COLD SPRING HARBOR SYMPOSIA ON QUANTITATIVE BIOLOGY

VOLUME LVII

COLD SPRING HARBOR SYMPOSIA ON QUANTITATIVE BIOLOGY

VOLUME LVII

The Cell Surface

COLD SPRING HARBOR LABORATORY PRESS
1992

COLD SPRING HARBOR SYMPOSIA ON QUANTITATIVE BIOLOGY
VOLUME LVII

International Standard Book Number 0-87969-063-1 (cloth)
International Standard Book Number 0-87969-064-X (paper)
International Standard Serial Number 0091-7451
Library of Congress Catalog Card Number 34-8174

Printed in the United States of America

COLD SPRING HARBOR SYMPOSIA ON QUANTITATIVE BIOLOGY

Founded in 1933 by
REGINALD G. HARRIS
Director of the Biological Laboratory 1924 to 1936

Previous Symposia Volumes

I (1933) Surface Phenomena
II (1934) Aspects of Growth
III (1935) Photochemical Reactions
IV (1936) Excitation Phenomena
V (1937) Internal Secretions
VI (1938) Protein Chemistry
VII (1939) Biological Oxidations
VIII (1940) Permeability and the Nature of Cell Membranes
IX (1941) Genes and Chromosomes: Structure and Organization
X (1942) The Relation of Hormones to Development
XI (1946) Heredity and Variation in Microorganisms
XII (1947) Nucleic Acids and Nucleoproteins
XIII (1948) Biological Applications of Tracer Elements
XIV (1949) Amino Acids and Proteins
XV (1950) Origin and Evolution of Man
XVI (1951) Genes and Mutations
XVII (1952) The Neuron
XVIII (1953) Viruses
XIX (1954) The Mammalian Fetus: Physiological Aspects of Development
XX (1955) Population Genetics: The Nature and Causes of Genetic Variability in Population
XXI (1956) Genetic Mechanisms: Structure and Function
XXII (1957) Population Studies: Animal Ecology and Demography
XXIII (1958) Exchange of Genetic Material: Mechanism and Consequences
XXIV (1959) Genetics and Twentieth Century Darwinism
XXV (1960) Biological Clocks
XXVI (1961) Cellular Regulatory Mechanisms
XXVII (1962) Basic Mechanisms in Animal Virus Biology
XXVIII (1963) Synthesis and Structure of Macromolecules
XXIX (1964) Human Genetics
XXX (1965) Sensory Receptors
XXXI (1966) The Genetic Code
XXXII (1967) Antibodies
XXXIII (1968) Replication of DNA in Microorganisms
XXXIV (1969) The Mechanism of Protein Synthesis
XXXV (1970) Transcription of Genetic Material
XXXVI (1971) Structure and Function of Proteins at the Three-dimensional Level
XXXVII (1972) The Mechanism of Muscle Contraction
XXXVIII (1973) Chromosome Structure and Function
XXXIX (1974) Tumor Viruses
XL (1975) The Synapse
XLI (1976) Origins of Lymphocyte Diversity
XLII (1977) Chromatin
XLIII (1978) DNA: Replication and Recombination
XLIV (1979) Viral Oncogenes
XLV (1980) Movable Genetic Elements
XLVI (1981) Organization of the Cytoplasm
XLVII (1982) Structures of DNA
XLVIII (1983) Molecular Neurobiology
XLIX (1984) Recombination at the DNA Level
L (1985) Molecular Biology of Development
LI (1986) Molecular Biology of *Homo sapiens*
LII (1987) Evolution of Catalytic Function
LIII (1988) Molecular Biology of Signal Transduction
LIV (1989) Immunological Recognition
LV (1990) The Brain
LVI (1991) The Cell Cycle

Front Cover (Paperback): Ablation of neurons at the optic chiasm prevents growth of axons of retinal ganglion cells into the optic chiasm. (For details, see Reichardt et al., p. 426, this volume.)

Back Cover (Paperback): Agrin-transfected cells induce clustering of AChRs. (For details, see Campanelli et al., p. 467, this volume.)

All Cold Spring Harbor Laboratory Press publications may be ordered directly from Cold Spring Harbor Laboratory Press, 10 Skyline Drive, Plainview, New York 11803. Phone: 1-800-843-4388 (Continental U.S. and Canada); (516)349-1930 (All other locations). FAX: (516)349-1946.

Symposium Participants

ADIE, ELAINE, Dept. of Biochemistry, University of Glasgow, Scotland

AGER, ANN, Dept. of Cell and Structural Biology, University of Manchester, United Kingdom

AKESON, RICHARD, Dept. of Basic Science Research, Children's Hospital Research Foundation, Cincinnati, Ohio

ALTSCHULER, DANIEL, Dept. of Cell Biology, Burroughs Wellcome Co., Research Triangle Park, North Carolina

AMANN, EGON, Pharma Research Laboratories, Hoechst Japan, Ltd., Kawagoe-Saitama, Japan

ANDERSON, KATHRYN, Dept. of Molecular and Cell Biology, University of California, Berkeley

ANDERSON, RICHARD, Dept. of Cell Biology and Neuroscience, University of Texas Southwestern Medical Center, Dallas

APPLEBURY, MEREDITH, Eye Research Laboratories, University of Chicago, Illinois

ARTAVANIS-TSAKONAS, SPYRIDON, Dept. of Molecular and Developmental Neurobiology, Boyer Center for Molecular Medicine, Yale University, New Haven, Connecticut

ASHKENAZI, AVI, Dept. of Molecular Biology, Genentech, Inc., South San Francisco, California

ATTISANO, LILLIAN, Dept. of Cell Biology, Memorial Sloan-Kettering Cancer Center, New York, New York

AXEL, RICHARD, Howard Hughes Medical Institute, Columbia University, New York, New York

BAUER, MARGARET, Dept. of Medical Microbiology and Immunology, University of Wisconsin, Madison

BEKKERS, AUGUST, Dept. of Enzymology and Protein Engineering, Institute of Biomembranes, Utrecht, The Netherlands

BELL, GREGORY, Dept. of Medicine, University of California, San Francisco

BERSTEIN, GABRIEL, Dept. of Pharmacology, University of Texas Southwestern Medical Center, Dallas

BEVILACQUA, MICHAEL, Dept. of Cellular and Molecular Medicine, Howard Hughes Medical Institute, University of California, San Diego

BIRNBAUMER, LUTZ, Dept. of Cell Biology, Baylor College of Medicine, Houston, Texas

BIRNBAUMER, MARIEL, Dept. of Cell Biology, Baylor College of Medicine, Houston, Texas

BLOBEL, GUNTER, Howard Hughes Medical Institute, Rockefeller University, New York, New York

BLOOM, THEODORA, *Nature*, London, United Kingdom

BONHOEFFER, FRIEDRICH, Abteilung Physikalische Biologie, Max Planck Institute, Tubingen, Germany

BOOTH, PAUL, Dept. of Cellular Biochemistry, Life Technologies Inc., Gaithersburg, Maryland

BORST, JANNIE, Dept. of Immunology, Netherlands Cancer Institute, Amsterdam, The Netherlands

BOSSY, BLAISE, Howard Hughes Medical Institute, University of California, San Francisco

BOURNE, HENRY, Department of Pharmacology, University of California, San Francisco

BRANDLEY, BRIAN, Glycomed Inc., Alameda, California

BRONSON, RICHARD, Dept. of Obstetrics and Gynecology, State University of New York, Stony Brook

BROWN, ABRAHAM, Lederle Laboratory, Pearl River, New York

BROWN, MICHAEL, Center for Genetic Disease, University of Texas Southwestern Medical Center, Dallas

BROWNING, JEFFREY, Dept. of Cell Biology, Biogen, Cambridge, Massachusetts

BRUGGE, JOAN, Dept. of Microbiology, University of Pennsylvania School of Medicine, Philadelphia

BURLINGAME, A.L., Dept. of Pharmaceutical Chemistry, University of California, San Francisco

BURNETT, J. PAUL, Dept. of Molecular and Cell Biology Research, Lilly Research Laboratories, Lilly Corporate Center, Indianapolis, Indiana

BURRIDGE, KEITH, Dept. of Cell Biology and Anatomy, University of North Carolina, Chapel Hill

BYTH, KATE, Dept. of Immunology and Pathology, Cambridge University, United Kingdom

CAMPBELL, BOB, Ares Advanced Technology, Randolph, Massachusetts

CARAS, INGRID, Dept. of Immunobiology, Genentech, Inc., South San Francisco, California

CARMELIET, PETER, Whitehead Institute, Cambridge, Massachusetts

CARPEN, OLLI, Dept. of Pathology, University of Helsinki, Finland

CATTERALL, CATHERINE, Dept. of Inflammation Biology, CellTech Ltd., Slough, Berkshire, United Kingdom

CHAN, PO-YING, Dept. of Cellular Interactions, Bristol-Myers Squibb PRI, Seattle, Washington

CHANG, CHIA-PING, School of Medicine, University of California at San Diego, La Jolla

CHANG, XIO-JIA, Dept. of Cell Adhesion Proteins, Genetics Institute, Cambridge, Massachusetts

CHARDIN, PIERRE, CNRS-Institut de Pharmacologie, Valbonne, France

CHEN, MANN-JY, Dept. of Microbiology and Immunology, Jefferson Cancer Institute, Philadelphia, Pennsylvania

CHERESH, DAVID, Dept. of Immunology, Scripps Clinic and Research Foundation, La Jolla, California

CHU, KETING, Dept. of Microbiology and Immunology, University of California, San Francisco

CLARK, CHRIS, Dept. of Genetics, Stanford University, Stanford, California

COUEZ, DOMINIQUE, Dept. of Cellular Biochemistry and Immunology, SmithKline Beecham, King of Prussia, Pennsylvania

COUGHLIN, SHAUN, Department of Medicine, Cardiovascular Research Institute, University of California, San Francisco

CRESSWELL, PETER, Dept. of Immunobiology, Howard Hughes Medical Institute, Yale University School of Medicine, New Haven, Connecticut

DANIELS, ALEJANDRO, Dept. of Pharmacology, Burroughs Wellcome Co., Research Triangle Park, North Carolina

DEMARTINO, JULIE, Dept. of Biochemical and Molecular Pathology, Merck Research Laboratories, Rahway, New Jersey

DEVREOTES, PETER, Dept. of Biological Chemistry, Johns Hopkins Medical School, Baltimore, Maryland

DICKESON, KENT, Dept. of Biotechnology and Microbiology, Wyeth-Ayerst Research, Radnor, Pennsylvania

DILIBERTO, EMANUEL, Dept. of Medical Biochemistry, Burroughs Wellcome Co., Research Triangle Park, North Carolina

DILORENZO, TERESA, Dept. of Otolaryngology, Long Island Jewish Medical Center, New Hyde Park, New York

DRICKAMER, KURT, Dept. of Biochemistry and Molecular Biophysics, Columbia University, New York, New York

DUDLEY, DAVID, Signal Transduction, Parke-Davis Pharmaceutical Corp., Ann Arbor, Michigan

EDELMAN, GERALD, Dept. of Developmental and Molecular Biology, Rockefeller University, New York, New York

EINSTEIN, RICHARD, Dept. of Molecular Genetics, SmithKline Beecham, King of Prussia, Pennsylvania

ENDEMANN, GERDA, Scios, Inc., Mountain View, California

FARRELL, FRANCIS, Dept. of Cell Biology, Burroughs Wellcome Co., Research Triangle Park, North Carolina

FAULDS, DARYL, Dept. of Cell Biology and Immunology, Berlex Biosciences, South San Francisco, California

FEATHERSTONE, CAROL, *Trends in Cell Biology*, Cambridge, United Kingdom

FIRTEL, RICHARD, Dept. of Biology, Center for Molecular Genetics, University of California at San Diego, La Jolla

FISKIN, MAX, Dept. of Microbiology, Kansas University Medical Center, Kansas City

FLEISSNER, ERWIN, Div. of Science and Mathematics, Hunter College, City University of New York, New York

FRANKE, WERNER, Institute of Cell and Tumor Biology, German Cancer Research Center, Heidelberg, Germany

FRASER, IAN, Dunn School of Pathology, Oxford University, Oxford, United Kingdom

FREMEAU, BOB, Dept. of Neurobiology, Duke University Medical Center, Durham, North Carolina

FURTH, MARK, Dept. of Molecular and Cell Biology, Regeneron Pharmaceuticals, Inc., Tarrytown, New York

GALAKATOS, NICHOLAS, Pharmaceuticals, Ciba-Geigy Corp., Summit, New Jersey

GARBERS, DAVID, Dept. of Pharmacology, Howard Hughes Medical Institute, University of Texas Southwestern Medical Center, Dallas

GEIGER, BENJAMIN, Dept. of Chemical Immunology, Weizmann Institute of Science, Rehovot, Israel

GERARD, CRAIG, Dept. of Pediatrics, Children's Hospital, Boston, Massachusetts

GERTLER, FRANK, Dept. of Oncology, University of Wisconsin, Madison

GILMAN, ALFRED, Dept. of Pharmacology, University of Texas Southwestern Medical Center, Dallas

GINSBERG, MARK, Dept. of Immunology, Scripps Clinic and Research Foundation, La Jolla, California

GOLDSTEIN, JOSEPH, Dept. of Molecular Genetics, University of Texas Southwestern Medical Center, Dallas

GOMPERTS, MIRANDA, Wellcome CRL Institute, University of Cambridge, United Kingdom

GOODMAN, COREY, Dept. of Molecular and Cell Biology, Howard Hughes Medical Institute, University of California, Berkeley

GUREVICH, VSEVOLOD, Dept. of Pharmacology, Thomas Jefferson University, Philadelphia, Pennsylvania

HAAG, FRIEDRICH, Dept. of Clinical Immunology, University Hospital, Hamburg, Germany

HADINGHAM, KAREN, Dept. of Biochemistry, Merck Sharp and Dohme Research Laboratories, Harlow, Essex, United Kingdom

HARRISON, STEPHEN, Dept. of Biochemistry and Molecular Biology, Harvard University, Cambridge, Massachusetts

HATTORI, YUTAKA, Dept. of Genetics, National Cancer Center Research Institute, Chou-Ku, Tokyo, Japan

HAWKINS, PHILIP, Dept. of Biochemistry, AFRC Institute of Animal Physiology and Genetics Research, Cambridge, United Kingdom

HAYASHI, MASAO, Dept. of Biology, Ochanomizu University, Bunkyo-ku, Tokyo, Japan

HEDMAN, HAKAN, Unit for Applied Cell and Molecular Biology, University of Umea, Sweden

HEDRICK, LORA, Dept. of Oncology and Pathology, Johns Hopkins School of Medicine, Baltimore, Maryland

HEEMELS, THERESE, Dept. of Cellular Biochemistry, The Netherlands Cancer Institute, Amsterdam, The Netherlands

HEIDARAN, MOHAMMAD, National Cancer Institute, Bethesda, Maryland

HELPER, JOHN, Dept. of Pharmacology, University of Texas Southwestern Medical Center, Dallas

HEMLER, MARTIN, Dept. of Tumor Virology, Dana-Farber Cancer Institute, Boston, Massachusetts

HEMPERLY, JOHN, Dept. of Neurobiology, Becton Dickinson Co., Research Triangle Park, North Carolina

HEN, RENE, LGME de CNRS, Strasbourg, France

HENDRICKSON, WAYNE, Dept. of Biochemistry and Molecular Biophysics, Columbia University, New York, New York

HEPLER, JOHN, Dept. of Pharmacology, University of Texas, Dallas

HETZEL, DONALD, Corporate Research and Development, Becton Dickinson and Co., Franklin Lakes, New Jersey

HIGGINS, LINDA, Dept. of Cell Biology, Scios, Inc., Mountain View, California

HINCK, LINDSAY, Dept. of Molecular and Cell Physiology, Stanford University, Stanford, California

HOBART, PETER, Dept. of Molecular Genetics, Pfizer Central Research, Groton, Connecticut

HOLGERSSON, JAN, Dept. of Molecular Biology, Massachusetts General Hospital, Boston

HORVITZ, H. ROBERT, Dept. of Biology, Howard Hughes Medical Institute, Massachusetts Institute of Technology, Cambridge

HSU, YEN-MING, Dept. of Protein Chemistry, Biogen Research Corp., Cambridge, Massachusetts

HU, SYLVIA, Mammalian Cell Research, Amgen, Inc., Thousand Oaks, California

HUNTER, TONY, Dept. of Molecular Biology and Virology, Salk Institute for Biological Sciences, La Jolla, California

HUTCHCROFT, JILL, Dept. of Medicinal Chemistry and Pharmacology, Purdue University, West Lafayette, Indiana

HYNES, RICHARD, Dept. of Biology, Center for Cancer Research, Massachusetts Institute of Technology, Cambridge

IMBODEN, JOHN, Dept. of Medicine, University of California, San Francisco

IWASHITA, SHINTARO, Dept. of Cellular Biochemistry, Mitsubishi-Kasei Institute of Life Sciences, Machida-shi, Tokyo, Japan

IZUMO, SEIGO, Dept. of Molecular Medicine, Beth Israel Hospital, Boston, Massachusetts

JACKSON, TREVOR, Dept. of Biochemistry, AFRC Institute of Animal Physiology and Genetic Research, Cambridge, United Kingdom

JACOBSON, KEN, Dept. of Cell Biology and Anatomy, University of North Carolina, Chapel Hill

JAIN, NEERAJ, Centre for Cellular and Molecular Biology, Hyderabad, India

JAN, LILY, Howard Hughes Medical Institute, University of California, San Francisco

JANIS, RICHARD, Dept. of Corporate Molecular Biology, Abbott Laboratories, Abbott Park, Illinois

JAVAHERIAN, KASHI, Dept. of Molecular Biology, Repligen Corp., Cambridge, Massachusetts

JESSELL, THOMAS, Center for Neurobiology and Behavior, Columbia University College of Physicians & Surgeons, New York, New York

JONES, GARETH, Biological Spectroscopy, Serc Daresbury Laboratory, Warrington, Cheshire, United Kingdom

KANDEL, ERIC, Center for Neurobiology and Behavior, Columbia University College of Physicians & Surgeons, New York, New York

KAWAHARA, DOUGLAS, Advanced Cellular Biology, Becton Dickinson, San Jose, California

KIM, SUNG-HOU, Dept. of Chemistry, Lawrence Berkeley Laboratory, University of California, Berkeley

KIMBLE, JUDITH, Dept. of Molecular Biology, University of Wisconsin, Madison

KINTNER, CHRIS, Dept. of Molecular Neurobiology, The Salk Institute for Biological Sciences, San Diego, California

KLEIN, PETER, Dept. of Biochemistry and Molecular Biology, Harvard University, Cambridge, Massachusetts

KOLANUS, WALDEMAR, Dept. of Molecular Biology, Massachusetts General Hospital, Boston

KOSHLAND, DANIEL, Dept. of Molecular and Cellular Biology, University of California, Berkeley

KRAMER, HELMUT, Dept. of Biological Chemistry, School of Medicine, University of California, Los Angeles

KRIEGER, MONTY, Dept. of Biology, Massachusetts Institute of Technology, Cambridge

KUBLER, DIETER, Dept. of Pathochemistry, German Cancer Research Center, Heidelberg

KUHSTOSS, STUART, Natural Products Research, Eli Lilly & Co., Indianapolis, Indiana

KUROSAKI, TOMOHIRO, Dept. of Biochemical Genetics, Memorial Sloan-Kettering Cancer Institute, New York, New York

KUSHNER, PINKY DROSTEN, ALS Research Ctr., California Pacific Medical Center, San Francisco

LAMPUGNANI, MARIA GRAZIA, Dept. of Vascular Biology, Mario Negri Institute, Milan, Italy

LANG, RICHARD, G.W. Hooper Research Foundation, University of California, San Francisco

LANGNER, KLAUS-DIETER, Preclinical Research, Behringwerke AG, Marburg/Lahn, Germany

LAROSA, GREGORY, Dept. of Molecular Biology, Repligen Corp., Cambridge, Massachusetts

LASKY, LAURENCE, Genentech, Inc., South San Francisco, California

LEAHY, DAN, Dept. of Biochemistry, Columbia University, New York, New York

LEE, CHONG-CHOU, Dept. of Developmental Biology, National Institutes of Health, Bethesda, Maryland

LEE, MONG-HONG, Dept. of Biological Sciences, University of Pittsburg, Pennsylvania

LEFKOWITZ, ROBERT, Department of Medicine, Duke University Medical Center, Durham, North Carolina

LEHTO, VELI-PEKKA, Dept. of Pathology, University of Oulu, Finland

LEVI, GIOVANNI, Physiopathologie du Developement, C.N.R.S., Paris, France

LEVIN, MARGARET, Dept. of Biological Sciences, University of Pittsburgh, Pennsylvania

LI, JIANXUN, Dept. of Signal Transduction, Rockefeller University, New York, New York

LI, MIN, Howard Hughes Medical Institute, University of California, San Francisco

LIMBIRD, LEE, Dept. of Pharmacology, Vanderbilt University School of Medicine, Nashville, Tennessee

LING, LEONA, Dept. of Molecular Biology, Biogen, Inc., Cambridge, Massachusetts

LIU, YAFANG, Dept. of Pharmacology, McGill University, Montreal, Quebec, Canada

LODISH, HARVEY, Whitehead Institute, Massachusetts Institute of Technology, Cambridge

LOMEDICO, PETER, Dept. of Molecular/Cellular Biology and Biochemistry, Hoffmann-LaRoche & Co., Ltd., Nutley, New Jersey

LOOR, FRANCIS, Preclinical Research, Sandoz Pharma, Ltd., Basel, Switzerland

LOWE, DAVID, Dept. of Molecular Biology, Genentech, Inc., South San Francisco, California

MAGEE, ANTHONY, Dept. of Eukaryotic Molecular Genetics, National Institute of Medical Research, Mill Hill, London, United Kingdom

Marco-Magazin, Marilyn, Protein Chemistry Unit, Sanofi ELF Bio Recherches, Labege-Innopole Voie, France

Massagué, Joan, Dept. of Cell Biology, Genetics Program, Memorial Sloan-Kettering Cancer Center, New York, New York

McAlmon, Karen, Dept. of Neonatology, Children's Hospital, Boston, Massachusetts

McCaffrey, Patricia, Dept. of Tumor Virology, Dana-Farber Cancer Institute, Boston, Massachusetts

McMahon, Andy, Dept. of Cell and Developmental Biology, Roche Institute of Molecular Biology, Nutley, New Jersey

McNeill, Helen, Dept. of Molecular and Cellular Physiology, Stanford University, Stanford, California

Miake-Lye, Ryn, Cell, Cambridge, Massachusetts

Miller, Linda, *Science*, Washington, D.C.

Miller, R. Tyler, Dept. of Internal Medicine, University of Texas Southwestern Medical Center, Dallas

Minami, Yasuhiro, Institute for Molecular and Cellular Biology, Osaka University, Suita-shi, Osaka, Japan

Moolenaar, Wouter, Dept. of Cellular Biochemistry, Netherlands Cancer Institute, Amsterdam

Moschuetzky, Heinz, Max Planck Institute, Freiburg, Germany

Murphy, Christine, Dept. of Pharmacology, University of Bath, Avon, United Kingdom

Murphy-Erdosh, Cynthia, Dept. of Neuroscience, Howard Hughes Medical Institute, University of California, San Francisco

Musil, Linda, Dept. of Anatomy and Cellular Biology, Harvard Medical School, Boston, Massachusetts

Muta, Tatsushi, Dept. of Biology, Faculty of Medicine, Kyushu University, Higashi-Ku, Fukuoka, Japan

Mynarcik, Dennis, Dept. of Endocrinology, State University of New York, Stony Brook

Na, Song-Qing, Dept. of Biology, Brandeis University, Waltham, Massachusetts

Nakanishi, Shigetada, Institute of Immunology, Faculty of Medicine, Kyoto University, Sakyo-ku, Kyoto, Japan

Napier, Mary, Dept. of Cardiovascular Research, Genentech, Inc., South San Francisco, California

Nastuk, Mary, Dept. of Biology, Worcester Foundation for Experimental Biology, Shrewsbury, Massachusetts

Nelson, W. James, Dept. of Molecular and Cell Physiology, Stanford University School of Medicine, Stanford, California

Neubert, Thomas, Dept. of Biochemistry, Howard Hughes Medical Institute, University of Washington, Seattle

Niessen, Carien, Dept. of Immunohematology, Central Laboratory of Blood Tranfusion Ser., Amsterdam, The Netherlands

Nigam, Sanjay, Dept. of Medicine, Harvard Medical School, Brigham & Women's Hospital, Boston, Massachusetts

Nitabach, Michael, Dept. of Biological Sciences, Columbia University, New York, New York

Nolte, Fredrick, Dept. of Immunology, University Hospital, First Medical Clinic, Hamburg, Germany

Nordmann, Rene, Preclinical Research, Sandoz Pharma, Ltd., Basel, Switzerland

North, Geoffrey, Current Biology, London, United Kingdom

Nüsslein-Volhard, Christiane, Max Planck Institut fur Entwicklungsbiologie, Tubingen, Germany

Offord, Jim, Dept. of Biotechnology, Parke-Davis, Ann Arbor, Michigan

Ohanjan, Vasken, Dept. of Medicine, University of Manchester, United Kingdom

Omer, Charles, Dept. of Cancer Research, Merck Research Laboratories, West Point, Pennsylvania

Ortega, Edwardo, Dept. of Geographic Medicine, New England Medical Center, Boston, Massachusetts

Ozenberger, Brad, American Cyanamid Co., Princeton, New Jersey

Pahlsson, Peter, Dept. of Pathology, University of Pennsylvania, Philadelphia

Pande, Gopal, Dept. of Pathology, Harvard Medical School, Boston, Massachusetts

Papkoff, Jackie, Dept. of Tumor Biology, Syntex Resarch, Palo Alto, California

Parada, Luis, Mammalian Genetics Section, Frederick Cancer Research & Development Center, Frederick, Maryland

Parker, Vann, Dept. of Mammalian Cell and Molecular Biology, Amgen, Inc., Thousand Oaks, California

Pausch, Mark, Agricultural Research Division, American Cyanamid Co., Princeton, New Jersey

Peach, Robert, Dept. of Cellular Interactions, Bristol-Myers Squibb Pharmaceuticals, Seattle, Washington

Pelletier, Anthony, Dept. of Cell Genetics,

Genentech, Inc., South San Francisco, California
PLOEGH, HIDDE, Netherlands Cancer Institute, Amsterdam
POULIOT, YANNICK, Dept. of Muscle Biochemistry, Montreal Neurological Institute, Montreal, Quebec, Canada
PREMONT, RICHARD, Dept. of Pharmacology, Mount Sinai School of Medicine, New York, New York
QWARNSTROM, EVA, Dept. of Pathology, University of Washington, Seattle
RADICE, GLENN, Center for Cancer Research, Massachusetts Institute of Technology, Cambridge
RALPH, LLOYD, Dept. of Pharmaceutics and Drug Delivery, Amgen, Inc., Thousand Oaks, California
RANDI, ANNA, Howard Hughes Medical Institute, Washington University School of Medicine, St. Louis, Missouri
REDDY, PRANHITHA, Dept. of Pharmacology, University of Washington, Seattle
REED, RANDALL, Dept. of Molecular Biology and Genetics, Johns Hopkins University, Baltimore, Maryland
REICHARDT, LOUIS, Howard Hughes Medical Institute, University of California, San Francisco
REYNOLDS, FREDERICK, Oncogene Science, Inc., Uniondale, New York
RIDLEY, ANNE, Dept. of Cell and Molecular Biology, Institute of Cancer Research, London, United Kingdom
ROBERTSON, MIRANDA, Garland Publishing, New York
RODER, JOHN, Dept. of Molecular Immunology, University of Toronto, Canada
RODRIGUES, MAURICIO, Dept. of Medical and Molecular Parasitology, New York University School of Medicine, New York, New York
ROMEO, CHARLES, Dept. of Molecular Biology, Massachussetts General Hospital, Boston
ROSENBERG, ROBERT, Dept. of Biology, Massachusetts Institute of Technology, Cambridge
ROSS, CHRISTOPHER, Dept. of Psychiatry and Neuroscience, Johns Hopkins University, Baltimore, Maryland
RUBIN, BEATRIX, Brain Research Institute, University of Zurich, Switzerland
RUBIN, GERALD, Dept. of Molecular and Cell Biology, Howard Hughes Medical Institute, University of California, Berkeley
RUOSLAHTI, ERKKI, La Jolla Cancer Research Foundation, La Jolla, California
SAAVEDRA, RAUL, Dept. of Skeletal Disorders, Harvard Medical School, Boston, Massachusetts
SANES, JOSHUA, Dept. of Anatomy and Neurobiology, Washington University School of Medicine, St. Louis, Missouri
SAWADA, IKUHISA, Dept. of Research and Planning, Yamanouchi Pharmaceutical Co., Ltd., Itabashi-ku, Tokyo, Japan
SCHELLER, RICHARD, Dept. of Biological Sciences, Stanford University, Stanford, California
SCHLESSINGER, JOSEPH, Dept. of Pharmacology, New York University Medical Center, New York, New York
SCHUESSLER, PETRA, Dept. of Microbiology and Hygiene, Institut fur Medizinische, Freiburg, Germany
SCHUMACHER, TON, Dept. of Cellular Biochemistry, The Netherlands Cancer Institute, Amsterdam
SEAMAN, WILLIAM, Dept. of Medicine and Microbiology/Immunology, University of California, San Francisco
SEED, BRIAN, Dept. of Molecular Biology, Harvard Medical School, Massachusetts General Hospital, Boston
SEEGER, ROBERT, Dept. of Hematology and Oncology, Childrens Hospital, Los Angeles, California
SHARFE, NIGEL, Dept. of Immunology and Pathology, Cambridge University, Cambridge, United Kingdom
SHARIF, MOHAMMED, Dept. of Medical Biological Chemistry, University of California, Davis
SHIH, AH-SHING, Dept. of Molecular Biology, Canji, Inc., San Diego, California
SHISHIDO, TOMOYUKI, Dept. of Signal Transduction, Rockefeller University, New York, New York
SHOWELL, HENRY, Dept. of Immunology and Infectious Disease, Pfizer Central Research, Groton, Connecticut
SIEH, MONICA, Howard Hughes Medical Institute, University of California, San Francisco
SIMON, MELVIN, Dept. of Biology, California Institute of Technology, Pasadena
SIMONS, KAI, Cell Biology Program, EMBL, Heidelberg, Germany
SIMPSON, PAT, Faculté de Médecine, Laboratoire de Génétique Moléculaire, Strasbourg, France

SINGH, SUNIL, Affymax Research Institute, Palo Alto, California

SMRCKA, ALAN, Dept. of Pharmacology, University of Texas Southwestern Medical Center, Dallas

SONG, ZHIWEI, Dept. of Biological Chemistry, University of Michigan, Ann Arbor

SPALDING, TRACY, Dept. of Physical Chemistry, National Institute for Medical Research, London, United Kingdom

SPRINGER, TIMOTHY, Center for Blood Research, Harvard Medical School, Boston, Massachusetts

STASSEN, FRANS, Research Dept., CIBA-GEIGY Pharmaceuticals, Summit, New Jersey

STEPHENS, LEN, Dept. of Biochemistry, AFRC Institute of Animal Physiology and Genetics Research, Cambridge, United Kingdom

STERNBERG, PAUL, Dept. of Biology, California Institute of Technology, Pasadena

STILLMAN, BRUCE, Cold Spring Harbor Laboratory, Cold Spring Harbor, New York

STROPP, UDO, Dept. of Biotechnology, Central Research, Bager-AG, Leverkusen, Germany

STRYER, LUBERT, Dept. of Cell Biology, Stanford University School of Medicine, Stanford, California

SUDHOF, THOMAS, Dept. of Molecular Genetics, University of Texas Southwestern Medical Center, Dallas

SUZUKI, SHINTARO, Dept. of Ophthalmology and Microbiology, Doheny Eye Institute, University of Southern California School of Medicine, Los Angeles

SWIMMER, CANDACE, ImmunoGen, Inc., Cambridge, Massachusetts

TAKEICHI, MASATOSHI, Dept. of Biophysics, Faculty of Science, Kyoto University, Sakyo-ku, Kyoto, Japan

TAKUMI, TORU, Whitehead Institute for Biomedical Research, Cambridge, Massachusetts

TAN, AGNES, Dept. of Cellular Biochemistry, The Netherlands Cancer Institute, Amsterdam

TANAKA, TOSHIAKI, Dept. of Tumor Immunology, Dana-Farber Cancer Institute, Boston, Massachusetts

TANG, WEI JEN, Dept. of Pharmacology, University of Texas Southwestern Medical Center, Dallas

TEIXEIRA, A., Dept. of Medical Oncology, Imperial Cancer Research Fund, London, United Kingdom

TEMPLE, GARY, Life Technologies, Inc., Gaithersburg, Maryland

THEIBERT, ANNE, Neurobiology Research Center, University of Alabama, Birmingham

THIELE, HEINZ-GUNTHER, Dept. of Immunology, University of Hamburg, Germany

THORNER, JEREMY, Dept. of Molecular and Cell Biology, University of California, Berkeley

TOLSTRUP, ANNE, Dept. of Molecular Biology, University of Aarhus, Denmark

TONEGAWA, SUSUMU, Center for Cancer Research, Massachusetts Institute of Technology, Cambridge

TONKS, NICHOLAS, Cold Spring Harbor Laboratory, Cold Spring Harbor, New York

TROPAK, MICHAEL, Dept. of Immunology and Neurobiology, Mt. Sinai Hospital Research Center, University of Toronto, Ontario, Canada

TSYGANKOV, ALEXANDER, Dept. of Molecular Biology, Bristol Meyers Squibb Research Institute, Princeton, New Jersey

TUCKER, GORDON, Institut de Recherches Servier, Suresnes, France

TYRRELL, DAVID, Dept. of Biochemistry, Glycomed, Inc., Alameda, California

UNANUE, EMIL, Dept. of Pathology, Washington University School of Medicine, St. Louis, Missouri

VAN DAMME, HENK, Dept. of Medical Biochemistry, University of Leiden, The Netherlands

VAN HUIJSDUIJNEN, R. HOOFT, Glaxo I.M.B., Geneva, Switzerland

VAN NOESEL, CAREL, Clinical Viro Immunology, Central Laboratory of Blood Transfusion, Amsterdam, The Netherlands

VAN SCHRAVENDIJK, MARIE ROSE, Dept. of Infectious Diseases, DNA Research Institute, Palo Alto, California

VARTICOVSKI, LYUBA, Dept. of Medical Research, Tufts University School of Medicine, Boston, Massachusetts

VICENTINI, LUCIA, Dept. of Pharmacology, University of Milan, Italy

VILLAFRANCA, J. ERNEST, Dept. of Protein Engineering, Agouron Pharmaceuticals, Inc., San Diego, California

VILLANO, JOHN, Dept. of Biochemistry, University of Texas Southwestern Medical Center, Dallas

VOGELI, GABRIEL, Dept. of Molecular Biology, Upjohn Co., Kalamazoo, Michigan

WALSH, FRANK, Dept. of Experimental Pathology, Guy's Hospital, London, United Kingdom

WANG, SY-SHI, Biomembrane Institute, Seattle, Washington

WARD, REBECCA, Research Dept., Genentech, Inc., South San Francisco, California

WARD, S., Dept. of Pharmacology and Pharmacy, University of Bath, Avon, United Kingdom

WATSON, STEVE, Dept. of Pharmacology, Oxford University, Oxford, United Kingdom

WATT, SUSANNE, Dept. of Medical Oncology, Imperial Cancer Research Fund, West Smithfield, United Kingdom

WATT, TERRI, Dept. of Osteobiology, Research and Development, Boehringer Mannheim, Indianapolis, Indiana

WEAVER, CHRISTINA, Dept. of Neuroscience, University of California, San Francisco

WEINMASTER, GERRY, Dept. of Neurobiology, Salk Institute, San Diego, California

WEIS, WILLIAM, Dept. of Biochemistry and Molecular Biophysics, Columbia University, New York, New York

WEISS, ARTHUR, Howard Hughes Medical Institute, University of California, San Francisco

WEISSMAN, IRVING, Dept. of Pathology, Lab. of Experimental Oncology, Stanford University Medical Center, Stanford, California

WESSENDORF, LISA, Dept. of Molecular Biology, Medical Research Council, Cambridge, United Kingdom

WESTMACOTT, DON, Dept. of Anti-Inflammatory Biology, Roche Products, Ltd., Hertfordshire, United Kingdom

WETZEL, RONALD, Dept. of Macromolecular Sciences, SmithKline Beecham, King of Prussia, Pennsylvania

WHEELER, GRANT, Center for Cancer Research, Massachusetts Institute of Technology, Cambridge

WIESCHAUS, ERIC, Dept. of Molecular Biology, Princeton University, Princeton, New Jersey

WILEY, DON, Dept. of Biochemistry and Molecular Biology, Harvard University, Cambridge, Massachusetts

WILKIE, THOMAS, Div. of Biology, California Institute of Technology, Pasadena

WILLERT, KARL, Dept. of Microbiology and Immunology, University of California, San Francisco

WILLIAMS, CAROL, Dept. of Immunology, Mayo Clinic, Rochester, Minnesota

WILLIAMS, LEWIS, Dept. of Cardiovascular Research, University of California, San Francisco

WITUNSKI, MICHAEL, Kiawah Island, South Carolina

WONG, WINNIE, Dept. of Biochemistry, BASF Bioresearch Corp., Cambridge, Massachusetts

WONG, YEE WAH, Dept. of Physiology and Biophysics, Columbia University, New York, New York

WRANA, JEFFREY, Dept. of Cell Biology and Genetics, Memorial Sloan-Kettering Cancer Center, New York, New York

WRIGHT, CLIFFORD, Dept. of Immunopathology, Parke Davis Pharmacology Research, Ann Arbor, Michigan

WU, MIN, Center for Cancer Research, Massachusetts Institute of Technology, Cambridge

XIE, MINGSHENG, Dept. of Physiology, Columbia University, New York, New York

XU, HUA, Dept. of Microbiology, University of California, San Francisco

YAMADA, KENNETH M., Lab. of Developmental Biology, National Institute of Dental Research, National Institutes of Health, Bethesda, Maryland

YAMAGATA, MASAHITO, Dept. of Anatomy and Neurobiology, Washington University, St. Louis, Missouri

YAMAMOTO, KYOHEI, Dept. of Cell Biology, Burroughs Wellcome Co., Research Triangle Park, North Carolina

YANCOPOULOS, GEORGE, Regeneron Pharmaceuticals, Inc., Tarrytown, New York

YANG, YU-CHUNG, Dept. of Medicine and Hematology/Oncology, Indiana University School of Medicine, Indianapolis

YOSHIHARA, YOSHIHIRO, Dept. of Neuroscience, Osaka Bioscience Institute, Suita, Osaka, Japan

YU, QIANG, Pulmonary Center, Boston University School of Medicine, Boston, Massachusetts

YUAN, JUNYING, Cardiovascular Research Center, Massachusetts General Hospital, Charlestown

ZIPURSKY, LARRY, Dept. of Biological Chemistry, University of California, Los Angeles

ZOLNIEROWICZ, STANISLAW, Friedrich Miescher Institute, Basel, Switzerland

First row: M. Brown; Picnic
Second row: D. Leahy, K. Drickamer, R. Anderson; H.R. Horvitz, B. Stillman, D. Luke III
Third row: A. Ager, G. Jones; C. Goodman, L. Reichardt
Fourth row: Picnic; T. Springer

First row: A. Ager, J. Witkowski, G. Jones, F. Walsh; M. Mathews, R. Hynes
Second row: B. Stillman, J. Kimble; H. Lodish, J. Goldstein
Third row: A. Tolstrup; M. Simon; R. Scheller

First row: J. Schlessinger, W. Moolenaar; R. Anderson, K. Simons
Second row: R. Miake-Lye, C. Kintner; S.-H. Kim, P. Sternberg, H. Bourne
Third row: Poster session

First row: W. Franke; F. Bonhoeffer, H.R. Horvitz; D. Kubler
Second row: R. Akeson, J. Sanes; D. Mynarcik, J. Whittaker
Third row: H. Lodish; J. Brugge; R. Reed
Fourth row: L. Lasky; C. Williams; J. Thorner, R. Hynes

Foreword

The biochemical processes that occur at the cell surface have profound effects on the well-being of single cells and whole organisms. Cell surface phenomena are principal areas of study in all disciplines of modern biological research, including physiology, development and differentiation, immunology, cancer, neurobiology, and cell growth and death. Moreover, the molecules that lie on or immediately subjacent to the cell surface comprise prime targets for alleviation or treatment of human diseases. Because of the importance of this research, it was decided to focus the 57th Cold Spring Harbor Symposium on the topic of The Cell Surface. There was, however, a considerable challenge to bring together a diverse group of scientists who study biology at the cell surface, but there was also comfort in the fact that common molecular themes appear in all species, from bacteria to man.

The very first Symposium, organized by Reginald Harris in 1933, was entitled Surface Phenomena, and in the intervening years, several other Symposia have dealt with many aspects of this broad topic. In 1992, given the breadth of the program, it was not possible to invite all the major scientists working in this field, and difficult decisions were necessary. In the organization of the meeting, I received enormous help and advice from Richard Hynes, a principal co-organizer, with additional help from Richard Axel and Cory Goodman. The meeting took place from May 27 to June 3, 1992. Following an introduction by Jim Watson, the first night speakers, Christiane Nüsslein-Volhard, Dan Koshland, Jr., Tony Hunter, and Gerald Edelman, provided a fascinating overview and a taste of things to come. Over the next six days, 324 participants witnessed a total of 82 oral presentations and 89 posters; the high standard set on the first night was maintained throughout the meeting. An insightful summary was presented by Mel Simon, who displayed an excellent grasp of all the science.

Essential funds to hold such a meeting were provided by the National Science Foundation; the U.S. Department of Energy, Office of Health and Environmental Research; and the following divisions of the National Institutes of Health: the National Cancer Institute, the National Institute of Neurological Disorders and Stroke, and the National Institute on Aging. Additional funding came from our corporate sponsors, who provide increasingly vital funding: Akzo Pharma International B. V.; Alafi Capital Company; American Cyanamid Company; Amgen Inc.; Applied Biosystems, Inc.; BASF Bioresearch Corporation; Becton Dickinson and Company; Boehringer Mannheim Corporation; Ciba-Geigy Corporation/Ciba-Geigy Limited; Diagnostic Products Corporation; The Du Pont Merck Pharmaceutical Company; Eastman Kodak Company; Genentech, Inc.; Glaxo; Hoffmann-La Roche Inc.; Johnson & Johnson; Kyowa Hakko Kogyo Co., Ltd.; Life Technologies, Inc.; Eli Lilly and Company; MetPath Inc.; Millipore Corporation; Monsanto Company; New England BioLabs, Inc.; Oncogene Science, Inc.; Pall Corporation; Perkin-Elmer Cetus Instruments; Pfizer Inc.; Recordati; Sandoz Research Institute; Schering-Plough Corporation; SmithKline Beecham Pharmaceuticals; Sumitomo Pharmaceuticals Co., Ltd.; Toyobo Co., Ltd.; The Upjohn Company; The Wellcome Research Laboratories, Burroughs Wellcome Co.; and Wyeth-Ayerst Research.

I thank the industrious staff in our Meetings Office, particularly Barbara Ward, Diane Tighe, and Micki McBride, for handling all the registrations, mailings, and housing arrangements in an expert and efficient manner. Herb Parsons and his staff provided wonderful audiovisual assistance. I am particularly indebted to Barbara Weinkauff, who handled much of the correspondence and many telephone calls during the organization of the meeting. Finally, the unfailing efforts of the Publications office, particularly Patricia Barker, Dorothy Brown, Mary Cozza, Joan Ebert, and Inez Sialiano, all under the guidance of Nancy Ford, made publication of this volume possible.

Bruce Stillman

Contents

Neurobiology

Membrane Traffic and Cytoskeletal Assembly

COLD SPRING HARBOR SYMPOSIA ON QUANTITATIVE BIOLOGY

VOLUME LVII

Receptors and Transmembrane Signaling

B.L. Stoddard, H.-P. Biemann, and D.E. Koshland, Jr.
Division of Biochemistry and Molecular Biology, Department of Molecular and Cell Biology, University of California, Berkeley, California 94720

Receptors were recognized as a cell-surface phenomenon long before they were recognized as protein molecules located in a membrane. This phenomenon was the ability of an organism or cell to recognize a specific stimulus from the outside environment and to use the information to initiate a signal transduction pathway, producing changes in the organism's behavior or metabolism that usually improve the organism's chances for survival or affect a cell's interactions with its immediate environment. An increased understanding of receptors led to the generalization that receptors are in fact protein molecules which could process information from outside a cell to specific signal transduction pathways within the cell. The crucial link in that process is the ability of a protein molecule to be influenced by a stimulus—a chemical, light, mechanical compression, or even heat—and to deliver that information across the membrane to the inside of the cell, where the signal is amplified and processed.

Sometimes this information transfer is achieved by a direct passage of chemicals, as in the case of ion channels. In most cases, however, information is transmitted via a conformational change occurring within the structure of the protein that spans the membrane. That process is still not completely understood, but important clues as to how it may occur are now emerging, and in this paper, we discuss some potential mechanisms for transmembrane signaling and the current knowledge of receptor structures as elucidated through X-ray and electron diffraction studies.

General Models for Signal Transduction

The current literature on receptors indicates that it may be appropriate to classify the mechanisms of transmembrane signaling into several groups, as shown in Figure 1. The first category might be called the *a*ssociation-*d*issociation models (labeled AD for conve-

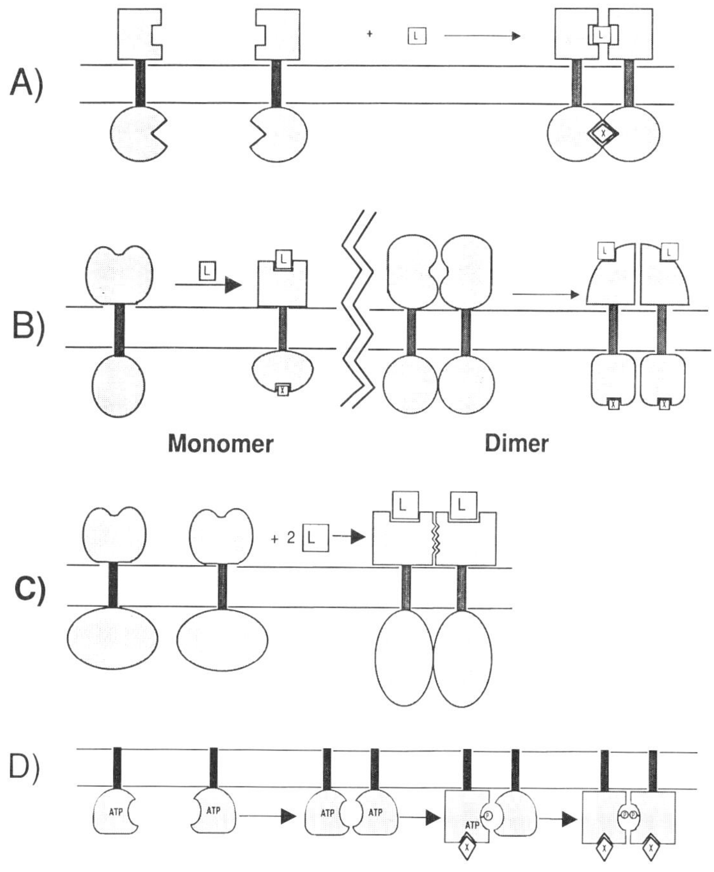

Figure 1. Mechanisms of transmembrane signaling by receptors. (*A*) Association-dissociation (AD) mechanism without conformational change. The ligand (L) binds to two subunits, inducing their association and generating a new site for the signaling molecule (X) in the cytoplasmic domain. (*B*) Binding-induced conformational change (BC) mechanism. Ligand induces a conformational change in the receptor that is transmitted from the ectodomain through the protein to the cytoplasmic domain. The receptor can be a monomer or an oligomer, but does not change its oligomeric state during the transmembrane signaling process. (*C*) Association-dissociation with binding-induced conformation change (ADBC) mechanism. An AD mechanism occurs, which generates a conformational change or a new site in the cytoplasmic domain as a result of binding-induced changes that lead to new attractions between subunits, which generates new sites within or across two subunits. (*D*) Association-dissociation inducing transphosphorylation between subunits (ADTP) mechanism. As a result of association of ectodomains, the cytoplasmic domains of the receptor are brought into close enough proximity to increase the rate of transphosphorylation, which in turn alters the signaling state of the receptor.

nience), and in one extreme form, as originally suggested for antibodies (Metzger 1974), it can involve a ligand that bridges across a dimer interface and creates a new complete binding site in the cytoplasmic domain of the receptor without any conformational change, enabling that new site to initiate chemical reactions which travel down the signal transduction pathway. Since the completion of the functional cytoplasmic site involves the juxtaposition of two half-sites, no conformational change within the protein would be required to produce a state of activation from a state of inactivation. It seems equally plausible that on some occasions a ligand in the recognition site on the exterior of the protein would induce a separation of subunits, generating two active monomers from one inactive dimer. This mechanism was initially proposed for the epidermal growth factor (EGF) receptor by Biswas et al. (1985). The essence of these two mechanisms is that no tertiary conformational changes need occur within the subunit in order to create a new active site.

The second model, which we can characterize as a *b*inding-induced *c*onformational change (the BC model), involves an induced conformational change that extends from the binding site on the outside of the membrane through the membrane to the inside (Mowbray and Koshland 1987; Thorsness et al. 1987). This mechanism requires no change in the association-dissociation state of the receptor and indeed can readily occur in a monomer. The BC model is likely to hold for the insulin receptor (IR) (Frattali et al. 1991), because its subunits are covalently cross-linked, as well as for the bacterial aspartate receptor, which is not observed to change oligomeric state upon binding (Foster et al. 1985). They rely on transmission of a conformational change within a subunit and/or changes in interactions between subunits. The sequence of signaling events on the inside of the cell is fairly well known in a number of cases. One type involves initiating a chain of phosphorylation through a cytoplasmic domain that is itself a kinase, and other types involve the activation of a G protein that activates an adenylyl cyclase cascade. For simplicity, we call the adenylyl cyclase cascade pathway and others the cytoplasmic signaling pathway.

There are subsets within each of these extreme receptor signaling mechanism possibilities, which are shown in Figure 1 and described where the individual cases are assessed. For example, a general mechanism in which both association-dissociation and a conformational change occur (which can be called for convenience the ADBC mechanism) seems very logical because that situation has been described for enzymes such as phosphorylase in which ligand-induced association-dissociation has been shown to be accompanied by a binding-induced conformational change. In an extreme case, dimerization could serve solely to modify the ligand-binding domains, causing them to transduce a conformational change through their own transmembrane domains to their cytoplasmic domains; no interactions between cytoplasmic domains would be involved in the transduction process. Evidence suggests multiple mechanisms by which ligand binding mediates receptor dimerization. In addition to the direct bridging mechanism shown in Figure 1A, ligands may bind independently to each ligand-binding domain and, via a conformational change, generate high-affinity recognition sites that mediate direct association of the external or cytoplasmic domains. Such an association mechanism could occur with or without ensuing conformational changes across the transmembrane domain or within the cytoplasmic domain.

There is clear evidence both for ligand-induced association-dissociation and for ligand-induced conformational changes. Receptors have also provided evidence for a new type of activation by association, i.e., the transphosphorylation of one subunit by the other generated by juxtaposition of the subunits. In this case, the association of receptors occurs first and the phosphorylation-induced conformational change occurs in a second step, referred to as the AD·TP mechanism (*a*ssociation/*d*issociation-*t*rans*p*hosphorylation) since it seems distinct from a case in which the ligand induces an initial conformational change causing association of the subunits (ADBC) and/or a conformational change that is transduced through the transmembrane domain (BC).

Our understanding of receptor molecules at the dynamic and functional level as introduced above has thus far outpaced our ability to visualize such molecules structurally. This is due to difficulties inherent in purifying and characterizing membrane-associated macromolecules to a point that is satisfactory for the application of such standard techniques as nuclear magnetic resonance (NMR), fluorescence, and even routine assay and chemical modification methodologies; in addition, growth of diffraction-quality crystals of membrane-bound receptors has proven to be an even more diffcult task. However, a select group of such molecules has recently provided atomic-resolution structural information of functional ligand-binding domain fragments and, in some cases, of the entire molecule. Such studies have allowed a direct correlation of structure with activity and potential signaling mechanisms for such molecules as the human growth hormone (hGH) receptor, the CD4 immunorecognition receptor, the aspartate receptor, and bacteriorhodopsin (which currently serves as a structural analogy for the many seven-transmembrane receptor molecules that have been identified and characterized). This review provides a broad overview of the current state of structural and dynamic information for a host of receptor species that bind soluble molecules, with emphasis on our understanding of transmembrane signaling mechanisms.

Because the structural tools are so difficult to apply to receptors, the precise mechanism for a particular receptor is probably not known for any one case, but it is valuable to consider the conceptual likelihoods based on current data because they help to create the incentive for additional tools. Already one such tool, "targeted disulfide cross-linking," has been developed

(Falke and Koshland 1987) and found to be useful in receptor studies (Milligan and Koshland 1988; Lynch and Koshland 1991). It thus augments such classic tools as X-ray crystallography, NMR, localized mutagenesis, and fluorescence, which frequently are difficult to apply to transmembrane receptors or give somewhat ambiguous results.

STRUCTURAL STUDIES OF RECEPTORS AND LIGANDS

Receptors are usually divided somewhat arbitrarily into overlapping groups or families of molecules on the basis of such traits as sequence homology, domain organization, proposed signaling mechanisms, or physiological function. Three-dimensional structural analysis of receptors has lagged far behind structural studies of soluble proteins, due to the difficulty of purifying and growing crystals of integral membrane proteins. However, diffraction studies of soluble fragments of some receptors and their ligands are providing atomic-resolution insights into the structural organization and patterns involved in receptor architecture and ligand recognition. In the first half of this paper, we summarize some important results of recent structural studies, which allow visualization of specific receptor species and important generalizations into entire families of these molecules. Current structural studies of receptors and their ligands (primarily receptors with one or two transmembrane domains) tell a recurring story of two structural strategies (α-helical bundles and immunoglobulin-like antiparallel β barrels) that have strong consequences on ligand binding and apparently on signaling mechanisms (Table 1).

The Aspartate Receptor

The aspartate receptor from the bacterial chemotaxis pathway shares no homology with the hematopoietic receptors and binds a number of ligands that are quite distinct from the hematopoietic factors discussed below. However, this molecule exhibits structural tendencies similar to the IR and the low-density lipoprotein (LDL) receptor (Russo and Koshland 1983). This receptor, which functions solely as a dimer, can bind a number of molecules, including aspartate, metals, aromatic rings, and periplasmic binding proteins (Koshland 1988). The receptor is composed of two identical subunits that possess single periplasmic and cytoplasmic domains and two transmembrane regions (Koshland 1988; Milligan and Koshland 1988). On the basis of circular dichroism (CD) data and the primary sequence, Moe and Koshland postulated the ligand-binding domain to be a four-helix bundle (see Youvan and Daldal 1986). The crystal structure of the periplasmic domain of the receptor has been solved to 2.0 Å resolution (Milburn et al. 1991) when the protein was expressed in recombinant form, cross-linked across the dimer interface at Cys-36, and truncated so that both transmembrane domains were removed, yielding a fully soluble protein. The structure of the ligand-binding domain is an antiparallel α-helical bundle, with helices up to 35 residues long and connected by short mobile loops. The structure is extremely close to the prediction mentioned above. The interhelical angles are approximately 20°, typical for packed helical bundles. The dimer interface is located along helix 1, which contacts the equivalent residues of helix 1′ in the second monomer. Recent evidence using targeted disulfide cross-linking (Lynch and Koshland 1992) shows that helices 1 and 1′ (along with 4 and 4′) continue through the membrane and comprise the receptor interface along the entire length of the molecule, into the cytoplasm. The transmembrane organization therefore consists of four α helices in which two (1 and 1′) are closer to each other than the others (4 and 4′). Minimal information is available on the structure of the cytoplasmic domain. A depiction of the receptor is shown in Figure 2A.

The aspartate-binding site has been located both crystallographically and through site-directed mutagenesis and is found at the dimer interface (Wolff and Parkinson 1988; Mowbray and Koshland 1990; Milburn et al. 1991). Residues from both monomers (Arg-64, 69′, 73′, Thr-154, and Tyr-149) contribute to aspartate binding. Interestingly, only one of two possible aspartate sites was found to be occupied in the crystal structure of the bound complex. Recent data in our laboratory suggest that half-of-sites binding and negative cooperativity between sites may contribute to the function of the receptor (H.-P. Biemann and D.E. Kosh-

Table 1. Some Classes of Receptors Yielding Structural Information

1 Transmembrane region per subunit	2 Transmembrane regions per subunit	7 Transmembrane regions per subunit
Hematopoietic receptors	bacterial chemotaxis receptors	bacteriorhodopsin
GHR	aspartate (Tar)	β-adrenergic receptor
IL-2,-3,-4,-6,-7 receptors	serine, leucine, indole (Tsr)	serotonin receptor
GM-CSF receptor	sugars (Trg)	thrombin receptor
IR	peptides (Tap)	
EGFR		
CD4 (immunorecognition receptor)		

A small number of the total known receptors are shown, grouped into a possible classification scheme based on the number of membrane-spanning regions. Structural studies of a small number of receptor- or membrane-bound protein species (hGHR, CD4, Tar, bacteriorhodopsin) allow important comparisons and conclusions about many related molecules.

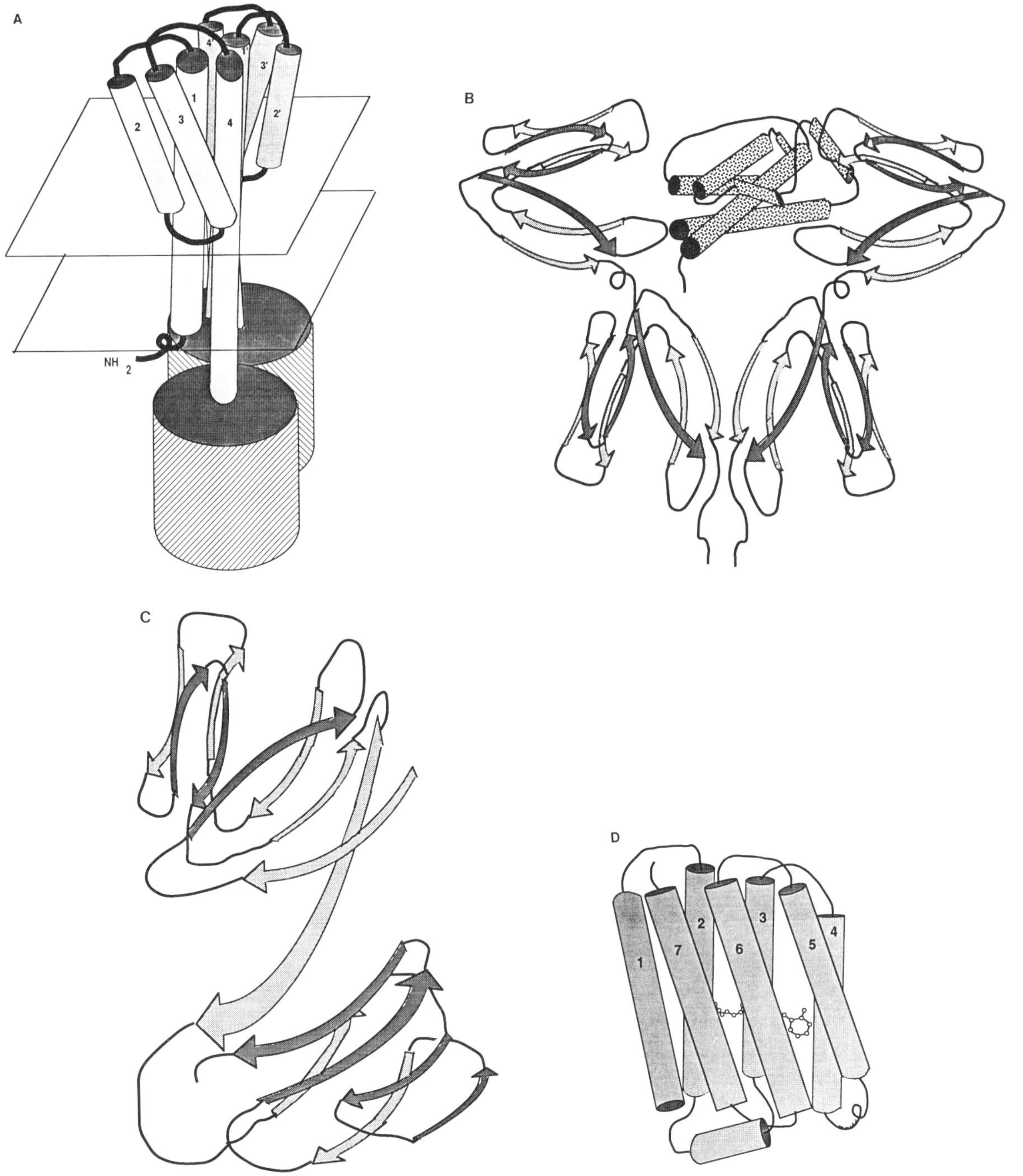

Figure 2. Schematic diagram of crystallographic structures described in this paper. (*A*) *E. coli* aspartate chemoreceptor. The receptor is a homodimer of two identical subunits, each of which is a four-helix bundle with short connecting loops. The aspartate-binding site(s) is located at the dimer interface and is shared between both subunits. Proposed signaling mechanisms all call upon conformational changes involving helices 1, 1′, 4, and 4′, which lead into a total of four transmembrane helices (two from each subunit). (*B*) hGHR. The ectodomain consists of a pair of β-barrel domains (which resemble Ig constant regions) linked by a single 4-residue helical turn. The domains are found at right angles to one another, giving the receptor monomer the approximate shape of the letter L. Two subunits are associated through their mutual assocations with a single growth hormone molecule, which binds at the end of all four barrels. The carboxy-terminal domains lead into the putative membrane-spanning regions of the receptor. (*C*) Human CD4 immunorecognition receptor. This membrane-bound molecule consists of four consecutive Ig-like domains, of which the first two (residues 1–182) have been solved crystallographically. The individual domains are similar to those observed for the GHR, but are linked by a single long β strand which is shared between both barrels and provides a long rigid structure of consecutive domains. The binding site for class II MHC is shared by both domains; the binding site for HIV gp120 is localized to a large area of domain 1 (residues 1–100). (*D*) Bacteriorhodopsin. This structure, solved by electron diffraction, currently serves as a model for the architecture of various seven-transmembrane receptors, such as the β-AR. The core of the protein serves as a site for chromophore, cofactor, or ligand binding and ion movement. Specific loops and regions on the cytoplasmic end of the molecule can serve as interaction points for effector molecules such as G proteins.

land, in prep.). A rotation of one subunit relative to the other has been suggested as one contribution to the transmembrane signal (Milburn et al. 1991). Other data, however, support the role of conformational changes within a single subunit as described elsewhere (Milligan and Koshland 1991).

Other difference maps of the receptor have revealed binding sites for platinum and 1,10 phenanthroline along the dimer interface (W.G. Scott et al., in prep.). Because chemotaxis receptors are capable of signaling in response to certain metals (cobalt, nickel) and aromatics (phenol), it is probable that these sites represent physiologically relevant binding sites and that signaling may proceed through similar means as that which is in response to the nearby binding of aspartate. In addition, mutagenic studies of the receptor have mapped out the binding sites of the receptor for maltose-binding protein (MBP); these residues are localized to two distinct regions located at the mobile loops farthest from the membrane and the immediately adjacent helices (Gardina et al. 1992). It would appear that MBP interacts with a large surface area of the receptor at a distance quite far from the membrane. Recent computational docking studies have indicated the mode of binding of MBP and the probable structure of the complex (Stoddard and Koshland 1992).

Human Growth Hormone Receptor

The complex of hGH to the serum-localized, soluble version of the extracellular domain of its receptor (hGHR) was solved to 2.8 Å resolution (De Vos et al. 1992). This receptor, upon binding its ligand, stimulates the growth and differentiation of various cell types. It belongs to the hematopoietic receptor superfamily, the members of which are all involved in cellular growth and differentiation. The receptor has extracellular and cytosolic domains and a single transmembrane sequence, similar to many receptor tyrosine kinases (Ullrich and Schlessinger 1990). The protein sequence of the extracellular domain of the receptor is similar to that of the receptors for interleukins 2, 3, 4, 6, and 7; colony-stimulating factors (CSF); and erythropoietin.

The extracellular domain of hGHR occurs naturally in the serum and binds hGH with approximately the same affinity as the intact receptor (Fuh et al. 1990). The crystal structure of the receptor fragment complexed to the growth hormone shows a complex of one molecule of hGH binding to a receptor dimer. This agrees with biochemical evidence (Cunningham et al. 1991) and is rather interesting, since this binding strategy is produced by the same residues from the two receptor monomers binding two different surfaces and sets of residues on the monomeric hormone ligand, which is itself a four-helix bundle.

The extracellular fragment of the receptor consists of two domains (residues 1–123 and 128–238) linked by a single 4-residue segment in an α-helical conformation. Each domain contains seven β strands which form a barrel of two antiparallel β sheets (with four and three strands, respectively). The two barrels lie approximately at right angles to one another, so that the individual receptor subunits form an "L" shape. The structure of the individual barrel domains of the receptor is similar to immunoglobulin constant domains, but with an exchange of strands between the two sheets as compared to the immunoglobulin fold. The topology of the growth hormone receptor domains is identical to domain D2 of CD4 (discussed below) and domain D2 of chaperone protein PapD (Holmgren and Branden 1989). The domains each contain three disulfide bridges that link strands of β sheet. Two of these link adjacent strands within each of the single β sheets, and the third stabilizes the barrel structure of the domain by linking strands from opposing β sheets. The second β domain is unusual in that it does not contain a pentapeptide sequence (Trp-Ser-X-Trp-Ser) near the carboxyl terminus, which is conserved among the other members of the hematopoietic superfamily.

The two receptor ectodomains in the soluble fragment are related by a twofold symmetry axis that runs approximately perpendicular to the helical axes of the bound hormone. The carboxy-terminal β-barrel domains from the two separate subunits are parallel to one another, with each having its carboxyl terminus pointing away from the hormone in the direction where the membrane would be. The hormone is bound at the top of these β domains, lying perpendicular to the β-sheet axes, and is also bound by the amino-terminal β domains from the two separate monomers, which are also roughly perpendicular to the helices of the growth hormone.

As a result of complex formation, two separate areas of the hormone surface are buried in the protein-receptor interface. The first interface is composed of helices 1 and 4 and the first connecting region and comprises about 1230 Å^2. The second binding site involves helices 1 and 3 and comprises approximately 900 Å^2. In comparison, the total surface area buried in the dimer interface of the receptor itself is about 500 Å^2 on each monomer. The residues of each receptor subunit that participate in hormone binding are largely the same (despite the nonsymmetrical hormone-binding surfaces involved). The structures of the binding determinants from the two receptor molecules are also similar, with an rms difference in Cα after superposition of 1.0 Å.

The majority of the binding interactions between the hormone and the receptor are apolar, primarily mediated through hydrophobic van der Waals contacts. In addition, a total of 13 possible salt bridges and hydrogen bonds are observed. Hormone binding appears to promote association at the base of the carboxy-terminal domain adjacent to the membrane, between the 3-stranded sheets of the β-barrel structure. It has been proposed on the basis of the crystal structure and additional biochemical evidence (Cunningham et al. 1991) that a sequential mechanism for receptor dimerization and signaling is followed in which the second receptor subunit binds to a complex of hGH and the first receptor molecule. Dimerization would then be supported by the extra binding energy acquired through the 500 $Å^2$ receptor interface.

The conclusions reached on receptor architecture and signaling for hGH are likely to be generalizable to the entire hematopoietic superfamily of receptors. These molecules all have limited sequence homology in the extracellular domain and a conserved set of four cysteine residues that lie within the first barrel. Bazan (1990) has previously proposed that each of these receptors contains two 100-residue domains, folded in an immunoglobulin-like barrel. For all of these receptor sequences, the residues that line up with positions involved in growth hormone binding are not conserved, implying specificity at the binding site for unique ligands.

CD4 Domain 2

The cell-surface glycoprotein CD4 is expressed on certain T lymphocytes, including helper T cells and major histocompatibility complex (MHC)-specific cytotoxic T cells. Interactions between CD4 and class II MHC are critical for immune response through the T-cell receptor. In addition, CD4 is the receptor for the viral glycoprotein gp120 of the human immunodeficiency virus (HIV), providing the virus a portal for fusion with the cell membrane. CD4 (55 kD) consists of an extracellular portion (residues 1–371), a single transmembrane segment (372–395), and a short cytoplasmic tail (396–433) that may transmit a signal to an associated tyrosine kinase, $p56^{lck}$.

The extracellular domain of CD4 consists of four consecutive immunoglobulin-like subunits (residues 1–100, 100–186, 186–290, and 290–370, respectively) with some limited homology with the extracellular domains of hGHR. Thus, this protein is another good case study of the structural theme of Ig-like domain structure in membrane-bound receptor and recognition factor species. The complete extracellular domain, like the intact membrane-spanning protein, appears to be monomeric (Sayre and Reinherz 1985; Kwong 1990). The known contact points to HIV gp120 and class II MHC molecules extend over domains 1 and 2 of CD4 (Clayton et al. 1988; Peterson and Seed 1988; Lamarre 1989). The structure of recombinant fragments of these CD4 extracellular domains (residues 1–182 or 183) was solved independently in the laboratories of Harrison and Hendrickson (Ryu et al. 1990; Wang et al. 1990). This protein, like the hGHR, also folds into a pair of antiparallel β-barrel domains (9 and 7 strands, respectively), each of which resembles an immunoglobulin constant domain (Fig. 2C). However, unlike hGHR, the two domains pack closely against one another, with a large hydrophobic interface. A long 15-residue β strand passes from one domain to the next and participates in the formation of both barrels; the axes of the two β-barrel cores are almost parallel to one another. Thus, although the architecture of structural domains in CD4 is similar to that of hGHR, the relative alignment of the domains is not. The total length of the full CD4 extracellular region (four consecutive β-barrel domains), from electron microscopy and hydrodynamic measurements and by extension of the crystal structures, is a stiff rod about 125 Å long and 25 Å–30 Å wide.

Although the structure of CD4 complexed to class II MHC or gp120 has not been solved, mutational analysis has allowed detailed mapping of the respective binding sites. Substitutions that interfere with MHC binding occur on both domains, primarily on the long β strand in domain 2 and on a pair of loops in domain 1. This implies that a very long extended region of contact must form upon binding of CD4 and MHC, and that intact CD4 projects beyond the outer antigen-presenting domain of the MHC molecule to interact with constant regions of the molecule. One current model suggests that the extended CD4 molecule, when bound to MHC, encourages the formation of a ternary complex with a T-cell receptor and a subsequent immune response to the presented antigen.

In contrast, the residues that map out the binding site of HIV gp120 localize along an edge of domain 1 and a pair of β strands, comprising a surface about 25 Å long. It is theorized that gp120 possesses a deep groove that binds the CD4 ridge and that the protein interacts with areas of CD4 to both sides of the primary binding interactions.

Atomic-level Structural Information from Bacteriorhodopsin: A Model for Seven-transmembrane Receptors

Although this paper concentrates on the receptors possessing one or two transmembrane regions, a survey of structural information about membrane-bound receptors would not be complete without at least a brief mention of receptors possessing larger numbers of transmembrane domains. As in the case of receptors containing only one or two transmembrane regions, nature has once again centered itself evolutionarily around a highly successful structural motif; in this case, the strategy is the bundling of seven amphipathic α helices in the membrane to create a hydrophilic core that may serve as a binding site for chromophores or cofactors, an ion channel, or a ligand-binding site. Of these, the adrenergic receptors are the most well characterized (Dohlman et al. 1991). However, integral membrane proteins are very difficult to crystallize for X-ray studies, and those with seven-transmembrane regions especially so, leading to little current structural information on these receptors. However, the structure of a different protein with similar organization, bacteriorhodopsin, has been solved through the use of electron diffraction methods and two-dimensional crystals (Henderson et al. 1990). This protein, from *Halobacterium halobium*, is a light-driven proton pump consisting of seven-transmembrane regions and an associated retinal chromophore. The protein occurs naturally as two-dimensional crystals. A three-dimensional map of the structure, at 3.5 Å resolution parallel to the membrane plane but lower in the perpendicular direction, was solved by using electron cryomicroscopy to obtain electron diffraction patterns and high-resolution micrographs.

A complete atomic model of the protein was built using residues 8–225. The transmembrane structure consists of seven transmembrane regions as predicted by sequence analysis, with the helices arranged sequentially in a clockwise fashion and a central channel that serves as the binding site for the retinal chromophore (which is complexed as a Schiff base with Lys-216) and as the route for proton delivery upon photon absorption. Twenty-one side chains, contributed by all seven helices, surround the retinal, and 26 residues from five helices form the proton channel. Ten of these residues are common to both functions. The retinal environment includes four tryptophans, which form a sandwich around the chromophore. The structure suggests that pK changes in the Schiff base of the bound chromophore act as the mechanism by which light energy is converted to proton pumping. Several residues, including three aspartates which line the channel of the protein (85, 96, and 212), are indicated as participating in this event. An arginine residue (82) affects the pK_a of one or two aspartate residues but does not itself participate in proton pumping. Henderson et al. therefore propose a minimal mechanism for proton pumping involving three aspartates, the Schiff base, and structured water molecules in the channel.

With respect to receptor species sharing structural similarity with bacteriorhodopsin, the importance of the seven-helix channel in positioning residues for chromophore binding, interactions with solvent, and proton uptake and dissociation serves as a model for the intricacies expected for signal transduction. For example, genetic analysis of the β-adrenergic receptor (β-AR) has revealed that the ligand-binding domain of the receptor, like that of rhodopsin, involves residues within the membrane-bound core of the protein. A model for ligand binding has been proposed in which agonist or antagonist is bound to an aspartate side chain in the third transmembrane helix, two serine residues in transmembrane helix 5, and a phenylalanine in helix number 6 (Strader et al. 1987). This pattern of binding is quite similar to retinal binding in bacteriorhodopsin, although the signaling mechanism of β-AR is quite different from the proton shuttle mechanism displayed by the light-reactive bacteriorhodopsin.

Prediction of Receptor Structures

Given the difficult nature of crystallization of membrane proteins such as receptors, and the very small number of currently available structures as detailed in this paper, it is no surprise that much effort is being invested toward the prediction of receptor structure and interactions. Predictions are based on information derived from genetic homology with proteins of known structure, sequence analysis (which allows the prediction of transmembrane incorporation and sometimes secondary structure), correlation with spectroscopic and hydrodynamic evidence, and energy dynamics calculations and binding simulations (to predict ligand-binding interactions and possibly mechanisms of signal transduction).

All four of the receptor and receptor ligand-binding domain structures summarized here (hGHR, CD4, Tar, and bacteriorhodopsin) were subjected to predictive schemes prior to structural solution, with a reasonable degree of success. G.R. Moe, of our laboratory (see Youvan and Daldal 1986), originally proposed that the structure of the ligand-binding domain of the Tar aspartate chemoreceptor was a four-helix bundle, based on a combination of evidence from CD measurements and from an analysis of the primary sequence of the periplasmic 212 residues of the receptor. Since CD indicated that approximately 90% of the periplasmic domain fragment is in an α-helical conformation, net diagrams were used to search for amphiphilic regions characteristic of bundled helix structural organization. The regions assigned to the four α helices in the prediction matched the actual helices in the crystal structure (Milburn et al. 1991). Knowledge of the typical packing interactions in such a structure and simple energy minimization considerations allowed construction of a left-twist bundle of antiparallel helices with a tilt between helices of approximately 18°, again closely approximating the final structure.

Once the structure of a receptor domain has been solved, predictive algorithms involving sequence substitution for related receptor species, energy minimization, and computational binding searches have been shown to yield accurate information about the function of many receptor species. The crystal structure of the *Salmonella typhimurium* aspartate chemoreceptor, for example, has provided high-quality models of the same receptor from *Escherichia coli*, as well as chemoreceptor domains for serine, leucine, and indole (the serine chemoreceptor Tsr). Docking studies of such computationally derived models have provided a clear picture of the binding sites and interactions involved in the receptor-mediated chemotactic response to MBP (Stoddard and Koshland 1992) and serine, leucine, and indole (C.J. Jeffery and D.E. Koshland, in prep.).

Bazan (1990) proposed that the ectodomain of hGHR was composed of a pair of Ig-like β-barrel folds; only the precise nature of strand connectivity and the tertiary fold leading to the overall shape of the receptor monomer remained unclear. Sequence similarity with a large number of related hematopoietic receptors (IL-2, -3, -4, -6, -7 receptors, CSF receptors, etc.) indicates that the same strategy of receptor architecture and ligand binding (β barrels in the receptor ectodomain complexed to ligands with α-helical bundle structures) should occur in many other systems. This is supported by the crystal structures of several ligands to these receptors, including CSF (Diederichs et al. 1991) and apolipoprotein E (Wilson et al. 1991), both of which show a fold identical to hGH and binding surfaces that map out similarly to hGH.

In an extension of the types of studies described above, one of the more intriguing hypotheses regarding currently unknown receptor structure concerns the extracellular domains of EGF and IR (Bajaj et al. 1987). Alignment and analysis of the sequences of EGFR from human and *Drosophila* and IR from humans indicate that these receptors each contain two large, related domains composed of short α helices followed by turns of conserved length and then short β strands; sequential repetition of α-helix-turn-β motif strongly implies either an antiparallel β-sheet structure or, more likely, an α/β-barrel structure. Since the α/β-barrel fold has been observed for more soluble protein structures than any other fold and is known to support ligand binding at the end of the barrel core (Farber and Petsko 1990), such a fold would not seem unreasonable for a large family of ligand-binding receptor domains. What is clear is that such predictive algorithms, when used on species such as EGFR, provide insight into possible relationships between putative structural themes in such receptors and functional studies, as described in the second half of this paper.

DYNAMIC STUDIES OF RECEPTORS AND LIGANDS

The Aspartate Receptor

From targeted disulfide cross-linking data and studies on truncated and intact receptors, the dynamics of the aspartate receptor suggest a BC mechanism, since no evidence for association-dissociation has been obtained from hydrodynamic or cross-linking studies (Foster et al. 1985; Milligan and Koshland 1988). Moreover, a dimeric receptor cross-linked between residues 36 of one subunit and 36′ of the other subunit showed normal transmembrane methylation and signaling, a situation in which monomer-dimer exchange is impossible (Falke and Koshland 1987). The ultracentrifuge and gel filtration experiments showed that dimer-tetramer associations did not occur (Milligan 1991; Milligan and Koshland 1991). Hence, the only solution to the transmembrane signaling must be some transmembrane conformational change. Targeted cross-linking studies (Falke and Koshland 1987) have shown that the receptor displays a great amount of global flexibility and that aspartate binding induces large changes in the receptor's dynamic behavior and conformation.

On the basis of these studies, a piston-like model had been proposed (Mowbray and Koshland 1987) to explain how a single movement of a transmembrane domain could change the juxtaposition of the cytoplasmic domain to the membrane and thus activate or deactivate it. This type of model was given further support by a series of experiments with truncated receptors (Milligan and Koshland 1991). In those experiments, one full subunit was cross-linked to another (called the 1 + 1 receptor), to a subunit which was missing the cytoplasmic domain (called the 1 + 2/3 receptor), and to a subunit missing both the cytoplasmic domain and the transmembrane regions (called the 1 + 1/3 receptor). All these chimeric species display signaling through changes in methylation rates upon aspartate binding, suggesting strongly that juxtaposition of cytoplasmic domains was unnecessary in this case, and some mechanism for an intra-subunit shift between the two transmembrane domains was needed to explain the transmembrane signaling. Such an effect could be explained by a relative shift between intrasubunit transmembrane helices by any of the mechanisms shown in Figure 3, of which the piston mechanism seems most likely, since continuous α-helices seem built to withstand deformation by up-and-down piston-like motions rather than sidewise bending mechanisms, which are called upon with mechanisms involving "scissors" or "see-saw" motion of helices, or rotation or simple translation of helices toward or away from each other. Moreover, a correlation with movements seen in the tryptophan repressor, which has structural analogies to the aspartate receptor (Lynch and Koshland 1992), adds plausibility to a piston-like effect.

On the other hand, the X-ray structure of the soluble fragment of the ectodomain shows the binding of a single aspartate molecule with very little conformational change within individual subunits. This anomaly can be explained in several ways. The soluble subunits, lacking their transmembrane appendages, might be released from physical constraints normally imposed by

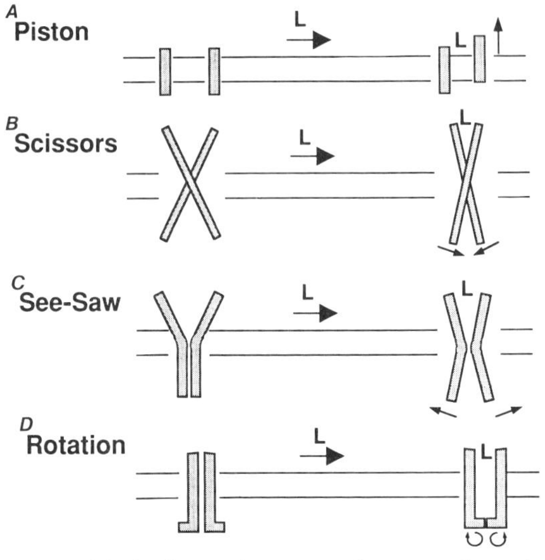

Figure 3. Mechanisms of transmembrane conformational change. (*A*) Piston model: Transmembrane domains move perpendicular to the membrane. (*B*) Scissors model: Helical domains move in scissors-type action as a result of ligand pulling two helices closer together in ectodomain, causing sites in signaling domain to move closer together also. (*C*) See-saw model: Transmembrane helices are curved and associated at a distinct pivot point so that closer juxtaposition in ectodomain causes them to move farther apart in cytoplasmic domain. (*D*) Rotation-translation model: Binding of ligand causes rotation of subunits or helices or a slight movement toward or away from each other without association-dissociation.

association with the lipid bilayer and therefore exist largely in the aspartate-bound conformation prior to ligand binding, thus showing little further conformational change upon binding aspartate. Such an effect is already known in the case of hemoglobin, where the very similarly structured myoglobin has high O_2 affinity and little conformational change on binding O_2. In that case, the monomeric myoglobin appears to be free of the constraints of the multi-subunit hemoglobin. The difference in subunit orientation between the bound and unbound forms of the protein might be explained simply by differences in crystal lattice contacts in the two crystal forms.

Alternative signaling mechanisms involve "scissors" or "seesaw"-type motions in which the two subunits rotate relative to each other, affecting contacts or conformations of the cytoplasmic domains. A hint of such intersubunit motion is evident when the crystal structures of the aspartate bound and unbound periplasmic domains are compared. Evidence against this type of mechanism includes the observations that (1) intersubunit disulfide cross-links in or near the transmembrane domains of the full-length receptor (36–36′ and 18–18′) still allow the receptor to signal and (2) two complete subunits are not needed because the 1 + 1/3 and 1 + 2/3 receptors retain function. A final description of how the signal is propagated through the transmembrane domains must await further experiments, but a strong possibility seems to be a major intrasubunit rearrangement accompanied by intersubunit conformational effects.

It is intriguing that although the mechanism of the aspartate receptor involves no association-dissociation, the aspartate does bind across the subunit interface, possibly bringing the subunits closer, or into a different relationship, in analogy with the AD mechanisms described above. It is also interesting that the insulin receptor shows a similar mechanism and that a hybrid of the aspartate ectodomain and the insulin cytoplasmic domain can still signal, although weakly (Moe et al. 1989).

Epidermal Growth Factor Receptor

EGF stimulates mitogenesis in epithelial cells by binding this 170-kD receptor and activating its tyrosine-specific autophosphorylation and exogenous kinase activities. The EGF receptor (EGFR) has been shown to undergo association-dissociation and probably transduces a tertiary conformational change through the membrane. Dimerization is proposed to result from activation of cryptic recognition sites on the ligand-binding domains that mediate direct association. EGF binding to a soluble extracellular fragment of the receptor has been shown to alter the fragment's intrinsic fluorescence (Greenfield et al. 1989). Thus, the ligand-binding domains of EGFR are likely to undergo ligand-induced conformational changes leading to association in much the same way as glycogen phosphorylase *b* oligomerizes upon binding AMP (Barford and Johnson 1992). Support for this mechanism comes from the observation that the receptor dimer has higher affinity for ligand than does the monomer. Ullrich and Schlessinger (1990) have proposed that EGF binds at the interface of two subdomains of the EGFR ectodomain. This interaction could cause tertiary structure changes that unmask recognition sites.

A multitude of observations support the proposal that dimerization of EGFR is an essential activating step: (1) EGF has been shown to cause receptor dimerization in vitro and in vivo (Yarden and Schlessinger 1987a; Honegger et al. 1990). (2) Aggregation induced by antibodies to the ectodomain is accompanied by autophosphorylation. (3) Immobilization on a solid support prevents stimulation by EGF (Yarden and Schlessinger 1987b). (4) EGF activation of solubilized receptor depends on receptor concentration (Yarden and Schlessinger 1987b). (5) Modifications of the transmembrane domain have been made and do not block signaling, suggesting that the transmembrane domain is simply a structural linker (Kashles et al. 1988). (6) Intermolecular autophosphorylation occurs upon dimerization, suggesting that dimerization activates the receptor by simply juxtaposing its kinase domains (Honegger et al. 1990). (7) Soluble or membrane-bound receptor fragments containing the ligand-binding domain, but not the kinase domain, act as dominant-negative inhibitors of signaling, suggesting that an intact dimer is necessary for function (Basu et al. 1989; Kashles et al. 1991). (8) The affinity of EGF for the dimer is higher than for the monomer. (9) A chimeric EGFR that has its cytoplasmic domain swapped with the IR does signal (Lammers et al. 1990), supporting the view that aggregation of kinase domains enabling their interaction is the important step in signaling.

Although dimerization thus seems to be necessary for signaling, Koland and Cerione (1988) have provided evidence that, under certain detergent conditions, isolated monomeric EGFR will autophosphorylate upon ligand addition, suggesting that a signal can be transduced across the transmembrane domain of a monomer and that this could play a role in the activation process in vivo. Thus, EGFR seems to fit in the ADBC category.

Other Single Transmembrane Domain Receptors

Similar data supporting association-dissociation mechanisms exist for the platelet-derived growth factor (PDGF), CSF, fibroblast growth factor (FGF), and hGH receptors. PDGF and its receptor have been extensively studied (for review, see Williams 1989). PDGF is composed of two disulfide cross-linked 15-kD subunits. One dimeric PDGF ligand binds two PDGFRs. Two highly homologous PDGF subunits termed A and B have been identified. They are found associated in homodimers and heterodimers. Upon activation, the receptor becomes an activated tyrosine kinase and associates with several cytoplasmic proteins that help mediate its strong mitogenic effects in mesenchymal cells. The primary structure of the receptor exhibits five Ig-like domains in the extracellular portion, a transmembrane domain, and two kinase domains in the cytoplasmic portion. A "kinase insert" region has been implicated in the association of the PDGFR with Src homology 2 (SH2)-containing proteins.

Because PDGF is bivalent, it is likely to mediate dimerization of PDGFR by directly bridging two receptors via the ligand-bridging mechanism without the necessity of direct receptor-receptor interactions or conformational changes. Evidence that PDGF induces receptor dimerization in vitro and in vivo has been presented by Kelly et al. (1991), who showed that upon dimerization, transphosphorylation of the ligand-binding domains occurs. A soluble PDGFR ligand-binding domain fragment served as a dominant-negative inhibitor (Duan et al. 1991) as was shown for the EGFR, again supporting the notion that only full-length homodimers can signal and that heterodimeric complexes of a single full-length monomer with a ligand-binding fragment do not function. These types of experiments support the notion that cytoplasmic domain interactions are a crucial result of dimerization for this receptor.

The CSF-1 receptor is structurally similar to the PDGFR. Lee and Nienhuis (1990) showed that dimerization of the cytoplasmic domains activates receptor kinase activity using a chimeric receptor. The ligand-binding domain was replaced with glycophorin A. Antiglycophorin A antibodies caused receptor aggregation and activation.

The FGF receptor family contains several related receptors with 2–3 Ig-like domains in the ligand-binding region, a transmembrane domain, and a cytoplasmic tyrosine kinase domain. The activated receptor has been shown to be a dimer that undergoes transphosphorylation (Bellot et al. 1991).

The structure of hGHR ligand-binding domain has been solved in a complex with ligand (De Vos et al. 1992). As described above, the structure indicates that the ligand activates the receptor by dimerization. No cytoplasmic activity has yet been identified for this receptor. If, like the CD4 receptor, it signals via contacts with a cytoplasmic protein, this could be accomplished by a conformational change on the cytoplasmic domains (ADBC mechanism, Fig. 1C) and/or by bringing together two domains to generate a shared site, similar to what occurs between the ligand-binding domains (AD mechanism, Fig. 1A).

Not all dimeric receptors are homodimers. The high-affinity receptor for nerve growth factor (NGF) has recently been identified as a heterodimer comprising the 75-kD NGFR and the 140-kD *trk* proto-oncogene product. Each of these transmembrane receptors had separately been shown to bind NGF, a dimeric ligand with low affinity. They manifest high-affinity binding as a heterodimer and form a heterotrimeric complex with ligand, shown by cross-linking experiments (Hempstead et al. 1991).

The Insulin Receptor

IR must signal via a BC model. It is a covalently cross-linked oligomer in the presence and absence of insulin. This receptor is of particular interest because of its role in glucose homeostasis and diabetes. Several instructive receptor mutants have been isolated from patients suffering from non-insulin-dependent diabetes mellitus. The receptor is an $\alpha_2\beta_2$ heterotetramer wherein each α ligand-binding subunit is disulfide cross-linked to a transmembrane β-kinase subunit (Rosen, 1987). Each of these $\alpha\beta$ half-receptors is analogous to the monomeric receptors described above. They are further cross-linked into the $\alpha_2\beta_2$ form, but reduction conditions can liberate free $\alpha\beta$ half-receptors. Isolated α subunits bind insulin, and soluble cytoplasmic fragments are active kinases. If the ligand-binding domains are removed proteolytically, the resulting receptor is constitutively kinetically active (Shoelson et al. 1988). Thus, the heterotetrameric configuration can be said to form an inhibitory context for the kinase domains that is relieved when ligand binds and transduces tertiary and/or quaternary conformational changes through the membrane. It should be noted that higher-order IR oligomers that are associated with greater kinase activity than the $\alpha_2\beta_2$ form have been observed in the presence of insulin (Kubar and Van Obbergen 1989). Although the $\alpha_2\beta_2$ form signals without a change in oligomeric state, it may be that amplification of the signal is mediated by oligomerization.

Clues to the basis of IR signaling come from many quarters. Florke et al. (1990) have shown that the receptor undergoes a conformational change upon binding insulin. Sucrose density sedimentation analysis revealed that, upon binding insulin, the solubilized IR's sedimentation coefficient decreases from 9.55 to 7.95. Evidence that insulin binding causes quaternary changes between the two $\alpha\beta$ half-receptors is provided by studies of insulin-binding behavior. Gu et al. (1988) and Sweet et al. (1987) have shown that the receptor exhibits negatively cooperative insulin binding. Additionally, Wilden et al. (1989) and Böni-Schnetzler (1986) have demonstrated that insulin mediates reassociation of $\alpha\beta$ half-receptors into heterotetramers and that this association is accompanied by kinase activation.

The apparent communication between the two α subunits suggests that a concerted conformational change may occur, altering the interaction between the half-receptors and thereby causing conformational changes in the cytoplasmic domain, and/or interactions between them. On the other hand, Mortensen et al. (1991) have provided evidence that signaling can occur within an isolated $\alpha\beta$ half-receptor. They reported insulin-responsive autophosphorylation and phosphorylation of exogenous substrates in a Triton-solubilized preparation of $\alpha\beta$ subunits purified by sucrose density gradient centrifugation without reassociation into heterotetramers. This work argues that insulin stimulation of IR kinase activity can occur as an intramolecular process through a single transmembrane domain.

Although it is not clear whether signals are transmitted through conformational changes in single $\alpha\beta$ subunits or between two $\alpha\beta$ subunits, much evidence suggests that intermolecular autophosphorylation is necessary for kinase activation toward exogenous substrates. Treadway et al. (1991) used insulin/insulin-like growth factor-1 hybrids (which are fully functional when wild type) in which one $\alpha\beta$ subunit lacks kinase activity as the result of a point mutation. Autophosphorylation only occurred on the mutant β chain, indicating that intermolecular autophosphorylation is the rule. These mutants, which were incompletely autophosphorylated, did not exhibit kinase activity toward exogenous substrates, implicating the premature termination of an intra-subunit transphosphorylation cascade as the cause.

Additional results pertaining to the likely mechanism of IR signaling include the finding that several mutations made in the IR transmembrane domain do not impede signaling (Frattali et al. 1991) and that a chimeric IR with the cytoplasmic domain of the EGFR still signals, although with higher basal activity (Lammers et al. 1990). These findings argue for a simplistic activation mechanism that is not dependent on precise amino acid sequences in these domains. For instance, insulin binding may bring the transmembrane domains closer to each other, thereby juxtaposing kinase domains and enabling their interaction through transphosphorylation.

Downstream Signaling and Interactions with Effector Molecules

Cytoplasmic domains of receptors employ a variety of mechanisms to couple with intracellular signaling pathways. They associate transiently and stably with other membrane and cytoplasmic proteins; they also interact enzymatically with substrates with or without stably associating. These substrates interface with diverse pathways, including second-messenger pathways, kinase cascades, and transcriptional regulators. When receptors are activated by physiological agonists, drugs, or mutations, the wide-ranging effects can include modulation of Na^+/H^+ exchange, free calcium levels, glucose transport, lipid metabolism, cytoskeletal organization, cell adhesion, gene expression, and parameters of malignant transformation.

Most of the receptors discussed in this paper have tyrosine kinase domains that are activated upon ligand binding. Tyrosine kinase activity has been shown to mediate signaling in several cases (Ullrich and Schlessinger 1990). IR and EGFR enhance their kinase activities by ligand-induced autophosphorylation. Once autophosphorylated, the IR remains active, even if insulin dissociates from the receptor, until the autophosphorylation is reversed. Autophosphorylation of EGFR at a proposed autoinhibition site at the carboxyl terminus may relieve its inhibitory action and enable interaction of the catalytic site with substrates. Evidence described elsewhere in this paper supports the role of autophosphorylation in activating IR kinase activity. The correlation is not definitive, however. Maddux and Goldfine (1991) reported that insulin and ATP binding causes carboxy-terminal conformational changes that may be sufficient for downstream signaling without the occurrence of autophosphorylation. Morrison and Pessin (1987) inhibited in vitro autophosphorylation with high substrate concentrations and still detected insulin-stimulated kinase activity. Additionally, autophosphorylation can occur under certain conditions in the unliganded receptor and is not sufficient for kinase activation. Thus, insulin binding may provide a signaling impulse that does not involve autophosphorylation.

Relevant substrates of the IR kinase include the mitogen-activated protein (MAP) kinase specific for serine and threonine and a phosphatidylinositol kinase (Yonezawa et al. 1990). Other serine/threonine kinases that are implicated in mitogenic regulation have been shown to be activated by tyrosine kinase receptors. They include raf, protein kinase C (PKC), and S6 kinase. The IR activates raf, presumably via an intermediary kinase termed raf kinase kinase (Lee et al. 1991). PKC is activated by PDGFR and activates transcription via transcription factors AP-1 and NFκB. S6 kinase is activated by IR and PDGFR, apparently through the serine/threonine-specific MAP kinase. Both S6 and MAP kinases have recently been found to exist in phosphorylated, active forms in the nucleus of growth-factor-stimulated cells, where they evidently influence transcription of *fos* and other genes (Chen et al. 1992). The *ras* protein is now directly implicated in this pathway of tyrosine kinase receptor signal transduction. Wood et al. (1992) showed that a dominant inhibitory mutant of c-Ha-*ras* suppressed NGF activation of MAP kinase, S6 kinase, and raf kinase. These two papers thus establish steps along a continuous pathway from the plasma membrane to the nucleus.

Many receptors with and without kinase domains act by associating with cytoplasmic kinases and other enzymes, such as phosphatidylinositol kinase, phospholipase C-γ, and GTPase activating protein, that lie in pathways of lipid-derived second messengers and G proteins (Ullrich and Schlessinger 1990). Separate autophosphorylated sites on the carboxyl termini of tyrosine kinase receptors have recently been shown individually to mediate association with different SH2-containing proteins like GAP and phospholipase C. The CD4 receptor on T cells signals through the cytoplasmic *lck* tyrosine kinase that is a member of the *src* family. Cell adhesion receptors lack intrinsic kinase domains, but have recently been shown to activate the $p125^{FAK}$ tyrosine kinase (J. Brugge, pers. comm.).

The adrenergic receptors, members of the seven-transmembrane family, couple to phospholipase C and adenylyl cyclase via G proteins (Dohlman et al. 1991).

Strader et al. (1989) have reviewed the evidence localizing ligand- and G-protein-binding sites. The β_2-adrenergic receptor interacts with G proteins via the third intracellular loop that lies between helices 5 and 6. This loop contains a sequence that has been directly implicated in G-protein interaction and forms a putative amphipathic α helix. Such sequences are found in both α_2 and β_2 adrenergic receptors and resemble the mastaparan peptide known to form a helix in solution, and capable of directly activating G_o. Because three of the four transmembrane side chains that interact with ligand in the hydrophobic core of the β_2 receptor lie in helices 5 and 6, direct allosteric communication between the ligand-binding domain and the effector loop is suggested. As might be predicted, G_s binding has been shown to enable the receptor to adopt its high-affinity ligand-binding state. The mechanisms whereby ligand and effector sites communicate through helices 5 and 6 remain to be elucidated.

Feedback

Receptor signaling levels are carefully modulated by the cell. Several examples exist of direct covalent modifications of receptor cytoplasmic domains by cellular enzymes that diminish receptor signaling. Such desensitization pathways appear to mediate feedback inhibition. The likely purposes of feedback pathways include (1) limiting signal intensity or duration and (2) allowing habituation to a wide range of agonist concentrations in order to retain responsiveness (Koshland 1988).

Bacteria rely on their chemotactic receptors to respond to gradients of attractant and repellent molecules in their aqueous milieu. After aspartate binds its receptor, the receptor modulates flagellar function via several intermediary phosphoproteins. At the same time, the receptor undergoes a conformational change that stimulates methylation of several glutamates on the receptor's cytoplasmic domain by a methyltransferase. Additionally, the activated receptor inhibits a methylesterase by indirectly causing it to be phosphorylated. The resulting increased methylation of the receptor dampens receptor signaling. The aspartate-induced methylation is slightly delayed relative to the signal to the flagella so that the response to an increase in aspartate is dampened soon after the flagella are modulated. In this way, the bacterium compares its current environment with its recent past (its "memory") and responds only to gradient changes, not absolute concentrations (Koshland 1988).

PKC down-regulates the kinase activities of IR and EGFR (see Ullrich and Schlessinger 1990 for references). The PKC phosphorylation site on the EGFR is Thr-654, which lies within 20 amino acids of the transmembrane domain. Evidence that PKC down-regulates these receptors in vivo comes from a variety of experiments (Friedman et al. 1984; Bollag et al. 1986; Takayama et al. 1988). PKC is implicated in circular feedback pathways because of evidence that activation of either receptor may activate the kinase (King and Cooper 1986; Nair et al. 1988). Because PKC is also activated by such agonists as PDGF, bombesin, and adrenergic ligands, it may mediate heterologous receptor desensitization, enabling cells to process multiple inputs in a sophisticated manner. Long-term treatment of cells with phorbol esters down-regulates PKC and sensitizes cells to various agonists. Given that PKC restrains signaling by receptors, it has been proposed that phorbol esters' mitogenic effects on cells in culture may be assisted by down-regulation of PKC and sensitization of cells to agonists in the medium (Biemann and Erikson 1990).

An elaborate set of desensitization pathways has been reported for the adrenergic receptors. In one such mechanism, β-AR is phosphorylated on its third intracellular loop (the proposed effector domain) by the cAMP-dependent protein kinase. In a more complex pathway, β-AR causes its own desensitization when it activates a heterodimeric G protein by releasing the $\beta\gamma$ complex from the α subunit. The $\beta\gamma$ complex then binds to the β-AR kinase (βARK), and translocates it to β-AR (Lefkowitz et al., this volume). βARK then phosphorylates β-AR at its extreme carboxyl terminus, but this phosphorylation alone does not inhibit receptor signaling. The receptor is inhibited after phosphorylation by β arrestin, a protein whose inhibitory effect is increased tenfold toward βARK-phosphorylated receptors (Lohse et al. 1992). This system is evidently replicated for rhodopsin.

CONCLUSIONS

Since the full explanation of the mechanism of transmembrane signaling for a single receptor species has yet to be elucidated, it is obviously too early to be very sure of any model. However, the accumulated literature suggests that transmembrane signaling is likely to be made up of some or all of the models shown in Figures 1 and 3. Certainly association-dissociation, binding-induced conformational changes, and combinations of those two mechanisms have been observed in soluble proteins, and the fragmentary evidence from receptor experiments, as well as simple logic, indicate that they are present in the phenomenon of transmembrane signaling. It is clear that both BC and AD mechanisms exist and that conformational mechanisms of the type shown in Figure 3 are highly probable. The conformational changes for the one- and two-transmembrane receptors provide a convenient clue to the changes in conformation induced by ligand binding in the seven-transmembrane receptors. It will be the work of the future to examine the individual cases in more detail to decide which and how many of these permutations are actually used.

ACKNOWLEDGMENT

The authors acknowledge financial support from the National Institutes of Health (grant DK-0976).

REFERENCES

Bajaj, M., M.D. Waterfield, J. Schlessinger, W.R. Taylor, and T. Blundell. 1987. On the tertiary structure of the extracellular domains of the epidermal growth factor and insulin receptors. *Biochim. Biophys. Acta* **916:** 220.

Barford, D. and L.N. Johnson. 1992. The molecular mechanism for the tetrameric association of glycogen phosphorylase promoted by protein phosphorylation. *Protein Sci.* **1:** 472.

Basu, A., M. Raghunath, S. Bishayee, and M. Das. 1989. Inhibition of tyrosine kinase activity of the epidermal growth factor (EGF) receptor by a truncated receptor form that binds to EGF: Role for interreceptor interaction in kinase regulation. *Mol. Cell. Biol.* **9:** 671.

Bazan, J.F. 1990. Structural design and molecular evolution of a cytokine receptor superfamily. *Proc. Natl. Acad. Sci.* **87:** 6934.

Bellot, F., G. Crumley, J.M. Kaplow, J. Schlessinger, M. Jaye, and C.A. Dionne. 1991. Ligand-induced transphosphorylation between different FGF receptors. *EMBO J.* **10:** 2849.

Biemann, H.-P. and R.L. Erikson. 1990. Abnormal protein kinase C downregulation and reduced substrate levels in non-phorbol ester-responsive 3T3-TNR9 cells. *Mol. Cell. Biol.* **10:** 2122.

Biswas, R., M. Basin, A. Sen-Majumdar, and M. Das. 1985. Intrapeptide autophosphorylation of the epidermal growth factor receptor: Regulation of kinase catalytic function by receptor dimerization. *Biochemistry* **24:** 3795.

Bollag, G.E., R.A. Roth, J. Beaudoin, D. Mochly-Rosen, and D.E. Koshland, Jr. 1986. Protein kinase C directly phosphorylates the insulin receptor in vitro and reduces its protein-tyrosine kinase activity. *Proc. Natl. Acad. Sci.* **83:** 5822.

Boni-Schnetzler, M., J.B. Rubin, and P.F. Pilch. 1986. Structural requirements for the transmembrane activation of the insulin receptor kinase. *J. Biol. Chem.* **261:** 15281.

Chen, R.-H., C. Sarnecki, and J. Blenis. 1992. Nuclear localization and regulation of erk-and rsk-encoded protein kinases. *Mol. Cell. Biol.* **12:** 915.

Clayton, L.K., R.E. Husseyu, R. Steinbrich, H. Ramachandran, Y. Hasain, and E.L. Reinherz. 1988. Substitution of murine for human CD4 residues identifies amino acids critical for HIV gp120 binding. *Nature* **335:** 363.

Cunningham, B.C., M. Ultsch, A.M. De Vos, M.G. Malkerrin, K.R. Clauser, and J.A. Wells. 1991. Dimerization of the extracellular domain of the human growth hormone receptor by a single hormone molecule. *Science* **254:** 821.

De Vos, A.M., M. Ultsch, and A.A. Kossiakoff. 1992. Human growth hormone and extracellular domain of its receptor: Crystal structure of the complex. *Science* **255:** 306.

Diederichs, K., T. Boone, and P.A. Karplus. 1991. Novel fold and putative receptor binding site of granulocyte-macrophage colony-stimulating factor. *Science* **254:** 1779.

Dohlman, H.G., J. Thorner, M.G. Caron, and R.J. Lefkowitz. 1991. Model systems for the study of seven-transmembrane-segment receptors. *Annu. Rev. Biochem.* **60:** 653.

Duan, D.-S.R., M.J. Pazin, L.J. Fretto, and L.T. Williams. 1991. A functional soluble extracellular region of the platelet-derived growth factor (PDGF) β-receptor antagonizes PDGF-stimulated responses. *J. Biol. Chem.* **266:** 413.

Falke, J.J. and D.E. Koshland. 1987. Global flexibility in a sensory receptor: A site-directed cross-linking approach. *Science* **237:** 1596.

Farber, G.K. and G.A. Petsko. 1990. The evolution of barrel enzymes. *Trends Biochem. Sci.* **15:** 228.

Florke, R.-R., H.W. Klein, and H. Reinauer. 1990. Structural requirements for signal transduction of the insulin receptor. *Eur. J. Biochem.* **191:** 473.

Foster, D.L., S.L. Mowbray, B.K. Jap, and D.E. Koshland. 1985. Purification and characterization of the aspartate chemoreceptor. *J. Biol. Chem.* **260:** 11706.

Frattali, A.L., J.L. Treadway, and J.E. Pessin. 1991. Evidence supporting a passive role for the insulin receptor transmembrane domain in insulin-dependent signal transduction. *J. Biol. Chem.* **266:** 9829.

Friedman, B., A.R. Frackelton, Jr., A.H. Ross, J.M. Connors, H. Fujiki, T. Sugimura, and M.R. Rosner. 1984. Tumor promoters block tyrosine-specific phosphorylation of the epidermal growth factor receptor. *Proc. Natl. Acad. Sci.* **81:** 3034.

Fuh, G., M.G. Mulkerrin, S. Bass, N. McFarland, M. Brochier, J.H. Bourell, D.R. Light, and J.A. Wells. 1990. The human growth hormone receptor: Secretion from *Escherichia coli* and disulfide bonding pattern of the extracellular binding domain. *J. Biol. Chem.* **265:** 3111.

Gardina, P., C. Conway, M. Kossman, and M. Manson. 1992. Aspartate and maltose-binding protein interact with adjacent sites in the Tar chemotactic signal transducer of *Escherichia coli*. *J. Bacteriol.* **174:** 1528.

Greenfield, C., I. Hiles, M.D. Waterfield, M. Federwisch, A. Wollmer, T.L. Blundell, and N. McDonald. 1989. Epidermal growth factor binding induces a conformational change in the external domain of its receptor. *EMBO J.* **8:** 4115.

Gu, J.-L., I.D. Goldfine, J.R. Forsayeth, and P. De Meyts. 1988. Reversal of insulin-induced negative cooperativity by monoclonal antibodies that stabilize the slowly dissociating ("Ksuper") state of the insulin receptor. *Biochem. Biophys. Res. Commun.* **150:** 694.

Hempstead, B.L., D. Martin-Zanca, D.R. Kaplan, L.F. Parada, and M.V. Chao. 1991. High-affinity NGF binding requires coexpression of the trk proto-oncogene and the low affinity NGF-receptor. *Nature* **350:** 678.

Henderson, R., J.M. Baldwin, T.A. Ceska, F. Zemlin, E. Beckman, and K.H. Downing. 1990. Model for the structure of bacteriorhodopsin based on high resolution electron cryo-microscopy. *J. Mol. Biol.* **213:** 899.

Holmgren, A. and C.-I. Branden. 1989. Crystal structure of chaperone protein PapD reveals an immunoglobulin fold. *Nature* **342:** 248.

Honegger, A.M., A. Schmidt, A. Ullrich, and J. Schlessinger. 1990. Evidence for epidermal growth factor (EGF)-induced intermolecular autophosphorylation of the EGF receptors in living cells. *Mol. Cell. Biol.* **10:** 4035.

Kashles, O., Y. Yarden, R. Fischer, A. Ullrich, and J. Schlessinger. 1991. A dominant negative mutation suppresses the function of normal epidermal growth factor receptors by heterodimerization. *Mol. Cell Biol.* **11:** 1454.

Kashles, O., D. Szapary, F. Bellot, A. Ullrich, J. Schlessinger, and A. Schmidt. 1988. Ligand-induced stimulation of epidermal growth factor receptor mutants with altered transmembrane regions. *Proc. Natl. Acad. Sci.* **85:** 9567.

Kelly, J.D., B.A. Haldeman, F.J. Grant, M.J. Murray, R.A. Seifert, D.F. Bowen-Pope, J.A. Cooper, and A. Kaslauskas. 1991. Platelet-derived growth factor (PDGF) stimulates PDGF receptor subunit dimerization and intersubunit trans-phosphorylation. *J. Biol. Chem.* **266:** 8987.

King, C.S. and J.A. Cooper. 1986. Effects of protein kinase C activation after epidermal growth factor receptor phosphorylation. *J. Biol. Chem.* **261:** 10073.

Koland, J.G. and R.A. Cerione. 1988. Growth factor control of epidermal growth factor receptor kinase activity via an intramolecular mechanism. *J. Biol. Chem.* **263:** 2230.

Koshland, D.E., Jr. 1988. Chemotaxis as a model second-messenger system. *Biochemistry* **27:** 5829.

Kubar, J. and E. Van Obbergen. 1989. Oligomeric states of the insulin receptor: Binding and autophosphorylation properties. *Biochemistry* **28:** 1086.

Kwong, P.D., S.-E. Ryu, W.A. Hendrickson, R. Axel, R.M. Sweet, G. Folena-Wasserman, P. Hensley, and R.W. Sweet. 1990. Molecular characteristics of recombinant

human CD4 as deduced from polymorphic crystals. *Proc. Natl. Acad. Sci.* **87:** 6423.
Lamarre, D., A. Ashkenazi, S. Fleury, D.H. Smith, R.-P. Sekaly, and D.J. Capon. 1989. The MHC-binding and gp120 binding functions of CD4 are separable. *Science* **245:** 743.
Lammers, R., E. Van Obbergen, R. Ballotti, J. Schlessinger, and A. Ullrich. 1990. Transphosphorylation as a possible mechanism for insulin and epidermal growth factor receptor activation. *J. Biol. Chem.* **265:** 16886.
Lee, A.W.-M. and A.W. Nienhuis. 1990. Mechanism of kinase activation in the receptor for colony-stimulating factor 1. *Proc. Natl. Acad. Sci.* **87:** 7270.
Lee, R.-M., U.R. Rapp, and P.J. Blackshear. 1991. Evidence for one or more Raf-1 kinase kinase(s) activates by insulin and polypeptide growth factors. *J. Biol. Chem.* **266:** 10351.
Lohse, M.J., S. Andexinger, J. Pitcher, S. Trukawinski, J. Codina, J.-P. Faure, M.G. Caron, and R.J. Lefkowitz. 1992. Receptor-specific desensitization with purified proteins. Kinase dependence and receptor specificity of β-arrestin and arrestin in the β_2-adrenergic receptor and rhodopsin systems. *J. Biol. Chem.* **267:** 8558.
Lynch, B.A. and D.E. Koshland, Jr. 1991. Disulfide cross-linking studies of the transmembrane regions of the aspartate sensory receptor of *Escherichia coli. Proc. Natl. Acad. Sci.* **88:** 10402.
———. 1992. Structural similarities between the aspartate receptor of bacterial chemotaxis and the Trp repressor of *E. coli*: Implications for transmembrane signalling. *FEBS Lett.* **307:** 3.
Maddux, B.A. and I.D. Goldfine. 1991. Evidence that insulin plus ATP may induce a conformational change in the β subunit of the insulin receptor without inducing receptor autophosphorylation. *J. Biol. Chem.* **266:** 6731.
Metzger, H. 1974. Effect of antigen binding on the properties of antibody. *Adv. Immunol.* **18:** 169.
Milburn, M.V., G.G. Prive, D.L. Milligan, W.G. Scott, J. Yeh, J. Jancarik, D.E. Koshland, Jr., and S.H. Kim. 1991. Three-dimensional structures of the ligand-binding domain of the bacterial aspartate receptor with and without a ligand. *Science* **254:** 1342.
Milligan, D.L. 1991. "Structure and dynamics of the aspartate chemoreceptor." Ph.D. thesis, University of California, Berkeley.
Milligan, D.L. and D.E. Koshland, Jr. 1988. Site-directed cross-linking: Establishing the dimeric structure of the aspartate receptor of bacterial chemotaxis. *J. Biol. Chem.* **263:** 6268.
———. 1991. Intrasubunit signal transduction by the aspartate chemoreceptor. *Science* **254:** 1651.
Moe, G.R., G.E. Bollag, and D.E. Koshland, Jr. 1989. Transmembrane signaling by a chimera of the *Escherichia coli* aspartate receptor and the human insulin receptor. *Proc. Natl. Acad. Sci.* **86:** 5683.
Morrison, B.D. and J.E. Pessin. 1987. Insulin stimulation of the insulin receptor kinase can occur in the complete absence of β subunit autophosphorylation. *J. Biol. Chem.* **262:** 2861.
Mortensen, E.R., J.G. Drachman, and G. Guidotti. 1991. The $\alpha\beta$ monomer of the insulin receptor has hormone-responsive tyrosine kinase activity. *Biochem. J.* **273:** 49.
Mowbray, S.L. and D.E. Koshland, Jr. 1987. Additive and independent responses in a single receptor. Aspartate and maltose stimuli on the Tar protein. *Cell* **50:** 171.
———. 1990. Mutations in the aspartate receptor of *Escherichia coli* which affect aspartate binding. *J. Biol. Chem.* **265:** 15638.
Nair, G.P., M.L. Standaert, R.J. Pollet, D.R. Cooper, and R.V. Farese. 1988. Effects of insulin and phorbol esters on diacylglycerol generation and synthesis and hydrolysis of phosphatidylcholine in BC3H-1 myocytes. *Biochem. Biophys. Res. Commun.* **154:** 1345.
Peterson, A. and B. Seed. 1988. Genetic analysis of monoclonal antibody and HIV binding sites on the human lymphocyte antigen CD4. *Cell* **54:** 65.
Rosen, O.M. 1987. After insulin binds. *Science* **237:** 1452.
Russo, A.F. and D.E. Koshland. 1983. Separation of signal transduction and adaptation functions of the aspartate receptor in bacterial sensing. *Science* **220:** 1016.
Ryu, S.-E., P.D. Kwong, A. Truneh, T.G. Porter, J. Arthos, M. Rosenberg, X. Dai, N.-H. Xuong, R. Axel, R.W. Sweet, and W.A. Hendrickson. 1990. Crystal structure of an HIV-binding recombinant fragment of human CD4. *Nature* **348:** 419.
Sayre, P. and E.L. Reinherz. 1985. Structural invariance of T4 molecules from T cell clones of different antigen and major histocompatibility complex specificities. *Eur. J. Immunol.* **15:** 291.
Shoelson, S.E., M.F. White, and R.C. Kahn. 1988. Tryptic activation of the insulin receptor. Proteolytic truncation of the α-subunit releases the β-subunit from inhibitory control. *J. Biol. Chem.* **263:** 4852.
Stoddard, B.L. and D.E. Koshland, Jr. 1992. Prediction of the structure of a protein-receptor complex using a binary docking method. *Nature* **358:** 774.
Strader, C.D., I.S. Sigal, and R.A.F. Dixon. 1989. Structural basis of β-adrenergic receptor function. *FASEB J.* **3:** 1825.
Strader, C.D., I.S. Sigal, R.B. Register, M.R. Candelore, E. Rands, and R.A.F. Dixon. 1987. Identification of residues required for ligand binding to the β-adrenergic receptor. *Proc. Natl. Acad. Sci.* **84:** 4384.
Sweet, L.J., B.D. Morrison, and J.E. Pessin. 1987. Isolation of functional $\alpha\beta$ heterodimers from the purified human placental $\alpha_2\beta_2$ heterotetrameric insulin receptor complex. A structural basis for insulin binding heterogeneity. *J. Biol. Chem.* **262:** 6939.
Takayama, S., M.F. White, and C.R. Kahn. 1988. Phorbol ester-induced serine phosphorylation of the insulin receptor decreases its tyrosine kinase activity. *J. Biol. Chem.* **263:** 3440.
Thorsness, P.E., S.L. Mowbray, and D.E. Koshland, Jr. 1987. Transmission of regulatory conformational changes through proteins. *Cold Spring Harbor Symp. Quant. Biol.* **52:** 623.
Treadway, J.L., B.D. Morrison, M.A. Soos, K. Siddle, J. Olefsky, A. Ullrich, D.A. McClain, and J.E. Pessin. 1991. Transdominant inhibition of tyrosine kinase activity in mutant insulin/insulin-like growth factor hybrid receptors. *Proc. Natl. Acad. Sci.* **88:** 214.
Ullrich, A. and J. Schlessinger. 1990. Signal transduction by receptors with tyrosine kinase activity. *Cell* **61:** 203.
Wang, J., Y. Yan, T.P.J. Garrett, J. Liu, D.W. Rodgers, R.L. Garlick, G.E. Tarr, Y. Husain, E.L. Reinherz, and S.C. Harrison. 1990. Atomic structure of a fragment of human CD4 containing two immunoglobulin-like domains. *Nature* **348:** 411.
Wilden, P.A., B.D. Morrison, and J.E. Pessin. 1989. Relationship between insulin receptor subunit association and protein kinase activation: Insulin-dependent covalent and Mn/Mg ATP-dependent noncovalent association of $\alpha\beta$ heterodimeric insulin receptors into an $\alpha_2\beta_2$ heterotetrameric state. *Biochemistry* **28:** 785.
Williams, L.T. 1989. Signal transduction by the platelet-derived growth factor receptor. *Science* **243:** 1564.
Wilson, C., M.R. Wardell, K.H. Weisgraber, R.W. Mahley, and D.A. Agard. 1991. Three-dimensional structure of the LDL receptor-binding domain of human apolipoprotein E. *Science* **252:** 1817.
Wood, K.W., C. Sarnecki, T.M. Roberts, and J. Blenis. 1992. *ras* mediates nerve growth factor receptor modulation of three signal-transducing protein kinases: MAP kinase, Raf-1, and RSK. *Cell* **68:** 1041.
Wolff, C. and J.S. Parkinson. 1988. Aspartate taxis mutants of the *Escherichia coli* Tar chemoreceptor. *J. Bacteriol.* **170:** 4509.
Yarden, Y. and J. Schlessinger. 1987a. Epidermal growth

factor induces rapid, reversible aggregation of the purified epidermal growth factor receptor. *Biochemistry* **26:** 1443.
———. 1987b. Self-phosphorylation of the epidermal growth factor receptor: Evidence for a model of intermolecular allosteric activation. *Biochemistry* **26:** 1434.
Yonezawa, K., G. Endeman, K.S. Kovacina, J.E. Chin, C. Stover, and R.A. Roth. 1990. Substrates of the insulin receptor kinase. *Adv. Second Messenger Phosphoprotein Res.* **24:** 266.
Youvan, D.C. and F. Daldal, eds. 1986. *Current communications in molecular biology: Microbial energy transduction: Genetics, structure, and function of membrane proteins*, p. 163. Cold Spring Harbor Laboratory, Cold Spring Harbor, New York.

A Model for Transmembrane Signaling in a Bacterial Chemotaxis Receptor

S.-H. Kim, G.G. Privé,* J. Yeh, W.G. Scott, and M.V. Milburn†
Department of Chemistry and Lawrence Berkeley Laboratory, University of California, Berkeley, California 94720

Bacteria in general can sense the increase in the concentration of attractants or decrease in repellent concentration in their environment and move toward attractants or away from the repellents by suppressing tumbling motions of their flagella (Berg and Brown 1972; Adler 1975; Macnab 1987; Koshland 1988). These bacteria have several types of receptors that can sense a wide variety of chemicals and physical properties (including aspartate, serine, cysteine, certain dipeptides, malate, citrate, maltose, glucose, galactose, ribose, phenol, Co^{++}, Ni^{++}, pH, and temperature), and their signals are integrated into a single intracellular pathway that ultimately regulates the mechanics of the flagellar motor.

Present addresses: *Molecular Biology Institute, University of California, Los Angeles, California 90024; † Glaxo, Inc., 5 Moore Drive, Research Triangle Park, North Carolina 27790.

Analogous to many transmembrane receptors, such as the receptors of epidermal growth factor, platelet-derived growth factor, insulin, and low-density lipoprotein, most bacterial chemotaxis receptors known so far in *Escherichia coli* and *Salmonella typhimurium* are composed of three domains: a ligand-binding domain, a transmembrane domain (consisting of two presumed helices in the case of the chemotaxis receptors), and a cytoplasmic domain (Fig. 1A). The signaling in the cytoplasm is accomplished by a fast process of phosphate transfer between several pairs of cytoplasmic proteins counterbalanced by phosphatases (Borkovich et al. 1989; Bourret et al. 1989), finally activating the rotor of each flagellum. This signaling is further modulated by a slower methylation and demethylation process of the cytoplasmic domains as an adaptation process (Russo and Koshland 1983). Most proteins in the signaling and adaptation pathways have been iden-

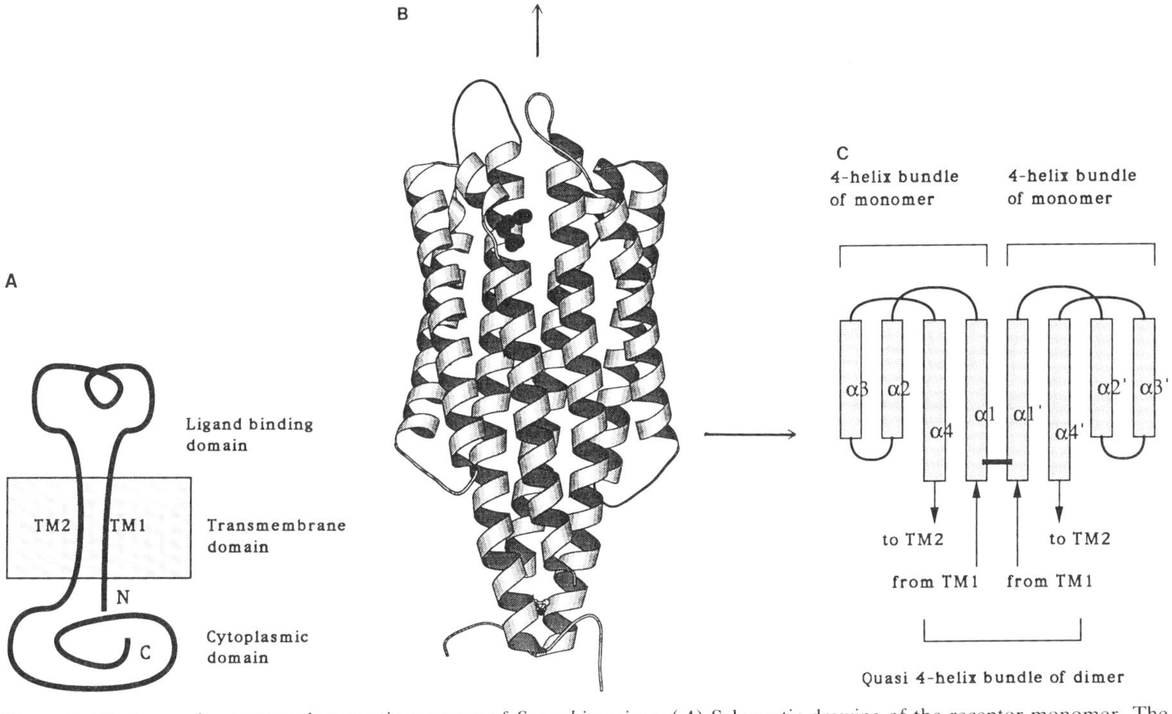

Figure 1. Features of aspartate chemotaxis receptor of *S. typhimurium*. (*A*) Schematic drawing of the receptor monomer. The first and second transmembrane regions are indicated by TM1 and TM2, respectively. (*B*) Backbone structure of the disulfide-bonded ligand domain dimer. The local twofold axis (shown as a vertical arrow) relating two monomers runs vertically through the middle of the dimer. One bound aspartate is shown. The crystal structure of the apo protein without the bound ligand is very similar to this. The virtual pivot axis around which one monomer rotates with respect to the other on aspartate binding is shown as a horizontal arrow. (*C*) Topological structure of the ligand domain dimer of the receptor. Two types of four-helical bundles are indicated.

tified, cloned, and expressed, thus making the bacterial chemotaxis system one of the best-characterized signal transduction systems. Although the biochemistry and regulation of these proteins have been extensively studied, the mechanism of transmembrane signaling is not known. We present a simple, hypothetical mechanism for transmembrane signaling based on the crystal structures of a ligand-binding domain of a bacterial chemotaxis receptor with and without a bound ligand.

METHODS

The crystal structures used in the analysis are those of a mutant of the ligand-binding domain of the aspartate receptor from *S. typhimurium* with and without aspartate (Milburn et al. 1991). The crystal structure of the apo form of the domain has since been refined at 2.2 Å resolution to an R factor of 18.9% (W. Scott et al., in prep.) and that of the aspartate complex at 2.0 Å to an R factor of 19% (G. Privé et al., in prep.). Both structures have been refined by simulated annealing refinement using the computer program package XPLOR (Brünger 1990). Preparation of the proteins, crystallization conditions, structure determinations, and description of the structures have been published previously (Milburn et al. 1991), and a brief summary is given below.

Crystal structure of the apo form of the ligand-binding domain. The ligand-binding domain of the aspartate receptor from *S. typhimurium* (beginning with Met-25 and ending with Arg-188) was produced from a multicopy plasmid. The proteins exported to the periplasm were extracted by mild osmotic lysis, precipitated by ammonium sulfate, and purified to homogeneity by anion exchange chromatography. The purified protein has been shown to bind aspartate with an affinity similar to intact receptors (Milligan 1991). Because crystals of the wild-type protein diffracted poorly, a disulfide-linked dimer of a Cys-36 mutant (asparagine at residue 36 was changed to cysteine) was crystallized and used for high-resolution studies. The corresponding intact receptor dimer containing a Cys-36 intermolecular cross-link has similar biological activity as the wild-type receptor. However, this cross-linked receptor appears to be somewhat activated when compared to the wild-type receptor (Milligan 1991).

High-quality single crystals of the Cys-36 mutant of the ligand-binding domain in the absence of aspartate were obtained by the sitting drop technique. The crystallization solutions contained 10 mg/ml protein, 0.05 M HEPES buffer (pH 7.4), 0.75 M Li_2SO_4, and 0.5 mM Cu(II)[1,10-phenanthroline]$_3$ (added to maintain the disulfide cross-link), and were equilibrated against 0.1 M HEPES and 1.5 M Li_2SO_4 at the same pH at 21°C. The space group of the crystals is $P6_522$ with cell dimensions $a = b = 80.03$ Å, $c = 155.6$ Å. One half of the Cys-36 cross-linked dimer resides in an asymmetric unit.

Multiple isomorphous replacement (MIR) with anomalous scattering phase refinement was carried out to 3.0 Å resolution. The MIR phase set was then solvent-flattened (Wang 1985), and phases were extended to 2.7 Å resolution with the calculated molecular envelope of the protein at 3.0 Å resolution. The resulting electron density map clearly revealed a cluster of four α helices. A total of 154 amino acids were assigned to the electron density starting with Met-25 and ending at Phe-180. The remaining 8 residues were disordered. The structure was further refined to an R factor of 18.9% using 2.2 Å resolution data (W. Scott et al., in prep.).

The ligand-binding domain of the aspartate receptor consists of two monomers, with each monomer being a four-helix bundle as shown in Figure 1B. This is consistent with an earlier biochemical observation that the functional form of the intact aspartate receptor or the ligand-binding domain is a dimer (Milligan and Koshland 1988). The first long helix spans the length of the periplasmic domain from residue 32 to residue 76, yielding a continuous 13-turn α helix with 45 amino acids. The longest dimension of the molecule along the axis of the helix bundle is about 80 Å, whereas the diameter is roughly 20 Å. Two monomers, A and B, are related by an exact crystallographic twofold axis that runs along the long axis of the dimer near the two amino-terminal helices (Fig. 1B,C). This results in a close association between $\alpha 1$ and $\alpha 1'$ helices from both monomers. With $\alpha 4$ and $\alpha 4'$ nearby, these four helices form a quasi-four-helix bundle in the core of the dimer (Fig. 1C). The dimer interface shows many nonpolar interactions between $\alpha 1$ and $\alpha 1'$, and some between $\alpha 4$ and $\alpha 4'$.

Crystal structure of the aspartate-complexed ligand domain. The same Cys-36 mutant protein described above was complexed with aspartate and crystallized from a solution containing 10 mg/ml protein, 0.5 mM aspartate, 50 mM HEPES buffer (pH 7.4), 0.75 M Li_2SO_4, and 0.5 mM Cu(II)[1,10-phenanthroline]$_3$ equilibrated against a reservoir of 1.5 M Li_2SO_4 and 0.1 M HEPES (Milburn et al. 1991). The protein crystallized in space group $P3_221$ with unit cell dimensions $a = b = 80.38$ Å, $c = 91.35$ Å. The structure was solved by molecular replacement with Patterson Correlation refinement (Brünger 1990) using a monomer from the unliganded structure as a probe model. Residues 25–34 and 76–86, corresponding to the amino terminus and the first loop region of the apo structure, respectively, were not included in the search model. Two independent and unambiguous solutions were found, corresponding to a Cys-36 cross-linked dimer in an asymmetric unit. The structure has been refined at 2.0 Å resolution to an R factor of 19% (G. Privé et al., in prep.).

The molecular symmetry of the dimer generates two ligand-binding sites, both at the dimer interface near the top of the molecule and distal from the membrane (Fig. 1B). Each ligand-binding site is composed of *two half-sites*, one from each monomer. Of the two ligand-binding sites, one is fully occupied and shows clear electron density for an aspartate. The second site shows weaker density that may be due to a disordered aspartate or a sulfate ion, which is present in high concentration in the crystals. Thus, it appears that only one of two sites is occupied fully, despite the fact that aspartate is present approximately in equimolar concentration to the monomers. This concentration is 100 to 1000 times the dissociation constant measured at the salt concentration of the crystallization condition (H. Biemann, unpubl.).

Modeling of the transmembrane domain. On the basis of the dimer structure of the ligand-binding domain, we have constructed a model joining the ligand-binding and transmembrane domains. The putative transmembrane (TM) regions of the protein sequence (residues 7–30 of TM1 and residues 191–209 of TM2) are quite hydrophobic in character and are predicted to form two α helices which we have modeled to continue uninterrupted from the amino- and carboxy-terminal long helices (α1 and α4, respectively). The attached transmembrane helices were then energy-minimized using the program XPLOR v2.1 (Brünger 1990) to reduce sterically unfavorable side-chain contacts and to optimize the interaction between two TM1 helices in a fashion similar to the $\alpha 1/\alpha 1'$ dimer interactions in the ligand domain structure. The resulting model is depicted in Figure 2. In the model, four α helices, two from each monomer, form a very long quasi-four-helix bundle and connect the ligand-binding region and the cytoplasmic domain, which are separated by a distance of more than 100 Å. This continuity of the quasi-four-helix bundle is used to provide the mechanical link for signal transmission in our model.

Structural comparison. To assess conformational differences between any pair of compared monomers, domains, or structural elements, we first superposition all or parts of the structures by a least-squares method. Then we calculate either distances between any two corresponding Cα atoms, or root-mean-square (r.m.s.) distances of all compared Cα atom pairs. In all comparative calculations, only Cα atoms in helices are used: Cα atoms for residues in the first loop region, residues 149–154 near the amino group of aspartate, the hydrophobic amino-terminal residues, and the disordered carboxy-terminal residues are not used in any calculations. In some cases, Cα atoms of residues 25–53 are further removed to eliminate any possible artifact due to the disulfide bond in the Cys-36 mutant. In some comparisons, structural features of monomer A of the apo protein are compared to monomer A of the aspartate complex after the two A monomers have been superpositioned (when inquiring about any conformational changes within the monomers) or after two B monomers have been superpositioned (when intermonomer changes are studied).

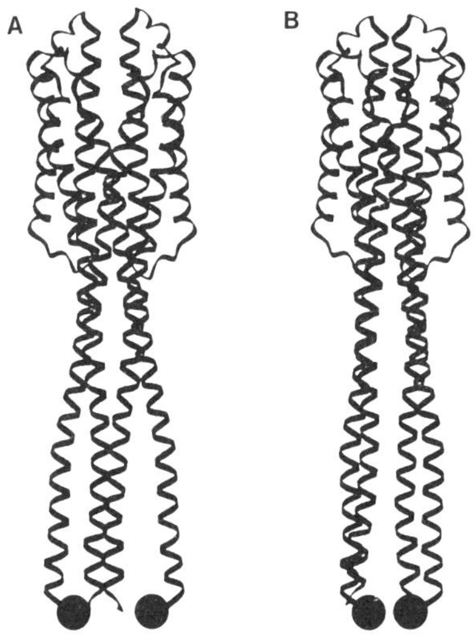

Figure 2. A molecular model containing the crystal structure of the ligand domain of apo receptor and four presumed transmembrane helices connected to the core four-helix bundle of the ligand domain. Filled circles represent the point where the cytoplasmic domain of the receptor starts. (*B*) Same as *A* but for the aspartate complex.

RESULTS

Conserved Monomer Conformation

The current prevailing model for the transmembrane signaling in the aspartate receptor is the "piston" model, in which a signal is transmitted by the change in relative positions of two transmembrane helices within a monomer (Fig. 3). A least-squares fit of the Cα atoms between the apo monomer and the monomers of the aspartate-complexed dimer reveal that the monomer structures remain largely unchanged on ligand binding except for loop L1, residues 149–154 near the amino group of aspartate, the hydrophobic amino-terminal residues, and the disordered carboxy-terminal residues. After residues 25–53 are further removed to eliminate possible artifact from the disulfide bond, the r.m.s. difference is 0.63 Å when A monomer of the apo protein is compared with A monomer of the aspartate complex, and 0.38 Å when B monomers of the two proteins are compared. These values are comparable to the r.m.s. difference of 0.78 Å obtained between Cα atoms of two noncrystallographically related monomers (A and B monomers) in the aspartate complex (Table 1).

We then examined the possible changes of α1 with respect to α4 on ligand binding within the monomers by superpositioning α1 and α4 of an apo monomer to those of the A or B monomer of the aspartate complex. The calculated r.m.s. difference was 0.60 Å for one monomer and 0.49 Å for the other. These values are again about the same magnitude as the corresponding r.m.s. distance of 0.74 Å between the two pairs of helices in the aspartate complex which are related by a noncrystallographic pseudo-twofold axis. These comparisons strongly suggest that the monomer conformation in the apo structure is practically the same as that in the aspartate complex.

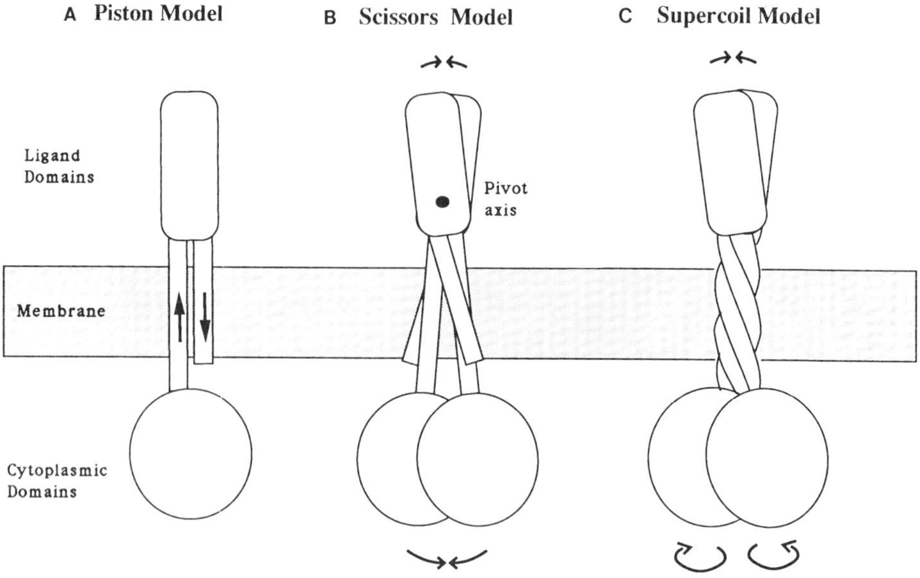

Figure 3. Three models for the mechanism of transmembrane signal transmission in aspartate chemotaxis receptor of *S. typhimurium*. (*A*) Piston model where ligand binding alters the relative position/orientation of two transmembrane helices (TM1 and TM2) within a monomer. (*B*) Scissors model where an attractant binding brings two ligand domains closer by rotation of one monomer with respect to the other around a pivot axis. (*C*) Supercoil model where an attractant binding unwinds the negatively supercoiled, central quasi-four-helical bundle of dimer, resulting in the change of the relative orientation of the two cytoplasmic domains.

Pivot Rotation of Monomer

The most significant difference between the apo and complex structures was found to be the relative orientation of one monomer with respect to the other. The effect of aspartate binding is to bring the residues in the half pockets from both monomers closer together. By pulling the upper part of the monomers closer to each other, the aspartate locks the dimer into a tighter complex and neutralizes the strong positive electrostatic potential in the binding pocket. This closing movement also results in a reduction of water-accessible area at the interface by 20%.

As mentioned earlier, the r.m.s. differences between two superimposed A monomers of the apo and complex structures is 0.63 Å, and the corresponding value for the B monomers is 0.38 Å. However, the r.m.s. difference between two A monomers after two B monomers are superpositioned is 1.38 Å, and that between two B monomers after two A monomers are superpositioned is 1.14 Å (Table 1), indicating that the primary conformational change upon ligand binding is in the relative position and orientation of one monomer with respect to the other in a dimer.

The simplest description of the conformational changes consistent with the above observation is a 4.5° rotation between two monomers about a virtual pivot axis which is roughly perpendicular to the twofold axis of the dimer (Fig. 1B). This pivot rotation occurs despite the disulfide cross-link of the two monomers near one end of the molecule (Fig. 2B). The fact that the ligand binds at the dimer interface triggering this rota-

Table 1. Conformational Differences between Apo and Aspartate Complex of the Aspartate Receptor Ligand Domain

Superpositioned	r.m.s. distance calculated	r.m.s. (Å)
Apo A/complex A	apo A/complex A	0.63
Apo B/complex B	apo B/complex B	0.38
Complex A/complex B	complex A/complex B	0.78
Apo A/complex A	apo B/complex B	1.14
Apo B/complex B	apo A/complex A	1.38
α1, α4 (apo A)/α1, α4 (complex A)	α1, α4 (apo A)/α1, α4 (complex A)	0.60
α1, α4 (apo B)/α1, α4 (complex B)	α1, α4 (apo B)/α1, α4 (complex B)	0.49
α1, α4 (complex A)/α1, α4 (complex B)	α1, α4 (complex A)/α1, α4 (complex B)	0.74

Two monomers in the apo dimer (apo A and apo B) are related by a crystallographic twofold axis, but those in the aspartate complexed dimer (complex A and complex B) are related by a pseudo twofold axis. In all calculations, only Cα atoms of helices are used. Thus excluded in comparisons are the hydrophobic amino-terminal residues, the first loop residues and residues 149–154 near the "amino pocket," and the disordered carboxy-terminal residues. Residues 25–53 are further excluded to remove any possible artifact due to Cys-36 cross-link.

tional motion also underscores the importance of the dimeric form for signaling.

It is possible that the high-salt environment of crystallization mimics the presence of aspartate, thus reducing the conformational difference to this small rotational difference.

DISCUSSION

The current prevailing piston model (Fig. 3A) is based on the monomeric form of the receptor in which one transmembrane helix moves in relation to the other *within the monomer* (Fig. 3A) (for review, see Koshland 1988). The strongest evidence in support of this model is given in a recent experiment of Milligan and Koshland (1991): The Cys-36 mutant of the *S. typhimurium* receptor was used to covalently cross-link two monomers, thus restricting the relative motion of the two monomers and stabilizing hybrid dimers in which one of the monomers was truncated to remove the cytoplasmic domain. In this study, the methylation rate of the cytoplasmic domain, which measures the extent of adaptation, was taken as an indirect measure of signal transmission. These investigators showed that in all cases, there was a preferential methylation in the aspartate-liganded form. This differential methylation of the Cys-36 cross-linked dimer, in which the movement of the two monomers is restricted and the second cytoplasmic domain is absent, was presented as support for the monomeric piston model of signal transmission.

Three points can be made to suggest that the interpretation of the results may be more complicated than was given. First, the Cys-36 *inter*-monomer cross-linked mutant behaves differently from the wild type in that the hydrodynamic properties of the wild-type dimer differ from those of the cross-linked dimer as measured by gel filtration (Milligan 1991) and also in that the methylation rates of the hybrid dimers are low compared to those of the intact wild-type dimer (Falke et al. 1988; Milligan and Koshland 1991). Second, the *intra*-monomer disulfide mutant with cross-links between residues Cys-36 and Cys-183 *within* a monomer, which should restrict a piston type of motion, shows similar preferential methylation behavior of the liganded form (Falke et al. 1988) as the *inter*-monomer, Cys-36 cross-linked dimer. Third, and giving the strongest evidence against this model, is that the crystal structures of the Cys-36 cross-linked ligand domain dimer with and without the ligand (Milburn et al. 1991) show little conformational change (see below and Results section above) within the monomer that could lead to the relative movement of the two transmembrane helices (TM1 and TM2) as required by the piston model.

Scissors and Supercoil Models

To estimate the effect of the pivot rotation of the ligand monomers on the cytoplasmic domains, we have built two models (Fig. 2); one for the ligand-bound receptor and the other for the apo receptor, by joining the four helices of the ligand domain ($\alpha 1$, $\alpha 4$, $\alpha 1'$, and $\alpha 4'$) to the four presumed transmembrane helices. The models were energy-minimized as described in the Methods section (Milburn et al. 1991). These models are consistent with biochemical evidence of cross-linking ligand domain and transmembrane helices; the $\alpha 1$ and TM1 of one monomer is in contact with those of the other monomer in a dimer: $\alpha 1$ helix is contiguous to TM1, and there is little contact between TM1 of one monomer and TM2 of the other monomer (Falke et al. 1988; Lynch and Koshland 1991; Pakula and Simon 1992).

As shown in Figure 2 and schematically in Figure 3B and 3C we propose that a small relative rotation of the periplasmic domains could cause a scissoring (Fig. 3B) or supercoil unwinding (Fig. 3C) motion of the transmembrane helices. Such motions would bring changes in the relative position and/or orientation of the two cytoplasmic domains, thus conveying the ligand-bound state of the periplasmic domain. The former assumes that the transmembrane helices have no mutual affinity, and the latter presumes that they interact with each other to form supercoils. Either mechanism or a combination of two would result in a change in the relative position and orientation of the two cytoplasmic domains.

Reverse Signaling

Simple mechanical models of the transmembrane signaling proposed here also imply that the signaling can travel in the reverse direction: Modification of the cytoplasmic domains or interaction of the domains with other proteins in the cytoplasm could cause rearrangement of the transmembrane helices or the quasi-four-helix bundle to transmit the information from the cytoplasmic side to the ligand domain side. This would affect the affinity of the ligand domains for their ligands in a subtle way. Such two-way signaling may have functional utility in this or other receptor systems in regulating the response to the environment depending on the cellular states.

"Frozen Dynamic" States of a Fuzzy Switch

It could be argued that the "open" and "closed" conformations of the apo and the complex structures are partly due to the fact that the two structures crystallized in two different lattices and both under high-salt condition, and they represent only one each of many conformations possible by the apo and the complex molecules, respectively, in solution. However, it is reasonable to assume that, due to many stabilizing hydrogen-bonding interactions between the ligand and the residues in the ligand domain interface as observed in the crystal structure of the complex, there may be primarily one conformation for the ligand-bound form. In these models, the unliganded receptor monomer is assumed to be in a dynamic motion (resembling, for example, a dynamically scissoring motion around a virtual pivot axis or a dynamically winding/unwinding motion around a vertical supercoil axis) relative to the other monomer within the dimeric receptor (Fig. 3A). This dynamic flexibility is "frozen" or quenched when the ligand binds to the protein, resulting in a partially fixed closed or unwound conformation. Such dynamic motion of the apo receptor monomers has been implicated in cross-linking experiments of cysteine-containing mutants (Falke and Koshland 1987), and the observed decrease in subunit exchange rate when a ligand binds to the wild-type receptor dimers (Milligan and Koshland 1988) is consistent with this concept.

This concept also provides a simple suggestion in describing three states of the receptor population. We make a starting assumption that the flagella motors are rotating counterclockwise in the absence of any signals from receptors (Bourret et al. 1989). In the absence of attractants or repellents, the receptors are assumed to be in dynamic motion generating signals for tumbling (clockwise rotation of flagella) stochastically, thus resulting in a random mixture of swimming and tumbling motions (Fig. 4). In this mode, a receptor acts as a "fuzzy switch," which randomly fires tumble signals. In the presence of attractant, the model assumes an equilibrium between two populations: a conformationally dynamic population with no bound ligand and a conformationally "frozen" population with bound attractants (A state). This suppresses tumble signals causing biased random motion toward attractants. In the presence of repellents, however, the model assumes an equilibrium between the conformationally dynamic population with no bound ligand and the frozen population with bound repellents (R state), which sends tumble signals. Both equilibria would shift when the concentration of the stimulant changes. This type of fuzzy signaling may be operative at other steps of the signaling pathway. The signal initiated at the receptor can then be amplified by, for example, multiple turnover of one or more downstream signal transfer steps, thus increasing the sensitivity of the response to a small change of the environment. Combined with the slower methylation steps of cytoplamic domains, temporal comparison of environment further provides a positive bias to the random motion of the cells (Segall et al. 1986). Implicit in this model is that reduction in dynamic mobility by cross-linking or antibodies (raised against the ligand domain) would increase the population of the dimer in a frozen conformation, either the attractant (A) or repellent (R) signal state.

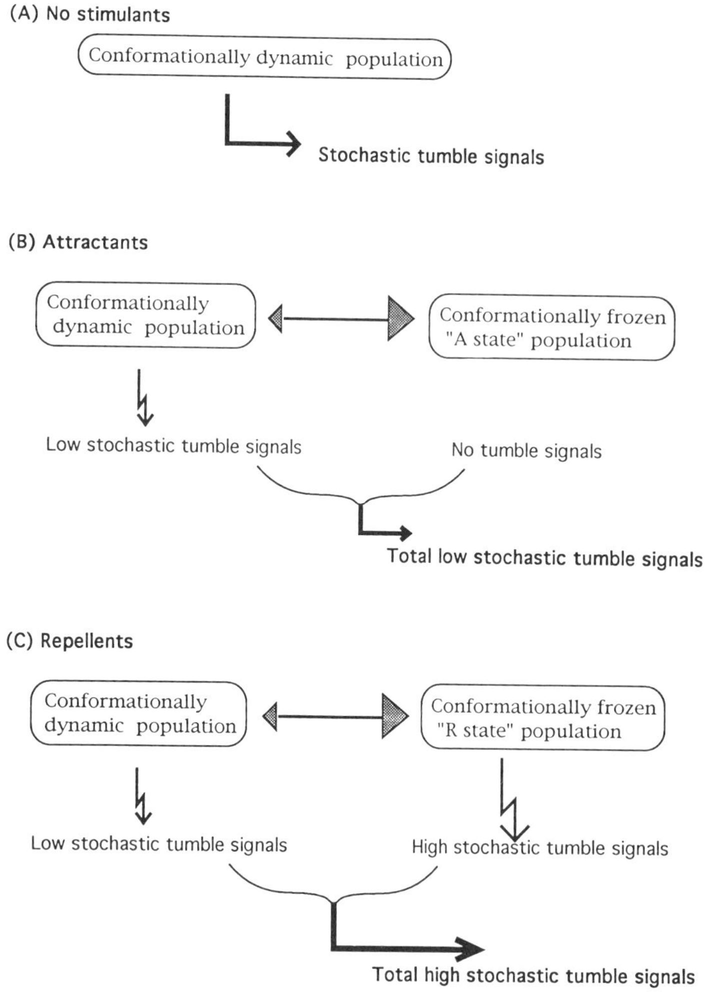

Figure 4. Three states of signaling in the aspartate receptor. (*A*) In the absence of stimulants; (*B*) in the presence of attractants; (*C*) in the presence of repellents.

α-Helices in Membrane

All models discussed assume the contiguity of the $\alpha 1$ helix with TM1 and the $\alpha 4$ helix with TM2, and that each α helix is a relatively rigid structural element, more stable conformationally than β sheets or other protein secondary structures. However, in crystal structures of proteins, long α helices are often bent slightly. Careful analysis of these studies (Barlow and Thornton 1988) showed that for these bent helices, one finds asymmetric interaction of the helix with surrounding parts of the molecule; if one side of the helix has more favorable interactions than the other side, the helix bends in that direction. For the scissors and supercoil unwinding models, we further assume that the flexibility of the helices in the long quasi-four-helix bundle of a receptor dimer is further reduced by the supercoiling of the helices, either two-by-two or all four together, in a manner similar to the supercoiling of the helices in the crystal structure.

One can also argue that within the membrane, where the environment is more homogeneously hydrophobic and lipid molecules move with fluidity, an α helix or supercoiled helices may bend less than on the surface of a soluble protein. Motions of helices *within membrane,* as suggested in the scissors or supercoil models (but not in the piston model), may be energetically much less costly than *on the surface* of an aqueous, water-soluble protein in which α helices usually face a hydrophobic core on one side and an aqueous environment on the other. Thus, although a conformational change in the ligand domain transmitted through the membrane may be partly dampened due to some limited amount of bending of supercoiled helices, we suggest that such bending is not extensive enough to negate the conformational change of transmembrane helices induced by the type of motion proposed in all three models.

Dimeric Receptors in Two States

The crystal structures discussed above suggest a two-step process of signaling. In the first step, receptor monomers are in equilibrium with their dimers in the absence of ligands (although the equilibrium is more toward the dimers). Each monomer has two separate half-sites with weak affinity for the ligand, whereas the dimers have two complete binding sites with strong affinity. We assume also that these receptor dimers with empty binding sites do not transmit the signal to the cytoplasmic domains.

The second step in this process is a conformational change of the dimers on ligand binding caused by the motions such as scissoring and supercoil unwinding.

There is increasing evidence that many receptors function as dimers. A recent crystal structure shows that human growth hormone binds to the dimer of its receptor (Cunningham et al. 1991; De Vos et al. 1992). It is also interesting to note that many mammalian transmembrane receptors such as the epidermal growth factor receptor are postulated to be activated by an association mechanism (Yardin and Schlessinger 1987; Ullrich and Schlessinger 1990). Although significant differences exist between this class of receptors and the aspartate receptor, the former having one transmembrane helix and the latter two per monomer, both do feature an association of monomers, and a conformational change of the helical bundle formed by dimers may be a common feature of signal transfer in both classes of receptor. The fact that the chimeras of the ligand plus transmembrane domain of the aspartate receptor and the cytoplasmic domain of the insulin receptor (Moe et al. 1989) or envZ protein (Utsumi et al. 1989) can transduce aspartate-dependent signal is further indication that there may be common mechanisms of signal transduction in many different types of receptors.

ACKNOWLEDGMENTS

This work has been supported by grants from the National Institutes of Health (AI-30725), the National Science Foundation (BBS-87-20137), Keck Foundation, and the Department of Energy. We gratefully acknowledge numerous discussions with Daniel E. Koshland, Jr. on the pros and cons of piston model, Berkley Lynch and Hans Biemann for discussions about biochemical properties of the aspartate receptors, Gerald Hazelbauer, Mike Manson, Phil Matzumura, and Steve Block for various aspects of chemotaxis, and Andy Pakula and Mel Simon for providing us with their cysteine cross-linking results prior to publication. The crystallographic works cited (Milburn et al. 1991) were done in collaboration with the laboratory of D. Koshland, Jr.

REFERENCES

Adler, J. 1975. Chemotaxis in bacteria. *Annu. Rev. Biochem.* **44:** 341.

Barlow, D.J. and J.M. Thornton. 1988. Helix geometry in proteins. *J. Mol. Biol.* **201:** 601.

Berg, H.C. and D.A. Brown. 1972. Chemotaxis in *Escherichia coli* analyzed by three-dimensional tracking. *Nature* **239:** 500.

Borkovich, K.A., N. Kaplan, J.F. Hess, and M.I. Simon. 1989. Transmembrane signal transduction in bacterial chemotaxis involves ligand-dependent activation of phosphate group transfer. *Proc. Natl. Acad. Sci.* **86:** 1208.

Bourret, R.B., J.F. Hess, K.A. Borkovich, A.A. Pakula, and M.I. Simon. 1989. Protein phosphorylation in chemotaxis and two-component regulatory systems of bacteria. *J. Biol. Chem.* **264:** 7085.

Brünger, A.T. 1990. X-PLOR version 2.1. Yale University, New Haven, Connecticut.

Cunningham, B.C., M. Ultsch, A.M. De Vos, M.G. Mulkerrin, K.R. Clauser, and J.A. Wells. 1991. Dimerization of the extracellular domain of the human growth hormone receptor by a single hormone molecule. *Science.* **254:** 821.

De Vos, A.M., M. Ultsch, and A.A. Kossiakoff. 1992. Human growth hormone and extracellular domain of its receptor: Crystal structure of the complex. *Science* **255:** 306.

Falke, J.J. and D.E. Koshland, Jr. 1987. Global flexibility in a sensory receptor: A site-directed cross-linking approach. *Science* **237:** 1596.

Falke, J.J., A.F. Dernburg, D.A. Sternberg, N. Zalkin, D.L. Milligan, and D.E. Koshland, Jr. 1988. Structure of a bacterial sensory receptor: A site-directed sulfhydryl study. *J. Biol. Chem.* **263:** 14850.

Koshland, D.E., Jr. 1988. Chemotaxis as a model second-messenger system. *Biochemistry* **27:** 5829.

Lynch, B.A. and D.E. Koshland, Jr. 1991. Disulfide cross-linking studies of the transmembrane regions of the aspartate sensory receptor of *Escherichia coli*. *Proc. Natl. Acad. Sci.* **88:** 10402.

Macnab, R.B. 1987. Motility and chemotaxis. *In* Escherichia coli *and* Salmonella typhimurium: *Cellular and molecular biology* (ed. F.C. Neidhardt), vol. 1, p. 732. American Society for Microbiology, Washington, D.C.

Milburn, M.V., G.G. Privé, D.L. Milligan, W.G. Scott, J. Yeh, J. Jancarik, D.E. Koshland, Jr., and S.-H. Kim. 1991. Three-dimensional structures of the ligand-binding domain of the bacterial aspartate receptor with and without a ligand. *Science* **254:** 1342.

Milligan, D.L. 1991. "Structure and dynamics of the aspartate chemoreceptor." Ph.D. thesis, University of California, Berkeley.

Milligan, D.L. and D.E. Koshland, Jr. 1988. Site-directed cross-linking: Establishing the dimeric structure of the aspartate receptor of bacterial chemotaxis. *J. Biol. Chem.* **263:** 6268.

———. 1991. Intra-subunit signal transduction by the aspartate chemoreceptor. *Science* **254:** 1651.

Moe, G.R., G.E. Bollag, and D.E. Koshland, Jr. 1989. Transmembrane signaling by a chimera of the *E. coli* aspartate receptor and the human insulin receptor. *Proc. Natl. Acad. Sci.* **86:** 5683.

Pakula, A. and M.I. Simon. 1992. Determination of transmembrane protein structure by disulfide cross-linking: The *E. coli* tar receptor. *Proc. Natl. Acad. Sci.* **89:** 4144.

Russo, A.F. and D.E. Koshland, Jr. 1983. Separation of signal transduction and adaptation functions of the aspartate receptor in bacterial sensing. *Science* **220:** 1016.

Segall, J.E., S.M. Block, and H.C. Berg. 1986. Temporal comparisons in bacterial chemotaxis. *Proc. Natl. Acad. Sci.* **83:** 8987.

Ullrich, A. and J. Schlessinger. 1990. Signal transduction by receptors with tyrosine kinase activity. *Cell* **61:** 203.

Utsumi, R., R.E. Brisette, A. Rampersaud, S.A. Forst, K. Oosawa, and M. Inouye. 1989. Activation of bacterial porin gene expression by a chimeric signal transducer in response to aspartate. *Science* **245:** 1246.

Wang, B.C. 1985. Resolution of phase ambiguity in macromolecular crystallography. *Methods Enzymol.* **115:** 90.

Yarden, Y. and J. Schlessinger. 1987. Epidermal growth factor induces rapid, reversible aggregation of the purified epidermal growth factor receptor. *Biochemistry* **26:** 1443.

Receptor Protein Tyrosine Kinases and Phosphatases

T. Hunter,* R.A. Lindberg,*† D.S. Middlemas,* S. Tracy,* and P. van der Geer*
*Molecular Biology and Virology Laboratory, The Salk Institute, P.O. Box 85800, San Diego, California 92186;
†Department of Biology, University of California, San Diego, La Jolla, California 92093

There are many mechanisms for transducing signals across the plasma membrane via cell-surface receptors (see Table 1). Two of these directly involve protein phosphorylation in the sense that the cell-surface receptors in question have the intrinsic ability to catalyze protein phosphorylation or dephosphorylation (Fig. 1). In the first category are the receptor protein kinases, which display specificity for phosphorylating either tyrosine or serine residues in their targets. A large number of growth factor receptors are receptor protein tyrosine kinases (PTKs), and in these cases ligand binding leads to activation of a cytoplasmic catalytic domain and the consequent tyrosine phosphorylation of a series of target proteins, which propagate a cytoplasmic signal. Examples of this type of receptor are the epidermal growth factor (EGF), plasma-derived growth factor (PDGF), and insulin receptors. Receptor protein serine kinases have only recently been discovered. So far there is no direct evidence that ligand binding activates the cytoplasmic catalytic domain, but it is assumed that this is the case. Examples of this type of receptor are the transforming growth factor β (TGFβ) and activin receptors. In the second category are receptor-like molecules possessing intrinsic protein tyrosine phosphatase (PTP) activity. In addition to the receptors that signal directly by phosphorylation/dephosphorylation, many G-protein-coupled receptors activate protein kinases indirectly via second messengers generated through G-protein activation of a signal-generating enzyme. Likewise, binding of ligands to the receptors of the hematopoietic cytokine family commonly stimulates protein tyrosine phosphorylation. However, this must occur indirectly, because these receptors lack a PTK catalytic domain. One possibility is that these receptors work as a bimolecular couple where a nonreceptor PTK is noncovalently associated with the cytoplasmic tail of a surface receptor lacking its own catalytic domain. A paradigm for such a bimolecular couple is the CD4/CD8 system (Fig. 1). The short cytoplasmic tails of CD4 and CD8 are noncovalently associated with Lck, a Src family PTK, and cross-linking of CD4 or CD8 leads to activation of the associated Lck PTK. There are more than 30 nonreceptor PTKs, and many of these may be coupled to surface receptors lacking their own catalytic domains in a similar fashion.

This paper focuses on the receptor PTKs and PTPs. Currently, more than 40 distinct vertebrate receptor PTK genes are known (Table 2). These enzymes are characterized by a large amino-terminal glycosylated external domain, which may have one or more of several distinct protein motifs (Fig. 2), and which provides a ligand-binding domain. There is a single membrane-spanning region, which is followed by a spacer region, a PTK catalytic domain, and a variable length carboxy-terminal tail. Receptor PTKs can be subdivided into families based on catalytic domain sequence similarity and on overall structure. The seven main families are typified by the EGF, PDGF, insulin, fibroblast growth factor (FGF), and nerve growth factor (NGF) receptor

Table 1. Primary Transmembrane Signal Transduction Mechanisms

1. Receptor protein kinases
   Ligand binding leads to activation of intrinsic protein kinase catalytic domain and phosphorylation of cytoplasmic substrates. Receptors can be either protein tyrosine kinase (e.g., EGF, PDGF, and insulin receptors) or protein serine kinase (e.g., TGF-β and activin receptors)
2. Receptors coupled to protein kinases
   Protein tyrosine kinases noncovalently associated with surface receptors are activated by binding of ligand to receptor
3. Protein tyrosine phosphatase receptors
   Ligand binding may regulate intrinsic phosphatase catalytic domain
4. Receptors coupled to G proteins
   Ligand binding leads to activation of specific G proteins. Activated G-protein subunits modulate activity of specific targets (e.g., adenylyl cyclase, phospholipases C and A_2, ion channels, cGMP phosphodiesterase)
5. Ion channels
   Ligand binding leads to channel opening and influx or efflux of ions
6. ANF receptor
   Ligand binding leads to activation of intrinsic guanylyl cyclase catalytic domain
7. Cytokine receptors
   These receptors are all related in structure but have an unknown signaling mechanism (e.g., IL-2, IL-3, IL-4, IL-5, IL-6, IL-7, GM-CSF, G-CSF, EPO, GH, PL, LIF, and CNTF receptors)
8. Others
   There is a major class of intracellular receptors belonging to the steroid receptor family whose ligands are sufficiently lipophilic to cross the plasma membrane and then associate with receptors inside the cell

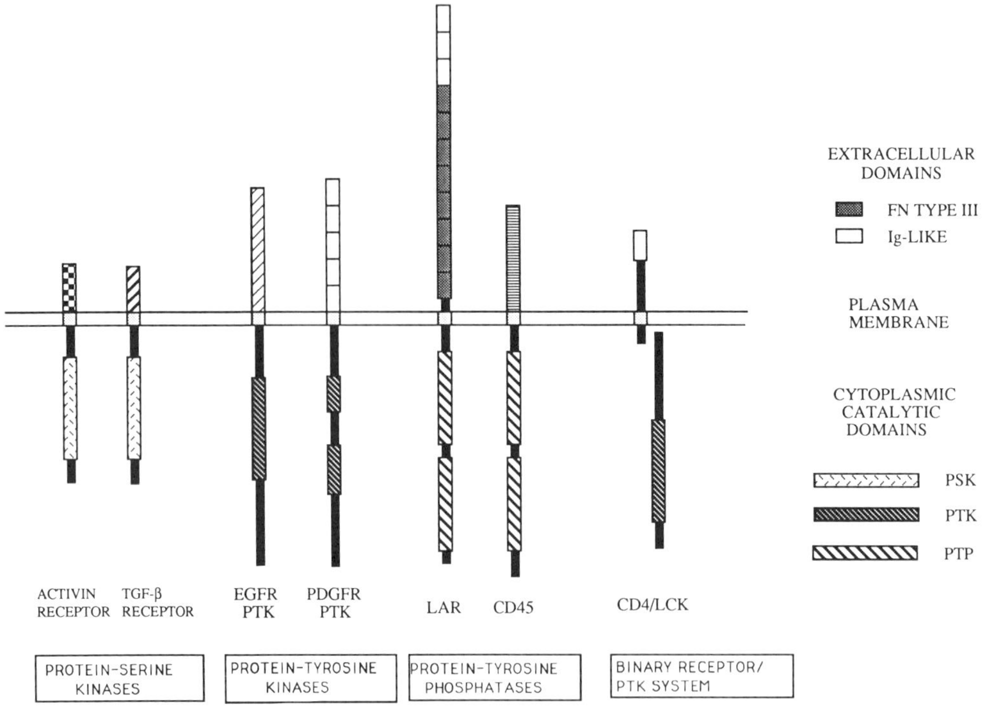

Figure 1. Transmembrane signal transduction mechanisms directly involving protein phosphorylation and dephosphorylation. Functional motifs present in extracellular and intracellular domains are represented by filled boxes as defined on the figure.

PTKs and the ECK and Ark receptor-like PTKs. The families can be grouped by similarity within the defined catalytic domain (Fig. 3), and sequences outside the catalytic domain show striking similarity as well. The spacing between the transmembrane domain and the catalytic domain, and the length of the carboxy-terminal tail, are usually similar within a family. In addition, the structural motifs and the cysteine residues are usually conserved within the external domain.

Binding of ligand to the external domain of receptor PTKs results in activation of the cytoplasmic catalytic domain as evidenced by autophosphorylation and the phosphorylation of cytoplasmic substrates. The most plausible model for activation is that ligand binding either stabilizes or induces receptor dimerization. Dimerization juxtaposes the two catalytic domains, which then leads to a conformational change that allows transphosphorylation of specific tyrosines in the cytoplasmic

Table 2. Growth Factor Receptor Protein Tyrosine Kinases

EGF receptor (cErbB protein)	PDGF receptor A-type (*patch* locus)
cErbB2 (Neu) protein	PDGF receptor B-type
cErbB3 protein	CSF-1 receptor (cFms protein)
X-*mrk* protein	SCF receptor (cKit protein-*W* locus)
Tyro2 protein (HER4?)	VEGF receptor (Flt1 protein)
	Flk1 (KDR) protein
Insulin receptor	Flk2 (Flt3) protein
IGF-1 receptor	Flt4 protein
Insulin receptor-related kinase	
cRos protein	ECK protein
Ltk protein	EPH protein
Tyro10 protein	Elk protein
	EEK protein
HGF receptor (Met protein)	Erk protein
cSea protein	Cek4/Mek4/HEK protein
	Cek5/Nuk protein
FGF receptor 1 (Flg, Cek1)	Sek protein
FGF receptor 2 (Bek, K-*sam*, Cek3)	
FGF receptor 3	AXL/UFO/Ark protein
FGF receptor 4	Tyro3 protein
FLG2 protein (Cek2)	
	Tie protein
NGF receptor (Trk protein)	RET protein
TrkB protein (BDNF/NT-3/NT-4/5 receptor)	Klg protein
TrkC protein (NT-3 receptor)	

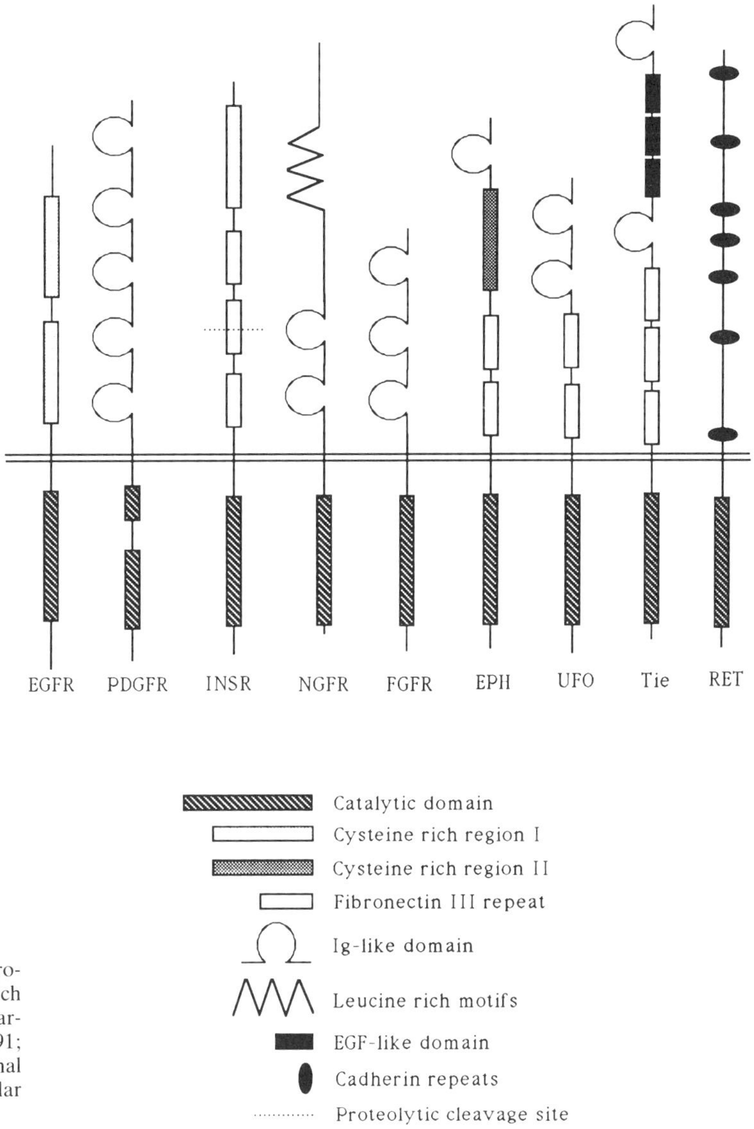

Figure 2. Structural topology of receptor protein tyrosine kinases. Compiled from Ullrich and Schlessinger 1990; O'Bryan et al. 1991; Partanen et al. 1992; Schneider and Schweiger 1991; and R. Schneider (pers. comm.). Functional motifs present in extracellular and intracellular domains are defined on the figure.

domain, which may lie within the catalytic domain or else in the carboxy-terminal tail. Such intermolecular autophosphorylation is a prerequisite for the ligand-bound receptor dimer to phosphorylate substrate proteins. This model is supported by both biochemical and genetic evidence. First, in most cases receptor dimerization is readily detectable by cross-linking following ligand binding. Second, the expression of kinase-inactive mutants of a particular receptor has a dominant negative effect on signaling via the wild-type receptor in the same cell, consistent with the formation of dimers containing one wild-type and one mutant receptor, where the functional receptor molecule cannot be activated for substrate phosphorylation because it is not transphosphorylated by its partner. Receptor autophosphorylation sites are believed to play an important role in substrate selection by providing binding sites for proteins that contain SH2 domains. Although not all receptor PTK substrates contain SH2 domains, such domains have been identified in a reasonable number of proteins that are implicated in mitogenic signaling whose tyrosine phosphorylation is stimulated upon treatment of cells with growth factors whose receptors are PTKs. For example, phospholipase-C-γ1 (PLC-γ1), the p85 subunit of phosphatidylinositol (PI)3 kinase, GAP120, Vav, and Nck are in this category. Tyrosine phosphorylation of PLC-γ1 stimulates its phospholipase activity, but the effect of tyrosine phosphorylation on the other proteins is not so clear-cut.

To investigate receptor PTK structure and function and to elucidate receptor-mediated signal pathways, we have been investigating three receptor PTK systems. Two of these involve receptor PTKs that we isolated in cDNA library screens that were initiated several years ago in an effort to learn more about the spectrum of receptor PTKs encoded by the mammalian genome. In these screens, cDNA libraries were probed with degenerate oligonucleotides corresponding to two short sequence motifs that are highly conserved in the catalytic domains of all PTKs. To enrich for receptor PTK

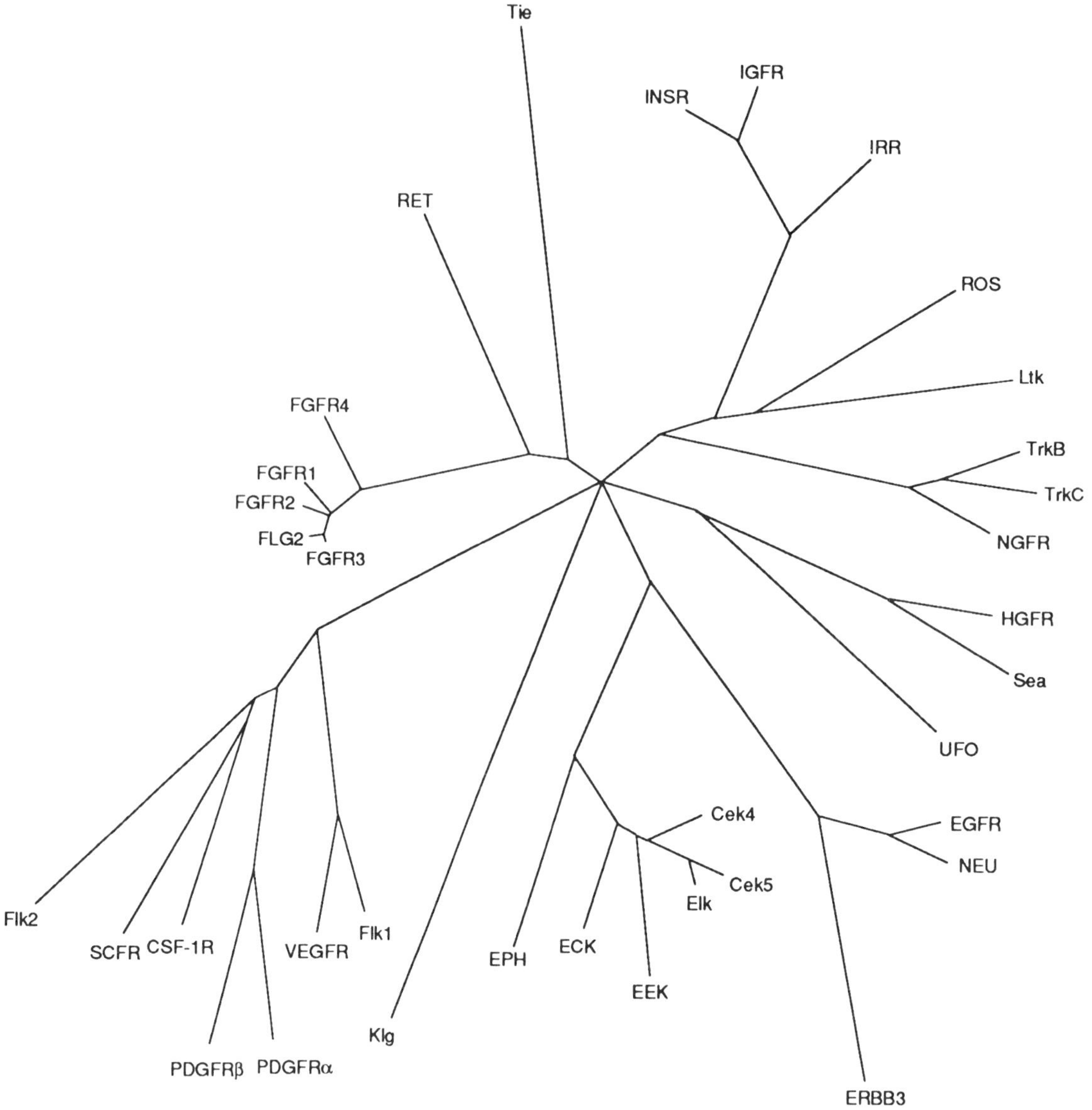

Figure 3. Molecular phylogeny of 35 receptor protein tyrosine kinases. The catalytic domain sequences for the receptors were obtained from the protein kinase database and analyzed as described by Hanks and Quinn (1991). The receptor protein tyrosine kinases shown are distinct vertebrate gene products (i.e., none are orthologs). The receptors are named as in the database or as the receptor for a known ligand. See Table 2 for a complete list of receptor protein tyrosine kinases.

cDNAs, one of these oligonucleotides was designed to match the consensus HRDLAARN, which is conserved among receptor PTKs and differs from the equivalent sequence in many of the nonreceptor PTKs. A cDNA for the ECK receptor PTK was isolated from a HeLa cell cDNA library (Lindberg and Hunter 1990). The *ECK* gene is expressed primarily in cells of epithelial origin and in tissues with high concentrations of epithelial cells, and this prompted the name, ECK, for *e*pithelial *c*ell *k*inase. Although ECK has a receptor-like structure, as yet the ligand for ECK has not been identified. The second receptor PTK, TrkB, was identified in an adult rat cerebellar library in a search for receptor PTKs that might be involved in neuronal function (Middlemas et al. 1991). TrkB is closely related to the Trk receptor PTK, originally identified in a mutant form as a human tumor oncogene (Martin-Zanca et al. 1986). Recently, neurotrophin-3 (NT-3) and brain-derived neurotrophic factor (BDNF) have been identified as ligands for TrkB and have been used to study TrkB function. Finally, we have studied the colony-stimulating factor 1 (CSF-1) receptor PTK, which is expressed primarily in cells of the myeloid lineage and is required for myeloid progenitor proliferation and differentiation, as well as for the biological activities of mature macrophages.

The ECK Receptor-like Protein Tyrosine Kinase

The ECK family of receptor-like PTKs has grown to at least eight distinct members (Table 2). For six of them, complete catalytic domain sequences have been reported, and we have placed them in a phylogenetic tree (Fig. 3). These six clearly form a family separate from other PTKs, both by the criterion of catalytic domain similarity and by overall structure. The cytoplasmic domain of the ECK receptor PTK family has a medium length carboxy-terminal tail, and the catalytic domain lacks an insert. Within the catalytic domain, the ECK family PTKs are most closely related to the EGF receptor PTK family. The ECK family receptor ligand-binding domain has a cysteine-rich region, which is not obviously related to cysteine-rich regions in other receptors, and two fibronectin III repeats (Fig. 2). A ligand has not yet been found for any of the ECK family receptors, so the biology of these receptors is unknown. Studies of the expression patterns of the ECK family receptors have shown that the four that cluster together in the phylogenetic tree (Cek4, Cek5, Elk, and EEK) are expressed most highly in brain (Chan and Watt 1991; Lhoták et al. 1991; Pasquale 1991; Sajjadi et al. 1991). In contrast, ECK and EPH are expressed more widely, although these genes are not commonly coexpressed; EPH is most highly expressed in lung, liver, kidney, and testes (Hirai et al. 1987), and ECK is most highly expressed in skin, lung, intestine, and ovary (Lindberg and Hunter 1990). On the basis of the close relationship of the external domains of these receptors, it seems likely that their ligands may be part of a related family, as is the case for ligands of the FGF and NGF receptor PTK families. Whether the functions of the ECK receptor PTK family members are similar remains to be seen.

ECK is expressed in many tissues and cells of different embryonic origins but is most highly expressed in epithelial cells, consistent with its original isolation from HeLa cells, which are of epithelial origin (Lindberg and Hunter 1990). For instance, skin, gut mucosa, and kidney tubule epithelial cells all express high levels of ECK. However, detectable but low levels of ECK are also found in non-epithelial cells. The *ECK* gene is expressed as a single 4.7-kb mRNA, both in human cell lines and in rodent tissues and cell lines. To characterize the ECK protein, polyclonal rabbit antisera were raised against a TrpE fusion protein containing 101 residues of the ECK carboxy-terminal tail. These antisera were used to demonstrate that in HeLa and A-431 cells, ECK is expressed as a 125-kD precursor, which is slowly converted into a 130-kD mature form, presumably as the protein is transported to the cell surface. In polarized MDCK epithelial cell monolayers, ECK has been localized to the basolateral membrane by immunofluorescence staining (C.M. Isacke and R.A. Lindberg, unpubl.). This implies that ECK may be a receptor for a humoral factor rather than an external factor. In MDCK cells a significant fraction of ECK staining was detected in intracellular membranes, consistent with the fact that the maturation of ECK appears to occur slowly.

Our attempts to study the ECK receptor PTK signal transduction pathways have been hampered by the lack of an ECK ligand to stimulate its PTK activity in intact cells. Both the 125-kD and 130-kD forms of ECK are phosphorylated on tyrosine when immunoprecipitates are incubated with labeled ATP, and both generate identical tryptic phosphopeptide maps containing six major peptides. So far, none of these autophosphorylation sites has been mapped. The ability of ECK to phosphorylate exogenous substrates has also not been tested.

To determine whether ECK has transforming potential and to search for a ligand, we expressed human ECK from a retroviral vector in several mouse and rat cell lines. For this purpose the entire coding region of the ECK cDNA was inserted into a vector that expresses the cDNA from cytomegalovirus promoter and contains a neomycin resistance marker (Scharfmann et al. 1991). NIH-3T3 cells were infected with this retrovirus and selected for G418 resistance. The selected cells were tested for ECK expression. Seventeen clonal cell lines were established, and they all expressed ECK at approximately equal levels as tested by immunoblotting anti-ECK immunoprecipitates with anti-ECK serum. The levels were 10- to 20-fold higher than the endogenous murine Eck levels. In addition, the expressed human ECK receptors were functional as determined by immune complex protein kinase assays. We tested whether overexpression of ECK from this vector would cause cell transformation in a focus-forming assay, but there was no increase in foci over control levels. Moreover, the cells expressing human ECK were not morphologically transformed. In the same assay the murine CSF-1 receptor expressed from the same vector induced foci in a CSF-1-independent manner. Exogenous CSF-1 is not required because NIH-3T3 cells secrete sufficient CSF-1 to elicit autocrine transformation. Overexpression of EPH in NIH-3T3 cells was also reported to have no effect on focus formation or morphology, but these cells were tumorigenic in nude mice (Maru et al. 1990). The most likely explanation for this result is that cell transformation is ligand-dependent and that the ligand for EPH is not present in calf serum or secreted by NIH-3T3 cells but is present in the mouse and drives tumor growth. We have not yet tested whether ECK-expressing cells are tumorigenic.

The TrkB Receptor Protein Tyrosine Kinase

The TrkB cDNA was originally isolated from a rat cerebellar library (Middlemas et al. 1991) in a search for PTKs that might be involved in cellular signal transduction in neurons. Northern analysis of various adult rat tissues indicates that the *trkB* gene is expressed predominantly in the nervous system. TrkB is a member of the growing Trk receptor PTK family, which presently consists of Trk, TrkB, and TrkC (Klein et al. 1989; Martin-Zanca et al. 1989; Lamballe et al. 1991; Middlemas et al. 1991). Like TrkB, both Trk and TrkC are expressed predominantly in the nervous system, with the expression of Trk being very restricted (Martin-Zanca et al. 1990). The Trk family receptor PTKs have a similar overall structure. Their external domains are characterized by a series of leucine-rich motifs (LRMs) near the amino terminus, and two immunoglobulin (Ig)-like domains near the transmembrane region (Schneider and Schweiger 1991). In addition, nearly all of the cysteines are conserved. The cytoplasmic domains are unusual in that they have a very short carboxy-terminal tail with less than 10 residues lying beyond the catalytic domain. Within the catalytic domain the Trk family PTKs are most closely related to the insulin receptor PTK family.

The discovery that NGF is a ligand for the Trk receptor (Kaplan et al. 1991a,b; Klein et al. 1991a) prompted several groups to investigate whether the NGF-related neurotrophins, BDNF and NT-3, might be ligands for the TrkB receptor. In collaboration with Louis Parada (Frederick Cancer Research Center) and Karoly Nikolics (Genentech), we found that BDNF and NT-3 are ligands for the TrkB receptor (Soppet et al. 1991). Other groups obtained similar findings at the same time (Klein et al. 1991b; Squinto et al. 1991). In our experiments, full-length TrkB was expressed in rodent fibroblasts, and the expressed TrkB bound both NT-3 and BDNF, as shown by Scatchard analysis and chemical cross-linking and by the ability of these neurotrophins to stimulate tyrosine phosphorylation of TrkB. TrkB-expressing fibroblasts respond mitogenically to NT-3 and BDNF and are transformed when TrkB and NT-3 or BDNF are coexpressed (Klein et al. 1991b). Human neurotrophin-5 (NT-5)/*Xenopus* neurotrophin-4 (NT-4) (which probably correspond to homologous gene products) has also recently been shown to be a ligand for the TrkB receptor (Berkemeier et al. 1991; Klein et al. 1992).

Northern analysis showed that the *trkB* gene, in contrast to the *ECK* gene, is expressed in a surprisingly complex fashion with as many as nine distinct RNAs being detected, ranging in size from 0.9 kb to 10.7 kb (Klein et al. 1990; Middlemas et al. 1991). To delineate the structural differences in the coding sequences of these mRNAs, Northern analysis of rat cerebral RNA was carried out with a variety of *trkB* probes corresponding to 5′-untranslated, coding, and 3′-untranslated regions. This analysis showed that most *trkB* mRNAs contain extracellular domain sequences, whereas only a subset of the mRNAs contain catalytic domain sequences (Middlemas et al. 1991). These results suggested some of the mRNAs encode a carboxy-terminally truncated receptor lacking the catalytic domain. mRNAs of 9.0 kb and 4.8 kb could encode a full-length receptor PTK, whereas mRNAs of 7.5 kb, 7.0 kb, and 2.4 kb could encode a carboxy-terminally truncated receptor lacking the catalytic domain.

To characterize the products of these mRNAs, we identified cDNA clones that encoded the extracellular region, but not the catalytic domain, by Southern hybridization of a cDNA sublibrary consisting of 38 *trkB* cDNAs with the relevant *trkB* probes. The complete nucleotide sequence of two cDNA clones, *trk*B.T1 and *trk*B.T2, indicated they encode two distinct carboxy-terminally truncated TrkB proteins, which have the complete extracellular region and transmembrane domain, but which differ in their short cytoplasmic tails (Fig. 4A) (Middlemas et al. 1991). The TrkB.T1 protein product is predicted to have a 23-amino-acid cytoplasmic tail in which the final 11 amino acids differ from the corresponding region of the full-length receptor PTK, whereas the predicted cytoplasmic domain of TrkB.T2 is 21 amino acids, the last 9 of which differ from the corresponding region in TrkB.FL and TrkB.T1. The mRNAs encoding truncated receptors presumably arise from alternative splicing. Our analysis suggests that the mRNA for TrkB.T1 is much more abundant than that for TrkB.T2.

Both the full-length TrkB receptor PTK ($gp145^{TrkB}$) and truncated receptors ($gp95^{TrkB}$) are expressed in mouse brain (Klein et al. 1990) and in the adult rat brain, where we have analyzed expression of TrkB by immunoblotting anti-TrkB immunoprecipitates with a polyclonal rabbit anti-TrkB antiserum raised against the extracellular domain of TrkB expressed in *Escherichia coli* (Fig. 4B). Klein et al. (1990) have shown that there is a differential expression of mRNAs for full-length TrkB and the truncated TrkB protein corresponding to rat TrkB.T1 in mouse brain. The mRNA for the truncated form is expressed at high levels in the ventricles and choroid plexus, whereas the mRNA for the full-length TrkB is expressed in cerebral cortex (Klein et al. 1990). However, there are circumstances under which mRNAs for both full-length and truncated TrkB proteins can be expressed in the same region of the brain (see below). We do not know where the mRNA for TrkB.T2 is expressed nor whether the expression of the various TrkB proteins follows that of their mRNAs. We are currently raising antibodies to peptides corresponding to the unique cytoplasmic tails of TrkB, TrkB.T1, and TrkB.T2 to characterize the expression of these proteins in different regions of the brain in more detail and to determine where they are expressed in neurons.

The function of truncated TrkB receptors, which consist of the entire extracellular region and transmembrane domain and a short cytoplasmic tail, remains unknown. A number of possibilities exist. The truncated TrkB receptors could act as clearance receptors or could be responsible for concentrating the cognate neurotrophins in particular areas of the brain for binding to the full-length TrkB receptor. If there are neurons where the full-length and truncated forms of TrkB are coexpressed, then one would expect that the truncated form would have a dominant negative effect on signaling via the full-length TrkB. An intriguing possibility was raised that the truncated receptors could be neuronal cellular adhesion molecules (Middlemas et al. 1991). It is interesting to note that the extracellular region of TrkB contains several structural motifs identified by detailed comparative sequence analysis (Schneider and Schweiger 1991). The amino-terminal region exhibits an apparent sequence homology with the *Drosophila toll* gene product, which contains LRMs and is involved in cell adhesion events.

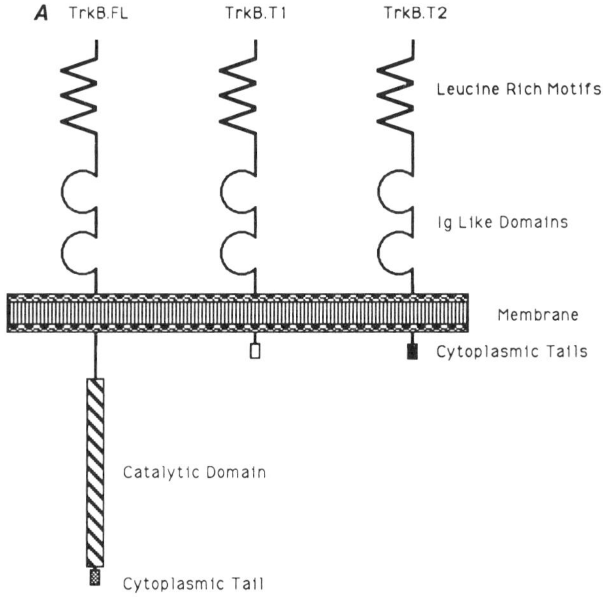

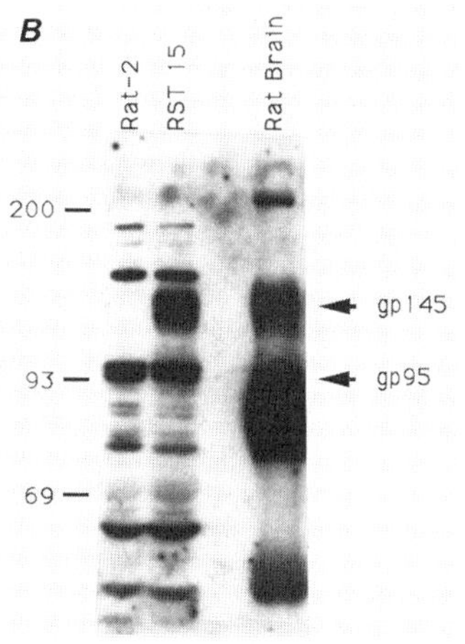

Figure 4. The structure and expression of full-length and truncated TrkB proteins. (*A*) Schematic of the full-length TrkB receptor PTK and two distinct truncated forms, TrkB.T1 and TrkB.T2. (*B*) Immunoblot analysis of TrkB proteins expressed in adult rat brain. Rat-2 cells and RST 15 cells (Rat-2 cells expressing TrkB) were lysed in boiling SDS gel loading buffer. A rat brain was homogenized in RIPA buffer and immunoprecipitated with polyclonal rabbit anti-TrkB antiserum. The Rat-2 and RST 15 whole-cell lysates and the brain immunoprecipitates were analyzed by SDS-PAGE on a 7.5% gel, followed by transfer to Immobilon-P and probing with anti-TrkB serum and detection with ^{125}I protein A.

TrkB contains three LRMs between two cysteine clusters. There are also two Ig-like domains near the transmembrane domain. It is possible that the LRMs could be involved in cellular adhesion whereas the Ig-like repeats could be involved in binding of the neurotrophins.

Although several neurotrophins are candidates for the actual physiological ligand for the TrkB receptor, recent experiments suggest that BDNF may be a physiological ligand for TrkB receptor. In a collaboration with Håkan Persson (Karolinska Institute) and Olle Lindvall (University of Lund), we have found that mRNAs for both BDNF and TrkB are elevated specifically and rapidly in the dentate gyrus following seizures induced by hippocampal kindling of adult rats. This treatment leads to the induction of mRNAs for both full-length and truncated TrkB proteins. Immunohistochemical staining with anti-TrkB antibodies indicates that a similar increase occurs at the protein level, and both full-length and truncated TrkB proteins can be detected by immunoblotting samples from this region of the brain. In contrast to TrkB, changes in mRNAs for Trk and TrkC were not observed in the hippocampus after kindling-induced seizures in rats. One hypothesis arising from these observations is that BDNF and TrkB are elevated in response to brain insults to afford neuroprotection via an autocrine/paracrine mechanism. We are currently investigating whether the full-length TrkB protein that is produced in response to kindling in the dentate gyrus contains P.Tyr, as would be expected if BDNF were secreted and then bound to TrkB and activated the TrkB receptor PTK.

Activation of the TrkB receptor PTK elicits a mitogenic response in fibroblasts (Klein et al. 1992), yet in neuronal cells this signal induces differentiation or promotes cell survival (Glass et al. 1991). To understand why different cells respond differently to the activation of TrkB, we have undertaken to identify the immediate substrates for the TrkB receptor PTK in different cell types. For this purpose, we have expressed TrkB in Rat-2 cells and in rat B50 cells, which are derived from a neuroblastoma and still retain some neuronal properties. We have begun by testing whether PLC-γ1 might be a substrate for the TrkB receptor PTK, and we have found that PLC-γ1 is rapidly phosphorylated on tyrosine residues in response to NT-3 treatment of TrkB-expressing Rat-2 cells. PLC-γ1 also becomes associated with the TrkB receptor upon NT-3 stimulation, presumably via specific association of one or both of the SH2 domains in PLC-γ1 with autophosphorylation sites on TrkB. We have found that there are several tyrosine residues phosphorylated in response to NT-3 treatment of TrkB-expressing Rat-2 cells, and we are in the process of trying to identify and mutate these tyrosines and to identify the tyrosine phosphorylation sites involved in PLC-γ1 activation. Such mutants could be used to delineate the importance of PLC-γ1 in cellular signal transduction pathways responsible for survival and differentiation of neurons.

The CSF-1 Receptor Protein Tyrosine Kinase

CSF-1 is a peptide growth factor that plays an important role in regulating survival, differentiation, and renewal of macrophages (Stanley et al. 1983). CSF-1 is also expressed in the uterus during early pregnancy and is thought to be involved in the regulation of placental trophoblast proliferation (Pollard et al. 1987). Responsiveness to CSF-1 depends on the expression of its receptor on the cell surface. The receptor for CSF-1 is encoded by the c-*fms* gene (originally identified as the cellular homolog of the McDonough feline sarcoma virus [SM-FeSV] v-*fms* oncogene), and is a ligand-stimulated PTK (Sherr et al. 1985). The CSF-1 receptor is in the PDGF receptor PTK family, which is characterized by the presence of multiple Ig domains in the extracellular domain and an insert within the catalytic domain (Fig. 2). The CSF-1 receptor is most closely related to the PDGF α and β receptors (Yarden et al. 1986; Matsui et al. 1989) and cKit (Qiu et al. 1988), the receptor for the stem-cell factor (Fig. 3), which all have five Ig domains in the extracellular ligand-binding domain and an insert that varies in length between 70 and 100 amino acid residues in the catalytic domain. Overall, this insert, which is now commonly known as the kinase insert, is poorly conserved between the different receptors in this family, but there are short stretches that are conserved between the different family members. Human, feline, and murine CSF-1 receptor cDNAs have been sequenced, and we will use the residue numbers for the murine sequence (Rothwell and Rohrschneider 1987), unless otherwise noted.

In our studies of the CSF-1 receptor PTK signal transduction, we have focused on the identification of tyrosine phosphorylation sites within the receptor. For this purpose we have used BAC1.2F5 cells, an SV40-immortalized murine macrophage line that requires CSF-1 for growth (Morgan et al. 1987), and an antiserum raised against a bacterially expressed fragment of the murine CSF-1 receptor. In these cells the CSF-1 receptor is synthesized as a 130-kD glycosylated precursor, which matures into a 160-kD protein that is present on the cell surface. Immunoprecipitation of CSF-1 receptors from ^{32}P-labeled BAC1.2F5 cells, followed by tryptic digestion and separation of the digestion products in two dimensions on thin-layer cellulose plates, has shown that the 160-kD mature form of CSF-1 receptor is phosphorylated on a large number of residues in resting CSF-1-deprived cells (van der Geer and Hunter 1990). Phosphoamino acid analyses on the whole receptor and on the individual phosphopeptides have shown that in resting cells all the phosphate is present on serine residues. The function of phosphorylation on serine residues in the CSF-1 receptor is largely unknown. Little work has been done on the characterization of any of these phosphorylation sites thus far, but we have preliminary evidence that the carboxy-terminal tail of the CSF-1 receptor is highly phosphorylated on serine residues in vivo.

CSF-1 binding stimulates receptor autophosphorylation (Downing et al. 1988; Roussel et al. 1988). CSF-1 receptor autophosphorylation in BAC1.2F5 cells treated at 37°C is not very easy to detect, but this is most likely because receptors are internalized and degraded very rapidly in response to ligand. In fact, two independent mechanisms for CSF-1 receptor down-regulation have been described, one of which is dependent on the presence of protein kinase C (Downing et al. 1989). CSF-1-dependent autophosphorylation is more readily detected if the BAC1.2F5 cells are stimulated with CSF-1 at 4°C. Although the time course is delayed, autophosphorylation and substrate phosphorylation take place normally at 4°C, whereas receptor internalization and degradation do not take place at this temperature (Sengupta et al. 1988). We showed initially that Tyr-706, which is located in the CSF-1 receptor kinase insert, is the major CSF-1-stimulated autophosphorylation site in BAC1.2F5 cells (van der Geer and Hunter 1990). Rohrschneider and co-workers showed that in addition to Tyr-706, Tyr-697 and Tyr-807 are also phosphorylated in response to ligand binding when CSF-1 is used to treat FDC-P1 and BALB/c-3T3 cells that are ectopically expressing CSF-1 receptors (Tapley et al. 1990). Tyr-973 (equivalent to Tyr-969 in the human CSF-1 receptor), four residues from the carboxyl terminus of the receptor, is deleted in vFms (Coussens et al. 1986), and mutation of Tyr-969 to phenylalanine accentuates CSF-1 receptor transforming activity (Roussel et al. 1987). It was therefore hypothesized that phosphorylation of this residue may be important for negative regulation of CSF-1 receptor kinase activity, analogous to what has been found for the Src PTK (Cooper et al. 1986; Cartwright et al. 1987; Kmiecik and Shalloway 1987; Piwnica-Worms et al. 1987). Thus far, however, there is no evidence for phosphorylation of Tyr-969. We have some evidence that a tyrosine residue in the carboxyl terminus can be phosphorylated in vitro in an autophosphorylation reaction. However, tryptic phosphopeptide maps of CSF-1 receptors autophosphorylated in vitro show a number of tryptic phosphopeptides (Tapley et al. 1990; van der Geer and Hunter 1990), only some of which have been identified. It is not clear whether all the sites that are phosphorylated in vitro are also phosphorylated in vivo.

We recently identified Tyr-721 in the murine CSF-1 receptor as the binding site for the p85 subunit of PI3 kinase (Reedijk et al. 1992). Tyr-721 was identified as a likely candidate for this binding site based on the similarity in the sequence directly downstream from this residue with that in other PI3 kinase binding sites (Cantley et al. 1991). To establish Tyr-721 as the binding site, Tyr-721 was mutated to phenylalanine, and bacterially expressed wild-type and mutant kinase inserts were used to study the interaction of the kinase insert with PI3 kinase (Reedijk et al. 1992). Using in vitro binding experiments, we showed that PI3 kinase from Rat-2 fibroblast lysates binds specifically to the CSF-1 receptor kinase insert in a manner dependent on tyrosine phosphorylation of the kinase insert. Mutation of Tyr-697 and Tyr-706 to phenylalanine had no effect on PI3 kinase binding, whereas mutation of Tyr-721 abolished binding completely (Reedijk et al. 1992). These results were confirmed using in vivo experiments. A Phe-697/706 double mutant CSF-1 receptor expressed in 208F rat fibroblasts has wild-type ability to bind PI3 kinase in response to CSF-1. In contrast, no PI3 kinase binding to the receptor was observed when 208F fibroblasts expressing Phe-721 mutant receptor were exposed to CSF-1. It has proven difficult to demonstrate that Tyr-721 is phosphorylated in the intact cell in response to CSF-1 binding, but we believe this is because only a small percentage of CSF-1 receptor molecules are phosphorylated at this site, which would be consistent with the observation that only a small fraction of CSF-1 receptor molecules associate with PI3 kinase (Downing et al. 1991). We are currently examining whether the Phe-721 mutant CSF-1 receptor shows any defects in cellular responses. The possibility exists that simple binding of PI3 kinase to an appropriate phosphotyrosine-containing sequence can stimulate its activity (Backer et al. 1992). If this were true, one would predict that the inability of the Phe-721 CSF-1 receptor to bind PI3 kinase would mean that CSF-1 cannot elicit increases in 3′ phosphoinositides in Phe-721 CSF-1 receptor-expressing cells. However, we have not yet formally tested this prediction.

We also showed using in vitro binding experiments that the carboxy-terminal SH2 domain of p85 binds with a higher affinity to the CSF-1 receptor kinase insert domain than the amino-terminal SH2 domain (Reedijk et al. 1992); similar results have been obtained using the PDGF receptor (Klippel et al. 1992). Under the same conditions, the GAP SH2 domains did not bind to the phosphorylated CSF-1 receptor kinase insert, supporting the current idea that different SH2 domains bind specifically to different phosphorylated tyrosine residues. This idea was based on the fact that different tyrosine-phosphorylated growth factor receptor PTKs bind different combinations of SH2-containing signaling molecules. Experiments using phosphotyrosine-containing peptides indicate that the specificity is indeed determined by residues surrounding the phosphorylated tyrosine residue (Fantl et al. 1992).

Wild-type murine CSF-1 receptors are able to transform NIH-3T3 fibroblasts, as measured by focus formation or growth in soft agar (Taylor et al. 1989). This is most likely because these receptors bind the CSF-1 produced by the NIH-3T3 cells, leading to autocrine stimulation. Deletion of the kinase insert does not appear to affect the transforming activity of CSF-1 receptors in NIH-3T3 fibroblasts (Taylor et al. 1989). In contrast, expression of murine CSF-1 receptors in Rat-2 fibroblasts does not lead to transformation, probably because these cells make very little if any CSF-1 (Reedijk et al. 1990; van der Geer and Hunter 1991). However, CSF-1 receptor-expressing Rat-2 fibroblasts do acquire responsiveness to CSF-1, and stimulation with CSF-1 results in immediate-early gene induction, changes in cell morphology, and an increase in growth rate (Reedijk et al. 1990; van der Geer and Hunter 1991). Rat-2 fibroblasts expressing kinase insert deletion mutant receptors are unable to grow in soft agar or to change morphology in response to CSF-1 (Reedijk et al. 1990). To study the role of individual tyrosine phosphorylation sites, we have mutated Tyr-697, Tyr-706, and Tyr-721 in the kinase insert to phenylalanine (or glycine) either singly or in combination. None of these mutations affects the PTK activity of the CSF-1 receptor as measured by in vitro phosphorylation assays. However, mutation of Tyr-706 affects the ability of the CSF-1 receptor to induce the expression of immediate-early genes and the ability to induce changes in cell morphology (the phenotypes of autophosphorylation site mutants are summarized in Table 3). CSF-1 treatment of Rat-2 fibroblasts expressing Phe-706 mutant CSF-1 receptors elicits a change in cell morphology and induces the expression of c-*jun* RNA, but, in contrast to wild-type CSF-1 receptor-expressing Rat-2 cells, c-*fos* RNA is not induced. Gly-706 mutant receptors induce c-*fos* RNA expression at least partially in response to CSF-1, but lack the ability to increase the levels of c-*jun* RNA or to change cell morphology (van der Geer and Hunter 1991). These data suggest that two independent signal transduction pathways are activated by this part of the protein (Fig. 5).

What signal pathways lead to the induction of c-*fos* and c-*jun*? The c-*fos* promoter is complex and possesses several elements through which it can be induced, including the serum response element (SRE) and the *sis*-inducible element (SIE). The major regulatory element in the c-*jun* promoter is a TPA response element. How do the mutations in the kinase insert affect these induction pathways? Since phenylalanine is a reasonably good structural analog of tyrosine, the Phe-706 mutation may simply prevent phosphorylation of Tyr-706, and the failure of the Phe-706 mutation to stimulate c-*fos* expression may mean that it does not activate

Table 3. Response to CSF-1 of Rat-2 and NIH-3T3 Cells Expressing Different Mutant CSF-1 Receptors

	Expressed in Rat-2 fibroblasts				Expressed in NIH-3T3 cells	
Receptors	increased growth rate	change in morphology	early response gene induction	c-*myc* induction	focus formation	c-*myc* induction
Wt	+ + +	+ + +	Wt	–	+ +	+ +
M614	–	–	–	n.d.	–	n.d.
F706	+ + +	+ + +	disturbed	–	+ +	n.d.
G706	+ + +	+/–	disturbed	–	+ +	n.d.
F807	+/–	+/–	Wt	–	–	–
G807	–	–	–	–	–	n.d.

(Wt) Wild-type receptors; (n.d.) not determined.

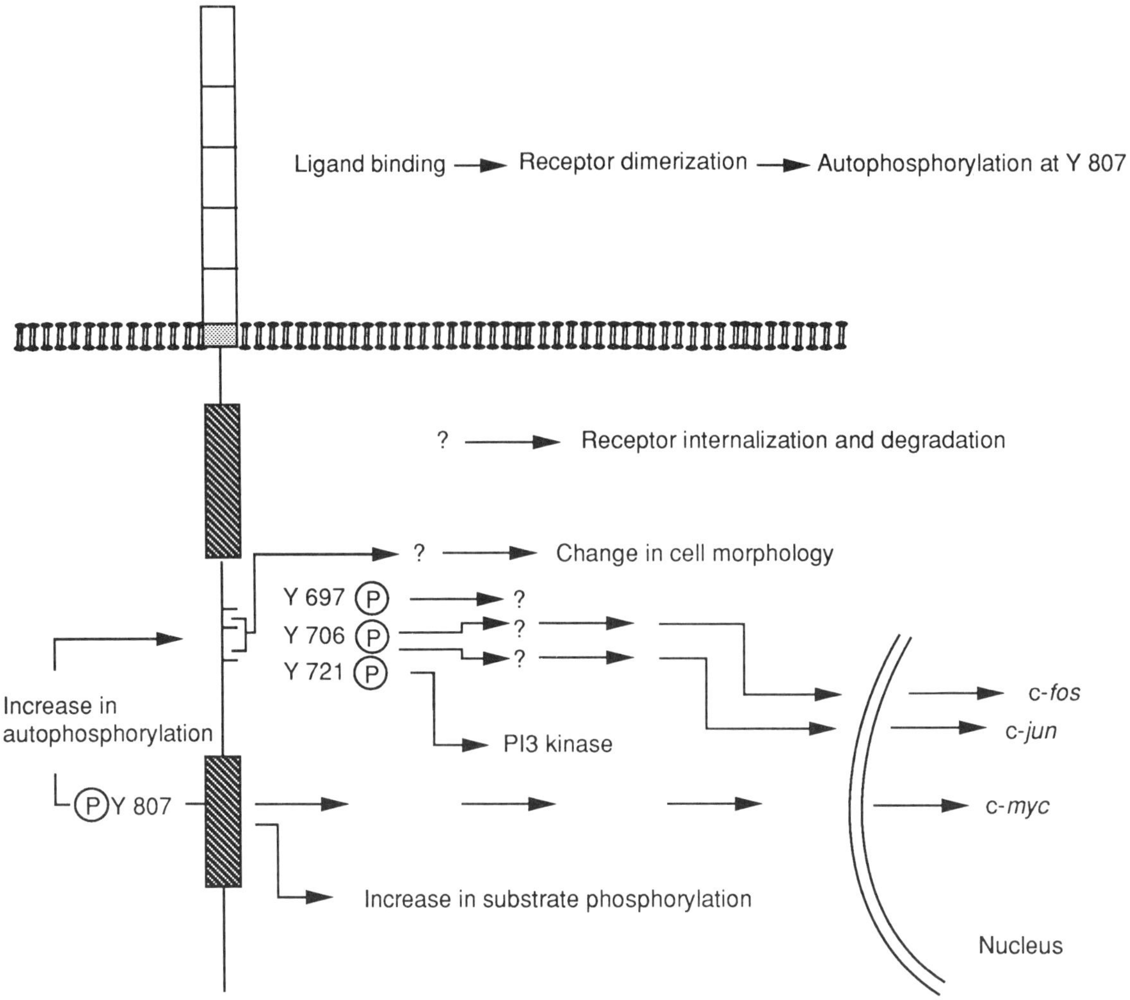

Figure 5. Signal transduction pathways initiated by CSF-1 binding to its receptor. CSF-1 binding to its receptor induces receptor dimerization followed by autophosphorylation at Tyr-807, which may precede the phosphorylation at Tyr-697, Tyr-706, and Tyr-721 in the kinase insert. The function of phosphorylation on Tyr-697 and Tyr-706 is not well understood. Mutational analysis of Tyr-706 indicates that this area of the receptor connects with two independent signal transduction pathways, one of which induces transcription of c-*jun* and the second of which leads to transcription of c-*fos*. Mutation of either Tyr-807 to phenylalanine or Tyr-706 to glycine effects the change in morphology seen when Rat-2 fibroblasts are grown in CSF-1. Tyr-721 is the site for association with PI3 kinase. It is not known what signals are involved in receptor internalization and degradation. It is known that kinase inactive receptors are internalized in response to CSF-1 but that degradation in the lysosomes depends on kinase activity and the presence of a kinase insert.

a pathway leading to the SRE or SIE, because a critical SH2-containing substrate cannot bind and therefore cannot be phosphorylated. On the other hand, the Gly-706 mutation may not only preclude phosphorylation of Tyr-706, but may also perturb the whole structure of this region. Therefore, it is harder to interpret the inability of the Gly-706 mutant to induce c-*jun* expression. One possibility is that the glycine mutation also affects the ability of substrates to bind to P.Tyr697 and/or P.Tyr721, and that one of these is responsible for activation of a pathway leading to the TRE. Of course, this would not explain why the Gly-706 mutant is still able to induce c-*fos* expression, if this requires phosphorylation at Tyr-706. The phenotypes of the Tyr-697 and Tyr-721 mutants have not been carefully analyzed in Rat-2 cells yet, but these mutants have been shown to be transforming in NIH-3T3 cells, which is not unexpected considering the results with expression of the kinase insert deletion mutant in NIH-3T3 cells (Taylor et al. 1989).

The CSF-1 receptor also autophosphorylates on Tyr-807. This residue is conserved in all PTKs studied thus far (Hanks and Quinn 1991), with the exception of CSK, the Src Tyr-527 regulatory PTK (Nada et al. 1991). Mutation of Tyr-807 to glycine abolishes CSF-1 receptor PTK activity (van der Geer and Hunter 1991). The Phe-807 mutant shows a two- to threefold decrease in autophosphorylation in response to CSF-1 in vivo and has reduced PTK activity in vitro (van der Geer and Hunter 1991). The Phe-807 receptor mutant does not transform NIH-3T3 cells and cannot induce an increase in growth rate and a change in morphology in response to CSF-1 in Rat-2 fibroblasts (Table 3). Phosphorylation of Tyr-807 has also been found to be important for the induction of c-*myc* RNA expression. Wild-type human CSF-1 receptors expressed in NIH-3T3 fibroblasts induce expression of c-*myc* RNA in response to CSF-1, whereas the Phe-809 mutant receptor (equivalent to the murine Phe-807 mutant) is deficient in induction of c-*myc* RNA expression, growth in soft agar, and growth in serum-free medium in response to CSF-1 (Roussel et al. 1991). These responses to CSF-1 are restored by overexpression of c-*myc* (Roussel et al. 1991). These data indicate that in NIH-3T3 cells, induction of c-*myc* expression is essential but not sufficient for CSF-1-dependent growth in soft agar and serum-free medium.

In contrast to NIH-3T3 cells, stimulation of Rat-2 fibroblasts, expressing wild-type CSF-1 receptors, does not lead to a substantial increase in c-*myc* RNA levels, although there is a strong growth response to CSF-1. We have now found that c-*myc* expression is induced by CSF-1 in rat 208F fibroblasts expressing wild-type CSF-1 receptors. Thus, it appears that the ability of CSF-1 to induce c-*myc* expression depends on the cell type, and that c-*myc* expression is not required in all cell types to increase their growth rate in response to CSF-1.

Signal transduction by the CSF-1 receptor is initiated by ligand binding and receptor dimerization (Li and Stanley 1991). Dimerization allows receptor autophosphorylation through an intermolecular mechanism (Fig. 5). Autophosphorylation on Tyr-807, located within the catalytic domain and conserved in all PTKs, may result in an increase in PTK activity, which is then followed by phosphorylation of other tyrosines (Fig. 5). Although precise kinetic analysis of autophosphorylation in response to ligand has yet to be done for the CSF-1 receptor, analysis of autophosphorylation site mutants in the CSF-1 receptor and other PTKs indicates that mutation of this conserved tyrosine residue affects PTK activity (Ellis et al. 1986; Weinmaster and Pawson 1986; Kmiecik and Shalloway 1987; Piwnica-Worms et al. 1987; Kazlauskas et al. 1991). It is thought that phosphorylation of other tyrosines within the receptor allows the receptor to interact with signaling molecules containing SH2 domains. In addition to a role in regulating receptor PTK activity, phosphorylation of Tyr-807 may be important for induction of c-*myc* expression (Roussel et al. 1991). However, it is not clear whether the effects on c-*myc* RNA expression are a direct effect of mutation of Tyr-807 or whether mutation of Tyr-807 results in a reduction of autophosphorylation at a secondary site that is important for c-*myc* RNA induction (Fig. 5).

The PDGF receptor, which is closely related to the CSF-1 receptor, binds the p85 subunit of PI3 kinase, GAP, PLC-γ1, and the cytoplasmic PTKs, cSrc, cYes, and Fyn (Kaplan et al. 1990a; Kazlauskas and Cooper 1990; Kazlauskas et al. 1990; Kypta et al. 1990; Morrison et al. 1990). It is thought that these signaling molecules bind specific phosphotyrosine residues in the PDGF receptor. In fact, the binding sites for the p85 subunit of PI3 kinase and for GAP have been identified (Fantl et al. 1992; Kashishian et al. 1992). PI3 kinase binds to P.Tyr740 and P.Tyr751, and GAP binds to P.Tyr771, all located within the PDGF receptor kinase insert (Fantl et al. 1992; Kashishian et al. 1992). Analysis of a GAP-binding site mutant in the PDGF receptor indicates that association of GAP is not essential for mitogenesis, at least not in epithelial cells, but analysis of mutations in the PI3 kinase-binding sites has yielded conflicting results (Fantl et al. 1992; Kazlauskas et al. 1992).

There is no evidence that either GAP or PLC-γ binds to the CSF-1 receptor in response to ligand binding, and so far PI3 kinase is the only protein reported to bind to the activated CSF-1 receptor. Analysis of the cellular responses of cells expressing Phe-721 CSF-1 receptors is currently being undertaken to learn more about the role of this site in receptor signaling. It is possible that proteins in addition to PI3 kinase bind to this site. The exact functions of the other CSF-1 receptor autophosphorylation sites are also unclear. Mutation of Tyr-706 in the kinase insert yields subtle phenotypes, indicating that this site is important for the induction of immediate-early gene expression and changes in cell morphology in response to CSF-1. It appears that the region around Tyr-706 may connect to two independent signaling pathways (Fig. 5). By analogy with other autophosphorylation sites, we anticipate that phosphorylation of these sites leads to interaction of the receptor with other as yet unidentified signaling molecules. We are currently analyzing what proteins can bind to the sequences containing P.Tyr-697, P.Tyr-706, and P.Tyr-721 to gain more insight into signal transduction pathways dependent on these sites.

Protein Tyrosine Phosphatases

Although it was clear from the start that phosphotyrosine in cellular protein turned over rapidly in the cell, little was known about protein phosphatases involved in dephosphorylation of protein tyrosine phosphate until recently (Hunter 1989; Alexander 1990; Fischer et al. 1991). The first PTP identified, PTP1B, was purified from human placenta on the basis of its ability to dephosphorylate phosphotyrosine residues specifically (Tonks et al. 1988). Sequence analysis of this protein revealed a domain homologous to a region duplicated in the previously cloned leukocyte common antigen, CD45 (Charbonneau et al. 1989). Many other PTPs have since been identified by their relatedness to these first two PTPs. As is the case for the PTKs, the PTP family can be divided into two main groups: those that span the membrane and those that are entirely cytoplasmic. The membrane-spanning group are reminiscent of receptor PTKs (Fig. 6). The receptor-like PTPs have an amino-terminal, glycosylated extracellular domain, a single membrane-spanning region, and a cytoplasmic region, which typically includes two catalytic domains. LAR (Streuli et al. 1988), PTPβ, PTPδ (Krueger et al. 1990), and PTPμ (Gebbink et al. 1991)

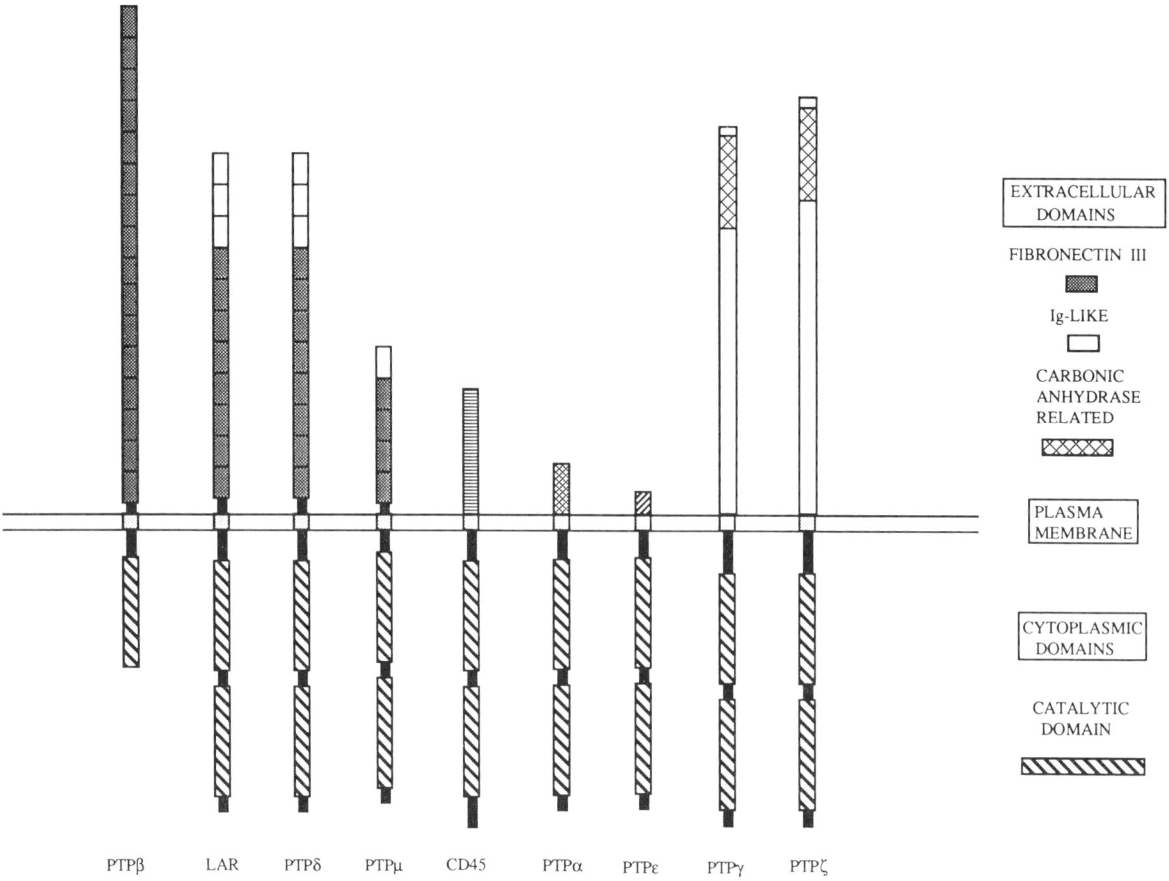

Figure 6. Schematic representation of vertebrate receptor-like protein tyrosine phosphatases. Functional motifs present in extracellular and intracellular domains are represented by filled boxes as defined on the figure.

possess large external domains that include Ig domains and fibronectin type III repeats, suggesting a function as cell adhesion molecules. PTPγ (Kaplan et al. 1990b; Krueger et al. 1990) and PTPζ (Krueger et al. 1990) have long external domains, with a region near the amino terminus related to carbonic anhydrase; the significance of this domain is unknown. Other receptor-like PTPs such as CD45 (Thomas et al. 1987), PTPα, and PTPϵ (Krueger et al. 1990) have shorter external domains. PTPβ is unique in having only a single catalytic domain (Krueger et al. 1990). A number of receptor-like PTPs have been identified in *Drosophila*, and all of these have Ig and fibronectin type III repeats in their external domains (e.g., DPTP, DLAR [Streuli et al. 1989], $DPTP_{10D}$, $DPTP_{99A}$ [Tian et al. 1991; Yang et al. 1991]).

Although no ligand has been conclusively identified for any of these receptor-like PTPs, it has been proposed that these proteins bind a ligand either directly or in conjunction with an accessory molecule. In principle, ligand binding to a receptor PTP could either activate or inhibit the activity of the cytoplasmic PTP domains. Alternatively, ligand binding might redistribute the receptor PTP in the plane of the membrane, bringing it into the vicinity of potential substrates, as has been proposed in the case of CD45 and the Lck PTK (Fischer et al. 1991). One promising avenue for the identification of a ligand for receptor-like PTPs is a genetic suppressor analysis of mutations in the *Drosophila* PTP genes.

The nonreceptor PTPs are also diverse in structure. They all possess a single catalytic domain and a wide variety of regulatory or subcellular localization domains. PTP1B and TCPTP have a regulatory carboxy-terminal domain that mediates attachment to the endoplasmic reticulum or cytoskeleton and inhibits activity. Proteolytic cleavage of this domain results in the liberation of the protein into the cytosol and PTP activation (Cool et al. 1990; Frangioni et al. 1992; Woodford-Thomas et al. 1992). The noncatalytic regions of PTP MEG1 (Gu et al. 1992) and PTPH1 (Yang and Tonks 1992) contain domains related to the amino-terminal domains of the cytoskeleton-associated proteins, band 4.1, talin, and ezrin. Recently, two distinct nonreceptor PTPs have been identified that contain two SH2 domains (Matthews et al. 1992; Plutzky et al. 1992; Shen et al. 1991; Yi et al. 1992). One of these is closely related to the *Drosophila corkscrew* gene product, which is in a signaling pathway downstream from the *torso* receptor PTK (Perkins et al. 1992). Since SH2 domains have been shown to bind specifically to phosphorylated tyrosine residues, the presence of SH2 domains has led to the speculation that these sequences mediate the association of these PTPs with potential tyrosine-phosphorylated substrates. Finally, there are two nonreceptor PTPs containing PEST sequences (Matthews et al. 1992) and another containing a domain related to a cytoplasmic retinol-binding protein (Gu et al. 1992).

We initiated an investigation of PTPs for several reasons. First, we hoped to gain insight into the spectrum of PTPs expressed by the mammalian genome. Second, we wanted to understand how PTP activity could be regulated. Finally, we wanted to determine what role particular PTPs have in mitogenic signal transduction pathways initiated by receptor PTKs. Toward these ends, we began by aligning the sequences for the PTPs, PTP1B, CD45, and LAR, which had been reported at that time (Thomas et al. 1987; Streuli et al. 1988; Charbonneau et al. 1989). Degenerate oligonucleotide primers were synthesized corresponding to the conserved regions of the catalytic domains. These primers were used to amplify a specific 199-bp fragment from NIH-3T3 cell RNA in a PCR reaction. The resulting fragment was used to probe an NIH-3T3 cDNA library, and a clone encoding a protein with a receptor-like structure similar to CD45 was isolated. The sequence of both the mouse and human homologs have since been reported by other investigators. The clones have been referred to as LRP (Matthews et al. 1990), HRPTPα (Kaplan et al. 1990b; Krueger et al. 1990), and R-PTPα (Sap et al. 1990). We have chosen to call our clone PTPα to avoid complicating the nomenclature. PTPα encodes a 793-amino-acid protein which includes a 19-residue signal sequence, a 123-residue serine/threonine-rich extracellular domain lacking cysteine residues but containing eight potential *N*-linked glycosylation sites, a transmembrane domain, two catalytic domains, and a short carboxy-terminal extension (Fig. 7).

PTPα RNA is widely expressed in cell lines and mouse tissues, with the highest levels being found in heart, kidney, and thymus. A polyclonal rabbit antiserum was raised against a peptide corresponding to the carboxy-terminal 19 amino acids of the predicted protein coupled to keyhole limpet hemocyanin. This antiserum immunoprecipitated a protein of 140 kD from [^{35}S]methionine/cysteine-labeled NIH-3T3 cells. Immunoprecipitation of p140 was blocked by preincubating the antiserum with the peptide antigen. This protein is considerably larger than the predicted size (88 kD). However, comigration of the major [^{35}S]methionine/cysteine-labeled tryptic peptides from the in-vivo-labeled p140 with cloned PTPα, which had been transcribed and translated in vitro, confirmed that p140 is PTPα. We presume that PTPα is modified by *N*-linked glycosylation at the available sites, and possibly may also contain *O*-linked sugars, and that these modifications account for the size difference.

PTPα immunoprecipitated from growing ^{32}P-labeled NIH-3T3 cells contained significant levels of ^{32}P, and phosphoamino acid analysis and tryptic phosphopeptide mapping showed that PTPα is phosphorylated predominantly on serine, with one major and several more minor phosphopeptides being present. Low levels of phosphotyrosine and phosphothreonine were also detected in PTPα, but represented less than 5% of the total phosphate content. The fact that PTPα is a phosphoprotein suggested that it might be regulated by phosphorylation in a manner analogous to the receptor PTKs. For this reason, we examined PTPα for changes in phosphorylation following stimulation of serum-starved NIH-3T3 cells with either TPA, PDGF, EGF, or whole serum. There was a rapid and reproducible, albeit small, increase in the total amount of phosphate, particularly in response to TPA and PDGF, although there was no change in the pattern of phosphopeptides with any of these factors.

To determine if the increase in phosphorylation following TPA stimulation was due directly to protein kinase C (PKC), the entire cytoplasmic portion of PTPα was expressed in bacteria as a GST fusion protein containing a thrombin-cleavable linker, purified on glutathione–Sepharose beads, cleaved with thrombin, and phosphorylated in vitro by purified PKC. The phosphopeptides derived from the in-vitro-phosphorylated protein comigrated with the major phosphopeptide and two of the minor phosphopeptides derived from the in-vivo-phosphorylated endogenous PTPα.

We then sought to identify the sites of PKC phosphorylation in PTPα. In other examples of transmembrane proteins that are phosphorylated by PKC, these sites lie in close proximity to the cytoplasmic face of the plasma membrane. With this in mind, we examined the PTPα sequence for possible sites of PKC phosphorylation (Ser/Thr.hydrophobic.Arg), taking into consideration the known mobility of the three phosphopeptides and the predicted mobilities of the candidate phosphorylation sites. Ser-180 was the best candidate, and therefore we mutated Ser-180 to alanine. The S180A mutant PTPα was expressed in bacteria as a GST fusion protein, purified, and phosphorylated in parallel with the wild-type PTPα. Comparison of the phosphotryptic maps from mutant and wild-type PTPα showed that the two peptides corresponding to two minor peptides were missing in the mutant, but the major peptide remained. We believe that the two minor peptides both contain Ser-180, and that the more negatively charged, less hydrophilic peptide was generated by cyclization of the amino-terminal glutamine of the tryptic peptide containing Ser-180. To identify the site of phosphorylation in the major peptide, we carried out further analysis including cyanogen bromide (CNBr) cleavage of the in-vitro-labeled wild-type protein. Both tryptic phosphopeptides were contained in a single ~7-kD CNBr fragment. This, coupled with the fact that the phosphoserine in the major peptide was localized to position three by Edman degradation, indicates that the major site of PKC phosphorylation is Ser-204. We are currently confirming this assignment by site-directed mutagenesis of Ser-204 to alanine.

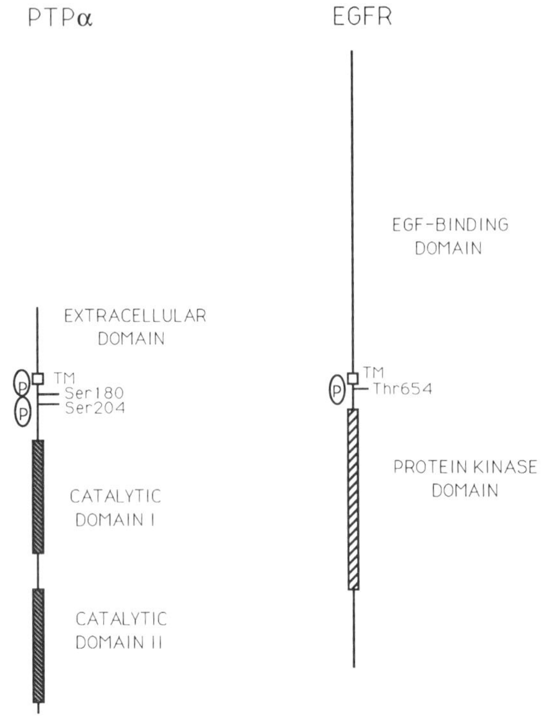

Figure 7. PKC phosphorylates PTPα at two sites near the transmembrane domain. The figure shows a comparison of the structures of PTPα and the EGF receptor indicating the location of the PKC phosphorylation sites.

Phosphorylation of the EGF receptor by PKC at Thr-654 (Hunter et al. 1984), which is in a location analogous to Ser-180 and Ser-204, reduces EGF-dependent signaling and contributes to the down-regulation of EGF receptor PTK activity following activation of PKC by EGF receptor-stimulated PI turnover (Fig. 7) (Cochet et al. 1984). This precedent suggests that phosphorylation of PTPα at two sites near the transmembrane domain by PKC could be functionally important. Consistent with this notion, J. den Hertog et al. (pers. comm.) have found that the in vitro phosphatase activity of PTPα is increased two- to threefold following TPA stimulation of cells expressing exogenous copies of PTPα and that the increase in activity is dependent on TPA-induced phosphorylation. A stimulatory effect of serine phosphorylation by an unknown protein kinase on the activity of the CD45 receptor-like PTP cells has also been noted (Ostergaard and Trowbridge 1991).

The fact that PTPα is phosphorylated at Ser-180 and Ser-204 in quiescent NIH-3T3 cells, where PKC activity is low, suggests that a protein kinase(s) other than PKC can also phosphorylate these sites. Such phosphorylation may maintain a basal level of PTPα activity. If PTPα activity is stimulated via PKC-mediated phosphorylation in the intact cell, then this could provide a mechanism for down-regulating the activity of ligand-activated receptor PTKs, such as the PDGF receptor, that induce PI turnover, by dephosphorylating either the receptor autophosphorylation sites or substrates phosphorylated by the receptor PTK.

To understand more about the function of PTPα, we are currently assessing the effect of expression of the phosphorylation site mutants of PTPα on signaling via PKC and mitogenic receptor PTKs. In this regard, an understanding of the substrate specificity of PTPα will be important. For this purpose we are also attempting to localize PTPα by immunolocalization to compare this with the known locations of PTKs and phosphotyrosine-containing proteins. The putative ligand-binding domain of PTPα is relatively small, and it is not certain that this domain is large enough for binding of a soluble ligand, especially since it is highly glycosylated. Attempts are being made to identify proteins that might interact with the external domain of PTPα by looking for coprecipitating proteins. It is possible that PTPα might associate with a ligand-binding subunit, as is the case for the IL-6 receptor.

SUMMARY AND CONCLUSIONS

It is clear that the number of receptor PTKs and PTPs encoded by a typical vertebrate genome is rather large. Although the signal pathways activated by the receptor PTKs may in many cases be common, specificity is provided by the ligand-binding domain and the availability of ligand. In addition, the precise spectrum of substrates that bind to and are phosphorylated by each receptor PTK can differ based on the number and nature of the autophosphorylation sites and on the repertoire of SH2-containing proteins and other substrates expressed in each cell type. It is also clear that receptor PTKs can activate multiple independent signaling pathways and that the output of these pathways can be integrated to provide a specific cellular response. The role of receptor PTPs in such integrated signaling networks is not yet obvious. In some cases, they may activate nonreceptor PTKs, whereas in other cases, they may counteract the effects of activated receptor and nonreceptor PTKs by dephosphorylating the PTKs themselves or their substrates. We know very little about the substrate specificity of PTPs, but in part this must be dictated by their subcellular location. It is possible that there are specific pairs of receptor PTKs and PTPs, which act in concert at the cell surface to activate and down-regulate specific signal pathways. Progress in understanding the function of receptor PTPs will depend on identifying ligands for receptor PTPs and then determining how ligand binding influences their activity.

ACKNOWLEDGMENTS

We thank Ann-Marie Quinn for her help in maintaining the protein kinase database and in generating the receptor protein tyrosine kinase phylogenetic tree. This work was supported by U.S. Public Health Service grants CA-14195 and CA-39870.

REFERENCES

Alexander, D.R. 1990. The role of phosphatases in signal transduction. *New Biol.* **2:**1049.

Backer, J.M., M.G. Myers, Jr., S.E. Shoelson, D.J. Chin, X.-J. Sun, M. Miralpeix, P. Hu, B. Margolis, E.Y. Skolnik, J. Schlessinger, and M.F. White. 1992. Phosphatidylinositol 3′-kinase is activated by association with IRS-1 during insulin stimulation. *EMBO J.* **11:** 3469.

Berkemeier, L.R., J.W. Winslow, D.R. Kaplan, K. Nikolics, D.V. Goeddel, and A. Rosenthal. 1991. Neurotrophin-5: A novel neurotrophic factor that activates *trk* and *trkB*. *Neuron* **7:** 857.

Cantley, L.C., K.R. Auger, C. Carpenter, B. Duckworth, A. Graziani, R. Kapeller, and S. Soltoff. 1991. Oncogenes and signal transduction. *Cell* **64:** 281.

Cartwright, C.A., W. Eckhart, S. Simon, and P.L. Kaplan. 1987. Cell transformation by $pp60^{c\text{-}src}$ mutated in the carboxy-terminal regulatory domain. *Cell* **49:** 83.

Chan, J. and V. Watt. 1991. *eek* and *erk*, new members of the *eph* subclass of receptor protein-tyrosine kinases. *Oncogene* **6:** 1057.

Charbonneau, H., N.K. Tonks, S. Kumar, C.D. Diltz, M. Harrylock, D.E. Cool, E.G. Krebs, E.H. Fischer, and K.A. Walsh. 1989. Human placenta protein-tyrosine-phosphatase: Amino acid sequence and relationship to a family of receptor-like proteins. *Proc. Natl. Acad. Sci.* **86:** 5252.

Cochet, C., G. Gill, J. Meisenhelder, J.A. Cooper, and T. Hunter. 1984. C-kinase phosphorylates the epidermal growth factor receptor and reduces its epidermal growth factor-stimulated tyrosine protein kinase activity. *J. Biol. Chem.* **259:** 2553.

Cool, D.E., N.K. Tonks, H. Charbonneau, E.H. Fischer, and E.G. Krebs. 1990. Expression of a human T-cell protein tyrosine-phosphatase in baby hamster kidney cells. *Proc. Natl. Acad. Sci.* **87:** 7280.

Cooper, J.A., K.L. Gould, C.A. Cartwright, and T. Hunter. 1986. Tyr 527 is phosphorylated in $pp60^{c\text{-}src}$: Implications for regulation. *Science* **231:** 1431.

Coussens, L., C. Van Beveren, D. Smith, E. Chen, R.L. Mitchell, C.M. Isacke, I. Verma, and A. Ullrich. 1986. Structural alteration of viral homologue of receptor proto-oncogene *fms* at carboxyl terminus. *Nature* **320:** 277.

Downing, J.R., C.W. Rettenmier, and C.J. Sherr. 1988. Ligand-induced tyrosine kinase activity of the colony-stimulating factor 1 receptor in a murine macrophage cell line. *Mol. Cell. Biol.* **8:** 1795.

Downing, J.R., M.F. Roussel, and C.J. Sherr. 1989. Ligand and protein kinase C downmodulate the colony-stimulating factor 1 receptor by independent mechanisms. *Mol. Cell. Biol.* **9:** 2890.

Downing, J.R., S.A. Shurtleff, and C.J. Sherr. 1991. Peptide antisera to the human colony-stimulating factor 1 receptor detect ligand-induced conformational changes and a binding site for phosphatidylinositol 3-kinase. *Mol. Cell. Biol.* **11:** 2489.

Ellis, L., E. Clauser, D.O. Morgan, M. Edery, R.A. Roth, and W.J. Rutter. 1986. Replacement of insulin receptor tyrosine residues 1162 and 1163 compromises insulin-stimulated kinase activity and uptake of 2-deoxyglucose. *Cell* **45:** 721.

Fantl, W.J., J.A. Escobedo, G.A. Martin, C.W. Turck, M. del Rosario, F. McCormick, and L.T. Williams. 1992. Distinct phosphotyrosines on a growth factor receptor bind to specific molecules that mediate different signaling pathways. *Cell* **69:** 413.

Fischer, E.H., H. Charbonneau, and N.K. Tonks. 1991. Protein tyrosine phosphatases: A diverse family of intracellular and transmembrane enzymes. *Science* **253:** 401.

Frangioni, J.V., P.H. Beahm, C.A. Jost, and B.G. Neel. 1992. The nontransmembrane tyrosine phosphatase PTP-1B localizes to the endoplasmic reticulum via its 35 amino acid C-terminal sequence. *Cell* **68:** 545.

Gebbink, M.F.B.G., I. van Etten, G. Hateboer, R. Suijkerbuijk, R.L. Beijersbergen, A.G. van Kessel, and W.H. Moolenaar. 1991. Cloning, expression and chromosomal localization of a new putative receptor-like protein tyrosine phosphatase. *FEBS Lett.* **289:** 1.

Glass, D.J., S.H. Nye, P. Hantzopoulos, M.J. Macchi, S.P. Squinto, M. Goldfarb, and G.D. Yancopoulos. 1991. TrkB mediates BDNF/NT-3-dependent survival and proliferation in fibroblasts lacking the low affinity NGF receptor. *Cell* **66:** 405.

Gu, M., I. Warshawsky, and P.W. Majerus. 1992. Cloning and expression of a cytosolic megakaryocyte protein-tyrosine-phosphatase with sequence homology to retinaldehyde-binding protein and yeast SEC14p. *Proc. Natl. Acad. Sci.* **88:** 5867.

Hanks, S.K. and A.M. Quinn. 1991. Protein kinase catalytic domain sequence database: Identification of conserved features of primary structure and classification of family members. *Methods Enzymol.* **200:** 38.

Hirai, H., Y. Maru, K. Hagiwara, J. Nishida, and F. Takaku. 1987. A novel putative tyrosine kinase receptor encoded by the *eph* gene. *Science* **238:** 1717.

Hunter, T. 1989. Protein-tyrosine phosphatases: The other side of the coin. *Cell* **58:** 1013.

Hunter, T., N. Ling, and J.A. Cooper. 1984. Protein kinase C phosphorylation of the EGF receptor at a threonine residue close to the cytoplasmic face of the plasma membrane. *Nature* **311:** 480.

Kaplan, D., D. Martin-Zanca, and L.F. Parada. 1991a. Tyrosine phosphorylation and tyrosine kinase activity of the *trk* proto-oncogene product induced by NGF. *Nature* **350:** 158.

Kaplan, D.R., B. Hempstead, D. Martin-Zanca, M.V. Chao, and L.F. Parada. 1991b. The *trk* proto-oncogene product: A signal transducing receptor for nerve growth factor. *Science* **252:** 554.

Kaplan, D.R., D.K. Morrison, G. Wong, F. McCormick, and L.T. Williams. 1990a. PDGF-β-receptor stimulates tyrosine phosphorylation of GAP and association of GAP with a signaling complex. *Cell* **61:** 125.

Kaplan, R., B. Morse, K. Huebner, C. Croce, R. Howk, M. Ravera, G. Ricca, M. Jaye, and J. Schlessinger. 1990b. Cloning of three human tyrosine phosphatases reveals a multigene family of receptor-linked protein-tyrosine-phosphatases expressed in brain. *Proc. Natl. Acad. Sci.* **87:** 7000.

Kashishian, A., A. Kazlauskas, and J.A. Cooper. 1992. Phosphorylation sites in the PDGF receptor with different specificities for binding GAP and PI3 kinase *in vivo*. *EMBO J.* **11:** 1373.

Kazlauskas, A. and J.A. Cooper. 1990. Phosphorylation of the PDGF receptor β subunit creates a tight binding site for phosphatidylinositol 3 kinase. *EMBO J.* **9:** 3279.

Kazlauskas, A., D.L. Durden, and J.A. Cooper. 1991. Functions of the major tyrosine phosphorylation site of the PDGF receptor β subunit. *Cell Regul.* **2:** 413.

Kazlauskas, A., C. Ellis, T. Pawson, and J.A. Cooper. 1990. Binding of GAP to activated PDGF receptors. *Science* **247:** 1578.

Kazlauskas, A., A. Kashisian, J.A. Cooper, and M. Valius. 1992. GTPase activating protein and phosphatidylinositol 3-kinase bind to distinct regions of the platelet-derived growth factor receptor β subunit. *Mol. Cell. Biol.* **12:** 2534.

Klein, R., D. Conway, L.F. Parada, and M. Barbacid. 1990. The *trk*B protein kinase gene codes for a second neurogenic receptor that lacks the catalytic kinase domain. *Cell* **61:** 647.

Klein, R., F. Lamballe, S. Bryant, and M. Barbacid. 1992. The *trk*B tyrosine protein kinase is a receptor for neurotrophin-4. *Neuron* **8:** 947.

Klein, R., L.F. Parada, F. Coulier, and M. Barbacid. 1989. *trk*B, a novel tyrosine protein kinase receptor expressed during mouse neural development. *EMBO J.* **8:** 3701.

Klein, R., S. Jing, V. Nanduri, E. O'Rourke, and M. Barbacid. 1991a. The *trk* proto-oncogene encodes a receptor for nerve growth factor. *Cell* **65:** 189.

Klein, R., V. Nanduri, S. Jing, F. Lamballe, P. Tapley, S. Bryant, C. Cordon-Cardo, K.R. Jones, L.F. Reichardt, and M. Barbacid. 1991b. The *trk*B tyrosine kinase is a receptor for brain-derived neurotrophic factor and neurotrophin-3. *Cell* **66:** 395.

Klippel, A., J.A. Escobedo, W.J. Fantl, and L.T. Williams. 1992. The C-terminal SH2 domain of p85 accounts for the high affinity and specificity of phosphatidylinositol 3-kinase to phosphorylated platelet-derived growth factor β receptor. *Mol. Cell. Biol.* **12:** 1451.

Kmiecik, T.E. and D. Shalloway. 1987. Activation and suppression of pp60$^{c\text{-}src}$ transforming ability by mutation of its primary sites of tyrosine phosphorylation. *Cell* **49:** 65.

Krueger, N.X., M. Streuli, and H. Saito. 1990. Structural diversity and evolution of human receptor-like protein tyrosine phosphatases. *EMBO J.* **9:** 3241.

Kypta, R.M., Y. Goldberg, E.T. Ulug, and S.A. Courtneidge. 1990. Association between the PDGF receptor and members of the *src* family of tyrosine kinases. *Cell* **62:** 481.

Lamballe, F., R. Klein, and R. Barbacid. 1991. *trk*C, a new member of the *trk* family of tyrosine protein kinases, is a receptor for NT-3. *Cell* **66:** 967.

Lhoták, V., P. Greer, K. Letwin, and T. Pawson. 1991. Characterization of *elk*, a brain-specific receptor tyrosine kinase. *Mol. Cell. Biol.* **11:** 2496.

Li, W. and E.R. Stanley. 1991. Role of dimerization and modification of the CSF-1 receptor in its activation and internalization during the CSF-1 response. *EMBO J.* **10:** 277.

Lindberg, R.A. and T. Hunter. 1990. cDNA cloning and characterization of *eck*, an epithelial cell receptor protein-tyrosine kinase in the *eph/elk* family of protein kinases. *Mol. Cell. Biol.* **10:** 6316.

Martin-Zanca, D., M. Barbacid, and L.F. Parada. 1990. Expression of the *trk* proto-oncogene is restricted to the sensory cranial and spinal ganglia of the neural crest origin in mouse development. *Genes Dev.* **4:** 683.

Martin-Zanca, D., S.H. Hughes, and M. Barbacid. 1986. A human oncogene formed by the fusion of truncated tropomyosin and protein kinase sequences. *Nature* **319:** 743.

Martin-Zanca, D., R. Oskam, G. Mitra, T. Copeland, and M. Barbacid. 1989. Molecular and biochemical characterization of the human *trk* proto-oncogene. *Mol. Cell. Biol.* **9:** 24.

Maru, Y., H. Hirai, and F. Takaku. 1990. Overexpression confers an oncogenic potential upon the *eph* gene. *Oncogene* **5:** 199.

Matsui, T., M. Heidaran, T. Miki, N. Popescu, W. La Rochelle, M. Kraus, J. Pierce, and S. Aaronson. 1989. Isolation of a novel receptor cDNA establishes the existence of two PDGF receptor genes. *Science* **243:** 800.

Matthews, R.J., E.D. Cahir, and M.L. Thomas. 1990. Identification of an additional member of the protein-tyrosine-phosphatase family: Evidence for alternative splicing in the tyrosine phosphatase domain. *Proc. Natl. Acad. Sci.* **87:** 4444.

Matthews, R.J., D.B. Bowne, E. Flores, and M.L. Thomas. 1992. Characterization of hematopoietic intracellular protein tyrosine phosphatases: Description of a phosphatase containing an SH2 domain and another enriched in proline-, glutamic acid-, serine-, and threonine-rich sequences. *Mol. Cell. Biol.* **12:** 2396.

Middlemas, D.S., R.A. Lindberg, and T. Hunter. 1991. *trk*B, a neural receptor protein-tyrosine kinase: Evidence for a full-length and two truncated receptors. *Mol. Cell. Biol.* **11:** 143.

Morgan, C., J.W. Pollard, and E.R. Stanley. 1987. Isolation and characterization of a cloned growth factor dependent macrophage cell line, BAC1.2F5. *J. Cell. Physiol.* **130:** 420.

Morrison, D.K., D.R. Kaplan, S.G. Rhee, and L.T. Williams. 1990. Platelet-derived growth factor (PDGF)-dependent association of phospholipase C-gamma with the PDGF receptor signaling complex. *Mol. Cell. Biol.* **10:** 2359.

Nada, S., M. Okada, A. MacAuley, J.A. Cooper, and H. Nakagawa. 1991. Cloning of a complementary DNA for a protein-tyrosine kinase that specifically phosphorylates a negative regulatory site of p60$^{c\text{-}src}$. *Nature* **351:** 69.

O'Bryan, J.P., R.A. Frye, P.C. Cogswell, A. Neubauer, B. Kitch, C. Prokop, R. Espinosa, B.M. Le, H.S. Earp, and E.T. Liu. 1991. *axl*, a transforming gene isolated from primary human myeloid leukemia cells, encodes a novel receptor tyrosine kinase. *Mol. Cell. Biol.* **11:** 5016.

Ostergaard, H.L. and I.S. Trowbridge. 1991. Negative regulation of CD45 protein tyrosine phosphatase activity by ionomycin in T cells. *Science* **253:** 1423.

Partanen, J., E. Armstrong, T.P. Makela, J. Korhonen, M. Sandberg, R. Renkonen, S. Knuutila, K. Huebner, and K. Alitalo. 1992. A novel endothelial cell surface receptor tyrosine kinase with extracellular epidermal growth factor homology domains. *Mol. Cell. Biol.* **12:** 1698.

Pasquale, E.B. 1991. Identification of chicken embryo kinase 5, a developmentally regulated receptor-type tyrosine kinase of the Eph family. *Cell Regul.* **2:** 523.

Perkins, L.A., I. Larsen, and N. Perrimon. 1992. *corkscrew* encodes a putative tyrosine phosphatase that functions to transduce the terminal signal from the receptor tyrosine kinase *torso*. *Cell* **70:** 225.

Piwnica-Worms, H., K.B. Saunders, T.M. Roberts, A.E. Smith, and S.H. Cheng. 1987. Tyrosine phosphorylation regulates the biochemical and biological properties of pp60$^{c\text{-}src}$. *Cell* **49:** 75.

Plutzky, J., B.G. Neel, and R.D. Rosenberg. 1992. Isolation of a *src* homology 2-containing tyrosine phosphatase. *Proc. Natl. Acad. Sci.* **89:** 1123.

Pollard, J.W., A. Bartocci, R. Arceci, A. Orlofsky, B.M. Ladner, and E.R. Stanley. 1987. Apparent role of the macrophage growth factor, CSF-1, in placental development. *Nature* **330:** 484.

Qiu, F., P. Ray, K. Brown, P.E. Barker, S. Jhanwar, F.H. Ruddle, and P. Besmer. 1988. Primary structure of *c-kit*: Relationship with the CSF-1/PDGF receptor kinase family-oncogenic activation of v-*kit* involves deletion of extracellular domain and C terminus. *EMBO J.* **7:** 1003.

Reedijk, M., X. Liu, and T. Pawson. 1990. Interactions of phosphatidylinositol kinase, GTPase-activating protein (GAP), and GAP-associated proteins with the colony-stimulating factor-1 receptor. *Mol. Cell. Biol.* **10:** 5601.

Reedijk, M., X. Liu, P. van der Geer, K. Letwin, M.D. Waterfield, T. Hunter, and T. Pawson. 1992. Tyr721 regulates specific binding of the CSF-1 receptor kinase insert to PI 3′-kinase SH2 domains: A model for SH2-mediated receptor–target interactions. *EMBO J.* **11:** 1365.

Rothwell, V.M. and L.R. Rohrschneider. 1987. Murine c-*fms* cDNA: Cloning, sequence analysis and retroviral expression. *Oncogene Res.* **1:** 311.

Roussel, M.F., J.L. Cleveland, S.A. Shurtleff, and C.J. Sherr. 1991. *Myc* rescue of a mutant CSF-1 receptor impaired in mitogenic signalling. *Nature* **353:** 361.

Roussel, M.F., J.R. Downing, R.A. Ashmun, C.W. Rettenmier, and C.J. Sherr. 1988. Colony-stimulating factor 1-mediated regulation of a chimeric c-*fms*/v-*fms* receptor containing the v-*fms*-encoded tyrosine kinase domain. *Proc. Natl. Acad. Sci.* **85:** 5903.

Roussel, M.F., T.J. Dull, C.W. Rettenmier, P. Ralph, A. Ullrich, and C.J. Sherr. 1987. Transforming potential of the c-*fms* proto-oncogene (CSF-1 receptor). *Nature* **325:** 549.

Sajjadi, F.G., E.B. Pasquale, and S. Subramani. 1991. Identification of a new *eph*-related receptor tyrosine kinase gene from mouse and chicken that is developmentally regulated and encodes at least two forms of the receptor. *New Biol.* **3:** 769.

Sap, J., P. D'Eustachio, D. Givol, and J. Schlessinger. 1990.

Cloning and expression of a widely expressed receptor tyrosine phosphatase. *Proc. Natl. Acad. Sci.* **87:** 6112.

Scharfmann, R., J.H. Axelrod, and I.M. Verma. 1991. Long-term in vivo expression of retrovirus-mediated gene transfer in mouse fibroblast implants. *Proc. Natl. Acad. Sci.* **88:** 4626.

Schneider, R. and M. Schweiger. 1991. A novel modular mosaic of cell adhesion motifs in the extracellular domains of the neurogenic *trk* and *trk*B tyrosine kinase receptors. *Oncogene* **6:** 1807.

Sengupta, A., W.-K. Liu, Y.G. Yeung, D.C.Y. Leung, A.R. Frackelton, and E.R. Stanley. 1988. Identification and subcellular localization of proteins that are rapidly phosphorylated on tyrosine in response to colony-stimulating factor 1. *Proc. Natl. Acad. Sci.* **85:** 8062.

Shen, S.H., L. Bastien, B.I. Posner, and P. Chretien. 1991. A protein-tyrosine phosphatase with sequence similarity to the SH2 domain of the protein-tyrosine kinases. *Nature* **352:** 736.

Sherr, C.J., C.W. Rettenmier, R. Sacca, M.F. Roussel, A.T. Look, and E.R. Stanley. 1985. The c-*fms* proto-oncogene product is related to the receptor for the mononuclear phagocyte growth factor, CSF-1. *Cell* **41:** 665.

Soppet, D., E. Escandon, J. Maragos, D.S. Middlemas, S.W. Reid, J. Blair, L.E. Burton, B.R. Stanton, D.R. Kaplan, T. Hunter, K. Nikolics, and L.F. Parada. 1991. The neurotrophic factors brain-derived neurotrophic factor and neurotrophin-3 are ligands for the *trk*B tyrosine kinase receptor. *Cell* **65:** 895.

Squinto, S.P., T.N. Stitt, T.H. Aldrich, S. Davis, S.M. Bianco, P. Masiakowski, M.E. Furth, D.M. Valenzuela, P.S. DiStefano, and G.D. Yancopoulos. 1991. *trk*B encodes a functional receptor for brain-derived neurotrophic factor and neurotrophin-3 but not nerve growth factor. *Cell* **65:** 885.

Stanley, E.R., L.J. Guilbert, R.J. Tushinski, and S.H. Bartelmez. 1983. CSF-1–A mononuclear phagocyte lineage-specific hemopoietic growth factor. *J. Cell. Biochem.* **21:** 151.

Streuli, M., N.X. Krueger, A.Y. Tsai, and H. Saito. 1989. A family of receptor-linked protein tyrosine phosphatases in humans and *Drosophila*. *Proc. Natl. Acad. Sci.* **86:** 8698.

Streuli, M., N.X. Krueger, L.R. Hall, S.F. Schlossman, and H. Saito. 1988. A new member of the immunoglobulin superfamily that has a cytoplasmic region homologous to the leukocyte common antigen. *J. Exp. Med.* **168:** 1553.

Tapley, P., A. Kazlauskas, J.A. Cooper, and L.R. Rohrschneider. 1990. Macrophage colony-stimulating factor-induced tyrosine phosphorylation of c-*fms* proteins expressed in FDC-P1 and BALB/c3T3 cells. *Mol. Cell. Biol.* **10:** 2528.

Taylor, G.T., M. Reedijk, V. Rothwell, L. Rohrschneider, and T. Pawson. 1989. The unique insert of the cellular and viral *fms* protein tyrosine kinase domain is dispensable for enzymatic and transforming activities. *EMBO J.* **8:** 2029.

Thomas, M.L., P.J. Reynolds, A. Chain, and Y. Ben-Neriah. 1987. B-cell variant of mouse T200 (Ly-5): Evidence for alternative mRNA splicing. *Proc. Natl. Acad. Sci.* **84:** 5360.

Tian, S.-S., P. Tsoulfas, and K. Zinn. 1991. Three receptor-linked protein-tyrosine phosphatases are selectively expressed on central nervous system axons in the *Drosophila* embryo. *Cell* **67:** 675.

Tonks, N.K., C.D. Diltz, and E.H. Fischer. 1988. Purification of the major protein-tyrosine-phosphatases of human placenta. *J. Biol. Chem.* **263:** 6722.

Ullrich, A. and J. Schlessinger. 1990. Signal transduction by receptors with tyrosine kinase activity. *Cell* **61:** 203.

van der Geer, P. and T. Hunter. 1990. Identification of tyrosine 706 in the kinase insert as the major colony-stimulated factor 1 (CSF-1)-stimulated autophosphorylation site in the CSF-1 receptor in a murine macrophage cell line. *Mol. Cell. Biol.* **10:** 2991.

———. 1991. Tyrosine 706 and 807 phosphorylation site mutants in the murine colony-stimulating factor-1 receptor are unaffected in their ability to bind or phosphorylate phosphatidylinositol-3 kinase but show differential defects in their ability to induce early response gene transcription. *Mol. Cell. Biol.* **11:** 4698.

Weinmaster, G.A. and T. Pawson. 1986. Protein kinase activity of FSV (Fujinama sarcoma virus) $P130^{gag\text{-}fps}$ shows a strict specificity for tyrosine residues. *J. Biol. Chem.* **261:** 328.

Woodford-Thomas, T.A., J.D. Rhodes, and J.E. Dixon. 1992. Expression of a protein-tyrosine phosphatase in normal and v-*src*-transformed mouse 3T3 fibroblasts. *J. Cell Biol.* **117:** 401.

Yang, Q. and N.K. Tonks. 1991. Isolation of a cDNA clone encoding a human protein-tyrosine phosphatase with homology to the cytoskeletal-associated proteins band 4.1, ezrin, and talin. *Proc. Natl. Acad. Sci.* **88:** 5949.

Yang, X., K.T. Seow, S.M. Bahri, S.H. Oon, and W. Chia. 1991. Two *Drosophila* receptor-like tyrosine phosphatase genes are expressed in a subset of developing axons and pioneer neurons in the embryonic CNS. *Cell* **67:** 661.

Yarden, Y., J.A. Escobedo, W.-J. Kuang, T.L. Yang-Feng, T.O. Daniel, P.M. Tremble, E.Y. Chen, M.E. Ando, R.N. Harkins, U. Francke, V.A. Fried, A. Ullrich, and L.T. Williams. 1986. Structure of the receptor for platelet-derived growth factor helps define a family of closely related growth factor receptors. *Nature* **323:** 226.

Yi, T., J.L. Cleveland, and J.N. Ihle. 1992. Protein tyrosine phosphatase containing SH2 domains: Characterization, preferential expression in hematopoietic cells, and localization to human chromosome 12p12-p13. *Mol. Cell. Biol.* **12:** 836.

The *Trk* Family of Tyrosine Kinases: Receptors for NGF-related Neurotrophins

L.F. Parada, P. Tsoulfas, L. Tessarollo, J. Blair, S.W. Reid, and D. Soppet
Molecular Embryology Section, NCI-Frederick Cancer Research and Development Center, ABL-Basic Research Program, Frederick, Maryland 21702-1201

The intricate architecture of the vertebrate nervous system is composed of diverse sets of highly specialized cells. This diversity arises initially from a simple neuroepithelium, which is formed during gastrulation when the underlying mesoderm interacts with dorsal embryonic ectoderm to induce formation of the neural plate. The earliest structuring of the brain becomes evident in the anterior neural tube with the formation of the prosencephalon, mesencephalon, and myelencephalon. Proliferative centers along the ventricular zones generate nascent cells that migrate radially toward defined locations in the mantle layer, where they differentiate into various specialized cells, including assorted neuronal subtypes, astrocytes, and oligodendrocytes.

The peripheral nervous system (PNS) stems from a group of cells, the neural crest, located at the apical ridges of the early neural tube (Le Douarin and Smith 1988). These migratory cells are multipotential, capable of giving rise to many nonneural cell types, as well as to the peripheral and autonomic nervous system.

As can be surmised, a multiplicity of cellular functions, including proliferation, pathfinding, migration, differentiation, appropriate synaptic connections, and survival, must be accounted for at the molecular level. Recent progress has led to the identification of multiple genes with potential roles in neural development, thus kindling hope that the molecular basis underlying the development and function of the nervous system may be charted in the coming years (Jessell 1988; Anderson 1989; Sanes 1989; Keynes and Lumsden 1990).

The nerve growth factor (NGF)-related neurotrophins are polypeptide growth factors whose function in neural development is well documented (Barde 1991). NGF, the prototype neurotrophin, functions in short distance tropic interactions of cranial ganglia (Davies and Lumsden 1984) but is principally studied for its role in the maintenance of cell survival of sensory cranial, spinal, and sympathetic neurons (Levi-Montalcini 1987).

During embryogenesis, successful neuromuscular and interneuronal connections over long distances are facilitated by an initial overproduction of neurons followed by competition for limiting quantities of neurotrophins emitted by the cells which are the targets of innervation. Those neurons that successfully make functional synapses survive due to the presence of neurotrophin, whereas those failing to make appropriate contacts do not receive neurotrophin and thus undergo programmed cell death (Oppenheim 1981). This property of NGF has been experimentally documented in vivo by use of blocking anti-NGF antibodies in the embryo, which results in virtual sympathectomy of the embryos due to enhanced death of the NGF-dependent neurons (Levi-Montalcini and Booker 1960), and by administration of exogenous NGF, which results in enlarged sensory ganglia due to reduced cell death (Levi-Montalcini et al. 1974; Hamburger et al. 1981).

During the past 2 years, several advances have greatly enhanced our understanding of neurotrophin action and function. Through the work of Barde, Thoenen, and their colleagues (Barde et al. 1987; Leibrock et al. 1989), brain-derived neurotrophic factor (BDNF) was isolated and sequenced. Comparison of NGF and BDNF sequences revealed extensive homology between the two genes and provided a strategy for a directed search of related genes. This fruitful approach has resulted in the cloning of additional members of the NGF neurotrophin family: NT-3 (Ernfors et al. 1990; Hohn et al. 1990; Jones and Reichardt 1990; Kaisho et al. 1990; Maisonpierre et al. 1990); NT-4 (Hallböök et al. 1991; Ip et al. 1992); and NT-5 (Berkemeier et al. 1991). A second advance has been the identification of the *trk* receptor tyrosine kinase (RTK) as a receptor for NGF in rat pheochromocytoma PC12 cells and in sensory neurons (Kaplan et al. 1991a,b). This work has led to the rapid association of the receptors encoded by the *trk*-related loci, *trkB* and *trkC*, with the additional known neurotrophins (Lamballe et al. 1991; Soppet et al. 1991; Squinto et al. 1991; P. Tsoulfas et al.; D. Soppet et al.; both in prep.).

METHODS

Materials. WGA-Sepharose 6B and Protein A-Sepharose 4B were from Pharmacia (Piscataway, New Jersey). Alkaline phosphatase-conjugated anti-mouse and anti-rabbit antibodies, BCIP, and nitro-blue tetrazolium were from Promega Biotec (Madison, Wisconsin). Peptides were generated at NCI-FCRDC. Antisera were prepared at NCI-FCRDC or by Cocalico Biologicals (Reamstown, Pennsylvania). Dulbecco's modified Eagle's medium (DMEM), Leibovitz L-15 medium, trypsin, calf serum, horse serum, and prestained molecular-weight markers were from GIBCO-BRL (Gaithersburg, Maryland). HPLC-purified NGF, BDNF, NT-3, and hNT-5 were supplied by Genentech (South San Francisco, California). xNT-4 was a gift from Drs. C. Ibáñez and H. Persson. Protease inhibitors and orthovanadate were from Sigma (St. Louis, Missouri).

Antisera and antibodies. Rabbit polyclonal antisera were raised against KLH-conjugated synthetic peptides corresponding to amino acids YNGQPLRESKIIHMDYYQEGEVS (peptide and antiserum 510) and peptide and antiserum 442 and 443 (Soppet et al. 1991). Antiserum 510 detects $gp150^{trkC}$ and $gp150^{trkC14}$.

Dissection of tissues and identification of TrkC proteins. Female outbred mice were sacrificed, and different regions of the brain, as well as other organs, were dissected and lysed in ice-cold lysis buffer, followed by WGA-Sepharose precipitation and Western transfer analysis as described previously (Soppet et al. 1991).

Transient transfections. Transfections were performed as described previously by Loeb et al. (1991). For PC12 cell transfections, the cells were grown on collagen-coated dishes, resuspended in PC12 medium by gentle trituration, and plated at 10–20% density on 10-cm collagen-coated dishes. The following day, cells were washed 4 times with DMEM and then 5 ml of DMEM containing 3 μg/ml insulin, 100 μg lipofectin (GIBCO-BRL), and 50 μg pMEX*neo*, pMEX*neotrkC*, pMEX*neotrkB*, or pMEX*neotrkC*14 was added. The lipofectin mixture was replaced with fresh PC12 medium after 8 hours. The following day, cells were fed with PC12 medium with or without 10 ng/ml BDNF, NT-3, or NT-5. Three days after treatment, the plates were scored for cells exhibiting processes more than 2 cell diameters in length.

In situ hybridization and probes. In situ hybridization was performed by standard methods on 5-μm paraffin sections as described previously (Tessarollo et al. 1992). The probes utilized were described in detail elsewhere: *trk* (Martin-Zanca et al. 1990); *trkB* (Klein et al. 1990b); *trkC* (P. Tsoulfas et al.; L. Tessarollo et al., both in prep.).

RESULTS AND DISCUSSION

Trk Gene Family Structure

We have cloned *Trk* genes from several vertebrate species (e.g., mouse, rat, and chicken). Figure 1 summarizes our current understanding of the various *Trk* family gene products and provides an idea of the inherent complexity of these three loci. The murine *trk* locus encodes a single major 3.2-kb mRNA species that encodes a glycoprotein of approximately 140 kD ($gp140^{prototrk}$; Martin-Zanca et al. 1989) bearing all the hallmarks of an RTK. The *trkB* locus is more complex, exhibiting multiple transcripts ranging from 2.0 kb to 11.0 kb and encoding at least three distinct receptors (Klein et al. 1989, 1990a; Middlemas et al. 1991). One receptor, $gp145^{trkB}$, is an RTK colinear with the *trk* receptor. cDNAs encoding at least two different *trkB* receptors that lack the catalytic tyrosine kinase domain have been described (*trk*BTK-T1 and *trk* BTK-T2; Fig. 1), and evidence for the existence of at least one of the truncated isoforms in vivo has been reported (Klein et al. 1990a; Soppet et al. 1991). The *trkC* gene encodes two major transcripts and a minimum of five receptors

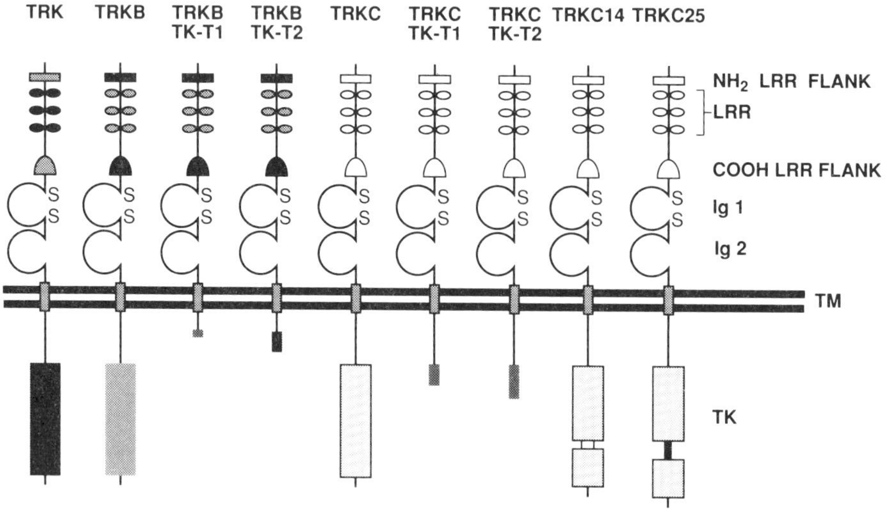

Figure 1. Schematic representation of different members and related isoforms of the *Trk* gene family: The amino-terminal sequences of the proteins contain the flank-LRR-flank structural motif (Rothberg et al. 1990; Krantz et al. 1991), followed by two tandem repeats of immunoglobulin (Ig)-like domains (Williams and Barclay 1988). (TM) Transmembrane domain; (TK) tyrosine kinase domain.

(P. Tsoulfas et al., in prep.), including three RTKs that differ by novel inserts in the catalytic domain and two truncated receptors (see Fig. 1). One *trkC* product, gp150^{trkC}, is fully colinear with gp140^{trk} and gp145^{trkB}, exhibiting extensive homology throughout the catalytic domain and lesser homology in the extracellular ligand-binding domain. Two additional receptors, gp150^{trkC14} and gp150^{trkC25}, contain additional amino acids at a novel site within the catalytic tyrosine kinase domain encoding inserts of 14 and 24 amino acids, respectively (P. Tsoulfas et al., in prep.). Truncated *trkC*, gp120$^{trkCTK\text{-}T1}$ and ^{T2}, differ from the truncated *trkB* receptors, since these molecules retain the juxtamembrane domain of the full receptors prior to diverging with a relatively long carboxyl terminus (see Fig. 1) (P. Tsoulfas et al., in prep.). We have not detected the existence of *trk* truncated isoforms. However, given the existence of truncated isoforms in *trkB* and *trkC*, coupled with the relatively limited expression exhibited by *trk*, we cannot rule out the existence of a truncated *trk* molecule that is tightly restricted in its expression.

Trk Gene Family Function

gp140^{prototrk} is expressed in rat PC12 cells, which differentiate in response to NGF by extending neurites and acquiring many biochemical characteristics of sympathetic neurons (Greene and Tischler 1976). These cells proved essential for the initial observation that gp140^{prototrk} became phosphorylated in response to NGF (Kaplan et al. 1991a) and subsequently to the demonstration that gp140^{prototrk} functions as an NGF receptor (Hempstead et al. 1991; Kaplan et al. 1991b; Klein et al. 1991). Subsequently, *trkB* has been shown to interact with BDNF and NT-3 (Soppet et al. 1991; Squinto et al. 1991), NT-4 (Ip et al. 1992), and NT-5 (Berkemeier et al. 1991; D. Soppet et al., in prep.). Furthermore, *trkC* has been identified as encoding the receptor for NT-3 (Lamballe et al. 1991; P. Tsoulfas et al., in prep.). A summary of these data is presented in Table 1.

Critical experiments using the PC12-derived *nnr* cell lines, which were selected for having lost responsiveness to NGF, demonstrated the functional interaction of *trk* receptor with NGF. Transient expression of gp140^{prototrk} in *nnr* cells reconstituted NGF response as assayed by neurite outgrowth (Loeb et al. 1991). These data were further extended by the overexpression of gp140^{prototrk} in PC12 cells. Incubation of these cells with NGF results in accelerated and enhanced responsiveness compared to normal PC12 cells (Hempstead et al. 1992).

The absence of PC12-like cell lines that respond to other neurotrophins has made direct assessment of the functional connection between neurotrophin(s) binding to *trkB* or *trkC* receptors more difficult. However, the use of PC12 and *nnr* cells in transient and stable transfection experiments allows assignment of BDNF as a functional receptor for *trkB* in a neurite outgrowth assay (Squinto et al. 1991; D. Soppet et al., in prep.). Similarly, NT-3 is a functional ligand for gp150^{trkC} in a neurite extension assay (P. Tsoulfas et al., in prep.).

Table 1. Summary of Interactions between *Trk* and Neurotrophin Gene Family Members

A. Trk *receptors associate directly with neurotrophins*

Receptor	Neurotrophin	References
gp 140^{prototrk}	NGF	Kaplan et al. (1991b); Klein et al. (1991)
		Meakin et al. (1992)
gp 145^{trkB}	BDNF	Soppet et al. (1991); Squinto et al. (1991)
	NT-3	Soppet et al. (1991); Squinto et al. (1991)
gp 150^{trkC}	NT-3	Lamballe et al. (1991); P. Tsouflas et al. (in prep.)
gp 150^{trkC14}	NT-3	P. Tsouflas et al. (in prep.)

B. Trk *receptor-mediated responses to neurotrophins*

Receptor	Neurotrophin	Phosphorylation	Binding	Neurite outgrowth	
				PC12	*nnr*-5
gp 140^{prototrk}	NGF	+ + +[a]	+ + +[a]	+ + +	+ + +
	NT-3	+[b]	n.d.	–	–
	NT-5	+ +[c]	n.d.	+ +	n.d.
gp 145^{trkB}	BDNF	+ + +[a]	+ + +[b]	+ + +	+ + +
	NT-3	+ + +[a]	+ + +[b]	–	–
	NT-5	+ + +[a]	n.d.	+ + +	+ + +
gp 150^{trkC}	NT-3	+ + +[a]	+ + +[b]	+ + +	n.d.
gp 150^{trkC14}	NT-3	+ + +[a]	+ + +[b]	–	–

Part A lists with references the demonstration of direct association between specific *Trk*-encoded receptors and neurotrophins. Part B details some of the biochemical and biological responses of *Trk*-neurotrophin interaction. The range of response is indicated by (+ + +) for a marked response to (–) for no detectable response. n.d. indicates not done.

[a]Assays performed in NIH-3T3 cells as well as PC12 cells and/or *nnr*-5 derivatives.

[b]Assays performed in NIH-3T3 cells alone.

[c]Assays performed in PC12 cells and/or *nnr*-5 derivatives.

trkC expression in these cells has led to further insights regarding the function of the receptors with differing inserts in the kinase domain. Thus, it appears that in PC12 cells the response of gp150^{trkC} (no insert) to NT-3 is different from that of gp150^{trkC14} (14-amino-acid insert; Table 1).

Further use of PC12 and *nnr* cells to assay receptor function should allow identification and study of differential signal transduction pathways for neurotrophins that bind and activate a particular receptor but which do not elicit similar biological responses.

Trk Family Expression

The results described above demonstrate the physical and functional interaction of *Trk* receptors with the different neurotrophins. However, it is also clear that the neurite outgrowth assay has limitations when considering the possible breadth of neurotrophin and *Trk* receptor functions in vivo (Barde 1991). Ultimately, it is fundamental to determine the expression of these receptors and their ligands in vivo to better predict and understand the scope of the potential interactions between these molecules. For this reason, we have studied the expression of the *Trk* gene family during murine embryogenesis and in the adult brain (Klein et al. 1990b; Martin-Zanca et al. 1990; L. Tessarollo et al., in prep.). A summary of the expression for all three *Trk* family receptors is presented in Figure 2.

The *trk* gene is the most transcriptionally restricted of the gene family members, with expression confined primarily to neural-crest-derived sensory neurons (Fig. 2) (Martin-Zanca et al. 1990). Figure 3 shows in situ hybridization of a 17.5-day embryo sagittal section. *trk* expression is confined to the trigeminal, superior, jugular, and dorsal root ganglia (DRG). Additional cranial ganglia and central nervous system (CNS) structures do not exhibit *trk* expression. We also observe *trk* transcripts in subsets of CNS neurons, particularly in the basal forebrain and in the mesencephalic nucleus of the trigeminal ganglion (Holtzman et al. 1992; L.F. Parada, unpubl.).

trkB transcripts are found throughout the nervous system during development and in the adult (Klein et al. 1990b; L. Tessarollo et al., in prep.). Figure 4 exhibits a series of adjacent sagittal sections from a 13.5-day embryo. Panel A highlights *trkB* expression in the spinal cord and in the sympathetic plexuses ventral to the spinal cord. Panel E, more parasagittal, shows punctate expression of *trkB* transcripts in DRG (see also panel G), in the acoustico-cochlear ganglia (panel I), and in the trigeminal ganglion and its facial afferents (Klein et al. 1989; 1990a,b). Thus, *trkB* expression is prevalent in many compartments of the nervous system. We also detect *trkB* transcripts in regions that may not reflect neural cells, including developing limb and regions outlining the forming vertebrae and ribs (Fig. 4).

trkC expression overlaps considerably with that of *trkB*, although *trkC* transcripts can be found in neural tissues where neither *trk* nor *trkB* transcripts have been previously documented and, additionally, in nonneuronal tissues (Fig. 2) (L. Tessarollo et al., in prep.). During embryonic development, the *trkC* gene is expressed throughout the CNS and PNS (Figs. 2 and 5). We observe *trkC* transcripts in the neural-crest-derived enteric ganglia (Fig. 5C), a novel site of *Trk* gene expression, and in nonneural tissues like the mesenchyme underlying the endothelium of large arteries and in the brown adipose tissue that forms behind the cervi-

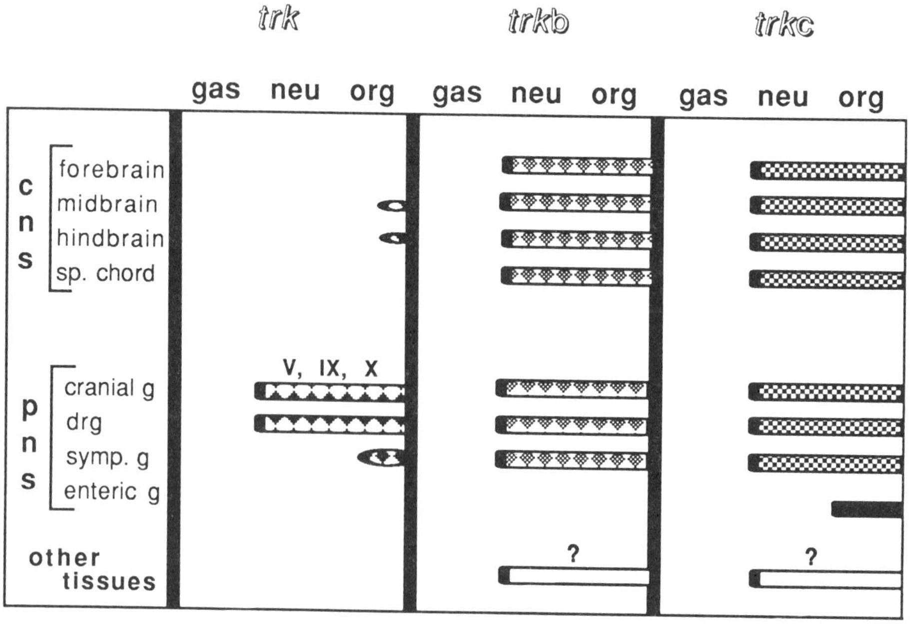

Figure 2. Summary of *Trk* gene family expression during postimplantation embryonic development. (gas) Gastrulation (day 6.5–9.0); (neu) day 9.0–11.5; (org) day 11.5–17.5; (V) trigeminal ganglion; (IX) superior ganglion; (X) jugular ganglion.

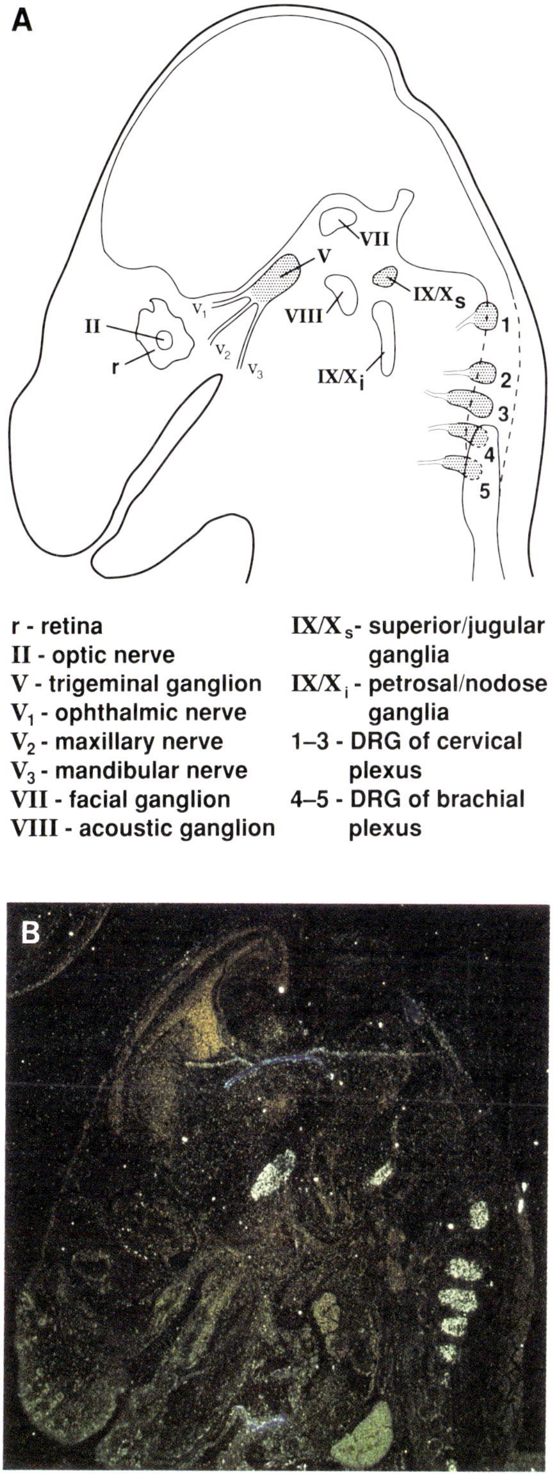

Figure 3. *trk* expression in a 17.5-day embryo sagittal section. (*A*) Diagram indicating various neural structures. (*B*) In situ hybridization darkfield view using a *trk*-specific probe (Martin-Zanca et al. 1990).

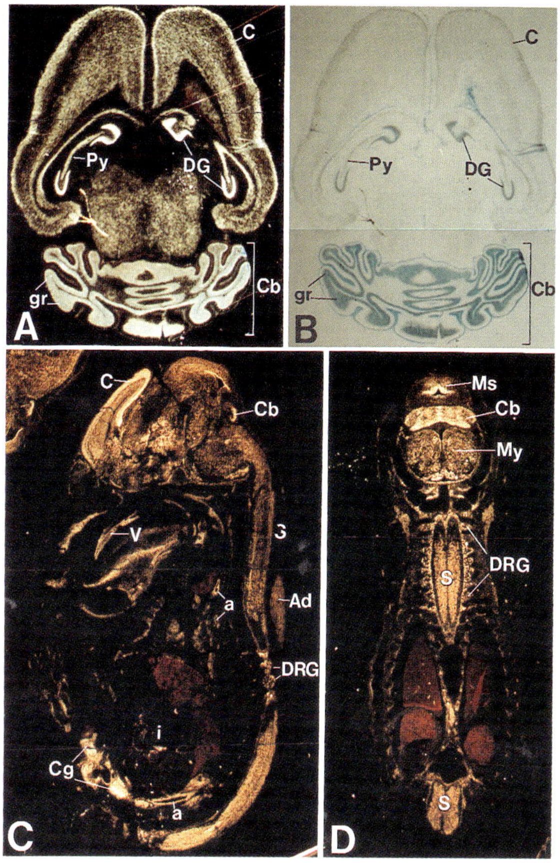

Figure 5. *trkC* expression in adult brain and in 17.5-day fetuses. (*A*) Darkfield and (*B*) lightfield midcoronal section through adult brain. (DG) Dentate gyrus; (Py) pyramidal neurons of the hippocampus; (Cb) cerebellum; (gr) granular cell layer; (C) cortex. (*C*) Sagittal and (*D*) frontal sections through 17.5-day fetuses. (C) Cortex; (Cb) cerebellum; (S) spinal cord; (V) trigeminal nerve tracts; (a) arteries; (Ad) brown adipose tissue; (i) intestine; (Cg) coeliac ganglion; (Ms) mesencephalon; (My) myelencephalon.

cal spinal cord. In the forming brain, *trkC* transcripts are most abundant in the cortex of the telencephalon (Fig. 5A) and in the metencephalon (cerebellum; Fig. 5B). This expression is maintained in the adult CNS, where *trkC* transcripts are found in the cerebral cortex primarily in the superficial and inner layers but not in intermediate layers (Fig. 5A). Like *trkB* (Klein et al. 1990b), *trkC* transcripts are also abundant in the pyramidal cells of the hippocampus and in the dentate gyrus (Fig. 5A) (L. Tessarollo et al., in prep.). Similarly, *trkC* transcripts are maintained in the mature cerebellum, where expression is abundant in the granular and Purkinje cell layer.

Several parallels can be drawn between the expression of the various *Trk* loci. Comparison of *Trk* gene expression in trigeminal and dorsal root ganglia suggests that all three genes are not expressed in a single neuron. *trk* transcripts are only found in neurons, whereas the *trkB* and *trkC* genes are expressed in neurons and also in nonneuronal cells within the ganglia as well as in Schwann cells along the axonal tracts (see Klein et al. 1989, 1990b; L. Tessarollo et al., in prep.).

trkB and *trkC* transcripts are coexpressed in many neurons, including pyramidal cells in Ammon's horn, granule cells of the dentate gyrus, and the Purkinje cells of the cerebellum (Klein et al. 1990b; L. Tessarol-

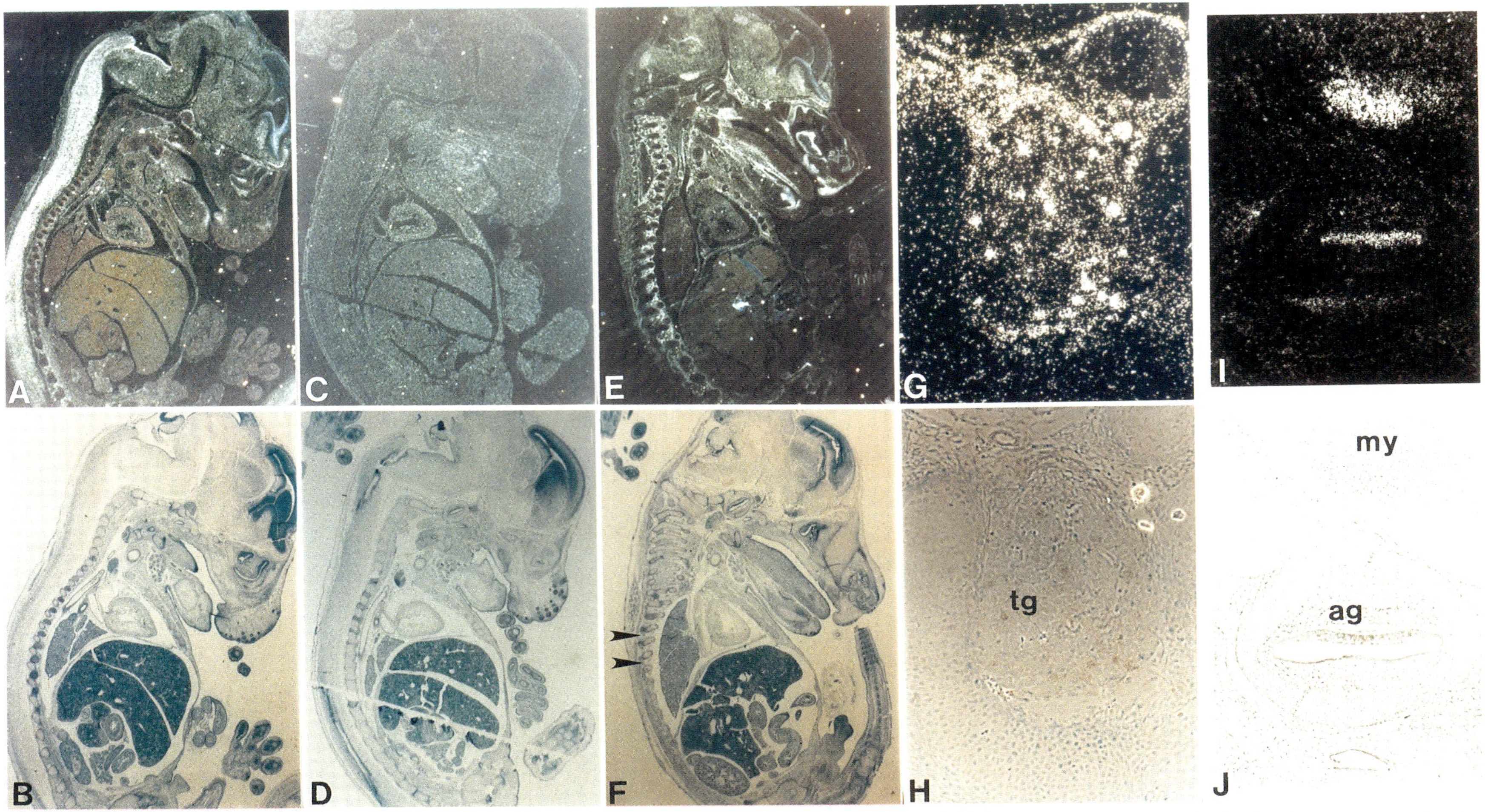

Figure 4. *trkB* expression in day-14.0 fetuses. (*A,C,E,G*, and *I*) Darkfield; (*B,D,F*, and *J*) lightfield sagittal sections. Panels *C* and *D* show a section hybridized with a sense probe. Panels *G* and *H* show the dorsal root ganglion (tg), *I* and *J* show the myelencephalon (my) and the acoustic ganglion (ag). Arrowheads in *F* indicate DRGs.

lo et al., in prep.). To understand how the various *Trk* receptors and neurotrophins might regulate nervous system function, it will be essential to chart fully expression of the cognate receptor/ligands during all stages of embryonic development and in the adult.

Trk Family Proteins

To examine the products of the various *Trk* loci, and to verify the existence of the proteins predicted by cDNA isolation and sequencing, we have generated receptor-specific antisera as well as antisera that recognize multiple receptors. These antibodies have permitted the demonstration of direct receptor/ligand interaction in recent studies (Kaplan et al. 1991a,b; Soppet et al. 1991; P. Tsoulfas et al., in prep.). Figure 6 shows a Western analysis assaying Trk proteins from *trk*-transfected NIH-3T3 cell lines. Antiserum 443, which recognizes the conserved carboxyl terminus of *Trk*-encoded proteins, readily detects gp140^{prototrk}, gp145^{trkB}, and gp150^{trkC} (Fig. 6, panel A). The antisera 442 and 510 have been generated from peptides corresponding to the less well conserved extracellular domains of gp145^{trkB} and gp150^{trkC}, respectively. Antiserum 442 specifically recognizes gp145^{trkB} (Fig. 6, panel B), whereas antiserum 510 recognizes gp150^{trkC} (Fig. 6, panel C) and gp150^{trkC14} (data not shown). Similar analysis of tissue extracts has demonstrated the presence of gp140^{prototrk} in embryonic DRG (Kaplan et al. 1991b); of gp145^{trkB}, gp110^{trkBTK}, gp150^{trkC}, and gp120^{trkCTK} in brain (Fig. 7A,B) (Klein et al. 1990a; Soppet et al. 1991; P. Tsoulfas et al., in prep.); and gp150^{trkC14} in brain (D. Soppet et al., unpubl.). Although *trkB* and *trkC* transcripts are present in many tissues (Klein et al. 1989; 1990a), we have not detected these proteins in tissues outside the nervous system (see also P. Tsoulfas et al., in prep.). This may reflect the greater sensitivity of transcript detection methods relative to the sensitivity of our antisera.

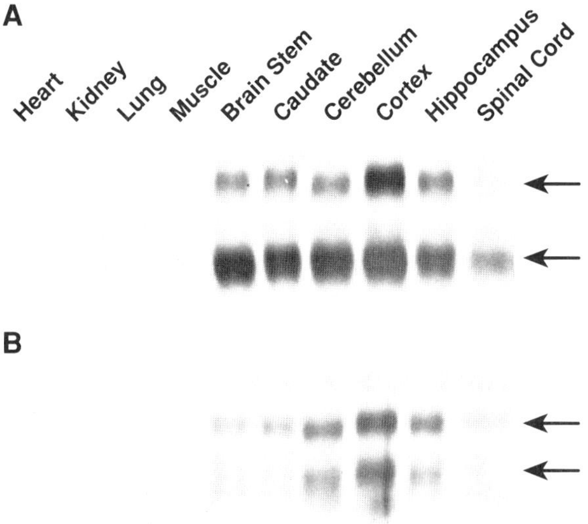

Figure 7. *trkB* and *trkC* gene product expression in mouse tissues. Western analysis of mouse tissues with *trkB*-specific antiserum 442 (*A*) or *trkC*-specific antiserum 510 (*B*). The lanes contain equivalent amounts of protein obtained from the tissues indicated at the top of the figure. The position of the 145–150-kD full-length (upper arrow in each panel) and 110–120-kD truncated forms (lower arrow in each panel) of the proteins are indicated on the right side of the figure.

SUMMARY

Neurotrophins are known to have important functions in the survival of embryonic and adult subpopulations of neurons. The identification of *Trk* family RTKs as receptors for NGF-related neurotrophins indicates phosphotyrosine-mediated signal transduction as a principal mechanism for neurotrophin signaling. Previous *trk* and *trkB* expression studies (Klein et al. 1989, 1990b; Martin-Zanca et al. 1990) and more recent studies with *trkC* (L. Tessarollo et al., in prep.) provide important clues about function. Thus, *trkB* and *trkC* expression in motor neurons and in many nonneuronal

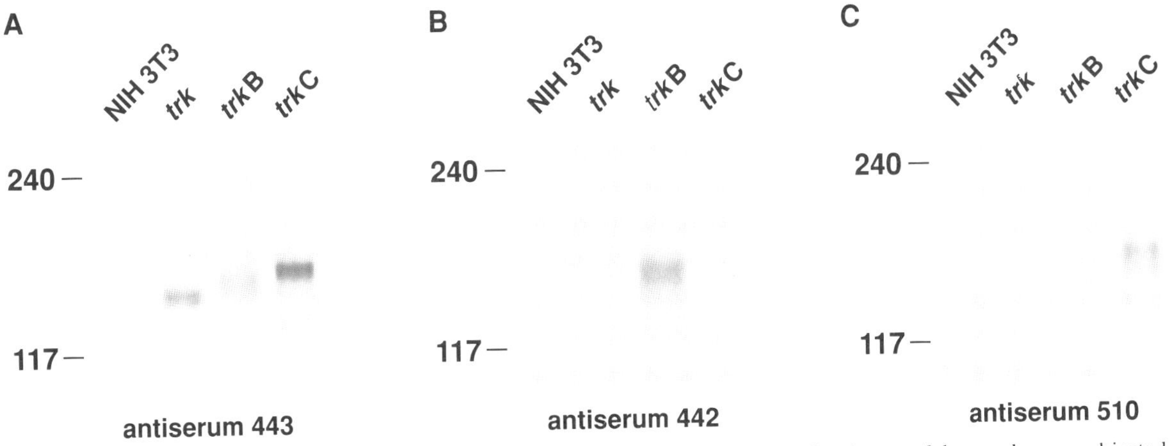

Figure 6. NIH-3T3 cells or stable NIH-3T3 transfectants for *trk*, *trkB*, or *trkC* as indicated at the top of the panels were subjected to Western analysis with antiserum 443 that recognizes *trk*, *trkB*, and *trkC* (*A*), *trkB*-specific antiserum 442 (*B*), or *trkC*-specific antiserum 510 (*C*). The apparent molecular weights are indicated on the left.

cells suggests that these cells are targets for neurotrophin action in vivo, even though this has not been demonstrated in the classic in vitro survival assays. Expression of *trkB* and *trkC* in nonneuronal cells implies that these receptors may act in additional aspects of organogenesis and development.

Current approaches to assay *Trk* receptor and neurotrophin function will be complemented by further studies in the living organism. Transgenic approaches aimed at ectopic expression and at interfering with normal receptor function should provide additional insights. Finally, reverse genetic approaches using targeted mutation of *Trk* receptors in embryonic stem cells (Stanton et al. 1992) will allow assessment of critical receptor requirements and provide powerful reagents for studying nervous system development and function.

ACKNOWLEDGMENTS

We are grateful to Johnna Maragos for excellent technical assistance, and we thank members of the Parada laboratory and our various colleagues and collaborators, particularly David Kaplan, for their many helpful discussions and encouragement. We thank Cindy Fitzpatrick for preparation of the manuscript and Richard Frederickson for assistance with art work. Research was sponsored by the National Cancer Institute, DHHS, under contract N01-CO-74101 with ABL. The contents of this publication do not necessarily reflect the views or policies of the Department of Health and Human Services, nor does mention of trade names, commercial products, or organizations imply endorsement by the U.S. Government.

REFERENCES

Anderson, D.J. 1989. The neural crest cell lineage problem: Neuropoiesis? *Neuron* **3:** 1.

Barde, Y.-A. 1991. The nerve growth factor family. *Prog. Growth Factor Res.* **2:** 237.

Barde, Y.-A., A.M. Davies, J.E. Johnson, R.M. Lindsay, and H. Thoenen. 1987. Brain derived neurotrophic factor. *Prog. Brain Res.* **71:** 185.

Berkemeier, L.R., J.W. Winslow, D.R. Kaplan, K. Nikolics, D.V. Goeddel, and A. Rosenthal. 1991. Neurotrophin-5: A novel neurotrophic factor that activates *trk* and *trk*B. *Neuron* **7:** 857.

Davies, A.M. and A.G.S. Lumsden. 1984. Relation of target encounter and neuronal death to nerve growth factor responsiveness in the developing mouse trigeminal ganglion. *J. Comp. Neurol.* **253:** 13.

Ernfors, P., C.F. Ibáñez, T. Ebendal, L. Olson, and H. Persson. 1990. Molecular cloning and neurotrophic activities of a protein with structural similarities to nerve growth factor: Developmental and topographical expression in the brain. *Proc. Natl. Acad. Sci.* **87:** 5454.

Greene, L.A. and A.S. Tischler. 1976. Establishment of a noradrenergic clonal line of rat adrenal pheochromocytoma cells which respond to nerve growth factor. *Proc. Natl. Acad. Sci.* **73:** 2424.

Hallböök, F., C.F. Ibáñez, and H. Persson. 1991. Evolutionary studies of the nerve growth factor family reveal a novel member abundantly expressed in *Xenopus* ovary. *Neuron* **6:** 845.

Hamburger, V., J.K. Brunso-Bechtold, and J. Yip. 1981. Neuronal death in the spinal ganglia of the chick embryo and its reduction by nerve growth factor. *J. Neurosci.* **1:** 60.

Hempstead, B., D. Kaplan, D. Martin-Zanca, L.F. Parada, and M. Chao. 1991. High affinity NGF binding requires co-expression of the *trk* protooncogene product and the low affinity NGF receptor. *Nature* **350:** 678.

Hempstead, B.L., S.J. Rabin, L. Kaplan, S. Reid, L.F. Parada, and D.R. Kaplan. 1992. Overexpression of the *trk* tyrosine kinase rapidly accelerates nerve growth factor-induced differentiation. *Neuron* **9:** 883.

Hohn, A., J. Leibrock, K. Bailey, and Y.-A. Barde. 1990. Identification and characterization of a novel member of the nerve growth factor brain-derived neurotrophic factor family. *Nature* **344:** 339.

Holtzman, D., I. Yiwen, L.F. Parada, S. Kinsman, C. Chen, J. Valletta, J. Zhou, J. Long, and W. Mobley. 1992. gp140^{trk} mRNA marks NGF-responsive forebrain neurons: Evidence that *trk* expression is induced by NGF. *Neuron* **9:** 465.

Ip, N.Y., C.F. Ibáñez, S.H. Nye, J. McClain, P.F. Jones, D.R. Gies, L. Belluscio, M.M. LeBeau, R. Espinosa III, S.P. Squinto, H. Persson, and G.D. Yancopoulos. 1992. Mammalian neurotrophin-4: Structure, chromosomal localization, tissue distribution, and receptor specificity. *Proc. Natl. Acad. Sci.* **89:** 3060.

Jessell, T.M. 1988. Adhesion molecules and the hierarchy of neural development. *Neuron* **1:** 3.

Jones, K.R. and L.F. Reichardt. 1990. Molecular cloning of the human gene that is a member of the nerve growth factor family. *Proc. Natl. Acad. Sci.* **87:** 8060.

Kaisho, Y., K. Yashimura, and K. Nakahama. 1990. Cloning and expression of a cDNA encoding a novel human neurotrophic factor. *FEBS Lett.* **266:** 187.

Kaplan, D.R., D. Martin-Zanca, and L.F. Parada. 1991a. Tyrosine phosphorylation and tyrosine kinase activity of the *trk* protooncogene product induced by NGF. *Nature* **350:** 158.

Kaplan, D.R., B.L. Hempstead, D. Martin-Zanca, M.V. Chao, and L.F. Parada. 1991b. The *trk* proto-oncogene product: A signal transducing receptor for nerve growth factor. *Science* **252:** 554.

Keynes, R. and A. Lumsden. 1990. Segmentation and the origin of regional diversity in the vertebrate central nervous system. *Neuron* **4:** 1.

Klein, R., D. Conway, L.F. Parada, and M. Barbacid. 1990a. The *trk*B tyrosine kinase gene codes for a second neurogenic receptor that lacks the catalytic domain. *Cell* **61:** 647.

Klein, R., D. Martin-Zanca, M. Barbacid, and L.F. Parada. 1990b. Expression of the tyrosine kinase receptor gene *trkB* is confined to the murine embryonic and adult nervous system. *Development* **109:** 845.

Klein, R., L.F. Parada, F. Coulier, and M. Barbacid. 1989. *Trk*B, a novel tyrosine protein kinase receptor expressed during mouse neural development. *EMBO J.* **8:** 3701.

Klein, R., S. Jing, V. Nanduri, E. O'Rourke, and M. Barbacid. 1991. The *Trk* proto-oncogene encodes a receptor for nerve growth factor. *Cell* **65:** 189.

Krantz, D.D., R. Zidovetzki, B.L. Kagan, and L.S. Zipursky. 1991. Amphipathic β structure of a leucine-rich repeat peptide. *J. Biol. Chem.* **266:** 16801.

Lamballe, F., R. Klein, and M. Barbacid. 1991. *trk*C a new member of the *trk* family of tyrosine protein kinases, is a receptor for neurotrophin-3. *Cell* **66:** 967.

Le Douarin, N.M. and J. Smith. 1988. Development of the peripheral nervous system from the neural crest. *Annu. Rev. Cell Biol.* **4:** 375.

Leibrock, J., F. Lottspeich, A. Hohn, M. Hofer, B. Hengerer, P. Masiakowski, H. Thoenen, and Y.-A. Barde. 1989. Molecular cloning and expression of brain-derived neurotrophic factor. *Nature* **341:** 149.

Levi-Montalcini, R. 1987. The nerve growth factor 35 years later. *EMBO J.* **6:** 1145.

Levi-Montalcini, R. and B. Booker. 1960. Destruction of the sympathetic ganglia in mammals by an antiserum to a nerve-growth protein. *Proc. Natl. Acad. Sci.* **46:** 384.

Levi-Montalcini, R., L. Revoltella, and P. Calissano. 1974. Microtubule proteins in the nerve growth factor-mediated response (interaction between the nerve growth factor and its target cells). *Recent Prog. Horm. Res.* **30:** 635.

Loeb, D.M., J. Maragos, D. Martin-Zanca, M.V. Chao, L.F. Parada, and L.A. Greene. 1991. The *trk* proto-oncogene rescues NGF responsiveness in mutant NGF-nonresponsive PC12 cell lines. *Cell* **66:** 961.

Maisonpierre, P.C., L. Belluscio, S. Squinto, N.Y. Ip, M.E. Furth, R.M. Lindsay, and G.D. Yancopoulos. 1990. A neurotrophin-3: A neurotrophic factor related to NGF and BDNF. *Science* **247:** 1446.

Martin-Zanca, D., M. Barbacid, and L.F. Parada. 1990. Expression of the *trk* proto-oncogene is restricted to the sensory cranial and spinal ganglia of neural crest origin in mouse development. *Genes Dev.* **4:** 683.

Martin-Zanca, D., R. Oskam, G. Mitra, T. Copeland, and M. Barbacid. 1989. Molecular and biochemical characterization of the human *trk* protooncogene. *Mol. Cell. Biol.* **9:** 24.

Meakin, S.O., U. Suter, C.C. Drinkwater, A.A. Welcher, and E.M. Shooter. 1992. The rat *trk* protooncogene product exhibits properties characteristic of the slow nerve growth factor receptor. *Proc. Natl. Acad. Sci.* **89:** 2374.

Middlemas, D.S., R.A. Lindberg, and T. Hunter. 1991. *Trk*B, a neural receptor protein-tyrosine kinase: Evidence for a full-length and two truncated receptors. *Mol. Cell. Biol.* **11:** 143.

Oppenheim, R.W. 1981. Neuronal cell death and some related regressive phenomena during neurogenesis: A selective historical review and progress report. In *Studies in developmental neurobiology: Essays in honor of Victor Hamburger* (ed. W.M. Cowan), p. 74. Oxford University Press, New York.

Rothberg, S.M., R.J. Jacobs, C.S. Goodman, and S. Artavanis-Tsakonas. 1990. Slit: An extracellular protein necessary for development of midline glia and commissural axon pathways contains both EGF and LRR domains. *Genes Dev.* **4:** 2169.

Sanes, J.R. 1989. Extracellular matrix molecules that influence neural development. *Annu. Rev. Neurosci.* **12:** 491.

Soppet, D., E. Escandon, J. Maragos, D.S. Middlemas, S.W. Reid, J. Blair, L.E. Burton, B.R. Stanton, D.R. Kaplan, T. Hunter, K. Nikolics, and L.F. Parada. 1991. The neurotrophic factor and neurotrophin-3 are ligands for the *trk*B tyrosine kinase receptor. *Cell* **65:** 895.

Squinto, S.P., T.N. Stitt, T.H. Aldrich, S. Davis, S.M. Bianco, C. Radziejewski, D.J. Glass, P. Masiakowski, M.E. Furth, D.M. Valenzuela, P.S. DiStefano, and G.D. Yancopoulos. 1991. *Trk*B encodes a functional receptor for brain-derived neurotrophic factor and neurotrophin-3 but nerve growth factor. *Cell* **54:** 885.

Stanton, B.R., A.S. Perkins, L. Tessarollo, D.A. Sassoon, and L.F. Parada. 1992. Loss of N*myc* function results in embryonic lethality and failure of the epithelial component of the embryo to develop. *Genes Dev.* (in press).

Tessarollo, L., L. Nagarajan, and L.F. Parada. 1992. C-*ros*: The vertebrate homolog of the *sevenless* tyrosine kinase receptor is tightly regulated during organogenesis in mouse embryonic development. *Development* **115:** 11.

Williams, A.F. and A.N. Barclay. 1988. The immunoglobulin superfamily-domains for cell surface recognition. *Annu. Rev. Immunol.* **6:** 381.

Cloning and Analysis of Neurotrophic Factor Receptors Using Function-based Strategies

D.J. Glass, D.R. Gies, T. Stitt, S. Davis, P. Hantzopoulos, N.Y. Ip, M. Goldfarb, and G.D. Yancopoulos

Regeneron Pharmaceuticals, Inc., Tarrytown, New York 10591-6707

"Neurotrophic factors" were proposed to exist following the observation, now decades old, that neurons die during normal development. This process of neuronal death, known as programmed or naturally occurring neuronal death, is thought to somehow be involved in the proper wiring of the nervous system. A variety of studies revealed that the amount of neuronal death observed was proportional to the amount of target innervated, leading to the notion that neurons compete for limiting amounts of target-derived neurotrophic activity (Hamburger and Levi-Montalcini 1949; Oppenheim 1981). The ability to assay for neurotrophic activity in vitro allowed the isolation of neurotrophic factors, the first of which was nerve growth factor (NGF) (Levi-Montalcini 1987). The subsequent purification and molecular cloning of brain-derived neurotrophic factor (BDNF) revealed that it shared about 50% amino acid identity with NGF (Leibrock et al. 1989) and inspired homology-based searches that have yielded additional NGF/BDNF relatives. Members of the NGF/BDNF family have been collectively referred to as the neurotrophins and include neurotrophin-3 (NT-3) (Ernfors et al. 1990; Hohn et al. 1990; Jones and Reichardt 1990; Kaisho et al. 1990; Maisonpierre et al. 1990; Rosenthal et al. 1990), neurotrophin-4 isolated from *Xenopus laevis* (xNT-4) (Holböök et al. 1991), and a mammalian neurotrophin that has alternatively been proposed to represent either the mammalian counterpart of xNT-4 (Ip et al. 1992a) or a novel member of this family (Berkemeier et al. 1991) and is thus designated here as NT-4/5. Purification approaches have also resulted in the identification and molecular cloning of a factor, named ciliary neurotrophic factor (CNTF) because it supports ciliary neuron survival, that bears no homology with the neurotrophins and thus represents a distinct class of neurotrophic factor (Adler et al. 1979; Lin et al. 1989; Stockli et al. 1989).

Until very recently, it was thought that neurotrophic factors were fundamentally different from other cytokines and growth factors, mainly because the actions of neurotrophic factors on neurons seemed so distinct from those of other factors on cells outside of the nervous system. The neurotrophic factor field has, however, undergone dramatic changes with the identification of the receptors for the neurotrophins and for CNTF. Interestingly, these two classes of neurotrophic factors use completely different receptors and signaling mechanisms to keep neurons alive, but these receptors and signaling mechanisms are not, as might have been expected, particular to neurons. The neurotrophins utilize a family of receptor tyrosine kinases, known collectively as the Trks, that are very similar to those used by traditional growth factors such as fibroblast growth factor (FGF) and epidermal growth factor (EGF) (Cordon-Cardo et al. 1991; Glass et al. 1991; Kaplan et al. 1991a,b; Klein et al. 1991a,b; Lamballe et al. 1991; Soppet et al. 1991; Squinto et al. 1991b). CNTF, on the other hand, actually shares receptor components with a family of distantly related hematopoietic cytokines, including leukemia inhibitory factor (LIF) and interleukin-6 (IL-6) (Davis et al. 1991; Ip et al. 1992b). We describe here novel function-based approaches that we have used to clone molecularly and to understand further the receptors for both the neurotrophins and for CNTF. These approaches should be applicable to the study of other ligands and their receptors.

METHODS

Cell culture, transfections, and assays. NIH-3T3 cells were cultured, utilized in MTT (3-[4,5-dimethylthiazol-2-yl]-2,5-diphenyl tetrazolium bromide) dye assays for cell growth, and transfected as defined previously (Glass et al. 1991; Nye et al. 1992). To construct PC12 cells stably expressing TrkB, a full-length rat *trkB* cDNA (Squinto et al. 1991b) subcloned in the pcDNAI*neo* expression vector (Invitrogen) was transfected into PC12 cells (using the PC12 isolate provided by E. Shooter, Stanford University) (10^7 cells per 10-cm dish) by incubating for 24 hours in 5 ml of OptiMEM medium (GIBCO BRL) containing 25 mg of plasmid DNA and 100 mg of Lipofectin (GIBCO BRL); cells were then rinsed and placed in fresh medium. Five days after transfection, cells were selected in 0.4 mg/ml G418. Resistant colonies were assayed for differentiation in the presence of 100 ng/ml BDNF, and one clone designated PC12/*trkB* was selected. PC12 cells were maintained and assayed as described previously (Squinto et al. 1991b).

Factors. Mouse 2.5S NGF was obtained from Bioproducts for Science (Indianapolis, Indiana). Purified recombinant human BDNF and NT-3 (>99% purified) were produced from Chinese hamster ovary (CHO) cells and kindly provided by Amgen/Regeneron Partners. Purified recombinant human NT-4 (>99% purified) was prepared from *Escherichia coli* (J. Fandl et al., unpubl.). Basic FGF was purchased from R&D Systems, Inc. (Minneapolis, Minnesota).

Genomic DNA preparation and Southern blotting. Genomic DNA preparation and Southern blotting were carried out as described previously (Maisonpierre et al. 1990), using probes specific for either the *trkA* or *trkB* genes.

Immunoprecipitations and immunoblotting. To detect changes in tyrosine phosphorylation upon addition of neurotrophins, fibroblasts were starved for 2 hours in serum-free Dulbecco's modified Eagle's medium (DMEM) and then treated for 5 minutes with the factors as indicated. Cells were then treated as described recently (Glass et al. 1991).

RESULTS

Function-based Cloning of the Receptors for NGF and BDNF

The finding that members of the *trk* gene family were expressed only in neurons led to the suggestion that the Trk proteins served as receptors for ligands specifically acting on neurons. The neurotrophins were potential candidates for such ligands, and subsequent studies revealed that TrkA is a receptor for NGF (Cordon-Cardo et al. 1991; Kaplan et al. 1991a,b; Klein et al. 1991a), that TrkB can serve as a receptor for both BDNF and NT-3 (Glass et al. 1991; Klein et al. 1991b; Soppet et al. 1991; Squinto et al. 1991b), and that TrkC is a receptor for NT-3 (Lamballe et al. 1991); neurotrophin/Trk interactions are discussed at greater length below.

Prior to the identification of the Trks as receptors for the neurotrophins, we initiated a function-based strategy designed to clone neurotrophin receptors. We reasoned that although the neurotrophins were known to have effects only on neurons, and despite the fact that these effects dramatically differed from those of traditional growth factors on nonneuronal cells, ectopic expression of a functional neurotrophin receptor in nonneuronal cells might bestow a selectable neurotrophin response on such cells. This reasoning was largely based on the observation that FGF was essentially indistinguishable from NGF in its ability to arrest growth and induce neuronal differentiation of PC12 rat pheochromocytoma cells (Togari et al. 1985), a traditional model system used to study NGF actions (Greene and Tischler 1976; Greene 1978). We therefore proposed that just as FGF could promote NGF-like responses in PC12 cells bearing FGF receptors, introduction of neurotrophin receptors into fibroblasts would allow them to display FGF-like responses (i.e., proliferative responses) to the neurotrophins. This proposal assumed that NGF and FGF receptor activation resulted in rather similar initial signals that were interpreted in dramatically different ways (i.e., growth arrest/neuronal differentiation vs. growth promoting) in the neurons used to study NGF as opposed to the fibroblasts utilized for FGF analysis. This assumption was supported by reports that NGF could promote mitogenic responses in adrenal precursor cells (Lillien and Claude 1985) and that the neurotransmitter serotonin could act as a mitogen on fibroblasts ectopically expressing a serotonin receptor (Julius et al. 1989).

On the basis of the above reasoning, we decided to utilize a variant of the NIH-3T3 fibroblast cell line in an attempt to clone neurotrophin receptors. This NIH-3T3 variant has the useful property that its survival, as well as its proliferation, is absolutely dependent on FGF under defined media conditions (Zhan et al. 1987). Transfection of these cells by genomic DNA, followed by removal of the FGF from the media, resulted in selection of a transfected clone that expressed a novel fgf gene (*fgf-5*) and had thus bypassed its normal requirement for exogenous FGF by establishing an autocrine loop involving the introduced *fgf-5* gene (Zhan et al. 1988). We transfected these NIH-3T3 cells with human placental genomic DNA, replaced the FGF in the defined media with a combination of both NGF and BDNF, and thus attempted to select clones that expressed functional neurotrophin receptors that would allow for survival and growth by NGF or BDNF instead of FGF (Fig. 1). Similar genomic transfections were also performed with NIH-3T3 cells ectopically expressing a receptor component, known as either p75 or the

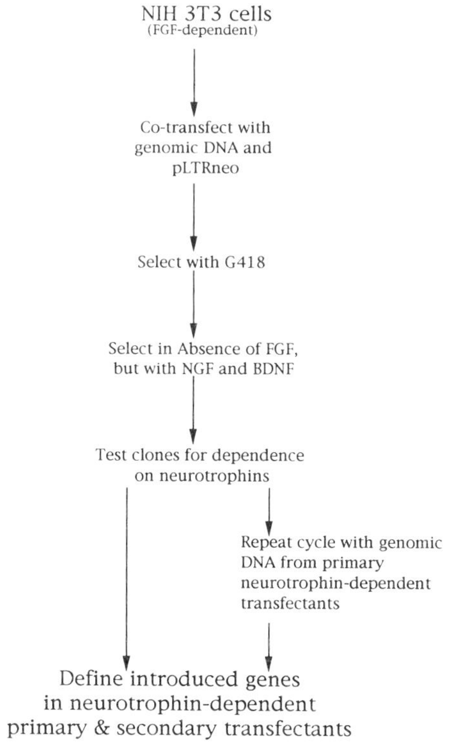

Figure 1. Flowchart of strategy used to clone neurotrophin receptors by selecting genomic DNA transfectants of NIH-3T3 fibroblasts in which NGF or BDNF can replace the normal FGF requirement.

low-affinity nerve growth factor receptor (hereafter called $p75^{LNGFR}$), that was thought to be a required accessory component of functional neurotrophin receptors (Hempstead et al. 1989; Rodriguez-Tebar et al. 1990); as we describe below, $p75^{LNGFR}$ is not absolutely required for signal transduction by neurotrophin receptors.

In the transfection outlined in Figure 1, 20 plates of NIH-3T3 fibroblasts (at a density of 1×10^6 cells/plate) were cotransfected with genomic human placental DNA and the pLTR*neo* plasmid encoding G418 resistance. Cells were then split (1:2) into G418-containing media to select for cells that had taken up the pLTR*neo* plasmid. Once G418-resistant colonies were established (>2000/plate), the cells were switched into serum-free defined media lacking FGF but containing both BDNF (5 nm) and NGF (5 nM). A total of five colonies survived after 1 week of selection in defined media containing BDNF and NGF. These five colonies were isolated and expanded, and each was then replated in either defined media alone or defined media supplemented with either 5 nM BDNF or 5 nM NGF. Three of these isolated primary transfectants were found to grow in the unsupplemented defined media, suggesting that

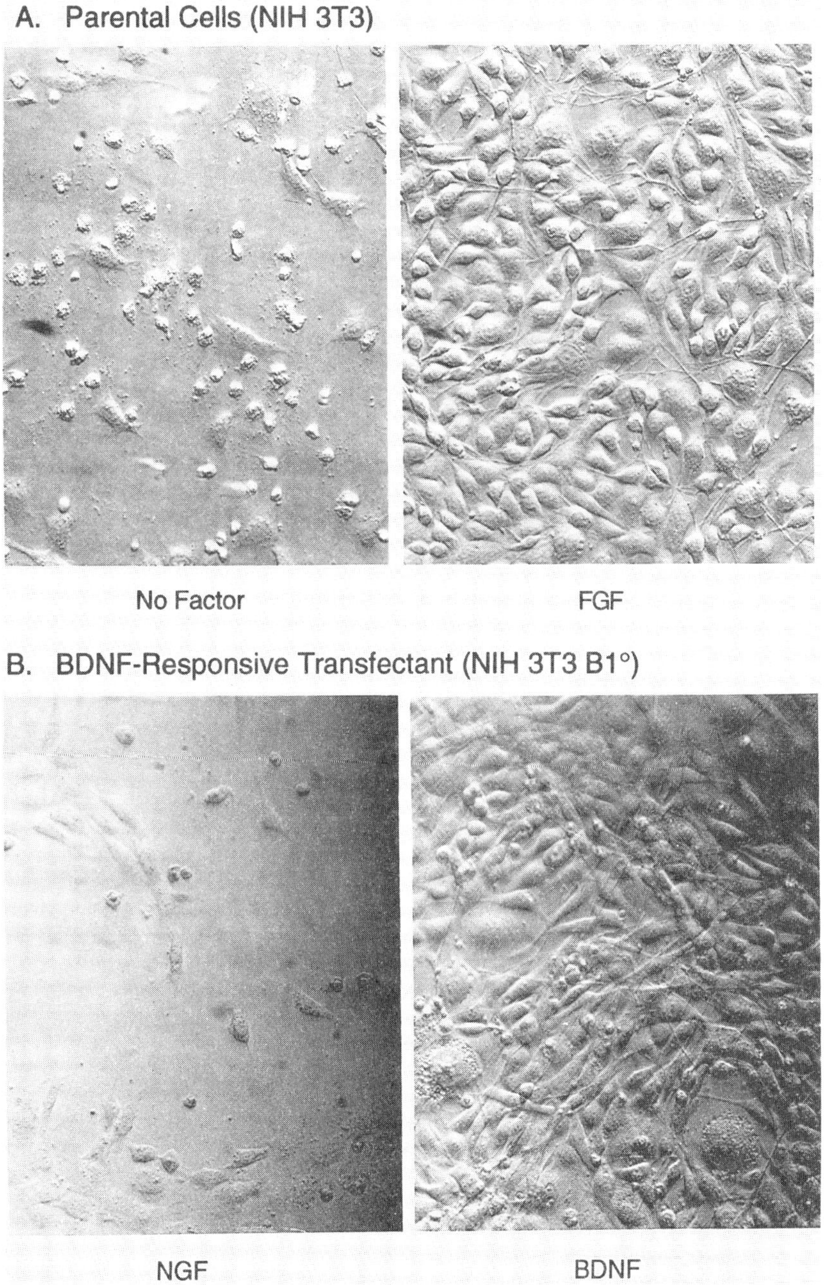

Figure 2. Photomicrographs comparing FGF effects on parental NIH-3T3 cells (*A*) with those of BDNF on a genomic DNA transfectant of NIH-3T3 cells (*B*) selected for its ability to survive in the presence of this neurotrophin.

they depended on a survival mechanism not involving the exogenously provided neurotrophins (as would be expected, for example, if the colony was expressing an FGF acting via an autocrine process). Of the other two primary transfectants, one was found to be dependent on BDNF for survival (3T3-B1°) and the other was dependent on NGF (3T3-N1°) (see Fig. 2 for comparison of FGF growth and survival promoting effects on parental NIH-3T3 cells [panel A] with those of BDNF on 3T3-B1° [panel B]). Repetition of the human placental genomic DNA transfections yielded NGF- and BDNF-responsive primary transfectants with frequencies similar to those in the experiment described above (data not shown). As has previously been seen in cloning experiments using genomic DNA transfections, secondary transfectants could be generated at substantially higher frequencies (Fig. 1).

The generation of NGF- and BDNF-responsive NIH-3T3 clones by the transfection process described above suggested that we had introduced genomic DNA encoding functional neurotrophin receptors into these cells. Subsequent to the identification of the Trks as receptors for the neurotrophins, we decided to exploit Trk-specific reagents to examine whether the NGF- and BDNF-responsive transfectants arose due to introduction of human *trk* genes. Southern analysis revealed that the primary NGF-responsive transfectant 3T3-N1°, as well as secondary NGF-responsive transfectants derived from this primary transfectant, contained a novel *trkA*-hybridizing fragment (not present in parental NIH-3T3 cells) apparently representing the introduced human *trkA* gene (Fig. 3A). Similarly, both the primary and secondary BDNF-responsive transfectants contained a novel *trkB*-hybridizing fragment (not present in parental NIH-3T3 cells) apparently representing the introduced human *trkB* gene (Fig. 3B). Analysis of neurotrophin-induced tyrosine phosphorylations in the various transfectants revealed that the NGF-responsive primary and secondary transfectants displayed NGF-induced tyrosine phosphorylation of a TrkA-sized protein that could be immunoprecipitated using a Trk-specific polyclonal antiserum (Fig. 4A), whereas the BDNF-responsive primary and secondary transfectants displayed BDNF-induced tyrosine phosphorylation of a TrkB-sized protein that could be immunoprecipitated using the same Trk-specific polyclonal antiserum (Fig. 4B); note that parental NIH-3T3 cells did not display inducible tyrosine phosphorylations in response to NGF or BDNF (data not shown).

Altogether, these data demonstrate that our strategy did indeed allow for the cloning of genes encoding functional receptors for NGF and BDNF. This ap-

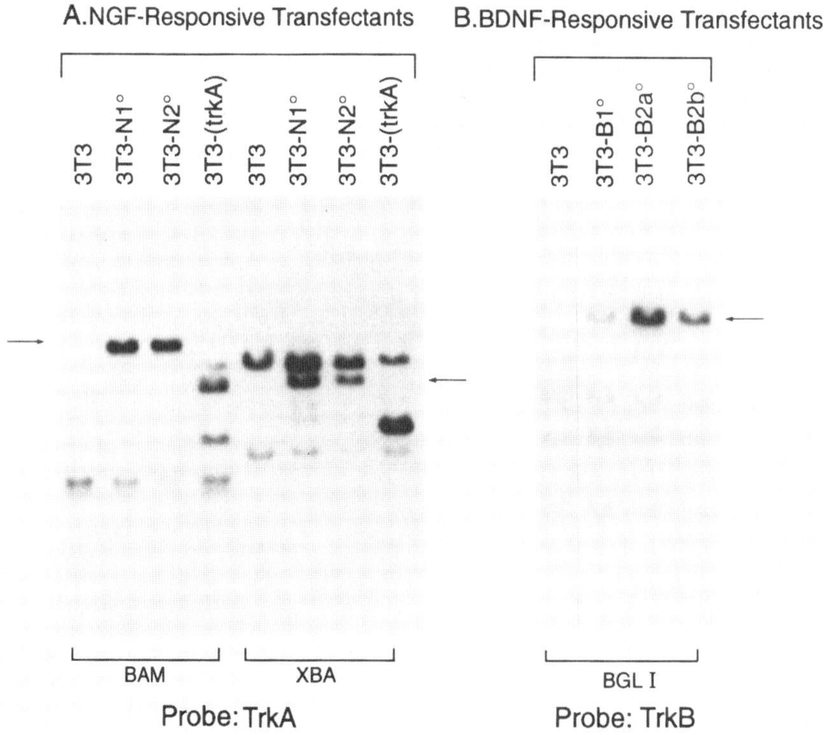

Figure 3. Southern blot analysis demonstrating that NGF (*A*) and BDNF (*B*) responsive genomic DNA transfectants arose due to the introduction of *trkA* (*A*) and *trkB* (*B*) genes, respectively. Restriction fragments shared by parental NIH-3T3 cells as well as transfectants are derived from endogenous *trkA* (*A*) or *trkB* (*B*) genes. Arrows indicate novel restriction fragments representing introduced *trk* genes found in primary and secondary transfectants but not in parental NIH-3T3 cells. (3T3) Parental NIH-3T3 cells; (3T3-N1°) primary NGF-responsive transfectant; (3T3-N2°) secondary transfectant derived from this primary transfectant; (3T3-[trkA]) NIH-3T3 subclone containing a transfected *trkA* expression vector used as a positive control for the *trkA* probing; (3T3-B1°) primary BDNF-responsive transfectant; (3T3-B2a° and 3T3-B2b°) two independent secondary clones derived from this primary transfectant.

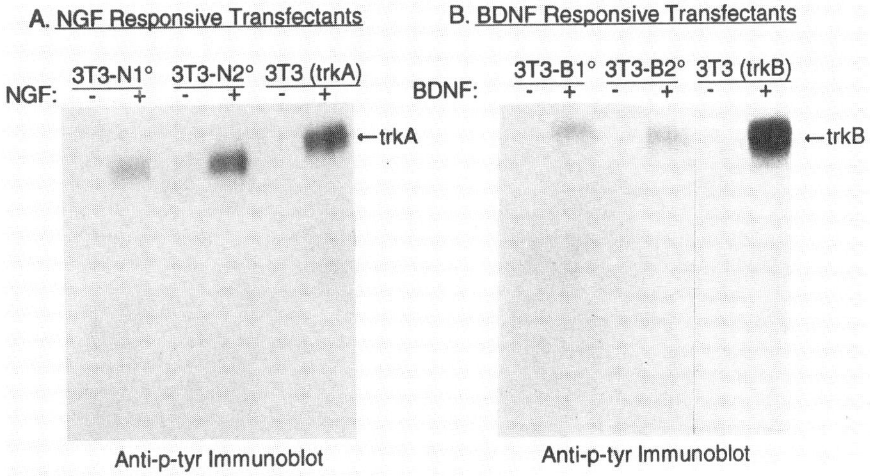

Figure 4. Phosphotyrosine immunoblotting reveals that NGF (*A*) and BDNF (*B*) selected genomic DNA transfectants contain appropriate Trk proteins that can be inducibly tyrosine phosphorylated in response to these neurotrophins. Cell lines are the same as those denoted as in Fig. 3.

proach did not depend on any prior knowledge of these receptors, nor was it biased toward any particular receptor. That this strategy yielded TrkA as the receptor for NGF and TrkB as the receptor for BDNF suggests that they are in fact the only functional receptors for these ligands; however, additional primary transfectants should be generated to verify this point. The powerful approach used here should also be of use in future attempts to identify the unknown receptors for available ligands.

Use of Trk-expressing NIH-3T3 Cells to Study the Interactions between the Neurotrophins and Their Receptors

The expansion of the neurotrophin family and the identification of the Trk receptors have been accompanied by several discrepancies concerning the total number of neurotrophins and their receptor interactions. For example, it remains unclear whether a recently isolated mammalian neurotrophin (NT-4/5) represents the mammalian counterpart of xNT-4 (as suggested by Ip et al. 1992a) or whether it is instead a novel member of the neurotrophin family that should be designated NT-5 (as proposed by Berkemeier et al. 1991). The reasons for this controversy involve the fact that NT-4/5 is considerably less related to xNT-4 than the other mammalian neurotrophins are to their *Xenopus* counterparts and that differences exist in the reported bioactivities of mammalian NT-4/5. Berkemeier et al. (1991) report that mammalian NT-4/5 can efficiently activate TrkA receptors as well as TrkB, whereas Ip et al. (1992a) claim that NT-4/5 mimics xNT-4 (Klein et al. 1992) in being a specific activator of TrkB. Issues concerning the bioactivities and evolutionary history of mammalian NT-4/5 have important implications: If this neurotrophin is not the homolog of xNT-4, then a mammalian neurotrophin corresponding to xNT-4 remains to be identified. Conflicts concerning the reported bioactivities of mammalian NT-4/5 reflect similar, although somewhat less dramatic, discrepancies about the biological actions of the other neurotrophins. Although most studies agree that NGF primarily utilizes the TrkA receptor and that BDNF primarily uses the TrkB receptor, there is disagreement whether (Rosenthal et al. 1990) or not (Ernfors et al. 1990; Squinto et al. 1991b) NT-3 is an efficient activator of TrkA receptors capable of inducing neuronal differentiation in TrkA-expressing PC12 cells. Discrepancy also exists as to whether NT-3 is as efficient as BDNF (Soppet et al. 1991) or is considerably less so (Glass et al. 1991; Klein et al. 1991b; Squinto et al. 1991b) in activating TrkB receptors. Finally, an important issue concerning the interaction between neurotrophins and the Trk receptors involves the role of $p75^{LNGFR}$, a molecule that can bind with similar affinities to all members of this neurotrophin family (Rodriguez-Tebar et al. 1990; Squinto et al. 1991b; Holböök et al. 1991). There are major disagreements as to whether $p75^{LNGFR}$ is required to convert Trk receptors to a high-affinity form (Hempstead et al. 1989, 1991; Kaplan et al. 1991b; Klein et al. 1991a; Lamballe et al. 1991), which is of importance primarily because such a conversion would determine the concentration at which the neurotrophins mediate their actions.

In an attempt to address the discrepancies concerning neurotrophin/Trk interactions, we utilized NIH-3T3 cells expressing each of the Trks in quantitative assays using highly purified preparations of the four mammalian neurotrophins. Furthermore, to understand whether $p75^{LNGFR}$ or other neuronally specific accessory molecules play a role in altering neurotrophin responsiveness, we compared the dose dependence of Trk-mediated signal initiation in NIH-3T3 cells as opposed to a neuronal-like cell line, the PC12 rat pheochromocytoma cell line (Greene and Tischler 1976; Greene 1978). PC12 cells normally express TrkA and $p75^{LNGFR}$. They may also express other neuron-specific accessory molecules, and they have served as the traditional model system for the study of NGF differentiative and survival actions. Analysis of the growth-promoting effects of each of the neurotrophins on Trk-expressing NIH-3T3 cells reveals that TrkA primarily mediates responses to NGF (with an EC_{50} between 1 and 10 ng/ml) and TrkC mediates responses to NT-3 (also with an EC_{50} between 1 and 10 ng/ml), whereas TrkB can mediate responses to BDNF, NT-4/5, or NT-3 (Fig. 4A,B,C). Although BDNF and NT-4/5 are nearly indistinguishable in this assay (with EC_{50} value of 1–10 ng/ml), NT-3 is about 50-fold less potent (with an EC_{50} of ~ 100 ng/ml). Remarkably, parental PC12 cells mirror the TrkA-expressing NIH-3T3 cells in their specificity and dose dependency (also displaying an EC_{50} of 1–10 ng/ml) for NGF (Fig. 4D). PC12 cells stably expressing a transfected TrkB-encoding plasmid gain the neurotrophin responses and dose dependencies expected from the assays with TrkB-expressing NIH-3T3 cells, except that they display markedly blunted responses to NT-3 (Fig. 4E).

Contrary to the findings of one previous study (Berkemeier et al. 1991), our results conclusively demonstrate that mammalian NT-4/5 parallels xNT-4 in its ability to interact functionally with TrkB but not with TrkA or TrkC, consistent with the suggestion that NT-4/5 represents the mammalian counterpart of xNT-4. We have also been able to show that despite a previous report (Rosenthal et al. 1990), NT-3 cannot efficiently activate TrkA or act on parental PC12 cells. Furthermore, in contrast to earlier results (Soppet et al. 1991), we have been able to verify that NT-3 is quantitatively a much less efficient activator of TrkB than is BDNF. Remarkably, our findings reveal mostly similar dose dependencies in Trk-expressing fibroblasts as compared to PC12 cells, arguing that accessory molecules (such as $p75^{LNGFR}$) expressed in neuronal-like cells do not markedly affect the ability of the Trk receptors to interact with, and initiate signaling in response to, their primary ligands. Only NT-3 responses mediated via TrkB seem to be notably different, depending on the cell in which the TrkB is expressed. Thus, NT-3 is much more effective at eliciting responses in TrkB-expressing NIH-3T3 cells than in TrkB-expressing PC12 cells. The basis for these cell-specific differences displayed by TrkB remain unclear but may involve processing events or accessory molecules (such as $p75^{LNGFR}$) restricted to neuron-like cells. Thus, although $p75^{LNGFR}$ does not appear to be capable of converting the Trk kinases into higher-affinity receptors (Hempstead et al. 1991; Kaplan et al. 1991b) that would be capable of responding to lower concentrations of the neurotrophins, it (or other neuronal accessory molecules) may be involved in modulating Trk-mediated responses in other ways. Although signal initiation due to Trk receptor activation seems, in general, to be quite similar in fibroblasts and neuronal cells, the different downstream signaling pathways available in these distinct cell types presumably allow for the very different biological responses (i.e., cell growth vs. growth inhibition and neuronal differentiation) they ultimately display (Nye et al. 1992).

Molecular Cloning of a CNTF Receptor Subunit Leads to the Elucidation of a CNTF Receptor Complex

In contrast to the wealth of data concerning neurotrophin receptors and their signaling pathways, up until very recently, little was known about the CNTF receptor and the signaling pathways it initiated. We decided to develop a reagent that would allow us to first identify and then molecularly clone CNTF-binding proteins. Toward this end, we genetically engineered the CNTF gene so that we could recombinantly produce an "epitope-tagged" CNTF. We reasoned that we could use this tagged CNTF as a bridge between a cell bearing receptors for CNTF and a detection scheme employing a monoclonal antibody specific for the epitope tag. Such detection schemes did indeed permit efficient detection of cells bearing specific CNTF receptors (Squinto et al. 1991a) and ultimately allowed for the cloning of a CNTF-binding protein when used to screen cDNA libraries transiently expressed in COS cells (Davis et al. 1991). As expected, this CNTF-binding protein (hereafter called CNTFRα) was notably expressed only within the nervous system, except for expression in skeletal muscle. Surprisingly, however, CNTFRα displayed no homology with the neurotrophin receptors, but it was related to one of the two receptor components (IL-6Rα) utilized by a hematopoietic cytokine, interleukin-6 (IL-6). Furthermore, CNTFRα was attached to the cell surface via a glycosyl-phosphatidylinositol linkage and thus lacked an intracytoplasmic domain, suggesting that, like IL-6Rα, it required a second receptor component for signal transduction. Subsequent studies have revealed that CNTF and IL-6 are distantly related members of a family of factors that also includes leukemia inhibitory factor (LIF)/cholinergic differentiation factor (CDF) and oncostatin M (OSM) (Bazan 1991; Rose and Bruce 1991) and that these related factors all utilize multicomponent receptors involving shared signal transducing components (Gearing et al. 1992; Ip et al. 1992b). In particular, the receptor complexes for these factors all include gp130 (Gearing et al. 1992; Ip et al. 1992b), the protein initially identified as the signal-transducing second component of the IL-6 receptor complex (Taga et al. 1989; Hibi et al. 1990; Murakami et al. 1991). CNTF and LIF/CDF (but not IL-6) also share an additional gp130-related receptor component (hereafter referred to as LIFRβ) (Ip et al. 1992b), which was initially identified as a LIF-binding protein (Gearing et al. 1991). It has been proposed that these receptor complexes all require homo- or heterodimerization of "β" subunits to initiate signal transduction, whereas the "α" components primarily serve to regulate the ligand specificity of the complex (Ip et al. 1992b). Thus, although IL-6 utilizes the IL-6Rα component in triggering homodimerization of gp130 (Fig. 5), evidence

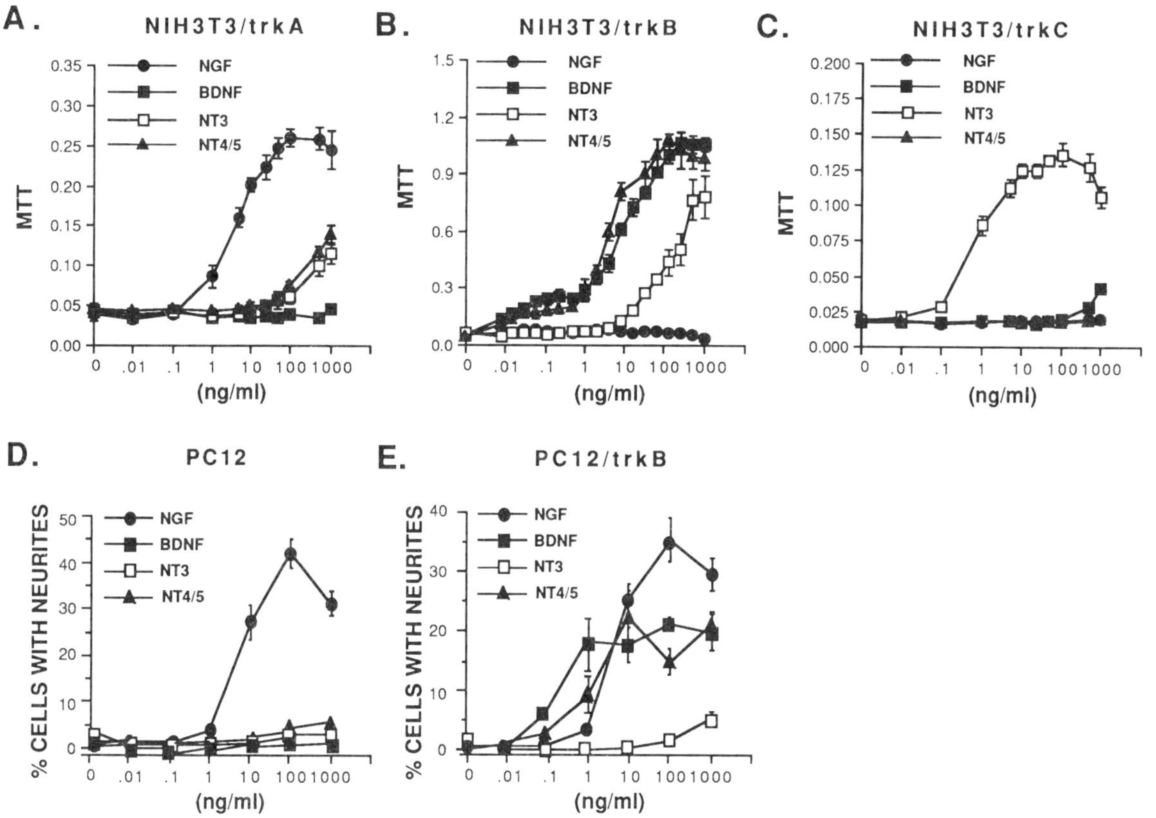

Figure 5. Specific activities of each of the neurotrophins for the Trks in fibroblasts and PC12 cells. NIH-3T3 cells expressing each of the Trks (*A–C*) were treated with various concentrations of each of the purified mammalian neurotrophins as indicated and assayed for cell growth via an MTT assay. Parental PC12 cells (*D*) or PC12 cells stably expressing TrkB (*E*) were treated with various concentrations of each of the purified mammalian neurotrophins and assayed for differentiation.

suggests that LIF may signal by inducing gp130/LIFRβ heterodimerization in the absence of an α component (Fig. 5) (Gearing et al. 1992; Ip et al. 1992b). Interestingly, addition of CNTFRα to the components comprising a functional LIF receptor complex (i.e., gp130 and LIFRβ) is sufficient to create a functional CNTF receptor complex (Fig. 5) (Ip et al. 1992b). Thus, the broad distributions of both gp130 and LIFRβ would allow for the widespread actions of LIF throughout the body, and the limited distribution of CNTFRα (predominately to the nervous system) would restrict the actions of CNTF to a subset of all LIF-responsive cells. Therefore, CNTFRα appears to be the only component of the CNTF receptor complex that is particular to the nervous system and that uniquely characterizes cells bearing functional CNTF receptors (Ip et al. 1992b). From an evolutionary viewpoint, it is remarkable that generally acting cytokines with important roles during hematopoiesis share overlapping receptor components with a neurotrophic factor, CNTF, whose actions seem to be primarily limited to the nervous system.

DISCUSSION

We have described function-based strategies that we have used to clone molecularly the receptors for the neurotrophins and for CNTF. Cloning the neurotrophin receptors involved an approach that assumed that they would mediate functional responses when ectopically expressed in nonneuronal cells. This assumption allowed for the cloning of the neurotrophin receptors in the absence of any additional information (structural, biochemical, or otherwise) about these receptors. Although this cloning was preempted by the correlative assignment of the Trk family as functional neurotrophin receptors, it represents an alternate solution to a long-standing receptor problem in neurotrophic factor biology that may be generally applicable to the cloning of other receptors. Furthermore, the establishment of Trk-expressing fibroblasts displaying growth responses to the neurotrophins has provided the means to reexamine a number of controversies concerning the bioactivities and receptor interactions of the neurotrophins (summarized in Fig. 6). Most notably, we have been able to show that mammalian NT-4/5 parallels xNT-4 in its bioactivities, consistent with the suggestion that it is indeed the mammalian counterpart of xNT-4. We have also demonstrated that all of the Trks are capable of mediating neurotrophin responses (to their primary ligands) with similar dose dependencies in fibroblasts but not in PC12 cells, arguing that accessory molecules (such as $p75^{LNGFR}$) found in neuronal cells do not markedly alter the ability of the Trks to initiate signaling. However, a neuron-like en-

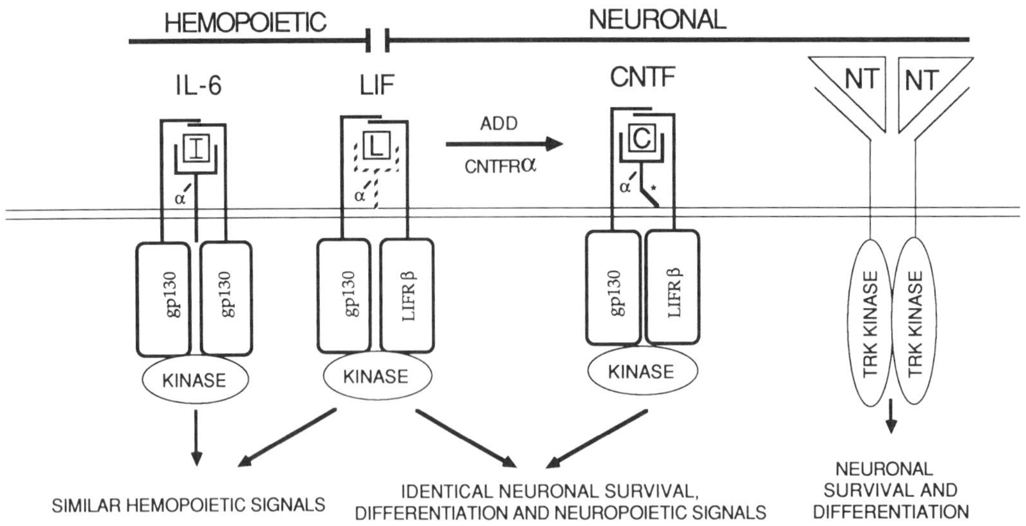

Figure 6. Schematic representations of models for IL-6, LIF, CNTF, and neurotrophin receptor complexes. In all cases, complex formation is apparently ligand-dependent and results in activation of a receptor-associated tyrosine kinase activity. Cytokine factors are represented as squares and a "generic" neurotrophin is depicted as triangle dimers; α subunits are known to exist for the IL-6 and CNTF receptor complexes and are thus depicted with solid lines (asterisk adjacent to CNTFRα/ membrane junction indicates GPI-linkage); a potential LIFRα component is indicated by a dashed line. Although depicted as membrane-bound, α subunits may also function as soluble cofactors.

vironment does seem to restrict the Trks (i.e., TrkB) in their ability to respond to their "nonpreferred" neurotrophin ligands (i.e., NT-3) (Fig. 7). Our findings with the neurotrophins and their receptors still leave a number of unresolved questions. It remains to be understood why NT-4/5 displays a much higher degree of evolutionary divergence than do the other three neurotrophins. It also remains a mystery as to why BDNF and NT-4/5 can both act as preferred ligands for TrkB. There may be physiologically important ways in which TrkB receptors can be made to distinguish between these two neurotrophins or the two factors may display differing in vivo availabilities that make it advantageous to maintain two distinct neurotrophins with similar Trk specificities. Alternatively, there may be yet another Trk receptor that displays differing specificities for BDNF as opposed to NT-4. In any event, defining the precise specificities and interactions of the neurotrophins for their receptors is absolutely required for understanding the roles of these factors in vivo.

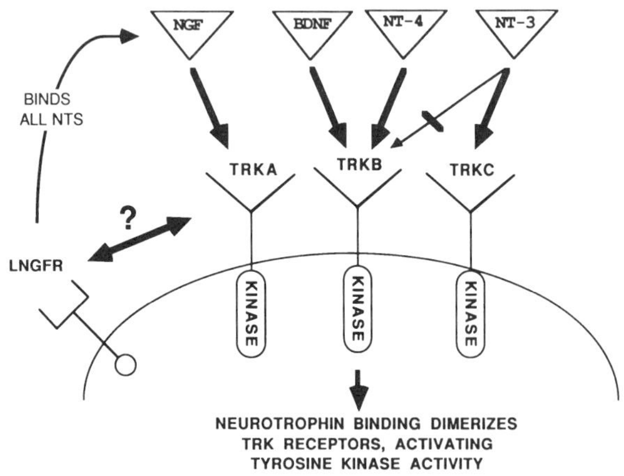

Figure 7. Schematic summary of neurotrophin interactions with their receptors. "Preferred" interactions are indicated by bold lines, whereas the "nonpreferred" interaction of NT-3 with TrkB is indicated by thin line; dash in this line denotes that this interaction is apparently blunted in neuronal environment. The possibility of an interaction between the Trks and $p75^{LNGFR}$ is indicated.

Our approach to cloning the CNTF receptor was much different from that used to clone the receptors for the neurotrophins. In the case of CNTF, we molecularly cloned a CNTF-binding protein based on its ability to bind CNTF, as well as a tagged version of CNTF. The homology of this cloned receptor component with the IL-6Rα component led to studies that have elucidated the structure of the CNTF receptor complex and have revealed that it shares components with the receptor complexes of hematopoietic cytokines that bear a distant relationship to CNTF. The CNTF receptor represents the first example in which a receptor for one factor (LIF) can be converted to serve as the receptor for another factor (CNTF) by the addition of a single conversion component (CNTFRα).

The CNTF receptor complex is remarkably unlike the receptors used by the neurotrophins, demonstrating that these two classes of neurotrophic factors utilize completely different receptor systems to keep neurons alive. However, there are certain unmistakable similarities shared by these very different receptor systems. In both cases, either the receptor (each of the Trks) or a required receptor component (CNTFRα) is almost exclusively restricted to the nervous system in its expression, largely limiting the actions of these factors to neurons. Furthermore, in both cases, homo- or heterodimerization of receptor components (either the Trks or gp130/LIFRβ) results in the activation of a tyrosine kinase activity responsible for initiating signal transduction (see Fig. 5). In the case of the CNTF receptor, this kinase activity is not intrinsic to the dimerizing components but apparently depends on yet another receptor-associated component (Fig. 5).

Perhaps the most surprising aspect of the receptor studies we have described is that although CNTF and the neurotrophins use strikingly different receptor systems, each of them has its analogs outside the nervous system. Thus, what makes neurotrophic factors so apparently special (i.e., in their abilities to promote neuronal survival and differentiation) cannot be that they initiate signals very different from those generated by traditional cytokines and growth factors but rather that they initiate these signals in neuronal cells and not in nonneuronal cells. A major remaining question is how initial signals very similar to those that often result in growth promotion in nonneuronal cells can be so uniquely interpreted within the context of a postmitotic neuron.

ACKNOWLEDGMENTS

We thank Drs. L.S. Schleifer, M.E. Furth, and the entire Discovery group at Regeneron for their enthusiastic support and insightful scientific input. We particularly thank Dr. A. Acheson for fibroblast growth assays, Dr. M.J. Macchi for tissue-culture expertise, and K. Mattsson, L. Defeo, and J. Ciliberti for excellent technical assistance. We also thank Amgen/Regeneron Partners for providing purified recombinant BDNF and NT-3, and Dr. E. Shooter for his isolate of the PC12 cell line.

REFERENCES

Adler, R., K.B. Landa, M. Manthorpe, and S. Varon. 1979. Cholinergic neuronotrophic factors: intraocular distribution of soluble trophic activity for ciliary neurons. *Science* **204:** 1434.

Bazan, J.F. 1991. Neuropoietic cytokines in the hematopoietic fold. *Neuron* **7:** 197.

Berkemeier, L.R., J.W. Winslow, D.R. Kaplan, K. Nikolics, D.V. Goeddel, and A. Rosenthal. 1991. Neurotrophin-5: A novel neurotrophic factor that activates trk and trkB. *Neuron* **7:** 857.

Cordon-Cardo, C., P. Tapley, S. Jing, V. Nanduri, E. O'Rourke, F. Lamballe, K. Kovary, R. Klein, K.R. Jones, L.F. Reichardt, and M. Barbacid. 1991. The *trk* tyrosine protein kinase mediates the mitogenic properties of nerve growth factor and neurotrophin-3. *Cell* **66:** 173.

Davis, S., T.H. Aldrich, D.M. Valenzuela, V. Wong, M.E. Furth, S.P. Squinto, and G.D. Yancopoulos. 1991. The receptor for ciliary neurotrophic factor. *Science* **253:** 59.

Ernfors, P., C.F. Ibanez, T. Ebendal, L. Olson, and H. Persson. 1990. Molecular cloning and neurotrophic activities of a protein with structural similarities to nerve growth factor. *Proc. Natl. Acad. Sci.* **87:** 5454.

Gearing D.P., C.J. Thut, T. VandenBos, S.D. Gimpel, P.B. Delaney, J. King, V. Price, D. Cosman, and M.P. Beckmann. 1991. Leukemia inhibitory factor receptor is structurally related to the IL-6 signal transducer, gp130. *EMBO J.* **10:** 2839.

Gearing, D.P., M.R. Comeau, D.J. Friend, S.D. Gimpel, C.J. Thut, J. McGourty, K.K. Brasher, J.A. King, S. Gillis, B. Mosley, S.F. Ziegler, and D. Cosman. 1992. The IL-6 signal transducer, gp130: An oncostatin M receptor and affinity converter for the LIF receptor. *Science* **255:** 1434.

Glass, D.J., S.H. Nye, P. Hantzopoulos, M.J. Macchi, S.P. Squinto, M. Goldfarb, and G.D. Yancopoulos. 1991. TrkB mediates BDNF/NT-3 dependent survival and proliferation in fibroblasts lacking the low affinity NGF receptor. *Cell* **66:** 405.

Greene, L.A. 1978. Nerve growth factor prevents death and stimulates neuronal differentiation of clonal PC12 pheochromocytoma cells in serum-free medium. *J. Cell Biol.* **78:** 747.

Greene, L.A. and A.S. Tischler. 1976. Establishment of a noradrenergic clonal line of rat adrenal pheochromocytoma cells which respond to nerve growth factor. *Proc. Natl. Acad. Sci.* **73:** 2424.

Hamburger, V. and R. Levi-Montalcini. 1949. Proliferation, differentiation and degeneration in the spinal ganglia of the chick embryo under normal and experimental conditions. *J. Exp. Zool.* **111:** 457.

Hempstead, B.L., L.S. Schleifer, and M.V. Chao. 1989. Expression of functional nerve growth factor receptors after gene transfer. *Science* **243:** 373.

Hempstead, B.L., D. Martin-Zanca, D.R. Kaplan, L.F. Parada, and M.V. Chao. 1991. High-affinity NGF binding requires coexpression of the *trk* proto-oncogene and the low-affinity NGF receptor. *Nature* **350:** 678.

Hibi, M. M. Murakami, M. Saito, T. Hirano, T. Taga, and T. Kishimoto. 1990. Molecular cloning and expression of an IL6 signal transducer, gp130. *Cell* **63:** 1149.

Hohn, A., J. Leibrock, K. Bailey, and Y.-A. Barde. 1990. Identification and characterization of a novel member of the nerve growth factor/brain derived neurotrophic factor family. *Nature* **344:** 339.

Holböök, F., C.F. Ibáñez, and H. Persson. 1991. Evolutionary studies of the nerve growth factor family reveal a novel member abundantly expressed in *Xenopus* ovary. *Neuron* **6:** 845.

Ip, N.Y., C.F. Ibáñez, S.H. Nye, J. McClain, P.F. Jones, D.R. Gies, L. Belluscio, M.M. Le Beau, R. Espinsosa III, S.P. Squinto, H. Persson, and G.D. Yancopoulos. 1992a. Mammalian neurotrophin-4: Structure, chromosomal localization, tissue distribution, and receptor specificity. *Proc. Natl. Acad. Sci.* **89:** 3060.

Ip, N.Y., S.H. Nye, T.G. Boulton, S. Davis, T. Taga, Y. Li, S.J. Birren, K. Yasukawa, T. Kisimoto, D.J. Anderson, N. Stahl, and G.D. Yancopoulos. 1992b. CNTF and LIF act on neuronal cells via shared signaling pathways that involve the IL-6 signal transducing receptor component gp130. *Cell* **69:** 1121.

Jones, K.R. and L.F. Reichardt. 1990. Molecular cloning of a human gene that is a member of the nerve growth factor family. *Proc. Natl. Acad. Sci.* **87:** 8060.

Julius, D., T.J. Livelli, T.M. Jessel, and R. Axel. 1989. Ectopic expression of the serotonin 1c receptor and the triggering of malignant transformation. *Science* **244:** 1057.

Kaisho, Y., K. Yoshimura, and K. Nakahama. 1990. Cloning and expression of a cDNA encoding a novel human neurotrophic factor. *FEBS Lett.* **266:** 187.

Kaplan, D.R., D. Martin-Zanca, and L.F. Parada. 1991a. Tyrosine phosphorylation and tyrosine kinase activity of the *trk* proto-oncogene product induced by NGF. *Nature* **350:** 158.

Kaplan, D.R., B.L. Hempstead, D. Martin-Zanca, M.V. Chao, and L.F. Parada. 1991b. The *trk* proto-oncogene product: A signal transducing receptor for nerve growth factor. *Science* **252:** 554.

Klein, R., F. Lamballe, S. Bryant, and M. Barbacid. 1992. The *trkB* tyrosine protein kinase is a receptor for neurotrophin-4. *Neuron* **8:** 947.

Klein, R., S.Q. Jing, V. Nanduri, E. O'Rourke, and M. Barbacid. 1991a. The *trk* proto-oncogene encodes a receptor for nerve growth factor. *Cell* **66:** 189.

Klein, R., V. Nanduri, S.Q. Jing, F. Lamballe, P. Tapley, S. Bryant, C. Cordon-Cardo, K.R. Jones, L. Reichardt, and M. Barbacid. 1991b. The *trkB* tyrosine kinase is a receptor for brain-derived neurotrophic factor and neurotrophin-3. *Cell* **66:** 395.

Lamballe, F., R. Klein, and M. Barbacid. 1991. *trk*C, a new member of the trk family of tyrosine protein kinases, is a receptor for neurotrophin-3. *Cell* **66:** 967.

Leibrock, J., F. Lottspeich, A. Hohn, M. Hofer, B. Hengerer, P. Masiakowski, H. Thoenen, and Y.-A. Barde. 1989. Molecular cloning and expression of brain-derived neurotrophic factor. *Nature* **341:** 149.

Levi-Montalcini, R. 1987. The nerve growth factor: Thirty-five years later. *EMBO J.* **6:** 1145.

Lillien, L.E. and P. Claude. 1985. Nerve growth factor is a mitogen for cultured chromaffin cells. *Nature* **317:** 632.

Lin, L.-F.H., D. Mismer, J.D. Lile, L.G. Armes, E.T. Butler, III, J.L. Vannice, and F. Collins. 1989. Purification, cloning and expression of ciliary neurotrophic factor (CNTF). *Science* **246:** 1023.

Maisonpierre, P.C., L. Belluscio, S. Squinto, N.Y. Ip, M.E. Furth, R.M. Lindsay, and G.D. Yancopoulos. 1990. Neurotrophin-3: A neurotrophic factor related to NGF and BDNF. *Science* **247:** 1446.

Murakami, M., M. Narazaki, K. HibiYasukawa, M. Hamaguchi, T. Taga, and T. Kishimoto. 1991. Critical cytoplasmic region of the interleukin 6 signal transducer gp130 is conserved in the cytokine receptor family. *Proc. Natl. Acad. Sci.* **88:** 11349.

Nye, S.H., S.P. Squinto, D.J. Glass, T.N. Stitt, P. Hantzopoulos, M.J. Macchi, N.S. Lindsay, N.Y. Ip, and G.D. Yancopoulos. 1992. K-252a and staurosporine selectively block autophosphorylation of neurotrophin receptors and neurotrophin-mediated responses. *Mol. Biol. Cell.* **3:** 677.

Oppenheim, R.W. 1981. Neuronal cell death and some related regressive phenomena during neurogenesis; a selective historical review and progress report. In *Studies in developmental neurobiology*. (ed. W.M. Cowan), p. 74. Oxford University Press, England.

Rodrigeuz-Tebar, A., G. Dechant, and Y.-A. Barde. 1990. Binding of brain-derived neurotrophic factor to the nerve growth factor receptor. *Neuron* **4:** 487.

Rose, T.M. and G. Bruce. 1991. Oncostatin M is a member of a cytokine family that includes leukemia-inhibitory factor, granulocyte colony-stimulating factor, and interleukin 6. *Proc. Natl. Acad. Sci.* **88:** 8641.

Rosenthal, A., D.V. Goeddel, T. Nguyen, M. Lewis, A. Shih, G.R. Laramee, K. Nikolics, and J.W. Winslow. 1990. Primary structure and biological activity of a novel human neurotrophic factor. *Neuron* **4:** 763.

Soppet, D., E. Escandon, J. Maragos, D.S. Middlemas, S.W. Reid, J. Blair, L.E. Burton, B.R. Stanton, D.R. Kaplan, T. Hunter, K. Nikolics, and L.F. Parada. 1991. The neurotrophic factors brain-derived neurotrophic factor and neurotrophin-3 are ligands for the *trkB* tyrosine kinase receptor. *Cell* **65:** 895.

Squinto, S.P., T.H. Aldrich, R.M. Lindsay, D.M. Morrissey, N. Panayotatos, S.M. Bianco, M.E. Furth, and G.D. Yancopoulos. 1991a. Identification of functional receptors for ciliary neurotrophic factor on neuronal cell lines and primary neurons. *Neuron* **5:** 757.

Squinto, S.P., T.N. Stitt, T.H. Aldrich, S. Davis, S.M. Bianco, C. Radziejewski, D.J. Glass, P. Masiakowski, M.E. Furth, D.M. Valenzuela, P.S. DiStefano, and G.D. Yancopoulos. 1991b. *trkB* encodes a functional receptor for brain-derived neurotrophic factor and neurotrophin-3 but not nerve growth factor. *Cell* **65:** 885.

Stockli, K.A., F. Lottspeich, M. Sendtner, P. Masiakowski, P. Carroll, R. Gotz, D. Lindholm, and H. Thoene. 1989. Molecular cloning, expression and regional distribution of rat ciliary neurotrophic factor. *Nature* **342:** 920.

Taga, T., M. Higi, Y. Hirata, K. Yamasaki, K. Yasukawa, T. Matsuda, T. Hirano, and T. Kishimoto. 1989. Interleukin-6 triggers the association of its receptor with a possible signal transducer, gp130. *Cell* **58:** 573.

Togari, A., G. Dickens, H. Kuzuya, and G. Guroff. 1985. The effect of fibroblast growth factor on PC12 cells. *J. Neurosci.* **5:** 307.

Zhan, X., B. Bates, X. Hu, and M. Goldfarb. 1988. The human FGF-5 oncogene encodes a novel protein related to fibroblast growth factors. *Mol. Cell. Biol.* **8:** 3487.

Zhan, X., A. Culpepper, M. Reddy, J. Loveless, and M. Goldfarb. 1987. Human oncogenes detected by a defined medium culture assay. *Oncogene* **1:** 369.

Interactions of PDGF and FGF Receptors with Cytoplasmic Signaling Molecules

K.G. Peters,†‡ J.A. Escobedo,*†‡ W.J. Fantl,†‡ and L.T. Williams*†‡
*Howard Hughes Medical Institute, †Cardiovascular Research Institute, ‡Department of Medicine, University of California, San Francisco, California 94143-0130

There has been significant progress in understanding the initial steps of signal transduction by the group of growth factor receptors that are membrane-spanning tyrosine kinases. These receptor tyrosine kinases phosphorylate themselves on tyrosines in response to ligand binding. During the last 3 years, an important function of autophosphorylation has become apparent: Individual phosphotyrosine residues of receptors appear to serve as highly selective binding sites that interact with specific cytoplasmic signaling molecules. These signaling molecules in turn appear to mediate the pleiotropic responses of cells to growth factors. We have focused on studies of the platelet-derived growth factor β (PDGFβ) receptor and fibroblast growth factor (FGF) receptor.

When activated by PDGF, the intracellular region of the PDGFβ receptor binds several signaling molecules. These include phosphatidylinositol 3-kinase (PI3 kinase) (Coughlin et al. 1989; Kazlauskas and Cooper 1990; Cantley et al. 1991; Escobedo et al. 1991; Fantl et al. 1992), phospholipase C-γ (PLC-γ) (Kumijan et al. 1989; Meisenhelder et al. 1989; Wahl et al. 1989; Morrison et al. 1990), GTPase activating protein (GAP) (Molloy et al. 1989; Kaplan et al. 1990; Kazlauskas et al. 1990; Fantl et al. 1992), and the c-*src*-encoded protein (Kypta et al. 1990). Studies of the interaction between the PDGFβ receptor and PI3 kinase led to the first observation of a signaling molecule that interacts directly with a receptor and also showed that a signaling molecule binds to a specific phosphotyrosine rather than to random phosphotyrosines in the cytoplasmic domain of a receptor (Coughlin et al. 1989; Escobedo et al. 1991; Fantl et al. 1992). Through receptor mutagenesis studies, it was shown that the PI3 kinase binds to two distinct sites on the PDGFβ receptor (Tyr-708 and Tyr-719). In similar studies, it was found that GAP binds to a third site (Tyr-739). As described below, PLC-γ binds to two additional sites in the carboxyl terminus of the PDGF receptor and to a single tyrosine in the FGF receptor.

Studies on the interactions of signaling molecules with tyrosine-phosphorylated receptors have been facilitated by the use of in vitro assays in which the interaction can be directly measured. Using this system, we found that phosphotyrosine-containing peptides as short as 5 amino acids representing receptor sequences can selectively block the interaction of signaling molecules with receptor tyrosine kinases (RTKs). The binding of peptides to signaling molecules is highly specific and occurs at low concentrations of peptides, but only when the peptide is phosphorylated on its key tyrosine residue. For example, the peptide phosphotyrosine - valine - proline - methionine - leucine, which represents a PDGF receptor PI3 kinase binding site, blocks PI3 kinase binding to the PDGFβ receptor but does not block binding of two other signaling molecules, GAP and PLC-γ. Correspondingly, the tyrosine-phosphorylated peptide representing the GAP binding site on the PDGFβ receptor blocks GAP binding but has no effect on the binding of PI3 kinase or PLC-γ.

This approach of using tyrosine-phosphorylated peptides to map binding sites for signaling molecules is especially useful in the case of a signaling molecule that has more than one binding site on the receptor. For example, PI3 kinase binds to two sites, at Tyr-708 and Tyr-719. By mutagenesis, it was extremely difficult to map the binding site because elimination of a single tyrosine did not eliminate the interaction of the receptor with PI3 kinase. However, by using a peptide representing either site, it was possible to block binding of PI3 kinase to the receptor because the peptide binds directly to the SH2 region of PI3 kinase that interacts with phosphotyrosines. Even though there were two separate sites on the receptor, PI3 kinase binding was blocked.

Mapping Intracellular Signaling Cascades by Selective Elimination of Binding Sites on Receptor Tyrosine Kinases

Mutating tyrosines on the PDGF receptor that are responsible for binding to specific signaling molecules provides an approach for assessing the roles of signaling molecules and cellular responses to PDGF. When the PI3 kinase binding sites on the PDGF receptor were mutated and the mutant receptor was expressed in murine epithelial cells, PDGF-stimulated PI3 kinase activation was abrogated even though inositol phosphate accumulation, calcium influx, and GAP tyrosine phosphorylation were normal (Fantl et al. 1992). Interestingly, we have recently shown in experiments done in collaboration with Dr. Yoshimi Kaziro that this mutant receptor, unlike wild-type PDGF receptor, was unable to increase the amount of GTP bound by $p21^{ras}$ and is similarly defective in activating Raf-1 kinase and MAP kinase (not shown). These experiments provide a tentative explanation for our previous observation (Fantl et al. 1992) that the PDGF receptor with mutated PI3 kinase binding sites was defective in eliciting a mitogenic response even though it could mediate increases in calcium, phosphatidylinositol (PI) turnover, and GAP phosphorylation. This mutant receptor does not activate the $p21^{ras}$ or Raf-1 kinase, molecules known to be important for mitogenesis. The results suggest that PI3 kinase is important for $p21^{ras}$ activation by PDGF. In other experiments using receptor mutants defective in phosphorylating GAP, we found $p21^{ras}$ activation by PDGF (reflected by an increase in GTP bound to $p21^{ras}$) to be normal. Therefore, it seems likely that PI3 kinase activated $p21^{ras}$ by enhancing protein guanine nucleotide exchange activity, rather than by reducing GAP activity.

Identification of a PLC-γ Binding Site on the FGF Receptor

The FGF receptor, like the PDGF receptor, is a tyrosine kinase. In contrast to the PDGF receptor, however, the FGF receptor does not phosphorylate GAP and activates PI3 kinase minimally, if at all. It has been shown previously that activation of the FGF receptor leads to stimulation of PLC-γ and the hydrolysis of PI. Therefore, we and other workers have assumed that this pathway of PLC-γ activation was important in the mitogenic response to FGF. Tyr-766 in the carboxy-terminal tail of the FGF receptor is highly conserved among FGF receptors from several species and has been previously identified as a binding site for PLC-γ (Mohammadi et al. 1991). To further establish Tyr-766 as the major binding site for PLC-γ on the FGF receptor, we synthesized several peptides encompassing Tyr-766 and its flanking sequences and compared these to other tyrosine-phosphorylated peptides in blocking the binding of PLC-γ to the FGF receptor. We found that a 9-amino-acid peptide that was phosphorylated on a tyrosine corresponding to Tyr-766 specifically blocked the association of PLC-γ with the FGF receptor at a peptide concentration of 10 μM. The unphosphorylated version of this peptide had little or no activity in blocking this interaction. A scrambled phosphopeptide containing the same amino acid composition in a different sequence also failed to block PLC-γ association with the FGF receptor. These results suggest that Tyr-766, when phosphorylated, was directly involved in the binding of PLC-γ to the receptor and that Tyr-766 and its flanking sequences constituted the major site of PLC-γ association.

To confirm the importance of Tyr-766 in the interaction between FGF receptor and PLC-γ, this tyrosine was mutated to phenylalanine and the mutant receptor was expressed in L6 myoblasts that are devoid of endogenous FGF receptors. FGF treatment of cells expressing wild-type FGF receptors enhanced tyrosine phosphorylation of PLC-γ (Fig. 1). However, in a cell line expressing the mutant (Y/F766) FGF receptor, there was little if any phosphorylation of PLC-γ stimulated by FGF. The mutant receptor also failed to associate with PLC-γ and did not elicit an increase in PI turnover in response to FGF (Fig. 2). These results further establish that Tyr-766 was necessary for ligand-induced association of PLC-γ with the FGF receptor in vivo and show that association with the receptor is required for FGF-induced phosphorylation and activation in PLC-γ.

FGF, like many other growth factors, stimulates a rapid increase in cytosolic calcium concentration. This

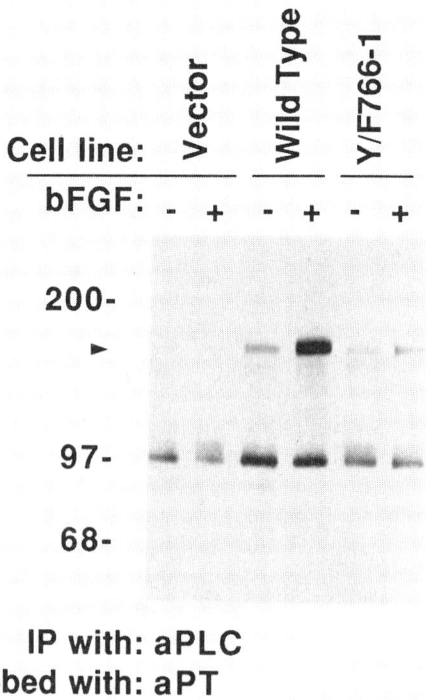

Figure 1. The Y/F766 FGF receptor mutant fails to mediate tyrosine phosphorylation of PLC-γ in response to FGF stimulation. Anti-PLC-γ (aPLC) immunoprecipitates made from unstimulated (−) or FGF-stimulated (+) L6 cells expressing either the wild-type or the mutant (Y/F766-1) FGF receptor were probed with an antiphosphotyrosine antibody (aPT). The arrowhead indicates tyrosine-phosphorylated PLC-γ.

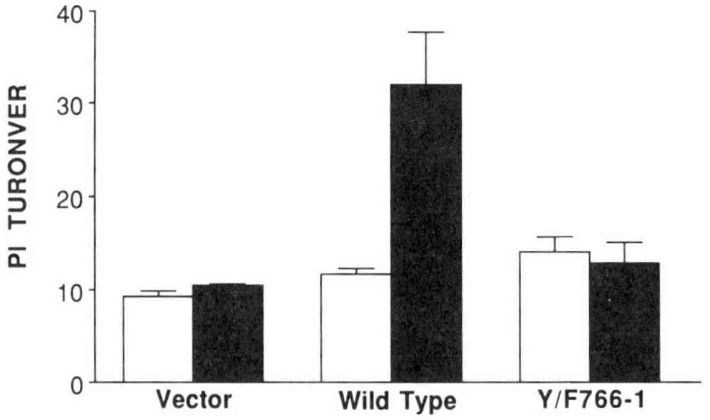

Figure 2. The Y/F766 FGF receptor mutant fails to mediate PI turnover in response to FGF stimulation. Cells expressing the wild-type or the mutant (Y/F766-1) FGF receptor were stimulated with FGF, and PI turnover was measured as accumulation of inositol-2-phosphate (CPM/cell). (Open bars) inositol phosphate accumulation in unstimulated cells; (solid bars) inositol phosphate accumulation in FGF-stimulated cells. Error bars indicate S.E.M. of triplicate samples.

process occurs in two phases, first by the transient release of calcium from intracellular stores and then later by sustained entry of calcium from the extracellular space. It has been shown by other investigators that the release of calcium from intracellular stores is probably mediated by inositol triphosphate. Since cytosolic calcium has been suggested as an important mediator of growth factor signals, we collaborated with Dr. Harlan Ives and Dr. Emily Wilson to test the ability of the Y/F766 FGF receptor to mobilize calcium. In the cells expressing the wild-type FGF receptor, FGF stimulated both the early and late phases of calcium mobilization (Fig. 3). However, in cells expressing the Y/F766 mutant receptor, FGF failed to stimulate an increase in intracellular calcium either from intracellular stores or from the extracellular space. The cells expressing either wild-type or mutant receptor had a normal calcium response to PDGF. From these studies, we conclude that PI turnover is required for FGF-induced calcium immobilization.

The most significant studies on the Y/F766 mutant FGF receptors were on the ability of this receptor to mediate FGF-stimulated mitogenesis. Despite the fact that this mutant receptor was defective in stimulating PI turnover and calcium immobilization, it was able to mediate a full FGF-stimulated mitogenic response (Fig. 4). The results of thymidine incorporation experiments were confirmed by cell count experiments, indicating that the receptor mediated the effects of FGF in all phases of the cell cycle. These studies on the FGF receptor have led to three important conclusions. First, they have shown that elimination of a single tyrosine at the interaction site of a signaling molecule with the receptor causes a loss in the ability of the receptor to phosphorylate and activate the signaling molecule. Second, activation of PLC-γ is required for the increase in calcium. Third, these studies led to the surprising conclusion that neither FGF-stimulated PI turnover nor FGF-stimulated calcium elevation was required for the mitogenic response to FGF in these cell lines.

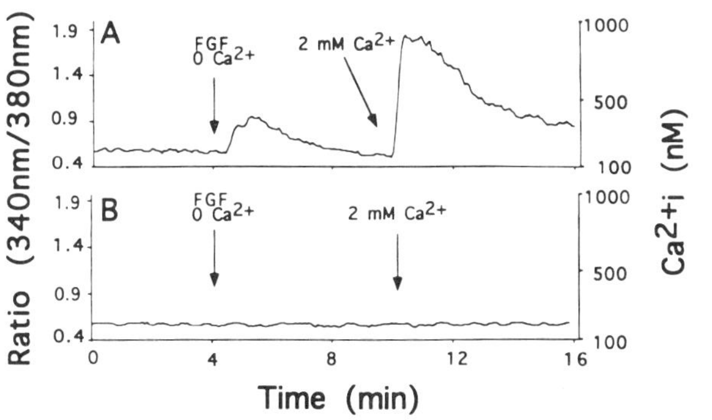

Figure 3. The Y/F766 FGF receptor does not mediate Ca^{++} mobilization from intracellular stores or from the extracellular space in response to FGF stimulation. To determine Ca^{++} mobilization from intracellular stores, basic FGF (bFGF) was added to L6 myoblasts expressing the wild-type FGF receptor (*A*) or the Y/F766 FGF receptor (*B*) in Ca^{++}-free buffer containing 10 mM EGTA (first arrow). After 10 min, the buffer was changed to Ca^{++}-containing buffer without EGTA (second arrow) to assess the contribution of Ca^{++} entry to calcium mobilization by the wild-type and Y/F766 mutant FGF receptors.

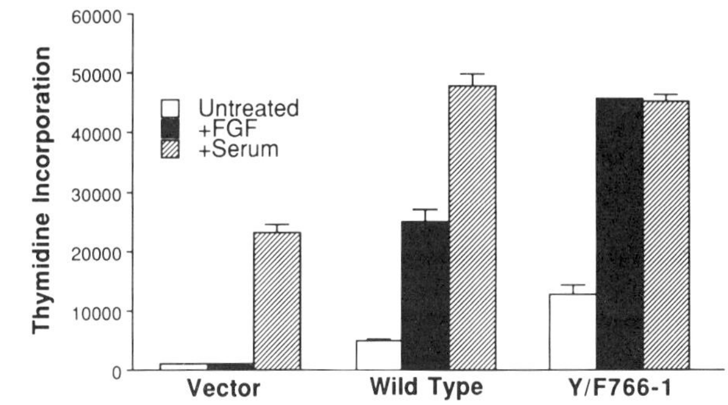

Figure 4. The Y/F766 FGF receptor mediates DNA synthesis in response to FGF stimulation. [^{3}H]Thymidine incorporation was measured in the L6 cells expressing the wild-type or the mutant (YF766-1) FGF receptor after stimulation with bFGF. (Open bars) Unstimulated; (solid bars) bFGF-stimulated; (hatched bars) serum-stimulated. Error bars indicate S.E.M. of triplicate samples.

In subsequent studies, we have found that the Y/F766 mutant receptor stimulates activity of MAP and Raf-1 kinases normally. Therefore, PLC-γ is not required for this pathway. Since this mutant receptor is capable of stimulating mitogenesis, the MAP kinase and Raf-1 kinases are strong candidates as mediators of FGF-stimulated mitogenesis. Future work in this area will be required to elucidate the roles of these pathways.

Activation of PLC-γ by PDGF Receptors

We recently found that mutations of Tyr-977 and Tyr-989 in the carboxy-terminal tail of the PDGF receptor caused a loss of the ability of the PDGF receptor to bind PLC-γ. This mutant receptor appears to be normal in interacting with other signaling molecules. When the mutant receptor was expressed in an epithelial cell line, PLC-γ was phosphorylated on tyrosines after PDGF addition to cells, even though there was no detectable association between PLC-γ and the receptor. However, this mutant receptor was incapable of stimulating PI hydrolysis. Therefore, association of PLC-γ with the receptor appeared to be required for PDGF-stimulated PLC-γ activity, perhaps because the association localizes PLC-γ to a site near its substrate or because the association allows the "correct" site on PLC-γ to be phosphorylated. The results of experiments testing the mitogenic response of the mutant receptor to PDGF were surprising. In contrast to the result with the FGF receptor that was defective in activating PLC-γ, the PDGF receptor mutant that failed to activate PLC-γ was unable to mediate a mitogenic response to PDGF. Therefore, PDGF-stimulated PLC-γ appears to play a role in mitogenesis in response to this growth factor. It is important to note that the experiments on PDGF and FGF receptors were done in different "host cells". This was necessitated by the requirement that the host cell lack endogenous receptors. However, our tentative conclusion is that the PDGF receptor requires PLC-γ stimulation for its mitogenic effect, whereas the FGF receptor relies more heavily on other pathways.

CONCLUSION

The experiments described in this paper focus on the signaling pathways stimulated by PDGF and FGF receptors. By mutating the receptor phosphotyrosines that bind specific signaling molecules, we selectively disrupted individual pathways. We found that PDGF receptors defective in binding and activating PI3 kinase were also defective in activating $p21^{ras}$, Raf-1 kinase, and MAP kinase, even though they mediated PI turnover, calcium elevation, and GAP phosphorylation normally. These mutants are known to be incapable of stimulating mitogenesis. We tentatively conclude that PI3 kinase plays an important role in activation of $p21^{ras}$ which, in turn, is known to regulate Raf-1 kinase and MAP kinase (de Vries-Smits et al. 1992; Thomas et al. 1992; Wood et al. 1992). Of course, this interpretation is subject to the usual caveats regarding studies of mutant proteins. Additional work will be required to determine how PI3 kinase might be important for $p21^{ras}$ activation.

In separate experiments, we were able to use point mutations of FGF and PDGF receptors to eliminate the activation of PLC-γ by these receptors. Surprisingly, the FGF receptor mutant that could not activate PLC-γ or elicit an increase in calcium had full activity in stimulating mitogenesis. Therefore, neither stimulation of PLC-γ nor increase in cytosolic calcium is required for FGF-induced mitogenesis. Other pathways, perhaps those involving $p21^{ras}$ or Raf-1 kinase, must be explored in future work on signal transduction by FGF receptors.

ACKNOWLEDGMENT

We thank Betty Cheung for her help in typing this manuscript.

REFERENCES

Cantley, L., K. Auger, C. Carpenter, B. Duckworth, A. Graziani, R. Kapeller, and S. Sultoff. 1991. Oncogenes and signal transduction. *Cell* **64:** 281.

Coughlin, S., J. Escobedo, and L. Williams. 1989. Role of phosphatidylinositol kinase in PDGF receptor signal transduction. *Science* **243:** 1191.

de Vries-Smits, A., B. Burgering, S. Leevers, C. Marshall, and J. Bos. 1992. Involvement of $p21^{ras}$ in activation of extracellular signal-regulated kinase 2. *Nature* **357:** 602.

Escobedo, J., D. Kaplan, W.M. Kavanaugh, C. Turck, and L. Williams. 1991. A phosphatidylinositol-3 kinase binds to platelet-derived growth factor receptors through a specific receptor sequence containing phosphotyrosine. *Mol. Cell. Biol.* **11:** 1125.

Fantl, W., J. Escobedo, G. Martin, C. Turck, M. Del Rosario, F. McCormick, and L. Williams. 1992. Distinct phosphotyrosines on a growth factor receptor bind to specific molecules that mediate different signaling pathways. *Cell* **69:** 413.

Kaplan, D., D. Morrison, G. Wong, F. McCormick, and L. Williams. 1990. PDGF β-receptor stimulates tyrosine phosphorylation of GAP and association of GAP with a signaling complex. *Cell* **61:** 125.

Kazlauskas, A. and J. Cooper. 1990. Phosphorylation of the PDGF receptor β subunit creates a tight binding site for phosphatidylinositol 3 kinase. *EMBO J.* **9:** 3279.

Kumijan, D., M. Wahl, S.-G. Rhee, and T. Daniel. 1989. Platelet-derived growth factor (PDGF) binding promotes the physical association of PDGF receptor with phospholipase-C. *Proc. Natl. Acad. Sci.* **86:** 8232.

Kypta, R., Y. Goldberg, E. Ulug, and S. Courtneidge. 1990. Association between the PDGF receptor and members of the src family of tyrosine kinases. *Cell* **62:** 481.

Meisenhelder, J., P.-G. Suh, S.-G. Rhee, and T. Hunter. 1989. Phospholipase C-γ is a substrate for the PDGF and the EGF receptor protein-tyrosine kinases in vivo and in vitro. *Cell* **57:** 1109.

Mohammadi, M., A. Honegger, D. Rotin, R. Fischer, F. Bellot, W. Li, C. Dionne, M. Jaye, M. Rubinstein, and J. Schlessinger. 1991. A tyrosine-phosphorylated carboxy-terminal peptide of the fibroblast growth factor receptor (flg) is a binding site for the SH2 domain of phospholipase C-γ1. *Mol. Cell. Biol.* **11:** 5068.

Molloy, C., K. Bottaro, M. Fleming, J. Gibbs, and S. Aaronson. 1989. PDGF induction of tyrosine phosphorylation of the GTPase activating protein. *Nature* **342:** 711.

Morrison, D., D. Kaplan, S.-G. Rhee, and L. Williams. 1990. Platelet-derived growth factor (PDGF)-dependent association of phospholipase C- γ with the PDGF receptor signaling complex. *Mol. Cell. Biol.* **10:** 2359.

Thomas, S., M. DeMarco, G. D'Arcangelo, S. Halegoua, and J. Brugge. 1992. Ras is essential for nerve growth factor- and phorbol ester-induced tyrosine phosphorylation of MAP kinases. *Cell* **68:** 1031.

Wahl, M., N. Olashaw, S. Nishibe, S.-G. Rhee, W. Pleger, and G. Carpenter. 1989. Platelet-derived growth factor induces rapid and sustained tyrosine phosphorylation of phospholipase C-γ in quiescent Balbc/3T3 cells. *Mol. Cell. Biol.* **9:** 2934.

Wood, K., C. Sarnecki, T. Roberts, and J. Blenis. 1992. Ras mediates nerve growth factor receptor modulation of three signal-transducing protein kinases: MAP kinase, Raf-1, and RSK. *Cell* **68:** 1041.

Role of SH2-containing Proteins in Cellular Signaling by Receptor Tyrosine Kinases

J. Schlessinger, M. Mohammadi, B. Margolis, and A. Ullrich*
New York University Medical Center, Department of Pharmacology, New York, New York 10016
**Max-Planck-Institut für Biochemie, Munich, Germany*

After the discoveries of nerve growth factor (NGF) and epidermal growth factor (EGF), it became clear that polypeptide growth factors play an important role in the control of cell growth and differentiation. Many growth factors have been characterized and shown to stimulate pleiotropic responses by binding to and activating cell-surface receptors with protein tyrosine kinase activity (Ullrich and Schlessinger 1990). All receptor tyrosine kinases exhibit a conserved molecular architecture containing a ligand-binding domain attached to a cytoplasmic region via a single transmembrane domain. The various cytoplasmic portions of receptor tyrosine kinases contain, in addition to the conserved catalytic domain, distinct regulatory sequences with tyrosine autophosphorylation sites and serine/threonine phosphorylation sites for different kinases. It was concluded that growth-factor-induced tyrosine phosphorylation is indispensable for signal transduction pathways essential for mitogenesis and transformation, because it was shown that tyrosine kinase activity is required for stimulation of both early and delayed responses such as phospholipase C and phospholipase A2 activities, Ca^{++} rise, S-6 phosphorylation, and c-*myc*, c-*fos* expression, and other responses leading to DNA synthesis or oncogenesis (Ullrich and Schlessinger 1990).

Receptor tyrosine kinases also play an important role in the control of cell differentiation. It was shown that the same receptor tyrosine kinase able to inhibit cell proliferation and induce differentiation when expressed in the context of one cell type is able to stimulate mitogenesis when expressed in another cellular environment. For example, activation of either NGF or fibroblast growth factor (FGF) receptors promotes the differentiation and survival of pheochromocytoma cells (PC12) and certain types of neuronal cells. However, when expressed in fibroblasts, activation of these two receptors leads to mitogenesis and even transformation (Togari et al. 1985; Damon et al. 1990; Kaplan et al. 1991; Klein et al. 1991; Wagner 1991). Hence, identification of signaling proteins that are activated by receptor tyrosine kinases and elucidation of the mechanisms that regulate their activity will reveal important clues concerning the processes controlling cell differentiation, mitogenesis, and transformation.

It is now well established that growth-factor-induced receptor dimerization is the mechanism responsible for stimulation of the catalytic activity of receptor tyrosine kinases and for the process of autophosphorylation (Schlessinger 1988). Tyrosine autophosphorylation sites have been identified and shown to be localized in the carboxy-terminal tail of EGF, FGF, and platelet-derived growth factor (PDGF) receptors and in the kinase insert regions of PDGF or colony-stimulating factor-1 (CSF-1) receptors (Kazaluskas and Cooper 1989, 1990; Ullrich and Schlessinger 1990; Margolis et al. 1990b; Kashishian et al. 1992; Mohammadi et al. 1991; Reedijk et al. 1992; Rotin et al. 1992b). The tyrosine-phosphorylated regions in growth factor receptors and other cellular proteins function as high-affinity binding sites for cellular proteins such as $pp^{60c\text{-}src}$, phospholipase Cγ (PLCγ), GTPase-activating protein of *ras* (GAP), and phosphatidylinositol (PI)3-kinase-associated protein p85 (Stahl et al. 1988; Suh et al. 1988; Trahey et al. 1988; Vogel et al. 1988; Meisenhelder et al. 1989; Margolis et al. 1989,1990c; Wahl et al. 1989; Burgess et al. 1990; Kypta et al. 1990; Escobedo et al. 1991; Otsu et al. 1991; Skolnik et al. 1991). These and other signaling proteins are able to recognize specifically different growth factor receptors, and the association between these proteins is strictly dependent on tyrosine phosphorylation of the receptor molecules (Margolis et al. 1990a). Nonphosphorylated growth factor receptors do not bind PLCγ, GAP, p85, and the other signaling proteins (Koch et al. 1991; Margolis 1992). Moreover, it was demonstrated that short tyrosine-phosphorylated peptides in growth factor receptors and other proteins such as middle T antigen represent high-affinity binding sites for association with target proteins (Cantley et al. 1991). The association between the tyrosine-phosphorylated regions in growth factor receptors and PLCγ, GAP, and other signaling proteins is mediated via a conserved region of approximately 95 amino acids termed *src homology 2* domains (SH2) (Koch et al. 1991; Heldin 1991; Margolis 1992). Hence, SH2 domains represent recognition motifs for specific tyrosine-phosphorylated peptide sequences. SH2 domains are frequently accompanied by another conserved region of approximately 45 amino acids termed SH3 domains, whose function is still poorly understood. High-affinity binding sites for the SH2 domain of PLCγ were identified in the EGF and FGF receptors (Margolis et al. 1990c; Mohammadi et al. 1992; Rotin et al. 1992b). The EGF receptor contains three high-affinity binding sites for the SH2 domains of PLCγ at Y992, Y1068, and Y1173 (Rotin et al. 1992b), whereas the FGF receptor contains one site at Y766 (Mohammadi et al. 1991).

EXPERIMENTAL PROCEDURES

Site-directed mutagenesis was performed according to the manufacturer's protocol (Amersham). A cDNA encoding the full-length human FGF receptor (*flg*) (Dionne et al. 1990) was subcloned in the m13MP19 replicative form using *Bam*HI-*Hin*dIII. A point mutation which changed Tyr-766 to phenylalanine was introduced using the following oligonucleotide: 5′-CCTCCAACCAGGAGTTCCTGGACCTGTCCATG-3′. The cDNAs encoding either wild-type or mutant FGF receptors were subcloned into the eukaryotic expression vector pMJ30 under the control of the adenovirus major late promotor and cytomegalovirus enhancer. L-6 rat myoblasts lacking endogenous FGF receptors were transfected with 0.5 μg of pSV2*neo* and 20 μg of wild-type or mutant FGF receptor expression vectors utilizing the calcium phosphate precipitation method. Clones were isolated after 2–3 weeks' selection in G418 (GIBCO) and screened for expression of FGF receptor by immunoprecipitation with anti-*flg* 1A and immunoblotting with anti-*flg* 3B antibodies. Binding experiments with ^{125}I-labeled acidic FGF (aFGF) followed by Scatchard analysis showed that the L-6 cells expressing wild-type FGF receptors possess approximately 4.5×10^5 receptors/cell and that Y766F cells possess approximately 4.2×10^5 receptors/cell. Transfected L-6 cells were stimulated with aFGF (100 ng/ml) for 5 minutes at 37°C. Cells were washed two times with phosphate-buffered saline (PBS) and scraped into 0.5 ml of lysis buffer (20 mM HEPES at pH 7.5, 150 mM NaCl, 10% glycerol, 1% Triton X-100, 1.5 mM $MgCl_2$, 1 mM EGTA, 10 μg/ml aprotonin, 10 μg/ml leupeptin, 1 mM phenylmethylsulfonyl fluoride [PMSF], 50 mM sodium fluoride, 10 mM sodium pyrophosphate, 200 μM sodium orthovanadate). The lysates were subjected to immunoprecipitation with anti-FGF receptor (anti-*flg* 1A) antibodies immobilized on protein A–Sepharose beads. The immunocomplexes were washed three times with HNTG (20 mM HEPES at pH 7.5, 150 mM NaCl, 0.1% Triton X-100, 10% glycerol), 3× Laemmli buffer was added, and the samples were boiled for 5 minutes. Following analysis by SDS-PAGE, the proteins were transferred electrophoretically to nitrocellulose membranes and immunoblotted with either anti-*flg* 3B (1:100) or antiphosphotyrosine (1:100) antibodies. The blots were treated with ^{125}I-labeled protein A and analyzed by autoradiography. The rabbit anti-FGF receptor serum (anti-*flg* 3B) was generated against a synthetic peptide from the kinase-insert region (residues 580–586) of human FGF receptor (*flg*). Anti-*flg* 1A serum was generated against a synthetic peptide derived from the carboxy-terminal tail of human FGF receptor (*flg* residues 808–822).

Immunoprecipitation experiments. Serum-starved L-6 myoblasts expressing either the wild-type or Y766F receptor mutant were stimulated with (+) or without (−) aFGF (100 ng/ml) for 5 minutes at 37°C and lysed, and the soluble lysates were immunoprecipitated with anti-FGF receptor (anti-*flg* 3B) antibodies. Following SDS-PAGE, the samples were immunoblotted with either anti-PLCγ or anti-FGF receptor antibodies (anti-*flg* 1A). In parallel, lysates from aFGF-treated and unstimulated cells were subjected to immunoprecipitation analysis with anti-PLCγ antibodies, analyzed by SDS-PAGE, and immunoblotted with either anti-PLCγ or antiphosphotyrosine antibodies. PLCγ rabbit antiserum was raised against a synthetic carboxy-terminal peptide of rat PLCγ (residues 1256–1274).

Determination of PI hydrolysis. L-6 myoblasts expressing either the wild-type or the Y766F receptor mutant were labeled with [^{3}H]myoinositol (2 μCi/ml) in Dulbecco's modified Eagle's medium (DMEM) containing 0.5% fetal bovine serum (FBS) for 24 hours and incubated in DMEM containing 20 mM LiCl for 20 minutes prior to addition of aFGF (100 ng/ml) for an additional 30 minutes at 37°C. Cells were extracted with 5% perchloric acid, and inositol phosphate formation was measured according to published procedures (Margolis et al. 1990c).

Thymidine incorporation. L-6 cells were seeded in 96-well plates (2×10^4 cells/well), and 24 hours later, the medium was changed to DMEM containing 0.1% FBS. After 48 hours of serum starvation, the medium was replaced by medium containing 200 ng/ml aFGF or, as a control, 10% fetal calf serum. After stimulation for 24 hours, cells were incubated with [^{3}H]thymidine (0.5 μCi/ml) for 16 hours at 37°C. The cells were washed with PBS, trypsinized, and harvested using a PHD Cell Harvester (Cambridge Technologies, Inc.), and the amount of incorporated [^{3}H]thymidine was quantitated by liquid scintillation counting (LKB).

RESULTS AND DISCUSSION

In response to stimulation by PDGF, EGF, FGF, and other growth factors, PLCγ becomes phosphorylated on tyrosine, serine, and threonine residues (Margolis et al. 1989; Meisenhelder et al. 1989; Wahl et al. 1989). Detailed analysis utilizing transfected cells expressing various receptor mutants combined with mutational analysis of PLCγ provided detailed information concerning the mechanism of PLCγ phosphorylation and activation. Figure 1 depicts a schematic diagram describing the steps involved in growth factor stimulation of PI hydrolysis. The binding of EGF or other growth factor receptors to their specific cell-surface receptors leads to stimulation of receptor tyrosine autophosphorylation. This creates a specific binding site for PLCγ that binds via its SH2 domains to at least three tyrosine autophosphorylation sites in the carboxy-terminal tail of EGF receptor (Y992, Y1068, and Y1173). The binding of the SH2 domains was shown to be crucial for efficient tyrosine phosphor-

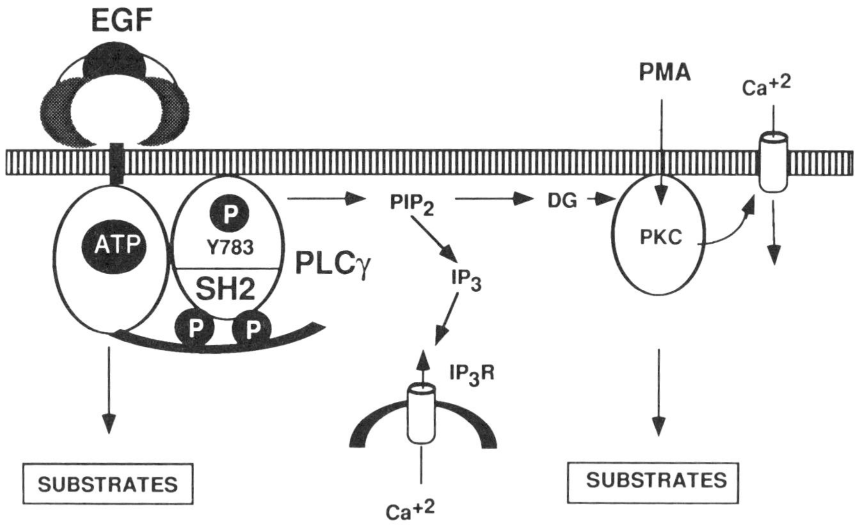

Figure 1. A model for the mechanism of activation of PI hydrolysis by EGF and other growth factors. EGF binding is followed by tyrosine autophosphorylation of five tyrosine residues in the carboxy-terminal tail of the EGF receptor. PLCγ binds to the tyrosine-phosphorylated tail (at positions Y992, Y1068, and Y1173), facilitating tyrosine phosphorylation of PLCγ on several sites. Phosphorylation of Y783 is essential for activation of phospholipase activity generating diacylglycerol and IP_3, which in turn activate PKC and release Ca^{++} from internal stores, respectively.

ylation of PLCγ by increasing the affinity (decreasing the K_m) for tyrosine phosphorylation by the EGF receptor kinase (Rotin et al. 1992a). EGF, PDGF, and other receptor tyrosine kinases phosphorylate PLCγ on several tyrosine residues both in vitro and in vivo. Elimination of PLCγ Y783 by site-directed mutagenesis prevented PDGF-induced stimulation of PI hydrolysis, indicating that phosphorylation on Y783 is essential for stimulation of PLC activity (Nishibe et al. 1990; Kim et al. 1991). Activation of PLCγ by tyrosine phosphorylation leads to the generation of two second messengers, diacylglycerol (DG) and IP_3, that are responsible for activation of PKC and release of Ca^{++} from internal stores, respectively (Fig. 1).

PI Hydrolysis Is Not Required for FGF-induced Mitogenesis

We have identified a tyrosine autophosphorylation site in the carboxy-terminal tail of FGF receptor that functions as a high-affinity binding site for the SH2 domain of PLCγ (Mohammadi et al. 1992). L-6 myoblasts devoid of endogenous FGF receptors were transfected with eukaryotic expression vector encoding either wild-type FGF receptor or FGF receptor in which Tyr-766 was replaced by a phenylalanine residue (Y766F). Positive clones were selected and screened for expression of FGF receptor by immunoprecipitation experiments and by binding experiments with ^{125}I-labeled aFGF. The stably transfected cell lines expressing either the wild-type or Y766F FGF receptor mutant were treated with aFGF, lysed, and subjected to immunoprecipitation with either anti-FGF receptor or anti-PLCγ antibodies, followed by SDS-PAGE and immunoblotting with either antiphosphotyrosine or anti-PLCγ antibodies. The results presented in Figure 2 show that the wild-type FGF receptor is able to stimulate tyrosine phosphorylation of PLCγ, whereas tyrosine phosphorylation of PLCγ was not detected in cells expressing the Y766F mutant. These results indicate that autophosphorylation at Y766 is essential for tyrosine phosphorylation of PLCγ. We have also shown that the association between PLCγ and the FGF receptor requires phosphorylation at Y766 (Fig. 2) (Mohammadi et al. 1992). We next examined the capacity of aFGF to stimulate PI hydrolysis in the transfected cell lines. Addition of aFGF to the L-6 transfectants expressing wild-type receptor was followed by a large increase in PI hydrolysis, whereas no stimulation of PI hydrolysis was detected in cells expressing the Y766F receptor mutant (Fig. 3). These results indicate that elimination of Y766 of FGF receptor abolishes association between FGF receptor and PLCγ and prevents tyrosine phosphorylation of PLCγ and PI hydrolysis. Nevertheless, the elimination of PI hydrolysis did not influence the capacity of aFGF to stimulate DNA synthesis in the transfected L-6 myoblasts. aFGF was able to stimulate DNA synthesis when added to the L-6 transfectants expressing either wild-type or Y766F mutant receptors (Fig. 3). It therefore appears that PI hydrolysis is not required for FGF-induced mitogenesis in L-6 myoblasts. This notion is consistent with studies showing that other growth factors such as CSF-1 are able to stimulate DNA synthesis without activation of PI hydrolysis (Downing et al. 1989). It is therefore possible that PI hydrolysis is not required for mitogenesis stimulated by growth factor receptors with tyrosine kinase activity. Nevertheless, FGF-induced PI hydrolysis might play a role in non-mitogenic effects of FGF. FGF-induced PI hydrolysis might participate in the control of cell shape as it has been shown that phosphatidylinositol phosphate-2

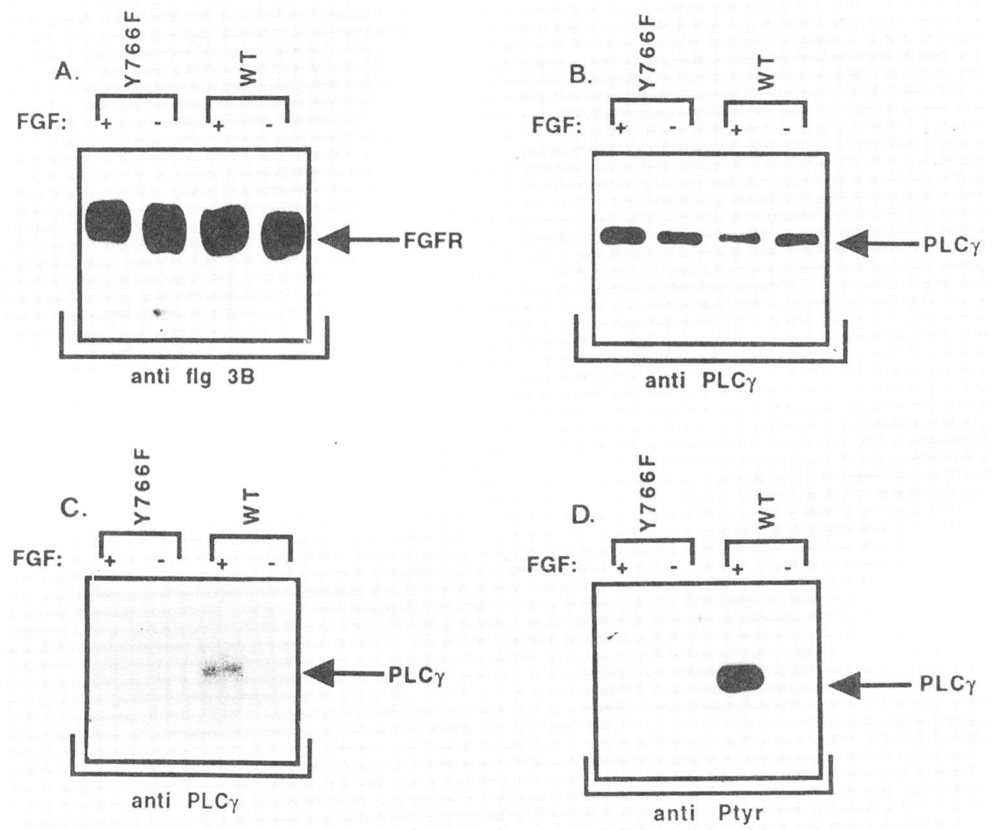

Figure 2. Point mutation at Y766 of the FGF receptor eliminates association between FGF receptor and PLCγ and tyrosine phosphorylation of PLCγ. Transfected L-6 myoblasts expressing either the wild-type or Y766F FGF receptor mutant were treated with aFGF and subjected to immunoprecipitation and immunoblotting analysis with different antibodies as described in experimental procedures. (*A*) Immunoprecipitation with anti-FGF receptor antibodies followed by immunoblotting with anti-FGF receptor antibodies (anti-*flg* 3B); (*B*) immunoprecipitation with anti-PLCγ antibodies followed by immunoblotting with anti-PLCγ antibodies; (*C*) immunoprecipitation with anti-FGF receptor antibodies followed by immunoblotting with anti-PLCγ antibodies; (*D*) immunoprecipitation with anti-PLCγ antibodies followed by immunoblotting with anti-phosphotyrosine antibodies.

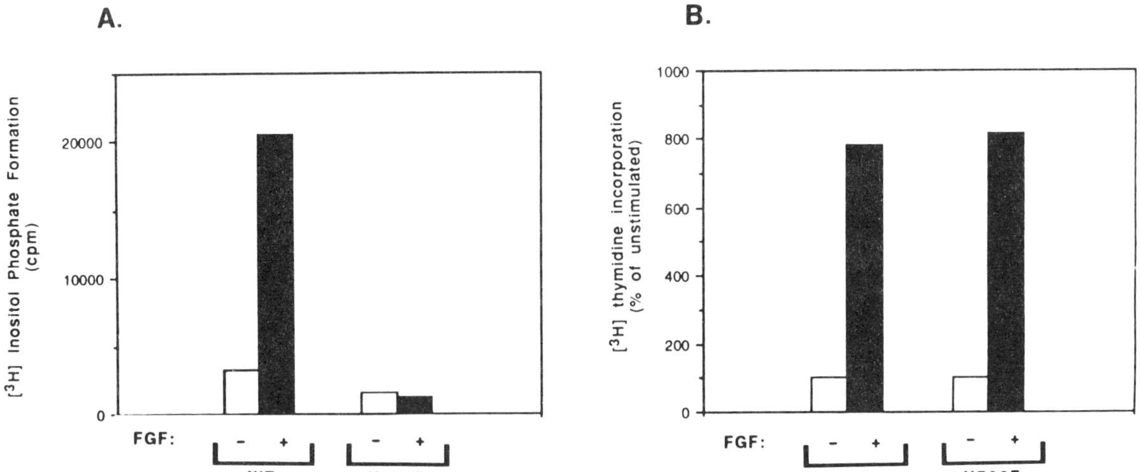

Figure 3. FGF-induced PI hydrolysis is not required for stimulation of DNA synthesis in L-6 myoblasts. Transfected L-6 myoblasts expressing either the wild-type or Y766F FGF receptor mutant were treated with aFGF and analyzed for stimulation of PI hydrolysis (*A*) or thymidine incorporation (*B*). (*A*) Inositol phosphate accumulation in unstimulated (□) or aFGF-stimulated cells (■); (*B*) thymidine incorporation in unstimulated (□) or aFGF-stimulated cells (■).

(PIP_2) is able to bind to the actin-binding protein profilin and thus be involved in the control of actin assembly (Goldschmidt-Clermont et al. 1991). PI hydrolysis may also play a role in the control of cell differentiation. It was shown that FGF induces the formation of ventral mesoderm in early amphibian mesoderm, and since PI hydrolysis is crucial for this process, it is possible that FGF-induced mesoderm induction is mediated via FGF-induced PI hydrolysis (Berridge et al. 1989; Paterno et al. 1989; Masalanski et al. 1992).

SH2-containing Proteins Control the Specificity of Receptor Signaling Pathways

We have developed an expression/cloning method called CORT by using the tyrosine-phosphorylated carboxy-terminal tail of EGF receptor as a probe for cloning of proteins that bind to the activated EGF receptor (Skolnik et al. 1991; Margolis et al. 1992b). Using CORT or other more conventional approaches, we have identified numerous SH2-containing proteins. Figure 4 depicts a list of SH2-containing proteins that appear to play a role in signaling pathways activated by growth factor receptors with tyrosine kinase activities and by cytoplasmic tyrosine kinases. On the basis of their primary structures, it is possible to divide these proteins into two main classes. Type I defines molecules that contain, in addition to the SH2 (and usually also SH3) domains, distinct enzymatic activities such as phospholipase activity (PLCγ), tyrosine kinase activity ($pp60^{c\text{-}src}$), and putative GDP/GTP exchanger (*vav*) (Katzav et al. 1989; Adams et al. 1992; Bustelo et al. 1992; Margolis et al. 1992a). It was shown that the SH2 domains of this class of signaling molecules are responsible for their interaction with tyrosine-phosphorylated receptors and other tyrosine-phosphorylated proteins.

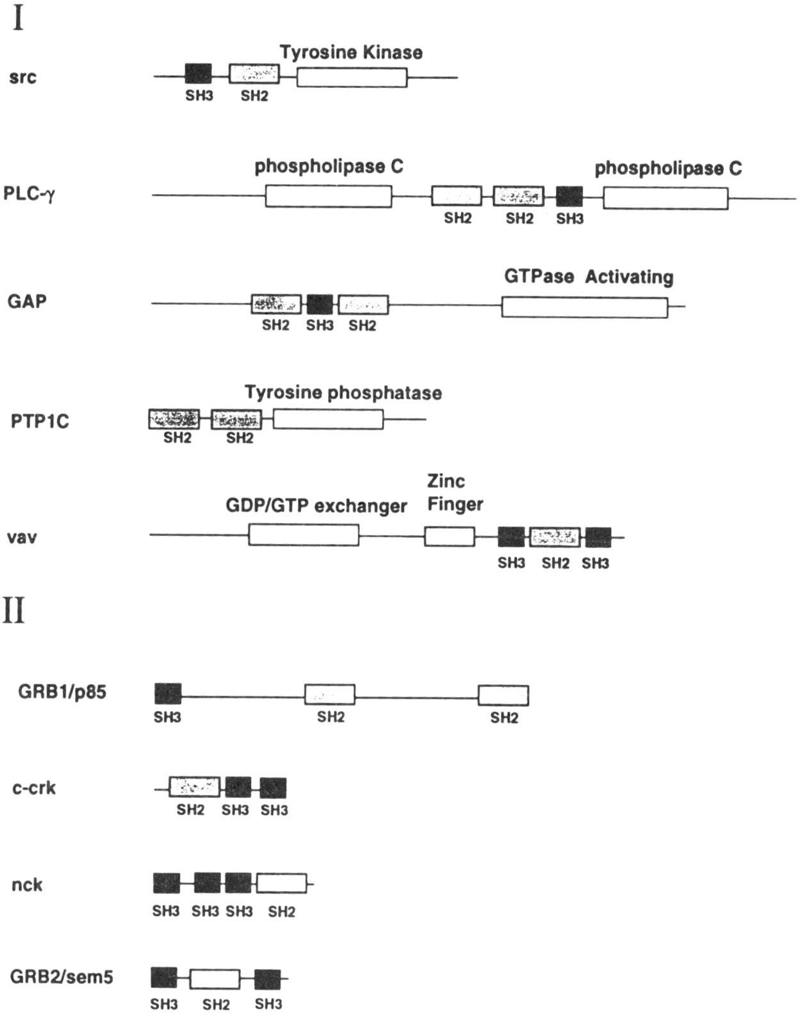

Figure 4. Signaling proteins containing SH2 domains. (I) Proteins that in addition to SH2 domains contain distinct enzymatic activities. (*src*) Tyrosine kinase activity; (PLCγ) phospholipase C activity; (GAP) GTPase-activating protein of *ras*; (PTP1C) putative tyrosine phosphatase activity (Shen et al. 1991); (*vav*) putative GDP/GTP exchanger activity. (II) SH2-containing proteins that serve as adapters or regulatory subunits of downstream signaling proteins. (GRB1/p85) Regulatory subunit of PI3 kinase; (c-*crk*) cellular homolog of v-*crk* oncogene; (*nck*) a protein phosphorylated by EGF and PDGF receptor; (GRB2/*sem-5*) protein involved in EGF receptor signal transduction in human and nematode, respectively.

It is thought that upon tyrosine phosphorylation or by virtue of their proximity and interaction with other proteins, these SH2-containing proteins are able to exert their distinct enzymatic activities. PLCγ can be considered as a prototype for this class of signaling proteins, and as mentioned above, the binding of the SH2 domain of PLCγ is followed by tyrosine phosphorylation that is essential for enhancement of its enzymatic activity (Kim et al. 1991; Rotin et al. 1992a).

The second class of SH2-containing proteins including *crk*, *nck*, and growth factor receptor bound (GRB2)/*sem-5* (Mayer et al. 1988; Lehmann et al. 1990; Clark et al. 1992; Lowenstein et al. 1992), are composed of virtually only SH2 and SH3 domains (Fig. 4). These proteins are thought to function as adapters or regulatory components of their specific catalytic subunits. For example, it is thought that the PI3-kinase-associated p85 is the regulatory subunit of p110, which is likely to function as the catalytic subunit of PI3 kinase (Escobedo et al. 1991; Otsu et al. 1991; Skolnik et al. 1991; Hu et al. 1992; McGlade et al. 1992). The oncogene product of the avian CT10 virus $p47^{gag\text{-}crk}$ is composed of only SH2 and SH3 domains. v-*crk* was shown to be associated with tyrosine-phosphorylated proteins and its expression leads to enhancement in total cellular tyrosine phosphorylation. Moreover, mutations in the SH2 domain of v-*crk* decrease tyrosine phosphorylation and abolish virus-induced transformation (Mayer et al. 1988, 1992; Mayer and Hanafusa 1990). GRB2 protein is another member of the type II SH2-containing proteins. GRB2 contains one SH2 domain flanked by two SH3 domains and has a wide tissue and cell distribution (Lowenstein et al. 1992). Like other signaling proteins, GRB2 associates with EGF and PDGF receptors in a ligand-dependent manner, both in vitro and in living cells. The association between GRB2 and growth factor receptors is mediated by the SH2 domain, is strictly dependent on receptor tyrosine autophosphorylation, and involves a direct interaction between GRB2 and the tyrosine-phosphorylated receptors. Nevertheless, both in vitro and in living cells, GRB2 does not appear to be phosphorylated on tyrosine, serine, or threonine residues at physiological levels of expression to any significant extent. GRB2 behaves like the PI3-kinase-associated p85 protein (Hu et al. 1992). Recent studies suggest that p85 is not tyrosine-phosphorylated at physiological levels of expression and that the binding of p85 to the tyrosine-phosphorylated PDGF receptor via its SH2 domain is essential for activation of PI3 kinase (Hu et al. 1992; McGlade et al. 1992). Only upon overexpression could PDGF-induced tyrosine phosphorylation of p85 be detected (Hu et al. 1992). Recent studies indicate that GRB2 represents the human homolog of the *Caenorhabditis elegans* signal transduction protein called *sem-5* (Clark et al. 1992; Lowenstein et al. 1992). It has been shown that *sem-5* and two other genes called *let-23* (EGF-receptor-like) and *let-60* (*ras*-like) lie along the same signal transduction pathway controlling *C. elegans* vulval induction (Aroian et al. 1990; Horvitz and Sternberg 1991; Clark et al. 1992). The structural similarity between GRB2 and *sem-5* is also reflected in the analogous functions of these two proteins. We have recently shown that GRB2 is able to rescue *sem-5* mutations in *C. elegans* (M.J. Stern et al., unpubl.) and that *sem-5* protein is able to bind specifically to the tyrosine-autophosphorylated human EGF receptor (E.R. Daly, unpubl.). On the basis of the genetics of *C. elegans* and the properties of GRB2, it is possible to propose a model for the role of GRB2/*sem-5* in signal transduction pathways mediated by growth factor receptors (Fig. 5). We have shown that GRB2 and *sem-5* are able to bind the tyrosine-autophosphorylated EGF receptor. It is therefore likely that *sem-5* will bind tyrosine-phosphorylated *let-23* via its SH2 domain according to the scheme presented in Figure 1 for the binding of PLCγ to the EGF receptor. Since loss-of-function mutations in *let-60* cause a phenotype similar to that of mutations in either *let-23* and *sem-5*, and since activated *ras* can rescue *let-23* and *sem-5* mutations, it is reasonable to assume that *let-60*/*ras* functions downstream from the EGF receptor and GRB2 and that GRB2 is involved in regulation of *ras* activity. Moreover, intact GRB2 protein and Ha-*ras* cooperate to stimulate DNA synthesis when microinjected into rat fibroblasts (Lowenstein et al. 1992). Hence, GRB2 and its as yet unidentified associated protein(s) may control growth-factor-induced *ras* signaling by stimulating the activity of a GDP/GTP exchange factor, by inhibiting a *ras* GTPase activity, or both.

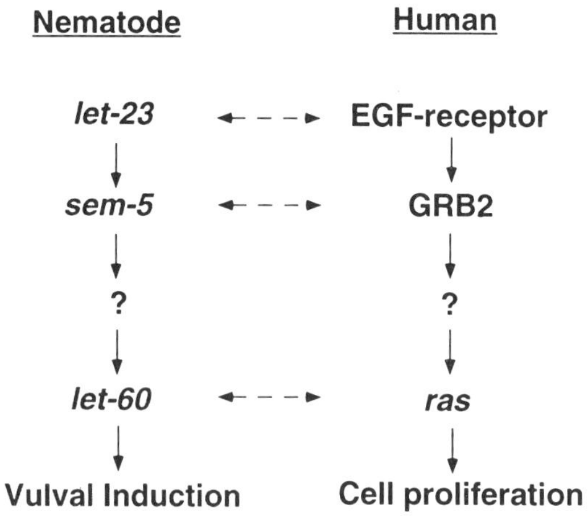

Figure 5. A model for the role of GRB2 and *sem-5* in *ras* signaling. In response to EGF stimulation, GRB2 (or *sem-5*) binds to the EGF receptor (or *let-23*) via its SH2 domain. *ras* (or *let-60*) acts downstream, leading to either cell proliferation or vulval induction. GRB2 (or *sem-5*) may control growth-factor-induced *ras* (*let-60*) signaling by stimulating a GDP/GTP exchange factor, by inhibiting a *ras* GTPase activity, or both.

ACKNOWLEDGMENTS

This work was supported by a grant from SUGEN (J.S.) and from HFSPO (J.S. and A.U.). B.M. is a Lucille P. Markey Scholar and a Kaplan Cancer Scholar.

REFERENCES

Adams, J.M., H. Houston, J. Allen, T. Lints, and R. Harvey. 1992. The hematopoetic expressed vav protooncogene shares homology with the dbl GDP-GTP exchange factor, bcr gene and a yeast gene CDC24 involved in cytoskeletal organization. *Oncogene* **7:** 618.

Aroian, R.V., M. Koga, J.E. Mendel, Y. Ohshima, and P.W. Sternberg. 1990. The *let-23* gene necessary for *Caenorhabditis elegans* vulval induction encodes a tyrosine kinase of the EGF receptor subfamily. *Nature* **348:** 693.

Berridge, M.J., C.P. Downes, and M.R. Hanley. 1989. Neural and developmental actions of Lithium: A unifying hypothesis. *Cell* **59:** 411.

Burgess, W.H., C.A. Dionne, J. Kaplow, R. Mudd, R. Frieser, A. Zilberstein, J. Schlessinger, and M. Jaye. 1990. Characterization and cDNA cloning of phospholipase C-γ, a major substrate for heparin-binding growth factor 1 (acidic fibroblast growth factor)-activated tyrosine kinase. *Mol. Cell. Biol.* **10:** 4770.

Bustelo, X.R., J.A. Ledbetter, and M. Barbacid. 1992. Product of vav proto-oncogene defines a new class of tyrosine protein kinase substrates. *Nature.* **356:** 68.

Cantley, L.C., K. Auger, C. Carpenter, B. Duckworth, A. Graziani, R. Kapeller, and S. Soltoff. 1991. Oncogenes and signal transduction. *Cell* **64:** 281.

Clark, S.G., M.J. Stern, and H.R. Horvitz. 1992. *C. elegans* cell-signaling gene *sem-5* encodes a protein with SH2 and SH3 domains. *Nature* **356:** 340.

Damon, D.H., P.A. D'Amore, and J.A. Wagner. 1990. Nerve growth factor and fibroblast growth factor regulate neurite outgrowth and gene expression in PC12 cells via both protein kinase C- and cAMP-independent mechanism. *J. Cell Biol.* **110:** 1333.

Dionne, C.A., G. Crumley, F. Bellot, J.M. Kaplow, G. Searfoss, M. Ruta, W.H. Burgess, M. Jaye, and J. Schlessinger. 1990. Cloning and expression of two distinct high affinity receptors cross reacting with acidic and basic fibroblast growth factors. *EMBO J.* **9:** 2685.

Downing, J.R., B. Margolis, A. Zilberstein, R. Ashum, A. Ullrich, C. Sherr, and J. Schlessinger. 1989. Phospholipase Cγ, a substrate for PDGF-receptor kinase is not phosphorylated on tyrosine during the mitogenic response to CSF1. *EMBO J.* **8:** 3345.

Escobedo, J.A., S. Navankasattusas, W.M. Kavanaugh, D. Milfay, V.A. Fried, and L.T. Williams. 1991. cDNA cloning of a novel 85 kd protein that has SH2 domains and regulates binding of PI3-kinase to the PDGF β-receptor. *Cell* **65:** 75.

Goldschmidt-Clermont, P.J., J.W. Kim, L.M. Machesky, S.G. Rhee, and T.D. Pollard. 1991. Regulation of phospholipase C-γ1 by profilin and tyrosine phosphorylation. *Science* **251:** 1231.

Heldin, C.H. 1991. SH2 domains: Elements that control protein interactions during signal transduction. *Trends Biol. Sci.* **16:** 450.

Horvitz, H.R. and P.W. Sternberg. 1991. Multiple intercellular signaling systems control the development of the *Caenorhabditis elegans* vulva. *Nature* **351:** 535.

Hu, P., B. Margolis, E.Y. Skolnik, R. Lammers, A. Ullrich, and J. Schlessinger. 1992. Interaction of phosphatidylinositol 3-kinase-associated p85 with epidermal growth factor and platelet-derived growth factor receptors. *Mol. Cell Biol.* **12:** 981.

Kaplan, D.R., B.L. Hempstead, D. Martin-Zanca, M.V. Chao, and L.F. Parada. 1991. The *trk* proto-oncogene product: A signal transducing receptor for nerve growth factor. *Science* **252:** 554.

Kashishian, A., A. Kazlauskas, and J.A. Cooper. 1992. Phosphorylation sites in the PDGF receptor with different specificities for binding GAP and PI3 kinase in vivo. *EMBO J.* **11:** 1373.

Katzav, S., D. Martin-Zanca, and M. Barbacid. 1989. vav, a novel human oncogene derived from a locus ubiquitously expressed in hematopoietic cells. *EMBO J.* **8:** 2283.

Kazlauskas, A. and J. Cooper. 1989. Autophosphorylation of the PDGF receptor in the kinase insert region regulates interactions with cell proteins. *Cell* **58:** 1121.

———.1990. Phosphorylation of the PDGF receptor β subunit creates a tight binding site for phosphatidylinositol 3 kinase. *EMBO J.* **9:** 3279.

Kim, H.K., J.W. Kim, A. Zilberstein, B. Margolis, J.G. Kim, J. Schlessinger, and S.G. Rhee. 1991. PDGF stimulation of inositol phospholipid hydrolysis requires PLC-γ1 phosphorylation on tyrosine residues 783 and 1254. *Cell* **65:** 435.

Klein, R., S. Jing, V. Nanduri, E. O'Rourke, and M. Barbacid. 1991. The *trk* proto-oncogene encodes a receptor for nerve growth factor. *Cell* **66:** 189.

Koch, C.A., D. Anderson, M.F. Moran, C. Ellis, and T. Pawson. 1991. SH2 and SH3 domains: Elements that control interactions of cytoplasmic signaling proteins. *Science* **252:** 668.

Kypta, R.M., Y. Goldberg, E.T. Ulug, and S.A. Courtneidge. 1990. Association between the PDGF receptor and members of the src family of tyrosine kinases. *Cell* **62:** 481.

Lehmann, J., G. Riethmuller, and J. Johnson. 1990. Nck, a melanoma cDNA encoding a cytoplasmic protein consisting of the src homology units SH2 and SH3. *Nucleic Acids Res.* **18:** 1048.

Lowenstein, E., R. Daly, A.G. Batzer, W. Li, B. Margolis, R. Lammers, A. Ullrich, E.Y. Skolnik, D. Bar-Sagi, and J. Schlessinger. 1992. The SH2 and SH3 domain containing protein 6RB2 links receptor tyrosine kinases to *ras* signaling. *Cell* **70:** 431.

Margolis, B. 1992. Proteins with SH2 domains: Transducers in the tyrosine kinase signaling pathway. *Cell Growth Differ.* **3:** 73.

Margolis, B., F. Bellot, A. Honegger, A. Ullrich, J. Schlessinger, and A. Zilberstein. 1990a. Tyrosine kinase activity is essential for the association of phospholipase C-γ with the epidermal growth factor receptor. *Mol. Cell. Biol.* **10:** 435.

Margolis, B., P. Hu, S. Katzav, W. Li, J.M. Oliver, A. Ullrich, A. Weiss, and J. Schlessinger. 1992a. Tyrosine phosphorylation of vav protooncogene product containing SH2 and SH3 domain and transcription factors motifs. *Nature* **356:** 71.

Margolis, B., O. Silvennoinen, F. Comoglio, C. Roonprapunt, E. Skolnik, A. Ullrich, A. Weiss, and J. Schlessinger. 1992b. High efficiency expression/cloning of EGF-receptor binding proteins with SH2 domains. *Proc. Natl. Acad. Sci.* (in press).

Margolis, B., N. Li, A. Koch, M. Mohammadi, D.R. Hurwitz, A. Zilberstein, A. Ullrich, T. Pawson, and J. Schlessinger. 1990b. The tyrosine phosphorylated carboxyterminus of the EGF receptor is a binding site for GAP and PLC-γ. *EMBO J.* **9:** 4375.

Margolis, B., S.G. Rhee, S. Felder, M. Mervic, R. Lyall, A. Levitski, A. Ullrich, A. Zilberstein, and J. Schlessinger. 1989. EGF induces tyrosine phosphorylation of phospholipase C-II: A potential mechanism for EGF receptor signaling. *Cell* **57:** 1101.

Margolis, B., A. Zilberstein, C. Franks, S. Felder, S. Kremer, A. Ullrich, S.G. Rhee, K. Skorecki, and J. Schlessinger. 1990c. Effect of phospholipase Cγ overexpression on PDGF-induced second messengers and mitogenesis. *Science* **248:** 607.

Maslanski, J.A., L.A. Leshko, and W.B. Busa. 1992. Lithium-sensitive production of inositol phosphates during amphibian embryonic mesoderm induction. *Science* **256:** 243.

Mayer, B.J. and H. Hanafusa. 1990. Association of the v-crk oncogene product with phosphotyrosine-containing proteins and protein kinase activity. *Proc. Natl. Acad. Sci.* **87:** 2638.

Mayer, B.J., M. Hamaguchi, and H. Hanafusa. 1988. A novel viral oncogene with structural similarity to phospholipase C. *Nature* **332:** 272.

Mayer, B.J., P.K. Jackson, R.A. Van Etten, and D. Baltimore. 1992. Point mutations in the abl SH2 domain coordinately impair phosphotyrosine binding in vitro and transforming activity in vivo. *Mol. Cell. Biol.* **12:** 609.

McGlade, C.J., C. Ellis, M. Reedjik, D. Anderson, G. Mbamalu, A. Reith, G. Panayotou, P. End, A. Bernstein, A. Kazlauskas, M. Waterfield, and T. Pawson. 1992. SH2 domains of signaling proteins control specificity of binding to growth factor receptors. *Mol. Cell. Biol.* **12:** 991.

Meisenhelder, J., P.G. Suh, S.G. Rhee, and T. Hunter. 1989. Phospholipase C-γ is a substrate for the PDGF and EGF receptor protein tyrosine kinases in vivo and in vitro. *Cell* **57:** 1109.

Mohammadi, M., C.A. Dionne, W. Li, N. Li, T. Spivak, A.M. Honegger, M. Jaye, and J. Schlessinger. 1992. A point mutation at tyr 766 of FGF-receptor eliminates phosphatidyl hydrolysis without affecting mitogenesis. *Nature* (in press).

Mohammadi, M., A. Honegger, D. Rotin, R. Fischer, F. Bellot, W. Li, C. Dionne, M. Jaye, M. Rubinstein, and J. Schlessinger. 1991. A tyrosine-phosphorylated carboxy-terminal peptide of the fibroblast growth factor receptor (*flg*) is a binding site for the SH2 domain of phospholipase C-γ1. *Mol. Cell. Biol.* **11:** 5068.

Nishibe, S., M. Wahl, S. Hernandez-Sotomayor, N. Tonks, S.G. Rhee, and G. Carpenter. 1990. Increase of the catalytic activity of phospholipase C-γ1 by tyrosine phosphorylation. *Science* **250:** 1253.

Otsu, M., I. Hiles, I. Gout, M. Fry, F. Ruiz-Larrea, G. Panayotou, A. Thompson, R. Dhand, J. Hsuan, N. Totty, A. Smith, S. Morgan, A. Courtneidge, P. Parker, and M. Waterfield. 1991. Characterization of two 85 kd proteins that associate with receptor tyrosine kinases, middle-T/$pp60^{c\text{-}src}$ complexes, and PI3-kinase. *Cell* **65:** 91.

Paterno, G.D., L.L. Gillepsie, M.S. Dixon, J.M.W. Slack, and J.K. Heath. 1989. Mesoderm inducing properties of Int-2 and kFGF: Two oncogene encoded growth factors related to FGF. *Development* **106:** 79.

Reedijk, M., X. Liu, P. van der Geer, K. Letwin, M.D. Waterfield, T. Hunter, and T. Pawson. 1992. Tyr721 regulates specific binding of the CSF-1 receptor kinase insert to PI 3′-kinase SH2 domains: A model for SH2-mediated receptor-target interactions. *EMBO J.* **11:** 1365.

Rotin, D., A.M. Honegger, B. Margolis, A. Ullrich, and J. Schlessinger. 1992a. Presence of SH2 domains of phospholipase $C\gamma_1$ enhances substrate phosphorylation by increasing the affinity toward the EGF-receptor. *J. Biol. Chem.* **267:** 9678

Rotin, D., B. Margolis, M. Mohammadi, R.M. Daly, G. Daum, N. Li, E. Fischer, W. Burgess, A. Ullrich, and J. Schlessinger. 1992b. SH2 domains prevent tyrosine dephosphorylation of the EGF receptor: Identification of tyr 992 as the high-affinity binding site for SH2 domains of phospholipase Cγ. *EMBO J.* **11:** 559.

Schlessinger, J. 1988. Signal transduction by allosteric receptor oligomerization. *Trends Biochem. Sci.* **13:** 443.

Shen, S.-H., L. Bastien, B.I. Posner, and P. Chretien. 1991. A protein-tyrosine phosphatase with sequence similarity to the SH2 domain of the protein-tyrosine kinases. *Nature* **352:** 736.

Skolnik, E.Y., B. Margolis, M. Mohammadi, E. Lowenstein, R. Fischer, A. Drepps, A. Ullrich, and J. Schlessinger. 1991. Cloning of PI3 kinase-associated p85 utilizing a novel method for expression/cloning of target proteins for receptor tyrosine kinases. *Cell* **65:** 83.

Stahl, M., C. Ferenz, K. Kelleher, R. Kriz, and J. Knopf. 1988. Sequence similarity of phospholipase C with the non-catalytic region of src. *Nature* **332:** 269.

Suh, P.G., S.H. Ryu, K.H. Moon, H.W. Suh, and S.G. Rhee. 1988. Inositol phospholipid-specific phospholipase C: Complete cDNA and protein sequences and sequence homology to tyrosine kinase-related oncogene products. *Proc. Natl. Acad. Sci.* **85:** 5419.

Togari, A., G. Dickens, H. Kuzuya, and G. Guroff 1985. The effect of fibroblast growth factor on PC12 cells. *J. Neurosci.* **5:** 307.

Trahey, M.G., G. Wong, R. Halenback, B. Rubinfield, G. Martin, M. Ladner, C.M. Long, W.J.. Crosier, K. Watt, K. Koths, and F. McCormick. 1988. Molecular cloning of two GAP complementary DNA from human placenta. *Science* **242:** 1696.

Ullrich, A. and J. Schlessinger. 1990. Signal transduction by receptors with tyrosine kinase activity. *Cell* **61:** 203.

Vogel, U.S., R.A. Dixon, M.D. Schaber, R.E. Diehl, M.S. Marshall, E.M. Scolnick, J.S. Sigal, and J.D. Gibbs. 1988. Cloning of bovine GAP and its interaction with oncogenic ras. *Nature* **335:** 90.

Wagner, J.A. 1991. The fibroblast growth factors: An emerging family of neural growth factors. *Curr. Top. Microbiol. Immunol.* **165:** 95.

Wahl, M.I., S. Nishibe, P.-G. Suh, S.G. Rhee, and G. Carpenter. 1989. Epidermal growth factor stimulates tyrosine phosphorylation of phospholipase C-II independently of receptor internalization and extracellular calcium. *Proc. Natl. Acad. Sci.* **86:** 1568.

Phosphatidylinositol-3 Kinase and Growth Regulation

S.P. Soltoff, C.L. Carpenter, K.R. Auger, R. Kapeller,
B. Schaffhausen,* and L.C. Cantley

*Departments of Physiology and of *Biochemistry, Tufts University School of Medicine, Boston, Massachusetts 02111*

Phosphatidylinositol-3 kinase (PI3 kinase) is a lipid kinase that is activated in the cascade of signal transduction events involved in the initiation of cell growth. It catalyzes the production of a family of phosphoinositides (PI(3)P, PI(3,4)P_2, and PI(3,4,5)P_3) with phosphate at the D-3 position on the inositol ring. Thus, this enzyme differs from PI4 kinase and PI(4)P5-kinase, which are involved in the production of PI(4,5)P_2, the substrate for phospholipase C (PLC). PLC hydrolyzes PI(4,5)P_2 to diacylglycerol and inositol-1,4,5-trisphosphate ($InsP_3$), which elevates intracellular free calcium concentration by releasing calcium from $InsP_3$-sensitive intracellulular stores. However, the PLCs that have been characterized do not hydrolyze the products of the PI3 kinase (Serunian et al. 1989; Lips et al. 1989). Recent work from several laboratories has demonstrated that ligand-activated protein tyrosine kinases form tight complexes with their cytosolic targets, and these complexes remain intact through immunoprecipitation protocols (for review, see Cantley et al. 1991). PI3 kinase was the first cytosolic enzyme found to be associated with activated protein tyrosine kinases. Within the last several years, and continuing to date, a large number of studies have demonstrated that this enzyme nearly ubiquitously associates with protein tyrosine kinases, including growth factor receptor and nonreceptor kinases. Some of these studies are reviewed here. PI3 kinase is only one of several signal transducing proteins that are recruited from the cytosol to the inner face of the plasma membrane in response to the binding of growth factors to their receptors. Other proteins include Ras GAP (GTPase activating protein), PI-specific PLCγ, and c-Raf. The recruitment of these proteins and their involvement in the transduction of cell proliferation has been reviewed recently (Cantley et al. 1991). PI3 kinase has been purified and consists of a heterodimer of 85-kD and 110-kD subunit proteins. The specific role(s) of the D-3 phosphoinositides is not yet known, but activation of PI3 kinase correlates well with the ability of several tyrosine kinases to initiate mitogenesis or cellular transformation. In addition to its involvement in signal transduction events in cell growth, PI3 kinase activity is stimulated by ligands that bind to receptors in terminally differentiated cells and receptors that are involved in the initiation of cellular differentiation. The present report is meant primarily to serve as a review of past and present studies by Cantley and co-workers and recent findings by other investigators. More comprehensive reviews on this and related subjects have been published recently (Carpenter and Cantley 1990; Cantley et al. 1991).

PI3 Kinase Activity Associates with Protein Tyrosine Kinases

Phosphatidylinositol kinase activity was observed to coimmunoprecipitate with polyomavirus middle T antigen (middle T)-specific or pp60$^{c\text{-}src}$-specific antisera of NIH-3T3 cells infected with polyomavirus; immunoprecipitates from mutants of middle T that were defective in transforming activity had reduced levels of PI kinase activity (Whitman et al. 1985; Kaplan et al. 1986). In these early experiments, lipid kinase assays were done by using PI or PI phosphates (PIPs) as substrates, and the products were separated using thin-layer chromatography (TLC) techniques. Subsequent studies demonstrated that there were two different PI kinases in fibroblasts: type-1 kinase (PI3 kinase) and type-2 kinase (a PI4 kinase). Type-1 kinase had a low K_m for ATP (10 μM), was resistant to inhibition by adenosine, and was inhibited by nonionic detergents such as Nonidet P-40 (NP-40). Type-2 kinase activity had a slightly higher K_m for ATP (34 μM), was inhibited by adenosine (K_i = 20 μM adenosine in the presence of 40 μM ATP), and was enhanced by NP-40 detergent (Whitman et al. 1987). Type-1 kinase was the enzyme that was observed to associate with the middle T–pp60$^{c\text{-}src}$ complex and to immunoprecipitate with the platelet-derived growth factor (PDGF) receptor in antiphosphotyrosine (anti-Tyr) immunoprecipitates or in wheat-germ agarose precipitates of PDGF-treated quiescent fibroblasts (Whitman et al. 1987). Two different PI4 kinases were separated from bovine brain (Endemann et al. 1987) and other tissues (for review, see Carpenter and Cantley 1990). One enzyme had properties similar to those of the type-2 PI kinase described above, and the other was designated type 3.

An initial identification of one of the subunits of the PI3 kinase protein was made when it was observed that an enhanced PI kinase activity and the phosphorylation on tyrosine of an 85-kD protein were jointly observed in 3T3 fibroblasts treated with PDGF or transformed by middle T (Kaplan et al. 1987). The monophosphate product of PI3 kinase was identified as PI(3)P and was shown to be distinct from PI(4)P, the product of PI4 kinase (Whitman et al. 1988). In this study, the PI kinases were immunoprecipitated from polyomavirus-transformed fibroblasts, and lipid kinase assays were performed using [^{32}P]ATP as the phosphate donor and PI as the substrate in the presence and absence of NP-40 detergent. The two PIPs migrated similarly on TLC, but when the PIP spots were extracted and deacylated, two distinct products, PI(3)P and PI(4)P, were separated using an anion-exchange high-performance liquid chromatography (HPLC) system.

A family of D-3 phosphorylated lipid products was generated in PDGF-treated smooth muscle cells (Auger et al. 1989). In these studies, lipid kinase assays were performed on anti-Tyr immunoprecipitates of PDGF-treated cells using PI, PI(4)P, and PI(4,5)P_2 as substrates. The resulting products, PI(3)P, PI(3,4)P_2, and PIP_3, were separable by TLC. The deacylated mono- and diphosphate lipids were distinguished from PI(4)P and PI(4,5)P_2 when analyzed on HPLC using a SAX (strong anion exchange) column, and the product of PI(4,5)P_2 phosphorylation was identified as PIP_3, later determined to be PI(3,4,5)P_3. Moreover, when quiescent cells were labeled in vivo with *myo*-[^{3}H]inositol, PDGF treatment produced a rapid rise in PI(3,4)P_2 and PIP_3. Only small changes in PI(4)P, PI(3)P, and PI(4,5)P_2 were observed. This study, and others that followed, demonstrated that growth factors stimulated PI3 kinase to produce a family of novel lipids. Unlike PI(4)P and PI(4,5)P_2, the novel D-3-phosphorylated lipids were not substrates for PLCβ, PLCγ, or PLCδ isolated from rat liver and bovine brain (Serunian et al. 1989). Significant levels of PI(3,4)P_2 and PIP_3 were produced in vivo by subconfluent NIH-3T3 fibroblasts and middle-T-antigen-transformed NIH-3T3 cells at high density; NIH-3T3 cells at high density and cells transfected with a transformation-defective mutant of middle T did not produce detectable levels of these products (Serunian et al. 1990). The rapidity with which the D-3 lipids are produced in response to growth factors and the association of these lipids with growth and transformation suggest that the lipids themselves serve as molecules that are involved in transmitting a growth signal to cells.

PI3 kinase was purified to homogeneity from rat liver and was found to be a heterodimer consisting of 85-kD (p85) and 110-kD (p110) subunits (Carpenter et al. 1990). The purified enzyme had characteristics similar to those of the type-1 kinase activity described previously (Whitman et al. 1987), and the 85-kD subunit was the same protein as that which was tyrosine-phosphorylated and associated with the PDGF receptor and the middle T–pp60$^{c\text{-}src}$ complex (Kaplan et al. 1987). The purified PI3 kinase used PI, PI(4)P, and PI(4,5)P_2 as substrates. The apparent K_m was 4 μM for PI(4,5)P_2, 9 μM for PI(4)P, and 60 μM for PI; the apparent K_m for ATP was 30–60 μM, depending on the lipid substrate (Carpenter et al. 1990).

The cDNA for p85 was cloned, sequenced, and expressed and was found to bind to tyrosine-phosphorylated receptor and nonreceptor tyrosine kinases (Escobedo et al. 1991b; Otsu et al. 1991). Since expressed p85 did not have PI3 kinase activity, but an antibody raised against p85 did immunoprecipitate PI3 kinase activity (Otsu et al. 1991), it was suggested that p110 was the catalytic subunit. p85 may function as a recognition protein to direct PI3 kinase to activated tyrosine kinases (see below).

The SH2 Domain of the p85 Subunit Binds to Receptor Tyrosine Kinases

As mentioned above, PI3 kinase is one of several proteins that have been found to associate with receptor tyrosine kinases upon ligand-induced activation. In addition to the PDGF (β) receptor (PDGFR), PI3 kinase activity associates with the colony-stimulating factor-1 receptor (CSF-1R) (Varticovski et al. 1989); epidermal growth factor (ErbB) receptor (EGFR) (Bjorge et al. 1990); *neu* (ErbB2) receptor (Peles et al. 1992); insulin receptor (Endenmann et al. 1990; Ruderman et al. 1990); insulin-like growth factor-1 (IGF-1) receptor (Kapeller et al. 1991); and gp140^{trk} (Soltoff et al. 1992), the nerve growth factor (NGF) receptor tyrosine kinase (see below). In addition to pp60$^{v\text{-}src}$ and middle T–pp60$^{c\text{-}src}$, PI3 kinase also immunoprecipitates with the following nonreceptor tyrosine kinases: pp130$^{gag\text{-}fps}$ (Fukui et al. 1989); pp160$^{gag\text{-}abl}$ (Varticovski et al. 1991); pp59$^{c\text{-}fyn}$ (Gutkind et al. 1990); and middle T/pp62$^{c\text{-}yes}$ (Fukui et al. 1989).

A number of receptor and nonreceptor protein tyrosine kinases, including the PDGFR, CSF-1R, and middle T, contain tyrosine residues that appear to require phosphorylation in order for PI3 kinase to bind (for review, see Cantley et al. 1991; Escobedo et al. 1991a). These proteins have a highly conserved consensus sequence of phospho-Tyr-Met/Val-X-Met, and this may serve as a recognition motif for p85 binding (below). Phosphopeptides based on this sequence block the in vitro association between PI3 kinase and the PDGFR (Escobedo et al. 1991a) or polyomavirus middle T–pp60$^{c\text{-}src}$ (Auger et al. 1992). Insulin receptor substrate-1 (IRS-1), which binds PI3 kinase and which is phosphorylated on tyrosine by the insulin receptor, also has this motif (Sun et al. 1991). Tyrosine kinases contain multiple tyrosine phosphorylated sites, and there appear to be distinct binding sites for different cytosolic enzymes that are recruited (below).

In common with other proteins that are recruited to protein tyrosine kinases and become phosphorylated on tyrosine, p85 has SH2 (*src* homology 2) and SH3 domains. These regions are made up of highly conserved stretches of amino acids that differ from the kinase domain of *src* and are noncatalytic. The SH2 region serves as a recognition site for the recruitment of cytosolic proteins to activated protein tyrosine kinases, and the SH3 domain may be involved in targeting to cytoskeletal proteins (Koch et al. 1991). p85 has two SH2 domains and one SH3 domain. p85 has an SH2 domain at the carboxyl terminus (C-SH2) and one in the middle of the protein closer to the amino terminus (N-SH2). Investigators recently have focused attention on the binding of the SH2 groups of p85 to various protein tyrosine kinases that are known to interact with PI3 kinase. Two important questions that have been investigated are (1) which SH2 domain of p85 binds to the tyrosine phosphorylated site and (2) whether the SH2 domains of different cytosolic proteins bind to the same phosphotyrosine-containing site. In some of these studies, GST (glutathione *S*-transferase) fusion proteins expressing various portions of p85 were constructed, and their binding to tyrosine kinase proteins was studied.

The binding of p85 to the EGFR and the PDGFR (Hu et al. 1992) and the binding of a GST fusion protein containing either SH2 domain to these receptors was enhanced in ligand-treated cells (Hu et al. 1992; McGlade et al. 1992). This suggests that receptor autophosphorylation is required for recruitment of PI3 kinase to the receptor and is consistent with early studies which indicated that PI3 kinase activity was associated with activated protein tyrosine kinases. A correlation was found between the PI3 kinase activity (in anti-Tyr immunoprecipitates) and the amount of p85 that associated with EGFR and PDGFR (in anti-p85 blots of anti-receptor immunoprecipitates): Lipid kinase activity and the p85-receptor complex were both greater for cells exposed to PDGF than for those exposed to EGF, although the cells contained eight- to tenfold more EGFRs than PDGFRs (Hu et al. 1992). Thus, no correlation was observed between PI3 kinase activity and receptor density. Both SH2 domains of p85 bound to the EGFR as well as the PDGFR, and GST-C-SH2 bound with an affinity similar to that of GST-N-SH2 (Hu et al. 1992; McGlade et al. 1992). However, other investigators have reported that a C-SH2 fusion protein bound to the PDGFR with much higher affinity than did a protein containing N-SH2 (Klippel et al. 1992). The C-SH2 protein did not bind to a mutant PDGFR that did not contain the PI3 kinase-binding site, and the C-SH2 protein appeared to bind to a specific site, since it did not compete with the binding of PLCγ1 and Ras GAP (other cystosolic proteins that have SH2 domains) to the PDGFR (Klippel et al. 1992). Point mutations of different tyrosine residues on the PDGFR effected selectively either the binding of PI3 kinase or GAP, demonstrating that these proteins bound to distinct tyrosine phosphorylated sites (Fantl et al. 1992; Kazlauskas et al. 1992). (The selective prevention of PI3 kinase binding also prevented the stimulation of DNA synthesis by PDGF, consistent with other studies in which the activation of PI3 kinase is associated with cell proliferation [Fantl et al. 1992]). In contrast, a fusion protein containing the SH2 domains of Ras GAP, although not those of $pp60^{src}$, blocked the association of PI3 kinase activity with the IGF-1 receptor (Yamamoto et al. 1992). The N-SH2 domain of GAP did not bind to the region of the CSF-1 receptor to which p85 bound, and the C-SH2 region of p85 complexed more efficiently to this region than did N-SH2 (Reedjik et al. 1992). Interestingly, the specificity of the binding of the SH2 domains of p85 with middle T was observed to be dependent on the particular assay that was used (Yoakim et al. 1992). When assayed in solution by the binding to phosphorylated middle T, fusion proteins containing either SH2 domain bound to middle T, but C-SH2 was more efficient than N-SH2 at competing with p85 for binding to middle T. However, when the fusion proteins were run on a gel, transferred to nitrocellulose, and probed with phosphorylated middle T, middle T bound to a protein containing N-SH2 but not C-SH2. Although the importance of SH2 domains in directing PI3 kinase from the cytosol to kinases has now been established, recent studies indicate that the particulars regarding the specificity of the two SH2 domains for different tyrosine kinases must be further characterized. Future efforts by investigators will resolve whether one or both SH2 groups are involved in binding and whether SH3 groups are jointly involved in targeting PI3 kinase to activated tyrosine kinases.

PI3 Kinase Activity Is Activated by Differentiation and Nonproliferative Stimulating Factors

In addition to the activation of PI3 kinase by the mitogenic factors mentioned above, $PI(3,4)P_2$ and/or $PI(3,4,5)P_3$ are produced in some cells in response to agonists that activate differentiated cellular functions or cellular differentiation per se. $PI(3,4,5)P_3$ was produced transiently in human neutrophils exposed to fMetLeuPhe. Its appearance peaked about 1 minute after exposure and returned to unstimulated levels after about 3 minutes (Traynor-Kaplan et al. 1989). This time course was parallel to the stimulation of the rate of O_2^- production, a biochemical index of neutrophil activation. $PI(3,4)P_2$ also was produced in these cells in response to this agonist (Traynor-Kaplan et al. 1989). Thrombin-treated human platelets responded with an increase in $PI(3,4)P_2$, which reached a maximum level after 5 minutes of exposure (Nolan and Lapetina 1990). In cultured Leydig tumor cells labeled with $[^3H]$inositol, EGF stimulated the in vivo production of $PI(3,4)P_2$, which reached a maximum level about 15 minutes after exposure to EGF (Pignataro and Ascoli 1990), and PI(3)P was produced from PI in anti-Tyr immunoprecipitates from EGF-treated cells (Miller and Ascoli 1990). EGF is not a mitogen in these cells, but it affects their hormonal responsiveness. Thus, in neutrophils, platelets, and Leydig tumor cells, agonists that stimulate PI3 kinase cause cell activation or affect cellular functions other than growth.

In PC12 cells, a sympathetic nerve-like cell derived from a rat pheochromocytoma, PI3 kinase was activated by NGF (Soltoff et al. 1992). NGF, which was the first peptide growth factor to be purified and characterized (see Levi-Montalcini 1987), causes these cells to differentiate and stimulates neurite outgrowth. Thus, since it does not cause these cells to divide, it is somewhat of a misnomer to call NGF a "growth factor." NGF binds to $gp140^{trk}$ and $p75^{NGFR}$, two distinct proteins required for binding to a high-affinity site. gp140[trk] is the protein product of the *trk* proto-oncogene (Hemstead et al. 1991; Kaplan et al. 1991; Klein et al. 1991) and has an intrinsic tyrosine kinase activity that is activated by NGF binding (Kaplan et al. 1991; Klein et al. 1991). p75[NGFR] has no catalytic activity. NGF stimulates the production of PI(3,4)P_2 and PI(3,4,5)P_3 in PC12 cells labeled in vivo with [^{32}P]orthophosphate, and anti-gp140[trk] and anti-Tyr immunoprecipitates from NGF-treated cells had increased PI3 kinase activity compared to untreated cells (Soltoff et al. 1992). In these cells, EGF also causes the in vivo production of PI(3,4)P_2 and PI(3,4,5)P_3, and PI3 kinase activity was enhanced in anti-Tyr immunoprecipitates of EGF-treated cells (S.P. Soltoff, unpubl.). These studies suggest that PI3 kinase is activated by peptides that promote cell differentiation as well as those that stimulate cell proliferation. Since there is a strong correlation between cellular transformation and proliferation with the activation of PI3 kinase, the observation that both proliferative and differentiative factors activate PI3 kinase in the same system obfuscates the connection between mitogenesis and PI3 kinase activation.

PI3 Kinase Studies: Current and Future Efforts

Several questions related to the activation of PI3 kinase may be raised: (1) Does PI3 kinase become phosphorylated? (2) Is the specific activity of PI3 kinase changed when it is recruited to the plasma membrane? (3) What fraction of PI3 kinase is involved in growth factor responses? The basis of these questions is a consideration of whether the receptor tyrosine kinases activate PI3 kinase upon the binding of its SH2 groups to the receptor or whether its recruitment to the plasma membrane is sufficient, serving to direct this enzyme to a location where its substrates (phospholipids) reside. PLCγ1, another SH2-containing cytosolic protein recruited to activated tyrosine kinase receptors, is activated by its tyrosine phosphorylation in intact cells or in vitro (Nishibe et al. 1990; Kim et al. 1991). Thus, the issue is whether PLCγ1 serves as a model for all cytosolic proteins recruited to growth factor receptors. In one study, it was reported that tyrosine phosphorylation of p85 was not detectable unless both p85 and PDGFR were overexpressed (Hu et al. 1992). However, NGF stimulated the tyrosine phosphorylation of p85 in PC12 cells and NIH-3T3 cells that expressed gp140[trk], and PDGF stimulated the tyrosine phosphorylation of p85 in the NIH-3T3 cells (Soltoff et al. 1992). The prevalence of this finding in other cells in response to different stimuli must be determined. These findings still beg the question of whether the tyrosine phosphorylation of p85 affects the specific activity of PI3 kinase. The PI3 kinase activity assayed in anti-p85 immunoprecipitates of PC12 cells is not substantially altered by the exposure of these cells to growth factors and is approximately 500–1000 times greater than the activity found in anti-Tyr immunoprecipitates of unstimulated cells and approximately 25–50 times the activity found in anti-Tyr immunoprecipitates of NGF-treated cells (S.P. Soltoff, unpubl.). This suggests that the cytoplasmic PI3 kinase is present in an active form and will produce the novel phospholipids once it is recruited to the membrane. It remains possible that phosphorylation of PI3 kinase by receptor tyrosine kinases is required to alter its activity in more subtle ways. Now that the appropriate reagents are available, questions regarding the recruitment of PI3 kinase may now be addressed using immunochemical techniques. It will be of great interest to determine the localization of PI3 kinase in response to various growth, differentiating, and stimulatory factors. Of course, the specific roles of the different D-3 lipids remains the basic unanswered question.

ACKNOWLEDGMENTS

Work in the authors' laboratory was supported by National Institutes of Health grants GM-36624 and GM-41890 (to L.C.C.), and by the Whitaker Health Sciences Fund (S.P.S.).

REFERENCES

Auger, K.R., C.L. Carpenter, S.E. Shoelson, H. Piwnica-Worms, and L.C. Cantley. 1992. Polyoma virus middle T antigen-pp60*src* complex associates with purified phosphatidylinositol 3-kinase in vitro. *J. Biol. Chem.* **267:** 5408.

Auger, K.R., L.A. Serunian, S.P. Soltoff, P. Libby, and L.C. Cantley. 1989. PDGF-dependent tyrosine phosphorylation stimulates production of novel polyphosphoinositides in intact cells. *Cell* **57:** 167.

Bjorge, J.D., T.-O. Chan, M. Antczak, H.-J. Kung, and D.J. Fujita. 1990. Activated type I phosphatidylinositol kinase is associated with the epidermal growth factor (EGF) receptor following EGF stimulation. *Proc. Natl. Acad. Sci.* **87:** 3816.

Cantley, L.C., K.R. Auger, C. Carpenter, B. Duckworth, A. Graziani, R. Kapeller, and S. Soltoff. 1991. Oncogenes and signal transduction. *Cell* **64:** 281.

Carpenter, C.L. and L.C. Cantley. 1990. Phosphoinositide kinases. *Biochemistry* **29:** 11147.

Carpenter, C.L., B.C. Duckworth, K.R. Auger, B. Cohen, B.S. Schaffhausen, and L.C. Cantley. 1990. Purification and characterization of phosphoinositide 3-kinase from rat liver. *J. Biol. Chem.* **265:** 19704.

Endemann, G., S.N. Dunn, and L.C. Cantley. 1987. Bovine brain contains two types of phosphatidylinositol kinase. *Biochemistry* **26:** 6845.

Endemann, G., K. Yonezawa, and R.A. Roth. 1990. Phosphatidylinositol or an associated protein is a substrate for the insulin receptor tyrosine kinase. *J. Biol. Chem.* **265:** 396.

Escobedo, J.A., D.R. Kaplan, W.M. Kavanaugh, C.W. Turck, and L.T. Williams. 1991a. A phosphatidylinositol-3 kinase binds to platelet-derived growth factor receptors through a specific receptor sequence containing phosphotyrosine. *Mol. Cell. Biol.* **11:** 1125.

Escobedo, J.A., S. Navankasattusas, W.M. Kavanaugh, D. Milfay, V.A. Fried, and L.T. Williams. 1991. cDNA cloning of a novel 85 kd protein that has SH2 domains and regulates binding of PI3-kinase to the PDGF β-receptor. *Cell* **65:** 75.

Fantl, W.J., J.A. Escobedo, G.A. Martin, C.W. Turck, M. del

Rosario, F. McCormick, and L.T. Williams. 1992. Distinct phosphotyrosines on a growth factor receptor bind to specific molecules that mediate different signaling pathways. *Cell* **69:** 413.

Fukui, Y., S. Kornbluth, S.-M. Jong, L.-H. Wang, and H. Hanafusa. 1989. Phosphatidylinositol kinase type I activity associates with various oncogene products. *Oncogene Res.* **4:** 283.

Gutkind, J.S., P.M. Lacal, and K.C. Robbins. 1990. Thrombin-dependent association of phosphatidylinositol-3 kinase with $p60^{c\text{-}src}$ and $p59^{fyn}$ in human platelets. *Mol. Cell. Biol.* **10:** 3806.

Hempstead, B., D. Kaplan, D. Martin-Zanca, L.F. Parada, and M. Chao. 1991. High affinity NGF binding requires co-expression of the *trk* proto-oncogene product and low affinity NGF receptor. *Nature* **350:** 678.

Hu, P., B. Margolis, E.Y. Skolnik, R. Lammers, A. Ullrich, and J. Schlessenger. 1992. Interaction of phosphatidylinositol 3-kinase-associated p85 with epidermal growth factor and platelet-derived growth factor. *Mol. Cell. Biol.* **12:** 981.

Kapeller, R., K.S. Chen, M. Yoakim, B.S. Schaffhausen, J. Backer, M.F. White, L.C. Cantley, and N.B. Ruderman. 1991. Mutations in the juxtamembrane region of the insulin receptor impair activation of phosphatidylinositol 3-kinase by insulin. *Mol. Endocrinol.* **5:** 769.

Kaplan, D.R., D. Martin-Zanca, and L.F. Parada. 1991. Tyrosine phosphorylation and tyrosine kinase activity of the *trk* proto-oncogene product induced by NGF. *Nature* **350:** 158.

Kaplan, D.R., M. Whitman, B. Schaffhausen, D.C. Pallas, M. White, L. Cantley, and T.M. Roberts. 1987. Common elements in growth factor stimulation and oncogenic transformation: 85 kd phosphoprotein and phosphatidylinositol kinase activity. *Cell* **50:** 1021.

Kaplan, D.R., M. Whitman, B. Schaffhausen, L. Raptis, R.L. Garcea, D. Pallas, T.M. Roberts, and L. Cantley. 1986. Phosphatidylinositol metabolism and polyoma-mediated transformation. *Proc. Natl. Acad. Sci.* **83:** 3624.

Kazlauskas, A., A. Kashishian, J.A. Cooper, and M. Valius. 1992. GTPase-activating protein and phosphatidylinositol 3-kinase bind to distinct regions of the platelet-derived growth factor receptor β subunit. *Mol. Cell. Biol.* **12:** 2534.

Kim, H.K., J.W. Kim, A. Zilberstein, B. Margolis, J.G. Kim, J. Schlessinger, and S.G. Rhee. 1991. PDGF stimulation of inositol phospholipid hydrolysis requires PLC-γ1 phosphorylation on tyrosine residues 783 and 1254. *Cell* **65:** 435.

Klein, R., S. Jing, V. Nanduri, E. O'Rourke, and M. Barbacid. 1991. The *trk* proto-oncogene product encodes a receptor for nerve growth factor. *Cell* **65:** 189.

Klippel, A., J.A. Escobedo, W.J. Fantl, and L.T. Williams. 1992. The C-terminal SH2 domain of p85 accounts for the high affinity and specificity of the binding of phosphatidylinositol 3-kinase to phosphorylated platelet-derived growth factor β receptor. *Mol. Cell. Biol.* **12:** 1451.

Koch, C.A., D. Anderson, M.F. Moran, C. Ellis, and T. Pawson. 1991. SH2 and SH3 domains: Elements that control interactions of cytoplasmic signaling proteins. *Science* **252:** 668.

Levi-Montalcini, R. 1987. The nerve growth factor: Thirty five years later. *Science* **237:** 1154.

Lips, D.L., P.W. Majerus, F.R. Gorga, A.T. Young, and T.L. Benjamin. 1989. Phosphatidylinositol 3-phosphate is present in normal and transformed fibroblasts and is resistant to hydrolysis by bovine brain phospholipase C II. *J. Biol. Chem.* **264:** 8759.

McGlade, C.J., C. Ellis, M. Reedijk, D. Anderson, G. Mbamalu, A.D. Reith, G. Panayotou, P. End, A. Bernstein, A. Kazlauskas, M.D. Waterfield, and T. Pawson. 1992. SH2 domains of the p85α subunit of phosphatidylinositol 3-kinase regulate binding to growth factor receptors. *Mol. Cell. Biol.* **12:** 991.

Miller, E.S. and M. Ascoli. 1990. Anti-phosphotyrosine immunoprecipitation of phosphatidylinositol 3′ kinase activity in different cell types after exposure to epidermal growth factor. *Biochem. Biophys. Res. Comm.* **173:** 289.

Nishibe, S., M.I. Wahl, S.M.T. Hernandez-Sotomayor, N.K. Tonks, S.G. Rhee, and G. Carpenter. 1990. Increase of the catalytic activity of phospholipase C-γ1 by tyrosine phosphorylation. *Science* **250:** 1253.

Nolan, R.D. and E.G. Lapetina. 1990. Thrombin stimulates the production of a novel polyphosphoinositide in human platelets. *J. Biol. Chem.* **265:** 2441.

Otsu, M., I. Hiles, I. Gout, M.J. Fry, F. Ruiz-Larrea, G. Panayotou, A. Thompson, R. Dhand, J. Hsuan, N. Totty, A.D. Smith, S.J. Morgan, S.A. Courtneidge, P.A. Parker, and M.D. Waterfield. 1991. Characterization of two 85 kd proteins that associate with receptor tyrosine kinases, middle T/$ppc60^{c\text{-}src}$ complexes, and PI3-kinase. *Cell* **65:** 91.

Peles, E., R. Lamprecht, R. Ben-Levy, E. Tzahar, and Y. Yarden. 1992. Regulated coupling of the Neu receptor to phosphatidylinositol 3′-kinase and its release by oncogenic action. *J. Biol. Chem.* **267:** 12266.

Pignataro, O.P. and M. Ascoli. 1990. Epidermal growth factor increases the labeling of phosphatidylinositol-3,4-bisphosphate in MA-10 Leydig tumor cells. *J. Biol. Chem.* **265:** 1718.

Reedjik, M., X. Liu, P. van der Geer, K. Letwin, M.D. Waterfield, T. Hunter, and T. Pawson. 1992. Tyr721 regulates specific binding of the CSF-1 receptor kinase insert to PI 3′-kinase SH2 domains: A model for SH2-mediated receptor-target interactions. *EMBO J.* **11:** 1365.

Ruderman, N.B., R. Kapeller, M.F. White, and L.C. Cantley. 1990. Activation of phosphatidylinositol 3-kinase by insulin. *Proc. Natl. Acad. Sci.* **87:** 1411.

Serunian, L.A., K.R. Auger, T.M. Roberts, and L.C. Cantley. 1990. Production of novel polyphosphoinositides in vivo is linked to cell transformation by polyoma middle T antigen. *J. Virol.* **64:** 4718.

Serunian, L.A., M. Haber, T. Fukui, J.W. Kim, S.G. Rhee, J.M. Lowenstein, and L.C. Cantley. 1989. Polyphosphoinositides produced by phosphatidylinositol 3-kinase are poor substrates for phospholipases C from rat liver and bovine brain. *J. Biol. Chem.* **264:** 17809.

Soltoff, S.P., S.L. Rabin, L.C. Cantley, and D.R. Kaplan. 1992. Nerve growth factor promotes the activation of phosphatidylinositol 3-kinase and its association with the *trk* tyrosine kinase. *J. Biol. Chem.* **267:** 17472.

Sun, X.J., P. Rothenberg, C.R. Kahn, J.M. Backer, E. Araki, P.A. Wilden, D.A. Cahill, B.J. Goldstein, and M.F. White. 1991. Structure of the insulin receptor substrate IRS-1 defines a unique signal transduction protein. *Nature* **352:** 73.

Traynor-Kaplan, A.E., B.L. Thompson, A.L. Harris, P. Taylor, G.M. Omann, and L.A. Sklar. 1989. Transient increase in phosphatidylinositol 3,4-bisphosphate and phosphatidylinositol trisphosphate during activation of human neutrophils. *J. Biol. Chem.* **264:** 15668.

Varticovski, L., G.Q. Daley, P. Jackson, D. Baltimore, and L.C. Cantley. 1991. Activation of phosphatidylinositol 3-kinase in cells expressing *abl* oncogene variants. *Mol. Cell. Biol.* **11:** 1107.

Varticovski, L., B. Druker, D. Morrison, L. Cantley, and T. Roberts. 1989. The colony stimulating factor-1 receptor associates with and activates phosphatidylinositol-3-kinase. *Nature* **342:** 699.

Whitman, M., D. Kaplan, T. Roberts, and L. Cantley. 1987. Evidence for two distinct phosphatidylinositol kinases in fibroblasts. *Biochem. J.* **247:** 165.

Whitman, M., C.P. Downes, M. Keeler, T. Keller, and L. Cantley. 1988. Type I phosphatidylinositol kinase makes a novel inositol phospholipid, phosphatidylinositol-3-kinase. *Nature* **332:** 644.

Whitman, M., D.R. Kaplan, B. Schaffhausen, L. Cantley, and T.M. Roberts. 1985. Association of phosphatidylinositol kinase activity with polyoma middle-T competent for transformation. *Nature* **315:** 239.

Yamamoto, K., D. Altschuler, E. Wood, K. Horlick, S. Jacobs, and E.G. Lapetina. 1992. Association of phosphorylated insulin-like growth factor-I receptor with the SH2 domains of phosphatidylinositol 3-kinase p85. *J. Biol. Chem.* **267:** 11337.

Yoakim, M., W. Hou, Y. Liu, C.L. Carpenter, R. Kapeller, and B.S. Schaffausen. 1992. Interactions of polyomavirus middle T with the SH2 domains of the pp85 subunit of phosphatidylinositol 3-kinase. *J. Virol.* **66:** (in press).

The Type II TGF-β Receptor Signals Diverse Responses in Cooperation with the Type I Receptor

J.L. WRANA, J. CÁRCAMO,* L. ATTISANO, S. CHEIFETZ,
A. ZENTELLA,* F. LÓPEZ-CASILLAS,* AND J. MASSAGUÉ*
*Howard Hughes Medical Institute, and Cell Biology and Genetics Program,
Memorial Sloan-Kettering Cancer Center, New York, New York 10021*

Transforming growth factor-β (TGF-β) is prototypic of a large family of cytokines conserved from insects through humans. TGF-β is multipotential, being able to control cell proliferation, differentiation, adhesion, migration, extracellular matrix production, and other activities (for review, see Lyons and Moses 1990; Massagué 1990; Roberts and Sporn 1990). Regulation of these functions is important in tissue development and repair processes. The ability of TGF-β to inhibit progression through the cell cycle and to control cell commitment to terminal differentiation provides unique opportunities to elucidate the molecular events that support these important processes. TGF-β and related cytokines initiate their effects by interacting with membrane receptors, some of which have been molecularly defined during the past year (for review, see Massagué 1992) but whose signaling mechanisms remain practically unknown.

One approximation to the identification of TGF-β receptors is based on tagging membrane proteins by chemical cross-linking to bound radiolabeled ligand. Conducted in a variety of cell types, these assays visualize various proteins that bind the three mammalian TGF-β isoforms, β1, β2, and β3, with high affinity and specificity (Fig. 1) (Massagué and Like 1985; Cheifetz et al. 1987, 1990; Segarini et al. 1989). The most widely distributed of these receptors are two ubiquitous glycoproteins of 53 kD and 70 kD, named TGF-β receptors I and II, respectively, according to their size (Cheifetz et al. 1987). Similar experiments conducted with two other groups of TGF-β-related factors, the activins and the bone morphogenetic proteins (BMPs), indicate that they too bind a set of specific type I and II receptors (Mathews and Vale 1991; Paralkar et al. 1991; Attisano et al. 1992).

Receptor I can be distinguished from other TGF-β receptors by its molecular size and sensitivity to reducing agents (Cheifetz and Massagué 1991), but its primary structure is not known. Recent molecular cloning studies indicate that the human type II TGF-β receptor (Lin et al. 1992), as well as several type II activin receptors (Mathews and Vale 1991; Attisano et al. 1992) and the related *daf-1* gene product of *Caenorhabditis elegans* (Georgi et al. 1990), are transmembrane proteins with a short cysteine-rich extracellular region and a cytoplasmic region that contains a protein kinase domain (Fig. 2A). One important feature that distinguishes these four receptors from the large group of cytokine receptors with tyrosine kinase activity is that their predicted kinase activity is toward serine and threonine residues. The ability to autophosphorylate on serine and threonine has indeed been demonstrated with a recombinant TGF-β receptor II kinase fragment (Lin et al. 1992). On the basis of their similarity with the tyrosine kinase receptors in overall structural configuration (Williams 1989; Ullrich and Schlessinger 1990), the type II receptors are regarded as central components of the signaling systems that mediate the action of TGF-β family members.

Other TGF-β-binding membrane proteins of known

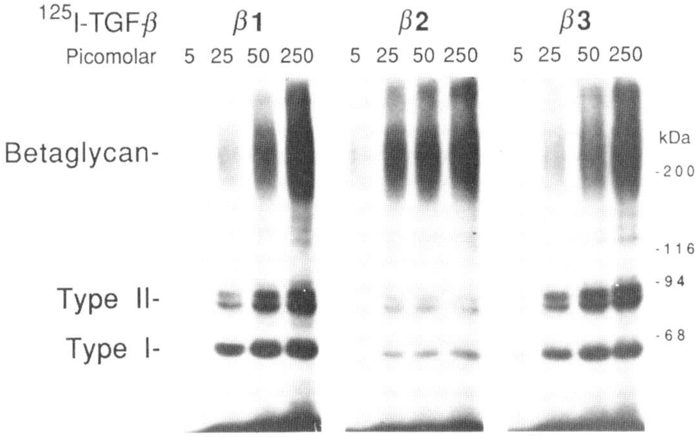

Figure 1. TGF-β-binding proteins in Mv1Lu mink lung epithelial cells. TGF-β receptor types I and II and betaglycan are shown as affinity-labeled products in reducing SDS-polyacrylamide electrophoresis gels. Samples were obtained by incubating and cross-linking subconfluent Mv1Lu cell monolayers with the indicated concentrations of ^{125}I-labeled TGF-β1, -β2, or -β3. In these cells, all three TGF-β isoforms bind equally well at low concentrations to receptors I and II. A subset of receptors I and II are saturated by higher concentrations of TGF-β1 or -β3 but not TGF-β2 (Cheifetz et al. 1990). TGF-β2 has a somewhat higher affinity for betaglycan than TGF-β1 or -β3.

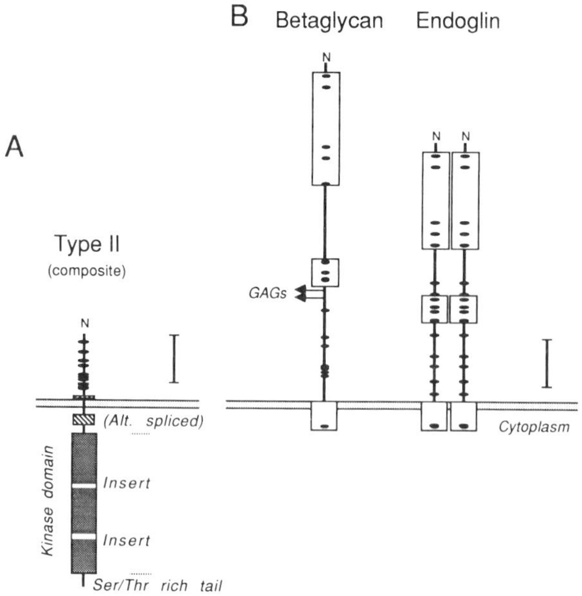

Figure 2. Schematic structures of the type II receptor, betaglycan, and endoglin as deduced from their cDNAs. (*A*) The composite serine/threonine kinase receptor structure shown as shared by type II receptors for TGF-β (TβR-II) and activin (ActR-II and ActT-IIB) (Mathews and Vale 1991; Attisano et al. 1992; Lin et al. 1992). They contain a relatively short (130–170 amino acids) extracellular region rich in cysteine (*ovals*), a transmembrane region, and a cytoplasmic region with a predicted serine/threonine kinase domain followed by a short serine/threonine-rich tail. The two inserts in the kinase domain are characteristic of this kinase subfamily. The juxtamembrane regions of ActR-IIB contain two alternatively spliced segments that generate four receptor isoforms. (*B*) Betaglycan and endoglin have large extracellular regions and short cytoplasmic tails. They show amino acid sequence homology (*boxes*) in two domains of the extracellular region and throughout the transmembrane and cytoplasmic regions. The probable sites for attachment of the heparan sulfate and chondroitin sulfate glycosaminoglycan (GAG) chains in betaglycan are indicated by arrows. Endoglin is a disulfide-linked homodimer. Bars, 100-amino-acid scale.

primary structure are betaglycan, also referred to as TGF-β receptor III, and endoglin (Fig. 2B). Betaglycan is a membrane proteoglycan that binds TGF-β through its 100-kD core protein, has a short cytoplasmic tail with no obvious signaling motif, and is widely distributed in mesenchymal, neural, and epithelial cells (Cheifetz et al. 1987; Segarini and Seyedin 1988; Andres et al. 1989; Massagué et al. 1990; López-Casillas et al. 1991; Wang et al. 1991). Endoglin is a disulfide-linked protein homodimer of 95-kD subunits that are structurally related to betaglycan, particularly in the transmembrane and cytoplasmic regions, and are present at highest levels in certain vascular endothelial cell types (Gougos and Letarte 1990; Cheifetz et al. 1992). The structure of these molecules, their absence from some cell types that respond to TGF-β, and the properties of certain TGF-β-resistant cell mutants (see below) provide evidence suggesting that betaglycan and endoglin may not be directly involved in signaling TGF-β responses.

Our current work focuses on the structural and functional analysis of receptors for TGF-β and related factors. To approach this problem, we have made use of a panel of TGF-β-resistant cell mutants. Some of these mutants, all isolated by being resistant to growth inhibition by TGF-β, have lost TGF-β binding to receptors I and II (Boyd and Massagué 1989; Laiho et al. 1990a). Mutants that fail to bind TGF-β via receptors I and II lack not only the growth inhibitory response, but also the early gene responses typically induced by TGF-β, suggesting that these receptors may mediate multiple TGF-β effects. This mutant cell system can be used to determine the individual contribution of each receptor component to specific TGF-β responses. Furthermore, it provides a system to analyze the functional role of defined domains in these receptors.

Using this approach, the present study addresses the question of whether the human TGF-β receptor II can mediate two distinct TGF-β responses, cell cycle arrest and elevated expression of extracellular matrix proteins. We find that both responses are restored when receptor II is introduced into cells devoid of active endogenous receptors. Furthermore, restoration of these responses by receptor II correlates with the reappearance of receptor I. These results suggest that TGF-β mediates its various responses through a signaling receptor system in which receptor II may interact with receptor I. This receptor model represents a significant departure from the mode of operation of tyrosine kinase receptors.

EXPERIMENTAL PROCEDURES

Cell lines and transfections. The Mv1Lu mink lung cell line (CCL-64, American Type Culture Collection) and its derivatives were maintained in minimal essential medium (MEM) (GIBCO) supplemented with 10% fetal bovine serum (GIBCO), nonessential amino acids (GIBCO), 100 units/ml penicillin, 100 μg/ml streptomycin, and 25 ng/ml of fungizone.

The human TβR-II expression vector was generated by cloning the *Kpn*I/*Not*I fragment of hTβR-II (Lin et al. 1992) into the pMEP vector (Invitrogen). Cells expressing hTβR-II were generated by transfection with lipofectin (GIBCO/BRL) as described by the manufacturer. Briefly, actively growing cells in a 100-mm dish were incubated overnight in serum-free medium containing 25 μg of DNA and 80 μg of lipofection reagent, and then washed into media containing 10% fetal bovine serum, antibiotics, and fungizone.

To select for stable transfections, cells grown for 2 days were subcultured into several 150-mm dishes and grown in the presence of 300 μg/ml of hygromycin (400 units/mg; Calbiochem) for 2 weeks with a medium change every 3–4 days. Individual colonies were ring-cloned, and all cells were maintained in the continuous presence of hygromycin. When necessary, expression of hTβR-II was induced by incubation of cells in the presence of 50 μM $ZnCl_2$ for at least 5 hours.

Receptor affinity-labeling assays. TGF-β1 and -β2 purified from porcine platelets were purchased from R&D Systems (Minneapolis, Minnesota). TGF-β3 purified to homogeneity from serum-free medium conditioned by Chinese hamster ovary cells expressing a transfected human TGF-β3 cDNA was provided by K. Iwata (Oncogene Science, Manhasset, New York). TGF-βs were iodinated by the chloramine-T method as described previously (Cheifetz et al. 1988). Affinity-labeling of confluent cell monolayers with ^{125}I-labeled TGF-βs and disuccinimidyl suberate (Pierce) was carried out as described previously (Massagué 1987). The concentration of iodinated ligand was 50 or 100 pM unless otherwise indicated. Labeled cell monolayers were solubilized with buffer containing Triton X-100 (Massagué 1987), and soluble extracts were electrophoresed on SDS-polyacrylamide gels under reducing conditions.

Growth inhibition assays. Cells grown in 24-well plates were labeled with 0.5 μCi/ml of ^{125}I-iododeoxyuridine (^{125}I-IdU, specific activity 2200 Ci/mmole, DuPont-New England Nuclear) for the last 2 hours of incubation. Cells were washed three times with cold phosphate-buffered saline and fixed with 95% methanol for 1 hour at 4°C. The cells were then washed once with phosphate-buffered saline and extracted with 0.2 N NaOH for 0.5 hour at 4°C. The extracts were collected and counted in a γ-counter. Each assay was carried out in triplicate, and values varied by less than 5%.

Fibronectin protein assay. For metabolic labeling, cells were treated as described in the figure legends and labeled with 50 μCi/ml of [^{35}S]methionine in methionine-free MEM for 2 hours. [^{35}S]methionine-labeled cell culture medium was heated in electrophoresis sample buffer containing dithiothreitol and subjected to SDS-PAGE. Gels were fixed and fluorographed with Enlightning (New England Nuclear).

RESULTS

The Mv1Lu cell line is highly sensitive to TGF-β, responding to picomolar concentrations of this factor with a complete cell cycle arrest in late G_1 phase and elevated expression of a set of early response genes that include various transcription factors and extracellular matrix components (Like and Massagueé 1986; Laiho et al. 1990b, 1991b). Previously, we isolated a panel of TGF-β-resistant cell mutants by selecting chemically mutagenized Mv1Lu cells for growth in the presence of TGF-β. The TGF-β-resistant cell clones that grew under these conditions fall into three classes according to their respective TGF-β receptor phenotypes (Boyd and Massagué 1989; Laiho et al. 1990a). One class of mutants fail to bind TGF-β through receptor I (Fig. 3), and are therefore named R (for "receptor") mutants. A second class of mutants, named DR (for "double receptor"), fail to bind TGF-β through receptors I and II (Fig. 3). The third class of mutants, named S (for "signaling"), bind through all three receptors but do not respond to TGF-β and are presumed to have mutations that affect downstream components (although mutations disrupting the signaling domain of the receptor would also display an S phenotype). No mutants were isolated that had lost betaglycan.

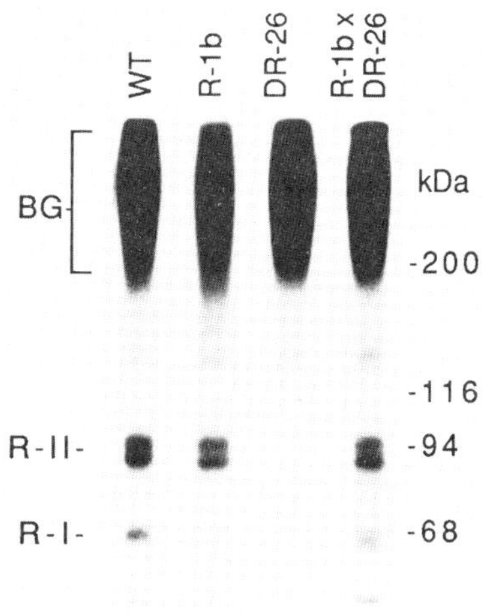

Figure 3. TGF-β receptor phenotypes of wild-type Mv1Lu cells, R and DR TGF-β-resistant mutants, and RxDR hybrids. (Modified from Laiho et al. 1991a.)

Although selection of these mutants was based on their ability to escape the growth inhibitory effect, they are also devoid of other responses to TGF-β (Boyd and Massagué 1989; Laiho et al. 1990a). Their resistance to TGF-β is recessive with respect to the wild-type phenotype, as demonstrated with cell hybrids with mutant and wild-type cells (Laiho et al. 1991a). Another property of this system is that R mutants complement DR mutants. Somatic cell hybrids between R and DR cells recover TGF-β responsiveness as well as the wild-type TGF-β receptor phenotype (Laiho et al. 1991a). Thus, the RxDR hybrid clones express receptor I even though this receptor is not detectable by affinity-labeling with ^{125}I-labeled TGF-β1 in either of the parent mutant cell lines (Fig. 3).

One interpretation of the selective loss of TGF-β receptors I and II in the mutants is that these receptors mediate a common set of TGF-β responses rather than mediating distinct responses separately. Furthermore, the concomitant loss of these receptors in DR mutants and the recovery of receptor I in RxDR hybrids suggests that receptors I and II may act cooperatively and with some degree of functional interdependence.

To begin to test this model, we expressed the recently cloned human TGF-β receptor II (TβR-II) (Lin et al. 1992) in representative R and DR cell clones. To this end, the hTβR-II cDNA was placed under the control of an inducible metallothionein promoter and stably transfected into mutant clones R-1B and DR-27. The resulting transfectants showed elevated ^{125}I-labeled TGF-β-binding activity in a Zn^{++}-inducible manner (data not shown).

To determine which cell-surface TGF-β-binding

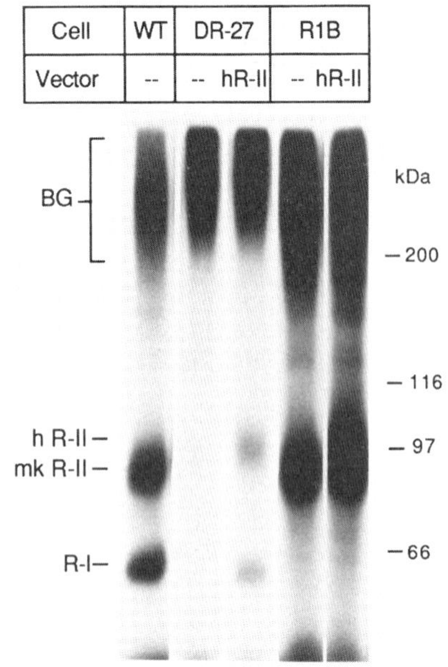

Figure 4. Affinity-labeled TGF-β receptor profiles in R and DR cell mutants stably transfected with the human TβR-II. The human receptor II can be distinguished from the endogenous mink receptor II by its slower electrophoretic mobility. Note the reappearance of a putative receptor I in DR transfectants but not in R transfectants. All transfectant cells were induced with Zn^{++} 18 hr before these assays.

species were present in these transfectants, cells were affinity-labeled by cross-linking to receptor-bound ^{125}I-labeled TGF-β1. This assay showed that hTβR-II was expressed in DR-27 transfectants as well as in R-1B transfectants (Fig. 4). The hTβR-II product could be distinguished from the endogenous mink TGF-β receptor II by the differences in electrophoretic mobility of their respective affinity-labeled complexes. The hTβR-II labeled product migrated as a 95-kD complex, as is characteristic of human receptor II in other cell types (Massagué et al. 1990), whereas the endogenous labeled receptor II in Mv1Lu cells migrated characteristically as an 85-kD species (Fig. 4). This difference is probably due to the presence of three extracellular *N*-glycosylation sites in hTβR-II and only two such sites in mink TβR-II (L. Attisano et al., unpubl.).

Interestingly, the DR-27 transfectants showed a TGF-β-labeled species that migrated in the position corresponding to receptor I (Fig. 4). The recovery of this putative receptor I in DR-27 transfected with hTβR-II was reminiscent of the expression of receptor I previously observed in hybrids between DR cells and R cells (Fig. 3) (Laiho et al. 1991a). Thus, DR-27 transfectants had an apparently wild-type TGF-β receptor phenotype with expression of an exogenous receptor II and a putative endogenous receptor I in addition to betaglycan. In contrast to these DR-27 transfectants, R-1B cells transfected with hTβR-II did not express detectable receptor I (Fig. 4).

Stable expression of hTβR-II in DR-27 cells restored their ability to respond to TGF-β1 with inhibition of

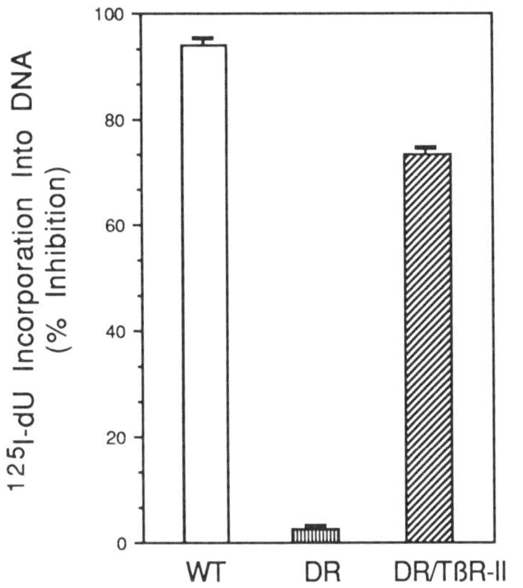

Figure 5. Restoration of the growth inhibitory response in DR cells by transfection of human TGF-β receptor II. DNA synthesis assays were conducted with monolayers of wild-type Mv1Lu cells (WT), DR-27 mutants, and DR-27 stably transfected with hTβR-II, incubated for 24 hr with or without 50 pM TGF-β1. Data are presented as the percent inhibition of ^{125}I-labeled deoxyuridine incorporation in TGF-β-treated cells relative to controls that did not receive TGF-β.

^{125}I-labeled deoxyuridine incorporation into DNA (Fig. 5). This parameter is a measure of the ability of TGF-β to arrest cell cycle, which in wild-type Mv1Lu cells occurs in late G_1 phase (Laiho et al. 1990b). Stable expression of hTβR-II failed to restore this response in R-1B cells, even after incubation of cells with Zn^{++} to induce maximal hTβR-II levels (data not shown).

Likewise, expression of hTβR-II restored the ability to respond to TGF-β with elevated expression of fibronectin in DR-27 cells, but not in R-1B cells (not shown) (Fig. 6). This parameter is a manifestation of the ex-

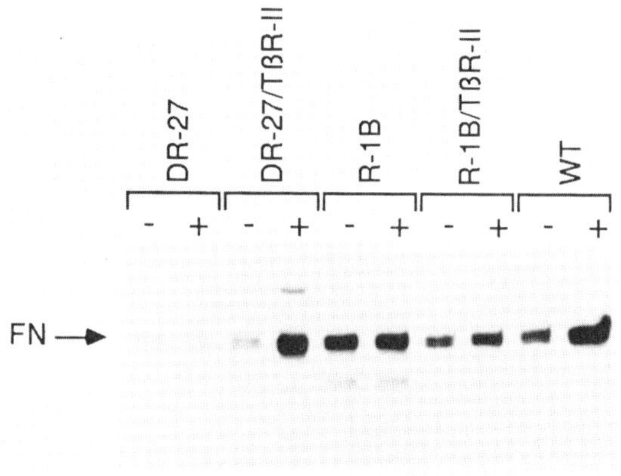

Figure 6. Restoration of the extracellular matrix response in DR cells by transfection of human TGF-β receptor II. Wild-type Mv1Lu cells (WT), DR-27 mutants, and DR-27 stably transfected with hTβR-II were incubated for 24 hr with or without 50 pM TGF-β1. Fibronectin synthesis was assayed by metabolically labeling cells with [^{35}S]methionine and detecting fibronectin secreted into the media by SDS-PAGE and fluorography.

tracellular matrix response observed in epithelial and mesenchymal cells when they are treated with TGF-β (Ignotz and Massagué 1986). These observations indicated that hTβR-II can mediate multiple TGF-β responses, and its ability to do so correlates with the reappearance of a TGF-β-binding species that has properties of TGF-β receptor I.

DISCUSSION

One of the most salient features of the biology of TGF-β and related cytokines is their multifunctionality. This property has raised the general question of whether the responses elicited by TGF-β are mediated by a common receptor complex or result instead from activation of independent receptors, each specifying an independent set of responses. The latter model, "independent receptors mediate independent responses," is inspired by the fact that TGF-β binds with high affinity and specificity to various coexisting membrane proteins (Cheifetz et al. 1987, 1992; Segarini et al. 1989; Massagué 1992). The better-known and more widely distributed of these proteins are TGF-β receptors I and II, the proteoglycan betaglycan, and endoglin. In addition, various less well characterized, but certainly distinct, TGF-β-binding proteins have been identified in some cell types (Massagué 1992).

Other evidence argues against this model and favors the alternative "one receptor-multiple responses" model. That evidence is based in part on the properties of TGF-β-resistant cell mutants in the present studies. In these cells, receptors I and II are implicated in mediation of TGF-β responses, since TGF-β binding to receptors I and II suffers a selective loss in many of the mutant cell clones. Concomitant with the loss of functional receptors I and II, these cells lost the growth inhibitory response (Boyd and Massagué 1989; Laiho et al. 1990a) and early responses of genes encoding *jun*B, plasminogen activator inhibitor-1, fibronectin, and thrombospondin (Laiho et al. 1991a; M. Ohtsuki and J. Massagué, unpubl.). These responses to TGF-β are typical of many other cell types, the vast majority of which also express receptors I and II (Massagué 1992).

We have begun to test this model by ectopically expressing the human TGF-β receptor II in TGF-β receptor-defective cell mutants. We show that receptor II restores diverse TGF-β responses to these mutants. One of these responses is growth inhibition, which in Mv1Lu cells occurs by arrest of cell progression through the G_1 phase of the proliferative cycle in a process that involves inhibition of retinoblastoma protein phosphorylation (Laiho et al. 1990b). Another response restored by receptor II is elevated fibronectin expression, which is representative of an extracellular matrix response separate from the growth inhibitory effect.

These results demonstrate that TGF-β receptor II can mediate diverse effects. The question is whether or not this serine/threonine kinase receptor can signal alone as the tyrosine kinase receptors do upon activation by their ligands (Williams 1989; Ullrich and Schlessinger 1990). The possibility that receptor II acts in concert with receptor I as part of a central TGF-β signaling receptor system, rather than as an isolated receptor protein, is supported by various lines of evidence. First, the R class of TGF-β-resistant mutants lack detectable receptor I but show normal binding of TGF-β to receptor II. Yet, these cells show the same degree of TGF-β resistance as mutants lacking TGF-β binding to receptor II do (Boyd and Massagué 1989). Second, mutants lacking binding to receptor II show a parallel loss of TGF-β binding to receptor I; no mutants were isolated that showed normal binding to receptor I but lacked receptor II (Laiho et al. 1990a). Third, restoration of TGF-β responses in hybrids between R and DR mutants is accompanied by reappearance of receptor I, perhaps a cryptic receptor I of DR mutants that is unable to bind ligand in the absence of receptor II (Laiho et al. 1991a).

This hypothesis is strengthened by two observations from the present studies. First, expression of the human TGF-β receptor II in DR mutants restores expression of this receptor as well as TGF-β binding to a protein that has properties of TGF-β receptor I. Restoration of TGF-β binding to receptor I in DR mutants correlated with restoration of TGF-β responsiveness. Second, no receptor I was rescued by expression of hTβR-II in R mutants, and receptor II in these cells was unable to signal TGF-β responses. These results raise the possibility that receptor I might depend on receptor II for ligand binding on the cell surface, whereas receptor II might depend on receptor I for signaling.

These results give weight to the notion that TGF-β may signal its many effects through a complex system involving receptors I and II. The evidence in support of this model is preliminary at the moment, but suggestive. Approaches to directly examine the nature of the interaction between these TGF-β receptor components are becoming available. If proven correct, the complex nature of this receptor system would represent a significant departure from what is currently known about transmembrane signaling by cytokine receptors.

ACKNOWLEDGMENTS

This work was supported by the National Cancer Institute (grant CA-34610). A.Z., F.L.-C., and J.C. are Associate Investigators and J.M. is an Investigator of the Howard Hughes Medical Institute. L.A. and J.W. are postdoctoral fellows of the Medical Research Council of Canada. L.A. is the recipient of a fellowship from the James Olson Fund.

REFERENCES

Andres, J.L., K. Stanley, S. Cheifetz, and J. Massagué. 1989. Membrane-anchored and soluble forms of betaglycan, a polymorphic proteoglycan that binds transforming growth factor-β. *J. Cell Biol.* **109:** 3137.

Attisano, L., J.L. Wrana, S. Cheifetz, and J. Massagué. 1992. Novel activin receptors: Distinct genes and alternative mRNA splicing generate a repertoire of serine/threonine kinase receptors. *Cell* **68:** 97.

Boyd, F.T. and J. Massagué. 1989. Growth inhibitory response to transforming growth factor-β linked to expression of a 53 kDa cell surface TGF-β receptor. *J. Biol. Chem.* **264:** 2272.

Cheifetz, S. and J. Massagué. 1991. Isoform-specific TGF-β binding proteins with membrane attachments sensitive to phosphatidylinositol-phospholipase C. *J. Biol. Chem.* **266:** 20767.

Cheifetz, S., H. Hernandez, M. Laiho, P. tenDijke, K.K. Iwata, and J. Massagué. 1990. Distinct transforming growth factor-β receptor subsets as determinants of cellular responsiveness to three TGF-β isoforms. *J. Biol. Chem.* **265:** 20533.

Cheifetz, S., T. Bellón, C. Calés, S. Vera, C. Bernabeu, J. Massagué, and M. Letarte. 1992. Endoglin is a component of the TGF-β receptor system in human endothelial cells. *J. Biol. Chem.* (in press).

Cheifetz, S., J.A. Weatherbee, M.L.-S. Tsang, J.K. Anderson, J.E. Mole, R. Lucas, and J. Massagué. 1987. The transforming growth factor-β system: A complex pattern of crossreactive ligands and receptors. *Cell* **48:** 409.

Georgi, L.L., P.S. Albert, and D.L. Riddle. 1990. *daf-1*, a *C. elegans* gene controlling dauer larva development, encodes a novel receptor protein kinase. *Cell* **61:** 635.

Gougos, A. and M. Letarte. 1990. Primary structure of endoglin, an RGD-containing glycoprotein of human endothelial cells. *J. Biol. Chem.* **265:** 8361.

Ignotz, R.A. and J. Massagué. 1986. Transforming growth factor-β stimulates the expression of fibronectin and collagen and their incorporation into the extracellular matrix. *J. Biol. Chem.* **261:** 4337.

Laiho, M., F.M.B. Weis, and J. Massagué. 1990a. Concomitant loss of transforming growth factor-β receptor types I and II in cell mutants resistant to TGF-β. *J. Biol. Chem.* **265:** 18518.

Laiho, M., J.A. DeCaprio, J.W. Ludlow, D.M. Livingston, and J. Massagué. 1990b. Growth inhibition by TGF-β linked to suppression of retinoblastoma protein phosphorylation. *Cell* **62:** 175.

Laiho, M., F.M.B. Weis, F.T. Boyd, R.A. Ignotz, and J. Massagué. 1991a. Responsiveness to transforming growth factor-β restored by complementation between cells defective in TGF-β receptors I and II. *J. Biol. Chem.* **266:** 9108.

Laiho, M., L. Rönnstrand, J. Heino, J.A. DeCaprio, J.W. Ludlow, D.M. Livingston, and J. Massagué. 1991b. Control of JunB and extracellular matrix protein expression by transforming growth factor-$\beta 1$ is independent of SV40 T antigen-sensitive growth inhibitory events. *Mol. Cell. Biol.* **11:** 972.

Like, B. and J. Massagué. 1986. The antiproliferative effect of type β transforming growth factor occurs at a level distal from receptors for growth-activating factors. *J. Biol. Chem.* **261:** 13426.

Lin, H.Y., X.-F. Wang, E. Ng-Eaton, R.A. Weinberg, and H.F. Lodish. 1992. Expression cloning of the TGF-β type II receptor, a functional transmembrane serine/threonine kinase. *Cell* **68:** 775.

López-Casillas, F., S. Cheifetz, J. Doody, J.L. Andres, W.S. Lane, and J. Massagué. 1991. Structure and expression of the membrane proteoglycan betaglycan, a component of the TGF-β receptor system. *Cell* **67:** 785.

Lyons, R.M. and H.L. Moses. 1990. Transforming growth factors and the regulation of cell proliferation. *Eur. J. Biochem.* **473:** 187467.

Massagué, J. 1987. Identification of receptors for type β transforming growth factor. *Methods Enzymol.* **146:** 174.

———. 1990. The transforming growth factor-β family. *Annu. Rev. Cell Biol.* **6:** 597.

———. 1992. Receptors for the TGF-β family. *Cell* **69:** 1067.

Massagué, J. and B. Like. 1985. Cellular receptor for type β transforming growth factor. Ligand binding and affinity labeling in human and rodent cell lines. *J. Biol. Chem.* **260:** 2636.

Massagué, J., S. Cheifetz, F.T. Boyd, and J.L. Andres. 1990. TGF-β receptors and TGF-β binding proteoglycans: Recent progress in identifying their functional properties. *Ann. N.Y. Acad. Sci.* **593:** 59.

Mathews, L.S. and W.W. Vale. 1991. Expression cloning of an activin receptor, a predicted transmembrane kinase. *Cell* **65:** 973.

Paralkar, V.M., R.G. Hammonds, and A.H. Reddi. 1991. Identification and characterization of cellular binding proteins (receptors) for recombinant human bone morphogenetic protein 2B, an initiator of bone differentiation cascade. *Proc. Natl. Acad. Sci.* **88:** 3397.

Roberts, A.B. and M.B. Sporn. 1990. The transforming growth factor-betas. In *Peptide growth factors and their receptors* (ed. M.B. Sporn and A.B. Roberts), p. 419. Springer-Verlag, Heidelberg.

Segarini, P.R. and S.M. Seyedin. 1988. The high molecular weight receptor to transforming growth factor-β contains glycosaminoglycan chains. *J. Biol. Chem.* **263:** 8366.

Segarini, P.R., D.M. Rosen, and S.M. Seyedin. 1989. Binding of TGF-β to cell surface proteins varies with cell type. *Mol. Endocrinol.* **3:** 261.

Ullrich, A. and J. Schlessinger. 1990. Signal transduction by receptors with tyrosine kinase activity. *Cell* **61:** 203.

Wang, X.-F., H.Y. Lin, E. Ng-Eaton, J. Downward, H.F. Lodish, and R.A. Weinberg. 1991. Expression cloning and characterization of the TGF-β type III receptor. *Cell* **67:** 797.

Williams, L.T. 1989. Signal transduction by the platelet-derived growth factor receptor. *Science* **243:** 1564.

Protein Tyrosine Phosphatases: The Problems of a Growing Family

N.K. Tonks, Q. Yang, A.J. Flint, M.F.B.G. Gebbink, B.R. Franza, Jr.,* D.E. Hill,† H. Sun, and S. Brady-Kalnay

*Cold Spring Harbor Laboratory, Cold Spring Harbor, New York 11724; *Freeman Laboratory of Cancer Cell Biology, Cold Spring Harbor Laboratory, Cold Spring Harbor, New York 11724; †Applied Biotechnology, Inc., Cambridge, Massachusetts 02138*

Protein tyrosine phosphorylation is now recognized as an important component of the control of many fundamental aspects of cellular function, including growth and differentiation, cell cycle and cytoskeletal integrity. In vivo, the net level of phosphorylation of tyrosyl residues in a target substrate reflects the balance between the competing action of kinases and phosphatases. We are examining physiological roles for protein tyrosine phosphorylation, pursuing the problem from the perspective of the enzymes that catalyze the dephosphorylation reaction, the protein tyrosine phosphatases (PTPases). The PTPases have, until recently, been somewhat neglected relative to the protein tyrosine kinases (PTKs). However, considerable progress has been made in identifying new members of the PTPase family, and it appears that they constitute a novel class of signal transducing molecules that rival the PTKs in their structural diversity and complexity.

One of the principal reasons that the study of PTPases has lagged behind that of the kinases is a technical one—the requirement for a suitably purified phosphorylated protein substrate with which to assay phosphatase activity. In the case of tyrosine phosphorylation, which accounts for less than 0.1% of phosphate linked to protein (Hunter and Sefton 1980), the low abundance of modified proteins raises problems in obtaining sufficient quantities for their utilization in routine assays. To overcome this problem, a number of artificial substrates have been utilized, most notably a reduced carboxamidomethylated and maleylated (RCM) derivative of lysozyme. The major PTPase activity in human placenta extracts was purified to homogeneity (Tonks et al. 1988a,b) using RCM lysozyme as substrate and a thiophosphorylated derivative of the protein bound to Sepharose as an affinity support.

Determination of the amino acid sequence of the purified enzyme, termed PTP1B, illustrated two important points (Charbonneau et al. 1988). First, PTP1B was not structurally related to the serine/threonine phosphatases. Thus, whereas the protein kinases are derived from a common ancestor, the protein phosphatases appear to have evolved in separate families. Second, a striking level of sequence identity was detected between PTP1B and the tandem intracellular domains in the leukocyte common antigen, CD45. CD45 is an abundantly expressed, hematopoietic cell-surface protein that exists in multiple forms due to alternative mRNA splicing of 3 exons encoding sequences at its amino terminus (for review, see Thomas 1989). Its structure can be described in terms of a PTPase-related intracellular segment, a single transmembrane segment, and an extracellular domain that is highly glycosylated, is rich in cysteinyl residues, and possesses the hallmarks of a ligand-binding motif. This observation was exciting because it suggested that CD45 can be considered as a prototype of transmembrane, receptor-like PTPases with the potential to trigger signaling events at the cell membrane through the ligand-modulated dephosphorylation of tyrosyl residues in proteins. This concept is supported by the demonstration of intrinsic PTPase activity in CD45 (Tonks et al. 1988c).

These observations have set in motion a rapidly expanding area of research involving many laboratories. At the present time, molecular biological approaches involving polymerase chain reaction (PCR) and low-stringency screening have identified 31 PTPases, excluding species homologs, as full-length sequences (for detailed review, see Charbonneau and Tonks 1992). Many additional partial sequences remain to be added to the list. The unique feature that defines the PTPase family is the sequence [I/V]H<u>C</u>XAGXXR[S/T]G. Within this motif, the cysteinyl residue is absolutely essential for activity, since the catalytic mechanism proceeds via the generation of a cysteine-phosphate intermediate (Guan and Dixon 1991). The structural features of several members of the PTPase family are illustrated in Figure 1.

Receptor-linked PTPases

With the exception of HPTPβ and a related molecule in *Drosophila*, DPTP10D, the transmembrane PTPases are characterized by the presence of two, tandemly repeated PTPase domains. The advantage of such an arrangement is unclear, although it should be remembered that oligomeric enzymes have the inherent capacity for cooperative interactions between their catalytic centers. However, it remains a point of controversy as to whether both domains are active (Streuli et al. 1990; Wang and Pallen 1991). For LAR and CD45, the idea has been proposed that domain 2 may

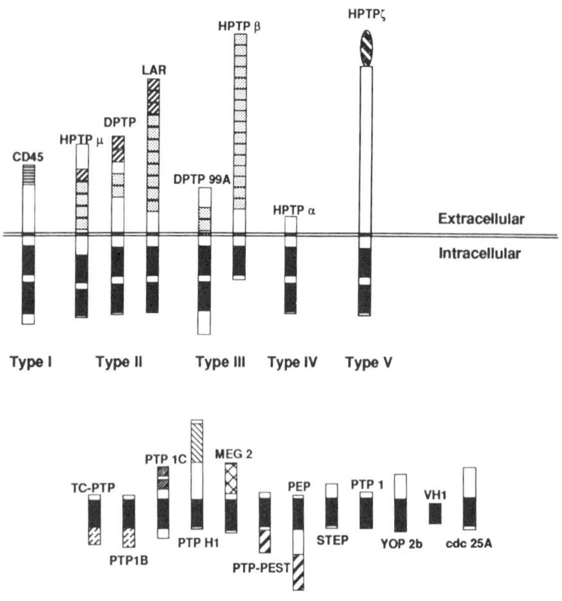

Figure 1. Members of the PTPase family. Conserved catalytic domains are shown in black. PTPs can be categorized as transmembrane, receptor-like, or nontransmembrane molecules. The receptor-like species can be subdivided into five types based on the structure of their extracellular segments. Type I represents the CD45 family, multiple isoforms of which arise from differential splicing of a primary mRNA transcript of a single gene; three exons encoding sequences at the extreme amino terminus (horizontal lines) are differentially expressed. Type II contains Ig-like (diagonal lines) and tandem FN type III-like repeat domains (stippled); this category includes LAR, DLAR, DPTP, and HPTPμ. Type III bears multiple FN type III-like repeats. Some type III isoforms such as HPTPβ have only one internal PTP domain. Type IV isoforms such as α and ϵ have small glycosylated extracellular segments. Type V possesses an amino-terminal motif with homology to carbonic anhydrase. Multiple nontransmembrane forms have also been identified. Many of the nonreceptor PTPs bear noncatalytic segments that are structurally related to other well-characterized proteins. The positions and relative sizes of these noncatalytic domains are shown as boxes containing distinct symbols; noncatalytic regions that have similar sequences are designated with identical patterns. The noncatalytic segments that have been identified include two SH2 domains in PTP1C, band 4.1 homology domains in PTPH1, an apparent lipid-binding domain in MEG2, and segments containing PEST sequences in PEP and PTP-PEST. In TC-PTP and PTP1B, the carboxy-terminal noncatalytic segments appear to play a role in modulating activity and controlling subcellular localization. PTP1, STEP, and Yop2b have noncatalytic sequences that are apparently unrelated to sequences in the databases. The protein from vaccinia virus, VH1, is much smaller than the other PTPs and presumably encodes only essential sequences within the catalytic domain. (Reprinted, with permission, from Tonks et al. 1993.)

itself not express activity but rather may function in modifying the specificity of domain 1 (Streuli et al. 1990). In support of this, inspection of the sequence of domain 2 in at least two PTPases, HPTPζ and HPTPγ, illustrates that the catalytically essential cysteinyl residue is replaced by aspartate. In those enzymes in which the essential cysteine has been retained in domain 2, there is the possibility that each domain may respond differently to potential ligands, display different substrate specificity, or be subjected to different modes of regulation. In the latter regard, it is interesting to note the unique 19-residue insert in domain 2 of CD45 that possesses potential sites of phosphorylation by casein kinase 2 (Charbonneau et al. 1988).

In contrast to the similarity among intracellular segments, the transmembrane PTPases can be readily distinguished by the structural diversity within their extracellular segments. At least five subtypes can be distinguished in this manner, as described in Figure 1. It should be stressed that these molecules must be considered putative receptors; no ligand has yet been identified which, upon binding to the extracellular segment of a transmembrane PTPase, alters the activity of the intracellular catalytic domains. The search for potential ligands remains a major challenge. A recent report suggested that CD45 may interact with a B-cell surface protein, CD22, although the report did not address the consequences for CD45 activity of this binding interaction (Stamenkovic et al. 1991). That this interaction is specific for the 180-kD form of CD45 also suggests that if in fact CD22 is a ligand for CD45, it is not the only one. As discussed below, the structural similarity between the type II receptor-like PTPases and adhesion molecules of the immunoglobulin superfamily suggests that these molecules may participate in homophilic binding interactions. In addition, the 280-residue motif at the amino terminus of PTPζ that is structurally related to carbonic anhydrase may form a binding pocket for low M_r ligands (Krueger and Saito 1992).

It is unclear what effect ligand binding will have on PTPase activity. Purified preparations of several receptor-like PTPases, including CD45 (Tonks et al. 1990), display considerable basal activity in vitro apparently in the absence of added ligand, raising the possibility that they are constitutively active in vivo. It is possible that ligand binding either inhibits activity or limits mobility of the enzyme on the membrane, thus restricting the spectrum of substrates with which it may interact.

PTPases and Cell Adhesion

The type II receptor-like PTPases are characterized by the presence of multiple immunoglobulin (Ig) and fibronectin (FN) type III-like repeats in the extracellular segment, which is characteristic of cell adhesion molecules such as N-CAMs (for review, see Edelman 1988). The first such PTPase identified was leukocyte common antigen-related (LAR) (Streuli et al. 1988), the 1234-residue extracellular segment of which contains 3 Ig-like and 8 FN III domains. LAR undergoes a posttranslational proteolytic cleavage to generate a mature protein that is a complex of two noncovalently associated subunits (Streuli et al. 1992; Yu et al. 1992). The 150-kD E subunit is derived from the N-CAM-like extracellular segment, whereas the 85-kD P subunit comprises the intracellular and transmembrane segments as well as a short stretch of extracellular sequence that interacts with the E subunit. Interestingly, in HeLa cells, the E subunit is shed in a cell-density-dependent manner; however, the effects of dissociation of E and P subunits on activity remain to be ascertained. In *Drosophila*, a role for DLAR and the other type II enzymes DPTP10D and DPTP99A (Hariharan et al. 1991; Tian et al. 1991; Yang et al. 1991) in cell adhesion and development has been suggested. In situ hybridization and immunocytochemistry data indicate that expression of these PTPases is restricted to the central nervous system, where the timing and pattern of their appearance on axons is consistent with a role in regulating neurite outgrowth and pathfinding during development.

We are particularly interested in a novel, receptor-like PTPase, termed RPTPμ, the sequence of which was reported by Gebbink et al. (1991). The cDNA predicts a protein of ~ 162 kD in which the extracellular segment includes one Ig-like and four FN III-like domains. There is a single transmembrane domain, and the intracellular segment bears two PTPase domains. This molecule is also characterized by the presence of a 158-residue juxtamembrane sequence that is some 70 residues longer than the equivalent segment in other receptor PTPases. We have also isolated cDNA for the human homolog of this molecule from a HeLa cell library (Q. Yang and N.K. Tonks, unpubl.). We have noted an intriguing similarity between the sequence of the juxtamembrane segment of this PTPase and that of the intracellular segment of the cadherin family of cell adhesion molecules (Fig. 2).

Cadherins are adhesion molecules that participate in the mechanisms of coordinated association of cells during tissue development and morphogenesis (for review, see Magee and Buxton 1991; Takeichi 1991). Essentially all cells that associate into tissues express cadherins, and reduction in the level of expression of E-cadherin has been linked with tumor invasiveness and metastasis (Frixen et al. 1991). Cadherins are concentrated in adherens-type junctions, where they are linked to cortical actin bundles. Four subclasses of cadherins have been well characterized at a structural level: L-CAM (liver cell adhesion molecule) in chickens, and E(epithelial)-, N(neural)-, and P(placental)-cadherins in mammals. However, the repertoire of cadherins is also now expanding rapidly (Napolitano et al. 1991; Suzuki et al. 1991). Their structure can be described in terms of three segments. The extracellular segment is

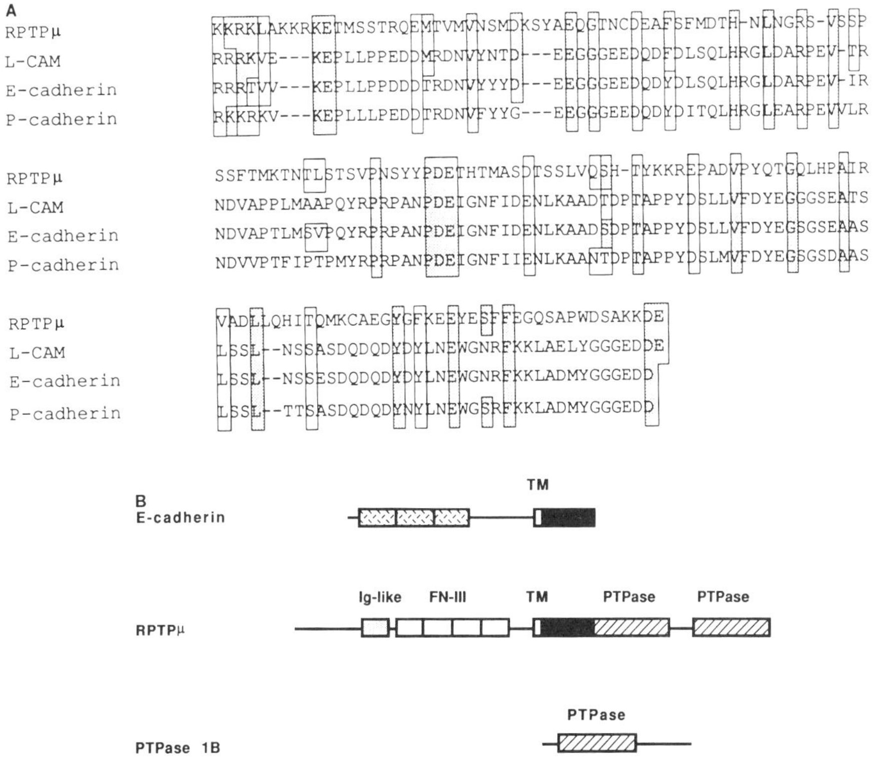

Figure 2. (*A*) Alignment of amino acid residues in the juxtamembrane domain of PTPμ with the conserved cytoplasmic domains of cadherins. Identities are highlighted in shaded boxes and conservative substitutions in open boxes. (*B*) Diagram illustrating the structure of PTPμ as well as its relationship to the cadherins and the low M_r PTPase, PTP1B. Diagonal lines denote PTPase catalytic domains. The segment of homology between PTPμ and the cadherins is represented in black. The transmembrane segments also share 8 identities. Using the ALIGN program, the mutation data matrix for scoring, and a gap penalty of 10 to compare residues 742–917 in PTPμ (Gebbink et al. 1991) with 553–722 in L-CAM (Gallin et al. 1987) and 555–724 in E-cadherin (Ringwald et al. 1987), scores of 7.3 and 7.5, respectively, were obtained. These alignment scores are expressed in units of standard deviation from the average background scores of 100 randomly generated sequences. Scores greater than 5 indicate homology.

glycosylated and comprises at least three repeated structural motifs containing stretches of acidic residues involved in binding Ca^{++}. Cadherin binding is homophilic and Ca^{++}-dependent; binding specificity is governed by the amino-terminal 113 residues of the extracellular segment. There is a single transmembrane domain and an intracellular segment of ~150 residues. It is the intracellular segment that is most highly conserved, displaying ~90% identity among the cadherins. Extensive analysis by site-directed mutagenesis has shown that deletions in the intracellular segment of E- and N-cadherin, particularly the carboxy-terminal 72 amino acids, abolish adhesion despite the presence of an intact extracellular segment (Nagafuchi and Takeichi 1988). The intracellular segment interacts with cortical actin indirectly, a process that is mediated by the binding of cytoplasmic proteins termed catenins. Catenin α (102 kD) displays homology with vinculin (Nagafuchi et al. 1991), catenin β (92–94 kD) displays homology with plakoglobin (McCrea et al. 1991), and catenin γ (~80 kD) remains to be characterized. Deletions of the carboxy-terminal sequences of cadherin also abolish the binding to catenins. Thus, the association of the cadherin cytoplasmic domain, the domain that displays homology with PTPμ, with catenins is essential for the interaction of cadherins with the cytoskeleton and their adhesive function.

What is the meaning of this homology between PTPμ and cadherins? We propose that this PTPase is directed to cytoskeletal targets through interaction between the juxtamembrane domain and molecules functionally similar to the catenins, thus localizing it to points of interface between the plasma membrane and the cytoskeleton. Interestingly, the adherens junctions have recently been identified as major sites of tyrosine phosphorylation within the cell and sites at which members of the *src* family of tyrosine kinases are concentrated (Tsukita et al. 1991). Furthermore, treatment of chick lens cells with inhibitors of PTPase activity leads to a pronounced accumulation of phosphotyrosine at adherens junctions, implying a ready reversibility of the phosphorylation reaction and a rapid turnover of tyrosine phosphate at these sites (Volberg et al. 1991).

The consensus view appears to be that although interaction with the cytoskeleton is crucial for the cell adhesion function of cadherins, this is not the case for members of the Ig-superfamily of adhesion molecules, such as N-CAM. In fact, the smallest form of N-CAM, termed ssd, lacks transmembrane and cytoplasmic segments and is attached to the external face of the membrane by a phosphatidylinositol glycan anchor (Edelman 1988). Nevertheless, there are reports of the triggering of signal transduction pathways involving G-protein-mediated changes in the levels of second messengers such as Ca^{++} by N-CAM and N-cadherin (Schuch et al. 1989; Doherty et al. 1991). PTPμ may provide a link between cell adhesion and signal transduction pathways involving reversible tyrosine phosphorylation. We also suggest that the cadherin-related juxtamembrane domain of PTPμ may regulate the interaction of this Ig-superfamily molecule with the cytoskeleton in addition to controlling its potential adhesive function in a manner analogous to the cadherins. The identification of the proteins with which PTPμ interacts should provide new insights into the role of reversible tyrosine phosphorylation in control of the cytoskeleton and cell adhesion.

Nontransmembrane PTPases

Generally, the PTPases are multidomain proteins bearing at least one catalytic domain and additional, noncatalytic, distinct structural motifs that appear to be important in regulation, including targeting to particular subcellular locations. Whereas the receptor-like PTPases have the potential for modulation of activity by binding of a ligand to the extracellular segment, several nontransmembrane PTPases, including two recently identified in our laboratory, further illustrate this general principle.

The structure of PTPH1 can be described in terms of three segments (Yang and Tonks 1991): an amino-terminal domain of ~320 residues that displays homology with the similarly located domains in band 4.1, ezrin, and talin; a central segment bearing several putative sites for phosphorylation by $p34^{cdc2}$ and casein kinase 2; and a carboxy-terminal catalytic domain. Intrinsic activity in PTPH1 has now been demonstrated (N.K. Tonks et al., unpubl.). In erythrocytes, band 4.1 promotes the association of actin and spectrin and, through its amino-terminal domain, binds to the transmembrane protein glycophorin in a phosphatidylinositol bisphosphate-dependent manner. Similar interactions between talin and β-integrin at focal adhesion plaques have also been proposed. Thus, it appears that this amino-terminal domain defines a family of proteins that are targeted to the interface between the plasma membrane and the cytoskeleton, suggesting that PTPH1 will display a similar subcellular location. Complementary DNA for a similar, but distinct, PTPase termed MEG01 has been isolated by Gu et al. (1991).

PTP-PEST is a widely expressed PTPase comprising an amino-terminal catalytic domain, in which intrinsic PTPase activity has also been demonstrated, and a carboxy-terminal segment that is rich in proline, glutamic acid, serine, and threonine residues and is characterized by the presence of PEST sequences (Yang et al. 1993). PEST sequences are a characteristic of proteins with short intracellular half-lives (for review, see Dice 1987). Such potential susceptibility to rapid degradation could be thought of as "terminal" targeting or targeting to destruction. Interestingly, like many PEST-sequence-containing proteins, PTP-PEST is the product of an inducible gene. In human rhabdomyosarcoma A204 muscle cells, stimulation with insulin leads to an increase of approximately fourfold in PTP-PEST mRNA levels. Possible links between PTP-PEST and signal transduction through the insulin receptor are under investigation. Another PEST-sequence-containing PTPase, PEP, was isolated recently by Matthews et al. (1992). It is a distinct gene product ~200 residues longer than PTP-PEST.

Recently, it has become apparent that there exists a new subfamily of PTPases which possess the conserved sequence motif surrounding the essential cysteinyl residue ([I/V]HCXAGXXR[S/T]G) but display little structural similarity elsewhere in the molecule to the other members of the family (Fig. 3). The founder member of this subfamily is the VH1 protein from the poxvirus vaccinia (Guan et al. 1991). Unlike the PTPases described thus far, which are absolutely specific for tyrosyl residues in proteins, VH1 displays dual specificity, dephosphorylating both phosphotyrosyl and phosphoseryl residues in vitro. Additional members now include the growth-factor-inducible murine immediate-early gene product 3CH134 (Charles et al. 1992) and its human homolog, CL100 (Keyse and Emslie 1992); the cell cycle control element $p80^{cdc25}$, which is the phosphatase that dephosphorylates and activates $p34^{cdc2}$ at the G_2/M phase transition of the eukaryotic cell cycle (Strausfeld et al. 1991); CDC14 from *Saccharomyces cerevisiae* (Wan et al. 1992); and an ORF from *Autographa californica* nuclear polyhedrosis virus (Tilakaratne et al. 1991).

Although it was believed that the exclusive provenance of tyrosine phosphorylation was eukaryotic cells, several reports have indicated the presence of phosphotyrosine in prokaryotes (Dadssi and Cozzone 1990; Atkinson et al. 1992). In addition, Dixon's group has ascertained that the essential virulence determinant of the bacterium *Yersinia*, the causative agent of the plague or Black Death, is a PTPase termed Yop2b, the activity of which is essential for virulence (Guan and Dixon 1990; Bliska et al. 1991). The Yop2b protein is encoded by the YopH gene located on a 70-kb megaplasmid. Rather than endogenous substrates, its targets are tyrosine phosphorylated proteins in the infected eukaryotic host cells. Thus, it has been suggested that the YopH gene may have been acquired through lateral gene transfer from a eukaryotic host, as suggested for retroviral PTKs. However, more recently, in the cyanobacterium *Nostoc commune*, a chromosomally encoded ORF termed IphP has been identified which, like VH1, displays sequence identity with the PTPases essentially only in the segment surrounding the active-site cysteine. Not only does the IphP gene encode an active phosphatase which, like VH1, displays dual specificity,

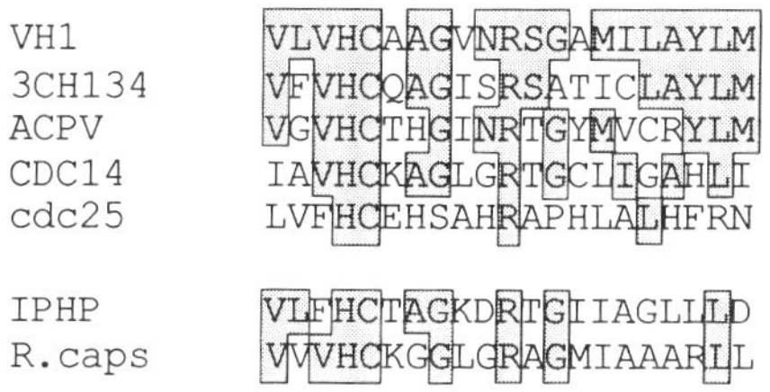

Figure 3. The VH1-related subfamily of PTPases.

but also a number of potential endogenous substrates have been identified (M. Potts et al., in prep.). We have also observed a PTPase-related ORF in the bacterium, *Rhodobacter capsulatus* (Protein Identification Resource Database, No. S19740). Such observations suggest that tyrosine phosphorylation may have arisen as a regulatory modification of proteins much earlier in evolutionary time than previously anticipated.

Regulation of PTPase Activity

Initial measurements of the activity of PTP1B and CD45 indicated a very high V_{max}, 10- to 1000-fold higher than that of the PTKs, and a high affinity for substrate in vitro (Tonks et al. 1988b, 1990). Thus, the PTPases have the potential to represent formidable barriers to the action of PTKs in vivo, suggesting that mechanisms must exist to harness their activity. We are studying such control mechanisms for several of the PTPases. As discussed above, many PTPases possess structural motifs suggestive of regulation through intracellular targeting. Superimposed on control mechanisms mediated through structural features of the enzyme is the potential for regulation by phosphorylation of the phosphatase.

There have been preliminary reports of phosphorylation of PTPases in vivo. For example, CD45 becomes phosphorylated on tyrosyl residues following stimulation of Jurkat T cells with phytohemagglutinin or anti-CD3 antibodies (Stover et al. 1991). Changes in phosphoserine levels in CD45 have also been reported; it is phosphorylated in response to phorbol ester stimulation of peripheral T cells (Autero and Gahmberg 1987), whereas stimulation of various T-cell lines with ionomycin, to mobilize intracellular Ca^{++}, leads to its dephosphorylation (Ostergaard and Trowbridge 1991). Our studies in this direction currently focus on PTP1B. Although it was originally purified as a monomeric catalytic subunit of 37 kD, isolation of cDNA for PTP1B predicted a longer protein of 435 amino acids (~50 kD) with an extension of 114 amino acids at the carboxyl terminus. This carboxy-terminal segment serves a regulatory function. The extreme carboxy-terminal 35 residues are both necessary and sufficient for targeting to membranes of the endoplasmic reticulum (ER) (Frangioni et al. 1992). The preceding 88 amino acids are predominantly hydrophilic and bear several putative sites for phosphorylation by serine/threonine kinases. We have demonstrated that PTP1B from HeLa cells is phosphorylated in vivo (Fig. 4) on seryl residues. In view of the fact that activation of the serine/threonine kinase $p34^{cdc2}$ occurs at the G_2/M phase transition of the cell cycle (for review, see Murray 1992), the possibility that $p34^{cdc2}$ was responsible for this phosphorylation was considered. Indeed, there are at least two sites in PTP1B that are phosphorylated in mitosis, of which one, S386, is a substrate for $p34^{cdc2}$ both in vitro and in vivo. There is a second site, S352, phosphorylation of which accounts for the pronounced retardation in electrophoretic mobility of PTP1B de-

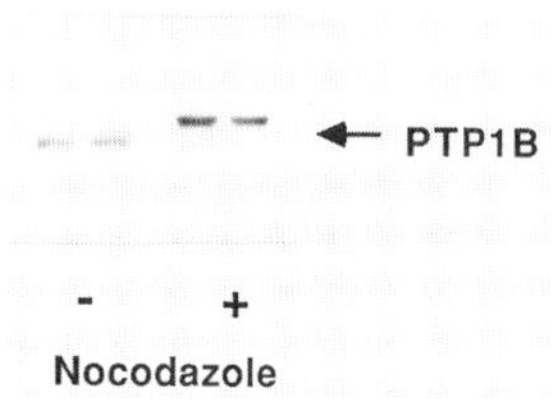

Figure 4. PTP1B is phosphorylated in a cell-cycle-dependent manner in HeLa cells. PTP1B was immunoprecipitated from ^{32}P-labeled HeLa cells using a monoclonal antibody FG6 and subjected to SDS polyacrylamide gel electrophoresis. The figure highlights the section of the gel containing PTP1B, identified by autoradiography.

tected in mitotic cells (Fig. 4). This site is not phosphorylated by p34^{cdc2} in vitro or in vivo, and the identity of the kinase involved remains to be established. The precise role of this mitotic phosphorylation is also unclear. There appears to be no gross change in intracellular localization of PTP1B coupled with the transition from G_2 to M phase, since the enzyme remains in a membrane fraction. However, a more detailed immunocytochemical examination is yet to be completed. When the activity of PTP1B from mitotic cells is measured in vitro with artificial substrates, a small inhibition is detected relative to the activity of this enzyme in asynchronous populations. It is possible that a more profound effect will be observed with a physiologically relevant substrate. The identity of the physiological substrates for PTP1B is unknown. Its localization raises the possibility of a role in controlling the dynamic changes in ER structure and association with the cytoskeleton in interphase cells (Lee and Chen 1988) or in the assembly/disassembly of this subcellular compartment with the cell cycle. The demonstration of ER localization of Ltk (Bauskin et al. 1991) also suggests that this tyrosine kinase or its substrates may be susceptible to the action of PTP1B in vivo.

In addition to mitotic phosphorylation, PTP1B is phosphorylated on S378 in vitro by PKC and on the same site in HeLa cells in vivo in response to TPA stimulation. There was no discernible effect of this modification on activity in vitro. Thus, PTP1B represents a point of convergence for the action of at least three serine/threonine kinases, suggestive of an important role in signal transduction in vivo. Such phosphorylation of a PTPase may be indicative of a new tier of control of the level of cellular phosphotyrosine and a novel mode of interaction between steps in signaling pathways involving serine/threonine and tyrosine phosphorylation.

PERSPECTIVES

Although research into the PTPases is largely in an exploratory phase where progress is primarily being made in the identification of new isoforms, the predicted structures of these enzymes are clearly consistent with important and diverse roles in controlling signal transduction in vivo. Several lines of evidence now dispel the notion that the PTPases are merely constitutively active housekeeping enzymes that function solely to antagonize the action of PTKs. First, the generation of various CD45$^-$ cell lines indicates that CD45 is essential for antigen-induced T-cell proliferation responses, for coupling of the T-cell receptor to the generation of phosphatidylinositol second messengers, and for the rapid increases in protein tyrosine phosphorylation associated with triggering of the T-cell receptor (for review, see Trowbridge 1991). It is suggested that these effects are mediated by the dephosphorylation and activation of members of the *src* family of PTKs by CD45. Similarly, CD45 has also been implicated in signaling through the B-cell receptor. Second, a key element in the control of the eukaryotic cell cycle is the dephosphorylation and activation of p34^{cdc2} by cdc25 (for review, see Murray 1992). Third, it has recently been shown that in rat embryo fibroblasts overexpressing the transmembrane enzyme RPTPα, there is a persistent activation of c-*src* with concomitant cellular transformation (Zheng et al. 1992). Finally, the *Drosophila* gene *corkscrew*, which encodes an SH2-domain-containing PTPase, functions in concert with D-*raf* to transduce positively the signal from *torso*, a receptor PTK, in embryonic cell fate determination (Perkins et al. 1992). Such studies suggest that PTPases have the potential to serve as an "on-switch," acting positively in triggering signal transduction events.

Evidence is accumulating in several laboratories, including our own (McGuire et al. 1991), to implicate a dysfunction in the action of PTPases in type II diabetes. Study of the PTPases may also yield important information regarding various neoplasias. It has already been demonstrated that overexpression of PTP1B can confer on 3T3 cells resistance to transformation by oncogenic *neu* (Brown-Shimer et al. 1992) and can at least partially antagonize the action of v-*src* (Woodford-Thomas et al. 1992). In addition, there are suggestions that some PTPases may be the products of tumor suppressor genes (LaForgia et al. 1991), i.e., deletion of a PTPase may contribute to the elevation of phosphotyrosine in some transformed cells. One would anticipate that further study of the structure, function, and mode of regulation of members of the PTPase family will not only furnish us with a more complete understanding of the physiological significance of tyrosine phosphorylation, but may also offer targets for pharmacological intervention in several human diseases.

ACKNOWLEDGMENTS

N.K.T. is a Pew Scholar in the Biomedical Sciences and acknowledges support from the National Cancer Institute (CA-53840), ICOS Corporation, and the Hansen Foundation. B.R.F.'s laboratory is supported by a grant (CA-40512) from the National Cancer Institute. A.J.F. is supported by a Cancer Research Institute/F.M. Kirby Foundation fellowship, M.F.B.G.G. by the Dutch Cancer Society, and H.S. by a Damon Runyon-Walter Winchell Cancer Research Fund fellowship. We thank Carol Marcincuk for typing the manuscript.

REFERENCES

Atkinson, M., C. Allen, and L. Sequeira. 1992. Tyrosine phosphorylation of a membrane protein from *Pseudomonas solanacearum*. *J. Bacteriol.* **174:** 4356.

Autero, M. and C.G. Gahmberg. 1987. Phorbol diesters increase the phosphorylation of the leukocyte common antigen CD45 in human T cells. *Eur. J. Immunol.* **17:** 1503.

Bauskin, A.R., I. Alkalay, and Y. Ben-Neriah. 1991. Redox regulation of a protein tyrosine kinase in the endoplasmic reticulum. *Cell* **66:** 685.

Bliska, J.B., K. Guan, J.E. Dixon, and S. Falkow. 1991. Tyrosine phosphate hydrolysis of host proteins by an essential *Yersinia* virulence determinant. *Proc. Natl. Acad. Sci.* **88:** 1187.

Brown-Shimer, S., K.A. Johnson., D.E. Hill, and A.M. Bruskin. 1992. Effect of protein tyrosine phosphatase 1B expression on transformation by the human *neu* oncogene. *Cancer Res.* **52:** 478.

Charbonneau, H. and N.K. Tonks. 1992. 1002 protein phosphatases? *Annu. Rev. Cell Biol.* **8:** 463.

Charbonneau, H., N.K. Tonks, K.A. Walsh, and E.H. Fischer. 1988. The leukocyte common antigen (CD45): A putative receptor-linked protein tyrosine phosphatase. *Proc. Natl. Acad. Sci.* **85:** 7182.

Charles, C.H., A.S. Abler, and L.F. Lau. 1992. cDNA sequence of a growth factor-inducible immediate early gene and characterization of its encoded protein. *Oncogene* **7:** 187.

Dadssi, M. and A.J. Cozzone. 1990. Evidence of protein tyrosine kinase activity in the bacterium *Acinetobacter calcoaceticus*. *J. Biol. Chem.* **265:** 20996.

Dice, J.F. 1987. Molecular determinants of protein half lives in eukaryotic cells. *FASEB J.* **1:** 349.

Doherty, P., S.V. Ashton, S.E. Moore, and F.S. Walsh. 1991. Morpho-regulatory activities of NCAM and N-cadherin can be accounted for by G protein-dependent activation of L- and N-neuronal calcium channels. *Cell* **67:** 21.

Edelman, G.M. 1988. Morphoregulatory molecules. *Biochemistry* **27:** 3533.

Frangioni, J.V., P.H. Beahm, V. Shifrin, C.A. Jost, and B.G. Neel. 1992. The nontransmembrane tyrosine phosphatase PTP1B localizes to the endoplasmic reticulum via its 35 amino acid C-terminal sequence. *Cell* **68:** 545.

Frixen, U.H., J. Behrens, M. Sachs, G.. Eberle, B. Voss, A. Warda, D. Lochner, and W. Birchmeier. 1991. E-cadherin-mediated cell-cell adhesion prevents invasiveness of human carcinoma cells. *J. Cell Biol.* **113:** 173.

Gallin, W.J., B.C. Sorkin, G.M. Edelman, and B.A. Cunningham. 1987. Sequence analysis of a cDNA clone encoding the liver cell adhesion molecule, L-CAM. *Proc. Natl. Acad. Sci.* **84:** 2808.

Gebbink, M.F.B.G., I. Van Etten, G. Hateboer, R. Suijkerbuijk, R.L. Beijersbergen, A.G. Van Kessel, and W.H. Moolenaar. 1991. Cloning, expression and chromosomal localization of a new putative receptor-like protein tyrosine phosphatase. *FEBS Lett.* **290:** 123.

Gu, M.J., D. York, I. Warshawsky, and P.W. Majerus. 1991. Identification, cloning and expression of a cytosolic megakaryocyte protein tyrosine phosphatase with sequence homology to cytoskeletal protein 4.1. *Proc. Natl. Acad. Sci.* **88:** 5867.

Guan, K. and J.E. Dixon. 1990. Protein tyrosine phosphatase activity of an essential virulence determinant in *Yersinia*. *Science* **249:** 5553.

———. 1991. Evidence for protein-tyrosine-phosphatase catalysis proceeding via a cysteine-phosphate intermediate. *J. Biol. Chem.* **266:** 17026.

Guan, K., S.S. Broyles, and J.E. Dixon. 1991. A Thr/Ser protein phosphatase encoded by vaccinia virus. *Nature* **350:** 359.

Hariharan, I.K., P.T. Chuang, and G.M. Rubin. 1991. Cloning and characterization of a receptor-class phosphotyrosine phosphatase gene expressed on central nervous system axons in *Drosophila melanogaster*. *Proc. Natl. Acad. Sci.* **88:** 11266.

Hunter, T. and B.M. Sefton. 1980. The transforming gene product of Rous sarcoma virus phosphorylates tyrosine. *Proc. Natl. Acad. Sci.* **77:** 1311.

Keyse, S.M. and E.A. Emslie. 1992. Oxidative stress and heat shock induce a human gene encoding a protein tyrosine phosphatase. *Nature* **359:** 644.

Krueger, N.X. and H. Saito. 1992. A human transmembrane protein tyrosine phosphatase, PTPζ, is expressed in brain and has an N terminal receptor domain homologous to carbonic anhydrases. *Proc. Natl. Acad. Sci.* 89: 7417.

LaForgia, S., B. Morse, J. Levy, G. Barnea, L.A. Cannizzaro, F. Li., P.C. Nowell, L. Boghosian-Sell, J. Glick, A. Weston, C.C. Harris, H. Drabkin, D. Patterson, C.M. Croce, J. Schlessinger, and K. Huebner. 1991. Receptor protein tyrosine phosphatase γ is a candidate tumor suppressor gene at human chromosome region 3p21. *Proc. Natl. Acad. Sci.* **88:** 5036.

Lee, C. and L.B. Chen. 1988. Dynamic behavior of endoplasmic reticulum in living cells. *Cell* **54:** 37.

Magee, A.I. and R.S. Buxton. 1991. Transmembrane molecular assemblies regulated by the greater cadherin family. *Curr. Opin. Cell Biol.* **3:** 854.

Matthews, R.J., D.B. Bowne, E. Flores, and M.L. Thomas. 1992. Characterization of haematopoietic intracellular protein tyrosine phosphatases: Description of a phosphatase containing an SH2 domain and another enriched in proline, glutamic acid, serine and threonine rich sequences. *Mol. Cell. Biol.* **12:** 2396.

McCrea, P., C.W. Turck, and B. Gumbiner. 1991. A homolog of the *Armadillo* protein in *Drosophilia* "Plakoglobin" associated with E cadherin. *Science* **254:** 1359.

McGuire, M.C., R.M. Fields, B.L. Nyomba, I. Raz, C. Bogardus, N.K. Tonks, and J. Sommercorn. 1991. Abnormal regulation of protein tyrosine phosphatase activities in skeletal muscle of insulin-resistant humans. *Diabetes* **40:** 939.

Murray, A.W. 1992. Creative blocks: Cell cycle check points and feedback controls. *Nature* **359:** 599.

Nagafuchi, A. and M. Takeichi. 1988. Cell binding function of E-cadherin is regulated by the cytoplasmic domain. *EMBO J.* **7:** 3679.

Nagafuchi, A., M. Takeichi, and S. Tsukita. 1991. The 102 kD cadherin-associated protein: Similarity to vinculin and post transcriptional regulation of expression. *Cell* **65:** 849.

Napolitano, E.W., K. Venstrom, E.F. Wheeler, and L.F. Reichardt. 1991. Molecular cloning and characterization of B-cadherin, a novel chick cadherin. *J. Cell Biol.* **113:** 893.

Ostergaard, H.L. and I.S. Trowbridge. 1991. Negative regulation of CD45 protein tyrosine phosphatase activity by ionomycin in T cells. *Science* **253:** 1423.

Perkins, L.A., I. Larsen, and N. Perrimon. 1992. *corkscrew* encodes a putative protein tyrosine phosphatase that functions to transduce the terminal signal from the receptor tyrosine kinase *torso*. *Cell* **70:** 225.

Ringwald, M., R. Schuh, D. Vestweber, H. Eiskelter, F. Lottspeich, J. Engel, R. Dolz, F. Jahnig, J. Epplen, S. Mayer, C. Muller, and R. Kemler. 1987. The structure of cell adhesion molecule uvomorulin. Insights into the molecular mechanism of Ca^{2+}-dependent cell adhesion. *EMBO J.* **6:** 3647.

Schuch, U., M.J. Lohse, and M. Schachner. 1989. Neural cell adhesion molecules influence second messenger systems. *Neuron* **3:** 13.

Stamenkovic, I., D. Sgroi, A. Aruffo, M.S. Sy, and T. Anderson. 1991. The B lymphocyte adhesion molecule CD22 interacts with leukocyte common antigen CD45RO on T cells and α2-6 sialyltranferase, CD75, on B cells. *Cell* **66:** 1133.

Stover, D.R., H. Charbonneau, N.K. Tonks, and K.A. Walsh. 1991. Protein tyrosine phosphatase CD45 is phosphorylated transiently on tyrosine upon activation of Jurkat T cells. *Proc. Natl. Acad. Sci.* **88:** 7704.

Strausfeld, U., J.C. Labbé, D. Fesquet, J.C. Cavadore, A. Pacard, K. Sadhu, P. Russell, and M. Dorée. 1991. Dephosphorylation and activation of a $p34^{cdc2}$/cyclin B complex *in vitro* by human CDC25 protein. *Nature* **351:** 242.

Streuli, M., N.X. Krueger, L.R. Hall, S.F. Schlossman, and H. Saito. 1988. A new member of the immunoglobulin superfamily that has a cytoplasmic region homologous to the leukocyte common antigen. *J. Exp. Med.* **168:** 1523.

Streuli, M., N.X. Krueger, T. Thai, M. Tang, and H. Saito. 1990. Distinct functional roles of the two intracellular phosphatase-like domains of the receptor-linked protein tyrosine phosphatases LCA and LAR. *EMBO J.* **9:** 2399.

Streuli, M., N.X. Krueger, P.D. Ariniello, M. Tang, J.M. Munro, W.A. Blattler, D.A. Adler, C.M. Disteche, and H. Saito. 1992. Expression of the receptor-linked protein tyrosine phosphatase LAR: Proteolytic cleavage and shedding of the CAM-like extracellular region. *EMBO J.* **11:** 897.

Suzuki, S., K. Sano, and H. Tanihara. 1991. Diversity of the cadherin family: Evidence for 8 new cadherins in nervous tissue. *Cell Regul.* **2:** 261.

Takeichi, M. 1991. Cadherin cell adhesion receptors as a morphogenetic regulator. *Science* **251:** 1451.

Thomas, M.L. 1989. The leukocyte common antigen family. *Annu. Rev. Immunol.* **7:** 339.

Tian, S.S., P. Tsoulfas, and K. Zinn. 1991. Three receptor-linked protein phosphatases are selectively expressed on central nervous system axons in the *Drosophila* embryo. *Cell* **67:** 675.

Tilakaratne, N., S.E. Hardin, and R.F. Weaver. 1991. Nucleotide sequence and transcript mapping of the *Hin*dIIIF region of the *Autographa californica* nuclear polyhedrosis virus genome. *J. Gen. Virol.* **72:** 285.

Tonks, N.K., C.D. Diltz, and E.H. Fischer. 1988a. Purification of the major protein tyrosine phosphatases of human placenta. *J. Biol. Chem.* **263:** 6722.

———. 1988b. Characterization of the major protein tyrosine phosphatases of human placenta. *J. Biol. Chem.* **263:** 6731.

———. 1990. CD45, an integral membrane protein tyrosine phosphatase. *J. Biol. Chem.* **265:** 10674.

Tonks, N.K., H. Charbonneau, C.D. Diltz, E.H. Fischer, and K.A. Walsh. 1988c. Demonstration that the leukocyte common antigen CD45 is a protein tyrosine phosphatase. *Biochemistry* **27:** 8695.

Tonks, N.K., A.J. Flint, M.F.B.G. Gebbink, H. Sun, and Q. Yang. 1993. Signal transduction and protein tyrosine dephosphorylation. *Adv. Messenger Phosphoprotein Res.* **28:** (in press).

Trowbridge, I.S. 1991. CD45: A prototype for transmembrane protein tyrosine phosphatases. *J. Biol. Chem.* **266:** 23517.

Tsukita, S., K. Oishi, T. Akiyama, Y. Yamanashi, T. Yamamoto, and S. Tsukita. 1991. Specific proto-oncogenic tyrosine kinases of *src* family are enriched in cell-to-cell adherens junctions where the level of tyrosine phosphorylation is elevated. *J. Cell Biol.* **113:** 867.

Volberg, T., B. Geiger, R. Dror, and Y. Zick. 1991. Modulation of intercellular adherens-type junctions and tyrosine phosphorylation of their components in RSV-transformed cultured chicken cells. *Cell Regul.* **2:** 105.

Wan, J., H. Xu, and M. Grunstein. 1992. CDC14 of *Saccharomyces cerevisiae*. *J. Biol. Chem.* **267:** 11274.

Wang, Y. and C.J. Pallen. 1991. The receptor-like protein tyrosine phosphatase HPTPα has two active catalytic domains with distinct substrates specificities. *EMBO J.* **10:** 3231.

Woodford-Thomas, T.A., J.D. Rhodes, and J.E. Dixon. 1992. Expression of a protein tyrosine phosphatase in normal and v-*src*-transformed mouse 3T3 fibroblasts. *J. Cell Biol.* **117:** 401.

Yang, Q. and N.K. Tonks. 1991. Isolation of a cDNA clone encoding a human protein tyrosine phosphatase with homology to the cytoskeletal-associated proteins band 4.1, ezrin and talin. *Proc. Natl. Acad. Sci.* **88:** 5949.

Yang, Q., D. Co, J. Sommercorn, and N.K. Tonks. 1993. Cloning and expression of PTP-PEST: A novel, human, nontransmembrane protein tyrosine phosphatase. *J. Biol. Chem.* (in press).

Yang, X., K.T. Seow, S.M. Bahri, S.H. Oon, and W. Chia. 1991. Two *Drosophila* receptor-like tyrosine phosphatase genes are expressed in a subset of developing axons and pioneer neurons in the embryonic CNS. *Cell* **67:** 661.

Yu, Q., T. Lenardo, and R.A.Weinberg. 1992. The N-terminal and C-terminal domains of a receptor tyrosine phosphatase are associated by non-covalent linkage. *Oncogene* **7:** 1051.

Zheng, X.M., Y. Wang, and C.J. Pallen. 1992. Cell transformation and activation of $pp60^{c\text{-}src}$ by overexpression of a protein tyrosine phosphatase. *Nature* **359:** 336.

The Erythropoietin Receptor: Dimerization, Activation, and Tumorigenesis

H.F. LODISH,*† D. HILTON,* G. LONGMORE,*‡
S.S. WATOWICH,* AND A. YOSHIMURA*
*Whitehead Institute for Biomedical Research, Cambridge, Massachusetts 02142; † Department of Biology, Massachusetts Institute of Technology, Cambridge, Massachusetts 02139; ‡ Division of Hematology-Oncology, Department of Medicine, Brigham and Women's Hospital, Harvard Medical School, Boston, Massachusetts 02115

Blood cells continuously originate from low numbers of self-renewing pluripotent stem cells that generate progenitor cells committed to one or a few hematopoietic lineages. Control of these differentiation pathways is determined, in part, through specific growth factors binding to their cognate receptor(s) (Metcalf 1989). Abrogation or alteration of this control network may lead to the development of leukemia.

Erythropoietin (EPO) is a serum glycoprotein hormone required for the survival, proliferation, and differentiation of committed erythroid progenitor cells (Krantz 1991). We isolated by expression cloning the murine erythropoietin receptor (EPO-R) cDNA (D'Andrea et al. 1989b); it has sequence homology with other cytokine receptors (D'Andrea et al. 1989a). Conserved structural features of the cytokine receptor superfamily include four similarly spaced exoplasmic cysteine residues, as well as a motif, WSXWS, located in the exoplasmic domain close to the membrane-spanning region (Cosman et al. 1990). The EPO-R and other members of the cytokine receptor family do not contain kinase-related or nucleotide-binding consensus sequences in their cytoplasmic domains, and the intracellular signaling pathways they initiate following ligand binding have yet to be defined.

Although little is known of the mechanisms by which cytokine receptors transduce their signal, dimerization of the receptors is thought to play a role. The receptors for interleukins 2, 3, 5, and 6, as well as granulocyte-macrophage colony stimulating factor, contain at least two different subunits (Hatakeyama et al. 1989; Taga et al. 1989; Hayashida et al. 1990; Devos et al. 1991; Kitamura et al. 1991), whereas the ligand-binding subunits of the granulocyte colony stimulating factor receptor, prolactin receptor, and growth hormone receptor form homodimers (Fukunaga et al. 1990; Cunningham et al. 1991; Rozakis-Adcock and Kelly 1991). Dimerization has been postulated to yield high-affinity receptors and also to provide the first step in the signal transduction pathway (Cunningham et al. 1991; Nicola and Metcalf 1991). Importantly, the crystal structure of the growth hormone receptor/growth hormone complex has recently been determined (De Vos et al. 1992). The exoplasmic domain contains two subdomains, each of which is built of seven β strands. In dimerization of the growth hormone receptor, two receptor subunits bind to different sites on a single growth hormone molecule. The receptor dimers are stabilized by growth hormone as well as by interactions between the membrane-proximal subdomains of the exoplasmic segments of the two subunits (De Vos et al. 1992).

Expression of the cloned EPO-R cDNA in the IL-3-dependent pro-B-cell line Ba/F3 allows the cells to grow in response to EPO, demonstrating that the EPO-R can functionally transmit a growth signal (Li et al. 1990). The recent demonstration that the mutation Arg-129 to cysteine (R129C) results in a constitutively active form of the EPO-R, often called cEPO-R, is provocative because it implicates the formation of aberrant inter- or intramolecular disulfide bonds in the process of receptor activation (Yoshimura et al. 1990b). We investigated the role of the novel cysteine residue in the constitutively active receptor and the possibility that this receptor may have an altered disulfide bonding pattern, using both biochemical and mutagenesis approaches. A cysteine at residue 129 is required for EPO-independent signaling. Analysis of several mutants of the EPO-R has revealed that all constitutively active mutants, but not the wild-type receptor or EPO-dependent mutants, form disulfide-linked homodimers in the endoplasmic reticulum (ER) and that a fraction of these dimers are transported to the plasma membrane.

Friend virus induces an acute erythroleukemia in adult mice (Friend 1957). A replication-defective spleen focus-forming virus (SFFV) and a replication-competent Friend murine leukemia virus (F-MLV) comprise the Friend virus complex. The SFFV *env* gene encodes a recombinant/deletion membrane glycoprotein, gp55 (Troxler et al. 1977), that is directly involved in leukemogenesis (Linemeyer et al. 1981; Ruta et al. 1983; Li et al. 1986, 1987; Aizawa et al. 1990). The gp55 protein binds to and activates the EPO-R, resulting in factor-independent cellular proliferation (Li et al. 1990; Ruscetti et al. 1990). Infecting mice with a SFFV-encoding EPO, rather than gp55, also results in erythrocytosis and splenomegaly similar to early Friend disease (Hoatlin et al. 1990). Activation of the EPO-R by gp55 may contribute to the early "preleukemic" polyclonal erythroblastosis (Hankins and Troxler 1980) observed in Friend disease. Consistent with the multistep nature of carcinogenesis (for review, see Weinberg 1989), it is apparent that multiple oncogenic events, acting at differing cellular sites, are necessary to achieve the full leukemic phenotype (for review, see Ben-David and Bernstein 1991; Sawyers et al. 1991; VanLohuizen and Berns 1990).

To determine whether the activated EPO-R could contribute to the development of leukemia, we generated a SFFV in which gp55 was replaced with the activated EPO-R, R129C. This virus, when injected into adult mice, results in polycythemia and splenomegaly. From infected mice we have isolated clonal hematopoietic growth-factor-independent cell lines expressing EPO-R R129C, which are arrested at the proerythroblast stage of erythroid differentiation. These cells, when injected back into mice, result in rapid development of erythroleukemia. This is the first demonstration of a truly oncogenic point mutation in any member of the cytokine receptor superfamily. As shown here, EPO-R R129C does not transform fibroblasts; this explains why oncogenic mutations in other members of this receptor superfamily have not yet been described.

MATERIALS AND METHODS

Mutagenesis techniques. The constitutive mutant of the EPO-R (R129C; also called cEPO-R), the carboxy-terminal truncated form of the receptor (tEPO-R), and the constitutive, truncated form of the receptor (tEPO-R-R129C) were isolated by a retroviral transduction system (Yoshimura et al. 1990b). The remaining mutant EPO-Rs (Fig. 1) were generated by polymerase chain reaction, using synthetic oligonucleotide primers encoding the desired amino acid substitutions. Mutant EPO-R cDNAs were subcloned into the mammalian expression vector pXM (Yang et al. 1986) and into M13 mp18 and mp19 vectors. Sequences of the mutant cDNAs were confirmed by the dideoxy chain terminating method, using synthetic oligonucleotides as primers.

Cell culture conditions, transfections, and mice. The wild-type and mutant EPO-R cDNAs in pXM were introduced into Ba/F3 cells by electroporation, and stable transformants were cloned, as described previously (Yoshimura et al. 1990b). IL-3-dependent, EPO-dependent, and factor-independent clones of Ba/F3 cells (Palacios and Steinmetz 1985) were maintained as described previously (Yoshimura et al. 1990b). NIH-3T3 subclone P1, kindly provided by M. Kamps (Whitehead Institute), was maintained in Dulbecco's modified Eagle medium (DME) supplemented with 10% heat-inactivated calf serum (CS). An NIH-3T3 cell line expressing v-*abl* was provided by R. Van Etten and P. Jackson (Whitehead Institute). The retroviral packaging cell lines ψ-cre and ψ-crip (Danos and Mulligan 1988) were provided by the Mulligan laboratory (Whitehead Institute) and maintained in DME supplemented with 10% CS. HCD57 cells are an erythropoietin-dependent erythroleukemia cell line derived from serial passages of cells isolated from the liver of newborn mice infected with Friend MLV (Ruscetti et al. 1990). They were kindly provided by D. Hankins (Bethesda, Maryland) and were maintained in Iscoves modified Dulbecco's medium (IMDM) supplemented with 20% FCS, 10^{-4} M β-mercaptoethanol, and 0.8 units/ml EPO (Amgen). BALB/c mice were obtained from Jackson Labs, Maine. NIH/Swiss mice were obtained from the National Cancer Institute. All mice

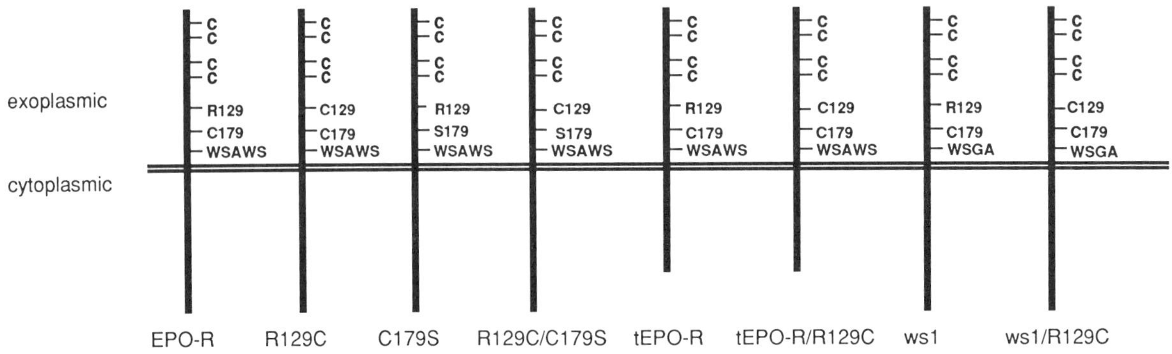

Growth in:	EPO-R	R129C	C179S	R129C/C179S	tEPO-R	tEPO-R/R129C	ws1	ws1/R129C
EPO	+	+	+	+	+	+	-	-
No Factor	-	+	-	+	-	+	-	-
Oligomerization	-	+	-	+	-	+	-	-

Figure 1. Schematic diagram of mutant EPO-Rs. Mutants of the EPO-R cDNA were generated or isolated as described in Materials and Methods. The receptors were assayed for their ability to confer EPO-dependent or factor-independent growth in Ba/F3 cells as described previously (Li et al. 1990; Yoshimura et al. 1990b). Assays for oligomerization were performed as indicated in Fig. 2 and by Watowich et al. (1992). Residues 129 and 179 are indicated; the four conserved exoplasmic cysteine residues (Cys-27, Cys-37, Cys-65, and Cys-81) and the conserved WSAWS sequence (residues 207–211) are not numbered.

were female and 4–6 weeks of age at the time of virus injection.

Metabolic labeling and immunoprecipitation. Ba/F3 cell lines expressing wild-type or mutant EPO-Rs were metabolically labeled with [^{35}S]methionine and cysteine (^{35}S-Express, NEN Research Products, Boston, Massachusetts). Cell lysates were prepared in buffer containing 1% Triton X-100, 0.5% sodium deoxycholate, 0.1% sodium dodecyl sulfate (SDS), 150 mM NaCl, 50 mM Tris-HCl (pH 7.4), 200 mM iodoacetamide, and 2 mM phenylmethylsulfonyl fluoride. Lysates were incubated with antipeptide antibodies raised against amino- or carboxy-terminal peptides of the EPO-R (Yoshimura et al. 1990a), followed by incubation with Protein A–agarose beads (Boehringer Mannheim, Indianapolis, Indiana). Proteins were eluted in gel sample buffer (1% SDS, 10% glycerol, 80 mM Tris-HCl [pH 6.8]) with or without 1% β-mercaptoethanol.

Gel electrophoresis and immunoblot analysis. One- and two-dimensional gel electrophoresis was carried out using 7.5% polyacrylamide-SDS gels. For two-dimensional gels, polypeptides were separated first under nonreducing conditions, then following reduction with 5% β-mercaptoethanol. Immunoblotting was performed; the filters were incubated with antisera raised to EPO-R peptides, followed by ^{125}I-labeled Protein A (Amersham Corporation, Arlington Heights, Illinois).

Biotinylated-EPO binding. Biotinylated erythropoietin (bEPO) was prepared (Wognum et al. 1990), and surface EPO-Rs were isolated as described previously, except that cross-linking of the bEPO/receptor complex was not performed (Yoshimura and Lodish 1992).

Plasmid construction and generation of viruses. *Kpn*I-*Xho*I fragments of EPO-R (D'Andrea et al. 1989b) and cEPO-R (i.e., R129C; Yoshimura et al. 1990b) cDNAs were subcloned into the compatible sites of the mammalian expression vector pXM, generating the plasmids pXM EPO-R and pXM cEPO-R. A *ras*-containing plasmid pUCJE was a kind gift from Robert Weinberg's laboratory (Whitehead Institute). The *Kpn*I-*Eco*RI DNA fragments, encompassing the coding sequences for EPO-R and cEPO-R, from pXM EPO-R or pXM cEPO-R, were isolated. The 5′ *Kpn*I end was blunted with T4 DNA polymerase (N.E. Biolabs) and ligated to *Bcl*I linkers (N.E. Biolabs). These EPO-R, 5′-*Bcl*I-*Eco*RI-3′ pieces were subcloned into the *Bam*HI-*Eco*RI sites of the retroviral vector pSFF (Bestwick et al. 1988). The plasmid pSFF was derived from a molecular clone (pL2-6K) of Friend SFFV (Li et al. 1987). The *Bam*HI-*Eco*RI fragment of the *env* gene in pL2-6K has been deleted. The resultant plasmids pSFF EPO-R and pSFF cEPO-R were transfected into a mixture of ψ-cre and ψ-crip packaging cells by calcium phosphate precipitation (Graham et al. 1980), and viruses containing the EPO-R (SFFV EPO-R) or cEPO-R (SFFV cEPO-R) cDNA sequence were obtained from the culture supernatant after 3 weeks of cocultivation, according to a previously described method (Bestwick et al. 1988). Rauscher-MLV was produced from R-wt/NIH cells (a kind gift from J.-P. Li, Whitehead Institute). For viral infection of cell lines, 5×10^5 to 1×10^6 cells were mixed with 2 ml of culture supernatant from virus-producing cells. Polybrene (8 μg/ml; Sigma) was added to facilitate the infection. After 4 hours of incubation at 37°C, fresh medium was added, and the cells were allowed to grow for 48 hours prior to analysis or removal of growth factors. To infect mice, freshly isolated culture supernatants from virus-producing cells were filtered through a 0.45 micron sterile filter, and 0.3 ml was injected via the tail vein.

Establishment of cell lines. Spleens from infected mice were suspended in Hanks balanced salt solution containing 25 mM HEPES (pH 7.2) (HBSS/H) on ice and teased apart with forceps. Large debris was allowed to settle and was discarded. The cells were washed twice in HBSS/H and plated in IMDM containing 20% FCS, EPO (2 units/ml), and IL-3 (100 units/ml). Cells were allowed to grow for 4 days, at which time they were transferred to plastic plates, in fresh culture medium, for 4 days. Nonadherent cells were collected and allowed to grow for another 4 days in regular tissue culture flasks in fresh culture medium containing EPO and IL-3. Following this, the cells were washed twice with serum-free medium and resuspended in IMDM containing 20% FCS but no added growth factors. Individual cells from this pool of factor-independent cells were cloned in agar; cells were plated in 60-mm petri dishes containing 5 ml of IMDM with 20% FCS, 5×10^{-5} M β-mercaptoethanol, and 0.35% SeaPlaque agarose, in the absence of any added growth factor. Clones were picked after 10 days.

Leukemogenicity assays. Growth-factor-independent clones isolated from the spleens of the diseased animals were washed once in serum-free medium and once in HBSS/H. Cells were resuspended in HBSS/H and 1×10^6 to 5×10^6 cells were injected into the tail veins or subcutaneously into the posterior flank of syngeneic mice that had received 0.5 Gy (500 rads) of gamma irradiation one day prior to cell challenge. Mice injected intravenously had weekly blood smears examined and spun hematocrits determined.

Fibroblast transformation assay. Transfection of NIH.3T3.P1 cells was carried out with a calcium phosphate technique (Roussel et al. 1984). Briefly, 50% confluent cells in 35-mm diameter culture dishes were transfected with 0.1–5.0 μg of plasmid DNA. After 18 hours of incubation, the cells were trypsinized and seeded into three 60-mm dishes. The cells were fed every 3 days with complete medium containing 5% FCS. On day 14, foci of transformed cells were identified and scored under the microscope. On day 21, the plates were washed and stained with 3% crystal violet in 40% methanol. Cell lines expressing EPO-R and cEPO-R were independently established by transfecting NIH.3T3.P1 cells with pSFF EPO-R and pSFF cEPO-R, respectively. Individual clones were isolated, and those expressing the EPO-R were identified through immunoblotting of extracted proteins. Their ability to form colonies in semisolid medium in either the presence or the absence of added EPO was determined (Freshney 1987).

RESULTS

A Constitutively Active and Dimeric Erythropoietin Receptor

The substitution Arg-129 to cysteine introduces a sixth cysteine residue into the EPO-R exoplasmic domain and confers constitutive (hormone-independent) activity (Fig. 1, mutant R129C; also called cEPO-R) (Yoshimura et al. 1990b). Mutation of residue 129 to serine, glutamic acid, or proline yields a wild-type, not a constitutive, phenotype (data not shown). Since a cysteine residue at position 129, rather than the loss of an arginine, was crucial for constitutive activity, we suspected that the new cysteine may form an intramolecular disulfide bond, possibly with Cys-179. To test this hypothesis, Cys-179 was mutated to a serine residue either in the wild-type receptor (C179S) or in the constitutive mutant (R129C/C179S). Neither the EPO-responsiveness of the wild-type receptor nor the constitutive activity of the R129C receptor was affected by this mutation (Fig. 1), demonstrating that Cys-179 is not involved in EPO-induced activation of the wild-type receptor or in constitutive activation of the R129C mutant.

To determine if the pattern of intramolecular or intermolecular disulfide bonding in R129C was different from that in the wild-type receptor, we immunoprecipitated metabolically labeled receptors and analyzed them by reducing and nonreducing SDS-PAGE. Both the newly synthesized wild-type EPO-R and R129C migrate with an apparent molecular weight of

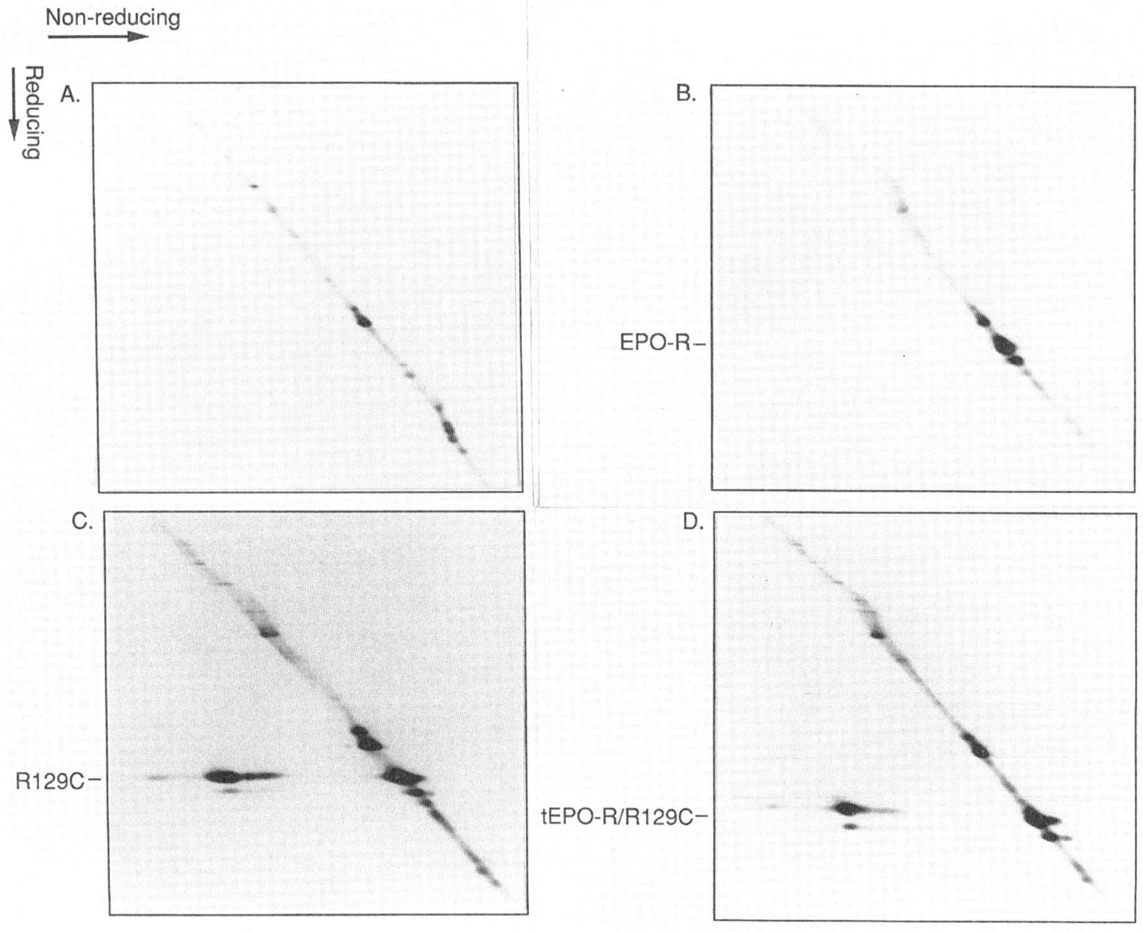

Figure 2. R129C and tEPO-R/R129C form disulfide-linked homodimers. Parental Ba/F3 cells (*A*) or cells expressing the wild-type EPO-R (*B*), R129C (*C*), or tEPO-R/R129C (*D*) were pulse-labeled with [^{35}S]methionine/cysteine for 10 min at 37°C and then chased for 2 hr at 18 °C. The EPO-Rs were immunoprecipitated and separated under nonreducing conditions in the first dimension, followed by reduction and electrophoresis in the second dimension. The positions of monomeric EPO-Rs are indicated.

64,000 under reducing conditions (Yoshimura et al. 1990b). Under nonreducing conditions, the newly synthesized wild-type EPO-R migrates with an apparent molecular weight of 64,000, whereas the constitutive receptor migrates as a monomer of $M_r \sim 64{,}000$ as well as an oligomer of $M_r \sim 160{,}000$. The constitutive, truncated EPO-R (tEPO-R/R129C) also forms disulfide-linked oligomers that migrate faster than the R129C oligomers, as expected for an oligomeric species composed of truncated receptor molecules (Watowich et al. 1992).

The identity of the polypeptides present in the EPO-R immunoprecipitates from parental Ba/F3 cells and cells expressing the wild-type receptor, R129C, and tEPO-R/R129C, was determined by two-dimensional gel electrophoresis. The only polypeptides that were generated by reduction of the disulfide-linked oligomers were monomers of R129C (Fig. 2C) and the faster-migrating monomers of tEPO-R/R129C (Fig. 2D). This demonstrates that the disulfide-linked species are EPO-R homo-oligomers and eliminates the possibility that the constitutive EPO-R forms a hetero-oligomer with another polypeptide of similar size. The ability of R129C/C179S, but not the wild-type receptor or C179S, to form disulfide-linked oligomers (data not shown) suggests that only Cys-129 is available for forming interchain disulfide bonds; thus, it is likely that the R129C oligomers are homodimers. The wild-type EPO-R migrated identically before and after reduction (Fig. 2B).

We assayed several other mutants of the EPO-R for their ability to form disulfide-linked oligomers and found that all constitutively active mutants of the EPO-R form disulfide-linked oligomers, whereas all hormone-responsive or inactive forms of the receptor do not (Fig. 1). Inactive receptor mutants include ws1 and ws1/R129C, which lack three residues, Ala-Trp-Ser, from the conserved WSXWS region and which have a Gly-Ala substitution in their place (Yoshimura et al. 1992). In addition, the ws1/R129C mutant contains the R129C mutation. When expressed in Ba/F3 cells, these mutant receptors are unable to transmit an EPO growth signal, since the cells will only grow in the presence of IL-3 (Fig. 1).

To determine if the disulfide-linked homodimers are present on the cell surface, we used biotinylated EPO (bEPO) to isolate cell-surface receptors (Yoshimura and Lodish 1992). Both R129C monomers and disulfide-linked dimers were found on the plasma membrane (Fig. 3, lane 3). Similarly, cells synthesizing R129C/C179S (Fig. 3, lane 5) and tEPO-R/R129C (data not shown) expressed both monomers and dimers on the surface. Following reduction, oligomers of the constitutive receptor were not detectable, and only monomers were seen (Fig. 3, lanes 8 and 10). Cells synthesizing hormone-responsive forms of the receptor (wild-type EPO-R, C179S, tEPO-R) did not express detectable levels of surface disulfide-linked dimers; as expected, only monomeric species are found on the plasma membrane (Fig. 3, lanes 2 and 4, data not shown).

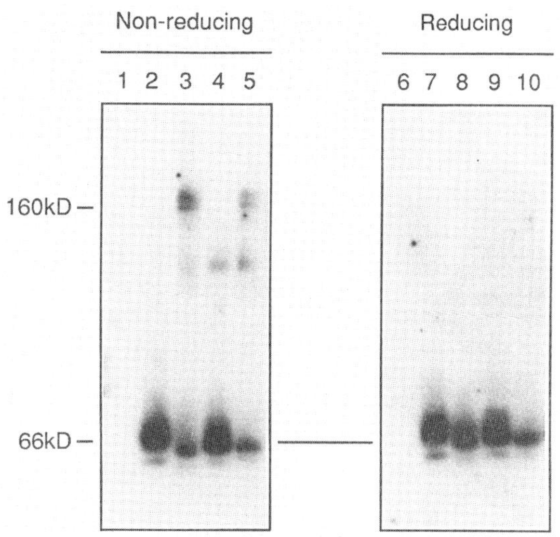

Figure 3. Disulfide-linked dimers of the constitutive EPO-R are found on the cell surface. Parental Ba/F3 cells (lanes *1,6*) or cells expressing the wild-type EPO-R (lanes *2,7*), R129C (lanes *3,8*), C179S (lanes *4,9*), or R129C/C179S (lanes *5,10*) were incubated with 10 nM bEPO for 4–6 hr at 4°C. The cells were washed and then were lysed in buffer containing 0.5% NP-40 and 200 mM iodoacetamide. Surface receptors that had bound bEPO were isolated on streptavidin-agarose beads, then separated by nonreducing and reducing gel electrophoresis and assayed by immunoblotting. The positions of cell-surface monomeric EPO-Rs (66 kD) and surface disulfide-linked dimers (160 kD) are indicated.

Infection of Mice with a Virus Expressing cEPO-R Results in Erythrocytosis and Splenomegaly

To test whether cEPO-R is oncogenic in vivo and could contribute to the development of leukemia, we engineered a SFFV in which the pathogenic *env* gene (*gp55*) was replaced with the cDNA encoding the constitutively activated erythropoietin receptor (Fig. 4).

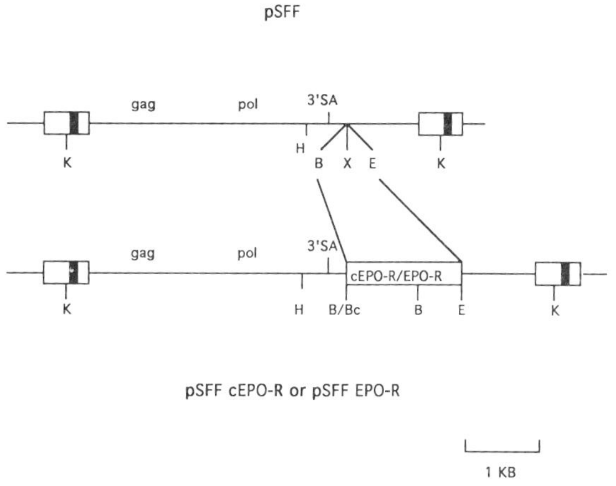

Figure 4. Plasmids used for the generation of viruses. The *Kpn*I-*Eco*RI fragment from pXM cEPO-R or pXM EPO-R, encompassing the coding region of cEPO-R or EPO-R cDNAs, was subcloned into the *Bam*HI-*Eco*RI site of the truncated *env* gene of the plasmid pSFF. Unique restriction sites are shown: B, *Bam*HI; Bc, *Bcl*I; E, *Eco*RI; H, *Hin*dIII; K, *Kpn*I; X, *Xho*I.

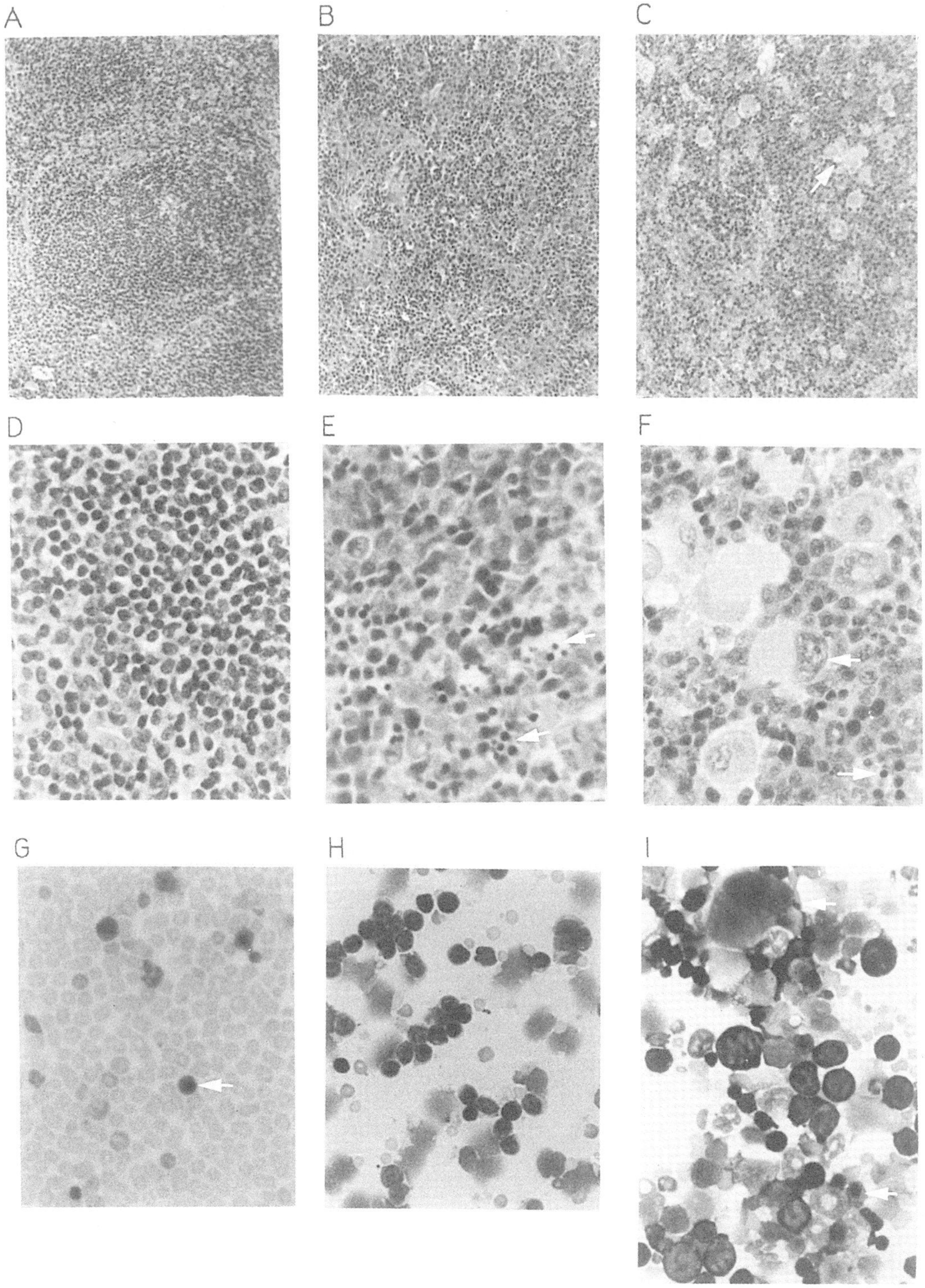

Figure 5. Pathological analysis of mice injected with SFFV cEPO-R virus. (*A*) Histological section from uninfected spleen demonstrating normal follicular architecture. (*B,C*) Histological section from spleens of mice A1 and P1, respectively. (*D*) Histological section of uninfected splenic nodule (white pulp) demonstrating normal lymphocytic morphology. (*E*) Histological section from the spleen of mouse A1. Arrow illustrates focus of erythropoiesis. (*F*) Histological section from the spleen of mouse D1. Upper arrow locates a megakaryocyte, lower arrow a focus of erythropoiesis. (*G*) Peripheral blood smear from mouse A1 showing extensive reticulocytosis and presence of an immature erythroid precursor (arrow). (*H*) Cytospin preparation of splenic cells from an uninfected mouse. (*I*) Cytospin preparation of splenic cells from A1. Upper arrow identifies a megakaryocyte, lower arrow a group of erythroid cells. All blood smears and cytospin preparations were treated with Wright-Giemsa stain. Histological sections were stained with hematoxylin and eosin. Magnifications: *A–C*, 2925×; *D–I*, 11,700×.

Control experiments showed that the virus, called SFFV cEPO-R, was of substantial titer and biologically functional: Infection of IL-3-dependent, pro-B Ba/F3 cells (Palacios and Steinmetz 1985), or EPO-dependent erythroleukemic HCD57 cells (Ruscetti et al. 1990), efficiently conferred factor-independent growth (data not shown). This showed, consistent with previous experiments (Yoshimura et al. 1990b), that cEPO-R can induce factor-independent proliferation of certain non-erythroid as well as erythroid cells.

Initial animal experiments utilized a viral complex generated by rescuing the defective SFFV cEPO-R virus from clones of NIH-3T3 cells by superinfecting these cells with a replication-competent Rauscher MLV (R-MLV) helper virus. Five weeks after injecting the resulting viral supernate into four adult BALB/c mice, one mouse (A1) developed erythrocytosis (hematocrit 65, normal 49) and splenomegaly (1.2 g, normal 0.1 g). Blood smears revealed extensive reticulocytosis, evidence of immature erythroid precursors in the blood, and occasional blast-like cells (Fig. 5G). The splenic architecture was entirely distorted with regression of the white pulp and infiltration with large basophilic blast-like cells (Fig. 5B,E; compare to normal panels A and D). Within the spleen, cells in various stages of erythroid differentiation were evident, as was an increase in the number of megakaryocytes.

Southern blot analysis of splenic DNA digested with *Kpn*I, which cuts only in the LTRs of the SFFV cEPO-R provirus, revealed proviral integration (data not shown). Importantly, gp55 was not expressed; this excludes the likelihood of a recombinational event generating wild-type SFFV that could cause the observed disease (Longmore and Lodish 1991).

In subsequent experiments, we infected mice with viruses from culture supernatants of SFFV cEPO-R-producing cells mixed with culture supernatants from R-MLV-producing cells in a 7:3 ratio, respectively. Injection of adult NIH/Swiss mice with this viral complex resulted in Friend-like disease in all mice (Table 1). The latency of disease was again 5 weeks (range 4–6 weeks). Histological examinations of blood and spleens (Fig. 5C,F) were similar to those seen with mouse A1. Five mice infected with R-MLV alone and five mice infected with a SFFV expressing wild-type EPO-R (SFFV EPO-R) did not develop polycythemia or splenomegaly. We conclude that the polycythemia and splenomegaly observed in infected mice was directly related to the SFFV expressing cEPO-R.

Characterization of Leukemic, Growth-factor-independent Cell Lines Established from Diseased Spleens

Growth-factor-independent cell lines were derived from the spleens of mice A1 and D1. Most analyses were carried out on 4 of 13 cell lines derived from the spleen of mouse A1. Histologically, the growth-factor-independent cell lines appear similar to the basophilic blast-like cells seen in the diseased spleens (data not shown).

Table 2 summarizes the characteristics of growth-factor-independent cell lines isolated from the spleen of a mouse infected with SFFV cEPO-R. All cell lines examined express GATA-1, a DNA-binding protein present at all stages of erythroid development yet also present in megakaryocytes and mast cells (Orkin 1990). These cell lines express globin but do not express the

Table 1. Mice Infected with SFFV cEPO-R

	Hematocrit (spleen weight, gm)			Splenic protein expression	
Mouse	week 4	week 5	week 6	cEPO-R	gp55
C1	47	48	48 (0.1)	–	–
C2	50	48	49 (0.1)	–	–
D1	58 (0.9)			+	–
D2	76 (1.3)			+	–
D3	58 (0.8)			+	–
D4	48	63 (1.1)		+	–
D5	50	61 (0.9)		+	–
D6	47	68 (0.8)		n.d.	n.d.
D7	48	56 (0.9)		n.d.	n.d.
D8	49	64 (1.2)		n.d.	n.d.
D9	49	48	64 (1.2)	n.d.	n.d.
D10	48	47	70 (1.5)	n.d.	n.d.

On the day of infection, culture supernatants from SFFV cEPO-R and R-MLV virus-producing cells were mixed 7:3, respectively, and filtered through a 0.45 μm sterile filter. From this mixture, 0.3 ml was injected via tail vein into 4–6-week-old NIH/Swiss female mice. Weekly the mice were examined for evidence of splenomegaly, spun hematocrits were determined, and blood smears were examined. Upon detection of an elevation in hematocrit or abnormality on blood smear, animals were sacrificed and organs were examined, weighted, and analyzed. Mice C1 and C2 are control animals that did not receive viral injections. At 6 weeks these animals were sacrificed and analyed. Splenic protein expression was determined by subjecting 250 μg of protein extract to PAGE under reducing conditions, transferring the proteins to nitrocellulose, and immunoblotting with antipeptide antisera to the EPO-R or with a monoclonal antibody to gp55. n.d. indicates not determined.

Table 2. Characterization of Cell Lines Established from the Spleen of Mouse A1

Cytokine-independent proliferation	+
Leukemogenic	+
SFFV cEPO-R proviral integration	+
cEPO-R proviral mRNA	+
EPO-R R129C protein	+
gp55 protein	−
spi-1 activation	−
p53 gene rearrangement and inactivation	+
GATA-1 mRNA	+
Globin mRNA	+
Band 3 mRNA	−
B220 protein	−
Thy-1 protein	−
MAC-1 protein	−

erythroid anion transporter Band 3, or markers of B cells, T cells, or macrophages (Longmore and Lodish 1991). Thus, the cell lines isolated appear to be arrested at the proerythroblast stage of erythroid differentiation.

Demonstration That Factor-independent Cell Lines Are Leukemogenic

Substantial evidence indicates that two of the cell lines tested (A1.4 and A1.5) are directly leukemogenic (Fig. 6). To determine whether the growth-factor-independent, clonal, proerythroblast cell lines isolated were fully transformed and leukemogenic, syngeneic mice were injected intravenously with 1×10^6 to 5×10^6 cells

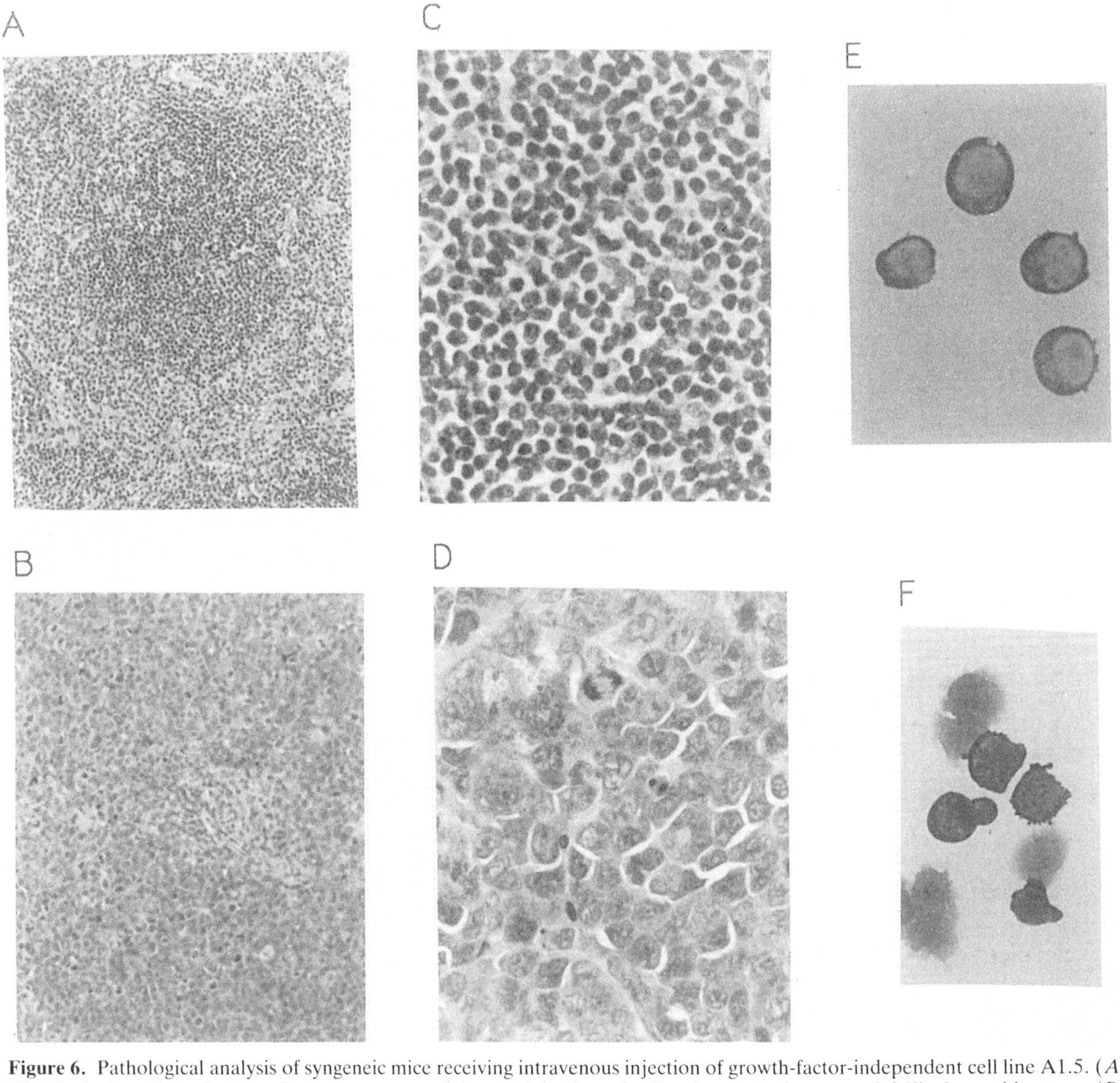

Figure 6. Pathological analysis of syngeneic mice receiving intravenous injection of growth-factor-independent cell line A1.5. (*A*) Histological section of spleen from a mouse receiving no injection of cells, demonstrating normal follicular architecture. (*B*) Histological section of spleen from mouse injected intravenously with 5×10^6 cells from cell line A1.5. The spleen was analyzed 2 weeks after injection. (*C*) Histological section of a normal splenic follicular nodule. (*D*) Histological section from the spleen from mouse described in *B*. (*E*) Cytospin preparation of the growth-factor-independent cell line A1.5, in culture, prior to injection into syngeneic mice. (*F*) Cytospin preparation of bone marrow from a mouse injected, intravenously, with 5×10^6 cells from cell line A1.5. Bone marrow was harvested 2 weeks after injection of A1.5 cells. All cytospin preparations were treated with Wright-Giemsa stain. Histological sections were stained with hematoxylin and eosin. Magnifications: *A* and *B*, 2925×; *C–F*, 11,700×.

of clone A1.4 or A1.5. After 2 weeks, mice receiving intravenous injections of cells became severely anemic (hematocrit 20–24, normal 49) without evidence of reticulocytosis. The total white blood cell and platelet counts were reduced. They also had significant splenomegaly (0.4–0.6 g, normal 0.1 g). Histological sections of the spleen and cytospin preparations from the bone marrow show total replacement of these organs with basophilic, leukemic blast-like cells with a high mitotic index (Fig. 6B,D,F; compare to the normal, panels A,C). Within the bone marrow and spleen there was no evidence of erythroid differentiation past the proerythroblast stage. Grossly, the bone marrow appeared white. Examination of cytospin preparations from the bone marrow of these mice (Fig. 6F) revealed complete replacement of the marrow cavity with erythroblastic appearing cells similar to the injected cells (Fig. 6E). Immunoblot analysis of proteins extracted from the leukemic spleens and bone marrow detected cEPO-R expression without evidence for gp55 expression (Longmore and Lodish 1991).

Importantly, mice injected intravenously with A1.4 or A1.5 cells did not, at any time, develop erythrocytosis, which would be indicative of infection with a SFFV cEPO-R virus secreted by the injected clones. Rather, they became severely anemic, and their bone marrow compartments were completely replaced with cells morphologically and biochemically similar to A1.4 or A1.5. Other evidence (Longmore and Lodish 1991) supports the clonal nature of the erythroleukemia induced by the injected cell lines and demonstrates that the erythropoietin receptor expressed in diseased spleens and growth-factor-independent cell lines retains the activating mutation.

cEPO-R Does Not Transform Fibroblasts

Expression of cEPO-R in the immortalized hematopoietic cell line BaF3 resulted in a fully transformed phenotype including cytokine-independent growth and tumorigenicity (Yoshimura et al. 1990b). Activated colony-stimulating factor 1(CSF-1) receptors when expressed in NIH-3T3 fibroblasts result in morphological transformation and anchorage-independent growth (Roussel et al. 1988). The CSF-1 receptor (c-*fms* proto-oncogene) contains a tyrosine kinase domain within its cytoplasmic domain and, as such, is different from the EPO receptor family of receptors, where the mode of signal transduction remains unknown. Thus, we were interested to know if the activated erythropoietin receptor R129C would transform nonhematopoietic cells. As shown in Table 3, there was no evidence of transformed foci of NIH-3T3 cells following transfection with plasmids expressing cEPO-R. Furthermore, isolated clones of NIH-3T3 cells expressing cEPO-R failed to grow in agar even in the presence of excess erythropoietin. As a control, NIH-3T3 cells expressing the transforming oncogene v-*abl* did grow without added growth factor. Isolated NIH-3T3 cells expressing EPO-R also did not proliferate in response to erythropoietin (data not shown), suggesting that the EPO-R does not transduce a proliferative signal in fibroblasts.

DISCUSSION

Dimerization of the EPO-R

The EPO-R can be activated by two different, hormone-independent mechanisms: (1) interaction with the gp55 glycoprotein of SFFV (Li et al. 1990) and (2) a point mutation, Arg-129 to cysteine, in the extracellular domain (Yoshimura et al. 1990b). The presence of a cysteine residue at position 129, and not the loss of an arginine, appears to be required for constitutive activity, since substitution of Arg-129 with serine, proline, or glutamic acid does not alter the ability of the EPO-R to confer EPO-responsive growth in Ba/F3 cells. This requirement for a cysteine residue led to the hypothesis

Table 3. Fibroblast Transformation by cEPO-R

Plasmid	Transformed foci per μg input DNA[a]	Colony formation in agar[b]	
		no additions	+EPO
pUCJE	5×10^3	n.d.	n.d.
pXM EPO-R	<10	n.d.	n.d.
pXM cEPO-R	<10	n.d.	n.d.
None	<10	0	0
v-*abl*	n.d.	+	+
pSFF EPO-R	<10	0	0
pSFF cEPO-R	<10	0	0

[a] NIH-3T3 P1 cells (one 35-mm diameter plate) were transfected by the calcium phosphate procedure with the indicated plasmid (0.5–5.0 μg DNA per plate.) 18 hr after transfection, cells were trypsinized and split into three 60-mm diameter culture dishes. Foci of morphologically transformed cells were scored under a microscope after 2 weeks. Data represent averages from two separate experiments. pUCJE is an expression plasmid containing the *ras* oncogene.

[b] Individual cells expressing EPO-R or cEPO-R were expanded and seeded in 0.35% agar in complete serum containing medium (no addition) or in complete medium containing serum plus 100 units/ml EPO. Cells were plated at 10^2/ml, 10^3/ml, and 10^4/ml. Colonies were counted 2 weeks after plating. NIH-3T3 cells transformed with the v-*abl* oncogene served as positive controls for colony growth in agar, whereas nontransfected NIH-3T3 cells served as negative controls. + indicates colonies for >20% of input cells. n.d. indicates not determined.

that the constitutive receptor may form novel intramolecular or intermolecular disulfide bonds, which alter the conformation of the receptor and render it constitutively active.

By analogy with the intramolecular disulfide-bonding pattern of the growth hormone receptor (De Vos et al. 1992), it could be predicted that the first and second cysteine residues (Cys-27 and Cys-37) of the EPO-R form a disulfide bond, as do the third and fourth (Cys-65 and Cys-81), leaving Cys-179 unpaired. Since the C179S mutant is functionally wild type, Cys-179 is not essential for normal receptor function, including ligand binding. The phenotype of mutant R129C/C179S demonstrates that Cys-179 is also not required for constitutive activity of the receptor; therefore, in the R129C mutant, Cys-179 is unlikely to pair with Cys-129 (see Fig. 1). Since the constitutively active R129C receptor (cEPO-R) binds EPO with an affinity similar to that of the wild-type receptor (Watowich et al. 1992), it is unlikely that intramolecular disulfide bond rearrangements have occurred. We conclude that EPO-R R129C and EPO-R R129C/C179S form disulfide-linked homodimers through Cys-129.

The formation of disulfide-linked dimers is correlated with constitutive activity. All constitutively active mutants of the EPO-R (R129C, tEPO-R/R129C, and R129C/C179S) form disulfide-linked dimers, whereas hormone-responsive (wild-type EPO-R, C179S, and tEPO-R) or inactive forms (ws1, ws1/R129C) do not (Fig. 1). The dimers appear to assemble in the ER (data not shown), and a small proportion of them reach the cell surface (Fig. 3), where they bind EPO with a single affinity of 700 pM (Watowich et al. 1992). The presence of Cys-129 is necessary but not sufficient for oligomerization and constitutive activation, since the mutant ws1/R129C, containing both Cys-129 and a mutation in the WSXWS region, fails to oligomerize and cannot deliver a proliferation signal in Ba/F3 cells. The failure of this mutant to form disulfide-linked dimers also indicates that covalent dimerization is not simply due to ER retention of misfolded receptors.

Conformational changes brought about by receptor oligomerization in response to ligand binding are likely to activate the tyrosine kinase receptors (Ullrich and Schlessinger 1990), and dimerization also appears to play a role in cytokine receptor signaling (Cunningham et al. 1991). Our results have revealed a strong correlation between the ability of mutant EPO-Rs to induce hormone-independent cell proliferation and their ability to form disulfide-linked dimers. Disulfide-linked dimerization of the R129C mutant may induce a conformational change in the receptor, mimicking the hormone-bound form of the wild-type receptor and rendering the receptor active in the absence of EPO. Preliminary evidence suggests that the wild-type receptor is capable of forming noncovalent homodimers, although we do not yet know the role of ligand binding in dimerization.

The crystal structure of the growth hormone receptor/growth hormone complex is known (De Vos et al. 1992). Sequence comparison between the EPO-R and the growth hormone receptor reveals that residue 129 of the EPO-R would fall in the membrane-proximal subdomain of the growth hormone receptor, at the dimer interface. Thus, Cys-129 is likely to be present at the dimer interface of the disulfide-linked R129C receptors and, by extrapolation, Arg-129 or neighboring residues may play a role in the noncovalent dimerization of the wild-type receptor.

Induction of Erythroleukemia by the cEPO-R

The mutant cEPO-R (EPO-R R129C) transduces a proliferative signal in the absence of any added growth factor; expression of cEPO-R in IL-3-dependent Ba/F3 cells (Yoshimura et al. 1990b) or EPO-dependent HCD57 cells results in deregulation of cellular proliferation manifest as growth-factor-independence. To test whether this activating mutation in the erythropoietin receptor can contribute to the development of leukemia, we generated a SFFV in which the pathogenic *env* gene, *gp55*, was replaced with the cDNA encoding cEPO-R. Mice injected with this virus (SFFV cEPO-R) developed polycythemia and splenomegaly. From the spleen of infected mice, we isolated factor-independent cells arrested at the proerythroblast stage of development, which produced leukemia when injected into syngeneic mice.

We did not detect any evidence of SFFV *env* gp55 expression in the spleens of mice infected with SFFV cEPO-R or the leukemic cell lines isolated from those spleens (Longmore and Lodish 1991). This excludes the likelihood of a recombinational event generating wild-type SFFV that could cause the observed disease. Only cEPO-R expressed from the integrated SFFV cEPO-R provirus was seen on Northern blot analysis; there was no detectable expression of the endogenous EPO-R (Longmore and Lodish 1991). Mice injected only with the helper virus R-MLV used in these experiments did not develop any hematologic disease. Similarly, mice injected with a SFFV expressing the wild-type EPO-R (SFFV EPO-R) did not develop disease. Thus, this disease is traceable to the introduction and expression of the cEPO-R gene in a target cell population. Table 2 summarizes the characteristics of growth-factor-independent cell lines isolated from the spleen of a mouse infected with SFFV cEPO-R.

Early in the course of Friend virus infection, there is a polyclonal erythroblastosis, without any increase in the self-renewal capacity or abrogation of differentiation potential of the cell (Hankins and Troxler 1980). After 4–6 weeks, clonal, growth-factor-independent, leukemogenic cells evolve (Mager et al. 1981; Wendling et al. 1981), which are arrested at the proerythroblast stage of differentiation. Activation of the EPO-R by gp55 is thought to be responsible for the early erythroblastosis, which could allow for subsequent genetic events that would result in the promotion of the leukemic clone, altered in its differentiation program. The role of gp55 in the later stages of Friend-virus-induced erythroleukemia is unclear. We have shown that the structurally activated cEPO-R can supplant gp55 as an oncogenic agent in Friend-virus-induced erythroleukemia. Presumably, cEPO-R mimics gp55 in promoting uncontrolled erythroblastosis, allowing further genetic events to occur that would promote the evolution of the full leukemic phenotype. As with gp55, the role, if any, of cEPO-R in the later stages of the disease is unclear.

Later events thought to contribute to the evolution of immortal, leukemic transformation in SFFV-induced erythroleukemia include insertional activation of the putative oncogene *spi-1* (Moreau-Gachelin et al. 1990) and/or the inactivation of the suppressor oncogene *p53* (for review, see Lane and Benchimol 1990; Ben-David and Bernstein 1991). Since the leukemic cell lines described in this paper were derived from infection of mice with a modified SFFV, SFFV cEPO-R, we expected to find activation of *spi-l* gene expression. However, individual leukemic clones derived from this pool did not express *spi-1*, nor was there evidence for proviral integration in the *spi-1* genomic locus. Thus, in contrast to erythroleukemia induced by Friend virus, activation of *spi-1* is unnecessary to achieve the full leukemic phenotype. Both alleles of the *p53* gene of the leukemic clone A1.5 were rearranged, and this resulted in inactivation of *p53* gene expression (Longmore and Lodish 1991). Presumably, the *p53* gene rearrangement in cell line A1.5 occurred in vivo and conferred a selective growth advantage, a situation similar to leukemic clones derived from infections with Friend virus. Whether or not *p53* inactivation alone is sufficient to generate the fully transformed leukemic phenotype in these erythroleukemic cells remains to be determined.

Following infection of mice with SFFV cEPO-R, there were elevations only in the number of circulating erythroid cells (erythrocytosis), and only erythroleukemic cell lines were isolated from the spleen. One would expect that factor-independent proliferation of cells expressing cEPO-R, as a result of SFFV cEPO-R infection, would be limited to those that can transduce a proliferative signal from the cEPO-R. Yet the disease caused by SFFV cEPO-R has only erythroid manifestations. The cEPO-R is capable of conferring factor-independent growth on a line of IL-3-dependent pre-B cells, Ba/F3 (Li et al. 1990; Yoshimura et al. 1990b); an IL-2-dependent cytotoxic T-cell line, CTLL-2 (data not shown); and an EPO-dependent erythroid cell line, HCD57. Thus, our results suggest that either the SFFV LTR restricts expression of cEPO-R to red cell precursors, or in vivo cEPO-R is only capable of transducing a signal in erythroid cells.

That cEPO-R did not transform fibroblasts results from the inability of EPO-R to transduce a proliferative signal in these cells. Presumably, fibroblasts lack some molecule(s), inherent to erythroid cells, necessary for signal transduction by members of the cytokine receptor superfamily. This may explain why oncogenic activations in members of the cytokine receptor superfamily have not been observed using fibroblasts in transformation assays.

ACKNOWLEDGMENTS

We thank Drs. Drorit Neumann and Jing-Po Li for many helpful discussions. We also thank Dr. K. Shimosaka and Mr. T. Hirano (Kirin Brewery Co., Yokohama, Japan) for their generous gift of pure, recombinant EPO. This work was supported by fellowships from the Anna Fuller Fund (710) and the American Cancer Society (PF-3546) to S.S.W., a fellowship from the Nakatomi Health Science Foundation (Saga, Japan) to A.Y., a National Institutes of Health/National Institute on Aging Physician Scientist Award AG-00294 to G.D.L., a postdoctoral fellowship from the Lucille P. Markey charitable trust to D.J.H., and National Institutes of Health grant HL-32262 to H.F.L.

REFERENCES

Aizawa, S., Y. Suda, Y. Furuta, T. Yagi, N. Takeda, N. Watanabe, M. Nagayoshi, and Y. Ikawa. 1990. Env-derived gp55 gene of Friend spleen focus-forming virus specifically induces neoplastic proliferation of erythroid progenitor cells. *EMBO J.* **9:** 2107.

Ben-David, Y. and A. Bernstein. 1991. Friend virus-induced erythroleukemia and the multistage nature of cancer. *Cell* **66:** 831.

Bestwick, R.K., S.L. Kozak, and D. Kabat. 1988. Overcoming interference to retroviral superinfection results in amplified expression and transmission of cloned genes. *Proc. Natl. Acad. Sci.* **85:** 5404.

Cosman, D., S.D. Lyman, R.L. Idzerda, M.P. Beckman, L.S. Park, R.G. Goodwin, and C.J. March. 1990. A new cytokine receptor superfamily. *Trends Biochem. Sci.* **15:** 265.

Cunningham, B.C., M. Ultsch, A.M. De Vos, M.G. Mulkerrin, K.R. Clauser, and J. Wells. 1991. Dimerization of the extracellular domain of the human growth hormone receptor by a single hormone molecule. *Science* **254:** 821.

D'Andrea, A.D., G.D. Fasman, and H.F. Lodish. 1989a. Erythropoietin receptor and interleukin-2 receptor β chain: A new receptor family. *Cell* **58:** 1023.

D'Andrea, A.D., H.F. Lodish, and G.G. Wong. 1989b. Expression cloning of the murine erythropoietin receptor. *Cell* **57:** 277.

Danos, O. and R.C. Mulligan. 1988. Safe and efficient generation of recombinant retroviruses with amphotropic and ecotropic host ranges. *Proc. Natl. Acad. Sci.* **85:** 6460.

De Vos, A.M., M. Ultsch, and A.A. Kossiakoff. 1992. Human growth hormone and extracellular domain of its receptor: Crystal structure of the complex. *Science* **255:** 306.

Devos, R., G. Plaetinck, J. Van der Heyden, S. Cornelis, J. Vandekerckhove, W. Fiers, and J. Tavernier. 1991. Molecular basis of a high affinity murine interleukin-5 receptor. *EMBO J.* **10:** 2133.

Freshney, R.I., ed. 1987. *Culture of animal cells: A manual of basic technique*, p. 140. A.R. Liss, New York.

Friend, C. 1957. Cell-free transmission in adult Swiss mice of a disease having the character of leukemia. *J. Exp. Med.* **105:** 307.

Fukunaga, R., E. Ishizaka-Ikeda, and S. Nagata. 1990. Purification and characterization of the receptor for murine granulocyte colony-stimulating factor. *J. Biol. Chem.* **265:** 14008.

Graham, F.L., S. Bacchetti, R. McKinnon, C. Stanners, B. Cordell, and H.M. Goodman. 1980. Transformation of mammalian cells with DNA using the calcium technique. In *Introduction of macromolecules into viable mammalian cells* (ed. B. Basergay et al.), p. 3. A.R. Liss, New York.

Hankins, W.D. and D. Troxler. 1980. Polycythemia and anemia-inducing erythroleukemia viruses exhibit differential erythroid transforming effects in vitro. *Cell* **22:** 693.

Hatakeyama, M., M. Tsudo, S. Minamoto, T. Kono, T. Doi, T. Miyata, M. Miyasaka, and T. Taniguchi. 1989. Interleukin-2 receptor β chain gene: Generation of three receptor forms by cloned human α and β chain cDNAs. *Science* **244:** 551.

Hayashida, K., T. Kitamura, D.M. Gorman, K.-I. Arai, T. Yokota, and A. Miyajima. 1990. Molecular cloning of a second subunit of the receptor for human granulocyte-macrophage colony-stimulating factor (GM-CSF): Reconstitution of a high-affinity GM-CSF receptor. *Proc. Natl. Acad. Sci.* **87:** 9655.

Hoatlin, M.E., S.L. Kozak, F. Lilly, A. Chakraborti, C.A. Kozak, and D. Kabat. 1990. Activation of the erythropoietin receptors by Friend viral gp55 and by erythropoietin and down-modulation by the murine Fv-2r resistance gene. *Proc. Natl. Acad. Sci.* **87:** 9985.

Kitamura, T., N. Sato, K.-I. Arai, and A. Miyajima. 1991. Expression cloning of the human IL-3 receptor cDNA reveals a shared β subunit for the human IL-3 and GM-CSF receptors. *Cell* **66:** 1165.

Krantz, S. 1991. Erythropoietin. *Blood* **77:** 419.

Lane, D.P. and S. Benchimol. 1990. p53: Oncogene or anti-oncogene. *Genes Dev.* **4:** 1

Li, J.P., R.K. Bestwick, C. Machida, and D. Kabat. 1986. Role of a membrane glycoprotein in Friend virus erythroleukemia: Nucleotide sequences of nonleukemogenic mutant and spontaneous revertant viruses. *J. Virol.* **57:** 534.

Li, J.P., R.K. Bestwick, C. Spiro, and D. Kabat. 1987. The membrane glycoprotein of Friend spleen focus-forming virus: Evidence that the cell surface component is required for pathogenesis and that it binds to a receptor. *J. Virol.* **61:** 2782.

Li, J.P., A.D. D'Andrea, H.F. Lodish, and D. Baltimore. 1990. Activation of cell growth by binding of Friend spleen focus-forming virus gp55 glycoprotein to the erythropoietin receptor. *Nature* **343:** 762.

Linemeyer, D., S. Ruscetti, E. Scolnick, L. Evan, and D.H. Duesberg. 1981. Biological activity of the spleen focus-forming virus encoded by a molecularly cloned subgenomic fragment. *Proc. Natl. Acad. Sci.* **78:** 1401.

Longmore, G.D. and H.F. Lodish. 1991. An activating mutation in the murine erythropoietin receptor induces erythroleukemia in mice: A cytokine receptor superfamily oncogene. *Cell* **67:** 1089.

Mager, D., T.W. Mak, and A. Bernstein. 1981. Quantitative colony method for tumorigenic cells transformed by two distinct strains of Friend leukemia virus. *Proc. Natl. Acad. Sci.* **78:** 1703.

Metcalf, D. 1989. The molecular control of cell division, differentiation commitment, and maturation in haemopoietic cells. *Nature* **339,:** 27.

Moreau-Gachelin, F., D. Ray, N.J. de Both, M.J.M. vab der Feltz, P. Tambourin, and A. Tavitian. 1990. Spi-1 oncogene activated in Rauscher and Friend murine virus-induced acute erythroleukemias. *Leukemia* **4:** 20.

Nicola, N.A. and D. Metcalf. 1991. Subunit promiscuity among hemopoietic growth factor receptors. *Cell* **67:** 1.

Orkin, S.H. 1990. Globin gene regulation and switching: Circa 1990. *Cell* **63:** 665.

Palacios, R. and M. Steinmetz. 1985. IL-3-dependent mouse clones that express B-220 surface antigen, contain Ig genes in germ-line configuration, and generate B lymphocytes in vivo. *Cell* **41:** 727.

Roussel, M.F., J.R. Downing, C.W. Rettenmier, and C.J. Sherr. 1988. A point mutation in the extracellular domain of the human CSF-1 receptor (c-*fms* proto-oncogene product) activates its transforming potential. *Cell* **55:** 979.

Roussel, M.F., C.W. Rettenmier, A.T. Look, and C.J. Sherr. 1984. Cell surface expression of v-*fms*-coded glycoproteins is required for transformation. *Mol. Cell. Biol.* **4:** 1999.

Rozakis-Adcock, M. and P.A. Kelly. 1991. Mutational analysis of the ligand-binding domain of the prolactin receptor. *J. Biol. Chem.***266:** 16472.

Ruscetti, S.K., N.J. Janesch, A. Chakraborti, S.T. Sawyer, and W.D. Hankins. 1990. Friend spleen focus-forming virus induces factor independence in an erythropoietin-dependent erythroleukemia cell line. *J. Virol.* **64:** 1057.

Ruta, M., R. Bestwick, C. Machida, and D. Kabat. 1983. Loss of leukemogenicity caused by mutations in the membrane glycoprotein structural gene of Friend spleen focus-forming virus. *Proc. Natl. Acad. Sci.* **80:** 4704.

Sawyers, C.L., C.T. Denny, and O.N. Witte. 1991. Leukemia and the disruption of normal hematopoiesis. *Cel* **64:** 337.

Taga, T., M. Hibi, Y. Hirata, K. Yamasaki, K. Yasukawa, T. Matsuda, T. Hirano, and T. Kishimoto. 1989. Interleukin-6 triggers the association of its receptor with a possible signal transducer, gp130. *Cell* **58:** 573.

Troxler, D.H., J.K. Boyars, W.P. Parks, and E.M. Scolnick. 1977. Friend strain of spleen focus-forming virus: A recombinant between mouse type C ecotropic viral sequences and sequences related to xenotropic virus. *J. Virol.* **22:** 361.

Ullrich, A. and J. Schlessinger. 1990. Signal transduction by receptors with tyrosine kinase activity. *Cell* **61:** 203.

VanLohuizen, M. and A. Berns. 1990. Tumorigenesis by slow-transforming retroviruses — An update. *Biochim. Biophys. Acta* **1032:** 213.

Watowich, S.S., A. Yoshimura, G.D. Longmore, D.J. Hilton, Y. Yoshimura, and H.F. Lodish. 1992. Homodimerization and constitutive activation of the erythropoietin receptor. *Proc. Natl. Acad. Sci.* **89:** 2140.

Weinberg, R.A., ed. 1989. Oncogenes and multistep carcinogenesis. In *Oncogenes and the molecular origins of cancer*, p. 307. Cold Spring Harbor Laboratory Press, Cold Spring Harbor, New York.

Wendling, F., F. Moreau-Gachelin, and P. Tambourin. 1981. Emergence of tumorigenic cells during the course of Friend virus leukemias. *Proc. Natl. Acad. Sci.* **73:** 3614.

Wognum, A.W., P.M. Lansdorp, R.K. Humphries, and G. Krystal. 1990. Detection and isolation of the erythropoietin receptor using biotinylated erythropoietin. *Blood* **76:** 697.

Yang, Y.-C., A.B. Ciarletta, P.A. Temple, M.P. Chung, S. Kovacic, J.S. Witek-Giannotti, A.C. Leary, R. Kriz, R.E. Donahue, G.G. Wong, and S.C. Clark. 1986. Human IL-3 (multi-CSF): Identification by expression cloning of a novel hematopoietic growth factor related to murine IL-3. *Cell* **47:** 3.

Yoshimura, A. and H.F. Lodish. 1992. In vitro phosphorylation of the erythropoietin receptor and an associated protein, pp130. *Mol. Cell. Biol.***12:** 706.

Yoshimura, A., A.D. D'Andrea, and H.F. Lodish. 1990a. Friend spleen focus-forming virus glycoprotein gp55 interacts with the erythropoietin receptor in the endoplasmic reticulum and affects receptor metabolism. *Proc. Natl. Acad. Sci.* **87:** 4139.

Yoshimura, A., G. Longmore, and H.F. Lodish. 1990b. Point mutation in the exoplasmic domain of the erythropoietin receptor resulting in hormone-independent activation and tumorigenicity. *Nature* **348:** 647.

Yoshimura, A., T. Zimmers, D. Neumann, G. Longmore, Y. Yoshimura, and H.F. Lodish. 1992. Mutations in the WSXWS motif of the erythropoietin receptor abolish processing, ligand-binding, and activation of the receptor. *J. Biol. Chem.* **267:** 11619.

Regulation of Protein Tyrosine Kinase Activation by the T-cell Antigen Receptor ζ Chain

A. WEISS,*†‡ A.C. CHAN,† M. IWASHIMA,*† D. STRAUS,*† AND B.A. IRVING§
*Howard Hughes Medical Institute, Departments of †Medicine, of ‡Microbiology and Immunology, and of §Physiology, University of California, San Francisco, California 94143

Resting antigen-specific T lymphocytes in the G_0 stage of the cell cycle encounter antigenic peptides bound to major histocompatibility complex (MHC) molecules on the surfaces of antigen-presenting cells. These T cells are activated to proliferate and exhibit a variety of differentiated functions. This complex cellular interaction involves a large number of plasma membrane molecules on the T cell and the antigen presenting cell. However, the T-cell antigen receptor (TCR) must play a central role in directing the initiation of antigen-specific responses.

The TCR is a complex oligomeric structure composed of the products of six genes, all transmembrane proteins, that are associated on the plasma membrane in an obligatory manner (for review, see Weiss 1991). The Ti αβ chains, derived from immunoglobulin-like genes, are responsible for the recognition of peptide antigen and the associated MHC molecule. Associated with the Ti heterodimer are the invariant CD3 chains (γ, δ, and ϵ), as well as a homo- or heterodimer consisting of the ζ chain and/or a related molecule. The association of the Ti, CD3, and ζ chains is mediated by sequences contained in and immediately adjacent to the extracellular face of the plasma membrane (Cosson et al. 1991; Manolios et al. 1991; Tan et al. 1991). Whereas the stoichiometry of the components of the TCR remains unclear, recent progress has been made in understanding the function of the CD3 and ζ (or related) chains. These invariant chains of the TCR, which do not encode intrinsic enzymatic activities, serve to couple the Ti ligand-binding subunit of the TCR to intracellular signal transduction pathways.

Stimulation of the TCR by antigen or by anti-TCR monoclonal antibodies activates a protein tyrosine kinase (PTK) as manifested by the tyrosine phosphorylation of several cellular proteins (Samelson et al. 1986; Hsi et al. 1989; June et al. 1990a). Stimulation of the TCR also activates the phosphatidylinositol (PI) signal transduction pathway (Imboden and Stobo 1985). Activation of phospholipase C (PLC) catalytic activity is responsible for the generation of the second messengers of the PI pathway, leading to an observed rise in cytoplasmic free calcium ($[Ca^{++}]_i$) and activation of protein kinase C (Berridge and Irvine 1989). Kinetic and PTK inhibitor studies suggested that the activation of PLC depends on the activity of the TCR-regulated PTK (June et al. 1990a,b). Recent studies demonstrating that stimulation of the TCR induces the tyrosine phosphorylation of the PLC γ1 isozyme account for the dependence of the PI second-messenger pathway on the activation of the TCR-regulated PTK (Park et al. 1991; Weiss et al. 1991). Thus, the activation of a PTK by the TCR appears to represent a proximal event in TCR signal transduction.

How the TCR activates kinase activity and which PTK is regulated by the TCR are not clear. On the basis of recent studies with chimeric receptors, the cytoplasmic domains of the TCR ζ and the CD3 ϵ chains, at least, appear to be able to interact with a TCR-regulated PTK (Irving and Weiss 1991; Romeo and Seed 1991; Letourneur and Klausner 1992). However, the mechanism by which these TCR chains activate PTK activity or which PTK is activated has not been determined. The current sudies were undertaken to define how the structural complexity of the TCR relates to its signal transduction function and to identify intracellular components of the TCR-regulated signaling pathways.

METHODS

Cell lines and antibodies. Jurkat, J.CaM mutants, and the β-chain-deficient mutant, J.RT3-T3.5, were described previously (Ohashi et al. 1985; Goldsmith and Weiss 1987; Goldsmith et al. 1988,1989). The monoclonal antibodies or heterosera were used at saturating concentrations as described previously (Goldsmith and Weiss 1987; Irving and Weiss 1991; Chan et al. 1991).

Plasmid constructions, mutagenesis, and transfection. The cDNA encoding the CD8/ζ chimera was constructed using the polymerase chain reaction (PCR) as reported recently (Irving and Weiss 1991). The CD8/ζ truncations described in Figure 4 were also generated using PCR amplifications. All PCR-derived products were sequenced prior to subcloning into the expression vector SFFV*neo*. Cells were transfected with the appropriate plasmid constructs by electroporation (Irving and Weiss 1991). Stable G418-resistant clones were screened for chimera expression by immunofluorescence. Clones expressing similar levels of chimeric proteins were used for comparative studies.

Immunoprecipitation and Western blot analysis. Cell lysates were prepared from unstimulated or stimulated cells as indicated for 2 minutes at 37°C (Irving and

Weiss 1991). Immunoprecipitates of the ζ chain were isolated from 10 × 10^7 cells using an anti-ζ heteroserum (no. 387) provided by L. Samelson and R. Klausner and resolved by SDS-PAGE and Western blotting. Tyrosine-phosphorylated proteins were detected using the 4G10 anti-phosphotyrosine monoclonal antibody and developed with alkaline-phosphatase-conjugated goat anti-mouse IgG.

Phospholipase C assays. To detect PLC activity of tyrosine phosphoproteins, immunoprecipitates were prepared from lysates of 2 × 10^7 cells stimulated as indicated for 60 seconds with the 4G10 anti-phosphotyrosine monoclonal antibody. Tyrosine phosphoproteins were eluted with 10 mM phenylphosphate and assessed for calcium-dependent PLC activity as described previously (Weiss et al. 1991). Results are expressed as the total inositol phosphates (InsP) generated in 15 minutes at 37°C. To assess changes in PLC catalytic activity, PLC γ1 was immunoprecipitated from lysates prepared of 10^7 cells 1 minute after stimulation as indicated using PLC γ1-specific monoclonal antibodies bound to Sepharose beads. The catalytic activity of the PLC bound to the beads was assessed for calcium-dependent PLC activity during a 15-minute assay at 37°C as described above, with the exception that these reaction mixtures contained 0.05% Triton X-100.

RESULTS

The Cytoplasmic Domain of the ζ Chain Is Sufficient to Activate Early and Late Signaling Events Associated with TCR Stimulation

To define the function of the individual invariant chains of the TCR, several laboratories constructed chimeric molecules incorporating the cytoplasmic domains of either the ζ chain or the CD3 ε chain (Irving and Weiss 1991; Romeo and Seed 1991; Letourneur and Klausner 1992). The ζ chain is of particular interest because it not only is a component of the TCR, but also is a structural component of the IgG Fc receptor in natural killer (NK) cells (Anderson et al. 1989; Lanier et al. 1989). Moreover, a ζ homolog, the γ chain, forms a disulfide-linked dimer within the oligomeric IgE Fc receptor on mast cells and basophils (Ravetch and Kinet 1991). All of these receptors initiate similar signal transduction events, including the induction of PTK and PLC activities.

Our laboratory constructed a CD8/ζ fusion protein in which the cytoplasmic domain of ζ was fused to the extracellular and transmembrane domains of the CD8 α chain (Irving and Weiss 1991). The CD8/ζ chimera or the wild-type CD8 α chain was expressed in the human-T-cell leukemic line Jurkat or a TCR-deficient mutant of Jurkat, J.RT3-T3.5, which fails to express functional

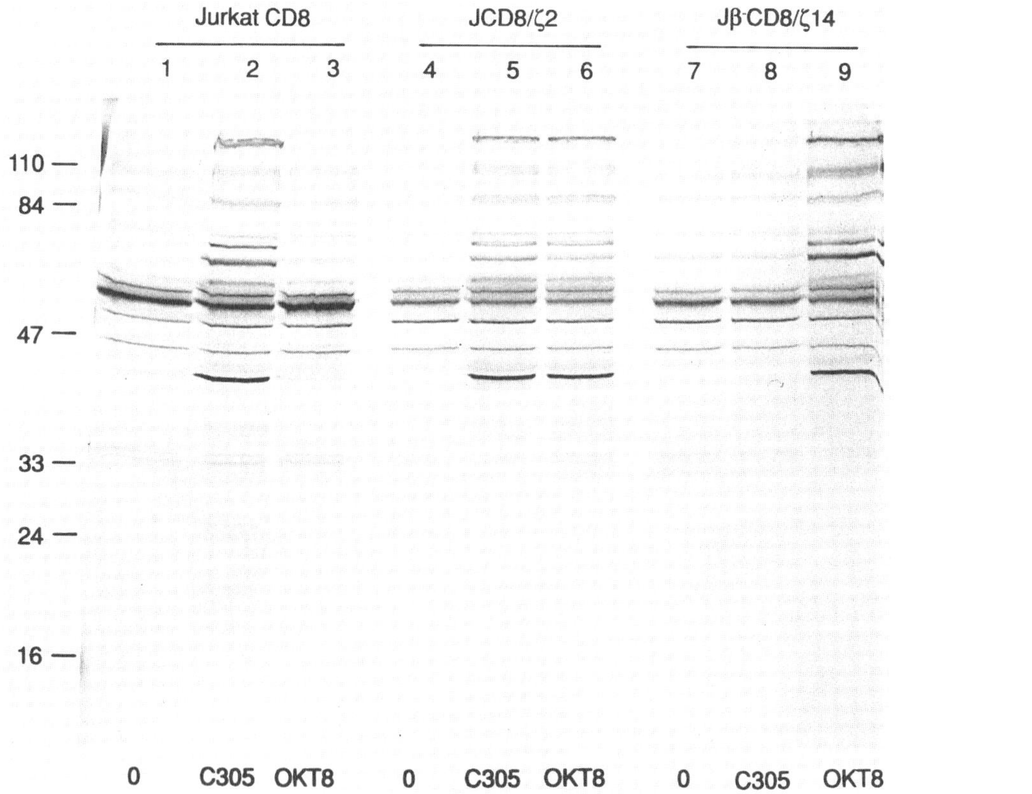

Figure 1. Stimulation of the CD8/ζ chimera induces new tyrosine phosphoproteins independently of the TCR and in a pattern indistinguishable from those induced by the TCR. Lysates were prepared from Jurkat cells expressing CD8 α (Jurkat CD8; lanes *1 – 3*), Jurkat expressing the CD8/ζ chimera (JCD8/ζ2; lanes *4 – 6*) or the Jurkat-derived Ti β-chain-deficient mutant expressing the CD8/ζ chimera (Jβ-CD8/ζ14; lanes *7 – 9*) following stimulation with the indicated monoclonal antibodies (C305 is an anti-Ti Vβ, and OKT8 is reactive with CD8). Lysates were analyzed by Western blotting and probed with an anti-phosphotyrosine monoclonal antibody. (Reprinted, with permission, from Irving and Weiss 1991 [copyright held by Cell Press].)

β-chain transcripts and, as a result, fails to transport other chains of the TCR to the plasma membrane (Ohashi et al. 1985). Both the CD8/ζ fusion protein or the CD8 α chain could be expressed in Jurkat as detected by immunofluorescence and immunoprecipitation (Irving and Weiss 1991). Moreover, the CD8/ζ fusion protein could be expressed in the TCR β-chain-deficient cell independently of any of the chains of the TCR, including the endogenous ζ chain. Thus, the cytoplasmic domain of the ζ chain could be freed of its obligatory interaction with the TCR by eliminating its transmembrane domain and expressing it as a fusion protein linked to the extracellular and transmembrane domains of CD8.

We examined the ability of the CD8/ζ chimera to activate the TCR-regulated PTK pathway. Signal transduction by the TCR can be induced by mimicking the interaction of the TCR with antigenic peptide/MHC by using anti-TCR monoclonal antibodies. Unlike the TCR, stimulation of CD8 α expressed in the TCR$^+$ Jurkat line with anti-CD8 monoclonal antibody induced no appreciable change in cellular tyrosine phosphoproteins (Fig. 1, compare lanes 2 and 3). In marked contrast, stimulation of the CD8/ζ chimera with anti-CD8 monoclonal antibody in the TCR$^+$ (lane 6) or in the TCR β-chain-deficient mutant (lane 9) induced a pattern of cellular tyrosine phosphoproteins that was indistinguishable from that induced by stimulation of the TCR with an anti-Ti Vβ monoclonal antibody in the same cells (lanes 5 and 8, respectively). These results demonstrate that the cytoplasmic domain of the ζ chain is sufficient to activate the TCR-regulated PTK pathway, and it can do so in a manner that is independent of the involvement of other TCR chains.

Stimulation of the CD8/ζ chimera could also activate PLC as manifested by an increase in inositol phosphates and elevation of $[Ca^{++}]_i$ (Irving and Weiss 1991). Moreover, as previously demonstrated for the intact TCR (Park et al. 1991; Weiss et al. 1991), stimulation of the CD8/ζ chimera induced the tyrosine phosphorylation of a PLC isozyme. This was detected by the recovery of PLC activity from immunoprecipitates of tyrosine phosphoproteins isolated from CD8/ζ-stimulated cells (Fig. 2A). Western blot analysis revealed that the tyrosine phosphorylation of PLC γ1 was induced following stimulation of the chimera (data not shown). Such tyrosine phosphorylation of PLC γ1 correlated with an increase in the PLC catalytic activity in a mixed micelle assay (Fig. 2B). Finally, late manifestations of T-cell activation associated with these early signal transduction events, such as expression of the activation antigen CD69 and production of interleukin-2, were also induced by stimulation of the CD8/ζ chimera (Irving and Weiss 1991). Thus, at least qualitatively, signal transduction by the CD8/ζ chimera mimics that induced by the intact oligomeric TCR.

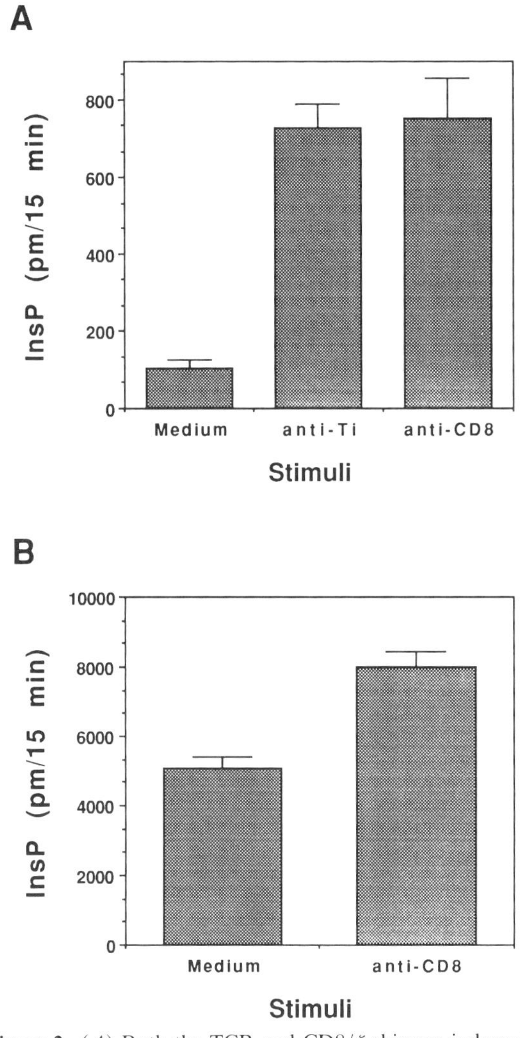

Figure 2. (*A*) Both the TCR and CD8/ζ chimera induce the recovery of PLC activity in tyrosine phosphoprotein immunoprecipitates. The TCR or CD8/ζ chimera on J.CD8/ζ.2 cells was stimulated as indicated. Tyrosine phosphoproteins were immunoprecipitated from lysates and assessed for associated PLC activity. (*B*) Increased PLC γ1 catalytic activity following CD8/ζ chimera stimulation. PLC γ1 was immunoprecipitated from unstimulated cells or cells stimulated via the CD8/ζ chimera with anti-CD8 monoclonal antibody. Calcium-dependent PLC catalytic activity was assessed in a mixed micelle assay.

The Cytoplasmic Domain of the TCR ζ Chain Is Associated with a PTK and a 70-kD Tyrosine Phosphoprotein (ZAP-70) following TCR Stimulation

The results obtained with the CD8/ζ chimera suggest that the cytoplasmic domain of ζ interacts with intracellular signal transduction components and can regulate their activities. Since the most proximal step in TCR signal transduction appears to be the activation of a PTK, manifested by the induction of tyrosine phosphorylation of cellular proteins, we examined the possible association of the ζ chain with tyrosine phosphoproteins.

Immunoprecipitates of the TCR ζ chain or the CD8/ζ chimera were isolated following stimulation of the TCR or the chimera and analyzed by Western blot analysis for tyrosine phosphoproteins. Stimulation of either of these signal-transduction-competent molecules resulted in the appearance of a 70-kD tyrosine

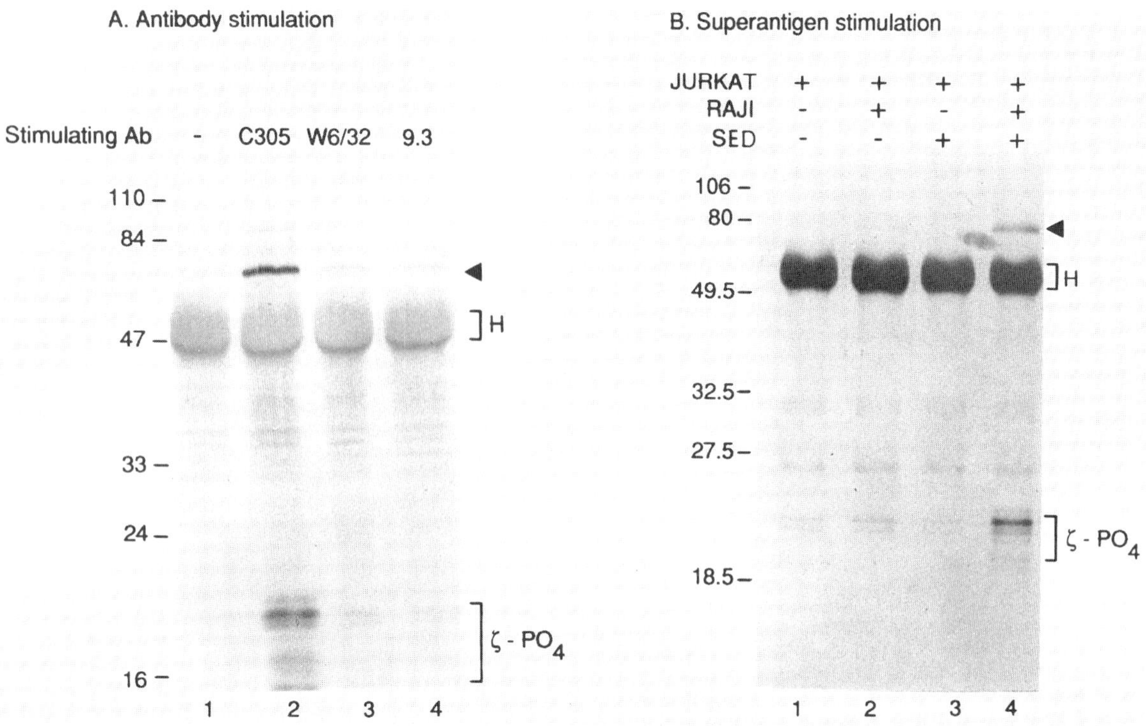

Figure 3. (*A*) A 70-kD tyrosine phosphoprotein (ZAP-70) is associated with the ζ chain following stimulation of the TCR. Jurkat cells were stimulated with medium (lane *1*), anti-Ti Vβ (C305; lane *2*), anti-HLA (W6-32; lane *3*), or anti-CD28 (9.3; lane *4*). Immunoprecipitates of the ζ chain were isolated from cell lysates and analyzed by Western blot analysis for tyrosine phosphoproteins using the 4G10 monoclonal antibody. (*B*) ZAP-70 is also induced following stimulation of Jurkat cells with the superantigen staphylococcal enterotoxin D (SED). Jurkat cells were stimulated as indicated with the antigen-presenting cell Raji in the presence or absence of SED. Immunoprecipitates of the ζ chain were analyzed in Western blots for tyrosine phosphoproteins. (*H*) Ig heavy chain; (ζ-PO4) phosphorylated ζ chain. Arrowhead is pointing to ZAP-70. (Reprinted, with permission, from Chan et al. 1991.)

phosphoprotein in immunoprecipitates of the isolated ζ chain (Fig. 3) or the CD8/ζ chimera (Chan et al. 1991). This protein, termed ZAP-70 for ζ-associated protein of 70 kD, was induced within 15 seconds of stimulation. Metabolic labeling studies suggested that the association or stability of the association is increased by TCR stimulation. In cells expressing both the TCR and CD8/ζ, the association of ZAP-70 either with TCR-associated ζ or with the chimera was determined by which receptor was stimulated (Chan et al. 1991). This specificity of ZAP-70 association with the stimulated receptor is similar to the specificity of induced tyrosine phosphorylation of the ζ chain or CD8/ζ chimera (Irving and Weiss 1991). Thus, ZAP-70 is a 70-kD tyrosine phosphoprotein that associates with the cytoplasmic domain of ζ following TCR stimulation.

Since the TCR and CD8/ζ induce the activation of a PTK, we examined the possibility that PTK activity is associated with the cytoplasmic domain of ζ. Immunoprecipitates of ζ or CD8/ζ were isolated before or after TCR or CD8/ζ stimulation and subjected to in vitro kinase assays by adding [γ-^{32}P]ATP. Tyrosine phosphorylation of ζ or CD8/ζ could be induced in vitro in immunoprecipitates isolated from unstimulated cells (Chan et al. 1991). However, this PTK activity, manifested by ζ or CD8/ζ phosphorylation, was markedly increased following TCR or chimera stimulation. One prominent in vitro substrate noted in the immunoprecipitates isolated from stimulated cells was a 70-kD protein, the presence of which correlated with the appearance of ZAP-70. Immunoprecipitates of several other cell-surface proteins did not contain PTK activity. Thus, the cytoplasmic domain of the ζ chain is constitutively associated with a PTK, and kinase activity is increased following receptor stimulation.

The identity of the PTK directly associated with the cytoplasmic domain of ζ is of great interest. Two members of the *src* family of PTKs, *fyn* and *lck*, have been implicated in TCR signal transduction (Samelson et al. 1990; Abraham et al. 1991; Cooke et al. 1991; Glaichenhaus et al. 1991). An alternative candidate for the TCR-activated PTK is ZAP-70. The appearance of ZAP-70 in TCR ζ or CD8/ζ immunoprecipitates correlates with receptor-mediated activation of the PTK pathway in intact cells. Its appearance also correlates with the increased in vitro PTK activity found in immunoprecipitates of ζ or CD8/ζ isolated from stimulated cells. We have recently preparatively purified CD8/ζ-associated ZAP-70 protein and obtained partial amino acid sequence (data not shown). This permitted the isolation of cDNA clones encoding ZAP-70. Preliminary analysis of the ZAP-70 sequence reveals that it is a novel protein that has homology with several known PTKs. Thus, it is tempting to speculate that

ZAP-70 may be the most proximal TCR-regulated PTK. However, the precise function and regulation of ZAP-70 are not yet clear. One method to assess the functional importance of ZAP-70 in TCR signal transduction is to determine whether the domain of TCR ζ required for ZAP-70 association is also required for signaling function.

Identifying a Functional Domain within the TCR ζ Chain Cytoplasmic Domain

To define the sequences of the cytoplasmic domain of ζ that interact with ZAP-70 and are required for function, sequential truncations were made in the cytoplasmic domain of the CD8/ζ chimera (Fig. 4A). These truncated CD8/ζ chimeric molecules were transfected into Jurkat cells. The abilities of these chimeras to associate with ZAP-70, to activate the PTK pathway, and to activate the PI pathway (as assessed by increases in $[Ca^{++}]_i$) were examined (Fig. 4B). Although there was a modest decrease in all three assays with the first truncation (T83), there was a coordinated loss of all three functional activities with truncation T34. This concomitant loss in ZAP-70 association, PTK activation, and PLC activation suggests the functional importance of ZAP-70 in TCR signal transduction pathways.

The sequence of ζ remaining in the most truncated chimera that retained functional activity suggested a functional sequence motif based on a previous observation made by Reth (1989), who identified a common motif when he compared the cytoplasmic sequences from a number of antigen-receptor-associated molecules (Fig. 5). This motif is triplicated within ζ (see Fig. 4A). Single copies of this motif are contained within the cytoplasmic domains of each of the CD3 chains. Thus, the TCR has as many as ten copies of this motif in the oligomeric receptor. This motif is also contained within the β and γ chains of the IgE Fc receptor on mast cells and within mb-1 and B29, the molecules associated with membrane immunoglobulin on B cells. Recent studies have suggested that the three motifs within ζ and the CD3 ϵ motif are functional (Letourneur and Klausner 1992; Romeo et al. 1992). In our own preliminary studies, we have linked single copies of the ζ1 and the ζ2 motifs to the transmembrane/cytoplasmic junction of CD8 and found that they are able to activate a PTK, interact with ZAP-70, and activate the PI pathway (data not shown). Moreover, three copies of the ζ1 motif appear to function more efficiently in these assays. Thus, this sequence motif may represent a common means by which many distinct antigen receptors couple to intracellular signaling machinery.

A

```
                  T12         T22              T34
                   ↓           ↓                ↓
RVKFSRSAEPPAYQQGQNQLYNELNLGRREEYDVLDKR

        T46        T56
         ↓          ↓
RGRDPEMGGKPRRKNPQEGLYNELQKDKMAEAYSEIGM

       T83
        ↓
KGERRRGKGHDGLYQGLSTATKDTYDALHMQALPPR
```

B

Construct	ZAP-70 Assn	PTK	$[Ca^{2+}]_i$
CD8/ζ	++++	++++	++++
CD8/ζ T83	++	++	++
CD8/ζ T56	++	++	++
CD8/ζ T46	++	++	++
CD8/ζ T34	-	-	-
CD8/ζ T22	-	-	-
CD8/ζ T12	-	-	-

Figure 4. (*A*) Schematic representation of the truncations of the CD8/ζ chimera. Shown is the cytoplasmic sequence of ζ only and the position and designations of the truncations. Shown underlined are the activation motifs identified within the ζ sequence. (*B*) Summary of the functional activities of the CD8/ζ truncations. Shown is the ability of each CD8/ζ truncated chimera, relative to the intact chimera, to associate with ZAP-70, activate the PTK pathway as assessed by Western blot analysis, and induce an increase in $[Ca^{++}]_i$ when stimulated.

hζ1	N	Q	L	Y	N	E	L	N	L	G	R	R	E	E		Y	D	V	L	D
hζ2	E	G	L	Y	N	E	L	Q	K	D	K	M	A	E	A	Y	S	E	I	G
hζ3	D	G	L	Y	D	G	L	S	T	A	T	K	D	T		Y	D	A	L	H
CD3γ	D	Q	L	Y	D	P	L	K	D	R	E	D	D	Q		Y	S	H	L	N
CD3ε	N	P	D	Y	E	P	I	T	K	G	Q	R	D	L		Y	S	G	L	N
CD3δ	D	Q	V	Y	Q	P	L	R	D	R	D	D	A	Q		Y	S	H	L	G
rIgE FcRγ	D	A	V	Y	T	G	L	N	T	R	N	Q	E	T		Y	E	T	L	G
rIgE FcRβ	D	R	L	Y	E	E	L		H	V	Y	S	P	I		Y	S	A	L	E
hMB-1	E	N	L	Y	E	G	L	N	L	D	D	C	S	M		Y	E	D	I	S
mB29	D	H	T	Y	E	G	L	N	I	D	Q	T	A	T		Y	E	D	I	V

Figure 5. A similar sequence motif is present within cytoplasmic domains of the invariant chains of several hematopoietic antigen receptors. Shown are the human (h), rat (r), or murine (m) sequences.

A Genetic Approach toward Identifying TCR-associated Signal Transduction Components

As an alternative to the direct biochemical approach toward studying TCR signal transduction, we previously isolated mutants of the Jurkat T-cell line that were deficient in their abilities to increase $[Ca^{++}]_i$ following TCR stimulation (Goldsmith and Weiss 1987; Goldsmith et al. 1988, 1989). These mutants were isolated using a combination of growth selection (TCR stimulation of Jurkat cells induces growth arrest) and cell sorting for cells that failed to increase $[Ca^{++}]_i$ in response to TCR stimulation. We hoped that this would lead to the unbiased identification of components of the signal transduction pathway.

Three mutants, J.CaM1 – 3, that express high levels of apparently normal TCRs and define distinct complementation groups have been isolated. Stimulation of the TCR on these cells fails to induce increases in $[Ca^{++}]_i$, inositol phosphates, or phosphorylation of PLC γ1 (Goldsmith and Weiss 1987; Goldsmith et al. 1988, 1989; Weiss et al. 1991). All of these cells express PLC γ1. A heterologous receptor, the human muscarinic receptor subtype 1, does activate the PI pathway in these three cells (Goldsmith et al. 1989), although this presumably occurs via the activation of PLC β (Bernstein et al. 1992). Thus, these studies lend further support to the notion that PLC γ1 tyrosine-phosphorylation is responsible for activation of the PI pathway by TCR stimulation. Moreover, these results suggest that the defect in each of the three J.CaM mutants is proximal to the tyrosine phosphorylation of PLC γ, although a structural mutation in PLC γ1 cannot be excluded.

We have attempted to assess the functional status of the TCR in the three mutants by expressing the CD8/ζ chimera in these cells. To date, only J.CaM1 and J.CaM3 have been successfully transfected with this functional chimeric receptor. In neither of these cells was the CD8/ζ chimera able to induce an increase in $[Ca^{++}]_i$ (data not shown). Thus, at least in the case of J.CaM1 and J.CaM3, the defect appears to be distal to the TCR but proximal to the tyrosine phosphorylation of PLC γ1.

It was important to assess the status of the PTK pathway in the J.CaM mutants since the activation of a PTK is essential for activation of PLC. Western blots of whole-cell lysates prepared from Jurkat and the three mutants following stimulation with anti-Ti Vβ or with phorbol ester were probed with an anti-phosphotyrosine monoclonal antibody (Fig. 6). These studies revealed that stimulation of the TCR on J.CaM1 and J.CaM3 fails to activate the PTK as no increase in the tyrosine phosphorylation of proteins was observed. In contrast, a PTK activity was induced in J.CaM2. However, in contrast to Jurkat, there appears to be a selective defect in the pattern of tyrosine phosphoproteins induced. Thus, the tyrosine phosphorylation of pp36-38, pp42-44, and pp150 was not observed in J.CaM2 but was readily detected in Jurkat. The failure to induce pp150 is compatible with the fact that this protein is PLC γ1 and that PLC activity is not induced in this cell. The pp42-44 doublet probably represents microtubule-associated protein-2 (MAP-2) kinase (data not shown), which was tyrosine-phosphorylated following phorbol ester treatment in J.CaM2. The failure to induce MAP-2 kinase tyrosine phosphorylation in response to TCR stimulation is consistent with the notion that the tyrosine phosphorylation of this protein is dependent on the activation of protein kinase C via the PI pathway (Nel et al. 1990). The identity of pp36-38 is not known. The results obtained with J.CaM2 suggest that the tyrosine phosphorylation of PLC γ1 may not result from a direct interaction between the TCR-activated PTK and PLC γ1. Thus, these results suggest that the defects in J.CaM1 and J.CaM3 result from the failure to activate the TCR-regulated PTK and that the defect in J.CaM2 is downstream from the TCR-regulated PTK.

Since the defects in J.CaM1 and J.CaM3 result in the failure to activate the TCR-associated PTK, we examined the possibility that these cells might be deficient in one of the *src* PTKs implicated in the TCR pathway. In preliminary studies, J.CaM1 has been found to be deficient in functional *lck* kinase activity that appears to result, in part, from defectively spliced mRNA (data not shown). Transfection of wild-type *lck* into J.CaM1 restores the ability of the TCR to activate the PTK and

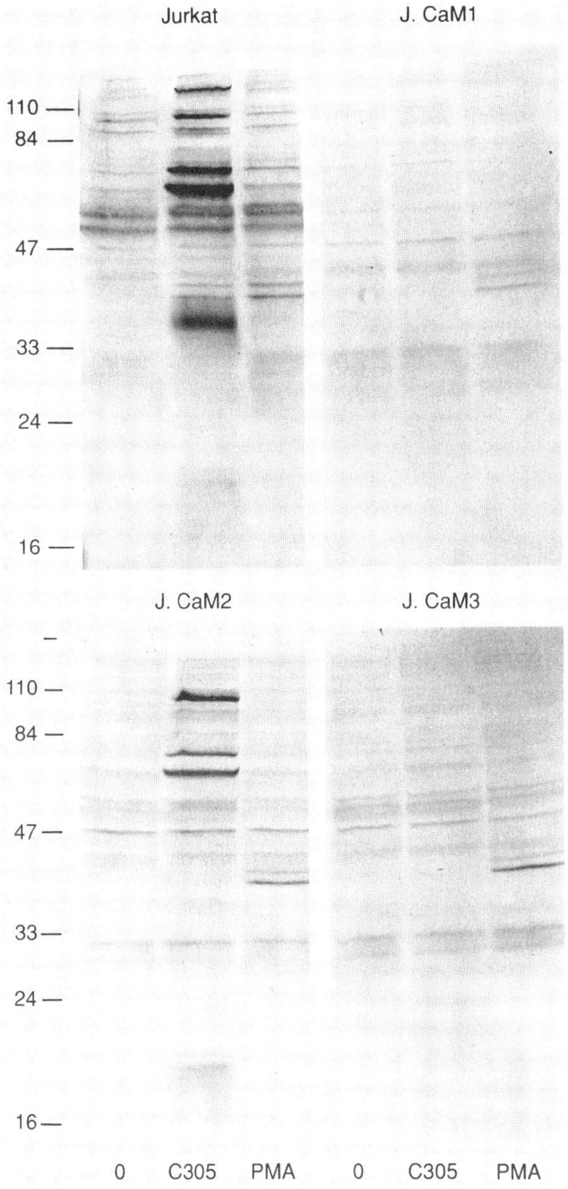

Figure 6. PTK activation in the J.CaM mutants. Shown is a Western blot analysis of tyrosine phosphoproteins in lysates obtained from Jurkat, J.CaM1, J.CaM2, and J.CaM3 cells following stimulation with medium, anti-Ti Vβ (C305), or phorbol myristate acetate (PMA).

PI pathways. J.CaM3 has functional *lck* kinase activity. Both J.CaM1 and J.CaM3 have functional *fyn* PTK activity. Preliminary studies suggest that J.CaM3 does express ZAP-70 mRNA. Thus, the defect in J.CaM3 remains unclear. However, the results with J.CaM1 clearly demonstrate that *lck* plays a critical role in TCR-mediated activation of a PTK.

DISCUSSION

The complex structure of the TCR has likely evolved to most efficiently carry out its functions of antigen recognition and signal transduction. From the studies with the CD8/ζ fusion protein, the ζ-chain dimer appears to function to couple the TCR to the intracellular signal transduction machinery. Our studies demonstrate that the cytoplasmic domain of ζ interacts with and regulates the activity of a PTK that may represent ZAP-70. A sequence motif, triplicated within ζ, appears to mediate the interaction with ZAP-70. This motif is also sufficient to construct a functional signaling chimeric receptor. These observations suggest that the interaction of ζ with ZAP-70 may be one of the means by which the TCR regulates subsequent signal transduction events. However, the observations made with the J.CaM mutants suggest that other proteins play critical roles in regulating TCR signal transduction. Indeed, the studies with J.CaM1 demonstrate that the *lck* PTK plays a role in regulating the activity of the TCR-activated PTK pathway.

Although our studies with the CD8/ζ chimera demonstrate that the cytoplasmic domain of ζ is sufficient to couple the TCR to the intracellular signaling machinery in a manner indistinguishable from that of the intact oligomeric TCR, it does not appear to be the only subunit capable of coupling to intracellular signaling molecules. Recent studies with chimeric molecules in which the cytoplasmic domain of CD3 ε was linked to a heterologous transmembrane molecule demonstrated that the CD3 ε chain is also able to couple to a PTK-regulated signaling pathway that can lead to T-cell activation (Letourneur and Klausner 1992). The notion that the CD3 chains are able to function in signal transduction was also supported in a recent study in which reconstituted TCRs, lacking any cytoplasmic ζ sequence, were still functionally competent in T-cell hybridomas to induce interleukin-2 production (Wegener et al. 1992). The TCR apparently has several independent subunits capable of signal transduction. The purpose of incorporating these independent functional subunits within the oligomeric TCR is not clear. All of the CD3 chains contain a single copy of the functional motif that is triplicated in ζ. It is possible that these subunits function redundantly and couple the TCR to the same PTK. This redundancy may serve as an amplification mechanism since the TCR must be responsive to relatively low concentrations of antigenic peptide (estimated to be a few hundred molecules or less on an antigen presenting cell). The CD8 chimera expressing three copies of the ζ1 motif appears to induce signaling events more efficiently than the chimera with only one copy (A. Weiss et al. unpubl.). This would also be consistent with the 100-fold increased concentration of monoclonal antibody required to stimulate a chimera with the cytoplasmic domain of CD3 ε that has only one copy of the motif compared to a chimera containing the cytoplasmic domain of ζ, which has three copies of the motif (Letourneur and Klausner 1992). Alternatively, the CD3 and ζ subunits may interact with distinct PTKs, since the sequences containing the motifs within these chains differ. Subtle differences were observed between the cellular tyrosine phosphoproteins induced following stimulation of the

CD3 ϵ or ζ chimeras (Letourneur and Klausner 1992). This is in contrast to our studies demonstrating an identical pattern of induced cellular tyrosine phosphoproteins following stimulation of the intact TCR versus the CD8/ζ chimera (see Fig. 2, lanes 5 and 6). However, it is equally possible that these distinct subunits interact functionally as the result of a conformational change following ligand binding. Such an interaction might promote the interaction and activation of PTKs that are bound and sequestered by the different subunits of the TCR. This might be somewhat analogous to the transphosphorylation and activation that occurs following ligand binding by PTK growth factor receptors (Ueno et al. 1991).

This sequence motif was shown in these studies to be able to interact with ZAP-70, a putative PTK, and to confer functional signal transduction capabilities when expressed in a chimeric receptor. This appears to be a conserved means by which several hematopoietic antigen receptors couple to intracellular PTKs. It is likely that these motifs expressed in distinct receptors in different cells interact with the same or related molecules. Each motif contains two tyrosines, and tyrosine phosphorylation of the receptor-associated chains is a prominent feature observed with stimulation of all of the receptors containing this motif. This raises interesting questions. Is tyrosine phosphorylation a means by which signaling molecules or PTKs are recruited to the activated receptor? If so, how does the PTK responsible for this phosphorylation recognize a stimulated receptor? An alternative hypothesis is that the unphosphorylated motif interacts with the most proximal signaling PTK. Receptor stimulation may involve homologous or heterologous dimerization and transphosphorylation of the motif-associated PTK leading to the activation of the kinase. Therefore, tyrosine phosphorylation of the motif may instead represent a mechanism of homologous or heterologous desensitization in which motif structure is distorted and preventing PTK-motif association. This is consistent with the hyporesponsiveness of TCRs on freshly isolated thymocytes in which the ζ chain is found to be constitutively phosphorylated (Nakayama et al. 1991). A clearer picture of the function of these motifs will emerge as their interactions with the relevant PTK can be studied in further detail.

One candidate for the PTK that is directly regulated by TCR stimulation is ZAP-70. Its tyrosine phosphorylation and coprecipitation with the ζ chain is observed within 15 seconds of TCR stimulation (Chan et al. 1991). It is not clear whether ZAP-70 is constitutively associated with ζ or whether the association is induced. Small amounts of ζ-associated ZAP-70 are observed in metabolic labeling studies and in ζ or chimera immunoprecipitates from unstimulated cells subjected to in vitro kinase assays. Increased amounts of ζ-associated ZAP-70 are observed following receptor stimulation. This may represent recruitment of additional ZAP-70 into the TCR complex or stabilization of a preexisting interaction. An increase in PTK activity is observed in vivo and in vitro with tyrosine phosphorylation and TCR association of ZAP-70. Moreover, the association of ZAP-70 with the CD8/ζ truncation chimeras correlates precisely with the signal transduction function of those chimeras. If ZAP-70 is indeed shown to be a PTK, how is its activity regulated? Transphosphorylation through the formation of homodimers or heterodimers with another PTK (possibly *lck*) seems to be the most likely model. A more precise role for ZAP-70 will emerge when purified preparations of the recombinant molecule are available and immunologic reagents are produced.

Another PTK, *lck*, required for TCR signal transduction function was identified as a result of our isolation of J.CaM TCR signal transduction mutants. Whereas a role for *lck* in TCR signaling function had previously been suggested (Abraham et al. 1991; Glaichenhaus et al. 1991), this was primarily in the context of its association with CD4 and CD8 during antigen recognition. Thus, during antigen recognition, the TCR and CD4 (or CD8) function as coreceptors interacting with different surfaces of MHC molecules. This is likely to promote the colocalization of *lck* with the TCR cytoplasmic domains and associated molecules. Under these circumstances, *lck* could interact with a TCR-associated PTK and regulate its function (by phosphorylation). However, our studies suggest that *lck* plays an important role in regulating TCR signal transduction function independent of the colocalization of the TCR and CD4 or CD8. This is relevant to the situations in which the TCR can recognize antigen in the absence of CD4 or CD8 participation as in memory T cells or in T cells expressing the $\gamma\delta$ TCR.

Supporting a role for *lck* in TCR signal transduction are studies with CD45 tyrosine-phosphatase-deficient mutants (Pingel and Thomas 1989; Koretzky et al. 1990, 1991). In these cells, there is not a general increase in basal tyrosine phosphorylation of cellular phosphoproteins, and stimulation of the TCR fails to activate the PTK or PI pathways. However, one critical substrate of CD45 appears to be *lck*. The carboxy-terminal tyrosine, Y505, is hyperphosphorylated in CD45-deficient cells (Ostergaard et al. 1989; M. Sieh and A. Weiss, unpubl.). This is a negative regulatory site for *lck* in transformation assays (Abraham and Veilette 1990). CD45 and *lck* have been coimmunoprecipitated, suggesting that they can interact in vivo (Schraven et al. 1991). However, it should be noted that it is not yet clear whether *lck* is the only relevant CD45 substrate in vivo responsible for the observed defect in TCR signal transduction.

The signaling impairment that results from the deficiency of *lck* in J.CaM1 and the signaling defect in CD45-deficient cells suggest that *lck* plays a critical role in TCR signal transduction. One hypothesis would be that *lck* is the PTK that is directly activated by the TCR. However, with few exceptions (Burgess et al. 1991), *lck* is not found in TCR immunoprecipitates that contain substantial kinase activity (Chan et al. 1991). Furthermore, stimulation of *lck* kinase activity by CD4 cross-linking does not induce the same pattern of cellular tyrosine phosphoproteins observed following TCR stimulation. Finally, the signaling phenotype of J.CaM1 is leaky since it can be stimulated by some anti-CD3 monoclonal antibodies to increase $[Ca^{++}]_i$ and weakly induce a selected subset of tyrosine phosphoproteins (Goldsmith and Weiss 1987, 1988; D. Strauss and A. Weiss, unpubl.). Thus, it is possible to bypass the *lck* defect partially. A more reasonable hypothesis is that *lck* regulates the activity of the TCR-associated PTK. This could occur during colocalization of CD4 (or CD8) and the TCR during antigen recognition or, as the case of J.CaM1 suggests, during TCR stimulation in the absence of CD4 or CD8 participation. Further experiments to define the critical domains of *lck* and the molecules with which it interacts are in progress.

These studies illustrate the complexity of the structure and function of TCR as it relates to signal transduction. Considerable progress has been made by simplifying the structure of the TCR into functional domains by constructing chimeric molecules. As these domains and the molecules that they interact with are identified, the challenge will be to understand the function and interactions of these components within the context of the whole oligomeric TCR.

ACKNOWLEDGMENTS

We thank Theresa Kadlecek for her excellent technical assistance. We are also grateful to Drs. Dan Littman and Roger Perlmutter for providing cDNAs used in these studies. This work was supported in part by National Institutes of Health grant GM-39553.

REFERENCES

Abraham, N. and A. Veilette. 1990. Activation of p56lck through mutation of a regulatory carboxy-terminal tyrosine residue requires intact sites of autophosphorylation and myristylation. *Mol. Cell. Biol.* **10:** 5197.

Abraham, N., M.C. Miceli, J.R. Parnes, and A. Veillette. 1991. Enhancement of T-cell responsiveness by the lymphocyte-specific tyrosine protein kinase $p56^{lck}$. *Nature* **350:** 62.

Anderson, P., M. Caligiuri, J. Ritz, and S.F. Schlossman. 1989. CD3-negative natural killer cells express ζ TCR as part of a novel molecular complex. *Nature* **341:** 159.

Bernstein, G., J.L. Blank, A.V. Smrcka, T. Higashijima, P.C. Sternweis, J.H. Exton, and E.M. Ross. 1992. Reconstitution of agonist-stimulated phosphatidylinositol 4,5-bisphosphate hydrolysis using purified m1 muscarinic receptor, $G_{q/11}$, and phospholipase C-β1. *J. Biol. Chem.* **267:** 8081.

Berridge, M.J. and R.F. Irvine. 1989. Inositol phosphates and cell signalling. *Nature* **341:** 197.

Burgess, K.E., A.D. Odysseos, C. Zalvan, B.J. Druker, P. Anderson, S.F. Schlossman, and C.E. Rudd. 1991. Biochemical identification of a direct physical interaction between the CD4: $p56^{lck}$ and Ti(TCR)/CD3 complexes. *Eur. J. Immunol.* **21:** 1663.

Chan, A.C., B.A. Irving, J.D. Fraser, and A. Weiss. 1991. The ζ chain is associated with a tyrosine kinase and upon T-cell antigen receptor stimulation associates with ZAP-70, a 70-kDa tyrosine phosphoprotein. *Proc. Natl. Acad. Sci.* **88:** 9166.

Cooke, M.P., K.M. Abraham, K.A. Forbush, and R.M. Perlmutter. 1991. Regulation of T cell receptor signaling by a *src* family protein-tyrosine kinase ($p59^{fyn}$). *Cell* **65:** 281.

Cosson, P., S.P. Lankford, J.S. Bonifacino, and R.D. Klausner. 1991. Membrane protein association by potential intramembrane charge pairs. *Nature* **351:** 414.

Glaichenhaus, N., N. Shastri, D.R. Littman, and J.M. Turner. 1991. Requirement for association of $p56^{lck}$ with CD4 in antigen-specific signal transduction in T cells. *Cell* **64:** 511.

Goldsmith, M.A. and A. Weiss. 1987. Isolation and characterization of a T-lymphocyte somatic mutant with altered signal transduction by the antigen receptor. *Proc. Natl. Acad. Sci.* **84:** 6879.

———. 1988. Early signal transduction by the antigen receptor without commitment to T cell activation. *Science* **240:** 1029.

Goldsmith, M.A., P.E. Dazin, and A. Weiss. 1988. At least two non-antigen-binding molecules are required for signal transduction by the T cell antigen receptor. *Proc. Natl. Acad. Sci.* **85:** 8613.

Goldsmith, M.A., D.M. Desai, T. Schultz, and A. Weiss. 1989. Function of a heterologous muscarinic receptor in T cell antigen receptor signal transduction mutants. *J. Biol. Chem.* **264:** 17190.

Hsi, E.D., J.N. Siegel, Y. Minami, E.T. Luong, R.D. Klausner, and L.E. Samelson. 1989. T cell activation induces rapid tyrosine phosphorylation of a limited number of cellular substrates. *J. Biol. Chem.* **264:** 10836.

Imboden, J.B. and J.D. Stobo. 1985. Transmembrane signalling by the T cell antigen receptor: Perturbation of the T3-antigen receptor complex generates inositol phosphates and releases calcium ions from intracellular stores. *J. Exp. Med.* **161:** 446.

Irving, B. and A. Weiss. 1991. The cytoplasmic domain of the T cell receptor ζ chain is sufficient to couple to receptor-associated signal transduction pathways. *Cell* **64:** 891.

June, C.H., M.C. Fletcher, J.A. Ledbetter, and L.E. Samelson. 1990a. Increases in tyrosine phosphorylation are detectable before phospholipase C activation after T cell receptor stimulation. *J. Immunol.* **144:** 1591.

June, C.H., M.C. Fletcher, J.A. Ledbetter, G.L. Schieven, J.N. Siegel, A.F. Phillips, and L.E. Samelson. 1990b. Inhibition of tyrosine phosphorylation prevents T cell receptor-mediated signal transduction. *Proc. Natl. Acad. Sci.* **87:** 7722.

Koretzky, G., J. Picus, T. Schultz, and A. Weiss. 1991. The tyrosine phosphatase CD45 is required for both T cell antigen receptor and CD2 mediated activation of a protein tyrosine kinase and interleukin 2 production. *Proc. Natl. Acad. Sci.* **88:** 2037.

Koretzky, G.A., J. Picus, M.L. Thomas, and A. Weiss. 1990. Tyrosine phosphatase CD45 is essential for coupling T cell antigen receptor to the phosphatidylinositol pathway. *Nature* **346:** 66.

Lanier, L.L., G. Yu, and J.H. Phillips. 1989. Co-association of CD3ζ with a receptor (CD16) for IgG Fc on human natural killer cells. *Nature* **342:** 803.

Letourneur, F. and R.D. Klausner. 1992. Activation of T cells by a tyrosine kinase activation domain in the cytoplasmic tail of CD3ϵ. *Science* **255:** 79.

Manolios, N., F. Letourneur, J.S. Bonifacino, and R.D. Klausner. 1991. Pairwise, cooperative and inhibitory interactions describe the assembly and probable structure of the T-cell antigen receptor. *EMBO J.* **10:** 1643.

Nakayama, T., L.E. Samelson, Y. Nakayama, T.I. Munitz, M. Sheard, C.H. June, and A. Singer. 1991. Ligand-stimulated signaling events in immature $CD4^+CD8^+$ thymocytes expressing competent T-cell receptor complexes. *Proc. Natl. Acad. Sci.* **88:** 9949.

Nel, A.E., C. Hanekom, A. Rheeder, K. Williams, S. Pollack, R. Katz, and G.E. Landreth. 1990. Stimulation of map-2 kinase activity in T lymphocytes by anti-CD3 or anti-Ti monoclonal antibody is partially dependent on protein kinase C. *J. Immunol.* **144:** 2683.

Ohashi, P., T. Mak, P. Van den Elsen, Y. Yanagi, Y. Yasunobu, A. Calman, C. Terhorst, J. Stobo, and A. Weiss. 1985. Reconstitution of an active surface T3/T-cell antigen receptor by DNA transfer. *Nature* **316:** 606.

Ostergaard, H.L., D.A. Shackelford, T.R. Hurley, P. Johnson, R. Hyman, B.M. Sefton, and I.S. Trowbridge. 1989. Expression of CD45 alters phosphorylation of the lck-encoded tyrosine protein kinase in murine lymphoma T-cell lines. *Proc. Natl. Acad. Sci.* **86:** 8959.

Park, D.J., H.W. Rho, and S.G. Rhee. 1991. CD3 stimulation causes phosphorylation of phospholipase C-γ1 on serine and tyrosine residues in a human T cell line. *Proc. Natl. Acad. Sci.* **88:** 5453.

Pingel, J.T. and M.L. Thomas. 1989. Evidence that the leukocyte-common antigen is required for antigen-induced T lymphocyte proliferation. *Cell* **58:** 1055.

Ravetch, J. and J.-P. Kinet. 1991. Fc receptors. *Annu. Rev. Immunol.* **9:** 457.

Reth, M. 1989. Antigen receptor tail clue. *Nature* **338:** 383.

Romeo, C. and B. Seed. 1991. Cellular immunity to HIV activated by CD4 fused to T cell or Fc receptor polypeptides. *Cell* **64:** 1037.

Romeo, C., M. Amiot, and B. Seed. 1992. Sequence requirements for induction of cytolysis by the T cell antigen/Fc receptor ζ chain. *Cell* **68:** 889.

Samelson, L.E., A.F. Phillips, E.T. Luong, and R.D. Klausner. 1990. Association of the fyn protein-tyrosine kinase with the T-cell antigen receptor. *Proc. Natl. Acad. Sci.* **87:** 4358.

Samelson, L.E., M.D. Patel, A.M. Weissman, J.B. Harford, and R.D. Klausner. 1986. Antigen activation of murine T cells induces tyrosine phosphorylation of a polypeptide associated with the T cell antigen receptor. *Cell* **46:** 1083.

Schraven, B., H. Kirchgessner, B. Gaber, Y. Samstag, and S. Meuer. 1991. A functional complex is formed in human T lymphocytes between the protein tyrosine phosphate CD45, the protein tyrosine kinase $p56^{lck}$ and pp32, a possible common substrate. *Eur. J. Immunol.* **21:** 2469.

Tan, L., J. Turner, and A. Weiss. 1991. Regions of the T cell antigen receptor a and β chains that are responsible for interactions with CD3. *J. Exp. Med.* **173:** 1247.

Ueno, H., H. Colbert, J.A. Escobedo, and L.T. Williams. 1991. Inhibition of PDGF β receptor signal transduction by coexpression of a truncated receptor. *Science* **252:** 844.

Wegener, A.-M.K., F. Letourneur, A. Hoeveler, T. Brocker, F. Luton, and B. Malissen. 1992. The T cell receptor/CD3 complex is composed of at least two autonomous transduction modules. *Cell* **68:** 83.

Weiss, A. 1991. Molecular and genetic insights into T cell antigen receptor structure and function. *Annu. Rev. Genet.* **25:** 487.

Weiss, A., G. Koretzky, R. Schatzman, and T. Kadlecek. 1991. Stimulation of the T cell antigen receptor induces tyrosine phosphorylation of phospholipase C γ1. *Proc. Natl. Acad. Sci.* **88:** 5484.

Activation of Immune System Effector Function by T-cell or Fc Receptor Intracellular Domains

C. Romeo, W. Kolanus, M. Amiot, and B. Seed

Department of Molecular Biology, Massachusetts General Hospital, Boston, Massachusetts 02114

It has recently become apparent that a number of the most prominent receptors in the immune system share common structural features and mediate similar activation pathways. These receptors include the T-cell and B-cell antigen receptors and several myeloid, natural killer cell, and mast cell Fc receptors. This paper focuses on the common attributes of some T-cell and Fc receptor chains and an Fc receptor subtype originally thought to function by a different mechanism.

The most abundant form of the T-cell antigen receptor (TCR) consists of at least six distinct polypeptide chains (Weiss and Stobo 1984; Ohashi et al. 1985; Berkhout et al. 1988; Sussman et al. 1988): the α/β antigen-binding chains, the three polypeptides of the CD3 complex, and the ζ chain (Fig. 1). If any of the chains are absent, efficient surface expression of the remaining members of the complex cannot be achieved. Until recently, the function of the invariant polypeptides in the complex was not known, although the structure of the ζ chain suggested a role in signal transduction. A 32-kD type I integral membrane homodimer, ζ has a 9-residue extracellular domain with no sites for N-linked glycan addition and a 112-residue (murine) or 113-residue (human) intracellular domain (Weissman et al. 1988a,b). The monomers are joined by a disulfide bond located near the beginning of the transmembrane domain, which also bears an aspartic acid residue not typically encountered in membrane-spanning sequences (Weissman et al. 1988a,b). An isoform of ζ called η (Baniyash et al. 1988a; Orloff et al. 1989) arises from an alternate mRNA splicing pathway (Jin et al. 1990) and is present in reduced amounts in cells expressing the antigen receptor. ζ is thought to be the limiting polypeptide for surface expression of the complete TCR complex (Sussman et al. 1988), and early evidence supported the notion that it played an important role in cellular activation programs triggered by receptor recognition of ligand (Weissman et al. 1989; Frank et al. 1990).

The idea that ζ was important for signal transduction gained support when the ζ chain was found to associate with the low-affinity IgG receptor, FcγRIIIA (see Fig. 1), on human natural killer cells (Lanier et al. 1989; Anderson et al. 1990). On murine natural killer cells, it appears that FcγRIIIA associates with a different polypeptide with similar structural features, the FcR γ chain (Ra et al. 1989a). The FcR γ chain was first described as part of the high-affinity IgE receptor complex, FcϵRI, present on mast cells and basophils (Fig. 1) where it associates with a ligand-recognition polypeptide, the α chain, and another polypeptide, the β chain, of undefined function (Blank et al. 1989; Ra et al. 1989b). γ has a homodimeric structure with a very short extracellular domain and cytoplasmic sequence motifs similar to those of ζ (Blank et al. 1989; Kinet 1989). The close similarity between ζ and γ was strengthened by the report that γ is expressed by a mouse T cell line, CTLL, in which it forms homodimers as well as γ-ζ and γ-η heterodimers (Orloff et al. 1990).

The B-cell antigen receptor (BCR) represents a transmembrane form of the immunoglobulin molecule, in which the immunoglobulin chains noncovalently as-

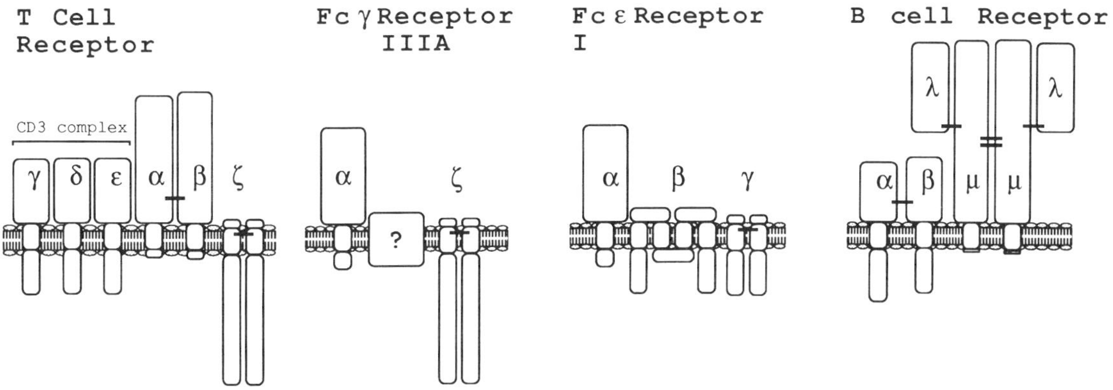

Figure 1. Schematic diagrams of multichain immunological receptor complexes. Intrachain disulfide linkages are shown as horizontal black bars. The subunit stoichiometry of the individual complexes is not precisely known. The FcγRIIIA receptor can accommodate other subunits (Kurosaki et al. 1992), so that possibility is indicated by a "? subunit". The B-cell antigen receptor shown represents the IgM form.

sociate with a disulfide-linked α-β heterodimer (Fig. 1), also known as the mb-1 (BCR-α) and B29 (BCR-β) proteins (Reth 1992). Efficient surface expression of the membrane-bound form of IgM requires the coexpression of both the α and β chains. Antigen binding and receptor cross-linking of the BCR complex induces intracellular signals that, along with T-cell help, stimulate B cells to divide and differentiate into antibody-secreting plasma cells or memory B cells. Although the role of the BCR chains in signal transduction is uncertain, Reth (1989) noted some interesting shared sequence elements in the cytoplasmic tails of ζ, FcR γ, CD3 γ and δ, and the BCR α and β chains, which suggested a shared functional role for these chains.

To study the function of ζ, we created chimeric molecules in which alternative extracellular domains were fused to the transmembrane and intracellular domains of ζ. It seemed possible that if endowed with an extracellular domain bearing sites for N-linked glycan addition, ζ chimeras might find their way to the cell surface. If expressed at the cell surface as an independent receptor chimera, ζ could be manipulated directly through its newly acquired extracellular domain.

RESULTS

Creation of ζ, η, and γ Chimeras

Human ζ (Weissman et al. 1988b) and γ (Küster et al. 1990) cDNAs were isolated by polymerase chain reaction from libraries prepared from the HPB-ALL tumor cell line (Aruffo and Seed 1987) and from human natural killer cells, and η cDNA (Jin et al. 1990) was isolated from a murine thymocyte library. ζ, η, and γ cDNAs were joined to the extracellular domain of an engineered form of CD4 or CD16 possessing a *Bam*HI site just upstream of the membrane-spanning domain (Zettlmeissl et al. 1990; A. Aruffo and B. Seed, unpubl.), which was joined to the *Bam*HI site naturally present in the ζ and η cDNAs at a similar location a few residues upstream of the membrane-spanning domain. To form the fusion protein with γ, a *Bam*HI site was engineered into the sequence at the same approximate location (Fig. 2). The gene fusions were introduced

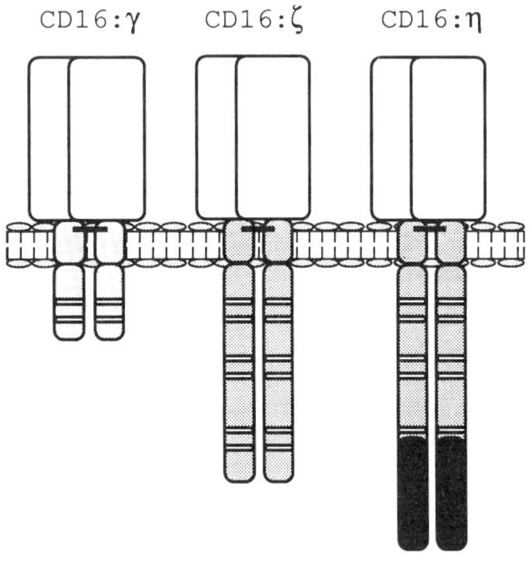

Figure 2. Schematic representation of the receptor chimeras. The extracellular portion of the phosphatidylinositol-linked form of CD16 was joined to the dimeric forms of FcR γ, TCR/FcR ζ, or TCR η just external to the transmembrane domain. The η sequence is identical to the ζ sequence at the amino terminus but diverges at residue 122 of the mature peptide (Jin et al. 1990). The open bars in the cytoplasmic domain represent conserved tyrosine-containing motifs.

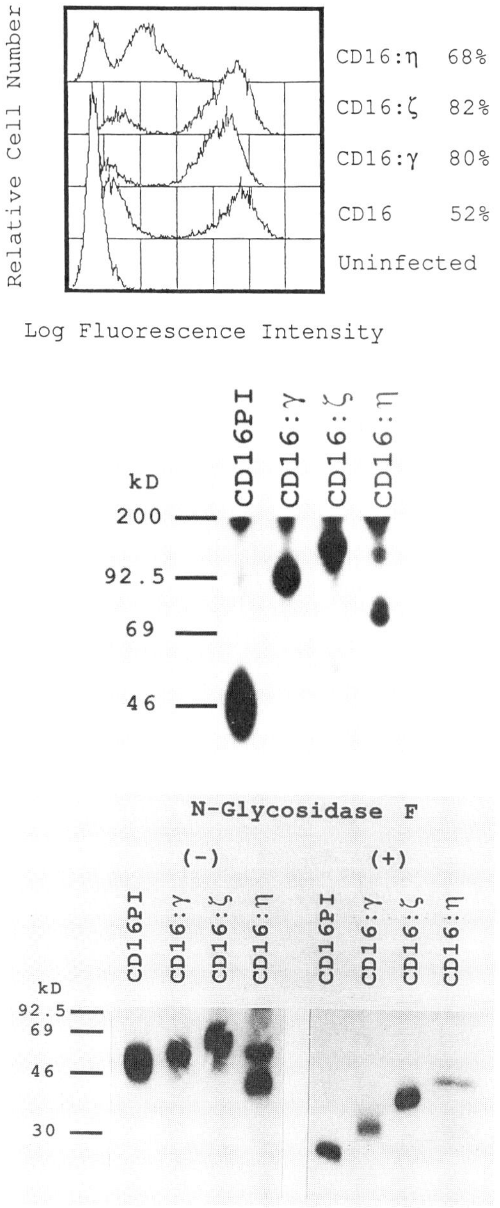

Figure 3. Characterization of the chimeras. (*Top*) Flow cytometric analysis of surface expression of CD16, CD16:γ, CD16:ζ, and CD16:η in CV1 cells. The percentage of cells positively reacting with monoclonal anti-CD16 antibody is shown to the right. (*Middle*) Immunoprecipitation of surface-labeled (^{125}I) CD16, CD16:γ, CD16:ζ, and CD16:η. Samples were run without reducing agent. Molecular mass standards in kilodaltons are indicated at the left. (*Bottom*) Samples are the same as in the middle panel run under reducing conditions and untreated (−) or treated (+) with *N*-glycosidase F to remove N-linked sugars.

into a vaccinia virus expression plasmid bearing the *Escherichia coli gpt* gene as a selectable marker (M. Amiot and B. Seed, unpubl.) and inserted into the genome of the vaccinia WR strain by homologous recombination and selection for growth in mycophenolic acid (Boyle and Coupar 1988; Falkner and Moss 1988). Flow cytometric analysis showed that the vaccinia recombinants direct the abundant production of CD16:ζ and CD16:γ fusion proteins at the cell surface, whereas the expression of CD16:η is substantially weaker (Fig. 3, top). The latter finding is consistent with a recent report that transfection of an η cDNA expression plasmid into a murine hybridoma cell line gave substantially less expression than transfection of a comparable ζ expression plasmid (Clayton et al. 1990). Immunoprecipitation of cells infected with the vaccinia recombinants revealed that the fusion proteins form covalent dimers, unlike the naturally occurring CD16 antigen (Fig. 3, middle).

Immunoprecipitation of surface-labeled cells expressing CD16:η yielded two bands, both corresponding to molecular masses lower than expected from the comparable values for the ζ chimera. However, digestion of the different chimeras with peptide:*N*-glycosidase F (Fig. 3, bottom) yielded peptide molecular masses (42, 38, and 32 kD for CD16:η, ζ, and γ, respectively) that accorded fairly well with the molecular masses predicted by the nucleotide sequences (42, 37, and 29 kD). The more rapidly migrating CD16:η form appears to lack significant glycosylation, as its molecular mass changes little upon digestion with *N*-glycosidase F. Because CD16:ζ and CD16:η are identical up to the last 21 residues of ζ, or the last 63 residues of η, the weaker cell-surface expression and apparent underglycosylation of CD16:η must be mediated by sequences at the carboxyl terminus of the cytoplasmic tail.

Calcium Mobilization in Response to Chimera Cross-linking

T-cell lines were infected with the vaccinia recombinants, and the relative cytoplasmic free calcium ion concentration was measured following cross-linking of the extracellular domains with antibodies. Both spectrofluorimetric (bulk population) and flow cytometric (single-cell) measurements were performed, with cells loaded with the dye Indo-1 (Grynkiewicz et al. 1985; Rabinovitch et al. 1986). Figure 4 (top) shows an analysis of data collected from cells of the Jurkat human T-cell leukemia line infected with vaccinia recombinants expressing the CD16:ζ fusion protein. Cross-linking of the chimeras reproducibly increased intracellular calcium, whereas similar treatment of cells expressing nonchimeric CD16 had little or no effect. When the chimera was expressed in mutant cell lines lacking antigen receptor, either REX33A (Breitmeyer et al. 1987; Sancho et al. 1989) or Jurkat mutant JRT3.T3.5 (Weiss et al. 1984), a strong response to

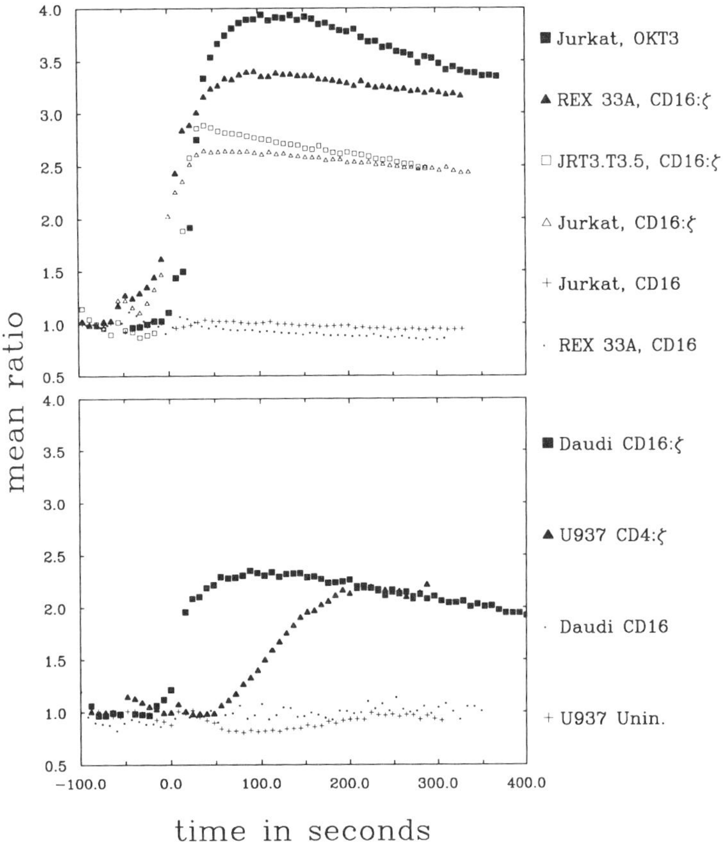

Figure 4. Calcium mobilization by receptor chimeras. (*Top*) Flow cytometric analysis of calcium mobilization following receptor cross-linking of CD16:ζ chimeras in TCR-positive or -negative cell lines. The mean ratio of violet to blue fluorescence of indo-1-loaded cells (a measure of relative calcium ion concentration) is shown for cell populations treated with antibodies at time 0. (■) Response to anti-CD3 antibody OKT3 of Jurkat (T-cell leukemia) cells; (▲) response to anti-CD16 antibody 3G8 cross-linking of CD16:ζ in the REX 33A TCR$^-$ mutant; (□) response to CD16:ζ cross-linking in the Jurkat TCR$^-$ mutant line JRT3.T3.5; (△) response to CD16:ζ cross-linking in Jurkat cells; (+) response to nonchimeric CD16 in Jurkat cells; (dots) response to nonchimeric CD16 in the REX 33A cell line. (*Bottom*) Analysis of calcium mobilization in Daudi cells, a B-cell (Burkitt) lymphoma, infected with vaccinia recombinants expressing CD16:ζ (■) or native CD16 (dots), or in U937 (histiocytic lymphoma) cells line expressing CD4:ζ following anti-CD4 cross-linking (▲), or of uninfected cells following similar treatment (+).

CD16 antibody cross-linking was seen. Infection with recombinants expressing CD16:ζ did not restore the anti-CD3 antibody response, showing that the fusion protein did not act by rescuing intracellular CD3 complex chains (data not shown).

Cellular activation by ζ chimeras could also be detected in other hematopoietic cell types. Figure 4 (bottom) shows calcium ion mobilization after cross-linking of CD16:ζ in Daudi cells, a Burkitt (B-cell) lymphoma cell line, or following cross-linking of CD4:ζ in U937 cells (from a histiocytic lymphoma cell line, thought to represent a neoplastic promonocyte). However, the capacity to mobilize calcium in response to cross-linking is not ubiquitously distributed, as neither fibroblasts, the K562 erythroleukemia cell line, nor the KG-1 (neoplastic myeloid precursor) cell line show this effect (data not shown).

Induction of Cytolysis by Chimeric Receptors

To evaluate the ability of the chimeras to redirect cell-mediated immunity, we developed a cytolysis assay in which cytotoxic T lymphocytes (CTLs) infected with vaccinia recombinants expressing CD16 chimeras specifically kill hybridoma cells expressing membrane-bound anti-CD16 antibodies. This assay is an extension of a hybridoma cytotoxicity assay originally developed to analyze effector mechanisms of cells bearing Fc receptors (Graziano and Fanger 1987a,b; Fanger et al. 1989; Shen et al. 1989). Expression of CD16:ζ in CTLs allows the armed CTL to kill 3G8 (anti-CD16; Fleit et al. 1982) hybridoma cells, whereas CTLs expressing the phosphatidylinositol-linked form of CD16 are inactive (Romeo et al. 1992). CTLs armed with CD16:ζ also do not kill hybridoma cells expressing an irrelevant antibody.

Chimeras between CD4 and ζ, η, or γ were also shown to be capable of directing CTLs to recognize and kill cells expressing the human immunodeficiency virus (HIV) envelope protein gp120 (Romeo and Seed 1991). CD4 is thought to interact with a nonpolymorphic region of the major histocompatibility complex (MHC) class II antigen (Gay et al. 1987; Sleckman et al. 1987); however, we could detect no specific hydrolysis directed by CD4:ζ against the Raji B cell line, which expresses abundant class II antigen (Romeo and Seed 1991).

Deletion and Point Mutation Analysis of ζ

To identify the ζ sequences necessary for cytolysis, a series of deletion mutants were prepared in which successively more of the ζ intracellular domain was removed from the carboxyl terminus (Fig. 5a). Most of the intracellular domain of ζ could be removed with little consequence for cytolytic potential; the full-length chimera CD16:ζ was essentially equal in efficacy to the chimera deleted to residue 65, CD16:ζD66* (Fig. 5b; here as elsewhere, mutations in ζ are represented by the "original residue-location-mutant residue" convention, so that D66*, for example, denotes replacement of Asp-66 by a termination codon). A substantial de-

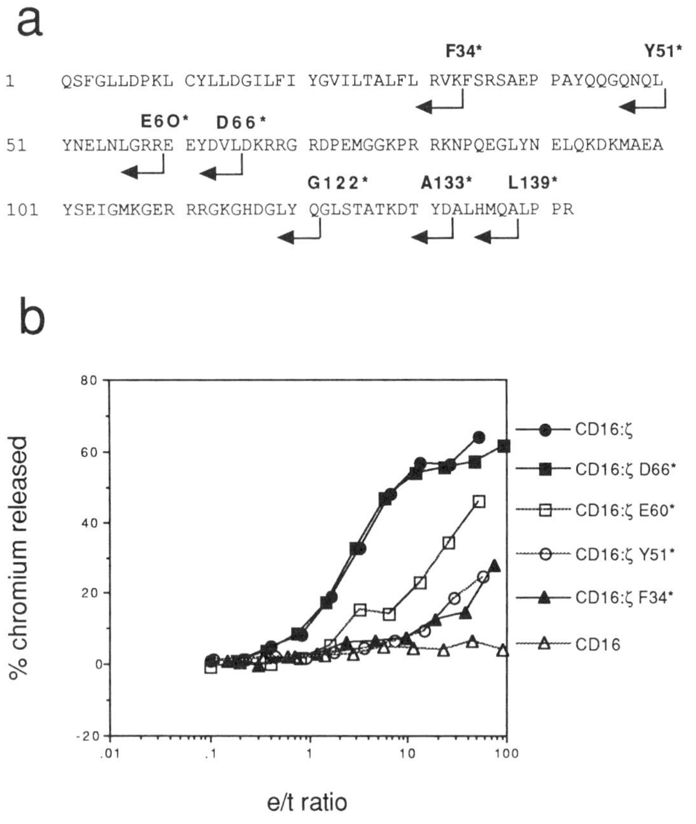

Figure 5. Deletion analysis of cytolytic potential. (*a*) Locations of the deletion endpoints. An asterisk, for example in D66*, denotes replacement of Asp-66 by a termination codon. (*b*) Cytolysis assay of undeleted CD16:ζ and four salient deletions. Hybridoma cells expressing surface antibody to CD16 were loaded with ^{51}Cr and incubated with increasing numbers of human CTLs infected with vaccinia recombinants expressing CD16:ζ chimeras. The percentage of chromium released, as a measure of specific cytolysis, is plotted as a function of the effector (CTL) to target (hybridoma) cell ratio (e/t). (Reprinted, with permission, from Romeo et al. 1992.)

crease in cytotoxicity was observed on deletion to ζ residue 59 (chimera CD16:ζE60*), and further deletion to residue 50 resulted in slightly less activity. However, complete loss of activity was not observed even when the intracellular domain was reduced to a three-residue transmembrane anchor chimera CD16:ζF34* (Fig. 5).

To eliminate the possible contribution of ζ transmembrane residues, the transmembrane and first 17 cytoplasmic residues of ζ were replaced by sequences encoding the membrane-spanning and first 14 or first 17 cytoplasmic residues of the CD5 or CD7 antigens, respectively (Fig. 6a). The resulting tripartite fusion proteins CD16:5:ζ(48–65) and CD16:7:ζ(48–65) do not form disulfide-linked dimers as do the simpler CD16:ζ chimeras because they lack the cysteine residue in the ζ transmembrane domain (Romeo et al. 1992). Both tripartite chimeras are able to mobilize calcium in Jurkat and TCR-negative cell lines and to mount a cytolytic response in CTLs (Fig. 6b and data not shown). However, truncation of the ζ portion to residue 59 in chimera CD16:7:ζ(48–59) abrogates the ability of tripartite fusion to direct calcium responsiveness in TCR-positive or -negative Jurkat cells or cytolysis in mature CTLs (Fig. 6b and data not shown).

To examine the contributions of individual residues within the ζ(48–65) motif, we prepared a number of mutant variants by site-directed mutagenesis and evaluated their ability to mediate receptor-directed killing (Fig. 6c). A number of relatively conservative substitutions (i.e., replacing acidic residues with their cognate amides) that spanned residues 59 to 63 yielded a moderate compromise of cytolytic efficacy. In general, the variants retained the ability to mobilize calcium. Collectively, however, these residues comprise an important submotif inasmuch as their deletion eliminates cytolytic activity. Conversion of Tyr-62 to either Phe-62 or Ser-62 eliminated both the cytotoxic and calcium responses. At the amino terminus of the 18-residue segment, replacement of Tyr-51 with Phe-51 abolished both calcium mobilization and cytolytic activity, whereas substitution of Leu-50 with Ser-50 eliminated the calcium response while only partially impairing cytolysis. At present, we suspect that the inability of the Leu-50/Ser-50 mutant to mobilize calcium in short-term flow cytometric assays does not fully reflect its ability to mediate a substantial increase in free intracellular calcium ion over the longer time span of the cytolysis assay.

ζ Contains Multiple Trigger Motifs

Within the cytoplasmic tail of ζ, one can identify three sequence elements containing a two-tyrosine motif similar to the one described above. To investigate the potential role of redundant sequence elements, the intracellular domain of ζ was divided into three segments, spanning residues 33–65, 71–104, and 104–138. Each of these segments was attached to a CD16:CD7 chimera by means of a *Mlu*I site introduced just distal to the basic membrane-anchoring sequences of the intracellular domain of CD7 (Fig. 7a). Comparison of the cytolytic efficacy of the three elements showed they were essentially equipotent (Romeo et al. 1992). Sequence comparison (Fig. 7b) shows that the second motif bears 11 residues between tyrosines, whereas the first and third motifs bear 10.

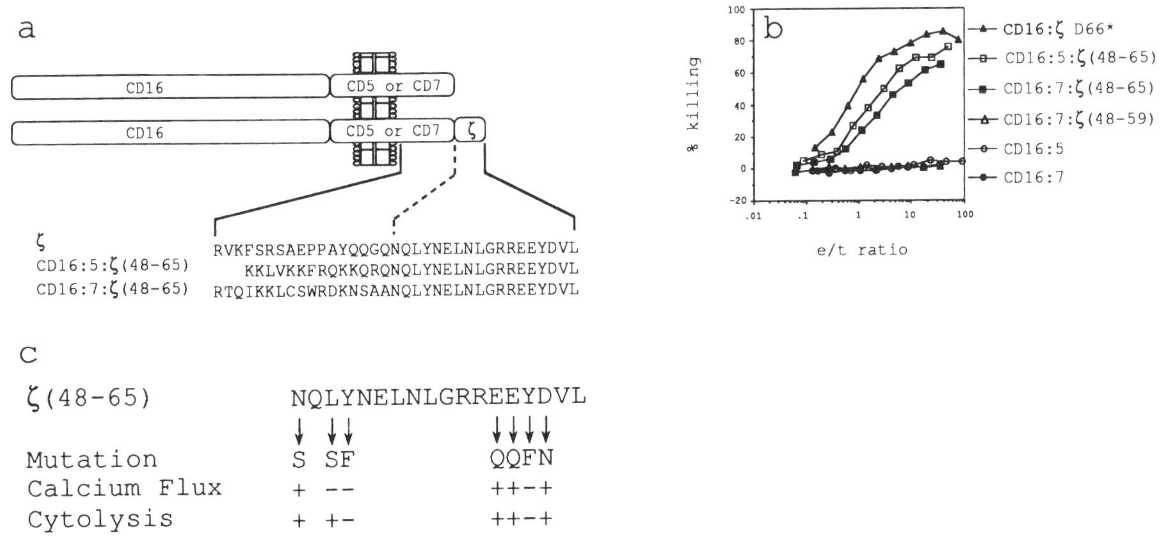

Figure 6. Elimination of the potential for transmembrane interactions reveals a short ζ segment capable of mediating cytolysis. (*a*) Schematic diagram of the monomeric bipartite and tripartite chimeras. At the top is the CD16:(CD5 or CD7) construct truncated after residue Gln-416 and Ala-193 of CD5 and CD7, respectively. Below are the CD16:CD5:ζ and CD16:CD7:ζ constructs and related controls. The peptide sequences of the intracellular domains are shown below. (*b*) Cytolytic activity mediated by tripartite fusion proteins. (▲) CD16:ζD66*; (□) CD16:5:ζ(48–65); (■) CD16:7:ζ(48–65); (△) CD16:7:ζ(48–59); (○) CD16:5; (●) CD16:7. (*c*) Contribution of individual amino acids to the activity of the 18-residue motif. Calcium ion mobilization and cytolytic potential are indicated for the amino acid substitutions shown. (*b* Reprinted, with permission, from Romeo et al. 1992.)

a

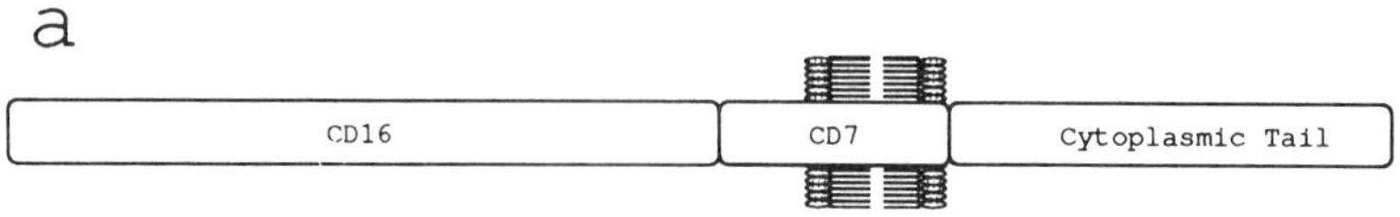

b

Cytoplasmic Tail Residues	Motif Residues	Cytolytic Function
ζ (33-65)	...NQL**Y**NE**L**NEGRR-EE**Y**DV**L***	Yes
ζ (71-104)	...EGL**Y**NE**L**QKDKMAEA**Y**SE**I***	Yes
ζ (104-138)	...DGL**Y**QG**L**STATK-DT**Y**DA**L**HMQA*	Yes
CD3-γ (119-160)	...DQL**Y**QP**L**KDRED-DQ**Y**SH**L**QGNQLRRN*	Yes
CD3-δ (109-150)	...DQV**Y**QP**L**RDRDD-AQ**Y**SH**L**GGNWARNK*	Yes
CD3-ε (134-185)	...NPD**Y**EP**I**RKGQR-DL**Y**SG**L**NQRRI*	Yes
FcR-γ (30-68)	...DGV**Y**TG**L**STRNQ-ET**Y**ET**L**KHEKPPQ*	Yes
BCR-α (135-192)	...ENL**Y**EG**L**NLDDC-SM**Y**ED**I**SRGLQGT**Y**QD**V**GNLHIGDAQLEKP*	Yes
FcγRII-A (236-311)	...NND**Y**ET**A**DGG**Y**MT**L**NPRAPTDDDKNI**Y**LT**L**PPNDHVNSNN*	Yes
CD28 (165-202)	...RSRLLHSD**Y**MN**M**TPRRPGPTRKH**Y**QP**Y**APPRDFAA**Y**RS*	No

Figure 7. Alignment of putative trigger motifs. (*a*) Schematic diagram of chimeras formed by appending cytoplasmic domains to the transmembrane domain of a CD16:CD7 chimera. (*b*) The cytoplasmic domains utilized along with their sequence locations are indicated at the left. The sequences of the putative tyrosine containing trigger motifs are shown in the middle with a single gap, represented by a dash, inserted to preserve the alignment of the tyrosines. An asterisk follows the carboxy-terminal residue of the peptide. Cytolytic function is indicated on the right.

Distribution and Function of the Trigger Motif

Two-tyrosine sequence motifs similar to those in the three functional elements of the ζ cytoplasmic domain are also found in the CD3 γ, δ, and ε chains, in the B-cell receptor α and β chains (mb-1 and B29, respectively), and in the Fc receptor β and γ chains (Kinet 1989; Reth et al. 1991; Wegener et al. 1992). In all cases tested, cytoplasmic tails containing these two-tyrosine motifs are capable of triggering the cytolytic effector program (Fig. 7b). The only absolutely conserved residues are the two tyrosines, and in all cases except one, the spacing between these tyrosines is ten residues. Both tyrosines also have conserved aliphatic amino acids precisely spaced three residues to their carboxy-terminal sides. The presence of hydrophilic and charged amino acids within the motif is likely important, although the exact nature and placement of these residues is variable. The BCR-α sequence is unique in that it contains an additional Y-X-X-Aliphatic element with the same ten-residue intratyrosine spacing. A fourth tyrosine residue (not shown) in the BCR-α cytoplasmic tail is embedded within a highly acidic environment located five residues from the amino-terminal tyrosine depicted in Figure 7b.

Recent studies have shown that similar CD8:ζ chimeras (Irving and Weiss 1991), IL–2Rβ(Tac):CD3 ε chimeras (Letourneur and Klausner 1992), and a T-cell receptor complex consisting of CD3 γ, δ, and ε but lacking ζ (Wegener et al. 1992), have the capacity to activate T-cell helper function. This suggests that the role of the trigger motif is to couple ligand-recognition chains to a common or related activation pathway in a variety of immune cells.

Expanded Role for Trigger Motifs

In the course of this work, it became apparent that another class of Fcγ receptors, FcγRII-A and -C, possesses the potential for triggering cytolytic T-cell effector programs (Kolanus et al. 1992). As shown in Figure 7b, structural similarities exist between the cytoplasmic domain of FcγRII-A and the ζ family of trigger motifs, and the FcγRII-A cytolytic function maps to this tyrosine containing sequence. The Y-X-X-Aliphatic element is present in three copies but with different intratyrosine spacings. FcγRII-A can be found on the surfaces of monocytes, macrophages, granulocytes, and platelets. The presence of this receptor on granulocytes and platelets extends the known range of trigger motif function to these cell types.

Accessory Receptors on T Cells Lack Autonomous Function

The ability to independently activate immune cells should not be concluded to be a general property of immune cell-surface receptors. Notable exceptions include the group of T-cell surface molecules often referred to as coreceptors or accessory molecules that are believed to influence or provide additional important signals for T-cell activation. This group of accessory receptors includes CD2, CD4, CD8, and CD28. The CD2 cytoplasmic tail is devoid of tyrosine residues, and a CD16:CD7:CD2(219–331) chimera was found not to trigger cytolysis (data not shown). The CD4 and CD8 cytoplasmic tails associate with the nonreceptor tyrosine kinase lck but by themselves do not transmit an activation signal in mature T cells (for review, see

Janeway 1992). The CD28 cytoplasmic tail contains four tyrosines but does not activate cytolytic function (Fig. 7b).

DISCUSSION

The T-cell and B-cell antigen receptors and the Fc receptors, FcγRIII-A, FcεRI, and FcγRII-A and -C, share many important features. All are capable of recognizing and responding to the ligand to generate activation signals that result in the execution of cellular effector programs. Although the nature of the ligand recognition apparatus and the ensuing effector program engaged are cell type-specific, common underlying features are present that serve to unify receptor function.

The T-cell antigen receptor recognizes antigen-derived peptides bound to MHC proteins. Upon receptor engagement, helper T lymphocytes synthesize and secrete cytokines, such as interleukin-2 (IL-2) and IL-4, that are important for activating the immune system, whereas CTLs kill their receptor-bound target cells. Immunoglobulin on the surface of a B cell can bind antigen directly and with T-cell help can go on to generate a humoral immune response by secreting soluble antibody. Antibody bound to antigen on target cells can in turn activate effector cells that bear receptors for the antibody Fc portion. Fc-receptor-initiated responses include phagocytosis, antibody-dependent cellular cytotoxicity, and the release of cellular mediators, such as histamine or leukocyte chemoattractants.

In the case of Fc receptor/antibody interactions, the physiological concentration of immunoglobulin in the serum is at least two orders of magnitude higher than the equilibrium binding constant of the Fc receptor for its antibody ligand. Therefore, the Fc receptors are always occupied by ligand without activating effector programs. It is the engagement of multivalent antibody/antigen complexes that leads to receptor aggregation and triggering of the signal transduction pathway (Ravetch and Kinet 1991). Since the T- and B-cell antigen receptors use common elements for triggering, it appears likely that receptor clustering is the primary mechanism by which they initiate signal transduction as well.

Ligand recognition by all of these receptors appears to be coupled to signal transduction by a common trigger element. The trigger element is a short peptide sequence containing at least two conserved tyrosine motifs. The tyrosine motif is characterized by the amino acid sequence Y-X-X-Aliphatic, where the aliphatic residue is predominately a leucine but can be an isoleucine. The spacing between the two tyrosines is conserved in the ζ family of trigger motifs that appear to be evolutionarily related (Wegener et al. 1992). We have uncovered a novel trigger element in an unrelated Fc receptor, FcγRII-A, that contains tyrosine motifs characteristic of the ζ trigger family but with a different spacing between the tyrosines. No known enzymatic activity has been described for the cytoplasmic tails containing this motif, and it appears likely that cellular activation is mediated through interactions between the motif and additional nonreceptor proteins (see Weiss et al., this volume). Interestingly, all of the receptors utilizing the trigger motif initially follow a similar path of activation marked by increases in tyrosine phosphorylation, generation of inositol triphosphates, and mobilization of intracellular calcium.

It is tempting to speculate that tyrosine phosphorylation of the motif is an important primary event in receptor signaling. In the case of T-cell receptor function this is unlikely for several reasons: (1) The kinetics of ζ phosphorylation are delayed relative to the kinetics of other early activation events, such as tyrosine phosphorylation of other proteins, generation of inositol triphosphates, and release of intracellular calcium (June et al. 1990); (2) the steady-state level of ζ phosphorylation following activation is around 5%, thus tyrosine-phosphorylated ζ represents a minor component of the signaling apparatus (Baniyash et al. 1988b); and (3) η and other variants of ζ, such as deletion and point mutants, are not tyrosine-phosphorylated following receptor triggering (Clayton et al. 1991; Koyasu et al. 1992). Rather, it seems reasonable to speculate that ζ phosphorylation plays a role in modulating TCR signaling, for example, through processes that might affect receptor coupling to the cytoskeleton or receptor internalization.

The triggering of cell-type-specific effector function described above can be distinguished from the diverse signals that either influence responsiveness to central activating signals or mediate activation pathways that do not engage effector function, e.g., as found in adhesion or chemotaxis. At the present time, no molecules apart from members of the trigger family have been shown to autonomously activate T cells, the most extensively characterized of immune system effector cells (Yokoyama and Shevach 1989). Aggregation of phospholipid-linked proteins (Kroczek et al. 1986; Yeh et al. 1987; Balk and Terhorst 1989), or treatment of CD2 with pairs of monoclonal antibodies (Meuer et al. 1984), can activate effector function in T cells, but this activation depends on the presence of one or more members of the trigger family (Schmitt-Verhulst et al. 1987; Moingeon et al. 1992). In less extensively characterized cell types, there is relatively little information as to the indispensability of the trigger molecules. In no small part, this scarcity of information can be attributed to the absence of suitable cell lines that both possess effector function and are amenable to genetic and biochemical analyses.

The use of T cells as surrogates for dissecting trigger motif function provides an alternative to studying receptor function in certain less amenable cell types such as monocytes, macrophages, granulocytes, and platelets. The common and central role that trigger motifs play in initiating signal transduction makes it likely that lessons learned from one type of immune cell will be broadly applicable to other cell types.

ACKNOWLEDGMENTS

We thank Sandro Aruffo for many of the *Bam*HI extracellular domain cassettes, Bernard Moss and Charlie Rice for generously providing recombinant vaccinia viruses and vectors, Franco Harris and Jim Kurnick for supplying the WH3 line, and members of the laboratory for discussion and criticism. C.R. was the recipient of grant 1309-9-RG from the American Foundation for AIDS Research. Other support was provided by an award to the Massachusetts General Hospital from Hoechst AG.

REFERENCES

Anderson, P., M. Caligiuri, C. O'Brien, T. Manley, J. Ritz, and S.F. Schlossman. 1990. Fc gamma receptor type III (CD16) is included in the zeta NK receptor complex expressed by human natural killer cells. *Proc. Natl. Acad. Sci.* **87:** 2274.

Aruffo, A. and B. Seed. 1987. Molecular cloning of a CD28 cDNA by a high efficiency COS cell expression system. *Proc. Natl. Acad. Sci.* **84:** 8573.

Balk, S. and C. Terhorst. 1989. Relationship between the T-cell receptor-T3 complex and Thy-1. *Immunol. Ser.* **45:** 411.

Baniyash, M., P. Garcia-Morales, J.S. Bonifacino, L.E. Samelson, and R.D. Klausner. 1988a. Disulfide linkage of the zeta and eta chains of the T cell receptor. Possible identification of two structural classes of receptors. *J. Biol. Chem.* **263:** 9874.

Baniyash, M., P. Garcia-Morales, E. Luong, L.E. Samelson, and R.D. Klausner. 1988b. The T cell antigen receptor zeta chain is tyrosine phosphorylated upon activation. *J. Biol. Chem.* **263:** 18225.

Berkhout, B., B. Alarcon, and C. Terhorst. 1988. Transfection of genes encoding the T cell receptor-associated CD3 complex into COS cells results in the assembly of the macromolecular structure. *J. Biol. Chem.* **263:**8528.

Blank, U., C. Ra, L. Miller, K. White, H. Metzger, and J.-P. Kinet. 1989. Complete structure and expression in transfected cell of high affinity IgE receptor. *Nature* **337:** 187.

Boyle, D.B. and B.E.H. Coupar. 1988. A dominant selectable marker for the construction of recombinant poxviruses. *Gene* **65:** 123.

Breitmeyer, J.B., S.O. Oppenheim, J.F. Daley, H.B. Levine, and S.F. Schlossman. 1987. Growth inhibition of human T cells by antibodies recognizing the T cell antigen receptor complex. *J. Immunol.* **138:** 726.

Clayton, L.K., A. Bauer, Y.J. Jin, L. D'Adamio, S. Koyasu, and E.L. Reinherz. 1990. Characterization of thymus-derived lymphocytes expressing $Ti\alpha-\beta CD3\gamma\delta\epsilon\zeta-\zeta$, $Ti\alpha-\beta CD3\gamma\delta\epsilon\eta-\eta$ or $Ti\alpha-\beta CD3\gamma\delta\epsilon\zeta-\zeta/\zeta-\eta$ antigen receptor isoforms: Analysis by gene transfection. *J. Exp. Med.* **172:** 1243.

Clayton, L.K., L. D'Adamio, F.D. Howard, M. Sieh, R.E. Hussey, S. Koyasu, and E.L. Reinherz. 1991. CD3 eta and CD3 zeta are alternatively spliced products of a common genetic locus and are transcriptionally and/or post-transcriptionally regulated during T-cell development. *Proc. Natl. Acad. Sci.* **88:** 5202.

Falkner, F.G. and B. Moss. 1988. *E. coli gpt* gene provides dominant selection for vaccinia virus open reading frame expression vectors. *J. Virol.* **62:** 1849.

Fanger, M.W., L. Shen, R.F. Graziano, and P.M. Guyre. 1989. Cytotoxicity mediated by human Fc receptors for IgG. *Immunol. Today* **10:** 92.

Fleit, H.B., S.D. Wright, and J.C. Unkeless. 1982. Human neutrophil Fc gamma receptor distribution and structure. *Proc. Natl. Acad. Sci.* **79:** 3275.

Frank, S.J., B.B. Niklinska, D.G. Orloff, M. Merćep, J.D. Ashwell, and R.D. Klausner. 1990. Structural mutations of the T cell receptor ζ chain and its role in T cell activation. *Science* **249:** 174.

Gay, D., P. Maddon, R. Sekaly, M.A. Talle, M. Godfrey, E. Long, G. Goldstein, L. Chess, R. Axel, J. Kappler, and P. Marrack. 1987. Functional interaction between the human T-cell protein CD4 and the major histocompatibility complex HLA-DR antigen. *Nature* **328:** 626.

Graziano, R.F. and M.W. Fanger. 1987a. Human monocyte-mediated cytotoxicity: The use of Ig-bearing hybridomas as target cells to detect trigger molecules on the monocyte cell surface. *J. Immunol.* **138:** 945.

———. 1987b. Fc gamma RI and Fc gamma RII on monocytes and granulocytes are cytotoxic trigger molecules for tumor cells. *J. Immunol.* **139:** 3536.

Grynkiewicz, G., M. Poenie, and R.Y. Tsien. 1985. A new generation of Ca^{2+} indicators with greatly improved fluorescence properties. *J. Biol. Chem.* **260:** 3440.

Irving, B.A. and A. Weiss. 1991. The cytoplasmic domain of the T cell receptor zeta chain is sufficient to couple to receptor-associated signal transduction pathways. *Cell* **64:** 891.

Janeway, C.A., Jr. 1992. The T cell receptor as a multicomponent signalling machine: CD4/CD8 coreceptors and CD45 in T cell activation. *Annu. Rev. Immunol.* **10:** 645.

Jin, Y.J., L.K. Clayton, F.D. Howard, S. Koyasu, M. Sieh, R. Steinbrich, G.E. Tarr, and E.L. Reinherz. 1990. Molecular cloning of the $CD3\eta$ subunit identifies a $CD3\zeta$-related product in thymus-derived cells. *Proc. Natl. Acad. Sci.* **87:** 3319.

June, C.H., M.C. Fletcher, J.A. Ledbetter, and L.E. Samelson. 1990. Increases in tyrosine phosphorylation are detectable before phospholipase C activation after T cell receptor stimulation. *J. Immunol.* **144:** 1591.

Kinet, J.-P. 1989. Antibody-cell interactions: Fc receptors. *Cell* **57:** 351.

Kolanus, W., C. Romeo, and B. Seed. 1992. Lineage-independent activation of immune system effector function by myeloid Fc receptors. *EMBO J.* (in press).

Koyasu, S., D.J. McConkey, L.K. Clayton, S. Abraham, B. Yandava, T. Katagiri, P. Moingeon, T. Yamamoto, and E.L. Reinherz. 1992. Phosphorylation of multiple CD3 zeta tyrosine residues leads to formation of pp21 in vitro and in vivo. Structural changes upon T cell receptor stimulation. *J. Biol. Chem.* **267:** 3375.

Kroczek, R.A., K.C. Gunter, R.N. Germain, and E.M. Shevach. 1986. Thy-1 functions as a signal transduction molecule in T lymphocytes and transfected B lymphocytes. *Nature* **322:** 181.

Kurosaki, T., I. Gander, U. Wirthmueller, and J.V. Ravetch. 1992. The beta subunit of the Fc epsilon RI is associated with the Fc gamma RIII on most cells. *J. Exp. Med.* **175:** 447.

Küster, H., H. Thompson, and J.-P. Kinet. 1990. Characterization and expression of the gene for the human Fc receptor γ subunit. Definition of a new gene family. *J. Biol. Chem.* **265:** 6448.

Lanier, L.L., G. Yu, and J.H. Phillips. 1989. Co-association of $CD3\zeta$ with a receptor (CD16) for IgG Fc on human natural killer cells. *Nature* **342:** 803.

Letourneur, F. and R.D. Klausner. 1992. Activation of T cells by a tyrosine kinase activation domain in the cytoplasmic tail of CD3 epsilon. *Science* **255:** 79.

Meuer, S.C., R.E. Hussey, M. Fabbi, D. Fox, O. Acuto, K.A. Fitzgerald, J.C. Hodgdon, J.P. Protentis, S.F. Schlossman, and E.L. Reinherz. 1984. An alternative pathway of T-cell activation: A functional role for the 50 kd T11 sheep erythrocyte receptor protein. *Cell* **36:** 897.

Moingeon, P., J.L. Lucich, D.J. McConkey, F. Letourneur, B. Malissen, J. Kochan, H.C. Chang, H.R. Rodewald, and E.L. Reinherz. 1992. CD3 zeta dependence of the CD2 pathway of activation in T lymphocytes and natural killer cells. *Proc. Natl. Acad. Sci.* **89:** 1492.

Ohashi, P.S., T.W. Mak, P. Van den Elsen, Y. Yanagi, Y. Yoshikai, A.F. Calman, C. Terhorst, J.D. Stobo, and A. Weiss. 1985. Reconstitution of an active surface T3/T-cell antigen receptor by DNA transfer. *Nature* **316:** 606.

Orloff, D.G., S.J. Frank, F.A. Robey, A.M. Weissman, and R.D. Klausner. 1989. Biochemical characterization of the eta chain of the T-cell receptor. A unique subunit related to zeta. *J. Biol. Chem.* **264:** 14812.

Orloff, D.G., C. Ra, S.J. Frank, R.D. Klausner, and J.-P. Kinet. 1990. Family of disulphide-linked dimers containing the ζ and η chains of the T-cell receptor and the γ chain of Fc receptors. *Nature* **347:** 189.

Ra, C., M.-H.E. Jouvin, and J.-P. Kinet. 1989a. Complete structure of the mouse mast cell receptor for IgE (FcϵRI) and surface expression of chimeric receptors (rat-mouse-human) on transfected cells. *J. Biol. Chem.* **264:** 15323.

Ra, C., M.-H.E. Jouvin, U. Blank, and J.-P. Kinet. 1989b. A macrophage Fcγ receptor and the mast cell receptor for IgE share an identical subunit. *Nature* **241:** 752.

Rabinovitch, P.S., C.H. June, A. Grossmann, and J.A. Ledbetter. 1986. Heterogeneity among T cells in intracellular free calcium responses after mitogen stimulation with PHA or anti-CD3. Simultaneous uses of Indo-1 and immunofluorescence with flow cytometry. *J. Immunol.* **137:** 952.

Ravetch, J.V. and J.P. Kinet. 1991. Fc receptors. *Annu. Rev. Immunol.* **9:** 457.

Reth, M. 1989. Antigen receptor tail clue. *Nature* **338:** 383.

———. 1992. Antigen receptors on B lymphocytes. *Annu. Rev. Immunol.* **10:** 97.

Reth, M., J. Hombach, J. Wienands, K.S. Campbell, N. Chien, L.B. Justement, and J.C. Cambier. 1991. The B-cell antigen receptor complex. *Immunol. Today* **12:** 196.

Romeo, C. and B. Seed. 1991. Cellular immunity to HIV activated by CD4 fused to T cell or Fc receptor polypeptides. *Cell* **64:** 1037.

Romeo, C., M. Amiot, and B. Seed. 1992. Sequence requirements for induction of cytolysis by the T cell antigen/Fc receptor zeta chain. *Cell* **68:** 889.

Sancho, J., T. Chatila, R.C. Wong, C. Hall, R. Blumberg, B. Alarcon, R.S. Geha, and C. Terhorst. 1989. T-cell antigen receptor (TCR)- alpha/beta heterodimer formation is a prerequisite for association of CD3-zeta 2 into functionally competent TCR.CD3 complexes. *J. Biol. Chem.* **264:** 20760.

Schmitt-Verhulst, A.M., A. Guimezanes, C. Boyer, M. Poenie, R. Tsien, M. Buferne, C. Hua, and L. Leserman. 1987. Pleiotropic loss of activation pathways in a T-cell receptor alpha-chain deletion variant of a cytolytic T-cell clone. *Nature* **325:** 628.

Shen, L., R.F. Graziano, and M.W. Fanger. 1989. The functional properties of Fc gamma RI, II and III on myeloid cells: A comparative study of killing of erythrocytes and tumor cells mediated through the different Fc receptors. *Mol. Immunol.* **26:** 959.

Sleckman, B.P., A. Peterson, W.K. Jones, J.A. Foran, J.L. Greenstein, B. Seed, and S.J. Burakoff. 1987. Expression and function of CD4 in a murine T cell hybridoma. *Nature* **328:** 351.

Sussman, J.J., J.S. Bonifacino, J. Lippincott-Schwartz, A.M. Weissman, T. Saito, R.D. Klausner, and J.D. Ashwell. 1988. Failure to synthesize the T cell CD3-ζ chain: Structure and function of a partial T cell receptor complex. *Cell* **52:** 85.

Wegener, A.-M.K., F. Letourneur, A. Hoeveler, T. Brocker, F. Luton, and B. Malissen. 1992. The T cell receptor/CD3 complex is composed of at least two autonomous transduction molecules. *Cell* **68:** 83.

Weiss, A. and J.D. Stobo. 1984. Requirement for the coexpression of T3 and the T cell antigen receptor on a malignant human T cell line. *J. Exp. Med.* **160:** 1284.

Weiss, A., R.L. Wiskocil, and J.D. Stobo. 1984. The role of the T3 surface molecules in the activation of human T cells: A two-stimulus requirement for IL2 production reflects events occurring at a pre-translational level. *J. Immunol.* **133:** 123.

Weissman, A.M., M. Baniyash, D. Hou, L.E. Samelson, W.H. Burgess, and R.D. Klausner. 1988a. Molecular cloning of the zeta chain of the T cell antigen receptor. *Science* **239:** 1018.

Weissman, A.M., S.J. Frank, D.G. Orloff, M. Mercep, J.D. Ashwell, and R.D. Klausner. 1989. Role of the zeta chain in the expression of the T cell antigen receptor: Genetic reconstitution studies. *EMBO J.* **8:** 3651.

Weissman, A.M., D. Hou, D.G. Orloff, W.S. Modi, H. Seuanez, S.J. O'Brien, and R.D. Klausner. 1988b. Molecular cloning and chromosomal localization of the human T-cell receptor ζ chain: Distinction from the molecular CD3 complex. *Proc. Natl. Acad. Sci.* **85:** 9709.

Yeh, E.T., H. Reiser, J. Daley, and K.L. Rock. 1987. Stimulation of T cells via the TAP molecule, a member in a family of activating proteins encoded in the Ly-6 locus. *J. Immunol.* **138:** 91.

Yokoyama, W.M. and E.M. Shevach. 1989. T cell activation via cell-surface antigens other than the CD3/T cell receptor complex. *Year Immunol.* **4:** 110.

Zettlmeissl, G., J.P. Gregersen, J.M. Duport, S. Mehdi, G. Reiner, and B. Seed. 1990. Expression and characterization of human CD4:immunoglobulin fusion proteins. *DNA Cell Biol.* **9:** 347.

G-Protein-coupled Receptors: Regulatory Role of Receptor Kinases and Arrestin Proteins

R.J. LEFKOWITZ, J. INGLESE, W.J. KOCH, J. PITCHER
H. ATTRAMADAL, AND M.G. CARON
Departments of Medicine and Biochemistry, Howard Hughes Medical Institute, Duke University Medical Center, Durham, North Carolina 27710

Receptors that couple to guanine nucleotide regulatory proteins comprise an extraordinarily diverse collection of functional entities. They mediate responses to stimuli as varied as neurotransmitters, hormones, chemoattractants, cytokines, odorants, and even photons of light (Dohlman et al. 1991). It has become increasingly clear that the functional and regulatory properties of these receptors are remarkably conserved, mirroring the substantial conservation of their structure and topographical features. Activation of such receptors with agonists or other stimuli leads to conformational changes that render the receptors "active." At least two major pathways exist along which the activated receptors may proceed. One pathway involves interaction with G protein, activation of a biochemical effector, and generation of a physiological response. The other pathway involves interaction with proteins other than the G proteins, leading to other fates not involving activation of physiological responses. Perhaps the best-studied of such pathways is that by which the activated receptors interact with proteins that lead to receptor desensitization. As with the process of receptor activation and interaction with G protein, with such desensitization mechanisms, substantial conservation of properties can be seen from one system to another. In this paper, we briefly review recently developed information dealing with the biochemical mechanisms by which activated G-protein-coupled receptors become desensitized. These mechanisms involve two newly appreciated families of proteins, receptor kinases and the arrestin proteins.

Figure 1 portrays in schematic form the known components of two model G-protein-coupled signal transducing systems, the adenylyl-cyclase-coupled β_2-adrenergic receptor and the visual light perception system. It also shows the proteins involved in the desensitization pathway. There are remarkable analogies and homologies between these two systems and others as well. In each case, a receptor protein, for example, the β_2-adrenergic receptor or rhodopsin, is present to initially interact with the stimulus. Many members of this receptor gene family have now been cloned, and their properties are well described in the literature (Dohlman et al. 1991). They all appear to have seven membrane-spanning domains interconnected by various hydrophilic extracellular and intracellular loops. In general, it appears to be the membrane-spanning domains that form the ligand-binding pocket, or in the case of rhodopsin, a pocket that nestles the chromophore retinal, which is covalently attached to the seventh transmembrane domain. The regions of the receptor that interact with G proteins comprise those segments of the intracellular hydrophilic loops which are found in closest proximity to the plasma membrane. Regions in the third cytoplasmic loop appear to be particularly important. Heterotrimeric G proteins, G_s in the case of the β-receptor system and transducin (G_t) in the case of the visual system, interact with the stimulus-modified receptors. The characteristics of these ubiquitous signal transducing proteins have been reviewed elsewhere (Bourne et al. 1991). They are heterotrimeric, consisting of α-GTP binding and hydrolyzing subunits and a $\beta\gamma$ complex that helps to anchor the α subunit in the membrane and that performs a number of other only recently appreciated signal transducing functions in its own right.

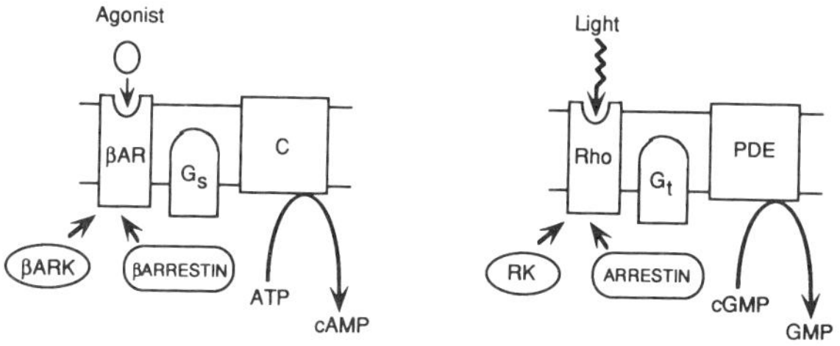

Figure 1. Schematic depicting the known signal transducing components of the adenylyl cyclase/β_2-adrenergic receptor (βAR) system and the visual light transduction system.

The third component of such signaling systems is an effector enzyme or ion channel, such as adenylyl cyclase or a cGMP phosphodiesterase. Such effectors often regulate intracellular levels of second messengers. In the case of cAMP, this goes on to activate the cAMP-dependent protein kinase. In the case of retinal phototransduction, cGMP is responsible for gating the sodium channel, which controls electrical signaling from the photon-activated retina.

Figure 1 also depicts the major components of a pathway that functions to desensitize or deactivate the stimulated receptors. In each of the cases shown, a cytoplasmic protein kinase, rhodopsin kinase (Lorenz et al. 1991; Palczewski et al. 1992) in the case of the retinal system, the β-adrenergic receptor kinase

(βARK) (Benovic et al. 1986) in the case of the β receptor, recognizes the activated conformation of the receptor. They phosphorylate their receptor substrates on multiple serine and threonine residues located in the distal portion of the carboxy-terminal cytoplasmic tail. These phosphorylation reactions lead to the binding of a second cytosolic protein, termed arrestin (Wilden et al. 1986) in the retinal system and βarrestin (Hausdorff et al. 1990) in the β-receptor system. These proteins bind to the phosphorylated forms of the receptors and appear to then sterically interdict their interaction with their G proteins thus leading to functional deactivation of the system.

All these components have been cloned and sequenced. To date, two forms of βARK are known (Benovic et al. 1989, 1991) as is rhodopsin kinase (Lorenz et al. 1991). All of these are typical protein kinases with centrally located conserved protein kinase catalytic domains. These three enzymes appear to form a distinct subfamily of serine/threonine kinases and share about 40–50% sequence identity within their catalytic domains. The noncatalytic regions of the proteins show much less similarity. Similarly, two forms of βarrestin known as βarrestin-1 (Lohse et al. 1990) and βarrestin-2 (Attramadal et al. 1992), as well as authentic retinal arrestin (Shinohara et al. 1987), have been cloned. They share 60–70% sequence identity.

RESULTS AND DISCUSSION

As mentioned, all of the above-described components of the β-adrenergic receptor and rhodopsin-mediated signaling and desensitizing systems have been cloned, and all have also been purified. It can be documented that the components described above, and no others, are sufficient to reconstitute in vitro both the pathway of signaling and the pathway of desensitization. For example, consider the data shown in Figure 2. Purified retinal components, rhodopsin and G_t, were reconstituted into phospholipid vesicles and phosphorylated by recombinant βARK1 purified from overexpressing Sf9 insect cells. It can be seen in panel A that arrestin is potent in desensitizing the phosphorylated rhodopsin, but the βarrestins are considerably weaker. Panel B shows that when the analogous reconstitution experiment is done with β_2-adrenergic receptor, purified from overexpressing Sf9 cells, and G_s, the addition of βarrestin-1 or βarrestin-2 leads to a progressive dose-related inhibition of signaling through the phosphorylated receptor and now retinal arrestin is much weaker. Half-maximal inhibition in either system appears to require about a 1:1 to a 2:1 ratio of βarrestin to receptor protein. These data document that there is marked system specificity in the effects of βarrestin versus arrestin to desensitize the receptors.

Recently, we have developed polyclonal antibodies that recognize the two distinct forms of βARK (Arriza et al. 1992) and the two distinct forms of βarrestin (Attramadal et al. 1992). Using these polyclonal antibodies, it has been documented that all four proteins are found throughout the central nervous system (Arriza et al. 1992; Attramadal et al. 1992). These data strongly suggest that the function of the βARK/βarrestin system is not limited to desensitizing only

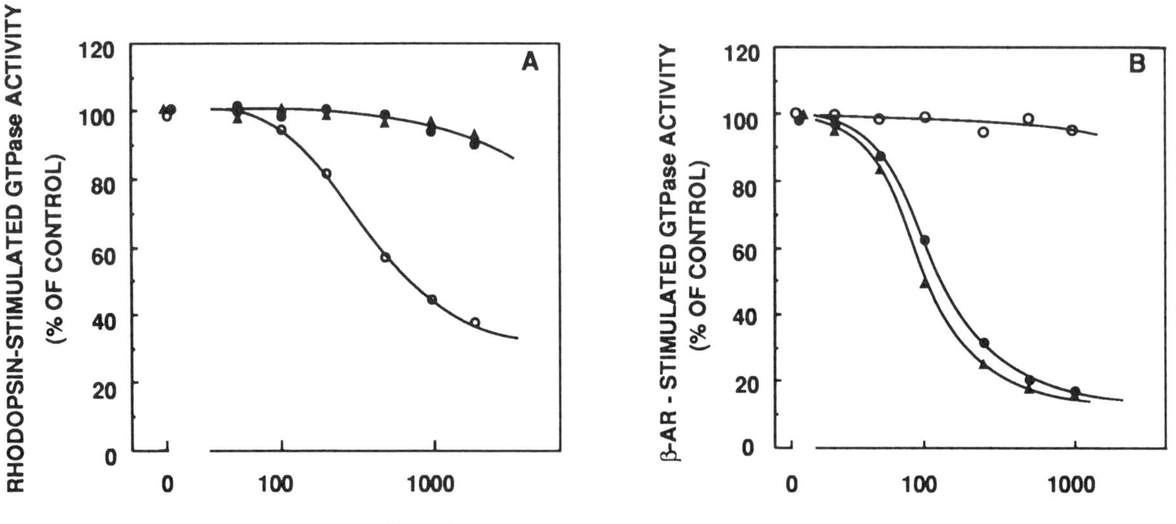

Figure 2. Inhibition of reconstituted signal transduction systems by βarrestin-1, βarrestin-2, and arrestin. (*A*) Inhibition of rhodopsin-stimulated G_t GTPase activity by purified βarrestin-1, βarrestin-2, and arrestin. The assay was performed with 200-fmoles of phosphorylated rhodopsin reconstituted into phospholipid vesicles with 100 fmoles-G_t and the addition of increasing amounts of arrestin (○), βarrestin-1 (●), or βarrestin-2 (▲) as indicated. The activities are calculated as percent control activity, i.e., light-dependent GTPase activity in the absence of any arrestin proteins. (*B*) Inhibition of β_2-adrenergic receptor stimulated G_s GTPase activity by purified arrestins. Receptor-stimulated G_s GTPase turnover rates were measured as the difference between isoproterenol (50 μM)- and propranolol (50 μM)-stimulated G_s GTPase activities. The assay was performed using 100 fmoles phosphorylated β_2-adrenergic receptor, 50 fmoles G_s, and increasing amounts of βarrestin-1 (●), βarrestin-2 (▲), or arrestin (○). (Adapted from Attramadal et al. 1992.)

β-adrenergic or even adrenergic receptors. Rather, this information suggests that many different neurotransmitter receptor systems may be regulated by βARKs and βarrestins. The exact nature of these systems, however, remains to be determined.

Another interesting finding was revealed by immunoelectron microscopy (Arriza et al. 1992; Attramadal et al. 1992). These studies revealed, as shown in Figure 3, that both forms of βARK and both forms of βarrestin are largely synaptic in their distributions. They can be found either postsynaptically, as shown in Figure 3, or presynaptically. These results suggest that the βARK/βarrestin system is crucially involved in mediating rapid regulation of synaptic neurotransmitter receptors. Moreover, these data confirm speculations that the βARK/βarrestin system might be of particular importance in synaptic regulation. These speculations were based on the known high expression of the βARK and βarrestin proteins in neuronal tissues, as well as the fact that the βARK system functions predominantly in the presence of high levels of receptor occupancy. Such high levels of receptor occupancy are classically observed at neurotransmitter synapses, as opposed to at peripheral receptor sites.

An interesting and characteristic property of the known members of the G-protein-coupled receptor kinase family is that although they are cytosolic enzymes, they rapidly translocate to the plasma membrane when the relevant receptor system is stimulated. Nonetheless, the mechanisms involved in this signal-orchestrated compartmentalization of the kinases has, until recently, been completely unknown. When amino acid sequences of the βARKs and rhodopsin kinase became known as a result of their molecular cloning, it became clear that rhodopsin kinase contained at its carboxyl terminus a classic CAAX box (Lorenz et al. 1991). Such carboxy-terminal sequences, CVLS in this case, are known to direct a series of three sequential post-translational modifications of proteins. These are, first, a prenylation of the cysteine via formation of a thioether bond with either a farnesyl (C15) or geranylgeranyl (C20) isoprenoid. The exact nature of the carboxy-terminal sequence directs which prenyl group is added. In the case of RK, CVLS directs farnesylation (Inglese et al. 1992b). Prenylation is followed by proteolysis of the three terminal amino acid residues, which is then followed by carboxyl methylation of the now carboxy-terminal cysteine. Although the full biological import of such prenylation reactions is not currently known, in the most-studied case, that of the *ras* proto-oncogene, it has been found that prenylation with the farnesyl moiety is required both for membrane association of Ras and for Ras-mediated cellular transformation. Accordingly, we have developed the hypothesis that the stimulus-mediated translocation of rhodopsin kinase might require the carboxy-terminal prenyl moiety.

To test this notion, we prepared three forms of rhodopsin kinase, as shown in Figure 4 (Inglese et al. 1992a). These included a cysteine to serine mutation which should block prenylation, the wild type which is farnesylated, and a third form in which the terminal serine is changed to leucine which should be geranylgeranylated. All the enzyme forms were transfected and expressed transiently in COS. Each expressed kinase was analyzed and found to undergo the correct isoprenyl modification. The three enzyme forms have been tested for their biological activities as shown in Figure 5. It can be seen that, whereas the two prenylated forms of the enzyme, including the wild-type farnesylated form, are about equally active, the nonprenylated form is markedly reduced in activity.

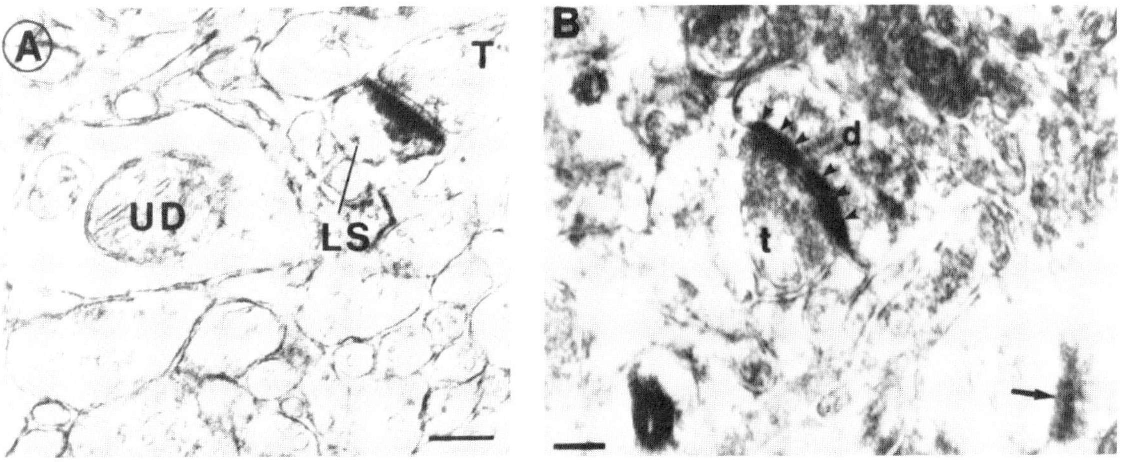

Figure 3. Ultrastructural localization of βarrestin and βARK at synaptic junctions. (*A*) The flocculent electron-dense material reflecting immunoperoxidase reactivity for βarrestin-1 is closely associated with the postsynaptic specialization of a spine (LS). LS is postsynaptic to an unlabeled terminal (T). (UD) Unlabeled dendrite lacking immunoperoxidase reactivity. Bar, 0.2 μm. (Adapted from Attramadal et al. 1992.) (*B*) βARK2 immunoreactive axon terminal (t). The postsynaptic densities (small arrowheads) formed here by the junction of terminal (t) with dendrite (d) are more electron-dense than other postsynaptic densities (arrow), suggesting the localization of βARK2 within postsynaptic densities. Bar, 0.25 μm. (Adapted from Arriza et al. 1992.)

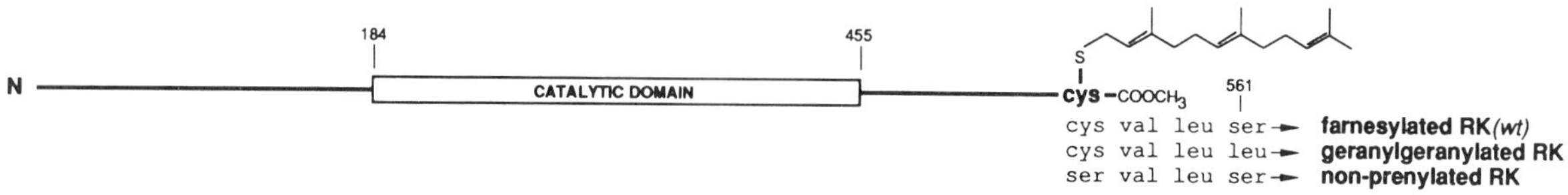

Figure 4. Schematic depicting rhodopsin kinase constructs. Illustration shows the carboxyl terminus of the processed form of the wild-type (wt) enzyme indicating the *S*-farnesylation and α-carboxyl methylation of the carboxy-terminal cysteine, and below this the CAAX motifs (where C is cysteine, A is an aliphatic amino acid, and X is the carboxy-terminal amino acid) of the preprocessed wt and mutant forms of the constructed rhodopsin kinases. (Adapted from Inglese et al. 1992a.)

To further test the involvement of isoprenyl groups in this translocation process, we developed a simple assay for translocation of the kinase. In this assay, enzyme preparations are incubated with rod outer segments containing rhodopsin in the presence or absence of light. The particulate receptors are then centrifuged, and the amount of kinase remaining in the supernatant and absorbed to the membranes is measured either enzymatically or by Western blotting.

This assay was applied to the three forms of rhodopsin kinase described above; the results are shown in Figure 6 (Inglese et al. 1992a). Each form displayed characteristic properties. The wild-type farnesylated (C15) form was largely in the soluble phase but translocated almost quantitatively to the membranes when the system was illuminated. The geranylgeranylated form bearing the more hydrophobic C20 geranylgeranyl group was almost entirely associated with membrane-bound receptors even in the absence of light and showed little further translocation when illuminated. Finally, the nonprenylated form was in the soluble phase and could not translocate even in the presence of light. These data then clearly indicate the essential requirement of the prenyl group for stimulus-mediated translocation of the enzyme to its receptor substrate. Moreover, they indicate that only the C15 moiety supports stimulus-driven translocation. In the presence of the C20 group, the enzyme is membrane-associated under nonstimulated conditions.

Inspection of the sequence of the two forms of βARK which have been cloned indicates that they do not contain CAAX boxes and accordingly will not be prenylated in vivo. Nonetheless, like rhodopsin kinase, they translocate in a stimulus-driven way from cytosol to membrane-bound receptor. How might this occur? We speculated that prenyl groups might also somehow be involved in the process. To test the feasibility of this notion, we prepared two mutant forms of βARK in which the sequence was altered to contain a CAAX box, CVLS to direct farnesylation in vitro, and CVLL to direct geranylgeranylation in vitro. Experiments with these mutant forms of βARK indicated that, as with rhodopsin kinase, the nonprenylated form of the enzyme (which is in this case the wild-type form) was the least active of the three with the two prenylated forms being about threefold more active (Inglese et al. 1992a). Moreover, the prenylated forms show significantly greater tendency to associate with membrane-

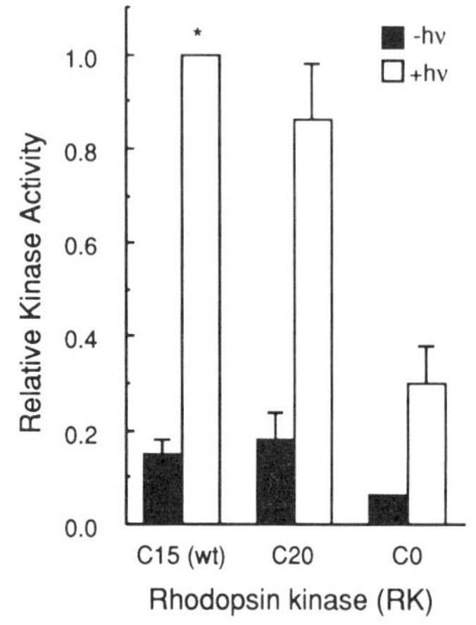

Figure 5. Relative activities of expressed recombinant rhodopsin kinases. Rhodopsin-containing rod outer segments were incubated and phosphorylated with COS-7 cell extracts expressing wild-type farnesylated rhodopsin kinase (C15), mutant geranylgeranylated rhodopsin kinase (C20), and non-isoprenylated rhodopsin kinase (C0). The activity values are normalized to the wild-type kinase (asterisk). The values shown are the mean ± S.E.M. of five separate experiments. (Adapted from Inglese et al. 1992a.)

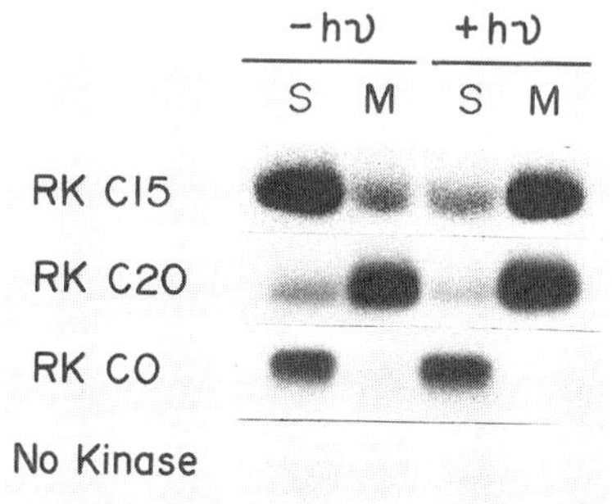

Figure 6. Rhodopsin kinase translocation-phosphorylation assay. COS-7 cell extracts (soluble fractions) of wt rhodopsin kinase (C15), geranylgeranylated rhodopsin kinase (C20), and non-isoprenylated rhodopsin kinase (C0) as described in Figs. 4 and 5 were incubated in the absence of ATP with rhodopsin-containing rod outer segments. The samples were either kept in the dark or briefly exposed (3 min) to light and centrifuged to separate supernatants (S) from membrane (M) fractions. Kinase activity on light-activated rhodopsin was then assessed in each fraction by the addition of [γ-^{32}P]ATP and rhodopsin-containing rod outer segments to membrane-depleted supernatants. The "no-kinase" lanes represent vector-only transfected COS-7 cell extract activity. (Adapted from Inglese et al. 1992a.)

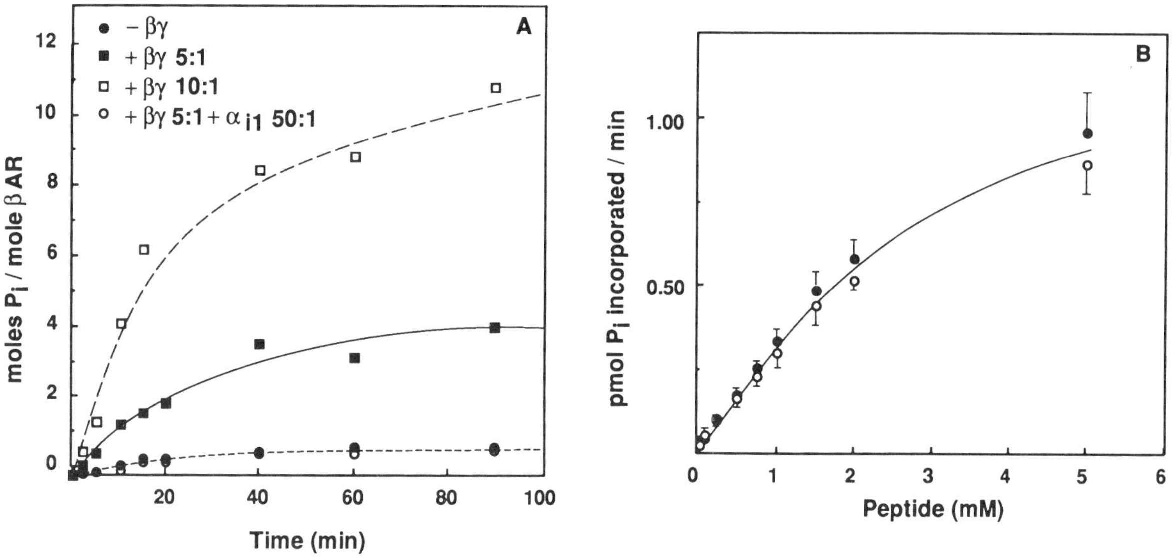

Figure 7. Effects of $\beta\gamma$ on βARK activity. (*A*) Enhancement of βARK-mediated β_2-AR phosphorylation by bovine brain $\beta\gamma$. Reconstituted β_2-AR (20 nM) was incubated with βARK (30 nM) in 20 mM Tris-HCl (pH 8.0), 2 mM EDTA, 10 mM $MgCl_2$, 1 mM dithiothreitol containing 100 μM ATP (~3000 cpm/pmole) and (−) isoproterenol (100 μM) in a total volume of 25 μl (○). Alternatively, 150 nM $\beta\gamma$ (molar ratio $\beta\gamma$:βARK, 5:1) (■), 300 nM $\beta\gamma$ (molar ratio $\beta\gamma$:βARK, 10:1) (□), or 150 nM $\beta\gamma$, 1.5 μM α_{i1} (molar ratio $\beta\gamma$:βARK:α_{i1}, 5:1:50) (●) was included in the phosphorylation incubation. Phosphorylation stoichiometries were determined assuming that all reconstituted receptor was accessible to kinase. (*B*) βARK phosphorylation of peptide substrates in the presence of $\beta\gamma$ subunits. The synthetic peptide RRREEEEESAAA, at the concentrations indicated, was incubated with purified βARK (25 nM) in 20 mM Tris-HCl (pH 8.0), 2 mM EDTA, 10 mM $MgCl_2$, 1 mM dithiothreitol, 100 μM [γ-^{32}P]ATP (250–500 cpm/pmole), 10 μM protein kinase inhibitor, and BSA (0.5 mg/ml), in a final volume of 25 μl at 30°C for 10 min. Phosphorylation reactions were performed in the absence (○) or presence of 255 nM $\beta\gamma$ subunits (molar ratio $\beta\gamma$:βARK, 10:1) (●). The results are the mean ± S.E.M. for three separate experiments. (Adapted from Pitcher et al. 1992.)

bound receptors. These findings indicate that prenyl groups have the *potential* to function as membrane anchors for βARK as well as rhodopsin kinase.

βARK is not prenylated in vivo, however, Nonetheless, a recent finding by Haga and Haga (1992) suggests a novel mechanism. These workers found that the $\beta\gamma$ subunits of heterotrimeric G proteins activate a muscarinic receptor kinase that has properties similar to βARK. As shown in Figure 7A, we were able to confirm this observation and noted a greater than tenfold stimulation of recombinant βARK1 by bovine brain $\beta\gamma$ subunits (Pitcher et al. 1992). As shown in Figure 7, the ability of $\beta\gamma$ to stimulate the enzyme is only observed when the substrate is receptor. When a model peptide substrate is used, $\beta\gamma$ had minimal, if any, effects on the kinase activity. These results suggest that $\beta\gamma$ does not simply directly activate βARK, but rather may have some other mechanism of action.

Since the γ subunit of $\beta\gamma$ is prenylated in vivo with a geranylgeranyl group, we speculated that perhaps $\beta\gamma$ subunits might "donate" prenyl groups to βARK, thus allowing it to anchor in the membrane during receptor phosphorylation. This hypothesis is particularly appealing since $\beta\gamma$ subunits are only free, i.e., dissociated from α and available to perform such a putative function, upon receptor stimulation. Thus, this proposed mechanism would provide a basis for a stimulus-dependent $\beta\gamma$-mediated translocation of βARK to the membrane-bound receptors. To test this notion, we examined the ability of $\beta\gamma$ subunits to promote the association of βARK to membrane-bound receptors. For convenience, in these assays we used rod outer segments. The data in Figure 8 show that $\beta\gamma$ does in fact translocate βARK from the soluble to the membrane-bound phase (Pitcher et al. 1992). In this system, the process is not light-stimulated. However, when the receptor component (i.e., rhodopsin) was heat-inactivated, about half of the $\beta\gamma$ effect was lost (see Fig. 8C). These data suggest that $\beta\gamma$ may associate not only with the βARK, but also with the receptor component of the system. However, it should be borne in mind that in vivo this pathway would be entirely "receptor dependent" simply because receptor stimulation is required to dissociate $\beta\gamma$ into a form in which it is capable of binding and translocating βARK.

In other experiments, we have been able to show that bovine brain $\beta\gamma$ binds tightly to an immobilized fusion protein in which the last 220 amino acids of βARK have been fused to glutathione-*S*-transferase (GST). Interestingly, under our experimental conditions, $\beta\gamma$ subunits derived from the visual G protein transducin, which are much weaker in stimulating βARK activity, are not able to bind to these fusion protein columns (Pitcher et al. 1992). This suggests the possibility of specificity in the actions of $\beta\gamma$ subunits from different G proteins to interact with and translocate different βARK isoenzymes.

In summary, we have documented the crucial involvement of prenyl groups in the translocation and receptor targeting of two different G-protein-coupled

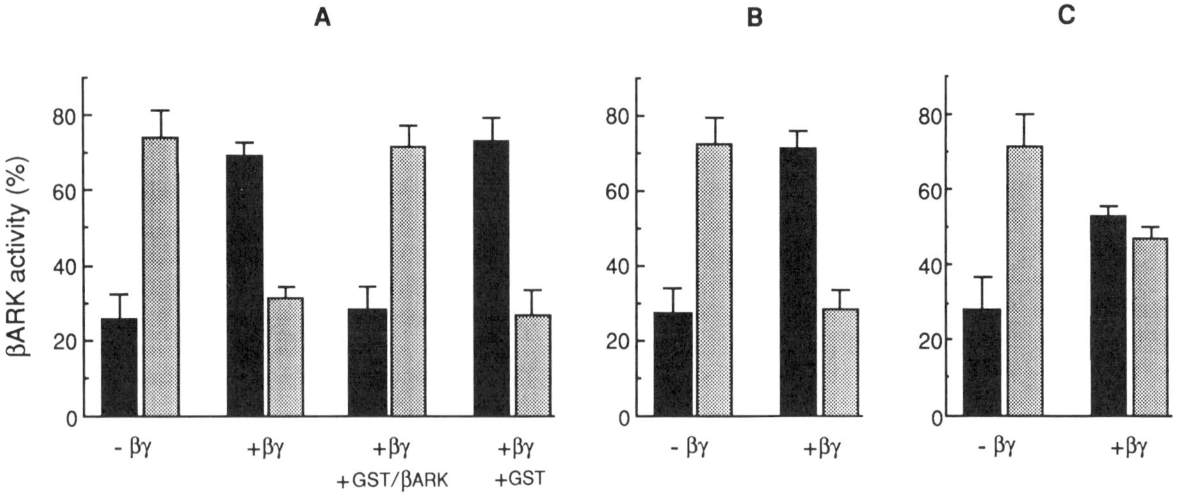

Figure 8. Membrane targeting of the βARK/$\beta\gamma$ complex. Purified βARK (15 nM) was incubated with 10 μl of native (*A* and *B*) or heat-inactivated (*C*) ROS membranes (final concentration of rhodopsin ~5 μM) in buffer A in a final volume of 50 μl. Incubations were performed at 22°C for 3 min under bright illumination (*A* and *C*) or in the dark (*B*). $\beta\gamma$ subunits purified from bovine brain (0.6 μM), GST-βARK fusion protein (3 μM), or GST (3 μM) were included in the incubations as indicated. Samples were subsequently centrifuged at 350,000*g* for 5 min. After centrifugation, the supernatants were rapidly removed, supplemented with ROS membranes (10 μl) and [γ-^{32}P]ATP (final concentration 100 μM, ~3000 cpm/pmole) and incubated under illumination at 30°C for 5 min. The sedimented ROS membranes were resuspended in buffer A containing 100 μM [γ-^{32}P]ATP and incubated under identical conditions. The $\beta\gamma$ subunits (0.3 μM) were added to both the supernatant and sedimented fractions of samples that lacked $\beta\gamma$ during the primary incubation. The total radioactivity incorporated into rhodopsin in both the supernatant and pellet fractions for a given sample was taken as 100%. Data shown are the means ± S.E.M. for three separate experiments. (Adapted from Pitcher et al. 1992.)

receptor kinases, rhodopsin kinase and βARK. In the case of rhodopsin kinase, the prenyl group is intrinsic to the enzyme. When the system is activated by light, the enzyme recognizes the conformationally altered form of rhodopsin, which it binds and phosphorylates. By virtue of its prenyl group, it is able to form a more stable association with the membrane-bound receptor. In the case of βARK, the paradigm appears to be more complicated. Here, the enzyme itself is not prenylated. Rather, upon receptor stimulation, activation with dissociation of G-protein $\beta\gamma$ subunits occurs, and these $\beta\gamma$ subunits (which are themselves prenylated) bind specifically to sequences located toward the carboxyl terminus of βARK. This βARK/$\beta\gamma$ complex is then able to interact with membrane-bound receptors, leading to their phosphorylation. Interaction of βARK with $\beta\gamma$ presumably facilitates the stable association or translocation of the enzyme to the membrane. Moreover, different forms of βARK might be translocated to different receptor substrates by virtue of specificity inherent in the interaction of $\beta\gamma$ with the enzymes. This latter point, however, is completely speculative at present.

These novel findings concerning the role of prenyl moieties in the stimulus-directed targeting of these regulatory G-protein-coupled receptor kinases to their receptor substrates provide new insights into the mechanisms by which prenylation may function to direct trafficking of enzymes to their crucial cellular substrates. Moreover, they suggest additional avenues of experimental approach to the problem of developing inhibitors for βARK-like enzymes toward the ultimate goal of attenuating desensitization reactions and prolonging and augmenting the therapeutic efficacy of drugs that function by stimulation of G proteins.

ACKNOWLEDGMENT

This work was supported in part by grant HL-16037 from the National Institutes of Health.

REFERENCES

Arriza, J.L., T.M. Dawson, R.B. Simerly, L.J. Martin, M.G. Caron, S.H. Snyder, and R.J. Lefkowitz. 1992. The G protein-coupled receptor kinases βARK1 and βARK2 are widely distributed at synapses in rat brain. *J. Neurosci.* **12:** 4045.

Attramadal, H., J.L. Arriza, C. Aoki, T.M. Dawson, J. Codina, M.M. Kwatra, S.H. Snyder, M.G. Caron, and R.J. Lefkowitz. 1992. βArrestin2—A novel member of the arrestin/βarrestin gene family. *J. Biol. Chem.* **267:** 17882.

Benovic, J.L., R.H. Strasser, M.G. Caron, and R.J. Lefkowitz. 1986. β-adrenergic receptor kinase: Identification of a novel protein kinase which phosphorylates the agonist-occupied form of the receptor. *Proc. Natl. Acad. Sci.* **83:** 2797.

Benovic, J.L., A. DeBlasi, W.C. Stone, M.G. Caron, and R.J. Lefkowitz. 1989. β-Adrenergic receptor kinase: Primary structure delineates a multigene family. *Science* **246:** 235.

Benovic, J.L., J.J. Onorato, J.L. Arriza, W.C. Stone, M. Lohse, N.A. Jenkins, D.J. Gilbert, N.G. Copeland, M.G. Caron, and R.J. Lefkowitz. 1991. Cloning, expression, and chromosomal localization of β-adrenergic receptor kinase 2. A new member of the receptor kinase family. *J. Biol. Chem.* **266:** 14939.

Bourne, H.R., D.A. Sanders, and F. McCormick. 1991. The GTPase superfamily: Conserved structure and molecular mechanism. *Nature* **349:** 117.

Dohlman, H.G., J. Thorner, M.G. Caron, and R.J. Lefkowitz. 1991. Model systems for the study of seven-transmembrane-segment receptors. *Annu. Rev. Biochem.* **60:** 653.

Haga, K. and T. Haga 1992. Activation by G protein $\beta\gamma$ subunits of agonist- or light-dependent phosphorylation of muscarinic acetylcholine receptors and rhodopsin. *J. Biol. Chem.* **267:** 2222.

Hausdorff, W.P., M.G. Caron, and R.J. Lefkowitz. 1990. Turning off the signal: Desensitization of β-adrenergic receptor function. *FASEB J.* **4:** 2881.

Inglese, J., W.J. Koch, M.G. Caron, and R.J. Lefkowitz. 1992a. The role of isoprenylation in the regulation of signal transduction via the G protein-coupled receptor kinases. *Nature* (in press).

Inglese, J., J.F. Glickman, W. Lorenz, M.G. Caron, and R.J. Lefkowitz. 1992b. Isoprenylation of a protein kinase. Requirement of farnesylation/α-carboxyl methylation for full enzymatic activity of rhodopsin kinase. *J. Biol. Chem.* **267:** 1422.

Lohse, M.J., J.L. Benovic, J. Codina, M.G. Caron, and R.J. Lefkowitz. 1990. β-Arrestin: A protein that regulates β-adrenergic receptor function. *Science* **248:** 1547.

Lorenz, W., J. Inglese, K., Palczewski, J.J. Onorato, M.G. Caron, and R.J. Lefkowitz. 1991. The receptor kinase family: Primary structure of rhodopsin kinase reveals similarities to the β-adrenergic receptor kinase. *Proc. Natl. Acad. Sci.* **88:** 8715.

Palczewski, K., G. Rispoli, and P.B. Detwiler. 1992. The influence of arrestin (48K protein) and rhodopsin kinase on visual transduction. *Neuron* **8:** 117.

Pitcher, J.A., J. Inglese, J.B. Higgins, J.L. Arriza, P.J. Casey, C. Kim, J.L. Benovic, M.M. Kwatra, M.G. Caron, and R.J. Lefkowitz. 1992. $\beta\gamma$ Subunits of G proteins target the β-adrenergic receptor kinase to membrane bound receptors. *Science* (in press).

Shinohara, T., B. Dietzschold, C.M. Craft, G. Wistow, J.J. Early, L.A. Donoso, J. Horwitz, and R. Tao. 1987. Primary and secondary structure of bovine retinal S antigen (48 KDa protein). *Proc. Natl. Acad. Sci.* **84:** 6975.

Wilden, U., S.W. Hall, and H. Kuhn. 1986. Phosphodiesterase activation by photoexcited rhodopsin is quenched when rhodopsin is phosphorylated and binds the intrinsic 48-kDa protein of rod outer segments. *Proc. Natl. Acad. Sci.* **83:** 1174.

Regulation of Mammalian Adenylyl Cyclases by G-Protein α and $\beta\gamma$ Subunits

W.-J. TANG, J.A. IÑIGUEZ-LLUHI, S. MUMBY, AND A.G. GILMAN
Department of Pharmacology, University of Texas Southwestern Medical Center, Dallas, Texas 75235

Overview of the G-Protein Family

Members of a family of heterotrimeric guanine nucleotide-binding regulatory proteins (G proteins) are responsible for transmission of information from many membrane-bound receptors to their intracellular effectors (for reviews, see Gilman 1987; Kaziro et al. 1991; Simon et al. 1991). There are several such pathways; the best-studied are those for dual (stimulatory and inhibitory) regulation of the rate of cAMP synthesis (e.g., by β-adrenergic and muscarinic agonists) and for light-stimulated hydrolysis of cGMP in retinal rods and cones. It is now appreciated that the general features of signal transduction pathways that have been elucidated for adenylyl cyclases and the visual system—that information flows from extracellularly oriented receptor to G protein to intracellular effector—are shared by a large number of transmembrane signaling systems that are triggered by hormones, neurotransmitters, autacoids, odorants, or physical stimuli.

Physiological events that are controlled by G proteins range (mechanistically) from those in which the direct interaction of G protein with effector has been demonstrated conclusively, to those in which a G protein is clearly involved but the mechanism is uncertain, to those in which control by a G protein is likely but unproven. Direct interactions between G proteins and effectors have been demonstrated for stimulatory regulation of adenylyl cyclase (Northup et al. 1983; May et al. 1985), retinal cGMP-specific phosphodiesterase (Stryer 1986), and phospholipase C-β1 (Smrcka et al. 1991; Taylor et al. 1991) by the α subunits of G_s, G_t (transducin), and G_q/G_{11}, respectively. There is clear evidence for the involvement of G proteins in the inhibitory regulation of adenylyl cyclase (Katada et al. 1984), the modulation of activity of voltage-sensitive Ca^{++} channels in skeletal and cardiac muscle (Brown and Birnbaumer 1988) and neurons (Hescheler et al. 1987), and the gating of K^+ channels in atrial muscle cells (Yatani et al. 1987, 1988), neurons (Van Dongen et al. 1988), and elsewhere (Brown and Birnbaumer 1990). However, in these cases, mechanisms remain to be determined. Although pathways that do not involve low-molecular-weight second messengers have been demonstrated, it is not yet known if the G protein-effector interaction is direct or if there are other macromolecular components that serve as primary targets for the G proteins involved. Finally, there are G proteins whose roles are reasonably obscure (e.g., G_{12}, G_{13} [Strathmann and Simon 1991]), pathways where mechanisms are quite uncertain (regulation of phospholipases A_2 and D [Jelsema and Axelrod 1987]), and processes where heterotrimeric G proteins have been implicated but evidence is incomplete (regulation of protein traffic through Golgi [Stow et al. 1991]).

The G-protein family contains a relatively large number of closely related members. The properties of the relevant polypeptides (including those deduced from cDNA cloning) are summarized in Tables 1 and 2. The subunits of heterotrimeric G proteins are designated α, β, and γ in order of decreasing mass. Differences in the identity of the α subunit are currently utilized to distinguish the various oligomers. Sixteen genes that encode at least 20 G-protein α subunits have been reasonably well characterized to date. These proteins are currently grouped into four major subclasses; members of one subclass are less than 50% identical to members of other subclasses (Strathmann and Simon 1991); sequence identities within a subclass range from 60% to greater than 90%. This scheme of classification is meaningful, in that it also encompasses important functional similarities. The α subunits contain a single, high-affinity (nM) binding site for guanine nucleotides and catalyze the hydrolysis of GTP to GDP. In many cases, this subunit also contains a site or sites for NAD-dependent ADP-ribosylation catalyzed by bacterial toxins (cholera toxin and/or pertussis toxin, depending on the α subunit). The α subunits of the G_i family (G_i, G_o, G_t, G_z) are probably all modified post- or co-translationally by the addition of amide-linked myristic acid at an amino-terminal glycine residue. This covalent modification is essential for membrane localization of these α subunits, presumably in part because myristoylation greatly increases the affinity of α for the G-protein $\beta\gamma$ subunit complex (Linder et al. 1991). $G_{s\alpha}$ is not myristoylated, and it is doubtful that members of the G_q and G_{12} subfamilies of α subunits are myristoylated (based on their amino-terminal amino acid sequences, which do not conform to those known to be required by protein *N*-myristoyl transferase). G-protein α subunits interact with and regulate effectors. It was thought that this important role was the exclusive province of α subunits, but this concept now appears to require significant modification (see below).

$G_{s\alpha}$ (expressed ubiquitously) and G_{olf} (expressed largely in olfactory neuroepithelium) are best known as direct activators of adenylyl cyclase. There are four splice variants of $G_{s\alpha}$, which differ primarily in the inclusion or exclusion of 15 contiguous amino acid

Table 1. Properties of Mammalian G-Protein α Subunits

Family/ subunit	M_r (kD × 10^{-3})	A.A. identity[a] (%)	Toxin[b]	Tissue distribution	Representative receptors[c]	Effector/role
G_s						
$\alpha_{s(S)}(2X)$[d]	44.2	100	CTX	ubiquitous	βAR, glucagon, TSH, others	↑adenylyl cyclase ↑Ca^{++} channels ↓Na^{+} channels
$\alpha_{s(L)}(2X)$[d]	45.7	—	CTX	ubiquitous		
α_{olf}	44.7	88	CTX	olfactory neuroepithelium	odorant	↑adenylyl cyclase
G_i						
α_{i1}	40.3	100	PTX	nearly ubiquitous		↑K^{+} channels ↓Ca^{++} channels ↓adenylyl cyclase (?) ↑phospholipase C (?) ↑phospholipase A_2 (?)
α_{i2}	40.5	88	PTX	ubiquitous	M_2Cho, α_2AR, others	
α_{i3}	40.5	94	PTX	nearly ubiquitous		
α_{oA}[d]	40.0	73	PTX	brain, others	Met-Enk, α_2AR, others	
α_{oB}[d]	40.1	73	PTX	brain, others		
α_{t1}	40	68	CTX,PTX	retinal rods	rhodopsin	↑cGMP-specific phosphodiesterase
α_{t2}	40.1	68	CTX,PTX	retinal cones	cone opsin	
α_g	40.5	67	CTX (?), PTX	taste buds	taste (?)	?
α_z	40.9	60	—	brain, adrenal, platelets	M_2Cho (?), others (?)	↓adenylyl cyclase (?), others (?)
G_q						
α_q	42	100	—	nearly ubiquitous	M_1Cho, α_1AR, others	↑phospholipase C-β1 others (?)
α_{11}	42	88	—	nearly ubiquitous		
α_{14}	41.5	79	—	lung, kidney, liver	?	
α_{15}	43	57	—	B cells, myeloid cells	?	?
α_{16}	43.5	58	—	T cells, myeloid cells	?	↑phospholipase C-β1
G_{12}						
α_{12}	44	100	—	ubiquitous	?	?
α_{13}	44	67	—	ubiquitous	?	?

[a]% Amino acid identity: Comparison is with the first-listed member of each family.
[b]Cholera toxin (CTX) and pertussis toxin (PTX) catalyze the ADP-ribosylation of an arginine residue (CTX) and a cysteine residue (PTX), respectively, of the indicated α subunits.
[c]Receptor abbreviations: βAR, β-adrenergic; M_2Cho, M_2-muscarinic cholinergic; α_2AR, α_2-adrenergic; Met-Enk, met-enkephalin; M_1Cho, M_1-muscarinic cholinergic; α_1AR, α_1-adrenergic.
[d]Splice variants. $\alpha_{s(S)}$ = short forms of α_s, and $\alpha_{s(L)}$ = long forms of α_s.

Table 2. Properties of Mammalian G-Protein β and γ Subunits

Subunit	M_r (kD × 10^3)	A.A. identity[a] (%)	Tissue distribution	Effector/role
β				
β_1	37.3	100	ubiquitous	required for G_α-receptor interaction
β_2	37.3	90	nearly ubiquitous	
β_3	37.2	83	nearly ubiquitous	inhibition of G_α activation
β_4	37.2	89	nearly ubiquitous	
				modulate activation of certain adenylyl cyclases by $G_{s\alpha}$ or calmodulin
				support of agonist-induced receptor phosphorylation and desensitization
γ				
γ_1	8.4	100	retina, other (?)	↑phospholipase C
γ_2	7.9	38	brain, adrenal, other (?)	↑K^+ channels (?)
γ_3	8.5	36	brain, testis	
γ_4	(?partial)	(34)	(kidney, retina (?))	↑phospholipase A_2 (?)
γ_5	7.3	25	liver, other (?)	
γ_6	7.5	35	brain, other (?)	

[a]% Amino acid identity: Comparison is with the first-listed member of each family.

residues encoded by exon 3 of the $G_{s\alpha}$ gene; there is as yet no known functional significance of this alternative splicing. $G_{s\alpha}$ can also activate voltage-sensitive Ca^{++} channels in cardiac and skeletal muscle (Brown and Birnbaumer 1990) and may have effects on Na^+ channels as well (Schubert et al. 1989). G_s is activated by receptors that stimulate adenylyl cyclase activity; β-adrenergic receptors are prototypical.

The β and γ subunits of G proteins are tightly associated with each other and have never been resolved under nondenaturing conditions. The $\beta\gamma$ complex is necessary for G protein-receptor interactions (Fung 1983; Florio and Sternweis 1985) and appears to play an important role in anchoring α to the membrane (Sternweis 1986) (see below). γ subunits contain prenyl modifications at their carboxyl terminus—the C20 (geranylgeranyl) prenyl modification is characteristic of the predominant brain γ subunits (Maltese and Robishaw 1990; Mumby et al. 1990; Yamane et al. 1990), whereas retinal γ contains a C15 (farnesyl) prenyl group (Fukada et al. 1990; Lai et al. 1990). The carboxy-terminal cysteine residue that is prenylated is also carboxymethylated (Yamane et al. 1990). The relative roles of α and $\beta\gamma$ subunits in modulating the activities of effector molecules has been a subject of considerable debate. There is genetic evidence for such a role for $\beta\gamma$ in the pheromone response pathway in budding yeast (Blumer and Thorner 1991). We have recently described striking effects of $\beta\gamma$ on adenylyl cyclase activity (see below). This system appears to constitute the first in which interactions between $\beta\gamma$ and an effector can be studied biochemically.

The $\beta\gamma$ subunit complexes of many purified G-protein oligomers can be interchanged functionally, and such exchange may also occur in vivo. Nevertheless, it is now clear that there are multiple forms of each of these subunits. Four closely related β-subunit genes have been described, and cDNAs that encode novel γ subunits are being cloned with increasing frequency. Thus, the combinatorial possibilities are sobering (minimally, 20 α/4 β/5–10 γ). The functional significance of this molecular heterogeneity remains to be defined.

Purification and reconstitution of the components involved in the regulation of adenylyl cyclase and cGMP phosphodiesterase and analysis of mutants with defects in the components of the adenylyl cyclase system have allowed detailed studies of the molecular mechanisms by which these enzymes are regulated (Gilman 1987; Kaziro et al. 1991; Simon et al. 1991). The essential protein-protein and protein-ligand interactions that characterize the system are diagramed in Figure 1. In the basal state, G proteins exist in their oligomeric form with GDP tightly bound to the α subunit. The spontaneous rate of dissociation of GDP is extremely slow (< 0.1/min) and is rate limiting for the overall functioning of the pathway. The interaction of the G protein with the appropriately liganded receptor stimulates the rate of dissociation of GDP, presumably as a result of a conformational change that results in an "opening" of the guanine nucleotide-binding site. However, in the presence of the relatively high intracellular concentration of GTP, the "open" guanine nucleotide site is rapidly filled.

The binding of GTP to H · R · G (see Fig. 1) has two important consequences. The first is that the affinity of H for R (and thus R for G) is decreased. Dissociation of H from R and, therefore, R from G · GTP allows R to recycle and thus function catalytically in the activa-

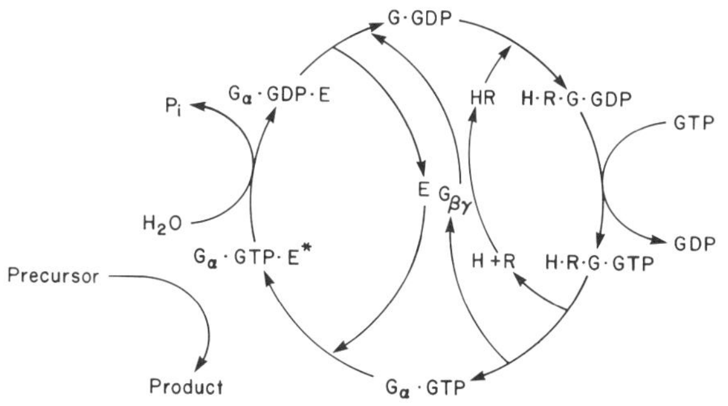

Figure 1. Mechanism of G-protein-regulated transmembrane signaling. The interactions of hormone (H), receptor (R), G protein (G), and effector (E) are shown. (Reprinted, with permission [Annual Reviews, Inc.], from Gilman 1987.)

tion of G. The second major consequence of binding of GTP to G is the "activation" of the G protein such that it can interact fruitfully with an appropriate effector. Such activation also greatly reduces the affinity of $G_{\alpha} \cdot GTP$ for $\beta\gamma$, and the resulting dissociation of subunits is a crucial component of the mechanism of activation. The $G_{\alpha} \cdot GTP$ complex is a direct regulator of adenylyl cyclase and the cGMP phosphodiesterase. The $\beta\gamma$ subunit complex also appears to play an important role in regulation of the activity of certain types of adenylyl cyclase (see below).

Deactivation of the G protein, which results in termination of signal transduction, is primarily due to the GTPase activity of $G_{\alpha} \cdot GTP$. The rate constant for hydrolysis of GTP by $G_{s\alpha}$ and $G_{i\alpha}$ is about 10 per minute at 30°C. Thus, the lifetime of $G_{\alpha} \cdot GTP$ is several seconds — sufficient for substantial amplification due to prolonged activation of the effector. To be emphasized is the crucial balance between the important rate constants in these systems. The basal rate of dissociation of GDP is very slow compared to the K_{cat} for the GTPase. Thus, in the basal state, essentially all of the G protein is inactive. When an appropriate agonist-receptor complex is present, the rate of dissociation of GDP is about equal to the K_{cat} for the GTPase. A significant fraction of the G protein can thus accumulate in the active form. Finally, the very slow *absolute* value of K_{cat} for the GTPase is important for significant signal amplification.

Membrane-bound Adenylyl Cyclases

Although there is an extensive descriptive literature on the regulatory properties of mammalian adenylyl cyclases, most of the detailed molecular information has been acquired only recently, since the initial cloning of a cDNA that encodes one of these proteins (Krupinski et al. 1989). Four such cDNAs have now been described, and the proteins are arbitrarily designated types I–IV (Krupinski et al. 1989; Bakalyar and Reed 1990; Feinstein et al. 1991; Gao and Gilman 1991). Type I adenylyl cyclase is a $G_{s\alpha}$- and Ca^{++}/calmodulin-activated form of the enzyme that is expressed predominantly in brain. Type II adenylyl cyclase, also found predominantly in brain, is insensitive to calmodulin. Expression of the type III protein is largely or wholly confined to the olfactory neuroepithelium; it too is activated by both $G_{s\alpha}$ and Ca^{++}/calmodulin. Type IV adenylyl cyclase resembles type II, although it is expressed in several peripheral tissues as well; however, it may not be the predominant cyclase at such sites.

Each of these four adenylyl cyclases appears to have the same interesting topology with regard to the plasma membrane. Briefly, each protein is hypothesized to contain a short, cytoplasmic amino-terminal domain, six transmembrane spans, a roughly 40-kD cytosolic domain, six additional transmembrane spans, and, finally, an additional 35-kD cytosolic domain at the carboxyl terminus (Fig. 2). This structure (topographically) is highly reminiscent of those of a variety of membrane-bound channels and transporters—particularly the cystic fibrosis transmembrane conductance regulator and the P glycoprotein (the product of the multidrug-resistance gene). The sequences of the adenylyl cyclases can be divided into domains for more detailed comparisons. As shown in Figure 2, regions C_{1a} and C_{2a}, comprising the bulk of each of the two large cytoplasmic domains, are the most conserved portions of the four adenylyl cyclases; the similarities between type II and type IV adenylyl cyclase are particularly high in these regions (76% and 79% amino acid identity). The C_{1a} and C_{2a} domains within each adenylyl cyclase are also similar to each other (roughly 50%), and they are highly homologous with the catalytic domains of a series of membrane-bound and heterodimeric soluble guanylyl cyclases (Chinkers and Garbers 1991). There is little sequence similarity among adenylyl cyclases in the regions designated N_1, C_{1b}, and C_{2b}. The length of the amino-terminal portions of the proteins (N_1) is quite variable. The carboxy-terminal region designated C_{2b} is present only in the type I and type III proteins. In general, sequence homologies

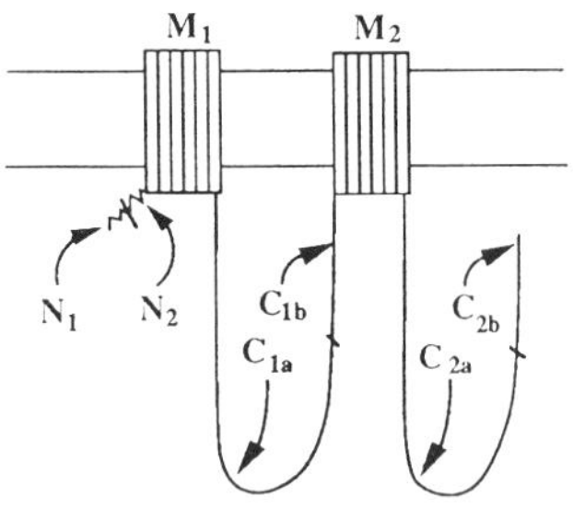

Types compared	% Identity							
	N_1	N_2	M_1	C_{1a}	C_{1b}	M_2	C_{2a}	C_{2b}
I / II	NH	18	33	58	NH	NH	52	-
I / III	NH	35	NH	54	NH	NH	54	NH
I / IV	NH	28	35	58	17	28	56	-
II / III	NH	35	19	54	27	25	56	-
II / IV	NH	56	52	76	31	43	79	-
III / IV	NH	33	28	54	32	NH	51	-

Figure 2. Comparisons of different adenylyl cyclases. The structural model of mammalian adenylyl cyclase is shown at the left. N denotes the amino-terminal region of the protein; the N_2 domain represents only 18 amino acid residues adjacent to the M_1 region. M_1 and M_2 are the first and second membrane-spanning domains. C_1 and C_2 are the predicted cytoplasmic domains. NH (nonhomologous) indicates that the percentage identity score is less than or equal to 15. A dash indicates that comparison is not applicable. (Modified from Gao and Gilman 1991.)

are also modest among the hydrophobic domains, although the type II and type IV proteins again show the greatest similarities. Two major questions arise from these structural considerations. Why do the adenylyl cyclases have structures that resemble those of channels and transporters (even though there is little similarity of sequence among the M_1 and M_2 domains of the adenylyl cyclases)? Why do the enzymes have a quasi-duplicated catalytic domain?

All of these adenylyl cyclases have been expressed in insect (Sf9) cells by inserting their cDNAs into recombinant baculovirus under the control of the strong polyhedron promoter. The type I and type II enzymes are the best characterized to date. Expression of either of these proteins increases adenylyl cyclase activity in Sf9 cell membranes by about 50- to 100-fold to a value of 10–20 nmole/min/mg membrane protein (~0.1% of membrane protein). The activity of the recombinant type I protein is increased markedly by addition of activated $G_{s\alpha}$ or Ca^{++}/calmodulin; as mentioned, the type II enzyme is insensitive to calmodulin. The recombinant type I protein can be purified to near homogeneity by forskolin-Sepharose chromatography, and it has an appropriate specific activity (about 5 μmole/min/mg) (Tang et al. 1991).

In initial attempts to define the domain structure of adenylyl cyclase, we expressed the amino-terminal and carboxy-terminal halves of the type I protein independently and coincidentally in Sf9 cells (Fig. 3). Protein accumulated when either the front or back halves were synthesized alone, but no activity was detected. However, a considerable amount of activity was recovered when the two constructs were expressed together, and this activity was stimulated by $G_{s\alpha}$, calmodulin, and forskolin. Thus, it seems possible that the catalytic activity of adenylyl cyclase might be controlled by regulating an interaction between the two homologous halves, and other experiments have been designed to test this hypothesis.

Charged amino acid residues were identified that were conserved not only among the four mammalian adenylyl cyclases and the membrane-bound guanylyl cyclases, but that were also identical in the homologous portions of the C_{1a} and C_{2a} domains of each adenylyl cyclase. Each residue of six such pairs of amino acids was mutated to alanine, and each of the twelve mutant proteins was expressed in Sf9 cells (Fig. 4A). Of interest, the catalytic activity of seven of these twelve proteins was compromised severely, although each protein was expressed (immunoblotting) and showed some evidence of native conformation (capacity to bind $G_{s\alpha}$). Thus, mutations in either the C_{1a} or C_{2a} domains are capable of obliterating catalytic activity. Most interesting, however, is that complementation is observed when catalytically inactive point mutants are expressed coincidentally with truncation mutants of the M_1C_1 or the M_2C_2 type (Fig. 4B). For example, catalytic activity is restored when a protein with a point mutation in the C_{2a} domain is coexpressed with M_2C_2, indicating formation of a complex between the full-length mutant protein and the truncation mutant. These results are thus again suggestive of the possibility that interactions between C_1 and C_2 are crucial for regulation of adenylyl cyclase activity and further suggest the possibility that adenylyl cyclase may function as a homodimer. However, the latter possibility is not favored because of failure to observe complementation when proteins with point mutations in C_{1a} and C_{2a} are coexpressed (Fig. 4C). It is hoped that assessment of the hydrodynamic properties of the protein will provide an unambiguous answer to the question of dimerization.

What is the significance of different species of adenylyl cyclase? Sensitivity to calmodulin and expression in specialized tissues provides a part of the explanation. However, we have also recently discovered another striking and unexpected difference between these forms—heir response to the G-protein βγ subunit complex (Tang and Gilman 1991). The adenylyl cyclases that were characterized originally with regard to differential effects of G-protein subunits (cyclases from S49 cells and platelets) are activated by $G_{s\alpha}$ and are largely insensitive to βγ. Thus, we were surprised to find prominent regulation of adenylyl cyclase activity by βγ when the novel cyclase clones were expressed

Virus / Adenylyl Cyclase Activity (nmol/min/mg)	Control virus	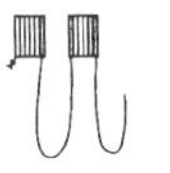 rAC-I	 $INM_1C_{1\text{-}s}$	 IM_2C_2	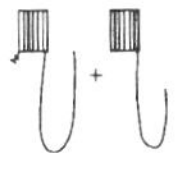 $INM_1C_{1\text{-}s} + IM_2C_2$
Basal	0.02	1.5	0.02	0.04	0.8
GTPγS + $rG_{s\alpha\text{-}s}$	0.03	4.5	0.05	0.05	2.3
Ca^{+2} + CaM	0.03	4.7	0.03	0.03	1.3
Forskolin	0.1	13.0	0.13	0.15	3.5

Figure 3. Adenylyl cyclase activity of truncated constructs. Cells were infected with recombinant baculovirus encoding the indicated construct and were harvested after 60 hr. Membranes were assayed for adenylyl cyclase activity without activators (basal) or with 50 nM GTPγS-$rG_{s\alpha}$, 50 μM Ca^{++} + 160 nM calmodulin, or 100 μM forskolin. (Modified from Tang et al. 1991.)

A.

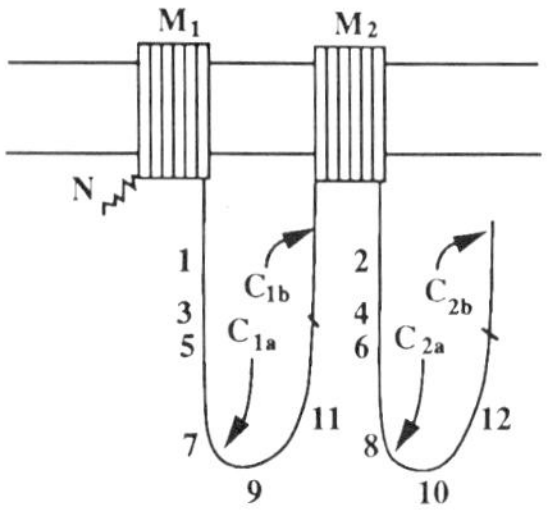

Virus	Adenylyl Cyclase Activity (nmol/ min/ mg)	$G_{s\alpha}$ binding (%)
wild type	5.1	100
1. D338A	0.1	34
2. D908A	0.1	17
3. R348A	3.2	113
4. K921A	0.4	60
5. K350A	2.3	90
6. K923A	0.1	92
7. H369A	1.7	87
8. H950A	0.1	36
9. R398A	0.1	84
10.R979A	0.1	20
11.Y418A	3.9	119
12.Y999A	0.2	23

B.

Virus	-	INM_1C_1	IM_2C_2	IIM_2C_2
-	-	0.1	0.2	0.1
D338A	0.2	0.5	0.1	0.1
D908A	0.2	0.2	1.0	1.8

C.

Virus	-	D338A	R398A
-	-	0.1	0.2
H950A	0	0.1	0.3
K923A	0.1	0.1	0.3

Figure 4. (*A*) Characterization of point mutants of adenylyl cyclase. Pairs of point mutants were constructed by changing highly conserved residues to alanine (relative positions are shown in the model of adenylyl cyclase at the top). (*B*) The ability of point mutants to be partially rescued by coexpression of truncated mutants. INM_1C_1 represents the NM_1C_1 construct from type I adenylyl cyclase; IM_2C_2 and IIM_2C_2 represent M_2C_2 constructs from type I and type II adenylyl cyclase, respectively. (*C*) Failure of coexpression of point mutants with mutations in the C_1 and C_2 domains to complement each other. In all panels, Sf9 cells were infected with the indicated baculovirus or pair of viruses and harvested after 48 hr. Membranes were assayed for adenylyl cyclase activity with 10 mM $MgCl_2$ and 100 μM forskolin (*A* and *C*) or with 10 mM $MgCl_2$, 5 mM $MnCl_2$, and 100 μM forskolin (*B*). In *A*, membranes were also assayed for their relative capacity to bind 5 nM GTPγS-r$G_{s\alpha}$.

and examined in isolation (Fig. 5). Type I adenylyl cyclase is inhibited by $\beta\gamma$. This observation by itself was not terribly surprising, since it had been documented by Katada and associates with a partially purified preparation of calmodulin-stimulated adenylyl cyclase from brain (Katada et al. 1987). However, these investigators proposed that the inhibitory effect was indirect and was mediated by interaction between $\beta\gamma$ and calmodulin. We have shown that this mechanism is incorrect. There is neither a stoichiometric nor a competitive relationship between $\beta\gamma$ and calmodulin when $\beta\gamma$ acts to inhibit type I adenylyl cyclase activity. In addition, $\beta\gamma$ can inhibit GTPγS · $G_{s\alpha}$-stimulated activity as well. In contrast, $\beta\gamma$ *stimulates* type II and type IV adenylyl cyclase activity strikingly in a conditional fashion—when GTPγS · $G_{s\alpha}$ is also present. Thus, when one compares type I and type II adenylyl cyclase expressed in the same cells and assayed under identical conditions (GTPγS · $G_{s\alpha}$ plus $\beta\gamma$), their specific activities differ by more than 20-fold; their activities are the same when only GTPγS · $G_{s\alpha}$ is present as an activator.

The data of Figure 5 are misleading with regard to the concentrations of G-protein subunits required to exert the effects described. Recombinant (*Escherichia coli*-derived) $G_{s\alpha}$ was used in these experiments, and the affinity of this protein for adenylyl cyclase is about 20-fold less than that of native $G_{s\alpha}$ (presumably because of the absence of an as yet unidentified covalent modification). When native $G_{s\alpha}$ is utilized, it activates adenylyl cyclase at 10–1000 pM concentrations, whereas the effects of $\beta\gamma$ are observed in the 1–100 nM range (Tang and Gilman 1991). This seems to make sense, since one would otherwise assume that type I adenylyl cyclase would be simultaneously activated by $G_{s\alpha}$ and inhibited by $\beta\gamma$ released from the G_s oligomer if the subunits were effective at similar concentrations. Rather, we view this mechanism as a means for cross-talk and interaction between signaling pathways. Thus, adenylyl cyclase is activated by G_s-linked pathways but can be inhibited or further activated conditionally and synergistically (depending on the type of cyclase in question) when other pathways are activated simulta-

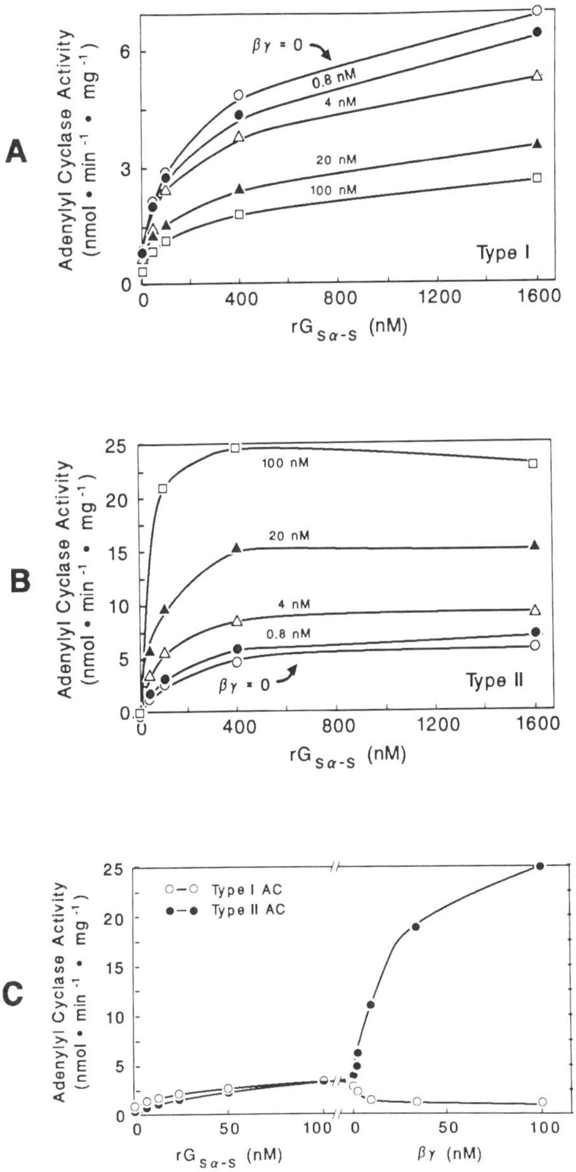

Figure 5. Effects of GTPγS-rG$_{s\alpha}$ and bovine brain $\beta\gamma$ on type I (*A*) and type II (*B*) adenylyl cyclase activity in membranes from Sf9 cells infected with recombinant baculovirus encoding the designated type of adenylyl cyclase. (*C*) Comparison of the effect of GTPγS-rG$_{s\alpha}$ (*left*) and bovine brain $\beta\gamma$ in the presence of 100 nM GTPγS-rG$_{s\alpha}$ (*right*) on type I and type II adenylyl cyclase. (Reprinted, with permission, from Tang and Gilman 1991 [copyright AAAS].)

neously. Concentrations of G_i and G_o proteins in the brain greatly exceed those of G_s, and pathways that utilize these G proteins would presumably be good candidates to release sufficient concentrations of $\beta\gamma$ to meet these needs. Rall and co-workers have observed highly synergistic effects of different neurotransmitters on cAMP synthesis in brain (Sattin et al. 1975). It will be of interest to see if conditional activation of adenylyl cyclase by G-protein $\beta\gamma$ subunits is the explanation for these phenomena.

The G-Protein $\beta\gamma$ Subunit Complex

The effects of G-protein $\beta\gamma$ subunits just described have heightened our interest in these polypeptides. As mentioned, there is substantial molecular diversity in both β and γ. Sequences of four distinct mammalian β (von Weizsäcker et al. 1992) and five γ (Fisher and Aronson 1992) subunits have been reported to date. Although $\beta\gamma$ subunit complexes are, to at least a certain extent, interchangeable among different α subunits, the molecular diversity of β and γ begs the question of functional significance. However, it has been very difficult to search for such differences, since preparations of $\beta\gamma$ (such as those used above) are often heterogeneous, and their composition is poorly defined. In addition, little is known about the functional importance of the fact that γ subunits are prenylated and processed at their carboxyl termini in a manner analogous to that described in particular for $p21^{ras}$.

To begin to approach these questions, we have expressed two different G-protein β subunits and three γ subunits both singly and in combinations in Sf9 cells (Iñiguez-Lluhi et al. 1992). Although expression of individual subunits is achieved in all cases, $\beta\gamma$ subunit activity (for example, support of pertussis toxin-catalyzed ADP-ribosylation of $G_{i\alpha 1}$) is detected only when β and γ are expressed concurrently. Of the six combinations of $\beta\gamma$ tested, only one ($\beta_2\gamma_1$) failed to generate a functional complex. Each of the other five complexes has been purified by subunit-exchange chromatography using $G_{o\alpha}$-agarose as the chromatographic matrix. When assayed for their ability to inhibit calmodulin-stimulated type I adenylyl cyclase activity or to potentiate $G_{s\alpha}$-stimulated type II adenylyl cyclase, recombinant $\beta_1\gamma_1$ is tenfold less potent than the other complexes examined. Differences were also detected in the ability of distinct $\beta\gamma$ complexes to support the ADP-ribosylation of $G_{i\alpha 1}$. Thus, G-protein α subunits and effectors can distinguish among different species of $\beta\gamma$. We suspect that these initial results are just the tip of the iceberg and that more profound specificity will be encountered when more widely divergent β and γ subunits are available.

We have also begun to probe the functional significance of the prenylation of γ. Two approaches were taken to address the possible role of such prenylation in dictating the cellular distribution of G proteins. Prenylation of γ was prevented either by site-directed mutagenesis or by inhibiting the synthesis of mevalonate, the precursor of cellular isoprenoids, with compactin (Muntz et al. 1992). When β or γ subunits were transiently expressed in COS cells, the proteins were found in the membrane fraction by immunoblotting. Replacement of Cys-68 of a γ_2 subunit with serine (γ_2C68S) prevented prenylation of the mutant protein and association of the protein with membranes. Interestingly, β subunits were also found in the cytoplasm when coexpressed with the nonprenylatable mutant γ subunit, and the cytosolic β and γ subunits were associated. Similarly, when Neuro 2A cells were treated with

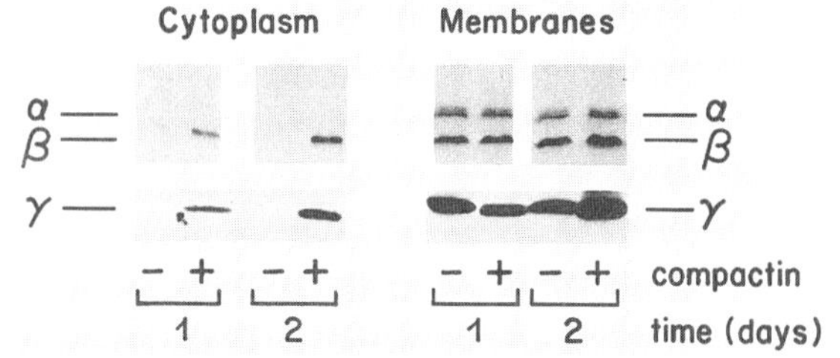

Figure 6. Effect of compactin on the distribution of G-protein subunits in Neuro 2A cells. Cells cultured in the presence or absence of 20 μM compactin were harvested and fractionated after 1 or 2 days. Proteins in the cytoplasmic and membrane fractions were resolved by SDS-PAGE and were subjected to immunoblotting to detect endogenous $G_{i\alpha}$, β, and γ_2 subunits. (Reprinted, with permission, from Muntz et al. 1992.)

compactin to inhibit protein prenylation, a fraction of endogenous β and γ was distributed in the cytoplasm (Fig. 6). These results indicate that prenylation facilitates association of γ subunits with membranes, that the cellular location of γ influences the distribution of β, and that prenylation is not an absolute requirement for interaction of β and γ.

To investigate these questions further, we have coexpressed the β_1 subunit with the γ_2C68S mutant subunit in Sf9 cells (Iñiguez-Lluhi et al. 1992). Virtually all of the mutant γ subunit is found in the cytosol, and there is a substantial distribution of the β subunit to the cytosol as well. (When wild-type β and γ are coexpressed, all of the β subunit is found in the particulate fraction.) Although gel filtration studies indicate that the cytosolic β_1 and γ_2C68S proteins are indeed associated in a tight complex, $\beta\gamma$ subunit activity (support of ADP-ribosylation of $G_{i\alpha 1}$) is not detectable. The failure of $\beta_1\gamma_2$C68S to support ADP-ribosylation of $G_{i\alpha 1}$ could be due to a very low affinity of this complex for G-protein α subunits or to the formation of a complex that is not a substrate for ADP-ribosylation by pertussis toxin. We thus compared the ability of $\beta_1\gamma_2$ and $\beta_1\gamma_2$C68S to interact with $G_{o\alpha}$-agarose (Fig. 7). The wild-type complex binds stably to immobilized $G_{o\alpha}$ and is eluted when the α subunit is activated with Al^{+++}, Mg^{++}, and F^-. In contrast, the mutant complex flows

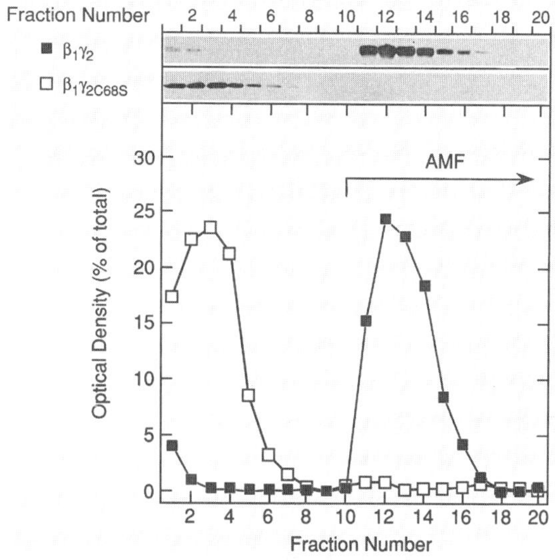

Figure 7. $G_{o\alpha}$-agarose chromatography of $\beta_1\gamma_2$ and $\beta_1\gamma_2$C68S. A cholate extract of membranes from cells expressing $\beta_1\gamma_2$ (■) and a cytosolic fraction from cells expressing $\beta_1\gamma_2$C68S (□) were applied to $G_{o\alpha}$-agarose columns. The columns were washed with detergent-containing buffer (fractions 1–10) and were then eluted with buffer containing Al^{+++}, Mg^{++}, and F^- to activate the $G_{o\alpha}$ and cause subunit dissociation. Aliquots of the fractions were subjected to SDS-PAGE and immunoblotting with a β subunit-specific antibody (shown in the figure) or with a γ subunit-specific antibody (not shown). (Reprinted, with permission, from Iñiguez-Lluhi et al. 1992.)

through the column. Thus, prenylation of γ and/or subsequent carboxy-terminal processing is an important determinant of the affinity of G-protein $\beta\gamma$ subunits for α.

Similar results were obtained when we examined the ability of partially purified $\beta_1\gamma_2$C68S to regulate adenylyl cyclase activity (Fig. 8). High concentrations of $\beta_1\gamma_2$C68S do not inhibit calmodulin-stimulated type I adenylyl cyclase activity, nor do they potentiate $G_{s\alpha}$-stimulated type II adenylyl cyclase activity. Furthermore, the mutant complex fails to diminish the effects of wild-type $\beta\gamma$ when the mutant is present in 100-fold excess.

Lipid modifications of proteins have often been

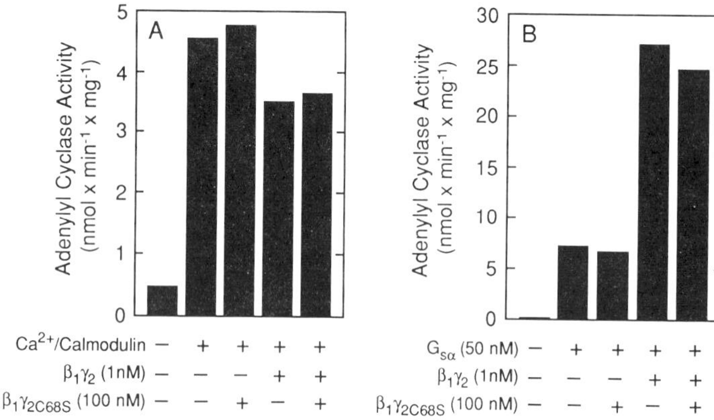

Figure 8. $\beta_1\gamma_2$C68S fails to regulate adenylyl cyclase activity. Sf9 cell membranes containing type I (*A*) or type II (*B*) adenylyl cyclase were assayed for adenylyl cyclase activity in the absence or presence of activators (50 μM $CaCl_2$/100 nM calmodulin for type I; 50 nM GTPγS-r$G_{s\alpha}$ for type II) and $\beta\gamma$ (1 nM $\beta_1\gamma_2$; 100 nM $\beta_1\gamma_2$C68S), as indicated. (Reprinted, with permission, from Iñiguez-Lluhi et al. 1992).

found to be essential determinants for localization of proteins to cellular membranes. However, this is not invariably so, and individual lipids do not specify the membrane to which such modified proteins are distributed. It is thus clear that lipid modifications of proteins provide more than a simple hydrophobic anchor for association of the protein with a membrane. There are now several examples where lipid modifications have been shown to be significant determinants of the affinities of protein-protein interactions. There is a "receptor" for the interaction of Src with the plasma membrane with specificity for both amino-terminal amino acid sequences in Src and for myristate (Resh and Ling 1990). Myristoylation of G-protein α subunits increases their affinity for $\beta\gamma$ (Linder et al. 1991) and, in an interestingly reciprocal fashion, prenylation of γ enhances the affinity of $\beta\gamma$ for α. Prenylation of $\beta\gamma$ also appears to have an important effect on the affinity of this protein complex for adenylyl cyclase. A unique aspect of lipid modifications of proteins may be their ability to target protein-protein interactions within a membrane environment.

ACKNOWLEDGMENTS

This work was supported by National Institutes of Health grant GM-34497, American Cancer Society grant BC-551, The Perot Family Foundation, the Lucille P. Markey Charitable Trust, and the Raymond and Ellen Willie Chair of Molecular Neuropharmacology. J.I.-L. is a recipient of a Howard Hughes Medical Institute predoctoral training award.

REFERENCES

Bakalyar, H.A. and R.R. Reed. 1990. A specialized adenylyl cyclase may mediate odorant detection. *Science* **250:** 1403.

Blumer, K.J. and J. Thorner. 1991. Receptor-G protein signaling in yeast. *Annu. Rev. Physiol.* **53:** 37.

Brown, A.M. and L. Birnbaumer. 1988. Direct G protein gating of ion channels. *Am. J. Physiol.* **254:** H401.

———. 1990. Ionic channels and their regulation by G protein subunits. *Annu. Rev. Physiol.* **52:** 197.

Chinkers, M. and D.L. Garbers. 1991. Signal transduction by guanylyl cyclases. *Annu. Rev. Biochem.* **60:** 553.

Feinstein, P.G., K.A. Schrader, H.A. Bakalyar, W.-J. Tang, J. Krupinski, A.G. Gilman, and R.R. Reed. 1991. Molecular cloning and characterization of a Ca^{2+}/calmodulin insensitive adenylyl cyclase from rat brain. *Proc. Natl. Acad. Sci.* **88:** 10173.

Fisher, K.J. and N.N. Aronson, Jr. 1992. Characterization of the cDNA and genomic sequence of a G protein γ subunit (γ_5). *Mol. Cell. Biol.* **12:** 1585.

Florio, V.A. and P.C. Sternweis. 1985. Reconstitution of resolved muscarinic cholinergic receptors with purified GTP-binding proteins. *J. Biol. Chem.* **260:** 3477.

Fukada, Y., T. Takao, H. Ohguro, T. Yoshizawa, T. Akino, and Y. Shimonishi. 1990. Farnesylated γ-subunit of photoreceptor G protein indispensable for GTP-binding. *Nature* **346:** 658.

Fung, B.K.-K. 1983. Characterization of transducin from bovine retinal rod outer segments. I. Separation and reconstitution of the subunits. *J. Biol. Chem.* **258:** 10495.

Gao, B. and A.G. Gilman. 1991. Cloning and expression of a widely distributed (type IV) adenylyl cyclase. *Proc. Natl. Acad. Sci.* **88:** 10178.

Gilman, A.G. 1987. G proteins: Transducers of receptor-generated signals. *Annu. Rev. Biochem.* **56:** 615.

Hescheler, J., W. Rosenthal, W. Trautwein, and G. Schultz. 1987. The GTP-binding protein, G_o, regulates neuronal calcium channels. *Nature* **325:** 445.

Iñiguez-Lluhi, J.A., M.I. Simon, J.D. Robishaw, and A.G. Gilman. 1992. G protein $\beta\gamma$ subunits synthesized in Sf9 cells: Functional characterization and the significance of prenylation of γ. *J. Biol. Chem.* (in press).

Jelsema, C.L. and J. Axelrod. 1987. Stimulation of phospholipase A_2 activity in bovine rod outer segments by the $\beta\gamma$ subunits of transducin and its inhibition by the α subunit. *Proc. Natl. Acad. Sci.* **84:** 3623.

Katada, T., K. Kusakabe, M. Oinuma, and M. Ui. 1987. A novel mechanism for the inhibition of adenylate cyclase via inhibitory GTP-binding proteins. *J. Biol. Chem.* **262:** 11897.

Katada, T., G.M. Bokoch, J.K. Northup, M. Ui, and A.G. Gilman. 1984. The inhibitory guanine nucleotide-binding regulatory component of adenylate cyclase. Properties and function of the purified protein. *J. Biol. Chem.* **259:** 3568.

Kaziro, Y., H. Itoh, T. Kozasa, M. Nakafuku, and T. Satoh. 1991. Structure and function of signal-transducing GTP-binding proteins. *Annu. Rev. Biochem.* **60:** 349.

Krupinski, J., F. Coussen, H.A. Bakalyar, W.-J. Tang, P.G. Feinstein, K. Orth, C. Slaughter, R.R. Reed, and A.G. Gilman. 1989. Adenylyl cyclase amino acid sequence: Possible channel- or transporter-like structure. *Science* **244:** 1558.

Lai, R.K., D. Perez-Sala, F.J. Canada, and R.R. Rando. 1990. The γ subunit of transducin is farnesylated. *Proc. Natl. Acad. Sci.* **87:** 7673.

Linder, M.E., I.-H. Pang, R.J. Duronio, J.I. Gordon, P.C. Sternweis, and A.G. Gilman. 1991. Lipid modifications of G protein subunits. Myristoylation of $G_{o\alpha}$ increases its affinity for $\beta\gamma$. *J. Biol. Chem.* **266:** 4654.

Maltese, W.A. and J.D. Robishaw. 1990. Isoprenylation of C-terminal cysteine in a G-protein γ subunit. *J. Biol. Chem.* **265:** 18071.

May, D.C., E.M. Ross, A.G. Gilman, and M.D. Smigel. 1985. Reconstitution of catecholamine-stimulated adenylate cyclase activity using three purified proteins. *J. Biol. Chem.* **260:** 15829.

Mumby, S.M., P.J. Casey, A.G. Gilman, S. Gutowski, and P.C. Sternweis. 1990. G protein γ subunits contain a 20-carbon isoprenoid. *Proc. Natl. Acad. Sci.* **87:** 5873.

Muntz, K.H., P.C. Sternweis, A.G. Gilman, and S.M. Mumby. 1992. Influence of γ subunit prenylation on association of guanine nucleotide-binding regulatory proteins with membranes. *Mol. Biol. Cell* **3:** 49.

Northup, J.K., M.D. Smigel, P.C. Sternweis, and A.G. Gilman. 1983. The subunits of the stimulatory regulatory component of adenylate cyclase. Resolution of the activated 45,000-dalton (α) subunit. *J. Biol. Chem.* **258:** 11369.

Resh, M.D. and H.-P. Ling. 1990. Identification of a 32K plasma membrane protein that binds to the myristylated amino-terminal sequence of $p60^{v-src}$. *Nature* **346:** 84.

Sattin, A., T.W. Rall, and J. Zanella. 1975. Regulation of cyclic adenosine 3′, 5′-monophosphate levels in guinea-pig cerebral cortex by interaction of *alpha* adrenergic and adenosine receptor activity. *J. Pharmacol. Exp. Ther.* **192:** 22.

Schubert, B., A.M.J. VanDongen, G.E. Kirsch, and A.M. Brown. 1989. β-Adrenergic inhibition of cardiac sodium channels by dual G-protein pathways. *Science* **245:** 516.

Simon, M.I., M.P. Strathmann, and N. Gautam. 1991. Diversity of G proteins in signal transduction. *Science* **252:** 802.

Smrcka, A.V., J.R. Hepler, K.O. Brown, and P.C. Sternweis.

1991. Regulation of polyphosphoinositide-specific phospholipase C activity by purified G_q. *Science* **251:** 804.

Sternweis, P.C. 1986. The purified α subunits of G_o and G_i from bovine brain require $\beta\gamma$ for association with phospholipid vesicles. *J. Biol. Chem.* **261:** 631.

Stow, J.L., J.B. de Almeida, N. Narula, E.J. Holtzman, L. Ercolani, and D.A. Ausiello. 1991. A heterotrimeric G protein, $G\alpha_{i-3}$, on Golgi membranes regulates the secretion of a heparan sulfate proteoglycan in LLC-PK_1 epithelial cells. *J. Cell Biol.* **114:** 1113.

Strathmann, M.P. and M.I. Simon. 1991. $G\alpha 12$ and $G\alpha 13$ subunits define a fourth class of G protein α subunits. *Proc. Natl. Acad. Sci.* **88:** 5582.

Stryer, L. 1986. Cyclic GMP cascade of vision. *Annu. Rev. Neurosci.* **9:** 87.

Tang, W.-J. and A.G. Gilman. 1991. Type-specific regulation of adenylyl cyclase by G protein $\beta\gamma$ subunits. *Science* **254:** 1500.

Tang, W.-J., J. Krupinski, and A.G. Gilman. 1991. Expression and characterization of calmodulin-activated (type-I) adenylyl cyclase. *J. Biol. Chem.* **266:** 8595.

Taylor, S.J., H.Z. Chae, S.G. Rhee, and J.H. Exton. 1991. Activation of the $\beta 1$ isozyme of phospholipase C by α subunits of the G_q class of G proteins. *Nature* **350:** 516.

Van Dongen, A.M.J., J. Codina, J. Olate, R. Mattera, R. Joho, L. Birnbaumer, and A.M. Brown. 1988. Newly identified brain potassium channels gated by the guanine nucleotide binding protein G_o. *Science* **242:** 1433.

von Weizsäcker, E., M.P. Strathmann, and M.I. Simon. 1992. Diversity among the beta subunits of heterotrimeric GTP-binding proteins: Characterization of a novel beta-subunit cDNA. *Biochem. Biophys. Res. Comm.* **183:** 350.

Yamane, H.K., C.C. Farnsworth, H. Xie, W. Howald, B.K.-K. Fung, S. Clarke, M.H. Gelb, and J.A. Glomset. 1990. Brain G protein γ subunits contain an all-*trans*-geranylgeranyl-cysteine methyl ester at their carboxyl termini. *Proc. Natl. Acad. Sci.* **87:** 5868.

Yatani, A., J. Codina, A.M. Brown, and L. Birnbaumer. 1987. Direct activation of mammalian atrial muscarinic potassium channels by GTP regulatory protein G_K. *Science* **235:** 207.

Yatani, A., R. Mattera, J. Codina, R. Graf, K. Okabe, E. Padrell, R. Iyengar, A.M. Brown, and L. Birnbaumer. 1988. The G protein-gated atrial K^+ channel is stimulated by three distinct $G_{i\alpha}$-subunits. *Nature* **336:** 680.

Detection of Coincident Signals by G Proteins and Adenylyl Cyclase

H.R. Bourne, K.D. Lustig, Y.H. Wong, and B.R. Conklin
Department of Pharmacology, Cardiovascular Research Institute, and Cell Biology Program, University of California School of Medicine, San Francisco, California 94143

Biological signals often participate in combinatorial regulation—that is, two or more simultaneous signals induce a response different from that elicited by either signal alone. Among the molecular machines that detect coincidence of two or more types of input signal, well-documented examples include the T-cell antigen receptor, neural NMDA receptors, $p34^{cdc2}$, and many (perhaps all) protein complexes that regulate gene transcription. Recent work suggests that it is time to add hormone-sensitive adenylyl cyclase to this list.

Until recently, we have assumed that all signals mediated by heterotrimeric G proteins, including hormonal stimulation of cAMP synthesis, follow pathways in which each regulatory element suffices by itself to activate the next one (Freissmuth et al. 1989; Birnbaumer 1990; Bourne et al. 1990): (1) A hormone (or sensory stimulus) binds to and activates a cell-surface receptor; (2) the hormone-receptor complex activates the appropriate G protein, promoting replacement of GDP by GTP on Gα, which in turn causes active GTP-bound Gα to dissociate from the $\beta\gamma$ complex; (3) Gα (or sometimes, $\beta\gamma$) binds to and activates the appropriate effector enzyme (e.g., adenylyl cyclase) or ion channel. The hormone$\rightarrow$G protein$\rightarrow$effector pathways are not always simple, of course. An individual G protein can serve as a target for more than one receptor (although not, so far as we know, for two receptors simultaneously), and a G protein may also regulate more than one class of effector (Brown and Birnbaumer 1990); even so, accurate transmission of a message by the G protein depends only on the specificity with which each element in the pathway binds to and activates the next. Now, however, recent studies of adenylyl cyclase show how detection of two or more coincident input signals by an effector molecule may confer additional specificity.

The story begins with the discovery that adenylyl cyclases constitute a surprisingly large family of proteins. Since molecular cloning of the first adenylyl cyclase gene (Krupinski et al. 1989) in 1989, researchers in the laboratories of Reed and Gilman have cloned three additional adenylyl cyclase genes (Bakalyar and Reed 1990; Feinstein et al. 1991; Gao and Gilman 1991); four more have been found but not yet reported (A.G. Gilman and J. Krupinski, pers. comm.). Biochemical characterization of partially purified recombinant adenylyl cyclases has begun to reveal a rationale for this profusion of genes: Although the GTP-bound α subunit of $G_s(\alpha_s \cdot GTP)$ stimulates all adenylyl cyclases (ACs), the individual isozymes respond in different ways to other regulators. cAMP synthesis by AC-I, for example, is stimulated by Ca^{++}-calmodulin and inhibited by G protein $\beta\gamma$ subunits (Tang et al. 1991); in contrast, AC-II (Tang and Gilman 1991) and AC-IV (Gao and Gilman 1991) are *stimulated* by $\beta\gamma$, but their activities are not affected by Ca^{++}-calmodulin.

Most importantly for our purposes, stimulation of AC-II by $\beta\gamma$ is strictly conditional, requiring coincidence of two input signals: $\beta\gamma$ alone has no effect on AC-II; $\beta\gamma$ stimulates *only* when AC-II is simultaneously activated by $\alpha_s \cdot GTP$ (Tang and Gilman 1991). Stimulation by $\beta\gamma$ is large (up to fivefold over that produced by α_s) and occurs at physiologically relevant concentrations of $\beta\gamma$ (5–10 nM). As Tang and Gilman (1991) recognized, the response of AC-II to $\beta\gamma$ in vitro suggests a mechanism for cooperation between G_s and other G proteins in stimulating cAMP synthesis.

Stimulation of AC-II by G_i-coupled Receptors

To explore regulation of AC-II in vivo, we used a human embryonic kidney cell line, called HEK-293. Their capacity for transient but simultaneous expression of multiple cDNAs has made HEK-293 cells especially useful for studying interactions among recombinant signaling proteins. For example, expression of recombinant, mutationally activated G protein α subunits produces constitutive stimulation of the appropriate responses in these cells: α_s increases cAMP synthesis (Wong et al. 1991), α_q increases production of inositol phosphates (Conklin et al. 1992), and $\alpha_{i1\text{-}3}$ and α_z inhibit cAMP synthesis (Wong et al. 1991, 1992). In addition, hormonal stimulation and inhibition of cAMP synthesis can be studied in HEK-293 cells using recombinant receptors (not present in the untransfected cells) to trigger both pathways (Wong et al. 1991, 1992): An appropriate agonist, human chorionic gonadotropin (hCG), triggers a recombinant lutropin receptor (LHR) to stimulate cAMP synthesis, utilizing the endogenous G_s and adenylyl cyclase molecules of HEK-293 cells. This stimulation is inhibited by agonists for exogenous G_i-coupled receptors—quinpirole, for the dopamine D_2 receptor (D_2-R), and N6-(2-phenylisopropyl) adenosine (PIA) for the adenosine A_1 receptor (A_1-R). Pertussis toxin blocks the effects of D_2-R or A_1-R agonists (Wong et al. 1992), presumably by preventing endogenous G_i proteins from interacting with the D_2-R or A_1-R.

To explore implications of the Tang and Gilman experiments (Tang and Gilman 1991), we studied regulation of recombinant AC-II expressed in HEK-293 cells. We predicted that recombinant AC-II in these cells would respond to stimulation by $\alpha_s \cdot$ GTP, but that (unlike the endogenous enzyme) $\alpha_s \cdot$ GTP-activated AC-II would synthesize cAMP more rapidly in cells treated with agonists that increase the concentration of free $\beta\gamma$ subunits. To test this prediction, we transfected the cells with AC-II cDNA in combination with cDNAs encoding other proteins: To provide $\alpha_s \cdot$ GTP, we expressed a mutant α_s (α_s-Q227L), with an amino acid substitution that causes constitutive activation of cAMP synthesis, owing to decreased hydrolysis of GTP (Graziano and Gilman 1989; Masters et al. 1989); as an alternative, in some experiments, we transfected the LHR, through which hCG could promote elevation of endogenous $\alpha_s \cdot$ GTP. To allow agonist-stimulated release of free $\beta\gamma$, cells were cotransfected with two G_i-coupled receptors, the D_2-R and the A_1-R.

The results were exactly as predicted (Federman et al. 1992): Agonist stimulation of either the D_2-R or the A_1-R increased cellular cAMP approximately threefold, but only when α_s was simultaneously active, and only in cells expressing AC-II; in the absence of AC-II, as described above, the G_i-coupled receptor agonists inhibited cAMP accumulation. Pertussis toxin blocked the stimulation of AC-II produced by D_2-R and A_1-R agonists, in keeping with the idea that the corresponding receptors specifically activate G_i proteins.

To test the interpretation that $\beta\gamma$ mediates the stimulation of cAMP synthesis by D_2-R and A_1-R agonists, we sought to reduce the concentration of free $\beta\gamma$ by coexpressing an α subunit that binds $\beta\gamma$ tightly but does not interact with adenylyl cyclase. We chose the α subunit (α_t) of transducin (G_t) from retinal rod cells (Medynski et al. 1985), which is known to interact with nonretinal $\beta\gamma$ (Kanaho et al. 1984): Expression of α_t completely blocked the ability of D_2-R and A_1-R agonists to augment cAMP synthesis in cells coexpressing the appropriate G_i-coupled receptors, α_s-Q227L, and AC-II (Federman et al. 1992). We presume that the inhibitory effect of α_t was due to binding and sequestration of $\beta\gamma$ subunits released by activation of G_i proteins.

What about other α subunits that bind $\beta\gamma$? α_q blocks stimulation of AC-II by G_i-coupled receptors, but α_{i2} does not (B.R. Conklin, unpubl.). One explanation of this difference is that α_i, but not α_t or α_q, can interact with G_i-coupled receptors. Thus, despite its ability to bind $\beta\gamma$, α_{i2} may fail to sequester free $\beta\gamma$ because $\alpha_i \cdot \beta\gamma$ complexes rapidly dissociate, owing to the ability of agonist-activated D_2- or A_1-receptors to catalyze replacement of GDP on G_i with GTP.

If α_{i2} fails to sequester $\beta\gamma$ because $\alpha_{i2} \cdot \beta\gamma$ complexes are rapidly dissociated by activation of G_i-coupled receptors, then one might predict that α_t will also lose its ability to sequester $\beta\gamma$ when stimulated by its normal receptor, rhodopsin, the photon detector in outer segments of retinal rod cells. Earlier evidence indicated that photoactivated rhodopsin (hereafter called R*) can activate G_i proteins (Kanaho et al. 1984), as well as G_t, although the reciprocal is not true—that is, G_i-coupled receptors cannot activate G_t (Cerione et al. 1986). Accordingly, we expressed bovine rhodopsin in HEK-293 cells, supplied the chromophore, 11-*cis*-retinal, and measured cAMP accumulation in the dark (rhodopsin inactive) or in the light (which produced R*). We tested the following predictions in HEK-293 cells: (1) In the absence of either AC-II or mutationally activated α_s, R* would not stimulate cAMP synthesis; (2) instead, in the absence of AC-II, R* would inhibit cAMP synthesis in a pertussis-toxin-sensitive fashion; (3) R* would augment $\alpha_s \cdot$ GTP-dependent AC-II activity by releasing $\beta\gamma$ from endogenous G_i proteins; (4) pertussis toxin would block this effect of R*; (5) because R* rapidly dissociates $\alpha_t \cdot \beta\gamma$ complexes, coexpression of α_t would not block this stimulatory effect of R*.

Each of these predictions proved correct (K.D. Lustig et al., in prep.). The first two of these five results simply demonstrate that R* can promote G_i-mediated inhibition of the endogenous adenylyl cyclase of HEK-293 cells. The third and fourth results show that, in the presence of active α_s, R* can activate AC-II, presumably by releasing free $\beta\gamma$ from G_i. The last result conforms nicely to the idea that only a stable GDP-bound α subunit can effectively sequester $\beta\gamma$ and that activation of appropriate receptors destabilizes $\alpha \cdot$ GTP $\cdot\, \beta\gamma$ by promoting exchange of GTP for bound GDP.

Stimulation of AC-II by G_q-coupled Receptors and Phorbol Esters

We next turned to effects of receptors coupled to G_q, the best-characterized member of a G-protein family that is structurally and functionally distinct from both the G_s and G_i-G_t protein families. G_q interacts with receptors different from those that activate G_s, G_i, and G_t, and regulates the activity of a different effector enzyme, phosphatidylinositol phospholipase C (PI-PLC); both these differences are thought to be conferred by the distinctive structure of α_q, which, in its GTP-activated state, binds to and stimulates PI-PLC (Smrcka et al. 1991; Taylor et al. 1991). In HEK-293 cells, an endogenous G_q-like protein mediates stimulation of PI-PLC by agonists acting on transfected recombinant bombesin receptors or muscarinic M1-acetylcholine receptors (m1-AChRs). Application of the $\beta\gamma$-sequestration "protocol" to G_q-coupled receptors in HEK-293 cells led to an informative and interesting surprise (K.D. Lustig et al., in prep.).

The ability of excess recombinant α_q to prevent G_i-mediated stimulation of AC-II led us to expect two results: (1) Agonist stimulation of G_q-coupled receptors would also stimulate cAMP synthesis in cells expressing AC-II and α_s-Q227L; (2) if such agonists acted by releasing $\beta\gamma$ from G_q, then coexpression of α_t would block the increase in cAMP synthesis. The first expectation was fulfilled, but the second was not—that is, in the presence of the appropriate recombinant receptors, bombesin and carbachol (an m1-AChR agonist) elevated α_s-Q227L-stimulated AC-II activity approximately threefold, but α_t had no effect. The difference in effectiveness of α_t on responses mediated by G_i and G_q could be explained in either of two ways. First, we can imagine that α_t has a high affinity for the $\beta\gamma$ subunits of G_i but fails to bind and sequester $\beta\gamma$ subunits released from G_q. We have no data to confirm or refute this possibility. It implies functional differences between $\beta\gamma$ subunits derived from G_i and G_q; such functional heterogeneity might reflect the structural heterogeneity of β and γ polypeptides, each of which belongs to a growing protein family encoded by four or more genes (Fong et al. 1987; Robishaw et al. 1989; Gautam et al. 1990; Levine et al. 1990; Simon et al. 1991).

An alternative explanation is that α_t fails to prevent the effects of bombesin and carbachol because these effects are not mediated by release of $\beta\gamma$ from G_q, but by a product of PI-PLC, such as inositol trisphosphate or diacylglycerol. If so, then phorbol esters, which mimic diacylglycerol as stimulators of protein kinase C, might reproduce the effects of agonists that stimulate G_q-coupled receptors. This turned out to be the case: Phorbol esters rapidly augmented cAMP accumulation in HEK-293 cells coexpressing α_s-Q227L and AC-II; like agonists acting on G_i- and G_q-coupled receptors, phorbol esters had little effect on cAMP synthesis in the absence of either $\alpha_s \cdot$GTP or AC-II.

From these results we conclude that AC-II is subject to two different conditional stimulators, $\beta\gamma$ and (probably) protein kinase C; both require simultaneous stimulation of AC-II by $\alpha_s \cdot$GTP. The two pathways account for effects of agonists acting on G_i- and G_q-coupled receptors, respectively. Stimulatory effects of phorbol esters and G_i-coupled receptors are partially additive, suggesting that protein kinase C and $\beta\gamma$ activate AC-II by distinct mechanisms.

So far, we have obtained no direct biochemical evidence that either stimulation of G_q-coupled receptors or treatment with phorbol esters leads to phosphorylation of AC-II or other substrates (e.g., Gα or G$\beta\gamma$ subunits) that play a part in the cAMP responses we have measured. Thus, our results are consistent with, but do not prove, the proposition that agonists for G_q-coupled receptors elevate activity of $\alpha_s \cdot$GTP-stimulated AC-II via protein kinase C.

Possible Physiological Significance

Taken together, the in vitro experiments (Tang and Gilman 1991) and our results in HEK-293 cells suggest that AC-II, and perhaps other adenylyl cyclase isozymes, can serve to detect coincident signals, mediated by either G_i or G_q, only when a third pathway, mediated by G_s, is simultaneously activated. Is there evidence that adenylyl cyclase serves as a coincidence detector in a mammalian organ or tissue, or that such a phenomenon could have physiological significance? The evidence, obtained in studies of the central nervous system, the predominant site of AC-II expression (Feinstein et al. 1991), suggests that the answer to both questions is yes.

Measurements of cAMP accumulation in brain slices have produced several puzzling results, dating back almost 20 years, that can now be explained by the idea that adenylyl cyclases detect coincident signals. Thus isoproterenol, an agonist at β-adrenoceptors, which stimulate G_s, increases cAMP content of cortical brain slices; this effect of isoproterenol is substantially augmented by simultaneous exposure to agonists that act on G_i- and G_q-coupled receptors, and also by treatment with phorbol esters (Sattin et al. 1975; Daly et al. 1980; Karbon and Enna 1985; Sugden and Klein 1988).

Recent electrophysiological experiments from the laboratory of Rodrigo Andrade (pers. comm.) suggest that detection by adenylyl cyclase of coincident signals from G_s- and G_i-coupled receptors may serve as a physiologically relevant mechanism for modulating synapses of pyramidal cells in the CA1 region of the hippocampus. Andrade observed that simultaneous application of agonists that act on G_i-coupled receptors markedly augments an electrophysiological effect of agonists that act on G_s; physiologically, the simultaneous stimulation of G_i- and G_s-coupled receptors would be expected to enhance synaptic transmission. In the absence of G_s stimulation, stimulation of G_i-coupled receptors had no such effect; this is in accord with the parallel requirement for activated α_s seen with $\beta\gamma$ stimulation of AC-II activity in vitro. Indeed, in the absence of G_s stimulation, agonists for G_i-coupled receptors actually inhibit transmission at the synapse Andrade studied.

PERSPECTIVE

We have reviewed evidence that at least one adenylyl cyclase isozyme, AC-II, can detect either of two coincident signals generated in cells simultaneously with activation of G_s. The coincident signal may be mediated either by $\beta\gamma$ subunits released from G_i or by activation of protein kinase C. Biochemical and electrophysiological evidence suggests that detection of coincident signals by adenylyl cyclase may have physiological significance in the central nervous system.

Many questions remain, including some mentioned above: We do not know in molecular detail how $\beta\gamma$ activates AC-II. We have not firmly established that protein kinase C mediates the effects on AC-II of G_q-coupled receptors or of phorbol esters. We have not identified the relevant substrate of protein kinase C. The question of structural versus functional heterogeneity of $\beta\gamma$ subunits remains to be explored.

If our interpretation of these results is correct, what are the implications for other regulatory processes mediated by G proteins? If AC-II can detect two signals in coincidence with $\alpha_s \cdot$GTP, how many different combinations of signals are detected by other adenylyl cyclase isozymes? Are there isozymes for which the conditional relationship between $\alpha_s \cdot$GTP and other regulators is "symmetrical"? In such a case, neither $\alpha_s \cdot$GTP nor a second signal would be able to activate the isozyme by itself, although the two signals together could do so. Do PI-PLC, ion channels, or other effectors of G-protein-mediated signals also act as coincidence detectors? Answers to these questions will further illuminate the rich versatility of signaling pathways initiated by receptors coupled to G proteins.

ACKNOWLEDGMENTS

Work from this laboratory discussed here is supported by grants from the National Institutes of Health and March of Dimes. We thank Rodrigo Andrade, Roger Nicoll, and Alfred G. Gilman for generous instruction and advice.

REFERENCES

Bakalyar, H.A. and R.R. Reed. 1990. Identification of a specialized adenylyl cyclase that may mediate odorant detection. *Science* **250:** 1403.

Birnbaumer, L. 1990. G proteins in signal transduction. *Annu. Rev. Pharmacol. Toxicol.* **30:** 675.

Bourne, H.R., D.A. Sanders, and F. McCormick. 1990. The GTPase superfamily. A conserved switch for diverse cell functions. *Nature* **348:** 125.

Brown, A.M. and L. Birnbaumer. 1990. Ionic channels and their regulation by G protein subunits. *Annu. Rev. Physiol.* **52:** 197.

Cerione, R.A., J.W. Regan, H. Nakata, J. Codina, J.L. Benovic, P. Gierschik, R.L. Somers, A.M. Spiegel, L. Birnbaumer, R.J. Lefkowitz, and M.G. Caron. 1986. Functional reconstitution of the α_2-adrenergic receptor with guanine nucleotide regulatory proteins in phospholipid vesicles. *J. Biol. Chem.* **261:** 3901.

Conklin, B.R., O. Chabre, Y.H. Wong, A.D. Federman, and H.R. Bourne. 1992. Recombinant $G_q\alpha$. Mutational activation and coupling to receptors and phospholipase C. *J. Biol. Chem.* **267:** 31.

Daly, J.W., W. Padgett, Y. Nimitkitpaisan, C.R. Creveling, D. Cantacuzene, and K.L. Kirk. 1980. Fluoronorepinephrines: Specific agonists for the activation of alpha and beta adrenergic-sensitive cyclic AMP generating systems in brain slices. *J. Pharmacol. Exp. Ther.* **212:** 383.

Federman, A.D., B.R. Conklin, K.A. Schrader, R.R Reed, and H.R. Bourne. 1992. Hormonal stimulation of adenylyl cyclase through G_i-protein $\beta\gamma$ subunits. *Nature* **356:** 159.

Feinstein, P.G., K.A. Schrader, H.A. Bakalyar, W.-J. Tang, J. Krupinski, A.G. Gilman, and R.R. Reed. 1991. Molecular cloning and characterization of a Ca^{2+}/calmodulin insensitive adenylyl cyclase from rat brain. *Proc. Natl. Acad. Sci.* **88:** 10173.

Fong, H.K.W., T.T. Amatruda III, B.W. Birren, and M.I. Simon. 1987. Distinct forms of the β subunit of GTP-binding regulatory proteins identified by molecular cloning. *Proc. Natl. Acad. Sci.* **84:** 3792.

Freissmuth, M., P.J. Casey, and A.G. Gilman. 1989. G proteins control diverse pathways of transmembrane signaling. *FASEB J.* **3:**2125.

Gao, B. and A.G. Gilman. 1991. Cloning and expression of a widely distributed (type IV) adenylyl cyclase. *Proc. Natl. Acad. Sci.* **88:** 10178.

Gautam, N., J. Northup, H. Tamir, and M.I. Simon. 1990. G protein diversity is increased by associations with a variety of γ subunits. *Proc. Natl. Acad. Sci.* **87:** 7973.

Graziano, M.P. and A.G. Gilman. 1989. Synthesis in *Escherichia coli* of GTPase-deficient mutants of G_{s_α}. *J. Biol. Chem.* **264:** 15475.

Kanaho, Y., S.C. Tsai, R. Adamik, E.L. Hewlett, J. Moss, and M. Vaughan. 1984. Rhodopsin-enhanced GTPase activity of the inhibitory GTP-binding protein of adenylate cyclase. *J. Biol. Chem.* **259:** 7378.

Karbon, E.W. and S.J. Enna. 1985. Characterization of the relationship between γ-aminobutyric acid B agonists and transmitter-coupled cyclic nucleotide-generating systems in rat brain. *Mol. Pharmacol.* **27:** 53.

Krupinski, J., F. Coussen, H.A. Bakalyar, W.-J. Tang, P.G. Feinstein, K. Orth, C. Slaughter, R.R. Reed, and A.G. Gilman. 1989. Adenylyl cyclase amino acid sequence: Possible channel- or transporter-like structure. *Science* **244:** 1558.

Levine, M.A., P.M. Smallwood, P.T. Moen, Jr., L.J. Helman, and T.G. Ahn. 1990. Molecular cloning of β_3 subunit, a third form of the G protein β-subunit polypeptide. *Proc. Natl. Acad. Sci.* **87:** 2329.

Masters, S.B., R.T. Miller, M.H. Chi, F.-H. Chang, B. Beiderman, N.G. Lopez, and H.R. Bourne. 1989. Mutations in the GTP-binding site of G_{s_α} alter stimulation of adenylyl cyclase. *J. Biol. Chem.* **264:** 15467.

Medynski, D.C., K. Sullivan, D. Smith, C. Van Dop, F.-H. Chang, B.K.-K. Fung, P.H. Seeburg, and H.R. Bourne. 1985. Amino acid sequence of the alpha subunit of transducin deduced from the cDNA sequence. *Proc. Natl. Acad. Sci.* **82:** 4311.

Robishaw, J.D., V.K. Kalman, C.R. Moomaw, and C.A. Slaughter. 1989. Existence of two γ subunits of the G proteins in brain. *J. Biol. Chem.* **264:** 15758.

Sattin, A., T.W. Rall, and J. Zanella. 1975. Regulation of cyclic adenosine 3′,5′-monophosphate levels in guinea-pig cerebral cortex by interactions of *alpha* adrenergic and adenosine receptor activity. *J. Pharmacol. Exp. Ther.* **192:** 22.

Simon, M.I., M.P. Strathmann, and N. Gautam. 1991. Diversity of G proteins in signal transduction. *Science* **252:** 802.

Smrcka, A.V., J.R. Hepler, K.O. Brown, and P.C. Sternweis. 1991. Regulation of polyphosphoinositide-specific phospholipase-C activity by purified G_q. *Science* **251:** 804.

Sugden, D. and D.C. Klein. 1988. Activators of protein kinase C act at a postreceptor site to amplify cyclic AMP production in rat pinealocytes. *J. Neurochem.* **50:** 149.

Tang, W.-J. and A.G. Gilman. 1991. Type-specific regulation of adenylyl cyclase by G protein $\beta\gamma$ subunits. *Science* **254:** 1500.

Tang, W.-J., J. Krupinski, and A.G. Gilman. 1991. Expression and characterization of calmodulin-activated (Type I) adenylylcyclase. *J. Biol. Chem.* **266:** 8595.

Taylor, S.J., H.Z. Chae, S.G. Rhee, and J.H. Exton. 1991. Activation of the β1 isozyme of phospholipase C by α subunits of the G_q class of G proteins. *Nature* **350:** 516.

Wong, Y.H., B.R. Conklin, and H.R. Bourne. 1992. G_z-mediated hormonal inhibition of cyclic AMP accumulation. *Science* **255:** 339.

Wong, Y.H., A. Federman, A.M. Pace, I. Zachary, T. Evans, J. Pouysségur, and H.R. Bourne. 1991. Mutant α subunits of G_{i2} inhibit cyclic AMP accumulation. *Nature* **351:** 63.

Thrombin Receptor Structure and Function

S.R. Coughlin,*† R.M. Scarborough,§ T.-K.H. Vu,* and D.T. Hung*‡
*Cardiovascular Research Institute, and Departments of †Medicine and ‡Laboratory Medicine, University of California, San Francisco, California 94143-0524; §COR Therapeutics, Inc., South San Francisco, California 94080

How do proteases that interact with the cell surface cause intracellular signaling events? One answer to this question was provided by the recent cloning and characterization of a functional thrombin receptor reviewed herein (Liu et al. 1991; Rasmussen et al. 1991; Vu et al. 1991a,b; Coughlin et al. 1992; Hung et al. 1992a,b,c,d; Scarborough et al. 1992).

Thrombin is a multifunctional serine protease generated at sites of vascular injury. Its varied actions on both soluble protein substrates and cells may be viewed as orchestrating not only hemostatic responses, but also perhaps inflammatory and proliferative responses to wounding. In addition to cleaving fibrinogen to the fibrin monomer to form the fibrous meshwork of blood clots, thrombin is a potent activator of a number of cell types. First and foremost, thrombin is the most potent stimulator of platelet aggregation (Davey and Luscher 1967; Shuman 1986), a thrombin activity probably critical in arterial thrombosis in vivo (Hanson and Harker 1988; Eidt et al. 1989; Fitzgerald and Fitzgerald 1989; Heras et al. 1989; Jang et al. 1989). A variety of thrombin actions on other cells types have been defined largely in vitro. Thrombin is chemotactic for monocytes (Bar-Shavit et al. 1983) and mitogenic for lymphocytes and for mesenchymal cells, including vascular smooth muscle cells (Chen and Buchanan 1975; Chen et al. 1976; McNamara et al. 1992), and it has a number of effects on the vascular endothelium. These include stimulating endothelial production of prostacyclin (Weksler et al. 1978), platelet-activating factor (Prescott et al. 1984), plasminogen activator-inhibitor (Gelehrter and Sznycer-Laszyk 1986), and the potent smooth muscle cell mitogen platelet-derived growth factor (Daniel et al. 1986). Thrombin also induces neutrophil adherence to the vessel wall (Zimmerman et al. 1986), probably by stimulating the expression of P-selectin on the endothelial surface (Hattori et al. 1989). Whether these disparate actions of thrombin are important mediators of inflammatory and proliferative responses in vivo remains to be defined. Identification of the receptor responsible for thrombin activation of platelets and other cells held the possibility of providing reagents for addressing such issues and of providing a new target for antithrombotic therapy and possibly antiproliferative or anti-inflammatory therapies.

Theories of possible mechanisms by which thrombin might activate platelets and other cells have included classical receptor occupancy, proteolytic cleavage of a receptor, and combinations thereof (Berndt et al. 1985; Shuman 1986). Because thrombin is a protease, identification of a functional thrombin receptor held out the possibility of identifying a novel mechanism of receptor activation.

Identification of the Thrombin Receptor by Expression Cloning

Traditional ligand-binding approaches to identify a functional thrombin receptor identified proteins that bind thrombin but have no apparent signaling function (Okamura et al. 1978; Gronke et al. 1987). In fact, because modified thrombins that lack both agonist and antagonist activities do bind to platelets in a manner indistinguishable from that of wild-type thrombin (Davey and Luscher 1967; Phillips 1974; Tollefsen et al. 1974; Martin et al. 1975; Workman et al. 1977), it has not been possible to conclude that the sites identified in binding studies are related to the functional thrombin receptor. (In retrospect, it is possible that classical binding approaches using native thrombin might never have succeeded because thrombin does cleave its receptor and, in doing so, removes one of the domains that mediate thrombin binding, a kind of Heisenberg uncertainty principle in action.) Because of the undefined nature of thrombin-receptor interaction, we adopted an approach that followed thrombin-induced responses. Expression cloning in Xenopus oocytes expressing exogenous mRNA or cRNA was employed to isolate a functional human thrombin receptor cDNA (Vu et al. 1991a). This strategy was based on studies by Nakanishi and colleagues who successfully used expression cloning in *Xenopus* oocytes to isolate a cDNA for the substance K receptor (Masu et al. 1987). The amino acid sequence deduced from the thrombin receptor cDNA revealed a novel member of the seven-transmembrane domain receptor family. Closest relatives included receptors for small peptides, substance P, and substance K (Vu et al. 1991a), suggesting that the thrombin receptor is peptide-receptor-like. This analogy was a useful hint at the outset of the structure/function work described below.

Thrombin Receptor Is Activated by a Novel Proteolytic Mechanism

Inspection of the receptor's amino acid sequence was revealing in light of a large body of literature on thrombin action. Thrombin cleaves peptides after arginines

(Muszbek and Laki 1984). We therefore examined sequences surrounding extracellular arginines in the receptor sequence and noted that the receptor's relatively long (100-residue) extracellular amino-terminal extension contained the putative thrombin cleavage site LDPR/S (single-letter amino acid code; with a slash / representing the point of cleavage after Arg-41 [R41] in the receptor; Figs. 1 and 2). This site resembled the known thrombin cleavage site found in the thrombin-activated zymogen protein C (LDPR/I) (Stenflo and Fernlund 1982). Carboxyl to this putative cleavage site lies a receptor domain resembling the carboxyl tail of the polypeptide hirudin, a leech-derived anticoagulant that binds thrombin with remarkable avidity. The carboxyl tail of hirudin is known to bind to the portion of thrombin's extended substrate recognition surface that recognizes residues carboxyl to the substrate's cleavage site (Figs. 1 and 2) (Jakubowski and Maragonore 1990; Rydel et al. 1990). Thus, the presence of a hirudin-like sequence carboxyl to the putative thrombin cleavage site within the thrombin receptor's amino-terminal extension suggested a model for thrombin-receptor interaction and implied that receptor proteolysis at the putative R41/S42 cleavage site might be important in receptor activation (Fig. 2).

We evaluated the importance of the R41/S42 site by changing the arginines in the receptor's amino-terminal extension to alanines, a maneuver designed to render each site uncleavable by thrombin. These mutant receptors were expressed in *Xenopus* oocytes, and their ability to respond to thrombin was assessed. The putative cleavage site mutant (R41A) could not be activated by thrombin, whereas the other mutants, R46A and R70A, were fully activatable. As an independent means of rendering the putative thrombin cleavage site uncleavable, Ser-42 at the site's P1′ position was changed to a proline. Like the R41A mutant, the S42P mutant receptor could not be activated by thrombin. These data strongly suggested that the receptor's putative thrombin cleavage site was critical for thrombin-induced receptor action (Vu et al. 1991a).

```
                         AGONIST          ANION-BINDING EXOSITE
               CLEAVAGE  PEPTIDE          BINDING
                   SITE  DOMAIN           DOMAIN

HUMAN TR38-60  ... LDPR/SFLLRN/PND-K/YEPFWE-DEE..
MOUSE TR38-62  ... VNPR/SFFLRN/PSENT/FELVPLGDEE..
                                        || |
HIRUDIN 55-65                      ...DFEEIPEEY'LQ-COO⁻
```

Figure 1. Functional domains within the thrombin receptor's extracellular amino-terminal extension. The cleavage recognition sequence (LDPR), thrombin cleavage site, agonist peptide domain, and anion-binding exosite binding domain (hirudin-like domain) within the human thrombin receptor are shown (standard single letter amino acid code). These are aligned with the murine thrombin receptor sequence and the anion-binding exosite binding sequence in hirudin's carboxyl tail. Specific residues important for thrombin-receptor interaction (Vu et al. 1991b) align with hirudin carboxyl tail residues that bind thrombin's anion-binding exosite (Rydel et al. 1990). A model for the interaction of these domains with thrombin is shown in Fig. 2.

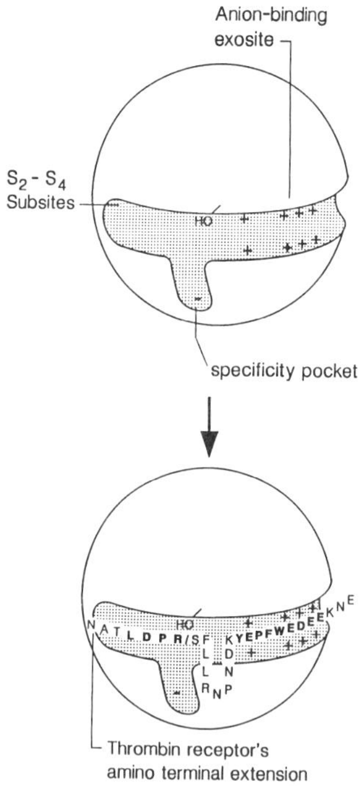

Figure 2. Model for thrombin-receptor interaction. Thrombin has an extended substrate-binding surface (represented by the dark canyon running laterally) that recognizes residues both amino and carboxyl to its substrate's cleavage site (a substrate's P1 arginine side chain would sit in thrombin's specificity pocket cartooned as a vertical cleft). Structure/function studies (Liu et al. 1991; Vu et al. 1991b) suggest that the receptor's hirudin-like domain (YEPFWEDEE) binds thrombin's anion-binding exosite. Y52, E53, and F55 appear to be important in this regard (Vu et al. 1991b). The receptor's cleavage site (LDPR/S) interacts with thrombin's S1–S4 subsites. Receptor-blocking antibodies (Hung et al. 1992c) and receptor-peptide-based thrombin inhibitors (Hung et al. 1992c) have been produced based on this model. (Similar figures were published in Vu et al. 1991b.)

How might proteolysis at this site lead to receptor activation? The favored hypothesis was that proteolysis at this site would unmask a new amino terminus that might serve as a peptide ligand for the receptor (Vu et al. 1991a). Precedent for proteolytic unmasking of a "ligand" existed in trypsinogen activation to trypsin (Bode et al. 1978), discussed in more detail below. Indeed, a peptide mimicking the new receptor amino terminus that would be revealed by receptor cleavage at the R41 position was a full agonist for the cloned receptor expressed in oocytes. Not only could the wild-type receptor be activated, but the R41A receptor, unactivatable by thrombin, responded fully to the new amino terminus "agonist peptide" as well. Thus, the agonist peptide could bypass the requirement for proteolysis and activate the receptor directly (Vu et al. 1991a).

The studies described above suggest that thrombin receptor activation proceeds by a novel mechanism depicted in Figure 3. Thrombin cleaves its receptor after Arg-41 in the receptor's amino-terminal extension, exposing a new amino terminus that functions as a

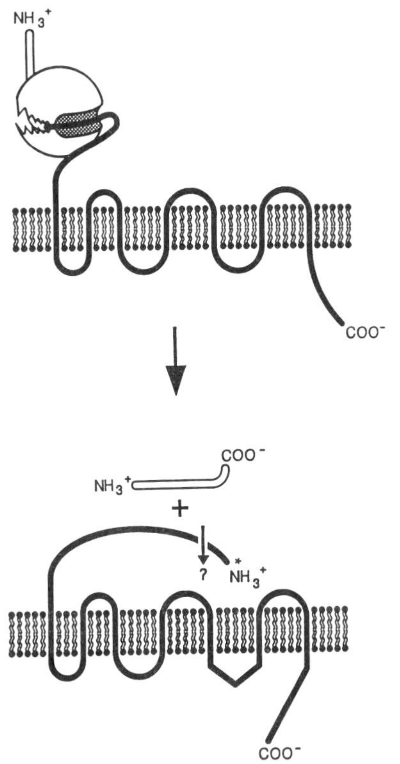

Figure 3. Model of thrombin receptor activation. Thrombin binds its receptor via the receptor's amino-terminal extension as detailed in Fig. 2; whether additional receptor domains may participate in thrombin binding is unknown. After binding to the amino-terminal extension, thrombin cleaves the receptor at the LDPR/S cleavage site (junction between open and filled receptor segments), releasing an inactive fragment of the receptor's amino terminus (open fragment) and exposing a new amino terminus. This newly unmasked amino terminus then functions as a tethered peptide ligand, binding to an as yet undefined pocket (small arrow and ?) to effect receptor activation (Vu et al. 1991a). Recent studies by Scarborough et al. (1992) suggest that structures at the extreme amino terminus (asterisk at the NH^{3+} on filled segment) of the receptor's agonist peptide domain are critical for its agonist function (see Fig. 4). (Similar figures were published in Vu et al. 1991b.)

tethered peptide ligand for the receptor. The new amino terminus then binds to an as yet undefined receptor site, effecting receptor activation (Vu et al. 1991a). This model predicted that receptor proteolysis was sufficient for receptor activation and that the thrombin cleavage recognition sequence would be the primary determinant of receptor specificity. Accordingly, replacing the receptor's thrombin cleavage site (LDPR/S) with that for enterokinase (DDDDK/S) completely switched the receptor's specificity, creating a functional "enterokinase receptor" that was fully activatable by enterokinase but unresponsive to thrombin (Vu et al. 1991b). These results strongly suggest that receptor proteolysis that unmasks the amino terminus beginning at position 42 (SFLL...) is sufficient for receptor activation (Vu et al. 1991b). No special "hormone-like" properties of thrombin need to be invoked for activation of this receptor.

Additional studies have shown the receptor's hirudin-like sequence can indeed bind thrombin's anion-binding exosite (Liu et al. 1991; Vu et al. 1991b). Detailed mutagenesis studies have defined the specific residues important in this interaction and strengthened the analogy between the receptor's anion-binding exosite binding domain and hirudin's carboxyl tail (Vu et al. 1991b). These relationships are illustrated in Figures 1 and 2.

Structure Activity Relationships for the Receptor's Agonist Peptide Domain Reveal a Mechanism for the Receptor's Proteolytic Switch

The unusual mechanism by which the thrombin receptor is activated demands that the function of the receptor's agonist peptide domain remain cryptic in the uncleaved receptor but be expressed upon receptor cleavage. Recent studies that define the structural requirements for agonist peptide function suggest a possible mechanism for this "proteolytic switch" (Scarborough et al. 1992). The five amino-terminal residues of the agonist peptide domain (SFLLR) are sufficient to specify agonist activity (Scarborough et al. 1992; Vassallo et al. 1992; Vouret-Craviari et al. 1992). Within this peptide, the protonated amino group at its extreme amino terminus (Scarborough et al. 1992) and the phenyl side chain of Phe-2 are vital for agonist function (Fig. 4) (Scarborough et al. 1992; Vassallo et al. 1992). Leu-4 and Arg-5 play lesser roles. Removing the agonist peptide's protonated amino group nearly abolished agonist function, and masking the agonist peptide's protonated amino group by acetylating its amino terminus or extending it to mimic the receptor's uncleaved state (i.e., LDPRSFLLR...) completely abolished agonist function. These studies suggested that the protonated amino group participates in specific interactions that mediate agonist peptide binding and receptor activation and that the agonist function of the receptor's agonist peptide domain changes from its cryptic state in the uncleaved receptor to its active state in the cleaved receptor in part by creation of the agonist peptide's protonated amino group upon receptor proteolysis (Scarborough et al. 1992). The analogy made above with trypsinogen activation may be extended to generate hypotheses regarding the nature of agonist peptide interaction with the receptor. The new amino-terminal protonated amino group unmasked upon trypsinogen cleavage to trypsin effects trypsin activation in part by forming an intramolecular salt bridge with the carboxylate side chain of trypsin residue Asp-194 (Bode et al. 1978). The protonated amino group of catecholamines causes $\beta 2$ adrenergic receptor activation in part by interacting with the carboxylate of Asp-113 within the $\beta 2$ receptor's third transmembrane domain (Strader et al. 1991) . Whether analogous mechanisms are utilized by the thrombin receptor agonist peptide domain's amino-terminal protonated amino group to effect thrombin receptor activation remains to be explored.

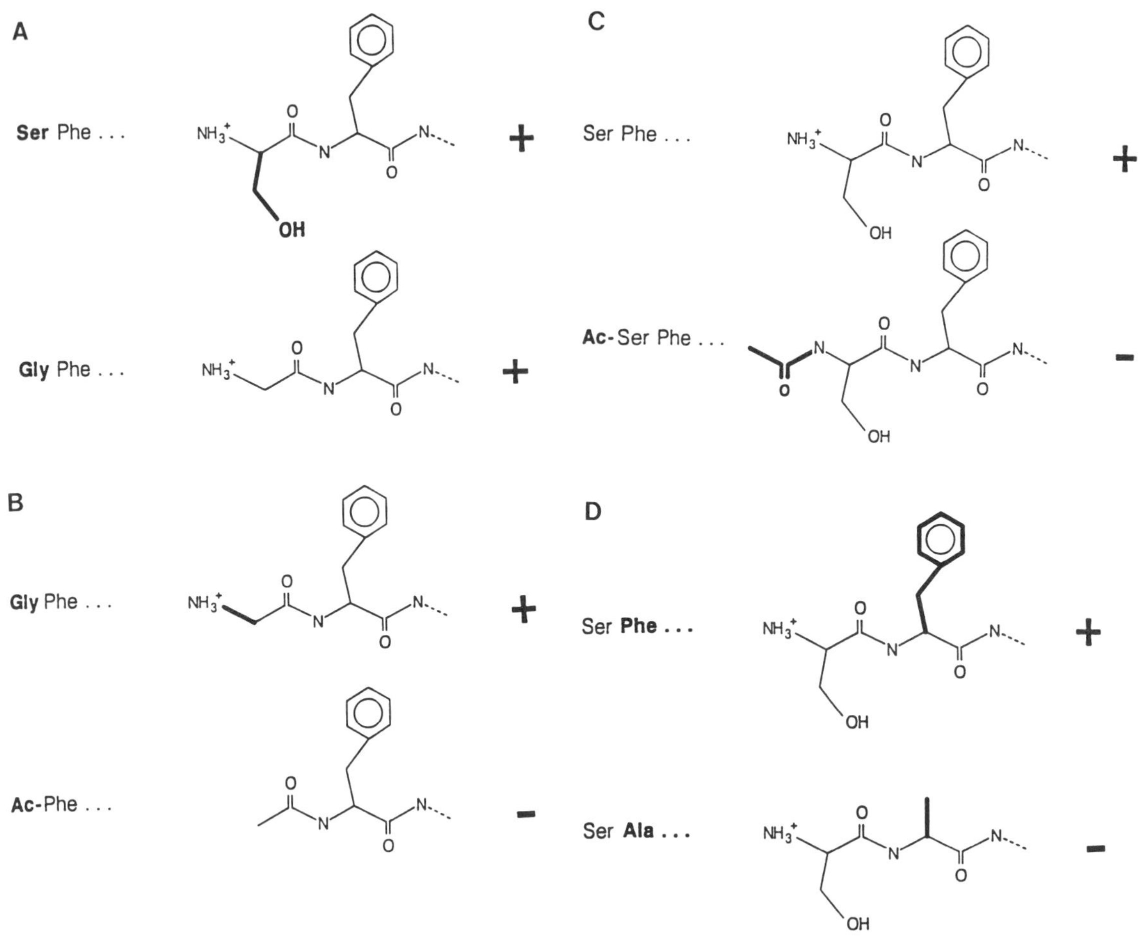

Figure 4. Structural requirements for agonist peptide function suggest a model for the proteolytic switch. Structures at the amino-terminal end of the thrombin receptor's agonist peptide domain (SFLLRN) are critical for its agonist function (Scarborough et al. 1992). In particular, the protonated amino group of Ser-1 created by receptor cleavage was vital for agonist function, as was the side chain of Phe-2. These conclusions are based on structure/function studies using peptides representing variations of the receptor's agonist peptide domain SFLLRN.., as cartooned in the figure. (+) A functional agonist peptide; (−) peptide with markedly reduced or absent agonist function. Substituting a glycine for Ser-1 had little effect on agonist function, suggesting the side chain of residue 1 is unimportant (*A*). In contrast, substituting an acetyl group for residue 1 (in essence removing the protonated amino group present in the glycine construct) markedly reduced agonist function (*B*). Blocking the amino-terminal protonated amino group also abolished agonist function (*C*), as did extending the amino terminus with the native receptor sequence (e.g., LDPRSFLLRN...). Substituting an alanine for Phe-2 abolished agonist function (*D*). No other individual residues within the agonist peptide SFLLRN were critical; Leu-4 and Arg-5 play relatively important roles. These data suggest that the protonated amino group contributed by Ser-1 functions in the context of Phe-2's phenyl side chain and other agonist determinants to effect receptor binding and activation. Because the protonated amino group is created by receptor cleavage, this model provides a mechanism for the agonist peptide to remain cryptic in the uncleaved receptor and to be unmasked by receptor proteolysis.

Unanswered Questions

Many questions surrounding the cellular and structural biology of the thrombin receptor remain unanswered. Where does the agonist peptide domain unmasked by receptor proteolysis bind and how does this binding event lead to a transmembrane signal that allows the receptor to talk to intracellular signaling molecules? Which signaling pathways does the cloned thrombin receptor activate? Thrombin is known to both activate phosphoinositide hydrolysis and inhibit adenylyl cyclase in responsive cells. Recent transfection studies with the cloned receptor suggest that the cloned thrombin receptor is able to speak to both pathways, probably via two distinct G proteins (Hung et al. 1992b). Confirmation of these findings by reconstitution experiments remains to be accomplished, as does examining the role of the cloned receptor in other signaling events.

The unique proteolytic mechanism by which the thrombin receptor becomes activated produces a ligand that is tethered to the receptor and cannot diffuse away. How is the receptor turned off? The agonist peptide can desensitize the cloned thrombin receptor in transfectants (D.T. Hung and S.R. Coughlin, unpubl.) and the native thrombin receptor in HEL cells (Brass

1992); thus, at least one specific mechanism independent of proteolysis by thrombin exists to shut off this receptor. By analogy with other seven-transmembrane receptors, it is probable that several mechanisms including receptor phosphorylation by a β adrenergic receptor kinase-like enzyme (Lohse et al. 1989) and by other protein kinases regulate thrombin receptor desensitization.

The unique proteolytic mechanism of thrombin receptor activation also recalls a long-standing question in thrombin biology. How does one get a concentration-dependent response if thrombin is acting as an enzyme to activate its receptor? That is, should not all receptors ultimately be cleaved and activated even at low thrombin concentrations, such that a response would always be maximal? Regardless of whether we can adequately explain it, the cloned receptor does confer concentration-dependent responses to transfected cells (Hung et al. 1991a). In fact, the designed enterokinase receptor confers enterokinase-dependent reponses to transfected cells (Hung et al. 1991b). The latter suggests that the ability of the receptor to confer concentration-dependent responses is "downstream" from receptor proteolysis and activation. The thrombin receptor is rapidly desensitized by thrombin, and thus part of the answer to this question may be related to a balance between the rate of thrombin receptor activation and the rate of desensitization.

The thrombin receptor is in essence a peptide receptor that contains its own ligand. Where did this receptor and its unusual activation mechanism come from? Is it an evolutionary specialization of a previous peptide receptor? Or is it representative of a more primitive heritage from which other peptide receptors and ligands may have evolved? Will this clone sire a family of structurally related thrombin or protease receptors? Or is this general mechanism used by more distantly related receptors for other proteases? How many of the assorted cell-activation functions of thrombin does the cloned receptor account for? The agonist peptide described above, which activates the cloned thrombin receptor directly and independent of thrombin and its protease activity, will be useful in this regard. Specifically, thrombin responses mimicked by the agonist peptide are likely mediated by the unusual mechanism described above via the cloned receptor or a highly related receptor. Thrombin responses not mimicked by the agonist peptide may be mediated by a different mechanism. These and a host of other basic questions remain to be answered.

ACKNOWLEDGMENTS

This work was supported in part by National Institutes of Health grants HL-44907 and HL-43821 and by University of California Tobacco-related Disease Research Program grant no. 2RT-19 and was done during the tenure of an award to S.R.C. from the American Heart Association and SmithKline Beecham.

REFERENCES

Bar-Shavit, R., A. Kahn, G.D. Wilner, and J.W. Fenton II. 1983. Monocyte chemotaxis: Stimulation by specific exosite region in thrombin. *Science* **220:** 728.

Berndt, M.C., C. Gregory, G. Dowden, and P.A. Castaldi. 1985. Thrombin interactions with platelet membrane proteins. *Ann. N.Y. Acad. Sci.* **485:** 374.

Bode, W., P. Schwager, and R. Huber. 1978. The transition of bovine trypsinogen to a trypsin-like state upon strong ligand binding. *J. Mol. Biol.* **118:** 99.

Brass, L.F. 1992. Homologous desensitization of HEL cell thrombin receptors. *J. Biol. Chem.* **267:** 6044.

Chen, L.B. and J.M. Buchanan. 1975. Mitogenic activity of blood components. I. Thrombin and prothrombin. *Proc. Natl. Acad. Sci.* **72:** 131.

Chen, L.B., N.N.H. Teng, and J.M. Buchanan. 1976. Mitogenicity of thrombin and surface alterations on mouse splenocytes. *Exp. Cell Res.* **101:** 41.

Coughlin, S.R., T.-K.H. Vu, D.T. Hung, and V.I. Wheaton. 1992. Perspectives: Characterization of the cloned platelet thrombin receptor: Issues and opportunities. *J. Clin. Invest.* **89:** 351.

Daniel, T.O., V.C. Gibbs, D.F. Milfay, M.R. Garavoy, and L.T. Williams. 1986. Thrombin stimulates c-sis gene expression in microvascular endothelial cells. *J. Biol. Chem.* **261:** 9579.

Davey, M.G. and E.F. Luscher. 1967. Actions of thrombin and other coagulant and proteolytic enzymes on blood platelets. *Nature* **216:** 857.

Eidt, J.F., P. Allison, S. Nobel, J. Ashton, P. Golino, J. McNatt, L.M. Buja, and J.T. Willerson. 1989. Thrombin is an important mediator of platelet aggregation in stenosed canine coronary arteries with endothelial injury. *J. Clin. Invest.* **84:** 18.

Fitzgerald, D.J. and G.A. Fitzgerald. 1989. Role of thrombin and thromboxane A2 in reocclusion following coronary thrombolysis. *Proc. Natl. Acad. Sci.* **86:** 7585.

Gelehrter, T.D. and R. Sznycer-Laszyk. 1986. Thrombin induction of plasminogen activator-inhibitor in cultured human endothelial cells. *J. Clin. Invest.* **77:** 165.

Gronke, R.S., B.L. Bergman, and J.B. Baker. 1987. Thrombin interaction with platelets: Influence of a platelet protease nexin. *J. Biol. Chem.* **262:** 3030.

Hanson, S.R. and L.A. Harker. 1988. Interruption of acute platelet-dependent thrombosis by the synthetic antithrombin PPACK. *Proc. Natl. Acad. Sci.* **85:** 3184.

Hattori, R., K.K. Hamilton, R.D. Fugate, R.P. McEver, and P.J. Sims. 1989. Stimulated secretion of endothelial vWF is accompanied by rapid redistribution to the cell surface of the intracellular granule membrane protein GMP-140. *J. Biol. Chem.* **264:** 7768.

Heras, M., J.H. Chesebro, W.J. Penny, K.R. Bailey, L. Badimon, and V. Fuster. 1989. Effects of thrombin inhibition on the development of acute platelet-thrombus deposition during angioplasty in pigs. *Circulation* **79:** 657.

Hung, D.T., T.-K.H. Vu, N.A. Nelken, and S.R.Coughlin. 1992a. Thrombin-induced events in non-platelet cells are mediated by the novel proteolytic mechanism defined by the cloned platelet thrombin receptor. *J. Cell Biol.* **116:** 827.

Hung, D.T., T.-K.H. Vu, Y. Wong, and S.R. Coughlin. 1992b. The cloned platelet thrombin receptor couples to at least two distinct effectors to stimulate both phosphoinositide hydrolysis and inhibit adenylyl cyclase. *J. Biol. Chem.* (in press).

Hung, D.T., T.-K.H. Vu, V.I. Wheaton, K. Ishii, and S.R. Coughlin. 1992c. The cloned platelet thrombin receptor is necessary for thrombin-induced platelet activation. Blocking antiserum to the thrombin receptor's hirudin-like domain. *J. Clin. Invest.* **89:** 1350.

Hung, D.T., T.-K.H. Vu, V.I. Wheaton, I.F. Charo, N.A.

Nelken, N. Esmon, C.T. Esmon, and S.R. Coughlin. 1992d. "Mirror image" antagonists of thrombin-induced platelet activation based on thrombin receptor structure. *J. Clin. Invest.* **89:** 444.

Jakubowski, J.A. and J.M. Maragonore. 1990. Inhibition of coagulation and thrombin-induced platelet activities by a synthetic dodecapeptide modeled on the carboxy-terminus of hirudin. *Blood* **75:** 399.

Jang, I.-K., H.K. Gold, A.A. Ziskind, R.C. Leinbach, J.T. Fallon, and D. Collen. 1989. Prevention of platelet-rich arterial thrombosis by selective thrombin inhibition. *Circulation* **81:** 219.

Liu, L., T.-K.H. Vu, C.T. Esmon, and S.R. Coughlin. 1991. The region of the thrombin receptor resembling hirudin binds to thrombin and alters enzyme specificity. *J. Biol. Chem.* **266:** 16977.

Lohse, M.J., R.J. Lefkowitz, M.G. Caron, and J.L. Benovic. 1989. Inhibition of β-adrenergic receptor kinase prevents rapid homologous desensitization of β2-adrenergic receptors. *Proc. Natl. Acad. Sci.* **86:** 3011.

Martin, B.M., R.D. Feinman, and T.C. Detwiler. 1975. Platelet stimulation by thrombin and other proteases. *Biochemistry* **14:** 1308.

Masu, Y., K. Nakayama, H. Tamaki, Y. Harada, M. Kuno, and S. Nakanishi. 1987. cDNA cloning of bovine substance K receptor through oocyte expression system. *Nature* **329:** 836.

McNamara, C.A., I.J. Sarembok, L.W. Gimple, J.W. Fenton, S.R. Coughlin, and G.K. Owens. 1992. Thrombin stimulation of smooth muscle cell proliferation is mediated by a prolteolytic, receptor-mediated mechanism. *J. Clin Invest.* (in press).

Muszbek, L. and K. Laki. 1984. Interaction of thrombin with proteins other than fibrinogen (thrombin-susceptible bonds). In *The thrombin* (ed. R. Machovich), p. 83. CRC Press, Boca Raton, Florida.

Okamura, T., M. Hasitz, and G.A. Jamieson. 1978. Platelet glycocalicin: Interaction with thrombin and role as thrombin receptor on the platelet surface. *J. Biol. Chem.* **253:** 3435.

Phillips, D.R. 1974. Thrombin interaction with human platelets: Potentiation of thrombin-induced aggregation and release by inactivated thrombin. *Thromb. Diath. Haemorrh.* **32:** 207.

Prescott, S.M., G.A. Zimmerman, and T.M. McIntyre. 1984. Human endothelial cells in culture produce platelet-activating factor when stimulated by thrombin. *Proc. Natl. Acad. Sci.* **81:** 3534.

Rasmussen, U.B., V. Vouret-Craviari, S. Jallet, Y. Schlesinger, G. Pages, A. Pavirani, J.-P. Lecocq, J. Pouysségur, and E. Van Obberghen-Schilling. 1991. cDNA cloning and expression of a hamster α-thrombin receptor coupled to Ca^{2+} mobilization. *FEBS Lett.* **288:** 123.

Rydel, T.J., K.G. Rabichandran, A. Tulinsky, W. Bode, R. Huber, C. Roitsch, and J.W. Fenton II. 1990. The structure of a complex of recombinant hirudin and human alpha-thrombin. *Science* **249:** 277.

Scarborough, R.M., M. Naughton, W. Teng, D.T. Hung, J. Rose, T.-K.H. Vu, V.I. Wheaton, C.W. Turck, and S.R. Coughlin. 1992. Tethered ligand agonist peptides: Structural requirements for thrombin receptor activation reveal mechanism of proteolytic unmasking of agonist function. *J. Biol. Chem.* **267:** 13146.

Shuman, M.A. 1986. Thrombin-cellular interactions. In *Ann. N.Y. Acad. Sci.* **485:** 228.

Stenflo, J. and P. Fernlund. 1982. Amino acid sequence of the heavy chain of bovine protein C. *J. Biol. Chem.* **257:** 12180.

Strader, C.D., T. Gaffney, E.E. Sugg, M.R. Candelore, R. Keys, A.A. Patchett, and R.A.F. Dixon. 1991. Allele-specific activation of genetically engineered receptors. *J. Biol. Chem.* **266:** 5.

Tollefsen, D.M., J.R. Feagler, and P.W. Majerus. 1974. The binding of thrombin to the surface of platelets. *J. Biol. Chem.* **249:** 2646.

Vassallo, R.R., T. Kiever-Emmons, K. Cichowski, and L.F. Brass. 1992. Structure-function relationships in the activation of platelet thrombin receptors by receptor-derived peptides. *J. Biol. Chem.* **267:** 6081.

Vouret-Craviari, V., E. Van Obberghen-Schilling, U.B. Rasmussen, A. Pavirani, J.-P. Lecocq, and J. Pouysségur. 1992. Synthetic thrombin receptor peptides activate G-protein-coupled signalling pathways but are unable to induce mitogenesis. *Mol. Biol. Cell.* **3:** 95.

Vu T.-K.H., D.T. Hung, V.I. Wheaton, and S.R. Coughlin. 1991a. Molecular cloning of a functional thrombin receptor reveals a novel proteolytic mechanism of receptor activation. *Cell* **64:** 1057.

Vu, T.-K.H., V.I. Wheaton, D.T. Hung, and S.R. Coughlin. 1991b. Domains specifying thrombin-receptor interaction. *Nature* **353:** 674.

Weksler, B.B., C.W. Ley, and E.A. Jaffe. 1978. Stimulation of endothelial cell prostacyclin production by thrombin, trypsin, and the ionophore A23187. *J. Clin. Invest.* **62:** 923.

Workman, E.F., Jr., G.C. White II, and R.L. Lundblad. 1977. Structure-function relationships in the interaction of alpha-thrombin with blood platelets. *J. Biol. Chem.* **252:** 7118.

Zimmerman, G.A., T.M. McIntyre, and S.M. Prescott. 1986. Thrombin stimulates neuthrophil adherence by an endothelial cell-dependent mechanism. *Ann. N.Y. Acad. Sci.* **485:** 349.

Guanylyl Cyclases: Ligands and Functions

T.D. Chrisman, S. Schulz, and D.L. Garbers
Howard Hughes Medical Institute, Department of Pharmacology, University of Texas Southwestern Medical Center, Dallas, Texas 75235

cGMP was discovered as a component of urine in 1963 (Ashman et al. 1963). Subsequently, guanylyl cyclases (Chinkers and Garbers 1991), responsible for formation of cGMP; phosphodiesterases (Beavo 1988), responsible for degradation of cGMP; specific effectors of guanylyl cyclases and phosphodiesterases (Yuen and Garbers 1992); and molecular targets for cGMP action (Stryer 1986; Edelman et al. 1987; Beavo 1988) were identified. cGMP is clearly accepted as a second messenger, even though its impact on cell behavior has not been elucidated in most cases.

We discuss here guanylyl cyclases, in particular those associated with the plasma membrane. These enzymes appear to span the membrane once and to serve as receptors for specific peptide ligands. Three members of this family have been cloned by our laboratory and others from mammalian tissues (Chang et al. 1989; Chinkers et al. 1989; Lowe et al. 1989; Schulz et al. 1989, 1990). Named in order of cloning, they are designated guanylyl cyclase-A (GC-A), guanylyl cyclase-B (GC-B), and guanylyl cyclase-C (GC-C); a model showing general features of their structure is given in Figure 1. Each appears to contain a single transmembrane domain and three other recognizable regions based on primary amino acid sequence. The amino-terminal extracellular domain (domain I) confers ligand selectivity and is the most variable among the receptor subtypes. Domain II consists of approximately 20 amino acid residues based on hydropathy plots and is the transmembrane segment. The cytoplasmic portion of the receptor represents about half of the total protein and has been divided into two domains on the basis of amino acid sequence and function. Domain III is homologous to the conserved catalytic domains of protein kinases and is more closely related to the protein tyrosine kinases than to the protein serine kinases (Chinkers and Garbers 1991); there is as yet no evidence that these receptor-cyclases possess protein kinase activity. Insight into the putative regulatory role of this domain comes from studies with GC-A showing that ATP is required for atrial natriuretic peptide (ANP) stimulation of the enzyme (Kurose et al. 1987; Song et al. 1988; Chinkers et al. 1991) and of domain-III-deleted forms of the enzyme that appear constitutively active, no longer requiring ATP or ANP for activity (Chinkers and Garbers 1989). Koller et al. (1992) have shown that partial deletion (as opposed to complete deletion) of domain III in GC-A or GC-B does not result in constitutive activation but does prevent peptide-ligand stimulation of the enzyme. Goraczniak et al. (1992) have reported that mutations within the sequence G^{503}-R-G-S-N-Y-G^{509}, a region similar to the conserved G-X-G-X-X-G motif of protein kinases, is the site of ATP binding and regulation by the nucleotide. Domain IV at the carboxyl terminus is the guanylyl cyclase catalytic domain as shown recently by bacterial expression of this region of GC-A (Thorpe and Morkin 1990). Prior to this, sequence analysis had shown this domain to be conserved in both subunits of heterodimeric forms of guanylyl cyclase (Chinkers and Garbers 1991) and in two internally homologous domains of bovine brain adenylyl cyclase (Krupinski et al. 1989).

METHODS

Porcine seminal plasma was provided by T. Berger (University of California at Davis). The sperm-poor

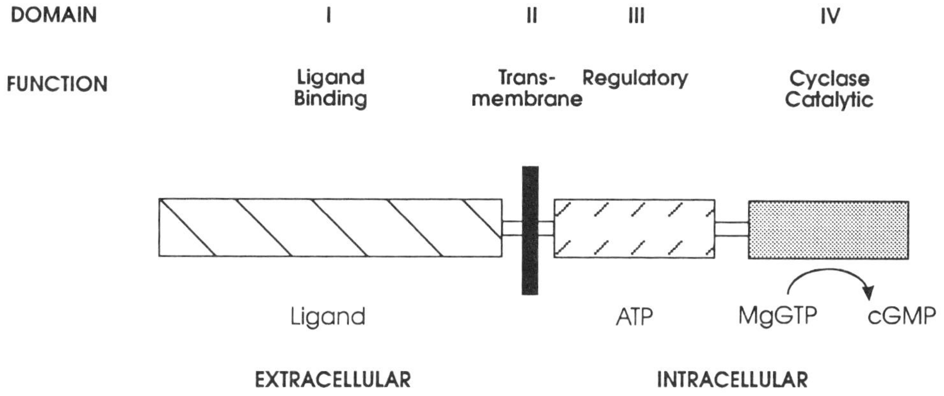

Figure 1. Topological model of guanylyl-cyclase-linked cell-surface receptors.

fraction was heated at 97°C for 15 minutes and clarified by centrifugation at 30,000*g* for 30 minutes. The supernatant fluid was fractionated by reversed-phase chromatography on a preparative C_{18} column (Vydac).

Bovine tracheas, obtained from a local slaughterhouse, were split longitudinally, and the mucosal surfaces were gently scraped with a spatula. Mucosal scrapings were resuspended in ice-cold phosphate-buffered saline (PBS), and clumps of cells were dispersed through a 10-ml pipet. The material was pelleted at 500*g* for 5 minutes (4°C) and then resuspended in ice-cold PBS. Light microscopy showed that the pellet consisted of 70–80% ciliated cells and 15–30% nonciliated cells. Following a wash in PBS, the pellet was resuspended in 9-volume Dulbecco's modified Eagle's medium (DMEM), 20 μg leupeptin and aprotinin, and 0.1 mM isobutyl methylxanthine (37°C, 95% O_2/5% CO_2). Aliquots (1 ml) of the cell suspension were incubated for 10 minutes at 37°C with shaking, and 10 μl of PBS or heat-stable enterotoxin (ST) in PBS was then added. The cells were incubated for an additional 10 minutes and 1 ml of 1 N perchloric acid (PCA) was then added. cGMP content was analyzed as described previously (Schulz et al. 1990).

Rat seminal vesicles were dissected free of connective and surrounding tissue, minced, and placed in cold PBS for 30 minutes. Tissue was removed, and the fluid was acidified by adding acetic acid to a final concentration of 1 N, heated at 97°C for 15 minutes, and centrifuged at 30,000*g* for 30 minutes. The supernatant fluid was applied to a PrepSep C_{18} column (Fisher Scientific), and the column was washed with 5 ml of 0.1% trifluoroacetic acid (TFA). Bound material was eluted with 2 ml of 60% acetonitrile/0.1% TFA and dried under vacuum. The dried material was dissolved in 0.1% TFA and fractionated as described in Figure 3.

Human embryonic 293 cells expressing rat GC-B were prepared by cloning the full-length cDNA for rat GC-B (Schulz et al. 1989) into the mammalian expression vector pRc/CMV (Invitrogen) to produce pRc-GC-B. The 293 cells were transfected at 50% confluency. Crude membranes were prepared from these cells as described recently (Potter and Garbers 1992). ^{125}I YCNP (iodination of YCNP was done as described in Koller et al. 1991) binding to the membranes was carried out for 2 hours at 37°C in DMEM. Specific binding was considered that competed for by 10 μM CNP (C-type natriuretic peptide).

The NIH-3T3 (ATCC CRL 1658) and Balb/3T3 (clone A31, ATCC CCL 163) fibroblast cell lines were grown to confluency in six-well plates. The cells were incubated in 1 ml of DMEM (containing 0.1 mM isobutyl methylxanthine) in the absence or presence of the indicated volumes of boiled porcine seminal plasma or ANP or CNP. Incubations were carried out for 10 minutes, and cGMP contents were determined as described above.

Confluent GC-B/293 cells in 12-well plates were used for cell-stimulation studies. The cells were incubated in the presence of 0.25 mM isobutyl methylxanthine in the presence or absence of the indicated fractions for 10 minutes at 37°C. Reactions were stopped by adding an equal volume of 1 N PCA, and cGMP contents were determined as described above. Northern and polymerase chain reaction (PCR) analyses were carried out as described by Schulz et al. (1992). For Northern analyses of GC-B and GC-C, the cDNA encoding the extracellular domain of rat GC-B (Schulz et al. 1989) or rat GC-C (Schulz et al. 1990) was used as a probe.

MAMMALIAN GUANYLYL CYCLASES AND THEIR PEPTIDE LIGANDS

Three structurally similar peptides that promote natriuresis have been isolated from vertebrates. Two of these, atrial natriuretic peptide (A-type natriuretic peptide, ANP) and brain natriuretic peptide (B-type natriuretic peptide, BNP), are most abundant in heart tissue (Brenner et al. 1990), and the third, C-type natriuretic peptide (CNP), was recently discovered in brain tissue (Sudoh et al. 1990). All three have a highly conserved 17-member disulfide ring with variations between species and peptide forms occurring principally outside this ring. The peptides cause natriuresis, diuresis, and smooth muscle relaxation when injected into rats (Sudoh et al. 1988, 1990); ANP is the most potent and CNP is the least effective (Sudoh et al. 1990).

Three natriuretic peptide receptors have been identified in mammals and have been cloned and expressed (Fuller et al. 1988; Chang et al. 1989; Chinkers et al. 1989; Lowe et al., 1989; Schulz et al. 1989). GC-A, GC-B, and ANP clearance receptor (ANP-CR) contain homologous extracellular domains but vary in ligand selectivity (Chinkers and Garbers 1991). ANP-CR is unique among the three in several respects: It does not appear to discriminate between natriuretic peptides effectively, it is the most abundant, and it does not contain the cytoplasmic kinase-like (III) or cyclase catalytic (IV) domains, possessing instead a short (37-amino-acid residue) cytoplasmic domain (Scarborough et al. 1986, 1988; Fuller et al. 1988). It has been proposed to serve a clearance function for circulating natriuretic peptides (Maack et al. 1987) but may also couple to other transducing systems (Anand-Srivastava et al. 1990). GC-A and GC-B have been shown to be receptors selective for ANP and CNP, respectively, although at high micromolar concentrations, the ligands effectively activate either receptor (Koller et al. 1991; Ohyama et al. 1992).

GC-A and ANP

The tissue distribution and some known physiological roles of GC-A and ANP are the subjects of several recent reviews (Brenner et al. 1990; Chinkers and Garbers 1991; Yuen and Garbers 1992). GC-A and ANP appear to have major roles in the regulation of blood volume and therefore blood pressure: GC-A expression is particularly evident in those tissues involved in hemodynamic regulation, e.g., kidney and adrenal gland (Wilcox et al. 1991), and ANP is stored and released from atrial myocytes (Debold 1982; Brenner et al. 1990) in response to increased central volume. In addition to the endocrine role of ANP released from the atria, GC-A and ANP may be involved in the "local" regulation of specific tissue functions by autocrine/paracrine mechanisms. Ritter et al. (1991) showed that primary cultures of kidney cells synthesize and release ANP into the medium. These authors raised the possibility that GC-A/ANP may also function as an autocrine/paracrine system in the kidney.

Biochemical studies of GC-A in overexpressing cells have provided some insight into the regulation of the enzyme receptor by ANP. The work of Kurose et al. (1987) showed that ANP was more effective in stimulating crude membrane preparations of guanylyl cyclase in the presence of millimolar amounts of ATP and that this effect of ATP was not attributed to GTP sparing. Chinkers and Garbers (1989) using expressed wild-type and kinase-domain-deleted mutants of GC-A showed that the kinase domain was required for ANP stimulation of the cyclase since the mutant was constitutively active and there were no effects of ANP or ATP.

GC-B and CNP

There have been few studies of the tissue distribution of GC-B and CNP. Primarily on the basis of two reports, both GC-B and its ligand have been considered to be confined almost exclusively to the central nervous system (CNS) and to be involved in CNS function. In situ studies (Wilcox et al. 1991) suggested that GC-B is found principally in those tissues derived from neural crest cells, and a survey of immunoreactive CNP (ir-CNP) detected this peptide only in CNS (Arimura et al. 1990; Kojima et al. 1990), although a more recent study reported very low amounts of ir-CNP in rat small intestine (Komatsu et al. 1991). CNP has not been found in circulating form (Komatsu et al. 1991), suggesting that it functions as an autocrine/paracrine factor or as a neurotransmitter. Recent work in our laboratory shows that both GC-B and CNP are found in a variety of tissues outside the CNS and supports a concept that GC-B and CNP function as an autocrine and/or paracrine system.

In preliminary studies, we found that boiled porcine seminal plasma markedly increased cGMP levels in several cell lines, the most dramatic elevations (greater than 200-fold) being seen in the mouse Balb/3T3 cell line (Table 1). In later work, the 3T3 cells were shown to contain GC-B. Moreover, seminal plasma also sharply elevated cGMP in cells overexpressing GC-B but not in control cells (human embryonic kidney cells) or in cells overexpressing GC-A. The factor was found to be heat-stable and to compete with ^{125}I YCNP for specific binding to membranes prepared from the GC-B overexpressing cells. Fractionation of the seminal plasma by reversed-phase chromatography resolved multiple peaks of (binding) activity as shown (arrows) in Figure 2. Further purification of the activity eluting at approximately 30% acetonitrile (fractions 63–65), followed by amino acid sequence analysis, showed it to be identical to pro-CNP (103 amino acids) predicted from a rat brain cDNA clone as determined by Kojima et al. (1990), although the existence of this form of the peptide in brain has not yet been reported. Although the other peaks of activity have not been purified further, use of an immunoassay for CNP showed the same "activity" profile as that shown in Figure 2. Therefore, porcine seminal plasma appears to contain multiple forms of CNP, presumably reflecting differential or incomplete processing of pro-CNP. Minamino et al. (1990) have reported the existence of two forms of CNP in porcine brain: the 22-amino-acid form (CNP-

Table 1. Elevations of cGMP Levels in 3T3 Fibroblast Cell Lines by Porcine Seminal Plasma and Natriuretic Peptides

Variable	pmole cGMP/well	
Seminal plasma, μl/ml medium	NIH-3T3	Balb/3T3
0	0.015	0.02
0.5	0.03	0.14
1	0.07	0.48
2	0.21	2.10
5	0.56	4.50
CNP, 2×10^{-8} M	2.40	9.30
ANP, 2×10^{-8} M	1.40	0.55

The indicated volumes of heated porcine seminal plasma were diluted in DMEM (containing 0.1 mM methyl isobutylxanthine) to a final volume of 1 ml. Other conditions were described in Methods. Porcine seminal plasma was heated to 97°C for 15 min.

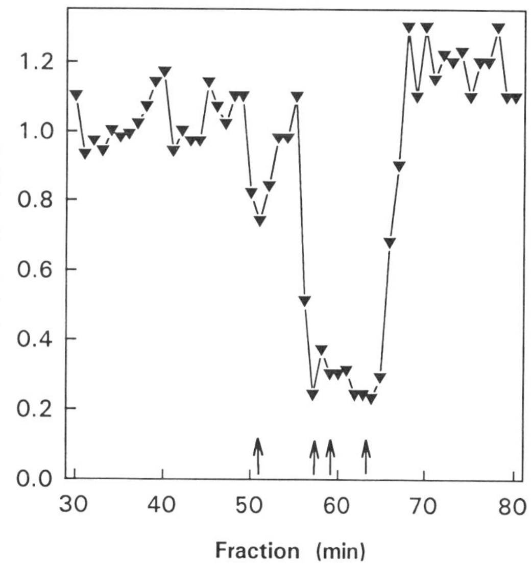

Figure 2. Fractionation of porcine seminal plasma by reversed-phase chromatography. Porcine seminal plasma was prepared as described in Methods and pumped onto the C_{18} column at 3 ml/min. Fractions were eluted with a 25–35% (fractions 30–80) acetonitrile gradient at 3 ml/min. Aliquots (1 μl) of the indicated fractions were analyzed for competition with ^{125}I YCNP binding to membranes of GC-B overexpressing cells.

22) first descibed by Sudoh et al. (1990) and a 53-amino-acid form (CNP-53) both derived from pro-CNP. Although the relative potencies of the different forms of ir-CNP have not been rigorously determined, subsequent studies (see Fig. 3) suggest that they are not equipotent in stimulating GC-B.

Quantitation of CNP in seminal plasma, epididymal fluid, and seminal vesicle fluid showed very high levels of the peptide in the porcine reproductive tract. Seminal plasma contained approximately 1 μM CNP, and seminal vesicle fluid contained about 70 μM CNP, whereas CNP was almost undetectable in the epididymis, suggesting that the seminal vesicle, rather than the testis, is the primary source of CNP in the ejaculate. The physiological significance of high concentrations of CNP in the ejaculate is unknown, although these amounts will stimulate both GC-A and GC-B.

We have identified ir-CNP in various tissues from several species. Fractionation of rat seminal vesicle fluid by reversed-phase chromatography (Fig. 3) or extracts of bovine tracheal mucosa (not shown) or *Xenopus laevis* skin (not shown) established the presence of multiple forms of ir-CNP in relatively high amounts as compared to that in porcine brain. Some forms markedly stimulated GC-B and others were only marginally active. These findings demonstrate that CNP as measured by immunologic assay and by "biologic" assay is widely distributed and not confined to the CNS as previously thought.

The widespread occurrence of CNP raised questions concerning the validity of models limiting the expression of its receptor, GC-B, to the CNS. Northern analysis of mRNA from several species (Table 2) suggests that, like its ligand, GC-B is found in tissues other than the brain. Consistent with these results, we have found in collaborative studies with Michael J. Welsh at the University of Iowa that CNP (10^{-8} M to 10^{-6} M) sharply elevates cGMP levels in cultured human airway epithelial cells, whereas even at high concentrations (micromolar), ANP has only small effects. The absence of CNP in the systemic circulation and the presence of GC-B and CNP in many tissues suggest that GC-B/CNP constitutes an autocrine/paracrine system perhaps regulating specialized functions such as epithelial fluid secretion in the airway.

Table 2. Tissue Distribution of GC-B and C-type Natriuretic Peptide

System	GC-B[a]	CNP[a]
Nervous		
central	+[1]	+[2,b]
adrenal medulla	+[1]	+[2,b]
retina	+[3]	n.d.
Respiratory		
lung	n.d.	+[3,b,c]
trachea	+[3]	+[3,b,c]
epithelial cell	+[3]	n.d.
Reproductive		
seminal fluid	n.d.	+[3,b,c]
seminal vesicle	n.d.	+[3,b,c]
epididymus	n.d.	±[3,b,c]
uterus	+[3]	+[3,b,c]

Immunoreactive CNP (ir-CNP) was determined by radioimmunoassay (Peninsula Laboratories). GC-B stimulation was determined as in Methods. n.d. indicates not determined.

[a]References: (1) Wilcox et al. (1991); (2) Sudoh et al. (1990); (3) this laboratory.

[b]ir-CNP.

[c]GC-B stimulation.

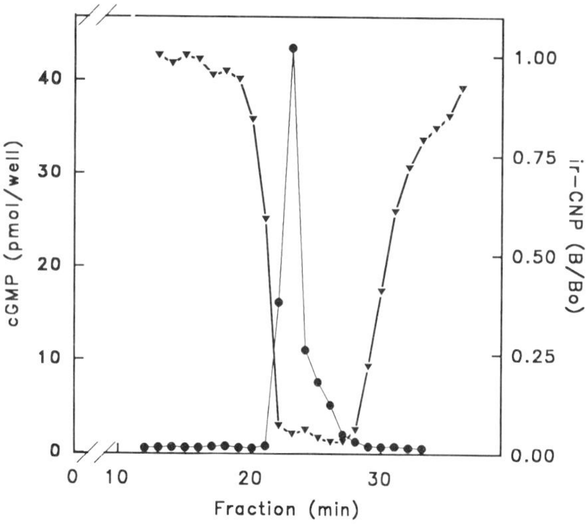

Figure 3. Fractionation of rat seminal vesicle extract by reversed-phase chromatography. Approximately 100 μg of protein (see Methods) was applied to a 4.6-mm × 250-mm C_{18} column (Vydac), and fractions were eluted with a 20–40% acetonitrile gradient (fractions 13–33) at 1 ml/min. Fractions were analyzed for immunoreactive CNP (ir-CNP; ▲) by radioimmunoassay according to the manufacturer's instructions (Peninsula Laboratories); Bo is the amount of ^{125}I YCNP bound in the absence of added sample. Measurement of cGMP-elevating activity (●) in cells overexpressing GC-B was described in Methods.

GC-C and Heat Stable Enterotoxin

GC-C was the third cell-surface peptide receptor of the family to be cloned and expressed (Schulz et al. 1990). Initially discovered in mammalian small intestine, this receptor is the target of bacterial heat-stable enterotoxins (ST) that cause a profound diarrhea in humans and domestic animals. Localized on the apical membrane of the small intestine epithelial cell (Field et al. 1978), the extracellular receptor faces the intestinal lumen, making it accessible to bacterial toxins and presumably to endogenous ligands.

Rat and human GC-C have been cloned from small intestine (Schulz et al. 1990; de Sauvage et al. 1991) or from cells derived from the intestine (Singh et al. 1991), and like GC-A and GC-B, it is a single transmembrane protein having both ligand-binding and guanylyl cyclase catalytic domains. These results suggest that earlier reports contending that the ST receptor is only indirectly associated with guanylyl cyclase (Kuno et al. 1986; Waldman et al. 1986) were based on proteolytic cleavage of the intestinal extracts. The receptor from both species has a topology similar to that of the natriuretic peptide receptors, GC-A and GC-B, but it is clearly different in the extracellular domain (I) based on amino acid sequence and is not stimulated by any known natriuretic peptides. The cyclase catalytic domain (IV) of GC-C is 55% identical with GC-A and GC-B; the kinase domain (III) is more divergent, being only 39% and 35% identical with GC-A and GC-B, respectively. GC-C contains an extended carboxyl tail not found in the natriuretic peptide receptors; the significance of this extension is not known, but it has been postulated to anchor the receptor to the cytoskeleton (Schulz et al. 1990).

Initial studies on the tissue distribution of ST-stimulated GC-C found it only in mammalian small intestine (Guerrant et al. 1980), whereas in the North American opossum, a marsupial, ^{125}I-labeled ST binding has been reported in the epithelia of many tissues (Krause et al. 1990). We have reexamined the tissue distribution of GC-C in the rat and the cow, and as demonstrated by Northern analysis, PCR, or ST elevations of cGMP, GC-C is found in several tissues (Table 3). Figure 4 shows the effects of ST on cGMP levels in intact cells isolated from bovine tracheal mucosa. The sensitivity to ST is identical to that found in bovine intestinal cell preparations (not shown) and is similar to that of rat GC-C overexpressed in COS-7 cells (Schulz et al. 1990) and in the T84 human colonic epithelial cell line (Guarino et al. 1987). These observations suggest that GC-C or closely related receptor-guanylyl cyclases are widespread in mammals as well as in other species.

Recently, Currie et al. (1992) isolated a 15-amino-acid peptide, guanylin, from rat small intestine that elevates cGMP in T84 cells (a human colonic epithelial cell line) and competes with ^{125}I-labeled ST binding to T84 membranes. Since GC-C appears to be the only ST receptor in this cell line (Singh et al. 1991; deSauvage et al. 1991), guanylin represents the first known endogenous ligand for GC-C. As shown by Currie et al. (1992) and in Table 3, guanylin mRNA is also found in

Table 3. Tissue Distribution of GC-C and Guanylin

Tissue	GC-C[a]	Guanylin[a]
Small intestine	+[1,2,b,c]	+[2,3,c]
Trachea	+[2,b,c]	—
Airway epithelium	+[2,b,c]	n.d.
Adrenal gland	+[2,c]	+[2,c]
Brain	+[2,c]	—
Heart	—	—
Kidney	—	+[2,3,c]
Liver	—	—
Olfactory mucosa	+[2,c]	n.d.
Testis	—	—
Uterus/oviduct	—	+[2,c]

Northern analyses and ST-stimulation were done as described in Methods. All tissues were from the rat, except airway epithelium (human) and trachea (bovine). n.d. indicates not determined.

[a]References: (1) Field et al. (1978); (2) this laboratory; (3) Currie et al. (1992).

[b]ST stimulation.

[c]RNA detection.

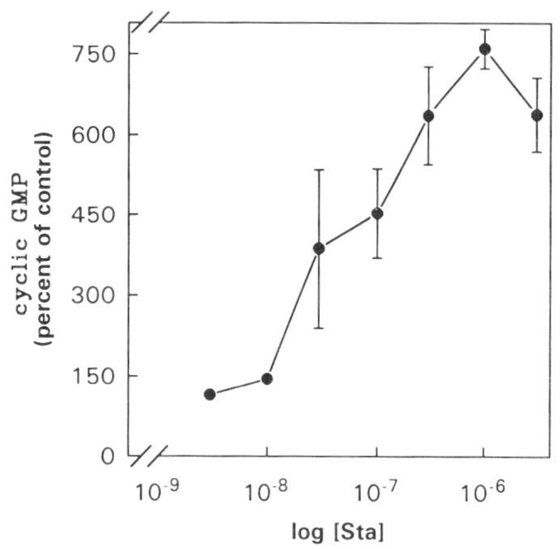

Figure 4. ST-dependent accumulation of cGMP in bovine tracheal cells. Each point represents the average of duplicate determinations plus or minus the range of the replicates. The data are presented as the percentage of control (PBS) with the average control value being 19 (range: 17–20) fmoles cGMP.

other tissues including the adrenal gland, kidney, and uterus/oviduct. These findings suggest that GC-C/guanylin will have regulatory roles in tissues other than the intestine, possibly in an autocrine/paracrine mechanism, as appears to exist for the receptor/ligand in intestine. It must be cautioned, however, that high guanylin concentration (micromolar) appears to be required for stimulation of GC-C, and therefore, the normal receptor for guanylin remains to be established.

Cell-surface receptor-guanylyl cyclases and their respective ligands have emerged as a unique signaling system combining structural and regulatory features of two other signaling systems, the adenylyl cyclase–G protein coupled receptors and the growth factor receptors. Unlike the adenylyl cyclases, which have multiple membrane-spanning domains and are linked indirectly to their cell-surface receptors via G proteins, the receptor-guanylyl cyclases contain, within a single protein, ligand-binding (I), regulatory (III), and catalytic (IV) domains. Although structurally similar to many growth factor receptors (Aaronson 1991), guanylyl cyclases have not yet been shown to possess intrinsic protein tyrosine/serine kinase activity.

In terms of physiology, the abundance of the guanylyl cyclase receptors in secretory epithelia and the known effects of their ligands to promote ion and water movement across the epithelium suggest that one role of this signaling system is the regulation of fluid volume and/or composition. The similarity to protein tyrosine kinase receptors raises the possibility that the guanylyl cyclases have other regulatory roles (separate from cGMP metabolism), although such evidence is lacking.

REFERENCES

Aaronson, S.A. 1991. Growth factors and cancer. *Science* **254:** 1146.

Anand-Srivastava, M.B., M.R. Sairam, and M. Cantin. 1990. Ring deleted analogs of atrial natriuretic factor inhibit adenylate cyclase/cAMP system: Possible coupling of clearance atrial natriuretic factor receptors to adenylate cyclase/cAMP signal transduction system. *J. Biol. Chem.* **265:** 8566.

Arimura, J.J., N. Minamino, K. Kangawa, and H. Matsuo. 1990. Isolation and identification of C-type natriuretic peptide in chicken brain. *Biochem. Biophys. Res. Commun.* **174:** 142.

Ashman, D.F., R. Lipton, M.M. Melicow, and T.D. Price. 1963. Isolation of adenosine 3′,5′-monophosphate and guanosine 3′,5′-monophosphate from rat urine. *Biochem. Biophys. Res. Commun.* **11:** 330.

Beavo, J.A. 1988. Multiple isozymes of cyclic nucleotide phosphodiesterase. *Adv. Second Messenger Phosphoprotein Res.* **22:** 1.

Brenner, B.M., B.J. Ballermann, M.E. Gunning, and M.L. Zeidel. 1990. Diverse biological actions of atrial natriuretic peptide. *Physiol. Rev.* **70:** 665.

Chang, M.S., D.G. Lowe, M. Lewis, R. Hellmiss, E. Chen, and D.V. Goedel. 1989. Differential activation by atrial and brain natriuretic peptides of two receptor guanylyl cyclases. *Nature* **341:** 68.

Chinkers, M. and D.L. Garbers. 1989. The protein kinase domain of the ANP receptor is required for signaling. *Science* **245:** 1392.

———. 1991. Signal transduction by guanylyl cyclases. *Annu. Rev. Biochem.* **60:** 553.

Chinkers, M., S. Singh, and D.L. Garbers. 1991. Adenine nucleotides are required for activation of rat atrial natriuretic peptide receptor/guanylyl cyclase expressed in a baculovirus system. *J. Biol. Chem.* **266:** 4088.

Chinkers, M., D.L. Garbers, M.S. Chang, D.G. Lowe, H. Chin, D.V. Goeddel, and S. Schulz. 1989. Molecular cloning of a new type of cell surface receptor: A membrane form of guanylyl cyclase is an atrial natriuretic peptide receptor. *Nature* **338:** 78.

Currie, M.G., K.F. Fok, J. Kato, R.J. Moore, F.K. Hamra, K.L. Duffin, and C.E. Smith. 1992. Guanylin: An endogenous activator of intestinal guanylate cyclase. *Proc. Natl. Acad. Sci.* **89:** 947.

Debold, A.J. 1982. Atrial natriuretic factor in the rat heart. Studies on isolation and properties. *Proc. Soc. Exp. Biol. Med.* **170:** 133.

deSauvage, F.T., T.R. Camerato, and D.V. Goeddel. 1991. Primary structure and functional expression of the human receptor for *Escherichia coli* heat-stable enterotoxin. *J. Biol. Chem.* **266:** 17912.

Edelman, A.M., D.K. Blumenthal, and E.G. Krebs. 1987. Protein serine/threonine kinases. *Annu. Rev. Biochem.* **56:** 567.

Field, M., W.J. Laird, and P.L. Smith. 1978. Heat-stable enterotoxin of *Escherichia coli*: In vivo effects on guanylyl cyclase activity, cyclic GMP concentration, and ion transport in small intestine. *Proc. Natl. Acad. Sci.* **75:** 2800.

Fuller, F., J.G. Porter, A.E. Arfsten, J. Miller, J.W. Schilling, R.M. Scarborough, J.A. Lewicki, and D.B. Schenk. 1988. Atrial natriuretic peptide clearance receptor: Complete sequence and functional expression of cDNA clones. *J. Biol. Chem.* **263:** 9395.

Goraczniak, R.M., T. Duda, and R.K. Sharma. 1992. A structural motif that defines the ATP-regulatory module of guanylate cyclase in atrial natriuretic factor signaling. *Biochem. J.* **282:** 533.

Guarino, A., M. Cohen, M. Thompson, K. Dharmsathaphorn, and R. Gianella. 1987. T84 cell receptor binding and guanyl cyclase activation by *Escherichia coli* heat-stable toxin. *Am. J. Physiol.* **253:** G775.

Guerrant, R.L., J.M. Hughes, B. Chang, D.L. Robertson, and F. Murad. 1980. Activation of intestinal guanylate cyclase by heat-stable enterotoxin of *Escherichia coli*: Studies of specificity, potential receptors and intermediates. *J. Infect. Dis.* 142: 220.

Kojima, M., N. Minamino, K. Kangawa, and H. Matsuo. 1990. Cloning and sequence analysis of a cDNA encoding a precursor for rat C-type natriuretic peptide (CNP). *FEBS Lett.* **219:** 209.

Koller, K.J., F.J. deSauvage, D.G. Lowe, and D.V. Goeddel. 1992. Conservation of the kinaselike regulatory domain is essential for activation of the natriuretic peptide receptor guanylyl cyclases. *Mol. Cell. Biol.* **12:** 2581.

Koller, K.J., D.G. Lowe, G.L. Bennett, N. Minamino, K. Kangawa, H. Matsuo, and D.V. Goeddel. 1991. Selective activation of the B natriuretic peptide receptor by C-type natriuretic peptide (CNP). *Science* **252:** 120.

Komatsu, Y., K. Nakao, S.-I. Suga, Y. Ogawa, M. Mukoyama, H. Arai, G. Shirakami, K. Hosoda, O. Nakagawa, N. Hama, I. Kishimoto, and H. Imura. 1991. C-type natriuretic peptide (CNP) in rats and humans. *Endocrinology* **129:** 1104.

Krause, W.J., R.H. Freeman, and L.R. Forte. 1990. Autoradiographic demonstration of specific binding sites for *E. coli* enterotoxin in various epithelia of the North American opossum. *Cell Tissue Res.* **260:** 387.

Krupinski, J., F. Coussen, H.A. Bakalyar, W.J. Tang, P.G. Feinstein, K. Orth, C. Slaughter, R.R. Reed, and A.G. Gilman. 1989. Adenylyl cyclase amino acid sequence: Possible channel- or transporter-like structure. *Science* **244:** 1558.

Kuno, T., Y. Kamisaki, S.A. Waldman, J. Gariepy, G. Schoolnik, and F. Murad. 1986. Characterization of the receptor for heat-stable enterotoxin from *Escherichia coli* in rat intestine. *J. Biol. Chem.* **261:** 1470.

Kurose, H., T. Inagami, and M. Ui. 1987. Participation of adenosine 5′-triphosphate in the activation of membrane-bound guanylate cyclase by atrial natriuretic factor. *FEBS Lett.* **219:** 375.

Lowe, D.G., M.S. Chang, R. Hellmiss, E. Chen, S. Singh, D.L. Garbers, and D.V. Goeddel. 1989. Human atrial natriuretic peptide receptor defines a new paradigm for messenger signal transduction. *EMBO J.* **8:** 1377.

Maack, T.M., M. Suzuki, F.A. Almeida, D. Nussenzvueig, R.M. Scarborough, G.A. McEnroe, and J.A. Lewicki. 1987. Physiological role of silent receptors of atrial natriuretic factor. *Science* **238:** 675.

Minamino, N., Y. Makino, H. Tateyama, K. Kangawa, and H. Matsuo. 1991. Characterization of immunoreactive human C-type natriuretic peptide in brain and heart. *Biochem. Biophys. Res. Commun.* **179:** 535.

Ohyama, Y., K. Miyamoto, Y. Saito, N. Minamino, K. Kangawa, and H. Matsuo. 1992. Cloning and characterization of two forms of C-type natriuretic peptide receptor in rat brain. *Biochem. Biophys. Res. Commun.* **183:** 743.

Potter, L.R. and D.L. Garbers. 1992. Dephosphorylation of the guanylyl cyclase-a receptor (GC-A) causes dephosphorylation. *J. Biol. Chem.* **267:** 14531.

Ritter, D., P. Needleman, and J.E. Greenwald. 1991. Synthesis and secretion of an atriopeptin-like protein in rat kidney cell culture. *J. Clin. Invest.* **87:** 208.

Scarborough, R.M., G.A. McEnroe, A. Arfsten, L.-L. Kang, K. Schwartz, and J.A. Lewicki. 1988. D-amino acid-substituted atrial natriuretic peptide analogs reveal novel receptor recognition requirements. *J. Biol. Chem.* **263:** 16818.

Scarborough, R.M., D.B. Schenk, G.A. McEnroe, A. Arfsten, L.-L. Kang, K. Schwartz, and J.A. Lewicki. 1986. Truncated atrial natriuretic peptide analogs: Comparison between receptor binding and stimulation of cyclic GMP accumulation in cultured vascular smooth muscle cells. *J. Biol. Chem.* **261:** 12960.

Schulz, S., T.D. Chrisman, and D.L. Garbers. 1992. Cloning and expression of guanylin: Its existence in various mammalian tissues. *J. Biol. Chem.* **267:** 16019.

Schulz, S., C.K. Green, P.S.T. Yuen, and D.L. Garbers. 1990. Guanylyl cyclase is a heat-stable enterotoxin receptor. *Cell* **63:** 941.

Schulz, S., S. Singh, R.A. Bellet, G. Singh, D.J. Tubb, H. Chin, and D.L. Garbers. 1989. The primary structure of a plasma membrane guanylyl cyclase demonstrates diversity with this new receptor family. *Cell* **58:** 1155.

Singh, S., G. Singh, J.-M. Heim, and R. Gertzer. 1991. Isolation and expression of a guanylate cyclase-coupled heat-stable enterotoxin receptor cDNA from a human cell line. *Biochem. Biophys. Res. Commun.* **179:** 1455.

Song, D.L., K.P. Kohse, and F. Murad. 1988. Brain natriuretic factor: Augmentation of cellular cyclic GMP, activation of particulate guanylate cyclase, and receptor binding. *FEBS Lett.* **232:** 125.

Stryer, L. 1986. Cyclic GMP cascade of vision. *Annu. Rev. Neurosci.* **9:** 87.

Sudoh, T., K. Kangawa, N. Minamino, and H. Matsuo. 1988. A new natriuretic peptide in porcine brain. *Nature* **332:** 78.

Sudoh, T., N. Minamino, K. Kangawa, and H. Matsuo. 1990. C-type natriuretic peptide (CNP): A new member of natriuretic peptide family identified in porcine brain. *Biochem. Biophys. Res. Commun.* **168:** 863.

Thorpe, D. and E. Morkin. 1990. The carboxyl region contains the catalytic domain of the membrane form of guanylate cyclase. *J. Biol. Chem.* **265:** 14717.

Ueda, S., N. Minamino, M. Aburaya, K. Kangawa, S. Matsukara, and H. Matsuo. 1991. Distribution and characterization of immunoreactive porcine C-type natriuretic peptide. *Biochem. Biophys. Res. Commun.* **175:** 759.

Waldman, S.A., T. Kuno, Y. Kamisaki, L.Y. Chang, J. Gariepy, P. O'Hanley, G. Schoolnik, and F. Murad. 1986. Intestinal receptor for heat-stable enterotoxin of *Escherichia coli* is tightly coupled to a novel form of particulate guanylate cyclase. *Infect. Immun.* **51:** 320.

Wilcox, J.N., A. Augustine, D.V. Goeddel, and D.G. Lowe. 1991. Differential regional expression of three natriuretic peptide receptor genes within primate tissues. *Mol. Cell. Biol.* **11:** 3454.

Yuen, P.S.T. and D.L. Garbers. 1992. Guanylyl cyclase-linked receptors. *Annu. Rev. Neurosci.* **15:** 193.

Lysophosphatidic Acid: A Novel Phospholipid with Hormone- and Growth Factor-like Activities

W.H. Moolenaar, R.L. van der Bend, E.J. van Corven, K. Jalink,
T. Eichholtz, and W.J. van Blitterswijk

Division of Cellular Biochemistry, The Netherlands Cancer Institute, 1066 CX Amsterdam, The Netherlands

Phospholipids are currently attracting a great deal of interest in studies on cellular signaling and growth control. Whereas a major function of phospholipids is to form bilayer membranes, the breakdown products of several plasma membrane phospholipids appear to act as signal molecules, i.e., as intracellular second messengers or as agonists that modulate cell function. Nature's simplest phospholipid, lysophosphatidic acid (LPA or monoacylglycerol-3-phosphate), is of particular interest in that it not only serves a critical precursor role in de novo lipid biosynthesis, but, more interestingly, also shows remarkable hormone- and growth-factor-like activities when added to appropriate target cells, apparently by activating its own G-protein-coupled receptor(s). Although LPA's lipid precursor role has been known for many decades, the possibility that this simple phospholipid may have an evolutionary, well-conserved role as a "first messenger" has been appreciated for only a few years. Understanding the multiple biological and biochemical activities of LPA, as well as its normal physiological functions, is obviously an important research goal. The intent of this paper is to briefly summarize our recent data on the multiple actions of LPA, with particular emphasis on its site of action and the underlying signaling events.

RESULTS AND DISCUSSION

LPA is rapidly produced in thrombin-activated platelets (Watson et al. 1985; Gerrard and Robinson 1989) and plasma-derived growth factor (PDGF)-stimulated fibroblasts (Fukami and Takenawa 1992) and is detectable in serum (Tokumura et al. 1986; T. Eichholtz et al., in prep.), suggesting that LPA may be secreted into the extracellular environment following its formation; direct proof for that notion awaits further studies. The production of LPA in activated cells is likely to be secondary to the formation of phosphatidic acid through the action of a specific phospholipase A_2, as schematically illustrated in Figure 1, but the existence of alternative metabolic routes leading to LPA formation cannot be excluded. Although exogenous LPA is partially metabolized by cells, the available evidence strongly suggests that the biological activity of LPA is not attributable to one of its metabolites (van der Bend et al. 1992a).

Exogenous LPA stimulates short-term, hormone-like effects in many cell types (Table 1) and, in fibroblasts, long-term DNA synthesis and cell division (Moolenaar et al. 1992; van Corven et al. 1989, 1992). Naturally occurring LPA (1-oleoyl) stimulates thymidine incorporation in fibroblasts with a half-maximal effect observed at about 15 μM and a saturating response at about 100 μM (van Corven et al. 1989), concentrations that are significantly higher than those required for evoking the early events listed in Table 1. The discrepancy between these dose-response relationships remains unexplained at present. As with polypeptide growth factors, the mitogenic response to LPA requires long-term presence of the stimulus. When LPA is removed from the culture medium several hours after stimulation, the cells fail to enter S phase. The growth-stimulating activity does not require the presence of peptide growth factors: Neither insulin nor epidermal growth factor (EGF) was found to act in a synergistic fashion with LPA (van Corven et al. 1989). In this respect, LPA acts differently from mitogenic peptides such as serotonin and vasopressin, which fail to stimulate cell proliferation unless synergizing growth factors are present (Rozengurt 1986).

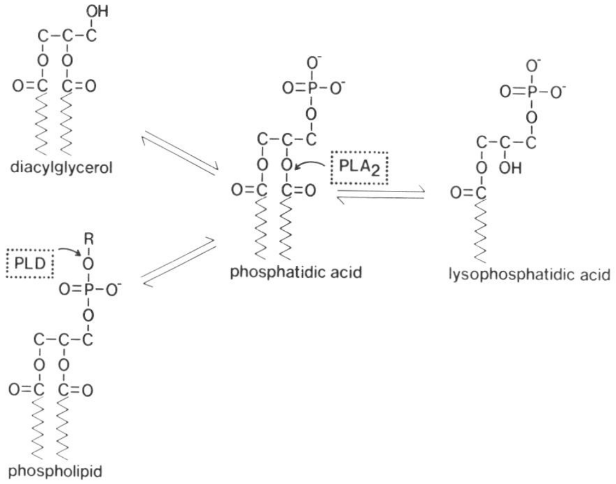

Figure 1. Proposed generation of LPA from newly formed PA during cell activation. Diacylglycerol is rapidly formed via receptor-linked activation of phospholipase C and is phosphorylated by diacylglycerol kinase to yield phosphatidic acid (PA). PA may also be formed through activation of phospholipase D (PLD) acting on PC and PE (R denotes phospholipid headgroup). PA may be hydrolyzed by a phospholipase A_2 (PLA_2) to LPA. The contribution of other pathways to generating LPA remains to be explored.

Table 1. Early Cellular Responses to LPA

Cell type	Response
Many cell types	Ca^{++} mobilization; arachidonate release; inhibition of adenylyl cyclase
Neuroblastoma cells	Ca^{++} mobilization; morphological changes
HT-29 colon carcinoma cells	Cl^- secretion
MDCK epithelial cells	Cl^- secretion
Mouse embryonic stem cells	Ca^{++} mobilization
Smooth muscle	Ca^{++} mobilization; contraction
Platelets	Ca^{++} mobilization; aggregation
Xenopus oocytes	Ca_i^{++}-dependent Cl^- current
Dictyostelium discoideum	chemotaxis

References can be found in Moolenaar et al. (1992).

Signal Transduction Pathways and Receptor Identification

We have identified several "classic" signal transduction pathways in the action of LPA (van Corven et al. 1989; Moolenaar et al. 1992). These are (1) GTP-dependent activation of phospholipase C (PLC), (2) release of arachidonic acid presumably as a result of phospholipase A_2 activation, (3) activation of phospholipase D (van der Bend et al. 1992b), and (4) pertussis-toxin-sensitive inhibition of adenylyl cyclase. In addition, studies on neuronal cells have revealed that LPA causes rapid changes in cytoskeletal organization (K. Jalink et al., in prep.), a process that is not attributable to the above second-messenger cascades, as discussed below. Together, the available evidence strongly suggests that LPA activates one or more specific G-protein-coupled receptors at the cell surface. In a recent study, van der Bend et al. (1992c) succeeded in identifying a putative high-affinity LPA receptor. A ^{32}P-labeled LPA analog containing a photoreactive fatty acid, diazirine-LPA, labels a membrane protein of apparent molecular mass 38–40 kD in various LPA-responsive cell types, with labeling being most prominent in neuronal cells and brain homogenates. Labeling is specific, in that only unlabeled LPA but not other phospholipids inhibit incorporation of diazirine-LPA into the 38–40 kD protein band; half-maximal inhibition is observed at approximately 10–20 nM unlabeled LPA. The 38–40 kD protein presumably represents a specific LPA cell-surface receptor mediating at least part of the multiple cellular responses to LPA (for further details, see van der Bend et al. 1992c). The putative LPA receptor is not detectable in human neutrophils, which are biologically unresponsive to LPA (Jalink et al. 1990). Clearly, the biochemical identification of a candidate LPA receptor is an important result, on which further experiments will be based.

As for the early signals evoked by LPA: A great variety of cell types respond to LPA by triggering the immediate breakdown of inositol phospholipids as measured by the formation of inositol triphosphate (1,4,5) and diacylglycerol (van Corven et al. 1989; Jalink et al. 1990). LPA-induced activation of PLC results in rapid but transient rise in Ca^{++}, which is primarily caused by the release of intracellularly stored Ca^{++}. Concomitantly, protein kinase C (PKC) is activated as shown by the phosphorylation of an endogenous 80-kD protein substrate (van Corven et al. 1989). The kinetics and shape of the LPA-induced Ca^{++} transient are virtually indistinguishable from those elicited by bradykinin and other hormones. Furthermore, LPA-induced Ca^{++} mobilization is subject to homologous desensitization, a common feature of many signaling systems where agonist-induced attenuation of cellular responsiveness is thought to have an important regulatory role (Benovic et al. 1988). From studies on permeabilized cells, it appears that the response is GTP-dependent, suggesting the involvement of a G protein (that is insensitive to pertussis toxin in human fibroblasts; van Corven et al. 1989). Thus, the early phosphoinositide-breakdown response to LPA has all the hallmarks of a G-protein-coupled receptor event.

In addition to activating PLC, LPA is known to inhibit cAMP accumulation in intact cells (van Corven et al. 1989). When adenylyl cyclase is prestimulated by either forskolin or certain receptor agonists (isoproterenol, prostaglandin E_1, etc.), addition of LPA decreases cAMP accumulation by 60–70% within 10 minutes. This decrease in cAMP is dose-dependent and is completely blocked by pretreating the cells with pertussis toxin, indicating that LPA acts through the G_i protein that inhibits adenylyl cyclase. Figure 2 summarizes the various early responses to LPA as identified in fibroblasts. Note the lack of correlation between early phosphoinositide hydrolysis and late initiation of DNA synthesis, supporting the notion that, contrary to common belief, activation of PLC is neither required nor sufficient for induction of a mitogenic response (Moolenaar 1991).

Changes in Neuronal Cell Shape: A Novel Signaling Mechanism

In a recent series of experiments, we have used the neuronal cell lines N1E-115 and NG108-15 as a model system to explore the neurobiological actions of LPA. These cells constitute a convenient model system to investigate early signaling events in the proliferative response. After serum starvation, these cells stop growing and subsequently begin to acquire various differentiated properties of mature neurons, including the formation of long neurites. Addition of 1-oleoyl-LPA (1 μM) to growth-factor-starved N1E-115 or NG108-15 cells causes rapid and dramatic changes in cell shape,

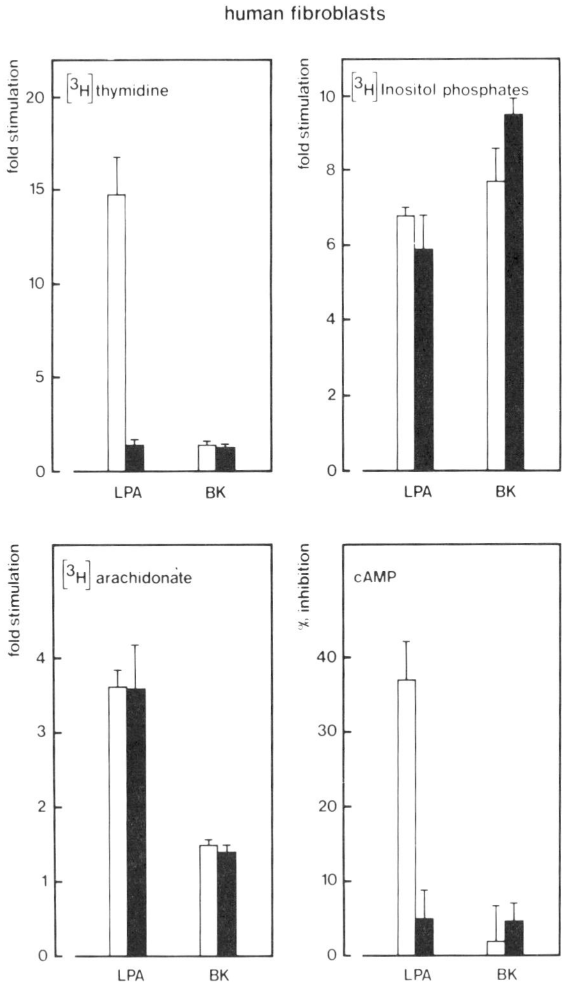

Figure 2. Mitogenic response and generation of second messengers induced by LPA and bradykinin (BK) in human fibroblasts in the presence (black bars) or absence (open bars) of 100 ng/ml pertussis toxin. For further details, see van Corven et al. (1989).

reminiscent of those observed during mitosis (K. Jalink et al., in prep.). Virtually every flattened cell starts to round up as early as 5–10 seconds after LPA addition, and rounding is complete within approximately 1 minute. Almost simultaneously, growth cones begin to collapse and developing neurites retract. The effects of LPA on neuronal cell shape are dose-dependent, with some rounding detectable at doses as low as 10 nM and maximal responses at 0.5–1 μM. In the continuous presence of LPA, cells maintain their rounded shape for 5–10 minutes. At 10–20 minutes after the addition of LPA, however, most cells gradually resume a flattened morphology, but neurite outgrowth remains suppressed for at least a few hours. Remarkably, a second application of LPA to such respread cells leaves the shape unaltered. LPA-induced cell rounding is thus subject to homologous desensitization, as one would expect for a receptor-mediated process. Further evidence for the morphological response's being receptor-mediated includes the findings that (1) the drug suramin, which blocks LPA-receptor interaction (van der Bend et al. 1992c), inhibits LPA-induced shape changes; (2) other, related lipids have no effect; (3) microinjected LPA fails to mimic the action of extracellularly applied LPA; and (4) the morphological response to LPA is fully mimicked by a synthetic peptide agonist of the cloned G-protein-coupled thrombin receptor, whereas LPA and thrombin clearly act through separate receptors (Jalink and Moolenaar 1992; K. Jalink et al., in prep.).

We have inferred that LPA- and thrombin-induced changes in neuronal cell shape are mediated by "contraction" of cortical actin filaments with no direct involvement of microtubules. What signaling mechanism is responsible for this unprecedented action of LPA and thrombin receptor agonists? In N1E-115 cells, as in many other cells, LPA stimulates phosphoinositide hydrolysis leading to calcium mobilization and activation of PKC. Yet, the following findings indicate that the PLC-Ca^{++}-PKC pathway is not responsible for the observed changes in cell shape: (1) Phosphoinositide-hydrolyzing neurotransmitters, Ca^{++} ionophores, and PKC-activating phorbol ester all fail to mimic LPA in inducing cell rounding; (2) LPA-induced shape changes are not prevented by down-regulating PKC by long-term treatment of the cells with phorbol ester; and (3) cells depleted of internal Ca^{++} (by addition of ionophore in Ca^{++} -free medium) show a normal morphological response to LPA despite the absence of a Ca^{++} transient. Similarly, bacterial toxin-sensitive G proteins, adenylyl cyclase, or cyclic nucleotides have no apparent role in mediating LPA-induced shape changes, since LPA action is neither inhibited nor mimicked by treatment of the cells with pertussis or cholera toxin, cAMP and cGMP analogs, or forskolin (10 μM). Finally, the involvement of phospholipase A_2 can be ruled out because potent inducers of this enzyme, such as bradykinin and ionomycin, free arachidonic acid, and prostaglandins have no effect on cell shape.

Taken together, these results suggest that LPA-induced cell rounding is not mediated by known G-protein-linked second-messenger cascades. By analogy with the cell rounding induced by active protein tyrosine kinases (Chinkers et al. 1981; Jove and Hanafusa 1987) and microinjected cdc2 kinase (Lamb et al. 1990), it seems plausible to assume that specific phosphorylations / dephosphorylations of certain actin-binding proteins (Pollard and Cooper 1986; Stossel 1989) may be responsible for the morphological effects of LPA. In support of this, we observed that LPA-induced cell rounding is blocked by both microinjected vanadate and exogenously added pervanadate, which is a membrane-permeable form of vanadate (Fantus et al. 1989), a widely used inhibitor of protein tyrosine phosphatases. Furthermore, brief preincubation with such (nonspecific) protein kinase inhibitors as genistein (50 μM), quercetin (50 μM), and staurosporine (1 μM) similarly inhibits cell rounding.

We have begun to analyze the possible role of the Src

protein tyrosine kinase in the rapid neuronal shape changes. It was found that both LPA and thrombin induce a small but statistically significant increase in Src kinase activity in N1E-115 cells as measured in an immunocomplex kinase assay (K. Jalink et al., in prep.). Although we cannot draw conclusions yet about cause-effect relationships, the results obtained are intriguing and deserve further study.

Activation of p21^{ras}

The product of the *ras* proto-oncogene (p21^{ras}) is a focal point of signal transduction by growth factor receptors, notably those with intrinsic tyrosine kinase activity. Given the mitogenic potency of LPA, we have examined the activation state of the Ras protein in Rat-1 fibroblasts following treatment with LPA. As illustrated in Figure 3, LPA causes the rapid accumulation of the GTP-bound, active form of Ras within 1 minute. The kinetics of activation is transient: The response subsides after about 10 minutes. This remarkable response to LPA is similar to that elicited by EGF, although the extent of p21^{ras} activation by LPA is somewhat weaker than observed with EGF (Fig. 3) (E.J. van Corven et al., in prep.). Like the other early events in LPA action, p21 · GTP accumulation conforms to all the criteria of a receptor-mediated process (E.J. van Corven et al., in prep.). Future experiments should reveal whether activation of p21^{ras} is uniquely observed with the putative LPA receptor, or else is a more common event in the action of certain G-protein-coupled receptors.

FUTURE PROSPECTS

The discovery that LPA exerts profound effects on many, but not all, cell types suggests that these low-abundance lipids may have a previously unrecognized role as a lipid mediator, whose function as a signaling molecule is seemingly well conserved during eukaryotic evolution. It seems plausible to assume that LPA (which is fairly well water-soluble) is secreted by cells in a manner similar to the secretion of platelet-activating factor, prostaglandins, and other lipid agonists. The released LPA may then activate target cells in a paracrine or autocrine fashion. Phospholipid labeling studies will allow a critical test of this hypothesis. It is premature to speculate further on the normal physiological (or perhaps pathophysiological) roles that LPA may play.

An obvious challenge for future studies is to establish the amino acid sequence of the putative LPA cell-surface receptor, either through protein purification or, perhaps more straightforward, via expression cloning in COS cells. Another challenge is to delineate the unique signal transduction events, involving activation of both the *src* and *ras* proto-oncogene products, that are responsible for reversal of the differentiated phenotype in neuroblastoma cells and initiation of DNA synthesis and cell division in fibroblasts and possibly other cell types.

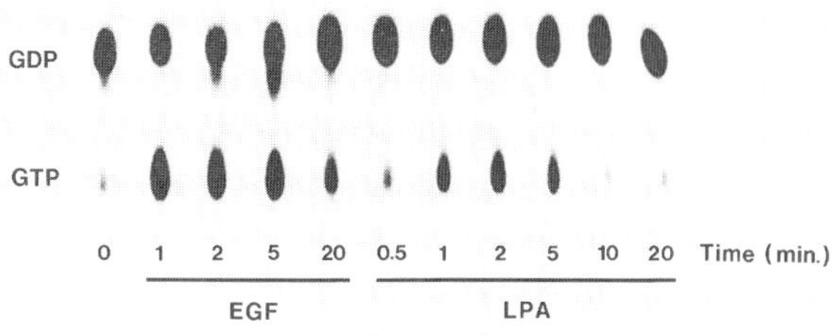

Figure 3. Activation of p21^{ras} by LPA and EGF in Rat-1 cells. Phosphate-labeled cells were treated with EGF (50 ng/ml) or LPA (10 μM) for varying time periods as indicated. p21^{ras} was immunoprecipitated using monoclonal 259, and radiolabeled GDP and GTP were separated by thin-layer chromatography (E.J. van Corven et al., in prep.).

ACKNOWLEDGMENTS

Research related to this review was supported by the Netherlands Cancer Foundation and by the Netherlands Organization for Scientific Research (NWO).

REFERENCES

Benovic, J.L., M. Bouvier, H.G. Caron, and R.J. Lefkowitz. 1988. Regulation of adenylate cyclase-coupled β-adrenergic receptors. *Annu. Rev. Cell Biol.* **4:** 405.

Chinkers, M., J.A. McKanna, and S. Cohen. 1981. Rapid rounding of human A431 cells by EGF. *J. Cell Biol.* **88:** 422.

Fantus, I.G., S. Kadota, G. Deragon, B. Foster, and B.I. Posner. 1989. Pervanadate mimics insulin action in rat adipocytes via activation of the insulin receptor tyrosine kinase. *Biochemistry* **28:** 8864.

Fukami, K. and T. Takenawa. 1992. Formation of phosphatidic acid in PDGF-stimulated 3T3 cells is a potential mitogenic signal. *J. Biol. Chem.* **267:** 10988.

Gerrard, J.M. and P. Robinson. 1989. Identification of the molecular species of lysophosphatidic acid produced when platelets are stimulated with thrombin. *Biochim. Biophys. Acta* **1001:** 282.

Jalink, K. and W.H. Moolenaar. 1992. Thrombin receptor activation causes rapid and dramatic changes in neural cell shape independent of classic second messengers. *J. Cell Biol.* **118:** 411.

Jalink, K., E.J. van Corven, and W.H. Moolenaar. 1990. Lysophosphatidic acid, but not phosphatidic acid, is a potent Ca^{2+}-mobilizing stimulus for fibroblasts. *J. Biol. Chem.* **265:** 12232.

Jove, R. and H. Hanafusa. 1987. Cell transformation by the viral *src* oncogene. *Annu. Rev. Cell Biol.* **3:** 31.

Lamb, N.J.C., A. Fernandez, A. Watsin, J.-C. Labbé, and J.-C. Cavadore. 1990. Microinjection of cdc2 kinase induces marked changes in cell shape, cytoskeletal organization, and chromatin structure in mammalian fibroblasts. *Cell* **60:** 151.

Moolenaar, W.H. 1991. G-protein-coupled receptors, phosphoinositide hydrolysis and cell proliferation. *Cell Growth Differ.* **2:** 350.

Moolenaar, W.H., K. Jalink, and E.J. van Corven. 1992. Lysophosphatidic acid: A bioactive phospholipid with growth factor-like properties. *Rev. Physiol. Biochem. Pharmacol.* **119:** 47.

Pollard, T.D. and J.A. Cooper. 1986. Actin and actin-binding proteins. *Annu. Rev. Biochem.* **55:** 987.

Rozengurt, E. 1986. Early signals in the mitogenic response. *Science* **234:** 161.

Stossel, T.P. 1989. From signal to pseudopod. How cells control cytoplasmic actin assembly. *J. Biol. Chem.* **264:** 18261.

Tokumura, A., K. Harada, K. Fukuzawa, and H. Tsukatani. 1986. Involvement of phospholipase D in the production of lysophosphatidic acid in rat plasma. *Biochim. Biophys. Acta* **875:** 31.

van Corven, E.J., A. Groenink, K. Jalink, T. Eichholtz, and W.H. Moolenaar. 1989. Lysophosphatidate-induced cell proliferation: Identification and dissection of signaling pathways mediated by G proteins. *Cell* **59:**45.

van Corven, E.J., A. van Rijswijk, K. Jalink, R. van der Bend, W.J. van Blitterswijk, and W.N. Moolenaar. 1992. Mitogenic action of lysophosphatidic acid and phosphatidic acid on fibroblasts. *Biochem. J.* **281:** 163.

van der Bend, R.L., J. de Widt, E.J. van Corven, W.H. Moolenaar, and W.J. van Blitterswijk. 1992a. Metabolic conversion of the biologically active phospholipid, lysophosphatidic acid, in fibroblasts. *Biochim. Biophys. Acta* **1125:** 110.

———. 1992b. The biologically active phospholipid, lysophosphatidic acid, induces phosphatidylcholine breakdown via activation of phospholipase D. *Biochem J.* **285:** 235.

van der Bend, R.L., J. Brunner, K. Jalink, E.J. van Corven, W.H. Moolenaar, and W.J. van Blitterswijk. 1992c. Identification of a putative membrane receptor for the bioactive phospholipid, lysophosphatidic acid. *EMBO J.* **11:** 2495.

Watson, S.P., R.T. McConnell, and E.G. Lapetina. 1985. Decanoyl lysophosphatidic acid induces platelet aggregation through an extracellular action: Evidence against a second messenger role for lysophosphatidic acid. *Biochem. J.* **232:** 61.

G-Protein-linked Signaling Pathways Mediate Development in *Dictyostelium*

R.L. JOHNSON,* R. GUNDERSEN,*‡ D. HERELD,* G.S. PITT,* S. TUGENDREICH,* C.L. SAXE,†§ A.R. KIMMEL,† AND P.N. DEVREOTES*

**Department of Biological Chemistry, Johns Hopkins University School of Medicine, Baltimore, Maryland 21205; †Laboratory of Cellular and Developmental Biology, National Institute of Diabetes and Digestive and Kidney Diseases (6/B1-12), National Institutes of Health, Bethesda, Maryland 20892*

cAMP Signaling during *Dictyostelium* Development

Dictyostelium is a simple model organism for developmental studies. Under conditions of growth, single amoebae comprise a homogeneous population of cells, but when deprived of nutrients, cells aggregate to form a multicellular structure consisting of several different cell types (Fig. 1). Development is first detected when amoebae migrate to central points within a monolayer and form aggregates of many thousands of cells. Cells become tightly adherent within the aggregate to create a mound, upon which a single tip forms. This structure rises, falls over, and forms a migrating pseudoplasmodium or "slug" with the tip forming the anterior end. Slugs are about a millimeter in size and crawl about the surface. This migratory phase, which can be bypassed, is influenced by multiple factors, such as light, temperature, and pH. Culmination results in a sorocarp or fruiting body, which consists of a stalk of dead vacuolated cells supporting an encased spore mass. Exposure of spores to nutrients causes amoebae to emerge and completes the life cycle (Raper 1940; Bonner 1944).

Dictyostelium is a relevant model system for study because its simple program has developmental features in common with more complex organisms such as the vertebrate embryo. Early aggregation involves long-distance signaling when central cells of a territory coordinate the aggregation of many thousands of cells from distances greater than a centimeter. At the mound stage, the tip acts as an organizer to maintain the anterior/posterior polarity of the migrating slug. Grafting of an additional tip onto a slug creates a second axis of polarity, resulting in two slugs that yield independent fruiting bodies (Raper 1940). Unlike more complex systems, such as the frog or mouse embryo, *Dictyostelium* can be cultured in large amounts and is haploid, which permits dissection of the developmental pathway using biochemical, genetic, and molecular biological techniques. Our studies have focused on the signal transduction pathways during development, primarily those utilizing extracellular and intracellular cAMP.

At several points within the developmental program, extracellular cAMP is required as a primary messenger. During early development, aggregation centers spontaneously emerge when the central cells within each territory secrete cAMP. The signal relay results in concentric or spiral cAMP waves that propagate outward at 6-minute intervals. Surrounding cells respond by moving up the cAMP gradient toward the aggregation center and secreting additional cAMP, which relays the signal outward to the territory boundaries (Tomchick and Devreotes 1981). As the cAMP wave passes, chemotaxis and secretion subside, and the cells become refractory to further stimulation. After several minutes of recovery, the cells regain sensitivity and a new wave of cAMP reinitiates the signaling cycle. After the pas-

Present addresses: ‡Department of Biochemistry, Microbiology, and Molecular Biology, University of Maine, Orono, Maine 04469; §Department of Anatomy and Cell Biology, Emory University School of Medicine, Atlanta, Georgia 30322.

Figure 1. The prominent stages of *Dictyostelium* development. From left to right: Single amoebae during growth (about 15 μm); aggregation territories (1–2 cm); tipped mound consisting of tightly packed cells (0.1–0.5 mm); migrating slug (1–2 mm); fruiting body (1–2 mm).

sage of about 30 waves, cells have moved into a single aggregate (Devreotes 1982). cAMP is also important to induce a set of early genes that are required for aggregation. Again, intermittent stimulation appears necessary, since constant exposure to cAMP inhibits the expression of these genes (Gerisch 1987).

During the mound and slug stages, cell-type differentiation occurs. Prestalk cells sort to the tip of the mound, whereas the remainder is mostly composed of prespore cells. Specific marker genes are characteristic of each cell type and require at least two morphogens, cAMP and differentiation inducing factor 1 (DIF-1), for induction. Prespore genes, which are expressed in cells constituting the posterior three quarters of the slug, are induced by micromolar levels of cAMP. In contrast, prestalk A genes require both cAMP and DIF-1 for proper induction, whereas prestalk B genes, a second class, need only DIF-1, and their expression is blocked by cAMP (Gerisch 1987; Williams et al. 1989).

Consistent with the requirement for cAMP at several points in the developmental program, cell-surface cAMP-binding sites have been detected throughout development. Although the largest number of sites is measured during aggregation when cell-cell signaling is occurring, a substantial number are also found during the slug stage (Schaap and Spek 1984). Pharmacological studies have indicated that chemotaxis (Van Haastert and Kien 1983), cell-cell signaling (Theibert et al. 1986), and cAMP-induced gene expression during early and late development (Schaap and Van Driel 1985; Gomer et al. 1986; Haribabu and Dottin 1986; Oyama and Blumberg 1986; Kimmel 1987) are mediated via similar cell-surface receptors. Moreover, much of the cAMP binding during development can be inhibited by GTPγS, which suggests that G proteins are coupled to these receptors (Schenk et al. 1991).

A cAMP Receptor Controls Early Development

A cAMP receptor, cAR1, appears to mediate the developmental events during aggregation. In early development, cAR1 protein expression coincides with that of cell-surface cAMP-binding sites. cAR1 is present at low amounts during growth; becomes induced upon starvation, with peak expression during cell aggregation; and then subsides rapidly (Klein et al. 1987). cAR1 is the major protein photoaffinity labeled with 8-N_3-[^{32}P]cAMP and thus accounts for most of the cAMP binding during aggregation (Klein et al. 1985). In the basal state, cAR1 migrates at 40 kD, but upon cAMP stimulation, it shifts in mobility to 43 kD and acquires at least 4 moles of phosphate per mole of receptor (Klein et al. 1987). The dose response and kinetics of phosphorylation correlate with that of adaptation of cAMP-induced cell shape changes, adenylyl cyclase activity, and myosin phosphorylation (Vaughan and Devreotes 1988).

cAR1 appears to couple to a specific Gα subunit, Gα2. Early studies showed that the addition of guanine nucleotides to membranes from aggregation-competent cells converted the high-affinity cAMP-binding sites to low affinity, indicating the involvement of G proteins (Janssens et al. 1986; Van Haastert et al. 1986). In the mutant frigid A (*fgdA*; Coukell et al. 1983), the guanine nucleotide effects on cAMP binding are greatly reduced, and basal and cAMP-stimulated GTPase activity is lowered (Kesbeke et al. 1988). *fgdA* alleles, which do not enter the developmental program, have defects in the Gα2 locus, which encodes a G-protein α subunit (Kumagai et al. 1989). Furthermore, disruption of the Gα2 gene by homologous recombination results in the *fgdA* phenotype and these cells ($g\alpha 2^-$ cells) have cAMP-binding sites of a single low affinity which are not influenced by GTP (Kumagai et al. 1991). Finally, the expression of Gα2 parallels that of cAR1 during early development (Pupillo et al. 1989).

Analysis of the cAR1 amino acid sequence reveals a topology similar to that of G-protein-coupled receptors found in vertebrates, such as the β-adrenergic receptor and rhodopsin (Fig. 2, top). The proposed structure of cAR1 contains an extracellular amino terminus, followed by seven membrane-spanning regions and ending with a long cytoplasmic carboxyl terminus (Klein et al. 1988). Present within the carboxy-terminal domain are four serine-rich regions, which are the sites of ligand-induced phosphorylation (D. Hereld et al., in

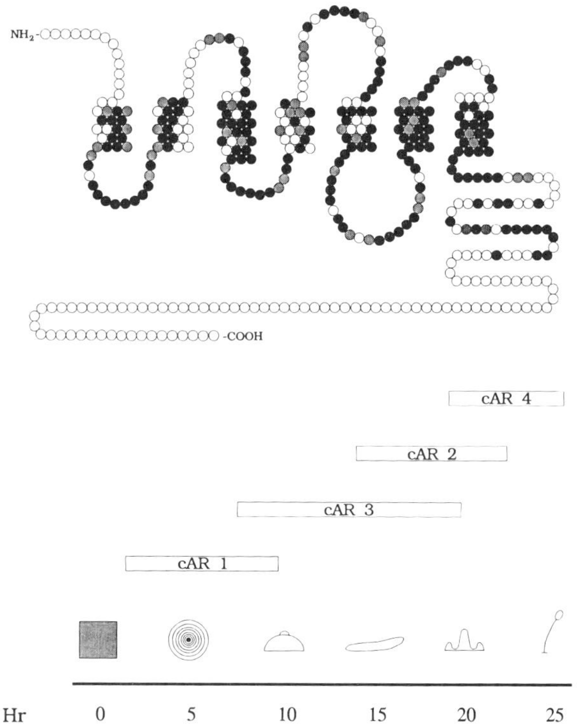

Figure 2. Proposed structure (*top*) and developmental expression (*bottom*) of cAR subtypes. (*Top*) Open circles represent amino acid residues unique to cAR1; gray and black circles indicate homologous and identical residues, respectively, to the three other cAMP receptors. (*Bottom*) The cAR subtypes are expressed sequentially during development but with some overlap. Open bars indicate the developmental expression of the major RNA transcript for each subtype.

prep.). Expression of a cAR1 cDNA in growing *Dictyostelium* cells, when few endogenous receptors are present, results in 30- to 50-fold higher levels of cell-surface cAMP binding and confirms that this receptor is a cAMP-binding protein (Klein et al. 1988).

cAR1 is needed for normal development. Similar to the $g\alpha2^-$ cells, cells that lack cAR1 by gene disruption ($car1^-$ cells) remain as a monolayer of single amoebae. These cells have almost no detectable surface cAMP-binding sites, and although they are chemotactic to folic acid, the $car1^-$ cells no longer respond to normal levels of cAMP. When developed on agar plates, the expression of early genes in the $car1^-$ cells is delayed and late gene expression is blocked. Expression of cAR1 from plasmid restores normal development, demonstrating that the defects result from the lack of this receptor (Sun and Devreotes 1991).

A Family of cAMP Receptor Subtypes

Recently, three additional cAMP receptors, cAR2 (C. Saxe et al., in prep.), cAR3 (R. Johnson et al., in prep.), and cAR4 (J. Louis et al., in prep.), have been identified. These share almost 60% amino acid identity with cAR1 in the transmembrane and loop regions (144 identical residues over 248 total, Fig. 2, top). However, the amino-terminal and carboxy-terminal regions of each receptor vary in amino acid length and composition. The sequence alignment of the four cARs is contiguous throughout the conserved region except for a 2–4-amino-acid gap in the first extracellular loop. This contrasts with the adrenergic and muscarinic receptor families, which have great variation in the third cytoplasmic loop (O'Dowd et al. 1989). The cARs are highly homologous at the nucleotide level as well; each contains an intron of varying size within a conserved position in the third putative transmembrane domain, and their nucleotide identity is similar to that seen at the amino acid level. From this evidence, it appears that these receptors have diverged recently during the evolution of cellular slime molds.

Although the cAR receptor family is highly related at the sequence level, the RNA of each is expressed at different times in development (Fig. 2, bottom). cAR1 is present mainly during early aggregation (Klein et al. 1987), although an additional transcript is induced later at much lower levels (Saxe et al. 1991a). These mRNA forms are transcribed from different promoters, but both encode an identical protein (J. Louis et al., in prep.). cAR3, which follows next, is induced at late aggregation with maximal expression occurring at the mound stage and continuing through later development at reduced levels. cAR2 is enriched in prestalk cells and is expressed initially at the mound stage but predominately at the slug stage (Saxe et al. 1991b). cAR4 is also expressed in prestalk cells and appears at the latest stages of development (J. Louis et al., in prep.).

The presence of four cAMP receptors during development is intriguing. The other receptor subtypes do not appear to act in place of cAR1 since the $car1^-$ cells are blocked in development. Instead, each cAR may mediate specific functions at various developmental stages. The high degree of sequence identity suggests that the differences among the receptors are subtle. To determine important distinctions between three subtypes, we expressed cAR1, cAR2, and cAR3 in growing *Dictyostelium* cells and investigated their biochemical properties.

Characterization of Three cAR Subtypes

Full-length clones of cAR1, cAR2, and cAR3 were ligated into an expression vector (Klein et al. 1988; Johnson et al. 1992) that utilizes an actin promoter for high levels of expression during growth and early development (Knecht et al. 1986). The constructs were stably transformed into wild-type AX-3 cells or the $car1^-$ cells, and clones that expressed high levels of each subtype were characterized further. Cellular membranes immunoblotted with an antibody recognizing a conserved epitope present in all the cARs (M. Caterina et al., in prep.) show that each subtype is expressed at comparable levels during growth (Fig. 3, top). cAR1, cAR2, and cAR3 migrate at 40 kD, 39 kD, and 62 kD, respectively, which are near the sizes calculated from their amino acid sequences. Growing cells transformed with the parent vector contain only low levels of cAR1 (Klein et al. 1988) and no detectable amounts of cAR2 and cAR3. This permits the characterization of each subtype without interference from the endogenous receptors present in development.

Using [^{3}H]cAMP as a ligand, the affinity of each cAR subtype was examined under physiological conditions. cAR1 and cAR3 are expressed at levels of 3×10^5 to 5×10^5 sites per cell and have different affinities for cAMP. cAR1 has two affinity states of 30 nM and 300 nM, which match those of the endogenous cAMP-binding sites expressed during early aggregation (Green and Newell 1975; Van Haastert 1985). cAR3 has two affinities as well, consisting of 20 nM and 500 nM. The majority of binding sites measured for cAR1 and cAR3 were of the lower affinity and, similar to the β-adrenergic receptor, the high-affinity state probably results from the coupling to a G protein (De Lean et al. 1980).

Little or no cAMP binding was detected on the cAR2 cells under physiological conditions. Previous studies on the endogenous receptors in developed wild-type cells have shown that the addition of 3 M ammonium sulfate to the binding mixture greatly increases the affinity for cAMP (Janssens and Van Driel 1984; Van Haastert 1985). Under these conditions, more than 2×10^5 binding sites of 11 nM affinity were measured in the cAR2 cells. Under physiological conditions, cAR2 appears to be a functional receptor, since it can mediate several cAMP-induced responses in vivo (see below). The dose dependency of these responses indicates a cAMP affinity of about 5 μM (J. Milne and P. Devreotes, in prep.), which is beyond the detection limit of the cAMP-binding assay (Johnson et al. 1992).

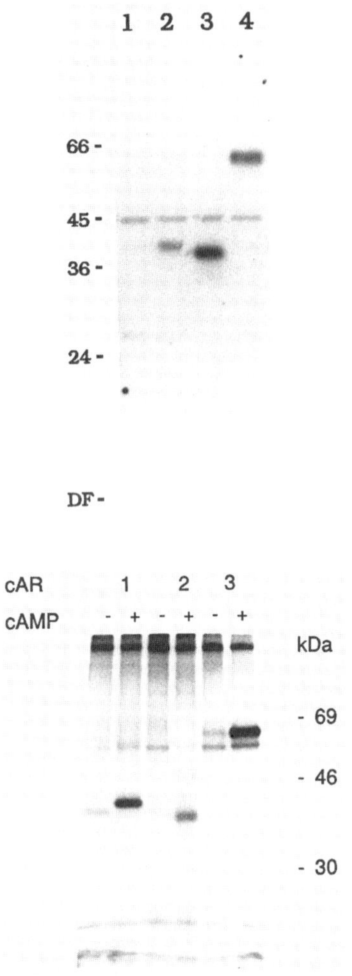

Figure 3. Immunoblot (*top*) and in vivo phosphorylation (*bottom*) of growing *Dictyostelium* cells expressing cAR subtypes from plasmid. (*Top*) Cells transformed with either the parent, cAR1, cAR2, or cAR3 expression vectors (lanes *1–4*, respectively) were examined for protein expression by immunoblotting with a cAR common antiserum. cAR1, cAR2, and cAR3 migrate at 40 kD, 39 kD, and 62 kD, respectively. (*Bottom*) Cells were labeled in vivo with $[^{32}P]H_3PO_4$ and then stimulated in the presence or absence of 1 μM cAMP. The pellets from detergent-extracted cells were then size-fractionated by SDS-PAGE.

When stimulated with cAMP, cAR1 becomes phosphorylated on multiple serines in its carboxy-terminal domain. The kinetics and dose response of this reaction correlate with the adaptation of several ligand-stimulated responses (Vaughan and Devreotes 1988; D. Hereld et al., in prep.). Both cAR2 and cAR3 contain multiple serines and threonines in their carboxy-terminal regions, two of which are conserved among all three subtypes. In vivo labeling experiments show that, similar to cAR1, both cAR2 and cAR3 become highly phosphorylated upon cAMP stimulation in growing cells (Fig. 3, bottom). In the basal state, cAR2 has little or no detectable phosphorylation and appears as a 1–2-kD smaller size, whereas cAR3 has some basal phosphorylation but does not have a detectable shift in apparent molecular weight. Since cAR1 is phosphorylated at multiple sites (Klein et al. 1987), and since comparable levels of each receptor are being expressed, both cAR2 and cAR3 must also acquire multiple phosphates. Amino acid analysis of cAR3 shows that threonine residues are phosphorylated in the basal state, whereas upon cAMP stimulation, serine residues are targeted (data not shown). Although each cAR is phosphorylated, the kinetics of phosphorylation for each varies. cAR1 phosphorylation is half-maximal at 45 seconds and reaches steady state by 5 minutes (Vaughan and Devreotes 1988). The time courses for cAR2 and cAR3 are slower; the half-time of cAR3 phosphorylation is near 4 minutes and reaches a plateau by 10 minutes, whereas that of cAR2 appears to be slower yet (data not shown).

When expressed simultaneously in a cell, cAR1 and cAR2 are differentially responsive to cAMP. cAR2 cells were developed to induce the endogenous cAR1, so that both subtypes are simultaneously present. Following stimulation with a range of cAMP concentrations, cell extracts were prepared and immunoblotted with cAR1 or cAR2 antisera (Fig. 4). Consistent with earlier studies on wild-type cells, the shift in cAR1 mobility begins at 1 nM and is complete at 100 nM cAMP (Vaughan and Devreotes 1988). cAR2, however, does not respond at these low concentrations; its mobility begins to shift at 100-fold higher cAMP concentrations (100 nM), and the transition is incomplete

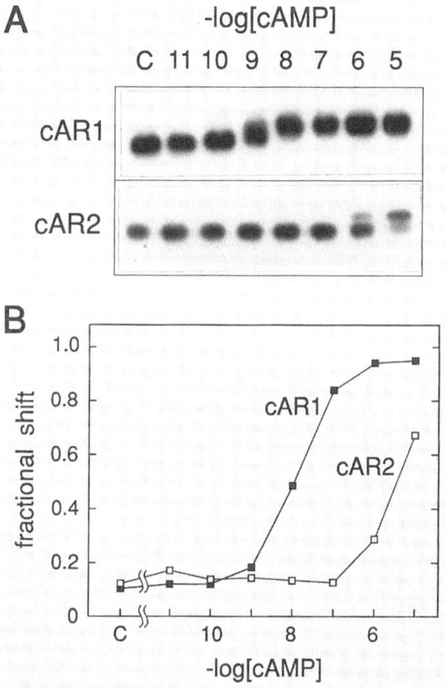

Figure 4. cAMP dose response of cAR1 and cAR2 mobility shift. Cells expressing both cAR1 and cAR2 were stimulated with the indicated concentrations of cAMP (shown in molar units) for 15 min. Unstimulated controls (lane *C*) are also shown. Cell extracts were prepared and immunoblotted with cAR1 or cAR2 antisera.

even at 10 μM. The dose dependency of this response for the two subtypes is consistent with the respective cAMP affinities and is independent of the presence of the other subtype. Since the mobility shift is a reflection of receptor phosphorylation, this result suggests that each receptor may activate its own specific kinase or, perhaps more likely, that only occupied receptors are targets for phosphorylation.

Receptor/G Protein Interactions: A Common Signaling Pathway?

The highly conserved intracellular loops of the cARs suggest that these receptors couple to the same G protein or class of G proteins. In vertebrates, structure/function experiments on rhodopsin and the β-adrenergic receptor have shown the cytoplasmic regions nearest the transmembrane domains are important in receptor/G protein coupling (Strader et al. 1987; O'Dowd et al. 1988; Weiss et al. 1988; Konig et al. 1989; Franke et al. 1990). Chimeras between the α_2- and β_2-adrenergic receptors demonstrate that the third loop is needed for G_s coupling (Koblika et al. 1988), whereas studies on β-adrenergic and muscarinic receptor chimeras indicate that both the second and third loops are required to switch G-protein specificity (Wong et al. 1990). Since these two intracellular loops have about 70% identical and 90% homologous residues among the cARs, one might expect these subtypes to activate the same G proteins.

cAR2 or cAR3 was expressed in the *car1*$^-$ cells and examined for the ability to act in place of cAR1. Neither of the two subtypes was able to restore development, since both transformants remained as a monolayer of amoebae. However, cAR2 and cAR3 could mediate several cAMP-induced responses that are absent in the parent *car1*$^-$ cells. cAMP stimulation of developed wild-type cells causes a G-protein α subunit, Gα2, to shift in mobility from 40 kD to 43 kD, which correlates with the phosphorylation of one or more serine residues (Gundersen and Devreotes 1990). The shift of Gα2, which is transient, peaks after 1–2 minutes of stimulation and returns to basal levels after 15 minutes. As shown in Figure 5, this response is receptor dependent, since in the *car1*$^-$ cells, Gα2 does not shift following ligand stimulation. However, the expression of either cAR1, cAR2 (data not shown), or cAR3 from plasmid restores the ability to shift Gα2 with the normal kinetics. Each subtype can also mediate cAMP-stimulated Ca^{++} uptake, which is absent in the parent *car1*$^-$ cells (J. Milne and P. Devreotes, in prep.). We cannot determine whether the other subtypes activate adenylyl cyclase, since the *car1*$^-$ cells retain this activity, although with altered kinetics (Pupillo et al. 1992).

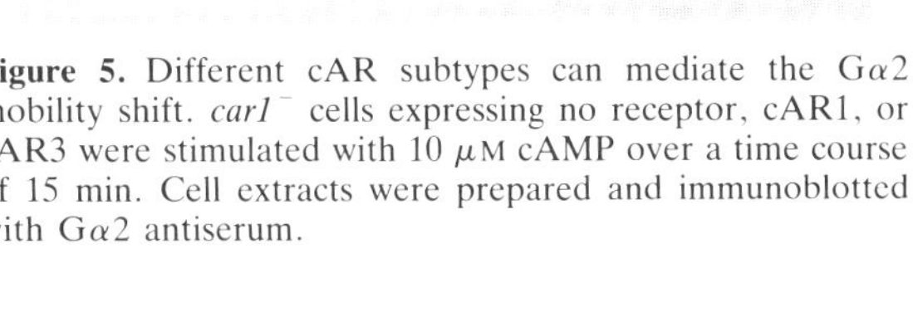

Figure 5. Different cAR subtypes can mediate the Gα2 mobility shift. *car1*$^-$ cells expressing no receptor, cAR1, or cAR3 were stimulated with 10 μM cAMP over a time course of 15 min. Cell extracts were prepared and immunoblotted with Gα2 antiserum.

The above results suggest that the cAR family may use a common signaling pathway. If these receptors have the same ligand and produce the same signal, how might they mediate different developmental processes? Aside from the differences in temporal and possible cell-type expression, the cARs may be differentially regulated by their cAMP affinities and rates of phosphorylation. The subtypes have different activation thresholds, as reflected in different dose responses for the induction of phosphorylation. The duration of the excited state may vary between the subtypes, as reflected in the differences in phosphorylation kinetics. The time course of cAR1 phosphorylation correlates well with adenylyl cyclase adaptation, suggesting that it mediates this process (Vaughan and Devreotes 1988). The longer half-times required for cAR2 and cAR3 to reach steady-state levels of phosphorylation indicate that these subtypes may deliver a prolonged signal. Hence, developmental changes may be mediated by variation in the duration of the signal as opposed to activation of different signals.

Gα2 appears to be a likely candidate for coupling to the three cARs, since the Gα2 protein is expressed throughout development (data not shown). However, an additional seven G α subunits have been identified that have distinct patterns of developmental expression. Gα1, Gα3, and Gα6 are expressed primarily during growth and early development (Pupillo et al. 1989; Firtel 1991; Wu and Devreotes 1991; M. Pupillo et al., in prep.). Gα4 and Gα5 are expressed in later development (Hadwiger et al. 1991), and Gα7 and Gα8 are more broadly expressed throughout development (Wu and Devreotes 1991). These additional α subunits could potentially couple to one or more of the cARs.

Both cAR1 and Gα2 have been implicated in regulating adenylyl cyclase activity. The dose response of adenylyl cyclase activation (Theibert et al. 1986) and adaptation (Dinauer et al. 1980) reflects that of the endogenous cAR1 in aggregation-competent wild-type cells. Furthermore, in the *car1*$^-$ cells, the kinetics and magnitude of adenylyl cyclase activation are altered, and the response no longer adapts after cAMP stimulation (Pupillo et al. 1992). The *fgdA* mutants, which express functional cAMP receptors but not Gα2, do not have cAMP-stimulated adenylyl cyclase activity in vivo. However, GTPγS can stimulate this activity in cellular lysates of *fgdA*, indicating that although Gα2 is part of this signaling pathway, a second GTP-binding protein may be involved (Kesbeke et al. 1988). To further understand the cAMP signaling pathway, the adenylyl cyclase involved in early aggregation was identified.

Identification of *Dictyostelium* Adenylyl Cyclases

Since the adenylyl cyclase activity in aggregation state cells is stimulated by GTP, we predicted that this enzyme would be similar to those identified from mammals and *Drosophila*, which interact with G proteins. Degenerate oligomers, corresponding to conserved regions of a putative catalytic domain of these cyclases, were used in the polymerase chain reaction, and two *Dictyostelium* adenylyl cyclases were identified. One of these, ACA, is expressed during early aggregation. Translation of the 4221-bp sequence results in a 1407-amino-acid protein (Pitt et al. 1992). When ACA is displayed on a hydropathy plot (Kyte and Doolittle 1982), two large hydrophilic regions and twelve hydrophobic, membrane-spanning domains are apparent (Fig. 6, top panel). The proposed topology of ACA is similar to that of the mammalian adenylyl cyclases, which display a duplicated structure consisting of two sets of six transmembrane regions, each followed by an intracellular cytoplasmic domain (Fig. 6, bottom panel). The proposed cytoplasmic domains, C_1 and C_2, share about 50% similarity between each other or to the respective domains in the mammalian Type I adenylyl cyclase (Pitt et al. 1992).

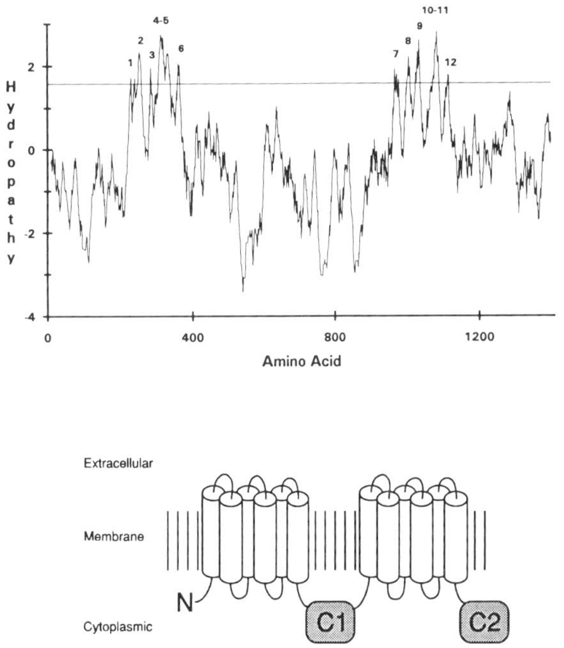

Figure 6. Hydropathy plot and proposed structure of the aggregation adenylyl cyclase (ACA). (*Top* panel) When analyzed on a hydropathy plot, the amino acid sequence of ACA contains three major hydrophilic regions separated by two large hydrophobic areas, each composed of six potential transmembrane domains. (*Bottom* panel) Proposed model of ACA shows a twelve-transmembrane domain protein with intracellular amino and carboxyl termini. Carboxy-terminal to each hydrophobic region are homologous cytoplasmic domains (C_1 and C_2).

The regulation of ACA mRNA was examined throughout development (Fig. 7). ACA appears as a single transcript of 5.8 kb that is not present during growth but is induced during development with peak expression during early aggregation (Pitt et al. 1992). This pattern of RNA expression coincides with that of adenylyl cyclase activity measured during early development (Klein 1976). During late aggregation, ACA levels drop, but expression increases again during later development, suggesting that this adenylyl cyclase may be important at multiple stages of development.

Evidence that ACA is the adenylyl cyclase that mediates cell-cell signaling during aggregation was obtained by targeted gene disruption. Cells that lack ACA (*aca*$^-$ cells), like those that lack cAR1 or Gα2, are blocked in development and remain as single amoebae. Although the *aca*$^-$ cells are chemotactic to cAMP, they have no measurable adenylyl cyclase activity upon Mn^{++} or GTPγS stimulation (Pitt et al. 1992). However, treatment of these cells with exogenous cAMP can induce both early and late cAMP-stimulated gene expression (P. Schaap et al., in prep.). Hence, the defect in the *aca*$^-$ cells lies not in responding to, but in sending, a cAMP signal.

Development can be restored in the *aca*$^-$ cells by expressing ACA from an actin promoter (*aca*$^-$/*ACA* cells). In addition, expression of ACG (*aca*$^-$/*ACG* cells), a topologically distinct adenylyl cyclase expressed only during spore germination, is also able to rescue the developmental block. The amino acid sequence of ACG suggests that it is composed of a large extracellular domain, followed by a single transmembrane region, and ending with a single intracellular catalytic domain. ACA and ACG are similar only in the catalytic regions, where they share about 60% homology.

Unlike the *aca*$^-$/*ACA* cells, the *aca*$^-$/*ACG* cells have adenylyl cyclase activity that is constitutive and not stimulated by GTPγS. In developed wild-type cells, cAMP stimulation induces an activation of adenylyl cyclase that peaks in 1 minute and then subsides rapidly (Dinauer et al. 1980). The *aca*$^-$/*ACA* cells have a fourfold higher activity that peaks after 2 minutes of ligand stimulation and requires a longer time to decay

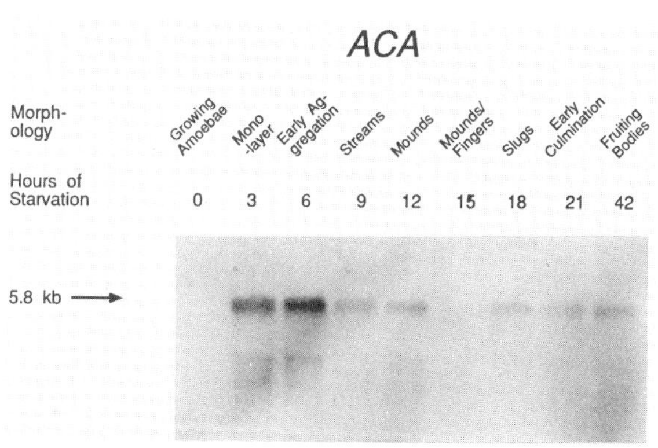

Figure 7. Developmental expression of ACA RNA. Total RNA was isolated at 3-hr time points throughout development and hybridized with a radiolabeled fragment of ACA.

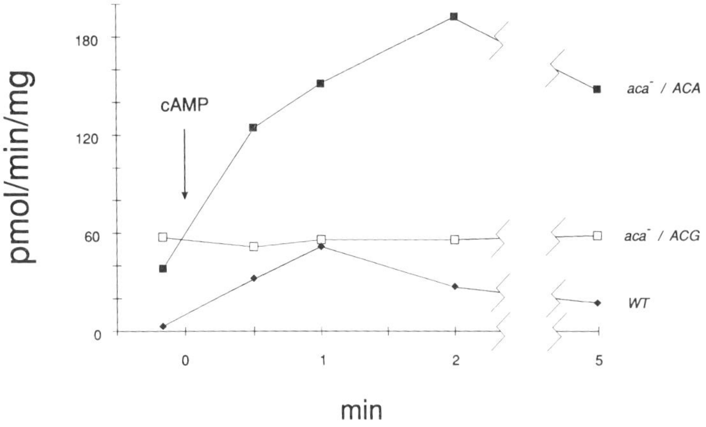

Figure 8. Time course of cAMP-stimulated adenylyl cyclase activity in wild-type and complemented *aca*⁻ cells. Intact cells were stimulated with cAMP for the indicated times and lysed; the lysates were assayed for adenylyl cyclase activity for 1 min. *aca*⁻ cells have no adenylyl cyclase activity (data not shown). Wild-type (◆), *aca*⁻/*ACA* (■), and *aca*⁻/*ACG* (□) responses are indicated.

(Fig. 8). In contrast, the *aca*⁻/*ACG* cells have a high level of basal activity, which is not modulated by the addition of cAMP. The evidence suggests that ACG does not couple to either a cAMP receptor or a G-protein α subunit when expressed during growth or early development (Pitt et al. 1992).

The topological structure of ACA and mammalian adenylyl cyclases is similar to that of the superfamily of twelve transmembrane transporters, which suggests that these cyclases may translocate cAMP across the membrane (Krupinski et al. 1989). However, the ability of ACG to complement the *aca*⁻ cells indicates that cAMP is secreted from these cells despite the fact that ACG possesses a single transmembrane domain. Furthermore, in vivo labeling studies using [^{3}H]adenine show that [^{3}H]cAMP is constitutively secreted from the *aca*⁻/*ACG* cells and accumulated in the media. Hence, ACA may not be required for cAMP export (Pitt et al. 1992).

PERSPECTIVES

The identification and gene disruption of cAR1, Gα2, and ACA, genes that are expressed during early aggregation, have shown that they are critical to the proper development of *Dictyostelium*. Their discovery has led to the isolation of additional related genes, which expands the role of receptor/G-protein signaling throughout the developmental program. What processes these additional cAR subtypes and G-protein α subunits are mediating is just beginning to be realized. The creation of null mutants will allow the determination of each gene's function during development. The involvement of G-protein/receptor signaling in *Dictyostelium* development suggests that this type of signaling system may also be involved in the development of more complex organisms, such as the vertebrate embryo.

ACKNOWLEDGMENTS

This work was supported in part by American Cancer Society grant NP-60833 and National Institutes of Health grants GM-28007 and GM-34933 to P.N.D. and by Biomedical Research Support grant SO-7 RR-05364 from Emory University to C.L.S. D.H. is a Damon Runyon-Walter Winchell Fellow.

REFERENCES

Bonner, J.T. 1944. A descriptive study of the development of the slime mold *Dictyostelium discoideum*. *Am. J. Bot.* **31:** 175.

Coukell, M.B., S. Lappano, and A.M. Cameron. 1983. Isolation and characterization of cAMP unresponsive (frigid) aggregation-deficient mutant of *Dictyostelium discoideum*. *Dev. Genet.* **3:** 283.

De Lean, A., J.M. Stadel, and R.J. Lefkowitz. 1980. A ternary complex model explains the agonist-specific binding properties of the adenylyl cyclase-coupled β-adrenergic receptor. *J. Biol. Chem.* **255:** 7108.

Devreotes, P. N. 1982. Chemotaxis. In *The development of* Dictyostelium discoideum (ed. W. F. Loomis), p. 117. Academic Press, New York.

Dinauer, M.C., S.A. MacKay, and P.N. Devreotes. 1980. Cyclic 3′, 5′-AMP relay in *Dictyostelium discoideum* cells. III. The relationship of cAMP synthesis and secretion during the cAMP signaling response. *J. Cell Biol.* **86:** 537.

Firtel, R.A. 1991. Signal transduction pathways controlling multicellular development in *Dictyostelium*. *Trends Genet.* **7:** 381.

Franke, R.R., B. Konig, T.P. Sakmar, H.G. Khorana, and K.P. Hofmann. 1990. Rhodopsin mutants that bind but fail to activate transducin. *Science* **250:** 123.

Gerisch, G. 1987. Cyclic AMP and other signals controlling cell development and differentiation in *Dictyostelium*. *Annu. Rev. Biochem.* **56:** 853.

Gomer, R.H., D. Armstrong, B.H. Leichtling, and R.A. Firtel. 1986. cAMP induction of prespore and prestalk gene expression in *Dictyostelium* is mediated by the cell-surface cAMP receptor. *Proc. Natl. Acad. Sci.* **83:** 8624.

Green, A.A. and P.C. Newell. 1975. Evidence for the existence of two types of cAMP binding sites in aggregating cells of *Dictyostelium discoideum*. *Cell* **6:** 129.

Gundersen, R.E. and P.N. Devreotes. 1990. In vivo receptor-mediated phosphorylation of a G protein in *Dictyostelium*. *Science* **248:** 591.

Hadwiger, J.A., T.M. Wilkie, M. Strathmann, and R.A. Firtel. 1991. Identification of *Dictyostelium* Gα genes expressed during multicellular development. *Proc. Natl. Acad. Sci.* **88:** 8213.

Haribabu, B. and R.P. Dottin. 1986. Pharmacological characterization of cyclic AMP receptors mediating gene regulation in *Dictyostelium discoideum*. *Mol. Cell. Biol.* **6:** 2402.

Janssens, P.M.W. and R. Van Driel. 1984. *Dictyostelium discoideum* cell membranes contain masked chemotactic receptors for cyclic AMP. *FEBS Lett.* **176:** 245.

Janssens, P.M.W., J.C. Arents, P.J. M.Van Haastert, and R.

Van Driel. 1986. Forms of the chemotactic adenosine 3′, 5′-cyclic phosphate receptor in isolated *Dictyostelium discoideum* membranes and interconversions induced by guanine nucleotides. *Biochemistry* **25:** 1314.

Johnson, R.L., P.J.M. Van Haastert, A.R. Kimmel, C.L. Saxe, B. Jastorff, and P.N. Devreotes. 1992. The cyclic nucleotide specificity of three cAMP receptors in *Dictyostelium. J. Biol. Chem.* **267:** 4600.

Kesbeke, F., B.E. Snaar-Jagalska, and P.J.M. Van Haastert. 1988. Signal transduction in *Dictyostelium fgd* A mutants with a defective interaction between surface cAMP receptors and a GTP-binding regulatory protein. *J. Biol. Chem.* **107:** 521.

Kimmel, A. 1987. Different molecular mechanisms for cAMP regulation of gene expression during *Dictyostelium* development. *Dev. Biol.* **122:** 163.

Klein, C. 1976. Adenylate cyclase activity in *Dictyostelium discoideum* amoebae and its change during differentiation. *FEBS Lett.* **68:** 125.

Klein, P., A. Theibert, D. Fontana, and P.N. Devreotes. 1985. Identification and cyclic AMP-induced modification of the cyclic AMP receptor in *Dictyostelium discoideum. J. Biol. Chem.* **260:** 1757.

Klein, P., R. Vaughan, J. Borleis, and P. Devreotes. 1987. The surface cyclic AMP receptor in *Dictyostelium. J. Biol. Chem.* **262:** 358.

Klein, P., T.J. Sun, C.L. Saxe, A.R. Kimmel, R.L. Johnson, and P.N. Devreotes. 1988. A chemoattractant receptor controls development in *Dictyostelium discoideum. Science* **241:** 1467.

Knecht, D.A., S.M. Cohen, W.F. Loomis, and H.F. Lodish. 1986. Developmental regulation of *Dictyostelium discoideum* actin gene fusions carried on low-copy and high-copy transformation vectors. *Mol. Cell. Biol.* **6:** 3973.

Koblika, B.K., T.S. Koblika, K. Daniel, J.W. Regan, M.G. Caron, R.J. Lefkowitz. 1988. Chimeric α_2-, β_2-adrenergic receptors: Delineation of domains involved in effector coupling and ligand binding specificity. *Science* **240:** 1310.

Konig, B., A. Arendt, J.H. McDowell, M. Kahlert, P.A. Hargrave, and K.P. Hofmann. 1989. Three cytoplasmic loops of rhodopsin interact with transducin. *Proc. Natl. Acad. Sci.* **86:** 6878.

Krupinski, J., F. Coussen, H.A. Bakalyar, W. Tang, P.G. Feinstein, K. Orth, C. Slaughter, R.R. Reed, and A.R. Gilman. 1989. Adenylyl cyclase amino acid sequence: Possible channel- or transporter-like structure. *Science* **244:** 1558.

Kumagai, A., J.A. Hadwiger, M. Pupillo, and R.A. Firtel. 1991. Molecular genetic analysis of two Gα protein subunits in *Dictyostelium. J. Biol. Chem.* **266:** 1220.

Kumagai, A., M. Pupillo, R. Gundersen, R. Miake-Lye, P.N. Devreotes, and R.A. Firtel. 1989. Regulation and function of Gα protein subunits in *Dictyostelium. Cell* **57:** 265.

Kyte, J. and R.F. Doolittle. 1982. A simple method for displaying the hydropathic character of a protein. *J. Mol. Biol.* **157:** 105.

O'Dowd, B.F., R.J. Lefkowitz, and M.G. Caron. 1989. Structure of the adrenergic and related receptors. *Annu. Rev. Neurosci.* **12:** 67.

O'Dowd, B.F., M. Hnatowich, J.W. Regan, W.M. Leader, M.G. Caron, and R.J. Lefkowitz. 1988. Site-directed mutagenesis of the cytoplasmic domains of the human β_2-adrenergic receptor. *J. Biol. Chem.* **263:** 15985.

Oyama, M. and D.D. Blumberg. 1986. Interaction of cAMP with the cell-surface receptor induces cell-type-specific mRNA accumulation in *Dictyostelium discoideum. Proc. Natl. Acad. Sci.* **83:** 4819.

Pitt, G.S., N. Milona, J. Borleis, K.C. Lin, R.R. Reed, and P.N. Devreotes. 1992. Structurally distinct and stage-specific adenylyl cyclase genes play different roles in *Dictyostelium* development. *Cell* **69:** 305.

Pupillo, M., R. Insall, G.S. Pitt, and P.N. Devreotes. 1992. Multiple cAMP receptors are linked to adenylyl cyclase in *Dictyostelium. Mol. Biol. Cell* (in press).

Pupillo, M., A. Kumagai, G.S. Pitt, R.A. Firtel, and P.N. Devreotes. 1989. Multiple α-subunits of guanine nucleotide-binding proteins in *Dictyostelium. Proc. Natl. Acad. Sci.* **86:** 4892.

Raper, K.B. 1940. Pseudoplasmodium formation and organization in *Dictyostelium discoideum. J. Elisha Mitchell Sci. Soc.* **56:** 241.

Saxe, C.L., R. Johnson, P.N. Devreotes, and A.R. Kimmel. 1991a. Multiple genes for cell surface cAMP receptors in *Dictyostelium discoideum. Dev. Genet.* **12:** 6.

———.1991b. Expression of a cAMP receptor gene of *Dictyostelium* and evidence for a multigene family. *Genes Dev.* **5:** 1.

Schaap, P. and W. Spek. 1984. Cyclic-AMP binding to the cell surface during development of *Dictyostelium discoideum. Differentiation* **27:** 83.

Schaap, P. and R. Van Driel. 1985. Induction of post-aggregative differentiation in *Dictyostelium discoideum* by cAMP. *Exp. Cell Res.* **159:** 388.

Schenk, P.W., S. Van Es, F. Kesbeke, and B.E. Snaar-Jagalska. 1991. Involvement of cyclic AMP cell surface receptors and G-proteins in signal transduction during slug migration of *Dictyostelium discoideum. Dev. Biol.* **145:** 110.

Strader, C.D., R.A.F. Dixon, A.H. Cheung, M.R. Candelore, A.D. Blake, and I. S. Sigal. 1987. Mutations that uncouple the β-adrenergic receptor form G_s and increase agonist affinity. *J. Biol. Chem.* **262:** 16439.

Sun, T.J. and P.N. Devreotes. 1991. Gene targeting of the aggregation stage cAMP receptor cAR1 in *Dictyostelium. Genes Dev.* **5:** 572.

Theibert, A., M. Palmisano, B. Jastorff, and P. Devreotes. 1986. The specificity of the cAMP receptor mediating activation of adenylate cyclase in *Dictyostelium discoideum. Dev. Biol.* **114:** 529.

Tomchik, K.J. and P.N. Devreotes. 1981. Adenosine 3′, 5′-monophosphate waves in *Dictyostelium discoideum*: A demonstration by isotope dilution-fluorography. *Science* **212:** 443.

Van Haastert, P.J.M. 1985. The modulation of cell surface cAMP receptors from *Dictyostelium discoideum* by ammonium sulfate. *Biochem. Biophys. Acta.* **845:** 254.

Van Haastert, P.J.M. and E. Kien. 1983. Binding of cAMP derivatives to *Dictyostelium discoideum* cells. *J. Biol. Chem.* **258:** 9636.

Van Haastert, P.J.M., R.J.W. De Wit, P.M.W. Janssens, F. Kesbeke, and J. DeGoede. 1986. G protein-mediated interconversions of cell-surface cAMP receptors and their involvement in excitation and desensitization of guanylate cyclase in *Dictyostelium discoideum. J. Biol. Chem.* **261:** 6904.

Vaughan, R.A. and P.N. Devreotes. 1988. Ligand-induced phosphorylation of the cAMP receptor from *Dictyostelium discoideum. J. Biol. Chem.* **263:** 14538.

Weiss, E.R., D.J. Kelleher, and G.L Johnson. 1988. Mapping sites of interaction between rhodopsin and transducin using rhodopsin antipeptide antibodies. *J. Biol. Chem.* **263:** 6150.

Williams, J.G., K.A. Jermyn, and K.T. Duffy. 1989. Formation and anatomy of the prestalk zone of *Dictyostelium. Development* (suppl.) **107:** 91.

Wong, S.K.F., E.M. Parker, and E.M. Ross. 1990. Chimeric muscarinic cholinergic: β-adrenergic receptors that activate Gs in response to muscarinic agonists. *J. Biol. Chem.* **265:** 6219.

Wu, L. and P.N. Devreotes. 1991. *Dictyostelium* transiently expresses eight distinct G-protein α-subunits during its developmental program. *Biochem. Biophys. Res. Comm.* **179:** 1141.

Molecular Genetic Analysis of Signal Transduction Pathways Controlling Multicellular Development in *Dictyostelium*

A.B. Cubitt, F. Carrel, S. Dharmawardhane, C. Gaskins, J. Hadwiger, P. Howard, S.K.O. Mann, K. Okaichi, K. Zhou, and R.A. Firtel

Department of Biology, Center for Molecular Genetics, University of California, San Diego, La Jolla, California 92093-0634

Understanding how cells within a multicellular organism communicate and coordinate cell growth and differentiation is a fundamental goal of modern biology. Cellular communication within an organism is mediated by the extracellular secretion of factors or by direct cell-cell contacts. In both cases, receptors provide the specificity of information transfer by their ability to discriminate between different signal molecules, by their distribution within the organism, and by their ability to interact with and activate specific downstream effectors. Receptors mediate signal transduction via a limited number of distinct second-messenger systems. However, it is becoming increasingly clear that signal transduction processes interact in a complex fashion within the cell to coordinate cell growth and differentiation.

In the cellular slime mold *Dictyostelium discoideum*, starvation triggers a series of developmental processes that lead to the formation of a multicellular organism. Although *Dictyostelium* represents a developmental system that is simpler than the metazoan system, it is now apparent that the basic mechanisms of signal transduction and developmental progression are similar, if not identical, in *Dictyostelium* and more complex developmental systems. However, in metazoan organisms, the developmental role and/or biochemistry of specific signal transduction pathways can be difficult to elucidate because they can play important roles in controlling cell-fate decisions in a number of tissues and also regulate cell growth. In *Dictyostelium*, however, growth and differentiation are distinct stages in the life cycle, which enables one to isolate mutations within genes that are required for development but are not essential for vegetative growth and to express genes with potentially lethal phenotypes from developmentally specific promoters. The relatively large number of cell-type- and stage-specific promoters that have been cloned and characterized in *Dictyostelium* enables assessment of the role of specific signal transduction molecules when ectopically expressed in particular cells. Furthermore, studies in *Dictyostelium* have the advantage that the cells can be grown and developed synchronously in large numbers and can be transformed at high efficiency, allowing one to obtain sufficient material for biochemical studies.

Although the developmental life cycle of *Dictyostelium* is essentially a continuous process, it can be conveniently divided into four stages: aggregation, mound formation, slug formation, and culmination. The formation of a multicellular organism in *Dictyostelium* is initiated when cells are deprived of a food source (Devreotes 1989; Firtel et al. 1989; Firtel 1991). The process of starvation triggers the induction of developmentally specific early genes required for aggregation and restricts the expression of vegetative specific genes (Gomer and Firtel 1987; Kimmel 1987; Mann and Firtel 1987, 1989). This leads to the autonomous secretion of pulses of cAMP that attract nearby cells and create an aggregation mound. Surrounding cells respond by chemotactically moving up the cAMP gradient and by relaying the signal through receptor-mediated activation of adenylyl cyclase (AC). In addition, cAMP also activates guanylyl cyclase (GC) and phospholipase C (PLC) activities and a slower, cAMP-mediated influx of Ca^{++} (see Europe-Finner and Newell 1987a,b; Devreotes 1989; Firtel et al. 1989; Janssens and van Haastert 1987; van Haastert et al. 1989; Firtel 1991; Milne and Coukell 1991). Extracellular cAMP pulses control the repression of a class of genes induced upon starvation that may be essential for the initial periods of early development and the induction of aggregation-stage-specific genes (Gerisch et al. 1975; Mann and Firtel 1987,1989; Mann et al. 1988). Biochemical and genetic studies have suggested that rises in intracellular cAMP concentrations mediate actin and myosin mobilization and control chemotaxis (see Newell et al. 1987), whereas PLC activation has been postulated to mediate pulse-induced gene expression and may also be essential in controlling association of actin with the membrane during chemotaxis (Mann and Firtel 1989; Firtel 1991; Shariff and Luna 1992). These intracellular second-messenger pathways rapidly adapt after cells are activated by oscillatory pulses of cAMP. High continuous concentrations of cAMP cause the cells to adapt and eventually lead to down-regulation of the signaling pathways. Kinetically, the adaptation of GC and PLC is distinct from the adaptation of AC. The former pathways adapt within 10 and 7 seconds, respectively, and AC activation adapts after 60–90 seconds. The adaptation of the signaling pathways during aggregation is essential for directed movement of cells toward the aggregation center (Dinauer et al. 1980). As a cAMP wave moves outward, cells nearest to the center are activated as the wave reaches these cells, resulting in an inward chemotactic movement up the cAMP gradient. As the wave moves through these cells and activates cells further from the center, the initial cells closer to the center do not respond, having already adapted the GC and PLC pathways. All of the above pathways are mediated by cell-surface cAMP receptors coupled through G proteins. Genes encoding four highly homologous cAMP receptors (cARs) containing seven putative transmembrane domains and eight G-protein α subunits (Gα) have been cloned (Klein et al. 1988; Pupillo et al. 1989; Hadwiger et al. 1991; Saxe et al. 1991a,b; Wu and Devreotes 1991; C. Gaskin et al., in prep.).

During mound formation, cellular differentiation begins and a subpopulation of cells arises in the tip of the developing aggregate that is thought to coordinate the differentiation of the remaining cells in the rapidly developing multicellular organism (Firtel 1991; Kimmel and Firtel 1991; Williams 1991; Schaap 1991; Siegert and Weijer 1991). Approximately 12 hours after the onset of starvation, the aggregation mound undergoes a gross morphological change, elongating to form an extended finger that falls over, producing a migrating pseudoplasmodium or slug. The tip of the slug coordinates slug migration and appears to be important for cell differentiation (Williams et al. 1989a,b). Although the molecular mechanisms whereby this is achieved have not been completely characterized, there is evidence to support the idea that gradients of cAMP and differentiation-inducing factor (DIF) are formed and maintained within the slug (see Firtel 1991; Kimmel and Firtel 1991; Schaap 1991; Williams 1991). During the multicellular stages, continuous fluxes of extracellular cAMP can mediate induction of cell-type-specific genes and are also thought to mediate spatial patterning, suggesting that at this stage, essential second-messenger pathways needed for prestalk and prespore gene expression do not adapt. The four cARs and coupled Gα subunits have been proposed to mediate some of these pathways. cAR1 is preferentially expressed during aggregation and cAR1 null cells lack cAMP-mediated activation of C and GC (Sun et al. 1990; Sun and Devreotes 1991). The roles of specific G proteins and effector pathways in controlling aggregation and cellular differentiation are discussed in subsequent sections.

It is important to note that up until the last 6 hours of the developmental cycle of *Dictyostelium*, cell differentiation is essentially completely reversible. Differentiated cells are capable of reinitiating cell growth when reexposed to a food source (Whittingham and Raper 1960), and if the tip of a slug is separated from the posterior, both parts of the slug are able to redifferentiate to form smaller, but normally patterned, slugs containing prestalk and prespore zones that go on to form complete culminants (Raper 1940). Thus, cell differentiation is maintained by spatial factors within the organism that generate the correct relative proportions of cell types independently of cell number. This process is mediated by transcriptional control of cell-type-specific genes regulated by signal transduction pathways acting under the direction of extracellular signals. Slug migration can continue for a variable length of time, depending on the environmental conditions, but ends when the decision to culminate is made. During this time, the slug undergoes dramatic cell relocalization and specialization to eventually form a mature fruiting body containing spore cells held several millimeters above the substratum by a vacuolated stalk (see Loomis 1982; Devreotes 1989; Firtel 1991; Kimmel and Firtel 1991; Williams 1991).

In this paper, we describe some of the recent approaches our laboratory has undertaken to understand how signal transduction from the cell surface regulates developmental progression in this system. To study the mechanisms controlling these processes, we have investigated the developmental pattern of expression and cell-type localization of a number of key intracellular signal transduction molecules. We have created null, overexpressor, and dominant negative mutations of these signal relay molecules in order to characterize which signal transduction events are occurring at particular stages and in specific cell types. In addition, we have initiated studies to isolate and clone the transcription factors regulated in response to developmental cues, enabling us to build a complete picture of developmental regulation in this organism. These approaches will enable us to characterize when and where programmed changes in gene expression occur and what specific transduction mechanisms cooperate to produce these changes. We have concentrated here on our laboratory's analysis of the role of cell-surface interactions in controlling developmental programs in *Dictyostelium*, with emphasis on the function and regulation of G proteins and the coupled effector and downstream pathways and the potential role of PLC in mediating developmental functions. The potential role of tyrosine phosphorylation in controlling developmental programs is also discussed.

Analysis of Guanine-nucleotide-binding Proteins in Signal Transduction

Heterotrimeric G proteins have been demonstrated to be essential intermediates in transducing information across the plasma membranes from the receptor to the effector in a variety of tissues (Gilman 1987, Johnson and Dhanasekaran 1989; Bourne et al. 1990, 1991; Simon et al. 1991). In many systems, it seems clear that the Gα subunits are primarily responsible for transducing this information; however, there is increasing evidence to support the idea that $\beta\gamma$ may also play a direct role in this process (Tang and Gilman 1991; Tang et al. 1991; Federman et al. 1992). Since many of the details of the cellular interactions involved in multicellular development and coordination in *Dictyostelium* have not been biochemically identified, G proteins provide excellent tools to probe selectively the role of particular communication systems in the developmental process.

At the present time, eight developmentally regulated Gα subunits have been identified in *Dictyostelium* (Fig. 1) (Kumagai et al. 1989,1991; Pupillo et al. 1989; Hadwiger et al. 1991; Wu and Devreotes 1991; C. Gaskins et al., in prep.). Each has a distinct, although overlapping, pattern of expression. Sequence comparison shows that each is approximately 50% identical to any other *Dictyostelium* Gα subunit and Gα subunits from other systems. Several, including Gα4 and Gα8, have amino acid changes in domains that are highly conserved between all Gα subunits. We have previously reported the effects of gene disruption and overexpression of the Gα subunits Gα1 and Gα2 (Kumagai et al. 1989,1991). *gα1* null cells aggregate normally and do not show a growth or developmental abnormality, whereas overexpressor strains, in which Gα1 is expressed from the actin 15 promoter, have large multinucleate cells and aberrant culminants (Kumagai et al. 1989; S. Dharmawardhane and R.A. Firtel, in prep.). *gα2* null cells do not aggregate and lack cAMP-mediated activation of AC, GC, and PLC (Kumagai et al. 1989,1991; Okaichi et al. 1992) but do display cAMP-mediated influx of Ca^{++} (Milne and Coukell 1991). Biochemical studies have shown that Gα2 couples to cAR1 during the aggregation phase of development (Kumagai et al. 1991) which is in agreement with the cAR1 null phenotype (Sun et al. 1990; Sun and Devreotes 1991). Although the analysis of *gα2* null cells suggests that Gα2 is directly involved in activating GC and PLC, GTPγS can activate AC in membranes from *gα2* null cells, suggesting that AC is not directly activated by this Gα subunit in vivo (Kesbeke et al. 1988).

To extend our understanding of the role of these G proteins, we have constructed dominant activating mutations of these proteins. In Ras and a number of Gα subunits, mutations of specific residues within the GTP-binding and GTP hydrolysis domains result in the constitutive activation of that protein. In Ha-*ras*, mutations at position 12 (G12V) or at position 61 (Q61L) result in a reduction of intrinsic GTPase activity, causing *ras* to remain in its active conformation (Bourne et al. 1991). In Gαs, amino acid substitutions in the equivalent positions (positions 43 and 227) also result in the activation of the G protein (Landis et al. 1989; Master et al. 1989; Woon et al. 1989; Gupta et al. 1991). We have used site-directed mutagenesis to construct analogous mutations in Gα1 and Gα2 corresponding at positions G45V and Q206L in Gα1 and at positions G40V and Q208L in Gα2. We have expressed these proteins in wild-type and *gα1* and *gα2* null backgrounds, respectively, to determine the influence of these mutations on the activation of second-messenger systems downstream from Gα2 and on the developmental progression of *Dictyostelium* (Okaichi et al.

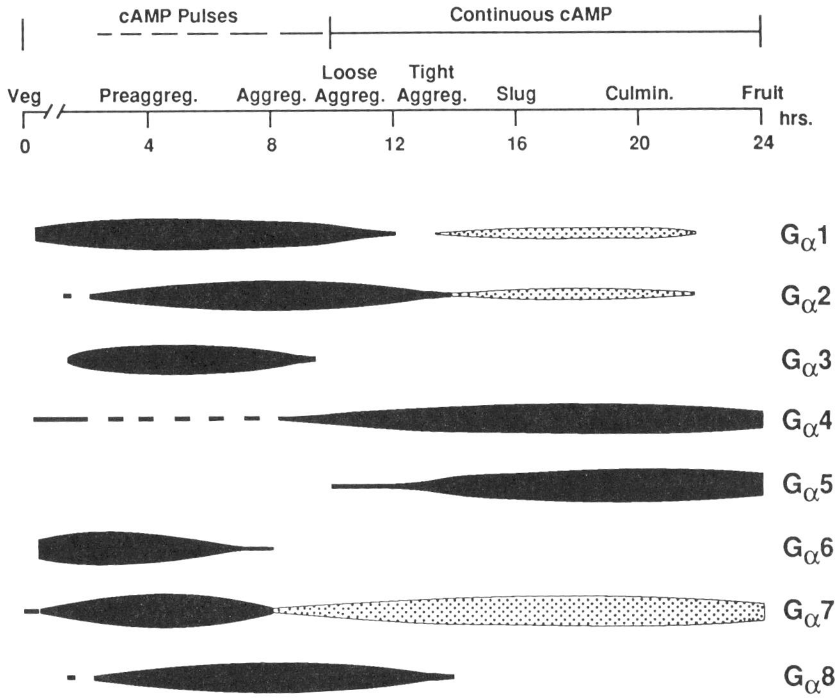

Figure 1. Developmental kinetics of Gα protein expression. The pattern of expression of the eight cloned Gα subunits is shown relative to the morphological differentiation time scale and the patterns of extracellular cAMP signal (see text). Gα1, Gα2, and Gα7 are expressed from two independent promoters. The expression patterns of both the early promoter (black) and late promoter (stippled) are shown. Gα1 and Gα2 are from Pupillo et al. (1989); Gα3 is taken from Wu and Devreotes (1991); Gα4 and Gα5 are from Hadwiger et al. (1991); Gα6, Gα7, and Gα8 are from C. Gaskins et al. (in prep.); see also Wu and Devreotes (1991).

1992; S. Dharmawardhane and R.A. Firtel, in prep.). Gα1 (G45V) expressed in wild-type or *gα1* null cells results in a defective aggregation pattern and a 4–5-day delay in culmination. Both G45V and Q206L mutations result in fruiting bodies with thick stalks, suggesting that Gα1 has a direct role in regulating culmination (S. Dharmawardhane and R.A. Firtel, in prep.).

Expression of either Gα2(G40V) or Gα2(Q208L) in wild-type KAx-3 cells results in a substantial delay in development (Okaichi et al. 1992). In cells expressing Gα2(G40V), aggregation occurs after approximately 10 hours and produces smaller mounds. Small slugs develop after 20 hours and culmination occurs by 30 hours to produce undersize fruiting bodies. Expression of Gα2(Q208L) results in a more severe suppression of development. Aggregation does not occur until 20 hours, and only 20–40% of the cells eventually form mounds. Culmination is delayed up to 36 hours and produces extremely small fruiting bodies.

To characterize the biochemical basis for the observed morphological phenotypes, the kinetics of activation of GC, AC, and PLC by cAMP were examined in KAx-3 cells expressing the mutated proteins and overexpressing the wild-type gene. In cells developed for 6 hours, expression of Gα2(G40V) or Gα2(Q208L) results in an almost complete inhibition of GC activation (Fig. 2) but causes a less severe reduction in the activation of AC in response to receptor activation.

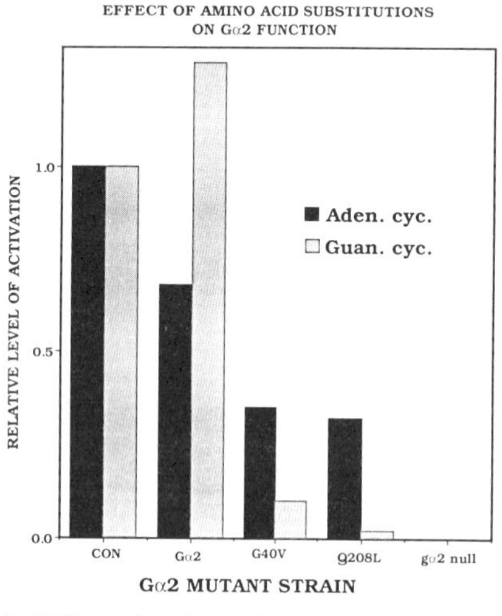

Figure 2. Effect of amino acid substitutions in Gα2 on the activation of AC and GC. Wild-type *Dictyostelium* cells, strain KAx-3, were transformed with Gα2 expression vectors, expressing wild-type Gα2 and Gα2 carrying amino acid substitutions at position 40 (Gα2[G40V]) and 208 (Gα2[Q208L]) from the cloned Gα2 promoter, and stable transformants were isolated. Cells were plated overnight on nonnutrient agar (16 hr, 6°C). Cells were harvested, and cAMP activation of AC and GC were assayed. The kinetics of activation were similar for all strains. The maximum levels of activation, as determined by the maximum level of cAMP or cGMP produced, were compared. The value for control KAx-3 cells (CON) was given a value of 1.0. (gα2) Cells transformed with wild-type Gα2; (*gα2*) *gα2* null cells. In all Gα2 transformants, the Gα2 protein was overexpressed about tenfold compared to the level found in control cells (for details, see Okaichi et al. 1992).

Previous results by Newell and co-workers and van Haastert and co-workers (Europe-Finner and Newell 1987a,b; van Haastert et al. 1989) showed a small, transient activation of PI-PLC (phosphatidylinositol-specific PLC) during the aggregation stage. In our hands, the level of activation varies between 25% and 50%, making a quantitative examination of the effects of Gα2 mutations on PI-PLC difficult to quantitate. Examination of the developmental kinetics of Ca^{++}-stimulated PLC activities during *Dictyostelium* development indicates that levels are highest during vegetative growth and decrease as development proceeds. Analysis of cAMP-receptor-mediated activation of PI-PLC activity shows a more reproducible and extended activation (twofold over 40 sec) in vegetative cells (Fig. 3). As a result, we examined the effect of Gα2 mutations on receptor-activated PI-PLC in vegetative cells. Cells expressing either of the mutated Gα2 proteins fail to exhibit increased inositol-1,4,5-triphosphate (Ins[1,4,5]P_3) levels in response to cAMP. Moreover, *gα2* null cells exhibit no cAMP-mediated activation of PI-PLC, similar to the results seen with GC, suggesting that GC and PI-PLC are regulated similarly by Gα2. Cells expressing Gα2(G40V) or Gα2(Q208L) displayed normal kinetics of cAMP receptor induction for the first 6 hours of starvation and similar total cAMP binding sites compared to KAx-3 cells and KAx-3 cells expressing wild-type Gα2. These results suggest that the lack of cAMP activation was not due to effects on receptor number.

Interestingly, overexpression of wild-type Gα2 causes an enhancement of cAMP-mediated activation of GC at all time points between vegetative growth and the aggregation stage, suggesting that Gα2 protein may be rate-limiting in the activation of GC activation. In contrast, overexpression of wild-type Gα2 results in a reproducible reduction in the level of AC activation (see Fig. 2), suggesting that excess Gα2 subunit was preferentially inhibitory for this pathway. These observations on the effect of Gα2 suggest possible mecha-

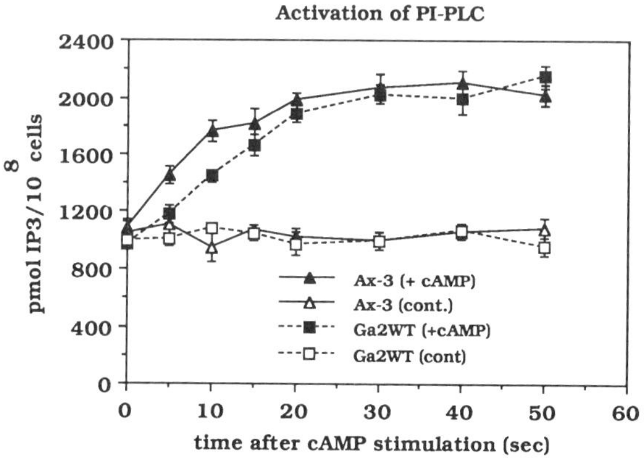

Figure 3. Kinetics of activation of PI-PLC in response to cAMP in strains expressing mutant Gα2 proteins in vegetative cells. Log-phase vegetative cells were harvested, washed, and activated with cAMP. Time points were taken and total Ins(1,4,5)P_3 was assayed. In *gα2* null cells, there is no cAMP activation of PI-PLC (for details, see Okaichi et al. 1992).

nisms by which AC is activated (see below). Cells transformed with wild-type Gα2 expression vector displayed kinetics of cAMP-mediated activation of PI-PLC activity similar to that of control cells.

Our analysis therefore showed that expression of the putative dominant-activating mutations Gα2(G40V) or Gα2(Q208L) in *Dictyostelium* results in the inhibition of multicellular development and signal transduction by cAMP, in contrast to those observed in mammalian cells in which Gαs and Gαi containing these mutations cause constitutive activation of the protein and downstream effectors (see above). In *Dictyostelium*, the apparently opposite effect may indicate that signal transduction mediated by Gα2 is under tight feedback control. In this organism, both GC and PI-PLC rapidly adapt at the aggregation stage (within 7–10 sec), whereas AC adapts within 60–90 seconds. A possible model to explain the apparent dominant negative phenotype is that these mutations may produce a low-level constitutive activation of the PI-PLC and GC pathways, leading to a constitutive adaptation and the inability to be effectively activated by cAMP. *gα2* null cells expressing the Gα2(G40V) protein do not develop at 20°C, but they will undergo development if initially plated at 4°C overnight before being transferred to normal conditions (20°C). At 4°C, GC does not adapt (van Haastert 1987), which may allow the pathways to function and permit development to proceed past a stage in which the adaptation would inhibit further development. Understanding how this adaptation process functions is an important future goal of our research.

Previous results showed that AC could be activated in membranes from *gα2* null cells by GTPγS (Kesbeke et al. 1988), indicating that the Gα2 subunit did not directly activate AC. This suggests that AC is either activated by another Gα subunit or via a different mechanism, possibly by Ca^{++} or by released $\beta\gamma$ subunits. The inhibition of AC activation by overexpressing the wild-type Gα2 subunit, which simultaneously leads to a potentiation of GC activation, is consistent with the model that AC is activated by released $\beta\gamma$ subunits. In this model, GTPγS would activate AC in *gα2* null membranes by releasing $\beta\gamma$ subunits from one or more of the G proteins containing other Gα subunits known to be present during the aggregation stage. Overexpressing Gα2 might therefore inhibit AC by acting as a sink for released $\beta\gamma$ subunits. A model for the role of Gα2 and other molecules in controlling signal transduction processes at the cell surface during aggregation is shown in Figure 4.

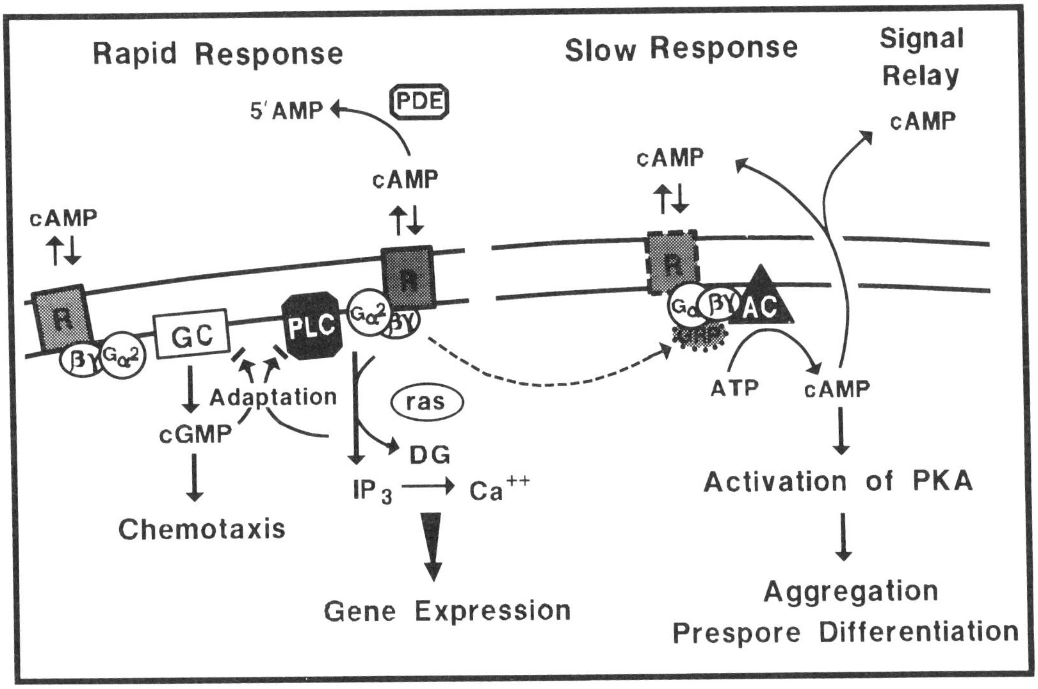

Figure 4. Model of function of cell-surface signaling pathways during aggregation. The responses are divided into a slow pathway involving the activation of AC and the rapid response pathways (GC and PLC). cAMP is secreted/released into the extracellular medium where it can feed back and activate unoccupied receptors on the same cell or activate receptors on adjacent cells, whereby it relays the cAMP signal. As described in the text, present evidence suggests that Gα2 directly activates GC and PLC. Studies with dominant-activating mutants of Gα2 indicate a negative feedback inhibition (adaption) of both of these pathways. Evidence also suggests that $\beta\gamma$ subunits released from the G protein containing Gα2 in response to cAMP may be a direct activator of AC. In addition, a number of other proteins, including GRP (see Theibert and Devreotes 1986), are essential for the in vivo activation of AC by cAMP. Activation of cAMP-dependent protein kinase A (PKA) is shown as a possible intracellular function and is known to also be essential for the in vivo activation of AC. Moreover, PKA null cells cannot activate prespore gene expression (Mann and Firtel 1991; Mann et al. 1992). Coupling of the two pathways is indicated by a dashed arrow. *ras* has been suggested to be involved in the PLC pathway, which is thought to be linked to the activation of gene expression. As in other organisms, *ras* is associated with the plasma membrane but is shown here for cartoon purposes only. Extracellular phosphodiesterase (PDE) clears the cAMP, thus allowing the receptors to deadapt. See text for other references and details.

Role of G Proteins in Controlling Late Development

Gα2 is also expressed during late development from a promoter distinct from the aggregation-stage promoter (F. Carrel and R.A. Firtel, in prep.). To examine the spatial pattern of *Gα2* expression, the isolated *Gα2* late promoter was dissected away from the aggregation-stage promoter and fused to the *Escherichia coli lacZ* reporter gene and transformed into wild-type *Dictyostelium* cells. Staining of β-galactosidase expression during development revealed that expression is first seen during the slug stage and is localized to the anterior of the prestalk region. During culmination, expression remained localized to the tip of the developing fruiting body and in the region of the newly forming stalk cells. Identical patterns of expression were also obtained with the full-length promoter at these times in development. To try to determine if Gα2 plays an essential role during this stage of development, we constructed a putative dominant negative mutation of Gα (Gα2[G206T]). The equivalent mutation in Gαs produces a Gαs subunit that, when expressed in wild-type cells, leads to an inhibition of receptor-mediated activation of AC in mammalian cells (Osawa and Johnson 1991). To investigate the role of Gα2 during culmination specifically, we expressed the altered G protein from the wild-type Gα2 promoter and from the *ecmA* promoter, a strong promoter with a spatial pattern of expression similar to that of the late Gα2 promoter (Williams et al. 1989a,b). Expression of the mutant Gα2 subunit from the aggregation-stage promoter resulted in an inhibition of aggregation (F. Carrel and R.A. Firtel, in prep.). When expressed from the *ecmA* promoter, development was normal until the slug stage. Culmination, however, was abnormal, with an aberrant stalk morphogenesis. Expression of the wild-type Gα2 gene from the *ecmA* promoter and the Gα2(G206T) mutation from the *SP60* promoter (a prespore specific promoter) showed no aberrant development, suggesting that the effect of *ecmA*/Gα2(G206T) expression was specific. These results demonstrate that this mutation can block Gα2 function during aggregation and culmination and suggest that Gα2 has an essential role during culmination, possibly in regulating stalk morphogenesis. Our analysis of Gα1 has suggested that it may have a role in culmination and aggregation (see above). During aggregation and culmination, our analysis suggests that Gα1 may be acting in concert with Gα2 in controlling essential developmental pathways. Further biochemical analysis on strains expressing these mutant Gα subunits should lead to a further understanding of their effects on common effector pathways.

Previous analyses have shown that the induction of prestalk- and prespore-specific gene expression during the formation of the mound is regulated by extracellular cAMP acting via cell-surface receptors, suggesting that G-protein-mediated signaling pathways are essential during the multicellular stages (see Firtel et al. 1989; Firtel 1991; Kimmel and Firtel 1991; Schaap 1991; Williams 1991). Of the eight Gα subunits identified to date, two, Gα4 and Gα5, are preferentially expressed during these stages (Hadwiger et al. 1991). As shown in Figure 1, Gα4, although expressed at a low level in vegetative cells, is induced at high levels at the time of mound formation. Gα5 shows little expression prior to the tipped aggregate stage, whereupon it is induced and expressed through the remainder of development. The developmental role of these two Gα subunits has been examined by creating null mutations via homologous recombination and by overexpressing the proteins from their cloned promoters. In the case of Gα4, both the null and overexpressing strains have distinct morphological phenotypes, suggesting that the level of Gα4 expression is critical in maintaining normal developmental progression (see Fig. 5) (Hadwiger and Firtel 1992). In wild-type cells, aggregates form an apical tip that elongates and produces a finger-like structure. This then falls onto the substratum, forming the migrating slug. In *gα4* null cells, however, cells aggregate normally, form a tip, and start to produce a first finger. Only the apical portion continues to elongate, producing a thin projection that in some cases becomes "knotted," whereas the basal region remains more rounded. The overexpressing strain aggregates normally but is delayed in forming a tight aggregate, which lifts off of the substratum to form a club-like structure (see Fig. 5). In the *gα4* null strains, prespore/spore development seems to be blocked at the stage of tip elongation since prespore-specific genes are only expressed at a very low level and no mature spores are produced ($<0.02\%$ of wild-type cells). Interestingly, the overexpressor cells show a 25-fold reduction in total number of spores produced, suggesting that spore formation is dependent on a finely balanced level of Gα4 expression and other signal transduction pathways.

Analysis of cAMP-induced cell-type-specific gene expression in Gα4 mutant strains showed that the cAMP-induced prestalk gene *pst-cath/CP2* is expressed at almost wild-type levels, whereas expression of the *Dictyostelium ras* homolog, *DdrasD*, is reduced. cAMP-induced prestalk and prespore genes can be induced in single-cell shaking culture by exogenous cAMP, and analysis of this regulation provides insight into cell-autonomous regulation of cAMP-mediated gene expression. Under these conditions, the prespore genes *SP60* and the prestalk-enriched gene *pst-cath/CP2* are expressed at normal levels seen during multicellular development, whereas *DdrasD* is overexpressed severalfold. In the *gα4* null cells, *pst-cath/CP2* was expressed to wild-type levels, *DdrasD* expression was reduced, and *SP60* was expressed to only a low level. In the Gα4 overexpressor line, development was delayed, expression of both *DdrasD* and *SP60* was severely delayed, and expression of *pst-cath/CP2* showed only a small delay. The results indicate that cAMP-regulated cell-type-specific gene expression is specifically effected in Gα4 mutants. The *gα4* null cells can be complemented with a low-copy-number Gα4-expression vector

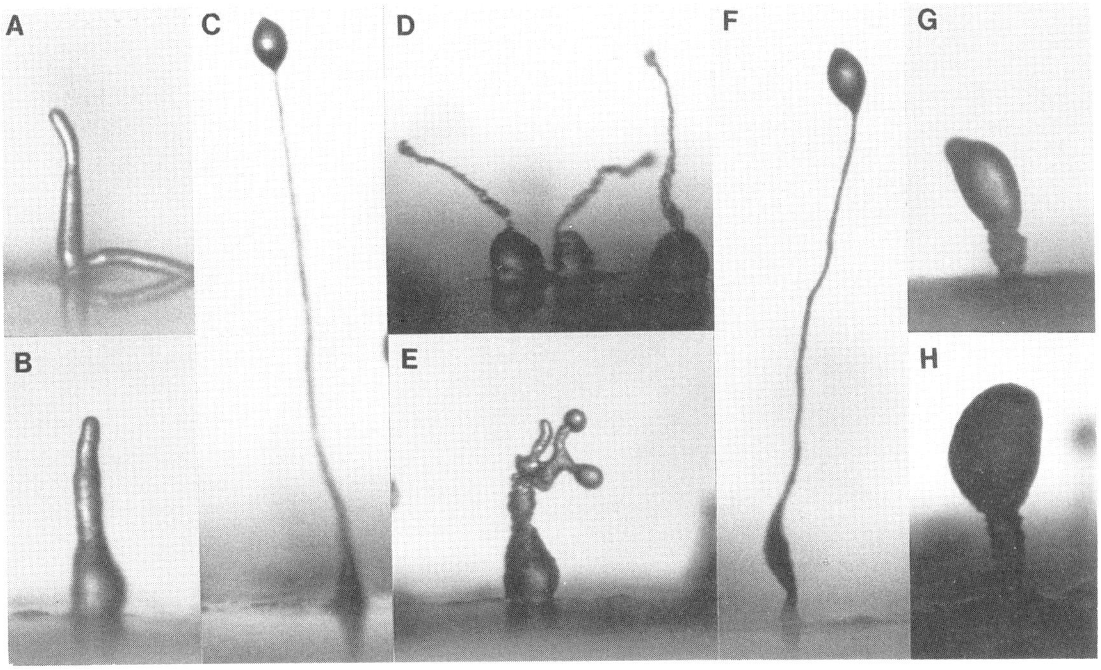

Figure 5. Developmental morphology of *gα4* null, Gα4 overexpressor, and wild-type cells. Logarithmically grown cells were washed and plated for development on nonnutrient agar. (*A*) Wild-type cells at the finger stage (14 hr after starvation); (*B*) *gα4* null cells at finger stage (15 hr after starvation); (*C*) wild-type cells at fruiting body stage (>26 hr after starvation); (*D,E*) *gα4* cells at final morphological stage (photo taken at 36 hr); (*F*) *gα4* null cells complemented with a low copy of the Gα4 expression vector at the fruiting body stage (photo taken at 36 hr); (*G, H*) Gα4-overexpressing cells at the final morphological stage (photo taken at 36 hr) (for details, see Hadwiger and Firtel 1992).

and show normal morphological differentiation, and a wild-type number of spores are produced.

To examine the spatial pattern of Gα4 expression, the cloned *Gα4* promoter was fused to the *E. coli lacZ* reporter gene and transformed into wild-type *Dictyostelium* cells. When the resulting transformants were stained for β-galactosidase expression during development, staining was seen in a subpopulation of cells, comprising an estimated 15% of the total cells. These cells are scattered or randomly distributed throughout the organism, with a slightly higher concentration in the posterior region in the slug stage. The spatial pattern of staining was similar to that of anterior-like cells that eventually form the upper and lower caps of the fruiting body (Williams et al. 1989a,b). At the present time, the function of these cells within the slug is not known. However, the restriction of Gα4 expression to these cells as suggested by the *lacZ* studies, taken together with the results from gene disruption and overexpression studies, suggests that anterior-like cells play an essential role in regulating morphogenesis and gene regulation. Gα4 may be expressed in other cells at a level below the limit of detection of our methods and therefore may serve a function in other cell types.

To determine whether Gα4 functions autonomously in cells, we examined the developmental potential of *gα4* null and Gα4 overexpressor lines in chimeric cell populations. In such experiments, two mutant strains or wild-type and mutant strains are mixed in various proportions, and the developmental potential of both the wild-type and mutant cells is examined. Chimeras of *gα4* null cells or the Gα4-overexpressing cells mixed with wild-type cells at a 50:50 ratio show a morphologically normal-looking fruiting body, suggesting that wild-type cells contribute an intercellular molecule that is deficient in the mutant strains. In addition, when the overexpressor and the *gα4* null cells were mixed, they also produced a normal-looking fruiting body, suggesting that the intercellular signaling deficiencies are different in the two mutants. Analysis of spore formation showed that the Gα4-overexpressing cells in the chimera with either wild-type cells or *gα4* null cells produced spores to a level similar to that of wild-type cells. The number of *gα4* null cells producing spores in either chimera is substantially raised, although the level of spore formation is approximately 3–4% of that of wild-type cells. To determine the spatial distribution of the *gα4* null and wild-type cells, each strain was individually marked with an *Act15/lacZ* promoter construct, which expresses β-galactosidase in all cell types, and mixed with unmarked cells of the other strain. The chimeras were allowed to develop and then stained to examine the location of the cell types within the maturing fruiting body. Analyses of these chimeras indicate that *gα4* null cells are found in all of the cell types of the mature fruiting body but appear to be present at a slightly lower level in the spore mass and a slightly higher relative level in the stalk.

Our analysis of Gα4 suggests that it has both cell-autonomous and non-cell-autonomous functions. The

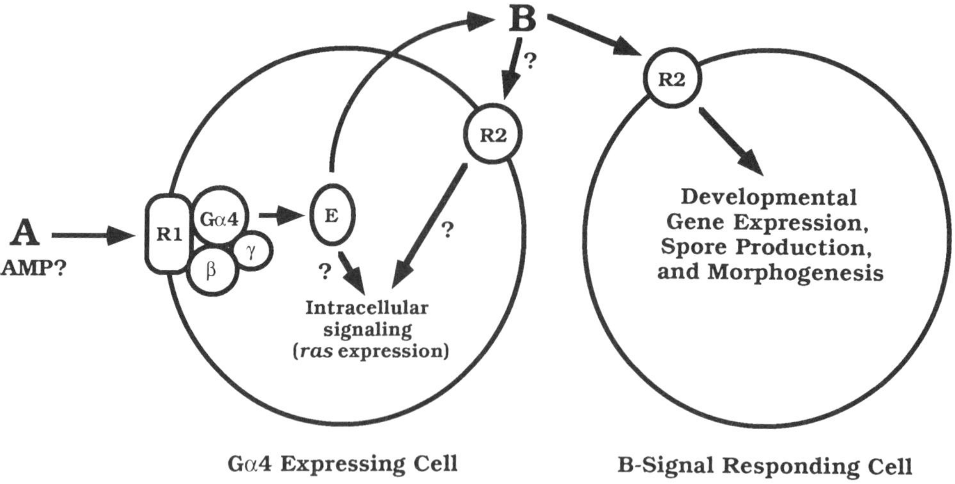

Figure 6. A model of Gα4 function. The Gα4 subunit, in conjunction with Gβ and Gγ subunits, is likely to be involved in the transduction of the signal "A" from a cell surface receptor "R" to an intracellular effector "E". Two candidates for R are the putative cAMP receptors, cAR2 and cAR3, both of which contain seven-transmembrane domains, characteristic of G-protein-coupled receptors, and are expressed during the multicellular stages of development (Saxe et al. 1991a,b). The downstream effector stimulated by the Gα4 subunit might be directly or indirectly responsible for the production of the second intercellular signal "B" that promotes proper morphology and spore production in cells that do not necessarily express the *Gα4* gene. The nature of this second intercellular signal is not known; however, two *Dictyostelium* morphogens, cAMP and DIF, are potential candidates for this signal (see text). In addition to DIF and cAMP, there might be other morphogens not yet identified that could be important for Gα4 function. In addition to producing this second intercellular signal, the Gα4 signal transduction pathway also appears to function cell-autonomously by regulating the expression of genes such as *Ddras* (see above discussion) and affecting plaque growth rates on bacterial plates (for details, see Hadwiger and Firtel 1992).

ability to participate in the formation of a wild-type fruiting body in chimeras with either wild-type or overexpressing cells combined with a partial complementation of the inability of the *gα4* null cells to produce spores suggests that Gα4 may be involved in an extracellular signaling process in which Gα4-producing cells are required for producing either an extracellular soluble factor or a cell-surface molecule that directly interacts with responding cells. Such a signal would appear to be essential for spore cell differentiation. In addition, other data, including the analysis of *DdrasD* expression in shaking culture (see Hadwiger and Firtel 1992), suggest that Gα4 may also have a cell-autonomous function with respect to gene expression. Our results therefore suggest that the Gα4 expression in anterior-like cells is important in controlling spatial patterning during *Dictyostelium* multicellular morphogenesis and prespore differentiation. A model for Gα4 function is outlined in Figure 6.

Gα5 expression initiates shortly after that of *Gα4*. *LacZ* expression studies indicate that Gα5 shows a spatial expression pattern similar to that of Gα4 and suggest that they are expressed in the same cells (J.A. Hadwiger and R.A. Firtel, in prep.). *gα5* null cells show no detectable morphological phenotype but are heterochronic, showing a delayed development starting at the tipped aggregate stage, the time of Gα5 expression. Gα5-overexpressing cells show accelerated development starting at the same time, suggesting that Gα5 plays a role in late development. The spatial pattern suggests that anterior-like cells have an essential cell-cell signaling role controlling spatial regulation during late development.

Characterization of Phospholipase-mediated Signal Transduction

Although we have made rapid advances in identifying and characterizing the presence of multiple cAMP receptor subtypes and Gα subunits, these advances are only the first stage in understanding the control of developmental regulation in this system. To increase our understanding of the role of individual signal transduction pathways during development, we have characterized the potential role of lipid-based signal transduction pathways. In mammalian cells, arguably the best-characterized lipid-based signal transduction system is that utilizing the hydrolysis of the inositol lipids (Berridge 1987). However, in *Dictyostelium*, measurements of inositol phosphates have been limited by the presence of large amounts of higher inositol phosphates that make direct measurements of Ins(1,4,5)P_3 in whole cells difficult (van Haastert et al. 1989; Stephens and Irvine 1990; Stephens et al. 1990). To overcome this problem, we have established methods to efficiently introduce [^{3}H]inositol into cells and to measure inositol phosphate formation free from high levels of [^{3}H]inositol-labeled compounds (Cubitt and Firtel 1992). In addition, we have assessed the potential importance of other lipid-based signal transduction pathways by measuring the hydrolysis of palmitate-labeled phospholipids during development using a 1,2-diacylglycerol (DG) mass assay.

[^{3}H]Inositol was readily taken up in vegetatively growing cells upon electroporation (see Cubitt and Firtel 1992). To measure inositol phosphate formation free from a high background of contaminants, experiments were carried out in membranes prepared from inositol-labeled cells. Under these conditions, a readily measurable membrane-bound PLC activity was activated by calcium. Half-maximal inositol phosphate formation occurred at a free calcium concentration of 2×10^{-7} M and was stimulated within 5 seconds of calcium addition. High-performance liquid chromatography (HPLC) analysis of the inositol phosphates formed revealed that calcium stimulated a rapid increase in Ins(1,4,5)P_3 and inositol 1,4-bisphosphate (Ins[1,4]P_2) after 5 seconds. At later times, additional peaks appeared, consistent with the subsequent metabolism of Ins(1,4,5)P_3 in the cell. To gain some insight into the potential developmental role of inositol-lipid-specific PLC activity, calcium-activated inositol phosphate formation was measured in membranes prepared from cells throughout development. Calcium-activated inositol phosphate formation was maximal in vegetative cells and decreased steadily until about 15 hours (Fig. 7). After this time, there was a small increase in inositol phosphate formation around 21 hours, although the magnitude of this effect was much smaller than that observed in vegetative cells and for the first 6 hours of development.

In many systems, signal transduction mediated by the formation of DG can occur by mechanisms other than inositol lipid hydrolysis. For example, it is clear that in mammalian cells, phosphatidylcholine (Exton 1990) and phosphatidylethanolamine (Hii et al. 1991) are hydrolyzed in response to receptor activation, and these lipids may be much larger sources of DG during sustained receptor activation than the inositol lipids (Exton 1990). To determine if other lipids undergo hydrolysis in *Dictyostelium*, cells were labeled with [^{3}H]palmitate, and phospholipid degradation was examined. Examination of phospholipids by two-dimensional thin-layer chromatography revealed that most classes of phospholipids undergo calcium-activated hydrolysis. The greatest relative decrease in palmitate labeling occurred in cardiolipin and phosphatidylethanolamine, which were labeled to the greatest extent under these conditions. Phosphatidylinositol and phosphatidylserine displayed smaller changes, and the levels of phosphatidylcholine decreased only slightly. These results demonstrate that there are high levels of calcium-activated phospholipase activities directed against a range of phospholipid substrates and that phosphatidylinositol hydrolysis is a relatively minor component of the total phospholipid hydrolysis under these conditions.

Although [^{3}H]palmitate labeling provides a useful method for assessing relative phospholipid hydrolysis, it is a poor method for determining the route or absolute level of phospholipid hydrolysis throughout the developmental cycle of *Dictyostelium*. This is because it is difficult to determine the specific activity of each lipid pool and to ensure that the pool-specific activity does not change during development when lipids undergo extensive remodeling (Ellingson 1974; Hase 1982; MacDonald and Weeks 1985). To overcome these problems, we have used a DG mass assay to measure DG levels during development. To assess the potential developmental role of phospholipid hydrolysis in *Dictyostelium* development, it is necessary to determine when, and to what degree, DG levels are activated. Figure 8 shows the developmental change in DG levels. To achieve this, cells were allowed to undergo multicellular development, and DG measurements were made (Fig. 8). Upon starvation, DG levels initially dropped, as compared to vegetative cell levels for the first 4–6 hours. After this time, DG levels started to increase gradually as aggregation and mound formation occurred, to reach values similar to those in vegetative cells. DG levels then decreased slightly before increasing again about 18 hours at the start of culmination. Between 18 and 22 hours, DG levels increased rapidly (threefold) as cells underwent culmination. By 24 hours, DG levels started to decrease. Total cellular phospholipid levels, in contrast, decreased progressively as development proceeded (twofold), resulting in

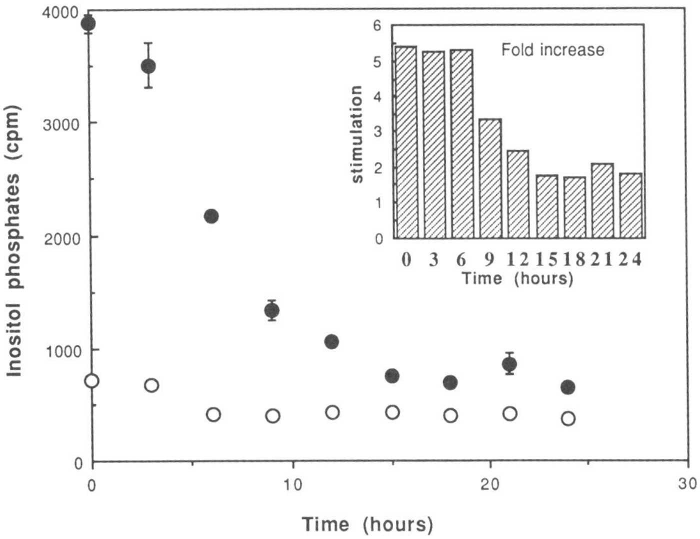

Figure 7. Developmental regulation of inositol-lipid-specific PLC. Cells were labeled with [^{3}H]inositol by electroporation and, after recovery and washing, plated for development as described in Cubitt and Firtel (1992). At the appropriate times in development, cells were washed from plates and membranes were prepared. Membranes were resuspended in assay buffer and incubated in the presence of 2×10^{-5}M free calcium or EGTA alone (control) for 20 min at 25°C. Reactions were terminated by the addition of 1 ml of ice-cold $CHCl_3$/MeOH/HCl, and neutral lipids were separated and measured as described in the above reference. Results represent the means of triplicate determinations plus or minus standard deviation. Symbols without error bars represent points in which the standard deviation was smaller than the size of the symbols. Data represent typical results from an experiment repeated independently three times (for details, see Cubitt and Firtel 1992).

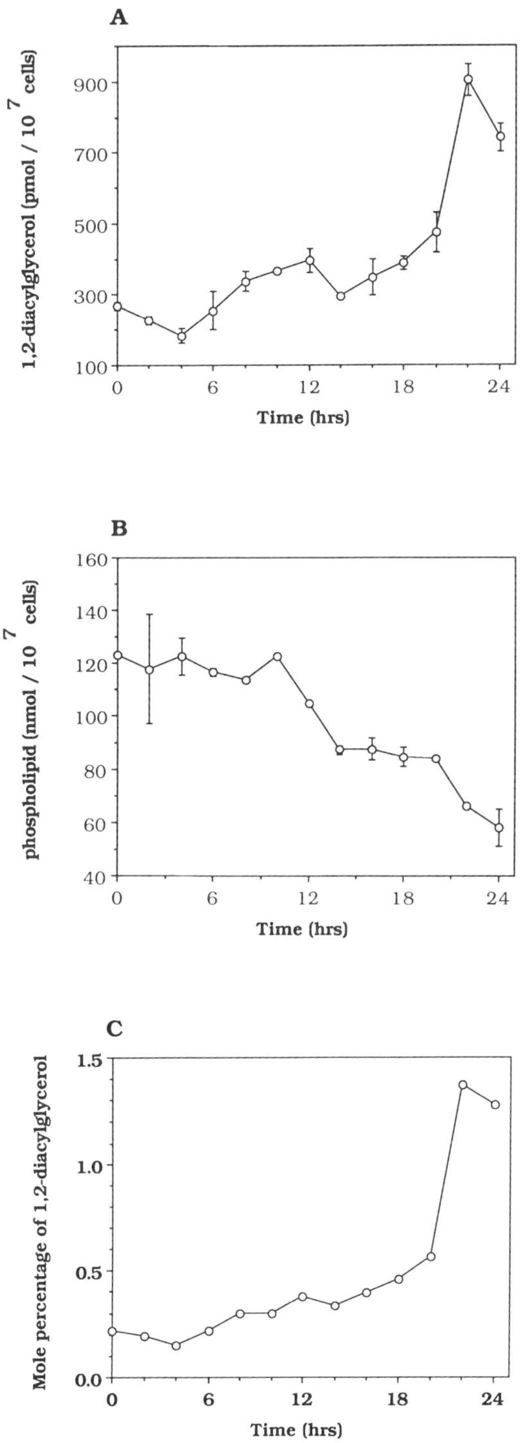

Figure 8. Measurement of DG (*A*) and total lipid phosphorus levels (*B*). (*C*) Mole percentage of DG (DG levels normalized for phospholipid content) in KAx-3 cells during development on filters. Cells (4×10^7) were washed and plated on filters under standard conditions in MES buffer and developed for 24 hr. Samples of cells were removed every 2 hr, and DG and total lipid phosphorus were measured as described recently (A. Cubitt and R.A. Firtel, in prep.). Data represent results from a typical experiment repeated three times. Points represent the means of duplicate samples plus or minus the experimental variation. The midpoint of aggregation is at 8 hr and the start of culmination is at 18 hr. Note the very rapid and large change at culmination.

larger mole percentage increases in DG during development (see Fig. 8B,C).

To examine directly the possibility that extracellular cAMP regulates DG formation, experiments were carried out in the presence or absence of cAMP pulsing in starved cells in shaking culture. Under these conditions, there was a significant increase in DG levels in cAMP-pulsed cells compared to the control nonpulsed cells. Maximal increases in DG were obtained in the *Synag* strain *Agip53* cells (aggregation-deficient cells that lack the ability to produce cAMP in response to receptor activation but can respond to cAMP) with 20 nM pulses of cAMP measured after 11 hours of development (see Fig. 9). To determine if this effect was mediated directly through cAMP receptors, *Agip53* cells were pulsed for 4 hours to initiate development but were subsequently incubated without further additions of cAMP under either fast- or slow-shake conditions. The results from such experiments demonstrated that DG levels increase in the absence of cAMP pulsing when cell-cell contacts are disrupted, compared to cells in which cell-cell interactions are sustained, suggesting that DG levels during aggregation are mediated by mechanisms other than cAMP and implying that cAMP activates DG formation indirectly. *Agip53* cells incubated in the absence of cAMP pulsing fail to exhibit increases in DG when incubated under fast- or slow-shake conditions, demonstrating that the mechanism of DG formation is developmentally regulated and was confirmed using *gα2* and *car1* null strains that are unable to proceed through development. In both of these cell lines, cAMP pulsing failed to elicit increases in DG, although DG levels decreased progressively as the incubation continued.

The results suggest a defined, developmental control of DG changes with a decrease upon starvation and developmentally controlled increases at the time of aggregation and again at culmination. The increases during aggregation are not directly activated by cAMP but do require developmental pathways that are controlled through cAMP-mediated pathways. Mutations lacking cAMP receptor pathways lack the increase in DG during aggregation. The mechanism controlling DG level increase during culmination is not well understood, nor is it known whether this is a general change or whether it is restricted to a specific cell type. Interestingly, recent results indicate that DG can mediate actin polymerization at the plasma membrane and thus may be involved in mediating cell movement (Shariff and Luna 1992). The maximum increases in DG levels observed take place at aggregation and culmination, times of increased cell movement, suggesting that regulated changes in DG levels may have an important role in controlling these pathways.

Regulation of Protein Phosphorylation during Development: Role of Phosphotyrosine Phosphatases and cAMP-dependent Protein Kinase

Protein phosphorylation represents a common mediator for the transduction of intracellular signals

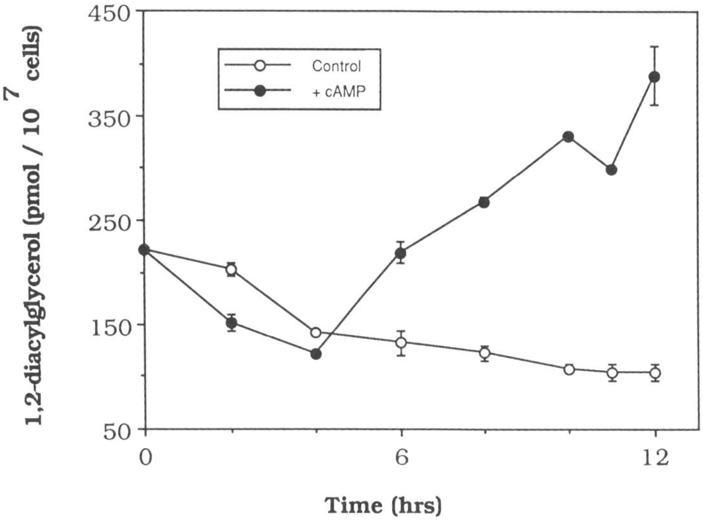

Figure 9. Effect of cAMP pulsing on DG levels in the *Synag* strain *Agip53* cells. Cells were washed twice and resuspended in MES buffer at a density of 4×10^6/ml and pulsed with 20 nM cAMP every 6 min. Aliquots of cells were removed every 2 hr for DG and total lipid phosphorus measurement. Control cells received no cAMP. Data represent results from a typical experiment repeated three times. Points represent the means of duplicate samples plus or minus experimental variation.

into biological responses. Considerable evidence indicates that many effects of intracellular mediators such as calcium, cAMP, cGMP, and DG are regulated by changes in protein phosphorylation in eukaryotes. It is now well established that the majority of protein phosphorylation occurs on serine and threonine residues, whereas phosphorylation on tyrosine occurs only on a minority of proteins (Sefton et al. 1980). Nevertheless, tyrosine phosphorylation (pTyr) serves as an important regulatory mechanism mediating the actions of both growth and differentiation pathways in many cells (Koch et al. 1991). In *Dictyostelium*, two developmentally regulated tyrosine protein kinases have been identified and cloned (Tan and Spudich 1990).

To characterize the pattern of pTyr during developmental progression in *Dictyostelium*, changes in protein pTyr were monitored throughout development by Western blot analysis (Fig. 10) probing with an antiphosphotyrosine-specific antibody. Distinct changes in the degree of pTyr were observed in a number of proteins as development progressed (P. Howard et al., in prep.). These specific, developmentally induced changes in pTyr suggest that this signaling pathway plays a significant regulatory role during development. To evaluate this role further, we have cloned and characterized a protein phosphotyrosine phosphatase (PTP1). Sequence comparisons with other protein tyrosine phosphatases reveal an amino acid identity of between 38% and 50% in the catalytic domain. Hydrophilicity analysis suggests that the protein does not contain a transmembrane-spanning domain and is probably intracellular, although it does contain an amino-terminal myristylation sequence, suggesting that the protein may be tethered to the plasma membrane. When the PTP1-encoded protein is expressed in *E. coli*, the purified protein has pTyr phosphatase activity, indicating that the protein does encode a phosphotyrosine phosphatase. There is a 99-amino-acid insert in the catalytic domain of unknown function. RNA blots show that the gene expresses two RNAs at very low levels during growth. Expression of both transcripts starts to rise prior to aggregation and mRNA levels reach maximum at approximately 10 hours and continue through the remainder of development. At 10 hours, two additional transcripts are seen that also remain through the remainder of development. Interestingly, the gene has a distinct spatial pattern of expression during development. This was examined using a *lacZ* reporter construct containing 900 bp of the 5′ upstream flanking region of PTP1. Analysis of the β-galactosidase staining pattern in cells expressing this construct showed that PTP1 is expressed in prestalk and anterior-like cells.

ptp1 null cells created by a homologous recombination and disruption of the *PTP1* gene show normal morphological development; however, development is accelerated, with the time period to the formation of the mounds decreased by 2 hours. In cells overexpressing PTP1, the developmental consequences of PTP1 overexpression are dependent on the copy number. At high copy number, overexpression of PTP1 leads to a

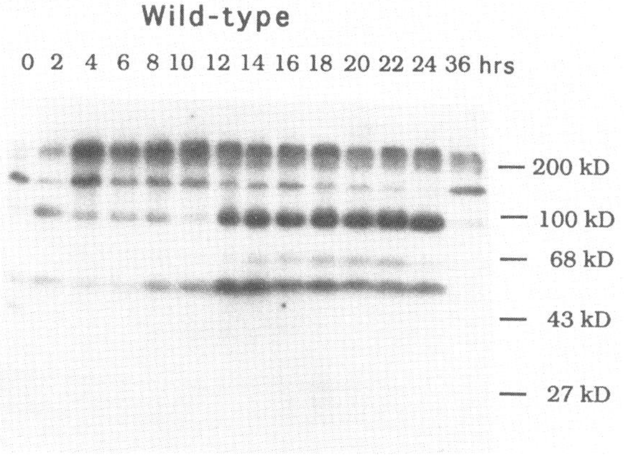

Figure 10. Developmental kinetics of protein tyrosine phosphorylation during *Dictyostelium* development. *Dictyostelium* cell lysates were sized by SDS-PAGE, blotted, and probed with an anti-phosphotyrosine (pTyr) antibody. Specific developmental changes in the pTyr-containing protein pattern can be seen. Aggregation initiates at about 6 hr, a tight aggregate is formed at about 11 hr, a slug at about 16 hr, and culmination is initiated by 18 hr (P. Howard et al., in prep.)

substantial delay, with cells not aggregating until 36–48 hours. Very few cells continue to proceed through development. Strains carrying lower copy numbers of the vector are able to form aggregates but develop an aberrant, multitipped phenotype, and culmination is very aberrant (Fig. 10). *ptp1* cells have an enhanced level of pTyr, whereas overexpression of PTP1 decreases pTyr. Analysis of the pTyr pattern in wild-type cells shows distinct developmental changes (Fig. 10). When this pattern is compared to that seen in *ptp1* null and PTP1 overexpressor lines, we observe changes in the anti-pTyr antibody staining pattern of a subset of pTyr-containing proteins, suggesting that these proteins may be specific substrates of PTP1. Moreover, the timing of major changes between the wild-type and PTP1 overexpressor and *ptp1* null cells corresponds to the time of PTP1 expression. Our analysis indicates that pTyr plays an essential role in *Dictyostelium* development. Although it is too early to assign specific developmental functions to particular pTyr-containing proteins that are affected in the mutants, we expect that at least some of the proteins whose pTyr expression pattern is altered in the PTP1 mutant stains play a role in the observed phenotypes. Finally, we have identified two additional developmentally regulated phosphotyrosine phosphatases in *Dictyostelium*. These genes will also be examined for their role in regulating pTyr-mediated pathways in this system.

cAMP-dependent Protein Kinase

Intracellular signal transduction pathways mediated by extracellular cAMP have been shown to be essential components of the cellular differentiation program in *Dictyostelium*. Activation of AC and the release of cAMP in the extracellular medium regulate chemotaxis during aggregation and are thought to be essential in controlling morphogenesis during multicellular development. Until recently, the potential role of intracellular cAMP, which generally acts via cAMP-dependent protein kinase (PKA) in eukaryotes, was not known. *Synag* mutants that are unable to activate AC and do not have a cAMP-mediated rise in intra- or extracellular cAMP can aggregate in response to exogenous cAMP pulses and thus can form coaggregates with wild-type cells. Many of these mutants go on to differentiate normally, suggesting that their only defect is in the ability to form the initial aggregate, because they are unable to relay a cAMP signal. These results also imply that cAMP-mediated increases in intracellular cAMP are not essential for development to proceed. Most genes that are induced or repressed by cAMP pulses during early development show normal regulation when *Synag* mutants are provided with exogenous pulses of cAMP (Mann et al. 1988). An exception is the repression of M4-1, which appears to require intracellular cAMP signaling (Kimmel and Carlisle 1986). Initially, as an approach to examine the possible role of PKA in development directly, we expressed in *Dictyostelium* the wild-type murine regulatory subunit R and mutant regulatory subunits that are unable to bind cAMP (Firtel and Chapman 1990). The latter would be predicted to act as a dominant negative regulator of PKA activity, binding to the catalytic subunit but not releasing it in response to cAMP. Expression of the dominant negative murine R subunit in wild-type *Dictyostelium* produced cells that do not aggregate but do activate cAMP pulse-induced gene expression. These results are consistent with the observations of Simon et al. (1989), who showed that overexpression of the *Dictyostelium* regulatory subunit results in a similar aggregation-minus phenotype.

To examine the role of intracellular regulation by PKA further, we used polymerase chain reaction (PCR) amplification to clone *Dictyostelium* serine/threonine protein kinases using oligonucleotide primers that showed specificity to conserved regions of the PKA/PKC subfamily. One cloned gene, *DdPK3*, shows approximately 50% sequence identity to both PKA and PKC within the catalytic domain (Mann and Firtel 1991; the same gene was also cloned by Burki et al. 1991). Subsequent sequence and biochemical analyses indicate that *DdPK3* encodes the *Dictyostelium* PKA catalytic subunit (Mann et al. 1992). Amino acid residues have been identified in the mammalian and yeast PKA catalytic subunits that are essential for interaction with the PKA-specific inhibitor PKI and the PKA regulatory subunit. These residues are all conserved in DdPK3. In addition, direct biochemical analysis has shown the presence of PKA activity in crude extracts of wild-type *Dictyostelium* cells and in extracts of *E. coli* expressing DdPK3 with the absence of PKA activity in *Ddpk3* null cells (Mann et al. 1992).

The role of DdPK3 in *Dictyostelium* development is being studied by examining the phenotypes of *Dictyostelium* strains in which the *DdPK3* gene has been disrupted (*Ddpk3* null cells) or in which DdPK3 is overexpressed from cell-type-specific promoters in a wild-type background. *Ddpk3* null cells do not aggregate, as expected from our results, with strains overexpressing the murine dominant negative regulatory subunit. These cells will coaggregate with wild-type cells but are sloughed off from the coaggregates, whereas the wild-type cells proceed through development. This phenotype suggests an impairment in cAMP relay during aggregation as well as in defects in later development. Analysis of the signaling pathways indicated that cAMP-receptor-mediated activation of GC is normal; however, the activation of AC is reduced, and the time course of activation is extended (S.K.O. Mann and R.A. Firtel, unpubl.). These observations suggest that PKA acts in a feedback pathway that is important for the adaptation of AC (the kinetics are extended) and for the activation itself. Most striking is the observation that *Ddpk3* null cells demonstrate normal activation of aggregation-stage pulse-induced genes and cAMP-induced prestalk genes but do not activate prespore genes. This indicates that the PKA catalytic subunit is essential for prespore gene induction and that either cAMP induction of prestalk genes occurs by mechanisms independent of each other or PKA is essential for the pathway that determines prespore specificity. Expression of DdPK3 from the actin-15 promoter rescues the aggregation-deficient phenotype, but morphological development stops at the first finger stage, even though biochemical assays showed that the total level of PKA in the whole organism was similar to that of control cells at the same time. These results suggest that the spatial pattern of DdPK3 activity is altered in these organisms and that PKA activity is essential for later development. We have further investigated the role of PKA activity by expressing DdPK3 downstream from the *SP60* prespore-specific promoter, the *ecmA* promoter (expressed in prestalk A cells), and the *ecmB* promoter (expressed in prestalk B cells). In all cases, distinct morphological phenotypes are obtained (see Fig. 11). For both *ecmA*/*DdPK3* and *ecmB*/*DdPK3* fusions, much of the morphological development is arrested at the tipped aggregate stage, although in some cases, apical mounds are formed and some prespore gene expression is obtained (S.K. Mann and R.A. Firtel, in prep.).

These results, combined with the results of the null mutant described above, indicate that PKA activity plays an essential role in late development. Interestingly, *Synag* cells that cannot activate AC can proceed through development in the presence of wild-type cells and produce spores as well as stalk cells. This suggests that some of the functions of PKA are independent of the rise in intracellular cAMP. These experiments, however, do not address the potential requirement of intracellular cAMP, which may be essential for activating PKA. The *Synag* strains examined have basal AC and thus a level of intracellular cAMP. It is known that the PKA catalytic and regulatory subunits dissociate in crude extracts, possibly indicating that the association between the subunits is weak in vivo. It is therefore possible that the free catalytic subunit exists inside cells and it is this activity that is essential. Williams and co-workers have shown that overexpression of a dominant negative mutant of the *Dictyostelium* regulatory subunit from the *ecmA* promoter inhibits stalk cell differentiation and culmination (Harwood et al. 1992). They have suggested that the results indicate that a rise in intracellular cAMP is essential for initiating culmination and the subsequent terminal differentiation of stalk and spore cells. Although these results demonstrate that this process requires PKA (as does the induction of prespore gene expression), measurements of an actual rise in cAMP at culmination remain controversial, possibly indicating that the primary inductive signal is mediated by other mechanisms.

DISCUSSION AND PERSPECTIVE

In the studies reviewed here, we have attempted to integrate our analysis of several different signal transduction pathways controlled through cell-surface interactions that are involved in the developmental regulation of *Dictyostelium*. Analyses of the effects of gene disruption and overexpression of Gα subunits have indicated potential roles of specific G proteins in regulating development. However, it is worth noting that we need a more complete understanding of the effect of overexpression of Gα subunits on the expression and function of $\beta\gamma$ in this system. Somewhat surprisingly, expression of dominant activating mutations of Gα2 results in a dominant negative phenotype. This result requires further investigation to understand how signal transduction pathways are regulated during aggregation, since it implies the existence of a negative feedback/adaptation pathway that is capable of completely blocking signal transduction mediated by cAMP. These pathways are clearly complex, as shown by a requirement for the PKA catalytic subunit for normal AC activation. Eight Gα subunits have been cloned. De-

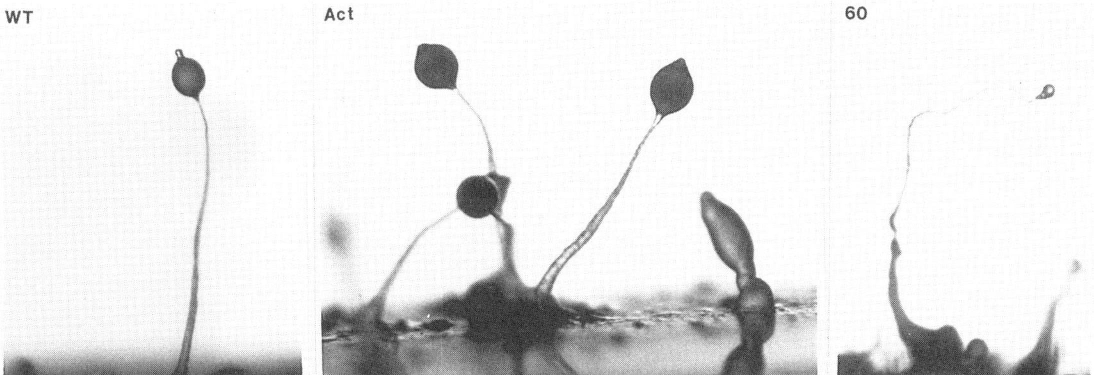

Figure 11. Phenotypes of cells overexpressing the PKA catalytic (PKAcat) subunit. (WT) Wild-type fruiting body; (Act) mature fruiting bodies of cells expressing the gene *DdPK3* encoding PKAcat from the *Act15* promoter on an integrating vector (note that fruiting bodies have a shorter, thicker stalk, larger base, and relatively larger spore mass compared to the wild-type cells); (60) fruiting body formed by cells expressing PKAcat from the *SP60* prespore promoter on an integrating vector (note abnormal structure). Most of the spores are at the bottom (see Mann et al. 1992; S.K.O. Mann and R.A. Firtel, in prep.).

velopmental phenotypes are known for a few, and in some cases (e.g., Gα7, Gα8), null cells or overexpressing cells have no observed phenotype (C. Gaskins et al., in prep.). It is possible that the phenotype is subtle and not detectable under laboratory conditions, that these Gα subunits are redundant, or that they do not have an essential function. Although it is clear that both Gα4 and Gα5 have important functions in controlling later morphogenesis, the downstream effectors and the upstream activators are not known. Further studies need to be directed at specifically defining the upstream activators and downstream effectors. These are only understood, in part, for Gα2 at the present time.

Our results also indicate that intracellular cAMP plays an essential role in regulating gene expression and morphogenesis. We have shown that cAMP-dependent protein kinase plays a role in regulating multicellular development in *Dictyostelium*. It will be interesting to define biochemically and at the molecular level the downstream pathways regulated by PKA. Further studies making use of AC null mutants are required to determine if PKA activity requires rises in intracellular cAMP.

Measurement of Ca^{++}-activated inositol-lipid-specific PLC activity during development revealed a significant membrane activity in vegetative cells and for the first 6 hours of development. Measurement of basal Ins(1,4,5)P_3 levels over the same time frame also revealed a progressive decrease as development proceeded. These results suggest that inositol lipid hydrolysis may be more important during vegetative growth and early development than at later stages. In comparison, DG levels start to increase around the time of aggregation and rise very rapidly during culmination. These results also suggest that inositol lipid hydrolysis alone is not responsible for increases in DG observed during development. We have observed Ca^{++}-activated hydrolysis of a number of phospholipid classes in membranes of *Dictyostelium*, and it is likely that these are also hydrolyzed in vivo to form DG. Understanding the mechanisms and regulation of DG formation is essential to elucidating the role played by DG formation during development. Regulation of DG formation during the aggregation phase was demonstrated to be indirectly activated by cAMP and to be directly influenced by cell-cell interactions. Cell-cell interactions have been established to be important mechanisms of cell regulation in other systems and have been demonstrated to influence prespore gene expression in *Dictyostelium* (Mehdy and Firtel 1985). A direct connection between DG formation, cell-cell contact, and prespore gene expression has not yet been established, although Ginsburg and Kimmel (1989) have shown that DG and Ins(1,4,5)P_3 can induce prespore gene expression. To be able to understand the role of DG changes, the enzymes responsible for DG production (either by phospholipid hydrolysis or lipid synthesis and the phospholipid substrates in the case of PLC/PLD activities) need to be identified and their roles examined using the molecular genetic approaches described here.

Tyrosine phosphorylation has been demonstrated to exert influences over various second-messenger systems, including the activation of phospholipase Cγ, phosphatidylinositol-3 kinase, and the *ras* GTase activating protein GAP. However, in many cases, the regulatory relationships between these activities seem complex since activation of *ras* or phospholipase Cγ (via DG) can influence tyrosine phosphorylation. The identification of the substrates and their function in the cell biology of *Dictyostelium* will be helpful in understanding the role of regulated tyrosine phosphorylation. As with the phosphotyrosine phosphatase, the tyrosine kinases responsible for the phosphorylation and how they are regulated need to be identified. Interestingly, the morphological phenotype of the PTP1 multitipped overexpression phenotype is similar to that of the activating mutation of *DdrasD* (*DdrasD*[G12T]). Moreover, expression of the dominant activating *DdrasD* mutation (*DdrasD*[Q61L]) from prestalk and prespore promoters has a distinct morphological phenotype. Understanding how these pathways interact with the G-protein-activated pathways should provide important information for understanding how developmental signals are integrated in this system.

REFERENCES

Berridge, M.J. 1987. Inositol trisphosphate and diacylglycerol: Two interacting second messengers. *Annu. Rev. Biochem.* **56:** 159.

Bourne, H.R., D.A. Sanders, and F. McCormick. 1990. The GTPase superfamily: A conserved switch for diverse cell functions. *Nature* **348:** 125.

———.1991. The GTPase superfamily: Conserved structure and molecular mechanism. *Nature* **349:** 117.

Burki, E., C. Anjard, J.C. Scholder, and C. Reymond. 1991. Isolation of two genes encoding putative protein kinases regulated during *Dictyostelium discoideum* development. *Gene* **102:** 57.

Cubitt, A. and R.A. Firtel. 1992. Characterization of phospholipase activity in *Dictyostelium discoideum*. Identification of a calcium dependent polyphosphoinositidase specific phospholipase C. *Biochem. J.* **283:** 371.

Devreotes, P. 1989. *Dictyostelium discoideum*: A model system for cell-cell interactions in development. *Science* **245:** 1054.

Dinauer, M.C., T.L. Steck, and P.N. Devreotes. 1980. Cyclic 3′, 5′-AMP relay in *Dictyostelium discoideum*. Adaptation of the cAMP signaling response during cAMP stimulation. *J. Biol. Chem.* **86:** 554.

Ellingson, J.S. 1974. Changes in the phospholipid composition in the differentiating slime mold *Dictyostelium discoideum*. *Biochim. Biophys. Acta* **834:** 301.

Europe-Finner, G.N. and P.C. Newell. 1987a. Cyclic AMP stimulates accumulation of inositol triphosphate in *Dictyostelium*. *J. Cell Sci.* **87:** 221.

———. 1987b. GTP analogues stimulate inositol triphosphate formation transiently in *Dictyostelium*. *J. Cell Sci.* **87:** 513.

Exton, J.H. 1990. Signalling through phosphatidylcholine breakdown. *J. Biol. Chem.* **265:** 1.

Federman, A.D., B.R. Conklin, K.A. Schrader, R.R. Reed, and H.R. Bourne. 1992. Hormonal stimulation of adenyl cyclase through Gi-protein Bc subunits. *Nature* **356:** 159.

Firtel, R.A. 1991. Receptor/G protein-mediated signal transduction pathways controlling *Dictyostelium* development. *Trends Genet.* **7:** 381.

Firtel, R.A. and A.L. Chapman. 1990. A role for cAMP-dependent protein kinase in early *Dictyostelium* development. *Genes Dev.* **4:** 18.

Firtel, R.A., P.J.M. van Haastert, A.R. Kimmel, and P. Devreotes. 1989. G protein linked signal transduction pathways in development: *Dictyostelium* as an experimental system. *Cell* **58:** 235.

Gerisch, G., H. Fromm, A. Huesgen, and U. Wick. 1975. Control of cell-contact sites by cyclic AMP pulses in differentiating *Dictyostelium* cells. *Nature* **255:** 547.

Gilman, A.G. 1987. G proteins: Transducers of receptor-generated signals. *Annu. Rev. Biochem.* **56:** 615.

Ginsburg, G. and A.R. Kimmel. 1989. Inositol triphosphate and diacylglycerol can differentially modulate gene expression in *Dictyostelium*. *Proc. Natl. Acad. Sci.* **86:** 9332.

Gomer, R.H. and R.A. Firtel. 1987. Cell-autonomous determination of cell-type choice in *Dictyostelium* development by cell-cycle phase. *Science* **237:** 758.

Gupta, S.K., C. Gallego, J.M. Lowndes, C.M. Pleiman, C. Sable, B.J. Eisfelder, and G.L. Johnson. 1991. Analysis of the fibroblast transformation potential of GTPase-deficient *gip2* oncogenes. *Mol. Cel. Biol.* **11:** 190.

Hadwiger, J.A. and R.A. Firtel. 1992. Analysis of $G_\alpha 4$, a G protein subunit required for multicellular development in *Dictyostelium*. *Genes Dev.* **6:** 38.

Hadwiger, J.A., T.M. Wilkie, M. Strathmann, and R.A. Firtel. 1991. Identification of *Dictyostelium* $G\alpha$ genes expressed during multicellular development. *Proc. Natl. Acad. Sci.* **88:** 8213.

Harwood, A.J., N.A. Hooper, M.-N. Simon, D.M. Driscoll, M. Veron, and J.G. Williams. 1992. Culmination in *Dictyostelium* is regulated by cAMP-dependent protein kinase. *Cell* **69:** 615.

Hase, A. 1982. Changes in phospholipid composition during the development of *Dictyostelium discoideum*. *Arch. Biochem. Biophys.* **219:** 21.

Hii, C.S.T., Y.S. Edwards, and A.W. Murray. 1991. Phorbol ester-stimulated hydrolysis of phosphatidylcholine and phosphatidylethanolamine by phospholipase D in HeLa cells. *J. Biol. Chem.* **266:** 20238.

Janssens, P.M.W. and P.J.M. van Haastert. 1987. Molecular basis of transmembrane signal transduction in *Dictyostelium discoideum*. *Microbiol. Rev.* **51:** 396.

Johnson, G.L and N. Dhanasekaran. 1989. The G-protein family and their interaction with receptors. *Endocr. Rev.* **10:** 317.

Kesbeke, F., B.E. Snaar-Jagalska, and P.J.M. van Haastert. 1988. Signal transduction in *Dictyostelium fgd* A mutants with a defective interaction between surface cAMP receptor and a GTP-binding regulatory protein. *J. Cell Biol.* **197:** 521.

Kimmel, A.R. 1987. Different molecular mechanisms for cAMP regulation of gene expression during *Dictyostelium* development. *Dev. Biol.* **122:** 163.

Kimmel, A.R. and B. Carlisle. 1986. A gene expressed in undifferentiated vegetative *Dictyostelium* is repressed by developmental pulses of cAMP and reinduced during dedifferentiation. *Proc. Natl. Acad. Sci.* **83:** 2506.

Kimmel, A.R. and R.A. Firtel. 1991. Signal transduction pathways regulating development of *Dictyostelium discoideum*. *Curr. Opin. Genet. Dev.* **1:** 383.

Klein, P., T.J. Sun, C.L. Saxe, A.R. Kimmel, and P.N. Devreotes. 1988. A chemoattractant receptor controls development in *Dictyostelium discoideum*. *Science* **241:** 1467.

Koch, C.A., D. Anderson, M.F. Moran, C. Ellis, and T. Pawson. 1991. SH2 and SH3 domains: Elements that control interactions of cytoplasmic signaling proteins. *Science* **252:** 668.

Kumagai, A., J. Hadwiger, M. Pupillo, and R.A. Firtel. 1991. Molecular analysis of two $G\alpha$ protein subunits in *Dictyostelium*. *J. Biol. Chem.* **266:** 1220.

Kumagai, A., M. Pupillo, R. Gundersen, R. Miake-Lye, P.N. Devreotes, and R.A. Firtel. 1989. Regulation and function of $G\alpha$ protein subunits in *Dictyostelium*. *Cell* **57:** 265.

Landis, C., S. Masters, A. Spada, A. Pace, H. Bourne, and L. Vallar. 1989. GTPase inhibiting mutations activate the alpha chain of Gs and stimulate adenylyl cyclase in human pituitary tumors. *Nature* **340:** 692.

Loomis, W.F., ed. 1982. *The development of* Dictyostelium discoideum. Academic Press, New York.

MacDonald, J.I.S. and G. Weeks. 1985. The biosynthesis and turnover of lipid during the differentiation of *Dictyostelium discoideum*. *Biochim. Biophys. Acta* **834:** 301.

Mann, S.K.O. and R.A. Firtel. 1987. Cyclic AMP regulation of early gene expression in *Dictyostelium discoideum*: Mediation via the cell surface cyclic AMP receptor. *Mol. Cell. Biol.* **7:** 458.

———. 1989. Two-phase regulatory pathway controls cAMP receptor-mediated expression of early genes in *Dictyostelium*. *Proc. Natl. Acad. Sci.* **86:** 1924.

———. 1991. A developmentally regulated, putative serine-threonine protein kinase is essential for development in *Dictyostelium*. *Mech. Dev.* **35:** 89.

Mann, S.K.O., C. Pinko, and R.A. Firtel. 1988. cAMP regulation of early gene expression in signal transduction mutants of *Dictyostelium*. *Dev. Biol.* **130:** 294.

Mann, S.K.O., W.M. Yonemoto, S.S. Taylor, and R.A. Firtel. 1992. *DdPK3*, which plays essential roles during *Dictyostelium* development, encodes the catalytic subunit of cAMP-dependent protein kinase. *Proc. Natl. Acad. Sci.* (in press).

Master, S., T. Miller, M. Chi, F. Chang, B. Beiderman, N. Lopez, and H. Bourne. 1989. Mutations in the GTP-binding site of Gs alpha alter stimulation of adenylyl cyclase. *J. Biol. Chem.* **264:** 15467.

Mehdy, M.C. and R.A. Firtel. 1985. A secreted factor and cyclic AMP jointly regulate cell-type-specific gene expression in *Dictyostelium discoideum*. *Mol. Cell. Biol.* **5:** 705.

Milne, J.L. and M.B. Coukell. 1991. A Ca^{++} transport system associated with the plasma membrane of *Dictyostelium discoideum* is activated by different chemoattractant receptors. *J. Cell. Biol.* **112:** 103.

Newell, P.C., G.N. Europe-Finner, and N.V. Small. 1987. Signal transduction during amoebal chemotaxis of *Dictyostelium discoideum*. *Microbiol. Sci.* **4:** 5.

Okaichi, K., A.B. Cubitt, G.S. Pitt, and R.A. Firtel. 1992. Amino acid substitutions in the *Dictyostelium* Ga subunit Ga2 produce dominant negative phenotypes and inhibit the activation of adenylyl cyclase, guanylyl cyclase, and phospholipase C. *Mol. Biol. Cell.* **3:** (in press).

Osawa, S. and G.L. Johnson. 1991. A dominant negative G alpha s mutant is rescued by secondary mutation of the alpha chain amino terminus. *J. Biol. Chem.* **266:** 4673.

Pupillo, M., A. Kumagai, G.S. Pitt, R.A. Firtel, and P.N. Devreotes. 1989. Multiple α subunits of guanine nucleotide binding proteins in *Dictyostelium*. *Proc. Natl. Acad. Sci.* **86:** 4892.

Raper, K.B. 1940. Pseudoplasmodium formation and organization in *Dictyostelium discoideum*. *J. Elisha Mitchell Sci. Soc.* **56:** 241.

Saxe, C.L., III, R.L. Johnson, P.N. Devreotes, and A.R. Kimmel. 1991a. Expression of a cAMP receptor gene of *Dictyostelium* and evidence for a multigene family. *Genes Dev.* **5:** 1.

———. 1991b. Multiple genes for cell surface cAMP receptors in *Dictyostelium discoideum*. *Dev. Genet.* **12:** 6.

Schaap, P. 1991. Intercellular interactions during *Dictyostelium* development. In *Microbiology cell-cell interactions*, (ed. M. Dworkin), p. 147. American Society for Microbiology, Washington, D.C.

Sefton, B.M., T. Hunter, K. Beemon, and W. Eckhart. 1980. Evidence that the phosphorylation of tyrosine is essential for cellular transformation by Rous sarcoma virus. *Cell* **20:** 807.

Shariff, A. and E.J. Luna. 1992. Diacylglycerol-stimulated formation of actin nucleation sites at plasma membranes. *Science* **256:** 245.

Siegert, F. and C.J. Weijer. 1991. Analysis of optical density wave propagation and cell movement in the cellular slime mould *Dictyostelium discoideum*. *Physica D* **49:** 224.

Simon, M.I., M.P. Strathmann, and N. Gautam. 1991. Diversity of G proteins in signal transduction. *Science* **252:** 802.

Simon, M.-N., D. Driscoll, R. Mutzel, D. Part, J. Williams, and M. Veron. 1989. Overproduction of the regulatory subunit of the cAMP-dependent protein kinase blocks the differentiation of *Dictyostelium discoideum*. *EMBO J.* **8:** 2039.

Stephens, L.R. and R.F. Irvine. 1990a. Stepwise phosphorylation of myo inositol leading to myo inositol hexakisphosphate in *Dictyostelium*. *Nature* **346:** 580.

Stephens, L.R., R.R. Kay, and R.F. Irvine. 1990b. A myo-inositol D-3 hydroxykinase activity in *Dictyostelium*. *Biochem. J.* **272:** 201.

Sun, T.J. and P.N. Devreotes. 1991. Gene targeting of the aggregation stage cAMP receptor cAR1 in *Dictyostelium*. *Genes Dev.* **5:** 572.

Sun, T.J., P.J.M. van Haastert, and P.N. Devreotes. 1990. Surface cAMP receptors mediate multiple responses during development in *Dictyostelium*: Evidenced by antisense mutagenesis. *J. Cell Biol.* **110:** 1549.

Tan, J.L. and J.A. Spudich. 1990. Developmentally regulated protein-tyrosine kinase genes in *Dictyostelium discoideum*. *Mol. Cell. Biol.* **10:** 3578.

Tang, W. and A.G. Gilman. 1991. Type-specific regulation of adenylyl cyclase by G protein $\beta\gamma$ subunits. *Science* **254:** 1500.

Tang, W.J., J. Krupinski, and A.G. Gilman. 1991. Expression and characterization of calmodulin-activated (type I) adenylyl cyclase. *J. Biol. Chem.* **266:** 8595.

Theibert, A. and P.N. Devreotes. 1986. Surface receptor-mediated activation of adenylate cyclase in *Dictyostelium*: Regulation by guanine nucleotides in wild-type cells and aggregation deficient mutants. *J. Biol. Chem.* **261:** 15121.

van Haastert, P.J.M. 1987. Differential effect of temperature on cAMP-induced excitation, adaptation, and deadaptation of adenylate and guanylate cyclase in *Dictyostelium discoideum*. *J. Cell Biol.* **105:** 2301.

van Haastert, P.J.M., M.J. de Vries, L.C. Penning, E. Roovers, J. van der Kaay, C. Erneux, and M.M. van Lookeren Campagne. 1989. Chemoattractant and guanosine 5′-[γ-thio]triphosphate induce the accumulation of inositol 1,4,5-triphosphate in *Dictyostelium* cells that are labelled with [^{3}H]inositol by electroporation. *Biochem. J.* **258:** 577.

Whittingham, W.F. and K.B. Raper. 1960. Non viability of stalk cells in *Dictyostelium*. *Proc. Natl. Acad. Sci.* **46:** 643.

Williams, J.G. 1991. Regulation of cellular differentiation during *Dictyostelium* morphogenesis. *Curr. Opin. Genet. Dev.* **1:** 358.

Williams, J., K. Jermyn, and K. Duffy. 1989a. Formation and anatomy of the prestalk zone of *Dictyostelium*. *Development* **7:** 91.

Williams, J.G., K.T. Duffy, D.P. Lane, S.J. McRobbie, A.J. Harwood, D. Traynor, R.R. Kay, and K.A. Jermyn. 1989b. Origins of the prestalk-prespore pattern in *Dictyostelium* development. *Cell* **59:** 1157.

Woon, C., L. Heasley, S. Osawa, and G. Johnson. 1989. Mutation of glycine 49 to valine in the alpha subunit of Gs results in the constitutive elevation of cyclic AMP synthesis. *Biochemistry* **28:** 4547.

Wu, L. and P.N. Devreotes. 1991. *Dictyostelium* transiently expresses eight distinct G-protein α-subunits during its developmental program. *Biochem. Biophys. Res. Commun.* **179:** 1141.

Signaling between the Extracellular Matrix and the Cytoskeleton: Tyrosine Phosphorylation and Focal Adhesion Assembly

L.H. ROMER,* K. BURRIDGE,† AND C.E. TURNER‡
*Department of Pediatrics, Division of Critical Care, and †Department of Cell Biology and Anatomy, University of North Carolina, Chapel Hill, North Carolina 27599;
‡Department of Anatomy and Cell Biology, SUNY Health Science Center, Syracuse, New York 13210

Recent developments in several areas of cell and developmental biology support the concept that adhesion molecules serve receptor and signal transduction functions in addition to more classically defined structural roles (Takeichi 1988; Albelda and Buck 1990; Hynes 1992). Adhesion receptors, including integrins, cadherins, and members of the immunoglobulin superfamily, influence many aspects of cell behavior, including growth control, differentiation, and motility (Edelman 1986; Buck and Horwitz 1987; Takeichi 1991; Schwartz et al. 1991). The integrin family of adhesion receptors mediates the interaction of cells with many different extracellular matrix (ECM) proteins. Specificity is imparted to this adhesive interaction by the combination of particular α and β chains to form integrin heterodimers that may recognize single or several ECM proteins (Albelda and Buck 1990; Hemler 1990; Hynes 1992). Integrin subunits span the plasma membrane and, in general, have relatively short and highly conserved cytoplasmic domains. Information initiated by the interactions of these receptors with the ECM is communicated across the cell membrane and then modulates cell behavior and cytoskeletal organization. In cultured cells, interactions of transmembrane integrin receptors with the cytoskeleton occur at focal adhesions, areas of tight juxtaposition between the cell surface and the ECM (Izzard and Lochner 1976; Burridge et al. 1988). In vitro binding experiments suggest that direct interactions occur between integrins and the focal adhesion proteins talin and α-actinin (Horwitz et al. 1986; Otey et al. 1990). Signaling mechanisms that facilitate focal adhesion formation and the accompanying cytoskeletal reorganization are incompletely understood but under intense scrutiny. For example, protein kinase C has been found in focal adhesions (Jaken et al. 1989), and the protein kinase C inhibitors H7 and HA1004 inhibit focal adhesion formation in rat and human fibroblasts (Woods and Couchman 1992). Furthermore, phorbol myristyl acetate, which activates protein kinase C, induces focal adhesion formation in fibroblasts cultured on the cell-binding domain of fibronectin (Woods and Couchman 1992).

There is also evidence that tyrosine kinases may play a role in integrin-mediated signal transduction. Phosphorylated tyrosine residues have been localized in the focal adhesions of normal cells (Maher et al. 1985), although the responsible kinases have been elusive. The tyrosine kinase pp60$^{v\text{-}src}$ has been localized to residual focal adhesions in Rous sarcoma virus (RSV)-transformed cells (Rohrschneider 1980). Upon platelet aggregation, the related tyrosine kinase pp60$^{c\text{-}src}$ has been noted to associate with cytoskeletal elements (Horvath et al. 1992), and the tyrosine phosphorylation that follows platelet aggregation is dependent on the interaction of the integrin IIb/IIIa with the ECM protein fibrinogen (Ferrell and Martin 1989; Golden and Brugge 1989; Golden et al. 1990). In addition, in platelets activated with thrombin, the p21^{ras} GAP protein has been noted to be phosphorylated on tyrosine after close association with the Src family tyrosine kinases Fyn, Lyn, and Yes (Cichowski et al. 1992). In KB cells and NIH-3T3 cells, antibody clustering of cell-surface integrins and plating cells on fibronectin have been shown to be associated with an integrin-dependent rise in the tyrosine phosphorylation of 130-kD proteins that are distinct from the α and β integrins (Guan et al. 1991; Kornberg et al. 1991). The discovery of the novel tyrosine kinase pp125^{FAK} localized in the focal adhesions of normal fibroblasts suggests a potential link in integrin-mediated signal transduction between the ECM and the cytoskeleton (Schaller et al. 1992).

We describe here the identification of two focal adhesion constituents, pp125^{FAK} and paxillin, as tyrosine kinase substrates that are phosphorylated during integrin-mediated adhesion of rat embryo fibroblasts (REF52) and NIH-3T3 cells to ECM. Both paxillin and the tyrosine kinase pp125^{FAK} are pp60$^{v\text{-}src}$ substrates in RSV-transformed cells (Glenney and Zokas 1988; Kanner et al. 1990). In normal cells, the 70-kD phosphoprotein paxillin binds to vinculin (Turner et al. 1990). Paxillin is also a major tyrosine kinase substrate during both avian and mammalian embryonic development (Turner 1991). We blocked the adhesion-induced elevation in phosphotyrosine in pp125^{FAK} and paxillin with the tyrosine kinase inhibitor herbimycin A. The inclusion of herbimycin A in the culture medium also interfered with normal focal adhesion and stress fiber assembly. These data implicate tyrosine phosphorylation as an important element in the series of events that link cell-ECM adhesion to cytoskeletal organization.

METHODS

Cell culture. Rat embryo fibroblasts (REF52) were grown in Dulbecco's modified Eagle's medium, supplemented with 10% fetal calf serum, 50 units/ml penicillin, and 50 μg/ml streptomycin. Mouse NIH-3T3 cells were grown in the same medium with 10% calf serum. Cells were passaged 1–3 days prior to use. For adhesion experiments, the NIH-3T3 cells were incubated with cycloheximide at 25 μg/ml for 2 hours. The cells were trypsinized, washed in serum-free medium containing 1 mg/ml turkey egg white trypsin inhibitor (Sigma Chemical Co., St. Louis, Missouri), washed again in serum-free medium, and plated in serum-free medium containing 25 μg/ml cycloheximide. REF52 cells were not treated with cycloheximide. These cells were trypsinized, washed once in medium containing serum, and washed twice in serum-free medium before plating in serum-free medium. Culture dishes or coverslips were coated with human plasma fibronectin (New York Blood Center) at 0.1 mg/ml either for 90 minutes at 37°C or overnight at 4°C. Culture dishes were coated with poly-L-lysine (Sigma) at 0.5 mg/ml for 90 minutes at 37°C. In experiments using herbimycin A (Gibco BRL, Gaithersburg, Maryland), cells were preincubated with, and plated 24 hours later in the presence of, 875 nM herbimycin A dissolved in DMSO. These cells were plated on fibronectin-coated surfaces.

SDS-PAGE and Western blotting. Samples were electrophoresed on 7.5% or 10% SDS-polyacrylamide gels (Laemmli 1970), except that bisacrylamide was used at concentrations of 0.19% and 0.13%, respectively. For analysis of whole cells, the medium was removed and the cells were dissolved by scraping with a Teflon policeman in 150 μl of Laemmli sample buffer containing 1 mM Na orthovanadate. The samples were boiled for 2 minutes and the DNA was sheared by passing multiple times through a 26 gauge needle. After electrophoresis, gels were either transferred to nitrocellulose sheets (Towbin et al. 1979) or stained with Coomassie blue. For Western blotting with antiphosphotyrosine, the nitrocellulose sheets were blocked with 5–10% cold fish gelatin (Sigma) in Tris-buffered saline (100 mM NaCl, 0.1% azide, 50 mM Tris Cl [pH 7.6]) containing 0.05% Tween-20 and then incubated for 1 hour with either ^{125}I-labeled py20 or unlabeled PT66 (Sigma). The nitrocellulose was then washed for 1 hour. Blots using PT66 were then incubated with affinity-purified rabbit anti-mouse IgG (Jackson ImmunoResearch, West Grove, Pennsylvania), followed by washing and a final incubation with ^{125}I-labeled protein A (8×10^5 cpm/ml) (a gift from Dr. S. Earp). After a final wash, the nitrocellulose was exposed with an intensifying screen to X-ray film at −70°C for autoradiography. py20 was a gift from Dr. John Glenney (University of Kentucky) or was purchased from ICN Biochemical, Inc. (Irvine, California). py20 was labeled with ^{125}I (New England Nuclear) using iodogen (Pierce Chemical Co., Radford, Illinois) (Fraker and Speck 1978) and used at 2×10^6 to 4×10^6 cpm/ml.

Immunoprecipitation. Cells from 60-mm culture dishes were scraped in 1 ml of cold lysis buffer (150 mM NaCl, 2 mM EDTA, 1 mM Na orthovanadate, 10 μg/ml leupeptin, 50 mM Tris HCl [pH 7.6], 0.1% NaN_3, 0.1% Triton X-100, and 0.1% Na deoxycholate). Lysates were centrifuged at 100,000*g* for 30 minutes at 4°C to remove particulate material, and the supernatants were used for immunoprecipitation. For pp125^{FAK}, monoclonal antibody 2A7 (a gift from Dr. Tom Parsons, University of Virginia) was added and incubated at 4°C for 90 minutes. The antigen-antibody complexes were precipitated by the addition of rabbit anti-mouse IgG (Jackson ImmunoResearch) bound to protein A–Sepharose. The samples were then incubated with end-over-end mixing for 14 hours at 4°C. The beads were sedimented in a microfuge, washed five times with 800 μl of lysis buffer, and boiled in Laemmli sample buffer prior to analysis by SDS-PAGE.

For immunoprecipitation of proteins containing phosphotyrosine, antiphosphotyrosine antibody (py20) coupled directly to agarose beads (a gift from Dr. Glenney) was added to cell lysate supernatants and incubated for 90 minutes at 4°C with end-over-end mixing. The beads were sedimented and washed with lysis buffer. To release phosphotyrosine-containing proteins, the beads were incubated with 10 mM phenylphosphate in lysis buffer containing 0.1 mg/ml ovalbumin for 30 minutes on ice. The beads were sedimented, and paxillin was immunoprecipitated from the supernatant by the addition of anti-paxillin bound to agarose beads (Glenney and Zokas 1988; Turner et al. 1990). After end-over-end mixing for 60 minutes, the beads were pelleted and washed as described above. Bound protein was released by boiling in Laemmli sample buffer.

Microscopy. Cells on coverslips were stained with antibodies or phalloidin after formaldehyde fixation and Triton X-100 permeabilization as described previously (Turner et al. 1990). Anti-paxillin antibody was used as a hybridoma culture supernatant. Fluorescein-labeled, affinity-purified goat anti-rabbit IgG and rhodamine-labeled, affinity-purified donkey anti-mouse IgG were obtained from Chemicon International, Inc. (Temecula, California). Fluorescein-labeled, affinity-purified goat anti-mouse IgG was obtained from Jackson Immunoresearch. Rhodamine-conjugated phalloidin was obtained from Molecular Probes, Inc. (Eugene, Oregon). Coverslips were viewed and photographed on a Zeiss Axiophot microscope, using a 63× objective and either T-max 400 or 3200 film (Kodak).

RESULTS

Western Blotting of Phosphotyrosine Activity Induced by ECM Adhesion

Newly formed focal adhesions of cells plated on fibronectin contain elevated levels of phosphorylation on tyrosine (Maher et al. 1985; Burridge et al. 1992). Western blot analysis with anti-phosphotyrosine antibodies was used to investigate the proteins involved in this process. REF52 were plated on fibronectin in serum-free medium, extracted after 30 minutes using Laemmli sample buffer containing Na orthovanadate, and then electrophoresed on SDS gels, transferred to nitrocellulose and blotted with ^{125}I-labeled py20 (Fig. 1, lane 3). Major bands containing increased phosphotyrosine activity were seen at molecular masses of 130 kD, 115 kD, and 70 kD, as compared with extracts of cells kept in suspension (Fig. 1, lane 1). This increased tyrosine phosphorylation was not seen in cells that adhered to surfaces in an integrin-independent manner, as evidenced by the low-level phosphotyrosine profile seen in extracts of cells plated on poly-L-lysine, uncoated tissue culture plastic, and uncoated bacteriological plastic (Fig. 1, lanes 2, 4, and 5, respectively). Proteins with approximate molecular masses of 130 kD, 115 kD, and 70 kD show a progressively increasing pattern of tyrosine phosphorylation when cells are plated for increasing times on fibronectin but not on plastic (Burridge et al. 1992). In contrast, when cells were kept in suspension, the level of tyrosine phosphorylation decreased progressively until it was essentially undetectable by about 90 minutes (data not shown). An identical pattern of bands with increased tyrosine phosphorylation is seen in the Western blots of cell extracts after plating on other substrates to which adhesion is mediated by integrins, including vitronectin, laminin, the cell-binding domain of fibronectin, and the synthetic RGD-containing polymer Pronectin-F (Burridge et al. 1992). These observations provide evidence for integrin mediation of this adhesion-associated increase in tyrosine phosphorylation.

We have examined the pattern of tyrosine phosphorylation in several cell types in response to adhesion to fibronectin using Western blotting with both py20 and PT66 — monoclonal anti-phosphotyrosine antibodies from different sources. Data from experiments with NIH-3T3 cells are shown in Figure 2. Because these cells exhibited some spreading on uncoated plastic with associated increases in phosphotyrosine, it seemed likely that these cells secreted ECM protein(s) and then adhered to these substrates. Cycloheximide pretreatment for 2 hours prior to plating and the inclusion of cycloheximide in the medium at the time of plating were therefore used to inhibit synthesis of ECM components by these cells. Cycloheximide treatment was associated with reduced adhesion of the NIH-3T3 cells to uncoated plastic, but did not affect cell adhesion to the fibronectin substrate. Increased tyrosine phosphorylation was seen in bands at molecular masses of 130 kD, 115 kD, and 70 kD in cell extracts after 20 and 60 minutes of adhesion and spreading on fibronectin as compared with cells kept in suspension (Fig. 2, lanes 2, 3, and 1, respectively). Although present, the increase in phosphotyrosine at approximately 70 kD is not as

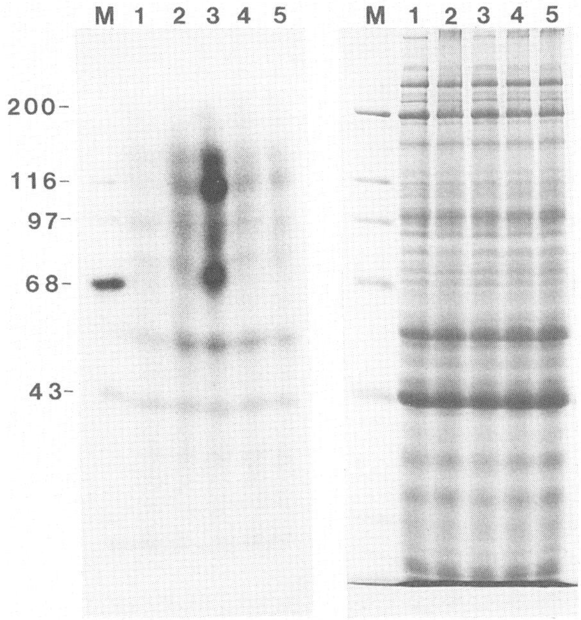

Figure 1. Western blot profiles of phosphotyrosine elevation during cell spreading on fibronectin. ^{125}I-labeled py20 immunoblots of REF52 lysates subjected to SDS-PAGE and transfer to nitrocellulose are shown (*left*) for cells kept in suspension (lane *1*), and plated for 30 min on polylysine (lane *2*), fibronectin (lane *3*), tissue culture plastic (lane *4*), and bacteriological plastic (lane *5*). Identically loaded lanes were stained with Coomassie blue (*right*). Molecular masses of marker proteins (M) are indicated in kilodaltons to the left of the autoradiograph.

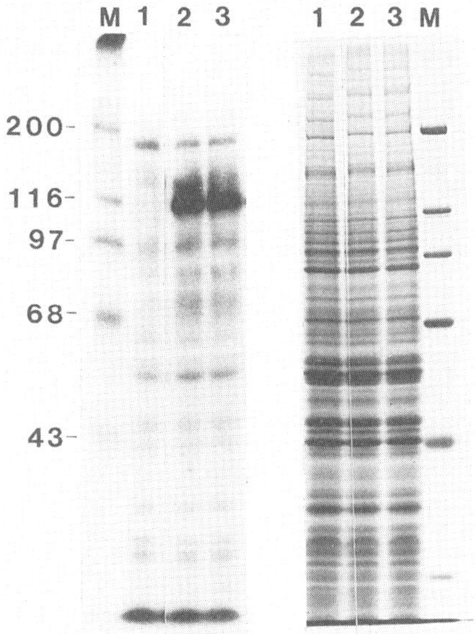

Figure 2. Changes in phosphorylation on tyrosine in protein extracts from NIH-3T3 cells spreading on fibronectin. Phosphotyrosine-containing proteins from NIH-3T3 mouse fibroblasts were blotted with ^{125}I-labeled py20 after separation with SDS-PAGE and immobilization on nitrocellulose. Elevated levels are seen at 130 kD, 115 kD, and 70 kD in samples from cells plated on fibronectin for 20 and 60 min (lanes *2* and *3*) as compared to cells kept in suspension (lane *1*). A Coomassie-blue-stained gel is shown (*right*) for comparison. Molecular masses of marker proteins (M) are displayed in kilodaltons to the left of the autoradiograph.

prominent in this blot as it is in the data from REF52. This may correlate with less tyrosine phosphorylation of the corresponding protein in NIH-3T3 mouse fibroblasts. We have noted elevated phosphotyrosine in proteins between 115 kD and 130 kD in other diverse cell types plated on fibronectin, including human umbilical vein and bovine adult venous endothelial cells, chicken embryo fibroblasts, BHK cells, and BALB/c-3T3 mouse fibroblasts, but these observations are not included here.

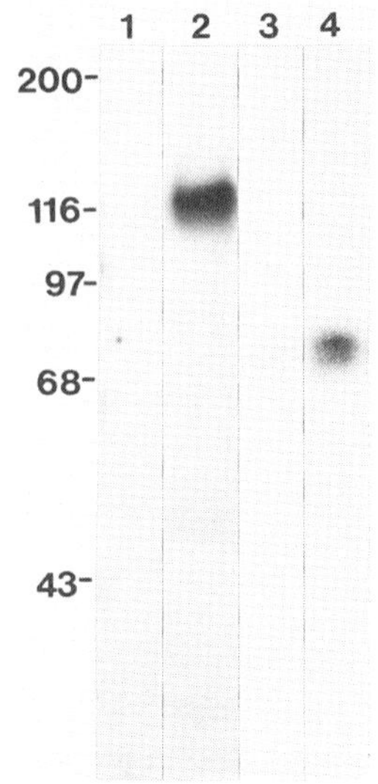

Figure 3. Identification of pp125^{FAK} and paxillin as proteins with elevated tyrosine phosphorylation in response to cell adhesion on fibronectin. This is a composite figure showing pp125^{FAK} (lanes *1*, *2*) and paxillin (lanes *3*, *4*) that have been immunoprecipitated from REF52 lysates of cells kept in suspension (lanes *1*, *3*) or plated on fibronectin (lanes *2*, *4*) for 60 min. Immunoprecipitates were blotted with ^{125}I-labeled py20 after SDS-PAGE and transfer to nitrocellulose. Molecular masses of marker proteins are shown to the left of the autoradiograph.

Identification of Adhesion-associated Tyrosine Kinase Substrates

Immunoprecipitation with monoclonal antibodies against focal adhesion proteins, followed by Western blotting with the anti-phosphotyrosine antibody py20, was used to identify the tyrosine phosphorylated proteins induced by adhesion to fibronectin. We did not pursue the possibility that any of these bands represented an integrin α or β chain, despite the fact that they themselves have molecular masses that would place them within the cluster of bands between 115 kD and 130 kD. Previous evidence indicated that this was unlikely (Guan et al. 1991; Kornberg et al. 1991). We then considered that the focal adhesion-associated tyrosine kinase pp125^{FAK} might be one component in the cluster of bands between 115 kD and 130 kD. Lysates from REF52 cells were prepared so as to compare cells plated on fibronectin-coated culture dishes for 60 minutes with cells kept in suspension for the same period of time. pp125^{FAK} was immunoprecipitated in parallel from both cell lysate samples with the addition of monoclonal antibody, and the resultant immunoprecipitates were analyzed by Western blotting with ^{125}I-labeled py20. Autoradiographs of the suspension and fibronectin-adhered samples are presented in lanes 1 and 2 of the composite in Figure 3. A nearly undetectable level of tyrosine phosphorylation was seen in the sample from REF52 kept in suspension (Fig. 3, lane 1), but a large amount of phosphorylation on tyrosine was seen in the pp125^{FAK} immunoprecipitated from cells adhered for 1 hour to the fibronectin substrate.

Previous work has identified the 70-kD focal adhesion phosphoprotein, paxillin, as a major tyrosine kinase substrate during embryonic development (Turner 1991). To determine whether paxillin was the 70-kD protein showing increased tyrosine phosphorylation in association with adhesion to fibronectin, py20 immunoprecipitates were prepared from REF52 kept in suspension or allowed to adhere to fibronectin for 60 minutes. Paxillin was then immunoprecipitated from the pools of phosphotyrosine-containing proteins that were released from py20 with phenylphosphate. Western blot analysis using ^{125}I-labeled py20 was performed after subjecting the final immunoprecipitates to SDS-PAGE and transferring to nitrocellulose. Autoradiographs of samples from suspension and fibronectin-adhered cells are presented in lanes 3 and 4 of the composite in Figure 3. In a pattern analogous to that seen with pp125^{FAK}, the phosphotyrosine content of the paxillin immunoprecipitated from cells kept in suspension was barely detectable, but cells that had been given 60 minutes to adhere to the fibronectin substrate showed a marked increase in the phosphotyrosine content of paxillin (Fig. 3, lanes 3 and 4).

The data therefore indicate that two adhesion-associated tyrosine kinase substrates in these cells are pp125^{FAK} and paxillin. We have also demonstrated increased phosphotyrosine content in paxillin immunoprecipitates from BALB/c-3T3 mouse fibroblasts after adhesion to fibronectin, as compared to paxillin immunoprecipitated from these cells kept in suspension (Burridge et al. 1992). Immunoprecipitation of [^{35}S]methionine-labeled REF52 extracts with both anti-paxillin and py20 antibodies has been employed to determine that 23% of paxillin is phosphorylated on tyrosine during the adhesion of these cells to fibronectin (Burridge et al. 1992). Compared to data from REF52, lower phosphotyrosine signal strength is seen in the paxillin immunoprecipitates from mouse fibroblasts and in the 70-kD band on Western blots of whole-cell extracts from other cell types. This is consistent with either a lower level of tyrosine phosphorylation on paxillin or an overall smaller paxillin content in other cell types as compared with REF52.

Effects of Tyrosine Kinase Inhibition on the Adhesion-induced Phosphorylation of Focal Adhesion Proteins

The tyrosine kinase inhibitor herbimycin A was employed to block the phosphotyrosine elevation seen with cell adhesion to fibronectin. After reconstitution with DMSO, herbimycin A was used at a working concentration of 875 nM both in cell culture medium used for the pretreatment of REF52 for 24 hours prior to plating, and in the serum-free medium used for plating the cells on fibronectin-coated surfaces. Herbimycin A (5 μM) has been used with rat vascular smooth muscle cells without any evidence of cytotoxicity (Weiss and Nuccitelli 1992). DMSO has been shown to induce the tyrosine phosphorylation of the epidermal growth factor receptor (Rubin and Earp 1983). DMSO-treated controls were used for all the experiments reported here and showed no change when compared with untreated controls (data not shown).

In general, REF52 treated with 875 nM herbimycin A did not spread as well on fibronectin-coated surfaces as untreated controls, although some heterogeneity was seen. Treated cells showed a marked decrement in all phosphotyrosine activity on Western blot analysis of whole-cell extracts (Fig. 4). Only trace amounts of phosphorylation on tyrosine are seen at approximately 115–130 kD in the herbimycin-A-treated sample (lane 1), as compared to the untreated control sample (lane 2). Immunoprecipitation has been used to demonstrate that herbimycin A treatment specifically inhibits the tyrosine phosphorylation of pp125^{FAK} (Burridge et al. 1992).

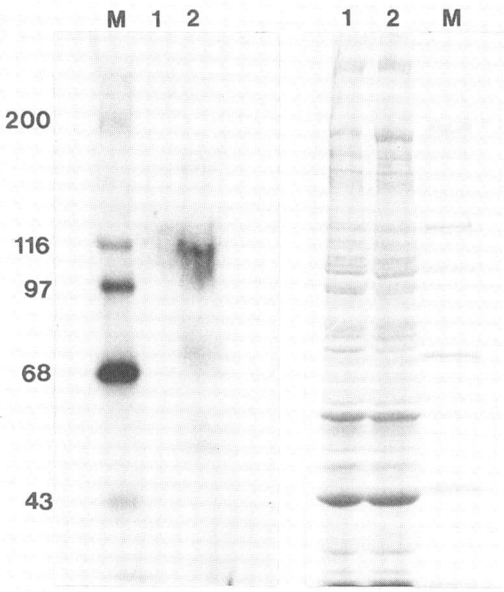

Figure 4. Herbimycin A inhibits adhesion-induced tyrosine phosphorylation. Lysates of REF52 plated for 1 hr in the presence (lane *1*) or absence (lane *2*, untreated control) of 875 nM herbimycin A were electrophoresed on polyacrylamide gels and either stained with Coomassie blue (*right*) or transferred to nitrocellulose and blotted with ^{125}I-labeled py20 to reveal phosphotyrosine-containing proteins (*left*). Note that herbimycin-A-treated cells were preincubated with the same concentration of the agent for 24 hr prior to plating. Molecular masses of marker proteins (M lanes) are indicated in kilodaltons to the left of the autoradiograph.

The distributions of pp125^{FAK} and paxillin were then examined in herbimycin-A-treated cells using immunofluorescence microscopy (Fig. 5). To allow comparison of focal adhesion staining between herbimycin-A-treated cells and untreated cells, only treated cells with the most normal morphology and spreading were selected for inclusion in these figures. Untreated controls exhibited focal adhesion staining patterns for pp125^{FAK} and paxillin (Fig. 5A,C). The pattern of pp125^{FAK} staining was noted to be significantly altered in many of the treated cells, with a ring of brightness at the cells' periphery and some diffuse perinuclear staining (Fig. 5B). Paxillin staining revealed a pattern of focal adhesions that was restricted to the cell periphery and markedly decreased in size as compared to untreated controls (Fig. 5D). py20 staining was also markedly altered from the control pattern in many herbimycin-A-treated cells. Whereas normal cells showed brightly staining peripheral focal adhesions, cells treated with herbimycin A showed only a rim of peripheral brightness that was not seen in untreated controls (Fig. 5E,F).

To determine whether focal adhesions were formed in the presence of herbimycin A, we examined talin staining patterns in these cells. Many herbimycin-A-treated cells showed a striking absence of focal adhesion formation when immunostained for talin (Fig. 6B). Instead, diffuse staining consistent with a cytoplasmic pool of free talin was seen. The cells in panels A and B of Figure 6 are the same as those in panels A and B of Figure 5. These cells were double-immunostained to document parallel changes in the staining patterns of talin and pp125^{FAK}. Rhodamine-conjugated phalloidin was used to visualize actin filaments in cells (Fig. 6C,D). Actin cytoskeletal organization was disordered in herbimycin-A-treated cells. An abundant pattern of peripheral actin filaments and stress fibers seen in normal REF52 was replaced by a sparse, spindly pattern of filaments in many cells treated with herbimycin A. A spectrum of effects was seen, ranging from cells with a somewhat reduced complement of nearly normal stress fibers to cells with only occasional thready fragments of filamentous actin and a perinuclear aggregation of what appeared to be actin clumps.

DISCUSSION

Adhesion receptors of the integrin superfamily facilitate cell-cell and cell-ECM interactions in animals as diverse as mammals and insects. An understanding of the signal transduction mechanisms involving integrins is just beginning to develop (for review, see Hynes 1992). Integrin-mediated cell-ECM adhesion and antibody-clustering of integrins have been noted to be associated with increased phosphotyrosine content of

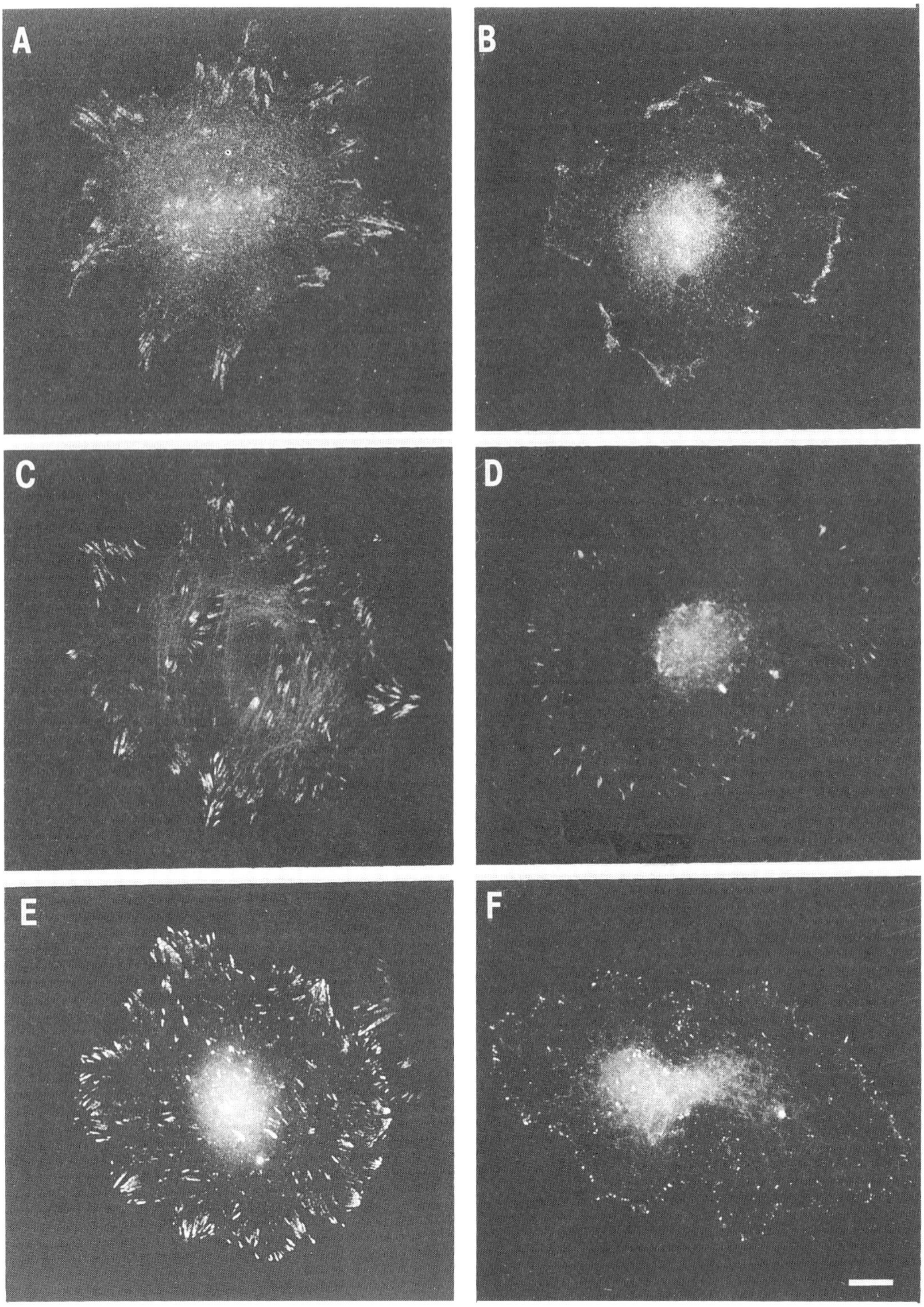

Figure 5. Patterns of cell staining for pp125^{FAK}, paxillin, and phosphotyrosine are altered by herbimycin A. REF52 were plated in the absence (*A, C, E*) or presence (*B, D, F*) of 875 nM herbimycin A on fibronectin-coated coverslips. Cells were stained for pp125^{FAK} (*A, B*), paxillin (*C, D*), or phosphotyrosine (*E, F*). Cells plated in the presence of herbimycin A had been pretreated with the same concentration of this agent for 24 hr. Bar, 10 μM.

intracellular proteins (Guan et al. 1991; Kornberg et al. 1991; Burridge et al. 1992). A scheme linking cell-ECM adhesion with intracellular events, including cytoskeletal assembly via changes in the tyrosine phosphorylation of intracellular proteins, is presented in Figure 7. The elevation in phosphotyrosine could result either from an inhibition of protein tyrosine phosphatases or from a stimulation of tyrosine kinases. Currently, we have no

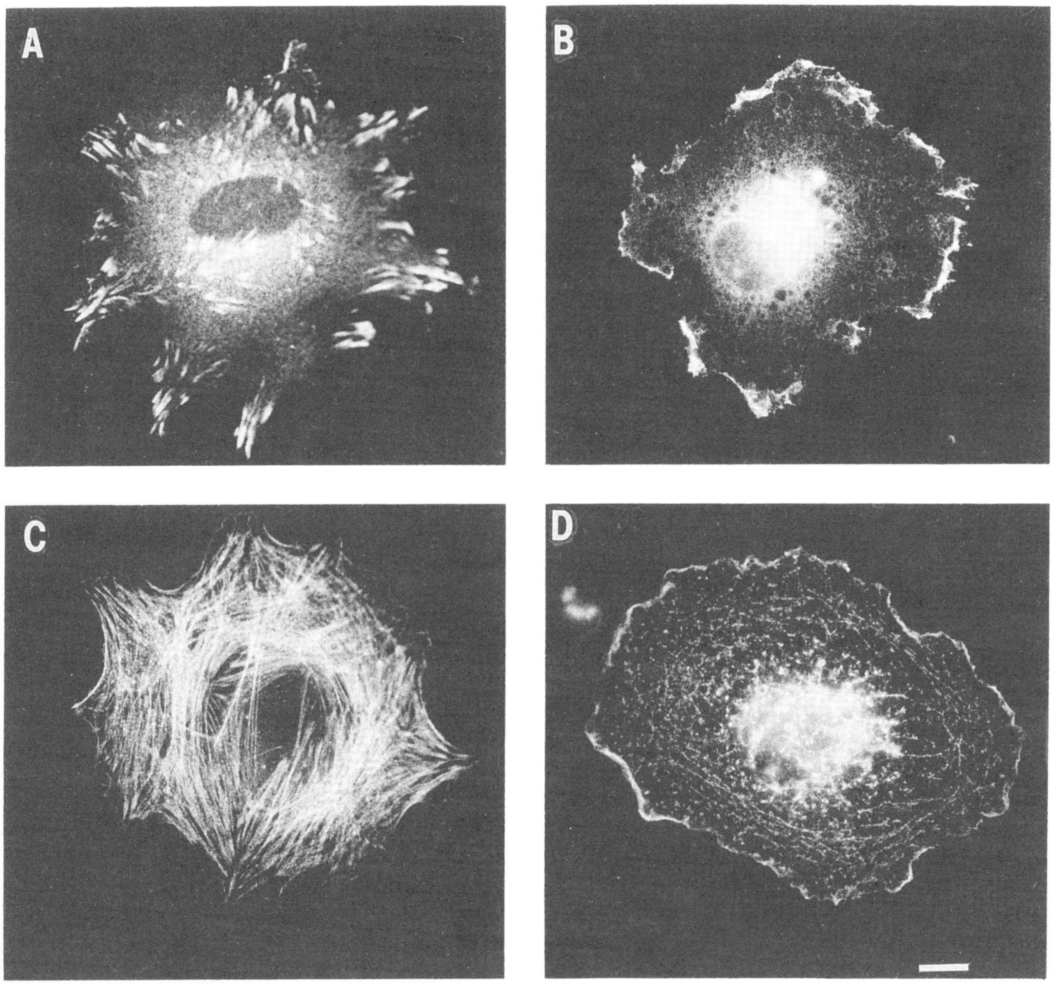

Figure 6. Focal adhesion formation and stress fiber organization are inhibited by herbimycin A. REF52 plated on fibronectin-coated coverslips for 1 hr in the absence (*A*, *C*) or presence (*B*, *D*) of 875 nM herbimycin A were immunostained for talin (*A*, *B*) and stained for actin using rhodamine-conjugated phalloidin (*C*, *D*). A double-labeling technique was used to document the colocalization of talin with pp125^{FAK} in control and treated cells (Figs. 5 and 6, panels *A* and *B*). Bar, 10 μM.

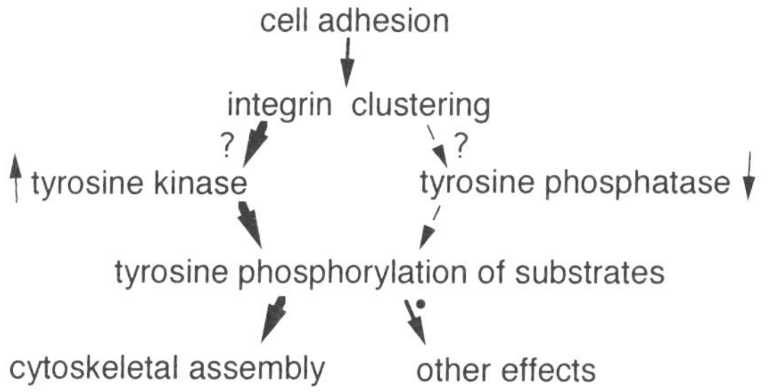

Figure 7. A scheme for tyrosine phosphorylation signal transduction in integrin-mediated cell adhesion. Cell adhesion and antibody binding to cell-surface integrins is associated with increased protein tyrosine phosphorylation (see also Kornberg et al. 1991; Guan et al. 1991). This effect may be mediated by either of two different biochemical pathways (?). Elevation in phosphotyrosine content is associated with cytoskeletal assembly and, perhaps, with other adhesion-induced events (·). See Discussion for further details.

information on tyrosine phosphatases in our system, but it is noteworthy that two tyrosine phosphatases have recently been identified that have sequence homology with the amino-terminal domain of talin (Gu et al. 1991; Yang and Tonks 1991). This domain of talin has been shown to target to focal adhesions when introduced into cells by microinjection (Nuckolls et al. 1990). The localization of these tyrosine phosphatases in cells and their role in focal adhesion function is under investigation. Our current data are consistent with the activation of tyrosine kinases in the series of events linking integrin-mediated ECM adhesion and cytoskeletal organization. We have identified two focal adhesion proteins as tyrosine kinase substrates that are phosphorylated during cell adhesion to ECM–paxillin and pp125^{FAK}. pp125^{FAK} is itself a tyrosine kinase that is concentrated in focal adhesions (Schaller et al. 1992). Others have also identified increased tyrosine phosphorylation in this kinase in association with integrin clustering and adhesion (Guan and Shalloway 1992; Lipfert et al. 1992; L. Kornberg et al., pers. comm.). It remains to be determined whether the elevated tyrosine phosphorylation found in pp125^{FAK} is associated with increased activity of this kinase and whether it is due to autophosphorylation or phosphorylation by some other kinase.

We have found that treatment of cells with herbimycin A inhibited the rise in tyrosine phosphorylation stimulated by cell adhesion to fibronectin. Herbimycin-A-treated cells also showed a decreased ability to form focal adhesions and to assemble stress fibers. These results contrast with previous work in which herbimycin A was found to promote the formation of stress fibers, a well-spread phenotype, and contact-inhibited growth behavior in rat kidney cells and chicken embryo fibroblasts transformed by RSV (Uehara et al. 1985, 1986; Suzukake-Tsuchiya et al. 1989). The fact that the inhibition of tyrosine kinase activity has very different consequences in transformed and normal cells may be explained by different overall levels of Src-family tyrosine kinases, or by the different activity of tyrosine kinases in the two cell systems. In transformed cells, the very high level of tyrosine kinase activity due to the unregulated expression of pp60$^{v\text{-}src}$ or related oncogene-encoded tyrosine kinases is associated with cytoskeletal disassembly, decreased adhesion, and cell rounding. The surfeit of aberrant tyrosine phosphorylations induced by pp60$^{v\text{-}src}$ and related kinases may disrupt interactions in focal adhesions by competitively inhibiting interactions that are dependent on specific tyrosine phosphorylations. Inhibition of this unregulated tyrosine kinase activity has the expected consequence of restoring a more normal phenotype, including the assembly of focal adhesions and stress fibers. Recent data from transfection experiments support this scenario (Woodford-Thomas et al. 1992). In these experiments, the phosphatase activity of v-*src*-transfected NIH-3T3 cells was increased tenfold above nontransfected levels with a resulting reversion of these transformed cells to a normal morphology.

In contrast to the cytoskeletal disruption associated with increased phosphotyrosine in RSV-transformed cells, in this study of normal cells, tyrosine phosphorylation is found in only a small subset of proteins and correlates with focal adhesion and stress fiber assembly as cells spread on a fibronectin substrate. This adhesion-associated increase in tyrosine phosphorylation may be the result of integrin-mediated autophosphorylation of pp125^{FAK} with the subsequent induction of paxillin phosphorylation, and these events may be independent of Src-family tyrosine kinase activity. Further work will be required to identify the specific tyrosine kinases and phosphorylated substrates that are critical in these events. Paxillin may be a tyrosine-phosphorylated substrate with a regulatory role in cytoskeletal organization. This is suggested by biochemical and embryological data. Paxillin binds to vinculin, and its presence in relatively small amounts in focal adhesions is consistent with a regulatory role (Turner et al. 1990). The fact that paxillin is a major tyrosine kinase substrate during embryonic development but has a very low phosphotyrosine content in adult tissues is also consistent with a role in the assembly of new cytoskeletal structures (Turner 1991). Phosphorylation on tyrosine in paxillin could generate a binding site that is recognized by an SH2 domain in another focal adhesion protein. Recently, an SH2 domain was identified in the actin-binding, focal adhesion component tensin (Davis et al. 1991). We are currently investigating whether there is an interaction between these two proteins that is dependent on tyrosine phosphorylation in paxillin. Such an association could play a role in regulating focal adhesion assembly.

We noted defective actin stress fiber assembly in herbimycin-A-treated cells. These findings are consistent with the presence of an impediment to some aspect of stress fiber formation in the presence of tyrosine kinase inhibition. Actin polymerization may be altered. The association between newly forming focal adhesions and stress fiber organization has been a subject of investigation (for review, see Burridge et al. 1988). Actin polymerization during stress fiber assembly has been tracked using the microinjection of rhodamine-conjugated actin (Wang 1984). These studies showed that stress fiber assembly proceeded from small discrete foci at the leading edges of spreading fibroblasts. This description is consistent with both the peripheral localization of new focal adhesion assembly (Izzard and Lochner 1980; DePasquale and Izzard 1987) and the sites of recently described talin-rich nodes that may represent focal adhesion precursors at the edge of spreading lamellipodia (DePasquale and Izzard 1991). These data support a role for new focal adhesions as sites for the nucleation of actin polymerization. Our observations indicate that phosphotyrosine activity in spreading cells is concentrated at peripheral sites consistent with newly forming focal adhesions (Burridge et al. 1992) and that tyrosine kinase inhibition interferes with the normal incorporation of focal adhesion constituents and with actin filament assembly.

Tyrosine kinase inhibitors have been used to study diverse processes, including growth factor receptor activities, interleukin receptor expression and transcription factor induction, and both PDGF- and thrombin-induced mitogenesis (Lyall et al. 1989; Levitski 1990; Bryckaert et al. 1992; Iwasaki et al. 1992; Otani et al. 1992; Weiss and Nuccitelli 1992). Although some protein tyrosine kinase inhibitors, such as genistein and erbstatin, have significant activity against protein kinases A and C (Akiyama et al. 1987; Bishop et al. 1990), the benzoquinonoid ansamycins (e.g., herbimycin) and tyrphostins have a high degree of specificity for tyrosine kinases (Uehara et al. 1988; Levitski 1990). Herbimycin A has been shown to block or reverse the transforming effects of Src-family tyrosine kinase oncogenes (including *src, yes, fps, ros*, and *abl*) but did not inhibit the effects of other oncogenes (including *raf, ras*, or *myc*) (Uehara et al. 1988). The mechanism of herbimycin A inhibition of pp60$^{v\text{-}src}$ tyrosine kinase activity is thought to include the binding of reactive sulfhydryl groups of pp60$^{v\text{-}src}$ to the benzoquinone moiety of the inhibitor and the blocking of the availability of an active site (Uehara et al. 1989; Fukazawa et al. 1990). Tyrphostins derive their high degree of specificity from their structure, which is engineered to mimic a tyrosine residue in a peptide chain (Levitski 1990). In our system, specific effects of tyrosine kinase inhibitors on the autophosphorylation and other activities of pp125^{FAK} will need to be explored. The use of a panel of tyrosine kinase inhibitors may enable us to understand the relative contributions of pp125^{FAK}, Src-family kinases, and others to the series of tyrosine phosphorylation events leading to the assembly of focal adhesions and stress fibers in response to cell-ECM adhesion.

Specific inhibitors of tyrosine phosphorylation may also be instrumental in defining the role of this signal transduction pathway in other adhesion-induced cellular events, including anchorage-dependent growth behavior. This phenomenon involves the adhesion of cells to ECM as a prerequisite for normal responses to growth factors by DNA replication and cell division. This requirement is frequently lost upon transformation. The basis for anchorage-dependence of normal growth has been an enigma, although various possible explanations have been advanced (for review, see Burridge 1986). Anchorage-dependence of growth has recently been correlated with the level of intracellular pH (Schwartz et al. 1989, 1990a,b, 1991). Nonadherent, anchorage-dependent cells were found to have a lower intracellular pH than the same cells adhering to a substrate. In addition, a more alkaline intracellular pH was correlated with stimulation of DNA synthesis and growth (Schwartz et al. 1989, 1990a). This effect has been shown to be triggered by integrin binding to ECM ligands (Schwartz et al. 1990a, 1991). Our finding of elevated tyrosine phosphorylation in response to adhesion suggests an alternative or additional explanation for anchorage-dependence of growth. Here is a biochemical response, the activation of a kinase, that may be critical in allowing cells to respond to growth factors. It should be possible to test this hypothesis in the future, and it will be important to determine the functions of these phosphotyrosine-containing proteins in growth control and cytoskeletal organization.

ACKNOWLEDGMENTS

We thank Drs. Tom Parsons, Lori Kornberg, Rudy Juliano, and Joan Brugge for valuable discussions. We are grateful to Dr. Tom Parsons for the gifts of antibodies against pp125^{FAK}, and to Dr. John Glenney for gifts of the monoclonal antibody py20. We express our appreciation to Natalie McLean for valuable and persevering technical assistance. This work was supported by National Institutes of Health grants GM-29860, HL-44918 (to K.B.), and GM-47607 (to C.T.).

REFERENCES

Akiyama, T., J. Ishida, S. Nakagawa, H. Ogawara, S. Watanabe, N. Itoh, M. Shibuya, and Y. Fukami. 1987. Genistein, a specific inhibitor of tyrosine-specific protein kinases. *J. Biol. Chem.* **262:** 5592.

Albelda, S.M. and C.A. Buck. 1990. Integrins and other cell adhesion molecules. *FASEB J.* **4:** 2868.

Bishop, W.R., J. Petrin, L. Wang, U. Ramesh, and R.J. Doll. 1990. Inhibition of protein kinase C by the tyrosine kinase inhibitor erbstatin. *Biochem. Pharmacol.* **40:** 2129.

Bryckaert, M.C., A. Eldor, M. Fontenay, A. Gazit, N. Osherov, C. Gilon, A. Levitski, and G. Tobelem. 1992. Inhibition of platelet-derived growth factor-induced mitogenesis and tyrosine kinase activity in cultured bone marrow fibroblasts by tyrphostins. *Exp. Cell Res.* **199:**255.

Buck, C.A. and A.F. Horwitz. 1987. Cell surface receptors for extracellular matrix molecules. *Annu. Rev. Cell Biol.* **3:** 179.

Burridge, K. 1986. Substrate adhesions in normal and transformed fibroblasts: Organization and regulation of cytoskeletal, membrane and extracellular matrix components at focal contacts. *Cancer Rev.* **4:** 18.

Burridge, K., C.E. Turner, and L.H. Romer. 1992. Tyrosine phosphorylation of paxillin and pp125^{FAK} accompanies cell adhesion to extracellular matrix: A role in cytoskeletal assembly. *J. Cell Biol.* **119:** (in press).

Burridge, K., K. Fath, T. Kelly, G. Nuckolls, and C. Turner. 1988. Focal adhesions: Transmembrane junctions between the extracellular matrix and the cytoskeleton. *Annu. Rev. Cell Biol.* **4:** 487.

Cichowski, K., F. McCormick, and J.S. Brugge. 1992. P21^{ras} GAP association with Fyn, Lyn, and Yes in thrombin-activated platelets. *J. Biol. Chem.* **267:** 5025.

Davis, S., M.L. Lu, S. Lin, J.A. Butler, B.J. Druker, T.M. Roberts, Q. An, and L.B. Chen. 1991. Presence of an SH2 domain in the actin-binding protein tensin. *Science* **252:** 712.

DePasquale, J.A. and C.S. Izzard. 1987. Evidence for an actin-containing cytoplasmic precursor of the focal contact and the timing of incorporation of vinculin at the focal contact. *J. Cell Biol.* **105:** 2803.

———.1991. Accumulation of talin in nodes at the edge of the lamellipodium and separate incorporation into adhesion plaques of focal contacts in fibroblasts. *J. Cell Biol.* **113:** 1351.

Edelman, G.M. 1986. Cell adhesion molecules in the regulation of animal form and tissue pattern. *Annu. Rev. Cell Biol.* **2:** 81.

Ferrell, J.E. and G.S. Martin. 1989. Tyrosine-specific protein phosphorylation is regulated by glycoprotein IIb-IIIa in platelets. *Proc. Natl. Acad. Sci.* **86:** 2334.

Fraker, P.J. and J.C. Speck. 1978. Protein and cell membrane iodinations with a sparingly soluble chloramide, 1,3,4,6-tetrachloro-3a, 6a-diphenyl-glycoluril. *Biochem. Biophys. Res. Commun.* **80:** 849.

Fukazawa, H., S. Mizuno, and Y. Uehara. 1990. Effects of herbimycin A and various SH-reagents on p60$^{v\text{-}src}$ kinase activity *in vitro*. *Biochem. Biophys. Res. Commun.* **173:** 276.

Glenney, J.R., Jr. and L. Zokas. 1988. Novel tyrosine kinase substrates from Rous sarcoma virus-transformed cells are present in the membrane skeleton. *J. Cell Biol.* **108:** 2401.

Golden, A. and J.S. Brugge. 1989. Thrombin treatment induces rapid changes in tyrosine phosphorylation in platelets. *Proc. Natl. Acad. Sci.* **86:** 901.

Golden, A., J.S. Brugge, and S.J. Shattil. 1990. Role of platelet membrane glycoprotein IIb-IIIa in agonist-induced tyrosine phosphorylation. *J. Cell Biol.* **111:** 3117.

Gu, M., J.D. York, I. Warshawsky, and P.W. Majerus. 1991. Identification, cloning, and expression of a cytosolic megakaryocyte protein-tyrosine phosphatase with sequence homology to cytoskeletal protein 4.1. *Proc. Natl. Acad. Sci.* **88:** 5867.

Guan, J.-L. and D. Shalloway. 1992. Regulation of pp125^{FAK} both by cellular adhesion and by oncogenic transformation. *Nature* (in press).

Guan, J.-L., J.E. Trevithick, and R.O. Hynes. 1991. Fibronectin/integrin interaction induces tyrosine phosphorylation of a 120 kDa protein. *Cell Regul.* **2:** 951.

Hemler, M. 1990. VLA proteins in the integrin family: Structure, functions and their role on leukocytes. *Annu. Rev. Immunol.* **8:** 365.

Horvath, A.R., L. Muszbek, and S. Kellie. 1992. Translocation of pp60^{c-src} to the cytoskeleton during platelet aggregation. *EMBO J.* **11:** 855.

Horwitz, A., K. Duggan, C. Buck, M.C. Beckerle, and K. Burridge. 1986. Interaction of plasma membrane fibronectin receptor with talin—A transmembrane linkage. *Nature* **320:** 531.

Hynes, R.O. 1992. Integrins: Versatility, modulation, and signaling in cell adhesion. *Cell* **69:** 11.

Iwasaki, T., Y. Uehara, L. Graves, N. Rachie, and K. Bomsztyk. 1992. Herbimycin A blocks IL-1-induced NF-kappa B DNA-binding activity in lymphoid lines. *FEBS Lett.* **298:** 240.

Izzard, C.S. and L.R. Lochner. 1976. Cell-to-substrate contacts in living fibroblasts: An interference reflection study with an evaluation of the technique. *J. Cell Sci.* **21:** 129.

———.1980. Formation of cell-to-substrate contacts during fibroblast motility: An interference reflection study. *J. Cell Sci.* **42:** 81.

Jaken, S., K. Leach, and T. Klauck. 1989. Association of type 3 protein kinase C with focal contacts in rat embryo fibroblasts. *J. Cell Biol.* **109:** 697.

Kanner, S.B., A.B. Reynolds, R.R. Vines, and J.T. Parsons. 1990. Monoclonal antibodies to individual tyrosine-phosphorylated protein substrates of oncogene-encoded tyrosine kinases. *Proc. Natl. Acad. Sci.* **87:** 3328.

Kornberg, L.J., H.S. Earp, C.E. Turner, C. Prockop, and R.L. Juliano. 1991. Signal transduction by integrins: Increased protein tyrosine phosphorylation caused by clustering β_1 integrins. *Proc. Natl. Acad. Sci.* **88:** 8392.

Laemmli, U.K. 1970. Cleavage of structural proteins during the assembly of the head of bacteriophage T4. *Nature* **227:** 680.

Levitski, A. 1990. Tyrphostins—Potential antiproliferative agents and novel molecular tools. *Biochem. Pharmacol.* **40:** 913.

Lipfert, L., B. Haimovich, M.D. Schaller, B.S. Cobb, J.T. Parsons, and J.S. Brugge. 1992. Integrin-dependent phosphorylation and activation of the protein tyrosine kinase pp125^{FAK} in platelets. *J. Cell Biol.* **119:** (in press).

Lyall, R.M., A. Zilberstein, A. Gazit, C. Gilon, A. Levitski, and J. Schlessinger. 1989. Tyrphostins inhibit epidermal growth factor (EGF)-receptor tyrosine kinase activity in living cells and EGF-stimulated cell proliferation. *J. Biol. Chem.* **264:** 14503.

Maher, P.A., E.B. Pasquale, J.Y.J. Wang, and S.J. Singer. 1985. Phosphotyrosine-containing proteins are concentrated in focal adhesions and intercellular junctions in normal cells. *Proc. Natl. Acad. Sci.* **82:** 6576.

Nuckolls, G.H., C.E. Turner, and K. Burridge. 1990. Functional studies of the domains of talin. *J. Cell Biol.* **110:** 1635.

Otani, H., J.P. Siegel, M. Erdos, and J.R. Gnarra. 1992. Interleukin (IL)-2 and IL-3 induce distinct but overlapping responses in murine IL-3-dependent 32D cells transduced with human IL-2 receptor beta chain: Involvement of tyrosine kinase(s) other than p56lck. *Proc. Natl. Acad. Sci.* **89:** 2789.

Otey, C.A., F.M. Pavalko, and K. Burridge. 1990. An interaction between α-actinin and the β_1 integrin subunit in vitro. *J. Cell Biol.* **111:** 721.

Rohrschneider, L.R. 1980. Adhesion plaques of Rous sarcoma virus-transformed cells contain the *src* gene product. *Proc. Natl. Acad. Sci.* **77:** 3514.

Rubin, R.A. and H.S. Earp. 1983. Dimethyl sulfoxide stimulates tyrosine residue phosphorylation of rat liver epidermal growth factor receptor. *Science* **219:** 60.

Schaller, M.D., C.A. Borgman, B.S. Cobb, R.R. Vines, A.B. Reynolds, and J.T. Parsons. 1992. pp125^{FAK}, a structurally unique protein tyrosine kinase associated with focal adhesions. *Proc. Natl. Acad. Sci.* **89:** 5192.

Schwartz, M.A., G. Both, and C. Lechene. 1989. Effect of cell spreading on cytoplasmic pH in normal and transformed fibroblasts. *Proc. Natl. Acad. Sci.* **86:** 4525.

Schwartz, M.A., C. Lechene, and D.E. Ingber. 1990a. Insoluble fibronectin activates the Na/H antiporter by clustering and immobilizing integrin $\alpha5\beta1$, independent of cell shape. *Proc. Natl. Acad. Sci.* **88:** 7849.

Schwartz, M.A., E.E. Rupp, J.V. Frangioni, and C.P. Lechene. 1990b. Cytoplasmic pH and anchorage-independent growth induced by V-Ki-ras, V-src and polyoma middle T. *Oncogene* **5:** 55.

Schwartz, M.A., D.E. Ingber, M. Lawrence, T.A. Springer, and C. Lechene. 1991. Multiple integrins share the ability to induce elevation of intracellular pH. *Exp. Cell Res.* **195:** 533.

Suzukake-Tsuchiya, K., Y. Moriya, and M. Hori. 1989. Induction by herbimycin A of contact inhibition in v-*src*-expressed cells. *J. Antibiot.* **42:** 1831.

Takeichi, M. 1988. The Cadherins: Cell-cell adhesion molecules controlling animal morphogenesis. *Development* **102:** 639.

———.1991. Cadherin cell adhesion receptors as a morphogenetic regulator. *Science* **251:** 1451.

Towbin, H., T. Staehelin, and J. Gordon. 1979. Electrophoretic transfer of proteins from polyacrylamide gels to nitrocellulose sheets: Procedure and some applications. *Proc. Natl. Acad. Sci.* **76:** 4350.

Turner, C.E. 1991. Paxillin is a major phosphotyrosine-containing protein during embryonic development. *J. Cell Biol.* **115:** 201.

Turner, C.E., J.R. Glenney, and K. Burridge. 1990. Paxillin: A new vinculin-binding protein present in focal adhesions. *J. Cell Biol.* **111:** 1059.

Uehara, Y., J. Fukazawa, Y. Murakami, and S. Mizuno. 1989. Irreversible inhibition of v-*src* tyrosine kinase activity by herbimycin A and its abrogation by sulfhydryl compounds. *Biochem. Biophys. Res. Commun.* **163:** 803.

Uehara, Y., M. Hori, T. Takeuchi, and H. Umezawa. 1985. Screening of agents which convert "transformed morphology" of Rous sarcoma virus-infected rat kidney cells to "normal morphology": Identification of an active agent as herbimycin and its inhibition of intracellular *src* kinase. *Jpn. J. Cancer Res.* **76:** 672.

———.1986. Phenotypic change from transformed to normal induced by benzoquinonoid ansamycins accompanies inactivation of p60^{src} in rat kidney cells infected with Rous sarcoma virus. *Mol. Cell. Biol.* **6:** 2198.

Uehara, Y., Y. Murakami, S. Mizuno, and S. Kawai. 1988. Inhibition of transforming activity of tyrosine kinase oncogenes by herbimycin A. *Virology* **164:** 294.

Wang, Y. 1984. Reorganization of actin filament bundles in living fibroblasts. *J. Cell Biol.* **99:** 1478.

Weiss, R.H. and R. Nuccitelli. 1992. Inhibition of tyrosine phosphorylation prevents thrombin-induced mitogenesis, but not intracellular free calcium release, in vascular smooth muscle cells. *J. Biol. Chem.* **267:** 5608.

Woods, A. and J.R. Couchman. 1992. Protein kinase C involvement in focal adhesion formation. *J. Cell Sci.* **101:** 277.

Woodford-Thomas, T.A., J.D. Rhodes, and J.E. Dixon. 1992. Expression of a protein tyrosine phosphatase in normal and v-*src*-transformed mouse 3T3 fibroblasts. *J. Cell Biol.* **117:** 401.

Yang, Q. and N.K. Tonks. 1991. Isolation of a cDNA clone encoding a human protein-tyrosine phosphatase with homology to the cytoskeletal-associated proteins band 4.1, ezrin and talin. *Proc. Natl. Acad. Sci.* **88:** 5949.

Mechanisms of Fibronectin and Integrin Function during Cell Adhesion and Migration

K.M. Yamada, S. Aota, S.K. Akiyama, and S.E. LaFlamme
Laboratory of Developmental Biology, National Institute of Dental Research, National Institutes of Health, Bethesda, Maryland 20892

Cells interact specifically with a variety of different extracellular molecules as they participate in embryonic development, tissue organization, wound healing, and tumor cell invasion (for review, see Hay 1991). Recent studies indicate that central elements of these dynamic interactions are specific recognition processes involving integrin receptors and a variety of extracellular ligands such as fibronectin (for general reviews, see Mosher 1989; Albelda and Buck 1990; Hynes 1990, 1992; Akiyama et al. 1990; Hemler 1990; McDonald and Mecham 1991; Ruoslahti 1991). For example, fibronectin along with other extracellular molecules serves as a migratory substrate for embryonic cells during gastrulation and neural crest cell migration; successful adhesion and migration of these cells depend on specific binding interactions between integrin receptors and recognition sites in fibronectin.

Biochemical characterization of these recognition sites and receptors has led to the development of synthetic peptide and antibody inhibitors that can be used experimentally to disrupt specific biological events in vitro and in vivo, blocking processes as diverse as cell migration and assembly of extracellular matrix. Complementary molecular biological approaches, especially mutagenesis, are now beginning to reveal the molecular strategies involved in these recognition events and in the subsequent interactions of integrins with the cytoskeleton. An underlying theme appearing in these studies is that cells often use overlapping but distinct binding mechanisms in each function. For example, the same peptide sequence is bound by different receptors depending on other sequences in the protein, certain integrin receptors can bind to more than one sequence, and the cytoplasmic domains of integrins can have complex cytoskeletal interactions. In each case, similar functions (at least at our current levels of resolution) can be mediated by more than one component of the system, and the final biological effect obtained often represents a sum of parallel cell-surface interactions. This overall strategy provides cells with considerable opportunities for subtle regulation of cell-matrix interactions by alterations in the composition of the extracellular matrix, e.g., by changes in alternative splicing, or of a cell's repertoire and spatial distribution of integrin receptors.

In this brief overview, we focus first on receptor recognition sites in the extracellular molecule fibronectin involved in cell adhesion, migration, and matrix assembly. We then turn to its receptors and their roles in these important cellular events in vitro and in vivo. We conclude by considering how the intracellular domains of integrin receptors may function and how their cytoplasmic interactions can be regulated by ligand occupancy.

FIBRONECTIN AND PEPTIDE INTERACTIONS WITH THE CELL SURFACE

Fibronectin is composed of three different types of repeating sequences, termed types I, II, and III, that are organized into distinct protease-resistant domains that contain binding sites for a variety of biological macromolecules, including heparin, collagen, fibrin, and cell-surface receptors (Fig. 1). Fibronectin contains at least two distinct cell adhesive regions; the best characterized region occupies the central part of the molecule. When isolated in the form of proteolytic fragments of 110, 75, or 37 kD, this central cell-binding domain retains full cell adhesive activity on a molar basis (Hayashi and Yamada 1983; Nagai et al. 1991).

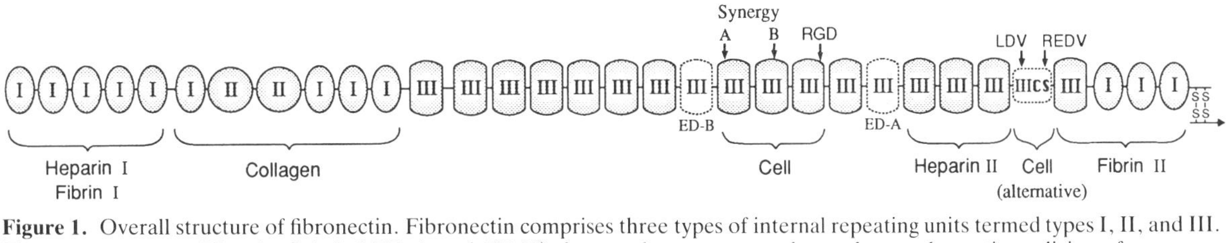

Figure 1. Overall structure of fibronectin. Fibronectin comprises three types of internal repeating units termed types I, II, and III. There are two type III units (labeled ED-A and ED-B) that can be present or absent due to alternative splicing of precursor mRNA. In addition, the human IIICS region can undergo alternative splicing of three subregions. Each fibronectin subunit chain is disulfide-bonded at the carboxy-terminal end. Binding domains are labeled along the bottom, including two domains for binding to heparin, two to fibrin, and one to collagen. The central cell-binding domain contains the RGD site, as well as subregions A and B of the synergy region; each is located in a different type III unit. The IIICS region provides an alternative cell-binding domain, which contains minimal recognition units with the peptide sequences LDV and REDV. See text for details and references.

Furthermore, radioactively labeled intact fibronectin and the 75-kD cell-binding fragment both bind directly to fibroblastic cells in suspension (Akiyama and Yamada 1985a; Akiyama et al. 1985) with similar affinities ($K_d = 0.4$–0.8 μM).

Initially, the cell adhesive signal of fibronectin was localized to an 11.5-kD fragment (Pierschbacher et al. 1981) and then to a short peptide region with the sequence Arg-Gly-Asp-Ser (Pierschbacher and Ruoslahti 1984; Yamada and Kennedy 1984). Small fragments and synthetic peptides containing this sequence can block fibronectin-mediated cell adhesive activity both in vitro (Pierschbacher and Ruoslahti 1984; Yamada and Kennedy 1984; Akiyama and Yamada 1985b) and in vivo (Boucaut et al. 1984b; Lash et al. 1987) if added in soluble form as a competitive inhibitor. In addition, deletion of the Arg-Gly-Asp (RGD) sequence by site-directed mutagenesis results in a loss of most cell adhesive activity from this central adhesion domain of fibronectin (Obara et al. 1988). However, the small 11.5-kD fibronectin fragment and RGD-containing synthetic peptides have very low cell adhesive activity, even though the RGD site is present (Pierschbacher and Ruoslahti 1984; Nagai et al. 1991). Direct binding studies reveal that the affinities of the 11.5-kD fragment and the RGD-containing synthetic peptides for cells are lower than those of intact fibronectin and the 75-kD fragment by one to two orders of magnitude (Akiyama et al. 1985). These studies all suggest that there is at least one region to the amino-terminal side of the 11.5-kD fragment that must also contribute substantially to the interaction of fibronectin with cells.

SYNERGY REGIONS IN FIBRONECTIN

To identify the polypeptide sequences besides RGD that are required for efficient interactions, a model system was developed using a cDNA fragment encoding the cell-binding domain of human fibronectin in the λgt11 expression system (Obara et al. 1987, 1988). Fibronectin expressed in a β-galactosidase fusion protein had full activity in adhesion assays. As expected, mutation of the RGDS sequence resulted in loss of nearly all cell adhesive function. In addition, however, deletion of an upstream region not including the RGDS site also caused major losses of activity (Obara et al. 1987).

To determine the sequences involved, a series of mutants deleting stepwise from the 5′ end of the fibronectin cDNA fragment were constructed (Obara et al. 1988; Aota et al. 1991), and the expressed fusion proteins were assayed for their cell adhesion activities (Fig. 2). Biological activity, as determined by the protein concentration required for 50% baby hamster kidney (BHK) cell spreading, showed a clear two-step decrease with progressive amino-terminal truncation. Although the three shortest constructs still contained the RGD sequence, they showed approximately 100-fold losses of activity, which is consistent with the previous

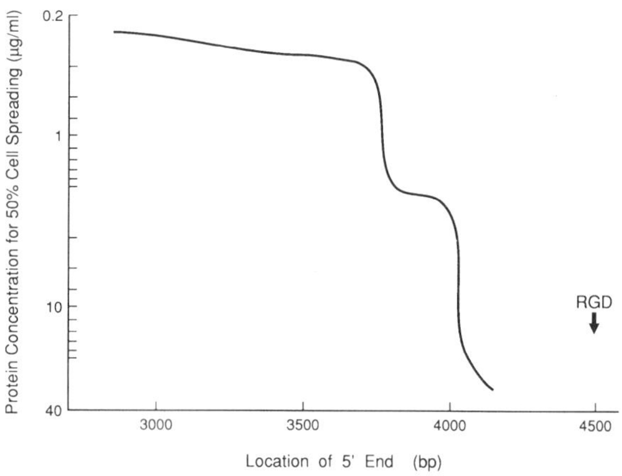

Figure 2. 5′-terminal deletion of the fibronectin central cell-binding domain and two-step loss of adhesive activity. Fibronectin cDNA encoding the cell-binding domain was truncated from the 5′ end, expressed in λgt11, purified, and then assayed for biological activity using a BHK cell-spreading assay. The protein concentrations required to produce spreading of 50% of the cells in each assay were plotted as a function of the 5′-terminal deletion points.

studies using proteolytic fragments (cf. Hayashi and Yamada 1983 and Akiyama et al. 1985). These results suggest that there are two subsites besides the RGD sequence important for biological activity.

To avoid possible artifacts derived from progressive shortening and varying sites of fusion with the exogenous polypeptide, a series of internal deletions designed to be one type III repeat unit in length were constructed (Aota et al. 1991). This study again demonstrated the existence of two important sites besides the RGD sequence, shown as A and B in Fig. 3. The B subregion was more important for cell-spreading activity, since its deletion resulted in greater than 20-fold losses in adhesive activity.

MONOCLONAL ANTIBODY ANALYSES OF FIBRONECTIN FUNCTION

As an alternative approach to this site-directed mutagenesis study, a set of monoclonal antibodies was produced against the cell-binding domain of fibronectin (Nagai et al. 1991). The epitopes of these antibodies were mapped using the internal deletion mutants described above as well as various small nonfused fibronectin polypeptide fragments expressed in *Escherichia coli* (Fig. 4). The inhibitory monoclonal antibody 3B8 bound to the center of the ninth repeat, far from the RGD sequence. Interestingly, its epitope mapped to the B site identified by the mutagenesis study (Aota et al. 1991). For example, 3B8 fails to bind to several mutants with single-amino-acid residue substitutions in this region. Two noninhibitory monoclonal antibodies, 13G12 and 12B8, which bind between the RGD sequence and the B subregion, can independently bind to the same type III repeat without competing with 3B8. This immunological study provides direct support for the conclusions based on site-directed mutagenesis.

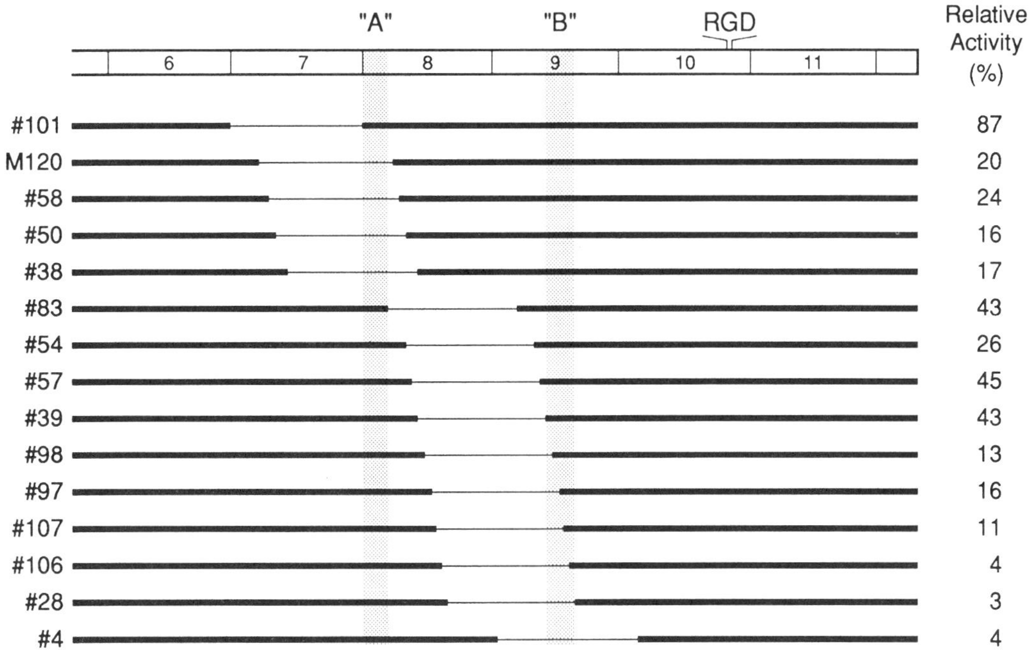

Figure 3. Internal deletion mutants and their cell-spreading activity. Using oligonucleotide-based site-directed mutagenesis, overlapped internal deletions (thin lines indicate deleted segments) designed to retain the type III repeating motif were introduced into the cell-binding domain of fibronectin. Their cell-spreading activities are shown relative to the full-length construct. Two subregions important for efficient cell-binding activity are shown by the gray boxes labeled *A* and *B*.

MINIMAL SEQUENCES, ALTERNATIVE SITES, AND OTHER POTENTIAL SYNERGY REGIONS

A 20-kD fibronectin fragment containing both the RGD and the more active of the two synergy subregions retains approximately 50% of the molar activity of the 110-kD fragment when used in soluble form as an inhibitor; the slight decrease in activity is consistent with the absence of the less-active synergy subregion from this fragment (S.K. Akiyama et al., in prep.). In contrast, the 20-kD fragment has less than 20% of the molar activity of the 110-kD fragment when coated directly on plastic for use as an adhesive substrate. Thus, the simple presence of both the RGD and synergy sites on an adhesive fragment is not sufficient to promote maximal cell adhesion. However, if a noninhibitory monoclonal antibody (MAb 13G12; Nagai et al. 1991) is used to immobilize the 20-kD fragment on plastic, this fragment then displays a molar activity identical to that of the larger 110-kD fragment in promoting cell adhesion. As expected, when this 20-kD fragment is coupled to an affinity column, it will bind the $\alpha_5\beta_1$ fibronectin receptor, whereas the 11.5-kD fragment lacking the B site does not (S.K. Akiyama et al., in prep.). These studies suggest that the manner in which small fragments are presented to cells can have major effects on adhesive activity, which will be important to consider in the future when using fragments to create biologically active adhesive substrates for potential medical applications.

Figure 4. Epitope mapping of anti-human fibronectin monoclonal antibodies. The epitopes of seven monoclonal antibodies (Nagai et al. 1991) were mapped using internally deleted mutants (Fig. 3) or short polypeptides expressed in *E. coli*. The thick bar indicates the range of each epitope, and the dashed portion indicates a region where binding sites do not overlap, because monoclonal antibodies mapping to this area can bind to fibronectin independently with no mutual inhibition (Nagai et al. 1991). Boldfaced lettering indicates that the antibody inhibits cell adhesion to fibronectin.

A second major region in fibronectin that mediates cell adhesion and migration in fibronectin is the IIICS (also termed the V) region (for references, see Humphries et al. 1987; Komoriya et al. 1991). Sequences in this region undergo alternative splicing, so that they are present in only certain isoforms of fibronectin. These sequences mediate cell-type-specific adhesion and migration dependent on the $\alpha_4\beta_1$ integrin, e.g., of neural crest cells and their derivatives, as well as certain lymphocytes. A key minimal amino acid sequence in this region is Leu-Asp-Val (LDV), although another spliced sequence containing Arg-Glu-Asp-Val has weak activity. Even though LDV is the key minimal sequence, decreasing the size of the full-length spliced sequence to this tripeptide entails a 20–25-fold loss of activity (Komoriya et al. 1991). This result strongly suggests that a "synergy" sequence is also involved in this site, which is recognized by the $\alpha_4\beta_1$ integrin, rather than $\alpha_5\beta_1$.

In the future, it will be interesting to determine whether other RGD-dependent cell adhesive proteins such as vitronectin also require additional information besides the RGD sequence. The precise mechanism of the synergistic enhancement in fibronectin still remains to be elucidated, i.e., direct binding of the RGD and synergy sites to the integrin receptor or indirect conformational effects of these subsites on the RGD sequence itself. Three-dimensional structural determination of the single type III repeat containing the RGD sequence, but lacking the synergistic region identified in this study, is successfully under way (Baron et al. 1992). The determination of the three-dimensional structure of a fully active cell-binding fragment of fibronectin containing both synergy and RGD domains will be of particular interest.

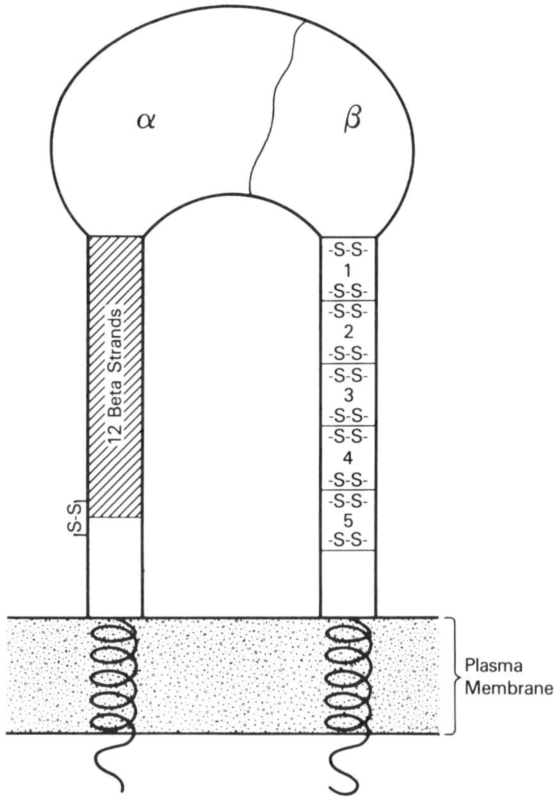

Figure 5. Schematic model of the $\alpha_5\beta_1$ fibronectin receptor. The α_5 subunit is at the left and the β_1 subunit at the right. The α_5 subunit contains 12 predicted β strands, and the β_1 subunit contains a series of intrachain disulfide bonds. Each subunit has a transmembrane segment and a short cytoplasmic tail. The external portion of this receptor is 20–22 nm long, with a head that is 12–15 nm in diameter. For details, see Nermut et al. (1988).

INTEGRINS

Roles and Structure

Integrins are heterodimeric cell-surface adhesion receptors, each consisting of an α subunit and a β subunit associated noncovalently in a 1:1 complex. More than 20 integrin dimers have been identified so far, which bind to a variety of adhesive proteins such as fibronectin, laminin, vitronectin, and collagen. Integrins play crucial roles in a wide variety of cellular interactions, ranging from simple attachment and spreading on adhesive substrates to more complex processes involved in migration and assembly of an extracellular matrix.

The overall structure of integrin receptors appears to be similar for the integrins examined to date. For example, the human $\alpha_5\beta_1$ fibronectin receptor consists of an ovoid head supported by two long, spindly legs (Fig. 5). All integrin subunits contain transmembrane domains and cytoplasmic tails, most of which are relatively short. As discussed below, the intracellular domains are thought to interact with cytoskeletal molecules.

Integrins in Cell Migration

Cell migration in vivo. Many steps in the embryonic development of multicellular organisms require the migration of cells to new sites. During morphogenesis, extracellular matrices composed of fibronectin and other adhesive macromolecules provide the substrates for cell adhesion and migration. In amphibians, fibronectin is initially found on the inner surface of the blastocoel roof at the early blastula stage, where it forms a three-dimensional meshwork covering the inner surface of the blastocoel (Boucaut and Darribère 1983). Integrin receptors are also present in blastulae and gastrulae of urodeles, particularly in fibronectin- and laminin-rich regions such as the blastocoel roof (Darribère et al. 1988). Synthetic peptides containing the RGDS sequence, Fab fragments of anti-integrin antibodies, and anti-fibronectin antibodies all individually inhibit migration of mesodermal cells across the inner blastocoel roof in a concentration-dependent manner when injected into the embryonic blastocoel cavity (Boucaut et al. 1984a,b; Darribère et al. 1988). Although effects on individual embryos can vary, the inhibition of integrin-mediated cell migration generally blocks gastrulation. Furthermore, intracellular microinjection of polyclonal antibodies or Fab fragments directed against the cytoplasmic domain of β_1 integrins can also disrupt matrix assembly and gastrulation above the cell progeny to which the antibodies segregate (Darribère et al. 1990). Such studies emphasize the importance of integrins in cell migration in development.

Neural crest cells are highly motile cells that originate in the neural tube and, after migration to various regions of the embryo, eventually differentiate into a wide range of tissues including melanocytes, facial bones, and peripheral nervous system. Neural crest cells migrate along fibronectin-rich pathways (Newgreen and Thiery 1980; Duband and Thiery 1982; Sternberg and Kimber 1986). RGD-containing synthetic peptides, anti-fibronectin antibodies, and anti-integrin antibodies individually inhibit cranial neural crest cell migration in vivo (Boucaut et al. 1984b; Bronner-Fraser 1985), resulting in the anomalous accumulation of neural crest cells within the lumen of the neural tube. These results indicate that fibronectin-integrin interactions are also important for correct migration of neural crest cells.

Cell migration in vitro. Cell migration on fibronectin can require both the RGD and synergy regions (Dufour et al. 1988; Nagai et al. 1991). For example, monoclonal antibodies to either site individually inhibit this process (Nagai et al. 1991). The migration of both normal (Akiyama et al. 1989) and oncogenically transformed human cells (Yamada et al. 1990) can be inhibited either by antifunctional antibodies that bind to the β_1 integrin polypeptide or by RGD-containing peptides (Albini et al. 1987; Akiyama et al. 1989; Yamada et al. 1990), although the inhibition by peptides is only partial. Inhibition of migration in these cases most probably occurs by disruption of β_1 integrin–fibronectin interactions.

The specific role of the α_5 integrin subunit in cell migration is less clear. Treatment of normal human fibroblasts and some transformed cells (such as 5637 bladder carcinoma cells) with anti-α_5 monoclonal antibodies has no inhibitory effect on migration (Akiyama et al. 1989; Yamada et al. 1990), whereas with certain other transformed cells (such as HT-1080 fibrosarcoma cells), there is partial or complete inhibition. The following mechanisms may explain these findings, which will require further investigation to evaluate: (1) the use of multiple, possibly redundant, integrin receptor systems for cell migration on fibronectin, (2) the rapid deployment of alternative fibronectin-binding integrins in migration to circumvent antibody inhibition, and (3) cell-type-specific use of different fibronectin-binding integrins in migration.

Individual human fibroblasts moving randomly in sparse culture on fibronectin substrates can become more migratory as the result of treatment with anti-α_5 antibodies (Akiyama et al. 1989). This unexpected result can be explained by the finding that the treatment of normal fibroblasts with the anti-α_5 monoclonal antibody results in a change of the distribution of the fibronectin receptor from a clustered to a diffuse pattern, accompanied by a loss of tight (focal) cell-substrate contacts. Both of these changes should promote a more migratory cellular phenotype (Chen et al. 1985; Duband et al. 1986, 1988). The random migration of human fibroblasts was, however, inhibited by a combination of anti-α_5 and anti-vitronectin receptor antibodies but not by either alone (S.S. Yamada et al., unpubl.); these cells therefore appear to use two redundant adhesion systems for migration.

In another cell system, Chinese hamster ovary (CHO) cell variants expressing reduced levels of the $\alpha_5\beta_1$ integrin could attach but not migrate on vitronectin substrates even though normal amounts of $\alpha_v\beta_3$ vitronectin receptor are expressed, suggesting that the $\alpha_5\beta_1$ and the $\alpha_v\beta_3$ receptors can act in a cooperative manner in cell migration (Bauer et al. 1992). CHO cell variants with 20% of the normal amount of $\alpha_5\beta_1$ integrin can attach well to but not migrate on fibronectin substrates (Schreiner et al. 1989; Bauer et al. 1992). These results suggest that "complex" processes such as migration may have more stringent requirements for high receptor number than the "simpler" process of cell attachment. It may also be possible that cell attachment and migration require different interactions of integrins with cytoskeletal or other intracellular components, a hypothesis that could be tested in the future.

Fibronectin and Integrins in Matrix Assembly

Fibroblasts in culture elaborate an extensive matrix of fibrillar fibronectin, a process that requires the $\alpha_5\beta_1$ integrin. Treatment of fibroblastic cell monolayers with anti-β_1 and anti-α_5 integrin monoclonal antibodies inhibits matrix assembly (Akiyama et al. 1989; Roman et al. 1989; Fogerty et al. 1990), although the exact extent of inhibition is a matter of controversy. Although long fibronectin fibrils are greatly disrupted after treatment with anti-α_5 or anti-β_1 integrin antibodies (more so with anti-β_1 than with anti-α_5 antibodies), cell-surface fibronectin is still present in the form of very short streaks or spots. Incomplete inhibition of matrix formation by anti-α_5 and anti-β_1 antibodies is consistent with the presence of other, as yet unidentified, "matrix assembly receptors."

The assembly of fibronectin into fibrils in vitro and in vivo appears to require multiple sites on fibronectin, including an amino-terminal region consisting of the first five type I repeats (McKeown-Longo and Mosher 1985; Quade and McDonald 1988), both the RGD and synergistic cell adhesive regions of fibronectin (McDonald et al. 1987; Nagai et al. 1991), and a site located at or near the junction of the ninth type I repeat and the first type III repeat of fibronectin that is identified by the monoclonal antibody L8 (Chernousov et al. 1991; Darribère et al. 1992). The precise function of each of these sites on fibronectin will be important to establish in the future.

Recombinant fibronectin polypeptides that lack the RGD sequence can still be incorporated into fibronectin fibrils (Schwarzbauer 1991), indicating that the RGD site is not required for fibril elongation. Conversely, the L8 site appears to function in fibril assembly or elongation but is not required for fibronectin binding to cells in vivo and in vitro (Chernousov et al. 1991; Darribère et al. 1992). The role of the amino-terminal region of fibronectin in matrix assembly is currently less clear. A 70-kD amino-terminal fragment of fibronectin can bind to cell monolayers (McKeown-Longo and Mosher 1985), as can a 29-kD amino-terminal fragment containing the first five type I repeats (Quade and McDonald 1988), and the 70-kD fragment can inhibit the binding of intact fibronectin (McKeown-Longo and Mosher 1985). Although the amino-terminal region of fibronectin does not mediate cell adhesion (McKeown-Longo and Mosher 1985), both anti-β_1 and anti-α_5 antibodies block ^{125}I-labeled 70-kD binding to cells, albeit only partially in the case of anti-α_5 antibodies (Fogerty et al. 1990). These results suggest that the $\alpha_5\beta_1$ integrin may play a role in the interaction between this region of fibronectin and cells.

REGULATION OF INTEGRIN FUNCTION

Integrin function in adhesion, migration, matrix assembly, and cytoskeleton organization depends on the quantity, affinity, and distribution of specific integrins. Although little is known about what mediates and regulates integrin spatial distribution and interactions with cytoskeletal proteins, such regulation is likely to be of central importance to integrin function. Biochemical studies have demonstrated that the β_1 intracellular domain can bind to both talin and α-actinin (Buck et al. 1986; Tapley et al. 1989; Otey et al. 1990). In intact cells, the ability of integrins to interact with the cytoskeleton has generally been assayed by their ability to concentrate in adhesion sites called focal contacts, where they colocalize with cytoskeletal proteins such as talin, vinculin, and α-actinin (Singer 1982; Chen et al. 1985; Damsky et al. 1985; Horwitz et al. 1986; Burridge et al. 1988). The β_1 intracellular domain is required for integrin localization to focal contacts, as shown by transfection studies using normal and mutant β_1 subunits in interspecies heterodimers (Solowska et al. 1989; Hayashi et al. 1990; Marcantonio et al. 1990). However, it was not clear from these studies whether the β_1 intracellular domain alone could interact with the cytoskeleton or whether, in analogy to ligand binding and especially talin binding (Buck et al. 1986), both the α and β subunits are required.

Analysis of Integrin Function Using Chimeras

The roles of individual integrin intracellular domains in receptor localization to adhesion sites and how their function may be regulated by ligand occupancy have been explored using novel model systems. To determine whether the α_5 and β_1 intracellular domains of the fibronectin receptor can function as independent domains, chimeric receptors were constructed containing the interleukin-2 (IL-2) receptor gp55 subunit (also known as the tac antigen) as the extracellular and membrane domains connected to either the α_5 or β_1 intracellular domain. A chimeric receptor lacking an intracellular domain served as a control (LaFlamme et al. 1992). When these chimeric receptors were transiently expressed in normal human fibroblasts and their expression was analyzed using antibodies to the IL-2 receptor, immunofluorescence microscopy demonstrated that the α_5 chimera and the receptor lacking an intracellular domain were expressed diffusely on the plasma membrane. In striking contrast, the β_1 chimera was concentrated in focal contacts (Fig. 6) and in other regions of the cell where ligand-occupied fibronectin receptors were concentrated, such as sites where endogenous fibronectin receptors bind extracellular fibronectin fibrils (LaFlamme et al. 1992). The α_5 intracellular domain does not therefore contain sufficient information to direct receptor localization. However, it is possible that the α_5 intracellular domain plays a supportive or regulatory role in receptor distribution or that it plays a more active role in receptor localization in other cell types. In contrast, the β_1 intracellular domain is alone sufficient to direct receptor localization to regions of the cell where ligand-occupied receptors are concentrated, such as to focal contacts.

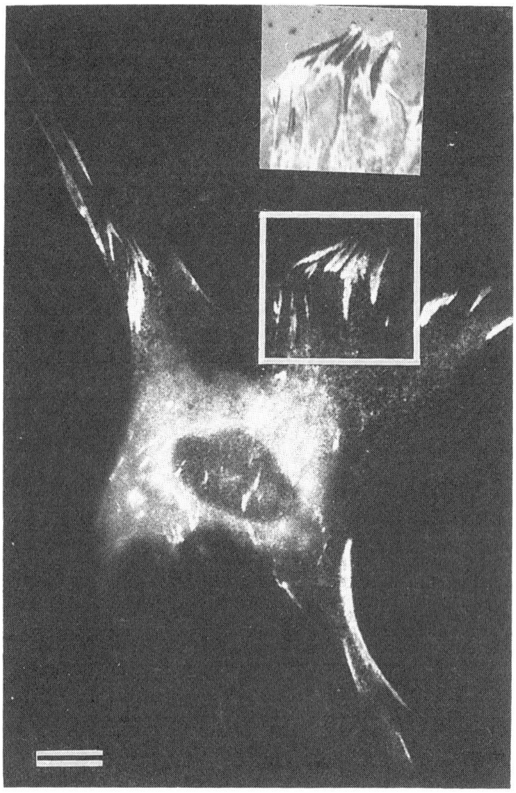

Figure 6. Expression of the β_1 chimera in normal human fibroblasts. Cells transfected with the β_1 chimera were immunostained using antibodies to the IL-2 receptor (LaFlamme et al. 1992). The inset shows the boxed region analyzed by interference reflection microscopy, where focal contacts are visualized as black streaks. Bar, 20 μm.

Biological Significance of β_1 Chimera Localization: The Role of Receptor Occupancy

The β_1 intracellular domain chimera cannot interact with substrate-bound ligand, such as fibronectin, yet it nevertheless concentrates in focal contacts formed by endogenous receptors. This result suggests that its localization is determined by an intracellular mechanism. Furthermore, this localization suggests that the

β_1 intracellular domain has an intrinsic ability to concentrate in focal contacts that must somehow be constrained in endogenous unoccupied receptors; otherwise, all types of β_1 integrins should concentrate in focal contacts regardless of the substrate ligand. Since the β_1 chimera mimicked a ligand-occupied receptor, at least in its ability to localize to focal contacts, it seemed possible that ligand occupancy might release this constraint on localization. To test this possibility, cells were allowed to spread and to form focal contacts in the absence of fibronectin on a laminin-coated substrate that had been blocked with heat-denatured bovine serum albumin (BSA) to prevent attachment of other adhesion factors. Under these conditions, the endogenous $\alpha_5\beta_1$ fibronectin receptor remains diffuse on the cell surface, and further ligand adsorption to the substrate is prevented. The ability of various soluble peptides and protein fragments to trigger the redistribution of these originally diffuse fibronectin receptors to focal contacts was compared. The cell-binding domain of fibronectin containing both the RGD site and the synergy region and high doses of a small pentapeptide containing the RGD site alone (GRGDS) produced redistribution of fibronectin receptors to focal contacts, whereas the controls could not (Fig. 7). An important control is to rule out effects due to adsorption of GRGDS or the 75-kD cell-binding domain of fibronectin to the BSA-blocked substrates. The substrates were preincubated with GRGDS or the 75-kD fragment under conditions similar to those used in the redistribution of the fibronectin and vitronectin receptors (LaFlamme et al. 1992) and then washed several times with serum-free medium. This treatment did not result in fibronectin receptor redistribution to focal contacts (Fig. 7). We therefore conclude that ligand occupancy, even with small peptide ligands, results in the redistribution of diffuse fibronectin receptors to focal contacts.

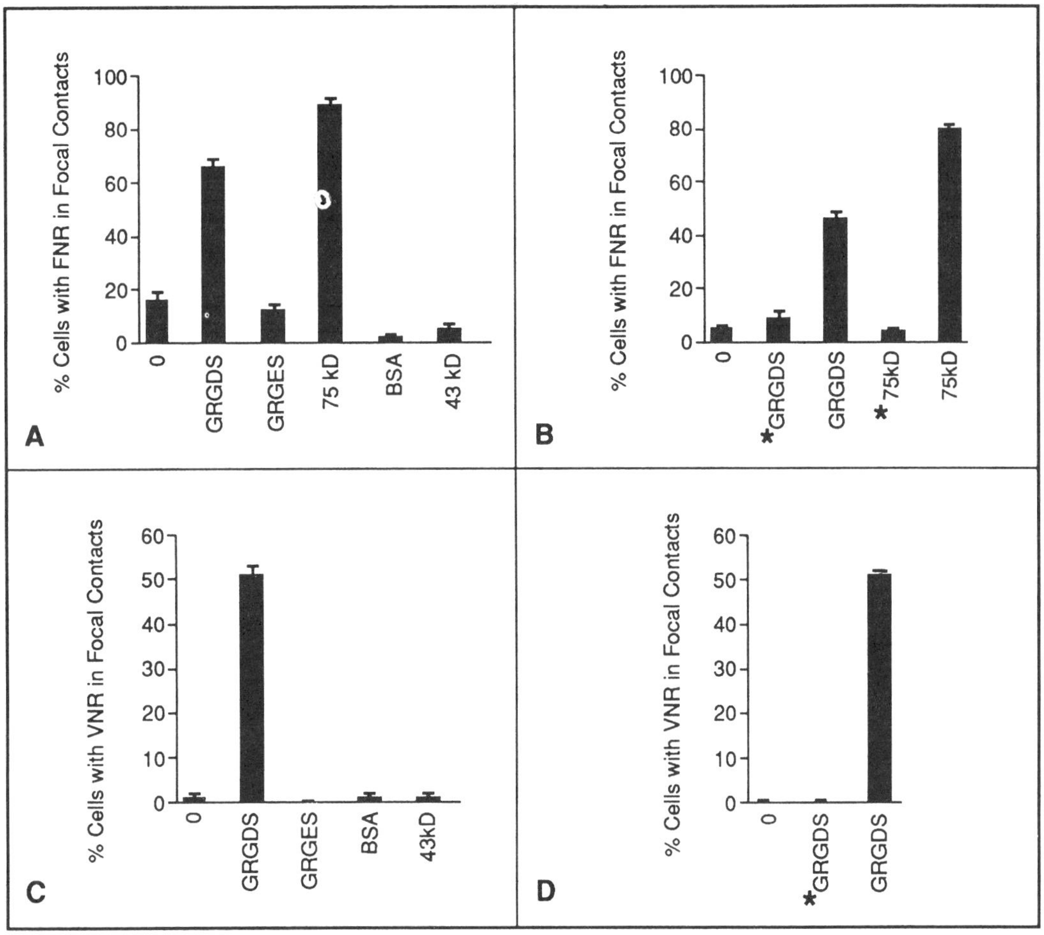

Figure 7. Redistribution of the fibronectin and vitronectin receptors by ligand occupancy. (*A*, *C*) Normal human fibroblasts pretreated as described previously on laminin-coated coverslips (LaFlamme et al. 1992) were incubated for an additional 60 min with serum-free medium alone or with either 100 μg/ml of the 75-kD cell-binding fragment of fibronectin (*A* and *B* only) or BSA or the 43-kD collagen-binding fragment of fibronectin, or 500 μg/ml GRGDS or GRGES. The percentage of cells with either fibronectin (FNR: *A* and *B*) or vitronectin receptors (VNR: *C* and *D*) in focal contacts was determined as described recently (LaFlamme et al. 1992). (*B*, *D*) Cells were treated as in *A* and *C*, except that the cells were incubated for 30 min after the addition of the peptides and protein fragments. In the control samples labeled *GRGDS and *75kD, the laminin-coated substrates were preincubated for 30 min with the indicated peptide or fragment and then washed well prior to the addition of cells.

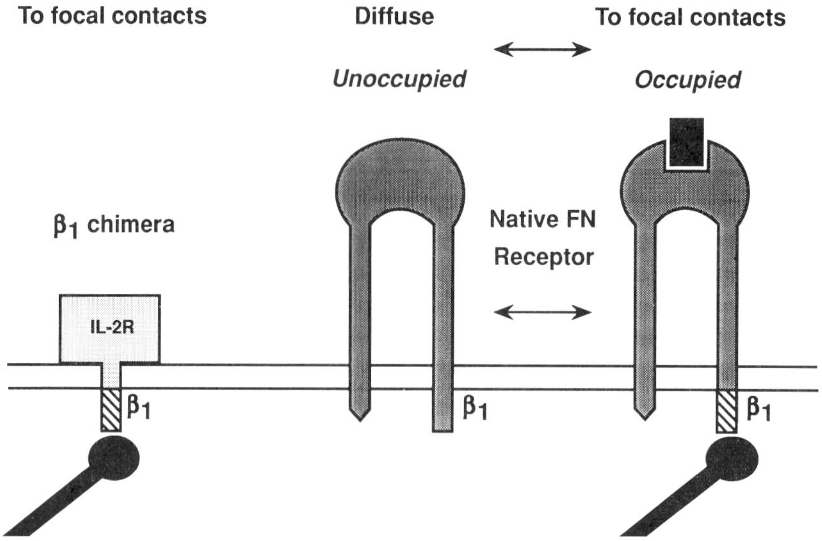

Figure 8. Schematic model for the regulation of fibronectin distribution receptor by ligand occupancy. The β_1 chimera is shown at focal contacts interacting with cytoskeletal protein(s) required for focal contact localization. The unoccupied fibronectin receptor would be distributed in a diffuse pattern on the cell surface because it cannot interact with the appropriate cytoskeletal protein(s). When the fibronectin receptor binds to its ligand fibronectin, a conformational change is proposed to occur in the receptor that then allows it to interact with the appropriate cytoskeletal protein(s) and to concentrate in focal contacts.

A simple model to explain this process (Fig. 8) involves a conformational change in the receptors that somehow unmasks or activates the β_1 intracellular domain. There is no direct evidence as yet that a conformational change in the fibronectin receptor occurs upon ligand occupancy; however, the distribution of the β_1 chimera to focal contacts and the redistribution of diffuse fibronectin receptors to focal contacts by ligand occupancy support the notion that the intracellular portions of occupied and unoccupied receptors are perceived differently by the cytoplasm and suggest that there is an intracellular mechanism involved in receptor distribution that is controlled by ligand occupancy.

Regulation of Localization of Other Integrin Receptors

Chimeric receptors containing the β_3 intracellular domain also concentrate in focal contacts (S.E. LaFlamme and K.M. Yamada, in prep.), suggesting that the β_3 intracellular domain also has an intrinsic ability to concentrate in focal contacts that must somehow be constrained in endogenous diffuse unoccupied receptors. An obvious question was whether, in analogy with the fibronectin receptor, ligand occupancy also releases this constraint on vitronectin receptors and targets them to focal contacts. Receptor occupancy with small peptide ligands also resulted in the redistribution of diffuse vitronectin receptors to focal contacts (see Fig. 7), a process that may also occur by a conformational change in the receptor that allows the β_3 intracellular domain to interact with the cytoskeleton.

The localization of the β_1 and β_3 chimeras to focal contacts supports the notion that the β_1 and β_3 intracellular domains have an intrinsic ability to concentrate in focal contacts that must somehow be constrained in endogenous unoccupied receptors. In addition, the function of these intracellular domains appears to be regulated by ligand occupancy, since occupancy of diffuse fibronectin and vitronectin receptors with even small peptide ligands releases this constraint and targets them to focal contacts. Therefore, besides the previously described *extracellular* mechanism regulating receptor distribution (Chen et al. 1986; Giancotti et al. 1986; Singer et al. 1988; Dejana et al. 1988; Akiyama et al. 1989; Fath et al. 1989; Roman et al. 1989; Carter et al. 1990), these findings suggest that there is also an *intracellular* mechanism that is regulated by ligand occupancy. Such a mechanism would provide a direct means of assembling adhesion sites and sites of fibronectin fibril formation.

The regulation of receptor distribution by ligand occupancy may be a characteristic of many integrin receptors. It will be interesting to determine whether this is a general characteristic of β_1 and β_3 integrins or whether the ability to regulate the function of β intracellular domains in this manner depends on the presence of a particular α subunit or ligand. Since the β_1 chimera can function like a ligand-occupied receptor in receptor distribution, it will be particularly interesting to determine whether the β_1 chimera can also function as a ligand-occupied receptor in other processes known to require integrin function, such as in the signal transduction pathways recently described to require integrin receptors (Guan et al. 1991; Kornberg et al. 1991; Schwartz et al. 1991), as well as in cell adhesion and migration using antibodies to the IL-2 receptor or IL-2 itself as a substrate.

Conversely, high levels of expression of the β_1 chimera might inhibit endogenous receptor function by competing with endogenous receptors for cytoskeletal proteins and possibly other cytoplasmic components required for normal receptor function. Preliminary evidence indicates that moderate to high levels of expression of the β_1 chimera inhibit endogenous $\alpha_5\beta_1$ fibronectin receptor localization to fibrils, as well as the function of this fibronectin receptor in matrix assembly, implying a central role for the β_1 intracellular domain in both of these processes (S.E. LaFlamme, unpubl.). These results are encouraging and suggest that the β_1 chimera and possibly other chimeras can be used as inhibitors to probe endogenous receptor function in vitro and possibly in vivo. These approaches would not only confirm the function of the corresponding heterodimeric receptors in processes such as cell adhesion, migration, matrix assembly, and signal transduction, but also establish a function for particular intracellular domains in these processes as well.

REFERENCES

Akiyama, S.K. and K.M. Yamada. 1985a. The interaction of plasma fibronectin with fibroblastic cells in suspension. *J. Biol. Chem.* **260:** 4492.

———. 1985b. Synthetic peptides competitively inhibit both direct binding to fibroblasts and functional biological assays for the purified cell-binding domain of fibronectin. *J. Biol. Chem.* **260:** 10402.

Akiyama, S.K., K. Nagata, and K.M. Yamada. 1990. Cell surface receptors for extracellular matrix components. *Biochim. Biophys. Acta* **1031:** 91.

Akiyama, S.K., E. Hasegawa, T. Hasegawa, and K.M. Yamada. 1985. The interaction of fibronectin fragments with fibroblastic cells. *J. Biol. Chem.* **260:** 13256.

Akiyama, S.K., S.S. Yamada, W.-T. Chen, and K.M. Yamada. 1989. Analysis of fibronectin receptor function with monoclonal antibodies: Roles in cell adhesion, migration, matrix assembly, and cytoskeletal organization. *J. Cell Biol.* **109:** 863.

Albelda, S.M. and C.A. Buck. 1990. Integrins and other cell adhesion molecules. *FASEB* **4:** 2868.

Albini, A., G. Allavena, A. Melchiori, F. Giancotti, H. Richter, P.M. Comoglio, S. Parodi, G.R. Martin, and G. Tarone. 1987. Chemotaxis of 3T3 and SV3T3 cells to fibronectin is mediated through the cell-attachment site in fibronectin and a fibronectin cell surface receptor. *J. Cell Biol.* **105:** 1867.

Aota, S., T. Nagai, and K.M. Yamada. 1991. Characterization of regions of fibronectin besides the arginine-glycine-aspartic acid sequence required for adhesive function of the cell-binding domain using site-directed mutagenesis. *J. Biol. Chem.* **266:** 15938.

Baron, M., A.L. Main, P.C. Driscoll, H.J. Mardon, J. Boyd, and I.D. Campbell. 1992. ^{1}H NMR assignment and secondary structure of the cell adhesion type III module of fibronectin. *Biochemistry* **31:** 2068.

Bauer, J.S., C.L. Schreiner, F.G. Giancotti, E. Ruoslahti, and R.L. Juliano. 1992. Motility of fibronectin receptor-deficient cells on fibronectin and vitronectin: collaborative interactions among integrins. *J. Cell Biol.* **116:** 477.

Boucaut, J.C. and T. Darribère. 1983. Fibronectin in early amphibian embryos. *Cell Tissue Res.* **234:** 135.

Boucaut, J.C., T. Darribère, H. Boulekbache, and J.P. Thiery. 1984a. Prevention of gastrulation but not neurulation by antibodies to fibronectin in amphibian embryos. *Nature* **307:** 364.

Boucaut, J.-C., T. Darribère, T.J. Poole, H. Aoyama, K.M. Yamada, and J.P. Thiery. 1984b. Biologically active synthetic peptides as probes of embryonic development: A competitive peptide inhibitor of fibronectin function inhibits gastrulation in amphibian embryos and neural crest cell migration in avian embryos. *J. Cell Biol.* **99:** 1822.

Bronner-Fraser, M. 1985. Alterations in neural crest migration by a monoclonal antibody that affects cell adhesion. *J. Cell Biol.* **101:** 610.

Buck, C.A., E. Shea, K. Duggan, and A.F. Horwitz. 1986. Integrin (the CSAT antigen): Functionality requires oligomeric integrity. *J. Cell Biol.* **103:** 2421.

Burridge, K., K. Fath, T. Kelly, G. Nuckolls, and C. Turner. 1988. Focal adhesions: Transmembrane junctions between the extracellular matrix and the cytoskeleton. *Annu. Rev. Cell Biol.* **4:** 487.

Carter, W.G., E.A. Wayner, T.S. Bouchard, and P. Kaur. 1990. The role of integrins $\alpha_2\beta_1$ and $\alpha_3\beta_1$ in cell-cell and cell-substrate adhesion of human epidermal cells. *J. Cell Biol.* **110:** 1387.

Chen, W.-T., E. Hasegawa, T. Hasegawa, C. Weinstock, and K.M. Yamada. 1985. Development of cell surface linkage complexes in cultivated fibroblasts. *J. Cell Biol.* **100:** 1103.

Chen, W.-T., J. Wang, T. Hasegawa, S.S. Yamada, and K.M. Yamada. 1986. Regulation of fibronectin receptor distribution by transformation, exogenous fibronectin, and synthetic peptides. *J. Cell Biol.* **103:** 1649.

Chernousov, M.A., F.J. Fogerty, V.E. Koteliansky, and D.F. Mosher. 1991. Role of the I-9 and III-1 modules of fibronectin in formation of an extracellular fibronectin matrix. *J. Biol. Chem.* **266:** 10851.

Damsky, C.H., K.A. Knudsen, D. Bradley, C.A. Buck, and A.F. Horwitz. 1985. Distribution of the CSAT cell-matrix antigen on myogenic and fibroblastic cells in culture. *J. Cell Biol.* **100:** 1528.

Darribère, T., K.M. Yamada, K.E. Johnson, and J.-C. Boucaut. 1988. The 140-kDa fibronectin receptor complex is required for mesodermal cell adhesion during gastrulation in the amphibian *Pleurodeles waltii*. *Dev. Biol.* **126:** 182.

Darribère, T., K. Guida, H. Larjava, K.E. Johnson, K.M. Yamada, J.P. Thiery, and J.-C. Boucaut. 1990. *In vivo* analyses of integrin β_1 subunit function in fibronectin matrix assembly. *J. Cell Biol.* **110:** 1813.

Darribère, T., V.E. Koteliansky, M.A. Chernousov, S.K. Akiyama, K.M. Yamada, J.P. Thiery, and J.-C. Boucaut. 1992. Three distinct regions of fibronectin are essential for matrix assembly *in vivo*. *Dev. Dynam.* (in press).

Dejana, E., S. Colella, G. Conforti, M. Abbadini, M. Gaboli, and P.C. Marchisio. 1988. Fibronectin and vitronectin regulate the organization of their respective arg-gly-asp adhesion receptors in cultured human endothelial cells. *J. Cell Biol.* **107:** 1215.

Duband, J.-L. and J.P. Thiery. 1982. Appearance and distribution of fibronectin during chick embryo gastrulation and neurulation. *Dev. Biol.* **94:** 337.

Duband, J.-L., S. Rocher, W.-T. Chen, K.M. Yamada, and J.P. Thiery. 1986. Cell adhesion and migration in the early vertebrate embryo: Location of the putative fibronectin receptor complex. *J. Cell Biol.* **102:** 160.

Duband, J.-L., G.H. Nuckolls, A. Ishihara, T. Hasegawa, K.M. Yamada, J.P. Thiery, and K. Jacobson. 1988. Fibronectin receptor exhibits high lateral mobility in embryonic locomoting cells but is immobile in focal contacts and fibrillar streaks in stationary cells. *J. Cell Biol.* **107:** 1385.

Dufour, S., J.-L. Duband, M.J. Humphries, M. Obara, K.M. Yamada, and J.P. Thiery. 1988. Attachment, spreading and locomotion of avian neural crest cells are mediated by multiple adhesion sites on fibronectin molecules. *EMBO J.* **7:** 2661.

Fath, K. R., C.-J. S. Edgell, and K. Burridge. 1989. The distribution of distinct integrins in focal contacts is determined by the substratum composition. *J. Cell Sci.* **92:** 67.

Fogerty, F.J., S.K. Akiyama, K.M. Yamada, and D.F. Mosher. 1990. Inhibition of binding of fibronectin to matrix assembly sites by anti-integrin ($\alpha_5\beta_1$) antibodies. *J. Cell Biol.* **111:** 699.

Giancotti, F. G., P. M. Comoglio, and G. Tarone. 1986. A 135,000 molecular weight plasma membrane glycoprotein involved in fibronectin mediated cell adhesion. Immuno-

fluorescence localization in normal and RSV-transformed fibroblasts. *Exp. Cell Res.* **163:** 47.
Guan, J.-L., J.E. Trevithick, and R.O. Hynes. 1991. Fibronectin/integrin interaction induces tyrosine phosphorylation of a 120-kDa protein. *Cell Regul.* **2:** 951.
Hay, E.D., ed. 1991. *Cell biology of extracellular matrix*, 2nd edition. Plenum Press, New York.
Hayashi, M. and K.M. Yamada. 1983. Domain structure of the carboxy-terminal half of human plasma fibronectin. *J. Biol. Chem.* **258:** 3332.
Hayashi, Y., B. Haimovich, A. Reszka, D. Boettiger, and A. Horwitz. 1990. Expression and function of chicken integrin β_1 subunit and its cytoplasmic domain mutants in mouse 3T3 cells. *J. Cell Biol.* **110:** 175.
Hemler, M.E. 1990. VLA proteins in the integrin family: Structures, functions and their role on leukocytes. *Annu. Rev. Immunol.* **8:** 365.
Horwitz, A., K. Duggan, C. Buck, M. C. Beckerle, and K. Burridge. 1986. Interaction of plasma membrane fibronectin receptor with talin—A transmembrane linkage. *Nature* **320:** 531.
Humphries, M.J., A. Komoriya, S.K. Akiyama, K. Olden, and K.M. Yamada. 1987. Identification of two distinct regions of the type III connecting segment of human plasma fibronectin that promote cell type-specific adhesion. *J. Biol. Chem.* **262:** 6886.
Hynes, R.O. 1990. *Fibronectins*. Springer Verlag, New York.
———. 1992. Integrins: Versatility, modulation, and signaling in cell adhesion. *Cell* **69:** 11.
Komoriya, A., L.J. Green, M. Mervic, S.S. Yamada, K.M. Yamada, and M.J. Humphries. 1991. The minimal essential sequence for a major cell type-specific adhesion site (CS1) within the alternatively spliced IIICS domain of fibronectin is leucine-aspartic acid-valine. *J. Biol. Chem.* **266:** 15075.
Kornberg, L.J., H.S. Earp, C.E. Turner, C. Prockop, and R.L. Juliano. 1991. Signal transduction by integrins: Increased protein tyrosine phosphorylation caused by clustering of β_1 integrins. *Proc. Natl. Acad. Sci.* **88:** 8392.
LaFlamme, S.E., S.K. Akiyama, and K.M. Yamada. 1992. Regulation of fibronectin receptor distribution. *J. Cell Biol.* **117:** 437.
Lash, J.W., K.K. Linask, and K.M. Yamada. 1987. Synthetic peptides that mimic the adhesive recognition signal of fibronectin: Differential effects on cell-cell and cell-substratum adhesion in embryonic chick cells. *Dev. Biol.* **123:** 411.
Marcantonio, E.E., J.-L. Guan, J.E. Trevithick, and R.O. Hynes. 1990. Mapping of the functional determinants of the integrin β_1 cytoplasmic domain by site-directed mutagenesis. *Cell Regul.* **1:** 597.
McDonald, J.A. and R.P. Mecham, eds. 1991. *Receptors for extracellular matrix*. Academic Press, New York.
McDonald. J.A., B.J. Quade, T.J. Broekelmann, R. LaChance, E. Hasegawa, and S. Akiyama. 1987. Both an aminoterminal matrix assembly domain and the carboxy-terminal cell adhesive domain are required for fibronectin assembly into a pericellular matrix by fibroblasts. *J. Biol. Chem.* **262:** 2957.
McKeown-Longo, P.J. and D.F. Mosher. 1985. Interaction of the 70,000-mol-wt amino-terminal fragment with the matrix-assembly receptor of fibroblasts. *J. Cell Biol.* **100:** 364.
Mosher, D.F. 1989. *Fibronectin*. Academic Press, New York.
Nagai, T., N. Yamakawa, S. Aota, S.S. Yamada, S.K. Akiyama, K. Olden, and K.M. Yamada. 1991. Monoclonal antibody characterization of two distant sites required for function of the central cell-binding domain of fibronectin in cell adhesion, cell migration, and matrix assembly. *J. Cell Biol.* **114:** 1295.
Nermut, M.V., N.M. Green, P. Eason, S.S. Yamada, and K.M. Yamada. 1988. Electron microscopy and structural model of human fibronectin receptor. *EMBO J.* **7:** 4093.
Newgreen, D. and J.P. Thiery. 1980. Fibronectin in early avian embryos: Synthesis and distribution along the migratory pathways of neural crest cells. *Cell Tissue Res.* **291:** 269.
Obara M., M.S. Kang, and K.M. Yamada. 1988. Site-directed mutagenesis of the cell-binding domain of fibronectin: Separable, synergistic sites mediate adhesive function. *Cell* **53:** 649.
Obara, M., M.S. Kang, S. Rocher-Dufour, A. Kornblihtt, J.P. Thiery, and K.M. Yamada. 1987. Expression of the cell binding domain of human fibronectin in *E. coli*: Identification of sequences promoting full to minimal adhesive function. *FEBS Lett.* **213:** 261.
Otey, C. A., F. M. Pavalko, and K. Burridge. 1990. An interaction between α-actinin and the β_1 integrin subunit in vitro. *J. Cell Biol.* **111:** 721.
Pierschbacher, M.D. and E. Ruoslahti. 1984. Cell attachment activity of fibronectin can be duplicated by small synthetic fragments of the molecule. *Nature* **309:** 30.
Pierschbacher, M.D., E.G. Hayman, and E. Ruoslahti. 1981. Location of the cell attachment site in fibronectin with monoclonal antibodies and proteolytic fragments of the molecule. *Cell* **81:** 259.
Quade, B.J. and J.A. McDonald. 1988. Fibronectin's amino-terminal matrix assembly site is located within the 29-kDa amino-terminal domain containing five type I repeats. *J. Biol. Chem.* **263:** 19602.
Roman, J., R.M. LaChance, T.J. Broekelman, C.J.R. Kennedy, E.A. Wayner, W.G. Carter, and J.A. McDonald. 1989. The fibronectin receptor is organized by extracellular matrix fibronectin: Implications for oncogenic transformation and for cell recognition of fibronectin matrices. *J. Cell Biol.* **108:** 2529.
Ruoslahti, E. 1991. Integrins. *J. Clin. Invest.* **87:** 1.
Schreiner, C.L., J.S. Bauer, Y.N. Danilov, S. Hussein, M.M. Sczekan, and R.L. Juliano. 1989. Isolation and characterization of Chinese hamster ovary cell variants deficient in the expression of fibronectin receptor. *J. Cell Biol.* **109:** 3157.
Schwartz, M.A., C. Lechene, and D.E. Ingber. 1991. Insoluble fibronectin activates the Na/H antiporter by clustering and immobilizing integrin $\alpha_5\beta_1$ independent of cell shape. *Proc. Natl. Acad. Sci.* **88:** 7849.
Schwarzbauer, J.E. 1991. Identification of the fibronectin sequences required for assembly of a fibrillar matrix. *J. Cell Biol.* **113:** 1463.
Singer, I.I. 1982. Association of fibronectin and vinculin with focal contacts and stress fibers in stationary hamster fibroblasts. *J. Cell Biol.* **92:** 398.
Singer, I.I., S. Scott, D.W. Kawka, D.M. Kazazis, J. Gailit, and E. Ruoslahti. 1988. Cell surface distribution of fibronectin and vitronectin receptors depends on substrate composition and extracellular matrix accumulation. *J. Cell Biol.* **106:** 2171.
Solowska, J., J.-L. Guan, E.E. Marcantonio, J.E. Trevithick, C.A. Buck, and R.O. Hynes. 1989. Expression of normal and mutant avian integrin subunits in rodent cells. *J. Cell Biol.* **109:** 853.
Sternberg J. and S.J. Kimber. 1986. Distribution of fibronectin, laminin and entactin in the environment of migrating neural crest cells in early mouse embryos. *J. Embryol. Exp. Morphol.* **91:** 267.
Tapley, P., A. Horwitz, C. Buck, K. Duggan, and L. Rohrschneider. 1989. Integrins isolated from Rous sarcoma virus-transformed chicken embryo fibroblasts. *Oncogene* **4:** 325.
Yamada, K.M. and D.W. Kennedy. 1984. Dualistic nature of adhesive protein function: Fibronectin and its biologically active peptide fragments can autoinhibit fibronectin function. *J. Cell Biol.* **99:** 29.
Yamada, K.M., D.W. Kennedy, S.S. Yamada, H. Gralnick, W.-T. Chen, and S.K. Akiyama. 1990. Monoclonal antibody and synthetic peptide inhibitors of human tumor cell migration. *Cancer Res.* **50:** 4485.

Functional Roles for Integrin α Subunit Cytoplasmic Domains

M.E. Hemler, P.D. Kassner, and B.M.C. Chan*
Dana-Farber Cancer Institute, Harvard Medical School, Boston, Massachusetts 02115

Cell-surface receptors in the integrin family mediate a wide variety of dynamic cell adhesion processes, such as those involved during complex tissue organizing events, immunolocalization of leukocytes, and tumor cell growth and metastasis. The integrin family comprises at least 20 distinct $\alpha\beta$ heterodimers, with $\alpha\beta$ subunit combinations as shown in Figure 1. This diversity of heterodimeric structures, many with multiple ligand-binding specificities, facilitates cell adhesion to a variety of extracellular matrix ligands, cell-surface prteins, and some serum and complement proteins (Hemler 1990; Larson and Springer 1990; Hynes 1992).

Whereas earlier attention was focused on identifying new integrin subunits and determining ligand-binding properties, more recent attention has turned toward studies of how integrins, spanning the cell membrane, act as a critical interface between the intracellular and extracellular environments. For example, as shown schematically in the top part of Figure 2, integrins bind to external ligands and translate this information into specific cellular responses, which lead to adhesion, spreading, or cell migration patterns, and ultimately, regulation of cell growth and differentiation.

Not only do integrins participate in this "outside in" signaling, but also they are subject to "inside out" signaling. A variety of different receptor-agonist interactions (e.g., see the bottom of Fig. 2) trigger an unknown intermediary mechanism (represented by a black box) which leads to an up-regulation of integrin function. By this means, cells can rapidly convert available receptors to become adhesive, without requiring new protein synthesis (Wright and Meyer 1986; Dustin and Springer 1989; Plow and Ginsberg 1989; Van Kooyk et al. 1989; Phillips et al. 1991; Hynes 1992).

Because integrins physically span the plasma membrane, it is expected that their cytoplasmic domains may play a key role in two-way signaling mechanisms as depicted in Figure 2. However, only a few experiments have addressed this issue. Expression of deleted forms of the integrin β_1 (Solowska et al. 1989) or β_2 (Hibbs et al. 1991) subunits, lacking cytoplasmic domains, resulted in diminished cell adhesion to immobilized lig-

*Present address: Department of Chemistry and Biochemistry, University of Guelph, Guelph, Ontario NIG 2WI Canada.

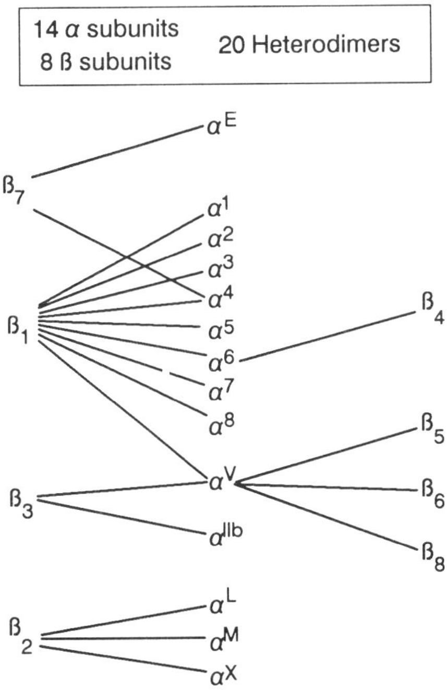

Figure 1. Subunit associations within the integrin family.

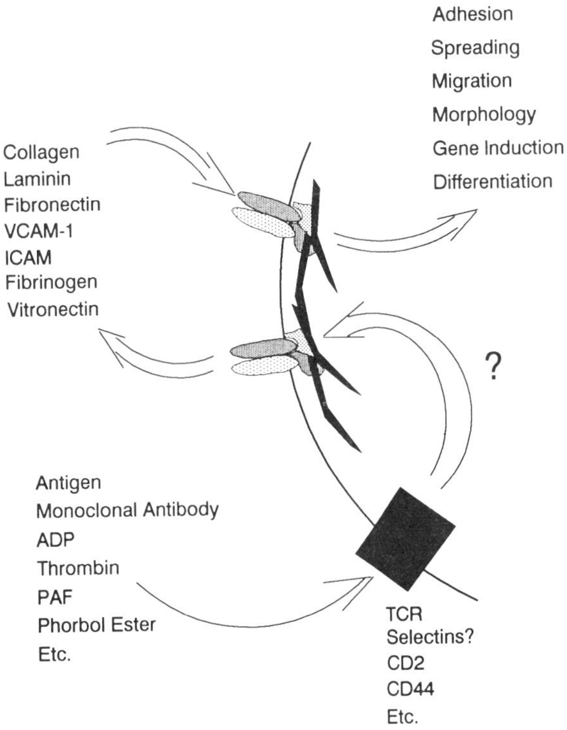

Figure 2. Schematic diagram of two-way signaling through integrins.

ands. The loss of β_2 integrin activity could not be overcome by phorbol ester treatment (Hibbs et al. 1991). In contrast, deletion of the β_3 cytoplasmic domain did not result in an alteration of $\alpha^{IIb}\beta_3$ activity expressed in CHO cells (O'Toole et al. 1991). Deletion of the integrin α^L cytoplasmic domain neither suppressed cell attachment functions, nor diminished the ability of LFA-1 to be functionally up-regulated (Hibbs et al. 1991). In contrast, both deletion of the cytoplasmic domain of α^{IIb} and swapping of the α^{IIb} cytoplasmic domain with that from α^5 caused an increase in the ligand-binding function of $\alpha^{IIb}\beta_3$. From these results, it is clear that there may be no simple set of rules applicable to all integrins. Furthermore, elucidation of the roles of integrin cytoplasmic domains is complicated by the fact that different cellular environments can yield variations in integrin functional activity (Larson et al. 1990; Chan and Hemler 1992).

Although not yet placed in the context of transmembrane signaling through integrins, both immunolocalization studies (Burn et al. 1988; Singer et al. 1988; Mueller et al. 1989) and direct biochemical methods (Horwitz et al. 1986; Tapley et al. 1989; Otey et al. 1990) have suggested that integrin cytoplasmic domains can associate with cytoskeletal proteins such as talin and α-actinin. Such associations are undoubtedly critical for integrin localization into focal contacts. Supporting this idea, deletion of the β_1 cytoplasmic domain (Solowska et al. 1989; Hayashi et al. 1990; Marcantonio et al. 1990) caused loss of localization to focal contacts, as did certain critical point mutations within β_1 (Reszka et al. 1992). In addition, expression of the β_1 cytoplasmic domain as part of a hybrid molecule is sufficient to drive the molecule into focal contacts (LaFlamme et al. 1992). Whereas the cytoplasmic and transmembrane domains of β_1 and β_3 are functionally interchangeable with respect to focal contact formation (Solowska et al. 1991), the cytoplasmic domain of β_5 subunit appears less able to localize there (Wayner et al. 1992). Integrin α^5 chain does not have inherent capability to localize to focal contacts, but a model has been proposed whereby it could act as a negative regulator of β_1 focal adhesion localization (LaFlamme et al. 1992).

Although focal contact formation has not directly been related to two-way signaling, it is intriguing that a focal adhesion kinase (FAK) has been discovered (Schaller et al. 1992), which is triggered through integrins (Guan et al. 1991; Kornberg et al. 1991). In future studies it will be interesting to see what role FAK plays in the two-way signaling mediated by integrins.

In our studies, we have initially focused on the roles of integrin α chains. Comparison of the known integrin α chain cytoplasmic domains (listed in Table 1) reveals that they are structurally diverse, aside from a consensus KXGFFKR motif found in most of the sequences. The lengths (counting from the first Lys) range from 15 amino acids (for α^1) to 76 amino acids (for α^7), and generally there is little amino acid similarity beyond the KXGFFKR motif. In addition, the recent discovery of alternatively spliced forms of α^3 (Tamura et al. 1991) and α^6 (Cooper et al. 1991; Tamura et al. 1991) cytoplasmic tails contributes to the overall sequence diversity. Although the α-chain cytoplasmic domains of different subunits are rather distinct when compared to one another, many are highly conserved when compared among different species. For example, the cytoplasmic tail of α^3 is identical in human and hamster, that from α^5 is identical in human and mouse, and α^4 and α^6 show only single differences when each is compared between human and mouse. An exception to this cross-species similarity is the nearly 50% dissimilarity seen between the cytoplasmic tails of human and mouse α^L.

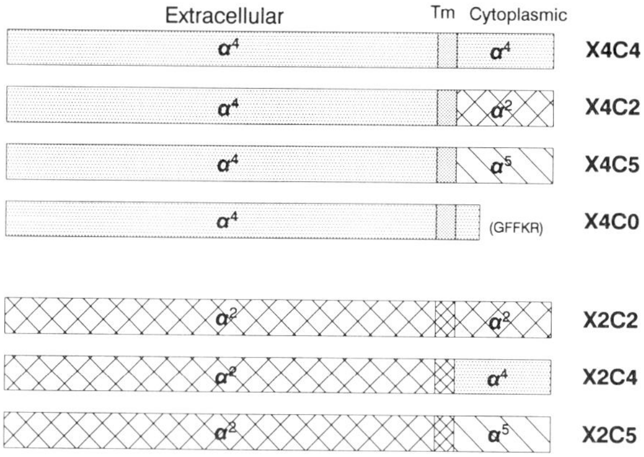

Figure 3. Schematic diagram of mutated integrin α chains.

The structural diversity seen among α-chain cytoplasmic domains predicts that each may play a distinct functional role. To test that hypothesis, we have constructed a series of α-chain chimeric molecules, as shown schematically in Figure 3. The α^2, α^4, and α^5 cytoplasmic tails are chosen for study because they all normally pair with the same β subunit (β_1), and thus physiologically relevant domain interactions are expected to be maintained.

RESULTS

Expression of Chimeric Molecules

Using the pF*neo* vector and techniques described elsewhere (Chan et al. 1992), the X2 series of constructs was expressed in both K562 cells (an erythroleukemic cell line) and RD cells (a rhabdomyosarcoma cell line). Regardless of which cytoplasmic tail was present, the different extracellular α^2 domain constructs were expressed equally well, as determined by flow cytometry and immunoprecipitation (Chan et al. 1992). Likewise, the X4 series of constructs containing either the α^4, α^2, or α^5 cytoplasmic domains were also expressed equally well in either K562 or MIP-101 cells (not shown). However, the X4CO construct (deleted just after the consensus GFFKR sequence) showed reduced expression in either K562 cells or MIP-101 cells (50–90% diminished compared to the other con-

Table 1. Comparison of Integrin α Subunit Cytoplasmic Domains

α Chain	Carboxy-terminal sequences	Reference
α1 ra	WKIGFFKRPLKKKMEK	Ignatius et al. (1990)
α2	WKLGFFKRKYEKMTKNPDEIDETTELSS	Takada and Hemler (1989)
α3	WKCGFFKRARTRALYEAKRQKAEMKSQPSETERLTDDY	Takada et al. (1991); Tsuji et al. (1991)
α3 ch	WKCGFFRRASTGAMYEAKGQKAEMRIQPSETERLTDDY	Hynes et al. (1989)
α3 ha	WKCGFFKRARTRALYEAKRQKAEMKSQPSETERLTDDY	Tsuji et al. (1990)
α3B ha	" " CDFFKPTRYYRIMPKYHAVRIREEERYPPPGSTLPTKKHWVTSWQIRDRYY	Tamura et al. (1991)
α4	WKAGFFKRQYKSILQEENRRDSWSYINSKSNDD	Takada et al. (1989)
α4 mo	WKAGFFKRQYRSILQEENRRDSWSYVNSKSNDD	Neuhaus et al. (1991)
α5	YKLGFFKRSLPYGTAMEKAQLKPPATSDA	Argraves et al. (1987); Fitzgerald et al. (1987)
α5 mo	YKLGFFKRSLPYGTAMEKAQLKPPATSDA	Holers et al. (1989)
α6	WKCGFFKRNKKDHYDATYHKAEIHAQPSDKERLTSDA	Tamura et al. (1990)
α6 mo	WKCGFFKRNKKDHYDATYHKAEIHTQPSDKERLTSDA	Cooper et al. (1991)
α6 ch	WKCGFFKRSKKDHYDATYHKAEIHAQPSDKERLTSDA	de Curtis et al. (1991)
α6B	" " CGFFKRSRYDDSVPRYHAVRIRKEEREIKDEKYIDNLEKKQWITKWNRNESYS	Tamura et al. (1991)
α6B mo	" " CGFFKRSRYDDSIPRYHAVRIRKEERE....	Cooper et al. (1991)
α7	WKLGFFKRAKHPEATVPQYHAVKILREDRQQFKEEKTGTIQRSNWGNSQWEG...	
	...SDAHPILAADWHPELGPGHPVSVTA	Schafer et al. (1989)
α8 ch	YKCGFFDRARPPQDDMADREQLTNNKTTDA	Bossy et al. (1991)
αIIb	WKVGFFKRNRPPLEEDDEEGE	Poncz et al. (1987)
αIIb ra	WKAGFFKRNRPPLEEEEEE–E	Poncz and Newman (1990)
αV	YRMGFFKRVRPPQEEQEREQLQPHENGEGNSET	Suzuki et al. (1987)
αV ch	YRMGFFKRVRPPQEEQEREQLQPHENGEGTSEA	Hynes et al. (1989)
αL	YKVGFFKRNLKEKMEAGRGVPNGIPAEDSEQLA–SGQEAGDPGCLKPLHEKDSESG	Larson et al. (1989)
αL mo	YKVGFFKRNLKEKMEADGGVPNGSPPEDTDPLAVPGEETKDMGCLEPSGRVTRTKA	Kaufmann et al. (1992)
αM	YKLGFFKRQYKDMMSEGGPPGAEPQ	Corbi et al. (1988); Arnaout et al. (1988); Hickstein et al. (1989)
αM mo	YKLGFFKRQYKDMMNEAAPQDAPPQ	Kaufmann et al. (1992)
αX	YKVGFFKRQYKEMMEEANGQIAPENGTQTPSPPSEK	Corbi et al. (1987)
ps2 dr	YKCGFFNRNRPTDHSQERQPLRNGYHGDEHL	Bogaert et al. (1988)

All sequences are of human origin unless otherwise indicated to be from mouse (mo), chicken (ch), rat (ra), hamster (ha), or *Drosophila* (dr). The conserved GFFKR sequence motif is underlined. The α3B and α6B are alternatively spliced forms, differing from the initially discovered α3 and α6 sequences. The mouse α6B sequence is not complete, and the human α7 sequence is continued on a second line.

structs). Thus, the absence of an α-chain cytoplasmic tail may interfere with chain assembly, processing, or heterodimer stability.

Role of Cytoplasmic Tails during VLA-2-dependent Cell Attachment to Immobilized Ligands

When the X2 series of constructs was expressed in RD cells, comparable VLA-2-mediated adhesion to collagen and laminin was observed for each construct, consistent with the expected ligand-binding properties of VLA-2. In a control experiment, no increased adhesion to fibronectin was observed (Table 2). From these results, it is clear that exchange of the α^2 cytoplasmic domains with those of α^4 and α^5 had no obvious effect on the relatively high level of adhesion mediated by VLA-2 in RD cells. In another experiment (not shown), K562 cells transfected with the X2 series of constructs showed only weak VLA-2-dependent adhesion to collagen or laminin. As described elsewhere (Chan and Hemler 1992), the host cell environment provided by K562 cells is not particularly conducive to high VLA-2 activity. Our results indicate that exchange of the α^2 cytoplasmic tail is not sufficient to overcome the relative inactivity of the VLA-2 heterodimer in K562 cells.

Role of Cytoplasmic Tails during VLA-4-dependent Cell Attachment to Immobilized Ligands

Each of the X4 series of constructs, expressed in either a carcinoma cell line (MIP-101) or K562 cells, was tested for ability to mediate adhesion to VLA-4 ligands. As indicated in Table 3, wild-type α^4 (X4C4) expressed in MIP-101 cells mediated moderate adhesion to the FN/40 fragment of fibronectin, to the CS1 peptide (sequence contained within FN/40), and to sVCAM-1. Despite exchange of the α^4 cytoplasmic tail with tails from α^2 or α^5, VLA-4-mediated adhesion to FN/40, CS1 peptide, or sVCAM-1 was essentially unchanged. Adhesion mediated by the X4CO construct in MIP-101 cells was very low and in some experiments was absent. It remains to be determined whether this is simply a reflection of the lower expression of that construct or is due to some functional deficiency caused by the absence of the α-chain cytoplasmic tail.

Table 2. VLA-2-mediated Adhesion to Immobilized ECM Proteins

Cell	Collagen	Laminin	Fibronectin
RD-pF	no	no	no
RD-X2C2	yes	yes	no
RD-X2C4	yes	yes	no
RD-X2C5	yes	yes	no

VLA-2-dependent adhesion was determined using an anti-α2 blocking antibody, and adhesion assays were carried out as described previously (Chan et al. 1992). Low levels of VLA-2-independent background adhesion to collagen and laminin were observed due to the activity of other integrins endogenously present in RD cells (primarily VLA-1 and VLA-6). Moderate adhesion to fibronectin was observed, but it also was VLA-2-independent and was entirely mediated by VLA-4 and VLA-5.

Table 3. VLA-4-mediated Adhesion to Its Immobilized Ligands

Cell	FN/40	CS1	sVCAM-1
MIP-pF	absent	absent	absent
MIP-X4C4	moderate	moderate	moderate
MIP-X4C5	moderate	moderate	moderate
MIP-X4C2	moderate	moderate	moderate
MIP-X4C0[a]	low	low	low
K-pF	(absent)[b]	absent	absent
K-X4C4	low	low	low
K-X4C5	low	low	low
K-X4C2	low	low	low
K-X4C0	absent/low[c]	absent/low	absent/low

[a]The low adhesion by MIP-X4X0 cells can possibly be explained by the low expression of that construct.

[b]Occasionally some VLA-4-independent adhesion to FN/40 was observed.

[c]Adhesion of K-X4C0 cells could not be determined accurately due to low and inconsistent expression levels.

In K562 cells, adhesion was rather low, but nonetheless similar for each chimeric construct (Table 3, bottom). Adhesion mediated by the X4CO construct was negligible, again perhaps partly due to lower expression levels. Together these results suggest that the cytoplasmic tails of α^2, α^4, and α^5 are interchangeable with respect to determining the level of constitutive VLA-2 or VLA-4 activity, whether it is high, low, or moderate.

Contribution of Cytoplasmic Tails to Post-ligand-binding Events

The processes of collagen gel contraction and random cell migration were chosen as examples of two widely differing events that happen subsequent to integrin-mediated cell adhesion to ligand. As shown in Table 4, the X2 series of constructs showed clear differences in their abilities to mediate these events. The C2 and C5 domains, but not the C4 domain, supported VLA-2-mediated collagen gel contraction by RD cells. Conversely, increased random migration on collagen- or laminin-coated surfaces occurred only when the C4 domain was attached to VLA-2 in place of the C2 domain. These widely differing responses provide clear indication that specific sequences within these different constructs play distinct roles in post-ligand-binding events. It is not yet clear if these are positive or negative roles.

To extend these results, migration mediated by the X4 series of constructs was also analyzed. When ex-

Table 4. Collagen Gel Contraction and Cell Migration Mediated by X2 Constructs Expressed in RD Cells

	Migration on		
Cells	collagen	laminin	Collagen gel contraction
RD-pF[a]	no	no	no
RD-X2C2	no	no	yes
RD-X2C4	yes	yes	no
RD-X2C5	no	no	yes

[a]A low level of background migration was seen using mock-transfected (RD-pF) cells. Migration and collagen gel contraction were measured as determined elsewhere (Chan et al. 1992).

Table 5. Comparison of X4 Constructs during VLA-4-dependent Migration of MIP-101 Cells

Cells	Migration on FN/40
MIP-pF	low
MIP-X4C2	low
MIP-X4C4	intermediate
MIP-X4C5	high
MIP-X4C0	?

Random migration of MIP-101 cells was carried out using plastic coated with 5 μg/ml of the FN/40 chymotryptic fragment of fibronectin using methods described previously (Chan et al. 1992).

pressed in K562 cells, adhesion and subsequent migration was too weak to yield useful information. However, when expressed in MIP-101 cells, the different X4 constructs yielded results (Table 5) that were only partially consistent with the results obtained using the X2 constructs in RD cells (Table 4). As shown, X4C5 supported the highest migration rate, X4C4 yielded an intermediate level, and X4C2 did not support migration above the background level seen with pF*neo* transfected cells.

Role of Cytoplasmic Tails during Cell Aggregation Triggered through VLA-4

Several antibodies directed against the α^4 chain or the β_1 chain of VLA-4 are able to trigger homotypic aggregation when added to various VLA-4-positive leukocyte cell lines, initially suggesting a novel ligand-binding function for VLA-4 (Bednarczyk and McIntyre 1990; Campanero et al. 1990). However, because later studies showed that this aggregation phenomenon does not seem to be mediated by any known VLA-4/ligand interaction (Pulido et al. 1991), it is hypothesized that VLA-4 could transmit a signal to trigger another effector pathway, such that VLA-4 itself does not mediate the aggregation.

To determine whether the cytoplasmic tail of α^4 has any specific role in triggering homotypic aggregation, we analyzed the X4 series of constructs expressed in K562 cells. As shown in Table 6, each of the chimeric constructs supported homotypic aggregation, either triggered by anti-α^4 or anti-β_1 antibodies, whereas untransfected K562 cells were not triggered to aggregate. In addition, the X4CO construct also supported homotypic aggregation, although at a lower level. This result is consistent with the lower level of expression of X4CO in K562 cells.

Table 6. Comparison of X4 Series Constructs during VLA-4-triggered Cell Aggregation

Cells	β-Triggered aggregation	α4-Triggered aggregation
K-pF	no	no
K-X4C2	yes	yes
K-X4C4	yes	yes
K-X4C4	yes	yes
K-X4C5	yes	yes
K-X4C0	yes	yes

Aggregation was induced by adding either MAb Li 1/2 (anti-β_1, Campanero et al. 1992), or MAb HP2/1 (anti-α4, Sánchez-Madrid et al. 1986) and then cells were visually inspected for the presence of aggregation after 3 hr.

DISCUSSION

The major conclusions from the studies mentioned here are (1) exchange of cytoplasmic tails does not seem to alter the cell-surface expression or cell adhesion properties of VLA-2 or VLA-4, (2) deletion of the α^4 cytoplasmic tail does not lead to increased constitutive VLA-4 activity, (3) exchange of cytoplasmic tails does have a dramatic effect on post-ligand-binding functions such as collagen-gel contraction and random cell migration, and (4) exchange or deletion or the α^4 cytoplasmic tail does not alter the aggregation-triggering properties of VLA-4.

As mentioned above, recent studies have suggested that the cytoplasmic domains of integrin α and β chains may play a role both in determining the constitutive activation state of integrins and in modulating integrin activity as a consequence of cell stimulation. Of particular interest was the report that deletion or swapping of the α^{IIb} cytoplasmic tail resulted in an increase in the affinity of $\alpha^{IIb}\beta_3$ for its ligand fibrinogen (O'Toole et al. 1991). To further explore the idea that integrin cytoplasmic domains might control integrin interactions with ligands, we have compared mutant and wild-type integrins expressed in multiple cell lines. Notably, we found that swapping of the cytoplasmic tails of α^2 or α^4 had no effect on the constitutive activity of VLA-2 or VLA-4, respectively. If the integrin was constitutively active (VLA-2 in RD cells, VLA-4 in MIP-101 cells), it remained active regardless of which tail was present. Similarly, if the integrin was constitutively inactive (VLA-2 or VLA-4 in K562 cells), it remained inactive. Deletion of the cytoplasmic tail of α^4 did not lead to increased constitutive activity of VLA-4 either in K562 cells or in MIP-101 cells. Together these results indicate that deletion or replacement of an α-chain cytoplasmic tail is not a universal mechanism for up-regulating the constitutive activity of integrins. This result is in contrast to results in which the cytoplasmic tail of α^{IIb} was deleted or exchanged (with α^5) and an increase in ligand-binding affinity was observed for $\alpha^{IIb}\beta_3$ expressed in CHO cells. We speculate that because we have utilized different integrins, different cellular environments, or a combination of both, a different mode of regulation of the constitutive activity of integrins may apply. In future studies, the cytoplasmic domain of the β_1 chain also needs to be analyzed to complement the current studies involving α chains.

Although we have not found that exchange of α-chain cytoplasmic domains influences the expression or adhesive activity of β_1 integrins, we have clearly demonstrated that two different post-ligand-binding events are dramatically dependent on cytoplasmic domains. In particular, we found that the α^2 and α^5 cytoplasmic domains are better suited for converting VLA-2-mediated adhesion to collagen into the stable exertion of physical force, such as occurs during collagen gel contraction. In contrast, the cytoplasmic domain of α^4 more effectively participated in the transient VLA-2-mediated adhesion that occurs during cell migration (Chan et al. 1992). Analysis of another integrin (VLA-4) expressed in another cell type (MIP-101) yielded similar results insofar as the cytoplasmic tail of α^4 was better able to support migration than the tail of α^2. However, the observation that the cytoplasmic tail of α^5 supported a high-level VLA-4-mediated MIP-101 cell migration, but not VLA-2-mediated migration of RD cells, again emphasizes that conclusions from one study cannot be extrapolated to apply to all integrins and all cell types. We hypothesize that the wide diversity of integrin cytoplasmic domain sequences, likely acting together with assorted unknown cellular factors found in different cellular environments, has the potential to provide an enormous variability in post-ligand-binding events. Because deletion or swapping of the α^4 cytoplasmic domain did not prevent VLA-4-triggered aggregation of K562 cells, we can essentially rule out a role for α^4 cytoplasmic domains in that process. Recent results have indicated that homotypic aggregation can be triggered by antibodies not only to α^4 (Bednarczyk and McIntyre 1990; Campanero et al. 1990), but also to β_1 (Campanero et al. 1992), to α^5 (Caixia et al. 1991), and to α^3 (J. Weitzman et al., unpubl.). This would suggest that perhaps β_1 is the essential component by which a pro-aggregation signal is delivered, leading to downstream activation of an unknown receptor or receptors that actually participate in the aggregation process. In other laboratories, monoclonal antibodies to a host of other cell-surface molecules have also been found to trigger aggregation (Kansas and Tedder 1992; Nakamura et al. 1992) that is mediated by unknown receptors. A hypothesis to be studied in future experiments is that perhaps antibody-triggering of β_1-containing integrins (VLA proteins) and a variety of other receptors feed into the same common downstream pathway.

In summary, these studies represent an initial attempt to analyze the role of integrin α-subunit cytoplasmic domains during ligand-binding and post-ligand-binding functions. Consistent with the widely differing primary structures of integrin cytoplasmic domains, we have begun to show that α-chain cytoplasmic domains can make widely differing contributions to post-ligand-binding events. Because we have not found a role for α-chain cytoplasmic tails in regulating cell-type-specific levels of constitutive activity, further studies will need to consider the cytoplasmic tail of the β_1 chain, or perhap other factors not acting through cytoplasmic domains. Finally, a recurring theme in these studies is that integrins can have distinct functional properties in different cellular environments. This consistent finding suggests that integrin cytoplasmic domains may interact with unknown cell-type-specific factors which remain to be discovered.

ACKNOWLEDGMENTS

This work was supported by National Institutes of Health grants GM-46526 and GM-38903, and by a Drug Discovery Award from Sandoz Company.

REFERENCES

Argraves, W.S., S. Suzuki, H. Arai, K. Thompson, M.D. Pierschbacher, and E. Ruoslahti. 1987. Amino acid sequence of the human fibronectin receptor. *J. Cell Biol.* **105:** 1183.

Arnaout, M.A., S.K. Gupta, M.W. Pierce, and D.G. Tenen. 1988. Amino acid sequence of the alpha subunit of human leukocyte adhesion receptor Mol (complement receptor type 3). *J. Cell Biol.* **106:** 2153.

Bednarczyk, J.L. and B.W. McIntyre. 1990. A monoclonal antibody to VLA-4 α chain (CDw49) induces homotypic lymphocyte aggregation. *J. Immunol.* **144:** 777.

Bogaert, T., N. Brown, and M. Wilcox. 1988. The *Drosophila* PS2 antigen is an invertebrate integrin which like the fibronectin receptor becomes localized to muscle attachments. *Cell* **51:** 929.

Bossy, B., E. Bossy-Wetzel, and L.F. Reichardt. 1991. Characterization of the integrin $\alpha 8$ subunit: A new integrin $\beta 1$-associated subunit, which is prominently expressed on axons and on cells in contact with basal laminae in chick embryos. *EMBO J.* **10:** 2375.

Burn, P., A. Kupfer, and S.J. Singer. 1988. Dynamic membrane-cytoskeletal interactions: Specific association of integrin and talin arises in vivo after phorbol ester treatment of peripheral blood lymphocytes. *Proc. Natl. Acad. Sci.* **85:** 497.

Caixia, S., S. Stewart, E. Wayner, W. Carter, and J. Wilkins. 1991. Antibodies to different members of the β_1 (CD29) integrins induce homotypic and heterotypic cellular aggregation. *Cell. Immunol.* **138:** 216.

Campanero, M.R., R. Pulido, M.A. Ursa, M. Rodriguez-Moya, M.O. De Landazuri, and F. Sanchez-Madrid. 1990. An alternative leukocyte homotypic adhesion mechanism, LFA-1/ICAM-1 independent, triggered through the human VLA-4 integrin. *J. Cell Biol.* **110:** 2157.

Campanero, M.R., A.G. Arroyo, R. Pulido, A. Ursa, M.S. De Matias, P. Sanchez-Mateos, P.D. Kassner, B.M.C. Chan, M.E. Hemler, A. Corbi, M.O. De Landazuri, and F. Sanchez-Madrid. 1992. Triggering of integrin-mediated intercellular adhesion through the common $\beta 1$ subunit of VLA heterodimers: Evidence for a role of $\alpha 2/\beta 1$ and $\alpha 4/\beta 1$ in cell-cell interactions. *Eur. J. Immunol.* (in press).

Chan, B.M.C. and M.E. Hemler. 1992. Multiple functional forms of the integrin VLA-2 derived from a single α^2 cDNA clone: Interconversion of forms induced by an anti β_1 antibody. *J. Biol. Chem.* (in press).

Chan, B.M.C., P.D. Kassner, J.A. Schiro, H.R. Byers, T.S. Kupper, and M.E. Hemler. 1992. Distinct cellular functions mediated by different VLA integrin α subunit cytoplasmic domains. *Cell* **68:** 1051.

Cooper, H.M., R.N. Tamura, and V. Quaranta. 1991. The major laminin receptor of mouse embryonic stem cells is a novel isoform of the $\alpha 6\beta 1$ integrin. *J. Cell Biol.* **115:** 843.

Corbi, A.L., T.K. Kishimoto, L.J. Miller, and T.A. Springer. 1988. The human leukocyte adhesion glycoprotein Mac-1 (Complement receptor type 3, CD11b) α subunit: Cloning, primary structure, and relation to the integrins, von Willebrand factor and factor B. *J. Biol. Chem.* **263:** 12403.

Corbi, A.L., L.J. Miller, K. O'Connor, R.S. Larson, and T.A. Springer. 1987. cDNA cloning and complete primary structure of the alpha subunit of a leukocyte adhesion glycoprotein, p150, 95. *EMBO J.* **6:** 4023.

De Curtis, I., V. Quaranta, R.N. Tamura, and L.F. Reichardt. 1991. Laminin receptors in the retina: Sequence analysis of the chick integrin alpha-6 subunit. *J. Cell Biol.* **113:** 405.

Dustin, M.L. and T.A. Springer. 1989. T-cell receptor cross-linking transiently stimulates adhesiveness through LFA-1. *Nature* **341:** 619.

Fitzgerald, L.A., M. Poncz, B. Steiner, S.C. Rall, Jr., J.S. Bennett, and D.R. Phillips. 1987. Comparison of cDNA-derived protein sequences of the human fibronectin and vitronectin receptor alpha-subunits and platelet glycoprotein IIb. *Biochemistry* **26:** 8158.

Guan, J.-L., J.E. Trevithick, and R.O. Hynes. 1991. Fibronectin/integrin interaction induces tyrosine phosphorylation of a 120 kDa protein. *Cell Regul.* **2:** 951.

Hayashi, Y., B. Haimovich, A. Reszka, D. Boettiger, and A. Horwitz. 1990. Expression and function of chicken integrin beta-1 subunit and its cytoplasmic domain mutants in mouse NIH 3T3 cells. *J. Cell Biol.* **110:** 175.

Hemler, M.E. 1990. VLA proteins in the integrin family: Structures, functions, and their role on leukocytes. *Annu. Rev. Immunol.* **8:** 365.

Hibbs, M.L., H. Xu, S.A. Stacker, and T.A. Springer. 1991. Regulation of adhesion to ICAM-1 by the cytoplasmic domain of LFA-1 integrin beta subunit. *Science* **251:** 1611.

Hickstein, D.D., M.J. Hickey, J. Ozols, D.M. Baker, A.L. Back, and G.J. Roth. 1989. cDNA sequence for the alpha-M subunit of the human neutrophil adherence receptor indicates homology to integrin alpha subunits. *Proc. Natl. Acad. Sci.* **86:** 257.

Holers, V.M., T.G. Ruff, D.L. Parks, J.A. McDonald, L.L. Ballard, and E.J. Brown. 1989. Molecular cloning of a murine fibronectin receptor and its expression during inflammation. *J. Exp. Med.* **169:** 1589.

Horwitz, A., K. Duggan, C. Buck, M.C. Beckerle, and K. Burridge. 1986. Interaction of plasma membrane fibronectin receptor with talin-a transmembrane linkage. *Nature* **320:** 531.

Hynes, R.O. 1992. Integrins: Versatility, modulation and signaling in cell adhesion. *Cell* **69:** 11.

Hynes, R.O., E.E. Marcantonio, M.A. Stepp, L.A. Urry, and G.H. Yee. 1989. Integrin heterodimer and receptor complexity in avian and mammalian cells. *J. Cell Biol.* **109:** 409.

Ignatius, M.J., T.H. Large, M. Houde, J.W. Tawil, A. Barton, F. Esch, S. Carbonetto, and L.F. Reichardt. 1990. Molecular cloning of the rat integrin $\alpha 1$ subunit: A receptor for laminin and collagen. *J. Cell Biol.* **111:** 709.

Kansas, G.S. and T.F. Tedder. 1992. Transmembrane signals generated through MHC class II, CD19, CD20, CD39, and CD40 antigens induce LFA-1-dependent and independent adhesion in human B cells through a tyrosine kinase-dependent pathway. *J. Immunol.* **147:** 4094.

Kaufmann, Y., E. Tseng, and T.A. Springer. 1992. Cloning of the murine lymphocyte function-associated molecule-1 α subunit and its expression in COS cells. *J. Immunol.* **147:** 369.

Kornberg, L.J., H.S. Earp, C.E. Turner, C. Prockop, and R.L. Juliano. 1991. Signal transduction by integrins: Increased protein tyrosine phosphorylation caused by clustering of β_1 integrins. *Proc. Natl. Acad. Sci.* **88:** 8392.

LaFlamme, S.E., S.K. Akiyama, and K.M. Yamada. 1992. Regulation of fibronectin receptor distribution. *J. Cell Biol.* **117:** 437.

Larson, R.S. and T.A. Springer. 1990. Structure and function of leukocyte integrins. *Immunol. Rev.* **114:** 181.

Larson, R.S., M.L. Hibbs, and T.A. Springer. 1990. The leukocyte integrin LFA-1 reconstituted by cDNA transfection in a nonhematopoietic cell line is functionally active and not transiently regulated. *Cell Regul.* **1:** 359.

Larson, R.S., A.L. Corbi, L. Berman, and T.A. Springer. 1989. Primary structure of the LFA-1 a subunit: An integrin with an embedded domain defining a protein superfamily. *J. Cell Biol.* **108:** 703.

Marcantonio, E.E., J. Guan, J.E. Trevithick, and R.O. Hynes. 1990. Mapping of the functional determinants of the integrin beta-1 cytoplasmic domain by site-directed mutagenesis. *Cell Regul.* **1:** 597.

Mueller, S.C., T. Kelly, M. Dai, H. Dai, and W.-T. Chen. 1989. Dynamic cytoskeleton-integrin associations induced by cell binding to immobilized fibronectin. *J. Cell Biol.* **109:** 3455.

Nakamura, S., J. Imanishi, and J. Minowada. 1992. R1-20, a novel monoclonal antibody reacting with a molecule distinct from integrin family, induces homotypic cell aggregation. *J. Immunol.* **149:** 30.

Neuhaus, H., M.C.-T. Hu, M.E. Hemler, Y. Takada, B. Holzmann, and I.L. Weissman. 1991. Cloning and expression of cDNAs for the α subunit of the murine lymphocyte-peyer's patch homing receptor. *J. Cell Biol.* **115:** 1149.

O'Toole, T.E., D. Mandelman, J. Forsyth, S.J. Shattil, E.F. Plow, and M.H. Ginsberg. 1991. Modulation of the affinity of integrin αIIbβ3 (GPIIb-IIIa) by the cytoplasmic domain of αIIb. *Science* **254:** 845.

Otey, C.A., F.M. Pavalko, and K. Burridge. 1990. An interaction between α-actinin and the $\beta 1$ integrin subunit in vitro. *J. Cell Biol.* **111:** 721.

Phillips, D.R., I.F. Charo, and R.M. Scarborough. 1991. GPIIb-IIIa: The responsive integrin. *Cell* **65:** 359.

Plow, E.F. and M.H. Ginsberg. 1989. Cellular adhesion: GPIIIb/IIIa as a prototypic adhesion receptor. *Prog. Hemostasis Thromb.* **9:** 117.

Poncz, M. and P.J. Newman. 1990. Analysis of rodent platelet glycoprotein IIb: Evidence for evolutionarily conserved domains and alternative proteolytic processing. *Blood* **75:** 1282.

Poncz, M., R. Eisman, R. Heidenreich, S.M. Silver, G. Vilaire, S. Surrey, E. Schwartz, and J.S. Bennett. 1987. Structure of the platelet membrane glycoprotein IIb: Homology to the alpha subunits of the vitronectin and fibronectin membrane receptors. *J. Biol. Chem.* **262:** 8476.

Pulido, R., M.J. Elices, M.R. Campanero, L. Osborn, S. Schiffer, A. Garcia-Pardo, R. Lobb, M.E. Hemler, and F. Sanchez-Madrid. 1991. Functional evidence for three distinct and independently inhibitable adhesion activities mediated by the human integrin VLA-4: Correlation with distinct $\alpha 4$ epitopes. *J. Biol. Chem.* **266:** 10241.

Reszka, A.A., Y. Hayashi, and A.F. Horwitz. 1992. Identification of amino acid sequences in the integrin β_1 cytoplasmic domain implicated in cytoskeletal association. *J. Cell Biol.* **117:** 1321.

Sánchez-Madrid, F., M.O. De Landazuri, G. Morago, M. Cebrian, A. Acevedo, and C. Bernabeu. 1986. VLA-3: A novel polypeptide association within the VLA molecular complex: Cell distribution and biochemical characterization. *Eur. J. Immunol.* **16:** 1343.

Schafer, I.A., A. Shapiro, M. Kovach, C. Lang, and R.B. Fratianne. 1989. The interaction of human papillary and reticular fibroblasts and human keratinocytes in the contraction of three-dimensional floating collagen lattices. *Exp. Cell Res.* **183:** 112.

Schaller, M.D., C.A. Borgman, B.S. Cobb, R.R. Vines, A.B. Reynolds, and J.T. Parsons. 1992. pp125^{FAK}, a structurally distinctive protein-tyrosine kinase associated with focal adhesions. *Proc. Natl. Acad. Sci.* **89:** 5192.

Singer, I.I., S. Scott, D.W. Kawka, D.M. Kazazis, J. Gailit, and E. Ruoslahti. 1988. Cell surface distribution of fibronectin and vitronectin receptors depends on substrate composition and extracellular matrix accumulation. *J. Cell Biol.* **106:** 2171.

Solowska, J., J.M. Edelman, S.M. Albelda, and C.A. Buck. 1991. Cytoplasmic and transmembrane domains of integrin β_1 and β_3 subunits are functionally interchangeable. *J. Cell Biol.* **114:** 1079.

Solowska, J., J.-L. Guan, E.E. Marcantonio, J.E. Trevithick, C.A. Buck, and R.O. Hynes. 1989. Expression of normal and mutant avian integrin subunits in rodent cells. *J. Cell Biol.* **109:** 853.

Suzuki, S., W.S. Argraves, H. Arai, L.R. Languino, M. Pierschbacher, and E. Ruoslahti. 1987. Amino acid sequence of the vitronectin receptor alpha subunit and comparative expression of adhesion receptor mRNAs. *J. Biol. Chem.* **262:** 4080.

Takada, Y. and M.E. Hemler. 1989. The primary structure of the VLA-2/collagen receptor $\alpha 2$ subunit (platelet GP Ia): Homology to other integrins and the presence of a possible collagen-binding domain. *J. Cell Biol.* **109:** 397.

Takada, Y., M.J. Elices, C. Crouse, and M.E. Hemler. 1989. The primary structure of the $\alpha 4$ subunit of VLA-4: Homology to other integrins and a possible cell-cell adhesion function. *EMBO J.* **8:** 1361.

Takada, Y., E. Murphy, P. Pil, C. Chen, M.H. Ginsberg, and M.E. Hemler. 1991. Molecular cloning and expression of the cDNA for the $\alpha 3$ chain of human $\alpha 3\beta 1$ (VLA-3), an integrin receptor for fibronectin, laminin and collagen. *J. Cell Biol.* **115:** 257.

Tamura, R.N., H.M. Cooper, G. Collo, and V. Quaranta. 1991. Cell type-specific integrin variants with alternative α chain cytoplasmic domains. *Proc. Natl. Acad. Sci.* **88:** 10183.

Tamura, R.N., C. Rozzo, L. Starr, J. Chambers, L.F. Reichardt, H.M. Cooper, and V. Quaranta. 1990. Epithelial integrin $\alpha 6\beta 4$: Complete primary structure of $\alpha 6$ and variant forms of $\beta 4$. *J. Cell Biol.* **111:** 1593.

Tapley, P., A. Horwitz, C. Buck, K. Duggan, and L. Rohrschneider. 1989. Integrins isolated from Rous sarcoma virus-transformed chicken embryo fibroblasts. *Oncogene* **4:** 325.

Tsuji, T., S. Hakomori, and T. Osawa. 1991. Identification of human galactoprotein b3, an oncogenic transformation-induced membrane glycoprotein, as VLA-3 α subunit: The primary structure of human integrin α_3. *J. Biochem.* **109:** 659.

Tsuji, T., F. Yamamoto, Y. Miura, K. Takio, K. Titani, S. Pawar, T. Osawa, and S. Hakomori. 1990. Characterization through cDNA cloning of galactoprotein b3 (Gap b3), a cell surface membrane glycoprotein showing enhanced expression on oncogenic transformation: Identification of Gap b3 as a member of the integrin superfamily. *J. Biol. Chem.* **265:** 7016.

Van Kooyk, Y., P. Van Dewiel-Van Kemenade, P. Weder, T.W. Kuijpers, and C.G. Figdor. 1989. Enhancement of LFA-1 mediated cell adhesion by triggering through CD2 or CD3 on T lymphocytes. *Nature* **342:** 811.

Wayner, E.A., R.A. Orlando, and D.A. Cheresh. 1992. Integrins $\alpha v\beta 3$ and $\alpha v\beta 5$ contribute to cell attachment to vitronectin but differentially distribute on the cell surface. *J. Cell Biol.* **113:** 919.

Wright, S.D. and B.C. Meyer. 1986. Phorbol esters cause sequential activation and deactivation of complement receptors on polymorphonuclear leukocytes. *J. Immunol.* **136:** 1759.

Ligand Binding to Integrins: Dynamic Regulation and Common Mechanisms

M.H. Ginsberg, T.E. O'Toole, J.C. Loftus, and E.F. Plow
Committee on Vascular Biology, The Scripps Research Institute, La Jolla, California 92037

Cell-cell and cell-substratum adhesive interactions regulate aggregation, cell polarity, and cell migration. In addition, such interactions can influence the differentiation of cells as manifested by changes in cell morphology, proliferation, and gene expression programs (Menko and Boettiger 1987; Ruoslahti and Pierschbacher 1987; Panayotou et al. 1989). This paper focuses on the integrins, a broadly distributed family (virtually every cell type expresses integrins, even erythrocyte precursors) of adhesion receptors. The existence of this family was first suspected in the mid 1980s (Ginsberg et al. 1985) because of the functional similarities among recognition specificities of cell adhesion receptors. This suspicion was heightened by the identification of gross structural (Pytela et al. 1986) and immunologic (Plow et al. 1986; Ginsberg et al. 1987) similarities among adhesion receptors. The cloning and sequencing of many of the members of this family (Tamkun et al. 1986; Argraves et al. 1987; Fitzgerald et al. 1987; Kishimoto et al. 1987; Loftus et al. 1987; Poncz et al. 1987; Suzuki et al. 1987; Takada and Hemler 1989; Takada et al. 1989) established its existence as a supergene family unambiguously. All integrins discovered to date are heterodimers of α and β subunits, which are products of separate genes and are mutually interdependent for correct processing and surface expression (O'Toole et al. 1989). When the organization of this family was first proposed in 1987 (Ginsberg et al. 1987; Hynes 1987), a simple vertical organization was envisioned consisting of three β subunits that were structurally related and formed a series of heterodimers with distinct α subunits. The α subunits were believed, at the time, to observe fidelity of pairing with specific β subunits. With the discovery that the α subunits appear to have evolved independently of β subunits (Takada and Hemler 1989), the discovery that novel β subunits (Cheresh et al. 1989b; Holzmann and Weissman 1989; Kajiji et al. 1989) can form heterodimers with existing α subunits, and finally the discovery that the α_1 subunit can form heterodimers with at least three different β subunits (Cheresh et al. 1989b; Vogel et al. 1990), more complicated schemes have been required to describe this family.

Given the broad biological utilization of cell adhesive interactions and given their effects on the differentiated state of cells, the conceptualization of the potential involvement of integrins in biology seems to be only limited by the imagination of the investigator. The following are a few examples of processes involving integrins: Members of the β_2 subfamily of integrins (CD11,18 contains LFA-1, MAC-1) play a central accessory role in T-cell-mediated cytolysis and in interaction of T cells with antigen-presenting cells (Springer 1985). Similarly, ligation of integrins (Van Noesel et al. 1988) may be an important cofactor for mitogenic responses triggered through the T-cell receptor. Integrins play a role in the homing of lymphocytes. This is exemplified by the accessory role of β_2 integrins in the adhesion of lymphocytes to endothelial cells (Prieto et al. 1988) and more recently by the demonstration that β_4-containing integrins play a role in the homing of B cells to Peyer's patches (Holzmann et al. 1989). Thus, changes in the integrin repertoire in lymphocytes, which are known to occur as a consequence of activation (Hemler et al. 1987), may control the targeting of these cells to specific inflammatory sites. Indeed, direct evidence for this has been found in rheumatoid synovial lymphocytes (Hemler et al. 1986). In inflammation, the interaction of polymorphonuclear leukocytes with endothelium, their extravasation, and subsequent function are dependent on integrins (VanderMeulen and Grinstein 1982; Nathan et al. 1989). In addition, cell-substratum interactions play a role in phagocytosis, and a novel leukocyte response integrin (Gresham et al. 1989) is involved in augmentation of the phagocytic response. The migratory and invasive property of pannus may be under the control of adhesion receptors such as integrins, and the $\alpha_v\beta_3$ integrin (vitronectin receptor) plays a role in bone resorption by osteoclasts (Davies et al. 1989). Finally, the occupancy of integrins controls expression of genes with direct relevance to tissue remodeling. For example, engagement of the $\alpha_5\beta_1$ fibronectin (Fn) receptor up-regulates the expression of metalloproteinases in synovial fibroblasts (Werb et al. 1989). From this brief litany, it is readily apparent that this supergene family is involved in a broad range of biological events.

Central to understanding integrin function is the basis of ligand recognition. Although this alone is a compelling intellectual challenge, an understanding of ligand recognition should facilitate development of agents that modulate integrin function. Indeed, animal studies have indicated the potential benefits of interference with integrin function (Coller et al. 1986; Humphries et al. 1986). The Arg-Gly-Asp (RGD) sequence was originally proposed to be the common recognition sequence for integrins (Ruoslahti and Pierschbacher 1986), although it is now apparent that the function of many integrins is not inhibited by synthetic peptides containing this sequence. Moreover, even among integrins that are inhibited by RGD peptides, there is specificity of protein recognition (Pytela et al. 1986). Although these findings might be due to differences in the conformation of the RGD sequence in different proteins, it is now clear that some of the ligands for integrins do not contain this sequence (Wayner et al. 1989). Nevertheless, the striking sequence identities among integrins, and in particular the remarkable conservation of disulfides (Phillips et al. 1988), strongly suggest that the overall structures of many of these receptors will be quite similar. It also suggests the possibility that there may be common features to ligand recognition by integrins.

We have chosen to utilize primarily β_3 integrins to address the ligand recognition issue because they were the first to be identified (Nurden and Caen 1974) and purified (Jennings and Phillips 1982), among the first to be cloned and sequenced (Loftus et al. 1986; Fitzgerald et al. 1987; Loftus et al. 1987; Poncz et al. 1987; Rosa et al. 1988), and the first to be expressed in completely recombinant form (O'Toole et al. 1989). Moreover, there are well defined and extensively studied synthetic peptide ligands for β_3 integrins (Plow et al. 1984,1985,1987; Gartner et al. 1985; Ginsberg et al. 1985; Haverstick et al. 1985; Ruggeri et al. 1986; Lam et al. 1987; Hautanen et al. 1989), which have a clear effect on biological function. Finally, β_3 integrins play a role in processes such as bone resorption (Davies et al. 1989), platelet aggregation (Phillips and Agin 1977), and endothelial cell attachment to certain substrates (Charo et al. 1987). A central feature of the function of all integrins is their dependence on divalent cations. When analyzed in depth with β_1 (Gailit and Ruoslahti 1988; Staatz et al. 1989), β_2 (Marlin and Springer 1987), and β_3 (Bennett and Vilaire 1979; Phillips and Baughan 1983; Ginsberg et al. 1986; Steiner et al. 1989), this dependence is met by millimolar concentrations of either magnesium or calcium. The concentrations of divalent cations may occupy cation-binding sites either in the integrins themselves or in the ligands. Our studies bear on this question, since we have identified a conformation-specific antibody (Frelinger et al. 1988) termed PMI-I, which recognizes a defined epitope (Loftus et al. 1987) in $\alpha_{IIb}\beta_3$ that is modulated by divalent cations. The modulation of this epitope correlates with the capacity of the receptor to bind fibrinogen (Fg) (Ginsberg et al. 1986). In addition, we have identified a patient with variant Glanzmann's thrombasthenia, whose $\alpha_{IIb}\beta_3$ does not bind macromolecular adhesive ligands (Ginsberg et al. 1986) or synthetic RGD peptide ligands (Ginsberg et al. 1990). $\alpha_{IIb}\beta_3$ from these individuals shows abnormal modulation of this divalent-cation-dependent epitope (Ginsberg et al. 1986). These data indicate that the relevant cation-binding site is within the integrin. A similar conclusion has been suggested by the data of Dransfield and Hogg (1990) for the leukocyte β_2 integrins.

A single point mutation has been identified in β_3 from affected individuals. The base change, which changes D_{119} to a Y, was introduced in the wild-type β_3 sequence by site-directed mutagenesis (Loftus et al. 1990). This substitution introduced a divalent-cation-binding defect similar to that observed in the patient's platelets, since the mutated receptor reacted with the PMI-I fully in the presence of divalent cations (Loftus et al. 1990). In contrast, maximal epitope expression of the wild-type receptor required the addition of divalent-cation-chelating reagents. The capacity of the mutant receptor to bind representative RGD ligands was examined utilizing the anti-LIBS1 monoclonal antibody and flow cytometry. This antibody recognizes an occupancy-dependent epitope on β_3 integrins that is not modulated by millimolar concentrations of divalent cations (Cembrowski and Mosher 1984; Tonnesen et al. 1984). As shown in Figure 1, RGD peptide ligands up-regulated the binding of anti-LIBS1 to wild-type $\alpha_{IIb}\beta_3$ and $\alpha_v\beta_3$, but failed to up-regulate the binding of this antibody to both receptors containing D_{119}Y. The ligand-binding defect in these receptors was further substantiated by their failure to bind to immobilized RGD peptide ligands (Fig. 2). Thus, the single-base change established the ligand-binding defective phenotype in both of these β_3 integrins.

Asp-119 is in striking proximity to the residue proposed to be cross-linked to the bound RGD peptide (Lys-125) (D'Souza et al. 1988). Moreover, when aligned with the other B subunits (Fig. 3), Asp-19 is absolutely conserved among integrins. Furthermore, Ser-121, Ser-123, Asp-126, and Asp-127 are also conserved. An oxygenated residue at the position corresponding to Ser-130 in β_3 is present in all B subunits, with the exception of the *Drosophila* lethal myospheroid gene product (MacKrell et al. 1988). The clustering of oxygenated residues could provide coordinating ligands for divalent cations. This idea is supported by the linear spacing of oxygenated residues that approximates the spacing of oxygenated calcium ligand residues of the loop of E-F hand sequences (Kretsinger 1980) contained in calmodulin and other divalent-cation-binding proteins. The occurrence of Met-124 in the place of an invariant glycine in the E-F loop structure suggests that this region will not have the geometry of the E-F loop, and thus mutational analysis will be required to assign function to the other oxygenated residues. Nevertheless, the proximity of the ligand-binding and putative cation-binding sites suggests the possibility of direct interactions between the bound cation and the ligand. Such a possibility has been suggested for thrombospondin binding to $\alpha_v\beta_3$ (Lawler et al. 1988) and for β_2 integrin function (Corbi et al. 1987). In addition, we have found that ligand binding to $\alpha_{IIb}\beta_3$ modulates the binding of the divalent-cation-sensitive PMI-I antibody (Frelinger et al. 1988), suggesting that ligand binding may perturb the interaction between bound divalent cation and the integrin.

In view of the remarkable conservation of the sequence around Asp-119 and the ubiquitous divalent cation requirement for integrin function, it seems likely

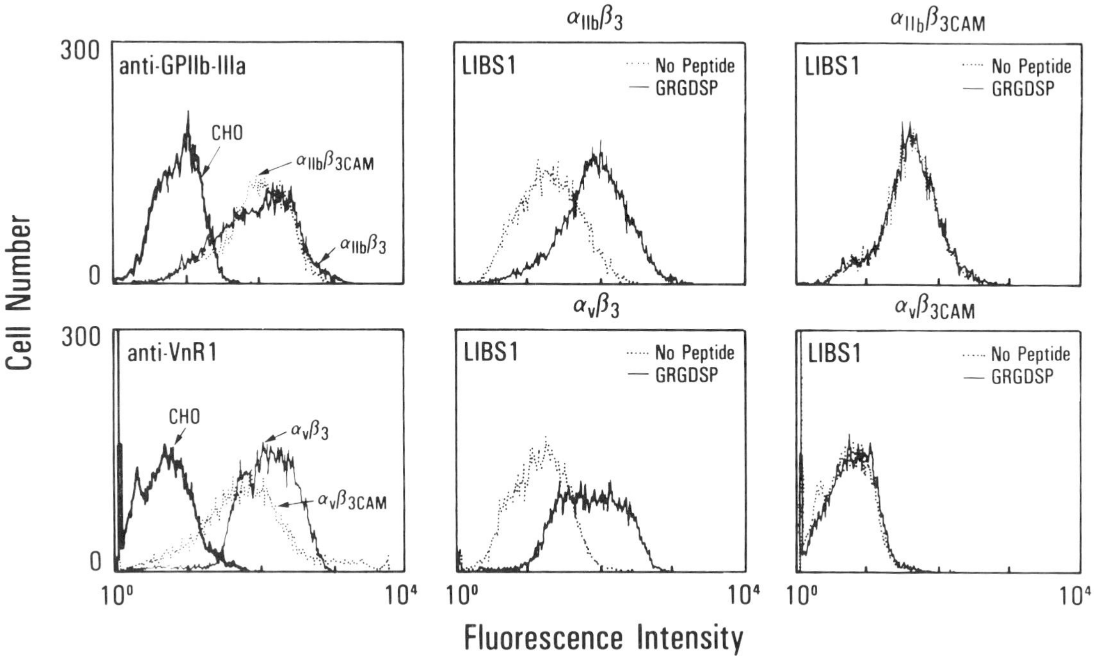

Figure 1. Flow cytometric analysis of the capacity of transfected cells to bind to RGD ligand. FACS was performed on stable CHO cell lines containing the indicated transfected integrins. The GRGDSP peptide was at a final concentration of 1 mM and was added together with the anti-LIBS1 antibody (0.1 μM) in the first incubation. Results are expressed as graphs of cell number (linear scale) on the ordinate versus fluorescence intensity (log scale) on the abscissa. The reporting antibodies are indicated at upper left of each panel. (Reprinted, with permission, from Loftus et al. 1990.)

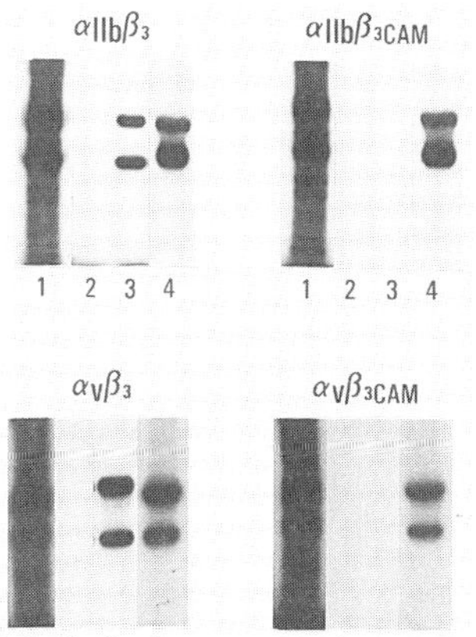

Figure 2. RGD affinity chromatography and immunoprecipitation of extracts of stably transfected cell lines. CHO cells stably expressing the indicated wild-type or mutant receptors were surface radioiodinated and insolubilized in octylglucoside-containing buffer. The extract was applied to a column of Sepharose 4B to which was coupled the KYGRGDS peptide. Unbound proteins were eluted with 30 ml of column buffer, and bound proteins were eluted by sequential elution with 3 ml of buffer containing GRGESP peptide (1 mg/ml), followed by 3 ml of buffer, and finally 3 ml of buffer containing GRGDSP (1 mg/ml). Fractions were analyzed by electrophoresis on 6.5% SDS-PAGE under nonreducing conditions followed by autoradiography. (Lane *1*) Starting material; (lane *2*) GRGESP eluted material; (lane *3*) GRGDSP eluted material; (lane *4*) immunoprecipitate starting material obtained with polyclonal anti-β_3 antibody. (Reprinted, with permission, from Loftus et al. 1990.)

that ligand interaction with divalent cations is a general feature of ligand binding to all integrins. Moreover, small peptide ligands that interact with certain integrins (RGDX [Ruoslahti and Pierschbacher 1986], LGGAKQAGDV [Plow et al. 1984], EILDVPST [Wayner et al. 1989]) all contain oxygenated residues capable of coordinating divalent cations. Furthermore, the D residue in RGDX was the most sensitive to conservative substitution (Ginsberg et al. 1985), and the D residue in LGGAKQAGDV was also central for activity (Rug-

Beta 1	D	L	S	Y	S	M	K	D	D	L	E	N
Beta 2	D	L	S	Y	S	M	I	D	D	L	R	N
Beta 3	D	L	S	Y	S	M	K	D	D	L	W	S
Beta 4	D	F	S	N	S	M	S	D	D	L	D	N
Beta 5	D	L	S	L	S	M	K	D	D	L	D	N
Beta 6	D	L	S	A	S	M	D	D	D	L	N	T
Avian (B1)	D	L	S	Y	S	M	K	D	D	L	E	N
Frog (B1)	D	L	S	F	S	M	K	D	D	L	E	N
l(1)mys	D	L	S	K	S	M	E	D	D	K	A	K
EF Loop	X		Y		Z	G		-X			-Z	

Figure 3. Alignment of the region of β_3 containing the Cam mutation with other integrin β subunits. Asp-119 is the first residue in the β_3 sequence. The consensus sequences for the E-F loop are shown on the bottom line with X, Y, Z, −X, and −Z, the five oxygenated residues which ligate the divalent cation. G is the invariant glycine, and I/V is the position of a hydrophobic residue. Circled are oxygenated residues with spacing identical to those of the loop. (Reprinted, with permission, from Loftus et al. 1990.)

geri et al. 1986). Thus, we hypothesize that a common feature of ligand binding to integrins is that oxygenated residues in the ligand interact with divalent cations occupying a divalent-cation-binding site of the integrins. This hypothesis also predicts that mutations affecting the aspartic acid residues corresponding to Asp-119 in other β subunits should inhibit ligand-binding function as should mutations affecting at least four other oxygenated residues.

The above-noted hypothesis explains one common feature of ligand recognition by integrins; however, as already emphasized, the integrins differ in ligand recognition specificity. It is clear that macromolecular ligands, such as Fn, may be recognized by multiple receptors on the cell surface, that the activation of cells can modulate the ligand recognition specificity of integrins (Plow and Ginsberg 1981), and that the cell of origin can also modulate this specificity (Kirchofer et al. 1990) with respect to macromolecular ligands. For this reason, we have elected to focus our attention on simple peptide ligands. Peptides based on the carboxyl terminus of the Fg γ chain (e.g., LGGAKQAGDV [L10]) interact with $\alpha_{IIb}\beta_3$ and inhibit the binding of RGD-containing peptides (Lam et al. 1987), which suggests a spatial relationship between the two peptide-binding sites. More recently, we have found that L10 failed to displace $\alpha_v\beta_3$ from an RGD affinity column (Lam et al. 1989). Similarly, Cheresh et al. (1989a) reported that endothelial cells containing $\alpha_v\beta_3$ failed to adhere to such γ-chain peptides. These experiments suggest that the ligand recognition specificity of integrins is controlled, in part, by the α subunit. Similarly, a wealth of data indicate that the macromolecular ligand recognition specificity of β_1 and β_2 integrins is also controlled by the α subunit.

To determine which regions of the α subunit might be involved in ligand specificity, we have chemically cross-linked a Fg γ-chain peptide to $\alpha_{IIb}\beta_3$. Cross-linking was saturable, activation-dependent, and divalent-cation-dependent. A combination of protein "restriction mapping" (D'Souza et al. 1990) and sequencing of peptide fragments placed the cross-linked region between amino acids 294 and 314 of the IIb heavy chain (D'Souza et al. 1990). Finally, a 12-residue peptide from this region binds Fg and inhibits Fg binding to $\alpha_{IIb}\beta_3$ (D'Souza et al. 1991), thus providing clear proof of its functionality.

A third potential interactive site was identified in β_3 comprising residues 211–222 by Charo et al. (1991). Synthetic peptides from this region inhibit Fg binding as do antibodies directed against such peptides. Recently, two independent spontaneous mutations at R_{214} in β_3 have been found to interfere with ligand-binding function (Bajt et al. 1992; Lanza et al. 1992). In addition, other potential interactive sites have been identified by hydropathic complementarity approaches or by mapping of monoclonal antibodies (Calvete et al. 1991a,b; Gartner et al. 1991). Thus, as might be expected from the heterodimeric nature of integrins, multiple contact points exist even for simple peptide ligands.

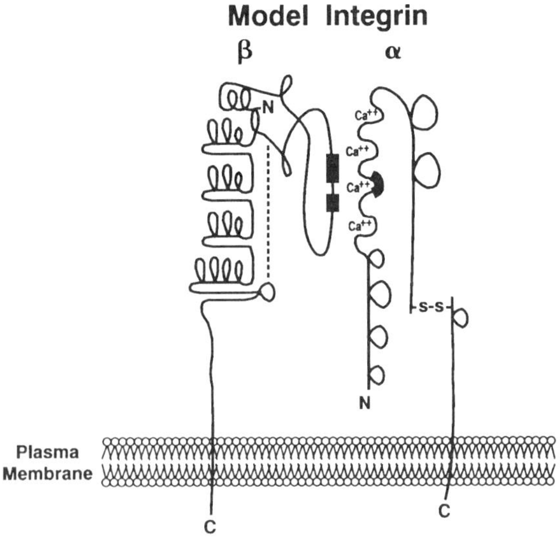

Figure 4. Schematic of a model integrin. This schematic model is based primarily on the structural features of $\alpha_{IIb}\beta_3$, the most intensively studied integrin. An indication of the disulfide-bonding pattern was proposed by Calvete et al. (1989, 1991c). Note that the amino terminus of β_3 is bonded to the central region (Cys-5-Cys-435). The dashed line indicates a proposed disulfide between Cys-405 and Cys-635. In α_{IIb}, the disulfides are all sequential. Two ligand contact sites in β_3 and one in α_{IIb} are indicated by dark bars. Note that both sets of ligand contact sites occupy disulfide-poor regions in the amino-terminal aspect of the subunits.

Some of those contact points are illustrated graphically in Figure 4.

A prominent feature of the integrins is the capacity of cells to change integrin affinity for ligands. This is a property of $\alpha_{IIb}\beta_3$ (Bennett and Vilaire 1979) and β_2 (Altieri and Edgington 1988) and apparently β_1 integrins (Shimizu et al. 1990). For example, in order for platelets to aggregate, the $\alpha_{IIb}\beta_3$ on the platelet surface must bind Fg (Peerschke et al. 1980), which requires prior platelet "activation." Similarly, for β_2 integrins to mediate cell-cell interactions (Dustin and Springer 1989) efficiently, leukocytes must be "activated." Soluble agonists such as thrombin or adenine nucleotides activate platelet and leukocyte integrins. These agonists probably do not act directly on the integrins but bind to cellular receptors, leading to engagement of "classical" signal transduction systems. G proteins usually transduce these signals, leading to activation of phospholipase C, changes in cytosolic calcium, and activation of cellular protein kinases. Furthermore, inhibitors of these signal transduction systems can inhibit agonist-induced activation of the binding function of integrins (Fujimoto et al. 1982; Shattil and Brass 1987; Dustin and Springer 1989). In an insightful study, Brass and Shattil (1987) dissected the mechanism of thrombin activation of platelet $\alpha_{IIb}\beta_3$ and found that G proteins mediate thrombin's effect. Nevertheless, following direct activation of protein kinase C (PKC), further G-protein requirement could not be demonstrated, showing that PKC was the most distal identified step. Integrin cytoplasmic domains are potential substrates for PKC. β_3 but not α_{IIb} is phosphorylated in response to activators of PKC. Hillery et al. (1991) showed that the stoichiometry of phosphorylation is low, suggesting that it cannot account for activation of $\alpha_{IIb}\beta_3$. There is also phosphorylation of neutrophil β_2 (Chatila et al. 1989), but certain agonists (Buyon et al. 1990; Merrill et al. 1990) can activate $\alpha_M\beta_2$ without its phosphorylation, and mutation of its phosphorylation site fails to block activation (Hibbs et al. 1991). Finally, macrophage α_6 becomes phosphorylated in response to activators of PKC (Shaw et al. 1990), but its causal relationship to affinity modulation is not established. Thus, phosphorylation of integrin cytoplasmic domains may have important functional consequences, but the phosphorylation of $\alpha_{IIb}\beta_3$ does not seem important for affinity modulation. The unanswered question is the mechanism by which these intracellular signal transduction events increase the affinity of ligand binding to integrins.

As noted above, activation refers to an increase in the ligand-binding affinity of an integrin in "activated" cells. The binding of large ligands such as Fg and Fn to $\alpha_{IIb}\beta_3$ seems absolutely cell-activation-dependent. In contrast, small Fg mimetic peptides and larger snake venom peptides bind without activation (Lam et al. 1987; Frelinger et al. 1988; Savage et al. 1990). Coller (1986) established that activation changes the size selectivity of macromolecular access to binding sites on $\alpha_{IIb}\beta_3$. His group also found that chymotrypsin activated $\alpha_{IIb}\beta_3$ without obvious proteolytic cleavage in the molecule (Peerschke and Coller 1984). These data suggested the hypothesis that activation did not affect the integrin, but it did affect the cellular microenvironment, thereby increasing access of macromolecules to the integrin. An alternate idea is that activation affects the conformation of the integrin, thereby influencing accessibility of macromolecules to the binding pocket. An IgM monoclonal antibody, PAC1, which binds activation-dependently at or near (Shattil et al. 1985; Taub et al. 1989; Ginsberg et al. 1990) the ligand-binding site of $\alpha_{IIb}\beta_3$ supports this idea.

To address this issue further, recombinant $\alpha_{IIb}\beta_3$ was expressed in a nonplatelet environment and was found in a resting state (O'Toole et al. 1990). Moreover, it could be activated in both platelets and in a nonplatelet environment by certain anti-integrin antibodies (O'Toole et al. 1990), which did not depend on signal transduction events. These antibodies appear to shift a conformational equilibrium between the resting and activated receptor (Frelinger et al. 1991), since they could activate purified soluble $\alpha_{IIb}\beta_3$ (Fig. 5). Differences in resonance energy transfer between fluorochromes coupled to monoclonal antibodies bound to $\alpha_{IIb}\beta_3$ in resting or activated platelets (Sims et al. 1991) also support a conformational mechanism.

Integrins generally possess short cytoplasmic domains that are topographically accessible to intracellular second messengers and therefore may mediate inside-out signal transduction. Moreover, integrin cytoplasmic domain sequences are highly conserved between species and yet there is less conservation between different α and β subunits in the same species (Marcantonio and Hynes 1988; Hemler 1990; Takada et al. 1991). This suggests that these polypeptides have indispensable and potentially unique roles. To test this hypothesis, we have examined the effect of alterations of the cytoplasmic domains of $\alpha_{IIb}\beta_3$ on its affinity state. Mutations of α_{IIb} and of β_3 ($\alpha_{IIb}\Delta991$, $\beta_3\Delta728$) putative cytoplasmic domains were expressed in Chinese hamster ovary (CHO) cells. As expected, the wild-type $\alpha_{IIb}\beta_3$ did not bind PAC1 unless it was first activated with monoclonal antibody (MAb) 62 (O'Toole et al. 1990). In sharp contrast, $\alpha_{IIb}\Delta991\beta_3$ bound PAC1 without activation, suggesting that it was

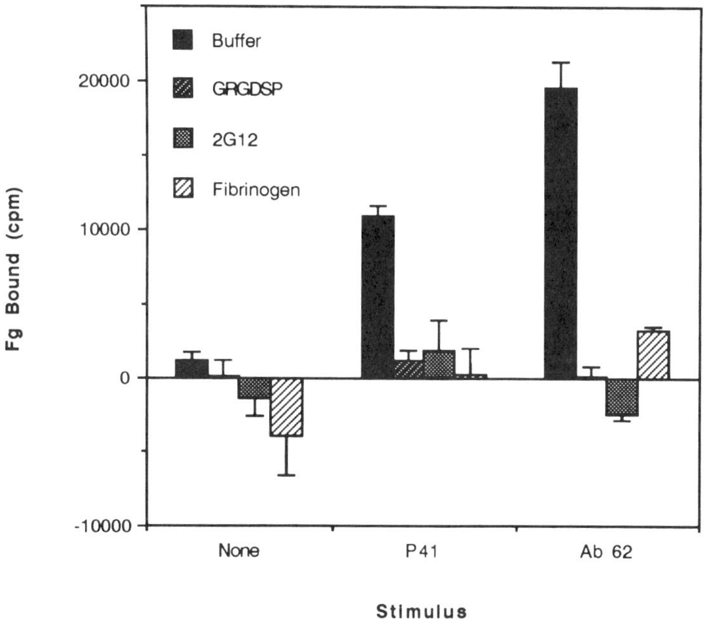

Figure 5. Activation of purified $\alpha_{IIb}\beta_3$. The binding of ^{125}I-labeled fibrinogen to purified $\alpha_{IIb}\beta_3$ captured in a microtiter well is described in the text and depicted above as Fg bound (cpm). Stimulated but not unstimulated receptor exhibits increased binding. Fibrinogen binding is inhibited with excess unlabeled ligand, the GRGDSP peptide, and the anti-$\alpha_{IIb}\beta_3$ antibody 2G12. (Reprinted, with permission, from O'Toole et al. 1990.)

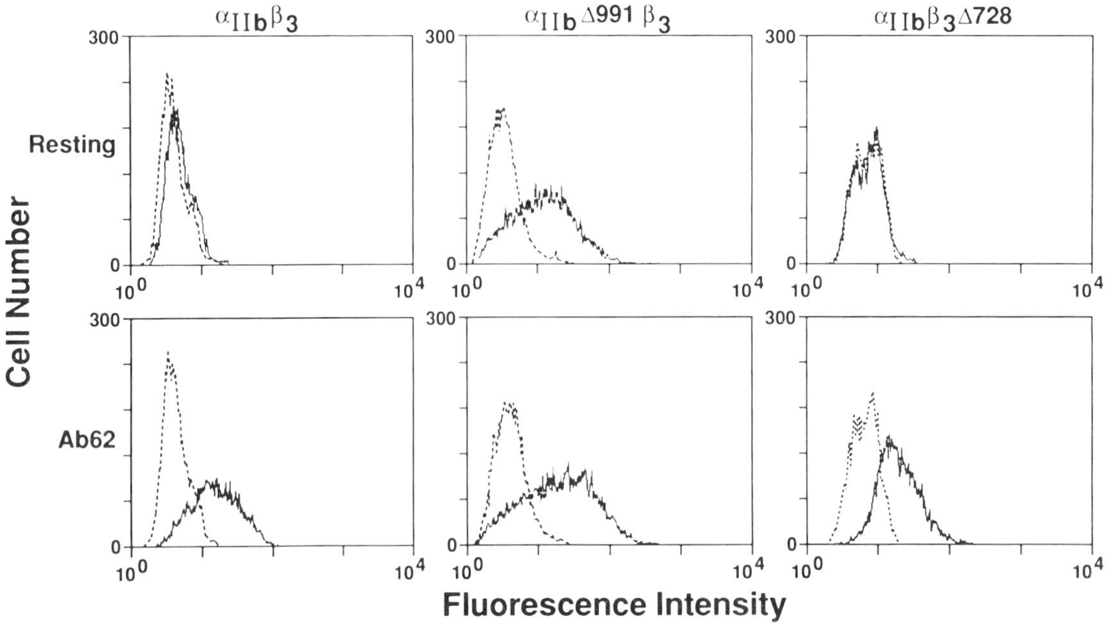

Figure 6. CHO cells expressing truncated α_{IIb} bind PAC1 in the absence of activators. The binding of PAC1 to CHO cells expressing $\alpha_{IIB}\beta_3$, $\alpha_{IIb}\Delta991\beta_3$, or $\alpha_{IIb}\beta_3\Delta728$ was measured by flow cytometry. Cells were harvested, stained with PAC1 in the presence or absence of 10 mM GRGDSP or 6 μM MAb 62, and analyzed by flow cytometry. Graphs of PAC1 binding in the presence of GRGDSP (dotted line) have been superimposed upon graphs of binding in its absence (solid line). (Reprinted, with permission, from O'Toole et al. 1991.)

constitutively active. PAC1 binding was specific, being blocked by 1 mM GRGDSP peptide (Fig. 6). The $\alpha_{IIb}\Delta991\beta_3$-bearing cells bound soluble ^{125}I-labeled Fg without addition of MAb 62 (O'Toole et al. 1991). An analysis of equilibrium binding of soluble Fg to $\alpha_{IIb}\Delta991\beta_3$ by the LIGAND (Munson and Rodbard 1980) program indicated a $K_a \pm$ S.E. of $1.25 \pm 0.13 \times 10^7$ M^{-1} (K_a = 80 nM), which is similar to the K_d of 110 nM observed for the activated wild-type receptor (O'Toole et al. 1990,1991). Moreover, Fg binding was inhibited by antibodies and EDTA, which inhibit Fg binding to platelet $\alpha_{IIb}\beta_3$. In contrast, the β_3 truncation did not bind Fg without activation. Furthermore, the cells bearing the α_{IIb} mutants underwent spontaneous aggregation upon addition of Fg (Fig. 7). In contrast, cells expressing the wild-type receptor only underwent Fg-dependent aggregation when the receptor was activated (O'Toole et al. 1991). Joining the α_5 cytoplasmic domain onto α_{IIb} (O'Toole et al. 1991), which produces a cytoplasmic domain eight residues longer than that of α_{IIb}, still results in a high-affinity receptor. These results suggest the hypothesis that specific structural elements in integrin α- or β-subunit cytoplasmic domains control its affinity state. A corollary of this hypothesis is that α- and β-subunit cytoplasmic domain splice variants (Silverstein et al. 1984; Wencel-Drake et al. 1984; van Kuppevelt et al. 1989; Altruda et al. 1990) may influence the integrin's affinity state. Three potential explanations for the capacity of α_{IIb} alteration to produce constitutively active receptors are depicted in Figure 8: (1) *Repressor*: A cytoplasmic repressor binds to the α-subunit cytoplasmic domain, and deletion of critical sequences results in loss of repressor binding. (2) *Intramolecular*: Interaction between the α- and β-subunit cytoplasmic domains holds the receptor in an inactive state; the mutation removes critical sequences required for this interaction. (3) *Activator*: The α subunit prevents the cytoplasmic face of the integrin from binding an activator element; removal of the relevant sequences of the α subunit thereby permits activator binding. Since the β_3 truncation did not activate $\alpha_{IIb}\beta_3$, explanation 1 is currently the favored hypothesis.

There is evidence of cell-type-specific elements in integrin activation pathways. For example, $\alpha_{IIb}\beta_3$ may be activated by thrombin in platelets but not in HEL cells, although HEL cells have a functional thrombin receptor (Brass et al. 1991). Similarly, in contrast to leukocytes, $\alpha_L\beta_2$ (LFA-1) is constitutively active when transfected into COS cells, and it cannot be further activated in these cells (Larson et al. 1990). In addition, $\alpha_5\beta_1$ and $\alpha_6\beta_1$ are constitutively active in platelets (Ill et al. 1985; Wayner et al. 1988), but lymphocytes (Shimizu et al. 1990) and macrophages (Shaw et al. 1990) require cellular activation to adhere optimally to Fn and laminin, respectively. These data suggest the second hypothesis—that the modulation of integrin affinity through cytoplasmic domains is controlled by cell-type-specific intracellular elements.

On the basis of the findings that mutations which alter the cytoplasmic domain of α_{IIb} result in constitutively active $\alpha_{IIb}\beta_3$, it is likely that spontaneous mutations of this type are pathogenic in humans. The α_{IIb} subunit is relatively platelet-specific (Ginsberg et al. 1987; Suzuki et al. 1987), and platelet aggregation requires only 50% functional $\alpha_{IIb}\beta_3$ (Kunicki et al. 1981), indicating that such mutations would be dominant. In a

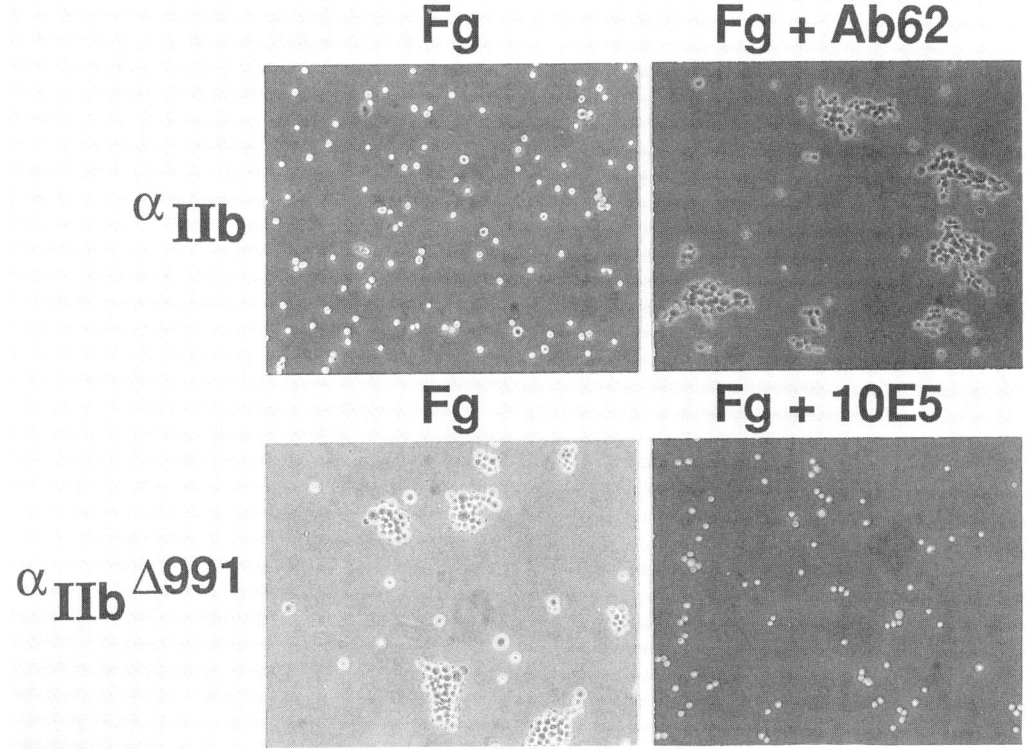

Figure 7. Agonist-independent aggregation of cells bearing $\alpha_{IIb}\Delta991\beta_3$. Cells expressing the indicated integrin α subunit were harvested with trypsin and EDTA, washed, and resuspended in Tyrode's solution at 10^7/ml (15). Cells (100 μl) were added to wells of a 24-well tissue-culture plate in the presence or absence of 6 μM MAb 62 and subjected to gyrorotation (100 rpm for 30 min). Fibrinogen (1 μM) was added and aggregation was stopped after 20 min by addition of 2% paraformaldehyde (150 μl). An inhibitory anti-$\alpha_{IIb}\beta_3$(MAb 10E5, 500 nM [16]) was added at the time of Fg addition to the $\alpha_{IIb}\Delta991\beta_3$ cells (*right, lower* panel). (Reprinted, with permission, from O'Toole et al. 1991.)

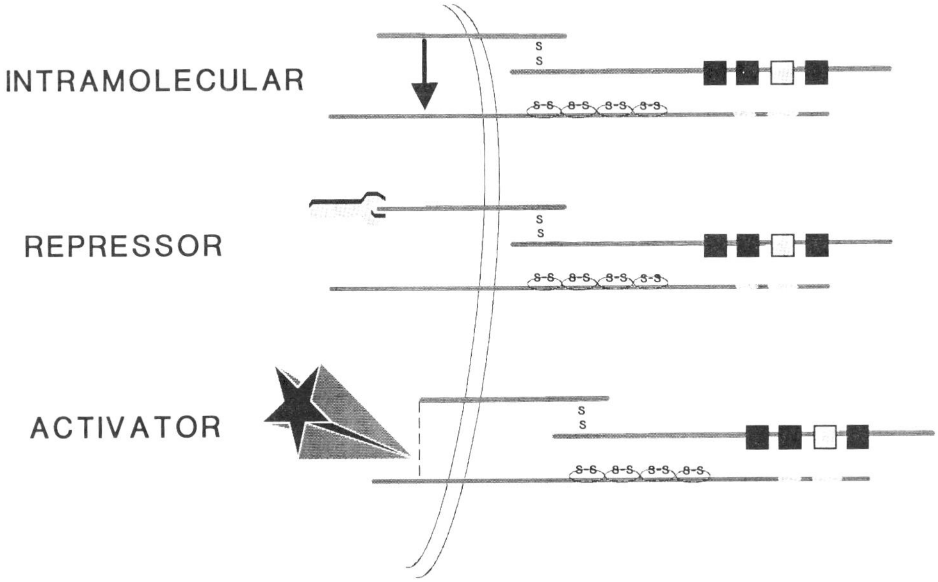

Figure 8. Three models to explain the activation of $\alpha_{IIb}\beta_3$ by changes in α_{IIb} cytoplasmic domain are depicted. *Intramolecular*: Interaction between the α and β cytoplasmic domains maintains the resting state. *Repressor*: Binding of a cytoplasmic repressor to the α cytoplasmic domain maintains the resting state. *Activator*: The α_{IIb} cytoplasmic domain blocks the binding of an activator to the integrin's cytoplasmic face.

hereditary form, the predicted disease might have features of congenital thrombocytopenia and a thrombotic tendency due to spontaneous platelet aggregation. Furthermore, somatic mutations such as might occur in cancers that express α_{IIb} (Chang et al. 1991) or in myeloproliferative syndromes could contribute to tumor-induced platelet aggregation (Grossi et al. 1988) or hyperaggregability of a subset of platelets and an acquired thrombotic tendency. An *Alu* repetitive sequence in intron 29 of α_{IIb} (Heidenreich et al. 1990) immediately precedes the exon containing the entire cytoplasmic domain. Since such repetitive sequences are potential "hot spots" for spontaneous recombination events, such occurrences might also produce changes in the α_{IIb} cytoplasmic tail either in the germ line or in somatic cells.

In summary, integrins mediate an extraordinary variety of biological processes by recognition of macromolecular ligands. In the present review, we have presented data supporting a hypothesis that a common aspect of integrin function involves the interaction of receptor-bound divalent cation with macromolecular and peptide ligands. Second, integrins differ in ligand recognition specificity even for small peptide ligands. We have discussed the evidence that this specificity is controlled in part by the α subunit and that the second putative divalent-cation-binding site may be critical in determining this specificity. Ultimately, the data presented here will have to be interpreted in light of the three-dimensional structure of the receptors. Furthermore, the changes in integrin affinity and specificity seen in different cell types (Kirchofer et al. 1990) or in a single cell type in response to changes in the cellular activation state (Plow and Ginsberg 1981) are in some cases due to conformational changes in the extracellular domain of the integrin. These changes represent a bona fide example of "inside-out" signal transduction and are likely to be mediated through the cytoplasmic domains of the integrins. In a sense, integrins are two-headed receptors that specifically recognize components of the extracellular and intracellular (cytoskeleton) matrix. Finally, it must also be emphasized that integrins may possess ligand-binding structures completely divorced from the ligand-binding site discussed here. For example, Miles et al. (1986) found that plasminogen binding to platelets was $\alpha_{IIb}\beta_3$-dependent but not divalent-cation-dependent. Moreover, the extent of plasminogen binding was the same in platelets expressing $\alpha_{IIb}\beta_3$ ($D_{119}Y$) and normal cells (Miles et al. 1986). Since these platelets contain $\alpha_{IIb}\beta_3$ with greatly reduced affinity for Fn and RGD peptides, these data directly indicate that the plasminogen recognition mechanism on $\alpha_{IIb}\beta_3$ is distinct from that for the RGD ligands.

ACKNOWLEDGMENTS

We gratefully acknowledge the able secretarial assistance of Ms. Eileen Hamlin and technical assistance of Ms. Jane Forsyth. This work was supported by National Institutes of Health grants HL-28235, HL-16411, AR-27214, and HL-39150. This is publication number 7481CVB.

REFERENCES

Altieri, D.C. and T.S. Edgington. 1988. The saturable high affinity association of factor X to ADP-stimulated monocytes defines a novel function of the Mac-1 receptor. *J. Biol. Chem.* **263:** 7007.

Altruda, F., P. Cervella, G. Tarone, C. Botta, F. Balzac, G. Stefanuto, and L. Silengo. 1990. A human integrin beta 1 subunit with a unique cytoplasmic domain generated by alternative mRNA processing. *Gene* **95:** 261.

Argraves, W.S., S. Suzuki, H. Arai, K. Thompson, M.D. Pierschbacher, and E. Ruoslahti. 1987. Amino acid sequence of the human fibronectin receptor. *J. Cell Biol.* **105:** 1183.

Bajt, M.L., M.H. Ginsberg, A.L. Frelinger III, M.C. Berndt, and J.C. Loftus. 1992. A spontaneous mutation of integrin $\alpha_{IIb}\beta_3$ (platelet GPIIb-IIIa) helps define a ligand binding site. *J. Biol. Chem.* **267:** 3789.

Bennett, J.S. and G. Vilaire. 1979. Exposure of platelet fibrinogen receptors by ADP and epinephrine. *J. Clin. Invest.* **64:** 1393.

Brass, L.F., D.R. Manning, A.G. Williams, M.J. Woolkalis, and M. Poncz. 1991. Receptor and G protein-mediated responses to thrombin in HEL cells. *J. Biol. Chem.* **266:** 958.

Buyon, J.P., S.G. Slade, J. Reibman, S.B. Abramson, M.R. Philips, G. Weissmann, and R. Winchester. 1990. Constitutive and induced phosphorylation of the alpha- and beta-chains of the CD11/CD18 leukocyte integrin family. Relationship to adhesion-dependent functions. *J. Immunol.* **144:** 191.

Calvete, J.J., A. Henschen, and J. Gonzalez-Rodriguez. 1989. Complete localization of the intrachain disulphide bonds and the *N*-glycosylation points in the alpha-subunit of human platelet glycoprotein IIb. *Biochem. J.* **261:** 561.

———. 1991a. Assignment of disulphide bonds in human platelet GPIIIa. A disulphide pattern for the beta-subunits of the integrin family. *Biochem. J.* **274:** 63.

Calvete, J.J., J. Arias, M.V. Alvarez, M.M. Lopez, A. Henschen, and J. Gonzalez-Rodriguez. 1991b. Further studies on the topography of human platelet glycoprotein IIb. *Biochem. J.* **273:** 767.

Calvete, J.J., J. Arias, M.V. Alvarez, M.M. Lopez, A. Henschen, and J. Gonzalez-Rodriguez. 1991c. Further studies on the topography of the N-terminal region of human platelet glycoprotein IIIa. Localization of monoclonal antibody epitopes and the putative fibrinogen-binding sites. *Biochem. J.* **274:** 457.

Cembrowski, G.S. and D.F. Mosher. 1984. Plasma fibronectin concentration in patients with acquired consumptive coagulopathies. *Thromb. Res.* **36:** 437.

Chang, S.Y., Y.Q. Chen, L.A. Fitzgerald, and K.V. Honn. 1991. Analysis of integrin mRNA in human and rodent tumor cells. *Biochem. Biophys. Res. Commun.* **176:** 108.

Charo, I.F., L.S. Bekeart, and D.R. Phillips. 1987. Platelet glycoprotein IIb-IIIa-like proteins mediate endothelial cell attachment to adhesive proteins and the extracellular matrix. *J. Biol. Chem.* **262:** 9935.

Charo, I.F., L. Nannizzi, D.R. Phillips, M.A. Hsu, and R.M. Scarborough. 1991. Inhibition of fibrinogen binding to GPIIb-IIIa by a GP IIIa peptide. *J. Biol. Chem.* **266:** 1414.

Chatila, T.A., R.S. Geha, and M.A. Arnaout. 1989. Constitutive and stimulus-induced phosphorylation of CD11/CD18 leukocyte adhesion molecules. *J. Cell Biol.* **109:** 3435.

Cheresh, D.A., S.A. Berliner, V. Vicente, and Z.M. Ruggeri. 1989a. Recognition of distinct adhesive sites on fibrinogen by related integrins on platelets and endothelial cells. *Cell* **58:** 945.

Cheresh, D.A., J.W. Smith, H.M. Cooper, and V. Quaranta. 1989b. A novel vitronectin receptor integrin ($\alpha_v\beta_x$) is responsible for distinct adhesive properties of carcinoma cells. *Cell* **57:** 59.

Coller, B.S. 1986. Activation affects access to the platelet receptor for adhesive glycoproteins. *J. Cell Biol.* **103:** 451.

Coller, B.S., J.D. Folts, L.E. Scudder, and S.R. Smith. 1986. Antithrombotic effect of a monoclonal antibody to the platelet glycoprotein IIb/IIIa receptor in an experimental animal model. *Blood* **68:** 783.

Corbi, A.L., L.J. Miller, K. O'Connor, R.S. Larson, and T.A. Springer. 1987. cDNA cloning and complete primary structure of the alpha subunit of a leukocyte adhesion glycoprotein, p150,95. *EMBO J.* **6:** 4023.

Davies, J., J. Warwick, N. Totty, R. Philp, M. Helfrich, and M. Horton. 1989. The osteoclast functional antigen, implicated in the regulation of bone resorption, is biochemically related to the vitronectin receptor. *J. Cell Biol.* **109:** 1817.

Dransfield, I. and N. Hogg. 1990. Regulated expression of Mg^{2+} binding epitope on leukocyte integrin alpha subunits. *EMBO J.* **8:** 3759.

D'Souza, S.E., M.H. Ginsberg, T.A. Burke, and E.F. Plow. 1990. The ligand binding site of the platelet integrin receptor GPIIb-IIIa is proximal to the second calcium binding domain of its alpha subunit. *J. Biol. Chem.* **265:** 3440.

D'Souza, S.E., M.H. Ginsberg, G.R. Matsueda, and E.F. Plow. 1991. A discrete sequence in a platelet integrin is involved in ligand recognition. *Nature* **350:** 66.

D'Souza, S.E., M.H. Ginsberg, T.A. Burke, S.C.-T. Lam, and E.F. Plow. 1988. Localization of an Arg-Gly-Asp recognition site within an integrin adhesion receptor. *Science* **242:** 91.

Dustin, M.L. and T.A. Springer. 1989. T-cell receptor cross-linking transiently stimulates adhesiveness through LFA-1. *Nature* **341:** 619.

Fitzgerald, L.A., B. Steiner, S.C. Rall, Jr., S.-S. Lo, and D.R. Phillips. 1987. Protein sequence of endothelial glycoprotein IIIa derived from a cDNA clone. Identity with platelet glycoprotein IIIa and similarity to "integrin." *J. Biol. Chem.* **262:** 3936.

Frelinger, A.L.,III, X. Du, E.F. Plow, and M.H. Ginsberg. 1991. Monoclonal antibodies to ligand-occupied conformers of integrin $\alpha_{IIb}\beta_3$ (glycoprotein IIb-IIIa) alters receptor affinity, specificity, and function. *J. Biol. Chem.* **266:** 17106.

Frelinger, A.L.,III, S.C.-T. Lam, E.F. Plow, M.A. Smith, J.C. Loftus, and M.H. Ginsberg. 1988. Occupancy of an adhesive glycoprotein receptor modulates expression of an antigenic site involved in cell adhesion. *J. Biol. Chem.* **263:** 12397.

Fujimoto, T., S. Ohara, and J. Hawiger. 1982. Thrombin-induced exposure and prostacyclin inhibition of the receptor for factor VIII/von Willebrand factor on human platelets. *J. Clin. Invest.* **69:** 1212.

Gailit, J. and E. Ruoslahti. 1988. Regulation of the fibronectin receptor affinity by divalent cations. *J. Biol. Chem.* **263:** 12927.

Gartner, T.K., R. Loudon, and D.B. Taylor. 1991. The peptides APLHK, EHIPA and GAPL are hydropathically equivalent peptide mimics of a fibrinogen binding domain of glycoprotein IIb/IIIa. *Biochem. Biophys. Res. Commun.* **180:** 1446.

Gartner, T.K., J.W. Power, E.H. Beachey, J.S. Bennett, and S.J. Shattil. 1985. The tetrapeptide analogue of the alpha chain and decapeptide analogue of the gamma chain of fibrinogen bind to different sites on the platelet fibrinogen receptor. *Blood* **66:** 305a.

Ginsberg, M.H., M.D. Pierschbacher, E. Ruoslahti, G.A. Marguerie, and E.F. Plow. 1985. Inhibition of fibronectin binding to platelets by proteolytic fragments and synthetic peptides which support fibroblast adhesion. *J. Biol. Chem.* **260:** 3931.

Ginsberg, M.H., A. Lightsey, T.J. Kunicki, A. Kaufman, G.A. Marguerie, and E.F. Plow. 1986. Divalent cation regulation of the surface orientation of platelet membrane glycoprotein IIb: Correlation with fibrinogen binding function and definition of a novel variant of Glanzmann's thrombasthenia. *J. Clin. Invest.* **78:** 1103.

Ginsberg, M.H., A.L. Frelinger III, S.C.-T. Lam, J. Forsyth, R. McMillan, E.F. Plow, and S.J. Shattil. 1990. Analysis of platelet aggregation disorders based on flow cytometric analysis of membrane glycoprotein IIb-IIIa with conformation-specific monoclonal antibodies. *Blood* **76:** 2017.

Ginsberg, M.H., J.C. Loftus, J.-J. Ryckwaert, M.D. Pierschbacher, R. Pytela, E. Ruoslahti, and E.F. Plow. 1987. Immunochemical and amino-terminal sequence comparison of two cytoadhesins indicates they contain similar or identical beta subunits and distinct alpha subunits. *J. Biol. Chem.* **262:** 5437.

Gresham, H.D., J.L. Goodwin, P.M. Allen, D.C. Anderson, and E.J. Brown. 1989. A novel member of the integrin receptor family mediates Arg-Gly-Asp-stimulated neutrophil phagocytosis. *J. Cell Biol.* **108:** 1935.

Grossi, I.M., J.S. Hatfield, L.A. Fitzgerald, M. Newcombe, J.D. Taylor, and K.V. Honn. 1988. Role of tumor cell glycoproteins immunologically related to glycoproteins Ib and IIb/IIIa in tumor cell-platelet and tumor cell-matrix interactions. *FASEB J.* **2:** 385.

Hautanen, A., J. Gailit, D.M. Mann, and E. Ruoslahti. 1989. Effects of modifications of the RGD sequence and its context on recognition by the fibronectin receptor. *J. Biol. Chem.* **264:** 1437.

Haverstick, D.M., J.F. Cowan, K.M. Yamada, and S.A. Santoro. 1985. Inhibition of platelet adhesion to fibronectin, fibrinogen, and von Willebrand factor substrates by a synthetic tetrapeptide derived from the cell-binding domain of fibronectin. *Blood* **66:** 946.

Heidenreich, R., R. Eisman, S. Surrey, K. Delgrosso, J.S. Bennett, E. Schwartz, and M. Poncz. 1990. Organization of the gene for platelet glycoprotein IIb. *Biochemistry* **29:** 1232.

Hemler, M.E. 1990. VLA proteins in the integrin family: Structures, functions, and their role on leukocytes. *Annu. Rev. Immunol.* **8:** 365.

Hemler, M.E., C. Huang, and L. Schwarz. 1987. The VLA protein family. Characterization of five distinct cell surface heterodimers each with a common 130,000 molecular weight beta subunit. *J. Biol. Chem.* **262:** 3300.

Hemler, M.E., D. Glass, J.S. Coblyn, and J.G. Jacobson. 1986. Very late activation antigens on rheumatoid synovial fluid T lymphocytes. Association with states of T cell activation. *J. Clin. Invest.* **78:** 696.

Hibbs, M.L., S. Jakes, S.A. Stacker, R.W. Wallace, and T.A. Springer. 1991. The cytoplasmic domain of the integrin lymphocyte function-associated antigen 1 β subunit: Sites required for binding to intercellular adhesion molecule 1 and the phorbol ester-stimulated phosphorylation site. *J. Exp. Med.* **174:** 1227.

Hillery, C.A., S.S. Smyth, and L.V. Parise. 1991. Phosphorylation of human platelet glycoprotein IIIa (GPIIIa). *J. Biol. Chem.* **266:** 14663.

Holzmann, B. and I.L. Weissman. 1989. Peyer's patch-specific lymphocyte homing receptors consist of a VLA-4-like α chain associated with either of two integrin β chains, one of which is novel. *EMBO J.* **8:** 1735.

Holzmann, B., B.W. McIntyre, and I.L. Weissman. 1989. Identification of a murine Peyer's patch-specific lymphocyte homing receptor as an integrin molecule with an α chain homologous to human VLA-4α. *Cell* **56:** 37.

Humphries, M.J., K. Olden, and K.M. Yamada. 1986. A synthetic peptide from fibronectin inhibits experimental metastasis of murine melanoma cells. *Science* **233:** 467.

Hynes, R.O. 1987. Integrins: A family of cell surface receptors. *Cell* **48:** 549.

Ill, C.R., E. Engvall, and E. Ruoslahti. 1985. Adhesion of

platelets to laminin in the absence of activation. *J. Cell Biol.* **99:** 2140.

Jennings, L.K. and D.R. Phillips. 1982. Purification of glycoproteins IIb and III from human platelet plasma membranes and characterization of a calcium-dependent glycoprotein IIb-III complex. *J. Biol. Chem.* **257:** 10458.

Kajiji, S., R.N. Tamura, and V. Quaranta. 1989. A novel integrin ($\alpha_E\beta_4$) from human epithelial cells suggests a fourth family of integrin adhesion receptors. *EMBO J.* **8:** 673.

Kirchofer, D., L.R. Languino, E. Ruoslahti, and M.D. Pierschbacher. 1990. $\alpha2$ $\beta1$ integrins from different cell types show different binding specificities. *J. Biol. Chem.* **265:** 615.

Kishimoto, T.K., K. O'Connor, A. Lee, T.M. Roberts, and T.A. Springer. 1987. Cloning of the B subunit of the leukocyte adhesion proteins: Homology to an extracellular matrix receptor defines a novel supergene family. *Cell* **48:** 681.

Kretsinger, R.H. 1980. Structure and evolution of calcium-modulated proteins. *CRC Crit. Rev. Biochem.* **8:** 119.

Kunicki, T.J., D. Pidard, J.-P. Cazenave, A.T. Nurden, and J.P. Caen. 1981. Inheritance of the human platelet alloantigen, P1A1, in type I Glanzmann's thrombasthenia. *J. Clin. Invest.* **67:** 717.

Lam, S.C-T., E.F. Plow, S.E. D'Souza, D.A. Cheresh, A.L. Frelinger III, and M.H. Ginsberg. 1989. Isolation and characterization of a platelet membrane protein related to the vitronectin receptor. *J. Biol. Chem.* **264:** 3742.

Lam, S.C-T., E.F. Plow, M.A. Smith, A. Andrieux, J.-J. Rychwaert, G.A. Marguerie, and M.H. Ginsberg. 1987. Evidence that Arginyl-Glycyl-Aspartate peptides and fibrinogen gamma chain peptides share a common binding site on platelets. *J. Biol. Chem.* **262:** 947.

Lanza, F., A. Stierle, D. Fournier, M. Morales, G. Andre, A.T. Nurden, and J. Cazenave. 1992. A new variant of Glanzmann's thrombasthenia (StrasbourgI). *J. Clin. Invest.* **89:** 1995.

Larson, R.S., M.L. Hibbs, and T.A. Springer. 1990. The leukocyte integrin LFA-1 reconstituted by cDNA transfection in a nonhematopoietic cell line is functionally active and not transiently regulated. *Cell Regul.* **1:** 359.

Lawler, J., R. Weinstein, and R.O. Hynes. 1988. Cell attachment to thrombospondin: The role of ARG-GLY-ASP, calcium, and integrin receptors. *J. Cell Biol.* **107:** 2351.

Loftus, J., J. Sorge, E.F. Plow, G.A. Marguerie, and M.H. Ginsberg. 1986. Molecular cloning of cDNA coding for a divalent cation regulated epitope in platelet membrane GPIIb. *Clin. Res.* **34:** 660A.

Loftus, J.C., T.E. O'Toole, E.F. Plow, A. Glass, A.L. Frelinger III, and M.H. Ginsberg. 1990. A β_3 integrin mutation abolishes ligand binding and alters divalent cation-dependent conformation. *Science* **249:** 915.

Loftus, J.C., E.F. Plow, A.L. Frelinger III, S.E. D'Souza, D. Dixon, J. Lacy, J. Sorge, and M.H. Ginsberg. 1987. Molecular cloning and chemical synthesis of a region of platelet GPIIb involved in adhesive function. *Proc. Natl. Acad. Sci.* **840:** 7114.

MacKrell, A.J., B. Blumberg, S.R. Haynes, and J.H. Fessler. 1988. The lethal myospheroid gene of *Drosophila* encodes a membrane protein homologous to vertebrate integrin beta subunits. *Proc. Natl. Acad. Sci.* **85:** 2633.

Marcantonio, E.E. and R.O. Hynes. 1988. Antibodies to the conserved cytoplasmic domain of the integrin B1 subunit react with proteins in vertebrates, invertebrates, and fungi. *J. Cell Biol.* **106:** 1765.

Marlin, S.D. and T.A. Springer. 1987. Purified intercellular adhesion molecule-1 (ICAM-1) is a ligand for lymphocyte function-associated antigen 1 (LFA-1). *Cell* **51:** 813.

Menko, A.S. and D. Boettiger. 1987. Occupation of the extracellular matrix receptor, integrin, is a control point for myogenic differentiation. *Cell* **51:** 51.

Merrill, J.T., S.G. Slade, G. Weissmann, R. Winchester, and J.P. Buyon. 1990. Two pathways of CD11b/CD18-mediated neutrophil aggregation with different involvement of protein kinase C-dependent phosphorylation. *J. Immunol.* **145:** 2608.

Miles, L.A., M.H. Ginsberg, and J.G. White. 1986. Plasminogen interacts with human platelets through two distinct mechanisms. *J. Clin. Invest.* **77:** 2001.

Munson, P.J. and D. Rodbard. 1980. LIGAND: A versatile computerized approach for characterization of ligand-binding systems. *Anal. Biochem.* **107:** 220.

Nathan, C., S. Srimal, C. Farber, E. Sanchez, L. Kabbash, A. Asch, J. Galit, and S.D. Wright. 1989. Cytokine-induced respiratory burst of human neutrophils: Dependence on extracellular matrix proteins and CD11/CD18 integrins. *J. Cell Biol.* **109:** 1341.

Nurden, A.T. and J.P. Caen. 1974. An abnormal platelet glycoprotein pattern in three cases of Glanzmann's thrombasthenia. *Br. J. Haematol.* **28:** 253.

O'Toole, T.E., J.C. Loftus, E.F. Plow, A. Glass, J.R. Harper, and M.H. Ginsberg. 1989. Efficient surface expression of platelet GPIIb-IIIa requires both subunits. *Blood* **74:** 14.

O'Toole, T.E., D. Mandelman, J. Forsyth, S.J. Shattil, E.F. Plow, and M.H. Ginsberg. 1991. Modulation of the affinity of integrin $\alpha_{IIb}\beta_3$ (GPIIb-IIIa) by the cytoplasmic domain of α_{IIb}. *Science* **254:** 845.

O'Toole, T.E., J.C. Loftus, X. Du, A.A. Glass, Z.M. Ruggeri, S.J. Shattil, E.F. Plow, and M.H. Ginsberg. 1990. Affinity modulation of the $\alpha_{IIb}\beta_3$ integrin (platelet GPIIb-IIIa) is an intrinsic property of the receptor. *Cell Regul.* **1:** 883.

Panayotou, G., P. End, M. Aumailley, R. Timpl, and J. Engel. 1989. Domains of laminin with growth-factor activity. *Cell* **56:** 93.

Peerschke, E.I. and B.S. Coller. 1984. A murine monoclonal antibody that blocks fibrinogen binding to normal platelets also inhibits fibrinogen interactions with chymotrypsin-treated platelets. *Blood* **64:** 59.

Peerschke, E.I., M.B. Zucker, R.A. Grant, J.J. Egan, and M.M. Johnson. 1980. Correlation between fibrinogen binding to human platelets and platelet aggregability. *Blood* **55:** 841.

Phillips, D.R. and P.P. Agin. 1977. Platelet membrane defects in Glanzmann's thrombasthenia. *J. Clin. Invest.* **60:** 535.

Phillips, D.R. and A.K. Baughan. 1983. Fibrinogen binding to human platelet plasma membranes identification of two steps requiring divalent cations. *J. Biol. Chem.* **258:** 10240.

Phillips, D.R., I.F. Charo, L.V. Parise, and L.A. Fitzgerald. 1988. The platelet membrane glycoprotein IIb-IIIa complex. *Blood* **71:** 831.

Plow, E.F. and M.H. Ginsberg. 1981. Specific and saturable binding of plasma fibronectin to thrombin-stimulated human platelets. *J. Biol. Chem.* **256:** 9477.

Plow, E.F., M.D. Pierschbacher, E. Ruoslahti, G.A. Marguerie, and M.H. Ginsberg. 1985. The effect of Arg-Gly-Asp-containing peptides on fibrinogen and von Willebrand factor binding to platelets. *Proc. Natl. Acad. Sci.* **82:** 8057.

———. 1987. Arginyl-Glycyl-Aspartic acid sequences and fibrinogen binding to platelets. *Blood* **70:** 110.

Plow, E.F., A.H. Srouji, D. Meyer, G.A. Marguerie, and M.H. Ginsberg. 1984. Evidence that three adhesive proteins interact with a common recognition site on activated platelets. *J. Biol. Chem.* **259:** 5388.

Plow, E.F., J. Loftus, E. Levin, D. Fair, D. Dixon, J. Forsyth, and M.H. Ginsberg. 1986. Immunologic relationship between platelet membrane glycoprotein GPIIb/IIIa and cell surface molecules expressed by a variety of cells. *Proc. Natl. Acad. Sci.* **83:** 6002.

Poncz, M., R. Eisman, R. Heidenreich, S.M. Silver, G. Vilare, S. Surrey, E. Schwartz, and J.S. Bennett. 1987. Structure of the platelet membrane glycoprotein IIb. Homology to the alpha subunits of the vitronectin and fibronectin membrane receptors. *J. Biol. Chem.* **262:** 8476.

Prieto, J., P.G. Beatty, E.A. Clark, and M. Patarroyo. 1988. Molecules mediating adhesion of T and B cells, monocytes and granulocytes to vascular endothelial cells. *Immunology* **63:** 631.

Pytela, R.P., M.D. Pierschbacher, M.H. Ginsberg, E.F. Plow, and E. Ruoslahti. 1986. Platelet membrane glycoprotein IIb/IIIa: Member of a family of Arg-Gly-Asp-specific adhesion receptors. *Science* **231:** 1559.

Rosa, J.-P., P.F. Bray, O. Gayet, G.I. Johnston, R.G. Cook, K.W. Jackson, M.A. Shuman, and R.P. McEver. 1988. Cloning of glycoprotein IIIa cDNA from human erythroleukemia cells and localization of the gene to chromosome 17. *Blood* **72:** 593.

Ruggeri, Z.M., R.A. Houghten, S.R. Russell, and T.S. Zimmerman. 1986. Inhibition of platelet function with synthetic peptides designed to be high-affinity antagonists of fibrinogen binding to platelets (platelet receptors/platelet aggregation/thrombosis). *Proc. Natl. Acad. Sci.* **83:** 5708.

Ruoslahti, E. and M.D. Pierschbacher. 1986. Arg-Gly-Asp: A versatile cell recognition signal. *Cell* **44:** 517.

———. 1987. New perspectives in cell adhesion: RGD and integrins. *Science* **238:** 491.

Savage, B., U.M. Marzec, B.H. Chao, L.A. Harker, J.M. Maraganore, and Z.M. Ruggeri. 1990. Binding of the snake venom-derived proteins applaggin and echistatin to the arginine-glycine-aspartic acid recognition site(s) on platelet glycoprotein IIb.IIIa complex inhibits receptor function. *J. Biol. Chem.* **265:** 11766.

Shattil, S.J. and L.F. Brass. 1987. Induction of the fibrinogen receptor on human platelets by intracellular mediators. *J. Biol. Chem.* **262:** 992.

Shattil, S.J., J.A. Hoxie, M. Cunningham, and L.F. Brass. 1985. Changes in the platelet membrane glycoprotein IIb-IIIa complex during platelet activation. *J. Biol. Chem.* **260:** 11107.

Shaw, L.M., J.M. Messier, and A.M. Mercurio. 1990. The activation dependent adhesion of macrophages to laminin involves cytoskeletal anchoring and phosphorylation of the alpha 6 beta 1 integrin. *J. Cell Biol.* **110:** 2167.

Shimizu, Y., G.A. VanSeventer, K.J. Horgan, and S. Shaw. 1990. Regulated expression and binding of three VLA (B1) integrin receptors on T cells. *Nature* **345:** 250.

Silverstein, R.L., L.L. Leung, P.C. Harpel, and R.L. Nachman. 1984. Complex formation of platelet thrombospondin with plasminogen. Modulation of activation by tissue activator. *J. Clin. Invest.* **74:** 1625.

Sims, P.J., M.H. Ginsberg, E.F. Plow, and S.J. Shattil. 1990. Effect of platelet activation on the conformation of the plasma membrane glycoprotein IIb-IIIa complex. *J. Biol. Chem.* **266:** 7345.

Springer, T.A. 1985. The LFA-1, Mac-1 glycoprotein family and its deficiency in an inherited disease. *Fed. Proc.* **44:** 2660.

Staatz, W.D., S.M. Rajpara, E.A. Wayner, W.G. Carter, and S.A. Santoro. 1989. The membrane glycoprotein Ia-IIa (VLA-2) complex mediates the Mg^{++}-dependent adhesion of platelets to collagen. *J. Cell Biol.* **108:** 1917.

Steiner, B., D. Cousot, A. Trzeciak, D. Gillessen, and P. Hadváry. 1989. Ca^{2+}-dependent binding of a synthetic Arg-Gly-Asp (RGD) peptide to a single site on the purified platelet glycoprotein IIb-IIIa complex. *J. Biol. Chem.* **264:** 13102.

Suzuki, S., W.S. Argraves, H. Arai, L.R. Languino, M.D. Pierschbacher, and E. Ruoslahti. 1987. Amino acid sequence of the vitronectin receptor alpha subunit and comparative expression of adhesion receptor mRNAs. *J. Biol. Chem.* **262:** 14080.

Takada, Y. and M.E. Hemler. 1989. The primary structure of the VLA-2/collagen receptor α^2 subunit (platelet GPIa): Homology to other integrins and the presence of a possible collagen-binding domain. *J. Cell Biol.* **109:** 397.

Takada, Y., M.J. Elices, C. Crouse, and M.E. Hemler. 1989. The primary structure of the α^4 subunit of VLA-4: Homology to other integrins and a possible cell-cell adhesion function. *EMBO J.* **8:** 1361.

Takada, Y., E. Murphy, P. Pil, C. Chen, M.H. Ginsberg, and M.E. Hemler. 1991. Molecular cloning and expression of the cDNA for α3-subunit of human $\alpha 3\beta 1$ (VLA-3), an integrin receptor for fibronectin, laminin and collagen. *J. Cell Biol.* **115:** 257.

Tamkun, J.W., D.W. DeSimone, D. Fonda, R.S. Patel, C.A. Buck, A.F. Horwitz, and R.O. Hynes. 1986. Structure of integrin, a glycoprotein involved in transmembrane linkage between fibronectin and actin. *Cell* **46:** 271.

Taub, R., R.J. Gould, V.M. Garsky, et al. 1989. A monoclonal antibody against the platelet fibrinogen receptor contains a sequence that mimics a receptor recognition domain in fibrinogen. *J. Biol. Chem.* **264:** 259.

Tonnesen, M.G., L.A. Smedly, and P.M. Henson. 1984. Neutrophil-endothelial cell interactions. Modulation of neutrophil adhesiveness induced by complement fragments C5a and C5a des arg and formyl-methionyl-leucyl-phenylalanine in vitro. *J. Clin. Invest.* **74:** 1581.

VanderMeulen, J. and S. Grinstein. 1982. Ca^{2+}-induced lysis of platelet secretory granules. *J. Biol. Chem.* **257:** 5190.

van Kuppevelt, T.H., L.R. Languino, J.O. Gailit, S. Suzuki, and E. Ruoslahti. 1989. An alternative cytoplasmic domain of the integrin beta 3 subunit. *Proc. Natl. Acad. Sci.* **86:** 5415.

Van Noesel, C., F. Miedema, M. Brouwer, M.A. De Rie, L.A. Aarden, and R.A.W. Van Lier. 1988. Regulatory properties of LFA-1 α and β chains in human T-lymphocyte activation. *Nature* **333:** 850.

Vogel, B.E., G. Tarone, F.G. Giancotti, and E. Ruoslahti. 1990. A novel fibronectin receptor with an unexpected subunit composition (alphaV B1). *J. Biol. Chem.* **265:** 5934.

Wayner, E.A., W.G. Carter, R.S. Piotrowicz, and T.J. Kunicki. 1988. The function of multiple extracellular matrix receptors in mediating cell adhesion to extracellular matrix: Preparation of monoclonal antibodies to the fibronectin receptor that specifically inhibit cell adhesion to fibronectin and react with platelet glycoproteins Ic-IIa. *J. Cell Biol.* **107:** 1881.

Wayner, E.A., A. Garcia-Pardo, M.J. Humphries, J.A. McDonald, and W.G. Carter. 1989. Identification and characterization of the T lymphocyte adhesion receptor for an alternative cell attachment domain (CS-1) in plasma fibronectin. *J. Cell Biol.* **109:** 1321.

Wencel-Drake, J.D., E.F. Plow, T.S. Zimmerman, R.G. Painter, and M.H. Ginsberg. 1984. Immunofluorescent localization of adhesive glycoproteins in resting and thrombin-stimulated platelets. *Am. J. Pathol.* **115:** 156.

Werb, Z., P.M. Tremble, O. Behrendtsen, E. Crowley, and C.H. Damsky. 1989. Signal transduction through the fibronectin receptor induces collagenase and stromelysin gene expression. *J. Cell Biol.* **109:** 877.

Role of α_V Integrins and Vitronectin in Human Melanoma Cell Growth

L.C. Sanders, B. Felding-Habermann, B.M. Mueller, and D.A. Cheresh
Department of Immunology, Scripps Research Institute, La Jolla, California 92037

The interaction of cells with the extracellular matrix (ECM) has been shown to facilitate many biological processes, such as cellular polarity, differentiation, proliferation, and migration (Hay 1981; Hynes 1981). A family of cell-surface receptors termed integrins play a fundamental role in promoting cell-matrix interactions as well as cell-cell interactions (Cheresh 1992; Hynes 1992). Consequently, these interactions can result in the transmission of signals from the extracellular environment to the intracellular compartment, providing a cellular mechanism that can convey various biological functions (Werb et al. 1989; Jaconi et al. 1991; Kornberg et al. 1991; Schwartz et al. 1991a,b; Leavesley et al. 1992).

Integrins are heterodimeric molecules composed of an α and a β subunit. Both subunits traverse the membrane and consist of large extracellular domains coupled to relatively small cytoplasmic tails. To date, 14 α and 8 β subunits have been described, making up at least 20 individual integrin receptors (Cheresh 1992; Hynes 1992). Depending on the α/β subunit pairing, the intact heterodimer can recognize one or multiple ligands present in the ECM or on the surface of opposing cells. In some cases, a given β subunit can associate with multiple α subunits. Alternatively, a specific α subunit can associate with a wide variety of β subunits, thus achieving diversity in the specificity and function of integrins. An example of the latter is the α_V integrin subunit, which can pair with multiple β subunits (β_1, β_3, β_5, β_6, β_8), each demonstrating differences in cell and/or tissue distribution and ligand specificity (Cheresh 1992). The most widely studied α_V integrins are $\alpha_V\beta_3$ and $\alpha_V\beta_5$, both of which mediate cell attachment to the ECM adhesive glycoprotein vitronectin (VN) (Cheresh et al. 1989; McLean et al. 1990; Ramaswamy and Hemler 1990; Smith et al. 1990; Wayner et al. 1991).

VN is a multifunctional adhesive glycoprotein known to promote the attachment and spreading of many different cell types in vitro. Cells can interact with VN either through its RGD domain, which involves integrins, or through its heparin-binding domain, which can bind proteoglycans on the cell surface (Reisfeld and Cheresh 1987; Tomasini and Mosher 1991). VN, like many other adhesive proteins, is produced in the liver and accumulates in the serum. Its regulatory role in the immune system and in fibrinolysis is well documented (Preissner 1989). VN not only exists in plasma, but has also been localized in the ECM of many different tissue types; these include human embryonic lung, fetal membrane, striated skeletal muscle, skin, and kidney (Hayman et al. 1983). Although the in vivo function of VN within the ECM is unclear, it has been suggested that it may be involved in cell adhesion and migration (Preissner 1989; Tomasini and Mosher 1991). Furthermore, increased VN deposits within skin ECM have been correlated with skin diseases such as discoid lupus erythematosus and dermatitis herpetiformis (Dahlback et al. 1989a,b). The observation that VN can be localized within human glioblastoma biopsy material suggests that VN may be involved in the progression of this highly lethal form of brain cancer (Gladson and Cheresh 1991). Recent immunohistological evidence indicates that VN accumulates in late-stage melanoma (C.L. Gladson and D.A. Cheresh, unpubl.). Interestingly, deposits of VN within the skin have been shown to increase with age, and additional evidence suggests that VN deposits may increase with exposure to the sun (Dahlback et al. 1989b). Both aging and sun exposure enhance the likelihood of contracting melanoma. Together, these observations implicate VN involvement in the progression of human melanoma.

It is not clear why more than one VN receptor exists or why multiple VN receptors are expressed by the same cell. However, it is tempting to suggest that each may send distinct signals to the cell, thereby potentiating separate biological functions in response to a VN substrate. Experimental evidence supports this contention. For example, two primary VN receptors, $\alpha_V\beta_3$ and $\alpha_V\beta_5$, contain identical α subunits with structurally distinct β subunits. Although β_3 and β_5 are 56% identical overall, their cytoplasmic tails are only 10–20% homologous. A structural motif unique to β_5 consists of a 10-amino-acid putative insert at the carboxyl terminus that has not been found on any other integrin β subunit (McLean et al. 1990; Ramaswamy and Hemler 1990). Although both $\alpha_V\beta_3$ and $\alpha_V\beta_5$ bind VN, recent evidence demonstrates that only $\alpha_V\beta_3$ is involved in cell spreading and is localized to focal contacts (Wayner et al. 1991). Cells containing only $\alpha_V\beta_5$ attached to VN but failed to spread, and $\alpha_V\beta_5$ was not detected in focal contacts by immunofluorescence (Wayner et al. 1991). Moreover, $\alpha_V\beta_3$ appears to promote migration of cells on VN, whereas $\alpha_V\beta_5$ does not (Leavesley et al. 1992). These data imply that the distinct cytoplasmic tails of β_3 and β_5 may differentially affect the biological responsiveness of a cell to a VN matrix.

Integrin-mediated cellular adhesion mechanisms have been implicated in the tumorigenicity and invasive properties of transformed cells (Humphries et al. 1986; Plantefaber and Hynes 1989; Ruoslahti and Giancotti 1989; Felding-Habermann et al. 1992; Seftor et al. 1992). Both in vitro and in vivo experiments suggest that a change in integrin expression occurs during malignant transformation (Plantefaber and Hynes 1989; Ruoslahti and Giancotti 1989). For example, overexpression of the $\alpha_5\beta_1$ integrin (the fibronectin receptor) has been correlated with decreased tumorigenicity, whereas expression of $\alpha_V\beta_3$ has been associated with increased tumorigenicity and invasiveness of human melanoma cells (Albelda et al. 1990; Giancotti and Ruoslahti 1990; Felding-Habermann et al. 1992; Nip et al. 1992; Seftor et al. 1992). Integrin $\alpha_V\beta_3$ is commonly found on melanoma cultured cell lines (Cheresh and Spiro 1987). In fact, $\alpha_V\beta_3$ integrin is highly expressed on human melanoma biopsy material from metastatic and vertically invasive primary lesions; however, it is not expressed on normal melanocytes, nevi of noninvasive primary melanoma (Albelda et al. 1990). In vitro invasion assays using reconstituted basement membrane have demonstrated an increase in melanoma cell invasiveness when soluble VN or antibodies to $\alpha_V\beta_3$ were present (Seftor et al. 1992). These observations suggest not only that $\alpha_V\beta_3$ is a marker of invasive melanoma, but also that it may play an important role in the malignant behavior of this neoplasm.

M21 cells, a human melanoma cell line derived from a metastatic lesion, express both $\alpha_V\beta_3$ and $\alpha_V\beta_5$. To establish the biological role of these receptors, a stable variant cell line lacking α_V mRNA was selected (M21-L) (Cheresh and Spiro 1987). Consequently, the M21-L cells fail to express $\alpha_V\beta_3$ and $\alpha_V\beta_5$ on their cell surface and do not attach to VN, fibrinogen, or von Willebrand factor (Cheresh and Spiro 1987). Therefore, the M21 and M21-L cells provide a model system to address the role of α_V integrins in the malignant phenotype of human melanoma. In this study, we assessed the role of α_V integrins and VN in the tumorigenicity of human melanoma cells in vivo. Specifically, we examined the growth properties of both M21 and M21-L cells for their ability to form tumors in athymic nude mice. Results presented here indicate that melanoma tumorigenicity depends on the expression of α_V integrins and the availability of VN in situ.

EXPERIMENTAL PROCEDURES

Cell propagation. M21 human melanoma cells and variants (M21-L, M21-L4, M21-L12) were propagated in culture using RPMI 1640 containing 10% fetal bovine serum as described previously (Cheresh and Spiro 1987). Cells grown in vivo were washed and resuspended in 0.1 phosphate-buffered saline (PBS) (pH 7.2), then injected subcutaneously into the flanks of 6-week-old female nude mice (Felding-Habermann et al. 1992). Coinjection of cells with VN and collagen occurred as described above with the addition of 500 μg/injection of protein.

Transfection of M21-L cells with α_V. M21-L cells were transfected with the plasmid pcDNA I/NEO (Invitrogen) containing α_V cDNA as described previously (Felding-Habermann et al. 1992). In brief, transfected cells were cultured in growth medium containing 500 μg/ml Geneticin (Sigma). From a panel of transfected subpopulations, cells expressing wild-type levels of α_V integrins (M21-L4) and nonexpressing control cells (M21-L12) were selected by repeated fluorescence-activated cell sorting using anti-α_V MAb LM142.

Cell adhesion assay. Cell adhesion experiments were performed as described previously (Wayner et al. 1991). In brief, ^{51}Cr-labeled cells were suspended in Hank's balanced salt solution containing 1% bovine serum albumin (BSA), 1 mM calcium, 1 mM magnesium, and 0.5 mM manganese. Cells were seeded in 48-well non-tissue-culture plastic microtiter plates, precoated with individual adhesive ligands (10 μg/ml). After adhesion, unbound cells were washed away, and adherent cells were harvested and counted. BSA-coated wells were used as a control.

Cell proliferation assay. M21 cells and variants were suspended in endothelial cell basal medium (EBM) (Clonetics Corporation) containing no serum or growth factors. Cells were seeded in 96-well non-tissue-culture plastic microtiter plates, precoated with adhesive ligands (10 μg/ml). At various time points after seeding, the increase in viable cell number was enumerated using a blue formazan colorimetric assay (Promega). BSA-coated wells were used as a control.

RESULTS AND DISCUSSION

To address the potential role of α_V integrins and VN in the malignant phenotype of human melanoma cells, a cell model system has been developed. Human melanoma cells (M21 cells) were selected by fluorescence-activated cell sorting for the lack of α_V expression. The resulting cell population (M21-L cells) fails to express α_V integrins (Fig. 1). Previous studies have demonstrated that M21-L cells fail to produce mRNA encoding α_V (Cheresh and Spiro 1987). These cells represent a stable population, since they have been passaged for several years and remain α_V negative. M21-L cells lacking α_V have lost expression of both $\alpha_V\beta_3$ and $\alpha_V\beta_5$ but maintain equivalent levels of β_1 integrins (Fig. 1). To examine whether the loss of α_V results in an altered adhesive phenotype, M21 or M21-L cells were allowed to attach to VN, fibronectin, types I and IV collagen, laminin, fibrinogen, and von Willebrand factor. As shown in Figure 2, M21-L cells readily attach to fibronectin, both collagens, and laminin, consistent with their expression of β_1 integrins. However, M21-L cells failed to attach to VN, fibrinogen, and von Willebrand factor, demonstrating their lack of α_V integrin expression (Fig. 2).

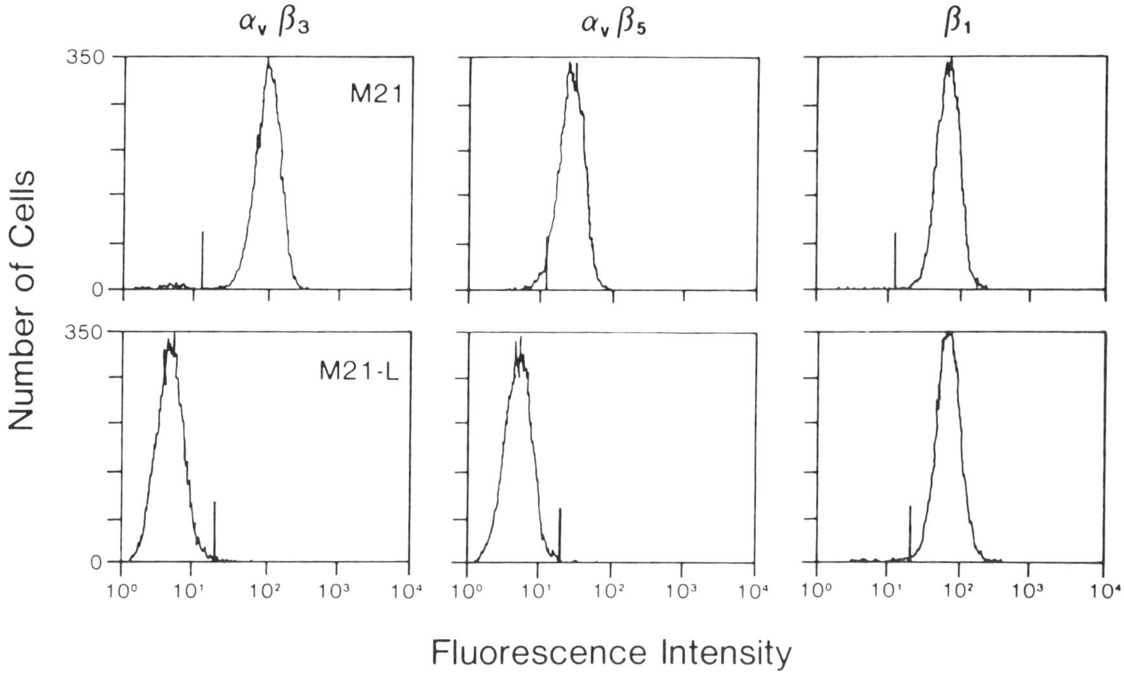

Figure 1. α_V integrin expression on M21 and M21-L cells. Flow cytometric analysis demonstrates differential expression of α_V integrins on M21 cells and M21-L cells. M21 cells express both $\alpha_V\beta_3$ and $\alpha_V\beta_5$, whereas M21-L cells express neither. M21 and M21-L express equivalent amounts of β_1 integrins.

To examine the growth properties of these cell lines, M21 and M21-L cells were injected subcutaneously into athymic nude mice. M21-L cells show a dramatic reduction in tumorigenicity compared to the M21 parental cell population (Fig. 3A). In vitro, however, these cells grow at the same rate, indicating that their growth properties are not intrinsically different (Fig. 3B). Moreover, when M21 or M21-L cells were cultured from their respective tumors, the cells maintained their respective α_V expression (data not shown). These data indicate that the reduction in tumor growth of M21-L cells is not due to in vivo selection of α_V-expressing cells.

To demonstrate a direct role for α_V integrins in human melanoma tumor growth, M21-L (α_V^-) cells were transfected with a cDNA encoding α_V (M21-L4 cells). This resulted in wild-type levels of $\alpha_V\beta_3$ and $\alpha_V\beta_5$ expression on the cell surface (Fig. 4). These cells regained the capacity to attach to VN, and tumorigenicity was fully restored (Figs. 5 and 6). A control cell population, M21-L12 cells (mock transfectants; α_V^-) failed to form significant tumors and were incapable of binding VN, behaving as the M21-L cells (Figs. 5 and 6). These results demonstrate that integrin α_V gene expression and the resulting adhesive phenotype are directly involved in the in vivo proliferation of human

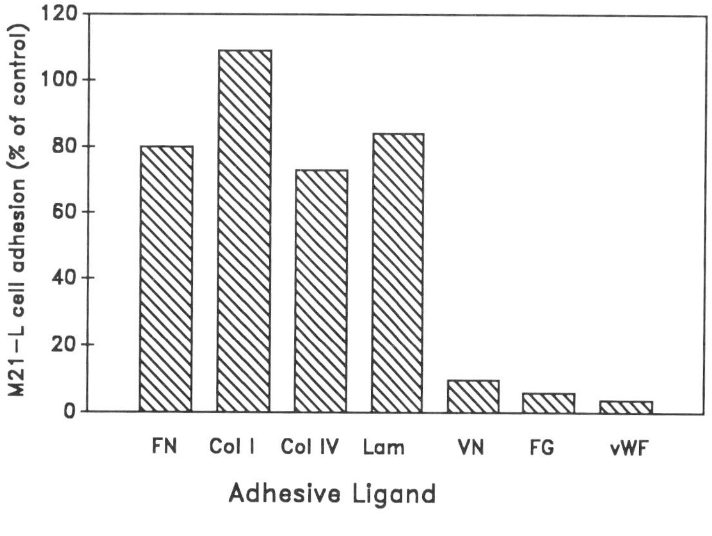

Figure 2. Adhesion of M21-L cells to various ligands. Attachment to various ligands demonstrates the different adhesive properties of M21-L (α_V^-) cells compared to M21 (α_V^+) cells. M21-L cell adhesion is plotted as percent of M21 cell adhesion. M21-L cells maintain their ability to adhere to fibronectin (FN), collagen type 1 (Col I), collagen type IV (Col IV), and laminin (Lam), presumably due to their equivalent β_1 expression. However, M21-L cells do not attach to VN, fibrinogen (FG), or von Willebrand factor (vWF), due to their loss of α_V integrin expression.

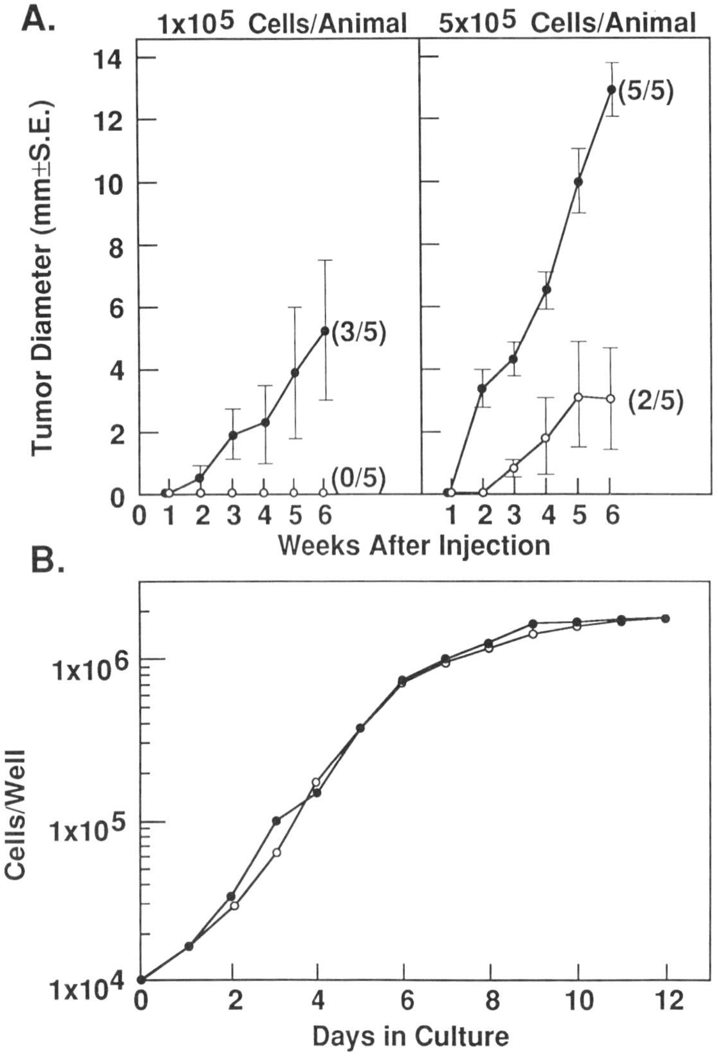

Figure 3. In vivo and in vitro growth properties of M21 and M21-L cells. (*A*) M21 human melanoma cells (●) or M21-L cells deficient in mRNA for α_V integrin (○) show different growth properties in vivo when injected subcutaneously into nude mice. (*B*) In vitro growth of M21 (●) and M21-L cells (○) was indistinguishable in complete growth medium containing 10% serum. (Reprinted, with permission, from Felding-Habermann et al. 1992 [copyright held by American Society for Clinical Investigation].)

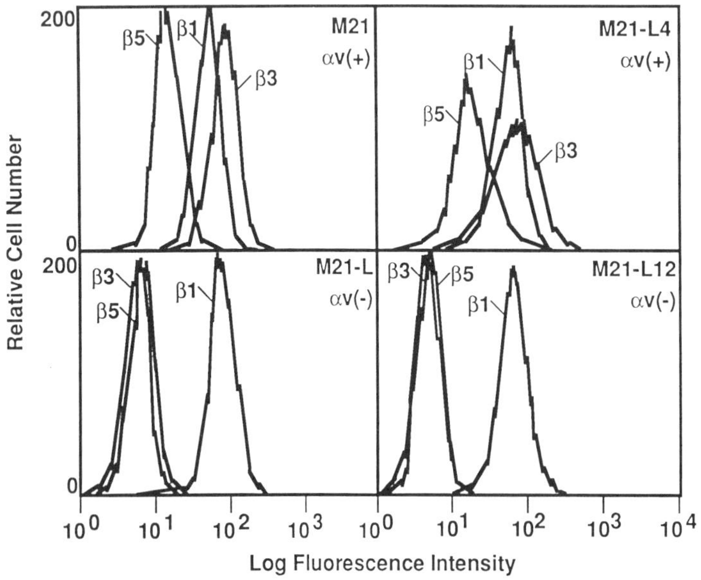

Figure 4. α_V and β_1 integrin expression on M21 cell variants. Flow cytometric analysis demonstrates differential expression of α_V integrins on M21 cell variants. M21 (wild type) and M21-L4 (transfected with cDNA for α_V) cells express $\alpha_V\beta_3$ and $\alpha_V\beta_5$; in contrast, M21-L (stable variant lacking α_V integrins) and M21-L12 (mock transfectant) cells do not. (Reprinted, with permission, from Felding-Habermann et al. 1992 [copyright held by American Society for Clinical Investigation].)

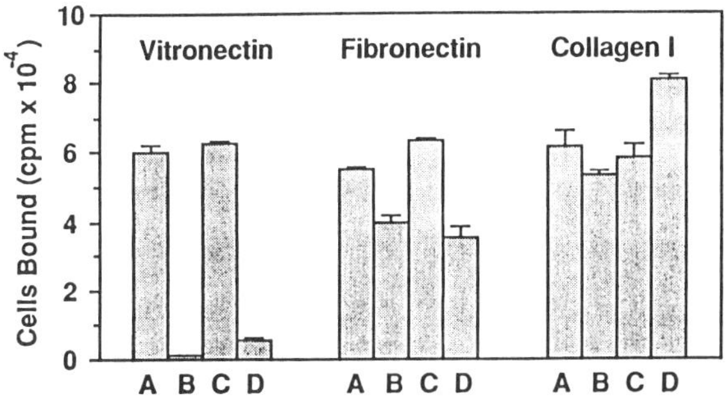

Figure 5. Adhesive properties of M21 cell variants. Attachment to immobilized ligands demonstrates the different adhesive properties of M21 cell variants. M21 and M21-L4 (α_V^+) cells adhere to VN, whereas M21-L and M21-L12 (α_V^-) cells do not. All four cell types attach to fibronectin and collagen consistent with their equivalent expression of β_1 integrins. (*A*) M21 cells; (*B*) M21-L cells; (*C*) M21-L4 cells, and (*D*) M21-L12 cells. (Reprinted, with permission from Felding-Habermann et al. 1992 [copyright held by American Society for Clinical Investigation].)

melanoma cells. In addition, these findings suggest that a ligand for α_V integrins such as VN may be capable of promoting the tumorigenicity of human melanoma cells.

In contrast to their in vivo growth differences, the M21 cells and adhesive variants proliferate identically in vitro, in the presence of serum (Fig. 3). However, in preliminary experiments when serum and exogenous growth factors are removed, these cells demonstrate anchorage-dependent proliferation (Fig. 7). Specifically, the α_V-expressing M21 and M21-L4 cells (positive α_V transfectants) proliferate in response to immobilized VN, demonstrating a mean doubling time of 96 hours. In contrast, M21-L and M21-L12 cells that lack α_V integrins fail to respond to immobilized VN (Fig. 7). However, all four cell types proliferate in response to collagen and fibronectin, which is likely due to their equivalent expression of β_1 integrins (Felding-Habermann et al. 1992). To our knowledge, this is the first demonstration of anchorage-dependent tumor cell proliferation. By establishing growth conditions in vitro that require α_V integrins, thereby mimicking the in vivo growth experiments, we are in a position to examine the role of α_V integrins in the proliferation of human melanoma cells. Our preliminary data suggest that VN, in vitro, can promote a proliferative signal to human melanoma cells by interacting with an α_V integrin. Further evidence supports a role for VN in this process in vivo as well. Specifically, when VN is coinjected with M21 cells subcutaneously into nude mice, a fourfold increase in proliferation was observed over cells injected without VN ($p < 0.05$) (Figs. 8 and 9). This increased proliferation was not observed when M21-L cells (α_V^-) were coinjected with VN, indicating that this response was dependent on α_V integrins (data not shown). Yet, both α_V^+ cells (M21 and M21-L4) and α_V^- cells (M21-L12) show increased tumorigenicity when coinjected with collagen, which is likely due to equivalent levels of β_1 integrins (data not shown). These observations and the fact that VN can be localized within the parenchyma of human melanoma biopsy material (C.L. Gladson and D.A. Cheresh, unpubl.) suggest a role for VN and α_V integrins in the malignant phenotype of human melanoma.

It is widely accepted that tumor cell-matrix interactions facilitate the progression of cancer, and several studies support the role of tumor cell-surface integrins in this process (for review, see Cheresh 1992). The role α_V integrins play in tumor cell proliferation and invasion have only recently been explored. For example, RGD-containing peptides, known to block the function of $\alpha_V\beta_3$ integrin, have been shown to inhibit melanoma tumor invasion in vitro and experimental metastases in a murine melanoma model system (Humphries et al. 1986; Gehlson et al. 1988). Similarly, an antibody to the β_3 subunit has been reported to inhibit growth of human melanoma cells in nude mice (Boukerche et al. 1989). In vitro invasion assays have shown melanoma cell invasiveness increases in the presence of soluble VN or antibodies to $\alpha_V\beta_3$ (Seftor et al. 1992). Integrin $\alpha_V\beta_3$ is commonly found on melanoma cell lines and is

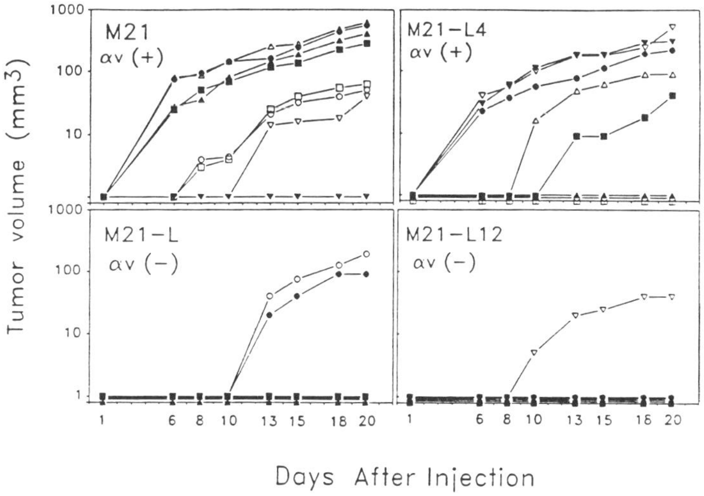

Figure 6. Tumorigenicity of M21 cell variants. M21 cell variants (1×10^6) were injected subcutaneously into the flanks of female nude mice. Tumors were allowed to proliferate for 20 days and the volume of each tumor was calculated ($w^2 \times 1/2$). M21 and M21-L4 (α_V^+) cells were significantly more tumorigenic than their α_V^- counterparts ($p < 0.01$ as determined by the Wilcoxon rank sum test). (Reprinted, with permission, from Felding-Habermann et al. 1992 [copyright held by American Society for Clinical Investigation].)

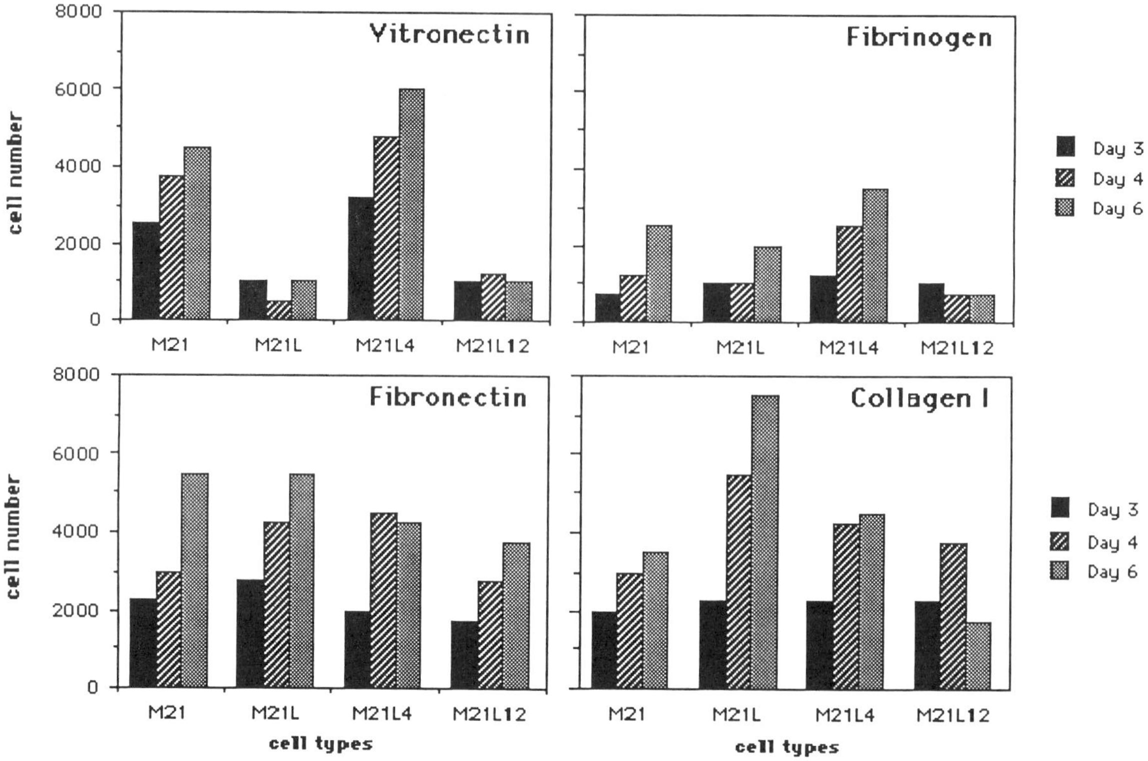

Figure 7. In vitro integrin-mediated anchorage-enhanced proliferation of human melanoma cells. M21 cell variants display different growth properties in response to immobilized adhesive ligands when grown in vitro, in the absence of serum and exogenous growth factors. M21 and M21-L4 (α_V^+) cells proliferate when grown in the presence of VN, whereas M21-L and M21-L12 (α_V^-) cells do not. Fibrinogen had a slight stimulative effect on the α_V^+ cells and no effect on the α_V^- cells. All four cell types showed proliferation in response to fibronectin and collagen, likely due to their identical β_1 integrin profile.

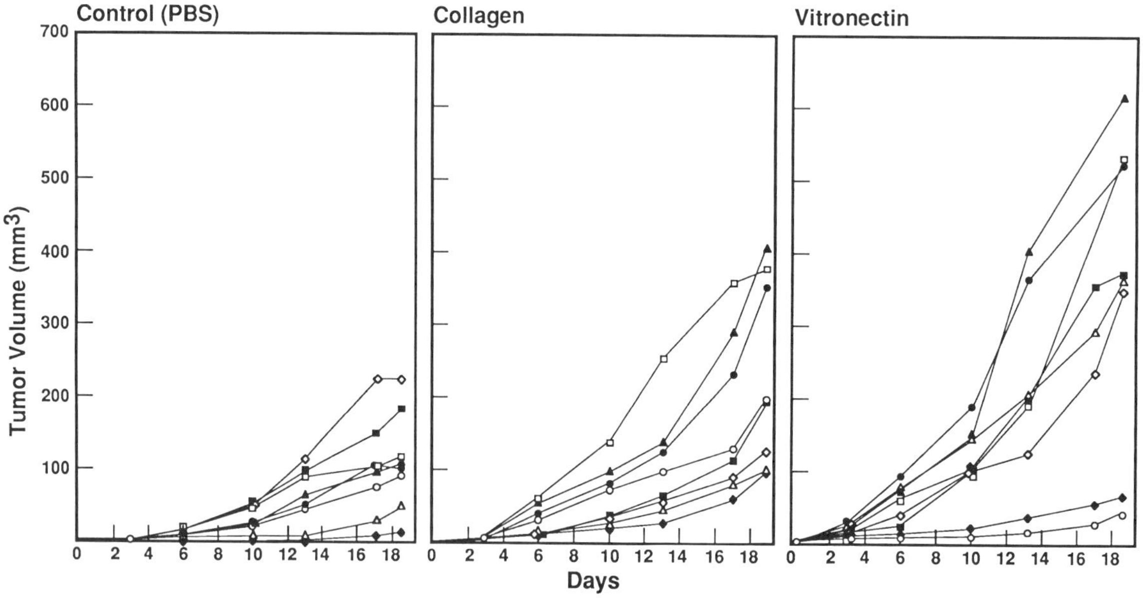

Figure 8. Enhancement of human melanoma tumorigenicity by extracellular matrix proteins. M21 cells (5×10^5) show enhanced tumorigenicity when subcutaneously coinjected with extracellular matrix molecules (500 μg/injection) into the flanks of nude mice. M21 (α_V^+) cells showed a three- to fourfold increase in tumor volumes when coinjected with VN and a twofold increase when coinjected with collagen, compared to cells injected with PBS alone.

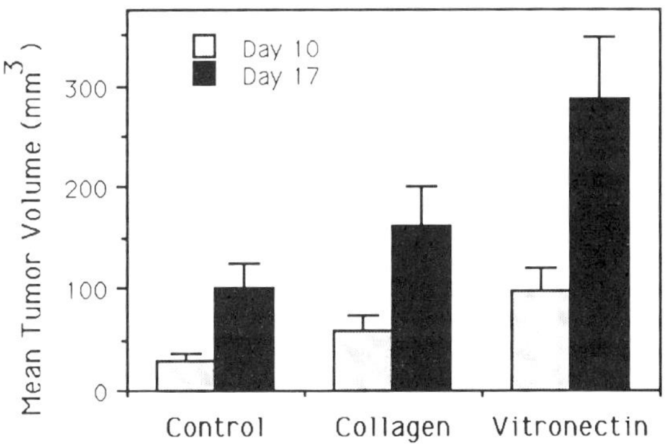

Figure 9. Extracellular matrix proteins enhance tumorigenicity of M21 cells. M21 tumor volumes were compared 10 and 17 days postinjection. M21 tumor volumes derived from cells coinjected with VN, when compared to tumor volumes obtained from cells injected with PBS alone, showed a significant difference at $p < 0.05$ (determined by the Wilcoxon rank sum test).

preferentially expressed on melanoma tissues, including vertically invasive primary and metastatic lesions, yet is absent on normal melanocytes, nevi, and horizontal primary melanoma lesions (Cheresh and Spiro 1987; Albelda et al. 1990). Metastatic melanoma, like many tumors, initiates at a primary site followed by cellular proliferation, and this can lead to invasion and metastasis. Our results indicate that specific growth conditions within the in vivo microenvironment favor cells that express α_V integrin(s). Consequently, it would appear that the expression of $\alpha_V\beta_3$ and/or $\alpha_V\beta_5$ is directly involved in the proliferative response of human melanoma in vivo. The mechanism by which this occurs is unclear; however, our preliminary results suggest the involvement of the extracellular adhesive ligand VN. Both α_V^+ cells (M21, M21-L4) and α_V^- cells (M21-L and M21-L12) grow at the same rate under typical tissue culture growth conditions — conditions that contain serum and growth factor. It is only when these cells are grown in non-serum conditions and in the presence of immobilized VN that they mimic the observed in vivo growth and thus demonstrate their anchorage dependence. These data strongly suggest that human melanoma cells have a distinct proliferative advantage in the presence of an adhesive ligand (i.e., VN) both in vivo and in vitro.

Nontransformed cells proliferate in an anchorage-dependent manner requiring adhesion to a matrix and stimulation by growth factors (End and Engel 1991). These extracellular signals stimulate an intracellular cascade of events leading to a proliferative response. A model is depicted in Figure 10 representing how these extracellular ligand-binding events may differ between normal and transformed cells. In this case, a normal cell requires signals from both growth factors and ECM molecules (anchorage-dependent growth); this synergy provides competency for normal cell division. However, a transformed cell is capable of responding to a signal potentiated by a growth factor (anchorage-independent growth) or an ECM molecule(s) (anchorage-dependent growth). This model may explain our observation that in the absence of serum and/or growth factors, binding of melanoma cells to VN through an α_V integrin stimulates proliferation. Consequently, VN acts as a mitogen for these cells. These observations provide evidence that the integrin-mediated cellular adhesion mechanism plays an important role in tumor cell proliferation.

ACKNOWLEDGMENTS

These studies were supported by National Institutes of Health grants CA-45726 and CA-50286. D.A.C. is a recipient of a Faculty Research Award from the American Cancer Society.

REFERENCES

Albelda, S.M., S.A. Mette, D.E. Elder, R. Stewart, L. Damjanovich, M. Herlyn, and C.A. Buck. 1990. Integrin distribution in malignant melanoma: Association of the β_3 subunit with tumor progression. *Cancer Res.* **50:** 6757.

Boukerche, H., O. Brethier-Vergnes, M. Bailly, J.F. Dore, L.K. Leung, and J.L. McGregor. 1989. A monoclonal antibody (LYP18) directed against the blood platelet glycoprotein IIb/IIIa complex inhibits human melanoma growth in vivo. *Blood* **74:** 909.

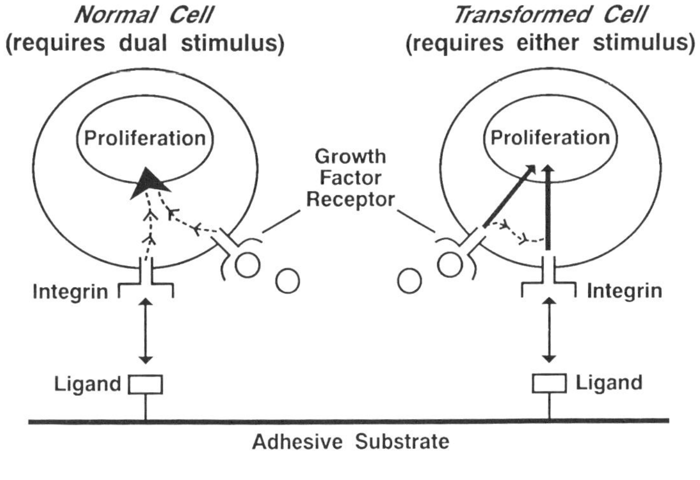

Figure 10. Cellular proliferation of normal versus transformed cells. A cellular proliferation model with emphasis on the extracellular signals required for cell proliferation. A normal cell requires stimulus from growth factors and ECM molecules, whereas a transformed cell requires stimulus from either.

Cheresh, D.A. 1992. Integrins: Structure, function and biological properties. *Adv. Mol. Cell Biol.* **7:** (in press).

Cheresh, D.A. and R.C. Spiro. 1987. Biosynthetic and functional properties of an RGD-directed receptor involved in human melanoma cell attachment to vitronectin, fibrinogen and von Willebrand factor. *J. Biol. Chem.* **262:** 17703.

Cheresh, D.A., J.W. Smith, H.M. Cooper, and V. Quaranta. 1989. A novel vitronectin receptor integrin ($\alpha v\beta x$) is responsible for distinct adhesive properties of carcinoma cells. *Cell* **57:** 59.

Dahlback, K., H. Lofberg, and B. Dahlback. 1989a. Vitronectin (S-protein of complement) colocalizes with Ig deposits and C9 neoantigen in discoid lupus erythematosus and dermatitis herpetiformis but not in bullous pemphigoid. *Br. J. Dermatol.* **120:** 725.

Dahlback, K., H. Lofberg, J. Alumets, and B. Dahlback. 1989b. Immunohistochemical demonstration of age-related deposition of vitronectin (S-protein of complement) and terminal complement complex on dermal elastic fibers. *J. Invest. Dermatol.* **92:** 727.

End, P. and J. Engel. 1991. Multidomain proteins of the matrix and cellular growth. In *Receptors for extracellular matrix* (ed. J.A. McDonald and R.P. Mecham), p. 79. Academic Press, New York.

Felding-Habermann, B., B.M. Mueller, C.A. Romerdahl, and D.A. Cheresh. 1992. Involvement of integrin αv gene expression in human melanoma tumorigenicity. *J. Clin. Invest.* **89:** 2018.

Gehlson, K.R., W.S. Argraves, M.D. Pierschbacher, and E. Ruoslahti. 1988. Inhibition of the in vitro tumor invasion by arg-gly-asp-containing peptides. *J. Cell Biol.* **106:** 25.

Giancotti, F.G. and E. Ruoslahti. 1990. Elevated levels of the alpha 5 beta 1 fibronectin receptor suppress the transformed phenotype of chinese hamster ovary cells. *Cell* **56:** 281.

Gladson, C.L. and D.A. Cheresh. 1991. Glioblastoma expression of vitronectin and the $\alpha v\beta 3$ integrin. *J. Clin. Invest.* **88:** 1924.

Hay, E.D. 1981. Collagen and embryonic development. In *Cell biology of extracellular matrix* (ed. E.D. Hay), p. 379. Plenum Press, New York.

Hayman, E.G., M.D. Pierschbacher, Y. Ohgren, and E. Ruoslahti. 1983. Serum spreading factor (vitronectin) is present at the cell surface and in tissues. *Proc. Natl. Acad. Sci.* **80:** 4003.

Humphries, M.J., K. Olden, and K.M. Yamada. 1986. A synthetic peptide from fibronectin inhibits experimental metastasis of murine melanoma cells. *Science* **233:** 467.

Hynes, R.O. 1981. Fibronectin and its relation to cellular structure and behavior. In *Cell biology of extracellular matrix* (ed. E.D. Hay), p. 295. Plenum Press, New York.

———.1992. Integrins: Versatility, modulation and signaling in cell adhesion. *Cell* **69:** 11.

Jaconi, M.E.E., J.M. Theler, W. Schlegel, R.D. Appel, S.D. Wright, and P.D. Lew. 1991. Multiple elevations of cytosolic-free Ca^{2+} in human neutrophils: Initiation by adherence receptors of the integrin family. *J. Cell Biol.* **112:** 1249.

Kornberg, L.J., H.S. Earp, C.E. Turner, C. Prockop, and R.L. Juliano. 1991. Signal transduction by integrins: Increased protein tyrosine phosphorylation caused by clustering of $\beta 1$ integrins. *Proc. Natl. Acad. Sci.* **88:** 8392.

Leavesley, D.I., G.D. Ferguson, E.A. Wayner, and D.A. Cheresh. 1992. Requirement of the integrin $\beta 3$ subunit for carcinoma cell spreading or migration on vitronectin and fibrinogen. *J. Cell Biol.* **117:** 1101.

McLean, J.W., D.J. Vestal, D.A. Cheresh, and S.C. Bodary. 1990. cDNA sequence of the human integrin $\beta 5$ subunit. *J. Biol. Chem.* **265:** 17126.

Nip, J., H. Shibata, D.J. Luskutoff, D.A. Cheresh, and P. Brodt. 1992. Human melanoma cells derived from lymphatic metastases use integrin $\alpha_V\beta_3$ to adhere to lymph node vitronectin. *J. Clin. Invest.* **90:** 1406.

Plantefaber, L.C. and R.O. Hynes. 1989. Changes in integrin receptors on oncogenetically transformed cells. *Cell* **56:** 281.

Preissner, K.T. 1989. The role of vitronectin as multifunctional regulator in the hemostatic and immune systems. *Blut* **59:** 419.

Ramaswamy, H. and M.E. Hemler. 1990. Cloning, primary structure and properties of a novel human integrin β subunit. *EMBO J.* **9:** 1561.

Reisfeld, R.A. and D.A. Cheresh. 1987. Human tumor antigens. *Adv. Immunol.* **40:** 323.

Ruoslahti, E. and F.G. Giancotti. 1989. Integrins and tumor cell dissemination. *Cancer Cells* **1:** 119.

Schwartz, M.A., C. Lechene, and D.E. Ingber. 1991a. Insoluble fibronectin activates the Na/H antiporter by clustering and immobilizing integrin $\alpha 5\beta 1$, independent of cell shape. *Proc. Natl. Acad. Sci.* **88:** 7849.

Schwartz, M.A., D.E. Ingber, M. Lawrence, T.A. Springer, and C. Lechene. 1991b. Multiple integrins share the ability to induce elevation of intracellular pH. *Exp. Cell Res.* **195:** 533.

Seftor, R.E.B., E.A. Seftor, K.R. Gehlsen, W.G. Stetler-Stevenson, P.D. Brown, E. Ruoslahti, and M.J.C. Hendrix. 1992. Role of the $\alpha v\beta 3$ integrin in human melanoma cell invasion. *Proc. Natl. Acad. Sci.* **89:** 1557.

Smith, J.W., D.J. Vestal, S.V. Irwin, T.A. Burke, and D.A. Cheresh. 1990. Purification and functional characterization of integrin $\alpha v\beta 5$: An adhesion receptor for vitronectin. *J. Biol. Chem.* **265:** 11088.

Tomasini, B.R. and D.F. Mosher. 1991. Vitronectin. *Prog. Hemostasis Thromb.* **10:** 269.

Wayner, E.A., R.A. Orlando, and D.A. Cheresh. 1991. Integrins $\alpha v\beta 3$ and $\alpha v\beta 5$ contribute to cell attachment to vitronectin but differentially distribute on the cell surface. *J. Cell Biol.* **113:** 919.

Werb, Z., P.M. Tremble, O. Behrendtsen, E. Crowley, and C.H. Damsky. 1989. Signal transduction through the fibronectin receptor induces collagenase and stromelysin gene expression. *J. Cell Biol.* **109:** 877.

Genetic Interactions with Integrins during Wing Morphogenesis in *Drosophila*

L.H.V. Wessendorf,* M. Wehrli, A. DiAntonio, and M. Wilcox†
MRC Laboratory of Molecular Biology, Cambridge, CB2 2QH United Kingdom

Cell adhesion is of critical importance during the development of a complex, multicellular organism. As cells divide and differentiate, they must also contact and selectively adhere to one another as they gradually create the shape and structures of the adult form. Additionally, such shaping is influenced by contacts with the space around the cells, the extracellular matrix, whereas the shape of the individual cells themselves is altered and maintained by the cytoskeleton. Integrins, as receptors for extracellular matrix components as well as cell-surface molecules, thus play an important role in these processes. Integrins are $\alpha\beta$ heterodimers that serve to connect components of the extracellular matrix (or certain cell-surface molecules) with the cytoskeleton on the inside of the cell via a transmembrane connection (for a recent review of integrins, see Hynes 1992). In *Drosophila*, the position-specific (PS) integrins are composed of two α chains, PS1α and PS2α, and a single β chain, PSβ (Wilcox et al. 1984; Bogaert et al. 1987; Leptin et al.1987, 1989). They were first isolated on the basis of their position-specific expression in the larval imaginal discs (Wilcox et al. 1981; Brower et al. 1984, 1985; Wilcox and Leptin 1985) and were subsequently shown to be expressed throughout development (Bogaert et al. 1987; Leptin et al. 1989; Zusman et al. 1990). Their effects range from their earliest involvement in germ band extension to later roles in eye and wing development. Whereas both extracellular ligands and intracellular contacts with cytoskeletal components have been well characterized for vertebrate integrins (for a current review, see Hynes 1992), little is known about integrin contacts on either side of the cell surface in *Drosophila*. Fibronectin, perhaps considered to be the "classical" integrin ligand, has not been found in flies, and although in vitro studies have shown that *Drosophila* PS integrins are capable of adhesive functions in assays with vertebrate vitronectin (Hirano et al. 1991), a fly vitronectin homolog has not been identified.

The existence of mutations in the PS integrin genes that result in viable as well as lethal phenotypes provides an opportunity for using a genetic approach to understand integrin function further and to gain knowledge about the specific molecules involved in integrin-dependent developmental processes in *Drosophila*. Viable mutations in both the PS2α and PSβ genes cause defects in wing development, including abnormalities in the wing surface itself and the angle at which the wing is held (Costello and Thomas 1981; De la Pompa et al. 1989; Wilcox et al. 1989; Wilcox 1990). We are investigating PS integrin function in *Drosophila* through a genetic analysis of their role in wing morphogenesis. In this way, we hope to gain insight into general interactions of integrins with other molecules, complementing and expanding the understanding of integrin function gained from vertebrate studies, and to understand better the functioning of cell adhesion in wing morphogenesis in the fly.

PS Integrin Structure and Function

The PS integrin family is comprosed of two α chains, PS1α and PS2α, and a β chain, PSβ. Both α chains form heterodimeric complexes with the single β chain (Wilcox et al. 1984; Leptin et al. 1987, 1989). Although the level of sequence and structural homology of all three subunits with vertebrate integrins is high (Bogaert et al. 1987; MacKrell et al. 1988; this paper), the two *Drosophila* α chains, PS1α and PS2α, share only as much sequence similarity with each other as with vertebrate α chains (see below), indicating that evolutionarily, the two fly α chains diverged quite early, before the separation of the arthropod and chordata lineages.

In addition to the structural diversity provided by the existence of the two α chains, further modulation of the PS2α chain structure is provided by alternative splicing of the transcript, which results in the addition of a 75-bp exon (Brown et al. 1989).

Cloning of the PS1α Gene

The *Drosophila* PS1α gene has now been cloned using oligodeoxynucleotide probes based on information from protein sequence analysis of immunopurified PS1α antigen (M. Wehrli, unpubl.). The PS1α cDNA of 4990 bp encodes an integrin α chain of 1146 amino acids. The protein sequence is more closely related to human integrin α chains α_3 (Takada et al. 1991) (32% identity) and α_6 (Tamura et al. 1990) (31.5% identity) of the β1 family than to the *Drosophila* PS2α integrin (21.7% identity). PS1α shares a conserved cysteine pattern with other integrin α chains; in addition, it has

Present address: *Department of Molecular and Cellular Biology, University of California, Berkeley, California 94720.
†Michael Wilcox passed away on 21 February, 1992. This paper is dedicated to his memory.

seven amino-terminal repeats similar to those found in other integrin α chains. The last of these repeats in PS1α encode three metal-binding domains, whereas most other integrin α chains have four such domains. There is further resemblance to other α chains in the transmembrane domain, and the cytoplasmic sequence GFFKR is conserved with all vertebrate integrin α chains. There are two potential cleavage sites in the region amino-terminal to the transmembrane domain that would allow the generation of a disulfide-linked heavy/light chain dimer, which has been described for isolated PS1α antigen (Wilcox et al. 1984; Leptin et al. 1987).

The PS1α gene contains eight exons dispersed over at least 37 kb. The overall structure of the gene is similar to that of the PS2α gene (Brown et al. 1989), with the bulk of the coding region being concentrated in large exons in the 3′ region. Thus, both structures are distinct from the vertebrate integrin genes α_X (Corbi et al. 1990) and α_{IIb} (Heidenreich et al. 1990), where the exons are fairly uniform in size and are not concentrated in the 3′ region. Analysis of intron positions in PS1α-coding regions shows that some coincide in PS1α and in PS2α. Furthermore, for every intron found in PS1α or PS2α, an intron can be found in the corresponding position in α_X or α_{IIb}, most often in both (but not the other way round). This conservation of intron positions between the fly and vertebrate genes suggests that integrin α chains originated from a common ancestral gene that was more closely related to the α_X and α_{IIb} gene structures than to PS1α or PS2α. The fly gene organization seen today could be the result of the elimination of particular introns from such an ancestral gene. Further comparison of the intron positions in PS1α, PS2α, α_X, and α_{IIb} with that of proposed protein domains fails to show any correlation of protein domains with exons, inconsistent with the model of exon shuffling proposed by Gilbert (1985).

In situ, the PS1α transcript is detected in embryonic development in the ectoderm and endoderm but not in the mesoderm. The transcript is first detected at the cellular blastoderm stage when more transcript is detected dorsally than ventrolaterally (Fig. 1). At gastrulation, the transcript shows a homogeneous distribution along the anteroposterior axis, and as development proceeds, accumulation of transcript is detected in a segmentally repeated manner. As morphological features become visible, the areas of high expression can be identified as the segment boundaries, where the presumptive tendon cells are found. PS1α antigen distribution (Leptin et al. 1989) follows the transcription pattern, with the delay expected for translation and processing. Thus, expression of the PS1α gene appears to be largely under transcriptional control.

Role of PS Integrins in Development

Functionally, the PS integrins are involved in a wide variety of developmental processes, ranging from their earliest known effects on germ band retraction to their progressively later roles in muscle attachment and development of the adult wings and eyes. Such diversity of function is reflected by their dynamic pattern of expression throughout development. Although all PS integrin subunits are transcribed from very early stages in development (Bogaert et al. 1987; Zusman et al. 1990; and above), protein expression on the cell surface is not seen until after the onset of gastrulation (Leptin et al. 1989; Wilcox 1990). This delay may be due to the fact that both α and β chains must be present for expression on the surface (i.e., there is no surface expression of an α chain without a β chain, and vice versa). Indeed, although the β chain is translated from maternal transcripts, translation of α chains may not occur until much later (Leptin et al. 1989). The PS1 complex appears earliest on the cell surface, during the germ band extension stage of gastrulation. At this point, the complex is expressed on the basolateral surface of the ectoderm, where it contacts the underlying mesoderm (Leptin et al. 1989). Null mutations in the zygotic PSβ gene, *lethal(1)myospheroid* (*mys*), have no effect on germ band retraction, but the absence of both maternal and zygotic transcripts often causes defects in this process (Wieschaus and Noell 1986; Leptin et al. 1989). Whereas PS1 complex expression is restricted to the ectoderm and endoderm of the embryo (Leptin et al. 1989; Wilcox 1990), PS2 complex expression occurs slightly later during germ band extension and is limited to the embryonic mesoderm (Bogaert et al. 1987). Initial PS2 protein expression on myoblasts becomes gradually more and more restricted to the basal surfaces of muscles as they develop, at points of attachment to the ectoderm (Bogaert et al. 1987). In the ectoderm, meanwhile, PS1 antigen becomes limited to tendon cells, the corresponding muscle attachment sites in this cell layer (Leptin et al. 1989). Thus, PS1 and PS2 appear to be directly juxtaposed on either side of the muscle attachments, and, as might be expected, null *mys* mutants have drastic defects in muscle attachment, where upon initial contraction, the muscles detach and round up (Leptin et al. 1989).

Later in development, the PS integrin complexes are expressed in a variety of tissues (Wilcox et al. 1981; Brower et al. 1984, 1985; Wilcox and Leptin 1985). They are again expressed in a complementary pattern in the epithelia of the wing imaginal disc. PS1 is expressed on the presumptive dorsal surface of the wing, and PS2 is expressed on the presumptive ventral surface of the wing (Wilcox et al. 1981; Brower et al. 1985; Wilcox 1990). Because these two surfaces must come together in an adhesive manner during wing morphogenesis (Waddington 1940; García-Bellido 1977), it seemed likely that the integrins would be involved in this process. Indeed, when mutations in the PS2α (*inflated*, *if*) and PSβ (*mys*) genes were examined, they showed defects in the apposition of the two wing surfaces, resulting in blistering and vein defects in the adult wing (Brower and Jaffe 1989; Wilcox et al. 1989; Zusman et al. 1990). Viable PSβ mutants have additional phenotypes, including jump muscle defects, and

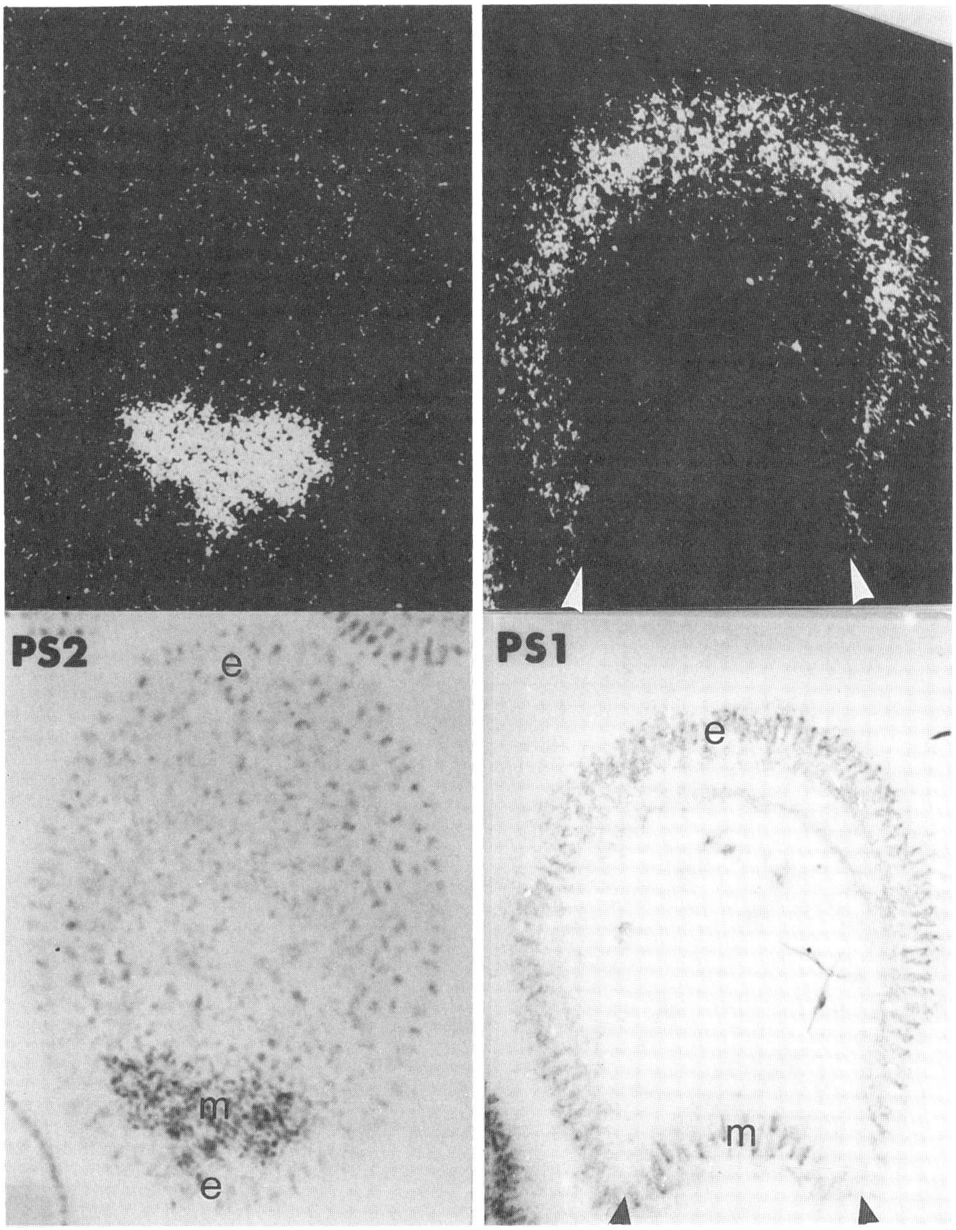

Figure 1. PS1α and PS2α transcription at the gastrulation stage. Cross sections of embryos at the gastrulation stage (stage 6 on the right, stage 7 on the left) hybridized with PS1α (*right*)- and PS2α (*left*)-specific probes, respectively. (*Top*) The two photographs show the detected transcript by dark-field microscopy, which visualizes bright silver grains generated by the radioactive probe. (*Bottom*) The same sections are shown in bright field. PS1α is expressed in the ectoderm (e) and not detected in the presumptive mesoderm (m). Arrowheads indicate the border between these two germ layers. PS1α is notably stronger expressed dorsally (which is up) than ventrolaterally. PS2α, in a somewhat later embryo, is detected in the invaginated mesoderm.

the wings of these flies are held at an abnormal angle to the body, presumably due to an effect on the indirect flight muscles (Costello and Thomas 1981; De la Pompa et al. 1989; Wilcox et al. 1989; Wilcox 1990).

A Genetic Approach to Understanding Integrin Function

Genetically, the severity and extent of integrin mutant wing phenotypes can be manipulated by changing the mutant background of the flies. This case is most clearly demonstrated by examining the relationship between the two alleles if^3, a mutation in the PS2α gene, and mys^{nj}, a mutation in the gene encoding PSβ. On their own, each mutant exhibits a different but mild wing defect, whereas put in the doubly homozygous situation, $if^3 mys^{nj}$, a drastic effect is seen. Now, both the frequency *and* severity of defects in the wing blade are dramatically increased, causing not only blistering, but also severe shrinking and crumpling of the wing blade (Wilcox et al. 1990). Such an interaction at the genetic level might be expected, since at the molecular level, PS2α and PSβ form a heterodimeric complex, and although a defect in one of the components may cause only a slight alteration of function, by mutating both members of the complex, complete failure may result. On the basis of the observation that an interaction at the molecular level (the $\alpha\beta$ heterodimeric complex) can be reflected at the genetic level (by the synergistic enhancement of phenotype seen in *if*^{3} *mys*^{nj}), we decided to use a genetic approach to search for other genes that might interact at the molecular level with the PS integrins.

An initial requirement for this study was a strategy that would allow us to distinguish between dominant and recessive genetic interactions. Dominant interactions occur when one copy of a recessive allele causes an effect, either enhancement or suppression, on the phenotype of another gene, whereas recessive interactions require two copies of the recessive allele. In the latter case, it can often be difficult to distinguish between an additive effect and an enhancement effect, since now two phenotypes are present; in the case of dominant interactions, the effect is clear. Additionally, we hoped that dominant synergistic interactions would indicate that the interacting gene would be involved at some level in integrin functioning.

As a model genetic system, we chose to analyze the role of integrins in the development of the adult fly wing (Wilcox 1990). The adult wing is an exquisitely patterned structure that offers many advantages for such an analysis. The morphogenetic processes that occur during the development of the wing from its undifferentiated, imaginal state to its adult form have been the subject of an intensive study pioneered by C.H. Waddington (1939, 1940). The wing is a large and detailed structure, exhibiting a distinct pattern of veins, so that even subtle developmental defects can be readily detected. In addition, since the wing is an expendable structure in the laboratory, the recovery of viable wing mutants should not be affected by any requirements of the wing for viability.

As previously mentioned, the integrins are expressed in the wing imaginal disc, on distinct epithelial regions that will eventually become apposed to give rise to the upper and lower surfaces of the wing in the adult. The events leading to the wing assuming its final, blade-like shape require an initial evagination and folding, followed by a series of attachments and detachments, during which the presumptive dorsal and ventral surfaces of the wing come together and pull apart, remaining in contact at basal attachment desmosomes via microtubule-containing structures known as transalar arrays (Milner and Muir 1987; Mogensen and Tucker 1988). During this time, integrins are associated with the basal junctions throughout their development (D. Fristrom, pers. comm.). Not surprisingly, then, it is in the apposition of the two wing surfaces that the *if*^{3} mutation in the gene encoding PS2α causes defects which eventually lead to the presence of blistering in 10–40% of adult wings (Wilcox et al. 1989). The *if*^{3} mutation most likely causes a regulatory defect that affects only the wing (Wilcox et al. 1989). Because *if*^{3} is both stubbornly recessive and relatively mild in its localized phenotype, we chose this allele for use in a genetic approach to identify other genes that might interact with integrins.

The genetic strategy was to test other known mutants for their ability to either enhance or suppress the phenotype of *if*^{3}. A similar genetic approach has been taken by Diaz-Benjumea and García-Bellido (1990), although not directed specifically toward integrin function. Their study has focused on characterizing interactions in double-mutant combinations of wing mutations from the different phenotypic classes (excess vein, loss of vein, etc.) and has revealed a high degree of synergistic interaction, although mostly in the doubly homozygous combination. However, because we were only interested in identifying interactions with integrins, and because of the difficulty of distinguishing between additive effects and enhancement in the case of recessive interactions, our study focused on the identification of dominant effects. In this way, we hoped to identify other components of the *Drosophila* integrin system, or modulators of the system.

From 35 second-site mutations analyzed, selected initially on the basis of phenotypes in the wing but supplemented with mutants displaying other nonwing phenotypes, we have now identified three known genes that act as dominant, synergistic effectors of the *if*^{3} phenotype (see Table 1), as well as mutations in two as yet uncharacterized genetic loci that also have a dominant effect. Recessive interactions (requiring the doubly homozygous situation) were observed with five of the mutations tested, whereas the remaining alleles exhibited no effect whatsoever. Of the known genes that we found to interact in a dominant fashion with *if*^{3}, *blistered* (*bs*) acts as a dominant enhancer, dramatically increasing the frequency of blistering usually observed with *if*^{3}, and two other genes, *slit* (*sli*) and *plexus* (*px*), act as dominant suppressors, reducing blistering nearly entirely (see Table 1). In the case of *px*, the effect is allele-specific; only one allele, *px*^{1}, acts as a dominant suppressor, and the remaining *px* alleles all act recessively to enhance the *if*^{3} phenotype.

Interacting Genes

***blistered* (bs).** *blistered*, which acts as a dominant enhancer of *if*^{3}, also shows a genetic effect on mutant alleles of *mys*, the gene encoding PSβ (M. Wilcox, unpubl.); however, this is not as strong as the interaction between *bs* and *if*^{3}. In a homozygous *mys*^{nj} background (held-out phenotype only), one copy of *bs* has no effect, but in the background of *mys*^{nj} over *mys*^{G}, a lethal allele, *bs* acts as a dominant enhancer, increasing the severity of the held-out effect dramatically, and to a lesser extent, enhancing wing vein defects. The interaction with *mys* is not as strong or as definitive as the interaction with *if*^{3}, reflecting the difference in the phenotypes being analyzed. Although

Table 1. Dominant Interactions between *if*^{3} (PS2α) and Other Mutants

	Wing defects (males)				Wing defects (females)			
	2W	1W	0W	%	2W	1W	0W	5
if^{3}	62	74	102	41	14	46	171	16
if^{3};*bs*^{2}/+	64	15	4	86	52	37	19	65
if^{3};*sli*^{IG107}/+	1	3	96	2.5	0	2	68	1.5
if^{3};*px*^{1}/+	3	8	142	5	3	5	160	3

The number of flies with blisters on two wings (2W), one wing (1W), or neither wing (0W) is shown for males and females. *if*^{3} is on the first chromosome, and *bs*, *sli*, and *px* are all on the second chromosome. Percentages shown were calculated based on the percentage of total wings showing blisters, rather than percentage of flies with blistered wings, to take into account differences in severity reflected by one-wing or two-wing blistering per fly.

the blistering frequency seen with *if*^{3} is simpler to quantitate, effects on the held-out phenotype seen with *mys*^{nj} can be only analyzed qualitatively. Since the *mys*^{nj} allele usually does not itself show effects in the wing blade, one might not expect to see genetic interactions in the wing blade.

Recently, *bs* has been implicated in genetic interactions with members of two other classes of mutant, those of the *Stubble-stubbloid* (*Sb-sbd*) locus and of the *Broad-Complex* (*BR-C*), as well as another, unrelated locus, *Enhancer of broad* (*E[br]*) (Gotwals and Fristrom 1991).
The *BR-C* has been shown to encode a family of transcription-factor-like zinc finger proteins (DiBello et al. 1991). Thus, *bs* exhibits genetic interactions with genes encoding vastly different types of molecules: the integrins, on one hand, and transcription factors, on the other. This indicates that its genetic interactions with the different classes most probably reflect different types of interactions or relationships at the molecular level. Although it is still unclear what kind of molecule *bs* encodes, such information will clarify not only whether it has a role in regulating integrin function, but also the molecular basis for its genetic interactions with these diverse classes of genes.

***slit* (sli).** The *sli* gene acts as a dominant suppressor of *if*^{3}. *sli* is known to encode a secreted molecule that contains both leucine-rich repeat and epidermal growth factor (EGF) repeat motifs in its extracellular domain, both of which are found in proteins that participate in cell-cell interactions and adhesive processes, suggesting a possible role for *sli* in such events (Rothberg et al. 1988, 1990). The *sli* protein is expressed predominantly in the midline neuroepithelium in the embryo, and null mutations in *sli* lead to defects in the embryonic central nervous system (CNS) (Rothberg et al. 1988, 1990). Interestingly, *sli* protein is also found near muscle attachment sites in the embryo (Rothberg et al. 1990), where PS integrins are also found. No expression or role of *sli* in the wing has been documented to date. Both the localization of *sli* at muscle attachment sites and the fact that *sli* encodes a protein with motifs suggestive of a role in cell interactions and/or adhesion may nevertheless be indicators that the observed genetic interaction between integrins and *sli* in the wing could be a result of an interaction between the two proteins at the molecular level.

***plexus* (px).** The fact that *px* exhibits an allele-specific interaction with *if*^{3}, where *px*^{1} acts as a dominant suppressor and other *px* alleles act as weak recessive enhancers, suggested that its gene product might interact closely at the molecular level with integrins, since different mutations could have such dramatically different effects at the genetic level. Additionally, *px* was observed to act as a dominant effector of other genes involved in cell-cell interaction and adhesion. At the genetic level, *px* acts as a dominant partial suppressor of *Ax* alleles of *Notch* (see Fig. 2), which are a result

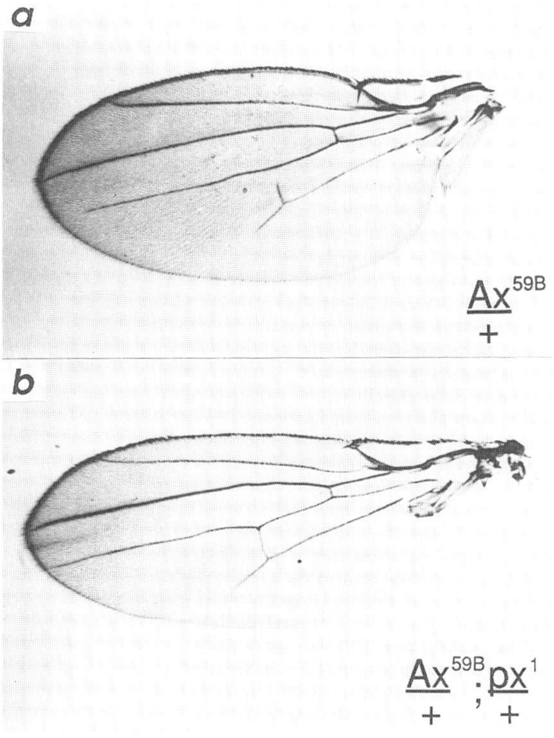

Figure 2. Dominant interaction between *Ax*^{598} and *px*^{1}. (*a*) Wings from *Ax*^{598}/ + flies; (*b*) wings from *Ax*^{598}/ + ; *px*^{1}/ + flies. Note the partial suppression by *px*^{1}/ + , particularly in vein 4 and the posterior crossvein. Similar results were obtained with other *Ax* and *px* alleles tested.

of point mutations in the EGF repeats of the *Notch* protein (Kelley et al. 1987). Conversely, *px* acts as a dominant enhancer of *Delta*, which also encodes an EGF repeat protein (Vassin et al. 1987; Kopczynski et al. 1988) and is a ligand for *Notch* (Fig. 3) (Rebay et al. 1991). Additional genetic interactions between *px* and other EGF-repeat-containing genes from *Drosophila* have also been observed, often of an allele-specific nature (L. Wessendorf, unpubl.). Recessive interactions have also been reported between *px* and *Notch* and *px* and *Delta* (Diaz-Benjumea and García-Bellido 1990). Interestingly, the putative gene encoding *px* has recently been cloned, and its gene product shows very weak sequence homology with the EGF-repeat-containing proteins, in particular to the neurogenic gene *Notch* (L. Wessendorf, unpubl.).

Although in all of these cases, *px* acts as a dominant effector of other phenotypes, another gene, *veinlet* (*ve*), acts as a dominant suppressor of the *px* phenotype itself (see Fig. 4). Interestingly, *ve* also acts as a reces-

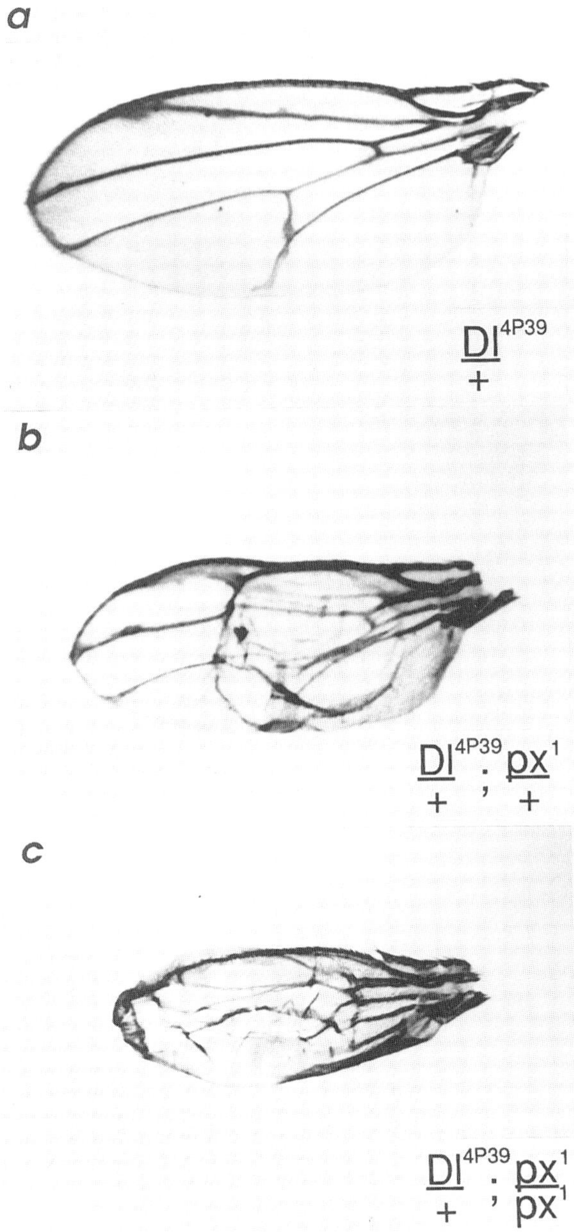

Figure 3. Interactions between Dl^{4P39} and px^1. (*a*) Wings from $Dl^{4P39}/+$ flies; (*b*) wings from $px^1/+$; $Dl^{4P39}/+$ flies (note the enhancement, indicated by the presence of large centrally located blisters, and resulting shrinkage of wing blade size); (*c*) wings from px^1; $Dl^{4P39}/+$ flies. Even stronger enhancement is seen in the homozygous *px* background. Similar interactions were observed with other *px* and *Dl* alleles tested.

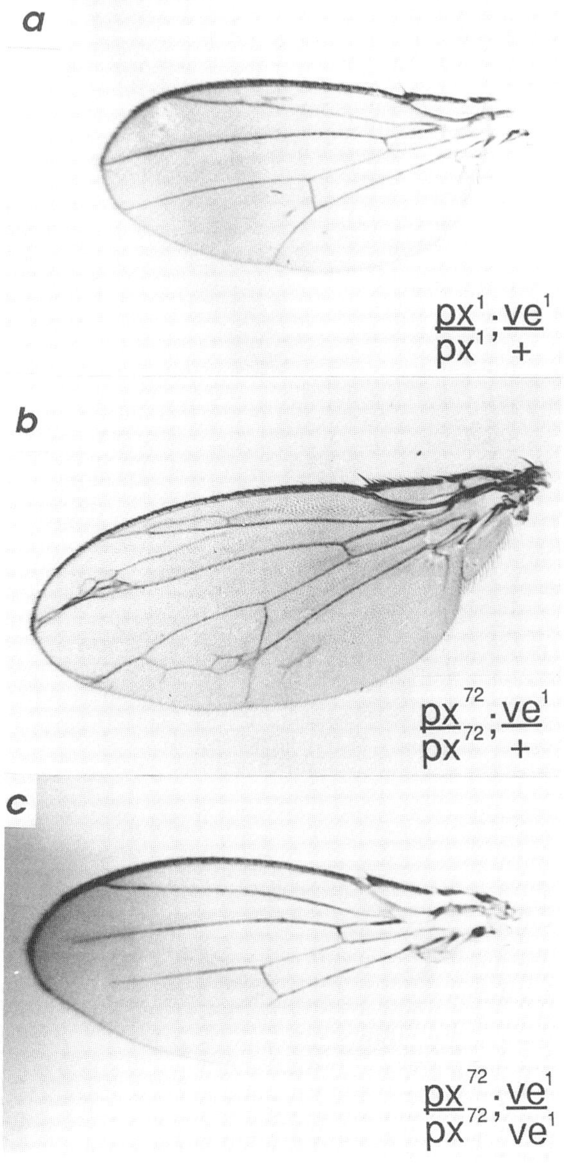

Figure 4. Allele-specific interactions between *px* and *ve*. (*a*) Wings from $px^1;ve^1/+$ flies, showing restoration of near wild-type vein pattern; (*b*) wings from $px^{72};ve^1/+$ flies (in this case, *px* phenotype is not suppressed by $ve^1/+$); (*c*) wings from $px^{72};ve^1$ flies; here, the *px* phenotype is suppressed, and vein truncations are due to the recessive ve^1 phenotype. A similar recessive interaction was noted by Diaz-Benjumea and García-Bellido (1990).

sive suppressor of the *if*^{3} phenotype (Wilcox 1990). Genetically, *ve* behaves as an allele of *rhomboid* (Diaz-Benjumea and García-Bellido 1990; Wilcox 1990).

The *rho* gene has been shown to encode a transmembrane protein, potentially having a role in cell interactions (Bier et al. 1990). Mutations in *rho* cause defects in both the peripheral and central nervous systems (Mayer and Nüsslein-Volhard 1988; Bier et al. 1990); on this basis, *rho* has been classified as a member of the spitz group of genes, characterized by their similar mutant phenotypes in the CNS (Mayer and Nüsslein-Volhard 1988). Alleles of another member of the spitz class, *Star* (*S*), also have a dominant suppressive effect on *px* (Lindsley and Grell 1968). Interestingly, these genes, which interact in the same manner with *px* in the wing, have been shown to be related in a different developmental process, that of CNS formation.

The roles of *Notch* and *Delta* in the very early development of the CNS, when neural versus epidermal cell fate decisions are being established, are also well documented (Poulson 1937; Lehman et al. 1983), and many *Notch* and *Delta* alleles cause CNS defects. Thus, it appears that many of the genes that interact genetically with *px* have earlier roles in the development of the CNS. Because it is known that these genes, at least in the cases of *Notch* and *Delta*, are also involved in the development of the wing (by their mutant wing phenotypes), it seems that the wing may serve as a model genetic system in which to identify interactions between genes that may function in similar developmental processes at other times during development.

These results have identified a network of genes that affect the *if*^{3} PS2α integrin mutant phenotype (see Fig. 5). It seems clear from the broad spectrum of molecules that can be observed to interact with the system in some way (from the *BR-C* of transcription factors to cell-surface molecules like Notch and Delta to secreted factors such as slit) that interactions at the genetic level can result from diverse types of interactions at the molecular level. Such a variety of molecular interactions might include specific effects at the level of gene expression or protein complex formation but may also result from direct although nonspecific effects on processes such as translation or protein processing. Additionally, it is possible that the products of these genes are not directly involved in a molecular interaction but that they are part of several parallel, unlinked components of the same process. Alternatively, the possibility exists that these genes are not related at all in a molecular sense and that the observed genetic interactions result from a "catastrophic addition" effect, where the process of wing morphogenesis is disabled in so many different, unrelated ways that a misleading phenotype is generated. However, the similar localization of *slit* and integrins, the possibility that *px* may contain a structural motif shared with other genes identified in this study, and the fact that many of the genes identified here have effects in another common aspect of development, that of the CNS, are all promising signs that such a genetic approach may yield clues about events at the molecular level.

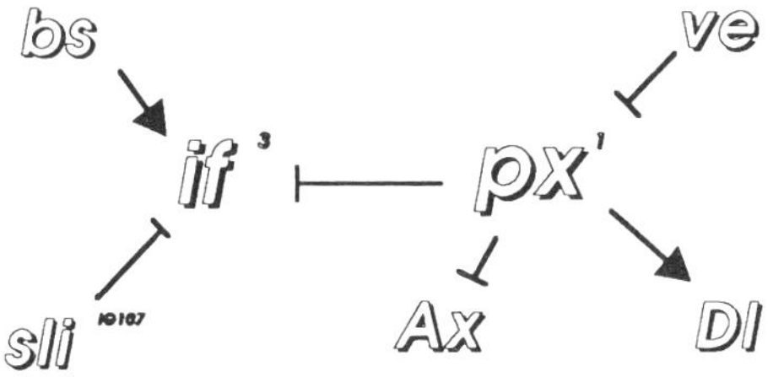

Figure 5. Dominant effectors influencing *if*^{3} and *px*^{1} phenotypes in the wing. Arrows indicate dominant enhancement effects, and bars indicate dominant suppression effects. For clarity and simplicity, only dominant interactions between alleles are shown. See text for discussion.

ACKNOWLEDGMENTS

This work was inspired and led by Michael Wilcox, and we will sadly miss his valuable expertise. Richard Smith and Julie Horsfield provided excellent technical assistance. We thank Peter Lawrence, Karen Heckingbottom, Jan Semenza, and Marcus Yaffee for helpful discussion of the manuscript. L.W. is supported by a Howard Hughes Medical Institute predoctoral fellowship, M.W. is supported by grants from the Schweizerizche Nationalfonds, Julius-Klaus-Stiftung, and Roche Research Foundation and A.D. was supported by a Herchel Smith Harvard Fellowship.

REFERENCES

Bier, E., L.Y. Jan, and Y.N. Jan. 1990. *rhomboid*, a gene required for dorsoventral axis establishment and peripheral nervous system development in *Drosophila melanogaster*. *Genes Dev.* **4:** 190.

Bogaert, T., N. Brown, and M. Wilcox. 1987. The *Drosophila* PS2 antigen is an invertebrate integrin that, like the fibronectin receptor, becomes localized to muscle attachments. *Cell* **51:** 929.

Brower, D.L. and S.M. Jaffe. 1989. Requirement for integrins during *Drosophila* wing development. *Nature* **342:** 285.

Brower, D.L., M. Piovant, and L.A. Reger. 1985. Developmental analysis of *Drosophila* position-specific antigens. *Dev. Biol.* **108:** 120.

Brower, D.L., M. Wilcox, M. Piovant, R.J. Smith, and L.A. Reger. 1984. Related cell-surface antigens expressed with positional specificity in *Drosophila* imaginal disks. *Proc. Natl. Acad. Sci.* **81:** 7485.

Brown, N.H., D.L. King, M. Wilcox, and F.C. Kafatos. 1989. Developmentally regulated alternative splicing of *Drosophila* integrin PS2α transcripts. *Cell* **59:** 185.

Corbi, A.L., J. Garcia-Aguilar, and T.A. Springer. 1990. Genomic structure of an integrin α subunit, the leukocyte p150,95 molecule. *J. Biol. Chem.* **265:** 2782.

Costello, W.J. and J.B. Thomas. 1981. Development of thoracic muscles in muscle-specific mutant and normal *Drosophila melanogaster*. *Soc. Neurosci. Abs.* **7:** 543.

De La Pompa, J.L., J.R. Garcia, and A. Ferrus. 1989. Genetic analysis of muscle development in *Drosophila melanogaster*. *Dev. Biol.* **131:** 439.

Diaz-Benjumea, F.J. and A. García-Bellido. 1990. Genetic analysis of the wing vein pattern of *Drosophila*. *Roux's Arch. Dev. Biol.* **198:** 33.

DiBello, P.R., D.A. Withers, C.A. Bayer, J.W. Fristrom, and G.M. Guild. 1991. The *Drosophila Broad-Complex* encodes a family of related proteins containing zinc fingers. *Genetics* **129:** 385.

García-Bellido, A. 1977. Inductive mechanisms in the process of vein formation in *Drosophila*. *Wilhelm Roux's Arch. Dev. Biol.* **182:** 93.

Gilbert, W. 1985. Genes-in-pieces revisited. *Science* **228:** 823.

Gotwals, P.J. and J.W. Fristrom. 1991. Three neighboring genes interact with the *Broad-Complex* and the *Stubble-stubbloid* locus to affect imaginal disc morphogenesis in *Drosophila*. *Genetics* **127:** 747.

Heidenreich, R., R. Eisman, S. Surrey, K. Delgrosso, J.S. Bennett, E. Schwartz, and M. Poncz. 1990. Organization of the gene for the platelet glycoprotein IIb. *Biochemistry* **29:** 1232.

Hirano, S., K. Ui, T. Miyake, T. Uemura, and M. Takeichi. 1991. *Drosophila* PS integrins recognize vertebrate vitronectin and function as cell-substratum adhesion receptors *in vitro*. *Development* **113:** 1007.

Hynes, R.O. 1992. Integrins: Versatility, modulation and signalling in cell adhesion. *Cell* **69:** 11.

Kelley, M.R., S. Kidd, W.A. Deutsch, and M.W. Young. 1987. Mutations altering the structure of epidermal growth factor-like coding sequences at the *Drosophila Notch* locus. *Cell* **51:** 539.

Kopczynski, C.C., A.K. Alton, K. Fetchel, P.J. Kooh, and M.A.T. Muskavitch. 1988. *Delta*, a *Drosophila* neurogenic gene, is transcriptionally complex and encodes a protein related to blood coagulation factors and epidermal growth factor of vertebrates. *Genes Dev.* **2:** 1723.

Lehmann, R., F. Jimenez, U. Dietrich, and J.A. Campos-Ortega. 1983. On the phenotype and development of mutants of early neurogenesis in *Drosophila melanogaster*. *Wilhelm Roux's Arch. Dev. Biol.* **192:** 62.

Leptin, M., R. Aebersold, and M. Wilcox. 1987. *Drosophila* position-specific antigens resemble the vertebrate fibronectin-receptor family. *EMBO J.* **6:** 1037.

Leptin, M., T. Bogaert, R. Lehmann, and M. Wilcox. 1989. The function of PS integrins during *Drosophila* embryogenesis. *Cell* **56:** 401.

Lindsley, D.L. and E.H. Grell. 1968. Genetic variations of *Drosophila melanogaster*. *Carnegie Inst. Washington Publ. No. 627.*

MacKrell, A.J., B. Blumberg, S.R. Haynes, and J.H. Fessler. 1988. The *lethal myospheroid* gene of *Drosophila* encodes a membrane protein homologous to vertebrate integrin β subunits. *Proc. Natl. Acad. Sci.* **85:** 2633.

Mayer, U. and C. Nüsslein-Volhard. 1988. A group of genes required for pattern formation in the ventral ectoderm of the *Drosophila* embryo. *Genes Dev.* **2:** 1496.

Milner, M.J. and J. Muir. 1987. The cell biology of *Drosophila* wing metamorphosis *in vitro*. *Roux's Arch. Dev. Biol.* **196:** 191.

Mogensen, M.M. and J.B. Tucker. 1988. Intermicrotubular actin filaments in the transalar cytoskeletal arrays of *Drosophila*. *J. Cell Sci.* **91:** 431.

Poulson, D.F. 1937. Chromosomal deficiencies and the embryonic development of *Drosophila melanogaster*. *Proc. Natl. Acad. Sci.* **23:** 133.

Rebay, I., R.J. Fleming, R.G. Fehon, L. Cherbas, P. Cherbas, and S. Artavanas-Tsakonas. 1991. Specific EGF repeats of Notch mediate interactions with Delta and Serrate: Implications for Notch as a multifunctional receptor. *Cell* **67:** 687.

Rothberg, J.M., D.A. Hartley, Z. Walther, and S. Artvanis-Tsakonas. 1988. *slit*: An EGF-homologous locus of *D. melanogaster* involved in the development of the embryonic central nervous system. *Cell* **55:** 1047.

Rothberg, J.M., J.R. Jacobs, C.S. Goodman, and S. Artavanis-Tsakonas. 1990. *slit*: An extracellular protein necessary for development of midline glia and commisural axon pathways contains both EGF and LRR domains. *Genes Dev.* **4:** 2169.

Takada, Y., E. Murphy, P. Pil, C. Chen, M.H. Ginsberg, and M.E. Hemler. 1991. Molecular cloning and expression of the cDNA for α_3 subunit of human $\alpha_3\beta_1$ (VLA-3), an integrin receptor for fibronectin, laminin and collagen. *J. Cell Biol.* **115:** 257.

Tamura, R.N., C. Rozzo, L. Starr, J. Chambers, L.F. Reichardt, H. Cooper, and V. Quaranta. 1990. Epithelial integrin $\alpha_6\beta_4$: Complete primary structure of α_6 and variant forms of β_4. *J. Cell Biol.* **111:** 1593.

Vassin, H., K.A. Bremer, E. Knust, and J.A. Campos-Ortega. 1987. The neurogenic gene *Delta* of *D. melanogaster* is expressed in neurogenic territories and encodes a putative transmembrane protein with EGF-like repeats. *EMBO J.* **6:** 3431.

Waddington, C.C. 1939. Preliminary notes on the development of the wings in normal and mutant strains of *Drosophila*. *Proc. Natl. Acad. Sci.* **25:** 299.

———.1940. The genetic control of wing development in *Drosophila*. *J. Genet.* **41:** 75.

Wieschaus, E. and E. Noell. 1986. Specificity of embryonic lethal mutations in *Drosophila* analyzed in germ-line clones. *Roux's Arch. Dev. Biol.* **195:** 63.

Wilcox, M. 1990. Genetic analysis of the *Drosophila* PS integrins. *Cell Differ. Dev.* **32:** 391.

Wilcox, M. and M. Leptin. 1985. Tissue-specific modulation of a set of related cell surface antigens in *Drosophila*. *Nature* **316:** 351.

Wilcox, M., D.L. Brower, and R.J. Smith. 1981. A position-specific cell surface antigen in the *Drosophila* wing imaginal disc. *Cell* **25:** 159.

Wilcox, M., A. DiAntonio, and M. Leptin. 1989. The function of the PS integrins in *Drosophila* wing morphogenesis. *Development* **107:** 891.

Wilcox, M., N. Brown, M. Piovant, R.J. Smith, and R.A.H. White. 1984. The *Drosophila* position-specific antigens are a family of cell surface glycoprotein complexes. *EMBO J.* **3:** 2307.

Zusman, S., R.S. Patel-King, C. ffrench-Constant, and R.O. Hynes. 1990. Requirements for integrins during *Drosophila* development. *Development* **108:** 392.

Toward a Genetic Analysis of Cell-Matrix Adhesion

R.O. Hynes, E.L. George, E.N. Georges,* J.-L. Guan,† H. Rayburn, and J.T. Yang
Howard Hughes Medical Institute and Center for Cancer Research, Department of Biology, Massachusetts Institute of Technology, Cambridge, Massachusetts 02139

The interactions of cells with their neighbors and with the extracellular matrix play crucial roles in development and in numerous physiological and pathological processes (Hay 1991; Hynes and Lander 1992). Studies of cell adhesion during the past decade and more have uncovered the existence of a large number of cell adhesion molecules. These include cell-surface receptors involved in cell-cell adhesion, such as the cadherins (Takeichi 1988, 1990, 1991), members of the immunoglobulin superfamily (Jessell 1988; Grumet 1991), and selectins (Bevilacqua et al. 1991; Lasky and Rosen 1992). Each of these families of receptors includes multiple related molecules. The same is true for cell-matrix adhesion, which is most frequently mediated by the family of adhesion receptors known as integrins (Hynes 1987, 1992; Albelda and Buck 1990; Hemler 1990).

Integrins are heterodimeric transmembrane receptors whose large extracellular domains interact with adhesive molecules of the extracellular matrix, or with counterreceptors on other cells, to mediate cell adhesion. Their smaller cytoplasmic domains interact with the cytoskeleton and are also involved in signal transduction processes (Shattil and Brugge 1991; Hynes 1992). The integrin family is very diverse (Table 1), and although each integrin shows selectivity in the ligands it binds, there is degeneracy of two kinds. First, many integrins bind several different ligands. Second, most extracellular matrix proteins can be recognized by multiple integrins. This is diagramed for two adhesive extracellular matrix molecules (laminin and fibrinogen) in Figure 1. Most cells express multiple integrins and are, therefore, able to interact with extracellular matrix molecules with considerable versatility. Further complexity is added by the fact that several integrin subunits can occur in alternatively spliced forms (see asterisks in Table 1). In the case of vertebrate integrins, this alternative splicing is only known to affect the cytoplasmic domains, where it is thought to modulate interactions with the cytoskeleton and/or signal transduction events. However, in *Drosophila*, at least two integrin subunits can be alternatively spliced in their extracellular domains, in regions of the molecules close to the ligand-binding and/or subunit interaction sites (Brown et al. 1989; G. Yee et al., in prep.). This raises the possibility that the alternative splicing may alter ligand-binding affinity or specificity.

Whereas some extracellular matrix proteins such as collagens and laminins are encoded by multigene families, alternative splicing also introduces important variations in several extracellular matrix molecules. Both fibronectin (FN) (Hynes 1990) and tenascin (Spring et al. 1989; Gulcher et al. 1991; Siri et al. 1991) are alternatively spliced to produce multiple isoforms. In the case of FN, one of the alternatively spliced segments includes a site for interaction with a specific integrin (Wayner et al. 1989; Guan and Hynes 1990; Mould et al. 1990, 1991). The alternative splicing of FN and the interaction of two different integrins with different sites in FN are diagramed in Figure 2.

Given this complexity of both integrins and their ligands, it is a challenging problem to decide which integrin-matrix interactions are involved and crucial in any given biological process. The most common approach has been to use blocking antibodies or peptides

Table 1. The Integrin Receptor Family

Subunits		Ligands and counterreceptors
	α_1	collagens/LM
	α_2	collagens/LM
	α_3^*	FN/LM/collagens
β_1^*	α_4	FN(V25)/V-CAM-1
	α_5	FN(RGD)
	α_6^*	LM
	α_7	LM
	α_8	?
	α_V	VN/FN (?)
	α_L	I-CAM-1/I-CAM-2
β_2	α_M	iC3b/FB/FX/I-CAM-1
	α_X	FB/iC3b?
β_3^*	α_{IIb}	FB/FN/VWF/VN/TSP
	α_V	VN/FB/VWF/TSP/FN/OP/collagen
β_4^*	α_6^*	LM??
β_5	α_V	VN
β_6	α_V	FN
$\beta_7(=\beta_P)$	α_4	FN(V25)/V-CAM-1?
β_8	α_V	?

Abbreviations: (C3bi) C3b component of complement, inactivated; (FB) fibrinogen; (FN) fibronectin; (FX) factor X; (I-CAM) intercellular adhesion molecule; (LM) laminin; (OP) osteopontin; (TSP) thrombospondin; (V-CAM-1) vascular cell adhesion molecule; (VN) vitronectin; (VWF) von Willebrand factor.

*Alternatively spliced.

Present addresses: *LGME/CNRS, U184 INSERM, Faculté de Médicine, 67000 Strasbourg, France; †Department of Pathology, Cornell University School of Veterinary Medicine, Ithaca, New York 14853.

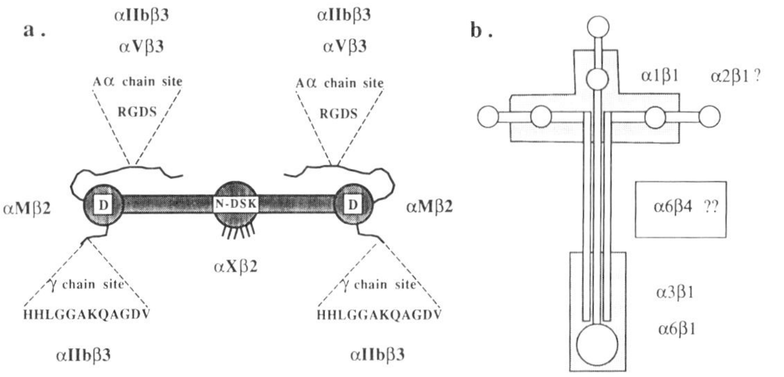

Figure 1. Multiple integrins binding to different parts of adhesive extracellular matrix glycoproteins. (*a*) Fibrinogen is recognized by four different integrins; both $\alpha_V\beta_3$ and $\alpha_{IIb}\beta_3$ recognize an RGD site in the Aα chain; $\alpha_{IIb}\beta_3$ in addition recognizes a different site at the carboxyl terminus of the γ chain; $\alpha_M\beta_2$ binds to fragment D and $\alpha_x\beta_2$ to the amino-terminal disulfide knot (DSK) region. (*b*) Laminin is recognized by two groups of integrins; $\alpha_1\beta_1$ (and probably the homologous $\alpha_2\beta_1$) binds to the cross region, whereas $\alpha_3\beta_1$ and $\alpha_6\beta_1$ (and probably $\alpha_7\beta_1$, which is closely related) bind near the end of the long arm. $\alpha_6\beta_4$ may or may not be an additional laminin receptor.

specific for particular integrins and/or matrix proteins. Although this approach has provided much information, it has its limits, especially for analyses of events in intact organisms. A complementary approach is to use genetic analyses to ablate specific components and dissect their functions in vivo. Because of its tractable genetics, *Drosophila* has been a primary organism for this approach to integrin function. Several groups have analyzed the effects of mutations in integrins on various developmental processes (Brower and Jaffe 1989; Leptin et al. 1989; Wilcox et al. 1989; Wilcox 1990; Zusman et al. 1992), and we have recently used this system to investigate the functional consequences of alternative splicing (Zusman et al. 1992) and of mutations in the cytoplasmic domain (Grinblat et al. 1992).

It would be advantageous to be able to apply genetic analyses to study the roles of integrin-matrix interactions in vertebrates, and we have initiated a program of research directed toward this aim. We have concentrated initially on FN, its alternatively spliced isoforms, and the subset of integrins that interacts with FN. We summarize here our progress to date.

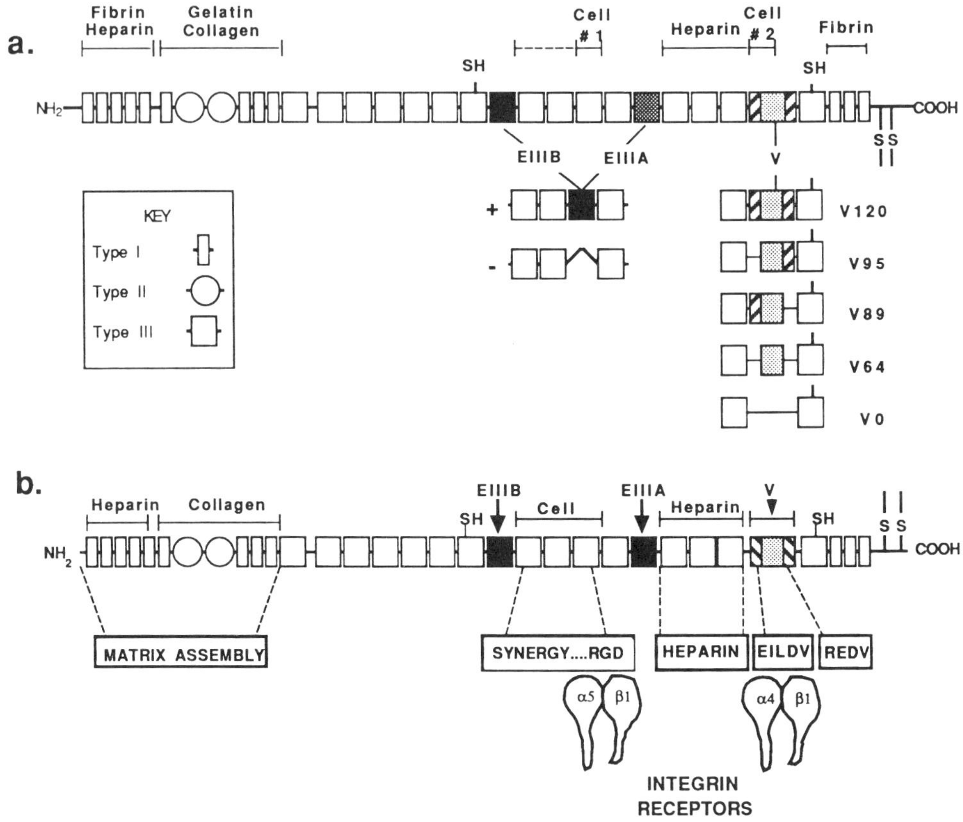

Figure 2. A single FN gene can generate up to 20 different FN isoforms. (*a*) These arise by alternative splicing at three positions designated EIIIB, EIIIA, and V. EIIIB and EIIIA are protein modules (type III repeats), which can be included or excluded by exon skipping. The V region can be fully included or partially or wholly excluded by complex splicing of a single exon; the pattern shown is that observed in humans. (*b*) Several integrins recognize FN (see Table 1). The best characterized are $\alpha_5\beta_1$, which recognizes an RGD sequence present in all isoforms of FN, and $\alpha_4\beta_1$, which recognizes a different sequence located in the alternatively spliced V25 segment of the V region. Both parts of the figure depict functional binding sites within each FN subunit.

RESULTS AND DISCUSSION

Distribution of FN Isoforms In Vivo

To provide context for the molecular genetic analyses that follow, we first provide a summary of descriptive studies of the expression of the different FN isoforms in vivo. As shown in Figure 2, there are three regions of alternative splicing of the FN transcript—EIIIB, EIIIA, and V (or IIICS)—which we refer to as B, A, and V for simplicity. Nucleic acid probes and antibodies specific for each of these segments can be prepared and have been used to define the patterns of expression of the different splice forms in embryos and adult tissues and during physiological and pathological responses. A number of generalizations can be made. First, it appears that early embryos express fibronectins (B^+ A^+ V^+) including all three splice segments (Norton and Hynes 1987; ffrench-Constant and Hynes 1989; DeSimone et al. 1992). This is particularly relevant in situations where extensive cell migration occurs, such as gastrulation, neural crest migration, and development of the heart (ffrench-Constant and Hynes 1988). In all these situations, FN is thought to be important in promoting cell migration (Hynes 1990). Second, as development of differentiated cell types proceeds, certain of the alternative segments are omitted in a cell-type-specific fashion (Vartio et al. 1987; Burton-Wurster et al. 1989; ffrench-Constant and Hynes 1989; Glukhova et al. 1989, 1990; Bennett et al. 1991; Pagani et al. 1991); Figure 3 diagrams a few examples. Finally, the pattern of alternative splicing can alter in response to various stimuli. Examples include wound healing (ffrench-Constant et al. 1989), myocardial infarction (Shekhonin et al. 1990; Knowlton et al. 1992), atherosclerosis (Glukhova et al. 1989, 1990; Mamuya and Brecher 1992), lung injury (Peters et al. 1988), and tumorigenesis (Carnemolla et al. 1989; Oyama et al. 1989). In each case, there appears to be an increase in the inclusion of the B and A segments, which are frequently present at low levels at steady state in adult tissues but increase in response to the various pathological situations. One can view this response as a

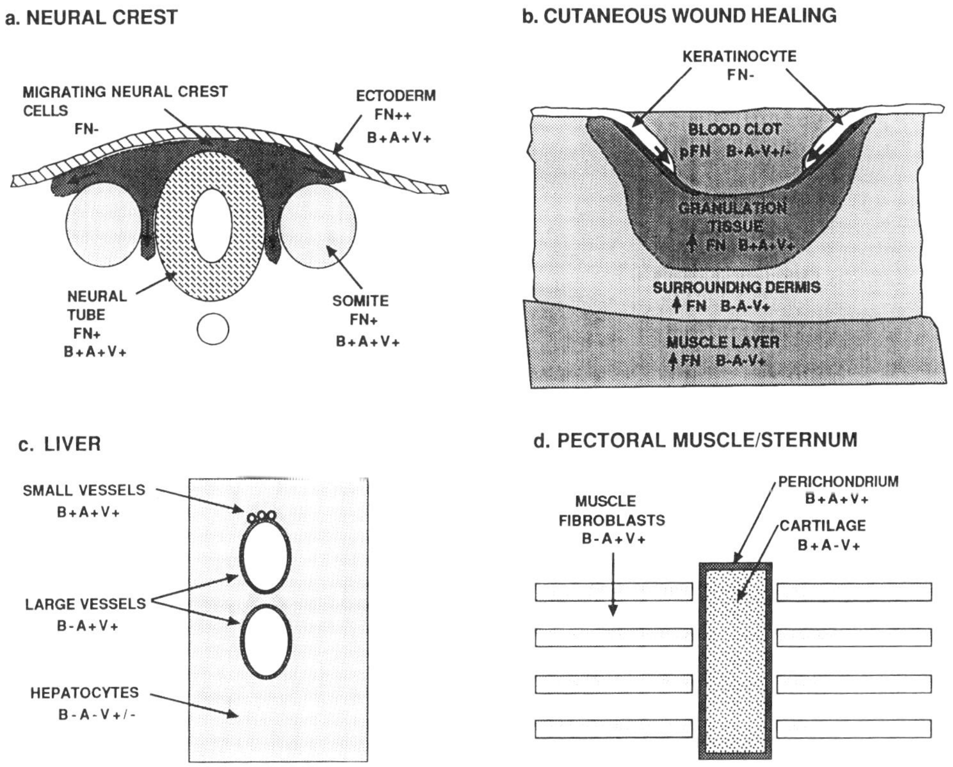

Figure 3. Patterns of alternative splicing of FN. Diagrams based on in situ hybridization (and some antibody) detection of the alternatively spliced segments. (*a*) FN expressed in early embryos and contributing to the matrix in which neural crest cells migrate includes all three splice segments (B^+ A^+ V^+). This FN is made by several cell types. This pattern of splicing is characteristic of other situations in early embryos where cell migration and proliferation are prominent (e.g., heart, area vasculosa, vasculogenesis). (*b*) In resting adult skin dermis, the predominant form of FN expressed is B^- A^- V^+. After wounding, the level of FN mRNA rises greatly, and the cells beneath the migrating keratinocytes produce B^+ A^+ V^+ FN. That is, their pattern of splicing reverts to that characteristic of early embryos (see *a*). (*c*) Liver is the site of synthesis of plasma FN (B^- A^- $V^{+/-}$, which is made by hepatocytes, circulates in the blood, and contributes to blood clots (see *b*), where it can serve as an important adhesive protein. In contrast, endothelial and other cells lining vessels in the liver express FN isoforms, which include the A, and sometimes also the B, segment. (*d*) The thorax illustrates different patterns of FN splicing by various differentiated cell types; cartilage includes the B segment but not the A; muscle fibroblasts express the inverse pattern, whereas perichondrial fibroblasts include all segments.

reversion to the pattern found in early embryos; this viewpoint highlights the fact that inclusion of the B and A segments correlates with proliferation and/or migration of cells.

Such descriptive analyses of the distribution of FN isoforms can only suggest hypotheses; they cannot test them. To do that, it is necessary to obtain the different isoforms in homogeneous form to test for differential functions or to manipulate the expression of the isoforms.

Molecular Genetic Analysis of FN Isoforms and Integrins In Vitro

Since all or most tissues express a mixture of different FN isoforms and since cells frequently form disulfide-bonded heterodimers of different isoforms expressed within the same cell (Schwarzbauer et al. 1987, 1989), it is essentially impossible to purify homogeneous FN isoforms from natural sources. Even plasma FN, which lacks the B and A segments, is 50% V^+, 50% V^- (Paul et al. 1986), and V^+/V^- heterodimers are the rule (Schwarzbauer et al. 1989). Since most functional studies on FN have been performed with plasma FN, they have not addressed the functions of the B or A segments. To circumvent this limitation, we have established cell lines expressing full-length FN isoforms encoded by cDNAs (Guan et al. 1990). By introducing the cDNA expression constructs into cells that do not synthesize FN, we were able to produce homogeneous FNs of each defined splicing pattern. Such recombinant-derived FN isoforms can then be used to test for functions dependent on the presence of specific splice segments.

In this way, we were able to demonstrate that lymphoid cells adhere specifically only to those FN forms containing the V segment (Guan and Hynes 1990; Guan et al. 1990). We were also able to show that this adhesion is mediated by the $\alpha_4\beta_1$ integrin, which recognizes a short sequence in a part of the V segment (V25) that can be alternatively spliced as a unit (Guan and Hynes 1990). These results conform with data from other laboratories using fragments of plasma FN and synthetic peptides (Wayner et al. 1989; Mould et al. 1990). The lymphoid cells used lack the $\alpha_5\beta_1$ integrin, which recognizes the RGD sequence present in all FNs. However, we introduced an α_5 cDNA construct into the cells by retroviral transduction and obtained cells expressing $\alpha_5\beta_1$, as well as $\alpha_4\beta_1$. Figure 4a shows that these cells adhere to all forms of FN, irrespective of the presence or absence of the V segment (or any other alternatively spliced segment), in marked contrast with the parent cell line, which expresses $\alpha_4\beta_1$ but not $\alpha_5\beta_1$ (Fig. 4b). This experiment demonstrates the use by cells of different FN receptor integrins to discriminate among different FN isoforms. Similar experiments to test for specific functions conferred by the B and A segments are under way but have, so far, failed to reveal a similarly clear distinction.

Molecular Genetic Analysis of FN Isoforms and Integrins In Vivo

To extend these analyses to the in vivo situation and initiate genetic analyses of the roles of different FN isoforms and different integrin receptors for FN, we have turned to molecular genetic approaches in mice. The development of methods for producing transgenic mice (Hanahan 1989) and, more recently, for generating targeted mutations by homologous recombination in embryonic stem (ES) cells (Capecchi 1989) have opened the way to production of mice mutant in genes of interest.

We have used the methods of Capecchi and coworkers (Thomas and Capecchi 1987; Mansour et al. 1988) to generate *null* mutations for FN and for α_4 and α_5 integrin subunits. In each case, clones were isolated from mouse genomic libraries that included the 5′ end of the gene. Sequencing of each of the clones located the exon encoding the translation initiation site and signal sequence. We then deleted a segment spanning the initiation codon, all or part of the signal sequence, and 5′-flanking sequences. In the case of the FN gene, these included the probable site of initiation of transcription as well as parts of the promoter (Patel et al. 1987). It is likely that the same is true for the segments deleted from the α_4 and α_5 integrin clones, but these genes have not been analyzed as extensively. The deleted segment in each case was replaced by a cassette comprising the *Escherichia coli* neomycin-resistance gene driven by the phosphoglycerate kinase (PGK) promoter and followed by a polyadenylation site. Each of the constructs contained 750–850 bp of 5′-flanking sequence from the gene and 5–6 kbp of genomic sequence 3′ of the neo^R insertion (Fig. 5a). In addition, at one or both ends of the construct, we inserted a second cassette comprising the herpes simplex virus thymidine kinase (TK) gene driven by the PGK promoter. The constructs were linearized and introduced into 1×10^8 to 4×10^8 D3 129/Sv ES cells (Doetschman et al. 1985) by electroporation at 25 μg/ml of DNA. The ES cells were plated for 1 day and then subjected to selection with G418 (200 μg/ml) and gancyclovir (2 μM) for 6–8 days. The G418 selects for cells expressing the neo^R cassette, whereas the gancyclovir selects against cells expressing the TK cassette (Mansour et al. 1988). Doubly resistant cells should include cells in which the neo^R cassette has inserted into the targeted gene by homologous recombination (Fig. 5a). Doubly resistant clones were screened for the appropriate recombinational event by Southern blotting or by PCR and, in each case, we obtained multiple clones with the desired insertional mutation. These were further tested by Southern blotting and grown up for injection into C57BL6 blastocysts (Hogan et al. 1986). Male chimeras with a high level of agouti coat color (derived from the 129/Sv ES cells) were bred with C57BL6 females and, in each case, we obtained transmission of ES cell-derived gametes. Screening of these agouti progeny by tail blotting revealed transmission of the

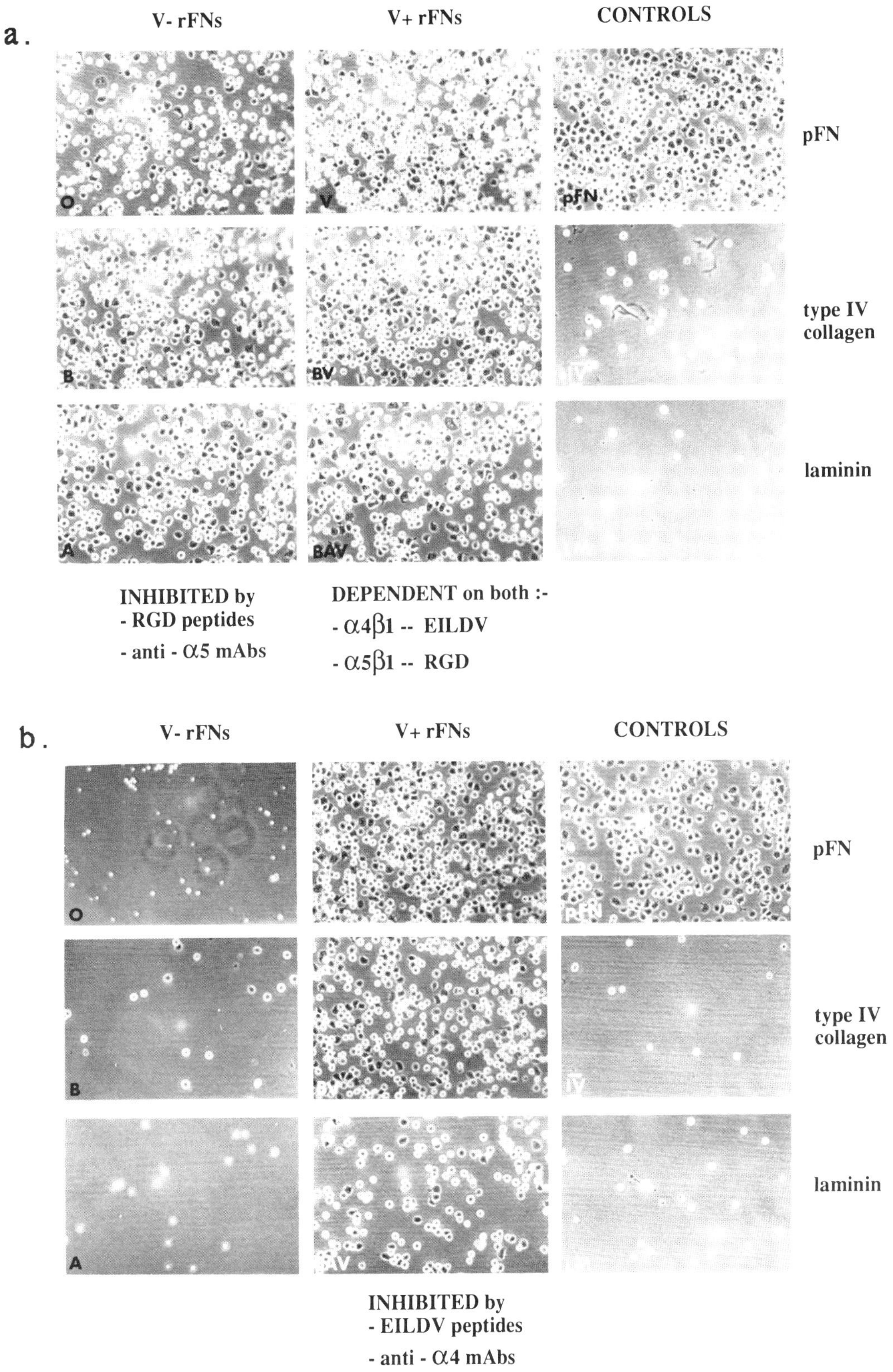

Figure 4. Cells expressing different integrins recognize different FN isoforms. (*a*) WEHI 231 cells transfected with α_5 integrin cDNA express both $\alpha_5\beta_1$ and $\alpha_4\beta_1$ and bind to all forms of recombinant full-length FN, but not to collagen or laminin. (*b*) The parent WEHI 231 cell line, which expresses $\alpha_4\beta_1$ but not $\alpha_5\beta_1$, adheres only to FN isoforms including the V segment. This binding can be inhibited by peptides from the V25 segment (containing the sequence EILDV) or by antibodies to α_4 or β_1. In contrast, adhesion of WEHI (α_5) cells to V^- isoforms of FN can be inhibited by peptides containing the RGD sequence or by antibodies to α_5 or β_1 (panel *a*).

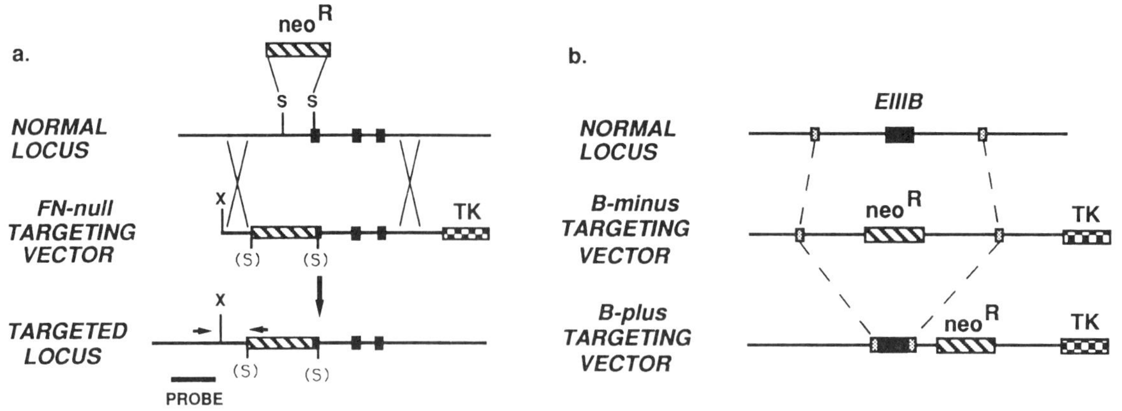

Figure 5. Targeting vectors for homologous recombination of the FN gene. (*a*) A segment flanked by two restriction sites (S) and containing part of the promoter and first exon is deleted and replaced by a selectable marker, the neomycin-resistance cassette (neo^R). DNA containing this replacement plus flanking sequences is fused to a counterselectable marker (TK) and transfected into ES cells. Homologous recombination with the wild-type locus inserts the neo^R marker into the genome but deletes the TK marker (see text). Thus, the targeted locus is defective (since it lacks the sites for initiation of transcription and translation) and can be selected on the basis of the neo^R insertion. Correct recombination can be confirmed by PCR analysis (arrows) and/or by Southern blotting using a probe outside the end (X) of the targeting vector. Exactly analogous strategies were used for the α_4 and α_5 genes. (*b*) The region of the gene encoding the EIIIB segment (in a single exon) is shown. Targeting vectors lacking the EIIIB exon or containing it fused to the adjacent exons were constructed. Each construct also contained the neo^R selectable marker in an adjacent intron and the counterselectable marker (TK) at one end. Insertion of these mutations into the gene can be obtained as depicted in *a*.

mutated alleles with approximately 50% frequency. Therefore, inactivation of the FN gene or of the α_4 or α_5 integrin genes produces recessive mutations, since the heterozygous mutant mice are viable and outwardly normal.

We have therefore been able to establish heterozygous F_1 lines for each of these presumptively *null* mutations. All three lines have been self-intercrossed. The FN and α_5 lines have been tested by tail blotting and, in both cases, it is clear that the mutations are embryonic lethals; no homozygous mutant progeny have been observed. Progeny of the α_4 line have not been tested as yet. These data clearly demonstrate the importance of FN and $\alpha_5\beta_1$ integrin for completion of embryonic development.

Analysis of embryonic lethal mutations requires sampling of embryos and/or fetuses at different stages of development. The embryos can then be examined, sectioned for histological and immunohistochemical analyses, and used to derive cultures for biochemical and cell biological analyses. By the latter procedures, we have confirmed that both the FN and α_5 mutations are true *null* mutations and that the respective proteins are not produced by homozygous mutant embryos. The morphological, histological, and immunohistological analyses are most advanced for the FN mutant embryos; some of the results are illustrated briefly here.

FN is first observed in the preimplantation blastocyst (Zetter and Martin 1978; R.O. Hynes et al., unpubl.) but is expressed at higher levels in postimplantation stages. Figure 6a illustrates a day-7.5 embryo at the primitive streak stage stained for FN. FN is concentrated at the interfaces between the three germ layers and is also present within the mesodermal layer, in agreement with published data (Wartiovaara et al. 1979). In contrast, day-7.5 embryos that are homozygous for the FN-*null* mutation express no detectable FN (Fig. 6b). However, it is of interest to note that the mutant embryo does contain three germ layers. That is, mesoderm forms in the absence of FN expression. This result contrasts with the result obtained in amphibian embryos injected with anti-FN antibodies (Boucaut et al. 1984). In such treated amphibian embryos, mesoderm fails to migrate into the blastocoel cavity. The ectoderm continues to expand but does not invaginate, and no mesoderm forms. However, when the equivalent antibody-blocking experiment is performed in avian embryos, mesoderm does enter from the primitive streak into the underlying space, although it does not migrate laterally (Harrisson et al. 1992). It appears that the result in the FN-*null* mouse embryo parallels that in the chicken embryo treated with anti-FN antibodies; "mesoderm" forms by delamination from the ectoderm at the primitive streak. Thus, FN is not required for delamination, a process presumably involving reduction in cell-cell adhesion. We are currently analyzing whether or not mesodermal migration is defective in the mutant embryos. Furthermore, in both these cases (mouse and chicken), it is unclear whether the cells that delaminate are true mesoderm. In each case, they are defined by position only, not by expression of particular mesodermal markers. Work is in progress to investigate this issue.

FN-*null* mouse embryos thus initiate the process of gastrulation in the absence of FN. However, from that point on they become progressively abnormal and are resorbed by day 10–11. Some folding of epithelial sheets occurs, but the embryos are disorganized and many structures fail to form normally, clearly indicating a requirement for FN in morphogenetic processes. Fur-

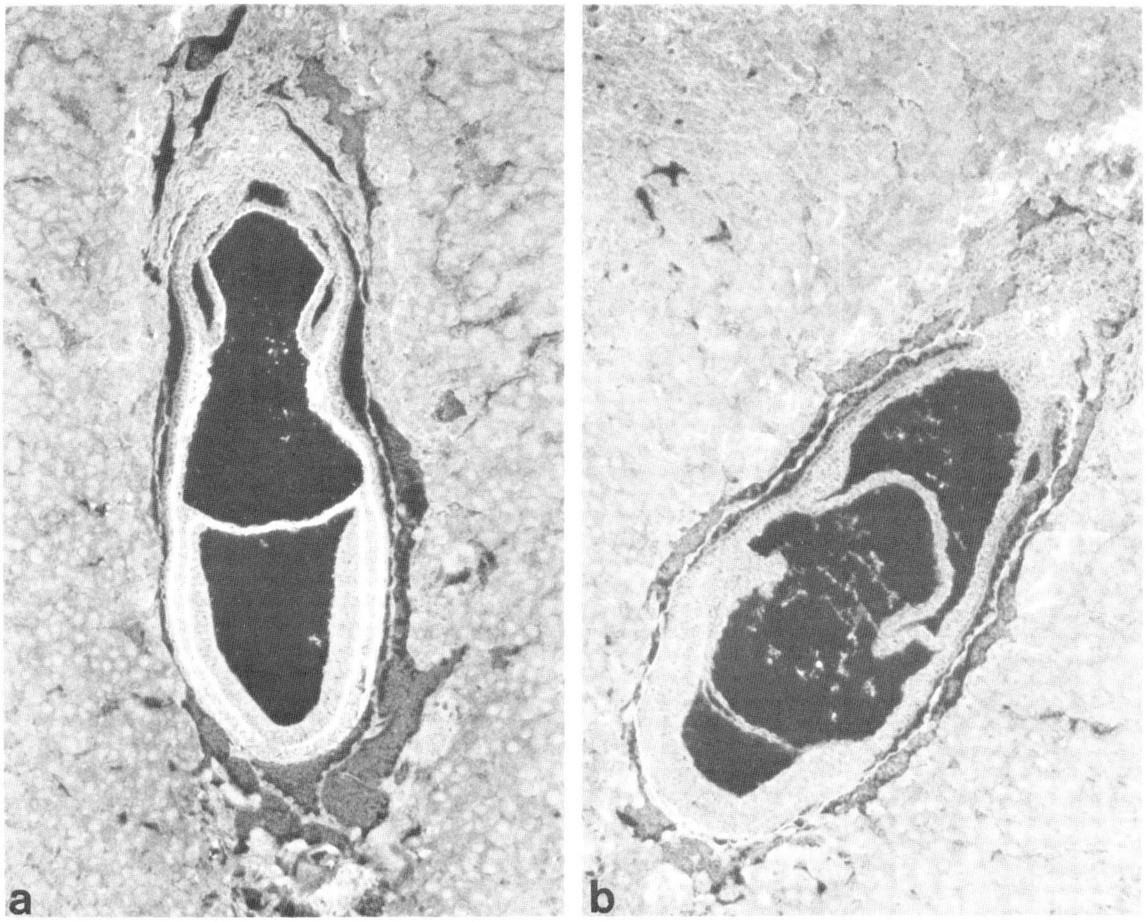

Figure 6. Comparison of wild-type and homozygous FN-*null* primitive streak stage embryos. (*a*) Wild-type embryos, stained for FN. FN is concentrated between the three germ layers and within the mesodermal layer. FN is also present in the chorion and between the ectodermal and mesodermal cell layers of the amnion. (*b*) Homozygous FN-*null* littermate, stained for FN. No FN is detectable in the embryo, yet three germ layers are present. As for the wild-type littermate, FN is detected in the maternal decidual tissue.

ther analysis of these embryos will provide further insights into the functions of FN during early development. The parallel analysis of the α_5 mutants will similarly define the functions of this integrin subunit.

Although the FN mutants clearly show the importance of FN for development, they provide no information about differential roles for the different isoforms. To address this issue, we have initiated two sets of experiments. In the first, we are attempting to generate more subtle mutations in the FN gene, which are intended to alter the pattern of splicing. In the second approach, we are attempting to generate lines of mice expressing FN transgenes of defined splicing pattern; the aim is to breed these transgenic lines with the FN-*null* line to ascertain which isoforms of FN can rescue the FN-*null* mutant phenotype and to what degree.

Both approaches have progressed to some extent, although neither is yet complete. Figure 5b illustrates one approach to the generation of FN mutations altered in splicing of the B segment, which lies in the center of the 70-kb FN gene. Two mutant constructs were made, one lacking the exon encoding the B segment and one in which a three-exon/two-intron segment of the gene was replaced by the equivalent segment of cDNA. Each construct also contained a neo^R cassette in an intron and a TK cassette at one end (Fig. 5b). ES cell clones with each of these mutations were obtained as described above, and chimeras were derived. Germ-line transmission was obtained in each case, and both heterozygous mutant lines were viable. However, in each case, homozygosity resulted in embryonic lethality. Abnormalities developed at about the same time as in the case of FN-*null* homozygotes, and immunohistochemical analyses indicated that little if any assembled FN was present in these embryos. These data raise the question of whether these mutations in the EIIIB region of the gene lead to more severe defects in FN expression than does simply altering the pattern of splicing. This is now under study. If these replacement constructs prove to be inadequate, an alternative approach will be to use the so-called "hit and run" method (Hasty et al. 1991), which involves insertion of a mutation and subsequent deletion of the adja-

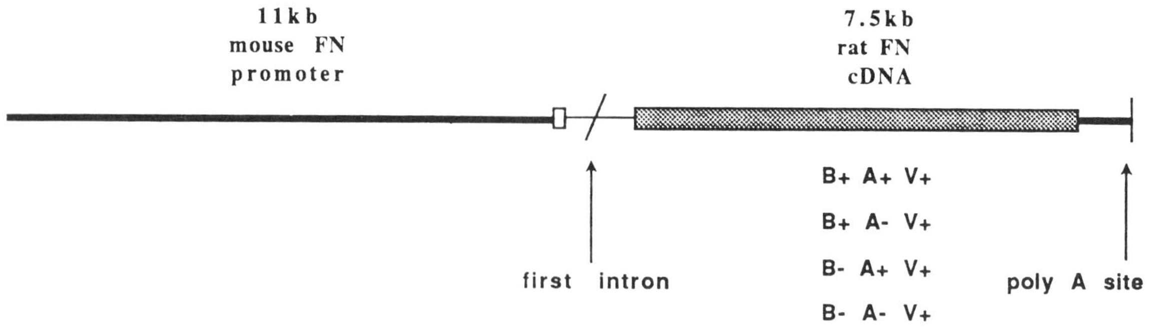

Figure 7. FN hybrid transgenes. Defined splicing patterns are listed below the rat cDNA portion of the transgene (stippled box). The first exon of the mouse FN gene is depicted by a white box. Fusion between mouse and rat sequences is in the first intron, depicted by a slanted line. Mice harboring these transgenes will first be analyzed for dominant effects of the transgene. Second, breeding with FN-*null* mutation mice will determine which FN splice variants are able to rescue the FN-*null* embryonic lethal phenotype.

cent wild-type segment. This technique has the potential advantage that no neo^R cassette is left in the mutated allele.

The transgenic approach we have used is to introduce a hybrid transgene comprising 11 kb of mouse FN promoter and 5′-flanking region fused within the first intron to a rat FN cDNA of defined splicing pattern (Fig. 7). Seven founder lines have been generated and are being tested for expression of rat FN. If these lines produce sufficient FN, they will be intercrossed with the FN-*null* mice to test for rescue of the embryonic lethality. Since the FN gene is so large (~70 kb, ~50 introns), we have omitted most of the introns. This may compromise expression. If the current FN transgenic lines prove inadequate, it will probably be necessary to include more of the introns and, perhaps, more flanking DNA.

CONCLUSIONS

Cell-matrix adhesion is complex, involving as it does multiple receptors and multiple ligands, both of which can sometimes undergo alternative splicing to generate isoforms that likely perform different functions. Genetic analyses represent one of the best available approaches to elucidation of the contributions of specific receptors or ligands to given functions in vivo. Such studies have been initiated in insects, where the genetics are relatively tractable. However, many of the roles of cell-matrix adhesion arise in important physiological and pathological processes such as hemostasis, thrombosis, inflammation, wound healing, and cancer, which need to be studied in mammalian systems. The development of methods for introducing genes and defined mutations in mice opens the way to analyses of cell-matrix adhesion by molecular genetics in a mammalian system. We have initiated such an approach to study the functions of FN and integrins that bind to it. The early results indicate that these molecules are indeed crucial for embryonic development (Table 2). The defects occurring in the mutants are beginning to provide further information about the roles of these molecules, and further analyses using transgenes and more subtle mutations should extend these studies further.

Table 2. Murine Mutations in Fibronectin and in Fibronectin Receptor Integrin Subunits

Mutation	Heterozygotes	Homozygotes
FN – *null*	viable, fertile	embryonic lethal
FN – B^-	viable, fertile	embryonic lethal
FN – B^+	viable, fertile	embryonic lethal
α_5 – *null*	viable, fertile	embryonic lethal
α_4 – *null*	viable	?

ACKNOWLEDGMENTS

We are grateful to Colleen Leslie for manuscript preparation. This work was supported by grants from the National Cancer Institute (RO1-CA-17007) and the National Heart Lung and Blood Institute (PO1-HL-41484) and by the Howard Hughes Medical Institute.

REFERENCES

Albelda, S.M. and C.A. Buck. 1990. Integrins and other cell adhesion molecules. *FASEB J.* **4:** 2868.

Bennett, V.D., K.M. Pallante, and S.L. Adams. 1991. The splicing pattern of fibronectin mRNA changes during chondrogenesis resulting in an unusual form of the mRNA in cartilage. *J. Biol. Chem.* **266:** 5918.

Bevilacqua, M., E. Butcher, B. Furie, B. Furie, M. Gallatin, M. Gimbrone, J. Harlan, K. Kishimoto, L. Lasky, R. McEver, J. Paulson, S. Rosen, B. Seed, M. Siegelman, T. Springer, L. Stoolman, T. Tedder, A. Varki, D. Wagner, I. Weissman, and G. Zimmerman. 1991. Selectins: A family of adhesion receptors. *Cell* **67:** 233.

Boucaut, J.C., T. Darribere, H. Boulekbache, and J.P. Thiery. 1984. Prevention of gastrulation but not neurulation by antibodies to fibronectin in amphibian embryos. *Nature* **307:** 364.

Brower, D.L. and S.M. Jaffe. 1989. Requirement for integrins during *Drosophila* wing development. *Nature* **342:** 285.

Brown, N.H., D.L. King, M. Wilcox, and F.C. Kafatos. 1989. Developmentally regulated alternative splicing of *Drosophila* integrin PS2 α transcripts. *Cell* **59:** 185.

Burton-Wurster, N., G. Lust, and R. Wert. 1989. Expression of the ED B fibronectin isoform in adult human articular cartilage. *Biochem. Biophys. Res. Commun.* **165:** 782.

Capecchi, M.R. 1989. Altering the genome by homologous recombination. *Science* **244:** 1288.

Carnemolla, B., E. Balza, A. Siri, L. Zardi, M.R. Nicotra, A. Bigotti, and P.G. Natali. 1989. A tumor-associated fibronectin isoform generated by alternative splicing of messenger RNA precursors. *J. Cell Biol.* **108:** 1139.

DeSimone, D.W., P.A. Norton, and R.O. Hynes. 1992. Identification and characterization of alternatively spliced fibronectin mRNAs expressed in early *Xenopus* embryos. *Dev. Biol.* **149:** 357.

Doetschman, T.C., H. Eistetter, M. Katz, W. Schmidt, and R. Kemler. 1985. The *in vitro* development of blastocyst-derived embryonic stem cell lines: Formation of visceral yolk sac, blood islands and myocardium. *J. Embryol. Exp. Morphol.* **87:** 27.

ffrench-Constant, C. and R.O. Hynes. 1988. Patterns of fibronectin gene expression and splicing during cell migration in chicken embryos. *Development* **104:** 369.

———. 1989. Alternatively splicing of fibronectin is temporally and spatially regulated in the chicken embryo. *Development* **106:** 375.

ffrench-Constant, C., L. Van De Water, H.F. Dvorak, and R.O. Hynes. 1989. Reappearance of an embryonic pattern of fibronectin splicing during wound healing in the adult rat. *J. Cell Biol.* **109:** 903.

Glukhova. M.A., M.G. Frid, B.V. Shekhonin, Y.V. Balabanov, and V.E. Koteliansky. 1990. Expression of fibronectin variants in vascular and visceral smooth muscle cells in development. *Dev. Biol.* **141:** 193.

Glukhova, M.A., M.G. Frid, B.V. Shekhonin, T.D. Vasilevskaya, J. Grunwald, M. Saginati, and V.E. Koteliansky. 1989. Expression of extra domain A fibronectin sequence in vascular smooth muscle cells is phenotype dependent. *J. Cell Biol.* **109:** 357.

Grinblat, Y., S. Zusman, G. Yee, F. Kafatos, and R.O. Hynes. 1992. The role of the cytoplasmic domain of the *Drosophila* PSβ integrin during development. *Development* (in press).

Grumet, M. 1991. Cell adhesion molecules and their subgroups in the nervous system. *Curr. Opin. Neurobiol.* **1:** 370.

Guan, J.-L. and R.O. Hynes. 1990. Lymphoid cells recognize an alternatively spliced segment of fibronectin via the integrin receptor $\alpha_4\beta_1$. *Cell* **60:** 51.

Guan, J.-L., J.E. Trevithick, and R.O. Hynes. 1990. Retroviral expression of alternatively spliced forms of rat fibronectin. *J. Cell Biol.* **110:** 833.

Gulcher, J.R., D.E. Nies, M.J. Alexakos, N.A. Ravikant, M.E. Sturgill, L.S. Marton, and K. Stefansson. 1991. Structure of the human hexabrachion (tenascin) gene. *Proc. Natl. Acad. Sci.* **88:** 9438.

Hanahan, D. 1989. Transgenic mice as probes into complex systems. *Science* **246:** 1265.

Harrisson, F., J. van Hoof, and J.-M. Foidart. 1992. Microinjection of anti-fibronectin antibodies in the chicken blastoderm: Inhibition of mesoblast cell migration but not of cell ingression at the primitive streak. *Development* (in press).

Hasty, P., R. Ramirez-Solis, R. Krumlauf, and A. Bradley. 1991. Introduction of a subtle mutation into the *Hox-2.6* locus in embryonic stem cells. *Nature* **350:** 243.

Hay, E.D., ed. 1991. *Cell biology of extracellular matrix*, 2nd edition. Plenum Press, New York.

Hemler, M.E. 1990. VLA proteins in the integrin family: Structures, functions and their role on leukocytes. *Annu. Rev. Immunol.* **8:** 365.

Hogan, B., F. Costantini, and E. Lacy. 1986. *Manipulating the mouse embryo. A laboratory manual.* Cold Spring Harbor Laboratory, Cold Spring Harbor, New York.

Hynes, R.O. 1987. Integrins: A family of cell surface receptors. *Cell* **4:** 549.

———. 1990. *Fibronectins.* Springer-Verlag, New York.

———. 1992. Integrins: Versatility, modulation, and signaling in cell adhesion. *Cell* **69:** 11.

Hynes, R.O. and A.D. Lander. 1992. Contact and adhesive specificities in the associations, migrations, and targeting of cells and axons. *Cell* **68:** 303.

Jessell, T.M. 1988. Adhesion molecules and the hierarchy of neural development. *Neuron* **1:** 3.

Knowlton, A.A., C.M. Connelly, G.M. Romo, W. Mamuya, C.S. Apstein, and P. Brecher. 1992. Rapid expression of fibronectin in the rabbit heart after myocardial infarction with and without reperfusion. *J. Clin. Invest.* **89:** 1060.

Lasky, L.A. and S. Rosen. 1992. The selectins: Carbohydrate-binding adhesion molecules of the immune system. In *Inflammation: Basic principles and clinical correlates* (ed. J. Gallin et al.). Raven Press, New York. (In press.)

Leptin, M., T. Bogaert, R. Lehmann, and M. Wilcox. 1989. The function of PS integrins during *Drosophila* embryogenesis. *Cell* **56:** 401.

Mamuya, W.S. and P. Brecher. 1992. Fibronectin expression in the normal and hypertrophic rat heart. *J. Clin. Invest.* **89:** 392.

Mansour, S.L., K.R. Thomas, and M.R. Capecchi. 1988. Disruption of the proto-oncogene *int-2* in mouse embryo-derived stem cells: A general strategy for targeting mutations to non-selectable genes. *Nature* **336:** 348.

Mould, A.P., A. Komoriya, K.M. Yamada, and M.J. Humphries. 1991. The CS5 peptide is a second site in the IIICS region of fibronectin recognized by the integrin $\alpha_4\beta_1$. Inhibition of $\alpha_4\beta_1$ function by RGD peptide homologues. *J. Biol. Chem.* **266:** 3579.

Mould, A.P., L.A. Wheldon, A. Komoriya, E.A. Wayner, K.M. Yamada, and M.J. Humphries. 1990. Affinity chromatographic isolation of the melanoma adhesion receptor for the IIICS region of fibronectin and its identification as the integrin $\alpha_4\beta_1$. *J. Biol. Chem.* **265:** 4020.

Norton, P.A. and R.O. Hynes. 1987. Alternative splicing of chicken fibronectin in embryos and in normal and transformed cells. *Mol. Cell Biol.* **7:** 4297.

Oyama, F., S. Hiroshashi, Y. Shimosato, K. Titani, and K. Sekiguchi. 1989. Deregulation of alternative splicing of fibronectin pre-mRNA in malignant human liver tumors. *J. Biol. Chem.* **264:** 10331.

Pagani, F., L. Zagato, C. Vergani, G. Casari, A. Sidoli, and F.E. Baralle. 1991. Tissue-specific splicing pattern of fibronectin messenger RNA precursor during development and aging in rat. *J. Cell Biol.* **113:** 1223.

Patel, R.S., E. Odermatt, J.E. Schwarzbauer, and R.O. Hynes. 1987. Organization of the rat fibronectin gene provides evidence for "exon shuffling" during evolution. *EMBO J.* **6:** 2565.

Paul, J.I., J.E. Schwarzbauer, J.W. Tamkun, and R.O. Hynes. 1986. Cell-type-specific fibronectin subunits generated by alternative splicing. *J. Biol. Chem.* **261:** 12258.

Peters, J.H., M.H. Ginsberg, C.M. Case, and C.G. Cochrane. 1988. Release of soluble fibronectin containing an extra type III domain (ED1) during acute pulmonary injury mediated by oxidants or leukocytes in vivo. *Am. Rev. Respir. Dis.* **138:** 167.

Schwarzbauer, J.E., C.S. Spencer, and C.L. Wilson. 1989. Selective secretion of alternatively spliced fibronectin variants. *J. Cell Biol.* **109:** 3445.

Schwarzbauer, J.E., R.S. Patel, D. Fonda, and R.O. Hynes. 1987. Multiple sites of alternative splicing of the rat fibronectin gene transcript. *EMBO J.* **6:** 2673.

Shattil, S.J. and J.S. Brugge. 1991. Protein tyrosine phosphorylation and the adhesive functions of platelets. *Curr. Opin. Cell Biol.* **3:** 869.

Shekhonin, B.V., S.B. Guriev, S.B. Irgashev, and V.E. Koteliansky. 1990. Immunofluorescent identification of fibronectin and fibrinogen/fibrin in experimental myocardial infarction. *J. Mol. Cell. Cardiol.* **22:** 533.

Siri, A., B. Carnemolla, M. Saginati, A. Leprini, G. Casari, F. Baralle, and L. Zardi. 1991. Human tenascin: Primary structure, pre-mRNA splicing patterns and localization of the epitopes recognized by two monoclonal antibodies. *Nucleic Acids Res.* **19:** 525.

Spring, J., K. Beck, and R. Chiquet-Ehrismann. 1989. Two contrary functions of tenascin: Dissection of the active sites by recombinant tenascin fragments. *Cell* **59:** 325.

Takeichi, M. 1988. The cadherins: Cell-cell adhesion molecules controlling animal morphogenesis. *Development* **102:** 639.

———. 1990. Cadherins: A molecular family important in selective cell-cell adhesion. *Annu. Rev. Biochem.* **59:** 237.

———. 1991. Cadherin cell adhesion receptors as a morphogenetic regulator. *Science* **251:** 1451.

Thomas, K.R. and M.R. Capecchi. 1987. Site-directed mutagenesis by gene targeting in mouse embryo-derived stem cells. *Cell* **51:** 503.

Vartio, T., L. Laitinen, O. Narvanen, M. Cutolo, L.E. Thornell, L. Zardi, and I. Virtanen. 1987. Differential expression of the ED sequence-containing form of cellular fibronectin in embryonic and adult human tissues. *J. Cell Sci.* **88:** 419.

Wartiovaara, J., I. Leivo, and A. Vaheri. 1979. Expression of the cell surface-associated glycoprotein, fibronectin, in the early mouse embryo. *Dev. Biol.* **69:** 247.

Wayner, E.A., A. Garcia-Pardo, M.J. Humphries, J.A. McDonald, and W.G. Carter. 1989. Identification and characterization of the T lymphocyte adhesion receptor for an alternative cell attachment domain (CS-1) in plasma fibronectin. *J. Cell Biol.* **109:** 1321.

Wilcox, M. 1990. Genetic analysis of the *Drosophila* PS integrins. *Cell Differ. Dev.* **32:** 391.

Wilcox, M., A. DiAntonia, and M. Leptin. 1989. The function of PS integrins in *Drosophila* wing morphogenesis. *Development* **107:** 891.

Zetter, B.R. and G.R. Martin. 1978. Expression of a high molecular weight cell surface glycoprotein (LETS protein) by preimplantation mouse embryos and teratocarcinoma stem cells. *Proc. Natl. Acad. Sci.* **75:** 2324.

Zusman, S., Y. Grinblat, G. Yee, F.C. Kafatos, and R.O. Hynes. 1992. Analyses of PS integrin functions during *Drosophila* development. *Development* (in press).

Glycosylation-dependent Cell Adhesion Molecule 1: A Novel Mucin-like Adhesion Ligand for L-Selectin

L.A. LASKY,* M.S. SINGER,‡ D. DOWBENKO,* Y. IMAI,‡ W. HENZEL,†
C. FENNIE,* S. WATSON,* AND S.D. ROSEN‡
*Departments of *Immunobiology and †Protein Chemistry, Genentech, Inc., South San Francisco, California 94080; ‡Department of Anatomy and Program in Immunology, University of California, San Francisco, California 94143-0452*

One of the most extraordinary aspects of the vertebrate immune system is the ability of circulating T and B lymphocytes to traffic efficiently through the peripheral lymphoid organs (Berg et al. 1989; Rosen 1989; Yednock and Rosen 1989; Butcher 1990). This mechanism evolved to ensure that circulating lymphocytes with diverse antigenic specificities would encounter foreign antigens that were sequestered by the afferent vessels of the lymphatic system into the lymph nodes. These encounters benefited the organism by allowing an enhanced probability of recognition of extrinsic antigens by the manifold antigen receptors (i.e., T-cell receptors and surface immunoglobulins) found on the surface of the circulating immune cells. Immune cells that were stimulated by antigens presented by antigen-presenting cells in the peripheral lymphoid organs would mount appropriate antibody- and cell-mediated responses, whereas immune cells that did not encounter an antigen that satisfied their antigenic specificities would flow back into the vascular system through the efferent lymphatic vessels and attempt to detect the appropriate antigen at other lymphoid sites. Since many antigens would, by this mechanism, be encountered when they were first experienced by the organism (i.e., at the level of the skin or in the gut), the ability to sequester and recognize these antigens in the lymph nodes allows a much earlier reaction against any pathogenic insult. The results of this recognition would, of course, be an embellished immune response directed against the particular offending antigen with subsequent protection and survival of the organism. It is therefore safe to say that the evolution of a highly efficient lymphocyte trafficking mechanism conferred an enormous selective advantage on the organism.

Early Investigations of Lymphocyte Homing

It was initially assumed that specific cell adhesive interactions were involved with the ability of lymphocytes to recirculate to the peripheral lymphoid organs. Experiments carried out by Gowans and colleagues in the 1960s demonstrated that lymphocytes taken from a particular lymphoid site tended to return, or "home" back to the same site when reinjected (Gowans and Knight 1964; Marchesi and Gowans 1964). These results led to the speculation that particular receptors, termed homing receptors, were expressed on the lymphocyte surface and that these homing receptors adhered to their cognate ligands that were found on the endothelial surface of a select lymphoid region (Butcher et al. 1980). This specific cell adhesive interaction would therefore allow the trafficking of subsets of cells to appropriate sites, such as peripheral or gut-associated lymph nodes.

Although these in vivo experiments were intriguing, they provided little information about the molecular nature(s) of the cell-surface component(s) involved in this trafficking response. The production of an in vitro cell-binding assay that appeared to mimic many of the adhesive interactions of lymph node homing in vivo allowed a number of contemporary techniques to be applied to the problem of lymphocyte trafficking (Woodruff et al. 1987). This assay essentially consisted of the analysis of the specific binding of lymphocytes to the postcapillary high endothelial venules (HEV) of frozen lymph node sections. In this way, various inhibitors of this binding phenomenon could be analyzed. For example, this technique allowed the analysis of a monoclonal antibody (termed MEL 14) directed against an approximately 90–110-kD murine leukocyte cell-surface component that blocked the binding of lymphocytes to the HEV of peripheral and mesenteric lymph nodes but not to the HEV of Peyer's patches (the gut-associated lymphoid tissue) (Gallatin et al. 1983). Since this antibody blocked peripheral, but not gut, lymphoid HEV binding, it was speculated that it recognized a lymphocyte homing receptor that was involved with the specific trafficking of these cells to peripheral lymphoid sites. Subsequent in vivo data further supported this contention by demonstrating that the MEL 14 antibody could completely inhibit the trafficking of lymphocytes to peripheral lymph nodes, but only marginally inhibit trafficking to the Peyer's patches (Hamann et al. 1991). Thus, it appeared that MEL 14 recognized a lymphocyte homing receptor specific for trafficking to the peripheral and mesenteric lymph nodes.

Concomitant with the experiments on MEL 14, Rosen, Stoolman, and colleagues were speculating on the possibility that the adhesion of lymphocytes to the HEV of peripheral lymph nodes was mediated by protein-carbohydrate interactions (Stoolman and Rosen 1983; Stoolman et al. 1984, 1987; Stoolman and Ebing 1989). Using the same frozen section cell-binding assay, these investigators discovered that the binding of lymphocytes to peripheral lymph node HEV could be inhibited by millimolar quantities of monomeric mannose-6-phosphate. They went on to show that micromolar quantities of a polymeric form of mannose-6-phosphate, termed polyphosphomannan ester (PPME), as well as a polymeric form of fucose-4-sulfate, termed fucoidin, were also efficient inhibitors of lymphocyte-HEV adhesion and that these adhesion-blocking carbohydrates bound to a lymphocyte surface component involved in HEV binding (Yednock et al. 1987a). The correlation between the inhibition caused by these carbohydrates and the previous work on MEL 14 antibody inhibition led these investigators to show that the approximately 90-kD lymphocyte surface antigen recognized by the MEL 14 antibody was equivalent to the carbohydrate-binding surface component involved with HEV adhesion (Yednock et al. 1987b). Further support for the involvement of endothelial carbohydrate ligands in HEV adhesion came from experiments that demonstrated that treatment of HEV sections or whole animals with sialidase, an enzyme that removes sialic acids from carbohydrates, resulted in a profound inhibition of lymphocyte binding to HEV (Rosen et al. 1985) and peripheral lymph node trafficking (Rosen et al. 1989; Imai et al. 1990). These data were thus consistent with the conclusion that the lymphocyte antigen recognized by MEL 14 was a novel type of adhesion molecule that mediated lymphocyte trafficking to peripheral lymph nodes by adhering to a unique carbohydrate ligand specifically expressed on HEV.

Molecular Analysis of the Homing Receptor: L-Selectin

The cloning of the cDNAs encoding the murine and human forms of the lymphocyte homing receptor revealed that this molecule represented a new type of cell adhesion glycoprotein (Bowen et al. 1989; Lasky et al. 1989; Siegelman et al. 1989; Tedder et al. 1989). Analysis of the structure of the homing receptor (Fig. 1) revealed that it contained a number of domains that were homologous to those found in other proteins. Perhaps most notable was the discovery of an amino-terminal domain that was related to a family of calcium-dependent carbohydrate-binding proteins or lectins (Drickamer 1988). These data were entirely consistent with the previous studies that indicated a role for protein-carbohydrate interactions in the binding of lymphocytes to HEV and strongly suggested that the lectin domain mediated, in large part, the adhesion of cells to this type of endothelium. Consistent with this hypothesis were studies that demonstrated that the MEL 14 antibody, which was previously shown to inhibit the adhesion of cells to HEV completely, mapped to an epitope contained within the amino-terminal lectin motif of the murine homing receptor (Bowen et al. 1990). Following the lectin domain was an epidermal growth factor (EGF)-like motif that was subsequently shown to be necessary for the appropriate function of the lectin domain, perhaps via a conformational interaction(s) between these domains. This motif was followed by two copies of a motif that was homologous to those found in proteins that interact with complement. Subsequent work demonstrated that these complement-binding protein-like domains are also required for appropriate lectin domain function, perhaps by their ability to mediate homing receptor oligomerization (Watson et al. 1991b). In summary, the structural analysis generated by the cloning of the homing receptor suggested that it was a mosaic protein that mediated cell adhesion through its amino-terminal lectin domain by the recognition of a carbohydrate presented by peripheral lymph node HEV.

The simultaneous description of the cDNAs encoding two other lectin-containing adhesion molecules, termed the endothelial-leukocyte adhesion molecule (ELAM) (Bevilacqua et al. 1989) and gmp140/protein activation-dependent granule external membrane protein (PADGEM) (Johnston et al. 1989), resulted in the description of a new lectin-dependent cell adhesion family (Stoolman 1989). ELAM and gmp140/

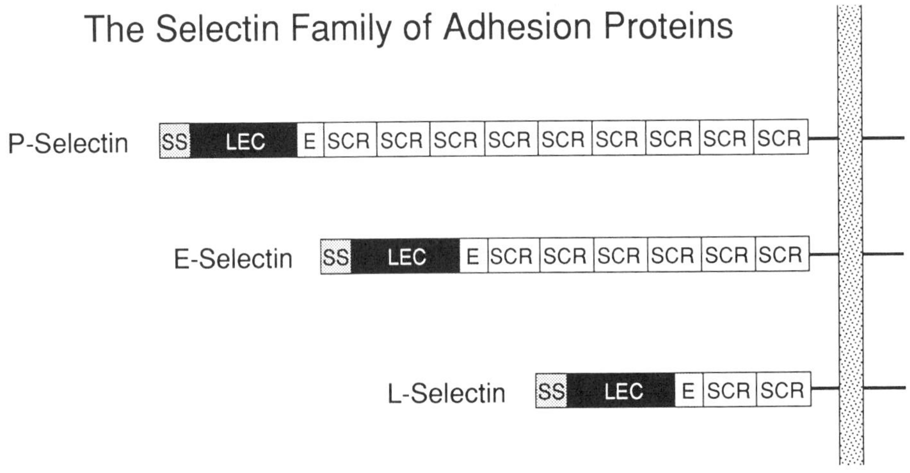

Figure 1. Domain structures of the three members of the selectin adhesion molecule family. Shown are the lectin (LEC), epidermal growth factor-like (E), and complement-binding protein-like (CBD) domains of L-, E-, and P-selectin.

PADGEM, which are expressed on endothelium activated by inflammatory or other mediates, were both found to have mosaic structures that were similar to the homing receptor (Fig. 1). All three proteins contained homologous amino-terminal lectin domains (~65% conserved), followed by conserved EGF-like motifs (~65% conserved) and various numbers of less conserved complement-binding protein-like domains (2 homing receptors; 6 ELAMs; 8 or 9 gmp140/PADGEMs). Subsequent work demonstrated that, analogous to the homing receptor, ELAM and gmp140/PADGEM both recognized sialyl Lewis x or a-like carbohydrate ligands on neutrophil, monocyte, and T-cell subset surfaces (Phillips et al. 1990; Walz et al. 1990). Because of the homologous nature of the protein structures and carbohydrate ligands, these various adhesion molecules were renamed the selectins (Bevilacqua et al. 1991): L-selectin (homing receptor), E-selectin (ELAM), and P-selectin (gmp140/PADGEM).

Biochemical Characterization of Endothelial Ligands for L-Selectin

Interest in the nature of the HEV ligand(s) recognized by L-selectin led Streeter and colleagues

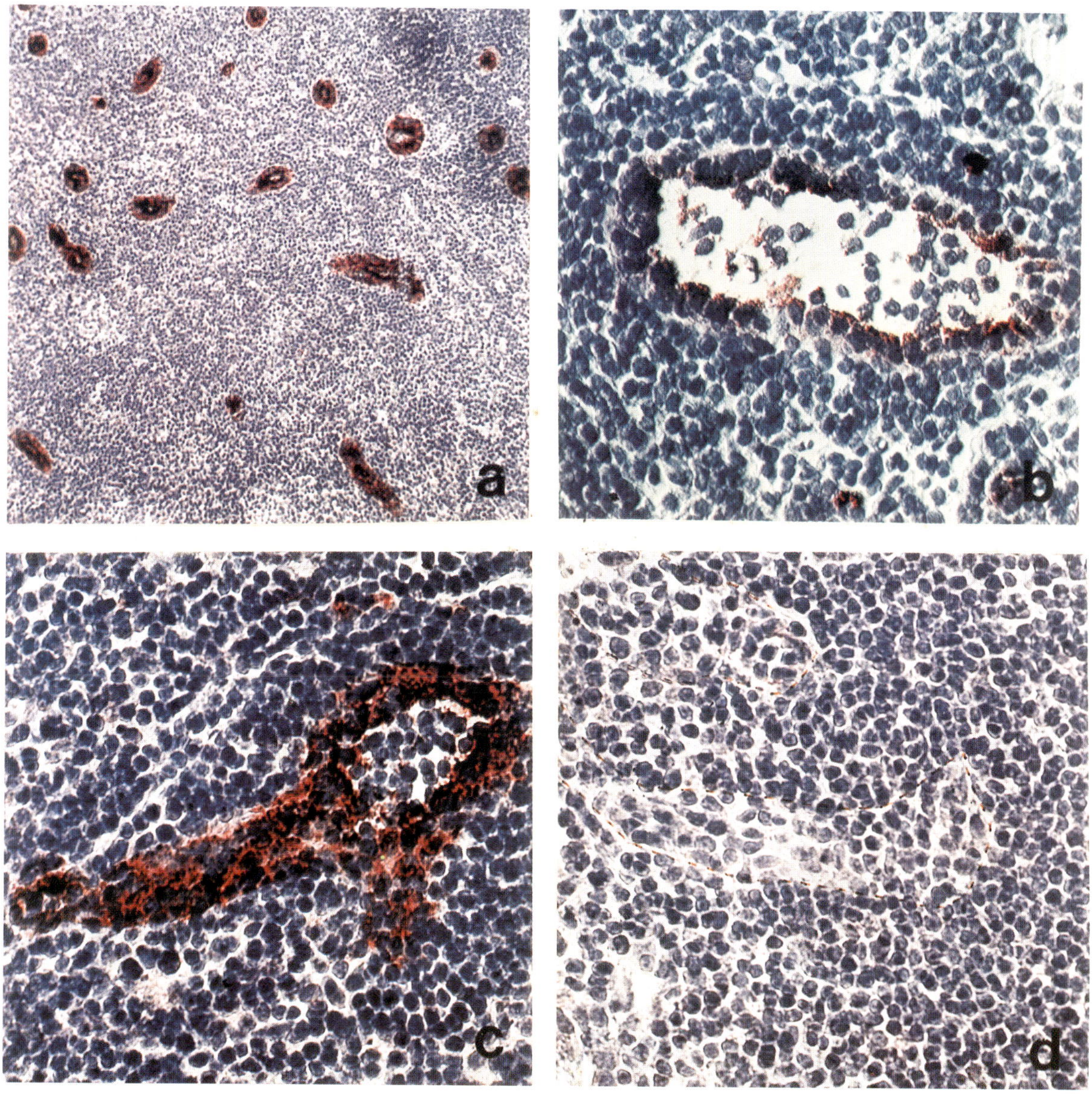

Figure 2. Immunohistochemical staining of peripheral lymph node high endothelial venules by an L-selectin-immunoglobulin chimera. Cryostat-cut sections of a murine peripheral lymph node were incubated with 3 μg per section of the L-selectin chimera. The sections were washed, incubated with a horseradish peroxidase (HRP)-conjugated anti-human immunoglobulin antibody, and processed for HRP activity as described previously (Watson et al. 1990). As shown, all of the peripheral lymph node HEVs are stained with the chimeric construct (*a–c*). Panel *d* shows the lack of staining of Peyer's patch high endothelial venules. (Reprinted, with permission, from Watson et al. 1990.)

(1988a,b) to develop a monoclonal antibody, termed MECA 79, against an antigen(s) that was relatively specific for the HEV of peripheral lymph nodes. This antibody was capable of inhibiting the L-selectin-mediated binding of cells to HEV in frozen section assays and was, presumably, directed against an adhesive ligand for L-selectin. Analysis of the protein antigens recognized by this antibody revealed a spectrum of molecular species. This result was interpreted to suggest that a diversity of endothelial ligands exist for the binding of lymphocytes via the lectin domain of L-selectin. In addition, the fact that this monospecific antibody recognized a diversity of proteins suggested that it recognized a posttranslational modification, perhaps carbohydrate in nature. Subsequent work demonstrated that antigens isolated from peripheral lymph nodes using the MECA 79 antibody could support L-selectin-mediated adhesion and that this adhesion was abolished by sialidase treatment of the isolated antigens (Berg et al. 1991). These results were consistent with the possibility that MECA 79 recognized an endothelial ligand(s) for L-selectin, although the molecular nature of this ligand(s) could not be accurately described.

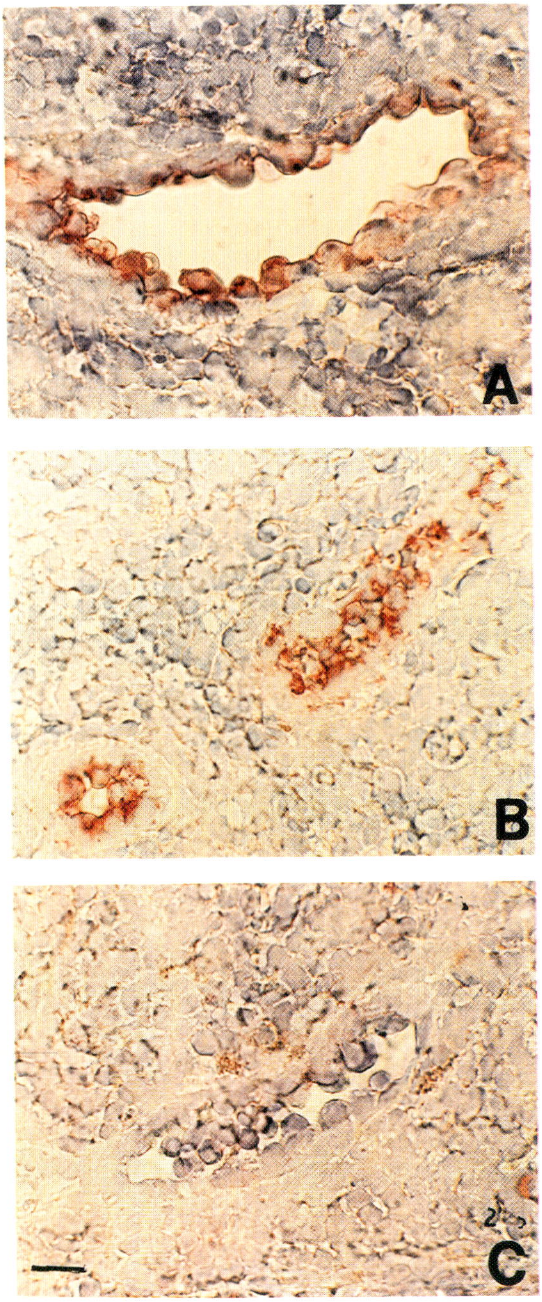

Figure 5. Immunohistochemical staining of peripheral lymph node HEVs with antisera against a peptide predicted from the L-selectin ligand cDNA sequence. Paraffin sections of peripheral lymph nodes were stained with antisera directed against a peptide spanning amino acids 141–151 of the amino acid sequence predicted from the L-selectin ligand cDNA clone. Panels *A* and *B* show staining with immune serum; panel *C* shows preimmune staining. Bar, 50 μm. (Reprinted, with permission, from Lasky et al. 1992.)

A second, potentially more specific reagent for the investigation of the HEV ligand(s) of L-selectin was generated by Watson et al. (1990, 1991b). These investigators took advantage of the built-in HEV ligand specificity of the L-selectin lectin domain by converting the extracellular portion of this glycoprotein into an antibody-like soluble protein. This "immunoglobulin chimera" was capable of inhibiting L-selectin-mediated lymphocyte adhesion to HEV in the frozen section assay. In addition, the chimera could be used as an immunohistochemical reagent to stain lymph node sections for the presence of an HEV ligand(s) (Fig. 2). These staining experiments revealed that the ligand(s) for L-selectin-mediated binding was luminally expressed and found only in peripheral and mesenteric, but not Peyer's patch, lymph nodes, in agreement with previous studies on L-selectin-dependent trafficking to these lymphoid sites. In addition, subsequent immunohistochemical studies demonstrated that the staining of the HEV by the chimera was abolished by treatment of sections with sialidase, consistent with previous cell-binding data and confirming the importance of sialic acid in the interaction between the L-selectin lectin domain and the HEV.

The immunohistochemical detection of an HEV ligand(s) for L-selectin using the immunoglobulin chimera set the stage for a concise biochemical analysis of the molecule(s). Early work by Ford and colleagues (Andrews et al. 1982) demonstrated that the HEV of peripheral lymph nodes specifically incorporated inorganic sulfate into a then unknown macromolecule that was apparently shed into the media during organ culture. Imai and colleagues (1991) took advantage of this early observation to demonstrate that much of the sulfate precursor was incorporated into two peripheral lymph-node-specific glycoproteins of 50 kD and 90 kD that could be precipitated by the L-selectin immunoglobulin chimera. The adhesive interaction between the chimera and these two sulfate-labeled HEV glycoproteins was mediated by the lectin domain of L-selectin: The interaction was calcium-dependent, it was inhibited by the MEL 14 antibody, it could be blocked by

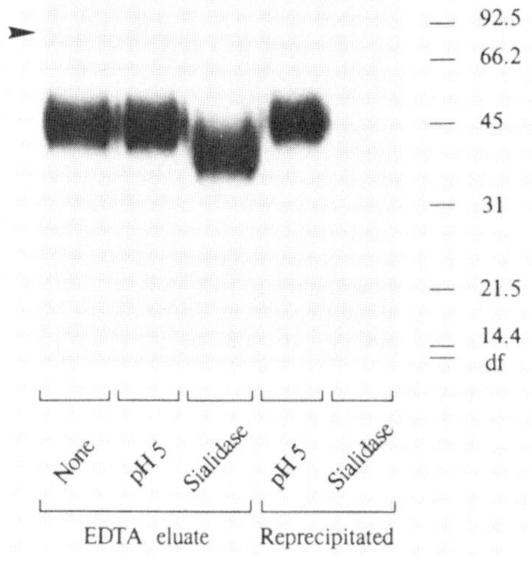

Figure 3. Sialidase treatment abolishes the binding of Sgp50 to the L-selectin-immunoglobulin chimera. [^{35}S]Sulfate-labeled 50-kD L-selectin ligand was prepared as described by Imai et al. (1991). Material was either treated with *Arthrobacter ureafaciens* sialidase (Sialidase) or mock-treated (pH 5). Separate aliquots of treated and mock-treated ligand were reprecipitated with the L-selectin chimera and analyzed by SDS-PAGE. As shown, the mock-treated ligand still interacts with the L-selectin chimera, whereas the smaller sialidase-treated material does not. (Reprinted, with permission, from Imai et al. 1990.)

various carbohydrates, and it was abolished by treatment with sialidase (e.g., see Fig. 3). The carbohydrate linkages on these L-selectin ligands were insensitive to the *N*-linked specific degradative enzyme *N*-glycanase, consistent with the interpretation that they were linked via *O* linkages to serine and threonine residues. In agreement with the work of Streeter, Berg, and colleagues (Streeter et al. 1988a; Berg et al. 1991), both glycoproteins contained the MECA 79 epitope. Interestingly, although previous work had indicated an approximately 90–110-kD MECA 79-reactive endothelial antigen, no evidence for a 50-kD species was provided (Berg et al. 1991). Finally, in a manner analogous to the carbohydrate ligands for E and P selectin, both the 50-kD and 90-kD L-selectin ligands were found to contain fucose, although the requirement for this sugar in L-selectin–carbohydrate binding has yet to be demonstrated. In conclusion, these data provided strong evidence that the 50-kD and 90-kD sulfated glycoproteins are tissue-specific endothelial ligands for L-selectin whose role is to mediate lymphocyte trafficking to peripheral lymph nodes.

Molecular Cloning of a cDNA Encoding the 50-kD L-Selectin Ligand

To characterize further the molecular nature of a ligand for L-selectin, Lasky et al. (1992) cloned a cDNA encoding the 50-kD sulfate-labeled glycoprotein. This group used a unique procedure to clone this cDNA that is of general use. The essential idea was to use the L-selectin-immunoglobulin chimera as an affinity matrix to purify sufficient 50-kD ligand for amino-terminal sequencing and oligonucleotide probe design. This procedure is of general use as long as the affinity of the ligand for the receptor-immunoglobulin chimera is sufficient for binding and purification. The conversion of a receptor into an antibody for cloning ligands is also of use in one of the currently available expression cloning formats as well. In this case, receptor-immunoglobulin can be coated onto matrices and used to pan cells expressing cDNAs encoding cell-surface ligands of interest. This form of the protocol could not, of course, have been used for the cloning of the 50-kD ligand, since a combination of both protein and glycosylation is required for appropriate ligand-receptor adhesive activity. The advantages of this procedure are manifold and include, most importantly, the ability to produce rapidly a reagent specific for a rare ligand without the need to resort to monoclonal antibody production.

The purification of the 50-kD L-selectin ligand consisted of a number of steps. The first step took advantage of previous work suggesting that the sulfated macromolecules produced by peripheral lymph node HEV are shed (Andrews et al. 1982; S. Watson, unpubl.). Thus, material shed in organ cultures from murine peripheral lymph node slices was isolated as the starting material. The next step took advantage of a novel property of the 50-kD ligand: its solubility after treatment with the organic solvent chloroform-methanol. Since most proteins are denatured by this treatment, this step resulted in a large degree of purification. The aqueous phase of this organic solvent treatment was further purified and concentrated by passage over a wheat-germ agglutinin column. The penultimate purification step consisted of binding the material eluted from wheat-germ agglutinin to an L-selectin-immunoglobulin column and eluting the column with the calcium chelating agent, EGTA. The entire purification process was followed by spiking the starting material with radioactive sulfate-labeled ligand. The material migrating on an SDS polyacrylamide gel in the ~50-kD region was eluted and subjected to amino-terminal microsequencing. This procedure resulted in an unambiguous sequence of 22 amino acids, from which an oligonucleotide probe was derived.

Screening of a murine peripheral lymph node library with the oligonucleotide probe derived from the purified 50-kD amino-terminal sequence resulted in the isolation of a single duplicate positive cDNA clone. Nucleotide sequencing of the clone revealed a short (~600 bp) sequence that encoded an open reading frame of 151 amino acids. The first 19 amino acids appeared to encode a hydrophobic signal sequence followed by the amino-terminal sequence predicted by protein microsequencing. No obvious transmembrane anchoring sequence was observed. However, the carboxy-terminal 19 amino acids encoded a highly significant amphipathic helix motif (see below). The most notable aspect of the coding sequence was its high

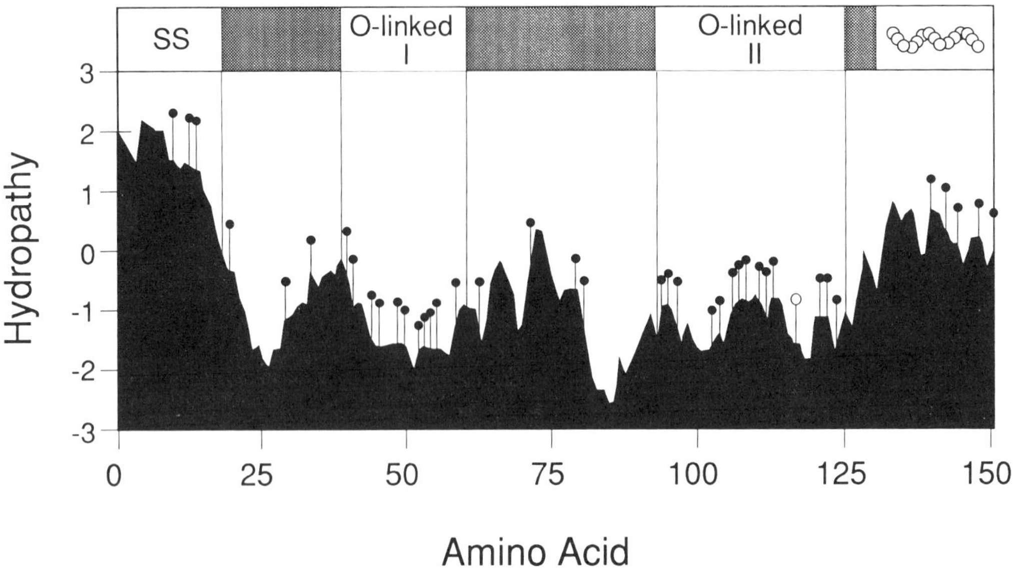

Figure 4. Domain structure of the protein predicted from the cDNA clone encoding the 50-kD sulfated L-selectin ligand. The figure illustrates a hydropathy plot (hydrophobic regions above the zero, hydrophilic regions below the zero) of the predicted amino acid sequence encoded by the L-selectin ligand cDNA clone. The dark balls correspond to serine or threonine residues (potential *O*-linked glycosylation sites), and the open ball corresponds to the single potential *N*-linked glycosylation site. The predicted domain structure of the ligand is shown above, with the signal sequence (SS), *O*-linked regions I and II, and carboxy-terminal amphipathic helical domains. (Reprinted, with permission, from Lasky et al. 1992.)

percentage of serine and threonine. The overall content of serine and threonine was ~30%, and the bulk of these residues was clustered into two domains containing more than 50% serine and threonine (*O*-linked regions I and II; Fig. 4). In addition, the native molecular weight of the encoded, processed protein was ~14,000. Since the molecular weight of the purified ligand was ~50,000, this suggests that the bulk (>70%) of the molecular weight of the protein is carbohydrate. Although the amino acid sequence of the 50-kD ligand was not related to any proteins in the protein sequence database, mucins are previously described proteins with high contents of serine and threonine, the majority of which contain *O*-linked carbohydrates (see below) (Sherbloom and Carraway 1980; Carraway and Hull 1991; Gendler et al. 1991; Gum et al. 1991; Perchet et al. 1991). Since the previous biochemical characterization of the ligand suggested that the bulk of the carbohydrate occurs on *O*-linked serine and threonine (Imai et al. 1991), these results suggested that the majority of the encoded serine and threonine residues contained *O*-linked carbohydrates and that the 50-kD ligand for L-selectin was mucin-like. Furthermore, it seemed inescapable that at least one of these *O*-linked carbohydrates was the sugar ligand recognized by the lectin domain of L-selectin. An analogous situation may be found in the case of the mucin-like mammalian egg sperm receptor, ZP3, which appears to bind sperm via a discrete set of *O*-linked carbohydrate chains (Florman and Wassarman 1985; Wassarman 1990).

Since previous work suggested that an L-selectin ligand would be expected to be tissue-specific to mediate regional lymphoid organ trafficking, mRNA expression analysis of various lymphoid and nonlymphoid organs was performed with the 50-kD ligand cDNA. In situ hybridization studies revealed that the mRNA encoding the L-selectin ligand was expressed only by the HEV cells in peripheral lymph nodes. Northern blot analysis showed that the mRNA encoding this ligand was expressed in a number of peripheral lymphoid sites, including the mesenteric lymph nodes and brachial, axillary, cervical, and popliteal lymph nodes. A more extensive tissue analysis revealed that the ligand was also expressed at reasonably high levels in lung lymphoid tissue and at very low levels in the gut-associated Peyer's patches (Bargatze et al. 1990; Hamann et al. 1991) but was not expressed in a number of nonlymphoid organs nor in spleen or thymus. Since these nonexpressing organs all contained high levels of endothelium, it may be concluded that the mRNA encoding this ligand is regionally expressed, and that this regional expression correlates with L-selectin-dependent trafficking pathways.

To characterize further the role of the 50-kD L-selectin ligand in regional lymphocyte trafficking, immunohistochemistry of peripheral lymph node HEV was performed using antisera directed against peptides predicted from the ligand cDNA sequence. As shown in Figure 5, the protein encoded by this cDNA appeared to be expressed on the luminal surface of the HEV. This result harkens back to the previous staining experiments with the L-selectin-immunoglobulin chimera (Fig. 2) (Watson et al. 1990) which demonstrated that the adhesive ligand recognized by this reagent was also expressed on the luminal surface of the HEV. This luminal expression would, of course, be expected for a glycoprotein whose role was to capture lymphocytes as they passed through the lumen of the HEV by a tethering interaction between the lectin domain of L-selectin and a carbohydrate ligand(s) ligated to a serine or threonine.

The sequence of the 50-kD L-selectin ligand, together with previous data on the biophysical characteristics of highly *O*-linked mucins, suggested a model for the structure of the 50-kD L-selectin ligand. Biophysical analyses of other mucins have demonstrated that they often form extended, relatively rigid, rodlike structures (Harding 1989; Jentoft 1990). The extended nature of these glycoproteins is, in large part, due to the inhibition of protein folding by the carbohydrate side chains that predominate throughout these molecules. In the same manner, it is highly likely that the two highly glycosylated regions, *O*-linked regions I and II, are also extended, relatively rigid, rod-like structures. As Figure 6 shows, we predict that these two domains will resemble "bottle brushes" with the highly extended amino acid backbones surrounded by *O*-linked carbohydrates. It is likely that these carbohydrate-rich domains act as scaffolds that present the *O*-linked carbohydrate ligands in a polyvalent fashion to the lectin domain of L-selectin. In this manner, the 50-kD L-selectin ligands can efficiently capture lymphocytes expressing L-selectin as they flow through the lumen of the HEV, thus directing regional lymphocyte trafficking and defining a novel mechanism for cell adhesion. The other, less highly glycosylated domains of the ligand are shown to be more highly compressed ("folded"). These domains may have other functions, such as a possible involvement in the intracellular trafficking of the ligand to the appropriate glycosylation pathway. Because the 50-kD ligand appears to act as a cell adhesion molecule by presenting polyvalent carbohydrates, we think that the previous, more cumbersome nomenclature (i.e., Sgp50 [Imai et al. 1991] or peripheral lymph node addressin [Streeter et al. 1988a]) should be replaced by the more descriptive term, *GLYCAM 1* (*GLY*cosylation-dependent *C*ell *A*dhesion *M*olecule 1).

Since the cDNA sequence did not reveal an obvious transmembrane anchoring motif, GLYCAM 1 may be associated with the cell surface by a relatively uncommon mechanism. Figure 6 illustrates two means by which GLYCAM 1 could associate with the luminal surface of the HEV. Panel A illustrates that GLYCAM 1 could associate via its carboxy-terminal amphipathic helices with a transmembrane protein. Such a mechanism appears to be the manner by which the mammalian egg sperm receptor, ZP3, associates with the hydrophobic ZP2 protein embedded within the egg cell surface. In the case of ZP3, this results in a thick glycocalyx that surrounds the egg with a high density of this sperm adhesion receptor. A similar type of thickened glycocalyx has been found in electron micrographs of peripheral lymph node HEV, and it is possible that much of this glycocalyx consists of loosely associated GLYCAM 1. An alternative type of membrane association is shown in panel B. In this case, GLYCAM 1 oligomerizes via its carboxy-terminal amphipathic helix such that the polar residues interact with each other, and the hydrophobic faces of the associated helices thermodynamically interact with the plasma membrane. In support of this hypothesis, gel filtration analysis has shown that GLYCAM 1 forms an oligomeric structure when shed from lymph node cultures (Y. Imai, unpubl.). However, an argument against this hypothesis is that transfection of GLYCAM 1 cDNA into kidney cells results in efficient secretion of the protein (C. Fennie, unpubl.)

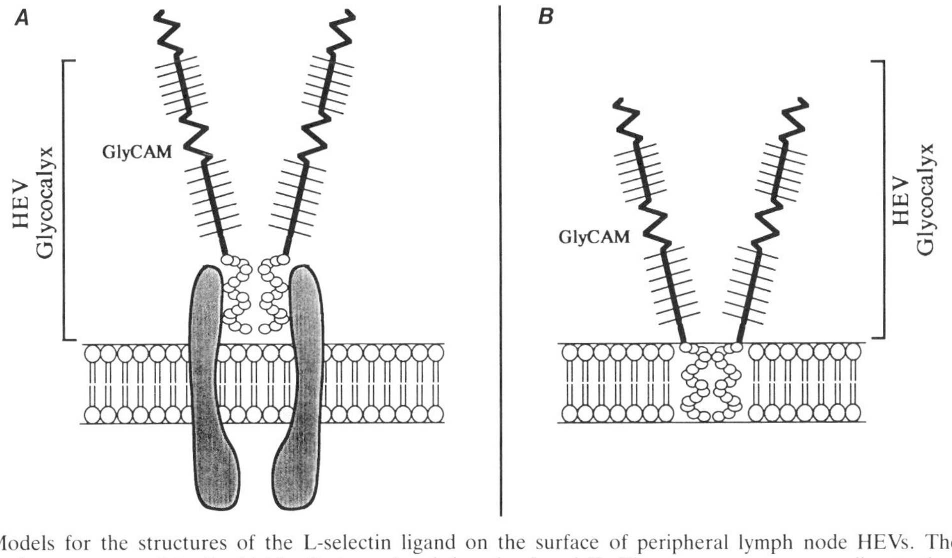

Figure 6. Models for the structures of the L-selectin ligand on the surface of peripheral lymph node HEVs. The extended brush-like regions correspond to the highly *O*-glycosylated domains I and II. The sequences corresponding to the less highly glycosylated amino-terminal and central domains are shown to be "compressed". Cell-surface attachment is accomplished either through an interaction with a transmembrane protein (*A*) or via an interaction between the amphipathic helix regions and the lipid bilayer. (Reprinted, with permission, from Lasky et al. 1992.)

SPECULATIONS AND FUTURE DIRECTIONS

The discovery of GLYCAM 1 demonstrates that cell adhesion may be accomplished by the tissue-specific expression of a scaffold for the presentation of carbohydrates to a selectin lectin domain. This tissue-specific expression of the appropriate carbohydrate ligand on the pertinent scaffold may thus direct lymphocytes to regional lymphoid sites by mediating adhesion between the HEV and cell-surface-localized L-secretin. These results suggest that this glycoprotein represents the fifth category of cell adhesion molecule (Springer 1990): the homotypically adherent cadherins, the integrins, their ligands, the immunoglobulin superfamily members, the selectins, and GLYCAM 1, a ligand for L-selectin. Even though the results described here are compelling, perhaps more exciting is that they suggest a number of interesting speculations and future directions to pursue.

Whereas the data described above are consistent with the ligand nature of GLYCAM 1, the relationships between this ligand and the 90-kD ligand remain to be determined. Comparative analyses of the 90-kD ligand versus GLYCAM 1 revealed that they interacted with the L-selectin immunoglobulin chimera in identical manners (Imai et al. 1991; Y. Imai, unpubl.). Initially, we suspected that the 90-kD ligand was a dimer of GLYCAM 1, but antisera directed against peptides from GLYCAM 1 do not cross-react with the 90-kD molecule. In addition, the 90-kD ligand, unlike GLYCAM 1, is not shed into the medium of lymph node organ cultures, suggesting a tighter interaction with the endothelial cell surface. Interestingly, there are also in vivo data demonstrating that GLYCAM 1 is shed into the serum (M. Brustein and S. Watson, unpubl.), suggesting a potentially important role for this shed form of the L-selectin ligand. Currently, we can only speculate on the physiological role of the 90-kD ligand versus GLYCAM 1. One possibility is that the 90-kD ligand is the transmembrane molecule which binds GLYCAM 1 to the endothelial cell surface (see Fig. 6A). It is not clear, however, why such a transmembrane anchoring protein would need to have L-selectin ligand activity to ensure its function. A second possibility is that shed GLYCAM 1 performs other physiological roles, such as lymphocyte activation or chemotaxis, whereas the 90-kD ligand is the actual HEV adhesive ligand for lymphocyte trafficking. An argument against this proposition is that staining of HEV with the antisera against GLYCAM 1 peptides reveals a luminal staining pattern (Fig. 5), as would be expected for an adhesive ligand such as GLYCAM 1. A final possibility is that both GLYCAM 1 and the 90-kD ligand act as adhesive molecules, but the 90-kD ligand may act as an "inflammation-specific" ligand, whereas GLYCAM 1 may be specific for peripheral lymph node HEV trafficking (Lewinsohn et al. 1987; Jutila et al. 1989; Ley et al. 1991; Von Andrian et al. 1991; Watson et al. 1991a). This might explain the relatively low levels of 90-kD ligand versus GLYCAM 1 in extracts of peripheral lymph nodes. Ultimately, the cloning and characterization of the cDNA encoding the 90-kD ligand will serve to answer many of these questions. In addition, the generation of a mouse that is homozygous for disruptive mutations in the GLYCAM 1 gene (a so-called "knock out" mouse) will further our understanding of the relative importance of GLYCAM 1 to lymph node trafficking.

A second interesting line of investigation is the nature of the carbohydrate ligand(s) that is specifically added to GLYCAM 1. The inclusion of sulfate in the carbohydrate side chains of GLYCAM 1 is an interesting and novel difference between these carbohydrates and the known carbohydrate ligands, sialyl Lewis x and a, for E- and P-selectin. That the selectins recognize similar carbohydrate ligands is suggested by the dependence on sialic acid for adhesion mediated by all three molecules, the inclusion of fucose in all three carbohydrate ligands (although the role of fucose in the L-selectin carbohydrate ligand has yet to be determined), and the demonstration that all three selectins can bind to sialyl Lewis x in vitro (C. Foxall, unpubl.). It will now be possible to test the importance of sulfate in the ligand activity of the GLYCAM 1 carbohydrate side chains through the combined use of peptide antisera directed against the protein portion of this glycoprotein ligand, inhibitors of sulfate addition, such as chlorate, and reactivity of sulfate-poor GLYCAM 1 with the L-selectin immunoglobulin chimera. In addition, these antipeptide antisera will be useful for following the biosynthetic pathway of GLYCAM 1 synthesis in the HEV. Thus, an analysis of where the appropriate carbohydrate modifications are accomplished to induce ligand activity can now be performed through a combination of peptide antisera and L-selectin immunoglobulin chimera assays. In addition, the production of large quantities of active GLYCAM 1 should now be possible. This may be accomplished by high-level expression of this glycoprotein in transfected endothelial cells. If these cells have the appropriate glycosyl and sulfate transferases, then stable cell lines expressing large quantities of active GLYCAM 1 may be obtained. This will allow for the characterization, by mass spectroscopy, of the active carbohydrates purified from the GLYCAM 1 produced by these stable transfectants.

An additional question that will be answered by transfection of cells and production of active GLYCAM 1 is the manner by which the appropriate carbohydrates are added to this glycoprotein and not to others. Although a considerable amount of information exists regarding the requirements of *N*-linked glycosylation, very little is known about how *O*-linked sites are chosen for carbohydrate addition. In the case of GLYCAM, not only is there a preponderance of *O*-linked carbohydrates, but there is also the question of how a highly specific type of carbohydrate, the ligand(s) for L-selectin, is added to this protein. Is there a protein-sequence-specific signal that allows for the intracellular trafficking of the GLYCAM polypeptide through a divergent glycosylation pathway or are there specific signals contained within the protein that are involved with the addition of the appropriate carbohydrate modifications when the protein passes through the constitutive secretion/glycosylation pathway? Are both *O*-linked regions used for L-selectin ligand addition or is one a structural "stalk" (Jentoft 1990), whereas the other acts as a scaffold for ligand presentation? Can the GLYCAM backbone be modified by other carbohydrates, such as sialyl Lewis x, and function as ligand for E- or P-selectin? What is the nature of the enzymes involved in carbohydrate ligand biosynthesis? Are the same sialyl and fucosyl transferases that are used for sialyl Lewis x synthesis used for synthesis of the L-selectin ligand? What is the nature of the undoubtedly unique sulfate transferase enzyme that adds sulfate residues to the carbohydrate chains of GLYCAM?

Questions regarding the cell biology of GLYCAM will need to be investigated. One important issue is the manner in which GLYCAM is bound to the cell surface. The apparently loose association of GLYCAM with the cell surface in vitro and in vivo suggests the possibility that the protein is a peripheral membrane protein. It will therefore be interesting to characterize any other, membrane-associated proteins that bind GLYCAM to the cell surface. In addition, the physiological relevance of shedding in vivo will require investigation. One possibility is that the shed form of the ligand binds to the lymphocyte surface and mediates a signal transduction pathway in the lymphocyte. Alternatively, the rapid shedding may merely be a means by which a high degree of cell adhesion regulation is accomplished. Another interesting aspect is the manner by which the ligand is luminally synthesized. Can this luminal synthesis by replicated in nonendothelial cell lines (i.e., in polarized epithelial cells)? What are the protein signals that mediate luminal expression? Is the carboxy-terminal amphipathic helix involved in this intracellular trafficking (Segrest et al. 1990)?

The potential role(s) of GLYCAM 1 in pathogenic inflammation will be a fruitful avenue of investigation with respect to the development of therapeutic inhibitors of L-selectin-mediated inflammatory responses. Previous data have demonstrated that L-selectin on neutrophil surfaces is involved with acute inflammatory responses and that these reactions are due to the mediation of neutrophil rolling by L-selectin (Lewinsohn et al. 1987; Jutila et al. 1989; Ley et al. 1991; Von Andrian et al. 1991; Watson et al. 1991a). It will be of great interest to determine if the ligand(s) involved in ectopic (i.e., nonperipheral lymph node) adhesion and rolling of neutrophils is the same as GLYCAM 1. Can GLYCAM 1 be induced in non-HEV cells by inflammatory mediators such as TNF (Spertini et al. 1991)? Will neutrophils roll and inflame acute sites in mice that have homozygous disruptions of the GLYCAM 1 gene? Will an active form of GLYCAM 1 be effective at inhibiting neutrophil rolling and influx into acute inflammatory sites? Can a form of GLYCAM expressing polyvalent sialyl Lewis x be used to inhibit inflammation mediated by the other selectins? It will also be important to examine the expression of GLYCAM in human pathogenic disease states. This will require the cloning of the human homolog of this glycoprotein and the production of antibodies directed against it for histochemical analyses of the endothelium in various disease states.

Finally, the manner by which the GLYCAM 1 gene is specifically regulated will prove to be of great interest. Initial data have demonstrated that GLYCAM mRNA is highly tissue-specific, consistent with the possibility that local environmental cues regulate the expression of the gene. In support of this contention, ligation of the afferent lymphatic vessels results in a virtually complete down-regulation of GLYCAM mRNA synthesis (Mebius et al. 1991; R. Mebius and D. Dowbenko, unpubl.), consistent with the hypothesis that localized factors either brought in or induced by afferent lymphatic flow are involved with GLYCAM 1 gene expression. Analysis of the genomic structure of the murine GLYCAM 1 gene has revealed that the coding sequence is disrupted into four separate exons and that these exons correlate, with some degree of accuracy, to functional domains of the protein. Interestingly, a series of glucocorticoid response elements has been found in the 5′ regulatory region of the GLYCAM gene (D. Dowbenko and P. Young, unpubl.). The biological relevance of this result is enhanced by the finding that GLYCAM mRNA is dramatically induced by lactating mammary gland tissue, but is at only low levels in postweaning breast tissue. This result suggests a potential role for this adhesion molecule in mammary gland immune function, perhaps as a mediator of lymphocyte trafficking to the breast. Finally, it will be very interesting to make transgenic animals containing various constructs derived from the GLYCAM gene in order to investigate the role(s) of genomic regulatory sites in tissue-specific gene expression.

CONCLUSION

GLYCAM 1 is the primary example of a new type of adhesion molecule that mediates lymphocyte trafficking by the tissue-specific presentation of carbohydrates to L-selectin. As the first example of such an adhesion molecule, it will provide a number of novel avenues for investigation. In addition, it may be that this glycoprotein is the initial example of a family of proteins whose role is to act as scaffolds for the presentation of carbohydrate ligands to lectin domains. Currently, there are at least three cell-surface mucins, the mammalian egg sperm receptor ZP3 (Wassarman 1990), leukosialin (CD43) (Cyster et al. 1990, 1991), and the early marker of hematopoietic cells, CD34 (Simmons et al. 1992), that may be additional examples of such adhesion molecules. Future analyses should provide added information about the generality of the use of these mucin-like molecules in cell adhesion events.

REFERENCES

Andrews, P., D. Milsom, and W. Ford. 1982. Migration of lymphocytes across specialized vascular endothelium. V. Production fo sulphated macromolecule by high endothelial cells in lymph nodes. *J. Cell. Sci.* **57:** 277.

Bargatze, R.F., P.R. Streeter, and E.C. Butcher. 1990. Expression of low levels of peripheral lymph node-associated vascular addressin in mucosal lymphoid tissues: Possible relevance to the dissemination of passaged AKR lymphomas. *J. Cell. Biochem.* **42:** 219.

Berg, E., M. Robinson, R. Warnock, and E.C. Butcher. 1991. The human peripheral lymph node vascular addressin is a ligand for LEC-CAM 1, the peripheral lymph node homing receptor. *J. Cell Biol.* **114:** 343.

Berg, E.L., L.A. Goldstein, M.A. Jutila, M. Nakache, L.J. Picker, P.R. Streeter, N.W. Wu, D. Zhou, and E.C. Butcher. 1989. Homing receptors and vascular addressins: Cell adhesion molecules that direct lymphocyte traffic. *Immunol. Rev.* **108:** 5.

Bevilacqua, M.P., S. Spengeling, M.A. Gimbrone, Jr., and B. Seed. 1989. Endothelial leukocytes adhesion molecule 1: An inducible receptor for neutrophils related to complement regulatory proteins and lectins. *Science* **243:** 1160.

Bevilacqua, M.P., E. Butcher, B. Furie, M. Gallatin, M. Gimbrone, J. Harlan, K. Kishimoto, L. Lasky, R. McEver, J. Paulson, S. Rosen, B. Seed, M. Siegelman, T. Springer, L. Stoolman, T. Tedder, A. Varki, D. Wagner, I. Weissman, and G. Zimmerman. 1991. Selectins: A family of adhesion receptors. *Cell* **67:** 233.

Bowen, B., C. Fennie, and L.A. Lasky. 1990. The Mel-14 antibody binds to the lectin domain of the murine peripheral lymph node homing receptor. *J. Cell Biol.* **110:** 147.

Bowen, B.R., T. Nguyen, and L.A. Lasky. 1989. Characterization of a human homologue of the murine peripheral lymph node homing receptor. *J. Cell Biol.* **109:** 421.

Butcher, E.C. 1990. Cellular and molecular mechanisms that direct leukocyte traffic. *Am. J. Pathol.* **136:** 3.

Butcher, E.C., R.G. Scollay, and I.L. Weissman. 1980. Organ specificity of lymphocyte migration: Medication by highly selective lymphocyte interaction with organ-specific determinants on high endothelial venules. *Eur. J. Immunol.* **10:** 556.

Carraway, K.L. and S.R. Hull. 1991. Cell surface mucin-type glycoproteins and mucin-like domains. *Glycobiology* **1:** 131.

Cyster, J.C., D.M. Shotton, and A.F. Williams. 1991. The dimensions of the T lymphocyte glycoprotein leukosialin and identification of linear protein epitopes that can be modified by glycosylation. *EMBO J.* **10:** 893.

Cyster, J., C. Somoza, N. Killeen, and A.F. Williams. 1990. Protein sequence and gene structure for mouse leukosialin (CD43), a T lymphocyte mucin without introns in the coding sequence. *Eur. J. Immunol.* **20:** 875.

Drickamer, K. 1988. Two distinct classes of carbohydrate-recognition domains in animal lectins. *J. Biol. Chem.* **263:** 9557.

Florman, H. and P. Wassarman. 1985. *O*-linked oligosaccharides of mouse egg ZP3 account for its sperm receptor activity. *Cell* **41:** 313.

Gallatin, W.M., I.L. Weissman, and E.C. Butcher. 1983. A cell surface molecule involved in organ-specific homing of lymphocytes. *Nature* **303:** 30.

Gendler, S.J., A.P. Spicer, E.N. Lalani, T. Duhig, N. Peat, J. Burchell, L. Pemberton, M. Boshell, and J.T. Papadimitriou. 1991. Structure and biology of a carcinoma-associated mucin, MUC1. *Am. Rev. Respir. Dis.* **144:** S42.

Geoffrey, J.S. and S.D. Rosen. 1989. Demonstration that a lectin-like receptor (gp90^{MEL}) directly mediates adhesion of lymphocytes to high endothelial venules of lymph nodes. *J. Cell Biol.* **109:** 2463.

Gowans, J. and E. Knight. 1964. The route of recirculation of lymphocytes in the rat. *Proc. R. Soc. Lond. B Biol. Sci.* **159:** 257.

Gum, J.R., Jr., J.W. Hicks, R.E. Lagace, J.C. Byrd, N.W. Toribara, B. Siddiki, F.J. Fearney, D.T.A. Lamport, and Y.S. Kim. 1991. Molecular cloning of rat intestinal mucin. *J. Biol. Chem.* **266:** 22733.

Hamann, A., D. Jablonski-Westrich, and H.-G. Thiele. 1991. Homing receptors reexamined: Mouse LECAM-1 (MEL-14 antigen) is involved in lymphocyte migration into gut-associated lymphoid tissue. *Eur. J. Immunol.* **21:** 2925.

Harding, S.E. 1989. The macrostructure of mucus glycoproteins in solution. *Adv. Carbohydr. Chem. Biochem.* **47:** 345.

Imai, Y., D.D. True, M.S. Singer, and S.D. Rosen. 1990. Direct demonstration of the lectin activity of gp90^{MEL}, a lymphocyte homing receptor. *J. Cell Biol.* **111:** 1225.

Imai, Y., M. Singer, C. Fennie, L. Lasky, and S. Rosen. 1991. Identification of a carbohydrate-based endothelial ligand for a lymphocyte homing receptor. *J. Cell Biol.* **113:** 1213.

Jentoft, N. 1990. Why are proteins *O*-glycosylated? *Trends Biochem. Sci.* **15:** 291.

Johnston, G.I., R.G. Cook, and R.P. McEver. 1989. Cloning of GMP-140, a granule membrane protein of platelets and endothelium: Sequence similarity to proteins involved in cell adhesion and inflammation. *Cell* **56:** 1033.

Jutila, M.A., L. Rott, E.L. Berg, and E.C. Butcher. 1989. Function and comparison of the neutrophil MEL-14 antigen in vivo: Comparison with LFA-1 and MAC-1. *J. Immunol.* 3318.

Lasky, L.A., M.S. Singer, T.A. Yednock, D. Dowbenko, C. Fennie, H. Rodriguez, T. Nguyen, S. Stachel, and S.D. Rosen. 1989. Cloning of a lymphocyte homing receptor reveals a lectin domain. *Cell* **56:** 1045.

Lasky, L.A., M. Singer, D. Dowbenko, Y. Imai, W. Henzel, C. Grimley, C. Fennie, N. Gillett, S. Watson, and S.D. Rosen. 1992. An endothelial ligand for L selectin is a novel mucin-like molecule. *Cell* **69:** 927.

Lewinsohn, D.M., R.F. Bargatze, and E.C. Butcher. 1987. Leukocyte-endothelial cell recognition: Evidence of a common molecular mechanism shared by neutrophils, lymphocytes, and other leukocytes. *J. Immunol.* **138:** 4313.

Ley, K., P. Gaehtgens, C. Fennie, M.S. Singer, L.A. Lasky, and S.D. Rosen. 1991. Lectin-like cell adhesion molecule 1 mediates leukocyte rolling in mesenteric venules in vivo. *Blood* **77:** 2553.

Marchesi, V. and J. Gowans. 1964. The migration of lymphocytes through the endothelium of venules in lymph nodes. *Proc. R. Soc. Lond. B Biol. Sci.* **159:** 283.

Mebius, R.E., P.R. Streeter, J. Brevé, A.M. Duijvestijn, and G. Kraal. 1991. The influence of afferent lymphatic vessel interruption on vascular addressin expression. *J. Cell Biol.* **115:** 85.

Phillips, M.L., E. Nudelman, F.C.A. Gaeta, M. Perez, A.K. Singhai, S.-I. Hakomori, and J.C. Paulson. 1990. ELAM-1 mediates cell adhesion by recognition of a carbohydrate ligand, sialyl-Le^{x}. *Science* **250:** 1130.

Porchet, N., J. Dufosse, J.P. Audie, V.G. Duperat, J.M. Perini, N.V. Cong, P. Degand, and J.P. Aubert. 1991. Structural features of the core proteins of human airway mucins ascertained by cDNA cloning. *Am. Rev. Resp. Dis.* **144:** S15.

Rosen, S.D. 1989. Lymphocyte homing: Progress and prospects. *Curr. Opin. Cell Biol.* **1:** 913.

Rosen, S.D., M.S. Singer, T.A. Yednock, and L.M. Stoolman. 1985. Involvement of sialic acid on endothelial cells in organ-specific lymphocyte recirculation. *Science* **228:** 1005.

Rosen, S.D., S.-I. Chi, D.D. True, M.S. Singer, and T.A. Yednock. 1989. Intravenously injected sialidase inactivates attachment sites for lymphocytes on high endothelial venules. *J. Immunol.* **142:** 1895.

Segrest, J.P., H. De Loof, J.G. Dohlman, C.G. Brouillette, and G.M. Anantharamaiah. 1990. Amphipathic helix

motif: Classes and properties. *Proteins Struct. Funct. Genet.* **8:** 103.
Sherbloom, A.P. and K.L. Carraway. 1980. A complex of two cell surface glycoproteins from ascites mammary adenocarcinoma cells. *J. Biol. Chem.* **255:** 12051.
Siegelman, M.H., M.V. de Rijn, and I.L. Weissman. 1989. Mouse lymph node homing receptor cDNA clone encodes a glycoprotein revealing tandem interaction domains. *Science* **243:** 1165.
Simmons, D.L., A.B. Satterthwaite, D.G. Tenen, and B. Seed. 1992. Molecular cloning of a cDNA encoding CD34, a sialomucin of human hematopoietic stem cells. *J. Immunol.* **148:** 267.
Spertini, O., F.W. Luscinskas, G.S. Kansas, J.M. Munro, J.D. Griffin, M.A.J. Gimbrone, and T.F. Tedder. 1991. Leukocyte adhesion molecule (LAM-1, L-selectin) interacts with an inducible endothelial ligand to support leukocyte adhesion. *J. Immunol.* **147:** 2565.
Springer, T.A. 1990. Adhesion receptors of the immune system. *Nature* **346:** 425.
Stoolman, L.M. 1989. Adhesion molecules controlling lymphocyte migration. *Cell* **56:** 907.
Stoolman, L.M. and H. Ebling. 1989. Adhesion molecules of cultured hematopoietic malignancies. A calcium-dependent lectin in the principle mediator of binding to the high endothelial venule of lymph nodes. *J. Clin. Invest.* **84:** 1196.
Stoolman, L.M. and S.D. Rosen. 1983. Possible role for cell surface carbohydrate-binding molecules in lymphocyte recirculation. *J. Cell Biol.* **96:** 722.
Stoolman, L.M., T.S. Tenforde, and S.D. Rosen. 1984. Phosphomannosyl receptors may participate in the adhesive interactions between lymphocytes and high endothelial venules. *J. Cell Biol.* **99:** 1535.
Stoolman, L.M., T.A. Yednock, and S.D. Rosen. 1987. Homing receptors on human and rodent lymphocytes-evidence for a conserved carbohydrate-binding specificity. *Blood* **70:** 1842.
Streeter, P.R., B.T.N. Rouse, and E.C. Butcher. 1988a. Immunohistologic and functional characterization of a vascular addressin involved in lymphocyte homing into peripheral lymph nodes. *J. Cell Biol.* **107:** 1853.
Streeter, P.R., E.L. Berg, B.T.N. Rouse, F. Bargatze, and E.C. Butcher. 1988b. A tissue-specific endothelial cell molecule involved in lymphocyte homing. *Nature* **331:** 41.
Tedder, T.F., C.M. Isaacs, T.J. Ernst, G.D. Demetri, D.A. Adler, and C.M. Disteche. 1989. Isolation and chromosomal localization of cDNAs encoding a novel human lymphocyte cell surface molecule, LAM-1. Homology with the mouse lymphocyte homing receptor and other human adhesion molecules. *J. Exp. Med.* **170:** 123.
Von Andrian, U., J.D. Chambers, L.M. McEvoy, R.F. Bargatze, K.E. Arfors, and E.C. Butcher. 1991. Two-step model of leukocyte-endothelial cell interaction in inflammation. *Proc. Natl. Acad. Sci.* **88:** 7538.
Walz, G., A. Aruffo, W. Kolanus, M. Bevilacqua, and B. Seed. 1990. Recognition by ELAM-1 of the sialyl-Le^x determinant on myeloid and tumor cells. *Science* **250:** 1132.
Wassarman, P.M. 1990. Profile of a mammalian sperm receptor. *Development* **108:** 1.
Watson, S., C. Fennie, and L.A. Lasky. 1991a. Neutrophil influx into an inflammatory site inhibited by a soluble homing receptor-IgG chimaera. *Nature* **349:** 164.
Watson, S., Y. Imai, C. Fennie, J.S. Geoffrey, S.D. Rosen, and L.A. Lasky. 1990. A homing receptor-IgG chimera as a probe for adhesive ligands of lymph node high endothelial venules. *J. Cell Biol.* **110:** 2221.
Watson, S., Y. Imai, C. Fennie, J. Geoffrey, M. Singer, S. Rosen, and L.A. Lasky. 1991b. The complement binding-like domains of the murine homing receptor facilitate lectin activity. *J. Cell Biol.* **115:** 235.
Williams, A.F. 1991. Out of equilibrium. *Nature* **352:** 473.
Woodruff, J.J., L.M. Clarke, and Y.H. Chin. 1987. Specific cell-adhesion mechanisms determining migration pathways of recirculating lymphocytes. *Annu. Rev. Immunol.* **5:** 201.
Yednock, T.A. and S.D. Rosen. 1989. Lymphocyte homing. *Adv. Immunol.* **44:** 313.
Yednock, T.A., L.M. Stoolman, and S.D. Rosen. 1987a. Phosphomannosyl-derivatized beads detect a receptor involved in lymphocyte homing. *J. Cell Biol.* **104:** 713.
Yednock, T.A., E.C. Butcher, L.M. Stoolman, and S.D. Rosen. 1987b. Receptors involved in lymphocyte homing: Relationship between a carbohydrate-binding receptor and the MEL-14 antigen. *J. Cell Biol.* **104:** 725.

Quantitative Determination of Selectin-Carbohydrate Interactions

R.M. Nelson,* A. Aruffo,† S. Dolich,* O. Cecconi,* G. Mannori,* and M.P. Bevilacqua*

*Howard Hughes Medical Institute, Cellular and Molecular Medicine and Department of Pathology, University of California, San Diego, La Jolla, California 92093-0669;
†Bristol Myers-Squibb Pharmaceutical Research Institute, Seattle, Washington 98121

The selectins are a recently described family of transmembrane glycoproteins that support adhesion of blood leukocytes to the vessel wall during inflammatory and immunological processes. The three known members have been designated E-, P-, and L-selectin for the cell type on which they were identified: endothelium, platelets, and leukocytes, respectively (Bevilacqua et al. 1991). Each contains an amino-terminal lectin-like domain, an epidermal growth factor (EGF) repeat, and a variable number of complement regulatory-like repeats (Bevilacqua et al. 1989; Camerini et al. 1989; Johnston et al. 1989; Lasky et al. 1989; Siegelman et al. 1989; Tedder et al. 1989). E-selectin is expressed on the surface of endothelial cells in response to bacterial endotoxin or inflammatory cytokines, such as interleukin-1 (IL-1) and tumor necrosis factor (TNF) (Bevilacqua et al. 1987, 1989). Assays in vitro using cultured endothelial monolayers and cells transfected with cDNA encoding E-selectin have shown that this molecule can support the adhesion of neutrophils, monocytes, and a subpopulation of T-lymphocytes (Bevilacqua et al. 1987, 1989; Graber et al. 1990; Hession et al. 1990; Carlos et al. 1991; Hakkert et al. 1991; Kuijpers et al. 1991; Luscinskas et al. 1991; Picker et al. 1991; Shimizu et al. 1991a,b; Leeuwenberg et al. 1992). P-selectin is found in storage granules of platelets and endothelial cells, from which it can be rapidly redistributed to the cell surface following stimulation by thrombin, histamine, and other mediators (Hsu-Lin et al. 1984; Stenberg et al. 1985; Bonfanti et al. 1989; McEver et al. 1989). Recent studies have demonstrated that P-selectin can mediate platelet and endothelial adhesive interactions with leukocytes (Larsen et al. 1989; Geng et al. 1990). L-selectin, normally found on the surface of most lymphocytes, supports their adhesion to high endothelial venules of peripheral lymph nodes (for review, see Stoolman 1989; Picker and Butcher 1992). It is also expressed by circulating polymorphonuclear leukocytes (PMN, neutrophils) and monocytes and can support their adhesion to cytokine-activated endothelium (Lewinsohn et al. 1987; Kishimoto et al. 1989; Smith et al. 1991; Spertini et al. 1991, 1992). Selectins may be especially important in leukocyte rolling on the vessel wall, a process that can precede firm attachment and extravasation during inflammation (Lawrence and Springer 1991; Ley et al. 1991; von Andrian et al. 1991).

The similarity of amino-terminal regions of selectins to the carbohydrate recognition domains found in C-type lectins (Drickamer 1988) suggested that their adhesive properties involved the recognition of carbohydrates. Intensive efforts to identify oligosaccharide ligands have ensued (for review, see Brandley et al. 1990; Rosen 1990; McEver 1991; Springer and Lasky 1991; Varki 1992b; Bevilacqua and Nelson 1993). Most studies on selectin-carbohydrate binding interactions have used cellular systems to facilitate the characterization of natural ligands. This work has resulted in the identification of sialyl Lewis x (sLe^x; Neu5Acα2-3Galβ1-4[Fucα1-3]GlcNAc), a structure found in abundance on PMN and monocytes (Symington et al. 1985; Ohmori et al. 1989; Munro et al. 1992), as a ligand of E-selectin (Goelz et al. 1990; Lowe et al. 1990; Phillips et al. 1990; Walz et al. 1990; Berg et al. 1991; Tiemeyer et al. 1991. P-selectin-dependent rosetting of activated platelets on leukocytes may involve similar oligosaccharides, since this process can be blocked by LNF-III, a pentasaccharide containing the Lewis x determinant (Larsen et al. 1990). It has also been demonstrated that sialic acid can be a component of P-selectin ligands (Corral et al. 1990; Moore et al. 1991) and that oligosaccharides containing sLe^x are recognized by this molecule (Polley et al. 1991; Zhou et al. 1991). More recently, murine L-selectin was also shown to interact with sLe^x-containing structures (Berg et al. 1992; Foxall et al. 1992). In addition, human E- and P-selectin and murine L-selectin have been shown to interact with molecules containing sLe^a (Neu5Acα2-3Galβ1-3[Fucα1-4]GlcNAc), a structural isomer of sLe^x (Berg et al. 1991, 1992; Handa et al. 1991; Tyrrell et al. 1991). These studies supported the hypothesis that oligosaccharide-binding specificities of the selectins are similar or identical. Finally, L- and P-selectin have been shown to interact with sulfated polysaccharides such as heparin, fucoidan, and dextran sulfate, and with 3-*O*-sulfate-galactosyl ceramide (sulfatides) (Yednock et al. 1987; Skinner et al. 1989; Imai et al. 1990; Aruffo et al. 1991; Handa et al. 1991). The molecular basis for binding specificity of selectins to oligosaccharides and polysaccharides has not yet been determined.

We have used cell adhesion assays and direct binding studies to quantitate the molecular interactions of selectins with oligosaccharide ligands. Our efforts have focused on the tetrasaccharides sLe^x, sLe^a, and related structures.

METHODS

Proteins and carbohydrates. Recombinant fusion proteins constructed from extracellular portions of human selectins and human immunoglobulin heavy-chain CH3, CH2, and hinge regions were prepared as described previously (Walz et al. 1990; Aruffo et al. 1991, 1992). These soluble selectin-Ig proteins contain the signal sequence, lectin-like domain, and EGF module along with 6, 2, and 2 of the complement regulatory modules for E-, P-, and L-selectin-Ig, respectively. Bovine serum albumin (BSA) neoglycoproteins with covalently linked sLe^x, sLe^a, Le^x, or Le^a were prepared as described previously (Pinto and Bundle 1983) and were kindly provided by Chembiomed Ltd. (Edmonton, Alberta). The oligosaccharides were synthesized with an aliphatic aglycone (8-methoxycarbonyloctyl; $-[CH_2]_8CO_2CH_3$) attached in a β-glycosidic linkage to the GlcNAc (Table 1) and linked to BSA through this structure (Lemieux et al. 1975). Incorporation of the aglycone glycoside forms of sLe^x, sLe^a, Le^x, and Le^a into the respective BSA neoglycoproteins (mole ratio) was 9, 13, 13, and 17. These and other oligosaccharides provided by Chembiomed Ltd. were unambiguously synthesized using a chemo-enzymatic route, purified by chromatographic methods with the final step on Biogel P-2, and freeze-dried in vacuo (P.V. Nikrad et al., in prep.). Most of the compounds were synthesized with the aglycone mentioned above, and all were assessed for purity by ^{1}H-nuclear magnetic resonance spectroscopy. Reducing tetrasaccharides sLe^x and sLe^a (sLe^x–OH and sLe^a–OH, see Table 1) were purchased from Oxford Glycosystems Ltd. (Rosedale, New York). Heparin, fucoidan, sulfatides (3-*O*-sulfate-galactosyl ceramide), and cerebrosides (galactosyl ceramide) were purchased from Sigma (St. Louis, Missouri). Sulfatides and cerebrosides were reconstituted with chloroform:methanol (1:1) at 10 mg/ml and stored at −20°C.

Cell adhesion assays. Human PMN were isolated from heparinized blood using a double density gradient of Histopaque 1077 and 1119 (Sigma, St. Louis, Missouri), essentially as described previously (English and Anderson 1974; Nelson et al. 1993). Contaminating red blood cells were removed via hypotonic lysis. Cell viability was assessed using trypan blue exclusion, and the cells were resuspended at a concentration of 2×10^6 cells/ml in phosphate-buffered saline (PBS) containing 0.5 mg/ml human serume albumin (HSA), 3 mM glucose, and 0.3 units aprotinin/ml (adhesion assay buffer). Human umbilical vein endothelial cells (Clonetics Corp., San Diego, California) were seeded in tissue culture wells that had been coated with 0.1% gelatin in water and allowed to grow to confluence in medium M199 (Gibco, Grand Island, New York) supplemented with 10% fetal calf serum (FCS), 5 units heparin/ml, and 15 μg endothelial cell growth supplement/ml (Sigma). Prior to adhesion assays, endothelial cell monolayers were washed once to remove cellular debris and activated with 200 units TNF-α/ml (Biogen Research Corp., Cambridge, Massachusetts) in M199 -10% FCS for 4 hours at 37°C (Bevilacqua et al. 1987).

Adhesion assays were performed in Nunclon MicroWell Terasaki plates (Nunc, Naperville, Illinois)

Table 1. Oligosaccharide Abbreviations and Structures

Abbreviations[a]	Structure[b]
sLacNAc	Neu5Acα2-3 Galβ1-4GlcNAcβ–OR
sLe^x	Neu5Acα2-3 Galβ1-4(Fucα1-3)GlcNAcβ–OR
sLe^xOCH_3	······ β–OCH_3
sLe^x–OH	······ GlcNAc
sLe^x(Glc)	······ Glc
$sLe^x(GlcN_3)$	······ $GlcN_3\beta$–OR
$sLe^x(GlcNH_2)$	······ $GlcNH_2\beta$–OR
sLe^x(GlcNHPr)	······ GlcNHPrβ–OR
$KDNLe^x$	Neu5OH ······ GlcNAcβ–OR
Fuc1-6sLacNAc	Neu5Acα2-3 Galβ1-4(Fucα1-6)GlcNAcβ–OR
sLe^c	Neu5Acα2-3 Galβ1-3GlcNAcβ–OR
$sLe^c(GlcN_3)$	······ $GlcN_3\beta$–OR
$sLe^c(GlcNH_2)$	······ $GlcNH_2\beta$–OR
sLe^c(GlcNHPr)	······ GlcNHPRβ–OR
sLe^a	Neu5Acα2-3 Galβ1-3(Fucα1-4)GlcNAcβ–OR
sLe^a–OH	······ GlcNAc
$sLe^a(GlcN_3)$	······ $GlcN_3\beta$–OR
$sLe^a(GlcNH_2)$	······ $GlcNH_2\beta$–OR
sLe^a(GlcNHPr)	······ GlcNHPrβ–OR
$2\text{-}6sLe^a$	Neu5Acα2-6 ······ GlcNAcβ–OR
2-6sLac	Neu5Acα2-6 Galβ1-4Glcβ–OR
2-6sGlcNAc	······ GlcNAcβ–OR
Sialyl Tn	······ GalNAcα–OR
A-trisacch.	GalNAcα1-3(Fucα1-2)Gal

[a]s = Neu5Ac; Pr = $C(O)CH_2CH_3$; KDN = keto-deoxy-nonulosonic acid.
[b]Dots indicate identity with the structure above; R = $(CH_2)_8CO_2CH_3$.

containing an endothelial cell monolayer or coated with recombinant selectin-Ig fusion proteins (R.M. Nelson et al., in prep.). To capture selectin-Ig molecules, wells were incubated with 10 μl of a 10 μg/ml solution of polyclonal goat anti-human IgG (γ-chain specific) antibody (Organon Teknika, Malverne, Pennsylvania) or protein A (Chemicon, Temecula, California) in 50 mM carbonate/bicarbonate buffer (pH 9.5) for 1 hour at room temperature or overnight at 4°C. Plates were washed twice with PBS and incubated with selectin-Ig proteins in PBS/1% HSA for 1 hour at room temperature. Wells coated with selectin-Ig fusion proteins or containing confluent endothelial cell monolayers were incubated with 5 μl of PBS or PBS containing carbohydrate for 30 minutes at 4°C prior to the addition of PMN. PMN (2×10^6 cells/ml) were added in 5 μl of adhesion assay buffer and allowed to adhere for 15–30 minutes at 4°C. Nonadherent cells were removed by washing with PBS. Adherent cells were fixed by adding 2.5% glutaraldehyde in PBS and were counted over a 0.24 mm^2 area in each of four replicate wells. Cell counts were averaged and expressed as adherent cells per mm^2. Adhesion of PMN to E- and P-selectin-Ig fusion proteins was inhibitable by specific monoclonal antibodies H18/7 and G559 (provided by R. McEver, University of Oklahoma), respectively. Adhesion to cytokine-activated endothelial cells was partially blocked by anti-E-selectin antibody H18/7 as described previously (Bevilacqua et al. 1987).

ELISA measuring selectin-carbohydrate interactions. Direct binding and competitive enzyme-linked immunosorbent assays (ELISA) were performed essentially as described previously (Nelson et al. 1993). Briefly, polystyrene 96-well microtiter ELISA plates were coated with unconjugated BSA, BSA neoglycoproteins, sulfatides, or cerebrosides and blocked with BSA. Test samples containing either purified selectin-Ig or selectin-Ig and carbohydrate were incubated in assay wells for 3 hours. Buffers contained 2 mM $CaCl_2$ unless otherwise indicated. After washing, bound selectin-Ig was detected using an antibody and colorimetric detection system. IC_{50} values were calculated by iteratively fitting the data to the equation: Fraction of maximal binding = IC_{50} divided by (IC_{50} + [oligosaccharide]), using nonlinear least-squares analysis software (Enzfitter, Elsevier-Biosoft, Cambridge, England).

RESULTS AND DISCUSSION

Synthetic oligosaccharides related to sLe^x and sLe^a were used to determine fine specificity of selectin recognition and to identify high-affinity ligands. We have focused primarily on E-selectin, with some comparisons to P- and L-selectin. We emphasized the use of solution-phase oligosaccharides to block cellular adhesion and direct molecular interactions and have chosen to highlight oligosaccharides that show inhibitory activity at concentrations $\leq$ 1 mM.

Solution-phase Oligosaccharides Block Leukocyte Adhesion

Cytokine-activated endothelial cells express E-selectin in abundance and thereby support leukocyte adhesion (Bevilacqua et al. 1987). sLe^x at a concentration of 1 mM had little effect on PMN adhesion to TNF-stimulated endothelial monolayers (Nelson et al. 1993) (0–30% blocking, 5 experiments). sLe^a was a more effective blocker, typically inhibiting adhesion by 30–50%. The nonsialylated trisaccharides Le^x and Le^a displayed no significant blocking activity at this concentration. To study selectin-dependent adhesion in a less biologically complex system, we used purified, recombinant selectin-Ig fusion proteins instead of endothelial monolayers. Selectin-Ig proteins were captured onto the surface of microtiter wells by binding to immobilized protein A or F(ab$'$)$_2$ fragments of anti-human Ig antibodies. This method allowed for systematic variation of the amount of selectin on the adhesive surface. As shown in Figure 1, increasing amounts of E-selectin-Ig supported increased adhesion of the promyelocytic cell line HL60. P-selectin-Ig was also found to support adhesion, but required more protein. Likewise, L-selectin-Ig bound HL60 cells, but was the least effective of the three selectins. The concentration-dependent effect of each selectin supports the concept that adhesion, in its simplest form, depends on the number of molecular interactions between participating surfaces. Differences among the selectins in adhesive activity for HL60 cells could be explained by recognition of the same ligand (e.g., sLe^x) with different affinities or by the involvement of different receptor-ligand interactions.

A series of related oligosaccharides (Table 1) was used to investigate the structural requirements for blocking adhesion. These studies demonstrated the importance of the sialic acid residue for recognition by E-selectin. For example, whereas sLe^a was found to be an effective blocker of PMN adhesion to E-selectin-coated plates (Fig. 2), Le^a was not (data not shown). In

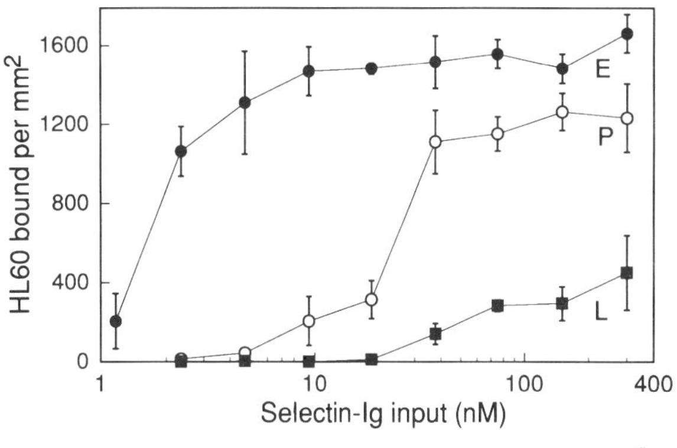

Figure 1. Adhesion of leukocytes to immobilized E-, P-, and L-selectin-Ig. Selectin-Ig fusion proteins at various concentrations were added to microtiter wells coated with recombinant protein A. After washing to remove unbound selectin-Ig, adhesion of HL60 cells was assessed as described in Methods.

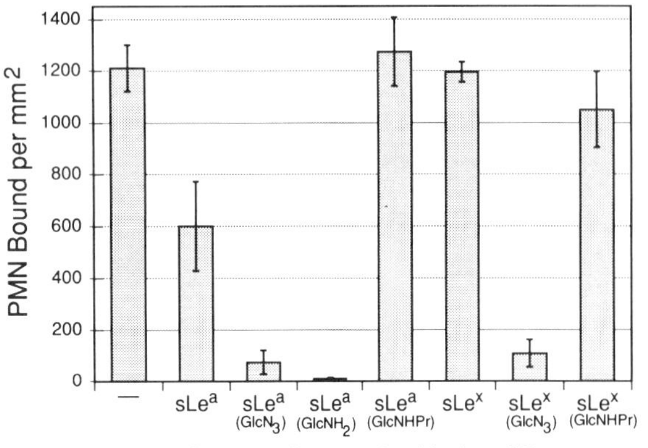

Figure 2. Inhibition of PMN adhesion to E-selectin-Ig by synthetic oligosaccharide derivatives of sLe^a and sLe^x. Solution-phase sLe^a, sLe^x, and amino-, azido-, and *N*-propionyl-substituted derivatives were tested for their ability to block PMN adhesion to immobilized E-selectin-Ig. Values shown represent the mean and standard deviation of cells counted in quadruplicate wells. Background adhesion to a nonspecific immunoglobulin fusion protein was subtracted from all values and was less than 5% of adhesion to E-selectin-Ig.

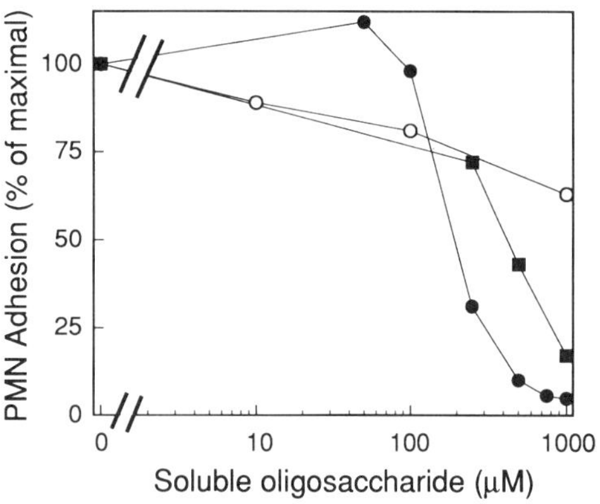

Figure 3. Concentration dependence of sLe^a, $sLe^x(GlcN_3)$, and $sLe^a(GlcNH_2)$ inhibition of PMN adhesion to E-selectin-Ig. Solution-phase sLe^a (○), $sLe^x(GlcN_3)$ (■), and $sLe^a(GlcNH_2)$ (●) at the indicated concentrations were tested for their ability to block PMN adhesion to immobilized E-selectin-Ig. Values represent the percentage of PMN bound in the absence of oligosaccharide competitors (% of maximal).

addition, a structural isomer of sLe^a with the sialic acid linked α2-6 (rather than α2-3) failed to block adhesion (< 10% blocking, 2 experiments). The fucose residue was also shown to be involved, in that a structure related to sLe^a but lacking fucose (sLe^c) was unable to block adhesion. In addition, sLacNAc (Neu5Acα2-3Galβ1-4GlcNAcβ–OR) and an α1-6 fucosylated derivative that differs from sLe^x only in fucose linkage (Fuc1-6sLacNAc) were ineffective.

We have recently identified two modifications of sLe^x and sLe^a that result in substantial increases in apparent binding affinities for E-selectin (Nelson et al. 1993). These are the addition of an 8-methoxycarbonyloctyl aglycone ($-[CH_2]_8CO_2CH_3$) attached in a β-glycosidic linkage to the reducing sugar, and the substitution of an amino or azido moiety for the *N*-acetyl group at carbon 2 of GlcNAc. The addition of this aglycone resulted in a modest enhancement of blocking activity in adhesion assays. Its effect is more clearly demonstrated in competitive binding assays described below. At a concentration of 1 mM, oligosaccharides related to sLe^x and sLe^a with amino and azido substitutions at carbon 2 of GlcNAc were found to be the most potent blockers of PMN adhesion to E-selectin (Fig. 2). Concentration-dependent effects of two of the most active compounds, $sLe^a(GlcNH_2)$ and $sLe^x(GlcN_3)$, are shown in Figure 3.

PMN adhesion to P-selectin-Ig was poorly blocked by soluble oligosaccharides related to sLe^x and sLe^a (not shown). Although several oligosaccharides (including nonsialylated and nonfucosylated structures) blocked PMN adhesion to P-selectin-Ig in certain assays, the effect was modest (typically <30% inhibition) and inconsistent. Neither azido nor amino substitutions at carbon 2 of GlcNAc greatly enhanced the ability of the compounds to block leukocyte adhesion to P-selectin-Ig.

Selectin-Ig Fusion Proteins Bind to Immobilized BSA-sLe^x and BSA-sLe^a

Soluble E- and P-selectin-Ig demonstrated concentration-dependent binding to immobilized BSA-sLe^x and BSA-sLe^a on microtiter plates (Fig. 4). Specific binding of E- and P-selectin-Ig was greatly diminished in the presence of 5 mM EGTA (Nelson et al. 1993), consistent with their function as C-type lectins (Drickamer 1988). Very little binding was detected on BSA-Le^x or BSA-Le^a. As noted in Methods, the mole ratios of oligosaccharide to BSA in the various BSA-neoglycoproteins were not identical, which may partly explain differences in selectin-Ig binding. However, as shown below in competition assays, differential binding may also reflect a difference in relative binding affinities. L-selectin-Ig demonstrated little or no binding to BSA-neoglycoproteins containing sLe^x, sLe^a, Le^x, or Le^a when assays were performed in the presence of 0.05% Tween-20 (Fig. 4); some binding was observed in the absence of detergent. A recent study has demonstrated that multimeric complexes containing murine L-selectin-Ig bound to immobilized sLe^x-glycolipid (Foxall et al. 1992). This interaction was inhibited by 50% in the presence of 5 mM sLe^x.

Consistent with previous reports (Imai et al. 1990; Aruffo et al. 1991; Handa et al. 1991), both L-selectin-Ig and P-selectin-Ig bound to immobilized 3-*O*-sulfategalactosyl ceramide (sulfatides) (Fig. 5) but not to galactosyl ceramide. This binding appeared to be largely calcium-independent (i.e., not abolished by EGTA) in the case of P-selectin-Ig. L-selectin-Ig binding was reduced approximately 50% in buffer containing 5 mM EGTA.

Solution-phase Oligosaccharides Compete with Solid-phase Carbohydrates for Binding to Selectins

We have examined the ability of solution-phase synthetic oligosaccharides to block the binding of E-

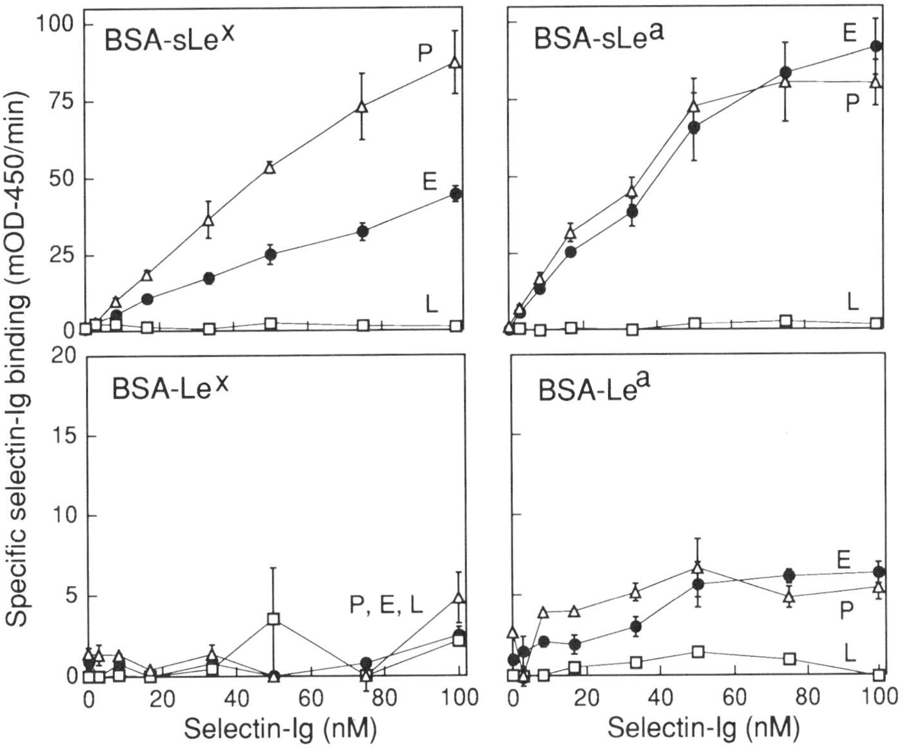

Figure 4. Specific binding of E-, P-, and L-selectin-Ig to immobilized BSA-neoglycoproteins. An ELISA was used to measure concentration-dependent binding to BSA-sLe^x, BSA-sLe^a, BSA-Le^x, and BSA-Le^a of E-selectin-Ig (●), P-selectin-Ig (△), and L-selectin-Ig (□). Bound E-selectin-Ig was detected using an enzyme-conjugated antibody and colorimetric substrate, as described in Methods. Specific binding to BSA-neoglycoproteins was calculated by subtraction of binding to unconjugated BSA at each concentration. Data points and error bars are the mean and standard deviation of measurements of rate of color generation from triplicate wells.

selectin-Ig and P-selectin-Ig to solid-phase BSA-neoglycoproteins, which we interpreted as competitive inhibition. Figure 6 shows the inhibitory effect of a series of structurally related oligosaccharides at 500 μM. The tetrasaccharides sLe^x and sLe^a with attached aliphatic aglycone blocked specific binding of 25 nM E-selectin-Ig

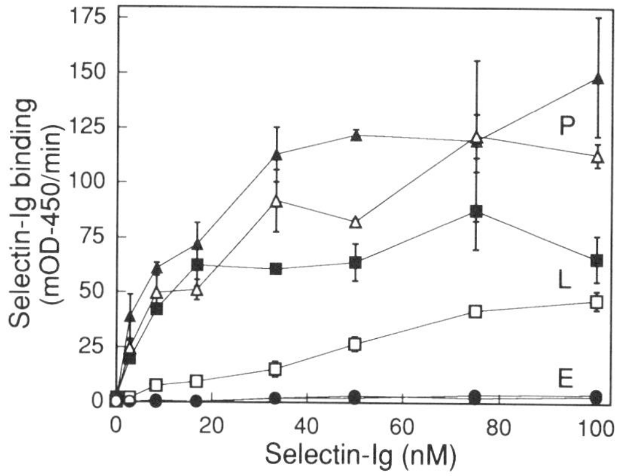

Figure 5. Specific binding of E-, P-, and L-selectin-Ig to immobilized sulfatides. An ELISA was used to measure concentration-dependent binding of selectin-Ig to 3-*O*-sulfate-galactosyl ceramide (sulfatides) in the presence of 2 mM $CaCl_2$ (filled symbols) or 5 mM EGTA (open symbols). Data points and error bars are the mean and standard deviation of measurements of rate of color generation from triplicate wells.

by more than 50%, with sLe^a exhibiting higher activity than sLe^x. In agreement with previous studies (Berg et al. 1991; Tyrrell et al. 1991; Nelson et al. 1993), and predicted by the results of adhesion studies described above, fucose and sialic acid were found to be important for E-selectin binding. First, the nonsialylated trisaccharide Le^x showed no blocking activity at this concentration, whereas Le^a demonstrated slight activity. Second, E-selectin-Ig binding was not inhibited by a structural isomer of sLe^a (2-6sLe^a) with the sialic acid linked α2-6 (rather than α2-3) to galactose. Third, a tetrasaccharide similar to sLe^x (Fuc1-6sLacNAc) with fucose linked α1-6 (rather than α1-4) had little or no inhibitory activity against E-selectin-Ig. Finally, our studies point to the importance of the *N*-acetyl group at carbon 5 of Neu5Ac. The oligosaccharide KDNLe^x (KDN = keto-deoxy-nonulosonic acid; for review of sialic acid diversity, see Varki 1992a), which contains a different type of sialic acid in which the *N*-acetyl moiety is replaced by a hydroxyl group, was significantly less active in E-selectin assays than was sLe^x. sLe^x and sLe^a with aliphatic aglycone substituents were more active than the reducing tetrasaccharides (Nelson et al. 1993). In addition, derivatives having amino or azido substitutions in place of the *N*-acetyl group of GlcNAc appeared to be especially potent.

To determine relative binding affinities, we compared inhibitory activities of selected oligosaccharides

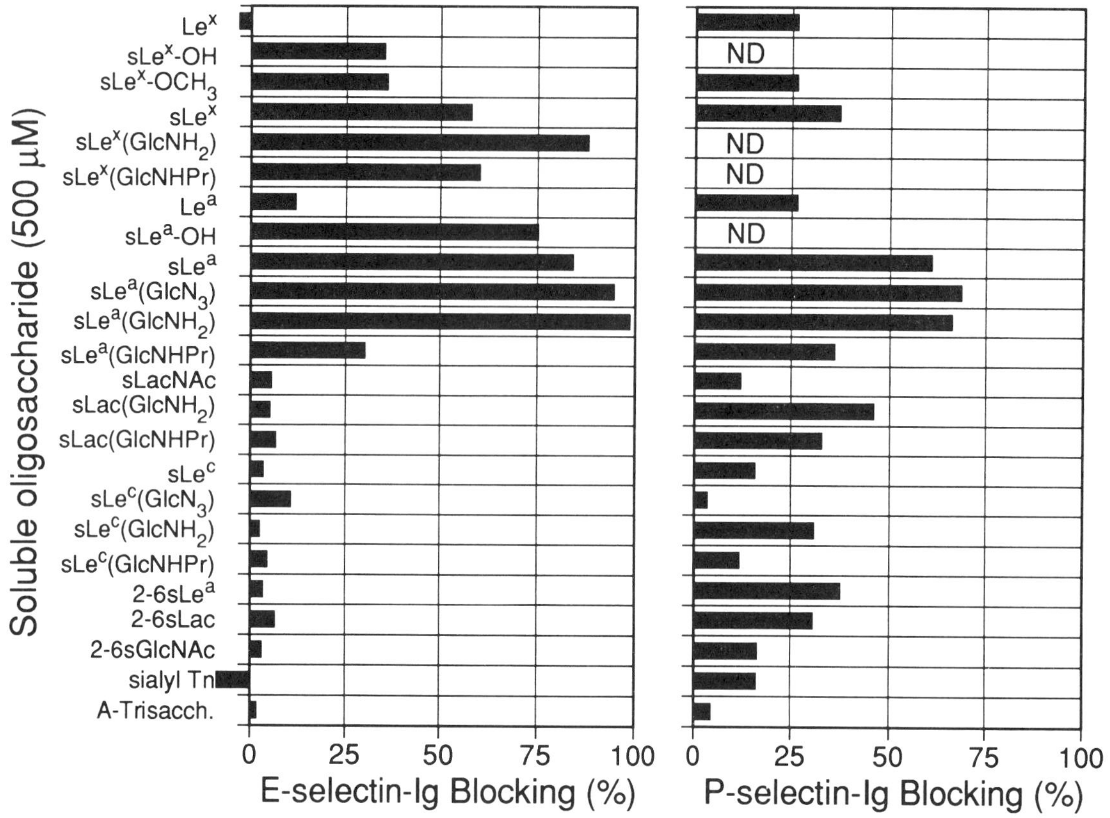

Figure 6. Inhibition of E- and P-selectin-Ig binding to immobilized BSA-sLe^a by sLe^a, sLe^x, and related oligosaccharides. Soluble oligosaccharides (500 μM) were co-incubated with 25 nM E-selectin-Ig or P-selectin-Ig in microtiter wells coated with BSA-sLe^a. Percent blocking is 100 minus percent binding in the absence of carbohydrate and was calculated from the mean of duplicate or triplicate measurements; ND indicates not done.

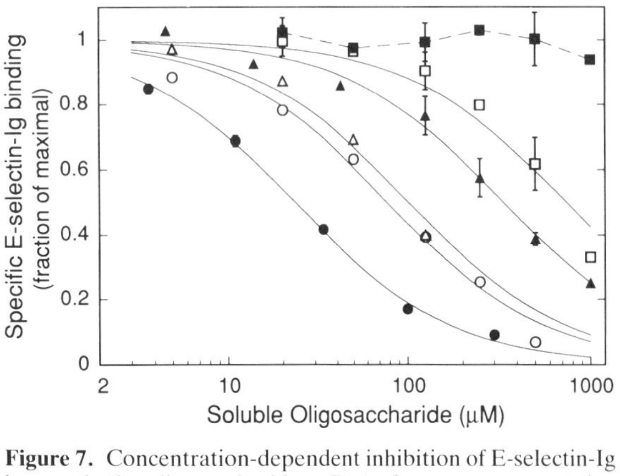

Figure 7. Concentration-dependent inhibition of E-selectin-Ig by synthetic oligosaccharides. Data from a representative competitive ELISA measuring the inhibition of specific E-selectin-Ig (25 nM) binding to BSA-sLe^a by six related oligosaccharides. Symbols and bars represent the mean and range of measurements made in duplicate wells. Key: Le^x (■); sLe^x–OH (□); sLe^x (▲); sLe^x(Glc) (△); sLe^x(GlcNH$_2$) (○); sLe^a(GlcNH$_2$) (●). Structural representations of oligosaccharides are shown in Table 1. Curves were drawn according to the calculated IC$_{50}$ values as described in Methods; the dashed line was drawn through the data points.

over a range of concentrations. A comparison of sLe^x without and with the 8-methoxycarbonyloctyl aglycone yielded IC$_{50}$ values ($\pm$ S.E.) of 740 $\pm$ 90 μM and 340 $\pm$ 20 μM, respectively (Fig. 7). This increase in activity may be caused by stabilization of the carbohydrate structure in a favorable configuration or may result from the aglycone contributing an additional binding component. The addition of hydrophobic sequences to certain mono- and oligosaccharides has been shown to increase their binding affinity for lectins. For example, the elderberry bark lectin, which binds Neu5Ac-Gal/GalNAc, shows increased affinity for glycosides with added β-linked nonpolar aglycones (Shibuya et al. 1987). More recently, monovalent sialosides substituted with aglycones terminating in aromatic structures have been shown to increase affinity for influenza hemagglutinin up to 65-fold (Toogood et al. 1991). The possibility of further enhancing E-selectin interactions with sLe^a or sLe^x by addition of other nonpolar groups deserves attention.

Titration of a tetrasaccharide related to sLe^x having the *N*-acetyl group of GlcNAc replaced by an amino group, sLe^x(GlcNH$_2$), yielded an IC$_{50}$ ($\pm$S.E.) of 77 $\pm$ 7 μM (Fig. 7). An analog of sLe^x synthesized with an *N*-propionyl (–NHC[O]CH$_2$CH$_3$) substitution at the same position (sLe^x[GlcNHPr]) was unchanged in activity relative to sLe^x (IC$_{50}$ = 360 $\pm$ 70 μM). Consistent with previous observations (Tyrrell et al. 1991), a compound similar to sLe^x having a terminal glucose rather than *N*-acetylglucosamine was more active than sLe^x itself. This tetrasaccharide, designated sLe^x(Glc), which has a hydroxyl group instead of the *N*-acetyl group on the GlcNAc of sLe^x, was found to have an IC_{50} of 100 ± 9 μM, a sevenfold increase in blocking activity over the reducing tetrasaccharide sLe^x–OH. The most potent blocker that we have identified to date (Nelson et al. 1993) is $sLe^a(GlcNH_2)$; analysis of data from the experiment shown in Figure 7 yielded an IC_{50} of 23 ± 1 μM.

A wide variety of oligosaccharides displayed some blocking activity against P-selectin-Ig (Fig. 6). These included some nonsialylated as well as some nonfucosylated compounds. None of the tested compounds appeared to be particularly potent competitors of immobilized BSA-sLe^a for binding to P-selectin-Ig. Concentration-dependent inhibition curves on the best blockers in Figure 6 yielded IC_{50} values in excess of ~300 μM. Interestingly, we have observed a low level of calcium-independent binding of P-selectin-Ig to BSA alone (Nelson et al. 1993). Taken together, these results may indicate that the BSA in the neoglycoprotein supplies an additional component of binding. It is noteworthy that studies on selectin recognition of oligosaccharides have been paralleled by investigations suggesting that certain proteins contribute to cellular ligands of L- and P-selectin (Siegelman et al. 1990; Imai et al. 1991; Kansas et al. 1991; Moore et al. 1991, 1992; Lasky et al. 1992). Antibodies that bind to the EGF repeat of L-selectin have been shown to block cellular adhesion but not the binding of certain carbohydrate ligands (Siegelman et al. 1990; Kansas et al. 1991). It has also been demonstrated that L-selectin can be used to affinity-purify two sulfated, fucosylated, and sialylated glycoproteins of 50 kD and 90 kD from murine lymph nodes (Imai et al. 1991). The 50-kD glycoprotein was cloned and found to be a mucin-like molecule containing two serine/threonine-rich domains, consistent with substantial *O*-linked glycosylation (Lasky et al. 1992). P-selectin has been shown to bind specifically to protease-sensitive sites on PMN and HL60 cells (Moore et al. 1991). Furthermore, it has been used to affinity-purify a glycoprotein from leukocyte extracts, with an M_r on polyacrylamide gels of approximately 120,000 under reducing conditions (Moore et al. 1992). Thus, both P- and L-selectin appear to bind with high affinity to a relatively small number of cell-surface glycoproteins. The function of the protein in these binding interactions remains to be determined.

The importance of the selectins in leukocyte adhesion and inflammation is now widely accepted. The ultimate goal of our studies is to identify anti-inflammatory therapeutics based on oligosaccharide ligands of the selectins. Our approach is centered on the use of quantitative bioassays to determine the inhibitory potency of candidate molecules. Future studies will involve the testing of additional synthetic oligosaccharides in vitro as well as evaluation of their therapeutic efficacy in vivo. It is anticipated that high-affinity carbohydrate ligands for selectins may prove useful in the treatment of a variety of human diseases involving inflammatory/immunological tissue damage, such as adult respiratory distress syndrome (ARDS), ischemic reperfusion injury, and arthritis.

ACKNOWLEDGMENTS

We thank the scientists at Chembiomed, Ltd. (Edmonton, Alberta), including A. P. Venot, M. A. Kasham, P. V. Nikrad, R. Ippolito, and F. Unger, for generously providing most of the oligosaccharides used in this study and for helpful discussions. A special thanks to A. P. Venot for his continued interest in this work. We thank R. McEver (University of Oklahoma Health Science Center) for antibody G559 against P-selectin, and M. Gimbrone (Brigham and Women's Hospital, Harvard Medical School, Boston, Massachusetts) for $F(ab')_2$ fragments of antibody H18/7 against E-selectin. We thank A. Corrigan for expert assistance in manuscript preparation. M.P.B. is an investigator in the Howard Hughes Medical Institute and a Pew scholar in the biomedical sciences.

REFERENCES

Aruffo, A., M.T. Dietsch, H. Wan, K.E. Hellstrom, and I. Hellstrom. 1992. Granule membrane protein 140 (GMP140) binds to carcinomas and carcinoma-derived cell lines. *Proc. Natl. Acad. Sci.* **89:** 2292.

Aruffo, A., W. Kolanus, G. Walz, P. Fredman, and B. Seed. 1991. CD62/P-selectin recognition of myeloid and tumor cell sulfatides. *Cell* **67:** 35.

Berg, E.L., J. Magnani, R.A. Warnock, M.K. Robinson, and E.C. Butcher. 1992. Comparison of L-selectin and E-selectin ligand specificities: The L-selectin can bind the E-selectin ligands sialyl-Le^x and sialyl-Le^a. *Biochem. Biophys. Res. Commun.* **184:** 1048.

Berg, E.L., M.K. Robinson, O. Mansson, E.C. Butcher, and J.L. Magnani. 1991. A carbohydrate domain common to both sialyl Le^a and sialyl Le^x is recognized by the endothelial cell leukocyte adhesion molecule ELAM-1. *J. Biol. Chem.* **23:** 14869.

Bevilacqua, M.P. and R.M. Nelson. 1993. Selectins. *J. Clin. Invest.* (in press).

Bevilacqua, M.P., S. Stengelin, M.A. Gimbrone, Jr., and B. Seed. 1989. Endothelial leukocyte adhesion molecule 1: An inducible receptor for neutrophils related to complement regulatory proteins and lectins. *Science* **243:** 1160.

Bevilacqua, M.P., J.S. Pober, D.L. Mendrick, R.S. Cotran, and M.A. Gimbrone, Jr. 1987. Identification of an inducible endothelial-leukocyte adhesion molecule. *Proc. Natl. Acad. Sci.* **84:** 9238.

Bevilacqua, M.P., E. Butcher, B. Furie, M. Gallatin, M. Gimbrone, J. Harlan, K. Kishimoto, L. Lasky, R. McEver, J. Paulson, S. Rosen, B. Seed, M. Siegelman, T. Springer, L. Stoolman, T. Tedder, A. Varki, D. Wagner, I. Weissman, and G. Zimmerman. 1991. Selectins: A family of adhesion receptors. *Cell* **67:** 233.

Bonfanti, R., B.C. Furie, B. Furie, and D.D. Wagner. 1989. PADGEM (GMP140) is a component of Weibel-Palade bodies of human endothelial cells. *Blood* **73:** 1109.

Brandley, B.K., S.J. Swiedler, and P.W. Robbins. 1990. Carbohydrate ligands of the LEC cell adhesion molecules. *Cell* **63:** 861.

Camerini, D., S.P. James, I. Stamenkovic, and B. Seed. 1989.

Leu-8/TQ1 is the human equivalent of the Mel-14 lymph node homing receptor. *Nature* **342:** 78.

Carlos, T., N. Kovach, B. Schwartz, M. Rosa, B. Newman, E. Wayner, C. Benjamin, L. Osborn, R. Lobb, and J. Harlan. 1991. Human monocytes bind to two cytokine-induced adhesive ligands on cultured human endothelial cells: Endothelial-leukocyte adhesion molecule-1 and vascular cell adhesion molecule-1. *Blood* **77:** 2266.

Corral, L., M.S. Singer, B.A. Macher, and S.D. Rosen. 1990. Requirement for sialic acid on neutrophils in a GMP-140 (PADGEM) mediated adhesive interaction with activated platelets. *Biochem. Biophys. Res. Commun.* **172:** 1349.

Drickamer, K. 1988. Two distinct classes of carbohydrate-recognition domains in animal lectins. *J. Biol. Chem.* **263:** 9557.

English, D. and B.R. Anderson. 1974. Single-step separation of red blood cells, granulocytes and mononuclear phagocytes on discontinuous density gradients of Ficoll-Hypaque. *J. Immunol. Methods* **5:** 249.

Foxall, C., S.R. Watson, D. Dowbenko, C. Fennie, L.A. Lasky, M. Kiso, A. Hasegawa, D. Asa, and B.K. Brandley. 1992. The three members of the selectin receptor family recognize a common carbohydrate epitope, the sialyl lewis-x oligosaccharide. *J. Cell Biol.* **117:** 895.

Geng, J.G., M.P. Bevilacqua, K.L. Moore, T.M. McIntyre, S.M. Prescott, J. M. Kim, G.A. Bliss, G.A. Zimmerman, and R.P. McEver. 1990. Rapid neutrophil adhesion to activated endothelium mediated by GMP-140. *Nature* **343:** 757.

Goelz, S.E., C. Hession, D. Goff, B. Griffiths, R. Tizard, B. Newman, G. Chi-Rosso, and R. Lobb. 1990. ELFT: A gene that directs the expression of an ELAM-1 ligand. *Cell* **63:** 1349.

Graber, N., T.V. Gopal, D. Wilson, L.D. Beall, T. Polte, and W. Newman. 1990. T cells bind to cy,okine-activated endothelial cells via a novel, inducible sialoglycoprotein and endothelial leukocyte adhesion molecule-1. *J. Immunol.* **145:** 819.

Hakkert, B.C., T.W. Kuijpers, J.F.M. Leeuwenberg, J.A. Van Mourik, and D. Roos. 1991. Neutrophil and monocyte adherence to and migration across monolayers of cytokine-activated endothelial cells: The contribution of CD18, ELAM-1, and VLA-4. *Blood* **78:** 2721.

Handa, K., E.D. Neudelman, M.R. Stroud, T. Shiozawa, and S. Hakomori. 1991. Selectin GMP-140 (CD62; PADGEM) binds to sialosyl-Le^a and sialosyl-Le^x, and sulfated glycans modulate this binding. *Biochem. Biophys. Res. Commun.* **181:** 1223.

Hession, C., L. Osborn, D. Goff, G. Chi-Rosso, C. Vassallo, M. Pasek, C. Pittack, R. Tizard, S. Goelz, and K. McCarthy. 1990. Endothelial leukocyte adhesion molecule 1: Direct expression cloning and functional interactions. *Proc. Natl. Acad. Sci.* **87:** 1673.

Hsu-Lin, S.C., C.L. Berman, B.C. Furie, D. August, and B. Furie. 1984. A platelet membrane protein expressed during platelet activation and secretion. Studies using a monoclonal antibody specific for thrombin-activated platelets. *J. Biol. Chem.* **259:** 9121.

Imai, Y., D.D. True, M.S. Singer, and S.D. Rosen. 1990. Direct demonstration of the lectin activity of $gp90^{mel}$, a lymphocyte homing receptor. *J. Cell Biol.* **11:** 1225.

Imai, Y., M.S. Singer, C. Fennie, L.A. Lasky, and S.D. Rosen. 1991. Identification of a carbohydrate-based endothelial ligand for a lymphocyte homing receptor. *J. Cell Biol.* **113:** 1213.

Johnston, G.I., R.G. Cook, and R.P. McEver. 1989. Cloning of GMP-140, a granule membrane protein of platelets and endothelium: Sequence similarity to proteins involved in cell adhesion and inflammation. *Cell* **56:** 1033.

Kansas, G.S., O. Spertini, L.M. Stoolman, and T.F. Tedder. 1991. Molecular mapping of functional domains of the leukocyte receptor for endothelium, LAM-1. *J. Cell Biol.* **114:** 351.

Kishimoto, T.K., M.A. Jutila, E.L. Berg, and E.C. Butcher. 1989. Neutrophil Mac-1 and MEL-14 adhesion proteins inversely regulated by chemotactic factors. *Science* **245:** 1238.

Kuijpers, T.W., B.C. Hakkert, M. Hoogerwerf, J.F.M. Leeuwenberg, and D. Roos. 1991. Role of endothelial leukocyte adhesion molecule-1 and platelet-activating factor in neutrophil adherence to IL-1-prestimulated endothelial cells. *J. Immunol.* **147:** 1369.

Larsen, E., T. Palabrica, S. Sajer, G.E. Gilbert, D.D. Wagner, B.C. Furie, and B. Furie. 1990. PADGEM-dependent adhesion of platelets to monocytes and neutrophils is mediated by a lineage-specific carbohydrate, LNF III (CD15). *Cell* **63:** 467.

Larsen, E., A. Celi, G.E. Gilbert, B.C. Furie, J.K. Erban, R. Bonfanti, D.D. Wagner, and B. Furie. 1989. PADGEM protein: A receptor that mediates the interaction of activated platelets with neutrophils and monocytes. *Cell* **59:** 305.

Lasky, L.A., M.S. Singer, T.A. Yednock, D. Dowbenko, C. Fennie, H. Rodriguez, T. Nguyen, S. Stachel, and S.D. Rosen. 1989. Cloning of a lymphocyte homing receptor reveals a lectin domain. *Cell* **56:** 1045.

Lasky, L.A., M.S. Singer, D. Dowbenko, Y. Imai, W.J. Henzel, C. Grimley, C. Fennie, N. Gillett, S.R. Watson, and S.D. Rosen. 1992. An endothelial ligand for L-selectin is a novel mucin-like molecule. *Cell* **69:** 927.

Lawrence, M.B. and T.A. Springer. 1991. Leukocytes roll on a selectin at physiologic flow rates: Distinction from and prerequisite for adhesion through integrins. *Cell* **65:** 859.

Leeuwenberg, J.F.M., T.M.A.A. Jeunhomme, and W.A. Buurman. 1992. Role of ELAM-1 in adhesion of monocytes to activated human endothelial cells. *Scand. J. Immunol.* **35:** 335.

Lemieux, R.U., D.R. Bundle, and D.A. Baker. 1975. The properties of a "synthetic" antigen related to the human blood-group Lewis a. *J. Am. Chem. Soc.* **97:** 4076.

Lewinsohn, D.M., R.F. Bargatze, and E.C. Butcher. 1987. Leukocyte-endothelial cell recognition: Evidence of a common molecular mechanism shared by neutrophils, lymphocytes, and other leukocytes. *J. Immunol.* **138:** 4313.

Ley, K., P. Gaehtgens, C. Fennie, M.S. Singer, L.A. Lasky, and S.D. Rosen. 1991. Lectin-like cell adhesion molecule 1 mediates leukocyte rolling in mesenteric venules *in vivo*. *Blood* **77:** 2553.

Lowe, J.B., L.M. Stoolman, R.P. Nair, R.D. Larsen, T.L. Berhend, and R.M. Marks. 1990. ELAM-1-dependent cell adhesion to vascular endothelium determined by a transfected human fucosyltransferase cDNA. *Cell* **63:** 475.

Luscinskas, F.W., M.I. Cybulsky, J.M. Kiely, C.S. Peckins, V.M. Davis, and M.A. Gimbrone, Jr. 1991. Cytokine-activated human endothelial monolayers support enhanced neutrophil transmigration via a mechanism involving both endothelial-leukocyte adhesion molecule-1 and intercellular adhesion molecule-1. *J. Immunol.* **146:** 1617.

McEver, R.P. 1991. Leukocyte interactions mediated by selectins. *Thromb. Haemostasis* 66: 80.

McEver, R.P., J.H. Beckstead, K.L. Moore, L. Marshall-Carlson, and D.F. Bainton. 1989. GMP-140, a platelet alpha-granule membrane protein, is also synthesized by vascular endothelial cells and is localized in Weibel-Palade bodies. *J. Clin. Invest.* **84:** 92.

Moore, K.L., A. Varki, and R.P. McEver. 1991. GMP-140 binds to a glycoprotein receptor on human neutrophils: Evidence for a lectin-like interaction. *J. Cell Biol.* **112:** 491.

Moore, K.L., N.L. Stults, S. Diaz, D.F. Smith, R.D. Cummings, A. Varki, and R.P. McEver. 1992. Identification of a specific glycoprotein ligand for P-selectin (CD62) on myeloid cells. *J. Cell Biol.* **118:** 445.

Munro, J.M., S.K. Lo, C. Corless, M.J. Robertson, N.C. Lee, R.L. Barnhill, D.S. Weinberg, and M.P. Bevilacqua. 1992. Expression of sialyl-Lewis X, an E-selectin ligand, in inflammation, immune processes, and lymphoid tissues. *Am. J. Pathol.* (in press).

Nelson, R.M., S. Dolich, A. Aruffo, O. Cecconi, and M.P. Bevilacqua. 1993. Higher affinity oligosaccharide ligand of E-selectin. *J. Clin. Invest.* (in press).

Ohmori, K., T. Yoneda, G. Ishihara, K. Shigeta, K. Hirashima, M. Kanai, S. Itai, T. Sasaoki, S. Arii, H. Arita, and R. Kannagi. 1989. Sialyl SSEA-1 antigens as a carbohydrate marker of human natural killer cells and immature lymphoid cells. *Blood* **74:** 255.

Phillips, M.L., E. Nudelman, F.C. Gaeta, M. Perez, A.K. Singhal, S. Hakomori, and J.C. Paulson. 1990. ELAM-1 mediates cell adhesion by recognition of a carbohydrate ligand, sialyl-Le^x. *Science* **250:** 1130.

Picker, L.J. and E.C. Butcher. 1992. Physiological and molecular mechanisms of lymphocyte homing. *Annu. Rev. Immunol.* **10:** 561.

Picker, L.J., T.K. Kishimoto, C.W. Smith, R.A. Warnock, and E.C. Butcher. 1991. ELAM-1 is an adhesion molecule for skin-homing T cells . *Nature* **349:** 796.

Pinto, B.M. and D.R. Bundle. 1983. Preparation of glycoconjugates for use as artificial antigens: A simplified procedure. *Carbohydr. Res.* **124:** 313.

Polley, M.J., M.L. Phillips, E. Wayner, E. Nudelman, A.K. Singhal, S-I. Hakomori, and J.C. Paulson. 1991. CD62 and endothelial cell-leukocyte adhesion molecule 1 (ELAM-1) recognize the same carbohydrate ligand, sialyl-lewis x. *Proc. Natl. Acad. Sci.* **88:** 6224.

Rosen, S.D. 1990. The LEC-CAMs: An emerging family of cell-cell adhesion receptors based upon carbohydrate recognition. *Am. J. Respir. Cell Mol. Biol.* **3:** 397.

Shibuya, N., I.J. Goldstein, W.F. Broekaert, M. Nsimba-Lubaki, B. Peeters, and W.J. Peumans. 1987. The elderberry (*Sambucas nigra L.*) bark lectin recognizes the Neu5Ac(α2-6)Gal/GalNAc sequence. *J. Biol. Chem.* **262:** 1596.

Shimizu, Y., S. Shaw, N. Graber, T.V. Gopal, K.J. Horgan, G.A. van Seventer, and W. Newman. 1991a. Activation-independent binding of human memory T cells to adhesion molecule ELAM-1. *Nature* **349:** 799.

Shimizu, Y., W. Newman, T.V. Gopal, K.J. Horgan, N. Graber, L.D. Beall, G.A. van Seventer, and S. Shaw. 1991b. Four molecular pathways of T cell adhesion to endothelial cells: Roles of LFA-1, VCAM-1, and ELAM-1 and changes in pathway hierarchy under different activation conditions. *J. Cell Biol.* **113:** 1203.

Siegelman, M.H., M. van de Rijn, and I.L. Weissman. 1989. Mouse lymph node homing receptor cDNA clone encodes a glycoprotein revealing tandem interaction domains. *Science* **243:** 1165.

Siegelman, M.H., I.C. Cheng, I.L. Weissman, and E.K. Wakeland. 1990. The mouse lymph node homing receptor is identical with the lymphocyte cell surface marker Ly-22: Role of the EGF domain in endothelial binding. *Cell* **61:** 611.

Skinner, M.P., D.J. Fournier, R.K. Andrews, J.J. Gorman, C.N. Chesterman, and M.C. Berndt. 1989. Characterization of human platelet GMP-140 as a heparin-binding protein. *Biochem. Biophys. Res. Commun.* **164:** 1373.

Smith, C.W., T.K. Kishimoto, O. Abbass, B. Hughes, R. Rothlein, L.V. McIntire, E. Butcher, and D.C. Anderson. 1991. Chemotactic factors regulate lectin adhesion molecule 1 (LECAM-1)-dependent neutrophil adhesion to cytokine-stimulated endothelial cells in vitro. *J. Clin. Invest.* **87:** 609.

Spertini, O., F.W. Luscinskas, M.A. Gimbrone, and T.F. Tedder. 1992. Monocyte attachment to activated human vascular endothelium *in vitro* is mediated by leukocyte adhesion molecule-1 (L-selectin) under nonstatic conditions. *J. Exp. Med.* **175:** 1789.

Spertini, O., F.W. Luscinskas, G.S. Kansas, J.M. Munro, J.D. Griffin, M.A. Gimbrone, Jr., and T.F. Tedder. 1991. Leukocyte adhesion molecule-1 (LAM-1, L-selectin) interacts with an inducible endothelial cell ligand to support leukocyte adhesion. *J. Immunol.* **147:** 2565.

Springer, T.A. and L.A. Lasky. 1991. Sticky sugars for selectins. *Nature* **349:** 196.

Stenberg, P.E., R.P. McEver, M.A. Shuman, Y.V. Jacques, and D.F. Bainton. 1985. A platelet alpha-granule membrane protein (GMP-140) is expressed on the plasma membrane after activation. *J. Cell Biol.* **101:** 880.

Stoolman, L.M. 1989. Adhesion molecules controlling lymphocyte migration. *Cell* **56:** 907.

Symington, F.W., D.L. Hedges, and S. Hakomori. 1985. Glycolipid antigens of human polymorphonuclear neutrophils and the inducible HL-60 myeloid leukemia line. *J. Immunol.* **134:** 2498.

Tedder, T.F., C.M. Isaacs, T.J. Ernst, G.D. Demetri, D.A. Adler, and C.M. Disteche. 1989. Isolation and chromosomal localization of cDNAs encoding a novel human lymphocyte cell surface molecule, LAM-1. *J. Exp. Med.* **170:** 123.

Tiemeyer, M., S.J. Swiedler, M. Ishihara, M. Moreland, H. Schweingruber, P. Hirtzer, and B.K. Brandley. 1991. Carbohydrate ligands for endothelial-leukocyte adhesion molecule 1. *Proc. Natl. Acad. Sci.* **88:** 11382.

Toogood, P.L., P.K. Galliker, G.D. Glick, and J.R. Knowles. 1991. Monovalent sialosides that bind tightly to influenza A virus. *J. Med. Chem.* **34:** 3138.

Tyrrell, D., P. James, N. Rao, C. Foxall, S. Abbas, F. Dasgupta, M. Nashed, A. Hasegawa, M. Kiso, D. Asa, J. Kidd, and B.K. Brandley. 1991. Structural requirements for the carbohydrate ligand of E-selectin. *Proc. Natl. Acad. Sci.* **88:** 10372.

Varki, A. 1992a. Diversity of the sialic acids. *Glycobiology* **2:** 25.

———. 1992b. Selectins and other mammalian sialic acid-binding lectins. *Curr. Opin. Cell Biol.* **4:** 257.

von Andrian, U.H., J.D. Chambers, L.M. McEvoy, R.F. Bargatze, K.-E. Arfors, and E.C. Butcher. 1991. Two-step model of leukocyte-endothelial cell interaction in inflammation: Distinct roles for LECAM-1 and the leukocyte β_2 integrins *in vivo*. *Proc. Natl. Acad. Sci.* **88:** 7538.

Walz, G., A. Aruffo, W. Kolanus, M. Bevilacqua, and B. Seed. 1990. Recognition by ELAM-1 of the sialyl-Le^x determinant on myeloid and tumor cells. *Science* **250:** 1132.

Yednock, T.A., E.C. Butcher, L.M. Stoolman, and S.D. Rosen. 1987. Receptors involved in lymphocyte homing: Relationship between a carbohydrate-binding receptor and the MEL-14 antigen. *J. Cell Biol.* **104:** 725.

Zhou, Q., K.L. Moore, D.F. Smith, A. Varki, R. McEver, and R.D. Cummings. 1991. The selectin GMP-140 binds to sialylated, fucosylated lactosaminoglycans on both myeloid and nonmyeloid cells. *J. Cell Biol.* **115:** 557.

Molecular Mechanisms of Complex Carbohydrate Recognition at the Cell Surface

W.I. Weis,† M.S. Quesenberry,*‡ M.E. Taylor,*§ K. Bezouška,*††
W.A. Hendrickson,† and K. Drickamer*
*Department of Biochemistry and Molecular Biophysics and †Howard Hughes Medical Institute, Columbia University, New York, New York 10032

Complex carbohydrates conjugated to lipids and proteins at cell surfaces, in the extracellular matrix, and in extracellular fluids can function in two fundamentally different ways. They can influence the intrinsic properties of the lipid or protein to which they are attached, or they can serve as recognition markers. In the latter case, lectins, or carbohydrate-recognition proteins, are generally required to mediate the biological activity of the saccharide. Endogenous animal lectins have been identified with increasing frequency. More than 50 of these have been characterized at the level of primary structure. These sequences fall into a remarkably small number of groups (Drickamer 1992). The largest and most diverse of these groups consists of Ca^{++}-dependent, or C-type, lectins (Drickamer 1988).

C-type animal lectins mediate several different types of interactions at the cell surface. Many of these proteins interact with endogenous saccharides. The asialoglycoprotein receptor is a prototype cell-surface C-type lectin that binds to saccharides attached to soluble proteins and mediates their endocytosis; the primary ligands for this receptor appear to be serum glycoproteins from which terminal sialic residues have been removed (Spiess 1990; Drickamer 1991). If both the lectins and their carbohydrate ligands are components of mammalian cell surfaces, saccharide recognition can form the basis for cell adhesion. For example, the selectin cell adhesion molecules mediate the initial interaction between leukocytes and endothelial cells (Stoolman 1989).

Instead of recognizing endogenous sugar structures, certain C-type animal lectins have high affinity for saccharides, which form part of the surface of bacterial and fungal pathogens. Mannose-rich structures are more abundant on these organisms than on mammalian cells, so their presence can be used as a crude way of distinguishing between self and non-self. Such antibody-independent discrimination appears to play an important role in innate immunity mediated by the mannose receptor on the surface of macrophages and by soluble mannose-binding proteins that initiate antibody-independent fixation of complement and act as opsonins (Stahl 1990).

The structures of several major subgroups of C-type animal lectins are summarized in Figure 1. Although the overall organization of these proteins varies widely, each contains one or more 130-amino-acid-long carbohydrate-recognition domains (CRDs). The primary structures of these domains share a characteristic pattern of 14 invariant residues and 17 residues that are conserved in character. The existence of the C-type CRDs as discrete protein domains with sugar-binding activity has been demonstrated by isolation of CRDs as protease-resistant fragments and by expression of truncated portions of the C-type lectins in vitro and in

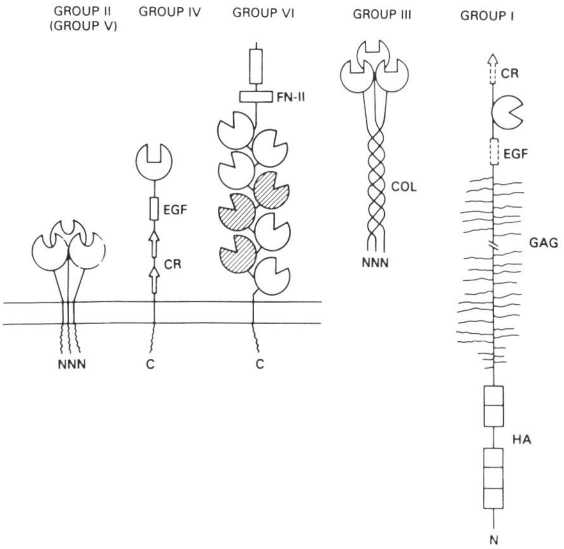

Figure 1. Summary of the structures of several groups of C-type animal lectins (see also Fig. 2). Representative structures for three groups of membrane-associated lectins are shown: *Group II*, the chicken hepatic lectin (homolog of the mammalian asialoglycoprotein receptor); *Group IV*, the selectin cell adhesion molecules; and *Group VI*, the macrophage mannose receptor. *Group III* lectins (collectins), such as mannose-binding protein, are found in extracellular fluids. *Group I* CRD-containing proteins are proteoglycans of the extracellular matrix. Other domains present in these molecules include *EGF*, epidermal growth factor-like repeats; *CR*, complement regulatory domains; *FN-II*, fibronectin type II repeats; *COL*, collagen-like sequences; *GAG*, glycosaminoglycan attachment sites; and *HA*, hyaluronic acid-binding domains.

Present addresses: ‡Department of Biology, Johns Hopkins University, 3400 North Charles Street, Baltimore, Maryland 20212; §Institute of Glycobiology, Department of Biochemistry, University of Oxford, South Parks Road, Oxford OX1 3UK, United Kingdom; ††Institute of Biotechnology, Charles University, Hlavova 8/2030, CS-128 40 Praha 2, Czechoslovakia.

bacteria (Loeb and Drickamer 1987; Quesenberry and Drickamer 1991; Weis et al. 1991a).

C-type CRDs are found at both the amino-terminal and carboxy-terminal ends of proteins, as well as at internal positions. It is particularly interesting that they can be found at the amino-terminal end of type II transmembrane proteins or at the carboxy-terminal end of type I proteins, since this places them at the most membrane-distal position in each case. C-type CRDs are associated with a variety of other protein domains, in various arrangements. For example, in some of the alternatively spliced forms of proteoglycan core proteins, the CRD is flanked by epidermal growth factor-like sequences and complement-regulatory repeats, whereas in the selectin cell adhesion molecules, domains of both of these types lie on the carboxy-terminal side of a CRD. In another group of C-type lectins, including the mannose-binding proteins discussed below, a CRD is juxtaposed with a collagen-like domain. These collagen-containing lectins are referred to as collectins. The macrophage mannose receptor has a unique domain organization, as it contains eight tandem CRD-like sequences in combination with a fibronectin type II repeat. It is the only known C-type lectin that contains more than one CRD within a single polypeptide.

Sequence similarity among the C-type CRDs indicates that these domains have a common evolutionary origin. The current state of our understanding of the evolution of C-type lectins is summarized in the first portion of this paper. The fact that the C-type CRDs are related in primary structure suggests that they are likely to be related in tertiary structure and in their mode of interacting with sugars. Analysis of the structure and function of a representative CRD is described. Finally, the role of multiple, clustered CRDs in recognition of complex oligosaccharides is considered.

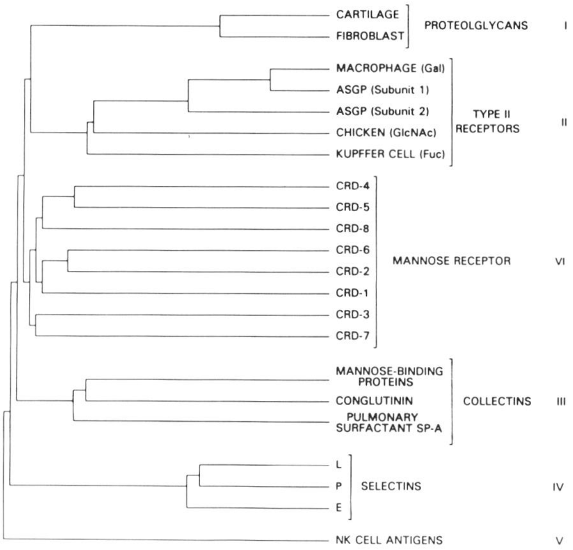

Figure 2. Dendrogram summarizing sequence similarity between various C-type CRDs. Similarities were determined on the basis of amino acid sequence comparisons (Bezouska et al. 1991; Kim et al. 1992). The CRDs of the mannose receptor have been segregated from the other proteins for clarity. The horizontal scale is arbitrary and should not be used to infer specific times of evolutionary divergence.

Evolution of C-type Animal Lectins

The degree of relatedness of the CRD portions of various animal lectins is summarized in Figure 2. Strikingly, CRDs in lectins that have similar domain organization are the CRDs that are most closely related to each other at the primary structure level. This observation suggests that the C-type lectins in each group are descended from a common ancestor. Since it is likely that association of CRDs with other functional domains is a result of exon shuffling, the shuffling events necessary to generate the array of domains that characterizes each of the groups of C-type lectins shown in Figure 1 probably occurred only once for each group. For example, the existence of a common ancestor for all of the collectins indicates that the CRD and the collagenous domain were only juxtaposed once in evolution, after which the individual group III lectins diverged from each other.

The genomic organization of many of the C-type animal lectins reflects the shuffling events that led to the current structures of the lectins. In each of the genes so far characterized for lectins in groups I–IV, the CRD-encoding segment is separated from the remainder of the gene by an intron. However, genes for lectins in groups I and II contain introns within the CRD-coding sequence, whereas those in groups III and IV do not (Bezouska et al. 1991). The conserved positions of the introns in the group I and II genes are consistent with the suggestion that they were present in the common ancestor for these CRDs.

Evolution of the macrophage mannose receptor is not so easily traced. On the basis of both sequence analysis and genomic organization, it appears that duplication of CRDs to form this receptor occurred at or before the time the precursors to the other lectins were generated by exon shuffling. In Figure 2, this conclusion is evident from the fact that many of the CRDs in the mannose receptor are less similar to each other than are the CRDs in widely divergent group I–V lectins. The positions of introns within and between the CRDs of the mannose receptor are also surprisingly complex (Kim et al. 1992).

Finally, it is noteworthy that the overall degree of sequence relatedness does not reflect the saccharide-binding characteristics of the C-type CRDs. For example, the chicken hepatic lectin binds primarily *N*-acetylglucosamine, whereas many of the other group II lectins bind galactose and *N*-acetylgalactosamine. Other lectins that bind *N*-acetylglucosamine are in group III. One explanation for this lack of concordance between overall sequences and sugar-binding activity is that relatively few amino acids in the CRD are directly involved in forming a sugar-binding site. Structural results discussed below are consistent with this conclusion.

With the identification of the sequence motif common to the C-type CRDs, it has been possible to predict, on the basis of primary structures, that proteins not initially identified as lectins might have sugar-binding activity. In two cases, the pulmonary surfactant apoprotein and aggrecan, the core protein of cartilage proteoglycan, this prediction has been experimentally verified (Haagsman et al. 1987; Halberg et al. 1988). However, it appears that the low-affinity IgE Fc receptor of lymphocytes, although it contains a CRD sequence motif and is certainly related in overall structure, probably binds the protein portion of IgE (Vercelli et al. 1989). A large number of type II transmembrane proteins that contain potential C-type CRDs have recently been implicated as possible receptors on natural killer cells (group V in Fig. 2) (Houchins et al. 1991; Wong et al. 1991). Saccharide-binding activity by these proteins has not yet been demonstrated.

C-type CRD Structure

The biological significance of serum mannose-binding protein is becoming increasingly evident, since it has been shown that it can fix complement in an antibody-independent manner (Ikeda et al. 1987) and forms part of the acute phase response (Ezekowitz et al. 1988). Complement fixation is presumed to be mediated by the collagenous domain of the mannose-binding proteins, by analogy with complement component C1q, in which Fc-binding domains are attached to collagenous stalks (Drickamer et al. 1986; Lu et al. 1990). The resulting hypothesis, that mannose-binding protein forms part of a primitive immune response, is supported by the finding that reduced serum levels of this protein, resulting from a mutation in the collagen-like amino-terminal domain, lead to susceptibility to recurrent severe infections (Sumiya et al. 1991).

As a first step toward understanding the mechanism by which C-type CRDs mediate saccharide-specific interactions, the three-dimensional structure of the C-type CRD from a rat mannose-binding protein has been deduced by X-ray diffraction studies of crystals obtained from bacterially expressed CRD. A fragment of the mannose-binding protein suitable for crystal structure analysis was obtained by limited proteolysis of the carboxy-terminal segment of the protein expressed in bacteria (Weis et al. 1991a). The proteolytic fragment corresponds to the minimum fragment of the mannose-binding protein that has sugar-binding activity when expressed in an in vitro translation system (Quesenberry and Drickamer 1991). Truncation of the mannose-binding protein results in formation of dimers rather than the trimeric building block of the native protein. As a result, crystals of the fragment can be used to deduce information about the CRD itself and its interaction with sugars but do not reveal the orientation of the CRD in its natural oligomeric configuration. Substitution of holmium ions for calcium made possible determination of the complete structure from data collected from a single crystal form, using the method of multiwavelength anomalous diffraction (Weis et al. 1991b).

The structure of the CRD protomer is an unusual fold containing extensive regions of non-regular secondary structure (Fig. 3a). All of the regular secondary structure elements, including two α helices and five β strands, are found in the lower two-thirds of the CRD. Two of the β strands (3 and 4) are extensively hydrogen-bonded in an antiparallel fashion, but there are no extended regions that would be described as β sheets. The upper one-third of the domain consists of

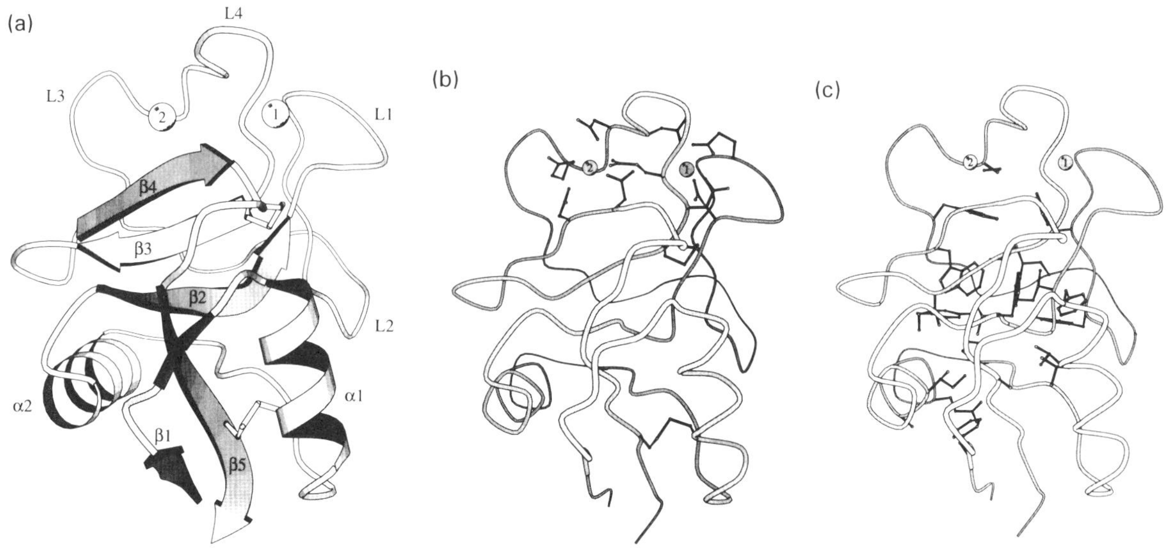

Figure 3. Structure of the C-type CRD from mannose-binding protein. (*a*) Ribbon diagram showing secondary structure elements of the mannose-binding CRD. Spheres 1 and 2 represent calcium ions. In *b* and *c*, the side chains of residues conserved in all of the C-type CRDs are shown, along with the α-carbon backbone of the entire polypeptide. Conserved calcium ion ligands and disulfide-bonded cysteines are shown in *b*, whereas other, largely hydrophobic side chains are shown in *c*.

less regular secondary structure and forms the binding sites for the two cations. The existence of two calcium-binding sites was expected from the second-order dependence of ligand binding on calcium concentration (Weis et al. 1991a).

The amino- and carboxy-terminal ends of the protein are paired as antiparallel β strands. This arrangement has important implications for the way CRDs in various proteins are arranged with respect to other modular protein domains. As noted above, C-type CRDs have been identified at the amino termini of selectins and at the carboxyl termini of endocytic receptors. Since the ends of the CRD are close to each other in space, the CRDs in these two classes of proteins may be projected away from the lipid bilayer in similar orientations, despite the different topology of these proteins in the membrane.

The CRD structure explains in large measure the pattern of conserved amino acids in the C-type CRD motif. In Figure 3, b and c, the positions of residues common to all C-type CRDs are highlighted. These residues form (1) two disulfide bonds, (2) ligation sites for two holmium (calcium) ions, (3) two hydrophobic cores, and (4) several surface turns. The importance of these residues in formation of the basic fold of this C-type CRD suggests that other domains in the family will have similar overall structures. Indeed, alignment of the sequences shows that most insertions and deletions can be accommodated on surface loops without disrupting the basic fold.

Structure of a CRD/Oligosaccharide Complex

A second crystal form of mannose-binding protein CRD, obtained in the presence of calcium ions and a mannose-containing glycopeptide from ovalbumin, has been analyzed (Weis et al. 1992). Terminal mannose residues on two branches of each oligosaccharide are observed to be ligated to CRDs that form part of two different dimers, thus cross-linking the dimers within the crystal. A model for interaction of one terminal mannose residue with a CRD, derived from the 1.7 Å resolution crystal structure, is presented in Figure 4.

The key feature of the mannose-CRD interaction is direct ligation of calcium ion 2 (Fig. 3a) by the equatorial 3- and 4-hydroxyl groups of mannose to form an 8-coordinate calcium ion complex. In the sugar-free, holmium-containing crystal, the cation is 7-coordinate, with water serving as the 7th ligand. Both sugar hydroxyls utilize their full bonding potential: One lone electron pair is used in a calcium coordination bond, the other lone pair accepts a proton from an asparagine side chain amide nitrogen, and the sugar hydroxyl proton forms a hydrogen bond with a carboxylate oxygen of glutamic acid. In each case, the remaining oxygens of the amino acid side chains involved in the hydrogen bonds are also coordinated to the calcium ion. This arrangement produces an intimately linked ternary complex of protein, calcium, and sugar. In contrast to other sugar-protein complexes that have been analyzed

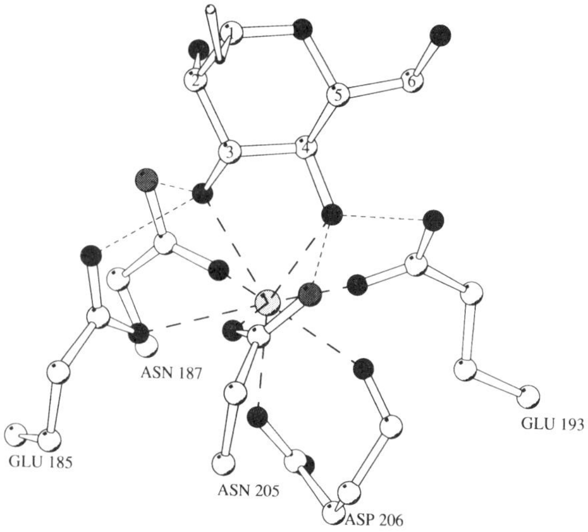

Figure 4. Structure of an oligosaccharide-CRD complex at calcium ion 2. The calcium ion is shown as a light gray sphere. White, dark gray, and black spheres represent carbon, nitrogen, and oxygen, respectively. Calcium coordination bonds are shown as thick dashed lines, and thin dashed lines represent hydrogen bonds. Numbers on the mannose carbon atoms represent ring positions. The α-glycosidic bond to the next sugar of the oligosaccharide (at carbon 1) has been cut off for clarity.

by X-ray crystallography (Quiocho 1986; Shaanan et al. 1991), there are few nonpolar van der Waals interactions in the oligosaccharide-CRD complex. The only significant interactions of this type occur between the exocyclic C6 of mannose and isoleucine 207, C4 of mannose and Cβ of histidine 189, and O2 of mannose and Cγ2 of histidine 189. The structure of the C-type lectin/sugar complex also contrasts with the structure of glycoside-ligated legume lectins, such as concanavalin A, since Ca^{++} and Mg^{++} are required to maintain structural integrity of the binding sites in these lectins but do not interact directly with the sugar (Derewenda et al. 1989).

The observed structure suggests that a primary determinant of specificity in saccharide binding is the presence of equatorial hydroxyl groups at positions 3 and 4 of the D-pyranose ring. Such groups are present in mannose, *N*-acetylglucosamine, and glucose, all of which are ligands for the mannose-binding protein. An additional ligand, L-fucose, probably binds via equatorial 2- and 3-hydroxyl groups. Superimposing the 2- and 3-hydoxyl groups of the L-fucose ring on the observed D-mannose structure produces the same interactions as those described above, without steric clashes. In contrast, D-galactose, which does not bind to mannose-binding protein, cannot form the same interactions. Specifically, it contains an axial rather than equatorial 4-hydroxyl group, and the equatorial 2- and 3-hydroxyl groups are unable to bind because they are twisted with the opposite handedness of the 3- and 4-hydroxyls of mannose.

The conclusion that the binding specificity of the mannose-binding protein lies in the limited network of hydrogen and coordination bonds shown in Figure 4 explains the broad range of monosaccharides that are bound with weak (~1 mM) affinity. Further binding specificity results from oligomerization, as discussed below.

Mutagenesis of a C-type CRD

Although the view of CRD-sugar interactions derived from crystallographic analysis is initially a static one, some idea of the natural dynamics of these domains in solution and at the cell surface can be obtained by changing the environmental conditions and by modifying the domain through site-specific mutagenesis. Both of these approaches have proven to be informative.

Mutagenesis of the mannose-binding CRD has been undertaken using a random cassette approach (Quesenberry and Drickamer 1992). In the absence of a way to develop a selection scheme in the bacterial expression system, it was necessary to utilize an efficient screening method to sort the mutants generated. The primary tool used for this purpose was a dot-blot filtration assay, in which aliquots of mutant CRDs were passed over nitrocellulose filters coated with yeast invertase. The high mannose structures on this glycoprotein are excellent ligands for the mannose-binding CRD. Bound CRDs can be visualized by staining with rabbit polyclonal antiserum followed by radioiodinated protein A. Since the random mutagenesis approach results in introduction of stop codons in some cases, mutants were also checked for production of full-length CRD by direct filtration through nitrocellulose, followed by staining with an antibody directed against the carboxyl terminus of the domain. Validation of this screening assay is shown in Figure 5.

Following characterization of mutants based on their ability to bind invertase at near physiological (2.5 mM) calcium ion concentrations, the changes made were identified by sequence analysis, and further characterization of the mutant phenotypes was undertaken. It was possible to purify most of the mutants at elevated calcium ion concentration on an affinity matrix containing high concentrations of mannose. This made possible determination of sugar-binding and calcium interaction properties of many of the domains. As might be expected, most of the amino acid residues that can be changed without affecting the activity of the CRD are located on the surface. In fact, changes made over a large portion of the surface are tolerated (Fig. 6). This observation is consistent with the structural studies suggesting that the interaction between the CRD and sugar ligands is restricted to a very small region of the molecule.

A further important outcome of these studies is confirmation of the importance of evolutionarily conserved residues. Most changes in the conserved calcium ion-liganding and hydrophobic core residues result in reduced affinity for invertase under physiological conditions. In many of these cases, the reduced activity can be attributed to loss of affinity for calcium ions. If the calcium concentration is increased sufficiently, the saccharide-binding activity of these mutant CRDs is indistinguishable from wild type in affinity and specificity. As might be expected, the residues changed in some of these calcium affinity mutants are calcium ion-liganding residues. More surprising is the identification of a group of calcium affinity mutants that map to the hydrophobic cores in the lower two-thirds of the domain. These residues are a considerable distance from the calcium ions themselves, suggesting that their effect on calcium affinity is indirect. The phenotype resulting

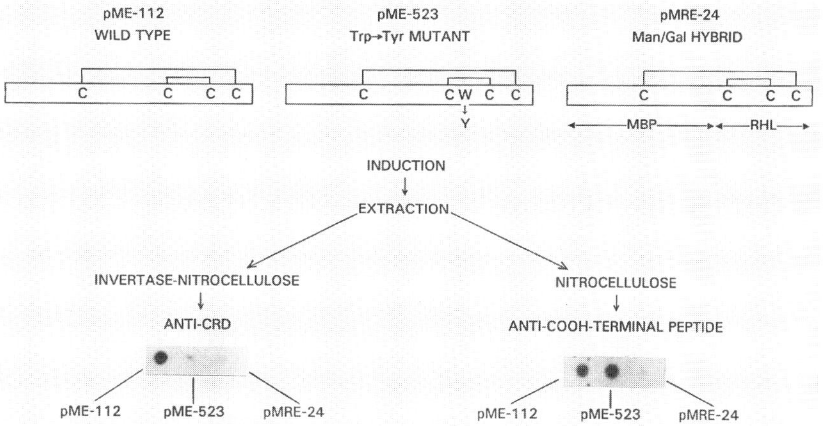

Figure 5. Strategy for characterization of mutant mannose-binding CRDs. (*Top*) Diagrams of wild-type and two mutant forms of the CRD. In the first mutant, an invariant tryptophan residue has been changed to tyrosine, whereas in the second, the 26 carboxy-terminal residues of the CRD have been substituted with the corresponding portion of the galactose-binding CRD from the rat asialoglycoprotein receptor (RHL). Both these proteins have been shown to lack sugar-binding activity. (*Bottom*) Following induction in the presence of calcium, active CRD was extracted from the cells and tested for activity by filtration through invertase-coated nitrocellulose followed by detection with polyclonal anti-CRD antiserum. Aliquots of the extract were also tested for expression of full-length protein by filtration through uncoated nitrocellulose and probing with antiserum against the carboxy-terminal peptide of the CRD. In both cases, bound antibodies were detected using radioiodinated protein A followed by autoradiography. (Reprinted, with permission, from Quesenberry and Drickamer 1992.)

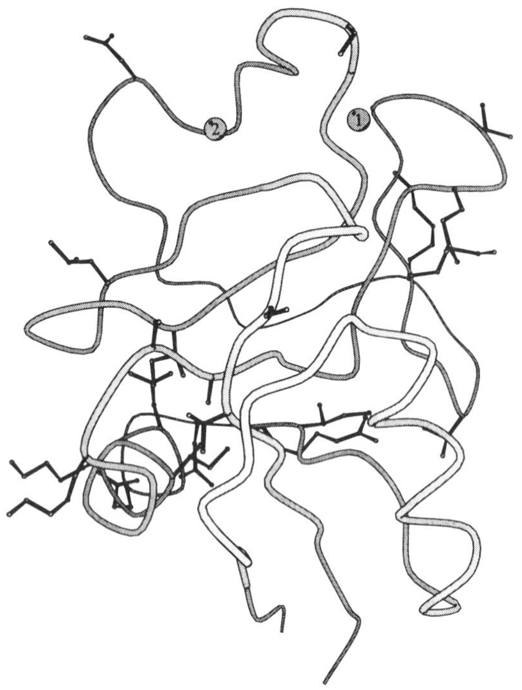

Figure 6. Examples of mutations that do not affect the activity of the mannose-binding CRD. Changing the side chains shown, most of which lie on the solvent-exposed surface, results in no change in the Ca^{++}- or sugar-binding activity of the CRD.

from these changes suggests that the precise geometry of the lower portion of the CRD is essential for positioning the irregular loops that form the calcium-binding sites.

An interesting consequence of the close interaction between the lower portion of the domain and the calcium ion-binding loops is that very slight changes in the lower portion of the domain may be utilized to effect biologically important changes in the activity of C-type CRDs. It is an important feature of the CRDs in endocytic receptors that they must be able to release their ligands in the slightly acidic (pH 5.5) environment of the endosomal compartment. It has been shown for the chicken hepatic lectin that this pH-mediated loss of sugar-binding activity is an indirect consequence of decreased affinity for calcium ions. This phenomenon may reflect direct titration of calcium-liganding carboxyl groups (that might have aberrantly high pKa values due to their local environment), or it might reflect partial titration of a number of side chains in the lower portion of the domain, resulting in a conformational change similar to that in the mutants. It is noteworthy that reducing the pH and removing calcium ions results in only very modest changes in the regular secondary structure of the CRD as observed by circular dichroism spectropolarimetry (G. V. Crichlow and K. Drickamer, unpubl.).

Finally, analysis of mutations in amino acid side chains involved in liganding calcium ion 2 and sugar ligand provide strong support for the bonding scheme shown in Figure 4. Of particular interest is the change Asn-187→Asp. In the wild-type CRD, the amide nitrogen of this side chain donates a hydrogen to one of the lone electron pairs on hydroxyl oxygen 3 of mannose. From the structure of the sugar-protein complex, it would be predicted that calcium coordination should be unaffected whereas sugar binding should be weakened or abolished by replacement of the side chain amide nitrogen with a carboxylate oxygen, since the carboxylate oxygen cannot serve as a proton donor. In addition, the presence of the partial negative charge on the carboxylate would be electrostatically unfavorable. Both of these predictions have been verified experimentally: This mutant CRD fails to bind to a mannose-containing affinity column, although the affinity for calcium measured by a proteolysis-resistance assay is indistinguishable from that of wild-type protein.

Oligomeric Structure and Oligosaccharide Recognition

The analysis of how a single CRD interacts with a terminal sugar residue reveals that the sugar is ligated in a relatively shallow indentation on the surface of the protein. Two consequences of the limited number of contacts between the sugar and the protein are that the CRD binds to diverse monosaccharides with only limited common structural features, and that the binding to monosaccharides is weak. In light of these characteristics of the mannose-binding and other C-type CRDs, the obvious question is how the receptors and other proteins that contain these CRDs achieve their exquisite specificity and high affinity for complex oligosaccharides.

The structure of the sugar ligands, particularly their branched nature, suggests that multiple interactions between individual monosaccharides within an oligosaccharide and one or more CRDs within a C-type lectin must be required to achieve high affinity and selectivity. In the case of truly independent binding events, the affinity for the multivalent ligand is the product of the individual affinities for monosaccharide. There are various ways in which the necessary multiple interactions may be achieved. One possibility is that there is a second saccharide-binding site in the CRD, which is not occupied in the crystal form that has been examined. Alternatively, multiple CRDs clustered together in an oligomer may each bind a single saccharide residue, so that the ultimate affinity derives from the clustering of CRDs. This last possibility is supported by the crystal structure analysis described above, in which a single oligosaccharide chain is observed to bind to different CRDs, although it must be noted that the specific disposition of CRDs in the crystal does not represent their arrangement in the intact lectin.

There is some reason to believe that each of these possibilities may be involved in explaining the ligand-binding characteristics of diverse C-type lectins. The fact that the selectins bind ligands with two distinct types of terminal residues (the sialic acid and fucose of sialyl Lewis-x antigen, for example [Feizi 1991]), yet contain only a single type of CRD, suggests that two binding sites may be present in the single CRD. Since all the ligands for calcium ion 2 in the mannose-binding CRD are conserved in the selectins, the terminal fucose residue could be accommodated in a manner exactly analogous to that described for the mannose-binding CRD described above. The second, sialic acid-binding, site might be associated with the other calcium ion or might be elsewhere in the structure.

For many of the other C-type animal lectins, there is evidence that clustering of CRDs plays an important role in achieving high affinity for oligosaccharides. As shown in Figure 1, many C-type lectins are homo- or hetero-oligomers. The structural basis for the formation of oligomers varies, although the core oligomer is often a trimer. In the collectins, collagen-like triple helical domains are an important factor in trimer formation, although the neck region connecting this collagenous stalk to the CRD is capable of stabilizing a trimer on its own (Weis et al. 1991a). The membrane anchor of the chicken hepatic lectin has been shown to play an essential role in oligomerization of this receptor (F. Verrey and K. Drickamer, in prep.), and a possible triple coiled-coil of α helices in the neck of the Kupffer cell fucose receptor would be a likely region for subunit interactions (Beavil et al. 1992).

The effect of oligomerization on ligand binding has been most extensively studied in the chicken hepatic lectin, the rat asialoglycoprotein receptor, and the macrophage mannose receptor. In the first case, it has been shown that a monomeric fragment of the protein and the hexameric, intact receptor bind *N*-acetylglucosamine with similar affinities, but the oligomer binds the multivalent ligand agalacto-α_1-acid glycoprotein with 25-fold greater affinity than does the monomeric domain (Loeb and Drickamer 1988). However, the fact that the monomeric CRD binds to clustered synthetic ligands better than it binds to monosaccharides suggests that there may be more than one sugar-binding site within the CRD also contributing to the overall valency of the receptor (Lee et al. 1989).

A more complex type of specificity may result from combinations of different types of CRDs in the asialoglycoprotein receptor. It has been demonstrated that the two different subunits of this receptor bind selectively to different terminal galactose residues on a triantennary glycopeptide (Rice et al. 1990). It is not known whether this selectivity is an intrinsic property of the different subunits or results from their combination in a specific orientation.

In contrast to the other proteins shown in Figure 1, clustering of C-type CRDs in the mannose receptor results from repetition of the CRD motif eight times within the single polypeptide (Taylor et al. 1990). Expression studies have demonstrated that, although CRD 4 alone is sufficient to mimic the binding of monosaccharides to the intact receptor, additional CRDs contribute to ligand binding (Fig. 7). CRDs 4–7 must be present to achieve high-affinity binding to a

DOMAINS	BINDING

1 2 3	−
4 5 6 7 8	+
4 5 6	+
4 5	+
4	+
5	−
5 6 7 8	+
5 6 7	+
6 7 8	−

Figure 7. Activity of multiple CRDs in the macrophage mannose receptor. Structure of the receptor is summarized at the top, with the activity of constructs containing subgroups of CRDs shown below. Activity was measured by retention on mannose-coupled Sepharose following in vitro transcription and translation. (Data from Taylor et al. 1992).

complex ligand (Taylor et al. 1992). The magnitude of this effect is dependent on the ligand tested, with the most significant effect being observed with a natural ligand, mannan from the outer wall of yeast. This observation suggests that the alternative forms of CRD clustering in the hepatic lectins and the mannose receptor may reflect a difference in their mode of binding to multivalent ligands, with the hepatic lectin structure providing optimal binding to clusters of terminal residues and the mannose receptor organization resulting in high-affinity binding to repeated polymers such as those found on the surfaces of many microorganisms.

CONCLUSIONS

It is possible to make tentative suggestions about why sugars have been utilized by nature in certain types of recognition events. One way in which use of sugars has an advantage over simple protein-protein interactions is that similar sugar structures can be appended to a variety of proteins using a common enzymatic machinery in the luminal compartment of cells. A second advantage of sugars is that in many cases the sugar and protein portions of a glycoprotein may be functionally orthogonal. That is, attachment of the saccharide often has little effect on the intrinsic activity of glycoprotein enzymes and hormones but is utilized as a tag to control turnover or clearance of the proteins. The common oligosaccharide structures can thus lead to clearance by a common mechanism for a variety of proteins with diverse functions (Drickamer 1991). In the context of the cell surface, the ability to put similar sugars on multiple protein and lipid anchors may serve as a convenient way to achieve high density of a common ligand on specific cell types, thus providing a broad target for initial low-affinity interactions between cells, such as those mediated by the selectins.

Although these ideas are speculative at the moment, they provide a conceptual framework in which to view the structural studies described above. They suggest that low-affinity, modest specificity resulting from the unique calcium/sugar interactions observed in the present studies may be an essential aspect of the function of this and related CRDs. In evolutionary terms, the limited degree of interaction between the sugar and the CRDs has allowed flexibility, so that modest changes in CRD structure result in development of significantly different monosaccharide-binding properties. This contrasts with the mannose 6-phosphate and thiol-dependent (S-type) β-galactoside-binding lectins, which seem to be locked into a more limited set of specificities, albeit with higher affinity.

The hypothesis that much of the specificity of animal lectins for complex oligosaccharides resides in the way C-type CRDs are arranged in clusters remains to be fully tested. An impediment to progress in this area has been the lack of structural information for oligosaccharides, so that the possible recognition surfaces of these structures can be identified and the basis for specificity in lectin recognition can be established. As structures are being determined at an increasing rate, progress on this front should increase.

ACKNOWLEDGMENTS

This work was supported by grants GM-42628 and GM-34102 from the National Institutes of Health. W.I.W. is a Life Sciences Research Foundation Fellow of the Howard Hughes Medical Institute. M.S.Q. was supported by training grant DK-07328 from the National Institutes of Health. K.D. is recipient of a Faculty Salary Award from the American Cancer Society.

REFERENCES

Beavil, A.J., R.L. Edmeades, H.J. Gould, and B.J. Sutton. 1992. α-Helical coiled-coil stalks in the low-affinity receptor for IgE (FcϵRII/CD23) and related C-type lectins. *Proc. Natl. Acad. Sci.* **89:** 753.

Bezouska, K., G.V. Crichlow, J.M. Rose, M.E. Taylor, and K. Drickamer. 1991. Evolutionary conservation of intron position in a subfamily of genes encoding carbohydrate-recognition domains. *J. Biol. Chem.* **266:** 11604.

Derewende, Z., J. Yariv, J.R. Helliwell, A.J. Kalb, E.J. Dodson, M.Z. Papiz, T. Wan, and J. Campbell. 1989. The structure of the saccharide-binding site of concanavalin A. *EMBO J.* **8:** 2189.

Drickamer, K. 1988. Two distinct classes of carbohydrate-recognition domains in animal lectins. *J. Biol. Chem.* **263:** 9557.

———. 1991. Clearing up glycoprotein hormones. *Cell* **67:** 1029.

———. 1992. Molecular structure of animal lectins. In *Molecular glycobiology* (ed. M. Fukuda), Oxford University Press, England. (In press.)

Drickamer, K., M.S. Dordal, and L. Reynolds. 1986. Mannose-binding proteins isolated from rat liver contain carbohydrate-recognition domains linked to collagenous tails. *J. Biol. Chem.* **261:** 6878.

Ezekowitz, R.A.B., L.E. Day, and G.A. Herman. 1988. A human mannose-binding protein is an acute phase reactant that shares sequence homology with other vertebrate lectins. *J. Exp. Med.* **167:** 1034.

Feizi, T. 1991. Cell-cell adhesion and membrane glycosylation. *Curr. Opin. Struct. Biol.* **1:** 766.

Haagsman, H.P., S. Hawgood, T. Sargeant, D. Buckley, R.T. White, K. Drickamer, and B.J. Benson. 1987. The major lung surfactant protein, SP 28-36, is a calcium-dependent, carbohydrate-binding protein. *J. Biol. Chem.* **262:** 13877.

Halberg, D.H., G. Proulx, K. Doege, Y. Yamada, and K. Drickamer. 1988. A segment of cartilage proteoglycan core protein has lectin-like activity. *J. Biol. Chem.* **263:** 9486.

Houchins, J.P., T. Yabe, C. McSherry, and F.H. Bach. 1991. DNA sequence analysis of NKG2, a family of related cDNA clones encoding type II integral membrane proteins on human natural killer cells. *J. Exp. Med.* **173:** 1017.

Ikeda, K., T. Sannoh, N. Kawasaki, T. Kawasaki, and I. Yamashina. 1987. Serum lectin with known structure activates complement through the classical pathway. *J. Biol. Chem.* **262:** 7451.

Kim, S.J., N. Ruiz, K. Bezouska, and K. Drickamer. 1992. Organization and characterization of the gene encoding the human macrophage mannose receptor. *Genomics* (in press).

Lee, R.T., K.G. Rice, N.B.N. Rao, Y. Ichikawa, T. Barthel, V. Piskarev, and Y.C. Lee. 1989. Binding characteristics of *N*-acetylglucosamine-specific lectin of the isolated chicken hepatocytes: Similarities to mammalian hepatic galactose/*N*-acetylgalactosamine-specific lectin. *Biochemistry* **28:** 8351.

Loeb, J.A. and K. Drickamer. 1987. The chicken receptor for endocytosis of glycoprotein contains a cluster of *N*-acetylglucosamine-binding sites. *J. Biol. Chem.* **262:** 3022.

———. 1988. Conformational changes in the chicken receptor for endocytosis of glycoproteins: Modulation of binding activity by Ca^{2+} and pH. *J. Biol. Chem.* **263:** 9752.

Lu, J., S. Thiel, H. Wiedemann, R. Timpl, and K.B.M. Reid. 1990. Binding of the pentamer/hexamer forms of a mannan-binding protein to zymosan activates the proenzyme $C1r_2C1s_2$ complex of the classical pathway of complement, without involvement of C1q. *J. Immunol.* **144:** 2287.

Quesenberry, M.S. and K. Drickamer. 1991. Determination of the minimum carbohydrate-recognition domain in two C-type animal lectins. *Glycobiology* **1:** 615.

———. 1992. Role of conserved and nonconserved residues in the Ca^{2+}-dependent carbohydrate-recognition domain of a rat mannose-binding protein: Analysis by random cassette mutagenesis. *J. Biol. Chem.* **267:** 10831.

Quiocho, F.A. 1986. Carbohydrate-binding proteins: Tertiary structures and protein-sugar interactions. *Annu. Rev. Biochem.* **55:** 287.

Rice, K.G., O.A. Weisz, T. Barthel, R.T. Lee, and Y.C. Lee. 1990. Defined stoichiometry of binding between triantennary glycopeptide and the asialoglycoprotein receptor of rat hepatocytes. *J. Biol. Chem.* **265:** 18429.

Shaanan, B., H. Lis, and N. Sharon. 1991. Structure of a legume lectin with an ordered *N*-linked carbohydrate in complex with lactose. *Science* **254:** 862.

Spiess, M. 1990. The asialoglycoprotein receptor: A model for endocytic transport receptors. *Biochemistry* **29:** 10008.

Stahl, P.D. 1990. The macrophage mannose receptor: Current status. *Am. J. Respir. Cell Mol. Biol.* **2:** 317.

Stoolman, L.M. 1989. Adhesion molecules controlling lymphocyte migration. *Cell* **56:** 907.

Sumiya, M., M. Super, P. Tabona, R.J. Levinsky, A. Takayuki, M.W. Turner, and J.A. Summerfield. 1991. Molecular basis of opsonic defect in immunodeficient children. *Lancet* **337:** 1569.

Taylor, M.E., K. Bezouska, and K. Drickamer. 1992. Contribution to ligand binding by multiple carbohydrate-recognition domains in the macrophage mannose receptor. *J. Biol. Chem.* **267:** 1719.

Taylor, M.E., J.T. Conary, M.R. Lennartz, P.D. Stahl, and K. Drickamer. 1990. Primary structure of the mannose receptor contains multiple motifs resembling carbohydrate-recognition domains. *J. Biol. Chem.* **265:** 12156.

Vercelli, D., B. Helm, P. Marsh, E. Padlan, R.S. Geha, and H. Gould. 1989. The B-cell binding site on human immunoglobulin E. *Nature* **338:** 649.

Weis, W.I., K. Drickamer, and W.A. Hendrickson. 1992. Structure of a C-type mannose-binding protein complexed with an oligosaccharide. *Nature* (in press).

Weis, W.I., G.V. Crichlow, H.M.K. Murthy, W.A. Hendrickson, and K. Drickamer. 1991a. Physical characterization and crystallization of the carbohydrate-recognition domain of a mannose-binding protein from rat. *J. Biol. Chem.* **266:** 20678.

Weis, W.I., R. Kahn, R. Fourme, K. Drickamer, and W.A. Hendrickson. 1991b. Structure of the calcium-dependent lectin domain from a rat mannose-binding protein determined by MAD phasing. *Science* **254:** 1608.

Wong, S., J.D. Freeman, C. Kelleher, D. Mager, and F. Takei. 1991. Ly-49 multigene family: New members of a superfamily of type II membrane proteins with lectin-like domains. *J. Immunol.* **147:** 1417.

Lymphocyte Homing Receptors

M.C.-T. Hu,* M.H. Siegelman,† B. Holzmann,§ D.T. Crowe,* and I.L. Weissman*
*Howard Hughes Medical Institute, Departments of Pathology and Developmental Biology, Stanford University School of Medicine, Stanford, California 94305; †Department of Pathology, University of Texas Southwestern Medical Center at Dallas, Texas 75235; §Institute for Medical Microbiology and Hygiene, Technical University of Munich, 8000 Munich 80, Germany

The immune response to foreign antigens takes place in lymphoid organs. Lymphocytes are largely mobile cells recirculating from the blood into and out of lymphoid organs (Gowans and Knight 1964; Howard et al. 1972; Gutman and Weissman 1973). Each lymphocyte is clonally predetermined to recognize a very restricted antigenic shape, and therefore lymphoid organs are required to bring together rare cells (in terms of antigen specificity) that must interact with each other. The recirculation of lymphocytes throughout the body plays a critical role in the normal function of the immune system by maximizing interactions of lymphocytes with antigen and by increasing collaborative interactions between many disparate cell types. Organ-specific lymphocyte homing is also important in augmenting the immune response in a tissue by enhancing circulation of the immunocompetent lymphocytes mostly in the type of tissue where they first encountered antigen. The specific portal of entry of both T and B lymphocytes (Howard et al. 1972; Gutman and Weissman 1973) from the blood stream into lymphoid organs was identified as specialized postcapillary venules bearing unusually high-walled endothelia and named high endothelial venules (HEVs) (Gowans and Knight 1964; Stamper and Woodruff 1976). The cell-surface molecules mediating the highly organ-specific binding of lymphocytes to HEV prior to transmigration through the vessel wall have been called "lymphocyte homing receptors" (Gallatin et al. 1983), and their complementary ligands on the endothelial surface have been named "addressins" (Streeter et al. 1988). Thus far, several functionally and antigenically distinct lymphocyte-HEV recognition systems governing the homing of lymphocytes to peripheral lymph nodes, mucosal lymphoid organs (Peyer's patches and appendix), skin, and inflamed synovium have been identified (for review, see Gallatin et al. 1986; Jalkanen et al. 1986b; Yednock and Rosen 1989).

The murine lymph-node-specific homing receptor is defined by antibody MEL-14 which recognizes a branched-chain ubiquitinated glycoprotein on the surface of lymphocytes with an apparent molecular mass of approximately 90 kD (for review, see Gallatin et al. 1986), and its cDNA has been cloned and well characterized (Lasky et al. 1989; Siegelman et al. 1989). Furthermore, a cDNA clone encoding the human homolog of this homing receptor has been isolated and characterized (Siegelman and Weissman 1989; Tedder et al. 1989). It is a member of a new class of adhesion receptors, now designated SELECTIN (Watson et al. 1990) or LEC-CAM (Stoolman 1989), which are clearly distinct from other major families of adhesion molecules such as the immunoglobulin gene superfamily, cadherins, and integrins. SELECTIN are putative multidomain proteins with contiguous homology units of lectin, epidermal growth factor (EGF), and complement regulatory short consensus repeats arranged in that order in their ectodomains (Bevilacqua et al. 1989; Johnston et al. 1989; Lasky et al. 1989; Siegelman et al. 1989).

The murine *l*ymphocyte *P*eyer's patch HEV *a*dhesion *m*olecule-1 (LPAM-1) involved in the organ-specific adhesion of lymphocytes to Peyer's patch HEV has been identified to be a member of the integrin family of adhesion receptors (Holzmann and Weissman 1989; Holzmann et al. 1989). LPAM-1 consists of an integrin α subunit (α_{4m}) homologous to human VLA-4 (very late antigen 4) α and a novel integrin β subunit (designated as β_p) that is invariantly involved in lymphocyte adhesion to Peyer's patch HEV (Holzmann and Weissman 1989). The functional role for the integrin α_4 subunit of LPAM-1 in homing to mucosal lymphoid organs had been shown indirectly by anti-α_{4m} monoclonal antibody R1-2 inhibition in HEV-binding assays in vitro in the mouse (Holzmann et al. 1989) and confirmed directly by lymphocyte migration studies in vivo in the rat (Issekutz 1991). Although α_{4m} can also associate with the integrin β_1 subunit to form LPAM-2 (Holzmann and Weissman 1989), it is now clear that $\alpha_{4m}\beta_1$ is not a lymphocyte-homing receptor to Peyer's patches (Hu et al. 1992). Furthermore, $\alpha_{4m}\beta_1$ is predominantly expressed on lymphoma cells that do not bind to Peyer's patch HEV, supporting that expression of $\alpha_{4m}\beta_1$ alone is not sufficient to mediate this adhesive event. In contrast, $\alpha_{4m}\beta_1$ does bind to vascular cell adhesion molecule-1 (VCAM-1) expressed on activated endothelia (Elices et al. 1990) and bone marrow and thymic stromal cell lines (Kina et al. 1991; Miyake et al. 1991).

Several other candidate molecules for adhesion to Peyer's patch HEV have also been identified. First, in humans, a lymphocyte surface glycoprotein(s) of approximately 90 kD (gp90^{Hermes}) defined by the Hermes series of monoclonal antibodies has been reported to be involved in lymphocyte organ-specific adhesion to lymph node, mucosal, or synovial HEVs (Jalkanen et al. 1986a, 1987). Recently, the Hermes antigen ($gp90^{Hermes}$) on lymphocytes has been shown to be identical to CD44 and human Pgp-1 antigens (Goldstein et al. 1989; Nottenburg et al. 1989; Picker et al. 1989). However, in mice, the homolog of Pgp-1 (Ly-24 or H-CAM) has not yet been demonstrated to be required in lymphocyte adhesion to HEV in lymph nodes or Peyer's patches. Perhaps this molecule can function as a nonorgan-specific accessory molecule for lymphocyte adhesion; it is a widespread primary cell-surface adhesion molecule (Carter and Wayner 1988; Kansas et al. 1989; Picker et al. 1989). Second, as an example of a nonorgan-specific adhesion molecule, the integrin LFA-1 has been shown by antibody-blocking studies to be involved in lymphocyte-HEV interactions (Hamann et al. 1988; Pals et al. 1988). The inhibitory effect of anti-LFA-1 antibodies is operative for lymph node and Peyer's patch HEV binding and other leukocyte-endothelial interactions (Butcher 1991; Lawrence and Springer 1991), suggesting that the role of the LFA-1 molecule in lymphocyte migration and homing is accessory. Finally, in the rat, a monoclonal antibody to HEBFpp defines a Peyer's patch adhesion molecule to be an approximately 80-kD single-chain protein (Chin et al. 1986), which appears to be distinct from LPAM-1. Whether a mouse homolog of this molecule operates concurrently with LPAM-1 to mediate adhesion of lymphocytes to Peyer's patch HEV remains to be established.

Structural Analysis of Lymph Node Homing Receptor cDNA

The analysis of the amino-terminal amino acid sequence (WTYHY) of the core protein of the mouse lymph-node-homing receptor (mLHR) led in our laboratory to preparation of a degenerate set of oligonucleotide probes that allowed isolation and characterization of a cDNA clone encoding the core protein of $gp^{90MEL-14}$ (Siegelman et al. 1989). Transfection of this mLHR cDNA into heterologous cells confirmed expression of the MEL-14 epitope and its role in specific adhesion to lymph node HEV. The distribution of expression of the hybridizing transcripts to this cDNA was largely restricted to hematolymphoid cells and tissues. Transcripts expressed in cell lines and tissues (except testis) in Northern blots exhibited identical patterns with bands at 1.5, 2.5, and 5.2 kb. Normal tissues showed a predominant lymphoid distribution, paralleling tissue-staining patterns for MEL-14. Thymus, spleen, and mesenteric lymph node were positive for the same transcripts found in cell lines, whereas liver, kidney, and brain showed no appreciable hybridization.

The mLHR cDNA clone sequence reveals a potentially highly glycosylated transmembrane protein with a short cytoplasmic tail. The complete nucleotide sequence of the cDNA clone contains a 54-bp 5′-untranslated region followed by an open reading frame of 1116 bp. The TGA stop codon at position 1170 is followed by 327 bp of 3′-untranslated region. The assignment of the initiation ATG codon is supported by the presence of flanking sequences GCC*ATG*G that match the consensus sequence of the translation initiation site (Kozak 1987). The reading frame encodes a protein with a bipartite hydrophilic hydrophobic signal sequence 38 amino acids in length, which is unusual for a eukaryotic signal sequence. Although hydropathy analysis confirms a generally hydrophobic signal sequence, the initial 15 residues are neutral to slightly hydrophilic. The mature protein begins with a tryptophan which is a highly unusual amino-terminal amino acid. The deduced mature protein is 334 amino acids in length, with a calculated molecular mass of 37.6 kD. This contrasts with the estimated molecular mass of the deglycosylated form of the protein, which is approximately 46 kD (van de Rijn et al. 1990); the extra approximately 8 kD has been attributed to a ubiquitin side chain, putatively located in the lectin domain (Siegelman et al. 1989). A hydrophobic transmembrane region encompassing amino acids from about 295 to 317 is followed by a cluster of positively charged residues and a hydrophilic cytoplasmic tail of 18 amino acids.

The deduced mature protein possesses ten potential asparagine-linked glycosylation sites, consistent with protein characterization studies showing extensive glycosylation on endoglycosidase F digestion (Siegelman et al. 1986; Van de Rijn et al. 1990) and tunicamycin inhibition experiments (Van de Rijn et al. 1990). Six of these are contained within the identical complement regulatory protein (CRP) repeat (see below). No clusters of serine, threonine, or both, characteristic of sites of *O*-linked glycosylation, are present in the deduced sequence. The mature protein contains 22 cysteine residues. Twelve of the cysteines are present in the CRP repeat structures, and an additional nine cysteines are concentrated in the 60 amino acids just preceding the repeat units in the EGF-like domain and carboxy-terminal lectin domain (see below), resulting in a highly cysteine-rich pretransmembrane region of 180 – 190 amino acids.

A striking feature of this protein is that it contains a tandem array of three kinds of domains that are the characteristic of the SELECTIN family. The three cell-surface glycoproteins comprising this new family of adhesion molecules are all transmembrane proteins with an unusual protein mosaic architecture. As shown in Figure 1, they each contain a tandem array of three kinds of domains: a calcium-dependent animal lectin domain, an EGF-like domain, and a variable number of repeat units, conforming to the short consensus repeats of CRP and other proteins (Bevilacqua et al. 1989; Johnston et al. 1989; Lasky et al. 1989; Siegelman et al. 1989). The lymph-node-homing receptor (LHR or L-SELECTIN) is present on lymphocytes and mediates the adherence of lymphocytes to endothelial cells in peripheral lymph nodes (Gallatin et al. 1983) and on neutrophils similarly involved in their extravasation at inflammatory sites (Lewinsohn et al. 1987). Endothelial leukocyte adhesion molecule-1 (ELAM-1

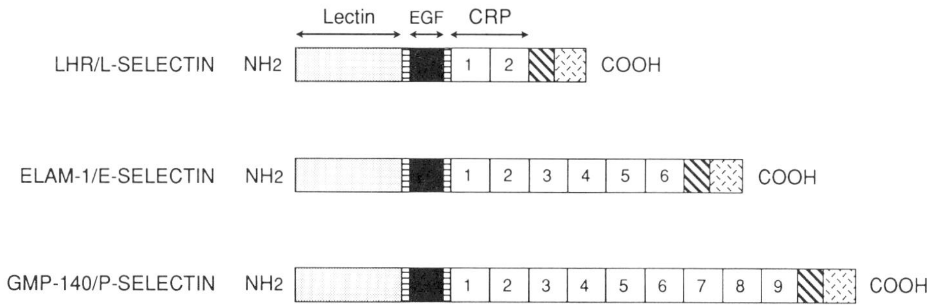

Figure 1. Comparison of the SELECTIN family of adhesion molecules. The lymph node homing receptor (LHR/L-SELECTIN) is compared to the other two members of the family, ELAM-1/E-SELECTIN and GMP-140/P-SELECTIN.

or E-SELECTIN) is a cell-surface receptor expressed on cytokine-induced endothelial cells (Bevilacqua et al. 1987) and mediates the binding of leukocytes. A granule membrane protein of platelets and endothelium (GMP-140, PADGEM, CD62, or P-SELECTIN) contained and stored within α granules of platelets and Weibel-Palade bodies of endothelial cells (Hsu-Lin et al. 1984; McEver and Martin 1984) is rapidly mobilized to the cell surface on activation and promotes binding of these platelets and endothelial cells to monocytes and neutrophils (Larsen et al. 1989; Hamburger and McEver 1990).

One type of carbohydrate-binding protein is the calcium-dependent (C-type) animal lectins; these are generally homologous to each other in a domain of 130–150 residues, a region containing the carbohydrate-binding activity in the asialoglycoprotein receptor, a type II membrane glycoprotein (for review, see Drickamer et al. 1989). The mLHR contains a sequence motif homologous to this type of lectin, in particular to the hepatic lectins of human and rat over the amino-terminal 118 amino acids, particularly preserved over a stretch of 45 amino acids (74–118) bearing similarity to the 50 carboxy-terminal residues of the binding domain in animal lectins. This region includes three invariant cysteines at 90, 109, and 116; Asn-Trp at 75–76; a characteristic Glu-Pro-Asn at 80–82; a Glu at 88, Cys-Val at 90–91; and the conserved Gly–Trp-Asn-Asp at 102–106. Only a highly conserved glycine (at 12) in the consensus sequence and present in other mammalian lectins is absent from this presumed carbohydrate-binding domain. A relative enrichment in positively charged residues and particularly lysines in this domain may have implications for sites of ubiquitination, carbohydrate binding, or both. The orientation of the lectin domain in mLHR is unusual, being inverted with respect to most other animal lectins that are oriented with the amino terminus anchored in the membrane and the carboxyl terminus containing the binding portion extracytoplasmically.

The EGF-like domain in mLHR consists of a single-copy homolog of the EGF repeat unit, which preserves many of the cysteine/glycine residues characteristic of this structure (118–155). All six consensus cysteines are present as well as Gly-147 and Gly-150, and Tyr-148. The relationship of these conserved residues is identical to that of human and bovine blood-clotting factors IX and X and the *Drosophila Notch* gene product. In addition, it is conserved with some other EGF-like repeat units in the four neurogenic loci in *Drosophila melanogaster*. These *Drosophila* molecules apparently mediate early cell-cell interaction events in embryogenesis controlling the developmental choice of ectodermal cells to either neural or epidermal pathways. Interestingly, this EGF-like domain shares homology with the cysteine-rich repeat units of the integrin β subunits (see below). The possible role of this EGF-like domain involved in receptor-ligand interactions is discussed below.

The third homology domain preserves the sequence motif of short consensus repeats present in complement regulatory and other proteins. Each motif is 62 amino acids in length, spans positions 156–217 and 218–279, and in the mouse is completely identical at the nucleotide and protein levels. The murine complement factor H, a serum protein with complement regulatory activity (Kristensen and Tack 1986), was identified as homologous over this region. In factor H, there are 20 contiguous homologous repeat motifs having approximately 10–31% homology with this domain of the mLHR. The same homologous repeat motif exists in a number of CRPs that bind the cleaved forms of C3 and C4 (for review, see Ahearn and Fearon 1989) and in other proteins such as the interleukin-2 (IL-2) receptor (Ishida et al. 1985; Leonard et al. 1985), the β_2-glycoprotein I (Lozier et al. 1984), and factor XIII (Ichinose et al. 1986). Every consensus sequence position is represented in the repeat motifs of mLHR. Except for a relative deletion of one residue between Cys-4 and Pro-7 of the consensus sequence, and an insertion of three residues between Pro-7 and Phe-30, the relative spacing of the remaining residues of the consensus sequence is completely preserved in the mLHR sequence. Two additional cysteines are also present in the mLHR repeats relative to all other such motifs. Comparison of the repeat units of the IL-2 receptor, β_2-glycoprotein I, and factor XIII and of the mLHR repeat structure shows greater similarity to complement regulatory proteins than to these noncomplement-related proteins. Although homologous repeat units are not unusual among cell-surface receptors and other proteins, the presence of a precisely identical repeat motif from nucleotide positions 633 to 818 and 819 to 1005 in the mLHR is distinctive and is not preserved in the homologous sequences in humans (Siegelman and Weissman 1989) or rats (M.H. Siegelman and I.L. Weissman, unpubl.). The finding of an identical rather than homologous repeat of the CRP-like sequence in mouse is striking. This implies a very recent gene segment duplication, rather than a reduction to two units from a multiunit ancestor gene segment by unequal crossing over.

Functional Role of the Lymph Node Homing Receptor in Lymphocyte Homing

The monoclonal antibody MEL-14 has been shown to be a strong inhibitor of functional binding between mLHR and lymph node HEV (Gallatin et al. 1983). It appears to recognize a novel epitope on the mLHR molecule composed of a combinatorial determinant between ubiquitin and the core polypeptide (Siegelman et al. 1986; St. John et al. 1986) and has been associated with the carbohydrate-binding domain (the lectin domain) of the molecule (Yednock et al. 1987). In vitro, mLHR-HEV recognition could be abrogated specifically by the monosaccharides mannose-6-phosphate and the structurally related fructose-1-phosphate, fucoidin, and PPME (a yeast phosphomonoester mannan fragment rich in mannose-6-phosphate) (for review, see Yednock and Rosen 1989). Furthermore, it has been shown that sialidase treatment of HEV ablates the binding of both lymphocytes and recombinant $gp^{90MEL-14}$ to tissue sections in vitro and the binding of lymphocytes to peripheral lymph node HEV in vivo (Yednock and Rosen 1989), suggesting that a form of sialic acid contributes to the mLHR-HEV recognition structure. Although recent studies on the carbohydrate ligands for ELAM-1 and GMP-140 (Goelz et al. 1990; Larsen et al. 1990; Lowe et al. 1990; Phillips et al. 1990; Walz et al. 1990; Tiemeyer et al. 1991) characterize the specific form of lectin-based carbohydrate binding to be directed to the human blood group antigen Lewis X (GMP-140) or sialyl Lewis X (ELAM-1), the minimal carbohydrate structures identified probably represent only the beginning of the search for the actual physiological ligands. It has not yet been demonstrated that these carbohydrate inhibitors represent the precise structures on the cells bearing the complementary ligand. The complete carbohydrate structures of the ligands still remain to be elucidated, and the identities of the protein or lipid molecules bearing carbohydrate ligands for E- and P-SELECTIN are as yet unknown. Recently, a ligand (or complementary adhesion molecule) for L-SELECTIN has been identified as a highly glycosylated, sulfate-containing mucoprotein called Gly-CAM 1 (Lasky et al., this volume).

Although the amino-terminal lectin domain of the mLHR clearly plays a crucial role in the adhesive function of these receptors, our evidence indicates that the EGF-like domain of mLHR may contribute to HEV binding directly or indirectly, possibly independent of the receptor's lectin activity (Siegelman et al. 1990). We demonstrated that the mLHR locus is inseparable in recombinant inbred mouse strains from the polymorphic murine lymphocyte cell-surface marker Ly-22. Anti-Ly-22 monoclonal antibody defines a new functional and polymorphic determinant on $gp^{90MEL-14}$ and, by primary structural analysis, localizes the Ly-22 epitope to the distal portion of the EGF-like domain. Because MEL-14 appears to recognize an epitope on mLHR composed of a combinatorial determinant between ubiquitin and the core polypeptide (Siegelman et al. 1986; St. John et al. 1986) and has been associated with the carbohydrate-binding domain of the molecule (Yednock et al. 1987), it was of interest to determine the relationship between Ly-22 and MEL-14 determinants. The results of antibody cross-blocking studies demonstrated that although preincubation of cells with MEL-14 antibody inhibited staining by MEL-14, preincubation with excess anti-Ly-22 antibody had no effect on subsequent MEL-14 staining, and vice versa. This suggested that separate epitopes on mLHR are recognized by MEL-14 and anti-Ly-22 antibodies. Since MEL-14 has been shown to inhibit binding of PPME (phosphomannan monoester) to lymphocytes (Yednock et al. 1987), we confirmed the expected inhibition of binding of fluoresceinated PPME to $EL\text{-}4^{MEL\text{-}14hi}$ cells pretreated with saturating levels of MEL-14. In contrast, preincubation with excess anti-Ly-22 had no detectable effect on subsequent staining with fluoresceinated PPME. Thus, it is clear that the Ly-22 epitope is distinct from both the MEL-14 determinant and the lectin activity of mLHR. To determine whether or not anti-Ly-22 had any effect on the binding of lymphocytes to lymph node HEV, we applied this antibody in a modified Stamper and Woodruff in vitro HEV-binding assay. The results demonstrated high-efficiency blocking of peripheral node lymphocytes to lymph node HEV by anti-Ly-22 antibodies, at least comparable to that of MEL-14 under the same conditions. It therefore appears that a monoclonal antibody directed against the EGF-like domain of mLHR impedes binding of lymphocytes to HEV with high efficiency.

Because anti-Ly-22 monoclonal antibody did not detectably impair the carbohydrate-binding activity of mLHR but blocked HEV binding, it seemed reasonable to suggest that the EGF-like domain (the Ly-22 antigen) may participate in the adhesion event, serving as at least part of the adhesive surface for the peripheral node HEV. Examples of a direct role of an EGF-like domain in protein-protein interactions exist (see below). Thus, we propose that at least two functional domains of mLHR, namely, lectin and EGF-like, may be required to contribute to the initial adhesion with the addressin molecules on lymph node HEV. It should be pointed out that there are alternative explanations for the data. For example, the EGF-like domain might play a role in regulating the conformation of the lectin domain or it might interact with other lymphocyte cell-surface molecules involved in adhesion to lymph node HEV, to name just two possibilities.

It should be noted that the standard frozen section HEV assay is likely a measure not only of the initial lumenal adhesive event of lymphocytes to high endothelial cells, but possibly of additional steps in the homing process before entry into the lymph node parenchyma proper occurs. However, we have little knowledge of the events following initial adhesion at the lumenal surface or what other molecules may be involved. Beyond the functional roles of these domains in mLHR, questions of the subsequent signals transduced through these or associated molecules, their ligands, or both on either of the two participating cells and the requirements for transendothelial migration provide ample areas to explore in the future. A model has been proposed that describes leukocyte extravasation as a multistep process requiring the activation of an adhesion molecule cascade (see the last section in the Peyer's patch homing receptor).

Cloning and Characterization of the cDNA Clone for β_p

Since the Peyer's patch homing receptor (LPAM-1) consists of an integrin α_{4m} subunit and a novel integrin β_p subunit (Holzmann and Weissman 1989), the cDNA clone encoding this β_p molecule had to be isolated in order to begin a functional analysis of this receptor. To isolate cDNA clones for the mouse β_p, a λ_{ZAP} cDNA expression library prepared from poly(A)$^+$ RNA isolated from TK1 cells was screened using anti-LPAM-1 polyvalent serum. This antiserum had been prepared by immunization with immunoaffinity-isolated LPAM-1 protein and specifically immunoprecipitated LPAM-1 from TK1 cell lysates (Holzmann and Weissman 1989), showing that it contains $\alpha_{4m}\beta_p$-specific antibodies; 2×10^6 recombinant phage were screened, and 54 positive phage were identified. After four rounds of purification, 17 of these phage were consistently positive and their cDNA inserts were subsequently excised in vivo into the pBluescriptSK$^-$ plasmid vector. Since no protein sequence data were available for β_p, independent methods were used to identify the desired clones. A comparison of the amino acid sequences of the human integrin β_1 (Argraves et al. 1987), β_2 (Kishimoto et al. 1987; Law et al. 1987), β_3 (Fitzgerald et al. 1987), chicken integrin β (Tamkun et al. 1986), and *Drosophila* integrin β (*mys*) (MacKrell et al. 1988) revealed at least two highly conserved regions among these β subunits. Thus, we synthesized two oligonucleotides that corresponded to these two most highly conserved integrin β-specific regions. Instead of using highly degenerate oligonucleotides, we designed each oligonucleotide with a unique nucleotide sequence according to the preference of codon usage in mammalian genes (Aota et al. 1988), and deoxyinosine was assigned in situations where codon usage has little preference. These two best-guess oligonucleotides were used as the opposing primers in polymerase chain reaction (PCR). Equal amounts of plasmid DNA from each positive clone were used as the template for the reaction. As a positive control, we used cDNAs synthesized from TK1 poly(A)$^+$ RNA for the reaction. As a negative template control, we used a cDNA clone pZ5/6 encoding the entire coding sequence for α_{4m} (Neuhaus et al. 1991). On the basis of an average 118-amino-acid residue distance between the two targeted amino acid sequences in the integrin β subunits, the predicted size of the amplified PCR product was 354 bp; 7 out of 17 positive cDNA clones yielded an estimated 350-bp PCR product comigrating with that of the positive control (data not shown). Moreover, the negative control cDNA did not yield any PCR product, suggesting strongly that the seven PCR-positive cDNA clones coded for β-like integrin subunit(s). Since the antiserum used for screening the TK1 library was specific for LPAM-1, these β-like cDNA clones likely encode β_p.

Instead of sequencing all seven β-like cDNAs, we developed a reverse Northern approach to verify the cDNA clones coding for β_p. We first prepared a DNA blot containing the cDNA inserts of the various clones. This DNA blot was then probed with the ^{32}P-labeled cDNA of poly(A)$^+$ RNA from RAW112 cells, which express α_{4m} and β_1 but not β_p. After removal of the RAW112-specific probe, this blot was reprobed with the ^{32}P-labeled cDNA of poly(A)$^+$ RNA from TK1 cells, which express α_{4m} and β_p but not β_1. The rationale of this sequential hybridization with cDNA library probes is that the cDNA clones encoding the β_p subunit should only hybridize to the TK1 ($\alpha_{4m}{}^+\beta_p{}^+$) library probe but not to the RAW112 ($\alpha_{4m}{}^+\beta_p{}^-$) library probe, whereas the cDNA clones encoding the α_{4m}-like subunit should hybridize to both library probes. This strategy allowed all of the candidate cDNA clones to be tested at once on a single DNA blot, whereas conventional Northern analysis would have required making probes from all of the candidate cDNA clones and hybridizing the RNA blot with one probe at a time. To exclude the vector DNA from the blot and improve the ratio of signal to noise for the hybridization, we amplified by PCR the cDNA inserts from four β-like cDNA clones containing inserts longer than 2 kb and one α-like cDNA clone (T29) as a control (Fig. 2A). The DNA blot containing these samples was first hybridized with the ^{32}P-labeled RAW112 library probe under high-stringency conditions. Following hybridization with the RAW112 library probe, all four β-like cDNA inserts were clearly negative, whereas the α-like cDNA inserts (T29) were obviously positive with a discrete band at approximately 2.6 kb (Fig. 2B). Probing of the same blot with an α_{4m}-specific probe (pZ5/6) revealed the same result (data not shown). When hybridized with the TK1 library probe, all four β-like cDNA inserts and the α-like cDNA insert were evidently positive with a discrete band in each lane, ranging from approximately 2.8 to 2.3 kb, indicating that all four β-like cDNA inserts solely hybridized to the TK1 library probe (Fig. 2C). The evidence, when taken together, strongly implicates these four β-like, TK1-specific cDNA clones as sequences that code for the novel β_p subunit.

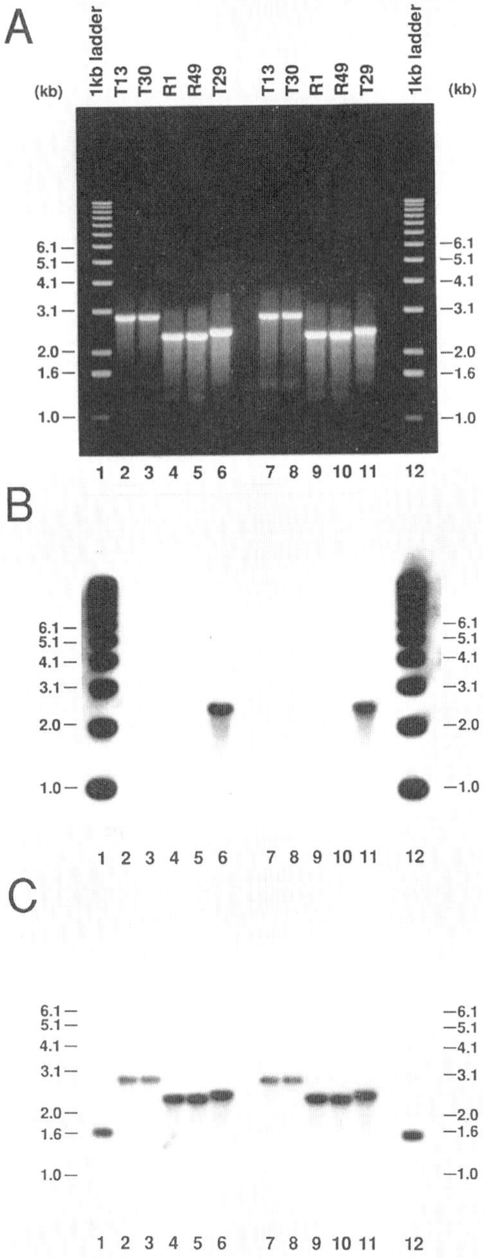

Figure 2. Reverse Northern analysis for β_p cDNA. (*A*) PCR amplification of four β-like cDNA inserts and one α-like cDNA insert was carried out with two sets of oligonucleotide primers, T3 and T7 primers (lanes *2–6*); M13 reverse and M13(−20) primers (lanes *7–11*), as duplicate experiments. Equal amounts (1/10 volume) of each resultant product were subjected to electrophoresis on a 1% agarose gel and transferred to a Zeta-probe nylon membrane (Bio-Rad). (*B*) DNA blot was hybridized with the ^{32}P-labeled RAW112 library probe and washed at high stringency. After autoradiography, the RAW112 library probe was removed from this blot. (*C*) Same DNA blot was hybridized with the ^{32}P-labeled TK1 library probe under high stringency conditions and autoradiographed.

To examine whether these β_p clones contain cDNA of the full-length RNA transcript, equal amounts of poly(A)$^+$ RNA from TK1 ($\alpha_{4m}{}^+\beta_p{}^+$), RAW112 ($\alpha_{4m}{}^+\beta_p{}^-$), and two control T cell lines TK5 and BW5147 ($\alpha_{4m}{}^-\beta_p{}^-$) were analyzed in a Northern analysis, using the β_p cDNA clone (T30) as a probe. A single band of approximately 2.7 kb was evident in the lane containing RNA from TK1 cells (lane 2) but was negative in RNAs from RAW112, TK5, and BW5147 cells (data not shown). Since the length of the T30 clone was approximately 2.7 kb, this result indicated that it contained a cDNA at or near the full-length RNA transcript of β_p. To confirm that a non-Peyer's patch homing B-cell lymphoma 38C13 cell line did not express β_p or β_1 (see below), equal amounts of poly(A)$^+$ RNAs from 38C13, TK1, and RAW112 were analyzed in Northern analyses, using either the β_p or mouse β_1 cDNA as a probe. The results show that 38C13 expressed neither β_p nor β_1 (data not shown).

Functional Role of β_p in Peyer's Patch-specific Lymphocyte Homing

The function of β_p as a unique component of the Peyer's patch-specific homing receptor was tested directly by expression of the β_p cDNA in a $\beta_p{}^-$ B-cell lymphoma. The 38C13 cell line (a cloned C3H/eb mouse B-cell lymphoma) was chosen because it expresses high levels of α_{4m} but not β_p or β_1 and does not bind to Peyer's patch HEV. Since 38C13 cells express high levels of L-SELECTIN and bind strongly to peripheral lymph node HEV, accessory molecules that are nonorgan-specific are likely to be present, and the organ-specific adhesion of the infected cells can be tested directly on both Peyer's patch and peripheral lymph node HEVs. To obtain 38C13 cells that stably express β_p (hereafter called 38-β_p cells), 38C13 cells were exposed to supernatants of an ecotropic helper cell line GP + E-86 (Markowitz et al. 1988), producing Lβ_pSN recombinant retroviruses. As a control, 38C13 cells were infected with retroviruses carrying a pLXSN vector that does not contain the β_p cDNA (these cells are called hereafter 38-LXSN). All vectors conferred G418 resistance to infected cells. Most of the infected cells expressed high levels of cell-surface β_p, whereas the parental 38C13 cells and the control LXSN retrovirus-infected cells were clearly negative for β_p surface expression. Furthermore, infection with these retroviruses did not change surface expression of other adhesive receptors, including L-SELECTIN, LFA-1, CD44 (Pgp-1), and surface immunoglobulin detected by the 38C13-specific anti-idiotypic antibody R7D4. These control stainings establish that quantitative differences in HEV binding (see below) between 38-β_p and 38C13 cells cannot be attributed to differences in expression levels of other known adhesion receptors. To examine whether β_p associated with α_{4m} to form heterodimers in 38-β_p cells, lysates from these cells were first immunoprecipitated with α_{4m}-specific monoclonal antibody R1-2 in the presence of Ca^{++}, and precipitates were analyzed with β_p-specific antisera by Western blotting. The immunoblot shows that a β_p polypeptide of about 130 kD is present in 38-β_p cells but not in the noninfected 38C13 or the control LXSN retrovirus-infected cells (data not shown).

One crucial question then was to test whether or not 38-β_p cells acquire a new Peyer's patch-specific adhesive phenotype. The binding capacity of 38-β_p cells for HEV in Peyer's patches or peripheral lymph nodes was examined in a modified in vitro HEV-binding assay (Stamper and Woodruff 1976; Butcher et al. 1979). The specific binding of 38-β_p cells to mouse Peyer's patch HEV is illustrated in Figure 3A. Although 38-β_p cells had no significant alteration in peripheral lymph node HEV-binding ability, they showed a highly significant increase (at least 20-fold) in Peyer's patch HEV-binding ability as compared with the noninfected 38C13 or the control-infected 38-LXSN cells (Fig. 3C). Furthermore, to assess the binding specificity of 38-β_p cells, monoclonal antibodies directed against α_{4m} (R1-2), $\beta_p(\beta_7)$ (M301) (Kilshaw and Murant 1991), mLHR (MEL-14), and a surface immunoglobulin anti-idiotype-positive staining control (R7D4) were tested for their blocking ability on cell adhesion to HEV in either Peyer's patches or peripheral lymph nodes. As shown in Figure 3C, preincubation of 38-β_p cells with anti-α_{4m} monoclonal antibody (R1-2) reduced their adhesion to Peyer's patch HEV by 80–90% but did not affect significantly their adhesion to peripheral lymph node HEV. Similarly, preincubation of these cells with anti-β_p monoclonal antibody (M301) (Kilshaw and Murant 1991) reduced their adhesion to Peyer's patch HEV by 40–50% but did not alter their ability to bind peripheral lymph node HEV. In contrast, preincubation of these cells with monoclonal antibody MEL-14 did not change their ability to bind Peyer's patch HEV but reduced their adhesion to peripheral lymph node HEV by 90–95%. Adhesion of these cells to Peyer's patch or peripheral lymph node HEV was not inhibited by treatment with the control monoclonal antibody (R7D4), which can bind strongly to 38-β_p cells. These results show that expression of β_p cDNA in 38C13 cells not only confers the β_p-specific antigenic epitopes in these cells, but also mediates their organ-specific adhesion to Peyer's patch HEV.

To confirm that β_p uniquely contributes to the homing function of all cells to Peyer's patch HEV, normal lymphocytes isolated from mesenteric lymph nodes were tested for binding to HEVs of Peyer's patches and peripheral lymph nodes. Most normal mesenteric lymph node lymphocytes express β_p and α_{4m} (Fig. 4A,B) and mLHR (Siegelman et al. 1989), and they bind equally well to Peyer's patch HEV and peripheral lymph node HEV (Fig. 4, lower panels). Preincubation of normal mesenteric lymphocytes with anti-β_p monoclonal antibody (M301) reduced their binding to Peyer's patch HEV by 70% but did not alter their ability to bind to peripheral lymph node HEV, whereas MEL-14 monoclonal antibody specifically blocked their

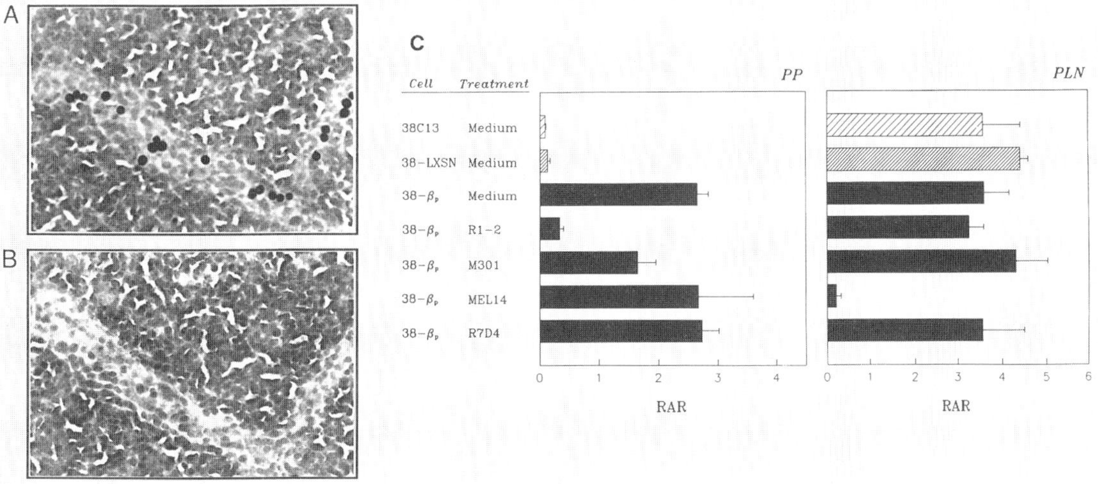

Figure 3. Adhesion of 38-β_p cells to Peyer's patch HEV is organ-specific and can be specifically inhibited by anti-α_{4m} and anti-$\beta_p(\beta_7)$ monoclonal antibodies. 38-β_p and 38C13 cells were incubated at 4°C for 30 min on fresh frozen sections of a mouse Peyer's patch, and cells bound to HEV were fixed to the section. (*A*) 38-β_p cells are the round dark circles lying above the plane of tissue. The outline of the HEV is delineated by a distinct, poorly stained basement membrane. Sections were stained with toluidine blue. (*B*) No 38C13 cells are bound to HEV. Magnification for all micrographs, 630 × (*C*) 38C13, 38-LXSN, and 38-β_p lymphoma cells were incubated with the indicated monoclonal antibodies for 30 min at 4°C and washed. Untreated or antibody-treated cells were mixed with an internal standard population of FITC-labeled mesenteric lymph node lymphocytes. Cell suspensions were incubated with mild rotation on freshly cut frozen sections of Peyer's patches (PP) or peripheral lymph nodes (PLN) for 30 min at 4°C. Relative adherence ratios (RAR) were calculated as described previously (Holzmann et al. 1989) and are presented as means with their standard deviations. The binding of untreated and unlabeled mesenteric lymph node lymphocytes, which were included in each experiment, defines an RAR unit.

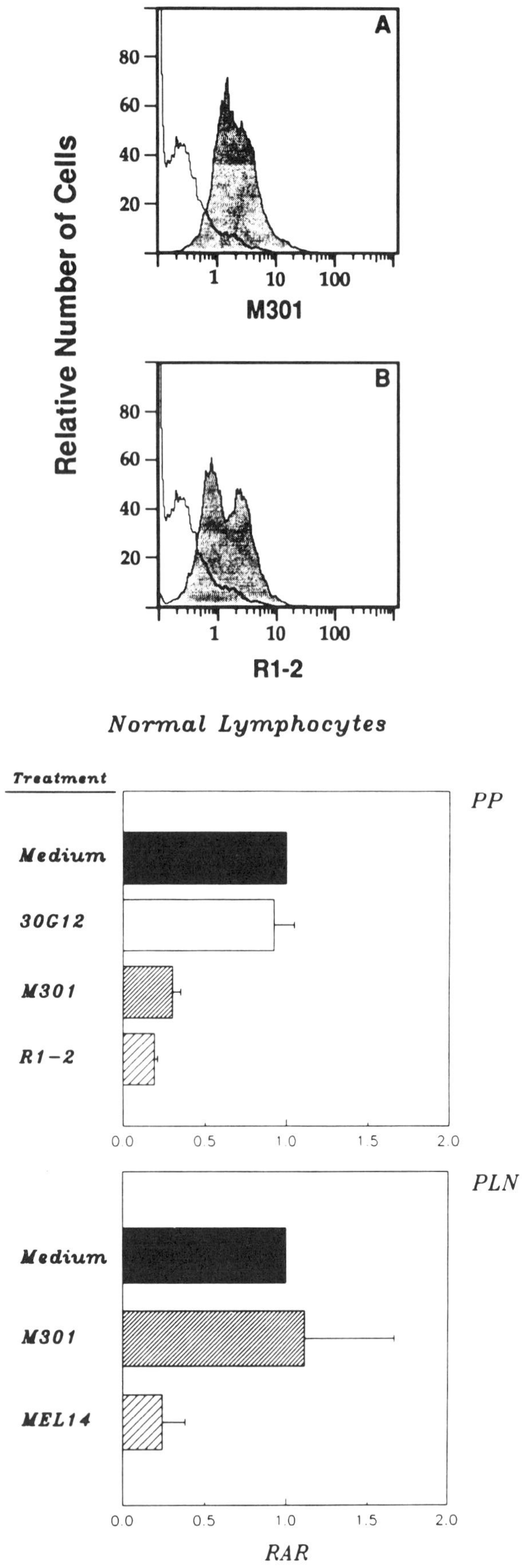

Figure 4. Adhesion of normal lymphocytes to Peyer's patch HEV can be specifically blocked by anti-β_p and anti-α_{4m} monoclonal antibodies (MAb). Expression of $\beta_p(\beta_7)$ and α_{4m} in normal lymphocytes was first analyzed by FACS. Normal lymphocytes isolated from mouse mesenteric lymph nodes were stained with either an isotype-matched control Mab R7D4 (unshaded histogram in each plot) or anti-$\beta_p(\beta_7)$ MAb M301 (*A*) or anti-α_{4m} MAb R1-2 (*B*) (shaded histogram in each plot), followed by FITC-labeled goat anti-rat IgG antibody, and analyzed by FACS. Normal lymphocytes were then incubated with either medium alone or the indicated monoclonal antibodies for 30 min at 4°C, washed, and mixed with an internal standard population of FITC-labeled mesenteric lymph node lymphocytes. An isotype-matched MAb 30H12 was used as a control antibody. Cell suspensions were incubated with frozen sections of Peyer's patches (PP) or peripheral lymph nodes (PLN), and RAR values were calculated as described in Fig. 3.

binding to peripheral lymph node HEV (Fig. 4, lower panels). Similarly, preincubation of these cells with the anti-α_{4m} monoclonal antibody R1-2 reduced their binding to Peyer's patch HEV by 80%, in accordance with the previous data (Holzmann et al. 1989). Thus, anti-β_p monoclonal antibody specifically blocked binding of normal lymphocytes to Peyer's patch HEV but did not inhibit their binding to lymph node HEV.

We have shown previously that TK50 lymphoma cells were the only Peyer's patch HEV-binding lymphoma cells thought to express $\alpha_{4m}\beta_1$ but not $\alpha_{4m}\beta_p$, as analyzed by coimmunoprecipitation with the anti-α_{4m} monoclonal antibody R1-2 (Holzmann and Weissman 1989). $\alpha_{4m}\beta_1$ was therefore proposed to be a possible receptor involved in lymphocyte adhesion to Peyer's patch HEV. However, $\alpha_{4m}\beta_1$ is predominantly expressed on lymphoma and/or leukemia cells that do not bind to Peyer's patch HEV, suggesting that expression of $\alpha_{4m}\beta_1$ alone is not sufficient to mediate this adhesive event. Surprisingly, FACS analysis with anti-β_p monoclonal antibody (M301) shows that TK50 lymphoma cells also expressed β_p (Fig. 5), indicating that all known Peyer's patch HEV-binding lymphoma cells express β_p. For instance, 38C13, RAW112 lymphoma cells, and the BW-α_{4m} transfectant express $\alpha_{4m}\beta_1$ but do not bind to Peyer's patch HEV, whereas TK1, TK23, TK40, and TK50 lymphoma cells and 38-β_p express $\alpha_{4m}\beta_p$ and bind to Peyer's patch HEV. It is not known, however, why β_p was not detected in TK50 cells by immunoprecipitation with the anti-α_{4m} monoclonal antibody R1-2 (Holzmann and Weissman 1989).

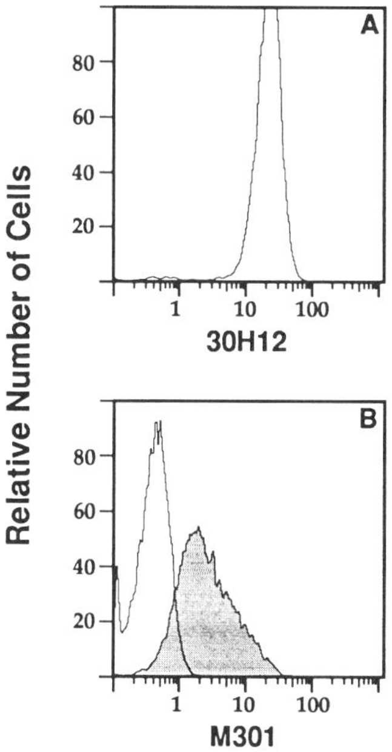

Figure 5. FACS analysis of $\beta_p(\beta_7)$ Expression in TK50 lymphoma cells. TK50 lymphoma (Thy1.2) were primary tumor cells and they were therefore propagated in SCID mice (Thy1.1). (*A*) TK50 lymphoma cells were isolated from mesenteric lymph node tumors from SCID mice and stained with biotin-labeled anti-Thy1.2 control MAb 30H12, followed by Texas-red-conjugated avidin, and analyzed by FACS. (*B*) Subsequently, TK50 (Thy1.2$^+$) cells were stained with either an isotype-matched control MAb R7D4 (unshaded histogram) or anti-β_7 MAb M301 (shaded histogram), followed by FITC-labeled goat anti-rat IgG antibody, and analyzed by FACS.

Structural Analysis of the Murine Integrin β_p Subunit

The nucleotide sequence of β_p contains a 32-bp 5′-untranslated region followed by a single open reading frame of 2418 bp encoding a polypeptide of 806 amino acids (Hu et al. 1992). The stop codon TGA at position 2451 is followed by a 178-bp 3′-untranslated region that contains the polyadenylation signal AATAAA at position 2600. The open reading frame starts with an ATG codon that is flanked by sequences, GCC*ATG*G, that match the consensus sequence of translation initiation site (Kozak 1987). Following the initiation codon ATG, there is a stretch of 19 mostly hydrophobic amino acids with the characteristics of a signal peptide (von Heijne 1986). Thus, the predicted mature β_p protein contains an extracellular region of 705 amino acids, a hydrophobic region of 21 amino acids (the transmembrane domain), and a cytoplasmic domain of 61 amino acids. The molecular mass of β_p estimated by SDS-PAGE is approximately 100 kD under nonreducing conditions and approximately 130 kD under reducing conditions (Holzmann et al. 1989). The predicted molecular mass of the mature β_p polypeptide after removal of the signal peptide is about 90 kD, which is less than the size estimated by SDS-PAGE. However, the actual difference may be due to anomalous migration of the glycosylated β_p subunit in SDS-PAGE. There are eight potential N-linked glycosylation sites having the consensus sequence N-X-S/T. All of these N-linked glycosylation sites are located on the extracellular portion of the polypeptide. The deduced amino acid sequences of mouse β_p and human β_7 (Yuan et al. 1990; Erle et al. 1991) show 86% identity over the entire cDNA, suggesting that mouse β_p might be the murine homolog of human β_7. There are three 100% conserved long stretches in the amino-terminal half of the extracellular domain, namely, residues 49–92, 149–212, and 251–309. However, there is a striking and very unusual difference between these two sequences. Although most of the cytoplasmic domain sequences are highly conserved, the very carboxy-terminal sequences are quite distinct (Fig. 6). To test whether the mouse genome contains sequences that are β_7-like in addition to β_p, genomic Southern analysis was performed. Southern blot analysis of mouse genomic DNA digested with six endonucleases (*Eco*RI, *Hin*dIII, *Bam*HI, *Xba*I, *Xho*I, and *Pst*I) showed that the pattern of restriction fragments hybridized to either the mouse β_p probe or the human β_7 probe was virtually identical (Hu et al. 1992). Similarly, human genomic DNA digested with the same enzymes as described above showed the same hybridization pattern of restriction fragments with either the mouse β_p probe or the human β_7 probe. Thus, these results demonstrate that β_p is indeed the murine homolog of human β_7. Origi-

```
mβp MVDSSTVLIFLLVLGGGQSELDTKITSSGEAAEWEDPDLSLQGSCQPVPSCQKCILSHPSCAWCKQLNFTASGEAEARRCARREELLARGCPAQELEEPRGRQEVLQDKPLSQGDRGEGATQLA  124
hβ7 ..ALPM..VL....SR.E....A..P.T.D.T..RN.H..ML.....A............................................LE.......Q......Q.....A.........  124

mβp PQRIRVTLRPGEPQKFRVRFLRAAGYPVDLYYLMDLSYSMKDDLERVRQLGHALLVRLQEVTHSVRIGFGSFVDKTVLPFVSTVPSKLHHPCPSRLERCQPPFSFHHVLSLTGDAQAFEREVGR  248
hβ7 ...V..........QLQ......E................................................................R....T......S.......................  248
            (R)                           (S)

mβp QNVSGNLDSPEGGFDAILQAALCQEQIGWRNVSRLLVFTSDDTFHTAGDGKLGGIFMPSDGRCHLDSNGVYTNSAEFDYPSVGQVAQALTAANIQPIFAVTGATLPVYQELRQLIPKSAVGELS  372
hβ7 .S...........................................................H.......L.SR.T..............S...........S.A.......SK...........  372
                                                                (S)

mβp EDSSNVVQLIMDAYDSLSSTVTLEHSPLPPGVSISFESHCKGPEKTEGEAGDRGQCNDVRVNQTVDFWVTLQATHCLPEAHVLRLWALGFSEELTVELHTVCDCNCGDAQPHAPYCSDGQGDLQ  496
hβ7 ..............N...........S.....H..Y..Q.E....R..K.E......H..I....T...S.........P.L...R........I.....L.....S.T..Q..H......H..  496
                                                                                                                         (A-)

mβp CGICSCAPGRLGQLCECSEADLSSPDLESGCRAPNGTGPLCSGKGRCQCGRCSCSGQSSGRLCECDDASCERHEGILCGGFGHCQCGVCHCHANHTGRACECSKSVDSCVSPEGGLCSGHGYCK  620
hβ7 ..V.........R.....V.E........................H..............H.....................R...........R........GDM...I...........R..  620
            (A)                                                                   (M)

mβp CNRCQCLDGYYGALCDQCLGCKSPCEQYRDCAECGAFGTGPLAANCSVVCADVNVTLTLAPNLDDGWCKERTIDNQLFFFLVEHAASG-IVLRVRPQEKGVDHTRAIILGCTGGIVAVGLGLVL  743
hβ7 ...................P...T..RH.........R.....T...TA..HT....A...I..........L...........DD.R.TV.........A...Q..V...V............  744
                                                                                           (A)                    (W)

mβp AYRLSVEIYDRREYRRFEKEQQQLNWKQDNNPLYKSAITTTVNPRFQGTNGRSPSLSLTREAD*  806
hβ7 ..............S..............S...........I.....EAD--..T.*        798
```

Figure 6. Comparison of the predicted amino acid sequences of mouse β_p and human β_7. The human β_7 sequence (hβ_7) (Erle et al. 1991) is displayed below the mouse β_p sequence (mβ_p). A dot indicates identity with the mouse sequence and a dash denotes a gap introduced to obtain optimal alignment. Nine different amino acid residues of the human β_7 sequence from another report (Yuan et al. 1990) are shown in parentheses. The numbers to the right of the sequence indicate the residue numbers for each protein.

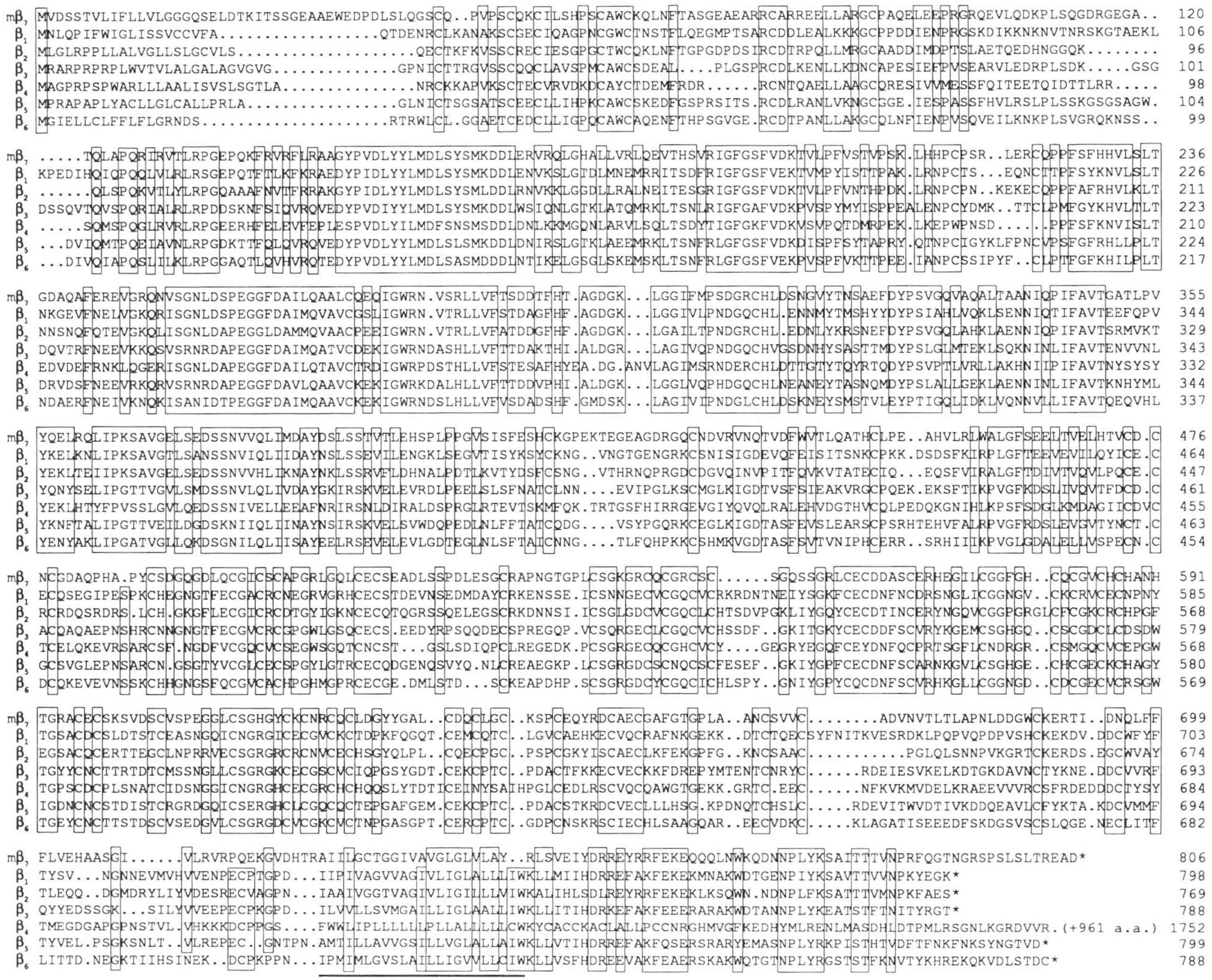

Figure 7. Alignment of mouse β_7 and other human integrin β subunits. The deduced amino acid sequence of mouse β_7 is compared with those of human integrin β_1 (Argraves et al. 1987), β_2 (Kishimoto et al. 1987; Law et al. 1987), β_3 (Fitzgerald et al. 1987; Rosa et al. 1988), β_4 (Hogervorst et al. 1990; Suzuki and Naitoh 1990), β_5 (Ramaswamy and Hemler 1990), and β_6 (Sheppard et al. 1990). The numbers along the right margin refer to the last amino acid number on each line for each mature integrin β subunit. The very large cytoplasmic domain (residues 768–1778) of the integrin β_4 subunit has been omitted since there are no corresponding sequences in the other integrin β subunits. Homologous residues (80% homology) among these integrin β subunits are boxed. Dashes denote gaps introduced to maximize homology, and asterisks indicate the termination codons of these protein sequences. The predicted transmembrane domains are underlined.

nally, we have referred to the LPAM-1 β chain as β_p; hereafter, we propose that only the β_7 nomenclature be used. The appellation LPAM-1 for $\alpha_4\beta_7$ remains appropriate, but the designation of LPAM-2 should be dropped in favor of its original name, VLA-4.

The β subunits of all known integrins are markedly similar (Fig. 7). The predicted amino acid sequence of the murine β_7 subunit is most related to human integrin β_2 (49% overall identity), and to a lesser extent, β_7 is related to β_1 (45% identity), β_5 (40% identity), β_3 (39% identity), β_6 (39% identity), and β_4 (34% identity). Generally speaking, a region of 242 amino acids (from residue 131 to 372 according to the β_7 sequence) located in the amino-terminal half of the mature protein is significantly conserved between β_7 and the other integrin β subunits (48–66% homology). Within this conserved region, there are two most highly conserved segments, residues 131–149 of β_7 (74–95% homology) and residues 233–248 of β_7 (75–94% homology). Although the gross structure of the β_7 subunit is similar to that of the other integrin β subunits, it has two unique features. One major difference is that only 54 cysteines present in the extracellular domain of the β_7 subunit are conserved with the other integrin β subunits, except the β_4 subunit, which lacks 6 of the 54 conserved cysteines. However, all 56 cysteines in the integrin β_1, β_2, β_3, β_5, β_6 subunits from human and the integrin β_1 subunits from mouse (Holers et al. 1989), chicken (Tamkun et al. 1986), *Xenopus* (DeSimone and Hynes 1988), and *Drosophila* (MacKrell et al. 1988) are present in the extracellular domain and fully conserved. On the other hand, there is a cysteine residue present in the predicted transmembrane domain of the β_7 subunit that is absent from all of the other integrin β subunits. The other major difference is that the β_7 subunit contains two extra hydrophobic regions near the transmembrane domain, residues 634–663 and 675–691 of the predicted mature protein, that are absent from all of the other known integrin β subunits. It is not known whether these two hydrophobic regions can form two membrane-spanning domains. If they do, the structure of the β_7 subunit will be very distinct from other known integrin β subunits. In general, the transmembrane domain is not significantly conserved and the cytoplasmic domain is weakly conserved among those β subunits. The cytoplasmic domain of the β_7 subunit is slightly longer than those of the other β subunits, except the β_4 subunit, which contains a very large ($\sim$ 1000 amino acids) cytoplasmic domain.

There are three tyrosine residues at positions 733, 738, and 758 in the cytoplasmic domain of β_7. Tyr-733 is located six amino acids downstream from a basic residue (arginine) and two amino acids downstream from an acidic residue (glutamic acid), and Tyr-758 is located seven amino acids downstream from a basic residue (lysine) and five amino acids downstream from an acidic residue (aspartic acid); these fit features characteristic for tyrosine kinase phosphorylation sites (Hunter and Cooper 1985). In particular, the sequences around Tyr-758 of the β_7 subunit are conserved with the β_1, β_3, and β_5 subunits, with a consensus sequence NPLYK, and show a degree of homology with tyrosine kinase phosphorylation sites in the EGF receptor, insulin receptor, and *neu* oncogene protein, as described previously (Tamkun et al. 1986). In the β_4 subunit, however, a tyrosine residue (residue 1668) with the homologous sequence NVPYK near the carboxyl terminus can be a potential site for tyrosine phosphorylation (Hogervorst et al. 1990; Suzuki and Naitoh 1990). Interestingly, the sequences near Tyr-758 of the β_7 subunit match the consensus sequence NPXY/F required for receptor-mediated endocytosis (Chen et al. 1990), which are also conserved in these integrin β subunits, except the β_4 subunit.

A striking feature of the β_7 subunit is the arrangement of its cysteine residues in the extracellular domain. Between amino acids 459 and 632, the β_7 subunit contains four tandem repeats of cysteine-rich motifs, with small gaps, each with eight cysteines (seven in the first repeat) in a conserved configuration (Fig. 8A). Comparison of these cysteine-rich motifs of β_7 to sequences in the NBRF database revealed many proteins that contain similar sequence motifs and the conserved arrangement of cysteines. The first protein described with this sequence motif is the EGF precursor, which contains ten homologous copies (Gray et al. 1983; Scott et al. 1983). It is generally believed that extracellular cysteines are disulfide-bonded. Since there are no disulfide links between α_{4m} and $\beta_p(\beta_7)$, these conserved cysteines within the EGF-like domain should be involved in intrachain disulfide bonds. The homologies per se and available functional data suggest that these domains share some common functional features (see below). Whether the mLHR and/or the β_7 EGF-like domains share structural homologies with other EGF/EGF-like sequences other than at the primary amino acid sequence level is yet to be determined. Another interesting feature of β_7 is that it contains one EF-loop consensus sequence (for review, see Strynadka and James 1989) at residues 159–170, which is conserved among all known integrin β subunits (Fig. 8B). Figure 8C illustrates the primary structure of β_7.

Several lines of evidence suggest that proteins containing one or more EGF-like domains are directly involved in protein-protein or receptor-ligand interactions.

1. EGF is a 53-residue, three-disulfide-bridged, single-chain polypeptide that binds to a specific cell-surface receptor having cytoplasmic tyrosine kinase activity that mediates its mitogenic and pleiotropic proliferative activities (for review, see Carpenter and Cohen 1979). Certain shorter synthetic peptides compete with authentic EGF for receptor-binding and mimic some of the functions of EGF, indicating that the minimal receptor-binding sequence lies within residues 20–31 (Komoriya et al. 1984). However, shorter synthetic peptides containing this minimal receptor-binding region have much less affinity for the receptor than authentic EGF (Heath and

A

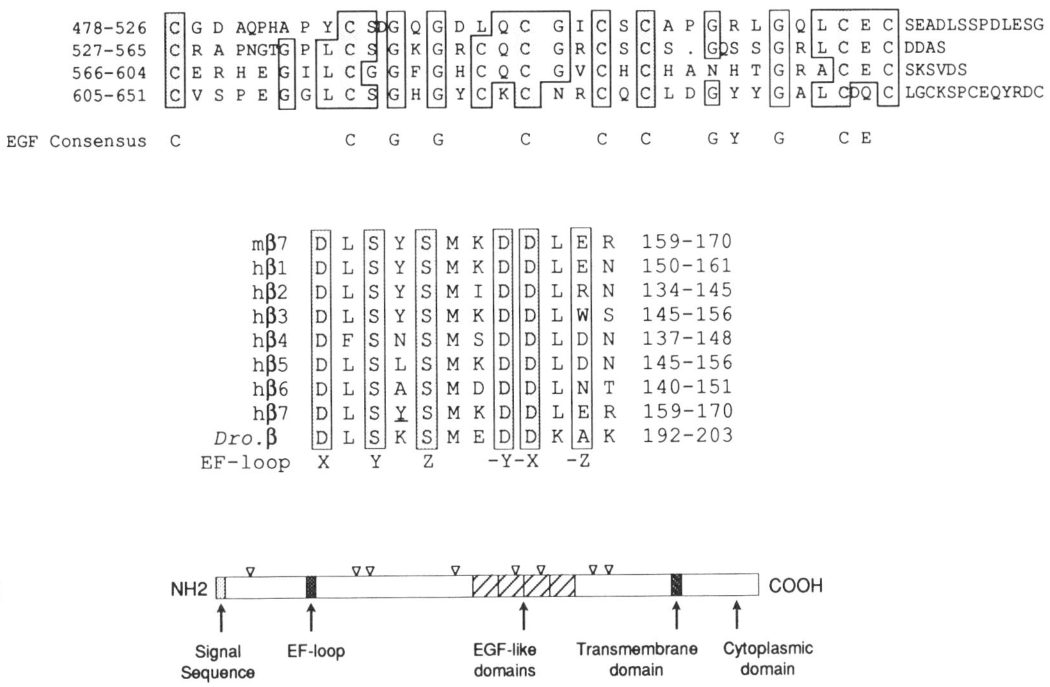

B

C

Figure 8. Structure of mouse β_7. (*A*) Homologous repeats of a cysteine-rich domain (EGF-like domain) in mouse β_7 were aligned. The numbers to the left of the amino acid sequence (one-letter code) are the residue numbers for β_7. Homologous residues are boxed and the cysteine residues are shaded. (*B*) Alignment of the region of mouse β_7 (mβ_7) containing the EF loop consensus sequence with other integrin β subunits. The aligned amino acid sequences (one-letter code) are from human β_1, β_2, β_3, β_4, β_5, β_6, β_7 and *Drosophila* integrin β (mys). The underlined tyrosine (Y) residue in human β_7 denotes that there is a discrepancy of this tyrosine residue in the β_7 sequence between two reports (Yuan et al. 1990; Erle et al. 1991). Oxygenated residues that align with those in the EF loop consensus sequence are boxed and the most conserved oxygenated residues are shaded. The residues that may form the putative calcium coordination octahedron are labeled by X, Y, Z, −Y, −X, −Z. (*C*) Schematic representation of the β_7 protein structure. The diagram summarizes the predicted domains that have been found in β_7, including the signal sequence, the EF loop, EGF-like domains, transmembrane domain, and cytoplasmic domain. Arrowheads indicate the potential *N*-glycosylation sites.

Merrifield 1986), suggesting that the tertiary structure of EGF contributes to its receptor binding affinity.

2. Urokinase-type plasminogen activator (uPA) binds to its receptor with high affinity through its amino-terminal region (residues 1–135), which contains an EGF-like domain (Stoppelli et al. 1985). Synthetic peptide studies show that the receptor-binding specificity locates within residues 18–32 of uPA, i.e., within the EGF-like domain (Appella et al. 1987).
3. The B1 chain of laminin appears to be involved in binding to collagen and to cell-surface receptors. The sequence of this molecule encodes several EGF-like, cysteine-rich repeats. Synthetic peptide studies have shown that residues 925–933 of this polypeptide in one of these EGF-like domains is mainly responsible for the receptor binding, cell attachment, and migration activities of laminin (Graf et al. 1987).
4. The LDL receptor contains three copies of EGF-like repeats in the extracellular domain. Structural analysis of the LDL receptor gene reveals that deletion of exons 7 and 8 (i.e., the first two copies of the EGF-like repeat) strongly interferes with the ligand-binding capacity and specificity of this receptor (Yamamoto et al. 1986). In addition to the binding of LDL, these EGF-like domains appear to allow the receptor to engage in repeated recycling without degradation (Davis et al. 1987).
5. Thrombomodulin is an endothelial cell-surface glycoprotein that contains six tandemly repeated EGF-like domains. A fragment of thrombomodulin containing the fifth and sixth EGF-like repeats has been proposed to act as the primary thrombin-binding site (Kurosawa et al. 1988).
6. Many other proteins containing one or more EGF-like domains are cell adhesion molecules, growth factors, receptors or receptor-like proteins, and proteins of the coagulation and fibrinolytic pathway, suggesting that the EGF-like domains are generally involved in protein-protein or receptor-ligand interactions.
7. As described above, anti-Ly-22 antibody strongly inhibits attachment of lymphocytes to lymph node HEV but does not block MEL-14 antibody binding to the adjacent lectin domain of mLHR, suggesting that this EGF-like domain is directly involved in the adhesion of lymphocytes to lymph node HEV (Siegelman et al. 1990).

Structural Analysis of the Murine Integrin α_4 cDNA

A murine cDNA clone encoding the integrin α_4 subunit (designated as α_{4m}) has been isolated and characterized (Neuhaus et al. 1991). The sequence of this cDNA encodes a peptide of 1039 amino acids and shows 84% homology with the human α_4 subunit (VLA-4 α), with an almost perfect conservation (31/32 amino acids) in the cytoplasmic domain. The primary structure of the α_{4m} subunit is very similar to that of human α_4 subunit (for review, see Hemler et al. 1990). Like other integrin α subunits, the α_{4m} contains a short carboxy-terminal cytoplasmic domain, a single transmembrane domain, and a large extracellular domain. The amino-terminal portion of the protein contains seven homologous repeats and three putative divalent cation binding sites of the general structure Asp-X-Asp-X-Asp-Gly-X-X-Asp or related sequences. These sites are among the most conserved between the α_{4m} and other α chains and may therefore be involved in the binding of integrin α and β chains. Although the integrin α subunits exhibit more sequence heterogeneity (17–24% homology) than the β subunits (see below), these putative divalent cation binding regions are among the most conserved sequences between the α_{4m} and other known α subunits, suggesting that they may be involved in the association of α and β subunits. Unlike the Leu-CAM and VLA-2 α subunits containing an additional amino acid sequence inserted between the amino terminus and the last divalent cation binding region (the I domain), α_{4m} does not contain an I domain. The positions of the characteristic cysteine residues are conserved between α_{4m} and human α_4. Unlike other integrin α subunits, a putative protease cleavage site is located in about the middle of the α_{4m} peptide sequence. This cleavage would yield fragments of approximately 80 kD and 70 kD, and cell-surface fragments of α_{4m} of approximately 84 and 62 kD have been shown by our previous immunoprecipitation data (Holzmann et al. 1989).

Tissue Distribution of $\alpha_4\beta_7$

The distribution of the expression of RNA species encoding the β_7 subunit is predominantly in lymphoid tissues, a finding that correlates well with the distribution of the α_{4m} subunit (Neuhaus et al. 1991). Although the integrin β_1 subunit is widely distributed on hematopoietic cells (Wayner et al. 1989), it is also expressed at a relatively high level in heart, kidney, brain, and lung (Holers et al. 1989). In contrast, by Northern analysis, we show no detectable β_7 transcripts in heart, muscle, and testis and very low levels in kidney, brain, liver, and lung (data not shown). The tissue distribution of other β subunits is still not completely elucidated. However, it has been documented that the β_2 integrin subunit is expressed exclusively on leukocytes (Kishimoto et al. 1989) and the β_3 integrin subunit is expressed primarily in platelets (Phillips et al. 1988). This implies that different cell types in different tissues (or organs) utilize different integrin molecules for adhesion function.

Possible Ligands for $\alpha_4\beta_7$

It has been shown very recently that phorbol-12-myristate-13-acetate (PMA)-stimulated TK1 cells can bind to the CS-1 region of fibronectin and VCAM-1, whereas unstimulated TK1 cells do not bind to either fibronectin or VCAM-1 (Rüegg et al. 1992). Thus, it has been proposed that the binding of unstimulated $\alpha_4\beta_7$ receptors to Peyer's patch HEV is very different from binding of PMA-stimulated $\alpha_4\beta_7$ to fibronectin and VCAM-1. The evidence for that proposal is mixed. First of all, $\alpha_4\beta_7$-positive TK1 cells can bind to Peyer's patch HEV constitutively in vitro without stimulation, whereas the same cells cannot bind to fibronectin and VCAM-1 without stimulation. Although binding of unstimulated TK1 cells to Peyer's patch HEV can occur at 4°C, i.e., this binding is not energy-dependent, the binding of PMA-stimulated TK1 cells to fibronectin and VCAM-1 in vitro occurred at 37°C (Rüegg et al. 1992) but not at 4°C. In addition, PMA stimulation significantly decreases binding of normal lymphocytes to Peyer's patch HEV (Bührer et al. 1992), whereas PMA stimulation markedly increases the binding of TK1 cells to fibronectin and VCAM-1, suggesting that PMA stimulation may alter the conformation of $\alpha_4\beta_7$ and/or the expression of other adhesion molecules. For instance, stimulation of T cells with ionomycin and PMA dramatically increased the expression of CD2 and CD44 (Geppert and Lipsky 1991; Bührer et al. 1992). Evidence against the role of $\alpha_4\beta_7$ in these interactions has been addressed with human cells, where it has been shown recently that $\alpha_4\beta_7$ does not play a major role in binding to fibronectin or VCAM-1 using PMA-stimulated cells (Chan et al. 1992). In contrast, $\alpha_4\beta_1$ (VLA-4) can bind to fibronectin and VCAM-1 constitutively without stimulation, and its binding affinity to fibronectin and VCAM-1 can be rapidly augmented upon stimulation with PMA (Shimizu et al. 1990; Chan et al. 1992). Thus, the binding specificity of unstimulated $\alpha_4\beta_7$ receptor to Peyer's patch HEV is distinct from the binding of PMA-stimulated $\alpha_4\beta_7$-positive cells to fibronectin and VCAM-1, but the role of $\alpha_4\beta_7$ other than adhesion in that interaction is not yet clear.

Recently, a model has been proposed that describes leukocyte extravasation as a multistep process requiring the activation of an adhesion molecule cascade. In this model, initial leukocyte binding and leukocyte rolling on endothelia is mediated by SELECTIN molecules, and then strong stabilization of binding is induced by activation of integrins through various mechanisms (Butcher 1991; Lawrence and Springer 1991). The data presented from our studies indicate that $\alpha_4\beta_7$ participates in one or more stages of the cascade (initial reversible binding, leukocyte activation, stable adhesion, and transmigration) and that it confers binding specificity for HEV in Peyer's patches. Although the binding capacity of $\alpha_4\beta_7$ for Peyer's patch HEV appears to be constitutively active, it is still possible that this function is dependent on cellular activation pathways (and not intrinsic to the molecule) and that it can be augmented by signals conferred to lymphocytes through other adhesion molecules.

CONCLUDING PERSPECTIVE

Our knowledge of lymphocyte traffic in the body has progressed remarkably in the last 10 years, primarily because of the identification and characterization of a number of organ-specific lymphocyte homing receptors, vascular addressins on HEV, and other cell adhesion molecules. The different ligand specificity, divalent cation requirements, and structure of these receptors and adhesion molecules suggest that each molecule contributes in a unique fashion to confer the final tissue-specific adhesive interaction. The dynamic role of lymphocyte homing receptors in deploying immunocompetent lymphocytes in lymphoid or nonlymphoid organs not only involves targeting and anchoring functions, but also may be involved in signaling pathways that transduce information from the particular extracellular environments to the interior of lymphocytes, influencing their proliferation and physiology. Searching for new lymphocyte homing receptors and their complementary ligands on HEV in lymphoid organs or tissues in nonlymphoid organs and studying structure-function relationships of these receptors and interactions with the cytoskeleton and signaling pathways should provide insights into the mechanisms governing the control of lymphocyte traffic. The use of soluble forms of these receptors to investigate and perhaps to control various inflammatory diseases and tumor metastasis processes might be exploited. Further advances in our understanding of the mechanisms that direct lymphocyte migration will be very useful from both scientific and clinical perspectives.

ACKNOWLEDGMENTS

We thank Libuse Jerabek and Lily Hu for the assistance with some mouse and cell culture work, respectively. This work was supported by the Howard Hughes Medical Institute and by National Institutes of Health grants AI-19512 and CA-42551. M.H. and D.C. are fellows of the Leukemia Society of America, M.S. is a fellow of Bristol-Myers Cancer Research and a McDonnell Scholar, and I.W. is an Investigator of the Howard Hughes Medical Institute.

REFERENCES

Ahearn, J., and D. Fearon. 1989. Structure and function of the complement receptors, CR1 (CD35) and CR2 (CD21). In *Advances in immunology* (ed. F. Dixon), p. 183. Academic Press, New York.

Aota, S., T. Gojobori, F. Ishibashi, T. Maruyama, and T. Ikemura. 1988. Codon usage tabulated from the GenBank genetic sequence data. *Nucleic Acids Res.* **16:** 315.

Appella, E., E.A. Robinson, S.J. Ullrich, M.P. Stoppelli, A. Corti, G. Cassani, and F. Blasi. 1987. The receptor-binding sequence of urokinase. *J. Biol. Chem.* **262:** 4437.

Argraves, W.S., S. Suzuki, H. Arai, K. Thompson, M.D. Pierschbacher, and E. Ruoslahti. 1987. Amino acid sequence of the human fibronectin receptor. *J. Cell Biol.* **105:** 1183.

Bevilacqua, M.P., S. Stengelin, M.A. Gimbrone, Jr., and B. Seed. 1989. Endothelial leukocyte adhesion molecule 1: An inducible receptor for neutrophils related to complement regulatory proteins and lectins. *Science* **243:** 1160.

Bevilacqua, M.P., J.S. Pober, D.L. Mendrick, R.S. Cotran, and M.A. Gimbrone, Jr. 1987. Identification of an inducible endothelial-leukocyte adhesion molecule. *Proc. Natl. Acad. Sci.* **84:** 9238.

Bührer, C., C. Berlin, D. Jablonski-Westrich, B. Holzmann, H.-G. Thiele, and A. Hamann. 1992. Lymphocyte activation and regulation of three adhesion molecules with supposed function in homing: LECAM-1 (MEL-14 antigen), LPAM-1/2 (α_4-integrin) and CD44 (Pgp-1). *Scand. J. Immunol.* **35:** 107.

Butcher, E.C. 1991. Leukocyte-endothelial cell recognition: Three (or more) steps to specificity and diversity. *Cell* **67:** 1033.

Butcher, E., R.G. Scollay, and I.L. Weissman. 1979. Lymphocyte adherence to high endothelial venules: Characterization of a modified in vitro assay, and examination of the binding of syngeneic and allogeneic lymphocyte populations. *J. Immunol.* **123:** 1996.

Carpenter, G., and S. Cohen. 1979. Epidermal growth factor. *Annu. Rev. Biochem.* **48:** 193.

Carter, W.G. and E.A. Wayner. 1988. Characterization of the class III collagen receptor, a phosphorylated, transmembrane glycoprotein expressed in nucleated human cells. *J. Biol. Chem.* **263:** 4193.

Chan, B.M.C., M.J. Elices, E. Murphy, and M.E. Hemler. 1992. Adhesion of vascular cell adhesion molecule 1 and fibronectin. Comparison of $\alpha_4\beta_1$ (VLA-4) and $\alpha_4\beta_7$ on the human B cell line JY. *J. Biol. Chem.* **267:** 8366.

Chen, W-J., J.L. Goldstein, and M.S. Brown. 1990. NPXY, a sequence often found in cytoplasmic tails, is required for coated pit-mediated internalization of the low density lipoprotein receptor. *J. Biol. Chem.* **265:** 3116.

Chin, Y.H., R.A. Rasmussen, J.J. Woodruff, and T.G. Easton. 1986. Monoclonal anti-HEBFpp antibody with specificity for lymphocyte surface molecules mediating adhesion to Peyer's patch high endothelium of the rat. *J. Immunol.* **136:** 2556.

Davis, C.G., J.L. Goldstein, T.C. Sudhof, R.G. Anderson, D.W. Russell, and M.S. Brown. 1987. Acid-dependent ligand dissociation and recycling of LDL receptor mediated by growth factor homology region. *Nature* **326:** 760.

DeSimone, D. and R.O. Hynes. 1988. *Xenopus laevis* integrins. Structural conservation and evolutionary divergence of integrin β subunits. *J. Biol. Chem.* **263:** 5333.

Drickamer, K., J.F. Mamon, G. Binns, and J.O. Leung. 1989. Primary structure of the rat liver asialoglycoprotein receptor. Structural evidence for multiple polypeptide species. *J. Biol. Chem.* **259:** 770.

Elices, M.J., L. Osborn, Y. Takada, C. Crouse, S. Luhowskyj, M.E. Hemler, and R.R. Lobb. 1990. VCAM-1 on activated endothelium interacts with the leukocyte integrin VLA-4 at a site distinct from the VLA-4/fibronectin binding site. *Cell* **60:** 577.

Erle, D.J., C. Ruegg, D. Sheppard, and R. Pytela. 1991. Complete amino acid sequence of an integrin β subunit (β_7) identified in leukocytes. *J. Biol. Chem.* **266:** 11009.

Fitzgerald, L.A., B. Steiner, S.C. Rall, Jr., S. Lo, and D.R. Phillips. 1987. Protein sequence of endothelial glycoprotein IIIa derived from a cDNA clone. Identity with platelet glycoprotein IIIa and similarity to "integrin." *J. Biol. Chem.* **262:** 3936.

Gallatin, M., T.P. St. John, M. Siegelman, R. Reichert, E.C.

Butcher, and I.L. Weissman. 1986. Lymphocyte homing receptors. *Cell* **44:** 673.
Gallatin, W.M., I.L. Weissman, and E.C. Butcher. 1983. A cell-surface molecule involved in organ-specific homing of lymphocytes. *Nature* **304:** 30.
Geppert, T.D., and P.E. Lipsky. 1991. Association of various T cell-surface molecules with the cytoskeleton. Effect of cross-linking and activation. *J. Immunol.* **146:** 3298.
Goelz, S.E., C. Hession, D. Goff, B. Griffiths, R. Tizard, B. Newman, G. Chi-Rosso, and R. Lobb. 1990. ELFT: A gene that directs the expression of an ELAM-1 ligand. *Cell* **63:** 1349.
Goldstein, L.A., D.F.H. Zhou, L.J. Picker, C.N. Minty, R.F. Bargatze, J.F. Ding, and E.C. Butcher. 1989. A human lymphocyte homing receptor, the Hermes antigen, is related to cartilage proteoglycan core and link proteins. *Cell* **56:** 1063.
Gowans, J. and E. Knight. 1964. The route of recirculation of lymphocytes in the rat. *Philos. Trans. R. Soc. Lond. B* **159:** 257.
Graf, J., Y. Iwamoto, M. Sasaki, G.R. Martin, H.K. Kleinman, F.A. Robey, and Y. Yamada. 1987. Identification of an amino acid sequence in laminin mediating cell attachment, chemotaxis, and receptor binding. *Cell* **48:** 989.
Gray, A., T. Dull, and A. Ullrich. 1983. Nucleotide sequence of epidermal growth factor cDNA predicts a 128,000-molecular weight protein precursor. *Nature* **303:** 722.
Gutman, G.A. and I.L. Weissman. 1973. Homing properties of thymus-independent follicular lymphocytes. *Transplantation* **16:** 621.
Hamann, A., D. Jablonski-Westrich, A. Duijvenstijn, E.C. Butcher, H. Baisch, R. Harder, and H.G. Thiele. 1988. Evidence for an accessory role of LFA-1 in lymphocyte-high endothelium interaction during homing. *J. Immunol.* **140:** 693.
Hamburger, S.A. and R.P. McEver. 1990. GMP-140 mediates adhesion of stimulated platelets to neutrophils. *Blood* **75:** 550.
Heath, W.F. and R.B. Merrifield. 1986. A synthetic approach to structure-function relationships in the murine epidermal growth factor molecule. *Proc. Natl. Acad. Sci.* **83:** 6367.
Hemler, M.E., M.J. Elices, C. Parker, and Y. Takada. 1990. Structure of the integrin VLA-4 and its cell-cell and cell-matrix adhesion functions. *Immunol. Rev.* **114:** 45.
Hogervorst, F., I. Kuikman, A.E.G. von dem Borne, Jr., and A. Sonnenberg. 1990. Cloning and sequence analysis of beta-4 cDNA: An integrin subunit that contains a unique 118 kd cytoplasmic domain. *EMBO J.* **9:** 765.
Holers, V.M., T.G. Ruff, D.L. Parks, J.A. McDonald, L.L. Ballard, E.J. Brown. 1989. Molecular cloning of a murine fibronectin receptor and its expression during inflammation. *J. Exp. Med.* **169:** 1589.
Holzmann, B. and I.L. Weissman. 1989. Peyer's patch-specific lymphocyte homing receptors consist of a VLA-4-like α chain associated with either of two integrin β chains, one of which is novel. *EMBO J.* **8:** 1735.
Holzmann, B., B.W. McIntyre, and I.L. Weissman. 1989. Identification of a murine Peyer's patch-specific lymphocyte homing receptor as an integrin molecule with an a chain homologous to human VLA-4α. *Cell* **56:** 37.
Howard, J.C., S.V. Hunt, and J.L. Gowans. 1972. Identification of marrow-derived and thymus-derived small lymphocytes in the lymphoid tissue and thoracic duct lymph of normal rats. *J. Exp. Med.* **135:** 200.
Hsu-Lin, S., C.L. Berman, B.C. Furie, D. August, and B. Furie. 1984. A platelet membrane protein expressed during platelet activation and secretion. Studies using a monoclonal antibody specific for thrombin-activated platelets. *J. Biol. Chem.* **259:** 9121.
Hu, M. C-T., D.T. Crowe, I.L. Weissman, and B. Holzmann. 1992. Cloning and expression of mouse integrin $\beta_p(\beta_7)$: A functional role in Peyer's patch-specific lymphocyte homing. *Proc. Natl. Acad. Sci.* **89:** (in press).
Hunter, T. and J.A. Cooper. 1985. Protein-tyrosine kinases. *Annu. Rev. Biochem.* **54:** 897.
Ichinose, A., B.A. McMullen, K. Fujikawa, and E.W. Davie. 1986. Amino acid sequence of the b subunit of human factor XIII, a protein composed of ten repetitive segments. *Biochemistry* **5:** 4633.
Ishida, N., H. Kanamori, T. Noma, T. Nikaido, H. Sabe, N. Suzuki, A. Shimizu, and T. Honjo. 1985. Molecular cloning and structure of the human interleukin 2 receptor gene. *Nucleic Acids Res.* **13:** 7579.
Issekutz, T.B. 1991. Inhibition of in vivo lymphocyte migration to inflammation and homing to lymphoid tissues by the TA-2 monoclonal antibody. A likely role for VLA-4 in vivo. *J. Immunol.* **147:** 4178.
Jalkanen, S., R.F. Bargatze, L.R. Herron, and E.C. Butcher. 1986a. A lymphoid cell surface glycoprotein involved in endothelial cell recognition and lymphocyte homing in man. *Eur. J. Immunol.* **16:** 1195.
Jalkanen, S., R.F. Bargatze, J. de los Toyos, and E.C. Butcher. 1987. Lymphocyte recognition of high endothelium: Antibodies to distinct epitopes of an 85-95KD glycoprotein antigen differentially inhibit lymphocyte binding to lymph node, mucosal, or synovial endothelial cells. *J. Cell Biol.* **105:** 983.
Jalkanen, S., R.A. Reichert, W.M. Gallatin, R.F. Bargatze, I.L. Weissman, and E.C. Butcher. 1986b. Homing receptors and the control of lymphocyte migration. *Immunol. Rev.* **91:** 39.
Johnston, G.I., R.G. Cook, and R.P. McEver. 1989. Cloning of GMP-140, a granule membrane protein of platelets and endothelium: Sequence similarity to proteins involved in cell adhesion and inflammation. *Cell* **56:** 1033.
Kansas, G.S., G.S. Wood, and M.O. Dailey. 1989. The family of cell surface glycoproteins defined by a putative anti-endothelial cell receptor antibody in man. *J. Immunol.* **142:** 3050.
Kilshaw, P.J. and S.J. Murant. 1991. Expression and regulation of $\beta_7(\beta_p)$ integrins on mouse lymphocytes: Relevance to the mucosal immune system. *Eur. J. Immunol.* **21:** 2591.
Kina, T., A.S. Majumdar, S. Heimfeld, H. Kaneshima, B. Holzmann, Y. Katsura, and I.L. Weissman. 1991. Identification of a 107-kD glycoprotein that mediates adhesion between stromal cells and hematolymphoid cells. *J. Exp. Med.* **173:** 373.
Kishimoto, T.K., K. O'Connor, A. Lee, T.M. Roberts, and T.A. Springer. 1987. Cloning of the β subunit of the leukocyte adhesion protein: Homology to an extracellular matrix receptor defines a novel supergene family. *Cell* **48:** 681.
Kishimoto, T.K., R.S. Larson, A.L. Corbi, M.L. Dustin, D.E. Staunton, and T.A. Springer. 1989. The leucocyte integrins. *Adv. Immunol.* **46:** 149.
Komoriya, A., M. Hortsch, C. Meyers, M. Smith, H. Kanety, and J. Schlessinger. 1984. Biologically active synthetic fragments of epidermal growth factor: Localization of a major receptor-binding region. *Proc. Natl. Acad. Sci.* **81:** 1351.
Kozak, M. 1987. An analysis of 5′-noncoding sequences from 699 vertebrate messenger RNAs. *Nucleic Acids Res.* **15:** 8125.
Kurosawa, S., D. Stearns, K. Jackson, and C. Esmon. 1988. A 10-kDa cyanogen bromide fragment from the epidermal growth factor homology domain of rabbit thrombomodulin contains the primary thrombin binding site. *J. Biol. Chem.* **263:** 5993.
Larsen, E., T. Palabrica, S. Sajer, G.E. Gilbert, D.D. Wagner, B.C. Furie, and B. Furie. 1990. PADGEM-dependent adhesion of platelets to monocytes and neutrophils is mediated by a lineage-specific carbohydrate, LNFIII (CD15). *Cell* **63:** 467.
Larsen, E., A. Celi, G.E. Gilbert, B.C. Furie, J.K. Erban, R. Bonfanti, D.D. Wagner, and B. Furie. 1989. PADGEM protein: A receptor that mediates the interaction of acti-

vated platelets with neutrophils and monocytes. *Cell* **59:** 305.
Lasky, L.A., M.S. Singer, T.A. Yednock, D. Dowbenko, C. Fennie, H. Rodriguez, T. Nguyen, S. Stachel, and S.D. Rosen. 1989. Cloning of a lymphocyte homing receptor reveals a lectin domain. *Cell* **56:** 1045.
Law, S.K.A., J. Gagnon, J.E.K. Hildreth, C.E. Wells, A.C. Willis, and A.J. Wong. 1987. The primary structure of the β subunit of the cell surface adhesion glycoproteins LFA-1, CR3 and p150,95 and its relationship to the fibronectin receptor. *EMBO J.* **6:** 915.
Lawrence, M.B. and T.A. Springer. 1991. Leukocytes roll on a selectin at physiologic flow rates: Distinction from and prerequisite for adhesion through integrins. *Cell* **65:** 859.
Leonard, W.J., J.M. Depper, M. Kanehisa, M. Kronke, N.J. Peffer, P.B. Svetlik, M. Sullivan, and W.C. Greene. 1985. Structure of the human interleukin-2 receptor gene. *Science* **230:** 633.
Lewinsohn, D.M., R.F. Bargatze, and E.C. Butcher. 1987. Leukocyte-endothelial cell recognition: Evidence of a common molecular mechanism shared by neutrophils, lymphocytes, and other leukocytes. *J. Immunol.* **138:** 4313.
Lowe, J.B., L.M. Stoolman, R.P. Nair, R.D. Larsen, T.L. Berhend, and R.M. Marks. 1990. ELAM-1-dependent cell adhesion to vascular endothelium determined by a transfected human fucosyltransferase cDNA. *Cell* **63:** 475.
Lozier, J., N. Takahashi, and F.W. Putnam. 1984. Complete amino acid sequence of human plasma β_2-glycoprotein I. *Proc. Natl. Acad. Sci.* **81:** 3640.
MacKrell, A.J., B. Blumberg, S.R. Haynes, and J.H. Fessler. 1988. The lethal myospheroid gene of *Drosophila* encodes a membrane protein homologous to vertebrate integrin β subunits. *Proc. Natl. Acad. Sci.* **85:** 2633.
Markowitz, D., S. Goff, and A. Bank. 1988. A safe packaging line for gene transfer: Separating viral genes on two different plasmids. *J. Virology* **62:** 1120.
McEver, R.P. and M.N. Martin. 1984. A monoclonal antibody to a membrane glycoprotein binds only to activated platelets. *J. Biol. Chem.* **259:** 9799.
Miyake, K., I.L. Weissman, J.S. Greenberger, and P.W. Kincade. 1991. Evidence for a role of the integrin VLA-4 in lympho-hemopoiesis. *J. Exp. Med.* **173:** 599.
Neuhaus, H., M.C.-T. Hu, M.E. Hemler, Y. Takada, B. Holzmann, and I.L. Weissman. 1991. Cloning and expression of cDNAs for the α subunit of the murine lymphocyte-Peyer's patch adhesion molecule. *J. Cell Biol.* **115:** 1149.
Nottenburg, C., G. Rees, and T. St. John. 1989. Isolation of mouse CD44 cDNA: Structural features are distinct from the primate cDNA. *Proc. Natl. Acad. Sci.* **86:** 8521.
Pals, S.T., A. den Otter, F. Miedema, P. Kabel, G.D. Keizer, R.J. Schepper, and C.J.L.M. Meijer. 1988. Evidence that leucocyte function-associated antigen-1 is involved in recirculation and homing of human lymphocytes via high endothelial venules. *J. Immunol.* **140:** 1851.
Phillips, D.R., I.F. Charo, L.V. Parise, and L.A. Fitzgerald. 1988. The platelet membrane glycoprotein IIb-IIIa complex. *Blood* **71:** 831.
Phillips, M.L., E. Nudelman, F.C. Gaeta, M. Perez, A.K. Singhal, S. Hakomori, and J.C. Paulson. 1990. ELAM-1 mediates cell adhesion by recognition of a carbohydrate ligand, sialyl-Le^x. *Science* **250:** 1130.
Picker, L.J., J. de los Toyos, M.J. Telen, B.F. Haynes, and E.C. Butcher. 1989. Monoclonal antibodies against the CD44 [In(Lu)-related p80], and Pgp-1 antigens in man recognize the Hermes class of lymphocyte homing receptors. *J. Immunol.* **142:** 2046.
Ramaswamy, H. and M.E. Hemler. 1990. Cloning, primary structure and properties of a novel human integrin β subunit. *EMBO J.* **9:** 1561.
Rosa, J.P., P.F. Bray, O. Gayet, G.I. Johnston, R.G. Cook, K.W. Jackson, M.A. Shuman, and R.P. McEver. 1988. Cloning of glycoprotein IIIa cDNA from human erythroleukemia cells and localization of the gene to chromosome 17. *Blood* **72:** 593.
Rüegg, C., A.A. Postigo, E.E. Sikorski, E.C. Butcher, R. Pytela, and D.J. Erle. 1992. Role of integrin $\alpha_4\beta_7/\alpha_4\beta_p$ in lymphocyte adherence to fibronectin and VCAM-1 and in homotypic cell clustering. *J. Cell Biol.* **117:** 179.
Scott, J., M. Urdea, M. Quiroga, R. Sanchez-Pescador, N. Fong, M. Selby, W. Rutter, and G.I. Bell. 1983. Structure of a mouse submaxillary messenger RNA encoding epidermal growth factor and seven related proteins. *Science* **221:** 236.
Sheppard, D., C. Rozzo, L. Starr, V. Quaranta, D.J. Erle, and R. Pytela. 1990. Complete amino acid sequence of a novel integrin β subunit (β_6) identified in epithelial cells using the polymerase chain reaction. *J. Biol. Chem.* **265:** 11502.
Shimizu, Y., G.A. Van Seventer, K.J. Horgan, and S. Shaw. 1990. Regulated expression and binding of three VLA (β_1) integrin receptors on T cells. *Nature* **345:** 250.
Siegelman, M.H. and I.L. Weissman. 1989. Human homologue of mouse lymph node homing receptor: Evolutionary conservation at tandem cell interaction domains. *Proc. Natl. Acad. Sci.* **86:** 5562.
Siegelman, M.H., M. van de Rijn, and I.L. Weissman. 1989. Mouse lymph node homing receptor cDNA clone encodes a glycoprotein revealing tandem interaction domains. *Science* **243:** 1165.
Siegelman, M.H., I.C. Cheng, I.L. Weissman, and E.K. Wakeland. 1990. The mouse lymph node homing receptor is identical with the lymphocyte cell surface marker Ly-22: Role of the EGF domain in endothelial binding. *Cell* **61:** 611.
Siegelman, M.H., M.W. Bond, W.M. Gallatin, T. St. John, H.T. Smith, V.A. Fried, and I.L. Weissman. 1986. Cell surface molecule associated with lymphocyte homing is a ubiquitinated branched-chain glycoprotein. *Science* **231:** 823.
Stamper, H.B., Jr. and J.J. Woodruff. 1976. Lymphocyte homing into lymph nodes: In vitro demonstration of the selective affinity of recirculation lymphocytes for high-endothelial venules. *J. Exp. Med.* **144:** 828.
St. John, T., W.M. Gallatin, M. Siegelman, H.T. Smith, V.A. Fried, and I.L. Weissman. 1986. Expression cloning of a lymphocyte homing receptor cDNA: Ubiquitin is the reactive species. *Science* **7231:** 845.
Stoolman, L.M. 1989. Adhesion molecules controlling lymphocyte migration. *Cell* **56:** 907.
Stoppelli, M.P., A. Corti, A. Soffientini, G. Cassani, F. Blasi, and R.K. Assoian. 1985. Differentiation enhanced binding of the amino-terminal fragment of human urokinase plasminogen activator to a specific receptor on U937 monocytes. *Proc. Natl. Acad. Sci.* **82:** 4939.
Streeter, P.R., E.L. Berg, B.T.N. Rouse, R.F. Bargatze, and E.C. Butcher. 1988. A tissue-specific endothelial cell molecule involved in lymphocyte homing. *Nature* **331:** 41.
Strynadka, N.C.J. and M.N.G. James. 1989. Crystal structures of the helix-loop-helix calcium-binding proteins. *Annu. Rev. Biochem.* **58:** 951.
Suzuki, S. and Y. Naitoh. 1990. Amino acid sequence of a novel integrin β_4 subunit and primary expression of the mRNA in epithelial cells. *EMBO J.* **9:** 757.
Tamkun, J.W., D.W. DeSimone, D. Fonda, R.S. Patel, C. Buck, A.F. Horwitz, and R.O. Hynes. 1986. Structure of integrin, a glycoprotein involved in the transmembrane linkage between fibronectin and actin. *Cell* **46:** 271.
Tedder, T.F., C.M. Isaacs, T.J. Ernst, G.D. Demetri, D.A. Adler, and C.M. Disteche. 1989. Isolation and chromosomal localization of cDNAs encoding a novel human lymphocyte cell surface molecule, LAM-1. Homology with the mouse lymphocyte homing receptor and other human adhesion proteins. *J. Exp. Med.* **170:** 123.
Tiemeyer, M., S.J. Swiedler, M. Ishihara, M. Moreland, H. Schweingruber, P. Hirtzer, and B.K. Brandley. 1991. Car-

bohydrate ligands for endothelial-leukocyte adhesion molecule 1. *Proc. Natl. Acad. Sci.* **88:** 1138.

van de Rijn, M., I.L. Weissman, and M.H. Siegelman. 1990. Biosynthesis pathway of $gp^{90MEL-14}$, the lymph node-specific homing receptor. *J. Immunol.* **145:** 1477.

von Heijne, G. 1986. A new method for predicting signal sequence cleavage sites. *Nucleic Acids Res.* **14:** 4683.

Walz, G., A. Aruffo, W. Kolanus, M. Bevilacqua, and B. Seed. 1990. Recognition by ELAM-1 of the sialyl-Le^x determinant on myeloid and tumor cells. *Science* **250:** 1132.

Watson, J., S. Kingsmore, G. Johnston, M.H. Siegelman, M. Le Beau, R. Lemons, N.S. Bora, T. Howard, I.L. Weissman, R. McEver, and M. Seldin. 1990. Genomic organization of the selectin family of leukocyte adhesion molecules on human and mouse chromosome 1. *J. Exp. Med.* **172:** 263.

Wayner, E.A., A. Garcia-Pardo, M.J. Humphries, J.A. MacDonald, and W.G. Carter. 1989. Identification and characterization of the T lymphocyte adhesion receptor for an alternative cell attachment domain (CS-1) in plasma fibronectin. *J. Cell Biol.* **109:** 1321.

Yamamoto, T., R.W. Bishop, M.S. Brown, J.L. Goldstein, and D.W. Russell. 1986. Deletion in cysteine-rich region of LDL receptor impedes transport to cell surface in WHHL rabbit. *Science* **232:** 1230.

Yednock T.A. and S.D. Rosen. 1989. Lymphocyte homing. *Adv. Immunol.* **44:** 313.

Yednock, T.A., E.C. Butcher, L.M. Stoolman, and S.D. Rosen. 1987. Receptors involved in lymphocyte homing: Relationship between a carbohydrate-binding receptor and the MEL-14 antigen. *J. Cell Biol.* **104:** 725.

Yuan, Q., W.-M. Jiang, G.W. Krissansen, and J.D. Watson. 1990. Cloning and sequence analysis of a novel β_2-related integrin transcript from T lymphocytes: Homology of integrincysteine-rich repeats to domain III of laminin B chains. *Int. Immunol.* **2:** 1097.

Extracellular Matrix/Growth Factor Interactions

E. RUOSLAHTI,* Y. YAMAGUCHI,* A. HILDEBRAND,* AND W.A. BORDER†

* *Cancer Research Center, La Jolla Cancer Research Foundation, La Jolla, California 92037;*
†*Division of Nephrology, University of Utah School of Medicine, Salt Lake City, Utah 84132*

Growth factors bind to extracellular matrices and onto cell surfaces, and this binding is mainly mediated by proteoglycans. Proteoglycans are proteins that carry an unusual carbohydrate, a glycosaminoglycan. The glycosaminoglycans are large carbohydrates that are composed of repeating disaccharide units and exist in four main forms: heparan sulfate and heparin, chondroitin sulfate and dermatan sulfate, keratan sulfate, and hyaluronic acid. The first three are protein-bound glycosaminoglycans in their natural form, and they all contain sulfate; hyaluronic acid is made as a free glycosaminoglycan and lacks sulfate (see Ruoslahti 1989; Kolset and Gallagher 1990). The glycosaminoglycan substitution depends on the ability of the protein to serve as an acceptor for xylosyltransferase, the enzyme that begins the synthesis of most types of glycosaminoglycans.

Proteoglycans have many different types of core proteins reflecting the varied locations and functions of these molecules (Ruoslahti 1989). Figure 1 shows examples of proteoglycan structures. The only common feature among proteoglycans may be the presence of the glycosaminoglycan. In this regard, the glycosaminoglycan substitution is similar to other types of glycosylation of proteins; it does not necessarily imply functional similarity. However, the diversity of proteoglycans is likely to be much lower than that of other types of glycoproteins, since cloning and sequencing analyses have repeatedly identified the same core proteins, or structurally very similar ones, from a variety of different sources. Thus, serglycin, a proteoglycan originally identified in a yolk sac tumor, has since been found to be the main proteoglycan of several other kinds of cells, notably various kinds of blood cells (Bourdon et al. 1985; Tantravahi et al. 1986; Alliel et al. 1988). Many of the other proteoglycans whose amino acid sequences are known fall into families of related proteoglycans. Thus, for example, the large proteoglycans versican and aggrecan belong to one family (Doege et al. 1987; Krusius et al. 1987; Zimmermann and Ruoslahti 1989), and the syndecans form another family (Marynen et al. 1989; Saunders et al. 1989; Carey et al. 1992), as do the decorin-type proteoglycans.

Many proteoglycans are constituents of the extracellular matrix; examples of such proteoglycans include the large proteoglycans versican, aggrecan, and perlecan, and the small proteoglycans decorin and fibromodulin. Some proteoglycans, such as the heparan sulfate proteoglycan, syndecan, have membrane-embedded core proteins. Finally, serglycin resides in the secretory granules of various types of hematopoietic cells.

The strong negative charge of glycosaminoglycans makes it possible for them to bind many substances, including some growth factors. However, the charge is not the whole basis of the binding; proteins can bind specifically to a given type of glycosaminoglycan, and the protein-binding site in a glycosaminoglycan can be a specific sugar sequence (Bray et al. 1989).

Many of the proteoglycans function as modulators of growth factors. The role of heparan sulfate proteoglycans as binders of fibroblast growth factors (FGFs) has been known for some time (see Burgess and Maciag 1989). The proteoglycan binding is thought to be important in protecting the FGFs from degradation and in creating a matrix-bound or cell-surface-bound reservoir of the growth factors. An active FGF/glycosaminoglycan complex can be generated from the reservoir by proteolysis of the proteoglycan core proteins or partial degradation of the heparan sulfate component (Ishai-Michaeli et al. 1990; Saksela and Rifkin 1990). Recently, it has been shown that the binding of basic FGF to its cell-surface signaling receptor requires prior binding of the growth factor either to the heparan sulfate side chains of a membrane heparan sulfate proteoglycan or to free heparan sulfate (heparin) chains (Rapraeger et al. 1991; Yayon et al. 1991).

The binding of growth factors by proteoglycans is not limited to the FGF/heparan sulfate proteoglycan interactions. Growth factors and cytokines that bind to heparin and heparan sulfate include granulocyte-macrophage colony stimulating factor and interleukin-3 (Roberts et al. 1988), pleiotrophin (Li et al. 1990; Merenmies and Rauvala 1990), and platelet factor 4, the prototype molecule in a family of a dozen or so growth factors and cytokines. Platelet factor 4 also binds avidly to chondroitin sulfate (Périn et al. 1988) and exists in platelet α granules complexed with the chondroitin sulfate proteoglycan serglycin (see Ruoslahti and Yamaguchi 1991).

The growth factor/proteoglycan interactions discussed above are mediated by the glycosaminoglycan components of the proteoglycans. At least one growth factor, transforming growth factor β (TGF-β), binds to proteoglycans through the core protein. One of the three receptors for TGF-β identified by cell-surface affinity labeling is a proteoglycan called betaglycan (Andres et al. 1989). The glycosaminoglycan component of betaglycan is not needed for the binding of

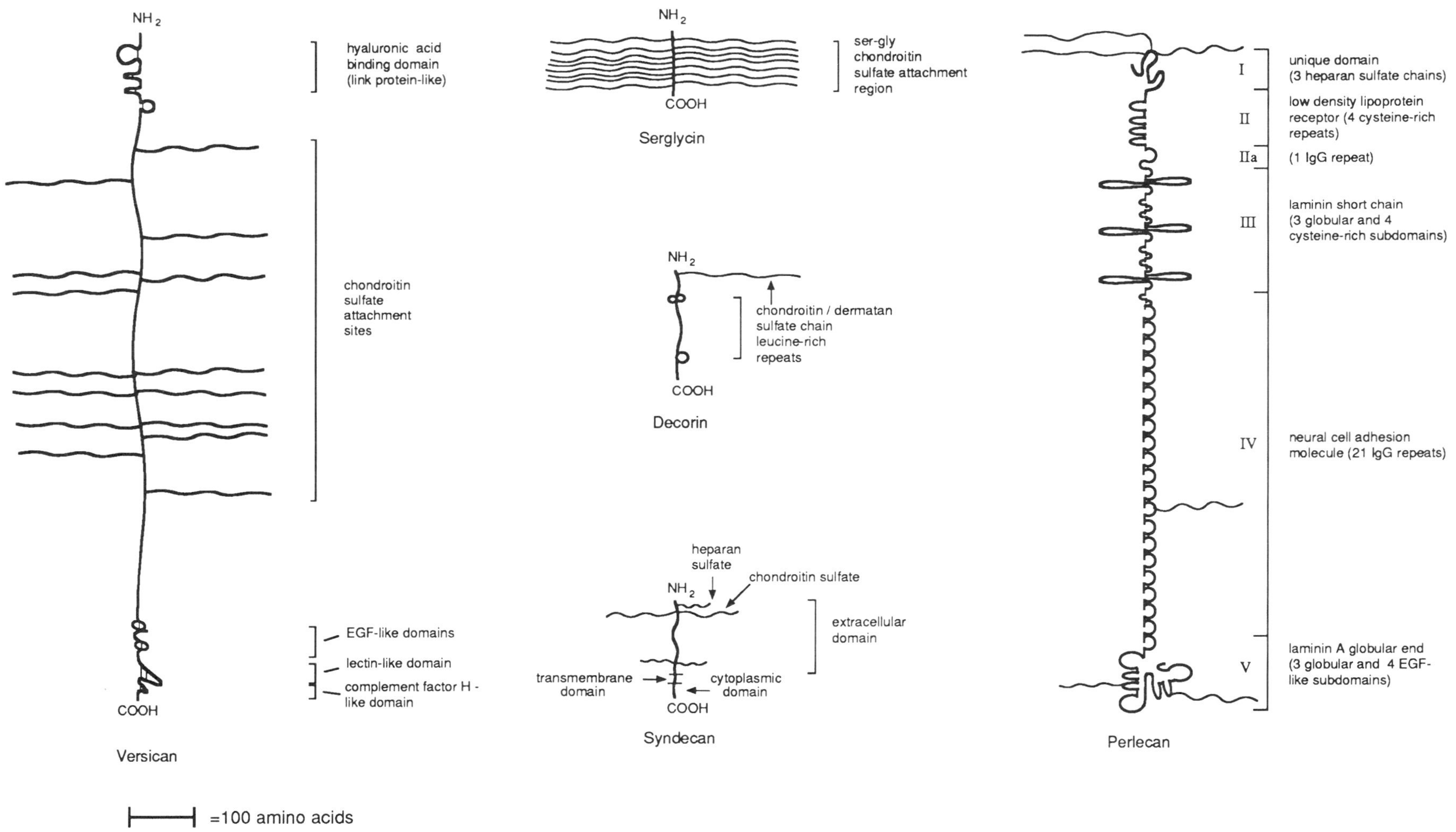

Figure 1. Schematic representation of proteoglycan structures. The four members of the decorin family each have a small core protein composed primarily of leucine-rich repeats and relatively few glycosaminoglycan chains. Decorin is depicted as a representative of this family and the structures of four other proteoglycans, versican, serglycin, syndecan, and perlecan, are shown for comparison.

TGF-β to this proteoglycan. Unlike the other two TGF-β receptors, betaglycan apparently does not function in signal transduction; its role is thought to be the binding of TGF-β for subsequent delivery to the signal transduction receptors. We have recently found that TGF-β is also bound by a group of proteoglycans, the prototype of which is the extracellular matrix proteoglycan, decorin.

RESULTS

The Decorin Family of Proteoglycans

The decorin-type proteoglycans comprise a family of proteoglycans characterized by small core proteins. The family includes at least four proteins, decorin, biglycan, fibromodulin, and lumican (Fig. 1)(Krusius and Ruoslahti 1986; Fisher et al. 1989; Oldberg et al. 1989; Blochberger et al. 1992). They each have an ~40-kD core protein that consists primarily of leucine-rich repeats of ~24 amino acids. This same leucine-rich motif was first identified in one of the plasma proteins, the leucine-rich α_2 glycoprotein (Takahashi et al. 1985), and it is present in a number of other proteins including the receptors for von Willebrand factor and lutropin, and the *Drosophila* developmental regulatory proteins Toll and chaoptin (see Patthy 1987; Ruoslahti 1988). The repeat is thought to be a general recognition and binding motif mediating protein-protein interactions.

Each of the members in the decorin family is a proteoglycan; i.e., they each carry covalently bound glycosaminoglycans attached to the core protein. In decorin and biglycan the glycosaminoglycan is of the chondroitin/dermatan sulfate type (chondroitin/dermatan sulfate consists of repeating disaccharide units of uronic acid linked to *N*-acetylgalactosamine). The difference between chondroitin sulfate and dermatan sulfate is that the uronic acid is a different isomer in the two. Fibromodulin and lumican are keratan sulfate proteoglycans (in keratan sulfate the repeating disaccharide is galactose-linked to *N*-acetylglucosamine with sulfate groups on the galactose moiety). Decorin contains one chondroitin sulfate chain attached to a specific serine residue near the amino terminus, and biglycan has two such chains. The ratio of chondroitin sulfate and dermatan sulfate in their side chains depends on the cell type from which the proteoglycan originates.

The tissue distributions of the decorin class vary from the almost universal presence of decorin and biglycan in various tissues to the preferential localizations of fibromodulin in the bone and lumican in the cornea. Decorin is bound to type I collagen in the extracellular matrix of tissues (Bianco et al. 1990) and is thought to regulate the size of the collagen fibrils (Vogel et al. 1984; Scott 1986). A similar function has been suggested for fibromodulin (Oldberg et al. 1989). Biglycan lacks the collagen-binding property and is localized on cell surfaces (Bianco et al. 1990). Little is known about the functional properties of lumican, but it is thought to be somehow involved in the collagen fibril alignment that makes the cornea transparent (Blochberger et al. 1992). Our recent results have shown that at least decorin and biglycan among these proteoglycans bind TGF-β.

TGF-β Binding by Decorin-type Proteoglycans

The initial observation that led to the discovery of the TGF-β-binding properties of decorin was made in a CHO cell expression system. The observation was that overexpression of decorin in the CHO cells inhibited their growth (Yamaguchi and Ruoslahti 1988). The basis of this effect turned out to be the ability of decorin to bind and neutralize TGF-β, which in these particular cells is a growth promoter (Yamaguchi et al. 1990).

The ability of decorin to bind TGF-β could be demonstrated in affinity chromatography, where decorin-Sepharose bound essentially all TGF-β activity from samples containing purified TGF-β1, whereas control Sepharose columns containing coupled albumin bound little of the activity. The TGF-β activity could be recovered by eluting with 8 M urea but not with 1 M NaCl. This latter result indicated that the TGF-β was binding to the protein core of decorin and not to the glycosaminoglycan, because urea typically disrupts protein-protein interactions, whereas protein-glycosaminoglycan interactions are reversed at high salt concentrations but persist in urea. That decorin was indeed binding TGF-β through the core protein was confirmed by testing decorin treated with chondroitinase to remove the glycosaminoglycan. Such decorin preparations were as active as the native proteoglycan in binding TGF-β.

Assays measuring the binding of radioactive TGF-β1 to decorin-coated microtiter wells were used to study the characteristics of the interaction between the two proteins. By measuring the ability of soluble decorin to compete for the binding of TGF-β1 to the insolubilized decorin, it was shown that the affinity of the interaction is in the range of $K_d = 10^{-8}$ (Yamaguchi et al. 1990). Purified biglycan also competed for the binding, whereas another chondroitin sulfate proteoglycan, aggrecan, had no effect. This result indicated that biglycan shares the ability of decorin to bind TGF-β, a result subsequently confirmed with bacterially produced decorin and biglycan fusion proteins (A. Hildebrand et al., in prep.). That the bacterially produced proteoglycan core proteins show the same TGF binding as the natural proteoglycans confirms the presence of the binding site in the protein part of the proteoglycan, since the bacterial products are devoid of glycosaminoglycan or other glycosylation.

The binding of TGF-β to decorin results in the neutralization of TGF-β activity (Yamaguchi et al. 1990). In the CHO cells, the proliferation of which is enhanced by TGF-β, decorin is a growth inhibitor, whereas in the mink lung epithelial cells commonly used to assay TGF-β activity, decorin counteracts the growth inhibitory activity of TGF-β. These results suggested that decorin and biglycan could be negative regulators of TGF-β activity. Data showing that TGF-β can stimulate the synthesis of decorin and biglycan lend further support to that idea.

Regulation of Decorin and Biglycan Synthesis by TGF-β

TGF-β up-regulates the synthesis of proteoglycans in cultured cells (Bassols and Massagué 1988). This effect appears to be one manifestation of the general stimulation of extracellular matrix synthesis by TGF-β. The matrix proteins whose synthesis is enhanced by TGF-β include fibronectin (Ignotz and Massagué 1986), tenascin (Pearson et al. 1988), and collagen (Ignotz and Massagué 1986; Varga et al. 1987), as well as integrin-type extracellular matrix receptors (Ignotz and Massagué 1987). The up-regulation of decorin and biglycan synthesis by TGF-β can be as high as 50-fold in mesangial cells from the kidney (Border et al. 1990a). Figure 2 illustrates the effect of TGF-β on decorin and biglycan production in mesangial cells. The synthesis of decorin, although enhanced in the mesangial cells, is not stimulated in all types of cells, whereas the synthesis of biglycan is (Schönherr et al. 1991). The marked enhancement of the synthesis of the two proteoglycans capable of neutralizing TGF-β by this growth factor suggests that these proteoglycans, in addition to being potentially capable of immobilizing TGF-β in tissues, can serve as negative regulators of TGF-β activity. That only the active, mature form of TGF-β binds to decorin, whereas the inactive precursor does not (A. Hildebrand et al., in prep.), lends further credence to this hypothesis. At this point, our results do not allow any conclusions as to the actual ability of tissue (collagen-bound) decorin to regulate TGF-β, but it has been shown that TGF-β also causes remarkable up-regulation of decorin and biglycan synthesis by TGF-β in vivo, and this up-regulation has been used as a marker of TGF-β activity in a disease involving matrix accumulation.

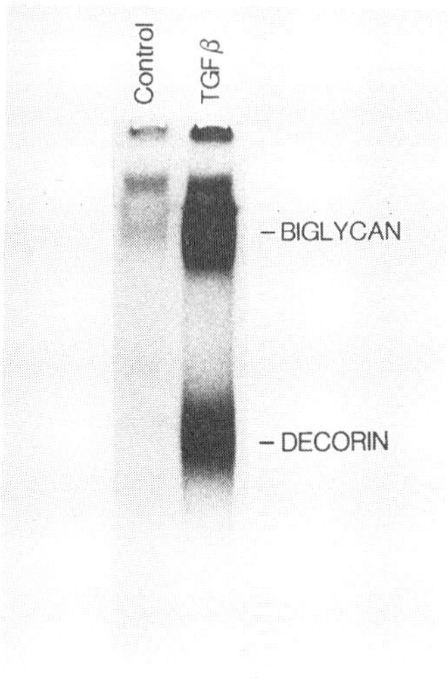

Figure 2. Stimulation of decorin and biglycan production by TGF-β1 in rat mesangial cells. Rat mesangial cells were cultured and treated with TGF-β1 as described previously (Border et al. 1990a).

TGF-β Is Responsible for the Obstruction of Glomerular Function by Matrix Accumulation

The accumulation of extracellular matrix in the glomeruli is the main pathological feature of glomerulonephritis, which is a kidney inflammation of suspected immunological origin. We have studied the underlying causes of the matrix accumulation in a rat model of glomerulonephritis induced by injecting the rats with an anti-thymocyte antibody (Border et al. 1990b). The anti-Thy-1 component of this antiserum damages the kidney mesangial cells, which express the Thy-1 antigen, setting an inflammatory process in motion. The kidneys of these rats show changes reminiscent of the human disease, including the accumulation of extracellular matrix in the glomeruli.

It was found that the matrix accumulation in the glomerulonephritic kidneys coincided temporally with increased expression of TGF-β1 mRNA, immunologically detectable TGF-β1 protein, and TGF-β activity (Border et al. 1990b; Okuda et al. 1990). The expression of TGF-β mRNA in glomerular preparations from the rat kidneys was first seen elevated on day 4 after the injection of the antiserum; it peaked between days 7 and 14 and then declined as the disease spontaneously subsided and the matrix resolved, which is what happens in this model. Since TGF-β is known to be a strong inducer of extracellular matrix synthesis, these results strongly suggested a role for TGF-β in the disease process. These kidneys also showed a strongly elevated decorin and biglycan production reminiscent of what was induced by TGF-β in the mesangial cells. A further indication of a role for TGF-β in the disease was obtained by examining the kidneys for the expression of two proteins, whose synthesis is known to be induced by TGF-β, the ED-A alternatively spliced form of fibronectin (Balza et al. 1988; Wang et al. 1991) and tenascin (Pearson et al. 1988). Immunofluorescence showed that both of these proteins were strongly expressed in the diseased glomeruli, whereas they were essentially undetectable in the normal glomeruli (T. Yamamoto et al., in prep.).

Evidence for a causal relationship between the TGF-β expression and the matrix accumulation was obtained by treating the glomerulonephritic rats with antibodies capable of neutralizing TGF-β1 activity (Border et al. 1990b). The neutralizing antibody suppressed the matrix accumulation and other manifestations of the disease, whereas normal rabbit serum had no effect. More recently, it has been possible to show that the natural inhibitor of TGF-β, decorin, when injected repeatedly into the rats developing glomerulonephritis, also protects them against the disease (Border et al. 1992).

DISCUSSION

Recent published data from our laboratories and the results presented here show that the tissue proteoglycans, decorin and biglycan, can bind TGF-β and neutralize its activity. Up-regulation of decorin and biglycan synthesis by TGF-β and the involvement of TGF-β in various disease processes characterized by extracellular matrix accumulation and fibrosis suggest that the proteoglycans may serve as physiological regulators of TGF-β activity and that they may provide pharmacological agents capable of controlling diseases associated with overproduction of TGF-β.

The decorin family of proteoglycans consists of at least four members; two of the four, decorin and biglycan, have been shown to bind to TGF-β, the other two remain to be tested. So far, we have not found any difference in the affinity of decorin and biglycan for TGF-β1, and they also compete for one another's binding. These results suggest that if there are differences between them with regard to TGF-β1 binding, the differences may relate to the tissue localization of the proteoglycan and, perhaps, the fate of the proteoglycan/TGF-β complex after the binding. It will also be interesting to see whether the three isoforms of TGF-β are each bound similarly by each of the proteoglycans.

TGF-β greatly enhances the synthesis of both decorin and biglycan by mesangial cells; in many other cell types, biglycan appears to be the proteoglycan whose expression is increased by TGF-β (Schönherr et al. 1991). TGF-β affects both the protein and glycosaminoglycan parts of these proteoglycans; more core protein is made and the glycosaminoglycan chains get longer (Bassols and Massagué 1988; Border et al. 1990a). Other proteoglycans and matrix proteins such as fibronectin and collagens are similarly up-regulated. A relatively specific increase is seen in the ED–A$^+$ alternatively spliced form of fibronectin and the matrix protein tenascin (Pearson et al. 1988; Wang et al. 1991), each of which is nearly undetectable in normal tissues such as the kidney, but increases in kidney diseases associated with elevated TGF-β levels (T. Yamamoto et al., in prep.). In addition to inducing increased synthesis of extracellular matrix, TGF-β also reduces its degradation by suppressing the synthesis of matrix-degrading enzymes (Overall et al. 1989). Since TGF-β induces an increase in the synthesis of the ED–A$^+$ fibronectin, tenascin, decorin, and biglycan, and since the increase can be as much as 50-fold (Border et al. 1990a), these proteins can serve as markers for TGF-β activity. Indeed, we have used each of these markers to monitor TGF-β responses in disease.

Increased production of TGF-β appears to be the underlying pathological factor in a number of diseases characterized by accumulation of extracellular matrix and fibrosis. We have found this to be the case in glomerulonephritis (Border et al. 1990b; Okuda et al. 1990). An elevation of TGF-β1 mRNA and TGF-β

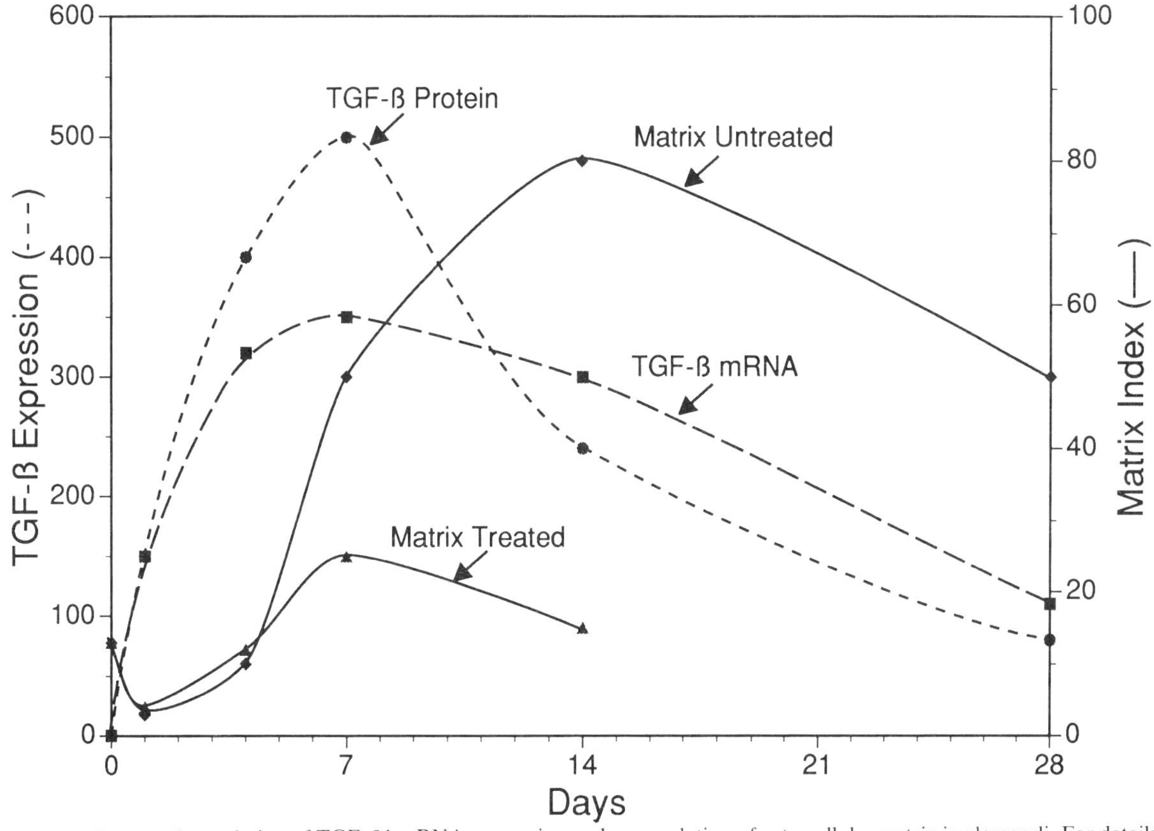

Figure 3. Temporal correlation of TGF-β1 mRNA expression and accumulation of extracellular matrix in glomeruli. For details see Okuda et al. (1990).

activity precedes matrix accumulation, and subsequent dissolution of the abnormal matrix follows a decline of TGF-β in this disease (Fig. 3). It was possible to reduce the matrix accumulation by in vivo administration of antibodies capable of neutralizing TGF-β1. Other workers have shown the same for skin scarring (Shah et al. 1992). Moreover, correlative evidence links TGF-β overproduction to lung fibrosis (Broekelman et al. 1991), liver cirrhosis (Castilla et al. 1991), myelofibrosis (Terui et al. 1990), and diabetic nephropathy (T. Yamamoto et al., in prep.). In addition, TGF-β may be a tumor promoter in tissues that have been wounded (Sieweke et al. 1990). We have termed the involvement of TGF-β in fibrotic conditions the "dark side" of TGF-β (Border and Ruoslahti 1992).

The factor or factors that induce the elevated production of TGF-β in injured tissue are not known, and it obviously will be very important to identify them. The results in the studies cited above indicate that the induction occurs primarily at the transcriptional level, since the level of TGF-β1 mRNA is greatly elevated in the injured tissues undergoing fibrosis. Once increased TGF-β production has been induced, it may become self-sustaining because TGF-β is capable of inducing its own synthesis (Van Obberghen-Schilling et al. 1988). This may give rise to a vicious circle of increasing TGF-β production that could potentially convert an acute injury into a chronic fibrotic process. Tissue components induced by TGF-β that are capable of neutralizing TGF-β, such as decorin and biglycan, may play a role in shutting down a TGF-β response.

ACKNOWLEDGMENTS

We thank Dr. Daniel Twardzik for providing purified TGF-β1. The original work by the authors and the preparation of this manuscript were supported by National Cancer Institute grants CA-42507, CA-28896, and Cancer Center Support grant CA-30199 to E.R., National Institute of Diabetes and Digestive and Kidney Diseases grant DK-43609 to W.A.B., and grants from Telios Pharmaceuticals, Inc., to E.R. and W.A.B.

REFERENCES

Alliel, P.M., J.-P. Périn, P. Maillet, F. Bonnet, J.-P. Rosa, and P. Jollès. 1988. Complete amino acid sequence of a human platelet proteoglycan. *FEBS Lett.* **236:** 123.

Andres, J.L., K. Stanley, S. Cheifetz, and J. Massagué. 1989. Membrane-anchored and soluble forms of betaglycan, a polymorphic proteoglycan that binds transforming growth factor-β. *J. Cell Biol.* **109:** 3137.

Balza, E., L. Borsi, G. Allemanni, and L. Zardi. 1988. Transforming growth factor β regulates the levels of different fibronectin isoforms in normal human cultured fibroblasts. *FEBS Lett.* **228:** 42.

Bassols, A. and J. Massagué. 1988. Transforming growth factor β regulates the expression and structure of extracellular matrix chondroitin/dermatan sulfate proteoglycans. *J. Biol. Chem.* **263:** 3039.

Bianco, P., L.W. Fisher, M.F. Young, J.D. Termine, and P. Gehron Robey. 1990. Expression and localization of the two small proteoglycans biglycan and decorin in developing human skeletal and non-skeletal tissues. *J. Histochem. Cytochem.* **38:** 1549.

Blochberger, T.C., J.-P. Vergnes, J. Hempel, and J.R. Hassell. 1992. cDNA to chick lumican (corneal keratan sulfate proteoglycan) reveals homology to the small interstitial proteoglycan gene family and expression in muscle and intestine. *J. Biol. Chem.* **267:** 347.

Border, W.A. and E. Ruoslahti. 1992. Transforming growth factor-β in disease: The dark side of tissue repair. *J. Clin. Invest.* **90:** 1.

Border, W.A., S. Okuda, L.R. Languino, and E. Ruoslahti. 1990a. Transforming growth factor-β regulates production of proteoglycans by mesangial cells. *Kidney Int.* **37:** 689.

Border, W.A., S. Okuda, L.R. Languino, M.B. Sporn, and E. Ruoslahti. 1990b. Suppression of experimental glomerulonephritis by antiserum against transforming growth factor β1. *Nature* **346:** 371.

Border, W.A., N.A. Noble, T. Yamamoto, J.R. Harper, Y. Yamaguchi, M.D. Pierschbacher, and E. Ruoslahti. 1991. Decorin, a natural inhibitor of transforming growth factor-β, protects against scarring in experimental kidney disease. *Nature* (in press).

Bourdon, M.A., Å. Oldberg, M. Pierschbacher, and E. Ruoslahti. 1985. Molecular cloning and sequencing analysis of a chondroitin sulfate proteoglycan cDNA. *Proc. Natl. Acad. Sci.* **82:** 13212.

Bray, B., D.A. Lane, J.M. Freyssinet, G. Pejler, and U. Lindahl. 1989. Anti-thrombin activities of heparin. Effect of saccharide chain length on thrombin inhibition by heparin cofactor II and by antithrombin. *Biochem. J.* **262:** 225.

Broekelmann, T.J., A.H. Limper, T.V. Colby, and J.A. McDonald. 1991. Transforming growth factor β1 is present at sites of extracellular matrix gene expression in human pulmonary fibrosis. *Proc. Natl. Acad. Sci.* **88:** 6642.

Burgess, W.H. and T. Maciag. 1989. The heparin-binding (fibroblast) growth factor family of proteins. *Annu. Rev. Biochem.* **58:** 575.

Carey, D.J., D.M. Evans, R.C. Stahl, V.K. Asundi, K.J. Conner, P. Garbes, and G. Cizmeci-Smith. 1992. Molecular cloning and characterization of *N*-syndecan, a novel transmembrane heparan sulfate proteoglycan. *J. Cell Biol.* **117:** 191.

Castilla, A., J. Prieto, and N. Fausto. 1991. Transforming growth factors β1 and α in chronic liver disease. *New Engl. J. Med.* **324:** 933.

Doege, K., M. Sasaki, E. Horigan, J.R. Hassell, and Y. Yamada. 1987. Complete primary structure of the rat cartilage proteoglycan core protein deduced from cDNA clones. *J. Biol. Chem.* **262:** 17757.

Fisher, L., J.D. Termine, and M.F. Young. 1989. Deduced protein sequence of bone small proteoglycan I (biglycan) shows homology with proteoglycan II (decorin) and several nonconnective tissue proteins in a variety of species. *J. Biol. Chem.* **264:** 4571.

Ignotz, R.A. and J. Massagué. 1986. Transforming growth factor-β stimulates the expression of fibronectin and collagen and their incorporation into the extracellular matrix. *J. Biol. Chem.* **261:** 4337.

———.1987. Cell adhesion protein receptors as targets for transforming growth factor-β action. *Cell* **51:** 189.

Ishai-Michaeli, R., A. Eldor, and I. Vlodavsky. 1990. Heparanase activity expressed by platelets, neutrophils, and lymphoma cells releases active fibroblast growth factor from extracellular matrix. *Cell Reg.* **1:** 833.

Kolset, S.O. and J.T. Gallagher. 1990. Proteoglycans in haemopoietic cells. *Biochim. Biophys. Acta* **1032:** 191.

Krusius, T. and E. Ruoslahti. 1986. Primary structure of an extracellular matrix proteoglycan core protein deduced from cloned cDNA. *Proc. Natl. Acad. Sci.* **83:** 7683.

Krusius, T., K.R. Gehlsen, and E. Ruoslahti. 1987. A fibroblast chondroitin sulfate proteoglycan core protein contains lectin-like and growth factor-like sequences. *J. Biol. Chem.* **262:** 13120.

Li, Y.-S., P.G. Milner, A.K. Chauhan, M.A. Watson, R.M. Hoffman, C.M. Kodner, J. Milbrandt, and T.F. Deuel. 1990. Cloning and expression of a developmentally regulated protein that induces mitogenic and neurite outgrowth activity. *Science* **250:** 1690.

Marynen, P., J. Zhang, J.J. Cassiman, H. Van den Berghe, and G. David. 1989. Partial primary structure of the 48- and 90-kilodalton core proteins of cell surface-associated heparan sulfate proteoglycans of lung fibroblasts. *J. Biol. Chem.* **264:** 7017.

Merenmies, J. and H. Rauvala. 1990. Molecular cloning of the 18-kDa growth-associated protein of developing brain. *J. Biol. Chem.* **265:** 16721.

Okuda, S., L.R. Languino, E. Ruoslahti, and W.A. Border. 1990. Elevated expression of transforming growth factor-β and proteoglycan production in experimental glomerulonephritis. Possible role in expansion of the mesangial extracellular matrix. *J. Clin. Invest.* **86:** 453.

Oldberg, Å., P. Antonsson, K. Lindblom, and D. Heinegård. 1989. A collagen-binding 59-kd protein (fibromodulin) is structurally related to the small interstitial proteoglycans PG-S1 and PG-S2 (decorin). *EMBO J.* **8:** 2601.

Overall, C.M., J.L. Wrana, and J. Sodek. 1989. Independent regulation of collagenase, 72 kDa progelatinase, and metalloendoproteinase inhibitor expression in human fibroblasts by transforming growth factor-β. *J. Biol. Chem.* **264:** 1860.

Patthy, L. 1987. Detecting homology of distantly related proteins with consensus sequences. *J. Mol. Biol.* **198:** 567.

Pearson, C.A., D. Pearson, S. Shibahara, J. Hofsteenge, and R. Chiquet-Ehrismann. 1988. Tenascin: cDNA cloning and induction by TGF-β. *EMBO J.* **7:** 2977.

Périn, J.-P., F. Bonnet, P. Maillet, and P. Jollès. 1988. Characterization and N-terminal sequence of human platelet factor proteoglycan. *Biochem. J.* **255:** 1007.

Rapraeger, A.C., A. Krufka, and B.B. Olwin. 1991. Requirement of heparan sulfate for bFGF-mediated fibroblast growth and myoblast differentiation. *Science* **252:** 1705.

Roberts, R., J. Gallagher, E. Spooncer, T.D. Allen, F. Bloomfield, and T.M. Dexter. 1988. Heparan sulfate bound growth factors: A mechanism for stromal cell mediated haemopoiesis. *Nature* **332:** 376.

Ruoslahti, E. 1988. Structure and biology of proteoglycans. *Annu. Rev. Cell Biol.* **4:** 229.

———. 1989. Proteoglycans in cell regulation. *J. Biol. Chem.* **264:** 13369.

Ruoslahti, E. and Y. Yamaguchi. 1991. Proteoglycans as modulators of growth factor activities. *Cell* **64:** 867.

Saksela, O. and D.B. Rifkin. 1990. Release of basic fibroblast growth factor-heparan sulfate complexes from endothelial cells by plasminogen activator-mediated proteolytic activity. *J. Cell Biol.* **110:** 767.

Saunders, S., M. Jalkanen, S. O'Farrell, and M. Bernfield. 1989. Molecular cloning of syndecan, an integral membrane proteoglycan. *J. Cell Biol.* **108:** 1547.

Schönherr, E., H.T. Järveläinen, L.J. Sandell, and T.N. Wight. 1991. Effects of platelet-derived growth factor and transforming growth factor-$\beta 1$ on the synthesis of a large versican-like chondroitin sulfate proteoglycan by arterial smooth muscle cells. *J. Biol. Chem.* **266:** 17640.

Scott, J.E. 1986. Proteoglycan-collagen interactions. *Ciba Found. Symp.* **124:** 104.

Shah, M., D.M. Foreman, and W.J. Ferguson. 1992. Control of scarring in adult wounds by neutralizing antibody to transforming growth factor β. *Lancet* **339:** 213.

Sieweke, M.H., N.L. Thompson, M.B. Sporn, and M.J. Bissell. 1990. Mediation of wound-related Rous sarcoma virus tumorigenesis by TGF-β. *Science* **248:** 1656.

Takahashi, N., Y. Takahashi, and F.W. Putnam. 1985. Periodicity of leucine and tandem repetition of a 24-amino acid segment in the primary structure of leucine-rich α_2-glycoprotein of human serum. *Proc. Natl. Acad. Sci.* **82:** 1906.

Tantravahi, R.V., R.L. Stevens, K.F. Austen, and J.H. Weis. 1986. A single gene in mast cells encodes the core peptides of heparin and chondroitin sulfate proteoglycans. *Proc. Natl. Acad. Sci.* **83:** 9207.

Terui, T., Y. Niitsu, K. Mahara, Y. Fujisaki, Y. Urushizaki, Y. Mogi, Y. Kohgo, N. Watanabe, M. Ogura, and H. Saito. 1990. The production of transforming growth factor-β in acute megakaryoblastic leukemia and its possible implications in myelofibrosis. *Blood* **75:** 1540.

Van Obberghen-Schilling, E., N.S. Roche, K.C. Flanders, M.B. Sporn, and A.B. Roberts. 1988. Transforming growth factor $\beta 1$ positively regulates its own expression in normal and transformed cells. *J. Biol. Chem.* **263:** 7741.

Varga, J., J. Rosenbloom, and S.A. Jimenez. 1987. Transforming growth factor β (TGFβ) causes a persistent increase in steady-state amounts of type I and type III collagen and fibronectin mRNAs in normal human dermal fibroblasts. *Biochem. J.* **247:** 597.

Vogel, K.G., M. Paulsson, and D. Heinegård. 1984. Specific inhibition of type I and type II collagen fibrillogenesis by the small proteoglycan of tendon. *Biochem. J.* **223:** 587.

Wang, A., D.S. Cohen, E. Palmer, and D. Sheppard. 1991. Polarized regulation of fibronectin secretion and alternative splicing by transforming growth factor β. *J. Biol. Chem.* **266:** 15598.

Yamaguchi, Y. and E. Ruoslahti. 1988. Expression of human proteoglycan in Chinese hamster ovary cells inhibits cell proliferation. *Nature* **336:** 244.

Yamaguchi, Y., D.M. Mann, and E. Ruoslahti. 1990. Negative regulation of transforming growth factor-β by the proteoglycan decorin. *Nature* **346:** 281.

Yayon, A., M. Klagsbrun, J.D. Esko, P. Leder, and D.M. Ornitz. 1991. Cell surface, heparin-like molecules are required for binding of basic fibroblast growth factor to its high affinity receptor. *Cell* **64:** 841.

Zimmermann, D. and E. Ruoslahti. 1989. Multiple domains of the large fibroblast proteoglycan, versican. *EMBO J.* **8:** 2975.

Mediation and Inhibition of Cell Adhesion by Morphoregulatory Molecules

G.M. Edelman
The Scripps Research Institute, La Jolla, California 92037

The cell surface plays a central role in regulating processes that lead to the formation of tissues and organs. These processes include the control of cell division, cell motion, adhesion, neurite extension, and embryonic induction. In the formation of a complex tissue, a remarkable set of coordinative events must occur to correlate mechanochemical changes in cell shape and mobility with cell interaction and gene expression. During the last 15 years, it has become apparent that three families of morphoregulatory molecules are centrally involved in such correlations. These include cell and substrate adhesion molecules and molecules involved in junctional complexes (Edelman et al. 1990).

Analyses of these molecules have revealed a remarkable set of properties and interactions. Two main themes have emerged: (1) Distinct place-dependent patterns of expression of these molecules can be seen during morphogenesis (Edelman 1988; Edelman and Crossin 1991; Takeichi 1991) and (2) the interactions of these molecules at the cell surface couple transmembrane control via the cytoskeleton (Edelman 1976) to specific external binding events (Nagafuchi and Takeichi 1988; Jaffe et al. 1990). To comprehend how these observations can be related to morphogenesis, it is important to understand the mechanisms of various binding events and signaling pathways affecting cell adhesion, shape, and motion. It is equally important to understand the means by which the genes for morphoregulatory molecules are expressed in a place-dependent manner during development.

In this paper, I touch briefly on some observations bearing on these two themes. I focus on certain CAMs (cell adhesion molecules) and on an extracellular matrix molecule, cytotactin. CAMs are responsible for cell-cell adhesion. Cytotactin is an example of a molecule whose interactions tend to inhibit adhesive events and lead to various alterations of cell shape and motion. A recent analysis of the promoter region of the gene for the neural cell adhesion molecule (N-CAM) and of that for cytotactin suggests that these promoters are targets for the products of homeobox-containing genes (Jones et al. 1992a,b). Although the significance of these observations is still not fully assessed, they may provide a basis (Edelman 1992) for connecting the regional expression of morphoregulatory molecules to their mechanochemical effects on tissue patterning.

CAMs

A number of reviews provide detailed accounts of cell adhesion (Edelman et al. 1990; Takeichi 1990; Edelman and Crossin 1991), and I therefore touch only briefly on some key properties of CAMs. Since the early descriptions of N-CAM and L-CAM (Thiery et al. 1977; Takeichi et al. 1981; Gallin et al. 1983), it has become evident that there are two main families of CAMs (Table 1). The N-CAM superfamily consists of various proteins having a series of extracellular domains homologous to immunoglobulins and to fibronectin type III repeats. The cytoplasmic domains of these molecules vary in length and structure as a result of alternative RNA splicing events. In some cases, such events lead to forms that are bound to the cell surface by phosphatidylinositol linkages. The binding of members of this family to yield cell-cell adhesion is divalent-cation-independent. The second main family of CAMs consists of so-called cadherins, the binding of which is divalent-cation-dependent (Brackenbury et al. 1977; Takeichi et al. 1981). These molecules share a common cytoplasmic domain but have extracellular domains of different binding specificities. No structural homologies are evident between the N-CAM and cadherin families. Different cells, however, express members of both families at their surfaces in particular combinations during development.

In most cases, CAM binding is homophilic, i.e., a given CAM binds to the identical CAM on an opposing cell (Hoffman and Edelman 1983). In general, therefore, the specificity of cell-cell interaction depends on the ability of each CAM to bind exclusively to itself. This is not universal, however: Certain CAMs bind to each other and to cell-surface molecules such as integrins in a heterophilic fashion (Grumet and Edelman 1984; Marlin and Springer 1987).

The binding functions of CAMs depend on direct or indirect interactions with the cytoskeleton. The redistribution of CAMs at the cell surface, the ability to bind extracellular ligands, and the coupling to various mechanisms governing cell shape, motion, and sorting all depend on such cytoskeletal interactions. The dependence of cytoskeletal linkage on the cytoplasmic domain of CAMs and the varying consequences of such a linkage (Fig. 1) has been shown by construction of deleted and chimeric molecules in transfection experiments (Nagafuchi and Takeichi 1988; Jaffe et al. 1990).

Table 1. Cell Adhesion Molecules

Neural	Immune	Miscellaneous	Invertebrates
N-CAM superfamily (calcium-independent)			
N-CAM	I-CAM-1	CEA	fasciclin II
Ng-CAM	I-CAM-2	NCA	amalgam
Nr-CAM	CD4	BGP-1	neuroglian
L1/NILE	LAR	poliovirus receptor	DLAR
Contactin/F11		PECAM-1/endo CAM	DPTP
F3		V-CAM	ApCAM
Tag-1/Axonin-1		OB-CAM	
MAG,SMP,DM-GRASP,SC1		PDGF R	
Po		MUC 18	
HT7		DCC	
Cadherin superfamily (calcium-dependent)			
L-CAM			
uvomorulin/E-cadherin	T-cadherin		EP-cadherin (X.)
N-cadherin/A-CAM	cadherins 4-11		XB-cadherin (X.)
P-cadherin	desmoglein/dgl		U-cadherin (X.)
B-cadherin	desmocollins I & II (dg2/3)		*fat* gene product (D.)
M-cadherin	PVA (Pemphigus vulgaris antigen)		
R-cadherin			

This table is a composite of the work from numerous laboratories. Individual references may be found in Edelman and Crossin (1991) and Takeichi (1990).

Evidence for the expression of specific combinations of CAMs at particular times and locations during development and tissue regeneration in a given species is extensive (Crossin et al. 1985; Edelman and Crossin 1991). This place-dependent expression begins as early in development as the two-cell stage. The tissue-specific combination of particular CAMs and the order of their expression in development appear to be critical in border formation, epithelial/mesenchymal transformation, and cell migration. Moreover, it has been shown that perturbation of CAM expression and binding leads to changes in morphology. Blockade of homophilic binding of N-CAM or neuron-glia CAM (Ng-CAM) by antibody fragments results, for example, in disruption of retinal structure and mapping (Buskirk et al. 1980), in alteration of patterns of neurite fasciculation (Hoffman et al. 1986), or in altered patterns of cell motion during cerebellar layering (Chuong et al. 1987). Similarly, superexpression or ectopic expression of cadherins after injection of their RNAs in embryos can lead to distorted morphologies (Kintner 1988).

The above brief survey indicates that a dynamic pattern of expression of CAMs at the cell surface affects the formation of epithelia and mesenchyme in a definite order that is characteristic of a given tissue. Moreover, CAM binding at the cell surface links the cytoskeleton to specific patterns of extracellular binding events. It is not difficult to see how such a process might alter tissue

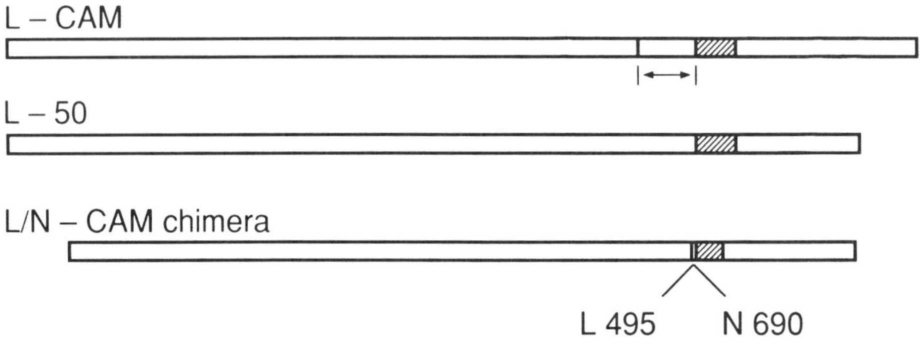

Figure 1. Chimeric and truncated molecules based on L-CAM. Schematic diagrams of L-CAM, L-50, and L/N-CAM are presented. All three molecules have identical extracellular domains with the exception of a 45-amino-acid segment, denoted by an arrow in the L-CAM construct, which is missing in the chimeric molecule. Hatched areas represent transmembrane domains. The junction between L-CAM and N-CAM amino acid sequences is indicated in the figure, with amino acids numbered as described previously (Cunningham et al. 1987; Gallin et al. 1987). Cells transfected with the L-50 cDNA (a form with the carboxy-terminal 50 amino acids missing) expressed the molecule but did not bind in short-term assays. Cells transfected with L-CAM cDNA bound in short-term assays and in long-term-sorting assays (Friedlander et al. 1989). They sorted from untransfected L cells and from cells transfected with the chimeric the L/N-CAM cDNA. In contrast, cells transfected with the L/N-CAM chimeric cDNA did not sort from untransfected cells, although the cells that expressed L-CAM or L-chimeric CAM bound equally well to each other and to themselves in short-term-binding assays. (Reprinted, with permission, from Jaffe et al. 1990.)

pattern by affecting cell shape and motion. Nevertheless, in most cases, the detailed mechanisms of these transmembrane signaling events remain to be worked out. A promising route to explore is the linkage of cadherins to cytoskeletal elements by intermediary cytoplasmic proteins, the catenins (Nagafuchi and Takeichi 1989; Ozawa et al. 1989).

Adhesion can obviously lead to cell patterning. It is perhaps less obvious, however, that such patterning also may depend on molecules that are counter-adhesive, i.e., that inhibit adhesion. A striking example is the extracellular matrix molecule, cytotactin.

Cytotactin: Counteracting Adhesion, Migration, and Process Extension

Cytotactin (Grumet et al. 1985) was independently discovered in a number of laboratories and has a variety of different names: tenascin (Chiquet-Ehrismann et al. 1986), hexabrachion (Erickson and Iglesias 1984), myotendinous antigen (Chiquet and Fambrough 1984), and $J1_{220/200}$ (Kruse et al. 1985). Its structure (Jones et al. 1988,1989; Spring et al. 1989) is remarkable in several respects. The molecule consists of six polypeptides linked at their amino termini by disulfide bonds to form a so-called hexabrachion. Each chain consists of a series of epidermal growth factor (EGF) repeats, followed by several fibronectin type III repeats (variable in number, depending on alternative RNA splicing), and finally by domains homologous to portions of fibrinogen.

Unlike fibronectin and laminin, cytotactin shows a place-dependent pattern of expression that is sharply restricted in time and place during development. In this respect, its pattern of expression resembles those observed for CAMs; indeed, it can be seen in patterns that alternate or correlate with those of CAMs at sites such as somites, the developing brain, and developing bones and joints. It is found in the extracellular matrix in the presence of other more generally widespread molecules such as collagen and fibronectin.

One of the striking properties of cytotactin is its effect on cell shape, process extension, and cell motion. Cells plated on substrates of cytotactin generally become rounded even when they adhere (Friedlander et al. 1988). Moreover, in cultured dorsal root ganglia, neurite extension (Fig. 2) into such regions is strongly inhibited. Several recent studies have indicated that when presented together with other molecules that support binding of the neuronal cell body, cytotactin can actually enhance neurite outgrowth from single cells (Wehrle and Chiquet 1990; Lochter et al. 1991).

An analysis of the detailed binding properties of the various domains of cytotactin (Fig. 3) has been recently carried out in my laboratory (Prieto et al. 1992). This work reveals a distinct pattern of cell attachment, flattening, or rounding that is characteristic of each type of domain. Thus, although EGF-like domains, fibrinogen-like domains, and certain fibronectin type III repeats discourage flattening and cell extension, other fibronectin-type-III-like domains support these properties.

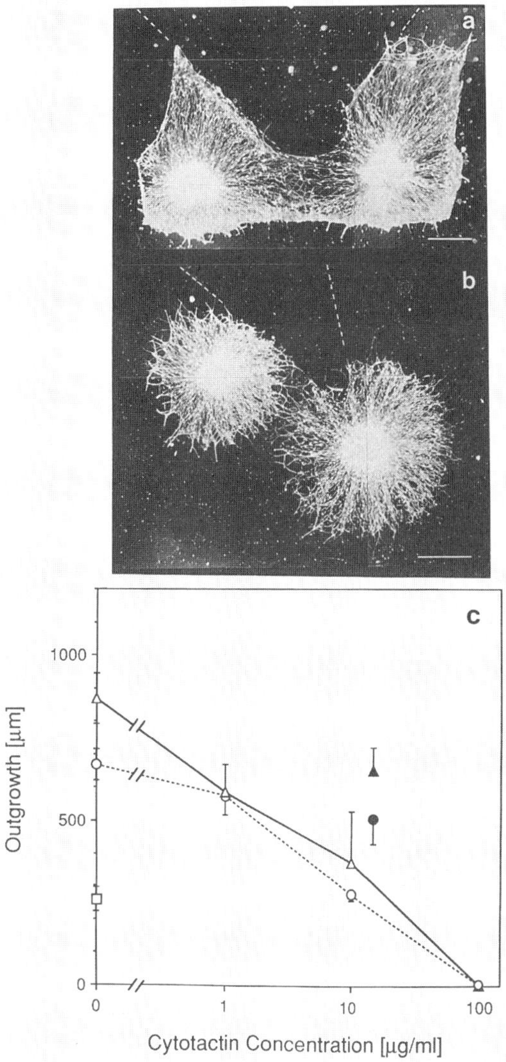

Figure 2. Effect of purified cytotactin on neurite outgrowth from dorsal root ganglia. (*a,b*) Neurite growth at a cytotactin boundary. Shown are micrographs of live explants. Dorsal root ganglia were explanted on a rectangular area incubated with 100 μg/ml laminin (*a*) or fibronectin (*b*) near a vertex of a triangular region on which 100 μg/ml cytotactin was incubated on top of laminin or fibronectin, before blocking with bovine serum albumin. Dashed lines (*a,b*) indicate the position of two sides of the triangles. (*a*) The bottom and sides of the outgrowth indicate the limits of the area coated with laminin. Note that neurites avoid the region coated with both cytotactin and laminin (*a*) or cytotactin and fibronectin (*b*). Bar, 1 mm. (*c*) Quantitation of neurite outgrowth on mixtures of cytotactin and laminin or fibronectin. Dorsal root ganglia were explanted on wells incubated with mixtures of the indicated concentrations of cytotactin and 10 μg/ml laminin (△) or fibronectin (○). (□) Outgrowth on bovine serum albumin. Neurite outgrowth in the presence of soluble cytotactin (10 μg/ml) was determined for ganglia grown in wells coated with 10 μg/ml laminin (▲) or fibronectin (●).

The possibility of a mosaic function composed of adhesive and counter-adhesive activities mediated by

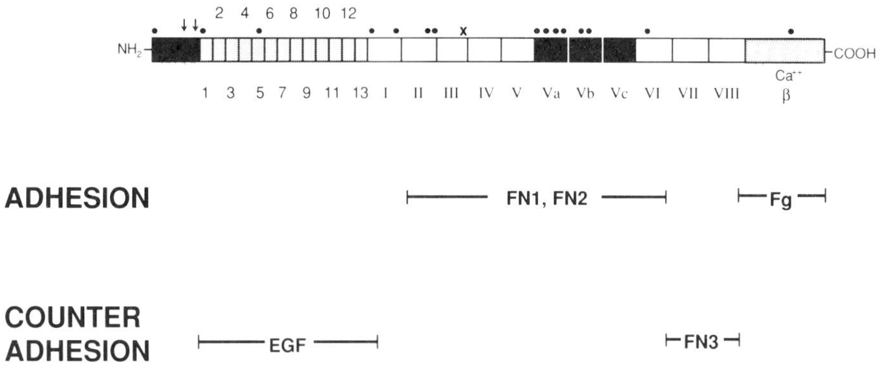

Figure 3. Summary of the cellular binding properties of different regions of cytotactin. Based on the linear sequence of chicken cytotactin (Jones et al. 1989) shown at the top, regions corresponding to the domains of the protein delineated below were prepared in bacterial expression vectors (A.L. Prieto et al., in prep.). The isolated domains were tested for their ability to support cell attachment (adhesive) or to prevent cell attachment and flattening on fibronectin (counter-adhesive).

the cytotactin molecule is raised by these findings. Both activities may actually operate in vivo. Although in the absence of other interactions the whole molecule is essentially counter-adhesive in tissue-culture experiments (Friedlander et al. 1988; Spring et al. 1989; Prieto et al. 1992), evidence of a central role in guiding cell migration has been found in cerebellar tissue slices (Chuong et al. 1987). Blockade of cytotactin binding by specific antibody fragments results in failure of migration of granule cells from the molecular layer into the internal granular layer. Moreover, the interactions of cytotactin with other matrix proteins such as fibronectin and proteoglycan (Hoffman and Edelman 1987) may modulate which of the cytotactin domains dominate the functional effects of the binding of the molecule to the cell surface.

A striking correlation between cell rounding and inhibition of cell division has been found when fibroblasts are bound to cytotactin in culture (Crossin 1991). The effects on cell division appear to be related to the inability of cytotactin-treated cells to alter their pH to more alkaline levels. Such effects have previously been shown to be due to the action of the Na^+/H^+ antiporter. Despite their coincidence, however, the cell rounding and antimitogenic effects of cytotactin binding appear to be independent and parallel events (Crossin 1991).

Clearly, cytotactin that shows place-dependent expression like CAMs is capable of modulating the cell surface and of affecting primary cellular processes such as shape change, migration, and division with resulting changes in cellular patterns. The early observations (Yahara and Edelman 1973; Edelman 1976) on concanavalin A as a global modulator of the cell surface affecting cell shape, cytoskeletal events, and cell division thus appear to have a natural counterpart in cytotactin (Crossin 1991). It is, of course, not at all certain that these different molecules act in the same fashion, and a detailed biochemical comparison of the two would be useful and revealing. In view of cytotactin's global modulatory effects on primary processes such as movement, shape, and cell division, an understanding of the genes controlling its expression would be as significant as an understanding of CAM expression.

Signaling and the Gene Control of Morphoregulatory Molecule Expression

A number of experiments indicate that signaling events between different tissues can regulate the expression of morphoregulatory molecules. This is particularly evident in the nervous system. Cutting a nerve to a muscle with subsequent Wallerian degeneration leads, for example, to a striking reexpression of N-CAM in the affected muscle. At the same time, N-CAM and Ng-CAM levels are reduced in the ventral horn on the side of the affected spinal segment, and they are increased in the adjoining dorsal root ganglia. These changes return to normal patterns only 200 – 300 days later, after Wallerian regulation is complete (Daniloff et al. 1986). Recent experiments on *Aplysia* indicate that simple conditioning is followed by sprouting events that are preceded by a downregulation of an N-CAM-like molecule on the surface of the neurons involved (Bailey et al. 1992; Mayford et al. 1992).

Similar activity-dependent signaling is seen for cytotactin in the whisker barrels of the somatosensory cortex in rodents. Cytotactin is normally expressed in the boundary of each whisker barrel (Crossin et al. 1989; Steindler et al. 1989). Ablating a row of whiskers (with destruction of their follicles) leads to a rearrangement of the cortical distribution of cytotactin to follow the fusion of the adjoining barrels which results from removal of input stimuli (Crossin et al. 1989). Thus, the expression of cytotactin in barrel borders depends somehow on the activity of axons mapping into the barrel.

To govern the local place-dependent expression of CAMs and of substrate-adhesion molecules (SAMs) such as cytotactin, specific signals must affect their local synthesis. This focuses attention on the nature of the genes that regulate the expression of morphoregulatory molecules. The discovery of an effective set of control genes affecting the transcription of such molecules would help explain how particular cellular patterns are actually constrained during morphogenesis. A particularly promising set of candidate genes are the homeobox-containing genes, for these genes and the proteins they specify have been shown to determine regional tissue patterns in a number of species (Kessel and Gruss 1990). Although the initial observations were carried out in *Drosophila* (Lewis 1978), it has been found that, in vertebrate species, a number of *Hox* genes arranged in four genomic complexes have a similar role (McGinnis and Krumlauf 1992). One of the striking features of such genes is that their arrangement in the genome reflects their effects along a particular developmental axis: Genes located 3′ affect more anterior structures, whereas those located in a more 5′ position affect more posterior structures.

Recent analyses of the promoter regions of the gene encoding N-CAM (Chen et al. 1990; Hirsch et al. 1990) and cytotactin (Jones et al. 1990) have, in fact, uncovered putative sequences for the binding of proteins encoded by homeobox-containing genes. A search was therefore carried out for evidence (Jones et al. 1992a,b) that the expression of such genes can regulate the transcription of reporter gene constructs driven by the promoters of N-CAM and cytotactin.

As shown in Figure 4, it was found that two *Xenopus laevis* genes, *Hox-2.5* and *Hox-2.4*, which are adjacent to each other in the genomic complex, can differentially modulate mouse N-CAM promoter activity after cotransfection of appropriate constructs into NIH-3T3 cells. A sequence necessary for expression was found in a 47-bp segment contained in the N-CAM promoter. When linked to a minimal promoter and a chloramphenicol acetyltransferase (CAT) reporter gene sequence, this segment enhanced basal transcription in 3T3 cells. This activity was greatly increased by cotrans-

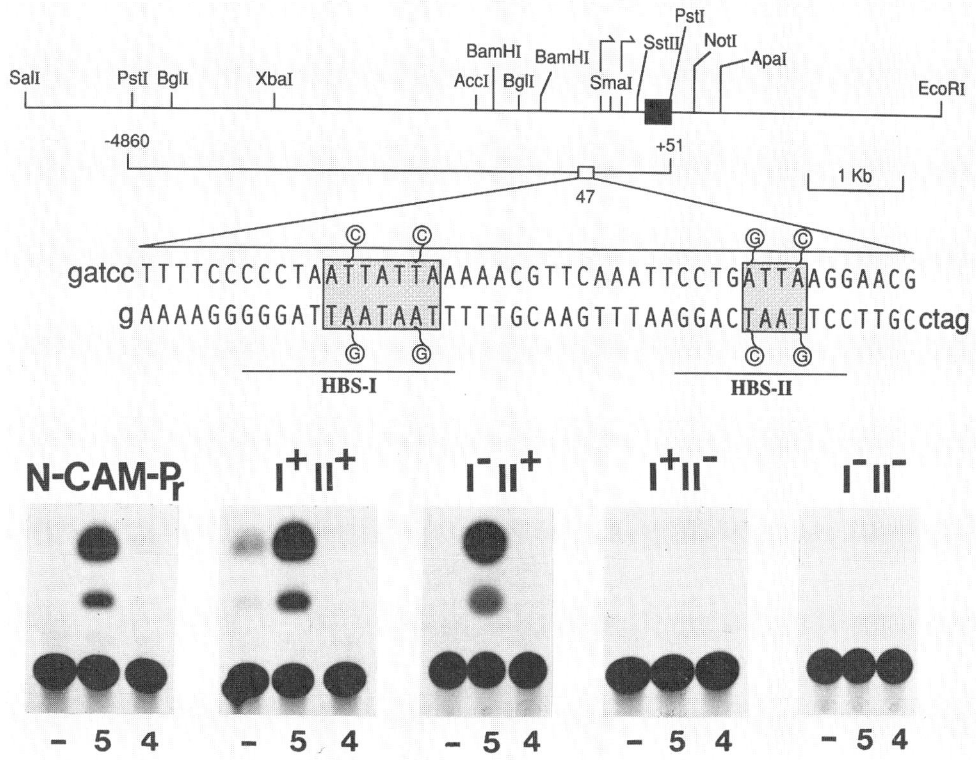

Figure 4. Cotransfection of NIH-3T3 cells with plasmids driving the expression of *Xenopus Hox-2.5* and *Hox-2.4* genes modulates the activity of a CAT reporter gene driven by the N-CAM promoter. A 4.9-kb *Pst*I fragment including the first exon (■) and the 4860 bp of 5′-flanking sequence from the mouse N-CAM gene was cloned upstream of a promoter-less CAT gene. This plasmid (N-CAM-Pr) gave no detectable activity when transfected without *Hox* gene expression plasmid (N-CAM-Pr, lane −). However, transfection with *Hox-2.5* activated CAT activity driven by the N-CAM promoter (N-CAM-Pr, lane *5*), whereas transfection with *Hox-2.4* had little or no effect (N-CAM-Pr, lane *4*). A 47-bp region (between 512 and 559 bp upstream of the ATG codon in the first exon) containing two potential homeodomain-binding sites (designated HBS-I and HBS-II) was cloned upstream of a minimal SV40 promoter and a downstream CAT gene. In addition, multiple base-pair substitutions were made in the TAAT sequence cores of HBS-I and HBS-II. These plasmids were tested in cotransfection experiments to determine whether the HBSs participate in the modulation of N-CAM promoter activity by *Hox* gene products and, if so, to determine whether specific HBSs mediated the response. The wild-type HBS sequence (HBS$I^{+}II^{+}$) had significant basal activity in NIH-3T3 cells. Cotransfection with *Hox-2.5* elevated expression from basal levels significantly, but cotransfection with *Hox-2.4* repressed the basal activity. Mutation of HBS-I (HBS-$I^{-}II^{+}$) had no effect on *Hox-2.5* activation. However, a mutation of either HBS-II (HBS-$I^{+}II^{-}$) or both HBSs (HBS-$I^{-}II^{-}$) completely abolished activation by *Hox-2.5*. These data suggest that *Hox-2.5* gene product activation of the N-CAM promoter is mediated by HBS-II and that *Hox-2.4* is a potent inhibitor of *Hox-2.5* activation.

fection with *Hox-2.5* and it was blocked by cotransfection by *Hox-2.4*. These results indicate that two neighboring *Hox* genes can differentially modulate transcription from the promoter of a CAM.

Similar cotransfection experiments using CAT reporter constructs and plasmids driving the expression of the mouse homeobox-containing gene, *Evx-1*, showed enhanced activity of the cytotactin promoter (Fig. 5). Deletion analysis localized the base sequences contributing to the activation to an 89-bp segment containing a phorbol 12-*O*-tetradecanoate-13-acetate response element (TRE/AP-1 element). Mutation of the TRE/AP-1 sequence abolished the ability of *Evx-1* to activate the promoter. The TRE/AP-1 element has been shown to be a target for transcription factors encoded by the *fos* and *jun* gene families. This raises the possibility that cytotactin promoter modulation by *Evx-1* may involve a growth factor signal transduction pathway. Cotransfection experiments (P. Gruss et al., unpubl.) with paired-box-containing genes (*Pax* genes) have also been shown to modulate the activity of both the N-CAM and cytotactin promoters, extending the observations to yet another family of homeobox-containing genes.

These findings with homeobox-containing genes, whose perturbation leads to regional anomalies (Kessel et al. 1990; LaMouellic et al. 1992), provide a clue to understanding the mechanism of place-dependent expression of morphoregulatory molecules. The results (Jones et al. 1992a,b) represent the first examples of

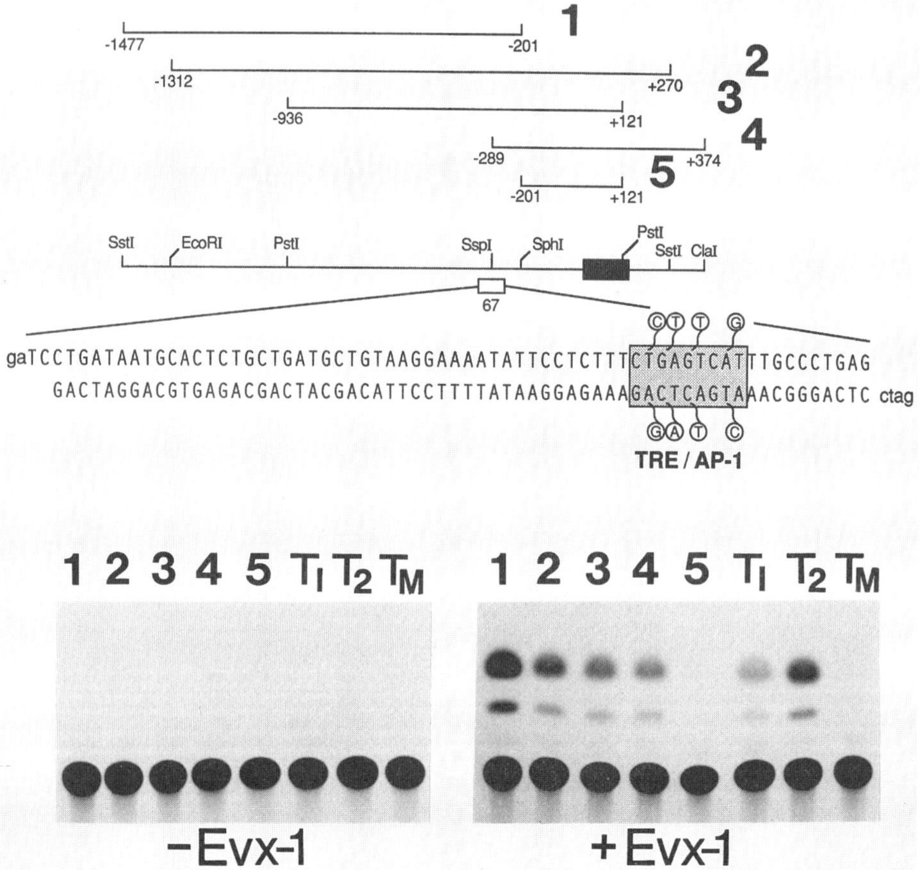

Figure 5. Sequences required for the activation of cytotactin promoter by a plasmid driving the expression of the homeobox-containing gene, *Evx-1*. Eight plasmids driving the expression of a CAT reporter gene were constructed and tested for their expression after cotransfection of NIH-3T3 cells with a plasmid driving the expression of *Evx-1*. A restriction map of part of the first intron, the first exon, and 1477 bp of 5′-flanking sequence of the chicken cytotactin gene is shown. Plasmids 1 – 5 (drawn above) containing the different segments of the promoter region shown were cloned upstream of a promoter-less CAT gene and tested for their activity in the absence (−*Evx-1*) or presence (+*Evx-1*) of a construct driving the expression of *Evx-1*. None of these reported constructs yielded detectable CAT expression without the added *Evx-1* plasmid (lanes *1 – 5*, −*Evx-1* panel). However, constructs 1 – 4 gave high levels of CAT activity, whereas construct 5 gave no activity (lanes *1 – 5*, + *Evx-1* panel). These data indicated that the region between −289 and −201 was required for *Evx-1* activation of the cytotactin promoter. This small segment of 89 bp was found to contain a TRE/AP-1 sequence (indicated by gray shaded box). Either one copy (T1) or two copies (T2) of a 67-bp segment from the cytotactin gene 5′-flanking sequence that included the TRE/AP-1 site were inserted upstream of an SV40 minimal promoter CAT gene cassette and tested for activation by *Evx-1*. Both constructs yielded significant CAT activity in the presence of cotransfected *Evx-1* expression plasmid. Mutation of the TRE/AP-1 element with multiple base-pair substitutions (indicated by circles) resulted in the loss of *Evx-1* activation. These data suggest that the TRE/AP-1 site is involved in the activation of the cytotactin promoter by *Evx-1*.

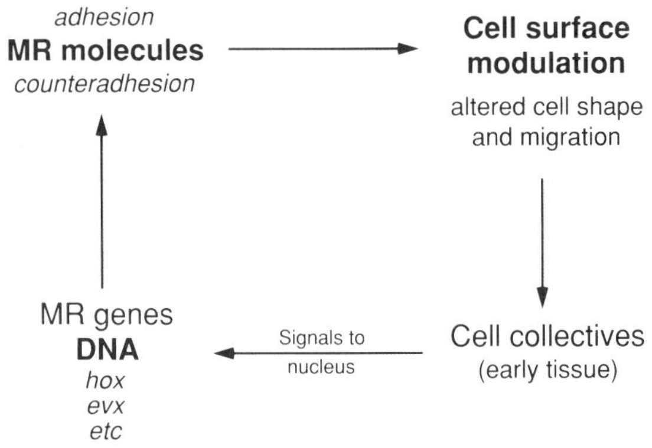

Figure 6. Hypothetical loop illustrating how place-dependent gene expression of homeobox-containing genes may lead to morphoregulatory (MR) molecule expression. This in turn leads to modulation of the cell surface and alteration of primary processes. The formation of cell collectives (by CAM and SAM interactions) changes nuclear signals for DNA expression, leading to expression of new MR molecules and new patterns. (Obviously, historegulatory genes are also involved: See Edelman 1992 for a description of the complete morphoregulator hypothesis.)

targets for these genes aside from other homeobox-containing genes. Of course, it will be necessary to confirm these observations in vivo before sound conclusions can be drawn. Nevertheless, if we assume that homeobox-containing genes, which are known to be expressed in local patterns, can affect the place-dependent expression of morphoregulatory molecules, we can begin to discern how the mechanochemical effects of these molecules might give rise to specific tissue patterns (Fig. 6).

Even if this hypothesis is confirmed, however, it will be necessary to show how the linkage of cells by particular combinations of CAMs in the presence of SAMs can lead to new signals to the nucleus to result in expression of different combinations of morphoregulatory molecules yielding new histology. Presumably, the specific alteration of signals depends on cell-cell interactions within a given particular cell collective and would be different for cells not so linked. Evidence for a role of CAMs and SAMs in such signaling is sparse (Doherty et al. 1991), but it is beginning to accumulate.

SUMMARY AND PROSPECT

The cell surface is obviously central to morphogenetic events. In 1976, I proposed (Edelman 1976) a series of hypotheses on the possible connections between transmembrane control of cell-surface receptors and the primary processes of development. The main suggestion was that interactions between the cytoskeleton and the cytoplasmic portions of particular receptors could act to transmit internal cellular states to the cell surface and that, reciprocally, the interactions of such receptors with various ligands could alter the state of the cytoskeleton and intracellular signals. The net result of such cell-surface modulation, as I called it, was to regulate cell-cell interaction and cell motion and shape and to contribute to the control of cell division.

As might be expected of that early stage in our knowledge of the cell surface, a few of my conjectures have turned out to be off the mark. But as I have hoped to illustrate briefly here, a number of the proposals of that paper have turned out to be fruitful, particularly those bearing on transmembrane control, on the primary processes of cell-cell adhesion, on cell shape changes, and on the guidance of cell migration by global counter-adhesive interactions. Further study of these various cell-surface interactions is likely to reveal a multilevel control system linking gene expression to cell-surface modulation. A major implication for morphogenesis is that, during development, these various events mediated by morphoregulatory molecules can lead to the selective formation of cell collectives. The resultant alterations in surface interactions at the boundaries of such collectives can lead to a series of new signals changing the particular combination of morphoregulatory molecules expressed at a given place. When the interactive components of this morphoregulatory system are more fully revealed, we shall be in a much better position to understand how tissue pattern and animal form arise and are inherited.

ACKNOWLEDGMENT

The work from the author's laboratory was supported by U.S. Public Health Service grants HD-09635, DK-04256, and AG-09326.

REFERENCES

Bailey, C.H., M. Chen, F. Keller, and E.R. Kandel. 1992. Serotonin-mediated endocytosis of apCAM: An early step of learning-related synaptic growth in *Aplysia*. *Science* **256:** 645.

Brackenbury, R., J.-P. Thiery, U. Rutishauser, and G.M. Edelman. 1977. Adhesion among neural cells of the chick embryo. I. An immunological assay for molecules involved in cell-cell binding. *J. Biol. Chem.* **252:** 6835.

Buskirk, D.R., J.-P. Thiery, U. Rutishauser, and G.M. Edelman. 1980. Antibodies to a neural cell adhesion molecule disrupt histogenesis in cultured chick retinae. *Nature* **285:** 488.

Chen, A., A. Reyes, and R. Akeson. 1990. Transcription initiation sites and structural organization of the extreme 5′ region of the rat neural cell adhesion molecule gene. *Mol. Cell. Biol.* **10:** 3314.

Chiquet, M. and D.M. Fambrough. 1984. Chick myotendinous antigen. I. A monoclonal antibody as a marker for tendon and muscle morphogenesis. *J. Cell Biol.* **98:** 1926.

Chiquet-Ehrismann, R., E.J. Mackie, C.A. Pearson, and T. Sakakura. 1986. Tenascin: An extracellular matrix protein involved in tissue interactions during fetal development and oncogenesis. *Cell* **47:** 131.

Chuong, C.-M., K.L. Crossin, and G.M. Edelman. 1987. Sequential expression and differential function of multiple adhesion molecules during the formation of cerebellar cortical layers. *J. Cell Biol.* **104:** 331.

Crossin, K.L. 1991. Cytotactin binding: Inhibition of stimulated proliferation and intracellular alkalinization in fibroblasts. *Proc. Natl. Acad. Sci.* **88:** 11403.

Crossin, K.L., C.-M. Chuong, and G.M. Edelman. 1985.

Expression sequences of cell adhesion molecules. *Proc. Natl. Acad. Sci.* **82:** 6942.

Crossin, K.L., S. Hoffman, S.-S. Tan, and G.M. Edelman. 1989. Cytotactin and its proteoglycan ligand mark structural and functional boundaries in somatosensory cortex of the early postnatal mouse. *Dev. Biol.* **136:** 381.

Cunningham, B.A., J.J. Hemperly, B.A. Murray, E.A. Prediger, R. Brackenbury, and G.M. Edelman. 1987. Neural cell adhesion molecule: Structure, immunoglobulin-like domains, cell surface modulation, and alternative RNA splicing. *Science* **236:** 799.

Daniloff, J.K., G. Levi, M. Grumet, F. Rieger, and G.M. Edelman. 1986. Altered expression of neuronal cell adhesion molecules induced by nerve injury and repair. *J. Cell Biol.* **103:** 929.

Doherty, P., S.V. Ashton, S.E. Moore, and F.S. Walsh. 1991. Morphoregulatory activities of N-CAM and N-cadherin can be accounted for by G protein-dependent activation of L- and N-type neuronal Ca^{2+} channels. *Cell* **67:** 21.

Edelman, G.M. 1976. Surface modulation in cell recognition and cell growth. *Science* **192:** 218.

———. 1988. *Topobiology: An introduction to molecular embryology.* Basic Books, New York.

———. 1992. Morphoregulation. *Dev. Dynam.* **193:** 2.

Edelman, G.M. and K.L. Crossin. 1991. Cell adhesion molecules: Implications for a molecular histology. *Annu. Rev. Biochem.* **60:** 155.

Edelman, G.M., B.A. Cunningham, and J.P. Thiery. 1990. *Morphoregulatory molecules.* John Wiley, New York.

Erickson, H.P. and J.L. Iglesias. 1984. A six-armed oligomer isolated from cell surface fibronectin preparations. *Nature* **311:** 267.

Friedlander, D.R., S. Hoffman, and G.M. Edelman. 1988. Functional mapping of cytotactin: Proteolytic fragments active in cell-substrate adhesion. *J. Cell Biol.* **107:** 2329.

Friedlander, D.R., R.-M. Mege, B.A. Cunningham, and G.M. Edelman. 1989. Cell sorting-out is modulated by both the specificity and amount of different cell adhesion molecules (CAMs) expressed on cell surfaces. *Proc. Natl. Acad. Sci.* **86:** 7043.

Gallin, W.J., G.M. Edelman, and B.A. Cunningham. 1983. Characterization of L-CAM, a major cell adhesion molecule from embryonic liver cells. *Proc. Natl. Acad. Sci.* **80:** 1038.

Gallin, W.J., B.C. Sorkin, G.M. Edelman, and B.A. Cunningham. 1987. Sequence analysis of a cDNA clone encoding the liver cell adhesion molecule, L-CAM. *Proc. Natl. Acad. Sci.* **84:** 2808.

Grumet, M. and G.M. Edelman. 1984. Heterotypic binding between neuronal membrane vesicles and glial cells is mediated by a specific cell adhesion molecule. *J. Cell Biol.* **98:** 1746.

Grumet, M., S. Hoffman, K.L. Crossin, and G.M. Edelman. 1985. Cytotactin, an extracellular matrix protein of neural and non-neural tissues that mediates glia-neuron interaction. *Proc. Natl. Acad. Sci.* **82:** 8075.

Hirsch, M.-R., L. Gaugler, H. Deagostini-Bazin, L. Bally-Cuif, and C. Goridis. 1990. Identification of positive and negative regulatory elements governing cell-type-specific expression of the neural cell adhesion molecule gene. *Mol. Cell. Biol.* **10:** 1959.

Hoffman, S. and G.M. Edelman. 1983. Kinetics of homophilic binding by E and A forms of the neural cell adhesion molecule. *Proc. Natl. Acad. Sci.* **80:** 5762.

———. 1987. A proteoglycan with HNK-1 antigenic determinants is a neuron-associated ligand for cytotactin. *Proc. Natl. Acad. Sci.* **84:** 2523.

Hoffman, S., D.R. Friedlander, C.-M. Chuong, M. Grumet, and G.M. Edelman. 1986. Differential contributions of Ng-CAM and N-CAM to cell adhesion in different neural regions. *J. Cell Biol.* **103:** 145.

Jaffe, S.H., D.R. Friedlander, F. Matsuzaki, K.L. Crossin, B.A. Cunningham, and G.M. Edelman. 1990. Differential effects of the cytoplasmic domains of cell adhesion molecules on cell aggregation and sorting-out. *Proc. Natl. Acad. Sci.* **87:** 3589.

Jones, F.S., G. Chalepakis, P. Gruss, and G.M. Edelman. 1992a. Activation of the cytotactin promoter by the homeobox-containing gene, Evx-1. *Proc. Natl. Acad. Sci.* **89:** 2091.

Jones, F.S., K.L. Crossin, B.A. Cunningham, and G.M. Edelman. 1990. Identification and characterization of the promoter for the cytotactin gene. *Proc. Natl. Acad. Sci.* **87:** 6497.

Jones, F.S., S. Hoffman, B.A. Cunningham, and G.M. Edelman. 1989. A detailed structural model of cytotactin: Protein homologies, alternative RNA splicing, and binding regions. *Proc. Natl. Acad. Sci.* **86:** 1905.

Jones, F.S., E.A. Prediger, D.A. Bittner, E.M. DeRobertis, and G.M. Edelman. 1992b. Cell adhesion molecules as targets for hox genes: N-CAM promoter activity is modulated by co-transfection with Hox 2.5 and 2.4. *Proc. Natl. Acad. Sci.* **89** 2086.

Jones, F.S., M.P. Burgoon, S. Hoffman, K.L. Crossin, B.A. Cunningham, and G.M. Edelman. 1988. A cDNA clone for cytotactin contains sequences similar to epidermal growth factor-like repeats and segments of fibronectin and fibrinogen. *Proc. Natl. Acad. Sci.* **85:** 2186.

Kessel, M. and P. Gruss. 1990. Murine developmental control genes. *Science* **249:** 374.

Kessel, M., R. Balling, and P. Gruss. 1990. Variation of cervical vertebrae after expression of a Hox-1.1 transgene in mice. *Cell* **61:** 301.

Kintner, C. 1988. Effects of altered expression of the neural cell adhesion molecule, N-CAM, on early development in *Xenopus* embryo. *Neuron* **1:** 545.

Kruse, J., G. Keilhauer, A. Faissner, R. Timpl, and M. Schachner. 1985. The J1 glycoprotein: A novel nervous system cell adhesion molecule of the L2/HNK-1 family. *Nature* **316:** 146.

LeMouellic, H., Y. Lallemand, and P. Brulet. 1992. Homeosis in the mouse induced by a null mutation in the *Hox-3.1* gene. *Cell* **69:** 251.

Lewis, E.B. 1978. A gene complex controlling segmentation in *Drosophila. Nature* **276:** 565.

Lochter, A., L. Vaughn, A. Kaplony, A. Prochiantz, M. Schachner, and A. Faissner. 1991. J1/tenascin in substrate-bound and soluble form displays contrary effects on neurite outgrowth. *J. Cell Biol.* **113:** 1159.

Marlin, S.D. and T.A. Springer. 1987. Purified intercellular adhesion molecule-1 (ICAM-1) is a ligand for lymphocyte function-associated antigen 1 (LFA-1). *Cell* **51:** 813.

Mayford, M., A. Barzilai, F. Keller, S. Schacher, and E.R. Kandel. 1992. Modulation of an N-CAM-related adhesion molecule with long-term synaptic plasticity in *Aplysia. Science* **256:** 638.

McGinnis, W. and R. Krumlauf. 1992. Homeobox genes and axial patterning. *Cell* **68:** 283.

Nagafuchi, A. and M. Takeichi. 1988. Cell binding function of E-cadherin is regulated by the cytoplasmic domain. *EMBO J.* **7:** 3679.

———. 1989. Transmembrane control of cadherin-mediated cell adhesion: A 94 kDa protein functionally associated with a specific region of the cytoplasmic domain of E-cadherin. *Cell Regul.* **1:** 37.

Ozawa, M., H. Baribault, and R. Kemler. 1989. The cytoplasmic domain of the cell adhesion molecule uvomorulin associates with three independent proteins structurally related in different species. *EMBO J.* **8:** 1711.

Prieto, A.L., C. Andersson-Fisone, and K.L. Crossin. 1992. Characterization of multiple adhesive and counteradhesive domains. in the extracellular matrix protein cytotactin. *J. Cell Biol.* (in press).

Spring, J., K. Beck, and R. Chiquet-Ehrismann. 1989. Two contrary functions of tenascin: Dissection of the active sites by recombinant tenascin fragment. *Cell* **59:** 325

Steindler, D.A., N.G.F. Cooper, A. Faissner, and M. Schachner. 1989. Boundaries defined by adhesion molecules during development of the cerebral cortex: The J1/tenascin glycoprotein in the mouse somatosensory cortical barrel field. *Dev. Biol.* **131:** 243.

Takeichi, M. 1990. Cadherins: A molecular family important in selective cell-cell adhesion. *Annu. Rev. Biochem.* **59:** 237.

———. 1991. Cadherin cell adhesion receptors as a morphogenetic regulator. *Science* **251:** 1451.

Takeichi, M., T. Atsumi, C. Yoshida, K. Uno, and T.S. Okada. 1981. Selective adhesion of embryonal carcinoma cells and differentiated cells by Ca^{2+}-dependent sites. *Dev. Biol.* **87:** 340.

Thiery, J.-P., R. Brackenbury, U. Rutishauser, and G.M. Edelman. 1977. Adhesion among neural cells of the chick embryo. II. Purification and characterization of a cell adhesion molecule from neural retina. *J. Biol. Chem.* **252:** 6841.

Wehrle, B. and M. Chiquet. 1990. Tenascin is accumulated along developing peripheral nerves and allows neurite outgrowth in vitro. *Development* **110:** 401.

Yahara, I. and G.M. Edelman. 1973. The effects of concanavalin A on the mobility of lymphocyte surface receptors. *Exp. Cell Res.* **81:** 143.

Cytoplasmic Control of Cadherin-mediated Cell-Cell Adhesion

M. Takeichi, S. Hirano, N. Matsuyoshi, and T. Fujimori
Department of Biophysics, Faculty of Science, Kyoto University, Kyoto 606, Japan

Cadherins constitute a multigene family of Ca^{++}-dependent cell-cell adhesion molecules with more than 15 members (Magee and Buxton 1991; Takeichi 1991). A role of the members of this molecular family is to bind cells very tightly through homophilic interactions. If cadherins are absent, cells can disperse freely. Many members of the family preferentially interact with the same molecules. As a result, cells selectively adhere to those expressing identical cadherin types (Nose et al. 1988; Friedlander et al. 1989; Miyatani et al. 1989). For this property, cadherins are thought to be important in specific cell-cell adhesion.

Cadherins are not simple adhesive substances such as glues, but their function is dynamically controlled by the cells. Cadherins are transmembrane proteins, and their carboxyl terminums is associated with cytoplasmic proteins, called α, β, and γ catenins (Ozawa et al. 1989). If portions of the cytoplasmic domain are deleted, these mutated cadherin molecules lose the ability to associate with the catenins and coincidentally lose their cell-binding function, although they can be expressed on the cell surface (Nagafuchi and Takeichi 1988, 1989; Ozawa et al. 1990; Ozawa and Kemler 1992). This finding indicated that the expression of cadherins on the cell surface is not sufficient for them to function, their activities requiring some support from the cytoplasm.

Cadherin-catenin complexes can be solubilized from the cell membrane by treatment with nonionic detergents. However, they tend to be resistant to such extraction at cell-cell junctions (Nagafuchi and Takeichi 1988; Nelson et al. 1990), suggesting that the cadherin-catenin complexes are tightly associated with cytoskeletal structures when they are in operation for cell-cell adhesion. Ultrastructural analyses revealed that cadherins are concentrated in a particular class of cell-cell junction, called zonula adherens, belt desmosome, intermediate junction, or intercellular adherens junction (Geiger and Ginsberg 1991). These junctions bear electron-dense cytoplasmic plaques that comprise a number of cytoskeletal proteins, vinculin, plakoglobin, β-actinin, radixin, actin filaments, and others (Geiger and Ginsberg 1991), and they also contain *src* family tyrosine kinases (Tsukita et al. 1991). The cadherin-catenin complexes are assumed to be interacting with these proteins directly or indirectly in the cell-cell junctions, and these interactions might be crucial for cadherins to function as cell adhesion molecules.

Recent molecular clonings revealed the primary structures of α-catenin and β-catenin. α-Catenin was found to be similar in part to vinculin (Herrenknecht et al. 1991; Nagafuchi et al. 1991), and β-catenin resembles plakoglobin or the *Drosophila armadillo* product (McCrea et al. 1991). As anti-*armadillo* protein antibodies can cross-react with β-catenin, it is thought that they are homologs (Peifer et al. 1992). Although we have accumulated such information on the structures of catenins, little is known about their function. In this paper, we describe our recent molecular and cytological analyses of the roles of catenins in cadherin function and discuss how cadherin-dependent cell-cell adhesion is regulated by the cytoplasmic proteins.

Identification of a Neural α-Catenin Subtype

α-Catenin was identified as a protein associated with E-cadherin, which is generally involved in epithelial cell-cell adhesion (Herrenknecht et al. 1991; Nagafuchi et al. 1991). We recently examined whether N-cadherin, which is expressed in various organs including the brain, is associated with any other specific catenins (Hirano et al. 1992). As an initial step of such study, we isolated N-cadherin–catenin complexes from a detergent extract of chicken embryonic brains and analyzed them by electrophoresis. At least two proteins coprecipitated with N-cadherin, and similar observations were made by other groups (Wheelock and Knudsen 1991; Ozawa and Kemler 1992). To generate monoclonal antibodies to the N-cadherin-associated proteins, we immunized rats with these molecular complexes and produced hybridomas. One monoclonal antibody NCAP1, thus obtained recognized two bands of 102 kD and 113 kD in immunoblots of cell lysates. We then screened chicken brain cDNA libraries with this antibody for cDNA clones of the antigens. Sequencing of the positive cDNA clones revealed that they encode a protein that is similar to, but not identical with, α-catenin, indicating that it is a subtype of α-catenin (Fig. 1). This novel protein was also similar to vinculin, as was α-catenin. Since this α-catenin-like molecule was identified as an N-cadherin-associated protein, we designated it αN-catenin. To distinguish this protein from the original α-catenin, we call the latter αE-catenin.

We next studied the tissue distribution of αN-catenin and compared it with that of αE-catenin. Previous immunoblot analysis demonstrated that αE-catenin is

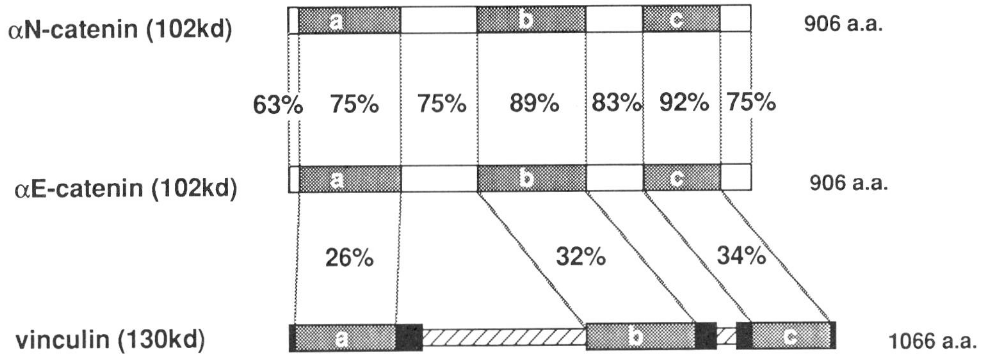

Figure 1. Schematic drawings of α-catenins and vinculin. Values represent the percentage of identities in amino acid sequences in the corresponding regions.

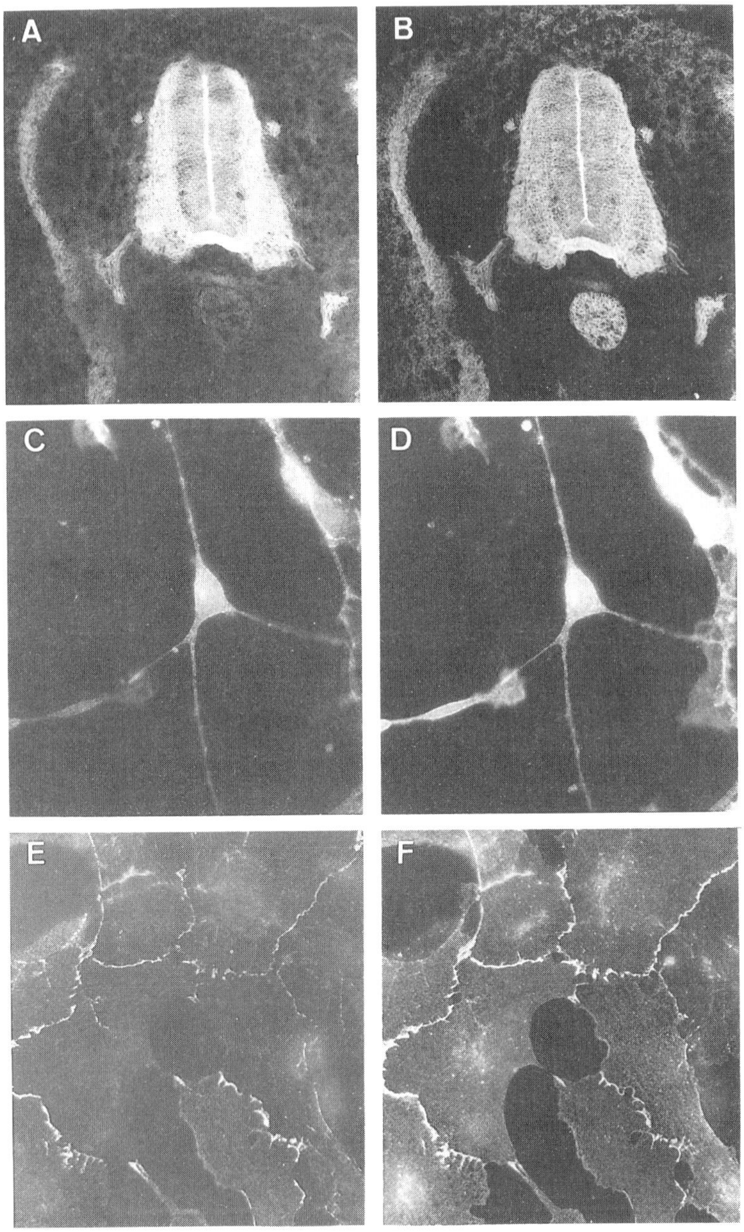

Figure 2. Double-immunofluorescence staining for αN-catenin and N-cadherin. (*A*, *B*) A transverse section of 10-day chicken embryo at the trunk region stained for αN-catenin (*A*) and N-cadherin (*B*). Note that the neural tube, myotome, and ventral roots are positive for both molecules, but the notochord stains only for N-cadherin. (*C–F*) A monolayer culture of 10-day embryonic brain cells stained for αN-catenin (*C*, *E*) and N-cadherin (*D*, *F*). (*C*, *D*) Neurons; (*E*, *F*) glias. Note the colocalization of the two molecules.

distributed in all the tissues examined, although its expression level was relatively low in the brain (Nagafuchi et al. 1991). In contrast, αN-catenin expression occurred predominantly in the brain, although many other tissues were not completely negative, suggesting that the tissue distributions of the two α-catenins are somehow complementary. Comparison of this pattern with that of N-cadherin distribution demonstrated that they were similar to each other with some exceptions. For example, although the heart showed a strong expression of N-cadherin, it did not express αN-catenin. We also immunostained sections of embryos and found that the neural tube and myotome expressed both N-cadherin and αN-catenin but the notochord expressed only αN-catenin (Fig. 2A,B). Such differential distributions of N-cadherin and αN-catenin were also found in other tissues. Thus, although αN-catenin generally colocalized with N-cadherin, there were other combinations.

To identify cell types expressing αN-catenin in the brain, cells of 10-day-old embryonic brains were cultured and stained for this molecule as well as for N-cadherin. Both neurons and glia expressed these molecules (Fig. 2C–F). They were distributed over the entire neuron surface and were concentrated at cell-cell junctions in glia.

Transfection of an α-Catenin-deficient Tumor Cell Line with αN-Catenin cDNA

To investigate the role of α-catenins, cell lines lacking these molecules should be useful, if they are available. An ideal cell line was recently found, the lung carcinoma PC9 (Kinjo et al. 1979). Cells of this line do

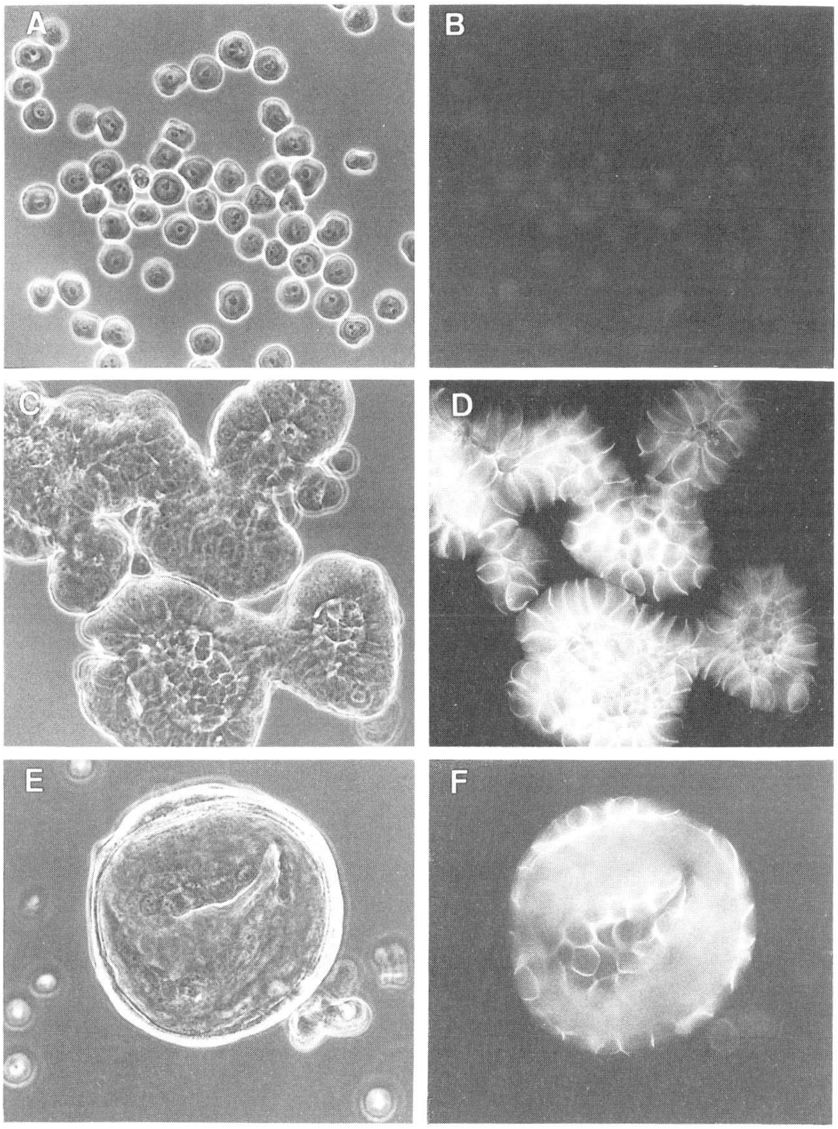

Figure 3. Effect of transfection of lung carcinoma PC9 cells with αN-catenin cDNA. (*A,B*) Parental PC9 cells that are deficient in α-catenins. These cells express E-cadherin and β-catenin. (*C–F*) Two examples of PC9 cell lines transfected with αN-catenin cDNA, displaying epithelial cell arrangements. Cells in *E* show a cystic form. This cyst was spherical, but part of the surface was dented during fixation of the sample. (*B,D,F*) Immunofluorescence staining for αN-catenin on the corresponding panels.

not express α-catenin, but they do express E-cadherin and β-catenin (Shimoyama et al. 1992). Interestingly, these cells grow as isolated cells or very loose aggregates in suspension and do not tightly adhere to each other (Fig. 3A). This adhesive property is unusual for cells expressing a cadherin, since such cells generally aggregate tightly. We therefore tested the possibility that the lack of α-catenin might have generated this unusual phenotype, by introducing exogenous αN-catenin cDNA into PC9 cells (Hirano et al. 1992).

Prior to this analysis, we tested whether αN-cadherin could bind with E-cadherin. L cells previously transfected with E-cadherin or N-cadherin were now transfected with αN-catenin cDNA. E-cadherin or N-cadherin from these cells was immunoprecipitated to analyze whether αN-catenin coprecipitated with these molecules. The results showed that αN-catenin was detected in both E-cadherin and N-cadherin immunoprecipitates. Thus, αN-catenin can bind with both of these cadherins.

When PC9 cells were transfected with αN-catenin cDNA, a striking change was observed in their association pattern. The transfected PC9 cells firmly adhered to each other forming compacted aggregates (Fig. 3C, E). Interestingly, many of these aggregates exhibited epithelium-like monolayer structures. In a most remarkable case, the cells formed cystic spheres, which resemble blastocysts, each composed of a shell of monolayer cell sheet and a fully expanded lumen. Such structures suggest that these cells have a polarity to transport substances in one direction, which is a key phenotype of the epithelia. This was rather surprising, since parental PC9 cells are completely round and isolated and do not exhibit such polarized properties. Considering that PC9 is of epithelial origin, they might have regained the original epithelial phenotype by expressing αN-catenin.

These transfected cell aggregates could be dissociated into single cells when they were incubated with anti-E-cadherin antibodies, indicating that the cadherin system is responsible for the formation of these structures. Thus, introducing only one species of cytoplasmic proteins into nonaggregating PC9 cells led them to undergo a dramatic change in their ability to form multicellular structures. These results suggest that αN-catenin is crucial not only for operation of the cadherin adhesion system, but also for designing an epithelium-type arrangement of cells. How can αN-catenin control the formation of such epithelial structures? Perhaps the E-cadherin–αN-catenin complexes interact with the cytoskeleton so as to induce a polarized distribution of themselves and then connect cells in a two-dimensional fashion. The mechanisms for such molecular interactions, however, remain to be elucidated.

Although we introduced αN-catenin into PC9 cells, their original αN-catenin might have been the αE type because of their epithelial origin (Kinjo et al. 1979). Probably both αE-catenin and αN-catenin can support E-cadherin function. We then have to question why these α-catenins have different tissue distributions. αN-catenin is predominantly expressed in the nervous system so that it might play some tissue-specific roles in neural cell-cell adhesion. To define such possible roles, we would need more specialized assay systems using neural cells, but not carcinoma cells.

Perturbation of Cadherin Action by v-*src*-mediated Tyrosine Phosphorylation

The function of adherens junctions, where cadherins are concentrated, might be linked with tyrosine phosphorylation. This was suggested by the findings that cell-cell junctions are phosphorylated at tyrosine residues (Maher et al. 1985; Maher and Pasquale 1988; Takata and Singer 1988) and that the isolated adherens junctions contain *src* family tyrosine kinases (Tsukita et al. 1991). Such kinases might be involved in signal transduction through the cell-cell junctions or in regulation of cell-cell adhesion. We examined the latter possibility by studying the cell-cell adhesion of v-*src*-transformed cells (Matsuyoshi et al. 1992).

The normal rat fibroblast line 3Y1 expresses P-cadherin. Both 3Y1 cells and 3Y1 cells transformed with v-*src* express the same amount of P-cadherin and can use this molecule for aggregation in suspension cultures. However, the aggregates of normal and v-*src*-transformed 3Y1 cells were morphologically distinct. Normal cells formed aggregates in which cells are tightly packed, showing a "compacted" appearance (Fig. 4A). Such compaction of cell aggregates generally depends on the action of cadherins, since anti-cadherin antibodies can inhibit it. In fact, anti-P-cadherin antibodies induced decompaction in 3Y1 cell aggregates. The v-*src*-transformed 3Y1 cells, however, failed to make such compacted aggregates. They formed loose aggregates in which individual cells remained round (Fig. 4B). This morphology resembled that of the normal 3Y1 aggregates whose P-cadherin was inhibited by antibodies. It thus seems that cadherins cannot normally function for compacting aggregates under v-*src* transformation, their functions probably being somehow modulated.

When these cells were cultured in collagen gel matrices at low densities, further remarkable differences were observed. Individual normal 3Y1 cells grew into dendritic colonies in which cells were firmly connected to each other (Fig. 4D). However, the transformed 3Y1 cells formed looser colonies, and they sometimes detached themselves from the main body of the colonies and freely migrated (Fig. 4E). Thus, in v-*src*-transformed 3Y1 cells, cadherins could not stably maintain cell-cell adhesion.

To test whether tyrosine phosphorylation mediated by v-*src* is directly involved in the above phenomenon, we examined the effect of a tyrosine kinase inhibitor, herbimycin A. When the aggregates of v-*src*-transformed 3Y1 cells were incubated with this antibiotic, they showed typical cadherin-dependent compaction (Fig. 4C). Thus, the abnormal adhesiveness of v-*src*-

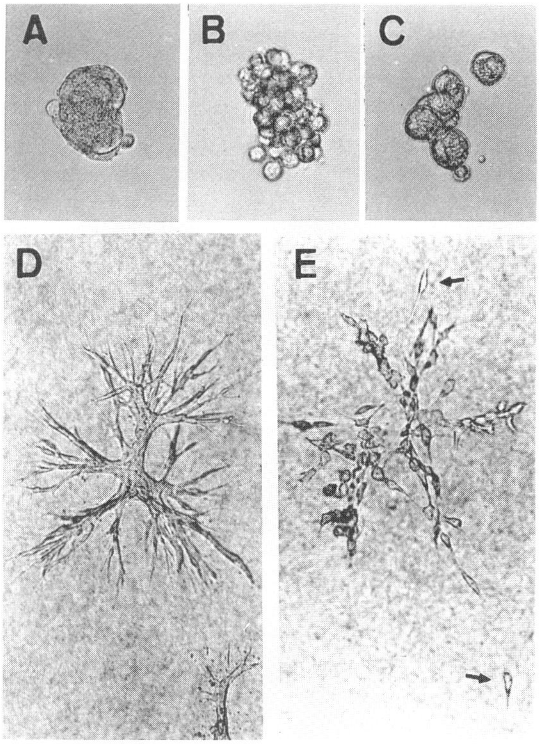

Figure 4. Aggregates of normal and v-*src*-transformed 3Y1 cells. (*A–C*) Cells aggregated in suspension cultures for 24 hr. (*A*) Normal cells; (B) v-*src*-transformed cells; (*C*) v-*src*-transformed 3Y1 cells in the presence of 0.2 μg of herbimycin A. (*D, E*) Colonies of normal (*D*) and v-*src*-transformed (*E*) 3Y1 cells grown in collagen gels. Arrows indicate cells which detached themselves from the colony.

transformed 3Y1 cells is apparently linked with enhanced tyrosine phosphorylation. We also tested the effect of a tyrosine phosphatase inhibitor, vanadate. The treatment of cells with this reagent is known to raise the level of tyrosine phosphorylation. When v-*src*-transformed 3Y1 cells were treated with vanadate, their cadherin-mediated aggregation was strongly inhibited, but the normal cells showed only a small response (Matsuyoshi et al. 1992). This result, along with the results of herbimycin A, suggests that increased tyrosine phosphorylation interferes with the cadherin adhesion system.

The uncompacted morphology of v-*src*-transformed 3Y1 cells is reminiscent of that of early mouse embryos. Early mouse embryos are composed of round blastomeres up to the early eight-cell stage and then undergo compaction at the late eight-cell stage. E-cadherin (uvomorulin) is responsible for this compaction it does not occur in the presence of anti-E-cadherin antibodies. However, the expression of E-cadherin is not sufficient for inducing the compaction. This adhesion molecule is expressed from the one-cell stage, but embryos cannot use it for compaction until the later stages. It is thought that some factors or physiological conditions are required for inducing E-cadherin-dependent compaction, which would appear only in later developmental stages. It was suggested that the activation of protein kinase C may be necessary for compaction (Winkel et al. 1990). Similar arguments would be possible for the 3Y1 system. The cadherin system may not be entirely functional when some particular components involved in this system are phosphorylated at tyrosine residues.

In attempts to search for molecules whose tyrosine phosphorylation affects cadherin function, we examined whether cadherins or catenins themselves are phosphorylated at tyrosine residues and in fact found that β-catenin is tyrosine-phosphorylated in v-*src*-transformed 3Y1 cells, whereas this molecule is little phosphorylated in normal 3Y1 cells (Fig. 5). Cadherins were also weakly tyrosine-phosphorylated only in the transformed cells. The phosphorylation of these proteins was, however, observed only when cells were preincubated with vanadate, suggesting that the phosphorylation accompanies rapid dephosphorylation under physiological conditions. We also found that the phosphorylation of β-catenin and cadherins at tyrosine residues is completely inhibited by herbimycin A. Thus, there was some correlation between the perturbation of cadherin action and the tyrosine phosphorylation of β-catenin and cadherins in the 3Y1 system.

It was reported that adherens junctions disappear when cells are transformed by v-*src* (Warren and Nelson 1987; Volberg et al. 1991). These observations suggest that cadherins cannot assemble to form particular junctions and distribute diffusely in the cell membrane of v-*src*-transformed cells. Possibly, individual

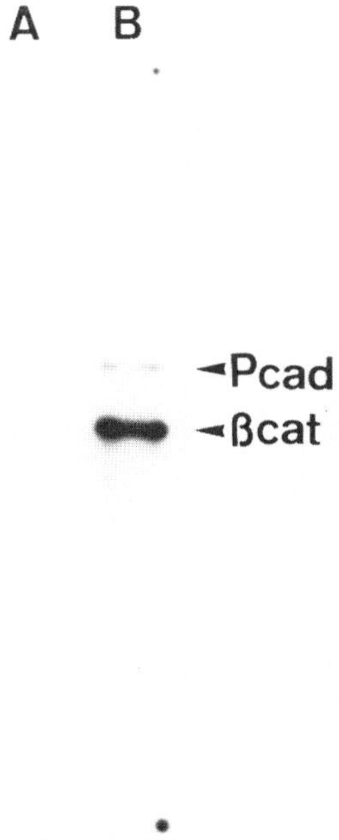

Figure 5. Immunoblot detection of phosphotyrosine in the cadherin-catenin complexes. P-cadherin was immunoprecipitated from normal (*A*) and v-*src*-transformed (*B*) 3Y1 cells, and the samples were blotted. Phosphotyrosine was then detected using anti-phosphotyrosine monoclonal antibody PY20. Cells were pretreated with 1 mM vanadate and 3 mM hydrogen peroxide for 20 min before lysis. (Pcad) P-cadherin; (βcat) β-catenin.

cadherin molecules have only low homophilic binding affinity, so that they might require coaggregation in the cell membrane to generate sufficient cell-cell binding energy. Such molecular assembly of cadherins might be prevented by v-*src*-mediated tyrosine phosphorylation, leading to relatively unstable cell-cell connections. Whether the tyrosine phosphorylation of β-catenin and cadherin is directly involved in such processes remains to be resolved.

Although the above findings were made using v-*src*-transformed cells, similar modulations of cadherin function may take place even in normal cells for the sake of physiological regulation of cell adhesion. As mentioned above, *src* family kinases are localized in cell-cell junctions of normal cells. One role of these enzymes might be to modulate cadherin-dependent adhesion. Cell-cell adhesion is probably a dynamic process, including detachment and readhesion of cells, and therefore ought to require fine regulatory mechanisms. Tyrosine kinases could be components of such mechanisms.

Inhibition of Epithelial Cell-Cell Adhesion by Overexpression of the N-Cadherin Cytoplasmic Domain

The experiments using PC9 cells described above revealed that α-catenin is a key cytoplasmic component for cadherin function. It is, however, still unknown whether other factors are also essential for the cadherin adhesion system. The cytoplasmic domains of cadherins are highly conserved in this family through the entire length, and the catenin-binding sites comprise only portions of them (Nagafuchi and Takeichi 1988, 1989; Ozawa et al. 1990). This raises the possibility that cytoplasmic factors other than the identified catenins are interacting with cadherins. Recently, Kintner (1992) found an interesting phenomenon: When *Xenopus* embryos were injected with the cytoplasmic domain of N-cadherin deleted in various positions, their tissues were dissociated. This finding suggests that the overexpressed cytoplasmic domains compete with endogenous cadherin molecules for cytoplasmic factors, leading to the inhibition of function of the endogenous cadherin molecules. Surprisingly, even the regions of the cytoplasmic domain that are not essential for catenin binding were also effective, suggesting the presence of some factors interacting with those regions.

We also carried out similar experiments using a keratinocyte line, PAM212, expressing E-cadherin and P-cadherin (T. Fujimori and M. Takeichi, in prep.). Cells of this line were transfected with cDNA encoding N-cadherin whose extracellular domain is largely deleted. This molecule was not functional for adhesion but could be inserted into the cell membrane. When we examined the morphology of the transfectants obtained, their cell-cell adhesions were found to be disrupted (Fig. 6). Cell aggregation experiments confirmed that the cadherin system itself was inhibited in these transfectants. This phenomenon was strictly correlated with the expression of the mutated N-cadherin, suggesting that the introduced cytoplasmic domains are competing with the endogenous E-cadherin or P-cadherin for some cytoplasmic components. Interestingly, we found that the endogenous E-cadherin in the transfectants coprecipitated with the same amount of α-catenin as that in untransfected cells. This finding suggests that depletion of α-catenin was not the main cause of the observed inhibition of the cadherin system. This in vitro system may thus be very useful for further study of the roles of cytoplasmic molecules in cadherin function.

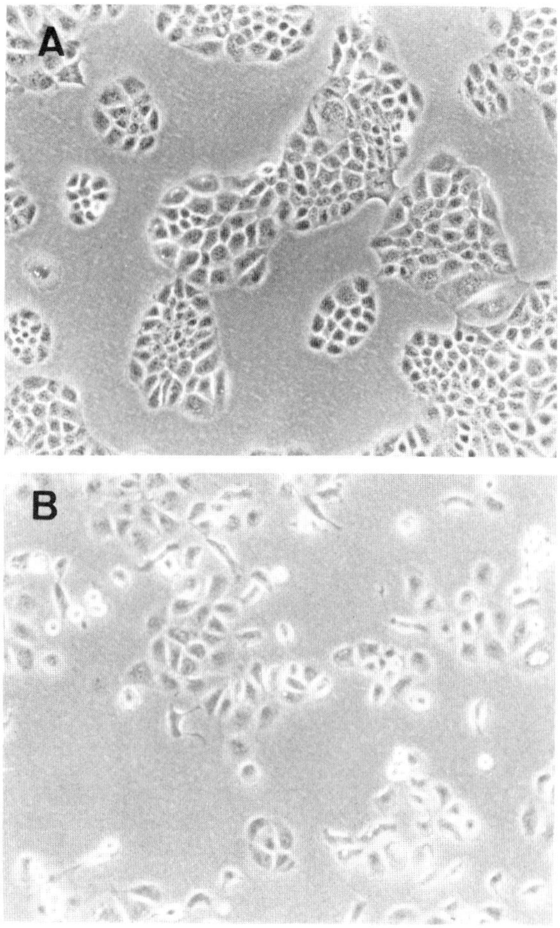

Figure 6. Effect of overexpression of the cytoplasmic domain of N-cadherin in PAM212 cells. (*A*) Parental cells; (*B*) a transfectant line. Note the disruption of cell-cell adhesion in *B*.

CONCLUDING REMARKS

As stated above, cell-cell adhesion is thought to be dynamic. For example, when cells enter mitosis, they round up by breaking their adhesion to neighboring cells and reattach to them after the completion of mitosis. In most tissues, there is constant relocation of cells responding to their turnover. During development, cell adhesion is more dynamic; morphogenetic movement generally accompanies disruption of preexisting cell-cell adhesions and establishment of new cell-cell connections. Such processes must require active regulation of cell adhesion and detachment. If cell-cell adhesion depended on purely physicochemical molecular interactions, such regulation would not be possible. The present studies demonstrated that cadherins, which are most important cell-cell bonds, require a group of cytoplasmic proteins for their function. This finding implies that these proteins are used for the regulation of cell-cell adhesion via modulation of cadherin activity. Tyrosine phosphorylation of catenins could be one course of action for such regulation.

Disrupting the regulatory mechanism of cadherin function may lead to abnormal or uncontrolled adhesion of cells. If this occurs in cancer cells, it might facilitate their invasion or metastasis because cells could more easily break down their mutual connections to move away from the primary tumor sites under such conditions. In fact, we showed in this work two examples of tumorigenic cells whose cadherin system is abnormal. In one case, cells were deficient in α-catenin, and, in the other case, v-*src*-mediated tyrosine phosphorylation partly suppressed cadherin action. In many human cancers, cells expressing cadherins do metastasize (Shimoyama and Hirohashi 1991). It is possible that such tumor cells might have some defects in the cadherin system at least temporally. Further analyses of the mechanisms of cytoplasmic control of cadherin action would thus be important to understand the molecular basis not only of normal morphogenetic behaviors of cells, but also of malignant cell behaviors that might be relevant to metastasis.

ACKNOWLEDGMENT

Original work described in this manuscript was supported by research grants from the Minister of Education, Science, and Culture of Japan.

REFERENCES

Friedlander, D.R., R.M. Mege, B.A. Cunningham, and G.M. Edelman. 1989. Cell sorting-out is modulated by both the specificity and amount of different cell adhesion molecules (CAMs) expressed on cell surfaces. *Proc. Natl. Acad. Sci.* **86:** 7043.

Geiger, B. and D. Ginsberg. 1991. The cytoplasmic domain of adherens-type junctions. *Cell Motil. Cytoskel.* **20:** 1.

Herrenknecht, K., M. Ozawa, C. Eckerskorn, F. Lottspeich, M. Lenter, and R. Kemler. 1991. The uvomorulin-anchorage protein α catenin is a vinculin homologue. *Proc. Natl. Acad. Sci.* **88:** 9156.

Hirano, S., N. Kimoto, Y. Shimoyama, S. Hirohashi, and M. Takeichi. 1992. Identification of a neural α-catenin as a key regulator of cadherin function and multicellular organization. *Cell* **70:** 293.

Kinjo, M., K. Oka, S. Naito, S. Kohga, K. Tanaka, S. Oboshi, Y. Hayata, and K. Yasumoto. 1979. Thromboplastic and fibrinolytic activities of cultured human cancer cell lines. *Br. J. Cancer* **39:** 2.

Kintner, C. 1992. Regulation of embryonic cell adhesion by the cadherin cytoplasmic domain. *Cell* **69:** 225.

Magee, A.I. and R.S. Buxton. 1991. Transmembrane molecular assemblies regulated by the greater cadherin family. *Curr. Opinion Cell Biol.* **3:** 854.

Maher, P.A. and E.B. Pasquale. 1988. Tyrosine phosphorylated proteins in different tissues during chick embryo development. *J. Cell Biol.* **106:** 1747.

Maher, P.A., E.B. Pasquale, J.Y.J. Wang, and S.J. Singer. 1985. Phosphotyrosine-containing proteins are concentrated in focal adhesions and intercellular junctions in normal cells. *Proc. Natl. Acad. Sci.* **82:** 6576.

Matsuyoshi, N., M. Hamaguchi, S. Taniguchi, A. Nagafuchi, S. Tsukita, and M. Takeichi. 1992. Cadherin-mediated cell-cell adhesion is perturbed by v-*src* tyrosine phosphorylation in metastatic fibroblasts. *J. Cell Biol.* **118:** 703.

McCrea, P.D., C.W. Turck, and B. Gumbiner. 1991. A homolog of the armadillo protein in *Drosophila* (plakoglobin) associated with E-cadherin. *Science* **254:** 1359.

Miyatani, S., K. Shimamura, M. Hatta, A. Nagafuchi, A. Nose, M. Matsunaga, K. Hatta, and M. Takeichi. 1989. Neural cadherin: Role in selective cell-cell adhesion. *Science* **245:** 631.

Nagafuchi, A. and M. Takeichi. 1988. Cell binding function of E-cadherin is regulated by the cytoplasmic domain. *EMBO J.* **7:** 3679.

———. 1989. Transmembrane control of cadherin-mediated cell adhesion: A 94 kDa protein functionally associated with a specific region of the cytoplasmic domain of E-cadherin. *Cell Regul.* **1:** 37.

Nagafuchi, A., M. Takeichi, and S. Tsukita. 1991. The 102 kd cadherin-associated protein: Similarity to vinculin and posttranscriptional regulation of expression. *Cell* **65:** 849.

Nelson, W.J., E.M. Shore, A.Z. Wang, and W.W. Hammerton. 1990. Identification of a membrane-cytoskeletal complex containing the cell adhesion molecule uvomorulin (E-cadherin), ankyrin, and fodrin in Madin-Darby canine kidney epithelial cells. *J. Cell Biol.* **110:** 349.

Nose, A., A. Nagafuchi, and M. Takeichi. 1988. Expressed recombinant cadherins mediate cell sorting in model systems. *Cell* **54:** 993.

Ozawa, M. and R. Kemler. 1992. Molecular organization of the uvomorulin-catenin complex. *J. Cell Biol.* **116:** 989.

Ozawa, M., H. Baribault, and R. Kemler. 1989. The cytoplasmic domain of the cell adhesion molecule uvomorulin associates with three independent proteins structurally related in different species. *EMBO J.* **8:** 1711.

Ozawa, M., M. Ringwald, and R. Kemler. 1990. Uvomorulin-catenin complex formation is regulated by a specific domain in the cytoplasmic region of the cell adhesion molecule. *Proc. Natl. Acad. Sci.* **87:** 4246.

Peifer, M., P. McCrea, K. J. Green, E. Wieschaus, and B. Gumbiner. 1992. The vertebrate adhesive junction proteins β-catenin and plakoglobin and the *Drosophila* segment polarity gene armadillo form a multigene family with similar properties. *J. Cell Biol.* **118:** 681.

Shimoyama, Y. and S. Hirohashi. 1991. Expression of E- and P-cadherin in gastric carcinomas. *Cancer Res.* **51:** 2185.

Shimoyama, Y., A. Nagafuchi, S. Fujita, M. Gotoh, M. Takeichi, S. Tsukita, and S. Hirohashi. 1992. Cadherin dysfunction in a human cancer cell line: Possible involvement of loss of α-catenin expression in reduced cell-cell adhesiveness. *Cancer Res.* **52:** (in press).

Takata, K. and S.J. Singer. 1988. Phosphotyrosine-modified proteins are concentrated at the membranes of epithelial and endothelial cells during tissue development in chick embryos. *J. Cell Biol.* **106:** 1757.

Takeichi, M. 1991. Cadherin cell adhesion receptors as a morphogenetic regulator. *Science* **251:** 1451.

Tsukita, S., K. Oishi, T. Akiyama, Y. Yamanashi, and T. Yamamoto. 1991. Specific proto-oncogenic tyrosine kinases of src family are enriched in cell-to-cell adherens

junctions where the level of tyrosine phosphorylation is elevated. *J. Cell Biol.* **113:** 867.
Volberg, T., B. Geiger, R. Dror, and Y. Zick. 1991. Modulation of intercellular adherens-type junctions and tyrosine phosphorylation of their components in RSV-transformed cultured chick lens cells. *Cell Regul.* **2:** 105.
Warren, S.L. and W.J. Nelson. 1987. Nonmitogenic morphoregulatory action of $pp60^{v\text{-}src}$ on multicellular epithelial structures. *Mol. Cell. Biol.* **7:** 1326.
Wheelock, M.J. and K.A. Knudsen. 1991. N-cadherin-associated proteins in chicken muscle. *Differentiation* **46:** 35.
Winkel, G.K., J.E. Ferguson, M. Takeichi, and R. Nuccitelli. 1990. Activation of protein kinase C triggers premature compaction in the four-cell-stage embryo. *Dev. Biol.* **138:** 1.

Cadherins and the Morphogenesis of Epithelial Tissues in *Xenopus* Embryos

C. Kintner
Molecular Neurobiology Laboratory, Salk Institute for Biological Studies, San Diego, California 92186

The development of the vertebrate nervous system begins when a portion of the embryo called the ectoderm is induced to form neural tissue rather than differentiating into epidermis. Ectoderm responds to induction in part by undergoing a number of morphological changes, including formation of the columnar epithelium of the neural plate, initiation of the elaborate movements of neurulation to form the neural tube, and segregation into the embryo away from the nonneural ectoderm (Schoenwolf and Smith 1990). In contrast, in the absence of induction, the ectoderm remains on the outside of the embryo and gives rise to a specialized epithelium comprising the epidermis. These types of morphological changes also occur as the ectoderm forms derivatives such as the lens and olfactory and otic placodes following other inductive interactions. How epithelial morphogenesis is mediated at the molecular level and how it is initiated by induction are the subjects of this paper.

In 1955, Townes and Holtfreter proposed that differential cell adhesion may be one of the key elements underlying epithelial morphogenesis (Townes and Holtfreter 1955). In their classic study, the epidermis and neural tube were isolated from early amphibian embryos and dissociated into single cells, and the two populations of cells were mixed together in explant culture. Over time, the two cell populations sorted out so that the epidermal cells reformed an epidermis with the correct polarity on the outside of the explant, and the neural cells reformed a neural tube with the correct polarity on the inside of the explant. This result suggested that cells belonging to the same tissue acquire the ability to recognize and adhere to each other and that the differential expression of these adhesion molecules in presumptive neural or epidermal cells would be the first step in tissue histogenesis during development.

The search for molecules capable of mediating cell adhesion led to the isolation and characterization of a family of cell-surface glycoproteins called the cadherins (for review, see Kemler et al. 1989; Takeichi 1991). Members of this family are proteins that span the membrane once and share extensive sequence similarity in regions on both sides of the membrane. Conserved regions in the cadherin extracellular domain bind homotypically and specifically, thereby linking adjacent cells expressing the same cadherin type. Conserved regions in the cadherin intracellular domain apparently bind to cytoplasmic components, thereby linking the cadherins to the cytoskeletal network. By binding on both sides of the membrane, the cadherins are thought to provide a transmembrane linkage that is required for specialized cell contacts in differentiated epithelial tissues. In line with this idea is the fact that the cadherins are found at sites of focal cell contact where they are localized to intermediate (or adherens)-type junctions (Boller et al. 1985; Volk and Geiger 1986). In addition, when cadherins are transfected in fibroblasts that are normally not adherent, the transfected cells become more epithelium-like in that they form focal contact sites where the cadherin protein is concentrated, partition the Na^+,K^+ ATPase to lateral surfaces, and have an increased frequency of dye coupling (Matsuzaki et al. 1990; McNeill et al. 1990). These morphological changes can be reversed by disrupting the adhesion between transfected cells or between epithelial cells in culture using blocking cadherin antibodies. Thus, these in vitro studies suggest that the cadherins, through the formation of intermediate junctions, play a key role in maintaining the morphology of differentiated epithelial tissues.

The pattern by which cadherins are expressed in developing epithelia suggests that these molecules could also play a role in morphogenesis. When new epithelial tissues form, they express different cadherin types (Takeichi 1991). A switch in cadherin type presumably results in differential adhesion, since each type tested when transfected into fibroblasts mediates specific adhesion by a homotypic binding mechanism. This specificity can be demonstrated by in vitro aggregation assays in which two cell lines expressing a different cadherin type (e.g., E-cadherin and P-cadherin) are mixed together. Over time, the cell mixture will sort into two distinct populations (Nose et al. 1988; Friedlander et al. 1989), indicating that boundaries of cadherin expression can mediate specific cell segregation. Thus, the differential adhesion that the cadherins impart to fibroblasts is similar in nature to that observed with embryonic cells by Townes and Holtfreter (1955).

As an in vivo model to study cadherin function, we have examined the role of the cadherins in early *Xenopus* embryos when the ectoderm undergoes morphogenesis in response to induction and forms epidermal and neural tissues. In this paper, I discuss experiments that were carried out to examine how induction regulates cadherin expression and how ectopic cadherin expression affects ectodermal morphology. I also discuss the properties of mutant cadherins where portions of the intracellular or extracellular domains have been deleted. The results from these studies are discussed in terms of how the cadherins regulate embryonic cell adhesion during the morphogenesis of epithelial tissues in *Xenopus* embryos.

METHODS AND EXPERIMENTAL PROCEDURES

Isolation of cDNAs encoding N-cadherin and E-cadherin. cDNAs encoding *Xenopus* cadherins were isolated by screening a stage-17 (early neurulae) cDNA library constructed in λgt10 (Kintner and Melton 1987) using the sequences of chicken N-cadherin cDNA encoding the cytoplasmic domain as a probe (Hatta et al. 1988). Phage clones were placed into groups by cross-hybridization, and representative clones with the largest insert from each group were subcloned into a plasmid vector and sequenced. This analysis identified a cDNA clone encoding *Xenopus* N-cadherin, one encoding a maternal cadherin (B. Gumbiner, pers. comm.) reported previously by Ginsberg et al. (1991), and one encoding a third *Xenopus* cadherin related to mouse E-cadherin. The assignment of this latter cadherin as E-cadherin was based on several criteria. First, RNA encoding this cadherin first appears in *Xenopus* embryos at the beginning of gastrulation, in agreement with the antibody-staining pattern reported for E-cadherin (Choi et al. 1990; Angres et al. 1991). Second, embryos injected with synthetic RNA transcripts encoding this cadherin and analyzed by Western analysis before gastrulation were found to express a polypeptide recognized by the E-cadherin antibodies.

RNase protection analysis of E-cadherin RNA. Embryos were obtained by in vitro fertilization and cultured as described previously (Kintner and Dodd 1991). Ectoderm was isolated from blastula embryos using standard procedures and, in some cases, treated with calcium-free medium in order to isolate and dissociate inner-ectodermal cells (Green and Smith 1990). Porcine Activin-A was a gift from Drs. Vaughn and Vale (Salk Institute). RNA was isolated from tissues and assayed for the presence of E-cadherin RNA using a probe whose sequence corresponds to the third and fourth EC domains. RNase protection probes for elongation factor-1α (EF-1α) and neural cell adhesion molecule (N-CAM) were used as described recently (Kintner and Dodd 1991).

Ectopic expression of cadherins in the ectoderm of* Xenopus *embryos by RNA injection. Synthetic RNAs were generated in vitro from cDNAs encoding N-cadherin and E-cadherin that were subcloned into the transcription vector SP64T (Krieg and Melton 1984). Synthetic RNAs dissolved in water at a concentration ranging from 0.05 to 0.2 μg/ml were injected into 2- to 4-cell embryos (10 nl/embryo). RNA was injected into embryos at the animal pole in order to direct the RNA into the ectoderm. In addition, the injected regions of the embryo were marked by coinjection of a small amount (0.5 ng/μl) of β-galactosidase RNA. Injected embryos were analyzed by X-gal staining and plastic sections as described previously (Detrick et al. 1990).

RESULTS

Several laboratories have generated antibody probes to the cadherin polypeptides expressed in early *Xenopus* embryos by raising monoclonal antibodies to cadherins isolated from the *Xenopus* A6 epithelial cell line (Choi and Gumbiner 1989; Angres et al. 1991). One group of these antibodies recognizes a *Xenopus* cadherin that was termed E-cadherin because its properties are similar to those of E-cadherin in the mouse (also called uvomorulin) and in the chick (also called L-CAM) (Levi et al. 1987; Choi and Gumbiner 1989). As is the case for E-cadherin in higher vertebrates, *Xenopus* E-cadherin is expressed primarily in differentiated epithelium such as epidermis. In mouse and chick embryos, however, E-cadherin is the primary adhesion molecule in embryos up to blastula stages, whereas *Xenopus* E-cadherin is not expressed in embryos until after gastrulation. This observation, and results using a pan-cadherin peptide antibody, suggested that another cadherin distinct from E-cadherin is present in *Xenopus* embryos during early stages of development (Choi et al. 1990). This putative cadherin was subsequently characterized by Hausen and colleagues using a monoclonal antibody that they also raised against cadherins isolated from the A6 cell line (Angres et al. 1991; Muller et al. 1992). This monoclonal antibody recognizes a cadherin, termed U-cadherin, that appears to be expressed in oocytes and eggs as well as the ectoderm and its early derivatives. This antibody also disrupts the reaggregation of cells dissociated from the ectoderm at early cleavage stages, indicating that U-cadherin mediates adhesion in early embryos (Angres et al. 1991).

To characterize further the various cadherin types present in *Xenopus* embryos, cDNA clones encoding *Xenopus* cadherins were isolated by screening a *Xenopus* stage-17 cDNA library using the sequence encoding the cytoplasmic domain of chick N-cadherin as a probe (Hatta et al. 1988). cDNA clones positive in this screen were subdivided into distinct classes by comparison of primary sequence data with the published sequences of mouse N-cadherin, E-cadherin, and P-cadherin (Boller et al. 1985; Hatta et al. 1988; Nose et al. 1988; Miyatani et al. 1989). This approach first identified a *Xenopus* homolog of N-cadherin, which by predicted amino acid sequence shows 85% identity with the published sequence of the mature mouse and chicken N-cadherin polypeptide (Detrick et al. 1990). A second class of cDNAs isolated in our screen encodes a cadherin polypeptide containing approximately 60% identity with mouse E-cadherin and chick L-CAM and 50% identity with mouse P-cadherin and N-cadherin. Subsequent analysis (see Methods and Experimental Procedures) indicated that this cadherin corresponds to the *Xenopus* E-cadherin identified by Gumbiner and colleagues using monoclonal antibodies (Choi and Gumbiner 1989). Finally, using a similar approach, Geiger and colleagues also isolated a cDNA clone corresponding to N-cadherin as well as a second class of cDNA clones whose sequence predicts a cadherin that is more related to mouse P- and E-cadherin than to N-cadherin (Ginsberg et al. 1991). The transcripts corresponding to this third cadherin were found to be expressed in oocytes and in early embryos through early tadpole stages. On the basis of these observations, it seems likely that this third cadherin is the same as the maternal cadherin that has been termed U-cadherin from antibody studies (see above) (Choi et al. 1990; Angres et al. 1991). In sum, cDNA cloning and antibody probes have provided evidence for three distinct cadherins expressed in early *Xenopus* embryos. The maternal cadherin is expressed ubiquitously at all early stages of development, E-cadherin is expressed in epidermal tissue but not neural tissue, and N-cadherin is expressed in neural tissue but not epidermis.

Expression of the Cadherins in the Ectoderm and Its Early Derivatives

The *Xenopus* N-cadherin cDNA was used as a probe to examine how the expression of N-cadherin in the ectoderm is regulated by induction. The developmental expression pattern of *Xenopus* N-cadherin is similar to that of N-cadherin in the early chick embryo (Hatta and Takeichi 1986; Hatta et al. 1987; Duband et al. 1988). *Xenopus* N-cadherin is not expressed at detectable levels in early cleavage-stage embryos but is first expressed around the time of gastrulation in the tissue anlagen that arise by mesodermal or neural induction (Detrick et al. 1990). N-cadherin is expressed in dorsal mesodermal derivatives, such as somitic mesoderm and notochord, and in neural tissue. To determine how the expression of N-cadherin relates to the formation of neural tissue following neural induction, the expression of N-cadherin RNA was assayed in isolated ectoderm after exposure to a neural inducer. Ectoderm was cultured alone or in contact with a heterologous neural inducer from the chick embryo, called Hensen's node, and then assayed for the expression of N-cadherin RNA (Kintner and Dodd 1991). In the absence of inducing tissue, the ectoderm does not express N-cadherin RNA. When combined with Hensen's node, N-cadherin RNA appears rapidly in the ectoderm during the same time course as that in normal embryos, such that N-cadherin RNA can be detected in induced ectoderm by the equivalent of stage 12 (Detrick et al. 1990). This stage corresponds to the point in development when gastrulation is complete, and the columnar epithelium of the neural plate first becomes apparent on the dorsal surface of the embryo. This expression of N-cadherin parallels that of N-CAM, which in *Xenopus* embryos is only expressed at significant levels in neural tissue and first appears in embryos in the neural plate at stage 12 (Jacobson and Rutishauser 1986; Kintner and Melton 1987; Levi et al. 1987). Thus, N-CAM and N-cadherin could be involved in the morphogenetic changes associated with early neural development that begin with the formation of the neural plate. In other respects, however, these molecules are expressed too late to participate in the early aspect of neural morphogenesis. This includes the specialized movements within dorsal midline cells, called notoplate cells, that are responsible for the axial elongation that gives the neural plate its characteristic shape (Gerhart and Keller 1986; Jacobson and Sater 1988).

cDNA probes for E-cadherin were used to determine the expression of E-cadherin RNA in the ectoderm before and after neural induction. Like other genes whose expression is associated with epidermogenesis in *Xenopus* embryos, the expression of E-cadherin in the ectoderm is initiated surprisingly early in development (Jamrich et al. 1987). Figure 1 shows the results of an RNase protection analysis of E-cadherin RNA carried out on RNA from total embryos or from cultured ectoderm isolated from embryos at the beginning of gastrulation. The results show that E-cadherin RNA is first detected in embryos and isolated ectoderm at late blastula stages (stage 10), indicating that the ectoderm initiates the expression of E-cadherin before gastrulation. Since total ectoderm was isolated for the analysis shown in Figure 1, it is possible that E-cadherin expression is only initiated within prospective epidermal regions and not neural regions of the ectoderm (Savage and Phillips 1989). To rule out this possibility, the ectoderm from blastula embryos was dissected into different regions along the prospective dorsal/ventral axis as diagramed in Figure 2. Dorsal ectoderm is fated to give rise to neural tissue and ventral ectoderm gives rise to epidermis. The dissected ectoderm was cultured

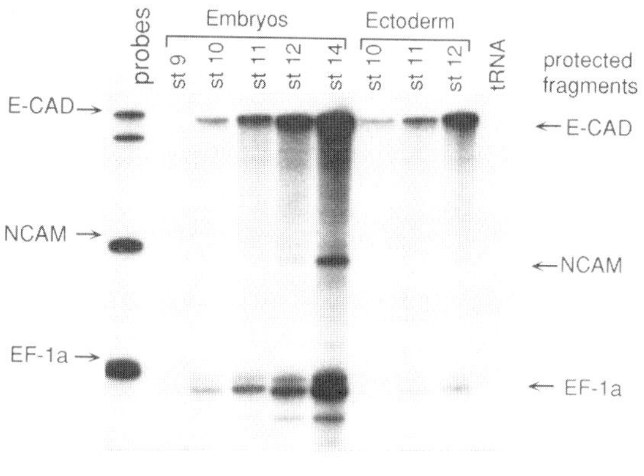

Figure 1. Expression of E-cadherin RNA in *Xenopus* embryos and the ectoderm. RNA was isolated from embryos, cultured at the designated stages (Nieuwkoop and Faber 1967), and assayed for the levels of E-cadherin, N-CAM, and EF-1α RNA using RNase protection analysis (Kintner and Dodd 1991). The left side designates the position of the probes, and the right side designates the position of the protected fragments. tRNA was used as a negative control. Note that E-cadherin is first detected at the beginning of gastrulation (stage 10). The levels of EF-1α RNA serve as a loading control, although at stages of development following the mid-blastula transition (stage 8), these levels change (Krieg et al. 1989).

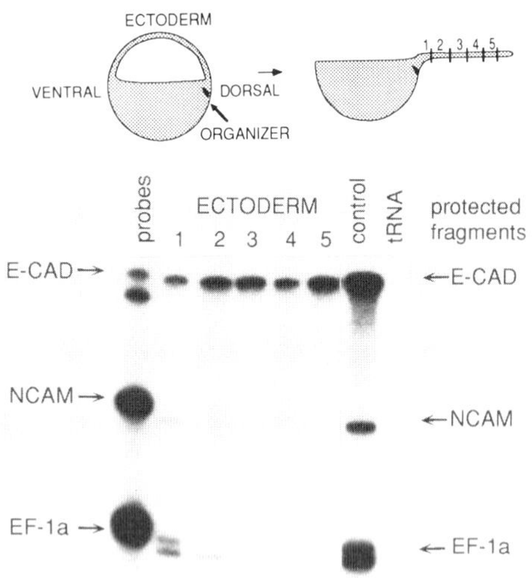

Figure 2. Expression of E-cadherin in presumptive neural and epidermal ectoderm. Ectoderm from blastula embryos was subdivided into five regions along the dorsal (presumptive neural) and ventral (presumptive epidermal) axes, as shown in the diagram at the top. Ectoderm was cultured for 4 hr, to the equivalent of stage 12, and subjected to RNase protection analysis for the expression of E-cadherin, N-CAM, and EF-1α. The positions of the probes and protected fragments are as described in Fig. 1. As a positive control, the same assay was carried out on RNA from stage-24 embryos, and as a negative control, on tRNA. Note that each region of the ectoderm expresses E-cadherin RNA at similar levels and that only region 1 appears to have formed neural tissue and to express N-CAM. This region of the ectoderm lies next to the organizer region of the blastula fate map (top diagram).

for 4 hours and then assayed for the expression of E-cadherin RNA. The results show that E-cadherin RNA is expressed at similar levels in both the dorsal (prospective neural) and ventral (prospective epidermal) ectoderm (Fig. 2).

For comparison, the RNA samples in Figures 1 and 2 were also assayed for the presence of N-CAM RNA. N-CAM, like N-cadherin, RNA is not expressed in the ectoderm until stage 12/14 when gastrulation is complete, and the neural plate appears on the dorsal surfaces of the embryos (Kintner and Melton 1987). Moreover, both N-CAM and N-cadherin RNAs do not appear to be expressed even transiently in isolated ectoderm (Fig. 1). In the experiment shown in Figure 2, the only region of blastula ectoderm that does express N-CAM after isolation from the rest of the embryo is the region located on the dorsal side of the embryo immediately adjacent to the so-called organizer region, which is known to be the source of signals underlying neural induction (Fig. 2, ectoderm sample 1). Thus, this ectoderm is likely to have been already induced to be neural tissue, and to express N-CAM, by planar signals from the organizer (Dixon and Kintner 1989). Together, the analyses of N-cadherin and E-cadherin RNAs suggest that (1) N-cadherin RNA is only expressed in the ectoderm following neural induction and does not occur in the portion of the ectoderm that gives rise to epidermal tissue and (2) E-cadherin is initially expressed at the beginning of gastrulation in both the prospective neural and epidermal portions of the ectoderm.

Since antibody studies have shown that E-cadherin is eventually expressed in the epidermis but not in the neural tube (Angres et al. 1991; Levi et al. 1991), the implication of the results described above is that induction turns off the expression of E-cadherin RNA in the portion of the ectoderm that forms neural tissue. Since a large fraction of the cells in the ectoderm do not come in contact with the inducing tissue when the ectoderm is induced using Hensen's node, the effects of induction on E-cadherin expression cannot be studied directly. In the case of mesodermal induction, this problem is overcome by using soluble inducers, such as Activin-A; the ectoderm can be dissociated into single cells in calcium-free medium, exposed to Activin-A, and then reaggregated into an explant (Green et al. 1990; Green and Smith 1990). Since the ectodermal cells are exposed to uniform concentrations of Activin-A, they are prevented from forming the epidermis and instead form mesodermal derivatives. We used this experimental paradigm to examine the expression of E-cadherin RNA when the ectoderm is prevented from forming the epidermis by mesodermal induction. Ectoderm was dissociated into single cells, exposed to Activin-A, and then assayed for the expression of E-cadherin by RNase protection. The results from these analyses (Fig. 3) show that dissociated ectodermal cells still express E-cadherin, but that Activin-A prevents this expression in a concentration-dependent manner.

These results indicate that mesodermal induction can prevent the expression of E-cadherin in the ectoderm. Interestingly, the ectoderm is known to lose the ability to respond to Activin-A at stage 10, or approximately

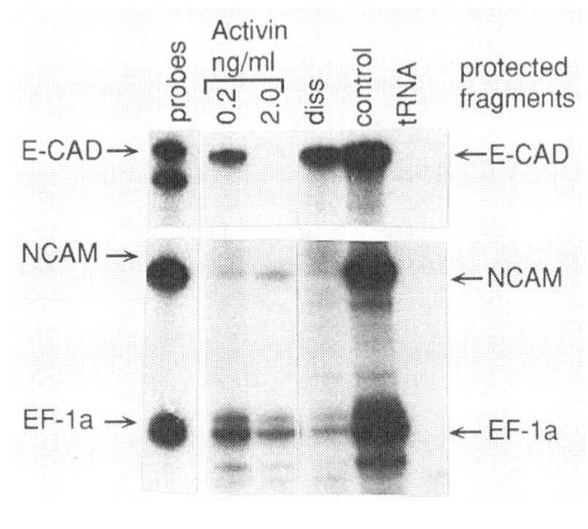

Figure 3. Expression of E-cadherin following mesodermal induction. Ectoderm was isolated from a blastula embryo and dissociated into single cells in calcium-free medium. Dissociated cells were treated with different concentrations of porcine Activin-A, or left untreated (diss), and then allowed to reaggregate by adding calcium to the medium. RNA was isolated from each sample and assayed for the expression of E-cadherin, N-CAM, and EF-1α. As a positive control, the same assay was carried out on RNA from stage-24 embryos, and as a negative control, on tRNA. Note that Activin-A at 2.0 ng/ml is capable of suppressing the expression of E-cadherin in dissociated ectodermal cells.

at the same stage when E-cadherin expression begins (Green et al. 1990; Kintner and Dodd 1991). Thus, Activin-A may prevent E-cadherin expression by directing ectoderm into a mesodermal fate before the onset of epidermogenesis. In contrast, the ectoderm can respond to neural induction until late gastrula stages, when the ectoderm already expresses E-cadherin (Sharpe and Gurdon 1990; Kintner and Dodd 1991). This observation indicates that neural induction can also extinguish the expression of E-cadherin by somehow reversing the course of epidermogenesis.

Cadherin Function in Ectodermal Morphogenesis

Analysis of N-cadherin and E-cadherin in *Xenopus* embryos indicates that the expression of these molecules correlates well with the morphological changes that occur in the ectoderm as it diverges into epidermal and neural tissue. To determine how the expression of these molecules relates to morphogenesis, we have carried out experiments in which N-cadherin or E-cadherin is ectopically expressed in the ectoderm before gastrulation. In *Xenopus*, ectopic expression is easily achieved by injecting synthetic mRNA into fertilized frog eggs, resulting in the expression of the injected RNA in the ectoderm within a few hours after injection. Moreover, since the early development of *Xenopus* embryos is so rapid, and synthetic RNAs are sufficiently stable, expression from injected RNA persists in the embryo through the stage at which the ectoderm forms both epidermal and neural tissues, or after about 14 hours of development.

Ectopic expression of N-cadherin has two obvious effects on the morphology of the ectodermal cell layer (Detrick et al. 1990; Fujimora et al. 1990). One of these effects was measured by generating mosaic embryos in which just one animal blastomere of an embryo at the 16-cell stage was injected with N-cadherin RNA (Detrick et al. 1990). The injected blastomere was marked by the coinjection of β-galactosidase RNA so that the descendants of the injected blastomere could be visualized at later stages by X-gal staining. Normally, when ectodermal cells derived from a given blastomere are marked, they remain together as a relatively coherent clone through gastrulation, although there is considerable mixing at the edge of the clone between cells derived from neighboring blastomeres (Wetts and Fraser 1989). Measuring the extent of the ectoderm cell mixing in mosaic embryos allows one to determine how the expression of a particular cell-surface molecule effects adhesion in an epithelial cell layer. When this analysis was carried out with N-cadherin, ectodermal cells expressing N-cadherin mixed very little with ectodermal cells from neighboring blastomeres relative to controls (Detrick et al. 1990). In some cases, N-cadherin expression suppressed cell mixing to the point where a sharp cell boundary arose between the descendants of the injected and neighboring blastomeres, indicating that ectodermal cell boundaries can be generated by the differential expression of N-cadherin. As a control, the same analysis was carried out with RNA encoding N-CAM, a member of the immunoglobulin-like cell adhesion molecule family (Edelman 1986). When a blastomere was injected with N-CAM RNA, its descendants now expressed N-CAM on their cell surfaces (Kintner 1988). Surprisingly, ectodermal cells expressing N-CAM mixed with cells derived from neighboring blastomeres as well as, if not better than, ectodermal cells injected with just β-galactosidase RNA (Detrick et al. 1990). This result indicates that N-cadherin is much more effective than N-CAM in generating cell boundaries in the ectodermal cell layer. The difference in the efficacy of these two molecules in this assay is more impressive in light of the fact that embryos injected with N-CAM RNA express considerably more N-CAM protein than embryos injected with N-cadherin RNA express N-cadherin (Kintner 1988; Detrick et al. 1990).

Not all the effects of ectopic N-cadherin expression in the ectoderm, however, could be attributed to increased cell adhesion. A second effect of N-cadherin expression on the morphology of the ectoderm was apparent in embryos injected with relatively large amounts of N-cadherin RNA. As these embryos proceeded into gastrulation, ectopic expression of N-cadherin produced a lesion in the ectodermal cell layer, eventually resulting in a complete breakdown in the integrity of the ectodermal epithelium (Detrick et al. 1990). To examine this defect further, embryos injected with N-cadherin RNA were analyzed at mid-gastrulation after embedding and sectioning in plastic (Fig. 4). The morphological appearance of the ectoderm in regions of high N-cadherin expression was not that of increased cell adhesion, but rather one of different morphology (compare A and B in Fig. 4). The cells were more spindle-shaped, and contacts between the cells were less pronounced. Notably, the epithelial cells in the outer ectodermal cell layer failed to maintain their normal tight apposition. These observations indicate that ectopic N-cadherin expression is not compatible with maintaining the ectoderm as an occluding epithelium at these stages of development.

The effects of ectopic E-cadherin expression on the ectoderm were also measured by injecting E-cadherin RNA into fertilized frog eggs. Preliminary analysis of these embryos indicates that the premature expression of E-cadherin at high levels is also not compatible with the ectoderm's maintaining its normal morphology. The embryos injected with E-cadherin RNA showed a lesion in the ectodermal cell layer that appeared near the start of gastrulation. Plastic sections of these embryos showed that the ectoderm in regions injected with E-cadherin RNA was altered in its morphology (compare A and C in Fig. 4). Interestingly, the change in morphology that occurred in the ectoderm in response to E-cadherin expression appeared to be different from that observed in response to N-cadherin expression. Specifically, a morphological change occurred primarily in the morphology of the inner ectodermal cells, which took on an appearance similar to that of the

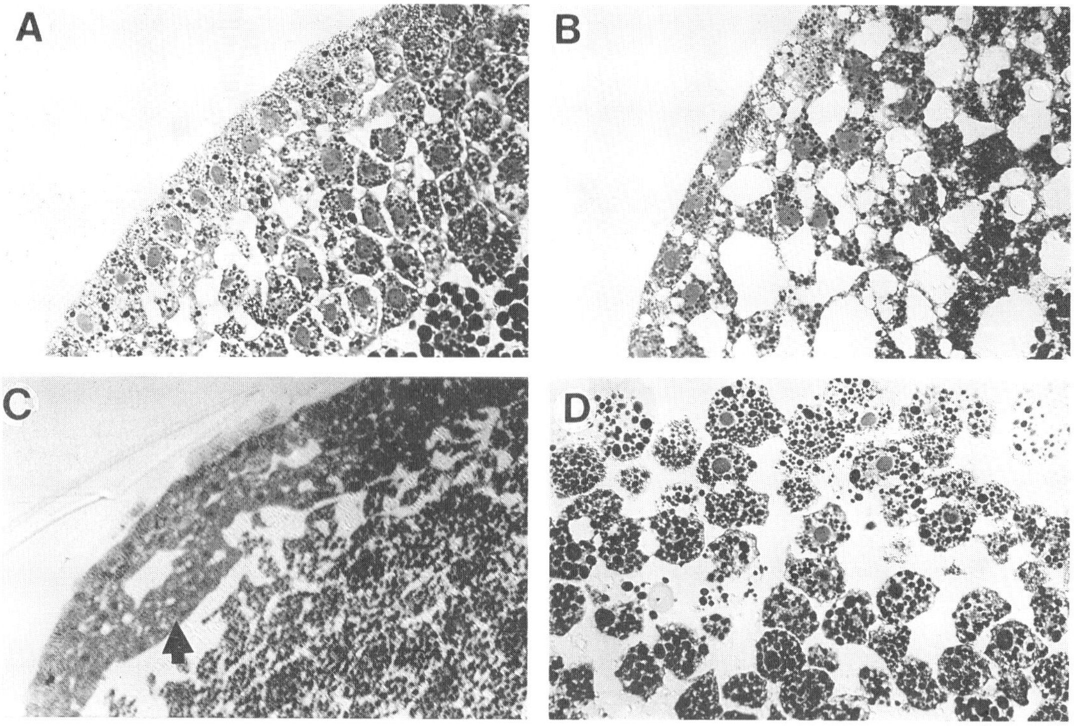

Figure 4. Effects of cadherin expression on ectodermal morphology. RNAs encoding N-cadherin (*B*), E-cadherin (*C*), or a truncated form of N-cadherin (*D*) were injected into the animal pole of fertilized eggs. At mid-gastrulation, embryos were fixed, embedded in plastics, sectioned, and stained with methylene blue. Shown in each case is a portion of the ectodermal cell layer. (*A*) Uninjected control embryo. The arrow in *C* points to the inner ectodermal cells that have taken on the appearance more appropriate for outer layer cells.

outer ectodermal cells (Fig. 4C). This effect on inner cells is consistent with the fact that, in epidermis, E-cadherin is normally expressed at high levels on the outer cells but only at very low levels on the inner cells. We speculate that the injection of E-cadherin RNA leads to the expression of E-cadherin at high levels within the inner ectodermal cell layer, resulting in changes in ectodermal morphology that are inappropriate for normal development.

Competition between Different Cadherin Types

The effects of ectopic N-cadherin and E-cadherin expression on the ectoderm suggest that these cadherin types are not functionally interchangeable with the maternal cadherin. Moreover, these effects suggest that the ability of E- or N-cadherin to alter the morphology of the ectoderm both in ectopic expression experiments and in normal development may reflect the ability of these cadherins to inhibit the maternal cadherin. One potential mechanism for inhibition is through the cadherin cytoplasmic domain, which is known to interact with intracellular components. The importance of these interactions to cadherin function has been examined extensively by introducing altered forms of E-cadherin into fibroblasts. Cell aggregation mediated by E-cadherin can be abolished by deleting 70 amino acids from the E-cadherin carboxyl terminus, even though this deleted form is still expressed on the cell surface (Nagafuchi and Takeichi 1988; Ozawa et al. 1989). This mutant protein also fails to interact with three intracellular polypeptides, called α-, β-, and γ-catenin, which are known to bind intact E-cadherin (Nagafuchi and Takeichi 1989; Ozawa et al. 1989, 1990). Recent studies show that α-catenin shares sequence similarity with vinculin (Nagafuchi et al. 1991) and that β-catenin is very similar to the product of the *Drosophila* segment polarity gene called *armadillo* and to a component of desmosomes called plakoglobin (McCrea et al. 1991). Since the interaction of the cadherin cytoplasmic domain with these cytoplasmic components appears to be essential for cadherin function, these interactions are likely to be a site where cadherin function is regulated. For example, cadherins coexpressed in the same cell could compete for binding to cytoplasmic components needed for function, and this competition would allow one cadherin type to be dominant over another.

To investigate the role of the cytoplasmic domain in the regulation of cadherin function, we generated a deleted form of N-cadherin that lacks most of the sequences encoding the extracellular domain (Kintner 1992). This deleted form of N-cadherin contains a signal peptide, a short extracellular domain, a transmembrane domain, and a complete intracellular domain. Embryos injected with RNA encoding the mutant cadherin show a very striking defect. The first external signs of a defect are apparent by mid-gastrulation stages when the ectodermal cell layer begins to disintegrate. At this point, the embryos stop gastrulating, as monitored by the closure of the blastopore, and the integrity of the ectodermal cell layer is completely lost. If the vitelline membrane is removed at this stage, the ectodermal cells completely dissociate from each other. Sections through injected embryos revealed that the ectodermal cells have lost all normal sites of cell contact (Fig. 4D). It should be emphasized that the effects on cell adhesion observed with the mutant cadherin occur at a stage when N-cadherin is not expressed at significant levels. Thus, the mutant effects cell adhesion by inhibiting the function of other adhesion molecules, most likely the maternal cadherin.

The effects of the mutant cadherin on cell adhesion were further analyzed in several ways (Kintner 1992). First, different regions of the cytoplasmic domain of the mutant were deleted to map the portion of the cytoplasmic domain that were capable of inhibiting adhesion when expressed in embryos. This analysis indicated that at least two regions were sufficient: one mapping next to the transmembrane domain and one mapping in the carboxy-terminal half of the cytoplasmic domain. Thus, several discrete areas of the cytoplasmic domain appear to be capable of generating a dominant-negative effect on cell adhesion. Second, the levels of the mutant protein required for the inhibition of cell adhesion were measured by titering the amounts of RNA encoding the mutant N-cadherin injected into embryos. This analysis indicated that the full effect on cell adhesion, in terms of dissociating the embryo into single cells, required a 10–50-fold excess of the mutant cadherin polypeptide over the endogenous maternal cadherin. It should be noted, however, that the mutant cadherin also had an effect on embryos when expressed at about the same level as the endogenous maternal cadherin. In these embryos, the tissues remained intact, but the mutant clearly affected the ability of the ectoderm and its derivative to form tissue with normal morphology. Thus, the mutant polypeptide needs only to be expressed at physiological levels to have an effect on cell adhesion. Third, the mutant cadherin was capable of inhibiting the binding of α-catenin to E-cadherin. This inhibition was demonstrated by injecting embryos with RNA encoding E-cadherin and with RNA encoding the mutant N-cadherin polypeptide. E-cadherin antibodies were then used to immunoprecipitate E-cadherin and associated proteins from extracts prepared from the injected embryos at the beginning of gastrulation. Western analysis of these immunoprecipitates indicated that E-cadherin was expressed equally well in embryos in the presence or absence of the mutant N-cadherin. Although α-catenin was associated with E-cadherin in the control embryos, no α-catenin was found associated with E-cadherin in the embryos injected with the mutant N-cadherin. Thus, the mutant cadherin may inhibit adhesion by preventing the binding of catenins to other cadherin types.

DISCUSSION

Regulation of Cadherin Expression by Embryonic Induction

The role of neural induction in diverting the ectoderm into neural and epidermal tissue is well established, although the molecular nature of the signaling mechanisms underlying neural induction is still unknown. In the case of the cadherins, induction has opposite effects in regulating the expression of N-cadherin, which is expressed in neural tissue, and E-cadherin, which is eventually expressed in the epidermis. Not surprisingly, the expression of N-cadherin transcripts is turned on relatively rapidly by neural induction. What is surprising is that the expression of E-cadherin RNA is initiated in all the ectoderm and is only later lost from neural tissue as a consequence of induction. Thus, neural induction is not acting on naive ectoderm but rather on ectoderm that has already initiated epidermal differentiation and the expression of E-cadherin. The fact that E-cadherin is already expressed in the ectoderm when induction occurs raises the possibility that changes in cell adhesion could regulate the induction process. For example, the ectoderm is known to lose the ability to respond to inducing tissues by the end of gastrulation, suggesting that some property of the ectoderm has changed that prevents induction from occurring. Changes in adhesion could be instrumental in this respect, since it is known that intimate and extended contact is required between inducing and responding tissues in order for induction to take place (Sharpe and Gurdon 1990). These observations indicate that differential adhesion needs to be examined as a potential factor in the regulation of inductive phenomena.

Cadherins and the Control of Epithelial Morphology

Ectopic expression of N-cadherin and E-cadherin in the ectoderm results in changes in the ectoderm morphology that reflect some of the changes that normally occur when the ectoderm forms neural and epidermal tissues. One such change is apparent as a decrease in cell mixing within the ectodermal cell layer, presumably reflecting the fact that these molecules mediate specific adhesion and are involved in the formation of tissue boundaries. A second change is an apparent transition of the ectoderm from the morphology of an embryonic epithelium to a morphology more appropriate for differentiated epithelium. One explanation of these latter changes is that they represent quantitative effects, i.e., by injecting RNA encoding E- or N-cadherin into embryos, the levels of cell adhesion in the ectoderm are raised to levels normally found in differentiated tissues, and these levels are incompatible with embryonic development. An alternative explanation is that the introduction of E- or N-cadherin into the ectoderm produces qualitative changes in line with the notion that the different cadherin types are somehow specialized to allow for the formation of different epithelial morphologies. For example, the ectoderm is an occluding epithelium that is relatively fluid, since the ectoderm must undergo rapid cell divisions during cleavage stages and then a rapid expansion in size during gastrulation as the ectoderm spreads to cover the outside of the embryo (Keller 1980). When gastrulation is complete, the ectoderm then forms a more static occluding epithelium comprising the epidermal cell layer. These changes in the properties of the ectoderm may depend in part on the expression of particular different cadherin types in that the maternal cadherin may have attributes appropriate for a fluid occluding epithelium and E-cadherin may be specialized for a differentiated epidermis. Similarly, the ectopic expression of N-cadherin also appears to recapitulate the transition that normally occurs when the ectoderm loses the morphology of an embryonic epithelium and takes on the appearance of a neuroepithelium (Schroeder 1970; Hartenstein 1989). Thus, N-cadherin may be compatible with a neuroepithelial morphology but not with the formation of an occluding epithelium such as the embryonic ectoderm or epidermis.

One model that might explain the different properties of the different cadherin types is based on the fact that cadherin function appears to be required for the formation of specialized junctions. For example, E-cadherin not only initiates the formation of adherens junctions, but also promotes (albeit indirectly) the formation of other specialized contacts such as tight and gap junctions (Gumbiner et al. 1988). Differences in epithelial cell morphology are likely to depend strongly on the complement of various junctional complexes. For example, freeze fracture and transmission electron microscopy have shown that tight junctions are dispersed and eventually lost as the ectoderm forms neuroepithelial cells (Decker and Friend 1974; Revel and Brown 1976). One possibility is that these changes in the levels of particular junctions are regulated by different cadherin types. These issues can be explored in more detail once more is known about how the junctional or cytoskeletal organization of epithelial cells changes during morphogenesis in relation to changes in the expression of N-cadherin or E-cadherin.

Regulation of Cadherin Function by the Cytoplasmic Domain

Intracellular interactions between the cytoplasmic tail and intracellular proteins have been proposed to link the cadherins to the cytoskeletal network as well as to regulate cadherin function (Nagafuchi and Takeichi 1989; Ozawa et al. 1990). The regulation of cadherin function through the cytoplasmic domain is likely to be particularly important during embryogenesis, when the appearance of new tissues such as epidermis and the neural tube depend in part on the expression of new cadherin types. We have examined the role of the cytoplasmic domain of N-cadherin in the regulation of cadherin function in embryos by expressing a form of N-cadherin that lacks extracellular sequences. This mutant has a very strong dominant-negative effect on adhesion in the ectodermal cell layer, suggesting that the cytoplasmic domain of N-cadherin is capable of inhibiting the maternal cadherin.

The mechanism by which the cytoplasmic domain of N-cadherin inhibits adhesion is unknown, but there are several possibilities that need to be considered. One possibility is that mutant cadherin is acting as a dominant-negative mutation (Herskowitz 1987). In other words, because the mutant cadherin lacks an extracellular binding domain but contains an intact cytoplasmic domain, it inhibits cadherin-mediated adhesion by competing for the binding of intracellular components. This model is supported by the fact that the cytoplasmic domains of different cadherin types share a high degree of sequence similarity and that, from this similarity, one would predict that the interactions of cytoplasmic components, such as the catenins described primarily for E-cadherin, will also occur with other cadherin types. Thus, in cases where cells express more than one cadherin, competition may occur between the different cadherins for the binding of intracellular components required for function.

Although a simple competition model would explain the effects of the mutant N-cadherin, there are other possibilities that should be kept in mind. For example, the extracellular deletion of N-cadherin may produce an altered protein that is analogous to similar deletions that have been made in receptor-configured tyrosine kinases. In these cases, deleting the ligand-binding region of a receptor kinase does not result in a dominant-negative mutant but rather constitutively activated forms of the receptor (Jarden and Ullrich 1988). In a similar manner, the extracellular deletion of N-cadherin could generate an activated form of the N-cadherin cytoplasmic domain if indeed this domain has signaling properties in addition to its interaction with the actin-based cytoplasmic network. Although such signaling properties are speculative at present, they would be in line with the notion that cell-cell contact results in intracellular signals that regulate tissue morphogenesis.

REFERENCES

Angres, B., A.H.J. Muller, J. Kellermann, and P. Hausen. 1991. Differential expression of two cadherins in *Xenopus laevis*. *Development* **111:** 829.

Boller, K., D. Vestweber, and R. Kemler. 1985. Cell-adhesion molecule uvomorulin is localized in the intermediate junctions of adult intestinal epithelial cells. *J. Cell Biol.* **100:** 327.

Choi, Y.-S. and B. Gumbiner. 1989. Expression of cell adhesion molecule E-cadherin in *Xenopus* embryos begins at gastrulation and predominates in the ectoderm. *J. Cell Biol.* **108:** 2449.

Choi, Y.-S., R. Sehgal, P. McCrea, and B. Gumbiner. 1990. A cadherin-like protein in eggs and cleaving embryos of *Xenopus laevis* is expressed in oocytes in response to progesterone. *J. Cell Biol.* **110:** 1575.

Decker, R.S. and D.S. Friend. 1974. Assembly of gap junctions during amphibian neurulation. *J. Cell Biol.* **62:** 32.

Detrick, R.J., D. Dickey, and C.R. Kintner. 1990. The effect of N-cadherin misexpression on morphogenesis in *Xenopus* embryos. *Neuron* **4:** 493.

Dixon, J.E. and C.R. Kintner. 1989. Cellular contacts required for neural induction in *Xenopus* embryos: Evidence for two signals. *Development* **106:** 749.

Duband, J.-L., T. Volberg, I. Sabanay, J.P. Thiery, and B. Geiger. 1988. Spatial and temporal distribution of the adherens-junction-associated adhesion molecule A-CAM during avian embryogenesis. *Development* **103:** 325.

Edelman, G.M. 1986. Cell adhesion molecules in the regulation of animal form and tissue pattern. *Annu. Rev. Cell Biol.* **2:** 81.

Friedlander, D.R., R.-M. Mege, B.A. Cunningham, and G.M. Edelman. 1989. Cell sorting-out is modulated by both the specificity and amount of different cell adhesion molecules (CAMS) expressed on cell surfaces. *Proc. Natl. Acad. Sci.* **86:** 7043.

Fujimora, T., S. Miyatani, and M. Takeichi. 1990. Ectopic expression of N-cadherin perturbs histogenesis in *Xenopus* embryos. *Development* **110:** 97.

Gerhart, J. and R. Keller. 1986. Region-specific cell activities in amphibian gastrulation. *Annu. Rev. Cell Biol.* **2:** 201.

Ginsberg, D., D. DeSimone, and B. Geiger. 1991. Expression of a novel cadherin (EP-cadherin) in unfertilized eggs and early *Xenopus* embryos. *Development* **111:** 315.

Green, J.B.A. and J.C. Smith. 1990. Graded changes in dose of a *Xenopus* activin A homologue elicit stepwise transitions in embryonic cell fate. *Nature* **347:** 391.

Green, J.B.A., G. Howes, K. Symes, J. Cooke, and J.C. Smith. 1990. The biological effects of XTC-MIF: Quantitative comparison with *Xenopus* bFGF. *Development* **108:** 173.

Gumbiner, B., B. Stevenson, and A. Grimaldi. 1988. The role of the cell adhesion molecule uvomorulin in the formation and maintenance of the epithelial junctional complex. *J. Cell Biol.* **107:** 1575.

Hartenstein, V. 1989. Early neurogenesis in *Xenopus*: The spatio-temporal pattern of proliferation and cell lineages in the embryonic spinal cord. *Neuron* **3:** 399.

Hatta, K. and M. Takeichi. 1986. Expression of N-cadherin adhesion molecules associated with early morphogenetic events in chick development. *Nature* **320:** 447.

Hatta, K., A. Nose, A. Nagafuchi, and M. Takeichi. 1988. Cloning and expression of a cDNA encoding a neural calcium-dependent cell adhesion molecule: Its identity in the cadherin gene family. *J. Cell Biol.* **106:** 873.

Hatta, K., S. Takagi, H. Fujisawa, and M. Takeichi. 1987. Spatial and temporal expression pattern of N-cadherin molecules correlated with morphogenetic processes of chicken embryos. *Dev. Biol.* **120:** 215.

Herskowitz, I. 1987. Functional inactivation of genes by dominant negative mutations. *Nature* **329:** 219.

Jacobson, A.G. and A.K. Sater. 1988. Features of embryonic induction. *Development* **104:** 341.

Jacobson, M. and U. Rutishauser. 1986. Induction of neural cell adhesion molecule (N-CAM) in *Xenopus* embryos. *Dev. Biol.* **116:** 524.

Jamrich, M., T.D. Sargent, and I.B. Dawid. 1987. Cell-type-specific expression of epidermal cytokeratin genes during gastrulation of *Xenopus laevis*. *Genes Dev.* **1:** 124.

Jarden, Y. and A. Ullrich. 1988. Growth factor receptors tyrosine kinases. *Annu. Rev. Biochem.* **57:** 443.

Keller, R. 1980. The cellular basis of epiboly: A SEM study of deep-cell rearrangement during gastrulation in *Xenopus laevis*. *J. Embryol. Exp. Morphol.* **60:** 210.

Kemler, R., M. Ozawa, and M. Ringwald. 1989. Calcium-dependent cell adhesion molecules. *Curr. Opin. Cell Biol.* **1:** 892.

Kintner, C.R. 1988. Effects of altered expression of the neural cell adhesion molecule, NCAM, on early neural development in *Xenopus* embryos. *Neuron* **1:** 545.

———. 1992. Regulation of embryonic cell adhesion by the cadherin cytoplasmic domain. *Cell* **69:** 225.

Kintner, C.R. and J. Dodd. 1991. Hensen's node induces neural tissue in *Xenopus* ectoderm. Implications for the action of the organizer in neural induction. *Development* **113:** 1495.

Kintner, C.R. and D.M. Melton. 1987. Expression of *Xenopus* N-CAM RNA is an early response of ectoderm to induction. *Development* **99:** 311.

Krieg, P.A. and D.A. Melton. 1984. Functional messenger RNAs are produced by SP6 in vitro transcription of cloned cDNAs. *Nucleic Acids Res.* **12:** 7057.

Krieg, P.A., S.M. Varnum, W.M. Wormington, and D.A. Melton. 1989. The mRNA encoding the elongation factor-1α is a major transcript at the mid-blastula transition. *Dev. Biol.* **133:** 93.

Levi, G., K.L. Crossin, and G. Edelman. 1987. Expression sequences and distribution of two primary cell adhesion molecules during embryonic development of *Xenopus laevis*. *J. Cell Biol.* **105:** 2359.

Levi, G., B. Gumbiner, and J.P. Thiery. 1991. The distribution of E-cadherin during *Xenopus laevis* development. *Development* **111:** 145.

Matsuzaki, F., R.-M. Mége, S.H. Jaffe, D.R. Friedlander, W.J. Gallin, J.I. Goldberg, B.A. Cunningham, and G.M. Edelman. 1990. cDNAs of cell adhesion molecules of different specificity induce changes in cell shape and border formation in cultured S180 cells. *J. Cell Biol.* **110:** 1239.

McCrea, P., C. Turck, and B. Gumbiner. 1991. A homolog of *Drosophila* armadillo (plakoglobin) associated with E-cadherin. *Science* **254:** 1359.

McNeill, H., M. Ozawa, R. Kemler, and W.J. Nelson. 1990. Novel function of the cell adhesion molecule uvomorulin as an inducer of cell surface polarity. *Cell* **62:** 309.

Miyatani, S., K. Shimamura, M. Hatta, A. Nagafuchi, A. Nose, M. Matsunaga, K. Hatta, and M. Takeichi. 1989. Neural cadherin: Role in selective cell-cell adhesion. *Science* **245:** 631.

Muller, A.H.J., B. Angres, and P. Hausen. 1992. U-cadherin in *Xenopus* oogenesis and oocyte maturation. *Development* **114:** 533.

Nagafuchi, A. and M. Takeichi. 1988. Cell binding function of E-cadherin is regulated by the cytoplasmic domain. *EMBO J.* **7:** 3679.

———. 1989. Transmembrane control of cadherin-mediated cell adhesion:A 94 kDa protein functionally associated with a specific region of the cytoplasmic domain of E-cadherin. *Cell Regul.* **1:** 37.

Nagafuchi, A., M. Takeichi, and S. Tsukita. 1991. The 102 kd cadherin-associated protein: Similarity to vinculin and posttranscriptional regulation of expression. *Cell* **65:** 849.

Nieuwkoop, P.D. and J. Faber. 1967. *Normal table of* Xenopus laevis. North-Holland, Amsterdam.

Nose, A., A. Nagafuchi, and M. Takeichi. 1988. Isolation of placental cadherin cDNA: Identification of a novel gene family of cell-cell adhesion molecules. *EMBO J.* **6:** 365.

Ozawa, M., H. Baribault, and R. Kemler. 1989. The cytoplasmic domain of the cell adhesion molecule uvomorulin associates with three independent proteins structurally related in different species. *EMBO J.* **8:** 1711.

Ozawa, M., M. Ringwald, and R. Kemler. 1990. Uvomorulin-catenin complex formation is regulated by a specific domain in the cytoplasmic region of the cell adhesion molecule. *Proc. Natl. Acad. Sci.* **87:** 4246.

Revel, J.-P. and S.S. Brown. 1976. Cell junctions in development, with particular reference to the neural tube. *Cold Spring Harbor Symp. Quant. Biol.* **40:** 443.

Savage, R. and C.R. Phillips. 1989. Signals from the dorsal blastopore lip region during gastrulation bias the ectoderm

toward a nonepidermal pathway of differentiation in *Xenopus laevis*. *Dev. Biol.* **133:** 157.
Schoenwolf, G.C. and J.L. Smith. 1990. Mechanisms of neurulation: Traditional viewpoint and recent advances. *Development* **109:** 243.
Schroeder, T.E. 1970. Neurulation in *Xenopus laevis*. An analysis and model based upon light and electron microscopy. *J. Embryol. Exp. Morphol.* **23:** 426.
Sharpe, C.R. and J.B. Gurdon. 1990. The induction of anterior and posterior neural genes in *Xenopus laevis*. *Development* **109:** 765.
Takeichi, M. 1991. Cadherin cell adhesion receptors as a morphogenetic regulator. *Science* **251:** 1451.
Townes, P.L. and J. Holtfreter. 1955. Directed movements and selective adhesion of embryonic amphibian cells. *J. Exp. Zool.* **128:** 53.
Volk, T. and B. Geiger. 1986. A-CAM: A 135kd receptor of intercellular adherens junctions. I. Immunoelectron microscopic localization and biochemical studies. *J. Cell Biol.* **103:** 1441.
Wetts, R. and S.E. Fraser. 1989. Slow intermixing of cells during *Xenopus* embryogenesis contributes to the consistency of the blastomere fate map. *Development* **105:** 9.

DCC: A Tumor Suppressor Gene Expressed on the Cell Surface

L. HEDRICK,*† K.R. CHO,*† J. BOYD,‡ J. RISINGER,‡ AND B. VOGELSTEIN†
*Department of Pathology and †The Oncology Center, The Johns Hopkins University School of Medicine, Baltimore, Maryland 21205; ‡Gene Expression Section, Laboratory of Molecular Carcinogenesis, National Institute of Environmental Health Sciences, National Institutes of Health, Research Triangle Park, North Carolina 27709

The development of human cancer has been proposed to be a multistep process (Foulds 1957; Nowell 1986). Colorectal tumorigenesis provides an excellent model system to test this multistep hypothesis at the molecular level (Vogelstein et al. 1988). Colorectal tumors progress through a continuum of histopathologically identifiable stages, proceeding from normal epithelium, through benign tumors called adenomas, to the malignant stage known as carcinomas (Sugarbaker et al. 1985). Several genes have now been identified which are altered during this progression. They include an oncogene (Ki-*ras*) (Bos et al. 1987; Forrester et al. 1987) and three tumor suppressor genes (*p53*, *DCC*, and *APC*) (Baker et al. 1989; Fearon et al. 1990; Groden et al. 1991; Kinzler et al. 1991), which are mutated in a preferred order during tumor progression. The tumor suppressor gene *DCC* (*d*eleted in *c*olorectal *c*arcinoma), located on the long arm of chromosome 18, encodes a cell-surface protein and is the focus of this paper. A brief description of its identification, cloning, and expression is provided, followed by recent results demonstrating the tissue localization of the DCC protein in normal colonic epithelium and in hyperplastic and neoplastic cells. Finally, preliminary results which suggest that DCC may be important in differentiation are presented.

Frequent and consistent loss of heterozygosity of specific chromosomes in human tumors has been associated with the presence of tumor suppressor genes contained within the region of loss (Knudson 1985; Friend et al. 1986; Baker et al. 1989). Based on this association, a search for potential tumor suppressor genes important in the development of colorectal cancer was pursued by analyzing all the non-acrocentric chromosomal arms in colorectal tumor DNA samples for loss of heterozygosity. This study revealed loss of heterozygosity on several chromosomal arms in a significant number of the colorectal tumors. In particular, the long arm of chromosome 18 was lost in approximately 45% of late adenomas and 75% of carcinomas (Vogelstein et al. 1988). To further delineate the critical region of loss, eight carcinomas, which had lost part but not all of 18q, were examined extensively with numerous probes spanning the length of the long arm of chromosome 18. One common region of deletion was centered at 18q21.3. Probes from genes and anonymous DNA segments that had been previously shown to map to this region were used to screen colorectal carcinoma DNA samples for possible alterations. One anonymous probe, p15-65, detected a homozygous deletion in one colorectal carcinoma. Homozygous deletions are rare and have been noted in tumors at loci encoding tumor suppressor genes (Wolf and Rotter 1985). These results strongly suggested that p15-65 was within, or at least very near, a potential tumor suppressor gene.

Consequently, p15-65 was used as a starting point for a bidirectional chromosomal walk that eventually encompassed 370 kb of contiguous genomic DNA. The walk generated 117 *Eco*RI fragments, which were used as probes in cross-species Southern blot analysis. Of the fragments, 24 hybridized to at least one of the species tested. Rat and human homologs from 7 of the evolutionarily conserved fragments were cloned and sequenced. The sequences were examined for the presence of open reading frames as well as for 5′ and 3′ splice donor and acceptor sites, which would indicate possible exons. A sensitive expression assay for mRNA was developed using the polymerase chain reaction to amplify cDNA. These studies revealed expression of two of the candidate exons in most normal tissues, including colonic mucosa. In contrast, 15 of 17 colorectal cancers failed to express the exons. The gene containing these exons was named *DCC*, and the results were consistent with the possibility that *DCC* represented a target tumor suppressor gene on chromosome 18q (Fearon et al. 1990). The cDNA was cloned, and the nucleic acid sequence of *DCC* was found to predict a 1447-amino-acid transmembrane protein with a unique cytoplasmic domain (K.R. Cho et al., unpubl.) and an extracellular portion with significant homology with neural cell adhesion molecules (N-CAM) and other cell-surface glycoproteins (Fig. 1). This was an intriguing finding because numerous studies of human cancers have revealed abnormalities in cell-cell relationships, including cell-cell and cell-extracellular matrix adhesion, loss of inhibition of growth by cell contact (Todaro and Green 1963), and disorganized tissue architecture (Cotran et al. 1989).

Somatic mutations of putative tumor suppressor genes in human tumors have been used as evidence for their importance in the development of the tumors analyzed (Baker et al. 1989). *DCC* is somatically altered in colorectal carcinomas by a variety of mecha-

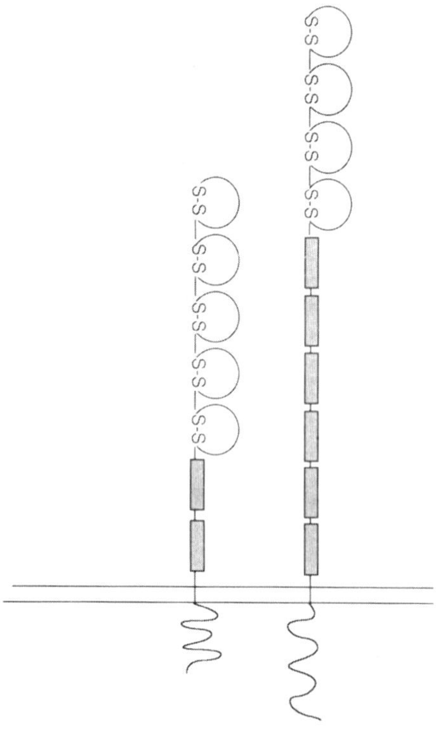

Figure 1. Schematic representation of the DCC protein and N-CAM. DCC contains four immunoglobulin-like C_2 domains and six fibronectin type III repeats, in contrast to N-CAM, which has five immunoglobulin-like C_2 domains and two fibronectin type III repeats. The extracellular domain of DCC is 1097 amino acids and the cytoplasmic domain is 326 amino acids in length.

nisms, including deletions, insertions, and point mutations (Fearon et al. 1990). The search for additional mutations is ongoing. An alternative approach to address *DCC*'s role in tumor suppression involves the definition of the biochemical and physiological properties of the protein. Initial results with this second approach are described below.

EXPERIMENTAL PROCEDURES

Polyclonal antibody preparation. Two polyclonal antibodies were raised against bacterial fusion proteins constructed in pATH vectors with nonoverlapping portions of the *DCC* cDNA. The fusion proteins were isolated from SDS-polyacrylamide gels and injected subcutaneously in rabbits. The antisera were affinity-purified using bacterial fusion proteins from constructs in pGEMEX expression vectors.

Immunoprecipitations, Western blots, and immunohistochemistry. Tissue culture cell lines were metabolically labeled with [^{35}S]methionine for 4 hours at 37°C. The cells were lysed in RIPA buffer under denaturing conditions. Approximately 20×10^6 counts were immunoprecipitated with affinity-purified anti-DCC antibodies and protein A–Sepharose beads. The samples were analyzed by SDS-PAGE. For Western blots, cell lines and tissues were solubilized in Laemmli's sample buffer, and 100–200 μg of protein was loaded per lane in SDS-polyacrylamide gels. The proteins were transferred to nitrocellulose with a semidry electroblot transfer apparatus. The filters were blocked with 10% goat serum and 10% nonfat dry milk in Tris-buffered saline and incubated with affinity-purified anti-DCC antibodies for 120 minutes at room temperature. The Amersham enhanced chemiluminescence system was employed for detection. Frozen tissue was cryostat-sectioned at 7-μm thickness for immunohistochemistry. The sections were fixed for 30 minutes at room temperature in 10% buffered formalin. Endogenous peroxidase activity was blocked by incubation in 0.3% H_2O_2, and signal intensity was increased using the antigen retrieval system from Biogenex. The sections were blocked with goat serum followed by incubation with 3 ng/μl of anti-DCC antibody. A secondary goat anti-rabbit biotinylated antibody, in conjunction with the Vectastain Elite reagent from Vector Laboratories, was used for detection.

Transfection of IMR-32 cell line. The IMR-32 cell line was derived from a human neuroblastoma and obtained from the American Type Culture Collection. A 1.0-kb fragment of *DCC* cDNA, surrounding the translation initiation site, was cloned into a eukaryotic expression vector in the antisense and sense orientations. IMR-32 cells were transfected with the plasmids by electroporation using a Bio-Rad Gene Pulser. A 0.4-cm cuvette containing 5×10^7 cells and 20 μg of purified plasmid in 1 ml of serum-free medium was pulsed at 0.2 kV and 250 μFD for 8.0 msec. The cells were then cultured in regular growth medium containing G418 (400 μg/ml). After approximately 15 days of selection, transfected clones were selected that contained the DCC-antisense vector, DCC-sense vector, or vector without insert. The clones were expanded and used for subsequent experiments in medium containing 100 μg/ml of G418. For morphological differentiation studies, the cells were cultured on Natrix extracellular matrix-coated dishes (Collaborative Research) in standard medium (with 100 μg/ml G418 for transfected cells) with or without nerve growth factor (NGF) (50 ng/ml). Fresh medium and NGF were added every 2 days, and neurite outgrowth in IMR-32 cells was generally apparent after 7–12 days of NGF treatment.

RESULTS AND DISCUSSION

Detection of DCC Gene Product

Our initial goal was to define cell lines and tissues that expressed detectable levels of the DCC gene product. Two rabbit polyclonal antibodies were raised against bacterial fusion proteins containing nonoverlapping epitopes of DCC, using selected portions of the *DCC* cDNA. One antibody was raised to a 1.65-kb portion of the extracellular domain, and the other was

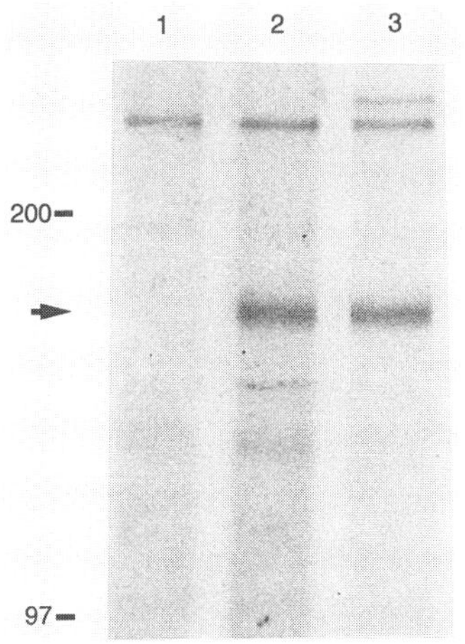

Figure 2. Immunoprecipitation of DCC from 577MF cells with two anti-DCC antibodies. The cells were metabolically labeled followed by lysis under denaturing conditions. (Lane *1*) Normal rabbit immunoglobulin, 6 ng/μl; (lane *2*) antiextracellular DCC antibody, 1.8 ng/μl; (lane *3*) anticytoplasmic DCC antibody, 1.2 ng/μl. The arrow indicates the 180-kD protein immunoprecipitated by both the extracellular and the cytoplasmic DCC antibodies. One of the higher-molecular-weight bands cross-reacts with normal immunoglobulin as well as the DCC antibodies. The cytoplasmic antibody immunoprecipitates an additional high-molecular-weight protein not recognized by the extracellular antibody.

raised to the entire cytoplasmic domain. In an initial screen of numerous cell lines by immunoprecipitation, *DCC* expression was detected at highest levels in one cell line, 577MF, a nonseminomatous testicular germ-cell tumor. Both antibodies precipitated a protein of approximately 180 kD from metabolically labeled 577MF cells, providing strong evidence that it represents the DCC gene product (Fig. 2). The sequence of the full-length *DCC* cDNA predicts an unmodified transmembrane molecule of approximately 153 kD. The sequence also suggests that *DCC* could be *N*-glycosylated (Fearon et al. 1990), possibly explaining the larger apparent molecular weight observed in 577MF. All of the colorectal carcinoma cell lines tested by immunoprecipitation were negative, consistent with previous RNA expression assays. Immunoprecipitation of 577MF cells labeled by cell-surface molecule iodination demonstrated that the DCC protein was present on the cell surface. To further investigate the cellular location of the DCC product, CHO cells transfected with a constitutive expression vector containing the full-length DCC cDNA were stained with DCC antibodies. The pattern of staining was consistent with a diffuse membrane localization of DCC. In addition, the signal showed increased intensity at cell-cell contact sites (Fig. 3). This pattern of staining has been reported in cell lines expressing molecules known to be involved in cell adhesion. These studies indicate that DCC is a cell-surface molecule and suggest that it may be concentrated at sites of cell-cell contact.

We then examined tissue distribution of DCC expression by Western blot analysis. Expression studies on normal tissues by reverse transcription polymerase chain reaction revealed that the highest level of expression was in brain. By Western blot analysis, a faint, but detectable, band was present at approximately 180 kD in human brain tissue lysates; however, DCC was not detected in any other tissues. We sought to increase the sensitivity of our expression assay by employing immunohistochemistry.

Tissue Localization of DCC Expression

The expression studies indicate that expression of DCC in cell lines and in tissues is extremely low. However, the previous analyses were performed with total cellular mRNA or total cellular protein. If DCC expression were locally expressed at high concentrations at either intra- or intercellular locations, it might be more readily detected by in situ methods. We tested

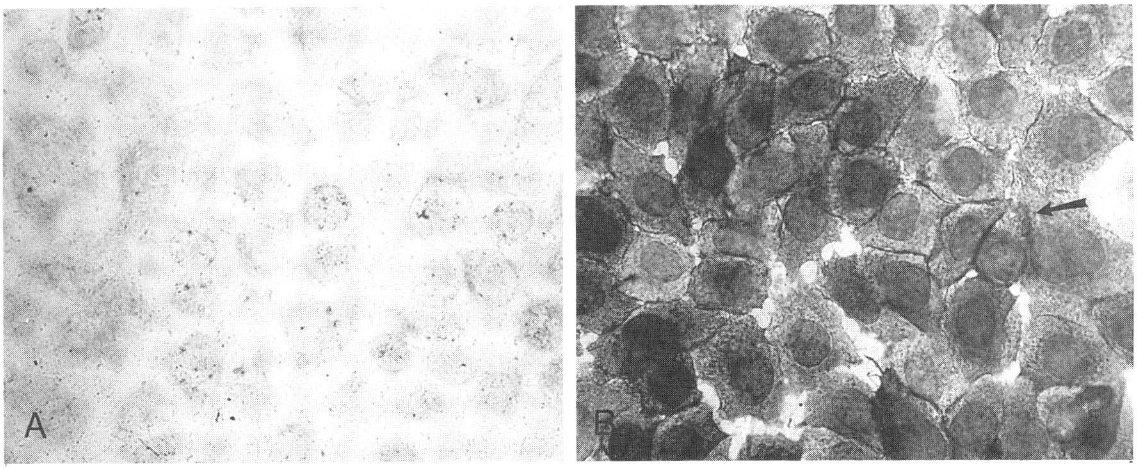

Figure 3. Immunocytochemistry of untransfected and transfected CHO cells. (*A*) Untransfected CHO cells incubated with 2 ng/μl anticytoplasmic DCC antibody; (*B*) transfected CHO cells expressing full-length DCC cDNA incubated with anticytoplasmic DCC antibody at 2 ng/μl. The arrow indicates increased signal intensity at cell-cell contact points. Magnification, 560×.

this hypothesis on tissue from the central nervous system, based on the earlier mRNA expression data. Staining of frozen human spinal cord sections demonstrated strong staining in axons at several levels of the spinal cord. DCC expression appeared to be present in both myelinated and nonmyelinated nerve cells. This technique was then applied to normal frozen colonic mucosal tissue sections. Normal colonic mucosa is composed of epithelial gland-like structures called crypts, surrounded by the supporting lamina propria, which contains fibroblasts and inflammatory cells. The colonic crypts contain two major differentiated cell types: absorptive cells, called enterocytes, responsible for fluid and electrolyte transport; and secretory cells, known as goblet cells, which produce and secrete mucin. The goblet cells are the most conspicuous cell type because of their large apical intracellular mucin accumulation. However, they only account for approximately 20% of the colonic crypt epithelium (Hafez et al. 1990). Immunohistochemical staining localized detectable DCC expression to a small subset of colonic epithelial cells, specifically the goblet cell population. Furthermore, the expression pattern of DCC within the goblet cells is specific. The signal is localized to the basolateral aspect of the cell and in the supranuclear cytoplasmic space, just below the apical mucin (Fig. 4). It was not possible to determine with certainty whether DCC was localized to the cell membrane in these sections, but the expression pattern is consistent with that possibility. In addition, the supranuclear staining may represent nascent protein in the Golgi apparatus, where it is known to be localized (Bloom and Fawcett 1975). To further evaluate the subcellular location of DCC, electron microscopic studies are in progress.

The restriction of DCC expression to a subset of differentiated epithelial cells in the colon raises many questions. From a standpoint of colorectal tumorigenesis, the pattern is interesting because colorectal tumors commonly lose cells with the goblet morphology and lose normal mucin production as they progress (Boland et al. 1982). The RNA expression data provide evidence that DCC expression is lost in colorectal carcinomas. With this in mind, we sought to determine the status of expression of the DCC gene product in tissue sections from colorectal tumors at different stages during its progression.

DCC Expression in Colorectal Tumor Progression

We examined five adenomas and six colorectal carcinomas by immunohistochemical analysis, using the cytoplasmic DCC antibody. The adenomas varied in size and in the amount of epithelial dysplasia, which were criteria used to determine stage. The early- and intermediate-stage adenomas appeared to retain expression of DCC, but the pattern of expression was abnormal. The positively staining cells showed a diffuse expression pattern with the majority of DCC in the apical aspect of the cells. Two late adenomas showed complete absence of DCC expression in the adenomat-

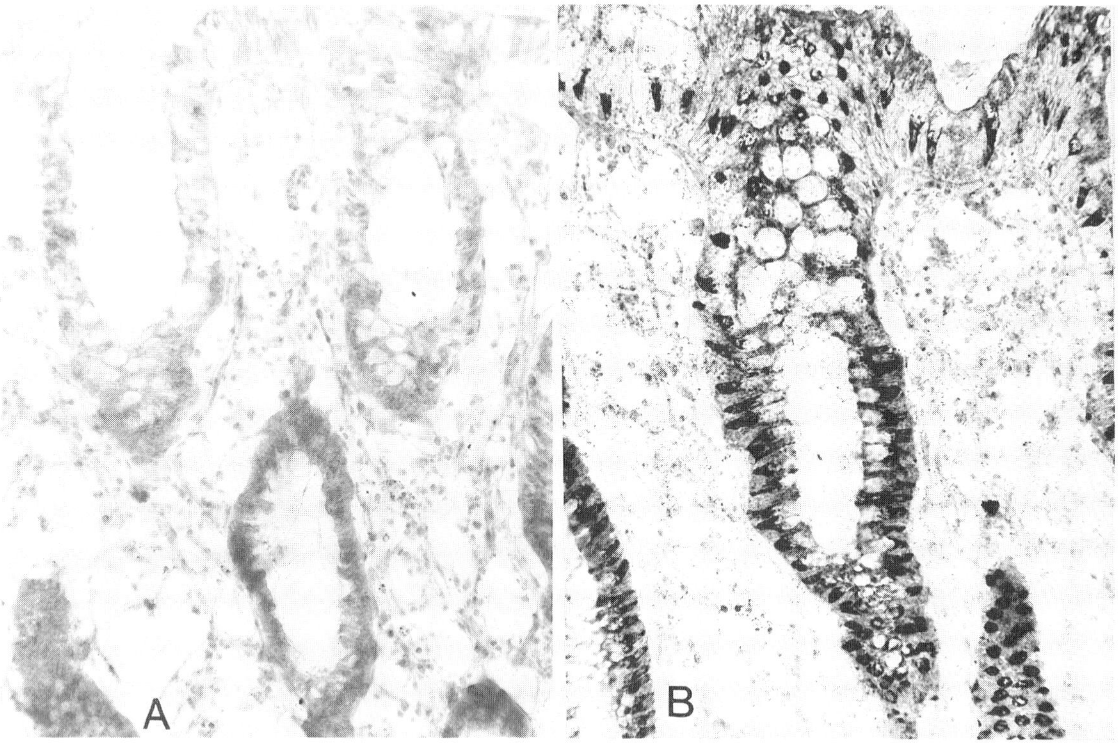

Figure 4. Immunohistochemistry of normal human colonic mucosa. (*A*) Normal colon incubated with 10 ng/μl of normal rabbit immunoglobulin; (*B*) normal colon incubated with 2 ng/μl of anticytoplasmic DCC antibody. Magnification, 212 ×.

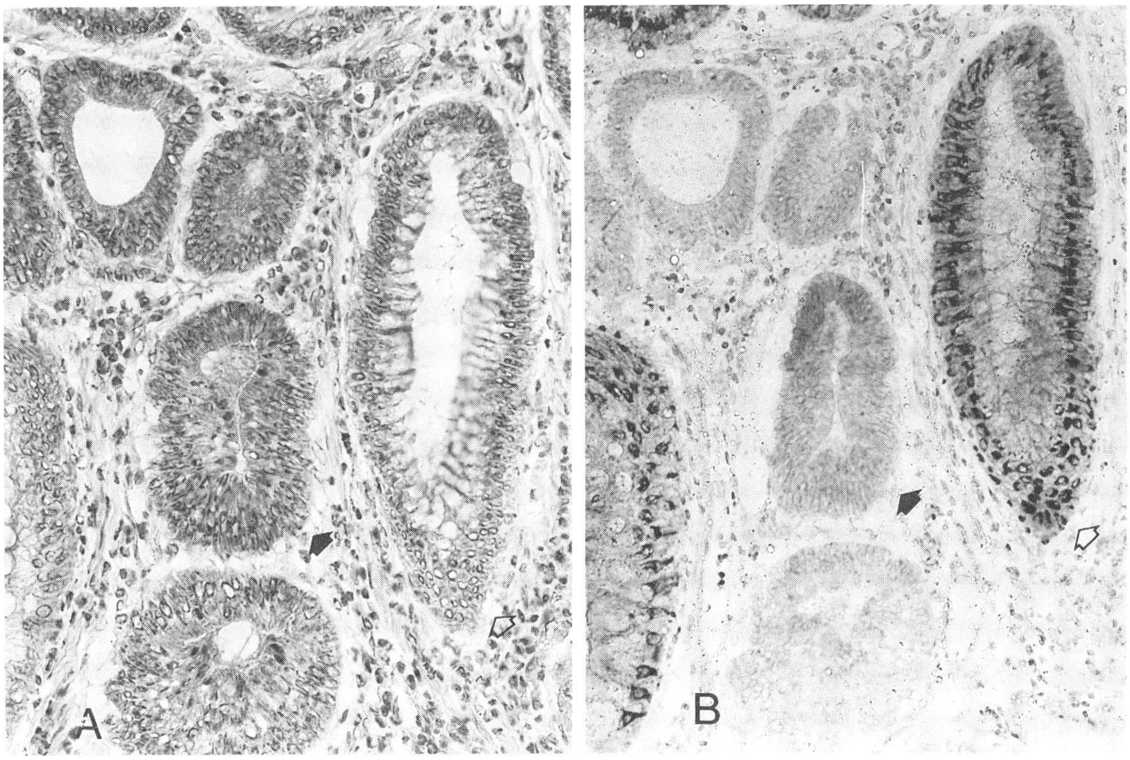

Figure 5. Immunohistochemistry of a colonic adenoma. (*A*) Hematoxylin and eosin-stained section showing both adenomatous and normal colonic glands; (*B*) anticytoplasmic DCC antibody demonstrating a complete lack of staining of the adenomatous glands with adjacent nonneoplastic glands staining normally. Open arrows indicate normal glands and closed arrows indicate adenomatous glands. Magnification, 212 × .

ous glands, whereas adjacent normal glands exhibited the normal staining pattern for DCC (Fig. 5). When these adenomas were stained with alcian blue, a histochemical stain for mucin, the adenomatous glands were negative. Four of the invasive carcinomas studied completely lacked detectable DCC expression. Two of the carcinomas were mucinous carcinomas, designated as such because they produce abundant extracellular mucin. These two carcinomas exhibited strong DCC staining, and like the adenomas, the staining pattern was abnormal with the neoplastic cells showing a strong diffuse staining pattern. These findings suggest that DCC expression is altered during colorectal tumor progression. It appears that in late-stage adenomas and in nonmucinous carcinomas, DCC expression is lost.

It is generally accepted that tumors are less well differentiated than the normal tissue in which they arise (Pignatelli and Bodmer 1988). The combination of the specific staining pattern of DCC in the goblet cells of normal colonic epithelium, and the loss of this differentiation in colonic tumors with the concomitant loss of DCC expression, raises the possibility that DCC is involved in differentiation. This possibility is supported by previous studies which show that differentiation is influenced by cell adhesion-molecule-mediated intercellular adhesion (Edelman 1988). However, the studies described above cannot determine whether DCC expression is a consequence of differentiation or whether it has a causative role in differentiation. To address this role, we turned to a tissue-culture cell line that expressed DCC and in which differentiation could be induced under appropriate conditions.

A Role in Differentiation

We knew from previous expression studies that DCC was expressed at relatively high levels in neuronal tissues. In addition, differentiation of some neuroblastoma cell lines in response to NGF is well documented (Waris et al. 1973). With this in mind, we tested several neuroblastoma cell lines for endogenous DCC expression. One cell line, IMR-32, expressed high levels of DCC mRNA and protein, readily detected by Northen blot, Western blot, and immunoprecipitation. Furthermore, when these cells were grown on an extracellular matrix, they differentiated in response to NGF, sending out long neuritic processes. Expression of DCC mRNA and protein was analyzed, by Northern and Western blots, after treatment with NGF. The level of DCC message reached its maximum level after 72–96 hours of NGF treatment, and the protein levels showed a twofold increase. These experiments provide further evidence that DCC expression is altered during differentiation but are unable to show a causative relationship. To directly test the possibility that DCC has an active role in differentiation, we used an antisense strategy to decrease the level of DCC expression and asked whether the cells would then differentiate in response to NGF. A 1.0-kb portion of the DCC cDNA surrounding the translation initiation site was cloned into a constitutive eukaryotic expression vector in the sense and antisense orientations. These constructs were used to transfect IMR-32 cells, and stable transfectants were selected with G418 and analyzed for the expression of DCC. Cells expressing the DCC cDNA in the sense orientation expressed normal endogenous DCC levels, as expected. In contrast, cells expressing the antisense cDNA segment expressed greatly reduced levels of DCC protein (Fig. 6). When these transfected cell lines were treated with NGF, cells expressing the sense clone responded with the typical extension of long neuritic processes. Cells expressing the antisense clone failed to elaborate neuritic processes, and, moreover, the cells continued to divide in the presence of NGF. These findings offer evidence that DCC plays a role in differentiation of IMR-32 cells in response to NGF.

In summary, the *DCC* gene is often deleted or mutated during the later stage of colorectal tumorigenesis, and its expression is altered. DCC has significant homology with N-CAM, and its product is expressed on the cell surface. DCC is selectively expressed in one type of differentiated cells in normal colonic epithelium, and its expression is lost during the progression of a subset of colorectal tumors. Finally, we have presented experiments which suggest that DCC has a causative role in differentiation of at least one cell line in culture. The possibility that tumor suppressor genes are involved in differentiation has been raised previously on the basis of chromosome transfer studies

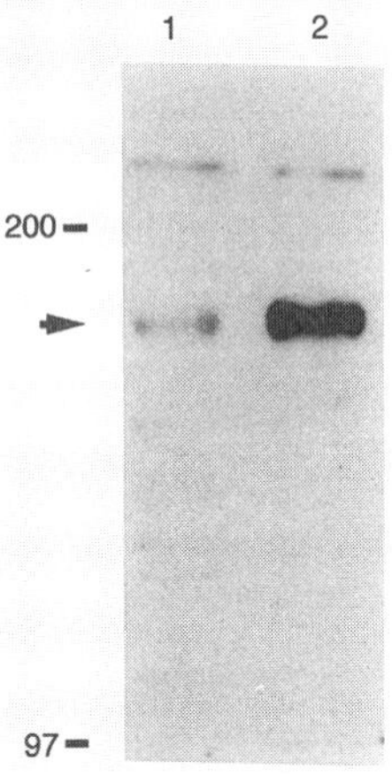

Figure 6. Western blot of antisense and sense IMR-32 transfected clones. (Lane *1*) 100 μg of protein from lysate of DCC-antisense transfected cells; (lane *2*) 100 μg of protein from lysate of DCC-sense transfected cells. DCC is indicated by the arrow. The additional high-molecular-weight band is a nonspecific cross-reacting band. The filter was incubated with anticytoplasmic antibody at 240 ng/ml in 10% nonfat dry milk followed by incubation with goat anti-rabbit HRP-conjugated secondary antibody. Signal was detected by enhanced chemiluminescence (ECL-Amersham).

(Peehl and Stanbridge 1982). The studies presented here raise the possibility that DCC functions as a tumor suppressor gene through induction of differentiation, a process that may be incompatible with continuous cellular proliferation.

ACKNOWLEDGMENTS

We thank Stanley Hamilton, Rahj Robinson, and William Fox for their help in collecting the frozen tissue samples. L.H. is a Howard Hughes Medical Institute physician research fellow. This work was supported by National Institutes of Health grant CA-43460 to B.V.

REFERENCES

Baker, S.J., E.R. Fearon, J.M. Nigro, S.R. Hamilton, A.C. Preisinger, J. M. Jessup, P. van Tuinen, D.H. Ledbetter, D.F. Barker, Y. Nakamura, R. White, and B. Vogelstein. 1989. Chromosome 17 deletions and p53 gene mutations in colorectal carcinomas. *Science* **244:** 217.

Bloom W. and D.W. Fawcett. 1975. Intestines. In *A textbook of histology*, p. 662. W.B. Saunders. Philadelphia.

Boland, C.R, C.M. Montgomery, and Y.S. Kim. 1982. Alterations in human colonic mucin occurring with cellular differentiation and malignant transformation. *Proc. Natl. Acad. Sci.* **79:** 2051.

Bos, J.L., E.R. Fearon, S.R. Hamilton, M.Verlaan-de Vries, J.H. van Boom, A.J. van der Eb, and B. Vogelstein. 1987. Prevalence of *ras* gene mutations in human colorectal carcinomas. *Nature* **327:** 293.

Cotran, R.S., V. Kumar, and S.L. Robbins. 1989. Neoplasia. In *Robbins pathologic basis of disease*, 4th edition, p. 244. W.B. Saunders, Philadelphia.

Edelman, G.M. 1988. Morphoregulatory molecules. *Biochemistry* **27:** 3533.

Fearon, E.F., K.R. Cho, J.M. Nigro, S.E. Kern, J.W. Simons, J.M. Ruppert, S.R. Hamilton, A.C. Preisinger, G. Thomas, K.W. Kinzler, and B. Vogelstein. 1990. Identification of a chromosome 18q gene that is altered in colorectal cancers. *Science* **247:** 49.

Forrester, K., C. Almoguera, H. Kyuhyung, W.E. Grizzle, and M. Perucho. 1987. Detection of high incidence of K-*ras* oncogenes during human colon tumorigenesis. *Nature* **327:** 298.

Foulds, L. 1957. Tumor progression. *Cancer Res.* **17:** 355.

Friend, S.H., R. Bernards, S. Rogelj, R.A. Weinberg, J.M. Rapaport, D.M. Albert, and T.P. Dryja. 1986. A human DNA segment with properties of the gene that predisposes to retinoblastoma and osteosarcoma. *Nature* **323:** 643.

Groden, J., A. Thliveris, W. Samowitz, M. Carlson, L. Gelbert, H. Albertsen, G. Josyln, J. Stevens, L. Spirio, M. Robertson, L. Sargeant, K. Krapcho, E. Wolff, R. Burt, J.P. Hughes, J. Warrington, J. McPherson, J. Wasmuth, D. Le Paslier, H. Abderrahim, D. Cohen, M. Leppert, and R. White. 1991. Identification and characterization of the familial adenomatous polyposis coli gene. *Cell* **66:** 589.

Hafez, M.M., D. Infante, S. Winawer, and E. Friedman. 1990. Transforming growth factor β_1 acts as an autocrine-negative growth regulator in colon enterocytic differentiation but not in goblet cell maturation. *Cell Growth Differ.* **1:** 617.

Kinzler, K.W., M.C. Nilbert, L.K. Su, B. Vogelstein, T.M. Bryan, D.B. Levy, K.J. Smith, A.C. Preseinger, P. Hedge, D. McKechnie, R. Finniear, A. Markham, J. Groffen, M.S. Boguski, S.F. Altschul, A. Horii, H. Ando, Y. Miyoshi, Y. Miki, I. Nishisho, and Y. Nakamura. 1991. Identification of FAP locus genes from chromosome 5q21. *Science* **253:** 661.

Knudson, A.J. 1985. Hereditary cancer, oncogenes, and antioncogenes. *Cancer Res.* **45:** 1437.

Nowell, P.C. 1986. Mechanisms of tumor progression. *Cancer Res.* **46:** 2203.

Peehl, D.M. and E.J. Stanbridge. 1982. The role of differentiation in the control of tumorigenic expression in human cell hybrids. *Int. J. Cancer* **30:** 113.

Pignatelli, M. and W.F. Bodmer. 1988. Genetics and biochemistry of collagen binding-triggered glandular differentiation in a human colon carcinoma cell line. *Proc. Natl. Acad. Sci.* **85:** 5561.

Sugarbaker, J.P., L.L. Gunderson, and R.E. Wittes. 1985. Colorectal cancer. In *Principles and practices of oncology*, 2nd edition (ed. V.T. Devita et al.), p. 800. J.B. Lippincott, Philadelphia.

Todaro, G.J. and H. Green. 1963. Quantitative studies of the growth of mouse embryo cells in culture and their development into established lines. *J. Cell Biol.* **17:** 299.

Vogelstein, B., E.R. Fearon, S.R. Hamilton, S.E. Kern, A.C. Preisinger, M. Leppert, Y. Nakamura, R. White, A.M.M. Smits, and J.L. Bos. 1988. Genetic alterations during colorectal-tumor development. *N. Engl. J. Med.* **319:** 525.

Waris, T., L. Rechardt, and P. Waris. 1973. Differentiation of neuroblastoma cells induced by nerve growth factor in vitro. *Experientia* **29:** 1128.

Wolf, D. and V. Rotter. 1985. Major deletions in the gene encoding the p53 tumor antigen cause lack of p53 expression in HL-60 cells. *Proc. Natl. Acad. Sci.* **82:** 790.

Intercellular Signaling during *Caenorhabditis elegans* Vulval Induction

P.W. STERNBERG, R.J. HILL, G. JONGEWARD, L.S. HUANG, AND L. CARTA
Howard Hughes Medical Institute and Division of Biology, California Institute of Technology, Pasadena, California 91125

The essential invariance of *Caenorhabditis elegans* development has allowed detailed analysis of the mechanisms of cell fate determination. A variety of studies suggest a major role for cell interactions in specifying cell fates during nematode development. The original observations of natural variability of certain cell fates and the cell fate changes after particular cell ablation experiments indicated that the fates of some cells during later embryonic and postembryonic development depend on cell interactions (Sulston and Horvitz 1977; Kimble and Hirsh 1979; Sulston et al. 1983). Blastomere interchange experiments have demonstrated that cell interactions are crucial to specification of fates during early embryogenesis (Priess and Thomson 1987; Schnabel 1991; Wood 1991). Molecular genetic analyses have revealed additional cell interactions and suggest that others will be uncovered. For example, mosaic analysis of the *her-1* gene indicates that sex determination involves cell communication (Hunter and Wood 1992). In addition, the pleiotropic effects of mutations in known signaling molecules, such as those discussed here, suggest hitherto undiscovered examples of intercellular signaling in *C. elegans* development.

The process of *C. elegans* vulval induction allows high-resolution analysis of how intercellular signals control cell fates during animal development (Horvitz and Sternberg 1991; Greenwald and Rubin 1992). The study of vulval induction has been based on the pioneering work of Horvitz and Sulston (1980; Sulston and Horvitz 1981), who demonstrated that mutants defective in vulval differentiation could be easily recognized and propagated. These mutants were mostly of two broad classes: vulvaless mutants that fail to generate a vulva and multivulva mutants that display excessive vulval differentiation (Fig. 1). In general, recessive vulvaless mutations define genes required to promote vulval differentiation, whereas recessive multivulva mutations define genes required to prevent excessive vulval differentiation. Additional genes have been identified by obtaining extragenic suppressors of existing mutations. Analyses of these mutations by developmental genetics and molecular biology have defined gene products that have a role in vulval induction.

Recent advances in *C. elegans* molecular genetics have allowed rapid progress in the analysis of vulval development as well as all aspects of *C. elegans* biology. Current methods that allow facile production of transgenic animals that carry multicopy transgenes composed of input DNAs (Mello et al. 1992) have been useful in several ways. Molecular cloning of a particular genetic locus now often relies on the demonstration that a genomic DNA clone can rescue the mutant phenotype in transgenic animals, thereby providing a functional test of whether a particular molecular clone corresponds to a genetic locus. Genomic clones for

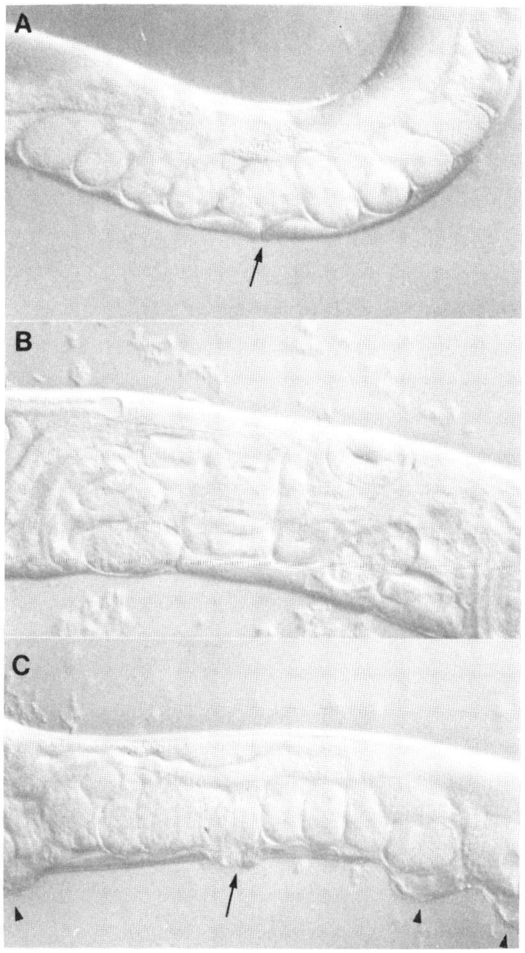

Figure 1. Nomarski photomicrographs of the vulval regions of adult hermaphrodites. (*A*) Wild-type hermaphrodite. The arrow indicates the vulva. (*B*) Vulvaless hermaphrodite *let-23(sy1)*. The animal is filled with embryos that were retained due to the lack of a vulva. (*C*) Multivulva hermaphrodite *lin-15(n309)*. The arrow indicates the vulva; arrowheads indicate nonfunctional pseudovulvae. Photographs were taken on Kodak technical pan film with a 40× objective.

most of the whole genome are available from the *C. elegans* genome project (Coulson et al. 1986, 1988; Sulston et al. 1992). *lacZ* fusion genes (Fire et al. 1990) have in some instances allowed rapid identification of cells that express particular genes. Construction of transgenes conferring novel dominant phenotypes has allowed tests of epistasis between genes with similar recessive phenotypes.

Our main goal in studying vulval induction has been to identify the inductive signal(s) and its receptor. In addition, we have sought to understand how the induction process is regulated. The regulation of the inductive signaling pathway might underlie the precision of the cell patterning that is central to normal vulval development.

Multiple Intercellular Signals Control VPC Fates

The vulva is formed from the progeny of three of the six vulval precursor cells (VPCs) that become specified to generate vulval cells by virtue of their position with respect to an inducing cell, the anchor cell of the gonad (Fig. 2). In the absence of the anchor cell, all six VPCs generate epidermal cells that fuse with the large epidermal syncytium called hyp7; this program of differentiation is the 3° fate. The induced VPCs, which are the three closest to the anchor cell (Fig. 3), are of two types, 1° and 2°, distinguished by the sets of progeny cells produced. For example, some of the progeny of the 1° VPC can attach to the anchor cell during morphogenesis, whereas the progeny of the 2° VPC cannot (Sternberg and Horvitz 1989). The simple problem of vulval induction is how the 1° and 2° VPCs are induced. The more complex problem is how the precise spatial pattern of 1° and 2° fates is established: 3°-3°-2°-1°-2°-3° from anterior to posterior. On the basis of cell ablation experiments coupled with genetic analyses, a three-signal model for VPC fate specification has been proposed (Fig. 3) (for review, see Horvitz and Sternberg 1991). In this view, VPCs are all initially equivalent but become specified to distinct fates by the combined action of three intercellular signals: an inductive signal from the anchor cell, a lateral signal among the VPCs, and a negative signal probably from hyp7.

The *inductive signal* is the major determinant of the extent of vulval differentiation and was defined by cell ablation experiments in which it was shown that the anchor cell of the somatic gonad was solely responsible for induction of the vulva (Kimble 1981). The inductive signal can act on each VPC individually, as single VPCs (in the absence of their five neighbors) can become 1°, 2°, or remain 3°, apparently depending on their distance from the anchor cell (Sternberg and Horvitz 1986). Thus, the signal may be distributed in a graded manner such that the distance of a VPC from the anchor cell determines its fate. Indirect evidence from a mutant that occasionally has a dorsally displaced gonad (and hence anchor cell) suggests that the inductive signal can act at a distance and is thus a secreted factor as induction still occurs in those animals with dorsally displaced anchor cells (Thomas et al. 1990). It is conceivable that the inductive signal acts via another tissue (e.g., hyp7, ventral cord motoneurons, or body wall muscle) and thus would not necessarily involve any diffusible signal. In this latter case, the genetic pathway described below is still valid, but the sites of action of some genes might then be different from the simple predictions.

The existence of a *negative signal* is suggested by observations made with mutations of the *lin-15* locus. As described below, *lin-15* is required to prevent VPCs from generating vulval cells in the absence of inductive signal. Genetic mosaic analysis indicates that *lin-15* does not act solely in the VPCs or anchor cell to repress vulval differentiation (Herman and Hedgecock 1990).

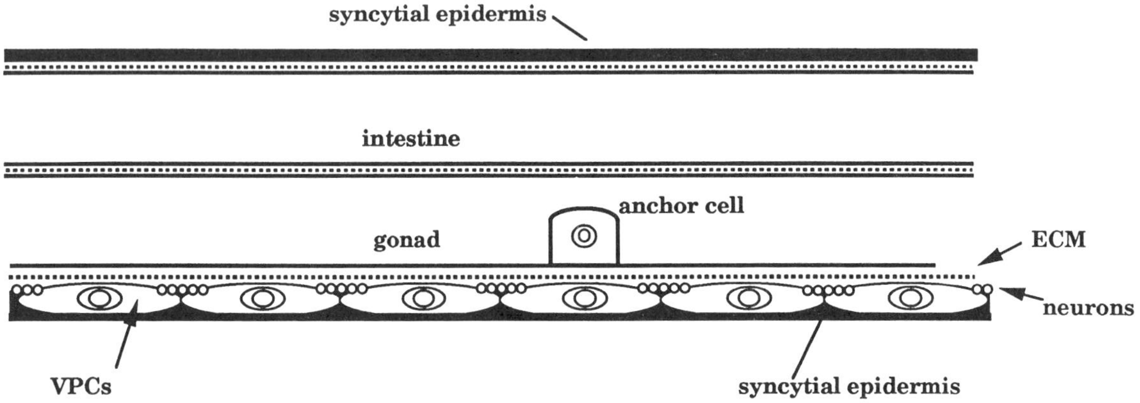

Figure 2. Cell biology of vulval induction. Schematic representation of a longitudinal section of a hermaphrodite at the time of induction. Six vulval precursor cells (VPCs) are arranged in an anterior to posterior line. The spindle-shaped VPCs are in contact with each other at their anterior and posterior ends. The anchor cell is located in the middle of the ventral portion of the developing gonad. Double circles within each cell represent the nucleus and nucleolus visible in Nomarski optics. The dashed line represents extracellular matrix (ECM) material between the gonad and the epidermis. The syncytial epidermis, called hyp7, is represented by black shading at the bottom. hyp7 would not actually be visible ventral to the VPCs in a section that includes VPC-VPC contact. The cell bodies of ventral cord motoneurons are represented by small circles. Based on the data of Sulston and Horvitz (1977) and L. Carta and P. Sternberg (in prep.).

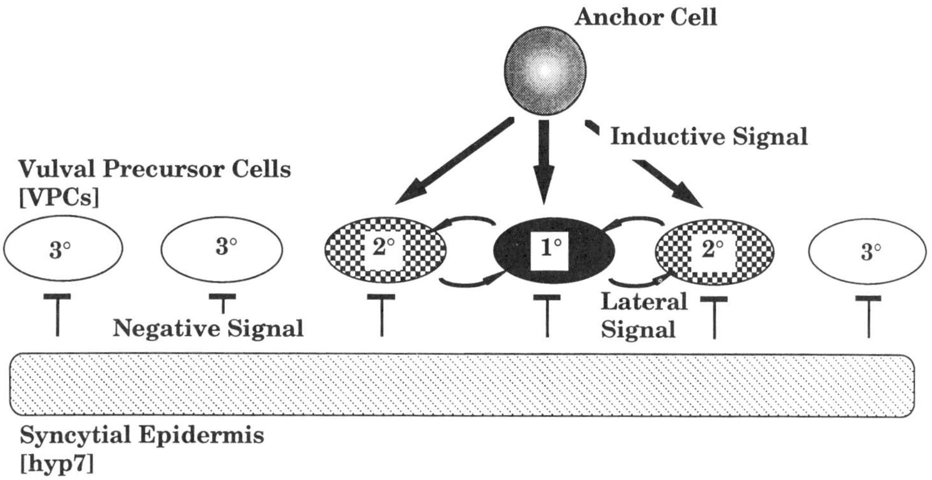

Figure 3. Three-signal model for cell intercellular signaling during vulval induction. See text for details. (Adapted from Horvitz and Sternberg 1991.)

Since *lin-15* acts in cells other than the anchor cell and VPCs, it defines a new signal.

The existence of the *lateral signal* has been proposed on the basis of three sets of observations. The VPC most proximal to the anchor cell prevents its immediate neighbors from also becoming 1°; in its absence, one of its neighbors becomes 1° (Sulston and White 1980; Sternberg 1988). This inhibition could have been passive, i.e., the 1° VPC simply prevents the inductive signal from reaching 2° VPCs, but analysis of the multivulva phenotype of *lin-15* mutants indicates that it is at least partially an active signaling process (Sternberg 1988). In a *lin-15* mutant, all six VPCs are either 1° or 2°. In the absence of neighbors in *lin-15* animals, a single VPC becomes 1°. If two neighboring VPCs are present, one becomes 1° and the other becomes 2°. Thus, signaling between the two VPCs specifies their fates in a given animal. This "lateral inhibition" phenomena by which a 1° VPC prevents its immediate neighbors from also being 1° can account for the alternating pattern of VPC fates seen in *lin-15* animals (e.g., 2°-1°-2°-1°-2°-1° or 1°-2°-2°-1°-2°-1°). *lin-12* is required for 2° fates, and in a mutant defective in both *lin-12* and *lin-15*, all VPCs are 1°. Thus, lateral signaling might work via *lin-12* to promote the 2° fate at the expense of the 1° fate (Greenwald et al. 1983; Sternberg and Horvitz 1989). The elegant mosaic analysis of Seydoux and Greenwald (1989) demonstrated that *lin-12* has such a role in other cells and acts on the receiving end of such a signaling process. *lin-12* is therefore an excellent candidate for being the receptor for a signal that stimulates 2° fate at the expense of 1° fate (Yochem et al. 1988).

Examination in the transmission electron microscope of longitudinal and transverse sections of hermaphrodites during the time of vulval induction (the early L3 stage, the third of four larval stages) suggests that the inductive signal occurs using a soluble signal, whereas the lateral signal could occur by a membrane-bound ligand (L. Carta and P. Sternberg, in prep.). The anchor cell is not in intimate contact with the VPCs during the early L3 when induction occurs. Thus, it is most likely that the inductive signal is soluble. Although it is conceivable that induction of 1° requires contact and that induction of 2° can act at a distance, the fact that VPCs can become 1° without apparent contact with the anchor cell during the time of induction suggests that this possibility is unlikely (Sternberg and Horvitz 1986; Thomas et al. 1990; L. Carta and P. Sternberg, unpubl.). Diffuse extracellular matrix material exists between the anchor cell and the VPCs that might contribute to the graded nature of vulval induction by binding and thus limiting spread of a soluble factor through the body cavity (pseudocoeleom). The VPCs appear to have access to the body cavity over limited portions of their cell bodies; this separation of the cell surfaces that could respond to the inductive signal might also contribute to the "graded" response by increasing the difference in the amount of signal received by adjacent cells.

In contrast to the apparent lack of contact between the anchor cell and each VPC, the VPCs contact each other over a limited extent of their cell surfaces, at their anterior and posterior tips (L. Carta and P. Sternberg, in prep.). Thus, signaling between the VPCs might occur via a membrane-bound ligand. This observation is consistent with cell ablation experiments in a *lin-15* mutant in which the lateral signal was defined (Sternberg 1988). In these experiments, lateral signaling occurred between cells that were in close proximity and most likely touching. This observation is also consistent with the proposed role of *lin-12* in mediating lateral signaling, since *lin-12* encodes a product likely to be a cell-surface receptor for a membrane-bound ligand (Yochem et al. 1988; see Artavanis-Tsakonas; Kimble et al.; both this volume).

Although the prospect of understanding how VPCs respond to and integrate multiple intercellular signals and thus become specified to one of three potential fates is intriguing, our recent work has focused on the simpler problem of how the anchor cell induces VPCs to generate vulval cells (1° or 2° fates) rather than to generate nonspecialized epidermis (3° fate). Below, we discuss two genes, *lin-3* and *let-23*, that are each required for vulval induction and that encode a candidate ligand and its receptor, respectively. After a brief discussion of additional positive-acting genes, we discuss two sets of negative regulators that somehow antagonize the action of the intercellular signaling and signal transduction pathway defined by the positive-acting genes.

lin-3 Encodes an Inductive Signal

Genetic and molecular experiments indicate that *lin-3* encodes an anchor cell signal that induces vulval formation. *lin-3* is necessary for induction of the vulva. Mutations that reduce *lin-3* activity result in a vulvaless phenotype (Horvitz and Sulston 1980; Sulston and Horvitz 1981; Ferguson and Horvitz 1985; Ferguson et al. 1987). In contrast to this reduction-of-function phenotype, transgenic animals carrying multiple copies of a *lin-3* genomic clone, and thus presumably overexpressing wild-type Lin-3, have a multivulva phenotype in which all six of the VPCs can have vulval fates (Hill and Sternberg 1992). Therefore, the level of *lin-3* activity may control the extent of vulval differentiation.

lin-3 encodes a putative signaling molecule (Hill and Sternberg 1992). The predicted *lin-3*-encoded protein has the architecture of growth factor precursors that are ligands for the mammalian epidermal growth factor (EGF) receptor. Specifically, the predicted *lin-3* product has a signal sequence, a single EGF domain, a transmembrane domain, and 181 residues in the putative cytoplasmic domain. An EGF domain is a motif consisting of six cysteine residues with semiconserved spacing (for review, see Carpenter and Wahl 1990). EGF-like domains are found in a wide variety of extracellular proteins, including the ligands for the mammalian EGF receptor subfamily of growth factor receptors. The precursors of ligands for the human EGF receptor include EGF, transforming growth factor-α (TGF-α), amphiregulin, heparin-binding EGF (HB-EGF), and vaccinia virus growth factor (VVGF) (Carpenter and Wahl 1990). The ligands for a second member of the EGF receptor subfamily (see below), encoded by the *neu* proto-oncogene, are the human heregulins and rat *neu* differentiation factor (NDF) (Holmes et al. 1992; Wen et al. 1992). Most of these precursors have a single EGF domain and are membrane-bound. The precursor to EGF has eight additional EGF-like domains; NDF and the heregulins also have an immunoglobulin-like domain amino-terminal to the EGF domain (Wen et al. 1992). Outside the EGF repeat, little similarity exists among all of these proteins as a class and with *lin-3*.

These human ligands can be processed from the membrane-bound precursor to a soluble form that contains an EGF domain and often other sequences. Some human ligands, such as TGF-α, can act in an unprocessed form (for review, see Massagué 1990). We have no direct evidence concerning whether or not Lin-3 is processed. There is, however, an alternative splice that adds 15 amino acids between the EGF repeat and the transmembrane domain. The presence of this extension might facilitate (or inhibit) processing of Lin-3. If so, alternative splicing might be one mode of regulation of this ligand.

Expression of *lin-3* in the anchor cell can stimulate vulval differentiation (Hill and Sternberg 1992). A *lin-3::lacZ* fusion was constructed that contains the extracellular and transmembrane domains of *lin-3*, including the EGF domain, and a cytoplasmic domain that consists mainly of the β-galactosidase protein encoded by the *Escherichia coli lacZ* gene. This transgene is bifunctional in vivo. It retains the ability to stimulate vulval differentiation and also specifically expresses β-galactosidase activity in the anchor cell at the time of vulval induction. Ablation experiments performed on animals carrying another *lin-3* transgene that contains an equivalent extent of genomic DNA confirm that *lin-3* can act as a signal. Ablation of the gonad, which generates the anchor cell, severely diminishes the ability of this transgene to stimulate vulval differentiation. These observations suggest that the detected expression of *lin-3* in the anchor cell can stimulate vulval differentiation.

We conclude from these results that *lin-3* encodes an inductive signal for vulval development. *lin-3* might encode the only inductive signal from the anchor cell to the VPCs or there might be multiple signals. We have not yet tested whether *lin-3* can replace the anchor cell for induction of vulval lineages. Since presumptive overexpression of *lin-3* in transgenic animals can cause all six VPCs to generate vulval cells, all six VPCs are able to respond to inductive signal in a given animal (Hill and Sternberg 1992). This observation supports the hypothesis that the anchor cell signal can act at a distance.

let-23 Might Be the Receptor for the *lin-3* Inductive Signal

let-23 genetics and structure indicate that it is an excellent candidate for a receptor of *lin-3* (Aroian et al. 1990; Aroian and Sternberg 1991). *let-23* mutations that reduce its activity result in a vulvaless phenotype (Ferguson and Horvitz 1985; Ferguson et al. 1987; Aroian and Sternberg 1991). *let-23* mutations block the ability of *lin-3* transgenes to stimulate excess vulval differentiation (Table 1) (Hill and Sternberg 1992). Thus, *lin-3* requires *let-23* for its action, and *let-23* mediates the inductive signal.

let-23 encodes a nematode homolog of the EGF receptor subfamily of human receptor tyrosine kinases (Aroian et al. 1990). This subfamily includes the EGF receptor, which is the product of the c-*erbB* proto-oncogene, c-*neu*/c-*erbB*-2/HER2, and c-*erbB-3*/HER3 (Carpenter and Wahl 1990). The tyrosine kinase sub-

Table 1. Interactions of Dominant Multivulva Phenotypes of Transgenic Animals with Chromosomal Mutations

	Multicopy transgene			
Chromosome	none	*lin-3*	*let-23*	*let-60*
+	WT	Muv	WT	Muv
lin-3	Vul	Muv	n.d.	Muv
let-23	Vul	Vul	WT	Muv
let-60	Vul	n.d.	n.d.	Muv

The interactions indicate that *lin-3* acts via *let-23*, which acts via *let-60*. (WT) wild type; (Muv) multivulva; (Vul) vulvaless; (n.d.) not determined. Multicopy *let-23* genes do not cause a dominant multivulva phenotype. The transgenic studies summarized here are from Aroian et al. (1990), Han and Sternberg (1990), and Hill and Sternberg (1992).

family is characterized by a cysteine-rich extracellular domain with characteristic spacing of cysteine residues, a transmembrane domain, a characteristic kinase domain, and a tyrosine-rich carboxy-terminal tail. Let-23 is not obviously more similar to any of the members of this human subfamily and appears to be single copy in the *C. elegans* genome. Thus, Let-23 may be a prototypical receptor. Similarly, Lin-3 is not obviously more similar to any of the ligands for this class of receptor. Thus, these *C. elegans* genes may be descended from a common ancestor to the more diverse mammalian families.

The mammalian receptors bind ligand with their extracellular domain, dimerize, *trans*-phosphorylate the carboxy-terminal cytoplasmic tail, and thus activate kinase activity or allow access to substrates (Ullrich and Schlessinger 1990; Hunter 1991; see Schlessinger et al.; Peters et al.; both this volume). Because of its overall similarity to the EGF receptor subfamily, *let-23* also is likely to function in this manner.

Other Positive-acting Signal Transduction Components

A number of other candidate positive-acting components of this pathway have been identified (Horvitz and Sternberg 1991). Genetic analysis of the *let-60* gene indicates that it acts like a switch to control whether VPCs have vulval or epidermal fates (Beitel et al. 1990; Han et al. 1990). *let-60* encodes a *C. elegans* Ras protein (Han and Sternberg 1990). Because of its high degree of sequence identity to biochemically characterized Ras proteins and because the genetic properties of *let-60* mutations are similar to the properties of analogous mutations in other Ras proteins, *let-60* is very likely to act like a molecular switch. Thus, *let-60 ras* likely exists in two general states: an active state bound to GTP and an inactive state bound to GDP. All five *let-60(gf)* mutations are in codon 13 (Beitel et al. 1990) and are thus analogous to activating mutations in other Ras proteins that disrupt the intrinsic GTPase activity of *ras* (Bourne et al. 1990, 1991). Dominant negative mutations of *let-60 ras* are by and large in residues analogous to dominant negative mutations in yeast and mammalian Ras proteins (Han and Sternberg 1991). Genetically, *let-60* acts after *lin-3* and *let-23* (Table 1 and Fig. 4) (Han et al. 1990). *lin-3*, *let-23*, and *let-60* have the same spectrum of pleiotropic effects on *C. elegans* development (Ferguson and Horvitz 1985; Aroian and Sternberg 1991; H. Chamberlin et al., unpubl.), suggesting that they can act together in multiple aspects of development.

The *sem-5* gene encodes a protein with two SH3 domains and one SH2 domain (Clark et al. 1992 and this volume), which are found in many proteins that interact with the mammalian EGF receptor. An SH2 domain confers the ability to bind to phosphotyrosine-

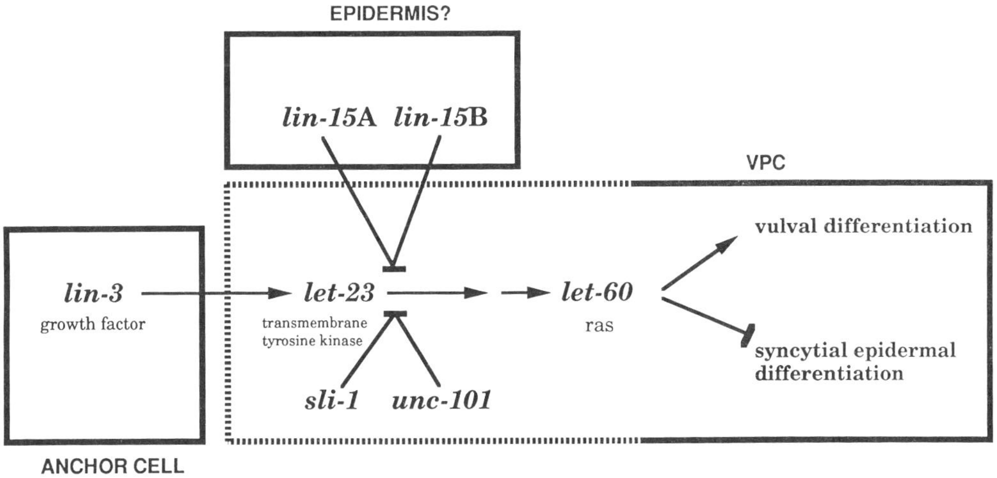

Figure 4. Proposed inductive signaling pathway and its negative regulation. *lin-3* encodes an EGF-domain-containing growth factor precursor. *let-23* encodes an EGF-receptor-like tyrosine kinase. *let-60* encodes a Ras protein. These three gene products act in series to promote vulval differentiation at the expense of epidermal differentiation. *lin-15* encodes two products, A and B, that both are required to antagonize the action of this pathway. Similarly, *sli-1* and *unc-101* encode products that are required to antagonize the action of this pathway. *lin-3* acts in the anchor cell; *lin-15* probably acts in the syncytial epidermis. *let-23* and *let-60* are proposed to act in the VPCs. We favor the hypothesis that *sli-1* and *unc-101* act in the VPCs. Since there is no direct evidence that *let-23* or *let-60* acts in the VPCs, we have enclosed this portion of the pathway in a dashed box.

containing peptides (for review, see Koch et al. 1991), which allows signal transduction either by activation due to phosphorylation by the receptor tyrosine kinase, or by favoring the interaction with other signaling molecules. Clark et al. thus propose that *sem-5* interacts with *let-23*. The order of action of *let-23* and *sem-5* has not yet been established. During vulval induction, *let-23*, *sem-5*, and *let-60* reduction-of-function mutations have similar vulvaless phenotypes, and thus likely act in a common pathway to control vulval differentiation in response to the *lin-3* inductive signal.

lin-2, *lin-7*, and *lin-10* are three other genes defined by recessive vulvaless mutations. These genes are distinct from *lin-3*, *let-23*, *let-60*, and *sem-5* (which we consider to represent the main trunk of the signaling pathway) in that putative null mutants are viable (Ferguson and Horvitz 1985; Kim and Horvitz 1990). Moreover, *lin-2*, *lin-7*, and *lin-10* are not absolutely required for vulval differentiation: Residual induction occurs in strains carrying putative null alleles (Horvitz and Sulston 1980; Sulston and Horvitz 1981; Ferguson and Horvitz 1985; Ferguson, et al. 1987). This residual differentiation depends on inductive signal (Sternberg and Horvitz 1989). Therefore, *lin-2*, *lin-7*, and *lin-10* might not be in the main trunk of the inductive signaling pathway. Another possibility is that they might mediate one of two partially redundant signal transduction pathways acting downstream from *let-23*. *lin-2*, *lin-7*, and *lin-10* might act to increase the efficacy of signal transduction by any of a variety of means. These three genes might be imagined to be positive regulators of other signal transduction components. A good working hypothesis is that these three genes act at the *let-23* step because, in general, *lin-2*, *lin-7*, or *lin-10* mutations have the genetic properties of partial reduction-of-function mutations in *let-23*, rather than of partial reduction-of-function mutations in *lin-3* or *let-60*. In particular, some *lin-2*, *lin-7*, *lin-10* mutations like some *let-23* mutations can display a hyperinduced phenotype in which excessive signal-dependent vulval differentiation occurs (Aroian and Sternberg 1991; G. Jongeward and P. Sternberg, in prep.; S. Kim, pers. comm.).

Negative Regulation by *lin-15* and Related Genes

A negative signaling pathway is defined by *lin-15* and a set of interacting genes. *lin-15* null mutants confer a multivulva phenotype: All six VPCs are 1° or 2° even in the absence of the inductive signal (Ferguson et al. 1987; Sternberg and Horvitz 1989). Even though vulval differentiation is independent of the gonadal signal in these *lin-15* mutant animals, VPCs are still responsive to the inductive and lateral signals (Sternberg 1988): The inductive signal can establish the pattern of 1° and 2° VPC fates in a *lin-15* mutant, and the lateral signal was defined in a *lin-15* mutant as the signal that promotes a 2° VPC fate at the expense of a 1° fate.

This multivulva phenotype was also found in animals defective in both *lin-8* and *lin-9* (Sulston and Horvitz 1981; Ferguson et al. 1987). However, animals defective only in either *lin-8* or *lin-9* alone are wild type (Table 2), suggesting that these two genes are functionally redundant (Horvitz and Sulston 1980). *lin-8* and *lin-9* are the founding members of two sets of genes, *A* and *B*, respectively (Ferguson and Horvitz 1989). When these authors mutagenized a *lin-8A* mutant and screened for multivulva mutants, alleles of *lin-35*, *lin-36*, *lin-37*, and *lin-15* were obtained that are defective in *B* function. When they mutagenized a *lin-9B* mutant, alleles of *lin-38* and *lin-15* were obtained that are defective in *A* function. As with *lin-8* and *lin-9*, any *A* or *B* allele by itself conferred no phenotype, and a combination of an *A* and a *B* mutation caused a multivulva phenotype. These and other experiments demonstrate that *lin-15* has three classes of mutations: those defective only in *A* function, those defective only in *B* function, and those that cause a multivulva phenotype. The multivulva mutations are genetically defined as defective in both *A* and *B* functions and hence are referred to as *AB* mutations. Moreover, *lin-15A* and *lin-15B* mutations complement: An animal carrying a *lin-15A* mutation in *trans* to a *lin-15B* mutation is wild type (Ferguson and Horvitz 1989).

Table 2. Examples of Interactions of *lin-15* and Related Genes

Genotype			
lin-8	*lin-9*	*lin-15*	Phenotype
+	+	+	wild type
A	+	+	wild type
+	B	+	wild type
A	B	+	multivulva
+	+	A	wild type
+	+	B	wild type
A	+	B	multivulva
+	B	A	multivulva
+	+	AB	multivulva

(+) Wild-type allele; (A) defective in an *A* activity; (B) defective in *B* activity; (AB) defective in both *A* and *B* activities.

It is not known how all of the genes with *A* and *B* functions interact. The *A* genes could act in a common pathway, and the *B* genes could act in a redundant pathway. Another possibility is that all *A* genes encode subunits of one protein complex, and all the *B* genes encode subunits of another protein complex (for further discussion, see Ferguson and Horvitz 1989). *lin-15A* and *B* could act at any step in this negative signaling pathway of *A* and *B* genes. For example, *lin-15A* could regulate the activity of one or more of the *A* genes or vice versa.

The molecular cloning of the *lin-15* locus confirms the genetic conclusion that *lin-15* encodes two functions that must be eliminated to result in a multivulva phenotype (L. Huang et al., in prep.; S. Clark and R. Horvitz, pers. comm.). A 15-kb genomic clone can complement *lin-15A*, *B*, and *AB* mutations in transgenic animals. This clone hybridizes to two distinct transcripts arranged tail to head. The more 5′ (upstream) transcript corresponds to *lin-15B* function: A genomic clone corresponding to only the upstream transcript rescues *lin-15B* mutants but not *lin-15A* mutants. The more 3′ (downstream) transcript corresponds to *lin-15A* function: A mutant defective in *lin-15A* function has a deletion of approximately 100 nucleotides in the downstream transcriptional unit. All *lin-15AB* alleles derived from wild-type strain N2 harbor deletions that disrupt both transcriptional units. These alleles were all obtained after ethylmethanesulfonate (EMS) mutagenesis, which typically makes transition mutations (Bejsovec and Anderson 1990). In contrast, only one of the *lin-15A* or *lin-15B* alleles has an obvious deletion, and thus it is likely that these mutations are point or pseudo-point mutations. However, deletions are not necessary to cause a multivulva phenotype. *lin-15AB* alleles obtained from mutagenesis of either *lin-15A* or *lin-15B* do not harbor deletions and are probably due to second tightly linked point mutations. The *lin-15* transcripts encode so far novel polypeptides, and thus additional experiments will have to be performed to deduce their biochemical functions.

The genetic mosaic analysis of Herman and Hedgecock (1990) demonstrates that *lin-15* cannot act only in the VPCs or the gonad but can act elsewhere in a distributed manner. A simple interpretation of their results is that *lin-15* acts in a negative signaling pathway that emanates from the hyp7 syncytial epidermis adjacent to the VPCs; this negative signal prevents vulval differentiation in the absence of the inductive signal (see Fig. 2). This analysis does not distinguish the sites of action of *lin-15A* and *B* because it was carried out with a *lin-15AB* mutant; thus, either the wild-type *lin-15A* or *lin-15B* functions (or both) could act in hyp7.

This negative signaling pathway, as represented by *lin-15*, acts to control the effect of *let-23* directly or indirectly. Identification of a *lin-15* null mutation allows a rigorous test of epistasis with *let-23* mutations. A severe *let-23* vulvaless mutation (one that does not, however, eliminate function of *let-23*; Aroian and Sternberg 1991) severely reduces the vulval differentiation of a *lin-15* null mutation (L. Huang and P. Sternberg, unpubl.), suggesting that *lin-15* normally acts as a negative regulator of *let-23* (Table 3). Since a *lin-15* mutation causes signal-independent vulval differentiation, the role of the *lin-15* pathway is to decrease somehow the basal (signal-independent) level of signal transduction by *let-23*. There are a number of ways in which this might be accomplished. First, the negative signaling pathway could affect *let-23* directly. For example, it could block the dimerization of Let-23 protein and thus lower basal kinase activity. Second, the negative signaling pathway could affect targets of *let-23*. For example, it might activate a protein tyrosine phosphatase that reverses the effect of *let-23* kinase or it might activate a negative regulator or *let-60 ras*, e.g., a GTPase-activating protein (GAP). Another possibility is that this pathway leads to activation of an SH2 protein that displaces factors necessary for signal transduction. Third, the negative signaling pathway could act upstream of *let-23*. For example, it might act to repress transcription of a component that activates *let-23* but not *lin-3*, since *lin-15* mutations are epistatic to *lin-3* mutations.

Table 3. The Extent of Vulval Differentiation Depends on the Inductive Signaling Pathway and *lin-15*

		Phenotype	
Genotype	AC	+	*lin-15*
+	+	wild type	multivulva
+	−	vulvaless	multivulva
let-23	+	vulvaless	vulvaless

(AC) Anchor cell; (+) present; (−) ablated.

Negative Regulation by *unc-101* and *sli-1*

In addition to negative regulatory activities revealed by the study of multivulva mutants, two other negative regulatory activities have been discovered by the study of extragenic suppressors of *let-23* vulvaless mutations. To identify negative regulators of the signaling pathway, two different partial reduction-of-function *let-23* mutations, which confer a vulvaless phenotype (Aroian and Sternberg 1991), were reverted (G. Jongeward and P. Sternberg, in prep.). We obtained several classes of mutations from these screens that phenotypically revert the vulvaless phenotype caused by the *let-23* mutation. One of the five existing *let-60(gf)* mutations was isolated, further supporting the hypothesis that *let-60* hyperactivity obviates the requirement for *let-23* in this pathway and that *let-60* is downstream from *let-23*. Two alleles of a known locus, *unc-101*, were isolated. Other suppressor mutations define a new X-chromosome locus, *sli-1* (*s*uppressor of *li*neage defect).

unc-101 and *sli-1* mutations differ in their properties. *unc-101* mutations have a variety of pleiotropic defects, including uncoordinated behavior (D. Riddle, pers. comm.), sensory ending defects of certain sensory neurons (D. Hall and E. Hedecock, pers. comm.), a defecation defect (Thomas 1990), abnormal male anatomy, and larval lethality. All existing *unc-101* alleles share this set of phenotypes. In contrast to *unc-101* mutations, *sli-1* mutations have no other known defects. However, *sli-1* mutations suppress other *let-23* mutant phenotypes that *unc-101* mutations apparently do not suppress.

sli-1 and *unc-101* mutations are recessive suppressors of partial reduction-of-function *let-23* mutations, but they do not suppress null alleles of *let-23*. Since *unc-101* and *sli-1* mutations can only suppress *let-23* mutations that retain some *let-23* activity, the normal function of these genes must be to antagonize the action of functional Let-23 protein. Because *sli-1* and *unc-101* mutations do not bypass the requirement for Let-23, they cannot be downstream negative regulators whose activities are antagonized by Let-23. *sli-1* and *unc-101* mutations are good suppressors of vulval induction defects of *lin-2*, *lin-7*, and *lin-10* but poor suppressors of *lin-3* or *let-60*. Although negative regulation by *sli-1* and *unc-101* could impinge on the signaling pathway at

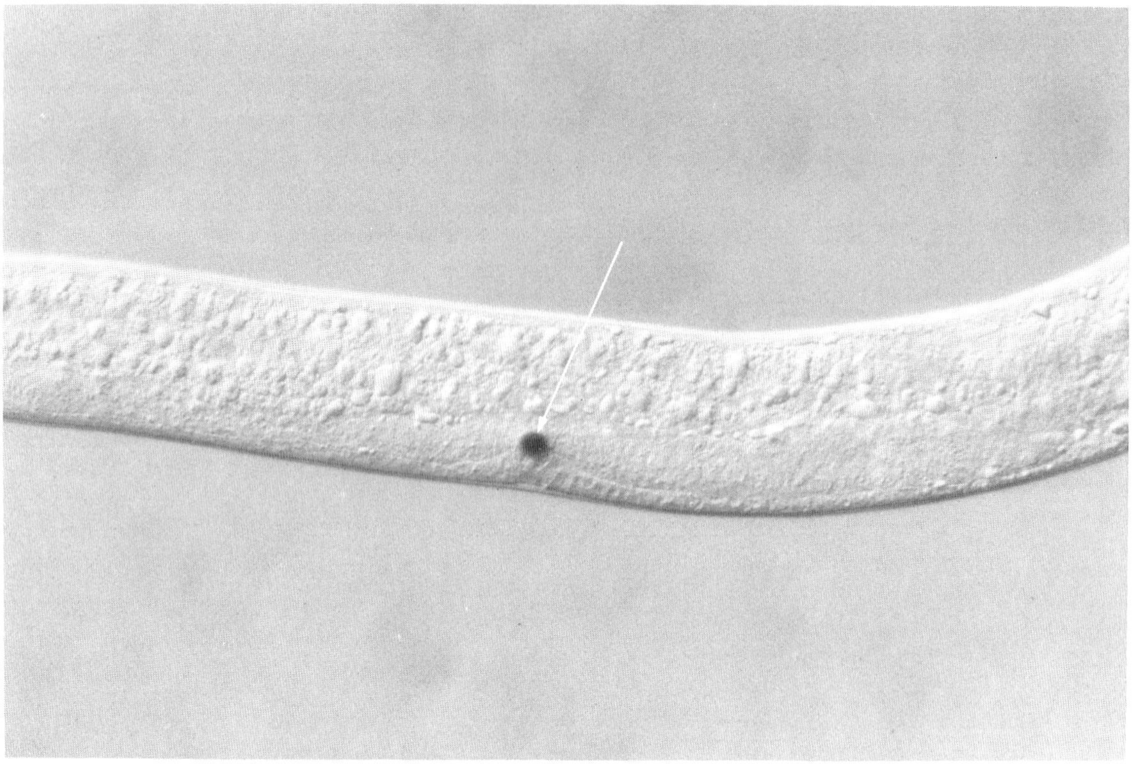

Figure 5. Nomarski photomicrograph of an L3 molt *lin-3::lacZ* transgenic hermaphrodite stained for β-galactosidase activity. Arrow indicates the anchor cell. The developing vulva is ventral (below in this micrograph) to the anchor cell. (For details, see Hill and Sternberg 1992.)

any step, these observations are most consistent with the targets of *sli-1* and *unc-101* being at the *let-23* step.

Mutations of either *unc-101* or *sli-1* in the absence of a *let-23* mutation confer no vulval induction defect (Table 4). However, animals defective in both *unc-101* and *sli-1* can be multivulval. Therefore, *sli-1* and *unc-101* must normally function during vulval induction. Vulval differentiation in the *unc-101; sli-1* double mutants is partially independent of inductive signal: Some vulval differentiation occurs in the absence of the anchor cell (G. Jongeward and P. Sternberg, in prep.). This partial independence of inductive signal also indicates that *sli-1* and *unc-101* do not act solely as negative regulators of *lin-3* in the anchor cell. It is possible that these genes prevent production of a ligand for *let-23* (*lin-3* or some other protein) in another tissue.

sli-1 and *unc-101* do not appear to interact strongly with the *A* and *B* classes of negative regulators. For example, strains carrying both an *unc-101* mutation and either a *lin-8A* or *lin-9B* mutation are not multivulval. Thus, there seem to be four separate negative regulatory activities. Although these might not all be direct, it is, however, conceivable that they all act on *let-23*. In addition to the possible mechanisms discussed above in the context of *lin-15*, *sli-1* and *unc-101* activities might be involved in downregulation (e.g., endoctyosis) or desensitization of Let-23 (e.g., serine/threonine phosphorylation) (for review, see Carpenter and Wahl 1990) because such negative regulation would not compensate for loss of receptor activity. It therefore would not be surprising if these four negative regulators were in fact all acting directly on *let-23*.

SUMMARY AND GENERAL CONCLUSIONS

On the basis of these results, we propose the following pathway (see Fig. 4). The inductive signal is en-

Table 4. Interactions of *let-23*, *unc-101*, and *sli-1*

let-23	*unc-101*	*sli-1*	Vulval differentiation	Gonad dependence
+	+	+	wild type	yes
sy1	+	+	< wild type	yes
sy1	−	+	≥ wild type	yes
sy1	+	−	≥ wild type	yes
+	−	+	wild type	yes
+	+	−	wild type	yes
+	−	−	≥ wild type	partial

coded by *lin-3* and acts on a receptor encoded by *let-23*. Activation of *let-23* leads to activation of *let-60 ras*, possibly by *sem-5*. *lin-15* and interacting genes act on or near *let-23* antagonistically to *lin-3* activation. *sli-1* and *unc-101* are distinct negative regulators of this signaling pathway.

Two types of observations indicate that the inductive signal is not likely to act by inhibiting *lin-15* activity. A *lin-15* mutant can be restored to approximately normal dependence on inductive signal by partially lowering *let-60 ras* activity (Han et al. 1990). In addition, the pattern of VPC fates is still responsive to the inductive signal in a *lin-15* mutant (Sternberg 1988). Similarly, *lin-3* is unlikely to act by inhibiting *sli-1* and *unc-101*. In hermaphrodites defective in both *sli-1* and *unc-101*, vulval differentiation still requires functional *let-23* (G. Jongeward and P. Sternberg, in prep.). Thus, with the caveat that the mutations used in this experiment might not completely eliminate function of these negative regulators, *sli-1* and *unc-101* are not in the main trunk of the inductive signaling pathway.

Vulval induction is of general interest in several ways. First, it provides a molecular genetic system to study the function of a particular class of intercellular signaling molecules: a growth factor, its receptor, and a Ras protein. The genetic identification of many pathways of negative regulation is expected to open up the molecular study of previously unidentified proteins, such as Lin-15A and B. Second, vulval induction is a paradigm for how multiple intercellular signals cooperate to specify the precise pattern of cell fates. The plethora of negative regulators might be involved in the integration of the multiple intercellular signals. Alternatively, they might simply reflect general cell mechanisms that impinge on many signaling pathways. If so, vulval induction provides a convenient in vivo assay with which to study these mechanisms.

ACKNOWLEDGMENTS

We thank Helen Chamberlin, Scott Clark, Ed Hedgecock, David Hall, Bob Horvitz, and Don Riddle for sharing their unpublished results and Paul Kayne, Bino Palmer, Kyria Tietze, and Junho Lee for critical reading of the manuscript. We are especially grateful to Craig Mello, Victor Ambros, Jim Kramer, and Andy Fire for continually sharing their expertise in the production of transgenic animals and Roger Hoskins, Denise Clark, and David Baillie for providing markers for the production transgenic animals. P.W.S. is an investigator of the Howard Hughes Medical Institute and a Presidential Young Investigator of the National Science Foundation. Our research reviewed here was supported by the Howard Hughes Medical Institute and grants to P.W.S. from the U.S. Public Health Service (HD-23690), the March of Dimes Birth Defects Foundation, and the Searle Scholars Program/Chicago Community Trust.

REFERENCES

Aroian, R.V. and P.W. Sternberg. 1991. Multiple functions of *let-23*, a *C. elegans* receptor tyrosine kinase gene required for vulval induction. *Genetics* **128:** 251.

Aroian, R.V., M. Koga, J.E. Mendel, Y. Ohshima, and P.W. Sternberg. 1990. The *let-23* gene necessary for *Caenorhabditis elegans* vulval induction encodes a tyrosine kinase of the EGF receptor subfamily. *Nature* **348:** 693.

Beitel, G., S. Clark, and H.R. Horvitz. 1990. The *Caenorhabditis elegans ras* gene *let-60* acts as a switch in the pathway of vulval induction. *Nature* **348:** 503.

Bejsovec, A. and P. Anderson. 1990. Functions of the myosin ATP and actin binding sites are required for *C. elegans* thick filament assembly. *Cell* **60:** 133.

Bourne, H.R., D.A. Sanders, and F. McCormick. 1990. The GTPase superfamily: A conserved switch for diverse cell functions. *Nature* **348:** 125.

———.1991. The GTPase superfamily: Conserved structure and molecular mechanism. *Nature* **349:** 117.

Carpenter, G. and M.I. Wahl. 1990. The epidermal growth factor family. *Handb. Exp. Pharmacol.* **95:** 69.

Clark, S.G., M.J. Stern, and H.R. Horvitz. 1992. *C. elegans* cell-signalling gene *sem-5* encodes a protein with SH2 and SH3 domains. *Nature* **356:** 340.

Coulson, A.R., J. Sulston, S. Brenner, and J. Karn. 1986. Toward a physical map of the genome of the nematode *Caenorhabditis elegans*. *Proc. Natl. Acad. Sci.* **83:** 7821.

Coulson, A., R. Waterston, J. Kiff, J. Sulston, and Y. Kohara. 1988. Genome linking with yeast artificial chromosomes. *Nature* **335:** 84.

Ferguson, E. and H.R. Horvitz. 1985. Identification and characterization of 22 genes that affect the vulval cell lineages of *Caenorhabditis elegans*. *Genetics* **110:** 17.

———.1989. The multivulva phenotype of certain *C. elegans* mutants results from defects in two functionally-redundant pathways. *Genetics* **123:** 109.

Ferguson, E.L., P.W. Sternberg, and H.R. Horvitz. 1987. A genetic pathway for the specification of the vulval cell lineages of *Caenorhabditis elegans*. *Nature* **326:** 259.

Fire, A., S. White-Harrison, and D. Dixon. 1990. A modular set of *lacZ* fusion vectors for studying gene expression in *Caenorhabditis elegans*. *Gene* **93:** 189.

Greenwald, I. and G.M. Rubin. 1992. Making a difference: The role of cell-cell interactions in establishing separate identities for equivalent cells. *Cell* **68:** 271.

Greenwald, I.S., P.W. Sternberg, and H.R. Horvitz. 1983. The *lin-12* locus specifies cell fates in *Caenorhabditis elegans*. *Cell* **34:** 435.

Han, M. and P.W. Sternberg. 1990. *let-60*, a gene that specifies cell fates during *C. elegans* vulval induction, encodes a ras protein. *Cell* **63:** 921.

———.1991. Analysis of dominant negative mutations of the *Caenorhabditis elegans let-60 ras* gene. *Genes Dev.* **5:** 2188.

Han, M., R. Aroian, and P.W. Sternberg. 1990. The *let-60* locus controls the switch between vulval and non-vulval cell types in *C. elegans*. *Genetics* **126:** 899.

Herman, R.K. and E.M. Hedgecock. 1990. The size of the *C. elegans* vulval primordium is limited by *lin-15* expression in surrounding hypodermis. *Nature* **348:** 169.

Hill, R.J. and P.W. Sternberg. 1992. The gene *lin-3* encodes an inductive signal for vulval development in *C. elegans*. *Nature* **358:** 470.

Holmes, W.E., M.X. Sliwkowski, R.W. Akita, W.J. Henzel, J. Lee, J.W. Park, D. Yansura, N. Abadi, H. Raab, G.D. Lewis, H.M. Shepard, W. Kuang, W.I. Wood, D.V. Goeddel, and R.L. Vandlen. 1992. Identification of heregulin, a specific activator of $p185^{erbB2}$. *Science* **256:** 1205.

Horvitz, H.R. and P.W. Sternberg. 1991. Multiple intercellular signalling systems control the development of the *C. elegans* vulva. *Nature* **351:** 535.

Horvitz, H.R. and J.E. Sulston. 1980. Isolation and genetic

characterization of cell-lineage mutants of the nematode *Caenorhabditis elegans*. *Genetics* **96:** 435.

Hunter, C.P. and W.B. Wood. 1992. Evidence from mosiac analysis of the masculinizing gene *her-1* for cell interactions in *C. elegans* sex determination. *Nature* **355:** 551.

Hunter, T. 1991. Cooperation between oncogenes. *Cell* **64:** 249.

Kim, S. K. and H. R. Horvitz. 1990. The *Caenorhabditis elegans* gene *lin-10* is broadly expressed while required specifically for the determination of vulval cell fates. *Genes Dev.* **4:** 357.

Kimble, J. 1981. Lineage alterations after ablation of cells in the somatic gonad of *Caenorhabditis elegans*. *Dev. Biol.* **87:** 286.

Kimble, J. and D. Hirsh. 1979. Post-embryonic cell lineages of the hermaphrodite and male gonads in *Caenorhabditis elegans*. *Dev. Biol.* **70:** 396.

Koch, C.A., D. Anderson, M.F. Moran, C. Ellis, and T. Pawson. 1991. SH2 and SH3 domains: Elements that control interactions of cytoplasmic signaling proteins. *Science* **252:** 668.

Massagué, J. 1990. Transforming growth factor-α. *J. Biol. Chem.* **265:** 21393.

Mello, C.C., J.M. Kramer, D. Stinchcomb, and V. Ambros. 1992. Efficient gene transfer in *C. elegans* after microinjection of DNA into germline cytoplasm: Extrachromosomal maintenance and integration of transforming sequences. *EMBO J.* **10:** 3959.

Priess, J.R. and J.N. Thomson. 1987. Cellular interactions in early *Caenorhabditis elegans* embryos. *Cell* **48:** 241.

Schnabel, R. 1991. Cellular interactions involved in the determination of the early *C. elegans* embryo. *Mech. Dev.* **34:** 85.

Seydoux, G. and I. Greenwald. 1989. Cell autonomy of *lin-12* function in a cell fate decision in *C. elegans*. *Cell* **57:** 1237.

Sternberg, P. W. 1988. Lateral inhibition during vulval induction in *Caenorhabditis elegans*. *Nature* **335:** 551.

Sternberg, P.W. and H.R. Horvitz. 1986. Pattern formation during vulval development in *Caenorhabditis elegans*. *Cell* **44:** 761.

———. 1989. The combined action of two intercellular signalling pathways specifies three cell fates during vulval induction in *C. elegans*. *Cell* **58:** 679.

Sulston, J. and H.R. Horvitz. 1977. Postembryonic cell lineages of the nematode *Caenorhabditis elegans*. *Dev. Biol.* **56:** 110.

———. 1981. Abnormal cell lineages in mutants of the nematode *Caenorhabditis elegans*. *Dev. Biol.* **82:** 41.

Sulston, J.E. and J.G. White. 1980. Regulation and cell autonomy during postembryonic development of *Caenorhabditis elegans*. *Dev. Biol.* **78:** 577.

Sulston, J.E., E. Schierenberg, J.G. White, and J.N. Thomson. 1983. The embryonic cell lineage of the nematode *Caenorhabditis elegans*. *Dev. Biol.* **100:** 64.

Sulston, J., Z. Du, K. Thomas, R. Wilson, L. Hillier, R. Staden, N. Halloran, P. Green, J. Thierry-Mieg, L. Qiu, S. Dear, A. Coulson, M. Craxton, R. Durbin, M. Berks, M. Metzstein, T. Hawkins, R. Ainscough, and R. Waterston. 1992. The *C. elegans* genome sequencing project: A beginning. *Nature* **356:** 37.

Thomas, J. H. 1990. Genetic analysis of defecation in *Caenorhabditis elegans*. *Genetics* **124:** 855.

Thomas, J.H., M.J. Stern, and H.R. Horvitz. 1990. Cell interactions coordinate the development of the *C. elegans* egg-laying system. *Cell* **62:** 1041.

Ullrich, A. and J. Schlessinger. 1990. Signal transduction by receptors with tyrosine kinase activity. *Cell* **61:** 203.

Wen, D., E. Peles, R. Cupples, S.V. Suggs, S.S. Bacus, Y. Luo, G. Trail, S. Hu, S.M. Silbiger, R.B. Levy, R.A. Koski, H.S. Lu, and Y. Yarden. 1992. Neu differentiation factor: A transmembrane glycoprotein containing an EGF domain and an immunoglobulin homology unit. *Cell* **69:** 559.

Wood, W.B. 1991. Evidence from reversal of handedness in *C. elegans* for early cell interactions determining cell fates. *Nature* **349:** 536.

Yochem, J., K. Weston, and I. Greenwald. 1988. *C. elegans lin-12* encodes a transmembrane protein similar to *Drosophila Notch* and yeast cell cycle gene products. *Nature* **335:** 547.

Genes Involved in Two *Caenorhabditis elegans* Cell-signaling Pathways

S.G. Clark, M.J. Stern,* and H.R. Horvitz
Howard Hughes Medical Institute, Department of Biology, Massachusetts Institute of Technology, Cambridge, Massachusetts 02139

Cell interactions are responsible for many aspects of animal development (see, e.g., Gurdon 1992). What are the signals, receptors, and signal transduction molecules that function as cells communicate with each other during development? Answers to this question are only beginning to emerge from studies in developmental biology (see, e.g., Greenwald and Rubin 1992; Hynes and Lander 1992; Jessell and Melton 1992). To what extent are particular intercellular signaling molecules shared among different sets of interacting cells? Are the same signaling molecules and pathways used in different cell types, at different locations, and at different times during development? Or does each group of interacting cells use a unique signaling pathway? Answers to these questions would provide fundamental insights into the molecular basis of animal development.

To address these issues, we are analyzing the role of cell interactions in the development of the nematode *Caenorhabditis elegans*. The complete development of *C. elegans* has been described at the level of single cells. For example, the *C. elegans* cell lineage is known to be essentially invariant and to generate a total of 959 somatic nuclei in the adult (Sulston and Horvitz 1977; Kimble and Hirsh 1979; Sulston et al. 1983). The invariance of *C. elegans* development reflects to a significant extent an invariance in cell interactions that regulate patterns of cell division, cell migration, cell differentiation, and cell death (for review, see Sulston 1988; Lambie and Kimble 1991). That reproducible cell interactions can be studied at the resolution of individual cells has greatly facilitated the analysis of intercellular signaling in *C. elegans*.

In this paper, we describe studies of how intercellular signaling controls two distinct aspects of *C. elegans* development: the migration of a pair of sex myoblasts and the induction of the vulva. The roles of individual cells in these signaling processes have been analyzed using a laser microbeam to kill specific cells and thereby reveal the ways in which these cells influence the fates of other cells. Mutations that perturb sex myoblast migration or vulval induction have identified genes that function in these as well as in other *C. elegans* cell signaling pathways. At least one of these genes acts in both sex myoblast migration and vulval induction. Genes involved in intercellular signaling in *C. elegans* are similar to genes that function during the development of other organisms, as well as to oncogenes associated with neoplastic growth.

Cell Interactions Coordinate the Development of the Egg-laying System

Sex myoblast migration and vulva induction are involved in the development of the egg-laying system of the adult *C. elegans* hermaphrodite. Egg laying by *C. elegans* requires the concerted action of the gonad, the vulva, the vulval and uterine muscles, and a pair of serotonergic motor neurons known as the hermaphrodite-specific neurons (HSNs) (Fig. 1). Eggs stored in

*Present address: Boyer Center for Molecular Medicine, Yale University, New Haven, Connecticut 06536.

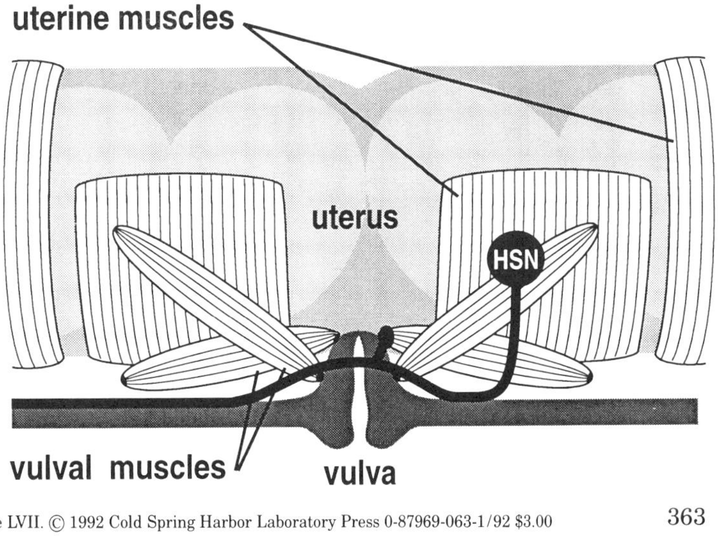

Figure 1. The *C. elegans* egg-laying system. Left lateral view of the egg-laying system, which consists of four components: the uterus, which stores eggs; the vulva, an opening that consists of specialized hypodermal cells and that connects the uterus to the external environment; the eight uterine and eight vulval muscles, which contract to squeeze the uterus and open the vulva, respectively; and the two HSNs, which innervate the egg-laying muscles. Only the four uterine and four vulval muscle cells and the one HSN neuron located on the left side of the animal are shown. Dorsal is up, and anterior is to the left. (Based on Sulston and Horvitz [1977] and White et al. [1986 and pers. comm.].)

the uterus are expelled through the vulva, an opening in the ventral hypodermis, by the contraction of the 16 uterine and vulval muscles, which squeeze the uterus and open the vulva. These egg-laying muscles are stimulated to contract by the HSNs. The proper functioning of all four components of the egg-laying system is essential for normal egg laying.

A cascade of cell interactions regulates the development of the egg-laying system. Specifically, the developing gonad attracts a pair of migrating myoblasts and thus determines the positions of the vulval and uterine muscles derived from these myoblasts (Thomas et al. 1990). The developing gonad also induces the formation of the vulva (for review, see Horvitz and Sternberg 1991). Finally, the developing vulval cells control the branching and synapse formation of the HSNs (Li and Chalfie 1990; Thomas et al. 1990; G. Garriga et al., in prep.). Below, we focus on how signals from the gonad control the development of the egg-laying muscles and the vulva.

The Gonad Attracts the Migrating Sex Myoblasts

The 16 vulval and uterine muscles involved in egg laying are derived from two precursor cells, the sex myoblasts or SMs (Fig. 2a) (Sulston and Horvitz 1977). The SMs are generated in the posterior ventral muscle quadrants during the first larval stage and are bilaterally symmetric in their positions. During the second larval stage, these cells migrate anteriorly to flank the precise center of the developing gonad. Then, during the third larval stage, the SMs undergo three rounds of cell division, and each produces four vulval and four uterine muscles.

The migration of the SMs is controlled in part by a signal from the somatic gonad (Thomas et al. 1990). If cells of the somatic gonad are killed using a laser microbeam, the SMs initiate migration normally but fail to assume their normal final positions flanking the precise center of the developing gonad; instead, the SMs in such animals are broadly distributed around their normal termination points (Fig. 2b). Thus, the gonad specifies the target site of the SM migration. That a gonadal signal can act at a distance to attract the migrating SMs was revealed by studies of a mutant defective in the gene *dig-1* (*dig*, *di*splaced *g*onad) (Fig. 2c). In *dig-1* animals, the gonad can be displaced anteriorly and dorsally (Thomas et al. 1990), apparently because of defects in adhesion between the developing gonad and its normal attachment site along the ventral hypodermis (M. Basson et al., unpubl.). The SMs in *dig-1* animals migrate to the position of the displaced gonad; if the somatic gonad is killed in a *dig-1* animal with a dorsal gonad, the SMs migrate anteriorly but fail to migrate to the dorsal side (Fig. 2d) (Thomas et al. 1990). These experiments indicate that a signal from the somatic gonad attracts the migrating SMs to their final positions.

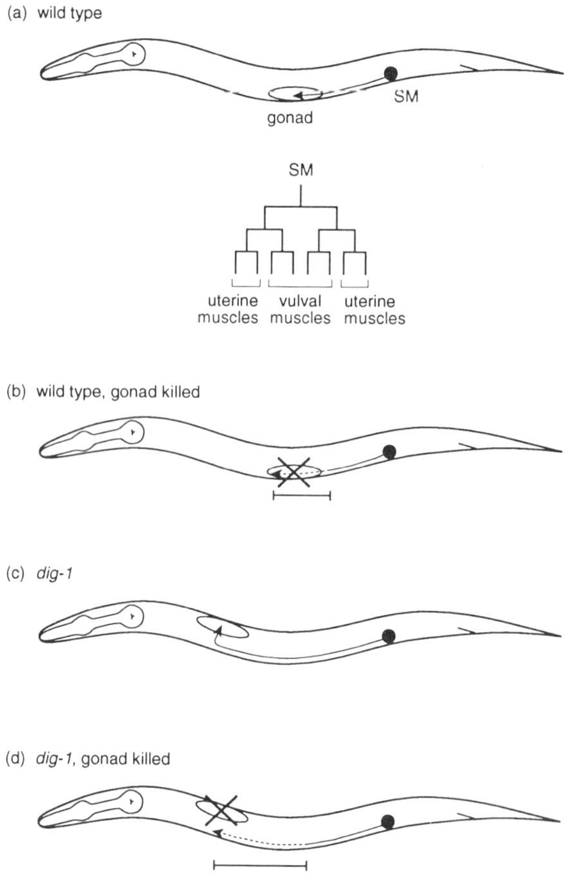

Figure 2. Development of the egg-laying muscles. (*a*) The vulval and uterine muscles are generated by the two sex myoblast (SM) cells (Sulston and Horvitz 1977). One SM is generated on each side of the late first-stage larva. During the second larval stage, each SM migrates anteriorly until it reaches a position flanking the center of the developing gonad. During the third larval stage, each SM undergoes three rounds of division, generating eight descendants that differentiate into four vulval and four uterine muscle cells. (*b*) In wild-type animals in which the gonad is killed with a laser microbeam, the SMs migrate anteriorly to a variable position along the ventral side (Thomas et al. 1990). The dotted line indicates that there is variability from animal to animal in the final positions of the SMs, and the bar below the animal indicates the extent of the variability. (*c*) In *dig-1* mutant animals, the gonad primordium can be displaced either anteriorly or both anteriorly and dorsally; the SMs migrate to the position of the displaced gonad (Thomas et al. 1990). (*d*) If the gonad is killed with a laser microbeam in a *dig-1* animal with a dorsally positioned gonad, the SMs do not migrate to a dorsal position, but rather remain ventral (Thomas et al. 1990). The dotted line indicates that there is variability from animal to animal in the final positions of the SMs, and the bar below the animal indicates the extent of the variability.

Genes Involved in Sex Myoblast Migration

Mutations in the genes *egl-15* and *egl-17* (*egl*, *eg*g-*l*aying defective) cause the premature termination of the SM migrations (Stern and Horvitz 1991). These genes were discovered because *egl-15* and *egl-17* mutants are defective in egg laying as a consequence of displaced vulval and uterine muscles, which fail to make their normal attachments to the uterus, vulva, and hypodermal wall (Trent et al. 1983; Stern and Horvitz 1991). Killing specific cells of the somatic gonad in *egl-15* and *egl-17* mutants with a laser microbeam allows the SMs to migrate further, revealing that in these mutants, the gonad stops the migration of the SMs (Stern and Horvitz 1991). The same gonadal cells that stop the SMs in *egl-15* and *egl-17* mutants attract the SMs in wild-type animals. These observations indicate that *egl-15* and *egl-17* mutations alter communication between the gonad and the SMs.

The *egl-15* gene is needed not only for SM migration, but also for early larval development (M.J. Stern and H.R. Horvitz, in prep.): *egl-15* mutations that affect SM migration only partially inactivate the gene, whereas complete loss-of-function *egl-15* mutations cause a general arrest of development at an early larval stage. The lethal phenotype caused by a temperature-sensitive lethal *egl-15* allele can be blocked by mutations in the gene *clr-1* (*clr*, *clear*, reflecting the clarity with which cell boundaries can be observed in these mutants; Hedgecock et al. 1990; M.J. Stern and H.R. Horvitz, in prep.). Interestingly, not only can *clr-1* mutations suppress the lethal phenotype caused by this *egl-15* mutation, but also some *egl-15* mutations can suppress the lethal phenotype caused by *clr-1* mutations. This finding suggested that both additional alleles of *egl-15* and alleles of additional genes like *egl-15* might be isolated as suppressors of *clr-1* mutations. Such suppressors were obtained and indeed led to the identification of new *egl-15* alleles, as well as of mutations in a number of newly discovered genes, including *sem-5* (*sem*, *sex muscle* abnormal) (Clark et al. 1992; M.J. Stern and H.R. Horvitz, in prep.). Like mutations in the *egl-15* and *egl-17* genes, mutations in the *sem-5* gene cause the posterior displacement of the sex myoblasts, apparently by perturbing communication between the gonad and the SMs. As discussed below, *sem-5* also functions in the intercellular signaling pathway responsible for induction of the vulval cell lineages.

The Gonad Induces Vulva Formation

In addition to controlling sex myoblast migration, the developing gonad also induces underlying hypodermal cells to divide and differentiate to form the vulva. If the gonad is killed with a laser microbeam, no vulva is formed (Sulston and White 1980). If the gonad is displaced anteriorly, as in *dig-1* mutants, the vulva is formed at a more anterior location (Thomas et al. 1990).

The role of cell interactions in controlling vulval development has been discussed recently in detail by Horvitz and Sternberg (1991). In brief, according to our current model, six cells—P3.p, P4.p, P5.p, P6.p, P7.p, P8.p (so-named because they are the posterior daughters of the cells P3, P4, etc.) —have the potential to generate vulval tissue. Each of these six cells can express any of three distinct fates, called 1°, 2°, and 3°; these fates are distinguished by their distinct patterns of cell divisions and by the progeny cell types they generate (Fig. 3a). A single cell of the somatic gonad, the anchor cell, normally induces the three closest precursor cells (P5.p, P6.p, and P7.p) to express the 2°, 1°, and 2° fates, respectively, which generate the 22 descendants that form the vulva (Fig. 3b). P3.p, P4.p, and P8.p, which are located further from the anchor cell, normally are uninduced and express the 3° fate, which generates nonvulval descendants that fuse with the hypodermal syncytium that envelops the animal. If the anchor cell is killed with a laser microbeam, all six of these cells express the nonvulval 3° fate (Fig. 3c). If the anchor cell is displaced with respect to P3.p–P8.p, as in *dig-1* animals, a corresponding shift occurs in which precursor cells are induced to express vulval cell lineages (Fig. 3d).

The inductive signal from the anchor cell has been proposed to function by counteracting a signal from the adjacent syncytial hypoderm, which inhibits the expression of vulval cell lineages by P3.p–P8.p (Fig. 3b) (Herman and Hedgecock 1990). If this inhibitory hypodermal signal is defective, all six precursor cells P3.p–P8.p express 1° and 2° fates, leading to the generation of multiple vulva-like structures (Fig. 3e). Additional interactions among the induced cells prevent adjacent P3.p–P8.p cells from both expressing a 1° lineage (Fig. 3b) (Sternberg 1988).

Genes Required for Vulval Induction

Many mutants have been isolated that display defects in the vulval cell lineages (Horvitz and Sulston 1980; Greenwald et al. 1983; Ferguson and Horvitz 1985, 1989). These mutants belong to two general phenotypic classes: vulvaless (Vul), in which no vulva is formed, and multivulva (Muv), in which multiple vulva-like structures are produced. In certain vulvaless mutants, all six cells P3.p–P8.p express the 3° cell fate (Fig. 3c). This phenotype is identical to that of animals lacking the anchor cell, suggesting that these mutants are defective in the signaling process required for vulval induction. In contrast, in many multivulva mutants, P3.p–P8.p all express either a 1° or a 2° cell fate, even in the absence of the inductive signal from the anchor cell (Fig. 3e). In these mutants, the inductive pathway seems to be activated independently of the inductive signal, presumably either because of a failure in the signal from the syncytial hypoderm that inhibits P3.p–P8.p from expressing vulval fates or because of an event that overcomes or bypasses this inhibition.

To identify additional genes involved in vulval induction, we and other investigators isolated suppressors of the multivulva phenotype caused by *lin-15* (*lin*, *lineage* abnormal) mutations (Beitel et al. 1990; Han et al. 1990; Aroian and Sternberg 1991; Clark et al. 1992; S.G. Clark and H.R. Horvitz, in prep.). Since *lin-15* mutations cause a multivulva phenotype even in the absence of an anchor cell, it seemed likely that mutations that prevented expression of the *lin-15* mutant phenotype would not act by blocking synthesis or re-

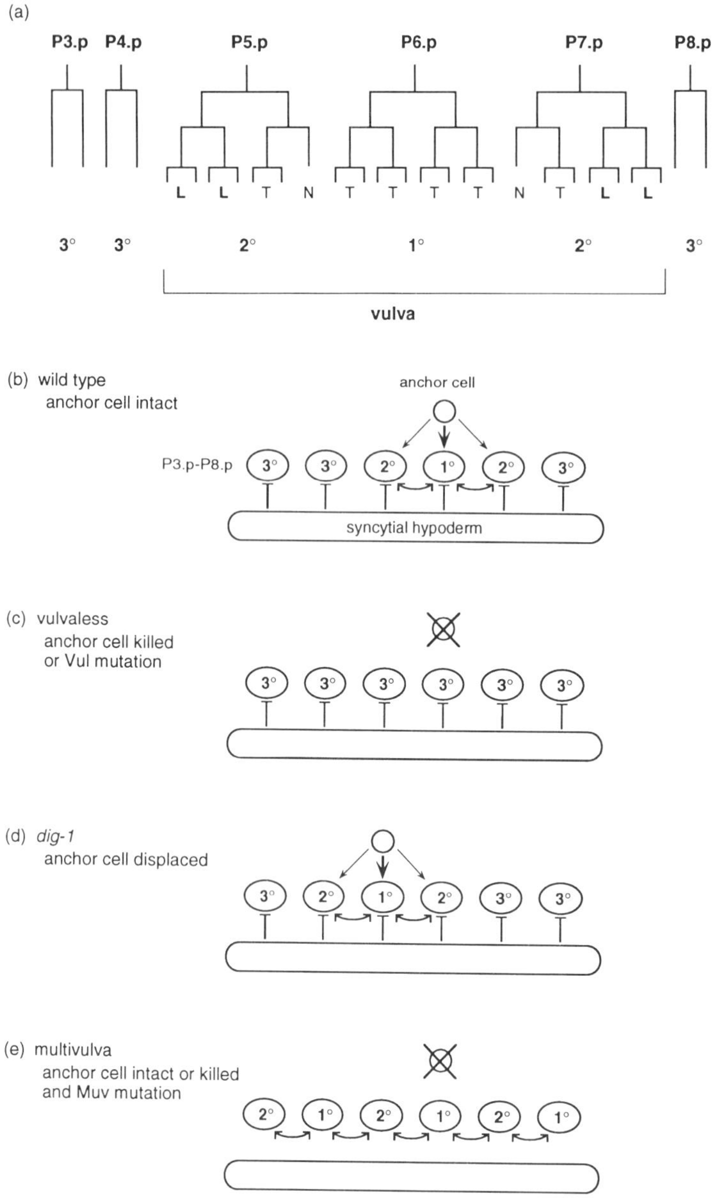

Figure 3. Models for vulval induction in wild-type, vulvaless, and multivulva animals. (*a*) Each P3.p–P8.p cell expresses one of three cell lineages, referred to as the 1°, 2°, and 3° cell fates. The cells P5.p–P7.p express the 1° and 2° cell fates, which generate the 22 descendants that form the vulva. P3.p, P4.p, and P8.p express the 3° cell fate, which generates two nonvulval progeny that join the syncytial hypoderm that envelops the animal. L, T, and N are distinct cell fates expressed by the descendants of P5.p, P6.p, and P7.p (see Horvitz and Sternberg 1991). (*b*) A signal (or signals) from the gonadal anchor cell induces vulval development by the cells P5.p–P7.p. P3.p, P4.p, and P8.p are uninduced and express the 3° cell fate. In addition, interactions among P3.p–P8.p prevent neighboring cells from both expressing the 1° cell fate, and interactions between the syncytial hypoderm and P3.p–P8.p prevent the uninduced cells P3.p, P4.p, and P8.p from expressing 1° and 2° cell fates. (*c*) In vulvaless (Vul) animals, all six cells P3.p–P8.p express the 3° cell fate, and no vulva is formed. Wild-type animals in which the gonadal anchor cell has been killed with a laser microbeam are vulvaless, as are mutant animals defective in genes that are needed for the functioning of the inductive signaling pathway (*lin-3*, *let-23*, *sem-5*, *let-341*, *let-60*, *lin-45*). (*d*) In *dig-1* animals, the anchor cell is displaced anteriorly, causing, for example, P4.p–P6.p to be induced and express the 1° and 2° cell fates. (*e*) In multivulva (Muv) animals, all six cells P3.p–P8.p express either the 1° or the 2° cell fate, and extra vulva-like structures are formed whether or not an anchor cell is present. Mutant animals can be Muv because of any of three defects: a failure in the signal from the syncytial hypoderm that normally prevents P3.p, P4.p, and P8.p from expressing vulval cell lineages (as in *lin-15* mutants and as depicted in this figure); the constitutive activation of a gene that functions within the signal transduction pathway (as in *let-60 ras* activation mutants); or the inactivation of a gene that functions within the signal transduction pathway to prevent the expression of vulval cell lineages (as in *lin-1* mutants). (Adapted from Beitel et al. [1990] and Horvitz and Sternberg [1991]; also, see text.)

lease of the inducing signal, but rather could act by blocking the signal transduction pathway that responds to that signal. Of the *lin-15* suppressor mutations we isolated in this way, some caused a vulvaless phenotype like that of animals that lack an anchor cell. The genes defined by these mutations were candidates for functioning in the inductive signaling pathway. Our experiments identified five such suppressor genes: *let-23, sem-5, let-60, let-341*, and *lin-45*.

let-23 Encodes a Receptor Tyrosine Kinase

One of these genes, *let-23* (*let*, *let*hal; severe mutations cause a larval lethal phenotype), was previously known to be required for vulval induction (Ferguson et al. 1987; Aroian and Sternberg 1991). *let-23* encodes a receptor tyrosine kinase similar to the epidermal growth factor (EGF) receptor (Aroian et al. 1990). Upon ligand binding, receptor tyrosine kinases are activated to form membrane-associated protein complexes and to phosphorylate specific tyrosine residues on intracellular proteins (Ullrich and Schlessinger 1990). On the basis of its genetic properties and the predicted structure of its protein product, the *let-23* gene probably acts within P3.p–P8.p and encodes the receptor of the inductive signal.

sem-5 Encodes a Protein with SH2 and SH3 Domains

A second gene, *sem-5*, acts not only in vulval induction, but also, as described above, in the regulation of sex myoblast migration (Clark et al. 1992). Of the six *sem-5* alleles, three were isolated as *lin-15* suppressors and, like *let-23* mutations, cause a noninduced, vulvaless phenotype (all six cells P3.p–P8.p express a 3° cell fate). Also like *let-23* mutations, these *sem-5* mutations

Table 1. *sem-5* and *clr-1* Gene Interactions

Genotype	Lethal (%)	Vul (%)	Wild type (%)
sem-5(n2019)	41	58	1
clr-1(e1745); sem-5(n2019)	42	42	16

The *clr-1(e1745ts)* mutation partially suppresses the vulvaless (Vul) phenotype but not the larval lethal (Lethal) phenotype caused by the *sem-5(n2019)* mutation, since the frequency of phenotypically wild-type individuals is greatly increased by the presence of the *clr-1* mutation. The *clr-1(e1745ts)* mutation also suppresses the Vul phenotype caused by the *sem-5(n1619)* mutation (M.J. Stern and H.R. Horvitz, in prep.). (Lethal) Animals that arrested during larval development and died as rod-like L1 or early L2 larvae; (Vul) vulvaless animals, which lacked a functional vulva; (Wild type) animals that had a functional vulva. For the experiments involving *sem-5(n2019)* and *clr-1(e1745ts); sem-5(n2019)* animals, 332 and 404 individuals were scored, respectively. Animals were raised at 20°C.

result in an incompletely penetrant rod-like larval lethal phenotype characteristic of many mutants defective in vulval induction. These *sem-5* mutations also cause a posterior displacement of the final positions of the sex myoblasts and suppress the clear and sterile phenotype of *clr-1* mutants. In contrast, the three *sem-5* mutations isolated as *clr-1* suppressors in the search for genes involved in sex myoblast migration do not result in larval lethality and cause only minimal defects in vulval induction. It appears that the search for *lin-15* suppressors, which was performed in a manner that allowed for the recovery of lethal mutations, led to the isolation of more severe *sem-5* alleles than did the search for *clr-1* suppressors. The *clr-1(e1745ts)* mutation also partially suppresses the vulvaless (but not the larval lethal) phenotype caused by *sem-5* mutations (Table 1). Although *clr-1* mutations do not cause vulval defects, this observation suggests that *clr-1* might function in vulval induction as well as in the regulation of sex myoblast migration.

The *sem-5* gene is predicted to encode a 228-amino-acid protein that consists almost entirely of one SH2 and two SH3 domains in the order SH3-SH2-SH3 (Clark et al. 1992). SH2 and SH3 src homology regions, first observed in the *src* family of protein tyrosine kinases (Sadowski et al. 1986), are present in many signaling proteins regulated by receptor and nonreceptor tyrosine kinases (Koch et al. 1991). Several SH2 domains are known to interact with phosphotyrosine-containing proteins (Moran et al. 1990; Koch et al. 1991; Matsuda et al. 1991). SH3 domains are suspected to mediate association with the cytoskeleton and cell membrane (Rodaway et al. 1989; Koch et al. 1991). Thus, *sem-5* might interact with phosphotyrosine-containing proteins to effect signal transduction. A recently identified human protein, GRB2 (growth factor receptor-bound protein 2), shows striking similarity to the Sem-5 protein (J. Schlessinger, pers. comm.). GRB2 associates with the intracellular domain of the activated EGF receptor, which supports the hypothesis that Sem-5 interacts directly with the Let-23 receptor tyrosine kinase.

Of the six *sem-5* mutations, one alters a highly conserved proline in the amino-terminal SH3 domain and another alters a highly conserved glycine in the carboxy-terminal SH3 domain (Fig. 4). The observations that mutations in either SH3 domain can decrease *sem-5* activity and that the two SH3 domains differ considerably in sequence indicate that these two domains are not redundant and that each is essential for normal function. Two other *sem-5* mutations alter adjacent codons in the SH2 domain, indicating that this domain is also needed for *sem-5* function, presumably for binding to phosphotyrosine-containing proteins.

let-60 Encodes a Ras Protein

A third gene identified by suppressors of the *lin-15* multivulva phenotype, *let-60*, acts as a genetic switch in the vulval inductive pathway (Beitel et al. 1990; Han et al. 1990). Mutations that decrease *let-60* activity confer a vulvaless phenotype, whereas mutations that increase *let-60* activity cause a multivulva phenotype. Thus, the activity of *let-60* specifies the state of the vulval inductive pathway: When *let-60* is active, the inductive pathway is active and a P3.p–P8.p cell will express a vulval 1° or 2° fate; when *let-60* is inactive, the inductive pathway is inactive and a P3.p–P8.p cell will express the nonvulval 3° fate. Presumably, in normal development, the anchor cell inductive signal activates *let-60*. Mutations that completely inactivate *let-60* result in a recessive rod-like larval lethal phenotype like that caused by severe *let-23* and *sem-5* mutations. Some *let-60* mutations, known as dominant-negative alleles, inactivate *let-60* and also antagonize wild-type *let-60*

```
                                                        L
                                                        ↑
sem-5 N   AEHDFQAGSPDELSFKRGNTLKVLNKDEDPHWYKAELDGNEGFIPSNYI
          |  ||      || ||||      ||| || |    |    |  ||||
sem-5 C   ALFDFNPQESGELAFKRGDVITLINKD DPNWWEGQLNNRRGIFPSNYV
                                                   ↓
                                                   R
```

Figure 4. *sem-5* SH3 domains. The amino (N)- and carboxy (C)-terminal SH3 domains of the Sem-5 protein differ in 28 of 49 positions (Clark et al. 1992). The mutation *n1619* causes a substitution of leucine for proline in the amino-terminal SH3 domain, and the mutation *n2195* causes a substitution of arginine for glycine in the carboxy-terminal SH3 domain. Vertical lines indicate amino acids that are identical in the two SH3 domains. The gap indicates that the amino-terminal domain has one extra amino acid in this region.

function in a mutant/ + heterozygote, causing a dominant vulvaless as well as a recessive larval lethal phenotype.

Molecular analysis indicated that *let-60* encodes a Ras protein in which 136 of its first 164 amino acids are identical to those encoded by the human N-*ras* gene (Han and Sternberg 1990). Ras proteins bind GDP and GTP, hydrolyze GTP, and act as switches in signal transduction pathways in yeast and mammalian cells (Barbacid 1987) as well as in insects (Simon et al. 1991). Mutations that activate *ras* in mammalian cells are oncogenic (Barbacid 1987).

All five of the independently isolated mutations that increase *let-60* activity and result in a multivulva phenotype cause a substitution of glutamic acid for glycine at codon 13 (Beitel et al. 1990). Codon 13 is a site altered in some oncogenic *ras* genes in mammals (Bos et al. 1985; Barbacid 1987). The dominant-negative *let-60* mutations alter residues in the GTP-binding region of the Let-60 Ras protein, and presumably for this reason eliminate *let-60* function and result in the recessive larval lethal phenotype; the dominant vulvaless phenotype caused by these mutations has been proposed to be a consequence of a competition by the inactive mutant Let-60 Ras protein for a positive activator needed for the function of the wild-type protein (Han and Sternberg 1991). Mutations that decrease *let-60 ras* activity and cause a recessive vulvaless phenotype can affect any of a number of codons (Beitel et al. 1990). Interestingly, mutations equivalent to some of these *let-60 ras* mutations do not affect the transforming activity of the mutationally activated v-Ha-*ras* oncogene (Sigal et al. 1986). For this reason, we proposed that the amino acids affected by these mutations might be important for the normal activation of Ras proteins but not for the function of Ras proteins once they have been activated (Beitel et al. 1990). This hypothesis has recently been tested by L. Howe et al. (pers. comm.), who constructed both c-Ha-*ras* and v-Ha-*ras* variants with mutations equivalent to those found in the *C. elegans let-60* gene. These researchers showed that substitutions at codon 66 like those in certain *let-60* mutants prevent the transforming activity of high concentrations of c-Ha-*ras* (which requires activation) but not that of v-Ha-*ras* (which is constitutively activated), supporting the idea that codon 66 of the Ras protein defines a site of interaction with a Ras-activating factor.

let-341 and *lin-45* Are Also Required for Vulval Induction

Two other genes identified in our screen for *lin-15* suppressors are *let-341* and *lin-45* (S.G. Clark and H.R. Horvitz, in prep.). The *let-341* gene had been defined previously in screens for essential genes located on the left arm of chromosome V by mutations that can cause a rod-like larval lethal phenotype similar to that caused by mutations in *let-23*, *let-60*, and *sem-5* (Johnsen and Baillie 1988,1991; Rosenbluth et al. 1988). The *let-341* mutations recovered in our screen are less severe and cause an incompletely penetrant larval lethal phenotype. Many of the surviving *let-341* animals develop into vulvaless adults in which the P3.p–P8.p cells are uninduced, indicating that this gene acts in vulval induction. Another mutation we found that results in an incompletely penetrant rod-like larval lethal and vulvaless phenotype is allelic to a mutation identified in a similar screen by Han et al. (1990). Together, these mutations define an additional gene required for vulval induction, *lin-45*.

A Genetic Pathway for Vulval Induction

Since vulvaless and multivulva mutants can have opposite effects on the process of vulval induction, double mutant combinations can be used to help order the genes defined by these mutants into a genetic pathway (Ferguson et al. 1987). For example, such a vulvaless–multivulva double mutant should be vulvaless if the process of inductive signaling is blocked by the vulvaless mutation after the point at which the pathway is activated by the multivulva mutation. Conversely, the double mutant should be multivulva if the block precedes the step that is activated.

The results of such genetic studies are summarized in Table 2. Vulvaless mutations in *let-23, sem-5, let-341, let-60*, and *lin-45* suppress the *lin-15* multivulva phenotype, consistent with the hypothesis that *lin-15* precedes these vulvaless genes in a genetic pathway for vulval induction. In contrast, *lin-15* multivulva mutations suppress vulvaless mutations in the gene *lin-3*. The *lin-3* gene is required for vulval induction, and mutations in this gene can cause an early rod-like larval lethal phenotype (Sulston and Horvitz 1981; Ferguson and Horvitz 1985; Ferguson et al. 1987). On the basis of the gene interaction data, *lin-15* does not precede *lin-3* in the pathway of vulval induction (see below for further discussion of this point).

let-60 multivulva mutations also distinguish between two classes of vulvaless mutations: *lin-45* mutations completely block the multivulva phenotype caused by the genetic activation of the *let-60* gene, whereas *let-23*, *let-341*, and *sem-5* mutations do not. In addition, animals carrying multiple copies of the wild-type *let-60* gene have a multivulva phenotype that is expressed even in the presence of vulvaless mutations in *lin-3* or *let-23*, indicating that an increase in *let-60* activity bypasses or reduces the need for *lin-3* and *let-23* in the vulval inductive pathway (Han and Sternberg 1990). In addition, an allele of *lin-45* that results in lethality causes lethality even in the presence of a *let-60* multivulva mutation (K. Kornfeld and H.R. Horvitz, unpubl.). Taken together, these observations suggest that *lin-3*, *let-23*, *sem-5*, and *let-341* act before *let-60* and *lin-45* might act after *let-60*.

Vulvaless mutations in all six of the genes *lin-3*, *let-23*, *let-60*, *sem-5*, *let-341*, and *lin-45* are suppressed by multivulva mutations in the gene *lin-1*. Since *lin-1* loss-of-function mutations cause all six of the cells P3.p–P8.p to express vulval cell lineages even in the

Table 2. Muv-Vul Double Mutant Phenotypes

	Muv mutants		
Vul mutants	*lin-15*	*let-60*	*lin-1*
lin-3	Muv	Muv	Muv
let-23	Vul	Muv	Muv
sem-5	Vul	Muv	Muv
let-341	Vul	Muv	Muv
let-60	Vul	—	Muv
lin-45	Vul	non-Muv	Muv

Double mutants containing a vulvaless and a multivulva mutation were constructed and examined using a dissecting microscope. The phenotypes of double mutants carrying *lin-3* and *lin-15*, *lin-3* and *lin-1*, and *let-23* and *lin-1* were described by Ferguson et al. (1987). The interaction between *lin-3* and *let-60* is based upon studies of a strain carrying a *lin-3* vulvaless mutation and an extrachromosomal array with multiple copies of the wild-type allele of *let-60*; the multivulva phenotype caused by this array suppresses the Lin-3 vulvaless phenotype, as reported by Han and Sternberg (1990). Although the *let-23(n1045cs)* and *lin-15(n309)* double mutant described previously (Ferguson et al. 1987) expresses aspects of both the vulvaless and multivulva phenotypes, a *let-23(n2020)* and *lin-15(n765ts)* double mutant is vulvaless; since *let-23(n2020)* causes a greater loss of gene function than does *let-23(n1045cs)* and since the *lin-15(n309)* and *lin-15(n765)* mutations seem to be equally severe under the conditions tested (S. Clark and R. Horvitz, unpubl.), we believe that this more recent experiment is more reliable. The other double mutant strains that were analyzed are: *let-23(mn23)* and *let-60(n1046gf)*, *sem-5(n1619)* and *lin-15(n765)*, *sem-5(n1619)* and *let-60(n1046gf)*, *sem-5(n1619)* and *lin-1(e1275)*, *let-341(n1613)* and *lin-15(n765)*, *let-341(n1613)* and *let-60(n1849gf)*, *let-341(n1613)* and *lin-1(e1275)*, *let-60(n1876lf)* and *lin-15(n765)*, *let-60(n1876lf)* and *lin-1(e1275)*, *lin-45(n2018)* and *lin-15(n765)*, *lin-45(n2018)* and *let-60(n1046gf)*, and *lin-45(n2018)* and *lin-1(e1275)* (S.G. Clark and H.R. Horvitz, in prep.), where "*gf*" is a gain-of-function allele and "*lf*" is a loss-of-function allele. Both multivulva (activation; *gf*) and vulvaless (inactivation; *lf*) alleles of *let-60* exist; no systematic analysis of *let-60* multivulva–vulvaless double mutants has been performed. (Muv) Multivulva phenotype; (Vul) vulvaless phenotype; (non-Muv) nonmultivulva phenotype in which animals had a nearly normal vulva but were often egg-laying-defective, a trait caused by weak vulvaless mutations (Ferguson et al. 1987).

absence of the inductive signal, *lin-1* normally acts to prevent the expression of vulval cell fates. Thus, these six vulvaless genes and the entire vulval inductive pathway act to prevent *lin-1* function.

A genetic pathway based on the phenotypes of double mutants is shown in Figure 5 (top). It should be noted that these gene interaction studies exclude rather than define orders of gene action. For example, the observation that a double mutant with a *lin-3* vulvaless and a *lin-15* multivulva mutation is multivulva implies that *lin-3* cannot act after *lin-15* in a linear pathway; *lin-3* could act before *lin-15* or, alternatively, *lin-3* and *lin-15* could act in parallel, as components of two independent pathways. The simplest way to depict the results of gene interaction studies is by the type of formal genetic pathway shown in Figure 5 (top), although it is important to note that any two apparently consecutive steps could function in parallel, rather than in series. These considerations are particularly relevant to the orders of action of *lin-3* and *lin-15* and of *let-60* and *lin-45*, as discussed below.

Genes Involved in Vulval Induction Act in Other Intercellular Signaling Pathways

The genes involved in the signaling pathway required for vulval induction also function in other intercellular signaling processes. The fates of two cells near the tail, P11 and P12, are controlled by cell interactions (Sulston and White 1980). In some *let-23* mutants, P12 expresses the fate normally expressed by P11, and in *lin-15* mutants, P11 expresses the fate normally expressed by P12 (Fixsen et al. 1985; Aroian and Sternberg 1991). Thus, as in vulval induction, in the control of P11 and P12 cell fates, the *let-23* and *lin-15* genes have opposite roles. In addition, at a low frequency, *let-341*, *let-60*, and *sem-5* mutations cause P12 to express a P11-like fate, suggesting that much if not all of the vulval inductive pathway might function in an intercellular signaling system that specifies the fates of these cells (S.G. Clark and H.R. Horvitz, in prep.).

Furthermore, as noted above, severe mutations in *lin-3*, *let-23*, *sem-5*, *let-341*, *let-60*, and *lin-45* all cause a characteristic rod-like larval lethality, suggesting that these genes act in a common process necessary for larval development. An activated *let-60 ras* mutation prevents the lethality conferred by mutations in at least the genes *let-23*, *sem-5*, and *let-341*, indicating that the activation of this *ras* gene bypasses the requirement of these genes for larval development as well as for vulval induction (Han et al. 1990; S.G. Clark and H.R. Horvitz, in prep.). These observations suggest that the larval lethality caused by *let-23*, *sem-5*, and *let-341* mutations also results from a defect in an intercellular signaling pathway controlled by these genes.

A Molecular Pathway for Vulval Induction

Knowledge of mutant phenotypes, the molecular structures, and the sites of expression and function of the genes involved in vulval induction can be combined with the results of gene interaction experiments to provide the basis for a model of the molecular pathway for vulval induction (Fig. 5, bottom). Recent studies have indicated that the Lin-3 protein is similar to the mammalian growth factors EGF and transforming growth factor-α (TGF-α), is expressed by the gonadal anchor cell, and seems likely to act as the inducing signal for vulval development (R.J. Hill and P.W. Sternberg, in prep.; also see Sternberg et al., this volume). Since *lin-3* acts before *let-23*, which encodes a receptor tyrosine kinase similar to the EGF receptor (Aroian et al. 1990; R.J. Hill and P.W. Sternberg, in prep.), it seems likely that the *lin-3* gene encodes an inducing signal that is released by the gonadal anchor cell and activates the Let-23 receptor on the surface of the responding P3.p–P8.p cells.

The role of *lin-15* is less clear. Preliminary data concerning the sequence of the *lin-15* gene have not revealed any clues concerning its molecular function (S.G. Clark and H. R. Horvitz, unpubl.; L. Huang et al., pers. comm.). Because *lin-15* loss-of-function mu-

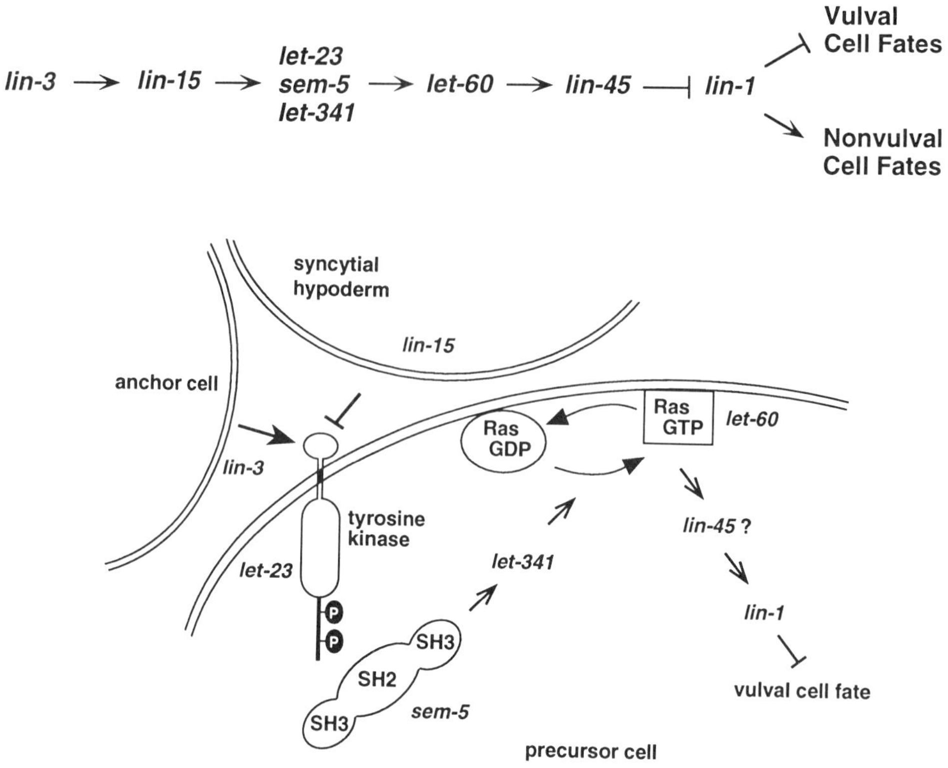

Figure 5. Models for genetic and molecular pathways for vulval induction. (*Top*) Gene interaction studies define three categories of genes that can mutate to a multivulva phenotype (*lin-15*, *let-60*, *lin-1*) and four categories of genes that can mutate to a vulvaless phenotype (*lin-3*, *let-23*/*sem-5*/*let-341*, *let-60*, *lin-45*) (see Table 2). As discussed in the text, these data can be most simply represented by the formal genetic pathway shown. (*Bottom*) The genetic and molecular data are consistent with the molecular model shown (see text).

tants are multivulva, the normal function of *lin-15* presumably is to prevent the activity of the inductive signaling pathway in uninduced precursor cells. Since the *lin-15* mulitvulva phenotype is expressed even after the gonad has been killed by a laser microbeam, *lin-15* cannot function (exclusively) in the gonad (Ferguson et al. 1987). Genetic mosaic studies confirm that *lin-15* acts outside the gonad and further indicate that this gene also acts outside P3.p–P8.p, leading to the hypothesis that *lin-15* functions in the syncytial hypoderm that envelops the animal (Herman and Hedgecock 1990). On the basis of these observations, Herman and Hedgecock (1990) proposed that (1) the syncytial hypoderm prevents P3.p–P8.p from expressing vulval cell lineages, (2) *lin-15* is needed for this inhibitory signal, and (3) the inductive signal from the anchor cell counteracts this inhibition. For these reasons, in the model in Figure 5 (bottom), we place *lin-15* function in the syncytial hypoderm acting in parallel to the signal specified by *lin-3* in the anchor cell. As discussed above, this pathway is consistent with the gene interaction data. A number of genes have been identified that can interact with *lin-15* in preventing expression of vulval cell lineages (Ferguson and Horvitz 1985, 1989), and future molecular characterization of these genes might help to reveal how *lin-15* functions.

The sites of synthesis and function of *let-23*, *sem-5*, *let-341*, *let-60*, *lin-45*, and *lin-1* are unknown. The simplest hypothesis is that all of these genes are expressed in and function in P3.p–P8.p. The order of action of *let-23*, *sem-5*, *let-341* also is unknown. The molecular structures of *let-23* and *sem-5* and the observation that the mammalian GRB2 protein binds to the mammalian EGF receptor suggest that the SH2 domain of the Sem-5 protein binds a tyrosine-phosphorylated form of the Let-23 receptor tyrosine kinase to effect signal transduction. If so, one possibility is that *let-341* acts after *sem-5* in the pathway of *let-60* activation. We have recently begun experiments to clone the *let-341* gene (J. Thomas and H.R. Horvitz, unpubl.). One plausible function for this gene, which is needed for the activation of the Let-60 Ras protein, is to encode a guanine nucleotide exchange factor; such exchange factors can convert an inactive GDP-bound form of Ras to an active GTP-bound form (West et al. 1990; Wolfman and Macara 1990; Jones et al. 1991). The signal transduction pathway involving *let-23*, *sem-5*, and *let-341* activates the *let-60 ras* gene, but where in this pathway *lin-45* acts is less clear. Although *lin-45* mutations completely block the multivulva phenotype caused by an activated *let-60 ras* mutation, this result could reflect a function for *lin-45* either downstream from or in parallel to *let-60*, as discussed above. Thus, one possibility is that *lin-45* mediates *ras* action. The entire pathway for vulval induction acts to prevent the inhibition of the expression of vulval cell fates by *lin-1*. The molecular cloning and characterization of both *lin-45* and *lin-1* are currently under way (A. Golden et al., pers. comm.; G. Beitel and H.R. Horvitz, unpubl.).

The Molecular Biology of Signal Transduction Is Highly Conserved

The molecular pathway of vulval induction involves a series of proteins similar to those involved in signal transduction in mammalian cells: Growth factors, receptor tyrosine kinases, SH2- and SH3-containing proteins, and Ras proteins have all been found to act in mammalian signal transduction pathways, as well as in oncogenic processes (for review, see Cantley et al. 1991). A receptor tyrosine kinase and a Ras protein, as well as a guanine nucleotide exchange factor and a GTPase-activating protein (both also associated with mammalian signal transduction; Bourne et al. 1990), have been identified in the signaling pathway that controls eye development in *Drosophila melanogaster* (Simon et al. 1991; also see Zipursky et al.; Simon et al.; both this volume).

It seems very likely that findings from each of these experimental systems are likely to be relevant to others. For example, we expect that regulators of other Ras proteins, such as a guanine nucleotide exchange factor and a GTPase-activating protein, also will be found to regulate the Let-60 Ras protein in the pathway for vulval induction. Conversely, proteins like Sem-5, such as the GRB2 protein (J. Schlessinger, pers. comm.), seem likely to function in mammalian and insect signal transduction and to transduce information between other receptor tyrosine kinases and Ras proteins. More generally, we hope that continued characterization of the pathways that regulate *C. elegans* sex myoblast migration and vulval induction will lead to the discovery of additional classes of molecules that function in signal transduction. Such molecules should include both those involved in transmitting information from the cell surface to regulatory proteins like Ras and those that are targets of Ras and other regulatory proteins and hence responsible for controlling and executing the cellular events that occur in response to intercellular signals.

Individual Proteins Can Act in Biologically and Molecularly Distinct Signal Transduction Pathways

Our studies of *C. elegans* sex myoblast migration and vulval induction reveal that individual proteins can act in distinct signal transduction pathways that control strikingly different biological processes. For example, the Sem-5 protein functions in both cell migration and inductive signaling. As discussed above, the Clr-1 protein might also act in both of these signal transduction pathways. In contrast, most of the molecules required for vulval induction do not seem to function in the control of sex myoblast migration (Fig. 6). Mutations in the genes *lin-3*, *let-23*, *let-341*, *let-60*, and *lin-45* do not affect sex myoblast migration, nor do mutations in the genes *egl-15* and *egl-17*, which function in sex myoblast migration, affect vulval induction. Thus, it seems that two partially overlapping sets of genes act to control the signal transduction processes responsible for sex myoblast migration and vulval induction. Certain proteins, such as the product of the *sem-5* gene, apparently can function as intermediates within otherwise distinct signal transduction pathways. Since the Let-23 receptor tyrosine kinase and the Let-60 Ras protein do not appear to function in sex myoblast migration, the Sem-5 protein seems likely to transduce information not only between these two molecules, but also between an unidentified receptor tyrosine kinase and an unidentified effector protein. The nature of the unknown molecules involved in sex myoblast migration and vulval induction and the degree to which these and other signal transduction pathways overlap are problems we hope to address in our future studies.

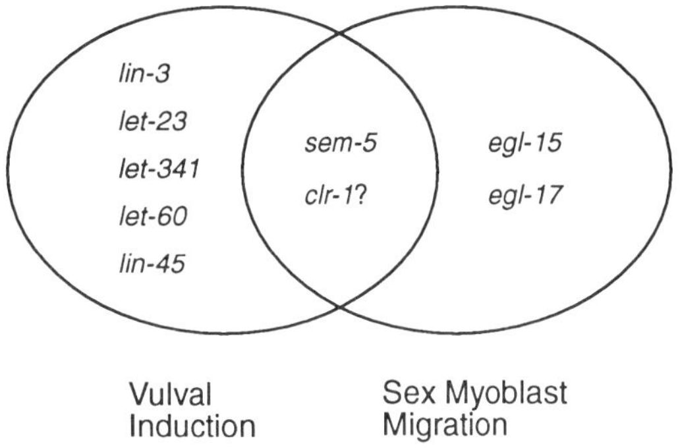

Figure 6. Genes involved in the control of sex myoblast migration and vulval induction define two partially overlapping sets. *lin-3*, *let-23*, *let-341*, *let-60*, and *lin-45* mutations cause defects in vulval induction (Ferguson et al. 1987; Beitel et al. 1990; Han et al. 1990; Aroian and Sternberg 1991; S.G. Clark and H.R. Horvitz, in prep.) but not in sex myoblast (SM) migration (M.J. Stern and H.R. Horvitz, in prep.). *egl-15* and *egl-17* mutations cause defects in SM migration but not in vulval induction (Stern and Horvitz 1991 and in prep.). *sem-5* mutations cause defects both in vulval induction and in SM migration (Clark et al. 1992). *clr-1* mutations partially suppress both the SM migration defects and the vulval induction defects caused by *sem-5* mutations (see Table 1; Clark et al. 1992; M.J. Stern and H.R. Horvitz, in prep.), suggesting that *clr-1* functions in both signal transduction pathways.

ACKNOWLEDGMENTS

We are grateful to Michael Basson, Greg Beitel, Erik Jorgensen, Josh Kaplan, Kerry Kornfeld, and Jeff Thomas for suggestions concerning this manuscript and to Erik Jorgensen for preparing Figure 1. This work was supported by U.S. Public Health Service research grants GM-24663 and GM-24943 and by the Howard Hughes Medical Institute. M.J.S. was supported by a postdoctoral fellowship from the American Cancer Society and by the Howard Hughes Medical Institute. H.R.H. is an Investigator of the Howard Hughes Medical Institute.

REFERENCES

Aroian, R.V. and P.W. Sternberg. 1991. Multiple functions of *let-23*, a *Caenorhabditis elegans* receptor tyrosine kinase gene required for vulval induction. *Genetics* **128:** 251.

Aroian, R.V., M. Koga, J.E. Mendel, Y. Ohshima, and P.W. Sternberg. 1990. The *let-23* gene necessary for *Caenorhabditis elegans* vulval induction encodes a tyrosine kinase of the EGF receptor subfamily. *Nature* **348:** 693.

Barbacid, M. 1987. *ras* genes. *Annu. Rev. Biochem.* **56:** 779.

Beitel, G.J., S.G. Clark, and H.R. Horvitz. 1990. *Caenorhabditis elegans ras* gene *let-60* acts as a switch in the pathway of vulval induction. *Nature* **348:** 503.

Bos, J.L., D. Toksoz, C.J. Marshall, M. Verlaan-de Vries, G.H. Veeneman, A.J. van der Eb, J.H. van Boom, J.W. Janssen, and A.C. Steenvoorden. 1985. Amino-acid substitutions at codon 13 of the N-*ras* oncogene in human acute myeloid leukaemia. *Nature* **315:** 726.

Bourne, H.R., D.A. Sanders, and F. McCormick. 1990. The GTPase superfamily: A conserved switch for diverse cell functions. *Nature* **348:** 125.

Cantley, L.C., K.R. Auger, C. Carpenter, B. Duckworth, A. Graziani, R. Kapeller, and S. Soltoff. 1991. Oncogenes and signal transduction. *Cell* **64:** 281.

Clark, S.G., M.J. Stern, and H.R. Horvitz. 1992. *C. elegans* cell-signaling gene *sem-5* encodes a protein with SH2 and SH3 domains. *Nature* **356:** 340.

Ferguson, E.L. and H.R. Horvitz. 1985. Identification and characterization of 22 genes that affect the vulval cell lineages of the nematode *Caenorhabditis elegans*. *Genetics* **110:** 17.

———.1989. The multivulva phenotype of certain *Caenorhabditis elegans* mutants results from defects in two functionally redundant pathways. *Genetics* **123:** 109.

Ferguson, E.L., P.W. Sternberg, and H.R. Horvitz. 1987. A genetic pathway for the specification of the vulval cell lineages of *Caenorhabditis elegans*. *Nature* **326:** 259.

Fixsen, W., P. Sternberg, H. Ellis, and H.R. Horvitz. 1985. Genes that affect cell fates during the development of *Caenorhabditis elegans*. *Cold Spring Harbor Symp. Quant. Biol.* **50:** 99.

Greenwald, I.S. and G.M. Rubin. 1992. Making a difference: The role of cell-cell interactions in establishing separate identities for equivalent cells. *Cell* **68:** 271.

Greenwald, I.S., P.W. Sternberg, and H.R. Horvitz. 1983. The *lin-12* locus specifies cell fates in *Caenorhabditis elegans*. *Cell* **34:** 435.

Gurdon, J.B. 1992. The generation of diversity and pattern in animal development. *Cell* **68:** 185.

Han, M. and P.W. Sternberg. 1990. *let-60*, a gene that specifies cell fates during *C. elegans* vulval induction, encodes a *ras* protein. *Cell* **63:** 921.

———.1991. Analysis of dominant-negative mutations of the *Caenorhabditis elegans let-60 ras* gene. *Genes Dev.* **5:** 2188.

Han, M., R.V. Aroian, and P.W. Sternberg. 1990. The *let-60* locus controls the switch between vulval and nonvulval cell fates in *Caenorhabditis elegans*. *Genetics* **126:** 899.

Hedgecock, E.M., J.G. Culotti, and D.H. Hall. 1990. The *unc-5, unc-6,* and *unc-40* genes guide circumferential migrations of pioneer axons and mesodermal cells on the epidermis in *C. elegans*. *Neuron* **4:** 61.

Herman, R.K. and E.M. Hedgecock. 1990. Limitation of the size of the vulval primordium of *Caenorhabditis elegans* by *lin-15* expression in surrounding hypodermis. *Nature* **348:** 169.

Horvitz, H.R. and P.W. Sternberg. 1991. Multiple intercellular signalling systems control the development of the *Caenorhabditis elegans* vulva. *Nature* **351:** 535.

Horvitz, H.R. and J.E. Sulston. 1980. Isolation and genetic characterization of cell lineage mutants of the nematode *Caenorhabditis elegans*. *Genetics* **96:** 435.

Hynes, R.O. and A.D. Lander. 1992. Contact and adhesive specificities in the associations, migrations and targeting of cells and axons. *Cell* **68:** 303.

Jessell, T.M. and D.A. Melton. 1992. Diffusible factors in vertebrate embryonic induction. *Cell* **68:** 257.

Johnsen, R.C. and D.L. Baillie. 1988. Formaldehyde-mutagenesis of the *eT1* balanced region in *Caenorhabditis elegans*: Dose-response curve and the analysis of mutational events. *Mutat. Res.* **201:** 137.

———.1991. Genetic analysis of a major segment [LGV(left)] of the genome of *Caenorhabditis elegans*. *Genetics* **129:** 735.

Jones, S., M.-L. Vignais, and J.R. Broach. 1991. The CDC25 protein of *Saccharomyces cerevisiae* promotes exchange of guanine nucleotides bound to Ras. *Mol. Cell. Biol.* **11:** 2641.

Kimble, J. and D. Hirsh. 1979. Post-embryonic cell lineages of the hermaphrodite and male gonads in *Caenorhabditis elegans*. *Dev. Biol.* **70:** 396.

Koch, C.A., D. Anderson, M.F. Moran, C. Ellis, and T. Pawson. 1991. SH2 and SH3 domains: Elements that control interactions of cytoplasmic signaling proteins. *Science* **252:** 668.

Lambie, E.J. and J. Kimble. 1991. Genetic control of cell interactions in nematode development. *Annu. Rev. Genet.* **25:** 411.

Li, C. and M. Chalfie. 1990. Organogenesis in *Caenorhabditis elegans:* Positioning of neurons and muscles in the egg-laying system. *Neuron* **4:** 681.

Matsuda, M., B.J. Mayer, and H. Hanafusa. 1991. Identification of domains of the v-*crk* oncogene product sufficient for association with phosphotyrosine-containing proteins. *Mol. Cell. Biol.* **11:** 1607.

Moran, M.F, C.A. Koch, D. Anderson, C. Ellis, L. England, G.S. Martin, and T. Pawson. 1990. Src homology region 2 domains direct protein-protein interactions in signal transduction. *Proc. Natl. Acad. Sci.* **87:** 8622.

Rodaway, A.R., M.J. Sternberg, and D.L. Bentley. 1989. Similarity in membrane proteins. *Nature* **342:** 624.

Rosenbluth, R.E., T.M. Rogalski, R.C. Johnsen, L.M. Addison, and D.L. Baillie. 1988. Genomic organization in *Caenorhabditis elegans*: Deficiency mapping on linkage group *V* (left). *Genet. Res.* **52:** 105.

Sadowski, I., J.C. Stone, and T. Pawson. 1986. A noncatalytic domain conserved among cytoplasmic protein-tyrosine kinases modifies the kinase function and transforming activity of Fujinami sarcoma virus P130gag-fps. *Mol. Cell. Biol.* **6:** 4396.

Sigal, I.S., J.B. Gibbs, J.S. DaLonzo, and E.M. Scolnick. 1986. Identification of effector residues and a neutralizing epitope of Ha-*ras*-encoded p21. *Proc. Natl. Acad. Sci.* **83:** 4725.

Simon, M.A., D.D. Bowtell, G.S. Dodson, T.R. Laverty, and G.M. Rubin. 1991. Ras1 and a putative guanine nucleotide exchange factor perform crucial steps in signaling by the sevenless tyrosine kinase. *Cell* **67:** 701.

Stern, M.J. and H.R. Horvitz. 1991. A normally attractive cell interaction is repulsive in two *C. elegans* mesodermal cell migration mutants. *Development* **113:** 797.

Sternberg, P.W. 1988. Lateral inhibition during vulval induction in *Caenorhabditis elegans*. *Nature* **335:** 551.

Sulston, J. 1988. Cell lineage. In *The nematode* Caenorhabditis elegans (ed. W.B. Wood et al.), p. 123-155. Cold Spring Harbor Laboratory, Cold Spring Harbor, New York.

Sulston, J.E. and H.R. Horvitz. 1977. Post-embryonic cell lineages of the nematode, *Caenorhabditis elegans*. *Dev. Biol.* **56:** 110.

———.1981. Abnormal cell lineages in mutants of the nematode *Caenorhabditis elegans*. *Dev. Biol.* **82:** 41.

Sulston, J.E. and J.G. White. 1980. Regulation and cell autonomy during postembryonic development of *Caenorhabditis elegans*. *Dev. Biol.* **78:** 577.

Sulston, J.E., E. Schierenberg, J.G. White, and J.N. Thomson. 1983. The embryonic cell lineage of the nematode *Caenorhabditis elegans*. *Dev. Biol.* **100:** 64.

Thomas, J.H., M.J. Stern, and H.R. Horvitz. 1990. Cell interactions coordinate the development of the *C. elegans* egg-laying system. *Cell* **62:** 1041.

Trent, C., N. Tsung, and H.R. Horvitz. 1983. Egg-laying defective mutants of the nematode *Caenorhabditis elegans*. *Genetics* **104:** 619.

Ullrich, A. and J. Schlessinger. 1990. Signal transduction by receptors with tyrosine kinase activity. *Cell* **61:** 203.

West, M., K. Hsiang-fu, and T. Kamata. 1990. A novel membrane factor stimulates guanine nucleotide exchange reaction of ras proteins. *FEBS Lett.* **259:** 245.

White, J., E. Southgate, J. Thomson, and S. Brenner. 1986. The structure of the nervous system of the nematode *Caenorhabditis elegans*. *Philos. Trans. R. Soc. Lond. B Biol. Sci.* **314:** 1.

Wolfman, A. and I.G. Macara. 1990. A cytosolic protein catalyzes the release of GDP from $p21^{ras}$. *Science* **248:** 67.

Signal Transduction Pathway Initiated by Activation of the *sevenless* Tyrosine Kinase Receptor

M.A. Simon, R.W. Carthew, M.E. Fortini, U. Gaul, G. Mardon, and G.M. Rubin
Howard Hughes Medical Institute, Department of Molecular and Cell Biology, University of California, Berkeley, California 94720

Transmembrane tyrosine kinase receptors play important roles in a variety of physiological and developmental processes. How the signals initiated by activation of such receptors effect changes in cell physiology is still poorly understood. The intracellular machinery used to interpret and implement the instruction(s) conveyed by the activation of tyrosine kinase receptors are being actively investigated in a wide variety of developing organisms and tissues, using both genetic and biochemical approaches. Biochemical studies with mammalian protein tyrosine kinase receptors (PTKs) have led to the identification of proteins that bind to or are phosphorylated by tyrosine kinases. Although some of these interactions suggest potential mechanisms of signal transmission, their role in vivo is still unclear.

The developing *Drosophila* eye has proven to be well suited for genetic, molecular, and cellular experiments aimed at understanding the mechanisms of signal transduction. The adult eye is made up of a simple array of approximately 800 20-cell units, called ommatidia. Each ommatidium contains eight photoreceptor cells, R1–R8, as well as four lens-secreting cone cells and eight other accessory cells. The cells that make up each ommatidium are thought to be recruited by a series of local cell-cell interactions, with differentiating cells instructing their immediate neighbors to adopt particular fates. Development of the *Drosophila* eye has been reviewed extensively (see Tomlinson 1988; Ready 1989; Banerjee and Zipursky 1990).

The R7 photoreceptor is the last of the eight photoreceptors to be recruited to the developing ommatidium (Fig. 1). Mutations in *sevenless* (*sev*; Tomlinson and Ready 1986) or *bride of sevenless* (*boss*; Reinke and Zipursky 1988; Krämer et al. 1991) have been shown to disrupt this recruitment. In each of these mutants, the presumptive R7 cell—the cell that finds itself in the pocket created by the differentiating R8, R1, and R6 cells—develops into a nonneuronal, lens-secreting cone cell. Thus, the presumptive R7 cell appears to face a simple choice between two alternative cell fates: It will develop into an R7 photoreceptor if it receives appropriate instructions, but it will adopt a nonneuronal cone cell fate if these instructions are disrupted by mutation of *sev* or *boss*. In this paper, we focus on recent progress in understanding this one inductive event, the recruitment of the R7 photoreceptor, and, in particular, the use of genetic analysis to reveal aspects of the intracellular pathway that transduces the signal initiated by activation of the receptor tyrosine kinase encoded by the *sev* gene.

sevenless: A Transmembrane Tyrosine Kinase Receptor with a Highly Specialized Role

The *sev*-encoded protein (Sev) is synthesized as a 280-kD glycoprotein precursor that is subsequently cleaved into 220-kD amino-terminal and 58-kD carboxy-terminal subunits that remain associated by noncovalent interactions (Simon et al. 1989). The carboxy-terminal subunit contains the transmembrane domain and a cytoplasmic domain highly homologous to known tyrosine kinases (Hafen et al. 1987). The Sev protein has been shown to have protein tyrosine kinase activity in vitro (Simon et al. 1989), and this activity appears to be essential for Sev function in vivo (Basler and Hafen 1988). Although Sev activity is only required in the R7 cell, Sev is expressed on the apical surface of many cells in the developing retina (Banerjee et al. 1987; Tomlinson et al. 1987) in a highly dynamic pattern that is controlled at the level of transcription of the *sev* gene. Temperature-inducible *hsp70-sev* fusion genes (Basler and Hafen 1989; Bowtell et al. 1989) and temperature-sensitive (Mullins and Rubin 1991) *sev* mutations have been used to show that Sev activity is required only to initiate R7 cell development and not for its subsequent differentiation or function. However, the Sev-mediated signal must be maintained for several hours to be effective (Fig. 1) (Mullins and Rubin 1991). A variety of genetic and molecular data indicate that *boss* encodes a cell-surface ligand on the R8 cell that activates the *sev*-encoded receptor on the surface of the R7 cell (see Zipursky et al., this volume).

Two features make the Sev pathway ideal for genetic approaches to understanding signaling by receptor tyrosine kinases. First, both the R7 cell and the Sev protein are dispensable for viability and fertility. Second, the functioning of this signaling pathway can be inferred from the presence of the R7 cell in a live, anesthetized fly.

Sina: A Nuclear Protein Required for R7 Cell Determination

Neither the Sev nor Boss proteins are required for viability or fertility; their loss appears to affect only eye

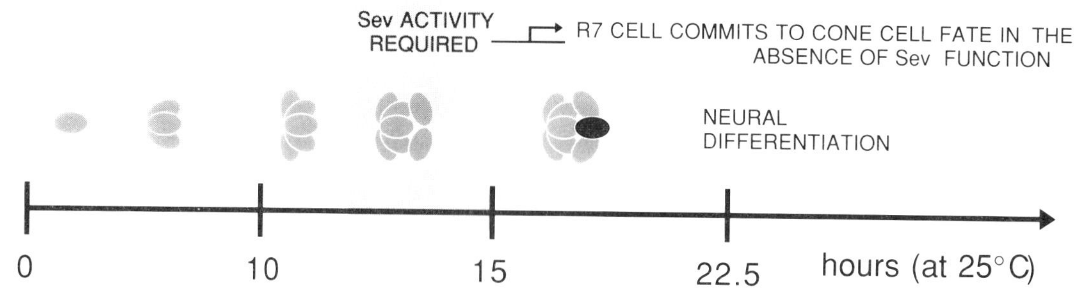

Figure 1. Summary of the time course of events involved in R7 cell recruitment. The time line is calibrated in hours of developmental time at 25°C. The developmental times given are approximate; the rate of development varies slightly between strains and is not uniform, being slower in the posterior portions of the eye disc (Basler and Hafen 1989). The times at which cells begin neural differentiation are indicated in the diagrams of assembling ommatidia shown above the time line, beginning with R8 and then adding R2 and R5, R3 and R4, R1 and R6, and finally R7 (darker shading). Also shown is the 6-hr period during which Sev activity appears to be required (Mullins and Rubin 1991). Finally, the time at which the presumptive R7 cell commits to an alternative developmental fate in the absence of Sev activity (Mullins and Rubin 1991) is shown.

development. We were therefore encouraged to seek mutations in other genes that would lead to the loss of the R7 cell and, like *sev* and *boss*, would not be required for viability. This effort led to the identification of the *sina* gene (Carthew and Rubin 1990). Loss of *sina* function results in the presumptive R7 cell developing as a cone cell, just as seen in *sev* and *boss* mutants. However, *sina* function is not limited to the eye, but appears to be required in other parts of the nervous system. The *sina* gene encodes a nuclear protein of unknown biochemical function. Sequence similarity to other proteins is limited to a small domain containing a cluster of cysteine residues (Brunk et al. 1991). Although the *sina* gene product is required for R7 determination, its role within the presumptive R7 cell remains unclear. Modification of Sina activity may be one end point for the signal transduction pathway initiated by Sev. Alternatively, Sina may function in parallel with the Sev pathway.

Sensitizing the System: A Way to Identify Particular Roles of Widely Utilized Proteins

The requirement of *sina* in developing tissues other than the eye emphasizes a limitation of the above approach. By requiring a homozygous mutant fly to survive to adulthood, where the presence or absence of an R7 cell can be assessed, we eliminate the possibility of identifying mutations in genes that also act in signal transduction pathways essential for earlier developmental stages. Given the widespread role of tyrosine kinases in development, we thought it likely that the vast majority of genes whose products act in the Sev-mediated signal transduction pathway would also act in earlier processes and thus be essential for viability.

To get around this problem, we created a situation in which Sev PTK activity was limiting and then screened for mutations that altered the strength of Sev signaling (Simon et al. 1991). We first carried out a systematic genetic screen for mutations that decrease the effectiveness of signaling by the Sev PTK, by looking for dominant enhancers of *sev* mutations. By adjusting the temperature at which flies carrying a temperature-sensitive allele of *sev* were grown, we could adjust Sev PTK activity to a level barely above the threshold necessary for R7 cell formation. Small reductions in the abundance or activity of other elements of the pathway might then be expected to lower signal strength sufficiently to cause a *sev* phenotype in flies grown at this threshold temperature. This sensitivity allowed us to identify genes encoding putative downstream elements of the pathway, by screening for genes in which inactivation of only one copy of the gene (which would be expected to reduce the level of gene product by half) resulted in the absence of the R7 cell. Since the other copy of the gene remained functional, we were able to identify these loci even though their functions were essential for viability. Seven genes were identified in this screen. The products of four of the seven loci also appear to be involved in transduction of signals from another PTK, the *Drosophila* homolog of the epidermal growth factor (EGF) receptor (Egfr), as judged from the ability of mutations in these loci to suppress *Ellipse*, a mutation that results in Egfr hyperactivity (Baker and Rubin 1989).

Ras Plays an Essential Role in the *sevenless* Signal Transduction Pathway

Our results suggest that four of the loci identified in our screen encode proteins that participate in the transduction of signals from at least two different PTKs. One of these four loci corresponds to the *Ras1* gene, which encodes a $p21^{ras}$ protein. Several previous studies have suggested a role for Ras activity in PTK function. Injection of monoclonal antibodies that inhibit Ras activity blocked the ability of PTKs to induce mitosis, and a dominant-negative *ras* gene was able to interfere with oncogenic transformation by several tyrosine-kinase-encoding oncogenes (see, e.g., Feig and Cooper 1988). Furthermore, recent studies of vulval development in the nematode *Caenorhabditis elegans* place the *ras* gene downstream from the PTK encoded by the *let-23* gene (see Sternberg et al.; Clark et al.; both this volume).

The fact that a twofold decrease in the level of Ras1 protein can give a dramatic effect on signaling by Sev suggests that Ras plays a critical role in this signaling pathway. To ask whether Ras1 activation alone is sufficient for Sev-mediated signaling, we used *sev* gene regulatory sequences to express wild-type and dominant activating *Ras1* alleles in those cells of the developing eye that normally express *sev*, R7, R3, R4, and the cone cells (Fortini et al. 1992). During normal eye development, R3 and R4 are thought to be unable to respond to Sev because of earlier cell-fate commitment, whereas the cone cells fail to adopt the R7 fate presumably because they do not contact R8, which alone expresses *boss*. Ubiquitous expression of *boss* in the developing eye disc induces the cone cell precursors to differentiate as ectopic R7 cells (Van Vactor et al. 1991). Similarly, expression of a truncated, constitutively active Sev protein under *sev* gene control results in *boss*-independent rescue of the normal R7 cell and transformation of cone cell precursors into supernumerary R7 cells (Basler et al. 1991). Identical results should thus be obtained with activated Ras1 under *sev* gene control if Ras1 activation can substitute for all of the Sev-mediated signal.

We (Fortini et al. 1992) found that a copy of the wild-type *Ras1* gene expressed under *sev* gene control has no apparent effect on eye development; seven independent P[*sevRas1*] transgenic lines all had normal external and internal adult eye morphology and still required *sev* activity for R7 formation. In contrast, 10 of 12 P[*sevRas1*$^{\mathrm{val12}}$] lines bearing the dominant activating *Ras1*$^{\mathrm{val12}}$ mutation exhibit various degrees of external eye roughening. Adult eye sections of the less severely affected lines reveal that most ommatidia contain as many as four or five supernumerary R7 cells. Moreover, R7 cell formation in P[*sevRas1*^{val12}] flies does not require either *sev* or *boss* gene function. The ability of *Ras1*$^{\mathrm{val12}}$ to reproduce the activated Sev phenotype and to bypass normal requirements for *sev* and *boss* suggests that Ras1 activation may be the primary consequence of ligand-induced Sev signaling during R7 cell determination. However, our results cannot exclude the possibility that endogenous Ras1 protein is not activated to the levels achieved by Ras1$^{\mathrm{val12}}$ and that normal Sev-mediated signaling requires additional pathways operating in parallel to the Ras1 pathway.

Because Ras proteins have been implicated in a variety of signaling processes, the consequences of Ras1 activation may be highly dependent on the cellular context. In this regard, the R3 and R4 photoreceptor precursors express *sev* but are not transformed into R7 cells by activated Sev or Ras1$^{\mathrm{val12}}$, perhaps because they are already committed to an alternative neural fate. One possibility is that Ras1 must be activated in all photoreceptor precursors for neural fate specification only and thus has already been activated in R3 and R4 by different means, perhaps involving other protein tyrosine kinases. Indeed, Ras1 is required for the development of all photoreceptors and interacts genetically with the locus encoding the *Drosophila* EGF receptor homolog (Simon et al. 1991).

Regulation of Ras Activity: Possible Links between Sev and Ras

The activity of Ras proteins is regulated by bound guanine nucleotides (for review, see Bourne et al. 1990, 1991); the GTP-bound state is active, and the GDP-bound state is inactive. The ratio of GTP: Ras to GDP: Ras is determined by two antagonistic reactions (see Fig. 2). An active GTP:Ras molecule is inactivated by the intrinsic GTPase activity of the Ras protein, a process that is greatly stimulated by RasGAP. An inactive GDP:Ras molecule is activated by the exchange of the bound GDP molecule for a GTP molecule, a reaction that is increased by guanine nucleotide exchange proteins.

How does the activation of the Sev receptor tyrosine kinase lead to the activation of Ras1 in the signaling pathway? In principle, activation of Ras could be accomplished by increasing the activity of an exchange protein or by decreasing the activity of RasGAP. Our genetic screens have led to the identification of the genes encoding both the exchange factor and RasGAP that appear to function in this pathway.

One of the four loci identified by Simon et al. (1991) as decreasing signal transduction initiated by either Sev or the Egfr encodes a protein with sequence similarity to the *Saccharomyces cerevisiae* CDC25 protein, which activates yeast RAS by facilitating exchange of GTP for GDP (Jones et al. 1991). A dominant allele of this locus, *Sos*, was isolated by Rogge et al. (1991) in an independent screen for dominant suppressors of *sev*. This gain-of-function mutation, *Sos*$^{\mathrm{J2}}$, does not bypass the requirement for *sev* function as it cannot suppress a *sev* null mutation. Rather, *Sos*$^{\mathrm{J2}}$ appears to increase the efficiency of transduction of a weak Sev-mediated signal.

If the similarity between the CDC25 protein and the Sos protein is indicative of similar activities, then our genetic results are consistent with a model in which the activation of the *sev* PTK by ligand binding stimulates the activity of the Sos protein. The activated Sos protein would then stimulate Ras1 protein activity by promoting the conversion of GDP:Ras1 to GTP:Ras1. Our experiments cannot exclude the possibility that the Sos protein might be a constitutive activator of Ras1 and that Sev could act solely by inhibiting the activity of a negative regulator such as RasGAP. Indeed, in a genetic screen for mutations that increase signal transduction, we isolated mutations in a gene, *Gap1*, which encodes a protein with sequence similarity to mammalian RasGap (Gaul et al. 1992). Our genetic analysis indicates that *Gap1* acts as a negative regulator of R7 cell determination, as loss of *Gap1*$^{+}$ activity leads to the formation of supernumerary, *sev*-independent R7 cells, and reduction in *Gap1*$^{+}$ activity increases the effectiveness of signaling by Sev. Analysis of mosaic

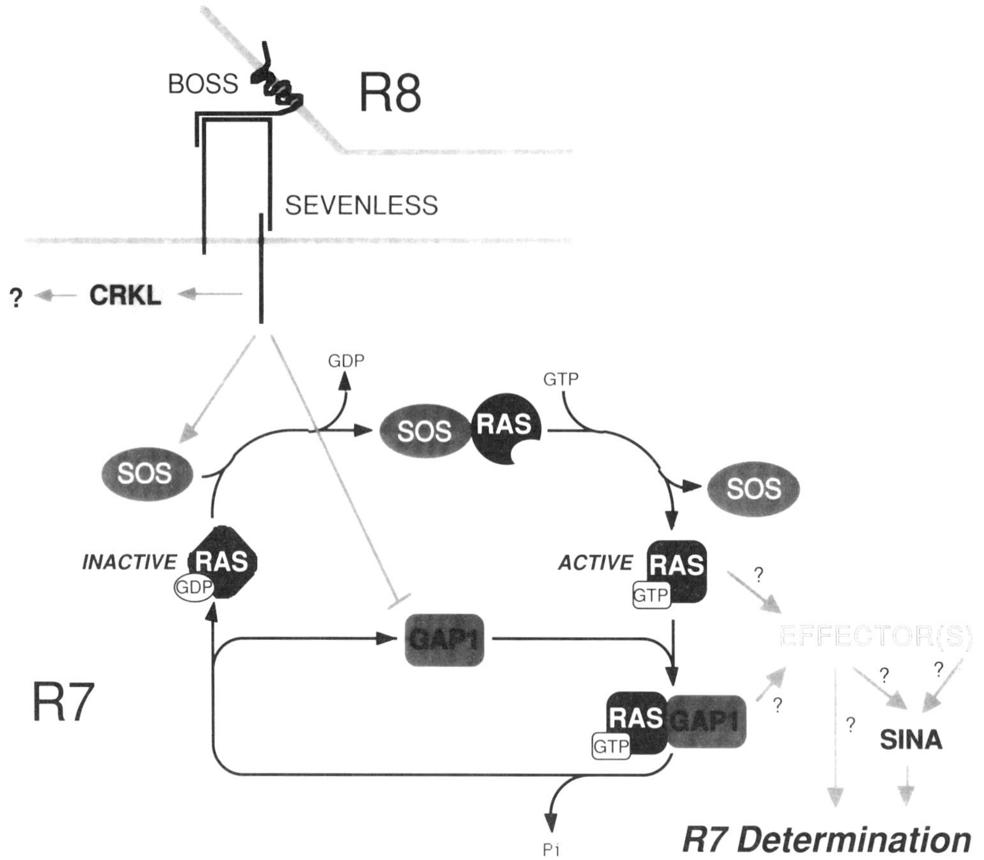

Figure 2. Model for the Sev-mediated signal transduction pathway (see text for details). Arrows shown in gray are based on inferences from genetic analysis.

individuals shows that *Gap1* acts cell-autonomously in the R7 and cone cell precursors, and the phenotypes of double mutants suggest that *Gap1* functions downstream from *sev*. Lack of *Gap1* product is sufficient to trigger R7 development, as the requirement for *sev*$^+$ function is eliminated by removal of *Gap1*$^+$ activity.

Biochemical studies will be required to determine if either Sos or Gap1 protein activity is actually regulated by Sev activity. The available data do, however, permit the conclusion that the level of either Sos or Gap1 protein activity can be a limiting step in the decision by the presumptive R7 cell to become an R7 cell. Regulation of Sos or Gap1 could be achieved directly through tyrosine phosphorylation of Sos or Gap1 protein by the Sev PTK. Alternately, activation by Sev could be indirect. If so, then the products of the remaining loci identified in the screen of Simon et al. (1991) are the prime candidates for the roles of intermediaries. A good candidate for such a biochemical link between Sev and Sos and Gap1 is the product of the *E(sev)2B* gene, which we have named *crk-like* (*crkl*) after the *crk* oncogene (Mayer et al. 1988), since it is composed of SH2 and SH3 domains. Similar proteins have been identified in *C. elegans* (Sem-5; see Clarke et al., this volume) and humans (GRP2; see Schlessinger et al., this volume). SH2 domains are known to interact directly with a number of tyrosine kinase receptors in a phosphorylation-dependent manner (for review, see Koch et al. 1991). Crkl shares 57% amino acid identity with Sem-5 and 64% identity with GRB2.

It has been proposed that the function of GAP proteins may not solely be to regulate Ras activity negatively, as their biochemical activity suggests, but that they may also function as downstream effectors of Ras-mediated signal transduction (McCormick 1989; Hall 1990). This hypothesis, which was originally based on the observation that GAP interacts with the Ras effector domain, has recently gained further support (for review, see Bollag and McCormick 1991; Lowy et al. 1991; Marshall 1991). Our genetic data argue against a role for Gap1 as the positive effector that transduces the Ras-mediated signal to downstream elements of the Sev pathway. Rather, our data implicate Gap1 as a negative regulator of Ras1 in the Sev signaling pathway. Loss-of-function mutations in Gap1 were isolated in three different mutagenesis screens with different bias, and all of these mutations resulted in the same phenotype characterized by the production of supernumerary, Sev-independent R7 cells. If Gap1 were the positive effector of Ras1, loss-of-function mutations in Gap1 would be expected to have the same phenotype as loss-of-function mutations in Ras1. The fact that loss-of-function mutations in Gap1 have an effect opposite to that of loss-of-function mutations in *Ras1* instead implicates *Gap1* as a negative regulator of the Sev signaling pathway. However, we cannot exclude the possibility that other GAP-like molecules present in the same cells as Gap1 act as effectors of Ras1.

The expression of Gap1 is spatially highly restricted, whereas that of Ras1 is not. For example, within the eye disc, Gap1 expression is restricted to the region posterior of the morphogenetic furrow, whereas Ras1 expression is found throughout the entire disc (Segal and Shilo 1986). This finding is in contrast to mammalian GAP and NF1, which appear to be expressed ubiquitously (see Bollag and McCormick 1991). The cell-type-specific expression and function of Gap1 is in contrast to the widespread expression and function of Ras1 and therefore suggests that distinct GTPase activating proteins may regulate Ras1 in different developmental pathways.

CONCLUDING REMARKS

The experimental advantages of being able to examine mutant phenotypes at single cell resolution in a developing tissue that is dispensable for viability and fertility ensure that new insights will continue to emerge from studies of the Sev-mediated signal transduction pathway. In particular, the utility of genetic screens for enhancers and suppressors of *sev* to identify components of the pathway has been established and has encouraged us to start similar endeavors from other points in the pathway. We believe that the success of our previous screens depended on our ability to create a situation where the pathway was "sensitized" by limiting Sevenless activity so that a twofold change in the abundance of another component of the pathway would produce a readily detectable change in phenotype. This ability to identify mutations in heterozygous individuals permits not only the screening of large numbers of individuals, but also the identification of genes with other roles during development whose total loss of function would be lethal. We have been able to individually adjust the activity of *Ras1*, *Gap1*, and *sina* to threshold levels and are currently using these "sensitized" backgrounds to identify other components of the pathway.

REFERENCES

Baker, N.E. and G.M. Rubin. 1989. Effect on eye development of dominant mutations in *Drosophila* homologue of the EGF receptor. *Nature* **340:** 150.

Banerjee, U. and S.L. Zipursky. 1990. The role of cell-cell interactions in the development of the *Drosophila* visual system. *Neuron* **4:** 177.

Banerjee, U., P.J. Renfranz, D.R. Hinton, B.A. Rubin, and S. Benzer. 1987. The *sevenless* protein is expressed apically in cell membranes of developing *Drosophila* retina; it is not restricted to cell R7. *Cell* **51:** 151.

Basler, K. and E. Hafen. 1988. Control of photoreceptor cell fate by the *sevenless* protein requires a functional tyrosine kinase domain. *Cell* **54:** 299.

———. 1989. Dynamics of *Drosophila* eye development and temporal requirements of *sevenless* expression. *Development* **107:** 723.

Basler, K., B. Christen, and E. Hafen. 1991. Ligand-independent activation of the *sevenless* receptor tyrosine kinase changes the fate of cells in the developing *Drosophila* eye. *Cell* **64:** 1069.

Bollag, G. and F. McCormick. 1991. Regulators and effectors of *ras* proteins. *Annu. Rev. Cell Biol.* **7:** 601.

Bourne, H.R., D.A. Sanders, and F. McCormick. 1990. The GTPase superfamily: A conserved switch for diverse cell functions. *Nature* **348:** 125.

———. 1991. The GTPase superfamily: Conserved structure and molecular mechanism. *Nature* **349:** 117.

Bowtell, D.D.L., M.A. Simon, and G.M. Rubin. 1989. Ommatidia in the developing *Drosophila* eye require and can respond to *sevenless* for only a restricted period. *Cell* **56:** 931.

Brunk, B.P., E.C. Martin, and P.N. Adler. 1991. *Drosophila* genes *posterior sex combs* and *suppressor two of zeste* encode proteins with homology to the murine *bmi-1* oncogene. *Nature* **353:** 351.

Carthew, R.W. and G.M. Rubin. 1990. *seven in absentia,* a gene required for specification of R7 cell fate in the *Drosophila* eye. *Cell* **63:** 561.

Feig, L. and G.M. Cooper. 1988. Inhibition of NIH 3T3 cell proliferation by a mutant ras protein with preferential affinity for GDP. *Mol. Cell. Biol.* **8:** 3235.

Fortini, M.E., M.A. Simon, and G.M. Rubin. 1992. Signaling by the *sevenless* protein tyrosine kinase is mimicked by Ras1 activation. *Nature* **355:** 559.

Gaul, U., G. Mardon, and G.M. Rubin. 1992. A putative Ras GTPase activating protein acts as a negative regulator of signaling by the *sevenless* receptor tyrosine kinase. *Cell* **68:** 1007.

Hafen, E., K. Basler, J.E. Edstroem, and G.M. Rubin. 1987. *sevenless,* a cell-specific homeotic gene of *Drosophila*, encodes a putative transmembrane receptor with tyrosine kinase domain. *Science* **236:** 55.

Hall, A. 1990. *ras* and GAP — Who's controlling whom? *Cell* **61:** 921.

Jones, S., M.-L. Vignais, and J.R. Broach. 1991. The CDC25 protein of *S. cerevisiae* promotes exchange of guanine nucleotides bound to ras. *Mol. Cell. Biol.* **11:** 2641.

Koch, A.A., D. Anderson, M.R. Moran, C. Ellis, and T. Pawson. 1991. SH2 and SH3 domains: Elements that control interactions of cytoplasmic signaling proteins. *Science* **252:** 668.

Krämer, H., R.L. Cagan, and S.L. Zipursky. 1991. Interaction of *bride of sevenless* membrane-bound ligand and the *sevenless* tyrosine-kinase receptor. *Nature* **352:** 207.

Lowy, D.R., K. Zhang, J.E. DeClue, and B.M. Willumsen. 1991. Regulation of $p21^{ras}$ activity. *Trends Genet.* **7:** 346.

Marshall, C.J. 1991. How does $p21^{ras}$ transform cells? *Trends Genet.* **7:** 91.

Mayer, B.J., M. Hanaguchi, and H. Hanafusa. 1988. A novel viral oncogene with phospholipase C. *Nature* **332:** 272.

McCormick, F. 1989. *ras* GTPase activating protein: Signal transmitter and signal terminator. *Cell* **56:** 5.

Mullins, M.C. and G.M. Rubin. 1991. Isolation of temperature-sensitive mutations of the tyrosine kinase receptor sevenless (*sev*) in *Drosophila* and their use in determining its time of action. *Proc. Natl. Acad. Sci.* **88:** 9387.

Ready, D.F. 1989. A multifaceted approach to neural development. *Trends Neurosci.* **12:** 101.

Reinke, R. and S.L. Zipursky. 1988. Cell-cell interaction in the *Drosophila* retina: The *bride of sevenless* gene is required in photoreceptor cell R8 for R7 cell development. *Cell* **55:** 321.

Rogge, R.D., C.A. Karlovich, and U. Banerjee. 1991. Genetic dissection of a neurodevelopmental pathway: *Son of sevenless* functions downstream of the *sevenless* and EGF receptor tyrosine kinases. *Cell* **64:** 39.

Segal, D. and B. Shilo. 1986. Tissue localization of *Drosophila melanogaster ras* transcripts during development. *Mol. Cell. Biol.* **6:** 2241.

Simon, M.A., D.D.L. Bowtell, and G.M. Rubin. 1989. Structure and activity of the *sevenless* protein: A protein kinase receptor required for photoreceptor development in *Drosophila*. *Proc. Natl. Acad. Sci.* **86:** 8333.

Simon, M.A., D.D.L. Bowtell, G.S. Dodson, T.R. Laverty, and G.M. Rubin. 1991. Ras1 and a putative guanine nucleotide exchange factor perform crucial steps in signaling by the *sevenless* protein tyrosine kinase. *Cell* **67:** 701.

Tomlinson, A. 1988. Cellular interactions in the developing *Drosophila* eye. *Development.* **104:** 183.

Tomlinson, A. and D.F. Ready. 1986. *Sevenless*: A cell-specific homeotic mutation of the *Drosophila* eye. *Science* **231:** 400.

Tomlinson, A., D.D.L. Bowtell, E. Hafen, and G.M. Rubin. 1987. Localization of the *sevenless* protein, a putative receptor for positional information, in the eye imaginal disc of *Drosophila*. *Cell* **51:** 143.

Van Vactor, D.L., Jr., R.L. Cagan, H. Krämer, and S.L. Zipursky. 1991. Induction in the developing compound eye of *Drosophila*: Multiple mechanisms restrict R7 induction to a single retinal precursor cell. *Cell* **67:** 1145.

Induction of the R7 Neuron in the *Drosophila* Compound Eye: The *bride of sevenless* and *sevenless* Interaction

S.L. Zipursky, H. Krämer, R. Cagan, A. Hart, and D. Van Vactor, Jr.
Howard Hughes Medical Institute, Department of Biological Chemistry, Molecular Biology Institute, University of California, Los Angeles, California 90024

Cell-cell interactions play a central role in development from simple unicellular organisms to complex higher vertebrates. In multicellular organisms, these interactions must be exquisitely controlled to ensure that the right numbers and positions of different cell types are established during embryogenesis. Little is known about the molecular and cellular mechanisms regulating these processes. Which cells produce cues and what types of molecules are they? What cells are competent to respond to these cues and what is the molecular structure of the receptors? What are the mechanisms that regulate the number of cells which respond to an extracellular signal? We have chosen to address these questions in the context of the development of one neuronal cell type, the R7 neuron, in the compound eye of *Drosophila melanogaster*.

The compound eye comprises a hexagonal array of simple eyes, or ommatidia, each containing an essentially invariant number of cells. These include eight different photoreceptor neurons (R1–R8) and a set of nonneuronal accessory cells such as cone cells that secrete the lens and a sheath of pigment cells that serves to isolate optically one ommatidium from its neighbors. Through the use of genetic mosaic analysis, Ready et al. (1976) and Lawrence and Green (1979) established that no strict cell lineages exist between any cells in the developing eye. This gave rise to the view that cell-cell interactions play a central role in regulating cell fate determination in the *Drosophila* eye. Recent studies using lineage tracers to assess the role of cell lineage during the development of the vertebrate nervous system have led to similar conclusions (see, e.g., Turner and Cepko 1987).

Pattern formation and cellular differentiation in the developing primordium of the compound eye, the eye imaginal disc, occur during the third and final stage of larval development (see schematic representation in Fig. 1). A wave of morphogenesis sweeps across this

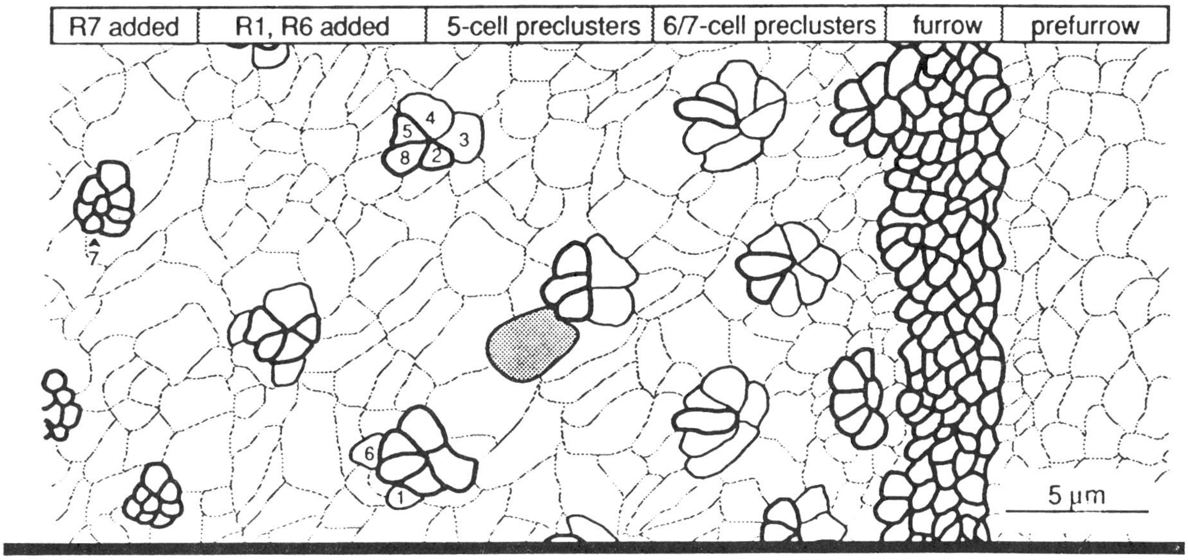

Figure 1. Development of the photoreceptor cell clusters in the developing compound eye of *D. melanogaster*: A tracing of the apical surface of the eye disc epithelium stained with cobalt sulfide is presented. This stain highlights the apical region of the disc epithelium. Pattern formation and differentiation in the primordium of the compound eye occurs as a wave proceeding from posterior to anterior. Anterior is to the right. The front of this wave is referred to as the morphogenetic furrow. Anterior to the furrow, cells are unpatterned and undifferentiated, whereas immediately posterior to the furrow, cells group into clusters and differentiate as neurons. The neurons differentiate in a stereotyped pattern with the R8 neuron differentiating first, followed by R2/R5, then R3/R4, R1/R6, and finally R7. The cells that are not part of the five-cell precluster undergo an additional round of cell division and give rise to the remaining R cells and later the pigment cells and cone cells. (Reprinted, with permission, from Cagan and Ready 1989.)

columnar epithelium from posterior to anterior. The front of this wave is marked by a depression (referred to as the morphogenetic furrow) in the apical surface of the disc epithelium. Anterior to the furrow, cells are unpatterned and undifferentiated. Immediately posterior to the furrow, cells form clusters and begin to differentiate into photoreceptor neurons. Tomlinson and Ready (1987) showed that there was a defined order of photoreceptor differentiation, with the R8 photoreceptor neuron differentiating first, followed by the pairwise differentiation of R2/R5, then R3/R4, followed by R1/R6, and finally R7. This sequential differentiation led to the proposal that photoreceptor cell identity is established as a consequence of a series of inductive events in which previously differentiating neurons induce pluripotent neighboring precursors to assume specific cell fates.

The first evidence for a specific cellular interaction between cells in the developing ommatidium was the demonstration through genetic mosaic analysis of the *bride of sevenless* (*boss*) gene. *boss* was shown to be required in the photoreceptor neuron R8, not for the development of R8 but rather for the development of a neighboring cell, the R7 photoreceptor neuron (Reinke and Zipursky 1988). The phenotype of *boss* mutations is identical to that of the previously identified *sevenless* (*sev*) gene in which the adult fly lacks the R7 photoreceptor neuron. This is a result of an incorrect developmental choice: The R7 precursor cell develops, not as a neuron but as a lens-secreting cone cell (Tomlinson and Ready 1986).

Molecular analysis by Hafen et al. (1987) established that *sev* encodes a receptor-like tyrosine kinase. This observation and the genetic requirement for *sev* in the R7 precursor cell led to the hypothesis that *sev* is a receptor for an inductive cue. The molecular analysis of the *boss* gene indicated that it too encoded a membrane protein with a large extracellular domain and seven transmembrane segments (Hart et al. 1990). Consistent with the genetic mosaic results, it was shown that *boss* is expressed specifically in the R8 photoreceptor neuron (Krämer et al. 1991). In this paper, we review recent studies from our laboratory that demonstrate both in vivo and in vitro that *boss* is a cell-surface-bound ligand of the *sev* receptor. In addition, we review developmental studies which show that multiple mechanisms act in concert to restrict the *boss* inductive cue to only one cell of the many cells within the developing ommatidium that express the *sev* receptor.

MATERIALS AND METHODS

Cell culture and aggregation assays. The procedures used to generate *boss*-expressing cells and the details of the *boss-sev* aggregation assays are described in detail in Krämer et al. (1991).

Immunohistology. Both electron and light microscopic analyses were carried out as decribed previously by Krämer et al. (1991), Van Vactor et al. (1991), and Cagan et al. (1992). A description of the epitopes to which different antibodies are directed is presented in Cagan et al. (1992). The XA12 marker is described in Hart et al. (1990). The *sev* and MAb22C10 antibodies were described by Basler and Hafen (1989b) and Zipursky et al. (1984).

Genetics. The isolation and molecular characterization of the $boss^{1}$ allele were described in Reinke and Zipursky (1988) and Hart et al. (1990). Other strains used have been described previously as follows: sev^{X1} and sev^{d2} alleles (Banerjee et al. 1987); $sev^{met2242}$ allele (Basler and Hafen 1989a); Rh4-*lacZ* line (Fortini and Rubin 1990); and the *nina E* allele used, $ninaE^{0117}$ (O'Tousa et al. 1985).

RESULTS

In Vitro Evidence That *boss* Is a Cell-surface Ligand for the *sev* Tyrosine Kinase

To determine whether *boss* and *sev* can directly interact, we assessed the ability of *boss*- and *sev*-expressing *Drosophila* S2 cell lines to aggregate. In one cell line, SL2-*sev*, a *sev* cDNA is expressed constitutively under the actin 5C promoter (Simon et al. 1989). In the other cell line, a full-length *boss* cDNA is expressed under the control of the inducible hsp70 promoter (Krämer et al. 1991). In suspension, both cell lines remain as single cells. However, upon mixing the S2*boss* and SL2*sev* cells, large aggregates are formed (Krämer et al. 1991). These aggregates contain approximately equal numbers of *sev*- and *boss*-expressing cells. Formation of aggregates is inhibited by antibodies directed against *boss* and *sev* (Fig. 2).

In Vivo Evidence That *boss* Is a Ligand for the *sev* Tyrosine Kinase

Antibodies were used to localize the expression of *boss* in the eye imaginal disc using both light and electron microscopy analyses. Staining was seen on the apical membranes, within a multivesicular body (MVB), and along the rough endoplasmic reticulum (RER) of the R8 cell (Fig. 3). In addition, strong staining was observed in an MVB in the R7 cell (Fig. 4). Since *boss* immunolocalization to the RER is only seen in R8 (Fig. 3E), the accumulation of *boss* in the MVB in R7 is due to transfer from R8. Using the *sev* null and the $sev^{met2242}$ alleles, in which the catalytically required lysine has been mutated to a methionine, internalization was shown to require the *sev* protein but not its catalytic activity. Furthermore, the *shibire* gene product, implicated in the process of endocytosis (Kosaka and Ikeda 1983), is necessary for the accumulation of *boss* within an MVB in R7. These observations are consistent with internalization of *boss* by *sev*-receptor-mediated endocytosis (Krämer et al. 1991).

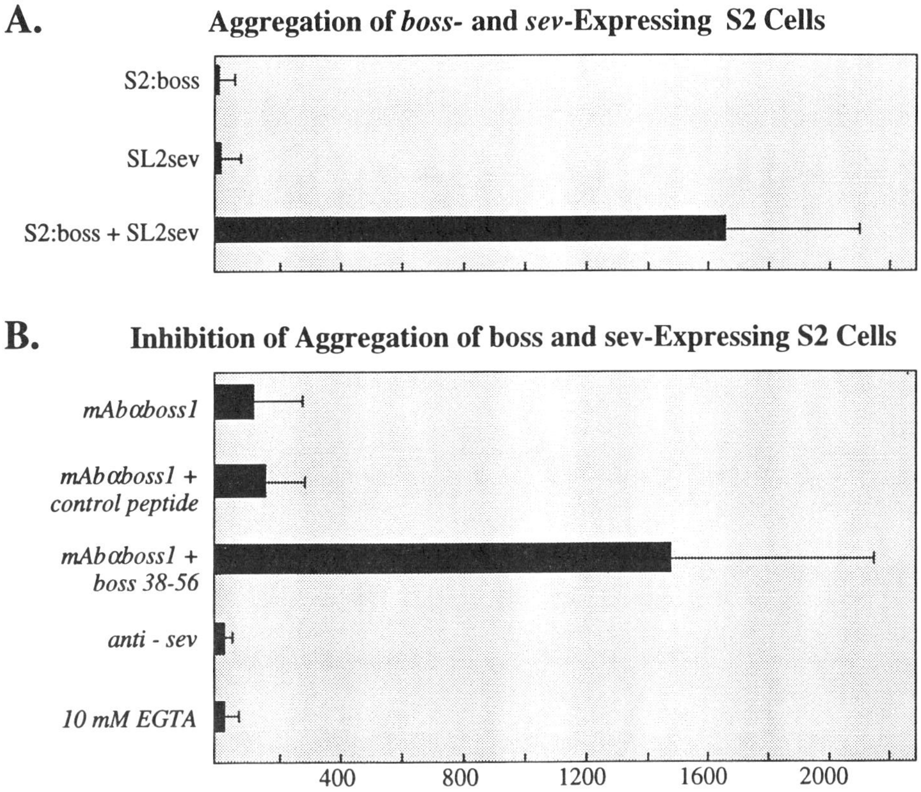

Figure 2. S2*boss* and SL-2*sev* cells form coaggregates. Cell lines expressing *boss* and *sev* were assessed for their ability to aggregate as described previously (Krämer et al. 1991). Aggregates larger than 50 μm in size were counted. Mixtures contained equal numbers of *boss*- and *sev*-expressing cells. The concentrations of other components were as follows: (1) Anti-*boss* 1 (i.e., the same as anti-*boss* NT1), 40 μg/ml; (2) control peptide and peptide corresponding to the amino-terminal region of the *boss* protein (*boss* 38–56), 30 μg/ml; (3) goat anti-*sev* either affinity-purified or as an antiserum was added at 1:100- and 1:200-fold dilutions, respectively. An unrelated goat serum showed no inhibition. (Reprinted, with permission, from Krämer et al. 1991 [copyright Macmillan].)

The Entire *boss* Integral Membrane Protein Is Internalized into *sev*-expressing Cells Both In Vivo and In Vitro

The internalization of *boss* into the R7 photoreceptor cell was initially demonstrated using an antibody to the amino terminus of *boss*. In addition, internalization of *boss* was also observed in tissue-culture cell aggregates formed between *sev*- and *boss*-expressing cells. The simplest model to account for this internalization is the proteolytic cleavage of the amino-terminal extracellular domain of *boss* and subsequent internalization of a soluble form of the amino terminus. To assess which part of *boss* is internalized into *sev*-expressing cells, we used a panel of antibodies directed toward different epitopes in both the intracellular and extracellular domains of *boss* (Cagan et al. 1992).

Both mouse monoclonal and rabbit antibodies were generated to five different regions of the *boss* protein, three of which are located in the amino-terminal extracellular domain and two in the carboxy-terminal cytoplasmic domain. Biochemical studies have verified the extracellular and cytoplasmic disposition of these epitopes.

Using antibodies to all five epitopes, internalization was observed both in tissue culture and in the R7 cell in the developing eye imaginal disc (Fig. 5) (Cagan et al. 1992). The requirements for carboxy-terminal epitope internalization were identical to those previously described for the amino-terminal epitope. Consistent with the internalization of the entire *boss* protein, biochemical studies revealed no evidence of specific proteolytic processing of the *boss* protein. These observations led to the rather surprising conclusion that the entire *boss* protein from its amino-terminal extracellular domain to the carboxy-terminal cytoplasmic domain is internalized into *sev*-expressing cells.

Ubiquitous Expression of the *boss* Protein Leads to Development of Ectopic R7 Cells

Many cells within the developing imaginal disc express the *sev* receptor (Banerjee et al. 1987). These include the R1, R3, R4, R6, R7 cells and the cone cell precursors (Tomlinson et al. 1987). Since *boss* is a cell-surface-bound ligand, the failure of the cone cell precursors to become R7 cells is likely due to the fact that these cells do not contact the R8 cell during normal development. We propose that the R cells, other than

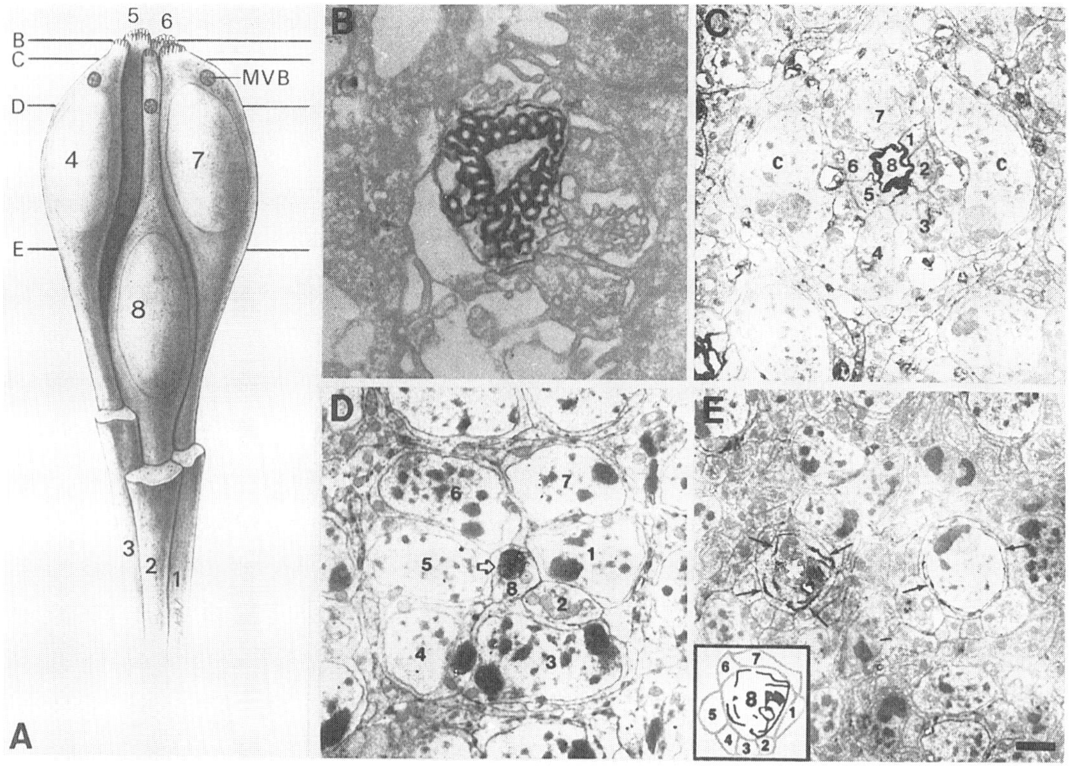

Figure 3. Expression of *boss* in the developing 8-cell cluster in the eye imaginal disc. (*A*) A cutaway diagram of a photoreceptor cell cluster. The apical surface of the disc epithelium is at the top of the diagram. The numbers indicate R cells, and the letters indicate different planes of section corresponding to those presented in panels *B–E*. (*B–E*) Stained for *boss* using anti-*boss* NT1. (*B*) A cross-sectional view of apical microvilli showing staining of the R8 cell; (*C*) staining of the plasma membrane of the R8 cell; (*D*) staining of a multivesicular body (open arrow) in the R8 cell; (*E*) staining of the RER of the R8 cell. (*Inset*) Tracing of the cluster marked by an asterisk in the electron micrograph. (Reprinted, with permission, from Krämer et al. 1991 [copyright Macmillan].)

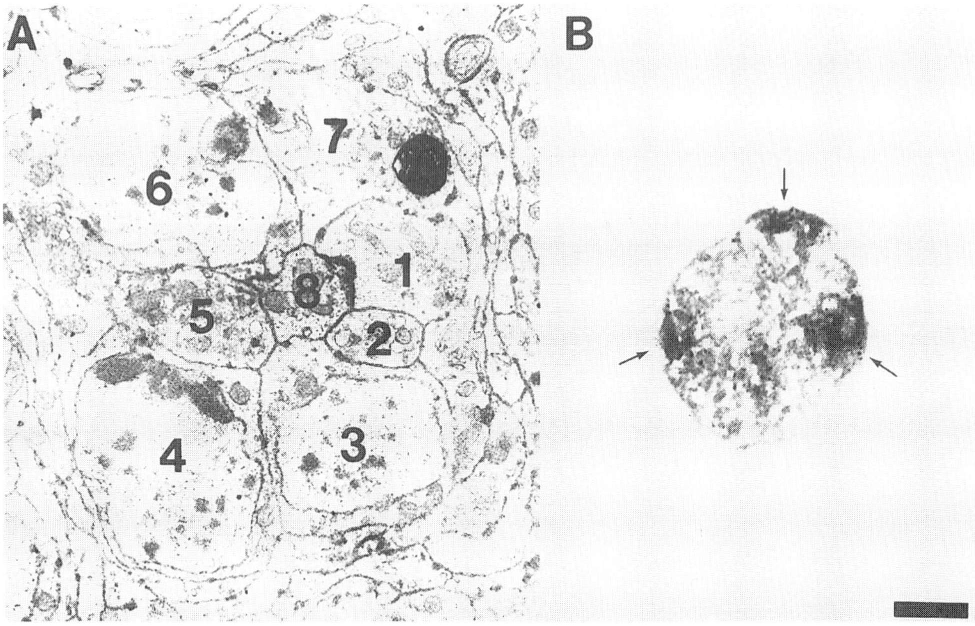

Figure 4. *boss* is localized to an MVB in the R7 cell. (*A*) At the 8-cell cluster stage during eye imaginal disc development, *boss* immunoreactivity as assessed with anti-*boss* NT1 is seen in a large-membrane-bound organelle referred to as a multivesicular body (MVB). The MVB appears to be similar to late endosomal compartments described in mammalian cells. It is within the MVB that the ligand and receptor complexes dissociate from one another. (*B*) A high-magnification view of an MVB stained with anti-*boss* NT1 . Arrows indicate localized *boss* immunoreactivity in the MVB. The MVB shown in this panel is not the same as that shown in panel *A*. Bar: (*A*) 1 μm; (*B*) 0.3 μm. (Reprinted, with permission, from Krämer et al. 1991 [copyright Macmillan].)

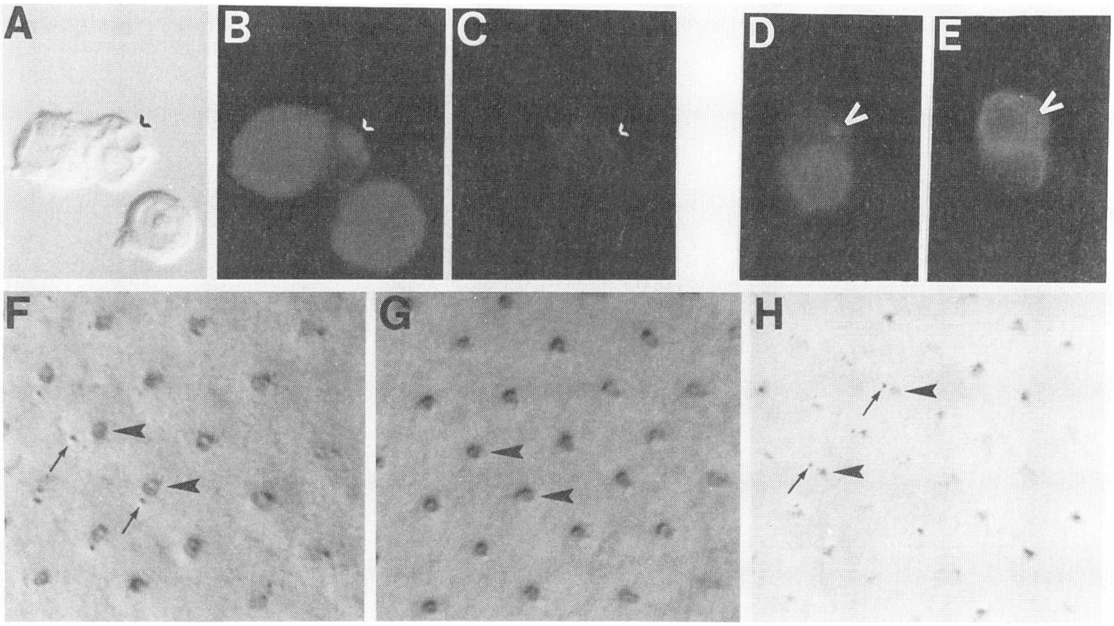

Figure 5. Internalization of *boss* occurs in tissue culture and requires the *sev* receptor. (*A,B*) A *boss*- and *sev*-expressing cell in contact. Arrowhead indicates the position of an organelle in the *sev*-expressing cell stained by both *boss* (*B*) and *sev* (*C*). *boss* immunoreactivity was detected with anti-*boss* CT1 that recognizes a carboxy-terminal epitope. (*D,E*) A pair of *boss* and *sev* cells stained with antibody to the amino terminus of *boss*; anti-*boss* NT1 (*D*) and anti-*sev* antibody (*E*). Arrowhead indicates the position of vesicular structure containing both *sev* and *boss* immunoreactivity. (*F*) Light microscopic view of the R8 cells and MVBs stained with anti-*boss* NT1 in a wild-type eye imaginal disc. (*G*) MVB staining is not observed in a sev^{x1} mutant eye disc. The sev^{x1} mutation is a protein null allele. (*H*) MVB staining is observed in the $sev^{met2242}$ allele. This mutant *sev* receptor contains a methionine at position 2242 replacing a lysine. The lysine is critical for catalytic activity and for R7 development. It is not required for internalization of *boss*. In panels F–H, the thin arrows indicate the *boss* immunoreactivity in the MVB and the arrowheads indicate staining in R8. (Reprinted, with permission, from Cagan et al. 1992 [copyright held by Cell Press].)

R7, that also express *sev* fail to assume an R7 cell fate due to their earlier commitment to an alternative fate. It was reasoned that earlier and ubiquitous expression of *boss* might lead to the induction of additional R7 cells.

To address the importance of the strict spatial and temporal expression pattern of *boss* in the R8 photoreceptor neuron, a *boss* cDNA was expressed under the control of the hsp70 promoter leading to expression in all cells in the developing eye imaginal disc. Ubiquitous *boss* expression resulted in the development of multiple R7 cells (Fig. 6) (Van Vactor et al. 1991). These cells were identified as R7 cells on the bases of morphological, molecular, and genetic criteria. The development of these additional R7 cells was shown to be strictly dependent on the *sev* receptor (Fig. 6C).

Cone Cell Precursors Are Transformed into R7 Cells in Response to Ectopic *boss* Expression

Using the R7-specific marker XA12, the development of ectopic R7 cells was assessed in the developing eye imaginal disc. Cone cell precursors were observed to express the marker in response to ubiquitous *boss* expression (Fig. 7). In contrast, none of the R1–R6 cells appear to be transformed into R7 cells. In addition, using the neuronal cytoplasmic marker MAb22C10 and electron microscopy, it was shown that early stages of R cell development appeared to be normal and confirmed the observation that the cone cell precursors assumed a neuronal identity in response to ectopic *boss* expression (Fig. 7C).

The R1–R6 Precursors Are Refractory to the *boss* Inductive Signal Due to Their Commitment to Alternative Fates

The R cells appear to be refractory to early expression of the *boss* cue. The simplest model to account for this observation is to propose that these cells have already been committed to alternative fates. To assess this, we set out to alter the early development of these cells using different mutant backgrounds. Would these cells blocked at an early stage of development remain competent to respond to the *boss* inductive cue?

The development of R7 neurons was assessed in two mutant backgrounds ($rough^{x63}$ and $Notch^{sp1}$) in which early stages of ommatidial assembly are aberrant (Tomlinson et al. 1988; Cagan and Ready 1989). In both mutant backgrounds, additional R7 cells develop (Van Vactor et al. 1991). Correlated with the appearance of these cells in the adult is the observation that in the eye imaginal disc, many cells internalize the *boss* ligand; in wild type, only the R7 precursor cell internalizes *boss*. In ro^{x63}, many ommatidia in the adult contained multiple R7 cells (two or three) as assessed by

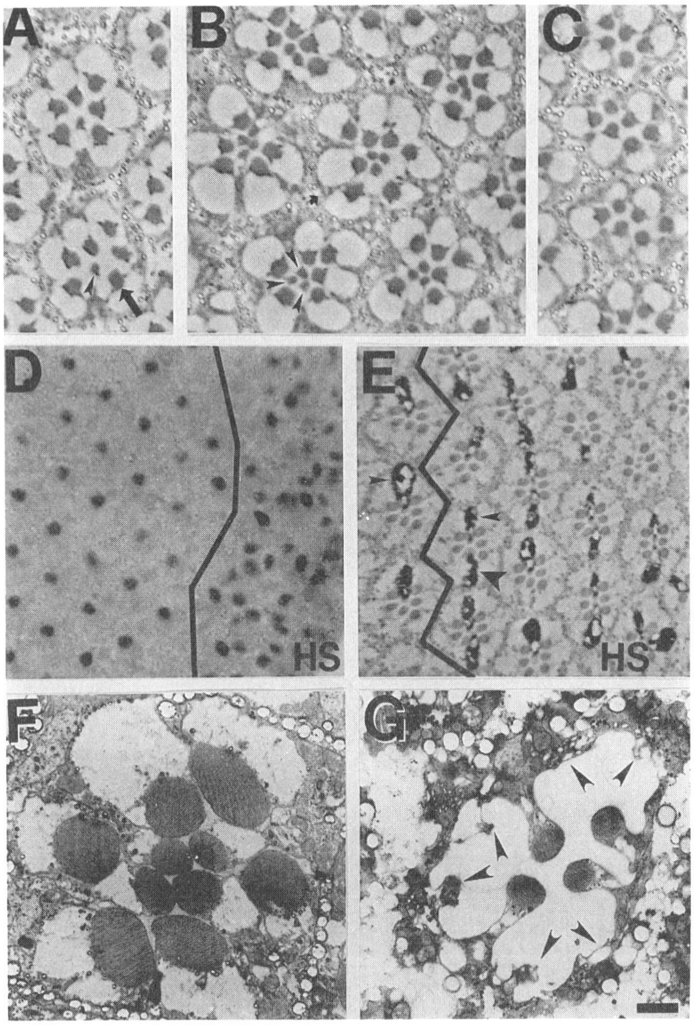

Figure 6. Ectopic *boss* expression under the hsp70 promoter leads to the development of additional R7 cells. (*A*) A light micrograph of a wild-type eye. Arrowhead indicates R7, and the thick arrow indicates a representative large rhabdomere-containing cell, R1. (*B*) Additional small rhabdomere-containing cells are seen in response to ectopic expression of *boss* in the HSB line. *boss* expression is ubiquitous in the eye disc in the absence of heat shock due to the chromosomal site of insertion of the hsp70 *boss* construct. Arrowheads indicate three small rhabodomere-containing cells in one ommatidium. Thick small arrow indicates an ommatidium with one additional large rhabdomere-containing cell. Ommatidia with additional large rhabdomere are seen infrequently. (*C*) The additional small rhabdomere-containing cells require *sev* gene function. No R7 cells are seen in HSB in a sev^{d2} mutant background. (*D*) Ectopic *boss* expression leads to an increase in the number of cells expressing the R7-specific enhancer trap, XA12. Also shown is a pupal eye disc at 50 hr postpupurium formation (ppf) stained for XA12 expression; pupae carrying the HSA construct were heat-pulsed at 15 hr ppf. Additional cells expressing XA12 are seen to the right of the line, indicating that within this region of the epithelium cells were competent to respond to the inductive cue presented by the heat shock (HS). (*E*) The HSA line carrying an R7-specific *lacZ* construct shows additional cells in the affected region HS (i.e., to the right of the jagged line). Small arrowheads indicate the position of the R7 cell, and large arrowheads show ectopic R7 cells. The R7-specific opsin used as a marker is only seen in 70% of wild-type ommatidia. (*F*) A transmission electron micrograph (TEM) of an HSB ommatidium containing four small rhabdomeres. (*G*) A TEM showing an HSB ommatidium in *ninaE* backround. Mutations in *ninaE*, encoding the R1–R6 opsin, lead to R1–R6 cell degeneration. Arrowheads indicate the degenerating rhabdomeres. Bar in *G* indicates the following values: (*A–C*) 5 μm; (*D*) 15 μm; (*E*) 10 μm; (*F, G*) 1.5 μm. (Reprinted, with permission, from Van Vactor et al. 1991 [copyright held by Cell Press].)

staining for *lacZ* driven by an R7-specific promoter (Rh4-lacZ). The conversion of R1–R6 precursors into R7 cells is supported by a concomitant reduction in the R1–R6 cells in ommatidia containing multiple R7 cells.

These data suggest that during ommatidial assembly, there are two classes of *sev*-expressing cells, namely, those that are competent to respond to the *boss* signal and those that are refractory to it (Fig. 8). Similar conclusions were reached by Basler et al. (1991) and Fortini et al. (1992) with activated forms of *sev* and *ras* expressed under the control of the *sev* enhancer.

DISCUSSION

This paper has reviewed recent progress in our understanding of cellular interactions that regulate cell fate in the developing compound eye. So far, one specific cellular interaction has been described, namely, the interaction between the signaling cell, the differentiating R8 neuron, and the responding cell, the R7 photoreceptor neuron. This interaction was defined through the use of genetic mosaic analysis which showed that the *boss* gene is required specifically in the R8 cell for the development of the neighboring R7 cell. Molecular and developmental studies described in this review support the view that *boss* plays a direct role in R7 induction through its interaction with the *sev* receptor.

The structure of the *boss* protein is unprecedented for a ligand of a receptor tyrosine kinase. Indeed, the structure of *boss*, with seven transmembrane segments and an amino-terminal extracellular domain, is similar to that of a large class of G-protein-coupled receptors. *boss* does not share statistically significant amino acid similarity with this class of receptors, but it does have a remarkably similar hydropathy profile. It is intriguing to consider the notion that in addition to serving as a ligand for *sev*, *boss* could in turn serve as a receptor for

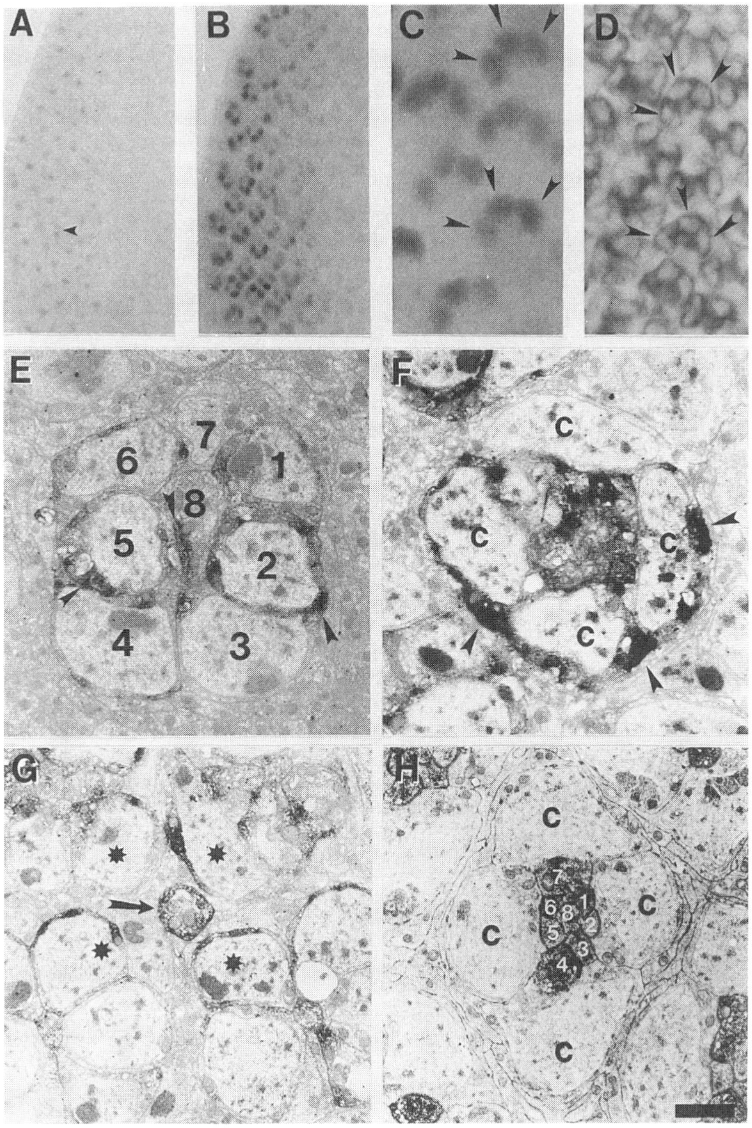

Figure 7. Developmental origin of additional R7 cells. (*A*) Wild-type eye disc viewed with the R7-specific enhancer trap, XA12. (*B*) Additional cells express XA12 in response to ectopic *boss* expression in HSB. (*C*) In the eye disc of HSB, XA12 is expressed in both the R7 cells (e.g., central arrow in each triplet) and the precursors to the anterior and posterior cone cells. (*D*) The MAb22C10 recognizes a cytoplasmic neuronal marker. In contrast to wild type, MAb22C10 is expressed in cone cells in the HSB mutant. Central arrowhead in each cluster indicates the R7 cell. Flanking arrows indicate the transformed cone cells. (*E*) Immunoelectron microscopy using MAb22C10 reveals staining of R cells in the 8-cell cluster in an HSB mutant. Arrowheads indicate cytoplasmic staining. The morphology of the 8-cell cluster looks normal. (*F*) In contrast to wild type, in HSB, MAb22C10 immunoreactivity is seen in cone cell precursors (arrowheads) as a result of ectopic boss expression. (*G*) Immunoelectron microscopy reveals that cells between clusters are occasionally transformed into neurons in HSB (arrow). Asterisks indicate the position of differentiating neurons in surrounding clusters. (*H*) Immunoelectron microscopy showing a wild-type cluster at the four-cone cell stage stained with MAb22C10. Note strong staining in the R cells and the absence of staining in the cone cell nuclei. Bar: (*A*, *B*) 30 μm; (*C*, *D*) 10 μm; (*E*–*H*) 2.5 μm. Abbreviations: (1–8), R1–R8; (c) cone cell.(Reprinted, with permission, from Van Vactor et al. 1991 [copyright held by Cell Press].)

the extracellular domain of *sev*. In light of this possibility, it is important to note that we have not seen any defects in the R8 neuron in a *sev* or *boss* mutant background. Hence, if *boss* were to function as a receptor during eye development, it would most likely function to enhance the inductive interaction between the R8 and R7 cells. One might imagine activation of *boss* leading to increased synthesis of the ligand and/or increasing the adhesive interactions between the R8 cell and the R7 cell. Alternatively, *boss* may function as receptor for a yet to be identified ligand. Indeed, *boss* is expressed in a number of sensory neurons where *sev* is not expressed.

It remains unclear whether the *boss*/*sev* interaction is the only intercellular communication required for R7 induction. Cagan and Ready (1989) using a *Notch* temperature-sensitive mutation previously showed that *Notch* is also required for R7 development. A number

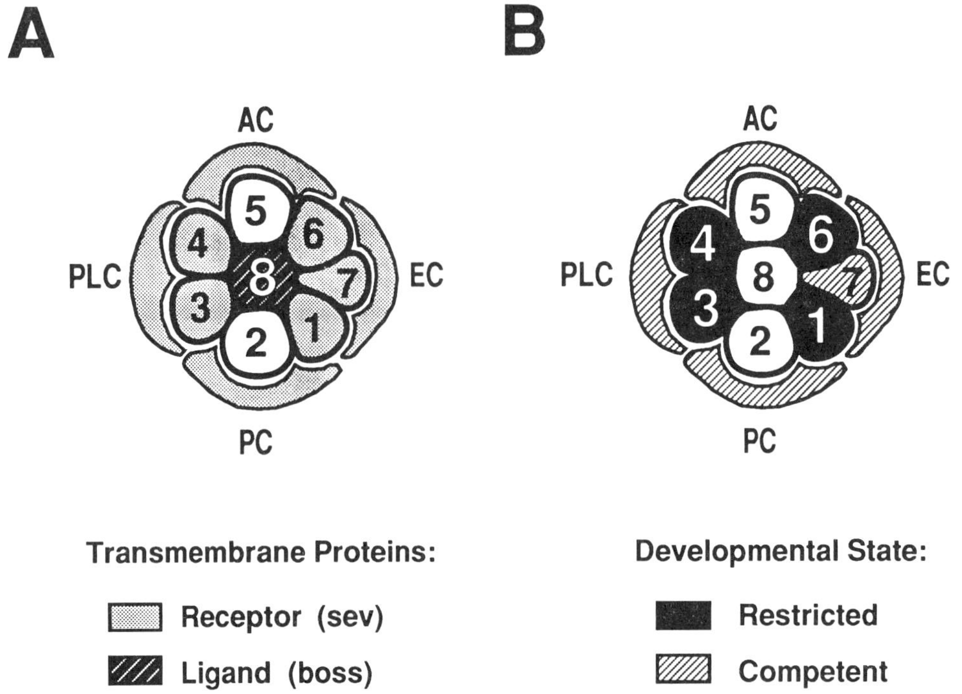

Figure 8. Multiple mechanisms limit R7 induction. (*A*) The *sev* receptor is expressed in cone cell precursors and the R1, R3, R4, R6, and R7 cells. Misexpression data provide an assessment of the developmental state of the *sev*-expressing cells. (*B*) The developmental state of the different *sev*-expressing cells is summarized. The cone cells are competent to assume an R7 cell fate if presented with the boss ligand. During normal development, these cells do not have access to the inductive cue. The R1, R3, R4, and R6 cells are refractory to *boss* expressed in R8 or expressed ubiquitously in the eye disc. However, mutations disrupt the acquisition of normal fates and leave these cells in a state responsive to the *boss* inductive cue. (Reprinted, with permission, from Van Vactor et al. 1991 [copyright held by Cell Press].)

of studies suggest that *Notch* functions as a receptor in a diverse set of intercellular interactions regulating cell fate. Similarly, the Delta protein, a ligand of *Notch*, shows a similar array of developmental phenotypes (M. Muskavitch, pers. comm.). This has led to the view that these transmembrane proteins may function as facilitators of cellular interactions upon which highly cell-type-specific interactions mediated by proteins such as *boss* and *sev* may be overlaid. The ability of ectopically expressed *boss* to induce additional R7 cells indicates that if the *boss/sev* interaction acts in concert with other molecular components, these additional proteins must be more generally expressed in the developing eye.

The original formulation of the model for inductive interactions in the developing eye invoked a combinatorial code for specifying unique cell fates (Tomlinson and Ready 1987). In the case of R7 development, it was proposed that interactions between the R1, R6, and R8 cells with the R7 precursor confer upon this cell its unique cell identity. The results of the ectopic expression of *boss* clearly establish that R7 cells can develop in a position other than that occupied by the R7 precursor, ruling out a strict requirement for this type of a combinatorial mechanism for establishing R7 cell fate.

A rather surprising finding was the observation that the entire *boss* protein, including the seven transmembrane segments, was internalized into the R7 precursor cell in the eye imaginal disc and into *sev*-expressing tissue-culture cells. In principle, one might imagine two different mechanisms for internalization of *boss*. In one model, small patches of the R8 membrane are detached and internalized into receptor-expressing cells via receptor-mediated endocytosis. Alternatively, protein transfer could occur via a partial phagocytic mechanism in which segments of *boss*-expressing microvilli are removed by the *sev*-expressing cells. We favor an endocytic mechanism largely for the reason that although we see many clathrin-coated vesicles and pits in unstained sections, we have not observed large double membrane vesicles as would be expected for a phagocytic mechanism (e.g., turnover of rod outer segment membranes in the overlying pigment epithelia of the vertebrate eye; Matsumoto et al. 1987). In the context of internalization of membrane proteins from adjacent cells, the studies by Kandel and colleagues (Bailey et al. 1992) on the serotonin-induced apCAM internalization provide a striking example of clathrin-mediated endocytosis of double membrane structures.

The spatial localization of *boss* to the R8 cell membrane plays a central role in regulating pattern formation in the developing eye imaginal disc. The *sev*-expressing cone cell precursors located one cell diameter away from R8 are competent to respond to the inductive cue as demonstrated by expressing *boss* ectopically. Interestingly, the *sev*-expressing R1, R3, R4, and R6 cells are refractory not only to normal *boss* expression, but also to *boss* expressed throughout disc development. Remarkably, correlated with the inability of these cells to respond to *boss* expressed on the adjacent R8 cell, these cells show no evidence of internalization of the *boss* ligand. By inhibiting the ability of these cells to assume their normal cell fate, these cells both respond to the inductive cue by becoming R7 cells and internalize *boss*. The mechanism by which commitment to alternative fates suppresses the ability of these cells to internalize the *boss* ligand is not known.

In summary, inductive interactions regulating development are widespread. The *boss*/*sev* interaction is one of only a few inductive ligand-receptor pairs regulating cell fate that have been described at the molecular level. As other inductive systems are dissected, it will be interesting to see whether the structure and regulation of receptors and inductive ligands share similarities with *boss* and *sev*.

ACKNOWLEDGMENTS

We thank the members of the Zipursky laboratory for helpful discussion during the course of these studies. This work was supported by the Howard Hughes Medical Institute, postdoctoral fellowships from the American Cancer Society (H.K.), and the National Institutes of Health (R.L.C.). This work was also supported by grants from the National Institutes of Health and the McKnight Foundation. S.L.Z. is an Investigator of the Howard Hughes Medical Institute.

REFERENCES

Bailey, C.H., M. Chen, F. Keller, and E. Kandel. 1992. Serotonin- mediated endocytosis of apCAM: An early step of learning-related synaptic growth in aplysia. *Science* **256:** 645.

Banerjee, U., P.J. Renfranz, D.R. Hinton, B.A. Rabin, and S. Benzer. 1987. The *sevenless*$^{+}$ protein is expressed apically in cell membranes of developing *Drosophila* retina; it is not restricted to cell R7. *Cell* **51:** 151.

Basler, K. and E. Hafen. 1989a. Ubiquitous expression of *sevenless*$^{+}$: Position-dependent specification of cell fate. *Science* **243:** 931.

———. 1989b. Dynamics of *Drosophila* eye development and temporal requirements for *sevenless* expression. *Development* **107:** 723.

Basler, K., B. Christen, and E. Hafen. 1991. Ligand-independent activation of the *sevenless* receptor tyrosine kinase changes the fate of cells in the developing *Drosophila* eye. *Cell* **64:** 1069.

Cagan, R.L. and D.F. Ready. 1989. *Notch* is required for successive cell decision in the developing *Drosophila* retina. *Genes Dev.* **3:** 1099.

Cagan, R.L., H. Krämer, A.C. Hart, and S.L. Zipursky. 1992. The *bride of sevenless* and *sevenless* interaction: Internalization of a transmembrane ligand. *Cell* **69:** 393.

Fortini, M.E. and G.M. Rubin. 1990. Analysis of *cis*-acting requirements of the Rh3 and Rh4 genes reveals a bipartite organization to rhodopsin promoters in *Drosophila melanogaster*. *Genes Dev.* **4:** 444.

Fortini, M.E., M.A. Simon, and G.M. Rubin. 1992. Signalling by the *sevenless* protein tyrosine kinase is mimicked by Ras 1 activation. *Nature* **355:** 559.

Hafen, E., K. Basler, J.E. Edstroem, and G.M. Rubin. 1987. *sevenless*, a cell-specific homeotic gene of *Drosophila*, encodes a putative transmembrane receptor with a tyrosine kinase. *Science* **236:** 55.

Hart, A.C., H. Krämer, D.L. Van Vactor, M. Paidhungat, and S.L. Zipursky. 1990. Induction of cell identity in the *Drosophila* retina: The bride of sevenless protein is predicted to contain a large extracellular domain and seven transmembrane segments. *Genes Dev.* **4:** 1835.

Kosaka, T. and K. Ikeda. 1983. Reversible blockage of membrane retrieval and endocytosis in the garland cell of temperature sensitive mutant of *Drosophila melanogaster, Shibire*. *J. Cell Biol.* **97:** 499.

Krämer, H., R.L. Cagan, and S.L. Zipursky. 1991. Interaction of *bride of sevenless* membrane-bound ligand and the *sevenless* tyrosine-kinase receptor. *Nature* **352:** 207.

Lawrence, P.A. and S.M. Green. 1979. Cell lineage in the developing retina of *Drosophila*. *Dev. Biol.* **71:** 142.

Matsumoto, B., D.M. Defoe, and J.C. Besharse. 1987. Membrane turnover in rod photoreceptors: Ensheathment and phagocytosis of outer segmental distal tips by pseudopodia of the retinal pigment epithelium. *Proc. R. Soc. Lond. B. Biol. Sci.* **230:** 339.

O'Tousa, J.E., W. Baehr, R.L. Martin, J. Hirsh, W.L. Pak, and M.L. Applebury. 1985. The *Drosphila ninaE* gene encodes an opsin. *Cell* **40:** 839.

Ready, D.F., T.E. Hanson, and S. Benzer. 1976. Development of the *Drosophila* retina, a neurocrystalline lattice. *Dev. Biol.* **53:** 217.

Reinke, R. and S.L. Zipursky. 1988. Cell-cell interaction in the *Drosophila* retina: The *bride of sevenless* gene is required in cell R8 for R7 cell development. *Cell* **55:** 321.

Simon, M., D.L.L. Bowtell, and G.M. Rubin. 1989. Structure and activity of the *sevenless* protein: A protein tyrosine kinase receptor required for photoreceptor cell development in *Drosophila*. *Proc. Natl. Acad. Sci.* **86:** 8333.

Tomlinson, A. and D.F. Ready. 1986. *sevenless*: A cell-specific homeotic mutation of the *Drosophila* eye. *Science* **231:** 400.

———. 1987. Neuronal differentiation in the *Drosophila* ommatidium. *Dev. Biol.* **120:** 366.

Tomlinson, A., B. Kimmel, and G. Rubin. 1988. *rough*, a *Drosophila* homeobox gene required in photoreceptors R2 and R5 for inductive interactions in the developing eye. *Cell* **55:** 71.

Tomlinson, A., D.D.L. Bowtell, E. Hafen, and G.M. Rubin. 1987. Localization of the *sevenless* protein, a putative receptor for positional information, in the eye imaginal disc of *Drosophila*. *Cell* **51:** 143.

Turner, D. and C. Cepko. 1987. A common progenitor for neurons and glia persists in rat retina late in development. *Nature* **328:** 131.

Van Vactor, D.L., Jr., R.L. Cagan, H. Krämer, and S.L. Zipursky. 1991. Induction in the compound developing eye of *Drosophila*: Multiple mechanisms restrict R7 induction to a single retinal precursor cell. *Cell* **67:** 1145.

Zipursky, S.L., T.R. Venkatesh, D.B. Twplow, and S. Benzer. 1984. Neuronal development in the *Drosophila* retina: Monoclonal antibodies as molecular probes. *Cell* **36:** 15.

Delta, *Notch*, and *shaggy*: Elements of a Lateral Signaling Pathway in *Drosophila*

P. Simpson, M. Bourouis, P. Heitzler, L. Ruel, M. Haenlin, and P. Ramain
Laboratoire de Génétique Moléculaire des Eucaryotes du CNRS, Faculté de Médecine, 67085 Cedex, France

In most invertebrates, the neural precursors for both central and peripheral nervous systems arise in a spaced pattern and are separated by intervening epidermal cells. Each precursor has a unique identity thought to be dependent on the position in the epithelium at which it is born (Taghert et al. 1984; Doe and Goodman, 1985; Doe et al. 1988a,b; Patel et al. 1989). Each precursor produces a defined set of specific neurons in a remarkably stereotyped array (Bate 1976). The generation of spaced precursors from a large area of the embryo may be a means of allowing the segregation of a large number of cells, each with a different positional identity. This mode of development means that an individual cell, the precursor, adopts a fate that is different from that of each of its neighboring cells. There is evidence, however, that at each site, more than one cell initially acquires the potential to become a neural precursor, and the question thus arises as to how the single, spaced cells are chosen. Such spacing patterns are quite common in biology.

For our studies of this phenomenon, we have chosen to concentrate on the sensory bristles of the peripheral nervous system of the adult notum of *Drosophila*. The precursors of the bristles arise in a manner similar to that of the neuroblasts of the central nervous system, but they present the advantage that they remain at their sites of origin within the epithelium. This gives rise to the pattern of bristles seen on the adult cuticle (Fig. 1). Areas of the notal epithelium from which the bristle precursors segregate are characterized by the expression of two genes, *achaete* (*ac*) and *scute* (*sc*) (Romani et al. 1989; Cubas et al. 1991; Skeath and Carroll 1991). In the absence of *ac* and *sc*, neural precursors do not form, and it is thought that the products of these two genes confer neural potential within the imaginal disc (Garcia-Bellido 1979; Ghysen and Dambly-Chaudière 1988). There are two types of bristles on each heminotum: 11 large macrochaetae that occupy specific sites and about 100 microchaetae that are uniformly spaced. The *ac* and *sc* genes are expressed in small clusters of cells, known as proneural clusters, at the site of formation of each macrochaete but display a later uniform expression over the area from which the microchaetae precursors segregate. If these proteins confer neural potential, this would mean that many more cells are initially capable of differentiating as neural cells than actually do so. In most clusters, a single sensory organ precursor segregates, after which the expression of *ac* and *sc* persists in the precursor itself, but disappears from the remaining cells of the group (Cubas et al. 1991; Skeath and Carroll 1991).

The Segregation of Spaced Neural Precursors Requires Cell Interactions

Studies on other insect species have led to the idea that developing neural cells have the ability to inhibit their neighbors. Laser ablation of neuroblasts in the grasshopper embryo results in an adjacent cell becoming a neuroblast (Doe and Goodman 1985). Similarly, the new bristles that arise at each molt on the epithelium of the bug *Rhodnius* always appear when division of epidermal cells has pushed the preexisting bristles sufficiently far apart (Wigglesworth 1940). This suggests that the neural precursors produce an inhibitory signal that prevents neighboring cells from developing as neural cells. This phenomenon has been called lateral inhibition. In the case of the bristle pattern in *Drosophila*, the endpoint of the signal would result in the repression of *ac* and *sc*. In *Drosophila*, at the time of their segregation, the precursors of the microchaetae are separated by one or two intervening epidermal cells (Garcia-Bellido and Merriam 1971; Hartenstein and Posakony 1989); subsequent division of the epidermal cells pushes the bristle precursors apart as development continues. The number of *ac/sc*-expressing cells within a proneural cluster from which a macrochaete will arise is between 20 and 30 (Cubas et al. 1991). However, the proteins reach significant levels in a small group of 4, 5, or 6 cells, and it is likely that only these cells are competent to initiate neural development (Heitzler and Simpson 1991; J. Modolell, pers. comm.). The cell interactions required for this process could therefore be contact-mediated, and it is not necessary to invoke a diffusible signal.

Notch, *Delta*, and *shaggy* Are Required for Lateral Inhibition

When the epithelium is mutant for *Notch* (*N*) or *Delta* (*Dl*), all cells with neural potential differentiate as neural precursors. This results in a field of adjacent microchaetae in the areas where microchaetae usually segregate in a spaced pattern and a tuft of six to seven macrochaetae at each site where there is usually a single one (Figs. 1a and 2) (Heitzler and Simpson

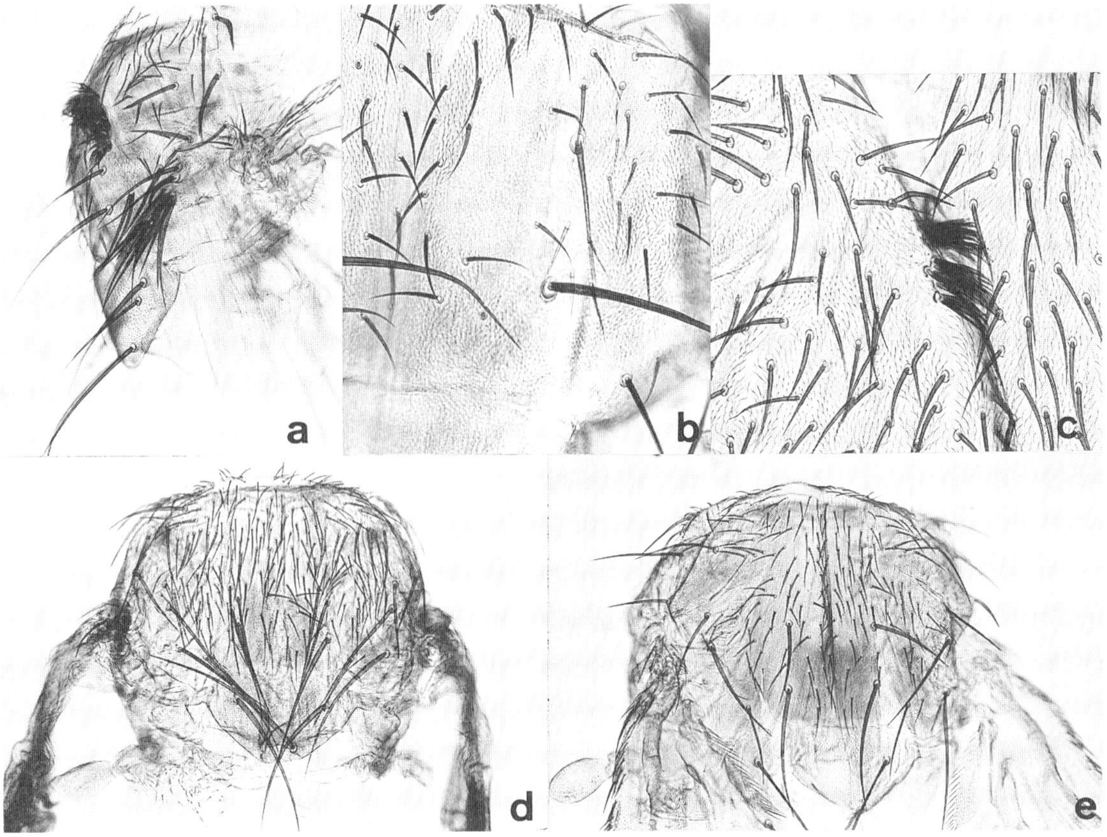

Figure 1. Mutant phenotypes. (*a*) Lateral view, thorax. Photograph of a clone mutant for Dl^{9P39}. A patch of adjacent microchaetae can be seen as well as a tuft of adjacent macrochaetae. (*b*) Clone mutant for Ax^{59b}. In the case of this strong allele, the mutant cells adopt the epidermal fate and thus bristles are not formed. The epidermal hairs have been marked with *multiple wing hairs*. (*c*) Clone mutant for both Ax^{SX1} and Dl^{9P39}. The double-mutant cells show the phenotype opposite to that of $Ax\ Dl^{+}$ cells and adopt the neural fate. Bristles are thus formed as in the case of Dl^{9P39} clones. (*d*) Phenotype of a sgg^{b12} mutant (normally lethal) that has been rescued by a heat-shock construct expressing SGG10. The bristle density remains higher than in the wild type. (*e*) Complete rescue of the bristle pattern of a sgg^{b12} mutant by a heat-shock construct expressing GSK-3β.

1991). The mutant phenotype for *shaggy* (*sgg*) is similar except that only about half of the competent cells become neural (Simpson and Carteret 1989). This phenotype suggests that these genes are required for the epidermal fate of the inhibited *ac/sc*-expressing cells. They are not needed for the differentiation of epidermal cells per se, however. In epithelia that are triply mutant for *ac*, *sc*, and either *N* or *sgg*, the cells differentiate as epidermis (Fig. 2) (Simpson and Carteret 1989). The *ac/sc* genes are thus epistatic over *N* and *sgg* and presumably also over *Dl*. The following is our interpretation of these results: In the absence of expression of the *ac/sc* genes, the default state of the epithelium is to differentiate as epidermis, and this will occur whether or not *N*, *Dl*, and *sgg* are functioning. Once *ac* and *sc* are expressed, then the cells are competent to develop as neural cells; in the absence of *N*, *Dl*, and *sgg*, they will differentiate sensory bristles. It therefore follows that in order to allow only some spaced cells to develop as neural cells and to prevent the others from doing so, *N*, *Dl*, and *sgg* are required.

N encodes an approximately 300-kD transmembrane protein, the extracellular domains of which contain 36 copies of a cysteine-rich motif displaying epidermal growth factor (EGF) homology and also three copies of another cysteine-rich motif, *Notch/lin-12* (Wharton et al. 1985; Kidd et al. 1986). The intracellular domain of *N* contains five copies of a 33-amino-acid repeat that is also found in *cdc10*, *SW16*, and *SW14* (genes controlling the cell cycle in yeast), *lin12* and *glp1* (genes controlling cell-fate decisions in the nematode that also show homology with the extracellular domain of *N*), and ankyrin (a cytoskeletal protein of the erythrocyte membrane). In the case of ankyrin, these repeats form a domain that interacts with integral membrane proteins and tubulin (Lux et al. 1990). A remarkably high conservation of all of these domains exists across species, pointing to an important functional role for each domain (Coffman et al. 1990; Ellisen et al. 1991; Weinmaster et al. 1991). *Dl* also encodes a transmembrane protein with several EGF-type repeats in the extracellular portion (Vässin et al. 1987; Kopczynski et al. 1988). However, *Dl* has only a small intracellular domain that is without homology with other known proteins. *sgg* encodes a family of at least four proteins, all of which contain a catalytic domain of

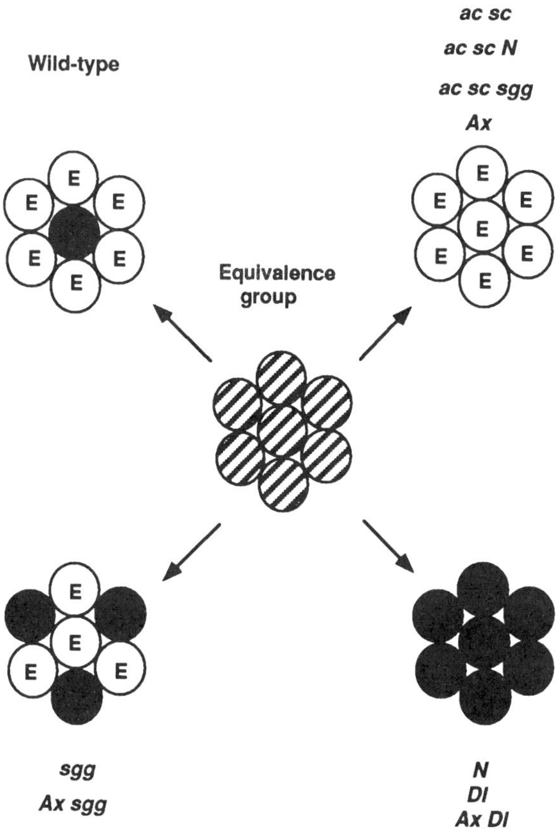

Figure 2. Scheme illustrating the phenotypes of mutants and mutant combinations referred to in the text. A central group of equipotential cells is shown. In the wild-type case, one cell adopts the neural fate (●) and the others become epidermal (E). In epithelia mutant for *N*, *Dl*, or *Ax Dl*, all cells of the group take up the neural fate. *Dl* is therefore epistatic over Ax. In epithelia mutant for *ac sc*, *ac sc N*, *ac sc sgg*, or *Ax*, all cells adopt the opposite epidermal fate. *ac* and *sc* are therefore epistatic over *N* and *sgg*. When mutant for *sgg* or *Ax sgg*, only about half of the cells of the group take up the neural fate, the others become epidermal. *sgg* is therefore epistatic over *Ax*.

kinase activity with putative serine-threonine specificity, as well as a region rich in glycine, alanine, and serine (Bourouis et al. 1990; Siegfried et al. 1990; our unpublished observations). The proteins from these three loci are therefore suitable candidates to be involved in cell-cell interactions. The question arises as to whether they constitute different elements of a signaling pathway. It has been shown in a cell-culture assay that the *N* and *Dl* proteins bind to one another (Fehon et al. 1990). Do *N* and *Dl* interact in a receptor-ligand relationship? If so, what mechanisms can be envisaged for generating intracellular signals, and does *sgg* act downstream in a signal transduction pathway?

Cells Mutant for *Notch*, but not *Delta*, Display Cell Autonomy

Lateral inhibition may therefore involve the production of a signal molecule by the presumptive neural precursors that would bind to a cellular receptor on the presumptive epidermal cells. Mutant cells defective in the receptor mechanism could never be inhibited and should always adopt the neural fate even when adjacent to wild-type cells in mosaic animals. Clones of cells of a mutant genotype in a background of wild-type cells can easily be generated in *Drosophila* by X-ray-induced mitotic recombination. Each epidermal cell secretes a fine nonsensory hair on the thorax, whereas each neural precursor produces (after two cell divisions) a sensory organ such as a bristle. Both bristles and hairs can be marked with suitable cuticle markers. Along the borders of marked clones, it is thus possible to make a cell by cell analysis and to study the epidermal or neural fate of adjacent mutant and wild-type cells (Heitzler and Simpson 1991).

Along the borders of clones mutant for *Dl*, where the mutant cells are in contact with wild-type cells, it was found that Dl^- cells are able to differentiate as epidermis (Fig. 3) (Heitzler and Simpson 1991). In the center of such clones where the mutant cells are not adjacent to wild-type cells, exclusively sensory bristles are formed. This argues that Dl^+ is not required for the reception mechanism. It may therefore be part of the signal. In contrast, cells mutant for *N* autonomously adopt the neural fate and cannot adopt the epidermal fate even when next to wild-type cells (Fig. 3) (Hoppe and Greenspan 1989; Heitzler and Simpson 1991). This suggests that N^- cells are insensitive to an inhibitory signal emanating from their wild-type neighbors and that *N* is thus required for the reception of the signal.

From earlier studies, it was postulated that the inhibitory signal is generated by the presumptive neural precursor. However, a study of the axonal projection patterns of the supernumerary macrochaetae in *sgg* mutant flies showed that the bristles have the same neuronal identity, so the cells of a proneural cluster have equal developmental potential and constitute an equivalence group (Kimble 1981; Simpson and Carteret 1990). Therefore, to begin with, they are all presumably synthesizing both *N* and *Dl*. Thus, it is first necessary to choose a single cell from among the group.

Mutant Cells Can Influence the Fate of Neighboring Wild-type Cells

Along the borders of clones mutant for *Dl*, the neural precursors are *always* of the wild-type genotype and the mutant cells *always* become epidermal (Fig. 3). Along the borders of the clones mutant for *N*, cells of the mutant genotype develop as neural cells, but their wild-type neighbors *always* become epidermal (Fig. 3). We also found that wild-type cells will adopt the epidermal fate if adjacent cells express a lower level of *N* activity than themselves but that they will produce neural precursors if adjacent cells express a higher level of *N* activity (Fig. 3) (Heitzler and Simpson 1991). These results indicate that the amount of these proteins has been measured by the cells prior to their choice of fates. It is therefore probable that *N* and *Dl* are also required for the mechanism that allows cells to choose between the two alternative fates. It appears that cells

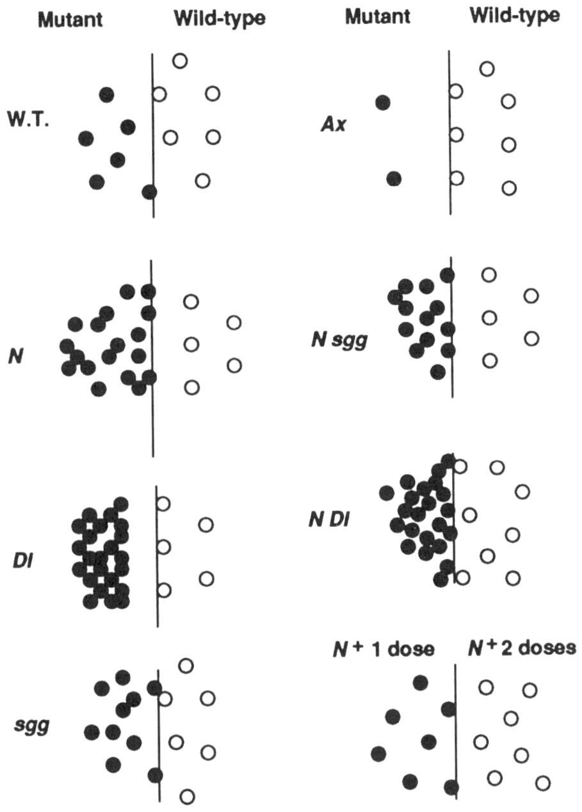

Figure 3. Scheme depicting the behavior of mutant and wild-type cells when juxtaposed along the borders of mutant clones. The solid lines represent the clone border. Only the bristles are drawn. (●) Mutant bristles; (○) wild-type bristles. Epidermal cells secrete hairs (not drawn) that are interspersed between the bristles. In the control case (W.T.) of marked but wild-type clones, marked and unmarked cells both form neural precursors along the border with equal frequency. This indicates a random choice of fate, epidermal or neural, for wild-type cells. Cells mutant for *N* form bristles in the center of the clone and also along the clone borders. They thus display cell autonomy and are not influenced by the presence of neighboring wild-type cells. However, the wild-type cells are themselves influenced by adjacent mutant *N* cells, and they only differentiate as epidermal cells along the border. The mutant *N* cells therefore force neighboring wild-type cells into the epidermal fate. Clones doubly mutant for *N* and *sgg* behave similarly. In contrast to *N*, cells mutant for *Dl* form bristles in the center of the clone where they are not in contact with wild-type cells but are able to differentiate as epidermis along the clone borders. They have thus been influenced by the wild-type neighbors and are said to behave nonautonomously. In this case, the only bristles to form along the clone border are of the wild-type genotype. Along the borders of clones mutant for *sgg*, or doubly mutant for *N* and *Dl*, both mutant and wild-type bristles form. Furthermore, adjacent mutant and wild-type bristles are found. In these cases, the wild-type cells have not affected their mutant neighbors, nor have the mutant cells affected the adjacent wild-type ones, and it would appear that signals traveling both ways between mutant and wild-type cells have been interrupted. Clones mutant for *Ax* differentiate few, if any, bristles and always adopt the epidermal fate along the clone borders. Wild-type cells are unaffected by *Ax* cells and can develop bristles along the border. This is in contrast to cells mutant for loss-of-function alleles, which force the wild-type cells to differentiate as epidermis. Clones of cells containing only a single copy of N^+ do not show a mutant phenotype. However, when juxtaposed to cells containing two copies of N^+, they produce all of the neural precursors along the border. There is thus a bias and cells with a smaller amount of N^+ will always be the ones to adopt the neural fate.

with less *N* protein than their neighbors have an increased ability to signal and will inhibit adjacent wild-type cells. Taken together, these results are consistent with the idea that *N* functions as a receptor and suggest a feedback mechanism whereby cells producing less receptor relative to their neighbors will be more efficient at signaling. In normal development, therefore, if one cell produced slightly less *N* product, perhaps through random physiological fluctuations, it would gain an early advantage, generate a greater signal, and inhibit adjacent cells. This means that what might, initially, be rather a small difference could be amplified into a stronger signal with time.

Cells mutant for *N* thus always inhibit neighboring wild-type cells (Fig. 3), and we postulate that the absence of the receptor led to the production of more signal as a result of a feedback loop. If the *Dl* protein were necessary for the signal, then cells doubly mutant for *N* and *Dl* should no longer be able to inhibit their neighbors. This was found to be the case (P. Heitzler and P. Simpson, in prep.). Along the borders of *N Dl* clones, both mutant and wild-type bristles develop (Fig. 3). Furthermore, mutant and wild-type bristles are sometimes found adjacent to one another. This observation is consistent with the view that these cells are defective in both the sending and receiving of the inhibitory signal: The mutant cell was not inhibited by the wild-type cell and the fate of the wild-type cell was not affected by the mutant one. These observations reinforce the case for *Dl* as the signaling molecule.

Gain-of-function Molecules of *Notch*

There is a special class of mutant alleles of *N* known as the *Abruptex* (*Ax*) alleles (Portin 1975). These are dominant and behave as gain-of-function alleles in that the mutant epithelium produces fewer bristles and thus the cells take up the epidermal fate (Figs. 1b and 2) (Palka et al. 1990; P. Heitzler and P. Simpson, in prep.). In homozygotes for some of the strongest mutants, no neural precursors develop at all. In mosaics, cells mutant for *Ax* behave autonomously, and when adjacent to wild-type cells, they differentiate as epidermal cells (Fig. 3). Furthermore, along the mosaic border, the cells that produce neural precursors are all wild type (Fig. 3). Once again, cells of one genotype have taken up one fate, and cells of the other genotype have taken up the other fate. These observations show that the *Ax* mutants behave in a fashion opposite to that of loss-of-function *N* alleles in every respect: They differentiate autonomously as epidermal cells rather than neural cells, and they fail to signal inhibition to their wild-type neighbors, unlike loss-of-function alleles that signal constitutively. The *Ax* mutant proteins therefore appear to behave as hyperactive *N* molecules.

The molecular lesions associated with seven *Ax* alleles have been determined, and each has a single-amino-acid change in one of the extracellular EGF-like repeats (Hartley et al. 1987; Kelley et al. 1987). Two lethal alleles, Ax^{59b} and Ax^{59d}, cause changes in a highly conserved cysteine residue required for the disulfide bonds between different EGF repeats necessary for the secondary structure. It is likely, then, that the conformation of the extracellular domain is altered in the *Ax* proteins. In view of the gain-of-function phenotype and the possible role of *N* as a receptor, there are two alternative ways to explain the hyperactivity of the *Ax* molecules: (1) The *Ax* molecules may be able to function constitutively in the absence of ligand-binding and this would cause the cells to develop as epidermis, and (2) these proteins may have an enhanced capacity to bind the ligand and thus transmit the signal more efficiently to the interior of the cell. If the *Dl* protein does indeed constitute the ligand, as postulated, then one should be able to distinguish between these possibilities by examining cells doubly mutant for *Ax* and *Dl*. If the *Ax* molecules act as constitutive receptors (i.e., they are ligand-independent), then such double mutant cells should be able to differentiate as epidermis. They are unable to do this, rather, *Ax Dl* cells develop as neural cells (Figs. 1c and 2) (P. Heitzler and P. Simpson, in prep). Therefore, these hyperactive *N* proteins require the *Dl* protein in order to function, and this argues in favor of the second alternative, i.e., they have an enhanced capacity for ligand binding. These experiments therefore provide further support for an interaction between the *N* and *Dl* proteins.

Another indication that the *Dl* and *N* proteins interact comes from studies of *Dl*-expressing and *N*-expressing cells in culture that have been shown to bind together (Fehon et al. 1990). Furthermore, these authors have demonstrated that only the EGF-like repeats 11 and 12 are required for the binding (Rebay et al. 1991). The *Ax* lesions cluster to a 226-amino-acid coding region spanning six EGF repeats (Kelley et al. 1987). This region includes repeats 23–29 and is therefore some distance away from repeats 11 and 12. *N* molecules, in which all of the extracellular parts of the protein are missing, except for EGF repeats 11 and 12, have been found to bind more effectively in the cell-culture assay (Rebay et al. 1991). Thus, one possible explanation for the behavior of the *Ax* proteins would be that they display a shape change that renders repeats 11 and 12 more accessible.

Competitive Protein-Protein Interactions?

Formally, when adjacent to wild-type cells, mutant *Ax* cells behave similarly to cells carrying more copies of N^+. We have previously argued that there is a correlation between the level of activity of *N*, whether due to a greater amount of wild-type protein or a hyperactive mutant *N* protein, and the cell's ability to signal. If the *Ax* molecules have an increased affinity for the *Dl* protein, then this enhanced binding has resulted in a lowered signaling capacity of the mutant cells. This has led us to the suggestion that, in addition to the binding of *N* molecules on one cell to *Dl* molecules on an adjacent cell, the *N* and *Dl* proteins on the surface of the *same* cell may interact. Binding between a cell's own *N* and *Dl* molecules would alter the availability of the proteins to interact with their counterparts on adjacent cells. Such competitive protein-protein interactions could provide a molecular basis for the feedback mechanism between the receptor and the signal (Fig. 4).

Cells carrying the null allele N^{55e11} and thought to be devoid of protein signal constitutively. In this case, it is simple to envisage that the absence of *N* would leave the *Dl* protein intact and free to bind to the *N* protein of adjacent cells. In contrast, in a cell carrying a greater number of *N* molecules relative to *Dl* molecules, *N* would bind to the *Dl* molecules within the cell's own membrane, as well as to those of adjacent cells. In this

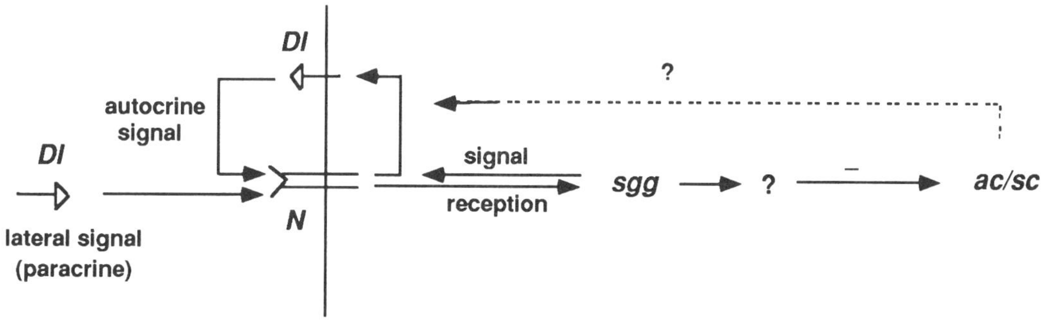

Figure 4. Scheme depicting the putative roles of *N*, *Dl*, and *sgg* during lateral inhibition. *Dl* is visualized as a signal protein that acts as a ligand for *N*, which may act as a receptor protein. The *N* and *Dl* proteins of opposing cells and also within the membrane of the same cell are postulated to bind together and generate a paracrine or an autocrine signal, respectively. *sgg* is thought to act downstream from *N* during signal transduction on the basis of the result of double-mutant *Ax sgg* cells that differentiate as neural cells. The end result of signal transduction is the cessation of *ac* and *sc* expression. *sgg* may act either directly on the *ac/sc* proteins or else on a transcription factor(s) whose role(s) is to modulate *ac/sc* expression. *sgg* may be upstream of *N* for the cell's ability to signal, since *sgg* can impair the cell's signaling ability via the *N* molecule.

fashion, this cell's ability to signal would be impaired since it would have little or no *Dl* protein left to bind to adjacent cells. In this context, there are several ways in which to explain the behavior of the *Ax* proteins that appear to have an enhanced affinity for the ligand. This may imply that they are altered in such a way as to be able only to bind *Dl* in certain configurations. One possibility is that the altered proteins may *only* bind the *Dl* molecules of the same cell, which would simultaneously reduce the availability of that cell's *Dl* molecules to interact with the receptor of neighboring cells and therefore to signal. Alternatively, perhaps they can no longer bind *Dl* molecules from the same cell but only those of surrounding cells. If activation of the receptor is only achieved by this means (i.e., a paracrine signal), then all the bound molecules of *Ax* (*N*) will transduce the signal. In this case, it would be necessary to invoke a different mechanism to account for the incapacity of these cells to signal.

In such a model, cells with the greatest number of bound *N* molecules become epidermal, whether they are bound to *Dl* proteins of the same or adjacent cells. This therefore raises the possibility that binding of the *N* and *Dl* molecules of the same cell could lead to activation of the receptor and that autocrine signaling might occur (Fig. 4). At the present time, we have no way of testing for an autocrine signal.

The signaling mechanism that is mediated by *N* and *Dl* allows the cells to choose between alternative fates, and eventually, one cell of a group adopts the neural fate and the others become epidermal. The decision to be a neural or an epidermal precursor remains labile for a considerable period of time, and following experimental intervention, cells can change their fate (Doe and Goodman 1985; Technau et al. 1988). We have suggested that small relative differences in the amounts of bound *N* proteins between cells could cause a bias that would eventually allow one cell to dominate. If this mechanism were to entail competitive protein-protein interactions between *N* and *Dl*, both between cells and within the same cell, then it is easy to imagine that the process would be rapidly reversible and could occur without the need for new synthesis and transport to the membrane of these large molecules. Once established, however, subsequent maintenance of the two developmental programs would probably involve new synthesis of *N* in the epidermal cells and *Dl* in the neural precursors, as well as the repression of *Dl* expression in the epidermal cells and of *N* expression in the neural cells. It has in fact been observed that *N* protein disappears from the bristle precursors in the developing wing (Fehon et al. 1991).

The Intracellular Domain of *Notch* Is Required for the Feedback between Signal and Reception

If the *N* and *Dl* molecules within the same cell membrane interact, it is possible that this would require the participation of the internal domain of the proteins. Binding of the two molecules would probably involve a certain amount of movement laterally within the membrane. In this context, it is interesting to note that the 33-amino-acid *SW16* repeats present in *N* are found in a number of other proteins including ankyrin. The ankyrins are integral membrane proteins that are thought to be involved in the movement of other proteins within the cell membrane (Lux et al. 1990). *SW16* repeats are also found in several mammalian transcription factors, where they are thought to be involved in protein-protein interactions in various cellular compartments (Haskill et al. 1991; Thompson et al. 1991). They could therefore control the association with other proteins and the functional activation of *N* as a signaling protein but perhaps also be involved in an association with *Dl*. For the moment, mutant alleles with lesions that map to the *SW16* repeats are not known, so further investigation of this possibility will have to await the construction of appropriately modified *N* molecules.

We have, however, studied the behavior of cells mutant for nd^1 and nd^2 whose lesions are in the intracellular domain of the *N* protein. Two lesions are associated with nd^1: a three-base-pair insertion at the 3′ side of the internal opa repeats (Wharton et al. 1985; Kidd et al. 1986) that results in the addition of an extra glutamine, and a missense mutation that results in a threonine to isoleucine change at amino acid position 2453 (Xu et al. 1990). The nd^2 mutant is due to the deletion of a single base pair at the 3′ end of the coding sequence, causing a frameshift changing the 14 carboxy-terminal amino acids into a new sequence of 23 amino acids (Xu et al. 1990). nd^1 animals display a hypomorphic loss-of-function phenotype associated with an enhanced signaling capacity, as is the case for protein null alleles. This suggests that the nd^1 proteins cannot interact with and sequester the *Dl* proteins. nd^2 animals show a variable phenotype in different regions of the thorax. Cells mutant for nd^2 show a very poor ability to signal (our unpublished observations), suggesting in this case that the *N* and *Dl* molecules are strongly bound together. These results argue in favor of a role of the intracellular domain of *N* for the interactions between receptor and signal molecules within the same cell membrane.

shaggy Is Required for the Feedback between Signal and Reception

Neuroepithelia mutant for *sgg* show a phenotype similar to that of *N* and *Dl*, although penetrance of the phenotype is incomplete even for known protein null alleles such as sgg^{D127} (Figs. 2 and 3) (Simpson and Carteret 1989; our unpublished observations). The *sgg* locus encodes a family of proteins with a catalytic domain of kinase activity. No clear transmembrane domains were found in these proteins, and antibody studies show they are mostly intracellular (our unpublished observations). In mosaics, *sgg* behaves autonomously in that the mutant cells can develop as neural cells when adjacent to wild-type cells (Fig. 3) (Heitzler and Simpson 1991). Along the mosaic border, however, both mutant and wild-type cells can form bristles, and occasionally a wild-type and a mutant bristle can be found adjacent to one another (Heitzler and Simpson 1991). Cells mutant for *sgg* therefore behave similarly to cells doubly mutant for *N* and *Dl*. This suggests that both the signaling and receiving capacities of *sgg* cells are impaired.

To investigate the role of *sgg* further, we have examined cells that are doubly mutant for *N* and *sgg*. *N sgg* cells develop, as expected, as neural cells, but like N^- sgg^+ cells, they constitutively inhibit their wild-type neighbors, i.e., along the mosaic borders, only mutant cells produce neural precursors, and the adjacent wild-type cells become epidermal (see Fig. 3; our unpublished results). This means that the inhibitory signal can function in the absence of *sgg*, and thus *sgg* is not absolutely required for signal production. Since N^- sgg^- cells constitutively signal inhibition to their neighbors but N^+ sgg^- cells produce only a poor signal, it follows that the absence of *sgg* decreases the cell's signaling ability only in the presence of *N* molecules. This argues that the *sgg* protein plays a role in the regulation of the cell's signal via the *N* molecule. We therefore conclude that *sgg* is somehow involved in the feedback between *N* and *Dl* (Fig. 4). If *sgg* should interact directly with the *N* protein, this will presumably involve part of the intracellular domain of *N*.

shaggy Is Required Downstream from *Notch* for the Transmission of the Inhibitory Signal

Although the mosaic analyses described above argue in favor of a role of *N* as a receptor, no mechanism for generating intracellular signals is known. The protein kinases produced at the *sgg* locus are obvious candidates to be involved in the transmission of the inhibitory signal to the interior of the cell. To determine whether *sgg* is absolutely required for the reception of the inhibitory signal, we studied the behavior of cells doubly mutant for *Ax* and *sgg*. Provided *Dl* is present, the gain-of-function *Ax* molecules of *N* always force the cells into the epidermal fate. Cells of the *Ax sgg* genotype display the phenotype of *sgg*, i.e., they differentiate as neural cells, albeit with incomplete penetrance (Fig. 2) (our unpublished observations). This strongly suggests that *sgg* is required downstream from *N* for signal transduction, since in its absence, cells carrying the *Ax* molecules no longer adopt the epidermal fate (Fig. 4).

shaggy Encodes a Family of at Least Four Protein Kinases

The study of the precise function of *sgg* has been hampered by the existence of multiple forms of the protein kinases. At least four classes of *sgg* proteins have been identified on Western blots (our unpublished observations). The known proteins all contain a central

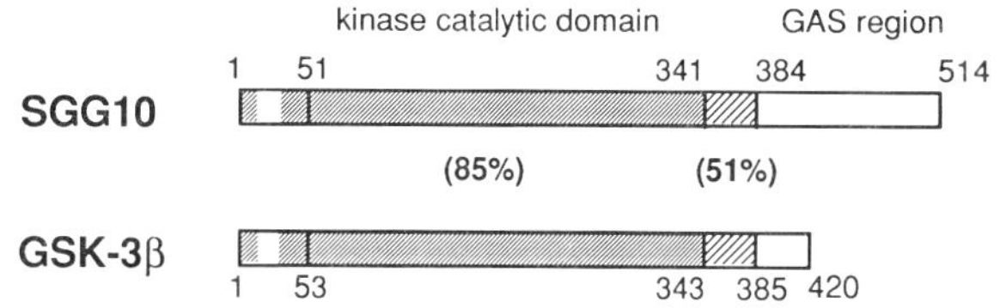

Figure 5. Extent of the homology between SGG10 and GSK-3β. The shaded areas show the extent of the regions of homologies. The SGG10 protein contains a central 290-amino-acid-long catalytic domain of kinase activity with an overall amino acid identity to the corresponding domain of GSK-3β of 85%. There are occasional stretches, 30 amino acids long, of complete conservation. There is a 130-amino-acid-long carboxyl terminus, the GAS region (see text). No such region is found in the GSK-3β sequence. The two domains are separated by a hinge segment. Note that homology extends outside the catalytic domain and is lowest in the hinge region. It is significant, however, that the homologies stop abruptly just after the hinge region, suggesting that the GAS region in SGG10 is a distinct additional domain.

kinase catalytic domain and a carboxy-terminal glycine, alanine, serine-rich region (GAS domain; Fig. 5) but differ in the length of the external amino-terminal or carboxy-terminal regions. One product, SGG10, is produced maternally and deposited in the eggs and is also found (albeit at reduced levels) throughout the rest of life cycle as a result of zygotic transcription. Three other products, SGG39, SGGX, and SGG46, appear zygotically at various times during embryogenesis, and their expression continues during larval life. Transgenic flies containing minigenes coding for these different proteins were constructed in order to test their capacities to rescue mutant flies (our unpublished results). SGG10 can rescue the zygotic lethality of *sgg* mutants, and these flies have a normal pattern of thoracic and wing bristles. Furthermore, the presence of a limited amount of SGG10 delivered via heat-shocked transgenic animals bearing the SGG10 sequences under the control of a heat-shock promoter led to the development of partial mutant *sgg* phenotypes, such as the presence of extra bristles on the thorax (see Fig. 1d). Only partial rescue was achieved with SGG39, and no rescue was possible with SGG46 (SGGX has not yet been tested). We are at present analyzing in more detail the *sgg* proteins present in the imaginal discs from which the adult notum is derived and are investigating the cellular location of the *sgg* proteins in epidermal and neural cells.

sgg proteins contain a particularly hydrophilic carboxy-terminal domain unusually rich in glycine, alanine, and serine residues (75% over the last 130 amino acids) and contain features found in some proteoglycan core proteins (Bourouis et al. 1990). Although there is no biochemical evidence in favor of a proteoglycan-like nature of the *sgg* products, this similarity raises the possibility that the *sgg* protein(s) may reside near the cell membrane. The possibility of an association of the *sgg* and *N* proteins is also being explored.

shaggy Is the *Drosophila* Homolog of the Mammalian Glycogen Synthase Kinase-3

The *sgg* proteins show a remarkably strong sequence homology with the glycogen synthase kinase-3 enzymes (GSK-3; more than 80% of identical amino acids in a region encompassing the kinase catalytic domain [Fig. 5]), mammalian proteins that have been mainly described with respect to their regulatory role on glycogen metabolism in skeletal muscle (Woodgett 1990, 1991). These proteins have more recently been shown to phosphorylate c-*jun* in vitro, hence inhibiting the ability of *jun* to bind DNA (Boyle et al. 1991). In a mammalian cell transfection assay, both the SGG10 and SGG39 proteins, but not SGG46, were able to downregulate c-*jun*-mediated *trans*-activation in a manner similar to that of the GSK-3 enzymes. This effect presumably results from the phosphorylating activity of the GSK-3/SGG proteins that prevented DNA-binding activity of nuclear Jun/AP-1 complexes (de Groot et al. 1992). We have also shown that one of the two forms of GSK-3 (GSK-3β), when introduced into *sgg* mutant flies, can rescue completely the mutant phenotype, albeit with a lower efficiency than SGG10 (Fig. 1e) (our unpublished results). This observation suggests that the *sgg* and GSK-3 enzymes can function similarly with respect to target substrates.

GSK-3 phosphoacceptor sites do not conform to a strict consensus. A minimal structural motif Ser-X X X-Ser(P) surrounded by proline residues has been proposed, and a requirement for a priming phosphorylation event seems to be the general rule (Fiol et al. 1990; Kemp and Pearson 1990; Woodgett 1991). The region of phosphoacceptor sites found in c-*jun* (TPPLSP) is not well conserved in *Drosophila jun*. Apparently, GSK-3 can also phosphorylate bacterially expressed c-*myb* and c-*myc* (Hughes et al. 1992), and thus an appealing possibility is that *sgg* modulates the activity of transcription factors required to activate the neural or the epidermal program.

Eventually, the inhibitory signal results in a cessation of *ac* and *sc* expression. These two genes, once activated, maintain their own expression through autoregulation (Martinez and Modolell 1991). *sgg* could thus act directly on *ac* and/or *sc* themselves and interrupt autoregulation. Furthermore, *ac* and *sc* are both expressed at the sites of bristle segregation in the imaginal epithelium, so if *sgg* were to act on only one of these two proteins, this would explain the partial penetrance of the mutant phenotype. This appears not to be the case, however, since the *sgg* mutant phenotype remains the same when either *ac* alone or *sc* alone are functional (our unpublished observations). We are currently examining the potential inhibitory action of *sgg* using a cell transfection assay and *trans*-activation of a reporter gene by *ac*/*sc*/*daughterless* heteromers.

An alternative possibility is that *sgg* acts negatively on a transcription factor(s) whose role is to switch on *ac* and *sc* at the appropriate sites in the epithelium. We have, in fact, obtained genetic evidence that the *sgg* mutant phenotype of neural hyperplasia depends on intact *cis*-regulatory sequences that govern the spatial expression of *ac* and *sc* (our unpublished results). We have recently isolated two putative transcription factors that may regulate *ac*/*sc* expression, and experiments are in progress to define further the role of these genes and their interactions with *sgg*.

CONCLUSIONS

Taken together, the observations described above argue in favor of roles for *N*, *Dl*, and *sgg* as separate elements in a cell-cell signaling pathway that enables cells to choose between epidermal and neural fates. There remain many unknowns. How is *N* activated after binding to *Dl*? Is the activity of *sgg* "induced" after activation of *N*? What are the substrates of *sgg*? The role of *N* is particularly intriguing. Our hypotheses imply considerable complexity in the functions attributed to this molecule. The protein would have to interact with the *Dl* molecules of the surrounding cells, modulate the activity of *Dl* within the same cell, and furthermore, after binding to *Dl*, initiate a signal transduction event inside the inhibited cell that would eventually lead to the cessation of *ac*/*sc* expression (see Fig. 4). The study of the behavior in mosaics of cells carrying *N* molecules bearing lesions in different domains of the molecule argues that in fact different functions can be attributed to different parts of this large protein. Further studies of this sort involving appropriately modified *N* proteins may enable a partial assignment of different domains to different functions. The existence of highly homologous *N* proteins in other organisms in which both extra- and intracellular domains are conserved points to a similar role for the *N* homolog in these species (Coffman et al. 1990; Ellisen et al. 1991; Weinmaster et al. 1991). The human homolog, *TAN-1*, is associated with a chromosomal translocation in some patients with T lymphoblastic leukemia (Ellisen et al. 1991). If this protein is involved in cell-cell interactions during the assignment of cell fates, then an abnormal, truncated, molecule could lead to abnormal cell communication and proliferation.

N and *Dl* appear not to be the only factors involved in helping cells to choose between epidermal and neural fates. As observed by Cubas et al. (1991), cells with a greater amount of *ac* and *sc* products relative to their neighbors stand a greater chance of becoming neural cells. It is the expression of the *ac* and *sc* proteins that provide the cells with neural potential in the first place. These proteins are expressed in clusters of cells, and the levels of protein are generally higher in the center of such clusters. In wild-type flies, it has been observed that the neural precursor very often originates from a cell near the center of the cluster (Cubas et al. 1991). Thus, either the amount of *ac* and *sc* affects the choice of the neural precursor in a manner independent of that of *N* and *Dl* or the *ac* and *sc* proteins, which are nuclear, could affect cell fate via the signaling mechanism mediated by the *N* and *Dl* proteins at the surface. Again, it is not known what happens downstream from *N* and how activation of *N* is signaled to the nucleus and results in the repression of *ac* and *sc*. However, a feedback along this signaling pathway could perhaps enable the levels of *ac* and *sc* in the nucleus to be perceived at the cell surface (Fig. 4).

There are many parallels between events leading to the segregation of neural precursors in *Drosophila* and the specification of cell fates in the vulva of the nematode. In the latter case, a diffusible inductive signal from the anchor cell is capable of specifying different fates among the equivalent target vulval precursor cells (VPC) cells, depending on the distance of the target cells from the source. This signal provides the nearest VPC cells with the possibility of taking up the 1° fate, whereas cells farther away from the source adopt the 2° fate (Sulston and Horwitz 1977; Sulston and White 1980; Kimble 1981; Sternberg and Horwitz 1986). This is then reinforced by a lateral signal, which is mediated by *lin-12*, an *N*-like protein, and which ensures that only one cell adopts the 1° fate (Greenwald et al. 1983; Sternberg 1988). It is thought that *lin-12* acts as a receptor for the lateral signal (Seydoux and Greenwald 1989). During selection of the neural precursor in *Drosophila*, it would seem that *ac* and *sc* provide an intrinsic "inductive signal" for the neural fate that can be considered as the 1° fate for these cells. This signal is also probably graded and may be stronger in the center of the cluster of equivalent cells. Lateral signaling involving *N* and *Dl* would then ensure that only a single cell adopts the 1° fate. In gain-of-function mutants of *lin-12*, isolated VPCs can adopt the 2° fate: Single VPCs, in the absence of the anchor cell, produce *lin-12* activity, and thus, *lin-12* is probably active in the 2° cell in an autocrine fashion (Sternberg and Horwitz 1989; Greenwald and Seydoux 1990).

lin-12 is involved in specifying cell fate in cells of different lineages. The *N*-mediated cell-signaling mechanism may also be used in many different tissues. *N* is required for the determination of different cell types in the ovary, the mesoderm, and the eye (Cagan and Ready 1989; Corbin et al. 1991; Ruohola et al. 1991). We have shown that $ac^- \ sc^- \ N^-$ cells, which become epidermal in the absence of neural potential, constitutively signal inhibition to wild-type neighbors in the same way as $ac^+ \ sc^+ \ N^-$ cells, showing that the signaling is not dependent on a neural fate (our unpublished observations). It is therefore possible that *N* could function in a similar signaling pathway in many different cell types, and the outcome of such cell interactions would depend on the developmental history of the cells and the regulatory proteins that they are expressing.

ACKNOWLEDGMENTS

We thank Cathie Carteret, Véronique Pantesco, and Claudine Ackerman for technical assistance. Our work is supported by the CNRS and the INSERM.

REFERENCES

Bate, C.M. 1976. Embryogenesis of an insect nervous system. 1. A map of the thoracic and abdominal neuroblasts in Locusta migratoria. *J. Embryol. Exp. Morphol.* **37:** 107.

Bourouis, M., P. Moore, L. Ruel, Y. Grau, P. Heitzler, and P. Simpson. 1990. An early embryonic product of the gene *shaggy* encodes a serine/threonine protein kinase related to the CDC28/cdc2$^+$ subfamily. *EMBO J.* **9:** 2877.

Boyle, W.J., T. Smeal, H.K. Libert, D.P. Angel, J.R. Woodgett, M. Karin, and T. Hunter. 1991. Activation of protein kinase C decreases phosphorylation of c-*jun* at sites that negatively regulate its DNA-binding activity. *Cell* **64:** 573.

Cagan, R.L. and D.F. Ready. 1989. Notch is required for successive cell divisions in the developing *Drosophila* retina. *Genes Dev.* **3:** 1099.

Corbin, V., A.M. Michelson, S.M. Abmayr, Y. Neel, E. Alcamo, T. Maniatis, and M. Young. 1991. A role for the *Drosophila* neurogenic genes in mesoderm differentiation. *Cell* **67:** 311.

Cubas, P., J.-F. de Celis, S. Campuzano, and J. Modolell. 1991. Proneural clusters of *achaete/scute* expression and the generation of sensory organs in the *Drosophila* imaginal wing disc. *Genes Dev.* **5:** 996.

Coffman, C., W. Harris, and C. Kintner. 1990. *Xotch*, the *Xenopus* homolog of *Drosophila Notch*. *Science* **249:** 1438.

de Groot, R., J. Auwerx, M. Bourouis, and P. Sassone-Corsi. 1992. Negative regulation of *jun/AP1*: Conserved function of glycogen synthase kinase-3 and the *Drosophila* kinase *shaggy*. *Oncogene* (in press).

Doe, C.Q. and C.S. Goodman. 1985. Early events in insect neurogenesis. II. Role of cell interactions and cell lineage in the determination of neuronal precursor cells. *Dev. Biol.* **111:** 206.

Doe, C.Q., D. Smouse, and C.S. Goodman. 1988a. Control of neuronal fate by the *Drosophila* segmentation gene *even-skipped*. *Nature* **333:** 376.

Doe, C.Q., Y. Hiromi, W.J. Gehring, and C.S. Goodman. 1988b. Expression and function of the segmentation gene *fushi tarazu* during *Drosophila* neurogenesis. *Science* **239:** 170.

Ellisen, L.W., J. Bird, D.C. West, A. Lee Soreng, T.C. Reynolds, S.D. Smith, and J. Sklar. 1991. TAN-1, the human homolog of the *Drosophila Notch* gene, is broken by chromosomal translocations in T lymphoblastic neoplasms. *Cell* **66:** 649.

Fehon, R.G., K. Johansen, I. Rebay, and S. Artavanis-Tsakonas. 1991. Complex cellular and subcellular regulation of *Notch* expression during embryonic and imaginal development of *Drosophila*: Implications for *Notch* function. *J. Cell Biol.* **113:** 657.

Fehon, R.G., P.J. Kooh, I. Rebay, C.L. Regan, T. Xu, M.A.T. Muskavitch, and S. Artavanis-Tsakonas. 1990. Molecular interactions between the protein products of the neurogenic loci *Notch* and *Delta*, two EGF-homologous genes in *Drosophila*. *Cell* **61:** 523.

Fiol, C.F., A. Wang, R.W. Roeska, and P. Roach. 1990. Ordered multisite protein phosphorylation. Analysis of glycogen synthase kinase-3 action using model peptide substrates. *J. Biol. Chem.* **265:** 6061.

Garcia-Bellido, A. 1979. Genetic analysis of the *achaete-scute* system of *Drosophila melanogaster*. *Genetics* **91:** 491.

Garcia-Bellido, A. and J. Merriam 1971. Parameters of wing imaginal disc development of *Drosophila melanogaster*. *Dev. Biol.* **24:** 61.

Ghysen, A. and C. Dambly-Chaudière. 1988. From DNA to form: The *achaete-scute* complex. *Genes Dev.* **2:** 495.

Greenwald, I. and G. Seydoux. 1990. Analysis of gain of function mutations of the *lin-12* gene of *Caenorhabditis elegans*. *Nature* **346:** 197.

Greenwald, I.S., P.W. Sternberg, and H.R. Horvitz. 1983.

The *lin-12* locus specifies cell fates in *Caenorhabditis elegans*. *Cell* **34:** 435.
Hartenstein, V. and J.W. Posakony. 1989. Development of adult sensilla on the wing and notum of *Drosophila melanogaster*. *Development* **107:** 389.
Hartley, D., T. Xu, and S. Artavanis-Tsakonas. 1987. The embryonic expression of the *Notch* locus of *Drosophila melanogaster* and the implications of point mutations in the extracellular EGF-like domain of the predicted protein. *EMBO J.* **6:** 3407.
Haskill, S., A.A. Beg, S.M. Tompkins, J.S. Morris, A.D. Yuochko, A. Sampson-Johannes, K. Mondal, P. Ralph, and A.S. Baldwin, Jr. 1991. Characterisation of an immediate early gene induced in adherent monocytes that encodes IκB-like activity. *Cell* **65:** 1281.
Heitzler, P. and P. Simpson. 1991. The choice of cell fate in the epidermis of *Drosophila*. *Cell* **64:** 1083.
Hoppe, P.E. and R.J. Greenspan. 1989. Local function of the *Notch* gene for embryonic ectodermal pathway choice in *Drosophila*. *Cell* **46:** 773.
Hughes, K., B.J. Pulverer, and J.R. Woodgett. 1992. Molecular characterisation of glycogen synthase kinase-3/ Factor A. *Adv. Protein Phosphat.* **6:** (in press).
Kelley, M.R., S. Kidd, W.A. Deutsch, and M.W. Young. 1987. Mutations altering the structure of epidermal growth factor-like coding sequences at the *Drosophila Notch* locus. *Cell* **51:** 539.
Kemp, B.E. and R.B. Pearson. 1990. Protein kinase recognition sequence motifs. *Trends Biochem.* **15:** 342.
Kidd, S., M.R. Kelley, and M.W. Young. 1986. Sequence of the *Notch* locus of *Drosophila melanogaster*: Relationship of the encoded protein to mammalian clotting and growth factors. *Mol. Cell. Biol.* **6:** 3094.
Kimble, J. 1981. Alterations in cell lineage following laser ablation of cells in the somatic gonad of *Caenorhabditis elegans*. *Dev. Biol.* **87:** 286.
Kopczynski, C.C., A.K. Alton, K. Fechtel, P.J. Kooh, and M.A.T. Muskavitch. 1988. *Delta*, a *Drosophila* neurogenic gene, is transcriptionally complex and encodes a protein related to blood coagulation factors and epidermal growth factor of vertebrates. *Genes Dev.* **2:** 1723.
Lux,, S.E., K.M. John, and V. Bennett. 1990. Analysis of cDNA for human erythrocyte ankyrin indicates a repeated structure with homology to tissue-differentiation and cell cycle control proteins. *Nature* **344:** 36.
Martinez, C. and J. Modolell. 1991. Cross-regulatory interactions between the proneural *achaete* and *scute* genes. *Science* **251:** 1485.
Palka, J., M. Scubiger, and H. Schwanniger. 1990. Neurogenic and antineurogenic effects from modifications at the *Notch* locus. *Development* **109:** 167.
Patel, N.H., B. Schafer, C.S. Goodman, and R. Holmgren. 1989. The role of segment polarity genes during *Drosophila neurogenesis*. *Genes Dev.* **3:** 890.
Portin, P. 1975. Allelic negative complementation at the *Abruptex* locus of *Drosophila melanogaster*. *Genetics* **81:** 121.
Rebay, I., R.J. Fleming, R.G. Fehon, L. Cherbas, P. Cherbas, and S. Artavanis-Tsakonas. 1991. Specific EGF repeats of *Notch* mediate interactions with *Delta* and *Serrate:* Implications for *Notch* as a multifunctional receptor. *Cell* **67:** 687.
Romani, S., S. Campuzano, E. Macagno, and J. Modolell. 1989. Expression of *achaete* and *scute* genes in *Drosophila* imaginal discs and their function in sensory organ development. *Genes Dev.* **3:** 997.
Ruohola, H., K.A. Bremer, D. Baker, J.R. Swedlow, L.Y. Jan, and Y.N. Jan. 1991. Role of neurogenic genes in establishment of follicle cell fate during oogenesis in *Drosophila*. *Cell* **66:** 433.
Siegfried, E., L.A. Perkins, T.M. Capaci, and N. Perrimon. 1990. Putative protein kinase product of the *Drosophila* segment polarity gene *zeste-white3*. *Nature* **345:** 825.
Seydoux, G. and I. Greenwald. 1989. Cell autonomy of *lin-12* function in a cell fate decision in *C. elegans*. *Cell* **57:** 1237.
Simpson, P. and C. Carteret. 1989. A study of *shaggy* reveals spatial domains of expression of *achaete-scute* alleles on the thorax of *Drosophila*. *Development* **106:** 57.
———. 1990. Proneural clusters: Equivalence groups in the epithelium of *Drosophila*. *Development* **110:** 927.
Skeath, J.B. and S.B. Carroll. 1991. Regulation of *achaete-scute* gene expression and sensory organ pattern formation in the *Drosophila* wing. *Genes Dev.* **5:** 984.
Sternberg, P.W. 1988. Lateral inhibition during vulval induction in *Caenorhabditis elegans*. *Nature* **335:** 551.
Sternberg, P.W. and H.R. Horvitz. 1986. Pattern formation during vulval development in *Caenorhabditis elegans*. *Nature* **335:** 551.
———. 1989. The combined action of two intercellular pathways specifies three cell fates during vulval induction in *C. elegans*. *Cell* **58:** 679.
Sulston, J.E. and H.R. Horvitz. 1977. Post embryonic cell lineages of the nematode *Caenorhabditis elegans*. *Dev. Biol.* **56:** 110.
Sulston, J.E. and J.G. White. 1980. Regulation and cell autonomy during post embryonic development of *Caenorhabditis elegans*. *Dev. Biol.* **78:** 577.
Taghert, P.H., C.Q. Doe, and C.S. Goodman. 1984. Cell determination and regulation during development of neuroblasts and neurones in grasshopper embryos. *Nature* **307:** 163.
Technau, G.M., T. Becker, and J.A. Campos-Ortega. 1988. Reversible commitment of neural and epidermal progenitor cells during embryogenesis of *Drosophila melanogaster*. *Roux's Arch. Dev. Biol.* **197:** 413.
Thompson, C.C., T.A. Brown, and S.L. McKnight. 1991. Convergence of Ets- and Notch-related structural motifs in a heteromeric DNA binding complex. *Science* **253:** 762.
Vässin, H., K.A. Bremer, E. Knust, and J.A. Campos-Ortega. 1987. The neurogenic gene *Delta* of *Drosophila melanogaster* is expressed in neurogenic territories and encodes a putative transmembrane protein with EGF-like repeats. *EMBO J.* **6:** 3433.
Weinmaster, G., V.J. Roberts, and G. Lemke. 1991. A homolog of *Drosophila Notch* expressed during mammalian development. *Development* **113:** 199.
Wharton, K.A., K.M. Johansen, T. Xu, and S. Artavanis-Tsakonas. 1985. Nucleotide sequence from the neurogenic locus *Notch* implies a gene product that shares homology with proteins containing EGF-like repeats. *Cell* **43:** 567.
Wigglesworth, V.B. 1940. Local and general factors in the development of "pattern" in *Rhodnius prolixus* (Hemiptera). *J. Exp. Biol.* **17:** 180.
Woodgett, J.R. 1990. Molecular cloning and expression of glycogen synthase kinase-3 /Factor A. *EMBO J.* **9:** 2431.
———. 1991. A common denominator linking glycogen metabolism, nuclear oncogenes and development. *Trends Biochem.* **16:** 177.
Xu, T., I. Rebay, R.J. Fleming, T.N. Scotgale, and S. Artavanis-Tsakonas. 1990. The *Notch* locus and the genetic circuitry involved in early *Drosophila* neurogenesis. *Genes Dev.* **4:** 464.

Regulation of Induction by GLP1, a Localized Cell Surface Receptor in *Caenorhabditis elegans*

J. Kimble,* S. Crittenden,* E. Lambie,*† V. Kodoyianni,*
S. Mango,* and E. Troemel*
*Department of Biochemistry and Laboratory of Molecular Biology, University of Wisconsin-Madison, Madison, Wisconsin 53706; †Dartmouth College, Hanover, New Hampshire 03755

During the development of the nematode *Caenorhabditis elegans*, cell interactions influence fundamental decisions of growth, differentiation, and pattern formation from early embryogenesis through adulthood (for review, see Lambie and Kimble 1991a). The powerful genetics and cellular simplicity of *C. elegans* make it an ideal organism to analyze the mechanisms by which cell interactions regulate cell fate. Specific cell interactions can be defined at the level of individual cells, and genes controlling those interactions can be identified and manipulated both genetically and molecularly. In this paper, we focus on the role of a single gene, *glp-1* (for *g*erm*l*ine *p*roliferation defective), in the control of cell fate by inductive signaling.

The activity of *glp-1* is required for at least two inductive events during *C. elegans* development. First, maternal *glp-1* is essential in the early embryo for induction of pharyngeal tissue (Austin and Kimble 1987; Priess et al. 1987). During normal development, two cells of the 4-cell embryo, called ABa and EMS, interact to generate a complete pharynx. Specifically, EMS (or its descendants) signals ABa (or its descendants) to make the anterior portion of the pharynx; this cell interaction occurs during the 4- to 28-cell stage of embryogenesis. In the absence of EMS, ABa does not generate pharyngeal tissue (Priess and Thomson 1987; Bowerman et al. 1992). Similarly, in the absence of *glp-1* activity, ABa makes no pharyngeal tissue (Austin and Kimble 1987; Priess et al. 1987). Therefore, a simple hypothesis is that *glp-1* directly mediates induction of ABa to form pharyngeal tissue. In addition to its postulated role in pharyngeal induction, maternal *glp-1* is required for hypodermal development and morphogenesis of the embryo. Second, zygotic *glp-1* is required for induction of germ-line proliferation during larval development and adulthood (Austin and Kimble 1987; Priess et al. 1987). During normal development, a single somatic cell, called the distal tip cell, signals the adjacent germ-line tissue to divide mitotically. In the absence of the distal tip cell, the germ line fails to proliferate: the few germ-line nuclei present leave mitosis, enter meiosis, and differentiate (Kimble and White 1981). Similarly, in the absence of *glp-1* gene activity, germ-line nuclei leave mitosis, enter meiosis, and differentiate (Austin and Kimble 1987; Priess et al. 1987). The simplest explanation of these results is that *glp-1* directly mediates the induction of germ-line proliferation by the distal tip cell.

The *glp-1* gene functions in the germ-line tissue rather than the distal tip cell and has been postulated to be part of the receiving mechanism in inductive signaling (Austin and Kimble 1987). Specifically, the germ line fails to proliferate in genetic mosaic animals whose distal tip cell has wild-type *glp-1*, but whose germ line lacks *glp-1*. In contrast, the germ line proliferates in mosaic animals whose germ line has wild-type *glp-1* and whose distal tip cell lacks *glp-1*. Therefore, *glp-1* functions in the germ line rather than the distal tip cell. A similar mosaic analysis cannot be done for maternal *glp-1* to clarify its role in embryogenesis; however, by analogy with the result obtained in the distal tip cell/germ-line mosaics, we predict that *glp-1* is required in ABa (or its descendants), which is the receiving cell rather than the signaling cell in pharyngeal induction.

The *glp-1* gene is a member of the LNG family, which is named for its founding members, *lin-12*, *Notch*, and *glp-1* (Austin and Kimble 1989; Yochem and Greenwald 1989). The *lin-12* and *Notch* genes, like *glp-1*, mediate cell interactions that regulate cell fate (Greenwald et al. 1983; Lehmann et al. 1983). Other members of the family include *Xenopus Xotch* (Coffman et al. 1990), rat *Notch* (Weinmaster et al. 1991), mouse *int-3* (Robbins et al. 1992), and human *TAN-1* (Ellisen et al. 1991). The developmental roles of the vertebrate LNG proteins are not understood, although lesions in TAN-1 and *int-3* protein are associated with lymphoid and mammary tumors, respectively (Ellisen et al. 1991; Robbins et al. 1992).

LNG proteins appear to be transmembrane proteins (Greenwald 1985; Wharton et al. 1985; Yochem et al. 1988; Austin and Kimble 1989; Yochem and Greenwald 1989; Coffman et al. 1990; Ellisen et al. 1991; Weinmaster et al. 1991). All possess one hydrophobic sequence at the amino terminus, which is thought to be a signal sequence, and a second hydrophobic sequence in the middle of the protein, which is thought to be a membrane-spanning domain. The *Notch* protein has been shown definitively to be a membrane protein with one transmembrane domain (Johansen et al. 1989; Kidd et al. 1989); the *glp-1* and *lin-12* proteins have been shown to be membrane associated (S. Crittenden and J. Kimble, unpubl.; see below), and by analogy with *Notch*, they, and all LNG proteins, are likely to be transmembrane proteins.

LNG proteins are composed of three types of repeated motifs (Greenwald 1985; Wharton et al. 1985;

Yochem et al. 1988; Austin and Kimble 1989; Yochem and Greenwald 1989; Coffman et al. 1990; Ellisen et al. 1991; Weinmaster et al. 1991). Here we focus on the *glp-1* protein, which we call GLP1. In the putative extracellular domain of GLP1 are ten EGF-like repeats and three cysteine-rich LNG repeats (other names for the LNG repeat include the *Notch* repeat [Wharton et al. 1985], the B-type cysteine-rich repeat [Kidd et al. 1989], or the LNR [*lin-12*/*Notch* repeat] motif [Yochem and Greenwald 1989]). The LNG repeats, which are unique to LNG proteins, are located just outside the single membrane-spanning domain; the tandem array of EGF-like repeats is found at the amino terminus. In the putative intracellular domain of GLP1 are six copies of the *cdc10*/SWI6 motif, which was originally discovered in two yeast regulatory genes (Aves et al. 1985; Breeden and Nasmyth 1987a,b). More recently, repeats of the *cdc10*/SWI6 motif have been found in the cytoskeletal protein ankyrin (Lux et al. 1990). All members of the LNG family have three LNG repeats extracellularly and six *cdc10*/SWI6 repeats intracellularly. The number of EGF-like repeats, however, is variable: GLP1 has ten EGF-like repeats, *lin-12* protein has 13 repeats, and *Notch* and the vertebrate homologs have 36 repeats. Therefore, there appears to be a *Notch* subfamily of proteins with 36 EGF-like repeats within the LNG superfamily. Nonetheless, the conservation of protein architecture among members of the LNG family, in addition to the similarity in their developmental roles, suggests that these proteins all function by a similar molecular mechanism.

In addition to being similar in structure and function, the *glp-1* and *lin-12* proteins are functionally interchangeable. Three lines of evidence support this idea. First is the phenotype of the *lin-12 glp-1* double mutant (Lambie and Kimble 1991b). Whereas both *lin-12* and *glp-1* single mutants usually reach adulthood (when derived from a *glp-1*/ + mother), the *lin-12 glp-1* double mutant invariably arrests in the first larval stage. Therefore, in *glp-1* and *lin-12* single mutants, the presence of the wild-type product of the other gene is essential for survival. This means that certain developmental processes can rely on either *glp-1* or *lin-12* function. Second is the similarity in the effects of certain dominant (d) mutations of *lin-12* and *glp-1* on vulval development (Greenwald et al. 1983; Mango et al. 1991). Normally, a cell in the somatic gonad, the anchor cell, signals to the underlying vulval precursor cells to form a vulva (Kimble 1981); in addition, the vulval precursor cells interact among themselves to ensure that only one vulva is made (Sulston and White 1980). In *lin-12(d)* and *glp-1(q35)* mutations, many pseudovulvae are made (Muv phenotype) in an anchor-cell-independent manner (Greenwald et al. 1983; Mango et al. 1991). Third is the effect of *lin-12* loss-of-function mutations on germ-line development (Seydoux et al. 1990). Normally, the germ line is organized such that mitotic cells are distal and meiotic cells are proximal (see Fig. 1). In *lin-12(lf)* mutants,

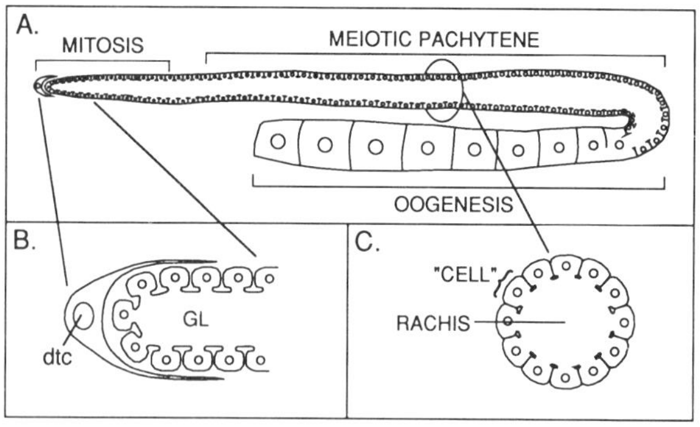

Figure 1. Anatomy of the adult hermaphrodite germ line. (*A*) Each ovotestis is a U-shaped structure with mitotic germ-line nuclei found distally, nuclei in meiotic pachytene found more proximally, and differentiating gametes found most proximally (Hirsh et al. 1976). (*B*) The distal tip cell sends out cytoplasmic processes that embrace the last few germ "cell" diameters in the germ-line tube (D. Hall, pers. comm.). (*C*) The germ-line tissue is a syncytium with most nuclei distributed around the periphery of the germ-line tube. Each germ-line nucleus is partially partitioned from a central core of cytoplasm, or rachis, by membranes; a germ "cell" refers to a unit composed of one nucleus plus its surrounding cytoplasm and membranes. For a review of gonadal anatomy, see Kimble and Ward (1988).

mitotic germ-line nuclei are present in an aberrant position at the proximal end of the gonad in addition to being found at the normal distal site. Seydoux and colleagues speculate that a signal from the anchor cell normally meant for LIN12, instead activates GLP1.

The functional overlap between *lin-12* and *glp-1* proteins implies that the major differences in the developmental roles of the *glp-1* and *lin-12* genes rely on differences in regulation of the two genes. The *glp-1* gene is primarily responsible for germ-line dependent interactions (embryonic induction of the pharynx depends on maternal *glp-1*, which is thought to be produced by the germ line). The *lin-12* gene, in contrast to *glp-1*, is primarily responsible for somatic interactions and specifically for signaling between cells of equivalent developmental potential (lateral signaling) (Greenwald et al. 1983). One simple mechanism to achieve this disparity in the roles of functionally similar proteins is to confine the expression of each to distinct tissues. By Northern blot analysis, it has been shown that *glp-1* RNA accumulates primarily in the germ line, whereas *lin-12* RNA is preferentially somatic (Austin and Kimble 1989). In addition, analysis of the *glp-1(q35)* mutation indicates that the somatic activity of *glp-1* may be inhibited posttranslationally (Mango et al. 1991). Therefore, two regulatory mechanisms have already been discovered that could explain the restriction of *glp-1* activity to the germ line.

In this paper, we address two fundamental questions about the function and regulation of GLP1. First, where is the GLP1 protein found during development? The current models for GLP1 function postulate that it acts as a receptor for an inductive signal. Is the location of GLP1 consistent with this model? Second, what are the critical functional domains in the GLP1 protein?

The conservation of sequence elements among the LNG proteins suggests that the repeated motifs may be critical. What are the molecular defects associated with *glp-1* mutants? Do the sites of the molecular lesions provide insight into the functional domains of GLP1 and LNG proteins generally?

MATERIALS AND METHODS

Methods used in the course of this work are described in detail elsewhere (Kodoyianni et al. 1992; S. Crittenden et al., in prep.). Briefly, for GLP1 protein localization, we used affinity-purified rat polyclonal antibodies raised against bacterially produced pGEX fusion proteins (Smith and Johnson 1988). Immunocytochemistry was done using germ lines extruded from the animal and with early embryos. Photographs were obtained on a BioRad MRC-600 laser scanning confocal microscope. For mapping mutations, lesions were localized roughly within the *GLP-1* gene, either by DNA-RNA mismatch (Myers et al. 1985) or by chemical cleavage of DNA-DNA hybrids (Cotton et al. 1988), and then the specific nucleotide changes were ascertained by DNA sequencing.

RESULTS

GLP1 Is Localized in Both Germ Line and Embryo

To learn the location of the GLP1 protein during development, we raised polyclonal antibodies against two fragments: One contained four EGF-like repeats (EGFL-7 to EGFL-10) and the other consisted of the three LNG repeats (S. Crittenden et al., in prep.). These antibodies recognize GLP1 in fixed tissues by immunocytochemistry and also in protein extracts by Western blot. We find the distribution of GLP1 to be restricted in both germ line and embryo.

Germ line. The germ line is a syncytial tissue: In the distal portion of the gonad, germ-line nuclei are arranged around a central core of cytoplasm and each nucleus is partially enclosed by membrane (Fig. 1). Within the germ-line syncytium, the most distal germ-line "cells" are mitotic, whereas more proximal "cells" have left the mitotic cell cycle and entered meiosis (Fig. 1) (Hirsh et al. 1976). We use the term germ "cell" to refer to a unit composed of one nucleus plus its surrounding cytoplasm and membranes (see Fig. 1C). GLP1 is associated with the plasma membranes of the distal-most germ-line "cells," which are in the mitotic cell cycle (Fig. 2). GLP1 is not detected in the distal tip cell. The germ-line staining of GLP1 is most intense at the distal end and becomes weaker more proximally. In the region where germ cells leave the mitotic cell cycle and enter meiosis, GLP1 becomes located within the cytoplasm of the syncytium (Fig. 2). No GLP1 is found in oocytes (data not shown). From these results, we make two conclusions. First, the localization of GLP1 to the mitotic region is consistent with the idea that GLP1 is a membrane receptor involved in induction of germ-line proliferation by the distal tip cell. Second,

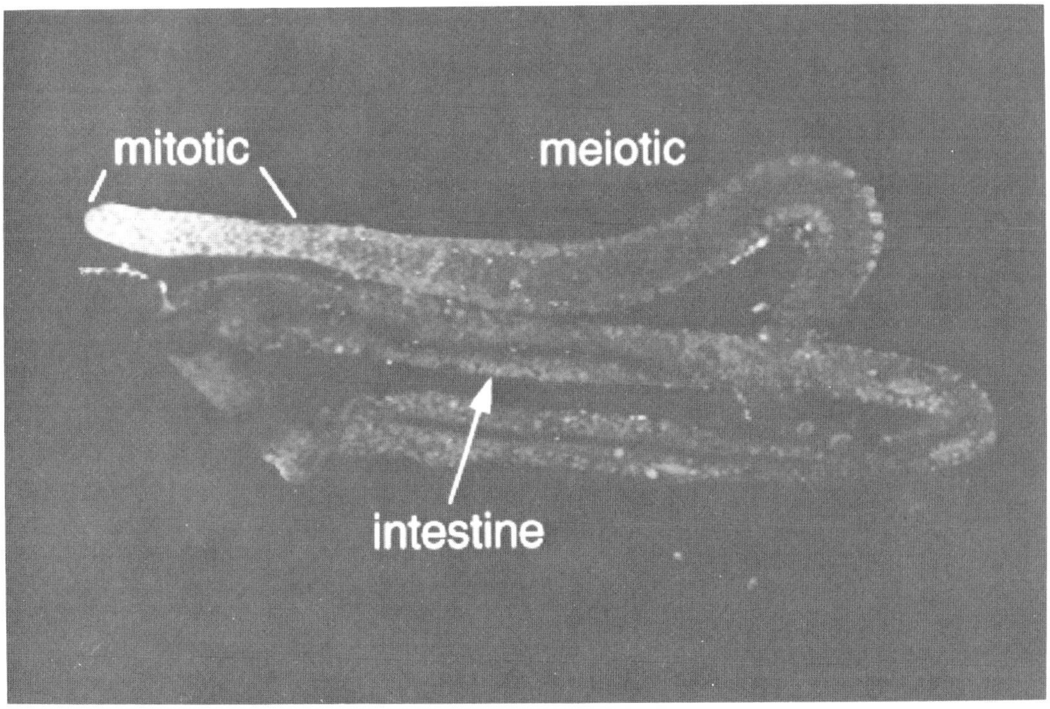

Figure 2. GLP1 is localized to the distal mitotic region of the germ line. An ovotestis and intestine were extruded from a wild-type hermaphrodite adult, stained with affinity-purified antibodies to the LNG region of GLP1, and examined using a BioRad MRC-600 confocal microscope. In the extruded gonad, GLP1 is associated with membranes of germ-line "cells" in the distal mitotic region. The level of GLP1 staining decreases as germ-line "cells" leave mitosis and enter meiosis. GLP1 is not detected in the intestine. For more detail, see S. Crittenden et al. (in prep.).

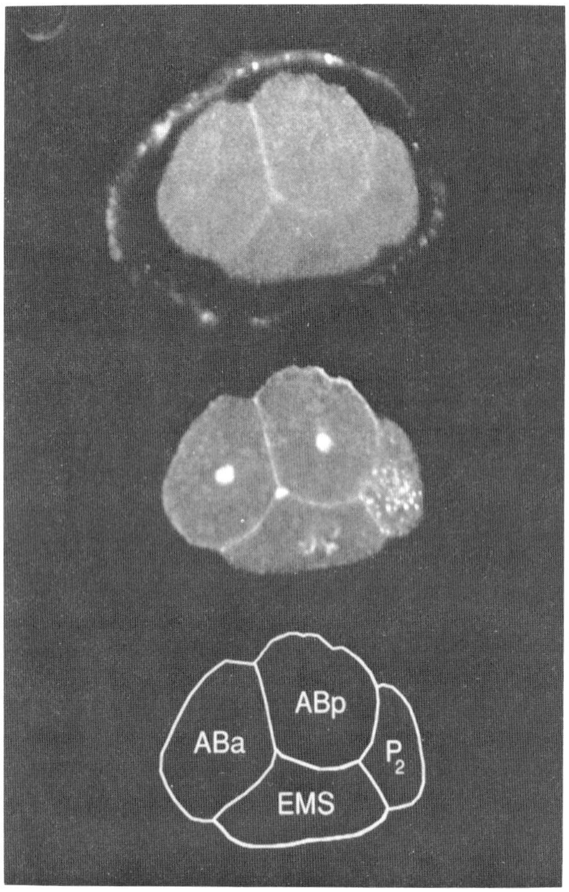

Figure 3. GLP1 is localized to ABa and ABp in the 4-cell embryo. Wild-type embryos were double stained with affinity-purified rat antibodies to the LNG region of GLP1 (fluorescein) and with mouse antibodies to actin, DNA, and P granules (rhodamine). These embryos were then examined using a BioRad MRC-600 confocal microscope. Anterior is left; dorsal is up. (*Top*) GLP1 staining is associated with membranes of the two daughters of AB, which are called ABa and ABp. GLP1 is found at regions of cell-cell contact including membranes between the two AB daughters and those between the AB daughters and the other two blastomeres at the 4-cell stage, EMS and P_2. (*Middle*) Actin staining reveals the boundaries of the blastomeres. DNA staining shows nuclei, and in this focal plane only the nuclei of ABa and ABp are seen. P granule staining is localized to P_2 and hence to the posterior end of the embryo. (*Bottom*) Line drawing of the embryo shown above with blastomeres labeled. For more detail, see S. Crittenden et al. (in prep.).

the failure to detect GLP1 in oocytes suggests that *glp-1* RNA rather than protein is contributed maternally to the embryo.

Early embryo. The embryo divides by an invariant pattern of cell divisions to generate a first larval stage worm (Sulston et al. 1983). The first division of the zygote is asymmetric, generating a larger blastomere, AB, and a smaller blastomere, P_1. AB gives rise primarily to hypodermis, nerves, and the anterior part of the pharynx; in contrast, P_1 gives rise primarily to muscle, intestine, germ line, and the posterior part of the pharynx. The daughters of AB, designated ABa and ABp, are equivalent in developmental potential (Priess and Thomson 1987). We find that GLP1 is associated with the plasma membranes of ABa and ABp at the 4-cell stage (Fig. 3); GLP1 is also found in all AB descendants through the 28-cell stage of embryogenesis (S. Crittenden et al., in prep.). These results bear on three aspects of GLP1 function in the embryo. First, the presence of GLP1 in ABa fulfills the prediction that GLP1 acts in the receiving cell of the inductive interaction between EMS and ABa (or their descendants). Second, the presence of GLP1 from the 4- to 28-cell stage corresponds remarkably well with the period of *glp-1(ts)* temperature sensitivity, supporting the idea that *glp-1* acts during this early phase of embryogenesis. Third, the presence of GLP1 in both ABa and ABp supports the proposal by Priess and Thomson (1987) that these two blastomeres are equivalent in developmental potential.

The Repeated Motifs in GLP1 Are Essential

To begin to delineate the functional domains of GLP1, we have characterized 17 *glp-1* mutations with varying phenotypes. Figure 4 summarizes the locations of the *glp-1* mutations within the GLP1 protein, their phenotypes, and their molecular defects. Our results are described below by phenotypic class.

In one phenotypic class are the putative null mutations (NULL, Fig. 4) (Austin and Kimble 1987; Kodoyianni et al. 1992; S. Crittenden et al., in prep.). All are nonconditional and have the same severe loss-of-function phenotype: a total failure of germ-line proliferation. The putative null alleles include two non-

← DOMAINS OF GLP1 →

TYPE OF MUTATION	EGF-like										LNG			*cdc10*/ SWI6						C-terminus
	1	2	3	4	5	6	7	8	9	10	1	2	3	1	2	3	4	5	6	
NONSENSE			↑ NULL								↑ NULL									↑ MX (cs)
MISSENSE	↑ E	↑(2) E		↑ E			↑ NULL					↑ NULL		↑(2) TS			↑(4) TS			
DELETION												[--------] NULL								

Figure 4. The molecular basis of *glp-1* mutations. *Domains of the* glp-1 *protein* (GLP1): Regions of GLP1 are represented from left to right as they are organized within GLP1, with the tandem array of EGF-like repeats found most amino-terminal and the unique carboxyl terminus found most carboxy-terminal. The transmembrane domain, located between the LNG repeats and *cdc10*/SWI6 (or ankyrin) repeats, has been omitted because no mutations mapped to this region. Individual repeats are numbered beginning from the most amino-terminal repeat. *Phenotypes of* glp-1 *mutations:* The various *glp-1* mutations have a variety of phenotypes: (NULL) phenotype similar to that of a null allele; (MX) mixed phenotype has both gain- and loss-of-function defects; (E) phenotype is more severe in embryo than germ line. *Number of* glp-1 *mutations:* An arrow with no parentheses signifies a single allele; when (*n*) is found next to an arrow, the number in parentheses indicates how many independently isolated alleles map to the specified repeat. See text for further explanation. (TS) Heat sensitive; (cs) cold sensitive. For more detail, see Kodoyianni et al. (1992).

sense mutations, two missense mutations, and one in-frame deletion (Fig. 1). No GLP1 is detected in mutants homozygous for the most amino-terminal nonsense mutation, *glp-1(q175)*; therefore, *glp-1(q175)* is likely to be a protein null. In contrast, GLP1 protein is detected in mutants homozygous for the in-frame deletion that removes most of the LNG repeats. Because GLP1 protein is present but activity is eliminated, the LNG repeats must be essential for GLP1 function. The presence or absence of GLP1 in the other null mutants has not yet been determined; yet the finding of a missense mutation in the seventh EGF-like repeat that leads to a null phenotype suggests that this EGF-like repeat is essential to GLP1 function.

A second phenotypic class includes six temperature-sensitive mutations (TS, Fig. 4) (Austin and Kimble 1987; Priess et al. 1987; Kodoyianni et al. 1992; S. Crittenden et al., in prep.). The effect of the *glp-1(ts)* mutations on germ-line proliferation is either as severe as that of the null mutations or nearly as severe. All six *glp-1(ts)* mutations are missense mutations located in the *cdc10*/SWI6 repeats, and GLP1 protein can be detected in *glp-1(ts)* homozygous mutants at restrictive temperature. Therefore, the *cdc10*/SWI6 repeats must be essential for GLP1 activity.

In a third phenotypic class are mutations with a more severe phenotype in the embryo than in the germ line (E, Fig. 4) (Austin and Kimble 1987; Priess et al. 1987; Kodoyianni et al. 1992). All four *glp-1(E)* mutations are missense mutations located in the amino-terminal EGF-like repeats. Three of the four are partial loss-of-function mutations and one (that in EGFL-1) may interfere with an embryo-specific function. The presence or absence of GLP1 in the *glp-1(E)* mutants has not yet been determined.

Finally, in a class by itself, is the allele *glp-1(q35)*, which has mixed gain- and loss-of-function character (MX, Fig. 1) (Austin and Kimble 1987; Mango et al. 1991; S. Crittenden et al., in prep.). This unique allele has a dominant multivulva (Muv) phenotype, which is typical of *lin-12* gain-of-function alleles, and a recessive germ-line proliferation defect, which is typical of *glp-1* loss-of-function alleles. *glp-1(q35)* is a nonsense mutation that is predicted to remove a unique stretch of amino acids from the carboxyl-terminus, leaving all of the repeated motifs intact. Examination of GLP1(q35) protein both by immunocytochemistry and on Western blots supports the notion that this protein is indeed present in truncated form. From our analysis of *glp-1(q35)*, we have proposed that the carboxy-terminal region deleted in GLP1(q35) protein may normally regulate the somatic expression of GLP1. Furthermore, we have provided evidence that the GLP1(q35) protein, when expressed at normal abundance, is functional and that the unique carboxy-terminal domain is not essential for its activity.

Positive Regulation of *glp-1*

The two partial loss-of-function *glp-1* mutations mapping to the second EGF-like repeat have a unique effect on germ-line development. The ovotestis of a wild-type hermaphrodite has about 1000 germ-line descendants, and that of a *glp-1(0)* hermaphrodite has only a few germ-line descendants (4–8). Intriguingly, the EGFL-2 missense mutants can possess either ovotestes with a virtually normal mitotic germ line or ovotestes with no mitotic germ line at all (Austin and Kimble 1987; Kodoyianni et al. 1992). Indeed, even within a single animal, one ovotestis can appear wild-type and the other Glp. This "all-or-none effect" on germ-line proliferation shows that *glp-1* is regulated independently in the two ovotestes and suggests that a positive feedback control can maintain *glp-1* in an active state.

DISCUSSION

GLP1 and the Germ Line

In the germ line, the distribution of GLP1 is restricted to the distal mitotic region of the syncytium (this paper; S. Crittenden et al., in prep.). GLP1 is found in membranes adjacent to the distal tip cell and also in membranes of mitotic germ cells that are not adjacent to the distal tip cell. Since the distal tip cell signals the germ line to remain in mitosis, this localization is consistent with a role of GLP1 in receiving that signal.

The mechanism by which GLP1 is localized in the germ-line syncytium is not understood. GLP1 germ-line localization might rely on transcriptional, post-transcriptional, or posttranslational controls. One simple model is that the distal tip cell could exert its control over germ-line proliferation by producing a ligand that binds and stabilizes GLP1. According to this model, the extent of GLP1 would be determined by the turnover of GLP1 plus its diffusion rate through the germ-line membrane. A gradient of GLP1 would thus be formed once GLP1 was released from the distal tip cell ligand. However, many alternatives exist, and there is no compelling reason to favor any one model at present.

The significance of GLP1 localization is not understood. It seems likely that the restriction of GLP1 to the mitotic region of the germ line reflects not only a requirement for its presence in that region, but also a requirement for its absence in more proximal cells. If this is true, the presence of Glp in more proximal germ-line nuclei might affect the differentiation of those germ-line nuclei or, as discussed in the next section, it might interfere with early embryogenesis.

GLP1 and the Early Embryo

In the early embryo, the distribution of GLP1 is restricted to the two anterior blastomeres, ABa and ABp, at the 4-cell stage, and to their descendants through the 28-cell stage (this paper; S. Crittenden et al., in prep.). Of particular importance to its role in pharyngeal induction, GLP1 is present in ABa membrane immediately adjacent to EMS membrane. Because ABa (or its descendants) is the receiving cell in pharyngeal induction and EMS (or its descendants) is the signaling cell, this location is consistent with the proposed role for GLP1 as a membrane receptor for the inducing signal.

In addition to its location in ABa, GLP1 is found in its developmentally equivalent sister ABp. This observation is intriguing in light of the similarities between *lin-12* and *glp-1*. As mentioned earlier, the *lin-12* and *glp-1* proteins overlap with each other functionally (Lambie and Kimble 1991b; Mango et al. 1991). The primary role of *lin-12* in development is to mediate lateral signaling—signaling that allows equivalent cells to adopt distinct fates (Greenwald et al. 1983). Therefore, the finding that GLP1 is present in two equivalent blastomeres that subsequently adopt individual fates suggests the possibility that GLP1 mediates lateral signaling between ABa and ABp (or their descendants), much as *lin-12* mediates lateral signaling between somatic precursor cells during postembryonic development. If this is true, the role of GLP1 in early embryogenesis may be much broader than previously considered.

The mechanism by which GLP1 is localized in the early embryo is likely to be posttranscriptional. This idea is based on our inability to detect GLP1 protein in oocytes, and on the embryonic requirement for maternal *glp-1*. Therefore, *glp-1* is likely to be contributed to embryos as maternal mRNA. Expression of that RNA might be regulated either by RNA localization or by localized activation of the RNA. One intriguing possibility is that the *glp-1* 3′UTR may control its temporal and spatial expression within the early embryo. In *Drosophila* and *Xenopus*, the regulated localization of maternal RNA is mediated by elements with the 3′UTRs of localized RNAs (Mowry and Melton 1992; MacDonald 1993). Alternatively, the localized appearance of GLP1 may result from posttranslational regulation, perhaps through the spatial control of protein stability.

The significance of the GLP1 restriction to AB descendants during early embryogenesis is not understood. As in the germ line, it seems likely that embryonic GLP1 restriction reflects not only a requirement for its presence in AB descendants, but also a requirement for its absence in the descendants of EMS and P_2.

GLP1 and the Control of Cell Fate

How does GLP1 mediate inductive interactions? By analogy with other membrane receptors, we postulate that the extracellular domain of GLP1 fields extracellular cues and that the intracellular domain acts in some manner to influence cell fate.The extracellular domain consists primarily of an amino-terminal array of EGF-like repeats and three LNG repeats, whereas the intracellular domain possesses six *cdc10*/SWI6 repeats and a unique carboxy-terminal tail. We find that each of twelve missense mutations with a loss-of-function phenotype maps to one or another of the tandemly repeated motifs, suggesting that the repeated motifs are essential to GLP1 function.

The specific functions of the various repeated motifs of GLP1 remain a matter of speculation. It seems likely that the intracellular domain is of particular importance for regulation of cell fate and that the *cdc10*/SWI6 repeats of GLP1 are the key structural elements mediating this function. A series of missense mutations within these repeats drastically reduce and perhaps eliminate GLP1 activity at restrictive temperature (Kodoyianni et al. 1992; S. Crittenden et al., in prep.). Therefore, the *cdc10*/SWI6 repeats are clearly critical to GLP1 function. Furthermore, a GLP1 protein with much of the unique carboxyl terminus removed, but with the *cdc10*/SWI6 repeats intact, remains functional. What might the *cdc10*/SWI6 repeats be doing? In several other proteins, *cdc10*/SWI6 repeats have been implicated in mediating protein-protein interactions (Lux et al. 1990; Haskill et al. 1991; LaMarco et al. 1991; Nolan et al. 1991; Thompson et al. 1991). Therefore, it is likely that the six *cdc10*/SWI6 repeats of LNG proteins are also important for protein-protein interactions. Furthermore, many *cdc10*/SWI6 repeat-containing proteins either bind or regulate a putative transcription factor: SWI4 and SWI6 of *Saccharomyces cerevisiae* (Andrews and Herskowitz 1989 a,b), *fem-1* of *C. elegans* (Spence et al. 1990), cactus of *Drosophila* (C. Nüsslein-Volhard, pers. comm.), IκB (Haskill et al. 1991), and the β1 subunit of GABP from mammals (Thompson et al. 1991).

We speculate that the *cdc10*/SWI6 repeats of GLP1, and indeed these repeats of all LNG proteins, may function by binding regulatory proteins that are critical for the specification of cell fate. The *cdc10*/SWI6 repeats of GLP1 might, for example, bind a protein and thereby prevent its entry into the nucleus. In the germ line of *C. elegans*, the bound factor might be a positive regulator of meiosis or a negative regulator of mitosis. According to this model, the LNG proteins could be thought of as the membrane receptor equivalents of cytoplasmic regulatory proteins that bear *cdc10*/SWI6 repeats (e.g., IκB). This model is invoked, in part, by analogy with tyrosine kinases, which can be found either as cytoplasmic proteins (e.g., c-*src*) or as membrane receptors (e.g., the EGF receptor).

ACKNOWLEDGMENTS

We thank Steve Paddock for his generous advice concerning use of the BioRad confocal microscope, and we thank Sean Carroll for permitting us to use that instrument, which is provided by the Howard Hughes Medical Institute. We are grateful to Leanne Olds for excellent assistance with technical illustrations. This research was supported by National Institutes of Health grant GM-31816 (to J.K.). V.K. and S.C. were supported by National Institutes of Health postdoctoral grants GM-14186 and GM-13720, respectively.

REFERENCES

Andrews, B.J. and I. Herskowitz. 1989a. The yeast SWI4 protein contains a motif present in developmental regulators and is part of a complex involved in cell-cycle dependent transcription. *Nature* **342:** 830.

———. 1989b. Identification of a DNA binding factor involved in cell-cycle control of the yeast *HO* gene. *Cell* **57:** 21.

Austin, J. and J. Kimble. 1987. *glp-1* is required in the germ line for regulation of the decision between mitosis and meiosis in *C. elegans*. *Cell* **51:** 589.

———. 1989. Transcript analysis of *glp-1* and *lin-12* homologous genes required for cell interactions during development of *C. elegans*. *Cell* **58:** 565.

Aves, S.J., B.W. Durckacz, A. Carr, and P. Nurse. 1985. Cloning, sequencing and transcriptional control of the *Schizosaccharomyces pombe cdc10* "start" gene. *EMBO J.* **4:** 457.

Bowerman, B., B.A. Eaton, and J.R. Priess. 1992. *skn-1*, a maternally expressed gene required to specify the fate of ventral blastomeres in the early *C. elegans* embryo. *Cell* **68:** 1061.

Breeden, L. and K. Nasmyth. 1987a. Cell cycle control of the yeast *HO* gene: *cis-* and *trans*-acting regulators. *Cell* **48:** 389.

———. 1987b. Similarity between cell-cycle genes of budding yeast and fission yeast and the *Notch* gene of *Drosophila*. *Nature* **329:** 651.

Coffman, C., W. Harris, and C. Kintner. 1990. *Xotch*, the *Xenopus* homolog of *Drosophila Notch*. *Science* **249:** 1438.

Cotton, R.G.H., N.R. Rodrigues, and R.D. Campbell. 1988. Reactivity of cytosine and thymine in single-base-pair mismatches with hydroxylamine and osmium tetroxide and its application to the study of mutations. *Proc. Natl. Acad. Sci.* **85:** 4397.

Ellisen, L.W., J. Bird, D.C. West, A.L. Soreng, T.C. Reynolds, S.D. Smith, and J. Sklar. 1991. TAN-1, the human homolog of the *Drosophila Notch* gene, is broken by chromosomal translocations in T lymphoblastic neoplasms. *Cell* **66:** 649.

Greenwald, I. 1985. *lin-12*, a nematode homeotic gene, is homologous to a set of mammalian proteins that includes epidermal growth factor. *Cell* **43:** 583.

Greenwald, I.S., P.W. Sternberg, and H.R. Horvitz. 1983. The *lin-12* locus specifies cell fates in *C. elegans*. *Cell* **34:** 435.

Haskill, S., A. A. Berg, S.M. Tompkins, J.S. Morris, A.D. Yurochko, A. Sampson-Johannes, K. Mondal, P. Ralph, and A.S. Baldwin. 1991. Characterization of an immediate-early gene induced in adherent monocytes that encodes the IkB-like activity. *Cell* **65:** 1281.

Hirsh, D., D. Oppenheim, and M. Klass. 1976. Development of the reproductive system of *Caenorhabditis elegans*. *Dev. Biol.* **49:** 200.

Johansen, K.M., Fehon, R.G., and S. Artavanis-Tsakonas. 1989. The *Notch* gene product is a glycoprotein expressed on the cell surface of both epidermal and neuronal precursor cells during *Drosophila* development. *J. Cell Biol.* **109:** 2427.

Kidd, S., M.K. Baylies, G.P. Gasic, and M.W. Young. 1989. Structure and distribution of the *Notch* protein in developing *Drosophila*. *Genes Dev.* **3:** 1113.

Kimble, J. 1981. Alterations in cell lineage following laser ablation of cells in the somatic gonad of *Caenorhabditis elegans*. *Dev. Biol.* **87:** 286.

Kimble, J. and S. Ward. 1988. Germ-line development and fertilization. In *The nematode* Caenorhabditis elegans (ed. W.B. Wood), p.191. Cold Spring Harbor Laboratory, Cold Spring Harbor, New York.

Kimble, J. and J.G. White. 1981. On the control of germ cell development in *Caenorhabditis elegans*. *Dev. Biol.* **81:** 208.

Kodoyianni, V., E. Maine, and J. Kimble. 1992. The molecular basis of loss-of-function mutations in the *glp-1* gene of *C. elegans*. *Mol. Biol. Cell.* (in press).

Lambie, E.J. and J. Kimble. 1991a. Genetic control of cell interactions in nematode development. *Annu. Rev. Genet.* **25:** 411.

———. 1991b. Two homologous regulatory genes, *lin-12* and *glp-1*, have overlapping function. *Development* **112:** 231.

LaMarco, K., C.C. Thomson, B.P. Byers, E.M. Walton, and S.L. McKnight. 1991. Identification of Ets- and *Notch*-related subunits in GA binding protein. *Science* **253:** 789.

Lehmann, R., F. Jimenez, U. Dietrich, and I.A. Campos-Ortega. 1983.. On the phenotype and development of mutants of early neurogenesis in *Drosophila melanogaster*. *Wilhelm Roux's Arch. Dev. Biol.* **192:** 62.

Lux, S.E., K.M. John, and V. Bennett. 1990. Analysis of cDNA for human erythrocyte ankyrin indicates a repeated structure with homology to tissue-differentiation and cell cycle control proteins. *Nature* **344:** 36.

MacDonald. 1993. The means to the ends: Localization of maternal messenger RNAs. *Semin. Dev. Biol.* (in press).

Mango, S.E., E.M. Maine, and J. Kimble. 1991. Carboxy-terminal truncation activates *glp-1* protein to specify vulval fates in *Caenorhabditis elegans*. *Nature* **352:** 811.

Mowry, K. and D. Melton. 1992. Vegetal messenger RNA localization directed by a 340 nucleotide RNA sequence element in *Xenopus* oocytes. *Science* **255:** 991.

Myers, R.M., Z. Larin, and T. Maniatis. 1985. Detection of single base substitutions by ribonuclease cleavage at mismatches in RNA:DNA duplexes. *Science* **230:** 1242.

Nolan, G.P., S. Ghosh, H.-C. Liou, P. Tempst, and D. Baltimore. 1991. DNA binding and IκB inhibition of the cloned p65 subunit of NFκB a *rel*-related polypeptide. *Cell* **64:** 961.

Priess, J.R. and J.N. Thomson. 1987. Cellular interactions in early *C. elegans* embryos. *Cell* 48:241.

Priess, J.R., H. Schnabel, and R. Schnabel. 1987. The *glp-1* locus and cellular interactions in early *C. elegans* embryos. *Cell* **51:** 610.

Robbins, J., B.J. Blondel, D. Gallahan, and R. Callahan. 1992. Mouse mammary tumor gene *int-3*: A member of the *notch* gene family transforms mammary epithelial cells. *J. Virology* **66:** 2594.

Seydoux, G., T. Schedl, and I. Greenwald. 1990. Cell-cell interactions prevent a potential inductive interaction between soma and germ line in *C. elegans*. *Cell* **61:** 939.

Smith, D.B. and K.S. Johnson. 1988. Single-step purification of polypeptides expressed in *Escherichia coli* as fusions with glutathione S-transferase. *Gene* **67:** 2140.

Spence, A.W., A. Coulson, and J. Hodgkin. 1990. The product of *fem-1*, a nematode sex-determining gene, contains a repeated motif found in cell cycle control proteins and receptors for cell-cell interactions. *Cell* **60:** 981.

Sulston, J.E. and J.G. White. 1980. Regulation and cell autonomy during postembryonic development of *Caenorhabditis elegans*. *Dev. Biol.* **78:** 577.

Sulston, J.E., E. Schierenberg, J.G. White, and J.N. Thomson. 1983. The embryonic cell lineage of the nematode *Caenorhabditis elegans*. *Dev. Biol.* **100:** 64.

Thompson, C.C., T.A. Brown, and S.L. McKnight. 1991. Convergence of Ets- and *Notch*-related structural motifs in a heteromeric DNA binding complex. *Nature* **253:** 762.

Weinmaster, G., V.J. Roberts, and G. Lemke. 1991. A homolog of *Drosophila Notch* expressed during mammalian development. *Development* **113:** 199.

Wharton, K.A., K.M. Johansen, T. Xu, and S. Artavanis-Tsakonas. 1985. Nucleotide sequence from the neurogenic locus *Notch* implies a gene product which shares homology with proteins containing EGF-like repeats. *Cell* **43:** 567.

Yochem, J. and I. Greenwald. 1989. *glp-1* and *lin-12*, genes implicated in distinct cell-cell interactions in *C. elegans*, encode similar transmembrane proteins. *Cell* **58:** 553.

Yochem, J., K. Weston, and I. Greenwald. 1988. The *Caenorhabditis elegans lin-12* gene encodes a transmembrane protein with overall similarity to *Drosophila Notch*. *Nature* **335:** 547.

Extracellular Morphogens in *Drosophila* Embryonic Dorsal-Ventral Patterning

K.V. Anderson, D.S. Schneider, D. Morisato, Y. Jin, and E.L. Ferguson
Genetics Division, Department of Molecular and Cell Biology, University of California, Berkeley, California 94720

Morphogens are molecules distributed in a concentration gradient that elicit different developmental fates at different concentrations. The graded distribution of a morphogen and a series of concentration-dependent responses of the cells exposed to a morphogen can combine to create a pattern of cell types in a defined spatial array. Several molecules active in early *Drosophila* development have been identified that fit this definition of a morphogen. The *bicoid* and *hunchback* proteins are two clear examples of morphogens that are distributed in gradients and define several different cell fates (Driever and Nüsslein-Volhard 1988; Struhl et al. 1992). The gradients of these morphogens depend on diffusion within the cytoplasm of the syncytial *Drosophila* embryo. Because other embryos do not have an early syncytial phase, these kinds of intracellular morphogens may not be of general use in development.

Gradients of extracellular morphogens could be used both in early embryogenesis and in later patterning processes in fields of cells. In biochemical terms, an extracellular morphogen would be a ligand for a membrane receptor, and different levels of receptor occupancy would trigger different cellular responses. We have defined two extracellular molecules with the properties of extracellular morphogens that organize two successive steps in dorsal-ventral patterning in the *Drosophila* embryo. These molecules apparently exert their effects on patterning by acting as localized ligands for broadly distributed membrane receptors. Our studies illustrate two different mechanisms for controlling the localized activity of these ligands that could be general mechanisms for generating gradients of extracellular morphogens.

Genetic analysis has defined a set of about 30 molecules necessary for the establishment of the dorsal-ventral pattern of the *Drosophila* embryo. Molecular analysis of these genes and their products has revealed that dorsal-ventral patterning depends on three sequential signal transduction pathways (Fig. 1). The first step in patterning, which has been most thoroughly studied by Schüpbach et al. (1991), occurs within the ovary and involves signaling from the oocyte to the surrounding follicle cells of the ovary. The next two steps in dorsal-ventral patterning take place after fertilization and make up the content of this paper.

The first patterning event after fertilization is defined by the 11 maternal dorsal group gene products (Fig. 2). In this step, a ventral extracellular cue activates a signal transduction cascade to create a gradient of the transcription factor *dorsal* in the nuclei of the early embryo. The next step in patterning defines cell identity in the dorsal half of the embryo. This process is specified by genes expressed in the zygote under the control of *dorsal*. Our data indicate that this step relies on graded activity of the transforming growth factor-β (TGF-β) family member *decapentaplegic* (*dpp*). In each of the two pathways that take place after fertilization, the data indicate that localization of the activity of an extracellular ligand is a crucial, highly regulated process necessary for the specification of the ordered pattern of cell types in the embryo.

The Dorsal Group Genes Encode Components of a Signal Transduction Pathway

Eleven different maternal effect genes, the dorsal group, are all required for the development of all lateral and ventral structures in the *Drosophila* embryo (Anderson and Nüsslein-Volhard 1984; Schüpbach and Wieschaus 1989). Genetic criteria indicate that the last gene in the dorsal group pathway is the *dorsal* gene (Anderson et al. 1985). The protein product of the *dorsal* gene is distributed in a gradient in the nuclei of the blastoderm embryo, being most concentrated in ventral nuclei, less concentrated in lateral nuclei, and absent from dorsal nuclei (Steward et al. 1988; Roth et al. 1989; Rushlow et al. 1989; Steward 1989). The *dorsal* protein is a transcription factor of the NF-κB/*rel* family (Steward 1987), and acts both as an activator of genes required for ventral development and as a repressor of genes required for dorsal development (Ip et al. 1991; Thisse et al. 1991).

The gradient of nuclear *dorsal* protein is established by regulated translocation of the *dorsal* protein from the cytoplasm to the nuclei of the blastoderm embryo (Roth et al. 1989; Rushlow et al. 1989; Steward 1989). In the absence of activity of any of the ten other dorsal group genes, the *dorsal* protein is present only in the cytoplasm. Thus, the dorsal group genes are all required to promote nuclear translocation of the *dorsal* protein, and they act in a dorsoventrally asymmetric manner so that *dorsal* protein is translocated into ventral, but not dorsal, nuclei.

The *Toll* gene acts genetically upstream of *dorsal* (Anderson et al. 1985). *Toll* is a transmembrane pro-

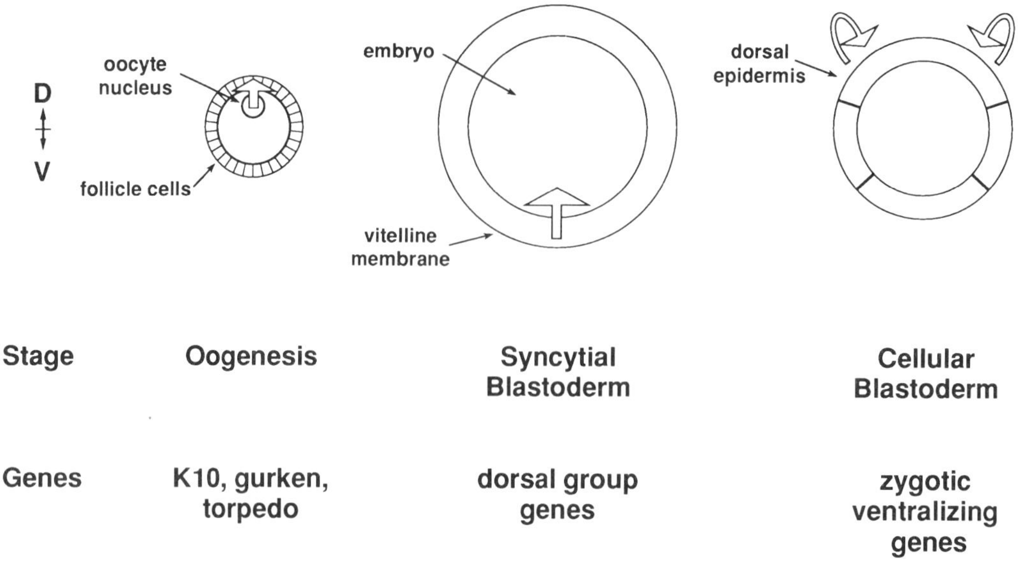

Figure 1. Three sequential intercellular communication steps define the dorsal-ventral pattern of the *Drosophila* embryo. Mutations in at least 30 genes have obvious effects on the dorsal-ventral pattern of the *Drosophila* larva. On the basis of their phenotypes, these mutations define three sequential steps in patterning. These steps are shown schematically here in cross-sectional view of the oocyte or embryo (dorsal up, ventral down). A group of genes that affect initial dorsal-ventral patterning of both the embryo and the eggshell act during oogenesis (Schüpbach et al. 1991). These genes, including fs(1)*K10*, *gurken*, and *torpedo*, appear to control a signaling pathway from the dorsally positioned oocyte nucleus to the nearest follicle cells, directing those cells to behave like dorsal follicle cells. After fertilization, at the syncytial blastoderm stage, the products of the maternally transcribed dorsal group genes are activated by an extracellular cue that is apparently localized in the ventral vitelline membrane (secreted by the follicle cells). Through a series of protein interactions, this ventral signal is used to define a gradient of the *dorsal* protein in the nuclei of the embryo. In response to the *dorsal* gradient, zygotic genes required for dorsal development are transcribed in the dorsal 40% of the embryo. The products of a number of these zygotic ventralizing genes are secreted to the extracellular space and act in a paracrine manner to organize pattern in the dorsal part of the embryo.

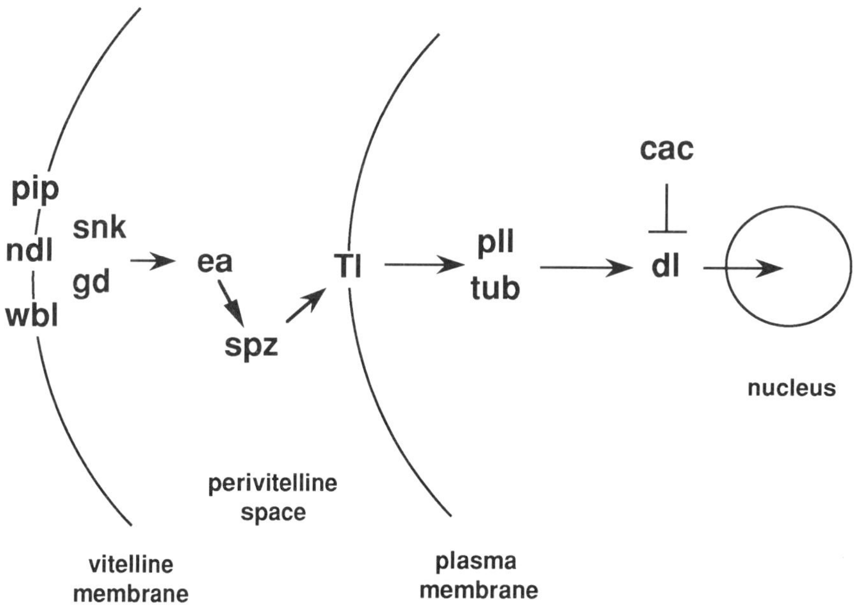

Figure 2. Signal transduction pathway defined by the dorsal group genes. The genetic order of action of the dorsal group genes defined by double-mutant analysis, combined with the cellular localization of cloned gene products, defines a signal transduction pathway. A signal in the ventral vitelline membrane, defined by the *pipe, nudel,* and *windbeutel* genes, apparently initiates the signaling chain. Two gene products related to the trypsin family of serine proteases, *snake* and *gastrulation defective*, act upstream of a third protease, *easter*. The product of the *spätzle* gene acts downstream from *easter* and is the gene product most directly required for activation of the transmembrane protein *Toll*. *Toll* apparently relays the signal to the cytoplasm to the *pelle* and *tube* proteins (P. Hecht and K. Anderson, unpubl.), which promote translocation of the *dorsal* protein from the cytoplasm to the nucleus. The *cactus* protein appears to keep the *dorsal* protein in the cytoplasm in the absence of the ventral activating signal (Roth et al. 1991).

tein found in the early embryonic plasma membrane and is uniform in concentration around the dorsal-ventral circumference (Hashimoto et al. 1988, 1991). The asymmetry of *dorsal* nuclear localization, and therefore of the dorsal-ventral pattern, originates with the genes upstream of *Toll* that appear to control the local ventral production of a *Toll* ligand. Four genes that are genetically upstream of *Toll* have been cloned, and all four encode products that are apparently synthesized in the embryo and secreted to the perivitelline space between the plasma membrane and the vitelline membrane, the innermost layer of the eggshell (Fig. 2) (DeLotto and Spierer 1986; Chasan and Anderson 1989; Konrad and Marsh 1990; D. Morisato and K.V. Anderson, unpubl.). The other three genes upstream of *Toll* are apparently active in the follicle cells that secrete the eggshell rather than in the oocyte itself (Manseau and Schüpach 1989; Stein et al. 1991). It is believed that these follicle-cell-dependent genes synthesize a localized component of the vitelline membrane, the inner layer of the eggshell, that triggers the dorsal group pathway on the ventral side of the embryo. From the genetic hierarchy and the location of these gene products, it seems that the information that controls *dorsal* nuclear localization originates outside the embryo and is relayed into the embryo via a signaling pathway whose components are encoded by the dorsal group genes.

Toll Appears to Be a Signal-transducing Receptor

The position of *Toll* in the genetic pathway, downstream from extracellular proteins and upstream of cytoplasmic components, together with its membrane localization, suggests that *Toll* acts as a receptor for a ventral extracellular signal that triggers a signal transduction cascade in the cytoplasm on the ventral side of the embryo. The *Toll* protein has features that are consistent with its acting as a ligand-binding, signal-transducing receptor (Hashimoto et al. 1988).

The *Toll* cytoplasmic domain is similar to the cytoplasmic domains of the mammalian and avian interleukin-1 receptors (IL-1R) (Fig. 3) (Schneider et al. 1991). Mutations that block *Toll* activity fall in regions conserved with the IL-1R (Schneider et al. 1991). Similarly, of a series of missense mutations in the IL-1R, only those changes at positions conserved with *Toll* block IL-1R activity (Heguy et al. 1992). However, since the mechanism used by the IL-1R to transmit a signal is not known, the homology does not clarify the mechanism of signal transduction by *Toll*.

The 777-amino-acid *Toll* extracellular domain is largely composed of 24-amino-acid leucine-rich repeats (LRRs) distributed in two blocks of repeats (Hashimoto et al. 1988). LRRs are found in a wide variety of proteins, both extracellular and cytoplasmic, and are believed to be involved in protein-protein interactions. Among the proteins that contain LRRs is a luteinizing hormone (LH) receptor, whose extracellular LH-binding domain is composed almost exclusively of LRRs (McFarland et al. 1989).

One recessive *Toll* allele, Tl^{rm9}, defines a site in the extracellular domain that should be important for ligand binding. The phenotype of this allele is unusual: Embryos laid by Tl^{rm9} females are lateralized (Anderson et al. 1985), which we interpret to mean that the receptor is activated to an intermediate level at all dorsal-ventral positions. The mutation responsible for the mutant phenotype of this allele is a threonine to asparagine change in the second block of leucine-rich repeats (Schneider et al. 1991; N. Machin and K.V. Anderson, unpubl.). In a bioassay developed by D. Stein for *Toll* ligand activity (Stein et al. 1991; see below), there is free extracellular ligand activity in both Tl^- and Tl^{rm9} embryos but none in wild-type, Tl^+, embryos. This suggests that the Tl^{rm9} protein binds ligand poorly and therefore that the second block of LRRs either is involved directly in ligand binding or is necessary for a conformation of the extracellular domain that allows ligand binding.

Truncated *Toll* proteins that stop before the transmembrane domain are dominant gain-of-function alleles that have no activity in *cis* but increase the activity of the wild-type transmembrane *Toll* protein in *trans* (Schneider et al. 1991). The shortest of these truncations stops two thirds of the way into the first block of LRRs, thereby defining an amino-terminal functional domain of *Toll*. We do not know how the truncated proteins act in *trans*. One possibility is that they have a low affinity for the *Toll* ligand, bind the ventrally produced ligand, and diffuse with the ligand to dorsal parts of the embryo because they are not tethered to the membrane. Once on the dorsal side, they release ligand, which then binds to and activates dorsally located transmembrane *Toll*. In this scenario, the amino-terminal region of the extracellular domain, in addition to the second block of LRRs, must include ligand-binding determinants.

Ventral Activation of the *easter* Protease Zymogen Is Crucial for Ventral Activation of *Toll*

Of the seven dorsal group genes that act upstream of *Toll*, we have been particularly interested in the *easter* gene because we have recovered a large number of dominant *easter* alleles that expand lateral pattern elements at the expense of dorsal structures (Chasan and Anderson 1989; Jin and Anderson 1990). The phenotypes of these alleles suggest that the activity of *easter* must normally be confined to ventral regions of the embryo to define the asymmetric pattern of the embryo.

The *easter* gene appears to encode a serine protease of the trypsin family and is predicted to be synthesized as an inactive zymogen that would require proteolytic processing for activity (Chasan and Anderson 1989). The *easter* protein is present in the perivitelline fluid and is apparently present in both ventral and dorsal parts of the embryo (Chasan et al. 1992). We have only

```
Toll, D.m.   K - F Q T E I K I W L Y A H N L L L W F V T E E D L D K D K K F D A F I S Y S H K - - D Q S F I E - D - Y L V P Q L E H G
Toll, D.v.   K - Y Q Q E I K I W L Y A H N L C M W F V T E A E L D K D K K F D A F I S Y S H K - - D Q S F V E - Q - H L V P Q L E H G
IL-1R: H     K I F K I D I V L W - Y R D S C Y D - F L P I K A S D - G K T Y D A Y I L Y P K T V G E G S T S D C D I F V F K V L P E V
IL-1R: M     K V F K V D I V L W - Y R D S C S G - F L P S K A S D - G K T Y D A Y I L Y P K T L G E G S F S D L D T F V F K L L P E V
IL-1R: Ch    K I F K I D I V L W - Y R S S C H P - L L G K K V S D - G K I Y D A Y V L Y P K N - R E S C L Y S S D I F A L K I L P E V

Toll, D.m.   P Q K - - - F Q L C V H E R D W L V G G H I P E N I M R S V A D S R R T I I V L S Q N F I K S E W - - A R L E F R A A H R
Toll, D.v.   P Q K - - - F T L C V H V R D W L V G G F I P E N I V R S V A D S R R T I I V L S P N F I K S D W - - A R M E F R A A H R
IL-1R: H     L E K Q C G Y K L F I Y G R D D Y V G E D I V E V I N E N V K K S R R L I I I L V R E T S G F S W L G G S S E E Q I A M Y
IL-1R: M     L E G Q F G Y K L F I Y G R D D Y V G E D T I E V T N E N V K K S R R L I I I L V R D M G G F S W L G Q S S E E Q I A I Y
IL-1R: Ch    L E R Q C G Y N L F I F G R N D L A G E A V I D V T D E K I H Q S R R V I I I L V P E P S C Y G I L E D A S E K H L A V Y

Toll, D.m.   S A L N E G R S R I I V I I Y S D I G D V E K L D E E L K A Y L K M N T Y L - K W - G D - - - - - - - - - P W F W D K L R
Toll, D.v.   A A L N E G R S R V I V I I Y S D I G D I E Q L D D E M K A Y L R M N T Y L - K W - G D - - - - - - - - - P W F W D K L R
IL-1R: H     N A L V Q D G I K V V L L E L E K I Q D Y E K M P E S I K - F I K Q K H G A I R W S G D F T Q G P Q S A K T R F W K N V R
IL-1R: M     N A L I Q E G I K I V L L E L E K I Q D Y E K M P D S I Q - F I K Q K H G V I C W S G D F Q E R P Q S A K T R F W K N L R
IL-1R: Ch    N A L I Q D G I K I I L I E L E K I E D Y A N M P E S I K - Y V K Q K Y G A I R W T G D F S E R S H S A S T R F W K K V R

Toll, D.m.   F A L P H R R P V G N I G N G A L I K T A L K G S T D D K L E L I K P S P V T P P L T T P P A E A T K N P L V A Q L N G V
Toll, D.v.   F A L P H R S P I G N S G N G A L I K S A L K G S T D D K L E L I K P S P V T P P L T T P P A R A T K N P L V A Q L N G G
IL-1R: H     Y H M P V Q R R S P S S K H Q L L S - - P A - - - T K E K L Q R E A H V P L G
IL-1R: M     Y Q M P A Q R R S P L S K H R L L T L D P V R D - T K E K L P A A T H L P L G
IL-1R: Ch    Y H M P S R K H G S S S G F H L S S

Toll, D.m.   T P H Q A I M I A N G K N G L T N L Y T P N G K S H - G N G H I N G A F I I N T N A K Q S D V
Toll, D.v.   T P H T A I M I A N G K N G L T N L Y A P N G K A H H G N G H I N G A F I I N T N A K Q S D V
```

Figure 3. Homology of the cytoplasmic domain of *Toll* to the cytoplasmic domain of the interleukin-1 (IL-1) receptor. The sequences of the cytoplasmic domains of the *Drosophila melanogaster* (D.m.) and *Drosophila virilis* (D.v.) (J.-L. Chen, I. Sulston and K.V. Anderson, unpubl.) *Toll* genes are aligned with the cytoplasmic domains of the human (H), murine (M), and avian (Ch) (Guida et al. 1992) IL-1 receptors; 57 of the first 184 amino acids are identical or conservative substitutions (boxed) in all five cytoplasmic domains.

been able to detect the zymogen form of the protein in either the perivitelline fluid or whole-embryo extracts (Chasan et al. 1992), suggesting that only a small fraction of the protein is processed to the active form.

We constructed a deletion form of *easter* that should be active without proteolytic processing. In this form of *easter*, *ea*ΔN, the prodomain of the protein was deleted, such that removal of the signal peptide should lead to the secretion of an active catalytic domain (Chasan et al. 1992). This form of *easter* promotes the development of lateral structures at all dorsal-ventral positions and thus appears to be constitutively active.

The *ea*ΔN mutant bypasses the requirement for five of the six other dorsal group genes that act upstream of *Toll* (Chasan et al. 1992). Unlike wild-type *easter*, *ea*ΔN is active in the absence of *pipe, nudel, windbeutel, snake*, or *gastrulation defective*. This suggests that these five genes normally act by promoting the activation of the *easter* zymogen. Both *snake* and *gastrulation defective* are also members of the trypsin family of serine proteases (DeLotto and Spierer 1986; Konrad and Marsh 1990), so it is possible that one of these two is the enzyme that activates *easter*. The *pipe, nudel*, and *windbeutel* genes could encode ventral products that act as localized cofactors in a zymogen activation complex.

The only gene upstream of *Toll* that is required for the production of lateral structures in the presence of *ea*ΔN is *spätzle* (Chasan et al. 1992). This places *spätzle* at the end of the extracellular pathway. We have cloned the *spätzle* gene and have found that it encodes a protein with an amino-terminal signal sequence but no other sequence homology with proteins in the database (D. Morisato and K.V. Anderson, unpubl.). Its position in the genetic pathway suggests that *spätzle* could encode the ligand that activates *Toll*.

Purification of a Protein with the Properties of a *Toll* Ligand

Stein defined a soluble activity present in perivitelline fluid that has the properties expected of a ligand that binds to and activates *Toll* (Stein et al. 1991). When perivitelline fluid containing this activity is injected into the perivitelline space of an embryo that lacks any of the genes upstream of *Toll*, it induces the development of ventral structures at the site of injection, indicating that it locally activates the *Toll* signal transduction pathway. The ligand-like activity was originally identified in the perivitelline fluid in embryos that have no *Toll* protein and is absent from Tl^{+} embryos, suggesting that if wild-type *Toll* is present, all of the ligand is bound to *Toll*.

We used the induction of ventral structures at the site of extracellular injection as a bioassay to purify from young wild-type embryos a protein that has this ligand-like activity (D.S. Schneider and K.V. Anderson, unpubl.). The activity, which is at least 10,000-fold purified, is an extremely stable, approximately 22-kD protein.

The one genetically defined candidate that could encode the *Toll* ligand was *spätzle*. To test whether the purified protein was encoded by the *spätzle* gene, we raised antisera to *spätzle* fusion proteins. Antisera against the *spätzle* protein recognize a 21–24-kD protein in the purified fractions, strongly suggesting that the *Toll* ligand is encoded by the *spätzle* gene (D.S. Schneider, D. Morisato, Y. Jin and K.V. Anderson, unpubl.). In this case, it appears that genetics and biochemistry have come together to identify a ligand for a membrane receptor and to define the function of this ligand in development.

Stein identified an activity in perivitelline fluid that rescues the *spätzle* mutant phenotype but does not define embryonic polarity by its site of injection (Stein and Nüsslein-Volhard 1992). Because it is soluble in the perivitelline fluid, it is likely that this *spätzle*-rescuing activity is uniformly distributed in the perivitelline fluid.

We can reconcile the existence of a uniformly distributed, non-polarity-inducing *spätzle* product with the localized, polarity-inducing activity of the purified protein if we suppose that *spätzle* must be proteolytically processed by *easter* to be active. Consistent with this hypothesis, we have found that the primary translation product of *spätzle* is larger than the purified ligand-like activity and that this protein is recognized only by antibodies to the carboxy-terminal half, and not the amino-terminal half, of *spätzle* (D. Morisato and K.V. Anderson, unpubl.). Further studies will be necessary to test whether *easter* processes *spätzle* from an inactive precursor to an active ligand and whether processed *spätzle* binds directly to *Toll*.

Local Receptor Activity Depends on Localized Zymogen Activation of the Protease That Cuts the Ligand, Rather Than on Local Synthesis of the Ligand

An extracellular morphogen can produce an asymmetric response in a field of cells if it causes local receptor activation. In theory, localized receptor activity could be caused by localized synthesis of any component of the cytoplasmic signaling pathway, of the receptor, or of the ligand. In the case of the dorsal group genes, it appears that none of these is locally transcribed or translated. Instead, receptor activation appears to depend on local proteolytic processing of the ligand precursor. Local processing, in turn, depends not on local synthesis of the processing enzyme but on local zymogen activation.

Ultimately, some component must be localized ventrally. The only genetically defined candidates for a localized ventral trigger are the three uncloned genes *nudel, pipe*, and *windbeutel*, all of which appear to be active in the follicle cells that secrete the eggshell, rather than in the embryo. We suppose that one (or more) of these proteins is ventrally localized in the vitelline membrane and that it serves as a nucleation center for the *easter* zymogen activation complex. The localization of this vitelline membrane protein is then relayed through a chain of protein interactions that ultimately leads to translocation of the *dorsal* protein to ventral nuclei.

We do not yet know where the gradient of activity that can be seen as the *dorsal* nuclear gradient originates. A number of possibilities exist. The vitelline membrane component that defines the *easter* activation complex could be distributed in a gradient in the vitelline membrane. Alternatively, that component could be tightly localized ventrally, and active *easter* could diffuse away from that site to generate a gradient. It is also possible that diffusion of processed *spätzle* creates the gradient or that the gradient arises only in the cytoplasm by diffusion of a signal from activated *Toll* within the cytoplasm of the syncytial embryo. At this point, an attractive model is that processed *spätzle* is a graded extracellular morphogen and that the six genes that act upstream of *spätzle* are necessary to allow production of the proper gradient.

Zygotically Active Genes Required for Patterning in the Dorsal Half of the Embryo

In response to the *dorsal* gradient, transcription of the mesoderm-defining genes *twist* and *snail* is activated ventrally. On the dorsal side of the embryo, a set of genes required for dorsal development is locally transcribed, as the result of repression by *dorsal* in ventral and lateral nuclei (Fig. 4). In the dorsal 40–50% of the embryo, there is no detectable nuclear *dorsal* protein, and genes required for dorsal development appear to be uniformly transcribed in this 40–50% of the embryo (Doyle et al. 1986; St. Johnston and Gelbart 1987; Shimmel et al. 1991). Thus, the patterning that occurs within this region cannot be directly determined by the *dorsal* gradient.

Among the seven zygotically transcribed genes required for normal patterning in the dorsal half of the embryo, the *decapentaplegic* (*dpp*) gene has the most extreme mutant phenotype (Irish and Gelbart 1987; Ferguson and Anderson 1992). Embryos that lack *dpp* are missing all dorsal and dorsolateral structures, and

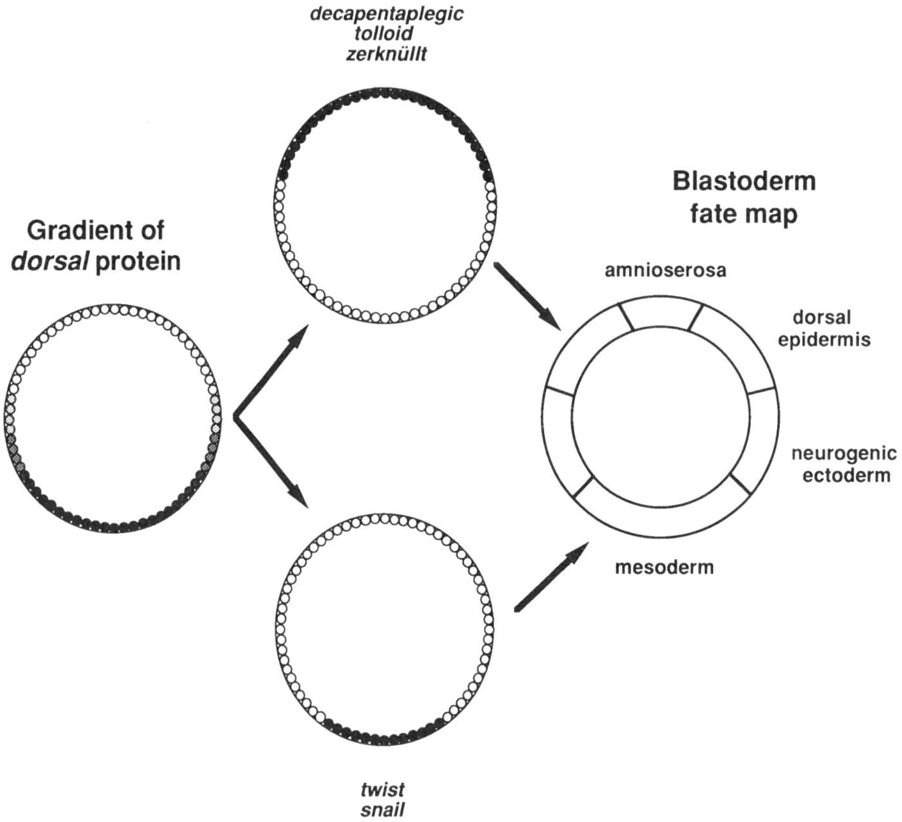

Figure 4. The *dorsal* gradient defines the domains of transcription of specific zygotic genes, which in turn define cell type in the embryo. The *dorsal* protein is graded in concentration in the nuclei in the ventral half of the blastoderm embryo. High levels of nuclear *dorsal* present in ventral nuclei activate the transcription of *twist* and *snail*, which are necessary for development of the mesoderm. High and intermediate levels of *dorsal* present in ventral and lateral nuclei repress the transcription of *decapentaplegic*, *tolloid*, and *zerknüllt*, limiting their transcription to the dorsal 40% of the embryo. *decapentaplegic*, *tolloid*, *zerknüllt*, and at least four other zygotically transcribed genes are necessary for the determination of the amnioserosa and the dorsal epidermis.

ventrolateral pattern elements are correspondingly expanded. *dpp* encodes a member of the TGF-β family (Padgett et al. 1987), most similar to the bone morphogenetic proteins BMP-2 and BMP-4 (Wozney et al. 1988), and presumably is secreted to the perivitelline space and acts by binding to a membrane receptor.

Strong alleles of *dpp* delete all dorsal structures, whereas weaker alleles cause a progressive loss of dorsal structures. The weakest alleles cause the loss of only the most dorsal pattern element, the amnioserosa, and the intermediate alleles delete the amnioserosa and some dorsal structures (Arora and Nüsslein-Volhard 1992; Ferguson and Anderson 1992). This allelic series, in which more *dpp* activity is required for dorsal than for dorsolateral structures, indicates that *dpp* is an essential component of a system of graded information that defines the dorsal pattern.

At least three of the six other genes required for dorsal development act by modifying the activity of *dpp*. This was demonstrated by increasing the number of copies of the wild-type *dpp* gene in embryos mutant for other genes (Ferguson and Anderson 1992). For instance, doubling the number of dpp^+ genes from two to four completely suppresses the phenotype of weak *tolloid* alleles and partially suppresses the phenotype of embryos that completely lack the *tolloid* gene. Since raising the gene dosage of *dpp* can compensate for the absence of *tolloid, tolloid* cannot be a component of the pathway downstream from *dpp* that allows reception and interpretation of the *dpp* signal. Instead, *tolloid* must act upstream of *dpp* to increase *dpp* activity. The pattern of *dpp* transcription is not altered in *tolloid* mutant embryos (Ray et al. 1991), so *tolloid* must modify *dpp* activity at a post-transcriptional level. The *tolloid* gene encodes a large secreted protein that is very similar to BMP-1 (Shimell et al. 1991), which was originally identified in a complex with BMP-2 (Wozney et al. 1988). Both *tolloid* and BMP-1 have an amino-terminal domain that resembles a crayfish metalloprotease, as well as several epidermal growth factor (EGF) repeats and repeats found in the complement proteases C1r and C1s. It seems likely that *tolloid* is a component of a complex that includes *dpp* and that *tolloid* acts by proteolytically cleaving a component of that complex, thereby elevating the activity of *dpp*.

Similar gene-dosage experiments showed that the *shrew* gene also acts by increasing *dpp* activity (Ferguson and Anderson 1992). As in *tolloid* mutants, the pattern of *dpp* transcription is not changed in *shrew* mutant embryos (Ray et al. 1991), indicating that this gene also regulates *dpp* activity post-transcriptionally.

The function of the *short gastrulation* (*sog*) gene in dorsal patterning had seemed complex, because although its only obvious phenotype was the absence of the most dorsal pattern element, the amnioserosa, genetic mosaic studies had shown that the gene was required in ventral, but not dorsal, cells (Zusman and Wieschaus 1985; Zusman et al. 1988). By changing the gene dosage of *dpp* in *sog* embryos, it became clear that *sog* inhibits *dpp* activity in the ventral part of the embryo: Although increasing the number of copies of *dpp* in a wild-type background does not cause a mutant phenotype, in *sog* embryos that have twice the number of copies of the *dpp* gene, dorsal structures expand into the regions normally occupied by ventrolateral structures (Ferguson and Anderson 1992). Thus, in the wild-type embryo, either *sog* blocks the diffusion of the *dpp* protein from its dorsal site of synthesis to more ventral regions or *sog* blocks the activity of *dpp* ventrally, preventing ventral cells from differentiating like dorsal cells.

A Gradient of *dpp* Activity

Because *dpp* is central to patterning in the dorsal half of the embryo and because patterning in this region depends on some graded activity, it seemed possible that *dpp* could act as a graded morphogen to organize the pattern in the dorsal half of the embryo. To test whether different levels of *dpp* activity can elicit different cell fates, we injected *dpp* transcripts at different concentrations into young embryos (E.L. Ferguson and K.V. Anderson, unpubl.). We found that different concentrations of *dpp* are sufficient to define at least three fates: High levels of *dpp* can drive all ectodermal cells to become the most dorsal pattern element, the amnioserosa; intermediate levels of *dpp* drive cells to become dorsal epidermis; and low levels of *dpp* are required to allow neurogenic ectoderm development. Thus, *dpp* can act as a morphogen to define a series of cell fates.

The *dpp* transcript is evenly distributed in the dorsal 40% of the embryo, and *dpp* RNA distribution does not change in any of the zygotic mutants (Ray et al. 1991). Because of the homology of *tolloid* and BMP-1, it seems likely that *tolloid* modifies *dpp* activity by interactions between the two proteins. Because *sog* acts in ventral embryonic cells, it is unlikely that it modifies translation or secretion of *dpp*, and it is more likely that it too acts by modifying the activity of the *dpp* protein. We suppose that extracellular interactions of *dpp* with the dorsally transcribed activators *tolloid* and *shrew* and the ventrally transcribed inhibitor *sog* lead to the production of a *dpp* activity gradient. Thus, in the case of the zygotically controlled patterning in the dorsal part of the embryo, the morphogen is an extracellular ligand that is produced in a broad domain. It is the interactions of the *dpp* ligand with other proteins that locally increase and reduce its ability to activate its receptor, thereby creating asymmetry within that field of cells.

Speculations on the Generality of These Patterning Processes

We believe our data indicate that in *Drosophila* dorsal-ventral patterning, morphogen gradients are established in the extracellular space that surrounds the early embryo. Thus, in contrast to the gradients that control anterior-posterior patterning in *Drosophila*, which arise from diffusion of proteins within the embryonic cytoplasm, the mechanisms that we believe define the *spätzle* and *dpp* morphogen gradients would not need to be restricted to embryos with an early syncytial stage. So there is no a priori reason these molecules used in *Drosophila* for early patterning could not be used in cellular embryos or in other patterning processes later in development.

Striking parallels exist between the activities of the two TGF-β family members, *dpp* and activin, in dorsal-ventral patterning in *Drosophila* and *Xenopus* embryos. In both cases, it has been shown that the factor can drive dorsal development and that small changes in factor concentration are sufficient to define a series of cell fates (Green and Smith 1990; E.L. Ferguson, and K.V. Anderson, unpubl.). It is not known whether the activity of an activin (or another TGF-β protein) is graded in the *Xenopus* embryo, and if such a gradient exists, it is not known how it arises. It is possible that an activin gradient forms in response to a maternally controlled gradient analogous to that controlled by *spätzle*. At this point, there is no way to evaluate the extent to which the same dorsal-ventral embryonic patterning pathways are used in *Drosophila* and vertebrates, but it seems likely that at least some aspects of patterning will be conserved.

REFERENCES

Anderson, K.V. and C. Nüsslein-Volhard. 1984. Information for the dorsal-ventral pattern of the *Drosophila* embryo is stored as maternal mRNA. *Nature* **311:** 223.

Anderson, K.V., G. Jürgens, and C. Nüsslein-Volhard. 1985. Establishment of dorsal-ventral polarity in the *Drosophila* embryo: Genetic studies on the role of the *Toll* gene product. *Cell* **42:** 779.

Arora, K. and C. Nüsslein-Volhard. 1992. Altered mitotic domains reveal fate map changes in *Drosophila* embryos mutant for zygotic dorsoventral patterning genes. *Development* **114:** 1003.

Chasan, R. and K.V. Anderson. 1989. The role of *easter*, an apparent serine protease, in organizing the dorsal-ventral pattern of the *Drosophila* embryo. *Cell* **56:** 391.

Chasan, R.,Y. Jin, and K.V. Anderson. 1992. Activation of the *easter* zymogen is regulated by five other genes to define dorsal-ventral polarity in the *Drosophila* embryo. *Development* **115:** 607.

DeLotto, R. and P. Spierer. 1986. A gene required for the specification of dorsal-ventral pattern in *Drosophila* appears to encode a serine protease. *Nature* **323:** 688.

Doyle, H., K. Harding, T. Hoey, and M. Levine. 1986. Transcripts encoded by a homeobox gene are restricted to dorsal tissues of *Drosophila* embryos. *Nature* **323:** 76.

Driever, W. and C. Nüsslein-Volhard. 1988. The *bicoid* protein determines position in the *Drosophila* embryo in a concentration-dependent manner. *Cell* **54:** 95.

Ferguson, E.L. and K.V. Anderson. 1992. Localized enhancement and repression of the activity of the TGF-β family member, *decapentaplegic* is necessary for dorsal-ventral pattern formation in the *Drosophila* embryo. *Development* **114:** 583.

Green, J.B.A. and J.C. Smith. 1990. Graded changes in dose of a *Xenopus* activin A homologue elicit stepwise transitions in embryonic cell fate. *Nature* **347:** 391.

Guida, S., A. Heguy, and M. Melli. 1992. The chicken IL-1 receptor: Differential evolution of the cytoplasmic and extracellular domains. *Gene* **111:** 239.

Hashimoto, C., S. Gerttula, and K.V. Anderson. 1991. Plasma membrane localization of the *Toll* protein in the syncytial *Drosophila* embryo: Importance of transmembrane signaling for dorsal-ventral pattern formation. *Development* **111:** 1021.

Hashimoto, C., K.L. Hudson, and K.V. Anderson. 1988. The *Toll* gene of *Drosophila*, required for dorsal-ventral embryonic polarity, appears to encode a transmembrane protein. *Cell* **52:** 269.

Heguy, A., C.T. Baldari, G. Macchia, J.L. Telford, and M. Melli. 1992. Amino acids conserved in interleukin-1 receptors (IL-1Rs) and the *Drosophila Toll* protein are essential for IL-1R signal transduction. *J. Biol. Chem.* **267:** 2605.

Ip, Y.T., R. Kraut, M. Levine, and C. Rushlow. 1991. The *dorsal* morphogen is a sequence-specific DNA-binding protein that interacts with a long-range repression element in *Drosophila*. *Cell* **64:** 439.

Irish, V.F. and W.M. Gelbart. 1987. The *decapentaplegic* gene is required for dorsal-ventral patterning of the *Drosophila* embryo. *Genes Dev.* **1:** 868.

Jin, Y. and K.V. Anderson. 1990. Dominant and recessive alleles of the *Drosophila easter* gene are point mutations at conserved sites in the serine protease catalytic domain. *Cell* **60:** 873.

Konrad, K. and L. Marsh. 1990. Abstract (28.93) from the 31st *Annual* Drosophila *Research Conference*.

Manseau, L. and T. Schüpbach. 1989. *cappuccino* and *spire*: Two unique maternal-effect loci required for both the anteroposterior and dorsoventral patterns of the *Drosophila* embryo. *Genes Dev.* **3:** 1437.

McFarland, K.C., R. Sprengel, H.S. Phillips, M. Köhler, N. Rosemblit, K. Nikolics, D.L. Segaloff, and P.H. Seeburg. 1989. Lutropin-choriogonadotropin receptor: An unusual member of the G protein-coupled receptor family. *Science* **245:** 494.

Padgett, R.W., R.D. St. Johnston, and W.M. Gelbart. 1987. A transcript from a *Drosophila* pattern gene predicts a protein homologous the transforming growth factor-β family. *Nature* **325:** 81.

Ray, R.P., K. Arora, C. Nüsslein-Volhard, and W.M. Gelbart. 1991. The control of cell fate along the dorsal-ventral axis of the *Drosophila* embryo. *Development* **113:** 35.

Roth, S., D. Stein, and C. Nüsslein-Volhard. 1989. A gradient of nuclear localization of the *dorsal* protein determines dorsoventral pattern in the *Drosophila* embryo. *Cell* **59:** 1189.

Roth, S., Y. Hiromi, D. Godt, and C. Nüsslein-Volhard. 1991. *cactus*, a maternal gene required for proper formation of the dorsoventral morphogen gradient in *Drosophila* embryos. *Development* **112:** 371.

Rushlow, C., K. Han, J.L. Manley, and M. Levine. 1989. The graded distribution of the *dorsal* morphogen is initiated by selective nuclear transport in *Drosophila*. *Cell* **59:** 1165.

Schneider, D.S., K.L. Hudson, T.-Y. Lin, and K.V. Anderson. 1991. Dominant and recessive mutations defines functional domains of *Toll*, a transmembrane protein required for dorsal-ventral polarity in the *Drosophila* embryo. *Genes Dev.* **5:** 797.

Schüpbach, T. and E. Wieschaus. 1989. Female sterile mutations on the second chromosome of *Drosophila melanogaster*. I. Maternal effect mutations. *Genetics* **121:** 101.

Schüpbach, T., R.J. Clifford, L.J. Manseau, and J.V. Price. 1991. Dorso-ventral signalling processes in *Drosophila* oogenesis. *Symp. Soc. Dev. Biol.* **49:** 163.

Shimell, M.J., E.L. Ferguson, S.R. Childs, and M.B. O'Connor. 1991. The *Drosophila* dorsal-ventral patterning gene *tolloid* is related to human bone morphogenetic protein-1. *Cell* **67:** 469.

Stein, D. and C. Nüsslein-Volhard. 1992. Multiple extracellular activities in *Drosophila* egg perivitelline fluid are required for establishment of embryonic dorsal-ventral polarity. *Cell* **68:** 429.

Stein, D., S. Roth, E. Vogelsang, and C. Nüsslein-Volhard. 1991. The polarity of the dorsoventral axis in the *Drosophila* embryo is defined by an extracellular signal. *Cell* **65:** 725.

Steward, R. 1987. *dorsal*, an embryonic polarity gene in *Drosophila*, is homologous to the vertebrate proto-oncogene, c-*rel*. *Science* **238:** 692.

———. 1989. Relocalization of the *dorsal* protein from the cytoplasm to the nuclei correlates with its function. *Cell* **59:** 1179.

Steward, R., S.B. Zusman, L.H. Huang, and P. Schedl. 1988. The *dorsal* protein is distributed in a gradient in early *Drosophila* embryos. *Cell* **55:** 487.

St. Johnston, R.D. and W.M. Gelbart. 1987. *decapentaplegic* transcripts are localized along the dorsal-ventral axis of the *Drosophila* embryo. *EMBO J.* **6:** 2785.

Struhl, G., P. Johnston, and P.A. Lawrence. 1992. Control of *Drosophila* body pattern by the *hunchback* morphogen gradient. *Cell* **69:** 237.

Thisse, C., F. Perrin-Schmitt, C. Stoetzel, and B. Thisse. 1991. Sequence-specific transactivation of the *Drosophila twist* gene by the *dorsal* gene product. *Cell* **65:** 1191.

Wozney, J.M., V. Rosen, A.J. Celeste, L.M. Mitsock, M.J. Whitters, R.W. Kriz, R.M. Hewick, and E.A Wang. 1988. Novel regulators of bone formation: Molecular clones and activities. *Science* **242:** 1528.

Zusman, S.B. and E.F. Wieschaus. 1985. Requirement for zygotic gene activity during gastrulation in *Drosophila melanogaster*. *Dev. Biol.* **111:** 359.

Zusman, S.B., D. Sweeton, and E.F. Wieschaus. 1988. *short gastrulation*, a mutation causing delays in stage-specific cell shape changes during gastrulation in *Drosophila melanogaster*. *Dev. Biol.* **129:** 417.

Adhesive Interactions That Regulate Development of the Retina and Primary Visual Projection

L.F. Reichardt, B. Bossy, I. de Curtis, K.M. Neugebauer,
K. Venstrom, and D. Sretavan
Program of Neurosciences, Department of Physiology and Howard Hughes Medical Institute, University of California, San Francisco, California 94143-0724

The ability of a nervous system to function properly requires that its connections be established precisely and reproducibly during development. This requires development of both a correct wiring pattern and proper signaling mechanisms at each synapse. To achieve the former, a neuron must establish appropriate connections with cells upstream and downstream in its signaling pathway. To meet the demands of the latter, neurons must acquire an identity that includes a selective repertory of receptors, ion channels, and neurotransmitters.

The generation of neural pathways is accomplished by directed growth of axons and dendrites. Growth cones are motile organelles at the tips of growing axons and dendrites that determine their directions and patterns of elongation. Growth cone behavior is determined by both intrinsic and extrinsic factors. The former regulate the competence of a neuron to respond to extrinsic signals. The latter determine whether particular signaling pathways are activated.

Cells and molecules that direct the establishment of individual axon pathways have been identified using both in vitro and in vivo assays. On the cellular level, particular cell types have been shown to promote, direct, or inhibit growth cone movements. These differences between cells must reflect specificity in the molecules expressed by these different cell types. Purified molecules include examples that promote neuronal process outgrowth, as well as others that actually inhibit such outgrowth. Molecules studied thus far include some that are believed to be soluble or diffusible in vivo, such as neurotrophins, neurotransmitters, and chemotropic factors. They also include others that are immobilized on cell surfaces or in extracellular matrices. Inhibitory molecules identified to date include neurotransmitters and cell-surface-associated glycoproteins and proteoglycans. Different populations of neurons exhibit different responses to individual promoters and inhibitors of neurite outgrowth, almost certainly reflecting differences in their repertories of receptors for these molecules (see, e.g., Kater and Guthrie 1990; Baier and Bonhoeffer 1992).

In vitro assays have been important for characterizing interactions of specific neurons with specific substrates or soluble factors. These assays make it possible to examine neuronal interactions with specific molecules whose expression patterns suggest importance in vivo. Assays in vitro have provided convincing evidence that some cells promote whereas others inhibit axonal growth. These assays have also identified specific neurotransmitters, neurotrophins, chemotropic factors, cell adhesion molecules, and extracellular matrix constituents as candidates to influence growth cone behavior by specific classes of neurons in vivo. To extend this approach, it is necessary to examine the behaviors of growth cones in their in vivo environment. Ideally, this needs to be done in real time in the presence or absence of reagents that inhibit individual molecular interactions. The retina and primary visual projection are comparatively accessible to studies in vivo. The retina develops as a thin sheet of tissue. Over much of its course, the primary visual projection is located on the surface of the embryonic nervous system. Living retinal axons growing in their native embryonic cellular environment can thus be visualized using videomicroscopy. These intact preparations are also accessible to perturbing reagents.

In this paper, we describe in vitro studies characterizing adhesive interactions of retinal neurons with cells and extracellular matrix constituents. Results indicate that retinal neurons utilize several cell adhesion molecules and extracellular matrix constituents for neurite growth. They utilize cell adhesion molecules and integrins as receptors. Both regulation of expression of their genes and modulation of receptor activity are shown to modulate these interactions. The axons of retinal ganglion cells form the visual projection. We describe the behavior of retinal ganglion cell growth cones in their in vivo environment, focusing on their behavior at an important pathway decision point, the optic chiasm. A population of early developing neurons whose somata are localized in the optic chiasm has been identified and shown to be important for normal development of the visual projection.

Adhesive Interactions That Mediate Axon Growth by Retinal Neurons In Vitro

Neuroepithelial cells and their neuronal derivatives in the retina and remainder of the central nervous system develop in contact with other cells and with numerous constituents of the extracellular matrix. Both cell-surface-associated adhesion molecules and extracellular matrix glycoproteins have been shown to affect dramatically development of cells in dissociated cell or explant culture in vitro and in embryos in vivo (for review, see Reichardt and Tomaselli 1991).

The Extracellular Matrix

Extracellular matrix proteins that are present during retinal or optic pathway embryogenesis and have been shown to have strong effects on development of retinal cells include isoforms of laminin, collagen IV, thrombospondin, and vitronectin (for review, see Reichardt and Tomaselli 1991; see also O'Shea and Dixit 1988; Neugebauer et al. 1991). Fibronectin has also been shown to promote neurite outgrowth (Akers et al. 1981). Restrictin and tenascin are two additional extracellular matrix proteins distributed in the retinal or retinal-tectal projection (Prieto et al. 1990; Rathjen et al. 1991). Most studies on neuronal interactions with extracellular matrix proteins have focused on the regulation of neurite outgrowth. Recent work indicates that the extracellular matrix may also be an important regulator of cell determination in the embryonic retina. Particularly dramatic are results suggesting that isoforms of laminin containing the S-laminin subunit regulate photoreceptor cell morphogenesis (Hunter et al. 1992).

Much of our work has focused on the effects of extracellular proteins that regulate neurite outgrowth and are candidates to promote in vivo development of axons and dendrites in the retina and retinal-tectal projection. These experiments have utilized either mixed populations of retinal neurons or purified retinal ganglion cells. Studies on the latter are particularly interesting, since all visual information is carried by their axons from the retina to the brain. Results, summarized in Table 1, show that retinal neurons and retinal ganglion cells interact with several constituents of the extracellular matrix, including laminin, thrombospondin, and vitronectin. Most of these proteins are present in the embryonic retina. Interactions with most, but not all, of these proteins stimulate neurite outgrowth.

Essentially all neurite-outgrowth-promoting interactions of neurons with extracellular matrix constituents appear to require integrin class receptors. To identify extracellular matrix receptors, it has been advantageous to use cells available in large quantities, such as neuronal cell lines, where methods such as affinity chromatography can complement antibody inhibition and immunocytochemical studies (cf. Ignatius and Reichardt 1988). Identified populations of primary embryonic neurons are present in comparatively small numbers, however, making biochemical analyses difficult. Therefore, our conclusions on receptors utilized by primary neurons rely primarily on antibody perturbations, analyses of immunoprecipitates, and immunocytochemistry. Results in Figure 1 show that integrins mediate the interactions of retinal neurons with most extracellular matrix glycoproteins. Integrin heterodimers containing the β_1 subunit mediate interactions with laminin, collagen IV, and fibronectin. Results examining interactions with thrombospondin and vitronectin are more complicated and are discussed below.

Laminin, one of the most potent extracellular matrix-associated neurite outgrowth factors, is present in the embryonic nervous system, and for this reason is the extracellular matrix molecule most intensely studied by neurobiologists. The $\alpha_6\beta_1$ heterodimer is a major candidate laminin receptor detected in immunoprecipitations from embryonic neural retina (de Curtis et al. 1991). Results in Figure 2 show that antibodies to α_6 or β_1 strongly inhibit retinal neuron adhesion to laminin. Thus, in the case of laminin, a single receptor, $\alpha_6\beta_1$, mediates a significant percentage of adhesion. Although many different classes of neurons interact with

Table 1. Effects of Extracellular Matrix Glycoproteins on Retinal Neurons

	Adhesion	Neurite outgrowth	Present in retina
Mixed neurons			
laminin	+ + +	+ + +	yes
fibronectin	+ +	+	–
collagen IV	+ + +	+ +	yes
collagen I	+ + +	+ +	–
vitronectin	+ + +	+ + +	yes
thrombospondin	+ + +	+ + +	yes
restrictin[a]	+ + +	–	yes
Retinal ganglion cells			
laminin	+ + +	+ +	
vitronectin[b]	n.d.	–	
fibronectin[b]	n.d.	+	
collagen I[b]	n.d.	+ + +	

Cell attachment and neurite outgrowth were measured in standard assays. Original data from this laboratory are published in Hall et al. (1987), Neugebauer et al. (1991), and de Curtis et al. (1991). n.d. indicates not determined.

[a]Data on restrictin have been published by Rathjen and co-workers (Rathjen et al. 1991).

[b]Data on vitronectin, fibronectin, and collagen I have been published by Carri et al. (1988), using organotypic cultures of embryonic (E6) retina from which the vast majority of neurites have been shown to be derived from retinal ganglion cells (Drazba and Lemmon 1990).

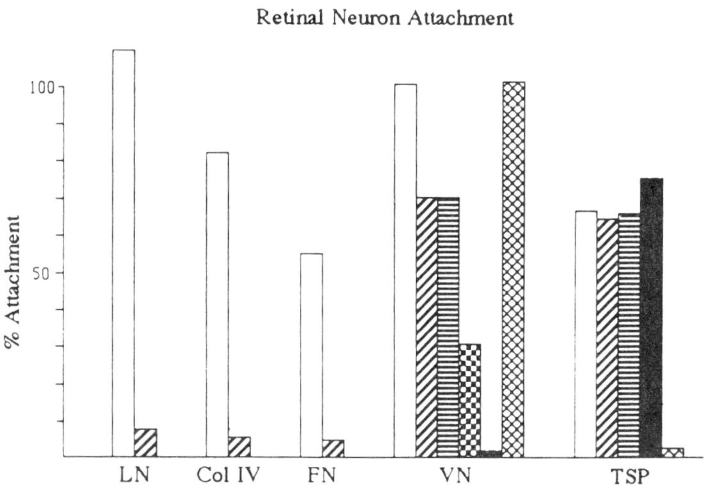

Figure 1. Effect of anti-integrin subunit-specific antibodies on cell attachment and neurite outgrowth. Cell attachment by E6 neural retinal cells was measured on laminin (LN), collagen IV (Col IV), fibronectin (FN), vitronectin (VN), and thrombospondin (TSP). Cell interactions with different matrix protein substrates were measured in control (□) or in the presence of anti-integrin β_1 (▨), anti-integrin α_V (▤), anti-integrin β_1 plus anti-integrin α_V (▩), the RGDS peptide (■), or heparin (▩).

laminin, they express and utilize different integrins that mediate interactions with this extracellular matrix protein. Sensory neurons, for example, have been shown to utilize $\alpha_1\beta_1$ and $\alpha_3\beta_1$, but not $\alpha_6\beta_1$ for this purpose (K.J. Tomaselli, unpubl.).

Vitronectin is a small extracellular matrix protein prominently expressed in the embryonic retina

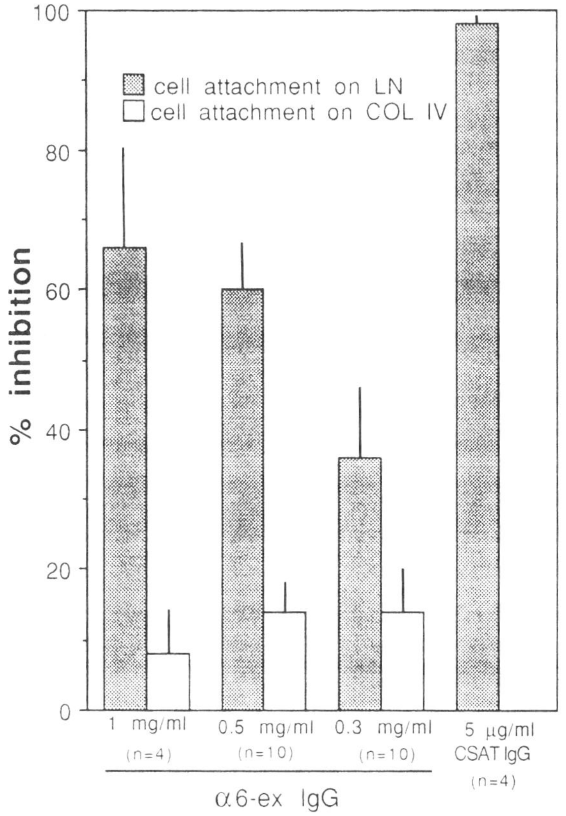

Figure 2. Cell attachment assay of E6 retinal neurons on LN or COL IV. Cells were prepared by trypsinization of E6 retinae, and the cell attachment assay was performed as described in Neugebauer et al. (1992). In every experiment, each sample was tested in triplicate. Bars indicate the standard error; *n* represents the number of experiments.

(Neugebauer et al. 1991). Results in Figure 1 show that integrins containing the α_{VN} and β_1 subunits mediate neuronal interactions with vitronectin. The α_V subunit is known to associate with multiple β subunits: β_1, β_3, β_5, β_6, and β_8 (cf. Moyle et al. 1991). Most of these heterodimers function as vitronectin receptors. The $\alpha_V\beta_3$, $\alpha_V\beta_5$, and $\alpha_V\beta_8$, but surprisingly not the $\alpha_V\beta_1$, heterodimers have been detected in embryonic neuroretina (Neugebauer et al. 1991; K. Venstrom, unpubl.). Several or all of these heterodimers seem likely to account for interactions of these neurons with vitronectin. The inhibitory effect of β_1 antibodies implies that a β_1-subunit-containing heterodimer is also important. This is probably not $\alpha_V\beta_1$, since this heterodimer cannot be detected in these cells. Data in Table 1 indicate that mixed retinal neurons, but not retinal ganglion cells, respond to vitronectin in neurite outgrowth assays. This difference is consistent with the distribution of the vitronectin receptor subunit, the integrin subunit α_{VN}, which is more strongly expressed in the retina than in the optic fiber layer or optic nerve (Carri et al. 1988; Neugebauer et al. 1991; B. Bossy and L.F. Reichardt, unpubl.).

Thrombospondin isoforms are prominently expressed in the embryonic central nervous system. Many types of neurons are able to interact with this extracellular matrix protein. These interactions may involve more than one class of receptors. Data in Figure 1 suggest that integrins do not mediate all adhesive interactions between retinal neurons and thrombospondin. Instead, adhesion is sensitive to heparin, suggesting that either proteoglycans or sulfated glycopeptides, each of which has been shown to exhibit heparin-sensitive binding to thrombospondin, may function as receptors (see also Kaesberg et al. 1989). Neurite outgrowth, however, is strongly inhibited by antibodies to the integrin β_1 subunit, more weakly inhibited by antibodies to the α_V subunit, and virtually eliminated by a combination of both antibodies (Neugebauer et al. 1991). The results suggest that integrins are particularly important in mediating motility-promoting interactions with thrombospondin. Preliminary results suggest that the integrin $\alpha_3\beta_1$ is one important neuronal receptor for this glycoprotein (M. DeFreitas, unpubl.).

In addition to mediating adhesive interactions with the better characterized extracellular matrix glycoproteins, integrins may mediate interactions of retinal neurons with additional molecules. The documented interactions of lymphocyte integrins with I-CAMs (Springer 1990) and of $\alpha_4\beta_1$ with V-CAM-1 (Hemler et al., this volume) provide precedents. Retinal neurons interact with the carboxy-terminal fragment of fibronectin, even though fibronectin is absent from most regions of the developing central nervous system. The only integrin known to interact with this domain of fibronectin is $\alpha_4\beta_1$, which has an alternate ligand, the cell adhesion molecule V-CAM-1. Recently, V-CAM-1 has been detected in the embryonic central nervous system (Rosen et al. 1992), suggesting that if retinal neurons do express $\alpha_4\beta_1$, V-CAM-1, not fibronectin, may be the physiologically relevant ligand.

A second integrin expressed by retinal neurons that is a candidate to interact with uncharacterized ligands in the embryonic retina is the $\alpha_8\beta_1$ heterodimer. cDNAs encoding the α_8 subunit were isolated by us in a search for novel integrins (Bossy et al. 1991). Antibodies were prepared to peptide sequences present in the deduced primary sequence. These were used to demonstrate that this α subunit associates with β_1 to form a heterodimer. The α_8 subunit has been expressed successfully in K562 cells, where it dimerizes with endogenous β_1 to form a surface heterodimer. The adhesive behavior of these cells expressing $\alpha_8\beta_1$ suggests that it is not a receptor for laminin, the collagens, or vitronectin. At present, its ligands are not identified. The expression pattern of α_8 in the embryonic chick retina and central nervous system is similar to that of adhesion molecules, such as Ng-CAM or F11, and suggests that it may interact with one of these molecules. As illustrated in Figure 3, anti-α_8 antibodies stain at both E6 and E11 the axonal fiber layer in the retina derived from retinal ganglion cells, but stain only weakly the cell bodies of these neurons. The ganglion cell axons are stained along their entire length, in the retina (Fig. 3A,C), optic nerve and optic chiasm (Fig. 3B), and optic tectum (Fig. 3E). Strong expression is seen in E3.5 embryos in the initial axons growing from the retina into the optic stalk (Fig. 3D). This integrin is also highly expressed in several additional axon tracts in the embryonic central nervous system (Bossy et al. 1991). Its strong localization to axons distinguishes it from all other integrin subunits examined to date in our laboratory.

Developmental Regulation of Integrin Receptors in the Retina

As development proceeds, individual neurons are exposed to different environments and exhibit changes in morphology, differentiation, and behavior of their axonal and dendritic processes. In particular, many different types of neurons have been shown to lose functional receptors for extracellular matrix proteins at times that correlate with the establishment of functional contacts with their targets. These include rat sensory neurons, which lose functional fibronectin receptors between E16 and P4 (Kawasaki et al. 1986); chick ciliary ganglion neurons, which lose functional laminin receptors between E8 and E14 (Tomaselli and Reichardt 1988); and chick retinal neurons and retinal ganglion cells, which lose functional fibronectin, laminin, vitronectin, and thrombospondin receptors between E6 and E12 (Cohen et al. 1986, 1989; Hall et al. 1987; de Curtis et al. 1991; Neugebauer et al. 1991).

Mechanisms by which receptor function is regulated have been examined in the retina, where evidence suggests that regulation occurs on more than one level. Analysis of integrin subunit expression in these cells indicates that there are striking temporal differences in levels of individual subunits, with some decreasing and others increasing (de Curtis et al. 1991). Focusing on the laminin receptor, $\alpha_6\beta_1$, which is important functionally on these neurons (Fig. 2), results in Figure 4 show that significant levels of α_6 mRNA are present in these neurons at both E6 and E12. Only marginally significant decreases in α_6 mRNA or protein are seen between E6 and E12, even though the neurons lose almost completely the abilities to adhere to or extend neurites on laminin during this interval of development. Immunocytochemistry, however, suggests that α_6 expression is decreased more significantly in the more developmentally advanced central sections of the E12 retina (I. de Curtis and L.F. Reichardt, in prep.). The results in these experiments suggest that laminin receptors continue to be expressed on the surfaces of neurons that have virtually lost the ability to interact with laminin. This result is consistent with a report that the K_d of retinal cell interactions with laminin is reduced more than tenfold between E6 and E11 (Cohen et al. 1989). Consequently, the activity of these receptors must be modulated. Additional evidence for this possiblity is presented below.

Retinal ganglion cells, the one population of retinal neurons that extend axons out of retina, are a comparatively minor cell population in the embryonic chick retina. These cells can be purified and identified by virtue of distinct colligative properties and antigenic repertories. Analysis of laminin receptor function in the purified cells reveals, as illustrated in Figure 4, that levels of α_6 mRNA are reduced dramatically between E6 and E12. Similar results have been obtained in measurements of α_6 protein (de Curtis et al. 1991). Some residual protein and mRNA was detected in these cells, but this may reflect contamination by other cells. The results suggest that laminin receptor activity in retinal ganglion cells is regulated primarily by control of mRNA levels.

It will be important to determine whether the loss of extracellular matrix receptor function seen in many populations of neurons is induced by target contact. In retinal ganglion cells, some evidence supports this possibility. After removal of their target, the optic tectum,

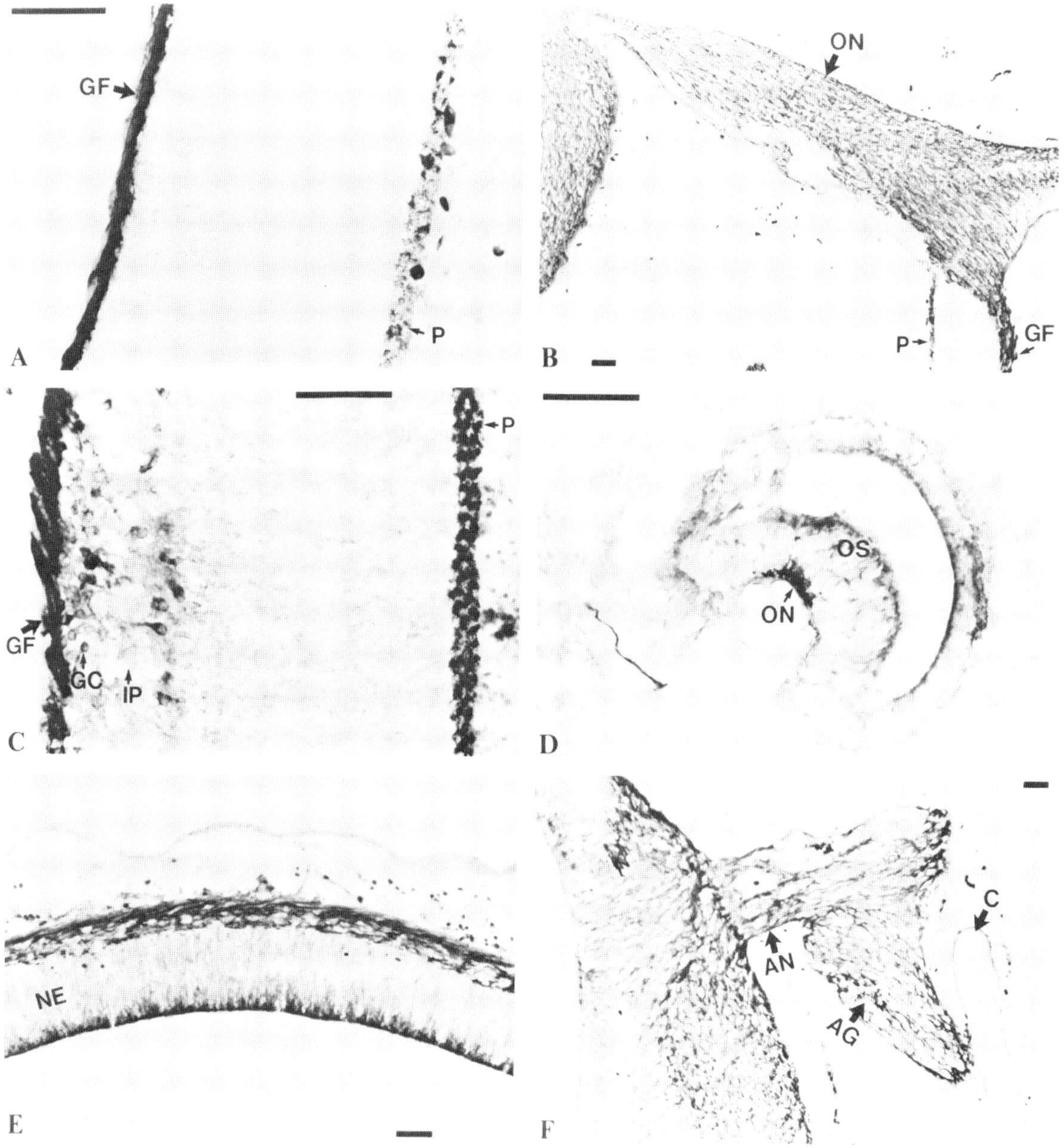

Figure 3. Indirect immunostaining of chick embryo sections with affinity-purified anti-α_8 antibodies. The α_8 antigen is visualized by the deposition of dark peroxidase reaction products. Erythrocytes have an endogenous peroxidase activity resulting in their nonimmune staining. (*A*) Section of a retina at E6. The ganglion cell fibers (GF) are immunoreactive. The pigmented cells (P) have an endogenous dark color that is present in controls (not shown). (*B*) Coronal section through the optic nerve at E6. The ventral side is at the top. The ganglion cell fibers (GF) of the retina, the optic nerve (ON), and the forebrain on the left are immunoreactive. The pigmented cells (P) have an endogenous dark color. (*C*) Section of a retina at E11. The ganglion cell fibers (GF), the ganglion cells (GC), and some cells in the inner plexiform layer (IP) are immunoreactive. The pigmented cells (P) have an endogenous dark color. (*D*) Section of the optic stalk at E3.5. The optic nerve (ON) on the ventral side of the optic stalk (OS) is strongly immunoreactive, as is an unidentified neuron on the left. (*E*) Coronal section of the dorsal part of the optic coronal tectum at E6. The circumferential fibers and the endfeet junctions on the ventricular side of the neuroepithelium (NE) are immunoreactive. (*F*) Coronal section through the acoustic ganglion at E6. The ventral side is at the bottom. The hindbrain at left, the acoustic nerve (AN), the acoustic ganglion (AG), and the side of the epithelium of a semicircular canal (C) in contact with the basal lamina are immunoreactive. Bars in *A–F*, 50 μm.

these cells are reported to remain able to interact with laminin at times when responsiveness has been lost in control animals (Cohen et al. 1989). For many years, it has been known that target-derived factors, such as neurotrophins, can regulate gene expression by mechanisms that depend on retrograde transport. Factors in optic tectum that regulate integrin expression have not yet been identified.

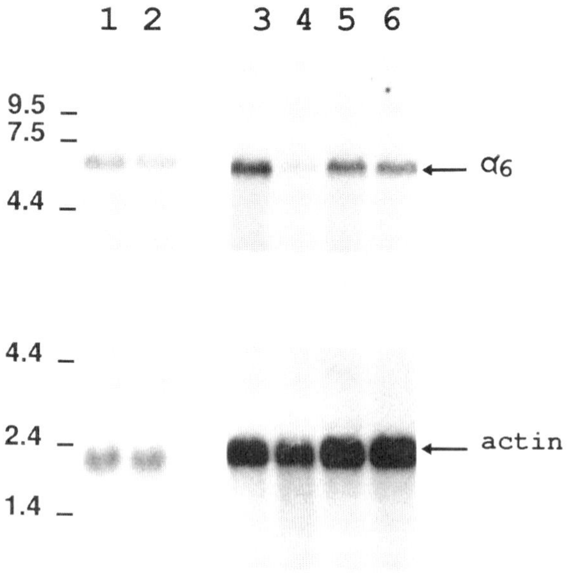

Figure 4. Distribution of α_6 mRNA in different retinal cellular fractions. Total RNA was extracted from total retinae (lanes *1* and *2*), from the retinal ganglion cell fraction (lanes *3* and *4*), and from the non-retinal ganglion cell fraction (lanes *5* and *6*) of E6 (lanes *1*, *3*, and *5*) and E12 (lanes *2*, *4*, and *6*) retinae, purified as described in de Curtis et al. (1991). Every sample was run in duplicate on an agarose gel and blotted onto nitrocellulose. The blots were incubated either with a ^{32}P-labeled double-stranded probe obtained using a gel-purified α_6 cDNA fragment (*top*) or with an actin probe (*bottom*). 20 μg (lanes *1* and *2*) or 10 μg (lanes *3*–*6*) total RNA was used for each lane.

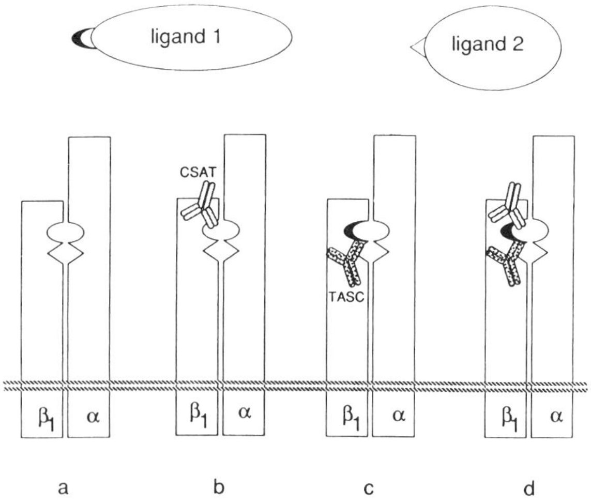

Figure 5. The effects of the activating antibody TASC and the inhibitory antibody CSAT on cell attachment suggest a general scheme for β_1-class integrin structure and function that features two distinct sites on the β_1 subunit involved in forming ligand-binding domains as well as a conformational change that leads to higher affinity binding at one of these sites. The α subunit depicted is a composite, allowing for all of the possible ligand-binding domains formed by $\alpha\beta$ dimers, and is not meant to represent any particular α subunit. (*a*) In the absence of antibodies, β_1-class integrins bind ligands "1" (at least LN, FN120, collagens, and FBG) and "2" (VN). (*b*) CSAT (stippled IgG) interferes with a site on the β_1 subunit required for the binding of ligand 1. (*c*) TASC (spotted IgG) binds a distinct site on the β_1 subunit that is required for binding to ligand 1. Occupancy of this site by TASC results in an activation of the site perturbed by CSAT as indicated by shading. This site now binds ligand 1 more effectively. (*d*) CSAT continues to recognize and perturb β_1 function in the presence of TASC, and thus even activated receptors are unable to bind ligand. Note that the antibodies are not drawn to scale but serve to indicate distinct sites on the β_1 subunit.

As discussed above, integrin activity also appears to be modulated by other factors in the retina. A monoclonal antibody named TASC has been isolated which binds an epitope in the chick integrin β_1 subunit and activates the latent ligand-binding activities of several β_1-subunit-containing integrin heterodimers (Neugebauer and Reichardt 1991; Neugebauer et al. 1991, 1992). A proposed scheme for TASC interactions with integrins is illustrated in Figure 5. Fab fragments of this antibody restore the ability of E12 retinal neurons to attach to laminin as rapidly as attachment can be measured (Neugebauer and Reichardt 1991). Endogenous factors modulating receptor function have not been identified.

Interactions of Retinal Neurons with Cellular Substrates In Vitro

In the developing primary visual system, retinal ganglion cell axons grow in contact with the endfeet of neuroepithelial astroglial precursors within the eye, optic nerve, and optic tectum. Descendants of these neuroepithelia have been shown to be effective substrates for axonal elongation (cf. Tomaselli et al. 1986). Neuronal responsiveness to astroglia is mediated only partially by integrins. In past work, we showed that neurite outgrowth over astroglia is promoted by adhesive interactions mediated by N-CAM, N-cadherin, and integrins (Neugebauer et al. 1988; Reichardt et al. 1990). Drazba and Lemmon (1990) showed that the adhesion molecule L1 is important in mediating outgrowth by retinal ganglion cells. Numerous additional cadherin and immunoglobulin superfamily adhesion molecules have been identified in the nervous system during the past few years, several of which are likely to modulate axonal growth in the primary visual projection (cf. Furley et al. 1990; Burns et al. 1991; Ranscht 1991; Tanaka et al. 1991). Experiments in vitro suggest that these molecules have apparently redundant functions on simple cellular substrates, such as astroglia. Inhibiting a single adhesive interaction is not sufficient to eliminate neurite outgrowth (cf. Neugebauer et al. 1988). Whether these molecules are redundant in vivo is not known.

The abilities of total populations of retinal neurons and of retinal ganglion cells to utilize a particular adhesive system to promote neurite outgrowth are developmentally regulated (cf. Neugebauer et al. 1988; Doherty et al. 1990, 1992). Within the retina, responsiveness to N-CAM is regulated differently by various cell types. Evidence indicates that the ability of retinal ganglion cells to utilize this adhesion molecule is lost, whereas other neurons in the retina exhibit enhanced responsiveness with increasing age (Neugebauer et al. 1988; Doherty et al. 1990). Differential splicing of exons containing extracellular or cytoplasmic polypeptide domains and regulation of polysialation have each been shown to regulate the ability of N-CAM to promote neurite outgrowth in vitro (cf. Doherty et al. 1990, 1992). Changes in polysialation have also been shown to influence axonal behavior in vivo (Landmesser et al. 1990). Thus, like integrin-mediated interactions, cell adhesion molecule-mediated neurite outgrowth appears to be regulated by both transcriptional and post-translational mechanisms. Again the significance of developmental regulation of these adhesive interactions is poorly understood.

Description of Growth Cone Behavior in the Optic Stalk and Optic Chiasm

During formation of the primary visual projection, axons of retinal ganglion cells exit the eye, traverse the neuroepithelial tube connecting the retina and central nervous system (the optic stalk), and grow over the ventral surface of the developing central nervous system to the midline, where the axons from both eyes intersect each other to form the optic chiasm (cf. Colello and Guillery 1990; Godement et al. 1990; Sretavan 1990). Individual axons exiting the optic chiasm either project across the midline into the contralateral optic tract or turn before crossing the midline to grow into the ipsilateral optic tract. This pattern of axon routing enables corresponding portions of the visual fields of each eye to be aligned for processing of information in the brain. Since retinal ganglion cell axons run superficially just underneath the pial surface of the brain, it is relatively accessible for visualization and manipulation. During development, the axons of retinal ganglion cells express cell adhesion molecules, such as L1/Ng-CAM, that make identification possible by immunocytochemistry. Alternatively, when injected into the eye, fluorescent dyes, such as diO, diI, and RITC, label all parts of retinal ganglion cells, including their growth cones and axons.

In rodents, previous work has shown that the adult pattern of axon routing is formed early in development (cf. Colello and Guillery 1990; Godement et al. 1990; Sretavan 1990). Retrograde labeling of ipsilaterally projecting axons has shown that they are derived from neurons in the ventral-temporal retina, whereas retrograde labeling of contralaterally projecting axons reveals that they are derived from retinal ganglion cells throughout the retina (cf. Sretavan 1990). Ipsilaterally and contralaterally projecting fibers develop simultaneously and are intermixed in the optic nerve before entering the optic chiasm. Results in Figure 6 show the behaviors in time lapse of several growth cones moving through the embryonic optic chiasm. All fibers grow initially toward the midline of the developing chiasm. Some cross and continue in the contralateral optic tract. Others turn before crossing the midline and enter

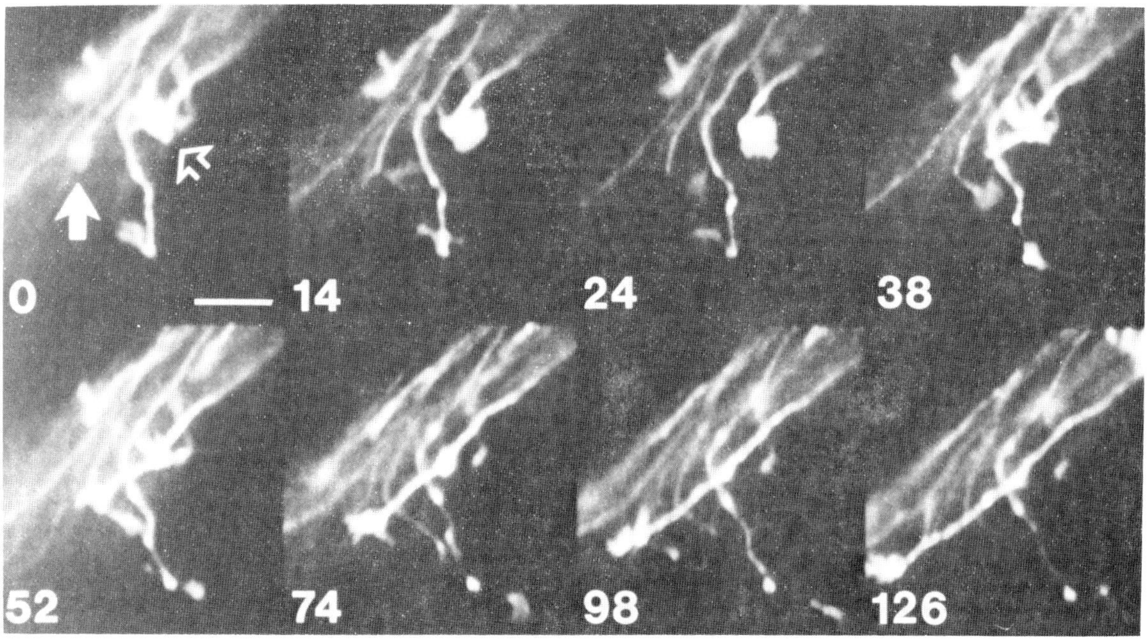

Figure 6. Time-lapse sequence illustrating the pathfinding behavior and trajectories of retinal ganglion cell axons at the optic chiasm from a one-eye embryo at E14. In each image, axons shown originate from the optic nerve at the upper right. The ipsilateral optic tract is toward the lower right, and the midline and the contralateral optic tract are toward the lower left. Time elapsed from the first image is shown at the bottom left corner in minutes. Bars, 25 μm. Both ipsilaterally (arrow) and contralaterally projecting axons (open arrow) in one-eye embryos navigate within the chiasm along the same trajectories as those of their normal counterparts, indicating that the pathfinding mechanism operating during normal development is very likely to be present in one-eye animals.

the ipsilateral optic tract. Results suggest strongly that the initial formation of the projection pattern depends on active pathfinding by retinal ganglion cell axons using guidance cues encountered by these growth cones in the vicinity of the midline of the chiasm.

In principle, the projection pattern of retinal ganglion cell axons at the optic chiasm could be established by either axon-axonal or axon-substrate interactions in the optic chiasm. However, early retinal axons appear to make appropriate pathway decisions in the absence of axons from the opposite eye (D. Sretavan and L.F. Reichardt, in prep.), implying that decisions made by these neurons do not depend on interactions with contralateral axons. Axon-axonal contacts clearly do influence growth cone behavior but do not appear to direct them into either optic chiasm. However, this does not eliminate the possibility that axon-axonal contacts may contribute to pathfinding by growth cones at later times (cf. Godement et al. 1990).

Two antigenically distinct populations of cells have been identified in the embryonic optic chiasm. Both populations are reported to be present at the time initial axons reach the commissure and are thus candidates to provide cues utilized in pathfinding. One is a "palisade" of radial glia that is recognized by several monoclonal antibodies, including SSEA-1 (Silver 1984; Mason et al. 1991; Colello and Guillery 1992). A second is a population of neurons, with cell somata in the chiasm, that are recognized by antibodies to L1, and other neuronal markers, such as MAP2 and β III tubulin (Feng and Sretavan 1991). These chiasm neurons extend along the lateral wall of the diencephalon. They may provide guidance information important in development of the chiasm.

To examine the function of cells present in the optic chiasm, we have identified antibodies that stain specifically the surfaces of chiasmic neurons. These antibodies do not recognize retinal ganglion cells, their axons, or other neurons in this portion of the embryonic central nervous system. An example of staining is shown in Figure 7. Injection of these antibodies with complement into E11.5–12 embryos results in complement-mediated lysis of the vast majority of these cells. The loss of cells persists during the following 4–5 days of embryonic development as assayed by staining with the same antibody. Comparison of projection patterns of retinal ganglion cell axons in control and experimental animals is presented in Figure 8. At E16 in controls, large numbers of axons labeled with the fluorescent dye, diI, have grown through the optic stalk, chiasm, and tract, generating a characteristic "X"-shaped projection. In animals lacking the endogenous neurons at the optic chiasm, the growth cones of retinal ganglion

Figure 7. Section through the ventral diencephalon illustrating positions of chiasm neurons. Specific staining with a monoclonal antibody illustrates an inverted "V" array of these neurons. Other experiments (not shown) demonstrate that these cells are labeled with antibodies to L1, MAP-2, and βIII-tubulin (Feng and Sretavan 1991).

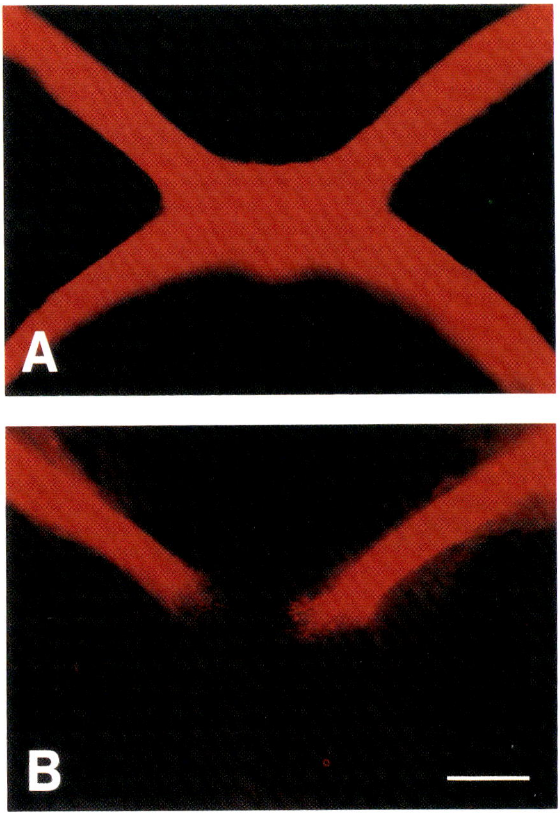

Figure 8. Ablation of neurons at the optic chiasm prevents growth of axons of retinal ganglion cells into the optic chiasm. (*A*) In control animals at E16, the axons of murine retinal ganglion cells, labeled by injection of diI in the eye, have grown through the optic chiasm forming an "X" characteristic of the adult visual projection. (*B*) Embryos were injected with monoclonal antibody and complement at E11.5. The axons of retinal ganglion cells were visualized at E16 by injection of diI into the eye. Although axonal outgrowth from the retina is not perturbed and normal-appearing optic nerves are present on both sides, the axons have not grown beyond the entrance to the optic chiasm. Other experiments at the light- and electron microscopy levels (not shown) showed that injection of antibody plus complement eliminated almost all chiasm neurons without detectably disrupting the underlying embryonic neuroepithelium. Injection of antibodies alone or injection of complement with control antibodies that did not label these cells did not eliminate these neurons or detectably perturb development of the normal pattern of axons of the visual projection at the optic chiasm. Bar, 200 μm.

cell axons appear to be paralyzed at their point of entry into the chiasm. The results argue that the presence of these neurons in the optic chiasm is essential for normal development of the visual projection. Similar experiments have not yet been done to examine the role of the palisade glial cells.

CONCLUSIONS

Understanding mechanisms of axonal guidance and growth cone motility requires characterization of axonal behavior in vivo, identification of cells and molecules capable of regulating in vivo behaviors, and direct experimental tests of resulting models.

Both within the embryonic retina and in the optic stalk, the growth cones of retinal ganglion cells are in contact with surfaces of neuroepithelial cells, extracellular matrix constituents secreted by these cells, and, at later stages, the axons of more developmentally advanced neurons. Thus, adhesive interactions relevant for development in this system must depend on ligands or receptors present on the neurons or their substrates in this pathway. Studies in vitro have identified several adhesion-promoting molecules that are candidates to mediate axonal outgrowth within the retina and in the retinal projection. These include a number of extracellular matrix proteins, including laminin; several integrin heterodimers, including receptors for laminin, collagen, thrombospondin, and vitronectin; and cell adhesion molecules such as L1/Ng-CAM, N-CAM, and N-cadherin. Many of these have been shown to mediate neurite growth over astroglia (see, e.g. Neugebauer et al. 1988; Drazba and Lemmon 1990). N-CAM, N-cadherin, L1/Ng-CAM, and integrins all appear to be functionally important. It is now timely to examine their functions in vivo.

Other investigators have shown that several cell adhesion molecules, including L1/Ng-CAM, N-cadherin, tag-1, and SC-1/DM-GRASP, when purified and coated on substrates, promote neurite outgrowth (Lagenauer and Lemmon 1987; Bixby and Jhabvala 1990; Furley et al. 1990; Burns et al. 1991; Tanaka et al. 1991). They thus can behave as ligands, analogous to extracellular matrix constituents. By utilizing species-specific antibodies to individual cell adhesion molecules, it has been possible to show that most of these can also function as neuronal receptors (cf. Neugebauer et al. 1988). Almost all neuronal cell adhesion molecules identified to date are able to interact with the same glycoprotein, i.e., homophilically. Both cadherins and immunoglobulin superfamily-class adhesion molecules also exhibit heterophilic interactions (cf. Volk et al. 1987; Kuhn et al. 1991). Similarly, most integrins are capable of interacting with more than one substrate. Thus, it is straightforward to identify a ligand or receptor in the retina or visual projection. Because both the retina and primary visual projection are accessible to experimental manipulation, it is possible to examine functions of individual ligands and receptors during development in vivo. It is much more difficult to prove that interactions detected in vitro are relevant in vivo.

Extensive work now shows that cell adhesion molecules and integrins are differentially expressed on neurons in the embryonic nervous system (cf. Bossy et al. 1991; Redies et al. 1992) and could thus in principle explain why different neurons make different pathway selections. At this point, the most direct experiment to test the significance of selective expression or regulation is to genetically ablate or ectopically express individual receptors or ligands, as has been done successfully in *Drosophila* (see, e.g., Grenningloh et al. 1991). Either would be helpful, for example, in determining whether the striking loss during development of retinal ganglion cell responsiveness to laminin is an important clue to observed changes in growth cone behavior in vivo.

Posttranslational modulation of cell-surface adhesion receptor function by extrinsic and intrinsic agents is likely to provide an important mechanism for regulating growth cone function. This paper suggests that integrins are regulated on neurons by such mechanisms, extending previous work on platelets and lymphocytes (cf. Springer 1990; Ginsberg et al., this volume). It will be interesting to determine whether cell adhesion molecule functions on neurons can be similarly modulated. At this time, to determine the significance of regulation of integrin activity it is necessary to identify endogenous regulators, analogous to the physiological substances, such as thrombin, that regulate integrins on platelets (cf. Coughlin et al., this volume). It will also be important to develop reagents able to monitor integrins in active and inactive states. Such reagents would make it possible to determine whether stimuli that direct growth cone behavior, e.g., contact with a guidepost cell, alter integrin activity locally. Antibodies able to bind selectively "inactive" or "activated" platelet integrin IIb/IIIa have been isolated (cf. Ginsberg et al., this volume), so it may be possible to isolate similar reagents to study other integrins.

Interactions in the retina and optic stalk that regulate the behaviors of retinal ganglion cell growth cones may prove to be similar to those mediating interactions between these neurons and cellular substrates in vitro. At the midline, the site of the future optic chiasm, however, growth cones encounter a more complex environment where they contact growth cones and axons from the contralateral eye, chiasmic neurons, and the midline palisade of specialized glia. Analyses of the behaviors of early retinal axons to enter this region show that they make decisions appropriate for developing an adult pattern of innervation (Colello and Guillery 1990; Godement et al. 1990; Sretavan 1990).

Analysis of growth cone behavior indicates that ipsilateral turning decisions are made by axons in the chiasm at positions close to, but on the ipsilateral side of, the midline. Mapping the positions of turning decisions suggests that local cues direct this turning behavior (Godement et al. 1990; Sretavan 1990; D. Sretavan and L.F. Reichardt, in prep.). Contralaterally and ipsilaterally projecting axons must differ in their responsiveness to these signals. The nature of the critical signals is not known. In principle, chemotropic factors, inhibitors of growth cone motility, or adhesion-promoting stimulators of neurite outgrowth could guide axons in this region (for review, see Dodd and Jessell 1988).

In an attempt to identify the function of one of the specialized cell types identified in the embryonic optic chiasm, the early developing neurons in this region were ablated with a cell-specific antibody and complement. The resulting paralysis of retinal ganglion cell growth cones at the entrance to the optic chiasm suggests that these cells have some critical function. This could be synthesis of one or more of the signals described above, such as a chemotropic factor or motility-enhancing adhesion molecule. The experiments do not provide evidence that these neurons give cues important in directing axons in an ipsilateral or contralateral direction, although they appear appropriately placed to influence that decision.

Further experiments should clearly attempt to delete the other populations of antigenically identifiable cell types in the chiasm. It will be useful to more carefully characterize the molecular environment at the optic chiasm. Finally, it will be important to examine interactions in vitro where the critical signals can be characterized.

REFERENCES

Akers, R.M., D.F. Mosher, and J.E. Lilien. 1981. Promotion of retinal neurite outgrowth by substratum-bound fibronectin. *Dev. Biol.* **86:** 179.

Baier, H. and F. Bonhoeffer. 1992. Axon guidance by gradients of a target-derived component. *Science* **255:** 472.

Bixby, J.L. and P. Jhabvala. 1990. Extracellular matrix molecules and cell adhesion molecules induce neurites through different mechanisms. *J. Cell Biol.* **111:** 2725.

Bossy, B., E. Bossy-Wetzel, and L.F. Reichardt. 1991. Characterization of the integrin α_8 subunit: A new integrin β_1-associated subunit, which is prominently expressed on axons and on cells in contact with basal laminae in chick embryos. *EMBO J.* **10:** 2375.

Burns, F.R., S. Von Kannen, L. Guy, J.A. Raper, J. Kamholz, and S. Chang. 1991. DM-GRASP, a novel immunoglobulin superfamily axonal surface protein that supports neurite extension. *Neuron* **7:** 209.

Carri, N.G., R. Perris, S. Johansson, and T. Ebendal. 1988. Differential outgrowth of retinal neurites on purified extracellular matrix molecules. *J. Neurosci. Res.* **19:** 428.

Cohen, J., J.F. Burne, J. Winter, and P. Bartlett. 1986. Retinal ganglion cells lose response to laminin with maturation. *Nature* **322:** 465.

Cohen, J., V. Nurcombe, P. Jeffrey, and D. Edgar. 1989. Developmental loss of functional laminin receptors on retinal ganglion cells is regulated by their target tissue, the optic tectum. *Development* **107:** 381.

Colello, R.J. and R.W. Guillery. 1990. The early development of retinal ganglion cells with uncrossed axons in the mouse: Retinal position and axon course. *Development* **108:** 515.

———. 1992. Observations on the early development of the optic nerve and tract in the mouse. *J. Comp. Neurol.* **317:** 357.

de Curtis, I., V. Quaranta, R.N. Tamura, and L.F. Reichardt. 1991. Laminin receptors in the retina: Sequence analysis of the chick integrin α_6 subunit; evidence for transcriptional and posttranslational regulation. *J. Cell Biol.* **113:** 405.

Dodd, J. and T.M. Jessell. 1988. Axon guidance and the patterning of neuronal projections in vertebrates. *Science* **242:** 692.

Doherty, P., J. Cohen, and F.S. Walsh. 1990. Neurite outgrowth in response to transfected NCAM changes during development and is modulated by polysialic acid. *Neuron* **5:** 209.

Doherty, P., C.E.C.K. Moolenaar, S.V. Ashton, R.J.A.M. Michalides, and F.S. Walsh. 1992. The VASE exon downregulates the neurite growth-promoting activity of NCAM140. *Nature* **356:** 791.

Drazba, J. and V. Lemmon. 1990. The role of cell adhesion molecules in neurite outgrowth on Muller cells. *Dev. Biol.* **138:** 82.

Feng, L.F. and D.W. Sretavan. 1991. A neuronal population is present at the optic chiasm prior to arrival of retinal ganglion cell axons. *Soc. Neurosci. Abstr.* **17:** 182.

Furley, A.J., S.B. Morton, D. Manalo, D. Karagogeos, J. Dodd, and T. Jessell. 1990. The axonal glycoprotein TAG-1 is an immunoglobulin superfamily member with neurite outgrowth-promoting activity. *Cell* **81:**157.

Godement, P., J. Salaun, and C.A. Mason. 1990. Retinal axon pathfinding in the optic chiasm: Divergence of crossed and uncrossed fibers. *Neuron* **5:** 173.

Grenningloh, G., E.J. Rehn, and C.S. Goodman. 1991. Genetic analysis of growth cone guidance in *Drosophila*: Fasciculin II functions as a neuronal recognition molecule. *Cell* **67:** 45.

Hall, D.E, K.M. Neugebauer, and L.F Reichardt. 1987. Embryonic neural retinal cell response to matrix molecules: Developmental changes and effects of CSAT antibody. *J. Cell Biol.* **104:** 623.

Hunter, D.D., M.D. Murphy, C.V. Olsson, and W.J. Brunken. 1992. S-laminin expression in adult and developing retinae: A potential cue for photoreceptor morphogenesis. *Neuron* **8:** 399.

Ignatius, M.J. and L.F. Reichardt. 1988. Identification and characterization of a neuronal laminin receptor: An integrin heterodimer that binds laminin in a divalent cation-dependent manner. *Neuron* **1:** 713.

Kaesberg, P.R., W.B. Ershler, J.D. Esko, and D.F. Mosher. 1989. Chinese hamster ovary cell adhesion to human platelet thrombospondin is dependent on cell surface heparan sulfate proteoglycan. *J. Clin. Invest.* **83:** 994.

Kater, S.B. and P.B. Guthrie. 1990. Neuronal growth cone as an integrator of complex environmental information. *Cold Spring Harbor Symp. Quant. Biol.* **55:** 359.

Kawasaki, Y., H. Horie, and T. Takenaka. 1986. The maturation-dependent change in fibronectin receptor density of mouse dorsal root ganglion neurons. *Brain Res.* **397:** 185.

Kuhn, T.B., E.T. Stoeckli, M.A. Condrau, F.G. Rathjen, and P. Sonderegger. 1991. Neurite outgrowth on immobilized axonin-1 is mediated by heterophilic interaction with L1 (G4). *J. Cell Biol.* **115:** 1113.

Lagenauer, C. and V. Lemmon. 1987. A L1-like molecule, the 8D9 antigen is a potent substrate for neurite extension. *Proc. Natl. Acad. Sci.* **84:** 7753.

Landmesser, L., L. Dahm, J. Tang, and U. Rutishauser. 1990. Polysialic acid as a regulator of intramuscular nerve branching during embryonic development. *Neuron* **4:** 655.

Mason, C.A., R. Blazaski, J. Dodd, and P. Godement. 1991. The midline of the mouse optic chiasm: Cellular composition and antigen expression during retinal axon growth. *Soc. Neurosci. Abstr.* **17:** 39.

Moyle, M., M.A. Napier, and J.W. McLean. 1991. Cloning and expression of a divergent integrin. *J. Biol. Chem.* **266:** 19650.

Neugebauer, K.M. and L.F. Reichardt. 1991. Cell surface regulation of β_1-integrin activity on developing retinal neurons. *Nature* **350:** 68.

Neugebauer, K.M., K. Venstrom, and L.F. Reichardt. 1992.

Adhesion of a chicken myeloblast cell line to fibrinogen and vitronectin through a β_1-class integrin. *J. Cell Biol.* **116:** 809.

Neugebauer, K.M., C.J. Emmett, K.A. Venstrom, and L.F. Reichardt. 1991. Vitronectin and thrombospondin promote retinal neurite outgrowth: Developmental regulation and role of integrins. *Neuron* **6:** 345.

Neugebauer, K.M., K.J. Tomaselli, J. Lilien, and L.F. Reichardt. 1988. N-cadherin, NCAM, and integrins promote retinal neurite outgrowth on astrocytes in vitro. *J. Cell Biol.* **107:** 1177.

O'Shea, K.S. and V.M. Dixit. 1988. Unique distribution of the extracellular matrix component thrombospondin in the developing mouse embryo. *J. Cell Biol.* **107:** 2737.

Prieto, A.L., F.S. Jones, B.A. Cunningham, K.L. Crossin, and G.M. Edelman. 1990. Localization during development of alternatively spliced forms of cytotactin mRNA by *in situ* hybridization. *J. Cell Biol.* **111:** 685.

Ranscht, B.. 1991. Cadherin cell adhesion molecules in vertebrate neural development. *Semin. Neurosci.* **3:** 285.

Rathjen, F.G., J.M. Wolff, and R. Chiquet-Ehrismann. 1991. Restrictin: A chick neural extracellular matrix protein involved in cell attachment co-purifies with the cell recognition molecule F11. *Development* **113:** 151.

Redies, C., H. Inuzuka, and M. Takeichi. 1992. Restricted expression of N- and R-cadherin on neurites of the developing chicken CNS. *J. Neurosci.* **12:** 3525.

Reichardt, L.F. and K.J. Tomaselli. 1991. Extracellular matrix molecules and their receptors: Functions in neural development. *Annu. Rev. Neurosci.* **14:** 531.

Reichardt, L.F., B. Bossy, S. Carbonetto, I. de Curtis, C. Emmett, D.E. Hall, M.J. Ignatius, T. Large, F. Lefcort, E. Napolitano, K.M. Neugebauer, and K.J. Tomaselli. 1990. Neuronal receptors that regulate axon growth. *Cold Spring Harbor Symp. Quant. Biol.* **55:** 341.

Rosen, G.D., J.R. Sanes, R. LaChance, J.M. Cunningham, J. Roman, and D.C. Dean. 1992. Roles for the integrin VLA-4 and its counter receptor VCAM-1 in myogenesis. *Cell* **69:** 1107.

Silver, J. 1984. Studies on the factors that govern directionality of axonal growth in the embryonic optic nerve and at the chiasm of mice. *J. Comp. Neurol.* **223:** 238.

Springer, T. 1990. Adhesion receptors of the immune system. *Nature* **346:** 415.

Sretavan, D. 1990. Specific routing of retinal ganglion cell axons at the mammalian optic chiasm during embryonic development. *J. Neurosci.* **10:** 1995.

Tanaka, H., T. Matsui, A. Agata, M. Tomura, I. Kubots, K.G. McFarland, B. Kohr, A. Lee, H.S. Phillips, and D.L. Shelton. 1991. Molecular cloning and expression of a novel adhesion molecule, SC1. *Neuron* **7:** 535.

Tomaselli, K.J. and L.F. Reichardt. 1988. Peripheral motoneuron interactions with laminin and Schwann cell-derived neurite-promoting molecules: Developmental regulation of laminin receptor function. *J. Neurosci. Res.* **21:** 275.

Tomaselli, K.J., L.F. Reichardt, and J.L. Bixby. 1986. Distinct molecular interactions mediate neuronal process outgrowth on extracellular matrices and non-neuronal cell surfaces. *J. Cell Biol.* **103:** 2659.

Volk, T., O. Cohen, and B. Geiger. 1987. Formation of heterotypic adherens-type junctions between L-CAM-containing liver cells and A-CAM-containing lens cells. *Cell* **50:** 987.

Second Messengers Underlying Cell-contact-dependent Axonal Growth Stimulated by Transfected N-CAM, N-cadherin, or L1

F.S. WALSH AND P. DOHERTY
Department of Experimental Pathology, UMDS, Guy's Hospital, London SE1 9RT, United Kingdom

During development, axons grow along highly stereospecific pathways to innervate and synapse with target tissue. In recent years, much effort has been devoted to attempting to understand these processes in molecular terms. A variety of studies (for review, see Doherty and Walsh 1989; Bixby and Harris 1991; Lumsden and Cohen 1991) have implicated a number of cellular components in regulating axonal growth and guidance. These are, first, soluble components such as neurotransmitters (Kater and Guthrie 1990) and chemotropic or trophic factors (Lumsden and Davies 1986; Heffner et al. 1990); second, components of the extracellular matrix such as fibronectin and a variety of laminin isoforms (Reichardt et al. 1990; Sanes et al. 1990); and third, an increasing number of membrane-associated glycoproteins that are members of different cell and substrate adhesion molecule families (Reichardt et al. 1990; Ranscht 1991; Rathjen and Jessell 1991; Walsh and Doherty 1991). In addition to positive cues for axonal growth, there are also inhibitory glycoproteins coexisting with these components (Keynes et al. 1991), although this class of protein is not as well characterized as the positive cues. A large number of observations regarding growth cone motility have been made in tissue culture systems; however, there is increasing information from in vivo model systems to back up these models developed from in vitro observations. Enormous progress has also been made in analyzing the structure and function of cell adhesion molecules (CAMs) that mediate axonal growth processes. It is now evident that members of three specific superfamilies of receptors are involved in these processes. These are (1) the integrins (Reichardt and Tomaselli 1991; Hynes 1992) that are the receptors for extracellular matrix components, (2) members of the immunoglobulin superfamily (Rathjen and Jessell 1991; Walsh and Doherty 1991), and (3) the cadherins (Takeichi 1991). We have been studying the structure and function of members of the immunoglobulin superfamily and the N-cadherin protein in axonal growth processes; in this paper, we compare and contrast different CAMs, using the neural cell adhesion molecule (N-CAM) as a model.

Diversity of N-CAM Isoforms

N-CAM was the first CAM to be studied in detail (Edelman 1985). It is known to bind in a Ca^{++}-independent homophilic (self) manner. A number of studies have assessed its role in axonal growth processes in vitro with different results from different model systems. Whereas a number of CAMs are active when isolated in a pure form, N-CAM is unusual in that most studies to date have been unable to reconstitute N-CAM-dependent neurite outgrowth with the purified protein. One possibility is that the protein is denatured during purification. However, the observation that N-CAM incorporated into liposomes can function in aggregation assays suggests a more complex situation. It is highly likely that N-CAM has to be mobile in a lipid bilayer in order to function efficiently as a neurite-outgrowth-promoting molecule. Most studies that have reported a positive role for N-CAM have utilized live cells (Bixby et al. 1987; Neugebauer et al. 1988; Seilheimer and Schachner 1988). Other studies showing a positive role for N-CAM in neurite outgrowth have utilized a molecular genetic approach. We have shown that 3T3 fibroblasts expressing human N-CAM via gene transfer express the protein at the cell surface and that most of the major isoforms are active in terms of neurite-outgrowth-promoting activity (Doherty et al. 1989, 1990b, 1991a,b, 1992a,b,c). However, to understand N-CAM function, it is important to discuss the diversity of isoforms that are found. The N-CAM gene is relatively large and is present as a single copy on human chromosome 11q23 and a syntenic region on mouse chromosome 9. In recent years, a large number of N-CAM cDNA clones have been isolated, and an almost complete picture of the intron/exon structure of the gene has been determined (Walsh and Doherty 1991). In terms of protein isoforms, there are three main classes resulting from different patterns of alternative splicing of the gene. These are transmembrane forms of 180 kD and 140 kD, glycosylphosphatidylinositol (GPI)-linked forms of about 120–125 kD, and soluble forms of about 110 kD. Because N-CAM is extensively glycosylated, these molecular sizes can vary quite considerably, especially if a certain type of carbohydrate unique to N-CAM, called polysialic acid (PSA), is present. To date, a total of 26 exons have been identified in the N-CAM gene in addition to the 5′ regulatory region. Alternative splicing in a cell- and tissue-specific manner generates the main membrane-associated forms of N-CAM. All isoforms utilize exons 1–10, which encode the five immunoglobulin-like domains and form the bulk of the extracellular polypeptide chain. Downstream from this, but still in the extracellular domain, are two regions that show some loose homology with fibronectin type III repeats. Although these have been reported in rodent and chicken cDNAs, the actual degree of homology found in human N-CAM with which we have worked is extremely low. Exon 15 contains the sequence information to attach N-CAM to the cell surface via a GPI anchor. Exon 15 encodes a 27-amino-acid region that is likely cleaved, although the actual site of cleavage has not yet been identified. This exon is used alternatively with exon 16. This exon encodes a 70-amino-acid sequence that crosses the plasma membrane, since it contains a region that is extremely hydrophobic and of predicted α-helical conformation. Downstream from exon 16 are three exons (17–19), which if used together, generate the largest N-CAM isoform of 180 kD. Exon 18 is alternatively used, and forms of N-CAM lacking the sequence encoded by this exon are of approximately 140 kD. Exon 18 is the largest coding exon in the N-CAM gene and inserts over 250 amino acids into the N-CAM sequence. A variety of studies have suggested that forms of N-CAM containing this sequence of about 180 kD are expressed in a highly restricted manner and may constitutively associate with the cytoskeleton (Pollerberg et al. 1985, 1986). In addition to these patterns of alternative splicing associated with membrane association and intracellular linkages, there are two regions of splicing in the N-CAM gene. The first of these is the MSD region (Walsh and Doherty 1991). This is a series of exons inserted between exons 12 and 13 in a region between the fibronectin repeats. The products of these exons are rich in amino acids serine and threonine, and in tissues such as skeletal muscle, they act as a site of *O*-linked carbohydrate attachment. The functional consequences of this are not clear, but the presence of *O*-linked carbohydrate may alter the linear structure of the N-CAM protein in this region, and as such, may act as a spacer region to increase the distance between the plasma membrane and the N-CAM regions important for homophilic interactions.

The final region of alternative splicing is within the fourth immunoglobulin domain, which is encoded by exons 7 and 8. A small exon of 30 base pairs has been identified in this region (Small and Akeson 1990). This inserts a 10-amino-acid peptide into the fourth immunoglobulin domain and switches the domain to a structure that more resembles an immunoglobulin variable domain. Another remarkable feature of the VASE exon is that its usage increases dramatically at later stages of brain development (see below). Studies by Small and Akeson (1990) have shown that as much as 50% of the N-CAM transcript in adult brain contains the VASE exon. As discussed below, usage of this exon has major consequences in terms of N-CAM's ability to act as a neurite-outgrowth-promoting molecule.

In addition to these changes in the primary amino acid sequence of N-CAM, there are also major changes in N-CAM glycosylation during development. N-CAM contains PSA, which is attached via an α2-8 linkage (Walsh and Doherty 1991). The PSA chains can be more than 200 residues in length, thus creating a large hydration sphere around the N-CAM protein. A number of studies have shown that PSA modulates N-CAM's adhesive function quite dramatically. Embryonic brain is high in N-CAM expressing PSA, whereas adult brain is rather low in PSA. As discussed below, it is of great interest that at times of greatest PSA expression, N-CAM can function very well as a neurite-outgrowth-promoting molecule, whereas at times of low PSA expression, N-CAM is poorly active in neurite outgrowth but is extremely active in adhesion assays. Temporarily, as PSA levels on N-CAM decrease, the usage of the VASE exon increases dramatically. These two components, namely PSA and usage of the VASE exon, appear to be major modulatory events in N-CAM's ability to function as a neurite-outgrowth-promoting molecule.

Cell Adhesion Molecule L1

Like N-CAM, the CAM L1 is a member of the immunoglobulin superfamily. Antibodies to L1 are active in inhibiting Schwann cell–neuron interactions (Bixby et al. 1988; Seilheimer and Schachner 1988) and also in inducing defasiculation of axon bundles. However, in addition to operating as a homophilic CAM, L1 appears to promote neurite outgrowth via heterophilic binding to a protein called axonin 1 (Kuhn et al. 1991; Zuellig et al. 1992), which appears to be a species homolog of the Tag 1 glycoprotein (Furley et al. 1990). A number of cDNA clones encoding mouse (Moos et al. 1988), rat (Miura et al. 1991), and human L1 (Hlavin and Lemmon 1991; Reid and Hemperly 1992) have recently been isolated and analyzed. There is a high degree of similarity between L1 across species, and, in general, the protein is 1260 amino acids in length. There are two isoforms of L1 that differ by use of an alternatively spliced exon that encodes four amino acids in the cytoplasmic domain. Both forms have been found in mouse, rat, and human L1, although the functional consequences of this insertion are not clear. There is no evidence of extensive alternative splicing in the L1 gene in the same manner as has been found for N-CAM. However, this is not the case in other L1-related genes such as NrCAM and BRAVO, where there appear to be a number of alternative spliced variants. Unlike N-CAM, the L1 glycoprotein appears to be able to operate as a neurite-outgrowth-promoting molecule when presented to neurons in a purified form. There is also increasing evidence that N-CAM and L1 may be physically associated in the plasma membrane (Simon et al. 1991), and a variety of in vivo studies in the neuromuscular system have suggested that PSA can modulate L1 function (Rutishauser 1991).

Cell Adhesion Molecule N-Cadherin

The cadherins comprise a family of transmembrane glycoproteins involved in Ca^{++}-dependent cell-cell interactions. Like N-CAM, cadherin interactions are generally homophilic. A number of cadherins have recently been identified, and each appears to be the product of separate genes (Takeichi 1991). The best-characterized cadherins are termed N-cadherin, E-cadherin, P-cadherin, and L-CAM. A large number of additional cadherins have been identified recently in a number of tissues (Takeichi 1991). There is extensive amino acid sequence similarity between individual types of cadherins in different species and between different cadherins in the same species. The cytoplasmic domain of cadherins that are believed to interact specifically with elements of the cytoskeleton are particularly highly conserved. The functional importance of the intracellular domain of N-cadherin has recently been studied by misexpression of N-cadherin cDNA clones. Kintner (1992) has developed a dominant negative mutation strategy to show that the intracellular domain of N-cadherin is extremely important for cadherin-dependent adhesion during embryogenesis. One of the sites of homophilic interactions has also been located to the amino-terminal region of about 113 amino acids. A cadherin cell adhesion recognition sequence has been located to this region, and peptides modeled around the HAV sequence appear to be functionally active in blocking N-cadherin-mediated interactions (Blaschuk et al. 1990; Nose et al. 1990; Doherty et al. 1991b). Other important sites of homophilic interaction in N-cadherin have not yet been identified.

Analysis of CAM Function in Neurite Outgrowth

A number of problems are associated with the molecular analysis of CAM function in neurite outgrowth. As discussed above, purified N-CAM does not promote neurite outgrowth. There are also a large number of N-CAM isoforms and glycosylation variants, and this makes it almost impossible to assess the role of individual isoforms of N-CAM using a classic biochemical approach. In addition, many primary cells such as Schwann cells or myotubes have a number of CAMs on their cell surface, which makes it difficult to assess the function of individual CAMs to the exclusion of others. An approach not limited by these problems is based on gene transfer technology. We have produced panels of NIH-3T3 cells that express a number of different N-CAM isoforms following cloning and transfection of full-length cDNAs in eukaryotic expression vectors. For comparison, we have also expressed the L1 glycoprotein and the N-cadherin protein in 3T3 cells. The parental 3T3 cells are particularly appropriate, because they do not express functionally active levels of N-CAM, L1, or N-cadherin. The model we have used to assess the neurite-outgrowth-promoting activity of these CAMs is based on a very simple experimental system. Monolayers of the transfected 3T3 cells or control cells are established and various populations of neurons are seeded at low cell density over a 16–48-hour period. Cultures are then fixed, and the pattern of neurite outgrowth is determined by identifying neurons via specific antigens, such as the Thy 1 glycoprotein, neurofilament, or GAP-43. The lengths of the neuronal processes are measured using standard computer-assisted morphometric analysis (Doherty et al. 1990a,b). We have now completed a series of studies comparing and contrasting different isoforms of N-CAM with the L1 and N-cadherin proteins, and at this stage, it is possible to make a number of generalizations about the mechanism of action of these CAMS. Table 1 shows a list of cDNA constructs that have been used to analyze the function of the different CAMs in neurite outgrowth. In addition, we show the different populations of neuronal cells used in these studies.

A Wide Variety of Neurons Respond to Transfected CAMs by Increased Neurite Growth

All the neurons listed in Table 1 respond to transfected N-CAM, N-cadherin, or L1 by increased neurite outgrowth. In the case of human and rat dorsal root

Table 1. Summary of the Transfected Cells and Probe Neurons Used to Study the Function of N-CAM, N-cadherin, and L1

Cellular monolayer
parental NIH-3T3 cells (do not support N-CAM-, N-cadherin-, or L1-dependent neurite outgrowth)
human N-CAM transfectants
115-kD secreted isoform
120-kD GPI isoform
120-kD + MSD1 region (37 a.a.'s)
140-kD TM isoform
140-kD + VASE exon (10 a.a.'s)
180-kD TM isoform
chick N-cadherin transfectants
human L1 transfectants
Probe neurons
rat/human DRG neurons (neurofilament protein assay)
chick retinal ganglion neurons at E6–E11
rat cerebellar granule neurons at PND1–PND9
rat hippocampal neurons at E17 and PND4

Confluent monolayers of control and transfected 3T3 cells are established and probe neurons are cultured on top of these for 16–48 hr. Cocultures are then fixed, and the extent of neuronal differentiation is quantitated by biochemical or direct morphometric analysis. (E6) Embryonic day 6; (PND1) postnatal day 1.

Table 2. Length of Longest Neurite on Transfectant Relative to Parental 3T3 Cell Monolayers

	Naive PC12 cells (48 hr)	Primed PC12 cells (16 hr)	Rat cerebellar neurons (PND1–PND6) (16 hr)
N-CAM 140 (clone 24)	214 ± 17 (11)	202 ± 12 (3)	178 ± 7 (6)
N-CAD (clone 1)	320 ± 25 (11)	184 ± 21 (3)	201 ± 16 (6)
L1 (clone 1)	n.d.	194 ± 10 (5)	173 ± 12 (5)

Naive PC12 cells, primed PC12 cells (treated with NGF for 3–4 days prior to the experiment), or rat cerebellar neurons were cultured on confluent monolayers of control and transfected 3T3 cells as indicated. The effect of the transfected CAM, measured as the percentage increase in the length of the longest neurite per cell on transfectants relative to control 3T3 cells is shown. Each value is the mean ± S.E. for the given number (*n*) of independent experiments. n.d. indicates not determined.

ganglion neurons, this was measured using an arbitrary biochemical measurement of neurite growth, namely, the level of immunoreactive neurofilament protein (Doherty et al. 1989). In subsequent studies, transfected N-CAM (Doherty et al. 1990 a,b), N-cadherin (Doherty et al. 1991b), and L1 (Williams et al. 1992) function has been assessed by directly measuring a number of morphometric parameters. The transfected CAMs appear to stimulate outgrowth from predetermined neurites, because the absolute length of neurites and the percentage of neurites greater than a given length consistently increased, whereas the absolute number of neurites per cell was relatively unaffected (see, e.g., Doherty et al. 1990b, 1992c). An increased number of branches per neurite for neurons grown over N-CAM-transfected cells (Doherty et al. 1990b) most probably simply reflected the fact that the primary neurites were longer. The absolute magnitude of the CAM response appears to be determined by several factors, including the length of the period of coculture, the age of the neurons, and the magnitude of basal outgrowth over the parental 3T3 cells. In general terms, all three CAMs can increase the length of the longest neurite by two- to threefold over a not inconsiderable length found on parental 3T3s. This has to be placed in the context of the fact that cells that express several CAMs (e.g., astrocytes) are only two- to threefold better than 3T3 fibroblasts in promoting neurite outgrowth from chick retinal ganglion cells (Doherty et al. 1990a). In some studies, we have made a direct comparison between neurite outgrowth in single clones of 3T3 cells expressing physiological levels of N-CAM (the 140-kD isoform), N-cadherin, or L1; these data are summarized in Table 2. It is shown that whereas all of the CAMs stimulate neurite outgrowth, none is consistently better than others at doing so. The specificity of the CAM responses could be demonstrated by "inhibition" by blocking antibodies. For example, antibodies to N-CAM can only block the N-CAM response, whereas antibodies to N-cadherin and L1 only block their respective responses. Furthermore, species-specific antibodies that bound exclusively to N-CAM or L1 in the neurons could also fully inhibit the respective responses, providing evidence that both operate via a homophilic binding mechanism (Doherty et al. 1990a; Williams et al. 1992).

Developmental Changes in Neuronal Responsiveness to CAMs

The ability of chick retinal ganglion cells (Doherty et al. 1990a), rat cerebellar neurons (Doherty et al. 1992b), and rat hippocampal neurons (Doherty et al. 1992c) to respond to transfected N-CAM (the 140-kD isoform) by increased neurite outgrowth is dramatically reduced over relatively short developmental periods (see Table 3). Broadly speaking, neurons that would normally be extending axons to their target respond, whereas those that have encountered their target and formed stable synapses do not. Failure to respond can-

Table 3. Developmental Profile of Neuronal Responsiveness to CAMs

		Level of response to		
Neuronal source		N-CAM	N-Cadherin	L1
RGC	E6	+++	+++	
	E9	–	+++++	n.d.
	E11	–	+++++	
Cerebellum	PND1	+++	++++	+++
	PND6	++++	++++	+++
	PND8	–	++++	+++
Hippocampus	E17	++++	+	
	PND4	+	++++	n.d.

Neurons from the chick retina (retinal ganglion cells, RGC), or the rat cerebellum and hippocampus were isolated at various developmental ages (as indicated) and their ability to respond to transfected N-CAM, N-cadherin and L1 was determined as in Table 2. The magnitude of the response varied from no significant effect (–) to a robust stimulation of neurite outgrowth (+++++). For details see text.

not be attributed to a general loss of CAM-dependent function, as the older neurons all show maintained or increased responsiveness to N-cadherin and/or L1. Thus, neuronal responsiveness to N-CAM appears to be subjected to more dynamic regulatory mechanisms than those controlling L1/N-cadherin responses, and this may reflect a target-dependent phenomenon. Specific loss of neuronal responsiveness to N-CAM could reflect either a loss of receptor (i.e., N-CAM) from the growth cone or a change of receptor function. The latter could be accomplished by either differential posttranslational processing of N-CAM and/or isoform switching.

PSA Is a Positive and Highly Specific Modulator of N-CAM-dependent Neurite Outgrowth

The PSA content of N-CAM is highly regulated during development of the nervous system. Recent studies have been facilitated by the availability of antibodies that can specifically recognize this relatively unusual carbohydrate structure. There is a general consensus that a high PSA content can be correlated with times and areas of relative plasticity in the developing and adult nervous system (see, e.g., Rutishauser et al. 1988; Walsh and Doherty 1991). Removal of PSA from N-CAM dramatically increases N-CAM-dependent adhesion in a variety of model systems (see, e.g., Hoffman and Edelman 1983), and more recently it has been suggested that PSA might inhibit general adhesion via a steric hindrance of membrane-membrane (Rutishauser et al. 1988) and membrane-substratum (Acheson et al. 1991) apposition. The above studies have largely been interpreted as suggesting that the primary purpose of PSA is to inhibit N-CAM function by perhaps preventing the *trans*-binding of N-CAM on one cell to N-CAM on a second cell. All of the younger neurons that respond to transfected N-CAM by extending longer neurites express PSA, and this can be specifically removed using the bacterial enzyme endoneuraminidase N (see, e.g., Rutishauser et al. 1988; Doherty et al. 1990a). We have found that although removal of PSA has been reported to increase adhesion, it substantially (by up to 80%) inhibits N-CAM-dependent neurite outgrowth. Furthermore, this is a very specific response in that integrin-, N-cadherin-, and L1-dependent neurite outgrowth are not *directly* affected by this treatment. (Table 4). These data cannot readily be explained, even by an inverse relationship between adhesion and neurite outgrowth, because we have only observed positive cooperative (Doherty et al. 1990b) and saturable (Doherty et al. 1992b) dose-response curves for N-CAM level and neurite outgrowth. Furthermore, these data are not obviously consistent with PSA operating by inhibiting the *trans*-binding of N-CAM, as this would be expected to inhibit N-CAM-dependent neurite outgrowth. The data do suggest that N-CAM-dependent neurite outgrowth and adhesion may be fundamentally different, and this can be explained by N-CAM-dependent activation of second-messenger pathways, rather than adhesion per se, underlying the neurite growth response (see below).

Table 4. PSA Specifically Promotes N-CAM-dependent Neurite Outgrowth

	% Inhibition of response		
Neuronal source	N-CAM	N-cadherin	L1
RGC (E6)	35–70	NE	n.d.
Hippocampus (E17)	~80	NE	n.d.
Cerebellum (PND4)	70–80	n.d.	NE

The effect of removal of PSA from neuronal N-CAM or CAM-dependent neurite outgrowth. Neurons were from the developing retina, hippocampus, and cerebellum at the time of greatest responsiveness to N-CAM (see Table 3). The neurons were then cultured on monolayers of control 3T3 cells and 3T3 cells expressing N-CAM, N-cadherin, and L1 (as in Table 2) in the presence and absence of a maximally active concentration of endoneuraminidase-N, which cleaves PSA from N-CAM. The results show the effects of this in terms of percentage inhibition of the respective CAM responses. (n.d.) Not determined; (NE) no significant effect. For details see text.

Although the above data indicate an important role for PSA in modulating N-CAM's ability to promote axonal growth, it is clear that removal of PSA does not totally inhibit this function (Doherty et al. 1990a). In addition, whereas N-CAM-dependent neurite outgrowth from chick retinal ganglion cells is down-regulated between embryonic day 6 (E6) and E9, PSA expression in the fiber tract is down-regulated between E10 and E14 (Schlosshauer et al. 1984) Similarly, the differences between PSA at E17 and postnatal day 4 (PND4) in the developing hippocampus are not substantial (Doherty et al. 1992c), and we cannot detect any obvious changes in PSA in the rat cerebellum between PND6 and PND8 (P. Doherty and S.V. Ashton, unpubl.). Because loss of neuronal responsiveness is not correlated with a loss of N-CAM from the growth cone (Doherty et al. 1990a), factors other than this and posttranslational processing are likely to contribute to this apparently general phenomenon.

N-CAM Isoforms and Neurite Outgrowth

The initial studies on N-CAM isoforms and neurite outgrowth suggested that if transfected N-CAM was linked to the membrane of the cellular substratum, the major determinant of neurite growth was the level rather than the type of N-CAM isoform (Doherty et al. 1990b). However, these studies compared the function of secreted, GPI-anchored (with and without the MSD1 domain, see above), and 140-kD transmembrane isoforms. More recently, we have found that modification of the cytoplasmic domain of the 140-kD isoform by use of exon 18 (creating the 180-kD isoform) substantially reduces its ability to promote neurite outgrowth following transfection and expression in 3T3 cells. This has been observed both with rat cerebellar granule cells as the probe neurons (Doherty et al. 1992a) and also with PC12 cells (P. Doherty, unpubl.). This was also shown to be a qualitative difference in N-CAM function, since it was observed with clones expressing both higher and lower levels of N-CAM

180-kD isoform as compared to the 140-kD positive control. Since the extracellular structures (as presented to the probe neurons) are identical between these isoforms, we conclude that the cytoplasmic domain of CAMs, even in the cellular substratum, can influence their function as neurite growth-promoting molecules. One possibility is that a reduced lateral mobility of N-CAM 180 in the cellular substratum (see Pollerberg et al. 1987) may also affect N-CAM's lateral mobility in the neuron (following *trans*-homophilic binding), and this may have direct consequences on the activation of

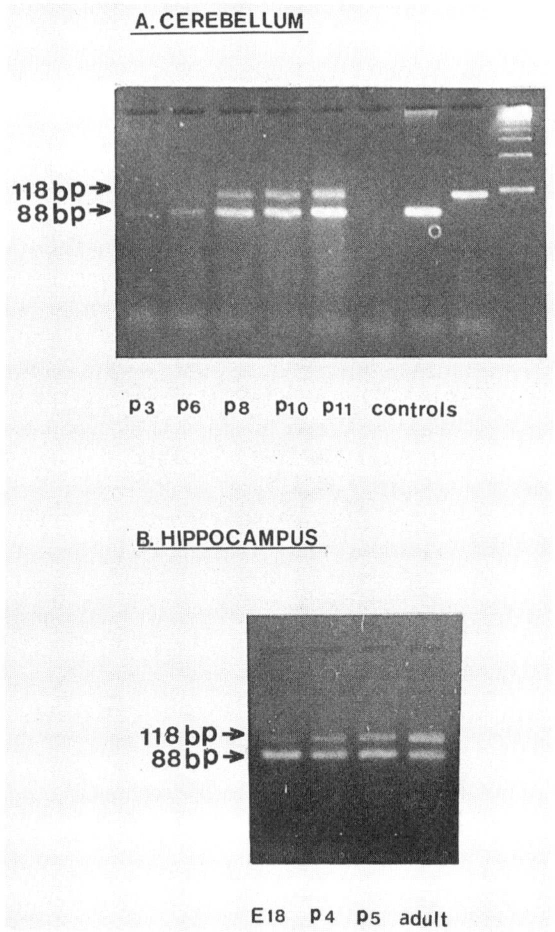

Figure 1. Developmental expression of the VASE exon in rat brain cerebellum and hippocampus. RNA was isolated from these regions at embryonic, postnatal, and adult ages and subjected to RT-PCR using the primer set outlined by Small and Akeson (1990). (*A*) Result of such an analysis in the cerebellum from postnatal day 3 to 11 (p3-p11). The control reactions were run with cDNA clones that either lacked or contained the VASE exon (Doherty et al. 1992b). These produced either an 88-bp band (lacking VASE) or a 118-bp band (including VASE). At the p3 and p6 timepoints, the main N-CAM transcripts are essentially lacking in VASE (88 bp), but at p8–p11, the usage of the VASE exon increases dramatically. (*B*) Result of a similar experiment in the hippocampus at embryonic day 18 (E18) to postnatal day 5 (p5) and adult brain. The results show that at E18 the major N-CAM transcripts present are low in VASE, but as the tissue develops, the usage of this exon increases dramatically such that the p5 samples are very near to the adult levels of about 50%.

the second-messenger pathways involved in CAM-dependent neurite outgrowth (see below).

Expression of N-CAM isoforms containing the product of the VASE exon show a reciprocal pattern of expression to N-CAM isoforms that contain the highest levels of PSA (Small and Akeson 1990). This observation suggests that VASE may have an opposite effect to PSA and may either promote N-CAM-dependent stability and/or inhibit N-CAM-dependent plasticity. We have directly tested this by again culturing cerebellar neurons and PC12 cells on monolayers of a number of independent clones of transfected 3T3 cells expressing overlapping levels of the N-CAM 140-kD isoform that differ only in the 10-amino-acid product of the VASE exon (spliced into the fourth Ig domain, see above). The results are clear and unequivocal; they demonstrate that use of VASE inhibits N-CAM's ability to promote neurite outgrowth without impairing its ability to support adhesion per se (Doherty et al. 1992b).

Recent evidence using the polymerase chain reaction suggests that the use of the VASE exon is the trigger that down-regulates neuronal responsiveness to N-CAM during normal development (Walsh et al. 1992). In the developing cerebellum, there is a rapid and dramatic increase in the ratio of N-CAM transcripts containing the product of the VASE exon over the PND6–PND8 period (see Fig. 1). There is also a similar change in the hippocampus over the E18–PND4 period. We would suggest that whereas the use of VASE might be the initial trigger, longer-term changes in N-CAM's PSA content could reinforce a change in N-CAM's function from a molecule that promotes plasticity (e.g., cell migration and axonal growth) to one that promotes and/or maintains stable connections between cells. In any case, it seems highly likely that two independent but complementary mechanisms have evolved to account for this bifunctionality. The demonstration of PSA in *Drosophila* (Roth et al. 1992) and alternative splicing of the N-CAM-related fasciclin II in insects (Grenningloh et al. 1991), and apCAM in *Aplysia* (Mayford et al. 1992) suggests an early evolution and subsequent conservation of these functions.

Activation of a CAM-specific Second Messenger Pathway in Neurons Accounts for Neurite Outgrowth but Not Adhesion

There is no direct correlation between N-CAM's ability to support adhesion and its ability to promote neurite outgrowth (see above); this has led us to suggest that recognition and signal transduction may be more important than adhesion per se for this response (Doherty et al. 1990a; see also Gundersen 1987). Studies, primarily on the giant synapses of *Aplysia* neurons, have shown that different neurotransmitters can promote or inhibit growth cone motility, and this can be correlated with their ability to modulate the level of intracellular calcium in the growth cone (for review, see Kater and Mills 1991). The binding of antibodies to

N-CAM and L1 on the surface of PC12 cells has been reported to stimulate a protein-kinase-C-dependent opening of L-type calcium channels (Schuch et al. 1989); however, no morphological consequences of this triggering were reported. We have tested a variety of pharmacological reagents for their ability to block CAM-dependent neurite outgrowth; these results are summarized in Table 5. In brief, antagonists of L- and N-type calcium channels can inhibit CAM (N-CAM, N-cadherin, and L1)-dependent neurite outgrowth from primary neurons and from PC12 cells. It is particularly evident from the L1 study (Williams et al. 1992) that L- or N-type antagonists can substantially inhibit the response on their own. However, similar results were also observed in a limited number of experiments for N-CAM/N-cadherin-dependent neurite outgrowth from the same neurons. The likeliest explanation of the current data is that a threshold level of calcium influx or cycling is required for the response, and that in some instances, flux through both types of calcium channel is required to reach this value. Thus, regulation at the level of calcium influx could contribute to the previously reported threshold effect of N-CAM on neurite outgrowth (Doherty et al. 1991b). A greater than maximal activation of a single pathway (i.e., calcium channel activation) would also readily explain the redundancy of individual CAMs apparent in some antibody perturbation of neurite outgrowth studies (see Bixby et al. 1987).

Is the calcium channel a true effector molecule involved in the transduction of the CAM recognition signal into a cellular response? One way to test this is to directly activate the pathway and determine if this can mimic the CAM response. We have found recently that potassium depolarization of PC12 cells can fully mimic CAM-dependent neurite outgrowth from PC12 cells and that this response can also be inhibited by N- and L-type calcium channel agonists (see Table 5) (Saffell et al. 1992). A similar response could also be induced by the calcium channel antagonist Bay K 8644. The clear implication of this study is that activation of this pathway following CAM binding may be more important for the neurite outgrowth response than adhesion per se.

Activation of calcium channels appears to be relatively exclusive to the CAM pathway. For example, neurite outgrowth that is dependent on integrin receptor function and nerve growth factor, or increased intracellular cAMP, is not affected in any way by the calcium channel antagonists (see Table 5). The precise mechanism involved in the activation of this CAM-specific pathway for neurite outgrowth remains to be determined. However, there are several important observations. First, protein kinase C does not appear to be central to the pathway, since activation of the enzyme does not mimic the response (Saffell et al. 1992) and inhibitors of the enzyme do not block the response (Doherty et al. 1991a). In addition, from the PC12 cell studies, it would appear that protein kinase A is not involved. These observations suggest that phosphorylation of the calcium channel is not a key signal in activation of the CAM-specific pathway. N-CAM-, N-cadherin-, and L1-dependent neurite outgrowth from primary neurons and PC12 cells can, however, be fully inhibited by pertussis toxin (see Table 5). This agent operates by ribosylating and inhibiting the function of a variety of heterotrimeric G proteins of the G_o/G_i families. The effects of pertussis toxin were again very specific to the CAM pathway (see Table 5) and the fact that the K^+ depolarization response is not affected places the G protein upstream of the calcium channels. To date there has been no report of a pertussis-toxin-sensitive G protein that can directly activate calcium channels; however, G_o may directly bind to and tonically inhibit calcium channels (Hescheler et al. 1987; Kasai 1991). Thus, the possibility that activation of the

Table 5. Effects of Calcium Channel Antagonists and Pertussis Toxin on Neurite Outgrowth Response

Response	Maximal inhibition by: L-type	N-type	both	pertussis toxin
PC12 cells (48 hr)				
control 3T3	no effect	no effect	no effect	no effect
N-CAM/N-CAD	55 ± 2% (10)	48 ± 4% (5)	92 ± 4% (5)	96 ± 1% (4)
K^+ depolarization	67 ± 7% (3)	28 ± 9% (3)	70 ± 8% (3)	no effect
NGF/collagen	no effect	no effect	no effect	no effect
↑cAMP/3T3	no effect	no effect	no effect	no effect
Primary neurons				
control 3T3[a,b]	no effect	no effect	no effect	no effect
N-CAM/N-CAD[a,b]	60–75% (4)	50–80% (4)	90–95% (3)	>95% (3)
L1[c]	91 ± 5% (6)	81 ± 8% (5)	99 ± 4% (5)	92 ± 8% (4)

Summary of the effect of calcium channel antagonists and pertussis toxin on the neurite outgrowth response induced by transfected CAMs (as indicated) or by K^+ depolarization (40 mM), treatment with NGF on a collagen substratum, or treatment with an agent that increases intracellular cAMP (cholera toxin, 1 ng/ml) for PC12 cells cultured on a 3T3 monolayer. L-type channels were blocked with either diltiazem, verapamil, or nifedipine (10 μM) and N-type by ω-conotoxin (0.25 μM). Pertussis toxin was used at 0.5–1 μg/ml. Studies in primary neurons were with [a]rat cerebellar granule cells or [b]rat hippocampal neurons. For the [c]L1 response, results with rat cerebellar granule cells and primed PC12 cells were identical and therefore pooled. For details see text.

CAM-specific pathway involves removal of a tonic inhibition should also be considered. In this context, it may be of some significance that G_o function can also be modulated by GAP-43 (Strittmater et al. 1990).

The precise mechanism whereby three independent CAMs can activate a common pathway remains to be determined. However, based on what is known about adhesion-dependent intercellular signaling in other cells, and in particular lymphocytes (for review, see Singer 1992), a three-stage model can be proposed. It is envisaged that initial apposition between elements of the growth cone and substratum would result in the clustering of CAMs via diffusional recruitment into the initial site of cell-cell contact. Since neurite outgrowth involves the making and breaking of contacts, this initial step would be dynamic and reversible. Rather than resulting in the generation of a stable adhesive force, this step could lead to the production of second messengers via activation of the CAM-specific pathway for neurite outgrowth. The fact that at least three CAMs can activate the same pathway raises the possibility of an adapter molecule that can interact with several CAMs and also the effector molecule(s). Coclustering of the adapter and/or effector may then account for a transient second-messenger signal arising in the growth cone.

In contrast to neurite outgrowth, cell adhesion is associated with the formation of stable adhesion plaques. This most probably involves linkage of the CAMs, via their cytoplasmic domains, to the underlying cytoskeleton (see, e.g., Nagafuchi and Takeichi 1988; Kintner 1992). Thus, we suggest that unique forms of individual CAMs may be adapted for adhesion or neurite growth as a consequence of their ability to stably, as opposed to transiently, cluster CAMs. We suggest that it is only the repeated transient clustering that results in prolonged activation of second-messenger pathways in neurons. In this context, PSA on neuronal N-CAM may favor neurite outgrowth at the expense of adhesion by inhibiting stable cluster formation; in contrast, use of the product of the VASE exon or exon 18 may have the opposite effect by promoting stable cluster formation.

CONCLUSIONS

The results of the present study have shown the potential of a transfection-based strategy for approaching the structure and function of CAMs. Using this approach, we have demonstrated that transfected N-CAM, N-cadherin, and L1 can promote neurite outgrowth from a wide variety of neurons. Of the three CAMs, neuronal responsiveness to N-CAM was subjected to the most dynamic regulation. In all three tested populations of neurons, N-CAM responsiveness was down-regulated with a time course suggesting that this phenomenon may be controlled by target innervation. The primary trigger for this may involve an increase in the proportion of neuronal N-CAM transcripts containing the product of the VASE exon; however, this change may also be reinforced by loss of PSA from neuronal N-CAM. In vitro studies support this by demonstrating that removal of PSA inhibits N-CAM-dependent neurite outgrowth and that N-CAM isoforms containing the product of the VASE exon are relatively poor substrates for neurite outgrowth. Evidence has also been obtained that all of the CAMs promote neurite outgrowth via the activation of a common CAM-specific second-messenger pathway in neurons. This may require the transient clustering of CAMs and effector molecules in the neuronal growth cone, and this may also be dependent on CAM motility in the substratum. In support of this, N-CAM isoforms that differ only in the size of their cytoplasmic domain (as a consequence of natural alternative splicing of the N-CAM gene) differ considerably in their ability to promote neurite outgrowth. This most likely relates to N-CAM's lateral diffusion properties in the cellular substratum. Overall, the primary purpose of PSA and isoform switching may be to change the function of N-CAM from a molecule that promotes plasticity (via activation of second-messenger pathways) to one that promotes stability (through direct adhesive interaction).

REFERENCES

Acheson, A., J.L. Sunshine, and U. Rutishauser. 1991. N-CAM polysialic acid can regulate both cell-cell and cell substrate interactions. *J. Cell Biol.* **114:** 143.

Bixby, J.L. and W.A. Harris. 1991. Molecular mechanisms of axon growth and guidance. *Annu. Rev. Cell. Biol.* **7:** 17.

Bixby, J.L., J. Lilien, and L.F. Reichardt. 1988. Identification of the major proteins that promote neuronal process outgrowth on Schwann cells in vitro. *J. Cell Biol.* **107:** 353.

Bixby, J.L., R.S. Pratt, J. Lilien, and L.F. Reichardt. 1987. Neurite outgrowth on muscle cell surfaces involves extracellular matrix receptors as well as Ca^{2+} dependent and independent cell adhesion molecules. *Proc. Natl. Acad. Sci.* **84:** 2555.

Blaschuk, O., R. Sullivan, S. David, and Y. Pouliot. 1990. Identification of a cadherin cell adhesion recognition sequence. *Dev. Biol.* **139:** 227.

Doherty, P. and F.S. Walsh. 1989. Neurite guidance molecules. *Curr. Opin. Cell Biol.* **1:** 1102.

Doherty, P., J. Cohen, and F.S. Walsh. 1990a. Neurite outgrowth in response to transfected NCAM changes during development and is modulated by polysialic acid. *Neuron* **5:** 209.

Doherty,P., S.V. Ashton, S.E. Moore, and F.S. Walsh. 1991a. Morphoregulatory activities of NCAM and N-cadherin can be accounted for by G-protein dependent activation of L- and N-type neuronal calcium channels. *Cell* **67:** 21.

Doherty, P., G. Rimon, D.A. Mann, and F.S. Walsh. 1992a. Alternative splicing of the cytoplasmic domain of neural cell adhesion molecule alters its ability to act as a substrate for neurite outgrowth. *J. Neurochem.* **58:** 2338.

Doherty, P., C.E.C.K. Moolenaar, S.V. Ashton, R.J.A.M. Michalides, and F.S. Walsh. 1992b. Use of the VASE exon down regulates the neurite growth promoting activity of NCAM140. *Nature* **356:** 791.

Doherty, P., L.H. Rowett, S.E. Moore, D.A. Mann, and F.S. Walsh. 1991b. Neurite outgrowth in response to transfected NCAM and N-cadherin reveals fundamental differences in neuronal responsiveness to CAMs. *Neuron* **6:** 247.

Doherty, P., S.D. Skaper, S.E. Moore, A. Leon, and F.S. Walsh. 1992c. A developmentally regulated switch in neu-

ronal responsiveness to NCAM and N-cadherin in the rat hippocampus. *Development* **115:** 885.
Doherty, P., M. Fruns, P. Seaton, G. Dickson, C.H. Barton, T.A. Sears, and F.S. Walsh. 1990b. A threshold effect of the major isoforms of NCAM on neurite outgrowth. *Nature* **343:** 464.
Doherty, P., C.H. Barton, G. Dickson, P. Seaton, L.H. Rowett, S.E. Moore, H.J. Gower, and F.S. Walsh. 1989. Neuronal process outgrowth of human sensory neurons on monolayers of cells transfected with cDNAs for five human N-CAM isoforms. *J. Cell. Biol.* **109:** 789.
Edelman, G.M. 1985. Cell adhesion and the molecular process of morphogenesis. *Annu. Rev. Biochem.* **54:** 135.
Furley, A.J., S.B. Morton, D. Mando, D. Karagogeous, J. Dodd, and T.M. Jessell. 1990. The axonal glycoprotein TAG1 is an immunoglobulin superfamily member with neurite outgrowth promoting activity. *Cell* **61:** 157.
Grenningloh, G., E.J. Rehm, and C.S. Goodman. 1991. Genetic analysis of growth cone guidance in *Drosophila*: Fasciclin II functions as a neuronal recognition molecule. *Cell* **67:** 45.
Gundersen, R.W. 1987. Response of sensory neurites and growth cones to patterned substrata of laminin and fibronectin in vitro. *Dev. Biol.* **121:** 423.
Heffner, C.D., A.G.S. Lumsden, and D.D.M. O'Leary. 1990. Target control of collateral extension and direction axon growth in the mammalian brain. *Science* **247:** 217.
Hescheler, J., W. Rosenthal, W. Trautwein, and G. Schultz. 1987. The GTP-binding protein, Go, regulates neuronal calcium channels. *Nature* **325:** 445.
Hlavin, M.L. and V. Lemmon. 1991. Molecular structure and functional testing of Human L1CAM: An interspecies comparison. *Genomics* **11:** 416.
Hoffman, S. and G.M. Edelman. 1983. Kinetics of neuronal binding by E and A forms of neural cell adhesion molecule. *Proc. Natl. Acad. Sci.* **80:** 5762.
Hynes, R.O. 1992. Integrins; Versatility, modulation, and signalling in cell adhesion. *Cell* **69:** 11.
Kasai, H. 1991. Tonic inhibition and rebound facilitation of a neuronal calcium channel by a GTP-binding protein. *Proc. Natl. Acad. Sci.* **88:** 8855.
Kater, S.B. and P.B. Guthrie. 1990. Neuronal growth cone as an integrator of complex environmental information. *Cold Spring Harbor Symp. Quant. Biol.* **55:** 359.
Kater, S.B. and L.R. Mills. 1991. Regulation of growth cone behaviour by calcium. *J. Neurosci.* **11:** 891.
Keynes, R.J., A.R. Johnson, and G.M.W. Cook. 1991. Contact inhibition of growth cone motility during neural development and regeneration. *Semin. Neurosci.* **3:** 321.
Kintner, C. 1992. Regulation of embryonic cell adhesion by the cadherin cytoplasmic domain. *Cell* **69:** 225.
Kuhn, T.B., E.T. Stoeckli, M.A. Condrau, F.G. Rathjen, and P. Sonderegger. 1991. Neurite outgrowth on immobilised Axonin-1 is mediated by a heterophilic interaction with L1 (G4). *J. Cell. Biol.* **115:** 1113.
Lumsden, A. and J. Cohen. 1991. Axon guidance in the vertebrate central nervous system. *Curr. Opin. Genet. Dev.* **1:** 230.
Lumsden, A.G.S. and A.M. Davies. 1986. Chemotropic effects of specific target epithelium in the developing mammalian nervous system. *Nature* **323:** 538.
Mayford, M., A. Brazilai, F. Keller, S. Schacher, and E.R. Kandel. 1992. Modulation of an NCAM-related adhesion molecule with long term synaptic plasticity in aplysia. *Science* **256:** 638.
Miura, M., M. Kobayashi, H. Asou, and K. Uyemura. 1991. Molecular cloning of cDNA encoding the rat neural cell adhesion molecule L1. *FEBS Lett.* **289:** 91.
Moos, M., R. Tacke, H. Scherer, D. Teplow, K. Friih, and M. Schachner. 1988. Neural adhesion molecule L1 as a member of the immunoglobulin superfamily with binding domains similar to fibronectin. *Nature* **334:** 701.
Nagafuchi, A. and M. Takeichi. 1988. Cell binding function of E-cadherin is regulated by the cytoplasmic domain. *EMBO J.* **7:** 3679.
Neugebauer, K.M., K.J. Tomaselli, J. Lilien, and L.F. Reichardt. 1988. N- cadherin, N-CAM and integrins promote retinal neurite outgrowth on astrocytes in vitro. *J. Cell. Biol.* **107:** 1177.
Nose, A., K. Tsuji, and M. Takeichi. 1990. Location of specificity determining sites in cadherin cell adhesion molecules. *Cell* **61:** 147.
Pollerberg, E., M. Schachner, and J. Daroust. 1986. Differentiation-state dependent surface mobility of two forms of the neural cell adhesion molecule. *Nature* **324:** 462.
Pollerberg, E.G., R. Sadoul, C. Goridis, and M. Schachner. 1985. Selective expression of the 180-KD component of the neural cell adhesion molecule N-CAM during development. *J. Cell. Biol.* **101:** 1921.
Pollerberg, G.E., K. Burridge, K.E. Krebs, S.R. Goodman, and M. Schachner. The 180-kD component of the neural cell adhesion molecule N-CAM is involved in cell-cell contacts and cytoskeleton-membrane interactions. *Cell Tissue Res.* **250:** 227.
Ranscht, B. 1991. Cadherin cell adhesion molecules in vertebrate neural development. *Semin. Neurosci.* **3:** 285.
Rathjen, F.G. and T.M. Jessell. 1991. Glycoproteins that regulate the growth and guidance of vertebrate axons: Domains and dynamics of the immunoglobulin/fibronectin type III subfamily. *Semin. Neurosci.* **3:** 297.
Reichardt, L.F. and K.J. Tomaselli. 1991. Extracellular matrix molecules and their receptors: Functions in neural development. *Annu. Rev. Neurosci.* **14:** 531.
Reichardt, L.F., B. Bossy, S. Carbonetto, I. DeCurtis, C. Emmett, D.E. Hall, M.J. Ignatius, F. Lefcort, E. Napolitano, T. Large, K.M. Nuegebauer, and K.J. Tomaselli. 1990. Neuronal receptors that regulate axon growth. *Cold Spring Harbor Symp. Quant. Biol. 55:* 341.
Reid, R.A. and J.J. Hemperly. 1992. Variants of human L1 cell adhesion molecule arise through alternate splicing of RNA. *J. Mol. Neurosci.* **3:** 127.
Roth, J., A. Kempf, G. Reuter, R. Schauer, and W. Gehring. 1992. Occurrence of sialic acid in *Drosophila melanogaster*. *Science* **256:** 673.
Rutishauser, U. 1991. Pleiotropic biological effects of the neural cell adhesion molecule (NCAM). *Semin. Neurosci.* **3:** 265.
Rutishauser, U., A. Acheson, A.K. Hall, D.M. Mann, and J. Sunshine. 1988. The neural cell adhesion molecule (N-CAM) as a regulator of cell-cell interactions. *Science* **240:** 53.
Saffell, J., F.S. Walsh, and P. Doherty. 1992. Direct activation of second messenger pathways mimic cell adhesion molecule-dependent neurite outgrowth. *J. Cell. Biol.* **118:** 663.
Sanes, J.R., D.D. Hunter, T.L. Green, and J.P. Merlie. 1990. S-Laminin. *Cold Spring Harbor Symp. Quant. Biol.* **55:** 419.
Schlosshauer, B., U. Schwartz, and U. Rutishauser. 1984. Topological distribution of different forms of neural cell adhesion molecule in the developing chick visual system. *Nature* **310:** 141.
Schuch, U., M.J. Lohse, and M. Schachner. 1989. Neural cell adhesion molecules influence second messenger systems. *Neuron* **3:** 13.
Seilheimer, B. and M. Schachner. 1988. Studies of adhesion molecules mediating interactions between cells of peripheral nervous system indicate a major role for L1 in mediating sensory neuron growth on Schwann cells in culture. *J. Cell Biol.* **107:** 341.
Simon, H., S. Klinz, T. Fahrig, and M. Schachner. 1991. Molecular association of the neural adhesion molecules L1 and N-CAM in the surface membrane of neuroblastoma cells is shown by chemical cross-linking. *Eur. J. Neurosci.* **3:** 634.
Singer, S.J. 1992. Intercellular communication and cell-cell adhesion. *Science* **255:** 1671.

Small, S.J. and R. Akeson. 1990. Expression of the unique NCAM VASE exon is independently regulated in distinct tissues during development. *J. Cell Biol.* **111:** 2089.

Takeichi, M. 1991. Cadherin cell adhesion receptors as a morphogenetic regulator. *Science* **251:** 1451.

Strittmatter, S.M., D. Valenzuela, T.E. Kennedy, E.J. Neer, and M.C. Fishman. 1990. Go is a major growth cone protein subject to regulation by GAP-43. *Nature* **344:** 836.

Walsh, F.S. and P. Doherty. 1991. Structure and function of the gene for neural cell adhesion molecule. *Semin. Neurosci.* **3:** 271.

Walsh, F.S., J. Furness, S.E. Moore, S.W. Ashton, and P. Doherty. 1992. Use of the NCAM VASE exon by neurons is associated with a specific down regulation of NCAM dependent neurite outgrowth in the developing cerebellum and hippocampus. *J. Neurochem.* (in press).

Williams, E.J., P. Doherty, G. Turner, R.A. Reid, J.J. Hemperly, and F.S. Walsh. 1992. Calcium influx into neurons can solely account for cell contact dependent neurite outgrowth stimulated by transfected L1. *J. Cell Biol.* (in press).

Zuellig, R.A., C. Rader, A. Schroeder, M.B. Kalousek, F. VonBohlen und Halbach, T. Osterwalder, C. Inan, E.T. Stoeckli, H. Affolter, A. Fritz, E. Haften, and P. Sonderegger. 1992. The axonally secreted cell adhesion molecule, axonin-1, primary structure, immunoglobulin-like and fibronectin- type-III-like domains and glycosyl-phosphatidylinositol anchorage. *Eur. J. Biochem.* **204:** 453.

Development of Neuromuscular Specificity in *Drosophila*

A. NOSE, D. VAN VACTOR, V. AULD, AND C.S. GOODMAN
Howard Hughes Medical Institute, Department of Molecular and Cell Biology, University of California, Berkeley, California 94720

The ability of motoneuron growth cones to find and recognize their correct muscles has been a model system for studies on the mechanisms of target recognition in both vertebrates and invertebrates. For example, in both chick (see, e.g., Tosney and Landmesser 1985) and zebrafish (see, e.g., Eisen et al. 1986) embryos, motoneuron growth cones make stereotyped pathway choices as they extend toward the regions where their appropriate target muscles arise. Similarly, in both grasshopper (Ball et al. 1985b) and *Drosophila* (see, e.g., Sink and Whitington 1990, 1991a; Halpern et al. 1991) embryos, identified motoneuron growth cones extend toward and innervate the appropriate target muscles in a highly specific fashion.

In vertebrates, much experimental evidence suggests that motoneuron growth cones make specific pathway choices (Tosney and Landmesser 1985). In addition, target muscles or other nearby mesodermal cells appear to express attractive guidance cues. For example, the growth cones of chick motoneurons normally grow toward the regions of their appropriate target muscles. If displaced experimentally, these growth cones often take circuitous routes to reach their same targets (Landmesser 1980; Lance-Jones and Landmesser 1981). Identified motoneurons in zebrafish establish cell-specific terminal fields primarily by directed outgrowth of branches on appropriate muscle fibers (Eisen et al. 1989; Liu and Westerfield 1990; Pike and Eisen 1990). In the *spt-1* mutation that eliminates trunk muscles, the pattern of motoneuron axon branches is seriously disrupted (Eisen and Pike 1991).

Muscle Pioneers Guide Development of Muscles and Motoneurons

Experimental analysis in insects suggests that, just as in vertebrates, so too in these simpler model systems, neuromuscular specificity involves a combination of pathway and target recognition. The advantage of these organisms is their relative simplicity of motoneurons and muscles, their segmentally repeated pattern, and in the case of *Drosophila*, the opportunity for genetic analysis. Thus, the goal of such studies is to understand neural specificity at the level of molecular mechanisms.

In the grasshopper embryo, skeletal muscles develop from muscle pioneers (Ho et al. 1983; Ball et al. 1985a; Ball and Goodman 1985a,b). These muscle pioneers appear early in development, enlarge relative to the surrounding mesodermal cells from which they arise, and extend growth-cone-like processes that find and establish the insertion sites of the muscles they prefigure. Smaller mesodermal cells cluster around and fuse with the muscle pioneers, which become multinucleate and later differentiate into muscle fibers. Motoneuron growth cones extend into the periphery before the differentiation of mature muscle fibers. They come in contact with several different muscle pioneers and display affinities for specific muscle pioneers in a highly selective fashion (Ball et al. 1985b). For example, in the grasshopper, when a specific muscle pioneer (for limb muscle 133a) is ablated prior to innervation, the motoneuron growth cone (Df) that would normally have innervated that muscle instead continues to extend further distally (Fig. 1) (Ball et al. 1985b).

In *Drosophila*, homologs to the muscle pioneers have been described (Leiss et al. 1988; Bate 1990; Dohrmann et al. 1990). These muscle precursors are thought to arise after fusion of muscle founder cells with surrounding myoblasts (Bate 1990). Just as in the grasshopper, these *Drosophila* muscle pioneers prefigure the larval muscles by extending growth-cone-like processes to the sites of muscle insertion (Bate 1990). They also serve as the host for the fusion of surrounding mesodermal cells before differentiating into mature muscle fibers and appear to provide guidance cues for motoneuron growth cones (Fig. 2).

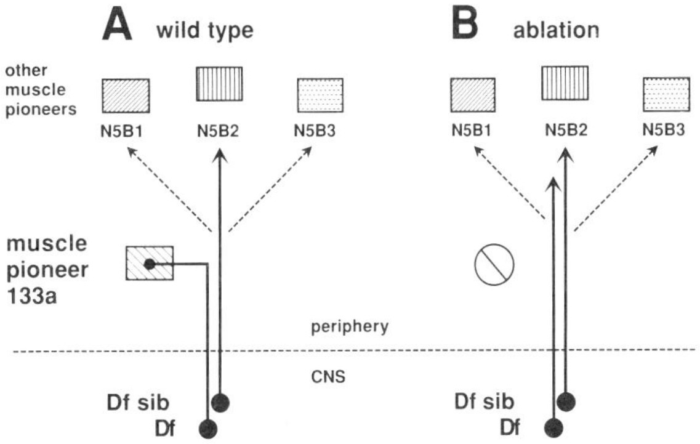

Figure 1. Experimental analysis of neuromuscular specificity in the grasshopper. In the grasshopper embryo, muscles develop from muscle pioneers (Ho et al. 1983). (*A*) Motoneuron growth cones come in contact with and display affinities for specific muscle pioneers in a highly selective fashion (Ball et al. 1985b). (*B*) When a specific muscle pioneer (for limb muscle 133a) is ablated prior to innervation, the motoneuron growth cone (Df) that would normally have innervated that muscle instead continues to extend further distally and follows the path taken by the growth cone of its sibling motoneuron (Df sib) along peripheral nerve 5B2 (Ball et al. 1985b).

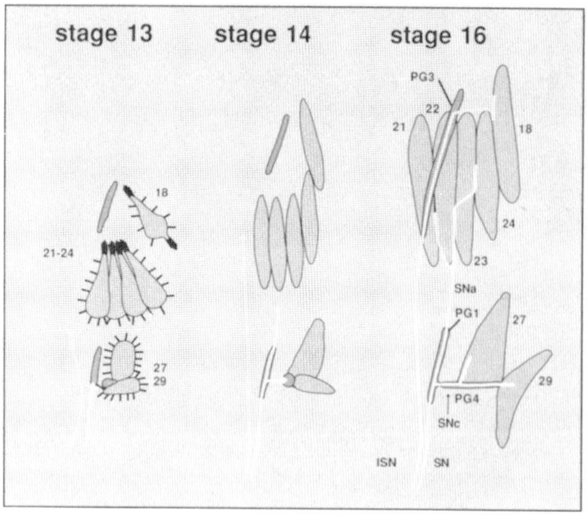

Figure 2. Development of muscles and motoneuron innervation in *Drosophila*. In the *Drosophila* embryo, muscle pioneers prefigure the larval muscles by extending growth-cone-like processes to the sites of muscle insertion (*left* panel) (Bate 1990). They also appear to provide guidance cues for motoneuron growth cones. Schematic timeline showing the development of a subset of muscles (21–24, 18, 27, and 29) from muscle pioneers, the outgrowth of the subset of motoneurons (white lines show axons and arrows show growth cones) that innervate these muscles (thick white lines in right panel show synapses), and the development of several peripheral glia (PG1, PG3, and PG4) along these motor pathways. The three panels show these events at stages 13–16 as revealed by their expression of the connectin protein (see text and Fig. 7). For simplicity, the only connectin-expressing muscle not shown is muscle 5. See text for details. (ISN) Intersegmental nerve; (SN) segmental nerve; (a,c) branches of the SN. (Adapted from Nose et al. 1992.)

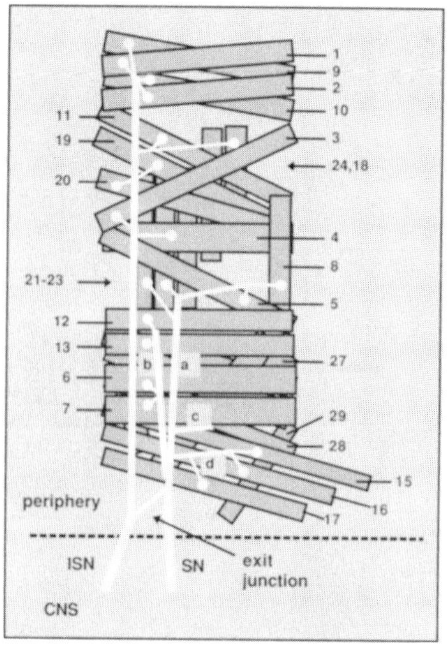

Figure 3. Pattern of muscles and peripheral nerves in *Drosophila*. Schematic diagram showing the pattern of 30 muscles in each abdominal hemisegment of a late *Drosophila* embryo (dorsal is up; ventral is down; numbers denote specific muscle fibers. Motoneurons exit the CNS in two peripheral nerve roots: the ISN or the SN. Immediately after exiting the CNS, a group of motoneurons (including RP1 and RP3) switch from the ISN to the SN at the exit junction. The ISN carries motoneurons (e.g., aCC, the three Us, RP2) to the most dorsal muscles (e.g., 1,2,9, and 10). The SN has four branches (a,b,c, and d) which carry motoneurons to more ventral and lateral muscles. (Adapted from Nose et al. 1992.)

Neuromuscular Specificity in *Drosophila*

Neuromuscular specificity in *Drosophila* is particularly well suited for studies on the molecular mechanisms of pathway and target recognition. The body wall musculature of *Drosophila* embryos and larvae is relatively simple and is arranged in a stereotyped array consisting of 30 individually identified muscle fibers in each abdominal hemisegment (Fig. 3). Most if not all of these muscle fibers are innervated by one or only a few motoneurons. For example, the RP1 motoneuron innervates only muscle fiber 13, and the RP3 motoneuron innervates muscle fibers 6 and 7 (Fig. 4A) (Sink and Whitington 1990,1991a; Halpern et al. 1991). The RP1 and RP3 growth cones initially contact and extend processes over a number of inappropriate muscles as well as contacting their correct targets. The processes to the inappropriate muscles are later withdrawn, generating the mature pattern of neuromuscular innervation (Sink and Whitington 1991a). The synapses from these motoneurons onto their specific muscle fibers always form at defined locations (Johansen et al. 1989a,b; Halpern et al. 1991).

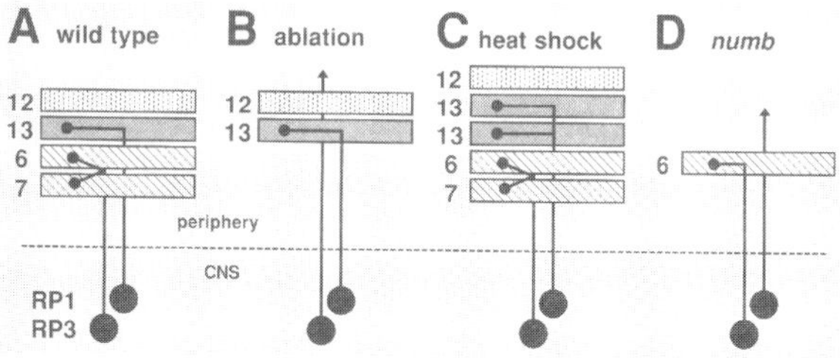

Figure 4. Experimental analysis of neuromuscular specificity in *Drosophila*. (*A*) The RP1 motoneuron normally innervates muscle fiber 13; the RP3 motoneuron normally innervates muscle fibers 6 and 7. (*B–D*) To test the specificity of motoneuron growth cones for specific muscle cell surfaces, a series of experiments were conducted that either duplicate or delete specific muscle fibers in *Drosophila*. (*B*) When muscle fibers 6 and 7 are surgically ablated, the RP3 growth cone arborizes over nontarget muscles in a variable and abnormal fashion (Sink and Whitington 1991b). (*C*) In specifically timed heat shock experiments, muscle fiber 13 can be duplicated; in these embryos, RP1, which now has two 13 targets, specifically innervates both 13 muscle fibers (Chiba and Keshishian 1991). (*D*) In *numb* mutant embryos, muscle fibers 12 and 13 are missing, and the RP1 growth cone extends abnormally (Chiba and Keshishian 1991).

Specific Glia Function in the Guidance of Motoneuron Growth Cones

In developing insect nervous systems, each neuromere has (on each side) two peripheral nerve roots in which motoneuron axons exit into the periphery and sensory axons enter into the central nervous system (CNS) (Figs. 2 and 3) (Bate 1982). One of these nerve roots forms near the boundary between adjacent neuromeres and is called the intersegmental nerve root (ISNR). The other nerve root forms near the middle of the neuromere and is called the segmental nerve root (SNR).

A pair of glia prefigure the ISNR (Figs. 5 and 6). The more medial of this pair of glia is located adjacent to where the first longitudinal pathways form; this conspicuous cell, called the segment boundary cell (SBC), was originally identified in the grasshopper (Bastiani and Goodman 1986) and later in *Drosophila* (Bastiani et al. 1985; Jacobs and Goodman 1989). In the grasshopper, the growth cones of the three Us and aCC, all motoneurons that innervate dorsal muscles, contact and extend out along the SBC. When the SBC is ablated, the aCC and U growth cones fail to exit the CNS, and the intersegmental nerve does not form (Fig. 5) (Bastiani and Goodman 1986). This experiment represents the best example in any organism of an identified glial cell being required for the formation of a specific axon pathway.

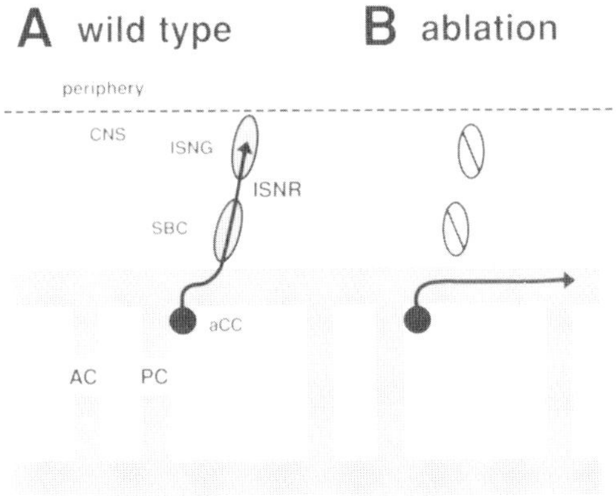

Figure 5. Role of glia during motoneuron pathfinding in the grasshopper. A pair of glia prefigure the intersegmental nerve root (ISNR) in the grasshopper embryo (the same is true in *Drosophila*; see Fig. 6). The more medial of this pair of glia is located adjacent to where the first longitudinal pathways form; this conspicuous cell is called the segment boundary cell (SBC) (Bastiani and Goodman 1986). (*A*) The growth cones of the three Us and aCC, all motoneurons that innervate dorsal muscles, contact and extend out along the SBC. Here aCC is shown extending out along the SBC and the next distal intersegmental nerve root glial cell (ISNG). (*B*) When the SBC and next ISNG are ablated, the aCC and U growth cones continue to extend posteriorly and the intersegmental nerve does not form (Bastiani and Goodman 1986).

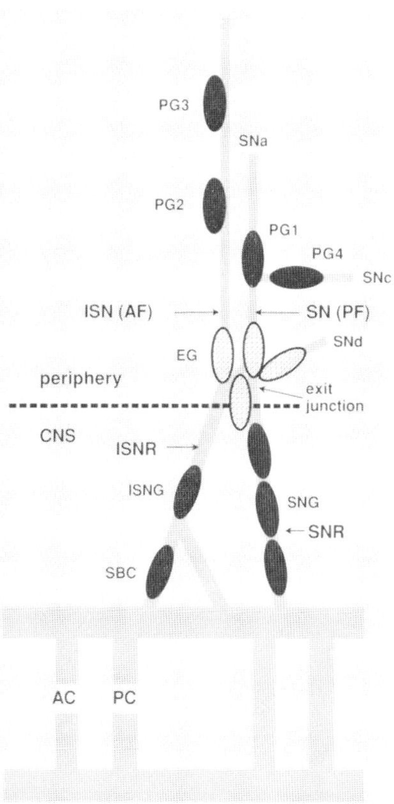

Figure 6. Pattern of glia along motor pathways in *Drosophila*. Schematic diagram showing the diversity and pattern of known glia along motor axon pathways in the *Drosophila* embryo. The figure shows the location of the intersegmental (ISG) and segmental (SG) nerve root glia along the intersegmental (ISNR) and segmental (SNR) nerve roots; the exit glia (EG) just outside the CNS, and the peripheral glia (PG) along the various branches of the intersegmental nerve (ISN; or anterior fascicle, AF) and segmental nerve (SN; or posterior fascicle, PF; branches SNa, SNc, and SNd) are shown. The more medial of the two intersegmental nerve root glia is the same as the segment boundary cell (SBC, see also Fig. 5). For further details, see text. (Adapted from Kläbt and Goodman 1991.)

Motoneuron Growth Cones Follow Specific Pathways in the Periphery

Motoneuron growth cones exit the CNS in either the ISNR or SNR. In *Drosophila*, of the roughly 34 motoneurons in each hemisegment, 9 that exit in the ISNR are well characterized (aCC, the 3 Us, RP1, RP2, RP3, RP4, and RP5; also in the ISNR are some of the axons of the 6 VUM neurons).

In the *Drosophila* embryo, the two major nerve roots initially come together just outside the CNS at a choice point containing four conspicuous glial cells (Fig. 6) (exit glia; Klämbt and Goodman 1991); the growth cones of motoneurons first exit the CNS and appear to extend along these glia. When these motoneuron pathways are initially formed, the exit glia sit just outside the CNS and where the major peripheral axon pathways form and where motor axons switch from one pathway to another. For example, the growth cones of the RP1, RP3, RP4, and RP5 motoneurons switch from the ISN to the SN in this exit junction and then form and extend in SN branch b (SNb) to innervate ventral muscles such as 6, 7, and 13. In total, there are five major peripheral pathways containing motoneuron axons: the intersegmental nerve (ISN or anterior fascicle) and the four major branches of the segmental nerve (posterior fascicle): SNa, SNb, SNc, and SNd. In general, dorsal muscles are innervated via the ISN, whereas groups of lateral and ventral muscles are innervated via one of the four branches of the SN.

The location of the exit glia and the complexity of the growth cone choices in this region lead to the suggestion that these glia may play a significant role in the patterning of the peripheral nerves just outside the CNS, and in the choices that motoneuron growth cones make of which pathway to follow. In this light, it is interesting to note that the exit glia display a remarkable pattern of molecular heterogeneity. Some enhancer trap lines label all four exit glia, some label only two, and others label only one (Auld et al. 1991; Klämbt and Goodman 1991).

Once the motoneuron growth cones leave the exit glia, they come into contact with other peripheral glia. Fredieu and Mahowald (1989) initially described some of these glia using a monoclonal antibody that labels three exit glia and three more distal peripheral glia. At present we can account for a total of four exit glia and at least four other peripheral glia (Fig. 6) (Klämbt and Goodman 1991; Nose et al. 1992).

In addition to glia and muscles, other cells are candidates to also be involved in the guidance of motoneuron growth cones. As motoneuron growth cones in the ISN first explore the periphery, they make dramatic contacts with cells of the tracheal system and later with afferent sensory axons (V. Hartenstein, pers. comm.; D. Van Vactor and C.S. Goodman, unpubl.). In addition, Bate and colleagues have described several twist-expressing mesodermal cells in more distal regions that associate with particular axon pathways (Bate et al. 1991). These, among other cell types, are candidates for playing instructive roles in the guidance of motoneurons to their target regions.

This interchange of motor axons and axon pathways just outside the *Drosophila* CNS at the location of the exit glia, and the more distal association of other glial cells and mesodermal cells at the locations of specific choice points, are reminiscent of the motoneuron plexus regions at the base of the developing limb buds in the chick embryo (Landmesser 1980; Lance-Jones and Landmesser 1981). In the plexus, chick motoneurons respond to positional information that routes them to their correct peripheral axon pathways, which lead to their correct target muscles. Therefore, it appears that in both invertebrates and vertebrates, nonmuscle cell types play a major role in motoneuron pathfinding.

Experimental Analysis of Neuromuscular Specificity

To test the specificity of motoneuron growth cones for specific muscle cell surfaces, a series of experiments were conducted that either duplicate or delete specific muscle fibers in *Drosophila* (Fig. 4B–D). For example, when muscle fibers 6 and 7 are surgically ablated, the RP3 growth cone arborizes over nontarget muscles in a variable and abnormal fashion (Sink and Whitington 1991b). Moreover, in *numb* mutant embryos where muscle fibers 12 and 13 are missing, the RP1 growth cone extends abnormally (Chiba and Keshishian 1991). Similarly, when muscle fiber 5 is missing (due to either genetic or laser lesioning), ectopic projections onto neighboring muscles are observed (Cash et al. 1992). In specifically timed heat shock experiments, muscle fiber 13 can be duplicated; in these embryos, RP1, which now has two 13 targets, specifically innervates both (Chiba and Keshishian 1991). Taken together, these observations of normal and manipulated embryos in both grasshopper (Fig. 1; see previous section) and *Drosophila* (Fig. 4) argue for a high degree of specificity in the ability of motoneuron growth cones to recognize particular muscles (Ball et al. 1985b; Chiba and Keshishian 1991; Sink and Whitington 1991b) and suggest that specific gene products active in target regions are responsible in part for determining the patterns of neuromuscular connectivity.

Specification of Muscles and Motoneurons

Experimental analysis in both vertebrates and invertebrates supports the hypothesis that subsets of muscles and motoneurons possess distinct molecular identities that allow motoneuron growth cones to find and recognize their appropriate muscle targets. In support of the hypothesis of muscle specification, four homeodomain proteins, *engrailed* in zebrafish and *S59*, *even-skipped*, and *apterous* in *Drosophila*, have been shown to be expressed in specific subsets of muscle fibers prior to innervation (Dohrmann et al. 1990; Hatta et al. 1990; Ball et al. 1991; C. Bourgouin et al., in prep.); mutations in at least one (*apterous* in *Drosophila*) lead to the absence of specific muscles (C. Bourgouin et al., in prep.).

In support of the notion of muscle and motoneuron cospecification, in at least one case, the muscles and the motoneurons that innervate them express the same homeodomain protein (Ball et al. 1991). *Drosophila even-skipped* (*eve*) is expressed in a subset of identified motoneurons (aCC, the three Us, and RP2) that innervate the *even-skipped*-expressing (dorsal-most) muscles (Ball et al. 1991). Loss of *eve* function results in the change of fate of at least two of these motoneurons (RP2 and aCC) (Doe et al. 1988). For example, in the absence of *eve*, the aCC growth cone extends posteriorly within the CNS, rather than projecting as normal out the intersegmental nerve and toward dorsal muscles (Doe et al. 1988). The expression of *eve* in both a subset of muscles and the motoneurons that innervate them suggests the possibility that this transcription factor might control the coordinated expression of recognition molecule(s) both pre- and postsynaptically.

Search for Neuromuscular Target Recognition Molecules

Two of the neural cell adhesion molecules (CAMs) uncovered by monoclonal antibody screens for CNS axonal glycoproteins (fasciclin III and fasciclin II) are also expressed in interesting patterns during motoneuron pathfinding and neuromuscular innervation. For example, fasciclin III (Patel et al. 1987), a homophilic CAM (Snow et al. 1989), is transiently expressed on the growth cones of several RP neurons (including RP1 and RP3) in the SNb peripheral nerve branch, and on the sites along muscle fibers 6 and 7 where RP3 innervates them (Fig. 7) (Halpern et al. 1991). By the time the synapse matures at stage 17, fasciclin III disappears from both the RP3 growth cone and muscle fibers 6 and 7 (Chang et al. 1991).

Fasciclin II is also expressed on many motoneuron growth cones including aCC (Grenningloh et al. 1991), and on several cells that the aCC growth cone contacts on its way to its dorsal muscle (D. Van Vactor and C.S. Goodman, unpubl.). In addition to their potential roles in motoneuron growth cone guidance, certain antibodies against these neural CAMs provide excellent reagents for following specific motoneuron growth cones (D. Van Vactor et al., unpubl.).

Another approach to identify molecules involved in neuromuscular target recognition is to screen a large number of enhancer trap lines (Harkins and Keshishian 1991; Nose and Goodman 1991). In principle, by screening a sizable fraction of the genome for patterns of gene expression (as reflected by the pattern of β-galactosidase expression), one can identify genes expressed by subsets of muscle fibers prior to or during innervation. In one screen of several thousand enhancer trap lines, 2 different lines were isolated that express β-galactosidase in small subsets of muscle fibers early in development (Harkins and Keshishian 1991); in another screen, more than 11,000 enhancer trap lines were examined, leading to the isolation of 5 lines that express β-galactosidase in small subsets of muscle fibers prior to and during innervation (Nose and Goodman 1991). Several of these lines are good candidates for genes encoding target recognition molecules that guide motoneuron growth cones.

Using this method (Nose et al. 1992), a novel homophilic CAM, called connectin, was discovered that is expressed on the surface of a small subset of muscle fibers during the period of motoneuron outgrowth and innervation. Interestingly, connectin is also expressed on the surface of the very motoneuron growth cones that innervate these muscle fibers (Fig. 8) and on the surface of a subset of peripheral glial cells

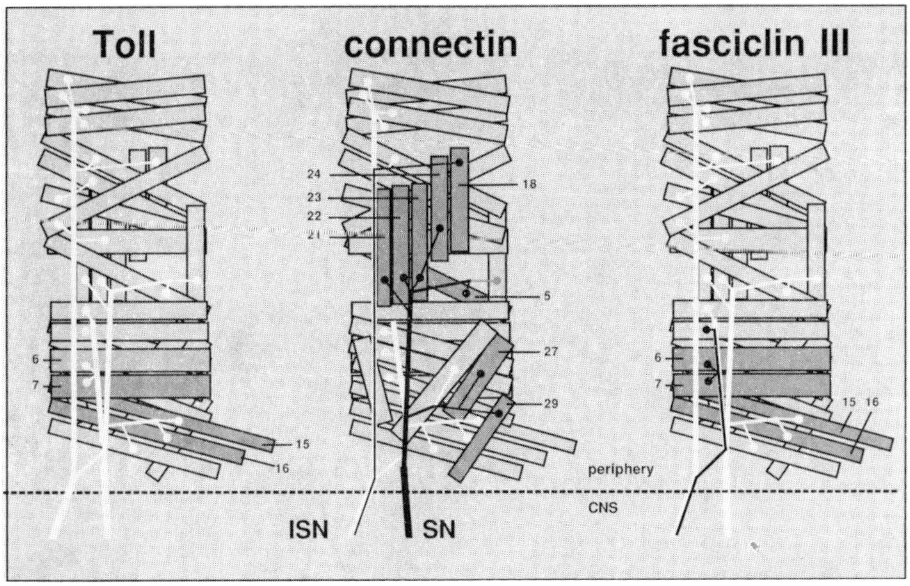

Figure 7. Surface glycoproteins are transiently expressed on subsets of muscles and motoneurons in *Drosophila*. Schematic diagrams showing expression of Toll, connectin, and fasciclin III (dark gray) on the surface of subsets of muscle fibers at stage 16, and expression of connectin and fasciclin III on subsets of motoneurons (black). *Left* and *right* diagrams are viewed from inside of embryo, *middle* diagram from outside in terms of layers of muscles. Lighter muscles are the nonexpressing muscles; white lines and circles are the nonexpressing motoneurons. Note the specificity of connectin-expressing motoneurons for connectin-expressing muscles. We have noticed only one exception. Connectin is sometimes seen on at least one axon (gray) which innervates muscle 8, a fiber that does not express connectin. It is known, however, that muscle 8 is jointly innervated by a motoneuron that also innervates muscle fiber 5 (Cash et al. 1992), a connectin-positive muscle fiber. Moreover, it is also possible that this is a transient aberrant nerve branch from motoneurons targeted to other connectin-positive muscles, since we do not see this connectin-positive axon to muscle 8 in all segments. (Adapted from Nose et al. 1992.)

Figure 8. Connectin is transiently expressed on a subset of muscles and the motoneurons that innervate these muscles. Photograph of a dissected embryo stained with anti-connectin antibodies and HRP immunocytochemistry, showing (from left to right) one thoracic (T1) and one abdominal (A1) segment as an example of segmental differences (dorsal is up). In the abdominal segment on the right, connectin is expressed on muscles 21–24 and 18 (note that muscle 18 is missing in thoracic segments) in the upper (dorsal) part of the photograph, and muscles 27 and 29 (out of focus) in the lower (more ventral) part of the photograph. At the very bottom of the photograph is the CNS, where subsets of motoneuron cell bodies express connectin, as do their axons exiting the CNS in the peripheral nerves. Note the connectin-positive motoneuron growth cone in the abdominal segment just extending over muscles 22 and 23 where it will synapse on them, and the boutons of a connectin-positive motoneuron synapsing on muscle 27 in the lower part of the photograph. Although connectin is expressed in a different but related pattern in thoracic versus abdominal segments, nevertheless, in each segment, the connectin expression is tightly linked between specific motoneurons and the muscles they innervate.

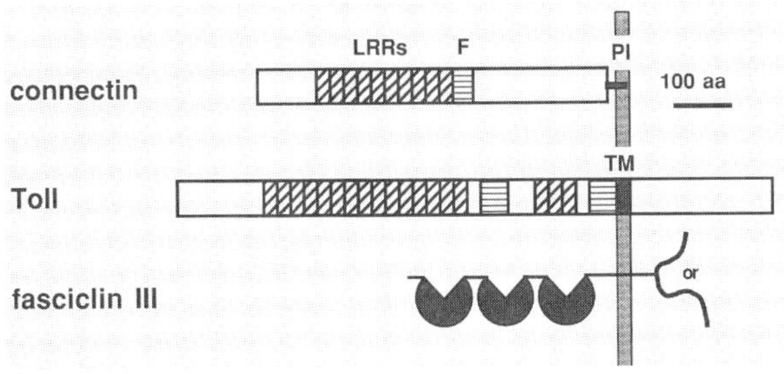

Figure 9. Schematic structure of muscle and motoneuron glycoproteins. Schematic diagram showing the domain structure of connectin, Toll, and fasciclin III. Each hatched rectangle represents one leucine-rich repeat (LRR). The horizontal lined areas denote carboxyl-flanking regions (F), and the solid box indicates signal peptide (connectin) and the transmembrane domain (in the case of Toll). The gray region represents the plasma membrane. Fasciclin III contains three immunoglobulin repeats in its extracellular domain and has two different cytoplasmic domains generated by alternative splicing. Connectin and fasciclin III can function as homophilic CAMs (Snow et al. 1989; Nose et al. 1992).

along the pathways leading to these muscles (Fig. 2). During synapse formation, the protein localizes to synaptic sites; afterward, it largely disappears.

cDNA sequence analysis reveals that connectin belongs to a family of CAMs and receptors with a repeating motif called the leucine-rich repeat (LRR) (Fig. 9) (see, e.g., Krantz and Zipursky 1990). Interestingly, another member of the LRR gene family, *Toll* (known for its role in dorsoventral pattern formation; Hashimoto et al. 1988), is also expressed on the surface of a different subset of muscle fibers prior to innervation (Nose et al. 1992). From a screen for genes downstream from *Ultrabithorax*, the *connectin* gene has been independently cloned and sequenced (Gould et al. 1990; A.P. Gould and R.A.H. White, in prep.).

Halpern et al. (1991) have shown that *Drosophila* fasciclin III, a homophilic CAM of the immunoglobulin gene superfamily (Snow et al. 1989; Grenningloh et al. 1990), is transiently expressed on a subset of muscles. We have shown that connectin and Toll, both members of the LRR gene family, are transiently expressed on different subsets of muscle fibers (Nose et al. 1992). These three examples provide strong evidence that individual muscle fibers are differentially labeled by specific CAMs.

Are the same recognition molecules deployed both on motoneuron growth cones and on the muscles they innervate? Fasciclin III is expressed on a subset of motoneurons (Halpern et al. 1991), one of which (RP3) innervates fasciclin-III-expressing muscles, whereas another (RP1) does not (in addition, most epidermal cells express fasciclin III; Patel et al. 1987). A more compelling example of pre- and postsynaptic specificity in the expression of a CAM is connectin, which is transiently expressed on the surface of eight muscles, the very motoneurons that innervate them (Fig. 8), and several glial cells along the pathways leading to them (Nose et al. 1992). This pattern of expression suggests a role for connectin in both pathway and target recognition during the generation of neuromuscular specificity.

If connectin does indeed help motoneuron growth cones recognize eight (21–24, 18, 5, 27, and 29) of the 30 muscles in each abdominal hemisegment, it must be only part of the mechanism that generates the final pattern of specificity. Although the connectin-expressing motoneurons innervate these muscles, they are able to distinguish among them. For example, certain motoneurons appear to selectively innervate one or two out of the cluster of five connectin-positive muscles (21–24, 18).

FUTURE PROSPECTS

We are still a long way from understanding how neuromuscular specificity is generated in insects. Nevertheless, there has been significant progress over the past decade in our understanding of muscle development, the function of glia, the patterns of motoneuron outgrowth, and the role of muscle pioneers in the formation of specificity. Moreover, recent studies have revealed three different cell adhesion/signaling molecules (Fig. 9) that are transiently expressed on subsets of muscle fibers during the period of innervation. One of these molecules, connectin, shows remarkable coexpression on the very motoneuron growth cones that innervate the connectin-positive muscles.

Clearly, using *Drosophila* as a model genetic organism, the most powerful approach to identify molecules involved in motoneuron pathfinding and neuromuscular target recognition is to screen for mutants that perturb these events. Such a mutant screen is now possible, given the availability of specific antibody probes and the knowledge about motoneurons, glia, other putative guideposts, and muscles described in this review. Using a monoclonal antibody that selectively stains most motoneuron growth cones and axons (D. Van Vactor et al., unpubl.), a large-scale F_2 mutant screen has been initiated (Van Vactor and Goodman 1992). Even though it is still in its infancy, a number of new mutants have been isolated that appear to perturb the projection of specific subsets of motoneuron growth cones.

There are many reasons to expect that knowledge gleaned from *Drosophila* will be relevant to our understanding of neural specificity and neuromuscular development in more complex organisms. The cellular and molecular mechanisms underlying growth cone guidance in insects appear remarkably similar to those used in vertebrates (see, e.g., Dodd and Jessell 1988; Harrelson and Goodman 1988; Klose and Bentley, 1989; McConnell et al. 1989). In vertebrate embryos such as chick (see, e.g., Tosney and Landmesser 1985) and zebrafish (see, e.g., Eisen et al. 1986), just as in insect embryos such as grasshopper (Ball et al. 1985b) and *Drosophila* (Sink and Whitington 1990; Halpern et al. 1991), motoneuron growth cones make stereotyped pathway choices as they extend toward the regions where their appropriate target muscles arise.

As various developmental questions have been explored at a molecular level, the mechanistic similarities between higher organisms and simpler model genetic systems such as *Drosophila* have become increasingly clear. We have no doubt that this relationship will equally apply to the principles underlying growth cone guidance and neuromuscular target recognition, and thus, the molecules and mechanisms revealed by molecular genetic analysis in *Drosophila* will be homologous to similar molecules and mechanisms in higher organisms.

ACKNOWLEDGMENTS

We are grateful to present and past members of the Goodman laboratory for their contributions of probes, results, and ideas that have provided the groundwork for the conclusions reviewed here. Research from the Goodman laboratory is supported by National Institutes of Health grants NS-18366 and HD-21294 and the Howard Hughes Medical Institute.

REFERENCES

Auld, V.J., C. Klämbt, R. Tsoo, and C.S. Goodman. 1991. Molecular characterization of embryonic glia and their role during growth cone guidance in *Drosophila*. *Soc. Neurosci. Abstr.* **17:** 742.

Ball, E.E. and C.S. Goodman. 1985a. Muscle development in the grasshopper embryo. II. Syncytial origin of the extensor tibiae muscle pioneers. *Dev. Biol.* **111:** 399.

———. 1985b. Muscle development in the grasshopper embryo. III. Sequential origin of the flexor tibiae muscle pioneers. *Dev. Biol.* **111:** 417.

Ball, E.E., R.K. Ho, and C.S. Goodman. 1985a. Muscle development in the grasshopper embryo. I. Muscles, nerves, and apodemes in the metathoracic leg. *Dev. Biol.* **111:** 383.

———. 1985b. Development of neuromuscular specificity in the grasshopper embryo: Guidance of motoneuron growth cones by muscle pioneers. *J. Neurosci.* **5:** 1808.

Ball, E.E., E.J. Rehm, N.H. Patel, and C.S. Goodman. 1991. Evolution of insect segmentation and homeotic genes and their function during neurogenesis. *Soc. Neurosci. Abstr.* **17:** 11.

Bastiani, M.J. and C.S. Goodman. 1986. Guidance of neuronal growth cones in the grasshopper embryo. III. Recognition of specific glial pathways. *J. Neurosci.* **6:** 3542.

Bastiani, M.J., C.Q. Doe, S.L. Helfand, and C.S. Goodman. 1985. Neuronal specificity and growth cone guidance in grasshopper and *Drosophila* embryos. *Trends Neurosci.* **8:** 257.

Bate, C.M. 1982. Pioneer neurons and axonal pathways. *Neurosci. Res. Program Bull.* **20:** 838.

———. 1990. The embryonic development of larval muscles in *Drosophila*. *Development* **110:** 791.

Bate, M., E. Rushton, and D.A. Currie. 1991. Cells with persistent *twist* expression are the embryonic precursors of adult muscles in *Drosophila*. *Development* **113:** 79.

Cash, S., A. Chiba, and H. Keshishian. 1992. Alternate neuromuscular target selection following the loss of single muscle fiber in *Drosophila*. *J. Neurosci.* **12:** 2051.

Chang, T.N., A. Chiba, D. Jay, and H. Keshishian. 1991. Fasciclin III expression during synaptic target recognition in *Drosophila*. *Soc. Neurosci. Abstr.* **17:** 214.

Chiba, A. and H. Keshishian. 1991. Growth cone choices of *Drosophila* motoneurons in response to mismatched targets. *Soc. Neurosci. Abstr.* **17:** 213.

Dodd, J. and T.M. Jessell. 1988. Axon guidance and the patterning of neuronal projections in vertebrates. *Science* **242:** 692.

Doe, C.Q., D. Smouse, and C.S. Goodman. 1988. Control of neuronal fate by the *Drosophila* segmentation gene *even-skipped*. *Nature* **333:** 376.

Dohrmann, C., N. Azpiazu, and N. Frasch. 1990. A new *Drosophila* homeo box gene is expressed in mesodermal precursor cells of distinct muscles during embryogenesis. *Genes Dev.* **4:** 2098.

Eisen, J.S. and S.H. Pike. 1991. The *spt-1* mutation alters segmental arrangement and axonal development of identified neurons in the spinal cord of the embryonic zebrafish. *Neuron* **6:** 767.

Eisen, J.S., P.Z. Myers, and M. Westerfield. 1986. Pathway selection by growth cones of identified motor neurons in live zebra fish embryos. *Nature* **320:** 269.

Eisen, J.S., S.H. Pike, and B. Debu. 1989. The growth cones of identified motoneurons in embryonic zebrafish select appropriate pathways in the absence of specific cellular interactions. *Neuron* **2:** 1097.

Fredieu, J.R. and A.P. Mahowald. 1989. Glial interaction with neurons during *Drosophila* embryogenesis. *Development* **106:** 739.

Gould, A.P., J.J. Brookman, D.I. Strutt, and R.A.H. White. 1990. Target of homeotic gene control in *Drosophila*. *Nature* **348:** 308.

Grenningloh, G., A. Bieber, J. Rehm, P.M. Snow, Z. Traquina, M. Hortsch, N.H. Patel, and C.S. Goodman. 1990. Molecular genetics of neuronal recognition in *Drosophila*: Evolution and function of immunoglobulin superfamily cell adhesion molecules. *Cold Spring Harbor Symp. Quant. Biol.* **55:** 327.

Grenningloh, G., E.J. Rehm, and C.S. Goodman. 1991. Genetic analysis of growth cone guidance in *Drosophila*: Fasciclin II functions as a neuronal recognition molecule. *Cell* **67:** 45.

Halpern, M.E., A. Chiba, J. Johansen, and H. Keshishian. 1991. Growth cone behavior underlying the development of stereotypic synaptic connections in *Drosophila* embryos. *J. Neurosci.* **11:** 3227.

Harkins, E.W. and H. Keshishian. 1991. Cellular and developmental characterization of two enhancer trap lines with expression in specific embryonic *Drosophila* muscles. *Soc. Neurosci. Abstr.* **17:** 214.

Harrelson, A.L. and C.S. Goodman. 1988. Growth cone guidance in insects: Fasciclin II is a member of the immunoglobulin superfamily. *Science* **242:** 700.

Hashimoto, C., K.L. Hudson, and K.V. Anderson. 1988. The *Toll* gene of *Drosophila*, required for dorsal-ventral embryonic polarity, appears to encode a transmembrane protein. *Cell* **52:** 269.

Hatta, K., T.F. Schilling, R.A. BreMiller, and C.B. Kimmel. 1990. Specification of jaw muscle identity in zebrafish: Correlation with *engrailed*-homeoprotein expression. *Science* **250:** 802.

Ho, R.K., E.E. Ball, and C.S. Goodman. 1983. Muscle pioneers: Large mesodermal cells that erect a scaffold for developing muscles and motoneurons in grasshopper embryos. *Nature* **301:** 66.

Jacobs, J.R. and C.S. Goodman. 1989. Embryonic development of axon pathways in the *Drosophila* CNS. I. A glial scaffold appears before the first growth cones. *J. Neurosci.* **9:** 2402.

Johansen, J., M.E. Halpern, and H. Keshishian. 1989a. Axonal guidance and the development of muscle fiber-specific innervation in *Drosophila* embryos. *J. Neurosci.* **9:** 4318.

Johansen, J., M.E. Halpern, K.M. Johansen, and H. Keshishian. 1989b. Stereotypic morphology of glutamatergic synapses on identified muscle cells of *Drosophila* larvae. *J. Neurosci.* **9:** 710.

Klämbt, C. and C.S. Goodman. 1991. The diversity and pattern of glia during axon pathway formation in the *Drosophila* embryo. *Glia* **4:** 205.

Klose, M. and D. Bentley. 1989. Transient pioneer neurons are essential for formation of an embryonic peripheral nerve. *Science* **245:** 982.

Krantz, D.E. and S.L. Zipursky. 1990. *Drosophila* chaoptin, a member of the leucine-rich repeat family, is a photoreceptor cell-specific adhesion molecule. *EMBO J.* **9:** 1969.

Lance-Jones, C. and L. Landmesser. 1981. Pathway selection by embryonic chick motoneurons in an experimentally altered environment. *Proc. R. Soc. Lond. B Biol. Sci.* **214:** 19.

Landmesser, L.T. 1980. The generation of neuromuscular specificity. *Annu. Rev. Neurosci.* **3:** 279.

Leiss, D., U. Hinz, A. Gasch, R. Mertz, and R. Renkawitz-Pohl. 1988. $\beta 3$ tubulin expression characterizes the differentiating mesodermal germ layer during *Drosophila* embryogenesis. *Development* **104:** 525.

Liu, D.W.C. and M. Westerfield. 1990. The formation of terminal fields in the absence of competitive interactions among primary motoneurons in the zebrafish. *J. Neurosci.* **10:** 3947.

McConnell, S.K., A. Ghosh, and C.J. Shatz. 1989. Subplate neurons pioneer the first axon pathway from the cerebral cortex. *Science* **245:** 978.

Nose, A. and C.S. Goodman. 1991. Connectin: A surface protein expressed on a subset of motoneuron axons and the target muscles they innervate in *Drosophila*. *Soc. Neurosci. Abstr.* **17:** 742.

Nose, A., V.B. Mahajan, and C.S. Goodman. 1992. Connectin: A homophilic cell adhesion molecule expressed on a subset of muscles and the motoneurons that innervate them in *Drosophila*. *Cell* **70:** 553.

Patel, N.H., P.M. Snow, and C.S. Goodman. 1987. Chracterization and cloning of fasciclin III: A glycoprotein expressed on a subset of neurons and axon pathways in *Drosophila*. *Cell* **48:** 975.

Pike, S.H. and J.S. Eisen. 1990. Identified primary motoneurons in embryonic zebrafish select appropriate pathways in the absence of other primary motoneurons. *J. Neurosci.* **10:** 44.

Sink, H. and P.M. Whitington. 1990. Location and connectivity of abdominal motoneurons in the embryo and larva of *Drosophila melanogaster*. *J. Neurobiol.* **22:** 298.

———.1991a. Pathfinding in the central nervous system and periphery by identified embryonic *Drosophila* motor axons. *Development* **112:** 307.

———.1991b. Early ablation of target muscles modulates the arborisation pattern of an identified embryonic *Drosophila* motor axon. *Development* **113:** 701.

Snow, P.M., A.J. Bieber, and C.S. Goodman. 1989. Fasciclin III: A novel homophilic adhesion molecule in *Drosophila*. *Cell* **59:** 313.

Tosney, K.W. and L.T. Landmesser. 1985. Growth cone morphology and trajectory in the lumbosacral region of the chick embryo. *J. Neurosci.* **5:** 2345.

Van Vactor, D. and C.S. Goodman. 1992. Genetic analysis of motoneuron pathfinding and neuromuscular connectivity. *Soc. Neurosci. Abstr.* **18:** 1273.

Rostrocaudal Differences among Muscles Revealed by a Transgene: Graded Expression at Low Copy Number

J.R. SANES,* M.J. DONOGHUE,*† M.C. WALLACE,† AND J.P. MERLIE†
*Departments of *Anatomy and Neurobiology and of †Molecular Biology and Pharmacology, Washington University School of Medicine, St. Louis, Missouri 63110*

The formation of a synapse requires the coordination of numerous interactions between the pre- and postsynaptic partners. The vertebrate skeletal neuromuscular junction has proven to be a useful model for studies of these interactions, because it is simple and accessible, and because a wealth of information is now available about its structure and function. Moreover, new synapses form readily in denervated adult muscle, permitting analysis of synaptogenesis uncomplicated by embryonic processes of morphogenesis and histogenesis. Axonal behavior is clearly guided by target-derived factors during reinnervation: For example, axons form synapses on denervated but not innervated muscle fibers, and they preferentially reinnervate original synaptic sites on denervated fibers, even though these sites occupy only ~0.1% of the muscle fiber surface. We and other workers have identified a set of cell membrane and extracellular matrix molecules that may be involved in these processes (for review, see Sanes et al. 1990).

To extend this work, we wanted to use the neuromuscular system to seek cell-surface molecules that mediate a more complex synaptogenetic phenomenon: the selection of an appropriate postsynaptic partner, from a myriad of candidates, by a growing axon. Accordingly, we undertook a series of electrophysiological experiments to determine whether reinnervation *is* selective in this respect in adult rodent muscle and, if so, whether a systematic set of synaptic preferences could be discerned. In fact, several prior studies had failed to demonstrate selective reinnervation of mammalian muscles: "native" and "foreign" axons were equally successful at establishing synapses in limb muscles following severance or transposition of their nerves (for references, see Wigston and Sanes 1985). However, studies on the peripheral autonomic nervous system by Purves et al. (1981; see also Purves and Lichtman 1985) suggested a new way to examine this issue. Purves had shown that neurons in sympathetic ganglia are selectively reinnervated by sympathetic preganglionic axons in adult mammals and that this selectivity appears to reflect a matching of pre- and postsynaptic partners from corresponding levels of the neuraxis. We therefore focused our attention on axial rather than limb muscles and sought matching based on the rostrocaudal positions of the synaptic partners rather than on their specific identities. Our main results (Fig. 1) (Wigston and Sanes 1982, 1985; Laskowski and Sanes 1987, 1988), as extended subsequently by Laskowski (Laskowski and High 1989; Laskowski et al. 1991), were the following: (1) For each of two thoracic muscles, the diaphragm and serratus anterior, the rostrocaudal axis of the spinal motor pool (the group of spinal motor neurons that supplies a muscle) is systematically mapped onto the rostrocaudal axis of the muscle (see also Landau et al. 1962). (2) These topographic maps of connectivity are established during embryogenesis. They are therefore likely to reflect cel

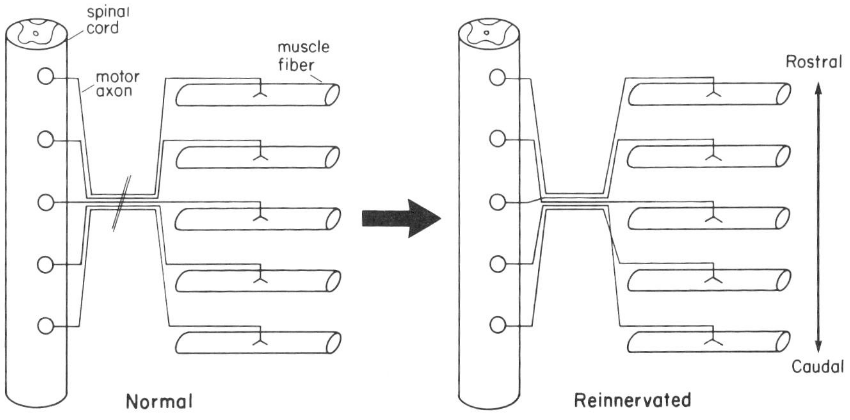

Figure 1. Muscle fibers in mammalian axial muscles are preferentially innervated and reinnervated by axons that arise from corresponding rostrocaudal levels of the spinal cord. Summary of electrophysiological studies detailed in Wigston and Sanes (1982, 1985), Laskowski and Sanes (1987, 1988), and Laskowski and High (1989).

  0-87969-063-1/92 $3.00

lular matching, although postnatal refinement of the maps might be activity- or usage-dependent (see Brown and Booth 1983; Bennett and Lavidis 1984). (3) The maps are restored, albeit imperfectly, when muscles are denervated and allowed to be reinnervated by their own multisegmental nerves. (4) The selectivity of reinnervation is particularly striking in neonates, but persists into adulthood. (5) Finally, intercostal muscles transplanted to an ectopic site in the neck are selectively reinnervated by positionally appropriate axons from the cervical sympathetic trunk: Muscles transplanted from the second and third thoracic interspaces (T2 and T3) are preferentially reinnervated by axons that arise from rostral thoracic levels of the spinal cord, T5 and T8 muscles are preferentially reinnervated by caudal axons, and T4 muscles receive an intermediate set of inputs. Thus, even when muscles are transplanted to a novel site and confronted with heterologous axons, a positional selectivity in synapse formation is expressed.

Together, these results suggest that a system of positional cues influences the formation of neuromuscular connections in the embryo and that this positional bias persists in adulthood (Fig. 1). The graded selectivity observed electrophysiologically (noted also in sympathetic ganglia; Purves et al. 1981) is consistent with the existence of molecular gradients that bias but do not rigidly determine synaptic preferences. Specifically, we speculated that adult muscles retain a positional memory of their rostrocaudal position that is stable to transplantation and that adult as well as embryonic muscles bear positional labels that affect axonal behavior. However, even if this speculation were valid, we had no way of knowing whether any individual molecule would be expressed in a gradient, how steep such a gradient might be, how far along the neuroaxis it might extend, or precisely where it would be localized in cellular terms. Moreover, none of the extracellular matrix and cell-surface molecules that are made by muscle and suspected to influence axonal behavior are known to vary in abundance, form, or distribution among muscles (J.R. Sanes, unpubl.). We were therefore disappointed but unsurprised when attempts to find positional labels (J. Covault et al., unpubl.) proved unsuccessful. What was needed was a more direct demonstration that a molecular gradient does exist in muscle and some indication of where the positional memory resides. Recently, we obtained new evidence on these points from a set of transgenic mice that had originally been generated and characterized for other purposes. In this paper, we describe these mice and what we have learned from them.

METHODS

Transgenic mice were generated as described by Hogan et al. (1986). Individual muscles were identified and dissected with the aid of Green's (1963) atlas. Chloramphenicol acetyltransferase (CAT) activity was assayed as described by Gorman (1985). Amounts of extract and times of incubation were adjusted to give 0.3%–50% conversion—within the linear range—but values are expressed as activity per 1 μg protein per 1-hour incubation, to facilitate comparisons among muscles. Protein concentrations were determined according to the method of Bradford (1976). DNA was prepared from cells and tissues as described by Davis et al. (1986), digested with appropriate restriction enzymes, and analyzed according to the method of Southern (1975) using probes that were labeled by random priming. RNase protection assays were performed as described by Melton et al. (1984). Primary cultures of adult muscle cells were prepared according to methods modified from Cossu et al. (1980). Cultured myoblasts were immortalized by infection with a recombinant retrovirus bearing a temperature-sensitive mutant SV40 T-antigen oncogene (Renfranz et al. 1991). Our modifications of these methods are detailed in Donoghue et al. (1991a,b, 1992a,b).

RESULTS AND DISCUSSION

Rostrocaudal Gradient of Transgene Expression

A few years ago, J. Kornhauser and J.P. Merlie generated transgenic mice using regulatory elements derived from the rat myosin light chain (MLC) 1f/3f gene. The transgene consisted of a 1.2-kb MLC1 promoter and a 0.9-kb downstream enhancer, placed 5′ and 3′, respectively, of a cassette containing a bacterial CAT gene and an intron and polyadenylation signal from the SV40 t-antigen gene. This construct had previously been shown to be expressed at much higher levels in myotubes than in fibroblasts or myoblasts in vitro (Donoghue et al. 1988). We hoped that expression would be muscle-specific in vivo as well, thus providing a vector that could be used to overexpress biologically interesting genes in muscle. Indeed, expression of this MLC1-CAT transgene proved to mimic expression of the endogenous gene in many respects: (1) Expression was high in adult hindlimb muscle, but negligible in all nonmuscle tissues tested. (2) Expression was developmentally regulated, increasing >10-fold between birth and adulthood. (3) Expression was muscle fiber type-specific, being higher in type II (fast twitch) than in type I (slow twitch) fibers, and higher in type IIB and IIX than in IIA fibers (Rosenthal et al. 1989; Donoghue et al. 1991a). However, further analysis of MLC1-CAT mice revealed an additional and unanticipated feature that has served as the starting point for our subsequent studies: Levels of CAT varied among muscles in a >100-fold, whole-body rostrocaudal gradient, such that muscles arising from successively more caudal somites expressed successively higher levels of CAT activity (Fig. 2a). CAT levels were higher in hindlimb muscles than in their forelimb serial homologs, but in general, the gradient was more prominent in axial than in limb musculature (Donoghue et al. 1991b). Particularly striking was the pattern of expression in the intercostals, a set of structurally and functionally similar muscles, each of which arises from a

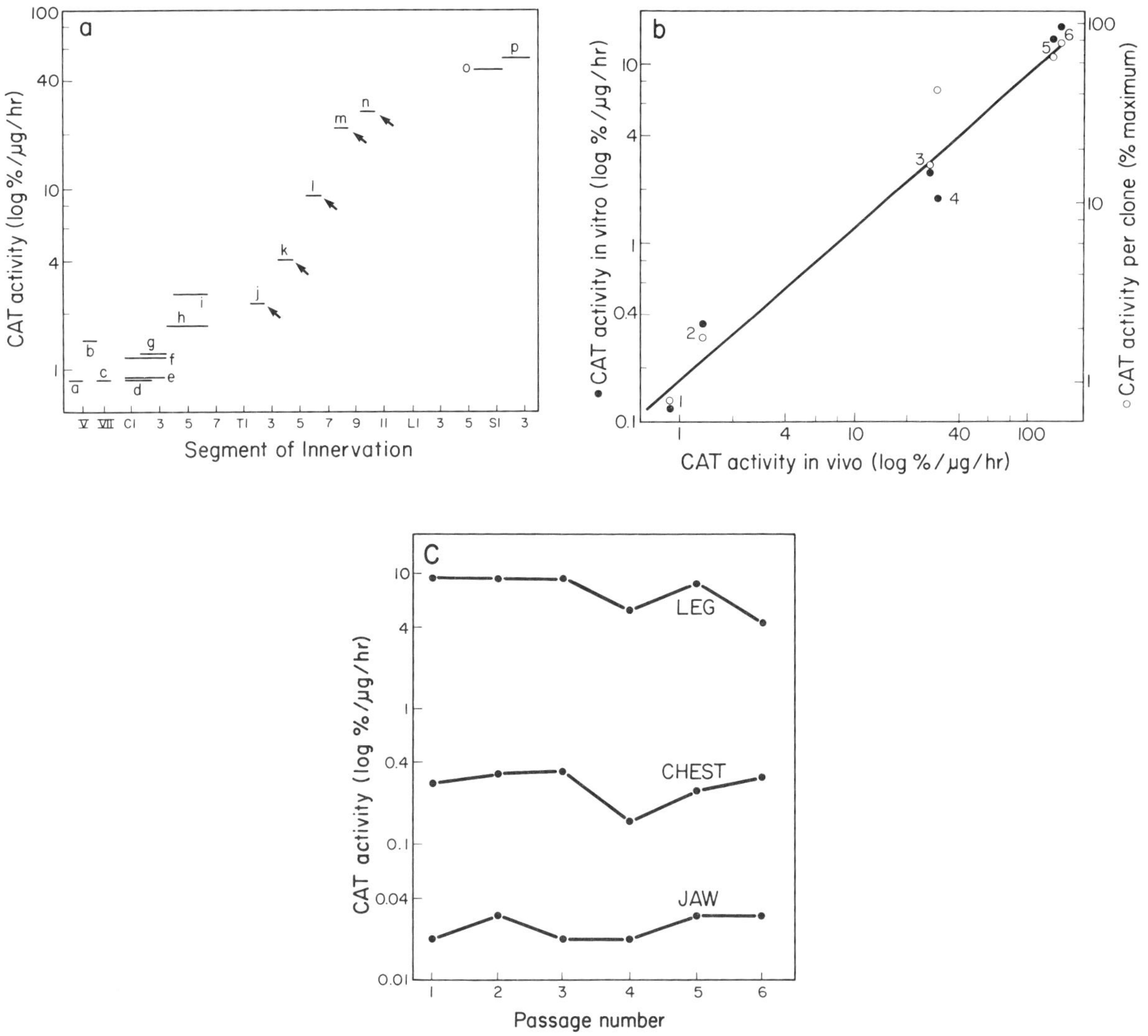

Figure 2. Transgene expression is positionally graded in muscles from MLC1-CAT mice. (*a*) CAT activities in axial muscles from adult MLC1-CAT mice, arranged by the cranial or spinal nerves that innervate them. Muscles shown are: a, masseter; b, anterior digastricus; c, posterior digastricus; d, sternohyoid; e, sternomastoid; f, omohyoid; g, cleidomastoid; h, diaphragm; i, clavotrapezius; j, intercostal 2; k, intercostal 4; l, intercostal 6; m, intercostal 8; n, intercostal 10; o, flexor caudalis; p, sacrococcygeus ventralis medialis. (Redrawn from Donoghue et al. 1991b.) (*b*) Relationship of CAT activities in cultured myotubes to those measured in adult muscles. Muscles from which cultures were prepared are: (1) masseter (jaw); (2) sternohyoid, omohyoid, sternomastoid, and cleidomastoid (neck); (3) latissimus dorsi and spinotrapezius (back); (4) pectoralis major and minor (chest); (5) tibialis anterior and extensor digitorum longus (hindlimb); and (6) gastrocnemius and plantaris (hindlimb). Closed circles show values from mass cultures and open circles show values from myogenic clones. (Redrawn from Donoghue et al. 1992b.) (*c*) CAT activities in successive passages of immortalized myogenic cell lines derived from jaw, chest, and leg muscles. (Redrawn from Donoghue et al. 1992b.)

single somite and is innervated through a single spinal nerve (see Wigston and Sanes 1985): The specific activity of CAT increased ~2-fold per two segments, or a total of ~10-fold across the ribcage.

The implication of this finding is that the MLC1-CAT transgene provides a marker of rostrocaudal position — the first so far described for adult muscle. Accordingly, we undertook a series of studies aimed at analyzing the control of transgene expression in MLC1-CAT mice. Our main results (detailed in Donoghue et al. 1991a,b, 1992b) and the conclusions we draw from them can be summarized as follows:

1. We measured the abundance of CAT mRNA in muscles that spanned the rostrocaudal axis. CAT mRNA levels were positionally graded and highly correlated (r = 0.97) with CAT activity levels. Thus, unless there are intermuscular differences in mRNA stability (a possibility addressed below), patterns of expression detected at the protein level can be used to monitor patterns of gene expression.
2. We assayed levels of CAT in individual muscles from three independently derived lines of MLC1-CAT mice. Southern analysis showed that each had inserted the transgene into a unique chromosomal

position. Rostrocaudal gradients of CAT activity of ~100-fold were detected in each line. Thus, graded expression does not depend on unique endogenous sequences near the site of integration (a so-called "enhancer trap") but is likely to be directed by sequences within the transgene itself.

3. To determine whether MLC-derived or reporter-associated sequences directed position-dependent expression, we analyzed nine other lines of transgenic mice. In three lines, the MLC-derived regulatory sequences from MLC1-CAT were linked to an acetylcholine receptor γ-subunit cDNA instead of to CAT. An RNase protection assay revealed rostrocaudal gradients of transgene mRNA abundance in all three lines. In other mice, the CAT/intron/polyadenylation signal cassette was linked to a promoter from the acetylcholine receptor α-subunit gene (three lines; Merlie and Kornhauser 1989); to a promoter from the skeletal α-actin gene (two lines; Petropoulos et al. 1989); or to regulatory elements from the muscle creatine kinase gene (one line; Johnson et al. 1989). Expression of CAT is muscle-specific in all these lines, and intriguing intermuscular differences in CAT levels were detected in some of them. However, in no case were levels graded in the striking fashion observed in MLC1-CAT mice. Thus, graded expression in the MLC1-CAT mice is likely to be directed by MLC-derived sequences rather than by sequences in the CAT cassette. Moreover, by showing that muscles do not vary systematically in their ability to process CAT RNA or protein, these results provide evidence that graded levels of CAT activity are likely to reflect graded transcriptional activation of the transgene.
4. Given that MLC-derived elements direct positionally graded expression of reporter genes, it seemed prudent to examine the possibility that the endogenous MLC 1f/3f gene was expressed in a hitherto undetected gradient. We therefore assayed MLC mRNA levels in a set of ten muscles that spanned the rostrocaudal axis. Probes were used that detected mRNAs encoding both MLC1f, which is produced from the promoter used in the transgene, and MLC3f, a second transcript that is initiated from an alternative promoter within the same gene. Neither mRNA was graded in abundance along the rostrocaudal axis. Thus, the position-dependence of MLC1-CAT expression is not evident in the endogenous MLC1f/3f gene.
5. Next, we devised a quantitative histochemical assay for CAT activity so that we could investigate the cellular basis of the graded expression. Using cryostat sections, we showed that CAT activity was confined to muscle fibers and that the average level of activity per fiber appeared to be continuously graded with rostrocaudal position. Within any single muscle, the transgene, like the endogenous gene (for references, see Donoghue et al. 1991a), was expressed at higher levels in fast-twitch than in slow-twitch fibers. However, the position-dependent variation in CAT level was apparent even when fibers of a single type were compared. Thus, fiber type and position affect CAT expression independently, and the positional gradient does not result from variations in fiber-type composition, from ectopic CAT expression in nonmuscle cells, or from discrete subpopulations of expressing or nonexpressing fibers.
6. Position-dependent expression of CAT might result from functional differences among muscles, such as muscle-specific patterns of synaptic activation or usage. Alternatively, if the gradient was established early in development, it might reflect an endogenous system of positional information. To distinguish these alternatives, we assayed CAT activity in positionally distinct muscles from embryos and neonates. Graded expression was evident by embryonic day (E)15, which was as early as muscles could be dissected. In situ hybridization using sectioned embryos has shown that levels of MLC1-CAT mRNA are also graded by E12.5 (Grieshammer et al. 1992). Thus, position-dependent transgene expression does not require mature patterns of synaptic connectivity or activity.
7. Our studies of transgene expression in vivo imply the existence of an endogenous positional memory in muscles. To learn whether the memory (as opposed to expression per se) resides within muscle cells themselves, and how stably it is maintained, we turned to tissue culture. We dissociated satellite cells (myoblasts) from adult muscles that spanned the rostrocaudal axis, cultured them until they fused to form myotubes, and then assayed the cultures for CAT. All cultures were prepared at identical cell density, and they all contained similar numbers of myotubes, amounts of protein, levels of acetylcholine receptors, and isoforms of myosin heavy chain. In contrast, specific activities of CAT varied ~100-fold among cultures, with values in vitro reflecting the rostrocaudal position of the muscle from which each culture was derived (closed circles in Fig. 2b). Similar results were obtained when individual myoblasts were stimulated to undergo several rounds of division in vitro before fusing to form clones of myotubes, which were subsequently recovered with cloning cylinders and assayed (open circles in Fig. 2b). Because each clone contained 100–1000 nuclei, the positional memory must have survived at least seven divisions in vitro. Moreover, because the clones were well separated, interactions between myogenic cells and accompanying nonmuscle cells were unlikely to have occurred in these cultures. Thus, positional differences in transgene expression do not appear to depend for their maintenance on axial gradients of diffusible substances, innervation, activity, nonmuscle cells, or extracellular matrix. Instead, muscle cells bear an intrinsic positional memory that is stable, heritable, and cell-autonomous.
8. As a final test of the stability of the positional memory, we immortalized myoblasts cultured from in-

dividual muscles by infecting them with a recombinant retrovirus that encodes a temperature-sensitive oncogene. The infected cells grew as myoblasts in rich medium at 33°C (the permissive temperature), but fused to form myotubes in serum-poor medium at 37°C (the nonpermissive temperature). The cells were then cloned at limiting dilution to generate clonal cell lines. When assayed as myotubes, lines from chest and back muscles contained manyfold higher levels of CAT than did lines from head and neck muscles, and lines from hindlimb muscles had severalfold higher CAT levels still. In addition, position-related differences in CAT activity were maintained through subcloning and over many passages in culture (Fig. 2c). Thus, the positional memory that regulates transgene expression is stable indefinitely and survives transformation. Moreover, the availability of cell lines that express a molecular correlate of their axial position permits a variety of cellular and molecular approaches to the study of positional information that would be infeasible in vivo.

Together, these results support the model sketched in Figure 3. We believe that the graded expression of CAT implies the existence of an endogenous regulator of transcription that differs among muscles, possibly itself in a gradient. The intermuscular differences arise within myoblasts during embryogenesis, and a positional memory of these differences is stably maintained throughout the animal's life. Why, given this system, is the endogenous MLC1f/3f gene not expressed in a gradient? One possible explanation is that the transgene incorporates only some of the regulatory elements present in the endogenous gene; for example, the endogenous gene might contain both "rostral" and "caudal" elements, with the former absent from the transgene. Alternatively, a caudal element may be masked or inactive in its native context but artifactually activated in the transgene. The first explanation seems contrived, but, in fact, an analogous case has been described recently in *Dictyostelium*: The SP60 gene is expressed throughout the prestalk region, and its promoter contains discrete elements that direct expression of a reporter in opposing (anterior-posterior and posterior-anterior) gradients (Haberstroh et al. 1991). In either case, our interest lies not in the MLC1f/3f gene but in the possibility that the position-dependent transcriptional regulatory system directs expression of endogenous genes that could be involved in morphogenesis or nerve-muscle interactions. It is our belief that the MLC1-CAT transgene will facilitate the search for such endogenous genes in two ways. First, we can use patterns of transgene expression to guide the selection of muscles for direct comparisons of endogenous components—e.g., by subtractive immunization, subtractive hybridization, or two-dimensional gel electrophoresis. Second, we can seek the "upstream" factors that control MLC1-CAT expression, in the hope that these will lead us in turn to endogenous "downstream" targets. Investigations of both types are currently in progress, but the remainder of this paper focuses on the second approach.

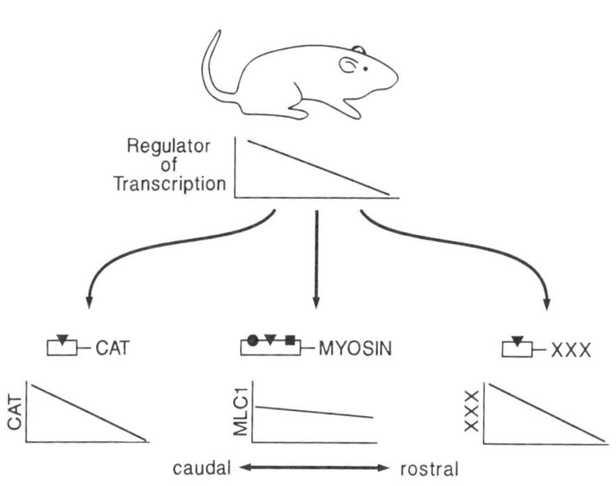

Figure 3. The gradient of MLC1-CAT transgene expression suggests the existence of an endogenous regulator of position-dependent gene expression. This, in turn, suggests that some endogenous genes are expressed in a gradient, even though the endogenous MLC1f/3f gene is apparently not. See text for explanation.

Regulation of Position-dependent Gene Expression

What molecular mechanisms might underlie position-dependent expression of the MLC1-CAT transgene? Two general possibilities are that some components of the transcriptional apparatus vary in abundance or activity along the rostrocaudal axis, or that the transgene itself differs among muscles. We refer to these modes of regulation, which are not mutually exclusive, as "*trans*" and "*cis*," respectively. We have begun to test a few specific *trans* and *cis* models of MLC1-CAT regulation.

Obvious candidates for *trans*-regulation are the myogenic factors. This is a group of four proteins, called myoD, myogenin, myf5, and MRF4 (also called herculin and myf6). All are structurally homologous, and all share the remarkable feature of being able to convert many types of nonmuscle cell into myoblasts following introduction by infection or transfection with an appropriate expression vector. All four factors are present in embryonic muscle, although each is regulated in a unique temporal pattern (Weintraub et al. 1991; Buckingham 1992). Finally, all are DNA-binding proteins that can serve as transcriptional activators of numerous muscle-specific genes, including the MLC1f/3f gene (Braun et al. 1990; Fujisawa-Sehara et al. 1992). We therefore asked whether the expression of any of these factors varies with rostrocaudal position.

Results of an RNase protection assay with a probe specific for MRF4 (Miner and Wold 1990) are shown in Figure 4. Levels of MRF4 mRNA were similar in head, neck, chest, forelimb, and hindlimb muscles. Similar results were obtained using probes specific for myogenin and myf5 (Miner et al. 1992): As previously reported, their mRNAs were less abundant than that

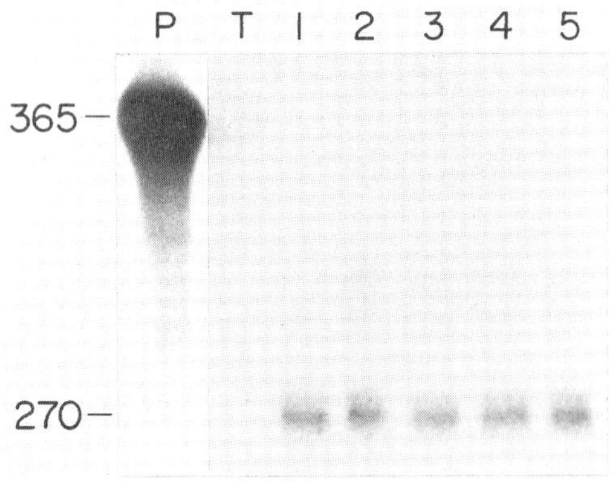

Figure 4. Levels of MRF4 mRNA vary little among adult muscles. RNase protection assay of RNA from muscle groups 1–5 listed in Fig. 5 legend. Undigested probe (P) and negative control (yeast tRNA; T) are also shown. The 365-nt MRF4 probe used (Miner and Wold 1990) protected a band of 270 nt.

for MRF4, but no significant intermuscular differences in abundance were detected (data not shown). Of course, it remains possible that these factors vary among muscles at the protein level, either in concentration or activity. In addition, our studies of myoD and of other putative transcriptional regulators, such as the Hox gene products (McGinnis and Krumlauf 1992), are still in progress.

More enlightening have been tests of one *cis* model: that the methylation state of the MLC1-CAT transgene differs among muscles. Methylation is the predominant covalent modification of vertebrate genomic DNA, and methylation patterns are heritable within somatic cells of an individual. Methylation has been implicated in several tissue-specific and developmentally regulated patterns of gene expression, in X-chromosome inactivation, and in parental imprinting; in most of these cases, expressed genes or alleles are hypomethylated, and nonexpressed forms are hypermethylated (for review, see Holliday et al. 1990; Razin and Cedar 1991; Surani 1991).

We used the isoschizomeric pair of restriction enzymes, *Msp*I and *Hpa*II, to analyze transgene methylation. Both enzymes recognize the sequence CCGG, which is frequently methylated at the second C, but *Msp*I is insensitive to methylation, whereas *Hpa*II cuts only unmethylated sites. The MLC1-CAT transgene contains five *Msp*I/*Hpa*II sites (Fig. 5a), and the methylation state of each of these can be analyzed by probing Southern blots of *Msp*I or *Hpa*II restriction digests with labeled fragments of the transgene and comparing the patterns of labeled bands in the two digests (for a more detailed explanation, see Donoghue et al. 1992a). Using this assay, we found that methylation of the MLC1-CAT transgene varies with body position in adult skeletal muscle. Consistent with the usual association of hypomethylation with high expression, methylation levels decreased as expression levels increased, in a rostrocaudal progression (Fig. 5b). Moreover, the MLC1-CAT transgene was more extensively methylated in rostrally derived myogenic cell lines, which express low levels of CAT, than in caudally derived myogenic cell lines, which express high levels of CAT (Figs. 2c and 5c). Thus, intermuscular differences in transgene methylation are related to body position both in vivo and in vitro.

A crucial question in studies of genomic methylation is whether methylation state is a determinant or consequence of expression. The availability of cell lines allowed us to address this issue by manipulating either expression or methylation and then assaying the effect of the manipulation on the other parameter. First, we took advantage of the observation that transgene expression, like expression of the endogenous MLC1f/3f gene, is myotube-specific: Proliferating myoblasts in rostral and caudal cell lines express negligible amounts of CAT, whereas myotubes express the transgene at levels that are manyfold higher and position-dependent (Donoghue et al. 1992a). We therefore compared transgene methylation in subconfluent myotubes and differentiated myotubes of rostral and caudal cell lines. In each of eight lines tested, the methylation state of myoblasts and myotubes was identical; examples are shown in Figure 5c. Thus, methylation state reflects the cell's history and/or origin rather than its current level of expression. Second, we treated rostrally derived (low expressing, hypermethylated) cells with the DNA demethylating agent, 5-azacytidine, then assayed transgene methylation and expression in the myotubes to which the treated cells gave rise. Treatment with 9 μM 5-azacytidine led to partial demethylation of the transgene (Fig. 5d) and to a ~100-fold increase in the specific activity of CAT (not shown; the increase varied from 90- to 160-fold among lines and was concentration-dependent). Taken together, these results suggest that methylation provides a genomic imprint of rostrocaudal body position that may serve as a component of the positional memory that mammalian cells retain into adulthood (Donoghue et al. 1992a). Further studies of the MLC1-CAT transgene are now required to learn how patterns of methylation are established, presumably in the embryo, and how methylation regulates expression.

Dependence of Graded Expression on Copy Number

One striking difference between the endogenous MLC1f/3f gene and the MLC1-CAT transgene is this: The MLC1f/3f gene is an ordinary single-copy gene (although a second, unlinked pseudogene is present in the mouse; Robert et al. 1984). In contrast, the transgene is present in multiple copies, which are arranged in a head-to-tail tandem array at a single insertion site, as is commonly found in transgenic mice generated by microinjection of linearized DNA (Palmiter and Brinster 1986). Each of the three lines we analyzed, lines 7, 52, and 63 (see above), bears about 40 transgene copies per genome (Rosenthal et al. 1989). Many studies of gene expression in cultured cells have shown that isolated regulatory sequences are more effective when

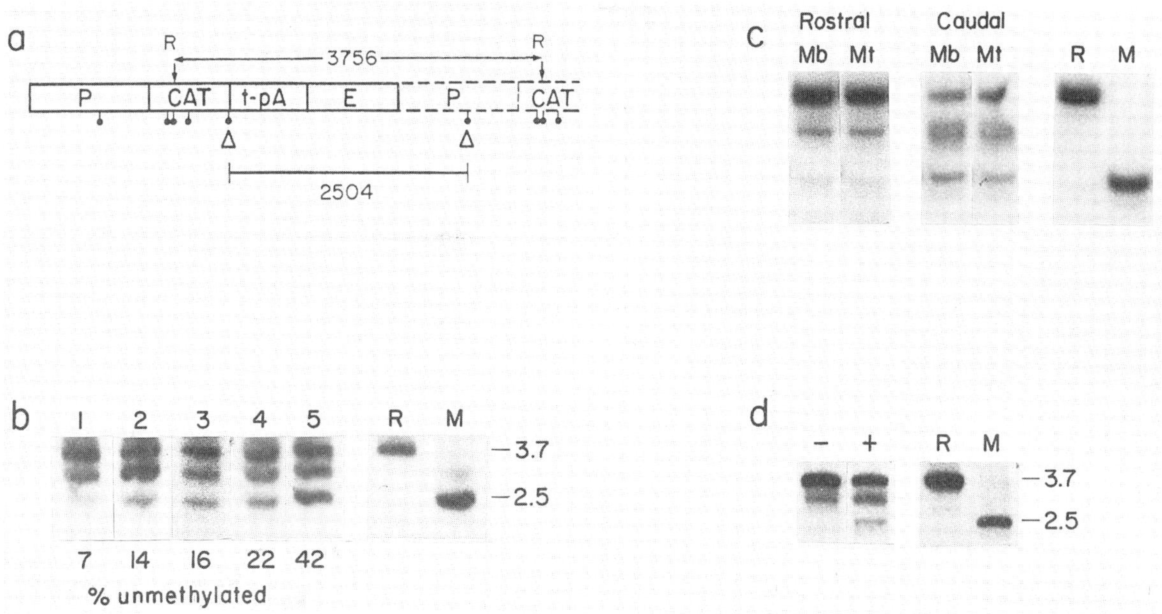

Figure 5. Position-dependent methylation of the MLC1-CAT transgene. (*a*) Diagram of the transgene. Lollipops show recognition sites for the methylation-sensitive restriction enzyme, *Hpa*II, and its methylation-insensitive isoschizomer, *Msp*I. Arrows show the single *Eco*RI site in each transgene copy. If the two sites marked by arrowheads are unmethylated, digestion with *Eco*RI plus either *Hpa*II or *Msp*I will produce a 2.5-kb fragment. If either or both sites are methylated, digestion with *Eco*RI plus *Msp*I still produces a 2.5-kb fragment, but digestion with *Eco*RI plus *Hpa*II produces a fragment of ~ 3 kb (if one site is methylated) or ~ 3.7 kb (if both sites are methylated). (*b*) Methylation analysis of DNA from the following muscle groups: (1) masseter, anterior digastricus and posterior digastricus (jaw); (2) omohyoid, sternohyoid, sternomastoid, clavotrapezius (neck); (3) pectoralis major and minor (chest); (4) triceps and biceps (forelimb); and (5) tibialis anterior and extensor digitorum longus (hindlimb). Lanes *1–5* show *Eco*RI plus *Hpa*II digests, revealing that the average number of unmethylated transgene copies per cell increases in a rostrocaudal gradient. Positions of 3.7-kb and 2.5-kb bands are shown in *Eco*RI (R) and *Eco*RI plus *Msp*I (M) digests, respectively. Values beneath each lane represent the density of the 2.5-kb band as a fraction of the total density of the 2.5-, 3.0-, and 3.7-kb bands. (*c*) Methylation analysis of DNA from myoblasts (Mb) and myotube (Mt)-rich cultures of rostrally (1-3A3, from jaw) and caudally (6-1B2, from leg) derived cell lines. Methylation state is position-dependent in myoblasts, which do not express the transgene, as well as in myotubes, which do. These lanes are of *Eco*RI plus *Hpa*II digests; R and M digests are shown to the right. (*d*) Demethylation of the transgene in a rostrally derived line by treatment with 5-azacytidine. This treatment caused a ~ 100-fold increase in CAT expression. "−" and "+" are *Eco*RI plus *Hpa*II digests of control and 5-azacytidine-treated cells, respectively. Positions of the 3.7-kb (methylated) and 2.5-kb (unmethylated) bands are shown by R and M digests, respectively, as in *b*.

introduced as concatamers than as monomers or dimers and that their effectiveness increases as a steep function of copy number (see, e.g., Schirm et al. 1987). By analogy, it seemed possible that a shallow, positional gradient of expression of the endogenous gene could be amplified by multimerization of the transgene. If binding were strictly cooperative, for example, a 1.15-fold gradient per copy could produce a >100-fold gradient from 40 copies ($1.15^{40} = 268$). Even if the contributions of multiple copies were only additive, the effect of high copy number could be significant (e.g., 2.5-fold × 40 = 100-fold). In this case, the transgene might provide an amplified position-dependent signal that could serve as a useful positional marker but would not mimic a steep gradient of expression of any endogenous gene. Moreover, the multiplicity of copies complicates the interpretation of results on the mechanism of position-dependent expression: We do not know how many of the transgene copies are transcriptionally active, whether it is the number of copies or the activity per copy that varies with position, or whether the hypomethylated copies are the active ones.

In light of these uncertainties, we generated new MLC1 CAT mice and sought lines with relatively few copies per genome. Although there is no established procedure for minimizing copy number, we took the precautions of decreasing the concentration of the DNA solution injected into each oocyte (~0.5 μg/ml instead of the ~10 μg/ml used in the original experiments; this corresponds to ~100 and ~2000 copies per oocyte, respectively) and of isolating the transgene from its plasmid with incompatible ends (*Acc*I at the 5′ end and *Xba*I at the 3′ end; an *Acc*I-*Acc*I fragment was used previously). Thirteen new MLC1-CAT transgenic founders were generated, as assessed by PCR analysis of tail DNA. Relative transgene copy number was estimated by Southern analysis of tail DNA, and CAT activity was measured in biopsies of thigh muscle. Of the 13 founder mice, 12 expressed the transgene. The founders with the lowest apparent copy number were bred, and DNA purified from livers of their offspring was subjected to a more quantitative test of copy number. The DNA was digested with *Eco*RI or *Hind*III, each of which cuts once within the transgene. Southern

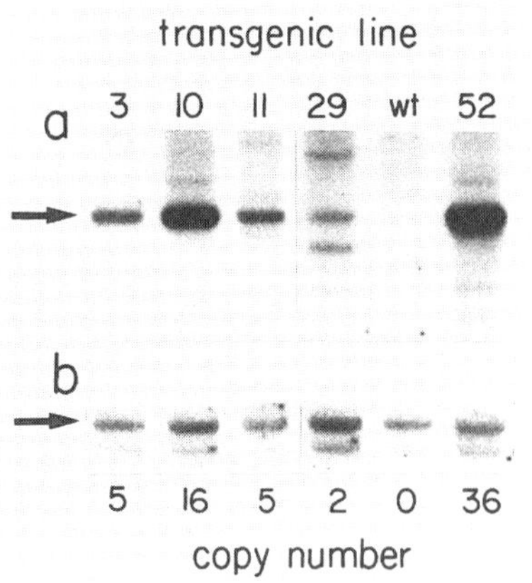

Figure 6. Identification of an MLC1-CAT mouse with few transgene copies. Liver DNA from the lines indicated and from a wild-type mouse (wt) was digested with *Hin*dIII and probed with the full transgene (*a*). The density of the 3.7-kb bands, which represents the complete transgene, was then determined. Faint bands, presumably representing flanking sequences, indicate that the transgene has inserted at a unique chromosomal position in each line. Subsequently, the blot was stripped and reprobed with a fragment of an endogenous single-copy gene, the acetylcholine receptor β subunit (*b*), to permit determination of absolute copy numbers (see text). Approximate numbers of transgene copies per diploid genome are shown beneath each lane.

blots of the digests were probed with the MLC1-CAT plasmid, and the intensity of the 3.7-kb transgene-sized band (see Fig. 5a) was determined by densitometry of autoradiographs (Fig. 6a). To correct for differences in DNA concentration or completeness of digestion among samples, the blots were reprobed with a fragment of the acetylcholine receptor β-subunit gene (Buonanno et al. 1989), which is known to be present in a single copy per haploid genome (Fig. 6b). Values for the transgene were then normalized to values for the endogenous gene and corrected for specific activity of the probes, exposure time, and probe-target ratio. Finally, these values were rounded to the nearest integer and increased by one, because the 3.7-kb band is generated from two tandem copies and therefore underestimates true copy number by one. Consistent with our previous results, line 52 contained about 40 copies of the transgene, and DNA from wild-type mice contained no 3.7-kb hybridizing band. Gratifyingly, three of the new lines contained only a few copies of the MLC1-CAT transgene per diploid genome—approximate copy numbers were 5, 5, and 2 for lines 3, 11, and 29, respectively (Fig. 6).

Next, we assayed CAT activity in individual muscles and muscle groups from animals of lines 3, 11, and 29. Specific activities of CAT were severalfold lower in the new lines than in those studied previously (compare Figs. 2a and 7), suggesting a weak relationship between

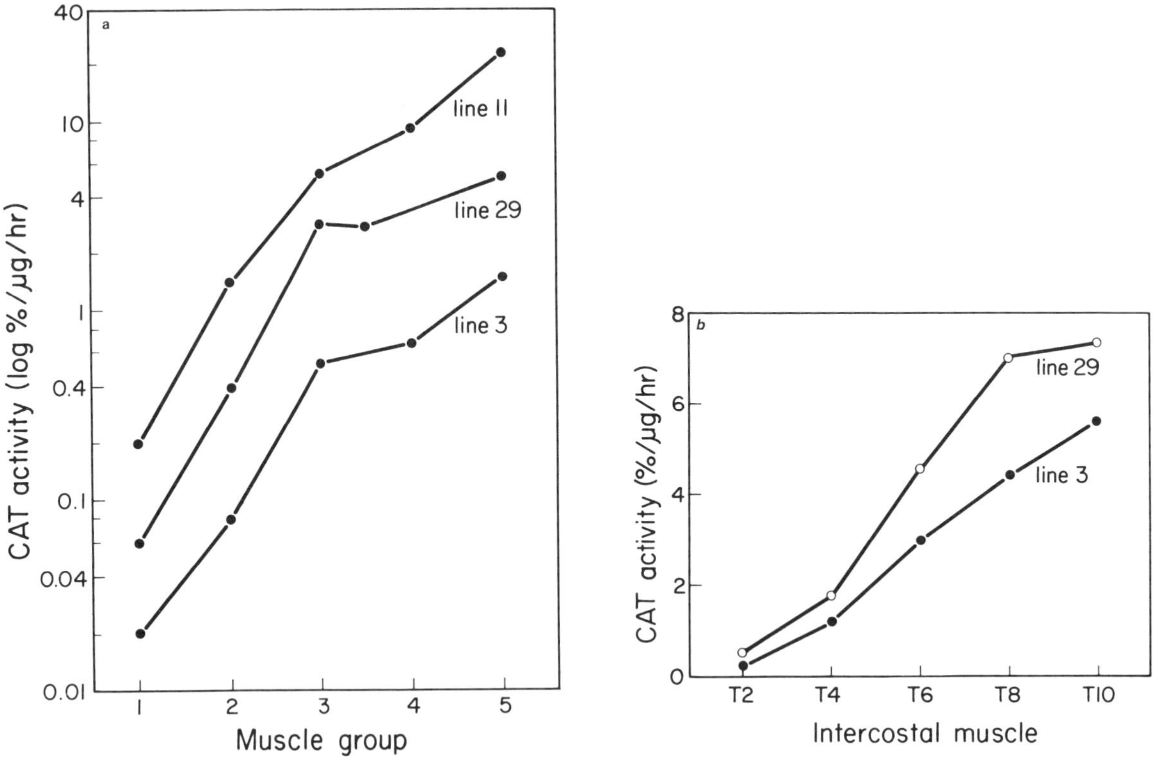

Figure 7. Position-dependent expression of the MLC1-CAT transgene in mice bearing ~2 (line 29) or ~5 (lines 7 and 11) transgene copies per genome. CAT activity was assayed in extracts of the 5 muscle groups listed in Fig. 5 legend (*a*), and in extracts of individual intercostal muscles (*b*). The gradient of expression is as marked in these mice as in mice that bear ~40 copies of the transgene (see Fig. 2).

transgene expression and copy number. Most important, however, CAT expression varied with rostrocaudal position in all three of the new lines. The gradient spanned the body, and was 75-, 115-, and 85-fold from jaw to hindlimb for lines 3, 11, and 29, respectively (Fig. 7a). Moreover, specific activities were graded along the intercostals, with differences of 23- and 14-fold between T2 and T10 for lines 3 and 29, respectively (Fig. 7b). These gradients of expression were similar in form, polarity, and magnitude to those documented earlier for lines 7, 52, and 63 (gradients of 104- to 212-fold from jaw to hindlimb and 8- to 13-fold along the intercostals; Donoghue et al. 1991b, 1992). Thus, the positional gradient of MLC1-CAT expression is not detectably copy number-dependent, and we can dismiss the possibility that graded expression represents a peculiarity of high transgene copy number. Instead, as few as 2–5 copies can direct expression in a ~100-fold gradient, and it is therefore reasonable to speculate that endogenous, single-copy genes exist that are expressed in steep gradients that span the rostrocaudal axis.

Finally, we examined the methylation state of the transgene in muscles from line 29 mice. Genomic DNA was prepared from jaw, chest, and hindlimb muscles and livers of adult mice and subjected to methylation analysis as described above. The sites indicated by arrowheads in Figure 5a (which flank a 2.5-kb fragment) were ~50% unmethylated in hindlimb muscle, more fully methylated in chest muscle, and more methylated still in jaw muscle (Fig. 8, lanes 1–3). Both sites were completely methylated in liver (Fig. 8, lane 4). Both the position-dependence of transgene methylation in muscle and the complete methylation in nonmuscle tissues are similar to the patterns observed in lines 52 and 63, which bear ~40 copies per genome (Fig. 5) (Donoghue et al. 1992a). In the previous study, we did not know whether partial methylation reflected methylation of all transgene copies in only some cells (cellular mosaicism) or methylation of only some transgene copies within individual cells. In the present case, at least some cellular mosaicism in methylation state must exist: Each cell contains only 2 transgene copies and therefore only one 2.5-kb or 3.7-kb transgene fragment, so variable ratios of the two fragments must result from differences among cells within a tissue. In this context, it will be interesting to investigate intercellular variations in transgene expression within individual muscles of line 29 mice. In any event, the position-dependence of transgene methylation in line 29 strengthens the argument that methylation of single-copy endogenous genes could provide a genomic imprint of body position. Moreover, these mice should be more suitable than the high-copy-number mice for detailed studies of the relationship of methylation to expression.

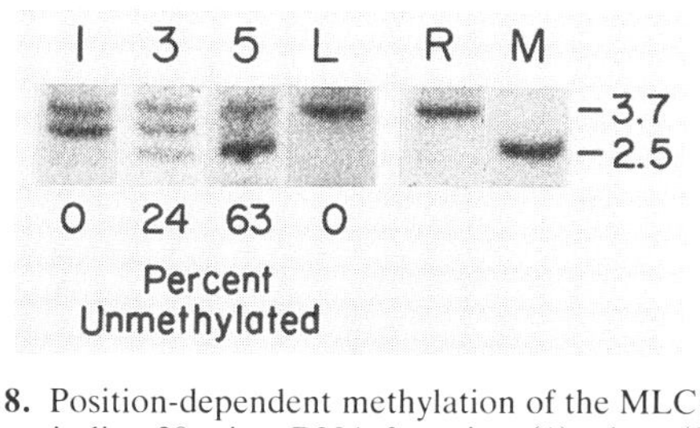

Figure 8. Position-dependent methylation of the MLC1-CAT transgene in line 29 mice. DNA from jaw (1), chest (3), and hindlimb (5) muscles and from liver (L) was digested with *Eco*RI plus *Hpa*II and probed with the complete transgene. The fraction of unmethylated transgene copies in muscle increases in a rostrocaudal gradient; the sites that flank the 2.5-kb band are fully methylated in liver DNA. Positions of the unmethylated (2.5 kb) and methylated (3.7 kb) bands are shown in the *Eco*RI (R) and *Eco*RI plus *Msp*I (M) digests, respectively. Methods and muscles are as in Fig. 5 legend.

CONCLUSIONS

This project began with the electrophysiological demonstration that muscles that arise from or occupy distinct rostrocaudal positions are preferentially innervated and reinnervated by axons from corresponding axial levels. This pattern suggested the existence of complementary sets of positional cues in muscles and nerves, which bias synapse formation in favor of positionally appropriate partners. Although no such cues have yet been found, the MLC1-CAT transgene provides direct evidence for the existence of position-dependent molecular differences among muscles and can serve as the starting point in the search for endogenous determinants of positional information. One complication with this model has been the possibility that the transgene's behavior results from its being present in numerous, tandemly repeated copies in the genome, in which case it might behave very differently from endogenous, single-copy genes. The new results presented here show, however, that gradients of transgene expression and methylation are demonstrable in animals that bear only a few copies of the transgene per genome. We are therefore encouraged to continue using the MLC1-CAT transgene in studies aimed at isolating endogenous genes that regulate position-dependent intercellular interactions in the mammalian neuromuscular system.

ACKNOWLEDGMENTS

We thank J. Alvarez, J. Cunningham, T. Hanley, Y. Johnson, R. Morris-Valero, and B. Patton for their participation in these experiments. This work was supported by grants from the Muscular Dystrophy Association and National Institutes of Health to J.R. Sanes and J.P. Merlie.

REFERENCES

Bennett, M.R. and N.A. Lavidis. 1984. Development of the topographical projection of motor neurons to a rat muscle accompanies loss of polyneuronal innervation. *J. Neurosci.* **4:** 2204.

Bradford, M.M. 1976. A rapid and sensitive method for the quantitation of microgram quantities of protein, utilizing

the principle of protein-dye binding. *Anal. Biochem.* **72:** 248.
Braun, T., E. Bober, B. Winter, N. Rosenthal, and H.H. Arnold. 1990. *Myf*-6, a new member of the human gene family of myogenic determinator factors: Evidence for a gene cluster on chromosome 12. *EMBO J.* **9:** 821.
Brown, M.C. and C.M. Booth. 1983. Postnatal development of the adult pattern of motor axon distribution in rat muscle. *Nature* **304:** 741.
Buckingham, M. 1992. Making muscle in mammals. *Trends Genet.* **8:** 144.
Buonanno, A., J. Mudd, and J.P. Merlie. 1989. Isolation and characterization and the β and ϵ subunit genes of mouse muscle acetylcholine receptor. *J. Biol. Chem.* **264:** 7611.
Cossu, G., B. Zani, M. Coletta, M. Bouchá, M. Pacifici, and M. Molinaro. 1980. *In vitro* differentiation of satellite cells isolated from normal and dystrophic mammalian muscles. A comparison with embryonic myogenic cells. *Cell Differ.* **9:** 357.
Davis, L.G., M.D. Dibner, and J.F. Battey. 1986. DNA preparation from cultured cells and tissue. In *Basic methods in molecular biology*, p. 47. Elsevier, New York.
Donoghue, M.J., J.D. Alvarez, J.P. Merlie, and J.R. Sanes. 1991a. Fiber type- and position-dependent expression of a myosin light chain-chloramphenicol acetyltransferase (CAT) transgene detected with a novel histochemical stain for CAT. *J. Cell Biol.* **115:** 423.
Donoghue, M.J., J.P. Merlie, N. Rosenthal, and J.R. Sanes. 1991b. A rostrocaudal gradient of transgene expression in adult skeletal muscle. *Proc. Natl. Acad. Sci.* **88:** 5847.
Donoghue, M.J., B.L. Patton, J.R. Sanes, and J.P. Merlie. 1992a. An axial gradient of transgene methylation in murine skeletal muscle: Genomic imprint of rostrocaudal position. *Development* (in press).
Donoghue, M., H. Ernst, B. Wentworth, B. Nadal-Ginard, and N. Rosenthal. 1988. A muscle-specific enhancer is located at the 3′ end of the myosin light-chain 1/3 gene locus. *Genes Dev.* **2:** 1779.
Donoghue, M.J., R. Morris-Valero, Y.R. Johnson, J.P. Merlie, and J.R. Sanes. 1992b. Mammalian muscle cells bear a cell-autonomous, heritable memory of their rostrocaudal position. *Cell* **69:** 67.
Fujisawa-Sehara, A., Y. Nabeshima, T. Komiya, T. Uetsuki, A. Asakura, and Y. Nabeshima. 1992. Differential transactivation of muscle-specific regulatory elements including the myosin light chain box by chicken myoD, myogenin, and MRF4. *J. Biol. Chem.* **267:** 10031.
Gorman, C.M. 1985. High efficiency gene transfer into mammalian cells. In DNA cloning (ed. D. Glover), vol. II, p. 143. IRL Press, Oxford.
Green, E.C. 1963. *The anatomy of the rat.* Hafner, New York.
Greishammer, U., D. Sassoon, and N. Rosenthal. 1992. A transgene target for positional regulators marks early rostrocaudal specification of myogenic lineages. *Cell* **69:** 79.
Haberstroh, L., J. Galindo, and R.A. Firtel. 1991. Developmental and spatial regulation of a *Dictyostelium* prespore gene: *cis*-acting elements and a cAMP-induced, developmentally regulated DNA binding activity. *Development* **113:** 947.
Hogan, B., F. Costantini, and E. Lacy. 1986. *Manipulating the mouse embryo: A laboratory manual.* Cold Spring Harbor Laboratory, Cold Spring Harbor, New York.
Holliday, R., F.R.S., M. Monk, and J.E. Pugh. 1990. DNA methylation and gene regulation. *Philos. Trans. R. Soc. Lond. Biol. Sci.* **326:** 177.
Johnson, J.E., B.J. Wold, and S.D. Hauschka. 1989. Muscle creatine kinase sequence elements regulating skeletal and cardiac muscle expression in transgenic mice. *Mol. Cell. Biol.* **9:** 3393.
Landau, B.R., K. Akert, and T.S. Roberts. 1962. Studies on the innervation of the diaphragm. *J. Comp. Neurol.* **119:** 1.
Laskowski, M.B. and J.A. High. 1989. Expression of nerve-muscle topography during development. *J. Neurosci.* **9:** 175.
Laskowski, M.B. and J.R. Sanes. 1987. Topographic mapping of motor pools onto skeletal muscles. *J. Neurosci.* **7:** 252.
———.1988. Topographically selective reinnervation of adult mammalian skeletal muscles. *J. Neurosci.* **8:** 3094.
Laskowski, M.B., A.S. Norton, and P.K. Berger. 1991. Branching patterns of the rat phrenic nerve during development and reinnervation. *Exp. Neurol.* **113:** 212.
McGinnis, W. and R. Krumlauf. 1992. Homeobox genes and axial patterning. *Cell* **68:** 283.
Melton, D.A., P.A. Krieg, M.R. Rebagliati, T. Maniatis, K. Zinn, and M.R. Green. 1984. Efficient in vitro synthesis of biologically active RNA and RNA hybridization probes from plasmids containing a bacteriophase SP6 promoter. *Nucleic Acids Res.* **12:** 7035.
Merlie, J.P. and J.M. Kornhauser. 1989. Neural regulation of gene expression by an acetylcholine receptor promoter in muscles of transgenic mice. *Neuron* **2:** 1295.
Miner, J.H. and B. Wold. 1990. Herculin, a fourth member of the *MyoD* family of myogenic regulatory genes. *Proc. Natl. Acad. Sci.* **87:** 1089.
Miner, J.H., J.B. Miller, and B.J. Wold. 1992. Skeletal muscle phenotypes initiated by ectopic *MyoD* in transgenic mouse heart. *Development* **114:** 853.
Palmiter, R.D. and R.L. Brinster. 1986. Germ-line transformation of mice. *Annu. Rev. Genet.* **20:** 465.
Petropoulos, C.J., M.P. Rosenberg, N.A. Jenkins, N.G. Copeland, and S.H. Hughes. 1989. The chicken skeletal muscle α-actin promoter is tissue specific in transgenic mice. *Mol. Cell. Biol.* **9:** 3785.
Purves, D. and J.W. Lichtman. 1985. *Principles of neural development.* Sinauer, Sunderland, Massachusetts.
Purves, D., W. Thompson, and J.W. Yip. 1981. Reinnervation of ganglia transplanted to the neck from different levels of the guinea pig sympathetic chain. *J. Physiol.* **313:** 49.
Razin, A. and H. Cedar. 1991. DNA methylation and gene expression. *Microbiol. Rev.* **55:** 451.
Renfranz, P.J., M.G. Cunningham, and R.D.G. McKay. 1991. Region-specific differentiation of the hippocampal stem cell line HiB5 upon implantation into the developing mammalian brain. *Cell* **66:** 713.
Robert, B., P. Daubas, A. Akimenko, A. Cohen, I. Garner, J.-L. Guenet, and M. Buckingham. 1984. A single locus in the mouse encodes both myosin light chains 1 and 3, a second locus corresponds to a related pseudogene. *Cell* **39:** 129.
Rosenthal, N., J.M. Kornhauser, M. Donoghue, K.M. Rosen, and J.P. Merlie. 1989. Myosin light chain enhancer activates muscle-specific developmentally regulated gene expression in transgenic mice. *Proc. Natl. Acad. Sci.* **86:** 7780.
Sanes, J.R., D.D. Hunter, T.L. Green, and J.P. Merlie. 1990. S-laminin. *Cold Spring Harbor Symp. Quant. Biol.* **55:** 419.
Schirm, S., J. Jiricny, and W. Schaffner. 1987. The SV40 enhancer can be dissected into multiple segments, each with a different cell type specificity. *Genes Dev.* **1:** 65.
Southern, E.M. 1975. Detection of specific sequences among DNA fragments separated by gel electrophoresis. *J. Mol. Biol.* **98:** 503.
Surani, M.A. 1991. Genomic imprinting: Developmental significance and molecular mechanism. *Genet. Dev.* **1:** 241.
Weintraub, H., R. Davis, S. Tapscott, M. Thayer, M. Krause, R. Benezra, T.K. Blackwell, D. Turner, R. Rupp, S. Hollenberg, Y. Zhuang, and A. Lassar. 1991. The myoD gene family: Nodal point during specification of the muscle cell lineage. *Science* **251:** 761.
Wigston, D.J. and J.R. Sanes. 1982. Selective reinnervation of adult mammalian muscles by axons from different segmental levels. *Nature* **299:** 464.
———.1985. Selective reinnervation of rat intercostal muscles transplanted from different segmental levels to a common site. *J. Neurosci.* **5:** 1208.

Agrin: A Synaptic Basal Lamina Protein That Regulates Development of the Neuromuscular Junction

J.T. CAMPANELLI,* M. FERNS,† W. HOCH,* F. RUPP,* M. VON ZASTROW,*
Z. HALL,† AND R.H. SCHELLER*
*Howard Hughes Medical Institute, Stanford University, Stanford, California 94305; †Department of Physiology, School of Medicine, University of California, San Francisco, California 94143-0444

Neurons communicate with each other by means of specialized intercellular contacts called synapses. The formation of synaptic contacts takes place during embryonic and early postnatal development and during regeneration in the peripheral nervous system. The cellular interactions that mediate the development of synaptic structures are not well understood, although it is likely that both the pre- and postsynaptic elements contribute to the regulation of this process (Dennis 1981; Schuetze and Role 1987). Although it is accepted that stable synaptic connections mediate nervous system function, it is becoming increasingly clear that normal brain function also involves the modulation of synaptic efficacy. Such plasticity, thought to be the molecular correlate of learning and memory, can be mediated by the emergence and disappearance of synaptic contacts in the adult (Bailey and Chen 1989). Thus, elucidation of the mechanisms involved in the development and regeneration of the synapse may yield insight into processes necessary for the plasticity that underlies learning and memory.

The most well characterized synapse is the neuromuscular junction, the contact made between nerve and muscle. The components of the neuromuscular junction include a presynaptic nerve terminal, a postsynaptic muscle cell membrane, and an intervening synaptic cleft occupied by the basal lamina: a form of extracellular matrix. All these components of the synapse are highly specialized with respect to nonsynaptic regions. The presynaptic nerve terminal is specialized to regulate the release of neurotransmitter in response to an influx of calcium ions triggered by an action potential; synaptic vesicles are aligned at a region of the nerve terminal known as the active zone, which also contains high concentrations of N-type calcium channels (Robitaille et al. 1990), as well as other components of the secretory apparatus. The membrane of the postsynaptic muscle fiber is specialized to respond to the secreted chemical messenger released by the nerve. In particular, the acetylcholine receptors (AChRs) are localized at densities of up to 10,000 per square micron beneath the nerve terminal (Fertuck and Salpeter 1974). Other postsynaptic specializations include a high density of the 43-kD protein, which is associated with the AChR, cytoskeletal specializations (Shear and Bloch 1985; Flucher and Daniels 1989), and muscle cell nuclei concentrated at synaptic sites that are specifically activated to express synaptic proteins (Changuex et al. 1990). A variety of components are highly concentrated in the synaptic basal lamina, including a unique complement of laminin and type IV collagen isoforms (Sanes et al. 1990), heparan sulfate proteoglycans (HSP) (Anderson and Fambrough 1983), and acetylcholinesterase (AChE) (Hall 1973; McMahan et al. 1978). The function of acetylcholinesterase is well documented; it terminates the postsynaptic response to presynaptically released acetylcholine by degrading the neurotransmitter. Other components of the synaptic basal lamina are believed to function in organizing the synaptic structure. These include a laminin B1 homolog (S-laminin) (Sanes et al. 1990), HSP (Anderson and Fambrough 1983; Bayne et al. 1984), and agrin (McMahan 1990). This exquisite organization raises the question: What are the molecular events that regulate this highly specialized, local differentiation of nerve and muscle?

The role of the nerve in directing postsynaptic differentiation has been well documented. Ingrowing motor neuron axons direct the accumulation of AChRs at the site of nerve-muscle contact (Anderson and Cohen 1977; Frank and Fischbach 1979). This clustering effect is specific for cholinergic neurons and is independent of neural activity (Cohen et al. 1979). Thus, neurally derived factors other than ACh influence the differentiation of postsynaptic structures.

A variety of factors can change the organization of the muscle cell membrane in ways that resemble changes that occur during development. Electrical activity in the muscle decreases expression of AChR subunits in non-synaptic nuclei and also stabilizes the postsynaptic apparatus (Lomo and Gunderson 1988). The neuropeptide CGRP, localized in at least some motor neuron presynaptic terminals, modulates the synthesis of AChR subunits (New and Mudge 1986), probably via a cAMP-mediated second-messenger pathway (Laufer and Changeux 1987). The protein ARIA modulates the synthesis of the α and ϵ subunits of the AChR (Harris et al. 1988; Martinou et al. 1991). Interestingly, when basic fibroblast growth factor (bFGF) is used to coat latex beads, high densities of receptor are localized at contact sites between the beads and the muscle fibers; this effect is specific for bFGF, since laminin, when similarly presented, is not effective (Peng et al. 1991). Finally, regeneration studies in which nerve and muscle are damaged and then allowed to regenerate show that molecules that reside in the synaptic basal lamina are capable of directing localized differentiation (Sanes et al. 1978; Burden et al. 1979). Thus, the control of synaptic differentiation may be a complex process involving multiple inductive interactions between nerve and muscle, some of which may be mediated by components of the basal lamina.

We have undertaken a molecular analysis of agrin, a synaptic basal lamina molecule implicated in regulating the formation of the postsynaptic apparatus of the neuromuscular junction (NMJ). Agrin has the ability to induce the formation of clusters of synaptic molecules on cultured myotubes (Godfrey et al. 1984; Wallace 1989), is present in the synaptic cleft of mature NMJs (Reist et al. 1987), and is produced by motor neurons at a time when synaptic contacts are being formed (Magill-Solc and McMahan 1988; Rupp et al. 1991). Thus, agrin fulfills many criteria expected of a neurally derived synaptic organizing molecule.

Identification of Agrin and cDNA Clones Encoding Agrin

Agrin's ability to induce the formation of AChR clusters on myotubes was first seen when extracts enriched in synaptic basal lamina from the *Torpedo* electric organ were applied to chick muscle fibers in culture (Godfrey et al. 1984). Monoclonal antibodies were produced that could remove clustering activity from these basal lamina extracts (Reist et al. 1987). These monoclonal antibodies, which stain synaptic basal lamina, were then used to prepare a highly purified agrin preparation containing a group of proteins having molecular masses of 150 kD, 135 kD, 95 kD, and 70 kD (Nitkin et al. 1987; Reist et al. 1987). The AChR clusters induced by the purified agrin were associated with other synaptic components, including acetylcholinesterase, butyrylcholinesterase, HSP, and the 43-kD protein, suggesting a more general role for agrin in regulating synaptic organization (Wallace 1986, 1989).

Several lines of evidence support the notion that motor neurons synthesize and secrete agrin. Ligation of frog sciatic nerve causes the accumulation of agrin immunoreactivity and clustering activity on the proximal side of the ligature (Magill-Solc and McMahan 1990). Furthermore, structures resembling Golgi stacks in spinal cord motor neuron cell bodies are immunoreactive (Magill-Solc and McMahan 1988). Although agrin staining is evident in myotubes cultured in the absence of nerve (Lieth et al. 1991), and in vivo in muscle masses prior to innervation (Fallon and Gelfman 1989), the data suggest that at least some of the synaptic agrin is derived from nerve. This supposition is supported by the use of species-specific anti-agrin antibodies and chimeric muscle-nerve cocultures in vitro (Reist et al. 1992).

To better understand the structure and function of agrin, the amino-terminal amino acid sequences of the four immunoreactive proteins from ray were determined (M. Smith et al., in prep.). The amino-terminal amino acid sequences of the 150-kD and 135-kD species are identical, as are the amino-terminal amino acid sequences of the 95-kD and the 70-kD proteins. Interestingly, the clustering activity is present only in the 95-kD and 150-kD-containing fractions (McMahan and Wallace 1989). Serum antibodies were generated to the affinity-purified agrin proteins and used to screen a cDNA expression library constructed using mRNA from ray electromotor neurons. Four cDNA clones were identified. Protein synthesized from one of these clones was able to bind antibodies that recognized the characteristic pattern of four agrin bands. This clone was sequenced and shown to contain an open reading frame that includes the amino-terminal amino acid sequence of the 95-kD and 70-kD proteins (M. Smith et al., in prep.). This marine ray clone was then used to screen chick and, in our laboratory, rat cDNA libraries. A set of overlapping clones that encode the full-length agrin protein were isolated and characterized in both rat (Rupp et al. 1991) and chick (Tsim et al. 1992).

Domain Structure of the Agrin Gene and Protein

The domain structure of the mammalian agrin gene and protein is illustrated in Figure 1 (Rupp et al. 1991, 1992). Most of the coding region is contained within a 16,000-nucleotide region of the mouse genome. The gene (AGRN/Agrn) has been assigned to chromosome 1 region pter-32 in human and to mouse chromosome 4 (Rupp et al. 1992). The agrin protein is initiated by a methionine followed by a hydrophobic domain which is predicted to serve as a cleaved signal sequence, consistent with an extracellular location for agrin. Nine domains contain a high degree of amino acid sequence identity with Kazal-type protease inhibitors. Interestingly, the region of significant homology with protease inhibitors is contained within the carboxyl half of each of the agrin repeats, whereas the amino half is strikingly similar to EGF repeats. Thus, the agrin protease inhibitor domains appear to have arisen by the fusion of an EGF motif with a protease inhibitor motif. This suggestion is strengthened by the intron/exon structure of the protease inhibitor repeats. The first seven such domains are each encoded on single distinct exons, whereas protease inhibitor domains eight and nine are split by an intron and are, therefore, encoded by two exons. The carboxy-terminal exon largely contains the homology with protease inhibitors, whereas the amino-terminal exon encodes an EGF repeat homolog. This organization suggests these domains have evolved by fusing an EGF repeat exon with a protease inhibitor exon. Although the function of the protease inhibitor domains remains to be demonstrated, they may serve to protect the agrin protein itself from degradation. It

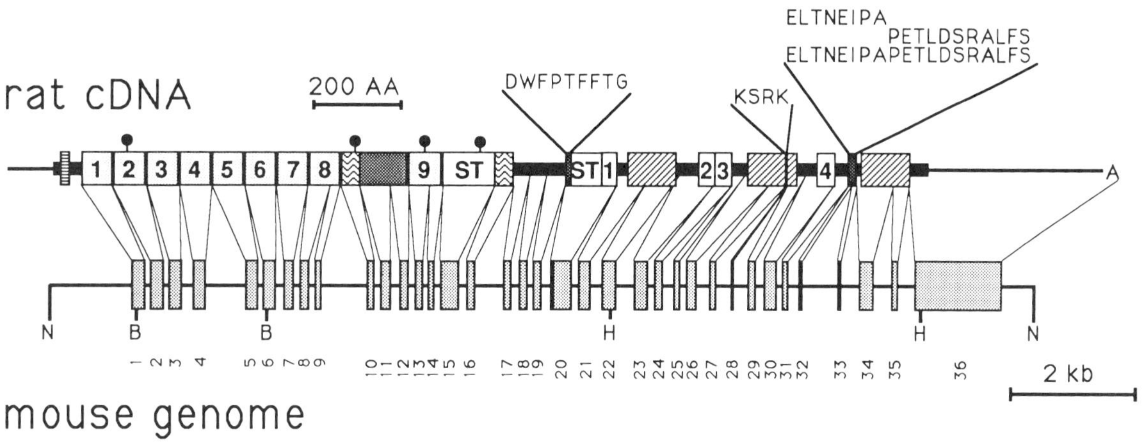

Figure 1. Domain organization of the rat agrin protein and the intron/exon organization of the mouse gene are illustrated. A set of overlapping cDNAs predicts a 1931-amino-acid reading frame flanked by in-frame stop codons. The initiator methionine is followed by a hydrophobic stretch of amino acids that is predicted to be a cleaved signal sequence ( ). Four potential *N*-linked glycosylation sites are located in the amino-terminal half of the protein ( ). Nine domains are homologous to Kazal-type protease inhibitors ( ). Two cysteine- and proline-rich sequences are not similar to other proteins in the database ( ). One region of agrin is similar to laminin domain III ( ), and three other portions of the agrin protein are similar to the laminin globular G domain ( ). Two domains are rich in serine and threonine and may serve as phosphorylation and/or *O*-linked glycosylation sites (ST). Four EGF repeats are located in the carboxy-terminal half of the agrin protein ( ). Three regions of the molecule differ in various cDNA clones as a result of alternative RNA splicing. The amino acid sequences of the alternatively spliced domains of agrin are written above the schematic of the rat protein ( ). The organization of 36 exons that encode most of the agrin protein is indicated along a region of 16 kb of mouse genomic DNA. The restriction enzyme sites indicated are as follows: (N = *Not*I; B = *Bam*HI; H = *Hin*dIII).

has been shown that agrin remains in the synaptic basal lamina after degeneration of pre- and postsynaptic structures (Reist et al. 1987). Alternatively, since development is a time of extensive tissue remodeling, the protease inhibitor domains may play a role in preferentially stabilizing molecules of the synaptic basal lamina.

Exons 10 and 16 are directly repeated sequences rich in cysteine and proline residues. Exons 11 and 12 encode a domain homologous to domain III of laminin. In laminin, this domain has been shown to bind nidogen, to have mitogenic activity on some cells, and to promote neurite attachment but not neurite outgrowth (Edgar et al. 1984; Panayotou at al. 1989). Two regions rich in serine and threonine (exon 15 and the majority of exon 20) may serve as *O*-linked glycosylation sites or protein phosphorylation domains. The carboxy-terminal half of the agrin protein contains four EGF repeats. These repeats not only contain the pattern of six cysteine residues, but also are identical to each other at 40% of their positions. These EGF repeats are not, however, well conserved with respect to the EGF repeats present in the protease inhibitor domains. Finally, the carboxy-terminal portion of the agrin protein also contains three repeated domains that share significant amino acid identity with repeats in the G domain of laminin and merosin. This region of laminin is the globular domain at the base of the cruciform and has been found to bind heparin (Sakashita et al. 1980; Ott et al. 1982) and presumably is involved in interactions with HSPs. Perhaps one or more of these domains of the agrin protein are involved in HSP binding as well.

Three regions of the agrin protein undergo alternative RNA splicing (Fig. 2). The first alternatively spliced exons are at the amino terminus of the second serine/threonine-rich domain. The donor site on the carboxyl terminus of exon 19 can be joined to either one of two acceptor sites on exon 20, separated by nine amino acids. An alternatively spliced region is also found within the second laminin G domain homolog. A four-amino-acid sequence, KSRK, is encoded on a separate exon (exon 28), which can be inserted into the second G domain homologous region. If this G domain is involved in heparin binding, insertion of these basic residues might effect agrin-proteoglycan interactions. A more complex pattern of splicing is observed carboxy terminal to the fourth EGF repeat. At this position, alternative splicing of eight- and eleven-amino-acid exons gives rise to four possibilities. Exon 31 can be spliced directly to exon 34, skipping the eight- and eleven-amino-acid exons. Alternatively, only the eight-amino-acid exon (exon 32), only the eleven-amino-acid exon (exon 33), or the eight- plus the eleven-amino-acid exons can be inserted at this site. Both the eight- and the eleven-amino-acid exons are acidic and contain proline residues, suggesting that they are present on the surface of the protein and adopt a nonhelical structure. A total of 16 forms of the agrin protein are possible from these combinations of alternative RNA splicing. It is not yet known which of these combinations occur in vivo, or how the splicing patterns may vary in different tissues and through development. However, the splicing carboxy-terminal to the fourth EGF repeat effects the ability of transfected agrin proteins to cluster AChR on cultured myotubes, as discussed in detail later in this paper.

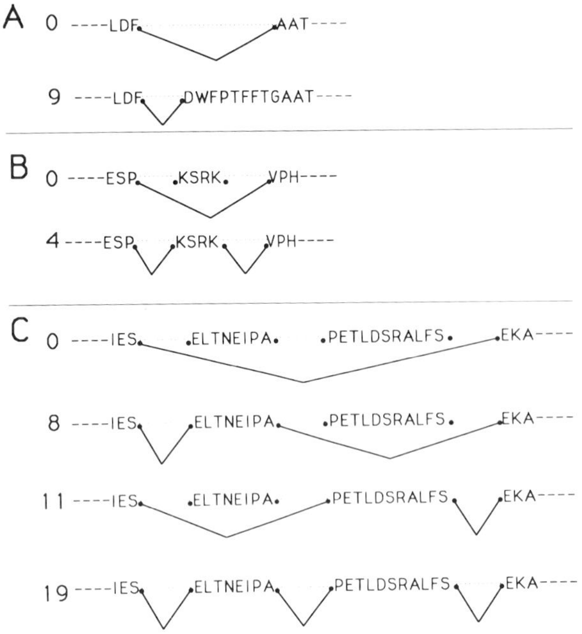

Figure 2. Three regions of the agrin gene are alternatively spliced. (*A*) Exon 19 can be spliced to either of two acceptor sequences on exon 20. The resulting proteins differ by 9 amino acids. (*B*) Exon 27 can be spliced to exon 28 or directly to exon 29, skipping exon 28. The resulting proteins differ by 4 amino acids. The sequences of two rat embryonic spinal cord cDNAs differ by the 12 nucleotides encoding the KSRK amino acids. (*C*) Exon 31 can be spliced directly to exon 34, skipping exons 32 and 33. Alternatively, exon 31 can be spliced to exon 32, then to exon 34, resulting in an 8-amino-acid insert. A further pattern is generated by splicing exon 31 to exon 33, then to exon 34, resulting in an 11-amino-acid insert. Finally, exons 31, 32, 33, and 34 can be joined, resulting in a 19-amino-acid insert.

Expression of Agrin

RNA hybridization and immunohistochemical techniques were used to determine the distribution of agrin transcripts and protein (Rupp et al. 1991; M. Ferns et al., in prep). Northern blots using RNA from brain and muscle demonstrate that the transcript is more abundantly expressed during embryonic development than in the adult (Fig. 3). The peak of expression in muscle is at embryonic day 16. After this time, the expression decreases dramatically, and in the adult is below the level of detection in this experiment. The agrin mRNA expression in brain peaks just prior to birth, considerably later than in muscle, and then dramatically declines but remains detectable in the adult brain.

The protein levels detected by Western blotting reflect generally the pattern of mRNA expression (Rupp et al. 1991). Antibodies were generated against the amino terminus (N antibody), middle (M antibody), and carboxyl terminus (L antibody) of the agrin molecule (Campanelli et al. 1991). Blots from embryonic brain, spinal cord, and muscle, reacted with the M antibody, reveal a 200-kD band (Fig. 4). In adult tissues, only a weak signal is detected in brain, and spinal cord and muscle protein levels are below the levels of

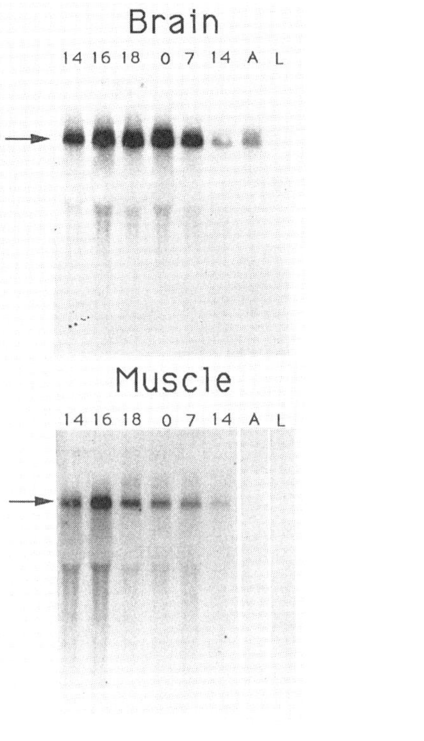

Figure 3. Developmental expression of agrin mRNA. Total RNA from embryonic days 14, 16, 18, birth (0), postnatal days 7, 14, and adult (A) were analyzed for agrin transcripts using Northern blotting. Liver RNA (L) is also included. The arrow indicates the position of the 8.5-kb transcript present in both brain (*upper* panel) and muscle (*lower* panel).

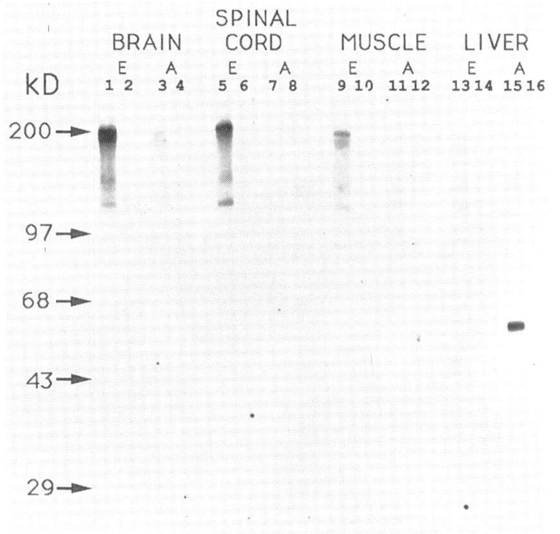

Figure 4. Agrin protein in embryonic and adult tissues. Protein from embryonic day 14 (E) and adult (A) tissues was fractionated on SDS acrylamide gels, blotted, and reacted with the agrin M antibody followed by ^{125}I-labeled goat anti-rabbit antibodies. Even-numbered lanes are reacted with the agrin M antibody and odd-numbered lanes are reacted with the agrin M antibody blocked with the fusion protein used to raise the antibody. A major band at 200 kD is observed in embryonic brain (lane *1*), spinal cord (lane *5*), and muscle (lane *9*). Adult liver (lane *15*) contains a 65-kD band of unknown origin.

detection afforded by these procedures. Little expression of the 200-kD proteins is detected in either embryonic or adult liver. A 68-kD band is detected in adult liver, but the nature of this band is not understood. Western blotting experiments with electric organ tissue also reveal a prominent 200-kD band. These experiments demonstrate that the major agrin species in vivo is approximately 200 kD—similar to the size of the protein predicted from analysis of cDNAs. The four lower-molecular-weight proteins purified from *Torpedo* are most likely the result of proteolysis that occurs during the isolation procedure.

To obtain a more detailed picture of the site of agrin expression, in situ hybridization and immunohistochemical techniques were employed (Fig. 5) (Rupp et al. 1991). In embryonic day 16 spinal cord (Fig. 5A), significant mRNA is detected in the ventricular zone, the spinal artery, and the dorsal root ganglia. However, the most notable feature of the hybridization to this tissue is the high level of expression seen in the ventral horn: the site of the motor neurons. Motor neuron cell bodies are immunoreactive with anti-agrin antibodies, as are the neuromuscular junctions (Rupp et al. 1991). In muscle, immunoreactivity is widespread in early embryonic muscle fibers and becomes progressively limited to synaptic sites as development proceeds (M. Ferns et al., in prep.). In the adult, the agrin immunoreactivity is highly localized to synaptic sites, and little or no staining is seen outside the synapse (Fig. 5B,C).

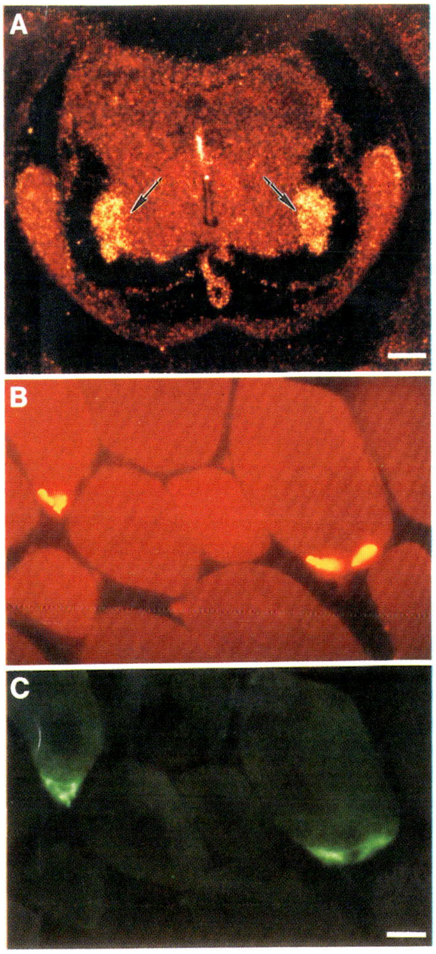

Figure 5. Localization of agrin mRNA and protein. (*A*) In situ hybridization to embryonic day 14–17 spinal cord. Intense labeling is observed in the motor neurons, as indicated by the arrows. Bar, 150 μm. (*B*) Identification of neuromuscular junctions using rhodamine-conjugated α-bungarotoxin to label AChRs. Two adult diaphragm muscle fibers are shown to have synapses in this section. (*C*) The agrin L antibody was reacted with the same tissue section shown in *B*, followed by a goat anti-rabbit antibody conjugated to fluorescein. Agrin is seen concentrated at synaptic sites in adult muscle. Bar, 6 μm.

Expression of Recombinant Agrins in Heterologous Cells

Recombinant agrin proteins have been expressed both stably in CHO cells and transiently using COS cells (Campanelli et al. 1991). Both systems yield comparable results. The transfected agrin gene product is approximately 210 kD and reacts with all three anti-agrin antibodies, demonstrating that full-length agrin protein is synthesized in the transfected cells. Stably transfected CHO cells have a perinuclear and reticular pattern of staining with the M antibody in the presence of saponin (Fig. 6A). In the absence of saponin, the immunoreactivity is patchy and concentrated between cells (Fig. 6B). The N and L antibodies have a similar staining pattern in the absence of saponin that is not altered by saponin permeabilization. The data suggest that the antibodies have different affinities for the agrin protein in different cellular compartments. Perhaps the molecule is not completely folded or is not assembled into a complex in the endoplasmic reticulum and Golgi, allowing accessibility to the major epitope recognized by the M antibody. The staining in the absence of saponin suggests the protein is present on the extracellular surface of the transfected cells. To investigate this issue in more detail, the cells were reacted with the M antibody, fixed, and reacted with secondary antibodies conjugated to gold particles. Labeling is observed along the cell surface (Fig. 6C) with a threefold higher density of labeling along the surface of projections from the cells (Fig. 6D). Gold particles can be seen up to 40 nm from the membrane associated with an electron-dense material characteristic of extracellular matrix. Finally, agrin-transfected CHO cells can be sorted by flow cytometry using the anti-agrin antibodies. We conclude from these studies that agrin is secreted and associates with the surface of the transfected cells, most likely via interactions with an extracellular matrix.

Functional Activity of Recombinant Agrin

Cells expressing agrin isoforms were cocultured with muscle fibers derived from primary myoblasts or myoblast cell lines (Fig. 7) (Campanelli et al. 1991; Ferns et al. 1992). As discussed above, the agrin protein may exist in as many as 16 different isoforms as a result of alternate RNA splicing. Thus far, we have limited our

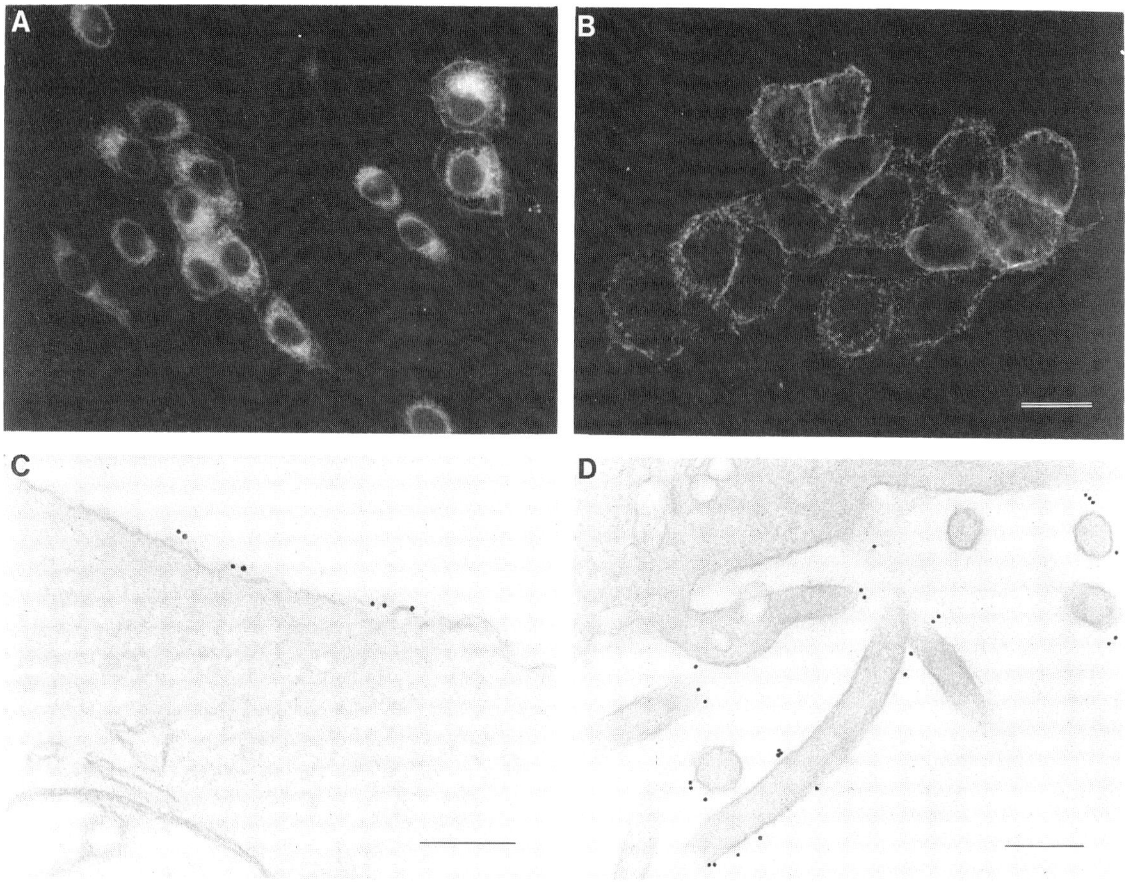

Figure 6. Transfected cells localize agrin on the extracellular surface. (*A*) Agrin$_0$-transfected CHO cells were permeabilized with saponin and stained with the agrin M antibody followed by a goat anti-rabbit antibody coupled to fluorescein. No staining of untransfected cells is observed. Immunoreactivity is perinuclear and reticular, characteristic of the endoplasmic reticulum and Golgi apparatus, as well as cell-surface associated. (*B*) Agrin$_0$-transfected cells that are not permeabilized show a patchy surface staining pattern. Agrin M antibody (shown) illustrates staining character seen with all three agrin antibodies in the absence of detergents. Bar, 14 μm (*A* and *B*). (*C*) Agrin$_0$-transfected cells were stained with the M antibody followed by gold (10 nm) conjugated goat-anti-rabbit antibody. The gold particles are observed along the cell surface. The particles are often observed associated with electron-dense material and up to 40 nm from the plasma membrane, suggestive of association with the extracellular matrix. Bar, 200 nm. (*D*) A threefold higher density of labeling is observed along projections from the surface of the transfected cells. Bar, 200 nm.

analysis of the AChR clustering activity of agrin to the splice variants generated at the position carboxy-terminal to the fourth EGF repeat. We refer to these isoforms as agrin$_0$, agrin$_8$, agrin$_{11}$, and agrin$_{19}$, depending on the inclusion of no exon, an 8-amino-acid exon, an 11-amino-acid exon, or an 8- plus 11-amino-acid exon, respectively. All these isoforms include both the 9- and 4-amino-acid exons discussed above.

Initial experiments focused on culturing agrin$_0$-transfected cells with primary myotubes (Campanelli et al. 1991). In these experiments, AChR aggregates were localized at contact sites between agrin-transfected cells and muscle fibers. The agrin$_0$ form is also capable of inducing AChR clusters at contacts made with myotubes derived from C2-C12 cells, a myoblast cell line (Ferns et al. 1992). A different result was obtained when agrin$_0$-transfected cells were cocultured with S27-derived myotubes. The S27 cell line is a genetic variant of the C2-C12 myoblasts selected for decreased incorporation of $^{35}SO_4$ into surface proteins (Gordon and Hall 1989). The S27 cells exhibit decreased incorporation of $^{35}SO_4$ and [^{3}H]glucosamine into proteoglycans and have reduced chain length heparan sulfate and chondroitin sulfate proteoglycans. Furthermore, the accumulation of extracellular matrix on the surface of S27 myotubes is reduced, and no spontaneous clusters of AChR are detected (H. Gordon et al., in prep.). Coculture of agrin$_0$-transfected cells with S27-derived myotubes did not result in the accumulation of AChR at contact sites.

At this point, we wondered if the other alternatively spliced forms of the protein might be active in clustering receptors on S27 cells. Agrin$_0$-, agrin$_8$-, agrin$_{11}$-, and agrin$_{19}$-transfected cells are all capable of inducing AChR aggregates at contact sites on C2-C12-derived myotubes, whereas only the agrin$_8$ and agrin$_{19}$ forms are active on S27-derived myotubes (Ferns et al. 1992). The agrin$_8$ and agrin$_{19}$ forms are also active in cluster-

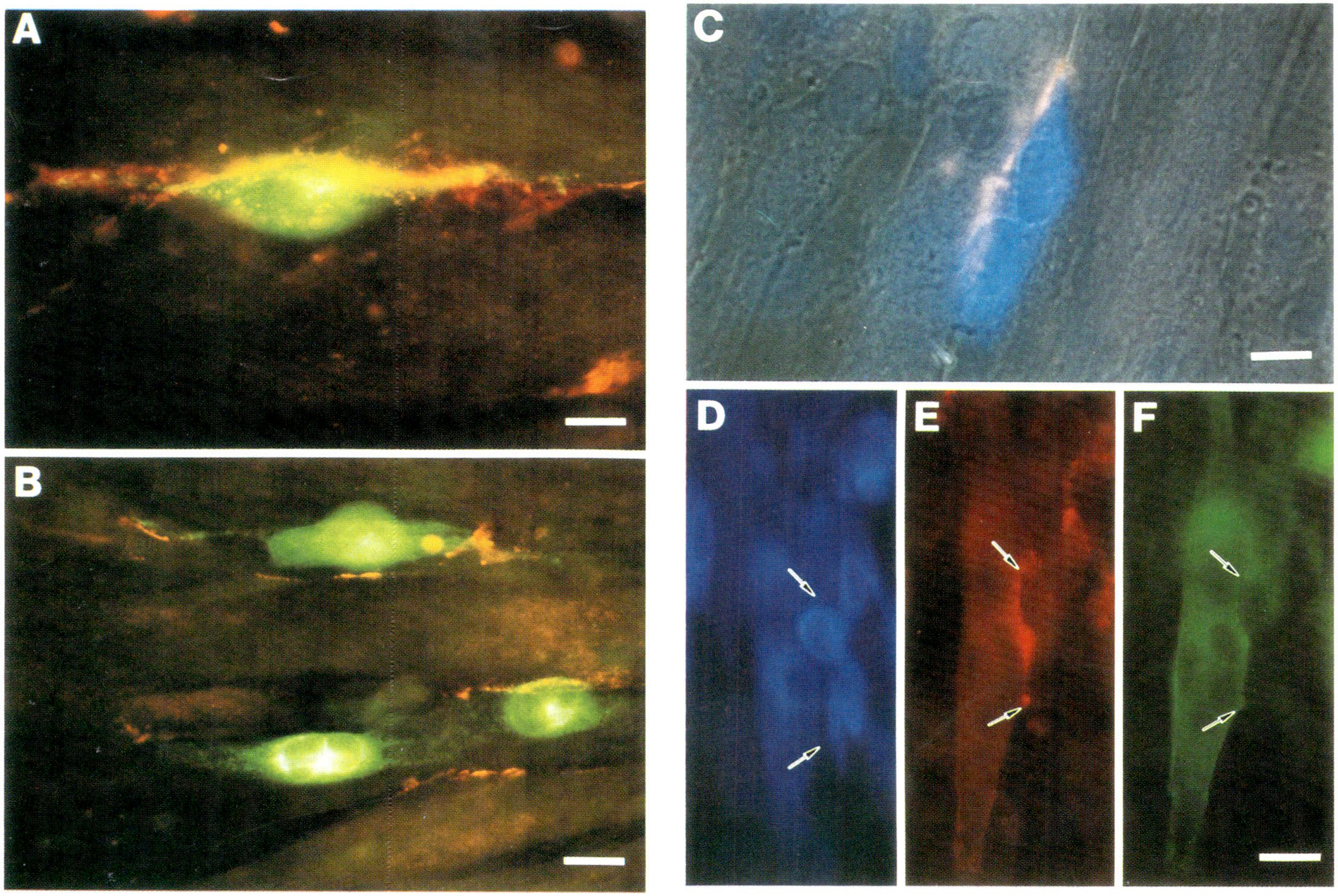

Figure 7. Agrin-transfected cells induce clustering of AChRs. (*A*) An $agrin_{19}$ transiently transfected COS cell cocultured with C2-C12 myotubes. The M antibody was reacted with the culture followed by a goat anti-rabbit antibody coupled to fluorescein (green). Rhodamine-conjugated α-bungarotoxin (red) labels AChRs. Bar, 6 μm. (*B*) An $agrin_{19}$ transiently transfected COS cell cocultured with S27 myotubes. Labeling is as in *A*. Bar, 9 μm. (*C*) An $agrin_0$ stably transfected CHO cell cocultured with C2-C12 myotubes. The M antibody was reacted with the culture followed by a goat anti-rabbit antibody conjugated to cascade blue (blue). AChRs are labeled with α-bungarotoxin conjugated to rhodamine (red). Bar, 14 μm. (*D-E*) An $agrin_8$-transfected CHO cell cocultured with a β_2-adrenergic receptor transfected C2-C12 myotube. The agrin M antibody was reacted with the culture followed by a goat anti-rabbit antibody conjugated to cascade blue (*D*; blue). α-Bungarotoxin conjugated to rhodamine (*E*; red) labels AChRs. Monoclonal antibody 12CA5, followed by goat anti-mouse fluorescein (*F*; green) labels adrenergic receptors. A high density of α-bungarotoxin is evident at the site of contact between the agrin-expressing CHO cells and the myotube (*E*), yet the adrenergic receptor is evenly distributed along the myotube surface (*F*). Bar, 22 μm.

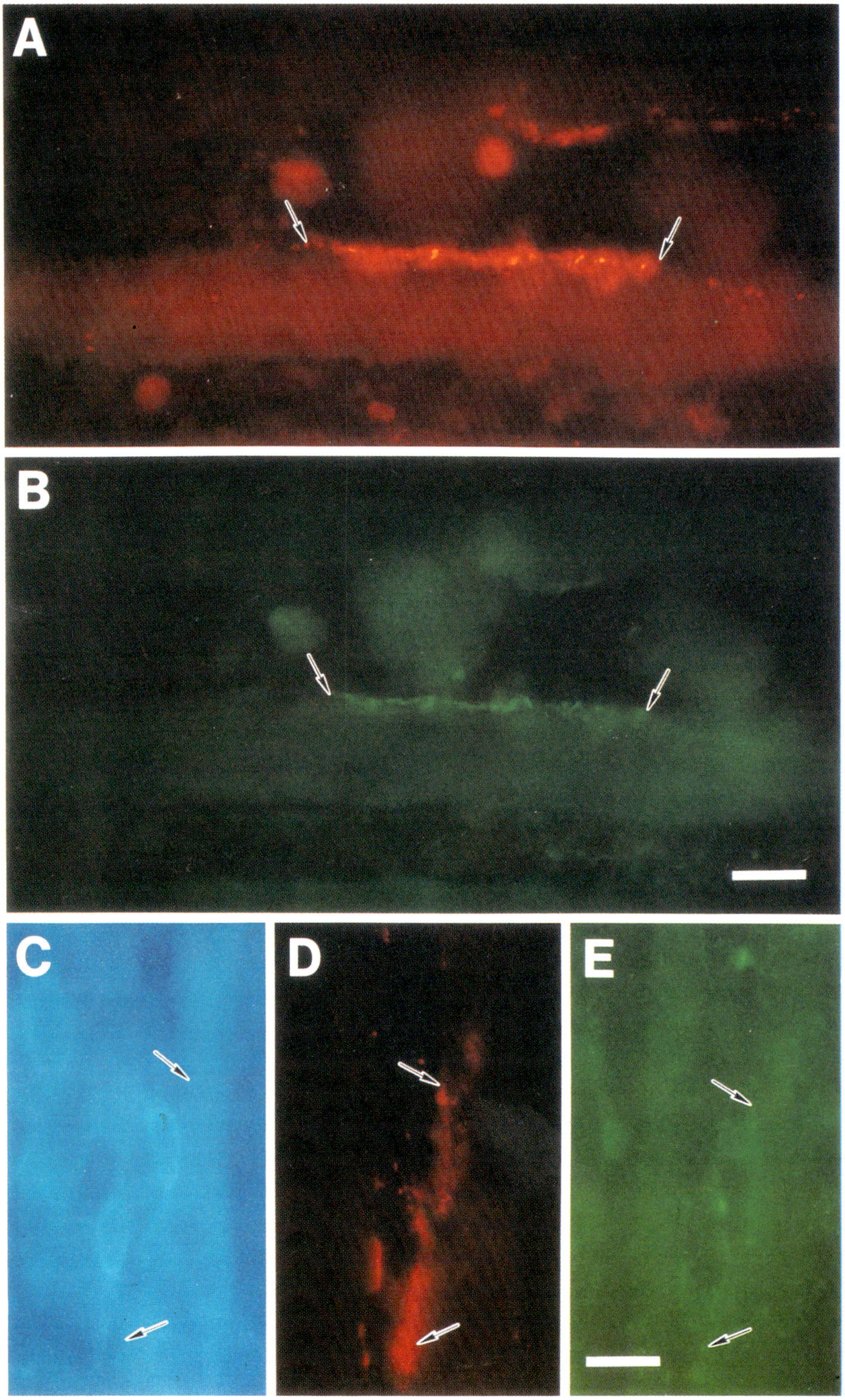

Figure 8. Agrin-induced AChR clusters may label with anti-phosphotyrosine antibodies. (*A, B*) $Agrin_8$-transfected COS cells cocultured with chick myotubes. α-Bungarotoxin conjugated to rhodamine (red) labels AChRs. A mouse monoclonal anti-phosphotyrosine antibody (PY20) followed by a goat anti-mouse antibody coupled to fluorescein (green) reveals phosphotyrosine immunoreactivity. Bar, 6 μm. (*C, D*) $Agrin_{19}$-transfected CHO cells cocultured with C2-C12 myotubes. The M antibody was reacted with the culture followed by a goat anti-rabbit antibody coupled to cascade blue (*C*; blue). α-Bungarotoxin coupled to rhodamine labels AChRs (*D*; red). Phosphotyrosine (*E*; green) visualized as in *B*, reveals phosphotyrosine immunoreactivity. Bar, 22 μm.

Table 1. RNA Splicing Regulates Agrin Activity

	Myotube type				
	rat primary	mouse C2	mouse S27	chick primary	C2 adrenergic
$Agrin_0$	+	+	–	–	–
$Agrin_8$	+	+	+	+	–
$Agrin_{11}$	+	+	–	– –/+	–
$Agrin_{19}$	+	+	+	+	–

ing AChR on primary chick myotubes, whereas the $agrin_{11}$ form is only very weakly active and the $agrin_0$ form is inactive. The data from these experiments are summarized in Table 1.

The relative levels of surface immunoreactivity for the $agrin_0$, $agrin_8$, $agrin_{11}$, and $agrin_{19}$ isoforms stably expressed in CHO cells were quantitated using fluorescence-activated cell sorting. Although the different lines express varying levels of cell-surface immunoreactivity, these differences do not correlate with the activity profile (Ferns et al. 1992). For instance, the $agrin_0$ form, which is inactive on S27 or chick myotubes, has a higher level of surface immunoreactivity than the $agrin_8$ form, which is highly active in all cases tested. Thus, the observed differences in activity are not the result of differing levels of expression of the different isoforms.

We then asked if other neurotransmitter receptor proteins cluster along with the AChR in our coculture system. Expression of epitope-tagged human α_2-C10- and β_2-adrenergic receptor cDNAs (von Zastrow and Kobilka 1992 and in prep.) were transfected into C2-C12 myoblasts. Flow cytometry of the transfected myoblasts showed that immunoreactive receptor protein was present on the cell surface. Immunocytochemical localization of adrenergic receptors on myotubes in the absence of detergents confirmed a cell-surface localization in the differentiated cells (Fig. 7F). As a further confirmation of the expression of transfected adrenergic receptors, α-receptor-transfected C2-C12 cells were shown to have 24-fold higher levels of specific yohimbine binding than nontransfected cells (360 vs. 15 fmole/mg). Similarly, β-receptor-transfected C2-C12 cells exhibited 6-fold higher levels of specific dihydroallprenolol binding than nontransfected cells (530 vs. 90 fmole/mg). Thus, C2-C12 cells can express functional α_2-C10-adrenergic and β_2-adrenergic receptor on their cell surface. Coculture of agrin-expressing CHO cells with adrenergic receptor-expressing myotubes resulted in the clustering of AChR at contact sites; however, these sites do not have clustered α- or β-adrenergic receptors (Fig. 7D–F). This result was obtained with the four agrin isoforms discussed above. From these experiments, we conclude that the clustering observed in our coculture experiments is specific for the AChR and not simply due to nonspecific aggregation of surface glycoproteins.

These experiments demonstrate a striking difference in the ability of the different spliced forms of agrin to specifically cluster AChR. The two most active forms include an 8- amino-acid exon, suggesting this sequence is important in regulating AChR clustering activity. The activity of $agrin_0$ and $agrin_{11}$ differs quite dramatically, depending on which muscle cells are used for the assay (Table 1). Since these forms are inactive on S27 myotubes, proteoglycans and/or an organized extracellular matrix are likely to mediate the action of at least some forms of agrin.

In the chick system, similar effects of splicing on clustering activity have been reported using truncated agrin molecules corresponding to the carboxy-terminal half of agrin (Ruegg et al. 1992). There, a comparison of alternative splice forms reveals that soluble agrin activity depends on the inclusion of an 11-amino-acid insertion at a splice site carboxy-terminal to the fourth EGF repeat. Sequence comparison reveals that the site of alternative splicing is conserved, and that 5 of the 11 amino acids are identical between rat and chick. Whether a chick exon corresponding to the rat 8-amino-acid splice variant exists remains to be determined. At the splice site in the second laminin G-like domain, the 4-amino-acid exon is perfectly conserved between species and in chick was found to be necessary for aggregating activity (Ruegg et al. 1992). The contribution of this exon to activity of full-length agrin in the rat system is currently being assessed.

Mechanism of Agrin-induced AChR Clustering

The accumulated data suggest that agrin is synthesized in motor neurons and secreted into the synaptic basal lamina. One activity of agrin is to initiate a redistribution of preexisting AChR, resulting in an accumulation beneath the presynaptic nerve terminal. The mechanism whereby agrin acts to cluster AChR at synaptic sites is unknown. Two recent sets of observations are potentially relevant. First, the four AChR subunits were expressed in either quail fibroblasts (Phillips et al. 1991) or oocytes (Froehner et al. 1990), resulting in dispersed AChR on the surface of these cells. If, however, the 43-kD protein is coexpressed along with the AChR subunits, the receptors are found in small patches on the surface of the heterologous expressing cells. These data suggest the 43-kD protein may play a role in forming the high concentrations of AChR at synaptic sites. Perhaps the 43-kD protein interacts with multiple AChR, resulting in a cross-linking of receptors. In addition, the 43-kD protein may function as a bridge between the receptor and the cytoskeleton. This raises the possibility that the clustering activity of agrin could be via regulation of the distribution or function of the 43-kD protein.

A second set of observations implicate tyrosine kinases in the mechanism of agrin action. AChRs clustered at synaptic sites in adult chick and rat are phosphorylated on several sites, including a tyrosine in the intracellular domain of the β subunit. When agrin is added to chick muscle fibers in culture, the induced patches of AChR also stain with anti-phosphotyrosine antibodies (Wallace et al. 1991). The phosphorylation occurs as the clusters are forming, consistent with a causal role in aggregation. However, in rat development, the AChR clusters are not associated with phosphotyrosine immunoreactivity until postnatal life, long after the initial formation of clusters (Qu et al. 1991). These data suggest that phosphorylation in rat may be playing a role later in development, perhaps in stabilizing the AChR clusters following synapse elimination.

We have examined the localization of phosphotyrosine immunoreactivity with AChR clusters induced by agrin-transfected cells (Fig. 8). Only weak phosphotyrosine immunoreactivity is observed at a fraction of the induced AChR clusters. At this point, we do not know if all clusters are weakly phosphorylated and only a few are labeled heavily enough to be detectable, or if only a fraction of the clusters contain phosphotyrosine.

CONCLUSIONS

In this paper, we have summarized our findings on the structure and function of the agrin gene and protein. Several points merit further discussion.

First, from analysis of the gene and cDNAs, we predict that the agrin protein is a secreted molecule of approximately 200 kD. This size is consistent with the sizes of the recombinant proteins when expressed in COS or CHO cells. Furthermore, this size is also consistent with Western blot data using antisera raised against recombinant rat agrin to probe proteins from mammalian brain, spinal cord, and muscle and also from ray electric organ, suggesting that the naturally occurring agrin molecule available for interaction with muscle fibers in vivo has a molecular mass of 200 kD.

Second, our expression studies of the agrin molecule in mammalian cell lines suggest that the agrin protein, although secreted, remains associated with the cell surface. Thus, in our coculture assays, the induction of high-density AChR aggregates is restricted to the site of contact between muscle cell and agrin-expressing cell. This situation is reminiscent of the formation of receptor clusters induced by neurites in vivo; these clusters are highly correlated with the sites of contact between nerve and muscle. This further supports the notion that the agrin produced by heterologous cells is not significantly different from the agrin present at the developing NMJ.

Third, the existence of multiple structural motifs suggests that the agrin protein may subserve multiple functions. Indeed, the agrin protein is capable of interacting with the cell surface of CHO and COS cells, as well as with the myotube surface. The result of the former interaction is accumulation of agrin on the cell surface, and the outcome of the latter interaction is the formation of AChR clusters. It is unlikely that these activities are identical, since the coexpression of agrin with the subunits of the AChR in COS cells does not result in the formation of AChR clusters (M. Yu and Z. Hall, unpubl.). The expression of a series of deleted agrin constructs would be a first step in analyzing the domains subserving these functions, as well as facilitating the discovery of other agrin functions. In this regard, a truncated form of chick agrin, with the amino-terminal half deleted, is secreted from transfected COS cells and retains clustering activity (Tsim et al. 1992). This suggests that the association of the full-length agrin molecule with the surface of transfected cells is an activity residing in the amino-terminal half of agrin, whereas clustering activity may reside in the carboxy-terminal half. It will be interesting to further assign specific functions to the predicted structural motifs.

Fourth, we have demonstrated that all the isoforms of agrin tested to date have the ability to direct the clustering of AChRs on C2-C12 and primary rat myotubes and that this clustering exhibits specificity for the AChR. Of the myotubes tested, these two cell types are most likely to reflect the cell-cell interactions occurring in the developing muscle. Although the cellular distribution of the agrin isoforms remains to be determined, all forms tested so far have been isolated from E17 spinal cord: a time when synaptic contacts between nerve and muscle are forming. Thus, at present, we cannot rule out a role for any of these isoforms in synaptic formation. A more detailed analysis of the temporal and spatial expression patterns of the different isoforms should prove informative in this respect.

Fifth, when tested on myotubes derived from different sources, the agrin isoforms differ in their ability to cluster AChRs. This is true when rat agrin is tested on chick myotubes or on a mouse muscle cell line deficient in proteoglycan synthesis (S27). Whether the chick/rat difference is a species difference is not known. In fact, it should be noted that the level of spontaneous AChR clustering exhibited by chick- and rat-derived myotubes is quite different, and as discussed above, the timing of cluster-associated phosphotyrosine staining differs between the two preparations.

The difference between C2-C12- and S27-derived myotubes may be a reflection of the mechanism of agrin-induced clustering. It is possible that the forms of agrin that are inactive on S27 cells ($agrin_0$ and $agrin_{11}$) require the presence of proteoglycans as a low-affinity receptor akin to the role suggested for proteoglycans in the activity of bFGF (Rapraeger et al. 1991; Yayon et al. 1991). The active forms ($agrin_8$ and $agrin_{19}$) may be of sufficient affinity for the clustering receptor that they are functionally independent of the putative low-affinity HSP receptor. Alternatively, the differences may suggest that the $agrin_8$ and $agrin_{19}$ forms each induce clustering by two independent mechanisms. For example, the amino-terminal half of the agrin molecule, which is identical in all of the agrin forms tested, may confer one mode of clustering activity that is dependent on proteoglycans, whereas the 8-amino-acid exon may confer a second independent clustering activity that is independent of proteoglycans. The $agrin_0$ and $agrin_{11}$ forms would thus have only one of the activities present in $agrin_8$ and $agrin_{19}$. The involvement of proteoglycans in the function of agrin is attractive, given the presence of domains homologous to the heparin-binding domain of laminin.

In summary, our isolation and analysis of the agrin gene and cDNAs have revealed a structurally diverse protein comprising a large variety of domains found in other molecules. We have begun functional tests of recombinant agrin and shown that the protein product displays AChR clustering activity. Taken together with our expression studies demonstrating that the agrin gene is expressed in motor neurons at the time that synaptic connections are being formed, it seems likely that agrin does mediate at least some of the nerve-induced changes that occur during postsynaptic muscle differentiation. Further experimentation should reveal more interesting properties of the agrin protein, which is likely to be as functionally diverse as its structure implies.

ACKNOWLEDGMENTS

We thank Sheree Harrison for excellent technical assistance. This work was supported by grants from the National Institutes of Health and Muscular Dystrophy Association to Z.H. J.T.C. was supported by National Institutes of Health training grant GM-07276, W.H. received a fellowship from Deutsche Forschungsgemeinschaft, and M.v.Z. is a Dana fellow in neuroscience.

REFERENCES

Anderson, M.J. and M.W. Cohen. 1977. Nerve-induced and spontaneous redistribution of acetylcholine receptors on cultured muscle cells. *J. Physiol.* **268:** 757.

Anderson, M.J. and D.M. Fambrough. 1983. Aggregates of acetylcholine receptors are associated with plaques of basal lamina heparan sulfate proteoglycan on the surface of skeletal muscle fibers. *J. Cell Biol.* **97:** 1396.

Bailey, C. and M. Chen. 1989. Time course of structural changes at identified sensory neuron synapses during long-term sensitization in Aplysia. *J. Neurosci.* **9:** 1774.

Bayne, E.K., M.J. Anderson, and D.M. Fambrough. 1984. Extracellular matrix organization in the developing muscle: Correlation with acetylcholine receptor aggregates. *J. Cell Biol.* **99:** 1486.

Burden, S., P.B. Sargent, and U.J. McMahan. 1979. Acetylcholine receptors accumulate at original synaptic sites in the absence of the nerve. *J. Cell Biol.* **82:** 412.

Campanelli, J.T., W. Hoch, F. Rupp, T. Kreiner, and R.H. Scheller. 1991. Agrin mediates cell contact-induced acetylcholine receptor clustering. *Cell* **67:** 909.

Changeux, J.P, C. Babinet, J.L. Bessereau, A. Bessis, A. Cartaud, J. Cartaud, P. Daubas, A. Devillers-Thiéry, A. Duclert, J.A. Hill, B. Jasmin, A. Klarsfeld, R. Laufer, H.O. Nghiêm, J. Piette, M. Roa, and A.M. Salmon. 1990. Compartmentalization of acetylcholine receptor gene expression during development of the neuromuscular junction. *Cold Spring Harbor Symp. Quant. Biol.* **55:** 381.

Cohen, M., M. Anderson, E. Zorychta, and P. Weldon. 1979. Accumulation of receptors at nerve muscle contacts in culture. *Prog. Brain Res.* **49:** 335.

Dennis, M.J. 1981. Development of the neuromuscular junction: Inductive interactions between cells. *Annu. Rev. Neurosci.* **4:** 43.

Edgar, D., R. Timpl, and H. Thoenen. 1984. The heparin-binding domain of laminin is responsible for its effects on neurite outgrowth and neuronal survival. *EMBO J.* **3:** 1463.

Fallon J. and C.E. Gelfman. 1989. Agrin-related molecules are concentrated at acetylcholine receptor clusters in normal and aneural developing muscle. *J. Cell Biol.* **108:** 1527.

Ferns, M., W. Hoch, J.T. Campanelli, F. Rupp, Z.W. Hall, and R.H. Scheller. 1992. RNA splicing regulates agrin-mediated acetylcholine receptor clustering activity on cultured myotubes. *Neuron* **8:** 1.

Fertuck, H.C. and M.M. Salpeter. 1974. Localization of acetylcholine receptor by ^{125}I-labeled alpha bungarotoxin binding at mouse motor end-plates. *Proc. Natl. Acad. Sci.* **71:** 1376.

Flucher, B.E. and M.P. Daniels. 1989. Distribution of Na^+ channels and ankyrin in neuromuscular junctions is complementary to that of acetylcholine receptors and the 43 kd protein. *Neuron* **3:** 163.

Frank, E. and G.D. Fischbach. 1979. Early events in neuromuscular junction formation in vitro. Induction of acetylcholine receptor clusters in the postsynaptic membrane and morphology of newly formed synapses. *J. Cell Biol.* **83:** 143.

Froehner, S.C., C.W. Luetje, P.B. Scotland, and J. Patrick. 1990. The postsynaptic 43K protein clusters muscle nicotinic acetylcholine receptors in *Xenopus* oocytes. *Neuron* **5:** 403.

Godfrey, E.W., R.M. Nitkin, B.G. Wallace, L.L. Rubin, and U.J. McMahan. 1984. Components of *Torpedo* electric organ and muscle that cause aggregation of acetylcholine receptors on cultured muscle cells. *J. Cell Biol.* **99:** 615.

Gordon, H. and Z. Hall. 1989. Glycosaminoglycan variants in the C2 muscle cell line. *Dev. Biol.* **135:** 1.

Hall, Z.W. 1973. Multiple forms of acetylcholinesterase and their distribution in endplate and non-endplate regions of rat diaphragm muscle. *J. Neurobiol.* **4:** 343.

Harris, D.A., D.L. Falls, R.M. Dill-Devor, and G.D. Fischbach. 1988. Acetylcholine receptor-inducing factor from chicken brain increases the level of mRNA encoding the receptor alpha subunit. *Proc. Natl. Acad. Sci.* **85:** 1983.

Laufer, R. and J.P. Changeux. 1987. Calcitonin gene-related peptide elevates cyclic AMP levels in chick skeletal muscle: Possible neurotrophic role for a coexisting neuronal messenger. *EMBO J.* **6:** 901.

Lieth, E., C.A. Cardasis, and J.R. Fallon. 1991. Muscle-derived agrin in cultured myotubes: Expression in the basal lamina and at induced acetylcholine receptor clusters. *Dev. Biol.* **148:** 1.

Lomo, T. and K. Gunderson. 1988. Trophic control of skeletal muscle membrane properties. In *Nerve muscle cell trophic communication* (ed. H.L. Fernandez and J.A. Donoso), p.62. CRC Press, Boca Raton, Florida.

Magill-Solc, C. and U.J. McMahan. 1988. Motor neurons contain agrin-like molecules. *J. Cell Biol.* **107:** 1825.

———. 1990. Synthesis and transport of agrin-like molecules in motor neurons. *J. Exp. Biol.* **153:** 1.

Martinou, J.C., D.L. Falls, G.D. Fischbach, and J.P. Merlie. 1991. Acetylcholine receptor-inducing activity stimulates expression of the ϵ-subunit gene of the muscle acetylcholine receptor. *Proc. Natl. Acad. Sci.* **88:** 7669.

McMahan, U.J. 1990. The agrin hypothesis. In *Cold Spring Harbor Symp. Quant. Biol.* **55:** 407.

McMahan, U.J. and B.G. Wallace. 1989. Molecules in basal

lamina that direct the formation of synaptic specializations at neuromuscular junctions. *Dev. Neurosci.* **11:** 227.

McMahan, U.J, J. Sanes, and L. Marshall. 1978. Cholinesterase is associated with the basal lamina at the neuromuscular junction. *Nature* **271:** 172.

New, H.V. and A.W. Mudge. 1986. Calcitonin gene-related peptide regulates muscle acetylcholine receptor synthesis. *Nature* **323:** 809.

Nitkin, R.M., M.A. Smith, C. Magill, J.R. Fallon, Y.M. Yao, B.G. Wallace, and U.J. McMahan. 1987. Identification of agrin, a synaptic organizing protein from *Torpedo* electric organ. *J. Cell Biol.* **105:** 2471.

Ott, U., E. Odermatt, J. Engel, H. Furthmayr, and R. Timple.1982. Protease resistance and conformation of laminin. *Eur. J. Biochem.* **123:** 63.

Panayotou, G., P. End, M. Aumailley, R. Timpl, and J. Engel. 1989. Domains of laminin with growth-factor activity. *Cell* **56:** 93.

Peng, H.B., L.P. Baker, and Q. Chen. 1991. Induction of synaptic development in cultured muscle cells by basic fibroblast growth factor. *Neuron* **6:** 237.

Phillips, W.P., C. Kopta, P. Blount, P.D. Gardner, J.H. Steinbach, and J.P. Merlie. 1991. ACh receptor-rich membrane domains organized in fibroblasts by recombinant 43-kilodalton protein. *Science* **251:** 568.

Qu, Z., E. Moritz, and R.L. Huganir. 1991. Regulation of tyrosine phosphorylation of the nicotinic acetylcholine receptor at the rat neuromuscular junction. *Neuron* **4:** 367.

Rapraeger, A.C., A. Krufka, and B.B. Olwin. 1991. Requirement of heparan sulfate for bFGF-mediated fibroblast growth and myoblast differentiation. *Science* **252:** 1705.

Reist, N.E., C. Magill, and U.J. McMahan. 1987. Agrin-like molecules at synaptic sites in normal denervated, and damaged skeletal muscles. *J. Cell Biol.* **105:** 2457.

Reist, N.E., M.J. Werle, and U.J. McMahan. 1992. Agrin released by motor neurons induces the aggregation of acetylcholine receptors at neuromuscular junctions. *Neuron* **8:** 865.

Robitaille, R., E. Adler, and M. Charlton. 1990. Strategic location of calcium channels at transmitter release sites of frog neuromuscular synapses. *Neuron* **5:** 773.

Ruegg, M.A., K.W.K. Tsim, S.E. Horton, S. Kröger, G. Escher, E.M. Gensch, and U.J. McMahan. 1992. The agrin gene codes for a family of basal lamina proteins that differ in function and distribution. *Neuron* **8:** 691.

Rupp, F., D.G. Payan, C. Magill-Solc, D.M. Cowan, and R.H. Scheller. 1991. Structure and expression of a rat agrin. *Neuron* **6:** 811.

Rupp, F., T.H. Özçelik, M. Linial, K. Peterson, U. Francke, and R.H. Scheller. 1992. Structure and chromosomal localization of the mammalian agrin gene. *J. Neurosci.* (in press).

Sakashita, S., E. Engvall, and E. Ruoslahti. 1980. Basement membrane glycoprotein laminin binds to heparin. *FEBS Lett.* **116:** 243.

Sanes, J.R, U.J. McMahan, and L. Marshall. 1978. Reinnervation of muscle fiber basal lamina after removal of myofibers. Differentiation axons at original synaptic sites. *J. Cell Biol.* **102:** 420.

Sanes, J.R., D.D. Hunter, T.L. Green, and J.P. Merlie. 1990. S-Laminin. *Cold Spring Harbor Symp. Quant. Biol.* **55:** 419.

Schuetze, S.M. and L.W. Role. 1987. Developmental regulation of nicotinic acetylcholine receptors. *Annu. Rev. Neurosci.* **10:** 403.

Shear, C.R. and R.J. Bloch. 1985. Vinculin in subsarcolemmal densities in chicken skeletal muscle: Localization and relationship to intracellular and extracellular studies. *J. Cell Biol.* **101:** 240.

Tsim, K.W.K., M.A. Ruegg, G. Escher, S. Kröger, and U.J. McMahan. 1992. cDNA that encodes active agrin. *Neuron* **8:** 677.

von Zastrow, M. and B.K. Kobilka. 1992. Ligand-regulated internalization and recycling of human β_2-adrenergic receptors between the plasma membrane and endosomes containing transferrin receptors. *J. Biol. Chem.* **267:** 3530.

Wallace, B.G. 1986. Aggregating factor from *Torpedo* electric organ induces patches containing acetylcholine receptors, acetylcholinesterase, and butyrylcholinesterase on cultured myotubes. *J. Cell Biol.* **102:** 783.

———. 1989. Agrin-induced specializations contain cytoplasmic, membrane, and extracellular matrix-associated components of the postsynaptic apparatus. *J. Neurosci.* **9:** 1294.

Wallace, B.G., Z. Qu, and R.L. Huganir. 1991. Agrin induces phosphorylation of the nicotinic acetylcholine receptor. *Neuron* **6:** 869.

Yayon, A., M. Klagsbrun, J.D. Esko, P. Leder, and D.M. Ornitz. 1991. Cell surface, heparin-like molecules are required for binding of basic fibroblast growth factor to its high affinity receptor. *Cell* **64:** 841.

Control of Floor Plate Identity and Function in the Embryonic Nervous System

A. Klar, T.M. Jessell, and A. Ruiz i Altaba
Howard Hughes Medical Institute, Department of Biochemistry and Molecular Biophysics, Center for Neurobiology and Behavior, Columbia University, New York, New York 10032

The patterning of the embryonic nervous system in vertebrates depends on cellular interactions that control the polarity of the neural plate, the identity and position of neural cell types, and the trajectory of developing axons. Within the region of the neural plate that gives rise to the spinal cord, many of these patterning events appear to be controlled by signals that derive from two midline cell groups, axial mesodermal cells of the notochord and an overlying group of neural plate cells that differentiate into the floor plate at the ventral midline of the neural tube (Kingsbury 1930; Jessell and Dodd 1992).

The differentiation of floor plate cells begins before neural tube closure and depends on inductive signals from the notochord (van Straaten et al. 1988; Placzek et al. 1990b; Schoenwolf and Smith 1990). The notochord and floor plate provide signals that contribute to the specification and allocation of cell types in the ventral spinal cord. Evidence for this has come primarily from experiments in chick embryos in which notochord or floor plate grafts placed next to the neural tube induce ectopic ventral neuronal types including motor neurons (Placzek et al. 1991; Yamada et al. 1991; Ericson et al. 1992). Inversely, prevention of floor plate differentiation by removal of the notochord results in the development of a spinal cord that lacks motor neurons and other ventral neurons (Hirano et al. 1991; Placzek et al. 1991; Yamada et al. 1991; Ericson et al. 1992). Similarly, in other vertebrates, the absence of the notochord and floor plate leads to abnormal neural patterning and the absence of ventral neuronal types (Bovolenta and Dodd 1991; Clarke et al. 1991; Ruiz i Altaba 1992). The floor plate also provides cues that orient the growth of axons to and across the ventral midline of the spinal cord. A secreted floor plate chemoattractant can orient the growth of commissural axons in vitro (Tessier-Lavigne et al. 1988; Placzek et al. 1990a,c), and these axons exhibit erratic behavior at the ventral midline of the spinal cord in embryos lacking a floor plate (Bovolenta and Dodd 1990,1991; Bernhardt et al. 1992).

The mechanisms involved in the differentiation of floor plate cells remain unknown. To address this problem, we have attempted to characterize molecules that are involved in the specification of floor plate cells and in mediating their signaling functions. One approach has been to identify transcription factors that are expressed selectively by midline cells during gastrulation, and this has led to the identification of *Pintallavis*, a member of the *HNF-3/fork head* family of transcription factors (Fig. 1) (Ruiz i Altaba and Jessell 1992). *Pintallavis* is expressed transiently by axial mesoderm and midline neural plate cells, and overexpression of this gene in frog embryos results in marked defects in neural development and patterning (Ruiz i Altaba and Jessell 1992).

A second approach has used subtractive hybridization techniques to identify genes expressed specifically in the floor plate. This has led to the identification of a gene, *F-spondin*, which encodes a secreted glycoprotein with adhesive properties (Klar et al. 1992). *F-*

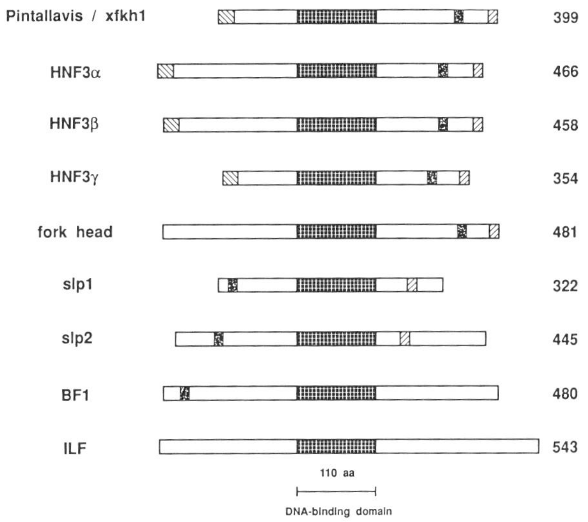

Figure 1. Schematic diagram showing the general structure of Pintallavis and related proteins. The position of the putative DNA-binding domain (defined on the basis of studies of the HNF-3 proteins) is marked. Other conserved domains are indicated by shaded boxes. Numbers on the right indicate the number of amino acids in each protein. Further details of each protein are provided for Pintallavis (Ruiz i Altaba and Jessell 1992), XFKH 1 (Dirksen and Jamrich 1992), HNF3 α, β, γ (Lai et al. 1990,1991), fork head (Weigel et al. 1989); BF-1 (Tao and Lai 1992), sloppy paired 1 and 2 (slp 1,2) (Grossniklaus et al. 1992), interleukin binding factor (ILF) (Li et al. 1991). Pintallavis and the HNF-3 proteins share similar amino-terminal sequences. The HPFSI motif found in the carboxy-terminal region of Pintallavis, the HNF-3 proteins, and fork head is also found in BF-1, slp-1, and slp-2, but in the amino-terminal region. The carboxy-terminal sequences of Pintallavis, the HNF-3 gene, and fork head are also found within the carboxyl half of the slp1 and slp2 proteins. The function of these three conserved sequences is not known.

spondin is expressed selectively by the floor plate at early stages of spinal cord development and promotes the adhesion of spinal cord neurons and the outgrowth of sensory axons in vitro (Klar et al. 1992). *F-spondin* also provides a marker with which to examine whether the expression of *Pintallavis* by midline neural plate cells is involved in their acquisition of floor plate properties. Deregulated expression of *Pintallavis* leads to the ectopic expression of *F-spondin* within the neural tube (A. Ruiz i Altaba et al., in prep.), suggesting that the expression of *Pintallavis* by cells at the midline of the neural plate in normal embryos may be involved in specifying floor plate identity or function.

RESULTS

Isolation and Characterization of *Pintallavis*

The notochord and floor plate exhibit a striking conservation in their signaling and inductive properties in early embryos (Hornbruch and Wolpert 1986; Wagner et al. 1990; Placzek et al. 1991; Yamada et al. 1991). In addition, both of these midline cell groups are unsegmented but are surrounded by intrinsically segmental tissues. The notochord is surrounded by the somites and the floor plate by neuromeres in the hindbrain and perhaps also in the diencephalon (Vaage 1969; Puelles et al. 1987; Keynes and Stern 1988; Lumsden and Keynes 1989; Fraser et al. 1990). The segregation of tissues into segmented and unsegmented structures may reflect an evolutionary conserved decision that could be controlled by similar molecules in different species. One candidate molecule for which such a function has been proposed is the *fork head* gene which, in flies, has been implicated in the repression of thoracic segmentation in anterior structures (Jürgens and Weigel 1988). *Fork head* encodes a nuclear protein that exhibits a domain that is similar to that of the DNA-binding domain of a class of transcription factors expressed in a variety of tissues including lung, liver, and other endodermal organs; the *HNF-3* genes (Fig. 1) (Weigel et al. 1989; Lai et al. 1990,1991; Weigel and Jackle 1990). This conservation prompted us to examine whether similar genes are expressed by midline cells in frog embryos at the time that the neural plate and notochord form. We have identified a gene, *Pintallavis*, which encodes a protein with a high degree of sequence similarity in the putative DNA-binding domain with the HNF-3 and fork head proteins (Ruiz i Altaba and Jessell 1992). Several other members of this family have recently been identified, including the mouse *BF-1* gene that is expressed in the embryonic forebrain (Tao and Lai 1992), an interleukin enhancer-binding protein (ILF), and the fly *sloppy paired* 1 and 2 genes (Fig. 1). *Pintallavis* exhibits 40–50% overall identity with *HNF-3*α, β, and γ and *fork head* but does not appear to be their *Xenopus* homolog. A closely related *Xenopus* gene, *XFKH1*, has been isolated (Dirksen and Jamrich 1992), the distribution of which is similar but not identical to that of *Pintallavis*, suggesting that they may represent the two copies of the same gene in the tetraploid *Xenopus* genome.

Embryonic Expression of *Pintallavis*

Pintallavis mRNA expression, analyzed by whole-mount in situ hybridization (Harland 1991), is first detected in $<30°$ of arc in the dorsal marginal zone of late blastulae (Fig. 2A) (Ruiz i Altaba and Jessell 1992), and by the time of appearance of the dorsal lip, expression of *Pintallavis* occupies a 60° arc in the dorsal marginal zone (Fig. 2A). At the onset of gastrulation, *Pintallavis* is expressed at high levels in cells undergoing involution at the dorsal midline (Fig. 2B). The dorsal expression of *Pintallavis* in the axial mesoderm is graded, with high levels detectable in posterior regions and low levels in prechordal mesoderm (Fig. 2C). Transverse sections of whole-mount labeled embryos at stage 12–12 1/2 show that *Pintallavis* mRNA is present at high levels in the notochord and in cells at the midline of the neural plate, overlying the notochord (Fig. 2C,D). The expression of *Pintallavis* in the midline of the neural plate appears to be restricted to deep layer cells. *Pintallavis* mRNA is also expressed in midline endodermal cells lining the archenteron, showing that *Pintallavis* is expressed by midline cells of all three germ layers.

In *Xenopus* embryos, the midline group of neural plate cells which expresses *Pintallavis* forms the notoplate (Jacobson 1981; Keller et al. 1991). Notoplate cells derive from the dorsal noninvoluting marginal zone and converge and extend to form a long strip at the midline of the neural plate. To determine whether *Pintallavis* expression in midline neural plate cells is intrinsic to the notoplate or depends on signals transmitted vertically from the underlying mesoderm, *Pintallavis* expression was examined in neural tissue in the absence of the underlying notochord. This can be achieved by the generation of exogastrulae in which notoplate extension and neural differentiation within the ectoderm still occur, although the mesoderm does not involute under the ectoderm (Holtfreter 1933; Kintner and Melton 1987; Ruiz i Altaba 1990,1992). In exogastrulae, *Pintallavis* is expressed in the mesoderm, but it is not detectable in the neural region (Fig. 3). This result suggests that the expression of *Pintallavis* by cells at the midline of the neural plate in normal embryos depends on inductive or maintenance signals from the underlying notochord. Consistent with this result, the lack of vertical signaling from the notochord leads to the absence of a floor plate (see below) and ventral neurons, such as those that express restrictin (Rathjen et al. 1991), in the neural tube of exogastrulae (Fig. 7) (see Ruiz i Altaba 1992).

Ectopic Expression of *Pintallavis*

The potential function of *Pintallavis* has been examined by injecting synthetic *Pintallavis* RNA into one cell of embryos at the 2-cell stage. A lineage tracer was

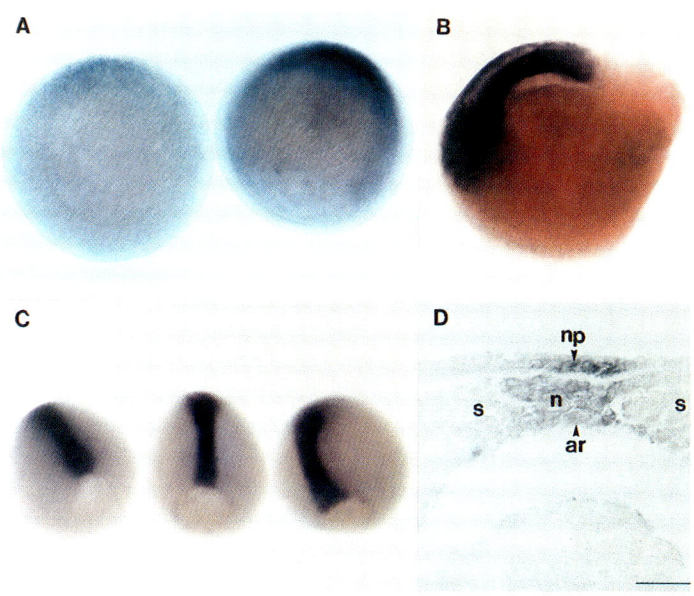

Figure 2. Expression of *Pintallavis* RNA in early *Xenopus* embryos as determined by whole-mount in situ hybridization. (*A*) *Left*: Ventral view of late blastula embryo (stage 8–8 1/2) showing expression of *Pintallavis* RNA in < 30° of arc. *Right*: Ventral view of early gastrula (stage 10) embryo showing expression of *Pintallavis* RNA in the dorsal marginal zone that will give rise to the axial mesoderm, notochord, and prechordal plate. In both cases dorsal side is up. (*B*) Lateral view of late gastrula (stage 12–12 1/2) embryo showing *Pintallavis* RNA localized to the notochord, overlying neural plate, and prechordal plate mesoderm. Anterior is to the left and dorsal side is up. (*C*) Dorsal view of three gastrula (stage 12 1/2) embryos showing midline expression of *Pintallavis*. Closing blastopore is to the bottom and anterior is up. (*D*) Cross-section of late gastrula-early neurula (stage 12 1/2) embryo labeled in whole mount showing expression of *Pintallavis* in the midline of the three germ layers. Dorsal side is to the top. (ar) Archenteron roof; (n) notochord; (np) neural plate; (s) somite. Scale bar for *A*, *B*, 300 μm; *C*, 500 μm; *D*, 100 μm.

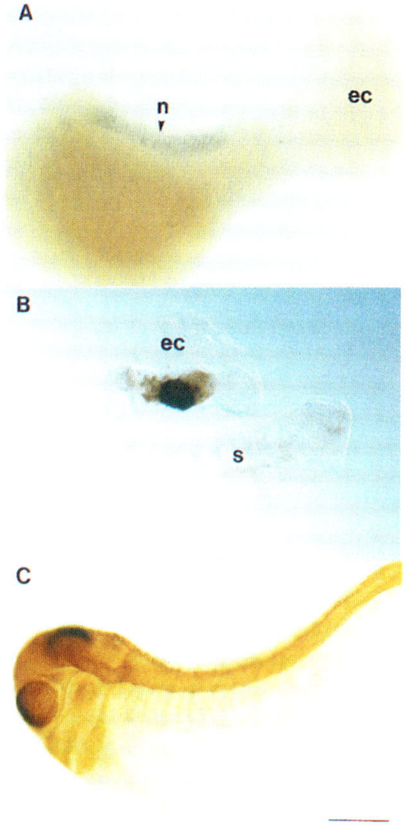

Figure 3. *Pintallavis* RNA is not expressed in neural ectoderm of *Xenopus* exogastrula embryos. (*A*) Localization of *Pintallavis* RNA by whole-mount in situ hybridization in a complete exogastrula embryo at stage 12–12 1/2. *Pintallavis* is expressed by the notochord (n) present in the mesodermal/endodermal component of the exogastrula but not in the ectodermal sac (ec). (*B*) Localization of Xen 1 neural antigen in a tadpole (stage 30–32) *Xenopus* embryo. Note labeling of the CNS, eye, cranial ganglia, and migrating neural crest. (*C*) Localization of Xen 1 neural antigen (Ruiz i Altaba 1992) in the ectodermal component of a complete exogastrula embryo (stage 36), similar to that in *A*. Scale bar for *A*, 250 μm; *B*, 400 μm; *C*, 200 μm.

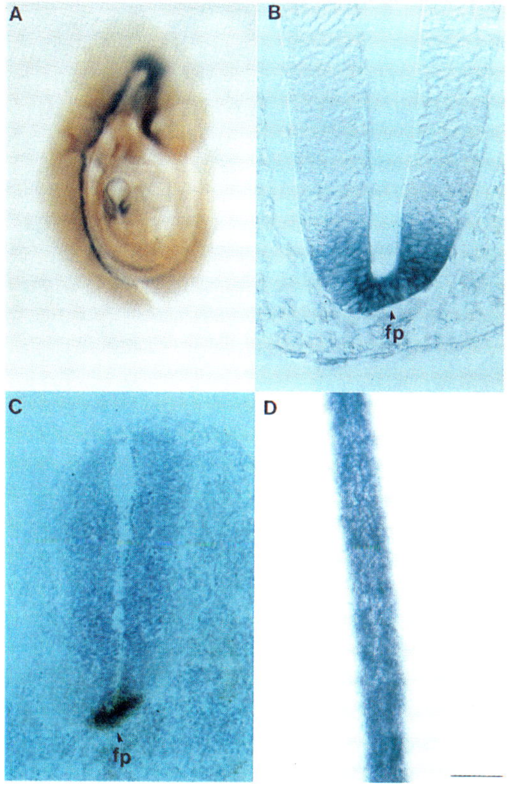

Figure 5. Localization of *F-spondin* RNA in rat embryos. (*A*) Restriction of *F-spondin* expression to the ventral neural tube at E10, as detected by whole-mount in situ hybridization. (*B*) Localization of *F-spondin* RNA to cells at the ventral midline of the neural tube of an E11 rat embryo, detected in a cryostat section of neural tube labeled previously by whole-mount in situ hybridization. (*C*) Cross-sections of the spinal cord at E12 showing localization of *F-spondin* RNA to the floor plate (fp), using a ^{35}S-labeled RNA probe. The section was counter-stained with cresyl violet after in situ hybridization. (*D*) Dorsal view of the ventral region of the dissected spinal cord of an E12 embryo showing restriction of *F-spondin* to midline fp cells, as detected by whole-mount in situ hybridization. Scale bar for *A*, 650 μm; *B*, 80 μm; *C*, *D*,120 μm.

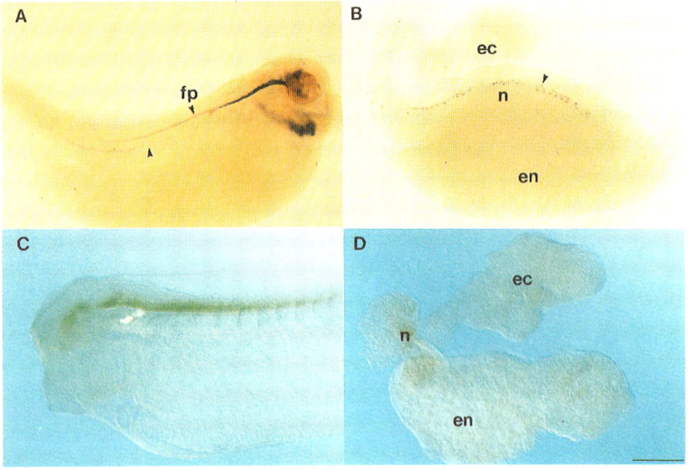

Figure 7. Localization of *F-spondin* RNA and restrictin protein in normal *Xenopus* embryos and in complete exogastrula. (*A*) Stage-38 embryo showing *F-spondin* RNA restricted to cells at the ventral midline of the neural tube (the floor plate, fp) and to cells in the hypochord (arrowhead), as detected by whole-mount in situ hybridization. Lateral cells in the midbrain and cells in the branchial arches also express *F-spondin*. (*B*) *F-spondin* is not expressed in the ectodermal (ec) component of complete exogastrula but is detected in hypochord (arrowhead) cells located adjacent to the notochord (n) and endoderm (en). (*C*) Localization of restrictin in the ventral CNS of tadpole stage *Xenopus* embryos, detected by immunoperoxidase labeling. (*D*) Lack of expression of restrictin in the ectodermal (ec) component of a complete exogastrula.

coinjected with *Pintallavis* RNA to identify the injected side of embryos. The injected halves of tadpole-stage embryos display a markedly shortened axis with a decrease in or absence of anterior neural structures (Ruiz i Altaba and Jessell 1992). The loss of anterior neural structures may result from expression of *Pintallavis* in anterior mesoderm, increasing the low level of *Pintallavis* RNA that is normally present in the prechordal plate, causing anterior mesodermal cells to acquire a more posterior character and thus eliminating their ability to induce forebrain structures. The repression of anterior neural differentiation may also result from the expression of *Pintallavis* in anterior neural plate cells, perhaps causing them to acquire posterior properties. In contrast to the decrease in size of forebrain structures, there was an increase in the amount of posterior neural tissue, which sometimes formed multiple tubular structures (Fig. 8). Although a marked perturbation of neural development was observed, there were no overt defects in mesodermal tissues. These results provide evidence that overexpression of *Pintallavis* results primarily in neural defects and suggest a role of *Pintallavis* in the induction and patterning of the neural axis.

The normal expression of *Pintallavis* by midline neural plate cells that differentiate into the floor plate raises the possibility that some of the defects in neural development caused by ectopic expression of *Pintallavis* result from the acquisition of floor plate properties by other neural tube cells. Examination of this possibility requires the identification of floor plate-specific

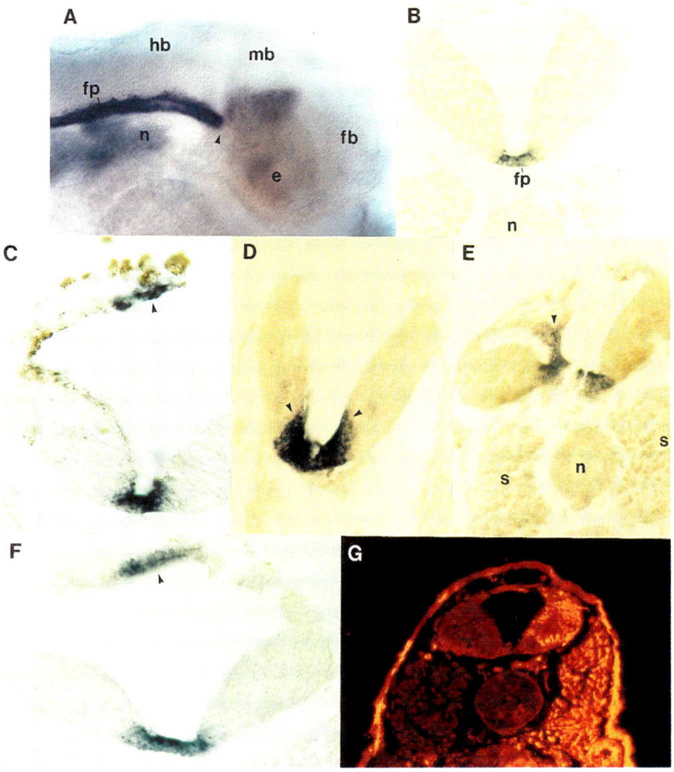

Figure 8. Ectopic expression of *F-spondin* RNA after injection of *Pintallavis* RNA into *Xenopus* embryos. (*A*) High-magnification view showing the localization of *F-spondin* RNA in the floor plate (fp) of the hindbrain (hb) and lower levels in the ventrolateral midbrain (mb) of a stage 36–38 *Xenopus* embryo detected by whole-mount in situ hybridzation. Note the sharp anterior boundary of *F-spondin* expression of the hindbrain-midbrain border (arrowhead). (e) Eye; (fb) forebrain; (n) notochord. (*B*) Cross-section through the spinal cord of a stage-36 *Xenopus* embryo showing the restriction of *F-spondin* RNA to floor plate (fp) cells, detected by whole-mount in situ hybridization. (*C–F*) Ectopic expression of *F-spondin* in the hindbrain of *Xenopus* tadpoles injected with *Pintallavis* RNA into one cell at the 2-cell stage. In *C* and *F*, *F-spondin* is expressed in cells adjacent to the dorsal midline. In *D*, there is a marked increase in the ventral domain of *F-spondin* expression. In *E*, the embryos display a large amount of neural tissue forming a secondary neural tube, and *F-spondin* is expressed ectopically at the junction of the tubular structures. *G* shows the distribution of the coinjected lineage tracer (rhodamine-lysine-dextran) in cells on one side of the neural tube of a stage-36 embryo. The lineage tracer is also detected in other tissues including somites and skin. Arrowheads describe the site of ectopic *F-spondin* expression. Scale bar for *A*, 150 μm; *B–F*, 15 μm; *G*, 25 μm.

genes that serve as markers with which to assess floor plate identity.

The *F-spondin* Gene Encodes a Floor Plate Adhesion Molecule

To identify molecules that could serve as markers of floor plate differentiation and that might also contribute to the signaling functions of the floor plate, we used subtractive hybridization methods to isolate genes that are expressed by the floor plate but not by dorsal spinal cord cells in E13 rat embryos (Klar et al. 1992). This screen led to the identification of a gene, *F-spondin*, whose expression is restricted to the floor plate during early neural development.

The *F-spondin* gene encodes a secreted protein which is separable into an amino-terminal domain containing 10 cysteine residues that exhibits no homology with other proteins and a carboxy-terminal domain which contains 6 repeats of a 55–59 residue domain (Fig. 4A). Domains similar to those present in the carboxyl half are also found in the adhesive glycoproteins thrombospondin I and II (Bornstein et al. 1991), in the adhesive circumsporozoite proteins of *Plasmodium* malarial parasites (Fig. 5B) (Rich et al. 1990), and in the Unc-5 protein, which appears to regulate axonal pathfinding in *Caenorhabditis elegans* (Culotti et al. 1991). Moreover, these domains, termed thrombospondin type 1 repeats (TSRs; Frazier 1991), have been shown to promote the adhesion of a variety of cell types and to mediate, in part, the neurite-outgrowth-promoting properties of thrombospondin (Neugebauer et al. 1991; O'Shea et al. 1991; Osterhout et al. 1992). The amino acid sequence VTCG that is contained within the TSRs of many of these proteins (Fig. 4) appears to be critical for cell adhesion (Rich et al. 1990; Prater et al. 1991).

Expression and Possible Functions of *F-spondin*

Analysis of the distribution of *F-spondin* mRNA in rat embryos by in situ hybridization shows that the gene is first expressed at 10.5 days postcoitum (E10.5) in cells located at the ventral midline of the neural tube (Fig. 5A) (Klar et al. 1992). The restricted expression of *F-spondin* mRNA in the floor plate is maintained at E11-E12 (Fig. 5B–D). By E13, however, *F-spondin* expression is also detectable in the ventral ventricular zone immediately above the floor plate (Klar et al. 1992). At later embryonic stages, expression spreads to the ventral and subsequently to the dorsal regions of the spinal cord.

The structural features of *F-spondin*, together with evidence that it is secreted from cells and binds to the cell surface, possibly in association with components of the extracellular matrix (Klar et al. 1992), raised the possibility that *F-spondin* can promote the adhesion of neural cells and the outgrowth of axons. To test this possibility, recombinant *F-spondin* containing a 10-amino-acid insert derived from the c-*myc* oncogene was purified using anti-*myc* antibodies from the medium of transfected COS-7 cells (Klar et al. 1992). A substrate of purified *F-spondin* promotes the adhesion of dorsal spinal cord neurons, many of which extend short neurites (Klar et al. 1992). Moreover, dorsal root ganglion neurons adhere and extend long neurites on a substrate of *F-spondin* (Fig. 6).

One potential in vivo function of F-spondin at the floor plate may be to guide axons at the ventral midline of the spinal cord. The growth cones of commissural axons cross the midline by extending between the basal surface of floor plate cells and the underlying basal lamina (Kuwada et al. 1990; Yaginuma et al. 1991; Bernhardt et al. 1992). F-spondin secreted by the floor plate may accumulate at high levels in association with the basal lamina under the floor plate, thus conferring it with adhesive properties that differ from those of basal lamina surrounding the remainder of the neuroepithelium. S-laminin, an isoform of the lamina B chain (Hunter et al. 1989), and other antigenic markers are restricted to the basal lamina under the floor plate (Sanes et al. 1990; Bovolenta and Dodd 1991), providing evidence for the specialization of floor plate basal lamina. The growth cones of commissural neurons may

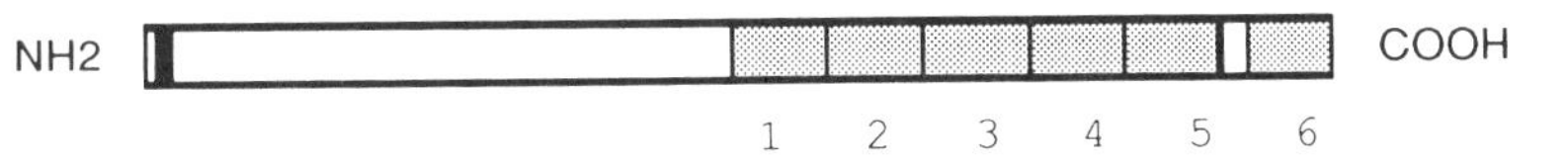


```
F-spondin            C----W--WS-CSVTCG-G---R-R-------------C---------C-----C
Thrombospondin         WS-WS-WS-CSVTCG-G---R-R-C-----------C---------C-----C
Thrombospondin II      WS-WS-WS-CSVTCG-G---R-R-C-----------C---------C-----C
Properdin               WS-W--WS-CSVTC--G---R-R-C-----------C---------C-----C
Complement COOH        W-CWS-WS-C----------R-R-C-----------C---------C
Complement NH2       C----WS-WS-C---C    ---R-R-------------C---------C-----C
Plasmodium cs                WS-CSVTCG-G---R-R
TRAP                         WS-CSVTCG-G---R-R---        ---C-------  C---- C
```

Figure 4. Structure and domain organization of rat F-spondin. (*Top*) Schematic representation of the domain structure of F-spondin. The stippled boxes and numbers show the thrombospondin type 1 repeats (TSRs). (*Bottom*) Comparison of the repeated F-spondin motif with the consensus TSRs found in thrombospondin I and II, region II of the *Plasmodium* circumsporozoite proteins (Rich et al. 1990), properdin (Goundis and Reid 1988), complement C6–C9 proteins (Haefliger et al. 1989), and thrombospondin-related anonymous protein (TRAP). (Modified from Klar et al. 1992).

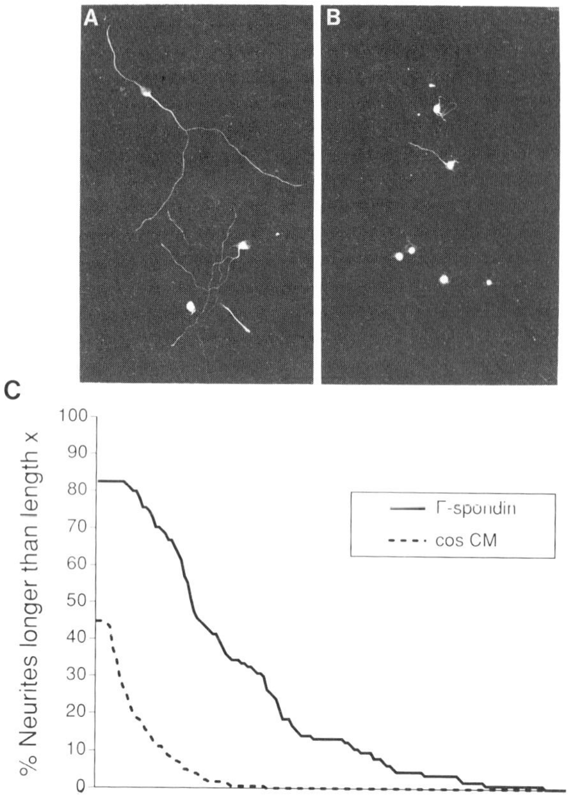

Figure 6. F-spondin promotes the extension of neurites in vitro. F-spondin^{myc} protein affinity-purified from transfected COS cells supernatant was immobilized onto a nitrocellulose substrate. (*A*) Neurons isolated from E14 rat dorsal root ganglia (DRG) extend long processes on F-spondin. (*B*) E14 DRG neurons do not extend on proteins isolated from control COS cell medium. In *A* and *B*, neurons were labeled with MAb 3A10 (Klar et al. 1992) to visualize neurites. (*C*) Quantitation of neurite outgrowth from E14 DRG neurons. The length of the longest neurite of each 3A10 positive neuron was measured. The percentage of neurons with neurites longer than a given length is plotted. (Modified from Klar et al. 1992.)

adhere preferentially to F-spondin or other basal lamina components, prompting a change in trajectory at the midline (Dodd et al. 1988). It is also possible that F-spondin is involved in general adhesive interactions of floor plate cells.

Regulation of *F-spondin* Expression by *Pintallavis*

***Isolation and expression of frog* F-spondin.** Independent of its potential function, the characterization of *F-spondin* and its restricted expression during early spinal cord development suggested that it could serve as a marker with which to examine the mechanisms that determine the identity of floor plate cells. To assess this, we used the rat *F-spondin* cDNA to isolate the frog *F-spondin* gene from a tadpole cDNA library. The frog and rat F-spondin proteins are highly conserved, exhibiting the same overall domain organization and 84% identity (A. Ruiz i Altaba et al., in prep.). The only region of extensive sequence divergence is the insert between the 5th and 6th TSRs. Indeed, the degree of conservation in the amino-terminal domain of the protein is greater than that in the carboxy-terminal domain which contains the TSRs, suggesting that the amino domain of the protein has an important role in the structure or function of the protein.

The pattern of expression of *F-spondin* mRNA during *Xenopus* development, determined by in situ hybridization, shows that the gene is first expressed by cells at the midline of the neural plate from (Nieuwkoop and Faber 1969) stage 22–23. At stage 34–36, the floor plate of the spinal cord and hindbrain expresses high levels of *F-spondin* mRNA (Fig. 7A) (A. Ruiz i Altaba et al., in prep.). Strikingly, there is no detectable expression in other regions of the spinal cord and hindbrain at any stage of development examined (up to stage 45). Thus, the gradual expansion in neural expression of *F-spondin* observed in the rat (Klar et al. 1992) does not occur in tadpoles. *F-spondin* is not expressed in the neural ectoderm of *Xenopus* exogastrulae (Fig. 8) (A. Ruiz i Altaba et al., in prep.), suggesting that floor plate differentiation and *F-spondin* expression require inductive signals from the notochord (Fig. 7B). There is a low level of expression of *F-spondin* mRNA in cells in the ventrolateral region of the midbrain and forebrain, but not at the ventral midline (not shown). Low levels of *F-spondin* mRNA are also detected in cells of the hypochord, in the caudal region of the notochord, and in branchial arches beginning about stage 36–38. These results establish that, in *Xenopus*, *F-spondin* is expressed selectively by floor plate cells over a prolonged period of spinal cord and hindbrain development.

F-spondin *expression in embryos injected with* Pintallavis *RNA*. Using *F-spondin* as a marker of floor plate cells in the hindbrain and spinal cord, we examined whether ectopic expression of *Pintallavis* in neural cells causes them to acquire floor plate properties. Synthetic RNA (Krieg and Melton 1984) derived from a full-length *Pintallavis* cDNA clone was injected into a single cell of 2-cell-stage embryos together with a lineage tracer (A. Ruiz i Altaba et al., in prep.). Injected embryos exhibited the same general phenotypes described above; in particular, a truncation of the embryo along the anteroposterior axis, a loss of anterior neural structures, and an expansion of the posterior neural tube.

The distribution of *F-spondin* in embryos injected with *Pintallavis* RNA was examined by in situ hybridization. In about 20% of embryos in which an overt neural phenotype was detected, *F-spondin* mRNA was expressed ectopically in the hindbrain, although not in the spinal cord (Fig. 8C–F) (A. Ruiz i Altaba et al., in prep.). Ectopic expression was detected only in two regions: in cells of the ventricular zone adjacent to and contiguous with the normal position of the floor plate (Fig. 8C,D) and in cells adjacent to the dorsal midline

(Fig. 8F). In some embryos, the number of ventral cells expressing *F-spondin* was up to fivefold greater than normal, and the region of labeled cells was observed to extend dorsally in the ventricular zone (not shown). Although the frequency of embryos exhibiting ectopic expression of *F-spondin* was low, the effect was specific. Injection of a frameshift mutant RNA which encodes a protein lacking the putative DNA-binding domain results in a pattern of *F-spondin* expression indistinguishable from that of uninjected embryos.

These findings show that deregulating the expression of *Pintallavis* in frog embryos results in the ectopic expression of *F-spondin* in cells of the embryonic hindbrain. The phenotype observed raises two major questions. First, why is the frequency of embryos exhibiting ectopic *F-spondin* expression low? Second, why is ectopic expression restricted to the hindbrain? The low frequency of ectopic *F-spondin* expression may relate to the time at which *Pintallavis* acts in neural cells. The normal pattern of expression of *Pintallavis* suggests that if the gene functions to control floor plate identity, it acts in midline neural plate cells. Studies on the stability of synthetic RNAs injected into *Xenopus* embryos have shown that exogenous RNA persists at high levels until the midblastula transition but declines dramatically at later stages, when the neural plate forms (Krieg and Melton 1984; Harland and Misher 1988; Ruiz i Altaba and Melton 1989). Thus, the levels of synthetic *Pintallavis* RNA present at the time that the neural plate cells may, in many embryos, be below the concentration required to drive *F-spondin* expression. Since hindbrain development precedes that of the spinal cord, a decrease in *Pintallavis* RNA levels with time could also explain why the phenotype is observed solely at hindbrain levels. Injection of higher amounts of *Pintallavis* RNA results in high toxicity or severely malformed embryos in which analysis of neural structures is difficult.

The detection of ectopic *F-spondin* expression only in cells of the ventral ventricular zone and the dorsal midline cannot be explained by a restriction of injected *Pintallavis* RNA to these cell types. Analysis of the distribution of the lineage tracer rhodamine-lysine-dextran (Gimlich and Cooke 1983) revealed that in only about 50% of embryos did all cells in the neural tube on the injected side express the lineage tracer. In the remaining embryos, the expression pattern was mosaic but usually included cells in the lateral region of the neural tube that do not express *F-spondin* after *Pintallavis* RNA injection. The lack of expression of *F-spondin* in the lateral region of the neural tube may indicate that these cells, which are primarily postmitotic neurons, express factors that inhibit the action of *Pintallavis*. Alternatively, cells in the ventral ventricular zone and dorsal midline may express additional factors that are required for *F-spondin* expression. Thus, *Pintallavis* expression is not sufficient to confer floor plate properties to all lateral neural plate cells.

The expanded domain of *F-spondin* expression in the ventral hindbrain could result from the proliferation of floor plate precursors or from a disruption of the normal convergent extension movements of cells at the midline of the neural plate (Keller and Danilchik 1988; Keller et al. 1991). A more interesting possibility is that during normal development, inductive signals from the notochord confer partial floor plate properties to neural plate cells that are located lateral to the group of cells that eventually undergo complete floor plate differentiation. The failure of these lateral cells to acquire a fully differentiated floor plate phenotype may occur because they do not express critical factors for floor plate differentiation or express these at levels below those required to stabilize or permit the progression of the full floor plate phenotype. The expression of *Pintallavis* in these cells may itself drive floor plate differentiation or enhance the expression of other factors to a level sufficient to stabilize floor plate differentiation, resulting in an expansion in the domain of *F-spondin* expression in the ventral hindbrain.

Cells at the dorsal midline of the neural tube form part of the roof plate and derive from the most lateral regions of the neural plate and are therefore unlikely to be influenced by notochord-derived signals. Thus, the ectopic expression of *F-spondin* in these cells cannot be the result of the proliferation or aberrant migration of floor plate cells. In these dorsal midline cells, ectopic expression of *Pintallavis* appears sufficient to drive expression of a floor plate-specific gene.

The lack of other floor plate markers in frog embryos leaves open the question of whether the ectopic expression of *F-spondin* in dorsal midline cells is indicative of a general acquisition of floor plate properties. In chick embryos, floor plate or notochord grafts placed next to the dorsal neural tube induce a dorsal floor plate and motor neurons while repressing dorsal neuronal differentiation (Yamada et al. 1991; Placzek et al. 1991). Ectopic expression of *Pintallavis* also represses the differentiation of Rohon-Beard primary sensory neurons (Hughes 1957) in the dorsal neural tube (Ruiz i Altaba and Jessell 1992), consistent with the possibility that dorsal midline cells that express *F-spondin* mRNA have acquired floor plate-like signaling properties.

The suppression of Rohon-Beard neurons is, however, evident in the anterior spinal cord where ectopic expression of *F-spondin* is not observed. This could reflect the fact that *F-spondin* expression defines a more advanced stage of floor plate differentiation than does the acquisition of signals that repress dorsal cell differentiation. It is also possible that the ectopic expression of *Pintallavis* in prospective neural cells during gastrulation inhibits the differentiation of Rohon-Beard neurons, which are known to be specified at this time (Lamborghini 1980). In this case, the induction of floor plate properties in dorsal midline cells could occur independently of the suppression of dorsal cell types. Additional markers are required to resolve this issue and also to determine whether ectopic expression of *Pintallavis* mimics the actions of a dorsal floor plate in inducing ventral neuronal types in adjacent dorsal regions of the neural tube.

Regulation of Floor Plate Properties by Cell Type-specific Transcription Factors

Patterning along the dorsoventral axis of the embryonic hindbrain and spinal cord involves the induction of midline neural plate cells that later give rise to the floor plate. Signals from the floor plate and notochord control the identity and distribution of neuronal types in the ventral region of the neural tube and the trajectory of axons that cross the ventral midline. The present studies provide an initial step in defining the molecular mechanisms that may underlie the control of pattern in this region of the central nervous system.

In vertebrate embryos, several classes of transcription factors have been implicated in the specification of cell fate. Homeobox genes which show restricted patterns of expression in developing embryos can change the fate of cells when ectopically expressed. These include *Xhox 3* (Ruiz i Altaba and Melton 1989), *X1HBox1* (Wright et al. 1989), *Hox1.1* (Kessel et al. 1990), and *Hox 1.4* (Wolgemuth et al. 1989). Helix-loop-helix genes can also affect cell fate. For example, *myoD* expression can initiate myogenesis in 10T 1/2 cells (Davis et al. 1987; Weintraub et al. 1991), and when injected into *Xenopus* embryos, can induce the expression of muscle-specific transcripts such as α-actin in non-muscle cells (Hopwood and Gurdon 1990). In frog embryos, however, cells that ectopically express *myoD* do not exhibit a complete change in their identity, suggesting that *myoD* and other regulatory genes may function to affect fate only in competent cells.

The *Pintallavis* gene encodes a putative DNA-binding transcription factor that may specify floor plate identity within the neural plate. The closest homologs of *Pintallavis* are the *HNF-3* genes, which function as transcription factors involved in the control of liver-specific gene expression (Fig. 1) (Lai et al. 1990,1991). Ectopic expression of *Pintallavis* can confer at least one floor plate-specific property to cells in other regions of the neural tube. The failure of all neural tube cells expressing *Pintallavis* to express floor plate properties, however, suggests that *Pintallavis* is insufficient, in itself, to confer floor plate properties to all cells or that it cannot override other regulatory signals in some cells. Defining the role of *Pintallavis* in cells that normally express the gene will require assays that interfere with its function. Inactivation of some transcription factors by antibody injection (Wright et al. 1989; Ruiz i Altaba et al. 1991) or by genetic ablation (Chisaka and Capecchi 1991; Lufkin et al. 1991; Mouellic et al. 1992) has indeed suggested a requirement for such genes in the control of cell fate and pattern in early vertebrate development.

The mechanism by which *F-spondin* expression is activated in *Pintallavis*-injected embryos remains unclear. *Pintallavis* expression by midline neural plate cells overlaps briefly with the expression of *F-spondin*; thus, it is possible that the activation of *F-spondin* transcription is regulated directly by *Pintallavis*. The maintenance of *F-spondin* transcription, however, cannot be dependent on *Pintallavis*, since *F-spondin* is expressed by floor plate cells long after *Pintallavis* expression ceases.

The characterization of F-spondin as a secreted glycoprotein involved in neuronal adhesion and neurite growth suggests that the signaling properties of the floor plate can be controlled by regulatory genes whose restricted expression is activated by notochord-derived inductive signals. There is increasing evidence that cell type-specific transcription factors involved in cell fate determination can regulate the expression of cell-surface or secreted proteins with adhesive or neurite-outgrowth-promoting activities. Studies in neural cell lines have provided evidence that the frog *Hox 2.4* and *2.5* genes can *trans*-activate the *NCAM* promoter (Jones et al. 1992b) and that the *Evx1* homeobox gene (Bastian and Gruss 1990), the mouse homolog of the frog *Xhox3* gene (Ruiz i Altaba and Melton 1990), can activate expression of cytotactin, an extracellular matrix glycoprotein (Jones et al. 1992a). The present experiments extend these studies by providing evidence in vivo for the regulation of the expression of a floor plate adhesion molecule by a cell type-specific transcription factor.

In addition to the expression of adhesion proteins, several other floor plate signaling functions have been identified. These include activities responsible for homeogenetic floor plate induction (Hatta et al. 1991; Yamada et al. 1991), dorsoventral patterning of neuronal types (Yamada et al. 1991), anteroposterior limb patterning (Wagner et al. 1990), and chemotropic guidance of commissural axons (Tessier-Lavigne et al. 1988). The identification of *Pintallavis* as a possible regulator of floor plate differentiation should permit the design of in vivo and in vitro assays to determine the contribution of this gene to the acquisition of these and other floor plate properties by cells at the midline of the neural plate.

ACKNOWLEDGMENTS

We thank C. Cox and M. Baldassare for help in the characterization of the frog and rat *F-spondin* genes, F. Rathjen for providing anti-restrictin antibodies, and Ira Schieren for help with the figures. T.M.J. is an Investigator and A.R.A. and A.K. are Research Associates of the Howard Hughes Medical Institute. This work was supported, in part, by National Institutes of Health Program Project grant 1PO1-NS-30532.

REFERENCES

Bastian, H. and P. Gruss. 1990. A murine even-skipped homologue Evx-1 is expressed during early embryogenesis and neurogenesis in a bychasic manner. *EMBO. J.* **9:** 1839.

Bernhardt, R.R., N. Nguyen, and J.Y. Kuwada. 1992. Growth cone guidance by floor plate cells in the spinal cord of zebrafish embryos. *Neuron* **8:** 869.

Bornstein, P., K. O'Rourke, K. Wikstrom, F.W. Wolff, R. Katz, P. Li, and V.M. Dixit. 1991. A second, expressed thrombospondin gene (Thbs2) exists in the mouse genome. *J. Biol. Chem.* **266:** 12821.

Bovolenta, P. and J. Dodd. 1990. Guidance of commissural

growth cones at the floor plate in the embryonic rat spinal cord. *Development* **109:** 435.

———. 1991. Perturbation of neuronal differentiation and axon guidance in the spinal cord of mouse embryos lacking a floor plate-analysis of Danforth's short-tail mutation. *Development* **113:** 625.

Chisaka, O. and M.R. Capecchi. 1991. Regionally restricted developmental defects resulting from targeted disruption of the mouse homeobox gene *hox 1.5*. *Nature* **350:** 473.

Clarke, J.D.N., N. Holder, S.R. Soffe, and J. Storm-Mathisen. 1991. Neuroanatomical and functional analysis of neural tube formation in notochordless *Xenopus* embryos: Laterality of the spinal cord is lost. *Development* **112:** 499.

Culotti, J.G., A. Spence, Y. Zhou, I. Scott, C. Leung-Hagesteijn, B. Stern, and E. Hedgecock. 1991. The unc-5 axon guidance gene of *C. elegans* has features of a cell adhesion receptor. *J. Cell Biol.* **115:** 122a.

Davis, R.L., H. Weintraub, and A.B. Lassar. 1987. Expression of a single transfected cDNA converts fibroblasts to myoblasts. *Cell* **51:** 987.

Dirksen, M.L. and M. Jamrich. 1992. A novel, activin-inducible blastopore lip-specific gene of *Xenopus laevis* contains a fork head DNA-binding domain. *Genes Dev.* **6:**599.

Dodd, J., S.B. Morton, D. Karagogeos, M. Yamamoto, and T.M. Jessell. 1988. Spatial regulation of axonal glycoprotein expression and subsets of embryonic spinal neurons. *Neuron* **1:** 105.

Ericson, J., S. Thor, T. Edlund, T.M. Jessell, and T. Yamada. 1992. Early stages of motor neuron differentiation revealed by expression of homeobox gene Islet-1. *Science* **256:** 1555.

Fraser, S., R. Keynes, and A. Lumsden. 1990. Segmentation in the chick embryo hindbrain is defined by cell lineage restrictions. *Nature* **344:** 431.

Frazier, W.A. 1991. Thrombospondins. *Curr. Opin. Cell Biol.* **3:** 792.

Gimlich, R.L. and J. Cooke. 1983. Cell lineage and the induction of second nervous systems in amphibian development. *Nature* **306:** 471.

Goundis, D. and K.B.M. Reid. 1988. Properdin, the terminal complement components, thrombospondin and the circumsporozoite protein of malaria parasites contain similar sequence motifs. *Nature* **335:** 62.

Grossniklaus, U., R. Kurth Pearson, and W.J. Gehring. 1992. The *Drosophila* sloppy paired locus encodes two proteins involved in segmentation that show homology to mammalian transcription factors. *Genes Dev.* **6:** 1030.

Haefliger, J.A., J. Tschopp, N. Vial, and D.E. Jennet. 1989. Complete primary structure and functional characterization of the sixth component of the human complement system. *J. Biol. Chem.* **264:** 18041.

Harland, R.M. 1991. In situ hybridization: an improved whole mount method for *Xenopus* embryos. *Methods Enzymol.* **36:** 675.

Harland, R. and L. Misher. 1988. Stability of RNA in developing *Xenopus* embryos and identification of a destabilizing sequence in TFIIIA messenger RNA. *Development* **102:** 837.

Hatta, K., C.B. Kimmel, R.K. Ho, and C. Walker. 1991. The cyclops mutation blocks specification of the floor plate of the zebrafish central nervous system. *Nature* **350:** 339.

Hirano, S., S. Fuse, and G.S. Sohal. 1991. The effect of the floor plate on pattern and polarity in the developing central nervous system. *Science* **251:** 310.

Holtfreter, J. 1933. Die totale exogastrulation, eine selbstablosung des ectoderms von entomesoderm. Entwicklung und functionelles verhalten nervenloser organe. *Wilhelm Roux' Arch. Entwicklungsmech. Org.* **129:** 669.

Hopwood, N. and J.B. Gurdon. 1990. Activation of muscle genes without myogenesis by ectopic expression of MyoD in frog embryo cells. *Nature* **347:** 197.

Hornbruch, A. and A. Wolpert. 1986. Positional signalling by Hensen's node when grafted to the chick limb bud. *J. Embryol. Exp. Morphol.* **94:** 257.

Hughes, A.F. 1957. The development of the primary sensory neurons in *Xenopus laevis* (Daudin). *J. Anat.* **91:** 323.

Hunter, D.D., V. Shah, J.P. Merlie, and J.R. Sanes. 1989. A laminin-like adhesive protein concentrated in the synaptic cleft of the neuromuscular junction. *Nature* **338:** 229.

Jacobson, A.G. 1981. Morphogenesis of the neural plate and tube. In *Morphogenesis and pattern formation* (ed. T.G. Connelly et al.), p. 233. Raven Press, New York.

Jessell, T.M. and J. Dodd. 1992. Floor plate-derived signals and the control of neural cell pattern in vertebrates. *Harvey Lect.* **86:** 87.

Jones, F.S., G. Chalepakis, P. Gruss, and G.M. Edelman. 1992a. Activation of the cytotactin promoter by the homeobox-containing gene *Evx-1*. *Proc. Natl. Acad. Sci.* **89:** 2091.

Jones, F.S., E.A. Prediger, D.A. Bittner, E.M. De Robertis, and G.M. Edelman. 1992b. Cell adhesion molecules as targets for *Hox* genes: Neural cell adhesion molecule promoter activity is modulated by cotransfection with *Hox-2.5* and *-2.4*. *Proc. Natl. Acad. Sci.* **89:** 2086.

Jürgens, G. and D. Weigel. 1988. Terminal versus segmental development in the *Drosophila* embryo: The role of the homeotic gene *fork head*. *Roux's Arch. Dev. Biol.* **197:** 345.

Keller, R. and M. Danilchik. 1988. Regional expression, pattern and timing of convergence and extension during gastrulation of *Xenopus laevis*. *Development* **103:** 193.

Keller, R., J. Shih, and P. Wilson. 1991. Cell motility, control and function of convergence and extension during gastrulation in *Xenopus*. In *Gastrulation* (ed. R. Keller), p. 101. Plenum Press, New York.

Kessel, M., R. Balling, and P. Gruss. 1990. Variations of cervical vertebrae after expression of a *Hox1.1* transgene in mice. *Cell* **61:** 301.

Keynes, R.J. and C.D. Stern. 1988. Mechanisms of vertebrate segmentation. *Development* **103:** 413.

Kingsbury, B.F. 1930. The developmental significance of the floor-plate of the brain and spinal cord. *J. Comp. Neurol.* **50:** 177.

Kintner, C.R. and D.A. Melton. 1987. Expression of the *Xenopus* N-CAM RNA in ectoderm is an early response to neural induction. *Development* **99:** 311.

Klar, A., M. Baldassare, and T.M. Jessell. 1992. *F-spondin*: A gene expressed at high levels in the floor plate encodes a secreted protein that promotes neural cell adhesion and neurite extension. *Cell* **69:** 95.

Krieg, P.A. and D.A. Melton. 1984. Functional messenger RNAs are produced by SP6 in vitro transcription of cloned cDNAs. *Nucleic Acids Res.* **12:** 7057.

Kuwada, J.Y., R.R. Bernhardt, and A.B. Chitnis. 1990. Pathfinding by identified growth cones in the spinal cord of zebrafish embryos. *J. Neurosci.* **10:** 1299.

Lai, E., V.R. Prezioso, W. Tao, W.S. Chen, and J.E. Darnell. 1991. Hepatocyte nuclear factor 3a belongs to a gene family in mammals that is homologous to the *Drosophila* homeotic gene fork head. *Genes Dev.* **5:** 416.

Lai, E., V.R. Prezioso, E. Smith, O. Litvin, R.H. Costa, and J.E. Darnell. 1990. HNF-3A, a hepatocyte-enriched transcription factor of novel structure is regulated transcriptionally. *Genes Dev.* **4:** 1427.

Lamborghini, J.E. 1980. Rohon-Beard cells and other large neurons in *Xenopus* embryos originate during gastrulation. *J. Comp. Neurol.* **189:** 323.

Li, C., C. Lai, D.S. Sigman, and R.B. Goynor. 1991. Cloning of a cellular factor, interleukin binding factor, that binds to NFAT-like motifs in the human immunodeficiency virus long terminal repeat. *Proc. Natl. Acad. Sci.* **88:** 7739.

Lufkin, T., A. Dierich, M. LeMeur, M. Mark, and P. Chambon. 1991. Disruption of the *Hox 1.6* homeobox gene results in defects in a region corresponding to its rostral domain of expression. *Cell* **66:** 1105.

Lumsden, A. and R. Keynes. 1989. Segmental patterns of neuronal development in the chick hindbrain. *Nature* **337:** 424.

Mouellic, H.L., Y. Lallemand, and P. Brulet. 1992. Homeosis in the mouse induced by a null mutation in the *Hox 3.1* gene. *Cell* **69:** 251.

Neugebauer, K.M., C.J. Emmett, K.A. Venstrom, and L.F. Reichardt. 1991. Vitronectin and thrombospondin promote retinal neurite outgrowth: Developmental regulation and role of integrins. *Neuron* **6:** 345.

Nieuwkoop, P. and J. Faber. 1969. *Normal table of* Xenopus laevis (*Daudin*). North Holland, Amsterdam.

O'Shea, K.S., L.-H.J. Liu, and V.M. Dixit. 1991. Thrombospondin and a 140 kd fragment promote adhesion and neurite outgrowth from embryonic central and peripheral neurons and from PC12 cells. *Neuron* **7:** 231.

Osterhout, D.J., W.A. Frazier, and D. Higgins. 1992. Thrombospondin promotes process outgrowth in neurons from the peripheral and central nervous system. *Dev. Biol.* **150:** 256.

Placzek, M., M. Tessier-Lavigne, T.M. Jessell, and J. Dodd. 1990a. Orientation of commissural axons in vitro in response to a floor plate derived chemoattractant. *Development* **110:** 19.

Placzek, M., M. Tessier-Lavigne, T. Yamada, T.M. Jessell, and J. Dodd. 1990b. Mesodermal control of neural cell identity: Floor plate induction by the notochord. *Science* **250:** 985.

———. 1990c. The guidance of developing axons by diffusible chemoattractants. *Cold Spring Harbor Symp. Quant. Biol.* **55:** 279.

Placzek, M., T. Yamada, M. Tessier-Lavigne, T.M. Jessell, and J. Dodd. 1991. Control of dorsoventral pattern in vertebrate neural development: Induction and polarizing properties of the floor plate. *Development* (suppl. 2), p. 105.

Prater, C.A., J. Plotkin, D. Jaye, and W.A. Frazier. 1991. The properdin-like type I repeats of a human thrombospondin contain a cell attachment site. *J. Cell Biol.* **112:** 1031.

Puelles, L., J.A. Amat, and M. Martinez-De-La-Torre. 1987. Segment-related, mosaic neurogenetic pattern in the forebrain and mesencephalon early chick embryos. I. Topography of AChE-positive neuroblasts up to stage HH18. *J. Comp. Neurol.* **266:** 247.

Rathjen, F.G., J.M. Wolf, and R. Chiquet-Ehrismann. 1991. Restrictin: A chick neural extracellular matrix protein involved in cell attachment co-purifies with the cell recognition molecule F11. *Development* **113:** 151.

Rich, K.A., F.W. George, J.L. Law, and W.J. Martin. 1990. Cell adhesive motif in region II of malarial circumsporozoite protein. *Science* **249:** 1574.

Ruiz i Altaba, A. 1990. Neural expression of the *Xenopus* homeobox gene Xhox3: Evidence for a patterning neural signal that spreads through the ectoderm. *Development* **8:** 595.

Ruiz i Altaba, A. 1992. Planar and vertical signals in the induction and patterning of the *Xenopus* nervous system. *Development* (in press).

Ruiz i Altaba, A. and T.M. Jessell. 1992. *Pintallavis*, a gene expressed in the organizer and midline cells of frog embryos: Involvement in the development of the neural axis. *Development* (in press).

Ruiz i Altaba, A. and D.A. Melton. 1989. Involvement of the *Xenopus* gene Xhox3 in pattern formation along the anterior-posterior axis. *Cell* **57:** ,317.

———. 1990. Axial patterning and the establishment of polarity in the frog embryo. *Trends Genet.* **6:** 57.

Ruiz i Altaba, A., T. Choi, and D.A. Melton. 1991. Expression of the Xhox3 homeobox protein in *Xenopus* embryos: Blocking its early function suggests the requirement of Xhox3 for normal posterior devepment. *Dev. Growth Differ.* **33:** 651.

Sanes, J.R., D.D. Hunter, T.L. Green, and J.P. Merlie. 1990. S-laminin. *Cold Spring Harbor Symp. Quant. Biol.* **55:** 419. Schoenwolf, G. and J.L. Smith. 1990. Mechanisms of neurulation: Traditional viewpoint and recent advances. *Development* **109:** 243.

Tao, W. and E. Lai. 1992. Telencephalon-restricted expression of BF-1, a new member of the HNF-3/fork head gene family, in the developing rat brain. *Neuron* **8:** 957.

Tessier-Lavigne, M., M. Placzek, A.G.S. Lumsden, J. Dodd, and T.M. Jessell. 1988. Chemotropic guidance of developing axons in the mammalian central nervous system. *Nature* **336:** 775.

Vaage, S. 1969. The segmentation of the primitive neural tube in chick embryos (*Gallus domesticus*). *Ergeb. Anat. Entwicklungsgesch.* **41:**1088.

van Straaten, H.W.M., J.W.M. Hekking, E.L. Wiertz-Hoessels, F. Thors, and J. Drukker. 1988. Effect of the notochord on the differentiation of a floor plate area in the neural tube of the chick embryo. *Anat. Embryol.* **177:** 317.

Wagner, M., C. Thaller, T.M. Jessell, and G. Eichele. 1990. Polarizing activity and retinoid synthesis in the floor plate of the neural tube. *Nature* **345:** 819.

Weigel, D. and H. Jackle. 1990. The fork head domain: A novel DNA binding motif of eukaryotic transcription factors? *Cell* **63:** 455.

Weigel, D., G. Jürgens, F. Küttner, E. Seifert, and H. Jäckle. 1989. The homeotic gene fork head encodes a nuclear protein and is expressed in the terminal regions of the *Drosophila* embryo. *Cell* **57:** 645.

Weintraub, H., R. Davis, S. Tapscott, M. Thayer, M. Krause, R. Benezra, T.K. Blackwell, D. Turner, R. Rupp, S. Hollenberg, C. Zhuang, and A.L. Lassar. 1991. MyoD: A nodal point during the specification of the muscle cell lineage. *Science* **251:** 761.

Wolgemuth, D.J.R., R.R. Behringer, M.P. Mostoller, R.L. Brinster, and R.D. Palmiter. 1989. Transgenic mice overexpressing the homeobox gene *Hox 1.4* exhibit abnormal gut development. *Nature* **337:** 464.

Wright, C.V.E., K.W.Y. Cho, J. Hardwicke, R.H. Collins, and E.M. DeRobertis. 1989. Interference with the function of a homeobox gene in *Xenopus* embryos produces malformations of the anterior spinal cord. *Cell* **59:** 81.

Yaginuma, H., S. Homma, L. Kunzi, and R.W. Oppenheim. 1991. Pathfinding by growth cones of commissural interneurons in the chick embryo spinal cord: A light and electric microscopic study. *J. Comp. Neurol.* **304:** 78.

Yamada, T., M. Placzek, H. Tanaka, J. Dodd, and T.M. Jessell. 1991. Control of cell pattern in the developing nervous system: Polarizing activity of the floor plate and notochord. *Cell* **64:** 635.

Neurexins

M. Geppert, Y.A. Ushkaryov, Y. Hata, B. Davletov, A.G. Petrenko, and T.C. Südhof
Department of Molecular Genetics and Howard Hughes Medical Institute, University of Texas Southwestern Medical Center, Dallas, Texas 75235

Structure and Function of the Synapse

Synapses are the primary points of communication between neurons in the brain. At a synapse, information is transferred between neurons in a unidirectional manner. Synaptic neurotransmission is very rapid, self-limiting, and highly localized. Synapses are abundant; a neuron can receive or make more than 100,000 synapses. More diffuse and longer-lasting communication processes between neurons are superimposed onto the synaptic signaling pathways. These "modulatory" signaling processes are often carried by neuropeptides that are released outside of synapses, whereas signaling at synapses is mediated by smaller molecules.

Morphologically, synapses contain several striking characteristics. The space between the plasma membranes of the pre- and postsynaptic cells slightly widens at the synapse and becomes filled with an undefined fuzzy material (Peters et al. 1991). Both pre- and postsynaptic plasma membranes are thickened, a phenomenon referred to as pre- and postsynaptic densities. Accumulations of uniformly sized 40–50-nm vesicles (called synaptic vesicles) are observed in the presynaptic nerve terminal. On the postsynaptic side, synapses are often located on spines emanating from the dendrites that seem to have their own intracellular membrane apparatus.

Information transfer at the synapse occurs by the release of signaling molecules, neurotransmitters, from the presynaptic neuron and their reception by the postsynaptic neuron. Neurotransmitters are actively taken up by synaptic vesicles and secreted by regulated synaptic vesicle exocytosis. They then diffuse across the synaptic cleft to react with postsynaptic receptors. One salient feature of neurotransmission is the directionality of the process. Synaptic vesicles, although present at high concentrations throughout the nerve terminal, only exocytose at the active zone of the synapse (the part of the presynaptic plasma membrane that is apposed to the postsynaptic membrane). Synaptic transmission is structurally and functionally organized along an axis extending from the presynaptic neuron to the postsynaptic neuron.

Synaptic vesicles play a central role in the function of the presynaptic nerve terminal, since they are the organelles from which neurotransmitters are actually released. Because of the relative ease with which they can be studied, many proteins of synaptic vesicles have been characterized structurally and functionally. By now, synaptic vesicles represent one of the best molecularly characterized organelle (for review, see Südhof and Jahn 1991). Synaptic vesicle proteins are synthesized in developing neurons before synapses form and are localized to vesicles that constitutively recycle throughout the neuron (Fletcher et al. 1991; Matteoli et al. 1992). On the presynaptic side, synaptogenesis consists of a transformation of this recycling pathway from a constitutive pathway with no specific localization to a regulated pathway restricted to a small part of the cell. It is likely that this transformation must involve cues from the postsynaptic cell acting via the plasma membrane.

Because signals at synapses are so restricted in space and time, the location of synapses on neurons with respect to each other and the wiring pattern produced by the synaptic connections are important determinants of brain function (Sheperd 1990). Synaptic connections in the brain probably form in a use-dependent manner and show a great specificity not only in terms of which neurons are connected, but also where on these neurons the connections occur. Furthermore, synapses are not static structures that are stable once formed. Instead, they are constantly remodeled during the life of an animal and are subject to several different types of plasticity.

Structurally and functionally, synapses can be thought of as a specialized and unusual form of cell-cell contact. It seems likely that in analogy to other cell-cell contacts, the formation and maintenance of synaptic connections may depend on cell-surface proteins that act as cell adhesion molecules. Because only neurons can make synapses and synapses are only a small part of a neuron, cell-surface receptors with a synapse function would have to be specific for the nervous system, would be different from the general molecules involved in less specific forms of cell-cell contact, and would have to participate in establishing specificity in the large diversity of synapses. As described below, we have identified a new family of brain-specific cell-surface receptors named neurexins that have many of the properties expected for proteins involved in establishing synaptic contacts.

α-Latrotoxin: Structure and Mode of Action

α-Latrotoxin is a component of venom from black widow spiders that causes massive neurotransmitter release from presynaptic nerve endings. Purified α-latrotoxin consists of a single polypeptide with a reported apparent relative molecular weight of 118,000 to 130,000 (Frontali et al. 1976; Valtorta et al. 1984; Ushkaryov and Grishin 1986). Cloning of α-latrotoxin yielded two unexpected results. First, the size of the toxin predicted from the cDNA sequence is almost 160,000. This value is considerably larger than the apparent molecular weight of the toxin on polyacrylamide gels, suggesting that the primary translation product of the toxin is proteolytically processed (Kiyatkin et al. 1990). Since the amino terminus of the toxin is known from peptide sequencing, it is likely that the carboxyl terminus of the toxin is trimmed off posttranslationally (Fig. 1, top). Second, the primary structure of α-latrotoxin reveals the presence of 20 copies of an internal repeat in the toxin (Fig. 1, bottom). These repeats are homologous to a widely distributed sequence motif present in a multitude of proteins. Because these repeats were first defined in *CDC10* (Breeden and Nasmyth 1987) but are present in the highest copy number in ankyrin (Lux et al. 1990), they are often referred to as either *CDC10* or ankyrin repeats.

α-Latrotoxin contains 20 copies of the *CDC10* repeats, almost as many as ankyrin (22 copies). The sequence conservation of these repeats is lower in α-latrotoxin than in ankyrin (Fig. 1, bottom). The structure of α-latrotoxin shows that the repeats are found in

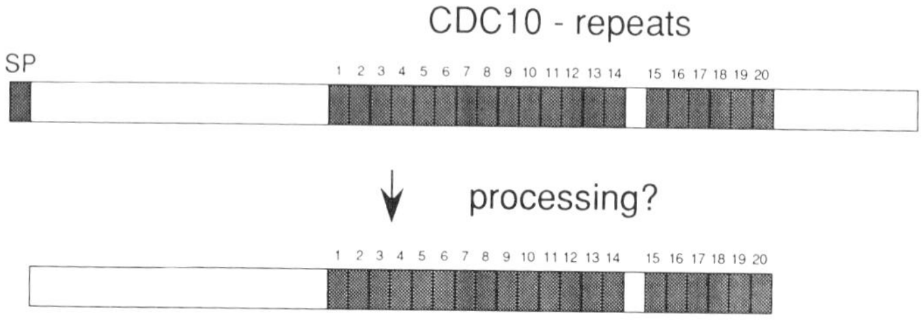


```
Latrotoxin  repeats

 1   H G R T V F H A A A K S G N D K I M F G L T F L A K S T E L N Q P D K      551
 2   K G Y T P I H V A A D S G N A G I V N L L I Q R G · V S I N S K T Y H      585
 3   F L Q T P L H L A A Q R G F V T T F Q R L M E S P E I N I N E R D K        619
 4   D G F T P L H Y A I R G G E R I L E A F L N Q I S · I D V N A K S N        652
 5   T G L T P F H L A I I K N D W P V A S T L L G S K K V D I N A V D E        687
 6   N N I T A L H Y A A I L G Y L E T T K Q L I N L K E I N A N V V S S P      721
 7   G L L S A L H Y A I L Y K H D D V A S F L M R S S N V N V N L K A L        755
 8   G G I T P L H L A V I Q G R K Q I L S L M F D I G · V N I E Q K T D        798
 9   E K Y T P L H L A A M S K Y P E L I Q I L L D Q G · S N F E A K T N        821
10   S G A T P L H L A T F K G K S Q A A L I L L N N E · V N W R D T D E        854
11   N G Q M P I H G A A M T G L L D V A Q A I I S I D A T V V D I E D K        883
12   N S D T P L N L A A Q N S H I D V I K Y F I D Q G · A D I N T R N K        921
13   K G L A P L L A F S K K G N L D M V K Y L F D K N · A N V Y I A D N        954
14   D G M N F F Y Y A V Q N G H L N I V K Y A M S E K · H K F E W S N T        987

15   A I C G P L H Q A A R Y G H L D I V K Y L V E E E F L S V D G S           1061
16   K T D T P L C Y A S E N G H F T V V Q Y L V S N G · A K V N H D C         1093
17   G G M T A I D K A I T K N H L Q V V Q F L A A N G · V D F R R K N S       1129
18   R G T T P F L T A V A E N A L H I A E Y L I R E K R Q D I N I N E Q N V  1163
19   D K D T A L H L A V Y Y K N L Q M I K L L I K Y G · I D V T I R N A       1196
20   Y D K T A L D I A I D A K F S N I V E Y L K T K S · G K F R R E Y K       1229

Consensus        - G - T P L H L A A - - G - - - V V - - L - - - - - · V N V N - - - -
                              Y   I             I I                   I D I D

Notch repeat     - G - T P L H L A A R - - - - - A V - - L L - - - - · A D - N A - D N
                                                   A                    N           D

Ankyrin repeat   - G - T P L H - A A - - G H - - - V - - L L - - G · A - - N - K T K
                                                                           D
```

Figure 1. (*Top*) Schematic diagram of the structure of α-latrotoxin. The published primary structure of α-latrotoxin translated from cDNA sequences (Kiyatkin et al. 1990) is probably incomplete at the 5′ end but still contains part of an amino-terminal hydrophobic sequence that may represent a signal peptide (SP). This is followed by a long sequence that has no homology with sequences in the current databanks and by 20 copies of an internal repeat that is similar to the *CDC10* repeats observed in many proteins. The carboxyl terminus of the protein may be cleaved during maturation because the apparent size of α-latrotoxin (120 kD) is much smaller than its predicted molecular mass (160 kD). This putative processing step would move the *CDC10* repeats to a carboxy-terminal position in the final polypeptide. (*Bottom*) Alignment of the *CDC10* repeats of α-latrotoxin with each other and with the consensus sequences derived from these repeats in ankyrin (Lux et al. 1990) and *Notch* (Wharton et al. 1985). Residues present in at least half of the α-latrotoxin repeats are shown in white on a black background and are used to derive the consensus sequence shown below the alignment.

the middle of the protein where they make up almost half of its total size. The putative carboxy-terminal processing of the toxin would place the repeats at the carboxyl terminus of the mature protein, possibly exposing them.

CDC10 repeats are found in a wide variety of proteins, from cell-surface receptors such as *Notch* (Wharton et al. 1985) and peripheral plasma membrane proteins such as ankyrin (Lux et al. 1990) to nuclear transcription factors such as SW14, SW16, and NF-κB (Breeden and Nasmyth 1987; Andrews and Herskowitz 1989; Kieran et al. 1990). The proteins that contain *CDC10* repeats share no functional properties, suggesting that the repeats function as a general protein-protein interaction domain. The presence of these repeats in α-latrotoxin is surprising because α-latrotoxin is an extracellular toxin, whereas in all other proteins the *CDC10* repeats are cytoplasmic. *CDC10* repeats may have undergone an evolutionary switch in α-latrotoxin and have been converted from a universal intracellular sequence motif to an extracellular function. Alternatively, it is also possible that α-latrotoxin acts intracellularly and needs to be translocated into the presynaptic cytosol for activity. As described below, little is known about the mechanism of action of α-latrotoxin, and so this question must await future elucidation.

The action of α-latrotoxin has been studied in greatest detail on the frog neuromuscular junction, but it has also been investigated in synaptosomes, PC12 cells, and other systems (for review, see Ceccarelli and Hurlbut 1980; Rosenthal and Meldolesi 1989). In all systems, α-latrotoxin causes massive neurotransmitter release. At the neuromuscular junction, this release is quantal and results in depletion of synaptic vesicles from the nerve terminal, suggesting that release occurs by synaptic vesicle exocytosis. In addition to neurotransmitter release, α-latrotoxin also induces ion fluxes and stimulates phosphoinositide hydrolysis in PC12 cells and synaptosomes (Grasso et al. 1980; Vincentini and Meldolesi 1984; Wanke et al. 1986).

Interestingly, the stimulation of neurotransmitter release by α-latrotoxin at the neuromuscular junction is independent of extracellular Ca^{++} (Misler and Hurlbut 1979) and apparently overcomes the block of neurotransmitter release caused by prior botulinum toxin treatment (Cull-Candy et al. 1976; Kao et al. 1976). The stimulation of phosphoinositide breakdown by α-latrotoxin is also Ca^{++}-independent; the activation of neurotransmitter release and of phosphoinositide breakdown requires a lag period of several minutes after toxin addition.

How does α-latrotoxin act? The presence of specific binding sites for the toxin in brain membranes suggests that it binds to a receptor (Tzeng and Siekevitz 1979). Indirect immunofluorescence experiments at the neuromuscular junction localized the binding sites to the presynaptic plasma membrane (Valtorta et al. 1984), and a high-affinity α-latrotoxin-binding protein has been purified from bovine brain (Petrenko et al. 1990, 1991). These results suggest that the first step in α-latrotoxin action consists in its binding to a specific receptor.

There are three hypotheses about the mechanism of action of α-latrotoxin after binding to the receptor: (1) The toxin directly activates the receptor, causing transduction of an activation signal across the plasma membrane. This hypothesis is compatible with the structure of the α-latrotoxin receptor (see below) but fails to explain the lag period between toxin addition and toxin action. (2) The toxin inserts into the membrane after receptor binding and generates nonspecific ion channels. This hypothesis is supported by the ability of purified toxin to induce ion channels in black lipid membranes (Finkelstein et al. 1976) but cannot explain the Ca^{++} independence of toxin action on the neuromuscular junction and PC12 cells. (3) The toxin is endocytosed after receptor binding and translocates into the cytosol, where it acts directly. This hypothesis is supported by the presence of *CDC10* repeats in the primary structure of α-latrotoxin and by the lag period of activation. However, translocation of such a large molecule would be unprecedented. At present, there are not sufficient data to discriminate between these different hypotheses.

Neurexins: A Family of Cell-surface Proteins Defined by the α-Latrotoxin Receptor

Purification of the α-latrotoxin receptor by affinity chromatography on immobilized α-latrotoxin revealed that it consists of two major components: A set of relatively high-molecular-weight proteins varying in size between 160,000 and 220,000, and a relatively low-molecular-weight protein of 29,000 (Petrenko et al. 1990, 1991). The high-molecular-weight components are immunologically related to each other (Fig. 2) and are each complexed with one or several low-molecular-weight proteins but not with each other (A.B. Petrenko et al., in prep.). α-Latrotoxin binding to the receptor causes its aggregation into high-molecular-weight complexes that shift the apparent size of the receptor on sucrose gradients from approximately 250,000 to more than 1,000,000 molecular weight (Petrenko et al. 1991, 1992; in prep.). This suggests that α-latrotoxin may cross-link the receptor.

Partial peptide sequences from the purified receptor were used to clone cDNAs encoding these sequences (Ushkaryov et al. 1992; A.B. Petrenko et al., in prep.). Sequencing of these clones revealed the presence of a novel gene family. Since the proteins encoded by these genes appear to be enriched in neurons and to represent cell-surface receptors, we call these proteins neurexins.

At least three genes for neurexins exist (Ushkaryov et al. 1992; Südhof et al., in prep.). Each gene contains two alternative promoters whose activity leads to two classes of transcripts. The two classes of transcripts differ in their 5′-coding regions but exhibit shared 3′-coding regions. The longer transcripts encode an

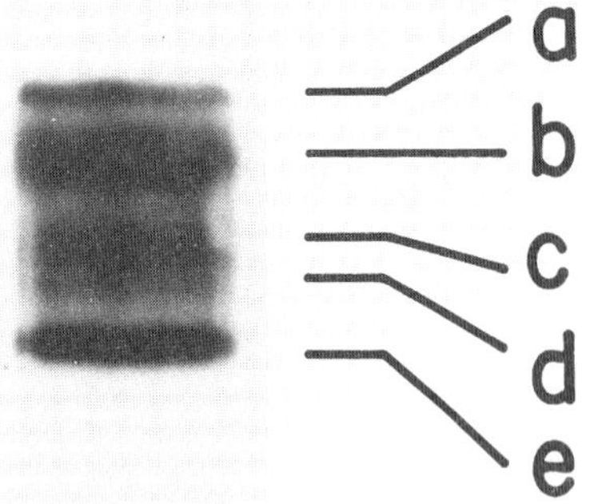

Figure 2. Immunoblot of the high-molecular-weight components of the purified α-latrotoxin receptor. α-Latrotoxin receptor was purified by affinity chromatography and sucrose gradient centrifugation and then immunoblotted with an antipeptide antibody to demonstrate the presence of multiple, immunologically related proteins with relative molecular weights ranging from 220,000 (*a*) to 160,000 (*e*). Immunoreactive bands were visualized using peroxidase-labeled secondary antibodies and enhanced chemiluminescence.

amino-terminal sequence of more than 1000 amino acids that is absent from the second class that encodes a shorter, different amino terminus. The longer forms of neurexins are referred to as α-neurexins and the shorter forms are referred to as β-neurexins.

Comparison of the peptide sequences derived from the purified α-latrotoxin receptor with the translated amino acid sequences of the neurexins revealed that all but one of the peptide sequences can be localized in the sequence of neurexin Iα. The peptide sequences obtained from the purified receptor account for more than 20% of the primary structure of neurexin Iα. Together with the reactivity of the purified α-latrotoxin receptor with several antibodies raised against parts of neurexin Iα, this suggests that the α-latrotoxin receptor constitutes a form of neurexin Iα (Ushkaryov et al. 1992).

Analysis of the primary structures of the neurexins reveals an organization into distinctive domains (Fig. 3) (Ushkaryov et al. 1992). The α-neurexins contain a typical signal sequence at the amino terminus, followed by three copies of a repeat. These three overall repeats (labeled A, B, and C in Fig. 3) are composed of a central epidermal growth factor (EGF) domain flanked by a left and right arm (labeled G_A and G_B in Fig. 3). These two arms of each overall repeat are weakly homologous to each other and to repeats found in the carboxyl termini of several extracellular matrix proteins. These proteins include laminin A, whose G-

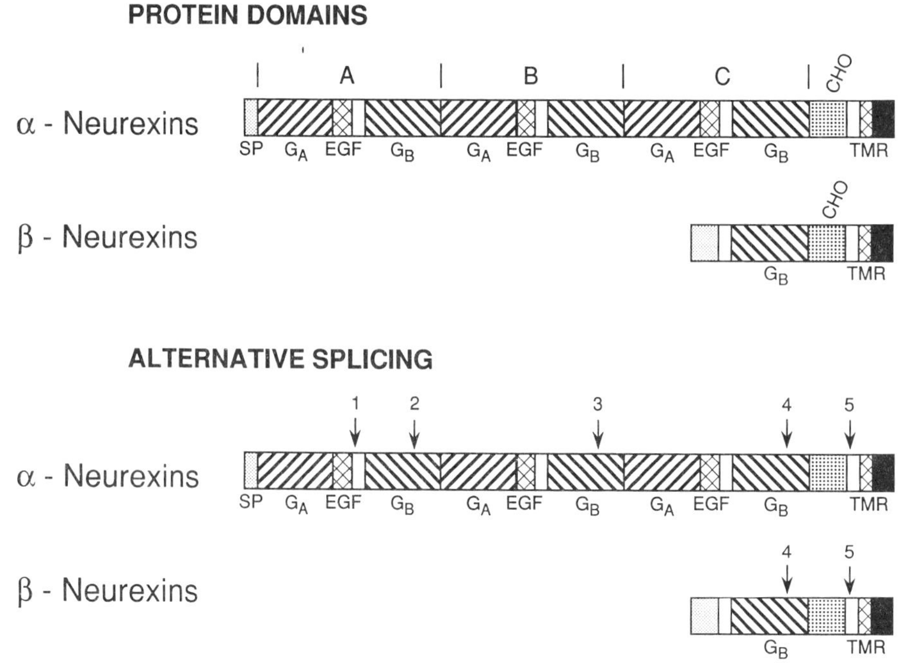

Figure 3. Structural organization and alternative splicing of neurexins (Ushkaryov et al. 1992). Each neurexin gene is transcribed from two promoters, resulting in families of proteins with divergent amino termini and variable carboxyl termini. Sequence analysis indicates the presence of three overall repeats in the α-neurexins (labeled A, B, and C). These three repeats are composed of central EGF domains flanked by right and left arms (hatched boxes labeled G_A and G_B). The right and left arms are weakly homologous to each other and to the G-domain repeats of laminin A and the corresponding sequence repeats found in merosin, agrin, perlecan, and slit. Whereas the α-neurexins contain a signal sequence that precedes the three repeats, the β-neurexins contain a hydrophobic amino-terminal sequence not typical of a classic signal sequence and then splice into the last half-repeat of the α-neurexins. After the repeats, the α- and β-neurexins contain a threonine- and serine-rich sequence typical of *O*-linked sugar regions (labeled CHO), followed by a transmembrane region and a cytoplasmic carboxyl terminus. Neurexins are subject to extensive alternative splicing that appears to be very similar in all neurexins. Three of the five alternative splice sites that are present in all neurexins are in the middle of the G_B repeats, whereas two lie between protein domains.

domain comprises five such repeats. In addition, the extracellular matrix proteins merosin, perlecan, agrin, and slit also contain one or several copies of these repeats.

Accordingly, α-neurexins contain six smaller repeats that are homologous to repeats found in laminin A and other extracellular matrix proteins and that are organized into three larger overall repeats in the α-neurexins. In contrast, the β-neurexins contain only a single copy of the smaller repeats corresponding to the right arm of the third overall repeat of the α-neurexins (Fig. 3). In the β-neurexins, this repeat is preceded by a hydrophobic sequence that is longer than a typical signal sequence. After the repeats, α- and β-neurexins contain a threonine- and serine-rich sequence that is similar to *O*-glycosylation domains found in cell-surface receptors such as the low-density lipoprotein (LDL) receptor (Südhof et al. 1985). Following this domain, each neurexin has a single transmembrane region and a short cytoplasmic tail. Despite the lack of a typical signal sequence for the β-neurexins, α- and β-neurexins probably have the same transmembrane orientation because both seem to be *O*-glycosylated to similar degrees.

The presence of at least three different homologous genes for neurexins, each driven by two independent promoters, creates a diversity of neurexins in the brain. In addition, all neurexins are extensively alternatively spliced. All α-neurexins contain at least five positions at which alternative splicing occurs, and β-neurexins contain at least two positions (Fig. 3). Most of the alternatively spliced sequences are small, with the exception of the alternatively spliced 100-amino-acid sequence at position 5 in neurexin II (Fig. 3). Several of the splice sites exhibit three or four different variants. Taking all splice sites into account and assuming that at least the widely separated splice sites are independent of each other, the three different currently known neurexin genes could potentially encode more than 200 different proteins. Although it is unknown whether the alternative splicing of the neurexins is functionally significant, the conservation of the alternative splice sites between different genes supports a functional role.

Northern blotting indicated that all neurexins are expressed in a brain-specific manner. The mRNAs for the different neurexins appeared to be differentially distributed between different parts of the nervous system. Using antibodies against two different neurexin epitopes, we found an exclusively neuronal stain that appeared to be concentrated on nerve endings, although a postsynaptic and extrasynaptic expression cannot currently be excluded (Fig. 4). Again, the staining observed with the neurexin antibodies was heterogeneous and differed between the two antibodies to different neurexins.

A surprising result of the purification of α-latrotoxin receptor was that synaptotagmin, a synaptic vesicle protein, copurified with the receptor during affinity chromatography (Petrenko et al. 1991). Synaptotagmin is an abundant synaptic vesicle protein that binds Ca^{++} and phospholipids, suggesting a role as the Ca^{++} trigger protein during synaptic vesicle exocytosis (Perin et al. 1990; Brose et al. 1992). Synaptotagmin copurified with the α-latrotoxin receptor during affinity chroma-

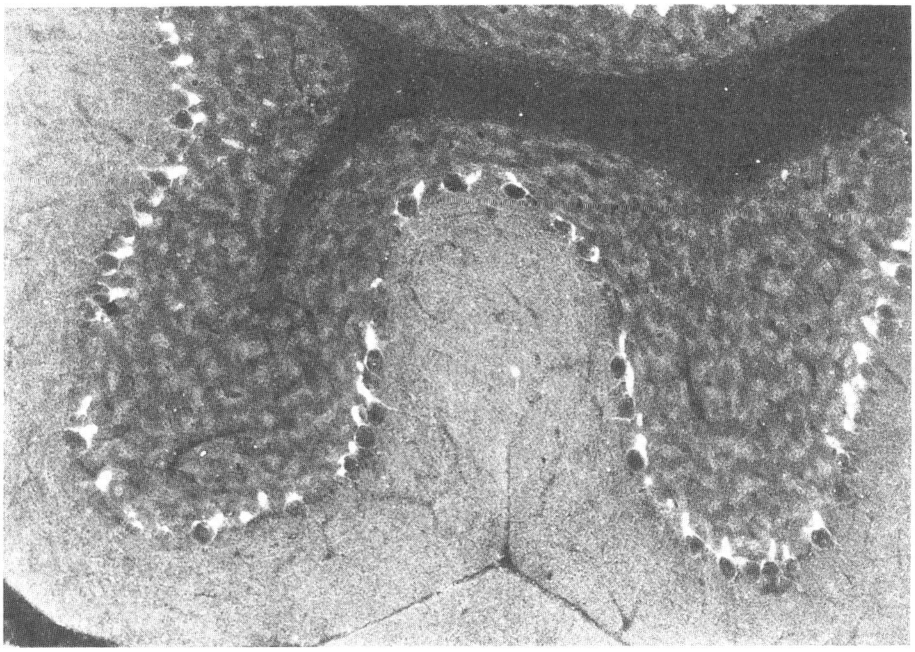

Figure 4. Immunocytochemical labeling of rat cerebellar cortex with an antibody to a peptide from the carboxyl terminus of neurexin II. Perfusion-fixed rat cerebellum was cut into 10-μm sections on a cryostat. Immunofluorescence labeling was carried out as described previously (Südhof et al. 1989). Although synaptic regions of the molecular layer are stained, the strongest signal is observed at the basket cell terminals around the base of the Purkinje cell neurons. No specific staining was observed using preimmune serum.

tography on immobilized α-latrotoxin because it directly interacted with the receptor; no interaction with α-latrotoxin was found. Since the α-latrotoxin receptor most probably represents a form of neurexin Iα, we tested the ability of different recombinant neurexins to bind to synaptotagmin. Recent experiments have shown that there may be a direct interaction between the cytoplasmic domains of all neurexins tested and synaptotagmin (Y.A. Hata et al., in prep.). Together, these results suggest that in addition to an extracellular function as cell-surface receptors, neurexins may also act as intracellular acceptor proteins for synaptic vesicles.

PERSPECTIVE

The neurexins have several interesting properties suggestive of a function as neuronal cell recognition molecules. Their structures resemble those of cell-surface receptors. The large extracellular domains of the α- and β-neurexins contain six copies or one copy, respectively, of a repeat similar to repeats found in several extracellular matrix proteins. These extracellular matrix proteins include laminin A (whose G-domain is composed of these repeats), agrin, slit, and perlecan (Sasaki et al. 1988; Rothberg et al. 1990; Noonan et al. 1991; Rupp et al. 1991; Kallunki and Tryggvason 1992; Murdoch et al. 1992). The G-domain of laminin A has been shown to represent its major heparin-binding site, and it may also interact with integrins (Gehlsen et al. 1992). In agrin, the region of the protein that contains the repeats homologous to the neurexins is alternatively spliced, and this alternative splicing regulates agrin activity (Ferns et al. 1992; Ruegg et al. 1992). These results suggest that in laminin A and agrin, the repeats homologous to neurexins may represent cell-interaction domains, raising the possibility that these repeats have similar functions in neurexins and that neurexins are cell adhesion molecules.

The neurexins differ from other proteins containing repeats homologous to the G-domain of laminin A. In the neurexins, these repeats occur in the context of transmembrane proteins anchored to the cell surface and account for a major portion of these proteins. In the extracellular matrix proteins, the repeats form only a small part of large secreted polypeptides that presumably become integrated into the extracellular matrix. It may be significant that all extracellular matrix proteins containing G-domain repeats have been directly or indirectly implicated in regulating axon guidance or synaptogenesis, although in each case, this activity seems to reside in the proteins when they are not part of an extracellular matrix.

With at least three genes, each driven by two alternative promoters, the neurexins form an unusually diverse family of proteins. Their diversity is enhanced by a high degree of alternative splicing that appears to be conserved between different neurexin genes. Although the exact number of different neurexin proteins is unknown, this number could easily exceed 200 forms. RNA blotting demonstrated that the expression of neurexins is specific to the nervous system. There was a differential distribution of different neurexins in different brain regions, and this heterogeneity extended to the patterns of alternative splicing.

The large number of neurexins and their differential distribution suggest that they may serve as marker proteins on the neuronal cell surface. For example, we found that PC12 cells express neurexins Iα and IIβ (Ushkaryov et al. 1992). PC12 cells did not express all splice forms of these two neurexins but only certain combinations. Coexpression of defined splice forms of neurexins could lead to tens of thousands of different combinations of neurexins expressed on the surfaces of neurons, with different types of neurons characterized by different neurexin combinations. A corollary of a function of the neurexins as neuronal marker proteins would be that there should be a reproducible pattern of expression of restricted splice forms of neurexins for a given population of neurons. This hypothesis is easily testable.

Antibodies against two different neurexin epitopes indicate that at least some neurexins are enriched in the nerve terminal. This localization would agree with the previously observed localization of the α-latrotoxin receptor to the presynaptic plasma membrane (Valtorta et al. 1984). Furthermore, the observed biochemical interaction of the neurexins with synaptotagmin, a synaptic vesicle protein with a probable function in synaptic vesicle exocytosis, would also suggest a presynaptic role. However, an exclusively presynaptic localization of the neurexins has not been firmly established. Further studies will be required to determine if the neurexins are simply enriched in the nerve terminal or if they are exclusively localized here, and if this is true for all neurexins or only for a subset.

The identity of a neuron is defined by its synaptic connectivity. Enrichment of a neuronal cell-surface protein on presynaptic nerve terminals would be compatible with a role as a recognition molecule that identifies a given neuron. However, other roles are equally possible: those of nerve-terminal cell adhesion molecules without a specificity function beyond marking synapses or those of receptors for unknown modulatory signals that are enriched on nerve terminals because modulation of nerve terminals modulates neurotransmission. These different proposed functions do not exclude each other, and each one implies a participation of neurexins in the working of a nerve terminal.

Neurons exhibit a high degree of topological specialization. They are divided into different cellular compartments that are functionally autonomous: dendrites, cell bodies, axons, and nerve terminals. Each of the different cell compartments of a neuron is characterized by different types of cell-cell interactions and by different intercellular signals. This suggests that the plasma membranes of the different cellular compartments will contain characteristic signaling molecules and cell adhesion proteins that are specific for each of these neuronal subcellular domains. Although many cell adhesion proteins with a widespread distribution have been characterized, few proteins with a restricted localization are known that could have such a compartment-specific function. The currently known characteristics of the neurexins suggest that they may fulfill such a function for the nerve terminal, in particular for the synapses formed by the nerve terminal.

REFERENCES

Andrews, B.J. and I. Herskowitz. 1989. The yeast SW14 protein contains a motif present in developmental regulators and is part of a complex involved in cell-cycle-dependent transcription. *Nature* **342:** 830.

Breeden, L. and K. Nasmyth. 1987. Similarity between cell-cycle genes of budding yeast and the *Notch* gene of *Drosophila*. *Nature* **329:** 651.

Brose, N., A.G. Petrenko, T.C. Südhof, and R. Jahn. 1992. Synaptotagmin: A calcium sensor on the synaptic vesicle surface. *Science* **256:** 1021.

Ceccarelli, B.F. and W.P. Hurlburt. 1980. Vesicle hypothesis of the release of quanta of acetylcholine. *Physiol. Rev.* **60:** 396.

Cull-Candy, S.C., H. Lundh, and S. Thesleff. 1976. Effects of botulinum toxin on neuromuscular transmission in the rat. *J. Physiol.* **241:** 353.

Ehrig, K., I. Leivo, W.S. Argraves, E. Ruoslahti, and E. Engvall. 1990. Merosin, a tissue-specific basement membrane protein, is a laminin-like protein. *Proc. Natl. Acad. Sci.* **87:** 3264.

Ferns, M., W. Hoch, J.T. Campanelli, F. Rupp, Z.W. Hall, and R.H. Scheller. 1992. RNA splicing regulates agrin-mediated acetylcholine receptor clustering activity on cultured myotubes. *Neuron* (in press).

Finkelstein, A., L.L. Rubin, and M.C. Tzeng. 1976. Black widow spider venom: Effect of the purified toxin on lipid bilayer membranes. *Science* **193:** 1009.

Fletcher, T.L., P.L. Cameron, P. De Camilli, and G. Banker. 1991. The distribution of synapsin I and synaptophysin in hippocampal neurons developing in culture. *J. Neurosci.* **11:** 1617.

Frontali, N., B. Ceccarelli, A. Gorio, A. Mauro, P. Siekevitz, M.C. Tzeng, and W.P. Hurlbut. 1976. Purification from black widow spider venom of a protein causing the depletion of synaptic vesicles at neuromuscular junctions. *J. Cell Biol.* **68:** 462.

Gehlsen, K.R., P. Sriramarao, L.T. Furcht, and P.N. Skubitz. 1992. A synthetic peptide derived from the carboxy terminus of the laminin A chain represents a binding site for the $\alpha_3\beta_1$ integrin. *J. Cell Biol.* **117:** 449.

Grasso, A., S. Alema, S. Rufini, and M.I. Senni. 1980. Black widow spider toxin-induced calcium fluxes and transmitter release in a neurosecretory cell line. *Nature* **283:** 774.

Kao, I., D.B. Crachman, and D.L. Price. 1976. Botulinum toxin: Mechanism of presynaptic blockade. *Science* **193:** 1256.

Kallunki, P. and K. Tryggvason. 1992. Human basement membrane heparan sulfate proteoglycan core protein: A 467-kD protein containing multiple domains resembling elements of the low density lipoprotein receptor, laminin, neural cell adhesion molecules, and epidermal growth factor. *J. Cell. Biol.* **116:** 559.

Kieran, M., V. Blank, F. Logeat, J. Vandekerckhove, F. Lottspeich, O. Le Bail, M.B. Urban, P. Kourilsky, P.A. Baeuerle, and A. Israel. 1990. The DNA binding subunit of NF-KB is identical to factor KBF1 and homologous to the rel oncogene product. *Cell* **62:** 1007.

Kiyatkin, N.I., I.E. Dulubova, I.A. Chekhovskaya, and E.V. Grishin. 1990. Cloning and structure of cDNA encoding α-latrotoxin from black widow spider venom. *FEBS Lett.* **270:** 127.

Lux, S.E., K.M. John, and V. Bennett. 1990. Analysis of cDNA for human erythrocyte ankyrin indicates a repeated structure with homology to tissue-differentiation and cell-cycle control proteins. *Nature* **344:** 36.

Matteoli, M., K. Takei, M.S. Perin, T.C. Südhof, and P. DeCamilli. 1992. Exoendocytotic recycling of synaptic vesicles in developing processes of cultured hippocampal neurons. *J. Cell. Biol.* **117:** 849.

Misler, S. and W.P. Hurlbut. 1979. Action of black widow spider venom on quantized release of acetylcholine at the frog neuromuscular junction: Dependence upon external Mg^{2+}. *Proc. Natl. Acad. Sci.* **76:** 991.

Murdoch, A.D., G.R. Dodge, I. Cohen, R.S. Tuan, and R.V. Iozzo. 1992. Primary structure of the human heparan sulfate proteoglycan from basement membrane (HSPG2/ Perlecan) *J. Biol. Chem.* **267:** 8544.

Noonan, D.M., A. Fulle, P. Valente, S. Cai, E. Horigan, M. Sasaki, Y. Yamada, and J.R. Hassel. 1991. The complete sequence of perlecan, a basement membrane heparan sulfate proteoglycan, reveals extensive similarity with laminin A chain, low density lipoprotein-receptor, and the neural cell adhesion molecule. *J. Biol. Chem.* **266:** 22939.

Perin, M.S., V.A. Fried, G.A. Mignery, R. Jahn, and T.C. Südhof. 1990. Phospholipid binding by a synaptic vesicle protein homologous to the regulatory region of protein kinase C. *Nature* **345:** 260.

Peters, A., S.L. Palay, and H.D. Webster. 1991. *The fine structure of the nervous system. Neurons and their supporting cells,* 3rd edition, Oxford University Press, England.

Petrenko, A.G., M.S. Perin, B. Davletov, Y.A. Ushkaryov, M. Geppert, and T.C. Südhof. 1991. Binding of synaptotagmin to the α-latrotoxin receptor implicates both in synaptic vesicle exocytosis. *Nature* **353:** 65.

Petrenko, A.G., V.A. Kovalenko, O.G. Shamotienko, I.N. Surkova, T.A. Tarasyuk, Y.A. Ushkaryov, and E.V. Grishin. 1990. Isolation and properties of the α-latrotoxin receptor. *EMBO J.* **9:** 2023.

Rosenthal, L. and J. Meldolesi. 1989. α-Latrotoxin and related toxins. *Pharmacal. Ther.* **42:** 115.

Rothberg, J.M., J. Jacobs, C.S. Goodman, and S. Artavanis-Tsakonas. 1990. Slit: An extracellular protein necessary for development of midline glia and commissural axon pathways contains both EGF and LRR domains. *Genes Dev.* **4:** 2169.

Ruegg, M.A., K.W.K. Tsim, S.E. Horton, S. Kröger, G. Escher, E.M. Gensch, and U.J. McMahan. 1992. The agrin gene codes for a family of basal lamina proteins that differ in function and distribution. *Cell* **8:** 691.

Rupp, F., D.G. Payan, C. Magill-Solc, D.M. Cowan, and R.H. Scheller. 1991. Structure and expression of a rat agrin. *Neuron* **6:** 811.

Sasaki, M., H.K. Kleinman, H. Huber, R. Deutzmann, and Y. Yamada. 1988. Laminin, a multidomain protein. *J. Biol. Chem.* **263:** 16536.

Sheperd, G.M., ed. 1990. *The synaptic function of the brain,* 3rd. edition. Oxford University Press, New York.

Südhof, T.C. and R. Jahn. 1991. Proteins of synaptic vesicles involved in exocytosis and membrane recycling. *Neuron* **6:** 665.

Südhof, T.C., J.L. Goldstein, M.S. Brown, and D.W. Russell. 1985. The LDL receptor gene: A mosaic of exons shared with different proteins. *Science* **228:** 815.

Südhof, T.C., A.J. Czernik, H. Kao, K. Takei, P.A. Johnston, A. Horiuchi, M. Wagner, S.D. Kanazir, M.S. Perin, P. DeCamilli, and P. Greengard. 1989. Synapsins: Mosaics of shared and individual domains in a family of synaptic vesicle phosphoproteins. *Science* **245:** 1474.

Tzeng, M.C. and P. Siekevitz. 1979. The binding interaction between α-latrotoxin from black widow spider venom and dog cerebral cortex synaptosomal membrane preparations. *J. Neurochem.* **33:** 263.

Ushkaryov, Y.A. and E.V. Grishin. 1986. Black widow spider

neurotoxin and its interaction with rat brain receptors. *Bioorg. Chem.* **12:** 71.

Ushkaryov, Y.A., A.G. Petrenko, M. Geppert, and T.C. Südhof. 1992. Neurexins: Synaptic cell surface proteins related to the α-latrotoxin receptor and laminin. *Science* (in press).

Valtorta, F., L. Maddedu, J. Meldolesi, and B. Ceccarelli. 1984. Specific localization of the α-latrotoxin receptor in the nerve terminal plasma membrane. *J. Cell Biol.* **99:** 124.

Vicentini, L. and J. Meldolesi. 1984. α-Latrotoxin of black widow spider venom binds to a specific receptor coupled to phosphoinositide breakdown in PC12 cells. *Biochem. Biophys. Res. Commun.* **122:** 538.

Wanke, E., A. Ferroni, P. Gattanini, and J. Meldolesi. 1986. α-Latrotoxin of the black widow spider venom opens a small, non-closing cation channel. *Biochem. Biophys. Res. Commun.* **134:** 320.

Wharton, K.A., K.M. Johansen, T. Xu, and S. Artavanis-Tsakonas. 1985. Nucleotide sequence from the neurogenic locus Notch implies a gene product that shares homology with proteins containing EGF-like repeats. *Cell* **43:** 567.

Elucidation of Biophysical and Biological Properties of Voltage-gated Potassium Channels

T.J. Baldwin, E. Isacoff, M. Li, G.A. Lopez, M. Sheng, M.L. Tsaur, Y.N. Jan, and L.Y. Jan
The Howard Hughes Medical Institute and Departments of Physiology and Biochemistry, University of California at San Francisco, San Francisco, California 94143-0724

The control of electrical potential and ionic fluxes across a cell-surface membrane is an essential element in governing numerous properties of all cell types. Potassium channels comprise a group of functionally heterogeneous polypeptides expressed in the surface membrane of almost all eukaryotic cells. They appear to be involved in a variety of cellular functions, including the regulation of membrane potential, secretion, and cell motility (for review, see Jan and Jan 1989; Ashcroft and Ashcroft 1990; Hille 1991). Voltage-gated potassium channels characteristically open in response to membrane depolarization and are selectively permeant to potassium ions. On the basis of their kinetic properties, these channels have been divided into the A-type potassium channels that inactivate rapidly after opening and the class of delayed rectifiers that show little inactivation.

In the nervous system, voltage-gated potassium channels have been implicated in regulating neuronal excitability and the strength of signaling between neurons (Kandel and Schwartz 1982). Potassium currents showing a wide range of voltage-dependent and kinetic properties have been characterized in neurons (Latorre et al. 1984; Rogawski 1985; Rudy 1988). Many varied combinations of these currents are found in different neurons and probably contribute to determining the excitability characteristics of individual cells.

The cloning of the *Shaker* gene in *Drosophila* and its homologs in various species (for review, see Jan and Jan 1990; Salkoff et al. 1992) provides defined molecular probes to elucidate the biophysical and biological properties of voltage-gated potassium channels. In addition to Shaker, three other subfamilies of voltage-gated potassium channels (Shab, Shal, and Shaw) have been defined based on amino acid sequence identity. Different members of the same subfamily share about 70% amino acid identity in the hydrophobic core domain, whereas members from different subfamilies share only about 40% amino acid identity.

We summarize here our studies on the structure-function relationship of potassium channels, using the electrophysiological characterization of mutant channels in *Xenopus* oocytes. In addition, we present the initial characterization of the regulation of expression of potassium channel genes and proteins in the adult rat brain.

EXPERIMENTAL PROCEDURES

The structure-function studies involved electrophysiological analysis of channels expressed in *Xenopus* oocytes. The methods for site-directed mutagenesis, heterologous expression, and voltage-clamp and patch-clamp recordings have been described previously (Isacoff et al. 1991; Lopez et al. 1991; Papazian et al. 1991). The initial characterization of regulation of expression of potassium channels in the brain includes cloning of potassium channel genes (Baldwin et al. 1991) and localizations of their mRNA and protein products in the adult rat brain (Sheng et al. 1992; Tsaur et al. 1992).

RESULTS

Involvement of the S4 Sequence in Sensing Voltage Changes

The intrinsic voltage dependence of voltage-gated potassium channels, similar to that of voltage-gated sodium channels and calcium channels, implies that the channel molecule contains charges or dipoles in the hydrophobic interior of the membrane (Hille 1991). The presence of a charged S4 sequence in each voltage-gated potassium channel polypeptide, and in each of the four internal repeats of the voltage-gated sodium or calcium channel α subunit, has led to the hypothesis that the S4 sequence spans the membrane and functions as the voltage sensor (Fig. 1) (Greenblatt et al. 1985; Kosower et al. 1985; Catterall 1986; Guy and Seetharamulu 1986; Noda et al. 1986). To test this S4 hypothesis, the seven basic residues of the S4 sequence of the Shaker B (ShB) potassium channel polypeptide, one of the alternative splice products of the *Shaker* gene, were replaced individually with either the neutral residue glutamine or a different basic residue (Papazian et al. 1991). These mutations alter specifically the voltage-dependent properties without affecting the voltage-independent kinetic properties or ion selectivity of the K^+ channel. The effects on the voltage dependence of activation, however, vary with the specific mutation and cannot be explained solely by considerations of the electrostatic interaction between the membrane potential and the S4 basic residues alone. These mutagenesis experiments suggest that structural considerations in

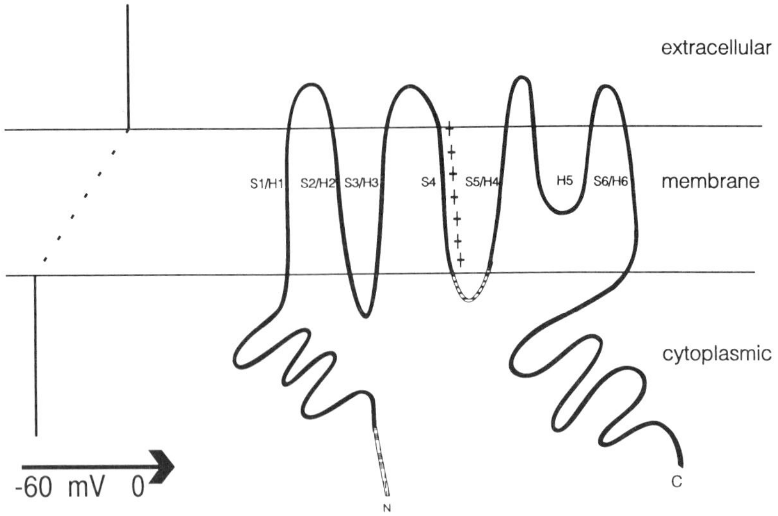

Figure 1. The intrinsic voltage sensitivity of a voltage-gated potassium channel is hypothesized to be contributed partly by the S4 sequence, which contains a basic residue at every third position. (*Right*) Model of the voltage-gated potassium channel polypeptide; (*left*) diagram indicating that much of the electric field generated by the membrane potential is in the hydrophobic interior of the membrane.

addition to electrostatic interactions are important in the involvement of the S4 sequence in sensing changes in membrane potential.

To test for structural constraints on the S4 sequence involving short-range hydrophobic interactions between the S4 sequence and its immediate surrounding, a further series of S4 mutants were generated, each with a highly conserved leucine residue replaced with alanine (Lopez et al. 1991). Each of these mutations in the S4 sequence causes a large shift of the voltage dependence curve of channel activation along the voltage axis (Fig. 2) but does not alter the voltage-independent properties. Since similar mutations in the other proposed membrane spanning segments do not cause such shifts of the voltage dependence curve (Lopez et al. 1991), the S4 sequence appears to be unique with respect to its close involvement in the voltage dependence of channel activation. The effects of the leucine to alanine mutations in S4 are unusually large; the free-energy difference (ΔG) between the closed and the open (activated) state is altered in individual mutants by 4–20 kcal/mole, i.e., these S4 mutations cause a change in the ΔG of 1–5 kcal/mole per subunit, assuming that a potassium channel is a tetramer (Fig. 3) (Tempel et al. 1987; MacKinnon 1991a). In comparison, leucine to alanine mutations in other proteins destabilize the protein by only 0–3 kcal/mole by creating a cavity within the protein; the greatest destabilizing effect results when no adjustment of structures surrounding the cavity is found in the mutant (Eriksson et al. 1992). The leucine to alanine mutations also have an additional destabilizing effect of 2 kcal/mole, so that the mutant protein denatures more easily due to the difference in hydrophobicity between

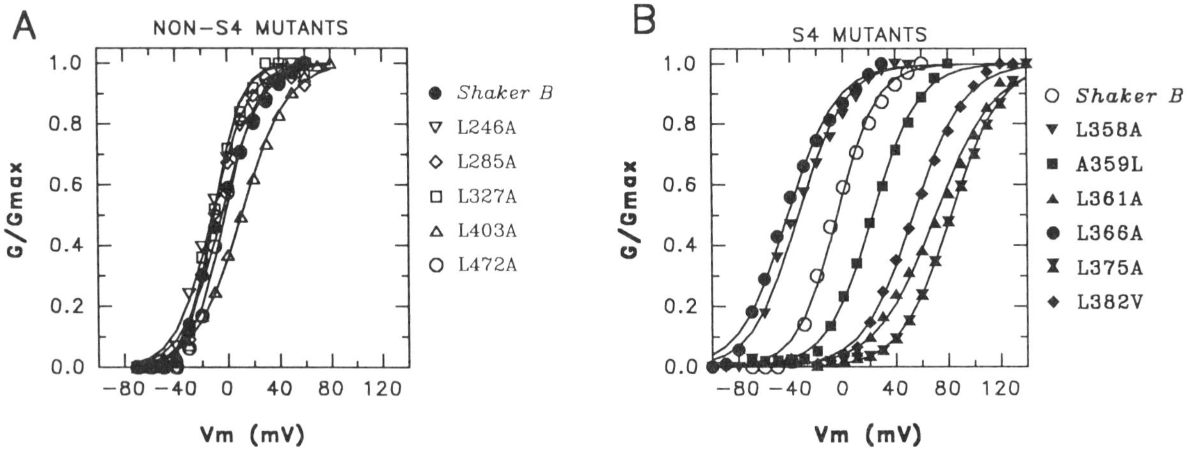

Figure 2. Mutations of highly conserved leucine residues in the S4 sequence (*B*) but not in the other putative membrane spanning segments (*A*) specifically alter the voltage-dependent properties of the ShB potassium channel, such as the normalized conductance-voltage curves shown in this figure. (Adapted from Lopez et al. 1991.)

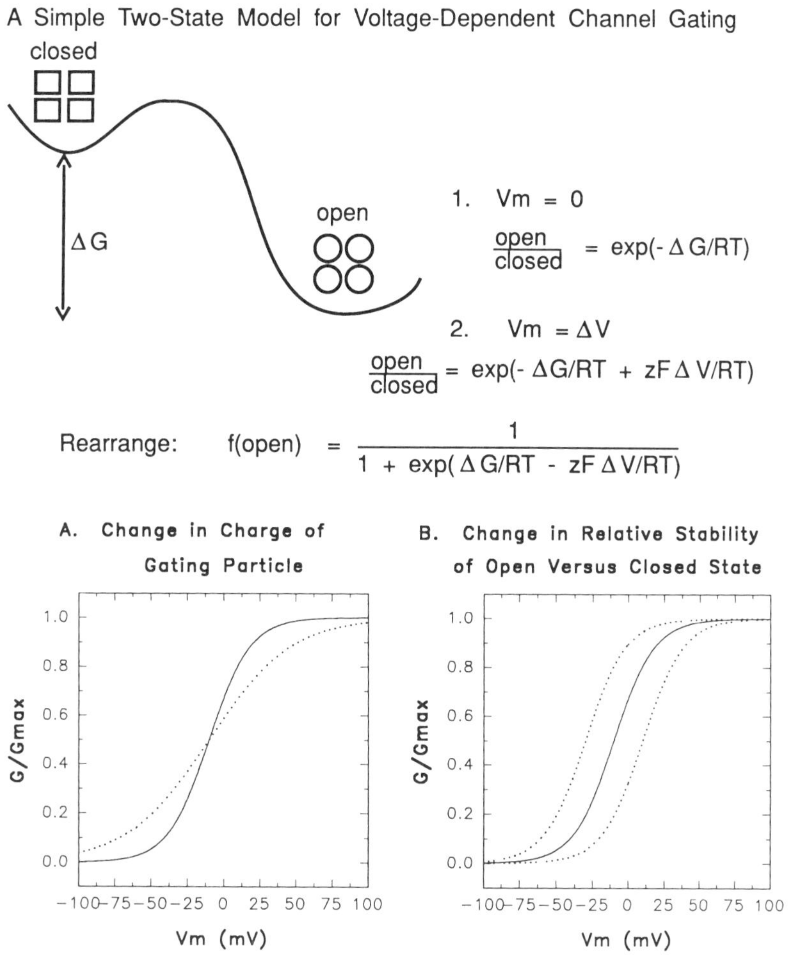

Figure 3. Voltage dependence of channel activation as illustrated by a simple two-state model. The fraction of channels that open (f[open]) at membrane potential $Vm = \Delta V$ depends on the charge of the intrinsic gating particle (z) and the difference in free energy between the open and the closed state (ΔG). A change in ΔG by a mutation, such as those shown in Fig. 2B, will shift the normalized conductance-voltage curve along the voltage axis.

leucine and alanine (Eriksson et al. 1992). This latter component due to the hydrophobicity difference would not alter ΔG between the closed and open state of a channel, since the channel in either state is not expected to approach a denatured state and to expose a buried leucine. Thus, a difference in the free energy of the channel between the closed and the open state may be altered by the leucine to alanine mutations if such mutations have differential effects on the packing and stability of the channel in the closed and open states. If the only effects of these mutations were to destabilize the channel protein by creating a cavity, the maximum alteration of ΔG should be no more than 3 kcal/mole; the voltage dependence curve would shift to the right if a larger cavity is created in the channel at the open state, and it would shift to the left if a larger cavity is created in the channel at the closed state. The observed largest shift of the voltage dependence curve to the right corresponds to an alteration of ΔG between the closed state and the open state by 5 kcal/mole. This implies that substitution of alanine for leucine at position 375 (and, to a lesser extent, at position 361) releases some strain and thereby stabilizes the closed state; the unusually large alteration of ΔG could be accounted for by the sum of the effects of stabilizing the closed state and the effects of destabilizing the open state by creating a cavity. Thus, these results indicate that the wild-type Shaker channel protein appears to be structurally constrained in a way that facilitates channel activation by depolarization.

Characterization of the Inactivation Gate and Its Receptor

Activation of the voltage-gated A-type potassium channel is followed by an inactivation process that is likely to be mediated via the "ball-and-chain" mechanism initially proposed for the inactivation of the sodium channel (Fig. 4) (Bezanilla and Armstrong 1977). The cytoplasmic inactivation gate for the ShB potassium channel has been associated with the amino terminus of the ShB polypeptide because deletions in the amino-terminal region remove fast inactivation, and

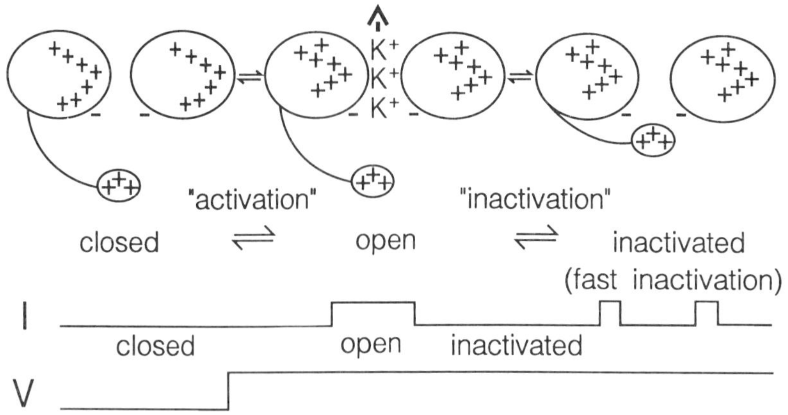

Figure 4. Ball-and-chain model of fast inactivation. After channel activation due to a change of membrane potential, presumably caused by movement of the charged gating particle, the channel opens allowing K^+ ions to flow through. The open channel becomes inactivated when the positively charged cytoplasmic inactivation gate interacts with a receptor at the cytoplasmic mouth of the pore, thereby blocking ion permeation.

cytoplasmic application of the ShB peptide corresponding to the first 20 amino acids of ShB restores inactivation (Hoshi et al. 1990; Zagotta et al. 1990). Similar observations have been made for the rat shal 1 (or Kv4.2) potassium channel, an A-type channel of the Shal subfamily (Baldwin et al. 1991; T.J. Baldwin and G.A. Lopez, unpubl.), suggesting that the involvement of the amino terminus as the cytoplasmic inactivation

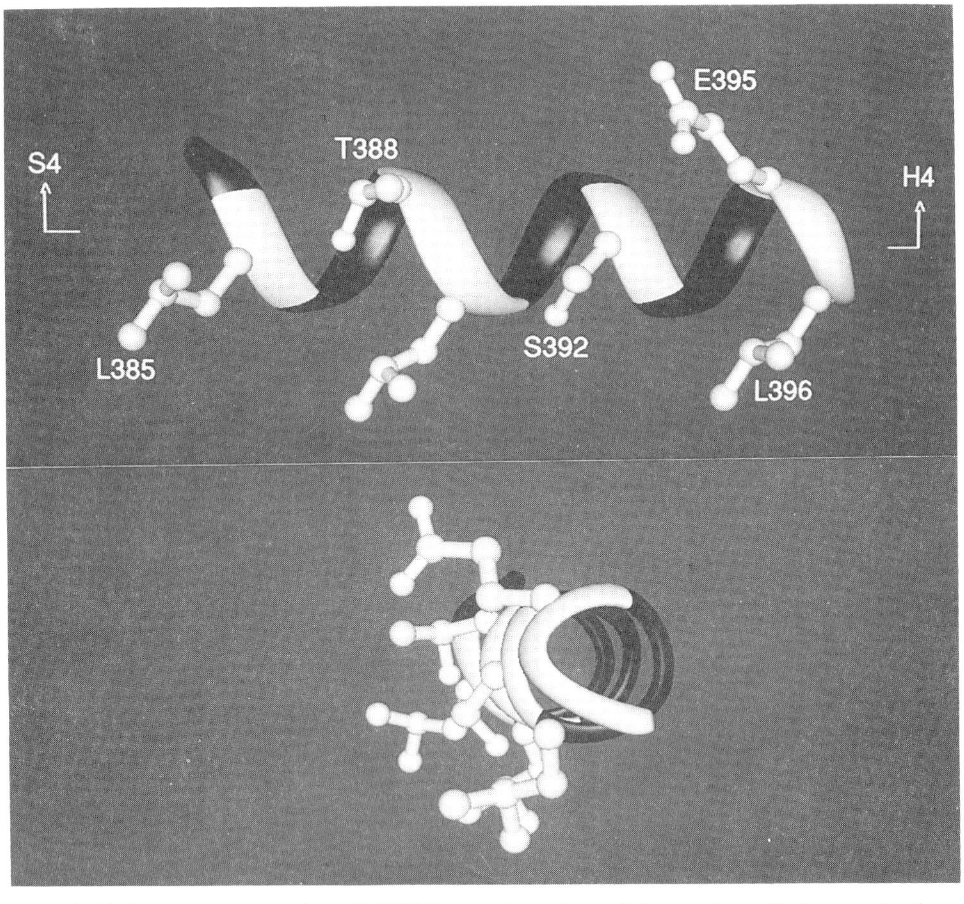

Figure 5. Hypothetical helical arrangement of the S4-S5(H4) loop which places all five residues affecting inactivation on one face of the helix. Mutations of any of these five residues (identified by the single letter code and the position in ShB) alter the affinity between the inactivation gate and its receptor. These mutations also reduce both inward and outward K^+ ion fluxes through a single channel. Thus, these five residues are near the permeation pathway and probably form part of the receptor for the inactivation gate.

gate may be a general mechanism across the superfamilies of voltage-gated potassium channels.

A search for components of the receptor for the cytoplasmic inactivation gate led to the identification of five amino acid residues in the sequence immediately following S4, i.e., the S4-S5 loop (Isacoff et al. 1991). Mutations of these residues either increase or decrease the affinity between the inactivation gate and the receptor. In particular, replacement of Leu-385 in ShB with residues of smaller side chain (valine or alanine) increases the affinity for the inactivation gate, indicating that the leucine residue in the wild-type channel destabilizes the interaction between the inactivation gate and its receptor. Mutations of these five residues also reduce the inward and outward potassium fluxes through a single channel, suggesting that these residues are in or near the permeation pathway. If the S4-S5 loop were folded into an α helix, then these five residues would be one face of this helix (Fig. 5). One possible scenario is that the cytoplasmic mouth of the potassium channel pore is lined with four such helices, which interact with both the permeant ion and the inactivation gate (Isacoff et al. 1991).

These five residues in ShB, which appear to interact with the cytoplasmic inactivation gate as well as potassium ions, are highly conserved among the four subfamilies of potassium channel polypeptides. Does this sequence conservation imply that these residues serve similar functions in different potassium channels? Cytoplasmic application of the ShB peptide to the mammalian DRK1 (or Kv2.1) channel, which shows little inactivation over hundreds of milliseconds and is considered a delayed rectifier potassium channel, causes fast inactivation (Isacoff et al. 1991). Moreover, when the corresponding leucine to alanine mutation that increases the affinity between the receptor and the inactivation gate of ShB is introduced into DRK1, it also enhances the inactivation induced by the ShB peptide (Fig. 6) (Isacoff et al. 1991). Thus, the DRK1 potassium channel not only contains a receptor that can interact with the ShB peptide, but this DRK1 receptor is likely to be similar to the receptor in ShB as well. Finally, even in the absence of the ShB peptide, the mutant DRK1 channel inactivates (Fig. 6), suggesting that DRK1 contains not only a receptor, but also a latent inactivation gate. Thus, a delayed rectifier that shows little inactivation may be structurally rather similar to an A-type potassium channel that inactivates; except in the former case, the affinity between the receptor and the inactivation gate is too weak to produce any significant inactivation.

Seeking Meaning for Potassium Channel Diversity Encountered In Vivo

A surprisingly large number of cDNAs encoding voltage-gated potassium channels have been isolated; at least 14 different potassium channel genes are reported to be expressed in the adult rat brain (Tempel et al. 1988; Frech et al. 1989; McKinnon 1989; Stühmer et al. 1989; Yokoyama et al. 1989; Chandy et al. 1990; Douglass et al. 1990; Grissmer et al. 1990; Grupe et al. 1990;

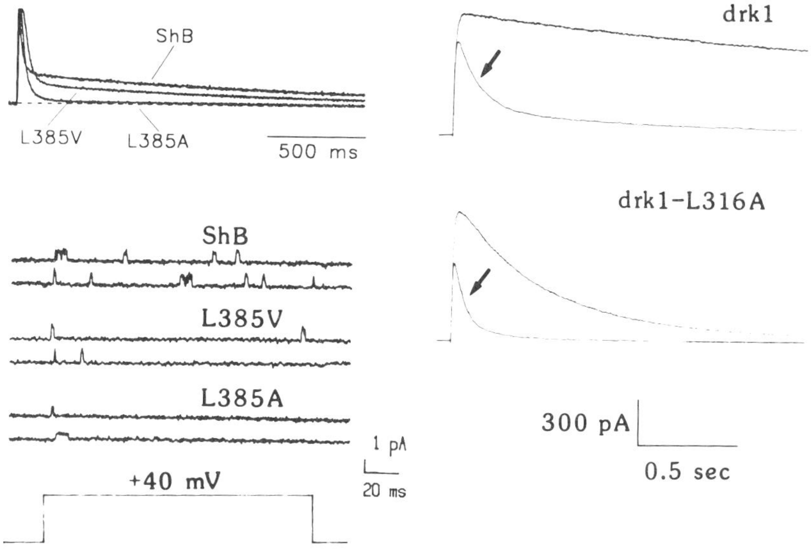

Figure 6. Conservation of the putative receptor for the inactivation gate in the ShB K^+ channel of the Shaker subfamily and the DRK1 K^+ channel of the Shab subfamily. Substitution of alanine for a leucine residue at equivalent positions of the two K^+ channel polypeptides (L385 of ShB and L316 of DRK 1) increases the affinity between the inactivation gate and its receptor. (*Left*) The L385A mutation of ShB hastens fast inactivation, so that the channel enters the inactivated state shortly after it opens and does not recover from inactivation over the duration of the voltage pulse. (*Right*) The DRK1 K^+ channels normally do not show fast inactivation, except in the presence of cytoplasmically applied ShB peptide that corresponds to the first 20 amino acids of ShB (arrow). The L316A mutation of DRK1 hastens the fast inactivation caused by the ShB peptide. Moreover, this mutant channel inactivates even in the absence of ShB peptide. Thus, the DRK1 channel appears to contain an intrinsic inactivation gate, except that its affinity for the receptor is too weak to cause any noticeable inactivation of the wild-type channel. (Adapted from Isacoff et al. 1991.)

T. McCormack et al. 1990; Swanson et al. 1990; Baldwin et al. 1991; Luneau et al. 1991; Pak et al. 1991a,b; Philipson et al. 1991; Roberds and Tamkun 1991; Schröter et al. 1991; Hwang et al. 1992). It is possible that the wide range of potassium channel properties observed in vivo are a consequence of the multitude of potassium channel gene products and their potential interactions to form novel heteromultimeric channels (Christie et al. 1990; Isacoff et al. 1990; K. McCormack et al. 1990; Ruppersberg et al. 1990; Covarrubias et al. 1991). Nevertheless, one wonders why so many genes exist that give rise to channels of apparently similar properties in expression systems such as the *Xenopus* oocyte. One possible explanation is that the expressions of different potassium channel genes are regulated differently in order to provide some characteristics of differential excitability and plasticity of individual neurons. Consistent with this notion, different potassium channel genes show overlapping but distinct patterns of mRNA expression in the brain (Drewe et al. 1992; Hwang et al. 1992; Rudy et al. 1992; Tsaur et al. 1992). Moreover, the level of expression of potassium channel mRNAs encoding RCK5 (Kv1.2), and rat shal 1 (Kv4.2) demonstrate a transient decrease in the granule cells of the dentate gyrus of the hippocampus in response to pentylenetetrazol-induced neuronal activity (Fig. 7) (Tsaur et al. 1992). A decrease of potassium channel expression in these excitatory neurons is expected to cause a further increase of neuronal excitability. Thus, the differential spatial and temporal patterns of regulation of potassium channel gene expression provide potential mechanisms for generating cell-type-specific and activity-dependent differences in neuronal excitability.

The characterization of the expression of some potassium channel proteins in the rat brain reveals another intriguing feature concerning potassium channel regulation. The proteins corresponding to the A-type K^+ channels RCK4 (Kv1.4) and rat shal 1 (Kv4.2) show distinct subcellular distributions in rat brain neurons. The RCK4 (Kv1.4) protein appears to be concentrated in axons and probably nerve terminals (presynaptic localization), whereas rat shal 1 (Kv4.2) appears to be targeted to dendrites (postsynaptic localization) (Fig. 8) (Sheng et al. 1992). Thus, as with other ion channels such as the Na^+ channel and acetylcholine receptor,

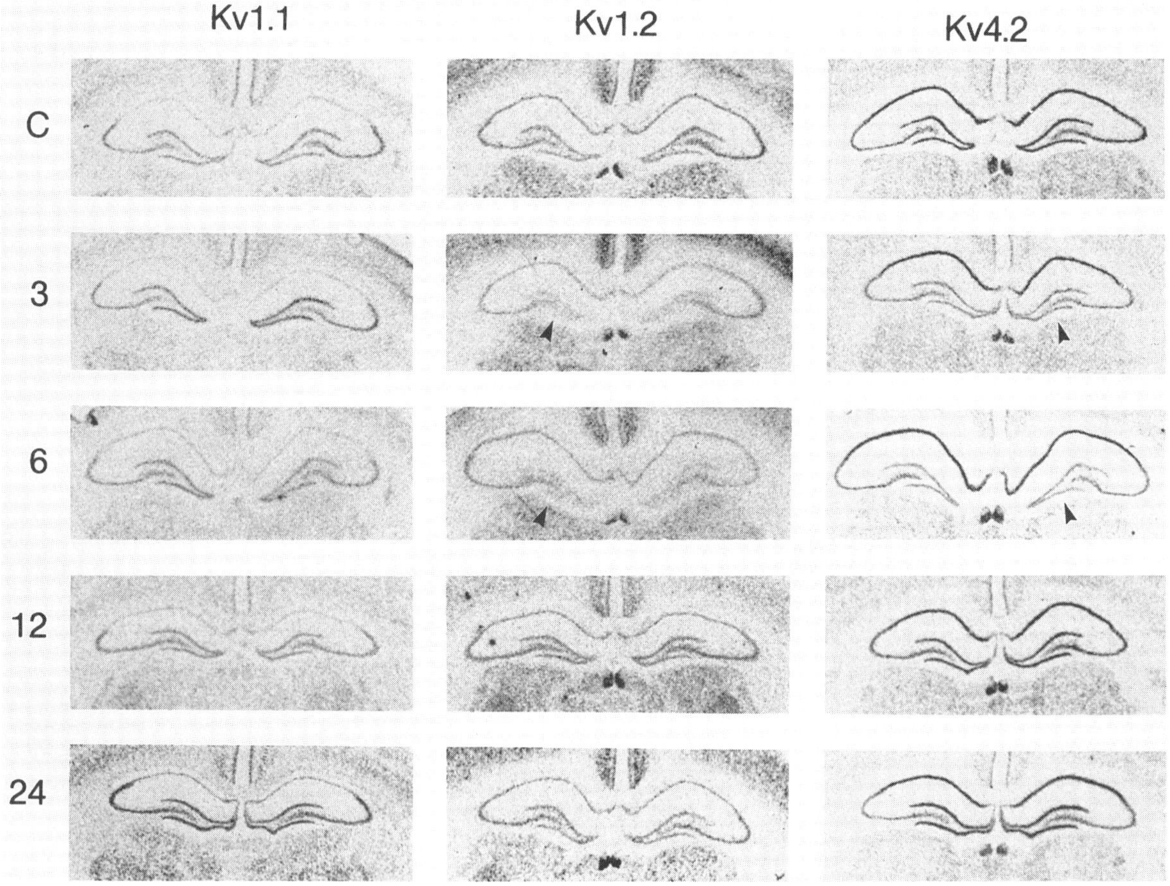

Figure 7. Neuronal activities induced by pentylenetetrazol injection cause a transient reduction of mRNA of Kv1.2 and Kv4.2, but not of Kv1.1, in the dentate gyrus of the hippocampal formation in the rat brain (arrowheads). This reduction is evident 3 hr and 6 hr after the treatment (second and third rows). At 12 hr or 24 hr after the treatment (fourth and fifth rows), the distributions of mRNA of these K^+ channel genes are similar to those in the control animals (first row). Note also that different K^+ channel genes show overlapping but distinct patterns of expression. (Adapted from Tsaur et al. 1992.)

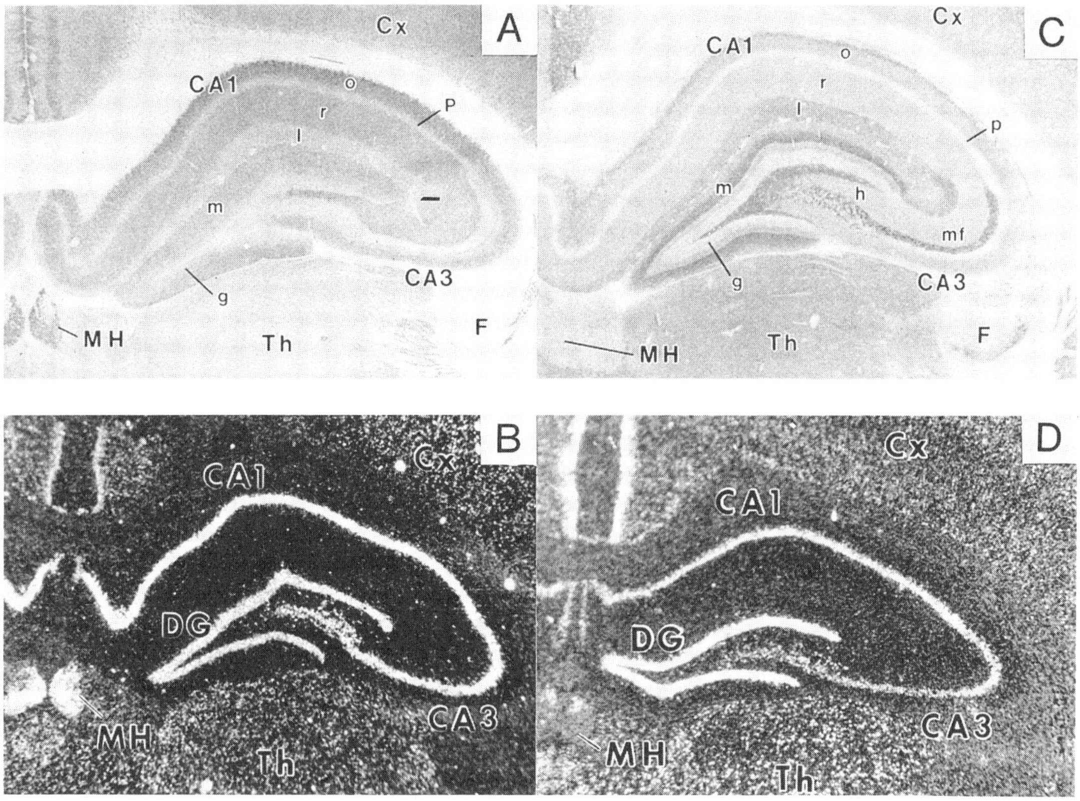

Figure 8. Segregation of two K^+ channel polypeptides to different parts of the neuron. The Kv4.2 protein (*A*) is present in dendrites and cell bodies, whereas the Kv1.4 protein (*C*) is localized in axons and possibly terminals. The differential subcellular distributions result in the complementary immunocytochemical staining patterns in the hippocampus of the rat brain (*A* and *C*), even though the expression patterns of Kv4.2 mRNA (*B*) and Kv1.4 mRNA (*D*) are fairly similar in the hippocampus. (CA1 and CA3) Fields of the Ammon's horn; (p) location of pyramidal cell bodies; (DG) dentate gyrus; (mf) mossy fiber tract; (g) stratum granulosum; (h) hilus; (m) stratum moleculare; (o) stratum oriens; (r) stratum radiatum; (l) stratum lacunosum-moleculare; (Cx) cortex; (MH) medial habenular nucleus; (F) fimbria; (Th) thalamus. (Adapted from Sheng et al. 1992.)

potassium channels show selective localization to specific domains of a cell. The segregation of potassium channel proteins implies subcellular specialization and provides an additional mechanism for modulating the excitability of particular subregions of a neuronal membrane.

PERSPECTIVE

The isolation of different K^+ channel genes has provided for the generation of specific molecular probes to examine both the biophysical characteristics of K^+ channels in heterologous expression systems and their functions in vivo. The former is well suited for mutational analysis and has allowed the identification of structural elements associated with specific channel functions (for review, see Miller 1990; MacKinnon 1991b; Stühmer 1991; Jan and Jan 1992). However, the molecular mechanisms for channel gating and potassium selectivity remain to be established. In addition, biochemical and crystallographic analyses of the channel protein will be essential for a complete understanding of potassium channel structure and function. As an initial step in the biochemical analysis, an element in the Shaker potassium channel amino-terminal hydrophilic domain has been found to be involved in subunit association (Li et al. 1992).

The modulation of potassium channel activity has been implicated in synaptic plasticity in several systems: for example, in the control of sensitization and learning in *Aplysia* (Kandel and Schwartz 1982; Byrne 1987; Abrams and Kandel 1988) and *Hermissenda* (Nelson et al. 1990). Thus, it is of considerable interest to determine whether regulation of potassium channel function or expression in the mammalian brain is involved in controlling excitability and synaptic efficacy. Understanding the function of specific potassium channels and how they are regulated by posttranslational modification such as phosphorylation in the brain represents a major challenge for the future. Furthermore, elucidation of defined roles for potassium channels in the mammalian brain might contribute to our understanding of certain disease states of the nervous system. Finally, although a large number of voltage-activated potassium channels have been characterized, we have merely scratched the surface of potassium channel diversity encountered in vivo. The characterization of new potassium channels, in particular those gated by second messengers such as Ca^{++} and G proteins, represents a major goal in elucidating the role potassium channels might play in governing plasticity within the brain.

ACKNOWLEDGMENTS

The studies discussed in this paper are supported by the National Institutes of Health, a National Institute of Mental Health grant awarded to the Silvio Conte Center for Neuroscience Research at the University of California at San Francisco, and Howard Hughes Medical Institute. M.L. is supported by a Helen Hay Whitney Foundation Fellowship and M.S. is supported by a Howard Hughes postdoctoral research fellowship for physicians. Y.N.J. and L.Y.J. are Hughes Investigators.

REFERENCES

Abrams, T.W. and E.R. Kandel. 1988. Is contiguity detection in classical conditioning a system or a cellular property? Learning in *Aplysia* suggests a possible molecular site. *Trends Neurosci.* **11:** 128.

Ashcroft, S.J.H. and F.M. Ashcroft. 1990. Properties and functions of ATP-sensitive K-channels *Cell. Signalling* **2:** 197.

Baldwin, T.J., M.-L. Tsaur, G.A. Lopez, Y.N. Jan, and L.Y. Jan. 1991. Characterization of a mammalian cDNA for an inactivating voltage-sensitive K^+ channel. *Neuron* **7:** 471.

Bezanilla, F. and C.M. Armstrong. 1977. Inactivation of the sodium channel. II. Gating current experiments. *J. Gen. Physiol.* **70:** 567.

Byrne, J.H. 1987. Cellular analysis of associative learning. *Physiol. Rev.* **67:** 329.

Catterall, W.A. 1986. Molecular properties of voltage-sensitive sodium channels. *Annu. Rev. Biochem.* **55:** 953.

Chandy, K.G., C.B. Williams, R.H. Spencer, B.A. Aguilar, S. Ghanshani, B.L. Tempel, and G.A. Gutman. 1990. A family of three mouse potassium channel genes with intronless coding regions. *Science* **247:** 973.

Christie, M.J., R.A. North, P.B. Osborne, J. Douglass, and J.P. Adelman. 1990. Heteropolymeric potassium channels expressed in *Xenopus* oocytes from cloned subunits. *Neuron* **4:** 405.

Covarrubias, M., A. Wei, and L. Salkoff. 1991. *Shaker*, *Shal*, *Shab* and *Shaw* express independent K^+ channel current systems. *Neuron* **7:** 763.

Douglass, J., P.B. Osborne, Y.-C. Cai, M. Wilkinson, M.J. Christie, and J.P. Adelman. 1990. Characterization and functional expression of a rat genomic DNA clone encoding a lymphocyte potassium channel. *J. Immunol.* **144:** 4841.

Drewe, J.A., S. Verma, G. Frech, and R.H. Joho. 1992. Distinct spatial and temporal expression patterns of K^+ channel mRNAs from different subfamilies. *J. Neurosci.* **12:** 538.

Eriksson, A.E., W.A. Baase, X.J. Zhang, D.W. Heinz, M. Blaber, E.P. Baldwin, and B.W. Matthews. 1992. Response of a protein structure to cavity-creating mutations and its relation to the hydrophobic effect. *Science* **255:** 178.

Frech, G.C., A.M.J. Vandongen, G. Schuster, A.M. Brown, and R.H. Joho. 1989. A novel potassium channel with delayed rectifier properties isolated from rat brain by expression cloning. *Nature* **340:** 642.

Greenblatt, R.E., Y. Blatt, and M. Montal. 1985. The structure of the voltage-sensitive sodium channel. Inferences derived from computer aided analysis of the *Electrophorus electricus* channel primary structure. *FEBS Lett.* **193:** 125.

Grissmer, S., B. Dethlefs, J.J. Wasmuth, A.L. Golding, and G.A. Gutman. 1990. Expression and chromosomal localization of a lymphocyte K^+ channel gene. *Proc. Natl. Acad. Sci.* **87:** 9411.

Grupe, A., K.L. Schröter, J.P. Ruppersberg, M. Stocker, T. Drewes, S. Beckh, and O. Pongs. 1990. Cloning and expression of a human voltage-gated potassium channel. A novel member of the RCK potassium channel family. *EMBO J.* **9:** 1749.

Guy, H.R. and P. Seetharamulu. 1986. Molecular model of the action potential sodium channel. *Proc. Natl. Acad. Sci.* **83:** 508.

Hille, B. 1991. *Ionic channels of excitable membranes.* Sinauer, Sunderland, Massachusetts.

Hoshi, T., W.N. Zagotta, and R.W. Aldrich. 1990. Biophysical and molecular mechanisms of *Shaker* potassium channel inactivation. *Science* **250:** 533.

Hwang, P.M., C.E. Glatt, D.S. Bredt, G. Yellen, and S.H. Snyder. 1992. A novel K^+ channel with unique localizations in mammalian brain: Molecular cloning and characterization. *Neuron* **345:** 530.

Isacoff, E.Y., Y.N. Jan, and L.Y. Jan. 1990. Evidence of the formation of heteromultimeric potassium channels in *Xenopus* oocytes. *Nature* **345:** 530.

———. 1991. Identification of a putative receptor for the cytoplasmic inactivation gate in the *Shaker* K^+ channel. *Nature* **353:** 86.

Jan, L.Y. and Y.N. Jan. 1989. Voltage-sensitive ion channels. *Cell* **56:** 13.

———. 1990. How might the diversity of potassium channels be generated? *Trends Neurosci.* **13:** 413.

———. 1992. Structural elements involved in specific K^+ channel functions. *Annu. Rev. Physiol.* **54:** 537.

Kandel, E.R. and J.H. Schwartz. 1982. Molecular biology of learning: Modulation of transmitter release. *Science* **218:** 433.

Kosower, E.M. 1985. A structural and dynamic molecular model for the sodium channel of *Electrophorus electricus*. *FEBS Lett.* **182:** 234.

Latorre, R., R. Corronado, and C. Vergara. 1984. K^+ channels gated by voltage and ions. *Annu. Rev. Physiol.* **46:** 485.

Li, M., Y.N. Jan, and L.Y. Jan. 1992. Specification of subunit assembly by the hydrophilic amino-terminal domain of the *Shaker* potassium channel. *Science* **257:** 1225.

Lopez, G.A., Y.N. Jan, and L.Y. Jan. 1991. Hydrophobic substitution mutations in the S4 sequence alter voltage-dependent gating in *Shaker* K^+ channels. *Neuron* **7:** 327.

Luneau, C.J., J.B. Williams, J. Marshall, E.S. Levitan, C. Oliva, J.S. Smith, J. Antanavage, K. Folander, R.B. Stein, and R. Swanson. 1991. Alternative splicing contributes to K^+ channel diversity in the mammalian central nervous system. *Proc. Natl. Acad. Sci.* **88:** 3932.

MacKinnon, R. 1991a. Determination of the subunit stoichiometry of a voltage-activated potassium channel. *Nature* **350:** 232.

———. 1991b. New insights into the structure and function of potassium channels. *Curr. Opin. Neurobiol.* **1:** 14.

McCormack, K., J.W. Lin, L.E. Iverson, and B. Rudy. 1990. *Shaker* K^+ channels subunits form heteromultimeric channels with novel functional properties. *Biochem. Biophys. Res. Commun.* **171:** 1361.

McCormack, T., E.C. Vega-Saenz de Mierra, and B. Rudy. 1990. Molecular cloning of a member of a third class of *Shaker*-family K^+ channel genes in mammals. *Proc. Natl. Acad. Sci.* **87:** 5227.

McKinnon, D. 1989. Isolation of a cDNA clone coding for a putative second potassium channel indicates the existence of a gene family. *J. Biol. Chem.* **264:** 8230.

Miller, C. 1990. Annus mirabilis of potassium channels. *Science* **252:** 1092.

Nelson, T.J., C. Collin, and D.L. Alkon. 1990. Isolation of a G protein that is modified by learning and reduces potassium currents in *Hermissenda*. *Science* **247:** 1479.

Noda, M., T. Ikeda, T. Kayano, H. Suzuki, H. Takeshima,

M. Kurasaki, H. Takahashi, and S. Numa. 1986. Existence of distinct sodium channel messenger RNAs in rat brain. *Nature* **320:** 188.

Pak, M.D., K. Baker, M. Covarrubais, A. Ratcliffe, and L. Salkoff. 1991a. A mouse brain homolog of the *Drosophila Shab* K^+ channel with conserved delayed rectifier properties. *J. Neurosci.* **11:** 869.

Pak, M.D., K. Baker, M. Covarrubias, A. Butler, A. Ratcliffe and L. Salkoff. 1991b. mShal, a subfamily of A-type K^+ channel cloned from mammalian brain. *Proc. Natl. Acad. Sci.* **88:** 4386.

Papazian, D.M., L.C. Timpe, Y.N. Jan, and L.Y. Jan. 1991. Mutations of the S4 sequence in the *Shaker* K^+ channel specifically alter its voltage dependence. *Nature* **349:** 305.

Philipson, L.H., R.E. Hice, K. Schaefer, J. LaMendola, G.I. Bell D.J. Nelson, and D.F. Steiner. 1991. Sequence and functional expression in *Xenopus* oocytes of a human insulinoma and islet potassium channel. *Proc. Natl. Acad. Sci.* **88:** 53.

Roberds, S.L. and M.M. Tamkun. 1991. Cloning and tissue-specific expression of five voltage-gated potassium channel cDNAs expressed in rat heart. *Proc. Natl. Acad. Sci.* **88:** 1798.

Rogawski, M.A. 1985. The A-current: How ubiquitous a feature of excitable cells is it? *Trends Neurosci.* **8:** 214.

Rudy, B. 1988. Diversity and ubiquity of K^+ channels. *Neuroscience* **25:** 729.

Rudy, B., C. Kentros, M. Weiser, D. Fruhling, P. Serodio, E. Vega-Saenz de Miera, M.H. Ellisman, J.A. Pollock, and H. Baker. 1992. Region-specific expression of a K^+ channel gene in brain. *Proc. Natl. Acad. Sci.* **89:** 4603.

Ruppersberg, J.P., K.H. Schröter, B. Sakmann, M. Stocker, S. Sewing, and O. Pongs. 1990. Heteromultimeric channels formed by rat brain potassium channel proteins. *Nature* **345:** 535.

Salkoff, L., K. Baker, A. Butler, M. Covarrubias, M.D. Pak, and A. Wei. 1992. An essential "set" of K^+ channels conserved in flies, mice and humans. *Trends Neurosci.* **15:** 161.

Schröter, K.H., J.P. Ruppersberg, F. Wunder, J. Rettig, M. Stocker and O. Pongs. 1991. Cloning and functional expression of a TEA-sensitive A-type potassium channel from a rat brain. *FEBS Lett.* **278:** 211.

Sheng, M., M.-L. Tsaur, Y.N. Jan, and L.Y. Jan. 1992. Subcellular segregation of two A-type K^+ channel proteins in rat central neurons. *Neuron* **9:** 271.

Stühmer, W. 1991. Structure-function studies of voltage-gated ion channels. *Annu. Rev. Biophys. Biophys. Chem.* **20:** 65.

Stühmer, W., J.P. Ruppersberg, K.H. Schröter, B. Sakmann, M. Stocker, and K.P. Giese. 1989. Molecular basis of functional diversity of voltage-gated potassium channels in mammalian brain. *EMBO J.* **8:** 3235.

Swanson, R., J. Marshall, J.S. Smith, J.B. Williams, M.B. Boyle. K. Folander, C.J. Luneau, J. Antanavage, C. Oliva, and S.A. Buhrow. 1990. Cloning and expression of cDNA and genomic clones encoding three delayed rectifier potassium channels in rat brain. *Neuron* **4:** 929.

Tempel, B.L., Y.N. Jan, and L.Y. Jan. 1988. Cloning of a probable potassium channel gene from mouse brain. *Nature* **332:** 837.

Tempel, B.L., D.M. Papazian, T.L. Schwarz, Y.N. Jan, and L.Y. Jan 1987. Sequence of a probable potassium channel component encoded at the *Shaker* locus of *Drosophila*. *Science* **237:** 770.

Tsaur, M.L., M. Sheng, D.H. Lowenstein, Y.N. Jan, and L.Y. Jan. 1992. Differential expression of K^+ channel mRNAs in rat brain and down-regulation in the hippocampus following seizures. *Neuron* **8:** 1055.

Yokoyama, S., K. Imoto, T. Kawamura, H. Higashida, N. Iwabe, T. Miyata, and S. Numa. 1989. Potassium channels form NG108-15 neuroblastoma-glioma hybrid cells—Primary structure and functional expression from cDNAs. *FEBS Lett.* **259:** 37.

Zagotta, W.N., T. Hoshi, and R.W. Aldrich. 1990. Restoration of inactivation in mutants of *Shaker* potassium channels by a peptide derived from ShB. *Science* **250:** 568.

Mechanisms of Sensitivity and Specificity in Olfaction

R.R. Reed
Howard Hughes Medical Institute, Department of Molecular Biology and Genetics, Johns Hopkins University School of Medicine, Baltimore, Maryland 21205

The ability of humans to detect more than 10,000 odorants is the result of complex anatomical, cellular, biochemical, and genetic specialization. The vertebrate olfactory system allows the detection of airborne molecules in the external environment with high sensitivity and converts this stimulus into the nerve impulses that are interpreted by the brain as distinct odors. Recent evidence from mammals as well as from aquatic animals suggests that similar mechanisms for the detection of odorants are utilized by all vertebrates (Nakamura and Gold 1987; Bruch and Teeter 1990, Firestein et al. 1990). The system is remarkably specific and highly sensitive (Amoore 1970), although in humans there has been reduced dependence on this modality.

Anatomy of the Olfactory Organ and Cellular Organization Contribute to Sensitivity

The remarkable sensitivity of the olfactory system is achieved at several levels. At the gross morphological level, the highly convoluted turbinates of the olfactory organ of the rat expose large amounts of surface area to contact odorants that are sniffed in through the nose. These turbinates are covered with a thick mucus layer that serves to solubilize the hydrophobic odorants and to provide the medium for the interaction of the stimulus with receptor molecules. Studies in several laboratories have demonstrated the existence of small extracellular proteins in the mucus that bind hydrophobic odorants with reasonably high affinity and may aid in the solubilization of odorant molecules (Pevsner et al. 1988a,b, 1990).

The cells responsible for detecting the odorant stimulus, the olfactory receptor neurons, reside in the neuroepithelium that lines the turbinates. The primary events in odorant detection occur in the cilia emanating from a single dendritic process that the cell extends to the surface. These cilial structures contribute to the sensitivity in a number of ways. First, they provide extensive membrane surface area at the presumed site of odorant recognition and transmembrane signaling. Second, their geometry results in the rapid accumulation of intracellular second messengers in proximity to the site of membrane-generated signal. Third, their electric properties provide for efficient propagation of the electrical signal generated in this region. The cell body of the sensory neuron sits centrally in the epithelium and extends a single long unbranched, unmyelinated axon that synapses in the olfactory bulb.

The axonal projections of the sensory neurons aggregate into large unmyelinated fasicles shortly after passing through the lamina propria. These bundles traverse the cribriform plate that separates the sensory epithelium from the olfactory bulb at the front of the brain. The fibers then make synapses in the glomerular structures of the bulb. There appears to be some molecular process that sorts the individual fibers to the correct glomerulus, although the mechanism responsible for this sorting is only now being addressed.

The olfactory epithelium contains two other cell types: neuroblast-like basal cells that are the progenitors of the mature sensory neurons and the sustentacular cells that have supportive and perhaps glial-like functions. Olfactory receptor neurons are unique among mature mammalian neurons in undergoing a continual process of degeneration and repopulation from the basal cell population throughout adult life (Monti-Graziadei and Graziadei 1979,1992). Understanding the mechanism by which olfactory neurons continually turn over yet maintain appropriate connections with the brain is one of the major challenges facing olfactory researchers. The sustentacular cells of the epithelium display apical microvilli and express a number of olfactory-specific proteins that are thought to play a role in the chemical modification and the subsequent inactivation and removal of odorant stimuli (Lazard et al. 1990, 1991).

Biochemical Cascades in Olfactory Signal Transduction

The initial events underlying odorant-induced signal transduction occur in the cilia of the olfactory epithelium. Odorant ligands bind to receptors leading to G-protein activation and interaction of the activated G protein with membrane-bound adenylyl cyclase. The activated adenylyl cyclase leads to increases in cAMP and opens a cyclic-nucleotide-responsive cation channel; this depolarizes the cell and initiates an action potential. A second pathway for odorant signaling involving inositol phosphates has been described in several species. The models for second-messenger involvement in odorant detection are based on considerable biochemical and molecular genetic evidence (Pace et al. 1985). Olfactory-specific or olfactory-enriched forms of many of the participants in these two cascades have now been identified: G_{olf} (Jones and Reed 1989), an adenylyl cyclase (Bakalyar and Reed 1990), a cyclic-nucleotide-gated ion channel (Dhallan et al. 1990), and a calcium/calmodulin-activated phosphodiesterase

(Borisy et al. 1992). Biochemical and electrophysiological evidence has implicated the second messenger, inositol 1,4,5-trisphosphate ($InsP_3$), in the transduction process (Restrepo et al. 1990). Generation of $InsP_3$ by phospholipase C in response to odorants appears to activate an $InsP_3$ receptor/channel complex in the cilial surface membrane. A receptor/channel complex localized to the ciliary membrane would lead to the influx of extracellular calcium. This calcium current could signal the presence of odorant by reducing the resting potential of the cell and thereby initiating an action potential. Alternatively, the influx of calcium could function by modulating the biochemical activity of other second-messenger enzymes. Interestingly, attempts to identify olfactory neuron-specific components that mediate the $InsP_3$ pathway, especially an olfactory-specific, pertussis-toxin-sensitive G protein that activates phospholipase C, have been unsuccessful (J. Davis and R. Reed, unpubl.).

The assumption that olfactory receptors would be members of the 7-transmembrane domain superfamily led Buck and Axel (1991) to identify and characterize 18 members of a large multigene family of receptors, consisting of perhaps 100 or more members, expressed in rat olfactory tissue. The localization of this family of receptors to the olfactory neurons was established by polymerase chain reaction (PCR) and confirmed by immunocytochemical techniques. Expression of this gene family is confined to olfactory neurons (A.M. Cunningham and R.R. Reed, unpubl.). The predicted size of this receptor family has been expanded to perhaps 500–1000 members by analysis of additional clones and found to exhibit regions of extensive variability within the transmembrane domains (Buck and Axel 1991; Levy et al. 1991), consistent with the hypothesis that this large multigene family functions in the remarkable sensory discrimination of diverse odorants.

Patterns of Receptor Gene Expression

The ability of olfactory neurons to respond differentially to odorant stimuli presumably results from different patterns of receptor gene expression in each receptor cell. The number of receptors present in each neuron can best be assessed by a PCR-based analysis of the receptor mRNAs present in a single cell. Such an analysis is confounded by the small size of the cells and the absence of coding region introns combined with the multicopy nature of the receptor family. The use of in situ hybridization to elucidate the pattern of receptor expression has revealed that individual receptor genes are expressed in only a small fraction of the mature receptor neurons.

The patterns of receptor expression within the olfactory epithelium have also been examined with antipeptide antibodies. An antibody generated against a conserved region juxtaposed to the seventh transmembrane span of the putative receptors is able to recognize essentially all olfactory neurons. Antibodies raised against an adjacent peptide sequence that is not present in the other cloned receptors have been used for immunocytochemical analysis. A fraction of the receptor cells are immunoreactive with this antibody, although the ability of this antibody to discriminate between receptor proteins of closely related sequence has not been addressed (A.M. Cunningham and R.R. Reed, unpubl.).

Genomic Organization of Olfactory Receptor Genes

There are 500–1000 DNA sequences in the rat, mouse, and human genomes that share extensive homology with the putative olfactory receptors. Preliminary analysis suggests that a fraction of these homologous sequences represent olfactory receptor genes with nucleotide insertion and deletion mutations that would render these gene products nonfunctional. For each of the receptor genes examined to date, the coding region is present within a single exon, although sequences characteristic of RNA splice donor and acceptor consensus sequences are observed upstream of the coding region in several genes from human and rat.

The distribution of the large family of putative olfactory receptors has been examined by fluorescent in situ hybridization (FISH) of metaphase chromosomes with recombinant genomic phage carrying receptor sequences. These initial studies revealed the presence of receptors on several chromosomes and suggested that multiple copies of receptor-related sequences might exist at each of these loci. A human genomic phage that hybridized strongly with receptor sequences from rat was characterized in detail and found to encode two olfactory receptors located approximately 10.2 kb apart in a head-to-tail orientation. This result, consistent with the data from FISH analysis, suggested that receptors were clustered at discrete sites in the genome.

An extensive analysis of receptors clustered on human chromosome 19, performed in collaboration with Anne Olson (Lawrence Livermore National Laboratory), has led to a long-range map of receptor organization on a portion of the short arm of human chromosome 19. The receptors that are close to each other in the chromosome are highly homologous in sequence. However, members of another receptor subfamily are interspersed with this cluster (Fig. 1). These observations suggest that evolution of the olfactory receptor gene family occurred by gene duplication and unequal chromosomal crossing-over. Moreover, the presence of receptors in islands of related sequences may have important implications for the mechanism which regulates the expression of members of this gene family in individual receptor cells.

Models for Receptor Gene Regulation

The mechanism for confining the expression of individual receptor genes to restricted subsets of olfactory neurons is likely to be complex. In principle, this

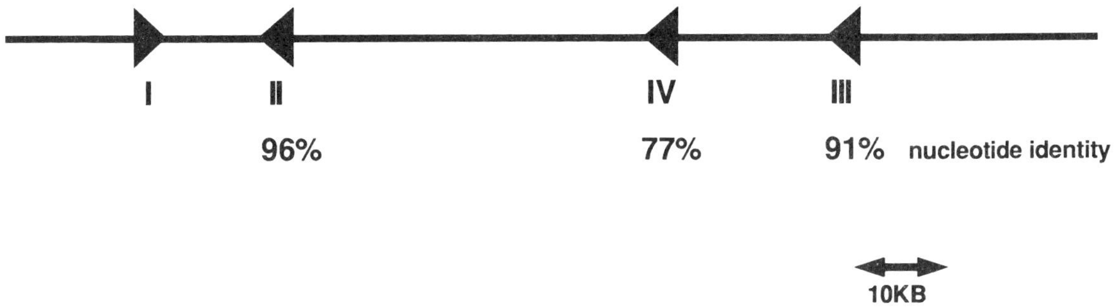

Figure 1. Genomic organization of an olfactory receptor subfamily on chromosome 19. Arrows indicate the direction of translation of each of the receptor genes. The nucleotide identity shared between each receptor (II, III, and IV) and receptor I are indicated. Receptor IV has a highly homologous partner approximately 120 kb to the left.

precise regulation could be achieved by hierarchical patterns of positive-acting transcription factors. The large number of differentially expressed genes would require an exquisitely intricate network of interacting sites and *trans*-acting proteins. The identification of the presumptive promoter sequences located immediately upstream of each of the receptor coding regions can best be addressed in transgenic mouse models by linking these receptor sequences to easily detected reporter proteins.

The data resulting from in situ hybridization studies of individual cloned receptors suggest that an olfactory neuron may express only a single receptor gene. Although there is no evidence of diversity arising in the olfactory system by a site-specific recombination event, DNA rearrangement could play a critical role in regulating expression. Specifically, the presence of a single expression site into which one of the dispersed "silent" genes was introduced into an "expression locus" by a gene conversion event could provide for the observed expression patterns and allow as few as two receptors to be expressed in each cell (Fig. 2). This model is a more complex form of the process responsible for yeast mating-type switching (Hicks et al. 1977). If these conversion events occurred in a stochastic fashion, cells expressing particular receptors would be randomly distributed throughout the olfactory epithelium.

Regardless of the mechanism that directs the expression of receptors in a subset of the sensory neurons, the problem of establishing and maintaining proper connections with the olfactory bulb remains formidable for the organism as well as for the investigator. It should be remembered that current models for second-messenger signaling in this system propose that the nature of the signal indicating the presence of an odorant, an action potential, is uniform in all of the neurons, i.e., no information regarding the chemical structure of the odorant detected is encoded in the signal. For the brain to interpret the signal, it must know which odorant activates that cell. Various models have been proposed to explain the way in which this is achieved (Shepherd 1985; Shepherd and Firestein 1991; Reed et al. 1992). One of the most attractive models is to have the information displayed as the receptor protein on all the extracellular membranes of the neuron. Proper connections would then be established by protein-protein interactions between the sensory neurons and the target neurons in the olfactory bulb.

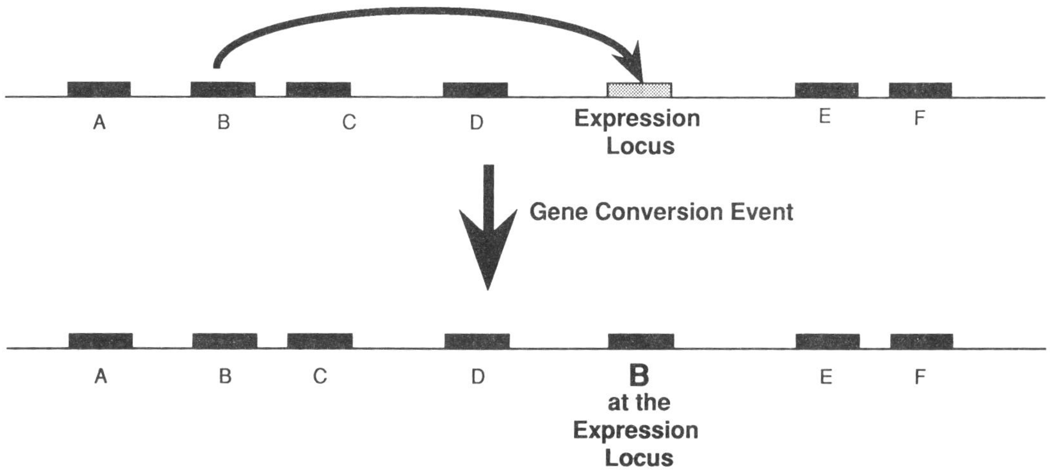

Figure 2. A simple model for the regulation of olfactory gene expression. The diagram illustrates conversion of sequences carried on the same chromosome as the expression locus, but conversion events from other chromosomes could easily be accommodated.

SUMMARY

The sensitivity and specificity of the mammalian olfactory system is the result of the contributions at the anatomical, cellular, and biochemical levels. The recent identification of a large gene family encoding putative olfactory receptors suggests that genetic diversity may play an important role in this system. Future challenges lie in understanding the mechanism that regulates expression of each of the receptors and establishes the complex neural processing network that interprets the primary neuronal responses.

REFERENCES

Amoore, J.E. 1970. *The molecular basis of odor.* C.C. Thomas, Springfield, Illinois.

Bakalyar, H.A. and R.R. Reed. 1990. Identification of a specialized adenylyl cyclase that may mediate odorant detection. *Science* **250:** 1403.

Borisy, F.F., G.V. Ronnett, A.M. Cunningham, D. Juilfs, J. Beavo, and S.H. Snyder. 1992. Calcium/calmodulin-activated phosphodiesterase expressed in olfactory receptor neurons. *J. Neurosci.* **12:** 915.

Bruch, R.C. and J.H. Teeter. 1990. Cyclic AMP links amino acid chemoreceptors to ion channels in olfactory cilia. *Chem. Senses* **15:** 419.

Buck, L. and R. Axel. 1991. A novel multigene family may encode odorant receptors: A molecular basis for odor recognition. *Cell* **66:** 175.

Dhallan, R.S., K.W. Yau, K.A. Schrader, and R.R. Reed. 1990. Primary structure and functional expression of a cyclic nucleotide-activated channel from olfactory neurons. *Nature* **347:** 184.

Firestein, S., G.M. Shepherd, and F.S. Werblin. 1990. Time course of the membrane current underlying sensory transduction in salamander olfactory receptor neurons. *J. Physiol.* **430:** 135.

Hicks, J.B., J.N. Strathern, and I. Herskowitz. 1977. The cassette model of mating-type interconversion. In *DNA insertion elements, plasmids and episomes* (ed. A. Bukhari et al.), p457. Cold Spring Harbor Laboratory, Cold Spring Harbor, New York.

Jones, D.T. and R.R. Reed. 1989. G_{olf}: An olfactory neuron specific-G protein involved in odorant signal transduction. *Science* **244:** 790.

Lazard, D., N. Tal, M. Rubinstein, M. Khen, D. Lancet, and K. Zupko. 1990. Identification and biochemical analysis of novel olfactory-specific cytochrome P-45011A and UDP-glucuronosyl transferase. *Biochemistry* **29:** 7433.

Lazard, D., K. Zupko, Y. Poria, P. Nef, J. Lazarovits, S. Horn, M. Khen, and D. Lancet. 1991. Odorant signal termination by olfactory UDP glucuronosyl transferase. *Nature* **349:** 790.

Levy, N.S., H.A. Bakalyar, and R.R. Reed. 1991. Signal transduction in olfactory neurons. *J. Steroid Biochem. Mol. Biol.* **39:** 633.

Monti-Graziadei, A.G. and P.P.C. Graziadei. 1979. Studies on neuronal plasticity and regeneration in the olfactory system: Morphologic and functional characteristics of the olfactory sensory neuron. In *Neural growth and differentiation* (ed. E. Meisami and M.A.B. Brazier), p. 373. Raven Press, New York.

———. 1992. Sensory reinnervation after partial removal of the olfactory bulb. *J. Comp. Neurol.* **316:** 32.

Nakamura, T. and G.H. Gold. 1987. A cyclic nucleotide-gated conductance in olfactory receptor cilia. *Nature* **325:** 442.

Pace, U., E. Hanski, Y. Salomon, and D. Lancet. 1985. Odorant-sensitive adenylate cyclase may mediate olfactory reception. *Nature* **316:** 255.

Pevsner, J., V. Hou, A.M. Snowman, and S.H. Snyder. 1990. Odorant-binding protein: Characterization of ligand binding. *J. Biol. Chem.* **265:** 6118.

Pevsner, J., R.R. Reed, P.G. Feinstein, and S.H. Snyder. 1988a. Molecular cloning of odorant-binding protein: Member of a ligand carrier. *Science* **241:** 336.

Pevsner, J., P.M. Hwang, P.B. Sklar, J.C. Venable, and S.H. Snyder. 1988b. Odorant binding protein and its mRNA are localized to lateral nasal gland implying a carrier function. *Proc. Natl. Acad. Sci.* **85:** 2383.

Reed, R.R., H.A. Bakalyar, A.M. Cunningham, and N.S. Levy. 1992. The molecular basis of signal transduction in olfactory sensory neurons. *J. Gen. Physiol.* **47:** 53.

Restrepo, D., T. Miyamoto, B.P. Bryant, and J.H. Teeter. 1990. Odor stimuli trigger influx of calcium into olfactory neurons of the channel catfish. *Science* **249:** 1166.

Shepherd, G.M. 1985. The olfactory system: The uses of neural space for a non-spacial modality. In *Contemporary spatial neurobiology* (ed. M. Correia and A.A. Perachio), p. 99. A.R. Liss, New York.

Shepherd G.M. and S. Firestein. 1991. Toward a pharmacology of odor receptors and the processing of odor images. *J. Steroid Biochem. Mol. Biol.* **39:** 583.

Molecular Biology of Smell: Expression of the Multigene Family Encoding Putative Odorant Receptors

A. Chess,* L. Buck,† M.M. Dowling,*§ R. Axel,*§ and J. Ngai*

§*Department of Biochemistry and Molecular Biophysics, and *Howard Hughes Medical Institute, Columbia University, New York, New York 10032; †Department of Neurobiology, Harvard Medical School, Boston, Massachusetts 02115*

Environmental stimuli are recognized by sensory neurons, and this information is transmitted to the brain where it is decoded to provide an internal representation of the external world. The olfactory sensory system of vertebrates can recognize and discriminate a large number of odorants of diverse molecular structure. Humans, for example, are capable of distinguishing among thousands of distinct odors. Although odorants often exhibit widely different structures, subtle changes in molecular structure can lead to striking differences in perceived odor. How do olfactory sensory neurons recognize the vast array of molecular structures that we define as odorants? The detection of distinct odorants presumably results from the association of odorous ligands with specific receptors on olfactory neurons. Olfactory perception may involve a small number of odorant receptors, each capable of interacting with a large number of odorants. Alternatively, a large number of receptors may associate with one or a small number of odorants. Olfactory perception therefore poses a problem in the recognition of multiple possible ligands, a problem also encountered in other biological systems, including the immune system. Unlike the immune system, however, the olfactory sensory system requires that the brain distinguish which receptors have been activated in order to identify different odorant stimuli.

Olfactory perception is initiated by the interaction of odors with receptors on sensory cells that line the posterior portion of the nasal cavity. These olfactory sensory neurons are bipolar: A specialized dendrite terminates in a collection of cilia on the mucosal surface, and an axon extends directly to the brain where it synapses on cells in the olfactory bulb, the first relay in the brain (for reviews, see Lancet 1986; Shepherd 1991). Odorants interact with specific receptors on olfactory cilia, leading to rapid and transient elevations in the intracellular second messengers, cAMP, or inositol triphosphate (IP_3) (Pace et al. 1985; Sklar et al. 1986; Boekhoff et al. 1990; Breer et al. 1990). Elevations in cAMP are thought to directly activate a cyclic nucleotide-gated cation channel (Nakamura and Gold 1987; Dhallan et al. 1990; Goulding et al. 1992), resulting in membrane depolarization and the generation of action potentials which are propagated along the sensory axon to the brain. In this manner, odorant-evoked elevations in cAMP are transduced into electrical signals in olfactory neurons. Although an IP_3-sensitive Ca^{++} channel has been identified in olfactory cilia (Restrepo et al. 1990), the mechanism by which elevations in IP_3 lead to changes in membrane potential remains unclear.

The elucidation of the logic underlying olfactory perception requires the characterization of the specific odorant receptors, the identification of the pattern of receptor expression in the olfactory neurons, and an analysis of the way in which these neurons connect to the brain. We have identified and cloned a large family of genes that are likely to encode odorant receptors in the rat (Buck and Axel 1991). This multigene family encodes seven-transmembrane-domain receptors and belongs to the superfamily of G-protein-coupled receptors. The size of the gene family in mammals (perhaps as large as 1000 genes) suggests that olfactory perception employs a large number of receptors, each capable of interacting with one or a small number of odorous ligands.

The interaction of odorous ligands with a large number of receptors on olfactory sensory neurons provides a mechanism for the specific recognition of a diverse array of odors. This specific binding at the level of the sensory neuron must then be decoded by the brain to determine which of the numerous receptors have been activated. The mechanisms employed to discern which receptors have been activated depend on the diversity of receptor expression in an individual sensory neuron. At one extreme, it is possible that a given sensory neuron may only express one type of odorant receptor. In such a model, the problem of distinguishing which receptors have been activated reduces to a problem of identifying which neurons have been activated. Perception of an odorant encountered by the organism would require the identification of that subset of cells responsive to a given odorant. The position of individual neurons in the olfactory epithelium or their projection patterns to the olfactory bulb may be used to identify which cells have been activated (for review, see Kauer et al. 1991; Shepherd 1991; for alternative models, see Bower 1991; Freeman 1991).

Other sensory systems spatially segregate afferent input from primary sensory neurons. The spatial segregation of information by the visual, auditory, and somatosensory systems is used to define the location of the stimulus in the external environment as well as to indicate the quality of the stimulus (for an overview, see Kandel et al. 1991). In contrast, olfactory processing does not extract spatial features of the odorant stimulus. Relieved of the necessity to encode information about the spatial localization of the sensory stimulus, the vertebrate olfactory system may use the spatial segregation of sensory input solely to encode the identity of the stimulus itself.

In initial experiments to discern the mechanism of processing of olfactory information, we have asked whether cells expressing a specific odorant receptor, and therefore responsive to specific odors, are spatially segregated in the olfactory epithelium (J. Ngai et al., in prep.). We have chosen the channel catfish as a model system for two major reasons: First, the repertoire of putative odorant receptors in the catfish is numerically far simpler than in mammals. It is therefore possible to examine the patterns of receptor expression of a significant fraction of the repertoire with a relatively small number of receptor probes. Second, the catfish olfactory epithelium is organized in a simple, regularly repeating structure, and this anatomical simplicity allows a straightforward analysis of the pattern of odorant receptor expression. We have therefore characterized the family of olfactory receptors expressed in the freshwater catfish. In situ hybridization indicates that the expression of specific receptor genes is restricted to olfactory sensory neurons such that each subfamily of genes is expressed in a small subpopulation of neurons. Analysis of tissue sections through the entire olfactory epithelium demonstrates no discernible pattern of receptor expression. These data suggest that in fish, neurons responsive to a given odorant are randomly distributed in the epithelium. If spatial segregation is employed to decode olfactory sensory information, then neurons expressing a given receptor are likely to converge on discrete loci within the olfactory bulb.

RESULTS

Experimental Strategy

The experimental design we employed to isolate genes encoding odorant receptors was based on three assumptions: First, the odorant receptors are likely to belong to the superfamily of receptor proteins that transduce intracellular signals by coupling to GTP-binding proteins. Second, the large number of structurally distinct odorous molecules suggests that the odorant receptors themselves should exhibit significant diversity and are therefore likely to be encoded by a multigene family. Third, expression of the odorant receptors should be restricted to the olfactory epithelium.

To identify molecules in the olfactory epithelium that resemble members of the seven-transmembrane-domain superfamily, homologs of this gene superfamily were amplified from rat olfactory epithelium cDNA using the polymerase chain reaction (PCR) (Buck and Axel 1991). We then asked whether any of the PCR products we obtained consisted of a mixture of DNA sequences, consistent with the amplification of members of a multigene family. We reasoned that restriction endonuclease digestion of an individual PCR product consisting of a single species of DNA would generate a set of DNA fragments whose molecular weight sum to the molecular weight of the original PCR product. On the other hand, if an individual PCR product consisted of several different DNA sequences (consistent with the amplification of members of a multigene family), restriction endonuclease digestion should generate a large array of fragments whose molecular weights sum to far greater than the molecular weight of the original PCR product. In this manner, we identified a multigene family in the rat that encodes a large number of seven-transmembrane-domain receptor proteins whose expression is restricted to the olfactory epithelium (Buck and Axel 1991). These properties suggest that this family is likely to encode a large number of distinct odorant receptors.

Rat Olfactory Receptor cDNAs

Complete nucleotide sequences were determined for the coding regions of ten of the most divergent cDNA clones initially isolated from this olfactory-specific gene family (Fig. 1) (Buck and Axel 1991). The deduced protein sequences of these cDNAs define a new multigene family that shares sequence and structural properties with the seven-transmembrane-domain superfamily of neurotransmitter and hormone receptors. This novel family, however, exhibits features different from any other member of the receptor superfamily thus far identified.

Each of the ten sequences contains seven hydrophobic stretches (19–26 amino acids) that represent potential transmembrane domains. These domains constitute the regions of maximal sequence similarity to other members of the seven-transmembrane-domain superfamily (O'Dowd et al. 1989a; Strader et al. 1989). On the basis of structural homologies with rhodopsin and the β-adrenergic receptors (O'Dowd et al. 1989a), it is likely that the amino termini of the olfactory proteins are located on the extracellular side of the plasma membrane and the carboxyl termini are located in the cytoplasm. In this scheme, three extracellular loops alternate with three intracellular loops to link the seven-transmembrane domains. Other structural features conserved with previously identified members of the superfamily include cysteine residues at fixed positions within the first and second extracellular loops, which are thought to form a disulfide bond. Finally, many of the olfactory proteins reveal a conserved cysteine within the carboxy-terminal domain that may serve as a palmitoylation site anchoring this domain to the membrane (O'Dowd et al. 1989b). These features, taken together with several short, conserved sequence motifs (see legend to Fig. 1), clearly define this new family as a member of the superfamily of genes encoding the seven-transmembrane-domain receptors.

```
                                        I                               II
F3       MDSSNRTRVSEFLLLGFVENKDLQPLIYGLFLSMYLVTVIGNISIIVAIISDPCLHTPMYFFLSNLSFVDICFISTTVPKML   82
F5       MSSTNQSSVTEFLLLGLSRQPQQQLLFLLFLIMYLATVLGNLLIILAIGTDSRLHTPMYFFLSNLSFVDVCFSSTTVPKVL   82
F6     MAWSTGQNLSTPGPFILLGFPGPRSMRIGLFLLFLVMYLLTVVGNLAIISLVGAHRCLQTPMYFFLCNLSFLEIWFTTACVPKTL   85
F12     MESGNSTRRFSSFFLLGFTENPQLHFLIFALFLSMYLVTVLGNLLIIMAIITQSHLHTPMYFFLANLSFVDICFTSTTIPKML   83
I3       MN--NQTFITQFLLLGLPIPEEHQHLFYALFLVMYLTTILGNLLIIVLVQLDSQLHTPMYLFLSNLSFSDLCFSSVTMPKLL   80
I7      MERRNHSGRVSEFVLLGFPAPAPLRVLLFFLSLLXYVLVLTENMLIIIAIRNHPTLHKPMYFFLANMSFLEIWYVTVTIPKML   83
I8       MN--NKTVITHFLLLGLPIPPEHQQLFFALFLIMYLTTFLGNLLIVVLVQLDSHLHTPMYLFLSNLSFSDLCFSSVTMLKLL   80
I9       MTRRNQTAISQFFLLGLPFPPEYQHLFYALFLAMYLTTLLGNLIIIILILLDSHLHTPMYLFLSNLSFADLCFSSVTMPKLL   82
I14      MTGNNQTLILEFLLLGLPIPSEYHLLFYALFLAMYLTIILGNLLIIVLVRLDSHLHMPMYLFLSNLSFSDLCFSSVTMPKLL   82
I15      MTEENQTVISQFLLLFLPIPSEHQHVFYALFLSMYLTTVLGNLIIIILIHLDSHLHTPMYLFLSNLSFSDLCFSSVTMPKLL   82

                        III                                    IV
F3     ----VNIQTQNNVITYAGCITQIYFFLLFVELDNFLLTIMAYDRYVAICHPMHYTVIMNYKLCGFLVLVSWIVSVLHALFQSLMM   163
F5     ----ANHILGSQAISFSGCLTQLYFLAVFGNMDNFLLAVMSYDRFVAICHPLHYTTKMTRQLCVLLVVGSWVVANMNCLLHILLM   163
F6     ----ATFAPRGGVISLAGCATQMYFVFSLGCTEYFLLAVMAYDRYLAICLPLRYGGIMTPGLAMRLALGSWLCGFSAITVPATLI   166
F12    ----VNIYTQSKSITYEDCISQMCVFLVFAELGNFLLAVMAYDRYVAXCHPLCYTVIVNHRLCILLLLLSWVISIFHAFIQSLIV   164
I3     ----QNMRSQDTSIPYGGCLAQTYFFMVFGDMESFLLVAMAYDRYVAICFPLHYTSIMSPKLCTCLVLLLWMLTTSHAMMHTLLA   161
I7     AGFIGSKENHGQLISFEACMTQLYFFLGLGCTECVLLAVMAYDRYVAICHPLHYPVIVSSRLCVQMAAGSWAGGFGISMVKVFLI   168
I8     ----QNIQSQVPSISYAGCLTQIFFFLLFGYLGNFLLVAMAYDRYVAICFPLHYTNIMSHKLCTCLLLVFWIMTSSHAMMHTLLA   161
I9     ----QNMQSQVPSIPYAGCLAQIYFFLFFGDLGNFLLVAMAYDRYVAICFPLHYMSIMSPKLCVSLVVLSWVLTTFHAMLHTLLM   163
I14    ----QNMQSQVPSISYTGCLTQLYFFMVFGDMESFLLVVMAYDRYVAICFPLRYTTIMSTKFCASLVLLLWMLTMTHALLHTLLI   163
I15    ----QNMQSQVPSIPFAGCLTQLYFYLYFADLESFLLVAMAYDRYVAICFPLHYMSIMSPKLCVSLVVLSWVLTTFHAMLHTLLM   163

                                    V                                         VI
F3     LALPFCTHLEIPHYFCEPNQVIQLTCSDAFLNDLVIYFTLVLLATVPLAGIFYSYFKIVSSICAISSVHGKYKAFSTCASHLSVV   248
F5     ARKSFCADNMIPHFFCDGTPLLKLSCSDTHLNELMILTEGAVVMVTPFVCILISYIHITCAVLRVSSPRGGWKSFSTCGSHLAVV   248
F6     ARLSFCGSRVINHFFCDISPWIVLSCTDTQVVELVSFGIAFCVILGSCGITLVSYAYIITTIIKIPSARGRHRAFSTCSSHLTVV   251
F12    LQLTFCGDVKIPHFFCELNQLSQLTCSDNFPSHLIMNLVPVMLAAISFSGILYSYFKIVSSIHSISTVQGKYKAFSTCASHLSIV   249
I3     ARLSFCENNVVLNFFCDLFVLLKLACSDTYINELMIFIMSTLLIIIPFFLIVMSYARIISSILKVPSTQGICKVFSTCGSHLSVV   246
I7     SRLSYCGPNTINHFFCDVSPLLNLSCTDMSTAELTDFVLAIFILLGPLSVTGASYMAITGAVMRIPSAAGRHKAFSTCASHLTVV   253
I8     ARLSFCENNVLLNFFCDLFVLLKLACSDTYVNELMIHIMGVIIIVIPFVLIVISYAKIISSILKVPSTQSIHKVFSTCGSHLSVV   246
I9     ARLSFCEDSVIPHYFCDMSTLLKVACSDTHDNELAIFILGGPIVVLPFLLIIVSYARIVSSIFKVPSSQSIHKAFSTCGSHLSVV   248
I14    ARLSFCEKNVILHFFCDISALLKLSCSDIYVNELMIYILGGLIIIIPFLLIVMSYVRIFFSILKFPSIQDIYKVFSTCGSHLSVV   248
I15    ARLSFCADNMIPHFFCDISPLLKLSCSDTHVNELVIFVMGGLVIVIPFVLIIVSYARVVASILKVPSVRGIHKIFSTCGSHLSVV   248

                               VII
F3     SLFYCTGLGVYLSSAANNSSQASATASVMYTVVTPMVNPFIYSLRNKDVKSVLKKTLCEEVIRSPPSLLHFFLVLCHLPCFIFCY   333
F5     CLFYGTVIAVYFNPSSSHLAGRDMAAAVMYAVVTPMLNPFIYSLRNSDMKAALRKVLAMRFPSKQ                       313
F6     LIWYGSTIFLHVRTSVESSLDLTKAITVLNTIVTPVLNPFIYTLRNKDVKEALRRTVKGK                            311
F12    SLFYSTGLGVYVSSAVVQSSHSAASASVMYTVVTPMLNPFIYSLRNKDVKRALERLLEGNCKVHHWTG                    317
I3     SLFYGTIIGLYLCPAGNNSTVKEMVMAMMYTVVTPMLNPFIYSLRNRDMKRALIRVICSMKITL                        310
I7     IIFYAASIFIYARPKALSAFDTNKLVSVLYAVIVPLFNPIIYCLRNQDVKRALRRTLHLAQDQEANTNKGSKIG              327
I8     SLFYGTIIGLYLCPSGDNFSLKGSAMAMMYTVVTPMLNPFIYSLRNRDMKQALIRVTCSKKISLPW                      312
I9     SLFYGTVIGLYLCPSANNSTVKETVMSLMYTMVTPMLNPFIYSLRNRDIKDALEKIMCKKQIPSFL                      314
I14    TLFYGTIFGIYLCPSGNNSTVKEIAMAMMYTVVTPMLNPFIYSLRNRDMKRALIRVICTKKISL                        312
I15    SLFYGTIIGLYLCPSANNSTVKETVMAMMYTVVTPMLNPFIYSLRNRDMKEALIRVLCKKKITFCL                      314
```

Figure 1. Protein sequences encoded by ten divergent rat cDNA clones. Ten divergent cDNA clones were subjected to DNA sequence analyses, and the protein sequence encoded by each was determined. Amino acid residues conserved in 60% or more of the proteins are shaded. The presence of seven hydrophobic domains (I–VII), as well as short conserved motifs shared with other members of the superfamily, demonstrates that these proteins belong to the seven-transmembrane-domain protein superfamily. Motifs conserved among members of the superfamily and the family of olfactory proteins include the GN in TM1 (transmembrane domain 1), the central W of TM4, the Y near the carboxy-terminal end of TM5, and the NP in TM7. In addition, the DRY motif carboxy-terminal to TM3 is common to many members of the G-protein-coupled superfamily. However, all the proteins shown here share sequence motifs not found in other members of this superfamily and are clearly members of a novel family of proteins.

Isolation and Sequence of Catfish Olfactory Receptor cDNAs

The isolation of cDNA clones encoding putative olfactory receptors from catfish sensory neurons was based on the assumption that these receptors share considerable homology with the family of rat olfactory receptors. Degenerate oligonucleotides from conserved regions of the rat receptors were therefore used in pairwise combinations as primers for PCR with catfish olfactory epithelium cDNA. The PCR products were then used as a probe to screen olfactory epithelium cDNA libraries under conditions of both low and high stringency. The nucleotide sequence of the coding regions of each of six divergent cDNA clones was determined (J. Ngai et al., in prep.). The deduced amino acid sequences of these catfish cDNAs again define a family of proteins that share sequence and structural properties with the seven-transmembrane-domain superfamily of neurotransmitter and hormone receptors (Fig. 2). Some sequence motifs are conserved in the olfactory receptors identified in both fish and mammals, whereas other motifs appear specific for the family of fish receptors. Overall, the individual fish re-

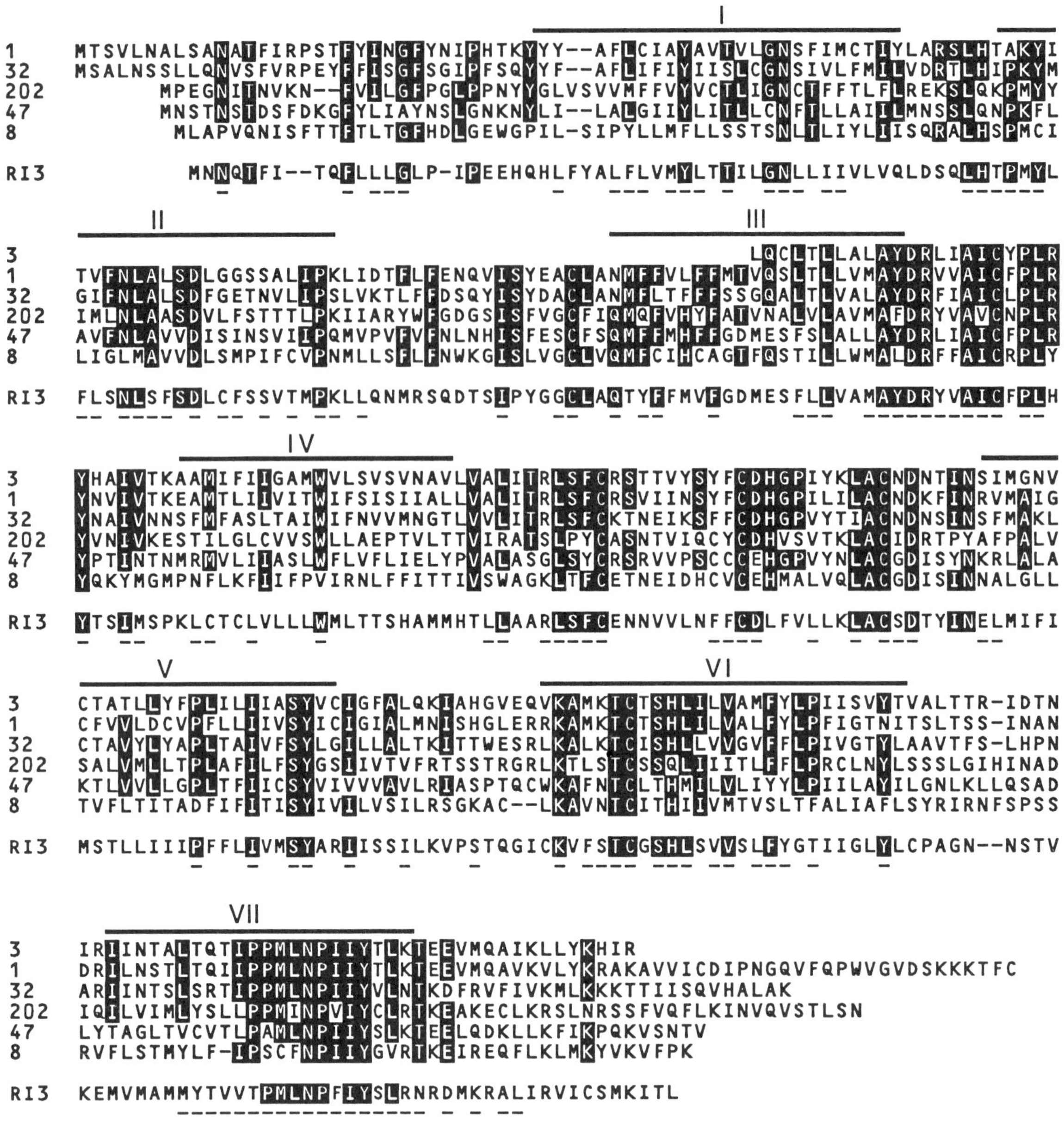

Figure 2. Protein sequences encoded by six catfish cDNA clones. Six divergent catfish olfactory receptor clones (clones 1, 32, 202, 3, 47, and 8) were subjected to DNA sequencing, and their predicted amino acid sequences are shown aligned to the protein sequence of rat cDNA clone I3. Catfish clone 3 is not full length. Residues in the catfish sequences are shaded if at least four out of six (or three out of five, where applicable) are identical at a given position. Residues in the rat I3 sequence identical to the conserved catfish residues are also shaded. Dashes under the rat I3 sequence correspond to positions where at least eight out of ten residues in the rat cDNAs shown in Fig. 1 are identical. The predicted positions of transmembrane domains I-VII are shown.

ceptors exhibit from 40% to 60% amino acid homology with one another and from 30% to 40% homology with representative rat sequences.

Although the rat and fish odorant receptor sequences are clearly members of the superfamily of seven-transmembrane-domain receptors, the olfactory receptors exhibit unique structural features that may be relevant to the proposed function of these proteins in odor recognition. Structure-function experiments involving in vitro mutagenesis suggest that adrenergic ligands interact with this class of receptor molecule by binding within the plane of the membrane (Kobilka et al. 1988; Strader et al. 1989). Not surprisingly, receptor families that bind the same class of ligands, such as the adrenergic or muscarinic acetylcholine receptor families, exhibit maximum sequence conservation (often >80%) within the transmembrane domains. In contrast, the family of receptors we have identified shows striking divergence within the third, fourth, and fifth transmembrane domains (Figs. 1 and 2). The divergence in potential ligand-binding domains is consistent with the idea that the family of molecules we have cloned is capable of associating with a large number of odorants of diverse molecular structures.

Size of the Multigene Families

The size of the receptor repertoire is likely to reflect the range of detectable odors and the degree of structural specificity exhibited by the individual receptors. We have performed Southern blot hybridization to genomic DNA to obtain an estimate of the size of the multigene families in both mammals and fish (Buck and Axel 1991; J. Ngai et al., in prep.). DNA probes were prepared from the coding regions of seven divergent classes of cDNAs from the rat. Stringency conditions were defined such that no cross-hybridization was observed between the different cDNA clones. These probes were then annealed to Southern blots of rat liver DNA after digestion with the restriction endonucleases *Eco*RI or *Hin*dIII. The olfactory receptor genes analyzed to date do not contain introns within the coding regions. This observation, together with the fact that most of the probes encompass only 400 nucleotides, suggests that the number of hybridizing bands closely approximates the number of receptor genes. Examination of the Southern blots reveals that all but one of the DNAs detect a relatively large, distinctive array of bands in genomic DNA (Fig. 3A). Clone I15 (probe 7), for example, detects about 17 bands with each restriction endonuclease, whereas clone F9 (probe 1) detects only about 5–7 bands with each enzyme. A single band is obtained with clone I7 (probe 5). These data suggest that each of the individual probes we have chosen is representative of subfamilies that range in size from a single member to as many as 17 members.

We detect a total of about 70 individual bands in this analysis, which could represent the presence of at least 70 different genes. It is probable that the total number of bands reflects only a minimal estimate of gene number, since it is unlikely that we have isolated representative cDNAs from all of the potential subfamilies and the hybridizations were performed under conditions of very high stringency. Additional screening of cDNA and genomic libraries continues to reveal new subfamilies of olfactory receptor cDNAs in mammals (Levy et al. 1991; Parmentier et al. 1992). These data suggest that the repertoire of odorant receptor genes in mammals may be extremely large and may consist of as many as 1000 genes.

If the odorant receptor repertoire reflects the range of detectable odors, we might expect that fish possess a far smaller number of odorant receptors than mammals. Extensive PCR and cloning experiments identify only 20 different receptor genes in the catfish, whereas similar experiments in mammals identify several hundred genes. Southern blotting of DNA from a single catfish with three divergent cDNA probes confirms this suggestion (Fig. 3B). The pattern of bands observed with Southern blotting is different with each of the three DNA clones. One of the cDNA clones (clone 3) detects only a single band with all enzymes, whereas the other two clones (clones 1 and 32) each detects at most 5 bands in the catfish genome. Under conditions of low stringency, cross-hybridization is observed between the individual cDNA probes, but few new bands are apparent in genomic blots. Using all six divergent cDNA clones, we detect only about 20 bands in the catfish genome. The number of olfactory receptor genes present within the catfish genome is therefore significantly smaller than that observed in mammalian DNA as a consequence of an apparently smaller number of subfamilies, as well as the smaller size of the individual subfamilies.

Subfamilies within the Multigene Family

Genomic blotting experiments, along with analysis of the sequences of the cDNA clones, demonstrate that considerable divergence exists between different family members. The multigene family in both fish and rat can be divided into subfamilies (operationally defined by cross-hybridization at high stringency), such that members of a given family share significant sequence conservation. One particularly clear example is provided by the receptor 32 subfamily we have characterized in the fish. A cDNA clone encoding receptor 32 detects 5 bands on Southern blots. Six different genes that define the receptor 32 subfamily were isolated from a genomic library constructed from kidney DNA from a single fish. Preliminary analysis suggests that at least three of these genes are linked within 30 kb of genomic DNA. The deduced amino acid sequences clearly demonstrate that the six genes define a discrete subfamily within the larger family of catfish olfactory receptor genes. Whereas only 40–60% amino acid conservation is observed among the six subfamilies, the six genes within the receptor 32 subfamily share more than 90% homology. Moreover, four of the sequences exhibit more than 95% conservation with one another.

The distribution of amino acid differences among the receptor 32 subfamily members is clearly nonrandom. More than 90% of the amino acid changes occur within transmembrane domains, although the transmembrane domains comprise less than 50% of the mass of receptor molecule. The ligands that activate the seven-transmembrane-domain receptors are thought to interact with the transmembrane domains within the plane of the membrane. The clustering of diversity within the transmembrane domains is therefore consistent with the suggestion that each of the subfamily members may interact with different odorous ligands. In such a model, the divergent subfamilies would encode receptors that bind odorants of widely differing molecular structures. Members of the individual subfamilies would then recognize subtle differences between more structurally related molecules.

Expression of the Gene Family

We have performed Northern blot analyses to demonstrate that expression of the members of this gene family is restricted to the olfactory epithelium. In mammals that may express hundreds of different receptor genes, the level of expression of a single receptor gene or subfamily in the mRNA population is extremely low.

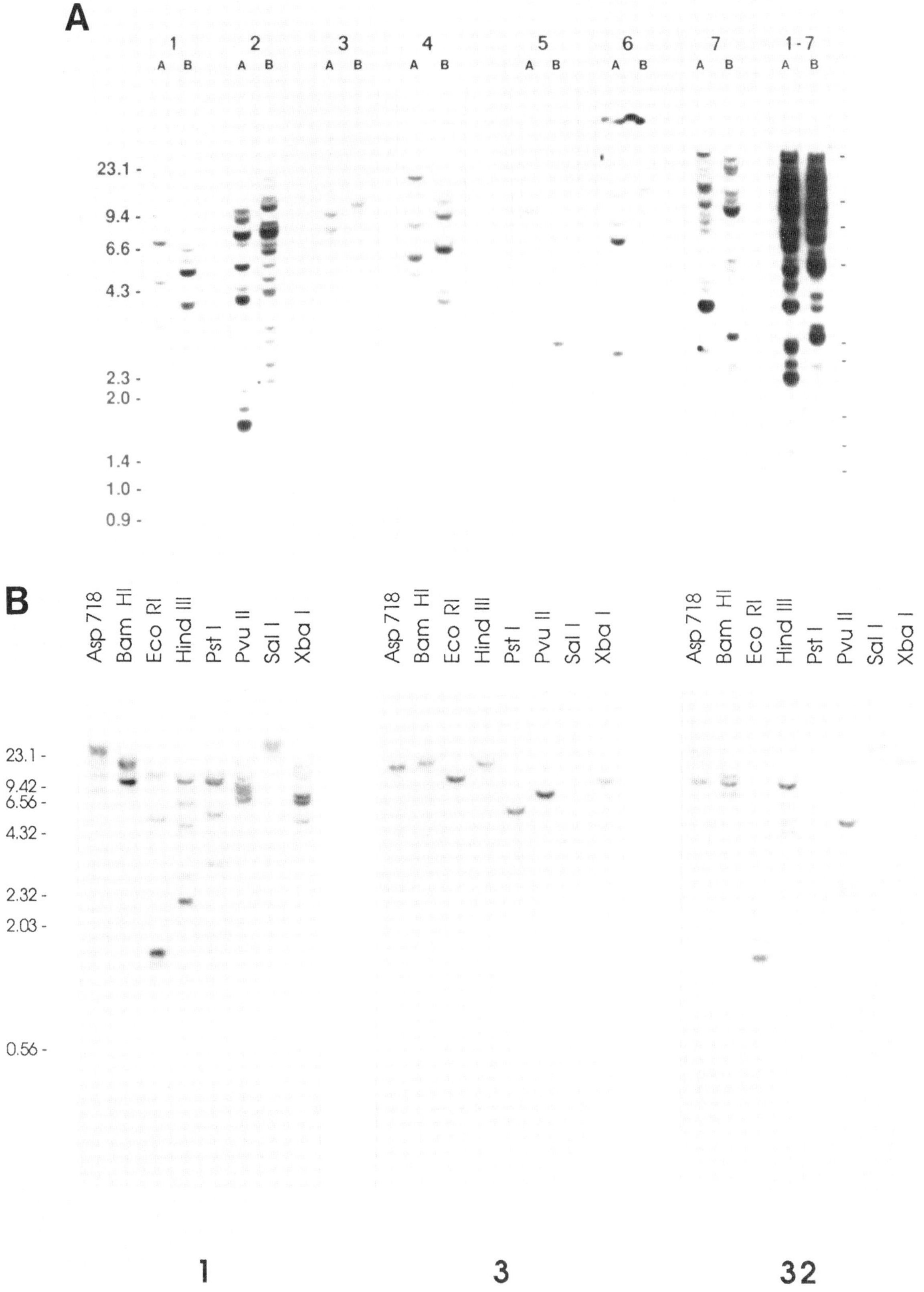

Figure 3. Southern blot analyses with divergent rat and catfish cDNAs. (*A*) Rat liver DNA (5 μg) was digested with *Eco*RI (A) or *Hin*dIII (B), electrophoresed in 0.75% agarose, blotted onto a nylon membrane, and hybridized with the ^{32}P-labeled probes indicated. The probes used were PCR-generated fragments of: (*1*) clone F9 (identical to F12 in Fig. 1); (*2*) F5; (*3*) F6; (*4*) I3; (*5*) I7; (*6*) I14; or (*7*) I15. The lane labeled "*1–7*" was hybridized to a mixture of the seven probes. The probes used showed either no cross-hybridization or only trace cross-hybridization with one another. Size markers on the left correspond to the four blots on the left (*1–4*), whereas the marker positions noted on the right correspond to the four blots on the right (*5–7* and "*1–7*"). (*B*) Catfish genome DNA (3 μg) was digested with the indicated restriction endonucleases, electrophoresed, blotted onto nitrocellulose, and hybridized with ^{32}P-labeled probes for receptors 1, 3, and 32.

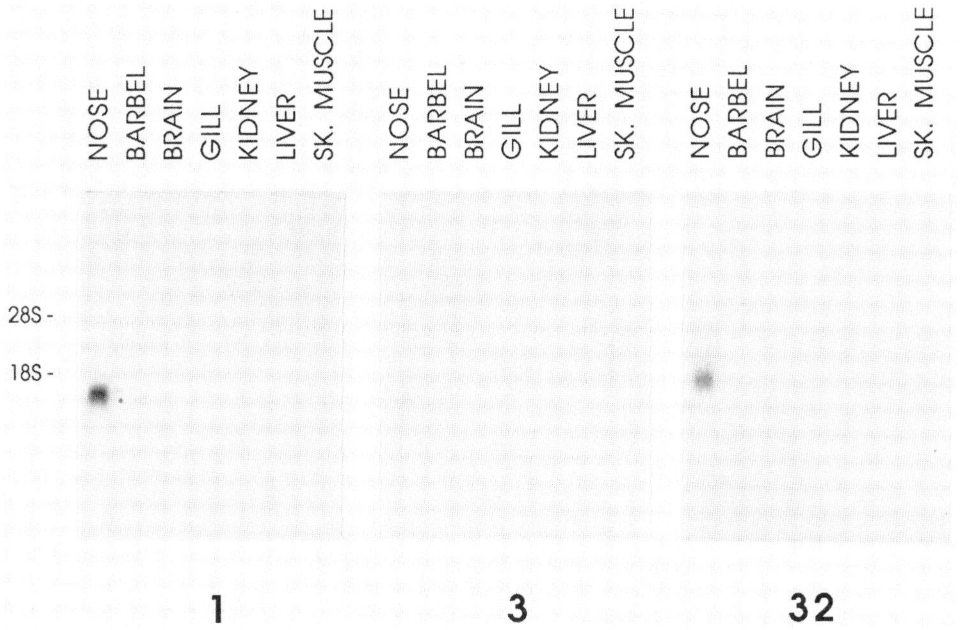

Figure 4. Northern blot analysis of catfish olfactory receptor expression. Poly(A)$^+$ RNA (2 μg) from catfish olfactory epithelium (nose), taste epithelium (barbel), brain, gill, kidney, liver, and skeletal muscle was electrophoresed on formaldehyde-agarose gels, blotted onto nitrocellulose, and hybridized with ^{32}P-labeled probes for receptors 1, 3, and 32. The positions of 28S and 18S ribosomal subunit migration are shown.

We therefore performed Northern blot analysis with a mixed probe consisting of the seven divergent cDNAs used above and detected 2 diffuse bands of about 5 kb and 2 kb in length in olfactory epithelium RNA (Buck and Axel 1991). No hybridization is observed to RNA from the brain or retina or other, non-neural tissues, including lung, liver, spleen, and kidney.

We have also used Northern blot hybridization, along with RNA in situ hybridization, to examine the patterns of expression of the family of catfish receptor cDNAs (J. Ngai et al., in prep.). In initial experiments, representative cDNAs from three of the six subfamilies were used as probe in Northern blots from a variety of fish tissues (Fig. 4). Each of the cDNA probes hybridizes to an RNA from olfactory epithelium with approximate sizes ranging from 1.5 kb to 20 kb. No hybridization was detected with RNA from barbel (taste epithelium), brain, gill, kidney, liver, or skeletal muscle.

Pattern of Receptor Expression in the Olfactory Epithelium

The identification of a large family of olfactory-specific receptors raises the question as to whether individual neurons express different receptors. If individual neurons express different complements of receptors, then the brain can identify distinct odors by determining which cells have been activated. One mechanism by which the brain can identify specific neurons would involve the spatial segregation of the sensory neurons or their projections.

What are the features of the vertebrate olfactory apparatus that might form the anatomical basis for a spatial map of olfactory information? Odorous stimuli are received from the environment by olfactory sensory neurons in the olfactory epithelium, which lines a portion of the nasal cavity. Each olfactory neuron projects a single dendritic process to the apical surface of mucosa and elaborates a number of sensory cilia. Each neuron extends a single unbranched axon; as the collection of axons emerge from the olfactory mucosal layer, they fasciculate to form the olfactory nerve. The axons of olfactory neurons form synapses on secondary neurons in discrete structures known as glomeruli within the olfactory bulb. Approximately 10^6 olfactory neurons project to $\sim 10^3$ glomeruli, but each olfactory neuron axon forms synapses within only a single glomerulus (Mori 1987).

The anatomy of the olfactory system therefore affords the opportunity for spatial segregation of afferent sensory input at all levels from the peripheral epithelium to the olfactory cortex (for review, see Kauer et al. 1991; Shepherd 1991). Spatial segregation could be apparent in the sensory epithelium, such that sensory neurons that express a given receptor may be localized in defined positions within the olfactory epithelium. This topographic localization would also be maintained in the projections of the sensory axon, such that neurons expressing the same receptor may project to the same glomeruli within the olfactory bulb. In a second model, sensory neurons expressing a given receptor may be randomly distributed rather than spatially segregated within the sensory epithelium. All neurons responsive to a given odorant would therefore project to discrete glomeruli, such that a topographic map is retained within the bulb.

We have used in situ hybridization to examine the expression and distribution of receptor mRNAs in the catfish olfactory epithelium. The catfish olfactory epithelium is organized in a simple, regularly repeating structure (see Caprio and Raderman-Little 1978), and this anatomical simplicity allows a straightforward spatial analysis of odorant receptor expression. The catfish olfactory epithelium is organized into a pair of bilateral structures known as olfactory rosettes. Each rosette consists of two parallel stacks of lamellae or leaflets that arise from a medial raphe. Each leaflet consists of two layers of epithelium sandwiching a central sheet of connective tissue. The olfactory sensory epithelium resides medially in each face of the lamella and is surrounded by nonsensory epithelium. Horizontal sections of an olfactory rosette show the pigment-containing central raphe with the numerous lamellae cut in cross section (Fig. 5A). Three major cell types comprise the pseudostratified olfactory epithelium: the regenerative, basal cells that lie closest to the basal lamina; the

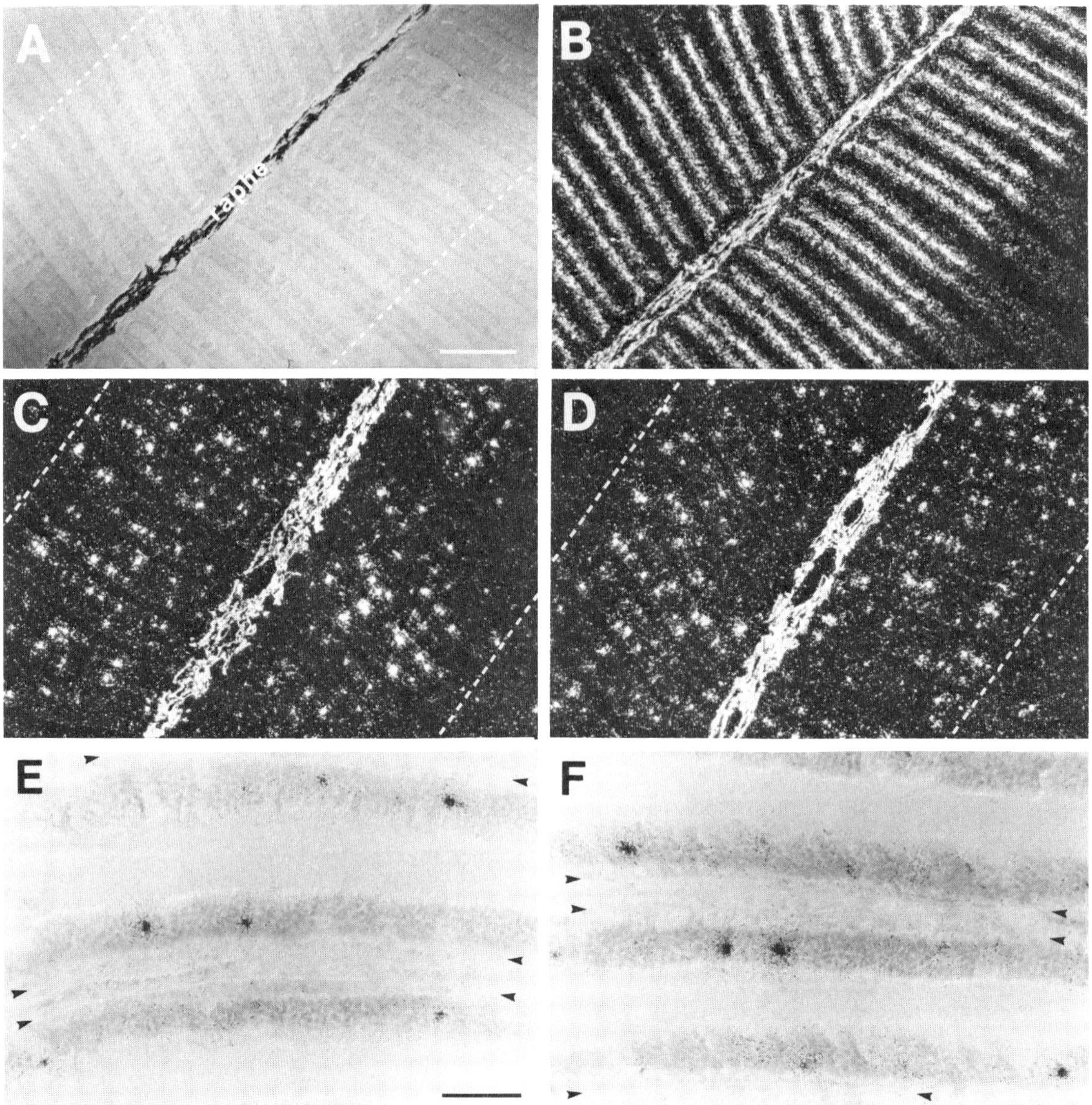

Figure 5. Localization of olfactory receptor and cyclic nucleotide-gated channel RNAs in the catfish olfactory epithelium by in situ hybridization. Catfish olfactory rosettes were sectioned and hybridized to ^{35}S-labeled RNA probes for the olfactory cyclic nucleotide-gated ion channel and olfactory receptors 1 and 32, as described previously (Goulding et al. 1992; J. Ngai et al., in prep.). (*A*) Low-power differential interference contrast (D.I.C.) micrograph of a horizonal section hybridized to an antisense probe for the cyclic nucleotide-gated ion channel. Multiple lamellae, each consisting of two epithelial layers, can be seen emanating from the medial pigmented raphe. Dashed lines show the lateral borders of the sensory neuroepithelium. (*B*) Dark-field micrograph corresponding to the area shown in *A*. Note the uniform labeling of the sensory neuroepithelium. (*C*, *D*) Dark-field micrographs of sections hybridized to antisense probes specific for receptor 1 (*C*) or receptor 32 (*D*). The punctate distribution of positive cells is found throughout the sensory neuroepithelium (delineated by dashed lines). (*E*, *F*) High-power, D.I.C. micrographs of sections probed with receptor 1 (*E*), or receptor 32 (*F*) antisense RNAs. Note localization over individual cells in the olfactory neuron cell body layer. Arrowheads indicate positions of basal laminae. Bar in *A*, 200 μm for *A*–*D*; bar in *E*, 50 μm for *E* and *F*.

sustentacular cells whose cell bodies reside most apically; and the olfactory neurons, whose cell bodies occupy an intermediate position in the epithelium.

In situ hybridizations were carried out by annealing individual ^{35}S-labeled antisense receptor probes with multiple serial sections of the entire neuroepithelium from single olfactory rosettes (J. Ngai et al., in prep.). The results with the three probes are qualitatively indistinguishable (two examples are shown in Fig. 5); in each case, we observe a striking punctate distribution of receptor RNA hybridization throughout the sensory neuroepithelium. Positive cells are found in all sections containing neuroepithelium; specific hybridization is never observed in the surrounding nonsensory epithelium (Fig. 5C,D). No specific signal is observed with the control sense probe. This dispersed, punctate pattern of hybridization with receptor probes is in sharp contrast to the pattern observed with a probe for the olfactory cyclic nucleotide-gated ion channel (Fig. 5B)(Goulding et al. 1992). RNA encoding this channel shows generalized expression in olfactory sensory cells throughout the epithelium.

Examination of the pattern of expression of receptor RNA at higher magnification reveals clusters of grains localized over individual olfactory neurons (Fig. 5E,F). Hybridization is only rarely observed in the basal cell layer. Each individual probe detects only 1–2% of the olfactory neurons. Our data indicate that specific receptors are localized to a small subset of olfactory neurons, suggesting that a single cell expresses only one or a few distinct odorant receptors. Each cell may therefore possess a unique identity as defined by the receptor it expresses.

Are olfactory neurons that express a given receptor spatially localized? Examination of multiple serial sections through an entire rosette indicates that the expression of a given odorant receptor RNA does not appear to segregate among a subset of lamellae. Moreover, no spatial segregation is apparent within the neuroepithelium of an individual leaflet. These data suggest that individual odorant receptors show no apparent spatial localization within the olfactory rosette. However, it is possible that more subtle nonrandom patterns may have escaped our detection. We therefore employed computer-assisted image processing techniques to reconstruct a single lamella from in situ hybridizations of serially sectioned rosettes. Analysis of the three-dimensional reconstruction of individual lamellae following in situ hybridization with probes specific for three subfamilies (receptors 1, 32, and 202) failed to demonstrate any discernible pattern of receptor expression. The results of our in situ hybridizations indicate that neurons expressing specific receptors appear to be randomly distributed within the olfactory epithelium.

DISCUSSION

How Large Is the Gene Family?

The vertebrate olfactory system can recognize and discriminate a large number of different odorant molecules. What properties of the odorant receptor permit the recognition of a vast array of ligands, and how does the brain distinguish which receptors have been activated? One approach to the understanding of olfactory discrimination involves the characterization of olfactory receptors and the neural pathways they activate. We have identified a large family of genes that are likely to encode odorant receptors in the rat. This multigene family encodes seven transmembrane receptors and therefore belongs to the superfamily of G-protein-coupled receptors. The size of the gene family in mammals (perhaps as large as 1000 genes) (Buck and Axel 1991; Levy et al. 1991; Parmentier et al. 1992) suggests that olfactory perception employs a large number of receptors, each capable of interacting with one or a small number of odorous ligands. Since the number of odorants detected by fish is thought to be far smaller than that detected by mammals, we anticipated that the repertoire of odorant receptors would be far smaller as well. Our studies are consistent with this model and suggest that the catfish genome may encode as few as 50 receptors (J. Ngai et al., in prep.).

The characterization of a large multigene family encoding putative odorant receptors indicates that the olfactory system utilizes a far greater number of receptors than the visual system. In color vision, for example, the discrimination of several hundred hues is accomplished by only three different photoreceptors (Rushton 1955,1965; Wald et al. 1955; Nathans et al. 1986). The photoreceptors each have different, but overlapping, absorption spectra that cover the entire spectrum of visible wavelengths. Discrimination of color results from comparative processing of the information from these three classes of photoreceptors in the brain. Whereas three photoreceptors can absorb light across the entire visible spectrum, our data suggest that a small number of odorant receptors cannot recognize the full spectrum of distinct molecular structures perceived by the mammalian olfactory system. Rather, olfactory perception probably employs an extremely large number of receptors each capable of recognizing a small number of odorous ligands.

Diversity of Olfactory Receptors and the Specificity of Odor Recognition

The olfactory proteins we have identified are members of the superfamily of seven-transmembrane-domain receptors. The predicted structural features of these receptors may render this family particularly well suited for the detection of a diverse array of distinct odorants. Experiments with other G-protein-coupled members of this class of receptors suggest that ligand binds to its receptor within the plane of the membrane such that the ligand contacts many, if not all, of the transmembrane helices (Kobilka et al. 1988; Strader et al. 1989). The family of olfactory proteins can be divided into several different subfamilies that exhibit significant sequence divergence within the transmembrane domains. This diversity may reflect the fact that the members of the family of odorant receptors must associate with odorants of widely different molecular structures. This pattern of diversity is also apparent upon analysis of the sequence differences among the members of a given subfamily. We have sequenced five members of the receptor 32 subfamily and observed that four of the five members share more than 95% amino acid homology. More than 90% of the amino acid differences between these genes cluster within transmembrane domains III and IV (J. Ngai et al., in prep.).

These observations suggest a model in which each of the individual subfamilies encodes receptors that bind distinct structural classes of odorant. Within a given subfamily, however, the sequence differences are far less extensive and are largely restricted to a small number of residues within putative ligand-binding domains. Thus, the members of a subfamily may recognize more subtle variations among odor molecules of a given structural class. It should be noted that such a model, unlike previous stereochemical models, does not necessarily predict that molecules with similar structures will have similar odors. The activation of distinct receptors with similar structures could therefore elicit different odor qualities.

Spatial Segregation and the Processing of Olfactory Information

Discrimination among the vast array of odors requires that the brain discern which of the numerous receptors have been activated. If individual olfactory sensory neurons express only one or a few receptor genes, then the problem of distinguishing which receptor has been activated can be reduced to a problem of distinguishing which neurons have been activated. Our data indicate that specific receptors are localized to a small subset of olfactory neurons, suggesting that a single cell expresses only one or a few distinct odorant receptors (J. Ngai et al., in prep.). Each cell may therefore possess a specific identity as defined by the receptors it expresses. The implication that individual neurons may express only a single receptor suggests that a topographic map defining the position of the specific neurons in the olfactory epithelium or their projections into the olfactory bulb may be employed to encode the quality of an olfactory stimulus.

We consider two models that invoke the spatial segregation of either olfactory neurons or their projections to identify those neurons activated by a given odorant (for reviews, see Kauer et al. 1991; Shepherd 1991). In one model, neurons expressing a given receptor are spatially localized within the epithelium and project to one or a few spatially segregated glomeruli within the olfactory bulb. In a second model, neurons expressing a given receptor may exhibit no spatial order and may be randomly distributed throughout the epithelium whereas their axons project to one or a few glomeruli that occupy spatially defined loci within the bulb. In either instance, exposure to a given odorant would result in the stimulation of a spatially restricted set of glomeruli such that individual odorants would be associated with specific topographic patterns of activity within the olfactory bulb. Distinguishing among these models will require an examination of the pattern of receptor expression in the neurons of the olfactory epithelium and the pattern of projections these neurons exhibit in the olfactory bulb.

Topography and Olfactory Coding in Fish

Several independent studies have demonstrated topographically defined patterns of activity in response to specific odorants in the fish olfactory bulb (Satou et al. 1979; Doving et al. 1980; Stacey and Kyle 1982; Demski and Dulka 1984). Electrophysiological recordings from olfactory bulb neurons in fish have demonstrated that individual glomeruli are differentially responsive to different odors (Macleod 1976). Is this topographic pattern in the olfactory bulb reflective of the spatial segregation of functional classes of neurons in the epithelium? Functional studies have failed to demonstrate any clear-cut segregation of odorant-induced activity in the fish olfactory epithelium (Thommesen 1983). Moreover, axon tracing experiments in fish show that individual glomeruli receive inputs from olfactory neurons dispersed throughout the olfactory epithelium (Riddle and Oakley 1991). Taken together, these data suggest that the fish olfactory bulb exhibits spatially organized units of activity, but this spatial order is not apparent within the sensory epithelium.

We have used probes representative of three distinct receptor subfamilies encompassing about 20 different genes and find no evidence for the restricted spatial localization of neurons expressing specific odorant receptors within the olfactory epithelium, with no segregation along anterior-posterior, dorsal-ventral, or medial-lateral axes. Furthermore, three-dimensional reconstruction techniques fail to demonstrate more subtle patterns of receptor expression. Neurons expressing specific receptors therefore appear to be randomly distributed within the olfactory epithelium (J. Ngai et al., in prep.).

Specific odors are therefore unlikely to elicit defined spatial patterns of activity within the olfactory epithelium. Rather, dispersed olfactory neurons with common receptors are likely to converge on common glomeruli in the olfactory bulb. Experiments combining the use of odorant-specific probes with axon tracing techniques should now permit a more direct test of this model.

Topography and Olfactory Coding in Higher Vertebrates

Mammals have evolved more complex strategies for encoding olfactory information than have fish: They are thought to recognize a far greater repertoire of odorants, and odorants elicit a far greater diversity of responses. This complexity may be reflected in differences in the organization of the olfactory systems between fish and higher vertebrates. For example, electrophysiological recordings in rats have shown that neurons in different regions of the olfactory epithelium exhibit different levels of responsiveness to specific odorants (Kauer and Moulton 1974; Thommesen and Doving 1977; Mackay-Sim et al. 1982). In contrast to fish, these observations suggest that neurons expressing specific odorant receptors are spatially segregated in the olfactory epithelium. In both fish and mammals, however, specific odorants appear to elicit spatially defined patterns of glomerular activity in the olfactory bulb (Adrian 1950; Pinching and Doving 1974; Stewart et al. 1979; Doving et al. 1980; Jourdan et al. 1980; Teicher et al. 1980; Jourdan 1982; Lancet et al. 1982; Royet et al. 1987; Mori et al. 1992).

The presence of odorant-specific spatial patterns in both the olfactory epithelium and olfactory bulb of higher vertebrates suggests the ordered projection of axons onto the bulb. Axon tracing in mammals demonstrates that defined subregions of the olfactory epithelium project to restricted areas in the olfactory bulb (Le Gros Clark 1951; Land 1973; Land and Shepherd 1974; Greer et al. 1981; Astic and Saucier 1986, 1988; Fujita et al. 1985; Mori et al. 1985; Saucier and Astic 1985). The projection from the olfactory epithelium to the olfactory bulb, however, lacks the point-to-point precision found in other sensory systems. Rather, specific glomeruli receive afferent inputs from neurons dispersed in fairly broad areas of the epithelium surface (Jastreboff et al. 1984; Astic et al. 1987), and neurons in the regions of the epithelium project to different glomeruli in the olfactory bulb. One particularly clear example of the segregation of functional classes of neurons and their projections is illustrated by the vomeronasal system. The vomeronasal organ is a distinct anatomical structure consisting of olfactory neurons found in the rostroventral part of the nasal cavity, quite distant from the main olfactory epithelium. The vomeronasal organ receives volatile chemical cues that govern sexual and nonsexual behaviors (Halpern 1987). This organ projects its primary sensory axons to the accessory olfactory bulb via the vomeronasal nerve. The accessory bulb in turn projects to more primitive regions of the cortex than the main olfactory bulb. Thus, the entire vomeronasal system is functionally distinct and anatomically segregated from the main olfactory system.

The data from higher vertebrates suggest that spatial segregation of functionally discrete classes of neurons is already apparent at the level of the olfactory epithelium and is maintained in the projections to the bulb. The availability of molecular probes for mammalian receptors should now permit a direct analysis of the patterns of expression of odorant receptors in the mammalian olfactory system. This may provide insight into the strategies employed in the more complex olfactory systems of mammals.

ACKNOWLEDGMENTS

This research was supported by the Howard Hughes Medical Institute (A.C., L.B., R.A., and J.N.), and by a grant from Columbia University for the M.D.-Ph.D. Program (Department of Biochemistry and Molecular Biophysics) (M.M.D.). J.N. was a Hughes Fellow of the Life Sciences Research Foundation for a portion of this work.

REFERENCES

Adrian, A.D. 1950. Sensory discrimination with some recent evidence from the olfactory organ. *Br. Med. Bull.* **6:** 330.

Astic, L. and D. Saucier. 1986. Anatomical mapping of the projection to the olfactory bulb in the rat. *Brain Res. Bull.* **16:** 445.

———. 1988. Topographical projection of the septal organ to the main olfactory bulb in rats: Ontogenetic study. *Brain Res.* **470:** 297.

Astic, L., D. Saucier, and A. Holley. 1987. Topographical relationships between olfactory receptor cells and glomerular foci in the rat olfactory bulb. *Brain Res.* **424:** 144.

Boekhoff, I., E. Tareilus, J. Strotmann, and H. Breer. 1990. Rapid activation of alternative second messenger pathways in olfactory cilia from rats by different odorants. *EMBO J.* **9:** 2453.

Bower, J.M. 1991. Piriform cortex and olfactory object recognition. In *Olfaction. A model system for computational neuroscience* (ed. J.L. Davis and H. Eichenbaum), p. 265. MIT Press, Cambridge, Massachusetts.

Breer, H., I. Boekhoff, and E. Tareilus. 1990. Rapid kinetics of second messenger formation in olfactory transduction. *Nature* **345:** 65.

Buck, L. and R. Axel. 1991. A novel multigene family may encode odorant receptors: A molecular basis for odor recognition. *Cell* **65:** 175.

Caprio, J. and R. Raderman-Little. 1978. Scanning electron microscopy of the channel catfish olfactory lamellae. *Tissue Cell.* **10:** 1.

Demski, L. S. and J.G. Dulka. 1984. Functional-anatomical studies on sperm release evoked by electrical stimulation of the olfactory tract in goldfish. *Brain Res.* **291:** 241.

Dhallan, R. S., K. Yau, K. Schrader, and R.R. Reed. 1990. Primary structure and functional expression of a cyclic nucleotide-activated channel from olfactory neurons. *Nature* **347:** 184.

Doving, K. B., R. Selset, and G. Thommesen. 1980. Olfactory sensitivity to bile-acids in salmonid fishes. *Acta. Physiol. Scand.* **108:** 123.

Freeman, W.L. 1991. Nonlinear dynamics in olfactory information processing. In *Olfaction. A model system for computational neuroscience* (ed. J.L. Davis and H. Eichenbaum), p. 225. MIT Press, Cambridge, Massachusetts.

Fujita, S.C., K. Mori, K. Imamura, and K. Obata. 1985. Subclasses of olfactory receptor cells and their segregated central projections demonstrated by a monoclonal antibody. *Brain Res.* **326:** 192.

Goulding, E.H., J. Ngai, R.H. Kramer, S. Colicos, R. Axel, S.A. Siegelbaum, and A. Chess. 1992. Molecular cloning and single-channel properties of the cyclic nucleotide-gated channel from catfish olfactory neurons. *Neuron* **8:** 45.

Greer, C.A., W.B. Stewart, J.S. Kauer, and G.M. Shepherd. 1981. Topographical and laminar localization of 2-deoxyglucose uptake in rat olfactory bulb induced by electrical stimulation of olfactory nerves. *Brain Res.* **217:** 279.

Halpern, M. 1987. The organization and function of the vomeronasal system. *Annu. Rev. Neurosci.* **10:** 325.

Jastreboff, P.J., P.E. Pedersen, C.A. Greer, W.B. Stewart, J.S. Kauer, T.E. Benson, and G.M. Shepherd. 1984. Specific olfactory receptor populations projecting to identified glomeruli in the rat olfactory bulb. *Proc. Natl. Acad. Sci.* **81:** 5250.

Jourdan, F. 1982. Spatial dimension in olfactory coding: A representation of the 2-deoxyglucose patterns of glomerular labeling in the olfactory bulb. *Brain Res.* **240:** 341.

Jourdan, F., A. Duveau, L. Astic, and A. Holley. 1980. Spatial distribution of [14C]2-deoxyglucose uptake in the olfactory bulbs of rats stimulated with two different odours. *Brain Res.*. **188:** 139.

Kandel, E.R., J.H. Schwartz, and T.M. Jessell. 1991. *Principles of neural science*, 3rd Edition. Elsevier, New York.

Kauer, J.S. and D.G. Moulton. 1974. Responses of olfactory bulb neurones to odour stimulation of small nasal areas in the salamander. *J. Physiol.* **243:** 717.

Kauer, J.S., S.R.. Neff, K.A. Hamilton, and A.R. Cinelli. 1991. The salamander olfactory pathway: Visualizing and modeling circuit activity. In *Olfaction. A model system for computational neuroscience* (ed. J.L. Davis and H. Eichenbaum), p. 43. MIT Press, Cambridge, Massachusetts.

Kobilka, B.K., T.S. Kobilka, K. Daniel, J.W. Regan, M.G. Caron, and R.J. Lefkowitz. 1988. Chimeric alpha-2-beta-2 adrenergic receptors: Delineation of domains involved in effector coupling and ligand binding specificity. *Science* **240:** 1310.

Lancet, D. 1986. Vertebrate olfactory reception. *Annu. Rev. Neurosci.* **9:** 329.

Lancet, D., C.A. Greer, J.S. Kauer, and G.M. Shepherd. 1982. Mapping of odor-related neuronal activity in the olfactory bulb by high-resolution 2-deoxyglucose autoradiography. *Proc. Natl. Acad. Sci.* **79:** 670.

Land, L.J. 1973. Localized projection of olfactory nerves to rabbit olfactory bulb. *Brain Res.* **63:** 153.

Land, L.J. and G.M. Shepherd. 1974. Autoradiographic analysis of olfactory receptor projections in the rabbit. *Brain Res.* **70:** 506.

Le Gros Clark, W.E. 1951. The projection of the olfactory epithelium on the olfactory bulb in the rabbit. *J. Neurol. Neurosurg. Psychiat.* **14:** 1.

Levy, N.S., H.A. Bakalyar, and R.R. Reed. 1991. Signal transduction in olfactory neurons. *J. Steroid Biochem. Mol. Biol.*. **39:** 633.

Mackay-Sim, A., P. Shaman, and D.G. Moulton. 1982. Topographic coding of olfactory quality: Odorant-specific patterns of epithelial responsivity in the salamander. *J. Neurophysiol.* **48:** 584.

Macleod, N. K. 1976. Spontaneous activity of single neurons in the olfactory bulb of the rainbow trout (*Salmo gairdneri*) and its modulation by olfactory stimulation with amino acids. *Exp. Brain Res.* **25:** 267.

Mori, K. 1987. Membrane and synaptic properties of identified neurons in the olfactory bulb. *Prog. Neurobiol.* **29:** 275.

Mori, K., N. Mataga, and K. Imamura. 1992. Differential specificities of single mitral cells in rabbit olfactory bulb for a homologous series of fatty acid odor molecules. *J. Neurophysiol.* **67:** 786.

Mori, K., S.C. Fujita, K. Imamura, and K. Obata. 1985. Immunohistochemical study of subclasses of olfactory nerve fibers and their projections to the olfactory bulb in the rabbit. *J. Comp. Neurol.* **242:** 214.

Nakamura, T. and G.H. Gold. 1987. A cyclic nucleotide-gated conductance in olfactory receptor cilia. *Nature* **325:** 342.

Nathans, J., D. Thomas, and D.S. Hogness. 1986. Molecular genetics of human color vision: The genes encoding blue, green and red pigments. *Science* **232:** 193.

O'Dowd, B.F., R.J. Lefkowitz, and M.G. Caron. 1989a. Structure of the adrenergic and related receptors. *Annu. Rev. Neurosci.* **12:** 67.

O'Dowd, B.F., M. Hnatowich, M.G. Caron, and R.J. Lefkowitz. 1989b. Palmitoylation of the human *B*-adrenergic receptor. Mutation of CYS^{341} in the carboxyl tail leads to an uncoupled non-palmitoylated form of the receptor. *J. Biol. Chem.* **264:** 7564.

Pace, U., E. Hanski, Y. Salomon, and D. Lancet. 1985. Odorant-sensitive adenylate cyclase may mediate olfactory reception. *Nature* **316:** 255.

Parmentier, M., F. Libert, S. Schurmans, S. Schiffmann, A. Lefort, D. Eggerick, C. Ledent, J. Perret, A. Grootegoed, and G. Vassart. 1992. Expression of members of the putative olfactory receptor germ cells. *Nature* **355:** 453.

Pinching, A.J. and K.B. Doving. 1974. Selective degeneration in the rat olfactory bulb following exposure to different odours. *Brain Res.* **82:** 195.

Restrepo, D., T. Miyamoto, B.P. Bryant, and J.H. Teeter. 1990. Odor stimuli trigger influx of calcium into olfactory neurons of the channel catfish. *Science* **249:** 1166.

Riddle, D.R. and B. Oakley. 1991. Evaluation of projection patterns in the primary olfactory system of rainbow trout. *J. Neurosci.* **11:** 3752.

Royet, J.P., G. Sicard, C. Souchier, and F. Jourdan. 1987. Specificity of spatial patterns of glomerular activation in the mouse olfactory bulb: Computer-assisted image analysis of 2-deoxyglucose autoradiograms. *Brain Res.* **417:** 1.

Rushton, W.A.H. 1955. Foveal photopigments in normal and colour-blind. *J. Physiol.* **129:** 41.

———.1965. A foveal pigment in the deuteranope. *J. Physiol.* **176:** 24.

Satou, M., M. Ichikawa, K. Ueda, and S.F. Takagi. 1979. Topographical relation between olfactory bulb and olfactory tracts in the carp. *Brain Res.* **173:** 142.

Saucier, D. and L. Astic. 1985. Analysis of the topographical organization of olfactory epithelium projections in the rat. *Brain Res. Bull.* **16:** 455.

Shepherd, G.M. 1991. Computation structure of the olfactory system. In *Olfaction. A model system for computational neuroscience* (ed. J.L. Davis and H. Eichenbaum), p. 3. MIT Press, Cambridge, Massachusetts.

Sklar, P.B., R.R.H. Anholdt, and S.H. Snyder. 1986. The odorant-sensitive adenylate cyclase of olfactory receptor cells: Differential stimulation by distinct classes of odorants. *J. Biol. Chem.* **261:** 15538.

Stacey, N.E. and A.L. Kyle. 1982. Effects of olfactory tract lesions on sexual and feeding behavior in the goldfish. *Physiol. Behav.* **30:** 621.

Stewart, W.B., J.S. Kauer, and G.M. Shepherd. 1979. Functional organization of the rat olfactory bulb analysed by the 2-deoxyglucose method. *J. Comp. Neurol.* **185:** 715.

Strader, C.D., I.S. Sigal, and R.A.F. Dixon. 1989. Structural basis of *B*-adrenergic receptor function. *FASEB J.* **3:** 1825.

Teicher, M.H., W.B. Stewart, J.S. Kauer, and G.M. Shepherd 1980. Sucking pheromone stimulation of a modified glomerular region in the developing rat olfactory bulb revealed by the 2-deoxyglucose method. *Brain Res.* **194:** 530.

Thommesen, G. 1983. Morphology, distribution, and specficity of olfactory receptor cells in salmonid fishes. *Acta Physiol. Scand.* **117:** 241.

Thommesen, G. and K.B. Doving. 1977. Spatial distribution of the EOG in the rat. A variation with odour stimulation. *Acta Physiol. Scand.* **102:** 205.

Wald, G., P.K. Brown, and P.H. Smith. 1955. Iodopsin. *J. Gen. Physiol.* **38:** 623.

Pharmacological and Genetic Approaches to the Analysis of Tyrosine Kinase Function in Long-term Potentiation

T.J. O'Dell,* S.G.N. Grant,* K. Karl,* P.M. Soriano,† and E.R. Kandel*

*Center for Neurobiology and Behavior, College of Physicians & Surgeons of Columbia University, Howard Hughes Medical Institute, New York, New York 10032; †Institute for Molecular Genetics, Howard Hughes Medical Institute, Baylor College of Medicine, Houston, Texas 77030

Long-term potentiation (LTP) in the CA1 region of the hippocampus is a stable enhancement of synaptic transmission that occurs following brief periods of high-frequency synaptic activity. The biophysical and biochemical events responsible for LTP have been studied extensively as a possible cellular mechanism underlying changes in synaptic efficacy involved in certain types of learning and memory in the vertebrate brain. Whereas it is known that coincident patterns of synaptic activity can induce LTP by virtue of their ability to elicit activation of *N*-methyl-D-aspartate (NMDA)-type glutamate receptors, which in turn leads to increases in levels of intracellular calcium, less is known about the subsequent biochemical steps responsible for the induction, expression, and maintenance of LTP (Nicoll et al. 1988; Bashir and Collingridge 1992). Several studies using protein kinase inhibitors have implicated activation of two calcium-sensitive serine-threonine kinases, protein kinase C (PKC), and calcium/calmodulin-sensitive kinase II (CaMKII), as necessary for the induction of LTP (Reymann et al. 1988; Malenka et al. 1989; Malinow et al. 1989). Hippocampal CA1 pyramidal cells contain high levels of PKC and CamKII, consistent with their involvement in LTP.

In addition to these serine-threonine kinases, hippocampal neurons express high levels of protein tyrosine kinase activity (Hirano et al. 1988). The proteins encoded by tyrosine kinase genes can be broadly defined into two structurally distinct categories; namely, receptor tyrosine kinases, which are membrane-spanning proteins linking an extracellular ligand-binding domain with the intracellular kinase domain, and nonreceptor tyrosine kinases, which can associate with the cytoplasmic side of the plasma membrane (for review, see Cooper 1989; Cantley et al. 1991). The receptor tyrosine kinases are signal transduction molecules for growth factors (e.g., epidermal growth factor [EGF], fibroblast growth factor [FGF], insulin) and neurotrophic factors (e.g., nerve growth factor [NGF], brain-derived neurotrophic factor [BDNF], neurotrophin-3 [NT-3]) and are expressed at varying levels throughout the nervous system and peripheral tissues. The nonreceptor tyrosine kinases may also play a role in transmembrane signal transduction pathways (Golden and Brugge 1989; Kypta et al. 1990; Klausner and Samelson 1991).

Whereas the functions of tyrosine kinases in the adult nervous system remain to be elucidated, the localization of tyrosine kinases to both the pre- and postsynaptic compartments suggests possible roles in both neurotransmitter release and receptor modulation. For several reasons, the nonreceptor tyrosine kinase *src* in particular has received much attention as a potentially important kinase in synaptic modulation. First, it is enriched in synaptic vesicle fractions (Pang et al. 1988) and can phosphorylate the synaptic vesicle protein synaptophysin in vitro (Yang et al. 1989; Barnekow et al. 1990). Second, *src* is localized to synaptic membranes and postsynaptic density (PSD), suggesting a role in modulation of neurotransmitter receptors or in transducing their signal at the postsynaptic level (Cudmore and Gurd 1991). Third, *src* has alternatively spliced forms (Brugge et al. 1987; Cartwright et al. 1987; Martinez et al. 1987; Pyper and Bolen 1990) that show higher kinase activity (Levy and Brugge 1989) and are exclusively expressed in brain (Cartwright et al. 1988; Ross et al. 1988; Sugrue et al. 1990).

These findings related to *src*, and the demonstrated interaction of tyrosine kinases with serine-threonine kinases in signal transduction pathways (Cantley et al. 1991), led us to investigate whether tyrosine kinases in general, and specifically, *src*, are involved in the flow of molecular information whereby activation of postsynaptic NMDA receptors leads to a long-lasting enhancement of synaptic transmission in the hippocampus.

As a first step in examining the role of tyrosine kinases in hippocampal physiology, we have investigated the effects of tyrosine kinase inhibitors on normal synaptic transmission, LTP, and neurotransmitter actions thought to involve second messengers. Next, we tested the hypothesis that *src* is involved in both short- and long-term (LTP) forms of synaptic plasticity, using mice that do not express a functional *src* protein. The *src* gene was disrupted using homologous recombination in embryonic stem cells to generate a null allele of the gene. The embryonic stem cell clones harboring the mutant *src* gene were used to derive chimeric mice that transmitted the mutation to their offspring, which were bred to produce mice homozygous for the *src* mutation (Soriano et al. 1991).

METHODS

Standard electrophysiological techniques were used to study 400-μm-thick transverse slices of guinea pig and mouse hippocampus placed in an interface-type recording chamber. The chamber was constantly perfused (1–3 ml/min) with an artificial cerebrospinal fluid (ACSF) containing (in mM) 124 NaCl, 4.4 KCl, 25 $NaHCO_3$, 1.0 $Na_2H_2PO_4$, 2.0 $CaCl_2$, 2.0 $MgCl_2$ (or $MgSO_4$), and 10.0 glucose. All experiments were done at 30°C. Synaptic potentials were evoked (0.02–0.04 Hz) using a fine wire, bipolar electrode placed in the stratum radiatum layer of the hippocampus to stimulate the presynaptic Schaffer-collateral/commissural fibers. The resulting excitatory postsynaptic potentials (EPSPs) were recorded either by using an extracellular microelectrode (5–10 Mohms, filled with 2.5 M NaCl) placed in the apical dendritic tree region of the CA1 pyramidal cells to record the field EPSPs, or by impaling individual CA1 pyramidal cell bodies with an intracellular microelectrode to record EPSPs intracellularly. The stimulation intensity was adjusted to give field EPSP amplitudes of approximately 1.0 mV and intracellularly recorded EPSPs of 4–7 mV. LTP was induced after recording control responses for 10–30 minutes by two, 1-second duration trains of 100 Hz stimulation separated by 20 seconds.

The in vitro kinase reactions were performed as follows: *src* was immunoprecipitated from hippocampus extracts using monoclonal antibody 327 (Oncogene Science) (Walaas et al. 1988) and used to phosphorylate poly (Glu, Tyr) 4:1 (Sigma) (Braun et al. 1984). Rat brain purified PKC (Lipidex) was used to phosphorylate [Ser]-PKC (19-31) peptide (Peninsula Laboratories) (Braun et al. 1984) in the presence of Ca^{++}, phosphatidyl serine, and TPA as described by Kikkawa et al. (1982). Heart protein kinase A (PKA) catalytic subunit (Sigma) phosphorylation of kemptide (Peninsula Laboratories) was carried out as described by Roskoski (1983). Rat brain Ca^{++}/calmodulin-dependent CaMKII phosphorylation of syntide-2 (gift of R.J. Colbran) was carried out as described by Hashimoto and Soderling (1987). All assays had a final concentration of 2% DMSO to dissolve inhibitors, and ATP was 10 μM; values are representative of two or more experiments.

RESULTS

Inhibition of *src* and the Specificity of Tyrosine Kinase Inhibitors

Genistein, an isoflavone, has been shown to inhibit tyrosine kinases at doses that have little effect on the serine-threonine kinases, cAMP-dependent PKA, phosphorylase kinase, and Ca^{++}/phospholipid-dependent PKC in vitro (Akiyama et al. 1987). In vivo studies show inhibition of tyrosine kinase activity but no inhibition of inositol phosphate production, Ca^{++} signaling, 1,2-diacylglycerol formation, phospholipase activity, or PKC activation and translocation (Dean et al. 1989; Mustelin et al. 1990). Recently, a more potent class of tyrosine kinase inhibitors, including lavendustin A and 2-hydroxy-5-(2,5-dihydroxybenzyl) aminobenzoic acid (compound 5), have been identified but less extensively characterized (Onoda et al. 1989). We have tested these inhibitors for their potency in inhibiting *src* tyrosine kinase activity isolated from the hippocampus and assessed their specificity by comparing their effect on CaMKII, PKC, and PKA. We have focused on both PKC and CaMKII, since their activities have been implicated as being necessary for the induction of LTP (Malenka et al. 1989; Malinow et al. 1989).

Because the neuronal form of the *src* protein tyrosine kinase $pp60^{c\text{-}src(+)}$ is highly expressed in hippocampal CA1 pyramidal cells (Ross et al. 1988; Sugrue et al. 1990) and has been suggested as having a role in synaptic plasticity, we immunoprecipitated $pp60^{c\text{-}src(+)}$ from hippocampi and assayed drug inhibition of tyrosine kinase activity (Table 1). Genistein was tenfold more potent in inhibiting *src* than PKC, CaMKII, or PKA, whereas an analog of genistein, genistin, did not significantly inhibit any of these kinases. Lavendustin A proved to be more potent on *src* than genistein and showed a wider range of specificity, with 200-fold less inhibition of PKC, CaMKII, and PKA. However, the derivative of lavendustin A, compound 5, was not as specific and showed potent inhibition of *src* and CaMKII, and less inhibition of PKC. To apply these in vitro specificity data to physiological experiments, we used genistein, genistin, and lavendustin A at concentrations that inhibit tyrosine kinase activity but not serine-threonine kinase activity and tested physiological events regulated by PKC or PKA in the presence of the

Table 1. Specificity of Tyrosine Kinase Inhibitors on Kinase Substrate Phosphorylation

	src	PKC	PKA	CamKII
Genistein	18	185	>200	>200[a]
Genistin	>200[a]	>200[a]	>200[a]	>200[a]
Lavendustin A	0.5	>100[a]	>100[a]	>100[a]
Compound B	0.5	70	>100[a]	0.2
Tryphostin RG50864	8	60	>100	65

Inhibitory concentration (IC_{50}) values (μM) determined by in vitro kinase assays.
[a]No inhibition of kinase activity was seen at this concentration.

tyrosine kinase inhibitors. In addition to the inhibitory activity on *src*, a nonreceptor tyrosine kinase, these inhibitors have also been reported to be similarly potent on other nonreceptor and receptor tyrosine kinases (Akiyama et al. 1987; Onoda et al. 1989).

Tyrosine Kinase Inhibitors Block LTP

To examine the possible role of protein tyrosine kinases in LTP, slices were bathed in 110 μM genistein, a concentration that substantially inhibits tyrosine kinase activity but has little effect on serine-threonine kinases (Table 1). This concentration of genistein had little effect on baseline synaptic transmission. Responses were 94.9 ± 13.1% (mean ±S.D., $n = 7$) of control after a 30-minute application. However, in the presence of genistein, high-frequency stimulation produced only a short-lasting potentiation (30–45 min), in contrast to the normal, long-lasting (> 1 hr) potentiation characteristic of LTP (Fig. 1A,B). To control for possible nonselective effects of genistein, we performed similar experiments with genistin, a structural analog of genistein that lacks antiprotein tyrosine kinase activity, and found that it did not block LTP (Fig. 1C). Although inhibition of protein tyrosine kinases prior to and during high-frequency stimulation blocked LTP, application of genistein 30 minutes after inducing LTP failed to inhibit established LTP (Fig. 1D). These find-

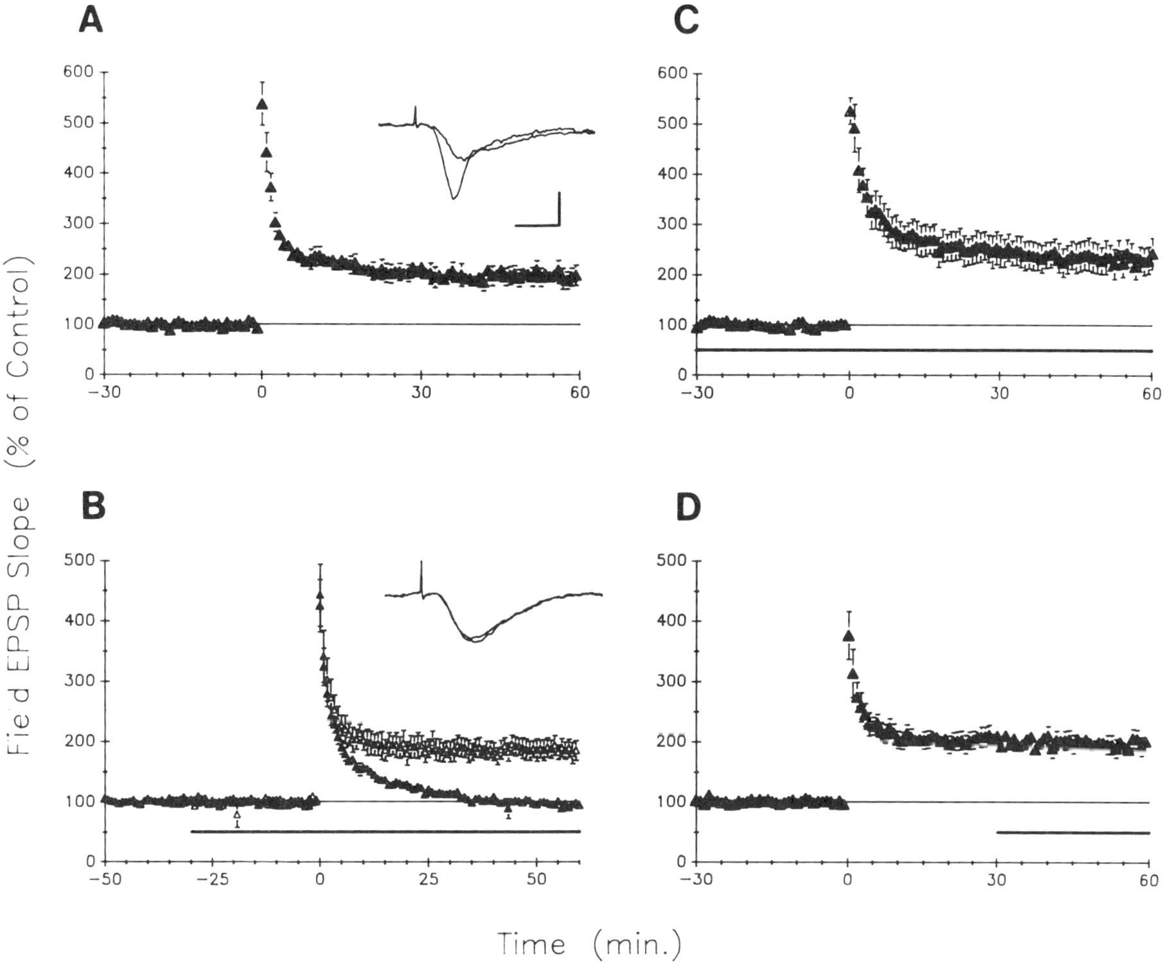

Figure 1. (*A*) Control LTP produced by high-frequency stimulation of the Schaffer collateral/commissural fibers in the stratum radiatum of the CA1 region of the hippocampus. Data (mean ± S.E.M.) from five individual slices are shown. The inset shows representative responses evoked 5 min before and 60 min after tetanic stimulation. Calibration bars are 1 mV, 6.5 msec. (*B*) Genistein blocks LTP. Bath application of 110 μM genistein for 30 min prior to tetanic stimulation results in only a decremental potentiation with responses returning to levels not significantly different from control levels ($p > 0.05$) by 30 min. Genistein was dissolved in DMSO and then diluted in ACSF to a final concentration of 110 μM. The concentration of DMSO required to solubilize this amount of genistein was 0.33%. The open triangles demonstrate that this concentration of DMSO does not block LTP ($n = 5$). (*C*) An analog of genistein that does not block tyrosine kinases does not block LTP. Slices ($n = 7$) were preincubated in 110 μM genistin (0.33% DMSO) for at least 30 min prior to eliciting LTP. (*D*) Application of genistein 30 min after tetanic stimulation does not block LTP. 110 μM genistein was added to the bath saline 30 min posttetanus and remained in the bath for the duration of the experiment. This concentration of genistein had no effect on the potentiated EPSPs in seven slices. No effect of genistein applied at 30 min posttetanus was observed in three slices where the amplitude of the potentiated EPSPs was monitored for at least 50 min in the presence of genistein (application times of 55 min in one slice and 70 min in two additional experiments).

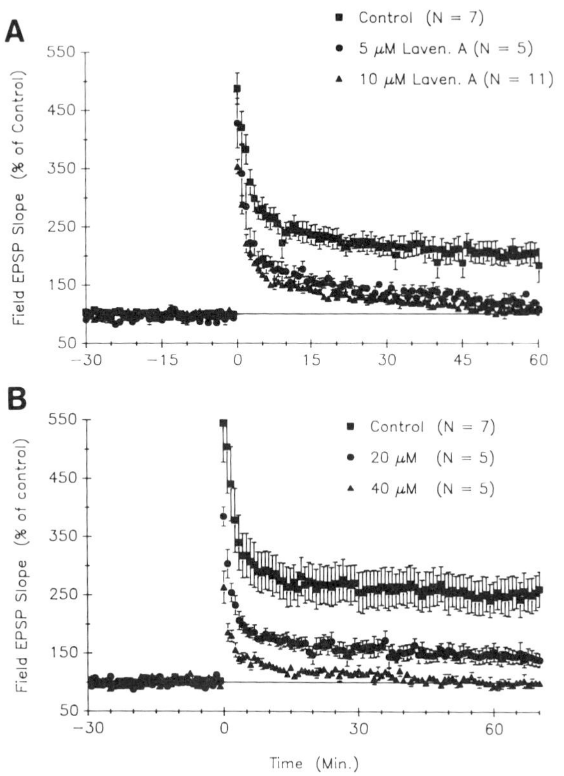

Figure 2. (*A*) Tetanic stimulation in the presence of lavendustin A produces only decremental potentiation. (*B*) Compound 5, a structural analog of lavendustin A, also inhibits LTP but is less potent. In both *A* and *B*, slices were bathed in the inhibitors for 0.5 to 4 hr prior to attempting to elicit LTP. The squares are control experiments in the absence of inhibitor.

ings suggest that whereas protein tyrosine kinases may be required for the induction of LTP, tyrosine kinase activity is not essential for the long-term maintenance and expression of LTP.

Other tyrosine kinase inhibitors had effects on LTP similar to that of genistein. Lavendustin A and its structural analog, compound 5, both blocked LTP when applied prior to tetanic stimulation and had little effect on established LTP (Fig. 2A,B). In in vitro kinase assays, lavendustin A is more potent than genistein in inhibiting c-*src* activity (Table 1) (O'Dell et al. 1991), a tyrosine kinase enriched in the CA1 region (Sugrue et al. 1990). Consistent with the block of LTP resulting from the ability of the inhibitors to block tyrosine kinases, lavendustin A was more potent than genistein in blocking LTP, with concentrations as low as 5 μM exerting pronounced effects.

Tyrosine Kinase Inhibitors Act Postsynaptically to Block LTP

Although the induction of LTP critically depends on biochemical processes in the postsynaptic cells, LTP is thought to depend, at least in part, on changes in transmitter release from the presynaptic terminals (Bliss et al. 1986; Bekkers and Stevens 1990; Malinow and Tsien 1990; Kullmann and Nicoll 1992). Thus, it is possible that the tyrosine kinase(s) involved in LTP could be located in either the pre- or postsynaptic sides of the synapse. Since bath application of kinase inhibitors would inhibit both pre- and postsynaptic kinases, the experiments described above do not indicate where tyrosine kinase activity is required for normal LTP. To determine whether the tyrosine kinases involved in the induction of LTP are located postsynaptically, we introduced lavendustin A into individual postsynaptic CA1 pyramidal cells via an intracellular microelectrode containing 1.0 mM lavendustin A. In these experiments, LTP was blocked in cells where tetanic stimulation was delivered 30 minutes after impaling a cell with a microelectrode containing lavendustin A, suggesting that postsynaptic tyrosine kinases are necessary for the induction of LTP (Fig. 3).

Tyrosine Kinase Inhibitors Do Not Appear to Affect PKA or PKC

The concentrations of genistein and lavendustin A that we find effective in blocking LTP are consistent with a selective action on tyrosine kinases as assessed using in vitro kinase assays to compare the ability of these compounds to inhibit the c-*src* nonreceptor tyrosine kinase and the serine-threonine kinases thought to be important in LTP, PKC, PKA, and CaMKII. As an additional test of specificity, we also examined the effects of the inhibitors on transmitter actions thought to be mediated by PKC and PKA.

Strong, depolarizing current pulses delivered via an intracellular recording electrode elicit a slow afterhyperpolarization (sAHP) in hippocampal neurons that is regulated by a number of neurotransmitters (Nicoll et al. 1990). Muscarinic receptor agonists that elicit breakdown of membrane phospholipids to form the second messengers diacylglycerol and IP_3 inhibit the sAHP (Nicoll et al. 1990), an effect that may be mediated by PKC activation, since it is mimicked by PKC activation with phorbol esters (Malenka et al. 1986b). β-Adrenergic receptor activation also inhibits the sAHP, although increases in intracellular cAMP levels and PKA activation are thought to mediate these effects (Madison and Nicoll 1982). We thus examined the effects of genistein on carbachol and norepinephrine (NE) inhibition of the sAHP to determine whether concentrations of genistein that block LTP interfered with the second-messenger systems that mediate the actions of these agonists. Genistein (110 μM) had no significant effect on the ability of carbachol and NE to inhibit the sAHP. Application of 10 μM carbachol and 10 μM NE-containing ACSF produced an average inhibition of the sAHP of 78.5 ± 7.2% (mean ±S.E.M., $n = 5$) and 73.9 ± 10.9 (mean ±S.E.M., $n = 5$), respectively. In the presence of genistein, the average inhibition of the sAHP by carbachol was 74.4 ± 7.6% (mean ±S.E.M., $n = 9$) and by norepinephrine was 78.9 ± 7.0 (mean ± S.E.M., $n = 5$). This suggests that the kinases thought to mediate the actions of these agonists (PKC

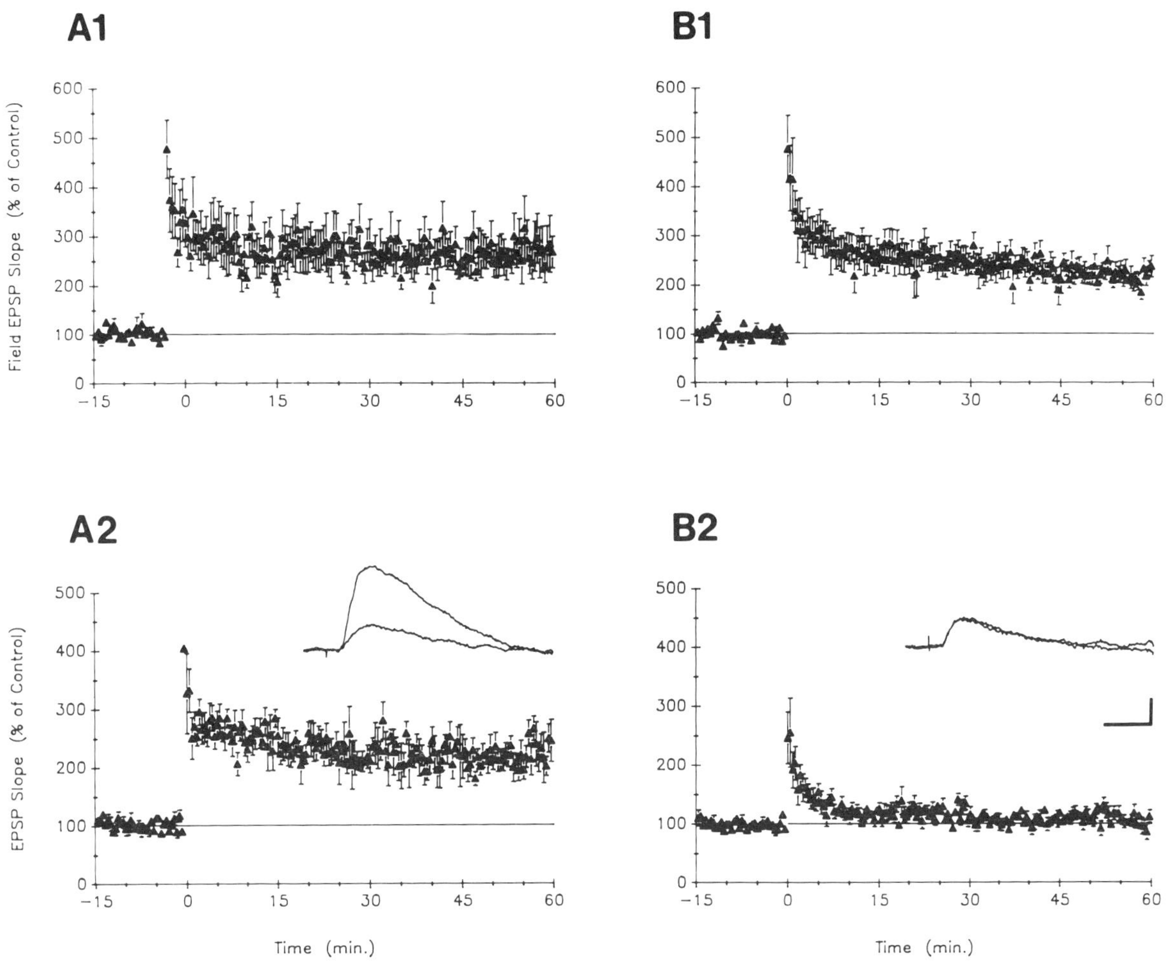

Figure 3. Postsynaptic injection of the tyrosine kinase inhibitor lavendustin A blocks LTP. EPSPs from a single postsynaptic cell were recorded with an intracellular electrode placed in the cell body of a CA1 pyramidal cell (*bottom* panels), while the responses of other, nearby cells were simultaneously monitored with an extracellular electrode placed in stratum radiatum (*top* panels). (*A1* and *A2*) Control experiments ($n = 8$) showing LTP of the field EPSPs (*A1*) and of the intracellular EPSPs (*A2*). In experiments with 1.0 mM lavendustin A in the extracellular electrode ($n = 8$), LTP of the intracellularly recorded EPSP was blocked (*B2*), while LTP was evident in surrounding cells in the same slices (*B1*). Traces are responses recorded 5 min pretetanus and 1 hr posttetanus in control experiments (*A2*) and in experiments with lavendustin A in the extracellular electrode (*B2*). Calibration bars are 5 mV and 18.5 msec.

and PKA) are not blocked by genistein. Surprisingly, even with very high concentrations of the PKC inhibitor H-7 (300 μM), we observed only a partial suppression of the effects of carbachol on the sAHP ($\sim 25\%$, $n = 7$), perhaps indicating that PKC only partially mediates this action of carbachol (Nicoll et al. 1990). Therefore, as an additional test for genistein inhibition of PKC activity, we examined the effects of genistein on the potentiation of synaptic transmission produced by directly activating PKC with phorbol esters.

Activation of PKC by phorbol esters produces a pronounced increase in synaptic transmission in the hippocampus (Malenka et al. 1986a). As shown in Figure 4, a drop of 10 μM phorbol 12,13-diacetate (PDAc)-containing ACSF onto the slice near the recording electrode resulted in a rapid increase in the amplitude of responses elicited by Schaffer collateral/commissural fiber stimulation. The PDAc-induced potentiation reached a maximal level 5–15 minutes following the application, and the amplitude of the field EPSPs remained significantly above control levels ($p < 0.01$) for the duration of the experiments. Concentrations of genistein that block LTP (110 μM) had no effect on the phorbol-ester-induced potentiation, suggesting that this concentration of genistein has little effect on PKC activity. Concentrations of H-7 that block LTP (300 μM) suppressed the response to PDAc; the PDAc-induced potentiation was significantly less ($p < 0.01$) in the presence of H-7 for the first 30 minutes following PDAc application. However, beyond this point the magnitude of the PDAc-induced potentiation was not significantly less in H-7-treated compared to control slices.

These physiological experiments, along with biochemical assays of inhibitor specificity, suggest that the concentrations of genistein and lavendustin A that block LTP do not appreciably inhibit PKC, PKA, and CaMKII. Although we cannot exclude other, un-

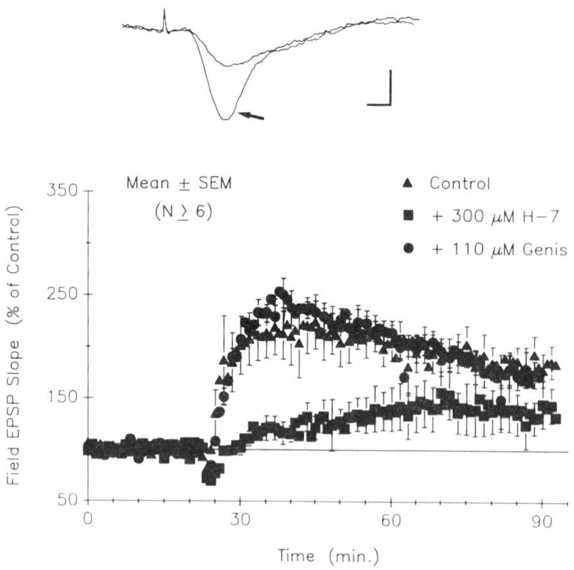

Figure 4. Concentrations of genistein (110 μM) that block LTP have no significant effect on the ability of phorbol esters to potentiate synaptic transmission. Different groups of slices were used as controls or preincubated for 1–4 hr in genistein (110 μM, 0.33% DMSO, $n = 6$) or H-7 (300 μM, 0.2% DMSO, $n = 8$). A drop of 10 μM phorbol-12,13-diacetate-containing ACSF was applied to the slice at approximately 24 min into the experiment. This elicited a transient depression of synaptic transmission that was rapidly followed by a potentiation that reached a maximal level 5–15 min later. Traces show potentiated EPSPs recorded 15 min after PDAc application (arrow) and a response recorded just prior to PDAc application. Calibration bars are 1.0 mV, 5.5 msec. Note that genistein had no effect on the phorbol-ester-induced potentiation (●). H-7 suppressed the response to PDAc over the first 30 min following PDAc application. However, beyond this point the magnitude of the PDAc-induced potentiation was not significantly less in H-7-treated compared to control slices. In preliminary experiments we were unable to inhibit the action of bath applied PDAc (10 μM) with H-7 (300 μM) in 0.01% DMSO. H-7 was only able to suppress the PDAc-induced potentiation using long preincubations in H-7 (1–4 hr) diluted in ACSF containing relatively high final concentrations of DMSO (0.2%) and using brief applications of PDAc.

known, actions of genistein and lavendustin A, the blockade of LTP by these inhibitors seems to be most readily explained by inhibition of tyrosine kinases. Although our experiments with protein tyrosine kinase inhibitors suggest that these kinases are part of the postsynaptic biochemical pathway that leads to LTP, several important questions remain unanswered. For instance, the relevant kinases and their substrates remain to be identified, and the nature of the interaction of protein tyrosine kinases with the serine-threonine kinases known to be involved in LTP remains to be determined. The experiments described next address the first of the questions: the identification of the specific tyrosine kinase(s) involved in LTP.

Is *src* Necessary for LTP?

Our initial efforts focused on the nonreceptor protein tyrosine kinase c-*src*, a likely candidate for the protein tyrosine kinase required for normal LTP since it is expressed in high levels in hippocampal neurons and is localized to both pre- and postsynaptic sites. We have electrophysiologically assessed synaptic transmission and plasticity in mice lacking the *src* tyrosine kinase to determine whether *src* may have a role in LTP.

To assess whether *src* might be involved in short-term forms of synaptic plasticity in the hippocampus, we compared paired-pulse facilitation of synaptic transmission in hippocampi from control and src^- mice. In the CA1 region of the hippocampus, single shocks delivered to the presynaptic fibers at a low frequency produce similar responses. If a second shock is applied soon after the first, the response elicited by the second stimulation pulse produces a much larger response. This phenomenon, known as paired-pulse facilitation, is a short-term form of synaptic plasticity (the response to the second stimulus returns to levels similar to the first as soon as the time interval between pulses is sufficiently increased) and is thought to predominantly reflect changes in transmitter release from the presynaptic terminals. The top trace in Figure 5A shows an example of paired-pulse facilitation in normal mouse

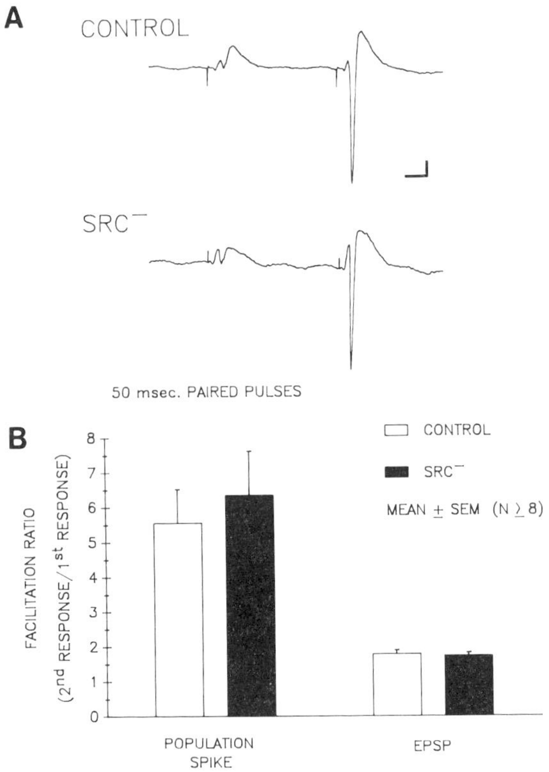

Figure 5. Paired-pulse facilitation is not altered in mice that lack the *src* tyrosine kinase. (*A*) Top trace is recording of the population spike elicited by two stimulations of the presynaptic fibers delivered at a time interval of 50 msec. Note the larger downward-going population spike elicited by the second stimulation pulse. The bottom trace is a similar recording taken from a slice from a src^- mouse. Calibration bars are 1 mV, 10.5 msec. (*B*) Summary of paired-pulse facilitation of the population spike and field EPSP in wild-type and src^- mice. Note that the *src* ablation has no effect on this form of synaptic plasticity.

hippocampus. Two pulses of stimulation, separated by 50 msec, were delivered to the presynaptic fibers in CA1. At this time interval there is a characteristic increase in the population spike elicited by the second stimulation pulse. The second trace in Figure 5A shows responses from a *src*⁻ mouse hippocampal slice using the same paired-pulse protocol. As shown in the summarized data in Figure 5B, where we measured paired-pulse facilitation of the population spike amplitude (recorded at the cell body layer of CA1) and of the field EPSP (recorded in stratum radiatum of CA1), this form of synaptic plasticity appears normal in the hippocampal slices from *src*⁻ mice.

We also examined LTP in slices from *src*⁻ and wild-type mice. Here, standard stimulation protocols (see Methods) were used to induce LTP, and changes in synaptic efficacy were monitored by measuring the amplitude of the population spike elicited by presynaptic fiber stimulation. As shown in Figure 6, in both wild-type and *src*⁻ mice, high-frequency stimulation produces a pronounced potentiation that persists for the duration of the experiment (1 hr). From these experiments we conclude that both short-term (paired-pulse facilitation) and long-term (LTP) forms of synaptic plasticity in the CA1 region of the hippocampus are normal in mice lacking the *src* tyrosine kinase.

DISCUSSION

Neurons in the vertebrate nervous system express high levels of protein tyrosine kinase activity, and within the brain some of the highest levels of tyrosine kinase activity are found in the hippocampus and cerebellum (Hirano et al. 1988), two structures thought to be important in learning and memory. At the cellular level, the localization of tyrosine kinase activity to synaptosomal and synaptic vesicle fractions has been taken to suggest a role for these kinases in transmitter release or synaptic plasticity. In the present experiments we used tyrosine kinase inhibitors to determine whether tyrosine phosphorylation is important in the biochemical events responsible for the NMDA receptor-dependent long-term potentiation of synaptic transmission in the CA1 region of the hippocampus. We found that genistein inhibits LTP following tetanic stimulation of the Schaffer collateral/commissural fibers only when present prior to tetanic stimulation, suggesting the tyrosine kinase activity is involved in the

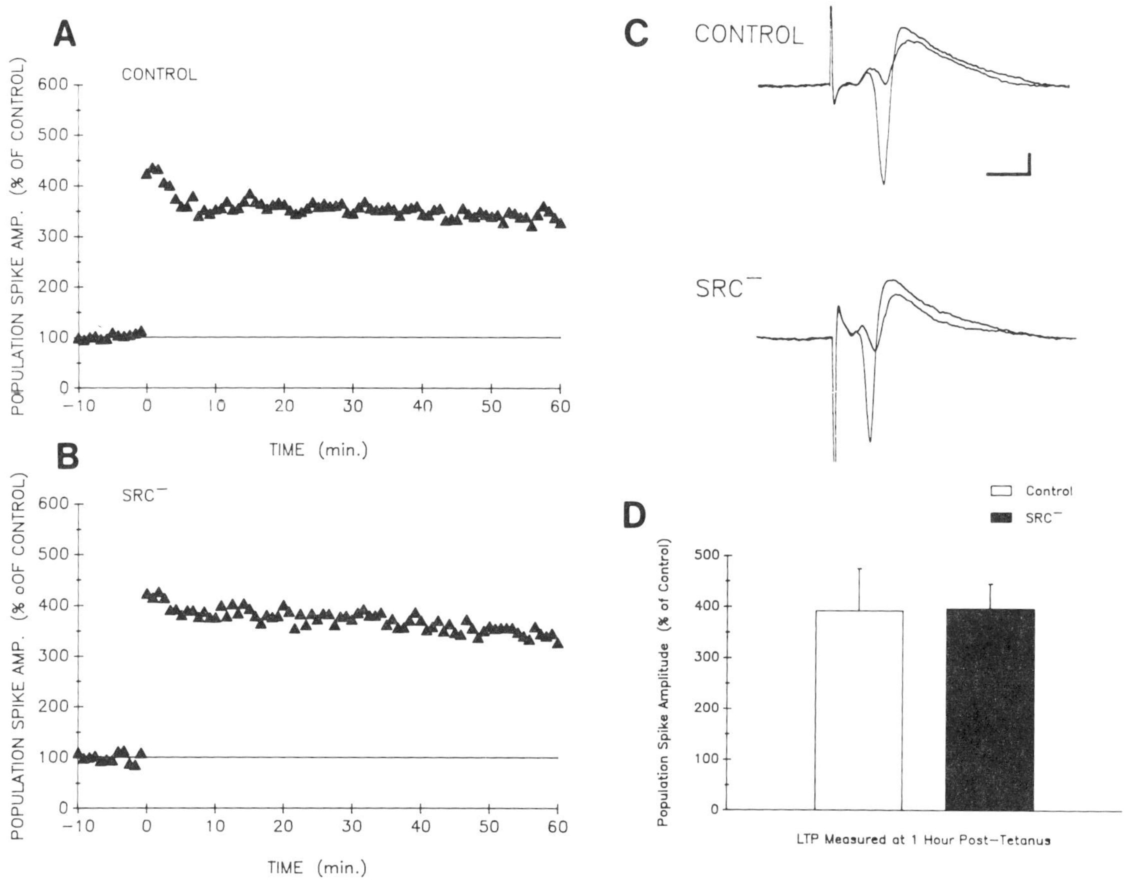

Figure 6. (*A,B*) Individual, representative experiments showing LTP in slices from a normal mouse (*A*) and a *src*⁻ mouse (*B*). In both cases, high-frequency stimulation (derived at time = 0) produced a pronounced potentiation. (*C*) Individual responses elicited before and 1 hr after high-frequency stimulation. Note the pronounced increase in the population spike amplitude elicited posttetanus. Calibration bars are 1 mV and 5.5 msec. (*D*) Summary of experiments comparing the magnitude of LTP measured 1 hr posttetanus in hippocampal slices from control and *src*⁻ mice (values are mean ±S.E.M., $n = 7$ slices for each).

events responsible for the induction of LTP but not in its maintenance and expression. Similar to other studies that have implicated serine-threonine kinases as having a role in LTP, the conclusions of this study rely on the specificity of the chemical inhibitors chosen. Although our data indicate that we have not inhibited serine-threonine kinases that modulate LTP, we cannot exclude an effect on some other critical molecule.

It is currently thought that the induction of LTP in the CA1 region of the hippocampus involves increases in protein phosphorylation in the postsynaptic CA1 pyramidal cells via the serine-threonine kinases PKC and CaMKII (Nicoll et al. 1988; Bashir and Collingridge 1992). Our data, which suggest that tyrosine kinase activity is also necessary for the induction of LTP, suggest that both tyrosine kinases and serine-threonine kinase activity act in a common pathway or cascade leading to enhanced synaptic transmission. What role might tyrosine kinases have in the network of kinase activity responsible for LTP?

In nonneuronal cells, tyrosine phosphorylation activates the gamma isozyme of phospholipase C (Nishibe et al. 1990). This suggests the possibility that tetanus-induced increases in tyrosine kinase activity may contribute to the activation of PKC during the induction of LTP. Such a mechanism could also eventually lead to the release of arachidonic acid, a potential retrograde messenger in LTP (Williams et al. 1989). Another possible role for tyrosine kinases in LTP is suggested by the recent findings that NMDA receptor activation in hippocampal primary neuronal cultures stimulates an increase in tyrosine phosphorylation of MAP kinase (Bading and Greenberg 1991) and that a MAP kinase-like protein, pp40, is similarly phosphorylated in response to phorbol esters in hippocampal slices (Stratton et al. 1989). Together these results suggest a pathway leading from NMDA receptor-mediated calcium influx to PKC that then activates a tyrosine kinase that phosphorylates MAP kinase. Rather than arising through a sequential cascade, the activation of MAP kinase by the different kinases may arise through parallel pathways. This view is appealing, since MAP kinase is a serine-threonine kinase that requires both serine-threonine and tyrosine phosphorylation for full activity (Anderson et al. 1990), and thus inhibition of either kinase pathway may prevent sufficient MAP kinase activation to induce LTP. Although somewhat speculative, the existence of such a convergence point for the biochemical processes involved in LTP is an attractive notion, since it could explain why inhibition of any single member of the network of kinases thought to be involved in LTP leads to complete, rather than partial, block of LTP.

The models presented above involve activation of tyrosine kinases following neurotransmitter receptor stimulation leading to the induction of LTP. An alternative mechanism, which is consistent with the tyrosine kinase inhibitor experiments, is that constitutive tyrosine kinase activity is required to maintain other signal transduction pathways responsive to LTP-inducing patterns of synaptic activity. For instance, several studies have recently found that growth factors that act through receptor-linked tyrosine kinases, such as EGF and FGF, can enhance the magnitude and duration of LTP without any apparent effects on baseline, unpotentiated synaptic transmission (Terlau and Seifert 1989, 1990; Abe et al. 1991; Ishiyama et al. 1991). Although at this stage it is unclear how these kinase pathways are activated or how they interact, the identification of the different kinases and biochemical techniques, perhaps using cultured hippocampal neurons, may lead to the identification of novel pathways from neurotransmitter receptors to regulation of synaptic plasticity.

Elucidating the role(s) tyrosine phosphorylation plays in LTP will require identification of the specific tyrosine kinases involved in LTP. Toward this end we have begun to study transgenic mice that lack nonreceptor tyrosine kinases. Our initial experiments with mice lacking the *src* tyrosine kinase indicate that *src* is not required for LTP. That *src* does not appear to be necessary for LTP does not, however, exclude it from having a function in LTP or other forms of synaptic plasticity. Other physiological protocols for the induction or modulation of synaptic efficacy may reveal a role for *src*. In addition to *src*, other related kinases in the same family, such as *fyn* and *yes*, as well as the nonreceptor tyrosine kinase *abl*, are expressed in the hippocampus, and these kinases may serve overlapping functions with *src* sufficient to prevent any defects in LTP. To explore the possible contribution of other kinases to LTP, we are now investigating LTP in mice deficient in *fyn*, *yes*, or *abl*. Our preliminary results suggest that whereas LTP appears normal in mice deficient in *yes* and *abl*, LTP is blunted in hippocampal slices from mice deficient in *fyn*. Clearly, such genetic approaches to examining the signal transduction pathways responsible for LTP hold promise, not just in the identification of the relevant tyrosine kinases and substrates in LTP, but also in providing valuable insights into the spectrum of biochemical, morphological, and behavioral changes accompanying LTP.

ACKNOWLEDGMENTS

We thank P. Stein for comments on an earlier draft, C. Lam for the figures, and A. Krawetz and H. Ayers for typing the manuscript. This work was supported in part by the National Institutes of Health and the Howard Hughes Medical Institute.

REFERENCES

Abe, K., F. Xie, and H. Saito. 1991. Epidermal growth factor enhances short-term and facilitates induction of long-term potentiation in rat hippocampal slices. *Brain Res.* **547:** 171.

Akiyama, T., J. Ishida, S. Nakagawa, H. Ogawara, S. Watanabe, N. Itoh, M. Shibiga, and Y. Fukami. 1987. Genistein, a specific inhibitor of tyrosine-specific protein kinases. *J. Biol. Chem.* **262:** 5592.

Anderson, N.G., J.L. Maller, N.K. Tonks, and T.W. Sturgill. 1990. Requirement for integration of signals from two distinct phosphorylation pathways for activation of MAP kinase. *Nature* **343:** 651.

Bading, H. and M. Greenberg. 1991. Stimulation of protein tyrosine phosphorylation by NMDA receptor activation. *Science* **253:** 912.

Barnekow, A., R. Jahn, and M. Schartl. 1990. Synaptophysin: A substrate for the protein tyrosine kinase pp60$^{c\text{-}src}$ in intact synaptic vesicles. *Oncogene* **5:** 1019.

Bashir, Z.I. and G.L. Collingridge. 1992. Synaptic plasticity: Long-term potentiation in the hippocampus. *Curr. Opin. Neurobiol.* **2:** 328.

Bekkers, J.M. and C.F. Stevens. 1990. Presynaptic mechanism for long-term potentiation in the hippocampus. *Nature* **346:** 724.

Bliss, T.V.P., R.M. Douglas, M.L. Errington, and M.A. Lynch. 1986. Correlation between long-term potentiation and release of endogenous amino acids from dentate gyrus of anaesthetized rats. *J. Physiol.* **377:** 391.

Braun, S., W.E. Raymond, and E. Racker. 1984. Synthetic tyrosine polymers as substrates and inhibitors of tyrosine-specific protein kinases. *J. Biol. Chem.* **259:** 2051.

Brugge, J., P. Cotton, A. Lustig, W. Yonemoto, L. Lipsich, P. Coussens, J.N. Barrett, D. Nonner, and T.W. Keane. 1987. Characterization of the altered form of the c-*src* gene produce in neuronal cells. *Genes Dev.* **1:** 287.

Cantley, L., K.R. Auger, C. Carpenter, B. Duckworth, A. Grazianai, R. Kapeller, and S. Soltoff. 1991. Oncogenes and signal transduction. *Cell* **64:** 281.

Cartwright, C.A., R. Simantov, W.M. Cowan, T. Hunter, and W. Eckhart. 1988. pp60$^{c\text{-}src}$ expression in the developing rat brain. *Proc. Natl. Acad. Sci.* **85:** 3348.

Cartwright, C.A., R. Simantov, P.L. Kaplan, T. Hunter, and W. Eckhart. 1987. Alterations in pp60$^{c\text{-}src}$ accompany differentiation of neurons from rat embryo striatum. *Mol. Cell. Biol.* **7:** 1830.

Cooper, J.A. 1989. The *src* family of protein-tyrosine kinases. In *Peptides and protein phosphorylation* (ed. B. Kemp and P.F. Alewood), p. 85. CRC Press, Boca Raton, Florida.

Cudmore, S.B. and J.W. Gurd. 1991. Postnatal age and protein tyrosine phosphorylation at synapses in the developing rat brain. *J. Neurochem.* **57:** 1240.

Dean, N., M. Kanemitsu, and A.L. Boynton. 1989. Effects of the tyrosine-kinase inhibitor genistein on DNA synthesis and phospholipid-derived second messenger generation in mouse 10T1/2 fibroblasts and rat liver T51B cells. *Biochem. Biophys. Res. Commun.* **165:** 795.

Golden, A. and J.S. Brugge. 1989. Thrombin treatment induces rapid changes in tyrosine phosphorylation in platelets. *Proc. Natl. Acad. Sci.* **86:** 901.

Hashimoto, Y. and T.R. Soderling. 1987. Calcium calmodulin-dependent protein kinase II and calcium phospholipid-dependent protein kinase activities in rat tissue assayed with a synthetic peptide. *Arch. Biochem. Biophys.* **252:** 418.

Hirano, A.A., P. Greengard, and R.L. Huganir. 1988. Protein tyrosine kinase activity and its endogenous substrates in rat brain: A subcellular and regional survey. *J. Neurochem.* **50:** 144.

Ishiyama, J., H. Saito, and K. Abe. 1991. Epidermal growth factor and basic fibroblast growth factor promote the generation of long-term potentiation in the dentate gyrus of anesthetized rats. *Neurosci. Res.* **12:** 403.

Kikkawa, U., Y. Takal, R. Minakuchi, S. Inohgra, and Y. Nishizuka. 1982. Calcium-activated, phospholipid-dependent protein kinase from rat brain. *J. Biol. Chem.* **257:** 13341.

Klausner, R.D. and L.E. Samelson. 1991. T cell antigen receptor activation pathways: The tyrosine kinase connection. *Cell* **64:** 875.

Kullmann, D.M. and R.A. Nicoll. 1992. Long-term potentiation is associated with increases in quantal context and quantal amplitude. *Nature* **357:** 240.

Kypta, R.M., Y. Goldberg, E.T. Ulug, and S.A. Courtneidge. 1990. Association between the PDGF receptor and members of the *src* family of tyrosine kinases. *Cell* **62:** 481.

Levy, J.B. and J.S. Brugge. 1989. Biological and biochemical properties of the c-*src* gene product overexpressed in chicken embryo fibroblasts. *Mol. Cell. Biol.* **9:** 3322.

Madison, D.V. and R.A. Nicoll. 1982. Noradrenaline blocks accommodation of pyramidal cell discharge in the hippocampus. *Nature* **299:** 636.

Malenka, R.C., D.V. Madison, and R.A. Nicoll. 1986a. Potentiation of synaptic transmission in the hippocampus by phorbol esters. *Nature* **321:** 175.

Malenka, R.C., D.V. Madison, R. Andrade, and R.A. Nicoll. 1986b. Phorbol esters mimic some cholinergic actions in hippocampal pyramidal neurons. *J. Neurosci.* **6:** 475.

Malenka, R.C., J.A. Kauer, D.J. Perkel, M.D. Mauk, P.T. Kelly, R.A. Nicoll, and W.N. Waxham. 1989. An essential role for postsynaptic calmodulin and protein kinase activity in long-term potentiation. *Nature* **349:** 554.

Malinow, R. and R.W. Tsien. 1990. Presynaptic enhancement shown by whole-cell recordings of long-term potentiation in hippocampal slices. *Nature* **346:** 177.

Malinow, R., H. Schulman, and R.W. Tsien. 1989. Inhibition of postsynaptic PKC or CamKII blocks induction but not expression of LTP. *Science* **235:** 862.

Martinez, R., B. Mathey-Prevot, A. Bernards, and D. Baltimore. 1987. Neuronal pp60$^{c\text{-}src}$ contains a six-amino acid insertion relative to its non-neuronal counterpart. *Science* **237:** 411.

Mustelin, T., K.M. Coogeshall, N. Isakov, and A. Altman. 1990. T cell antigen receptor-mediated activation of phospholipase C requires tyrosine phosphorylation. *Science* **247:** 1584.

Nicoll, R.A., J.A. Kauer, and R.C. Malenka. 1988. The current excitement in long-term potentiation. *Neuron* **1:** 97.

Nicoll, R.A., R.C. Malenka, and J.A. Kauer. 1990. Functional comparison of neurotransmitter receptor subtypes in mammalian central nervous system. *Physiol. Rev.* **70:** 513.

Nishibe, S., M.I. Wahl, S.M. Teresa Jermamdez-Sotomayor, N.K. Tonks, S.G. Rhee, and G. Carpenter. 1990. Increase of the catalytic activity of phospholipase C-γ 1 by tyrosine phosphorylation. *Science* **250:** 1253.

O'Dell, T.J., E.R. Kandel, and S.G.N. Grant. 1991. Long-term potentiation in the hippocampus is blocked by tyrosine kinase inhibitors. *Nature* **353:** 558.

Onoda, T., H. Iinuma, Y. Sadaki, M. Hamada, K. Isshiki, H. Naganawa, T. Takeuchi, K. Tatsuta, and K. Unezawa. 1989. Isolation of a novel tyrosine kinase inhibitor, lavendustin A, from *Streptomyces griseolavendus*. *J. Natl. Prod.* **52:** 1252.

Pang, D.T., J.K.T. Wang, F. Valtorta, F. Benfenati, and P. Greengard. 1988. Protein tyrosine phosphorylation in synaptic vesicles. *Proc. Natl. Acad. Sci.* **85:** 762.

Pyper, J.M. and J.B. Bolen. 1990. Identification of a novel neuronal c-*src* exon expressed in human brain. *Mol. Cell. Biol.* **10:** 2035.

Reymann, K.G., R. Brodemann, H. Kase, and H. Matthies. 1988. Inhibitors of calmodulin and protein kinase C block different phases of hippocampal long-term potentiation. *Brain Res.* **461:** 388.

Roskowski, T. 1983. Assays of protein kinases. *Methods Enzymol.* **99:** 3.

Ross, C.A., G.E. Wright, M.D. Resch, C.A. Pearson, and S.H. Snyder. 1988. Brain-specific *src* oncogene mRNA mapped in rat brain by in situ hybridization. *Proc. Natl. Acad. Sci.* **85:** 9831.

Soriano, P., C. Montgomery, R. Geske, and A. Bradley. 1991. Targeted disruption of the c-*src* proto-oncogene leads to osteopetrosis in mice. *Cell* **64:** 693.

Stratton, K.R., P.F. Worley, P.L. Huganir, and J.M. Baraban. 1989. Muscarinic agonists and phorbol esters increase tyrosine phosphorylation of a 40-kilodalton protein in hippocampal esters. *Proc. Natl. Acad. Sci.* **86:** 2498.

Sugrue, M.M., J.S. Brugge, D.R. Marshak, P. Greengard, and E.L. Gustafson. 1990. Immunocytochemical localization of the neuron-specific form of the c-*src* gene product, pp60$^{c\text{-}src^{+}}$ in rat brain. *J. Neurosci.* **10:** 2513.

Terlau, H. and W. Seifert. 1989. Influence of epidermal growth factor on long-term potentiation in the hippocampal slice. *Brain Res.* **484:** 352.

———. 1990. Fibroblast growth factor enhances long-term potentiation in the hippocampal slice. *Eur. J. Neurosci.* **2:** 973.

Walaas, S.I., A. Lustig, P. Greengard, and J.S. Brugge. 1988. Widespread distribution of the c-*src* gene product in nerve cells and axon terminals in the adult rat brain. *Mol. Brain Res.* **3:** 215.

Williams, J.H., M.L. Errington, M.A. Lynch, and T.V.P. Bliss. 1989. Arachidonic acid induces a long-term activity-dependent enhancement of synaptic transmission in the hippocampus. *Nature* **341:** 739.

Yang, X., R. Martinez, J. LeBeau, O. Wiestler, and G. Walter. 1989. Evolutionary expression of the neuronal form of the *src* protein in the brain. *Proc. Natl. Acad. Sci.* **86:** 4751.

α Calcium/Calmodulin Kinase II Mutant Mice: Deficient Long-term Potentiation and Impaired Spatial Learning

A.J. SILVA,* Y. WANG,† R. PAYLOR,‡ J.M. WEHNER,‡
C.F. STEVENS,† AND S. TONEGAWA*

*Howard Hughes Medical Institute at Center for Cancer Research and Department of Biology, Massachusetts Institute of Technology, Cambridge, Massachusetts 02139; †Howard Hughes Medical Institute at Salk Institute, La Jolla, California 92037; ‡Institute for Behavioral Genetics, University of Colorado, Boulder, Colorado 80309

Long-term potentiation (LTP) is a type of synaptic plasticity that has been widely studied as a candidate mechanism for some types of learning and memory (Bliss and Gardner Medwin 1973; Bliss and Lomo 1973; Schwartzkroin and Webster 1975; McNaughton et al. 1978). In some hippocampal synapses where LTP has been studied, the voltage-sensitive and glutamate-gated ion channel, the *N*-methyl-D-aspartate receptor (NMDAR) plays a critical role in the induction of LTP by regulating a calcium (Ca^{++}) current (Collingridge and Singer 1990). The Ca^{++} influx facilitated by an opening of NMDARs leads to an increase in the synaptic connection via a series of biochemical events that include activation of Ca^{++} and/or calmodulin-dependent kinases.

To date, the evidence supporting the linkage between LTP and learning and memory primarily comes from the analysis of rats in which the NMDAR is blocked by an antagonist, aminophosphonovaleric acid (APV) (Morris et al. 1986; Staubli Thibault et al. 1989; Davis et al. 1992). The interpretation of those results is difficult, however, because inhibiting NMDAR function also disrupts synaptic function (Bekkers and Stevens 1990b) and might therefore alter the character of information processing in the hippocampus. Thus, the deficits in learning might be due to this alteration in hippocampal synaptic function and not to the deficits in LTP. A second line of evidence linking LTP to learning and memory comes from the observation that induction of saturating levels of LTP in the hippocampus impairs the ability of rats to acquire new spatial information (Castro et al. 1989). However, these findings have also been given an alternative interpretation (Keith and Rudy 1990; Sutherland et al. 1991).

We have adopted a new strategy for the study of the mechanisms of mammalian memory, suitable for addressing the problem of LTP and learning: production of mice with mutations in individual enzymes likely to be involved in the regulation of candidate memory mechanisms, such as LTP. Such specific mutations can be made by using gene targeting (Capecchi 1989). As a first step in this program, we report here studies on a strain of mutant mice that do not express the α isoform of Ca^{++}/calmodulin-dependent protein kinase type II (α CaMKII). This enzyme is neural specific and is present presynaptically, as well as being richly represented adjacent to the postsynaptic membrane at synapses that express LTP (Erondu and Kennedy 1985; Schulman and Lou 1989). Calmodulin (CaM) loaded with Ca^{++} activates this enzyme and induces its autophosphorylation. Once autophosphorylated, the CaMKII holoenzyme no longer requires Ca^{++} or CaM for activity. This switch-like mechanism can maintain the enzyme in an active state beyond the duration of the activating Ca^{++} signal (Miller et al. 1988; Hanson et al. 1989) and has been invoked in learning models (Miller and Kennedy 1986; Lisman and Goldring 1988). Furthermore, pharmacological experiments have implicated this holoenzyme in the induction of LTP (Malenka et al. 1989; Malinow et al. 1989).

We find that although postsynaptic mechanisms seem normal in the CA1 hippocampal region of these mutant mice, they exhibit little or no LTP. These mice are therefore an ideal model to examine the association of hippocampal α CaMKII activity, LTP, and learning and memory processes. We examined whether α CaMKII mutant mice can learn to solve a complex spatial learning task, the Morris water tasks (Morris 1981), and found that these mutant mice have a pronounced deficit in spatial learning performance compared to normal wild-type littermates. Our results demonstrate that α CaMKII is important for spatial learning and support the hypothesis that LTP is the electrophysiological basis for certain types of learning processes.

EXPERIMENTAL PROCEDURES

Production of α CaMKII mutant mice. We transfected the linealized p23 plasmid into E14 embryonic stem (ES) cells (Thompson et al. 1989) by electroporation and isolated 150 (*neo*) colonies, of which two (E14-20 and E14-84) were shown by Southern blotting analysis to harbor a homologously integrated plasmid. We injected the E14-20 ES cells into C57BL/6J blastocysts and transferred the blastocysts into pseudo-pregnant mothers. Twelve male chimeric mice were born, and we bred them with BALB/c females. Southern blot analysis of tail DNA from the offspring of the chimeric males revealed that 11 of the 12 males transmitted the α CaMKII mutation to 20–40% of their offspring. Southern blot analysis of a litter from a cross between two mice heterozygous for the α CaMKII mutation identified mice with two mutant copies of the α CaMKII locus (mutation homozygous), as well as mice with one mutant copy (mutation heterozygous) and mice with normal α CaMKII loci (wild type).

In vitro phosphorylation assays. The brain was isolated quickly, dissected, weighed, and homogenized gently in 4 volumes of cold extraction buffer containing 10 mM Tris-HCl (pH 7.4), 1 mM EGTA, 0.5 mM DTT, 0.1 mM PMSF, 5 mg/l leupeptin, 20 mg/l soybean trypsin inhibitor. This homogenate was kept on ice and was typically diluted 1/10 to 1/200 in the same cold buffer. The assays were carried out after 2- to 20-fold dilution of the homogenate in cold water and were immediately followed by the addition of cold assay buffer to a final concentration of 50 μM Syntide 2, 25 mM HEPES (pH 7.4), 10 mM MgCl, 0.5 mM DTT, 15 μM ATP, and 50 μCi/ml of [γ-^{32}P]ATP. For the Ca^{++}/CaM-induced assays, we also added to this buffer 1.5 μM calmodulin and 2.0 mM $CaCl_2$. For the Ca^{++}/CaM-independent reactions, we added 0.5 mM EGTA to the assay buffer. After adding the assay, buffer reactions were briefly mixed and quickly placed at 30°C for 45 seconds. The reaction was terminated by spotting half of the reaction volume (25 μl) in perforated disks of phosphocellulose. These disks were then washed of nonincorporated [γ-^{32}P]ATP with 1% phosphoric acid and water. The radioactivity bound to the disks was counted, and the values were plotted. The phosphorylation results were derived from four independent experiments, each with at least three different concentrations of homogenates to check that substrate was not limiting, and with duplicates at each concentration point.

Electrophysiological procedures. All animal handling and tissue preparation were in accordance with a protocol approved by the Salk Institute and MIT Animal Use and Care Committee. Transverse hippocampal slices (~350 μm) were prepared from normal (wild type) or mutant mice (male or female, 1–4.5 months old, mostly 1.5–3 months old). Slices were then maintained in an incubation chamber for at least 1 hour at room temperature (24° ± 1°C). An individual slice was transferred to a submerged recording chamber, where it was held by a net made with flattened platinum wire and nylon threads and continuously perfused with artificial cerebrospinal fluid (ACSF) at a rate of ~2 ml/min. The temperature in the recording chamber was 30.5° ± 0.5°C. The ACSF, equilibrated with 95% O_2/5% CO_2, is composed of (in mM): NaCl 120, KCl 3.5, NaH_2PO_4 1.25, $NaHCO_3$ 26, $MgCl_2$ 1.3, $CaCl_2$ 2.5, picrotoxin (PCTX) 0.05. The solution for dissection has the same composition as regular ACSF except there is no PCTX and NaCl is replaced with equimolar sucrose (Aghajanian and Rasmussen 1989). CA3 region was usually removed to prevent epileptiform activity. The cell layer was visualized under an inverted microscope with phase contrast (Zeiss, Thornwood, New York). Extracellular field excitatory postsynaptic potentials (f-EPSPs) were recorded in the stratum radiatum of CA1 with electrodes (1–2 Mohms) filled with ACSF. Excitatory postsynaptic currents (EPSCs) were recorded in CA1 pyramidal neurons with the whole-cell patch-clamp mode, using electrode 3–4 Mohms; no fire polishing; soft glass (Drummond Sci. Co., Broomal, Pennsylvania), filled with (in mM) cesium gluconate 130, $CsCl_2$ 5, EGTA 0.5, $MgCl_2$ 1, Mg-ATP 2, GTP 0.2, NaCl 5, HEPES 10 (pH 7.25). The seal formed on cell bodies was typically 2–3 gigaohms, and the input resistance of cells was typically around 100 Mohms. Bipolar tungsten stimulating electrodes (Frederick Haer & Co., Brunswick, Maine) were positioned in Schaffer collateral-commissural afferents to evoke f-EPSPs (150–200 μm away) or to evoke EPSCs (50–100 μm away). The stimulus intensity was adjusted to evoke pretetanic responses of similar sizes for all the neurons or slices. The stimulus duration was 100 μsec. Recordings were performed with an Axopatch-1A (Axon Instruments, Inc., Foster City, California), filtered at 1–2 kHz, and sampled at 5–10 kHz. Data were collected and analyzed with programs written by C.F. Stevens in AxoBASIC/QuickBASIC. The data collected from normal other strains of mice were combined with the data from normal littermates, since they were indistinguishable. CNQX (6-cyano-7-nitroquinoxaline-2,3-dione) and D-APV (D-2-amino-5-phosphonovaleric acid) were from Cambridge Research Biochemicals (Valley Stream, New York), and PCTX was from Sigma (St. Louis, Missouri).

RESULTS

α CaMKII Mutant Mice

To produce mice with a mutation in the α CaMKII locus, we constructed the plasmid p23 (Fig.1A) which contains a 6.1-kb mouse genomic α CaMKII sequence that is disrupted by insertion of a neomycin-resistance gene (*neo*) from the plasmid pgkneo (Adra et al. 1987). The insertion is within the α CaMKII exon encoding most of the regulatory domain, and the inserted sequence replaced a 130-bp mouse genomic sequence flanked by a pair of *Sph*I sites, which includes the entire inhibitory domain and five amino acids in the amino end of the calmodulin-binding domain (Colbran et al. 1989). We have mutation homozygous mice that were born in June of 1990: The α CaMKII mutation appears not to affect long-term survival under laboratory conditions. In the experiments described in this paper, unless otherwise noted, we used as controls the wild-type littermates of the mutant mice. Northern blot analysis of total brain RNA confirmed that the α CaMKII mutant mice lacked α CaMKII mRNA or any truncated form of it, whereas the β CaMKII mRNA was present at the normal level (data not shown). Western blot analysis of forebrain tissue from the mutants and their normal littermates with the monoclonal antibody 6J9 (Erondu and Kennedy 1985), which recognizes α CaMKII, confirmed that α CaMKII is absent in the mutants but present in the mutation heterozygous and wild-type mice (Fig.1C). A similar analysis with a poly-

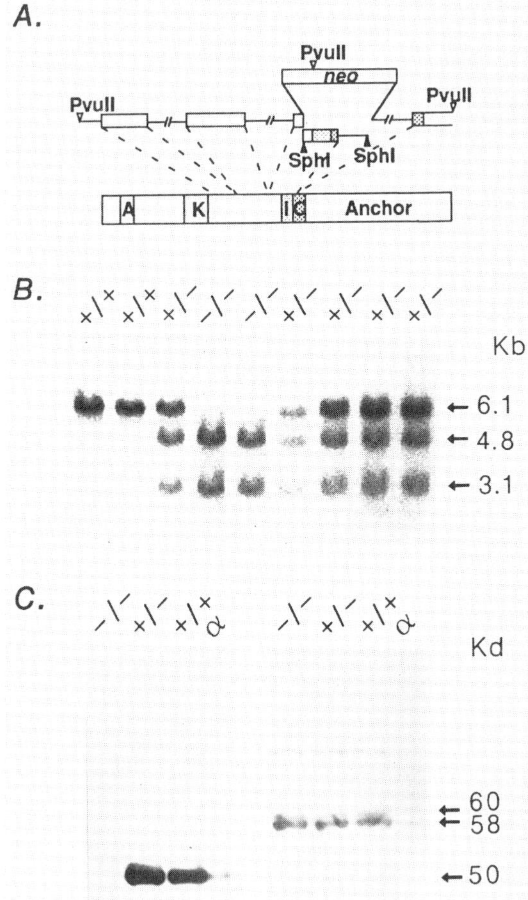

Figure 1. The α CaMKII mutant mice lack α CaMKII. (*A*) Schematic representation of the p23 construct used for the targeting experiment. The 6.1-kb *Pvu*II genomic fragment used in the construction of p23, as well as the *neo* and its insertion site, are shown on the upper part of this figure. We also show the deleted *Sph*I fragment and the functional domains of the α CaMKII enzyme in the lower tracings of this figure. The intron/exon boundaries and their correspondence to the functional domains of the protein are shown by dashed lines. (A) ATP-binding domain; (K) catalytic domain; (I) inhibitory domain; (C) CaM-binding domain; (Anchor) association/anchor domain. (*B*) Western analysis with a monoclonal antibody that recognizes α CaMKII (6g9; left panel), and with a polyclonal antibody that recognizes the β CaMKII, and β′ CaMKII ("Darcy"; right panel) of forebrain homogenates (5 mg) from a mutation homozygous (+ / −), mutation heterozygote (+/ −), and wild-type (+ / +). As a control we included purified rat α CaMKII protein (70 ng) next to the other samples in the Western procedure. Parallel to the autoradiographs, we show the m.w. of the protein bands detected (α CaMKII, β CaMKII, and β′ CaMKII, respectively, 50, 58, and 60 kD). An ANOVA with repeated measures confirms this observation. There was main effect of Strain ($F[1,23] = 22.778$, $p = .0001$) showing that overall, wild-type controls were better than the mutant mice. There was also a main effect of Trial Block ($F[5,115] = 43.371$, $p < .00001$) showing that overall, animals' latency to swim to the platform decreased during training. Finally, there was a significant Strain × Trial Block interaction ($F[5,115] = 4.528$, $p = 008$). Post hoc analysis of the interactions (Neuman-Keuls, $p < .05$) showed that the wild-type animals had significantly lower escape latencies compared to the mutant mice on trial block numbers 1, 2, 3, and 4, but were not different on trial blocks 5 and 6.

clonal antibody that recognized β CaMKII detected the kinase in all three types of mice (Fig.1C). Consistent with the complete absence of α CaMKII in the mutant mice is the reduced (45 ±4%) level of in vitro phosphorylation of α CaMKII-specific peptide substrate, synthide2, in the forebrain homogenates of these mice as compared to the control mice. In contrast, such difference in the in vitro phosphorylation levels was not observed with the homogenates of cerebellum, where the α isoform is known to be a minor CaMKII component (Erondu and Kennedy 1985; Burgin et al. 1990). We confirmed the specificity of our CaMKII assays by adding a peptide inhibitor of CaMKII activity and showing that it reduces phosphorylation levels by 95%.

Ca^{++} calmodulin-independent kinase II activity of the forebrain of mutant mice was also measured and shown not to be increased compared to control mice ($n = 4$). This observation, as well as the lack of any truncated RNA (see above), strongly suggests that the mutant mice do not harbor truncated α CaMKII enzyme, which retains the catalytic domain without the inhibitory domain.

General Observations on Brain Anatomy and Behavior

Observations at the light microscopic level reveal no obvious differences between the mutants and wild-type mice. We concentrated our survey on the hippocampus and neocortex, two structures with high expression of α CaMKII (Burgin et al. 1990). A comparison between coronal brain sections, stained with thionin, from mutants and wild-type mice ($n = 3$; Fig. 2) showed that the arrangement of cells in the major layers of the neocortex and hippocampus appear normal. Tangential sections of the somatosensory cortex stained with cytochrome oxidase or with cresyl violet revealed the same topographical organization of the barrel fields as described for normal mice (data not shown).

The behavior of the mutant mice appears remarkably normal: The mutant pups gain weight at the same rate as their normal littermates, which suggests that their suckling behavior is not deficient. The mutants whisk, sniff, mate, and are not ataxic, indicating that the absence of the α CaMKII does not result in generalized loss of neuronal function. However, they are clearly more jumpy than their wild-type littermates. For instance, unlike wild-type littermates, mutant mice try to avoid human touch long after weaning. Interestingly, this "nervousness" was previously reported for animals with hippocampal lesions (Douglas 1967).

Synaptic Function

Using whole-cell recording in hippocampal slices, we have studied synaptic currents evoked in CA1 pyramidal cells by stimulating Schaeffer collaterals. Our conclusions are based on studies of 37 neurons or slices from 14 homozygotes, 2 slices from 1 heterozygote, and 35 neurons or slices from 20 littermates or other strains.

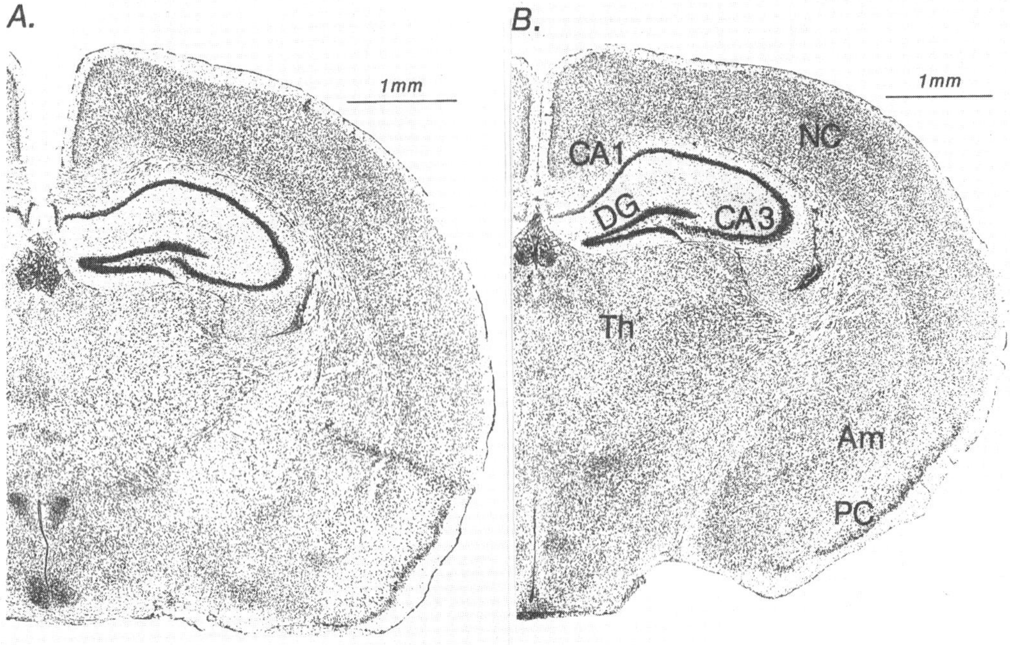

Figure 2. Coronal sections from a wild-type (*A*) and α CaMKII mutant (*B*) mouse brain. The frozen brain sections (14 μm) were thaw-mounted on slides coated with gelatin and stained with 1% thionin. The rostral/caudal levels of the two slices were matched for the hippocampus, but due to slightly different angles of sectioning, the ventral portion of the slice from the α CaMKII mutant mouse is posterior to the one from the wild-type mouse. (Am) Amygdala; (CA1 and CA3) pyramidal cell fields of the hippocampus; (DG) dentate gyrus; (NC) neocortex; (PC) pyriform cortex; (Th) thalamus.

Typical synaptic currents from mutant and normal hippocampal pyramidal neurons are illustrated in Figure 3, A and B. These experiments were carried out in normal magnesium (1.3 mM), and at the neuron's membrane potential (−50 mV) that reveals the NMDA as well as the non-NMDA components of the synaptic currents. Although the absolute amplitude of the synaptic currents depends on stimulus intensity, we used comparable stimuli in all experiments and found no apparent difference in the peak size and time course of synaptic currents from normal and mutant mice. In the presence of CNQX (10 μM; antagonist of non-NMDA receptor) and PCTX (50 μM; antagonist of $GABA_A$ receptor) and at holding potential −30 mV, the rise time of the NMDA component (average 4–10 traces) was 6.1 msec (±0.5 S.E.M.) for the normal mice ($n = 4$) and 6.1 msec (±0.4) for the mutant mice ($n = 4$); the decay time constant was 31.2 msec (±4.9) for the normal mice and 33.7 msec (±4.9) for the mutant mice. The rise time of the non-NMDA component (average 4–15 traces) was 1.5 msec (±0.2) for the normal mice ($n = 4$) and 1.6 msec (±0.1) for the mutant mice ($n = 6$); the decay time constant was 8.9 msec (±0.8) for the normal mice and 7.1 msec (±0.9) for the mutant mice at holding potential −80 mV. All above corresponding values were not significantly different (p values ranged from 0.2 to 0.9, t-test). Furthermore, the ratio of NMDA to non-NMDA components does not differ between normal and mutant neurons (Fig. 3C): For 8 normal neurons the NMDA/non-NMDA component ratio was 29.1% (±5), and for 8 mutant neurons the ratio was 23.8% (±3; $p > 0.3$, t-test).

The previous results demonstrated that, at one particular voltage, the NMDA component of synaptic currents is the same size at normal and mutant synapses. The Mg^{++} block of NMDAR channels that endows these channels with the voltage dependence so important for making LTP associative could, however, differ between normal and mutant synapses. For example, the action of protein kinase C has recently been shown to modify the Mg^{++} dissociate constant at the NMDAR regulatory site (Chen and Huang 1990), and some direct or indirect action of α CaMKII could possibly have an analogous effect. Were this the case, an essential difference in normal and mutant synaptic function would not have been revealed in the preceding experiments. We have investigated this point by directly measuring the voltage dependence of NMDAR in normal and mutant mice. As illustrated in Figure 4, the dependence of NMDAR channel conductance on voltage is not distinguishably different between normal and mutant mice. The activation curves from both normal mice ($n = 4$) and mutant mice ($n = 3$) are well fitted by the equation developed by Jahr and Stevens (1990a,b) (see legend to Fig. 4). In summary, we could detect no differences between NMDAR function in normal and mutant mice.

Historic effects, as revealed by the response to the second of a stimulus pulse pair, did appear to be one aspect of synaptic function that differs between normal

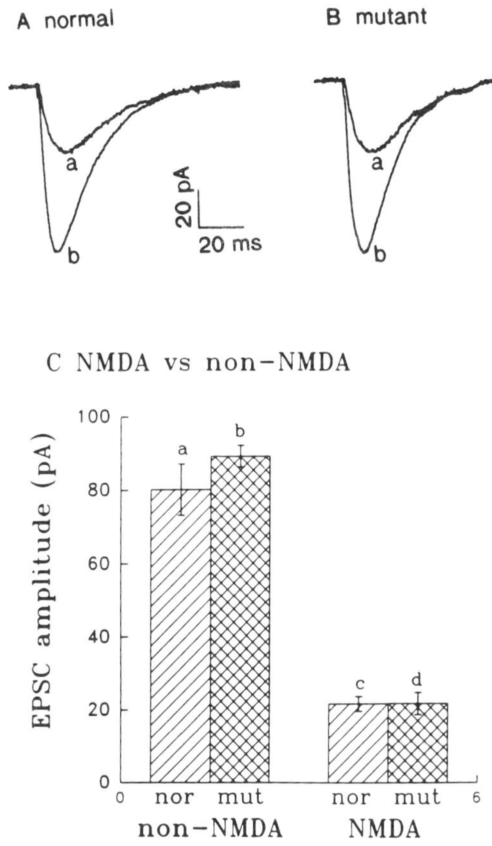

Figure 3. Synaptic currents from whole-cell recordings as a function of time for normal (*A*) and mutant (*B*) mice. The superimposed pair of traces in *A* and *B* show the total synaptic current that includes both the NMDA and non-NMDA components (*b*) and the NMDA component alone (*a*). The NMDA component was obtained by subtracting the non-NMDA component remaining after application of 50 μM D-APV (an antagonist of NMDA receptor; traces not shown) from the total current (shown in *b*). (*C*) Averages of EPSC amplitudes for non-NMDA and NMDA components from eight normal (nor) and eight mutant (mut) mice. The holding potential was −50 mV and the stimulus intensity was increased to reveal the NMDA (in addition to non-NMDA) components of the synaptic currents. The NMDA components were measured at 20–35 msec after the peak current (always judged by the traces mostly after D-APV or at −80 mV). There are no significant differences between *a* and *b* ($p > 0.2$, t-test) and between *c* and *d* ($p > 0.9$).

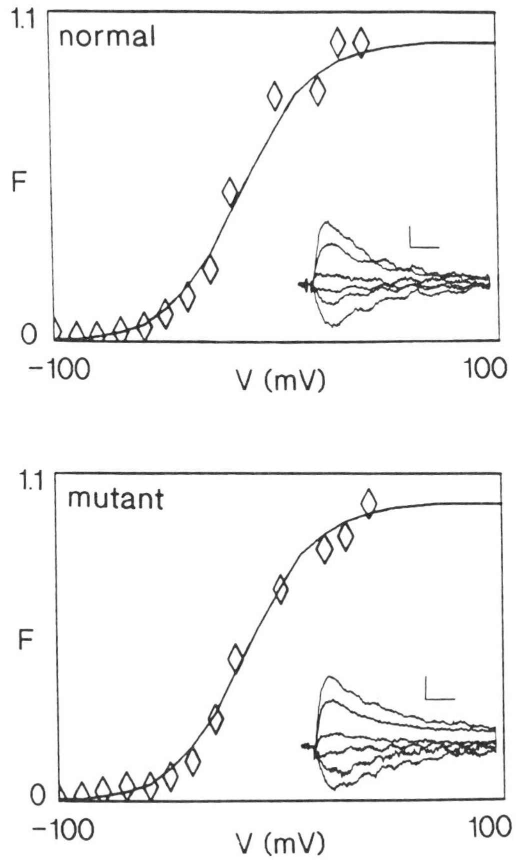

Figure 4. Voltage dependence of NMDA receptor channels for typical normal (*top*) and mutant (*bottom*) neurons in the presence of CNQX (10 μM) and PCTX (50 μM). The fraction (F) of the maximum conductance through the NMDA receptor channels is given as a function of membrane potential. The insets show sample traces of synaptic current as a function of times. The holding potentials (mV) for the sample traces are (from top to bottom) +40, +30, +20, −90, −30, −20 for normal neurons and +40, +30, +20, −70, −40, −20 for mutant neurons. The vertical calibration bars indicate 25 pA current. The horizontal bars indicate 10 and 15 msec for normal and mutant neurons, respectively. The smooth curves were fitted with the equation: $g(V) = 1/[1 + \exp(-0.062V)(C/3.57)]$, where g is conductance in pS, V is membrane potential in mV, and C is extracellular magnesium concentration in mM (Jahr and Stevens 1990a,b). Note that no free parameters are used in fitting the theoretical curve to the data.

and mutant mice. Figure 5 presents the average paired pulse results. Although the slices from both normal and mutant mice showed paired pulse facilitation, the degree of facilitation was less in mutant mouse slices at all interpulse intervals tested. The impaired pulse facilitation in the mutant mice may not be surprising in view of the fact that synapsin I, a molecule believed to be involved in making vesicles available for release, is regulated by CaMKII (Lin et al. 1990).

Deficient Long-term Potentiation

We have used two approaches to evaluate LTP in the mutant mice: field potential recording to survey properties of large populations of neurons, and whole-cell recording to investigate LTP under conditions that are most sensitive for detecting the presence of LTP (the effect is typically larger with whole-cell recordings than with field potentials) and to permit examination of the factors, such as the release probability, that participate in LTP.

Figure 6 shows the typical time course of synaptic strength in a normal and a mutant mouse following tetanic stimulation that typically induces LTP. As is apparent from the figure, the normal mouse exhibited an obvious increase in synaptic strength, whereas synaptic transmission in the mutant mouse was unaffected by the tetanic stimulation except for a usual

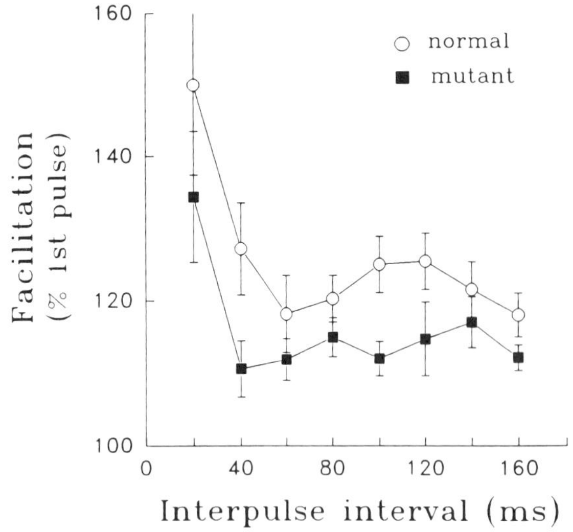

Figure 5. Paired pulse facilitation in normal and mutant mice. The size of the peak field potential for the second pulse of a pair (expressed as the percent of the first pulse size) as a function of the time interval between the pulses. Data are from 10 normal and 10 mutant slices. Error bars show the S.E.M.

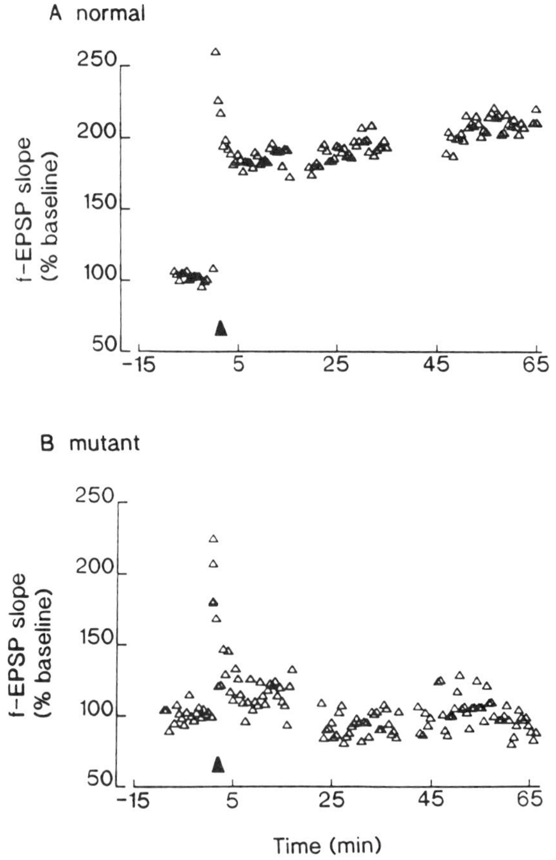

Figure 6. Synaptic strength, expressed as a percentage of the average pretetanus initial field potential rate of change (f-EPSP), as a function of time in minutes during the experiments for a typical normal (*A*) and mutant (*B*) slice. Testing stimuli were given once each 30 sec. Our standard tetanus, given at time = 0 (indicated by arrowhead), was 100 Hz, 3 trains, and 20 pulses for each train with an intertrain interval of 11 sec. In some mutant slices, a second tetanus (100 Hz, 200-msec duration, 5 trains, and 6-sec interval) was given.

amount of posttetanic potentiation. Of 11 slices from normal mice, 9 exhibited LTP (defined as an increase in synaptic strength of 20% or greater sustained for at least 30 minutes), whereas only 2 of 16 slices from mutant mice showed LTP (Table 1). These two slices were from different mutant mice. The distribution of synaptic strength 30–60 minutes after a tetanus for 11 normal and 16 mutant neurons is shown in Figure 7. Many mutant neurons actually showed some decrease in synaptic posttetanus period. The effect was not just an increased threshold for LTP because, in slices from mutant mice, a second tetanus with increased trains (group of pulses, see legend to Fig. 8) still failed to produce potentiation. Based on field potential recordings that report the characteristics of large neuronal populations, we conclude that most hippocampal synapses do not exhibit significant LTP in mice lacking α CaMKII.

With whole-cell recording, 9 of 10 neurons from normal mice exhibited LTP with an average increase in synaptic strength of 286% (±48) of baseline at 30–60 minutes following a tetanic stimulus. For the mutant mice, 2 of 12 cells exhibited LTP, and the average synaptic strength after tetanic stimulation for these two cells was 221% (±34); for the neurons from mutant mice that did not show LTP (less than 20% increase), the average was 98% (±8, $n = 10$) (Table 1; Fig. 8). These two cells were from two different mutant mice.

For normal cells, LTP is accompanied by a decrease in the number of "failures"—those stimuli that produce no postsynaptic response—and an increase in the quantal content as revealed by a standard quantal anal-

Table 1. Summary of the Changes of Synaptic Strength in Normal and Mutant Mice

Mouse	LTP *n*/ Total *n*	Potentiation (% baseline) mean ± S.E.M.	No LTP *n*/ total *n*	Change (% baseline) mean ± S.E.M.
Field EPSP				
normal	9/11	183 ± 18	2/11	114 ± 1
homozygote	2/16	168 ± 32	14/16	99 ± 3
heterozygote	1/2	141	1/2	115
EPSC				
normal	9/10	286 ± 48	1/10	96
homozygote	2/12	221 ± 39	10/12	98 ± 8

n represents the number of slices (field EPSP slope) or neurons (EPSC amplitude).

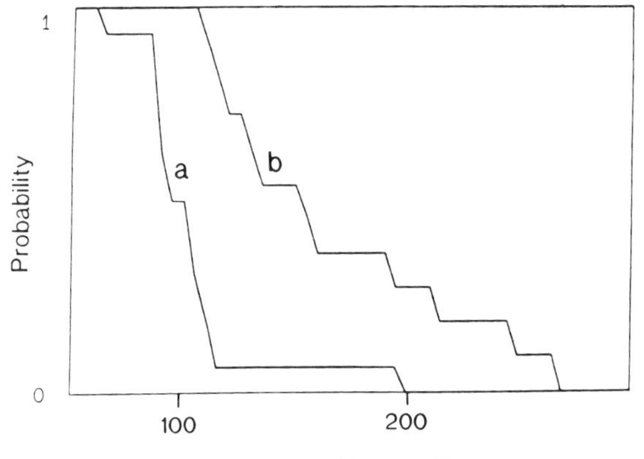

Figure 7. Cumulative probability as a function of initial field EPSP slope, expressed as percentage of the pretetanus average, and valued for neurons 30–60 min after tetanic stimulation. The 16 mutant slices are represented by graph a and 11 normal slices by graph b. The ordinate gives the fraction of the slices that exhibited a posttetanic field EPSP slope of the corresponding abscissa size or smaller.

ysis (Bekkers and Stevens 1990a; Malinow and Tsien 1990). The synapses from one mutant animal that did exhibit LTP exhibit the same properties: a decrease in failures following tetanic stimulation (from a pretetanic failure rate of 2.3% to a posttetanic rate of less than 0.07%) and an increase in quantal content that can account for the entire amount of potentiation.

We conclude, then, that mice lacking expression of α CaMKII are markedly deficient in LTP, but in those cases where LTP does occur, it appears to be indistinguishable from what is observed in normal animals.

The Morris Water Task

In the Morris water task, mice are placed in a round pool filled with opaque water. To escape the water, they must swim to a submerged platform. In the "visible-platform" version of the Morris task, a visually conspicuous white flag is placed on top of the submerged platform which is positioned in random locations during each trial. To solve this task and swim directly to the platform, it is sufficient for an animal to learn that the flag indicates the location of the platform. Hence, distal extramaze cues are irrelevant in this task. In the "hidden-platform" version of the Morris task, the escape platform is located in a fixed location within the pool. Since there are no immediate proximal cues indicating the platform's location, and an animal cannot see the platform through the water, it must learn the multiple spatial relationships between distal objects in the room surrounding the pool and the platform in order to locate and swim directly to it. There are five phases to the behavioral study. All animals were subjected to each phase in the order presented below.

Phase 1: Mutant mice can solve the visible-platform task. In the visible-platform task, the platform location alternated among four possible places within each block of trials (see legend to Fig. 9). Animals were

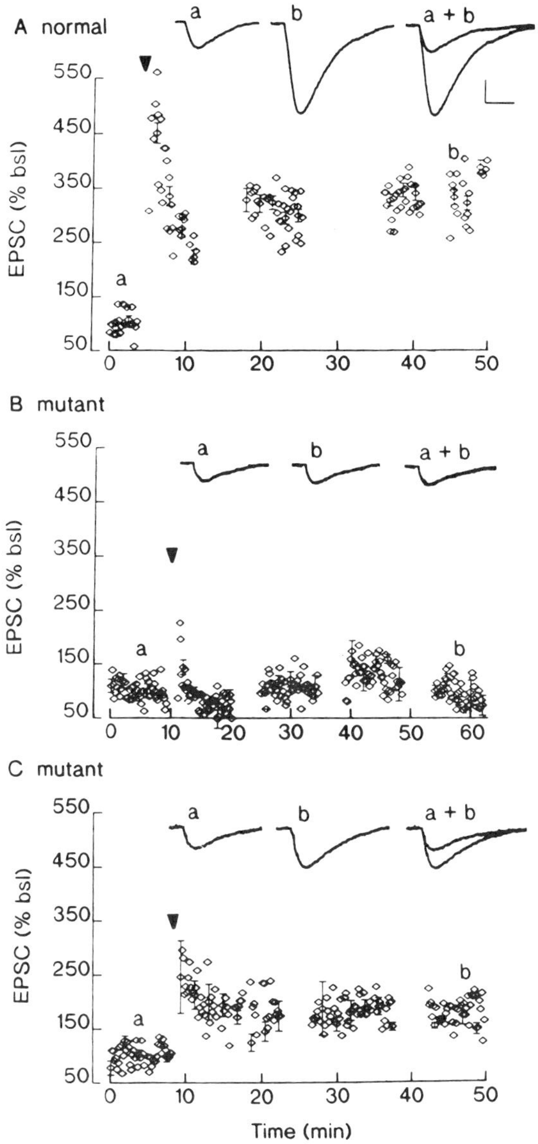

Figure 8. Peak amplitude of synaptic current (EPSC) from whole-cell recordings as a function of time. Amplitudes are expressed as percentage of mean amplitude before tetanic stimulation. (*A*) Typical normal cell, (*B*) typical mutant cell, and (*C*) one of two mutant cells that exhibited LTP. Traces a and b show typical recordings of synaptic current made at the times indicated by the same letters in the amplitude graphs immediately below. The last trace is the preceding pair superimposed. Testing stimuli were given once each 2 sec; the points plotted here are averages of 5 consecutive samples. Holding potential was either −60 or −70 mV (−60 mV for this figure). The tetanus was given at the time point indicated by arrowhead. The tetanus protocol was typically 50 Hz, 2–3 trains, 30 pulses for each train, and 10–11 sec between trains while holding membrane potential at −30 to −40 mV. (With higher frequency tetani, such as 100 Hz, the synaptic current was often depressed, particularly in the mutant neurons.) The amplitude calibration bar (vertical) is 20 pA and the time calibration bar (horizontal) is 10 msec. In many cases (including field potential recording), we tested different things between the times we collected data for synaptic strength; this accounts for the gaps in the graphs. In addition, two to seven different tetanus protocols were tested in some mutant neurons if the typical tetanus protocol did not induce LTP. These procedures made it difficult to average plots among neurons or slices precisely, so we preferred to show the representative plots.

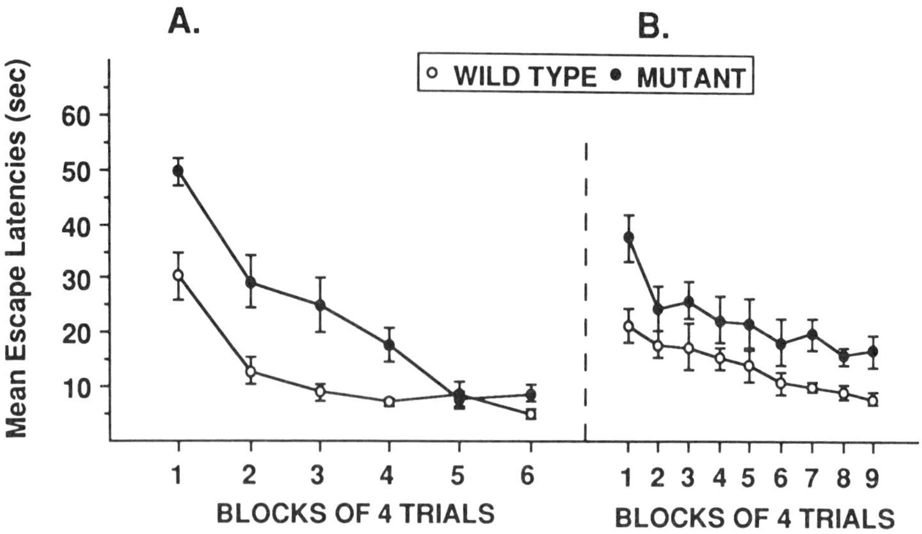

Figure 9. Mean escape latencies for animals in the Morris water task. (*A*) In the first phase of the experiment, wild-type controls ($n = 14$) and α CaMKII mutant mice ($n = 11$) were trained to navigate to a randomly located visible platform. The platform was rendered visible by attaching a small white flag to its top. Each animal was first pretrained to climb onto the platform and given a 15-sec practice swim to ensure that all animals could swim. On each trial, a subject was allowed to search the pool for 60 sec. Once a subject found the platform, it was allowed to remain there for 45 sec. Animals were given 12 trials a day, in blocks of 4 trials, on 2 consecutive days. α CaMKII mutant mice were initially impaired at locating the visible platform but overcame this deficit and learned to locate it just as rapidly as controls. An ANOVA with repeated measures confirms this observation. There was a main effect of Strain ($F[1,23] = 22.778$, $p = .0001$) showing that, overall, wild-type controls were better than the mutant mice. There was also a main effect of Trial Block ($F[5,115] = 43.371$, $p < .00001$) showing that overall, animals' latency to swim to the platform decreased during training. Finally, there was a significant Strain × Trial Block interaction ($F[5,115] = 4.528$, $p = .008$). Post hoc analysis of the interactions (Newman-Keuls, $p < .05$) showed that the wild-type animals had significantly lower escape latencies compared to the mutant mice on trial block numbers 1,2,3, and 4, but were not different on trial blocks 5 and 6. (*B*) All animals were then trained to find a hidden platform located in a fixed location. The top of the platform was 1 cm below the surface of the water. Wild-type and α CaMKII mutant mice were given either 3 or 5 days of training as described above (only 3 days are shown). Wild-type controls had lower escape latencies than the mutants. One wild-type animal floated on all trials, so its data were excluded from the analysis. Thus, 13 wild-type and 11 mutants were used. There was a main effect of Genotype ($F[1,22] = 11.576$, $p = .0024$) showing that, overall, wild-type controls had significantly lower escape latencies than the mutants. The main effect of Trial Block ($F[8,176] = 8.54$, $p < .00001$) was also significant, showing that animals improved during training. The Genotype × Trial Block interaction was not significant. Bars indicate the S.E.M.

tested for 2 consecutive days with three blocks of four trials per day, and the time required to reach the platform was recorded. Figure 9 shows that the α CaMKII mutant mice initially took longer than the wild-type mice to reach the platform, but by the end of training, they were locating it as rapidly as controls. The results show that although α CaMKII mutant mice appear to be initially impaired, they are able to overcome this deficit by training. Therefore, α CaMKII mutant mice (1) are able to learn to associate the flag with the escape platform, (2) are motivated to escape the water, and (3) have the coordinated motor skills needed to swim in water.

The exact nature of the initial impairment is not clear but is likely due to the fact that α CaMKII mutant mice initially responded differently from the wild-type animals when placed on the platform. Unlike wild-type mice, the α CaMKII mutant mice immediately jumped into the water. This "jumping response" occurred a number of times before the first trial, and thus by the time their jumping response habituated, they appeared fatigued. Therefore, they may have taken longer on the first day of "visible-platform" training simply because they were tired. On the second day, however, they did not show the jumping response and, hence, were not fatigued before their trials were given.

Phase 2: Training on the hidden-platform task. Seven to twelve days after the visible-platform task, animals were trained on the Morris hidden-platform task. Animals received the standard 3 days of training used previously for inbred strains of mice (Upchurch and Wehner 1988). As can be seen in Figure 9, wild-type controls quickly learned to locate the hidden platform, and by the end of the third day of training, they were navigating directly to it in less than 10 seconds. The α CaMKII mutant mice took longer than the wild-type mice to locate the hidden platform, but their performance improved during training. However, after performance reached plateau levels, the mutant mice still took approximately 20 seconds to locate the platform, approximately twice as long as the wild-type mice. To ensure that mice were at asymptotic levels of performance, some animals from each group ($n = 5$) were given an additional 2 days of training. Although wild-type animals were significantly better than the mutant mice at locating the platform during extended training, neither group improved significantly compared to day 3 of training. In fact, in all measures of performance during each phase of this study, extended training did not significantly alter the performance of wild-type or α CaMKII mutant mice.

Phase 3: Probe trial. To evaluate whether the mice indeed located the hidden platform by learning the multiple spatial relationships between distal cues and the hidden platform, we subjected them to a probe trial. In this trial, the platform was removed and mice were allowed to search the pool for 1 minute. If an animal located the platform during training by using distal cues, it should selectively search the place where the platform was located during training more than other places in the pool. Animals were given their probe trial immediately following the last training trial (i.e., following 3 or 5 days of training). The results are shown in Figure 10.

The wild-type mice spent a larger percentage of their time in the quadrant where the platform had been during training (training quadrant) than in the other three quadrants. They also crossed the exact spot where the platform had been located during training (training site) more often than equivalent locations in other quadrants. In contrast, α CaMKII mutant mice did not spend more time searching in the training quadrant than in the other quadrants; they spent an equal amount of time in all four quadrants. The mutant mice did not cross the training site more often than equivalent locations in the adjacent quadrants. They did cross the equivalent site in the opposite quadrant less frequently than the training site. This difference, however, may be an artifact of the testing protocol in which the animals always started the probe trial from positions in the opposite quadrant. α CaMKII mutant mice may have learned to swim directly away from the start location before implementing any search strategy. Since they were always started from places opposite to the training site, their "starting strategy" reduced the likelihood of crossing the opposite quadrant site early

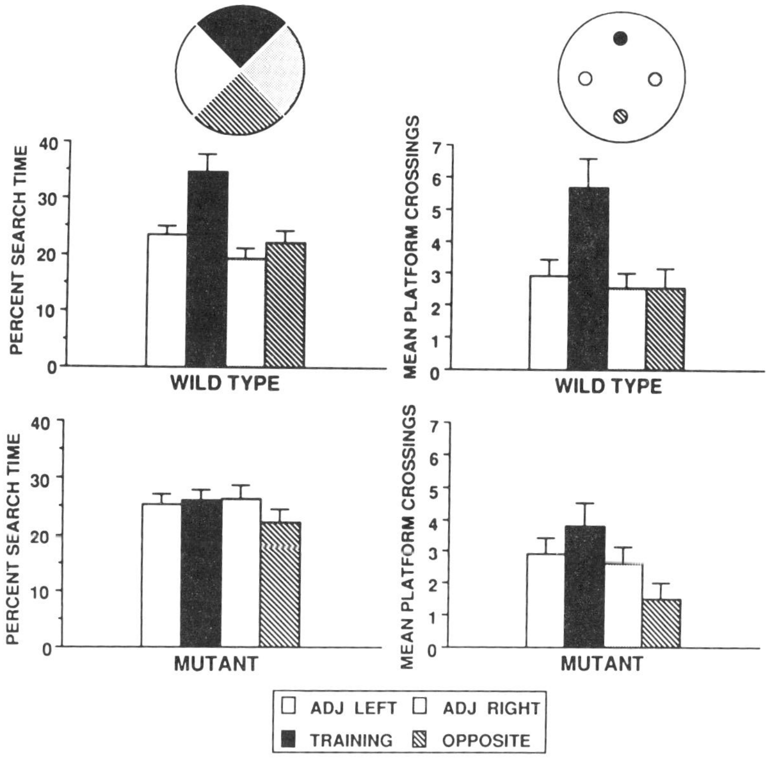

Figure 10. Data from animals given a probe trial after 3 or 5 days of training. Since there were no differences in performance for wild-type or α CaMKII mutant mice, the data were analyzed together. These data show that wild-type animals selectively searched the place where the platform had been located during training, whereas, in general, the α CaMKII mutants' search was not selective. The upper panel represents data from the wild-type animals. An ANOVA showed that wild-type mice spent significantly more time in the quadrant where the platform had been located during training than in the other three quadrants ($F[3,36] = 7.29$, $p = .006$; Newman-Keuls post hoc analysis: Training > Opposite, Adjacent Left, Adjacent Right, $p < .01$). Similarly, an ANOVA showed that wild-type mice crossed the exact site where the platform had been located significantly more often than equivalent locations in the other three quadrants ($F[3,36] = 13.673$, $p < .0001$; Newman-Keuls analysis: Training > Opposite, Adjacent Left, Adjacent Right, $p < .01$). The lower panel shows the data for the α CaMKII mutant mice. In contrast to wild-type animals, α CaMKII mutant mice did not selectively search any quadrant of the pool (ANOVA $F[3,30] = .689$, $p > .05$). The α CaMKII mutant mice also failed to cross the exact place where the platform had been located compared to both adjacent sites. However, they did cross the opposite site less often than the training site. ANOVA $F[3,30] = 3.85$, $p < .05$; Newman-Keuls analysis: Training = Adjacent Left and Adjacent Right, but Training > Opposite, $p < .05$). Above the panels on the right, we show a schematic drawing of the pool with the four virtual quadrants used in the analysis, and on the left we show the four platform sites. Bars indicate the S.E.M.

in the probe. In future studies, the probe trial start location should be randomly chosen.

These data indicate that the wild-type mice learned the spatial relationships between distal cues and the hidden platform (spatial strategy). In addition, these results suggest that the mutant mice improved their performance by developing an alternate strategy that does not depend on specific cues surrounding the pool. For example, mutant mice may have learned the distance between the wall of the pool and platform. Since the pool is circular and the wall is uniform, this strategy would not allow the mutant mice to distinguish among the four quadrants of the pool.

Phase 4: Random-platform task. To confirm that mutant mice are impaired at spatial learning, they were subjected to a "random platform" task. In this task, mice were given trials with the platform in its original location intermixed with trials where the platform was moved to any one of the seven other locations as shown in Figure 11. If the mutant mice were impaired in spatial learning, they should be able to find the platform in the new locations as readily as when it was in the original training site.

Figure 11 also shows that the wild-type mice took significantly less time to locate the platform when it was in its original location (training trials) compared to new locations (random trials). In contrast, the α CaMKII mice took just as long to locate the platform when it was in new locations as when it was in its training site. These results support the hypothesis that the mutant mice developed a "nonspatial" strategy to find the platform. Another observation which confirms the validity of our hypothesis is that on the random trials, wild-type mice crossed the training site significantly more often than the α CaMKII mutant mice (Fig. 12).

In agreement with the data shown in Figure 9B, mutant mice took approximately twice as long as the wild-type mice to locate the platform when it was in its training site. This is most probably a reflection of the strategy employed by the mutant mice, which does not allow as precise a localization of the platform as a spatial strategy does.

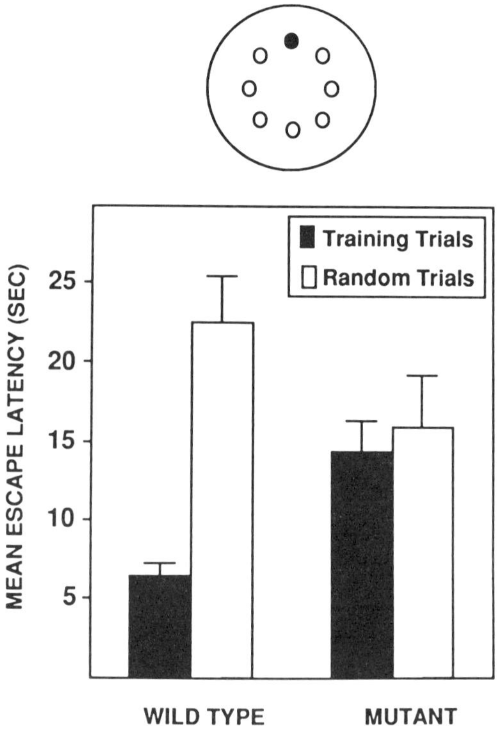

Figure 11. Latency analysis of the "random-platform" task. Two days after the animals were given 3 or 5 days of training and subsequent probe trials, they were given a block of four trials with the platform in its original training location. On the next block of four trials, the platform was in the original location for the first two trials but was in two new ("random") locations on the next two trials. On the final block of four trials, the platform was placed in the original spot on the first trial and then in different random locations on the next three trials. At the top of the panel, we show a schematic drawing of the locations of the platform in the training trials (●) and in the random trials (○). In each random trial, only one of the seven locations was used. Represented in the figure is the mean latency to find the platform on the seven trials when the platform was in its original location (black bars) compared to the five trials when it was in random locations (white bars). The wild-type controls clearly took less time to find the platform when it was in its original location compared to when it was in random locations (correlated T-test; $t[12] = -6.504$, $p < .001$). The α CaMKII mutant mice, however, were just as proficient at locating the platform in random locations as when it was in its original location (correlated T-test $t[10] = -.336$, $p > .05$). One α CaMKII mutant mouse appeared to have had a seizure, and it failed to swim normally on the day of random platform training. Thus, its data were not analyzed. Bars indicate the S.E.M.

Phase 5: " + " (Plus) maze task. The results described thus far suggest that the mutant mice are impaired in learning the spatial relationships between distal cues and the escape platform (i.e., true impaired spatial learning). However, an alternative explanation of the results is that they are impaired in another process(es), such as in the ability to see and attend to distal cues or to make an association between the distal environment and the escape platform. To exclude these latter possibilities, we tested the mice in a water-filled " + " maze (L. Baskall et al., in prep.).

The " + " maze is a four-armed clear Plexiglas maze filled with opaque water. An escape platform is placed in one arm of the maze with its top 1 cm below the surface of the water and is kept there throughout all trials. The + maze was placed directly on top of the Morris pool to ensure that the same distal cues were utilized. On each trial, a mouse was placed in one of the three arms that did not contain the platform and allowed to swim toward the intersection. At this point, the mouse must choose one of the three remaining arms to enter. If the mouse chose the correct arm containing the platform, it was allowed to climb onto the platform and then was removed from the maze and scored as a correct choice. When the animal swam into another arm of the maze that did not contain the platform, it was trapped in that arm and the trial was scored as a mistake. The arm used to start the animal on each trial was chosen pseudorandomly with the restriction that

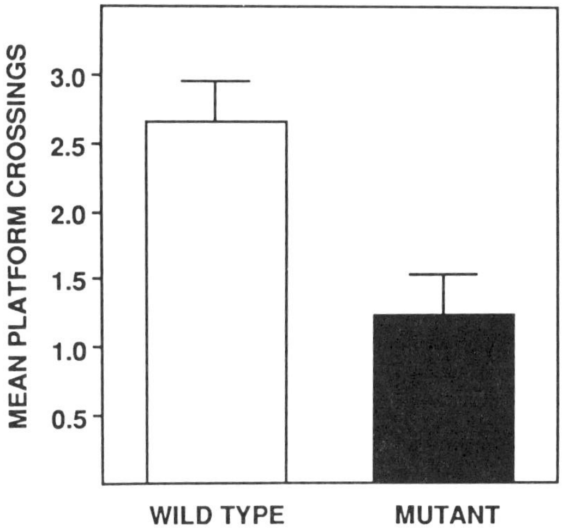

Figure 12. Platform-crossing analysis in the random-platform task. Represented in this figure are the mean number of times mice crossed the original training site on the five trials with the platform in random locations. Wild-type controls crossed the training site significantly more often than α CaMKII mutant mice on these five trials ($t[21] = 3.318$, $p < .01$). Bars indicate the S.E.M.

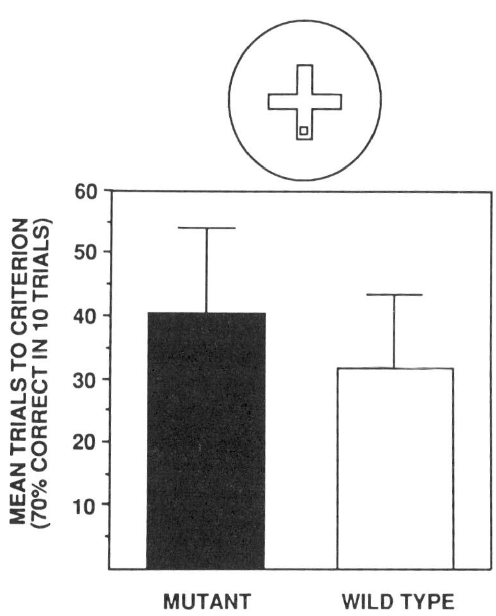

Figure 13. Analysis of performance on the " + " maze. The figure shows the mean number of trials to reach the learning criterion of 70% correct in 10 trials in a water-filled " + " maze. On each trial, an animal was placed facing the center of the maze in one of the three arms that did not contain the platform. An animal was allowed to swim into any of the arms, but if it chose an incorrect arm, it was trapped for 20 sec. If it chose the correct arm, it was allowed to remain on the platform for 10 sec. Each animal was given 15 trials on day 1, and 25 trials on each subsequent day. Each animal was given one trial at a time with an intertrial time of 1–3 min. These animals had been given extended training in the Morris task. The data indicate that both wild-type and α CaMKII mutant mice were able to learn the location of a hidden escape platform in the + maze. The difference between the two groups' performance was not significant ($t[8] = .469$, $p > .05$). A schematic representation of the + maze placed on top of the Morris pool and the location of the hidden platform are shown at the top of the panel. Bars indicate the S.E.M.

each arm which did not contain the platform was used in a block of three trials. Since on each trial there are three possible arms into which an animal can swim, random choices should result in the correct choice 33% of the time. Likewise, other predictable strategies such as always swimming left, right, or straight would also produce correct choices about 33% of the time, because the start position is rotated such that in each ten trials, each of the three start arms are used at least three times. The criterion used to assess whether mice had learned to locate the platform using a distal cue was 70% correct choices in ten trials.

The + maze task has several features in common with the hidden-platform version of the Morris task. Both tasks require subjects to navigate through water and locate a hidden platform by using the extramaze environment. Thus, in both tasks, animals must be able to see and attend to distal cues. The difference between the tasks is that the hidden-platform version of the Morris task requires an animal to learn multiple spatial relationships between distal cues and the escape platform. In the + maze, however, an animal always takes the same path to the platform, thus, it always has the same distal environment to swim toward for escape. Therefore, the maze can be solved by learning the single relationship between a particular distal environment and the escape platform.

Figure 13 shows that both the wild-type mice and the α CaMKII mutant mice learned to solve the task and that there was no significant difference in the number of trials to reach criterion between the two genotypes ($t[8] = 0.469$, $p < 0.05$). These results demonstrate that the impairment of the α CaMKII mutant mice in the Morris hidden-platform task was not due to an inability to see distal cues, attend to the extramaze environment, or learn a simple association between escape and distal environment.

DISCUSSION

In this study, we have constructed mutant mice that are deficient in α CaMKII by the gene targeting technique. The mutant mice exhibit no obvious neuroanatomical defects and retain intact postsynaptic mechanisms, including NMDAR function, but are strikingly deficient in their ability to produce LTP in the CA1 region of the hippocampus. Thus, these mice provided a highly suitable animal model for studying the relationship between LTP and learning processes. Indeed, we have shown that the mutant mice have remarkably specific learning impairments, which leads to the conclusion that α CaMKII has a prominent role in spatial learning, but that it is not essential for some types of nonspatial learning. Our results considerably strengthen the contention that the synaptic changes exhibited in LTP are the basis for spatial memory.

Our observations also extend earlier conclusions based on pharmacological experiments that CaMKII holoenzyme is involved in LTP (Malinow et al. 1989); we have now shown that the α form of this enzyme is involved in LTP. Because LTP was greatly diminished but apparently normal when present in the mutant mice, we favor the notion that α CaMKII is involved in a regulatory pathway for the processes responsible for LTP. In this view, the magnitude of potentiation would be altered in the mutants, but any LTP that appeared would have usual characteristics. We cannot rule out, of course, that α CaMKII is directly in the pathway for the production of LTP and that some other kinase (β CaMKII, for example) is substituting in the instances where mutant mice exhibited LTP.

A deficit in LTP could be caused by a decreased number of connections between CA3 and CA1 neurons in the mutant mice. Although we have no anatomical evidence relating to number of connections, our electrophysiological results make this possibility highly unlikely: (1) Comparable stimulus intensities in normal and mutant mice produced comparable-sized responses; (2) the total depolarization for normal and mutant preparations was similar because the response sizes were the same for both; and (3) for the whole-cell experiments, we did not rely on synaptic depolarization for unblocking NMDA receptors at activated synapses; rather the cells, which were voltage-clamped, were held at membrane potentials of −30 or −40 mV during the tetanus, voltages that produce LTP even for exceedingly small inputs.

The precise role of α CaMKII in LTP induction remains unknown. We hope that our α CaMKII mutant mice will be useful in the identification of this role. To date we have tested whether α CaMKII modulates the activities of protein kinase C, a kinase that has also been implicated in LTP induction (Malinow et al. 1989). We found that normal levels of activity of this kinase are present in homogenates from forebrain of mutant mice, suggesting that the targets of α CaMKII are elsewhere. Synapsin I was shown to be one target for this kinase (McGuiness et al. 1985), and our observation of diminished paired pulse facilitation raises the possibility that the effect might be a presynaptic one. We also tested the hypothesis that α CaMKII has the unique capacity to target the CaMKII holoenzyme to the membrane (Erondu and Kennedy 1985), where it can respond to localized Ca^{++} rises (Connor and Muller 1991; Guthrie et al. 1991). We found that the proportion of the CaMKII holoenzyme activity associated with the membrane is unaltered in the forebrain of the mutant mice, which does not support the hypothesis (A.J. Silva, unpubl.).

The most remarkable feature of the α CaMKII mutant mice is the apparent lack of widespread abnormalities. However, this does not mean that abnormalities are restricted to those described here. At the microscopic level, a high-resolution analysis may very well reveal fine anatomical differences between the mutant mice and the wild-type control. At the electrophysiological level, we have thus far analyzed only the hippocampal CA1 field cells. In light of the known distribution of α CaMKII, cells in the other fields of the hippocampus, as well as in amygdala and neocortex, should be examined for possible deficit. At the behavioral level, we suspect that the modulation of the acoustic startle response (Davis 1992) is impaired in these mice, since they seem to have an abnormally enhanced acoustic startle response. Hence, it is interesting that the amygdala is another prominent site of α CaMKII expression (Erondu and Kennedy 1985; Burgin et al. 1990) and that LTP in this part of the brain might be involved in the modulation of the acoustic startle response (Davis 1992). On the other hand, our behavioral work did show that α CaMKII is not essential for all learning. For instance, although it has previously been shown that α CaMKII is expressed in the basal ganglia (Erondu and Kennedy 1985), and that this structure is essential for learning the visible-platform version of the Morris task (Wishaw and Kolb 1984), our study indicated that the α CaMKII mutant mice can learn this task.

Finally, we have demonstrated that α CaMKII mutant mice are a very useful model for studying the relationship between a particular gene product (α CaMKII), LTP, and behavior. We view our results as a promising start in the application of the gene targeting technique for the study of mammalian learning and memory. We expect that other similarly constructed mice with mutations in judiciously chosen genes will be useful tools for studying mammalian behavior. In this regard, perhaps even more useful would be the mice with subtle rather than null mutations or mice with mutations directed to specific regions of the brain. Construction of these mutant mice may be feasible (Hasty et al. 1991; O'Gorman et al. 1991).

ACKNOWLEDGMENTS

We thank Alan Smith, Yoshi Ichikawa, Mary B. Kennedy, Martin Hooper, Neal Waxman, Paul Kelly, and Allan Collins for advice and/or various invaluable reagents. We also thank James R. Pauly, Gerland Schneider, and Reha Erzurumlu for invaluable help with the neuroanatomy. This research was supported by Howard Hughes Medical Institute (S.T. and C.F.S.), Human Frontier Science Program grant 76834 (S.T.), National Science Foundation grant BNS-882007 (J.M.W.), and National Institutes of Health grant 5R-01NS-12961-17 (C.F.S.). J.M.W. is the recipient of a Research Scientist Development Award (AA-001410).

REFERENCES

Adra, C.N., P.H. Boer, and M.W. McBurney. 1987. Cloning and expression of the mouse pgk-1 gene and the nucleotide sequence of its promoter. *Cell* **60:** 65.

Aghajanian, G.K. and K. Rasmussen. 1989. Intracellular studies in the facial nucleus illustrating a simple new method for obtaining viable motoneurons in adult rat brain slices. *Synapse* **3:** 331.

Bekkers, J.M. and C.F. Stevens. 1990a. Presynaptic mechanism for long-term potentiation in the hippocampus. *Nature* **346:** 724.

———. 1990b. Computational implications of NMDA receptor channels. *Cold Spring Harbor Symp. Quant. Biol.* **50:** 131.

Bliss, T.V.P. and T.J. Lomo. 1973. Long-lasting potentiation of synaptic transmission in the dentate area of the anaesthetized rabbit following stimulation of the perforant path. *J. Physiol.* **232:** 331.

Bliss, T.V.P. and A.R. Gardner Medwin. 1973. Long-lasting potentiation of synaptic transmission in the dentate area of the unanaesthetized rabbit following stimulation of the perforant path. *J. Physiol.* **232:** 357.

Burgin, K.E., M.N. Waxham, S. Rickling, S.A. Westgate, W.C. Mobley, and P.T. Kelly 1990. In situ hybridization histochemistry of Ca^{2+}/calmodulin-dependent protein kinase in developing rat brain. *J. Neurosci.* **10:** 1788.

Capecchi, M.R. 1989. Altering the genome by homologous recombination. *Science* **244:** 1288.

Castro, C.A., L.H. Silbert, B.L. McNaughton, and C.A. Barnes. 1989. Recovery of spatial learning deficits after decay of electrically induced synaptic enhancement in the hippocampus. *Nature* **342:** 545.

Chen, L. and L.Y.M. Huang. 1990. Protein kinase C reduces Mg^{2+} block of NMDA-receptor channels as a mechanism of modulation. *Nature* **356:** 521.

Colbran, R.J., C.M. Schworer, Y. Hashimoti, Y.L. Fong, D.P. Rich, M.K. Smith, and T.R. Soderling. 1989. Review article: Calcium/calmodulin-dependent protein kinase II. *Biochem. J.* **258:** 313.

Collingridge, G.L. and W. Singer. 1990. Excitatory amino acid receptors and synaptic plasticity. *Trends Pharmacol. Sci.* **11:** 290.

Connor, J.A. and W. Muller. 1991. Dendritic spines as individual neuronal compartments for synaptic Ca^{2+} responses. *Nature* **354:** 74.

Davis, M. 1992. The role of the amygdala in fear-potentiated startle: Implications for animal models of anxiety. *Trends Pharmacol. Sci.* **13:** 35.

Davis, S., S.P. Butcher, and R.G.M. Morris. 1992. The NMDA receptor antagonist D-2-amino-5-phosphonopentanoate (D2AP5) impairs spatial learning and LTP in vivo at intracerebral concentrations comparable to those that block LTP in vitro. *J. Neurosci.* **12:** 21.

Douglas, R.J. 1967. The hippocampus and behavior. *Psychol. Bull.* **67:** 416.

Erondu, N.E. and M.B. Kennedy. 1985. Regional distribution of type II Ca^{2+}/calmodulin-dependent protein kinase in rat brain. *J. Neurosci.* **5:** 3270.

Guthrie, P.B., M. Segal, and S.B. Kater. 1991. Independent regulation of calcium revealed by imaging dendritic spines. *Nature* **354:** 76.

Hanson, P.I., M.S. Kapiloff, L.L. Lou, M.G. Rosenfeld, and H. Schulman. 1989. Expression of a multifunctional Ca^{2+}/calmodulin-dependent protein kinase and mutational analysis of its autoregulation. *Neuron* **3:** 59.

Hasty, P., R. Ramirez-Solis, R. Krumlauf, and A. Bradley. 1991. Introduction of a subtle mutation into the Hox-2.6 locus in embryonic stem cells. *Nature* **350:** 243.

Jahr, C.E. and C.F. Stevens. 1990a. A quantitative description of NMDA receptor-channel kinetic behavior. *J. Neurosci.* **10:** 1830.

———. 1990b. Voltage dependence of NMDA- activated macroscopic conductances predicted by single-channel kinetics. *J. Neurosci.* **10:** 3178.

Keith, J.R. and J.W. Rudy. 1990. Why NMDA-receptor-dependent long-term potentiation may not be a mechanism of learning and memory: Reappraisal of the NMDA-receptor blockade strategy. *Psychobiology* **18:** 251.

Kennedy, M.B., M. Bennett, and N. Erondu. 1983. Biochemical and immunochemical evidence that the "major" postsynaptic density protein is a subunit of a calmodulin-dependent protein kinase. *Proc. Natl. Acad. Sci.* **80:** 7357.

Lin, J.W., M. Sugimori, R.R. Llinas, T.L. McGuinness, and P. Greengard. 1990. Effects of synapsin I and calcium/calmodulin-dependent protein kinase II on spontaneous neurotransmitter release in the squid giant synapse. *Proc. Natl. Acad. Sci.* **87:** 8257.

Lisman, J.E. and M.A. Goldring. 1988. Feasibility of long-term storage of graded information by the Ca/calmodulin-dependent protein kinase molecules of the post-synaptic density. *Proc. Natl. Acad. Sci.* **85:** 5320.

Malenka, R.C., J.A. Kauer, D.J. Perkel, M.D. Mauk, P.T. Kelly, R.A. Nicoll, and M.N. Waxham. 1989. An essential role for postsynaptic calmodulin and protein kinase activity in long-term potentiation. *Nature* **340:** 554.

Malinow, R. and R.W. Tsien. 1990. Letters. Presynaptic enhancement shown by whole-cell recordings of long-term potentiation in hippocampal. *Nature* **346:** 177.

Malinow, R., H. Schulman, and R.W. Tsien. 1989. Inhibition of postsynaptic PKC or CaMKII blocks induction but not expression of LTP. *Science* **245:** 862.

McGuiness, T., Y. Lai, and P. Greengard. 1985. Ca^{2+}/calmodulin-dependent protein kinase II isozymic forms from rat forebrain and cerebellum. *J. Biol. Chem.* **260:** 1969.

McNaughton, B.L., R.M. Douglas, and G.V. Goddard. 1978. Synaptic enhancement in fascia dentata: Cooperativity among coactive afferents. *J. Physiol.* **157:** 277.

Miller, S.G. and M.B. Kennedy. 1986. Regulation of brain type II Ca^{2+}/calmodulin-dependent protein kinase by autophosphorylation: A Ca^{2+}-triggered molecular switch. *Cell* **44:** 861.

Miller, S.G., B.L. Patton, and M.B. Kennedy. 1988. Sequences of autophosphorylation sites in neuronal type II CaM kinase that control Ca^{2+}-independent activity. *Neuron* **1:** 593.

Morris, R.G.M. 1981. Spatial localisation does not depend upon the presence of local cues. *Learn. Motiv.* **12:** 239.

Morris, R.G.M., E. Andersen, G.S. Lynch, and M. Baudry. 1986. Selective impairment of learning and blockade of long-term potentiation by an *N*-methyl-D-aspartate receptor antagonist, AP5. *Nature* **319:** 774.

O'Gorman, S., D.T. Fox, and G.M. Wahl. 1991. Recombinase-mediated gene activation and site-specific integration in mammalian cells. *Science* **251:** 1351.

Schulman, H. and L.L. Lou. 1989. Multifunctional Ca^{2+}/calmodulin-dependent protein kinase: Domain structure and regulation. *Trends Biochem. Sci.* **14:** 62.

Schwartzkroin, P.A. and K. Wester. 1975. Long-lasting facilitation of synaptic potential following tetanization in the in vitro hippocampal slice. *Brain Res.* **89:** 107.

Staubli Thibault, U., M. DiLorenzo, and G. Lynch. 1989. Antagonism of NMDA receptors impairs acquisition but not retention of olfactory memory. *Behav. Neurosci.* **103:** 54.

Sutherland, R.J., H.C. Dringenberg, J.M. Hoesing, and R.W. Skelton. 1991. Is LTE in the hippocampus necessary for place learning? *Soc. Neurosci. Abstr.* **17:** 483.

Thompson, S., A.R. Clarke, A.M. Pow, M.L. Hooper, and D.W. Melton. 1989. Germ line transmission and expression of a corrected *HPRT* gene produced by gene targeting in embryonic stem cells. *Cell* **56:** 313.

Upchurch, M. and J.M. Wehner. 1988. DBA/wlbg mice are incapable of chronologically-based learning in the Morris water task. *Pharmacol. Biochem. Behav.* **29:** 325.

Whishaw, I.Q. and B. Kolb. 1984. Decortication abolishes place but not cue learning in rats. *Behav. Brain Res.* **11:** 123.

Structure and Interactions of CD4

S.C. Harrison,*† J. Wang,* Y. Yan,* T. Garrett,*† J. Liu,*
U. Moebius,‡ and E. Reinherz‡§

*Harvard University Department of Biochemistry and Molecular Biology and †Howard Hughes Medical Institute, Cambridge, Massachusetts 02138; §Harvard Medical School Department of Medicine and ‡Dana Farber Cancer Institute, Boston, Massachusetts 02115

The CD4 glycoprotein is found on most thymocytes (Reinherz et al. 1980) and on a subset of T lymphocytes (Reinherz et al. 1979; Biddison et al. 1982; Krensky et al. 1982; Meuer et al. 1982). Contacts between CD4 and class II major histocompatibility complex (MHC) molecules transmit signals for thymocyte differentiation (Kruisbeek et al. 1985) and for T-cell activation (Eichmann et al. 1987). CD4 is also the receptor for human immunodeficiency virus (HIV) (Dalgleish et al. 1984; Klatzmann et al. 1984; Maddon et al. 1986).

The extracellular portion of CD4 (residues 1–375) contains four immunoglobulin (Ig)-like domains (Maddon et al. 1985). When expressed as a secreted protein in mammalian cell culture, CD4 (1–375) is a rod-like, monomeric molecule (Kwong et al. 1990; T. Kirchhausen et al., in prep.). Crystals of this fragment suggest a tendency to oligomerize under crystallization conditions (Kwong et al. 1990; J. Wang and S.C. Harrison, unpubl.), but it is not yet known whether the clusters found in these crystals bear any relationship to lateral contacts that might occur in the cell surface. Such contacts could in principle be part of a signaling process, because the cytoplasmic tail of CD4 binds noncovalently to a tyrosine kinase known as $p56^{lck}$ (Rudd et al. 1988; Veillette et al. 1988). The crystals of CD4 (1–375) are not very well ordered; those of a smaller two-domain fragment, CD4 (1–182), are far better ordered. We have determined the structure of this fragment (Wang et al. 1990); another group has done so independently (Ryu et al. 1990). We review the properties of CD4 (1–182) here, and we discuss experiments aimed at defining the parts of its surface that form specific contacts with class II molecules and with the HIV envelope glycoprotein.

METHODS

Crystals and structures. Two-domain fragments of human CD4 were produced in Chinese hamster ovary (CHO) cells (CD4 [1–182]) and in *Escherichia coli* (CD4 [1–183]). The bacterially expressed material was refolded as described previously (Garlick et al. 1990). Crystallization by hanging-drop vapor diffusion, using polyethylene glycol (PEG) as precipitant, yielded three crystal forms denoted A, B, and C. Structures of CD4 (1–182) and CD4 (1–183) were determined and refined using the C and A forms, respectively (Wang et al. 1990; T. Garrett et al.; Y. Yan et al.; both in prep.). The molecular packing in the B form has also been determined (J. Wang, unpubl.).

gp120-binding studies. CD4 and various mutant forms were expressed on the surface of stably transfected murine T-cell hybridoma MA5.8 cells. Binding of gp120 was determined at 37°C using indirect immunofluorescence (Moebius et al. 1992a).

Class II MHC-binding studies. CD4 and various mutant forms were expressed transiently in COS-1 cells; rosette formation with Raji cells, which express class II proteins, was determined 48 hours posttransfection (Clayton et al. 1989; Moebius et al. 1992b).

RESULTS AND DISCUSSION

Structure of CD4 Domains 1 and 2

The CD4 (1–182) fragment contains two abutting, immunoglobulin-like domains (Fig. 1) (Ryu et al. 1990; Wang et al. 1990). Domain 1 is very similar to a Vκ module. It has nine β strands, forming two β sheets. One sheet contains strands AGFCC′C″, and the other sheet contains strands BED. The residues at the heart of the domain — a pair of disulfide-bridged cysteines, a tryptophan, and various key hydrophobics — are exactly those characteristic of known immunoglobulin structures (Lesk and Chothia 1982). Some specific hallmarks of light-chain variable domains (Colman 1988; Hunkapiller and Hood 1989) are also present: a glycine in the AB loop; an Arg-Asp salt bridge linking the beginning of strand D with the EF loop; and a tyrosine two residues before the second cysteine, hydrogen-bonded to the carbonyl of the aspartic acid four residues before it (Fig. 2). Domain 1 differs from an IgV domain in two important ways: It lacks the features responsible for dimerization of variable modules in an immunoglobulin molecule (Chothia et al. 1985; Colman 1988), and it contains elaborations of the C′C″ and DE loops. The absence of dimerization structures is consistent with the observations summarized above in that any oligomeric association on the cell surface is probably mediated by membrane-proximal domains. The elaborations of the C′C″ and DE loops probably have a role in the specific function of CD4: its interaction with class II MHC molecules (see below). These elaborations can be described as follows. (1) The C′C″ loop, which corresponds to the CDR2 of light-chain

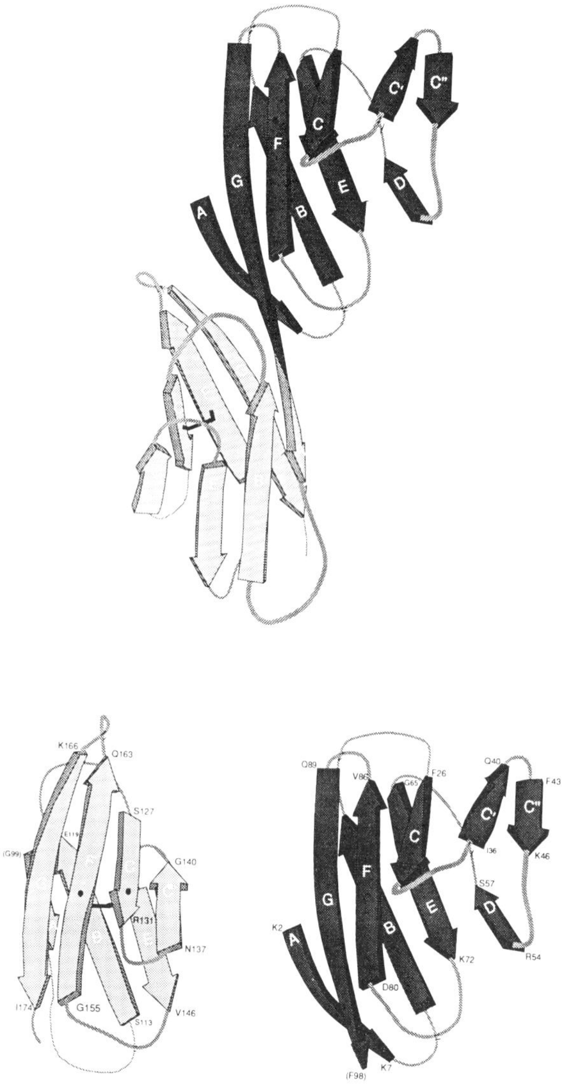

Figure 1. (*Top*) Richardson ribbon diagram, showing folded structure of the polypeptide chain in the first two domains of CD4. Letters label β strands. Domain 1 is darkly hatched; domain 2 is lightly hatched. In this, as in all other figures, the membrane-proximal part of the molecule is at the bottom of the illustration. (*Bottom*) Individual domains shown in corresponding orientations. The first and last residues of each β strand are indicated by single-letter amino acid code and sequence numbers.

variable domains, contains more residues than are usually present in the IgV structures. The prominently exposed side chain of Phe-43, just at the end of the C′C″ β turn, is a particularly striking feature of human CD4. (2) The D strand terminates "early" (at Ser-57) in order to accommodate a DE insertion that includes two irregular helical turns (residues 58–64). (3) The side chain of Trp-62, in the middle of this insertion, lies beneath the C′C″ loop, causing it to project away from the body of the domain. This tryptophanyl side chain effectively extends the hydrophobic core of the domain in order to support the protruding C′C″ corner. Because of its salience, we call this edge of the domain the "C′C″ ridge," but we also emphasize that the DE insertion is essential to its conformational prominence.

Domain 2 can be described as a truncated variable domain with a number of idiosyncratic features. There are seven strands: GFCC′ in one sheet and ABE in the other. The C″ and D strands are absent, and the C′ strand makes a "shortcut" from C to E. The conserved disulfide linking strands B and F across the barrel is absent; in its stead is a disulfide linking strands C and F within a sheet. The cysteine in the C strand occupies the location in sequence normally held by a tryptophan; the B-strand position that is usually cysteine is a leucine in domain 2; the F-strand cysteine is conserved. The absence of an intersheet disulfide appears to release some packing constraints. In comparison with the IgV domain and with domain 1, the two sheets of domain 2 are displaced relative to each other by nearly 5 Å in a direction roughly normal to the strand direction, and one sheet "rocks" with respect to the other about this axis by about 15°.

The contact between the domains is a lap joint formed by opposite ends of the two IgV-like structures. The last β strand of domain 1 continues uninterrupted to form the first strand of domain 2. A six-residue hydrophobic sequence exists at the boundary, with side chains alternately pointing into the cores of domains 1 and 2. The β-strand connector emerges from domain 1 between the AB and EF loops, and it enters domain 2 between the BC and FG loops. These loops in domain 2 splay outward, presenting their inside faces to the bottom part of domain 1. The interface thus generated is hydrophobic in character, with a total buried surface of about 550 Å^2 per domain. It therefore appears to be quite stable, and large-scale hinge motion is unlikely.

Organization of the Four-domain Extracellular Segment

The characteristic features of the domain 1/domain 2 interface appear to be repeated at the domain 2/domain 3 interface (T. Garrett et al., in prep). In particular, the AB and EF loops of domain 2 superimpose well on the homologous loops of domain 1, and Leu-109 and Leu-151, which face solvent in the two-domain fragment, correspond to hydrophobic residues in domain 1 that lie in the interface. The G strand emerges in a similar way from both domains, and residues 174–179 mimic residues 95–100 in their continuous hydrophobic character. We therefore suggest that domains 2 and 3 are connected by a hydrophobic overlap like the one at the interface between domains 1 and 2. We also suggest that a similar contact occurs between domains 3 and 4. A six-residue hydrophobic sequence is present at the domain boundary, and the AB and FG loops of domain 3, which can be assigned with confidence by working forward and backward from the linker sequences, also exhibit the key sequence characteristics of these loops in domains 1 and 2. The rod-like molecule resulting from such joints would have a somewhat zigzag or notched appearance, as shown in Figure

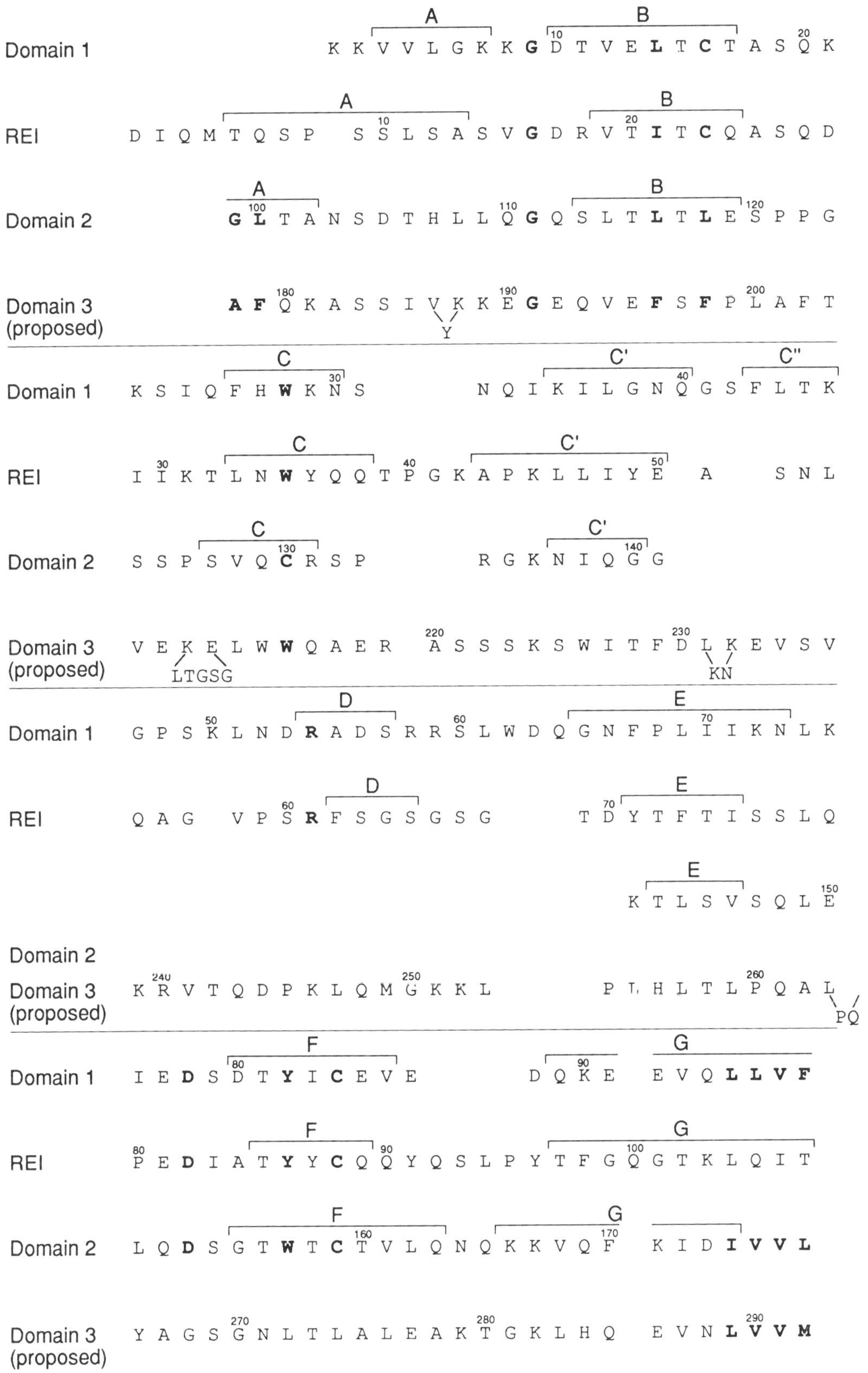

Figure 2. Aligned sequences of CD4 domains 1, 2, and 3, and the IgVκ domain of Bence-Jones protein REI (Epp et al. 1974). The alignments of CD4 domains 1 and 2 with each other and with REI are based on structural superpositions (Wang et al. 1990); the alignment of CD4 domain 3 is based on a proposed IgV organization, working forward and back from the hydrophobic interdomain linker sequences as discussed in the text and in T. Garrett et al. (in prep.). Key residues discussed in the text are shown in bold characters.

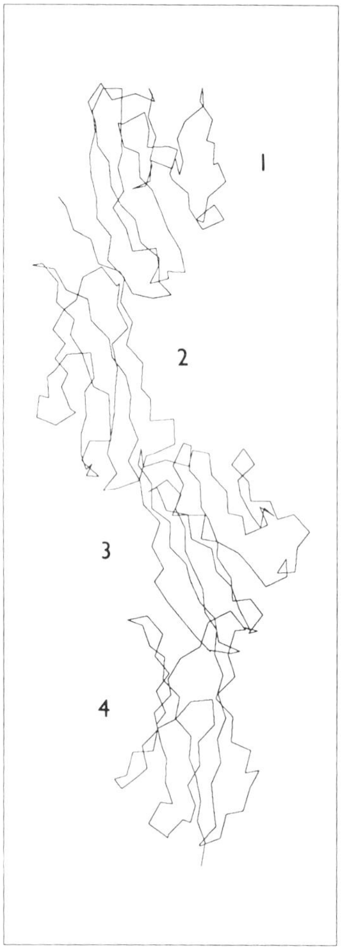

Figure 3. Proposed organization of CD4, domains 1–4, based on the assumption that the linkages between domains 2 and 3 and 3 and 4 are similar to the linkage between domains 1 and 2. Domain 3 has been represented by domain 1, and domain 4 is represented by domain 2. Because the actual number of residues in domains 3 and 4 does not correspond precisely to the numbers in domains 1 and 2, respectively, the representation is only schematic.

3, where domains 3 and 4 have been represented by appropriately rotated and translated copies of domains 1 and 2, respectively. It would be about 130 Å long, in good agreement with estimates from electron microscopy (T. Kirchhausen et al., in prep) and hydrodynamic parameters (Kwong et al. 1990).

Binding to HIV gp120

The interaction of CD4 with the HIV envelope glycoprotein has been mapped to the neighborhood of the C′C″ ridge by a number of laboratories, using various mutagenesis schemes (Clayton et al. 1988; Peterson and Seed 1988; Arthos et al. 1989; Sattentau et al. 1989; Ashkenazi et al. 1990; Brodsky et al. 1990). In most cases, altered CD4 is expressed on the cell surface, and binding of gp120 to the expressing cells is determined. If a true dissociation constant is to be measured, the determination must be made under equilibrium conditions. The dissociation rate of the CD4-gp120 complex is slow, however, and only when binding is studied above 30°C can equilibrium be established in a reasonable experimental interval. Because most published measurements are made at lower temperatures—in many cases, to avoid endocytic uptake of CD4—they yield, in effect, a relative associative rate constant (k_{on}), rather than a true equilibrium dissociation constant (K_D). To overcome this problem, we have used stably transfected murine T-cell lines, which do not down-regulate surface CD4 at 37°C, even in the presence of anti-CD4 monoclonal antibodies or gp120 (Moebius et al. 1992a). The escape from endocytic uptake is probably due to the presence of $p56^{lck}$ in these cells. The formation of a CD4-$p56^{lck}$ complex, mediated by the cytoplasmic tail of CD4, appears to inhibit entry of CD4 into coated vesicles, perhaps because the complex is anchored to cytoskeletal elements by the SH3 domain of $p56^{lck}$ (Pelchen-Matthews et al. 1991).

The data in Table 1 include on and off rates as well as K_D values for binding of gp120 to CD4 variants expressed on the surface of MA5.8 cells (Moebius et al. 1992a). The site-directed mutations were chosen to pose specific questions about the specificity and extent of the contact, given previous results implicating residues in the vicinity of the CDR2-like region. The key results can be summarized as follows: (1) The exposed Phe-43 is a critical determinant. Of the various hydrophobic substituents examined, only leucine produces

Table 1. gp120 Binding Properties of Selected CD4 Variants

Variant[a]	k_{on} $min^{-1}\mu M^{-1}$	k_{off} $(min^{-1} \times 10^{-3})$[b]	K_D (nM)[c]
Wild type	6.1 ± 2.4	12.5 ± 1.2	0.9
S19Y	5.9	13.1	1.2
H27A	5.5	13.3	1.6
K35A	1.8	173.0	45.0
G38A	3.2	28.0	9.4
Q40A	17.0	5.8	0.5
F43A	n.d.	n.d.	500.0
F43I	n.d.	n.d.	500.0
F43L	5.9	21.6	2.5
F43W	3.9	77.0	16.0
F43Y	1.7	53.0	32.0
T45A	7.7	9.6	0.6
K46A	2.0	33.0	19.0
P48Q	4.5	35.0	4.5
S49V	3.6	46.0	13.0
N52A	5.7	10.2	0.8
A55V	5.0	18.0	2.4
D56A	4.3	12.2	1.6
R59A	2.4	27.0	7.9
R59K	7.1	9.5	0.6
W62Y	n.d.	n.d.	152.0
E85Q	3.6	17.0	1.5
K29A/E85Q	3.7	50.0	6.7
Q89L	7.3	12.8	1.2

Data from Moebius et al. (1992a).

[a]Variants are indicated by residue number and single-letter code.

[b]k_{on}, k_{off} indicate kinetic rate constants for association and dissociation. n.d. indicates not determined.

[c]K_D indicates dissociation constant, measured at equilibrium (note that K_D is approximately equal to k_{off}/k_{on} in all cases).

less than a tenfold decrease in affinity, and the isoleucine mutant is essentially inactive. Phe-43 probably fits into a complementary hydrophobic pocket that cannot accommodate a β-branched residue. (2) Phe-43 is partly surrounded on the molecular surface by positively charged side chains (Lys-29, Lys-35, Lys-46, Arg-59), all of which influence gp120 binding. Acidic residues on gp120 may be required to neutralize this concentration of positive charge. It is interesting to note that the significant effect of changing Lys-35 to Ala-35 would not have been detected by measuring only relative k_{on}, because the main effect of eliminating the lysyl side chain is to accelerate dissociation. The result is in accord with what might be expected from a salt bridge at the rim of an extended site. The rate of association, probably largely diffusion-controlled, might be little affected by removal of the charged side chain, but the barrier to dissociation might be substantially lowered. (3) The conformation of the C′C″ turn, critical for positioning Phe-43, depends on the large Trp-62 side chain for "support." Tyr-62 mutant binds only weakly, showing that rearrangements in this region are deleterious. The extent of a local conformational change produced by the tyrosine substitution is hard to predict. The indole nitrogen of Trp-62 donates a hydrogen bond to the buried hydroxyl of Ser-57; Tyr-62 can do the same but with a perturbed geometry. A substantial rearrangement of the irregular helix in the DE loop might therefore accompany the mutation. (4) The "lower" part of the C′C″ ridge is involved in the contact but probably less intimately than the "upper" part. Changing Ser-49 to Val-49 is expected to perturb the C″D corner because the serine side chain points inward and hydrogen bonds to a buried water molecule. This mutation increases K_D about tenfold.

A model for the interface with gp120 consistent with these results involves most of the C′C″ ridge and a contact surface that extends as far as Lys-35 on one side and Arg-59 on the other (see Figs. 1 and 4). Specificity for the side chain of Phe-43 is quite marked; in contrast, the interaction around Pro-48 and Ser-49 may be less demanding. The total buried surface area at the interface is probably between 600 and 900 Å, slightly larger than an antibody-antigen interaction and comparable to other "strong" protein-protein contacts.

Comparison of structures from two crystal forms of two-domain CD4 shows that the gp120-binding region exhibits a somewhat unexpected flexibility (Y. Yan et al., in prep.). Different packing interactions in the two crystals produce somewhat different conformations for the C′C″ turn and the C″ strand as shown in Figure 4. There is a modest rearrangement of the peptide backbone in the C′C″ turn, leading to a rather more noticeable reorientation of Phe-43, and there is a shift of about 3 Å in the position of Pro-48. Crystal packing forces do not generally perturb a folded protein to this extent, and we conclude that significant "induced fit" will be possible in the CD4-binding site for gp120.

Even more extensive structural changes in the HIV envelope glycoprotein (gp120/gp41) appear to accom-

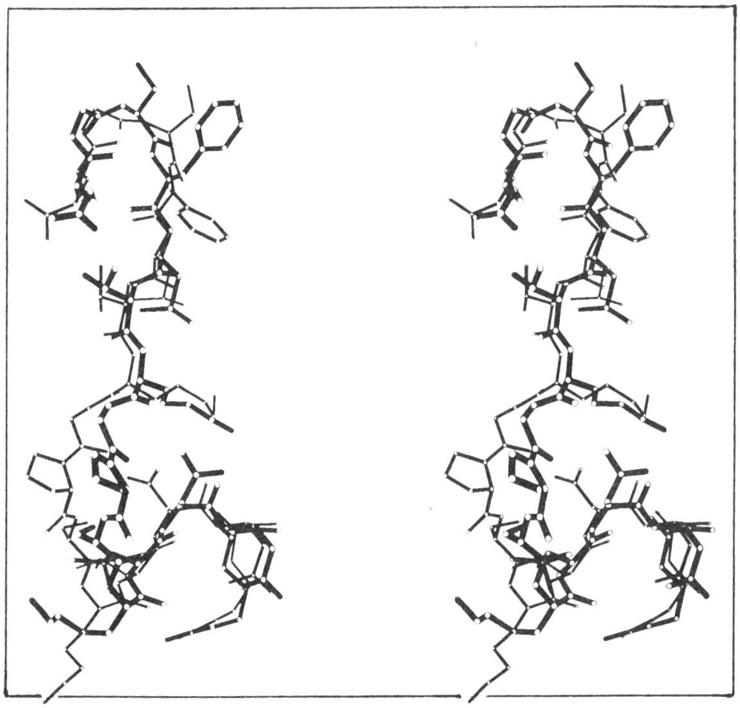

Figure 4. Comparison of the C′C″ ridge in the refined structures of the A (light lines) and C (heavy lines) crystal forms of two-domain CD4 (Y. Yan et al., in prep). Residues from Gln-40 to Arg-54 are shown in stereo. The flexibility of the polypeptide chain in this region allows crystal packing forces to dictate rather different detailed conformations. The view corresponds to looking from the right at the C′C″ ridge in Fig. 1. Phe-43 is at the top, and the C″D corner (with Pro-48 in the foreground) is at the bottom.

pany binding. Early experiments with laboratory strains of HIV showed that soluble CD4 can bind as tightly to virions as to free gp120 and that it can neutralize infectivity in cell culture (Deen et al. 1988; Fisher et al. 1988; Hussey et al. 1988; Traunecker et al. 1988). Subsequent studies with fresh patient isolates have demonstrated, however, that binding to virus not adapted to growth in transformed T cells is significantly weaker than binding to free gp120, released from virions or expressed as a recombinant molecule, derived from the same viral stocks (Daar et al. 1990; Ashkenazi et al. 1991; Brighty et al. 1991; Moore et al. 1992; Turner et al. 1992). Passage in culture of a CD4-resistant fresh isolate can convert it to CD4 sensitivity: The accompanying minor sequence changes in gp120 are not correlated from experiment to experiment, and the CD4 affinity of the free gp120 does not change (Turner et al. 1992). Related experiments also show that binding of CD4 to virions induces dissociation of gp120 ("shedding") (Moore et al. 1990, 1991; Hart et al. 1991). Patient isolates resistant to neutralization by soluble CD4 are relatively refractory to CD4-induced shedding (Moore et al. 1992). Dissociation of gp120 from the virion may therefore be important for neutralization of the virus. The HIV envelope glycoprotein, gp120/gp41, is an oligomeric (probably tetrameric) complex (Pinter et al. 1989; Schawaller et al. 1989; Doms et al. 1990). Binding of soluble CD4 induces a conformational change, as monitored by antibody reactivity and protease sensitivity (Sattentau and Moore 1991). It is reasonable to propose a model to account for all these observations, on the basis of the notion that CD4 binding induces both tertiary and quaternary structural changes in gp120/gp41 and that the interactions between gp120 and gp41 are weaker in the CD4-bound conformation (Fig. 5). The CD4 affinity is necessarily lower for the unbound gp120/gp41 conformation (this is a thermodynamic consequence of the assumption that CD4 binding triggers the change; compare T-state and R-state hemoglobin). Differences between patient isolates and laboratory-adapted strains could be differences in the number of bound CD4 ligands needed to induce the transition, in the kinetics of the transition, or both.

Binding to MHC Class II Proteins

An individual interaction of CD4 with a class II protein is relatively weak, and it has been difficult to detect except by following adhesion between cells expressing CD4 and cells expressing class II molecules (Doyle and Strominger 1987; Clayton et al. 1989). Results from different assay systems yield somewhat different conclusions concerning the effects of particular mutations on the interaction (Fleury et al. 1991; Moebius et al. 1992b). It is clear, however, that the class II interface is much more extended than the gp120 interface; it includes a larger surface of domain 1 as well as a significant contact in domain 2. The data summarized in Table 2 show that changing residues lying on opposite edges of domain 1 can reduce the apparent affinity for class II proteins and that there is a distinct overlap with the gp120-binding site. In particular, the prominent C′C″ turn and the DE insert appear to play a role. These are features of the domain that depart from the usual IgV structure, and it is plausible to suggest that they evolved in connection with specific functions of CD4. Class II proteins can be assumed to be very similar in structure to class I MHC molecules (Brown et al. 1988). Examination of the class I model shows a very deeply "notched" surface. The α_3 and β_2-microglobulin domains (which correspond to the β_2 and α_2 domains of class II chains) are asymmetrically placed with respect to the peptide-presenting " platform," creating a substantial hollow. The CD loop in the β_2 domain, implicated by recent mutagenesis experiments as part of the CD4 contact (Cammarota et al. 1992; König et al. 1992), lies within this hollow. The extensive contact surface on CD4 implies, however, that other regions (and probably other domains) on the class II molecule will also be involved. We propose that the zigzagged surface created by the offset of domains 1 and 2 in CD4 is complementary to the notched surface of class II proteins and that the contact is rather like a

(1)		(2)		(3)		(4)
$[gp120/gp41]_4$	$\xrightarrow{CD4}$	$[gp120/gp41]_4\ (CD4)_x$	—>	$\{gp120{:}gp41\}_4\ (CD4)_x$	—>	x gp120 · CD4 + (4 - x) gp120 + $\{gp41\}_4$
unbound state of env tetramer		binding of one or more CD4 (x≤4)		conformational change in env tetramer		dissociation of gp120 (with bound CD4 if occupied)
low affinity for CD4				high affinity for CD4		

Figure 5. A scheme for events accompanying the binding of CD4 to the HIV oligomeric envelope glycoprotein (here assumed to be a tetramer). State *3* might be the fusion-active configuration in actual viral infection (or syncytium formation), and state *4* might actually be a side reaction. Triggering of dissociation, state *4*, when virus is not bound to a cell surface is believed to be a key mechanism for neutralization of laboratory strains by soluble CD4. For simplicity, it is assumed that the proposed cooperative conformational change in the tetramer (from state *2* to *3*), which is triggered by binding of *x* CD4 ligands, leads to dissociation in state *4* of all gp120, whether liganded or not. The dotted line between gp120 and gp41 in the high-affinity state of the tetramer (braces) is meant to indicate weakened interaction between the two chains of the envelope glycoprotein.

Table 2. Class II MHC-binding Properties of Selected CD4 Variants

Variant[a]	Rosette formation[b]
Wild type	++
S19Y	+
H27A	++
K35A	+/−
G38A	++
Q40A	++
F43A	−
F43I	−
F43L	+
F43Y	+ I
F43W	++
T45A	++
K46A	−
P48Q	++
S49V	−
N52A	++
R59A	+/−
R59K	++
W62Y	−
Q89L	++
T17E/A18S	+
H27T/N30F/N32D/I34R	+/−
G99K, S104P, H107S	−

Data from Moebius et al. (1992b).
[a]Variants are indicated by residue number and single-letter code.
[b]Rosette formation with Raji B cells to COS-1 cells expressing CD4 scored as ++ (identical to wild-type CD4), + (modest reduction, over 50% wild-type level), +/− (substantial reduction, less than 50% wild-type level), and − (no detectable rosette formation).

handshake, where two concave surfaces partly surround each other.

ACKNOWLEDGMENTS

We thank D.C. Wiley, R. Fisher, J. Rosa, and L. Burkly for discussions. The work was supported by grants AI-30361 (to S.C.H.), GM-39589 (to Don C. Wiley), AI-27336 (to E.L.R.), and P30-AI-28691 (to U.M.); U.M. acknowledges an AIDS Stipendium from the Bundesministerium für Forschung und Technologie.

REFERENCES

Arthos, J., K.C. Deen, M.A. Chaikin, J.A. Fornwald, G. Sathe, Q.J. Sattentau, P.R. Clapham, R.A. Weiss, J.S. McDougal, C. Pietropaola, R. Axel, A. Truneh, P.J. Maddon, and R.W. Sweet. 1989. Identification of the residues in human CD4 critical for the binding of HIV. *Cell* **57:** 469.

Ashkenazi, A.L., L.G. Presta, S.A. Marsters, T.R. Camarato, K.A. Rosenthal, B.M. Fendley, and D.J. Capon. 1990. Mapping the CD4 binding site for human immunodeficiency virus by alanine-scanning mutagenesis. *Proc. Natl. Acad. Sci.* **87:** 7150.

Ashkenazi, A., D.H. Smith, S.A. Marsters, L. Riddle, T.J. Gregory, D.D. Ho, and D.J. Capon. 1991. Resistance of primary isolates of human immunodeficiency virus type 1 to soluble CD4 is independent of CD4-rgp120 binding affinity. *Proc. Natl. Acad. Sci.* **88:** 7056.

Biddison, W.E., P. Rao, M. Thalle, G. Goldstein, and S. Shaw. 1982. Possible involvement of the OKT4 molecule in T cell recognition of class II HLA antigens: Evidence from studies of cytotoxic T lymphocytes specific for SB antigens. *J. Exp. Med.* **156:** 1065.

Brighty, D.W., M. Rosenberg, I.S.Y. Chen, and M. Ivey-Hoyle. 1991. Envelope proteins from clinical isolates of human immunodeficiency virus type 1 that are refractory to neutralization by soluble CD4 possess high affinity for the CD4 receptor. *Proc. Natl. Acad. Sci.* **88:** 7802.

Brodsky, M.H., M. Warton, R.M. Myers, and D.R. Littman. 1990. Analysis of the site in CD4 that binds to the HIV envelope glycoprotein. *J. Immunol.* **144:** 3078.

Brown, J., T. Jardetsky, M.A. Saper, B. Samraoui, P.J. Bjorkman, and D.C. Wiley. 1988. A hypothetical model of the foreign antigen binding site of class II histocompatibility molecules. *Nature* **332:** 845.

Cammarota, G., A. Scheirle, B. Takacs, D.M. Doran, R. Knorr, W. Bannworth, J. Guardiola, and F. Sinigaglia. 1992. Identification of a CD4 binding site on the β_2 domain of HLA-DR molecules. *Nature* **356:** 799.

Chothia, C., J. Novotny, R. Bruccoleri, and M. Karplus. 1985. Domain association in immunoglobulin molecules. *J. Mol. Biol.* **186:** 651.

Clayton, L.K., M. Sieh, D.A. Pious, and E.L. Reinherz. 1989. Identification of human CD4 residues affecting class II MHC versus HIV-1 gp120 binding. *Nature* **339:** 548.

Clayton, L.K., R.E. Hussey, R. Steinbrich, H. Ramachandran, Y. Husain, and E.L. Reinherz. 1988. Substitution of murine for human CD4 residues by site directed mutagenesis identifies amino acids critical for HIV-gp120 binding. *Nature* **335:** 363.

Colman, P.M. 1988. Structure of antibody-antigen complexes: Implications for immune recognition. *Adv. Immunol.* **43:** 99.

Daar, E.S., X.L. Li, T. Moudgil, and D.D. Ho. 1990. High concentrations of recombinant soluble CD4 are required to neutralize primary human immunodeficiency virus type 1 isolates. *Proc. Natl. Acad. Sci.* **87:** 6574.

Dalgleish, A.G., P.C. Beverly, P.R. Clapham, D.H. Crawford, M.F. Greaves, and R.A. Weiss. 1984. The CD4 (T4) antigen is an essential component of the receptor for the AIDS retrovirus. *Nature* **312:** 763.

Deen, K.C., J.S. McDougal, R. Inacker, G. Folena-Wasserman, J. Arthos, J. Rosenberg, P.J. Maddon, R. Axel, and R.W. Sweet. 1988. A soluble form of CD4 (T4) protein inhibits AIDS virus infection. *Nature* **331:** 82.

Doms, R.W., P.L Earl, S. Chakrabarti, and B. Moss. 1990. Human immunodeficiency virus types 1 and 2 and simian immunodeficiency virus *env* proteins possess a functionally conserved assembly domain. *J. Virol.* **64:** 3537.

Doyle, C. and J.L. Strominger. 1987. Interaction between CD4 and class II MHC molecules mediates cell adhesion. *Nature* **330:** 256.

Eichmann, K., J.-I. Jönsson, I. Falk, and F. Emmrich. 1987. Effective activation of resting mouse T lymphocytes by crosslinking submitogenic concentrations of the T cell antigen receptor with either Lyt-2 or L3T4. *Eur. J. Immunol.* **17:** 643.

Epp, O., P. Colman, H. Fehlhammer, W. Bode, M. Schiffer, and R. Huber. 1974. Crystal and molecular structure of a dimer composed of the variable portions of the Bence-Jones protein REI. *Eur. J. Biochem.* **45:** 513.

Fisher, R.A., J.M. Bertonis, W. Meier, V.A. Johnson, D.S. Costopoulos, T. Liu, R. Tizard, B.D. Walker, M.S. Hirsch, R.T. Schooley, and R.A. Flavell. 1988. HIV infection is blocked *in vitro* by recombinant soluble CD4. *Nature* **331:** 76.

Fleury, S., D. Lamarre, S. Meloche, S.-E. Ryu, C. Cantin, W.A. Hendrickson, and R.P. Sekaly. 1991. Mutational analysis of the interaction between CD4 and class II MHC: Class II antigens contact CD4 on a surface opposite the gp120-binding site. *Cell* **66:** 1037.

Garlick, R.L., R.J. Kirschner, F.M. Eckenrode, W.G. Tarpley, and C.C. Tomich. 1990. *Escherichia coli* expression, purification and biological activity of a truncated soluble CD4. *AIDS Res. Hum. Retroviruses* **6:** 465.

Hart, T.K., R. Kirsch, E. Ellens, R.W. Sweet, D.M. Lambert,

S.R. Petteway, Jr., J. Leary, and P.J. Bugelski. 1991. Binding of soluble CD4 proteins to human immunodeficiency virus type 1 and infected cells induces release of envelope glycoprotein gp120. *Proc. Natl. Acad. Sci.* **88:** 2189.

Hunkapiller, T. and L. Hood. 1989. Diversity of the immunoglobulin gene superfamily. *Adv. Immunol.* **44:** 1.

Hussey, R.E., N.E. Richardson, M. Kowalski, N.R. Brown, H.-C. Chang, R.F. Siliciano, T. Dorfman, B. Walker, J. Sodrowski, and E.L. Reinherz. 1988. A soluble CD4 protein selectively inhibits HIV replication and syncytium formation. *Nature* **331:** 78.

Klatzmann, D., E. Champagne, S. Chamaret, J. Gruest, D. Guetard, T. Hercend, J.-C. Gluckman, and L. Montagnier. 1984. T lymphocyte T4 molecule behaves as the receptor for human retrovirus LAV. *Nature* **312:** 767.

König, R., L.-Y. Huang, and R.N. Germain. 1992. MHC class II interaction with CD4 mediated by a region analogous to the MHC class I binding site for CD8. *Nature* **356:** 796.

Krensky, A.M., C.S. Reiss, J.W. Mier, J.L. Strominger, and S.J. Burakoff. 1982. Long-term human cytolytic T cell lines allospecific for HLA-DR6 antigen are $OKT4^+$. *Proc. Natl. Acad. Sci.* **79:** 2365.

Kruisbeek, A.M., J.J. Mong, B.J. Fowlkes, J.A. Carmen, S. Bridges, and D.L. Longo. 1985. Absence of the $Lyt\text{-}2^-$, $L3T4^+$ lineage of T cells in mice treated neonatally with anti-I-A correlates with absence of intrathymic I-A-bearing antigen-presenting cell function. *J. Exp. Med.* **161:** 1029.

Kwong, P.D., S.-E. Ryu, W.A. Hendrickson, R. Axel, R.M. Sweet, G. Folena-Wasserman, P. Hensley, and R.W. Sweet. 1990. Molecular characteristics of recombinant human CD4 as deduced from polymorphic crystals. *Proc. Natl. Acad. Sci.* **87:** 6423.

Lesk, A.M. and C.J. Chothia. 1982. Evolution of proteins formed by β- sheets. II. The core of the immunoglobulin domains. *J. Mol. Biol.* **160:** 325.

Maddon, P.J., A.G. Dalgliesh, J.S. McDougal, P.R. Clapham, R.A. Weiss, and R. Axel. 1986. The T4 gene encodes the AIDS virus receptor and is expressed in the immune system and the brain. *Cell* **47:** 333.

Maddon, P.J., D.R. Littman, M. Godfrey, D.E. Maddon, L. Chess, and R. Axel. 1985. The isolation and nucleotide sequence of a cDNA encoding the lymphocyte protein T4: A new member of the immunoglobulin family. *Cell* **42:** 93.

Meuer, S.C., S.F. Schossman, and E.L. Reinherz. 1982. Clonal analysis of human cytotoxic T lymphocytes: $T4^+$ and $T8^+$ effector T cells recognize products of different major histocompatibility complex regions. *Proc. Natl. Acad. Sci.* **79:** 4395.

Moebius, U., L.K. Clayton, S. Abraham, S.C. Harrison, and E.L. Reinherz. 1992a. The HIV gp120 binding site on CD4: Delineation by quantitative equilibrium and kinetic binding studies of mutants in conjunction with a high resolution CD4 atomic structure. *J. Exp. Med.* **176:** 507.

Moebius, U., L.K. Clayton, S. Abraham, A. Diener, J.J. Yunis, S.C. Harrison, and E.L. Reinherz. 1992b. The HIV-gp120 binding C′C″ ridge of CD4 domain 1 is also involved in interaction with class II MHC molecules. *Proc. Natl. Acad. Sci.* (in press).

Moore, J.P., J.A. McKeating, W.A. Norton, and Q.J. Sattentau. 1991. Direct measurement of soluble CD4 binding to human immunodeficiency virus type 1 virions: gp120 dissociation and its implications for virus-cell binding and fusion reactions and their neutralization by soluble CD4. *J. Virol.* **65:** 1133.

Moore, J.P., J.A. McKeating, R.A. Weiss, Q.J. Sattentau, M.K. Gorny, and S. Zolla-Pazner. 1990. Dissociation of gp120 from HIV-1 virions induced by soluble CD4. *Science* **250:** 1139.

Moore, J.P., J.A. McKeating, Y. Huang, A. Ashkenazi, and D.D. Ho. 1992. Virions of primary human immunodeficiency virus type 1 isolates resistant to soluble CD4 (sCD4) neutralization differ in sCD4 binding and glycoprotein gp120 retention from sCD4-sensitive isolates. *J. Virol.* **66:** 235.

Pelchen-Matthews, A., J.E. Armes, G. Griffiths, and M. March. 1991. Differential endocytosis of CD4 in lymphocytic cells. *J. Exp. Med.* **173:** 575.

Peterson, A. and B. Seed. 1988. Genetic analysis of monoclonal antibody and HIV binding sites on the human lymphocyte antigen CD4. *Cell* **54:** 65.

Pinter, A., W.J. Honnen, S.A. Tilley, C. Bona, H. Zaghouani, M.K. Gorny, and S. Zolla-Pazner. 1989. Oligomeric structure of gp41, the transmembrane protein of human immunodeficiency virus type 1. *J. Virol.* **63:** 2674.

Reinherz, E.L., P. Kung, G. Goldstein, and S. Schlossman. 1979. A separation of functional subsets of human T cells by a monoclonal antibody. *Proc. Natl. Acad. Sci.* **76:** 4061.

Reinherz, E.L., P. Kung, G. Goldstein, R.H. Levey, and S. Schlossman. 1980. Discrete stages of intrathymic differentiation: Analysis of normal thymocytes and leukemic lymphoblasts of T lineage. *Proc. Natl. Acad. Sci.* **77:** 1588.

Rudd, C.E., J.M. Trevillyan, J.D. Dasgupta, L.L. Wong, and S.F. Schlossman. 1988. The CD4 receptor is complexed in detergent lysates to a protein-tyrosine kinase (pp58) from human T lymphocytes. *Proc. Natl. Acad. Sci.* **85:** 5190.

Ryu, S., P.D. Kwong, A. Truneh, T.G. Porter, J. Arthos, M. Rosenberg, X. Dai, N. Xuong, R. Axel, R.W. Sweet, and W.A. Hendrickson. 1990. Crystal structure of HIV-binding recombinant fragment of CD4. *Nature* **348:** 419.

Sattentau, Q. and J.P. Moore. 1991. Conformational changes induced in the human immunodeficiency virus envelope glycoprotein by soluble CD4 binding. *J. Exp. Med.* **174:** 407.

Sattentau, Q.J., J. Arthos, K. Deen, N. Hanna, D. Healey, P.C. Beverly, R. Sweet, and A. Truneh. 1989. Structural analysis of the human immunodeficiency virus-binding domain of CD4. *J. Exp. Med.* **170:** 1319.

Schawaller, M., G.E. Smith, J.J. Skehel, and D.C. Wiley. 1989. Studies with crosslinking reagents on the oligomeric structure of the *env* glycoprotein of HIV. *Virology* **172:** 367.

Traunecker, A., W. Luke, and K. Karjalainen. 1988. Soluble CD4 molecules neutralize human immunodeficiency virus type 1. *Nature* **331:** 84.

Turner, S., R. Tizard, J. DeMarinus, R.B. Pepinsky, J. Zullo, R. Schooley, and R. Fisher. 1992. Resistance of primary isolates of human immunodeficiency virus type 1 to neutralization by soluble CD4 is not due to lower affinity with the viral envelope glycoprotein gp120. *Proc. Natl. Acad. Sci.* **89:** 1335.

Veillette, A., M.A. Bookman, E.M. Horak, and J.B. Bolen. 1988. The CD4 and CD8 T cell surface antigen are associated with the internal membrane tyrosine-protein kinase p56lck. *Cell* **55:** 301.

Wang, J., Y. Yan, T.P. Garrett, J. Liu, D.W. Rodgers, R.L. Garlick, G. Tarr, E. Husain, E.L. Reinherz, and S.C. Harrison. 1990. Atomic structure of a fragment of human CD4 containing two immunoglobulin-like domains. *Nature* **348:** 411.

Structural Aspects of CD4 and CD8 Involvement in the Cellular Immune Response

W.A. HENDRICKSON,*† P.D. KWONG,† D.J. LEAHY,† S.-E. RYU,†§
H. YAMAGUCHI,† S. FLEURY,‡ AND R.-P. SÉKALY‡
*Howard Hughes Medical Institute and †Department of Biochemistry and Molecular Biophysics, Columbia University, New York, New York 10032; ‡Laboratoire d'Immunologie, Institut de Recherches Cliniques de Montréal, Montréal, Quebéc H2W 1R7, Canada

Thymus-derived lymphocytes (T cells) participate in the cellular immune response, and they do so when stimulated into action through interactions between T-cell receptor molecules on their surfaces and peptide antigens presented on target cell surfaces as complexes with molecules of the major histocompatibility complex (MHC). Mature T cells are of two major types, distinguished by the exclusive occurrence of either the CD4 or CD8 glycoproteins on their surfaces. The predominant CD4-bearing ($CD4^+CD8^-$) cells are the helper T cells, whereas the predominant CD8-bearing ($CD4^-CD8^+$) cells are the cytotoxic or killer T cells. The $CD4^+$ helper T cells are restricted to interact with class II MHC molecules, which occur on certain immune system cells such as macrophages and B cells, whereas the $CD8^+$ killer T cells interact with class I MHC molecules, which are found on all cell types. The differentiation of T cells to perform these distinctive roles probably also involves signaling events stimulated by interactions of CD4 and CD8 with MHC molecules.

The response to a viral pathogen illustrates the roles played by each of these components. Virus particles can be picked up, by surface immunoglobins on B cells, for example, and then endocytosed and processed into peptides for presentation on class II MHC molecules. Stimulation of helper T cells by these class-II-presented peptides leads to the production of cytokines that in turn stimulate the proliferation of lymphocytes selected for antibody production (B cells) or for cellular defense (T cells) specifically directed against the invading pathogen. Subsequent stimulation of appropriate cytotoxic T cells by peptide antigens from infected cells, presented on class I MHC molecules, then leads to the production of lytic factors that can eliminate the infected cell. The involvement of CD4 or CD8 is essential for the efficient stimulation of the respective responses. Thus, the molecules can properly be described as coreceptors in the cellular immune response. In addition, CD4 is subverted to become the receptor for infection by the human immunodeficiency virus (HIV) (Dalgleish et al. 1984; Klatzmann et al. 1984).

Relatively little is known about the molecular details of the events associated with CD4 and CD8 involvement in the immune response. Many of the interactions between these coreceptors and other molecules in the T-cell signaling processes are intrinsically weak and difficult to measure. The much stronger interaction between CD4 and the gp120 coat protein of HIV is better understood (Sweet et al. 1991). To further the characterization of these interactions, we are engaged in structural studies on CD4 and CD8. Results from crystal structure of recombinant soluble fragments of the human CD4 and CD8 molecules, taken together with mutational studies, are now beginning to provide a picture of the molecular interactions that underlie the specific stimulatory processes of the immune response.

Several important features that distinguish the CD4 and CD8 structures were already apparent from the amino acid sequences deduced from cDNA and genomic clones of CD4 (Maddon et al. 1985, 1987; Clark et al. 1987) and CD8 (Littman et al. 1985; Sukhatme et al. 1985; Norment and Littman 1988). As shown schematically in Figure 1, both the CD4 and CD8 molecules are single-pass transmembrane molecules with large, extracellular amino-terminal portions and short cytoplasmic tails. A domain with clear homology to immunoglobulin (Ig) variable domains is at the amino terminus in each case. There is also a clear similarity between the carboxy-terminal tails of CD4 and CD8, which in each case interact via a noncovalent bridge to the lymphocyte kinase $p56^{lck}$. Otherwise, however, CD4 and CD8 differ. Whereas CD4 is monomeric, CD8 is a disulfide-bridged dimer. Moreover, whereas the extracellular portion of CD4 comprises four distinct intron-separated domains (D1, D2, D3, and D4), with sparse *N*-linked glycosylation, the Ig-like domain of CD8 is followed by a 50-residue stretch that is heavily decorated by *O*-linked glycosylation. Interestingly, as shown in Figure 1, despite the large differences in chain length between CD4 and CD8, the amino-terminal domains can extend equally from the cell surface.

EXPERIMENTAL PROCEDURES

Protein preparation. Soluble forms of human CD4 and CD8α were produced in transfected Chinese hamster ovary (CHO) cells and purified as described previously (Kwong et al. 1990; Ryu et al. 1990; Leahy et al. 1992a). D3D4 fragments of CD4 were produced as

§Present address: Department of Biochemistry and Molecular Biology, Harvard University, Cambridge, Massachusetts 02138.

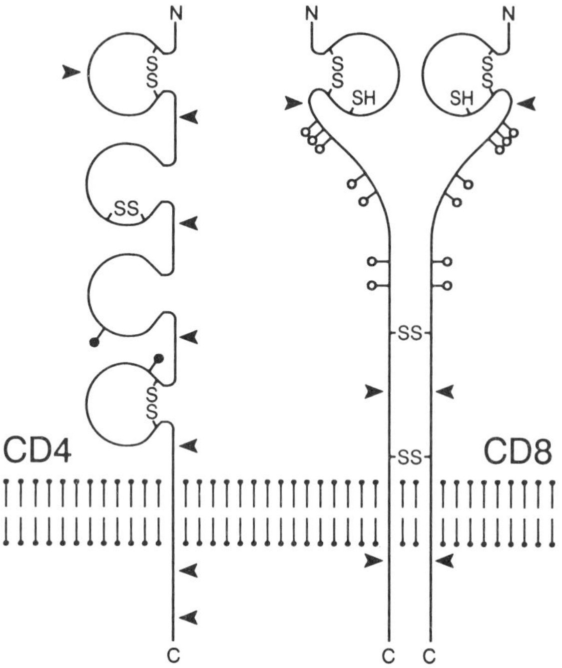

Figure 1. Schematic diagram of the CD4 and CD8 cell-surface topology. These are reductions from the human CD4 and human CD8α sequences. Polypeptide strands are represented by lines roughly proportionate to chain length. Cysteines are represented by S, sites of *N*-linked glycosylation are represented by filled lollipops, sites of *O*-linked glycosylation are represented by open lollipops, and intron positions are represented by arrowheads.

described by Healy et al. (1990). The Q425 monoclonal antibody was produced as described previously (Healy et al. 1990) and reduced to Fab fragments with standard procedures (Parham et al. 1982).

Diffraction analysis. Crystals of soluble fragments of CD4 and CD8 were grown and characterized and used for data measurements as described previously. Diffraction data from trigonal crystals of soluble CD4 were measured at the F1 beamline at the Cornell synchrotron, CHESS (Kwong et al. 1990; Ryu et al. 1990; Leahy et al. 1992a). Crystals of the D1D2 fragment of CD4 were also grown in a second crystal form, similar to one described by Charnow et al. (1990), under the same conditions used to obtain our first D1D2 crystals (Ryu 1991). Data from these crystals were measured on a Xuong-Hamlin detector system. Crystals of the Q425 Fab were grown by hanging-drop vapor diffusion from ammonium phosphate.

Functional assays. Experiments to assess the interaction between class II MHC molecules and mutant CD4 molecules were conducted in cellular assays of adhesion and IL-2 production as reported previously (Fleury et al. 1991).

RESULTS AND DISCUSSION

Crystals of Soluble CD4

The extracellular portion of human CD4, consisting of 370 amino acid residues and two *N*-linked glycosylations, can be expressed in soluble form by secretion from CHO cells (Deen et al. 1988). This recombinant protein could be crystallized into several polymorphic forms (Kwong et al. 1990). As purified, the CD4 glycoprotein is heterogeneous in sialation (Carr et al. 1989), but it is made homogeneous after neuraminidase treatment. Nevertheless, all crystals of this intact soluble CD4 (sCD4) are poorly ordered. Diffraction is typically limited to spacings no finer than 5 Å.

Despite the poor diffraction from these sCD4 crystals, some interesting features of the molecules can be deduced. First, it appears that CD4 oligomerizes at the high protein concentrations and low hydration needed for crystallization. All five distinctive forms that were characterized have multiples of two molecules per asymmetric unit. Moreover, all of these crystals, except for one with eight unique sCD4 molecules, have a lattice dimension of approximately 125 Å along a crystallographic twofold axis. These results suggest that the crystallized molecules may be tetramers having a length of 125 Å along a molecular dyad axis of symmetry.

The cause of the poor diffraction is not understood, but it may well be associated with internal flexibility of the molecules. The juncture between domains D2 and D3 is particularly sensitive to proteolytic cleavage (Ibegbu et al. 1989; Healy et al. 1990; Ryu 1991). Whether such flexibility is functionally significant is not clear, but the propensity to associate into oligomers could be important in transmembrane signaling (Ullrich and Schlessinger 1990; Hendrickson 1992). A model for cell-surface association of CD4 molecules consistent with the crystal properties is shown in Figure 2.

Structure of the D1D2 Fragment of CD4

Consistent with the proteolysis results, a recombinant construct of CD4 that corresponds to the first two domains (D1D2) forms a stable fragment that retains all antibody epitopes mapped to these domains and is fully active in binding to gp120 of HIV (Arthos et al. 1989). Crystals of this fragment diffract well, and the structure was determined at 2.3 Å resolution from the initial crystals (type I). More recently, a second crystal structure (type II) has also been determined, and further refinement has been carried out (Ryu 1991). The molecular structure in both cases is seen to be very similar, with a folded topology as shown in Figure 3.

Both domains of the D1D2 fragment are folded in variations of the domains found in immunoglobulins, but they are connected in a novel way (Ryu et al. 1990; Wang et al. 1990). The D1 domain has the topology of Ig variable domains, but it differs in having a half-length A strand, abbreviated CC′ and FG (CDR3-like)

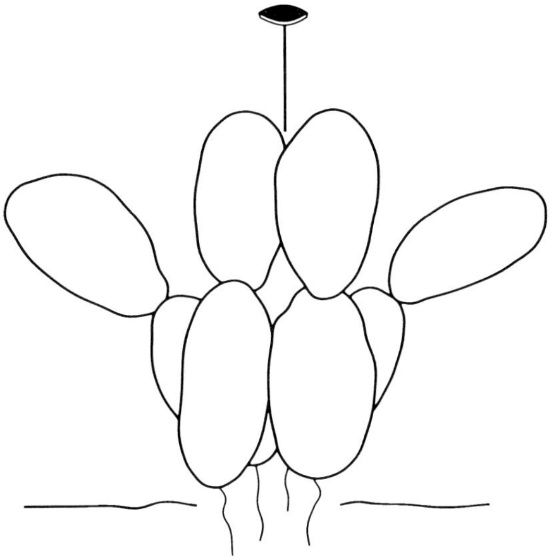

Figure 2. Possible mode of CD4 self-association. A CD4 oligomer is drawn here to be consistent with molecular dimensions and symmetry deduced from sCD4 crystal parameters, and with relative orientations that could obtain on a cell surface. The D1D2 portion is seen to be somewhat flexibly linked to the membrane proximal D3D4 portion. Self-associations in this model are mediated by the D3D4 moiety.

loops, and a prominently extended C′C″ (CDR2-like) loop. The CC′ and FG loops of Ig variable domains are responsible for the dimerization in antibodies, and in their shortened form in CD4 they cannot support this interaction. D1D2 is a monomer in crystals and in solution. The CDR2-like loop, which juts out prominently from the body of D1, is strongly implicated in the binding to HIV and features a solvent-exposed phenylalanine at the reverse turn of this loop.

The D2 domain has seven β strands, as in Ig constant domains, but these have a novel strand-switched topology that is also seen in the bacterial chaperonin Pap D (Holmgren and Brändén 1989), in the human growth hormone receptor (de Vos et al. 1992), and in fibronectin type III domains (Leahy et al. 1992b). The D2 domain also features an intra-sheet disulfide bridge rather than the usual cross-sheet bridge. D2 is also quite small with only about 75 ordered residues.

The D1 and D2 domains are intimately connected, with the last strand of D1 running directly into the first strand of D2. There is a substantial contact interface, and the relative disposition of these domains is the same in two very differently packed crystal structures. This gives the D1D2 fragment a rod-like shape (25 × 25 × 60 Å) and stiffness. Its length, both in sequence and morphology, is approximately half that of whole sCD4.

Structure of a Soluble CD8 Fragment

A soluble fragment of human CD8 chains was expressed in CHO cells and shown to form dimers (Leahy at al. 1992a). Small, but nevertheless useful, crystals were grown after deglycosylation and proteolytic truncation of these molecules. The resulting polypeptide chain comprises the Ig-like domain and about half of the glycosylated stalk region. The crystal structure, determined at 2.6 Å resolution (Leahy et al. 1992a), shows that the CD8 dimer is organized very much as in the variable domain tips of antibodies (Fig. 4). Here, in contrast with CD4, the dimerization loops are extended and utilized as in the immunoglobulins. The CDR2-like loop is also elongated in CD8, but its conformation is quite different from that in CD4.

CD8 α chains have three internal cysteine residues, two of which are in the positions on B and F strands that are normally disulfide-bridged in Ig domains. A biochemical analysis of murine CD8 $\alpha\beta$ heterodimers has, however, shown that the α-chain bridge in these molecules is instead between cysteines on the B and C strands (Kirszbaum et al. 1989). What we clearly observe in the CD8 homodimer structure is the usual Ig-like bridge, but the C strand thiol is disposed such that it would seem possible to isomerize to the B to C alternative. The distinction in disulfide bridging in these cases could reflect species differences, but it might also be due to an isomerization that distinguishes α_2 homodimers from $\alpha\beta$ heterodimers (Leahy et al. 1992a).

Portions of the CD8 chains extending beyond the Ig-like domain are disordered in the crystal structure although they are known chemically to be present. The results do not, however, distinguish between a flexibly attached rigid strut and a thoroughly "floppy" chain. The stalk region on the intact CD8 molecules is heavily complexed with *O*-linked oligosaccharides that terminate in sialyl groups. The resulting negative charges might be expected to repel the membrane surface and thereby to extend the stalk region and to raise the Ig-like domains away from the surface.

Interaction of CD4 with Class II MHC Molecules

A direct physical interaction between CD4 or CD8 molecules and their respective MHC partners, such that dissociation constants could be measured, has been difficult to demonstrate. This difficulty is not unexpected in light of the high polyvalency in cell-cell interactions, which would seem to require weak intrinsic molecular interactions in order to avoid uncontrolled adhesion. Less direct but nevertheless compelling evidence for these interactions is available, however, from suitably constructed cell biology experiments.

The cellular interaction between CD4 and class II MHC molecules can be seen in cell adhesion assays and in assays of IL-2 production in response to T-cell stimulation. Impairment of function in these assays can be observed when mutations are made that affect the interfaces of contact involved in the interactions. The interpretation of such experiments can, of course, be confused by indirect effects in this complex system, but knowledge of the structure of some of the components

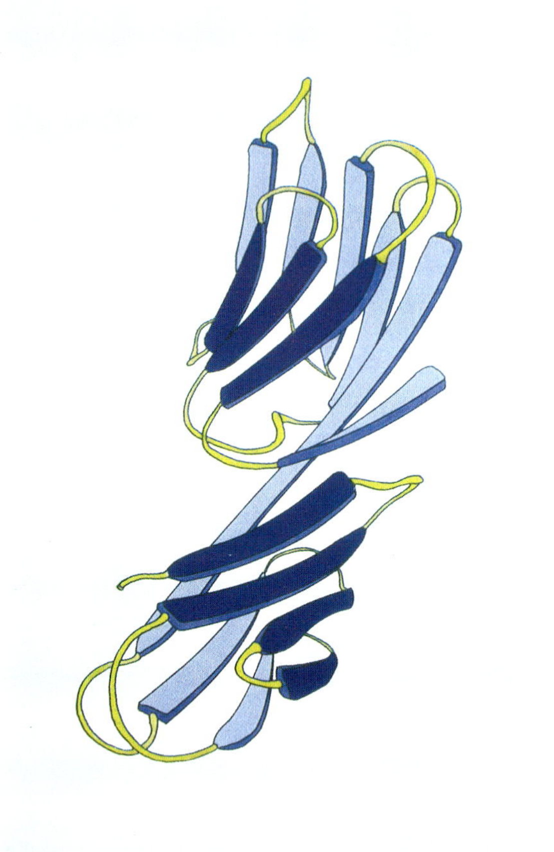

Figure 3. Ribbon diagrams of the D1D2 fragment of CD4. Purple ribbons represent the β strands, and yellow tubes represent loops. The D1 domain is above and the D2 domain is below. (Reprinted, with permission from Hendrickson and Yarmolinski, as published in Ryu et al. 1990.)

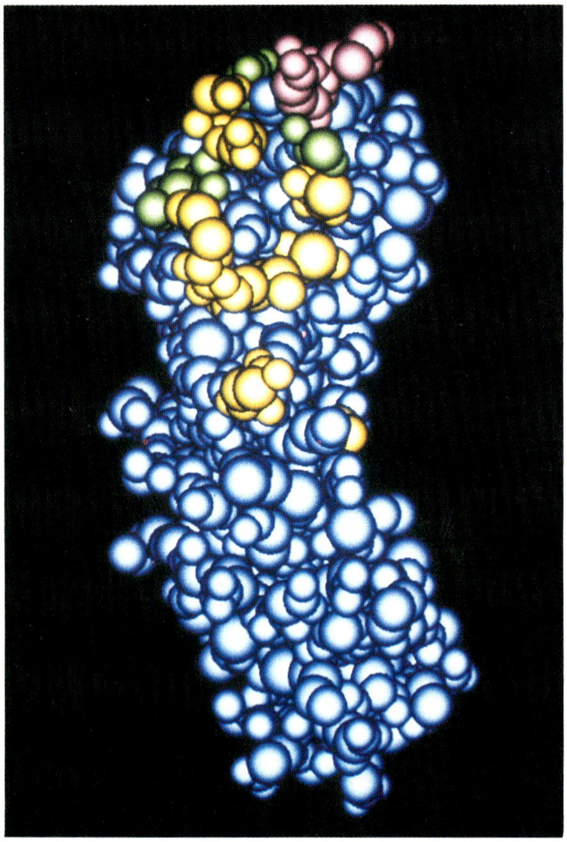

Figure 5. Space-filling drawing showing CD4 residues implicated in class II binding. The D1D2 fragment is shown oriented as in Fig. 5C (right) of Fleury et al. (1991). Amino acid residues implicated in MHC class II interactions are shown in yellow (K1, K2, K7, T15, T17, D88, Q89, and Q165). Those having partial effects are shown in green (S19, E87, and K90), and those tested but showing no effect are shown in pink (Q20, K21, and S23).

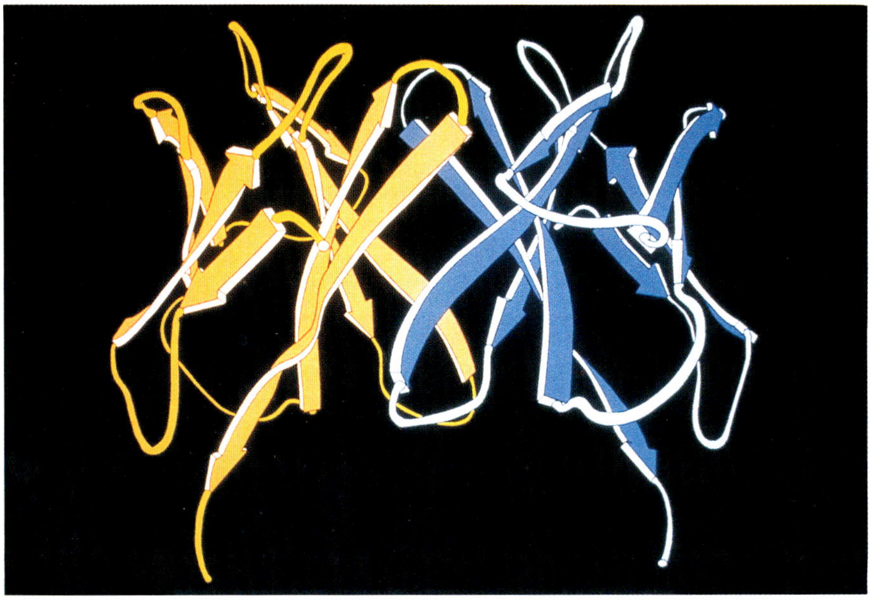

Figure 4. Ribbon diagram of the CD8α homodimeric Ig domains. Ribbons represent β strands with arrows indicating the chain direction; one protomer is shown in yellow and the other is in blue. The carboxyl termini at the bottom connect to the glycosylated flexible stalk that will extend to the transmembrane segment. (Reprinted, with permission, from Leahy et al. 1992a [copyright held by Cell Press].)

is helpful. Most of the early CD4 mutations examined in this way involved insertions and deletions (Clayton et al. 1989; LeMarre et al. 1989), which are prone to indirect effects.

A more recent study of 30 CD4 mutants, covering all four domains and including several single-site mutations, suggested that one face of the D1D2 unit of CD4 is involved in the class II interaction (Fleury et al. 1991). The cellular assay system used in this work (Gay et al. 1987) is designed to isolate the CD4-class II component of the interaction by using the interaction between the T-cell receptor on a murine T-cell hybridoma (3DT 52.5.8) and its class I target displayed on a murine fibroblast (DAP-3 D^d) to establish the basal level response, and then observing the signal enhancement from the interaction between human CD4 molecules expressed in the hybridoma and human class II molecules expressed in the fibroblast. The residues specifically implicated by the study of Fleury et al. (1991) are 19 and 89 from the ends of CDR1-like and CDR3-like loops in D1 and 165 from the FG loop of D2.

We have now extended the mutational analysis to several other residues in D1 that have their side chains exposed on the implicated surface. Preliminary results from an alanine scan, compiled in Table 1, further define the site of interaction and remain consistent with a contact surface along a side of D1 with some participation from D2. It is interesting to note that the replacement of Ser-19 by alanine has little effect, in contrast to the previously reported effect on replacement by the bulky tyrosine side chain. On the other hand, signaling function is completely disrupted when Gln-89 is replaced by either leucine or alanine. The apparent interaction surface is shown in Figure 5. Recent studies also show that murine as well as human

Table 1. Effect of Mutations in CD4 on its Interactions with MHC Class II

Residue[a]	Fractional solvent accessibility	Effect on IL-2 production[b]
WT	—	22
*K1	0.68	4
K2	0.52	1
*K7	0.66	1
T15	0.98	1
T17	0.75	17
*S19	1.03	19
*Q20	0.88	16
*K21	0.98	17
S23	0.68	16
E87	0.52	1
D88	0.71	2
Q89	0.89	2
*K90	0.80	14

[a]In each case, the indicated residue was mutated to alanine. Those residues marked by asterisks are conserved between mouse and human CD4.

[b]Values cited are the ratios between IL-2 production for the T cells when stimulated by lymphocytes bearing both mouse D^d (class I) and human DR (class II) MHC molecules, in comparison with control cells bearing only the class I molecules.

CD4 can interact productively with human class I molecules, which is consistent with the conservativ character of this surface (Table 1). Interestingly, th CDR2-like region, which appears not to be involved i the class II interaction based on analysis of a seve residue deletion (43–49) in this region (Fleury et a 1991), is poorly conserved between mouse and ma

The CD4-binding site, or sites, on class II MH molecules are only now beginning to be establishe Recent studies (Cammarota et al. 1992; König et 1992) identify a carboxylate-rich site on the $\beta 2$ dom that is homologous with the $\alpha 3$-domain site identi earlier on class I molecules (Salter et al. 1990).

Interaction of CD8 with Class I MHC Molecule

Although the actual nature of molecular intera between CD8 and class I molecules is unknown, s important features are beginning to emerge. Salt al. (1990) and Potter et al. (1989) have demonst that a loop on the $\alpha 3$ domain of class I molecule principal binding site for CD8. On the CD8 side, ers et al. (1991) have shown that mutations i CDR1-like and CDR2-like loops of CD8 had the est effect on class I binding in a cell adhesion These results help to limit possibilities for th lecular interaction.

As shown in Figure 6, the sizes and shapes o and class I components are compatible with a plementary association. The shape of the class molecules as derived from the crystal structure of man et al. (1987) features a notch that can acc date the CD8 dimer in an orientation that could an interaction between the CD8-binding loop o domain and the CDR1-like and CDR2-like l CD8. Interestingly, in a binding of this kind, o dimer might be able to bind simultaneously to t I molecules. It is also noteworthy that the fa CD8 Ig-like domain that relates to the putativ binding face of CD4 would be accessible i model for making additional interactions with I molecule.

Electrostatic as well as shape compleme likely to be important in the CD8-class I int The CD8-binding region on class I molecule four highly conserved carboxylate residues. the residues from the CDR-like loops on CD implicated in class I binding endow that surf strongly positive electrostatic potential. In possible that relatively nonspecific electrosta tions might account for the relatively weak a between these molecules. Electrostatic inter this are used in the rather promiscuous a made by electron-transfer proteins such as c c (Wendoloski et al. 1987).

Association of CD4 and CD8 with p

Both CD4 and CD8 associate with the kinase $p56^{lck}$, a tyrosine kinase similar to th

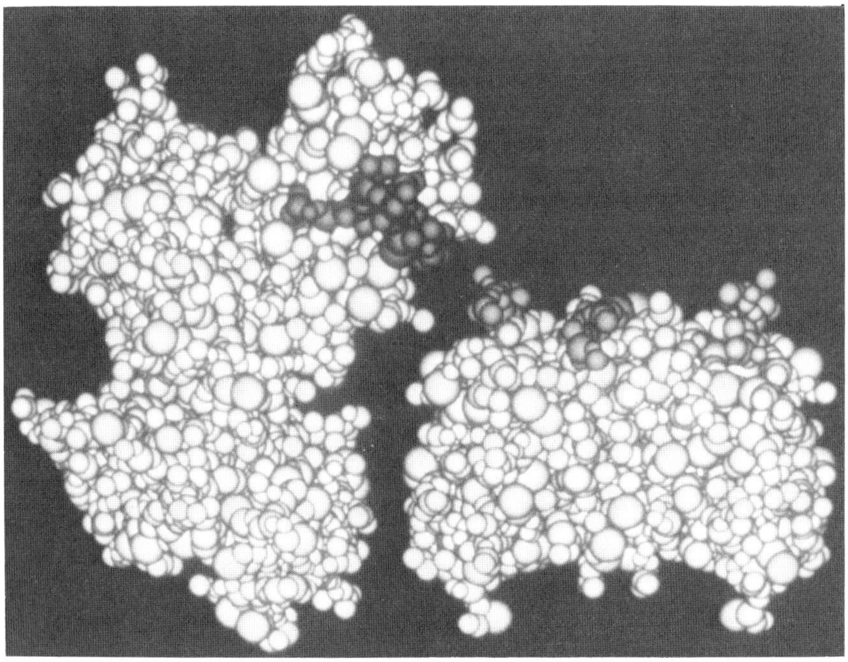

Figure 6. Juxtaposed space-filling representations of the CD8α dimer and class I HLA-A2 molecule. The CD8 coordinates are from Leahy et al. (1992), and the HLA coordinates are from Bjorkman et al. (1987). CD8 is shown on the left with its membrane-directed carboxyl termini above, and the class I molecule is on the right with its peptide groove on the top right and the membrane-directed α-chain carboxyl terminus at the bottom left. The CDR1-like and CDR2-like loops of CD8 that have been implicated in MHC binding are darkened, and the loop in α3 of HLA (residues 223–229) that has been implicated in CD8 binding is also darkened. (Adapted, with permission, from a color version in Leahy et al. 1992a [copyright held by Cell Press].)

protein (Rudd et al. 1988; Veillette et al. 1988). The resulting complexes then have a form similar to classic growth factor receptors (Ullrich and Schlessinger 1990), which itself suggests that CD4 and CD8 will be involved in T-cell activation. The short cytoplasmic tails of CD4 and CD8 have similar sequences that both feature several positively charged residues and a pair of conserved cysteine residues. The interaction of these tails with lck is to the amino-terminal region of lck, which is rich in negatively charged residues and also includes a pair of cysteine residues (Turner et al. 1990). The amino terminus of lck is myristylated and thereby anchored to the cytoplasmic face of the lymphocyte membrane, and this localization is essential for the association. Electrostatic complementation must be involved in the direct interaction between the lck and CD4 or CD8 terminal tails, and it has been suggested that zinc ions might form a bridge coordinating two cysteines from each chain (Turner et al. 1990); we are conducting experiments to test this hypothesis.

Interactions of CD4 with HIV

The hallmark of AIDS is the immune deficiency brought on by profound depletion of CD4$^+$ T cells. The complicity of CD4 itself in the infection of these cells by HIV has long been clear (Dalgleish et al. 1984; Klatzmann et al. 1984), and the more recent mutational and structural analyses have helped to define the molecular interaction. The virus uses CD4 as its cell-attachment receptor in a high-affinity interaction between the gp120 envelope glycoprotein and CD4 (nanomolar dissociation constant). The site on CD4 that is responsible for this binding is localized to the D1 domain (Arthos et al. 1989), and a vast number of single-site mutations within D1 further delimit the site (Arthos et al. 1989; Ashkenazi et al. 1990; Brodsky et al. 1990). The structure of the D1D2 fragment identifies the implicated residues with the CDR2-like region that juts out prominently from D1 (Ryu et al. 1990; Wang et al. 1990). All but four of the solvent-accessible residues found initially to impair gp120 binding when mutated are in the span from 41 to 59 (Ryu et al. 1990). More recent analyses further emphasize the importance of the 41 to 59 span. One exception is at a partially buried site, and its effect may be indirect; the negative effects from another two sites have not been confirmed in other experiments (Sweet et al. 1991). The exceptionally exposed Phe-43 at the CDR2-like β turn and Arg-59, which serves as a defining spacer for positioning the C′C″ hairpin of the CD2-like region, are particularly sensitive (Sweet et al. 1991). Unfortunately, little detail is known about the CD4-binding site on gp120.

Events following the initial binding of HIV to CD4 on the cell surface lead to fusion of the HIV envelope with the cell membrane and consequent infection. The CDR3-like loop of D1 is implicated in fusion by mutation experiments (Ashkenazi et al. 1990; Camerini and Seed 1990) and by antibody binding to this epitope (Truneh et al. 1991). The precise mechanism for fusion is unclear, but the binding-induced "stripping" of gp120 from the viral particle (Moore et al. 1990; Hart

et al. 1991) to expose gp41 molecules on the envelope membrane appears to be involved.

Interactions of Antibody Q425 with the D3D4 Fragment of CD4

Monoclonal antibodies with epitopes that map to the CDR2-like region of D1 inhibit the binding of gp120 to CD4 and block HIV infection (Sattentau et al. 1986); those that map to the CDR3-like loop of D1 have limited effect on gp120 binding but block fusion and infection (Truneh et al. 1991). Antibodies against other epitopes generally have little effect on HIV interactions, but a series of antibodies directed against certain portions of the D3 domain have the interesting property of blocking HIV infection without impairing the affinity of CD4 for gp120 (Healey et al. 1990). Indeed, if anything, monoclonal antibodies Q425 and Q428 actually enhance the binding of CD4 by gp120. These properties are particular to the epitope covered by Q425 and Q428, which is thought to be structurally near the D2 to D3 junction; other anti-D3 antibodies do not block infection. Our working hypothesis is that junctional flexibility within CD4 is needed for membrane fusion after HIV binding. It may also be important for normal signal transduction.

The Q425 antibody also affords us another approach to the structural analysis of CD4. We have produced Fab fragments of Q425 and used these in attempts to produce sCD4 cocrystals with hopes of blocking the internal flexibility that may be limiting CD4 crystal quality. Interestingly, divalent cations are required to form the complex. Unfortunately, however, although complex formation can be demonstrated by migration shifts in native gel electrophoresis, under the crystallization conditions, only either sCD4 crystals or Fab crystals could be obtained. The Fab crystals are suitable for analysis, and they should ultimately be of interest for comparison with a Fab-CD4 complex structure.

Attempts are now in progress to crystallize the complex between the Q425 Fab fragment and the D3D4 fragment of CD4. Sequence comparisons indicate that D3, despite its absence of disulfide bridge, may be structurally similar to D1 and that D4 is similar to D2 (Ryu et al. 1991). Furthermore, these comparisons are compatible with a tight association within the D3D4 fragment as is found within D1D2. In distinction, however, the dimerization loops that are missing in D1D2 would appear to be present in D3D4. Perhaps these are involved in the tendency of CD4 to oligomerize as it crystallizes and, conceivably, also in signaling oligomerization.

ACKNOWLEDGMENTS

We thank Richard Axel, Ray Sweet, and Quentin Sattentau for their pivotal roles in some of the experiments and for discussions. We thank Hans-Erik Aronson for help in producing Figure 5.

REFERENCES

Arthos, J., C.K. Deen, M.A. Chaikin, J.A. Fornwald, Q.J. Sattentau, P.R. Clapham, R.A. Weiss, J.S. McDougal, C. Pietropaolo, P.J. Maddon, A. Truneh, R. Axel, and R.W. Sweet. 1989. Identification of the residues in human CD4 critical for the binding of HIV. *Cell* **57:** 469.

Ashkenazi, A., L.G. Presta, S.A. Marsters, T.R. Camerato, K.A. Rosenthal, B.M. Fendly, and D.J. Capon. 1990. Mapping the CD4 binding site for human immunodeficiency virus by alanine-scanning mutagenesis. *Proc. Natl. Acad. Sci.* **87:** 7150.

Bjorkman, P.J., M.A. Saper, B. Samraoui, W.S. Bennett, J.L. Strominger, and D.C. Wiley. 1987. Structure of the human class I histocompatibility antigen, HLA-A2. *Nature* **329:** 506.

Brodsky, M.H., M. Warton, R.M. Myers, and D.R. Littman. 1990. Analysis of the site in CD4 that binds to the HIV envelope glycoprotein. *J. Immunol.* **144:** 3078.

Camerini, D. and B. Seed. 1990. A CD4 domain important for HIV-mediated syncytium formation lies outside the virus binding site. *Cell* **60:** 747.

Cammarota, G., A. Scheirle, B. Takacs, D.M. Doran, R. Knorr, W. Bannwarth, J. Guardiola, and F. Sinigaglia. 1992. Identification of a CD4 binding site on the β_2 domain of HLA-DR molecules. *Nature* **356:** 799.

Charnow, S.M., D.H.Peers, R.A. Byrn, M.G. Mulkerrin, R.J. Harris, W.-C. Wang, P.J. Bjorkman, D.J. Capon, and A. Ashkenazi. 1990. Enzymatic cleavage of a CD4 immunoadhesin generates crystallizable, biologically active Fd-like fragments. *Biochemistry* **29:** 9885.

Carr, S.A., M.E. Hemling, G. Folena-Wasserman, R.W. Sweet, K. Anumula, J.R. Barr, M.J. Huddleston, and P. Taylor. 1989. Protein and carbohydrate structural analysis of a recombinant soluble CD4 receptor by mass spectrometry. *J. Biol. Chem.* **264:** 21286.

Clark, S.J., W.A. Jeffries, A.N. Barclay, J. Gagnon, and A.F. Williams. 1987. Peptide and nucleotide sequences of rat CD4 (W3/25) antigen: Evidence for derivation from a structure with four immunoglobulin related domains. *Proc. Natl. Acad. Sci.* **84:** 1649.

Clayton, L.K., M. Sieh, D.A. Piou, and E.L. Reinherz. 1989. Identification of human CD4 residues affecting class II MHC versus HIV-1 gp 120 binding. *Nature* **339:** 548.

Dalgleish, A.G., P.C.L. Beverley, P.R. Clapham, D.H. Crawford, M.F. Greaves, and R.A. Weiss. 1984. The CD4 (T4) antigen is an essential component of the receptor for the AIDS retrovirus. *Nature* **312:** 763.

Deen, K.C., J.S. McDougal, R. Inacker, G. Folena-Wasserman, J. Arthos, M. Rosenberg, P.J. Maddon, R. Axel, and R. Sweet. 1988. A soluble form of CD4 (T4) protein inhibits AIDS virus infection. *Nature* **331:** 82.

DeVos, A.M., M. Ultsch, and A.A. Kossiakoff. 1992. Human growth hormone and extracellular domain of its receptor. *Science* **255:**306.

Fleury, S., D. Lamatte, S. Meloche, S.-E. Ryu, D. Cantin, W.A. Hendrickson, and R.-P. Sékaly. 1991. Mutational analysis of the interaction between CD4 and Class II MHC: Class II antigens contact CD4 on a surface opposite the gp120-binding site. *Cell* **66:** 1037.

Gay, D., P. Maddon, R. Sékaly, M.A. Talle, M. Godfrey, E. Long, G. Goldstein, L. Chess, R. Axel, J. Kappler, and P. Marrack. 1987. The T-cell accessory molecule CD4 recognises a monomorphic determinant on isolated Ia. *Nature* **328:** 626.

Hart, T.K., R. Kirsh, H. Ellens, R.W. Sweet, D.M. Lambert, S.R. Petteway, J. Leary, and P.J. Bugelski. 1991. Binding of soluble CD4 proteins to human immunodeficiency virus type 1 and infected cells induces release of envelope glycoprotein gp120. *Proc. Natl. Acad. Sci.* **88:** 2189.

Healey, D., L. Dianda, J.P. Moore, J.S. McDougal, M.J. Moore, P. Estess, D. Buck, P.D. Kwong, P.C.L. Beverley, and Q.J. Sattentau. 1990. Novel anti-CD4 monoclonal antibodies separate human immunodeficiency virus infection and fusion of CD4 cells from virus binding. *J. Exp. Med.* **172:** 1233.

Hendrickson, W.A. 1992. Modes of transduction. *Curr. Biol.* **2:** 57.

Holmgren, A. and C.-I. Brändén. 1989. Crystal structure of chaperone protein PapD reveals an immunoglobulin fold. *Nature* **342:** 248.

Ibegbu, C.C., M.S. Kennedy, P.J. Maddon, K.C. Deen, D. Hicks, R.W. Sweet, and J.S. McDougal. 1989. Structural features of CD4 required for binding to HIV. *J. Immunol.* **142:** 2250.

Kirszbaum, L., J.A. Sharpe, N. Goss, J. Lahnstein, and I.D. Walker. 1989. The α-2 chain of murine CD8 lacks an invariant Ig-like disulfide bond but contains a unique intrachain loop instead. *J. Immunol.* **142:** 3931.

Klatzmann, D., E. Champagne, S. Chamaret, J. Gruest, D. Guetard, T. Hercend, J.-C. Gluckman, and L. Montagnier. 1984. T-lymphocyte T4 molecule behaves as the receptor for human retrovirus LAV. *Nature* **312:** 767.

König, R., L.-Y. Huang, and R.N. Germain. 1992. MHC class II interaction with CD4 mediated by a region analogous to the MHC class I binding site for CD8. *Nature* **356:** 796.

Kwong, P.D., S.-E. Ryu, W.A. Hendrickson, R. Axel, R.M. Sweet, G. Folena-Wasserman, P. Hensley, and R.W. Sweet. 1990. Molecular characteristics of recombinant human CD4 as deduced from polymorphic crystals. *Proc. Natl. Acad. Sci.* **87:** 6423.

LaMarre, D., A. Ashkenazi, S. Fleury, D.H. Smith, R.P. Sékaly, and D.J. Capon. 1989. The MHC-binding and gp 120-binding functions of CD4 are separable. *Science* **245:** 743.

Leahy, D.J., R. Axel, and W.A. Hendrickson. 1992a. Crystal structure of a soluble form of the human T cell coreceptor CD8 at 2.6 Å resolution. *Cell* **68:** 1145.

Leahy, D.J., W.A. Hendrickson, I. Aukhil, and H.P. Erickson. 1992b. Structure of a fibronectin type III domain from tenascin phased by MAD analysis of the selenomethionyl protein. *Science* (in press).

Littman, D.R., Y. Thomas, P. Maddon, L. Chess, and R. Axel. 1985. The isolation and sequence of the gene encoding T8: A molecule defining functional classes of T lymphocytes. *Cell* **240:** 237.

Maddon, P.J., D.R. Littman, M. Godfrey, D.E. Maddon, L. Chess, and R. Axel. 1985. The isolation and nucleotide sequence of a cDNA encoding the T cell surface protein T4: A new member of the immunoglobulin gene family. *Cell* **42:** 93.

Maddon, P.J., S.M. Molinaux, D.E. Maddon, K.A. Zimmerman, M. Godfrey, F.W. Alt, L. Chess, and R. Axel. 1987. Structure and expression of the human and mouse T4 genes. *Proc. Natl. Acad. Sci.* **84:** 9155.

Moore, J.P., J.A. McKeating, R.A. Weiss, and Q.J. Sattentau. 1990. Dissociation of gp120 from HIV- 1 virions induced by soluble CD4. *Science* **250:** 1139.

Norment, A.M. and D.R. Littman. 1988. A second subunit of CD8 is expressed in human T cells. *EMBO J.* **7:** 3433.

Parham, P., M.J. Androlewicz, F.M. Brodsky, N.J. Holmes, and J.P. Ways. 1982. Monoclonal antibodies: Purification fragmentation and application to structural and functional studies of class I MHC antigens. *J. Immunol. Methods* **53:** 133.

Potter, T.A., T.V. Rajan, R.F. Dick II, and J.A. Bluestone. 1989. Substitution at residue 227 of H-2 class I molecules abrogates recognition by CD8-dependent, but not CD8-independent, cytotoxic T lymphocytes. *Nature* **337:** 73.

Rudd, C.E., J.M. Trevillyan, J.D. Dasgupta, L.L. Wong, and S.F. Schlossman. 1988. The CD4 receptor is complexed in detergent lysates to a protein-tyrosine kinase (pp56) from human T lymphocytes. *Proc. Natl. Acad. Sci.* **85:** 5190.

Ryu, S.-E. 1991. "Crystallographic studies on HIV-binding fragments of human CD4." Ph.D. thesis, Columbia University, New York, New York.

Ryu, S.-E., P.D. Kwong, A. Truneh, T.G. Porter, J. Arthos, M. Rosenberg, X. Dai, N. Xuong, R. Axel, R.W. Sweet, and W.A. Hendrickson. 1990. Crystal structure of an HIV-binding recombinant fragment of human CD4. *Nature* **348:** 419.

Salter, R.D., R.J. Benjamin, P.K. Wesley, S.E. Buxton, T.P.J. Garrett, C. Clayberger, A.M. Krensky, A.M. Norment, D.R. Littman, and P. Parham. 1990. A binding site for the T-cell co-receptor CD8 on the $\alpha3$ domain of HLA-A2. *Nature* **345:** 41.

Sanders, S.K., R.O. Fox, and P. Kavathas. 1991. Mutations in CD8 that affect interactions with HLA class I and monoclonal anti-CD8 antibodies. *J. Exp. Med.* **174:** 371.

Sattentau, Q.J., A.G. Dalgleish, R.A. Weiss, and P.C. Beverley. 1986. Epitopes of the CD4 antigen and HIV infection. *Science* **234:** 1120.

Sukhatme, V.P., K.C. Sizer, A.C. Vollmer, T. Hunkapiller, and J.R. Parnes. 1985. The T cell differentiation antigen Leu-2/T8 is homologous to immunoglobulin and T cell receptor variable regions. *Cell* **40:** 591.

Sweet, R.W., A. Truneh, and W.A. Hendrickson. CD4: Its structure, role in immune function and AIDS pathogenesis, and potential as a pharmacological target. *Curr. Opin. Biotechnol.* **2:** 622.

Truneh, A., D. Buck, D.R. Cassatt, R. Juszcak, S. Kassis, S.E. Ryu, D. Healey, R.W. Sweet, and Q.J. Sattentau. 1991. A region in domain 1 of CD4 distinct from the primary gp120 binding site is involved in HIV infection and virus-mediated fusion. *J. Biol. Chem.* **266:** 5942.

Turner, J.M., M.H. Brodsky, B.A. Irving, S.D. Levin, R.M. Perlmutter, and D.R. Littman. 1990. Interaction of the unique N-terminal region of tyrosine kinase $p56^{lck}$ with cytoplasmic domains of CD4 and CD8 is mediated by cysteine motifs. *Cell* **60:** 755.

Ullrich, A. and J. Schlessinger. 1990. Signal transduction by receptors with tyrosine kinase activity. *Cell* **61:**203.

Veillette, A., M.A. Bookman, E.M. Horak, and J.B. Bolen. 1988. The CD4 and CD8 T cell surface antigens are associated with the internal membrane tyrosine protein kinase $p56^{lck}$. *Cell* **55:** 301.

Wang, J., Y. Yan, T.P.J. Garrett, J. Liu, D.W. Rodgers, R.L. Garlick, G.E. Tarr, Y. Husain, E.L. Reinherz, and S.C. Harrison. 1990. Atomic structure of a fragment of human CD4 containing two immunoglobulin-like domains. *Nature* **348:**411.

Wendoloski, J.J., J.B. Matthew, P.C. Weber, and F.R. Salemme. 1987. Molecular dynamics of a cytochrome c-cytochrome b_5 electron transfer complex. *Science* **238:** 794.

Biochemical Anatomy of Antigen Processing

C.A. NELSON, C.V. HARDING, AND E.R. UNANUE
Washington University School of Medicine, Department of Pathology, St. Louis, Missouri 63110

We summarize here our recent studies dealing with the cellular and biochemical analyses of antigen presentation by the class II proteins of the major histocompatibility complex (MHC). Antigen presentation is the process through which immunogenic proteins are prepared for recognition by T lymphocytes. T lymphocytes of the CD4 subset respond to peptides from proteins that are processed in intracellular acidic vesicles of antigen-presenting cells (APCs). Such processed peptides become bound to class II MHC proteins, and together, they form a complex recognizable by CD4 T lymphocytes. CD8 T lymphocytes recognize peptides presented by class I MHC molecules. These peptides originate from proteins degraded in the cytosol, translocated into the endoplasmic reticulum, where they bind to nascent class I MHC molecules, and then transported as a bimolecular complex to the plasma membrane. Therefore, the MHC molecules serve as peptide carriers that rescue peptides from extensive catabolism, with the class I MHC sampling primarily cytosolic peptides and the class II MHC sampling primarily peptides from internalized proteins. At the same time, the peptide-MHC bimolecular complex represents the antigenic determinant that clonally distributed antigen-specific T-cell receptors engage. Consequently, the process of antigen presentation directs the activation of T cells within the immune system. Some of our early studies were summarized in a previous Cold Spring Harbor Symposium (Unanue et al. 1989).

In this paper, we detail first our recent studies that identify a cellular locus where protein antigens are processed in APCs and, second, our first attempts to characterize peptides extracted from class II MHC molecules of APCs. The majority of these studies have been published or are now in the process of being written (Harding et al. 1990, 1991a,b; Collins et al. 1991; Nelson et al. 1992).

The initial studies on antigen processing demonstrated that protein molecules required an intracellular processing step before recognition by CD4 T cells. This intracellular processing step was inhibited by small amines that neutralized the intracellular acid pH. This intracellular processing event was bypassed by using peptides instead of the whole protein molecules (Allen et al. 1984; Shimonkevitz et al. 1984). Indeed, peptides derived from protein antigen could be added to, and presented by, lightly fixed APCs. This indicated that fixed APCs must contain surface-bound class II MHC molecules that effectively bind peptides. (In many of these studies, the T cells that recognize the peptide-MHC complex are T-cell hybridomas; T-cell hybridomas are derived from the fusion of antigen-specific T cells with a marked thymoma, are easy to grow, and respond to their specific peptide-MHC complex by producing cytokines such as interleukin-2 [IL-2]).

Before presenting our data, it is pertinent to discuss the concept of determinant selection, or immunodominance. Determinant selection is a concept first identified by Alan Rosenthal and associates working with the T-cell response to insulin in guinea pigs (Rosenthal et al. 1977). In brief, it states that the different alleles of the MHC gene complex, a highly polymorphic gene, influence which portion of the protein antigen is recognized by the individual. Part of the explanation for immunodominance, i.e., why some peptides are selected over others, resides in the binding affinity of a given peptide for the MHC proteins. Indeed, peptides that do not bind, or bind weakly, are not presented by the MHC molecules (Babbitt et al. 1985, 1986; Buus et al. 1987). On the other hand, among the many peptides that do bind, only selected peptides are immunodominant.

Our own studies on antigen processing examined in detail the response to the protein hen egg-white lysozyme (HEL) in mice of the H-2^k genotype. We found that a great many of the T cells from mice immunized with HEL responded to a tryptic peptide from residues 46–61 (Allen et al. 1984, 1987). When the response to HEL (46–61) was examined in detail, it became apparent that the 10 mer from residues 52–61 was responsible for its binding affinity to the class II molecule I-A^k (see Table 1 for the sequence of 52–61). HEL (52–61) bound to isolated and purified I-A^k molecules with affinities from 10^{-6} to 10^{-7} M (Babbitt et al. 1985); the isolated complex alone was sufficient to trigger specific T-cell hybridomas (Babbitt et al. 1986). The 52–61 sequence contained some amino acid residues critical for binding to I-A^k and others contained residues critical for recognition by the T-cell receptor (Allen et al. 1987). The view emerged that the peptides derived from processing adopted a novel conformation as they interacted with class II MHC (Table 1).

The studies described in this paper focused on two major issues: (1) to establish the biochemical basis and intracellular site of processing and (2) to establish the nature of HEL peptides physiologically processed by the APCs. References are limited to pertinent studies that apply to our experiments and are not intended to be comprehensive of the entire literature.

Table 1. Sequence of I–A^k-associated Peptides

Fraction	Source	Peptide sequence	Position	Length	Yield ± S.D. (%)
1	HEL	D G S T <u>D Y G I L Q I N S R</u>	48–61	14	7.7 ± 1.8
2	HEL	D G S T <u>D Y G I L Q I N S</u>	48–60	13	3.3 ± 1.0
3	HEL	D G S T <u>D Y G I L Q I N S R</u> W	48–62	15	37.8 ± 6.7
4	HEL	D G S T <u>D Y G I L Q I N S R</u> W	48–62	15	33.8 ± 5.3
5	HEL	<u>D Y G I L Q I N S R</u> W W (C)	52–63(64)	12(13)	17.4 ± 8.8
6	hsp70	I I A N D Q G N R T T P S Y	28–41	14	
7	I–A^K β-chain	T P R R G E V Y T C H V E H P	165–179	15	
8	S30 ribosomal	K V H G S L A R A G K V R G Q T P K V A K Q	75–96	22	
		A G K V R G Q T P K V A K Q E K K K K K T	83–103	21	
9	Ryudocan	E P L V P L D N H I P E N A Q P G	84–100	17	

Underlined is the sequence from residues 52 to 61, previously identified as representing the immunodominant peptide of HEL in I–A^k-bearing mice. Last column refers only to the HEL peptides isolated from I-A^k.

MATERIALS AND METHODS

The APCs used in our studies are lines of B-cell hybridomas or lymphomas that express I-A^k. Also included are peritoneal macrophages isolated from CBA/J mice (H-2^k). The T-cell hybridomas show specificity for HEL (52–61) bound to I-A^k (Allen and Unanue 1984). The antigen presentation assays were described previously (Allen and Unanue 1984). Isolation of I-A^k, binding assays using synthetic peptides, and cellular fractionations are also found in our published studies (Luescher and Unanue 1990; Harding et al. 1991a). The liposomes are described in the studies from Huang's laboratory (Connor et al. 1984) and from our own laboratory (Harding et al. 1991a,b).

RESULTS

Cellular Studies

One of the issues that we recently examined is the number of peptide-MHC complexes in an APC required to trigger a T cell (Harding and Unanue 1990). This point becomes important in the context of how the immune system discriminates between self and non-self protein antigens. Peptides derived from autologous proteins show the same properties as foreign peptides to bind or not to bind a given MHC protein (Babbitt et al. 1985). Therefore, the issue becomes critical in determining how much competition there may be between self-proteins internalized by APCs and foreign proteins. In our experiments, we fed APCs a given amount of I^{125}-labeled HEL (52–61) for 4 hours and then examined two parameters: (1) the amount of peptide bound to I-A^k (i.e., aliquots of cells were lysed and I-A^k was immunoprecipitated with specific monoclonal antibodies) and (2) the response by a known T-cell hybridoma (i.e., after the incubation, the cells were washed, fixed, and incubated with the T-cell hybridoma, after which the amounts of IL-2 were measured). We measured the number of stimulatory complexes at two concentrations of peptides, 100 nM and 300 nM, and estimated values of 210 and 340 molecules, respectively, per APC. Such concentrations of peptide resulted in stimulation of the T-cell hybridomas. The amounts of MHC-containing peptides represented about 0.1% of the total in the APCs. We calculated that a concentration of endogenous competing peptide on the order of 10^{-4} M was needed to explain the low degree of occupancy by competition. Regardless of whether this is the explanation, the results show that T cells are remarkably sensitive and can be triggered by very few molecules. Similar results were also obtained by Demotz et al. (1990). This low level of engagement of antigen-specific receptor raises the issue of the contribution of the peptide-MHC and T-cell receptor to the physical interactions of both cells. The physical interactions could be explained by the observations that triggering the T-cell receptor results in activation of the adhesion molecules of T cells, like the lymphocyte functional antigen 1 (LFA-1) (Dustin and Springer 1989). We have recently confirmed this important observation from Springer's laboratory using the peptide-MHC complexes as ligand to trigger the T-cell receptor (Harding and Unanue 1991).

A molecule such as HEL requires two biochemical events for processing. Since HEL is a compact globular protein with four disulfide bonds, it must be denatured by reduction of its disulfides. Second, the denatured protein, of course, first must be catabolized to peptides that can bind to I-A^k. We believe that both of these events take place in late endocytic compartments compatible with lysosomes.

Incubation of HEL with lysosomal enzymes did not result in the efficient generation of bioactive peptides unless the protein was first denatured (Fig. 1). Our evidence points to the lysosomal compartment as the site where reduction takes place (Collins et al. 1991). We used a reagent developed by Bretscher and Lutter (1988), where a ^{125}I-labeled Gly-Tyr peptide was conjugated via disulfide linkage to either transferrin or α_2-macroglobulin labeled with I^{131}. The ratio of both I labels was followed in the cell and taken as indication of reducing activity. Although the transferrin conjugate was not reduced, the α_2-macroglobulin conjugate was affected after about 15–20 minutes of internalization in a high-density compartment. Thus, early endosomes did not show reducing properties by Percoll fractionation of the intracellular organelles (Fig. 2) and only late compartments in the endocytic pathway. Other workers have identified a transport system in lysosomes for

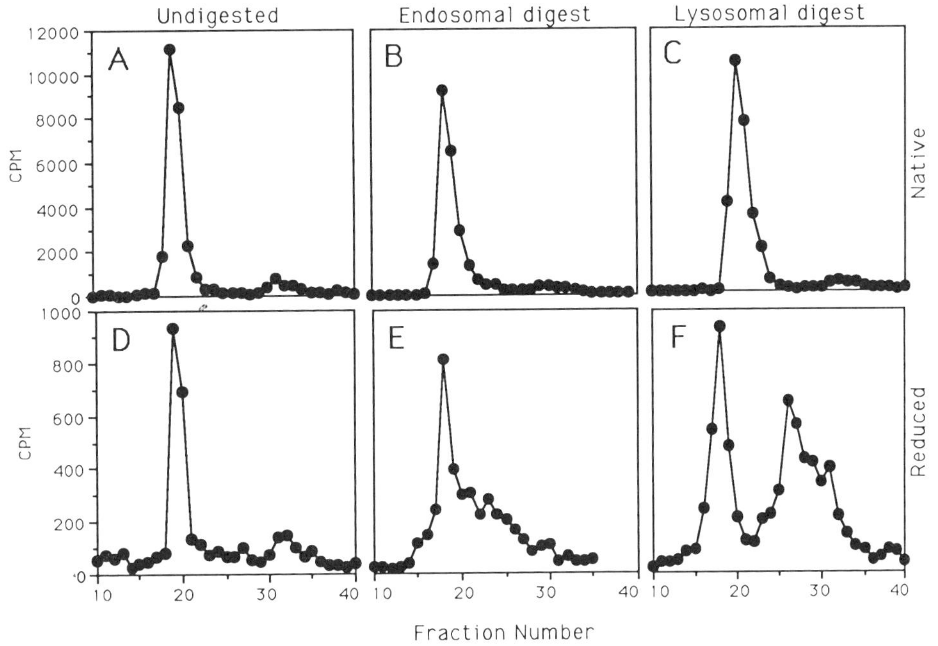

Figure 1. Catabolism of HEL requires reduction of its disulfides. Murine macrophages were isolated, and their intracellular vesicles were fractionated in Percoll gradients into endosomes and lysosomes. Each fraction was solubilized and incubated with HEL for 2 hr at 37°C at pH 5.5. One set was incubated in the presence of dithiothreitol. The figure shows the samples after gel filtration using Bio-Rad TSK-125 columns. (Reprinted, with permission, from Collins et al. 1991.)

cysteine that could act as the reducing agent (Pisoni et al. 1990). Finally, the reduction of the protein allows not only a higher propensity to proteolysis, but also the generation of T-cell epitopes in peptides containing cysteine linkages (Collins et al. 1991; Jensen 1991).

The site of generation of peptides in the APCs was first speculated to be the endosomal system. In endosomes, some limited proteolytic activity would generate the peptides that then become available to internal class II molecules. At a later stage, proteins would be extensively catabolized in lysosomes and lost to the immune system. Our experiments indicate that it is in late endocytic-lysosomal-type vesicles that immunogenic peptides are generated. Our approach was to encapsulate HEL in lysosomes of different chemical compositions (Connor et al. 1984) and to follow them in the APCs. Liposomes made of a mixture of dioleoylphosphatidylethanolamine and palmitoylhomocysteine were stable at neutral pH but were destabilized at a pH of 6.5 or lower, releasing their content in early endosomes. In contrast, liposomes made up of a mixture of dioleoylphosphatidylcholine or dioleoylphosphatidylserine were stable at wide ranges of pH and did not release their content until they reached lysosomes as a result of their enzymatic lysis. Thus, our system allowed us to examine the immunogenicity of HEL released in endosomes (and from there flowing to various compartments) or released selectively in lysosomes. HEL released in lysosomes was highly immunogenic, in fact, much more than similar amounts given as soluble HEL or contained in the pH-sensitive liposomes (Figs. 3 and 4). Similar results were obtained with three other protein antigens (Harding et al. 1991b).

These two sets of experiments indicate that lysosomal compartments are key vesicles for generating the immunogenic peptides. The issues arise therefore on where the peptides bind to class II MHC molecules. About three quarters of class II MHC molecules of the APCs are on the plasma membrane and the remaining are in intracellular compartments. In macrophages, the intracellular class II was localized by immunoelectron microscopy to various endocytic vesicles, in early endocytic vesicles as well as in a second type that contained some lysosomal proteins such as lysosomal associated membrane protein-1 (Lamp-1) and cathepsin D. In B cells, particularly in lymphoblastoid lines, Peters et al. (1991) identified a vesicle late in the endocytic pathway that contained class II MHC molecules and lysosomal enzymes. This late endosome or lysosome is morphologically very similar to the second type of vesicle of the macrophages and is a likely candidate for the site of coupling of peptide with class II MHC. This compartment may also represent the same site where the protein is degraded (and which was targeted in the liposome experiments described above) or it may be a site different from degradation, therefore implying that the peptides require transport from one site to another (i.e., from the denaturing/proteolytic compartment to the class II MHC coupling site). Regardless, the site of coupling is definitely intracellular. We examined this point directly in an experiment in which macrophages from one MHC strain were fed HEL and cocultured with a second set of macrophages of a different MHC allele that were not fed HEL. We then used T cells that recognized HEL only in the context of the MHC alleles of the second set. Such T cells were not triggered,

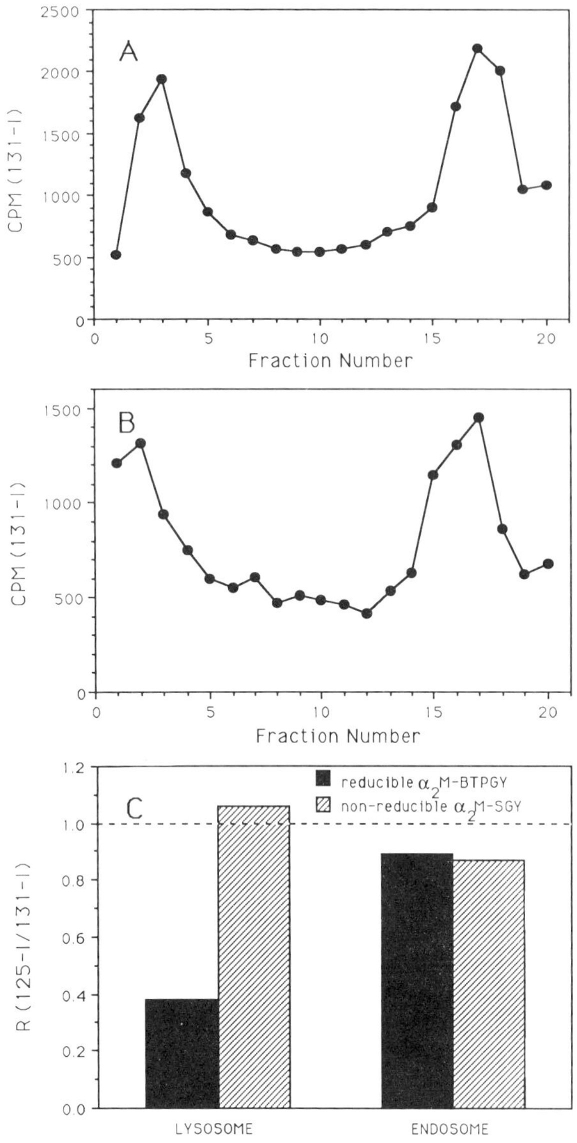

Figure 2. Disulfide-reducing activity is found in lysosomes. Macrophages were incubated for 30 min with α_2-macroglobulin linked to a reducible compound (*A*) and to a nonreducible conjugate (*B*). The conjugate consisted of ^{131}I-labeled α_2 macroglobulin and ^{125} I-labeled peptide made of glycine and tyrosine. The α_2-macroglobulin was labeled with I^{131}, and a peptide made of glycine and tyrosine was labeled with I^{125}. The profile of I^{131} is shown in *A* and *B*. In *C*, the cells were lysed, and their organelles were fractionated by Percoll sedimentation into endosomes and lysosomes. The ratio of each label was measured after precipitation by α_2-macroglobulin. Note that the protein loses its I^{125} label when it arrives in the lysosomes. (Reprinted, with permission, from Collins et al. 1991.)

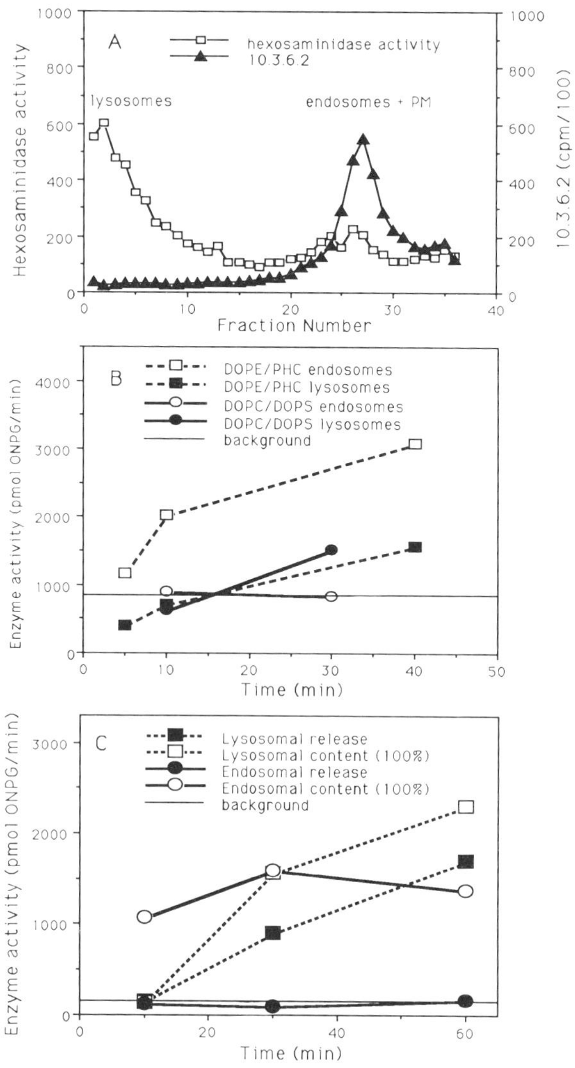

Figure 3. Release of lipsosome contents in different subcellular fractions. Macrophages were incubated with liposomes made as described in the text. Each incubation contained a mixture of the same liposomes of the same compositions; some contained FITC-dextran, and others contained anti-FITC IgG. The cells were then fractionated in endosomes and lysosomes. The degree of release was estimated by the isolation of free FITC or FITC as part of an immune complex. (*A*) Percoll fractionation profile, which also includes that of membranes incubated with anti-I-A^k antibody; (*C*) measurements of release and the amounts targeted to a given compartment of the acid-resistant liposomes.

indicating the absence of peptides released from one macrophage to another (Harding et al. 1990).

One issue that has been examined by several laboratories is whether the class II MHC molecules that bind peptides represent recently produced molecules or molecules that may be recycling between the plasma membrane and the cell interior. Although some evidence for recycling has been obtained, it is possible that this represents a small pathway that varies among different APCs. In our experience, the inhibition of protein synthesis reduced presentation of processed protein in macrophages, an indication that recently synthesized class II MHC molecules may represent the key molecules in coupling with peptides.

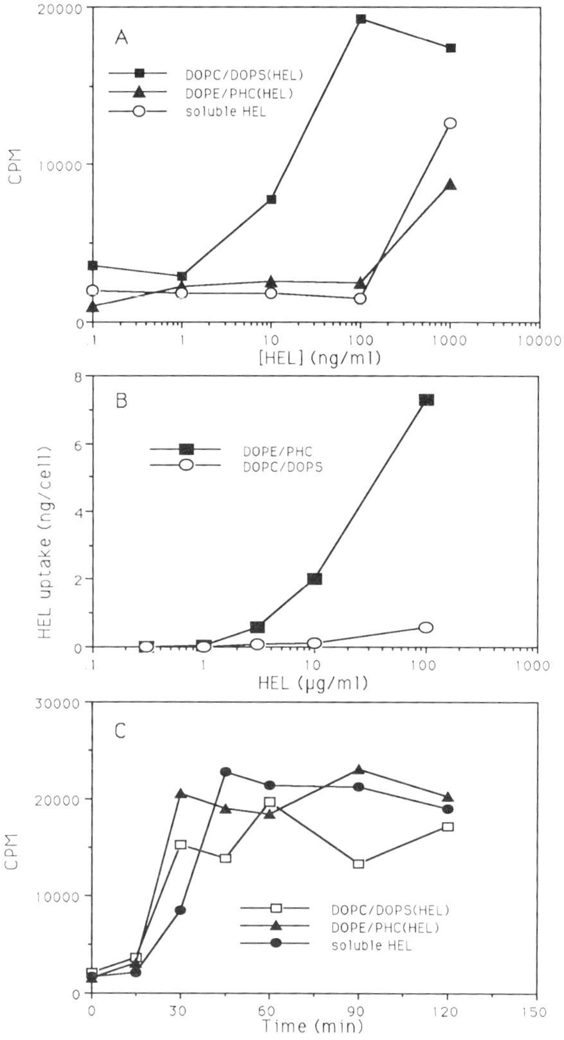

Figure 4. Processing and presentation of the liposome-encapsulated HEL. Macrophages were incubated with HEL encapsulated in the preparation indicated of liposomes or with soluble HEL, fixed after 2 hr in *A* and *B*, or at the indicated times in *C*. The amounts of I-A^k-peptide complex were assayed by the addition of a specific T-cell hybridoma. (*A,C*) Amounts of IL-2 secreted are depicted in the vertical axis. (*B*) Degree of uptake of HEL in each liposome; note the low degree with the acid-resistant preparations. (*C*) Kinetics of presentation.

Identification of Immunogenic Peptides Bound to MHC

One approach to understand determinant selection is to isolate and examine peptides bound to MHC proteins resulting from the processing of a protein antigen. The biochemical analysis of peptides from class I or class II MHC molecules has now been successfully attempted. Peptides extracted from class I molecules are of eight to nine amino acids in length and demonstrate binding motifs specific to a given allele (Van Bleek and Nathenson 1990; Falk et al. 1991; Jardetsky et al. 1991; Hunt et al. 1992). The peptides extracted from class II MHC molecules are more heterogeneous, in the limited analysis so far reported (Buus et al. 1988; Rudensky et al. 1991; Srinivasan et al. 1991). In our studies, we cultured APCs for 12 hours with HEL and then extracted and isolated the I-A^k molecule using detergent and specific monoclonal antibodies. The I-A^k molecule was then precipitated, and the peptides were extracted after elution by low pH treatment (Nelson et al. 1992). The peptide profile obtained from reverse-phased high-performance liquid chromatography (HPLC) on C18 columns is shown in Figure 5; panels A and B show five major peptides isolated after incubation with HEL. The amino acid sequence, determined

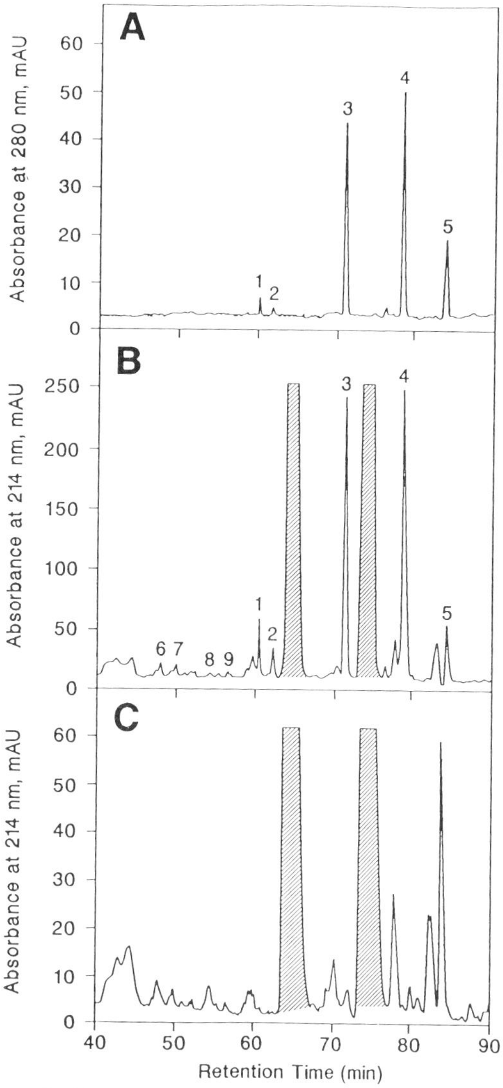

Figure 5. HPLC profiles of peptides eluted from purified I-A^k and fractionated on a C18 column. M12.C3.F6 cells (80 liters) were cultured in 1 mg/ml HEL, and the class II MHC was extracted and isolated. (*A*) Profile at 280 nm absorbance; (*B*) profile at 214 nm absorbance. (*C*) Acid-eluted peptides recovered from M12.C3.F6 cells cultured in media alone, at 214 nm absorbance. The shaded peaks in the 214-nm profiles represent the detergents (Mega 8/Mega 9) used in the purification. Sequence data were obtained for the numbered peaks (Table 1).

by Edman degradation, indicated that the five peptides contained all or part of the sequence from residues 52–61. The sequences of the five peptides are shown in Table 1. All of the peptides extended beyond the 52–61 sequence by several amino acids, either on the amino or carboxyl end. Recall that the 52–61 sequence binds to the allele-specific combining site.

Our results explain the immunodominance of the 10-mer sequence. Indeed, the 52–61 sequence was included in all five major peptides presented by I-A^k. Our results also indicate the size heterogeneity of peptides selected by the class II MHC molecules. Of the five isolated peptides, four were immunogenic for T-cell hybridomas specific for the 52–61 sequence.

We have speculated on the reasons behind the selection by the APCs of the 52–61 with additional residues on both ends (Nelson et al. 1992). The fragments were selected from the primary proteolytic process, the nature of which is not known at present and which would establish the chemical composition of the major peptides. The longer fragments appear to have certain advantages. Perhaps the peptides with longer ends represent fragments that have higher affinities for I-A^k, although preliminary data indicate that this is not the case.

Prior to these experiments, Luescher et al. (1990) had carried out an examination of I-A^k molecules bound to photoactivatable compounds attached to the 52–61 peptide or to longer sequences. Our purpose was to identify contact sites in the I-A^k molecule for the photoprobe. Our major results established that the photoprobe (iodo-salicylic acid) bound to 52–61 labeled a hydrophilic peptide of I-A^k; however, the same probe bound to 46–61 labeled a more hydrophobic site. This site was identified in a region in either the α_2 or β_2 domains of I-A^k. Somehow, the longer peptide had extended beyond the allele-specific site and established contact beyond it with the invariant domains α_2 or β_2.

The best evidence that we have at this time is that the longer peptides could help stabilize the α/β heterodimer. The class II MHC molecules are membrane-spanning molecules that show some conformational flexibility. It is possible that peptides help to stabilize such a heterodimer. Recent experiments from McConnell's (Dornmair and McConnell 1990) and Germain's (Germain and Hendrix 1991) laboratories indicated that class II MHC molecules could be found when electrophoresed in SDS-PAGE, under conditions where the protein was not boiled, in at least two states: a stable dimer where the α and β chains migrate together or a loose form where they dissociate. The amounts of stable versus loose forms varied among APCs of different strains. Our preliminary experiment indicates that longer peptides fed to APCs favor the formation of the stable molecules.

Finally, it is of interest to examine the molecules that are spontaneously found in the MHC isolated from APCs, i.e., the library of natural peptides normally being presented. That APCs are normally presenting natural peptides has not been well established using immunological (Lorenz and Allen 1988) and biochemical methods (cited above). The natural peptides extracted from APCs include peptides from proteins naturally found in the vesicular system (transferrin, invariant chain, class I or II molecules), as well as molecules that are normally in the cytosol. For example, we found two ribosomal proteins and an hsp70 peptide. This indicates that, as with the class I system, there is a mechanism for transport of cytosolic proteins to the class-II-MHC-bearing compartment. So far, very few peptides from extracellular proteins have been identified, but the limited studies have only used B-cell lines as APCs, cells where the rate of endocytosis is limited.

DISCUSSION

The scenario emerging from the analysis of antigen presentation is that the MHC is a peptide-binding system that samples the inside of the cell and transports peptides to the plasma membrane. At that site, the T lymphocyte system recognizes the complex and is activated, or not, depending on a wide range of cellular conditions. We have arrived at a stage where we can examine in quantitative, biochemical, and cellular terms many of the elements that establish the immunogenicity of a protein antigen. For the class II MHC system, which constitutes the work that we have summarized from our laboratory, it is largely the extracellular milieu that is being sampled. Of course, this represents a critical system to induce T-cell reactivities to foreign proteins that reach the circulation and, very importantly, to a range of pathogenic microorganisms such as intracellular pathogenic bacteria and viruses.

The analyses of the peptides derived from proteins that are internalized fit some of the predictions made from immunological studies: Peptides that have high affinity for class II MHC are selected by the APCs. In fact, in our system, determinant selection is well explained by the finding of a single peptide sequence, which is the major determinant recognized by CD4 T cells of mice immunized with the protein. At this time, cellular studies need to be done to define the precise intracellular stages and compartments where proteolysis of the protein and coupling to MHC take place. The biochemical analysis of the peptides tells us that peptides with ends that extend beyond the combining site of class II MHC are critical. This may indicate that longer peptides have an advantage by nonspecific binding areas that help to stabilize the class II molecule.

REFERENCES

Allen, P.M. and E.R. Unanue. 1984. Antigen processing and presentation by macrophages. *Am. J. Anat.* **170:** 483.

Allen, P.M., D.J. Strydom, and E.R. Unanue. 1984. Processing of lysozyme by macrophages: Identification of the determinant recognized by two T cell hybridmas. *Proc. Natl. Acad. Sci.* **81:** 2489.

Allen, P.M., G.R. Matsueda, R.J. Evans, J.B. Dunbar, Jr., G. Marshall, and E.R. Unanue. 198. Identification of the T-cell and Ia contact residues of a T-cell antigenic epitope. *Nature* **27:** 713.

Babbitt, B.P., P.M. Allen, G. Matsueda, E. Haber, and E.R. Unanue. 1985. Binding of immunogenic peptides to Ia histocompatibility molecules. *Nature* **317:** 359.
Babbitt, B.P., G. Matsueda, E. Haber, E.R. Unanue, and P.M. Allen. 1986. Antigen competition at the level of peptide-Ia binding. *Proc. Natl. Acad. Sci.* **83:** 4509.
Bretscher, M.S. and R. Lutter. 1988. A new method for detecting endocytosed proteins. *EMBO J.* **7:** 4087.
Buus, S., A. Sette, S.M. Colon, and H.M. Grey. 1988. Autologous peptides constitutively occupy the antigen binding site of Ia. *Science* **242:** 1045.
Buus, S., A. Sette, S.M. Colon, C. Miles, and H.M. Grey. 1987. The relation between major histocompatibility complex (MHC) restriction and the capacity of Ia to bind immunogenic peptides. *Science* **235:** 1353.
Collins, D.S., E.R. Unanue, and C.V. Harding. 1991. Reduction of disulphide bonds within lysosomes is a key step in antigen processing. *J. Immunol.* **147:** 4054.
Connor, J., M.B. Yatvin, and L. Huang. 1984. pH sensitive liposomes: Acid-induced fusion. *Proc. Natl. Acad. Sci.* **81:** 1715.
Demotz, S., H.M. Grey, and A. Sette. 1990. The minimal number of class II MHC-antigen complexes needed for T cell activation. *Science* **249:** 1028.
Dornmair, K. and H.M. McConnell. 1990. Refolding and reassembly of separate alpha and beta chains of class II molecules of the major histocompatibility complex leads to increased peptide-binding capacity. *Proc. Natl. Acad. Sci.* **87:** 4134.
Dustin, M.L. and T.A. Springer. 1989. T-cell receptor cross-linking transiently stimulates adhesiveness through LFA-1. *Nature* **341:** 619.
Falk, K., O. Rotzschke, S. Stevanovic, G. Jung, and H.-G. Rammensee. 1991. Allele-specific motifs revealed by sequencing of self-peptides eluted from MHC molecules. *Nature* **351:** 290.
Germain, R. and L.R. Hendrix. 1991. MHC class II structure, occupancy and surface expression determined by post-endoplasmic reticulum antigen binding. *Nature* **353:** 134.
Harding, C.V. and E.R. Unanue. 1990. Quantitation of antigen-presenting cell MHC class II/peptide complexes necessary for T-cell stimulation. *Nature* **346:** 574.
———. 1991. Modulation of antigen presentation and peptide-MHC-specific, LFA-1-dependent T cell-macrophage adhesion. *J. Immunol.* **147:** 767.
Harding, C.V., D.S. Collins, O. Kanagawa, and E.R. Unanue. 1991a. Liposome-encapsulated antigens engender lysosomal processing for class II MHC presentation and cytosolic processing for class I presentation. *J. Immunol.* **147:** 2860.
Harding, C.V., D.S. Collins, J.W. Slot, H.J. Geuze, and E.R. Unanue. 1991b. Liposome-encapsulated antigens are processed in lysozomes, recycled, and presented to T cells. *Cell* **64:** 393.
Harding, C.V., E.R. Unanue, J.W. Slot, A.L. Schwartz, and H.J. Geuze. 1990. Functional and ultrastructural evidence for intracellular formation of major histocompatibility complex class II-peptide complexes during antigen processing. *Proc. Natl. Acad. Sci.* **87:** 5553.
Hunt, D.F., R.A. Henderson, J. Shabanowitz, K. Sakaguchi, H. Michel, N. Sevilir, A.L. Cox, E. Appele, and V.H. Engelhard. 1992. Characterization of peptides bound to the class I MHC molecule HLA-A2.1 by mass spectrometry. *Science* **255:** 1261.
Jardetzky, T.S., W.S. Lane, R.A. Robinson, D.R. Masdden, and D.C. Wiley. 1991. Identification of self-peptides bound to purified HLA-B27. *Nature* **353:** 326.
Jensen, P.E. 1991. Reduction of disulfide bonds during antigen processing. Evidence from a thiol-dependent insulin determinant. *J. Exp. Med.* **174:** 1121.
Lorenz, R.G. and P.M. Allen. 1988. Direct evidence for functional self protein/Ia-molecule complexes in vivo. *Proc. Natl. Acad. Sci.* **85:** 5220.
Luescher, I.F. and E.R. Unanue. 1990. Purification and photoaffinity labeling of the I-A^k histocompatibility molecule. *J. Immunol. Methods* **135:** 233.
Luescher, I.F., D.L. Crimmins, B.D. Schwartz, and E.R. Unanue. 1990. The sites in the I-A^k histocompatibility molecule photoaffinity labeled by an immunogenic lysozyme peptide. *J. Biol. Chem.* **265:** 11177.
Nelson, C.A., R.W. Roof, D.W. McCourt, and E.R. Unanue. 1992. Identification of the naturally processed form of hen-egg white lysozyme bound to the murine MHC class II molecule, I-A^k. *Proc. Natl. Acad. Sci.* **89:** 7380.
Peters, P.J., J.J. Neefjes, V. Oorschot, H.L. Ploegh, and H.J. Geuze. 1991. Segregation of MHC class-II molecules from MHC class-I molecules in the Golgi complex for transport to lysosomal compartments. *Nature* **329:** 668.
Pisoni, R.L., T.L. Acker, K.M. Lisowski, R.M. Lemons, and J.G. Thoene. 1990. A cysteine-specific lysosomal transport system provides a major route for the delivery of thiol to human fibroblast lysosomes: Possible role in supporting lysosomal proteolysis. *J. Cell Biol.* **110:** 327.
Rosenthal, A.S., M.A. Barcinski, and J.T. Blake. 1977. Determinant selection is a macrophage dependent immune response gene function. *Nature* **267:** 156.
Rudensky, A.Y., P. Preston-Hurlburt, S.-C. Hong, A. Barlow, and C.A. Janeway, Jr. 1991. Sequence analysis of peptides bound to MHC class II molecules. *Nature* **353:** 622.
Shimonkevitz, R., S.M. Colon, S. Buus, and H.M. Grey. 1984. Antigen recognition by H-2 restricted T cells. II. A tryptic ovalbumin peptide that substitutes for processed antigen. *J. Immunol.* **133:** 2067.
Srinivasan, M., E.W. Marsh, and S.K. Pierce. 1991. Characterization of naturally processed antigen bound to major histocompatibility complex class II molecules. *Proc. Natl. Acad. Sci.* **88:** 7928.
Unanue, E.R., C.V. Harding, I.F. Luescher, and R.W. Roof. 1989. Antigen-binding function of class II MHC molecules. *Cold Spring Harbor Symp. Quant. Biol.* **54:** 383.
Van Bleek, G.M. and S.G. Nathenson. 1990. Isolation of an endogenous processed immunodominant viral peptide from the class I H-2^{kb} molecule. *Nature* **348:** 213.

MHC Products: Biosynthesis, Intracellular Traffic, and "Empty" Molecules

H.L. PLOEGH
Center for Cancer Research, Department of Biology, Massachusetts Institute of Technology, Cambridge, Massachusetts 02139

The B- and T-cells of the immune system possess two different classes of receptors by which antigen-specific recognition is achieved. A fundamental difference in antigen recognition is that the B-cell receptors for antigen, the immunoglobulins, interact with *intact* protein antigens, whereas their counterparts on T lymphocytes recognize short *fragments* of antigens—peptides of 8–20 residues in length—held tightly by the products of the major histocompatibility complex (MHC), which are cell-surface glycoproteins (Braciale and Braciale 1991; Neefjes et al. 1991). This latter attribute forces the attention of the T cells to focus on the cell surface. Because of the polymorphic nature of the MHC products, the architecture of the peptide-binding portions of MHC products shows considerable variation for the different alleles. Differences in residues that line the peptide-binding pockets dramatically influence the sets of peptides that can be bound by allelic products. However, not all polymorphic positions are directly involved in peptide binding, and some of them are placed such that they would interact with the T-cell receptor. A series of events that take place in the thymus ensure that T cells "learn" to recognize the MHC products on the thymic cells with which they interact in the course of T-cell development. The T cells thus instructed will not, as a rule, interact with cells from an individual that differs with respect to its MHC products. For example, a T cell from individual X, specific for the combination of an antigenic peptide and an MHC, cannot usually recognize the same antigenic peptide when it is presented by a different MHC molecule. Thus, the polymorphism of MHC products contributes not only to peptide-binding specificity, but also to establishing MHC restriction. These two aspects are intimately linked in development (Van Kaer et al. 1992).

A very useful classification has been the distinction between MHC class I and MHC class II products. Both types of products are heterodimers, as expressed at the cell surface. The former are anchored in the lipid bilayer by a single stretch of hydrophobic residues (although the possibility of phosphatidylinositol glycan linkage is an alternative used by some of the so-called non-classical MHC class I products), whereas the latter consist of two transmembrane subunits. Almost without exception, MHC class I products present their peptide to T-cell receptors on $CD8^+$ cells, and MHC class II molecules serve as the restricting elements for $CD4^+$ T cells. The structure and function of the CD4 and CD8 coreceptors are discussed by Hendrickson et al. and Harrison et al. (both this volume). A further key observation was the almost exclusive use of MHC class II molecules to present peptides derived from proteins that somehow gain access to the endocytic pathway (for review, see Braciale and Braciale 1991; Neefjes and Ploegh 1992a). In contrast, class I molecules combine with peptides that are derived from proteins synthesized by the class-I-expressing cell itself (Goldberg and Rock 1992). An interesting complication is the predominantly cytosolic disposition of the proteins that generate class-I-bound peptides, whereas the peptide-binding site on class I molecules itself is located on the luminal side (topologically equivalent with the outside of the cells) when peptide binding occurs (see below). The solution to this paradox is a specialized mechanism by which peptides, generated in the cytosol, are translocated across the ER membrane by a heterodimer of two multi-transmembrane segment-containing proteins, the products of the TAP-1 and TAP-2 genes (see also Click et al., this volume). These products belong to the family of ATP-binding cassette-containing transporters (DeMars and Spies 1992). To date there is no evidence either for involvement of ATP or for transport of short peptides (as opposed to much larger fragments).

Several questions have emerged from the general picture sketched above, two of which I address specifically. If there is indeed a dichotomy in the compartments sampled by MHC class I and class II molecules and hence the routes of antigen presentation (often referred to as the "endogenous" and "exogenous" pathways of antigen presentation, respectively), are these differences reflected by the intracellular pathways traveled by MHC products? The results obtained in several laboratories, and summarized below, show that this is indeed the case.

Implicit in the scheme of antigen presentation is the notion that MHC products are initially assembled in the absence of peptide and bind them only later. Does the cell discriminate between "empty" and peptide-filled molecules, and if so, what are the consequences? The available evidence suggests that peptide fulfills an essential structural role, in addition to serving as the nominal antigen recognized by T cells.

Trafficking of MHC Products

The trafficking of MHC products has been analyzed

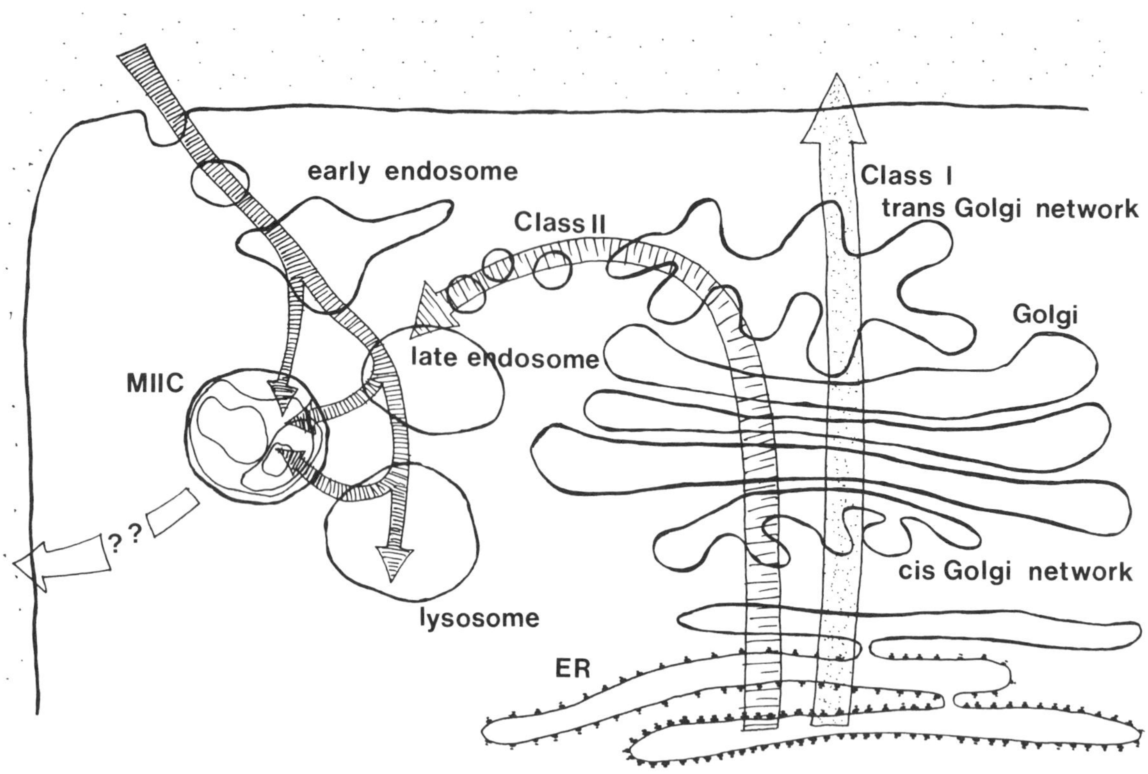

Figure 1. Schematic representation of the pathways traveled by MHC class I and MHC class II molecules. The major pathway traveled by class I molecules (stippled arrow), class II molecules (hatched arrow), and exogenously provided antigens (finely hatched arrow) are indicated. Multiple routes to the MHC class II compartment (MIIC) are available and may be used to different extents in different cell types. The exact point of entry of class II molecules must be considered under debate, but for human Epstein-Barr virus-transformed cells, a relatively late point of entry is favored. Following acquisition of peptide, the pathway traveled by loaded class II molecules to the cell surface is not known: It may involve a direct transfer from MIIC to the cell surface (open arrow with question mark) or, alternatively, parts of the endocytic pathway may be traveled in a retrograde fashion.

by a combination of biochemical and morphological methods (see Fig. 1 for schematic outline) (Teyton et al. 1990; Peters et al. 1991; Baas et al. 1992). MHC class I and class II proteins both start their journey in the ER where signal sequence cleavage (Dobberstein et al. 1979; Ploegh et al. 1979), cotranslational glycosylation, and covalent modifications such as disulfide bonds are completed. Subunit association likewise takes place early in the course of biosynthesis. For class I molecules, heavy chains associate rapidly with the light chain, β_2-microglobulin (β_2m) (Stam et al. 1986; Krimpenfort et al. 1987; Rein et al. 1987). As described below, the antigenic peptide should perhaps also be considered a class I subunit, as its association with class I molecules has important consequences for surface expression and stability of the complex. The class II molecules associate in the ER with a third subunit, a type II transmembrane protein referred to as the invariant chain or Ii (Cresswell 1992). The invariant chain, in particular its cytoplasmic tail, contains information necessary and sufficient to direct targeting of class II molecules to the appropriate compartments (Viville et al. 1993). Until the class I and class II products reach the trans-Golgi they travel together, and it is only in the trans-Golgi network (TGN) that they separate. By immunoelectron microscopy performed on human B-lymphoblastoid cell lines, areas of the TGN selectively labeled for MHC class I, MHC class II, and the mannose-6-phosphate receptor have been observed. In close proximity to the TGN, small profiles heavily labeled for MHC class II molecules are observed (Peters et al., 1991), but by far the most prominent accumulation is seen in compartments that can be recognized morphologically by having considerable amounts of internal membrane. These compartments have been referred to as MHC class II compartments, or MIIC (Peters et al. 1991). When analyzed for the presence of a series of cytochemical markers, these MIIC were positive for a number of lysosomal markers, including LAMP and cathepsin D, but were not stained by antibodies against the late endosomal marker, mannose-6-phosphate receptor. Using a variety of endocytic tracers, including cationized ferritin, colloidal-gold-conjugated bovine serum albumin (BSA), and horseradish peroxidase (HRP), again colocalization of the endocytic tracers and class II molecules was observed only relatively late in the endocytic pathway. We have not observed any colocalization of MHC class II molecules and transferrin-HRP, internalized via the transferrin receptor and delivered predominantly to early endosomes. Combined, these data argue in favor of an accumulation of MHC class II molecules late in the endocytic pathway. Biochemical experiments argue in favor of an intersection of the biosynthetic and endocytic pathways (Neefjes et al. 1990). Although the association of class II molecules with invariant chain ensures delivery to the endocytic pathway, its removal is required to allow class II molecules to bind peptide (Roche and Cresswell 1990; see also Click et al., this volume and references therein) and also to allow the egress of class II molecules from the endocytic pathway to the cell surface (Neefjes and Ploegh 1992b). Interference with the breakdown of invariant chain by exposing cells to lysosomotropic drugs, such as chloroquine or leupeptin, has multiple effects: Because class II molecules do not bind peptide when associated with invariant chain, the conformational transitions usually observed when class II molecules bind peptide no longer occur. Second, the continued presence of invariant chain leads to retention of the class II molecules in the endocytic pathway (Cresswell et al. 1990; Teyton et al. 1990). Because the proteases involved in invariant chain breakdown are in all likelihood the same as those involved in generating the antigenic peptides presented by class II molecules, the effects of lysosomotropic drugs on antigen presentation are complex.

Peptide-MHC Class I Interactions

Peptides are essential subunits that allow the MHC products that have bound them to function as restriction elements in intact cells. In the recent past, the use of mutant cell lines and advances in the production of large quantities of each of the MHC subunits have contributed to our understanding of how these molecules are put together (Townsend et al. 1989; Ljunggren et al. 1990; Bijlmakers and Ploegh 1993). On the one hand, the success in achieving reconstitution and even crystallization of MHC molecules (Elliott and Eisen 1990; Silver et al. 1991; Garboczi et al. 1992; Zhang et al. 1992) from their subunits produced with recombinant DNA technology (Parker et al. 1992) shows that assembly does not require the environment of the living cell. In this respect, MHC class I molecules behave as do other multisubunit structures that have been successfully reassembled in vitro. Thoughts on protein folding, based largely on refolding studies of unfolded polypeptides, suggest that the folding pathways under these in vitro conditions need not be identical to those observed in the living cell: The absence of chaperonins that might aid folding in the living cell and the likelihood of protein folding being well on its way prior to completion of the polypeptide chain are just two of the obvious differences. At present, there are no reports that analyze the refolding pathways and assembly of MHC heavy chain and β_2m in the presence of peptide.

In contrast, and perhaps paradoxically, studies performed on intact cells and in detergent extracts have resulted in rather detailed proposals (Townsend et al. 1990; Elliott 1991) for folding and assembly pathways of class I molecules in intact cells. The starting point for these studies was the discovery by Townsend et al. (1989) that the RMA-S cell line, selected in the laboratory of K. Karre for low expression of class I molecules, had a defect in the provision of peptides to class I molecules. By simply including in the tissue culture medium peptides known to be restricted by the H-2b products present on the RMA-S cell line, the expression of class I molecules at the cell surface could be dramatically enhanced. Biochemical analysis of biosynthetically labeled cells exposed to these peptides suggested that peptides induced assembly of class I molecules. To achieve this, peptides would have to be taken up by the cells and delivered to the class I assembly compartment. It could later be shown that cells indeed took up peptide, but that the observed effect of peptide on class I assembly could be explained by release of peptide at the time of lysis and subsequent binding to class I molecules (Townsend et al. 1990). (Let me clarify the term "assembly." Very few of the published experiments are likely to address this issue directly. The most commonly used technique for investigating the assembly and subsequent maturation of class I molecules is the pulse-chase experiment, in which the class I molecules are recovered by immunoprecipitation from detergent extracts. The properties that are scored for are the presence or absence of the epitope(s) recognized by the antibodies used, and the ability of the antibody-class I complex, including coimmunoprecipitating subunits, to survive the washing procedure inherent in immunoprecipitation protocols. If in these experiments both heavy and light chains are recovered in a complex, we may safely assume that assembly has taken place; however, the absence of the sought-after subunit cannot itself be construed as evidence for lack of assembly. We must therefore clearly distinguish between assembly, as in the putting together of the relevant subunits in the physical sense, and assembly, or rather the lack thereof, as judged from the inability to coimmunoprecipitate the relevant subunits.)

The early experiments suggested that unless cells had been fed peptide, no assembly of class I molecules took place. It has since become apparent that rather dilute detergent extracts, after prolonged incubation, will yield predominantly free subunits in immunoprecipitation (Elliott et al. 1991). Much of the experimental evidence that would support peptide-induced assembly relies on the ability of peptide, added to the extract before precipitation, to promote recovery of both heavy and light chains of class I molecules. This ability was exploited to construct an indirect binding assay, based on which a detailed kinetic scheme was put forward that described the possible routes of assembly. A prominent player in this scheme is the free class I heavy chain which (capable of binding peptide) could act as the nucleation point for recruitment of β_2m into the complex. Binding of peptide to free class I heavy chains was in fact demonstrated, but for it to bind, a far greater concentration of peptide was needed than that observed for a heavy chain (HC)-β_2m complex (Elliott et al. 1991). We do not know, even to an approximation, what the concentration of peptide in the class I assembly compartment could be; but given the much greater affinity of the HC-β_2m complex for peptide, the "empty" class I molecule is a most attractive candidate for the physiological intermediate. Recently, Margulies and co-workers (Ribaudo and Margulies 1992) and my own laboratory have succeeded in obtaining mouse class I molecules in an in vitro translation system that supports their assembly. Simultaneous translation of heavy chains and β_2m is required to observe assembly. None of the published reports substantiates peptide binding on the luminal side of the microsome (Bijlmakers et al. 1993; Levy et al. 1991a,b; Ribaudo and Margulies 1992), as opposed to subsequent binding in the microsomal detergent extract. The binding specificity of the class I molecules produced in vitro is indistinguishable from that obtained in intact cells. On the basis of these experiments, it is suggested that under conditions of in vitro translation, the intermediate capable of peptide binding is not the free heavy chain, but the empty heterodimer. It was also observed that if RMA-S cells were lysed at higher densities, and if incubation times were kept short, the preferential form in which class I molecules were recovered in fact consisted of properly formed empty molecules (Elliott et al. 1991). We had proposed earlier that such empty molecules were likely to be conformationally distinct from their "filled" counterparts, because they underwent an irreversible loss of epitopes following exposure to 37°C (Ljunggren et al. 1990; Schumacher et al. 1990). Given the notorious stability of class I complexes isolated from normal cells and the behavior of empty molecules, it seemed plausible to postulate a "receptive" or "open" conformation for empty molecules, which would transform into a closed state upon binding of peptide (Schumacher et al. 1990). This model would result in deviations from the kinetic parameters observed for peptide binding in the simple model where HC-β_2m + peptide is in equilibrium with the peptide-filled complex. The on rates observed for peptide binding, together with the equilibrium binding constants and observed off rates, indeed revealed such inconsistencies (Elliott 1991). The existence of conformationally distinct intermediates therefore remains an attractive proposition.

Assembly In Vitro of Human Class I Molecules

The first paper on in vitro translation of human class I heavy and light chains failed to show assembly (Dobberstein et al. 1979; Ploegh et al. 1979), in all likelihood because poly(A)$^+$ RNA contains only small amounts of class I mRNA, but surely other factors contributed as well. Recently two reports appeared dealing with in vitro translation of HLA-B27 (Levy et al. 1991a,b). These reports differ in several important aspects from the results obtained with mouse class I molecules and seem to diverge from what we presently know about the peptide-binding specificity of class I molecules. First, the experiments reported are performed under completely reducing conditions, a set of circumstances unlikely to support proper disulfide bond formation. For the synthesis in vitro of molecules capable of ligand binding, including three different mouse class I molecules (Ribaudo and Margulies 1992; Bijlmakers et al. 1993), without exception it was found necessary to include oxidized glutathione to promote disulfide bond formation. It is possible that HLA-B27 is an exception to this rule, but I consider this unlikely. A further point of interest concerns the use of the W6/32 antibody in the studies by Levy et al. (1991a,b). Of all anti-class I antibodies, this is probably the best characterized and the most widely used. The W6/32 antibody reacts with all human class I molecules synthesized by mutant cells that lack the TAP1/TAP2 peptide transporter (DeMars and Spies 1992), and such molecules have been shown to consist of HC-β_2m complexes devoid of peptide (with the exception of HLA-A2, which may be unusual in its ability to present signal peptides [Wei and Cresswell 1992]). Because the W6/32 antibody depends on the presence of β_2m, it follows that empty class I molecules react with W6/32. An increase in W6/32 reactivity upon addition of peptide has not been documented, as long as the molecules are not exposed to temperatures above 30°C. For HLA-B27, there is detailed information on the types of peptide it will bind. For 14 of 14 peptides sequenced, a length of 9 residues was observed, and all peptides had an arginine residue at position 2 (Madden et al. 1991). The peptide capable of inducing assembly in vitro had a length of 11 residues and contained a modified amino terminus (Levy et al. 1991a). It has, however, been suggested that a free amino terminus is essential for stable peptide binding, and the residues involved in accommodating the free peptide amino terminus are in fact quite conserved.

Peptide Mixtures as a Tool to Explore Peptide-Class I Interactions

The studies of Rammensee and co-workers (for review, see Rötzschke and Falk 1991) have changed the perception of how to approach MHC-peptide interactions. These workers immunopurified MHC class I products, extracted peptides from them, and subjected the peptides to Edman degradation. If each allelic class I product displays a unique architecture of the peptide-binding site (as has been shown so clearly by X-ray crystallographic analysis), then peptides bound by a particular class I allele are expected to share certain structural features. This turned out to be the case (Rötzschke and Falk 1991). Even though unique sequences were not readily identifiable, owing to the large heterogeneity of the peptide mixtures recovered, particular amino acids showed up very strongly at certain positions in the peptide when sequenced as a pool. Therefore, many of the peptides in these preparations must share that particular amino acid at that position, the so-called "anchor" position. For H-2K^b and spontaneous mutants closely related to it, it was shown by sequencing of radiolabeled, bound peptides that a few amino acid replacements in the peptide-binding pocket sufficed to dramatically change the profile of peptides bound (Van Bleek and Nathenson 1991). My laboratory has taken the approach of synthetic peptide mixtures to explore this issue (Deres et al. 1992; Schumacher et al. 1992).

In a procedure analogous to the synthesis of degenerate oligonucleotides, a number of different amino acids (as opposed to a single one) are incorporated at successive cycles of peptide synthesis, thus resulting in a large number of peptides produced in a single run. By using radiolabeled versions of such peptide mixtures, we could produce a "fingerprint" of the binding specificity of the H-2D^b molecule (Schumacher et al. 1992). We could also demonstrate that 2–3 amino acid replacements in the H-2D^b molecule, as they occur in natural D^b variants, changed this fingerprint. The peptide mixtures used were conceived such that they contained actual T-cell epitopes that could be recognized by virus-specific cloned cytotoxic T lymphocytes and allow class I molecules to select peptides (Schumacher et al. 1991) from this mixture. Even though a given peptide including the viral epitope was present only in one part per 864, such peptide mixtures could still sensitize uninfected target cells for lysis by these cytotoxic T cells (Guimezanes et al. 1992; Schumacher et al. 1992). This experiment shows that synthetic peptide mixtures can be used to search for T-cell epitopes, in those cases where the restricting class I molecules, but not the peptide or protein antigen, are known. The use of such peptide libraries is by no means restricted to the analysis of class I-peptide interactions: Other receptor-ligand or enzyme-substrate interactions could be explored using this approach.

REFERENCES

Baas, E.J., H.-M. van Santen, M.J. Kleijmeer, H.J. Geuze, P.J. Peters, and H.L. Ploegh. 1992. Peptide-induced stabilization and intracellular localization of empty HLA class I complexes. *J. Exp. Med.* **176:** 147.

Bijlmakers M.J. and H.L. Ploegh. 1993. Putting together an MHC class I molecule. *Curr. Opin. Immunol.* (in press).

Bijlmakers M.J., J.J. Neefjes, E.H.M. Wojcik-Jacobs, and H.L. Ploegh. 1993. The assembly of H2-K^b class I molecules translated *in vitro* requires oxidized glutathione and peptide. *Eur. J. Immunol.* **23:** (in press).

Braciale, T.J. and V.L. Braciale. 1991. Antigen presentation: Structural themes and functional variations. *Immunol. Today* **12:** 124.

Cresswell, P. 1992. Chemistry and functional role of the invariant chain. *Curr. Opin. Immunol.* **4:** 87.

Cresswell, P., R. Avva, J. Davis, C. Lamb, J. Riberdy, and P. Roche. 1990. Intracellular transport and peptide properties of HLA class II glycoproteins. *Semin. Immunol.* **2:** 273.

DeMars, R. and T. Spies. 1992. New genes in the MHC that encode proteins for antigen processing. *Trends Cell Biol.* **2:** 81.

Deres, K., T.N.M. Schumacher, K.-H. Wiesmüller, S. Stevanovic, G. Greiner, G. Jung, and H.L. Ploegh. 1992. Preferred size of peptides that bind to H-2K^b is sequence dependent. *Eur. J. Immunol.* **6:** 1603.

Dobberstein, B., H. Garoff, G. Warren, and J.L. Robinson. 1979. Cell-free synthesis and membrane insertion of mouse H-2d histocompatibility antigen and β_2-microglobulin. *Cell* **17:** 759.

Elliott, T.J. 1991. How do peptides associate with MHC class I molecules? *Immunol. Today* **12:** 386.

Elliott, T.J. and H.N. Eisen. 1990. Cytotoxic T lymphocytes recognize a reconstituted class I histocompatibility antigen (HLA-A2) as an allogeneic target molecule. *Proc. Natl. Acad. Sci.* **87:** 5213.

Elliott, T., V. Cerundolo, J. Elvin, and A. Townsend. 1991. Peptide-induced conformational change of the class I heavy chain. *Nature* **351:** 402.

Garboczi, D.N., D.T. Hung, and D.C. Wiley. 1992. HLA-A2-peptide complexes. Refolding and crystallization of molecules expressed in *Escherichia coli* and complexed with single antigenic peptides. *Proc. Natl. Acad. Sci.* **89:** 3429.

Goldberg, A.L. and K.L. Rock. 1992. Proteolysis, proteasomes and antigen presentation. *Nature* **357:** 375.

Guimezanes, A., T.N.M. Schumacher, H.L. Ploegh, and A. Schmitt Verhulst. 1992. A viral peptide can mimic an endogenous peptide for allorecognition of a class I MHC product. *Eur. J. Immunol.* **6:** 1651.

Krimpenfort, P., G. Rudenko, F. Hochstenbach, D. Gussow, A. Berns, and H.L. Ploegh. 1987. Crosses of two independently derived transgenic mice demonstrate functional complementation of the genes encoding heavy (HLA-B27) and light (β_2-microglobulin) chains of HLA class I antigens. *EMBO J.* **6:** 1673.

Levy, F., R. Larsson, and S. Kvist. 1991a. Translocation of peptides through microsomal membranes is a rapid process and promotes assembly of HLA-B27 heavy chain and β_2-microglobulin translated in vitro. *J. Cell Biol.* **115:** 959.

Levy, F., R. Gabathuler, R. Larsson, and S. Kvist. 1991b. ATP is required for *in vitro* assembly of MHC class I antigens but not for transfer of peptides across the ER membrane. *Cell* **67:** 265.

Ljunggren, H.-G., N.S. Stam, C. Öhlen, J.J. Neefjes, P. Höglund, M.-T. Heemels, J. Bastin, T.N.M. Schumacher, A. Townsend, K. Kärre, and H.L. Ploegh. 1990. Empty MHC class I molecules come out in the cold. *Nature* **346:** 476.

Madden, D.R., J.C. Gorga, J.L. Strominger, and D.C. Wiley. 1991. The structure of B27 reveals nonamer peptides bound in an extended conformation. *Nature* **353:** 321.

Neefjes, J.J. and H.L. Ploegh. 1992a. Intracellular transport of MHC class II molecules. *Immunol. Today* **13:** 179.

———. 1992b. Intracellular of endosomal proteolytic activities by leupeptin blocks surface expression of MHC class II molecules and their conversion to SDS-stable heterodimers (a, β, peptide) in endosomes. *EMBO J.* **11:** 411.

Neefjes, J.J., T.N.M. Schumacher, and H.L. Ploegh. 1991. Assembly and intracellular transport of major histocompatibility complex molecules. *Curr. Opin. Cell Biol.* **3:** 601.

Neefjes, J.J., V. Stollorz, P.J. Peters, H.J. Geuze, and H.L. Ploegh. 1990. The biosynthetic pathway of MHC class II but not class I molecules intersects the endocytic route. *Cell* **61:** 171.

Parker, K.C., M.L. Silver, and D.C. Wiley. 1992. An HLA-A2/β_2-microglobulin-peptide complex assembled from subunits expressed separately in *Escherichia coli. Mol. Immunol.* **29:** 371.

Peters, P.J., J.J. Neefjes, V. Oorschot, H.L. Ploegh, and H.J. Geuze. 1991. Segregation of MHC class II molecules from MHC class I molecules in the Golgi complex for transport to lysosomal compartments. *Nature* **349:** 669.

Ploegh, H.L., L.E. Cannon, and J.L. Strominger. 1979. Cell-free translation of the mRNAs for the heavy and light chains of HLA-A and HLA-B antigens. *Proc. Natl. Acad. Sci.* **76:** 2273.

Rein, R.S., G.H.A. Seemann, N. Stam, J.J. Neefjes, and

H.L. Ploegh. 1987. Association with β_2-microglobulin controls the expression of transfected human class I genes. *J. Immunol.* **138:** 1178.

Ribaudo, R.K. and D.H. Margulies. 1992. Independent and synergaic effects of disulfide bond formation, β-microglobulin, and peptides on class I MHC folding and assembly in an in vitro translation system. *J. Immunol.* **149:** 2935.

Roche, P.A. and P. Cresswell. 1990. Invariant chain-association with HLA-DR molecules inhibits immunogenic peptide binding. *Nature* **345:** 615.

Rötzschke, O. and K. Falk. 1991. Naturally-occurring peptide antigens derived from the MHC class-I-restricted processing pathway. *Immunol. Today* **12:** 447.

Schumacher, T.N.M., M.-T. Heemels, J.J. Neefjes, W.M. Kast, C.J.M. Melief, and H.L. Ploegh. 1990. Direct binding of peptide to empty MHC class I molecules on intact cells and in vitro. *Cell* **62:** 563.

Schumacher, T.N.M., M.L.H. de Bruijn, L.N. Vernie, W.M. Kast, C.J.M. Melief, J.J. Neefjes, and H.L. Ploegh. 1991. Peptide selection by MHC class I molecules. *Nature* **350:** 703.

Schumacher, T.N.M., G.M. van Bleek, M.-T. Heemels, K. Deres, K.W. Li, M. Imarai, L.N. Vernie, S.G. Nathenson, and H.L. Ploegh. 1992. Synthetic peptide libraries in the determination of T cell epitopes and peptide binding specificity of class I molecule. *Eur. J. Immunol.* **22:** 1405.

Silver, M.L., K.C. Parker, and D.C. Wiley. 1991. Reconstitution by MHC-restricted peptides of HLA-A2 heavy chain with β_2-microglobulin, in vitro. *Nature* **350:** 619.

Stam, N., H. Spits, and H.L. Ploegh. 1986. Monoclonal antibodies raised against denatured HLA-B locus heavy chains permit biochemical characterization of some HLA-C locus products. *J. Immunol.* **137:** 2299.

Teyton, L., D. O'Sullivan, P.W. Dickson, V. Lotteau, A. Sette, P. Fink, and P.A. Peterson. 1990. Invariant chain distinguishes between the exogenous and endogenous antigen presentation pathways. *Nature* **348:** 39.

Townsend, A., T. Elliott, V. Cerundolo, L. Foster, B. Barber, and A. Tse. 1990. Assembly of MHC class I molecules analyzed in vitro. *Cell* **62:** 285.

Townsend, A., C. Öhlen, J. Bastin, H.G. Ljunggren, L. Foster, and K. Kärre. 1989. Association of class I major histocompatibility heavy and light chains induced by viral peptides. *Nature* **340:** 443.

Wei, M.L. and P. Cresswell. 1992. HLA-A2 molecules in an antigen processing mutant cell contain signal sequence-derived peptides. *Nature* **356:** 443.

Van Bleek, G. and S. Nathenson. 1991. The structure of the antigen-binding groove of major histocompatibility complex class I molecules determines specific selection of self-peptides. *Proc. Natl. Acad. Sci.* **88:** 11032.

Van Kaer, L., P.G. Ashton-Rickardt, H.L. Ploegh, and S. Tonegawa. 1992. TAP1 mutant mice are deficient in antigen presentation, surface class I molecules, and $CD4^-8^+$ T cells. *Cell* **71:** 1205.

Viville, S., J.J. Neefjes, V. Lotteau, A. Dierich, M. Lemeur, H.L. Ploegh, C. Benoist, and D. Mathis. 1993. Mice lacking the MHC class II-associated invariant chain. *Cell* (in press).

Zhang, W., A.C.M. Young, M. Imarai, S.G. Nathenson, and J.C. Sacchetini. 1992. Crystal structure of the major histocompatibility complex class I H-2Kb molecule containing a single viral peptide: Implications for peptide binding and T-cell receptor recognition. *Proc. Natl. Acad. Sci.* **89:** 8403.

Transport and Expression of Class I MHC Glycoproteins in an Antigen-processing Mutant Cell Line

E.-M. Click,* K.S. Anderson, M.J. Androlewicz, M.L. Wei,† and P. Cresswell
Section of Immunobiology, Howard Hughes Medical Institute, Yale University School of Medicine, New Haven, Connecticut 06510

Class I major histocompatibility complex (MHC) molecules assemble in the endoplasmic reticulum (ER) from three components: a 44-kD transmembrane glycoprotein, encoded by one of a number of MHC-linked genes; the soluble 12-kD protein β_2-microglobulin (β_2m); and a small peptide of eight to ten amino acids. The trimeric complex so formed contains the peptide embedded in a binding cleft consisting of two parallel α helices overlaying a platform of eight β strands, contributed by the amino-terminal two domains of the glycoprotein (Bjorkman et al. 1987). Amino acid substitutions in the α helices and β strands encoded by different class I alleles generate variability in the binding site and consequent variability in the peptides bound. Sequence "motifs" in the spectrum of bound peptides are a discernible result of this binding-site polymorphism, with preferred amino acid residues found at two or three positions in the peptides associated with particular alleles (Rotzschke et al. 1990; Rotzschke and Falk 1991). These positions represent dominant interactions between the peptide and amino acid side chains in the binding cleft.

The peptides associated with class I MHC molecules are predominantly derived from cytosolic proteins (Jardetsky et al. 1991). Genes encoding proteins homologous to proteasome components are encoded in the MHC, leading to the suggestion that proteasomes, or a subset of them, are involved in generating the set of class-I-associated peptides (Brown et al. 1991; Glynne et al. 1991; Kelly et al. 1991; Ortiz-Navarette et al. 1991). Transport of these peptides into the ER is believed to be a property of a specific transporter molecule (for review, see Monaco 1992). Current evidence suggests that this molecule is a heterodimer of two multimembrane spanning proteins of the ATP-binding cassette (ABC) family of transporters (Kelly et al. 1992; Spies et al. 1992). The two subunits are encoded by genes in the MHC currently designated TAP.1 and TAP.2 (for *t*ransporters associated with *a*ntigen *p*rocessing) (Monaco 1992).

Much of the evidence for the preceding description comes from studies of mutant cell lines that are defective in class-I-restricted antigen processing. The mouse cell line RMA-S and the human cell line BM36.1 lack a functional TAP.2 molecule and are poorly lysed by virus-specific class-I-restricted T cells when virally infected (Townsend et al. 1989; Attaya et al. 1992; Kelly et al. 1992). The human cell line .134 is defective in TAP.1 expression and is similarly resistant to lysis (Spies et al. 1990; Spies and DeMars 1991). The human cell line .174 and its hybrid derivative T2 have a large homozygous deletion of the MHC that encompasses all of the functional class II genes as well as TAP.1, TAP.2, and the two genes encoding proteasome homologs (DeMars et al. 1985; Salter et al. 1985). These cells are totally resistant to lysis by class-I-restricted T cells upon viral infection, but, like the other mutant cell lines described, they can be killed by these same T cells if they are incubated with the viral peptide that binds the class I molecule to create the specific T-cell epitope (Cerundolo et al. 1990; Hosken and Bevan 1990).

All of the above mutant cell lines share the characteristic of very low cell-surface expression of class I MHC molecules. This presumably results from the lack of the peptide component of the tripartite class I glycoprotein complex. In the absence of one or both of the TAP gene products, these peptides are not transported into the ER. In the absence of the peptide, the partially assembled class I complexes are not expressed on the cell surface because, according to one view (Ljunggren et al. 1990), they are unstable or, according to another view (Alexander et al. 1990), they fail to exit the ER. A number of exceptions have been noted to the observation that class I molecules are absent from, or low on, the surface of these mutant cells. First, the endogenous HLA-A2 allele is expressed on .134, .174, and T2 at 20–50% of wild-type levels (DeMars et al. 1985; Salter and Cresswell 1986). Second, the products of a number of mouse class I alleles transfected into T2 are well expressed on the cell surface, in many cases at wild-type levels (Alexander et al. 1989). Recently, we and other investigators have shown that HLA-A2 in the T2 cell line binds a limited number of peptides at very high levels (Henderson et al. 1992; Wei and Cresswell 1992). These peptides are signal-sequence-derived peptides and are presumably accessible to HLA-A2 molecules because their entry into the ER does not require

Present addresses: *Department of Biochemistry, Duke University Medical Center, Durham, North Carolina 27706; †Department of Internal Medicine, Brigham and Women's Hospital, Boston, Massachusetts 02115.

the TAP proteins. However, H-2K^b and H-2D^P molecules in T2 are expressed on the surface despite the fact that they are virtually (H-2K^b) or completely (H-2D^P) devoid of bound endogenous peptides (Wei and Cresswell 1992).

We examine in detail here the biosynthesis in the T2 cell of HLA alleles (A3 and B7), which are poorly expressed, and the H-2D^P allele, which is well expressed on the cell surface. We also show that the transport defect extends to a soluble nonmembrane-associated variant of HLA-B7, and we present data on the relative stabilities to denaturation of class I molecules isolated from T2 and wild-type cells.

MATERIALS AND METHODS

Cell lines. T2 is a hybrid of the B-lymphoblastoid cell line (LCL) .174 (DeMars 1984) and the T-LCL CEM (Foley et al. 1965) selected for lack of cell-surface expression of class I MHC (Salter et al. 1985). HMy2.C1R (C1R; class-I-reduced) is a derivative of the B-LCL LICR-LON-HMy2 (Edwards et al. 1982) that expresses no detectable HLA-A or HLA-B locus protein products from the endogenous class I genes (Alexander et al. 1989). Transfected cell lines T2.A3, T2.B7, T2.D^P, C1R.A3, C1R.B7, and C1R.D^P have been described previously (Alexander et al. 1989; Wei and Cresswell 1992). T2 and C1R transfectants expressing a soluble variant of HLA-B7 (T2.B7s and C1R.B7s) were similarly derived (see below). Cells were grown in Iscove's modified Dulbecco's medium (IDMEM) (GIBCO), containing 5% iron/transferrin-supplemented calf serum (Hyclone Laboratories) or 10% fetal bovine serum (GIBCO).

Plasmid DNA. The hybrid gene HLA-B7 × Q10, obtained from B. Kahn-Perles, encodes a soluble class I molecule (HLA-B7s) consisting of the $\alpha1$ and $\alpha2$ domains of HLA-B7 and the $\alpha3$ and truncated transmembrane domains of the nonclassical mouse H-2 allele Q10^b (Kahn-Perles et al. 1989). Plasmid pSV2*neo* (ATCC no. 37149) was used to confer resistance to the antibiotic G418.

Electroporation. Electroporations were performed essentially as described previously (Alexander et al. 1989). T2 or C1R cells (20×10^6 in 0.5 ml of serum-free IDMEM) were mixed with HLA-B7 × Q10 plasmid DNA (20 μg, linearized with *Eco*RI; Boehringer Mannheim) and pSV2*neo* (2 μg, linearized with *Bam*HI; Boehringer Mannheim) and then pulsed with 250 V, 500 microFarads (C1R) or 210 V, 500 microFarads (T2).

Isolation of class I transfectants. G418-resistant clones were screened for synthesis of soluble HLA-B7 by SDS-PAGE of immunoprecipitates from metabolically labeled cells.

Metabolic labeling. Cells in log phase were preincubated and labeled in methionine-free Dulbecco's modified Eagle's medium (DMEM; GIBCO) containing 3% dialyzed fetal calf serum (FCS). For a 4-hour continuous labeling, 2×10^6 cells were incubated with 100 μCi [^{35}S]methionine (Trans-label, ICN). For pulse-chase experiments, 20×10^6 cells were pulse-labeled for 15 minutes with 625 μCi [^{35}S]methionine (1.25 mCi/ml) and then chased by the addition of 6 ml of chase medium (IDMEM containing 5% Hyclone plus 50-fold excess methionine and cysteine); 1-ml aliquots (3.3×10^6 cells) were withdrawn at various times and assayed as described above.

Antibodies. Monoclonal antibody 4E (anti-HLA-B locus; Yang et al. 1984) was provided by S.-Y. Yang. Monoclonal antibody 7-16.10 (anti-D^P; Harmon et al. 1983) was provided by J. Frelinger. Monoclonal antibody GAP.A3 was produced in this laboratory (Berger et al. 1982). Serum F is a rabbit antiserum that reacts with free or β_2m-associated class I heavy chains and also free β_2m (Cresswell and Ayres 1976). Rabbit anti-human IgG was purchased from ICN Biochemicals, and rabbit anti-β_2m was obtained from Boehringer Mannheim. R.Ip35N, a rabbit antipeptide antibody to the cytoplasmic domain of the p35 form of the invariant chain, has been described previously (Lamb and Cresswell 1992). R.A3e7 is a rabbit antipeptide antibody to the cytoplasmic domain of HLA-A3.

Cell lysis and immunoprecipitations. Labeled cells were extracted in 0.5 ml of 1% Triton X-100, 0.15 M NaCl, 0.01 M Tris (pH 7.5), containing 0.2 mM phenylmethylsulfonyl fluoride (PMSF) and 0.2 mM *N*-α-tosyl-L-lysyl-chloromethyl ketone (TLCK). Extracts were precleared with 2 μl of rabbit anti-human IgG serum and protein A–Sepharose beads. Specific immunoprecipitation was performed with 2 μl of ascites, containing monoclonal antibody, antiserum (2 μl), or DEAE-purified rabbit anti-β_2m IgG (20 μg) and protein A–Sepharose beads (Sigma). Labeled culture medium was similarly precleared and immunoprecipitated. Samples were subjected to two-dimensional gel analysis (Alexander et al. 1990) or SDS-PAGE (Laemmli 1970). Radioactively labeled proteins were visualized by fluorography using Enlightening (New England Nuclear-Dupont).

Endoglycosidase H digestion. Endoglycosidase H (Endo H) digestions were performed as described previously (Salter and Cresswell 1986). Endo H was obtained from ICN Biologicals.

Cell-surface biotinylation. Cells (8×10^6) were resuspended in 4 ml of 130 mM NaCl, 20 mM bicine (pH 7.5), containing 0.2 mM NHS-LC-biotin (Pierce, Rockford, Illinois), with or without 1% Triton X-100, and incubated for 30 minutes on ice. Glycine was added to 10 mM to stop the reaction, and aliquots were immunoprecipitated with R.Ip35N or with a mixture of GAP.A3 and R.A3e7. The HLA-A3 precipitates were separated by two-dimensional electrophoresis, with isoelectric focusing in the first dimension and SDS-PAGE in the second dimension (Alexander et al. 1989). R.Ip35N precipitates were separated by SDS-PAGE (Laemmli 1970). The gels were electroblotted onto nitrocellulose, blocked overnight, washed twice with phosphate-buffered saline (PBS)/0.05% Tween 20, and incubated with PBS/0.05% Tween 20/5 mg/ml bovine serum albumin (BSA) and 1:1000 Avidin D-HRP (Vector) for 60 minutes at room temperature. The blots were then washed five times with PBS/Tween/BSA and developed by incubation with 0.06 M Tris (pH 7.4), 0.04% NiCl, 0.01% H_2O_2, and 0.06 mg/ml diaminobenzidine (Sigma). The reaction was stopped by adding 2% sodium azide and washing in H_2O.

RESULTS

Transport of HLA-A3 and H-2D^P in the C1R and T2 Cell Lines.

HLA-A3 was shown to be virtually absent from the surface of T2.A3, according to reactivity with the monoclonal antibody GAP.A3, yet present on the surface of the control cell C1R.A3; however, the mouse allele H-2D^P is expressed on the surface of both cells (Wei and Cresswell 1992). To determine whether the difference in cell-surface expression is mirrored in the biosynthesis and intracellular processing of the two class I molecules, cells were pulse-labeled and chased, and immunoprecipitated class I glycoproteins were treated with Endo H. Analysis by SDS-PAGE (Fig. 1) showed the acquisition of Endo H resistance by the class I molecules in those cells demonstrating high cell-surface expression (C1R-A3, C1R-D^P, and T2.D^P). However, in T2.A3, which has negligible HLA-A3 on its cell surface, the HLA-A3 glycoprotein never became Endo-H-resistant and was virtually undetectable by 4 hours of chase.

Quantitation of the Endo-H-sensitive and -resistant forms of HLA-A3 at the various chase times relative to the total A3 in the pulse label (Fig. 2) shows that the half-life of the Endo-H-sensitive form of HLA-A3 is essentially the same in both T2 and C1R. The difference lies in the fact that in C1R, HLA-A3 is converted to an Endo-H-resistant form, whereas in T2, the epitope recognized by monoclonal antibody GAP.A3 disappears. This suggests that the HLA-A3 molecule is recognized by GAP.A3 when it is initially formed in the ER of either T2 or C1R, but the epitope is lost in T2. Binding of GAP.A3 requires the association of HLA-A3 with β_2m, presumably to stabilize a conformational epitope recognized by the antibody. The loss of the GAP.A3 epitope could reflect a conformational change in HLA-A3, e.g., β_2m dissociation, or the loss could be due to degradation of the HLA-A3 protein. To distinguish between these two possibilities, a similar pulse-chase analysis was performed precipitating with a reagent that recognizes both native and denatured forms of HLA-A3 (rabbit antiserum R.A3e7), as well as the conformational dependent reagent GAP.A3. Because the R.A3e7 antibody reacts with the HLA-A2 and B5

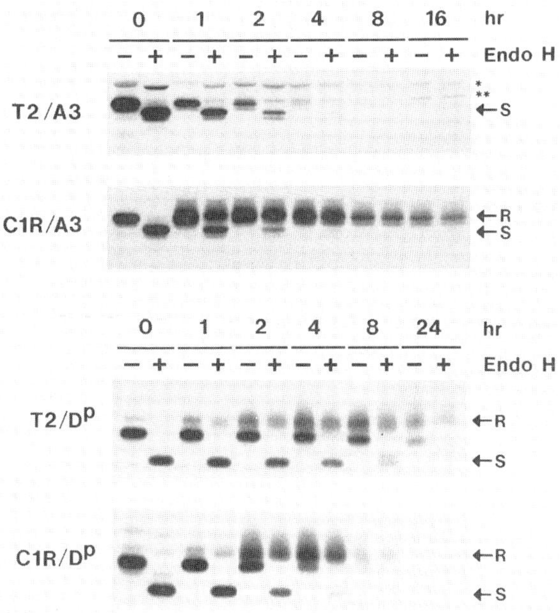

Figure 1. Kinetics of maturation of HLA-A3 and H-2D^P expressed in mutant (T2) and control (C1R) cell lines. Cells were pulse-labeled for 15 min with [^{35}S]methionine and then chased with excess unlabeled methionine. Aliquots were withdrawn at various times (1–24 hr) after initiation of chase, and cell lysates were immunoprecipitated with monoclonal antibody GAP.A3 (HLA-A3) or 7-16.10 (H-2D^P). Immunoprecipitated proteins were digested (+) or mock-treated (−) with Endo H and then separated by SDS-PAGE and visualized by fluorography. Arrows indicate Endo-H-sensitive (S) and -resistant (R) forms. Contaminant bands identified as immunoglobulin heavy chain (*) and actin (**) can be seen in T2/A3.

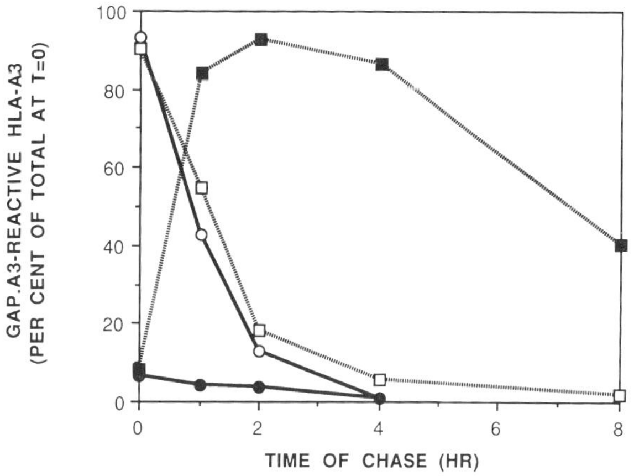

Figure 2. Quantitation of Endo-H-sensitive and -resistant forms of HLA-A3 expressed in T2 and C1R. The Endo-H-treated lanes for T2.A3 and C1R.A3 in Fig. 1 were scanned using an LKB-2400 densitometer, and the peak areas were integrated. For each cell line, the amount of Endo-H-sensitive (Endo H^S) or Endo H-resistant (Endo H^R) HLA-A3 at each time point is expressed as a fraction of the total HLA-A3 (Endo H^S + Endo H^R) at the beginning of chase (t = 0). (○) T2:A3-S; (●) T2:A3-R; (□) C1R:AS-S; (■) C1R:A3-R.

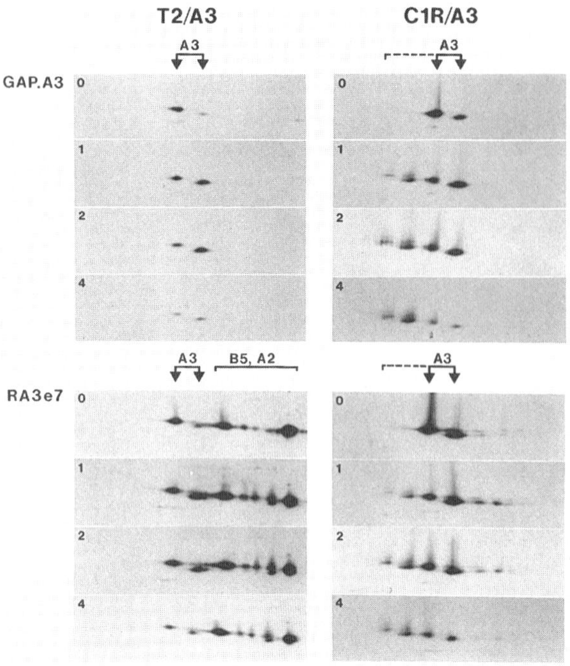

Figure 3. Rates of disappearance of GAP-A3-reactive and total HLA-A3 are identical in T2.A3. C1R.A3 and T2.A3 cells were pulse-labeled with [^{35}S]methionine for 15 min and chased for 1–4 hr with excess cold methionine. Cell lysates were precleared and divided into two aliquots. One was immunoprecipitated with GAP.A3, and the other half was precipitated with R.A3e7. Proteins were resolved by isoelectric focusing (IEF) in the first dimension followed by SDS-PAGE in the second dimension. Radioactive proteins were visualized by fluorography.

alleles endogenous to T2 in addition to HLA-A3, the immunoprecipitated class I molecules were analyzed by two-dimensional PAGE. These data showed a parallel loss of HLA-A3 with both reagents in T2.A3, suggesting that HLA.A3 is degraded in T2 prior to transport into the medial Golgi (Fig. 3).

To demonstrate further that HLA-A3 was not present on the surface of T2.A3, the surface proteins of C1R.A3 and T2.A3 were specifically biotinylated by a nonpermeable biotinylating reagent, NHS-LC-biotin. Surface class I molecules were then immunoprecipitated from detergent-extracted cells by a mixture of the antibodies R.A3e7 and GAP.A3, separated by two-dimensional PAGE, electroblotted onto nitrocellulose, and detected by incubation with avidin coupled to horseradish peroxidase (HRP). Figure 4 clearly shows that HLA-A3 is easily detectable on the surface of C1R.A3, whereas only HLA-A2 is detectable on T2.A3. As a control for the cell-surface specificity of biotinylation, extracts were in parallel immunoprecipitated with an antibody, RIp35N, which recognizes the intracellular p35 form of the class-II-associated invariant chain. After SDS-PAGE and electroblotting, p35 was not detectable unless detergent was present during the biotinylation (data not shown).

Transport of Soluble HLA-B7 in C1R and T2

The transport characteristics of soluble HLA-B7 were evaluated by radioactively labeling the transfected cells with [^{35}S]methionine for 4 hours and immunoprecipitating HLA-B7s from both detergent-extracted cells and supernatant. T2 and C1R transfectants expressing membrane-associated HLA-B7 were included as controls, and the immunoprecipitated HLA molecules were analyzed for their susceptibility to Endo H. The data (Fig. 5) clearly show that HLA-B7s is secreted by C1R.B7s in an Endo-H-resistant form, whereas secretion from T2.B7s is marginal. The soluble B7 molecules that are still cell-associated in C1R.B7s are Endo-H-sensitive, and the membrane-associated B7 molecules in C1R.B7 are endo-H-resistant. However, both the soluble and membrane forms of HLA-B7 that remain cell-associated in T2 are Endo-H-sensitive. Thus, the behavior, in terms of intracellular transport, of HLA-B7s in the two cell lines is essentially the same as that of membrane HLA-B7. In T2, both molecules remain Endo-H-sensitive, arguing that they fail to traverse the medial Golgi, whereas in C1R, both molecules traverse the medial Golgi, acquiring Endo H resistance. HLA-B7 is expressed on the plasma membrane in C1R, whereas HLA-B7s is secreted. Figure 6 shows that in the C1R.B7s cell line, the soluble HLA-B7 molecule is virtually all secreted by 8 hours of chase. In a similar experiment, HLA-B7s secreted from the T2 cell line was maximally 10% of the total synthesized,

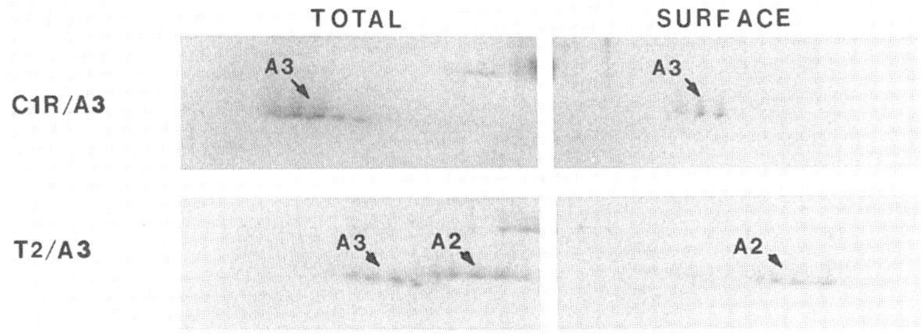

Figure 4. HLA-A3 is not detectable on the surface of T2.A3. The cell lines C1R.A3 and T2.A3 were biotinylated in the presence of 1% Triton X-100 (TOTAL) or in the absence of detergent (SURFACE). Cells were then lysed and immunoprecipitated with a mixture of GAP.A3 and R.A3e7. The precipitates were separated by two-dimensional gel electrophoresis, blotted onto nitrocellulose, and developed with avidin-HRP.

the vast majority being degraded with kinetics similar to those of HLA-A3 in T2.A3 (Fig. 1, and data not shown).

Instability of Class I Molecules Synthesized in T2

To examine the role of peptide occupancy on the stability of class I molecules, the resistance of the HLA-B7s molecule from C1R.B7s and T2.B7s to ethanol precipitation and resolubilization in detergent was examined. Labeled cell lysates or culture supernatants from C1R.B7s or T2.B7s were precipitated with ethanol and redissolved in extraction buffer containing 1% Triton X-100 before precipitation with antibody to β_2m (Fig. 7). There was little effect of ethanol precipitation on the amount of HLA-B7s precipitable from detergent extract or supernatant from C1R.B7s, indicating stable association of the B7s heavy chain with β_2m. However, ethanol precipitation of the T2.B7s detergent extract virtually eliminated HLA-B7s molecules precipitable with anti-β_2m antibody, arguing that the HLA-B7s molecules synthesized in T2 are unstable and that β_2m is easily dissociated.

DISCUSSION

From the data presented here, it is clear that class I glycoproteins in T2 associate with β_2m despite the absence of peptides. This is true both for the human alleles that are effectively retained in the cell and for the mouse alleles that are transported to the cell surface. It is also clear that retention in the ER does not depend on membrane attachment, since the behavior of the soluble HLA-B7s molecule is essentially identical to that of the membrane form of HLA-B7 (see Fig. 4). Previous experiments have shown that retention of human class I molecules in T2 is determined by the α1

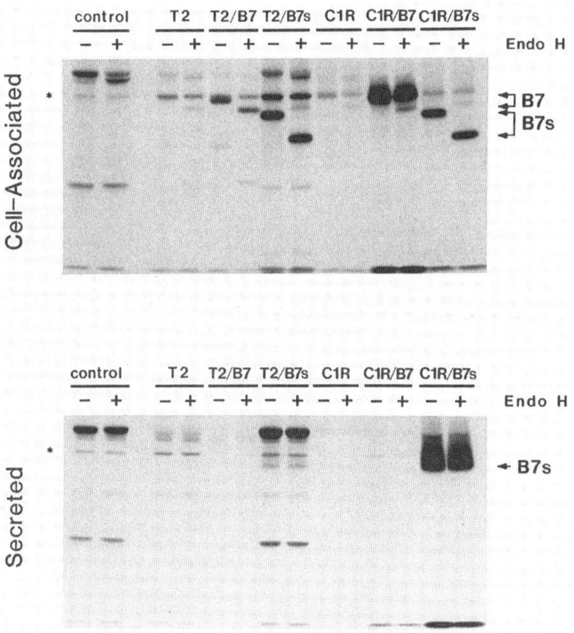

Figure 5. Impaired transport of soluble HLA-B7 expressed in T2. C1R and T2 transfectants expressing membrane and soluble HLA-B7 were continuously labeled for 4 hr with [^{35}S]methionine, and detergent extracts (*top*) and supernatants (*bottom*) were precipitated with the monoclonal antibody 4E. Immunoprecipitates were untreated (−) or treated (+) with Endo H. Control lanes show immunoprecipitates from T2.B7s cells and supernatant with a negative control monoclonal antibody. Asterisk indicates the actin contaminant.

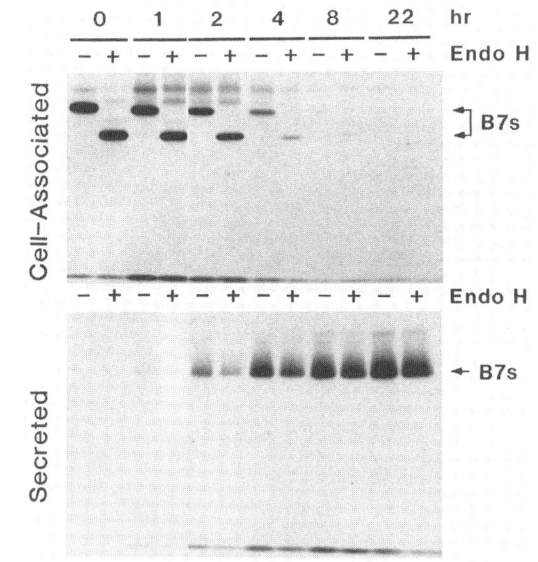

Figure 6. Pulse-chase analysis of soluble HLA-B7 in C1R.B7s. T2/B7s and C1R/B7s cells were pulse labeled with [^{35}S]methionine for 15 min and then chased with excess cold methionine for 0–7 hr as indicated. Cell lysates and culture medium were immunoprecipitated with antibody 4E and subjected to SDS-PAGE and fluorography. All gels were exposed identically to demonstrate relative incorporation of ^{35}S label.

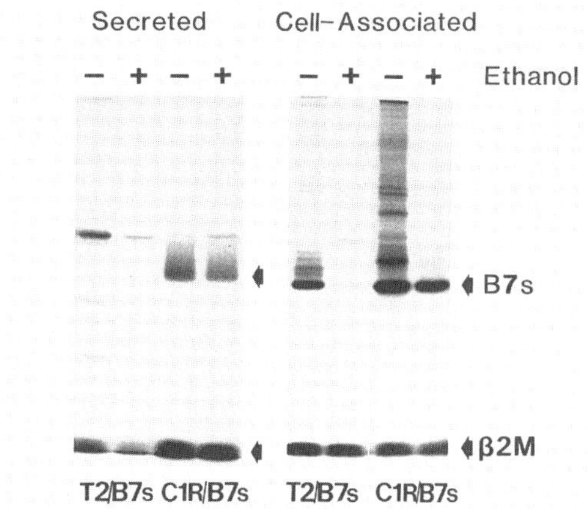

Figure 7. Instability of HLA-B7s synthesized in T2. T2.B7s and C1R.B7s were labeled for 4 hr with [^{35}S]methionine. Culture supernatants (*left*) and detergent extracts (*right*) were divided in half. One half was precipitated with ethanol overnight at −20°C and redissolved in extraction buffer (+), and the other half received no extra treatment (−). All samples were immunoprecipitated with rabbit anti-β_2m and subjected to SDS-PAGE and fluorography.

and/or $\alpha2$ domains (Alexander et al. 1990). The data presented here show that retention does not depend on signals transmitted from the $\alpha1$ and $\alpha2$ domains to the transmembrane or cytoplasmic regions of the class I molecule. Retention in the ER of the p35 form of the invariant chain, for example, is a function of the cytoplasmic domain of this molecule (Bakke and Dobberstein 1990; Lotteau et al. 1990).

Other molecules are retained in the ER as a result of misfolding. These can be mutant molecules that fail to achieve a normal conformation or individual subunits of heteromultimeric proteins that remain in the ER in the absence of association with other subunits (Rose and Doms 1988). Such molecules are frequently found in association with the chaperone immunoglobulin binding protein (BiP) (Bole et al. 1986), although it is unclear whether BiP plays an active role in retention. Although class I glycoprotein–β_2m complexes without bound peptide could be regarded as incomplete multimers, this does not appear to result in BiP association. Clearly, they are not grossly misfolded, since several conformation-dependent antibodies (e.g., the GAP.A3 reagent used in Fig. 1) react with them. In mouse cells, class I molecules have been shown to interact with an 88-kD protein in the ER prior to their transport through the Golgi apparatus (Degan and Williams 1991). A similar interaction of free class I heavy chains with a 90-kD protein has been shown in human Daudi cells, and it has been proposed that this protein (IP90) plays a role in the assembly of multimeric proteins similar to that proposed for BiP (Hochstenbach et al. 1992). However, association of such chaperone molecules with class I–β_2m complexes has not been demonstrated in human cells.

We previously hypothesized (Alexander et al. 1990; Wei and Cresswell 1992) that the retention of human class I glycoproteins in T2 is the property of a specific molecule that interacts with "empty" HLA molecules and prevents their egress from the ER. Once the class I molecules bind peptides, they are then free to transport. The mouse class I alleles studied, we proposed, are not bound by this retention molecule and so are expressed on the cell surface in the absence of peptides. To date, we have not been successful in identifying such a retention molecule. This may indicate that it does not exist or that it operates by a substoichiometric mechanism, perhaps involving retrieval from the *cis*-Golgi reticulum or salvage compartment, as proposed for the receptor that retrieves ER proteins containing the carboxy-terminal KDEL retention motif. Klausner and co-workers have, in fact, demonstrated the cycling of mouse class I glycoproteins between the ER and *cis*-Golgi reticulum in a mouse cell line defective in class-I-restricted processing (Hsu et al. 1991).

In the murine RMA-S cell line, the K^b and D^b alleles are poorly expressed on the cell surface, compared to the expression levels seen in transfected T2 cells. The levels of expression can be increased in RMA-S by incubation at reduced temperature, arguing that lack of expression is in fact a consequence of reduced stability at 37°C (Ljunggren et al. 1990). One could argue that the elevated expression of D^b and K^b on the surface of T2 compared to RMA-S occurs because H-2 molecules associated with human β_2m are more stable than those associated with mouse β_2m. However, transfection of human β_2m into RMA-S does not substantially increase its K^b and D^b expression at 37°C (K.S. Anderson, unpubl.), arguing that this is an insufficient explanation. It is clear, however, that complexes of β_2m with both human (Fig. 7) and mouse (Cerundolo et al. 1991) class I glycoproteins are much less stable than they are when peptides are associated.

To date, HLA-A2 remains the sole example of an HLA allele with high levels of expression on T2. This probably results from the binding of fragments of at least three different signal sequences: from IP-30, an interferon-γ-inducible protein (Henderson et al. 1992; Wei and Cresswell 1992); from calreticulin, an abundant ER protein (Henderson et al. 1992); and from the α-subunit of the signal sequence receptor (Wei and Cresswell 1992; T. Rappoport, pers. comm.). Calreticulin and the signal sequence receptor are universally expressed in mammalian cells, and IP-30 is synthesized constitutively in T2 (M. Androlewicz, unpubl.). Clearly, signal-sequence-derived peptides are readily accessible to class I molecules in T2, although in wild-type HLA-A2-positive cells, the normal pool of cytosolically derived peptides competes effectively with them for access to class I molecules (Henderson et al. 1992; Wei and Cresswell 1992). This argues that the binding of cytosolically derived peptides probably occurs in the rough ER, since that is where signal peptides are cleaved during protein translocation. The ER-associated proteolytic system (Lippincott-Schwartz et al. 1988), however, does not appear to generate effectively peptides that associate with class I molecules, arguing that it occurs at some other site than class I–peptide loading.

ACKNOWLEDGMENTS

We thank Ms. Nancy Dometios for expert preparation of the manuscript. This work was supported by grant AI-15775 from the National Institutes of Health and by the Howard Hughes Medical Institute. Matthew J. Androlewicz was supported by an award from Merck, Sharp and Dohme Research Laboratories, Rahway, New Jersey.

REFERENCES

Alexander, J., J.A. Payne, R. Murray, J.A. Frelinger, and P. Cresswell. 1989. Differential transport requirements of HLA and H-2 class I glycoproteins. *Immunogenetics* **29:** 380.

Alexander, J., J.A. Payne, B. Shigekawa, J. Frelinger, and P. Cresswell. 1990. The transport of class I major histocompatibility complex antigens is determined by sequences in the $\alpha1$ and $\alpha2$ protein domains. *Immunogenetics* **31:** 169.

Attaya, M., S. Jameson, C.K. Martinez, E. Hermal, C. Aldrich, J. Forman, K. Fischer Lindahl, M.J. Bevan, and J.J. Monaco. 1992. Ham-2 corrects the class I antigen-processing defect in RMA-S cells. *Nature* **355:** 647.

Bakke, O. and B. Dobberstein. 1990. MHC class II-associated

invariant chain contains a sorting signal for endosomal compartments. *Cell* **63:** 707.

Berger, A.E., J.E. Davis, and P. Cresswell. 1982. Monoclonal antibody to HLA-A3. *Hybridoma* **1:** 87.

Bjorkman, P.J., M.A. Saper, B. Samroui, W.S. Bennett, J. Strominger, and D.C. Wiley. 1987. Structure of the human class I histocompatibility antigen, HLA-A2. *Nature* **329:** 506.

Bole, D.G., L.M. Hendershot, and J.F. Kearney. 1986. Post-translational association of immunoglobulin heavy chain binding protein with nascent heavy chains in nonsecreting and secreting hybridomas. *J. Cell Biol.* **102:** 1558.

Brown, M.G., J. Driscoll, and J.J. Monaco. 1991. Structural and serological similarity of MHC-linked LMP and proteasome (multicatalytic proteinase) complexes. *Nature* **353:** 355.

Cerundolo, V., T. Elliott, J. Elvin, J. Bastin, H.-G. Rammensee, and A. Townsend. 1991. The binding affinity and dissociation rates of peptides for class I major histocompatibility complex molecules. *Eur. J. Immunol.* **21:** 2069.

Cerundolo, V., J. Alexander, K. Anderson, C. Lamb, P. Cresswell, A. McMichael, F. Gotch, and A. Townsend. 1990. Presentation of viral antigen controlled by a gene in the major histocompatibility complex. *Nature* **345:** 449.

Cresswell, P. and J.L. Ayres. 1976. HLA antigens: Rabbit antisera reacting with all A series or all B series specificities. *Eur. J. Immunol.* **6:** 82.

Degen, E. and D.B. Williams. 1991. Participation of a novel 88-kD protein in the biogenesis of murine class I histocompatibility molecules. *J. Cell Biol.* **112:** 1099.

DeMars, R. 1984. Mutations that dissect the D-region of a human B-lymphoblastoid cell line. *Dis. Markers* **2:** 175.

DeMars, R., C. Rudersdorf, C. Chang, J. Petersen, J. Strandtmann, N. Korn, B. Sidwell, and H. Orr. 1985. Mutations that impair a posttranscriptional step in expression of HLA-A and -B antigens. *Proc. Natl. Acad. Sci.***82:** 8183.

Edwards, P.A., C.M. Smith, A.M. Neville, and M.J. O'Hare. 1982. A human-human hybridoma system based on a fast-growing mutant of the ARH-77 plasma cell leukemia-derived line. *Eur. J. Immunol.* **12:** 641.

Foley, G.E., H. Lazarus, S. Farber, G.G. Uzman, B.A. Boone, and R.E. McCarthy. 1965. Continuous culture of human lymphoblasts from peripheral blood of a child with acute leukemia. *Cancer* **18:** 522.

Glynne, R., S.H. Powis, S. Beck, A. Kelly, L.-A. Kerr, and J. Trowsdale. 1991. A proteasome-related gene between the two ABC transporter loci in the class II region of the human MHC. *Nature* **353:** 357.

Harmon, R.C., N. Stein, and J.A. Frelinger. 1983. Monoclonal antibodies reactive with H-2 determinants. *Immunogenetics* **18:** 541.

Henderson, R.A., H. Michel, K. Sakaguchi, J. Shabanowitz, E. Appella, D.F. Hunt, and V.H. Engelhard. 1992. HLA-A2.1-associated peptides from a mutant cell line: A second pathway of antigen presentation. *Science* **255:** 1264.

Hochstenbach, F., V. David, S. Watkins, and M.B. Brenner. 1992. Endoplasmic reticulum resident protein of 90 kilodaltons associates with the T- and B-cell antigen receptors and major histocompatibility complex antigens during their assembly. *Proc. Natl. Acad. Sci.* **89:** 4734.

Hosken, N.A. and M.J. Bevan. 1990. Defective presentation of endogenous antigen by a cell line expressing class I molecules. *Science* **248:** 367.

Hsu, V.W., L.C. Yuan, J.G. Nuchtern, J. Lippincott-Schwartz, G.J. Hammerling, and R.D. Klausner. 1991. A recycling pathway between the endoplasmic reticulum and the Golgi apparatus for retention of unassembled MHC class I molecules. *Nature* **352:** 441.

Jardetsky, T.S., W.S. Lane, R.A. Robinson, D.R. Madden, and D.C. Wiley. 1991. Identification of self peptides bound to purified HLA-B27. *Nature* **353:** 326.

Kahn-Perles, B., C. Barra, J. Jehan, D. Fourel, M. Barad, J.L. Maryanski, and F.A. Lemonnier. 1989. Cytotoxic T lymphocyte recognition of secreted HLA class I molecules. *J. Immunol.* **142:** 3021.

Kelly, A., S.H. Powis, R. Glynne, E. Radley, S. Beck, and J. Trowsdale. 1991. Second proteasome-related gene in the human MHC class II region. *Nature* **353:** 667.

Kelly, A., S.H. Powis, L.-A. Kerr, I. Mockridge, T. Elliott, J. Bastin, A. Uchanska-Ziegler, J. Trowsdale, and A. Townsend. 1992. Assembly and function of the two ABC transporter proteins encoded in the human major histocompatibility complex. *Nature* **355:** 641.

Laemmli, U.K. 1970. Cleavage of structural proteins during the assembly of the head of bacteriophage T4. *Nature* **227:** 680.

Lamb, C.A. and P. Cresswell. 1992. Assembly and transport properties of invariant chain trimers and HLA-DR invariant chain complexes. *J. Immunol.* **114:** 3478.

Lippincott-Schwartz, J., J.S. Bonifacino, L.C. Yuan, and R.D. Klausner. 1988. Degradation from the endoplasmic reticulum: Disposing of newly synthesized proteins. *Cell* **54:** 209.

Ljunggren, H.-G., N.J. Stam, C. Ohlen, J.J. Neefjes, P. Hoglund, M.-T. Heemels, J. Bastin, T.N.M. Schumacher, A. Townsend, K. Karre, and H. Ploegh. 1990. Empty MHC class I molecules come out in the cold. *Nature* **346:** 476.

Lotteau, V., L. Teyton, A. Peleraux, T. Nilsson, L. Karlsson, S.L. Schmid, V. Quaranta, and P.A. Peterson. 1990. Intracellular transport of class II MHC molecules directed by invariant chain. *Nature* **348:** 600.

Monaco, J.J. 1992. A molecular model of MHC class-I-restricted antigen processing. *Immunol. Today* **13:** 173.

Ortiz-Navarrete, V., A. Seelig, M. Gernold, S. Frentzel, P.M. Kloetzel, and G.J. Hammerling. 1991. Subunit of the "20S" proteasome (multicatalytic proteinase) encoded by the major histocompatibility complex. *Nature* **353:** 662.

Rose, J.K. and R.W. Doms. 1988. Regulation of protein export from the endoplasmic reticulum. *Annu. Rev. Cell Biol.* **4:** 257.

Rotzschke, O. and K. Falk. 1991. Naturally-occurring peptide antigens derived from the MHC class-I-restricted processing pathway. *Immunol. Today* **12:** 447.

Rotzschke, O., K. Falk, K. Deres, H. Schild, M. Norda, J. Metzger, G. Jung, and H.-G. Rammensee. 1990. Isolation and analysis of naturally processed viral peptides as recognized by cytotoxic T cells. *Nature* **348:** 252.

Salter, R.D. and P. Cresswell. 1986. Impaired assembly and transport of HLA-A and -B antigens in a mutant T × B cell hybrid. *EMBO J.* **5:** 943.

Salter, R.D., D.N. Howell, and P. Cresswell. 1985. Genes regulating HLA class I antigen expression in T-B lymphoblast hybrids. *Immunogenetics* **21:** 235.

Spies, T. and R. DeMars. 1991. Restored expression of major histocompatibility class I molecules by gene transfer of a putative peptide transporter. *Nature* **351:** 323.

Spies, T., V. Cerundolo, M. Colonna, P. Cresswell, A. Townsend, and R. DeMars. 1992. Presentation of viral antigen by MHC class I molecules is dependent on a putative peptide transporter heterodimer. *Nature* **355:** 644.

Spies, T., M. Bresnahan, S. Bahram, D. Arnold, G. Blanck, E. Mellins, D. Pious, and R. DeMars. 1990. A gene in the human major histocompatibility complex class II region controlling the class I antigen presentation pathway. *Nature* **348:** 744.

Townsend, A., C. Ohlen, J. Bastin, H.-G. Ljunggren, L. Foster, and K. Karre. 1989. Association of class I major histocompatibility heavy and light chains induced by viral peptides. *Nature* **340:** 443.

Wei, M.L. and P. Cresswell. 1992. HLA-A2 molecules in an antigen-processing mutant cell contain signal sequence-derived peptides. *Nature* **356:** 443.

Yang, S.Y., Y. Morishima, N.H. Collins, T. Alton, M.S. Pollack, E.J. Yunis, and B. Dupont. 1984. Comparison of one-dimensional IEF patterns for serologically detectable HLA-A and -B allotypes. *Immunogenetics* **19:** 217.

Dedicated Transporters for Peptide Export and Intercompartmental Traffic in the Yeast *Saccharomyces cerevisiae*

K. Kuchler,* H.M. Göransson,† M.N. Viswanathan,‡ and J. Thorner‡

**Department of Molecular Genetics, University of Vienna, Vienna, A-1030, Austria; †Department of Microbiology, University of Umeå, Umeå, S-901 87, Sweden; ‡Department of Molecular and Cell Biology, Division of Biochemistry and Molecular Biology, University of California, Berkeley, California 94720*

In both prokaryotic and eukaryotic cells, there are proteins that reside in the plasma membrane which catalyze the export of specific peptides, proteins, and other macromolecules. Similar molecules are located in the membranes that delimit intracellular organelles and appear to mediate the import of polypeptides into these compartments. The dedicated transporters that are responsible for these transmembrane translocation events are all members of a very large superfamily of integral membrane proteins that share extensive sequence similarity and domain organization and also contain a highly conserved motif that confers on these transporters the ability to bind and hydrolyze ATP. Hence, this family of membrane-bound transporters has been called the ATP-binding cassette or "ABC" proteins (Higgins 1992) or the "traffic ATPases" (Ames et al. 1990).

One representative member of this class of specialized membrane proteins is the *STE6* gene product (Ste6p) of the yeast *Saccharomyces cerevisiae* (Kuchler et al. 1989; McGrath and Varshavsky 1989). Here we review the evidence that Ste6p is a member of the ABC family of membrane-bound transporters and the findings that demonstrate that Ste6p is responsible for the secretion of the peptide mating pheromone, **a**-factor, from haploid cells of the *MAT***a** genotype. We also present more recent data on the structure, localization, and function of Ste6p. Using Ste6p as the paradigm, we also address experimentally additional questions about the physiological role of this type of transporter, including (1) whether Ste6p can mediate the export of another peptide from yeast cells other than **a**-factor; (2) whether a mammalian homolog of Ste6p, the human *mdr1* gene product or "P-glycoprotein," first recognized because of its ability to confer multidrug resistance to animal cells in culture (Endicott and Ling 1989; Gottesman and Pastan 1993), can substitute for Ste6p in vivo; and (3) how many ABC-type transporters there might be in a eukaryotic cell, even in an organism, like *S. cerevisiae*, with a relatively small genome size (~15 Mb) (Olson 1991).

EXPERIMENTAL PROCEDURES

Yeast strains and culture conditions. Yeast strains used throughout this study were W303-1A (*MAT***a** *ura3-52 leu2-3,112 his3-11,15 trp1-1 ade2-1 can1-100*) and its isogenic *ste6Δ* derivative, WKK7 (*ste6Δ::HIS3*), constructed as described previously (Kuchler et al. 1989). Mating tester strains were RC757 (*MATα sst2-1 rme1 his6 met1 can1 cyh2*) (Chan and Otte 1982) and L1543 (*MATα lys9*) (Trueheart and Fink 1989). Synthetic medium (SD), supplemented with nutrients appropriate for maintenance of plasmids, or rich medium (YPD) was prepared as described previously (Sherman et al. 1986). Cells were grown routinely at 30°C.

Recombinant DNA manipulations. Plasmids pKK1, a multicopy plasmid carrying the *MF***a***1* gene, and pKK16, a multicopy plasmid carrying both the *MF***a***1* and *STE6* genes, were constructed as described previously (Kuchler et al. 1989). Multicopy (2 μm DNA-based) plasmids and low-copy-number (yeast centromeric DNA-containing, or CEN-based) plasmids expressing the human *mdr1(G185V)* cDNA from the *STE6* promoter and translational start site were constructed according to Kuchler and Thorner (1992a). Both *mdr1*$^{+}$ and *mdr1(G185V)* cDNAs were also expressed from the strong constitutive *ADH1* promoter (Ammerer 1983) in the multicopy yeast expression vector, pAD4M (gift from A.J. Brake, Chiron Corporation, Emeryville, California). Standard methods were used for DNA-mediated transformation of yeast (Ito et al. 1983) and *Escherichia coli* (Sambrook et al. 1989).

Immunological reagents. Polyclonal rabbit anti-Ste6p antibodies (serum 9384) were raised against an *E. coli* TrpE-Ste6 fusion protein constructed, expressed, and purified as described in detail elsewhere (Kuchler et al. 1993). Epitope tags encoding an antigenic determinant from the c-*myc* gene (Evan et al. 1985) were ligated in-frame at two different positions within the *STE6* coding sequence as described in detail by Kuchler et al. (1993). Hybridoma cell lines (generously provided by J.M. Bishop, University of California, San Francisco) that secrete the monoclonal antibody Myc-9E10, which recognizes the *c-myc* epitope employed in this work, were used to produce mouse ascites fluid that was used as the source of the monoclonal antibody without further purification. Polyclonal antibodies (serum 838) directed against the yeast H^{+}-

translocating plasma membrane ATPase (*PMA1* gene product) were generously provided by R. Serrano (Polytechnic University, Valencia, Spain) (Serrano et al. 1991). Polyclonal antibodies (pooled sera A, B, and C) directed against the glycolytic enzyme, yeast phosphoglycerate kinase (*PGK1* gene product), were prepared in this laboratory as described previously (Baum et al. 1978).

Preparation and analysis of membranes and membrane proteins. The methods used to prepare cell extracts, to partially purify plasma membranes by differential centrifugation, and to solubilize membrane protein from such fractions are described in detail elsewhere (Kuchler and Thorner 1992a; Kuchler et al. 1993). When membranes were to be used for photoaffinity-labeling (see below), cells were lysed, membranes were prepared, and proteins were extracted in buffers lacking 2-mercaptoethanol. When total extracts were analyzed by immunoblotting, cells were resuspended in PAGE sample buffer containing 5% SDS and 10% 2-mercaptoethanol before disruption of the cells by vortex mixing with glass beads. Membrane fractionation by sucrose gradient buoyant density sedimentation was conducted essentially as described previously by Bowser and Novick (1991). Protein concentration was measured according to the method of Lowry et al. (1951) in the presence of 1% SDS to solubilize membrane proteins, using bovine serum albumin as the standard. For immunoblot analysis, about 50–100 μg of protein was suspended in SDS-PAGE sample buffer, heated at 55°C for 10 minutes, and subjected to electrophoresis in a 7% SDS-PAGE gel (Dreyfuss et al. 1984). Transfer of proteins from SDS-PAGE gels to nitrocellulose filters was carried out by standard procedures (Towbin et al. 1979). Proteins on immunoblots were detected using a chemiluminescence detection system (Amersham) and conditions recommended by this manufacturer.

Radiolabeling of cells and immunoprecipitation. Total cellular proteins were radiolabeled either with $^{35}SO_4^{--}$, or with a mixture of ^{35}S-Cys and ^{35}S-Met, or with $^{32}PO_4^{---}$, and analyzed by SDS-PAGE and autoradiography, as described in detail elsewhere (Kuchler et al. 1989, 1993; Kuchler and Thorner 1992a).

Photoaffinity labeling. Ste6p in intact yeast membranes, and Ste6p partially purified by immunoprecipitation from yeast membranes that were solubilized in the presence of 1% Triton X-100, was subjected to photolysis in the presence of 8-azido-[α-^{32}P]ATP as described in detail elsewhere (Kuchler and Thorner 1992a; Kuchler et al. 1993).

Mating pheromone secretion. Export of **a**-factor was assessed by two independent methods, a diffusion ("halo") bioassay that measures the release of extracellular **a**-factor directly (Kuchler et al. 1989) and a qualitative mating assay that measures the ability of a patch of *MAT***a** cells to mate with a patch of an appropriate *MAT*α partner on an agar plate, a process that requires the production of extracellular **a**-factor (Kuchler and Thorner 1992a; Kuchler et al. 1993).

Indirect immunofluorescence. Immunofluorescence microscopy of haploid *MAT***a** cells was carried out following the procedures recommended by Pringle et al. (1991) with minor modifications. To prepare pheromone-treated cells for analysis by immunofluorescence, exponentially growing cells (A_{600nm} = 0.75-1) were exposed to α-factor (final concentration 7 μM) for 120 minutes prior to fixation with 5% formaldehyde. For higher resolution analysis of the distribution of Ste6p, a homozygous *MAT***a**/*MAT***a** diploid strain (FC180; gift from F. Cross, Rockefeller University, New York, New York) was transformed with plasmids expressing either c-Myc-tagged Ste6p or the corresponding vector control, YEp352, and examined by the same methods, both before and after pheromone treatment, as described in detail by Kuchler et al. (1993). Fluorescent images were recorded either photographically or via computer-based digital methods.

Polymerase chain reaction and DNA sequence analysis. Identification and isolation of *S. cerevisiae* genomic DNA sequences homologous to *STE6* and other genes encoding ABC-transporter proteins was accomplished using the polymerase chain reaction (PCR) technique. The primers used for amplification were designed on the basis of consensus sequences derived from an alignment of the most highly conserved segments, the so-called Walker motifs (Walker et al. 1982), in the ATP-binding domains of more than 30 presumptive eukaryotic ABC-type transporters. The oligonucleotide primers synthesized also reflected the codon usage bias of *S. cerevisiae* (Davis and Thorner 1987). The sense primer (corresponding to the amino acid sequence -VGSSGCGK-), was 5′-GT(C,T)GGTT-C(A,C,T) TC(A,C,T) GG(A,C,T) TG(C,T) GG(A,T) AA-3′. The antisense primer (corresponding to the amino acid sequence -DEATSALD-), was 5′-(A,G) TC (C,T)AAAGC(A,G,T) GA (A,G,T) GT (A,G,T) GC(C,T)TCATC-3′. The primers were incubated with genomic DNA in a thermal cycler in a volume of 100 μl using Ampli-Taq™ DNA polymerase and reaction buffer supplied by the manufacturer (Perkin-Elmer-Cetus). The reaction mixture was incubated, first, for 5 cycles under the following program: 95°C (30 sec), 72°C (90 sec), and 35°C (30 sec); then, incubation was continued for an additional 30 cycles in a similar way, except that annealing was carried out at 55°C instead of 35°C. Products were analyzed by electrophoresis in 2% agarose gels, and amplified DNA fragments of the expected size were purified, treated with T4 DNA polymerase, and ligated into the *Sma*I site of the pBluescript II KS vector. The ligation mixtures were used for DNA-mediated transformation of *E. coli* strain TG2. Plasmids were recovered by the rapid boiling procedure of Holmes and Quigley (1981) and analyzed by restriction digestion for the presence of inserts of the proper size. Insert-containing plasmids were subjected to nucleotide sequence analysis following standard procedures using bacteriophage T7 DNA polymerase (Sequenase™; United States Biochemical Corp.) under the conditions recommended by the supplier.

RESULTS

Yeast a-Factor Export Requires an ATP-dependent Transporter

Despite its deceptively simple life-style as a unicellular microorganism, *S. cerevisiae* has three distinct cell types. There are two haploid cell types, *MAT***a** and *MAT*α, which produce and respond to extracellular peptide hormones. These peptides bind to G-protein-coupled cell-surface receptors and initiate a signal transduction pathway that induces processes that permit the haploid cells to conjugate or "mate" to form the third cell type, the *MAT***a**/*MAT*α diploid cell (for review, see Sprague and Thorner 1992). These peptide hormones are referred to, therefore, as "mating pheromones." *MAT*α cells secrete α-factor pheromone, which acts on the *MAT***a** cells, and *MAT***a** cells produce **a**-factor pheromone, which acts on the *MAT*α cells.

Work from this laboratory demonstrated previously that one of the pheromones, α-factor (a tridecapeptide), is generated by *MAT*α cells, first as a 165-residue precursor, which enters the secretory pathway where it undergoes Asn-linked glycosylation and proteolytic processing to produce the mature bioactive peptide prior to its release via secretory vesicles (for review, see Fuller et al. 1988). In marked contrast, we discovered more recently that the pheromone produced by *MAT***a** cells, **a**-factor (a dodecapeptide), is generated by a completely different route. The **a**-factor is initially synthesized as a 36-residue precursor but does not enter the secretory pathway and is converted in the cytosol (or at the inner face of the plasma membrane) to its mature form prior to its release (Sterne and Thorner 1986, 1987; Sterne 1989). Active **a**-factor carries two posttranslational modifications (Anderegg et al. 1988). A farnesyl chain is attached in thioether linkage to the sulfhydryl group of the carboxy-terminal cysteine residue; the genes (*RAM1* and *RAM2*) that encode the subunits of the protein:prenyl transferase responsible for this modification have been identified (Schafer et al. 1989; for additional details, see Schafer and Rine 1992). In addition, the carboxyl group of the carboxy-terminal cysteine is methyl-esterified; the gene (*STE14*) that encodes the *S*-farnesyl-cysteine carboxyl:*S*-adenosylmethionine methyltransferase responsible for this modification also has been identified (Sterne-Marr et al. 1990). Thus, mature bioactive **a**-factor is actually a lipopeptide.

Most importantly, we demonstrated that export of **a**-factor does not require the normal secretory vesicle-based secretion pathway defined by the temperature-sensitive *sec* mutations (Schekman 1985), but rather requires the function of the *STE6* gene product (Kuchler et al. 1989; Sterne 1989). We showed further that the *STE6* gene encodes a molecule that is highly homologous in amino acid sequence and predicted topology to human Mdr1 and other members of the ABC-transporter superfamily (Kuchler et al. 1989; McGrath and Varshavsky 1989). Moreover, we found that the level and activity of Ste6p are, in fact, rate-limiting for the release of **a**-factor by *MAT***a** cells (Kuchler et al. 1989). Given its sequence homology with known transport proteins like Mdr1 and bacterial permeases, and its requirement for production of extracellular **a**-factor, we proposed that Ste6p is the membrane protein directly responsible for mediating the transmembrane translocation of the pheromone (Kuchler et al. 1989). We also proposed, by analogy, that the physiological role of Ste6p homologs in other organisms, in particular Mdr1, might be to catalyze the export of peptides and proteins that lack a classic amino-terminal hydrophobic leader sequence (Kuchler and Thorner 1990, 1992b). In fact, to date, Ste6p is the first and only eukaryotic Mdr-like transporter whose authentic and biologically relevant substrate has been unequivocally identified.

Ste6p Is an Integral but Unglycosylated Plasma Membrane Protein

The *STE6* gene product belongs, on the basis of sequence homology, to the superfamily of prokaryotic and eukaryotic ATP-dependent transporter proteins. Ste6p and its closest relatives, including human Mdr1 and human cystic fibrosis transmembrane conductance regulator (CFTR) (Riordan et al. 1989), have a distinct domain structure (Fig. 1). In the amino-terminal half of these proteins, there are six highly hydrophobic and potentially α-helical segments, each long enough to be able to span a lipid bilayer. These transmembrane segments (TMSs) are followed by the ABC domain that has been shown to contain the ATP-binding motif. The

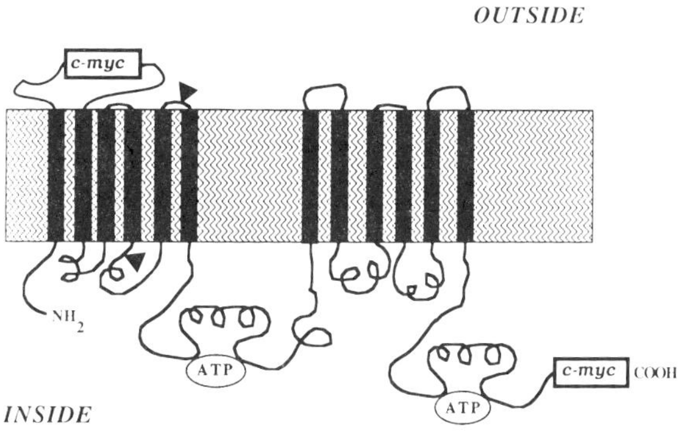

Figure 1. Predicted topology of Ste6p in the yeast plasma membrane. Solid line represents the polypeptide chain, vertical bars indicate transmembrane helices, and ovals depict consensus ATP-binding motifs. The approximate positions of insertion of epitope tags, c-Myc (□) and FLAG (▲), are indicated schematically.

carboxy-terminal half of the molecule contains a nearly identical repeat of the sequences in the amino-terminal portion. Hence, the overall structure of Ste6p, Mdr1, and CFTR could be abbreviated as $[(TMS)_6\text{-}ABC]_2$. A direct role for Ste6p in **a**-factor transport makes several predictions about the topology, membrane insertion, subcellular localization, and regulation of this protein. As one approach for addressing these issues, we generated and employed specific antibodies and other probes to characterize in more detail the structure, membrane association, biochemical properties, and physiological functions of Ste6p.

The *STE6* DNA sequence predicts an open reading frame of 1290 amino acids (calculated molecular weight 144,744) (Kuchler et al. 1989; McGrath and Varshavsky 1989). To determine if yeast cells actually produce a corresponding polypeptide, and to examine the relationship between the primary translation product and the native functional protein, polyclonal antibodies directed against a TrpE-Ste6 fusion protein were raised in rabbits. When *MAT***a** cells were radiolabeled with ^{35}S-cysteine and ^{35}S-methionine, one of the sera raised (9384), but not the corresponding preimmune serum, specifically and efficiently immunoprecipitated a protein from total cell extracts solubilized with SDS that had an apparent molecular weight of 145,000, in excellent agreement with the size of Ste6p deduced from its gene sequence (Kuchler et al. 1993). This 145-kD species was absent in extracts of a *MAT***a** *ste6Δ* mutant and markedly overproduced in *MAT***a** cells carrying the *STE6* gene on a multicopy plasmid, confirming that the 145-kD protein is indeed the product of the *STE6* gene. Because the polyclonal antiserum failed to detect Ste6p on immunoblots, several epitope-tagged versions of Ste6p were constructed (Fig. 1) and employed for detection of Ste6p by immunoblotting. Three epitope-tagged derivatives of Ste6p were fully functional, as judged by two independent assays for production of extracellular **a**-factor (Kuchler et al. 1993). One of these derivatives contained a 10-residue antigenic determinant (-EQKLISEEDL-) from the c-Myc oncoprotein (Evan et al. 1985) inserted in-frame between amino acids 70 and 71 of Ste6p. Another derivative had a longer 16-residue segment (-LEEQKLISEEDL LRKR-) containing the same determinant attached in-frame as the carboxy-terminal sequence, immediately following residue 1285 of Ste6p. The third derivative contained the c-Myc epitopes at both locations. In a similar way, a linker encoding the so-called "FLAG" epitope (-DYKDDDDK-) (Hopp et al. 1988) was inserted in-frame into the *STE6* coding sequence either at a unique *Eco*RI site (between residues 205 and 206), which lies within the second predicted internal hydrophilic loop, or at a unique *Bcl*I site (between residues 282 and 283), which lies within the third predicted external hydrophilic loop (Fig. 1). Like all three of the c-Myc-tagged versions, the Ste6p derivative containing the FLAG epitope in the third external loop was able to complement the *ste6Δ* mutation for **a**-factor secretion, as judged by restoration of a halo of growth inhibition in a lawn of a *MATα sst2* indicator strain, and was capable of promoting mating with an appropriately marked *MATα* partner, a process that also requires production of extracellular **a**-factor. Thus, in these four derivatives, the presence of the epitope insertion did not disrupt the proper folding, localization, or function of Ste6p. In contrast, the Ste6p derivative carrying the FLAG insertion within the second internal loop was completely nonfunctional (Kuchler et al. 1993).

The presence of even the single 10-residue c-Myc epitope was sufficient to permit sensitive detection of Ste6p on immunoblots using the anti-c-Myc monoclonal antibody 9E10 and a chemiluminescence detection system (ECL, Amersham). For example, the amount of Ste6p expressed from a *CEN* plasmid was readily detectable in 50 μg of protein from a total cell extract separated by gel electrophoresis (Kuchler et al. 1993). For either immunoprecipitation or immunoblotting of Ste6p, it was crucial that samples solubilized with SDS not be heated above 55°C; when boiled in the presence of SDS, Ste6p aggregated and did not enter polyacrylamide gels.

The most homologous counterpart to Ste6p in metazoans, Mdr1, is a glycoprotein in situ (termed P-glycoprotein) that carries a single N-linked mannose-rich oligosaccharide (Endicott and Ling 1989). The presumed transmembrane topology of Ste6p (Fig. 1) closely resembles that of Mdr1 and places one of the 14 consensus sites for addition of Asn-linked carbohydrate present in Ste6p within the first putative external hydrophilic loop and, hence, presumably exposed to the lumen of the compartments of the secretory pathway. The single Asn-linked chain in Mdr1 is attached to a site in a nearly identical position in this mammalian protein (Endicott and Ling 1989). However, treatment with tunicamycin, an inhibitor of Asn-linked protein glycosylation, had no detectable effect on the apparent molecular weight of Ste6p, as judged by its mobility upon SDS-PAGE, and solubilized Ste6p failed to bind to concanavalin A-agarose (Kuchler et al. 1993). Concanavalin A is a lectin that is capable of recognizing both the O- and the N-linked mannose residues that are found attached to authentic yeast glycoproteins (Kukuruzinska et al. 1987). Thus, native Ste6p appears not to carry any oligosaccharide chains and, therefore, glycosylation does not seem to be required for either the membrane delivery or function of Ste6p (Kuchler et al. 1993).

Hydropathy analysis of the deduced Ste6 polypeptide predicts 12 highly hydrophobic, potentially α-helical, putatively membrane-spanning segments (Fig. 1). To determine whether Ste6p is indeed a membrane-associated protein, subcellular fractionation by differential centrifugation of total cell-free extracts was performed and analyzed by immunoblotting. Ste6p was found exclusively in the membrane fraction; no immunoreactivity was found in the high-speed supernatant fraction nor, as expected, in any subcellular fractions prepared from a *ste6Δ* strain. To determine the nature of this membrane association, a variety of extraction conditions were employed as a means to solubilize Ste6p from the membrane fraction. Total membranes were suspended in different solutions and resedimented at 200,000*g* in an ultracentrifuge, and equivalent portions of the resulting supernatant solutions were analyzed by immunoblotting. Two treatments that normally release only peripherally bound proteins from membranes (0.1 M Na_2CO_3 at pH 11 and 2 M urea) did not release detectable amounts of Ste6p, as predicted if Ste6p is indeed an intrinsic membrane protein. The nonionic detergent Triton X-100 was nearly as efficient in extracting Ste6p from total membranes as the strong anionic detergent SDS, whereas CHAPSO (a zwitterionic detergent frequently used for solubilization of intrinsic membrane proteins) was moderately effective under the same circumstances. Thus, as expected for an integral membrane protein, only detergent treatment solubilized Ste6p (Kuchler et al. 1993).

Although total membranes were isolated in the presence of a serine protease inhibitor and EDTA, these precautions were not sufficient to completely prevent proteolysis of Ste6p during the time required for membrane isolation. When whole cells were suspended in SDS-containing sample buffer prior to lysis with glass beads, proteolytic degradation of Ste6p was virtually eliminated. Much more severe nonspecific breakdown of Ste6p was observed when cells were first converted to spheroplasts in an osmotically stabilizing medium by digestion of the cell wall with Zymolyase at 30°C for 30 minutes prior to membrane solubilization with SDS. It appears, therefore, that the bulk of the Ste6p in intact spheroplasts was susceptible to the proteases known to contaminate commercial preparations of Zymolyase (Scott and Schekman 1980), suggesting that Ste6p is located in the plasma membrane and that at least a portion of Ste6p is exposed at the exocellular surface.

Because Ste6p appeared to be an integral membrane protein, and because other mammalian and bacterial ATP-dependent transporters are located in the cytoplasmic membrane, it seems reasonable to presume that Ste6p is localized to the plasma membrane. To test this prediction rigorously, however, the intracellular distribution of Ste6p was analyzed by two independent methods for subcellular fractionation of membranes. First, the partitioning of Ste6p by differential centrifugation was compared to that of known marker proteins. Just like the plasma membrane-associated H^+-translocating ATPase (*PMA1* gene product), the bulk of the total cellular content of Ste6p sedimented at 12,000*g*; like Pma1p, the remainder of the total Ste6p sedimented at 100,000*g* (Fig. 2A). In marked contrast, none of the phosphoglycerate kinase (*PGK1* gene product) sedimented at 12,000*g*, whereas the majority of Pgk1p remained in the supernatant fraction even after sedimentation at 100,000*g* (Fig. 2A). To demonstrate that the Ste6p found in the 12,000*g* pellet fraction was membrane-associated rather than simply aggregated or associated with some other particulate material, flotation in a sucrose buoyant density gradient was employed (Fig. 2B). For this purpose, the material in the

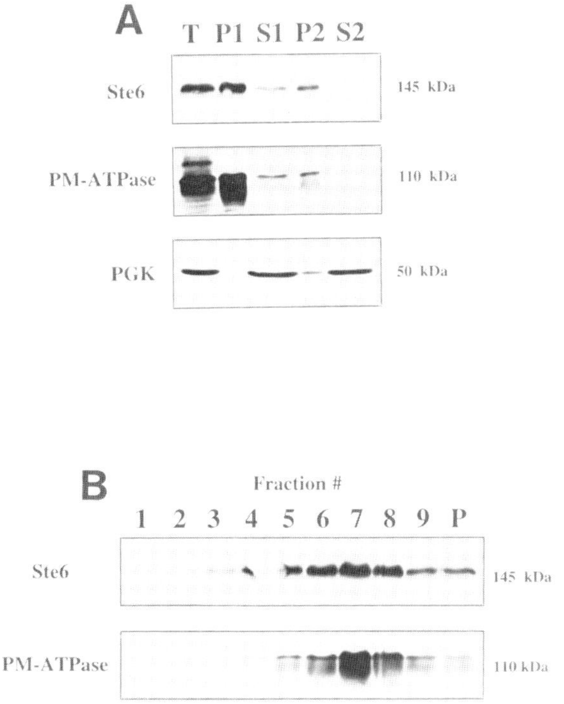

Figure 2. Ste6p cofractionates with an authentic plasma membrane marker protein. (*A*) Whole-cell homogenates were prepared and fractionated by differential centrifugation; the content of Ste6p, Pma1p (PM-ATPase), and phosphoglycerate kinase (PGK) was determined by SDS-PAGE and immunoblotting. The apparent molecular masses of these three species, based on their mobilities relative to a set of standard proteins, are indicated. (T) Total extract; (P1 and S1) pellet and supernatant fractions, respectively, of 12,000*g* sedimentation of T; (P2 and S2) pellet and supernatant fractions, respectively, of 100,000*g* sedimentation of S1. (*B*) A P1 fraction was prepared from a whole-cell homogenate and subjected to flotation in a sucrose buoyant density gradient according to the method of Bowser and Novick (1991); the content of Ste6p and Pma1p (PM-ATPase) in each fraction was determined by SDS-PAGE and immunoblotting. Fraction 1, top of the gradient (30% sucrose); Fraction 9, bottom of the gradient (55% sucrose); (P) insoluble material in the pellet at the bottom of the tube. (Reprinted, with permission, from Kuchler et al. 1993 [copyright held by Rockefeller University Press].)

12,000*g* pellet was suspended in 55% sucrose, overlaid with sucrose solutions of decreasing density, and subjected to prolonged high-speed centrifugation (Bowser and Novick 1991). As expected for an authentic membrane-associated protein, Ste6p moved from the dense 55% sucrose cushion to the lighter densities characteristic of membranes. Most significantly, the distribution of Ste6p was identical, fraction-for-fraction, to that observed for the plasma membrane ATPase (Fig. 2B).

More than 40 eukaryotic members of the superfamily of Ste6p-like transporters have been described to date. Like the other members of the family that are above 100,000 kD, Ste6p is predicted from its sequence to have two discrete nucleotide-binding folds (Fig. 1). Hydrolysis of ATP has been shown to energize the transport process mediated by the highly related transporter, Mdr1 (Hamada and Tsuruo 1988), and by other members of this superfamily (for review, see Higgins 1992; Kuchler and Thorner 1992b). As expected, Ste6p solubilized from membranes with Triton X-100 and partially purified by immunoprecipitation is an ATP-binding protein, as judged by affinity labeling of a 145-kD species using the photoactivatable ATP analog, 8-azido-ATP (Kuchler et al. 1993). This cross-linking was specific to Ste6p because no labeled species was detectable in cells not expressing Ste6p, because appearance of the photolabeled 145-kD product required UV-irradiation, and because labeling was effectively competed by the presence of excess unlabeled nucleotide.

Mating Pheromone Action Regulates the Level and the Localization of Ste6p

The *STE6* gene is only expressed in *MAT***a** cells (Wilson and Herskowitz 1984), and its transcription is significantly induced when *MAT***a** cells are exposed to α-factor pheromone (K. Kuchler et al., unpubl.). Transcriptional activation by α-factor requires a specific *cis*-acting sequence, the pheromone response element, found upstream of the promoter for *STE6* (Wilson and Herskowitz 1986) and the promoters for other pheromone-inducible genes (Van Arsdell and Thorner 1987; Van Arsdell et al. 1987). To determine if mating pheromone induction leads to a concomitant elevation of the level of Ste6p, *MAT***a** cells expressing epitope-tagged Ste6p from a *CEN* plasmid were treated with α-factor, and total extracts were prepared at various times and analyzed by immunoblotting. Treatment of *MAT***a** cells with α-factor increased the expression of Ste6p by at least an order of magnitude within 100 minutes after addition of pheromone; in contrast, in the same cells, the level of neither phosphoglycerate kinase nor calmodulin was detectably altered during the same time period (Kuchler et al. 1993). Increased expression of Ste6p in pheromone-treated cells also resulted in elevated **a**-factor secretion, consistent with our earlier findings using Ste6p artificially elevated by overproduction from a multicopy vector (Kuchler et al. 1989).

To examine whether Ste6p is a phosphoprotein, and whether its phosphorylation is stimulated by α-factor, cells overexpressing Ste6p were radiolabeled with $^{32}PO_4^{---}$ at steady state in the absence or presence of α-factor, and Ste6p was immunoprecipitated with the rabbit polyclonal anti-Ste6p serum. Even in the absence of pheromone treatment, Ste6p was detectably phosphorylated. In the presence of α-factor, the radioactivity incorporated into Ste6p was significantly increased (Kuchler et al. 1993); however, the increase in phosphate incorporation mirrored the increase in total Ste6p observed over the same time period. Therefore, it is likely that increased expression of Ste6p following pheromone induction, rather than increased phosphorylation of preexisting Ste6p molecules, accounts for the observed apparent increase in phosphorylation upon α-factor treatment.

As yet another means to determine its intracellular location, the distribution of Ste6p was examined by indirect immunofluorescence. In *MAT***a** haploid cells expressing epitope-tagged Ste6p from a multicopy plasmid, Ste6p was found throughout the plasma membrane in a patchy, rather nonuniform pattern, as indicated by uneven apparent surface staining (Kuchler et al. 1993). In addition, fluorescence was consistently observed in vesicle-like bodies that seemed to subtend the inner surface of the plasma membrane. To increase the resolution of this technique, *MAT***a**/*MAT***a** diploid cells expressing the epitope-tagged Ste6p were examined. When viewed by setting the focal plane in the center of these large diploid cells, the patchy rim-like staining and plasma membrane-associated vesicular bodies were more obvious (Kuchler et al. 1993). Essentially the same pattern of plasma membrane and vesicle-like staining, although less intense, was observed in cells expressing Ste6p from a low copy *CEN* vector, or when optical sections (0.5 μm) of the *MAT***a**/*MAT***a** diploids were observed by confocal microscopy. The staining in vesicle-like bodies could reflect newly synthesized Ste6p within the compartments of the secretory pathway en route to the cell surface. Alternatively, these structures could represent a special pool of secretory vesicles (Holcomb et al. 1988a) containing stored Ste6p, providing a reservoir from which Ste6p could be delivered to the plasma membrane in a rapid (and perhaps regulated) manner.

Given the striking increase in the level of Ste6p that occurred after pheromone stimulation, it was of interest to explore whether localization of Ste6p also was affected by α-factor treatment, because cytoskeletal elements and other intracellular structures undergo rather pronounced changes in organization when haploid cells are exposed to pheromone (Baba et al. 1989; Drubin 1991). As a result of these underlying changes, budding of *MAT***a** and *MAT*α cells is suppressed and the morphology of the cells becomes altered, such that the cell wall and plasma membrane grow asymmetrically to produce a prominent projection. The elongated mating cells fuse at the tips of their respective projections (for review, see Sprague and Thorner 1992). Indeed, a dramatic change in the surface distribution of Ste6p was found upon pheromone treatment. After exposure to α-factor, plasma membrane staining of Ste6p either in haploid *MAT***a** cells (Fig. 3) or in diploid *MAT***a**/*MAT***a** cells (Kuchler et al. 1993) was observed primarily along the surface of the projection, and most predominantly at its tip. The intracellular vesicle-like structures were present in the pheromone-treated cells, although the number and intensity of staining of these bodies appeared to be reduced, suggesting that the redistribution of Ste6p observed upon pheromone treatment may involve, at least in part, mobilization and plasma membrane fusion of the Ste6p-containing vesicles. Therefore, these vesicles may represent an exocytotic compartment previously unidentified in yeast.

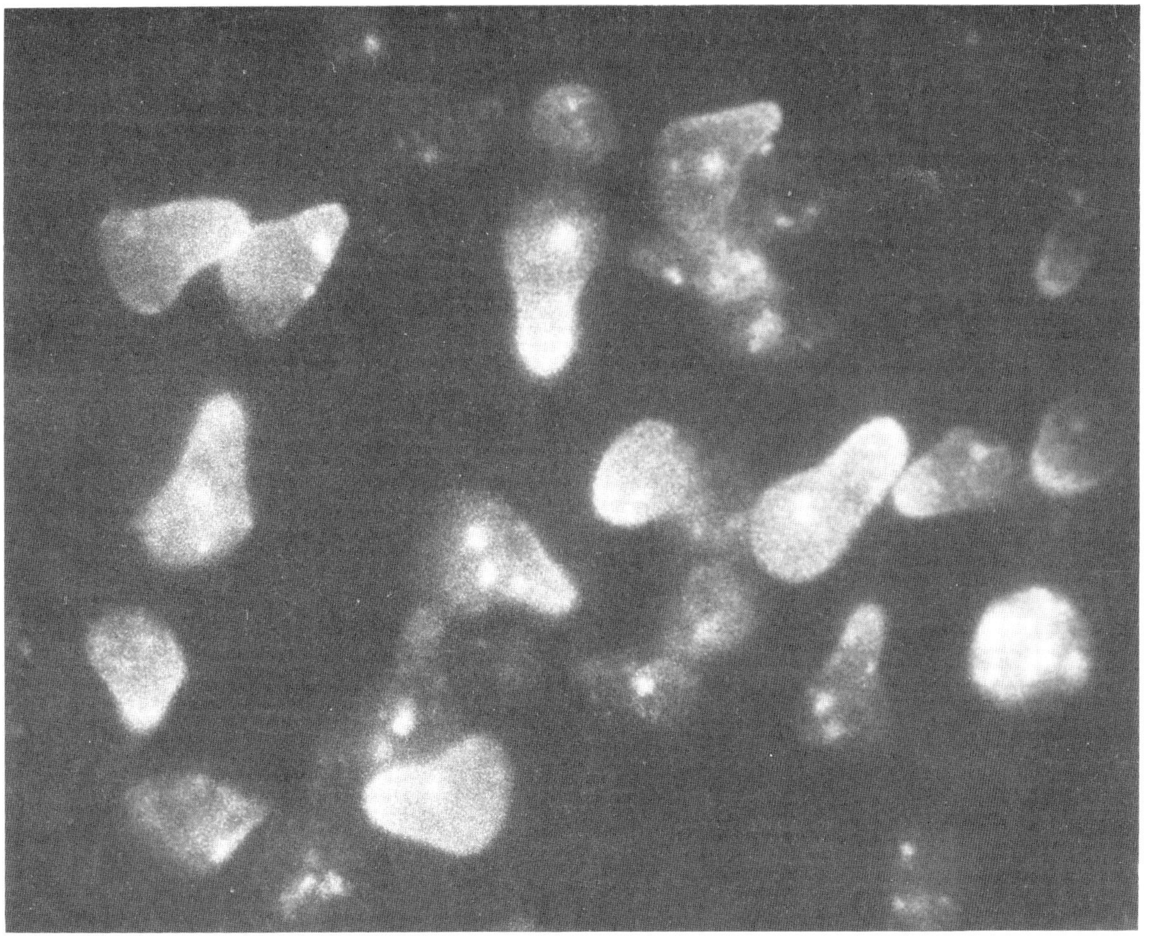

Figure 3. Ste6p localizes to the tip of the projection in pheromone-treated cells. Exponentially growing cultures of a *MAT***a** *ste6*Δ haploid strain (WKK7) carrying either a multicopy plasmid expressing Ste6p doubly tagged with the c-Myc epitope or the corresponding vector (YEp352) were split into two portions; one sample (−α-factor) was fixed with formaldehyde immediately and the other sample (+ α-factor) was treated with 7 μM α-factor for 120 min prior to formaldehyde fixation. After permeabilization, the fixed cells were applied to cover slips, decorated with anti-c-Myc antibody 9E10 followed by fluorescein isothiocyanate-conjugated sheep-anti-mouse immunoglobulin, stained with 4,6-diamino-2-phenyl-indole to reveal nuclear DNA (not shown), and viewed in a fluorescence microscope. Only cells expressing Ste6p after pheromone treatment are shown in this panel; prior to pheromone treatment, the same cells are spherical in shape and the fluorescence pattern displays a patchy distribution that completely surrounds the circumference of each cell (Kuchler et al. 1993). The control cells containing vector alone were not detectably stained either before or after pheromone treatment.

Ste6p Can Mediate the Export of Another Polypeptide Fused to a-Factor

The precursors to a number of peptides secreted by animal cells resemble pro-**a**-factor in that (1) the segment destined to be released as the mature bioactive peptide is situated at the carboxy-terminal end of the precursor; (2) the precursor lacks any obvious amino-terminal hydrophobic leader sequence (or any strikingly hydrophobic domain, for that matter); (3) the precursor does not become glycosylated; (4) proteolytic processing seems to occur in the cytosol and not within any compartment of the classic secretory pathway; and (5) in several instances, the precursor appears to undergo posttranslational modification with a lipophilic substituent prior to its release from cells. Notable examples are the precursors to the secreted mammalian peptides, pro-interleukin-1α (IL-1α; 271 residues) and pro-interleukin-1β (IL-1β; 269 residues).

When expressed in *S. cerevisiae*, mature IL-1β (153 residues) remains in the cytosol and is not glycosylated; however, when an amino-terminal hydrophobic leader is attached, the IL-1β is directed to the secretory pathway, resulting in the extracellular release of active and glycosylated cytokine (Baldari et al. 1987; Livi et al. 1991). We have expressed mature IL-1α (159 residues) in yeast, both alone and as a fusion protein to the amino terminus of either mature or pro-**a**-factor (K. Kuchler and J. Thorner, in prep.). Unlike mature IL-1α itself, the chimeric IL-1α::**a**-factor and IL-1α::pro-**a**-factor proteins associate nearly quantitatively with the plasma membrane fraction, presumably by virtue of the prenylation of the carboxyl terminus of the **a**-factor segment, and remain unglycosylated. Most intriguingly, we also observed export of both the IL-1α::pro-**a**-factor and IL-1α::**a**-factor fusion proteins, albeit at low levels, in a manner that depends strictly on the presence of a functional *STE6* gene. Although indirect, these findings suggest that export of this mammalian cytokine could be achieved via the action of a eukaryotic ABC-type transporter. The attachment of mature **a**-factor to the carboxyl terminus of IL-1α was as effective as attachment of pro-**a**-factor in targeting the chimeric protein to the membrane and in promoting its export. This observation suggests that the recognition element(s) for association of **a**-factor with its export machinery, which presumably includes Ste6p, is confined to the carboxyl terminus of **a**-factor and/or its modifications. The finding that the chimeric IL-1α::**a**-factor molecules associate exclusively with the plasma membrane, but with no other cellular membrane, implies that a receptor or binding protein for prenylated polypeptides may exist in the yeast plasma membrane. Evidence for such a "docking protein" for prenylated Ras has been obtained in animal cells (Hancock et al. 1991; P.J. Casey, pers. comm.).

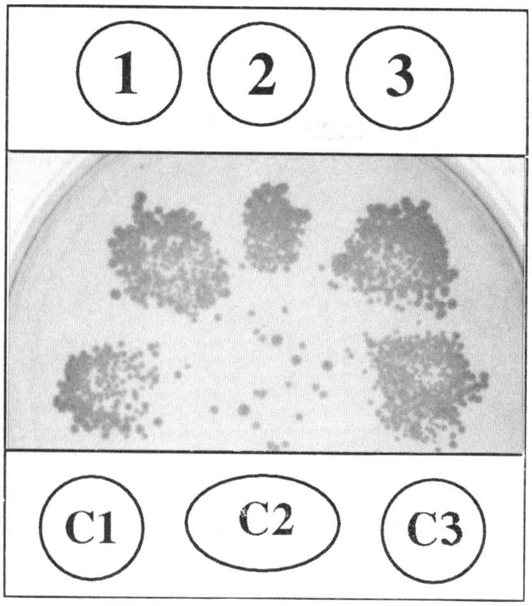

Figure 4. Human Mdr1 expressed in yeast can mediate **a**-factor export and thereby suppress the mating defect of a *MAT***a** *ste6*Δ mutant. Exponential phase cultures of the indicated *MAT***a** strains grown, when necessary, under selective conditions for the maintenance of plasmids were spotted onto a lawn of a *MAT*α partner of appropriate genotype on a YPD plate and, after incubation for 6 hr, the dish was replica-plated onto minimal medium selective for prototrophic diploid cells. (*1*, *2*, and *3*) Three independently derived colonies of the *MAT***a** *ste6*Δ strain transformed with p*ADH-mdr1*$^+$; (C1) *MAT***a** *STE6*$^+$; (C2) *MAT***a** *ste6*Δ transformed with p*ADH-mdr1(G198V)*; (C3) *MAT***a** *ste6*Δ transformed with pKK16 (which contains both the *STE6* and *MFA1* genes).

Human Mdr1 Expressed in Yeast Can Mediate a-Factor Export

Our finding that Ste6p is the **a**-factor transporter led us to propose (Kuchler et al. 1989; Kuchler and Thorner 1990) that other Mdr-like proteins in other systems (for review, see Kuchler and Thorner 1992b) might catalyze the transmembrane translocation of peptides or proteins. To test this hypothesis directly, we first used the *STE6* promoter and translational initiation signals to express a mutant *mdr1* cDNA encoding a multidrug-resistant Mdr1 variant (which carries a G185V mutation) in yeast cells (Kuchler and Thorner 1992a). We were able to demonstrate that this construct permitted copious expression of mammalian Mdr1 in yeast and that at least a portion of the Mdr1 molecules had come to be inserted into the plasma membrane in a properly folded state, as judged by their ability both to bind ATP and to confer upon the yeast cells a moderately elevated level of resistance to the antibiotic valinomycin, an Mdr1-directed drug that is capable of killing yeast (Kuchler et al. 1989). Nonetheless, we could not demonstrate that the drug-resistant version of human Mdr1 was capable of substituting for Ste6p to mediate the export of **a**-factor. Another group, however, subsequently showed independently that the normal form of the murine homolog of human Mdr1, mouse Mdr3, was able to complement a *ste6*Δ mutant, but a multidrug-resistant variant of Mdr3 was not (Raymond et al. 1992). Hence, we reexamined whether wild-type human Mdr1 (as opposed to the multidrug-resistant variant we tested initially) is capable of mediating **a**-factor export in place of Ste6p (K. Kuchler et al., in prep.). Indeed, when expressed from a strong constitutive promoter, normal Mdr1, but not the Mdr1(G185V) mutant, was able to mediate a readily detectable level of **a**-factor export, as judged by restoration of mating ability to a *MAT***a** *ste6*Δ mutant (Fig. 4) or by production of a halo of growth inhibition in a *MAT*α *sst2* indicator lawn. In addition, high-level expression of normal human Mdr1 conferred upon the yeast cells nearly complete resistance to the growth-inhibitory effects of even millimolar concentrations of valinomycin. Since the physiologically relevant substrate(s) of Mdr1 is still unknown (West 1990; Gottesman and Pastan 1993), our results suggest that the function of P-glycoproteins in mammalian cells may include the transmembrane export of endogenous peptides and/or proteins. Consistent with this view, in a separate and contemporaneous study, it was shown that the Mdr1 homolog in Chinese hamster ovary cells is capable of mediating the efficient extrusion of a growth-inhibitory synthetic peptide, *N*-acetyl-leucyl-leucyl-norleucinal (which blocks the cellular protease, calpain) (Sharma et al. 1992).

The Yeast Genome Encodes Multiple *STE6*-like Gene Products

In most organisms, multiple Mdr-like genes are present (Higgins 1992; Gottesman and Pastan 1993). For example, human Mdr2 is a close homolog of human Mdr1 and is expressed in a variety of normal human tissue types; yet, unlike Mdr1, overproduction of Mdr2 does not confer a multidrug-resistance phenotype on animal cells in culture. Furthermore, in addition to the class of ABC transporters of the $[(TMS)_6\text{-ABC}]_2$ type represented by Ste6p, Mdr1, and CFTR, there are also members of the superfamily that are essentially half the size, with a domain structure of either $(TMS)_6$-ABC or ABC-$(TMS)_6$. Such "half-sized" ABC transporters in humans include, for example, the *tap1* and *tap2* gene products, which are proteins required for the import of peptide antigens into the endoplasmic reticulum to permit association of such antigens with the class I major histocompatibility receptors of the human immune system (Monaco 1992), and Pmp70, a human peroxisomal membrane protein which, when defective, results in a fatal cerebro-hepato-renal dysfunction known as Zellweger's syndrome (Gartner et al. 1992).

With regard to the diversity of ABC transporters found in a single genome, *S. cerevisiae* is fast outpacing other eukaryotic organisms. The *ADP1* gene was discovered as part of the European effort to sequence chromosome III (Purnelle et al. 1991). The deduced Adp1 polypeptide is most similar in size, sequence, and domain arrangement to the *Drosophila* White (O'Hare et al. 1984) and Brown (Dreesen et al. 1988) proteins, which possess the ABC-$(TMS)_6$ topology and which appear to be involved in the uptake of precursors for the synthesis of the pteridine pigments in the secondary pigment cells of the fly compound eye. Larger Ste6/Mdr1-like transporters have also been identified in *Drosophila* (Wu et al. 1991). Deletions of *ADP1* cause no discernible phenotype and, hence, no function has been ascribed to it. The *S. cerevisiae PDR5* gene was originally identified as a locus on chromosome XV that confers elevated resistance to the protein synthesis inhibitor, cycloheximide, and simultaneous cross-resistance to an unrelated inhibitor, sulfometuron methyl (Leppert et al. 1990). Thus, phenotypically, Pdr5p displays one of the hallmarks of an Mdr-like ABC transporter, namely, the ability to confer a multidrug-resistance phenotype. Correspondingly, the deduced sequence of Pdr5p indicates that it is a large 1511-residue polypeptide with readily detectable homology with the ABC-transporter superfamily (Wang et al. 1992).

As one approach to determine whether additional *STE6*-like gene products are encoded in the yeast genome, we utilized PCR and primers based on sequences highly conserved in the nucleotide-binding folds of more than 30 members of the ABC-transporter superfamily, in organisms as evolutionarily distant as yeasts and humans. Among the PCR products we have characterized to date, we have identified at least two novel Ste6p-like genes in the *S. cerevisiae* genome (Göransson and Thorner 1992). We have designated these genes *SSH1* and *SSH2* (for *S*terile-*S*ix *H*omologs). The sequence of the nucleotide-binding domain of Ssh1p revealed that it is even more homologous to authentic Mdr-like proteins from other organisms than is Ste6p (Fig. 5); however, both the length of the open reading frame of the entire gene, which we isolated using the PCR product as a probe (H.M. Göransson and J. Thorner, in prep.), and the size of its transcript, as determined by hybridization to $poly(A)^+$ RNA, indicate that Ssh1p is a half-sized ABC transporter. *SSH1* resides on chromosome XVI, and an *ssh1Δ* mutation has no detectable effect on growth, mating, or sporulation under a wide variety of conditions tested and, hence, its function is unknown.

In contrast, the *SSH2* gene (Fig. 6), which also resides on chromosome XVI, appears to be the yeast

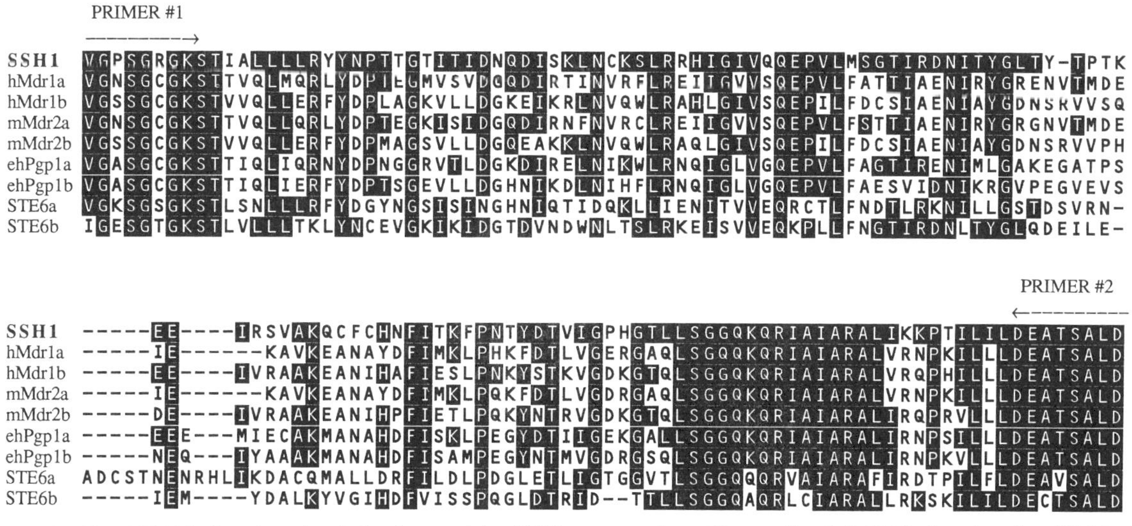

Figure 5. The ATP-binding domain of the *S. cerevisiae SSH1* gene product. The predicted ATP-binding domain of Ssh1p is aligned with each of the two ATP-binding domains in human Mdr1 (hMdr1a and hMdr1b); mouse Mdr2 (mMdr2a and mMdr2b); a P-glycoprotein homolog from *Entamoeba histolytica* (ehPgp1a and ehPgp1b); and *S. cerevisiae* Ste6p (STE6a and STE6b). Positions of the primers used for PCR amplification are indicated. Residues that are identical between Ssh1p and any of the other proteins are represented by white-on-black letters. The dashes represent single-residue gaps introduced to optimize the alignment.

```
         PRIMER #1
         --------->
SSH2     LGPNGCGKSSIQRIIAEIWPVYNKNGLLSIPSENNIFFIPQK-PYFS--RGGTLRDQ--LYTMSRMNSFDRGF-
rPmp70   CGPNGCGKSSLFRVLGELWPLF--GGHLTKPERGKLFYVPQR-PYMT---LGTLRDQ-VIYPDGKEDQKKKGI-
SpHmt1   VGESGGGKSTIMRILLRFFDVNSGSITIDDQDIRNVTLSSLRSSIGVVPQDSTLFNDTILYNIKYAKPSAT---
ScADP1   MGGSGAGKTTLLDILAMKRKTG-HVSGS--IKVNGISMDRKSFSKIIGFVDQDDFLLPTLTVFETVLNSAL---
DmBr     LGGSGAGKTTLLAAISQRLRGNLTGDVVLNGMAMERHQMTRISSFLPQFEINVKTFTAYEHLYFMSHFKMH---
DmWh     MGSSGAGKTTLLNALAFRSPQGIQVSPSGMRLLNGQPVDAKEMQARCAYVQQDDLFIGSLTAREHLIFQAMVRM
SSH1     VGPSGRGKSTIALLLLRYYNPTTGTITIDNQDISKLNCKSLRRHIGIVQQEPVLMSGTIRDNITYGLTYTPTK-
STE6a    VGKSGSGKSTLSNLLLLFYDGYNGSISINGHNIQTIDQKLLIENITVVEQRCTLFNDTLRKNILLGSTDSVRN-
STE6b    IGESGTGKSTLVLLLTKLYNCEVGKIKIDGTDVNDWNLTSLRKEISVVEQKPLLFNGTIRDNLTYGLQDEILE-

                                                                       PRIMER #2
                                                                      <---------
SSH2     RDKELVQILVEVKLDYLLKRGVGLTYLDAIADW------KDLLSGGEKQRVNFARIMFHKPLYVVLDEATSALD
rPmp70   SDQVLKGYLDNVQLGHILEREGG---WDSVQDW------MDVLSGGEKQRMAMARLFYHKPQFAILDECTSAVS
SpHmt1   ----NEEIYAAAKAAQIHDRILQF--PDGYNS--RVGERGLKLSGGEKQRVAVARAILKDPSIILLDEATSALD
ScADP1   -LRLPKALSFEAKKARVYKVLEELRIIDIKDRIIGNEF-DRGISGGEKRRVSIACELVTSPLVLFLDEPTSGLD
DmBr     -RRTTKAEK-RQRVADLLLA-VGLR--DAAHT------RIQQLSGGERKRLSLAEELITDPIFLFCDEPTTGLD
DmWh     ----PRHLTYRQRVARVDQVIQELSLSKCQHTIIGVPGRVKGLSGGERKRLAFASEALTDPPLLICDEPTSGLD
SSH1     ---------EEIRSVAKQCFCHNFITKFPNTYDTVIGPHGTLLSGGQKQRIAIARALIKKPTILILDEATSALD
STE6a    ADCSTNENRHLIKDACQMALLDRFILDLPDGLETLIGTGGVTLSGGQQQRVAIARAFIRDTPILFLDEAVSALD
STE6b    ---------IEMYDALKYVGIHDFVISSPQGLDTRIDT--TLLSGGQAQRLCIARALLRKSKILILDECTSALD
```

Figure 6. The ATP-binding domain of the *S. cerevisiae SSH2* gene product. The predicted ATP-binding domain of Ssh2p is aligned with the ATP-binding domains in the 70-kD rat peroxisomal membrane protein (rPmp70); a gene product that confers heavy-metal tolerance in *Schizosaccharomyces pombe* (SpHmt1); *S. cerevisiae* Adp1p (ScADP1); the Brown (DmBr) and White (DmWh) proteins of *Drosophila melanogaster*; *S. cerevisiae* Ssh1 (SSH1); and *S. cerevisiae* Ste6p (STE6a and STE6b). Positions of the primers used for PCR amplification are indicated. Residues that are identical between Ssh1p and any of the other proteins are represented by white-on-black letters. The dashes represent single-residue gaps introduced to optimize the alignment.

homolog of the human and rat peroxisomal membrane protein, Pmp70, which has been implicated in the import of the constituent proteins of the peroxisome (Kamijo et al. 1990; Gartner et al. 1992). Definitive genetic tests are currently under way to determine if *SSH2* function is required for the biogenesis of peroxisomes in *S. cerevisiae*, and reagents are being developed to determine if Ssh2p is a component of the peroxisomal membrane. A considerable number of additional PCR products were generated with the primers we used and remain to be characterized.

DISCUSSION

On the basis of the essential role of Ste6p in **a**-factor export (Kuchler et al. 1989) and its homology to known transport proteins (Kuchler et al. 1989; McGrath and Varshavsky 1989), we proposed previously that Ste6p is a transporter and that it directly mediates the secretion of **a**-factor by *MAT***a** cells. By analogy with Mdr1 (Endicott and Ling 1989) and with certain other polytopic membrane proteins (Hartmann et al. 1989), we suggested further that Ste6p is an integral plasma membrane-associated protein containing 12 membrane-spanning segments positioned with both its amino and carboxyl termini on the cytosolic face of the plasma membrane. The effects of insertion of the epitope tags we used for immunological detection provide some indirect support for this topological arrangement (Fig. 1). Neither of two different epitopes (c-Myc and FLAG) inserted in regions of the protein predicted to be on the external surface of the cell, nor a c-Myc epitope at the extreme carboxyl terminus, detectably perturbed the function of Ste6p, whereas one of the same epitopes (FLAG) inserted into a region of the protein predicted to face the cytosol was incompatible with the function of Ste6p. That a portion of Ste6p is exposed to the exterior of the cell was further supported by the observation that Ste6p is accessible to attack by the protease(s) present in the enzyme preparation standardly used for the generation of yeast spheroplasts. The application of various labeling and extraction procedures demonstrated that Ste6p is an intrinsic ATP-binding membrane protein with an apparent molecular weight (145,000) in excellent agreement with the size of the molecule deduced from its predicted amino acid sequence. Furthermore, both subcellular fractionation (Fig. 2) and immunofluorescence microscopy (Fig. 3) demonstrated that Ste6p is located in the plasma membrane.

Unlike both mammalian P-glycoprotein (Endicott and Ling 1989; Kane et al. 1990) and CFTR (Riordan et al. 1989), yeast Ste6p does not appear to carry any oligosaccharide moieties, even though Ste6p has a consensus site for addition of N-linked carbohydrate at a position precisely analogous to that found in mammalian Mdr1. This observation indicates that glycosylation is required neither for the function of Ste6p function nor for its intracellular transport. In this regard, it is noteworthy that other polytopic plasma membrane proteins of yeast, such as Pma1p, the H^+-translocating ATPase (Serrano et al. 1991), and Fur4p, the uracil permease (Silve et al. 1991), are not *N*-glycosylated, despite the presence of potential *N*-glycosylation sites. Likewise, human Mdr1 and mouse Mdr3 are not glycosylated when expressed in yeast in functional form (Kuchler and Thorner 1992a; Raymond et al. 1992); and, at least for Mdr1, glycosylation is apparently also not essential for its function even in animal cells (Beck and Cirtain 1982; Ling et al. 1983). In contrast, oligosaccharide addition appears to be important for the proper targeting of CFTR because certain mutations that result in incompletely glycosylated CFTR also lead to degradation of the protein in the endoplasmic reticulum (Cheng et al. 1990).

Although the majority of Ste6p protein was localized to the cell surface, indirect immunofluorescent staining also detected Ste6p in vesicle-like structures at or near the inner side of the plasma membrane (Fig. 3). The spatial distribution of *STE2* gene product, the receptor for α-factor mating pheromone on the surface of *MAT***a** cells (Blumer et al. 1988; Reneke et al. 1988), displays almost an identical pattern when examined by indirect immunofluorescence microscopy (Marsh and Herskowitz 1988; Jackson et al. 1991). The apparent vesicular staining for Ste6p could be an artifact of overexpression; however, this conclusion seems unwarranted because a similar staining pattern, albeit much weaker, was observed when Ste6p was expressed from a *CEN* plasmid instead of from a multicopy vector. The observed vesicle-like bodies could arise from some perturbation of the structure of Ste6p caused by the presence of the c-Myc epitopes, resulting in its mislocalization or aggregation within the secretory compartment. This explanation also seems unlikely because the epitope-tagged Ste6p derivatives examined immunocytologically were fully functional physiologically. Moreover, our laboratory has used c-Myc-epitope tagging to successfully localize several other yeast proteins, including Sst2 (H. Dohlman, unpubl.) and Mot1 (Davis et al. 1992), without detecting any vesicle-like staining whatsoever. Therefore, the vesicular bodies most likely represent newly synthesized Ste6p being delivered to the plasma membrane. In fact, many nascent plasma membrane proteins in yeast can also be found in intracellular vesicles isolated by biochemical fractionation (Holcomb et al. 1988b). On the other hand, it has been reported that a significant fraction of yeast Pma1p exists in an intracellular pool of membrane vesicles in which this H^+-ATPase is fully functional (Serrano et al. 1991). Thus, the vesicles that contain Ste6p may not simply represent way stations in its intracellular transport. Alternatively, the Ste6p-containing vesicles could function as an intracellular reservoir of this transport protein, which in response to an appropriate stimulus (presumably exposure to α-factor) can be mobilized for rapid insertion into the plasma membrane. A precedent for such a regulatory mechanism exists in mammalian adipocytes in the case of insulin-stimulated exocytosis of preexisting vesicles that contain a particular glucose transporter isotype (James et al. 1989). In fact, the regulated exocytotic insertion of plasma membrane transporters and channels appears to be a more general phenomenon (Al-Awqati 1989).

The mating pheromones (**a**-factor and α-factor) trigger a signal transduction cascade that induces a complex series of biochemical and morphological changes in their corresponding target cells (*MAT*α and *MAT***a** haploids, respectively). The responses induced result in a massive rearrangement of the cytoskeleton, leading to asymmetric growth and formation of a projection. The tip of this elongated projection is the actual site of conjugation where the two haploid cell types fuse to form a zygote. We found that both the level and the localization of Ste6p undergo a dramatic change when yeast cells are exposed to mating pheromone. As we have shown previously, elevated levels of Ste6p result in a dramatic increase in the production of extracellular **a**-factor (Kuchler et al. 1989). Moreover, both genes that encode the **a**-factor precursors, *MF***a***1* and *MF***a***2*, are also pheromone-inducible (Gething et al. 1985; Dolan et al. 1989). Hence, in response to receipt of an α-factor signal, *MAT***a** cells will be provoked into emitting a greatly elevated **a**-factor signal. Furthermore, because we found that the majority of Ste6p in α-factor-treated *MAT***a** cells localizes primarily to the tip of the growing projection (Fig. 3), most of the **a**-factor secreted will presumably be released from the tip of this projection, leading to a pronounced anisotropy in the gradient of **a**-factor surrounding the *MAT***a** cell. The ability of yeast cells to respond to pheromones, and to increase pheromone production in response to a stimulus, has been referred to as "courtship" and has been shown to be a prerequisite for efficient mating between haploid cells (Jackson and Hartwell 1990a). A given haploid cell selects for mating only the haploid cell of opposite mating type that produces the highest level of extracellular pheromone and that is able, in turn, to respond to pheromone by formation of its own projection, a process referred to as "partner discrimination" (Jackson and Hartwell 1990b). Thus, both the elevated production and the highly polarized localization of Ste6p in an α-factor-induced *MAT***a** cell help to provide a molecular explanation for the phenomena of courtship and partner discrimination.

Our results suggest that, in response to pheromone, a *MAT***a** haploid will release a high level of **a**-factor at a highly localized site on its cell surface, the projection tip. Furthermore, because **a**-factor is a lipopeptide, its hydrophobicity and poor diffusibility will help to maintain its high concentration at the projection tip. Presumably only the *MAT*α cell nearest the projection tip of the *MAT***a** cell will be exposed to a concentration of **a**-factor sufficiently potent to elicit the complementary biological response. Thus, we propose that Ste6p may be essential for efficient mating for two distinct reasons: Ste6p is required for **a**-factor secretion per se (Kuchler et al. 1989) and, in addition, Ste6p may be critical for establishment and/or maintenance of the polarity in **a**-factor secretion that is required for effective courtship and proper partner selection. This latter role for Ste6p may explain why exogenously added **a**-factor does not rescue efficiently the mating defect of *MAT***a** *mf***a***1 mf***a***2* mutants (Michaelis and Herskowitz 1988; Marcus et al. 1991). Similarly, during mammalian development, there are often circumstances where one cell induces a differentiative event in a neighboring cell or a tissue induces developmental changes that are confined to a single layer of overlying epithelial cells.

Some of the agents that appear to be responsible for such events, for example, IL-1α and IL-1β and acidic and basic fibroblast growth factors, do not appear to be secreted by the classic secretory pathway (Kuchler and Thorner 1990, 1992b). Given that Ste6p is highly homologous to mammalian Mdr proteins and given that we have shown here that at least one mammalian Mdr protein is capable of transporting a heterologous peptide (Fig. 4), our results suggest that the transporter-mediated release of a differentiative peptide may be a device for the highly localized delivery of a developmental signal that has been evolutionarily conserved from yeast to mammals.

Compelling evidence from a number of divergent systems (for review, see Kuchler and Thorner 1990, 1992b) indicates that members of the ABC-transporter superfamily are located in a variety of organellar membranes, as well as in the plasma membrane. Indeed, the identification of multiple *STE6*-related genes in *S. cerevisiae*, including the *SSH1* and *SSH2* loci that we have discovered (Figs. 5 and 6), suggests that ABC transporters may be considerably more abundant in eukaryotic cells than first anticipated. The remaining challenge is to devise appropriate methods and approaches to determine the relevant physiological roles of each of these molecules.

ACKNOWLEDGMENTS

This work was supported by an Erwin Schrödinger postdoctoral fellowship from the Fonds zur Förderung der wissenschaftlichen Forschung of the Austrian government and by a postdoctoral fellowship from the Austrian Chamber of Commerce (to K.K.), by a postdoctoral fellowship (ALTF 141-1990) from the European Molecular Biology Organization (to H.M.G.), by a President's Undergraduate Research Fellowship and by funds from the Undergraduate Research Assistants Program of the University of California (to M.N.V.), by National Institutes of Health research grant GM-21841 and by funds provided by the W.M. Keck Foundation (to J.T.), and by facilities provided by the Cancer Research Laboratory, University of California, Berkeley. The authors thank J. Michael Bishop of the University of California, San Francisco, for providing hybridoma lines; Stan Grell of the Cancer Research Laboratory for his expert assistance in the production and isolation of ascites fluid; Ramon Serrano for the gift of anti-Pma1 antiserum; Fred Cross for the gift of diploid strain FC180; and other members of our laboratory for helpful suggestions and useful discussions.

REFERENCES

Al-Awqati, Q. 1989. Regulation of membrane transport by endocytotic removal and exocytotic insertion of transporters. *Methods Enzymol.* **172:** 49.

Ames, G. F.-L., C. Mimura, and V. Shyamala. 1990. Bacterial periplasmic permeases belong to a family of transport proteins operating from *Escherichia coli* to human: Traffic ATPases. *FEMS Microbiol. Rev.* **75:** 429.

Ammerer, G. 1983. Expression of genes in yeast using the *ADC1* promoter. *Methods Enzymol.* **101:** 192.

Anderegg, R.J., R. Betz, S.A. Carr, J.W. Crabb, and W. Duntze. 1988. Structure of the *Saccharomyces cerevisiae* mating hormone **a**-factor: Identification of S-farnesyl cysteine as a structural component. *J. Biol. Chem.* **263:** 18236.

Baba, M., N. Baba, Y. Ohsumi, K. Kanaya, and M. Osumi. 1989. Three-dimensional analysis of morphogenesis induced by mating pheromone α-factor in *Saccharomyces cerevisiae*. *J. Cell Sci.* **94:** 207.

Baldari, C., J.A. Murray, P. Ghiara, C. Cesarini, and C.L. Galeotti. 1987. A novel leader peptide which allows efficient secretion of a fragment of human interleukin-1β in *Saccharomyces cerevisiae*. *EMBO J.* **6:** 229.

Baum, P., J. Thorner, and L. Honig. 1978. Identification of tubulin from the yeast *Saccharomyces cerevisiae*. *Proc. Natl. Acad. Sci.* **75:** 4962.

Beck, W.T. and M.C. Cirtain. 1982. Continued expression of vinca alkaloid resistance by CCRF-CEM cells after treatment with tunicamycin or pronase. *Cancer Res.* **42:** 184.

Blumer, K.J., J.E. Reneke, W.E. Courchesne, and J. Thorner. 1988. Functional domains of a peptide hormone receptor: The α-factor receptor (*STE2* gene product) of the yeast *Saccharomyces cerevisiae*. *Cold Spring Harbor Symp. Quant. Biol.* **53:** 591.

Bowser, R. and P. Novick. 1991. Sec15 protein, an essential component of the exocytotic apparatus, is associated with the plasma membrane and with a soluble 19.5 S particle. *J. Cell Biol.* **112:** 1117.

Chan, R.K. and C.A. Otte. 1982. Physiological characterization of *Saccharomyces cerevisiae* mutants super-sensitive to G1-arrest by α-factor and **a**-factor pheromones. *Mol. Cell. Biol.* **2:** 21.

Cheng, S.H., R.J. Gregory, J. Marshall, S. Paul, D.W. Souza, G.A. White, C.R. O'Riordan, and A.E. Smith. 1990. Defective intracellular transport and processing of CFTR is the molecular basis of most cystic fibrosis. *Cell* **63:** 827.

Davis, J.L., R. Kunisawa, and J. Thorner. 1992. A presumptive helicase (*MOT1* gene product) affects gene expression and is required for viability in the yeast *Saccharomyces cerevisiae*. *Mol. Cell. Biol.* **12:** 1879.

Davis, T.N. and J. Thorner. 1987. Isolation of the yeast calmodulin gene using synthetic oligonucleotide probes. *Methods Enzymol.* **139:** 248.

Dolan, J.W., C. Kirkman, and S. Fields. 1989. The yeast STE12 protein binds to the DNA sequence mediating pheromone induction. *Proc. Natl. Acad. Sci.* **86:** 5703.

Dreesen, T.D., D.J. Johnson, and S. Henikoff. 1988. The brown protein of *Drosophila melanogaster* is similar to the white protein and to components of active transport complexes. *Mol. Cell. Biol.* **8:** 5206.

Dreyfuss, G., S.A. Adam, and D.Y. Choi. 1984. Physical change in cytoplasmic messenger ribonucleoproteins in cells treated with inhibitors of mRNA transcription. *Mol. Cell. Biol.* **4:** 415.

Drubin, D. 1991. Development of cell polarity in budding yeast. *Cell* **65:** 1093.

Endicott, J.A. and V. Ling. 1989. The biochemistry of P-glycoprotein-mediated multidrug resistance. *Annu. Rev. Biochem.* **58:** 137.

Evan, G.I., G.K. Lewis, G. Ramsay, and J.M. Bishop. 1985. Isolation of monoclonal antibodies specific for human c-*myc* proto-oncogene product. *Mol. Cell. Biol.* **5:** 3610.

Fuller, R.S., R.E. Sterne, and J. Thorner. 1988. Enzymes required for yeast prohormone processing. *Annu. Rev. Physiol.* **50:** 345.

Gartner, J., H. Moser, and D. Valle. 1992. Mutations in the 70 kd peroxisomal membrane protein gene in Zellweger syndrome. *Nature Genet.* **1:** 16.

Gething, M.-J., ed. 1985. *Current communications in molecular biology: Protein transport and secretion*, p. 103. Cold Spring Harbor Laboratory, Cold Spring Harbor, New York.

Göransson, H.M. and J. Thorner. 1992. Eukaryotic Mdr1/P-glycoprotein homologues: Unconventional secretion processes mediated by a growing family of ATP-dependent membrane translocators. In *Protein synthesis and targeting in yeast* (ed. M.F. Tuite et al.), p. 339. Springer-Verlag, Berlin.

Gottesman, M.M. and I. Pastan. 1993. Biochemistry of multidrug resistance mediated by the multidrug transporter. *Annu. Rev. Biochem.* **62:** (in press).

Hamada, H. and T. Tsuruo. 1988. Characterization of the ATPase activity of the M_r 170,000 to 180,000 membrane glycoprotein (P-glycoprotein) associated with multidrug resistance in K562/ADM cells. *Cancer Res.* **48:** 4926.

Hancock, J.F., K. Cadwallader, H. Paterson, and C.J. Marshall. 1991. A CAAX or a CAAL motif and a second signal are sufficient for plasma membrane targeting of ras proteins. *EMBO J.* **10:** 4033.

Hartmann, E., T.A. Rapoport, and H. Lodish. 1989. Predicting the orientation of eukaryotic membrane-spanning proteins. *Proc. Natl. Acad. Sci.* **86:** 5786.

Higgins, C.F. 1992. ABC transporters: From microorganisms to man. *Annu. Rev. Cell Biol.* **8:** 67.

Holcomb, C.L., W.J. Hansen, T. Etcheverry, and R. Schekman. 1988a. Secretory vesicles externalize the major plasma membrane ATPase in yeast. *J. Cell Biol.* **106:** 641.

———. 1988b. Plasma membrane protein intermediates are present in the secretory vesicles of yeast. In *Molecular biology of intracellular protein sorting and organelle assembly*, p. 153. A.R. Liss, New York.

Holmes, D.S. and M. Quigley. 1981. A rapid boiling method for the preparation of bacterial plasmids. *Anal. Biochem.* **114:** 193.

Hopp, T.P., K.S. Prickettt, V.L. Price, R.T. Libby, C.J. March, D.P. Corretti, D.L. Urdal, and P.J. Conlon. 1988. A short polypeptide marker sequence useful for recombinant protein identification and purification. *BioTechnology* **6:** 1204.

Ito, H., Y. Fukuda, K. Murata, and A. Kimura. 1983. Transformation of intact yeast cells treated with alkali cations. *J. Bacteriol.* **153:** 163.

Jackson, C.L. and L.H. Hartwell. 1990a. Courtship in *Saccharomyces cerevisiae*: Both cell types choose mating partners by responding to the strongest pheromone signal. *Cell* **63:** 1039.

———. 1990b. Courtship in *Saccharomyces cerevisiae*: An early cell-cell interaction during mating. *Mol. Cell. Biol.* **10:** 2202.

Jackson, C.L., J.B. Konopka, and L.H. Hartwell. 1991. *S. cerevisiae* α pheromone receptors activate a novel signal transduction pathway for mating partner discrimination. *Cell* **67:** 389.

James, D.E., M. Strube, and M. Mueckler. 1989. Molecular cloning and characterization of an insulin-regulatable glucose transporter. *Nature* **338:** 83.

Kamijo, K., S. Taketani, S. Yokota, T. Osumi, and T. Hashimoto. 1990. The 70-kDa peroxisomal membrane protein is a member of the Mdr (P-glycoprotein)-related ATP-binding protein superfamily. *J. Biol. Chem.* **265:** 4534.

Kane, S.E., I. Pastan, and M.M. Gottesman. 1990. Genetic basis of multidrug resistance of tumor cells. *J. Bioenerg. Biomembr.* **22:** 593.

Kuchler, K. and J. Thorner. 1990. Membrane translocation of proteins without hydrophobic signal sequences. *Curr. Opin. Cell Biol.* **2:** 617.

———. 1992a. Functional expression of human *mdr1* cDNA in *Saccharomyces cerevisiae*. *Proc. Natl. Acad. Sci.* **89:** 2302.

———. 1992b. Secretion of peptides and proteins lacking hydrophobic signal sequences: The role of adenosine triphosphate-driven membrane translocators. *Endocr. Rev.* **13:** 499.

Kuchler, K., H.G. Dohlman, and J. Thorner. 1993. The **a**-factor transporter (*STE6* gene product) and cell polarity in the yeast *Saccharomyces cerevisiae*. *J. Cell Biol.* **120:** (in press).

Kuchler, K., R.E. Sterne, and J. Thorner. 1989. *Saccharomyces cerevisiae STE6* gene product: A novel pathway for protein export in eukaryotic cells. *EMBO J.* **8:** 3973.

Kukuruzinska, M.A., M.L.E. Bergh, and B.J. Jackson. 1987. Protein glycosylation in yeast. *Annu. Rev. Biochem.* **56:** 915.

Leppert, G., R. McDevitt, S.C. Falco, T.K. Van Dyk, M.B. Ficke, and J. Golin. 1990. Cloning by gene amplification of two loci conferring multiple drug resistance in *Saccharomyces*. *Genetics* **125:** 13.

Ling, V., N. Kartner, T. Sudo, L. Siminovitch, and J. R. Riordan. 1983. Multidrug resistance phenotype in Chinese hamster ovary cells. *Cancer Treat. Rep.* **67:** 869.

Livi, G.P., J.S. Lillquist, L.M. Miles, A. Ferrara, G.M. Sathe, P.L. Simon, C.A. Meyeres, J.A. Gorman, and P.R. Young. 1991. Secretion of *N*-glycosylated interleukin-1β in *Saccharomyces cerevisiae* using a leader peptide from *Candida albicans*: Effect of *N*-linked glycosylation on biological activity. *J. Biol. Chem.* **266:** 15348.

Lowry, O.H., N.J. Rosebrough, A.L. Farr, and R.J. Randall. 1951. Protein measurements with the folin phenol reagent. *J. Biol. Chem.* **193:** 265.

Marcus, S., G. A. Caldwell, D. Miller, C.-B. Xue, F. Naider, and J.M. Becker. 1991. Significance of C-terminal cysteine modifications to the biological activity of the *Saccharomyces cerevisiae* **a**-factor mating pheromone. *Mol. Cell. Biol.* **11:** 3603.

Marsh, L. and I. Herskowitz. 1988. From membrane to nucleus: The pathway of signal transduction in yeast and its genetic control. *Cold Spring Harbor Symp. Quant. Biol.* **53:** 557.

McGrath, J.P. and A. Varshavsky. 1989. The yeast *STE6* gene encodes a homologue of the mammalian multidrug resistance P-glycoprotein. *Nature* **340:** 400.

Michaelis, S. and I. Herskowitz. 1988. The **a**-factor pheromone of *Saccharomyces cerevisiae* is essential for mating. *Mol. Cell. Biol.* **8:** 1309.

Monaco, J.J. 1992. A molecular model of MHC class I-restricted antigen processing. *Immunol. Today* **13:** 173.

O'Hare, K., C.V. Murphy, R. Levis, and G.M. Rubin. 1984. DNA sequence of the *white* locus of *Drosophila melanogaster*. *J. Mol. Biol.* **180:** 437.

Olson, M.V. 1991. Genome structure and organization in *Saccharomyces cerevisiae*. In *The molecular and cellular biology of the yeast* Saccharomyces: *Genome dynamics, protein synthesis, and energetics* (ed. J.R. Broach et al.,), vol. 1, p. 1. Cold Spring Harbor Laboratory Press, Cold Spring Harbor, New York.

Pringle, J.R., A.E.M. Adams, D.G. Drubin, and B.K. Haarer. 1991. Immunofluorescence methods for yeast. *Methods Enzymol.* **194:** 565.

Purnelle, B., J. Skala, and A. Goffeau. 1991. The product of the YCR105 gene located on the chromosome III from *Saccharomyces cerevisiae* presents homologies to ATP-dependent permeases. *Yeast* **7:** 867.

Raymond, M., P. Gros, M. Whiteway, and D.Y. Thomas. 1992. Functional complementation of yeast *ste6* by a mammalian multidrug resistance *mdr* gene. *Science* **256:** 232.

Reneke, J.E., K.J. Blumer, W.E. Courchesne, and J. Thorner. 1988. The carboxy-terminal segment of the yeast α-factor receptor is a regulatory domain. *Cell* **55:** 221.

Riordan, J.R., J.M. Rommens, B.-S. Kerem, N. Alon, M. Rozmahel, Z. Grzelczak, J. Zielenski, S. Lok, N. Plavsic, J.-L. Chou, M.L. Drumm, M.C. Ianuzzi, F.S. Collins, and L.-C. Tsui. 1989. Identification of the cystic fibrosis gene: Cloning and characterization of complementary DNA. *Science* **245:** 1066.

Sambrook, J., E.F. Fritsch, and T. Maniatis. 1989. *Molecular cloning: A laboratory manual*, 2nd edition. Cold Spring Harbor Laboratory, Cold Spring Harbor, New York.

Schafer, W.R. and J.D. Rine. 1992. Protein prenylation: Genes, enzymes, targets, and functions. *Annu. Rev. Genet.* **26:** 209.

Schafer, W.R., R. Kim, R.E. Sterne, J. Thorner, S.-H. Kim, and J. Rine. 1989. Genetic and pharmacological suppression of oncogenic mutations in *RAS* genes of yeast and humans. *Science* **245:** 379.

Schekman, R. 1985. Protein localization and membrane traffic in yeast. *Annu. Rev. Cell Biol.* **1:** 115.

Scott, J.H. and R. Schekman. 1980. Lyticase: Endoglucanase and protease activities that act together in yeast cell lysis. *J. Bacteriol.* **142:** 414.

Serrano, R., C. Montesinos, M. Roldan, G. Garrido, C. Ferguson, K. Leonard, B.C. Monk, D.S. Perlin, and E. Weiler. 1991. Domains of yeast plasma membrane and ATPase-associated protein. *Biochim. Biophys. Acta* **1062:** 157.

Sharma, R.C., S. Inoue, J. Roitelman, R.T. Schimke, and R.D. Simoni. 1992. Peptide transport by the multidrug resistance pump. *J. Biol. Chem.* **267:** 5731.

Sherman, F., G.R. Fink, and J.A. Hicks. 1986. *Laboratory course manual for methods in yeast genetics*. Cold Spring Harbor Laboratory, Cold Spring Harbor, New York.

Silve, S., C. Volland, C. Garnier, R. Jund, M.R. Chevallier, and R. Haguenauer-Tsapis. 1991. Membrane insertion of uracil permease, a polytopic yeast plasma membrane protein. *Mol. Cell. Biol.* **11:** 1114.

Sprague, G.F., Jr. and J. Thorner. 1992. Pheromone response and signal transduction during the mating process of *Saccharomyces cerevisiae*. In *The molecular and cellular biology of the yeast* Saccharomyces: *Gene expression* (ed. E. Jones et al.), vol. 2, p. 657. Cold Spring Harbor Laboratory, Cold Spring Harbor, New York.

Sterne, R.E. 1989. "A novel pathway for peptide hormone biogenesis: Processing and secretion of the mating pheromone **a**-factor by *Saccharomyces cerevisiae*." Ph.D. thesis, University of California, Berkeley.

Sterne, R.E. and J. Thorner. 1986. Processing and secretion of a yeast peptide hormone by a novel pathway. *J. Cell Biol.* **103:** 189a.

———. 1987. Peptide hormone export by a novel pathway: Yeast **a**-factor mating pheromone is a lipopeptide. *J. Cell Biol.* **105:** 80a.

Sterne-Marr, R.E., L.C. Blair, and J. Thorner. 1990. *Saccharomyces cerevisiae STE14* gene is required for COOH-terminal methylation of **a**-factor mating pheromone. *J. Biol. Chem.* **265:** 20057.

Towbin, H., T. Staehelin, and J. Gordon. 1979. Electrophoretic transfer of proteins from polyacrylamide gels to nitrocellulose sheets: Procedure and some applications. *Proc. Natl. Acad. Sci.* **76:** 4350.

Trueheart, J. and G.R. Fink. 1989. The yeast cell-fusion protein FUS1 is *O*-glycosylated and spans the plasma membrane. *Proc. Natl Acad. Sci.* **86:** 9916.

Van Arsdell, S.W. and J. Thorner. 1987. Hormonal regulation of gene expression in yeast. In *Transcriptional control mechanisms* (ed. D. Granner et al.), p. 325. A.R. Liss, New York.

Van Arsdell, S.W., G.L. Stetler, and J. Thorner. 1987. The yeast repeated element *sigma* contains a hormone-inducible promoter. *Mol. Cell. Biol.* **7:** 749.

Walker, J.E., M. Sarate, M.J. Runswicke, and N.J. Gay. 1982. Distantly related sequences in the α- and β-subunits of ATP synthase, myosin, kinases, and other ATP requiring enzymes and a common nucleotide binding fold. *EMBO J.* **1:** 945.

Wang, M., E. Balzi, L. Van Dyck, J. Golin, and A. Goffeau. 1992. Sequencing of the yeast multidrug resistance *PDR5* gene encoding a putative pump for drug efflux. *Yeast* **8:** S528.

West, I.C. 1990. What determines the substrate specificity of the multi-drug resistance pump? *Trends Biochem. Sci.* **15:** 42.

Wilson, K.L. and I. Herskowitz. 1984. Negative regulation of *STE6* gene expression by the $\alpha 2$ product of *Saccharomyces cerevisiae*. *Mol. Cell. Biol.* **4:** 24207.

———. 1986. Sequences upstream of the *STE6* gene required for its expression and regulation by the mating type locus in *Saccharomyces cerevisiae*. *Proc. Natl. Acad. Sci.* **83:** 2536.

Wu, C.-T., M. Budding, M.S. Griffin, and J.M. Croop. 1991. Isolation and characterization of *Drosophila* multidrug resistance gene homologs. *Mol. Cell. Biol.* **11:** 3940.

Each Caveola Contains Multiple Glycosyl-phosphatidylinositol-anchored Membrane Proteins

Y.-S. YING, R.G.W. ANDERSON, AND K.G. ROTHBERG
Department of Cell Biology and Neuroscience, University of Texas Southwestern Medical Center, Dallas, Texas 75235

Cells take up a variety of essential, low-molecular-weight molecules and ions by passive diffusion through water-filled pores in the plasma membrane. This process works well when the extracellular concentration is sufficient to support inward movement. When the concentration outside the cell is too low for direct transport, the ion or molecule must be first concentrated to facilitate import. Prokaryotic and eukaryotic cells have developed special cell machinery for this purpose (Ames 1986; Anderson et al. 1992).

Gram-negative bacteria sometimes must take up essential nutrients against gradients as large as 10^5-fold (Ames 1986). This is accomplished by a periplasmic transport system that is composed of two parts: (1) a collection of high-affinity binding proteins that are free in the periplasmic space, which is enclosed by the inner and outer plasma membranes of the cell, and (2) an integral membrane protein complex that resides in the inner membrane. Each binding protein is specific for a class of low-molecular-weight molecules or ions. These ligands can gain access to the periplasm through nonspecific pores in the outer membrane. Once a ligand binds to its binding protein, the complex attaches to the integral membrane component and the ligand moves across the membrane into the cytoplasm. Ligand translocation is energy dependent (for review, see Ames 1986).

Eukaryotic cells also need a mechanism for facilitating the import of low-molecular-weight molecules and ions that are sparse in the environment. For example, the concentration of the essential vitamin 5-methyltetrahydrofolate in plasma is as much as 10^3-fold lower than the intracellular concentration (Anderson et al. 1992), which under some conditions (Matsue et al. 1992) is too low to support cell growth. Recently, we described an uptake pathway that shares some features with periplasmic transport in bacteria. This process is called potocytosis (Anderson et al. 1992). The best-studied example of a ligand internalized by potocytosis is 5-methyltetrahydrofolate.

Many cultured epithelial cells express a high-affinity receptor (K_d = 1 nM) for 5-methyltetrahydrofolate (Antony et al. 1981, 1987; McHugh and Cheng 1979; Kamen and Capdevila 1986) that has the same ability to bind folates as does a soluble folate-binding protein found in the plasma and milk of many animal species (Kane and Waxman 1989). The relationship between these two proteins was clarified when the cloning of the cDNA (Lacey et al. 1989) and subsequent biochemical analysis of the pure protein (Luhrs and Slomiany 1989) showed that the receptor is a glycosyl-phosphatidylinositol (GPI)-anchored membrane protein (Ferguson and Williams 1988; Low 1989). Most likely, much of the soluble folate-binding protein found in vivo is generated by cleavage of the membrane anchor.

The mechanism of action of this receptor was determined by studying 5-methyltetrahydrofolate uptake in folate-depleted MA104 cells (Kamen and Capdevila 1986). The folate concentration within these cells is ~ 1 μM, whereas the concentration in cells grown under physiological conditions (5–10 nM 5-methyltetrahydrofolate) is ~ 5–6 μM. Half of the membrane receptors are able to bind folate at 0°C; however, when the cells are incubated continuously at 37°C for 1 hour, all the receptors become occupied with ligand (Kamen et al. 1988). Folate bound to external receptors can be distinguished from folate bound to internal receptors because the former is released by a brief acid wash. Acid-labile receptors exchange with acid-resistant receptors once every hour at 37°C (Kamen et al. 1988). Therefore, folate receptors recycle.

Once the receptor/vitamin complex is internalized at 37°C, the vitamin is delivered to the cytoplasm of the cell. The arrival of 5-methyltetrahydrofolate in the cytoplasm is detected in two ways. First, folate bound to membranes can be separated from cytoplasmic folate by cell fractionation on Percoll gradients (Kamen et al. 1989). Second, once the vitamin arrives in the cytoplasm, it is polyglutamated. The polyglutamated forms are easily detected by high-pressure liquid chromatography. Both methods document that as soon as 5-methyltetrahydrofolate is internalized into the acid-resistant compartment, it is released from the receptor and transported into the cytoplasm.

Receptor-mediated accumulation of 5-methyltetrahydrofolate in the cytoplasm (Kamen et al. 1991) is inhibited by probenecid, even though binding to the receptor and internalization into the acid-resistant compartment are unaffected by the drug. Probenecid is a general inhibitor of organic anion transport (Henderson and Zevely 1985; Lipman et al. 1990). This suggests that within the acid-resistant compartment folate dissociates from the receptor, possibly in response to low pH (Kamen et al. 1988), and moves across the membrane through an organic anion carrier.

The location of folate receptors on the cell surface has been determined by light and electron microscopic immunocytochemistry (Rothberg et al. 1990a,b). Folate receptors are organized into dense clusters (Rothberg et al. 1990b). Up to 18% of these clusters are associated with a membrane invagination called a caveola. The majority of the caveolae contain receptor clusters even though they occupy less than 2% of the apical surface. Other potential routes of internalization, such as clathrin-coated pits, are devoid of receptors (Rothberg et al. 1990b). The bulk of the evidence indicates that caveolae internalize receptor-bound 5-methyltetrahydrofolate by forming a closed compartment near the cell surface that does not merge with other endocytic pathways.

The closed caveolae are equivalent to the periplasmic space in bacteria, and the GPI-anchored proteins perform the concentrative function of the periplasmic binding proteins. Open caveolae receive the ligand. After the caveolae close, the ligand is released from the receptor or is enzymatically converted to a new species of molecule. The volume of the closed caveolae is so small that large concentration gradients are easily generated. The concentrated molecule or ion then naturally flows down its gradient into the cell through an appropriate carrier in the membrane. The coupling of a GPI-anchored receptor to a caveola provides a high-affinity route of internalization for small molecules that allows cells to flourish under conditions where the extracellular concentration of essential molecules is too low to support growth (Matsue et al. 1992).

There are three cardinal features of potocytosis: (1) uptake mediated by a GPI-anchored membrane protein; (2) sequestration of ligand by caveolae; (3) delivery of the sequestered ligand to the cytoplasm through carriers in the caveola membrane. To meet these criteria, caveolae must contain multiple GPI-anchored membrane proteins. New lipid-linked membrane proteins continue to be identified (Ferguson 1992), often in the absence of any known function. Only in the case of alkaline phosphatase (Jemmerson et al. 1985) and the folate receptor (Rothberg et al. 1990b) is there convincing evidence for occupancy of caveolae by a native, GPI-anchored protein. We have carried out a detailed immunocytochemical analysis of two additional GPI-anchored proteins, Thy-1 and prion PrP(C). We show that both these proteins are clustered in association with caveolae, that each cluster contains multiple GPI-anchored proteins, and that this distribution occurs in a variety of cell types.

MATERIALS AND METHODS

Materials

Chemicals used for immunofluorescence, electron microscopy, and cell culture were obtained as described previously (Lacey et al. 1989; Rothberg et al. 1990a,b, 1992). Antibodies were obtained from the following sources: goat anti-rabbit IgG, goat anti-mouse IgG, goat anti-mouse IgG conjugated to fluorescein isothiocyanate (FITC), goat anti-rabbit IgG conjugated to FITC, and rabbit anti-rat IgG conjugated to FITC were from Zymed Laboratories, Inc. (South San Francisco, California); rabbit anti-rat IgG was from Sigma (St. Louis, Missouri); sheep anti-rabbit IgG was from Organon Teknika (Cappel, Durham, North Carolina); mouse anti-dinitrophenol (DNP) IgG was from Oxford Biomedical (Oxford, Michigan); goat anti-rabbit IgG conjugated to gold (10 and 15 nm), donkey anti-sheep IgG conjugated to gold (5 nm), and rabbit anti-goat IgG conjugated to gold (5 and 10 nm) were from Biocell (Ted Pella, Redding, California); mouse anti-placental alkaline phosphatase IgG was from Medix Biotech Inc. (Foster City, California); rabbit anti-alkaline phosphatase IgG was from Accurate Chemical and Scientific Corp. (Westbury, New York); mouse anti-folate receptor IgG was from Centocor (Malvern, Pennsylvania); rat anti-Thy-1 IgG was either from Gibco (Gaithersburg, Maryland) or from the tissue-culture supernatant of hybridoma cells 30-H12 obtained from American Type Tissue Culture (ATCC TIB-107); and rabbit anti-Prion PrP(C) (designated R073; Serban et al. 1990) was a kind gift from Dr. Stanley Prusiner.

Methods

Cell culture. JEG-3 cells, a chorionic carcinoma cell line (ATCC-HTB 36), and Neuro-2A, a neuroblastoma cell line (ATCC-CCL 131), were grown continuously as a monolayer in Dulbecco's modified Eagle's medium supplemented with 10% fetal bovine serum and 100 units/ml penicillin/streptomycin. NIH-3T3 (ATCC-CRL 1659) and 3T3-L1 (ATCC-CL 173) cells, mouse fibroblast cell lines, were grown in Dulbecco's modified Eagle's medium containing 4500 mg/l *d*-glucose, 10% calf serum, and 100 units/ml penicillin/streptomycin. Cells for each experiment were set up according to a standard protocol. On day 0, 0.5×10^4 to 2.5×10^4 cells were plated onto 35-mm dishes and grown for 3–5 days. Medium was changed after 2 days. For induction of 3T3-L1 cell differentiation to adipocytes, cells were plated at a higher density, 5×10^4 cells/dish, and grown for 5 days so that cells were postconfluent. Cells were then induced in complete medium containing insulin (10 μg/ml), dexamethasone (0.3 μM), and isobutyl methylxanthine (0.5 μM) for 2 days. The dexamethasone and isobutyl methylxanthine were removed and the cells were maintained in insulin (10 μg/ml). Cells were used 8 days after induction.

Immunocytochemistry: Indirect immunofluorescence. Cells were chilled to 4°C for 20 minutes directly out of the incubator and then rinsed with ice-cold buffer A (minimum essential medium containing 20 mM HEPES [pH 7.4] and 0.2% crystalline BSA) and incubated an additional 10 minutes at 4°C. Antibodies were diluted in buffer A as indicated in the figure legends. Cells were incubated with the primary antibodies for 60 minutes, rinsed with buffer A, and subsequently incubated with FITC-conjugated secondary antibody (25 μg/ml in buffer A) for 60 minutes, all at 4°C. After washing in buffer A, cells were quickly rinsed with Dulbecco's phosphate-buffered saline (DPBS) and fixed with 3% paraformaldehyde in DPBS for 30 minutes. Finally, cells were rinsed with DPBS and water before mounting in a 2.5% solution of 1,4 diazobicyclo (2,2,2) octane. Cells were observed and photographed using a Ziess photomicroscope III with appropriate filter package.

Immunocytochemistry: Immunoelectron microscopy. Cells were chilled, incubated with the appropriate primary antibody at the indicated temperature, and washed exactly as described above for the immunofluorescence protocol. In double-labeling experiments, the cells were incubated with a mixture of the two primary antibodies. The cells were extensively washed in buffer A at 4°C and incubated in the presence of the appropriate anti-IgG followed by gold-labeled anti-anti-IgG as described previously (Rothberg et al. 1990b). For double-labeling experiments, the secondary and tertiary antibodies were added sequentially. In the case of alkaline phosphatase, the secondary antibody was conjugated to DNP instead of to gold. The cells were then rinsed in buffer B (0.1 M phosphate buffer, 3 mM KCl, 3 mM $MgCl_2$ [pH 7.4]), fixed in 2% glutaraldehyde in buffer B for 60 minutes, rinsed in buffer B, and postfixed in 1% OsO_4 in buffer C (0.1 M phosphate buffer plus 1.5% potassium ferricyanide) for 60 minutes, all at 4°C. Cells were then dehydrated in ethanol, stained en bloc with 1% uranyl acetate, and embedded in Eponate. Alkaline-phosphatase-specific DNP was localized in thin sections according to the method of Pathak and Anderson (1989). Sections were stained with 4% uranyl acetate and 1% lead citrate and observed using a JEOL 100 CX electron microscope.

Transfection of 3T3-L1 cells with folate receptor. 3T3-L1 cells (2.0×10^5) were plated onto 100-mm dishes and cultured in 10 ml of Dulbecco's modified Eagle's medium with high glucose (4500 mg/l) containing 10% calf serum and 100 units/ml penicillin/streptomycin (Medium A). After 24 hours, the medium was replaced with 3 ml of Medium A containing 3 mg/ml of polybrene (Sambrook et al. 1989) and further cultured for 18 hours. 100 ng of the cDNA for the folate receptor (designated FRwt) was mixed with fresh Medium A before adding 3 mg/ml of polybrene, and the cells were incubated with the mixture for 6 hours at 37°C. The DNA was removed, 30% DMSO in Medium A was added for 4 minutes, and the cells were cultured in Medium A for 24 hours. The medium was then changed to Medium A containing 1 mg/ml of G418 (selection medium). The selection medium was changed every 2 days to remove dead cells. After 14 days, individual colonies were selected and expanded in 12-well plates. These expanded colonies were assayed for the presence of the folate receptor by indirect immunofluorescence and subcloned by limiting dilution in selection medium.

Quantitative methods. Electron microscope negatives were obtained by randomly photographing 50 regions per experimental treatment. Coded negatives were projected to a final magnification of either 90,720 × or 136,000 × with a Slide King II lantern projector (Besler, Linden, New Jersey). Surface length, number of gold particles, number of clusters, number of caveolae, and number of caveolae labeled were determined directly from the enlarged areas. The code was broken to identify the sample at the completion of the quantification.

RESULTS

Thy-1 and prion were the new GPI-anchored proteins chosen for this study because of the availability of well-characterized antibodies as well as cultured cells that express these proteins. The same immunocytochemical methods were used that had previously been developed for localizing the folate receptor in MA104 cells (Rothberg et al. 1990b), which allows one to directly compare the results of the two studies.

Thy-1 and Prion Are Clustered in Caveolae

Immunogold localization of surface antigens was routinely performed by incubating cells in the presence of primary and secondary antibodies followed by gold-labeled, tertiary antibody before fixation and processing for electron microscopy (Rothberg et al. 1990b). NIH-3T3 cells that had been incubated in the presence of anti-Thy-1 IgG at 4°C had numerous clusters of gold particles distributed over the surface (arrows, Fig. 1A). Many of these clusters were found within or near caveolae (arrowheads, Fig. 1B). Quantitative analysis of the gold particles (Table 1) showed that 78% of the gold was in clusters, 65% of the caveolae were labeled, and 13% of the gold clusters were in caveolae. When the antibody incubations were carried out at room temperature, the association with caveolae was more striking: 89% clustered, 77% of the caveolae labeled (with a twofold increase in the number of caveolae), and 24% of the clusters in caveolae. We did not see any evidence of gold clusters associated with clathrin-coated pits or other membrane specializations. No labeling was seen when an irrelevant monoclonal antibody (Rothberg et al. 1992) was substituted for the anti-Thy-1 IgG (data not shown).

Prion is a cellular derived protein that can cause neurodegenerative disease in animals and humans (Prusiner 1991). The cellular protein, designated PrP(C), is synthesized as a GPI-anchored membrane protein (Stahl et al. 1992) that can be converted to the infectious form (designated PrP[SC]) by an unknown, posttranslational process (Prusiner 1991). Initially we

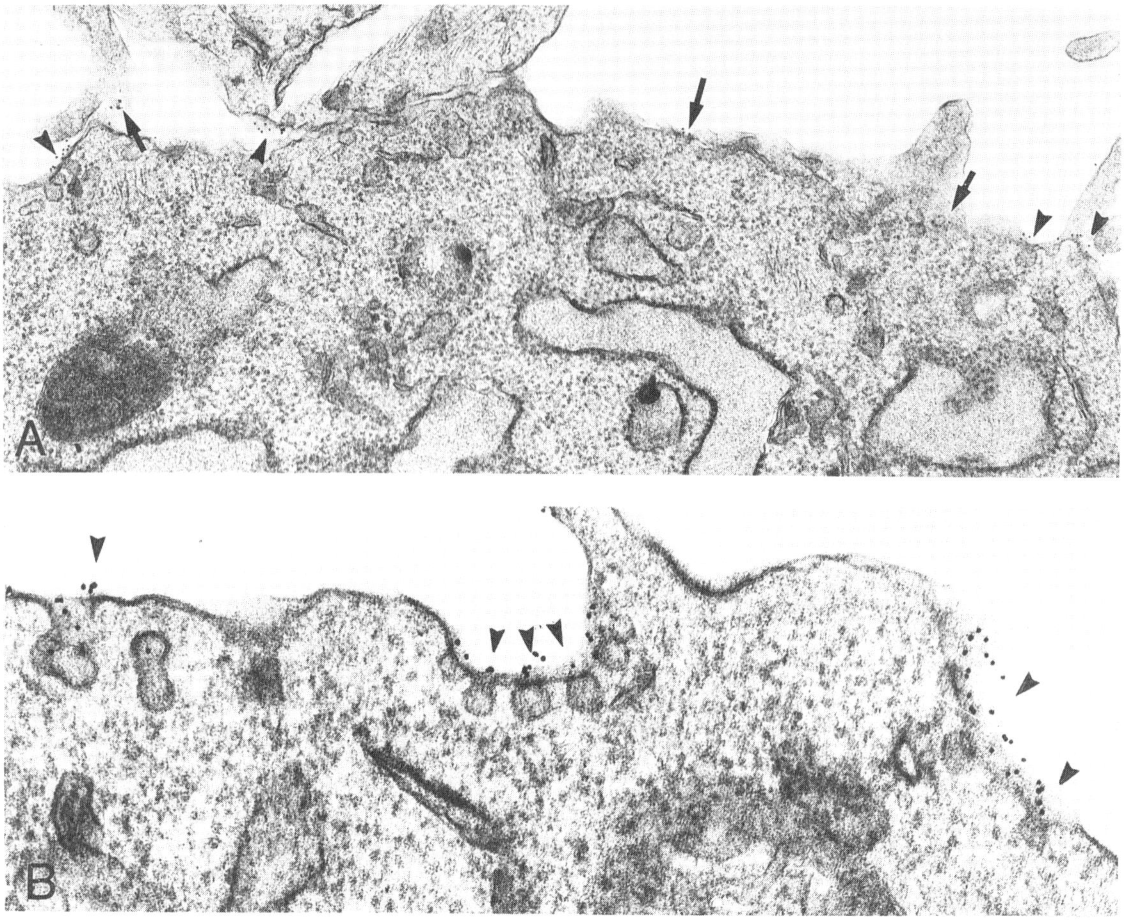

Figure 1. Immunogold localization of Thy-1 on the surface of 3T3 cells. Cells were chilled to 4°C and incubated in the presence of rat anti-Thy-1 IgG (20 μg/ml) followed by rabbit anti-rat IgG (25 μg/ml) and goat anti-rabbit IgG conjugated to gold (1/30 dilution in buffer A) as described previously. The cells were then fixed and processed for electron microscopy. (*A*) Arrows indicate clusters of gold particles, arrowheads indicate caveolae. (*B*) Arrows indicate gold within caveolae. Magnifications: *A*, 22,000 × ; *B*, 144,500 × .

Table 1. Quantitative Analysis of Immunogold Labeling

Cell	NIH-3T3	NIH-3T3	Neuro-2A
GPI-anchored protein	Thy-1	Thy-1	PrP(C)
Temperature	4°C	22°C	4°C
Gold clustered (%)	78	89	85
Caveolae labeled (%)	65	77	47
Clusters in caveolae (%)	13	24	12
Caveolae/100 μm	19.2	30.4	6.7
Gold/caveolae	4.8	6.9	4.6

The indicated cells were chilled to 4°C and incubated in the presence of either 20 μg/ml of anti-Thy-1 IgG or a 1/100 dilution of anti-PrP(C) serum in buffer A for 1 hr at 4°C. Cells were rinsed and processed to localize the primary antibody by indirect immunogold labeling as described in Fig. 5. Each set of cells was then processed for electron microscopy. Fifty photographs were taken of each sample. Quantitative analysis of the gold particle distribution was carried out on magnified images from the negatives as described previously. Total gold particles, number of caveolae, and length of membrane evaluated were (1) anti-Thy-1 at 4°C: 2440 particles, 79 caveolae, 411 μm; (2) anti-Thy-1 at 37°C: 4938 particles, 137 caveolae, 451 μm; (3) anti-PrP(C): 1683 particles, 96 caveolae, 673 μm. A cluster of gold is an aggregate of gold that contains more than three gold particles. The number of clusters per micrometer of membrane in nonimmune treated samples was below the limit of detection.

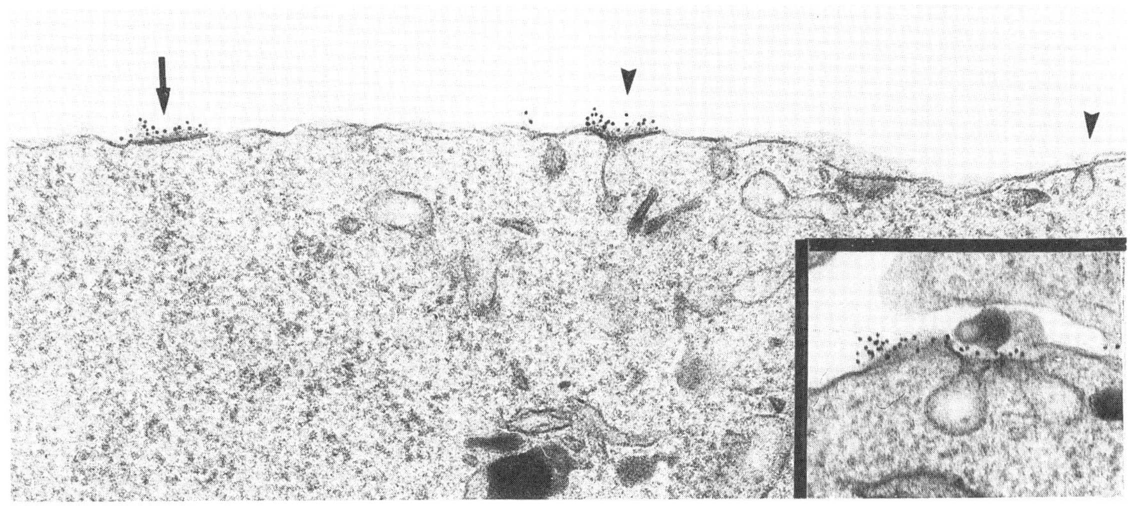

Figure 2. Immunogold localization of PrP(C) in Neuro-2A cells. Cells were chilled to 4°C and incubated in the presence of rabbit anti-PrP(C) serum (1/100 dilution in buffer A), goat anti-rabbit IgG (25 μg/ml), and rabbit anti-goat IgG conjugated to gold (1/30 dilution in buffer A) as described previously. Arrows indicate clusters of gold particles, arrowheads indicate caveolae. Magnification, 46,000 × ; inset, 62,000 × .

chose to localize PrP(C) in the neuroblastoma cell line Neuro-2A. Neuro-2A cells incubated in the presence of anti-PrP(C) serum had numerous clusters of gold particles on their surface (Fig. 2). Many of the clusters were associated with caveolae (Fig. 2 inset). A quantitative analysis of the gold distribution (Table 1) showed that 85% of the gold was in clusters, 47% of the caveolae were labeled, and 12% of the gold clusters were in caveolae. The density of caveolae in these cells was 3–4 times less than in NIH-3T3 cells (Table 1).

GPI Protein Is in Caveolae Regardless of Cell Type

We have now localized three different GPI-anchored membrane proteins—the folate receptor (Rothberg et al. 1990b), Thy-1, and PrP(C)—in three different types of cells: MA104, NIH-3T3, and Neuro-2A. The results suggest that a general property of GPI-anchored membrane proteins is their ability to cluster and associate with caveolae. This predicts that GPI-anchored proteins should maintain this organizational pattern during cell differentiation. Moreover, a foreign GPI-anchored protein introduced into a cell by transfection of a cDNA should also be clustered.

3T3-L1 cells can be induced to differentiate into adipocytes in culture (Kuri-Harcuch and Green 1981; Guller et al. 1988; Hauner 1990; Toscani et al. 1990; Sparks et al. 1991), which is accompanied by a ninefold increase in the number of caveolae (Fan et al. 1983). Figure 3 shows that Thy-1 was clustered on the surface of undifferentiated cells and that many of the clusters were associated with caveolae (arrowheads, Fig. 3A). The expression of Thy-1 seemed to decline in response to differentiation (Fig. 3B). There were fewer clusters but more caveolae, which resulted in a dramatic increase in the proportion of gold that was associated with caveolae (arrowheads, Fig. 3B). More than 40% of the anti-Thy-1 IgG-specific gold particles were located in caveolae.

3T3-L1 cells ordinarily do not express the folate receptor (Fig. 4A). Therefore, we transfected these cells with a full-length cDNA for the receptor, and positive cells were isolated by standard methods. Immunofluorescence showed that the receptor was organized into clusters (Fig. 4B) that were indistinguishable from clusters observed on the surface of MA104 cells (Rothberg et al. 1990b). Immunogold labeling experiments showed that many of these clusters were associated with caveolae (see Fig. 5C). Similar results were obtained when PAM-212 cells, a folate-receptor-negative, mouse keratinocyte cell line, were transfected with the folate receptor cDNA (Matsue et al. 1992).

Each Caveola Contains Multiple GPI-anchored Proteins

A single cell type may express many different GPI-anchored membrane proteins (Ferguson and Williams 1988; Low 1989; Ferguson 1992). The folate receptor, Thy-1, and PrP(C) were all found to be clustered, and in each case a high percentage (up to 89%) of caveolae were labeled. These results suggest that each cluster is a composite of many different species of GPI-anchored proteins. For this reason, we designed a double-labeling protocol to see if the folate receptor, Thy-1, and PrP(C) are coclustered on the surface of 3T3 cells.

NIH-3T3 cells express both Thy-1 and PrP(C). Double labeling of these cells with monospecific antibodies (5 nm = PrP(C); 15 nm = Thy-1) showed (Fig. 5) that each cluster of gold contained both sizes of gold particles and that caveolae were double labeled (arrows, Fig. 5A). We rarely found clusters of gold that

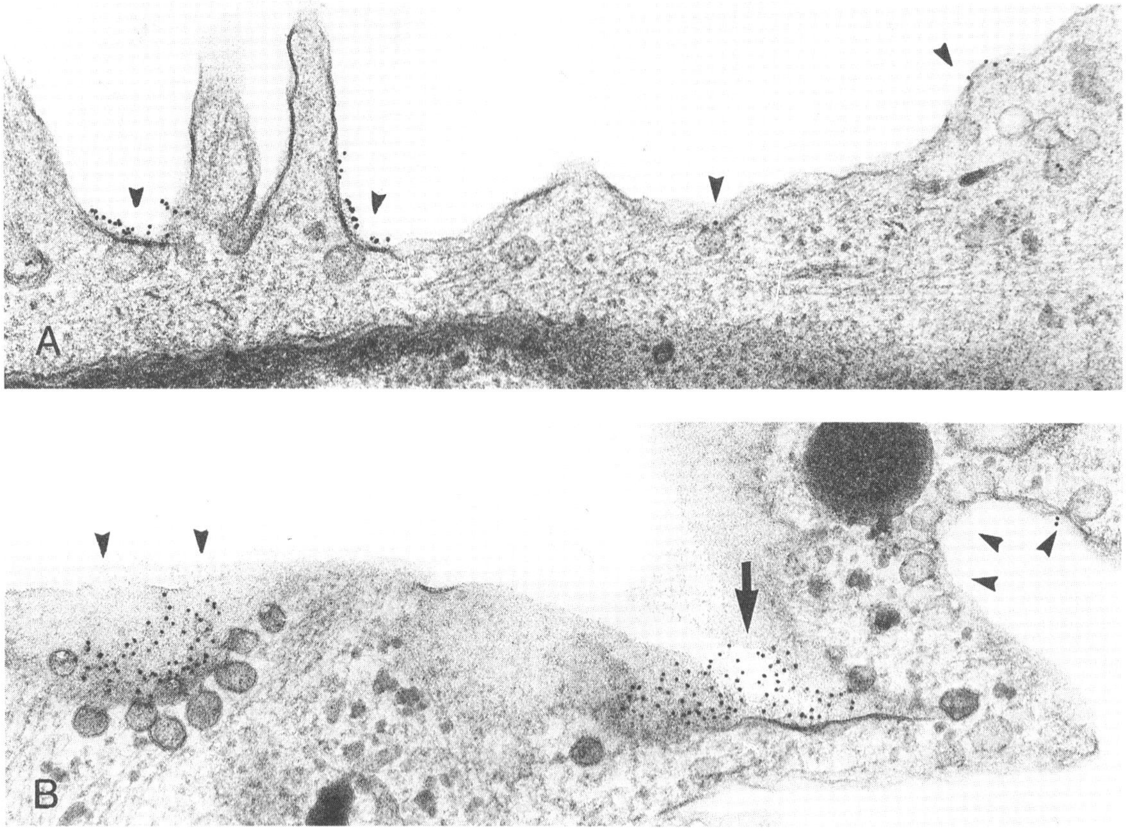

Figure 3. Localization of Thy-1 on the surface of undifferentiated (*A*) and differentiated (*B*) 3T3-L1 cells. Either untreated cells (*A*) or differentiated cells (*B*) were chilled to 4°C and processed to localize Thy-1 as described in Fig. 1. Arrows indicate clusters of gold particles, arrowheads indicate caveolae. Magnification, 55,000 ×.

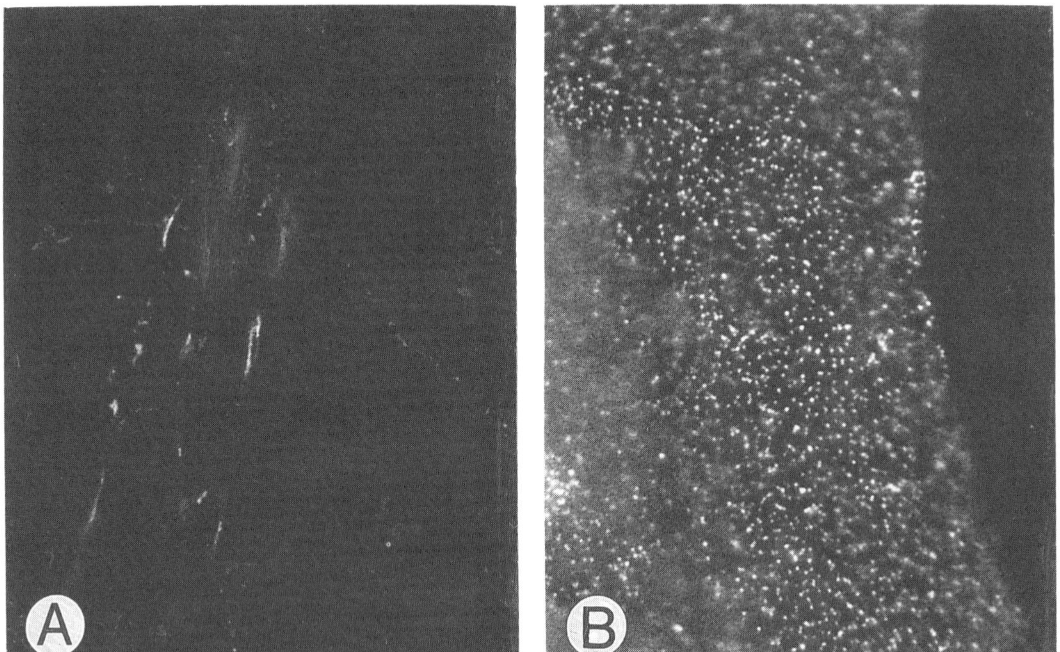

Figure 4. Immunofluorescence localization of folate receptor in normal (*A*) and transfected (*B*) 3T3-L1 cells. Cells were chilled to 4°C and incubated in the presence of rabbit anti-folate receptor IgG (50 μg/ml), followed by FITC-labeled anti-rabbit IgG (25 μg/ml) before fixing and processing as described previously. Magnification, 1375 ×.

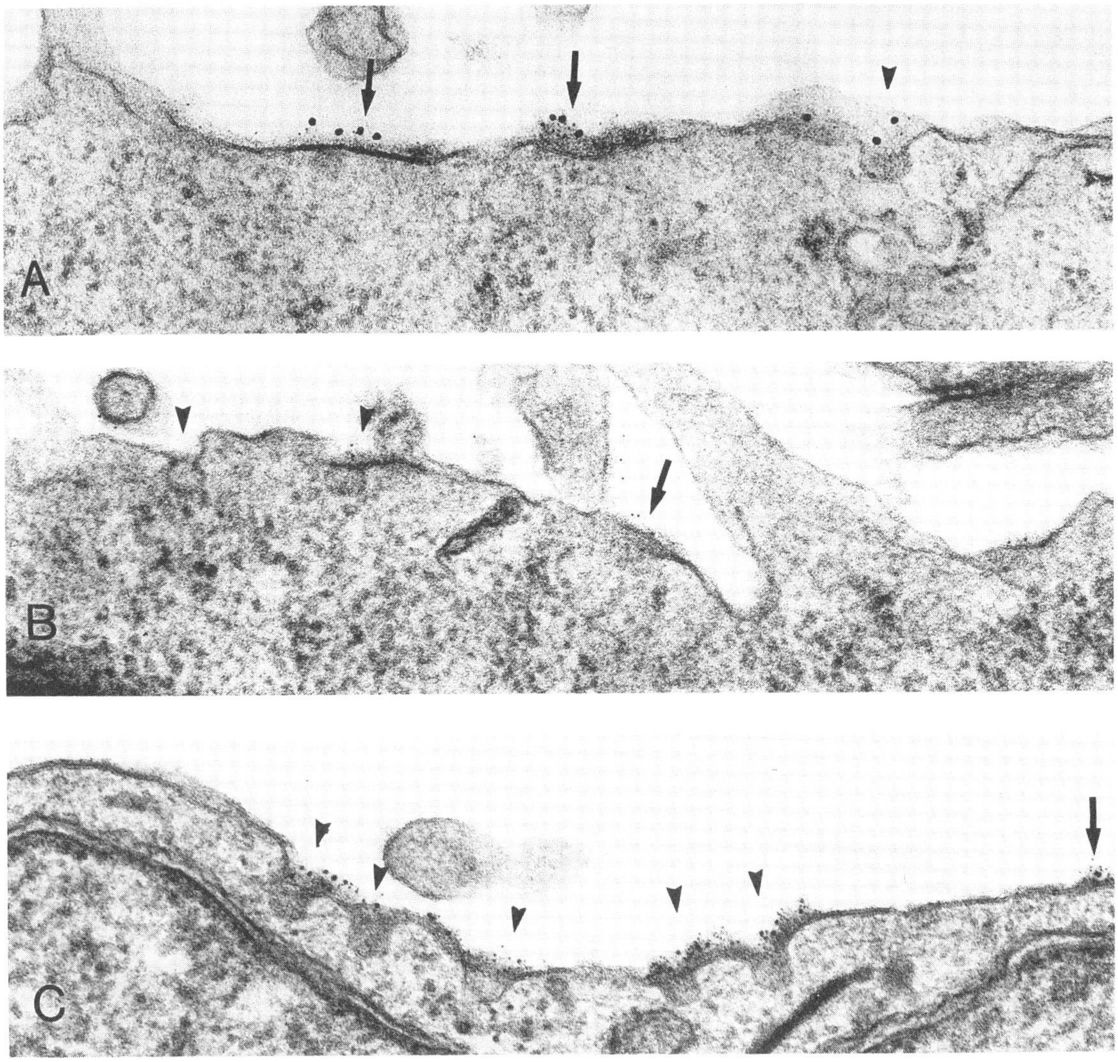

Figure 5. Immunogold colocalization of binding sites for either Thy-1 and PrP(C) (*A*), PrP(C) and non-immune IgG (*B*), or Thy-1 and the folate receptor (*C*) in normal (*A*,*B*) and transfected 3T3-L1 (*C*) cells. NIH-3T3 cells or 3T3-L1 cells that had been transfected with the cDNA for the folate receptor were chilled to 4°C and incubated in the presence of a mixture of two antigen-specific antibodies (20 μg/ml anti-Thy-1 IgG, 1/100 dilution of anti-PrP(C) serum, and 30 μg/ml of anti-folate receptor IgG) followed by the appropriate anti-IgG and gold-labeled anti-anti-IgG as described previously. 15-nm gold corresponds to Thy-1, and 5-nm gold labels either PrP(C) (*A*,*B*) or the folate receptor (*C*). Arrows indicate gold clusters, arrowheads indicate caveolae. Magnifications: *A*, 70,000 ×; *B*, 73,000 ×; *C*, 72,000 ×.

contained a single size gold particle. Only the small gold particles were seen when a nonimmune IgG was substituted for the anti-Thy-1 IgG (Fig. 5B).

3T3-L1 cells also express Thy-1 and PrP(C) (data not shown). Therefore, we colocalized Thy-1 and the folate receptor in 3T3-L1 cells that had been transfected with the folate receptor cDNA (Fig. 5C). Following an incubation with both anti-Thy-1 IgG and anti-folate receptor IgG, these cells were found to have both sizes of gold probes in each cluster (Fig. 5C). Many of these clusters were also found within caveolae. Therefore, in transfected 3T3-L1 cells, each cluster contains at least three different species of GPI-anchored membrane protein: Thy-1, PrP(C), and folate receptor.

As a final test, we used indirect immunofluorescence to localize both alkaline phosphatase and the folate receptor on the surface JEG-3 cells. Anti-folate receptor IgG bound in the usual punctate distribution found in all other cells that have been examined (Fig. 6A). Anti-alkaline phosphatase IgG appeared to be distributed on the surface with exactly the same pattern (Fig. 6B). Coclustering was confirmed by double labeling with two different sizes of gold-conjugated antibody (Fig. 6C; 10-nm gold = folate receptor and 5-nm gold = alkaline phosphatase). Each cluster contained both 5-nm and 10-nm gold particles. Labeled caveolae were also found to have both sizes of gold particles (arrowheads, Fig. 6D,E).

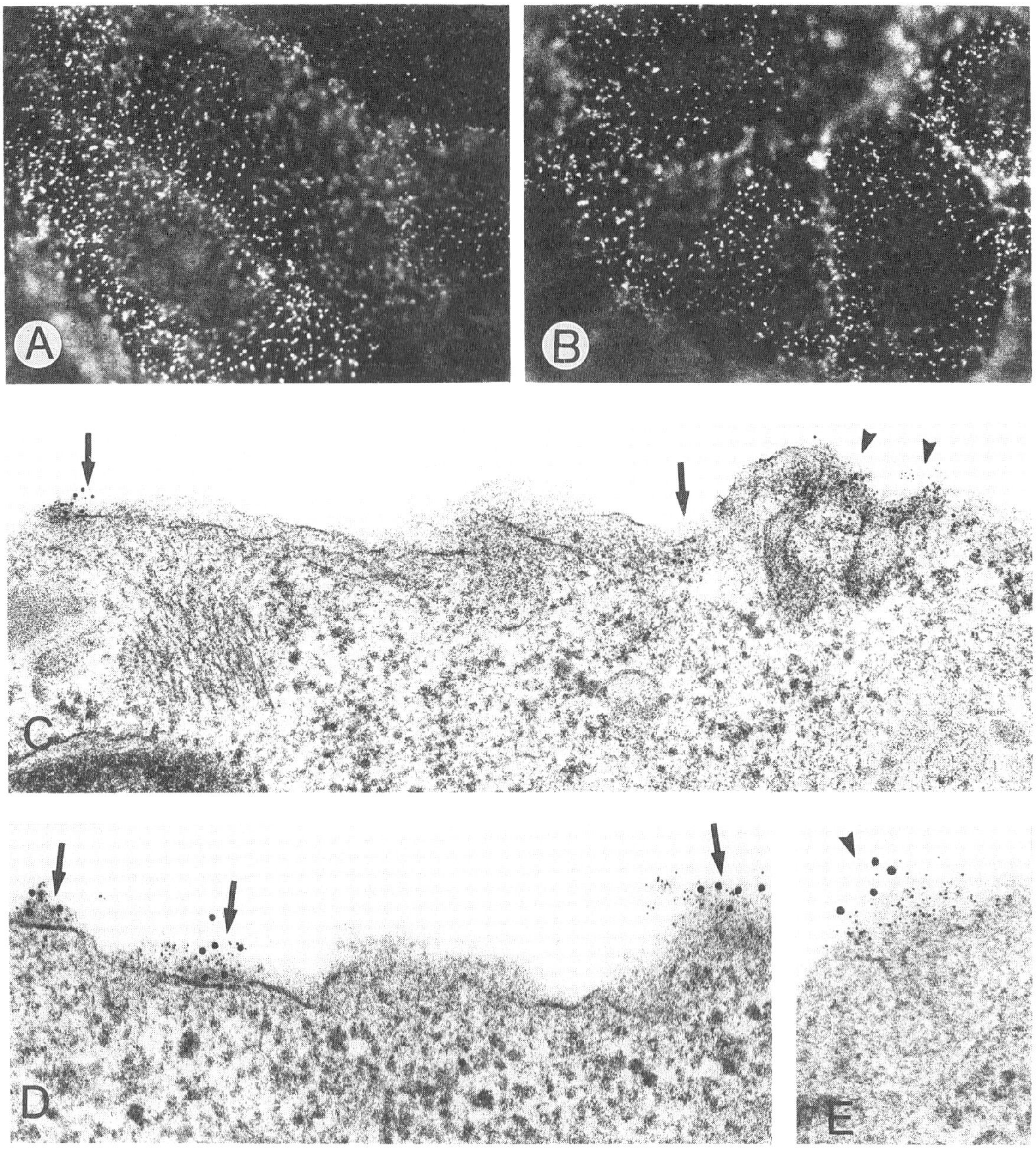

Figure 6. Indirect immunofluorescence localization of either the folate receptor (*A*) or alkaline phosphatase (*B*) and immunogold colocalization of the two antigens (*C–E*) in JEG-3 cells. Cells were chilled to 4°C and processed to localize the indicated antigens either by indirect immunofluorescence (*A,B*) or immunogold (*C–E*) methods using 50 μg/ml (*A*) or 100 μg/ml (*C–E*) of anti-folate receptor IgG and 30 μg/ml (*B–E*) of anti-alkaline phosphatase IgG. The cells were then incubated with a 1/30 dilution of anti-rabbit IgG conjugated to 5-nm gold (alkaline phosphatase) and 30 μg/ml of DNP-labeled anti-rabbit IgG. Samples were embedded and DNP was localized in thin sections using a 10-nm gold probe as described previously. (*C–E*) Arrows indicate the gold clusters, arrowheads indicate the caveolae. Magnifications: *A,B*, 1250 ×; *C*, 54,400 ×; *D,E*, 70,000 ×.

Caveolae Do Not Internalize Anti-Thy-1 IgG

Folate-receptor-specific immunogold label is rarely found directly within the lumen of caveolae in MA104 cells (Rothberg et al. 1990b). In contrast, anti-Thy-1 IgG-specific gold particles were often found within 3T3 cell caveolae (see arrowheads, Figs. 1B and 3). The smaller size of MA104 caveolae may make them less accessible to IgG probes such as anti-folate receptor IgG, which would account for why we never saw delivery of this immunoglobulin to endosomes (Rothberg et al. 1990b). If this is the case, then maybe 3T3 cell caveolae are able to endocytose anti-Thy-1 IgG and deliver it to endosomes. NIH-3T3 cells were incubated in the presence of anti-Thy-1 IgG for 2 hours at 37°C (Fig. 7). The cells were then fixed and processed for indirect immunofluorescence localization of either surface IgG (Fig. 7A) or internalized IgG (Fig. 7B). De-

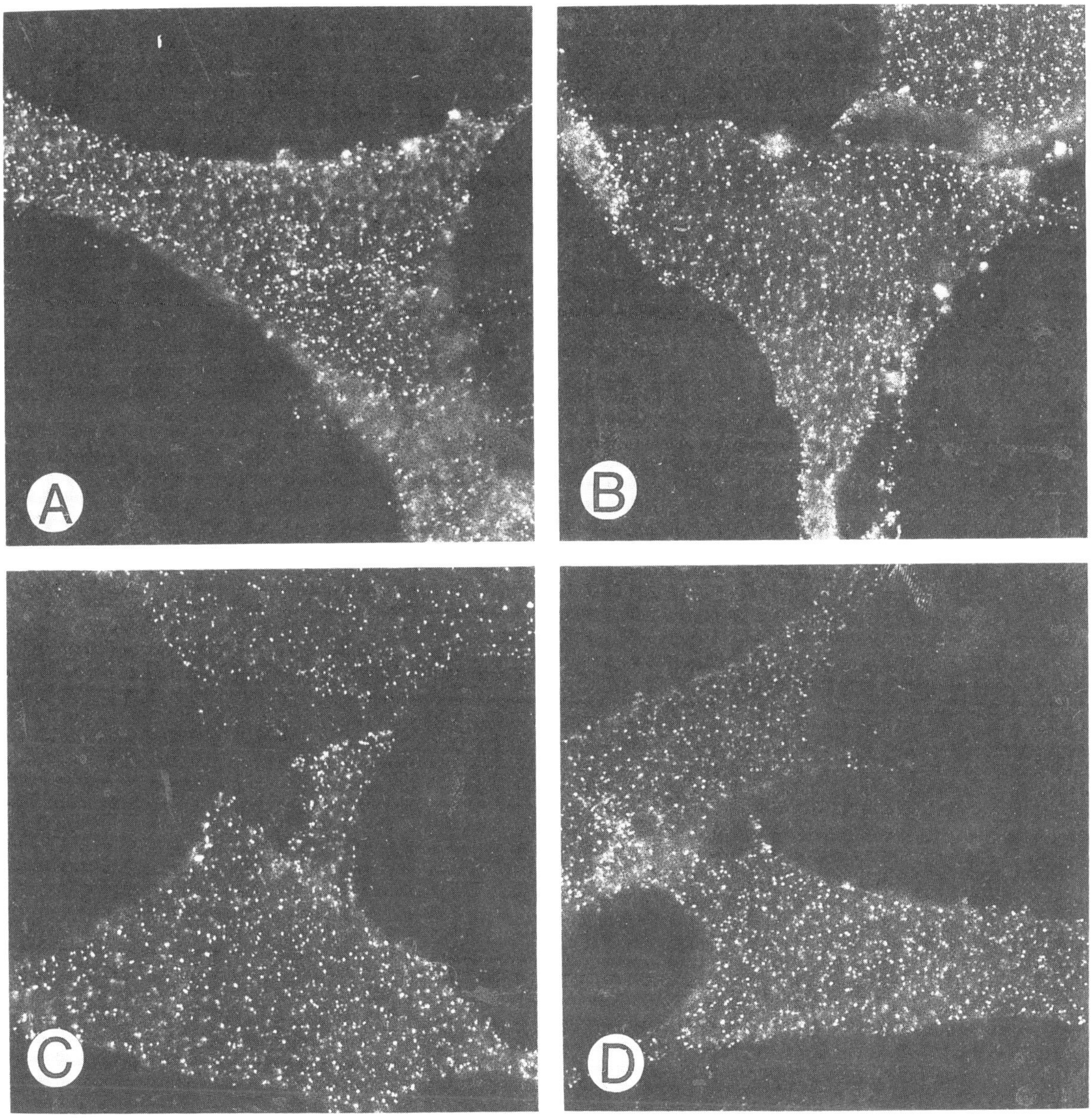

Figure 7. Distribution of surface-bound (*A,C,D*) and internalized (*B*) anti-Thy-1 IgG in cells incubated in the presence of anti-Thy-1 IgG. NIH-3T3 cells were incubated in the presence of 100 μg/ml of rat anti-Thy-1 IgG for either 2 hr at 37°C (*A,B*) or 1 hr at 4°C (*C,D*). One set of cells that had been incubated at 4°C was subsequently warmed to 37°C for 1 hr (*D*). Cells were then processed to localize either surface-bound (*A,C,D*) or internalized IgG (*B*) as described previously. Magnification, 1225×.

spite the lengthy incubation time, we found no evidence for internalization. There was no decline in the amount of IgG on the surface during the incubation, and cells did not accumulate detectable amounts of the IgG in internal vesicles (compare Fig. 7A with 7B). In another test, the cells were incubated with anti-Thy-1 IgG at 4°C, washed, and then warmed to 37°C for 1 hour to allow internalization. Staining of the cell surface with FITC-conjugated anti-rat IgG before (Fig. 7C) and after (Fig. 7D) warming to 37°C showed that there was no decline in the number of IgG binding sites during incubation.

DISCUSSION

Four different GPI-anchored, integral membrane proteins have now been localized on the cell surface using immunogold labeling techniques: the folate receptor (Rothberg et al. 1990b), alkaline phosphatase (Jemmerson et al. 1985), Thy-1, and prion PrP(C). In all cases, they are found to be highly clustered with a substantial fraction of the clusters within or near caveolae. Moreover, the fraction of clusters in caveolae increases with an increase in the number of observed caveolae. The formation of clusters appears to be an intrinsic property of the protein and its lipid anchor, because we found this behavior to be independent of cell type, state of cell differentiation, and whether or not the protein is native to the cell. Even a chimeric protein that has been engineered with a GPI anchor recognition site becomes associated with caveolae as a consequence of acquiring the lipid mooring (Keller et al. 1992).

The reproducible association of different GPI-anchored proteins with caveolae affirms the generality of the potocytosis hypothesis. A major function of the anchor is to direct the protein to caveolae. The anchor is also probably important for the clustering and sorting of the protein in the Golgi apparatus (van Meer et al. 1987; Lisanti et al. 1989; van Meer 1989; Rothberg et al. 1990b). We have proposed that there are at least three ways that GPI-anchored proteins are able to deliver molecules by potocytosis (Anderson et al. 1992): by binding a ligand such as folate; by enzymatically generating products that enter cells such as adenosine from adenosine monophosphate (Che et al. 1992); or by binding a circulating protein that carries a low-molecular-weight ligand such as thyroxin bound to transthyretin (Heltianu et al. 1989). Each interaction results in the uptake of a specific set of molecules through caveolae.

The clusters reflect the steady-state distribution of these proteins on the cell surface. Regardless of the protein, we consistently found that from 10% to 20% of the gold label was present as single particles scattered over the membrane. Most likely, single GPI-anchored proteins are constantly exchanging with the clusters in a dynamic equilibrium that favors the clustered configuration. The unclustered GPI-anchored protein may interact with other membrane proteins on the surface. The ability of cholesterol-binding drugs to uncluster GPI-anchored proteins, even after fixation with formaldehyde, indicates that these microdomains are quite fragile. This is why special care must be taken when applying immunocytochemical techniques not to uncluster or alter the distribution of these proteins during the experiment (Rothberg et al. 1990b). We recommend immunogold labeling, aldehyde fixation, and processing of samples strictly at 4°C.

Transmembrane proteins that internalize macromolecules by receptor-mediated endocytosis are clustered in clathrin-coated pits (Goldstein et al. 1985; Anderson 1991). Each coated pit is able to gather multiple, different proteins (Anderson 1991) that have in common a reverse-turn, three-dimensional conformation in their cytoplasmic domain (Bansal and Gierasch 1991). The physical basis for the clustering of GPI-anchored proteins appears to be quite different (Rothberg et al. 1990a). Nevertheless, each of these clusters also contains multiple proteins. If the folate receptor is typical, these proteins are densely packed in each domain (Rothberg et al. 1990b). A cluster, then, is a compact aggregate of glycosyl-phosphatidylinositol lipids attached to an array of different proteins that carry out specific uptake functions. The close apposition of these proteins within each cluster ($\sim$30,000 molecules/μm^2) affords an opportunity for protein-protein interactions that result in the chemical modification of select molecules within the group. Some of the enzymatic reactions known to be expressed by these proteins include hydrolase, phosphatase, and phosphotransferase activities (Ferguson and Williams 1988). This organizational scheme will permit complex chemical reactions to take place in each cluster, often within caveolae. These reactions may be crucial for controlling the import of essential, small molecules and ions, or the transmission of signals by potocytosis.

It is still not known whether caveolae pinch off from the membrane as do clathrin-coated pits during receptor-mediated endocytosis (Lin et al. 1991). If this is a normal part of the potocytosis cycle, then a ligand internalized by caveolae should appear in a vesicle that merges with an internal, vacuolar compartment. The origin of a ligand-containing vesicle is difficult to determine without a specific membrane marker such as a cytoplasmic coat. Our attempts to use caveolin, a protein component of the caveola coat (Rothberg et al. 1992), as a marker for vesicles derived from caveolae failed to find evidence for budding. We also were unable to detect internalization and delivery of anti-folate receptor IgG to endosomes in MA104 cells (Rothberg et al. 1990b). In the current study, we incubated NIH-3T3 cells in the presence of anti-Thy-1 IgG for an extended time but still failed to find evidence for significant accumulation in endosomes. Immunofluorescence is less sensitive than immunogold for detecting ligand internalization (Keller et al. 1992). The possibility that a small amount of anti-Thy-1 IgG is internalized nonspecifically by coated pits also cannot be ruled out (Tran et al. 1987; Montesano et al. 1982; Keller et al. 1992). Nevertheless, this experiment indicates that the quantity of IgG internalized by caveolae is far below what can be accomplished by receptors in clathrin-coated pits, which suggests that caveolae are not a major source of endocytic vesicles.

This study establishes that an intrinsic property of the GPI-anchored protein is an ability to cluster and associate with caveolae on the surfaces of cells. In this location, they are available for mediating the internalization of small molecules by potocytosis. A significant challenge for the future is to identify the molecules that are transported into cells by this pathway.

ACKNOWLEDGMENTS

We thank Drs. Stanley Prusiner and Albert Taraboulos for kindly providing the anti-PrP(C) IgG. We also are indebted to Drs. Stephen Lacey and Tim Ritter for assistance with the transfection experiments. This work was supported in part by grants from the National Institutes of Health (HL-20948 and GM-43169) and the Perot Foundation.

REFERENCES

Ames, G.F.-L. 1986. Bacterial periplasmic transport systems: Structure, mechanism, and evolution. *Annu. Rev. Biochem.* **55:** 397.

Anderson, R.G.W. 1991. Molecular motors that shape endocytic membrane. In *Intracellular trafficking of proteins* (ed. C.J. Steer and J.A. Hanover), p. 13. Cambridge University Press, London.

Anderson, R.G.W., B.A. Kamen, K.G. Rothberg, and S.W. Lacey. 1992. Potocytosis: Sequestration and transport of small molecules by caveolae. *Science* **255:** 410.

Antony, A.C., R.S. Kincade, R.S. Verma, and S.R. Krishnan. 1987. Identification of high affinity folate binding proteins in human erythrocyte membranes. *J. Clin. Invest.* **80:** 711.

Antony, A.C., C. Utley, K.C. Van Horne, and J.F. Kolhouse. 1981. Isolation and characterization of a folate receptor from human placenta. *J. Biol. Chem.* **256:** 9684.

Bansal, A. and L.M. Gierasch. 1991. The NPXY internalization signal of the LDL receptor adopts a reverse-turn conformation. *Cell* **67:** 1195.

Che, M., T. Nishida, Z. Gatmaitan, and I.M. Arias. 1992. A nucleoside transporter is functionally linked to ectonucleotidases in rat liver canalicular membrane. *J. Biol. Chem.* **267:** 9684.

Fan, J.Y., J.-L. Carpentier, E.V. Obberghen, C. Grunfeld, P. Gordon, and L. Orci. 1983. Morphological changes of the 3T3-L1 fibroblast plasma membrane upon differentiation to the adipocyte form. *J. Cell Sci.* **61:** 219.

Ferguson, M.A. 1992. Glycosyl-phosphatidylinositol membrane anchors: The tale of a tail. *Biochem. Soc. Trans.* **20:** 243.

Ferguson, M.A. and A.F. Williams. 1988. Cell-surface anchoring of proteins via glycosyl-phosphatidylinositol structures. *Annu. Rev. Biochem.* **57:** 285.

Goldstein, J.L., M.S. Brown, R.G.W. Anderson, D.W. Russell, and W.J. Schneider. 1985. Receptor-mediated endocytosis: Concepts emerging from the LDL receptor system. *Annu. Rev. Cell Biol.* **1:** 1.

Guller, S., R.E. Corin, D.C. Mynarcik, B.M. London, and M. Sonenberg. 1988. Role of insulin in growth hormone-stimulated 3T3 cell adipogenesis. *Endocrinology* **122:** 2084.

Hauner, H. 1990. Complete adipose differentiation of 3T3-L1 cells in a chemically defined medium: Comparison to serum-containing culture conditions. *Endocrinology* **127:** 865.

Heltianu, C., L. Dobrila, F. Anthoe, and M. Simionescu. 1989. Evidence for thyroxine transport by the lung and heart capillary endothelium. *Microvasc. Res.* **37:** 188.

Henderson, G.B. and E.M. Zevely. 1985. Inhibitory effects of probenecid on the individual transport routes which mediate the influx and efflux of methotrexate in L1210 cells. *Biochem. Pharmacol.* **34:** 1725.

Jemmerson, R., F.G. Klier, and W.H. Fishman. 1985. Clustered distribution of human placental alkaline phosphatase on the surface of both placental and cancer cells. *J. Histochem. Cytochem.* **33:** 1227.

Kamen, B.A. and A. Capdevila. 1986. Receptor-mediated folate accumulation is regulated by the cellular folate content. *Proc. Natl. Acad. Sci.* **83:** 5983.

Kamen, B.A., A.K. Smith, and R.G.W. Anderson. 1991. The folate receptor works in tandem with a probenecid-sensitive anion carrier. *J. Clin. Invest.* **87:** 1442.

Kamen, B.A., C.A. Johnson, M.T. Wang, and R.G.W. Anderson. 1989. Regulation of the cytoplasmic accumulation of 5-methyltetrahydrofolate in MA104 cells is independent of folate receptor regulation. *J. Clin. Invest.* **84:** 1379.

Kamen, B.A., M.T. Wang, A.J. Streckfuss, X. Peryea, and R.G.W. Anderson. 1988. Delivery of folates to the cytoplasm of MA104 cells is mediated by a surface membrane receptor that recycles. *J. Biol. Chem.* **263:** 13602.

Kane, M.A. and S. Waxman. 1989. Role of folate binding proteins in folate metabolism. *Lab. Invest.* **60:** 737.

Keller, G.-A., M.W. Siegel, and I.W. Caras. 1992. Endocytosis of glycophospholipid-anchored and transmembrane forms of CD4 by different endocytic pathways. *EMBO J.* **11:** 863.

Kuri-Harcuch, W. and H. Green. 1981. Suppression of the adipose conversion of 3T3 cells by acidified serum. *J. Cell. Physiol.* **108:** 455.

Lacey, S.W., J.M. Sanders, K.G. Rothberg, R.G.W. Anderson, and B.A. Kamen. 1989. cDNA for the folate binding protein correctly predicts anchoring to the membrane by glycosylphosphatidylinositol. *J. Clin. Invest.* **84:** 715.

Lin, H.C., M.S. Moore, D.A. Sanan, and R.G.W. Anderson. 1991. Reconstitution of coated pit budding from plasma membranes. *J. Cell Biol.* **114:** 881.

Lipman, B.J., S.C. Silverstein, and T.H. Steinberg. 1990. Organic anion transport in macrophage membrane vesicles. *J. Biol. Chem.* **265:** 2142.

Lisanti, M.P., I.W. Caras, M.A. Davitz, and E. Rodriguez-Boulan. 1989. A glycophospholipid membrane anchor acts as an apical targeting signal in polarized epithelial cells. *J. Cell. Biol.* **109:** 2145.

Low, M.G. 1989. The glycosyl-phosphatidylinositol anchor of membrane proteins. *Biochim. Biophys Acta* **988:** 427.

Luhrs, C.A. and B.L. Slomiany. 1989. A human membrane-associated folate binding protein is anchored by a glycosylphosphatidylinositol tail. *J. Biol. Chem.* **264:** 21446.

Matsue, H., K.G. Rothberg, A. Takashima, B.A. Kamen, R.G.W. Anderson, and S.W. Lacey. 1992. The folate receptor allows cells to grow in physiological concentrations of 5-methyltetrahydrofolate. *Proc. Natl. Acad. Sci.* **89:** 6006.

McHugh, M. and Y.C. Cheng. 1979. Demonstration of high affinity folate binder in human cell membranes and its characterization in cultured human KB cells. *J. Biol. Chem.* **254:** 11312.

Montesano, R., J. Roth, A. Robert, and L. Orci. 1982. Non-coated membrane invaginations are involved in binding and internalization of cholera and tetanus toxins. *Nature* **296:** 651.

Pathak, R.K. and R.G.W. Anderson. 1989. The use of dinitrophenol-IgG conjugates to detect sparse antigens by immunogold labeling. *J. Cytochem.* **37:** 69.

Prusiner, S.B. 1991. Molecular biology of prion diseases. *Science* **252:** 1515.

Rothberg, K.G., Y.-S. Ying, B.A. Kamen, and R.G.W. Anderson. 1990a. Cholesterol controls the clustering of the glycophospholipid-anchored membrane receptor for 5-methyltetrahydrofolate. *J. Cell Biol.* **111:** 2931.

Rothberg, K.G., Y.-S. Ying, J.F. Kolhouse, B.A. Kamen, and R.G.W. Anderson. 1990b. The glycophospholipid-linked folate receptor internalizes folate without entering the clathrin-coated pit endocytic pathway. *J. Cell Biol.* **110:** 637.

Rothberg, K.G., J.E. Heuser, W.C. Donzell, Y.-S. Ying, J.R. Glenney, and R.G.W. Anderson. 1992. Caveolin, a protein component of caveolae membrane coats. *Cell* **68:** 673.

Sambrook, J., E.F. Fritsch, and T. Maniatis. 1989. *Molecular cloning: A laboratory manual*, 2nd edition. Cold Spring Harbor Laboratory Press, Cold Spring Harbor, New York.

Serban, D., A. Taraboulos, S.J. DeArmond, and S.B. Prusiner. 1990. Rapid detection of Creutzfeldt-Jakob disease and scrapie prion proteins. *Neurology* **40:** 110.

Sparks, R.L., E.E. Strauss, A.I. Zygmunt, and T.E. Phelan. 1991. Antidiabetic AD4743 enhances adipocyte differentiation of 3T3 T mesenchymal stem cells. *J. Cell. Physiol.* **146:** 101.

Stahl, N., M.A. Baldwin, R. Hecker, K.M. Pan, A.L. Burlingame, and S.B. Prusiner. 1992. Glycosylinositol phos-

pholipid anchors of the scrapie and cellular prion proteins contain sialic acid. *Biochemistry* **31:** 5043.
Toscani, A., D.R. Soprano, and K.J. Soprano. 1990. Sodium butyrate in combination with insulin or dexamethasone can terminally differentiate actively proliferating Swiss 3T3 cells into adipocytes. *J. Biol. Chem.* **265:** 5722.
Tran, D., J.-L. Carpentier, F. Sawano, P. Gorden, and L. Orci. 1987. Ligands internalized through coated or non-coated invaginations follow a common intracellular pathway. *Proc. Natl. Acad. Sci.* **84:** 7957.
van Meer, G. 1989. Lipid traffic in animal cells. *Annu. Rev. Cell. Biol.* **5:** 247.
van Meer, G., E.H.K. Stelzer, R.W. Wijnaendts-van-Resandt, and K. Simons. 1987. Sorting of sphingolipids in epithelial (Madin-Darby canine kidney) cells. *J. Cell. Biol.* **105:** 1623.

Molecular Flypaper, Atherosclerosis, and Host Defense: Structure and Function of the Macrophage Scavenger Receptor

M. KRIEGER,* J.M. ABRAMS,*† A. LUX,* AND H. STELLER*†

Department of Biology and the Howard Hughes Medical Institute in the Department of Brain and Cognitive Sciences,† Massachusetts Institute of Technology, Cambridge, Massachusetts* 02139

Macrophages participate in host defense, development (Gordon et al. 1988), and pathology. For example, they engage in a variety of antibacterial, antimycobacterial, antiprotazoan, and antitumor reactions, including the generation of reactive oxygen and nitrogen intermediates and the secretion of proteases. They recognize and process antigens for T-cell presentation, release key mediators of inflammation (e.g., TNF), and are perhaps best characterized by their ability to endocytose and phagocytose a wide variety of molecules and cells.

Macrophage recognition of foreign substances is not limited to those materials that are identified and tagged by opsonins such as antibodies and complement (Gordon et al. 1988). In addition, macrophages have a capacity to differentiate independently between self and nonself (Ezekowitz et al. 1991; Janeway 1992; Taylor et al. 1992). This self-nonself discrimination may have predated the emergence of specific immunity mediated by lymphocytes and antibodies (Gordon et al. 1988; Janeway 1992). Janeway (1992) has proposed that the polyspecificity and self-nonself discrimination required of macrophage-like cells in primitive organisms could have been achieved with multiple receptors and/or receptors with multiple or broad ligand-binding specificity. We have suggested that macrophage scavenger receptors are attractive candidates for such host-defense discriminatory receptors (Freeman et al. 1990; Krieger 1992).

Development and Disease

In addition to host defense, macrophages may play an important role in the recognition of effete cells during the course of development and aging (Gordon et al. 1988). For example, we have recently observed that a large number of cells undergo programmed cell death during *Drosophila* embryogenesis and that the apoptotic cell corpses are phagocytosed by macrophages (J.M. Abrams et al., in prep.). Macrophages are also involved in the pathogenesis of several diseases, including fibrotic lung disease (Brain 1985) and atherosclerosis (Brown and Goldstein 1983; Goldstein and Brown 1989; Havel and Kane 1989; Steinberg et al. 1989), the principal source of one of the leading causes of death in industrialized nations, coronary heart disease. A hallmark of atherosclerosis is the formation of atherosclerotic plaques in artery walls. Plaques result from the accumulation underneath the endothelial lining of an artery of cells, cell debris, extracellular matrix, and cholesterol and cholesteryl esters. Macrophages and smooth muscle cells are major cellular components in plaques, where they are often filled with cholesteryl-ester-rich foam-like fat droplets that give these cells, called foam cells, a distinctive microscopic appearance. A growing plaque protrudes into the lumen of a coronary artery, narrowing the lumen and thus reducing blood flow. The reduced flow of blood can cause angina, and the reduction in the diameter of the lumen increases the likelihood of blockage by a clot and a consequent heart attack.

Atherosclerotic plaque formation apparently begins with the attachment of monocytes to the lumenal surface of the endothelium and is followed by their migration into the subendothelial space. There, the monocytes differentiate into macrophages and, when there are high-plasma low-density lipoprotein (LDL) levels, they accumulate massive amounts of lipoprotein cholesterol and become foam cells (Brown and Goldstein 1983; Steinberg et al. 1989). Several lines of evidence suggest that the well-known LDL receptor pathway described in the elegant studies of Brown, Goldstein, and colleagues (Goldstein and Brown 1989) is not required for, and may not normally be involved in, cholesterol accumulation during foam cell development. For example, in humans and animals with genetic defects in the gene for the LDL receptor, plasma LDL levels are abnormally high and plaque formation is accelerated.

A popular model for lipoprotein-cholesterol accumulation in macrophage foam cells is the macrophage scavenger receptor model (Brown and Goldstein 1983; Goldstein and Brown 1989; Steinberg et al. 1989). Macrophages, but not their precursor monocytes, express high levels of a receptor activity, also discovered by Brown, Goldstein, and colleagues (Goldstein et al. 1979), which can mediate the endocytosis of chemically modified LDL. For macrophages in culture, this endocytosis can mediate the conversion of the cells into lipid-rich cells that are strikingly similar in appearance to the macrophage/foam cells seen in atherosclerotic plaques. A receptor responsible for this in vitro foam cell formation is called the macrophage scavenger receptor because of its distinctive ligand-binding properties (see below). Chemical modifications that can convert LDL into a scavenger receptor ligand include oxidation and acetylation. Oxidized LDL (OxLDL) has been detected in atherosclerotic plaques, and the antioxidant probucol can inhibit plaque formation in an animal model of atherosclerosis (Carew et al. 1987; Kita et al. 1987; for review see, Steinberg et al. 1989). These findings, and the observation that scavenger receptors can mediate macrophage–foam cell formation in vitro, have focused attention on the properties of scavenger receptors and their possible role in atherogenesis.

Binding Properties of Scavenger Receptors

Macrophage scavenger receptors exhibit unusually broad ligand specificity (Goldstein et al. 1979; Brown et al. 1980; Brown and Goldstein 1983; Kodama et al. 1988, 1990; Rohrer et al. 1990; Nishikawa et al. 1990; Freeman et al. 1991; Hampton et al. 1991; Krieger 1992). Their ligands are polyanions, including chemically modified proteins such as acetylated LDL (AcLDL), OxLDL, and maleylated bovine serum albumin (M-BSA), but not native LDL or native BSA, and certain polyribonucleotides (poly I and poly G, but not poly C). Some polysaccharides (dextran sulfate, but not chondroitan sulfate) and phospholipids (phosphatidylserine, but not phosphatidylcholine) are also ligands. The only common feature of scavenger receptor ligands recognized thus far is that they comprise polyanionic monomers or macromolecular complexes; however, many polyanions are not ligands. The broad binding specificity of the scavenger receptor and its expression by macrophages suggest that it may play a role not only in atherogenesis, but also in other physiological and pathophysiological systems, e.g., macrophage-associated immune responses and inflammation (Freeman et al. 1990; Krieger 1992). Bacterial endotoxin (lipopolysaccharide) is a ligand for scavenger receptors, and scavenger receptors appear to play a quantitatively important role in the clearance of endotoxin from the circulation by the liver (Hampton et al. 1991). Thus, scavenger receptors presumably help protect the body from endotoxic shock during gram-negative bacterial sepsis. These observations support the proposal that scavenger receptors may be pattern recognition receptors for the identification and subsequent removal of pathogens (Freeman et al. 1990; Hampton et al. 1991; Janeway 1992; Krieger 1992).

To facilitate analysis of the functions of scavenger receptors, we and other investigators began purification and characterization of the scavenger receptor molecule. Wong et al. (1983) and Dressel, Via, and their colleagues (Via et al. 1985; Dressel et al. 1987) first reported the partial purification of ligand-binding proteins from cultured cells and tissues. From bovine lung membranes, we purified scavenger receptors approximately 240,000-fold (Kodama et al. 1988). These purified bovine scavenger receptors were composed of a set of related *N*-glycosylated proteins, including ~220-kD trimers, ~150-kD dimers, and ~77-kD monomers. The trimers and dimers could be converted into the monomers by reduction. We were able to detect ligand binding to the trimeric form but not to the dimeric or monomeric forms.

On the basis of partial amino acid sequencing, we cloned two scavenger receptor cDNAs, designated type I and type II, from a bovine lung cDNA library (Kodama et al. 1990; Rohrer et al. 1990). These were derived from alternatively spliced mRNA products of a single gene. Transfection of corresponding cDNA expression vectors into COS (Kodama et al. 1990; Rohrer et al. 1990) or Chinese hamster ovary (CHO) cells (Freeman et al. 1991; Penman et al. 1991) confers on them receptor activity exhibiting the distinctive, broad ligand-binding specificity characteristic of macrophage scavenger receptors. Type I and/or type II scavenger receptor expression has been detected in macrophages in vitro (Kodama et al. 1990) and in vivo (Matsumoto et al. 1990; Naito et al. 1991; Yla-Herttuala et al. 1991) and in stimulated cultured smooth muscle cells (Pitas 1990; M. Freeman, pers. comm.). Expression of scavenger receptor activity by endothelial cells has been observed previously (see, e.g., Nagelkerke et al. 1983; Pitas et al. 1985). Although some workers initially assumed that the same receptor proteins were responsible for this activity in macrophages and endothelial cells, recent studies have suggested that the scavenger receptor activity expressed in endothelial cells may be mediated by a different receptor protein with similar binding properties (Via et al. 1992; M. Freeman, pers. comm.).

Predicted Structure

The sequence of the type I bovine scavenger receptor cDNA predicts a 453-amino-acid protein with the following domains (see Fig. 1) (Kodama et al. 1990): (I) amino-terminal cytoplasmic (amino acid residues 1–50), (II) transmembrane (51–76), (III) spacer (77–108), (IV) α-helical coiled-coil (109–271), (V) collagenous (272–343), and (VI) carboxy-terminal cysteine-rich (344–453), designated SRCR (scavenger receptor cystein-rich). The SRCR domain helped to define a previously unrecognized family of remarkably well-conserved cysteine-rich protein domains (Freeman et al. 1990; Aruffo et al. 1991; Krieger 1992). The type I receptor has seven potential N-linked glycosylation sites in the spacer and α-helical coiled-coil domains and eight cysteines, six of which are in the SRCR domain.

The type II scavenger receptor is identical to the type I receptor, except that the 110-amino-acid SRCR domain is replaced by a 6-amino-acid carboxyl terminus (Fig. 1) (Rohrer et al. 1990). Despite its truncated carboxyl terminus, the type II scavenger receptor

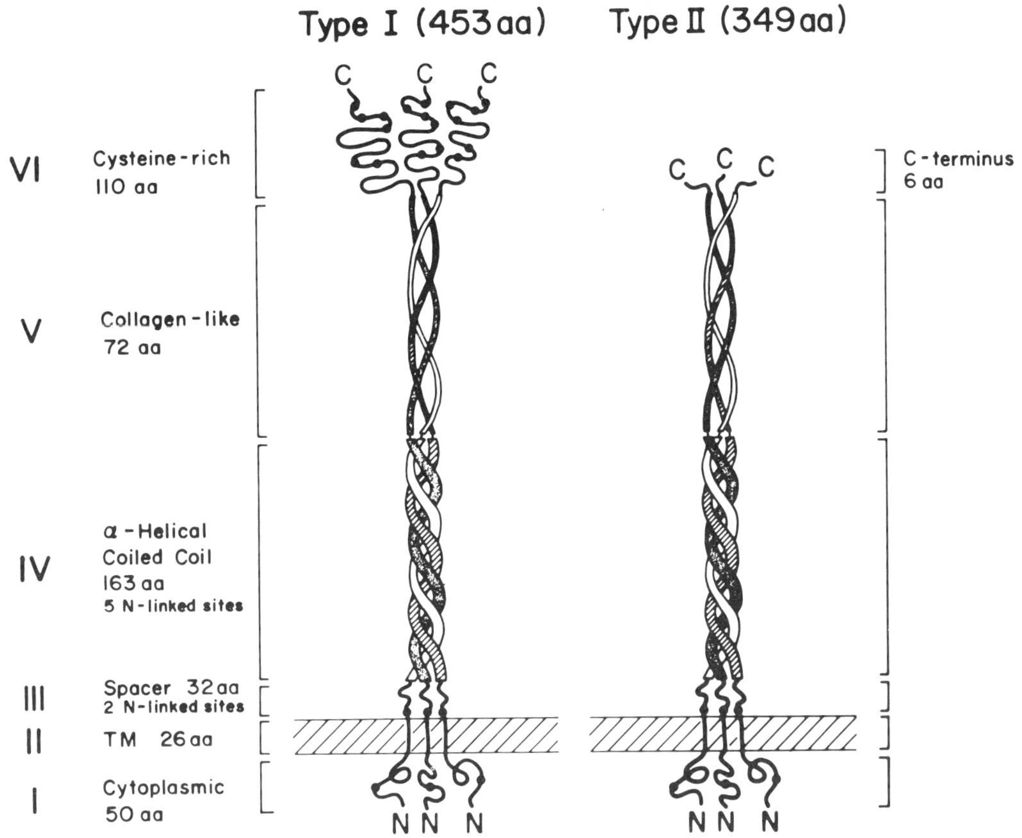

Figure 1. Models of the predicted structures of the type I and type II bovine scavenger receptors. The type I and II receptors share five identical domains: (I) amino-terminal cytoplasmic domain; (II) transmembrane (TM) domain; (III) spacer domain; (IV) α-helical coiled-coil domain comprising heptad repeats with aliphatic amino acids in the first and fourth positions; (V) collagen-like domain comprising 24 Gly-X-Y repeats. The 110-amino-acid carboxy-terminal cysteine-rich domain of the type I receptor is replaced by a 6-residue carboxyl terminus in the type II receptor. Small filled circles in the cytoplasmic, spacer, and cysteine-rich domains represent the approximate positions of cysteine residues. This model assumes that the α-helical coiled-coil domain IV forms a single triple-stranded left-handed superhelix that merges with the right-handed collagen-like triple helix of domain V to form a single long fibrous stalk. In this illustration, coiled-coil domains IV and V are overwound to emphasize their triple helical structures. (Reprinted, with permission, from Rohrer et al. 1990 [copyright Macmillan].)

mediates the endocytosis of modified LDL with essentially the same affinity as the type I receptor and with a similar broad ligand specificity (Rohrer et al. 1990; Freeman et al. 1991). Several studies suggest that the collagenous domain plays a critical role in ligand binding and probably represents the ligand-binding domain of these scavenger receptors (Kodama et al. 1991; S. Acton et al., in prep.). There is an attractive mechanism that may explain both the receptor's high affinity for ligands and its broad ligand specificity. The collagen triple helix may provide a selectively sticky surface that exhibits broad specificity at low affinity for monovalent interactions, and high-affinity binding may be a consequence of polyvalent ligand binding either to multiple sites on a single collagen triple helix or to multiple collagenous domains if the receptors aggregate into higher-order oligomers (Kodama et al. 1990). The scavenger receptor may be acting as a kind of molecular flypaper (Krieger 1992). On the basis of the sequence of the bovine receptor, cDNAs encoding the type I and type II scavenger receptors were cloned from murine (Freeman et al. 1990), human (Matsumoto et al. 1990), and rabbit (M. Freeman, pers. comm.) cDNA libraries. As expected, the sequences of the receptors from these four species are highly similar.

We have used metabolic labeling and immunochemical techniques to study the biosynthesis and posttranslational processing of bovine scavenger receptors expressed in CHO (Penman et al. 1991) and COS (S. Acton et al., in prep.) cells. Newly synthesized receptor subunits are inserted into the membrane of the endoplasmic reticulum (ER). In the lumen of the ER, the subunits undergo N-linked glycosylation and hydroxylation of prolines and/or lysines. In addition, the precursor monomer chains rapidly oligomerize into trimers, each comprising a monomer and a Cys-83 (spacer domain, see Fig. 1) disulfide-linked dimer. The precursor trimers are transported to and through the Golgi apparatus and *trans*-Golgi network, where their high-mannose N-linked oligosaccharides are converted into complex, endoglycosidase-H-resistant sugars. The mature receptor trimers are then transported to the cell surface, where they participate in the receptor-mediated endocytosis of ligands. (For a more-detailed review of the structure and intracellular processing of scavenger receptors, see Krieger 1992.)

Functions of Scavenger Receptors

The physiological and pathophysiological functions of the scavenger receptor have not been established with certainty. It seems likely, even though it has not yet been proven, that macrophage scavenger receptors are involved in foam cell formation during atherogenesis (Brown and Goldstein 1983; Goldstein and Brown 1989; Steinberg et al. 1989; Krieger 1992). We have found that CHO cells, which stably express bovine scavenger receptors, can be converted into lipid-laden foam-like cells by incubation in culture with modified LDL (Freeman et al. 1991). In addition, scavenger receptor mRNA and protein have been detected in atherosclerotic plaques (Matsumoto et al. 1990; Yla-Herttuala et al. 1991). Because of their broad binding specificity, scavenger receptors may be involved in other physiological systems. It seems likely that they participate in macrophage-associated host defense activities, for example, the clearance of pathogens such as endotoxin (Hampton et al. 1991), and possibly in the recognition of effete cells during development.

We have recently described the expression of scavenger-receptor-like endocytosis by macrophages in intact *Drosophila melanogaster* embryos and in primary embryonic cell cultures (Abrams et al. 1992). The *Drosophila* Schneider L2 cell line, but not the Kc cell line, also exhibited a scavenger-receptor-mediated endocytic pathway similar to that described for mammalian macrophages. The relationship between the structures of the mammalian and *Drosophila* scavenger receptors has not yet been established. Nevertheless, these findings suggest that scavenger receptors participate in important well-conserved functions and raise the possibility that they may be pattern recognition receptors that arose early in the evolution of host defense mechanisms. They have also established a new system for the investigation of receptor-mediated endocytosis in *Drosophila* and for the study of scavenger receptor function in disease, host defense, and development.

ACKNOWLEDGMENTS

We thank our many current and former colleagues and collaborators for their contributions to the studies reviewed here and Kathy Sweeney for assistance in preparing the manuscript. We also thank Mason Freeman for permitting us to cite results of his laboratory's unpublished studies of rabbit scavenger receptors. This work was supported by grants to M.K. from the National Institute of Health Heart, Lung, and Blood Institute (HL41484) and Arris Pharmaceutical Corporation. H.S. is an Assistant Investigator with the Howard Hughes Medical Institute, and J.A. was supported by an American Cancer Society fellowship.

REFERENCES

Abrams, J.M., A. Lux, H. Steller, and M. Krieger. 1992. Macrophages in *Drosophila* embryos and L2 exhibit scavenger receptor-mediated endocytosis. *Proc. Natl. Acad. Sci.* (in press).

Aruffo, A., M.B. Melnick, P.S. Linsley, and B. Seed. 1991. The lymphocyte glycoprotein CD6 contains a repeated domain structure characteristic of a new family of cell surface and secreted proteins. *J. Exp. Med.* **174:** 949.

Brain, J.D. 1985. Macrophages in the respiratory tract. In *Handbook of physiology: Section 3: The respiratory system* (ed. A.P. Fishman and A.B. Fisher), vol. 1, p. 447. American Physiological Society, Bethesda, Maryland.

Brown, M.S. and J.L. Goldstein. 1983. Lipoprotein metabolism in the macrophage: Implications for cholesterol deposition in atherosclerosis. *Annu. Rev. Biochem.* **52:** 223.

Brown, M.S., S.K. Basu, J.R. Falck, Y.K. Ho, and J.L. Goldstein. 1980. The scavenger cell pathway for lipoprotein degradation: Specificity of the binding site that mediates the uptake of negatively-charged LDL by macrophages. *J. Supramol. Struct.* **13:** 67.

Carew, T.E., D.C. Schwenke, and D. Steinberg. 1987. Antiatherogenic effect of probucol unrelated to its hypocholesterolemic effect: Evidence that antioxidants in vivo can selectively inhibit low density lipoprotein degradation in macrophage-rich fatty streaks and slow the progression of atherosclerosis in the Watanabe heritable hyperlipidemic rabbit. *Proc. Natl. Acad. Sci.* **84:** 7725.

Dressel, H.A., E. Friedrich, D.P. Via, H. Sinn, R. Ziegler, and G. Schettler. 1987. Binding of acetylated low density lipoprotein and maleylated bovine serum albumin to the rat liver: One or two receptors? *Embo J.* **6:** 319.

Ezekowitz, R.A., D.J. Williams, H. Koziel, M.Y. Armstrong, A. Warner, F.F. Richards, and R.M. Rose. 1991. Uptake of Pneumocystis carinii mediated by the macrophage mannose receptor. *Nature* **351:** 155.

Freeman, M., J. Ashkenas, D.J. Rees, D.M. Kingsley, N.G. Copeland, N.A. Jenkins, and M. Krieger. 1990. An ancient, highly conserved family of cysteine-rich protein domains revealed by cloning type I and type II murine macrophage scavenger receptors. *Proc. Natl. Acad. Sci.* **87:** 8810.

Freeman, M., Y. Ekkel, L. Rohrer, M. Penman, N.J. Freedman, G.M. Chisolm, and M. Krieger. 1991. Expression of type I and type II bovine scavenger receptors in Chinese hamster ovary cells: Lipid droplet accumulation and nonreciprocal cross competition by acetylated and oxidized low density lipoprotein. *Proc. Natl. Acad. Sci.* **88:** 4931.

Goldstein, J.L. and M.S. Brown. 1989. Familial hypercholesterolemia. In *The metabolic basis of inherited disease*, 6th edition. (ed. C.R. Scriver et al.), p.1215. McGraw-Hill, New York.

Goldstein, J.L., Y.K. Ho, S.K. Basu, and M.S. Brown. 1979. Binding site on macrophages that mediates uptake and degradation of acetylated low density lipoprotein, producing massive cholesterol deposition. *Proc. Natl. Acad. Sci.* **76:** 333.

Gordon, S., V.H. Perry, S. Rabinowitz, L.P. Chung, and H. Rosen. 1988. Plasma membrane receptors of the mononuclear phagocyte system. *J. Cell. Sci. Suppl.* **9:** 1.

Hampton, R.Y., D.T. Golenbock, M. Penman, M. Krieger, and C.R. Raetz. 1991. Recognition and plasma clearance of endotoxin by scavenger receptors. *Nature* **353:** 342.

Havel, R.J. and J.P. Kane. 1989. Disorders of the biogenesis and secretion of lipoproteins containing the B apolipoproteins. In *The metabolic basis of inherited disease*, 6th edition (ed. C.R. Scriver et al.), p. 1129. McGraw-Hill, New York.

Janeway, C.A. 1992. The immune system evolved to discriminate infectious nonself from noninfectious self. *Immunol. Today* **13:** 11.

Kita, T., Y. Nagano, M. Yokode, K. Ishii, N. Kume, A. Ooshima, H. Yoshida, and C. Kawai. 1987. Probucol prevents the progression of atherosclerosis in Watanabe heritable hyperlipidemic rabbit, an animal model for familial hypercholesterolemia. *Proc. Natl. Acad. Sci.* **84:** 5928.

Kodama, T., P. Reddy, C. Kishimoto, and M. Krieger. 1988. Purification and characterization of a bovine acetyl low density lipoprotein receptor. *Proc. Natl. Acad. Sci.* **85:** 9238.

Kodama, T., M. Freeman, L. Rohrer, J. Zabrecky, P. Matsudaira, and M. Krieger. 1990. Type I macrophage scavenger receptor contains alpha-helical and collagen-like coiled coils. *Nature* **343:** 531.

Kodama, T., T. Doi, A. Matsumoto, K. Higashino, Y. Kurihara, Y. Kawabe, S. Uesugi, T. Imanishi, H. Itakura, and Y. Yazaki. 1991. The C-terminal region of a collagen-like domain containing a cluster of basic amino acids mediate binding of modified low density lipoproteins by macrophage scavenger receptors *Circulation* **84:** 229.

Krieger, M. 1992. Molecular flypaper and atherosclerosis: Structure of the macrophage scavenger receptor. *Trends Biochem. Sci.* **17:** 141.

Matsumoto, A., M. Naito, H. Itakura, S. Ikemoto, H. Asaoka, I. Hayakawa, H. Kanamori, H. Aburatani, F. Takaku, H. Suzuki, Y. Kobari, T. Miyai, K. Takahashi, E.H. Cohen, R. Wydro, D.E. Housman, and T. Kodama. 1990. Human macrophage scavenger receptors: Primary structure, expression, and localization in atherosclerotic lesions. *Proc. Natl. Acad. Sci.* **87:** 9133.

Nagelkerke, J.F., K.P. Barto, and T.J. vanBerkel. 1983. In vivo and in vitro uptake and degradation of acetylated low density lipoprotein by rat liver endothelial, Kupffer, and parenchymal cells. *J. Biol. Chem.* **258:** 12221.

Naito, M., T. Kodama, A. Matsumoto, T. Doi, and K. Takahashi. 1991. Tissue distribution, intracellular localization, and in vitro expression of bovine macrophage scavenger receptors. *Am. J. Pathol.* **139:** 1411.

Nishikawa, K., H. Arai, and K. Inoue. 1990. Scavenger receptor-mediated uptake and metabolism of lipid vesicles containing acidic phospholipids by mouse peritoneal macrophages. *J. Biol. Chem.* **265:** 5226.

Penman, M., A. Lux, N.J. Freedman, L. Rohrer, Y. Ekkel, H. McKinstry, D. Resnick, and M. Krieger. 1991. The type I and type II bovine scavenger receptors expressed in Chinese hamster ovary cells are trimeric proteins with collagenous triple helical domains comprising noncovalently associated monomers and Cys83-disulfide-linked dimers. *J. Biol. Chem.* **266:** 23985.

Pitas, R.E. 1990. Expression of the acetyl low density lipoprotein receptor by rabbit fibroblasts and smooth muscle cells. Up-regulation by phorbol esters. *J. Biol. Chem.* **265:** 12722.

Pitas, R.E., J. Boyles, R.W. Mahley, and D.M. Bissell. 1985. Uptake of chemically modified low density lipoproteins in vivo is mediated by specific endothelial cells. *J. Cell. Biol.* **100:** 103.

Rohrer, L., M. Freeman, T. Kodama, M. Penman, and M. Krieger. 1990. Coiled-coil fibrous domains mediate ligand binding by macrophage scavenger receptor type II. *Nature* **343:** 570.

Steinberg, D., S. Parthasarathy, T.E. Carew, J.C. Khoo, and J.L. Witztum. 1989. Beyond cholesterol. Modifications of low-density lipoprotein that increase its atherogenicity. *N. Engl. J. Med.* **320:** 915.

Taylor, M.E., K. Bezouska, and K. Drickamer. 1992. Contribution to ligand binding by multiple carbohydrate-recognition domains in the macrophage mannose receptor. *J. Biol. Chem.* **267:** 1719.

Via, D.P., H.A. Dresel, S.L. Cheng, and A.M. Gotto, Jr. 1985. Murine macrophage tumors are a source of a 260,000-dalton acetyl-low density lipoprotein receptor. *J. Biol. Chem.* **260:** 7379.

Via, D., A. Fanslow, H.A. Dresel, D.K. Dennison, E. Levine, S. Wolle, B. Liu, L. Chan, and A.M. Gotto, Jr. 1992. Identification and density dependant regulation of the AcLDL receptor in normal and transformed bovine aortic endothelial cells (BAEC). *FASEB J.* **6:** A371.

Wong, H., A.M. Fogelman, M.E. Haberland, and P.A. Edwards. 1983. Identification of the scavenger receptor from rabbit aveolar macrophages. *Arteriosclerosis* **3:** 475a.

Yla-Herttuala S., M.E. Rosenfeld, S. Parthasarathy, E. Sigal, T. Sarkioja, J.L. Witztum, and D. Steinberg. 1991. Gene expression in macrophage-rich human atherosclerotic lesions. 15-lipoxygenase and acetyl low density lipoprotein receptor messenger RNA colocalize with oxidation specific lipid-protein adducts. *J. Clin. Invest.* **87:** 1146.

Biogenesis of Cell-surface Polarity in Epithelial Cells and Neurons

K. Simons, P. Dupree, K. Fiedler, L.A. Huber, T. Kobayashi, T. Kurzchalia, V. Olkkonen, S. Pimplikar, R. Parton, and C. Dotti
European Molecular Biology Laboratory, 6900 Heidelberg, Germany

In multicellular organisms, cells are differentiated into cell types differing in architecture and organized in such a way that they can perform their different functions. Epithelial cells and neurons are characterized by their stably polarized cell surfaces. Epithelia are cellular sheets that separate compartments of the organism. The cells in a simple epithelium are linked through junctional complexes to form a permeability barrier (Simons and Fuller 1985). These cellular sheets perform a variety of vectorial functions in transporting and secretory epithelia that reflect the polar organization of the epithelial cell. The plasma membrane of each cell is divided into two domains: an apical domain facing the external milieu and a basolateral domain which is in contact with the internal milieu and the blood supply. These plasma membrane domains have different lipid and protein compositions.

Neurons are also characterized by a cell-surface organization in which proteins are segregated and maintained in discrete functional domains. The nerve cell plasma membrane is much more complicated in its architecture than that of the epithelial cell surface. However, a fundamental dichotomy is the segregation of proteins into presynaptic and postsynaptic regions (Kelly 1988; Black and Baas 1989). In some neuronal cells, the cell surface can be roughly divided into two territories: the axonal domain and the somatodendritic domain comprising the cell body and the dendrites. The presynaptic terminals are usually axonal and the postsynaptic sites somatodendritic. A major question in nerve cell architecture is how the neuron segregates the axonal and the dendritic surface proteins from each other. Does this process have anything in common with the mechanisms that epithelial cells employ to establish their polarized cell-surface organization?

We have been using two experimental systems to study this question, MDCK cells and fetal rat hippocampal neurons. The MDCK cell is a dog kidney cell line that grows on permeable supports to form a polarized epithelial sheet (Fig. 1) that mimics the epithelial organization seen in vivo (Simons and Fuller 1985). Cultured fetal rat hippocampal neurons (Banker and Cowan 1977; Goslin and Banker 1991) undergo a sequence of developmental events after plating (Dotti et al. 1988). First, several short processes are formed, and one of these begins to grow rapidly to form the axon (Fig. 2). Subsequently, the remaining processes start to elongate to become the cell's dendrites. Finally, the axons form synapses with the dendrites and cell bodies of other neuronal cells in the culture. These mature neurons have been studied extensively and display the characteristics of hippocampal neurons in situ, including the segregation of axonal and dendritic marker proteins to their correct destinations (Banker and Waxman 1988; Goslin and Banker 1991).

Viral Glycoproteins as Tools to Study Cell-surface Polarity

Enveloped RNA viruses such as Semliki Forest virus (SFV), influenza virus, and vesicular stomatitis virus (VSV) have been employed as widely used tools to analyze membrane traffic in mammalian cells (Simons and Warren 1984; Roman and Garoff 1985). They were first used to elucidate the biosynthetic and endocytic pathways in fibroblasts. Later this approach was used successfully in epithelial MDCK cells as well (Rodriguez-Boulan and Sabatini 1978). Influenza virus hemagglutinin is routed to the apical surface of MDCK cells, whereas the spike glycoproteins of VSV and SFV are basolateral (Rodriguez-Boulan and Pendergast 1980; Fuller et al. 1985b; Roman and Garoff 1986). These proteins have therefore been used as model apical and basolateral proteins, respectively, to study polarized sorting and segregation in filter-grown MDCK cells.

The first question that we tried to answer was whether the virus strategy could also be applied to neuronal cells. When mature stage 5 hippocampal neurons (Fig. 2) were infected with influenza virus, VSV, or SFV, we found that the delivery of newly synthesized influenza virus hemagglutinin was mainly axonal, whereas the VSV and SFV glycoproteins were excluded from the axon and delivered to the surface of the cell body and the dendrites (Dotti and Simons 1990; Dotti et al. 1993). This polarized sorting was restricted to mature neurons. When stage 3 cells (Fig. 2) from which axons were growing were infected with these viruses, no overt polarity was observed. The viral glycoproteins were delivered to all the neurites (Dotti and Simons 1990; Kobayashi et al. 1992).

In MDCK cells, the site of sorting for apical and basolateral viral proteins has been localized to the trans-Golgi network (TGN), the exit compartment of the Golgi complex (Fuller et al. 1985a; Griffiths and Simons 1986; Hughson et al. 1988). The apical and the

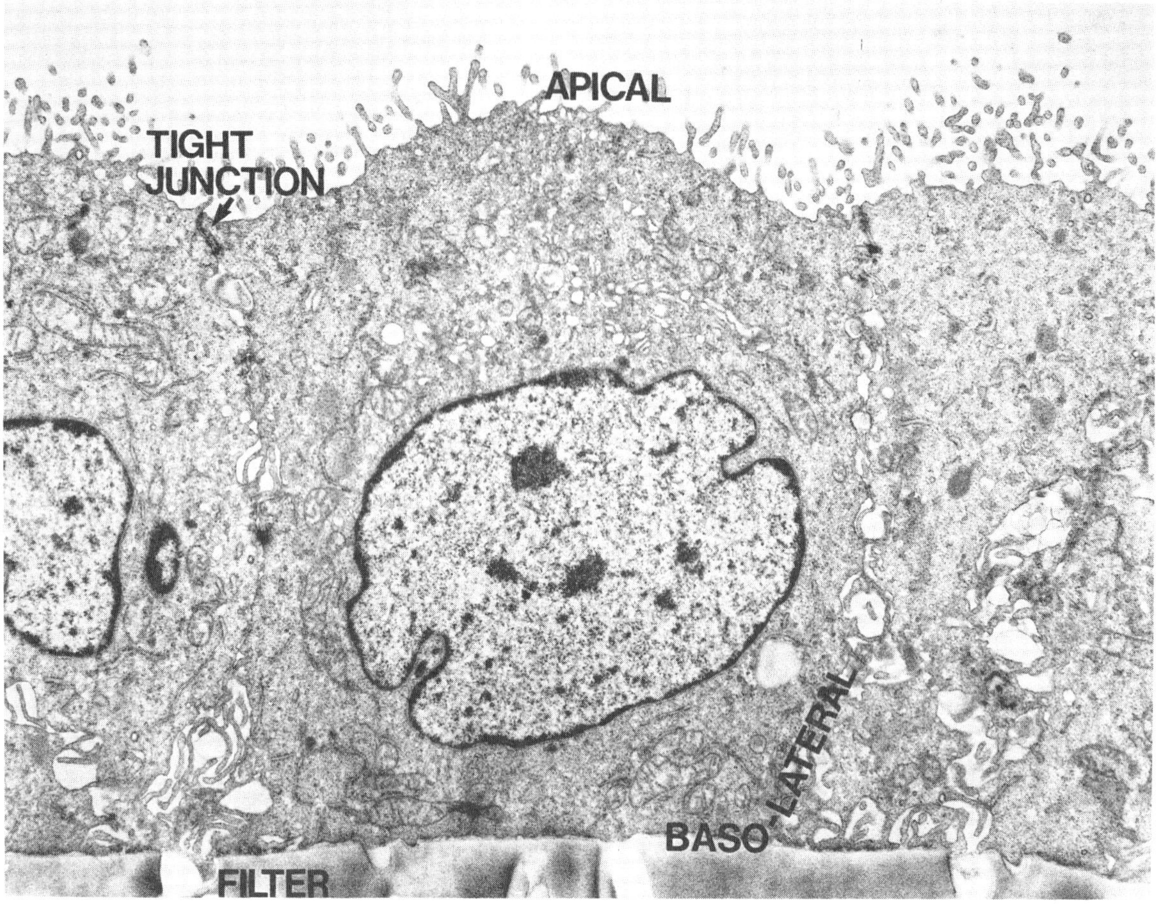

Figure 1. Electron micrograph of filter-grown MDCK cells.

basolateral viral glycoproteins move together from their site of synthesis to the TGN where two vesicle populations are formed: one for apical proteins and another for basolateral proteins (Bennett et al. 1988; Wandinger-Ness et al. 1990). These are targeted to their respective destinations, where they deliver their cargo (Simons and Wandinger-Ness 1990). There is another scenario for epithelial sorting, operating in hepatocytes (Hubbard et al. 1989). Both apical and basolateral proteins are first routed to the basolateral membrane, and then segregation occurs. The apical proteins are transported by transcytosis to their final destination, presumably after endocytosis into early basolateral endosomes (Mostov et al. 1992). In hepatocytes, the direct route from the TGN to the apical surface is evidently lacking.

Where the sorting occurs in stage 5 hippocampal neurons is not yet clear, but the experimental data are

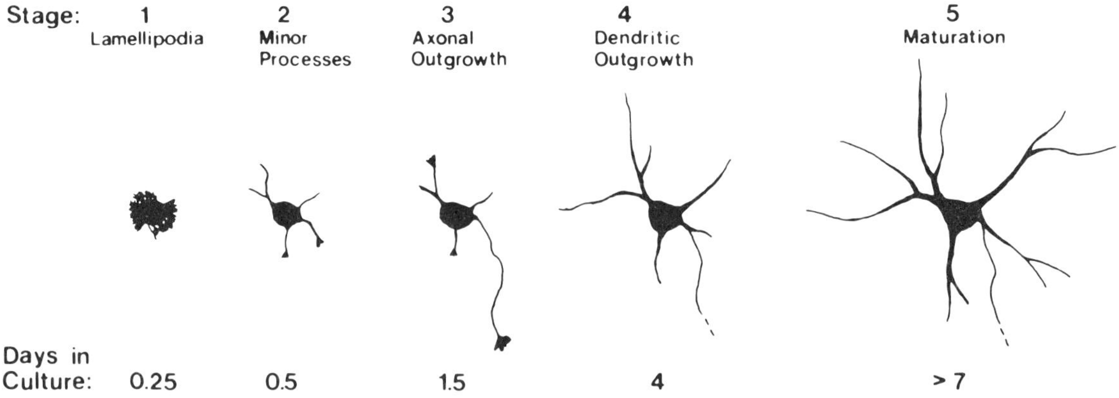

Figure 2. Differentiation of rat hippocampal neurons in culture. (Reprinted, with permission, from Dotti et al. 1988.)

consistent with the view that this occurs in the Golgi complex located in the cell body (Dotti and Simons 1990), as is the case in MDCK cells. From the existing data, one cannot exclude the possibility that a small fraction of the influenza virus hemagglutinin is first routed to the somatodendritic domain and then moved by transcytosis to the axon. However, this is unlikely because the hemagglutinin is poorly endocytosed (Roth 1989) and would therefore remain in the dendrites after delivery.

Sorting of Cellular Surface Proteins in MDCK Cells and Hippocampal Neurons

The results obtained with the sorting of viral glycoproteins in hippocampal neurons imply that the apical and the axonal domains are equivalent, whereas the basolateral surface would correspond to the somatodendritic domain. Recent studies have shown that glycophosphatidylinositol (GPI)-anchored proteins are sorted to the apical surface in most epithelial cells (Lisanti and Rodriguez-Boulan 1991). We therefore studied the distribution of the Thy-1 protein, a GPI-anchored protein in hippocampal neurons. This protein, which is apical when expressed in MDCK cells (Wilson et al. 1990), is mainly axonal in stage 5 hippocampal neurons (Dotti et al. 1991). GPI-anchored proteins are common in neurons and are known to be present on the axonal surface (Lisanti and Rodriguez-Boulan 1991). However, few studies have so far been carried out to see whether they are generally routed to the axon and excluded from the cell body and the dendrites.

One protein found ubiquitously in mammalian cells including epithelia and neurons is the transferrin receptor. This protein is basolateral in most epithelial cell types (Fuller and Simons 1986). The basolateral plasma membrane faces the blood supply and contains proteins responsible for nutrient uptake as well as for growth factor and hormonal regulation (Wandinger-Ness and Simons 1991). The transferrin receptor is somatodendritic in hippocampal neurons (Cameron et al. 1991; Parton et al. 1992), consistent with the equivalence of this plasma membrane domain with the basolateral surface, and reflecting the role of this neuronal surface domain in regulating cell growth and nutrient uptake.

Recent studies have explored the sorting of a typical neuronal cell-surface protein, the voltage-dependent sodium channel in MDCK cells. The form of the sodium channel primarily localized in axons was expressed after transfection in MDCK cells. This axonal protein was sorted to the apical surface (K.J. Angelides and B. Wible, pers. comm.). Altogether, these studies confirm the data obtained with the viral glycoproteins and demonstrate that there are shared features between the biosynthetic sorting pathways in hippocampal neurons and MDCK cells. It should be pointed out that not all proteins are routed either to the apical or to the basolateral domains in MDCK cells. The Na^+, K^+ ATPase is such a protein. Hammerton et al. (1991) have demonstrated that newly synthesized α subunits of the ATPase are routed in equal amounts to the apical and basolateral surfaces. The apical Na^+, K^+ ATPase is rapidly removed from the plasma membrane and presumably degraded, whereas the basolateral sodium pumps are stabilized in place by interactions with ankyrin and spectrin in the basolateral cytocortex. The steady-state concentration of the sodium pumps is therefore basolaterally disposed as in most epithelial cells. In retinal pigment epithelial cells, the Na^+, K^+ ATPase is apical (Gundersen et al. 1991). The sodium pump is composed of α and β subunits, both of which exist in several isoforms. The MDCK and the retinal pigment epithelial cells contain only the α_1 subunit. In the pigment epithelial cells, the ankyrin-spectrin skeletal network has a polarity opposite to that of MDCK cells and is found underlying the apical membrane linked to the Na^+, K^+ ATPase (Gundersen et al. 1991). What about hippocampal neurons? These express both the α_1 and the α_3 subunits. Both of these polypeptides were found to be nonpolarized and present on both axons and dendrites (Pietrini et al. 1991). The sodium pumps in hippocampal neurons may therefore be stabilized by ankyrin-spectrin interactions in both membrane territories.

Table 1 summarizes the comparison between MDCK

Table 1. Localization of Surface Proteins in MDCK Cells and Hippocampal Neurons

Protein	Location	Reference
VSV G protein	basolateral/dendritic	Rodriguez-Boulan and Pendergast (1980); C.G. Dotti et al. (in prep.); Dotti and Simons (1990)
SFV E protein	basolateral/dendritic	Roman and Garoff (1986); Dotti et al. (1993)
Transferrin receptor	basolateral/dendritic	Fuller and Simons (1986); Parton et al. (1992); Cameron et al. (1991)
Influenza virus hemagglutinin	apical/axonal	Rodriguez-Boulan and Pendergast (1980); Dotti and Simons (1990)
Thy-1 protein	apical/axonal	Wilson et al. (1990); Dotti et al. (1991)
Na-channel	apical/axonal	Angelides et al. (1988); K.J. Angelides and B. Wible (pers. comm.)
Na^+, K^+ ATPase	apical and basolateral/ axonal and dendritic	Hammerton et al. (1991); Pietrini et al. (1991)

cells and hippocampal neurons. The striking feature is how similarly, on one hand, apical and axonal proteins, and on the other, basolateral and somatodendritic proteins, are being handled. One amazing characteristic of the epithelial sorting mechanisms is their flexibility in performance in different epithelial cell types (cf. MDCK cells and hepatocytes). In addition to these differences in organization of the traffic routes, it has also become apparent that basolateral and apical delivery and retention are not fixed characteristics but can be modulated when appropriate. The LDL receptor is basolateral in most epithelia but is apical in the epithelium of the renal tubule (Pathak et al. 1990), and in MDCK cells, it is in both surface domains (Li et al. 1991). The mechanism responsible for this tissue specificity is not known. It could involve the interactions with the ankyrin-spectrin membrane skeleton as shown for Na^+, K^+ ATPase (Hammerton et al. 1991) or posttranslational modification of the receptor itself or modulation of the sorting machinery. If such modulation of the neuronal mechanisms were also possible, this might have important implications for handling of neuronal surface proteins that appear to be able to change their localization in different neuronal cell types, as evidenced, for instance, by the possibility of routing synaptic vesicle components to the dendrites (Pow and Morris 1989). This plasticity in sorting behavior should be kept in mind in future studies.

Sorting and Targeting Machinery

Only elucidation of the sorting signals and of the sorting and targeting machinery will clarify how similar epithelial and neuronal cells are in this respect on the molecular level. The sorting signals that determine apical and basolateral delivery are being identified (Mostov et al. 1992). The first signal for polarized sorting to be rigorously established was the GPI moiety which anchors the carboxyl terminus of GPI-linked proteins to the outer leaflet of the plasma membrane bilayer (Lisanti and Rodriguez-Boulan 1991). Replacing the transmembrane hydrophobic domain of a normally basolateral protein with a GPI anchor will redirect the protein to the apical surface.

The search for signals that direct non-GPI-anchored apical proteins and basolateral proteins to their destinations has recently led to the identification of peptide segments in the cytoplasmic "tails" of basolateral proteins that function as sorting signals (Mostov et al. 1992). The signals for apical proteins are still unknown. Apical proteins lacking their cytoplasmic tails are still apically sorted, thus the apical signal should be localized either in the transmembrane or, most likely, in the exoplasmic domains. "Anchorless" apical membrane proteins are usually secreted into the apical medium when transfected into MDCK cells (Corbeil et al. 1992; Vogel et al. 1992).

The mechanisms by which apical and basolateral proteins become sorted are not yet characterized. We have postulated a rafting mechanism to explain the delivery of GPI-anchored proteins to the apical surface. Not only GPI-linked proteins, but also glycosphingolipids are preferentially sorted to the apical membrane in MDCK cells (Simons and van Meer 1988). Hydrogen bonding may cause glycosphingolipids to cluster into microdomains, "rafts," that could include GPI-anchored proteins as well. The rafts could function as sorting platforms for further inclusion of non-GPI-anchored apical proteins in the TGN (Simons and Wandinger-Ness 1990). We assume that transmembrane sorting proteins bind to signals in the exoplasmic domains of the apical cargo proteins and that the cytosolic domains of the putative sorting proteins bind cytosolic proteins that lead to proper vesiculation and targeting (Fig. 3) of the released vesicles. These rafts with their cargo would then leave the TGN apical transport vesicles for delivery to the apical cell surface. Since GPI-anchored proteins are axonal, one would expect that glycosphingolipids are preferentially sorted to axons as well. However, this is not yet known.

To identify molecular machinery involved in the epithelial sorting process, an in vitro system has been developed that allows the isolation of TGN-derived, exocytic vesicles from MDCK cells (Bennett et al. 1988). This system was based on the ability of MDCK cells to form a fully polarized monolayer when grown on a permeable support. After infection with influenza virus or VSV and accumulation of the glycoproteins in the TGN by incubation at 20°C (Matlin and Simons 1983), the cells were mechanically perforated using a

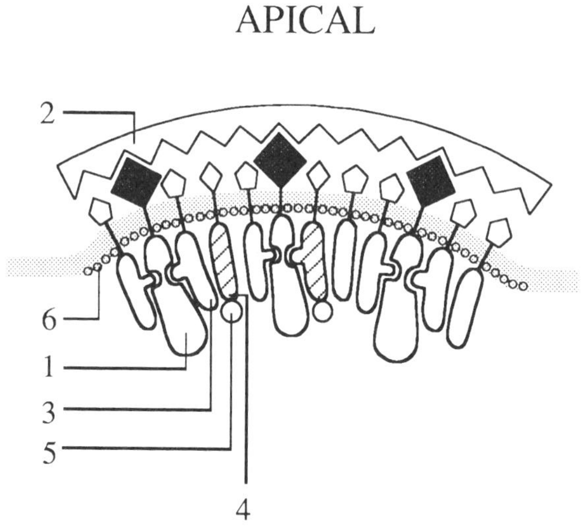

1. Apical Sorting Receptor.
2. Vesiculation and Apical Targeting Machinery.
3. Apical Membrane Protein.
4. Apical Secretory Protein Receptor.
5. Apical Secretory Protein.
6. Glycosphingolipids.

Figure 3. Hypothetical "apical" membrane patch in the trans-Golgi network of MDCK cells. This patch will vesiculate into an apical transport vesicle. GPI-linked proteins will be included in the patch as well (not indicated in the figure). (Modified from Simons and Wandinger-Ness 1990.)

nitrocellulose filter. Sealed vesicles containing influenza virus hemagglutinin or VSV G protein could be collected from the incubation medium. It was further possible to separate apical and basolateral vesicles by immunoisolation with antibodies against the cytoplasmically exposed domains of the glycoproteins (Wandinger-Ness et al. 1990). The protein patterns of the two populations of vesicles contained a number of proteins in common but also proteins specifically enriched in one or the other of the populations. Since there is no synthesis of cellular proteins in the conditions of viral infection used, the nonviral proteins in these vesicles are probably essential, perhaps recycled, components of the machinery involved in sorting, vesicle budding, movement, or fusion, and therefore interesting candidates for further analysis.

We have recently found that influenza virus hemagglutinin forms a large complex together with several integral membrane proteins in the apical transport vesicles, some of which are components of the basolateral vesicles as well. This complex formation allowed us to purify one of these proteins, which we call VIP21 (vesicle integral protein of 21 kD) (Kurzchalia et al. 1992). Using partial amino acid sequences, we constructed degenerate oligonucleotides that were employed to clone the MDCK cDNA for the protein. This cDNA encoded a 178-amino-acid protein containing a hydrophobic region between residues 102 and 134 that could form a transmembrane stretch. Antibodies were prepared against peptides from the amino-terminal and the carboxy-terminal regions of the protein (P. Dupree et al., in prep.). Using immunofluorescence and immunoelectron microscopy, VIP21 was localized to the Golgi complex, cytoplasmic vesicles, and the apical and basolateral plasma membrane domains. Interestingly, VIP21 appeared to be associated with non-clathrin plasma membrane invaginations. In addition, nonepithelial cells such as BHK21 cells expressed the protein. From our previous studies on the protein composition of immunoisolated apical and basolateral transport vesicles, we know that VIP21 is a major component of both apical and basolateral vesicles (Wandinger-Ness et al. 1990). The function of VIP21 is not yet known, but we assume that it is a communal constituent in post-Golgi sorting and targeting.

As shown in Figure 4A and B, VIP21 is also expressed in hippocampal neurons. Anti-VIP21 antibodies demonstrate a punctate labeling in both axons and dendrites. This vesicle protein is routed to both surface domains in neurons as well as in MDCK cells, reinforcing its generalized distribution in post-Golgi trafficking.

Recent studies demonstrate that *ras*-like GTP-binding proteins play an important role in membrane traffic. The first two to be characterized were ypt1p and sec4p in *Saccharomyces cerevisiae* (Gallwitz et al. 1983; Salminen and Novick 1987; Segev et al. 1988). The family of ypt1/sec4 proteins in mammalian cells are called Rab proteins (Touchot et al. 1987; Zahraoui et al. 1989; Chavrier et al. 1990). These proteins are localized to different segments of the biosynthetic and endocytic pathways and are thought to be involved in conferring specificity to transport vesicle budding or targeting (Bourne 1988; Pfeffer 1992). The present Rab catalog comprises 24 members (Chavrier et al. 1992). We have been searching for Rab proteins involved in apical or basolateral vesicle delivery in MDCK cells. So far, we have no candidate for apical delivery. However, Rab8p is enriched in basolateral transport vesicles (L. Huber et al., in prep.). Immunofluorescence microscopy has shown that Rab8p is in vesicular structures and on the basolateral plasma membrane. It is also present in nonepithelial BHK21 cells, where anti-Rab8p antibodies show a punctate labeling suggesting the presence of Rab8p in cytoplasmic vesicles. Interestingly, Rab8p is present in hippocampal neurons as well. The localization is limited to the cell body and to the dendrites (L. Huber et al., in prep.). Figure 4, C–E shows the punctate labeling seen with anti-Rab8p antibodies. There appears to be little, if any, Rab8p in axons. These findings suggest that the similarities in the handling of apical/axonal and of basolateral/dendritic proteins may reflect similar mechanisms for polarized sorting and targeting in MDCK cells and hippocampal neurons.

Role of Microtubules

The organization of the microtubule networks is completely different in MDCK cells and hippocampal neurons. In MDCK cells, the minus ends of the microtubules are mainly underneath the apical membrane, and the plus ends are toward the basal membrane. Above the nucleus there are also microtubules running parallel to the substratum with mixed polarity (Bacallao et al. 1989). In mature stage-5 hippocampal neurons, the axonal microtubules are uniformly oriented with their plus ends distal to the cell body, as has been found in axons of other neuronal cell types (Baas et al. 1988, 1989). The microtubules in the dendrites closer to the cell body are of mixed polarity, and the microtubules extending into the distal regions of the dendrites change into uniform polarity with their plus ends distally. The dendritic and the axonal microtubules differ in their complement of microtubule-associated proteins (Olmstead 1986). MAP2, for instance, is only on microtubules in the cell body and the dendrites, whereas the tau protein is axonal. There is no information as to whether MDCK cells also contain different populations of microtubules.

Several studies have shown that microtubules play a role in the delivery of apical as well as axonal proteins from the Golgi complex to the cell surface (Vale 1987; Rodriguez-Boulan and Nelson 1989). Disrupting the microtubules by treatment with nocodazole or colchicine slows down the delivery of apical proteins in MDCK cells, whereas there seems to be less effect in the basolateral direction (Breitfeld et al. 1990). The lack of protein synthetic machinery in the axon requires that axonal proteins have to be transported far away from the cell body. The fast axonal transport machinery

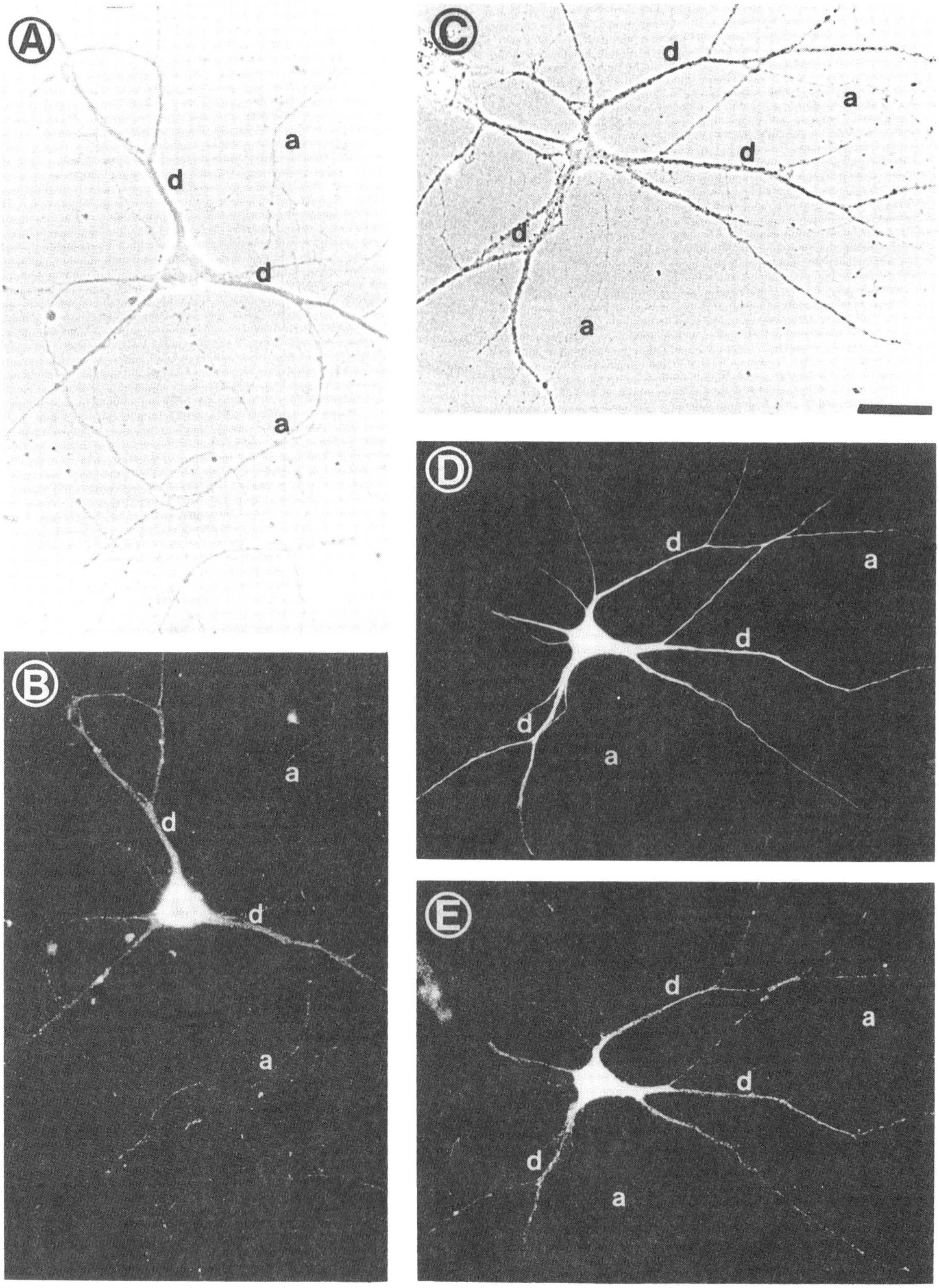

Figure 4. Localization of VIP21 (*A*,*B*) and Rab8 (*C*–*E*) in mature hippocampal neurons (stage 5). Anti-VIP21 (*B*) stains the cell body and dendrites (d) and axon (a) as shown in the corresponding phase-contrast micrograph (*A*). Anti-Rab8 (*E*) only stains the cell body and the dendrites (d) but not axons (a). *D* shows the same cell as in *E* stained with anti-MAP2 (specific dendritic marker), and *C* is the corresponding phase-contrast micrograph.

is known to employ anterograde motors of the kinesin type (Vale 1987). Transport of the viral glycoproteins from the cell body in mature hippocampal neurons is blocked by depolymerization of the microtubules in both the axonal and the dendritic directions (C.G. Dotti et al., in prep.). Most of the influenza virus hemagglutinin, and the VSV and SFV glycoproteins, remain in the cell body under these conditions. We assume that sorting into separate vesicles has occurred but that the vesicles cannot move into the axon or the dendrites when the microtubules are absent. The transport vesicles and the motor proteins involved in the polarized targeting process remain to be identified and characterized.

The Fence Function between the Two Surface Domains in MDCK Cells and Hippocampal Neurons

To maintain the compositional differences between the apical and the basolateral domains, mechanisms must exist that obstruct the randomization of the lipids and the proteins in the two membrane domains. Two mechanisms are known to prevent the mixing. One is the restriction of membrane protein mobility by anchoring to the underlying membrane skeleton (Nelson and Veshnock 1987), and the other is the existence of a localized barrier to diffusion at the tight junctions (Gumbiner 1987; Simons 1990). The former mechanism is unlikely to be of importance for GPI-anchored proteins, which are known to be mobile on the surface of MDCK cells (Lisanti and Rodriguez-Boulan 1991), and for lipids such as glycosphingolipids enriched in the apical membrane (van Meer et al. 1987). By fusing exogenous fluorescent lipids into the apical membrane of filter-grown MDCK cells, it could be demonstrated that the tight junction forms a fence for diffusion of the labeled lipids, but only in the outer (exoplasmic) leaflet of the bilayer. Lipids introduced into the inner (cytoplasmic) leaflet rapidly diffused to the basolateral side (van Meer and Simons 1986).

Neurons also require mechanisms to prevent mixing of the axonal proteins (and lipids) into the somatodendritic plasma membrane domain and vice versa. It is obvious that there is no equivalent to the tight junction at the axon hillock. Tight junctions are formed by close cell-cell contacts along the apex of neighboring epithelial cells. Is there a diffusion barrier between the somatodendritic and axonal domains in mature hippocampal neurons? Since GPI-anchored proteins cannot directly interact with the underlying membrane skeleton, one would have to assume that a "fence" is required to restrict mobility. Recent studies in our laboratory suggest that such a fence is present. Fluorescent lipids were fused into the axon of stage-5 hippocampal neurons by the same method that we used earlier to demonstrate the fence function of the tight junctions in MDCK cells (van Meer and Simons 1986). Fusion was induced by low pH and was mediated by the influenza virus hemagglutinin which was expressed on the axonal surface after viral infection. The fluorescent lipids were shown to be mobile on the axonal membrane by loss of fluorescence energy transfer and by photobleaching experiments (Kobayashi et al. 1992). Nevertheless, the fluorescent lipids did not move into the plasma membrane of the cell body or the dendrites. The barrier to diffusion operated in both leaflets of the bilayer and is therefore different from epithelial tight junctions. The structural components of this fence remain to be characterized.

OUTLOOK

Neurons and epithelial cells have a completely different cell architecture. How epithelial and neuronal cell types achieve their characteristic appearances is an enigma that will occupy molecular cell biologists for years to come (Thompson 1942). Morphogenesis involves a complex interplay between signal transduction, gene expression, molecular structure, membrane trafficking, dynamic cytoskeletal networks, and organelle biogenesis. How all these parameters are integrated to generate cell architecture is beyond available experimental capability.

The generation and maintenance of the polarized cell surfaces of epithelial cells and neurons are problems that are beginning to yield to experimental attack. We are glimpsing possible mechanisms, and these suggest that despite the differences in appearance, similarities prevail. Proteins and lipids seem to be sorted to their destinations by vesicular carriers. Some proteins directed to the apical surface in MDCK cells appear to be axonal in hippocampal neurons, whereas several basolateral proteins are somatodendritic. Diffusion barriers of different construction in neurons and epithelial cells prevent the mixing of the plasma membrane domains. These domains are further divided into subdomains in which proteins are anchored either by interactions with neighboring cells (or the extracellular matrix) or by linkage to the underlying membrane skeleton. Neurons form synaptic junctions between each other and with other cells. Junctions with completely different functions are found along the lateral membranes of neighboring epithelial cells. The basolateral membranes of various epithelial cell types are divided into subdomains by infoldings in which anion exchangers and sodium pumps are concentrated (Drenckhahn et al. 1985; Koob et al. 1987). These membrane proteins are known to interact with ankyrin and spectrin in the membrane skeleton (Rodriguez-Boulan and Nelson 1989; Bennett 1992). Subdomains have also been identified in neurons that appear to be organized by interactions with the underlying spectrin-based skeleton. For instance, voltage-dependent sodium channels are localized to distinct regions in neurons (Waxman and Ritchie 1985; Angelides et al. 1988; Black et al. 1990). These proteins interact with ankyrins, which in turn bind to spectrins (Srinivasan et al. 1988; Joe and Angelides 1991). There is a family of ankyrin isoforms that are localized to different regions in neurons (Bennett 1992). These isoforms may be responsible for organizing discrete subdomains within the somatodendritic and axonal plasma membrane domains.

This is a promising start. Only further work will unravel how far the similarities in the mechanisms responsible for polarizing the cell surfaces of neurons and epithelial cells extend.

ACKNOWLEDGMENTS

We thank Liane Meyn and Hilkka Virta for excellent technical assistance throughout these investigations.

REFERENCES

Angelides, K.J., L.W. Elmer, D. Loftus, and E. Elson. 1988. Distribution and lateral mobility of voltage-dependent sodium channels in neuronal cells. *J. Cell Biol.* **106:** 1911.

Baas, P.W., M.M. Black, and G.A. Banker. 1989. Changes in microtubule polarity orientation during the development of hippocampal neurons in culture. *J. Cell Biol.* **109:** 3085.

Baas, P.W., J.S. Deitch, M.M. Black, and G.A. Banker. 1988. Polarity orientation of microtubules in hippocampal neurons: Uniformity in the axon and nonuniformity in the dendrite. *Proc. Natl. Acad. Sci.* **85:** 8335.

Bacallao, R., C. Antony, C. Dotti, E. Karsenti, E.H.K. Stelzer, and K. Simons. 1989. The subcellular organization of Madin-Darby canine kidney cells during the formation of a polarized epithelium. *J. Cell Biol.* **109:** 2817.

Banker, G.A. and W.M. Cowan. 1977. Rat hippocampal neurons in dispersed cell culture. *Brain Res.* **126:** 397.

Banker, G.A. and A.B. Waxman. 1988. Hippocampal neurons generate natural shapes in cell culture. In *Intrinsic determinants of neuronal form and function* (ed. R.I. Lasek), p. 61. A.R. Liss, New York.

Bennett, V. 1992. Ankyrins. Adaptors between diverse plasma membrane proteins and the cytoplasm. *J. Biol. Chem.* **267:** 8703.

Bennett, M., A. Wandinger-Ness, and K. Simons. 1988. Release of putative exocytic transport vesicles from perforated MDCK cells. *EMBO J.* **7:** 4075.

Black, J.A., J.D. Kocsis, and S.G. Waxman. 1990. Ion channel organization of the myelinated fiber. *Trends Neurosci.* **13:** 48.

Black, M.M. and P.W. Baas. 1989. The basis of polarity in neurons. *Trends Neurosci.* **12:** 211.

Bourne, H.R. 1988. Do GTPases direct membrane traffic in secretion? *Cell* **53:** 669.

Breitfeld, P.P., W.C. McKinnon, and K.E. Mostov. 1990. Effect of nocodazole on vesicular traffic to the apical and basolateral surfaces of polarized MDCK cells. *J. Cell Biol.* **111:** 2365.

Cameron, P.L., T.C. Südhof, R. Jahn, and P. de Camilli. 1991. Colocalization of synaptophysin with transferrin receptors: Implications for receptor biogenesis. *J. Cell Biol.* **115:** 151.

Chavrier, P., K. Simons, and M. Zerial. 1992. The complexity of the Rab and Rho GTP-binding protein subfamilies revealed by a PCR cloning approach. *Gene* **112:** 261.

Chavrier, P., R.G. Parton, H.-P. Hauri, K. Simons, and M. Zerial. 1990. Localization of low molecular weight GTP-binding proteins to exocytic and endocytic compartment. *Cell* **62:** 317.

Corbeil, D., G. Boileau, G. Lemay, and P. Crine. 1992. Expression and polarized apical secretion in Madin-Darby canine kidney cells of a recombinant soluble form of neutral endopeptidase lacking the cytosolic and transmembrane domains. *J. Biol. Chem.* **267:** 2798

Dotti, C.G. and K. Simons. 1990. Polarized sorting of viral glycoproteins to the axon and dendrites of hippocampal neurons in culture. *Cell* **62:** 63.

Dotti, C.G., J. Kartenbeck, and K. Simons. 1993. Polarized distribution of the viral glycoproteins of vesicular stomatitis, fowl plague, and Semliki forest viruses in hippocampal neurons in culture: A light and electron microscopy study. *Brain Res.* (in press).

Dotti, C.G., R.G. Parton, and K. Simons. 1991. Polarized sorting of glypiated proteins in hippocampal neurons. *Nature* **349:** 158.

Dotti, C.G., C.A. Sullivan, and G.A. Banker. 1988. The establishment of polarity by hippocampal neurons in culture. *J. Neurosci.* **8:** 1454.

Drenckhahn, D., K. Schlüter, D.P. Allen, and V. Bennett. 1985. Colocalization of band 3 with ankyrin and spectrin at the basal membrane of intercalated cells in the rat kidney. *Science* **230:** 1287.

Fuller, S.D. and K. Simons. 1986. Transferrin receptor polarity and recycling accuracy in "tight" and "leaky" strains of Madin-Darby canine kidney cells. *J. Cell Biol.* **103:** 1767.

Fuller, S.D., R. Bravo, and K. Simons. 1985a. An enzymatic assay reveals that proteins destined for the apical and basolateral domains of an epithelial cell line share the same late Golgi compartments. *EMBO J.* **4:** 297.

Fuller, S.D., C.-H. von Bonsdorff, and K. Simons. 1985b. Cell surface influenza haemagglutinin can mediate infection by animal viruses. *EMBO J.* **4:** 2475.

Gallwitz, D., C. Donath, and C. Sander. 1983. A yeast gene encoding a protein homologous to the human c-*has/bas* protooncogene product. *Nature* **306:** 704.

Goslin, K. and G. Banker. 1991. Rat hippocampal neurons in low density culture. In *Culturing nerve cells* (ed. G. Banker and K. Goslin), p. 251. MIT Press, Cambridge, Massachusetts.

Griffiths, G. and K. Simons. 1986. The *trans* Golgi network: Sorting at the exit site of the Golgi complex. *Science* **234:** 438.

Gumbiner, B. 1987. Structure, biochemistry, and assembly of epithelial tight junctions. *Am. J. Physiol.* **22:** 749.

Gundersen, D., J. Orlowski, and E. Rodriguez-Boulan. 1991. Apical polarity of Na,K-ATPase in retinal pigment epithelium is linked to a reversal of the ankyrin-fodrin submembrane cytoskeleton. *J. Cell Biol.* **112:** 863.

Hammerton, R.W., K.A. Krzeminski, R.W. Mays, T.A. Ryan, D.A. Wollner, and W.J. Nelson. 1991. Mechanism for regulating cell surface distribution of Na^+, K^+-ATPase in polarized epithelial cells. *Science* **254:** 847.

Hubbard, A.L., B. Stieger, and J.R. Bartles. 1989. Biogenesis of endogenous plasma membrane proteins in epithelial cells. *Annu. Rev. Physiol* **51:** 755.

Hughson, E., A. Wandinger-Ness, H. Gausepohl, G. Griffiths, and K. Simons. 1988. The cell biology of enveloped virus infection of epithelial tissues. In *The molecular biology of infectious diseases: Centenary Symposium of the Pasteur Institute* (ed. M. Schwartz), p.75. Elsevier, Paris.

Joe, E.-H. and K.I. Angelides. 1991. Development, maintenance, and modulation of voltage-dependent sodium channel topography in nerve cells. *Curr. Top. Membr. Transp.* **39:** 229.

Kelly, R.B. 1988. The cell biology of the nerve terminal. *Neuron* **1:** 431.

Kobayashi, T., B. Storrie, K. Simons, and C.G. Dotti. 1992. A functional barrier to movement of lipids in polarized neurons. *Nature* (in press).

Koob, R., M. Zimmermann, W. Schoner, and D. Drenckhahn. 1987. Colocalization and coprecipitation of ankyrin and Na^+, K^+-ATPase in kidney epithelial cells. *Eur. J. Cell Biol.* **45:** 230.

Kurzchalia, T.V., P. Dupree, R.G. Parton, R. Kellner, H. Virta, M. Lehnert, and K. Simons. 1992. VIP21, a 21 KD membrane protein is an integral component of trans-Golgi-network-derived transport vesicles. *J. Cell Biol.* **118:** 1003.

Li, C., S. Stifani, W.I. Schneider, and M.I. Poznansky. 1991. Low density lipoprotein receptors on epithelial cell (Madin-Darby canine kidney) monolayers. *J. Biol. Chem.* **266:** 9263.

Lisanti, M.P. and E. Rodriguez-Boulan. 1991. Polarized sorting of GPI-linked proteins in epithelia and membrane microdomains. *Cell Biol. Int. Rep.* **15:** 1023.

Matlin, K.S. and K. Simons. 1983. Reduced temperature prevents transfer of a membrane glycoprotein to the cell surface but does not prevent terminal glycosylation. *Cell* **34:** 233.

Mostov, K., G. Apodaca, B. Aroeti, and C. Okamoto. 1992. Plasma membrane protein sorting in polarized epithelial cells. *J. Cell Biol.* **116:** 577.

Nelson, W.J. and P.I. Veshnock. 1987. Ankyrin binding to ($Na^+ + K^+$) ATPase and implications for the organization of membrane domains in polarized cells. *Nature* **328:** 533.

Olmstead, J.-B. 1986. Microtubule-associated proteins. *Annu. Rev. Cell Biol.* **2:** 421.

Parton, R.G., K. Simons, and C. Dotti. 1992. Axonal and dendritic endocytic pathways in cultured neurons. *J. Cell Biol.* **118:** 1003.

Pathak, R.K., M. Yokode, R.E. Hammer, S.L. Hofmann, M.S. Brown, J.L. Goldstein, and R.G.W. Anderson. 1990. Tissue-specific sorting of the human LDL receptor in polarized epithelia of transgenic mice. *J. Cell Biol.* **111:** 347.

Pfeffer, S.R. 1992. GTP-binding proteins in intracellular transport. *Trends Cell Biol.* **2:** 41.

Pietrini, G., M. Matteoli, G. Banker, and M.J. Caplan. 1991. Isoforms of the Na,K-ATPase are present in both axons and dendrites of hippocampal neurone in culture. *Proc. Natl. Acad. Sci.* **89:** 8414.

Pow, D.V. and J.F. Morris. 1989. Dendrites of hypothalamic magnocellular neurons release neurohypophysial peptides by exocytosis. *Neuroscience* **32:** 435.

Rodriguez-Boulan, E. and J. Nelson. 1989. Morphogenesis of the polarized epithelial cell phenotype. *Science* **245:** 718.

Rodriguez-Boulan, E. and M. Pendergast. 1980. Polarized distribution of viral envelope proteins in the plasma membrane of infected epithelial cells. *Cell* **20:** 45.

Rodriguez-Boulan, E. and D.D. Sabatini. 1978. Asymmetric budding of viruses in epithelial monolayers: A model system for study of epithelial polarity. *Proc. Natl. Acad. Sci.* **75:** 5071.

Roman, L.M. and H. Garoff. 1985. Revelation through exploitation: The viral model for intracellular traffic. *Trends Biochem. Sci.* **10:** 428.

———. 1986. Alteration of the cytoplasmic domain of the membrane-spanning glycoprotein p62 of Semliki forest virus does not affect its polar distribution in established lines of Madin-Darby canine kidney cells. *J. Cell Biol.* **103:** 2607.

Roth, M.G. 1989. Molecular biological approaches to protein sorting. *Annu. Rev. Physiol.* **51:** 797.

Salminen, A. and P.J. Novick. 1987. A *ras*-like protein is required for a post Golgi event in yeast secretion. *Cell* **49:** 527.

Segev, N., J. Mulholland, and D. Botstein. 1988. The yeast GTP-binding YPT1 protein and a mammalian counterpart are associated with the secretion machinery. *Cell* **52:** 915.

Simons, K. 1990. The epithelial tight junction: Occluding barrier and fence. In *Morphoregulatory molecules* (ed. G.M. Edelman et al.), p. 341. Wiley, New York.

Simons K. and S.D. Fuller. 1985. Cell surface polarity in epithelia. *Annu. Rev. Cell Biol.* **1:** 243.

Simons, K. and G. van Meer. 1988. Lipid sorting in epithelial cells. *Biochemistry* **27:** 6197.

Simons, K. and A. Wandinger-Ness. 1990. Polarized sorting in epithelia. *Cell* **62:** 207.

Simons, K. and G. Warren. 1984. Semliki forest virus: A probe for membrane traffic in the animal cell. *Adv. Protein Chem.* **36:** 79.

Srinivasan, Y., L. Elmer, J. Davis, V. Bennett, and K. Angelides. 1988. Ankyrin and spectrin associate with voltage-dependent sodium channels in brain. *Nature* **333:** 177.

Thompson, D.W. 1942. *On growth and form.* Cambridge University Press, England.

Touchot, N., P. Chardin, and A. Tavitian. 1987. Four additional members of the *ras* gene superfamily isolated by an oligonucleotide strategy: Molecular cloning of YPT-related cDNAs from a rat brain library. *Proc. Natl. Acad. Sci.* **84:** 8210.

Vale, R.D. 1987. Intracellular transport using microtubule-based motors. *Annu. Rev. Cell Biol.* **3:** 347.

van Meer, G., B. Gumbiner, and K. Simons. 1986. The function of tight junction in maintaining differences in lipid composition between the apical and basolateral cell surface domains of MDCK cells. *EMBO J.* **5:** 1455.

van Meer, G., E.H.K. Stelzer, R.W. Wijnaendts-van-Resandt, and K. Simons. 1987. Sorting of sphingolipids in epithelial (Madin-Darby canine kidney) cells. *J. Cell Biol.* **105:** 1623.

Vogel, L.K., M. Spiess, H. Sjostrom, and O. Noren. 1992. Evidence for an apical sorting signal on the ectodomain of human aminopeptidase-N. *J. Biol. Chem.* **267:** 2794.

Wandinger-Ness, A. and K. Simons. 1991. The polarized transport of surface proteins and lipids in epithelial cells. In *Intracellular trafficking of proteins* (ed. C.J. Steer and J.A. Hannover), p. 575. Cambridge University Press, England.

Wandinger-Ness, A., M.K. Bennett, C. Antony, and K. Simons. 1990. Distinct transport vesicles mediate the delivery of plasma membrane proteins to the apical and basolateral domains of MDCK cells. *J. Cell Biol.* **111:** 987.

Waxman, S.G. and J.M. Ritchie. 1985. Organization of ion channels in the myelinated nerve fiber. *Science* **228:** 1502.

Wilson, J.M., N. Fasel, and J.-P. Kraehenbuhl. 1990. Polarity of endogenous and exogenous glycosyl-phosphatidylinositol-anchored membrane proteins in Madin-Darby canine kidney cells. *J. Cell Sci.* **96:** 143.

Zahraoui, A., N. Touchot, P. Chardin, and A. Tavitian. 1989. The human Rab genes encode a family of GTP-binding proteins related to yeast YPT1 and SEC4 products involved in secretion. *J. Biol. Chem.* **264:** 12394.

Regulation of Epithelial Cell Polarity: A View from the Cell Surface

W.J. Nelson, R. Wilson,* D. Wollner, R. Mays, H. McNeill, and K. Siemers
Department of Molecular and Cellular Physiology, Stanford University School of Medicine, Stanford, California 94305-5426

Structural and Functional Polarity of Epithelial Cells

Polarized transporting epithelial cells are found in a variety of organs and tissues, where they form a closed monolayer of cells that separates two biological compartments in the body (Rodriguez-Boulan and Nelson 1989). This cell type regulates the ionic composition of these compartments by vectorial transepithelial transport of ions, solutes, and fluid. For example, in the kidney, polarized epithelial cells regulate the reabsorption of ions, solutes, and fluid from the ultrafiltrate to the blood supply (Rodriguez-Boulan and Nelson 1989); in the choroid plexus, ions and fluid are transported from the blood supply to the cerebral space to form the cerebral spinal fluid (Wright 1972). The regulation of either a reabsorptive or secretory function requires the development of restricted distributions of ion and solute transport proteins to structurally and functionally distinct domains of the plasma membrane, termed apical and basolateral, that face different biological compartments (Rodriguez-Boulan and Nelson 1989).

What are the mechanisms involved in regulating the restricted distributions of membrane proteins to different domains of the plasma membrane in this important cell type? Are there factors involved that are specific to polarized epithelial cells? Recent studies indicate that protein distribution in the cell is regulated by at least two sites: the Golgi complex and the plasma membrane. However, our thinking about protein sorting has been complicated by the fact that different epithelial cells appear to have different mechanisms for regulating cell-surface polarity of membrane proteins. For instance, in Madin-Darby canine kidney (MDCK) cells, which are derived from a kidney epithelium, proteins are sorted in the Golgi complex into different vesicle populations which are then vectorially delivered to either the apical or basolateral membrane domain (Simons and Fuller 1985; Rodriguez-Boulan and Nelson 1989). In contrast, hepatocytes (Bartles et al. 1987; Bartles and Hubbard 1988) and intestinal epithelial cells (Matter et al. 1990) deliver most, if not all, proteins directly to the basolateral membrane; apical proteins are then retrieved by endocytosis and then delivered to the apical membrane. Do these results indicate that each cell type has a unique mechanism for establishing cell-surface polarity, or are there mechanisms that are conserved between the different cell types? In addition, given that many of the proteins that have a restricted membrane distribution in polarized epithelial cells are also expressed in nonpolarized cells, what are the signals in epithelial cells that induce polarization of these proteins? Here, we discuss recent studies in MDCK cells indicating that in this cell type, protein sorting occurs not only at the level of the Golgi complex, as emphasized in previous studies, but also at the level of the plasma membrane, as indicated in hepatocytes and intestinal cells. The contributions of these different sites for protein sorting in regulating cell-surface polarity of membrane proteins are discussed in the context of the development of epithelial cell polarity.

Maintenance of Cell-surface Polarity: Protein Sorting in the Golgi Complex

The first studies to investigate the delivery of membrane proteins to different membrane domains in polarized MDCK cells used viral glycoproteins to show that apical (hemagglutinin of influenza virus) and basolateral (G protein of VSV) membrane proteins are delivered vectorially from the Golgi complex to the appropriate membrane domain (Rodriguez-Boulan and Sabatini 1978; Matlin et al. 1983; Rindler et al. 1984). Immunoelectron microscopy indicated that protein sorting occurred in the *trans*-Golgi network (TGN) (Rindler et al. 1984, 1985). Subsequent studies on the delivery of a variety of endogenous membrane proteins demonstrated that the viral glycoproteins utilized normal cellular pathways of protein trafficking between the TGN and different cell-surface domains of these cells (Le Bivic et al. 1990, 1991a).

Recent studies have begun to yield compelling evidence for the existence of different sorting signals that mediate protein delivery to either the apical or basolateral membrane (see Fig. 1) (for recent reviews, see Mostov et al. 1992; Wollner and Nelson 1992). One type of sorting signal for the direct delivery of proteins to the apical membrane involves anchorage of proteins to the lipid bilayer through glycosylphosphatidylinositol (GPI) (Brown et al. 1989; Lisanti et al.

*Present address: Department of Chemistry, Philadelphia College of Textiles and Science, Philadelphia, Pennsylvania 19144.

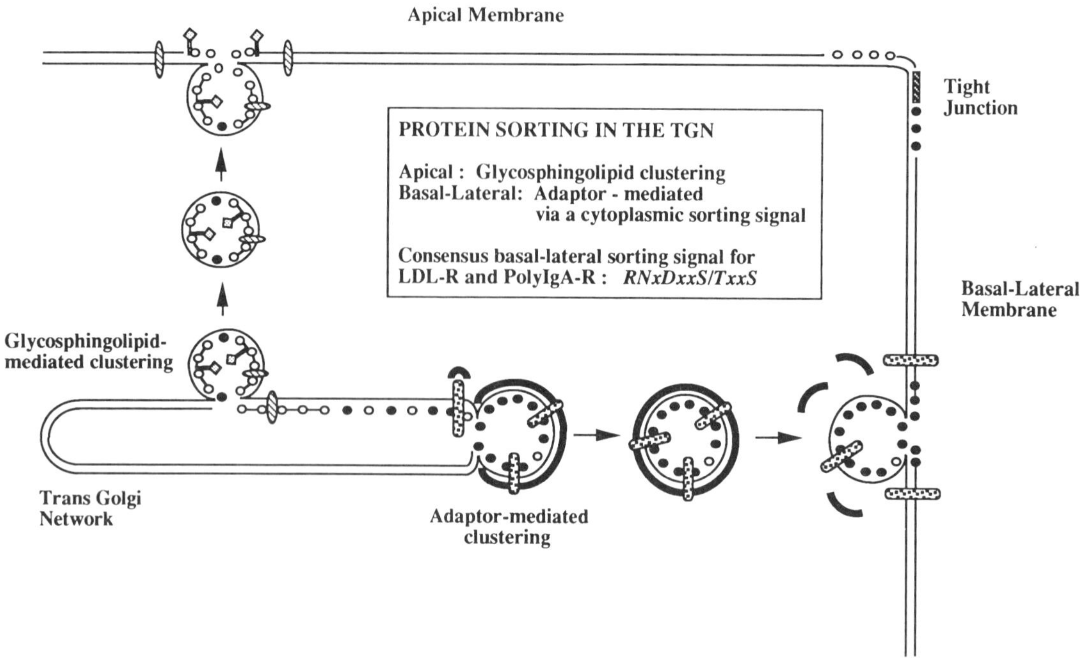

Figure 1. Protein sorting in the TGN. Schematic representation of sorting pathways in the TGN in a polarized MDCK epithelial cell. Both proteins and lipids are sorted in the TGN by clustering mechanisms. Glycosphingolipids and GPI-anchored proteins cluster to form apical vesicles, whereas basolateral membrane proteins are clustered through recognition of a cytoplasmic sorting signal by adapter proteins. Vesicles bud from the TGN and dock with the appropriate membrane domain.

1989). Since glycosphingolipids are also sorted in the TGN for delivery to the apical membrane (van Meer and Simons 1988), it has been proposed that GPI-anchored proteins in the TGN may be sorted through their incorporation with glycosphingolipids into glycosphingolipid rafts, or clusters (see Fig. 1) (Simons and Wandinger-Ness 1990; Brown and Rose 1992). Non-GPI linked apical proteins may contain sorting information in the ectodomain of the protein, which could be recognized by a putative apical sorting receptor that interacts with glycosphingolipids (Simons and Wandinger-Ness 1990).

There is also evidence that strongly indicates the existence of a positive sorting signal for basolateral membrane proteins in MDCK cells (for review, see Mostov et al. 1992; Wollner and Nelson 1992). The identification of a putative basolateral sorting signal was initiated in studies on "nonpolarized" fibroblasts by Lavarovits and Roth (1988), who showed that conversion of a cysteine to a tyrosine at position 543 in the cytoplasmic domain of the viral protein hemagglutinin (HA-Y_{543}) caused the mutated protein to be efficiently internalized through coated vesicles at the cell surface; by comparison, internalization of wild-type HA was very poor.

In polarized MDCK cells, wild-type HA is sorted from basolateral proteins in the TGN, perhaps by association with glycosphingolipids (Skibbens et al. 1989; Brown and Rose 1992), and then delivered directly to the apical membrane (Matlin and Simons 1984; Rindler et al. 1984, 1985). Significantly, Brewer and Roth (1991) showed that HA-Y_{543} was efficiently sorted to the basolateral membrane rather than being delivered to the apical membrane. Thus, in addition to conferring delivery to the basolateral membrane, the signal for endocytosis in HA was also dominant to the inherent sorting signal in the protein for delivery to the apical membrane.

At the same time, when Le Bivic et al. (1991b) analyzed the delivery of nerve growth factor receptor (NGFR) to the cell surface in MDCK cells, they found that wild-type NGFR, like HA, was delivered to the apical membrane and that NGFR was internalized from the apical membrane domain at a very low rate (6%). However, an internal deletion of 58 amino acids in the cytoplasmic domain of NGFR resulted in a mutant receptor that was efficiently delivered to the basolateral membrane. In addition, the truncated NGFR exhibited a greater than 100-fold increase in the amount of ligand-induced endocytosis compared to the wild-type receptor; analysis of this mutation revealed that a tyrosine residue (Y_{308}) was moved into close proximity to the transmembrane domain. Further evidence for a role of an endocytosis signal in basolateral protein sorting in the TGN was obtained in an analysis of IgG Fc receptor (FcR) sorting in MDCK cells (Hunziker and Mellman 1990; Hunziker et al. 1991). Two closely related receptors were sorted to different membrane domains and had different rates of endocytosis at the cell surface via coated vesicles. The receptor carrying an endocytosis signal (FcRII-B2) was delivered to the basolateral membrane, whereas the other receptor (FcRII-B1) was delivered to both membrane domains equally. Significantly, deletion of the endocytosis signal of FcRII-B2 resulted in efficient expression of the truncated protein on the apical membrane (Hunziker et al. 1991).

This result supported the hypothesis that a basolateral sorting signal may be similar to the signal for cell-surface endocytosis. However, analysis of low-density lipoprotein (LDL) and poly IgA receptors showed that these signals may not necessarily be the same. The LDL receptor has a well-characterized signal for endocytosis that contains a tyrosine or aromatic amino acid (Davis et al. 1987). Hunziker et al. (1991) found that mutations in the LDL receptor in which this critical tyrosine was replaced by alanine resulted in a receptor that was delivered efficiently to the basolateral membrane but was poorly internalized at the cell surface. Similarly, Casanova et al. (1991) found that a 14-amino-acid sequence next to the transmembrane domain of the poly IgA receptor was sufficient for delivery of the receptor to the basolateral membrane, but the truncated receptor was poorly internalized from the cell surface (Casanova et al. 1991), suggesting that it did not contain a signal for cell surface endocytosis.

Taken together, these recent results strongly suggest that there are positive sorting signals that specify protein delivery to either the basolateral membrane or the apical membrane domain of MDCK cells. In the absence of any sorting signal, proteins may be delivered to both membrane domains. For example, FcRII-B1 was delivered to both cell surfaces (Hunziker et al. 1991); Na/K-ATPase may also represent another protein that does not possess a strong sorting signal in MDCK cells (see below). These results imply that exclusive delivery to either surface requires some sorting information in the protein.

The sorting machinery that recognizes and distinguishes between an apical or basolateral membrane sorting signal is not known at present. However, given the nature of the putative sorting signals, it seems likely that different classes of proteins are positively sorted by a process of protein clustering in the plane of the lipid bilayer and the subsequent budding of separate vesicles from the TGN (see Fig. 1) (Mostov et al. 1992; Wollner and Nelson 1992). The clustering of proteins destined to form apical vesicles may be mediated by the aggregation of glycosphingolipids (see above). Adapter proteins, perhaps similar to the adaptins that are involved in clathrin-mediated coated pit assembly at the cell surface (Pearce and Robinson 1990), may recognize a basolateral signal in the cytoplasmic domain of some proteins and facilitate the clustering of these proteins into a budding vesicle in the TGN (see Fig. 1).

Maintenance of Cell-surface Polarity: Protein Sorting at the Plasma Membrane

Studies in polarized epithelial cells, other than MDCK cells, indicate that sorting of membrane protein to different membrane domains occurs at the level of the plasma membrane. In hepatocytes and intestinal epithelial cells, membrane proteins are delivered from the TGN to the basolateral membrane domain (sinusoidal membrane in hepatocytes) from where apical membrane proteins are efficiently retrieved and then transported to the apical membrane domain (Bartles et al. 1987; Bartles and Hubbard 1988; Matter et al. 1990). Do these results imply that hepatocytes and intestinal epithelial cells use a mechanism for sorting proteins at the plasma membrane that is different from that in the TGN of MDCK cells? Two recent studies indicate that the mechanism(s) may be similar. First, expression of rat liver dipeptidyl peptidase IV, which is delivered to the apical membrane via the basolateral membrane in hepatocytes, is delivered directly from the TGN to the apical membrane in transfected MDCK cells (Casanova et al. 1992). Second, deletion analysis of the LDL receptor in transgenic mice showed that a short cytoplasmic sequence, similar to that described for the poly IgA receptor (see above), specified targeting and accumulation of the truncated protein in the basolateral membrane of hepatocytes (Yokode et al. 1992). Significantly, deletions within this sequence resulted in the appearance of the mutated protein at the apical membrane (bile canaliculus): This is reminiscent of the results by Hunziker et al. (1991) on the targeting of the LDL receptor in transfected MDCK cells (see above). Taken together, these results indicate that the same sorting signals on these proteins are recognized by the sorting machinery in the TGN of MDCK cells and at the plasma membrane of hepatocytes/intestinal epithelial cells.

Development of Cell-surface Polarity: Analyzing the Contributions of Protein Sorting in the TGN and at the Plasma Membrane

In vivo, polarized epithelial cells arise through the conversion of nonpolarized mesenchymal precursor cells (Rodriguez-Boulan and Nelson 1989). This conversion requires the expression of proteins that induce specific cell-cell (E-cadherin) and cell-substratum (laminin A chain) contacts (Ekblom et al. 1986; Ekblom 1989) and the reorganization of constitutively expressed proteins such as Na/K-ATPase from a uniform distribution on the cell surface of nonpolarized precursor cells to a polarized distribution on the basolateral membrane domain in some cell types. We have used MDCK cells, an in vitro model of polarized epithelia that develop structural and functional polarity in stages similar to those that occur in vivo, to investigate mechanisms that regulate the cell-surface distribution of Na/K-ATPase (Hammerton et al. 1991).

To examine the generation of the cell-surface polarity of Na/K-ATPase, we induced E-cadherin-dependent cell-cell contact in confluent monolayers of MDCK cells and then visualized Na/K-ATPase distribution by indirect immunofluorescence using laser-scanning confocal microscopy (see Fig. 2) (Hammerton et al. 1991). At early time points after induction of cell-cell contact (<48 hr), we detected staining of α-subunit Na/K-ATPase at the apical surface and at sites of cell-cell

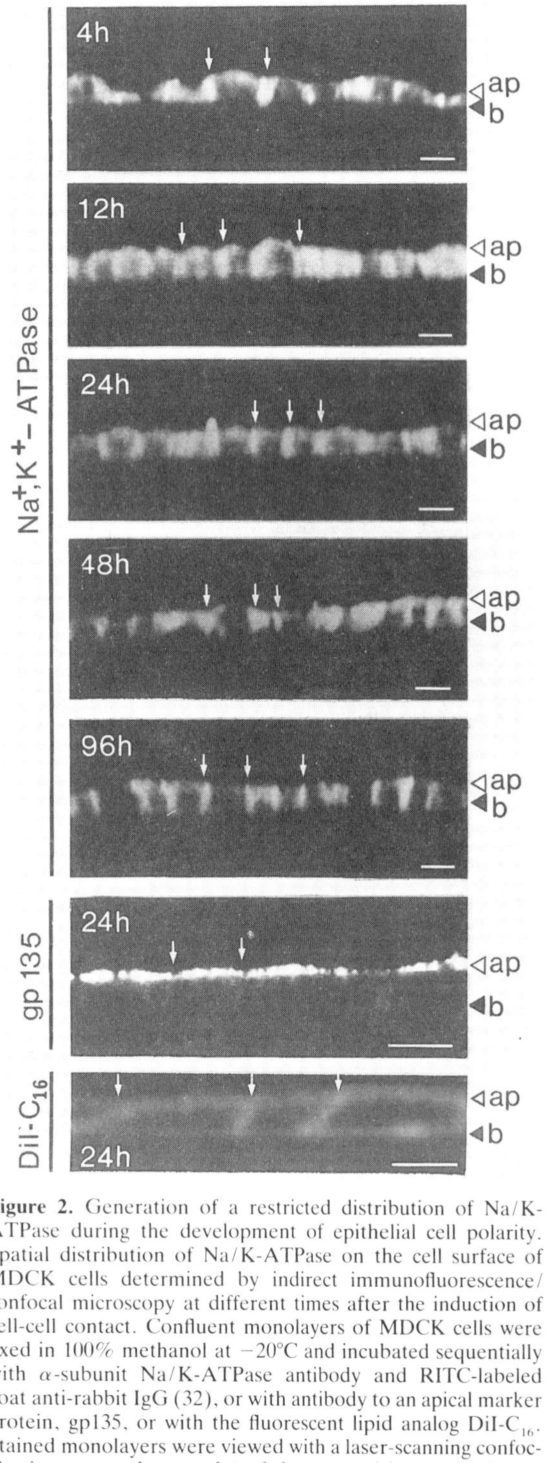

Figure 2. Generation of a restricted distribution of Na/K-ATPase during the development of epithelial cell polarity. Spatial distribution of Na/K-ATPase on the cell surface of MDCK cells determined by indirect immunofluorescence/confocal microscopy at different times after the induction of cell-cell contact. Confluent monolayers of MDCK cells were fixed in 100% methanol at −20°C and incubated sequentially with α-subunit Na/K-ATPase antibody and RITC-labeled goat anti-rabbit IgG (32), or with antibody to an apical marker protein, gp135, or with the fluorescent lipid analog DiI-C_{16}. Stained monolayers were viewed with a laser-scanning confocal microscope; the *x-z* plot of the scanned images is shown. Note that the magnification of the images for the gp135 and DiI-C_{16} staining are twice that of the Na/K-ATPase staining. (AP) Apical membrane; (B) basal membrane; (arrows) lateral membrane. Bar, 10 μm.

contacts (lateral membranes). At later time points (> 48 hr), the staining of lateral membranes was very prominent when compared, for example, with cells labeled briefly with the fluorescent lipid analog, DiI-C_{16}, which labels all cell surfaces equally. A low level of Na/K-ATPase staining was also detectable at the apical membrane at these later times, but the staining was less distinctive than that present on the apical membrane shortly after induction of cell-cell contact when compared, for instance, to the staining of gp135, a marker protein of the apical membrane (see Fig. 1) (Ojakian and Schwimmer 1988; Wang et al. 1990).

The cell-surface distribution of Na/K-ATPase was determined quantitatively by labeling cell-surface protein with the biotinylating reagent, sulfo-NHS-biotin, at different times after induction of cell-cell contact. This type of reagent is well suited for these studies (Sargiacomo et al. 1989). It is soluble in water, but the sulfo group, which confers an overall negative charge to the molecule, prevents its diffusion across the lipid bilayer of the plasma membrane. This ionic nature also impairs its diffusion across the cationic-selective tight junction that forms at the boundary between the apical and basolateral membranes. Thus, labeling can be confined to the extracellular surface of specific membrane domains, provided the integrity of the tight junction is maintained.

Biotinylated Na/K-ATPase was recovered initially (24–48 hr) from both the forming apical and basolateral membrane domains. Between 48 and 72 hours after cell-cell contact, however, we detected a 400–500% increase in the amount of Na/K-ATPase on the basolateral membrane. During the same period, the amount of biotinylated Na/K-ATPase detected on the apical membrane decreased by more than 65%. Ninety-six hours after the induction of cell-cell contact, more than 85% of the total cell-surface Na/K-ATPase was localized to the basolateral membrane domain (Hammerton et al. 1991).

We determined the subunit organization of Na/K-ATPase on the cell surface during the development of polarity by taking advantage of the ability of α-subunit-specific antibodies to coimmunoprecipitate an $\alpha\beta$-complex from whole-cell extracts. Analysis of cell-surface and newly synthesized Na/K-ATPase that had been biotinylated at either the apical membrane or basolateral membrane revealed the presence of an $\alpha\beta$-subunit complex in which both subunits were biotinylated. This result strongly indicates that Na/K-ATPase is delivered to, and resides in, both membrane domains as an $\alpha\beta$-subunit complex.

To determine whether direct targeting of newly synthesized Na/K-ATPase to the basolateral membrane contributed to the establishment of the restricted distribution of this protein, we quantitatively analyzed the delivery of newly synthesized Na/K-ATPase to both cell-surface domains (Hammerton et al. 1991). At all times following the induction of cell-cell contact, we detected delivery of approximately equal amounts of newly synthesized α-subunit Na/K-ATPase to the apical and basolateral membrane domains. During the course of the experiment, the total amount of newly synthesized α-subunit Na/K-ATPase delivered to the cell surface decreased. These results indicate that Na/

K-ATPase is not delivered directly to the basolateral membrane but is inserted into vesicles in the *trans*-Golgi network that are then transported to both the apical and basolateral membrane domains. We suggest that sorting of Na/K-ATPase in the Golgi complex and targeting to one membrane domain do not appear to play a role in the development or maintenance of the cell-surface polarity of this protein. This is in contrast to many other proteins that have been studied which are delivered directly to either the apical or basolateral membrane domain (Rodriguez-Boulan and Nelson 1989; Mostov et al. 1992). These proteins appear to be expressed specifically in polarized epithelial cells, whereas Na/K-ATPase is expressed in nonpolarized cells, as well as in polarized epithelial cells where it may be present on either the apical or basolateral membrane. Differences in the sorting of Na/K-ATPase and other proteins may be due to differences in their affinity for the protein-sorting machinery that is thought to reside in the TGN.

We sought to determine the kinetics of development of functional polarity of Na/K-ATPase by measuring the binding of the cardiac glycoside, ouabain, to the apical and basolateral membrane domains. Ouabain binds specifically to an extracellular domain of the α-subunit of Na/K-ATPase and blocks the E_2-P$\rightarrow E_2$ conformational change in the enzyme during K^+ transport across the lipid bilayer; hence, ouabain binding to Na/K-ATPase is evidence for an active pump (Jorgensen 1986). Addition of [^{3}H]ouabain to cell monolayers 3.5–4.5 hours after the induction of cell-cell contact showed binding of ouabain to both the forming apical membrane and the basolateral membrane, indicating the presence of active Na/K-ATPase on both cell-surface domains. The binding was specific since more than 98% was competitively inhibited by the addition of a 100-fold excess of unlabeled ouabain. By 12 hours after the induction of cell-cell contact, there was a decrease in the binding of ouabain to the apical membrane. By 24 hours, we detected little or no ouabain binding on the apical membrane, although Na/K-ATPase was still detected at this membrane domain by cell-surface biotinylation. At this time, more than 95% of cell-surface ouabain binding was restricted to the basolateral membrane domain (Hammerton et al. 1991). The lack of ouabain binding to the apical membrane may account for the fact that a previous study failed to detect with NAB-ouabain the delivery of newly synthesized Na/K-ATPase to that membrane domain (Caplan et al. 1986).

It was noteworthy that at times when little or no ouabain binding was detected on the apical membrane (>24 hr), substantial amounts of Na/K-ATPase were still detected at that membrane surface, and as newly synthesized protein that was delivered from the Golgi complex. We do not know whether loss of ouabain binding is due to selective inactivation of Na/K-ATPase in this membrane domain. However, the loss of ouabain binding on the apical membrane coincided with the formation of that domain, as determined by the kinetics of development of polarity of specific marker proteins. Significantly, the apical membrane of MDCK cells is characterized by the presence of high concentrations of glycosphingolipids and decreased membrane fluidity (van Meer and Simons 1988), which may influence Na/K-ATPase activity (Kimelberg and Papahadjopoulos 1972, 1974). Although the basis for the loss of Na/K-ATPase activity on the apical membrane is unknown, an important consequence is that the cell rapidly develops functional polarity of Na/K-ATPase in the absence of either cell-surface polarity of Na/K-ATPase protein or polarized delivery of newly synthesized Na/K-ATPase to the basolateral membrane domain.

Given that newly synthesized Na/K-ATPase is delivered to both the apical and basolateral membrane domains in approximately equal amounts, how is the polarized distribution of Na/K-ATPase on the basolateral membrane established? We investigated the possibility that preferential stabilization of newly synthesized protein on the basolateral membrane rather than the apical membrane was responsible for determining the unequal cell-surface distribution of Na/K-ATPase (Hammerton et al. 1991). We examined the half-time of residence of newly synthesized Na/K-ATPase in each membrane domain at various times after the induction of cell-cell contact. Our results showed a dramatic change in the residence time of newly synthesized Na/K-ATPase on different cell-surface domains. Shortly after cell-cell contact was established, Na/K-ATPase had a very short residence time (~3 hr) on both the forming apical and basolateral membrane domains, which is similar to the residence time of Na/K-ATPase on the cell surface of single MDCK cells. At subsequent times (24, 72, 144 hr), Na/K-ATPase that was delivered to the apical membrane continued to have a relatively short residence time (<2 hr). In contrast to the rapid loss of Na/K-ATPase from the apical membrane, the residence time of Na/K-ATPase delivered to the basolateral membrane gradually increased; by 144 hours after cell-cell contact, the residence time was more than 36 hours. Comparison of the residence time of Na/K-ATPase on the apical and basolateral membranes reveals that at 24, 72, and 144 hours after cell-cell contact, the residence time was up to 6-, 12-, and >36-fold longer on the basolateral membrane than on the apical membrane.

The average residence time for total cell-surface proteins detected by biotinylation during the chase period was similar on each membrane domain (12–14 hr) and remained relatively constant at different times after induction of cell-cell contact. Thus, the residence time for Na/K-ATPase on the apical surface was up to 14-fold less than that of total protein, indicating that Na/K-ATPase was being selectively removed. However, the residence time for Na/K-ATPase on the basolateral membrane became 3–4-fold longer than that of total protein, indicating that this pool of Na/K-ATPase was preferentially stabilized.

Taken together, these results indicate that generation of a restricted cell-surface distribution of Na/K-ATPase in this cell type is determined by stabilization and accumulation of protein on the basolateral membrane, rather than by direct delivery of newly synthesized protein from the Golgi complex to the basolateral membrane.

Is this mechanism for generating cell-surface polarity unique to Na/K-ATPase? Analysis of the development of cell-surface polarity of two other basolateral membrane proteins, Dg-1 and E-cadherin, showed that restricted distributions were established within 8 hours of the induction of cell-cell contact (Wollner et al. 1992). At that time, however, newly synthesized Dg-1 and E-cadherin were delivered predominantly to the forming apical membrane and not to the basolateral membrane. The residence time of these proteins in the apical membrane was very short ($t_{1/2}$ 30–60 and 120 min, respectively), compared to that in the basolateral membrane ($t_{1/2}$ 24 and 8 hr, respectively). Thus, development of the cell-surface polarity of these proteins appears to be regulated by the differential removal of these proteins from the apical membrane and their retention in the basolateral membrane, rather than by direct delivery of newly synthesized proteins to the forming basolateral membrane. In contrast to the situation with Na/K-ATPase, however, the delivery of both DG-1 and E-cadherin became vectorial between the TGN and the basolateral membrane (Wollner et al. 1992).

Selective Protein Retention on the Basolateral Plasma Membrane

The significant differences between the rates of removal of Na/K-ATPase, DG-1, and E-cadherin from the apical and basolateral membrane domains indicate that endocytosis plays an important role in the rapid development of cell-surface polarity. The rates of removal of these proteins ($t_{1/2}$ 1–2 hr) (Hammerton et al. 1991; Wollner et al. 1992) correspond well with a previous study that reported that less than 2 hours were required for constitutive internalization of surface markers from either the apical or basolateral plasma membrane domains of MDCK cells (Balcarova et al. 1984). This rate is significantly faster than the rate of removal of these proteins from the basolateral membrane ($t_{1/2}$ 24–36 hr). A rapid constitutive rate of removal of cell-surface proteins would act as a counterforce to the accumulation of proteins being delivered to that region of the membrane, whereas a decrease in the rate of removal would result in protein accumulation in the membrane.

A decrease in the rate of endocytosis of proteins could result from the formation of complexes with other membrane and cytoplasmic proteins. Previously, we and other investigators showed that induction of cell-cell contact results in the assembly of Dg-1 into a large protein complex containing other components of the membrane core domain of the desmosome, cytoplasmic plaque proteins (predominantly, desmoplakin I/II), and the cytokeratin intermediate filament network (Penn et al. 1987; Pasdar and Nelson 1988a,b, 1989). Assembly of the desmosome results in an increase in the insolubility of the component proteins in buffers containing Triton X-100 and an increase in their metabolic stability, which are consistent with the formation of a large protein complex. We suggest that a consequence of protein complex formation is a decrease in the rate of protein removal from the cell surface that would normally occur if the component proteins remained in an unassembled state. Significantly, in the absence of cell-cell contact, unassembled Dg-1 is rapidly removed from the cell surface ($t_{1/2} < 2$ hr); this coincides also with the rapid rate of loss of Dg-1 from the apical membrane where there is no cell-cell contact and hence no induction of desmosome assembly. It is also noteworthy that loss of cell-cell contact results in the rapid internalization and degradation of the desmosomal proteins and other cell adhesion proteins (Pasdar and Nelson 1988a,b, 1989).

Induction of cell-cell contact also results in the formation of protein complexes between the cell adhesion protein E-cadherin and components of the actin-based cytoskeleton (Ozawa et al. 1990; Nagafuchi et al. 1991). These interactions appear to occur specifically on the membrane at sites of cell-cell adhesion and not elsewhere on the membrane. Thus, specific linkage of E-cadherin to the actin cytoskeleton may also reduce the rate of removal of this protein from sites of cell-cell contact. On the other hand, we found that E-cadherin is rapidly removed from the apical membrane where there is no cell-cell contact and hence where complex formation with the actin cytoskeleton might not be induced.

There is also strong evidence to support a role for E-cadherin and the membrane-cytoskeleton in the stabilization and retention of Na/K-ATPase on the basolateral membrane domain (Nelson et al. 1990a). Previously, we showed that development of a restricted distribution of Na/K-ATPase on the basolateral membrane coincided both temporally and spatially with E-cadherin-dependent cell-cell contact and assembly of the membrane-cytoskeletal proteins, ankyrin and fodrin (Nelson and Veshnock 1986, 1987b). Significantly, Na/K-ATPase binds with high affinity to ankyrin in vitro (Koob et al. 1987; Nelson and Veshnock 1987a; Morrow et al. 1989), and a complex comprising Na/K-ATPase, ankyrin, and fodrin has been isolated from whole-cell extracts (see Fig. 3) (Nelson et al. 1990b). What is the connection between the induction of cell-cell contact through E-cadherin and the reorganization of Na/K-ATPase and membrane-cytoskeleton distribution? There is evidence for a direct linkage between Na/K-ATPase and E-cadherin through the membrane-cytoskeleton: (1) A complex comprising E-cadherin, ankyrin, and fodrin has been isolated from whole cells (Nelson and Hammerton 1989); (2) fibroblasts expressing exogenous E-cadherin show colocalization of E-cadherin, Na/K-ATPase, and fodrin at cell-cell contacts (McNeill et al. 1990); and (3) fibroblasts express-

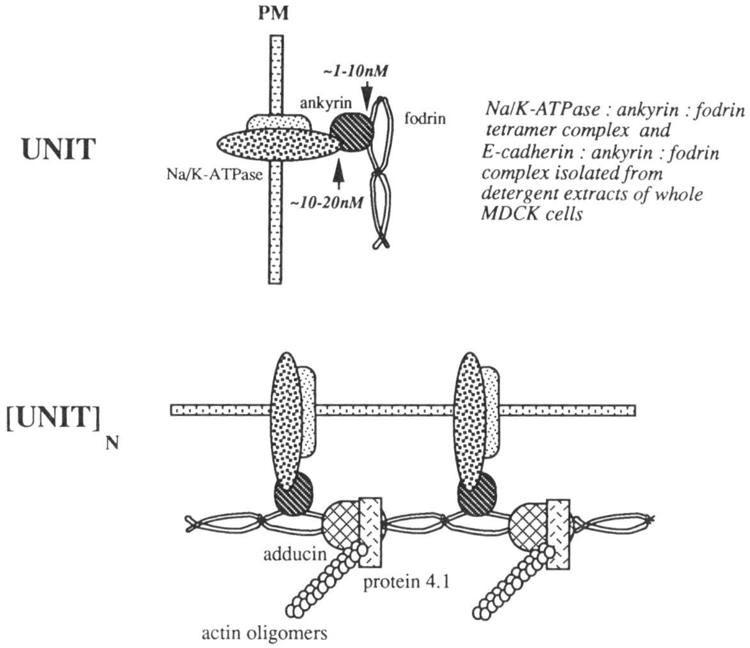

Figure 3. Unit complexes of the membrane-cytoskeleton. Schematic representation of protein-protein interactions between Na/K-ATPase, ankyrin, and fodrin (*top*) as determined by in vitro and in vivo studies (for details and references, see text). Shown are possible interactions between the Unit complexes of the membrane-cytoskeleton and other cytoskeleton proteins during the assembly of the Unit complexes into a detergent-insoluble cytoskeleton (*bottom*).

ing a truncated E-cadherin that lacks portions of the cytoplasmic domain do not develop these restricted distributions of either Na/K-ATPase or fodrin (McNeill et al. 1990). These results indicate that E-cadherin is at, or near the top of, a molecular cascade of events that initiate the assembly of the membrane-associated cytoskeleton at sites of cell-cell contact that leads to the reorganization of Na/K-ATPase distribution.

Mechanism for Generating Cell-surface Polarity of Na/K-ATPase in MDCK Cells

On the basis of the results of the present study and our previous observations, we propose a simple working model to determine how cells generate a restricted distribution of Na/K-ATPase on part of the basolateral membrane in this cell type (see Fig. 4). We suggest that specific cell-cell contacts, induced through E-cadherin,

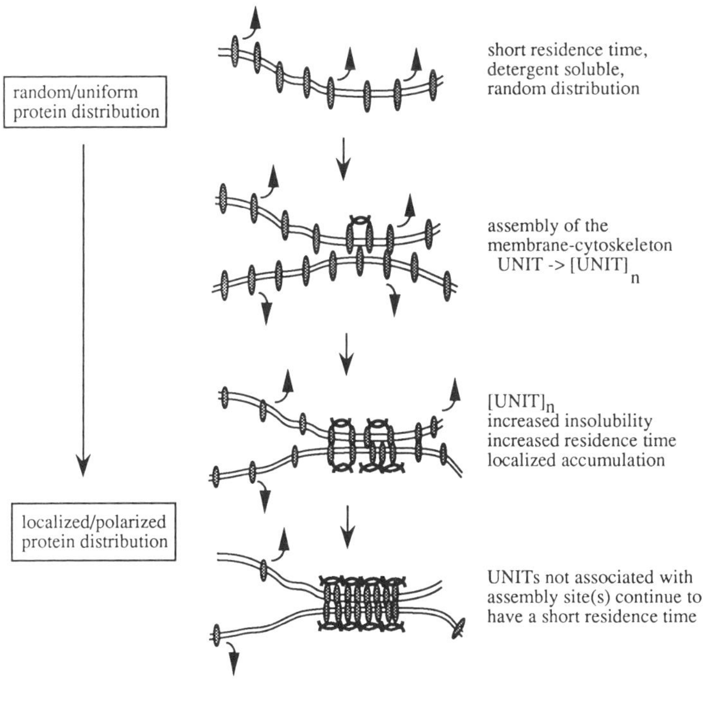

Figure 4. Assembly of the membrane-cytoskeleton. Schematic representation of the consequences of assembly of the membrane-cytoskeleton on the differential accumulation of membrane and cytoskeletal proteins in restricted regions of the plasma membrane (for details, see text).

act as nucleation sites on the cell surface for the assembly of the membrane-cytoskeleton. Homotypic interactions between E-cadherin molecules on adjacent cells may transduce a signal for localized assembly of the membrane-cytoskeleton; this might be similar to the effect of the addition of a ligand to the cell surface of lymphocytes, which also results in patching and capping of ligand-receptor complexes by the cytoskeleton (Bourguignon and Bourguignon 1984). As a consequence of the assembly of the membrane-cytoskeleton, other membrane proteins that are bound to the membrane-cytoskeleton (e.g., Na/K-ATPase) are also selectively incorporated into the forming cytoskeletal complex at cell-cell contacts. Since cell-cell contacts are not formed at either the apical or basolateral membranes, we predict that there is no signal for membrane-cytoskeleton assembly and hence Na/K-ATPase is rapidly removed (i.e., short residence time) from those regions of the membrane. The consequence of these events is the preferential accumulation of Na/K-ATPase at lateral membranes (cell-cell contacts) as shown directly by our immunofluorescence/confocal microscopy.

Multiple Cellular Sites and Mechanisms for Regulating the Cell-surface Distribution of Membrane Proteins in Polarized Epithelial Cells

We suggest that there are at least four important stages in the sorting process that give rise to a restricted distribution of membrane proteins at the cell surface (see Fig. 5) (Wollner and Nelson 1992): (1) sorting of apical and basolateral membrane proteins in the TGN into different populations of transport vesicles in MDCK (renal?) cells, (2) capture and docking of the appropriate transport vesicle at the cell surface, (3) retention of the protein in the appropriate membrane domain, and (4) retrieval of proteins from the incorrect domain and delivery to lysosomes or to the correct membrane domain.

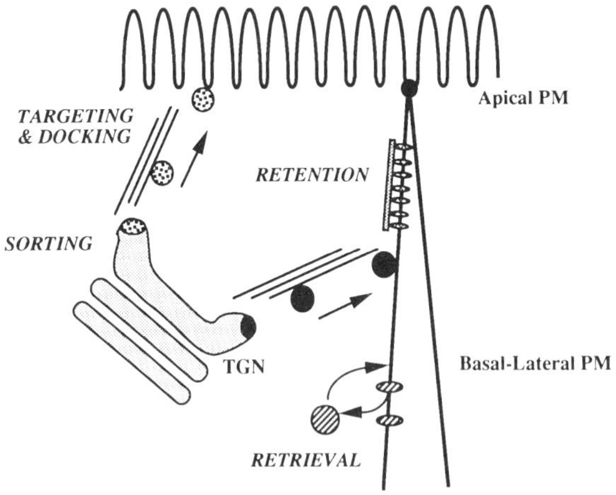

Figure 5. Multiple sites for sorting membrane proteins in the establishment and maintenance of epithelial cell polarity (for details, see text).

As summarized above, recent studies in MDCK cells have identified candidate signals for sorting of apical and basolateral membrane proteins, respectively, in the TGN. However, the machinery involved is poorly understood. In vitro reconstitution of vesicle budding and transport from the Golgi complex to the plasma membrane should provide an approach to this problem (Gravotta et al. 1990; Wandinger-Ness et al. 1990). The mechanisms involved in transport and docking of transport vesicles with the appropriate membrane domain are also poorly understood. An important recent advance has been the isolation of apical and basolateral transport vesicles from polarized MDCK cells. The detailed characterization of these vesicles should provide insight into candidate proteins involved in the sorting, transport, and docking processes. Isolated apical and basolateral vesicles have been shown to bind to microtubules in vitro (van der Sluijs et al. 1990), suggesting that microtubules may be important in delivery of vesicles, perhaps via different motor proteins (kinesin, dynein), to either the apical or basolateral membrane. The distribution and polarity of microtubules could provide a suitable spatial framework for the transport component of the sorting process (Salas et al. 1986; Rindler et al. 1987; Achler et al. 1989; Hunziker et al. 1990).

Finally, if an endocytosis-like signal proves to be responsible for the sorting of proteins to basolateral vesicles, activation of a retention system at the cell surface could be critical to block the recognition of that signal by the cell-surface endocytosis machinery, which would result in the rapid internalization of the protein and its loss from the cell surface. However, it is also important to consider that some internalized proteins may be recycled back to their original (basolateral) membrane in a retrieval process that would account for their apparent distribution at the basolateral membrane.

ACKNOWLEDGMENTS

Work from the authors' laboratory was supported by grants from the National Institutes of Health, National Science Foundation, and the March of Dimes Foundation. R.W. was supported by a postdoctoral fellowship from the National Kidney Foundation, D.W. was supported by a postdoctoral fellowship from the National Institutes of Health, and J.N. is the recipient of an established investigator award from the American Heart Association.

REFERENCES

Achler, C., D. Filmer, C. Merte, and D. Drenckhahn. 1989. Role of microtubules in polarized delivery of apical membrane proteins to the brush border of the intestinal epithelium. *J. Cell Biol.* **109:** 179.

Balcarova, S.J., S.E. Pfeiffer, S.D. Fuller, and K. Simons. 1984. Development of cell surface polarity in the epithelial Madin-Darby canine kidney (MDCK) cell line. *EMBO J.* **3:** 2687.

Bartles, J.R. and A.L. Hubbard. 1988. Plasma membrane protein sorting in epithelial cells. Do secretory pathways hold the key? *Trends Biochem. Sci.* **13:** 181.

Bartles, R.T., H.M. Feracci, B. Steiger, and A.L. Hubbard. 1987. Biogenesis of the rat hepatocyte plasma membrane in vivo: Comparison of the pathways taken by apical and basolateral proteins using subcellular fractionation. *J. Cell Biol.* **105:** 1241.

Bourguignon, L. and G.J. Bourguignon. 1984. Capping and the cytoskeleton. *Int. Rev. Cytol.* **87:** 195.

Brewer, C.B.R. and M.B. Roth. 1991. A single amino acid change in the cytoplasmic domain alters the polarized delivery of influenza virus hemagglutinin. *J. Cell Biol.* **114:** 413.

Brown, D.A. and J.K. Rose. 1992. Sorting of GPI-anchored proteins to glycolipid-enriched membrane subdomains during transport to the apical cell surface. *Cell* **68:** 533.

Brown, D.A., B. Crise, and J.K. Rose. 1989. Mechanism of membrane anchoring affects polarized expression of two proteins in MDCK cells. *Science* **245:** 1499.

Caplan, M.J., H.C. Anderson, G.E. Palade, and J.D. Jamieson. 1986. Intracellular sorting and polarized cell surface delivery of Na/K-ATPase, an endogenous component of MDCK cell basolateral plasma membranes. *Cell* **46:** 623.

Casanova, J.E., G. Apodaca, and K. Mostov. 1991. An autonomous signal for basolateral sorting in the cytoplasmic domain of the polymeric immunoglobulin receptor. *Cell* **66:** 65.

Casanova, J., Y. Mishumi, Y. Ikehara, A. Hubbard, and K. Mostov. 1992. Direct apical sorting of dipeptidyl peptidase IV in MDCK cells. *J. Biol. Chem.* **266:** 24428.

Davis, C.G., I.R. van Driel, D.W. Russell, M.S. Brown, and J.L. Goldstein. 1987. The low density lipoprotein receptor: Identification of amino acids in the cytoplasmic domain required for rapid endocytosis. *J. Biol. Chem.* **262:** 4075.

Ekblom, P. 1989. Developmentally regulated conversion of mesenchyme to epithelium. *FASEB J.* **3:** 2141.

Ekblom, P., D. Vestweber, and R. Kemler. 1986. Cell-matrix interactions and cell adhesion during development. *Annu. Rev. Cell Biol.* **2:** 27.

Gravotta, D., M. Adesnik, and D.D. Sabatini. 1990. Transport of influenza virus HA from the trans Golgi network to the apical surface of MDCK cells permeabilized in their basal-lateral membrane: Energy dependence and involvement of GTP-binding proteins. *J. Cell Biol.* **111:** 2893.

Hammerton, R.W., K.A. Krzeminski, R.W. Mays, T.A. Ryan, D.A. Wollner, and W.J. Nelson. 1991. Mechanism for regulating cell surface distribution of Na/K-ATPase in polarized epithelial cells. *Science* **254:** 847.

Hunziker, W. and I. Mellman. 1990. Expression of macrophage-lymphocyte Fc receptors in MDCK cells: Polarity and transcytosis differ for isoforms with or without coated pit localization domains. *J. Cell Biol.* **109:** 3291.

Hunziker, W., P. Vale, and I. Mellman. 1990. Differential microtubule requirements for transcytosis in MDCK cells. *EMBO J.* **9:** 3515.

Hunziker, W., C. Harter, K. Matter, and I. Mellman. 1991. Basolateral sorting in MDCK cells requires a distinct cytoplasmic domain determinant. *Cell* **66:** 907.

Jorgensen, P.L. 1986. Mechanism of the Na/K-ATPase pump: Protein structure and conformations of the pure Na/K-ATPase. *Biochim. Biophys. Acta* **694:** 27.

Kimelberg, H.K. and D. Papahadjopoulos. 1972. Phospholipid requirements for Na/K-ATPase activity: hesdgroup specificity and fatty acid fluidity. *Biochim. Biophys. Acta* **282:** 277.

———. 1974. Effects of phospholipid acyl chain fluidity, phase transitions, and cholesterol on Na/K-ATPase-stimulated adenosine triphosphatase. *J. Biol. Chem.* **249:** 1071.

Koob, R., M. Zimmerman, W. Schoner, and D. Dreukhahn. 1987. Co-localization and co-precipitation of ankyrin and Na/K-ATPase in kidney epithelial cells. *Eur. J. Cell Biol.* **45:** 230.

Lavarovits, J. and M. Roth. 1988. A single amino acid change in the cytoplasmic domain allows the influenza virus hemagglutinin to be endocytosed through coated pits. *Cell* **53:** 743.

Le Bivic, A., A. Quaroni, B. Nichols, and E. Rodriguez-Boulan. 1991a. Biogenetic pathways of plasma membrane proteins in Caco-2, a human intestinal epithelial cell line. *J. Cell Biol.* **111:** 1351.

Le Bivic, A., Y. Sambuy, K. Mostov, and E. Rodriguez-Boulan. 1990. Vectorial targeting of an endogenous apical membrane sialoglycoprotein and uvomorulin in MDCK cells. *J. Cell Biol.* **110:** 1533.

Le Bivic, A., Y. Sambuy, A. Patak, M. Patil, M. Chao, and E. Rodriguez-Boulan. 1991b. An internal deletion in the cytoplasmic tail reverses the apical localization of human NGF receptor in transfected MDCK cells. *J. Cell Biol.* **115:** 607.

Lisanti, M.P., I.W. Caras, M.A. Davit, and E. Rodriguez-Boulan. 1989. A glycophospholipid membrane anchor acts as an apical targeting signal in polarized epithelial cells. *J. Cell Biol.* **109:** 2145.

Matlin, K.S. and K. Simons. 1984. Sorting of an apical plasma membrane glycoprotein occurs before it reaches the cell surface in cultured epithelial cells. *J. Cell Biol.* **99:** 2131.

Matlin, K., D.F. Bianton, M. Pesonen, D. Louvard, N. Genty, and K. Simons. 1983. Transepithelial transport of a viral membrane glycoprotein implanted into the apical plasma membrane of Madin-Darby canine kidney cells. I. Morphological evidence. *J. Cell Biol.* **97:** 627.

Matter, K., M. Brauchbar, K. Bucher, and H.-P. Hauri. 1990. Sorting of endogenous plasma membrane proteins occurs from two sites in cultured human intestinal epithelial cells (Caco-2). *Cell* **60:** 429.

McNeill, H., M. Ozawa, R. Kemler, and W.J. Nelson. 1990. Novel function of the cell adhesion molecule uvomorulin as an inducer of cell surface polarity. *Cell* **62:** 309.

Morrow, J.S., C.D. Cianci, T. Ardito, A.S. Mann, and M. Kashgarian. 1989. Ankyrin links fodrin to the alpha subunit of Na/K-ATPase in Madin-Darby canine kidney cells and in intact renal tubule cells. *J. Cell Biol.* **108:** 455.

Mostov, K., G. Apodaca, B. Aroeti, and C. Okamoto. 1992. Plasma membrane protein sorting in polarized epithelial cells. *J. Cell Biol.* **116:** 577.

Nagafuchi, A., M. Takeichi, and S. Tsukita. 1991. The 102 kd cadherin-associated protein: Similarity to vinculin and posttranscriptional regulation of expression. *Cell* **65:** 849.

Nelson, W.J. and R.W. Hammerton. 1989. A membrane-cytoskeletal complex containing Na/K-ATPase, ankyrin, and fodrin in Madin-Darby canine kidney (MDCK) cells: Implications for the biogenesis of epithelial cell polarity. *J. Cell Biol.* **108:** 893.

Nelson, W.J. and P.J. Veshnock. 1986. Dynamics of membrane-skeleton (fodrin) organization during development of polarity in Madin-Darby canine kidney epithelial cells. *J. Cell Biol.* **103:** 1751.

———. 1987a. Ankyrin binding to Na + ,K + -ATPase and implications for the organization of membrane domains in polarized cells. *Nature* **328:** 533.

———. 1987b. Modulation of fodrin (membrane skeleton) stability by cell-cell contact in Madin-Darby canine kidney epithelial cells. *J. Cell Biol.* **104:** 1527.

Nelson, W.J., R.W. Hammerton, A.Z. Wang, and E.M. Shore. 1990a. Involvement of the membrane-cytoskeleton in the development of epithelial cell polarity. *Semin. Cell Biol.* **1:** 359.

Nelson, W.J., E.M. Shore, A.Z. Wang, and R.W. Hammerton. 1990b. Identification of a membrane-cytoskeletal complex containing the cell adhesion molecule uvomorulin (E-cadherin), ankyrin, and fodrin in Madin-Darby canine kidney epithelial cells. *J. Cell Biol.* **110:** 349.

Ojakian, G.K. and R. Schwimmer. 1988. The polarized distribution of an apical cell surface glycoprotein is maintained by interactions with the cytoskeleton of Madin-Darby canine kidney cells. *J. Cell Biol.* **107:** 2377.

Ozawa, M., M. Ringwald, and R. Kemler. 1990. Uvomorulin-catenin complex formation is regulated by a specific domain in the cytoplasmic region of the cell adhesion molecule. *Proc. Natl. Acad. Sci.* **87:** 4246.

Pasdar, M. and W.J. Nelson. 1988a. Kinetics of desmosome assembly in Madin-Darby canine kidney epithelial cells: Temporal and spatial regulation of desmoplakin organization and stabilization upon cell-cell contact. I. Biochemical analysis. *J. Cell Biol.* **106:** 677.

———. 1988b. Kinetics of desmosome assembly in Madin-Darby canine kidney epithelial cells: Temporal and spatial regulation of desmoplakin organization and stabilization upon cell-cell contact. II. Morphological analysis. *J. Cell Biol.* **106:** 687.

———. 1989. Regulation of desmosome assembly in epithelial cells: Kinetics of synthesis, transport, and stabilization of desmoglein I, a major protein of the membrane core domain. *J. Cell Biol.* **109:** 163.

Pearce, B.M.F. and M. Robinson. 1990. Clathrin, adaptors, and sorting. *Annu. Rev. Cell Biol.* **6:** 151.

Penn, E.J., C. Hobson, D.A. Rees, and A.I. Magee. 1987. Structure and assembly of desmosome junctions: Biosynthesis and turnover of the major desmosome components of Madin-Darby canine kidney cells in low calcium medium. *J. Cell Biol.* **105:** 2327.

Rindler, M.J., I.E. Ivanov, and D. Sabatini. 1987. Microtubule-acting drugs lead to the nonpolarized delivery of the influenza hemagglutinin to the cell surface of polarized Madin-Darby canine kidney cells. *J. Cell Biol.* **104:** 231.

Rindler, M.J., I.E. Ivanov, H. Plesken, and D.D. Sabatini. 1985. Polarized delivery of viral glycoproteins to the apical and basolateral membranes of Madin-Darby canine kidney cells infected with temperature sensitive viruses. *J. Cell Biol.* **104:** 231.

Rindler, M.J., I.E. Ivanov, M. Plesken, E. Rodriguez-Boulan, and D.D. Sabatini. 1984. Viral glycoproteins destined for apical or basolateral plasma membrane domains traverse the same Golgi apparatus during their intracellular transport in doubly infected Madin-Darby canine kidney cells. *J. Cell Biol.* **98:** 1304.

Rodriguez-Boulan, E. and W.J. Nelson. 1989. Morphogenesis of the polarized epithelial cell phenotype. *Science* **245:** 718.

Rodriguez-Boulan, E. and D.D. Sabatini. 1978. Asymmetric budding of viruses in epithelial monolayers: A model system for the study of epithelial cell polarity. *Proc. Natl. Acad. Sci.* **75:** 5071.

Salas, P.J., D.E. Misek, S.D. Vega, D. Gundersen, M. Cereijido, and E. Rodriguez-Boulan. 1986. Microtubules and actin filaments are not critically involved in the biogenesis of epithelial cell surface polarity. *J. Cell Biol.* **102:** 1853.

Sargiacomo, M., M. Lisanti, L. Graeve, A. Le Bivic, and E. Rodriguez-Boulan. 1989. Integral and peripheral protein composition of the apical and basolateral membrane domains in MDCK cells. *J. Membr. Biol.* **107:** 277.

Simons, K. and S.D. Fuller. 1985. Cell surface polarity in epithelia. *Annu. Rev. Cell Biol.* **1:** 243.

Simons, K. and A. Wandinger-Ness. 1990. Polarized sorting in epithelia. *Cell* **62:** 207.

Skibbens, J.E., M. Roth, and K.S. Matlin. 1989. Differential extractability of influenza virus hemagglutinin during intracellular transport in polarized epithelial cells and nonpolar fibroblasts. *J. Cell Biol.* **108:** 821.

van der Sluijs, P., M.K. Bennett, C. Antony, K. Simons, and T.E. Kreis. 1990. Binding of exocytic vesicles from MDCK cells to microtubules in vitro. *J. Cell Sci.* **95:** 545.

van Meer, G. and K. Simons. 1988. Lipid sorting in epithelial cells. *Biochemistry* **27:** 6197.

Wandinger-Ness, A., M. Bennett, C. Antony, and K. Simons. 1990. Distinct transport vesicles mediate the delivery of plasma membrane proteins to the apical and basolateral domains of MDCK cells. *J. Cell Biol.* **111:** 987.

Wang, A.Z., G.K. Ojakian, and W.J. Nelson. 1990. Steps in the morphogenesis of a polarized epithelium. I. Uncoupling the roles of cell-cell and cell substratum contact in establishing plasma membrane polarity in multicellular epithelial (MDCK) cysts. *J. Cell Sci.* **95:** 137.

Wollner, D.A. and W.J. Nelson. 1992. Establishing and maintaining epithelial cell polarity: Roles of protein sorting, delivery and retention. *J. Cell Sci.* **102:** 185.

Wollner, D.A., K.A. Krzeminski, and W.J. Nelson. 1992. Remodelling the cell surface distribution of membrane proteins during the development of epithelial cell polarity. *J. Cell Biol.* **116:** 889.

Wright, E.M. 1972. Mechanisms of ion transport across the choroid plexus. *J. Physiol.* **226:** 545.

Yokode, M., R. Pathak, R.E. Hammer, M.S. Brown, J.L. Goldstein, and R.G.W. Anderson. 1992. Cytoplasmic sequence required for basolateral targetting of LDL receptor in livers of transgenic mice. *J. Cell Biol.* **117:** 39.

Cytoplasmic Control of Cell Adhesion

B. GEIGER,* O. AYALON,* D. GINSBERG,* T. VOLBERG,*
J.L. RODRÍGUEZ FERNÁNDEZ,† Y. YARDEN,* AND A. BEN-ZE'EV†
*Departments of *Chemical Immunology and †Molecular Genetics and Virology, The Weizmann Institute of Science, Rehovot, 76100, Israel*

Cell adhesion involves receptor-mediated cell-surface interactions with various immobilized extracellular "ligands" present either on the surface of neighboring cells or on extracellular matrix networks. These interactions play a central role in cytoskeletal organization and tissue morphogenesis as well as in cell motility, growth, and differentiation (Vasiliev and Gelfand 1981; Edelman et al. 1990; Ben-Ze'ev 1991; Takeichi 1991). These short- and long-range effects are especially prominent in adherens-type junctions (AJ), a class of cell adhesions that are associated with the microfilament system through an elaborate network of interconnecting anchor molecules (Geiger 1982, 1989; Geiger et al. 1985, 1990; Geiger and Ginsberg 1991). During the last decade, many AJ constituents were discovered and shown to segregate into three major subcellular domains. The *junctional receptors* are transmembrane polypeptides belonging either to the integrin family, in extracellular matrix adhesions (Hynes 1987, 1992), or to the cadherin family, in cell-cell contacts (Takeichi 1988, 1990, 1991; Takeichi et al. 1989). The cell-type-specific diversity of these receptors has a profound apparent effect on the specificity of adhesive interactions.

The second junctional domain consists of *cytoskeletal anchor proteins*, molecules whose main function is to link (directly or indirectly) microfilaments to the cytoplasmic aspects of the junctional membrane. Some of these molecules (e.g., vinculin) are apparently present in all AJ, whereas others (e.g., talin or catenins) show preferential association with either matrix or intercellular junctions (Geiger et al. 1985; Geiger and Ginsberg 1991). The *cytoskeletal elements* of AJ, mostly actin and actin-associated proteins, appear to be ubiquitously present in all these junctions, irrespective of the nature of the particular adhesive surface components (Geiger et al. 1985; Geiger and Ginsberg 1991).

During the process of junction formation, these proteins are recruited from other subcellular domains and assemble in the contact site; the membrane receptors diffuse laterally and segregate into the newly formed junctions, and the plaque and cytoskeletal constituents bind to the cytoplasmic moieties of these receptors. It appears that the primary event which triggers this cascade of molecular interactions is the binding of the adhesion molecules to the polyvalent and immobilized extracellular ligands, largely as proposed in a Cold Spring Harbor Symposium more than a decade ago (Geiger 1982). In the present paper, we extend this model and show that the formation of AJ involves an *initiation* step, triggered by the extracellular interaction, followed by a *maturation* step, which depends on the binding of the surface receptors to the plaque and cytoskeletal domains and is apparently critical for the formation of stable adhesions. We show here that modulation of the maturation step has a dramatic effect on cell structure and may occur through changes in both the absolute and relative amounts of the different junctional molecules, as well as through their posttranslational modification (mostly phosphorylation).

MATERIALS AND METHODS

Molecular genetic techniques. Unless otherwise indicated, the molecular genetic techniques employed here were carried out according to the method of Sambrook et al. (1989), and most enzymes were purchased from Boehringer Mannheim (Germany). Details concerning the preparation of the various cDNA constructs encoding N-cadherin (Hatta and Takeichi 1986), integrin (Tamkun et al. 1986), vinculin (Bendori et al. 1989; Price et al. 1989), and the β_1 cadherin/integrin chimeric receptors (Geiger et al. 1992) are described in detail elsewhere. Transfection of cDNA (cloned into pECE or pJ4Ω expression vectors [Ellis et al. 1986; Shaulsky et al. 1991]) into Chinese hamster ovary (CHO) or 3T3 cells was carried out using the calcium phosphate method and transfectants selected with 600 μg/ml G-418 (Gibco, USA).

Immunochemistry. Antibodies used in the present study include monoclonal anti-N-cadherin (anti-A-CAM, clone ID-7.2.3; Volk and Geiger 1984), which specifically reacts with the chick protein. Anti-β_1-integrin was a rabbit antibody (R-363) prepared against the 36 carboxy-terminal amino acids of the β_1 chain of integrin (Marcantonio and Hynes 1988). Vinculin antibodies were either monoclonal (Sigma Immunochemicals, USA) or prepared in rabbits (Geiger 1979). Mouse monoclonal antibodies, reactive with phosphotyrosine (P-Tyr), were prepared by immunization of BALB/c mice with KLH-P-Tyr as described previously (Heffetz and Zick 1989). Affinity-purified rabbit anti-P-Tyr antibodies were generated as described previously (Heffetz and Zick 1989); rhodamine-labeled phalloidin was purchased from Sigma (St. Louis, Missouri). Antibodies to the *neu* oncogene, to *raf-1*, and to phosphatidylinositol (PI) 3′-kinase (85-kD subunit) were prepared as described previously (Peles et al. 1992a,b). Secondary fluorescent antibodies were purchased from Jackson (West Grove, Pennsylvania). Immunofluorescence microscopy was carried out using an Axiophot microscope (Zeiss, Germany) equipped for fluorescein and rhodamine examination.

Cultured cells and tumorigenicity assay. BALB/c-3T3 clone A31 cells and SVT2 cells (Aaronson and Todaro 1968), which are transformed by SV40, were grown in Dulbecco's modified Eagle's medium (DMEM) plus 10% calf serum (Gibco, Grand Island, New York). MDCK, MDBK, and MCF7 cell lines were originally obtained from the American Type Culture Collection. Chicken lens cells were prepared as described previously (Volk and Geiger 1986a,b).

For examination of tumorigenic activity, SVT2 cells and various transfectants were injected subcutaneously into syngeneic BALB/c mice at either 10^6 or 5×10^6 cells per animal. The appearance and size of tumors and the survival time were subsequently determined.

Immunoprecipitation and polyacrylamide gel electrophoresis. For immunoprecipitation, cells were labeled with 200 μC/ml of [^{35}S]methionine for 4 hours. The cells were lysed in RIPA buffer (1% NP-40, 0.5% deoxycholate, 0.1% SDS, 150 mM NaCl, 5 mM EDTA, 25 mM Tris [pH 8.0]). The lysates were clarified by centrifugation at 15,000*g*, and equal amounts of radioactive proteins (5×10^6 cpm) were incubated with the chicken vinculin-specific antibody. The immune complexes were precipitated with *Staphylococcus aureus* (The Enzyme Center, Boston, Massachusetts). The proteins were then separated on 8% SDS-containing polyacrylamide gels.

Two-dimensional gel electrophoresis was performed as described previously (Ben-Ze'ev 1990). [^{35}S]Methionine-labeled proteins (5×10^7 cpm) were separated by two-dimensional gel electrophoresis, transferred to nitrocellulose by electroblotting, and visualized by autoradiography. The positions of the transfected and the endogenous vinculins on the two-dimensional gels were identified by immunoblotting with the broad-range monoclonal anti-vinculin antibody, followed by alkaline phosphatase-linked anti-mouse IgG (Promega, Madison, Wisconsin). The levels of mouse and chicken vinculin, which were separated on the two-dimensional gels, were determined using a laser densitometer and computerized analysis of the images with the QUEST software system as described previously (Garrels 1989).

H_2O_2/vanadate pretreatment. Cultured cells on glass coverslips were washed with serum-free DMEM. H_2O_2 (2 mM) and sodium orthovanadate (1 mM) in serum-free medium were added to the cells for a period of 15 minutes, or as indicated, at 37°C in a humidified incubator. For the examination of long-term effects, cells were pretreated with H_2O_2/vanadate for 15 minutes, then washed once with 10% FCS-containing medium and further incubated in serum-containing medium for different periods of time. The treatment was terminated by washing the cells with PBS, followed by fixation.

RESULTS AND DISCUSSION

The common view of AJ structure and formation suggests that the assembly of the various junctional domains is controlled by the modulation of either the amount or availability of a particular junctional component that is rate-limiting during junction formation. However, as is shown below, independent changes in different junctional molecules may profoundly affect adhesive interactions. This is in line with the idea that AJ are highly dynamic structures (Kreis et al. 1982) and their formation is a highly cooperative process (Bendori et al. 1989; Geiger et al. 1990). Several examples for such perturbations are described below.

Effect of Quantitative and Qualitative Modulation of Cadherins on Junction Formation

A primary role in the induction of AJ formation is attributed to the junctional receptors. This was most convincingly demonstrated for cadherins by the capacity of transfection of mesenchymal cells with the respective cDNA clones to induce cell aggregation and acquisition of an epithelioid morphology. In fact, such transfections recently became the standard bioassay for adhesion molecule activity. We have confirmed these observations for different cadherins, including chick, human, and zebrafish N-cadherin and *Xenopus* EP-cadherin (Ginsberg et al. 1991; Salomon et al. 1992; S. Bitzur et al., unpubl.). Furthermore, examination of various EP-cadherin-transfected CHO clones pointed to a high variability in the amounts of the expressed protein, which had a profound effect on acquired morphology. At low levels of cadherin, morphological changes were hardly detected (Fig. 1, compare a to b); cells expressing moderate levels of the protein underwent an apparent epithelialization and spreading (Fig. 1c); and clones expressing relatively high amounts of EP-cadherin were significantly more condensed (Fig. 1d). This observation indicated not only that this cadherin has a morphogenetic capability, but also that the acquired phenotype may be greatly affected by the level of the expressed cadherin. This finding is consistent with other reports indicating that cells expressing different amounts of the same cadherin can sort out from each other in a mixed culture (Friedlander et al. 1989). These results also imply that other essential junctional components (catenins, for example) are either constitutively expressed in cells, irrespective of their cadherin levels, or the transfection with cadherin cDNA stimulates the expression of these components.

Immunocytochemical labeling of cadherin-transfected mesenchymal cells confirmed that the foreign protein was indeed associated with adhesion sites (Fig. 2b, compare to *neo*-transfected cells in Fig. 2a). The clear

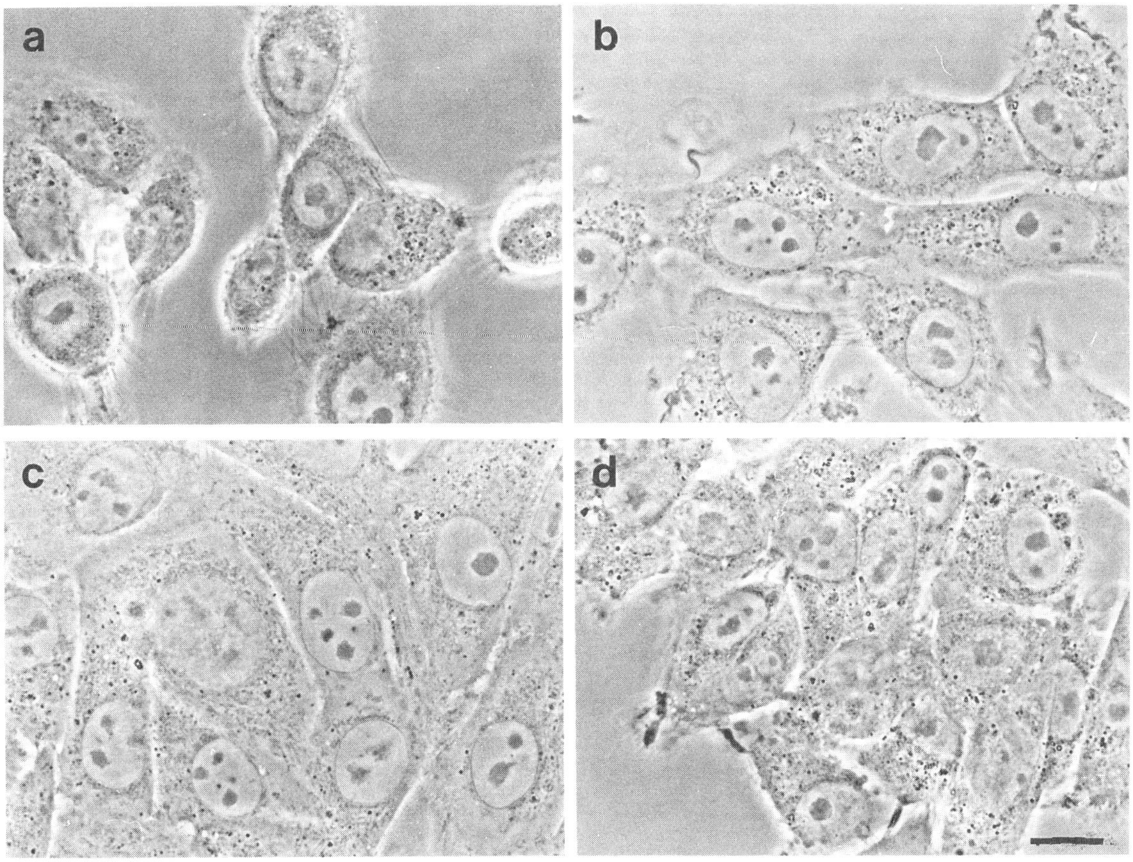

Figure 1. Phase-contrast microscopy of CHO cells transfected with cDNAs encoding the *neo*-resistance selective marker (*a*) or EP-cadherin (*b–d*). The cadherin-transfected clones that were selected for structural analysis expressed either low levels (*b*), moderate levels (*c*), or high levels (*d*) of EP-cadherin. Notice that at low levels of exogenous cadherin, the effect on cell morphology is very limited, whereas at higher EP-cadherin levels, the cells flatten down and form epithelioid sheets. At higher levels of EP-cadherin expression, cell clustering was noticed. Bar, 25 μm.

association of cadherins with AJ enabled a detailed examination of functional sites along these molecules. It has been shown, for example, that cadherins missing either amino-terminal or carboxy-terminal segments lose the capacity to mediate cell adhesion (Nagafuchi and Takeichi 1988; Ozawa et al. 1989). The requirement for an intact cytoplasmic domain is also demonstrated in Figure 2c, showing immunolabeling of CHO cells, transfected with a truncated N-cadherin cDNA, lacking most of the intracellular region. These findings confirm the notion that interactions of cadherins with the membrane-bound plaque are essential for an effective and long-term assembly of AJ. This property could be attributed to the capacity of the cytoskeletal and plaque components to recruit and stabilize the transmembrane receptors and thus promote multivalent, high-affinity interactions.

To directly challenge this hypothesis, we have transfected different cell types with a cDNA encoding a chimeric receptor consisting of the cytoplasmic and transmembrane regions of β_1-integrin and the extracellular domain of N-cadherin. A complete report on the distribution of the hybrid molecule is given elsewhere (Geiger et al. 1992), and only the major finding is outlined here. Immunolabeling of the transfected 3T3 cells with anti-chicken N-cadherin antibodies indicated that the primary sites of organization of the cadherin β_1-integrin hybrid molecule were cell-substrate focal contacts (Fig. 2e,f). In these sites its distribution was largely similar to that of chicken β_1-integrin transfected into the same cells (Fig. 2d), suggesting that adhesion molecules such as integrins may be recruited into existing adhesions due to interaction of their cytoplasmic domains with components of the junctional plaque. Recent studies with chimeric receptors bearing the integrin cytoplasmic domain confirm these findings (LaFlamme et al. 1992). Cells expressing relatively high levels of the chimeric protein exhibited increased intercellular adhesion and contained the chimeric protein both in cell-cell and cell-matrix AJ (Fig. 2e,f). This confirms that effective cadherin mediated adhesion requires an intact cytoplasmic domain, which may provide an anchorage to the cytoskeleton. This domain, however, may be substituted with another actin-associated region, such as the intracellular region of the integrin β_1 chain (Hayashi et al. 1990; Marcantonio et al. 1990; Otey et al. 1990).

On the basis of these observations, we propose that

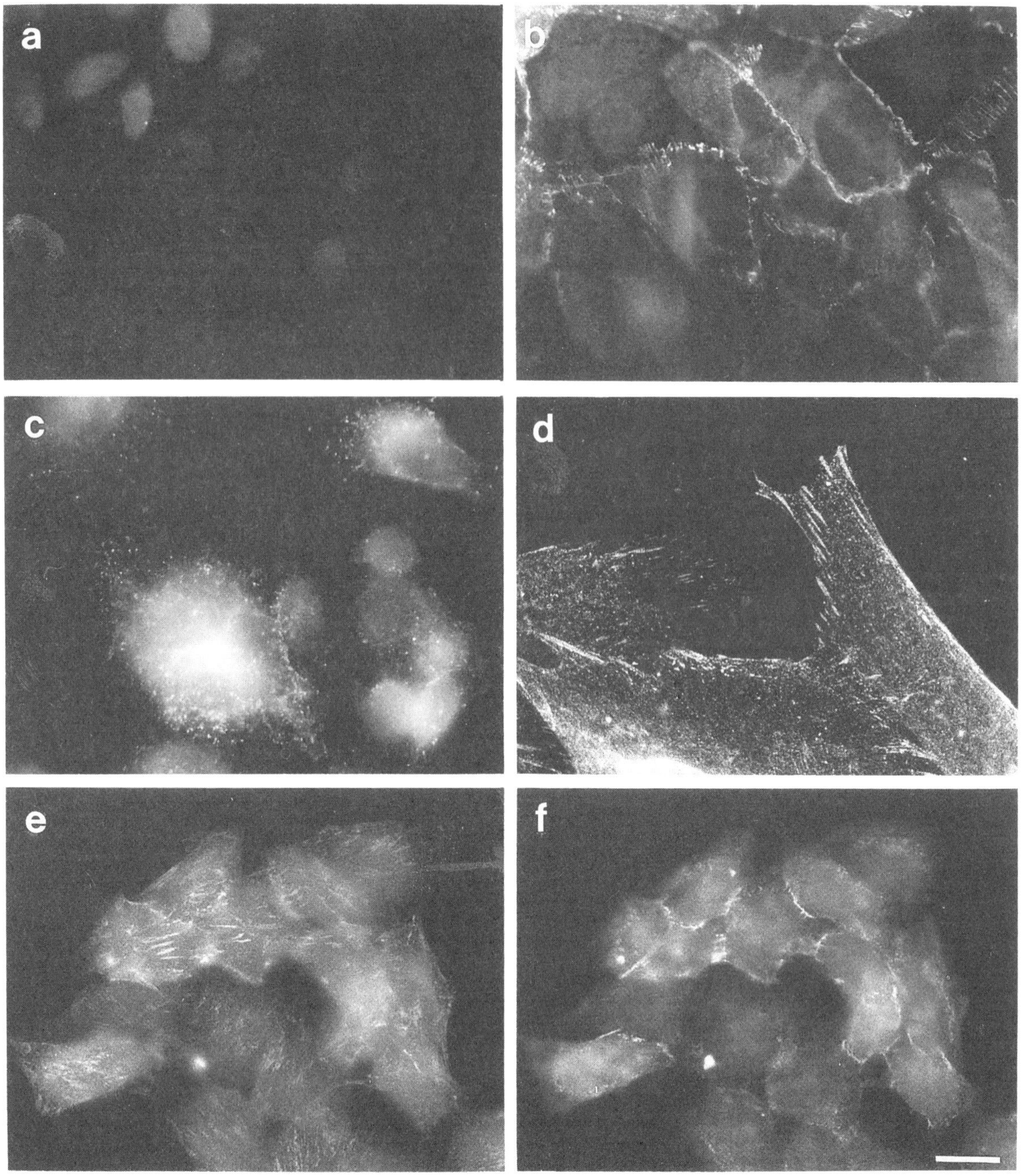

Figure 2. Immunofluorescent labeling of CHO cells (*a*) and several stably transfected CHO lines (*b–f*). The cDNAs used for transfection were all cloned into the pECE eukaryotic expression vector. They encode the following proteins: (*b*) full-length chicken N-cadherin; (*c*) truncated chicken N-cadherin, missing most of the cytoplasmic domain; (*d*) full-length chicken β_1-integrin; (*e,f*) a chimeric receptor consisting of the extracellular domain of chicken N-cadherin and the cytoplasmic and transmembrane domains of β_1-integrin, photographed at ventral (*e*) and dorsal (*f*) focal planes. The antibodies used for labeling included chicken-specific anti-N-cadherin (*a,b,c,e,f*) and anti-chicken β_1-integrin (*d*). Note that the intact cadherins and β_1-integrin display a mutually exclusive distribution. The truncated cadherin loses the capacity to mediate cell adhesion whereas the chimeric receptor becomes associated both with focal contacts formed with the matrix and with cell-cell adhesions. Bar, 10 μm.

the formation of AJ occurs in two distinct steps: *initiation* and *maturation*. The former is most likely triggered by the interaction of the adhesion receptor with the extracellular ligand, leading to the assembly of a membrane- and cytoskeleton-bound plaque. Consequently, AJ initiated by a single or a few adhesion molecules may contain enough plaque and cytoskeletal components to bind additional adhesion molecules. This is in line with the fact that the assembly of microfilaments and associated proteins is critical for stable adhesion and that the interaction between the cytoplasmic molecules of AJ (e.g., vinculin, talin, α-actinin, and actin) is a highly cooperative process with a capacity to efficiently recruit junctional molecules from diffusible pools throughout the cell and bind them to the contact site (Bendori et al. 1989; Geiger et al. 1990).

Effect of Modulation of Vinculin Levels on Cell Behavior

The studies described above pointed to the central role of the submembranal plaque of AJ both in the formation of membrane-microfilament interactions and in the stabilization of these adhesions. Thus, alterations in the quantities of plaque constituents, their availability, and their capacity to interact with the membrane and cytoskeletal proteins may have a profound effect on cell adhesion and the establishment of a stable junction. Of the various known plaque proteins, we have chosen vinculin, since this protein appears to play a pivotal role in the assembly of the junctional plaque (Bendori et al. 1989). In addition, the level of vinculin expression was shown to be modulated by a variety of environmental factors, such as alterations in substrate adhesiveness (Ungar et al. 1986; Bendori et al. 1987), induction of differentiation (Ben-Ze'ev and Amsterdam 1987), or growth stimulation (Bellas et al. 1991; Glück et al. 1992).

To experimentally determine the effect of modulation of vinculin levels on AJ formation, we transfected cDNA encoding the chicken vinculin into different cultured cells and examined clones stably expressing the foreign protein. The properties of these transfectants were described in detail elsewhere (Rodríguez Fernández et al. 1992a,b), and the major findings are briefly summarized below.

1. Expression of chick vinculin in 3T3 cells resulted in a

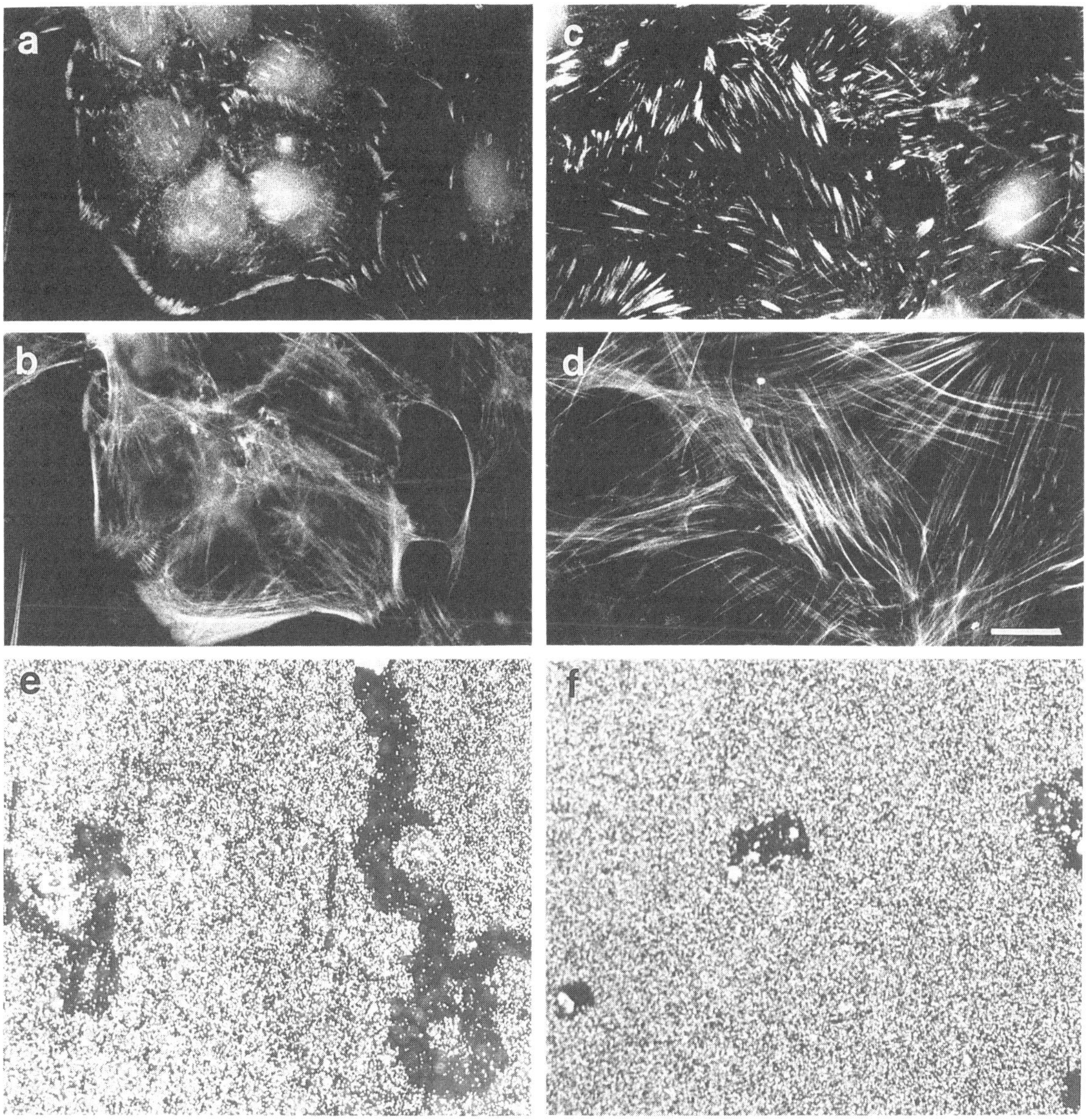

Figure 3. Double fluorescent labeling for vinculin (*a*,*c*) and actin (*b*,*d*) as well as dark-field visualization of phagokinetic tracks (*e*,*f*) of 3T3 cells transfected with chicken vinculin cDNA (*c*,*d*,*f*) or only with the *neo*-resistance selective marker (*a*,*b*,*e*). The level of the exogenous chicken vinculin in the former cells is about 20% relative to the endogenous protein. Notice that the vinculin transfectants have considerably larger focal contacts and stress fibers and that their motility is suppressed. Bar, 10 μm.

significant increase in the number and size of focal contacts (Fig. 3a,c) and stress fibers (Fig. 3b,d). It is noteworthy that although the transfected cells contained only 20% chicken vinculin (relative to the level of the endogenous protein), the apparent effect on focal contacts and stress fibers was conspicuous.

2. The motility of the vinculin-transfected cells was dramatically suppressed, as shown by examination of the phagokinetic tracks left behind moving cells (Fig. 3e,f) and by the rate of closure of a "wound" introduced in a confluent cell monolayer (Rodríguez Fernández et al. 1992a).
3. The level of expression of chicken vinculin in the transfected 3T3 cells was never higher than ~20% (relative to the endogenous protein), regardless of the vector system used or the mode of transfection.
4. Several transformed cell lines (e.g., the SV40-transformed SVT2 line and the rat adenocarcinoma line Bsp73-ASML [ASML]) contain diminished levels or no vinculin, respectively (Rodríguez Fernández et al. 1992b). Expression of transfected chicken vinculin in these cells resulted in a dramatic suppression of their capacity to form tumors in syngeneic hosts or in nude mice, as well as a decrease in their ability to grow in semisolid medium (Fig. 4 and the inserted table).
5. The level of expression of chicken vinculin in both transformed lines was considerably higher than that obtained in normal 3T3 cells. Vinculin levels (exogenous plus endogenous) within the transformed cells increase, therefore, to approximately the same level found in nontransformed cells. In fact, suppression of tumorigenicity was achieved only in cells expressing vinculin at levels comparable to those found in nontransformed 3T3 cells.
6. As previously shown, the level of vinculin expression in cells is markedly affected by environmental conditions (Ben-Ze'ev 1991). As shown in Figure 5, serum stimulation of quiescent 3T3 cells transfected with chicken vinculin leads to a four- to fivefold increase in the rate of synthesis of the endogenous protein. Interestingly, densitometric scanning of the autoradiographs shown in Figure 5 indicated that there was a small, yet significant and reproducible increase also in the level of the chick vinculin, whose transcription is controlled by the constitutive LTR promoter (Fig. 5). These and previous results suggest that although the regulation of vinculin levels occurs mainly at the level of transcription (Ben-Ze'ev et al. 1990; Bellas et al. 1991), posttranscriptional events may also contribute to the apparent changes. The mechanisms underlying these regulatory stages are still unclear, however. Given the fact that the cytoplasmic plaque has an important role in the establishment of stable contacts, it is conceivable that modulation of its formation might significantly affect cell adhesion. It still remains to be determined whether qualitative differences between the transfected chick vinculin and the endogenous mouse protein are responsible for the dramatic changes in cell behavior, or whether the levels of vinculin in the cells are tightly controlled and even small perturbations may lead to rather dramatic phenotypic changes.

Involvement of Tyrosine Phosphorylation in the Modulation of AJ

Cell adhesions are believed to be molecularly dynamic structures whose constituents are present in either junctional (immobile) or extrajunctional (diffusible) pools between which a dynamic equilibrium is maintained (Kreis et al. 1982, 1984). Shifts in this equilibrium in one direction or the other may thus lead either to extension of junctions and increased adhesiveness, or to junction disassembly. Such changes may occur due to modulation of the levels of junctional constituents, as discussed above, or to structural modifications of these components. In view of the apparent anti-adhesive effect of oncogenic tyrosyl kinases and the association of these enzymes with AJ, we have focused here on the effect of tyrosine phosphorylation on AJ.

The effect of tyrosine phosphorylation in normal cells is poorly understood, mainly because P-Tyr is hardly detectable in them. This is attributable not only to the low levels of tyrosine kinase activity, but mainly to the rapid dephosphorylation by P-Tyr-specific phosphatases. As previously shown, addition of vanadate and H_2O_2 (pervanadate) to cells results in a rapid increase in P-Tyr levels, and the staining of cells with anti-P-Tyr antibodies is predominantly apparent at intercellular AJ (Volberg et al. 1991, 1992). As demonstrated in Figure 6, the primary phosphorylation sites in a variety of epithelial cells are the intercellular junctions, with little or no labeling of cell-matrix adhesion plaques. In mesenchymal cells such as 3T3, focal contacts became heavily labeled following 5–15 minutes of pervanadate treatment. Moreover, staining of cells for actin, vinculin, and cadherin suggested that cell contacts were augmented due to the short pervanadate treatment. On the other hand, longer incubation of cells, following pervanadate stimulation, resulted in dramatic changes in the organization of the cytoskeleton and deterioration of the adhesion sites: In MDCK epithelial cells, intercellular junctions gradually disassembled concomitant with a marked enhancement of focal contact and stress fiber formation (Volberg et al. 1991). These effects were already noticed after 30 minutes of incubation, and a full effect was obtained after 2 hours (Fig. 7). In mesenchymal cells, where phosphorylation was typically associated with focal adhesions, long incubation (>30 min) with pervanadate usually resulted in massive cell detachment.

AJ as Signal Transduction Sites

Cell adhesion is known to have not only short-range effects on the structure of cells, but also long-range

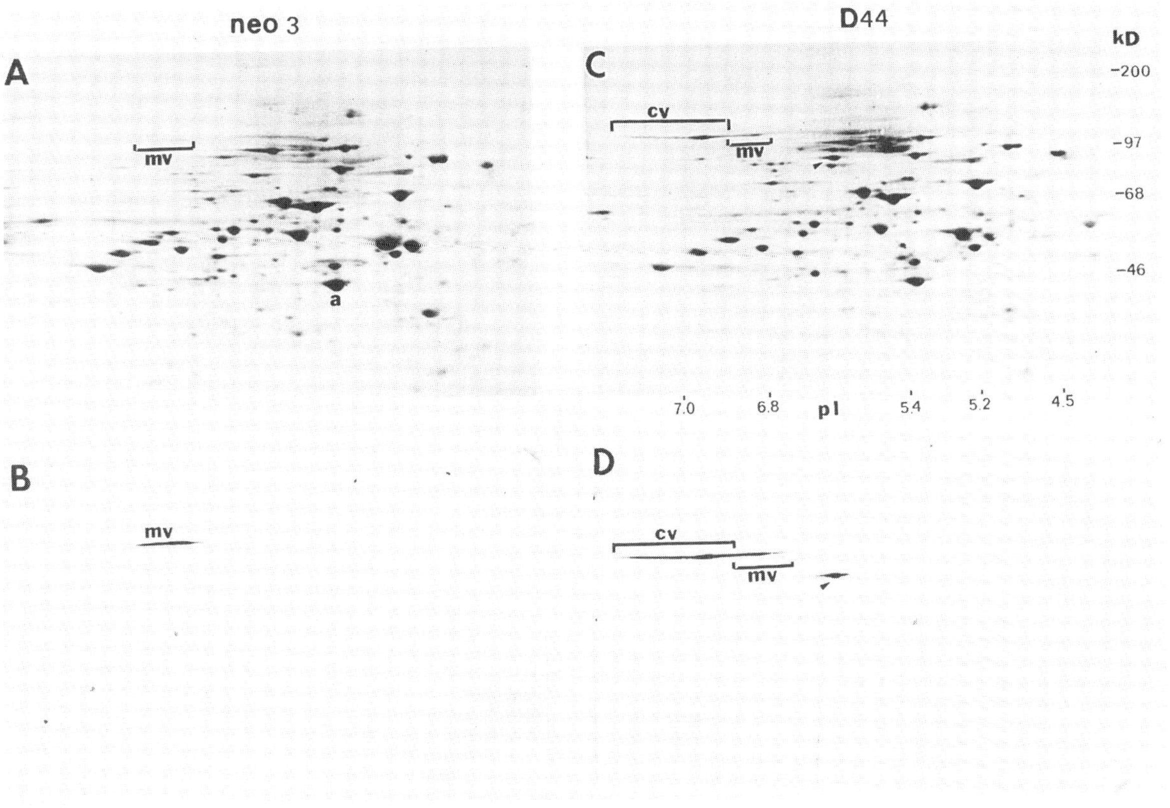

Tumorigenicity of SVT2 Clones Expressing Different Vinculin Levels

Clone	Vinculin Level* Chicken	Vinculin Level* Mouse	Tumor Incidence (%)
3T3	0	750	0/6 (0)
neo 1	0	ND	6/6 (100)
neo 3	0	140	6/6 (100)
SVT2	0	135	6/6 (100)
D 34	344	170	1/6 (16)
D 43	105	ND	5/6 (83)
D 41	80	ND	5/6 (83)
D 44	670	160	0/6 (0)

Figure 4. Identification and quantitation of chicken vinculin, expressed in transfected, SV40-transformed 3T3 cells. Cells of a neo^r clone (neo3) (*A,B*) and of a chicken vinculin-transfected clone (D44) (*C,D*) were labeled for 16 hr with [^{35}S]methionine and 2.5 $\times$ 10^7 cpm were analyzed by two-dimensional gel electrophoresis. The separated proteins were transferred to nitrocellulose by electroblotting, and their pattern was obtained by autoradiography (*A,C*). Vinculin was identified on the same blots with a broadly specific anti-vinculin monoclonal antibody, followed by alkaline phosphatase-coupled anti-mouse IgG (*B,D*). (mv) Mouse vinculin; (cv) chicken vinculin; (a) actin. Arrowheads in *C* and *D* point to a degradation product of cv. Notice that the cv level in clone D44 is four times higher than that of the endogenous mv level. The attached table summarizes the relative levels of chicken and mouse vinculins in 3T3 cells, as determined by quantitative computerized analysis of two-dimensional gels from [^{35}S]methionine-labeled total cell lysates. The numbers represent arbitrary OD units. ND indicates not done. SV40 transformed 3T3 cells (SVT2) and several transfectants expressing the *neo* resistance gene (neo 1,3) or different amounts of chicken vinculin (D34, D43, D41, and D44). Notice that the level of endogenous vinculin in the SV40-transformed cells is about 20% of its level in 3T3 cells and that in some of the transfectants (D44, for example) normal vinculin levels were restored. Also included in the table is the tumor incidence in syngeneic (BALB/c) mice injected with the various clones. As shown, both SVT2 cells and the *neo*-transfected clones produced tumors in all the injected animals, whereas those transfectants expressing nearly normal vinculin levels lost their tumorigenic capacity.

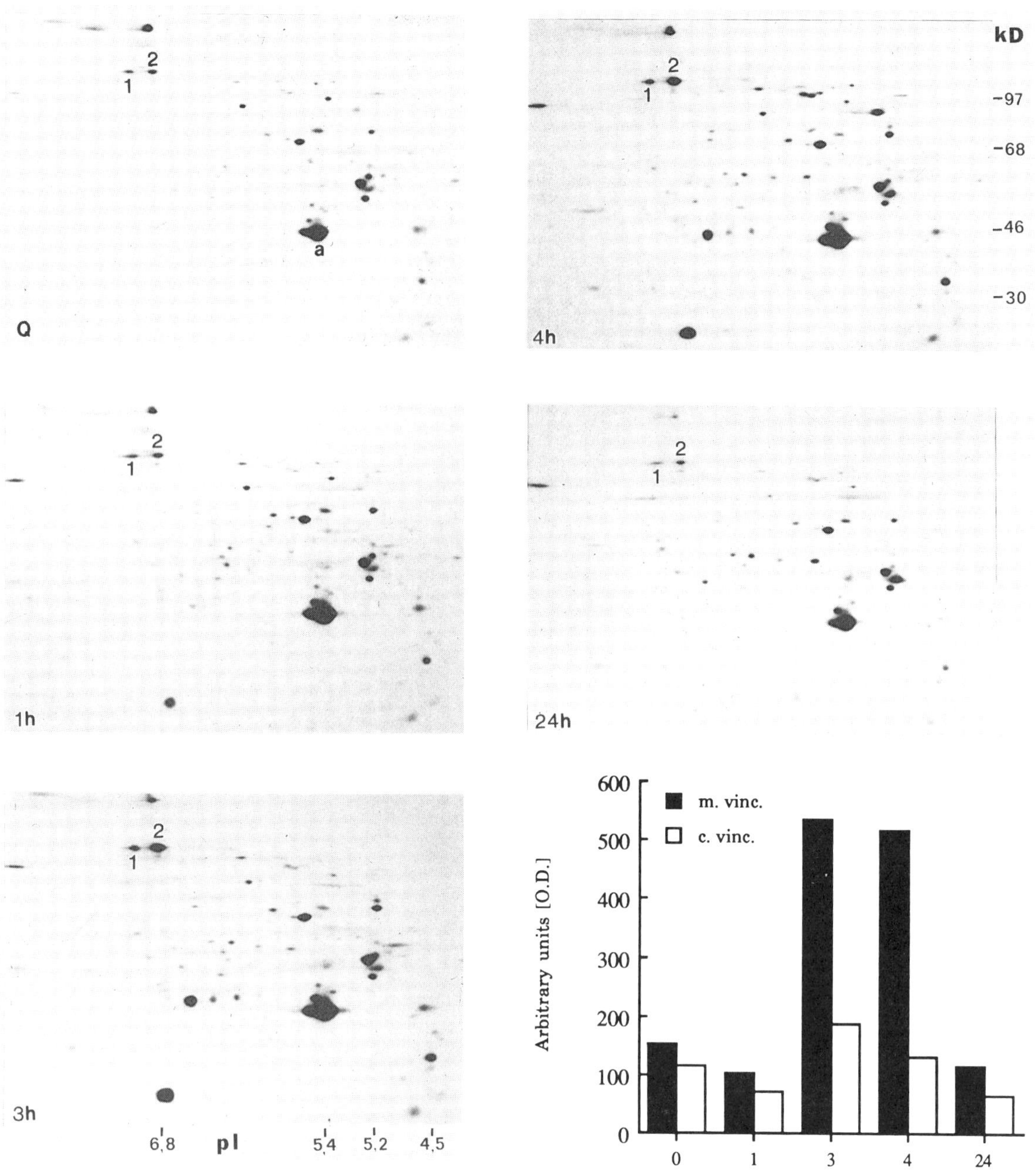

Figure 5. Induction of vinculin synthesis in quiescent 3T3 cells, stimulated with serum. Quiescent, chicken vinculin-transfected, 3T3 cells (*Q*) were treated with 20% calf serum for different times (1–24 hr), and the cells were pulse-labeled for 30 min with [^{35}S]methionine. Total cell radioactive proteins (5 × 10^5 cpm) were separated by two-dimensional gel electrophoresis followed by autoradiography. (*a*) Actin; (1) chicken vinculin; (2) mouse vinculin. The levels of chicken (c. vinc.) versus mouse (m. vinc.) were determined by computerized densitometric analysis of the two-dimensional gel autoradiographs. Notice that the expression levels of the endogenous mouse vinculin transiently increased four- to fivefold about 3–4 hr after serum stimulation. The levels of expression of the foreign (chicken) vinculin showed only a limited (yet consistent) increase at that time point.

effects on cell growth and differentiation. For example, matrix adhesions are generally considered to be essential for normal cell growth, whereas high cell density usually leads to growth arrest. This notion is not conceptually new, yet the molecular mechanism(s) responsible for adhesion-mediated signal transduction is still obscure.

The results discussed in the previous section, indicating that AJ are prominent tyrosine phosphorylation sites, and the increasing list of molecules associated

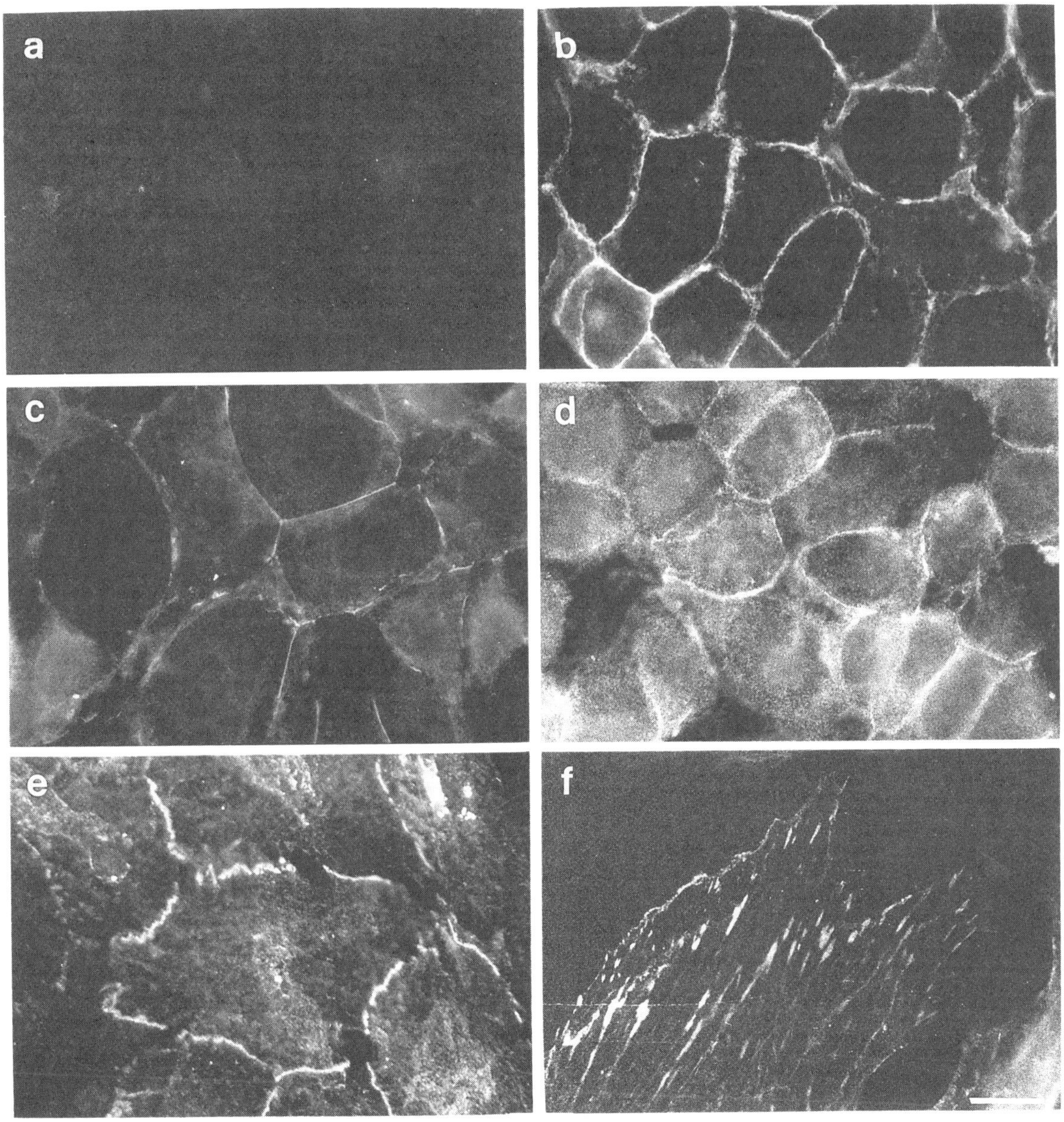

Figure 6. Distribution of P-Tyr-containing proteins in various cultured cells following 15-min treatment with vanadate and H_2O_2 (to suppress protein tyrosine phosphatase activity). For comparison, nontreated cells stained with anti-P-Tyr are shown in *a*. Cells presented here include MDCK (*a,b*), MDBK (*c*); MCF-7 (*d*); chicken lens cells (*e*), and 3T3 cells (*f*). Notice that in all the epithelial cells (*b–e*) the staining is found mostly in the cell-cell junction, whereas in 3T3 fibroblasts, which have poor cell-cell adhesion, prominent labeling is detected along stress fibers. Bar, 10 μm.

with the cytoplasmic domains of these adhesions (Geiger and Ginsberg 1991) allow us to propose a general model for the mechanism of long-range signaling. According to this hypothesis, AJ specifically bind, and locally immobilize, enzymes involved in signal transduction as well as their substrates. It is considered that the assembly of AJ is essential for bringing these entities together, and they thus play an indispensable role in the generation of signals (Fig. 8). For example, in close analogy with the mechanism of action of several growth factors, clustering of receptor-tyrosine kinases in newly formed junctions, due to their binding to membrane or plaque components, may lead to cross-phosphorylation and activation of these enzymes. Junctional kinases may also phosphorylate additional molecules that participate in signal transduction cascades or in the construction of the junction itself, thus leading to activation of signal transduction pathways or modulation of junction structure (see arrows in Fig. 8).

This junctional signaling model is presented here in a rather general form mostly because the molecules involved are still poorly characterized. To identify such molecules, we have recently stained cells with antibodies to several molecules that participate in different signal transduction pathways. The preliminary results,

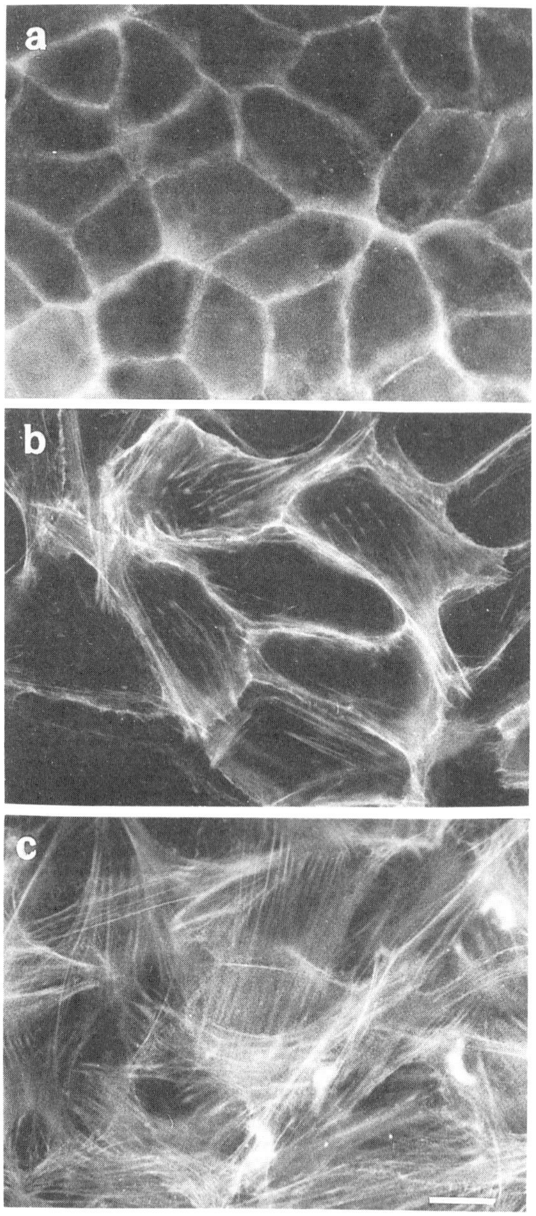

Figure 7. Effect of pervanadate treatment on the microfilament system in MDCK cells. Nontreated cells (*a*) as well as cells pretreated with pervanadate for 15 min and fixed 30 min (*b*) or 2 hr (*c*) later were fluorescently stained for actin. Notice that microfilaments, which were initially associated with cell-cell junctions, underwent dramatic reorganization manifested by detachment from cell-cell AJ and concomitant assembly into an elaborate network of stress fibers. Bar, 10 μm.

depicted in Figure 9, are highly encouraging, as they show that a variety of such molecules indeed localize in different types of AJ. These include the *neu* oncogene, *raf-1*, and PI 3′-kinase. These components join a long list of molecules that show an apparent association with AJ, including protein kinase C (Jaken et al. 1989), various kinases of the *src* family (Rohrschneider 1980; Gentry and Schneider 1984; Rohrschneider and Najita 1984; Tsukita et al. 1991), and phospholipase C (McBride et al. 1991). It is a major future challenge to

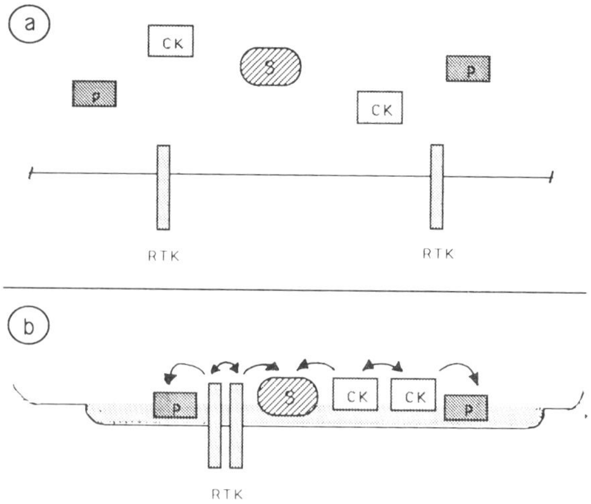

Figure 8. Hypothetical model depicting a possible mode of signal transduction in AJ. In nonattached cells (panel *a*) there are diffusible receptor tyrosine kinases (RTK) in the membrane; various cytoplasmic kinases (CK) capable of phosphorylating tyrosine, serine, or threonine residues, different substrates for these enzymes (s), and junctional plaque proteins (p). Upon the induction of AJ (panel *b*) and the construction of a submembrane plaque (shaded area along the membrane in *b*), these constituents may cluster in the nascent adhesion, thus potentiating a complex network of phosphorylation events (represented by the arrows). These modifications could lead to activation or inactivation of the respective kinases, modification of substrates which then activate signal transduction cascades, and changes in AJ structure (if modification of certain plaque proteins affect their binding properties).

identify the junctional molecules that bind these enzymes and the nature of the junctional substrates that are modified by them. Elucidation of these aspects may shed light on the mechanisms by which junctional signaling is integrated in the overall regulation of cell growth and differentiation.

CONCLUSION

The studies discussed in this paper addressed two major issues: (1) the short-range molecular events which take place at the cytoplasmic faces of AJ and (2) the possible nature of the long-range signals generated in them.

We propose that a key junctional domain which plays a pivotal role in both activities is the submembrane plaque. Through cooperative interactions of its components, the junctional plaque stimulates the extensive assembly of AJ components and mediates the membrane anchorage of microfilament bundles. In addition, the plaque may be involved in the recruitment of adhesion molecules, like integrins and cadherins, and their immobilization in nascent adhesions.

Moreover, we show that the AJ plaque contains several proteins which are involved in long-range signal transduction processes in cells. These include several protein kinases and other elements of signal transduction cascades. It is proposed that the assembly of AJ

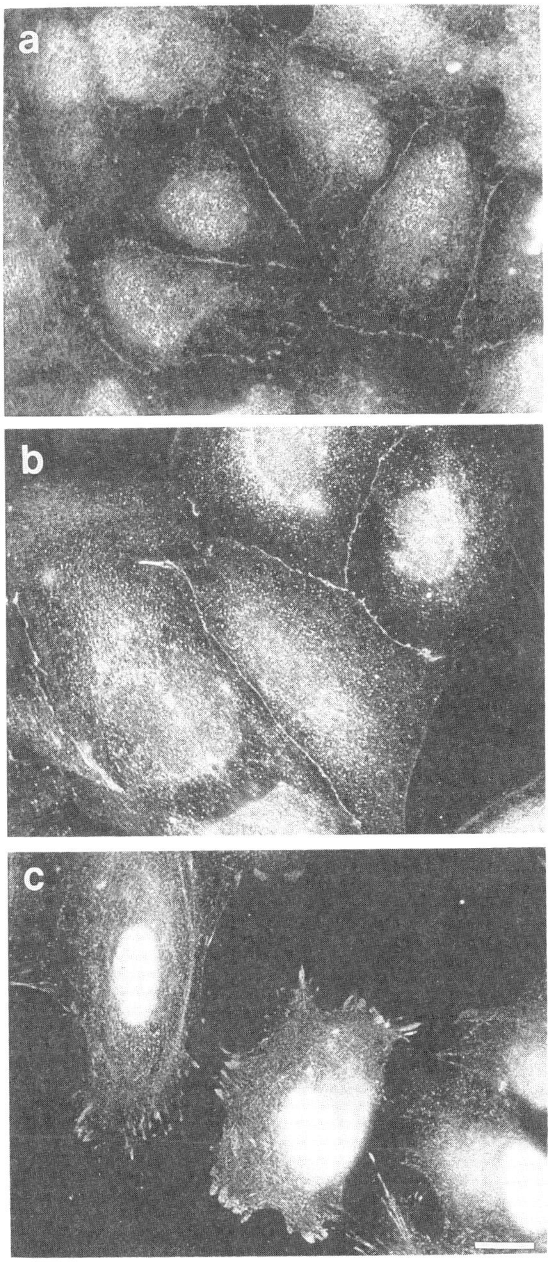

Figure 9. Distribution of the *neu* receptor, (*a*) *raf-1* and PI 3′-kinase (*c*) in cultured human endothelial cells. The former two proteins were primarily associated with the intercellular contacts of these cells, and the latter was predominantly associated with focal contacts. Bar, 10 μm.

and the local concentration of these proteins may be involved in the generation of long-range adhesion signals.

ACKNOWLEDGMENTS

This study was supported by grants from the National Council for Research and Development (Israel)-German Cancer Research Center (Heidelberg), German-Israeli binational research program, The Minerva Foundation, and the Leo and Julia Forchheimer Center for Molecular Genetics at the Weizmann Institute of Science. B.G. is the Erwin Neter Professor for Cell and Tumor Biology. I express my gratitude to Mrs. E. Majerowich for her excellent assistance in typing this manuscript and preparing it for publication.

REFERENCES

Aaronson, S. and G. Todaro. 1968. Development of 2T2-like lines from Balb/c mouse embryo cultures: Transformation susceptibility to SV40. *J. Cell Physiol.* **42:** 141.

Bendori, R., D. Salomon, and B. Geiger. 1987. Contact-dependent regulation of vinculin expression in cultured fibroblasts: A study with vinculin-specific cDNA clones. *EMBO. J.* **6:** 2897.

———. 1989. Identification of two distinct functional domains on vinculin involved in its association with focal contacts. *J. Cell Biol.* **108:** 2383.

Ben-Ze'ev, A. 1990. Application of two-dimensional gel electrophoresis in the study of cytoskeletal protein regulation during growth activation and differentiation. *Electrophoresis* **11:** 191.

———. 1991. Animal cell shape changes and gene expression. *BioEssays* **13:** 207.

Ben-Ze'ev, A. and A. Amsterdam. 1987. In vitro regulation of granulosa cell differentiation: Involvement of cytoskeletal protein expression. *J. Biol. Chem.* **262:** 5366.

Ben-Ze'ev, A., R. Reiss, R. Bendori, and B. Grodecki. 1990. Transient induction of vinculin gene expression in 3T3 fibroblasts stimulated by serum-growth factors. *Cell Regul.* **1:** 621.

Bellas, R.E., R. Bendori, and S.R. Farmer. 1991. Epidermal growth factor activation of vinculin and β_1-integrin gene transcription in quiescent Swiss 3T3 cells. *J. Biol. Chem.* **266:** 12008.

Edelman, G., B. Cunningham, and J.-P. Thiery. 1990. *Morphoregulatory molecules.* Wiley, New York.

Ellis, L., E. Clauser, D.O. Morgan, M. Endery, R.A. Roth, and W.J. Rutter. 1986. Replacement of insulin receptor tyrosine residues 1162 and 1163 compromises insulin-stimulated kinase activity and uptake of 2-deoxyglucose. *Cell* **45:** 721.

Friedlander, D.R., R.M. Mege, B.A. Cunningham, and G.M. Edelman. 1989. Cell sorting-out is modulated by both the specificity and amount of different cell adhesion molecules (CAMs) expressed on cell surfaces. *Proc. Natl. Acad. Sci.* **86:** 7043.

Garrels, J.I. 1989. The QUEST system for quantitative analysis of two-dimensional gels. *J. Biol. Chem.* **264:** 5269.

Geiger, B. 1979. A 130K protein from chicken gizzard: Its localization at the termini of microfilament bundles in cultured chicken cells. *Cell* **18:** 193.

———. 1982. Involvement of vinculin in contact-induced cytoskeletal interactions. *Cold Spring Harbor Symp. Quant. Biol.* **46:** 671.

———. 1989. Cytoskeleton-associated cell contacts. *Curr. Opin. Cell Biol.* **1:** 103.

Geiger, B. and D. Ginsberg. 1991. The cytoplasmic domain of adherens-type junctions. *Cell Motil. Cytoskeleton* **20:** 1.

Geiger, B., T. Volk, and T. Volberg. 1985. Molecular heterogeneity of adherens junctions. *J. Cell Biol.* **101:** 1523.

Geiger, B., D. Ginsberg, D. Salomon, and T. Volberg. 1990. The molecular basis for the assembly and modulation of adherens-type junctions. *Cell Differ. Dev.* **32:** 343.

Geiger, B., D. Salomon, D. Takeichi, and R.O. Hynes. 1992. A chimeric N-cadherin/β_1-integrin receptor which localizes to both cell-cell and cell-matrix adhesion. *J. Cell Sci.* (in press).

Gentry, L. and L. Rohrschneider. 1984. Common features of the *yes* and *src* gene products defined by peptide-specific antibodies. *J. Virol.* **51:** 539.

Ginsberg, D., D. DeSimone, and B. Geiger. 1991. Expression of a novel cadherin (EP-cadherin) in unfertilized eggs and early *Xenopus* embryos. *Development* **111:** 315.

Glück, U., J.L. Rodriguez Fernández, R. Pankov, and Z. Ben-Ze'ev. 1992. Regulation of adherens junction protein expression in growth-activated 3T3 cells and in regenerating liver. *Exp. Cell Res.* (in press).

Hatta, K. and M. Takeichi. 1986. Expression of N-cadherin adhesion molecules associated with early morphogenetic events in chick development. *Nature* **320:** 447.

Hayashi, Y., B. Haimovich, A. Reszka, D. Boettiger, and A. Horwitz. 1990. Expression and function of chicken integrin β_1 subunit and its cytoplasmic domain mutants in mouse NIH 3T3 cells. *J. Cell Biol.* **110:** 175.

Heffetz, D. and Y. Zick. 1989. H_2O_2 potentiates phosphorylation of novel putative substrates for the insulin receptor kinase in intact fao cells. *J. Biol. Chem.* **264:** 10126.

Hynes, R.O. 1987. Integrins: A family of cell surface receptors. *Cell* **48:** 549.

———. 1992. Integrins: Versatility, modulation and signaling in cell adhesion. *Cell* **69:** 11.

Jaken, S., K. Leach, and T. Klauck. 1989. Association of type 3 protein kinase C with focal contacts in rat embryo fibroblasts. *J. Cell Biol.* **109:** 697.

Kreis, T.E., B. Geiger, and J. Schlessinger. 1982. Mobility of microinjected rhodamine actin within living chicken gizzard cells determined by fluorescence photobleaching recovery. *Cell* **29:** 835.

Kreis, T.E., Z. Avnur, J. Schlessinger, and B. Geiger. 1984. Dynamic properties of cytoskeletal proteins in focal contacts. In *Molecular biology of the cytoskeleton* (ed. G.G. Borisy et al.), p. 45. Cold Spring Harbor Laboratory, Cold Spring Harbor, New York.

LaFlamme, S.E., S.K. Akiyama, and K.M. Yamada. 1992. Regulation of fibronectin receptor distribution. *J. Cell Biol.* **117:** 437.

Marcantonio, E.E. and R.O. Hynes. 1988. Antibodies to the conserved cytoplasmic domain of the integrin β_1 subunit react with proteins in vertebrates, invertebrates and fungi. *J. Cell Biol.* **106:** 1765.

Marcantonio, E.E., J.-L. Guan, J.E. Tevithick, and R.O. Hynes. 1990. Mapping of the functional determinants of the integrin β_1 cytoplasmic domain by site-directed mutagenesis. *Cell Regul.* **1:** 597.

McBride, K., S. Rhee, and S. Jaken. 1991. Immunocytochemical localization of phospholipase C-g in rat embryo fibroblasts. *Proc. Natl. Acad. Sci.* **88:** 7111.

Nagafuchi, A. and M. Takeichi. 1988. Cell binding function of E-cadherin is regulated by the cytoplasmic domain. *EMBO J.* **7:** 3679.

Otey, C.A., F.M. Pavalko, and K. Burridge. 1990. An interaction between α-actinin and the β_1 integrin subunit in vitro. *J. Cell Biol.* **111:** 721.

Ozawa, M., H. Baribault, and R. Kemler. 1989. The cytoplasmic domain of the cell adhesion molecule uvomorulin associates with three independent proteins structurally related in different species. *EMBO J.* **8:** 1711.

Peles, E., R. Lamprecht, R. Ben-Levy, E. Zahar, and Y. Yarden. 1992a. Regulated coupling of the Neu receptor to phosphatidylinositol 3′-kinase and its release by oncogenic activation. *J. Biol. Chem.* **267:** (in press).

Peles, E., S.S. Bacus, R.A. Koski, H.S. Lu, D. Wen, S.G. Ogden, R. Ben-Levy, and Y. Yarden. 1992b. Isolation of the Neu/HER-2 stimulatory ligand: A 44 kd glycoprotein that induces differentiation of mammary tumor cells. *Cell* **69:** 205.

Price, G.J., P. Jones, M.D. Davison, B. Patel, R. Bendori, B. Geiger, and D.R. Critchley. 1989. Primary sequence and domain structure of chicken vinculin. *Biochem. J.* **259:** 453.

Rodriguez Fernández, J.L., B. Geiger, D. Solomon, and A. Ben-Ze'ev. 1992a. Overexpression of vinculin suppresses cell motility in BALB/c 3T3 cells. *Cell Motil. Cytoskeleton* **22:** 127.

Rodriguez Fernández, J.L., B. Geiger, D. Solomon, H. Sabanay, M. Zöller, and A. Ben-Ze'ev. 1992b. Suppression of tumorigenicity in transformed cells transfected with vinculin cDNA. *J. Cell Biol.* (in press).

Rohrschneider, L. 1980. Adhesion plaques of Rous sarcoma virus-transformed cells contain the *src* gene product. *Proc. Natl. Acad. Sci.* **77:** 5514.

Rohrschneider, L. and L. Najita. 1984. Detection of the v-*abl* gene product at cell-substratum contact sites in Abelson murine leukemia virus-transformed fibroblasts. *J. Virol.* **51:** 547.

Salomon, D., O. Ayalon, R. Patel-King, R. Hynes, and B. Geiger. 1992. Extrajunctional distribution of N-cadherin in cultured human endothelial cells. *J. Cell Sci.* **120:** 7.

Sambrook, J., E.F. Fritsch, and T. Maniatis. 1989. *Molecular cloning: A laboratory manual.* Cold Spring Harbor Laboratory, Cold Spring Harbor, New York.

Shaulsky, G., N. Goldfinger, M. Tosky, A.J. Levine, and V. Rotter. 1991. Nuclear localization is essential for the activity of p53 protein. *Oncogene* **6:** 2055.

Takeichi, M. 1988. Cadherins: Key molecules for selective cell-cell adhesion. *IARC Sci. Publ.* 76.

———. 1990. Cadherins: A molecular family important in selective cell-cell adhesion. *Annu. Rev. Biochem.* **59:** 237.

———. 1991. Cadherin cell adhesion receptors as a morphogenetic regulator. *Science* **251:** 1451.

Takeichi, M., K. Hatta, A. Nose, A. Nagafuchi, and M. Matsunaga. 1989. Cadherin-mediated specific cell adhesion and animal morphogenesis. *Ciba Found. Symp.* **144:** 243.

Tamkun, J.W., D.W. DeSimone, D. Fonda, R.S. Patel, C. Buck, A.F. Horwitz, and R.O. Hynes. 1986. Structure of integrin, a glycoprotein involved in the transmembrane linkage between fibronectin and actin. *Cell* **42:** 271.

Tsukita, S., K. Oishi, T. Akiyama, Y. Yamanashi, T. Yamamoto, and S. Tsukita. 1991. Specific proto-oncogenic tyrosine kinases of *src* family are enriched in cell-to-cell adherens junctions where the level of tyrosine phosphorylation is elevated. *J. Cell Biol.* **113:** 867.

Ungar, F., B. Geiger, and A. Ben-Ze'ev. 1986. Cell contact- and shape-dependent regulation of vinculin syntheses in cultured fibroblasts. *Nature* **319:** 787.

Vasiliev, J. and I. Gelfand, eds. 1981. *Neoplastic and normal cells in culture.* Cambridge University Press, England.

Volberg, T., B. Geiger, R. Dror, and Y. Zick. 1991. Modulation of intercellular adherens-type junctions and tyrosine phosphorylation of their components in RSV-transformed cultured chick lens cells. *Cell Regul.* **2:** 105.

Volberg, T., Y. Zick, R. Dror, I. Sabanay, C. Gilon, A. Levitzki, and B. Geiger. 1992. The effect of tyrosine-specific protein phosphorylation on the assembly of adherens-type junctions. *EMBO J.* **11:** 1733.

Volk, T. and B. Geiger. 1984. A 135-kd membrane protein of intercellular adherens junctions. *EMBO J.* **3:** 2249.

———. 1986a. A-CAM: A 135-kD receptor of intercellular adherens junctions. I. Immunoelectron microscopic localization and biochemical studies. *J. Cell Biol.* **103:** 1441.

———. 1986b. A-CAM: A 135-kD receptor of intercellular adherens junctions. II. Antibody-mediated modulation of junction formation. *J. Cell Biol.* **103:** 1451.

Desmosomal Proteins: Mediators of Intercellular Coupling and Intermediate Filament Anchorage

W.W. Franke, S.M. Troyanovsky, P.J. Koch, R. Troyanovsky, B. Fouquet, and R.E. Leube

Division for Cell Biology, German Cancer Research Center, D-6900 Heidelberg, Federal Republic of Germany

The establishment and maintenance of intra- and intercellular architecture involves stable cell contact structures ("junctions") that appear as distinct plasma membrane domains characterized by close apposition of the surface membranes of neighboring cells and a dense layer ("plaque") on either cytoplasmic side. These plaques in turn can serve as anchoring plates for the association of specific kinds of cytoskeletal filaments. Two major kinds of such plaque-bearing intercellular junctions ("adhering junctions") can be distinguished by morphological (Farquhar and Palade 1963) and biochemical criteria (Franke et al. 1982; Geiger et al. 1983, 1985; Cowin et al. 1985b, 1986):

1. Desmosomes (maculae adhaerentes), mostly 0.1–0.5 μm in diameter, are involved in the anchorage of intermediate-sized filaments (IFs) and in the intercellular coupling—and hence tissue formation—of epithelial cells, but also of myocardial and meningeal cells and the cells of the dendritic reticulum of lymph node follicles. In epithelia, these plaques are usually associated with IFs of the cytokeratin category (for reviews, see Moll et al. 1982; Steinert and Roop 1988), whereas desmosomal plaques of myocardiac and Purkinje fiber cells of the heart are associated with desmin-containing IFs and those of certain meningeal and of dendritic reticulum cells with vimentin IFs (Cowin et al. 1985b; Moll et al. 1986).
2. Adhering cell junctions anchoring actin microfilaments (also referred to as intermediate junctions) occur in various sizes and morphological forms, from small puncta adhaerentia and streak-like fasciae adhaerentes to the cell-encircling belt of the zonula adhaerens that is typical of polar epithelial cells.

Biochemical analyses of isolated junctional structures and immunolocalization studies have shown that these two major types of adhering junctions differ markedly in their protein composition. So far, only one protein has been found in common to the plaques of both kinds of junctions, i.e., plakoglobin (Cowin et al. 1986; Kapprell et al. 1987; Franke et al. 1989). All other junctional components identified appear to be mutually exclusive. The microfilament-anchoring junctions are associated with plaques containing various actin-associating proteins such as vinculin, α-catenin, β-catenin, and α-actinin (see Geiger et al.; Takeichi et al.; both this volume) and are enriched in specific transmembrane glycoproteins such as E-cadherin (also called L-CAM or uvomorulin). Desmosomes, however, contain a different set of proteins. Figure 1 presents the typical pattern of major polypeptides identified in desmosomes isolated from a stratified epithelium, that of the bovine muzzle. From results of comparative studies of desmosomes in diverse cell types and tissues one can distinguish between *constitutive* and *accessory* desmosome-associated proteins. Constitutive proteins are present, in at least one subform, in desmosomes of all kinds of cells and include (1) nonglycosylated cytoplasmic proteins assembled in the plaque, such as desmoplakin I and plakoglobin, and (2) transmembrane glycoproteins such as desmoglein and desmocollins I and II. Desmosomes of certain epithelia contain some additional (accessory) proteins such as desmoplakin II, an alternative splice variant of desmoplakin I (Green et al. 1990; Schwarz et al. 1990; Virata et al. 1992), the basic "band 6 protein," desmocalmin, and desmoyokin (for review, see Schwarz et al. 1990).

The recent finding that the product of the developmental mutant gene *armadillo* of *Drosophila*, which is important for correct segmental polarity, is highly homologous to human plakoglobin (Peifer and Wieschaus 1990) and also to the β-catenins of the amphibian species *Xenopus laevis* (McCrea et al. 1991) and the mouse (R. Kemler et al., in prep.) has raised the question of the possible orthologous relationship of these polypeptides. However, we have recently cDNA-cloned the authentic plakoglobin of *X. laevis* and found that its sequence is closer to the human plakoglobin than to the β-catenin of the same species (Fig. 2) (Fouquet et al. 1992), indicating that plakoglobin and β-catenin are different proteins coexisting in the same species and probably also in some cell types. It remains to be examined, however, whether there are further genes of a larger gene family of plakoglobin-related proteins.

The determination of the amino acid sequences of the cDNA-cloned desmosomal glycoproteins desmoglein and desmocollins I and II of stratified epithelia has revealed that both kinds of proteins are members of the cadherin family of cell adhesion molecules (Fig. 3) (Koch et al. 1990, 1991a,b; Collins et al. 1991; Mechanic et al. 1991; Nilles et al. 1991; Parker et al. 1991; Wheeler et al. 1991; for review, see Magee and Buxton

		Localization	M_r value (kDa)	Molecular size (aa/mol.wt. x10^3)	Size of precursor (aa /mol. wt x10^3)	Constitutive	Genotypically different subforms
–Desmoplakin I		Cytoplasmic	250	2677 / 310	–	+	
–Desmoplakin II		Cytoplasmic	215	2078 / 238	–	–	
–Desmoglein		Transmembrane	165	994 / 106	1043 / 112	+	+
–Desmocollin I	–Type 1	Transmembrane	130	761 / 85	893 / 99.6	+	+
	–Type 2			774 / 85.5	n.d.		
–Desmocollin II	–Type 1	Transmembrane	115	707 / 79	839 / 93.5	+	+
	–Type 2			720 / 79.5	n.d.		
–Plakoglobin		Cytoplasmic	83	82	–	+	*
–Band 6–protein		Cytoplasmic	75	n.d.	–	–	

Figure 1. Major polypeptides identified in isolated desmosomes from bovine muzzle epithelium after separation by SDS-polyacrylamide gel electrophoresis and staining with Coomassie blue (for methods, see Cowin et al. 1985a,b, 1986) shown by photography (*left*; the fraction also contains some residual cytokeratins) and in schematic presentation, together with the specific topology, molecular weight, number of amino acid (aa) residues, and the size of the unprocessed precursor molecule. (n.d.) Not determined. The scheme also contains information whether the specific polypeptide is a *constitutive* desmosomal component and whether different subforms of the specific polypeptide are known (the asterisk in the case of plakoglobin indicates that there are closely related but somewhat larger proteins such as β-catenin which, however, are not isoforms of plakoglobin).

1991). The specific "desmosomal cadherins," however, are not identical in all cells, and three different forms of each desmoglein (Fig. 4) and desmocollin (Koch et al. 1991b, 1992; S.M. Troyanovsky et al., in prep.) have been distinguished so far, including examples of coexpressions as well as cell type-specific synthesis. Despite considerable amino acid sequence divergence in the various domains of these proteins (Fig. 4), the different subforms of desmogleins and desmocollins are readily identified by characteristic molecular features such as the extended carboxy-terminal cytoplasmic domain of all desmogleins with 2–6 repeating elements of a motif of 29 ± 1 amino acids and the coexisting alternative splice forms (I and II) of all desmocollins which result from a "mini-exon" containing a stop codon (Fig. 5).

From their molecular topology, the desmosomal cadherins with their large extracellular *N*-glycosylated domain on the outer surface and their cytoplasmic portion contributing to the plaque (Schmelz et al. 1986; Koch et al. 1990) are obvious candidates for cell adhesion molecules involved in at least three different functions: intercellular coupling, plaque assembly, and specific IF anchorage. Previous transfection experiments with E-cadherin cDNA deletion constructs indicated that the carboxy-terminal region of approximately 70 amino acids is involved in the attachment of actin microfilaments to adhering junctions of the *zonula*, *fascia*, and *punctum* types (Nagafuchi and Takeichi 1988; Ozawa et al. 1989, 1990; Nagafuchi et al. 1991; see also Matsuzaki et al. 1990). Therefore, we have performed a series of transfection experiments using deletion mutants and chimeric proteins to examine possible functional contributions of the cytoplasmic domain (tail) of the desmosomal cadherins.

In one series of such experiments, we combined the entire tails or parts thereof of desmoglein and desmocollin with a specific topogenic "harpoon" sequence securing the rapid and stable insertion of the chimeric protein (and thus the cadherin tails under examination) into a defined membrane structure. Two particularly successful targeting protein elements in our experiments are characterized by the existence of four transmembrane domains: connexin, the hallmark constituent protein of gap junctions (Paul 1986), as a plasma membrane-specific element; and synaptophysin, a protein accumulating in small cytoplasmic vesicles resembling synaptic vesicles of neuronal and neuroendocrine cell (Leube et al. 1989).

```
                *..  ..* *.****** ******** **.** ***.  * ..*.*  *    **  *******   *   *.*  .    ** *** **
  1           MEVMNLMEQPIKVTEWQQTYTYDSGIHSGANTCVPSVSSKGIMEEDE-ACGRQYTLKKTTTYTQGVPPSQGDLEYQMSTTARAKRVREAMC pghum
  1           MDLGDVVEMPMKVTEWQKTYTYDSGINSGINTSVPSLNGKMVIEDDSLAYPNSYTTVKTTTYTQQQNP---DMESHLN-MTRAQRVRAAMY pgxl
  1 MATQADLMELDMAMEPDRKAAVSHWQQQSYLDSGIHSGATTTAPSLSGKGNPEDEDVD-TNQVLYEWEQGFSQSFTQDQVADIDGQYAMTRAQRVRAAMF catbxl
                .*.   .  . *. **     **** **. *..*** **  .**. .    *          .*                ***********

    *    *    *  **.*** **.*.******.*****.***********.***.************.** ** *************.*** ***.***.*
 91 PGVSGEGQLALLATQVEGQATNLQRLAEPSQLLKSAIVHLINYQDDAELVTRALPELTKLLNDEDPVVVTKAAMIVNQLSKKEASRRALMGSPQLVAAVV pghum
 88 PETV-EDHSYLFSTQIEGQQTNVQKLAEPSQMLKSAIMHLINYQDDAELATRAIPELTKLLNDEDPMVVNKASMIVNQLSKKEASRKALMQSPQIVAAIV pgxl
100 PETLDEGMQIPSTQFDSAHPTNVQRLAEPSQMLKHAVVNLINYQDDAELATRAIPELTKLLNDEDQVVVNKAAVMVHQLSKKEASRHAIMRSPQMVSAIV catbxl
    ***. *    . .        ****.********* *.. ************************** .*****  ..* *********.*.* ***.* ***

    **** ***.******************** ****************************************************************
191 RTMQNTSDLDTARCTTSILHNLSHHREGLLAIFKSGGIPALVRMLSSPVESVLFYAITTLHNLLLYQEGAKMAVRLADGLQKMVPLLNKNNPKFLAITTD pghum
187 RTMQHTSDMDTARCTTSILHNLSHHREGLLSIFKSGGIPALVRMLSSPVESVLFYAITTLHNLLLYQEGAKMAVRLADGLQKMVPLLNKNNPKFLAITTD pgxl
200 RTMQNTNDVETARCTAGTLHNLSHHREGLLAIFKSGGIPALVKMLGSPVDSVLFYAITTLHNLLLHQEGAKMAVRLAGGLQKMVALLNKTNVKFLAITTD catbxl
    **** * *..*****    ************ ***********.** ***.*************** *********** ******.**** *.*******

    **************** ***** ******** ***********************************************************.
291 CLQLLAYGNQESKLIILANGGPQALVQIMRNYSYEKLLWTTSRVLKVLSVCPSNKPAIVEAGGMQALGKHLTSNSPRLVQNCLWTLRNLSDVATKQEGLE pghum
287 CLQLLAYGNQESKLIILGNGGPQGLVQIMRNYNYEKLLWTTSRVLKVLSVCPSNKPAIVEAGGMQALGKHLTSNSPRLVQNCLWTLRNLSDVATKQEGLD pgxl
300 CLQILAYGNQESKLIILASGGPQALVNIMRTYSYEKLLWTTSRVLKVLSVCSSNKPAIVEAGGMQALGLHLTDSSQRLVQNCLWTLRNLSDAATKQEGME catbxl
    ***.*************    **** ** *** * ******************* ***************** ***   * **************.******..

     ********** ******************** .*******   *** *** **** ***** ******.******* .**.******* **********
391 SVLKILVNQLSVDDVNVLTCATGTLSNLTCNNSKNKTLVTQNSGVEALIHAILRAGDKDDITEPAVCALRHLTSRHPEAEMAQNSVRLNYGIPAIVKLLN pghum
387 NVLKILVNQLSSDDVNVLTCATGTLSNLTCNNGRNKTLVTQSNGVESLIHTILRASDKDDIAEPAVCAMRHLTSRHQDAEVAQNSVRLHYGIPAIVKLLN pgxl
400 GLLGTLVQLLGSDDINVVTCAAGILSNLTCNNYKNKMMVCQVGGIEALVRTVLRAGDREDITEPAICALRHLTSRHQEAEMAQNAVRLHYGLPVVVKLLH catbxl
     .*  **  * ***.**.*** * ******** .** .* *  *.* *..*.*** *..** ***.**.********.**.*** ******.*..****

     * *************************** .* *********** *******.*  **********.******************.***.* .*******
491 QPNQWPLVKATIGLIRNLALCPANHAPLQEAAVIPRLVQLLVKAHQDAQRHVA-AGTQQPYTDGVRMEEIVEGCTGALHILARDPMNRMEIFRLNTIPLF pghum
487 PPYQWPLVKATIGLIRNLALCPANHAPLYDAGVIPRLVQLLVKSHQDAQRHAA-SGTQQPYTDGVKMEEIVEGCTGALHILARDPVNRMDIYKLNTIPLF pgxl
500 PPSHWPLIKATVGLIRNLALCPANHAPLREQGAIPRLVQLLVRAHQDTQRRTSIGGTQQQFVEGVRMEEIVEGCTGALHILARDIHNRIVIRGLNTIPLF catbxl
    **  ***.***.**************** . *.*********. *** **.     ****   .**.****************. **. *  *******

    ****** *******  ************** ******************* ****************.*********** .*. *************.
590 VQLLYSSVENIQRVAAGVLCELAQDKEAADAIDAEGASAPLMELLHSRNEGTATYAAAVLFRISEDKNPDYRKRVSVELTNSLFKHDPAAWEAAQSMIPI pghum
586 VQLLYSPVENIQRVSSGVLCELAQDKEAADTIDAEGASAPLMELLHSRNEGIATYAAAVLFRISEDKNADYRKRVSVELTNAIFRQDPAAWEAAQSMIPL pgxl
600 VQLLYSPIENIQRVAAGVLCDVAQDKEAAEAIEAEGATAPLTELLHSRNEGVATYAAAVLFRMSEDKPQDYKKRLSVELTSSLFRTEPMPWNEAADL-GL catbxl
    *******.******  ****..*******. *.****.*** *********.**********.****  **.**.*****  .** .*..*  *  .  .

    *.** *.*.  ** **  *.**.*.   *** .*..* ***         ****
690 NEPYGDDMDATYRPMYSSDVPLDPLEMHMDMDGDYPIDTYSDGLRPPYPTADHMLA*                                        pghum
686 NDPYSDEMEN-YRAMYPEDIPLEPMG--GDMDVEYAMDGYSDHPGRGHYADNHMMA*                                        pgxl
699 DIG-AQGEPLGYRQ---DDSSYRSFHAAGYGQDAMGMDSMMDHDMGGHHPGADYPVDGLPDLSHAQDLMDGLPPGDSNQLAWFDTDL*          catbxl
              **    .*        *       **   **   **        .
```

Figure 2. Amino acid sequence comparison of *Xenopus* plakoglobin (pgxl, from embryonic cDNA; see Fouquet et al. 1992), human plakoglobin (pghum, upper line; cf. Franke et al. 1989), and *Xenopus* β-catenin (catbxl, lower line, from McCrea et al. 1991). Identical amino acid residues between the two plakoglobins and with *Xenopus* β-catenin are marked by asterisks above and below the specific reference sequence; conservative amino acid exchanges are marked by dots. Horizontal bars represent gaps introduced to optimize alignment. The sequence between amino acid positions 162 and 171 of human plakoglobin presents a correction of the previously published sequence (Franke et al. 1989). Note that the *Xenopus* plakoglobin sequence is much more similar to that of the human plakoglobin than to that of the β-catenin of the same species (*Xenopus*). This shows that plakoglobin and β-catenin are different proteins coexisting in the same species and that the sequence of plakoglobin has been markedly conserved during evolution.

METHODS

Experimental strategies. For our transfection experiments we used plasmid constructs in which cDNA segments encoding various cytoplasmic portions of desmocollins or desmogleins were combined with the amino-terminal part, including all four transmembrane regions, of a cell type-specific gap junction protein, connexin 32 (cf. Paul 1986), isolated from rat liver cDNA. The resulting chimeric protein was thus designed to integrate into the plasma membrane and to form connexons clustering into gap junction structures which, however, expose the cytoplasmic tail of the specific desmosomal cadherin.

Constructions of plasmids. A complete cDNA encoding the rat liver connexin 32 cDNA was restricted with *Hin*dIII and *Eco*RV, and the resulting fragment was cloned into the appropriate sites of Bluescript. A fragment encoding the first 219 connexin amino acids,

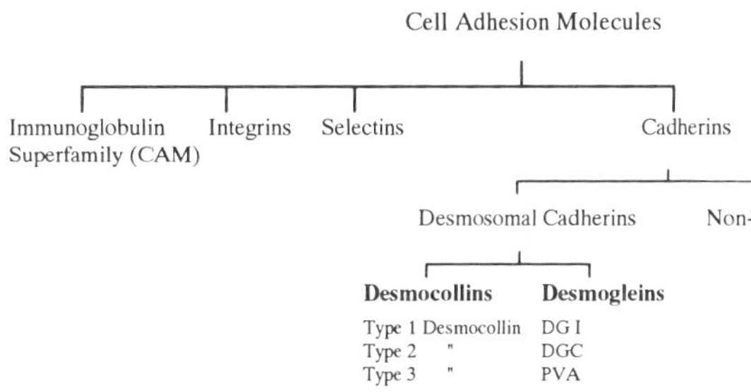

Figure 3. Among the various types of vertebrate adhesion molecules, the cadherins represent a relatively large "superfamily" (for review, see Takeichi 1990). This family of calcium-dependent homophilic adhesion molecules can be further divided into non-desmosomal and desmosomal cadherins, and the latter subgroup comprises the desmocollins (type-1, -2, and -3 desmocollin) and the desmogleins (DGI, epidermal desmoglein; DGC, "colon-type" desmoglein; PVA, pemphigus vulgaris antigen).

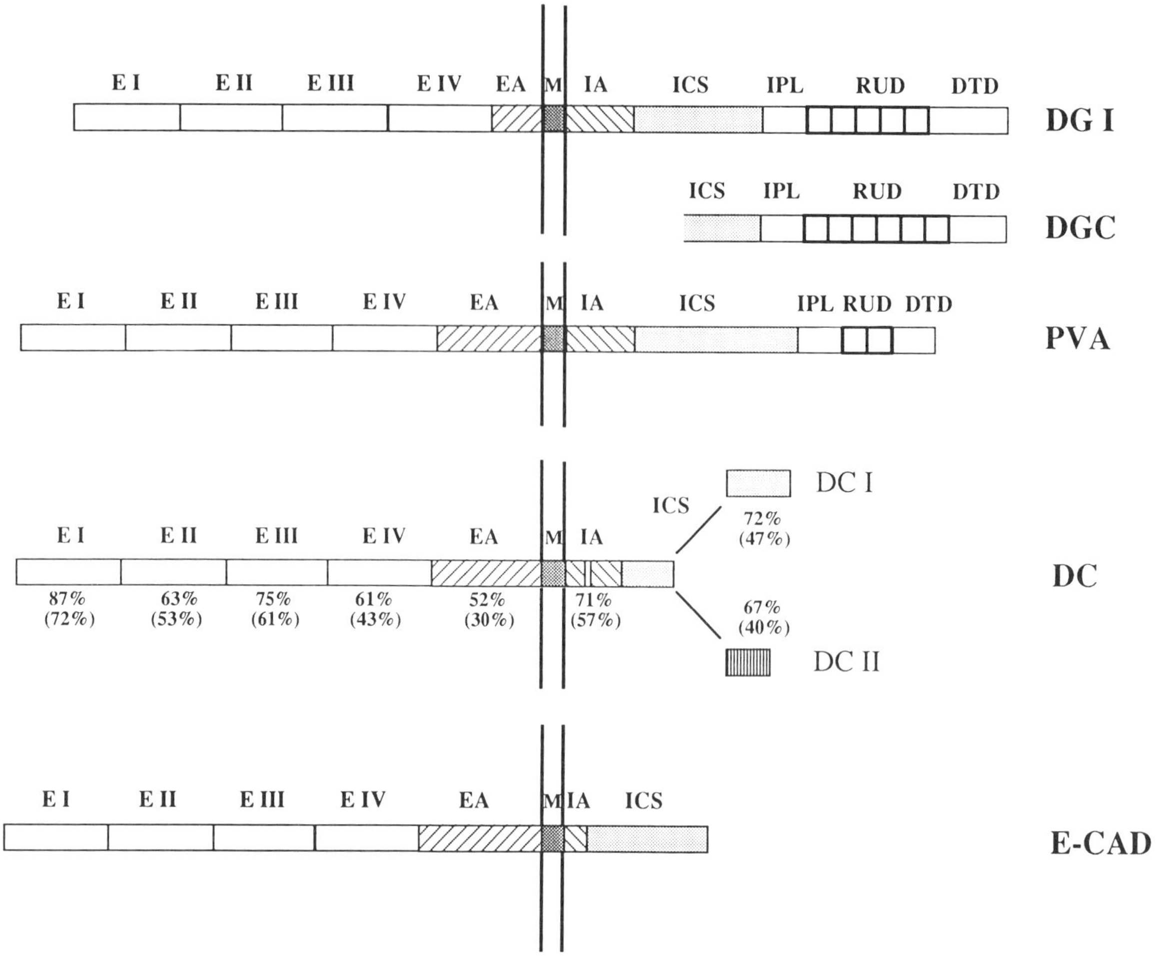

Figure 4. Schematic presentation of topology and domains of various cadherins. Epidermal desmoglein (DGI), colon-type desmoglein (DGC; only a partial sequence has been published yet), and the pemphigus vulgaris antigen (PVA; Amagai et al. 1991), one of the two desmocollins (DC) published with full sequence (Koch et al. 1992) with the two alternative splicing subforms I and II, and E-cadherin (E-CAD). The molecular domains are designated: EI–EIV (four extracellular repeating elements), EA (extracellular anchoring domain), M (membrane-spanning domain), IA (intracellular cadherin-typical segment), IPL (intracellular proline-rich linker region), RUD (repetitive unit domain, consisting of 2, 5, or 6 units), DTD (desmoglein-specific terminal domain). The homology in the individual domains between type-1 and type-2 desmocollin (DC) is indicated (identical amino acids per segment are given in brackets).

i.e., comprising all four transmembrane domains, was amplified with appropriate primers and inserted into the *Hin*dIII/*Nar*I site of plasmid pSR[5] encoding the tail portion of synaptophysin (cf. Leube et al. 1989). The *Nar*I site thus generated was then used for excision by treatment with *Nar*I/*Xba*I or *Nar*I/*Bam*HI, and the

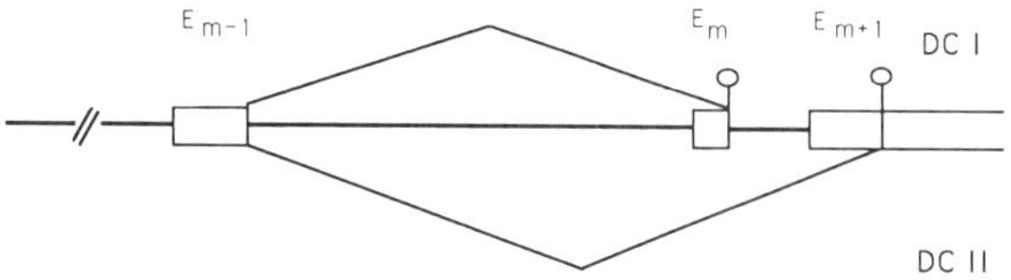

Figure 5. Schematic illustration of the gene structure encoding the cytoplasmic portion of desmocollin genes (S.M. Troyanovsky et al., in prep.) and the two splicing forms. The 48-bp mini-exon (E_m) and the flanking exons (E_{m-1} and E_{m+1}) are shown. Lines indicate the events leading to the formation of the specific splice variants DCI and DCII (cf. Fig. 4). Note that both exons E_m and E_{m+1} contain stop codons (♀).

synaptophysin tail segment was replaced by different, PCR-amplified fragments of cDNA clones encoding parts of the tail domain of desmocollins and desmogleins (Koch et al. 1990, 1991a,b, 1992). The largest construct used, for example, encoded rat connexin 32 fused with the last 460 amino acids of bovine desmoglein. The longest cytoplasmic portions of forms I and II of type 1 and type 2 desmocollin (Koch et al. 1991b, 1992) were integrated into the chimeric constructs at *Nar*I sites introduced in the second codon after the assumed end of transmembrane region.

The *Hin*dIII/*Bam*HI or *Hin*dIII/*Xba*I inserts were subcloned in expression vectors Lk 444, containing the neomycin resistance gene and the β-actin promoter, or pBEHpac18, containing the puromycin resistance gene and promoter elements of the SV40 early genes.

Transfections. Different cell lines were used for transient and stable transfections. Only results obtained from the desmosome-rich human epithelial cell line A-431, originally derived from a squamous cell carcinoma of the vulva (for desmosomes and IF proteins, see Moll et al. 1986; Leube et al. 1988), are described. Cells were transfected with CsCl-purified plasmid DNA using the Ca^{++}-phosphate coprecipitation method without glycerol shock (cf. Leube et al. 1989). Stably transfected cells were selected by additions of geneticin (1 mg/ml) or puromycin (4 μg/ml) to the culture medium.

Localization techniques and electron microscopy. Cells grown on glass coverslips were processed for immunofluorescence microscopy as described previously (Cowin et al. 1986; Leube et al. 1989), using the following primary antibodies: (1) rabbit antibodies specific for a sequence located in a cytoplasmic loop of rat liver connexin which does not cross-react with the endogenous connexin of A-431 cells (from Dr. V. Krutovskich, IARC, Lyon, France; (2) guinea pig antibodies specific for bovine desmocollin type 1 (Koch et al. 1991b) which do not cross-react with endogenous desmocollin of A-431 cells; (3) monoclonal murine antibody DG3.10 which reacts with a defined epitope located in one of the repeating units of the cytoplasmic portion of bovine and human desmoglein (Schmelz et al. 1986; Koch et al. 1990); (4) various murine monoclonal antibodies against desmoplakins I and II (Cowin et al. 1985a); (5) guinea pig antibodies against bovine desmoplakin; (6) murine monoclonal antibody PG5.1 against plakoglobin (Cowin et al. 1986); (7) murine monoclonal antibodies against human cytokeratins 8 (H1; Troyanovsky et al. 1989) and 18 (cf. Leube et al. 1988); (8) monoclonal cytokeratin antibody lu-5.

Conditions and secondary antibodies used for double-label immunofluorescence were essentially as described previously (see, e.g., Cowin et al. 1986; Leube et al. 1989). For electron microscopy, cells were fixed and processed essentially as described (Cowin et al. 1986; Leube et al. 1989). For immunoelectron microscopy, coverslips were fixed with 2% freshly prepared formaldehyde and permeabilized with 0.1% saponin. After incubation with primary antibodies for several hours, cells were washed several times with phosphate-buffered saline (PBS) and incubated overnight with immunoglobulins against the species of the specific primary antibody that had been attached to colloidal gold particles either 5 nm (Amersham-Buchler, Braunschweig, Germany) or 1 nm (Bio Trend, Cologne, Germany) in diameter. After further washes with PBS, samples were fixed with glutaraldehyde, washed, postfixed with 2% OsO_4 for 30 minutes, and processed as described (Leube et al. 1989; Troyanovsky et al., in prep.).

RESULTS

The expression of the gene constructs used for transfection of A-431 cells with wt-connexin 32 or with combined parts of connexin 32 and cadherins, as well as the stability of the resulting chimeric connexin-cadherin proteins, was monitored by SDS-polyacrylamide gel electrophoresis (S.M. Troyanovsky et al., in prep.). In addition, the correct topogenesis of the chimeric molecules was routinely examined by immunofluorescence microscopy and electron microscopy.

Plaque Assembly and IF Anchorage Potential of the Desmocollin Tail Segment

A-431 cells transfected with wt-connexin 32 or with the connexin-cadherin constructs encoding the amino-terminal part of connexin with all four transmembrane domains showed immunofluorescence predominantly in variously sized, mostly large, aggregate structures at plasma membrane sites connecting adjacent cells (Fig. 6a) and in certain "intracellular" dots. In cells transfected with constructs combining this amino-terminal part of connexin 32 with the cytoplasmic portion of desmocollin type 1, form I (Co-DcII), a variable proportion of the connexin-32-positive aggregates were also strongly positive for desmosomal plaque proteins such as desmoplakin (Fig. 4b) and plakoglobin which, of course, also occurred on the numerous small punctate arrays of the endogenous desmosomes (Fig. 4b) (cf. Moll et al. 1986).

Double-label immunofluorescence microscopy with cytokeratin antibodies (see Methods) showed that these novel cell boundary structures positive for both connexin 32 and desmosomal plaque proteins were also associated with IF bundles (data not shown). Colocalization of connexin 32 and desmosomal plaque proteins was not observed in cells transfected with chimeric constructs containing the shorter (cf. Figs. 4 and 5) tail portion of the corresponding desmocollin forms II (S.M. Troyanovsky et al., in prep.).

Electron microscopy of Co Dc1I-transfected cells revealed the organization of the novel structural entity generated (Fig. 7a–c). In addition to their normal desmosomes, these cells contained variously sized, often very large junctional aggregates in which the paracrystalline connexon packing typical of gap junctions was associated with a dense cytoplasmic plaque typical of desmosomes (Fig. 7a). In most cases, bundles of IFs were seen to associate with such gap junction plaques (Fig. 7 and insert). Occasionally, the new junction structures formed were asymmetric in that a plaque and IF bundle association were seen only on one side of the junction whereas on the other side, i.e., in the adjacent cell, the gap junction surface appeared free (Fig. 7b). Whether these asymmetric structures reflect mixtures of connexin 32 chimeric with endogenous connexins, resulting in connexins containing desmocollin tails in one cell but not in the other, remains to be examined.

Not infrequently, we noted roundish structures deep in the cytoplasm which, however, showed the same features of a double-membrane gap junction structure associated with a dense plaque at which IFs insert (Fig. 7c). This organization could represent gap junctions in finger-like invaginations or detached junctional arrays.

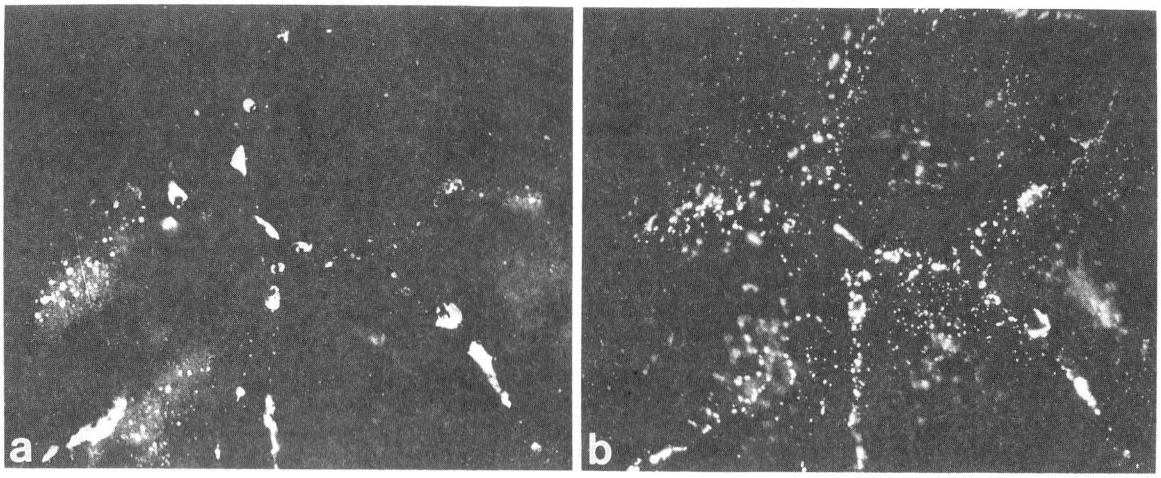

Figure 6. Double-label immunofluorescence microscopy of cells of a subline of human A-431 epithelial cells that has been stably transfected with a plasmid that encodes a chimeric protein comprising the first part of connexin 32, including the four transmembrane domains, and the cytoplasmic portion of desmocollin type 1, form I, after staining with connexin-32-specific (*a*) and desmoplakin (*b*) antibodies. Note that several of the connexin-positive aggregates at intercellular boundaries (*a*) are also positive for desmoplakin (*b*). Other structures are exclusively immunostained for either connexin (*a*) or desmoplakin (*b*). Magnification, 860 ×.

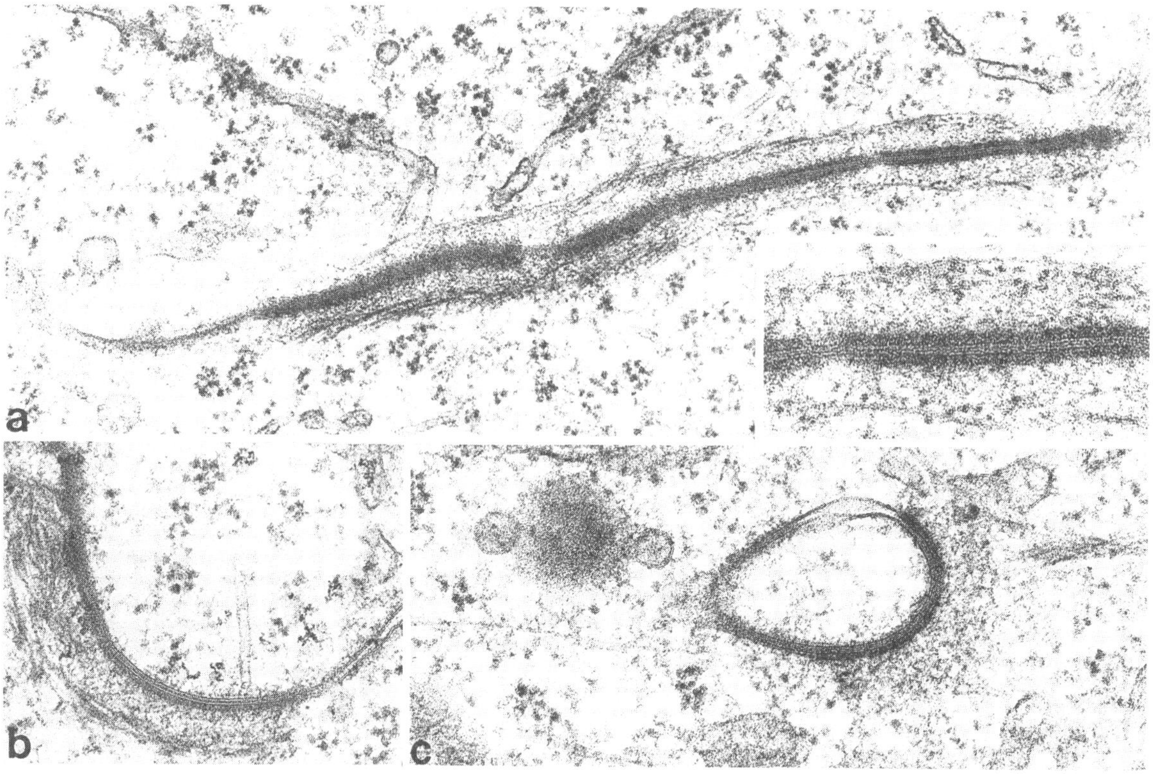

Figure 7. Electron micrographs of the cells of the same subline shown in Fig. 6. (*a*) Survey picture of the boundary of two adjacent cells showing a large region occupied by a novel kind of junction consisting of typical gap junction structure at the plasma membrane level, covered on either side by a densely stained plaque and associated with bundles of IFs which laterally abut on the plaque surface (for higher magnification, see insert in the lower right). (*b*) Asymmetric junction containing the connexin-desmocollin chimeric protein Co-DcII, showing a dense plaque and IF bundles attached in the bottom cell but not in the upper cell (here an accidental microtubule ends at the junctional membrane). (*c*) Plasma membrane junction structure similar to that shown in (*b*) but seemingly located deep in the cell. This arrangement may represent a fingerlike protrusion of one cell into a neighboring cell or a detached cell-surface portion included in the cytoplasm. Note IF bundles oriented toward the plaque covering these junction-type membranes. Magnifications: (*a*) 48,000 × ; (insert in *a*) 100,000 × ; (*c*,*d*) 65,000 ×.

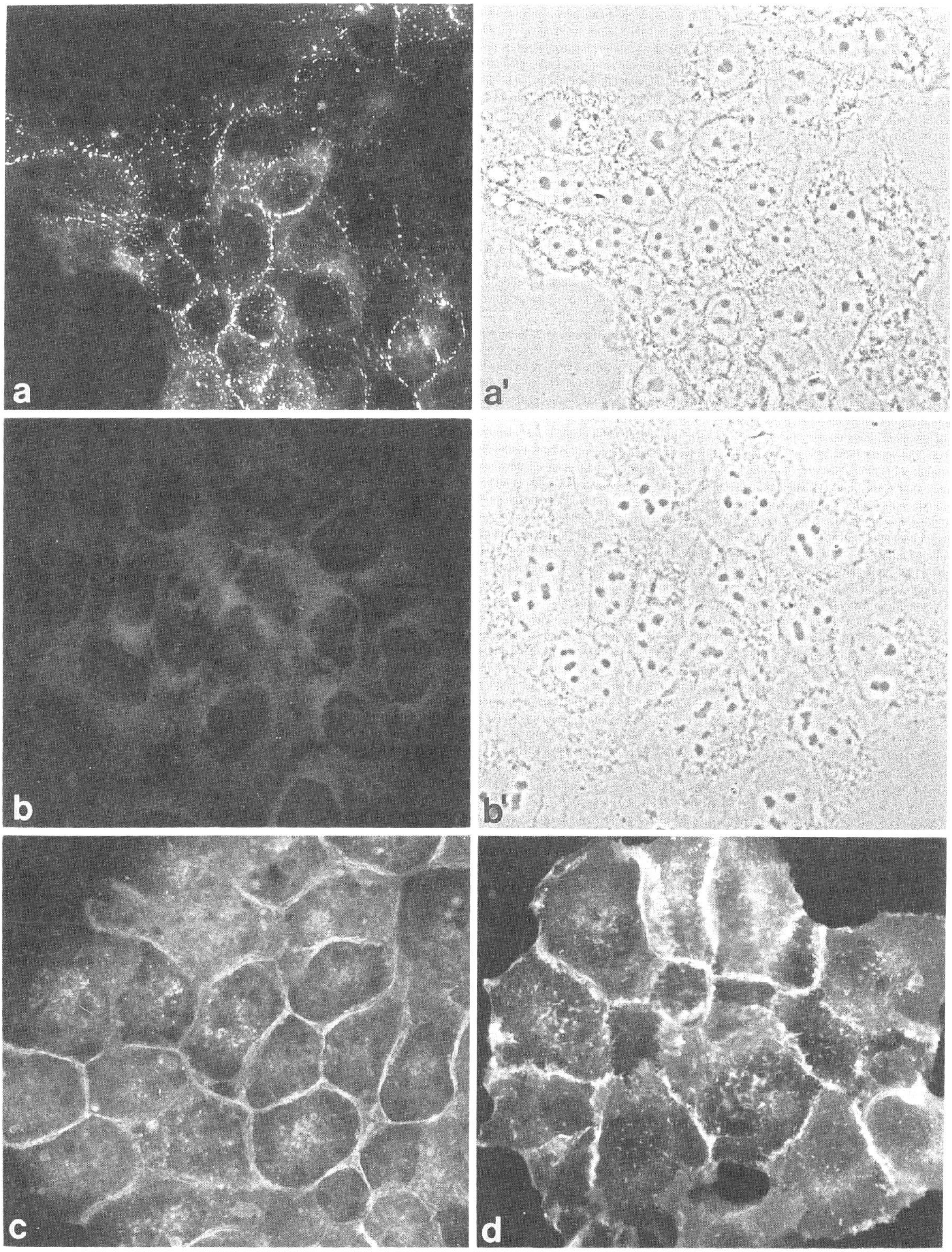

Figure 8. Immunofluorescence microscopy of normal A-431 cells (*a*,*a'*) and cells of an A-431 subline stably transfected with the construct Co-Dg carrying the cytoplasmic portion of bovine desmoglein DGI in combination with the first part of connexin 32 (*b*–*d*), after immunostaining with antibodies specific for desmoplakin (*a* and *b*; *a'* and *b'* are the corresponding phase contrast images), connexin 32 (*c*) and desmoglein antibody DG3.10 (*d*). Note numerous desmosomes in normal cells (punctate staining in *a*) but absence of desmosomal staining in the transfected cells (*b*). Immunostaining for the "foreign" connexin 32 (*c*) as well as for the endogenous and the "foreign" desmoglein (*d*) is concentrated at the plasma membrane of intercellular boundaries where it is, however, not clustered in punctate arrays but rather evenly spread. Magnification, 570 ×.

Immunoelectron microscopy proved that these novel junctional structures combining gap junctional and desmosomal features were positive for connexin 32 and for desmosomal marker proteins such as desmoplakin and also for cytokeratins (S.M. Troyanovsky et al., in prep.).

Dispersive Potential of Desmoglein Tail Domains

Completely different results were obtained in A-431 cells transfected with similar chimeric connexin 32 constructs in which the desmocollin tail was replaced by the cytoplasmic portion of desmoglein (Co-Dg). Desmosomes abundant in control cells were totally absent in cells stably transfected with Co-Dg constructs (Fig. 8a,b), showing a weak and diffuse desmoplakin immunostaining throughout the cytoplasm. The connexin-32-containing chimeric proteins synthesized in the transfected cells did not cluster into aggregates at intercellular boundaries but appeared rather evenly spread over the plasma membrane (Fig. 8c), similar to the general, nonpunctate plasma membrane immunostaining seen with desmoglein antibodies (Fig. 8d).

The disappearance of all desmosomes and the lack of formation of gap junction-like structures with desmosomal plaques and IFs attached was confirmed by electron microscopy. A-431 cell-derived sublines stably transfected with Co-Dg constructs had a peculiar altered morphology (Fig. 9a). Adjacent cells did not establish stable junctional structures, with the exception of occasional *puncta adhaerentia*-like small contact sites, but showed numerous microvillar protrusions interdigitated in the intercellular contact zone. Immunoelectron microscopy showed that both connexin 32 and desmoglein sequences (Fig. 9b) were not concentrated in clusters but evenly distributed along the surface membranes. Otherwise, these cells devoid of desmosomes grew and proliferated normally.

Double-label experiments with cytokeratin antibodiies showed that the masses of IF bundles present in A-431 cells had detached from the plasma membrane and accumulated in large whorls in the perinuclear cytoplasm (cf. S.M. Troyanovsky et al., in prep.).

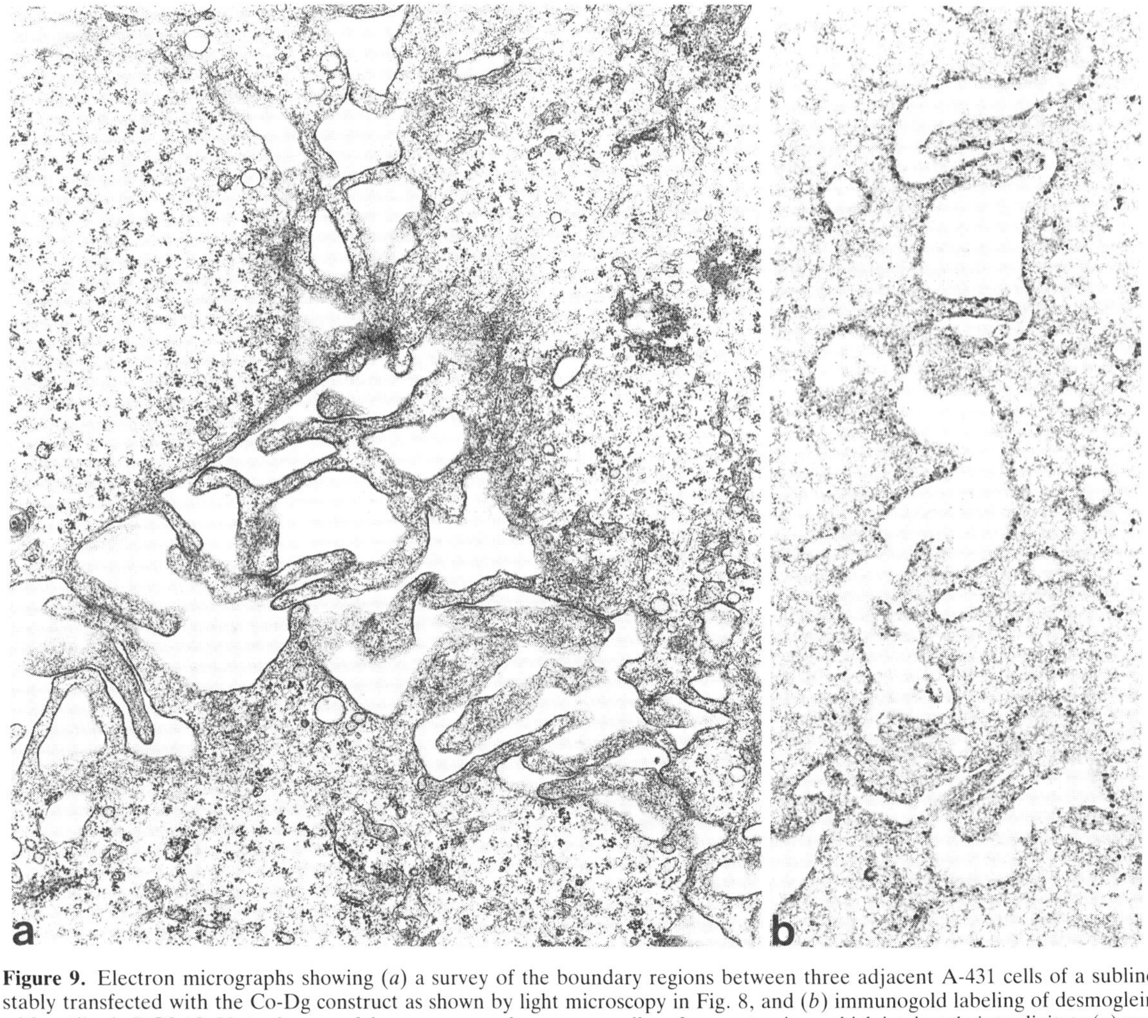

Figure 9. Electron micrographs showing (*a*) a survey of the boundary regions between three adjacent A-431 cells of a subline stably transfected with the Co-Dg construct as shown by light microscopy in Fig. 8, and (*b*) immunogold labeling of desmoglein with antibody DG3.10. Note absence of desmosomes and numerous cell-surface protrusions which intricately interdigitate (*a*) and colloidal immunogold particles along the entire plasma membrane (*b*). Magnification, 20,000 ×.

DISCUSSION

The desmosomal cadherins are transmembrane glycoproteins which are usually clustered in the desmosomal domains where their extracellular, glycosylated portion is involved in homotypic cell coupling and their cytoplasmic portion (tail) contributes to the plaque. Our transfection experiments using connexin-cadherin chimeric proteins have shown that the tail of desmocollin(s) contains sufficient information to direct the accumulation of cytoplasmic plaque proteins such as desmoplakins and plakoglobin and thus to provide a specific anchorage site for IFs. This plaque assembly and IF anchorage competence seems to require the carboxy-terminal portion of the form I of desmocollins, whereas the alternative carboxy-terminal element encoded by the mini-exon of splice form II (see Fig. 5) apparently can bind some desmoplakin but does not form a complete, IF-anchoring plaque. Using deletions and point mutations in chimeric protein combinations of the desmocollin tail with connexin or non-desmosomal cadherins, we are currently trying to identify the minimum sequence requirements for plaque assembly and IF bundle architecture.

Our observation that plakoglobin can also assemble on the surface of small cytoplasmic vesicles containing the desmocollin I tail in combination with the amino-terminal portion of synaptophysin (data not shown; cf. Leube et al. 1989; S.M. Troyanovsky et al., in prep.) suggests that cell-cell contact and the plasma membrane environment are not absolutely necessary for the assembly of this plaque protein. In the plasma membrane at least, the cytoplasmic portion of form I-desmocollin alone seems sufficient to induce and maintain the assembly of a complete plaque which then in turn can associate with IFs. This, however, does not exclude functional contributions of other desmosomal proteins, desmogleins included.

The marked dominant-negative effect which the cytoplasmic portion (tail) of desmoglein has exhibited when coupled to another transmembrane protein, connexin 32, and used for transfection of epithelial cells such as A-431 cells cannot yet be satisfactorily explained. Not only does the addition of desmoglein tail prevent the clustering of the connexin molecules into a gap junction structure, but it also results in the disassembly of all endogenous desmosomes and the release of the cytokeratin IF bundles from their desmosome anchorage at the plasma membrane. As a result, the addition of desmoglein tail portions seems generally to interfere with intercellular coupling and cytoskeletal organization.

The mechanism of this dispersive action of the cytoplasmic portion of desmoglein is not understood. Competition for certain binding proteins involved in desmosome formation and other processes is one possible hypothesis that has to be tested. To some extent, our observations with desmoglein resemble recent findings of an inhibition of E-cadherin-mediated cell adhesion by sequence elements located in the cytoplasmic portion of E-cadherin (Kintner 1992; see also Takeichi et al., this volume).

ACKNOWLEDGMENTS

We thank Drs. J. Kartenbeck and L. Eshkind, as well as Mrs. C. Grund, for valuable discussions and competent help, and Ms. E. Gundel and Mrs. A. Wohlfahrt for typing the manuscript. This work has been supported by the Deutsche Forschungsgemeinschaft.

REFERENCES

Amagai, M., V. Klaus-Kovtun, and J.R. Stanley. 1991. Autoantibodies against a novel epithelial cadherin in pemphigus vulgaris, a disease of cell adhesion. *Cell* **67:** 869.

Collins, J.E., P.K. Legan, T.P. Kenny, J. MacGarvie, J.L. Holton, and D.R. Garrod. 1991. Cloning and sequence analysis of desmosomal glycoprotein-2 and glycoprotein-3 (desmocollins)—Cadherin-like desmosomal adhesion molecules with heterogeneous cytoplasmic domains. *J. Cell Biol.* **113:** 381.

Cowin, P., H.-P. Kapprell, and W.W. Franke. 1985a. The complement of desmosomal plaque proteins in different cell types. *J. Cell Biol.* **101:** 1442.

Cowin, P., W.W. Franke, C. Grund, H.-P. Kapprell, and J. Kartenbeck. 1985b. The desmosome-intermediate filament complex. In *The cell in contact* (ed. G.M. Edelman and J.-P. Theiry), p. 427. Wiley, New York.

Cowin, P., H.-P. Kapprell, W.W. Franke, J. Tamkun, and R.O. Hynes. 1986. Plakoglobin: A protein common to different kinds of intercellular adhering junctions. *Cell* **46:** 1063.

Farquhar, M.G. and G.E. Palade. 1963. Junctional complexes in various epithelia. *J. Cell Biol.* **17:** 375.

Fouquet, B., R. Zimbelmann, and W.W. Franke. 1992. Identification of plakoglobin in oocytes and early embryos of *Xenopus laevis*: Maternal expression of a gene encoding a junctional plaque protein. *Differentiation* **51:** (in press).

Franke, W.W., M.D. Goldschmidt, R. Zimbelmann, H.M. Mueller, D.L. Schiller, and P. Cowin. 1989. Molecular cloning and amino acid sequence of human plakoglobin, the common junctional plaque protein. *Proc. Natl. Acad. Sci.* **86:** 4027.

Franke, W.W., R. Moll, D.L. Schiller, E. Schmid, J. Kartenbeck, and H. Müller. 1982. Desmoplakins of epithelial and myocardial desmosomes are immunologically and biochemically related. *Differentiation* **23:** 115.

Geiger, B., E. Schmid, and W.W. Franke. 1983. Spatial distribution of proteins specific for desmosomes and adhaerens junctions in epithelial cells demonstrated by double immunofluorescence microscopy. *Differentiation* **23:** 189.

Geiger, B., Z. Avnur, T. Volberg, and T. Volk. 1985. Molecular domains of adherens junctions. In *The cell in contact* (ed. G.M. Edelman and J.-P. Theiry), p. 461. Wiley, New York.

Green, K.J., D.A.D. Parry, P.M. Steinert, M.L.A. Virata, R.M. Wagner, B.D. Angst, and L.A. Nilles. 1990. Structures of the human desmoplakins. Implications for function in the desmosomal plaque. *J. Biol. Chem.* **265:** 2603.

Kapprell, H.-P., P. Cowin, and W.W. Franke. 1987. Biochemical characterization of the soluble form of the junctional plaque protein, plakoglobin, from different cell types. *Eur. J. Biochem.* **166:** 505.

Kintner, C. 1992. Regulation of embryonic cell adhesion by the cadherin cytoplasmic domain. *Cell* **69:** 225.

Koch, P.J., M.J. Walsh, M. Schmelz, and W.W. Franke. 1990. Identification of desmoglein, a constitutive desmosomal glycoprotein, as a member of the cadherin family of cell adhesion molecules. *Eur. J. Cell Biol.* **53:** 1.

Koch, P.J., M.D. Goldschmidt, M.J. Walsh, R. Zimbelmann, and W.W. Franke. 1991a. Complete amino acid sequence of the epidermal desmoglein precursor polypeptide and identification of a second type of desmoglein gene. *Eur. J. Cell Biol.* **55:** 200.

Koch, P.J., M.D. Goldschmidt, R. Zimbelmann, R. Troyanovsky, and W.W. Franke. 1992. Complexity and expression patterns of the desmosomal cadherins. *Proc. Natl. Acad. Sci.* **89:** 353.

Koch, P.J., M.D. Goldschmidt, M.J. Walsh, R. Zimbelmann, M. Schmelz, and W.W. Franke. 1991b. Amino acid sequence of bovine muzzle epithelial desmocollin derived from cloned cDNA: A novel subtype of desmosomal cadherins. *Differentiation* **47:** 29.

Leube, R.E., B. Wiedenmann, and W.W. Franke. 1989. Topogenesis and sorting of synaptophysin: Synthesis of a synaptic vesicle protein from a gene transfected into non-neuroendocrine cells. *Cell* **59:** 433.

Leube, R.E., B.L. Bader, F.X. Bosch, R. Zimbelmann, T. Achtstätter, and W.W. Franke. 1988. Molecular characterization and expression of the stratification-related cytokeratins 4 and 15. *J. Cell Biol.* **106:** 1249.

Magee, A.I. and R.S. Buxton. 1991. Transmembrane molecular assemblies regulated by the greater cadherin family. *Curr. Opin. Cell Biol.* **3:** 854.

Matsuzaki, F., R.M. Mége, S.H. Jaffe, D.R. Friedlander, W.J. Gallin, J.I. Goldberg, B.A. Cunningham, and G.M. Edelman. 1990. cDNAs of cell adhesion molecules of different specificity induce changes in cell shape and border formation in cultured S180 cells. *J. Cell Biol.* **110:** 1239.

McCrea, P.D., C.W. Tuck, and B. Gumbiner. 1991. A homolog of the *armadillo* protein in *Drosophila* (plakoglobin) associated with E-cadherin. *Science* **254:** 1359.

Mechanic, S., K. Raynor, J.E. Hill, and P. Cowin. 1991. Desmocollins form a subset of the cadherin family of cell adhesion molecules. *Proc. Natl. Acad. Sci.* **88:** 4476.

Moll, R., P. Cowin, H.-P. Kapprell, and W.W. Franke. 1986. Biology of disease. Desmosomal proteins: New markers for identification and classification of tumors. *Lab. Invest.* **54:** 4.

Moll, R., W.W. Franke, D.L. Schiller, B. Geiger, and R. Krepler. 1982. The catalog of human cytokeratins: Patterns of expression in normal epithelia, tumors and cultured cells. *Cell* **31:** 11.

Nagafuchi, A. and M. Takeichi. 1988. Cell binding function of E-cadherin is regulated by the cytoplasmic domain. *EMBO J.* **7:** 3679.

Nagafuchi, A., M. Takeichi, and S. Tsukita. 1991. The 102 kd cadherin-associated protein: Similarity to vinculin and posttranscriptional regulation of expression. *Cell* **65:** 849.

Nilles, I.A., D.A.D. Perry, E.E. Powers, B.D. Angst, R.M. Wagner, and K.J. Green. 1991. Analysis and expression of human desmoglein: Cadherin-like component of the desmosome. *J. Cell Sci.* **99:** 809.

Ozawa, M., H. Baribault, and R. Kemler. 1989. The cytoplasmic domain of the cell adhesion molecule uvomorulin associates with three independent proteins structurally related in different species. *EMBO J.* **8:** 1711.

Ozawa, M., M. Ringwald, and R. Kemler. 1990. Uvomorulin-catenin complex formation is regulated by a specific domain in the cytoplasmic region of the cell adhesion molecule. *Proc. Natl. Acad. Sci.* **87:** 4246.

Parker, A.E., G.N. Wheeler, J. Arnemann, S.C. Pidsley, P. Ataliotis, C.L. Thomas, D.A. Rees, A.I. Magee, and R.S. Buxton. 1991. Desmosomal glycoproteins II and III: Cadherin-like junctional molecules generated by alternative splicing. *J. Biol. Chem.* **266:** 10438.

Paul, D.L. 1986. Molecular cloning of cDNA for rat liver gap junction protein. *J. Cell Biol.* **103:** 123.

Peifer, M. and E. Wieschaus. 1990. The segment polarity gene *Armadillo* encodes a functionally modular protein that is the *Drosophila* homolog of human plakoglobin. *Cell* **63:** 1167.

Schmelz, M., R. Duden, P. Cowin, and W.W. Franke. 1986. A constitutive trans-membrane glycoprotein of M_r 165,000 (desmoglein) in epidermal and non-epidermal desmosomes. II. Immunolocalization and microinjection studies. *Eur. J. Cell Biol.* **42:** 184.

Schwarz, M.A., K. Owaribe, J. Kartenbeck, and W.W. Franke. 1990. Desmosomes and hemidesmosomes: Constitutive molecular components. *Annu. Rev. Cell Biol.* **6:** 461.

Steinert, P.M. and D.R. Roop. 1988. Molecular and cellular biology of intermediate filaments. *Annu. Rev. Biochem.* **57:** 593.

Takeichi, M. 1990. Cadherins: A molecular family important in selective cell-cell adhesion. *Annu. Rev. Biochem.* **59:** 237.

Troyanovsky, S.M., V.I. Guelstein, T.A. Tchipysheva, V.A. Krutosvkikh, and G.A. Bannikov. 1989. Patterns of expression of keratin 17 in human epithelia: Dependency of cell position. *J. Cell Sci.* **93:** 419.

Virata, M.L.A., R.M. Wagner, D.A.D. Parry, and K.J. Green. 1992. Molecular structure of the human desmoplakin I and II amino terminus. *Proc. Natl. Acad. Sci.* **89:** 544.

Wheeler, G.N., E.E. Parker, C.L. Thomas, P. Ataiotis, D. Poynter, J. Arnemann, A.J. Rutman, S.C. Pidsley, F.M. Watt, D.A. Rees, R.S. Buxton, and A.L. Magee. 1991. Desmosomal glycoprotein I, a component of intercellular desmosome junctions, is related to the cadherin family of cell adhesion molecules. *Proc. Natl. Acad. Sci.* **88:** 4796.

Role of the Zygotic Genome in the Restructuring of the Actin Cytoskeleton at the Cycle-14 Transition during *Drosophila* Embryogenesis

E.D. Schejter, L.S. Rose, M.A. Postner, and E. Wieschaus
Department of Molecular Biology, Princeton University, Princeton, New Jersey 08544

In many multicellular organisms, the early stages of embryonic development are characterized by rapid cleavage divisions, which increase the number of cells but do not contribute greatly to the developmental diversity of the embryo. In amphibian embryos, this early period of synchronous mitotic cycles gives way to the "mid-blastula transition," during which the cell cycle is lengthened, cell divisions become asynchronous, and cells acquire new characteristics, such as motile activity (Kirschner et al. 1985). In *Drosophila*, an analogous event occurs after the syncytial nuclei have undergone 13 synchronous divisions. At the onset of the "cycle-14 transition," the nuclei temporarily stop dividing, and during the period that follows, leading up to gastrulation, several important morphological changes take place. These include a clearing of yolk and lipid from the cortical cytoplasm, a restructuring of the cortical actin cytoskeleton into a contractile hexagonal network, an increase in plasma membrane, and finally, cellularization of the blastoderm.

An important characteristic of the mid-blastula transition in most organisms is that it represents a shift from dependence on maternally supplied gene products to an increasing reliance on gene products made in the embryo itself. This conclusion is based primarily on the strong burst of zygotic transcription associated with the transition. Similar observations made in *Drosophila* (Edgar and Schubiger 1986) have been complemented by genetic data produced in our laboratory (Merrill et al. 1988; Wieschaus and Sweeton 1988) and are summarized below. These results have demonstrated that the first overt requirements for zygotic gene activity are observed during the cycle-14 transition. This paper focuses primarily on one major feature of the transition, namely, blastoderm cellularization, and on the contribution of three zygotically expressed genes to this process.

Requirement for Zygotic Gene Activity during the Cycle-14 Transition

The earliest stages of embryonic development in *Drosophila* are characterized by rapid and synchronous rounds of nuclear division. These take place initially within the interior of the embryo, and later, following a concerted process of nuclear migration, at the embryonic cortex. Since these divisions are not associated with cytokinesis, at the onset of cycle 14, the embryo is a syncytium, with about 6000 nuclei at its surface. Development to this stage is fully driven by maternally supplied gene products, as was demonstrated by introduction of RNA synthesis inhibitors such as α-amanitin into early embryos (Arking and Parente 1980; Edgar et al. 1986). In these treated embryos, the normal course of development through the first 13 cleavage divisions is unaltered. When the embryos reach cycle 14, however, they fail to show the characteristic morphological changes associated with the cycle-14 transition. Most prominently, the embryo is completely blocked in cellularization.

This observation suggests that the events of the cycle-14 transition require the transcription of genes whose products function in the processes that bring about these morphological changes. Because α-amanitin-treated embryos develop normally up to cycle 14, these genes represent the earliest visible requirements for zygotic transcription in *Drosophila*. To identify these early-acting zygotic genes, we have studied the development of embryos homozygously deleted for cytologically defined chromosomal segments spanning the entire genome (Merrill et al. 1988; Wieschaus and Sweeton 1988). Several conclusions could be drawn from this work.

1. Large fractions of the genome could be eliminated with no deleterious effect on the morphological changes associated with the cycle-14 transition, indicating that such genes are rare.
2. Only seven loci that are required during the cycle-14 transition were identified, and these were mapped to specific chromosomal regions. One of the seven loci affects nuclear spacing and the cell cycle pause, another affects the clearing of the cortical cytoplasm, three loci affect the restructuring of the actin cytoskeleton, and two affect the production of cell membrane. The effects of RNA synthesis inhibition on the cycle-14 transition can be attributed to the summed requirements for these seven loci.
3. Complete disruption of the cycle-14 transition cannot be achieved by the elimination of any single chromosomal region. There thus seems to be no master regulatory locus responsible for all aspects of the cycle-14 transition. Instead, elimination of individual loci results in alterations of specific transition events.

4. When a particular aspect of the transition is blocked, the remaining cycle-14 program proceeds with its normal kinetics and morphology, arguing that the transition is composed of separate parallel processes.

Much of our recent work has focused on the three genes that affect the cellularization process by regulating the restructuring of the cortical actin cytoskeleton. Presented below are results that provide phenomenological and molecular detail to the means by which the zygotic genome contributes to this aspect of the cycle-14 transition.

Features of the Process of Blastoderm Cellularization

The cellularization process involves the invagination of membrane down from the surface and between the cortical nuclei, which are arranged in an hexagonal array. Actin and myosin are strongly concentrated at the inward tips of the advancing cell membranes. When viewed parallel to the surface, these cytoskeletal elements are found initially to form a hexagonal network over the entire embryo. As the cell membranes deepen, the network is broken into individual rings, which gradually constrict as the cells close off basally (Fullilove and Jacobson 1971; Warn and Magrath 1983; Katoh and Ishikawa 1989; Young et al. 1990). The inhibition of cell membrane formation by actin depolymerizing agents (Zalokar and Erk 1976; Foe and Alberts 1983) and by anti-myosin antibodies (Lutz and Kiehart 1987) strongly suggests that the actin-myosin network present at the cellularization front is functionally required for cellularization. On the basis of these findings and of our own observations of the structural changes and dynamics of the process, we favor a mechanistic explanation for cellularization that is closely modeled after the contractile ring mechanism thought to govern conventional cytokinesis (Schroeder 1990). In this model, contraction of the actin-myosin network provides the mechanical force that pulls membrane into the interior between individual nuclei during cellularization (Fig. 1). Although this contraction is taking place parallel to the embryonic surface, the linking up of the individual contractile units into a network over the entire embryo results in a net inward motion. Attachment of the network to the plasma membrane results in membrane invagination, by a process akin to the contraction of a net over an inflated balloon. Whether contraction of the actin-myosin network suffices to explain blastoderm cellularization, or whether additional processes are involved, is still a matter of debate (Warn and Robert-Nicoud 1990).

Cell membrane formation during cellularization follows upon the restructuring of the actin-based cytoskeleton directly underlying the plasma membrane (Simpson and Wieschaus 1990). The cortical actin cytoskeleton undergoes two major transitions during early embryonic development. The first transition takes place when the nuclei migrate from the interior into the cortex at cycle 10. At this time, the homogeneous layer of cortical actin becomes organized into cap-like structures above each nucleus, which undergo stereotypic expansion and reformation with each succeeding round of nuclear division (Karr and Alberts 1986). As the nuclei become more densely packed, the margins of the caps become juxtaposed. The second transition is associated with the events of cycle 14. At the onset of cycle 14, the caps reform, with the actin along the adjoining cap margins exhibiting a mesh-like organization. Shortly thereafter, this meshwork resolves into a hexagonal pattern of fine lines, giving rise to the contractile network described previously (Fig. 2). This restructuring of the actin-based cytoskeleton provides the juncture at which zygotic gene products first influence the cellularization process.

Requirements for Zygotic Gene Activity in Formation of the Cycle-14 Cortical Cytoskeleton

Three of the genetic loci that are required for cycle-14 functions specifically affect the restructuring of the cortical actin-based cytoskeleton (Fig. 2) (Merrill et al. 1988; Wieschaus and Sweeton 1988; Simpson and Wieschaus 1990; E. Schejter and E. Wieschaus, unpubl.). In embryos deficient for either the *nullo* or *serendipity-α* (*sry-α*) loci, many of the interfaces between adjacent cells are missing, resulting in the appearance of multinucleated cells. This defect can be attributed to an early failure in proper formation of the actin-based hexagonal array (Fig. 2E). Although the resolution of the actin meshwork in these embryos into aligned fibers is properly initiated, the resulting array is nonuniform. The actin lines vary in thickness, and many become discontinuous. In embryos missing the third locus, *bottleneck* (*bnk*), the newly resolved hexagonal network is rapidly converted into a series of uneven contractile rings, reminiscent of the appearance of the cellularization front during the late stages of the process (Fig. 2F). A variety of secondary features are observed in *bnk* embryos, including misplaced nuclei and multinucleated cells. In all three cases, disruption of the network is partial, and membrane extension proceeds. The resulting cellular blastoderms contain many aberrantly shaped and incomplete cells.

The cellular mechanisms responsible for the reorganization of the cortical actin cytoskeleton into a contractile hexagonal array are unknown. By analogy to a key element in a proposed model for conventional cytokinesis (White and Borisy 1983), one possible mechanism for the reshaping of a mesh-like array of actin into the fine lines of the hexagonal array is by the induction of contractility. If the actin fibers in the network are linked, then as the meshwork begins to contract, the tension will straighten out the fibers and force them into a parallel alignment. This model indicates that an inability to contract the network properly will result in an unresolved, mesh-like array in cycle 14.

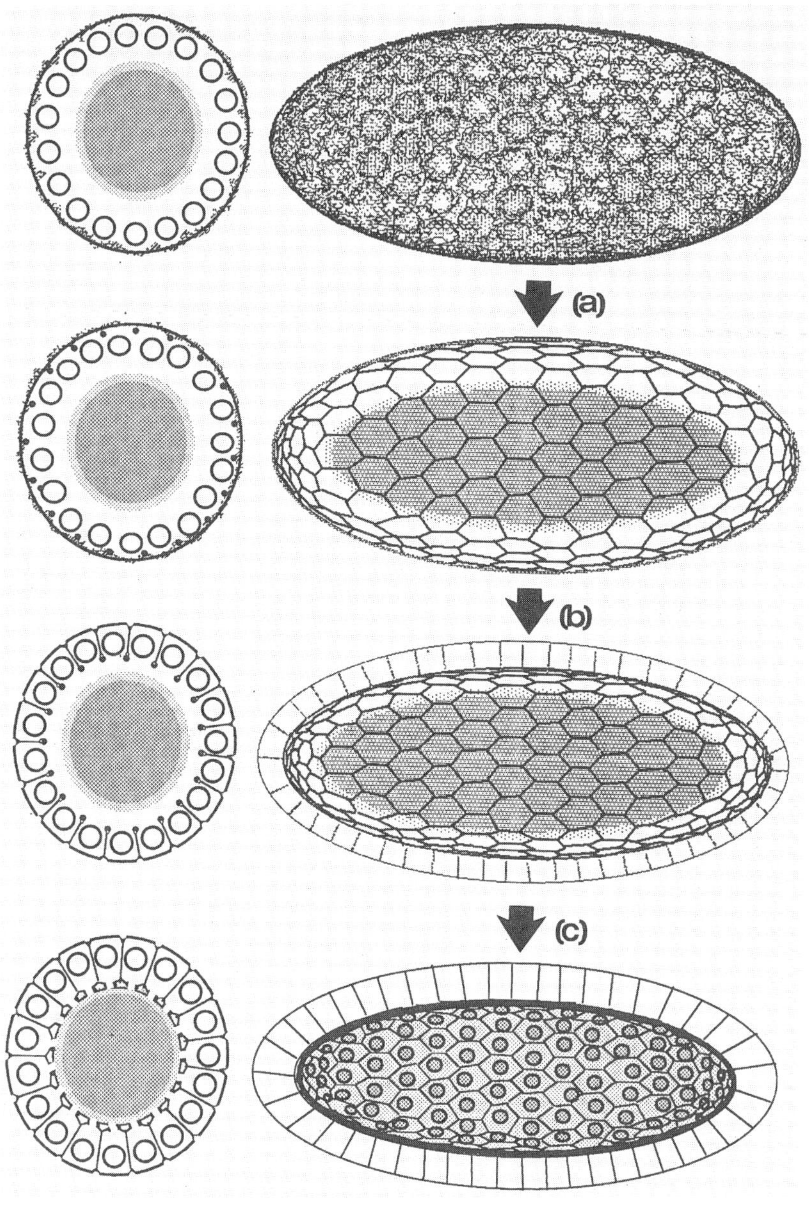

Figure 1. A schematic representation and mechanistic model of blastoderm cellularization. Lateral views of a cellularizing blastoderm are shown, along with illustrations of cross sections (left-hand column) at the corresponding stages. At the onset of cycle 14 (*top*), the actin-based cytoskeleton forms cap structures above all cortical nuclei (shown as circles in section). A period of transition follows (*a*; and see Fig. 2), during which a hexagonal actin array is formed. The array contracts (*b*) and, as a result, pulls membrane in between neighboring nuclei, through the cleared cortical cytoplasm, and toward the lipid and yolk-rich interior (stippled area). As the cellularization front passes the bases of the nuclei (*c*), the contractile network rounds up and, while the membranes continue to invaginate, eventually closes off the cells basally.

Since no such phenotype is produced in embryos lacking *nullo*, *sry*-α, or *bnk* or, for that matter, in embryos lacking any portion of their zygotic genome, we conclude that contractility is not dependent on zygotic gene activity. The function of the three genes appears to be required for maintaining the structural integrity of the actin-myosin network once contraction has initiated. In the proposed model, both factors that bridge the individual fibers of adjacent contractile units and factors that stabilize contacts within units would be required. In embryos lacking bridging factors, the array would begin to constrict but would break up into individual rings, as observed in *bnk* embryos. Lack of stabilizing contacts within contractile units would allow fibers to be pulled from one interface and accumulate at another, as seen in *nullo* and *sry*-α embryos.

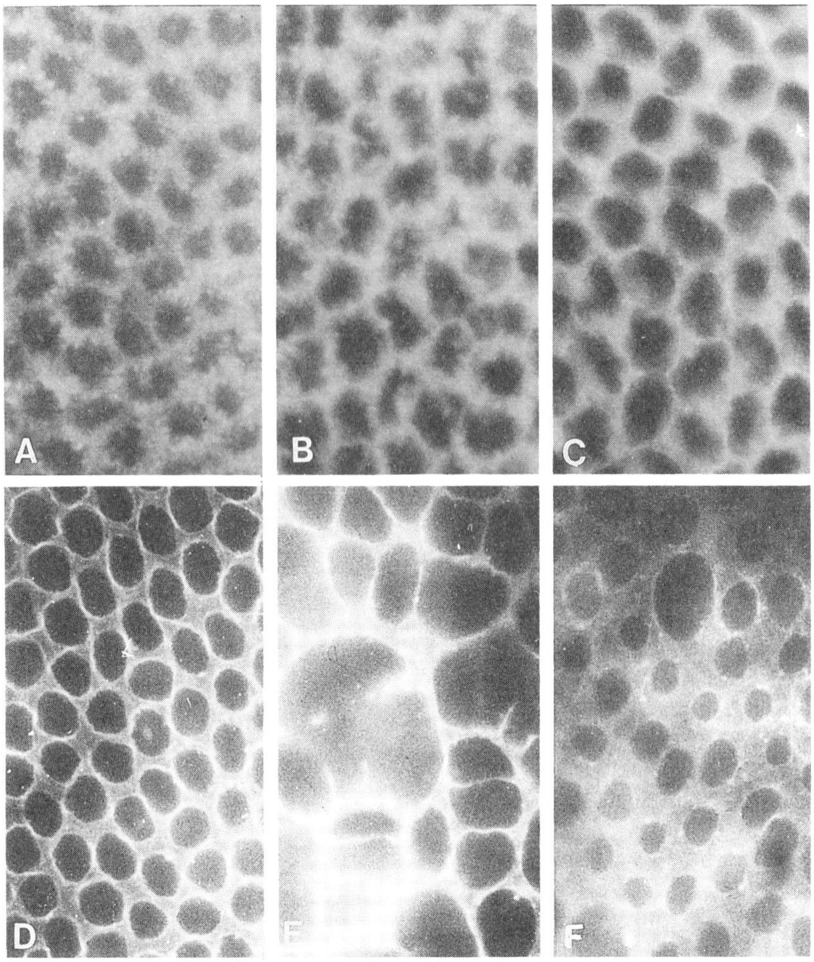

Figure 2. (*A–C*) Resolution of the hexagonal actin network in early cycle-14 embryos. At the base of the perinuclear caps, F-actin, visualized with rhodamine-labeled phalloidin, exhibits a mesh-like quality (*A*), which begins to sharpen within a few minutes into an array of thin lines (*B, C*). The final, sharp array lacks any mesh-like features (not shown). These initial stages are unaffected in the three mutants. (*D–F*) Contractile array in wild-type and mutant embryos. As cell membranes are extended, the array begins to round up (*D*) and will eventually close off the cells basally. In embryos lacking *nullo* or *serendipity-α* (*E*), the interfaces of the hexagonal array vary in thickness and are often fragmented and discontinuous, giving rise to multinucleated cells. The contractile array of *bottleneck* embryos early in cycle 14 (*F*) exhibits the stretched and contracted appearance normally found considerably later in wild-type embryos.

Molecular Features of *nullo*, *serendipity-α*, and *bottleneck*

All three of the genes zygotically required for proper formation of the contractile actin network have been cloned (Vincent et al. 1985; Schweisguth et al. 1990; Rose and Wieschaus 1992, E. Schejter and E. Wieschaus, unpubl.). All three genes are intron-less and encode blastoderm-specific transcripts. The protein products are relatively small (24–56 kD) and bear no obvious sequence similarities to previously described proteins. Both the *nullo* and *bnk* proteins are highly charged, with a large excess of basic residues. The *nullo* protein contains a potential amino-terminal myristoylation site. The inability to obtain point mutations in these genes, despite exhaustive rounds of mutagenesis, can perhaps be explained by the finding that the major structural feature of two of the proteins is that of excess charge. If this global feature is of functional significance, it is understandable why single primary sequence changes may have little effect.

A particularly striking feature of these genes is that they are transcribed exclusively during the late stages of the syncytial blastoderm. The earliest signs of transcription for all three coincide with the general onset of zygotic transcription during cleavage cycles 10 and 11. The kinetics of transcript accumulation and degradation differ somewhat among the three. *nullo* (Fig. 3) and *bnk* exhibit sharp peaks of transcription during the cycle-13 mitosis and very early in cycle 14, respectively, after which both transcripts completely disappear within a very short time. *sry-α* transcripts reach their expression peak later in cycle 14 and persist somewhat longer, but they too cannot be detected by the end of cellularization. No restrictions are observed on the spatial distribution of any of the three transcripts.

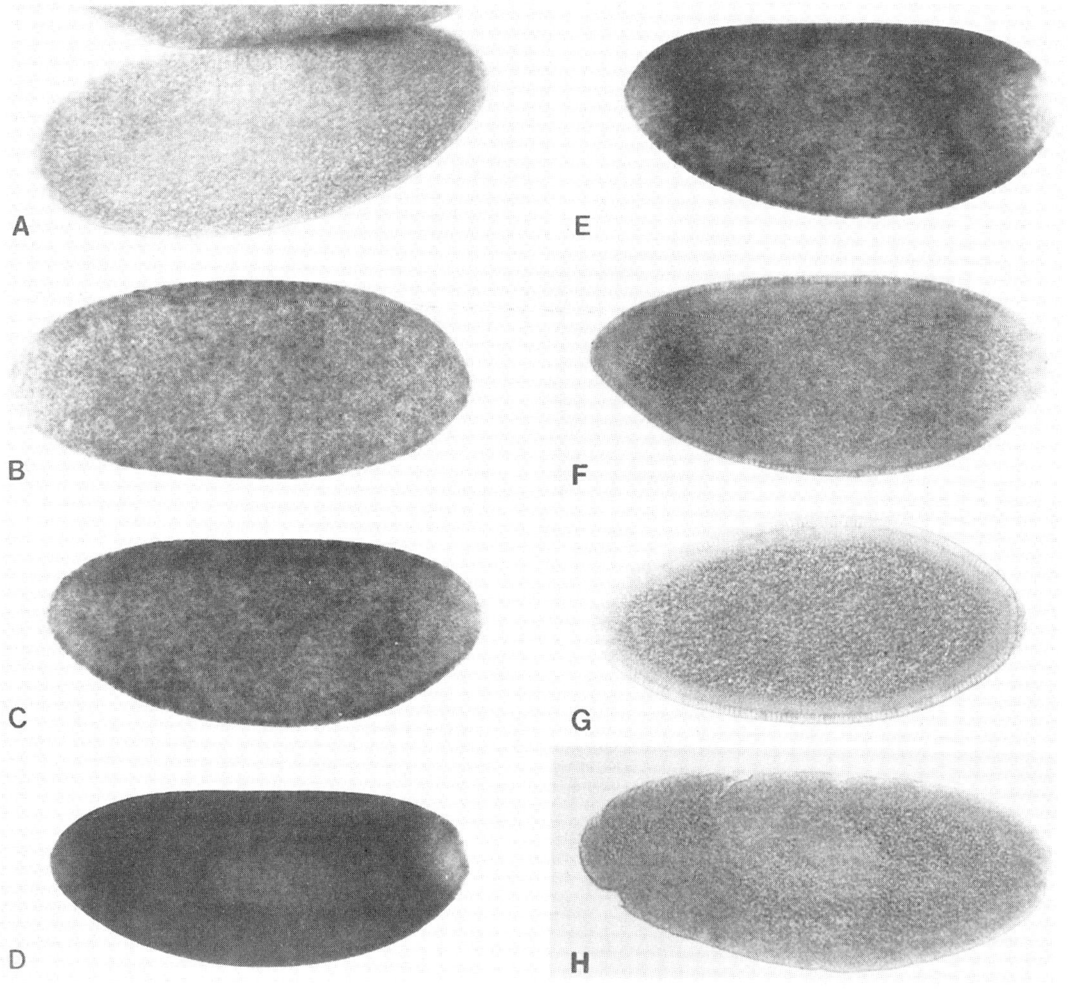

Figure 3. In situ hybridization to *nullo* transcripts in wild-type embryos during early embryogenesis. Anterior is to the left. (*A*) Cleavage-stage embryo; (*B*) cycle-12 embryo; (*C*) cycle-13 embryo, interphase; (*D*) cycle-13 embryo, mitosis; (*E*) cycle-14 embryo, early interphase, during formation of the contractile array; (*F*) cycle-14 embryo during the earliest stages of membrane invagination; (*G*) cycle-14 embryo membranes have invaginated to bases of nuclei; (*H*) extended germ-band-stage embryo. (Reprinted, with permission, from Rose and Wieschaus 1992.)

The Cycle-14 Transition and the Overt Requirement for Zygotic Gene Activity: How Are They Related?

The cycle-14 transition represents a critical stage in the early development of the *Drosophila* embryo. During this period and in its wake, several developmental pathways are initiated or altered in a profound way. It is during the cycle-14 transition that reliance on gene products derived from the zygotic genome first becomes apparent. Understanding the manner in which the shift to zygotic control is achieved is of importance to the elucidation of the developmental program of the early embryo.

A prominent feature of the cycle-14 transition is the arrest and eventual shift in the cell cycle program from rapid, synchronous, and ubiquitous mitotic cycles to a pattern of temporally and spatially restricted divisions. As in other organisms that experience a mid-blastula transition, this aspect has been shown to depend on the nucleo-cytoplasmic ratio (Edgar et al. 1986) and has been given a genetic and molecular description with the cloning of *string*, the *Drosophila* homolog of *cdc25* (Edgar and O'Farrell 1989). We have been studying in detail another feature of the cycle-14 transition: the restructuring of the cortical actin cytoskeleton in anticipation of blastoderm cellularization. In the course of these studies, we have been able to provide a detailed description of this event and to identify and characterize a number of zygotic gene products that are functionally required for proper execution of the process. The results of these studies, some of which are described above, enable us to speculate on the role zygotic gene activity plays in the cycle-14 transition.

One attractive possibility is that it is the presence or absence of at least some of the relevant zygotic gene products that brings about the cytoskeletal reorganization at the cycle-14 transition. According to this view, maternally provided reagents such as actin, myosin, and "contractility factors" may be poised to act to initiate cellularization but require a triggering event that is supplied by zygotic gene activity. Several features of the *nullo* gene illustrate how such an event may be regulated. Analysis of embryos genetically mosaic for *nullo* (Simpson and Wieschaus 1990) showed *nullo* function to be extremely nonautonomous, in a manner that suggested that the *nullo* product was highly diffusible and limited in dosage. A signal that is kept at a constant level throughout the embryo cortex, but that functions only after accumulation to some threshold level, would be well suited to orchestrate the cytoskeletal transition of cycle 14. Our analysis of the embryonic phenotypes produced following elimination of defined regions of the zygotic genome argues against the presence of a single genetic locus that may play this master triggering role. However, it is possible that it is the summed activity of several loci that constitute this central function. To test these notions, two complementary sets of experiments are now being planned. One approach seeks to eliminate combinations of the various zygotic genes simultaneously and to assess the phenotypic consequences. A second approach will attempt to produce high levels of the zygotic gene products prematurely, in an effort to determine whether the embryo can restructure its cortical cytoskeleton prior to cycle 14. The molecular cloning of the three genes involved in this process and the availability of antisera (Schweisguth et al. 1990; M. Postner and E. Wieschaus, unpubl.) to monitor their expression make these experiments feasible.

A second aspect worthy of consideration is the manner in which events of the cycle-14 transition control the expression of the gene products that are required during the transition. Although initiation of transcription and the accumulation of transcripts from all three genes follow a general pattern for early zygotic expression, the extremely rapid disappearance of the transcripts during cycle 14 is unusual. The maternal mutation *maternal haploid* produces haploid embryos in which the cycle-14 transition is altered. In these embryos, the proper nucleo-cytoplasmic ratio required for cell cycle arrest is not obtained following the 13th mitotic cycle, and an additional 14th round of nuclear division takes place (Edgar et al. 1986). Transcription initiation and transcript accumulation of *nullo* and *sry-α* in *maternal haploid* embryos follow their normal course, but during cycle 14, the transcript levels remain high, and it is only following the 14th mitosis that the transcripts disappear (Rose and Wieschaus 1992). These observations suggest that a program of RNA transcript degradation is initiated with the onset of the cycle-14 transition. We are interested in identifying regulatory sequences within the three genes that respond to this program, which may be of general significance. In this context, it is relevant to inquire why expression of the three genes is so severely restricted. By forcing expression at inappropriate developmental stages, it may be possible to gain insight into the actual functions of these zygotic gene products.

A final issue is that of the timing mechanisms employed by the embryo in bringing about the various aspects of the cycle-14 transition. As discussed previously, the arrest and shift in the cell cycle program are dependent on the nucleo-cytoplasmic ratio. We have proposed that restructuring of the cortical actin cytoskeleton may be governed by a "long-term" clock that responds to the synthesis and accumulation of zygotic gene products. From other observations, we would like to argue that a second timing mechanism is utilized as well. Under certain conditions, including those imposed by genetic gynandromorphs (Simpson and Wieschaus 1990), local patches of the embryo cortex display a slight asynchrony of cell cycle events. When this occurs, the regions that enter cycle 14 first also reorganize their cytoskeleton early. This suggests that the restructuring can only occur at a fixed point in the cell cycle, implying the presence of a "short-term," cell-cycle-dependent clock that participates in these events.

REFERENCES

Arking, R. and A. Parente. 1980. Effects of RNA inhibitors on the development of *Drosophila* embryos permeabilized by a new technique. *J. Exp. Zool.* **212:** 183.

Edgar, B.A. and P.H. O'Farrell. 1989. Genetic control of cell division patterns in the *Drosophila* embryo. *Cell* **57:** 177.

Edgar, B.A. and G. Schubiger. 1986. Parameters controlling transcriptional activation during early *Drosophila* development. *Cell* **44:** 871.

Edgar, B.A., C.P. Kiehle, and G. Schubiger. 1986. Cell cycle control by the nucleo-cytoplasmic ratio in early *Drosophila* development. *Cell* **44:** 365.

Foe, V.E. and B.M. Alberts. 1983. Studies of nuclear and cytoplasmic behavior during the five mitotic cycles that precede gastrulation in *Drosophila* embryogenesis. *J. Cell Sci.* **61:** 31.

Fullilove, S.L. and A.G. Jacobson. 1971. Nuclear elongation and cytokinesis in *Drosophila montana*. *Dev. Biol.* **26:** 560.

Karr, T.L. and B.M. Alberts. 1986. Organization of the cytoskeleton in early *Drosophila* embryos. *J. Cell Biol.* **102:** 1494

Katoh, K. and H. Ishikawa. 1989. The cytoskeletal involvement in cellularization of the *Drosophila melanogaster* embryo. *Protoplasma* **150:** 83.

Kirschner, M., J. Newport, and J. Gerhart. 1985. The timing of early developmental events in *Xenopus*. *Trends Genet.* **1:** 41.

Lutz, D.A. and D.P. Kiehart. 1987. Localized disruption of cortical organization following microinjection of cytoplasmic myosin antibody into pre-cellular *Drosophila* embryos. *J. Cell Biol.* **105:** 127a.

Merrill, P.T., D. Sweeton, and E. Wieschaus. 1988. Requirements for autosomal gene activity during precellular stages of *Drosophila melanogaster*. *Development* **104:** 495.

Rose, L.S. and E. Wieschaus. 1992. The *Drosophila* cellularization gene *nullo* produces a blastoderm specific transcript whose levels respond to the nucleo-cytoplasmic ratio. *Genes Dev.* (in press).

Schroeder, T.E. 1990. The contractile ring and furrowing in dividing cells. *Ann. N.Y. Acad. Sci.* **582:** 78.

Schweisguth, F., J. Lepesant, and A. Vincent. 1990. The membrane-associated *serendipity alpha* protein is required for the cellularization of the *Drosophila* embryo. *Genes Dev.* **4:** 922.

Simpson, L.D. and E. Wieschaus. 1990. Zygotic activity of the *nullo* locus is required to stabilize the actin-myosin network during cellularization in *Drosophila*. *Development* **110:** 851.

Vincent, A., H.V. Colot, and M. Rosbash. 1985. Sequence and structure of the serendipity locus of *Drosophila melanogaster*. A densely transcribed region including a blastoderm-specific gene. *J. Mol. Biol.* **186:** 149.

Warn, R.M. and R. Magrath. 1983. F-actin distribution during the cellularization of the *Drosophila* embryo visualized with FL-phalloidin. *Exp. Cell Res.* **143:** 103.

Warn, R.M. and M. Robert-Nicoud. 1990. F-actin organization during the cellularization of the *Drosophila* embryo as revealed with a confocal laser scanning microscope. *J. Cell Sci.* **96:** 35.

White, J.G. and G.G. Borisy. 1983. On the mechanisms of cytokinesis in animal cells. *J. Theor. Biol.* **101:** 289.

Wieschaus, E. and D. Sweeton. 1988. Requirements for X-linked zygotic gene activity during cellularization of early *Drosophila* embryos. *Development* **104:** 483.

Young, P.E., T.C. Pesacreta, and D.P Kiehart. 1990. Dynamic changes in the distribution of cytoplasmic myosin during *Drosophila* embryogenesis. *Development* **111:** 1.

Zalokar, M. and I. Erk. 1976. Division and migration of nuclei during early embryogenesis of *Drosophila melanogaster*. *J. Microsc. Biol. Cell.* **25:** 97.

Distinct Patterns of Actin Organization Regulated by the Small GTP-binding Proteins Rac and Rho

A.J. RIDLEY AND A. HALL
Institute of Cancer Research, Chester Beatty Laboratories, London SW3 6JB, United Kingdom

Many extracellular signals stimulate cellular responses that involve changes in the organization of actin microfilaments. For example, directional motility induced by chemoattractants requires precise spatial control of actin polymerization and depolymerization (Cooper 1991). Actin reorganization is also an early response to several growth factors, although it is not known what role such alterations play in cell growth control (Burridge et al. 1988). As yet, the signal transduction pathways linking transmembrane receptors to the actin cytoskeleton have not been delineated, but the targets are likely to include some of the many actin-binding proteins that can modulate actin polymerization and organization (Vandekerckhove 1990; Bretscher 1991).

The superfamily of Ras-related GTP-binding proteins consists of more than 40 members and can be divided into four subfamilies on the basis of sequence homology: Ras, Rho, Rab, and Arf, a grouping that also reflects similarities in function (Balch 1990; Hall 1990; Kahn et al. 1991). Ras proteins themselves have been shown to be essential for many receptor-mediated signal transduction pathways leading to either cell proliferation or differentiation. Members of the Rab and Arf subfamilies are involved in intracellular vesicle trafficking, and recent evidence indicates that the Rho subfamily regulates actin organization.

The mammalian Rho subfamily includes Rho (A, B, and C), Rac (1 and 2), CDC42 (CDC42Hs and G25K), rhoG, and TC10 (Hall 1990; Munemitsu et al. 1990; Shinjo et al. 1990; Vincent et al. 1992). Biochemical characterization of the mammalian Rho and Rac proteins has shown that, like Ras, they bind and hydrolyze GTP, and that their intrinsic GTPase activity can be stimulated by several GTPase activating proteins (GAPs) (Garrett et al. 1989, 1991; Diekmann et al. 1991). In addition, two proteins, rhoGDI and smgGDS, have been characterized that can alter the rate of nucleotide exchange on Rho and Rac (Fukumoto et al. 1990; Hiraoka et al. 1992).

Evidence that Rho proteins could be involved in regulating actin cytoskeletal organization came initially through the use of an exoenzyme, C3 transferase, produced by *Clostridium botulinum*, which ADP-ribosylates Rho proteins on amino acid Asn-41 (Aktories et al. 1989; Sekine et al. 1989). When C3 transferase is introduced into a variety of cell types, they lose their actin stress fibers and round up (Rubin et al. 1988; Chardin et al. 1989; Paterson et al. 1990). More direct evidence that Rho acts on the actin cytoskeleton came from microinjecting cells with recombinant Rho proteins or plasmids expressing *rho* cDNAs. Rho was found to induce a rapid reorganization of actin into stress fibers in fibroblasts (Paterson et al. 1990).

The close sequence homology (60%) between Rac and Rho suggested that Rac might also affect actin filament distribution, and we show here that recombinant Rac protein indeed stimulates actin reorganization to form membrane ruffles. In addition, we have analyzed early changes in the actin cytoskeleton following stimulation of quiescent serum-starved Swiss 3T3 fibroblasts by a variety of growth factors, and we have shown that Rac and Rho are selectively required for distinct patterns of actin reorganization.

MATERIALS AND METHODS

Lysophosphatidic acid (LPA), phorbol 12-myristate 13-acetate (PMA), and insulin were obtained from Sigma Chemical Company, bombesin from Calbiochem, platelet-derived growth factor (PDGF) from Amersham International, and epidermal growth factor (EGF) from Collaborative Research.

Cell culture and microinjection. Swiss 3T3 cells were maintained in Dulbecco's modified Eagle's medium (DMEM) containing 10% bovine fetal calf serum (FCS). For growth factor stimulation and microinjection, cells were seeded at 5×10^4/well in 15-mm-diameter wells containing 13-mm circular glass coverslips. To facilitate localization of microinjected cells, a cross was scored on coverslips with a glass pen. After 7–10 days, cells were starved in DMEM containing 0.2% $NaHCO_3$ for 16 hours. In each experiment, approximately 150 adjacent cells were microinjected in one quadrant of the cross. The efficiency of injection was determined by coinjecting mouse immunoglobulin at 0.5 mg/ml with the protein being assayed. The same batch of C3 transferase was used in all experiments described.

Immunofluorescence. For actin localization, cells were fixed in 3% paraformaldehyde for 20 minutes, permeabilized with 0.2% Triton X-100 for 5 minutes, and incubated with 0.1 μg/ml tetramethylrhodamine-isothiocyanate (TRITC)-labeled phalloidin (Sigma Chemical Co.) for 40 minutes. For localization of vinculin together with actin, cells were washed with ice-cold 2-[*N*-morpholino]ethane sulfonate (MES) buffer

(50 mM MES [pH 6.0], 5 mM $MgCl_2$, 3 mM EGTA), briefly permeabilized at 0°C for 1 minute with 0.5% Triton X-100 in MES buffer, washed with MES buffer, and fixed in 3% paraformaldehyde for 20 minutes. They were incubated with a 1:200 dilution of the mouse anti-vinculin antibody 284 (a gift from Michael Wilkinson) for 60 minutes, followed by 1:200 TRITC-conjugated goat anti-mouse IgG (Sigma Chemical Co.) together with 0.1 μg/ml fluorescein isothiocyanate (FITC)-labeled phalloidin. Cells were viewed with a Zeiss microscope and photographed with Kodak T-MAX 400 film.

Expression and purification of recombinant proteins. Normal and V14RhoA cDNAs and a V12 Ha-Ras cDNA were expressed under the control of a tryptophan promoter in *Escherichia coli*, and proteins were purified as described previously (Garrett et al. 1989; Paterson et al. 1990). Rac1 cDNAs were subcloned into the pGEX-2T vector (Smith and Johnson 1988) and purified to homogeneity as described previously (Ridley et al. 1992). Active protein concentrations were determined by a filter binding assay using [^{3}H]GDP or [^{3}H]GTP, as described previously (Hall and Self 1986).

RESULTS

RhoA Induces the Assembly of Stress Fibers and Focal Adhesions

By microinjecting purified recombinant proteins into fibroblasts, it has been shown that both RhoA and a constitutively activated mutant, V(Val)14RhoA, rapidly stimulate the formation of actin stress fibers (Paterson et al. 1990). The ends of stress fibers are normally associated with the plasma membrane at distinct areas known as focal adhesions. Here, integrin receptors are clustered, and their extracellular domains bind proteins such as fibronectin and collagen. At the intracellular side of focal adhesions, a number of cytoplasmic proteins are localized, including vinculin, talin, α-actinin, and tensin (Burridge et al. 1988). To determine whether Rho-induced stress fiber formation was accompanied by the clustering of such proteins into new focal adhesions, the localization of actin, vinculin, and talin was followed by immunofluorescence over a 2-hour time course after microinjection of V14RhoA. Confluent, serum-starved Swiss 3T3 cells had very few stress fibers or membrane ruffles, and filamentous actin was localized principally at the plasma membrane in a cortical actin network and also as punctate aggregates (Fig. 1a). Vinculin (Fig. 1c) and talin (not shown) were localized to the cytoplasm, with a slight clustering at the ends of the rare stress fibers. Microinjection of V14RhoA led within 5 minutes to the loss of punctate actin aggregates and a diffuse increase in polymerized actin. New stress fibers were apparent by 15 minutes and increased in number and diameter for up to 1 hour (shown at 30 min in Fig. 1b). Vinculin (Fig. 1d) and talin (not shown) progressively localized into new focal adhesions with a time course that paralleled that for stress fiber formation. Neither process significantly preceded the other within the limits of detection by immunofluorescence.

Rac1 Stimulates Actin Reorganization to Form Membrane Ruffles

Since Rac proteins are approximately 60% homologous to Rho, it is possible that they similarly induce actin reorganization. We therefore purified both recombinant Rac1 and Rac1 with a mutation of amino acid 12 from glycine (G) to valine (V), equivalent to the oncogenic V12 mutation in Ras. This mutation in Rac1 has previously been reported to decrease its intrinsic GTPase activity and also to render it insensitive to GTPase activating proteins (GAPs) (Diekmann et al. 1991). Microinjection of both V12Rac1 and normal Rac1 into serum-starved Swiss 3T3 cells stimulated rapid actin reorganization, but with a phenotype very distinct from that induced by Rho. Within 5 minutes, actin filaments accumulated at the plasma membrane forming distinct structures known as membrane ruffles. This response reached a maximum by 15 minutes (Fig. 1e) and was maintained for up to 4 hours after injection. V12Rac1 was approximately fourfold more potent than normal Rac1 in inducing this response, as estimated by determining the lowest concentration of each protein able to induce significant membrane ruffling upon microinjection.

In addition to the stimulation of membrane ruffling, a later response to Rac1 microinjection was the formation of new stress fibers, first detected between 20 and 30 minutes after injection (Fig. 1f). These fibers accumulated in number and thickness over 2 hours but never reached the density observed following microinjection of Rho. As a control, a mutant protein V12A35Rac1, with amino acid 35 changed from threonine to alanine (A), was injected. This amino acid is conserved in all Ras-related proteins and has been shown to be essential for both Ras and Rho function (Paterson et al. 1990; Bourne et al. 1991). No alteration in the actin cytoskeleton was observed between 15 minutes and 4 hours after microinjection of V12A35Rac1.

Growth Factor-stimulated Actin Reorganization

The rapid cellular responses to Rho and Rac suggest that they could normally function as components of signal transduction pathways leading to actin reorganization. To investigate this possibility, it was first necessary to characterize the actin cytoskeletal rearrangements induced in Swiss 3T3 cells by extracellular factors. Several growth factors for Swiss 3T3 cells have been reported to induce actin reorganization in other cell lines (Bockus and Stiles 1984; Kadowaki et al. 1986; Mellström et al. 1988), and in particular, we had observed that cells in the presence of serum contained abundant stress fibers, even when quiescent, whereas

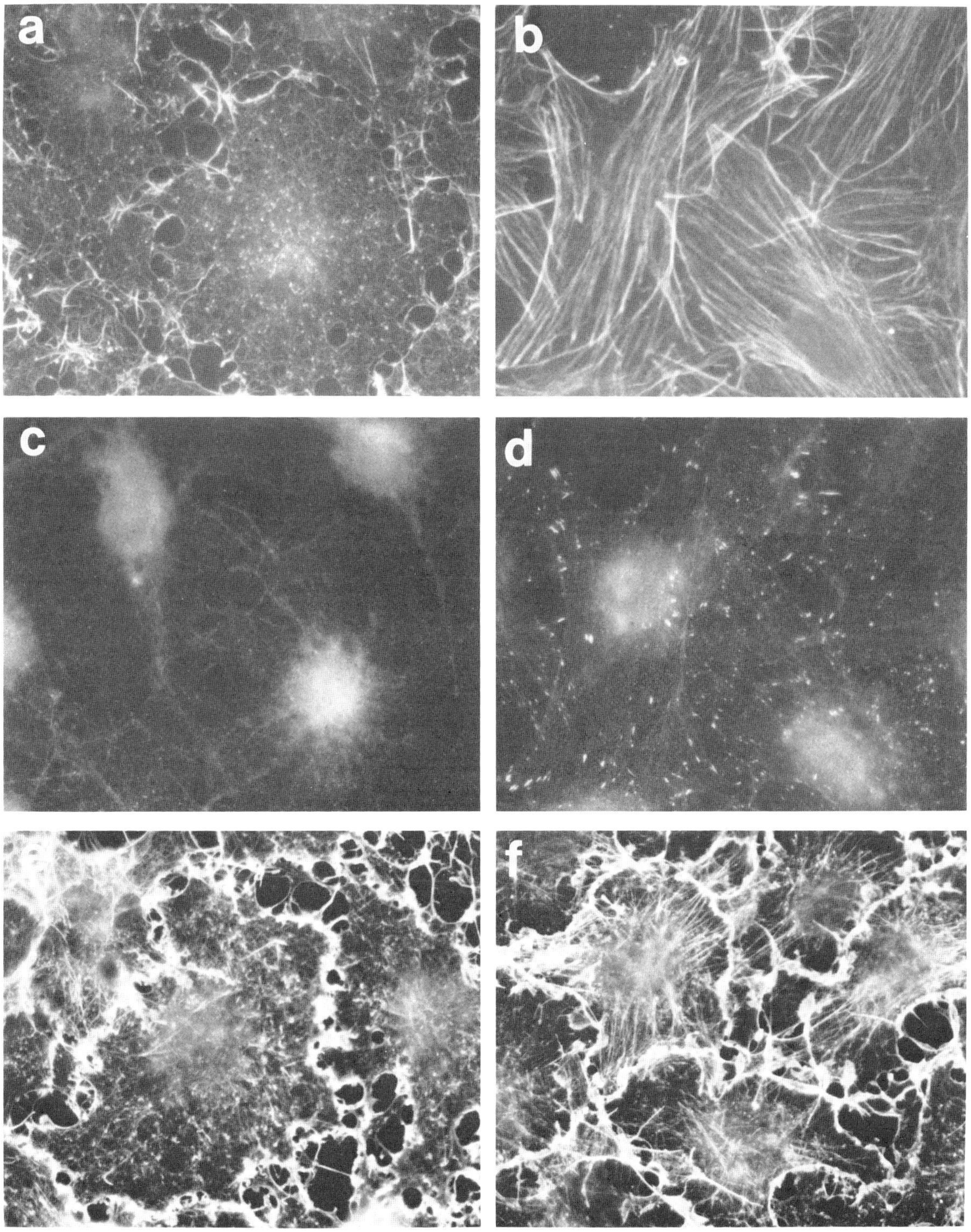

Figure 1. Microinjection of Rho induces assembly of focal adhesions and stress fibers, whereas Rac stimulates membrane ruffling. Purified recombinant proteins were microinjected at 300 μg/ml into confluent, serum-starved Swiss 3T3 cells. Cells were fixed either without injection (*a,c*), or 30 min after injection of V14RhoA (*b,d*), or 15 min (*e*) and 30 min (*f*) after injection of V12Rac1. Actin filaments were revealed using FITC-conjugated phalloidin (*a,b*) or TRITC-conjugated phalloidin (*e,f*), and focal adhesions were localized with a mouse anti-vinculin antibody (*c,d*). (*a–d*, Reprinted with permission, from Ridley and Hall 1992; *e* and *f*, reprinted, with permission, from Ridley et al. 1992 [both copyrights held by Cell Press].)

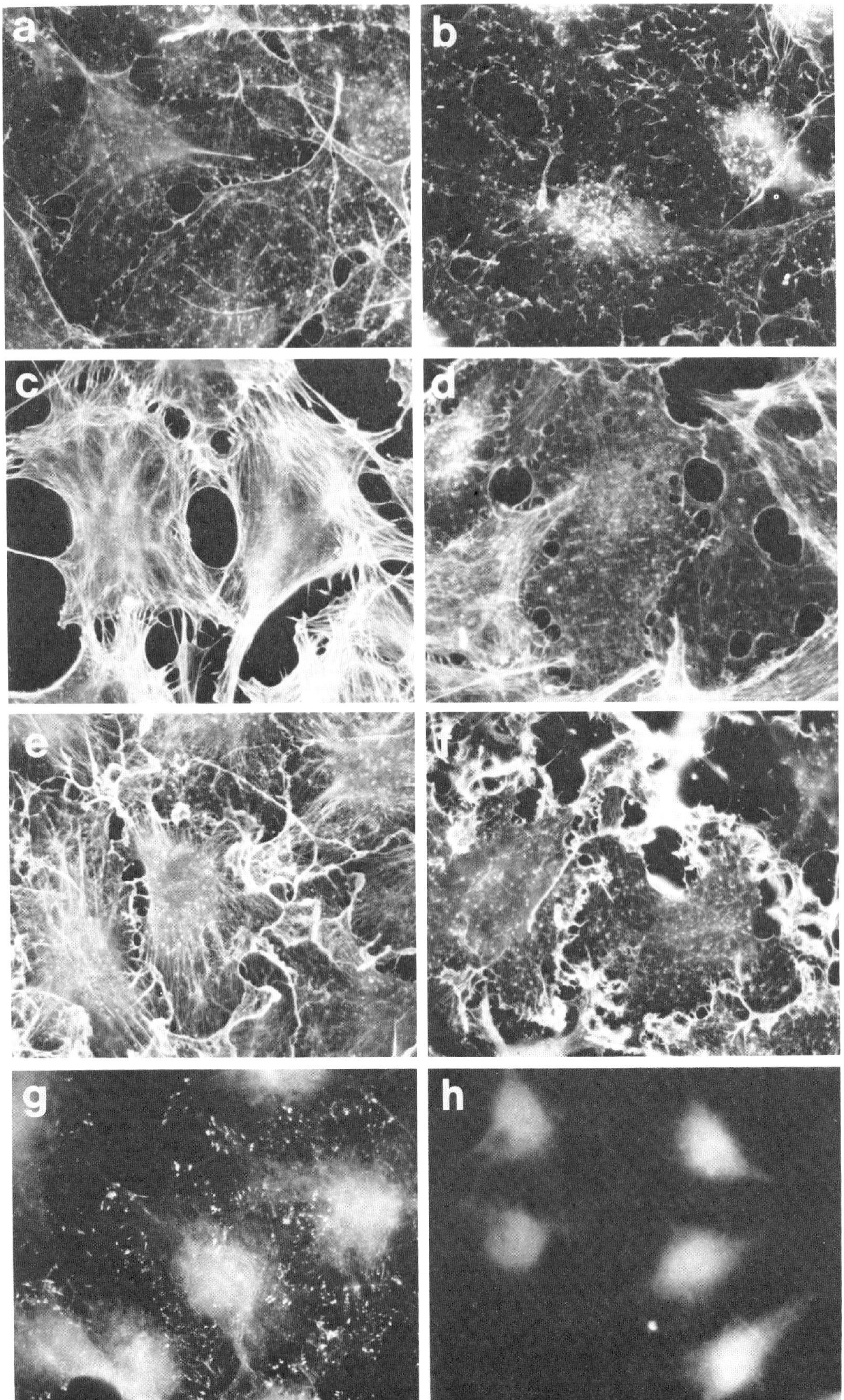

Figure 2. (*See facing page for legend.*)

Table 1. Growth Factor-induced Changes in the Actin Cytoskeleton of Serum-starved Swiss 3T3 Cells

Addition	Stress fibers 5 min	Stress fibers 30 min	Ruffling 10 min
FCS (0.2%)	+ + + +	+ + + +	+
LPA (20 ng/ml)	+ + + +	+ + + +	–
Bombesin (10 nM)	+ +	+ +	+ +
PDGF (3 ng/ml)	–	+ +	+ + +
EGF (10 ng/ml)	–	+ +	+ + +
Insulin (1 μg/ml)	–	+	+ +
PMA (100 nM)	–	–	+ + +

Cells were fixed at the indicated times, and actin filaments were revealed with TRITC-labeled phalloidin. The relative density of actin filaments was compared visually at 5 min for early stress fibers, at 10 min for actin in membrane ruffles, and at 30 min for late stress fibers.

after 16 hours of serum deprivation they contained very few. We have shown that readdition of serum rapidly stimulates stress fiber formation and that within 1 minute of addition, new actin polymerization is detectable. Further analysis of the active component in FCS responsible for this rapid actin cytoskeletal organization led to its identification as a phospholipid, most likely LPA, bound to serum albumin (Ridley and Hall 1992). LPA at 20 ng/ml stimulates stress fiber assembly with a time course indistinguishable from the response to 0.2% FCS (e.g., Fig. 2c). This response to LPA was very similar to that induced by microinjecting V14RhoA and was similarly accompanied by new focal adhesion formation, as determined through the localization of vinculin and talin to the ends of newly formed stress fibers (data not shown; Ridley and Hall 1992). LPA has been reported to stimulate other intracellular responses in cells with a similar potency and apparently acts through a specific transmembrane receptor (Van der Bend et al. 1992).

The reorganization of actin into membrane ruffles by Rac1 resembles that induced by a variety of extracellular ligands in several cell types, and in fibroblasts a rapid response is elicited by PDGF (Mellström et al. 1988). We therefore followed actin cytoskeletal rearrangements after addition of PDGF at 3 ng/ml to Swiss 3T3 cells. Within 2 minutes, accumulation of actin filaments in membrane ruffles was clearly discernible and reached a maximum by 5–10 minutes (Fig. 2e). A later response to PDGF was the formation of new stress fibers, first detectable around 10 minutes and gradually increasing to reach a maximum between 30 and 60 minutes. Over a 2-hour time course the density of stress fibers never reached that observed following addition of 20 ng/ml LPA, even though the membrane ruffling response was maximal at 3 ng/ml PDGF. The sequence of changes in the actin cytoskeleton induced by PDGF, first inducing membrane ruffling and later stress fiber formation, was identical to that induced by microinjected Rac proteins.

Other growth factors for Swiss 3T3 cells were also analyzed for their ability to induce actin reorganization (Table 1). Both EGF and insulin, which, like PDGF, act through transmembrane tyrosine kinase receptors, induced a similar pattern of responses to PDGF, although there were qualitative differences. Bombesin induced both membrane ruffling and early stress fiber formation resembling a combination of the responses to LPA and to PDGF.

Rho Proteins Are Required for Stress Fiber and Focal Adhesion Assembly

The similarity between the actin rearrangements induced by microinjecting Rho and by adding LPA or FCS suggested that Rho might itself be involved in mediating stress fiber formation in response to extracellular factors. To test this, the activity of endogenous Rho proteins was inhibited by introducing into cells the bacterial exoenzyme, C3 transferase, which ADP-ribosylates Rho proteins on amino acid Asn-41 and thereby inactivates them (Aktories et al. 1989; Sekine et al. 1989; Paterson et al. 1990). C3 is very specific for Rho proteins and is at least 400-fold less active on the closely related Rac and CDC42 proteins (Just et al. 1992; Ridley and Hall 1992). Swiss 3T3 cells were microinjected with C3 transferase and subsequently stimulated with various extracellular factors. The rapid stimulation of stress fiber assembly by LPA (Fig. 2c,d), FCS, and bombesin (data not shown) was completely inhibited in cells microinjected with C3 at concentrations above 125 μg/ml. The later increase in stress fibers in PDGF-stimulated cells was also inhibited by C3, whereas the membrane ruffling response in the

Figure 2. Rho is required for growth factor-induced stress fiber and focal adhesion formation. Actin filaments are shown in serum-starved Swiss 3T3 cells without injection (*a*), 20 min after injection with C3 transferase at 160 μg/ml (*b*), microinjected with C3 at 80 μg/ml and stimulated with LPA (*c*), microinjected with C3 at 160 μg/ml and stimulated with LPA (*d*), stimulated with PDGF (*e*), or microinjected with C3 at 160 μg/ml and stimulated with PDGF (*f*). Vinculin localization is shown in cells stimulated with LPA (*g*) or in cells microinjected with C3 at 160 μg/ml and stimulated with LPA. LPA at 20 ng/ml or PDGF at 3 ng/ml was added approximately 15 min after microinjected C3 transferase. Cells were fixed 10 min later and stained with TRITC-phalloidin to reveal actin filaments, or with a mouse anti-vinculin antibody to localize focal adhesions. (Adapted from Ridley and Hall 1992.)

same cells was not altered (Fig. 2e,f). Finally, C3 also prevented the formation of focal adhesions in response to growth factors, for example, LPA (Fig. 2g,h). These results suggest that Rho proteins are required specifically for growth factor-induced stress fiber and focal adhesion assembly, but not for membrane ruffle formation.

Rac Proteins Are Required for Growth Factor-induced Membrane Ruffling

Because Rho proteins are not required for growth factor-induced membrane ruffling, and exogenous Rac1 stimulates membrane ruffling, we next tested whether endogenous Rac proteins mediated the ruffling response. To inhibit Rac activity in cells, a dominant negative inhibitory Rac protein was introduced. Amino acid 17 was mutated from serine to asparagine (N) in both normal and V12Rac1, to form N17Rac1 and V12N17Rac1. This substitution was chosen because a Ha-Ras protein with an equivalent mutation, N17Ras, has been shown to act as a dominant inhibitor of endogenous Ras proteins, probably by competitively inhibiting their interaction with a nucleotide exchange factor (Farnsworth and Feig 1991). The effects of N17Rac1 and V12N17Rac1 were indistinguishable, and the following results refer to experiments performed with V12N17Rac1, although identical results were obtained with N17Rac1.

When V12N17Rac1 was introduced into serum-starved Swiss 3T3 cells, it had no effect on morphology or on the actin cytoskeleton (Fig. 3a). In addition, stress fiber formation in response to LPA was not altered in cells containing V12N17Rac1 (Fig. 3b,c). However, when V12N17Rac1-injected cells were stimulated with PDGF, membrane ruffling was inhibited (Fig. 4a,b). Membrane ruffling in response to PMA (Fig. 4c,d) and to other growth factors listed in Table 1 (data not shown) was also prevented by V12N17Rac1.

In addition to inhibiting membrane ruffling induced by PDGF, V12N17Rac1 prevented the later increase in stress fibers (Fig. 4a,b). As reported above, Rho proteins were also required for this process, and it is therefore likely that Rac functions upstream of Rho on signal transduction pathways leading from PDGF, but not LPA, to actin reorganization.

Ras-induced Actin Reorganization Is Dependent on Rho and Rac

Activated Ras proteins have previously been shown to stimulate membrane ruffling in fibroblasts (Bar-Sagi and Feramisco 1986), and we were interested to determine whether the Ras and Rac responses were interconnected. In serum-starved Swiss 3T3 cells, microinjection of V12 Ha-Ras stimulated actin reorganization into membrane ruffles within 15–30 minutes, and new stress fibers were also formed by 30 minutes. This response reached a maximum by 1 hour (Fig. 5a) and was considerably slower and less extensive than the

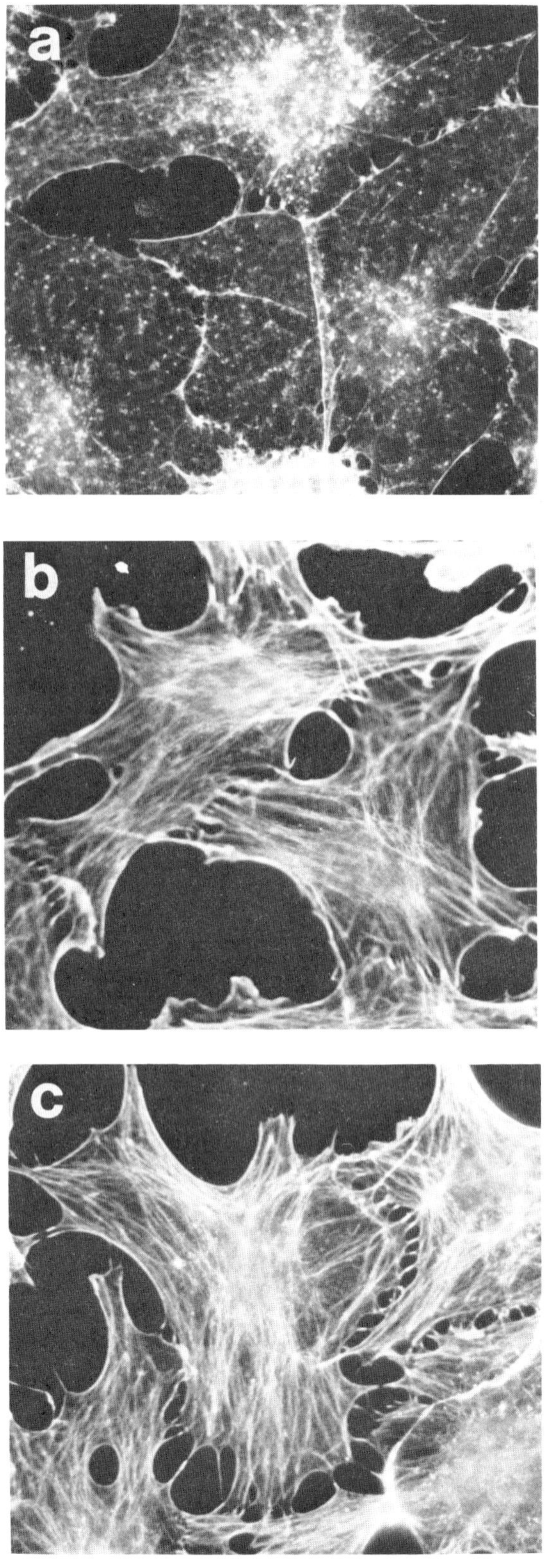

Figure 3. Rac is not required for LPA-induced stress fiber formation. Actin filaments are shown in serum-starved Swiss 3T3 cells 25 min after microinjection with V12N17Rac1 (*a*), stimulated with LPA for 10 min (*b*), or injected with V12N17Rac1 15 min before stimulating with LPA for 10 min (*c*). V12N17Rac1 was injected at 900 μg/ml, and LPA was added at 20 ng/ml. (Adapted from Ridley et al. 1992.)

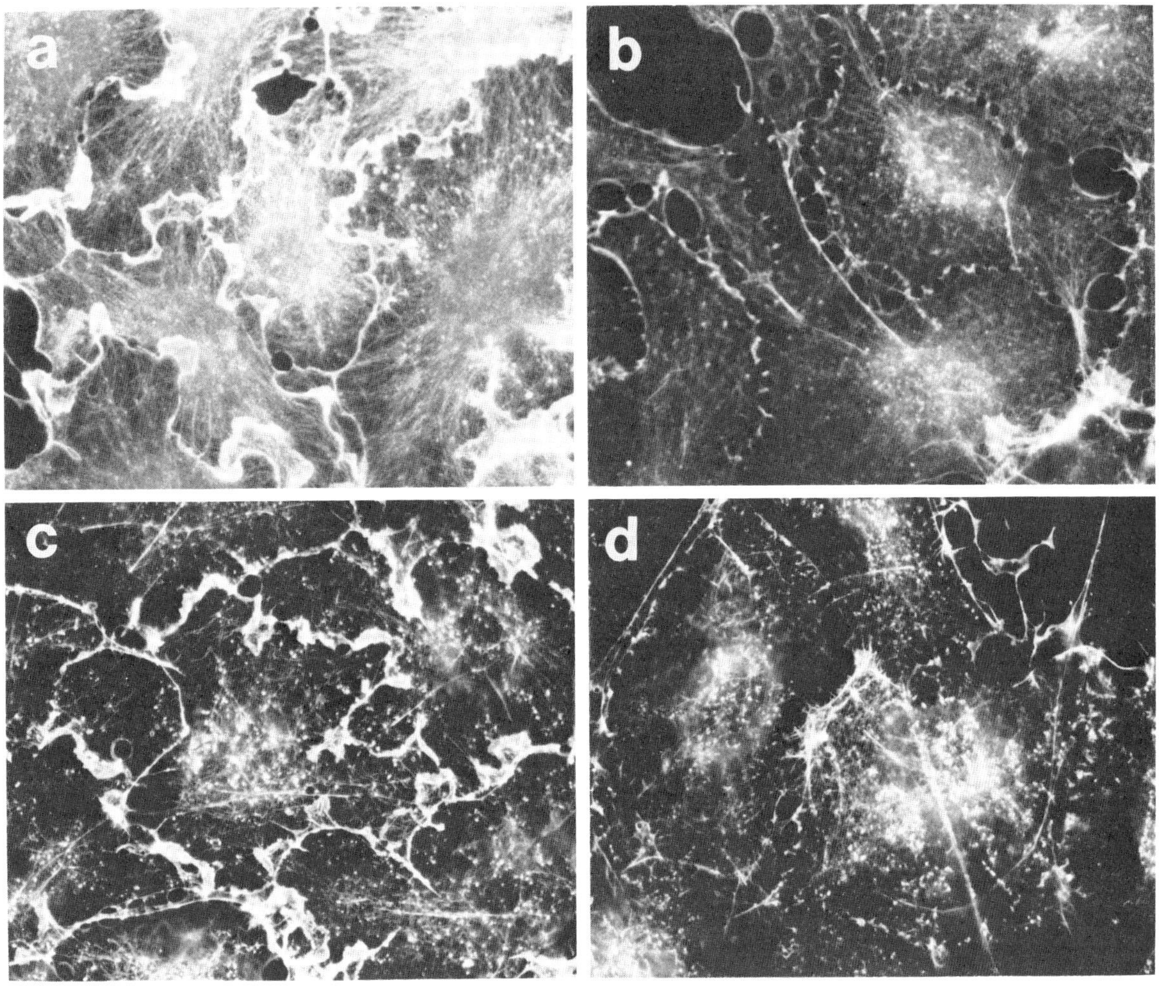

Figure 4. Rac is required for growth factor-induced membrane ruffling. Actin filaments are shown in uninjected cells (*a,c*) or in cells injected with V12N17Rac1 at 900 μg/ml (*b,d*). Approximately 15 min after injection, 3 ng/ml PDGF (*a,b*) or 100 nM PMA (*c,d*) was added for 10 min, then cells were fixed and stained with TRITC-phalloidin. (Adapted from Ridley et al. 1992.)

rapid response to V12Rac1 (Fig. 1e,f). When cells were comicroinjected with V12 Ha-Ras and V12N17Rac1, actin reorganization was completely inhibited (Fig. 5d), indicating that the response to V12 Ha-Ras was mediated by endogenous Rac proteins. Coinjection of C3 transferase with V12 Ha-Ras selectively inhibited stress fiber formation but not membrane ruffling (Fig. 5c), in concordance with the results with growth factors showing that Rho proteins are not required for membrane ruffling responses. Actin reorganization induced by V12 Ha-Ras was unaffected by coinjection of the inactive Rac protein V12A35Rac1, indicating that the inhibitory effect is not merely due to the presence of another small GTP-binding protein (data not shown). Finally, although V12 Ha-Ras stimulated actin reorganization, endogenous Ras proteins were not required either for the equivalent growth factor-induced alterations or for the responses to microinjected Rac1 or RhoA, as these were not altered in cells microinjected with the Ras-neutralizing antibody Y13-259 (data not shown; Ridley et al. 1992). This antibody preparation was clearly active, since it inhibited actin reorganization when coinjected with V12 Ha-Ras (Fig. 5b).

DISCUSSION

We have shown that the Ras-related GTP-binding proteins Rho and Rac regulate distinct patterns of actin reorganization in fibroblasts. Changes in two types of actin filament structure have been studied in serum-starved Swiss 3T3 cells stimulated with extracellular factors. First, an increase in stress fibers was observed, accompanied by new focal adhesions, within 2 minutes of adding either LPA or bombesin to cells. Other growth factors such as PDGF also stimulated stress fiber formation, but this response occurred significantly later than and was not as extensive as that induced by LPA. Stress fiber formation in response to all growth factors tested was dependent on Rho function. A second type of actin rearrangement induced by several growth factors including PDGF was the formation of membrane ruffles. Rac but not Rho function was essen-

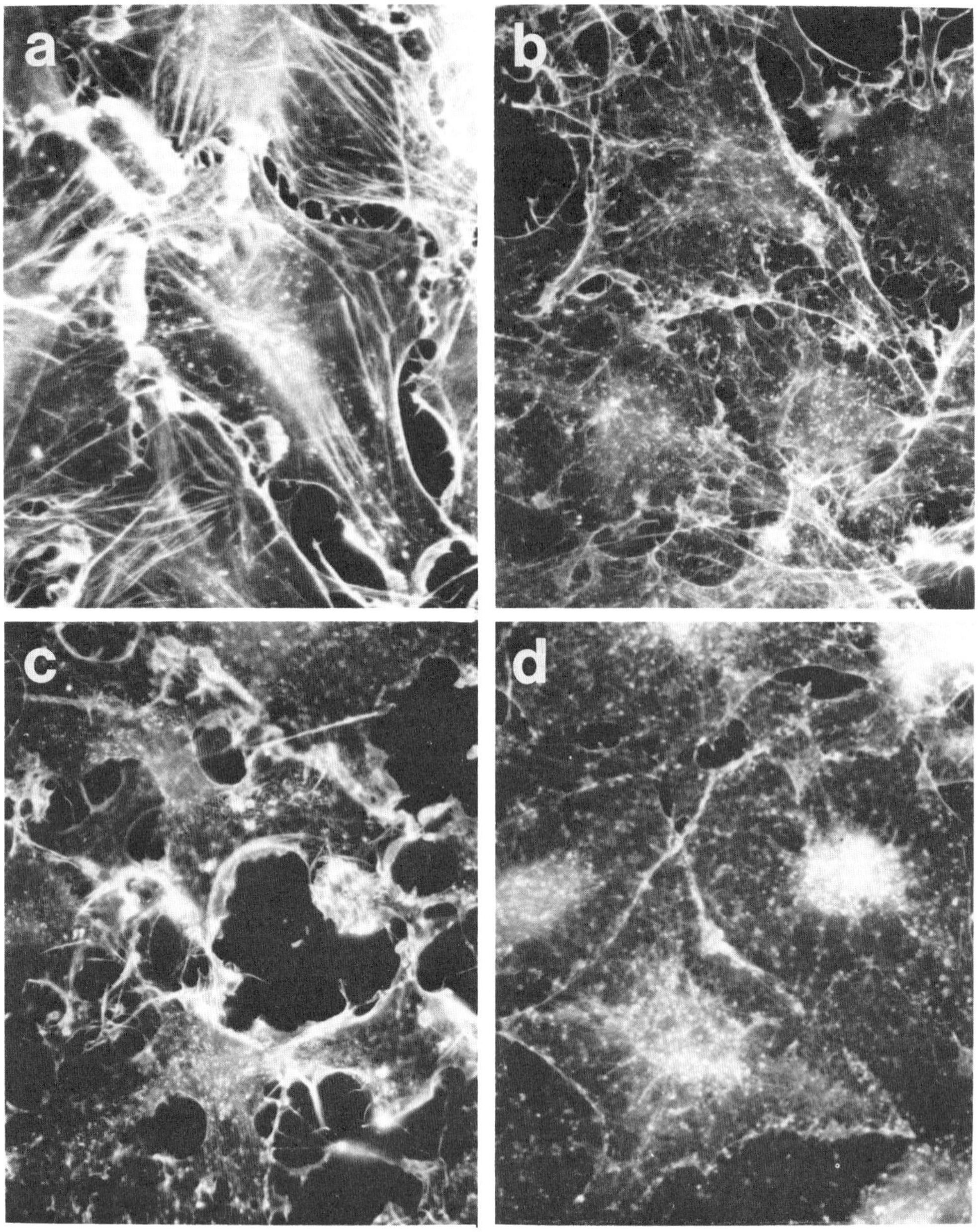

Figure 5. Actin reorganization induced by V12 Ha-Ras is mediated by Rac and Rho. Actin filaments are shown in serum-starved cells injected with V12 Ha-Ras alone (*a*), or together with Y13-259 anti-Ras antibody (*b*), C3 transferase (*c*), or V12N17Rac1 (*d*). V12 Ha-Ras was injected at 800 μg/ml, V12N17Rac1 at 400 μg/ml, C3 transferase at 160 μg/ml, and Y13-259 antibody at 10 mg/ml. Cells were fixed and stained with TRITC-phalloidin 1 hr after microinjection. (Adapted from Ridley and Hall 1992.)

tial for this process, and the two types of actin reorganization can therefore be separately regulated through distinct signal transduction pathways initiated by growth factor receptors. However, there can be interaction between the two processes, since Rho-dependent stress fiber formation induced by PDGF also requires Rac function, suggesting that in this case Rac activation occurs first and subsequently leads to Rho activation.

Another function for Rac proteins has been recently described: They have been characterized as essential components of the NADPH-dependent superoxide generating system, a process unique to phagocytic and some lymphocytic cells (Abo et al. 1991; Knaus et al. 1991). *rac2* mRNA is detectably expressed only in hematopoietic cells, whereas *rac1* is widely expressed (Didsbury et al. 1989; Shirsat et al. 1990; Moll et al. 1991), suggesting that Rac2 is specifically involved in superoxide generation, although both Rac1 and Rac2 can activate the process in vitro. Extracellular factors that stimulate superoxide production in neutrophils also induce membrane ruffling (Omann et al. 1987), and it is possible that Rac proteins play a coordinating role in regulating both processes.

Rho and Rac appear to act downstream from several different types of growth factor receptors as mediators of actin reorganization, just as Ras is essential for the mitogenic responses to many growth factors (Hall 1990). It is not yet known whether Rho or Rac function is required for mitogenesis, but it is interesting that endogenous Ras proteins are not essential for actin rearrangements, indicating that not all growth factor-induced responses are dependent on Ras. The constitutively activated V12 Ha-Ras protein does induce both membrane ruffling and new stress fiber assembly, but these responses are considerably delayed compared to the responses to activated Rho or Rac and are in fact mediated by endogenous Rho and Rac proteins. V12Ras has been shown to stimulate other signal transduction processes within 5 minutes of entering cells; for example, protein kinase C and MAP-kinase activation (Morris et al. 1989; Leevers and Marshall 1991). It is therefore probable that Rac and Rho are indirectly activated as a consequence of V12Ras-induced alterations in other intracellular signals.

A model consistent with our data placing Rho and Rac on growth factor-mediated signal transduction pathways is shown in Figure 6. In this model, growth factors lead to activation of Rho and/or Rac by stimulating an increase in the level of GTP-bound proteins. This can be achieved either by stimulating exchange of GDP for GTP, or by inhibiting GTP hydrolysis, or by a combination of both mechanisms. For Ras, both types of control have been shown to be important in different cell systems. For example, in NIH-3T3 cells, an increase in nucleotide exchange appears to be the predominant mechanism for increasing Ras-GTP levels in response to PDGF or FCS, whereas in T cells stimulated through the antigen receptor, the major control is through GAP activity (Downward et al. 1990; Zhang et al. 1992). Since Rho and Rac can be activated independently, it is likely that some of the factors regulating their activity show specificity toward one or the other protein. At present only one protein, smgGDS, has been identified with exchange factor activity in vitro toward Rho and Rac, and this is also active on a number of other Ras-related proteins (Hiraoka et al. 1992). The proto-oncogene product Dbl has been shown to act as an exchange factor for the Rac- and Rho-related protein CDC42, and the domain responsible for this activity has homology with similar domains in several other proteins, including Vav and Bcr (Hart et al. 1991; Hall 1992). It is not yet known whether these domains also have exchange factor activity toward specific members of the Rho family.

Several proteins that stimulate GTP hydrolysis on Rho-related proteins have been identified, and these show some selectivity in vitro. RhoGAP was the first to be identified and stimulates GTP hydrolysis on both Rho and Rac. It has sequence homology with a domain

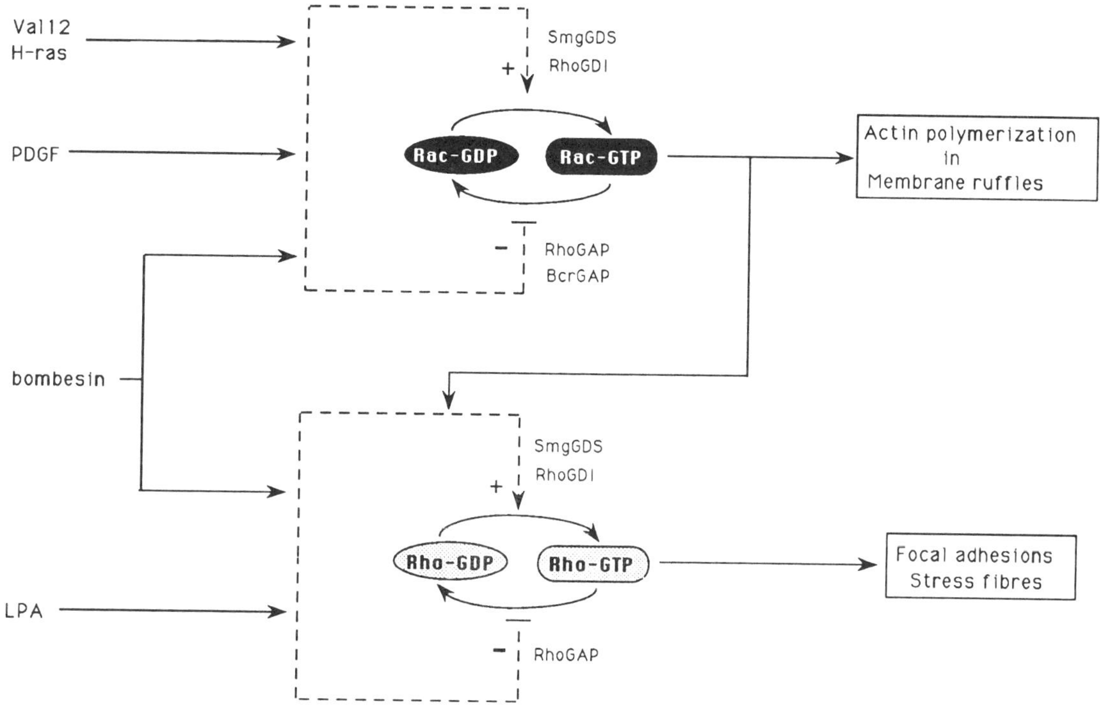

Figure 6. A model for Rho and Rac action. Growth factors that stimulate membrane ruffling, such as PDGF and bombesin, require Rac function. They presumably activate Rac by inducing an increase in the GTP-bound form, either by activating guanine nucleotide exchange proteins or by inhibiting GTPase activating proteins (GAPs) such as Bcr (the breakpoint cluster region gene product) or RhoGAP. Rac-GTP then stimulates membrane ruffling, by an as-yet-undefined mechanism. In addition, it induces Rho activation, which then stimulates stress fiber formation, accounting for the observed delay in this response to PDGF. Rho can also be activated independently of Rac by bombesin and LPA, and thus bombesin directly stimulates both Rho- and Rac-dependent responses. (Reprinted, with permission, from Ridley et al. 1992 [copyright held by Cell Press].)

of Bcr, which acts as a GAP for Rac and not Rho (Diekmann et al. 1991). Another protein with this RhoGAP domain, *n*-chimerin, also acts on Rac but not Rho and is specifically expressed in brain, suggesting that different cell types may use different proteins to control Rho or Rac activity. Two other proteins with RhoGAP domains have recently been cloned, p190 and p85 (Otsu et al. 1991; Settleman et al. 1992b), and p190 has been shown to have GAP activity on Rho, Rac, and cdc42 (Settleman et al. 1992a). Either could provide a direct link between growth factor receptors and Rho or Rac. p190 binds to RasGAP after stimulation of fibroblasts with growth factors, and RasGAP binds to several activated growth factor receptors through a SH2 domain-phosphotyrosine interaction (Ellis et al. 1990; Moran et al. 1991). p190 may therefore perform a dual function, interacting with both Ras-mediated and Rho/Rac-mediated responses. p85 is a subunit of phosphatidylinositol (PI) 3 kinase and also binds to activated growth factor receptors (Otsu et al. 1991).

It is possible that GAPs for Rho and Rac may act not only as regulators, but also as effectors of their function, as has been suggested for RasGAPs (McCormick 1989). Further analysis of GAPs may therefore provide insight into the mechanisms whereby Rho and Rac activate downstream responses. Rac and Rho share the common property of stimulating actin polymerization but differ in the pattern of actin filament organization induced. Increased actin polymerization in response to growth factors could be achieved simply by increasing the availability of actin monomers, which would then spontaneously polymerize (Stossel 1989; Cooper 1991). The concentration of actin monomers in cells is far in excess of that required for spontaneous polymerization in vitro, but these monomers are believed to be sequestered by actin-binding proteins and are thus not normally available for polymerization. Several actin-binding proteins, including profilin, also bind to phosphatidylinositols, which prevent their binding to actin. It has therefore been suggested that the availability of actin monomers could be modulated by alterations in the level of phosphatidylinositols induced by extracellular factors, thereby leading to increased actin polymerization (Stossel 1989). As yet there is no in vivo evidence to support this model, but it is possible that Rac and Rho may be involved in modulating PI turnover and thereby stimulate actin polymerization. In this respect, it is interesting that the p85 subunit of PI3 kinase has a RhoGAP-like domain; possibly Rho or Rac might regulate the production of PI(3,4,5)trisphosphate.

The different patterns of actin organization induced by Rac and Rho presumably involve distinct actin filament-binding and cross-linking molecules, and indeed, several of these proteins have been selectively localized to certain actin structures (Bretscher 1991). Rac and Rho might specifically recruit different subsets of actin-binding proteins to the sites of actin polymerization, thereby dictating the type of structure formed. Alternatively, the location of Rac- and Rho-induced actin nucleation sites themselves could dictate the type of structure formed. For example, if actin nucleation is associated with proteins which can then incorporate into focal adhesions, this may ensure that stress fibers are formed.

The identification of Rac and Rho as regulators of distinct types of actin organization provides a first link between extracellular factor receptors and the actin cytoskeleton. Their biochemical properties as GTP-binding proteins suggest that they are likely to be key regulatory molecules, and further study of their interactions with other molecules should illuminate the processes both of signal transduction and of actin organization.

ACKNOWLEDGMENTS

This work was supported by the Cancer Research Campaign (U.K.) and the Medical Research Council (U.K.). We are grateful to Caroline Johnston for the N17Rac1 mutant, to Klaus Aktories for purified C3 transferase, to Michael Wilkinson for the mouse anti-vinculin antibody V284, and to Keith Burridge for the mouse anti-talin antibody 8d4.

REFERENCES

Abo, A., E. Pick, A. Hall, N. Totty, C.G. Teahan, and A.W. Segal. 1991. The small GTP-binding protein, p 21rac1, is involved in the activation of the phagocyte NADPH oxidase. *Nature* **353:** 668.

Aktories, K., S. Braun, S. Rösener, I. Just, and A. Hall. 1989. The *rho* gene product expressed in *E. coli* is a substrate for botulinum ADP-ribosyltransferase C3. *Biochem. Biophys. Res. Commun.* **158:** 209.

Balch, W.E. 1990. Small GTP-binding domains in vesicular transport. *Trends Biochem. Sci.* **15:** 473.

Bar-Sagi, D. and J.R. Feramisco. 1986. Induction of membrane ruffling and fluid-phase pinocytosis in quiescent fibroblasts by *ras* proteins. *Science* **233:** 1061.

Bockus, B.J. and C.D. Stiles. 1984. Regulation of cytoskeletal architecture by platelet-derived growth factor, insulin and epidermal growth factor. *Exp. Cell Res.* **153:** 186.

Bourne, H.A., D.A. Sanders, and F. McCormick. 1991. The GTPase superfamily: Conserved structure and molecular mechanism. *Nature* **349:** 117.

Bretscher, A. 1991. Molecular aspects of microfilament structure and assembly. *Curr. Opin Struct. Biol.* **1:** 281.

Burridge, K., K. Fath, T. Kelly, G. Nuckolls, and C. Turner. 1988. Focal adhesions: Transmembrane junctions between the extracellular matrix and the cytoskeleton. *Annu. Rev. Cell. Biol.* **4:** 487.

Chardin, P., P. Boquet, P. Maduale, M.R. Popoff, E.J. Rubin, and D.M. Gill. 1989. The mammalian G protein rhoC is ADP-ribosylated by *Clostridium botulinum* exoenzyme C3 and affects actin microfilaments in Vero cells. *EMBO J.* **8:** 1087.

Cooper, J.A. 1991. The role of actin polymerization in cell motility. *Annu. Rev. Physiol.* **53:** 585.

Didsbury, J., R.F. Weber, G.M. Bokoch, T. Evans, and R. Snyderman. 1989. Rac, a novel *ras*-related family of proteins that are botulinum toxin substrates. *J. Biol. Chem.* **264:** 16378.

Diekmann, D., S. Brill, M.D. Garrett, N. Totty, J. Hsuan, C. Monfries, C. Hall, L. Lim, and A. Hall. 1991. Bcr encodes a GTPase activating protein for p21rac. *Nature* **351:** 400.

Downward, J., J.D. Graves, P.H. Warne, S. Rayter, and D.A. Cantrell. 1990. Stimulation of $p21^{ras}$ upon T-cell activation. *Nature* **346:** 719.

Ellis, C., M. Moran, F. McCormick, and T. Pawson. 1990. Phosphorylation of GAP and GAP-associated proteins by transforming and mitogenic tyrosine kinases. *Nature* **343:** 377.

Farnsworth, C.L. and L.A. Feig. 1991. Dominant inhibitory mutations in the Mg^{2+}-binding site of Ras^{H} prevent its activation by GTP. *Mol. Cell. Biol.* **11:** 4822.

Fukumoto, Y., K. Kaibuchi, Y. Hori, H. Fujioka, S. Araki, T. Ueda, A. Kikuchi, and Y. Takai. 1990. Molecular cloning and characterization of a novel type of regulatory protein (GDI) for the rho proteins, ras p21-like small GTP-binding proteins. *Oncogene* **5:** 1321.

Garrett, M.D., G.N. Major, N. Totty, and A. Hall. 1991. Purification and *N*-terminal sequence of the $p21^{rho}$ GTPase activating protein, rhoGAP. *Biochem. J.* **276:** 833.

Garrett, M.D., A.J. Self, C. van Oers, and A. Hall. 1989. Identification of distinct cytoplasmic targets for ras/R-ras and rho regulatory proteins. *J. Biol. Chem.* **264:** 10.

Hall, A. 1990. The cellular function of small GTP-binding proteins. *Science* **249:** 635.

———. 1992. Signal transduction through small GTPases—A tale of two GAPs. *Cell* **69:** 389.

Hall, A. and A.J. Self. 1986. The effect of Mg^{2+} on the guanine nucleotide exchange rate of $p21^{N\text{-}ras}$. *J. Biol. Chem.* **261:** 10963.

Hart, M.J., A. Eva, T. Evans, S.A. Aaronson, and R.A. Cerione. 1991. Catalysis of guanine nucleotide exchange on the CDC42Hs protein by the *dbl* oncogene product. *Nature* **354:** 311.

Hiraoka, K., K. Kaibuchi, S. Ando, T. Musha, K. Takaishi, A. Mizuno, L. Ménard, E. Tomhave, J. Didsbury, R. Snyderman, and Y. Takai. 1992. Both stimulatory and inhibitory GDP/GTP exchange proteins, smgGDS and rhoGDI, are active on multiple small GTP-binding proteins. *Biochem. Biophys. Res. Commun.* **182:** 921.

Just, I., C. Mohr, G. Schallehn, L. Menard, J.R. Didsbury, J. Vandekerckhove, J. van Damme, and K. Aktories. 1992. Purification and characterization of an ADP-ribosyltransferase produced by *Clostridium limosum*. *J. Biol. Chem.* **267:** 10274.

Kadowaki, T., S. Koyasu, E. Nishida, H. Sakai, F. Takadu, I. Yahara, and M. Kasuga. 1986. Insulin-like growth factors, insulin and epidermal growth factor cause rapid cytoskeletal reorganization in KB cells. *J. Biol. Chem.* **261:** 16141.

Kahn, R.A., F.G. Kern, J. Clark, E.P. Gelmann, and C. Rulka. 1991. Human ADP-ribosylation factors. *J. Biol. Chem.* **266:** 2606.

Knaus, U.G., P.G. Heyworth, T. Evans, T.J. Curnette, and G.M. Bokoch. 1991. Regulation of phagocyte oxygen radical production by the GTP-binding protein Rac2. *Science* **254:** 1512.

Leevers, S.J. and C.J. Marshall. 1992. Activation of extracellular signal-regulated kinase, ERK2, by p21*ras* oncoprotein. *EMBO J.* **11:** 569.

McCormick, F. 1989. *ras* GTPase activating protein: Signal transmitter and signal terminator. *Cell* **56:** 5.

Mellström, K., C.-H. Heldin, and B. Westermark. 1988. Induction of circular membrane ruffling on human fibroblasts by platelet-derived growth factor. *Exp. Cell Res.* **177:** 347.

Moll, J., G. Sansig, E. Fattori, and H. van der Putten. 1991. The murine *rac*1 gene: cDNA cloning, tissue distribution and regulated expression of *rac*1 mRNA by disassembly of actin microfilaments. *Oncogene* **6:** 863.

Moran, M., P. Polakis, F. McCormick, T. Pawson, and C. Ellis. 1991. Protein-tyrosine kinases regulate the phosphorylation, protein interactions, subcellular distribution, and activity of p21 ras GTPase-activating protein. *Mol. Cell. Biol.* **11:** 1804.

Morris, J.D.H., B. Price, A.C. Lloyd, A.J. Self, C.J. Marshall, and A. Hall. 1989. Scrape-loading of Swiss 3T3 cells with *ras* protein rapidly activates protein kinase C in the absence of phosphoinositide hydrolysis. *Oncogene* **4:** 27.

Munemitsu, S., M.A. Innis, R. Clark, F. McCormick, A. Ullrich, and P. Polakis. 1990. Molecular cloning and expression of a G25K cDNA, the human homolog of the yeast cell cycle gene. *Mol. Cell. Biol.* **10:** 5977.

Omann, G.M., R.A. Allen, G.M. Bokoch, R.G. Painter, A.E. Traynor, and L.A. Sklar. 1987. Signal transduction and cytoskeletal activation in the neutrophil. *Physiol. Rev.* **67:** 285.

Otsu, M., I. Hiles, I. Gout, M.J. Fry, F. Ruiz-Larrea, G. Panayotou, A. Thompson, R. Dhand, J. Hsuan, N. Totty, A.D. Smith, S.J. Morgan, S.A. Courtneidge, P.J. Parker, and M.D. Waterfield. 1991. Characterization of two 85 kd proteins that associate with receptor tyrosine kinases, middle-T/$pp60^{c\text{-}src}$ complexes, and PI3-kinase. *Cell* **65:** 91.

Paterson, H.F., A.J. Self, M.D. Garrett, I. Just, K. Aktories, and A. Hall. 1990. Microinjection of recombinant $p21^{rho}$ induces rapid changes in cell morphology. *J. Cell Biol.* **111:** 1001.

Ridley, A.J. and A. Hall. 1992. The small GTP-binding protein Rho regulates the assembly of focal adhesions and actin stress fibres in response to growth factors. *Cell* **70:** 389.

Ridley, A.J., H.F. Paterson, C.L. Johnston, D. Diekmann, and A. Hall. 1992. The small GTP-binding protein Rac regulates growth factor-induced membrane ruffling. *Cell* **70:** 401.

Rubin, E.J., D.M. Gill, P. Boquet, and M.R. Popoff. 1988. Functional modification of a 21-kilodalton G protein when ADP-ribosylated by exoenzyme C3 in *Clostridium botulinum*. *Mol. Cell. Biol.* **8:** 418.

Sekine, A., M. Fujiwara, and S. Narumiya. 1989. Asparagine residue in the rho gene product is the modification site for botulinum ADP-ribosyltransferase. *J. Biol. Chem.* **264:** 8602.

Settleman, J., C.F. Albright, L.C. Foster, and R.A. Weinberg. 1992a. Association between GTPase activators for Rho and Ras families. *Nature* **359:** 153.

Settleman, J., V. Narasimhan, L.C. Foster, and R.A. Weinberg. 1992b. Molecular cloning of cDNAs encoding the GAP-associated protein p190: Implications for a signaling pathway from Ras to the nucleus. *Cell* **69:** 539.

Shinjo, K., J.G. Koland, M.J. Hart, V. Narasimhan, D.I. Johnson, T. Evans, and R.A. Cerione. 1990. Molecular cloning of the gene for the human placental GTP-binding protein Gp (G25K): Identification of this GTP-binding protein as the human homolog of the yeast cell-division-cycle protein CDC42. *Proc. Natl. Acad. Sci.* **87:** 9853.

Shirsat, N.V., R.J. Pignolo, B.L. Kreider, and G. Rovera. 1990. A member of the *ras* gene superfamily is specifically expressed in T, B and myeloid hemopoietic cells. *Oncogene* **5:** 769.

Smith, D.B. and K.S. Johnson. 1988. Single-step purification of polypeptides expressed in *Escherichia coli* as fusions with glutathione S-transferase. *Gene* **67:** 31.

Stossel, T.P. 1989. From signal to pseudopod. *J. Biol. Chem.* **264:** 18261.

Vandekerckhove, J. 1990. Actin-binding proteins. *Curr. Opin. Cell Biol.* **2:** 41.

Van der Bend, R.L., J. Brunner, K. Jalink, E.J. van Corven, W.H. Moolenaar, and W.J. van Blitterswijk. 1992. Identification of a putative membrane receptor for the bioactive phospholipid, lysophosphatidic acid. *EMBO J.* **11:** 2495.

Vincent, S., P. Jeanteur, and P. Fort. 1992. Growth-regulated expression of rhoG, a new member of the *ras* homolog gene family. *Mol. Cell. Biol.* **12:** 3138.

Zhang, K., A.G. Papageorge, and D.R. Lowy. 1992. Mechanistic aspects of signalling through Ras in NIH 3T3 cells. *Science* **257:** 671.

Summary: The Cell Surface Regulates Information Flow, Material Transport, and Cell Identity

M.I. SIMON
Division of Biology 147-75, California Institute of Technology, Pasadena, California 91125

The 57th Symposium of the Cold Spring Harbor series was organized around the theme of the cell surface. Although each of the sessions dealt with cell-surface-related functions, the range of systems that were discussed and the variety in the points of view brought to bear have made this one of the most diverse of the Cold Spring Harbor Symposia. There were more than 80 speakers, 90 posters, and 340 participants. Subjects ranged from bacterial genetics to X-ray crystallography, embryonic development, carbohydrate chemistry, and neurobiology.

The diversity of subject matter at this 57th Symposium reflects our current view of the global role that the cell surface plays in integrating and regulating cell function and identity; it is in marked contrast to earlier pictures. Jim Watson in his introductory remarks to this year's meeting pointed out that from the beginning of the series in 1933, participants in the Cold Spring Harbor Symposia have been fascinated with questions involving cell-surface structure and function. The very first symposium was devoted to "Cell Surface Phenomena," and again in 1940, the 8th Cold Spring Harbor Symposium focused on "Cell Permeability and Cell Membranes." Figure 1A is a representation of the cell membrane taken from that symposium. The surface of the cell was seen as a barrier that acted as the demarcation between the extracellular world and a poorly organized cytoplasm. Twenty years ago this image changed with the introduction of a more refined and highly articulated model of the cell membrane (Fig. 1B) by Singer and Nicholson (1972). The picture included proteins that floated serenely, like icebergs in a sea that represented the lipid bilayer. Some of these polypeptides extended across the membrane while others acted as cell-surface receptors and still others were peripheral to the inner surface of the membrane and interacted with cytoplasmic components. Jon Singer (1992) realized that a key to membrane function lay in the proteins that traversed and/or were associated with the surface, although at the time there were very few well-characterized transmembrane proteins. In the last 20 years we have experienced a revolution in our ability to identify membrane proteins and to study their function. The development of new biochemical approaches and the introduction and refinement of a variety of techniques, including recombinant DNA technology as well as novel biophysical and analytical methods, have totally changed our view of the cell surface. Figure 1C illustrates the contemporary view as it was presented by Michael Brown in his Dorcas Cummings Lecture at this 57th Symposium. The cell surface is envisioned to be a dynamic organelle, extending from the extracellular matrix through to the cytoskeleton. Much of the activity at the surface is controlled by membrane proteins that regulate interactions with a variety of different kinds of ligands in the environment of the cell and transmit information to the cytoplasm. Conditions in the interior of the cell modulate the properties of these receptors. Surface proteins also function to move material in and out of the cell and to establish specific contacts that determine the differentiated identity of the cell. In addition, proteins on the cytoplasmic face of the membrane are linked to cell-

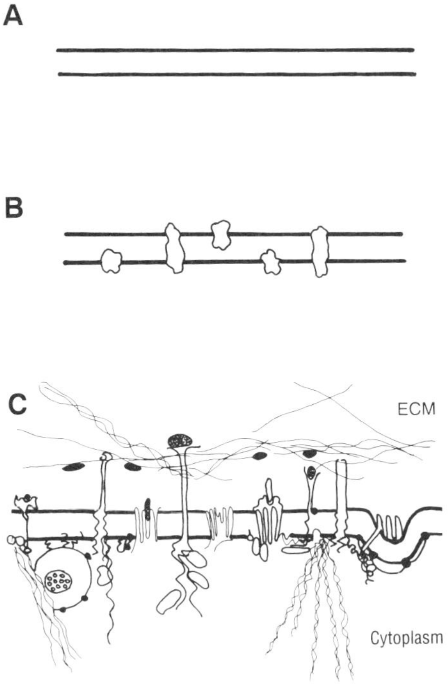

Figure 1. Historical representations of the cell surface. (*A*) Model of the cell membrane taken from the 8th Cold Spring Harbor Symposium, 1940. (*B*) 1972 version of the cell surface as modified from Singer and Nicholson (1972). (*C*) Current (1992) representation. ECM is the extracellular matrix. The protein components represented are described in the text.

All authors cited here without dates refer to papers in this volume.

surface receptors and seed the assembly of intracellular cytoskeletal elements that control the superstructure of the cell. Simultaneously, fusion of vesicles that have been assembled and transported to the surface renews and modifies the composition of the membrane while ligand-bound receptors are recycled and extracellular structures are generated. Thus, the surface continuously regulates the flow of both information and material into and out of the cell. It controls the structure and morphology of the cell and generates and responds to signaling molecules that affect the cell's ability to relate to its environment and to its neighbors. Ultimately, tissue morphology and complex neural patterning depend on instructive interactions at the cell surface.

The major problems that engaged this 57th Symposium centered around questions related to how cell-membrane-associated proteins work. That is, how do they act to acquire information from the environment? How is this information transduced into a form that the cell can use? How do cell-surface proteins direct cellular movements and cell-cell interactions? How is contact between the cell surface and the nucleus maintained, and how do cell-surface functions regulate the growth and development of complex organisms? In addition to processing information and transmitting signals, the cell surface controls the specific secretion and the distribution of membrane proteins. We are beginning to understand the mechanisms that regulate the timing of appearance and the spatial distribution of specific receptors, adhesion mediators, and specialized structures on the cell surface. Questions about the systems that control vesicle traffic, membrane fusion, transport selectivity, and the remodeling of cell-surface properties were discussed in the context of their function in the nervous system and in the immune system. We will summarize how the theme of information and material transfer across the cell membrane was dealt with at this symposium.

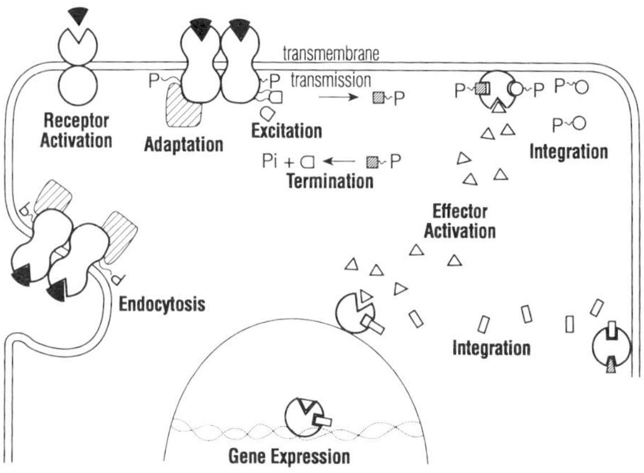

Figure 2. General characteristics of signal-transducing systems. Ligand binding to receptor activates the receptor. This can take the form of the stabilization of receptor dimerization and results in the transfer of information to the intracellular portion of the receptor, initiating an excitation process and an adaptation process. The excitation process generates active transducer molecules, and multiple excitation events may be integrated into a signal that activates an effector, which amplifies the process by generating many molecules of second messenger. These can in turn activate or inhibit a process that leads to a cellular response. The activation and translocation of a transcription factor is shown in the cartoon above. The adaptation process is shown as leading to eventual endocytosis of receptor.

SIGNAL TRANSDUCTION

Cells process information at the membrane by generating and regulating circuits of specific protein-protein interaction. These circuits act to convert a ligand-binding event at the surface into the modification of intracellular activity. In general, all information-processing circuits have certain functional correlates in common; Figure 2 outlines the general characteristics of these systems. Usually the circuit includes a cell-surface receptor protein that can bind extracellular ligand. The receptor is not a simple passive antenna-like component whose function is restricted to recognizing certain molecules; it has a number of other functions. Ligand recognition is accompanied by a change in the structure of the receptor protein, and this change is transmitted across the membrane to an intracellular component that acts as a signal generator. Two classes of signals are generated: excitation signals and adaptation signals. The excitation signal results from a cascade of events initiated by interaction of the intracellular portion of the receptor with another molecule, e.g., a GTP-binding protein or a substrate for endogenous receptor activity, that eventually leads to the stimulation of an effector. The effector function can result from the activation of a protein kinase, the generation of second-messenger molecules, the opening of an ion channel, the nucleation of the growth of cytoskeletal elements, or a variety of other intracellular changes. These changes are often coupled to reactions that amplify the initial event. Thus, for example, a single ligand-binding event can be converted into the generation of hundreds of cyclic nucleotide molecules or dozens of activated protein kinases. Furthermore, information from a variety of inputs can be integrated to reinforce or diminish specific effector activity; e.g., ion channel opening and protein kinase activation can be coordinated to generate a cellular response. The response may take the form of the activation of pathways that lead to the expression of specific genes, to the secretion of a protein or neurotransmitters, or to changes in cell membrane permeability. Finally, each of these circuits must also include a process that allows the excitation signal to be terminated so that the cell can remain responsive to additional extracellular events.

In addition to excitation signaling, most transmembrane systems incorporate a feedback mechanism that adapts or desensitizes the information-processing circuit to maintain its responsiveness to ambient levels of ligand. The adaptive signal often involves the modification of the sensitivity of the receptor to ligand or change in the receptor's ability to further generate excitation signals. The desensitization process may include the removal of the receptor from the cell surface. Receptor activity is thus down-regulated or modulated so that the sensitivity of the system can be reset (Fig. 2).

The remarkable flexibility inherent in protein-protein interactions and the modular design of the genes that encode many transmembrane proteins have permitted an enormous variety of evolutionary experiments with information-processing-circuit design. While maintaining the general features of each of the elements of the circuit (Fig. 2), information-processing pathways that function repetitively with time constants of milliseconds, and other pathways that integrate information over long periods of time and control a single event, have evolved. The extraordinary ability of cells to respond to changes in concentrations of a vast diversity of protein factors, signaling molecules, and environmental conditions is the result of the computational power of these intracellular circuits. The differentiated state of the cell determines the repertoire of signaling systems and how they are integrated to function in a particular cell. Conversely, in many cases, the induction of a particular differentiated state is the result of an instructional system that depends on signal transduction and information-processing circuits. The key to generalizations about the strategies that are used by different information-processing pathways lies in the analysis of the detailed activity of each of their component elements and how they carry out their specific functions. The 57th Cold Spring Harbor Symposium provided some of that detail.

Ligand Binding and Transmembrane Transmission

Dan Koshland introduced the question of receptor function with a review of the structure and possible mechanisms of action of a variety of receptors (Stoddard et al.). How does information about ligand binding get from the outside to the inside of the cell? In general, it is assumed that this transmission event involves a conformational change in the transmembrane receptor. One of the simplest forms of conformational change is the aggregation of individual subunits to form a multimeric structure (see Schlessinger et al.). Alternatively, the conformational change could be the result of ligand-induced stabilization of a specific configuration of the receptor protein. One straightforward approach to resolving this issue, i.e., the nature of the conformational change, is to determine the atomic structure of the receptor in the presence and the absence of bound ligand. Unfortunately, it has been impossible to crystallize many of the transmembrane proteins. Therefore, the approach has been to divide the problem by overproducing the extracellular portion of the protein (hoping eventually to overproduce the intracellular portion as well) and crystallizing it in the presence and absence of bound ligand. Together, Sung-Hou Kim (Kim et al.), Dan Koshland (Stoddard et al.), and their collaborators generated a beautiful picture of the ligand-binding portion of the aspartate receptor derived from *Escherichia coli*. The crystal structure shows a feature that is also found in other cell-surface receptors; that is, the protein appears to exist in the membrane as a homodimer where each of the monomers contributes portions of the ligand-binding site. X-ray crystallography gives us an extremely clear, highly resolved view of the structure of the ligand-binding site. It allows a rationalization of the behavior of mutants that affect ligand binding. However, it is much more difficult to determine the exact nature of the transmembrane transmission event. Koshland has used a series of functional assays with reconstituted receptors and has come to the conclusion that a transverse motion, i.e., like that of a pulley, is involved in transmembrane transmission. Analysis of the crystal structures by Sung-Hou Kim suggests that the conformational change involves the stabilization of a small rotational movement in the molecules that allows the two subunits to pivot around each other. Thus, the structural approach provides a picture of ligand binding; however, it is difficult to deduce the dynamic nature of subsequent events from the crystallographic data.

The dynamics of the transmembrane conformational change can be studied with voltage-gated potassium channels. Electrophysiology provides a millisecond record that reflects the opening and closing of the channel, and site-directed mutagenesis and heterologous expression of cloned genes allow the systematic variation of the structure of the channel. Lilly Jan and her co-workers (Baldwin et al.) showed that they could use this system to identify and modify regions of the channel that are integrally involved in sensing voltage changes in the membrane and in closing or inactivating the channel. These experiments generate models of the kinds of changes in protein structure that could regulate function. However, all the three-dimensional information in the system is the result of guesses about how the proteins fold and how they are disposed in the membrane. As Dr. Jan points out, the amino acid substitution data are consistent with these models, but structural information is required in order to generate a precise picture of function.

Once it is possible to frame the answer to a question at the atomic level of resolution, it is very difficult to move back to experimental paradigms that provide a lower-resolution picture. X-ray crystal structures taken together with data on the effects of specific mutations that define the residues involved in receptor interactions will eventually allow us to deduce the mode of ligand binding and the nature of some of the protein-protein interactions in various receptors and perhaps even in ion channel structures. A number of talks made this point at the meeting. Harrison et al. described the structure of CD4, a component of the T-cell receptor and part of the receptor for the HIV virus. They showed how they could rationalize the function of a variety of mutants that affect the binding of the GP120 ligand to this protein. Hendrickson and colleagues reported on the structures of soluble forms of CD4 and CD8. Again, the structures help rationalize the results of mutagenesis of CD4 and the effect of mutations on its relatively weak but critical interaction with class II major histocompatibility (MHC) protein. The structures also provide insight into the CD8 class I MHC interactions. Other receptor proteins include repeated copies of amino acid sequence homology motifs, such as the epidermal growth factor (EGF) family repeats or the fibronectin type III family repeats. These all presumably have homologous three-dimensional structures. The fibronectin type III repeats have been crystallized, and their structures are being determined. We will soon get an idea of the nature of the precise protein-protein interactions that are involved both in adhesive responses and in specific ligand binding to a variety of receptors. These studies may help to clarify how multiple motifs function to drive ligand binding. However, it is not yet clear that we will be able to deduce the dynamic properties required for signal transduction in multicomponent complexes from the snapshots provided by crystallography.

Serpentine Receptors

One of the most versatile and pervasive receptor structures is found in the serpentine or seven-pass membrane receptor family. These receptors provide the ultimate challenge to the crystallographer. For the most part, they are thought to be embedded in the membrane, leaving only a small portion of the polypeptide accessible to the aqueous medium on the extracellular surface. Ligand binding almost certainly stabilizes a specific transmembrane conformation of the receptor, leading to the activation of a complex cytoplasmic signal-transducing system. One of the most remarkable developments over the last few years has been the revelation of the nature of the receptor system involved in olfaction. The olfactory receptors are derived from a large family of genes with perhaps as many as 1000 members (Chess et al.; Reed). They include subfamilies of serpentine receptors that are presumably specific for different odorants. Axel and co-workers (Chess et al.), moving to a simpler system with a smaller family of receptors (the olfactory system of the catfish), found that there may be only one kind of receptor per cell. Although this finding may clear up the nature of the strategy used to resolve different inputs during odorant perception, the ability of the system to restrict expression to just one receptor family member per cell raises intriguing questions concerning the regulation of specific receptor gene expression. Furthermore, the mechanisms involved in mapping specific odorant receptor cells to different parts of the olfactory bulb, and the question of how regenerating olfactory neurons are reintegrated into the system, move the problem of the specificity of signal transduction from the level of intracellular interactions to that of intercellular specificity. Axel pointed out that there is evidence for spatial segregation of olfactory signal-processing elements in the mammalian olfactory bulb and precedent for spatially segregated pathways which project neurons that control subsets of odorant responses to different portions of the cortex. The olfactory system clearly provides a potential source of enormous diversity in receptor structure-function relationships. However, in order to understand how it is organized, we will be required to confront the questions of how nerve processes grow to appropriate regions of the brain and how the complex wiring of the nervous system is generated and regenerated.

As if 1000 odorant receptors were not enough, this family of serpentine receptors extends to cover the detection of an even larger number of ligands in a variety of cell types. The serpentine receptors detect hormones, neurotransmitters, light, taste, and a variety of "factors." In addition to its extraordinary flexibility that allows it to adapt to the requirements for binding a wide range of ligands, this structure also demonstrates various ways of triggering specific responses. Thus, for example, Coughlin described (Coughlin et al.) the thrombin receptor. This seven-pass membrane receptor provides a substrate for the proteolytic enzyme thrombin, which can digest the receptor and liberate a peptide that acts as a specific ligand. In addition to generating rapid initial responses to ligand binding, these receptors also control the growth response of cells. The pathway by which signals are transduced through the thrombin receptors is not at all clear. The thrombin receptor is just one of the many serpentine receptors that can affect cell growth. Moolenaar reminded us (Moolenaar et al.) that the cell membrane is a repository of second messengers and ligands that are liberated by the activation of a variety of enzymes. Thus, lysophosphatidic acid, which is found as a product of lipid breakdown, binds to a specific receptor, and this receptor can induce a variety of signals that regulate cell morphology and growth.

Networks and Circuitry

In contrast to the olfactory system, where there appear to be multiple receptors to accommodate a variety of ligands and to couple them to a single (or small number of different) intracellular signaling system(s), many other circuits include a series of receptor isoforms that respond to the same ligand. Thus, for example, there are five muscarinic receptors, eight serotonin receptors, more than nine adrenergic receptors, multiple dopamine receptors, and so on. Members of these subgroups all bind the same ligand but differ in the way that they couple to excitation and adaptation signal-generating systems. Figure 3 shows the general signal transduction pathway associated with serpentine receptors. It involves the activation by ligand of the receptor, which then induces the specific dissociation of a heterotrimeric G-protein complex. The α subunit binds GDP and the receptor catalyzes the exchange of GTP for GDP bound to the α subunit; this then releases the $\beta\gamma$ heterodimer. The free receptor can act as

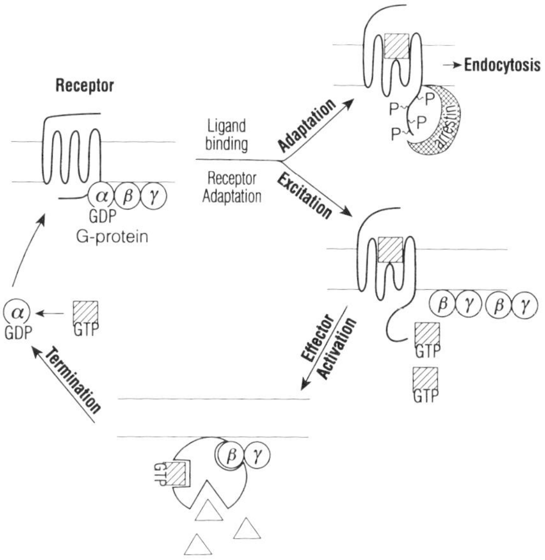

Figure 3. Schematic representation of the stages involved in signal transduction via the serpentine receptor-heterotrimeric G-protein system. Ligand binding activates the receptor's ability to catalyze the exchange of GDP on the α subunit of the heterotrimeric G protein for GTP, which stabilizes an alternate configuration of the α subunit and may dissociate the $\beta\gamma$ complex. The GTP-bound α subunit and perhaps the $\beta\gamma$ subunit as well interact with and activate or inhibit specific effectors. Hydrolysis of the GTP bound on the α subunit to GDP terminates the signal. Receptor adaptation often involves the initial phosphorylation of the receptor and its interaction with another signal-inhibiting molecule, arrestin.

substrate for a receptor kinase, which phosphorylates the receptor and initiates the desensitization process. There is also a great deal of diversity in the heterotrimeric G proteins. Genes encoding 16 different α subunits, 4 different β subunits, and more than 5 γ subunits have been found. The α subunits can be subdivided into at least four families that couple ligand binding to receptor to the activation of different effectors. Thus, for example, the GTP-bound form of the α_s family activates adenylyl cyclase and the α subunits of Gαq subfamily activate the β isoform of phospholipase C. The tremendous variety of serpentine receptors, the apparent redundancy of specific subfamilies of receptors, and the diversity in the heterotrimeric GTP-binding protein families suggest that there are highly sophisticated signal transduction mechanisms involving combinatorial associations of G-protein subunits that generate intracellular information-processing circuits and networks (Simon et al. 1991).

The heterotrimeric G-protein system was part of the subject of the 1988 Cold Spring Harbor Symposium. At that time, Henry Bourne was concerned that whereas effector-activating functions had been assigned to the α subunit that binds GTP, the $\beta\gamma$ heterodimer was left "twisting in the wind." A much more integrated reading of the function of the $\beta\gamma$ complex emerged from discussions at the 57th Symposium. A variety of new roles for the $\beta\gamma$ subunits and for the diversity of the α subunit of the heterotrimeric G-protein signaling system were presented. Tang et al. showed that there are multiple isoforms of adenylyl cyclase, which is one of the effectors activated by the G_s α subunit of the heterotrimeric G protein. While it is well accepted that the GTP-bound form of the G_s α protein can activate adenylyl cyclase, they found that the β and γ subunits can modulate the effect of the α subunit and, depending on the nature of the adenylyl cyclase isoform, $\beta\gamma$ can have a variety of different effects. The apparently bewildering complexity of receptors, heterotrimeric G proteins, and effectors may provide specific pathways composed of various isomorphic elements that interact to generate an appropriate response in different cells. Depending on the nature of the adenylyl cyclase isoform, the $\beta\gamma$ subunit can increase the activation by the α subunit or can moderate or markedly decrease activity.

Bourne and co-workers extended the notion of a role for $\beta\gamma$ subunits in the activation process. Using a series of transfection assays, they were able to demonstrate dramatic effects of the timing of the release or sequestration of $\beta\gamma$ subunits on the cellular response. They suggested that, because of the presence of multiple isoforms, a variety of different information-processing circuits could be generated, including those that were additive or subtractive or circuits that operated independently and ignored each other. The nature of the effector isoform and its potential to be modified by still other products of G-protein-activated pathways could generate a set of conditions that would require the coordinated or simultaneous application of signals, such as two different neurotransmitters, in order to generate a response.

The work by Lefkowitz and co-workers defined another element that contributes to information processing circuit design. They showed that $\beta\gamma$ subunits could play an important role in the feedback mechanism that is coupled to signal generation. They found that free $\beta\gamma$ subunits could mobilize the βARK (β-adrenergic receptor kinase) enzyme, i.e., the kinase that phosphorylates specific seven-pass membrane receptors and initiates the process that desensitizes their signaling ability. This finding demonstrates that the initial ligand-binding event generates two signals: one, the αGTP-activating signal and second, the $\beta\gamma$ signal, which may play a role in effector activation but may also play a role in modulating activity. Thus it seems that a variety of levels of control of signal transduction exist in a Rube Goldberg-like structure of interdependent events that generate a pathway. They provide stages at which feedback and interactions from different systems inside the cell can occur. This is precisely the kind of baroque structure that evolution can "tinker" with over eons, adding features and generating new layers that add a specific response.

Some effectors, G proteins, and receptors are confined by differentiation to single cell types where they have evolved and have been fine-tuned to carry on one job; the visual system and the olfactory system provide examples of these kinds of specializations. On the other hand, many cells have a large number of different receptors, combinations of $\beta\gamma$ and α G-protein subunits and effector isoforms. In these cells, differential gene expression, intracellular localization, and the specificity of protein-protein interactions act to maintain specific signaling circuits that can detect, integrate, or ignore a variety of chemical signals and generate appropriate responses.

Analysis of Function

How do we determine which circuit is activated in a particular cell type? Gilman (Tang et al.) calculated there may be as many as 5000 different combinations of components in the G-protein systems: How do we know which of these components is being used in a particular cell at a particular time? How do we resolve the enormous complexity that is generated by all of these isoforms? One approach to determining these relationships is to look for a biological system in which the complexity can be simplified. Just as Axel and co-workers (Chess et al.) turned to the catfish to analyze the olfactory system, Devreotes (Johnson et al.) and Firtel (Cubitt et al.) have turned to the slime mold *Dictyostelium discoideum* to analyze the diversity of serpentine receptors and G-protein signal-transducing systems. They clearly described the nature of eight different Gα proteins that have been cloned and sequenced and characterized with respect to the developmental cycle of *Dictyostelium*. In this system, the G proteins are expressed differentially during stages of development and their expression can be separated in time. In addition, there are four cAMP receptors whose expression is also differentially regulated, and two different adenylyl cyclases. Presumably, each of these G proteins has a special role at a given stage in development. The *D. discoideum* system is ideal for the analysis of signaling function. The genes can be eliminated by homologous recombination, and the knockouts can be studied. However, the result of these experiments is not simple. All eight of the Gα-subunit genes were eliminated one at a time. Firtel and Devreotes found that one of the knockouts showed a clear phenotype. Subtle effects on the elimination of development were found with two other Gα subunits. However, the other Gα proteins did not have a demonstrable effect on the function of *Dictyostelium* during development and differentiation. This kind of result, i.e., the apparent lack of phenotype when a gene that regulates an "important" function is eliminated, has been found in almost every genetic system that has been examined. Thus, for example, the recent sequencing of yeast chromosome III led to the knockout of a large number of gene functions. Again, the vast majority of these knockouts resulted in no apparent effect on growth. Similar results have been found in genetic studies of complex organisms such as *Caenorhabditis elegans* and *Drosophila melanogaster*. In the vast majority of cases, elimination of a transcript by gene mutation does not lead to a clear visible phenotype. This does not mean, of course, that there is *no* phenotype, it simply means that the systematic search for a phenotype has to be much more sophisticated.

The notion that gene knockouts that are apparently benign might manifest a specific phenotype when examined under the appropriate test conditions was clearly demonstrated in the presentations by Kandel, Tonagawa, and Hynes. Kandel's work (O'Dell et al.) with neural circuits in the sea slug, *Aplysia*, suggested that protein phosphorylation was involved in the signal transduction processes that accompany "learning." To test the involvement of protein kinases in complex brain functions such as memory and learning in mammals, he and his co-workers screened a number of lines of mice that carried gene knockouts of four different members of a family of protein tyrosine kinases. Physiological studies revealed no debilitating phenotype resulting from these mutations. However, they found that one of these lines of mice, where the gene for the fyn-kinase had been eliminated, showed a marked defect in long-term potentiation (LTP), which is thought to play a role in learning and memory. Susumu Tonagawa (Silva et al.) came at this problem from another direction. He and his co-workers also argued that they should be able to knock out specific genes that encoded an isoform of a critical signaling protein if the isoform was involved in neural function; the elimination of that isoform would not result in a developmentally lethal phenotype but could eliminate a specific pathway involved in a dispensable function. They described the results of knockouts of the calcium-calmodulin kinase II (CaMKII). Their experiments demonstrated a marked defect in LTP and a clearly defined defect in the ability of these animals to acquire three-dimensional information and to use this information in learning specific tasks. These experiments do not point to the specific role that kinase isoforms play in the network of interactions that regulate LTP. However, the data imply that the multiplicity of isoforms that one finds in all of these signaling systems underlies the tremendous capacity of the system to handle information. It further demonstrates our ability to analyze interesting biological functions, e.g., learning and memory, which involve information processing at the intercellular and intracellular levels. The key to the analysis lies in the multiplicity of isoform systems that initially appear to have redundant or overlapping activities. Using appropriate techniques, some of these are found to be associated with specialized information-processing pathways that regulate highly evolved functions required for development and behavior.

At a more basic level, the gene knockout experiments can be applied directly to try to understand the function of rudimentary cell-surface systems that are essential for tissue structure and development. The disruption of these functions results in the elimination of an essential step and has a dramatic effect on development or differentiation. These kinds of studies were clearly outlined in the work presented by Hynes and co-workers on fibronectin and integrin gene knockouts. They observed dramatic changes that resulted in catastrophic embryonic lethals, i.e., the mice could not get through the developmental process in the absence of a specific gene product. This system provides the ability to look at the diversity of the extracellular matrix proteins and to analyze their function. Isoforms are generated not only by duplicating genes, but also by introducing splice variants that appear with different developmental timing and may be involved in different functions. Hynes's work clearly provides a glimpse into the future, where the effects of the elimination and modification of these variants on extracellular matrix protein expression and structure can be studied in the context of the developing embryo. Ultimately, we will have to take many different approaches, including biochemical reconstitution, gene replacement, the use of dominant negative mutations, and genetic suppression, to sort out the roles that individual isoforms play in regulating cell-surface functions.

Complex Receptors and Receptor Complexes

Arguably, the most ubiquitous and well-studied of the signal-transducing systems are those that involve transient protein phosphorylation. The integral association of protein kinases and phosphatases with transmembrane receptors and their involvement in hormonal responses, growth control, and immune system function have focused a great deal of attention on these systems. More than 100 different tyrosine kinases have been identified, and many dozens of different serine-threonine kinases have been found. In addition, there is an emerging family of protein phosphatases that will almost certainly demonstrate the same level of diversity as the kinases. This diversity is the result of the combinatorial association of a series of functional domains: a ligand-binding region on the surface of the cell, a transmembrane component, and a signal-generating element in the cytoplasm. Almost all of the receptor tyrosine kinases (RTKs) demonstrate this modular arrangement. The introduction of specific mutations in domains of the RTK has been used to ascertain function. In the work described by Hunter (Hunter et al.), Schlessinger (Schlessinger et al.), Cantley (Soltoff et al.), and Williams (Peters et al.), the route that information follows from ligand binding through transmembrane transmission to the activation of the kinase and its autophosphorylating function was examined. The work summarized in this symposium demonstrated that a series of proteins, some of which are clearly substrates for the tyrosine kinase, contain specific homology regions, SH2 domains, that are able to bind to specialized autophosphorylation sites which act to dock these substrates to the receptor tyrosine kinase. In some cases, the receptor may then phosphorylate specific tyrosine residues on the substrate molecules. On the other hand, the simple sequestration of this protein may be sufficient to generate a signaling activity. The SH2 domain binds with nanomolar affinity to a specific tyrosine phosphate (Schlessinger et al; Peters et al.). This specificity provides another way of examining the role of substrates in receptor function. For example, by eliminating the site at which the SH2 domain of the substrate phospholipase C binds to the EGF or platelet-derived growth factor (PDGF) protein kinase domain, the phospholipase activation was eliminated, but mitogenic activity of the receptor was not affected. The activation of this particular substrate may be important for other cellular activities, but it does not appear to be important for regulating growth. On the other hand, similar experiments strengthen the notion that phosphatidyl inositol 3 (PI3) kinase is important for mitogenic activity. These results suggest that components of the PI3 kinase may be part of the pathway for information transfer from the cell surface to the nucleus; however, the precise role of the phosphorylated substrate could be quite complex. It might act directly to activate other kinases, or it could be involved in events that control receptor turnover. For example, it might be required at some stage in endocytosis. Introducing a break in the route of information transfer in these kinds of experiments only demonstrates that a pathway is coupled to function, but it does not tell us what part of the pathway (e.g., the excitation, adaptation, integration, amplification, or termination step) is associated with the mutation that was introduced. Nonetheless, the association of specific receptors with their substrates through SH2 domains provides a clear indication of a relationship between the two proteins. It has allowed the screening of expression libraries so that families of molecules that contain SH2 domains can be further identified (Schlessinger et al.). However, the problem entailed in defining signaling pathways by isolating proteins that bind to each other is that without the appropriate mutagenic experiments or diagnostic cellular systems, there is no way to show that this association truly represents a function required for a particular cellular pathway.

Alternative approaches in addition to the biochemical isolation and reconstitution of the pathway are necessary to provide functional information. Thus, one of the most satisfying series of results presented at the 57th Symposium came from the studies on morphogenesis in *Drosophila* and in *C. elegans*. Here the genetic paradigm is followed. It involves the isolation of mutants that are defective in specific pathways responsible for organogenesis; however, when the genes that encode these mutants are characterized, the same molecules that the biochemists have found associated with cell-surface receptor-signaling pathways (such as homologs of the EGF receptor tyrosine kinase and of the PDGF receptor tyrosine kinase pathways) appear. Thus, for example, when Schlessinger and co-workers use expression libraries and phosphorylated peptides to screen for genes that are receptor tyrosine kinase substrates and carry SH2 domains, they find a cDNA that encodes a protein called GRB2 which is homologous in sequence to the SEM-5 gene found by Horvitz (Clark et al.) and Sternberg (Sternberg et al.) as a product of the pathway that leads to vulval development in *C. elegans*. Many of the other SH2 domain proteins that have been fished out biochemically can be clearly assigned to the regulatory pathways generated by testing for genetic epistasis. This genetic approach allows the establishment of an information flow pathway that extends from appropriate ligands through receptor tyrosine kinases, SH2 domain proteins to the Ras-GTP-binding protein, and eventually to the induction of specific patterns of gene expression.

In *Drosophila*, the mutant screens designed by Rubin and his colleagues (Simon et al.) to uncover the components of the pathway that regulates cell-cell interaction during photoreceptor differentiation reveal many of the same players, including a homolog of the EGF receptor and of the SEM5 (GRB2) gene and the entire Ras pathway, including a gene that probably encodes a GDP-GTP exchanger and a gene that encodes a homolog of GAP, a protein that stimulates Ras to increase its GTPase activity. It is this congruence of work from orthogonal approaches that provides strong support for the notion that the proteins which have been isolated by their ability to bind tightly to each other are probably associated in the pathway for signal transduction. Of course, the dynamic interactions and kinetics of the events that occur during these transitions are central to the signal transduction process but cannot be deduced from these experiments. Furthermore, even if we assume that the receptor tyrosine kinase activates the Ras pathway, it is not clear whether this is the major or the only pathway that is sufficient to account for receptor function. Nor do we know where the signal pathway proceeds from the Ras-GDP-GTP cycle or precisely how it intersects with the RTK. Again, it is difficult to ascertain which of the signaling pathways is the excitation pathway and which of these pathways are involved in adaptation or desensitization events. In fact, some of the mutants that have been isolated can best be assigned functions that are important for the termination of signaling. Zipursky and co-workers have shown that the protein on a neighboring cell which appears to be critical for generating the signal required for photoreceptor formation is internalized by a mechanism that may involve the stimulation of specific phagocytosis by the developing photoreceptor. The termination of one stage of signaling is an essential part of the process.

Termination of Signaling

Endocytosis or receptor desensitization is obviously not the only way in which a signal can be terminated. Multiple terminating mechanisms exist along any signaling pathway. One of these involves protein tyrosine phosphatases. The phosphatases that appear to have receptor function include an extracellular region that may be composed of a variety of different kinds of ligand- or other protein-binding domains. These "receptors" generally include an undistinguished transmembrane domain and two repeats of what appear to be catalytic domains of the protein tyrosine phosphatase. This tripartite modular arrangement is characteristic of many receptors. It is probably safe to suggest that many ligand-binding transmembrane proteins can exist as monomers or dimers and may be activated by protein-protein interaction and dimerization. On the other hand, it is just as possible, for example, that the protein phosphatases exist in an unliganded form that is activated, and that once ligand is bound, the phosphatase is inactivated. If the function of the phosphatase is to remove phosphoryl groups from other receptors, then its inactivation would raise the levels of endogenous receptor activity and perhaps trigger a signal-transducing pathway. Very little is known about the precise nature and the specificity of the substrates for the protein phosphate phosphatases. The protein tyrosine phosphatases appear to be most efficient, however, since there is a great deal of serine and threonine phosphate found when proteins are analyzed at steady state, but only 0.1% of the protein phosphorylation found in the cell is on tyrosine (Hunter et al.). Thus, it is possible that the limiting factor in the system, and perhaps one of the major terminating signals, is the phosphatase function. Nick Tonks (Tonks et al.) presented evidence that some tyrosine phosphate phosphatases may themselves be regulated by serine-threonine phosphorylation, suggesting that there can be integrated cross-regulation of the different kinase systems. Tyrosine kinases may regulate serine-threonine kinase activity and serine-threonine kinases may regulate tyrosine kinases. There certainly are examples of kinases that can phosphorylate all three residues. Activation can also involve the removal of a resident phosphoryl group from a specific enzyme. Thus, for example, guanylyl cyclase is thought to be activated by phosphatase action (Chrisman et al.).

Vertical, Lateral, and Cytoplasmic Association

Many of the protein tyrosine kinases that act together with receptors are not included as part of the receptor molecule. Both Parada (Parada et al.) and Hunter (Hunter et al.), as well as Joan Massagué (Wrana et al.), presented evidence that protein tyrosine kinases and protein serine-threonine kinases can act in concert with other cell-surface proteins to modify receptor function and ligand recognition. Perhaps the most dramatic example of the formation of receptor complexes came from an examination of cytokine function. A variety of different cytokines may share a transmembrane component, but ligand-binding specificity can be determined by the introduction of an additional subunit that alters the ligand-binding properties of the complex (Glass et al.; Lodish et al.). The cytokines are but one example of a large group of receptors that acquire ligand specificity or other functions by the aggregation of a number of different proteins in the plane of the membrane to assemble an effective receptor. The T-cell receptor is one example of such a complex cluster. The presence both of this receptor and of a

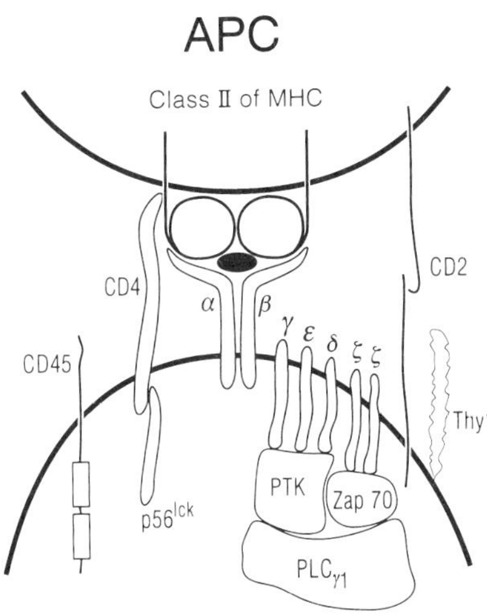

Figure 4. Recognition of the antigen-presenting cell (APC) by the complex of proteins that makes up the T-cell receptor. This rendition of the process is a modification of the scheme presented by Weiss et al. and by Romeo et al.

variety of indirectly associated transmembrane proteins is required for effective recognition of the MHC receptor on the antigen-presenting cell (Fig. 4). However, once an appropriate ligand is found and cell-cell interaction and recognition occur, there follows a complex series of signal-transducing events. A variety of cytoplasmic proteins interact with the activated complex receptor, including two molecules that are protein tyrosine kinases, ZAP-70 and LCK (Weiss et al.). In addition, there is a membrane-bound tyrosine phosphatase CD-45 which is also required for T-cell-receptor signal transduction. The active receptor also requires the assembly of intracellular components that provide a cascade of site-specific transient modifications, including phosphorylation and perhaps nucleotide binding, that results in the simultaneous information transfer to other parts of the cell. Modification of portions of this assemblage provides an opportunity to delineate the nature of its functional elements (Romeo et al.; Weiss et al.). It is perhaps the complexity of the assembly process that ensures both the recognition of the appropriate ligand or antigen-presenting cell and the coincident activation of a number of intracellular processes that key the cell's responsiveness. Thus, the assembly of the receptor acts as an encryption device that provides a fail-safe step which will not allow the signaling process to be completed until the appropriate elements are in place. In addition, the stages of assembly and the duration of effective contact with the antigen-presenting cell may affect the nature of the signal that is generated and its relative ability to direct T-cell function and differentiation. The formation of complex aggregates may also result in signal integration. In fact, the process is not unlike that involved in arranging a symposium. The appropriate invitation (ligand) leads to the assembly of a variety of participants. An accelerated exchange of information occurs, resulting in the modification, activation, or inhibition of information-processing pathways in individual participants. They send appropriate messages to different locations, thus amplifying and utilizing the encounter to effect changes in a variety of local activities. We might refer to these complex aggregates as signal-processing centers or, more simply, as symposons.

ADHESION AND CELL FUNCTION

The Integrins Form Information-processing Circuits

The interaction between the antigen-presenting cell and the T-cell-receptor complex demonstrates that signaling results from transient interactions between receptors assembled on the surface of different cells. These interactions lead to the clustering of an assemblage of membrane-associated proteins that generate a signal-processing center. The intracellular cluster may include tyrosine kinases, serine-threonine kinases, phosphatases, phospholipases, and other transducers that initiate changes in cell function and morphology. There are numerous other induction processes where signaling is mediated directly by cell-cell or cell-matrix interactions. Cell migration, axonal growth, morphogenesis, and a variety of differentiated states of the cell are maintained by these kinds of interactions. At this 57th Symposium it became clear that molecules traditionally associated with cell adhesion also act as complex receptors and participate in signal-transducing systems. The adhesion receptors exist in subfamilies in which diversity is generated by a large number of isoforms. These are differentially expressed and presumably are responsible for tissue-specific effects. The integrins, for example, constitute a relatively large family of heterodimeric proteins that specifically recognize portions of extracellular matrix proteins. Some of the diversity in ligand recognition is probably the result of combinatorial arrangements of the various integrin α and β subunits and their specific expression in different cell types. These receptors also have the familiar modular structure of transmembrane receptors. Their extracellular domains are involved in ligand recognition, whereas intracellular domains appear to be responsible for signal generation and signal transduction. The paradigms for exploring receptor function work as well with the integrins and cadherins as they do with other transmembrane protein systems. The expression of chimeric homologs, the ectopic overproduction of internal or external fragments of the molecule, and the specific mutagenesis of receptor domains have all been used to understand the function of these molecules. Since the integrins are heterodimeric and can exist in specific combinatorial associations, the analysis of their function is a bit more complex. Nonetheless, these paradigms have resulted in a clearer picture both of the ligand specificity of the different isoforms of the integrins and of the signaling capacity of different integrin subunits and integrin complexes. Yamada (Yamada et al.), Ginsberg (Ginsberg et al.), and Hemler (Hemler et al.) summarized the evidence that the binding of the integrin receptor to multiple sites on the extracellular matrix leads to a conformational change in the integrin that may be transmitted across the membrane, changing the relationship of the intracellular domains and leading to segregation of the integrin complex to regions of focal adhesion in the cell. Most interesting, however, is the evidence that intracellular conditions can in turn influence the ligand-binding properties of the receptor. There are a variety of systems in which this "inside out" signaling has been observed. The ability to change affinity for extracellular matrix substrates is particularly important for the regulation of cell migration. The discovery (Ginsberg et al.) that specific antibodies can recognize changes in integrin structure that correlate with affinity changes provides a powerful tool for exploring this process. The ability to modulate receptor binding by changing intracellular signaling conditions also emphasizes the integrative properties of the adhesion receptors. A variety of intracellular excitation signals, for example, those generated by GTP-binding proteins or by protein kinases, could change the affinity of certain integrins for their extracellular matrix substrate. These kinds of changes may be critical in controlling a variety of cellular responses.

Even though the cytoplasmic portion of the integrin receptors is not extensive, it can nonetheless be shown to participate in important signaling functions. The cytoplasmic region regulates the ability of the integrin to associate with and initiate the assembly of focal adhesions. The aggregation of integrin receptors in focal adhesion structures generates a cascade of events, including the activation of specific tyrosine kinases and perhaps phosphatases, that mediate the modification and assembly of components of the cytoskeleton. Burridge and co-workers (Romer et al.) have demonstrated the activation of FAK-125, a tyrosine kinase that phosphorylates specific cytoskeletal substrates, including paxelin and talin. Suddenly a number of the "usual suspects" begin to appear. There are a variety of proteins in the focal adhesion that contain SH2 domains, and the presence of the Src-kinase, which has an SH3 homology domain, may be responsible for interactions with cytoskeletal actin components.

The binding of the integrin receptor to extracellular matrix elements apparently initiates or stabilizes the formation of an active aggregate or focal adhesion that begins to organize cytoskeletal elements. Hemler and co-workers found that the ectopic expression of specific integrins can have different effects depending on the nature of the cell and its signaling state. This supports the suggestion that integrin activation depends on post-ligand-binding events that are determined by the intracellular state of the cell and by the other signals that are being simultaneously processed. Thus, the receptor-ligand-cytoplasmic component complex acts to integrate a variety of intracellular activities.

Another level of signal integration was revealed by the work of Ridley and Hall. They showed that the activated form of the small GTP-binding protein Rho could elicit the rapid formation of polymerized actin stress fibers in much the same way as a variety of growth factors. Another member of the subfamily of small GTP-binding proteins, Rac, when activated, induced a second form of actin polymerization manifested as membrane ruffling. Their experiments suggest that like Ras, the Rac and Rho proteins act downstream from the growth factor receptors. Rac and Rho are, of course, accompanied by a coterie of proteins essential for their function, an activity that stimulates nucleotide exchange, another that stimulates GTPase activity, and connections that provide "input" from the growth factor system and "output" to the actin polymerization system. This assemblage of activities provides ample opportunity for integrating signals from a variety of receptors or for generating cell-specific circuitry to regulate cytoskeletal reorganization.

Although the outlines of the signal-transducing processes are emerging, the details remain to be elucidated. The molecules that initially interact directly with the cytoplasmic portion of β-integrin are not known, and the specific process that leads to integrin receptor association with focal adhesions is unclear. The sequence of events that occur after ligand binding and the temporal order of events that lead to cytoskeletal restructuring are also still obscure. The precise connection of this system with other signal-transducing elements that regulate growth factor responses is not known.

Signaling by Other Components of Adhesion-Receptor Systems

In addition to producing immediate effects on cytoskeletal structure, the integrin receptors can contribute to the control of a number of the differentiated properties of the cell. Thus, for example, cells that lack the fibronectin-integrin receptor were found to lose their ability to generate tumors, and restoration of these receptors to the cells restored tumorigenicity. The extracellular matrix through its interaction with integrins and their signaling capacities can contribute to the control of cellular proliferation and oncogenesis. It is not clear whether the integrin signal alone is sufficient to induce cell proliferation; however, it can have dramatic effects on specific tumors such as melanomas (Sanders et al.). This point was cogently demonstrated by the description of the characterization of the tumor suppressor DCC gene (Hedrick et al.). When the gene was cloned and characterized, it was found to have many of the characteristics of a cell adhesion molecule (CAM), supporting the idea that these CAMs promote differentiation and in doing so generate signals that are incompatible with proliferation. Their loss or effects on their signaling capacity may allow the cell cycle to continue and thus stimulate proliferation. A clear, albeit auxiliary, role for the extracellular matrix in cell proliferation was demonstrated by Ruoslahti et al. They showed that a variety of extracellular matrix proteins such as the proteoglycans and byglycins can interact specifically with growth factors and bind transforming growth factor β (TGF-β) with very high affinity. These molecules may act to screen specific cells from the affects of TGF-β, since the binding of TGF-β by the proteoglycan blocks its activity. The proteoglycan may also serve to induce the synthesis of TGF-β and thus enhance the effects of TGF-β in a number of tissues. It may also stabilize a gradient of growth factor and guide morphogenesis. There are a variety of systems in which the extracellular matrix plays multiple roles. It binds to cell surfaces and anchors cells stimulating morphogenetic changes and differentiation, it modulates the adhesivity of cells to specific substrates restructuring the cytoskeletal and guiding cell migration, and it provides signals to specific cell types that control cell proliferation and growth.

The cadherins are another large family of proteins that interact with each other and that have been shown to be critically important for the integrity of specific tissues. Thus, blocking cadherin interaction between cells by using antibodies that are specific for the extracellular domain of the cadherin can lead to tissue disruption and to the dissociation of cell-cell interactions. Furthermore, cadherin function also requires activities that are regulated by its intracellular domain. Kintner overexpressed the intracellular domain of specific cadherin molecules in *Xenopus* and showed that the cytoplasmically expressed domain was titrated by another molecule that interfered with endogenous signal transduction. Furthermore, he and his co-workers showed that ectopic expression of different cadherin domains led to striking effects, suggesting that tissue-specific cadherins signaled by interacting with specific subsets of cytoplasmic molecules. In fact, Takeichi has shown (Takeichi et al.) that there are three intracellular polypeptides called α-, β-, and γ-catenin that bind specifically to cadherins. There may be different subfamilies of catenins that interact with specific cadherin signaling regions in the cytoplasm to develop and generate an intracellular signal. Takeichi et al. cloned the genes that encode the cytoplasmic catenin molecules and showed that the cadherins bind very tightly to specific catenins and are found localized to adherens junctions. There appear to be a specific α-catenin that interacts with N-cadherin and another that interacts with E-cadherin. These catenins show the same tissue distributions as the parent cadherin, suggesting that there is a family of isoforms of interacting proteins that are found in certain cell types and that generate appropriate signals which lead to cell-type-specific interactions. The catenins contribute to the control of the formation of cytoskeletal infrastructure. Enhanced tyrosine phosphorylation induced either by introducing an activated tyrosine kinase (v-Src) into the cell or by blocking tyrosine phosphatase activity resulted in the inhibition of cadherin-mediated cell aggregation. On the other hand, treatment of cells with herbamycin, which blocked the activity of v-Src, allowed the cells to revert to a cadherin-dependent compacted phenotype (Takeichi et al.). β-Catenin appears to be phosphorylated on tyrosine, thus it is possible that the modifications that occur when the catenins bind to the cytoplasmic domain of aggregated N-cadherins are the events that trigger intracellular signaling, and phosphorylation may further modify the nature of catenin-cadherin interaction. This sequence of events is consistent with the inhibition of aggregation observed when the cytoplasmic domain of the cadherin was introduced ectopically into cells. It may interfere with the assembly of other downstream molecules to block the signaling events that occur in adherens junction formation and that are necessary for the adhesion phenotype.

B. Geiger (Geiger et al.) further outlined the pathway initiated by the binding of adhesion receptors to polyvalent immobilized extracellular ligands, which induces the formation and stabilization of adherens junctions. These proteins interact with cytoplasmic components, and after a variety of steps, the complex participates in the assembly of a cytoskeletal scaffold. Geiger proposed that the adherens junction plaque itself is a signal-transducing machine that incorporates both tyrosine kinase and protein kinase activities. By antibody-staining techniques, they found that the junctions include the proteins Raf, PI3 kinase, protein kinase C, members of the Src family, and phospholipase C. Thus, again, the progression of events appears to involve an initial adhesion event which is transduced into the aggregation of a variety of signaling and cytoskeletal elements. However, after a flurry of phosphorylation and dephosphorylation induced in this junctional region, followed by the assembly of appropriate cytoskeletal components, the trail ends. How do signals from these aggregates get to the nucleus? How do the events that lead to cell adhesion or the loss of adhesion get translated into events that lead to cell proliferation, differentiation, or the expression of specific subsets of genes? Edelman outlined the pathways of induction mediated by adhesion resulting in cytoplasmic signaling that could lead to the activation of specific receptors and the formation of specific cell aggregates. However, the details of the way in which information generated by the reorganization of the cytoskeleton is translated into either growth-regulating cytoplasmic signals or into nuclear signals that would feed back on the process and drive cellular differentiation are not known. Some hints emerge from the apparent involvement of homeodomain-like regions in the control of expression of genes that encode CAMs, suggesting that their controlled expression is part of a pathway that regulates pattern formation in development.

The Importance of Modulating Stickiness

We obtained a better picture of how a combination of signaling processes mediated by different cell types can coordinate behavior and result in specific cellular localization when the role that some classes of adhesion molecules play in inflammation was described. Selectins recognize specific carbohydrates and are responsible for the interaction of leukocytes with endothelial cells. They participate in the localization of subsets of cells to a given tissue because of their ability to recognize complex carbohydrate molecules on the cell surface. These recognition events are critical in the "homing" behavior of subsets of lymphocytes to different tissues (Hu et al.). The molecular nature of some of the ligands and of the selectin receptors was described in this symposium (Weis et al.; Lasky et al.; R.M. Nelson et al.). The different responses mediated by selectin, integrin, and immunoglobulin-like (I-CAM) receptor interactions can be combined to explain the complex behavior observed in the extravasation of white cells from the bloodstream at an appropriate site. An inflammatory response results in the expression of a series of chemotactic factors that diffuse into the endothelial space and induce the appearance of certain selectin molecules on the extracellular surface. As neutrophils with appropriate complex carbohydrates on their surface pass through the vasculature, they interact via selectin specificity with the vascular endothelium and move from the bloodstream to roll along the surface of the endothelial cells. This slows down the motion of the neutrophils and allows them to respond to inflammatory factors via a variety of G-protein-mediated receptors that presumably modify the adhesive properties of β-integrins on the surface of the neutrophils, causing them to bind tightly to the endothelial cell wall. Then the white cell initiates a process of extravasation that allows it to move out of the vasculature to the site of inflammation. A consecutive series of events that are the result of an exchange of adhesive interactions and the interaction of other signaling molecules between two cells that occurs in a timed sequence leads to specific cell migration in response to a gradient of chemotactic factors released at the site of infection.

There are large families of different adhesive molecules. Each is involved in binding to a variety of different ligands, including the extracellular matrix and complex carbohydrates, as well as to each other. The diversity of isoforms is probably responsible for some of the diversity in ligand specificity, but it is also responsible for tissue-specific and cell-specific differences in the activities of a given type of adhesive molecule. It is clear, particularly in the case of the integrins, some of the CAMs, and the cadherins, that they participate in complex mechanisms involved in cytoplasmic signaling. The cadherins interact with a series of cytoplasmic molecules that help sequester and perhaps initiate the formation of complexes associated with the membrane. The integrins may also be sequestered to form complex aggregates. These complexes can include a number of intracellular effectors that modify the components of the cytoskeleton by phosphorylation and dephosphorylation or by generating appropriate second messengers leading to the formation of cytoskeletal structures emanating from the junctional regions. In addition, the process of assembly of the cytoskeleton may itself be responsible for generating other signals that affect cell growth and differentiation. The information transfer in these systems occurs in both directions—from the outside of the cell to the inside as well as from the inside to the outside of the cell. Thus, signaling events inside the cell can lead to the modification of the cytoplasmic portion of a number of adhesion receptors, which in turn modifies the adhesive or ligand-binding characteristics of the protein. These properties result in a great deal of flexibility in the response to environmental factors. Changes in adhesivity are probably also required during the cell cycle for cell division and growth. Furthermore, differentiated cells and proliferating cells probably have different adhesive requirements. The cell surface, cytoskeleton, and cell cycle are intrinsically linked to each other.

Neurite Outgrowth and Neural Connections

One of the most intriguing problems in understanding the role of adhesion receptors and their coordination with information transduction is the question of how specificity is established in the formation of neural connections. Axons can travel over long distances, make specific turns, and connect precisely to the correct target. Does the growth cone have a program that reads the surface like a missile homing in on its target by recognizing preprogramed landmarks? The interactions between the growing axon and substrate molecules, as well as a variety of cells along the way, play an important role in determining the direction of growth cone movement. A variety of experiments have established the involvement of CAMs, integrins, cadherins, and extracellular matrix molecules in enhancing neurite outgrowth. However, the program of messages required to properly guide axonal migration cannot be easily established from these kinds of experiments. Interactions analogous to the conversation between the site of inflammation, endothelial cells, and subsets of circulating neutrophils may be involved. Thus, for example, in studies of neurite outgrowth and neurite elongation, Walsh and Doherty showed that N-CAM, N-cadherin, and L1 all stimulate neurite outgrowth in primary neurons. However, these effects can be inhibited by treatment with pertussis toxin, which blocks signal transduction through the G-protein family. The precise mechanism by which this occurs is not at all clear. It may require the modulation of the avidity of adhesion receptors. During the time that there is growth cone movement and nerve elongation, some of the adhesion characteristics of certain receptors may be suppressed. Once a pathway has been established, interaction with other intracellular signals may modulate adhesiveness and strengthen or weaken the interactions of the neuron with its substrate. These kinds of changes in adhesive properties and the specific collaboration between the growth cone, underlying cells, and extracellular matrix proteins that have "anti-adhesive" as well as adhesive properties can be seen directly by video microscopy of growing retinal neurons (Reichardt et al.). The nature of the integrins or cadherins and CAMs that are present in the growing neuron, and the nature of many of the extracellular matrix components, can be identified by testing for interference of growth cone extension using appropriate antibodies. The combinatorial features of the integrins and the diversity of their families as well as the variety of extracellular matrix molecules and their modifications provide an enormous repertoire of possible interactions. Reichardt's work suggests that it may be possible to follow the dynamic interactions that occur during axon outgrowth, including the appearance and disappearance of specific receptors at different stages during axon elongation and the modulation of the adhesive properties of these receptors as a function of time. Although a preformed gradient of guidance molecules may play an important role in migration, the ability of cells to respond at a particular time to a specific signal is also important. Antibodies provide a means for eliminating cells that may be providing signals in the pathway and thus determining which of the adjacent cells are responsible for elaborating messages that result in changes of direction of the growing axon. However, it is not easy to put the story together. In addition to being able to follow the growth cone by video microscopy, we require methods to find out what correlated changes are occurring within the cell so that we can compare changes in adhesive properties of receptors, changes in second-messenger levels, and changes in specific gene transcription with neurite outgrowth and axon migration.

Another approach involves genetics (Nose et al.). In some organisms such as *Drosophila*, it is possible to screen for mutations that affect molecules which might be involved in processes required to generate appropriate neural migration. A variety of sophisticated genetic techniques, including enhancer traps, allow one to screen for molecules that are regulated and appear at a specific time in the migrating nerve or target cell and then disappear once synaptic connections have been stabilized. Specific molecules have been found that are required to reroute the growing axon. The *Drosophila* system can be used to misexpress the genes that encode these proteins and to show that they can function ectopically and lead to characteristic derangements in the structure of the nervous system. This experimental paradigm is extremely powerful; it is limited only by imagination in conceiving the genetic screening approaches that will give insight into the kinds of molecules involved and by the laborious nature of the process required to screen mutants, identify gene products, and associate them with a specific function in the assembly of neuromuscular interactions.

Once specific connections have been established between nerves or between nerves and muscles, the nerve cell may also be involved in redirecting the structure of its target cell and in building the synaptic structure required for appropriate function. The analysis of agrin (Campanelli et al.) suggests that it is one of the molecules that can induce the formation of clusters of acetylcholine receptors at the synaptic junction in myotubes. It is a multidomain complex molecule that has a variety of different spliced forms. These isoforms may have different functions and behave differently in terms of their ability to regulate receptor clustering and guide the restructuring of the synaptic region. Agrin itself may respond to intracellular signals and so its function may also be controlled by the state of the target cell. Neurexins (Geppert et al.) are receptors for latharotoxin, the toxin elaborated by the black widow spider. They are localized in the synapse and could also play a role in forming appropriate connections and in restructuring the surface of the axon and of the target cell. Interestingly, neurexins also exist as complex gene products with a large number of processed messages that could contribute to the diversity of the structures that they form and perhaps to their ability to stabilize specific synaptic connections.

SPACE, TIME, AND MORPHOGENESIS

At every stage in the process of axonal outgrowth, elongation, migration, and target selection, there appear to be a series of "hand-shakes" that require specific intracellular conditions in the migrating cell and certain extracellular molecules. These interact with receptors in an appropriately timed fashion leading to spatial specificity and regulating cell-specific gene expression. The control of floorplate identity and function in the embryonic nervous system is an example of a series of processes coupled in time that result in the redistribution in space of appropriate signaling molecules which can guide nerve growth and axon migration. Jessell and his co-workers (Klar et al.) are beginning to populate the sequence of morphological events that occur during spinal chord formation with molecules that mark different stages in the process. Thus, the induction of pintallavis gene expression (pintallavis may be a transcription factor) in midline neural plate cells appears to be linked in sequence to the secretion of F-spondin, which is an extracellular matrix protein necessary for neural development. F-spondin may be involved in regulating the movement of the growth cones of the commissural neurons and in the formation of the spinal cord. In addition to the pintallavis gene, there are presumably a variety of other transcription factors that act in a combinatorial fashion to regulate the events involved in the expression of the proteins that are required for differentiation. As the genes that are specifically expressed and their patterns of expression, both in time and in space, are characterized, it may be possible to assemble the inductive program involved in the coordinate regulation of patterning and neuronal guidance.

In some cases, differentiation is driven by specific factors that are induced, are present for a short period of time, and then disappear so that another program can be engaged. However, in other systems, it may be necessary for the differentiated cell to maintain a marker of differentiation. A critical question concerns the nature of the cell memory that allows certain cells to assume and maintain a specific identity over time so they can act as targets for one growing axon or another.

Sanes and his co-workers have shown that axial muscle cells that are innervated by spinal motor neurons maintain their specific identity and can be selectively reinnervated, not only in neonates but also in adults. This suggests that the appropriate targets remember and display their relative positions. Studying a transgenic mouse that contained an insert with a reporter gene expressed specifically in muscle cells, these workers found graded expression of the reporter in the muscles that span the rostrocaudal axis. These muscle cells apparently have an endogenous positional memory that controls the expression of this particular promoter. In fact, dissociated cultured myoblasts from adult muscles demonstrated this same graded activity. The positional differences do not depend on axial gradients of diffusible substances but are endogenous to each of the cells. Possibly, this gradient represents the imprinting of an embryonic condition that first was manifested during the differentiation of these cells, so they now maintain a constant marker of their initial position which is somehow inherited. There may be different patterns of methylation that control graded gene expression in the different cell types, and these patterns may be transmitted in a heritable fashion that marks the cells with a spatial coordinate.

All the mechanisms that one might imagine to have been involved in the process of neurite extension, axon migration, and synapse formation have been found to occur: i.e., the formation of diffuse inductive signals from underlying tissues, the formation of a gradient of surface markers, the elaboration of specific guiding molecules that change the directions of the outgrowing axon, signals that allow the selection of appropriate target cells, and molecules that reorganize the structure of the connection between the target cell and the nascent synapse. These changes reflect the ability both of the signaling system to respond to a variety of different molecules in the substratum and of adjacent cells to elaborate extracellular matrix substances or cell-surface receptors at the appropriate time. Furthermore, the finding that there may be an imprinting mechanism which generates a stable gradient of gene expression that could act as a reference point for the organization and maintenance of neural outgrowth patterns demonstrates the levels of complexity that exist in these guidance systems and their apparent relationship to the mechanisms that guide morphogenesis and differentiation. However, not all of these systems will be found in all organisms; for example, *Drosophila* does not show DNA methylation and there is little or no evidence for imprinting. On the other hand, a variety of marvelous mechanisms described by C. Nüsslein-Volhard (pers. comm.) demonstrate that cDNAs encoding molecules which are critical in morphogenesis can be spatially distributed in a gradient fashion by processes initiated in oogenesis. The gradient eventually leads to restriction of the developmental pathway followed by localized subsets of cells. The morphogenetic gradient need not be established intracellularly. Anderson showed how dorsal-ventral patterning could be the result of an asymmetry generated at a stage many steps removed from the cell-surface receptor or the intracellular transcription factor. The cell-surface receptor is uniformly distributed, the potential ligand is uniformly distributed, and the zymogen precursor of the enzyme that activates the ligand is uniformly distributed. It is only the system which presumably activates the zymogen and triggers the cascade that may lead to a gradient of processed ligand that needs to be localized in order to generate the gradient.

One of the most interesting systems described in *Drosophila* with homologs in a variety of other organisms is the notch-delta system. When a gradient of inductive ligand interacts with a patch of "equivalent" cells, a subset or a single cell may begin to differentiate. The interactions between the notch and delta receptors on this and adjacent cells drives the polarity that has been initiated so that one or a small number of initially equivalent cells differentiate along one pathway, whereas their neighbors are prevented from taking that path and are driven to an alternate fate. How does a molecule like notch carry out this function? How does it act in the determination of so many different cell types? An analogous function in nematodes (lin 12, Sternberg et al.) has also been found to play a similar role in organogenesis, and other signaling molecules that have similar functions have been described in the differentiation of the gametogenic system (Gip-1, Kimble et al.). Notch behaves as if it functions in integrating both cell-surface signals and the patterns of regulation of transcription factors. It must maintain a communication link between the nucleus, the cell surface, and neighboring cells and must have the ability to regulate multiple signaling systems. The analysis of the molecular mechanisms that underlie these activities will be fascinating.

Another system in which the timing of zygotic gene expression generates a massive spatial reorganization was presented by Weischaus and his co-workers (Schejter et al.). The network of fibers that will move the membrane to subdivide the multinucleate stage-14 *Drosophila* embryo into thousands of separate cells has been elaborated, and only the appropriate triggering event is required to initiate the processes. The synthesis and accumulation of three zygotic gene products are necessary, and experiments are currently under way to pinpoint the nature of the clock that synchronizes this event.

MEMBRANE TRAFFIC

Just as enormous subways, pipelines, sewers, and aqueducts hundreds of feet below the surface of a metropolitan city transport people and material, eliminate waste, and deliver energy, so beneath the cell surface there are intricate and highly specialized systems that populate, restructure, and maintain the various membrane surfaces that envelop different parts of the cell. In fact, the transport, localization, and recycling of cell-surface receptors is an integral part of the signal transduction process. This was beautifully illustrated by the work presented on the assembly of the class I and class II MHC receptors (C. Nelson et al.; Click et al.; Ploegh). Two separate pathways exist for the assembly of the class I or class II MHC. The peptides that are incorporated into the class II complex are derived through the lysosomal compartment from proteins that are for the most part of extracellular origin. The class I complex is assembled in the endoplasmic reticulum, incorporating peptides that were degraded in the cytosol and specifically transported to be incorporated into the class I complex and shipped to the surface of the cell. Two separate pathways intersect with two different complex protein degradation systems and sample different populations of potential antigens. They generate homologous but functionally separate receptors that will stimulate different subsets of lymphocytes. Underlying this process and a variety of other processes that involve the transport of peptides into specific compartments of the cell is a family of transporter molecules (Kuchler et al.). The diversity and number of different transporters in a single cell type and their localization in the cell hint at some of the complexity that exists in the apparatus for sorting of proteins and peptides onto different cellular locations.

The exquisite precision of the sorting processes was also clear in the work presented by Simons and by Nelson on the transport of different proteins to the specialized surfaces of polarized cells. A variety of mechanisms contribute to this specificity. Kai Simons (Simons et al.) introduced the discussion of the role that the Ras-like GTP-binding proteins play in the sorting of membrane vesicles. They are thought to be involved in the processes that address and deliver these vesicles. There are thus 24 different isoforms of Rab, and many of these presumably have their own nucleotide-exchange-stimulating activity and a GTPase-stimulating activity. Although this impressive assemblage could provide enough information-processing hardware to generate vesicle specificity, precisely how it functions is not at all clear. Nelson (W.J. Nelson et al.) also described a variety of processes for achieving the localization of proteins to specific regions of the cell surface. One relatively simple notion is the presence of a retention signal, i.e., another protein or structure unique to a membrane surface that prevents the recycling of the target protein. Random sorting and endocytosis would eventually localize the target protein in the region of the retention signal.

It is clear that there are highly segregated, specialized subregions of the membrane. These are maintained by junctional regions between cells and by interaction with junctional regions that anchor the membrane to the underlying cytoskeleton. There are probably also other mechanisms that generate specialized patches on the cell surface. Anderson and coworkers described one such specialized feature: the caveolae, a region where ligand-binding proteins that are linked to the outer layer of the membrane bilayer through a glycosyl-phosphatidylinositol anchor accumulate. This region may play a role in concentrating and transporting a variety of low-molecular-weight dilute nutrients against a concentration gradient. There are certainly other highly specialized patches and surfaces that keep certain proteins from freely diffusing across the entire cell surface. The problems associated with specific protein sorting and with cytoskeletal assembly and disassembly were just touched upon during the 57th Symposium. This is an area that deserves much closer scrutiny and a much longer visit.

PERSPECTIVES

In trying to characterize the 57th Symposium, I keep coming back to the word complexity. The complex mechanisms that handle enormous amounts of information and route and reroute materials provide the cell with its identity—the "persona" that allows it to relate to its neighbors, carry out highly specialized functions, or drive the growth and morphogenesis of the organism. This complexity provides an extraordinary challenge to our ability to analyze the function of individual information-processing pathways. Myriads of receptors, isomorphic families of transducers, and multiple effectors are differentially distributed in space in the same cell, and their activities are coordinated in time, so that in addition to understanding their individual circuitry, the rules that connect or separate them need to be clarified. In his discussion of cell polarity, Kai Simons concludes, "Morphogenesis involves a complex interplay between signal transduction, gene expression, molecular structure, membrane trafficking, dynamic cytoskeletal networks, and organelle biogenesis. How all these parameters are integrated to generate cell architecture is beyond available experimental capability."

Although this may currently be true, we also felt at the 57th Symposium the clear sense that the elusive "experimental capability" was not very far away. New tools and new approaches to the analysis of the kind of complexity we have seen are required and are almost at hand. The global genomic projects that Jim Watson and others have helped initiate will provide us with complete catalogs of isoform families, or at least the means to obtain them. Comparisons between organisms where high-resolution genetic techniques are available and those where genetic techniques are less discriminating, but where biochemistry and cell-free reconstitution are possible, will allow us to use the most appropriate techniques to arrive at insights and generalizations about information-processing circuitry. Physical techniques, including high-resolution X-ray crystallography, are beginning to explain the nature of ligand binding and specific protein-protein interaction. Video microscopy of living tissue provides a view of intracellular reactions separated in time and may give us an idea of the mechanisms involved in coordinating information in the cell in time and space. Nor should we take lightly the beneficial effects of this symposium. A variety of scientists, many of whom would never see each other because they segregate to their own parochial meetings, were brought together in the congenial atmosphere of Cold Spring Harbor and certainly went home with a new view of the processes and problems posed at the cell surface.

I thank the staff for making this meeting enjoyable as well as informative, and I thank Jim Watson for maintaining this marvelous tradition. Finally, I say good-bye to Barbara McClintock, who passed away a few months after this meeting. Just as it has so many times in the past, her presence at Cold Spring Harbor during the 57th Symposium made this an important and memorable occasion for me.

REFERENCES

Simon, M.I., M.P. Strathmann, and N. Gautam. 1991. Diversity of G proteins in signal transduction. *Science* **252:** 802.

Singer, S.J. 1992. The structure and function of membranes—A personal memoir. *J. Membr. Biol.* **129:** 3.

Singer, S.J. and G.L. Nicholson. 1972. The fluid mosaic model of the structure of cell membranes. *Science* **175:** 720.

Author Index

Subject Index

D

E

H

I

S